THE COGNITIVE NEUROSCIENCES

Michael S. Gazzaniga, *Editor-in-Chief*

Section Editors:
Emilio Bizzi
Ira B. Black
Colin Blakemore
Leda Cosmides
Stephen M. Kosslyn
Joseph E. LeDoux
J. Anthony Movshon
Steven Pinker
Michael I. Posner
Pasko Rakic
Daniel L. Schacter
John Tooby
Endel Tulving

RADFORD BOOK
MIT PRESS
BRIDGE, MASSACHUSETTS
DON, ENGLAND

Fourth printing, 1997

This book was set in Baskerville by Asco Trade Typesetting Ltd., Hong Kong and was printed and bound in the United States of America.

Library of Congress Cataloging-in-Publication Data

The Cognitive neurosciences / Michael S. Gazzaniga, editor-in-chief;
section editors, Emilio Bizzi ... [et al.].
p. cm.
"A Bradford book."
Includes bibliographical references and index.
ISBN 0-262-07157-6
1. Cognitive neuroscience. I. Gazzaniga, Michael S. II. Bizzi, Emilio.
[DNLM: 1. Brain—physiology. 2. Cognition—physiology. WL 300 C6765 1994]
QP360.5.C643 1994
153—dc20
DNLM/DLC
for Library of Congress 93-40288
CIP

CONTENTS

II NEURAL AND PSYCHOLOGICAL DEVELOPMENT

III SENSORY SYSTEMS

V ATTENTION

VI MEMORY

IX EMOTION

X EVOLUTIONARY PERSPECTIVES

XI CONSCIOUSNESS

PREFACE

At some point in the future, cognitive neuroscience will be able to describe the algorithms that drive structural neural elements into the physiological activity that results in perception, cognition, and perhaps even consciousness. To reach this goal, the field has departed from the more limited aims of neuropsychology and basic neuroscience. Simple descriptions of clinical disorders are a beginning, as is understanding basic mechanisms of neural action. The future of the field, however, is in working toward a science that truly relates brain and cognition in a mechanistic way. That task is not easy, and many areas of research in the mind sciences are not ready for that kind of analysis. Yet that is the objective.

A wide spectrum of effort is represented in this book. Our intention was to start at the molecular level and continue right up to the problem of human conscious experience. Achieving goals of this magnitude requires a critical approach and good scientific manners. It is easy to disparage attempts at explaining data beyond their internal constraints. To see how local phenomena go beyond conventional boundaries of discussion and to fit them into the understanding of a deeper problem is a difficult enterprise. After all, it is easier to ask questions about the procedural aspects of an experiment than its conceptual premises.

Cognitive neuroscience walks a thin line. It must be exacting and build its foundation on the best and most stringent of observations about the mysteries of nature. On the other hand, it has to explore, in an intelligent and probing and verifiable way, how primary data speak to the issues of how brain enables mind.

This volume has risen to the occasion. It is not complete. It is the beginning. It represents the taking stock of five years of attacking the problem of mind and brain. Each section grapples with the most fundamental problem of modern science—the problem of the explanatory gap. The gap here is the one between biologic process and the processes of mind. This problem and the science built up to understand how the brain enables mind has come to be called cognitive neuroscience.

To respond to this challenge, the section editors carefully picked contributors to their own topic and brought out of them their timely thoughts on this, the most difficult problem of modern science. For three incomparably beautiful weeks in Lake Tahoe, more than a hundred scientists lectured and interacted with twenty-five fellows of the Summer Institute. These fellows had the privilege of hearing the reports live, and their inspiring questions helped many of the authors refine their papers.

The work represented in this volume has been generously sponsored in the main by the McDonnell Foundation and the Pew Charitable Trusts. Private foundations play a crucial and pivotal role in launching new sciences. In the present instance, the vision of a truly new discipline was shared by Dr. John T. Bruer of the McDonnell Foundation. He has been singular in his support and enthusiasm. Besides working with his scientific advisory board (chaired by George A. Miller), Dr. Bruer has been active in enlisting the support of his own Foundation board as well as gaining support of other foundations such as the Pew Charitable Trust. In addition, most of the section editors have played major roles in this effort, as have the contributors. We hope we have put together a source book that will serve students of the mind sciences well for years to come.

There are many people to thank, but perhaps none more important than Flo Batt. She arrived to the project in the nick of time and pulled together both the meeting arrangements as well as many of the grueling editorial duties. Thanks also to the National Institute of Mental Health, which provided partial support for the fellows. Dr. Richard Nakamura took great pride in assisting this project. The Cognitive Neuroscience Institute also provided some support. The Institute has played a crucial role in the development of this field and has been generous in its support over the past thirteen years. Any royalties accruing from this effort will be awarded to the Institute to help in its future programs. Finally, The MIT Press has done a wonderful job with this large, complex book. Fiona Stevens had the vision, and many others worked on fulfilling the dream.

Michael S. Gazzaniga
Center for Neuroscience
University of California, Davis

THE COGNITIVE NEUROSCIENCES

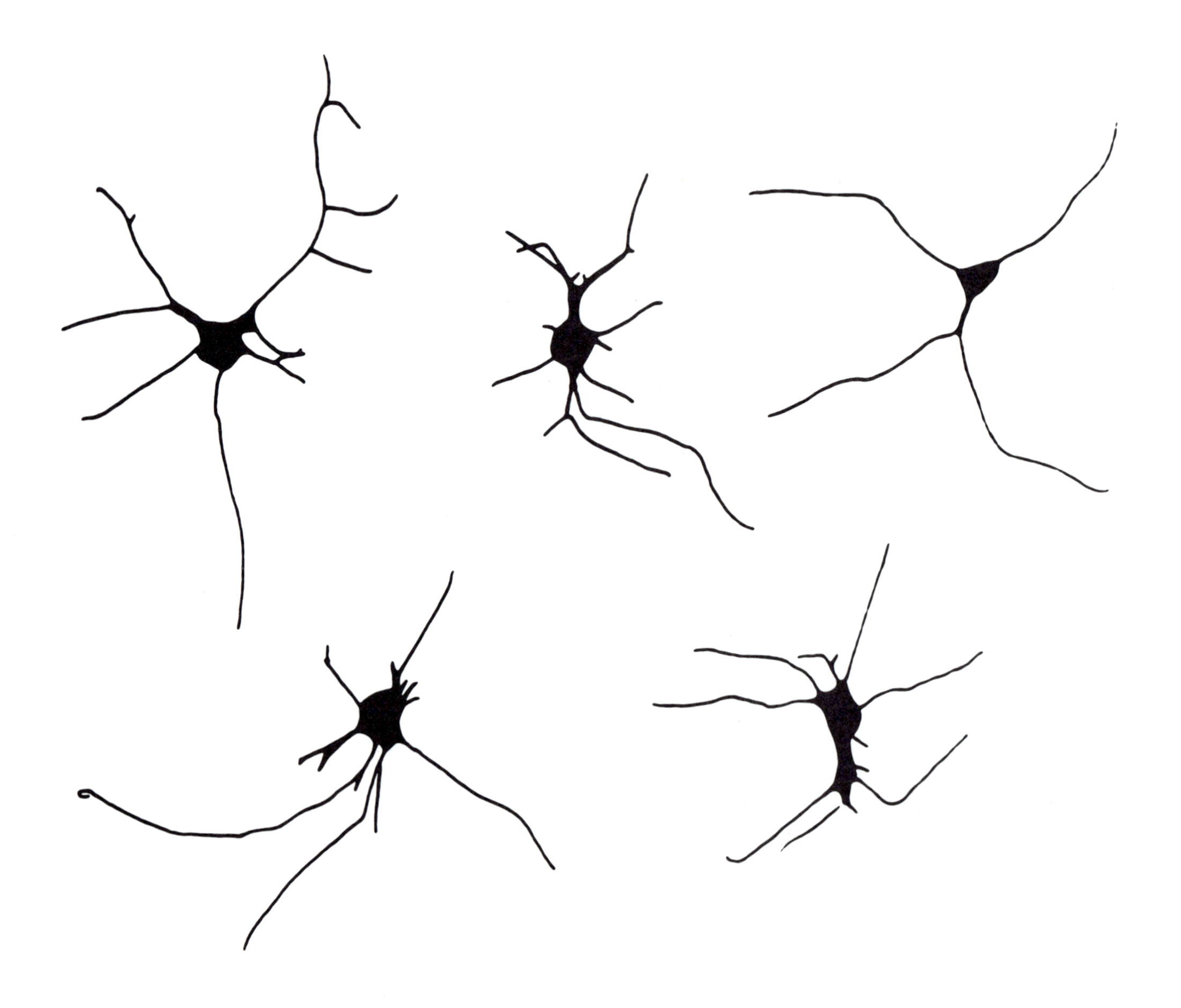

I MOLECULAR AND CELLULAR PLASTICITY

Camera lucida tracings of representative CaBP-positive cells demonstrate morphological maturation of Purkinje cells elicited by NGF (200 U/ml). (Reprinted from Cohen-Cory, Dreyfus, and Black, 1991, with permission.)

Introduction

IRA B. BLACK

IN THE PRESENT context, *plasticity* refers to brain mutability and flexibility, which underlies alteration of structure and function over time in response to environmental change. The molecular and cellular determinants of plasticity constitute the basic building-block mechanisms on which perception, memory, and cognition are based. Several preliminary observations may provide perspective and indicate the rudimentary state of our present understanding.

Literally thousands of individual plastic mechanisms have already been defined at the molecular and cellular levels. Environmental change alters gene action, the activity of enzymes that synthesize neurotransmitters, the function of neuronal growth factors, the growth of neuronal processes, the formation of neural connections, the strength of connections, and the very survival of neurons, for example. Yet, at this early stage, the molecular and cellular plastic mechanisms governing any specific cognitive process remain to be elucidated. Indeed, the conviction that cellular physicochemical mechanisms generate cognition is an article of faith for some observers. Despite these caveats and reservations, a number of important provisional statements are warranted.

Most generally, a large subset of molecular and cellular plastic mechanisms that govern cognitive function also govern function of nonneural cells and systems. For example, the regulation of gene expression, the alteration of cell-cell signaling, and the regulation of cell genesis and death are common to systems throughout the organism. Available evidence suggests that the brain uses mechanisms employed by other systems as well; the brain does not occupy a privileged, unique, or novel position in terms of component mechanisms employed. Rather, the highly complex systems of polarized brain neurons and glia mix and match, combine and permute the component mechanisms into novel patterns. Consequently, while a variety of model biological systems may provide insights into brain plastic mechanisms, cognition arises from the unique architecture of brain systems based on the novel organization of common plastic mechanisms.

Just as multiple organ systems use common plastic mechanisms, it may be anticipated that multiple cognitive processes share common underlying molecular and cellular mechanisms. Based on current knowledge, it would be unreasonable to assume that different forms of memory, or different perceptual processes, for example, use different molecular or cellular mechanisms. A search for *the* unique set of mechanisms that exclusively underlies a given form of memory may thus be ill-conceived.

Conversely, different systems, and therefore different brain functions, appear to use different specific plastic mechanisms. As discussed later, some systems implement long-term change by altering expression of trophic factor genes, whereas other systems alter expression of trophic receptor genes (Black, this volume). In this sense, then, plastic mechanisms may be system-specific, each system may follow its own rules with unique temporal profiles, novel magnitudes of response, and so on. Modularity of brain function may be based, in part, on system specificity of plastic mechanisms.

The chapters in this first part graphically illustrate the issues of specificity, selectivity, and commonality at multiple levels of analysis. Moreover, plasticity assumes a thematic coherence in the consideration of perceptual plasticity, underlying synaptic plasticity, which in turn is governed by specific transmitter, trophic, and growth mechanisms at the molecular level. Gilbert analyzes perceptual plasticity by examining visual fill-in: The visual system generates continuity from discontinuous contours, for example, leading to the generation of visual percepts. Fill-in is based on the activity of neurons projecting horizontally across the visual cortex, spanning an area much larger than a single receptive field. The horizontal neurons provide one structural basis for lateral integration across visual space, subserving fill-in over varying time scales.

Kaas has documented the plasticity of body image perception based on experience. For example, cortical motor maps are plastic, their representation depending on ongoing use. Kaas concludes that cortical representation is a time-based functional construct, in which order is derived from the time coincidence of stimuli, and that it is not a true absolute anatomical construct.

Singer considers the issue of synchrony of discharge among a dispersed set of neurons as generating an ensemble that represents a percept. Convergence of this pattern on a higher-order neuron provides an unambiguous description and representation of a perceptual object. Predictions arising from this "convergence-binding" model are critically examined.

Although the precise synaptic molecular mechanisms underlying the foregoing perceptual plasticity remain to be defined, Bailey and Kandel provide detailed descriptions of the synaptic plastic mechanisms available. Using the marine mollusk *Aplysia californica* as a model, Bailey and Kandel describe ultrastructural changes in synaptic active zone number and area with behavioral habituation or sensitization. Moreover, the changes in active zones were paralleled by changes in synapse number and neuritic arbor density. Changes in active zone and synapse numbers persist for weeks, paralleling the persistence of long-term memory in this system. Finally, the growth changes in active zone and synapse number may be elicited by the transmitter serotonin, which may be functioning as a growth factor in this context.

The relationships among depolarization, transmitter, and growth factor gene expression are explicitly explored by Black. This work has begun to define molecular mechanisms through which millisecond-to-millisecond stimuli are converted to quasi-permanent changes in the nervous system. In sum, depolarizing stimuli elicit nerve growth factor (NGF) gene expression in cultured hippocampal neurons. Because NGF evokes neurite outgrowth and transmitter and synapse changes over days to weeks, acute stimuli may elicit relatively long-term neural changes. In parallel studies, specific transmitter receptor subtypes that evoke responsiveness to trophic factor are defined.

Chalupa describes the extraordinary cytoarchitectonic specificity of retinal ganglion cell arbors to suggest a role for precise trophic interactions. In contrast, Shepherd analyzes signal processing in the olfactory system to derive general underlying principles.

In summary, this part of the text presents the phenomenon of plasticity from a remarkably diverse set of perspectives, including the perceptual, systems, cortical, synaptic, cellular, and molecular.

1 Trophic Interactions and Brain Plasticity

IRA B. BLACK

ABSTRACT Increasing evidence suggests that trophic interactions that influence brain ontogeny play a role throughout life, mediating processes as diverse as learning, memory, and regrowth after injury. This chapter describes recent insights suggesting that impulse activity regulates trophic interactions in the brain, converting millisecond-to-millisecond signaling into long-term changes in neural circuit architecture and function. For example, depolarizing stimuli regulate trophic factor gene expression, an effect mediated by excitatory transmitters. Responsiveness to trophic factors is also regulated by activity, as depolarization increases expression of trophic receptor genes. Further, different excitatory transmitter receptor subtypes subserve trophic or regressive effects, enabling a neural system to memorialize precisely excitatory or inhibitory experiences. A variety of open questions are delineated to define provisional future directions.

Emerging evidence suggests that mechanisms underlying developmental plasticity play roles throughout life. For example, processes as diverse as learning, memory, and regrowth after injury may be mediated by cellular and molecular mechanisms that also govern normal development of the nervous system. Commonality of mechanisms is exemplified by our increasing awareness of the multiple, critical functions served by brain growth and survival (trophic) factors.

Although growth and survival factors have been the focus of study in developmental biology for decades, their critical roles throughout life have been appreciated only recently. Yet growth factors appear to occupy a central functional niche, potentially integrating experience, impulse activity, synaptic pathway formation (and plasticity), and neural circuit architecture during maturity as well as development. Moreover, traditional distinctions among growth factors, trophic factors, and neurotransmitters are dissolving. A summary of the salient characteristics of the now-classical agent, nerve growth factor (NGF), may help provide a perspective for these new views.

The pioneering studies of Levi-Montalcini, Hamburger, and Cohen established that NGF is required for the survival and normal development of peripheral sensory and sympathetic neurons (Levi-Montalcini and Hamburger, 1951, 1953; Cohen and Levi-Montalcini, 1956; Cohen, 1960; Levi-Montalcini and Angeletti, 1968). However, even these initial studies indicated that, in addition to survival, NGF exerted other effects that could play roles in contexts different from development. For example, the factor elicits neurite outgrowth, and targets treated with NGF become innervated by increased numbers of terminals of responsive neurons (Olson, 1967; Levi-Montalcini and Angeletti, 1968). These experiments clearly suggested that NGF influences neural circuit architecture, at least during development.

The foregoing observations were complemented by the seminal demonstration that target NGF is transported from innervated target to perikarya of innervating neurons in a retrograde fashion (Hendry et al., 1974; Hendry, 1975a, b, 1976). Consequently, NGF may act as the messenger mediating retrograde communication of information from visceral target to innervating neuron in the peripheral nervous system. Because the density of target innervation correlates with the concentration of target NGF messenger RNA (mRNA) (Shelton and Reichardt, 1984), the elaboration of trophic factor by targets may regulate architecture of the pathway as well as communication between afferent neurons and particular targets.

The recent realization that trophic factors are synthesized in the brain stimulated interest in mechanisms of action and regulation. For example, the observation that NGF regulates the function of basal forebrain cholinergic neurons (Hefti, 1983; Hefti, Dravid, and Hartikka, 1984; Hefti et al., 1985; Martinez et al., 1985, 1987; Large et al., 1986) and the localization of NGF to

IRA B. BLACK Department of Neuroscience and Cell Biology, UMDNJ–Robert Wood Johnson Medical School, Piscataway, N.J.

hippocampal targets (Whittemore et al., 1986) suggested that trophic functions are critical centrally as well as peripherally. Because the septohippocampal system appears to play an important role in contextual-spatial memory (Eichenbaum, Otto, and Cohen, 1992), trophic and mental functions might somehow be related. Central targets might play roles analogous to those in the periphery, and central interactions might mediate mnemonic function.

The developing trophic field grew more intriguing and complex with the discovery that brain-derived neurotrophic factor (BDNF) and NGF exhibit sequence similarity (Leibrock et al., 1989). The inference that these were members of a neurotrophin gene family was quickly confirmed by the molecular cloning and sequencing of neurotrophins 3 and 4 (NT3, NT4) (Hallböök, Ibáñez, and Persson, 1991; Maisonpierre et al., 1990). Localization of NGF and BDNF, in particular, to the hippocampus (Hofer et al., 1990; Whittemore et al., 1986) extended the potential scope of target trophic interactions in the brain.

In turn, a critical issue concerns the mechanisms that regulate the elaboration of trophic factors by targets. Are factors simply produced constitutively by targets, independent of intercellular, environmental stimuli? Alternatively, is expression of trophic factors regulated, possibly allowing the extracellular environment to influence trophic interactions? The latter alternative, of course, may have far-reaching implications, potentially allowing experience to affect trophic interactions, synaptic pathway modulation, and neural communication. Further, if regulation extends beyond development into maturity, environmental influences may access neural pathway architecture throughout life. Hence, experience, through the mediation of intercellular signals, might alter pathway efficacy in a relatively long-lasting manner. As this may constitute an important form of plasticity and even contribute to such critical processes as learning and memory in some systems, trophic regulation has attracted increasing attention.

Having reviewed this background material briefly, we now can articulate some questions concerning the regulation of trophic expression in the brain. Most generally, are trophic gene expression and factor elaboration regulated processes in the brain? Does experience gain access to trophic interactions? More specifically, do pathway use and impulse activity play roles in trophic interactions? Can we define specific transmitter systems that govern trophic interactions? Is trophic receptor expression and, consequently, neuronal responsiveness regulated by experience? The answers to these and other related questions have been sought through a combination of in vivo and culture techniques.

Trophic gene expression and the balance between excitatory and inhibitory activity

Although trophic interactions in the brain have been intensely studied for nearly a decade, the relationship of trophic gene expression to neural activity has been approached only recently. Experiments performed in vivo initially raised the possibility that innervating neurons might regulate trophic expression in targets. For example, fimbrial transection elevated NGF expression in neonatal hippocampus, suggesting that innervation regulates target elaboration of the trophic factor (Whittemore et al., 1987).

Complementary studies demonstrated that limbic seizures increase hippocampal NGF gene expression (Gall and Isackson, 1989; Gall, Murray, and Isackson, 1990; Isackson et al., 1991). Electrolytic induction of limbic seizures increased NGF mRNA in neurons of the hippocampal dentate gyrus within 1 hour and in neocortical and olfactory forebrain neurons several hours thereafter. Results were quantitated by in situ hybridization and nuclease protection. NGF mRNA decreased from $1\frac{1}{2}$ to 15 hours after seizure onset, indicating that the elevation was transient, not long-lasting. These observations raised the possibility that increased impulse activity, in this case pathological, may increase NGF gene expression, either directly or indirectly. The in vivo nature of the paradigm and the widespread nature of limbic seizures precluded more detailed analyses. Other approaches have begun to define mediating mechanisms.

Using cultured hippocampus as neuronal dissociates in a fully defined medium or as explants, direct evidence that depolarization regulates NGF gene expression was obtained (Lu et al., 1991a). The depolarizing stimuli —either elevated K^+ (35 mM) or the Na^+ channel blocker veratridine—evoked a threefold increase in NGF mRNA (figure 1.1). The effect of veratridine was specifically blocked by tetrodotoxin, which antagonizes the channel-blocking actions of the alkaloid, suggesting that depolarization itself increased NGF gene expression. Moreover, using the hippocampal neuronal dissociates, regulation by intrinsic spontaneous impulse activity was analyzed. To increase spontaneous activity

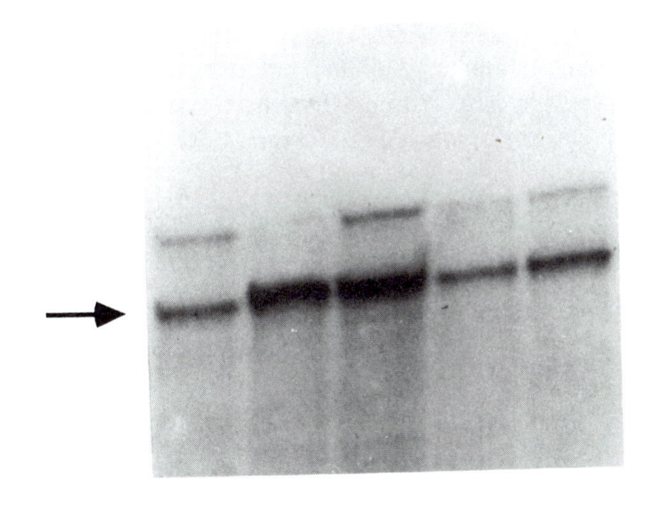

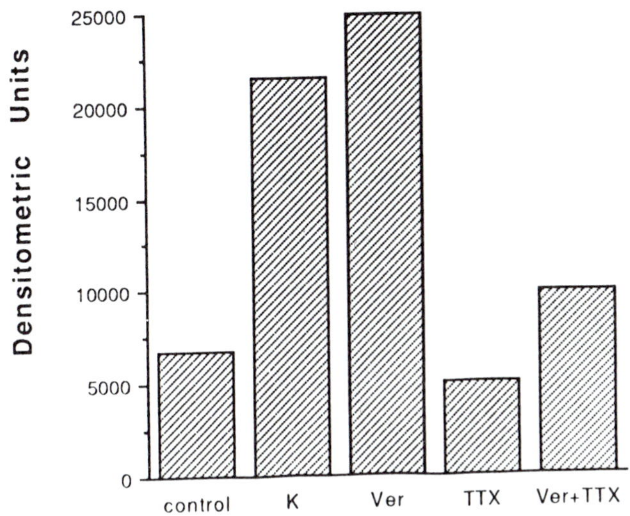

FIGURE 1.1 Regulation of NGF mRNA expression in hippocampal neurons by depolarization. E18 hippocampal neurons were cultured for 6 days in serum-free medium. These cultures were then treated with the following agents for an additional 2 days and harvested for NGF mRNA measurement: K^+, 35 mM; veratridine (Ver), 1 μM; tetrodotoxin (TTX), 1 μM. (Top) Autoradiogram with 40 μg of total RNA per lane. Arrow indicates the 411–base pair NGF mRNA fragment protected by RNase T2. (Bottom) Densitometric plot of the autoradiogram. The result represents three experiments with independent RNA preparations. (Reprinted from Lu et al., 1991a, by permission.)

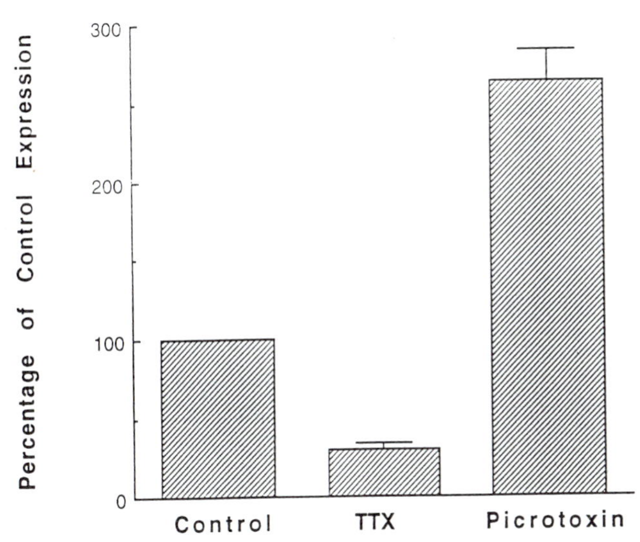

FIGURE 1.2 Regulation of NGF gene expression by spontaneous neuronal activity. E18 hippocampal neurons were cultured in serum-free medium. Control group was cultured for 8 days. Tetrodotoxin (TTX) group was cultured for 2 days and treated with 1 μM TTX for 6 days. Picrotoxin group was cultured for 8 days and treated with the gamma-aminobutyric acid antagonist picrotoxin (1 μM) for 6 hours on the last day. All groups were cultured for the same amount of time before harvest for NGF RNase protection. Autoradiograms were analyzed densitometrically, and numerical values were normalized to percentage of control. (Reprinted from Lu et al., 1991a, by permission.)

in the cultures, inhibitory gamma-aminobutyric acid interneurons were blocked with picrotoxin. Exposure to picrotoxin elicited nearly a threefold increase in NGF mRNA, suggesting that NGF gene expression may represent a balance between excitatory and inhibitory impulse activity (figure 1.2). These studies suggested that depolarizing influences and endogenous impulse activity directly regulate trophic gene expression. One natural next set of questions concerned the specific transmitters involved and the mediating receptors.

Lindholm and colleagues examined a variety of transmitters in cultured hippocampal dissociates (Zafra et al., 1990). Kainic acid, a glutamate receptor agonist, dramatically increased neuronal NGF mRNA. In contrast, *N*-methyl-D-aspartic acid (NMDA) or inhibitors of NMDA glutamate receptors exerted no effects. Further, the kainate effect was blocked by antagonists of non-NMDA receptors, implying that depolarization of hippocampal neurons by glutamate, through the mediation of non-NMDA receptors, regulates NGF gene expression. The same transmitter-receptor systems also elevated expression of BDNF, suggesting that multiple neurotrophins are regulated by impulse activity in the hippocampus (Zafra et al., 1990).

Recent studies have substantiated this contention and raised the possibility that trophins play roles in physiological processes involved in memory. Long-term potentiation (LTP)—enhanced efficacy of synaptic transmission, particularly in the hippocampus—consequent to prior impulse activity has attracted attention as one possible mechanism underlying memory function (Bliss and Lømo, 1973). Recent studies have implicated trophic interactions in LTP (Patterson et al., 1992). Induction of LTP in the hippocampal CA1 pyramidal cell layer increased BDNF and NT3 messages in CA1 neurons. The alterations were apparently cell-specific, as nonstimulated regions of the pyramidal cell layer and the dentate exhibited no alterations in mRNA. Moreover, NGF mRNA was unchanged. Because trophins are known to increase transmitter levels and induce axon terminal sprouting, Patterson and colleagues speculate that activity-induced increases in trophin expression may play a role in the relatively stable changes in synaptic transmission that occur in LTP.

Trophic responsiveness: Regulation of receptor gene expression

Whereas the foregoing studies indicate that impulse activity may regulate trophic factor gene expression, other evidence suggests that responsiveness to trophic stimulation also is critically regulated. Emerging evidence suggests that specific presynaptic transmitters regulate the responsiveness of receptive neurons to trophic factors by influencing expression of trophic receptor genes.

One initial indication of transmitter-trophic interactions was derived from an unexpected source, the cerebellum. A number of studies indicated that NGF protein (Large et al., 1986), NGF receptor protein (Taniuchi, Schweitzer, and Johnson, 1986; Eckenstein, 1988; Schatteman et al., 1988; Yan and Johnson, 1988), and NGF receptor mRNA (Buck et al., 1988; Schatteman et al., 1988) are transiently expressed in the cerebellum in several species. These observations raised the possibility that NGF regulates the development of some cerebellar populations. The discovery that Purkinje cells in the developing cerebellum in vivo express high- and low-affinity NGF receptors identified a key population that might be regulated by the trophic factor (Cohen-Cory, Dreyfus, and Black, 1989). This result was of particular interest, since the development of Purkinje cells was known to be regulated by multiple epigenetic factors (Altman, 1972; Ito, 1984; Altman and Bayer, 1985), including presynaptic innervation (Rakic and Sidman, 1973; Berry and Bradley, 1976; Sotelo and Arsenio-Nunes, 1976).

To elucidate mechanisms by which trophic and presynaptic stimulation might regulate Purkinje ontogeny, the effects of NGF and excitatory transmitters on survival and morphological maturation were studied in dissociated cell culture (Cohen-Cory, Dreyfus, and Black, 1991). Simultaneous exposure to depolarizing agents and NGF specifically enhanced Purkinje survival. NGF in combination with either high potassium or veratridine markedly increased Purkinje survival. Moreover, NGF together with the excitatory presynaptic transmitter aspartate (figure 1.3) or glutamate increased survival twofold. NGF also increased Purkinje cell size and promoted neurite elaboration (figure 1.4). These effects required simultaneous exposure to NGF and aspartate, glutamate, or pharmacological depolarizing agents. Enhanced survival and neurite elaboration were not elicited by exposure to trophic factor or transmitters alone. These results suggest a novel mechanism in which trophic factor and afferent stimulation interact to promote survival and morphogenesis of developing Purkinje cells. The effects on neurite

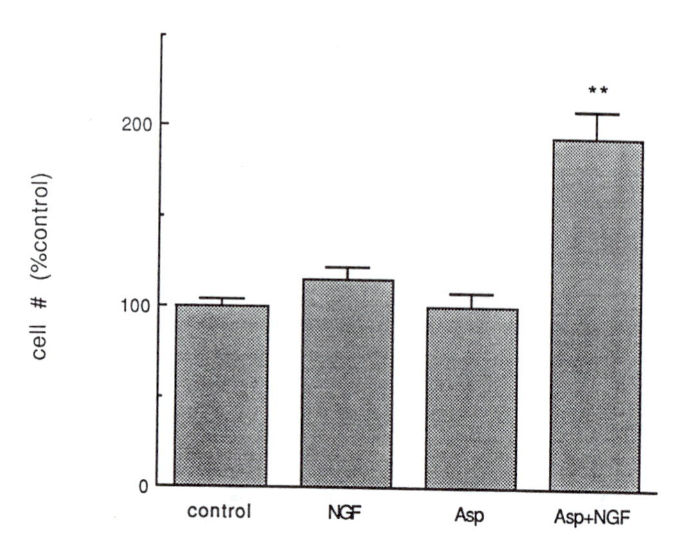

FIGURE 1.3 Effects of aspartate and NGF on Purkinje cell survival. E18 rat cerebellar cells were grown in culture, and immunostained for calcium-binding protein to identify Purkinje cells. Duplicates of two independent experiments were analyzed per condition. Doses of 10 μM aspartate (Asp) and 200 units/ml NGF were used for the time in culture. Values depicted (mean $\pm$ SEM) are as follows: Control, 100 $\pm$ 4.49%; NGF, 115.0 $\pm$ 6.72%; Asp, 99.75 $\pm$ 8.42%; Asp + NGF, 194.0 $\pm$ 15.70%. The actual number of cells in representative control cultures was 1688 $\pm$ 99.42. **, differs from control, NGF, and aspartate by $p < .05$. (Reprinted from Cohen-Cory, Dreyfus, and Black, 1991, by permission.)

growth were particularly provocative, as neurite growth may be a critical determinant of synaptic connections and circuit architecture. In this system, then, the combined action of transmitter and trophic agent may regulate neural system organization.

To begin investigating the molecular mechanisms by which trophic interactions and neural activity regulate Purkinje survival, cytoarchitecture, and function, the influence of excitatory signals on the expression of both NGF and the p75 glycoprotein (the low-affinity component of the NGF receptor) was characterized (Cohen-Cory et al., 1991, 1993). Glycoprotein p75 was used as an index of receptor expression and cellular responsiveness, because it is well defined and conveniently monitored. Expression of NGF and p75 mRNAs was examined in mixed, neuron-enriched, or pure glial cultures. Expression of the NGF gene was localized to proliferating glia, whereas expression of p75 was restricted to Purkinje cells. To ascertain whether presynaptic activity potentially modulates receptor expression, p75 gene expression was studied in cultures exposed to excitatory signals. Depolarization with high potassium, veratridine, or exposure to the excitatory transmitter aspartate resulted in a two- to threefold increase in the expression of both p75 protein and mRNA. The increases did not require the presence of glia, indicating a direct effect of excitatory signals on the neuronal population. Moreover, mRNA and receptor increased per neuron. It may be concluded that local glia provide trophic support for Purkinje cells, whereas impulse activity modulates Purkinje responsiveness by regulating expression of trophic receptor subunits. These observations are complemented by the recent finding that depolarizing influences increase expression of the high-affinity receptor, *trk*, in a neuronal cell line (Birren, Verdi, and Anderson, 1992). In summary, it is apparent that impulse activity may influence trophic interactions at the level of receptor expression, as well as regulating ligand expression.

Transmitter-trophic interactions exert both progressive and regressive actions

My colleagues and I sought to identify the excitatory amino acid (EAA) receptors involved in the regulation of Purkinje survival to begin defining the molecular triggers that potentially govern NGF receptor expression (Mount, Dreyfus, and Black, 1993). EAAs activate multiple ionotropic receptors as well as a G-protein-linked "metabotropic" receptor. Ionotropic receptors are composed of three major subtypes, each named for its preferred agonist (the NMDA, quisqualate/AMPA, and kainate receptors). Ionotropic and G-protein-linked receptors have been identified on Purkinje cells (Dupont, Gardette, and Crepel, 1987; Garthwaite, Yamini, and Garthwaite, 1987; Huang, Bredt, and Snyder, 1990; Krupa and Crepel, 1990; Linden et al., 1991; Llano et al., 1991; Masu et al., 1991; Sekiguchi, Okamoto, and Sakai, 1987). To ascertain how receptors modulate survival, Purkinje cells were cultured with NGF, and EAA receptor subtypes were selectively stimulated or antagonized.

Initially, we characterized the potential role of ionotropic EAA receptors by exposing cultures to the antagonists MK-801, D-APV, and DNQX. Each increased cell number, suggesting that endogenous ionotropic activity *decreased* survival. To determine whether G-protein-linked EAA receptor stimulation modulates survival, the metabotropic agonist ACPD was examined. Alone, ACPD had no effect on survival. However, simultaneous exposure to ACPD and NGF significantly elevated Purkinje cell number. Further, increased survival was blocked by the selective metabotropic antagonist L-AP3, which also reduced cell number in the absence of exogenous EAA. Thus, endogenous metabotropic stimulation is normally necessary for survival.

In sum, excitatory transmitters apparently shape development through a novel mechanism, simultaneously exerting trophic actions through G-protein-linked receptors and regressive influences through ionotropic receptors. The ultimate effect of an excitatory transmitter may be determined by the balance between metabotropic-trophic and ionotropic-regressive actions. In the case of Purkinje cells, aspartate, the putative EAA transmitter of innervating climbing fibers, regulates survival through trophic and regressive influences. In turn, the work cited earlier suggests that the metabotropic-trophic effect is mediated through the expression of NGF receptors. The detailed relationship among metabotropic, ionotropic, and NGF receptors and survival and death remains to be more fully characterized. Nevertheless, it is already apparent that plasticity represents a balance between progressive and regressive influences exerted through specific transmitter receptor subtypes. It is apparent, then, that exquisitely precise regulation may derive from the delicate balance

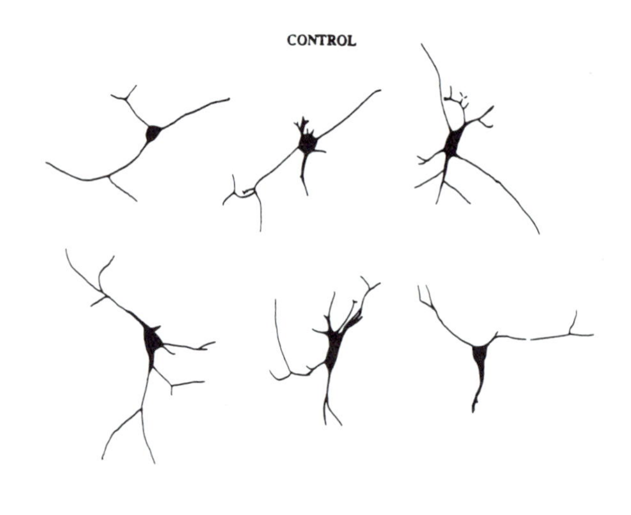
CONTROL

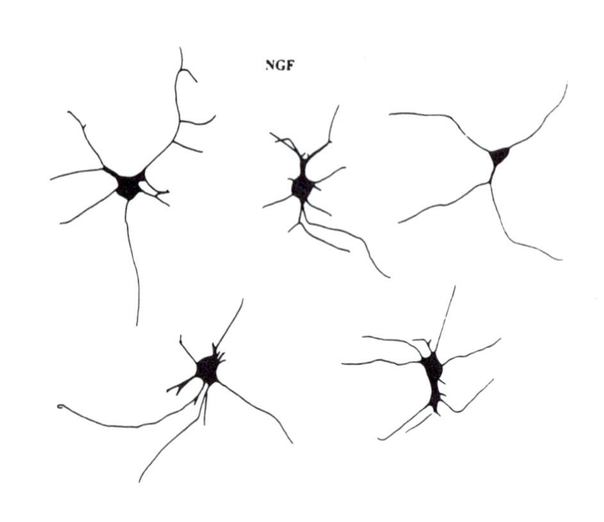
NGF

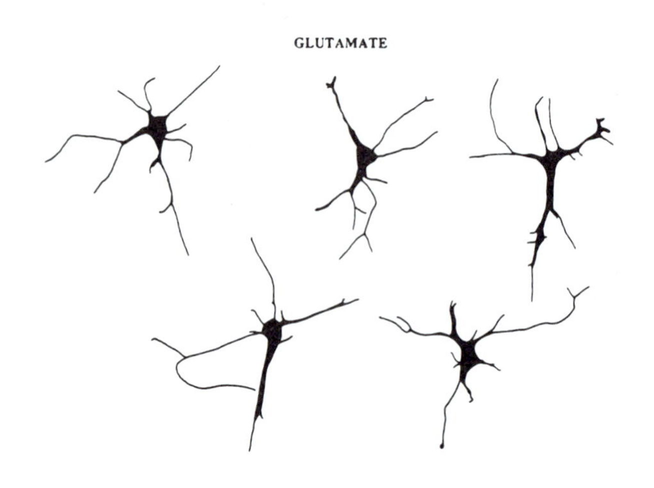
GLUTAMATE

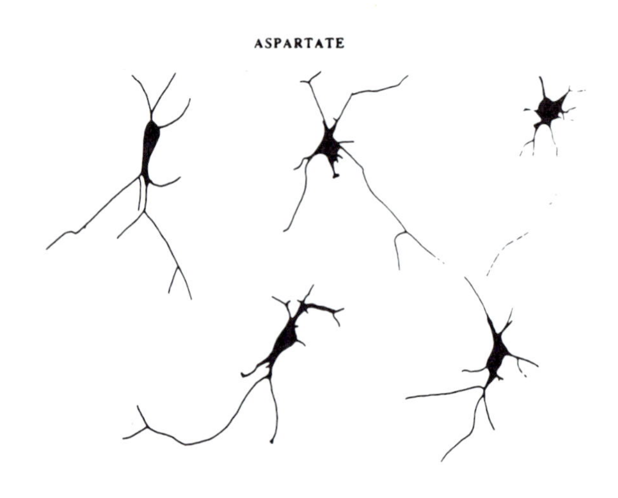
ASPARTATE

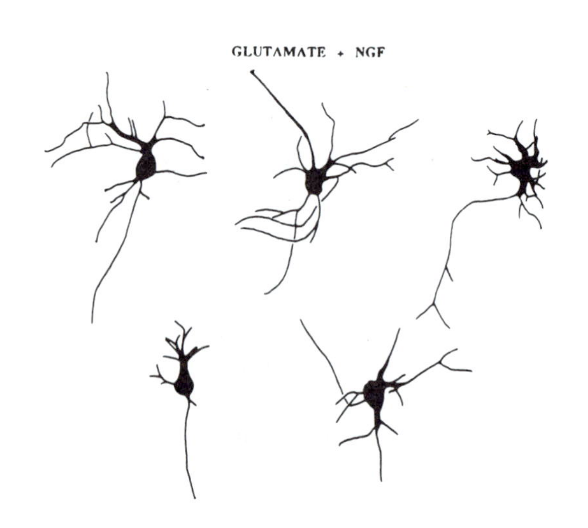
GLUTAMATE + NGF

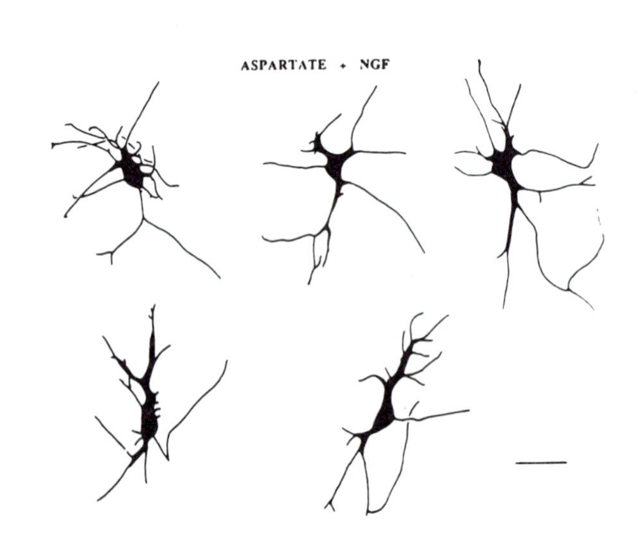
ASPARTATE + NGF

among excitatory, inhibitory, and modulatory actions of transmitters and trophic agents.

Some perspectives and future directions

The foregoing early work provides a view of the emerging complexities of trophic regulation. It is already apparent that impulse activity may influence trophic interactions at multiple loci in different cells and systems. In turn, environmental stimuli that elicit activity in any particular system have the potential to gain access to trophic regulation and to the long-term changes that result. In principle, then, virtually instantaneous, millisecond-to-millisecond impulses may be converted into quasipermanent neural circuit changes through distinct, definable molecular mechanisms.

Surprisingly perhaps, these changes may be either trophic or regressive, depending on the specific EAA receptor subtypes stimulated. Potentially, then, some stimuli may increase pathway efficacy, whereas others may reduce efficacy in some systems. Relationships among experience, impulse activity, and pathway efficacy may be exceedingly complex and fine-tuned.

What variables govern the balance between metabotropic-trophic and ionotropic-regressive influences? Is the final biological effect simply a reflection of relative expression of receptor subtype genes, or do other factors govern access of EAA to already expressed receptors? Is the expression of different receptor subtypes independently regulated at the transcriptional, translational, and posttranslational levels? Approaches to these and related questions may help elucidate the intimate relationship(s) between survival and death, between trophism and regression. More generally, we may inquire whether the coexistent potential for both increased and decreased function is widespread, exhibited by most neuronal systems. Finally, does the trophic-regressive balance contribute to the pathogenesis of disease, and does receptor subtype specificity suggest new therapeutic approaches?

In addition to the trophic-regressive balance, studies with the hippocampal neuronal cultures imply that a balance between excitatory and inhibitory transmitters also regulates trophic factor gene expression. Consequently, experience may increase or decrease pathway efficacy through excitatory or inhibitory transmitter influences. In sum, excitatory-inhibitory, trophic-regressive checks and balances may allow precise modulation of brain neural circuits.

Focusing on trophic effects alone, multiple unanswered questions demand attention. Several are listed here as examples to indicate future research directions. Though impulse activity and specific EAA transmitters can regulate both trophic factor and receptor gene expression, are factor and receptor genes coregulated in most systems, or does regulation occur at different levels in different systems? As proximate glia express trophic factor in the cerebellar dissociates (Cohen-Cory et al., 1993) as well as in other brain areas in vivo and in vitro (Lu et al., 1991b), should we now envisage a trophic neuron-glia-neuron loop, as well as the classic neuron-neuron synaptic circuit? With multiple trophic, tropic, and growth factors, in addition to the neurotrophins, how is biological specificity and selectivity conferred; what is the balance between specificity and redundancy; which factors act sequentially on defined populations; and what are the rules of sequential activity? Are we now in a position to move from environmental stimulus to impulse activity to trophic regulation to mental function and behavior, as implied by the example of LTP cited previously?

ACKNOWLEDGMENTS The author thanks Betty J. Wheeler for excellent technical assistance. This work was supported by National Institutes of Health grants P01-HD23315 from the National Institute of Child Health and Human Development; R01-NS10259 from the National Institute of Neurological Disorders and Stroke; the McKnight Foundation Award in Neuroscience; and a Juvenile Diabetes Foundation grant.

FIGURE 1.4 Morphological maturation of Purkinje cells elicited by excitatory neurotransmitters or NGF or both. Camera lucida tracings of representative CaBP-positive cells were obtained from each experimental condition. Note the marked enhancement in neurite elaboration and in cell size after treatment with glutamate plus NGF or aspartate plus NGF. L-Aspartate and L-glutamate were used at a 10-μM final concentration, and NGF was used at 200 units/ml. Scale bar, 40 μm. (Reprinted from Cohen-Cory, Dreyfus, and Black, 1991, by permission.)

REFERENCES

ALTMAN, J., 1972. Postnatal development of the cerebellar cortex of the rat: II. Phases in the maturation of the Purkinje cells and of the molecular layer. *J. Comp. Neurol.* 145:399–464.

ALTMAN, J., and S. A. BAYER, 1985. Embryonic development of the rat cerebellum: III. Regional differences in time of origin, migration and settling of Purkinje cells. *J. Comp. Neurol.* 231:42–65.

Berry, M., and P. Bradley, 1976. The growth of the dendritic trees in the cerebellum of the rat. *Brain Res.* 112:1–35.

Birren, S. J., J. M. Verdi, and D. J. Anderson, 1992. Membrane depolarization induces p140^{trk} and NGF responsiveness, but not p75^{LNGFR}, in MAH cells. *Science* 257:395–397.

Bliss, T. V. P., and T. Lømo, 1973. Long-lasting potentiation of synaptic transmission in the dentate area of the anaesthetized rabbit following stimulation of the perforant path. *J. Physiol.* 232:331–356.

Buck, C. R., H. J. Martinez, M. V. Chao, and I. B. Black, 1988. Differential expression of the nerve growth factor receptor gene in multiple brain areas. *Dev. Brain Res.* 44: 259–268.

Cohen, S., 1960. Purification of a nerve-growth promoting protein from the mouse salivary gland and its neurocytotoxic antiserum. *Proc. Natl. Acad. Sci. U.S.A.* 46:302–311.

Cohen, S., and R. Levi-Montalcini, 1956. A nerve growth–stimulating factor isolated from snake venom. *Proc. Natl. Acad. Sci. U.S.A.* 42:571–574.

Cohen-Cory, S., C. F. Dreyfus, and I. B. Black, 1989. Expression of high- and low-affinity nerve growth factor receptors by Purkinje cells in the developing rat cerebellum. *Exp. Neurol.* 105:104–109.

Cohen-Cory, S., C. F. Dreyfus, and I. B. Black, 1991. NGF and excitatory neurotransmitters regulate survival and morphogenesis of cultured cerebellar Purkinje cells. *J. Neurosci.* 11:462–471.

Cohen-Cory, S., R. C. Elliott, C. F. Dreyfus, and I. B. Black, 1991. Expression of the NGF receptor gene by cerebellar Purkinje neurons in culture is regulated by depolarizing influences. *Soc. Neurosci. Abstr.* 17:373.5.

Cohen-Cory, S., R. C. Elliott, C. F. Dreyfus, and I. B. Black, 1993. Depolarizing influences increase low-affinity NGF receptor gene expression in cultured Purkinje neurons. *Exp. Neurol.* 119:165–173.

Dupont, J. L., R. Gardette, and F. Crepel, 1987. Postnatal development of the chemosensitivity of rat cerebellar Purkinje cells to excitatory amino acids: An in vitro study. *Dev. Brain Res.* 34:59–68.

Eckenstein, F., 1988. Transient expression of NGF-receptor-like immunoreactivity in postnatal rat brain and spinal cord. *Brain Res.* 446:149–154.

Eichenbaum, H., T. Otto, and N. J. Cohen, 1992. The hippocampus—what does it do? *Behav. Neural Biol.* 57: 2–36.

Gall, C. M., and P. J. Isackson, 1989. Limbic seizures increase neuronal production of messenger RNA for nerve growth factor. *Science* 245:758–761.

Gall, C. M., K. Myrrat, and P. J. Isackson, 1990. Kainic acid-induced seizures stimulate increased expression of nerve growth factor mRNA in rat hippocampus. *Mol. Brain Res.* 9:113–123.

Garthwaite, G., B. Yamini, Jr., and J. Garthwaite, 1987. Selective loss of responsiveness to *N*-methyl-D-aspartate in rat cerebellum during development. *Dev. Brain Res.* 36: 288–292.

Hallböök, F., C. F. Ibáñez, and H. Persson, 1991. Evolutionary studies of the nerve growth factor family reveal a novel member abundantly expressed in *Xenopus* ovary. *Neuron* 6:845–858.

Hefti, F., 1983. Alzheimer's disease caused by a lack of nerve growth factor? *Ann. Neurol.* 13:109–110.

Hefti, F., A. Dravid, and J. Hartikka, 1984. Chronic intraventricular injections of nerve growth factor elevate hippocampal choline acetyltransferase activity in adult rats with septo-hippocampal lesions. *Brain Res.* 293:305–311.

Hefti, F., J. Hartikka, F. Eckenstein, H. Gnahn, R. Heumann, and M. Schwab, 1985. Nerve growth factor (NGF) increases choline acetyltransferase but not survival or fiber outgrowth of cultured fetal spetal cholinergic neurons. *Neuroscience* 14:55–68.

Hendry, I. A., 1975a. The retrograde trans-synaptic control of the development of cholinergic terminals in sympathetic ganglia. *Brain Res.* 86:483–487.

Hendry, I. A., 1975b. The response of adrenergic neurons to axotomy and nerve growth factor. *Brain Res.* 94:87–97.

Hendry, I. A., 1976. Control in the development of the vertebrate sympathetic nervous system. In *Review of Neuroscience*, vol. 2, S. Ehrenpreis and I. J. Kopin, eds. New York: Raven, pp. 149–193.

Hendry, I. A., K. Stöckel, H. Thoenen, and L. L. Iversen, 1974. The retrograde axonal transport of nerve growth factor. *Brain Res.* 68:103–121.

Hofer, M., S. R. Pagliusi, A. Hohn, J. Leibrock, and Y-A. Barde, 1990. Regional distribution of brain-derived neurotrophic factor mRNA in the adult mouse brain. *EMBO J.* 9:2459–2464.

Huang, P. M., D. S. Bredt, and S. H. Snyder, 1990. Autoradiographic imaging of phosphoinositide turnover in brain. *Science* 249:802–804.

Isackson, P. J., M. M. Huntsman, K. D. Murray, and C. M. Gall, 1991. BDNF mRNA expression is increased in adult rat forebrain after limbic seizures: Temporal patterns of induction distinct from NGF. *Neuron* 6:937–948.

Ito, M., 1984. *The Cerebellum and Neural Control.* New York: Raven.

Krupa, M., and F. Crepel, 1990. Transient sensitivity of rat cerebellar Purkinje cells to *N*-methyl-D-aspartate during development. A voltage clamp study in in vitro slices. *Eur. J. Neurosci.* 2:312–316.

Large, T. H., S. C. Bodary, D. O. Clegg, G. Weskamp, U. Otten, and L. F. Reichardt, 1986. Nerve growth factor gene expression in the developing rat brain. *Science* 234: 352–355.

Leibrock, J., F. Lottspeich, A. Hohn, M. Hofer, B. Hengerer, P. Masiakowski, H. Thoenen, and Y-A. Barde, 1989. Molecular cloning and expression of brain-derived neurotrophic factor. *Nature* 341:149–152.

Levi-Montalcini, R., and P. U. Angeletti, 1968. Nerve growth factor. *Physiol. Rev.* 48:534–569.

Levi-Montalcini, R., and V. Hamburger, 1951. Selective growth stimulating effects of mouse sarcoma on the sensory and sympathetic nervous system of the chick embryo. *J. Exp. Zool.* 116:321–362.

LEVI-MONTALCINI, R., and V. HAMBURGER, 1953. A diffusible agent of mouse sarcoma, producing hyperplasia of sympathetic ganglia and hyperneurotization of viscera in the chick embryo. *J. Exp. Zool.* 123:233–288.

LINDEN, D. J., M. H. DICKINSON, M. SMEYNE, and J. A. CONNOR, 1991. A long-term depression of AMPA currents in cultured cerebellar Purkinje neurons. *Neuron* 7:81–89.

LLANO, I., J. DREESSEN, M. KANO, and A. KONNERTH, 1991. Intradendritic release of calcium induced by glutamate in cerebellar Purkinje cells. *Neuron* 7:577–583.

LU, B., M. YOKOYAMA, C. F. DREYFUS, and I. B. BLACK, 1991a. Depolarizing stimuli regulate nerve growth factor gene expression in cultured hippocampal neurons. *Proc. Natl. Acad. Sci. U.S.A.* 88:6289–6292.

LU, B., M. YOKOYAMA, C. F. DREYFUS, and I. B. BLACK, 1991b. NGF gene expression in actively growing brain glia. *J. Neurosci.* 11:318–326.

MAISONPIERRE, P. C., L. BELLUSCIO, S. SQUINTO, N. Y. IP, M. E. FURTH, R. M. LINDSAY, and G. D. YANCOPOULOS, 1990. Neurotrophin-3: A neurotrophic factor related to NGF and BDNF. *Science* 247:1446–1451.

MARTINEZ, H. J., C. F. DREYFUS, G. M. JONAKAIT, and I. B. BLACK, 1985. Nerve growth factor promotes cholinergic development in brain striatal cultures. *Proc. Natl. Acad. Sci. U.S.A.* 82:7777–7781.

MARTINEZ, H. J., C. F. DREYFUS, G. M. JONAKAIT, and I. B. BLACK, 1987. NGF selectively increases cholinergic markers but not neuropeptides in rat basal forebrain in culture. *Brain Res.* 412:295–301.

MASU, M., Y. TANABE, K. TSUCHIDA, R. SHIGEMOTO, and S. NAKANISHI, 1991. Sequence and expression of a metabotropic glutamate receptor. *Nature* 349:760–765.

MOUNT, H. J., C. F. DREYFUS, and I. B. BLACK (1993). Purkinje cell survival is differentially regulated by metabotropic and ionotropic excitatory amino acid receptors. *J. Neurosci.* 13:3173–3179.

OLSON, L., 1967. Outgrowth of sympathetic adrenergic neurons in mice treated with a nerve-growth factor (NGF). *Z. Zellforsch.* 81:155–173.

PATTERSON, S. L., L. M. GROVER, P. A. SCHWARTZKROIN, and M. BOTHWELL, 1992. Neurotrophin expression in rat hippocampal slices: A stimulus paradigm inducing LTP in CA1 evokes increases in BDNF and NT-3 mRNAs. *Neuron* 9:1081–1088.

RAKIC, P., and R. L. SIDMAN, 1973. Organization of the cerebellar cortex secondary to deficit of granule cells in weaver mutant mice. *J. Comp. Neurol.* 152:133–162.

SCHATTEMAN, G. C., L. GIBBS, A. A. LANAHAN, P. CLAUDE, and M. BOTHWELL, 1988. Expression of NGF receptor in the developing and adult primate central nervous system. *J. Neurosci.* 8:860–873.

SEKIGUCHI, M., K. OKAMOTO, and Y. SAKAI, 1987. NMDA receptors on Purkinje cell dendrites in guinea pig cerebellar slices. *Brain Res.* 437:402–406.

SHELTON, D. L., and L. F. REICHARDT, 1984. Expression of the nerve growth factor gene correlates with the density of sympathetic innervation in effector organs. *Proc. Natl. Acad. Sci. U.S.A.* 81:7951–7955.

SOTELO, C., and M. L. ARSENTIO-NUNES, 1976. Development of Purkinje cells in the absence of climbing fibers. *Brain Res.* 111:389–395.

TANIUCHI, M., J. B. SCHWEITZER, and E. M. JOHNSON, 1986. Nerve growth factor receptor molecules in rat brain. *Proc. Natl. Acad. Sci. U.S.A.* 83:1950–1954.

WHITTEMORE, S. R., T. EBENDAL, L. LÄRKFORS, L. OLSON, A. SEIGER, I. STRÖMBERG, and H. PERSSON, 1986. Developmental and regional expression of ß nerve growth factor messenger RNA and protein in the rat central nervous system. *Proc. Natl. Acad. Sci. U.S.A.* 83:817–821.

WHITTEMORE, S. R., L. LÄRKFORS, T. EBENDAL, V. R. HOLETS, A. ERICSSON, and H. PERSSON, 1987. Increased ß-nerve growth factor messenger RNA and protein levels in neonatal rat hippocampus following specific cholinergic lesions. *J. Neurosci.* 7:244–251.

YAN, Q., and E. M. JOHNSON, 1988. An immunohistochemical study of the nerve growth factor (NGF) receptor in developing rats. *J. Neurosci.* 8:3481–3498.

ZAFRA, F., B. HENGERER, J. LEIBROCK, H. THOENEN, and D. LINDHOLM, 1990. Activity dependent regulation of BDNF and NGF mRNAs in the rat hippocampus is mediated by non-NMDA glutamate receptors. *EMBO J.* 9:3545–3550.

2 Molecular and Structural Mechanisms Underlying Long-term Memory

CRAIG H. BAILEY AND ERIC R. KANDEL

ABSTRACT Studies of the cellular basis of learning in *Aplysia* have demonstrated that aspects of memory storage are represented at the level of individual neurons by changes in the strength and structure of identified synaptic connections. Short-term memory, lasting minutes to hours, involves an alteration in synaptic effectiveness of preexisting synaptic connections as a result of the covalent modification of preexisting proteins by second-messenger cascades. By contrast, long-term memory, lasting days or weeks, is associated with the growth of new synaptic connections activated by altered gene expression and new protein synthesis. The structural changes can be induced by repeated presentation of modulatory neurotransmitters that appear to mimic the effects of growth factors during cellular differentiation and are associated with a rapid and transient modulation of NCAM-related cell adhesion molecules, believed to contribute to axonal outgrowth and guidance in the developing nervous system. Thus, the mechanisms underlying the growth of new synaptic connections initiated by experience in the adult may eventually be understood in the context of the activity-dependent refinement of synaptic connectivity during neural development.

Among the most interesting and important problems facing both cognitive psychologists and neurobiologists is an understanding of the mechanisms that underlie learning and memory. Modern behavioral and biological studies have shown that learning and memory are not a unitary process—not a single faculty of the mind—but a family of distinct processes, each with its own rules. In the most general sense, *learning* can be considered the process by which new information about the world is acquired and *memory* the process by which that knowledge is retained. Recent studies have demonstrated that learning can be divided into at least two general categories (Milner, Corkin, and Teuber, 1968; Weiskrantz, 1970; Tulving, 1991; Squire, 1992). *Explicit* or *declarative memory* is the conscious acquisition of knowledge about people, places, and things and is particularly well developed in the vertebrate brain. *Implicit* or *nondeclarative memory* is the nonconscious learning of motor skills and other tasks and includes simple associative forms such as classical conditioning and nonassociative forms such as sensitization and habituation. The two types of learning seem to involve different neural circuits in the brain. Explicit learning uniquely depends on temporal lobe and diencephalic structures—for example, the hippocampus, subiculum, and entorhinal cortex—whereas implicit learning does not depend on temporal lobe function but rather involves the same sensory, motor, or associational pathways used in the expression of the learning process. Thus, whereas explicit learning is most readily studied in mammals, implicit forms can be effectively studied in both nonmammalian vertebrates and higher invertebrates.

For both explicit and implicit learning tasks, the type of memory system employed is a determining factor in how the information is encoded, stored, and recalled. A striking feature that has emerged from cognitive studies in human beings as well as behavioral studies in experimental animals is that memory for both types of learning appears to be a single, graded continuous process, the duration of which is related to the number of training trials (Weiskrantz, 1970; Craik and Lockhart, 1972; Wickelgren, 1973). The pioneering studies of Ebbinghaus (1855/1963), based on the learning of nonsense syllables, first demonstrated that repetition of a task can

CRAIG H. BAILEY Center for Neurobiology and Behavior, College of Physicians and Surgeons of Columbia University and the New York State Psychiatric Institute, New York, N.Y.
ERIC R. KANDEL Howard Hughes Medical Institute and Center for Neurobiology and Behavior, College of Physicians and Surgeons of Columbia University and the New York State Psychiatric Institute, New York, N.Y.

increase both the strength and duration of the memory for that task. Consequently, long-term memory has often been considered a graded extension of short-term memory. However, experimental results from clinical studies in humans as well as a variety of studies on different animal models suggest that, like learning, memory is not a unitary process and can exist in at least two forms: a short-term form, which can last seconds, minutes, or hours, and a long-term form that can persist for days, weeks and, in the case of humans, even a lifetime (James, 1890; McGaugh, 1966; Atkinson and Shiffrin, 1968; Davis and Squire, 1984). Short-term memory can be dissociated from long-term memory in humans following specific clinical conditions (Squire and Zola-Morgan, 1988) and in experimental animals using inhibitors of protein synthesis (Agranoff, 1972; Barondes,1975; Flexner, Flexner, and Stellar, 1983).

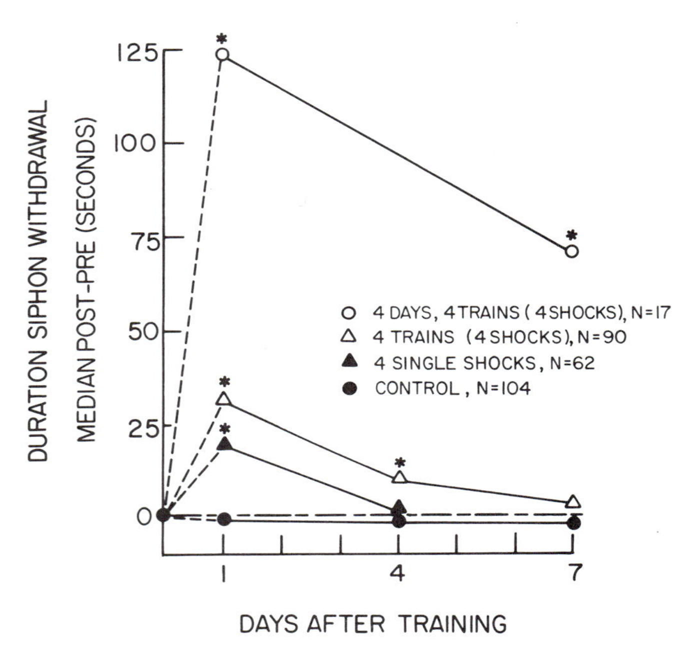

FIGURE 2.1 Behavioral long-term sensitization. A summary of the effects of long-term sensitization training on the duration of siphon withdrawal in *Aplysia californica*. The retention of the memory for sensitization is a graded function proportional to the number of training trials. Experimental animals received four single shocks for 1 day (filled triangles), four trains of shocks for 1 day (open triangles), or four trains of shocks per day for 4 days (open circles). Control animals were not shocked (filled circles). A pretest determined the mean duration of siphon withdrawal for all animals before training. Posttraining testing was carried out 1, 4, or 7 days after the last day of training. The asterisks indicate a significant difference between the duration of siphon withdrawal for the trained and control animals (Mann-Whitney U tests, $p <$ .01). N, number of animals per group. (Reprinted from Frost et al., 1985, by permission.)

One of the major surprises in the modern study of behavior has been the realization that learning and memory are probably universal features of the nervous system (Thorpe, 1956). In recent years, this realization has encouraged the use of several higher-invertebrate preparations where the advantages of a tractable central nervous system and identified neurons have facilitated the study of learning and memory at the cellular and molecular level (Kandel, 1976; Kandel and Schwartz, 1982; Alkon, 1984; Carew and Sahley, 1986; Byrne, 1987; Bailey and Kandel, 1993). One such model system has been the gill- and siphon-withdrawal reflex of *Aplysia*. In *Aplysia*, as is the case in other mollusks, the mantel cavity, a respiratory chamber housing the gill, is covered by a protective sheet, the mantel shelf, that terminates in a fleshy spout, the siphon. A tactile stimulus to the siphon mantel shelf in *Aplysia* activates siphon sensory neurons, which relay the information to interneurons and to gill and siphon motor neurons (Castellucci et al., 1970; Byrne, Castellucci, and Kandel, 1974, 1978). These reflexes exhibit several types of associative and nonassociative learning, including classical conditioning and habituation and sensitization. We focus here on sensitization of the gill- and siphon-withdrawal reflexes and, specifically, on the molecular and structural mechanisms that underlie their long-term form.

The memory for long-term sensitization in Aplysia *has a representation in the monosynaptic component of the reflex*

Sensitization is an elementary form of nonassociative learning, by which an animal learns about the properties of a single noxious stimulus. The animal learns to strengthen its defensive reflexes and to respond vigorously to a variety of previously neutral or indifferent stimuli after it has been exposed to a potentially threatening or noxious stimulus. In *Aplysia*, sensitization of the gill- and siphon-withdrawal reflex can be induced by a strong stimulus applied to another site, such as the neck or tail. This activates facilitatory interneurons, which synapse on the sensory neurons and strengthen the synaptic connection between the sensory neurons and their central target cells (Hawkins, Castellucci, and Kandel, 1981).

As in the case for other defensive withdrawal reflexes, the behavioral memory for sensitization of the gill- and siphon-withdrawal reflex is graded, and retention is proportional to the number of training trials. A single stimulus to the tail gives rise to short-term sensitization lasting minutes to hours. Repetition of the stimulus produces long-term behavioral sensitization that can last days to weeks (figure 2.1) (Pinsker et al., 1973; Frost et al., 1985).

The memory for both short- and long-term sensitization is represented on an elementary level by the monosynaptic connections between identified mechanoreceptor sensory neurons and their follower cells. Although this component accounts for only a part of the behavioral modification measured in the intact animal, its simplicity has allowed the reduction of the analysis of the short- and long-term memory of sensitization to the cellular and molecular level. For example, this monosynaptic pathway can be reconstituted in dissociated cell culture (Montarolo et al., 1986; Rayport and Schacher, 1986), where serotonin (5-HT), a modulatory neurotransmitter normally released by sensitizing stimuli (Glanzman et al., 1989; Mackey, Kandel, and Hawkins, 1989) can substitute for the shock to the neck or tail used during behavioral training in the intact animal. A single application of 5-HT produces short-term changes in synaptic effectiveness, whereas four or five applications of 5-HT over a period of $1\frac{1}{2}$ hours (or continuous application of 5-HT for $1\frac{1}{2}$ to 2 hours) produces long-term changes lasting 1 or more days (Montarolo et al., 1986).

Biophysical studies of this monosynaptic connection suggest that both the similarities and the differences in memory reflect, at least in part, intrinsic cellular mechanisms of the nerve cells participating in memory storage. Thus, studies of the connections between sensory and motor neurons in both the intact animal and on cells in culture indicate that the long-term changes are surprisingly similar to the short-term changes (Frost et al., 1985; Montarolo et al., 1986; Dale, Kandel, and Schacher, 1987). A component of the increase in synaptic strength observed during both the short- and long-term changes is due, in each case, to enhanced release of transmitter by the sensory neuron, accompanied by an increase in the excitability of the sensory neurons, attributable to the depression of a specific potassium channel (Klein and Kandel, 1980; Hochner, Schacher, and Kandel, 1986; Dale, Kandel, and Schacher, 1987; Scholz and Byrne, 1987; Dale, Schacher, and Kandel, 1988).

Despite these several similarities, the short-term cellular changes differ from the long-term process in two important ways. First, the short-term change involves only covalent modification of preexisting proteins and an alteration of preexisting connections. Both short-term behavioral sensitization in the animal and short-term facilitation in dissociated cell culture do not require ongoing macromolecular synthesis; the short-term change is not blocked by inhibitors of transcription or translation (Montarolo et al., 1986; Schwartz, Castellucci, and Kandel, 1971). By contrast, these inhibitors selectively block the induction of the long-term changes in both the semi-intact animal (Castellucci et al., 1989) and in primary cell culture (Montarolo et al., 1986). Most striking is the finding that the induction of long-term facilitation at this single synapse exhibits a critical time window in its requirement for protein and RNA synthesis characteristic of that necessary for learning in both vertebrate and invertebrate animals (Agranoff, 1972; Barondes, 1975; Davis and Squire, 1984; Flexner, Flexner, and Stellar, 1983; Montarolo et al., 1986). From a molecular perspective, these studies indicate that the long-term behavioral and cellular changes, which last more than 24 hours, require the expression of genes and proteins not required for the short term. Whereas the gene products required for short-term memory are preexisting and must be turned over relatively slowly, some gene products required for long-term memory must be newly synthesized. Second, the long-term process, but not the short-term process, involves a structural change. Bailey and Chen (1983, 1988a,b,c, 1989) have demonstrated that long-term sensitization training is associated with the growth of new synaptic connections by the sensory neurons onto their follower cells. This synaptic growth can be induced in the intact ganglion by the intracellular injection of adenosine 3′:5′-cyclic phosphate (cAMP), a second messenger activated by 5-HT (Nazif, Byrne, and Cleary, 1991a), and can be reconstituted in sensorimotor cocultures by repeated presentations of 5-HT (Glanzman, Kandel, and Schacher, 1990; Bailey et al., 1992b).

The finding that long-term memory involves a structural change and requires new macromolecular synthesis raises several interesting conceptual issues. First, it raises questions of how the covalent modifications of preexisting connections during short-term memory become transformed into a structural change. Are the structural changes dependent on, and perhaps induced by, the changes in gene and protein expression in the neurons involved? What molecules trigger the gene induction, and how does this activation lead to a structural change? Second, it provides a way of thinking about the stability of long-term memory. Is the stability of long-term memory achieved because of the relative stability of synaptic structures? If so, is the loss of memory with time reflected by a corresponding reduction in the number of synaptic connections? Finally, the finding that an alteration in the number of synaptic connections is the most reliable anatomical marker for the long-term process raises the question of how closely these structural mechanisms in the mature animal resemble the activity-dependent de novo synapse formation in the developing animal?

In this chapter, we shall address these issues by focusing on recent molecular studies of learning-related synaptic growth during long-term sensitization in *Aplysia*. Because structural changes are a signature of long-term memory processes, ranging in complexity from those produced by elementary forms of nonassociative learning in invertebrates to those produced by higher-order associated tasks in mammals, our hope is that principles derived from this approach may be applicable to more complex systems and, ultimately, to human behavior.

Long-term sensitization involves a persistent and transcriptionally dependent increase in protein phosphorylation

Both the similarities and differences between short- and long-term memory for sensitization are reflected in the cellular properties of the participating neurons. This localization makes it possible to focus on the biochemistry of these neurons to determine the molecular mechanisms that may underlie the two forms of information storage. Using such a biochemical approach, one can ask two questions: (1) What are the molecular bases for the similarity of long- and short-term memory, a similarity that gives memory its apparently graded and continuous properties? (2) What accounts for the difference in susceptibility to blockade by inhibitors of protein and RNA synthesis? That is, how do the short- and long-term processes differ biochemically?

The similarity between the short- and long-term processes suggested, at a first approximation, that the same molecular machinery involved in short-term facilitation was also involved in long-term facilitation. Bernier and coworkers (1982) demonstrated that short-term facilitation of sensory neurons is accompanied by an increase in the intracellular concentration of cAMP. This increase in cAMP after 5-HT application presumably activates a cAMP-dependent protein kinase as an early step in the cascade of events leading to facilitation. To test the possible involvement of the cAMP-dependent protein kinase, Castellucci and colleagues (1980) injected the purified catalytic subunit into sensory neurons and found that this produced a broadening of the action potential and facilitated transmitter release. Moreover, intracellular injection of a specific inhibitor of the cAMP-dependent protein kinase was found to block the facilitatory effects of 5-HT on sensory neurons (Castellucci et al., 1982; Ghirardi et al., 1992; Goldsmith and Abrams, 1991).

As the short-term effect involves cAMP-dependent protein phosphorylation, perhaps the long-term process also involves an activation of this cascade, but now the activation is autonomous and persists even in the absence of a stimulus. Sweatt and Kandel (1989) addressed this possibility by using an in vivo phosphorylation assay, based on quantitative two-dimensional gels of phosphoproteins from intact sensory neurons, to examine how the pattern of substrate phosphorylation present in the short term (following a single exposure to 5-HT) relates to that present in the long term (after repeated exposure to 5-HT). They found that long-term facilitation produced by repeated or prolonged exposure to 5-HT or cAMP induced increased phosphorylation of the same group of 17 substrate proteins, which are similarly phosphorylated during short-term facilitation (figure 2.2). However, whereas the increased phosphorylation of the substrate proteins after a single pulse of 5-HT or cAMP is transient and does not require new protein synthesis, the increased phosphorylation of the same substrates after repeated or prolonged exposure persists for at least 24 hours and is blocked by inhibitors of protein or RNA synthesis applied during the 5-HT application (see figure 2.2).

What underlies the persistent increase in protein phosphorylation? Does it represent depression of a phosphatase or enhanced activity of a kinase? If the latter, which kinase is involved? The fact that the long-term changes lead to phosphorylation of the same substrates

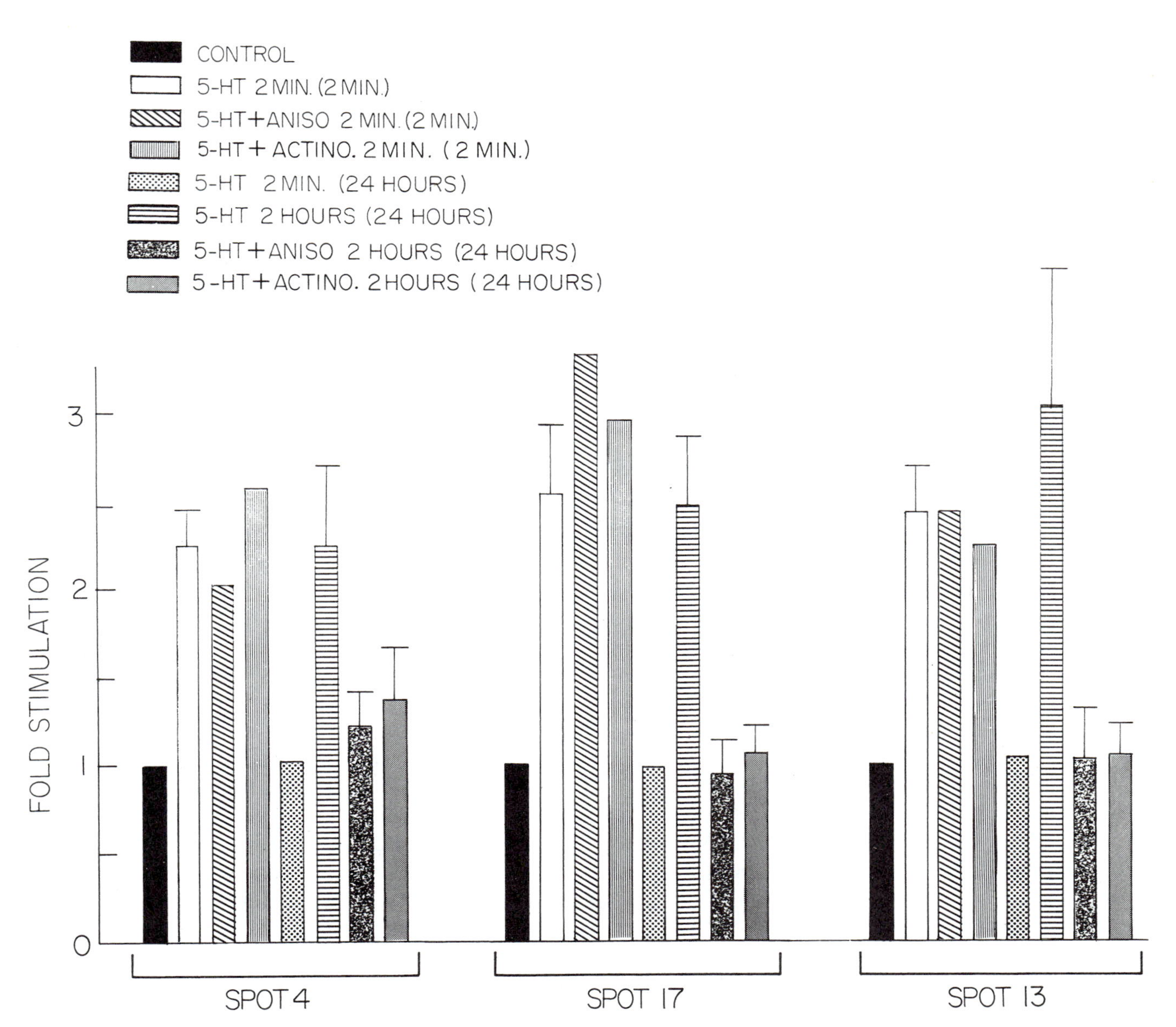

FIGURE 2.2 Short- and long-term effects of 5-HT on the phosphorylation of individual proteins. The incorporation of phosphate into the same set of 17 proteins was increased during both short- and long-term facilitation of isolated *Aplysia* sensory neurons. Representative data are shown for three proteins. The protein synthesis inhibitor anisomycin or the RNA synthesis inhibitor actinomycin D applied during the period of 5-HT application had no effect on phosphorylation increases associated with short-term facilitation but blocked the persistence of phosphorylation during long-term facilitation. (Reprinted from Sweatt and Kandel, 1989, by permission.)

involved in the short-term process suggests that cAMP-dependent kinase might also be involved in the long term. To explore the possible mechanisms underlying the persistent phosphorylation, Bergold and coworkers (1990) examined changes in the properties of the cAMP-dependent protein kinase (A kinase). The A kinase is activated during both short-term and long-term sensitization in *Aplysia*, and the amount of the regulatory subunit is lowered, as compared to the catalytic subunit, in the sensory neurons of long-term sensitized animals (Greenberg et al., 1987). Bergold and colleagues (1990) found that the facilitatory stimuli (5-HT or cAMP) also diminish the ratio of the regulatory to catalytic subunits in the sensory neurons and that this reduction in the regulatory subunit requires new protein synthesis. The work of Schwartz and his colleagues indicates that the down-regulation of the regulatory subunit does not occur at the level of transcription (Bergold et al., 1990, 1992) but rather involves proteolytic cleavage of the regulatory subunit by the ubiquitin pathway (Hegde, Chain, Schwartz, 1992). Ubiquitin itself and components of the ubiquitin proteolytic pathway might be among the new proteins made in sensory neurons during long-term sensitization.

Long-term sensitization is induced by the activation of cAMP-responsive genes

The transcriptional dependence of long-term facilitation has been further demonstrated by the examination of changes in the expression of specific gene products induced by 5-HT. Barzilai and associates (1989) focused on the $1\frac{1}{2}$-hour period of training and studied the incorporation of labeled amino acids [^{35}S]methionine into proteins in the sensory neurons (figure 2.3). They found that 5-HT produces three temporally discrete sets of changes in specific proteins that could be resolved on two-dimensional gels. First, at 30 minutes 5-HT induces a rapid and transient increase in the rate of synthesis of 10 proteins and a transient decrease in 5 proteins. These changes subside within 1 hour and are, in all cases, dependent on transcription. Second, at 3 hours, there is a transient increase in four proteins that is also dependent on transcription. Finally, at 24 hours, there is a sustained increase in the expression of two proteins that persists during the maintenance phase of long-term memory. The 15 early proteins induced by repeated exposure to 5-HT can also be induced by cAMP. These features—rapid induction, transcriptional dependence, and second-messenger mediation—suggest the possibility that this control might involve a gene cascade, whereby early regulatory proteins activate later effector genes. The early proteins induced by 5-HT and cAMP during the acquisition phase of long-term facilitation in *Aplysia* may therefore resemble the immediate early gene products induced in vertebrate cells by growth factors. In vertebrates, some of the immediate early genes encode regulators; others encode effector proteins (Sorrentino, 1989).

What turns on these early proteins and triggers the long-term process? In most systems studied thus far, the cAMP-dependent protein kinase is responsible for the activation of cAMP-induced gene expression by phosphorylation of cAMP-responsive element-binding proteins (CREBPs), which bind to an upstream enhancer sequence, the cAMP-responsive element (CRE) (Montminy et al., 1986). To determine whether a similar cascade is involved in the initiation of long-term facilitation, Dash, Hochner, and Kandel (1990) examined extracts of *Aplysia* sensory neurons and found that they contained proteins specifically bound to a mammalian CRE sequence. Then, by blocking the induction of long-term facilitation with injections of oligonucleotides containing the CRE sequence into the nucleus of the sensory neuron, they demonstrated that these CREB-like proteins were essential in activating the long-term process.

A recent study by Kaang, Kandel, and Grant (1993) has explored further the steps whereby cAMP regulates genes important for long-term facilitation by addressing two questions: (1) Can the facilitating transmitter 5-HT activate transcription in the sensory neurons? (2) Does 5-HT activate transcription through the PKA-mediated phosphorylation of CREB-related transcription factors that bind to CRE? Based on gene transfer into individual *Aplysia* sensory neurons (Kaang et al., 1992), these investigators found that repeated pulses of 5-HT result in the induction of a reporter plasmid driven by the CRE. The induction was graded and did not occur following a single pulse of 5-HT. In addition, 5-HT did not induce a reporter gene driven by a tPA response element (TRE). Moreover, coexpression of a reporter (GAL4-CAT) and a wild-type or mutant GAL4-CREB fusion protein revealed that the transcription induced by 5-HT requires that CREB be phosphorylated on SER^{119} by protein kinase A. These data, together with the earlier finding of Dash, Hochner, and Kandel (1990), indicate that the phosphorylation of CREB-like proteins and the consequent activation of transcription represents a key component of the molecular switch for extending the short-term process for presynaptic facilitation, which is independent of protein synthesis, into the long-term process, which requires gene expression.

The studies in *Aplysia* also provide evidence that a conventional chemical transmitter acting through a second-messenger pathway can produce synaptic actions of different time courses as a function of the number of times the transmitter is applied. A single pulse of 5-HT activates the cAMP-dependent protein kinase so

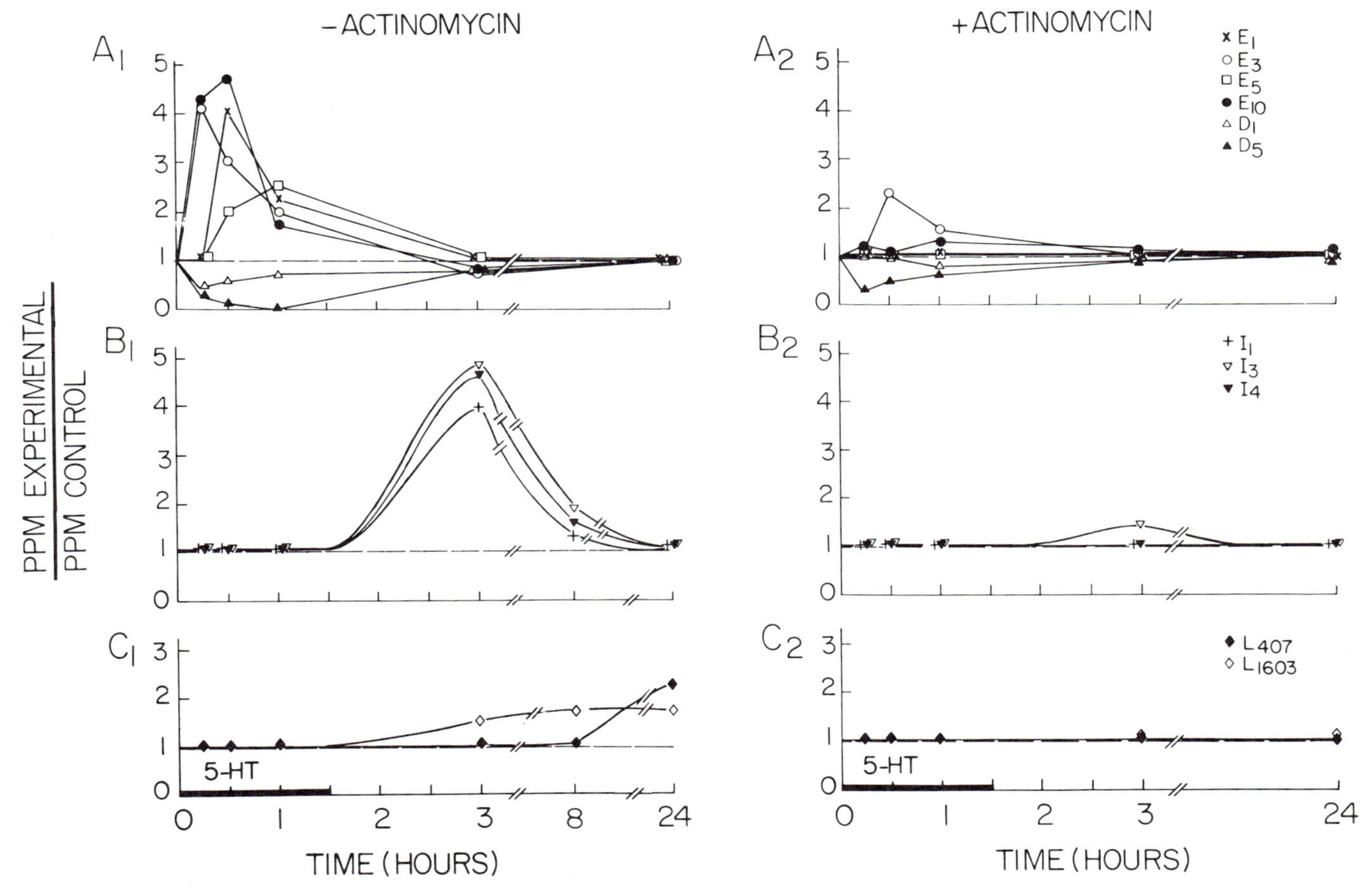

FIGURE 2.3 The effects of 5-HT and actinomycin D on the incorporation of [^{35}S]methionine into specific proteins of isolated *Aplysia* sensory neurons in primary cell culture. Each plot represents the ratio of the average parts per million (PPM) of 5-HT-treated clusters over the average PPM for the control for a specific protein at the indicated time points. (A_1) Early increases and decreases in the net synthesis of specific proteins. (B_1) Intermediate increases peaking at 3 hours. (C_1) Late increases in the net synthesis of specific proteins. (A_2, B_2, C_2) All the changes in protein expression were either partially or completely blocked by the addition of actinomycin D 1 hour before the application of 5-HT. (Reprinted from Barzilai et al., 1989, by permission.)

that it phosphorylates substrates in the cytoplasm, leading to an enhancement of transmitter release in a period of minutes. Repeated presentation of 5-HT or exposure to drugs that increase cAMP is capable of translocating the catalytic subunit to the nucleus (Bacskai et al., 1993), where it can phosphorylate CREB which, in turn, can activate CRE-driven genes to prolong the enhancement of transmitter release for 1 or more days. Thus, the distinction between short- and long-term facilitation derives in part from the ability of 5-HT to select, based on the number of stimuli presented, between a cellular program involving only the cytoplasmic actions of the cAMP pathway and a program that involves, in addition, the recruitment of nuclear transcription factors.

Long-term sensitization involves the growth of new sensory neuron synapses

The gene induction accompanying long-term facilitation in *Aplysia* seems to be a general property of several types of learning in both vertebrate and invertebrate nervous systems. In the mammalian hippocampus, a robust form of synaptic enhancement known as *long-term potentiation* (LTP), which shares several physiological properties with long-term facilitation in *Aplysia*, is believed to be a contributing component to the neural mechanisms for explicit forms of learning. Induction of LTP is associated with the increased expression of mRNA encoding both transcription factors (Cole et al., 1989; Wisden et al., 1990) and effector proteins potentially important for growth (Qian et al., 1993). These findings suggest that in mammals as well as in *Aplysia*, long-term changes in synaptic plasticity may be linked to gene induction and that one of the functions of this induction is to initiate synaptic growth. Indeed, an increasing body of morphological data emanating originally from studies on mammalian development, and more recently embodying results from studies on nonmammalian vertebrates and higher invertebrates, now

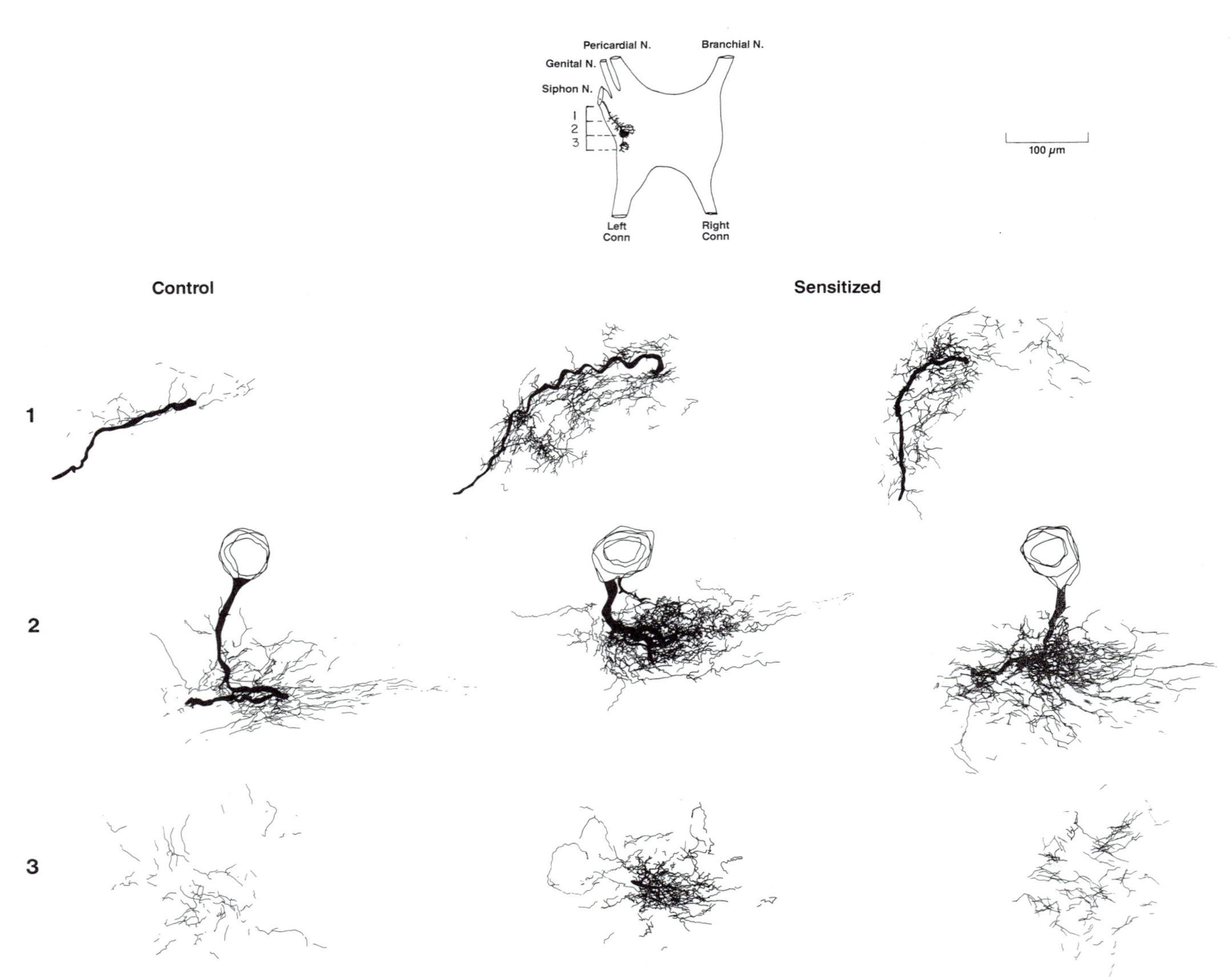

FIGURE 2.4 Serial reconstruction of sensory neurons from long-term sensitized and control animals. Total extent of the neuropil arbors of sensory neurons from one control and two long-term-sensitized animals are shown. In each case, the rostral (row 3) to caudal (row 1) extent of the arbor is divided roughly into thirds. Each panel was produced by the superimposition of camera lucida tracings of all horse radish peroxidase–labeled processes present in 17 consecutive slab-thick sections and represents a linear segment through the ganglion of roughly 340 μm. For each composite, ventral is up, dorsal is down, lateral is to the left, and medial is to the right. By examining images across each row (rows 1, 2, and 3), the viewer is comparing similar regions of each sensory neuron. In all cases, the arbor of long-term-sensitized cells is markedly increased compared to control. (Reprinted from Bailey and Chen, 1988a, by permission.)

suggests that long-term memory resembles a process of neuronal growth and differentiation (for review, see Bailey and Kandel, 1993).

In *Aplysia*, Bailey and Chen first demonstrated that long-term behavioral sensitization is accompanied by significant structural changes. By combining selective intracellular labeling techniques with complete serial reconstructions, they have shown that the memory for long-term sensitization is accompanied by a family of alterations at identified sensory neuron synapses. These changes reflect structurally detectable modifications at two different levels of synaptic organization: (1) alterations in focal regions of membrane specialization of the synapse (the number, size, and vesicle complement of sensory neuron active zones are larger in sensitized animals than in control animals) (Bailey and Chen, 1983, 1988b) and (2) a parallel but more pronounced and widespread effect involving modulation of the total number of presynaptic varicosities per sensory neuron (Bailey and Chen, 1988a). Sensory neurons from long-term sensitized animals exhibited a twofold increase in the total number of synaptic varicosities as well as an enlargement in the size of each neuron's axonal arbor (figure 2.4).

The increase in the number of sensory neuron varicosities and the enlarged neuropil arbor observed after behavioral training for long-term sensitization can also be induced in the intact ganglion by intracellular injection of cAMP (Nazif, Byrne, and Cleary, 1991a) and can be reconstituted in dissociated sensorimotor neuron cocultures by the repeated presentation of 5-HT that evokes long-term facilitation (Glanzman, Kandel, and Schacher, 1990). In culture, this increase can be correlated with long-term (24-hour) enhancement of the amplitude of the sensory–to–motor neuron synaptic potential and depends on the presence of an appropriate target cell, similar to the synapse formation that occurs during development (Glanzman, Kandel, and Schacher, 1989). The nature of this interaction between the presynaptic cell and its target, as well as its contribution to growth changes, are not known.

TRANSIENT AND ENDURING COMPONENTS OF THE STRUCTURAL CHANGE As a first step in examining the mechanisms that might give rise to the growth of new synaptic connections during long-term sensitization, Bailey and Chen (1989) compared the time course of each class of structural changes with the behavioral duration of the memory. By examining the morphology of sensory neuron synapses at 1–2 days, 1 week, and 3 weeks after the completion of behavioral training, these researchers found that not all the structural changes persisted as long as the memory. For example, the increased size and vesicle complement of sensory neuron active zones present 24 hours after the completion of training were found to have returned to control levels when tested 1 week later. These data indicate that, insofar as modulation of active zone size and associated vesicles is 1 of the mechanisms underlying long-term sensitization, it may represent only a transient structural component of the increased synaptic responsiveness underlying the long-term process. By contrast, the duration of changes in varicosity and active zone number, which persisted unchanged for at least 1 week and were only partially reversed at the end of 3 weeks, endured in parallel with the behavioral time course of memory, suggesting that only the increases in the number of sensory neuron synapses contribute to the maintenance of long-term sensitization (figure 2.5).

EARLY GENES AND INDUCTION OF THE STRUCTURAL CHANGE Which proteins might contribute to the growth of new sensory neuron synapses? A series of recent studies has begun to explore the roles of the early proteins in inducing structural change by attempting to characterize and identify each of the 15 proteins observed by Barzilai and coworkers (1989) to be specifically altered in expression during the acquisition of long-term sensitization. Six of these 15 early proteins have now been identified. Strikingly, the two proteins that increase (clathrin and tubulin) and the four proteins that decrease their level of expression (NCAM-related cell adhesion molecules) all seem to relate to structural changes. We now consider these in turn.

Mayford and colleagues (1992) first focused on four proteins, D1–D4, the expression of which is decreased in a transcriptionally dependent manner after the application of 5-HT or cAMP. These proteins range in size from 100 kD to 140 kD, have similar peptide maps, and cross-react selectively and specifically with two monoclonal antibodies (mAb) raised by Keller and Schacher (1990) against proteins in the neuropil of *Aplysia*. Using these mAbs, Mayford and coworkers (1992) cloned and sequenced cDNAs for these proteins and found that they encoded different isoforms of an immunoglobulin-

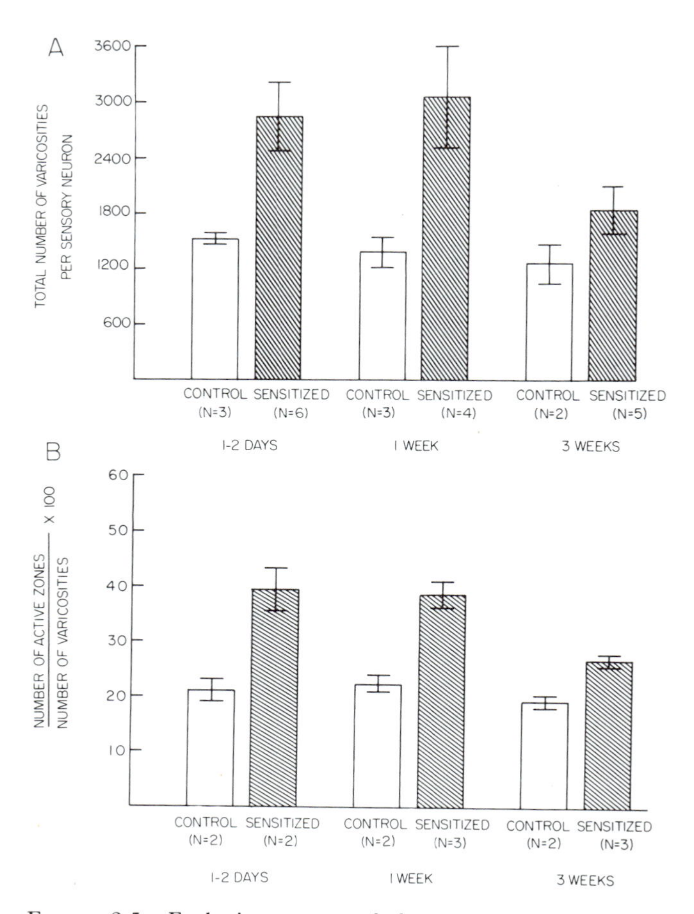

FIGURE 2.5 Enduring structural changes at sensory neuron synapses during long-term sensitization. (A) Duration of changes in the total number of varicosities per sensory neuron. (B) Duration of changes in the incidence of sensory neuron active zones. (Reprinted from Bailey and Chen, 1989, by permission.)

related cell adhesion molecule, designated apCAM, which shows greatest homology to NCAM in vertebrates and fasciclin II in *Drosophila.* Imaging of fluorescently labeled mAbs to apCAM shows that not only is there a decrease in the level of expression but that even preexisting protein is lost from the surface membrane of the sensory neurons within 1 hour after the addition of 5-HT. This transient modulation by 5-HT of cell adhesion molecules, therefore, may represent one of the early molecular steps required for initiating learning-related growth of synaptic connections. Indeed, blocking the expression of the antigen by mAb causes defasciculation, a step that appears to precede normal synapse formation in *Aplysia* (Glanzman, Kandel, and Schacher, 1989; Keller and Schacher, 1990).

SEROTONIN INITIATES THE INTERNALIZATION OF APCAMS BY A COORDINATED PROGRAM OF CLATHRIN-MEDIATED ENDOCYTOSIS What are the mechanisms by which 5-HT modulates apCAM, and what significance does this modulation have for the structural changes induced by 5-HT? Bailey and associates (1992a) addressed these questions by combining thin-section electron microscopy with immunolabeling using a gold-conjugated mAb specific to apCAM. They found that within 1 hour of its application, 5-HT led to a 50% decrease in the density of gold-labeled apCAM complexes at the surface membrane of the sensory neuron (figure 2.6). This down-regulation was particularly prominent at sites where the processes of the sensory neurons contact one another and was achieved by a heterologous, protein synthesis–dependent activation of the endosomal pathway, which led to internalization and apparent degradation of apCAM. As is the case for the down-regulation at the level of expression, the 5-HT-induced internalization of apCAM can be simulated by cAMP (Bailey, Chen, and Kandel, 1992). Concomitant with the down-regulation of apCAM, Hu and colleagues (1993) have recently demonstrated that, as part of this coordinated program for endocytosis, 5-HT and cAMP also induce in the sensory neurons increased formation of coated pits and coated vesicles and increased expression of the light chain of clathrin (apClathrin) (figure 2.7). Because the apClathrin light chain contains all the important functional domains of both LCa and LCb of mammalian clathrin believed to be essential for coated pit assembly and disassembly (Brodsky et al., 1991), the increase in clathrin may be an important component in activating the endocytic cycle required for the internalization of apCAM.

The ability of 5-HT to modify the structure of the surface and internal membrane systems of sensory neurons in *Aplysia*, by initiating a rapid, protein synthesis–dependent sequence of steps, bears a striking similarity to the ruffling of the cell surface and membrane remodeling induced in nonneuronal systems by epidermal growth factor and other well-characterized growth factors (Bretscher, 1989; Dadabay et al., 1991) or by nerve growth factor in PC12 cells (Connolly, Grun, and Grun, 1984). These similarities suggest that modulatory transmitters important for learning, such as serotonin, may serve a double function. In addition to producing transient regulation of the excitability of neurons, repeated or prolonged exposure of a modulatory transmitter can also produce an action comparable to that

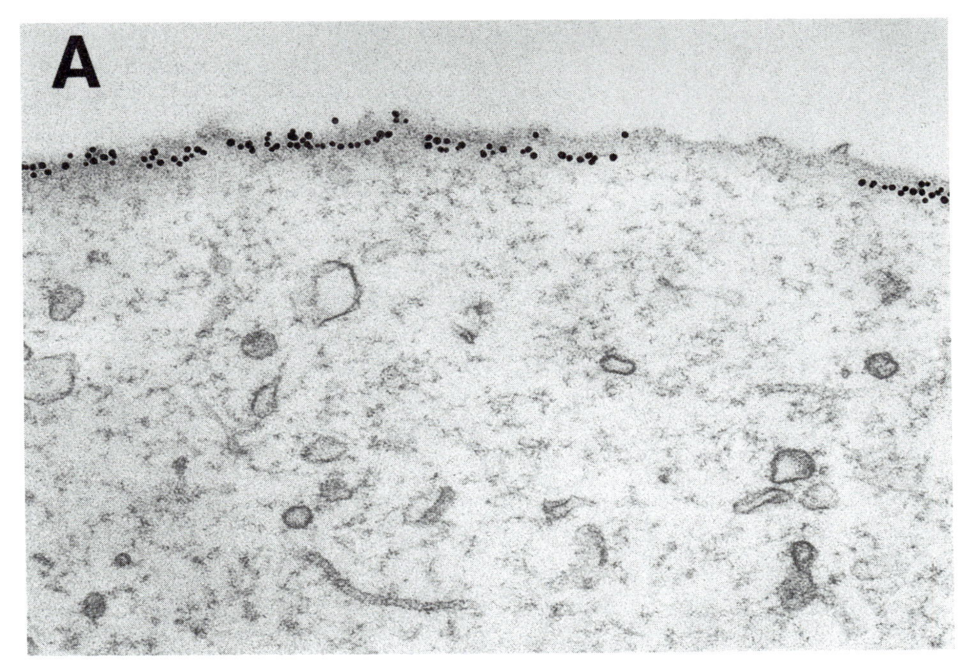

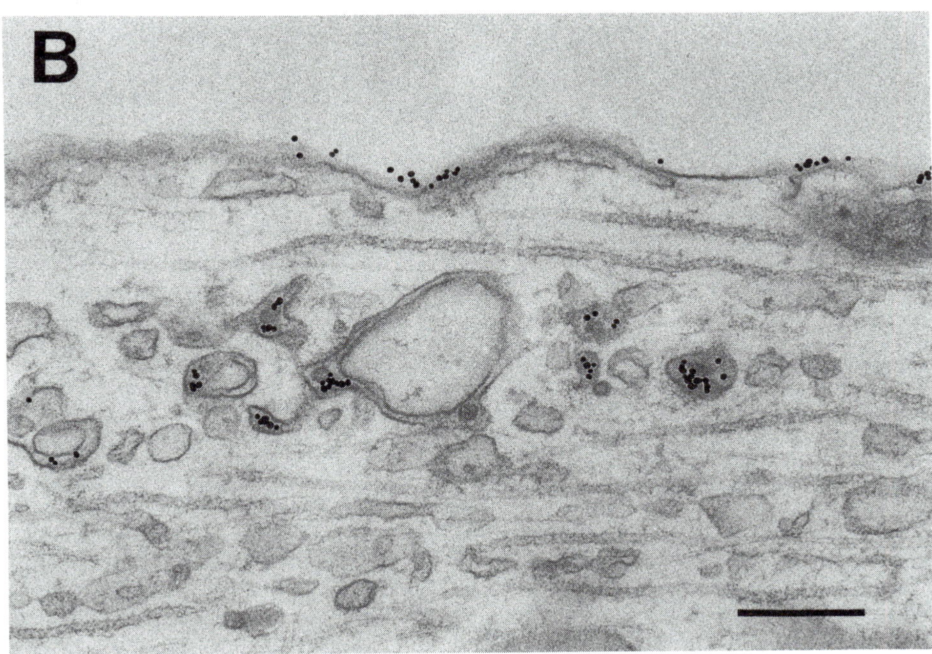

FIGURE 2.6 5-HT-induced down-regulation of apCAM at the sensory neuron membrane. (A) Prior to 5-HT treatment, the surface membrane has a uniform, linear appearance. Gold-conjugated mAbs to apCAM decorate the plasma membrane, and little gold is found inside the cell where the cytoplasm has a normal complement of cytoskeletal elements and membrane-bound vesicular profiles. (B) Three prominent changes in cellular architecture induced by a 1-hour exposure to 5-HT. First, there is an apparent ruffling of the cell surface as the plasma membrane now appears to be thrown up into a series of undulating contours. Second, there is activation of the endosomal pathway leading to a striking reorganization of the cytoplasm. 5-HT recruits a variety of polymorphic vesicular and tubular endosomal profiles. Finally, there is a significant redistribution of apCAM, resulting in a decrease in the number of gold complexes at the surface membrane and a concomitant increase in its internalization into a variety of endosomal subcomponents. Scale, 0.25 μm. (Reprinted from Bailey et al., 1992a, by permission.)

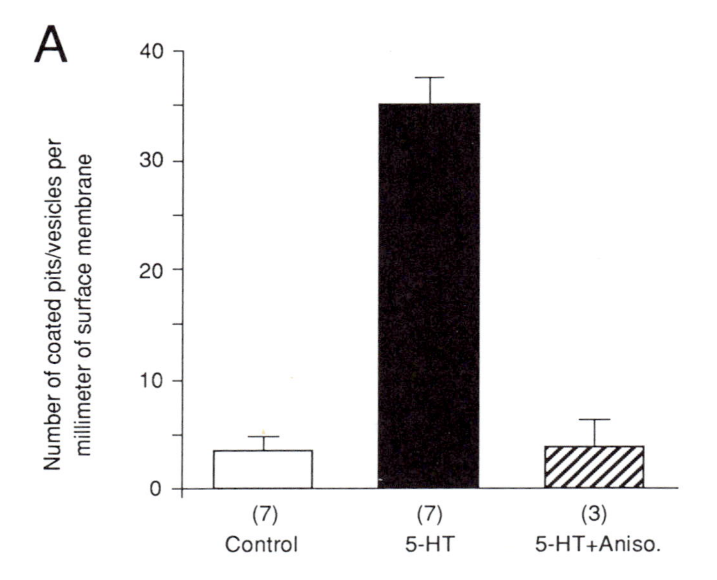

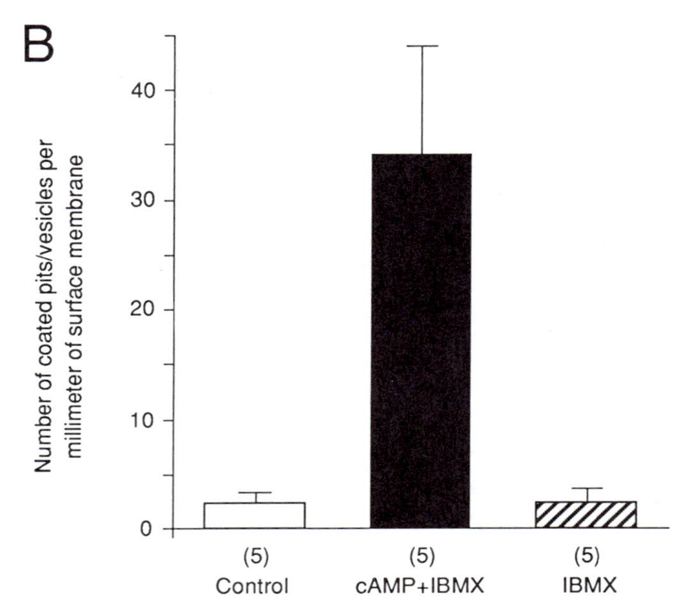

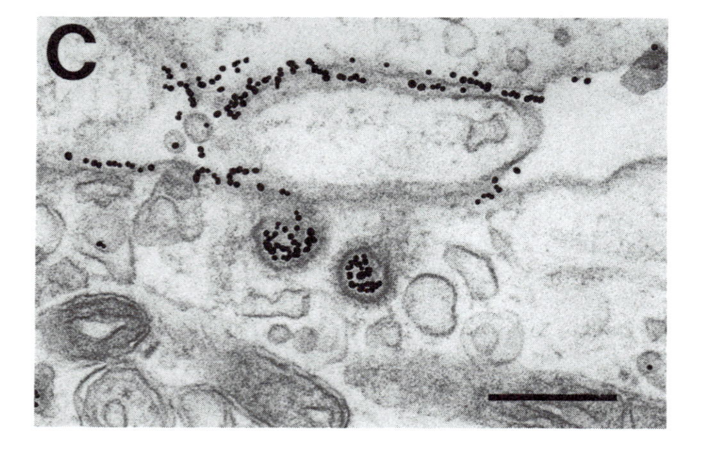

FIGURE 2.7 5-HT-induced activation of clathrin-mediated endocytic pathway in the sensory neuron. Exposure of sensory neurons to either 5-HT (A) or permeable analogs of cAMP (B) produce a 10-fold increase in the density of coated pits and coated vesicles compared to untreated control cells. This increase can be blocked by 10 μmol anisomycin (aniso.). IBMX, 3-isobutyl-1-methyl-xanthine. For (A) and (B), each bar represents the mean $\pm$ SEM. (C) The clustering of coated vesicles depicted in this micrograph is characteristic of 5-HT or cAMP-treated cells and may reflect an earlier synchronous burst of endocytic activity at the surface membrane. Note the heavy concentration of gold-conjugated mAbs to apCAM present in the two coated vesicles. This removal of cell adhesion molecules from patches of apparent adhesion may facilitate defasciculation of the axonal processes of the sensory neurons and is believed to represent one of the early steps in the structural program that underlies long-term facilitation. Scale, 0.25 μm. (Reprinted from Hu et al., 1993, by permission.)

of a growth factor, resulting in more persistent regulation of the architecture of the neuron.

Based on these findings, Bailey and colleagues (1992a) have suggested that the 5-HT–induced internalization of apCAM and consequent endocytically activated membrane remodeling may represent the first morphological steps in the structural program underlying long-term facilitation. According to this view, learning-related synapse formation is preceded by and perhaps requires endocytic activation, which can then serve a double function. First, the removal of cell adhesion molecules from the neuronal surface at sites of apposition may destabilize adhesive contacts and facilitate defasciculation, a process that may be important in disassembly. Second, the massive endocytic activation might lead to a redistribution of membrane components that favors synapse formation. The assembly of membrane components required for initial synaptic growth may involve insertion, by means of targeted exocytosis, of endocytic membrane retrieved from sites of adhesion and recycled to sites of new synapse formation. Synapse formation may require, in addition, the recruitment of new transport vesicles from the trans-Golgi network.

LATE GENES AND MAINTENANCE OF THE STRUCTURAL CHANGE The finding that both the behavioral and cellular changes of long-term sensitization require altered gene expression raises the question of whether the structural changes associated with the long-term process also depend on new macromolecular synthesis. This question has been addressed directly by examining in parallel the effects of inhibitors of protein and RNA synthesis on both the structural and functional changes that accompany long-term facilitation. The long-term functional changes in synaptic strength evoked by 5-HT are dependent on new macromolecular synthesis during the period of transmitter application (Montarolo et al., 1986). Utilizing sensorimotor cocultures, Bailey and colleagues (1992b) have shown that anisomycin and actinomycin D block the long-lasting increases produced by 5-HT in both synaptic potential and varicosity number. These results indicate that macromolecular synthesis is required for the expression of the long-lasting structural changes in the sensory cell and that this synthesis is coupled with the long-term functional modulation of sensory-to-motor neuron synapses. A similar dependence on new protein synthesis has been reported for the in vivo structural changes in pleural sensory neurons, where the cAMP-mediated increase in varicosity number can also be blocked by anisomycin (Nazif, Byrne, and Cleary, 1991b).

To establish a more precise correlation between the inhibition of macromolecular synthesis and the changes in synaptic structure that accompany long-term memory requires a better understanding of the learning-induced gene products and how their altered expression affects the phenotypical properties of the participating neurons. Recent progress along this line has been made by Kennedy and colleagues (Kennedy et al., 1988a,b, 1989), who have now cloned and identified two late proteins whose expression increases 24 hours after both behavioral training (Castellucci et al., 1988) and 5-HT application to isolated primary cultures of sensory neurons (Barzilai et al., 1989). One of these proteins is the *Aplysia* homolog of GRP78/BiP (Kennedy et al., 1989; Kuhl et al., 1992), which is believed to chaperone the correct folding and assembly of secretory and transmembrane proteins in the lumen of the endoplasmic reticulum (Pelham, 1988). The specific increase in *Aplysia* BiP expression occurs 3 hours after 5-HT application (Kennedy et al., 1989) and is coincident with the 3-hour increase in overall protein synthesis observed by Barzilai and associates (1989). This change in the specific expression of *Aplysia* BiP may be a response by the sensory neuron to meet the demands of the 3-hour general increase in protein synthesis. Furthermore, the increase in general protein synthesis may, in part, underlie the structural growth-related change in the sensory–to–motor neuron connections observed during the maintenance phase of long-term sensitization. Parallel findings by Brostrum and coworkers (1987) have shown that the application of cAMP or phorbol esters to GH3 cells, which induce an increase in total protein synthesis, similarly induces a coincident increase in BiP synthesis. This suggests that modulation of BiP expression may be a general response of cells to an increased posttranslational load on the endoplasmic reticulum due to increased protein synthesis.

Toward a working molecular model of learning-related synapse formation

Figure 2.8 illustrates a model of putative macromolecular events underlying long-term sensitization in an *Aplysia* sensory neuron. In this model, 5-HT, a modulatory transmitter released by facilitating interneurons that synapse presynaptically on the sensory neuron, acts to initiate separate memory processes with different durations. Short-term memory, which has a time course of minutes to hours, involves the covalent modification of pre-existing proteins. 5-HT binds to its transmembrane receptor, activating adenyl cyclase to generate cAMP and activate the A kinase. These transmembrane and cytoplasmic signaling mechanisms transduce the extracellular signal and initiate the modification of target proteins such as closure of the 5-HT-sensitive potassium channel. The changes are independent of new macromolecular synthesis, and their duration determines the time course of short-term memory retention.

In contrast to the short-term effects, repeated activation of the facilitatory interneurons triggers the induction of long-term sensitization. Unlike the covalent modification mechanisms, the acquisition of long-term memory that lasts more than 1 day depends, in this model, on the induction of new proteins (see figure 2.8, filled rectangle and triangle). This induction is initiated by the same second messengers involved in short-term memory that now modify transacting regulators of gene expression such as CREBP (see figure 2.8, open circles). These transcription factors initiate a cascade of regulated and effector gene expression. Early effectors may include the induction of a protease specific for the regulatory subunit of the A kinase, thereby generating persistent kinase activity, or clathrin light chains that may be involved in the removal of apCAM from the cell surface through activation of the endocytic pathway and, thus, may contribute to initial stages of the struc-

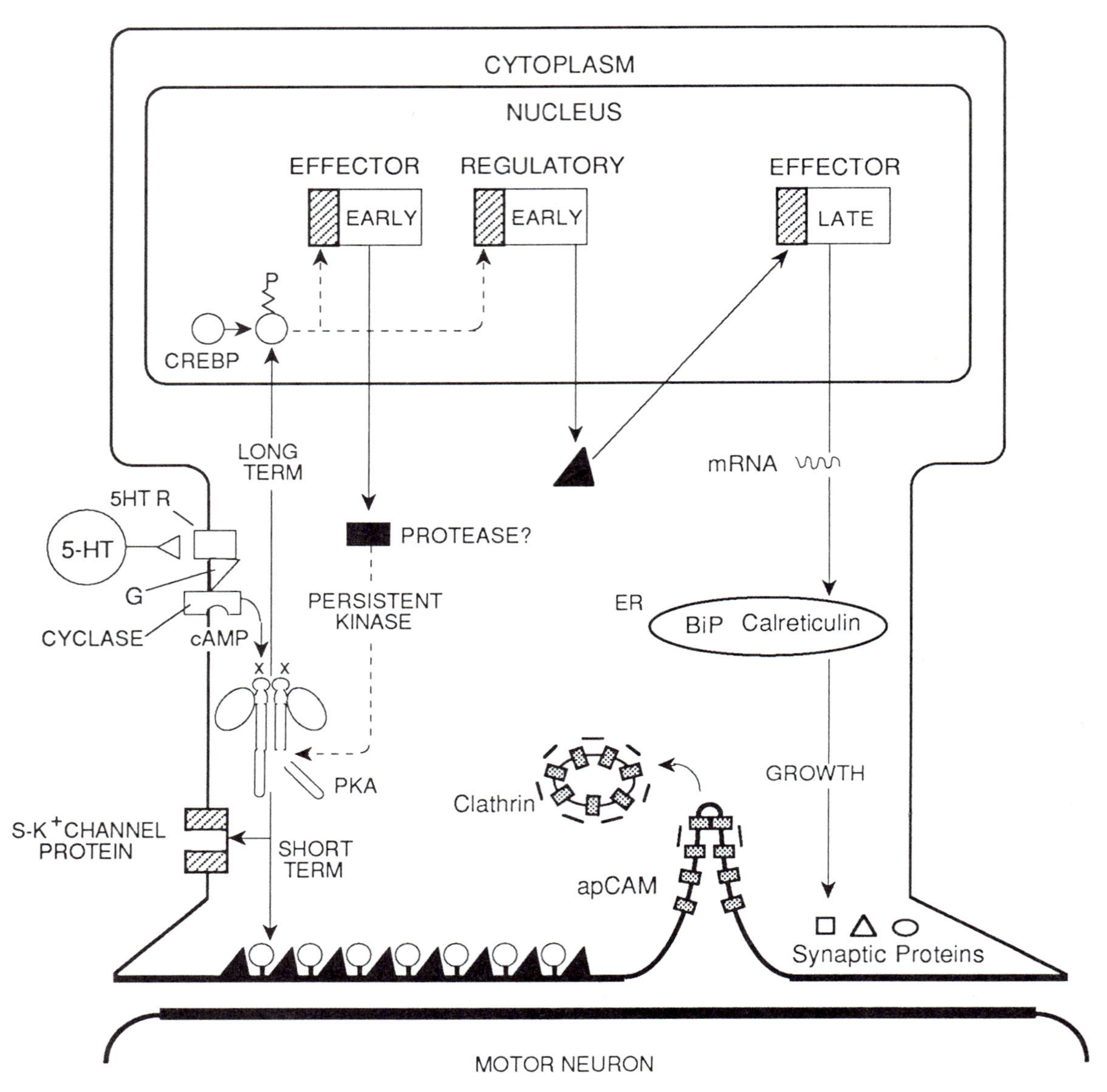

FIGURE 2.8 Model of the molecular and structural mechanisms underlying long-term sensitization (Reprinted from Kennedy, Hawkins, and Kandel, 1992, by permission.)

tural program involved in the growth of new synaptic connections. The induced regulatory genes may, in turn, initiate further rounds of transcriptional activation, generating a cascade of sequential gene expression affecting genes encoding proteins such as BiP, which may reflect a general response designed to meet the posttranslational demands of increased protein synthesis, as well as structural proteins necessary for the construction of the new synaptic arbors associated with long-term sensitization. Memory lasting hours is retained by the half-life of the effector proteins or by functional modifications, such as phosphorylation, of these proteins. Certain of the early effector proteins may also serve to reinforce and maintain the initial response —for example, by proteolytic activation of a protein kinase. Memory lasting days, weeks, or months (longer than the half-life of effector proteins) is initiated by the early regulatory genes whose protein products trigger the maintained expression of late effector genes, which may contribute to and stabilize the growth of new synaptic connections coincident with the maintenance phase of long-term memory.

Conclusions

Aspects of the neuronal changes that underlie memory storage are represented at the level of individual synapses. Short-term memory involves only the covalent modification of pre-existing proteins and an alteration in the strength of preexisting connections. By contrast, long-term memory is associated with an alteration in gene expression and the growth of new synaptic connections. Despite the association of morphological changes with different forms of learning and memory, surprisingly little is known about the mechanisms that underlie this behaviorally relevant structural plasticity. Recent molecular studies of the learning-related synaptic growth that accompanies long-term sensitization in *Aplysia* have begun to address this issue.

These studies indicate that long-term memory involves the flow of information from membrane receptors to the genome, as seen in other processes of cellular differentiation and growth. Such changes could reflect the recruitment by environmental stimuli of developmental processes that are latent or inhibited in the fully differentiated neuron. Indeed, recent evidence suggests that the growth changes accompanying long-term memory storage share features in common with synapse formation during development.

First, in both cases, the structural change exhibits a requirement for new protein and mRNA synthesis. These alterations in transcriptional and translational state can be initiated in the long-term process by the repeated or prolonged exposure to modulatory transmitters that, in this respect, appear to mimic the effects of growth factors and hormones during the cell cycle and differentiation. Thus, modulatory transmitters important for learning activate not only the cytoplasmic second-messenger cascades required for the short-term process but also a nuclear messenger system by which the transmitter can exert long-term regulation over the excitability and architecture of the neuron through changes in gene expression.

Second, the structural change is associated with a rapid and transient modulation of NCAM-related cell adhesion molecules that are present during the early development of the nervous system, where they appear to have a role in cell migration and process outgrowth. In the adult, the *Aplysia* NCAM seems normally to exert an inhibitory constraint on growth. As we have here described, by their ability to engage second-messenger systems that regulate gene expression, modulatory transmitters utilized for learning can initiate synaptic growth resembling the programs of cellular differentiation normally associated with the actions of growth factors. These cellular programs can include alterations in the display of cell surface adhesion molecules. Thus, aspects of the regulatory mechanisms underlying learning-related synaptic growth may eventually be understood in the context of the basic cell and molecular processes that govern synapse formation during development.

ACKNOWLEDGMENTS This work was supported in part by National Institutes of Health grants MH37134 and GM32099 to C.H.B. and by a grant from the Howard Hughes Medical Institute to E.R.K.

REFERENCES

Agranoff, B. W., 1972. *The Chemistry of Mood, Motivation and Memory*. New York: Plenum Press.

Alkon, D. L., 1984. Calcium-mediated reduction of ionic currents: A biophysical memory trace. *Science* 226:1037–1045.

Atkinson, R. C., and R. M. Shiffrin, 1968. Human memory: A proposed system and its control processes. In *The Psychology of Learning and Motivation*, vol. 2, K. W. Spence and J. T. Spence, eds. New York: Academic Press, p. 89.

Bacskai, B. J., B. Hochner, M. Mahaut-Smith, S. R. Adams, B.-K. Kaang, E. R. Kandel, and R. Tsien, 1993. Spatially resolved dynamics of cAMP and protein kinase A subunits in *Aplysia* sensory neuron. *Science* 260:222–226.

Bailey, C. H., and M. Chen, 1983. Morphological basis of long-term habituation and sensitization in *Aplysia*. *Science* 220:91–93.

Bailey, C. H., and M. Chen, 1988a. Long-term memory in *Aplysia* modulates the total number of varicosities of single identified sensory neurons. *Proc. Natl. Acad. Sci. U.S.A.* 85:2373–2377.

Bailey, C. H., and M. Chen, 1988b. Long-term sensitization in *Aplysia* increases the number of presynaptic contacts onto the identified gill motor neuron L7. *Proc. Natl. Acad. Sci. U.S.A.* 85:9356–9359.

Bailey, C. H., and M. Chen, 1988c. Morphological basis of short-term habituation in *Aplysia*. *J. Neurosci.* 8:2452–2459.

Bailey, C. H., and M. Chen, 1989. Time course of structural changes at identified sensory neuron synapses during long-term sensitization in *Aplysia*. *J. Neurosci.* 9:1774–1780.

Bailey, C. H., M. Chen, and E. R. Kandel, 1992. Early steps in learning-related synaptic growth: cAMP simulates the heterologous endocytosis of apCAMS induced by 5-HT in sensory neurons of *Aplysia*. *Soc. Neurosci. Abstr.* 18:941.

Bailey, C. H., M. Chen, F. Keller, and E. R. Kandel, 1992a. Serotonin-mediated endocytosis of apCAM: An early step of learning-related synaptic growth in *Aplysia*. *Science* 256:645–649.

Bailey, C. H., and E. R. Kandel, 1993. Structural changes accompanying memory storage. *Annu. Rev. Physiol.* 55:397–426.

Bailey, C. H., P. G. Montarolo, M. Chen, E. R. Kandel, and S. Schacher, 1992b. Inhibitors of protein and RNA synthesis block the structural changes that accompany long-term heterosynaptic plasticity in the sensory neurons of *Aplysia*. *Neuron* 9:749–758.

Barondes, S. H., 1975. Protein synthesis dependent and protein synthesis independent memory storage processes. In *Short-Term Memory*, D. Deutsch and J. A. Deutsch, eds. New York: Academic, pp. 379–390.

Barzilai, A., T. E. Kennedy, J. D. Sweatt, and E. R. Kandel, 1989. 5-HT modulates protein synthesis and the expression of specific proteins during long-term facilitation in *Aplysia* sensory neurons. *Neuron* 2:1577–1586.

Bergold, P. J., S. A. Beushausen, T. C. Sacktor, S. Cheley, H. Bayley, and J. H. Schwartz, 1992. A regulatory subunit of the cAMP-dependent protein kinase downregulated in *Aplysia* sensory neurons during long-term sensitization. *Neuron* 8:387–397.

Bergold, P. J., J. D. Sweatt, E. R. Kandel, and J. H. Schwartz, 1990. Protein synthesis during acquisition of long-term facilitation is needed for the persistent loss of regulatory subunits of the *Aplysia* cAMP-dependent protein kinase. *Proc. Natl. Acad. Sci. U.S.A.* 87:3788–3791.

Bernier, L., V. F. Castellucci, E. R. Kandel, and J. H. Schwartz, 1982. Facilitatory transmitter causes a selective and prolonged increase in adenosine 3′:5′-monophosphate in sensory neurons mediating the gill and siphon withdrawal reflex in *Aplysia*. *J. Neurosci.* 2:1682–1691.

Bretscher, A., 1989. Rapid phosphorylation and reorganization of ezrin and spectrin accompany morphological changes induced in A-431 cells by epidermal growth factor. *J. Cell. Biol.* 108:921–930.

Brodsky, F. M., B. L. Hill, S. L. Acton, I. Nathke, D. H. Wong, S. Ponnambalam, and P. Parham, 1991. Clathrin light chains: Arrays of protein motifs that regulate coated-vesicle dynamics. *Trends Biochem. Sci.* 16:208–213.

Brostrom, M. A., K.-V. Chin, C. Cade, D. Gmitter, and C. O. Brostrom, 1987. Stimulation of protein synthesis in pituitary cells by phorbol esters and cyclic AMP. *J. Biol. Chem.* 262:16515–16523.

Byrne, J. H., 1987. Cellular analysis of associative learning. *Physiol. Rev.* 67:329–439.

Byrne, J. H., V. Castellucci, and E. R. Kandel, 1974. Receptive fields and response properties of mechanoreceptor neurons innervating skin and mantle shelf of *Aplysia*. *J. Neurophysiol.* 37:1041–1064.

Byrne, J. H., V. Castellucci, and E. R. Kandel, 1978. Contribution of individual mechanoreceptor sensory neurons to defensive gill-withdrawal reflex in *Aplysia*. *J. Neurophysiol.* 41:418–431.

Carew, T. J., and C. L. Sahley, 1986. Invertebrate learning and memory: From behavior to molecules. *Annu. Rev. Neurosci.* 9:435–487.

Castellucci, V. F., H. Blumenfeld, P. Goelet, and E. R. Kandel, 1989. Inhibitor of protein synthesis blocks long-term behavioral sensitization in the isolated gill-withdrawal reflex of *Aplysia*. *J. Neurobiol.* 20:1–9.

Castellucci, V. F., E. R. Kandel, J. H. Schwartz, F. D. Wilson, A. C. Nairn, and P. Greengard, 1980. Intracellular injection of the catalytic subunit of cyclic AMP–dependent protein kinase simulates facilitation of transmitter release underlying behavioral sensitization in *Aplysia*. *Proc. Natl. Acad. Sci. U.S.A.* 77:7492–7496.

Castellucci, V. F., T. E. Kennedy, E. R. Kandel, and P. Goelet, 1988. A quantitative analysis of 2-D gels identifies proteins in which labeling is increased following long-term sensitization in *Aplysia*. *Neuron* 1:321–328.

Castellucci, V. F., A. Nairn, P. Greengard, J. H. Schwartz, and E. R. Kandel, 1982. Inhibitor of adenosine 3′:5′-monophosphate-dependent protein kinase blocks presynaptic facilitation in *Aplysia*. *J. Neurosci.* 2:1673–1681.

Castellucci, V. F., H. Pinsker, I. Kupfermann, and E. R. Kandel, 1970. Neuronal mechanisms of habituation and dishabituation of the gill-withdrawal reflex in *Aplysia*. *Science* 167:1745–1748.

Cole, A. J., D. W. Saffen, J. M. Barban, and P. F. Worley, 1989. Rapid increase of an immediate early gene messenger RNA in hippocampal neurons by synaptic NMDA receptor activation. *Nature* 340:474–476.

Connolly, J. L., S. A. Green, and L. A. Greene, 1984. Comparison of rapid changes in surface morphology and coated pit formation of PC12 cells in response to nerve growth factor, epidermal growth factor and dibutyryl cyclic AMP. *J. Cell Biol.* 98:457–465.

Craik, F. I. M., and R. S. Lockhart, 1972. Levels of processing: A framework for memory research. *J. Verbal Learn. Verbal Behav.* 11:671–684.

Dadabay, C. Y., E. Patton, J. A. Cooper, and L. J. Pike, 1991. Lack of correlation between changes in polyphosphoinositide levels and actin/gelsolin complexes in A431 cells treated with epidermal growth factor. *J. Cell Biol.* 112:1151–1156.

Dale, N., E. R. Kandel, and S. Schacher, 1987. Serotonin produces long-term changes in the excitability of *Aplysia* sensory neurons in culture that depend on new protein synthesis. *J. Neurosci.* 7:2232–2238.

Dale, N., S. Schacher, and E. R. Kandel, 1988. Long-term facilitation in *Aplysia* involves increases in transmitter release. *Science* 239:282–285.

Dash, P. K., B. Hochner, and E. R. Kandel, 1990. Injection of the cAMP-responsive element into the nucleus of Aplysia sensory neurons blocks long-term facilitation. *Nature* 345:718–721.

Davis, H. P., and L. R. Squire, 1984. Protein synthesis and memory: A review. *Psychol. Bull.* 96:518–559.

Ebbinghaus, H., 1963. *Memory: A Contribution to Experimental Psychology*. New York: Dover. (Original work published 1855.)

Flexner, J. B., L. B. Flexner, and E. Stellar, 1983. Memory in mice as affected by intracerebral puromycin. *Science* 141:57–59.

Frost, W. N., V. F. Castellucci, R. D. Hawkins, and E. R.

Kandel, 1985. Monosynaptic connections from the sensory neurons of the gill- and siphon-withdrawal reflex in *Aplysia* participate in the storage of long-term memory for sensitization. *Proc. Natl. Acad. Sci. U.S.A.* 82:8266–8269.

Ghirardi, M., O. Braha, B. Hochner, P. G. Montarolo, E. R. Kandel, and N. E. Dale, 1992. The contributions of PKA and PKC to the presynaptic facilitation of evoked and spontaneous transmitter release at depressed and nondepressed synapses in sensory neurons of *Aplysia*. *Neuron* 9:479–489.

Glanzman, D. L., E. R. Kandel, and S. Schacher, 1989. Identified target motor neuron regulates neurite outgrowth and synapse formation of *Aplysia* sensory neurons in vitro. *Neuron* 3:441–450.

Glanzman, D. L., E. R. Kandel, and S. Schacher, 1990. Target-dependent structural changes accompanying long-term synaptic facilitation in *Aplysia* neurons. *Science* 249: 799–802.

Glanzman, D. L., S. L. Mackey, R. D. Hawkins, A. M. Dyke, P. E. Lloyd, and E. R. Kandel, 1989. Depletion of serotonin in the nervous system of *Aplysia* reduces the behavioral enhancement of gill withdrawal as well as the heterosynaptic facilitation produced by tail shock. *J. Neurosci.* 9:4200–4213.

Goldsmith, B. A., and T. W. Abrams, 1991. Reversal of synaptic depression by serotonin at *Aplysia* sensory neuron synapses involves activation of adenylyl cyclase. *Proc. Natl. Acad. Sci. U.S.A.* 88:9021–9025.

Greenberg, S. M., V. F. Castellucci, H. Bayley, and J. H. Schwartz, 1987. A molecular mechanism for long-term sensitization in *Aplysia*. *Nature* 329:62–65.

Hawkins, R. D., V. F. Castellucci, and E. R. Kandel, 1981. Interneurons involved in mediation and modulation of gill-withdrawal reflex in *Aplysia*: II. Identified neurons produce heterosynaptic facilitation contributing to behavioral sensitization. *J. Neurophysiol.* 45:315–326.

Hegde, A. N., C. Chain, and J. H. Schwartz, 1992. Long-term memory in *Aplysia*: Molecular signals for the ubiquitin-mediated proteolysis of regulatory subunits of the cAMP-dependent protein kinase. *Soc. Neurosci. Abstr.* 18:810.

Hochner, B., S. Schacher, and E. R. Kandel, 1986. Action-potential duration and the modulation of transmitter release from the sensory neurons of *Aplysia* in presynaptic facilitation and behavioral sensitization. *Proc. Natl. Acad. Sci. U.S.A.* 83:8410–8414.

Hu, Y., A. Barzilai, M. Chen, C. H. Bailey, and E. R. Kandel, 1993. 5-HT and cAMP induce the formation of coated pits and vesicles and increase the expression of clathrin light chains in sensory neurons of *Aplysia*. *Neuron* 10:921–929.

James, W., 1890. *The Principles of Psychology* (vols. 1–2). New York: Holt.

Kaang, B.-K., E. R. Kandel, and S. G. N. Grant, 1993. Activation of cAMP-responsive genes by stimuli that produce long-term facilitation in *Aplysia* sensory neurons. *Neuron* 10:427–435.

Kaang, B.-K., P. J. Pfaffinger, S. G. N. Grant, E. R. Kandel, and Y. Furukawa, 1992. Overexpression of an *Aplysia* shaker K^+ channel gene modifies the electrical properties and synaptic efficacy of identified *Aplysia* neurons. *Proc. Natl. Acad. Sci U.S.A.* 89:1133–1137.

Kandel, E. R., 1976. *Cellular Basis of Behavior: An Introduction to Behavioral Neurobiology*. San Francisco: Freeman.

Kandel, E. R., and J. H. Schwartz, 1982. Molecular biology of an elementary form of learning: Modulation of transmitter release by cyclic AMP. *Science* 218:433–443.

Keller, F., and S. Schacher, 1990. Neuron-specific membrane glycoproteins promoting neurite fasciculation in *Aplysia californica*. *J. Cell Biol.* 111:2637–2650.

Kennedy, T. E., M. A. Gawinowicz, A. Barzilai, E. R. Kandel, and J. D. Sweatt, 1988a. Sequencing of proteins from two-dimensional gels using in situ digestion and transfer of peptides to polyvinylidene difluoride membranes: Application to proteins associated with sensitization in *Aplysia*. *Proc. Natl. Acad. Sci. U.S.A.* 85:7008–7012.

Kennedy, T. E., R. D. Hawkins, and E. R. Kandel, 1992. Molecular interrelationships between short- and long-term memory. In *Neuropsychology of Memory*, L. R. Squire and N. Butters, eds. New York: Guilford Press.

Kennedy, T. E., D. Kuhl, A. Barzilai, E. R. Kandel, and J. D. Sweatt, 1989. Characterization of changes in late protein and mRNA expression during the maintenance phase of long-term sensitization. *Soc. Neurosci. Abstr.* 15: 1117.

Kennedy, T. E., K. Wager-Smith, A. Barzilai, E. R. Kandel, and J. D. Sweatt, 1988b. Sequencing proteins from acrylamide gels. *Nature* 336:499–500.

Klein, M., and E. R. Kandel, 1980. Mechanism of calcium current modulation underlying presynaptic facilitation and behavioral sensitization in *Aplysia*. *Proc. Natl. Acad. Sci. U.S.A.* 77:6912–6916.

Kuhl, D., T. E. Kennedy, A. Barzilai, and E. R. Kandel, 1992. Long-term sensitization training in *Aplysia* leads to an increase in the expression of BiP, the major protein chaperone of the ER. *J. Cell Biol.* 119:1069–1076.

Mackey, S. L., E. R. Kandel, and R. D. Hawkins, 1989. Identified serotonergic neurons LCB1 and RCB1 in the cerebral ganglia of *Aplysia* produce presynaptic facilitation of siphon sensory neurons. *J. Neurosci.* 9:4227–4235.

Mayford, M., A. Barzilai, F. Keller, S. Schacher, and E. R. Kandel, 1992. Modulation of an NCAM-related adhesion molecule with long-term synaptic plasticity in *Aplysia*. *Science* 256:638–644.

McGaugh, J. L., 1966. Time-dependent processes in memory storage. *Science* 153:1351.

Milner, B., S. Corkin, and H. L. Teuber, 1968. Further analysis of the hippocampal amnestic syndrome: 14 year follow-up of H. M. *Neuropsychologia* 6:215–234.

Montarolo, P. G., P. Goelet, V. F. Castellucci, J. Morgan, E. R. Kandel, and S. Schacher, 1986. A critical period for macromolecular synthesis in long-term heterosynaptic facilitation in *Aplysia*. *Science* 234:1249–1254.

Montminy, M. R., K. A. Sevarino, J. A. Wagner, G. Mandel, and R. H. Goodman, 1986. Identification of a cyclic-

AMP-responsive element within the rat somatostatin gene. *Proc. Natl. Acad. Sci. U.S.A.* 83:6682–6686.

NAZIF, F. A., J. H. BYRNE, and L. J. CLEARY, 1991a. cAMP induces long-term morphological changes in sensory neurons of *Aplysia*. *Brain Res.* 539:324–327.

NAZIF, F. A., J. H. BYRNE, and L. J. CLEARY, 1991b. Long-term (24 hr) morphological changes induced by cAMP in pleural sensory neurons of *Aplysia* are dependent on de novo protein synthesis. *Soc. Neurosci. Abstr.* 17:1589.

PELHAM, H., 1988. Coming in from the cold. *Nature* 322:776–777.

PINSKER, H. M., W. A. HENING, T. J. CAREW, and E. R. KANDEL, 1973. Long-term sensitization of a defensive withdrawal reflex in *Aplysia*. *Science* 182:1039–1042.

QIAN, Z., M. E. GILBERT, M. A. COLICOS, E. R. KANDEL, and D. KUHL, 1993. Differential screening reveals that tissue-plasminogen activator is induced as an immediate early gene during seizure, kindling, and LTP. *Nature* 361: 453–457.

RAYPORT, S. G., and S. SCHACHER, 1986. Synaptic plasticity in vitro: Cell culture of identified *Aplysia* neurons mediating short-term habituation and sensitization. *J. Neurosci.* 6: 759–763.

SCHOLZ, K. P., and J. H. BYRNE, 1987. Long-term sensitization in *Aplysia*: Biophysical correlates in tail sensory neurons. *Science* 235:685–687.

SCHWARTZ, J. H., V. F. CASTELLUCCI, and E. R. KANDEL, 1971. Functions of identified neurons and synapses in abdominal ganglion of *Aplysia* in absence of protein synthesis. *J. Neurophysiol.* 34:939–953.

SORRENTINO, V., 1989. Growth factors, growth inhibitors and cell cycle control. *Anticancer Res.* 9:1925.

SQUIRE, L. R., 1992. Memory and the hippocampus: A synthesis from findings with rats, monkeys, and humans. *Psychol. Rev.* 99:195–231.

SQUIRE, L. R., and S. ZOLA-MORGAN, 1988. Memory: Brain systems and behavior. *Trends Neurosci.* 11:170–175.

SWEATT, J. D., and E. R. KANDEL, 1989. Persistent and transcriptionally-dependent increase in protein phosphorylation upon long-term facilitation of *Aplysia* sensory neurons. *Nature* 339:51–54.

THORPE, W. H., 1956. *Learning and Instincts in Animals*. Cambridge, Mass.: Harvard University.

TULVING, E., 1991. Concepts of human memory. In Memory: Organization and Locus of Change, L. R. Squire, N. M. Weinberger, G. Lynch, and J. L. McGaugh, eds. New York: Oxford University, pp. 3–32.

WEISKRANTZ, L., 1970. A long-term view of short-term memory in psychology. In *Short-Term Changes in Neural Activity and Behavior*, G. Horn and R. A. Hinde, eds. Cambridge.: Cambridge University Press.

WICKELGREN, W. A., 1973. The long and the short of memory. *Psychol. Bull.* 80:425–438.

WISDEN, W., M. L. ERRINGTON, S. WILLIAMS, S. B. DUNNETT, C. WATERS, D. HITCHCOCK, G. EVAN, T. V. P. BLISS, and S. P. HUNT, 1990. Differential expression of immediate early genes in hippocampus and spinal cord. *Neuron* 4:603–614.

3 The Nature and Nurture of Retinal Ganglion Cell Development

LEO M. CHALUPA

ABSTRACT This chapter summarizes the results of recent neurophysiological, immunocytochemical, and morphological studies that have documented several interesting ontogenetic features of mammalian retinal ganglion cells. These studies deal with the establishment of excitable membrane properties during prenatal life; the relation of neuropeptide content (somatostatin and neuropeptide Y) to distinct cell classes; and the remodeling of dendritic processes as a consequence of selective perturbations of the developing retina. Evidence is provided that the diversity of retinal ganglion cells is greater than that indicated on the basis of morphological and physiological criteria. Furthermore, certain attributes of retinal ganglion cells, previously believed to reflect retinal circuitry, appear to be intrinsic to these neurons, whereas others, such as the tangential extent of dendrites as well as dendritic stratification to ON and OFF sublaminae of the inner plexiform layer, are regulated by extrinsic factors.

Retinal ganglion cells provide the only means by which visual information is conveyed to the cerebrum. Given the paramount importance of vision for our experience of the world, it is not unexpected that these neurons have received intensive study since the very early days of brain research. Indeed, figure 3.1, reprinted from Polyak's monumental volume, *The Retina* (Polyak, 1941), shows a diagram by an eleventh-century Arabic neuroscientist depicting the two eyes and their projections to the visual centers of the brain. Readers who have been exposed to a course in neuroanatomy will undoubtedly discern that certain details in this drawing are not entirely accurate. It must be recognized, however, that this pioneering investigator of the visual system did not have the benefit of modern neuroanatomical tracing methods, nor is it likely that he received sufficient extramural funding for his efforts. Tremendous progress has been made since the time of this drawing, much of it within the last decade. It can be argued that we now know more about the anatomical and functional properties of retinal ganglion cells than we do about any other neurons of the mammalian brain.

A somewhat related, but quite separate body of literature, has exploited these retinal cells for studies dealing with developmental issues. A major reason for the popularity of the visual system in developmental studies is the relative accessibility of the retina to experimental manipulations. It is feasible, for instance, to make intraocular injections of various substances so as to influence the normal state of developing ganglion cells. Consequently, we now have a great deal of descriptive information regarding key features of ganglion cell development. In the domestic cat, the mammalian species commonly employed for visual development studies, it has been established when retinal ganglion cells are born (Walsh et al., 1983), when they first extend axons into the optic stalk (Williams et al., 1986), how the number of optic fibers fluctuates during early development (Williams et al., 1986), when the main retinorecipient targets are innervated (Shatz, 1983; Williams and Chalupa, 1982), and the type of restructuring that occurs in axonal arbors to attain the exquisite specificity that characterizes mature retinofugal pathways (Sretavan and Shatz, 1987).

This chapter provides an overview of recent studies dealing with some developmental features of retinal ganglion cells as well as certain related organizational properties of these neurons in the mature retina. For the most part, I shall focus on work carried out in my laboratory on the developing cat retina. One of the long-term objectives of this research program is to estab-

---

LEO M. CHALUPA Section of Neurobiology, Physiology, and Behavior, Division of Biology, and Department of Psychology, University of California, Davis, Calif.

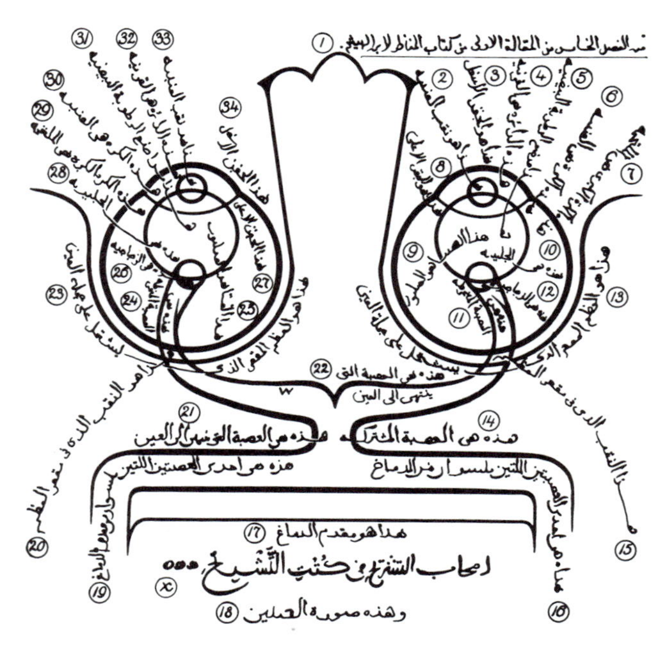

FIGURE 3.1 Diagram of the visual system from an eleventh-century manuscript by Ibn al-Haitham. (Reprinted from Polyak, 1941, by permission.)

lish which aspects of ganglion cell development are regulated by factors intrinsic to these neurons as opposed to those that reflect environmental influences. Because most of these studies deal with prenatal development, the term *environmental influences* denotes events occurring within the developing brain as well as those mediated later by visual experience. Recent reviews dealing with related aspects of visual system development are available elsewhere (Chalupa and White, 1990; Chalupa and Dreher, 1991; Chalupa, Skaliora, and Scobey, 1993).

Ontogeny of excitable membrane properties

Retinal ganglion cells have been the focus of several recent studies that have documented the importance of correlated activity for the stabilization and elimination of early neuronal connections (Hebb, 1949). The influence of neuronal activity on the development of the visual system was initially suggested by the monocular deprivation experiments of Wiesel and Hubel (1963, 1965). Definitive evidence for this concept, however, was provided subsequently by their students Stryker and Harris (1986), who showed that intraocular injections of the sodium voltage-gated channel blocker tetrodotoxin (TTX) prevents the formation of ocular dominance columns in cat visual cortex. Such injections silence all spontaneous as well as light-driven activity of ganglion cells. More recently, Shatz and Stryker (1988) extended this manipulation to fetal animals by demonstrating that infusions of TTX into the developing brain prevent the formation of ocular domains within the dorsal lateral geniculate nucleus. This important finding implied that fetal retinal ganglion cells are capable of firing sodium-mediated action potentials. Because most photoreceptors are not born until fairly late in development, and because light cannot activate the retina in utero, these TTX experiments further suggested that fetal ganglion cells are capable of discharging "spontaneous" action potentials. A heroic effort by Galli and Maffei (1988) showed that this is indeed the case. Recordings from retinal ganglion cells of rat embryos demonstrated spontaneous discharges (Galli and Maffei, 1988), with adjacent cells showing correlated firing patterns (Maffei and Galli-Resta, 1990). Periodic bursts of synchronous activity have also been reported in fetal cat and postnatal ferret retinas (Meister et al., 1991).

The foregoing observations raised a number of fundamental issues related to the functional development of retinal ganglion cells: When are developing retinal ganglion cells first capable of discharging action potentials? How does the excitability of these neurons change during ontogeny? What are the mechanisms (i.e., ionic conductances) underlying spike generation in maturing ganglion cells?

To address these and related questions, we have made whole-cell patch clamp recordings from acutely dissociated retinal ganglion cells of fetal and postnatal timed-pregnant cats (Skaliora, Scobey, and Chalupa, 1993). This approach makes it feasible to assess the electrophysiological properties of even the smallest neurons in the fetal retina. In these experiments, ganglion cells are identified by retrograde labeling following injections of a fluorescent tracer into retinorecipient areas.

To date, the findings have revealed a pronounced developmental increase in the proportion of ganglion cells capable of discharging action potentials. At embryonic day 30 (E30), which is 35 days before birth, only one third of the ganglion cells studied could generate spikes in response to current injections (figure 3.2A). In all cases, these very young neurons responded to maintained depolarization with only a single spike. With maturity, the proportion of excitable ganglion cells increased rapidly so that by E38, almost 80% of the cells

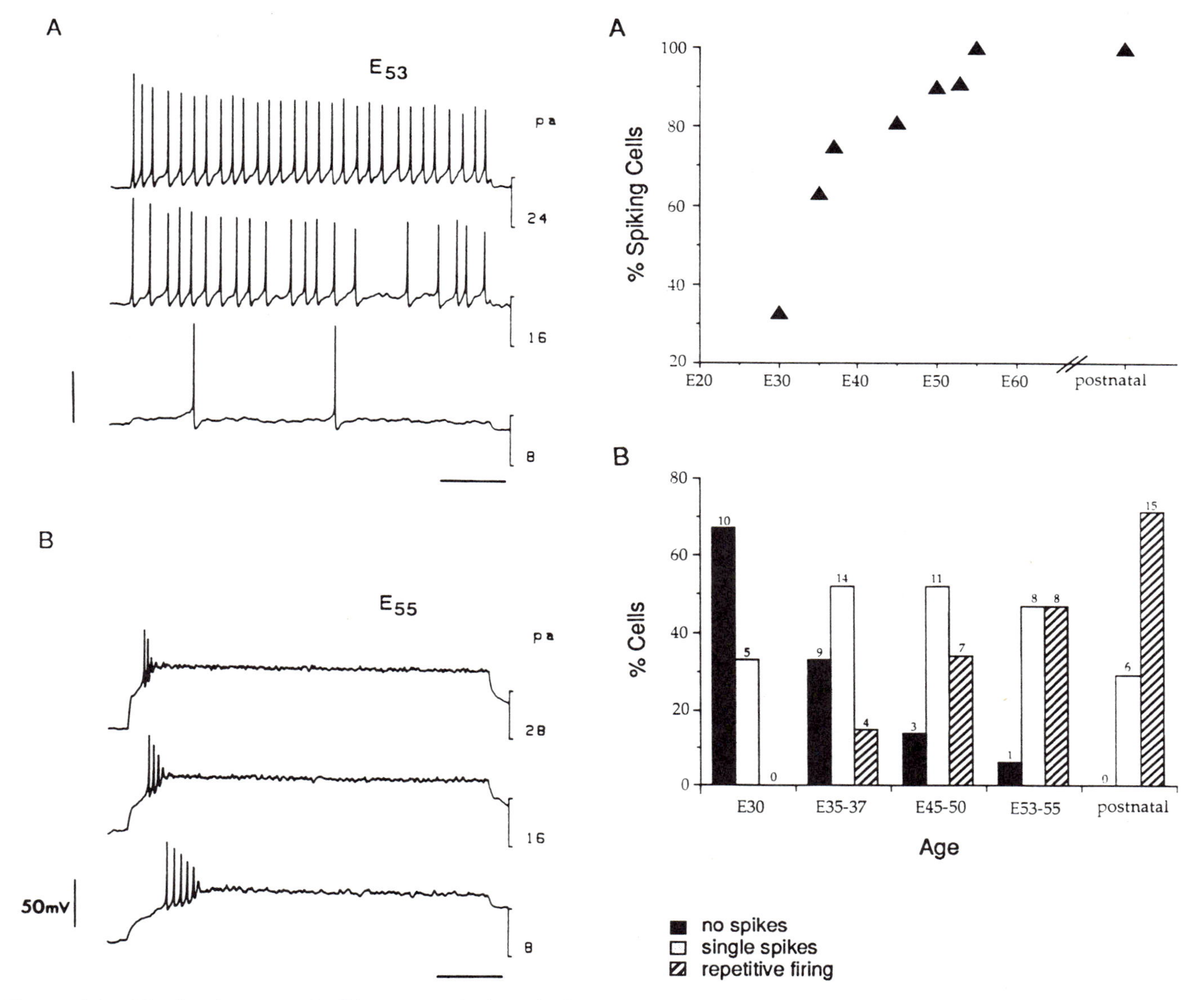

FIGURE 3.2 The development of spiking properties in retinal ganglion cells. (A) Progressive increase, as a function of age, in the percentage of cells capable of generating action potentials in response to injected depolarizing currents. (B) Cells that responded with only a single spike are differentiated from those capable of repetitive firing. Number of cells in each group is denoted over the bars. (Reprinted from Skaliora, Scobey, and Chalupa, 1993, by permission.)

FIGURE 3.3 Responses of fetal ganglion cells (at E53 and E55) to injections of depolarizing currents illustrating a sustained (A) and a transient (B) spiking pattern. The intensity of the stimulus is shown to the right of each trace. (Reprinted from Skaliora, Scobey, and Chalupa, 1993, by permission.)

fired action potentials, while at E55 all cells were capable of spiking. At the same time, there was also a progressive increase in the proportion of cells manifesting repetitive firings to maintained depolarization (figure 3.2B). By late fetal and postnatal ages, two strikingly different response patterns were evident. Individual cells responded to electrical stimulation with either single spikes or in a sustained manner, with the majority manifesting the latter pattern (figure 3.3). This finding was unexpected because it is believed that transient and sustained responses to light, recorded in ganglion cells of the intact retina, are due to differences in retinal circuitry (Werblin, 1977). However, our recordings from isolated ganglion cells suggest that these distinct firing patterns may reflect intrinsic differences in membrane properties among different classes of retinal ganglion cells (cf. Mobbs, Everett, and Cook, 1992). The resolution of this basic issue requires further study.

At all fetal ages, the spiking activity of retinal ganglion cells could be abolished by application of TTX. This indicates that action potentials are sodium-mediated very early in development. Furthermore, voltage-clamp recordings revealed several ontogenetic trends in sodium current properties that could contribute to the observed developmental changes in the spiking capability of these neurons (Skaliora, Scobey, and Chalupa, 1993). In particular, there is a twofold increase in the sodium current densities and a shift in voltage dependence of both activation and steady-state inactivation. This means that with maturity, sodium currents become significantly greater and activate at more negative potentials, with inactivation occurring at less negative potentials. Interestingly, these changes in sodium currents were found to be largely restricted to the period between E30 and E38, which coincides with the period of massive axon ingrowth and innervation of retinorecipient targets (Shatz, 1983; Williams and Chalupa, 1982). One intriguing implication of this finding is that target contact by ingrowing retinal ganglion cells may trigger changes in either the expression or the functional state of sodium voltage-gated channels of these neurons. Furthermore, since retinal ganglion cells can generate sodium-mediated action potentials as early as E38, the functional maturation of these cells cannot be the limiting factor for activity-mediated refinements in retinal projections, which begin approximately 1 week later (about E45; Williams and Chalupa, 1982; Shatz, 1983). Studies now in progress are concerned with documenting developmental changes in potassium and calcium conductances with the objective of providing a model to account for the ontogeny of excitable membrane properties in retinal ganglion cells.

Neuropeptide content of retinal ganglion cells

It has long been recognized that retinal ganglion cells are comprised of diverse classes of neurons (Cajal, 1893). A major advance in retinal research was achieved when it became apparent that morphologically defined classes of ganglion cells can be distinguished on the basis of projectional patterns as well as functional properties (Stone, 1983). Such structure-function correlates have been documented most thoroughly for cat and monkey ganglion cells, but it is reasonable to assume that this is a general principle of retinal organization.

Two classes of ganglion cells—alpha and beta, which are morphological designations (Boycott and Wässle, 1974) for the functionally delineated Y and X cells (Enroth-Cugell and Robson, 1966)—have been particularly well characterized in the cat. The alpha(Y) cells are the largest ganglion cells in the cat retina. Although they constitute only approximately 5% of the ganglion cell population, their axon terminals arborize extensively to innervate some 50% of the neurons within the dorsal lateral geniculate nucleus and the superior colliculus. As might be expected on the basis of their large size, alpha cells have relatively large receptive fields and provide the most rapid conduction of visual information to retinorecipient targets. Furthermore, alpha cells respond to light in a transient manner (Cleland, Levick, and Sanderson, 1973) and show nonlinear spatial summation to sinusoidal gratings drifted across the visual field (Enroth-Cugell and Robson, 1966). In contrast, beta(X) cells account for approximately 50% of the ganglion cell population, have the smallest receptive fields, respond to light in a sustained and highly linear manner, and project to the two main laminae (A and A1) of the lateral geniculate nucleus (Stone, 1983).

Most of the remaining population of retinal ganglion cells in the cat have been morphologically classified as gamma cells (Boycott and Wässle, 1974), which have been shown to correspond functionally to W cells (Saito, 1983). In terms of somal size, these are the smallest retinal cells and share the common property of slow conduction velocity. Gamma cells project primarily to the superior colliculus as well as to the C layers of the lateral geniculate nucleus. It is commonly assumed that further study will lead to a greater parceling of gamma cells because their functional properties are markedly heterogeneous (Stone and Fukuda, 1974; Rodieck, 1979; Stanford, 1987). Indeed, two smaller classes—termed *delta* (Boycott and Wässle, 1974; Wässle, Voigt, and Patel, 1987; Dacey, 1989) and *epsilon* (Leventhal, Keens, and Tork, 1980) cells—have been distinguished from the overall gamma group.

The relevance of the foregoing retinal typologies to cognitive neuroscience is that visual input can be construed as being conveyed to the brain by relatively distinct channels specialized for the processing of fairly selective aspects of the visual world. Although there is still considerable disagreement about the perceptual relevance of the information conveyed by distinct classes of ganglion cells (Shapley, 1990), the notion of visual parallel processing is now firmly established. Furthermore, certain salient aspects of the "channels," derived from the organization of retinal ganglion cells, continue to be preserved at higher levels of the visual system, including the multiple extrastriate areas of the visual cortex. The magnocellular and parvocellular streams of information processing in the primate visual system represent perhaps the best known example of this concept (Livingstone and Hubel, 1988).

Despite the detailed anatomical and physiological knowledge that has accumulated about retinal ganglion cells, remarkably little is known about the neurochemical properties of these neurons. There is some evidence, however, that glutamate may serve as the fast-action excitatory neurotransmitter of retinal ganglion cells (Montero, 1990; Nunes Cardozo, Buijs, and van der Want, 1991). Given the ubiquitous presence of glutamate within the central nervous system, this is not unexpected.

Several years ago, my colleagues and I began to examine the cytochemical properties of ganglion cells in the adult cat retina. Our major goal was to establish whether cytochemical markers could prove useful in distinguishing among different classes of cat retinal ganglion cells. This seemed a reasonable undertaking as studies dealing with other species had demonstrated that some retinal ganglion cells do contain neuropeptides (Kuljis and Karten, 1988; Brecha et al., 1987). Furthermore, because certain neuropeptides have been implicated in developmental processes (see, for example, Bulloch, 1987; Grimm-Jorgensen, 1987), we also wanted to examine the ontogeny of neuropeptide expression in fetal ganglion cells.

Initially, we demonstrated that somatostatin was present in two groups of cells (figure 3.4) within the

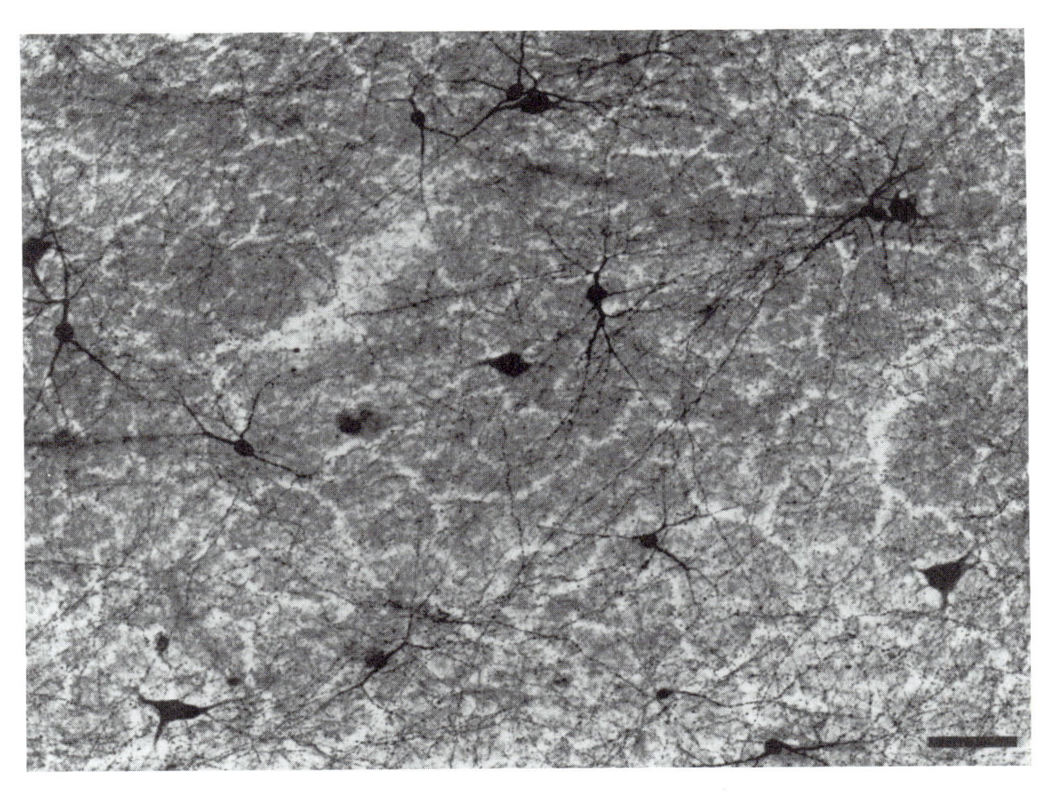

FIGURE 3.4 Somatostatin-immunoreactive cells in the ganglion cell layer of the adult cat. Two types of cells can be distinguished. The larger profiles, with limited staining of their dendritic processes, were shown to be alpha ganglion cells and the smaller, more darkly staining profiles with extensive processes are displaced amacrine cells. Scale bar, 100 μm. (Reprinted from White and Chalupa, 1991, by permission.)

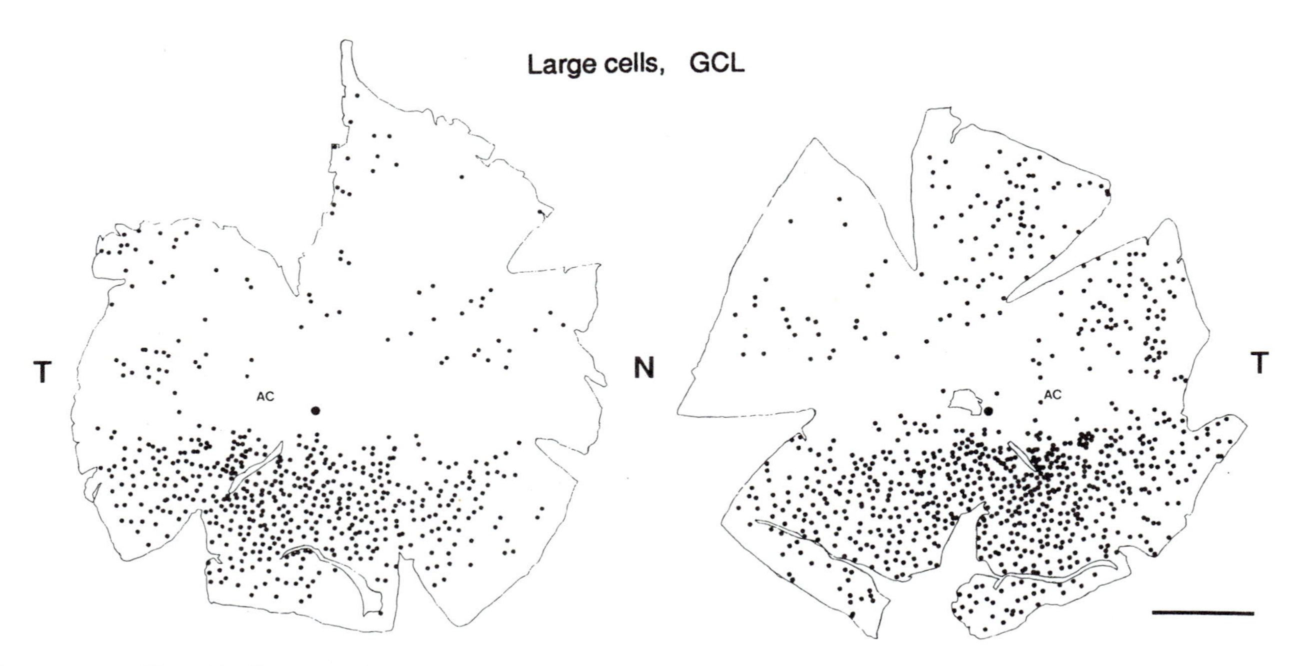

FIGURE 3.5 The distribution of somatostatin-immunoreactive alpha ganglion cells in the mature cat retina. (Reprinted from White et al., 1990, by permission.)

ganglion cell layer of the adult cat (White et al., 1990). It was established subsequently that one of these groups is composed of wide-field displaced amacrine cells and the other of alpha ganglion cells (White and Chalupa, 1991). Curiously, both cell types are preferentially distributed in the inferior retina. The unusual distribution of the somatostatin-immunoreactive alpha cells is shown in figure 3.5. Because alpha cells are distributed throughout the retina, this means that only a subgroup of alpha cells are somatostatin-immunoreactive. In other words, all somatostatin-immunoreactive ganglion cells are alpha cells, but not all alpha cells are somatostatin-immunoreactive. It was also shown that within a local region of the inferior retina, the somatostatin-immunoreactive alpha cells are arrayed with a greater regularity than the overall population of alpha cells.

As discussed previously, ganglion cells have been grouped into distinct classes on the basis of a cluster of common anatomical and functional properties, although there is sometimes disagreement as to which properties are salient for a given population of neurons (Stone, 1983). The localization of a neuroactive substance within a group of cells is not sufficient to invoke a new class. Nevertheless, the finding that somatostatin content distinguishes a subgroup of alpha ganglion cells suggests that the diversity of retinal ganglion cells may be greater than recognized previously on the basis of morphological and functional criteria.

The finding that somatostatin immunoreactivity identifies a subgroup of alpha cells prompted us to look for other neuropeptides in cat retinal ganglion cells. Recently, we discovered that an antibody for neuropeptide Y (NPY) recognizes a subpopulation of gamma-type ganglion cells (Hutsler, White, and Chalupa, 1993). These neurons project to the superior colliculus and to the C layers of the lateral geniculate nucleus (figure 3.6). Unlike the somatostatin-immunoreactive alpha cells, they are distributed across the entire retina, with the highest density at the area centralis (figure 3.7). Furthermore, within a local region of the retina, there was no indication that the NPY-immunoreactive gamma cells are arrayed in a regular mosaic pattern. These findings provide the first evidence that gamma ganglion cells in the cat retina can be subdivided on the basis of immunocytochemical properties.

The retinal ganglion cells we have identified on the basis of their neuropeptide content demonstrate several

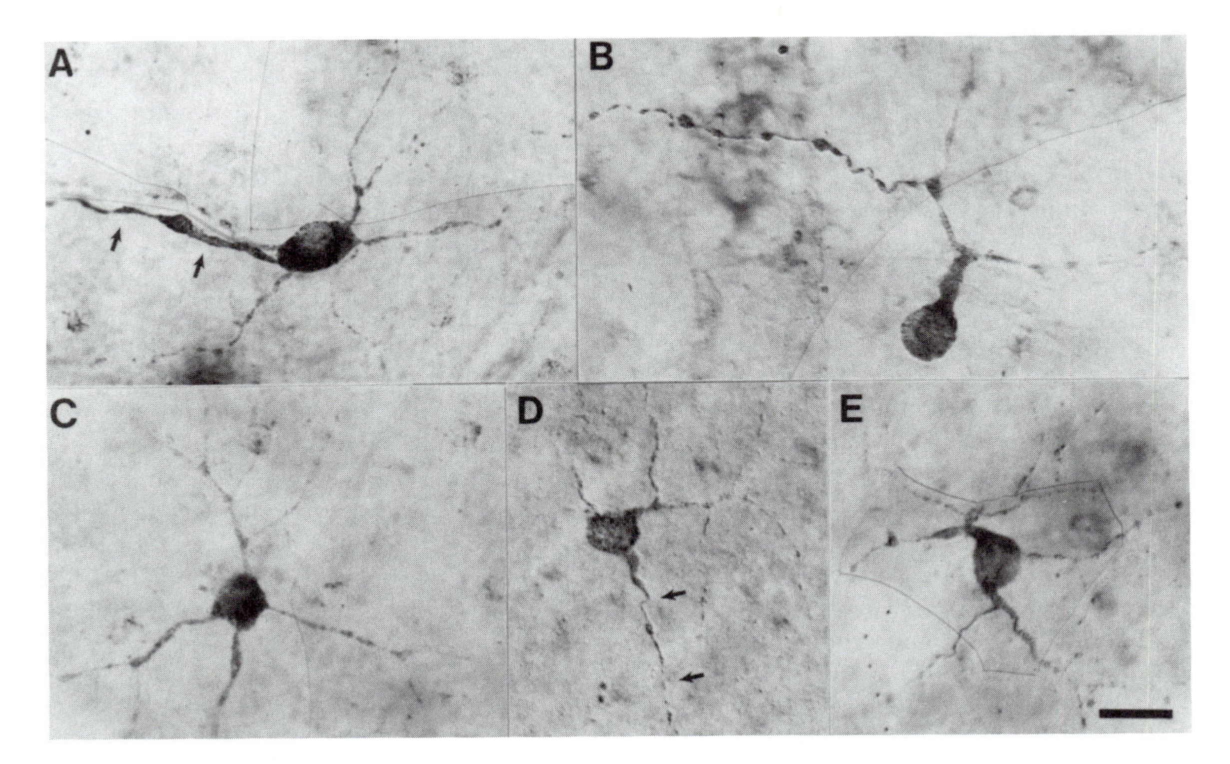

FIGURE 3.6 Examples of NPY-immunoreactive retinal ganglion cells in the mature cat. Calibration bar, 20 μm. (From Hutsler, White, and Chalupa, 1993.)

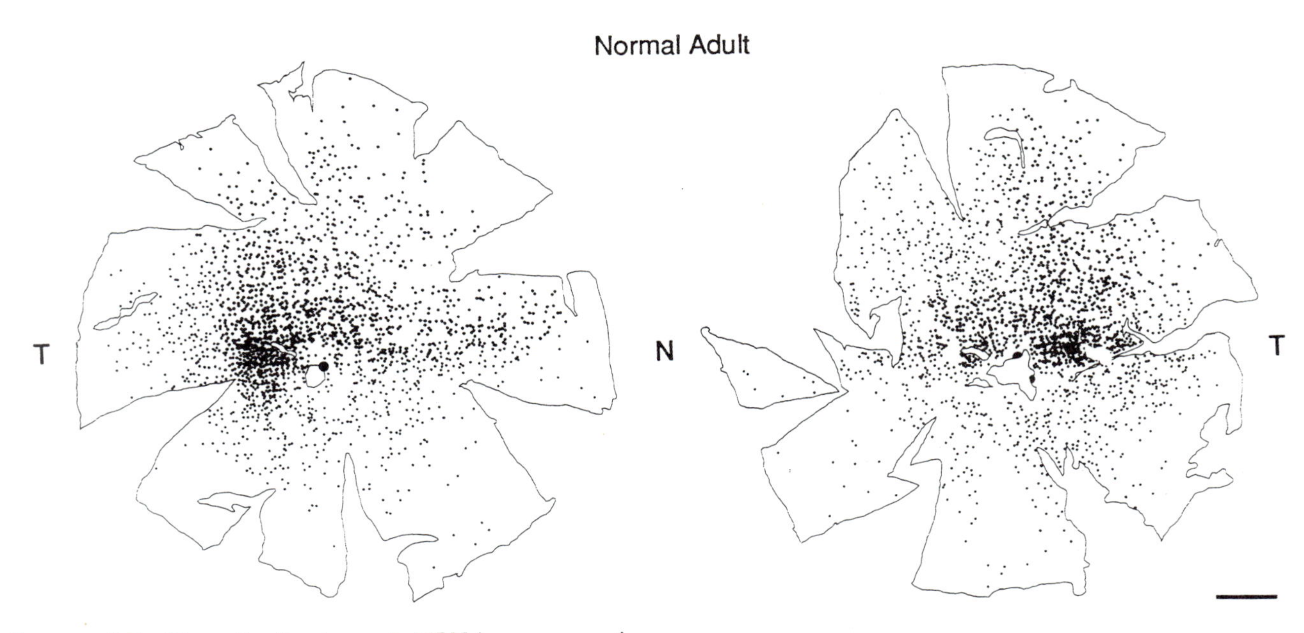

FIGURE 3.7 The distribution of NPY-immunoreactive gamma-type ganglion cells in the mature cat retina. T, temporal; N, nasal. Calibration bar, 3 mm. (From Hutsler, White, and Chalupa, 1993.)

unexpected properties. First, each peptide is contained in only a single retinal ganglion cell class, somatostatin in alpha cells and NPY in gamma cells. Second, in both cases, only a subgroup of the cells in each class have been found to be immunoreactive. Third, both the somatostatin-immunoreactive alpha cells and the NPY-containing gamma cells display unusual distributions. The former are largely confined to the inferior retina, whereas the latter are not arrayed in a regular pattern within local regions of the retina. The functional implications of these properties will be considered next.

It is commonly assumed that for a group of retinal ganglion cells to constitute a functionally meaningful class, the cells must provide complete coverage of the visual field and be distributed in an orderly mosaic pattern (e.g., Peichl, 1991; Wässle, Levick, and Cleland, 1975). This viewpoint is based on what has been shown to be the case for the most thoroughly characterized ganglion cell classes in the cat retina, the alpha and beta cells. Undoubtedly, regular mosaics and complete coverage of the retina are critically important attributes of ganglion cell functions mediating high-resolution spatial vision. However, this may not be a requirement for other aspects of visual function. If ganglion cells that contain neuropeptides convey either visual information that is not associated with high spatial resolution or information that is supplementary to high spatial resolution, this might account for the unusual distributional properties of the two ganglion cell subgroups identified on the basis of neuropeptide immunoreactivity. Although the functional roles of neuropeptides in visual processing remain to be established, it is likely that they act as modulators of visual activity rather than in the rapid transmission of high-resolution spatial vision. Within this framework, retinal ganglion cells could be considered as serving dual functions: the conveyance of object-derived information to the visual centers of the brain via fast-acting neurotransmitters and the modulation of visually related information at either central targets, or even within the retina, by their release of specific neuropeptides. As yet, however, very little is known about the effects of neuropeptides on the response properties of visual neurons. The immunocytochemical findings summarized earlier indicate that this issue warrants further study. It also remains to be established whether other specific subgroups of ganglion cells in the cat retina can be distinguished by their neurochemical content.

Ontogeny of neuropeptide expression

Immunocytochemical work on the mature retina raised a number of pertinent questions regarding the development of somatostatin and NPY in retinal cells: When do these neurons first express such immunoreactivities? Is there transient expression of these neuropeptides or are the mature numbers and distributions of these cells attained through a gradual incremental process? How does the development of neuropeptide content relate to what is known about other aspects of retinal development? Can such expression of immunoreactivity be modified by developmental manipulations?

The studies we have carried out indicate that the ontogeny of somatostatin immunoreactivity within the cat retina differs in two fundamental respects from that of NPY immunoreactivity (Hutsler and Chalupa, 1992; White and Chalupa, 1992). First, somatostatin immunoreactivity was observed early in development, with immunoreactive profiles present in the central region of the retina as early as E30 (White and Chalupa, 1992). In contrast, NPY-immunoreactive cells were not observed in the ganglion cell layer until nearly E50 (Hutsler and Chalupa, 1992). Second, there is clear evidence for the transient expression of somatostatin. Immunoreactive ganglion cells were found throughout the superior and inferior retina at E51. These neurons, which could be distinguished from alpha cells by their granular staining, disappeared completely by postnatal day 38. Immunoreactive alpha cells were first seen at 5 days after birth. These neurons did not achieve their mature number and staining characteristics until more than a month later. Furthermore, from the time that the immunoreactive alpha cells could be recognized, they were distributed mainly in the inferior retina, as is the case at maturity. In contrast, there was no indication of transient expression of NPY-immunoreactive ganglion cells; their number increased gradually to attain adult levels shortly after birth.

Because somatostatin immunoreactivity in alpha cells develops after birth, around the time of eye opening, we also sought to determine whether visual experience plays a role in the expression of this neuropeptide. It had been demonstrated previously that normal visual experience is required for the expression of certain other neuroactive substances in retinal neurons, including vasoactive intestinal polypeptide in the primate (Stone et al., 1988) and dopamine in the primate and chick

(Iuvone et al., 1989; Stone et al., 1989). We found, however, that neither intraocular injections of TTX nor dark rearing significantly alters somatostatin immunoreactivity in the developing cat retina.

The early appearance of somatostatin immunoreactivity as well as its transient expression suggests that this peptide may have a function in some aspect of retinal development. There is evidence in other systems that somatostatin can enhance process outgrowth (Bulloch, 1987; Grimm-Jorgensen, 1987) as well as influence the activity of choline acetyltransferase during a specific developmental period (Kentori and Vernadakis, 1990). Thus, it may prove fruitful to investigate the regulatory role of somatostatin in the development of the mammalian visual system. In contrast, the relatively late and restricted expression of NPY in the developing retina would appear to argue against such a function for this neuropeptide.

Development and remodeling of dendrites

The first retinal ganglion cells are born in the cat at approximately 19 days after conception (E19). Within a very short period of time (on the order of several hours), these neurons extend an axon into the optic stalk to begin the formation of the optic nerve (Williams et al., 1986). By comparison, the growth and differentiation of dendritic processes takes place considerably later, so that the differentiation of ganglion cell classes, based on dendritic morphologies, is not possible until nearly E50 (Maslim, Webster, and Stone, 1986; Ramoa, Campbell, and Shatz, 1988). Perhaps the most intriguing aspect of dendritic development is the remarkable remodeling that takes place during late fetal and early postnatal ages. Among the changes that have been documented are the following: the tangential growth of dendritic fields, the retraction of processes that initially ramified widely within the inner plexiform layer (IPL), and the elimination of dendritic spines and collateral branches (Dann, Buhl, and Peichl, 1988; Ramoa, Campbell, and Shatz, 1988). The significance of such regressive events is unclear. It is commonly assumed that these reflect the outcome of competitive interactions for presynaptic inputs. However, the results of a recent serial electron-microscopical study by Wong, Yamawaki, and Shatz (1992) suggests that the loss of exuberant spines does not reflect a presynaptic mechanism as dendritic spines on developing ganglion cells rarely receive synaptic contacts. As will be discussed later, this should not be interpreted as invalidating the idea that presynaptic mechanisms play a key role in the remodeling of other features of retinal ganglion cell dendrites.

In the adult cat, alpha and beta cell classes are subdivided into ON and OFF types depending on their responses to light increments or decrements. ON and OFF cells can also be distinguished morphologically on the basis of where the dendrites of these neurons stratify within the IPL. (The dendrites of OFF cells stratify distally, whereas ON cell dendrites stratify proximal to the cell body.) At maturity the dendrites of these two types of cells (ON and OFF) provide complete coverage of the retina, with limited overlap between dendrites of the same type. This mosaic pattern reflects a regular distribution of ON and OFF cells across the retinal surface. On the basis of this mature pattern of retinal organization, it has been suggested that during early development, interactions among neighboring ganglion cells of the same type regulate dendritic growth of individual neurons (Boycott and Wässle, 1974; Wässle and Reimann, 1978). This has been taken to mean that growing dendrites recognize the dendrites of type-specific cells and that such recognition results in the inhibition of subsequent dendritic growth.

This hypothesis predicts that the dimensions of dendritic fields of retinal ganglion cells should be modified by manipulations that result in a change in the density of ganglion cells during development. Several studies have demonstrated that decreasing ganglion cell density in neonatal animals (by local retinal lesions or transections of the optic tract) does result in an extension of dendritic arbors toward the region of cell depletion (Linden and Perry, 1982; Perry and Linden, 1982; Eysel, Peichl, and Wässle, 1985). A study carried out in my laboratory has shown further that increasing retinal ganglion cell density above normal values, by monocular enucleation during fetal development, results at maturity in a significant decrease in the size of dendritic fields (figure 3.8) (Kirby and Chalupa, 1986). Collectively, the foregoing observations indicate that dendritic growth is not controlled entirely by mechanisms intrinsic to individual ganglion cells. Rather, dendritic field size is regulated by cellular interactions within the developing retina.

What is less clear, however, is the basis of the cellular interactions that control dendritic growth. One possi-

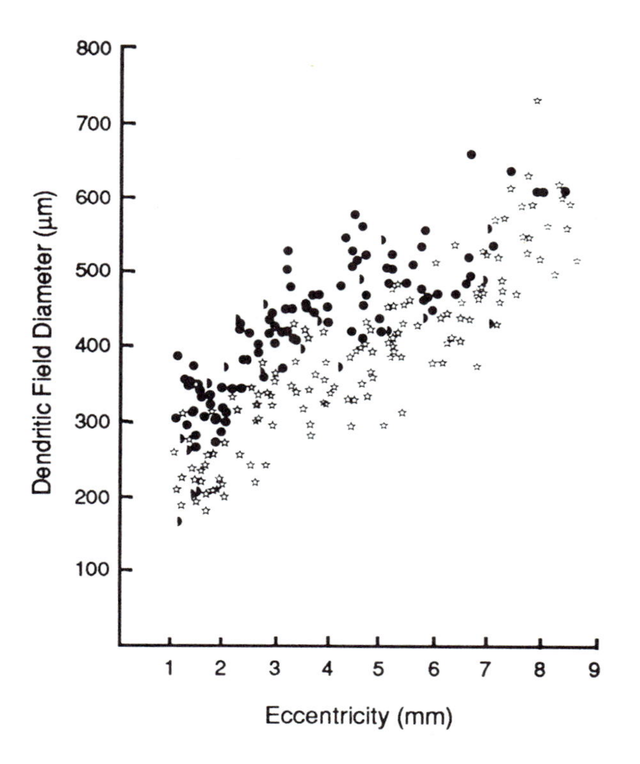

FIGURE 3.8 The size of dendritic fields of alpha ganglion cells is plotted as a function of eccentricity (distance from area centralis) for normal animals as well as those that were monocularly enucleated before or after birth. The fetal enucleates have a higher-than-normal density of ganglion cells. Correspondingly, the dendritic fields of ganglion cells in the fetal enucleates tend to be significantly smaller than normal at all eccentricities. (Reprinted from Kirby and Chalupa, 1986, by permission.)

bility is that there are indeed inhibitory interactions that restrict the growth of developing retinal ganglion cell dendrites. However, the changes in the size of dendritic fields that have been observed after manipulations that either increase or decrease the developing ganglion cell population could also be explained by competition for retinal afferents among ganglion cells. According to the afferent competition hypothesis (Perry and Linden, 1982), the size of dendritic fields in retinal ganglion cells is dependent on the availability of afferents (i.e., amacrine and bipolar cell inputs). Thus, manipulations that decrease ganglion cell density cause an increase in the size of dendritic fields because the number of available afferents is presumed to be greater than normal; conversely, increasing ganglion cell density results in smaller dendritic fields because the available number of afferents is less than normal. Surprisingly, the degree to which dendritic growth is dependent on afferent inputs or cell-specific interactions among ganglion cells remains unresolved. A recent report cites a further complication, claiming that there may also be non-class-specific interactions contributing to the development of retinal ganglion cell morphologies (Ault et al., 1993).

A potentially fruitful means for breaking out of this quagmire is to manipulate the state of developing retinal afferents (either their number or functional properties) without affecting the density of the ganglion cell population. Such an approach would provide a direct evaluation of the role of retinal afferents in the morphological development of retinal ganglion cells. A recent study in my laboratory has utilized such a strategy (Bodnarenko and Chalupa, 1993).

As mentioned previously, retinal ganglion cell dendrites initially ramify widely within the IPL prior to becoming restricted to ON and OFF sublaminae. In cat retina, this stratification process appears to coincide with the time period when bipolar cells are forming synaptic contacts with ganglion cells (Maslim and Stone, 1986). This temporal coincidence raised the possibility that the synaptic activity of bipolar cells might provide a signal regulating the restriction of exuberant dendritic processes. An opportunity to test this idea was provided by the fact that rod bipolar cells and ON-cone bipolar cells hyperpolarize in response to the application of the glutamate agonist 2-amino-4-phosphonobutyrate (APB) (Slaughter and Miller, 1981). Hyperpolarization of bipolar cells by APB prevents the release of glutamate by these retinal interneurons so that under dark-adapted conditions, all ganglion cell responses to light are blocked (Wässle et al., 1991).

Daily intraocular injections of APB in newborn cats resulted in a profound arrest of the stratification of retinal ganglion cell dendrites (Bodnarenko and Chalupa, 1993). As may be seen in figure 3.9, at P2, when the APB treatment was initiated, approximately 46% of the beta ganglion cells in the central region of the retina were still multistratified. This percentage decreased rapidly during the normal course of development so that by P13 only 12% or so of the dendrites were multistratified. In contrast, daily treatment with APB largely maintained the immature state of stratification, so that at P13 the treated retina contained nearly three times

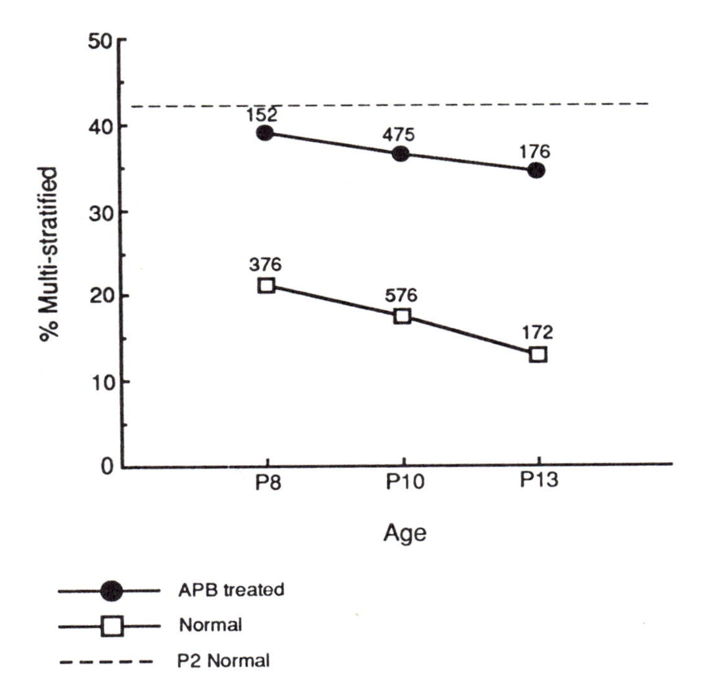

FIGURE 3.9 The incidence of multistratified beta ganglion cells in normal and APB-treated retinas. In all cases, intraocular injections of APB were initiated at P2. The numbers above each data point denote the number of cells that were analyzed. As may be seen, APB treatment largely arrested the stratification process. (From Bodnarenko and Chalupa, 1993)

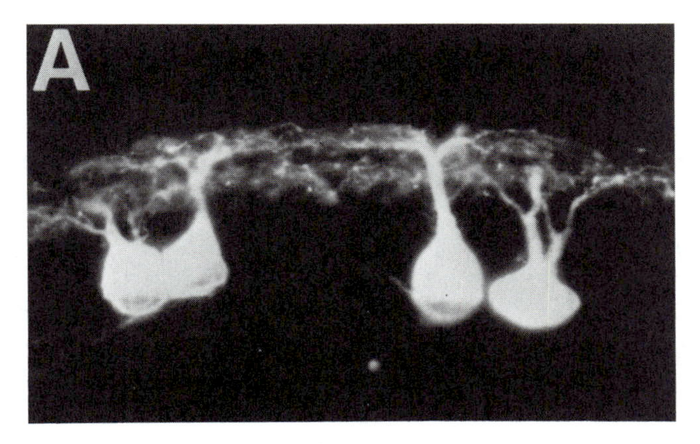

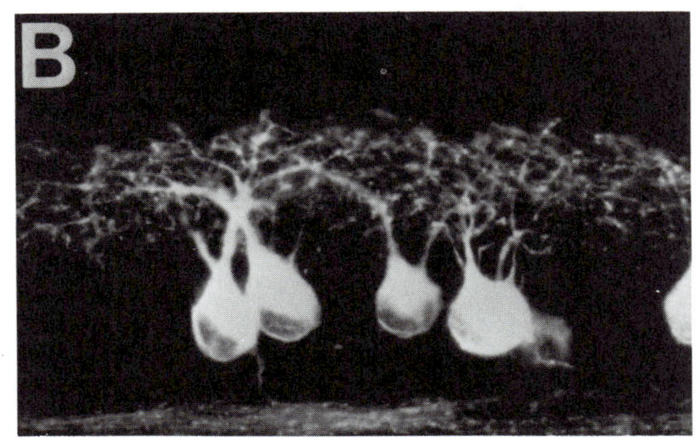

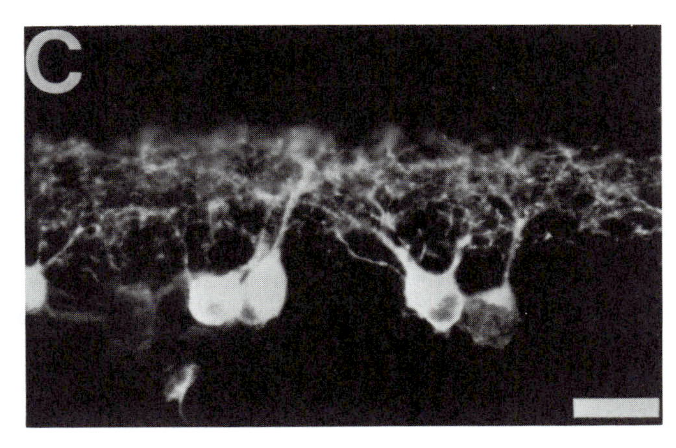

FIGURE 3.10 Examples of DiI-labeled beta ganglion cells in transverse sections of the normal retina (A), an APB-treated retina (B), and a normal P2 retina (C). Note the stratification of dendritic processes into two distinct sublaminae in the normal case and the multistratified appearance of the APB-treated retina and the P2 retina. Calibration bar, 20 μm. (From Bodnarenko and Chalupa, 1993)

the incidence of multistratified ganglion cells as is the case normally. Examples of the normal, APB-treated cells as well as those of a normal P2 animal are depicted in figure 3.10.

These findings have several important implications. First, they provide the first demonstration that activity plays a major role in the morphological restructuring of mammalian retinal ganglion cells. Second, the results suggest strongly that the key factor regulating the retraction of initially exuberant retinal dendrites is presynaptic glutamate-mediated activity. In this context, it should be noted that a previous study found that blockade of retinal ganglion cell action potentials by intraocular TTX application did not alter the dendritic stratification process (Wong, Herrmann, and Shatz, 1991). Thus, it is not the firing pattern of retinal ganglion cells but rather the synaptic input from bipolar cells that provides the signal for the stabilization or the retraction of dendritic processes. Third, our results also indicate that ON and OFF ganglion cells were affected by the APB treatments to the same degree. This means that glutamate-mediated activity alone cannot establish which ganglion cells will maintain dendrites within the ON or OFF sublaminae of the IPL. Some other factor must be responsible for the differentiation of ON cells from OFF cells. What this may be is unknown, but one possibility is that differences between ON cells and OFF cells are specified before the segregation of dendrites within the IPL.

The foregoing studies demonstrate clearly that dendritic morphologies of developing retinal ganglion cells can be modified, in both their tangential and vertical extents, by certain types of experimental interventions. Because dendritic morphology relates closely to the functional properties of retinal ganglion cells (Yang and Masland, 1992), it would be of considerable interest to examine the perceptual capabilities of animals with altered dendritic processes.

A more fundamental question is whether the normal retinal environment contributes to the generation of specific ganglion cell classes. In other words, would the morphological properties that differentiate alpha, beta, and gamma cells be expressed if these neurons were grown outside the retina? Recently, Montague and Friedlander (1991) showed that postnatal cat retinal ganglion cells can survive and grow in culture and that morphologically distinct classes can be differentiated. As indicated previously, retinal ganglion cell classes are recognizable in cat retina approximately 2 weeks before birth (Ramoa, Campbell, and Shatz, 1988). This means that the postnatal cells harvested by Montague and Friedlander showed a capability to re-express their class-specific morphological properties. It remains to be established whether distinct classes of ganglion cells would differentiate if they had been removed from the retina at an earlier developmental period, before the expression of class-specific features. Unfortunately, it has not proved feasible to maintain early fetal ganglion cells viably in culture. This may reflect the dependence of very young neurons on target-released tropic factors. A potentially fruitful strategy to address this issue is to grow ganglion cells within a retinorecipient nucleus. Such a study recently was initiated in my laboratory (Miguel-Hidalgo and Chalupa, 1993). Miguel-Hidalgo has injected into the superior colliculus dissociated retinal ganglion cells before the time of their morphological differentiation. These neurons were previously labeled with a retrograde fluorescent tracer to make their identification unequivocal. The results must be considered preliminary, but they show that many of the dissociated retinal ganglion cells can survive for several weeks within the superficial layers of the superior colliculus and that some exhibit remarkable growth and differentiation of dendritic processes after early removal from the retina (figure 3.11). Future studies with such transplants of dissociated neurons could provide greater insights concerning the contributions of the retinal milieu to the morphological, neurochemical, and electrophysiological properties of mammalian retinal ganglion cells.

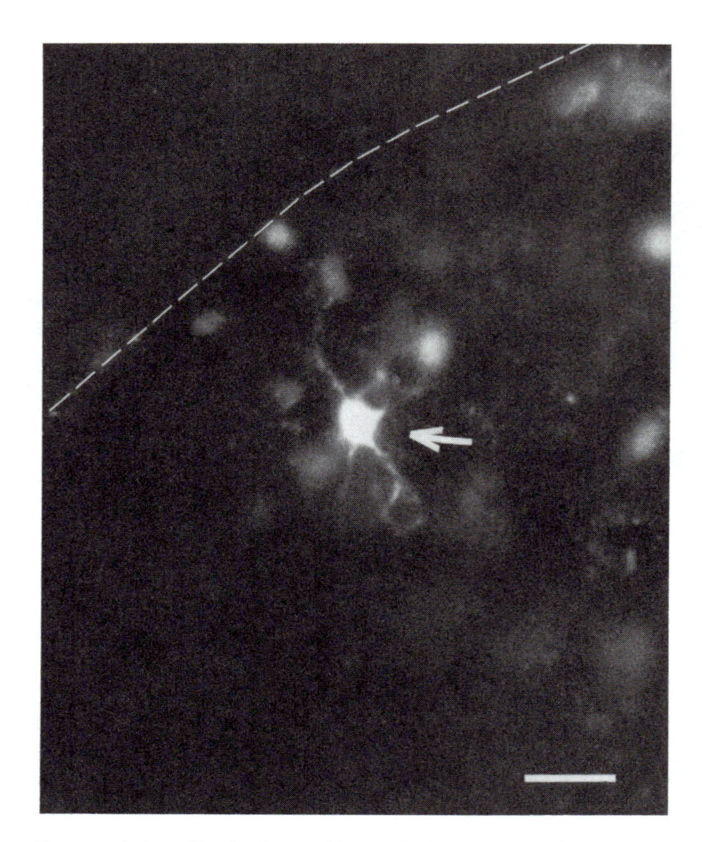

FIGURE 3.11 Retinal ganglion cells in the superficial layers of the superior colliculus. These neurons were initially labeled by the retrograde tracer fluorogold, dissociated from the retina, and subsequently injected into the superior colliculus of a littermate animal. Following a 2-week survival period, pronounced differentiation of dendritic processes is evident in the retinal ganglion cell denoted by the arrow. Dashed line indicates top of the superior colliculus. Calibration line, 100 μm.

ACKNOWLEDGMENTS This research was generously supported by a grant from the National Eye Institute of the National Institutes of Health (EYO3391). I have been fortunate to have among my laboratory staff many capable and dedicated individuals. In particular, I wish to express my thanks to Stefan Bodnarenko, Jeff Hutsler, Michael Kirby, José Javier Miguel-Hidalgo, Irini Skaliora, Cheryl A. White, and Robert W. Williams, whose research has provided the data on which this chapter is based. I also thank Cara Snider for her highly capable technical assistance.

REFERENCES

AULT, S. J., K. G. THOMPSON, Y. ZHOU, and A. G. LEVENTHAL, 1993. Selective depletion of beta cells affects the development of alpha cells in cat retina. *Visual Neurosci.* 10:237–245.

BODNARENKO, S. R., and L. M. CHALUPA, 1993. Stratification of On and Off ganglion cell dendrites is dependent on glutamate-mediated afferent activity in the developing retina. *Nature* 364:144–146.

BOYCOTT, B. B., and H. WÄSSLE, 1974. The morphological types of ganglion cells of the domestic cat's retina. *J. Physiol.* 240:397–419.

BRECHA, N., D. JOHNSON, J. BOLZ, S. SHARMA, J. G. PARNAVELAS, and A. R. LIEBERMAN, 1987. Substance P-immunoreactive retinal ganglion cells and their central axon projections in the rabbit. *Nature* 327:155–158.

BULLOCH, A. G. M., 1987. Somatostatin enhances neurite outgrowth and electrical coupling of regenerating neurons in *Helisoma*. *Brain Res.* 412:6–17.

CAJAL, S. R. 1893. La retine des vertebres. *La Cellule.* 9:17–257.

CHALUPA, L. M., and B. DREHER, 1991. High precision systems require high precision "blueprints": A new view regarding the formation of connections in the mammalian visual system. *J. Cogn. Neurosci.* 3:209–219.

CHALUPA, L. M., I. SKALIORA, and R. P. SCOBEY, 1993. Responses of isolated cat retinal ganglion cells to injected currents during development. In *Progress in Brain Research*, vol. 95, T. P. Hicks, S. Molochnikoff, and T. Ono, eds. Amsterdam: Elsevier, pp. 25–31.

CHALUPA, L. M., and C. A. WHITE, 1990. Prenatal development of visual system structures. In *Development of Sensory Systems in Mammals*, J. R. Coleman, ed. New York: Wiley, pp. 3–60.

CLELAND, B. G., W. R. LEVICK, and K. J. SANDERSON, 1973. Properties of sustained and transient ganglion cells in the cat retina. *J. Physiol. (Lond.)* 228:649–680.

DACEY, D. M., 1989. Dopamine-accumulating retinal neurons revealed by in vitro fluorescence display a unique morphology. *Science* 240:1196–1198.

DANN, J. F., E. H. BUHL, and L. PEICHL, 1988. Postnatal dendritic maturation of alpha and beta ganglion cells in cat retina. *J. Neurosci.* 8:1485–1499.

ENROTH-CUGELL, C., and J. G. ROBSON, 1966. The contrast sensitivity of retinal ganglion cells of the cat. *J. Physiol. (Lond.)* 187:517–519.

EYSEL, U., L. PEICHL, and H. WASSLE, 1985. Dendritic plasticity in the early postnatal feline retina: Quantitative characteristics and sensitive period. *J. Comp. Neurol.* 242:134–145.

GALLI, L., and L. MAFFEI, 1988. Spontaneous impulse activity of rat retinal ganglion cells in prenatal life. *Science* 242: 90–91.

GRIMM-JORGENSEN, Y., 1987. Somatostatin and calcitonin stimulate neurite regeneration of molluscan neurons in vitro. *Brain Res.* 403:121–126.

HEBB, D. O., 1949. *Organization of Behavior.* New York: Wiley.

HUTSLER, J. J., and L. M. CHALUPA, 1992. Prenatal development of neuropeptide Y immunoreactivity in the ganglion cell layer of the cat retina. *Soc. Neurosci. Abstr.* 18:1029.

HUTSLER, J. J., C. A. WHITE, and L. M. CHALUPA, 1993. Neuropeptide Y immunoreactivity identifies a group of gamma-type retinal ganglion cells in the cat. *J. Comp. Neurol.* 336:468–480.

IUVONE, P. M., M. TIGGES, A. FERNANDES, and J. TIGGES, 1989. Dopamine synthesis and metabolism in rhesus monkey retina: Development, aging, and the effects of monocular deprivation. *Visual Neurosci.* 2:465–471.

KENTORI, S ., and A. VERNADAKIS, 1990. Growth hormone–releasing hormone and somatostatin influence neuronal expression in developing chick brain: II. Cholinergic neurons. *Brain Res.* 512:297–303.

KIRBY, M. A., and L. M. CHALUPA, 1986. Retinal crowding alters the morphology of alpha ganglion cells. *J Comp. Neurol.* 251:532–541.

KULJIS, R. O., and H. J. KARTEN, 1988. Neuroactive peptides as markers of retinal ganglion cell populations that differ in anatomical organization and function. *Visual Neurosci.* 1:73–81.

LEVENTHAL, A. G., J. KEENS, and I. TORK, 1980. The afferent ganglion cells and cortical projections of the retinal recipient zone of the cat's "pulvinar complex." *J. Comp. Neurol.* 194:535–554.

LINDEN, R., and V. H. PERRY, 1982. Ganglion cell death within the developing retina: A regulatory role for retinal dendrites? *Neuroscience* 7:2813–2827.

LIVINGSTONE, M. S., and D. H. HUBEL, 1988. Segregation of form, color, movement, and depth: Anatomy, physiology, and perception. *Science* 240:740–749.

MAFFEI, L., and L. GALLI-RESTA, 1990. Correlation in the discharges of neighboring rat retinal ganglion cells during prenatal life. *Proc. Natl. Acad. Sci. U.S.A.* 87:2861–2864.

MASLIM, J., and J. STONE, 1986. Synaptogenesis in the retina of the cat. *Brain Res.* 373:35–48.

MASLIM, J., M. WEBSTER, and J. STONE, 1986. Stages in the structural differentiation of retinal ganglion cells. *J. Comp. Neurol.* 254:383–402.

MEISTER, M., R. O. L. WONG, D. A. BAYLOR, and C. J. SHATZ, 1991. Synchronous bursts of action potentials in ganglion cells of the developing mammalian retina. *Science* 252:939–943.

MIGUEL-HIDALGO, J. J., and L. M. Chalupa, 1993. Transplantation of dissociated retinal ganglion cells into the superior colliculus of postnatal rats. *Soc. Neurosci. Abstr.* 19:1741.

MOBBS, P., K. EVERETT, and A. COOK, 1992. Signal shaping by voltage-gated currents in retinal ganglion cells. *Brain Res.* 574:217–223.

MONTAGUE, P. R., and M. J. FRIEDLANDER, 1991. Morphogenesis and territorial coverage of isolated mammalian retinal ganglion cells. *J. Neurosci.* 11:1440–1457.

MONTERO, V. M., 1990. Quantitative immunogold analysis reveals high glutamate levels in synaptic terminals of retino-geniculate, cortico-geniculate, and geniculate cortical axons in the cat. *Vis. Neurosci.* 4:437–443.

Nunes Cardozo, B., R. Buijs, and J. van der Want, 1991. Glutamate-like immunoreactivity in retinal terminals in the nucleus of the optic tract in rabbits. *J. Comp. Neurol.* 309:261–270.

Peichl, L., 1991. Alpha ganglion cells in mammalian retinae: Common properties, species differences, and some comments on other ganglion cells. *Visual Neurosci.* 7:155–169.

Perry, V. H., and R. Linden, 1982. Evidence for dendritic competition in the developing retina. *Nature* 297:683–685.

Polyak, S., 1941. *The Retina.* Chicago: University of Chicago.

Ramoa, A. S., G. Campbell, and C. J. Shatz, 1988. Dendritic growth and remodeling of cat retinal ganglion cells during fetal and postnatal development. *J. Neurosci.* 8: 4239–4261.

Rodieck, R. W., 1979. Visual pathways. *Annu. Rev. Neurosci.* 2:193–225.

Saito, H., 1983. Pharmacological and morphological differences between X- and Y-type ganglion cells in the cat's retina. *Vision Res.* 23:1299–1308.

Shapley, R., 1990. Visual sensitivity and parallel retinocortical channels. *Annu. Rev. Psychol.* 41:635–658.

Shatz, C. J., 1983. The prenatal development of the cat's retinogeniculate pathway. *J. Neurosci.* 3:482–499.

Shatz, C. J., and M. P. Stryker, 1988. Prenatal tetrodotoxin infusion blocks segregation of retinogeniculate afferents. *Science* 242:87–89.

Skaliora, I., R. P. Scobey, and L. M. Chalupa, 1993. Prenatal development of excitability in cat retinal ganglion cells: Action potentials and sodium currents. *J. Neurosci.* 13:313–323.

Slaughter, M. M., and R. F. Miller, 1981. 2-Amino-4-phosphonobutyric acid: A new pharmacological tool for retina research. *Science* 211:182–185.

Sretavan, D. W., and C. J. Shatz, 1987. Axon trajectories and pattern of terminal arborization during the prenatal development of the cat's retinogeniculate pathway. *J. Comp. Neurol.* 255:386–400.

Stanford, L. R., 1987. W-cells in the cat retina: Correlated morphological and physiological evidence for two distinct classes. *J. Neurophysiol.* 57:218–244.

Stone, J., 1983. *Parallel Processing in the Visual System.* New York: Plenum.

Stone, J., and Y. Fukuda, 1974. Properties of cat retinal ganglion cells: A comparison of W-cells with X- and Y-cells. *J. Neurophysiol.* 37:722–748.

Stone, R. A., A. M. Laties, E. Raviola, and T. N. Wiesel, 1988. Increase in retinal vasoactive intestinal polypeptide after eyelid fusion in primates. *Proc. Natl. Acad. Sci. U.S.A.* 85:257–260

Stone, R. A., T. Lin, A. M. Laties, and P. M. Iuvone, 1989. Retinal dopamine and form-deprivation myopia. *Proc. Natl. Acad. Sci. U.S.A.* 86:704–706.

Stryker, M. P., and Harris, W. A., 1986. Binocular impulse blockade prevents the formation of ocular dominance columns in cat visual cortex. *J. Neurosci.* 6:2117–2133.

Walsh, C., E. H. Polley, T. L. Hickey, and R. W. Guillery, 1983. Generation of cat retinal ganglion cells in relation to central pathways. *Nature* 302:611–614.

Wässle, H., W. R. Levick, and B. G. Cleland, 1975. The distribution of the alpha type ganglion cells in the cat's retina. *J. Comp. Neurol.* 159:419–438.

Wässle, H., and H. J. Reimann, 1978. The mosaic of nerve cells in the mammalian retina. *Proc. R. Soc. Lond. [Biol.]* 200:441–461.

Wässle, H., T. Voigt, and B. Patel, 1987. Morphological and immunocytochemical identification of indoleamine-accumulating neurons in the cat retina. *J. Neurosci.* 7:1574–1585.

Wässle, H., M. Yamshita, U. Greferath, U. Grünert, and F. Müller, 1991. The rod bipolar cell of the mammalian retina. *Visual Neurosci.* 7:99–112.

Werblin, F. S., 1977. Regenerative amacrine cell depolarization and formation of on-off ganglion cell response. *J. Physiol. (Lond.)* 264:767–785.

White, C. A., and L. M. Chalupa, 1991. Subgroup of alpha ganglion cells in the adult cat is immunoreactive for somatostatin. *J. Comp. Neurol.* 304:1–13.

White, C. A., and L. M. Chalupa, 1992. Ontogeny of somatostatin immunoreactivity in the cat retina. *J. Comp. Neurol.* 317:129–144.

White, C. A., L. M. Chalupa, D. Johnson, and N. C. Brecha, 1990. Somatostatin-immunoreactive cells in the adult cat retina. *J. Comp. Neurol.* 293:134–150.

Wiesel, T. N., and D. H. Hubel, 1963. Single-cell responses in striate cortex of kittens deprived of vision in one eye. *J. Neurophysiol.* 26:1003–1017.

Wiesel, T. N., and D. H. Hubel, 1965. Comparison of the effects of unilateral and bilateral eye closure on cortical unit responses in kittens. *J. Neurophysiol.* 28:1029–1040.

Williams, R. W., M. J. Bastiani, B. Lia, and L. M. Chalupa, 1986. Growth cones, dying axons, and developmental fluctuations in the fiber population of the cat's optic nerve. *J. Comp. Neurol.* 246:32–69.

Williams, R. W., and L. M. Chalupa, 1982. Prenatal development of retinocollicular projections in the cat: An anterograde tracer transport study. *J. Neurosci.* 2:604–622.

Wong, R. O. L., K. Herrmann, and C. J. Shatz, 1991. Remodelling of retinal ganglion cell dendrites in the absence of action potential activity. *J. Neurobiol.* 22:685–697.

Wong, R. O. L., R. M. Yamawaki, and C. J. Shatz, 1992. Synaptic contacts and the transient dendritic spines of developing retinal ganglion cells. *Eur. J. Neurosci.* 4:1387–1397.

Yang, G., and R. H. Masland, 1992. Direct visualization of the dendritic and receptive fields of directionally selective retinal ganglion cells. *Science* 258:1949–1952.

4 The Reorganization of Sensory and Motor Maps in Adult Mammals

JON H. KAAS

ABSTRACT Sensory representations in developing mammals have long been known to be plastic and subject to environmental influences during certain critical periods. There is also a growing awareness that receptive field properties of neurons in cortical maps in adult mammals are dynamically maintained and vary with learning, experience, and conditions of stimulation. In addition, major perturbations in sensory activation, such as those produced by lesions of sensory surfaces, are capable of producing major reorganizations in sensory maps. Such changes have been demonstrated in a range of mammalian species, in auditory, somatosensory, visual, and motor systems, and at both cortical and subcortical levels in these systems. Some of the alterations in maps occur immediately after the sensory manipulation, suggesting the rapid potentiation of previously weak synaptic influences, whereas other modifications develop over days to weeks, time enough for changes in neurotransmitter and neuromodulator expression, synaptic turnover, and neuronal growth. Some types of reorganization clearly suggest the growth of new connections in the central nervous system. Adult plasticity in sensory and motor systems may be responsible for improvements in sensorimotor skills with practice, adjustments to sensory loss and impairments, recovery from central nervous system damage, and misperceptions and mislocalizations after damage to sensory systems.

We are all aware of changes in behavior that must reflect significant, persisting changes in brain microstructure. Most notably, we improve in sensory and motor skills with practice, adjust to visual and auditory impairments, and recover abilities lost immediately after brain damage. It also seems reasonable to suppose that alterations in sensory and motor systems themselves have much to do with these behavioral adjustments. Sensory and motor systems consist largely of interconnected orderly representations of sensory surfaces and body effectors, and such maps occupy most of the cortex in most mammals (Kaas, 1987). Thus, much of the brain of most mammals is directly involved in processing sensory information and in mediating motor performance. If these systems are mutable in adult mammals, much of the brain can be directly involved in behavioral change. However, the widespread belief was that sensory systems are highly plastic only during a short developmental time, the critical period (e.g., Fox, 1992; Hubel and Wiesel, 1970; see also Rauschecker, 1991). Early, innovative efforts to demonstrate plasticity in subcortical somatosensory representations of rats (e.g., Wall and Egger, 1971) and monkeys (Pollin and Albe-Fessard, 1979) were viewed with skepticism, few investigators were drawn into studies of adult plasticity, and a general acceptance of the validity of adult plasticity failed to emerge. This situation changed with investigations of adult plasticity in cortical representations in the early 1980s.

The clear advantage of studying sensory representations in cortex, rather than in subcortical structures, is that cortical maps often are large and easily accessible on the surface of the brain. Thus, it is possible to obtain detailed maps of normal organization using microelectrode mapping techniques (see Welker, 1976), to employ procedures that might alter the internal structure of normal representations, and then to compare detailed maps from experimental and control animals. As in the earlier experiments on reorganization in the spinal cord and brain stem, the initial experiments on plasticity in somatosensory cortex depended on producing major deprivations via nerve cuts or crushes. Such major deprivations, as for eye closures or nerve cuts in studies of developmental plasticity, were believed to be the most likely to produce map changes of magnitudes that could be effectively measured and documented.

JON H. KAAS Department of Psychology, Vanderbilt University, Nashville, Tenn.

This expectation was realized, and convincing changes in sensory, and now motor, maps have been demonstrated repeatedly in a number of laboratories (for review, see Kaas, 1991).

Although the existence of adult plasticity seems well established, important questions remain. First, what is the functional significance of map plasticity? The plasticity observed in the deprivation and overstimulation experiments appears similar to the more subtle changes in receptive field structure reported after manipulations of sensory stimuli (Pettet and Gilbert, 1992), changes in attention or emotional state (Desimone et al., 1990), and learning (Weinberger, David, and Lepun, 1993). The same neural mechanisms may be mediating map changes and some of these rapidly induced changes in receptive fields. Thus, experiments on the more dramatic forms of adult plasticity may help reveal the neural mechanisms of learning and perceptual modification, and changes in sensory and motor maps after deprivation might be those that mediate sensorimotor adjustments and learning. In addition, changes in maps might mediate recovery after damage to the nervous system. For example, after partial lesions of a cortical representation of the visual field—the middle temporal visual area (MT)—monkeys rapidly recover from a localized deficit in visual tracking (Newsome et al., 1985). Perhaps, as suggested by Yamasaki and Wurtz (1991), remaining parts of MT (or other areas) reorganize to recover and utilize the missing information. Finally, some reorganizations may not mediate recoveries, compensations, or learning, but may produce further malfunction by producing inappropriate responses to sensory stimuli. For example, mislocalization to an amputated arm of sensory tactile stimuli on the face in humans (Ramachandran, Rogers-Ramachandran, and Stewart, 1992) may be a result of the reorganization of somatosensory representations so that cortex normally activated by the arm is activated by receptors in the face, as can occur in monkeys with sensory loss (Pons et al., 1991). Thus, "arm" cortex may not respecify when activated by afferents from the face to help mediate face sensation but may persistently and incorrectly signal inputs from the missing arm. Tragically, similar types of reactivation without respecification could be responsible for the phenomena of phantom and thalamic pain (e.g., Melzack, 1990), where normal tactile stimuli become intensely painful. Because adult plasticity may relate to any or all of these possibilities, experiments on adult plasticity afford opportunities to make significant advances in understanding behavioral phenomena that currently seem mysterious.

Another issue is where does plasticity occur? Historically, there has been some reluctance to believe that primary sensory maps are capable of change in adults because stability in function would seem to depend on stability in representation. Instead, changes reflecting experience were thought to occur in higher-order association cortex. Although it is now clear that plasticity does occur in primary sensory areas, levels of plasticity may vary across levels in a system, and higher areas may be even more plastic. Studies of plasticity in primary sensory areas have not been motivated by the belief that these structures are the most likely place for placticity to occur but by practical concerns. The internal organizations of only primary and a few other sensory representations are known in enough detail, and can be revealed with enough fidelity, that modest changes in organization can be detected. Thus, there has been only limited progress in assessing the relative amounts of plasticity that occur at other levels in sensory systems, and more research is needed. In addition, comparisons across levels can be complicated by differences in the organization of sensory representations at successive levels, so that local changes in the activation pattern at one level may produce widespread effects at subsequent levels. It is important to determine whether changes in map structure observed at one level can be attributed to modifications relayed from another level.

Still another important issue is what mediates adult plasticity? Are the mechanisms similar to or different from those that permit developmental plasticity? Whereas altered growth and the formation of new connections can be important in developmental plasticity, regeneration and growth generally are not considered to be features of adult plasticity. Nevertheless, recent evidence suggests that axonal growth may be important.

The dynamic and changeable quality of sensory representations

Amputation of body parts and section or deactivation of sensory nerves (Metzler and Marks, 1979; Calford and Tweedale, 1988; 1991b; Byrne and Calford, 1991; Nicoletis et al., 1993a), retinal lesions (Chino et al., 1992), and electrical stimulation of cortex (Nudo, Jenkins, and Merzenich, 1990; Recanzone, Merzenich, and Dinse, 1992) all produce immediate changes in receptive fields of central neurons. These types of rapid changes seem very similar to the alterations produced by other more normal manipulations of neural pathways. For example, receptive field sizes and locations can be rapidly modified by directed attention (e.g., Desimone et al., 1990), associative sensory stimulation and learning (e.g., Delacour, Houcine, and Talbi, 1987; Diamond and Weinberger, 1989; Gonzalez-Lima and Aqudo, 1990; Scheich et al., 1992; Diamond, Armstrong-James, and Ebner, 1993; Weinberger, David, Lepun, 1993), stimuli falling outside the classical receptive field (e.g., Allman, Miezin, and McGuinness, 1985; Fiorani et al., 1992; Gilbert, 1992; Pettet and Gilbert, 1992), and repetitive stimulation (e.g., Bonds, 1991; Lee and Whitsel, 1992). The usual explanation for such rapid changes is the selective potentiation or weakening of various excitatory and inhibitory components of the multitude of synaptic inputs that affect the excitability of any central neuron. The connections within sensory hierarchies are highly divergent and convergent, and the traditional receptive field of a neuron reflects only a portion of its total synaptic input (see Snow et al., 1988; Roberts and Wells, 1990).

Although it is uncertain how the dynamics of synaptic influence shift from moment to moment in the situations just described, a number of possibilities have been discussed in the general context of the unmasking of ineffective synapses (Wall, 1977; Killackey, 1989). More specifically, it has long been obvious that selective denervation could remove feedforward activation of neurons with locally distributed inhibitory connections, thereby reducing the inhibition on neurons to the extent that they express previously subthreshold inputs. Indeed, when local inhibitory mechanisms are chemically suppressed, receptive fields of neurons in the altered cortex immediately enlarge (e.g., Dykes et al., 1984; Alloway, Rosenthal, and Burton, 1989; Jacobs and Donoghue, 1991).

Another possibility is that peripheral amputations and other traumatic manipulations, even in anesthetized animals, briefly activate widespread modulating systems that produce response enhancement. Intense, painful, and otherwise meaningful stimuli could activate the widely distributed noradrenergic and cholinergic brainstem projection systems, both of which have been associated with neuromodulation and potentiation (Armstrong-James and Fox, 1983; Foote and Morrison, 1987). The widespread release of excitatory neurotransmitters could enhance subthreshold activity, thereby enlarging receptive fields. The observation that amputations and denervations produce immediate expansions of receptive fields in both contralateral and ipsilateral somatosensory cortex (Calford and Tweedale 1990, 1991a,b) suggests the involvement of widespread systems. Both noradrenergic and cholinergic systems have been implicated as "permissive" factors in neuronal plasticity induced by learning and experience (Kasamatsu and Pettigrew, 1976; Bear and Singer, 1986; Ebner and Armstrong-James, 1990; however, see Tromblay et al., 1986) and in reorganizations of sensory maps after partial sensory deprivations (Juliano et al., 1990; Kano, Iino, and Kano, 1991; Lee, Weisskopf, and Ebner, 1991; Haniseh et al., 1992). Finally, the rapid strengthening or potentiation of synapses when the presynaptic and postsynaptic neurons are coactive, as suggested by Hebb (1949), is a mechanism of rapid change that may be mediated by *N*-methyl-D-aspartic acid (NMDA) receptors (Brown et al., 1988; Brown, Kairiss, and Keenan, 1990; Bear et al., 1990), perhaps in conjunction with neuromodulating substances.

The ability of all cortical maps to reorganize

In addition to the short-term dynamic changes in sensory representations, there are additional, often more dramatic alterations that take time to be expressed. These slowly developing changes may be dependent on mechanisms that require protein synthesis, structural modifications in neurons, and persisting changes in levels of neuromodulators and transmitters and their receptors.

Many of the early experiments demonstrated slowly developing reorganizations of primary somatosensory (area 3b) cortex of owl and squirrel monkeys. In these monkeys, the central fissure is shallow or absent, and most or all of the hand representation is accessible under visual guidance on the dorsolateral surface of the brain (figure 4.1A) The normal representation of the hand in area 3b is extremely orderly (figure 4.1B, C) in that the glabrous skin of the digits is consistently represented from thumb to little finger (digits 1–5) in a lateromedial cortical sequence, with the tips rostral and the bases caudal, followed by the pads of the palm (Merzenich et al., 1978; Kaas et al., 1979; Sur, Nelson, and Kaas, 1982). The dorsal skin of the hand activates an extremely small proportion of the hand representation, typically along the margins of the tissue devoted to the glabrous skin. Our original manipulation of the input

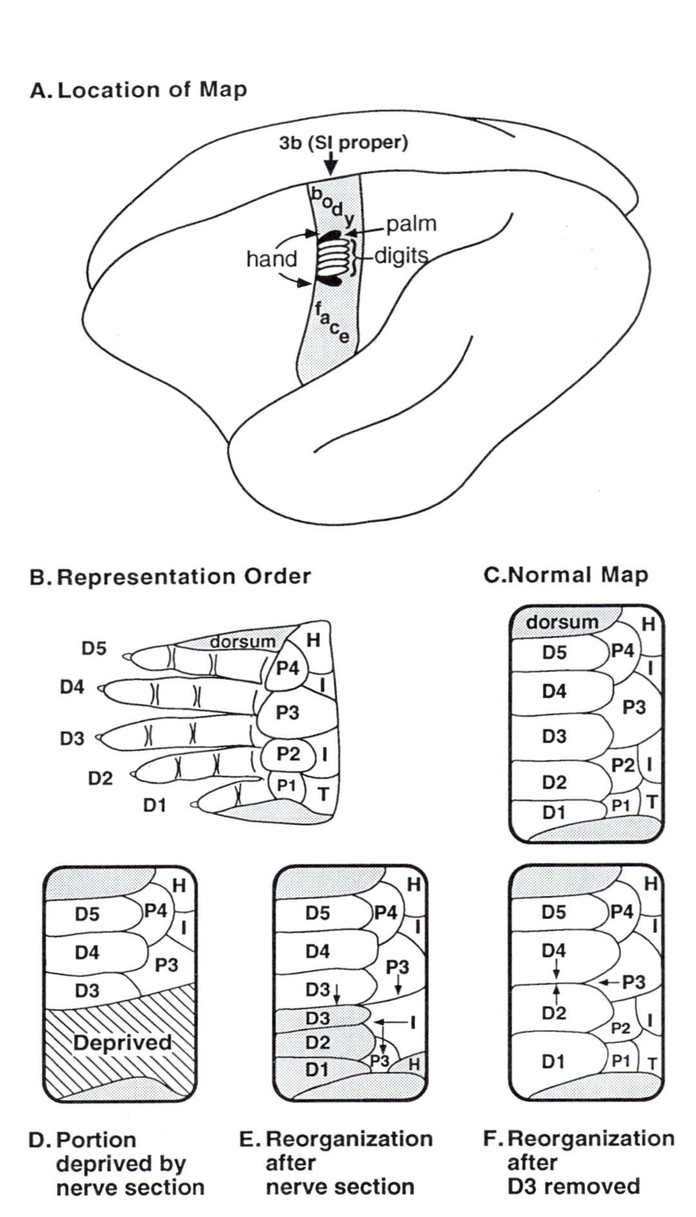

FIGURE 4.1 Reorganization of primary somatosensory cortex (area 3b) in owl monkeys. (A) Dorsolateral view of an owl monkey brain showing the location of the hand representation within area 3b. (B) Ventral surface of the hand split along the palm to reflect the somatotopic pattern of the hand representation. (C) Pattern of the hand representation in area 3b. (D) Portion of the hand representation deprived of normal activation by section of the median nerve. (E) Somatotopic pattern of the reorganized cortex months after median nerve section (based on Merzenich et al., 1983a). Most of the reactivation is from receptors on the dorsum of digits 1–3. (F) Reorganization after D3 removal (based on Merzenich et al., 1984). Digits and pads of the hand are traditionally numbered. Insular (I), hypothenar (H), and thenar (T) pads are indicated.

to this map was to cut and ligate the median nerve to the hand (Merzenich and Kaas, 1982; Kaas, Merzenich, and Killackey, 1983; Merzenich et al., 1983a; 1983b). This nerve subserves the thumb half of the glabrous hand, and cutting it removed all input from cutaneous receptors of the glabrous surfaces of digits 1, 2, and half of 3, and adjoining pads of the palm. The corresponding portion of the map in area 3b was thereby deprived of its normal source of activation (figure 4.1D). Immediately after the deactivation, some of the deprived cortex already could be activated by new inputs, but it took weeks to months before all of the deprived cortex was reactivated, mainly from inputs from the back of the hand (figure 4.1E). Thus, inputs from the back of the hand replaced inputs from the front of the hand, producing a greatly enlarged representation of the hairy surfaces of digits 1–3. In related experiments, Merzenich and colleagues (1984) demonstrated reorganization of the hand map after removing a digit (figure 4.1F). After several months of recovery, the part of area 3b normally devoted to the glabrous surface of the amputated third digit became activated by inputs from the glabrous surfaces of digits 2 and 4 (see Code, Eslin, and Juliano, 1992, for additional results). These experiments established the concept that neurons in the somatosensory cortex of adult mammals can acquire new receptive fields. Subsequently, a number of other manipulations have been shown also to alter map structure in area 3b of these and other monkeys (see Kaas, 1991).

Similar results have been obtained after deprivation from primary somatosensory cortex (S1) of other mammals. After removal of a digit in raccoons, S1 cortex formerly activated by that digit gradually becomes responsive to the adjoining digits (figure 4.2) (Welker and Saidenstein, 1959; Rasmusson, 1982; Kelahan and Doetsch, 1984; Turnbull and Rasmusson, 1991). The reactivation is particularly obvious in raccoons because they have a large hand representation that is exposed primarily on the cortical surface of the brain, and shallow sulci separate the gyral representations of the individual digits. Thus, the location of cortex normally representing any digit is apparent on visual inspection of the cortical surface, and abnormal organization is easily revealed. Reactivations of deprived portions of S1 after removing inputs have also been reported for: (1) rats, in which inputs from the saphenous nerve of the hind paw expanded their cortical territory to occupy partially the sciatic nerve terri-

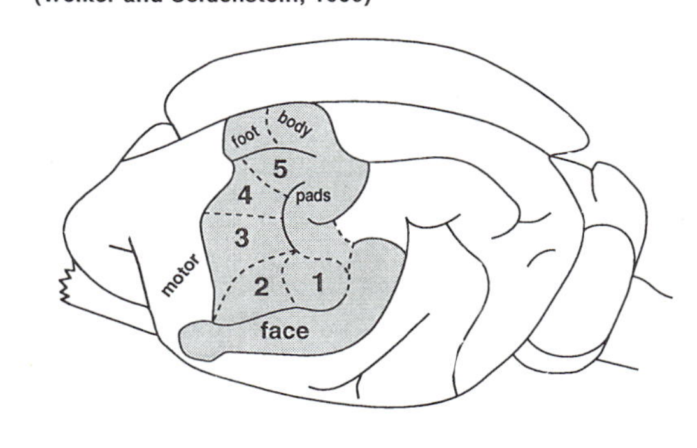

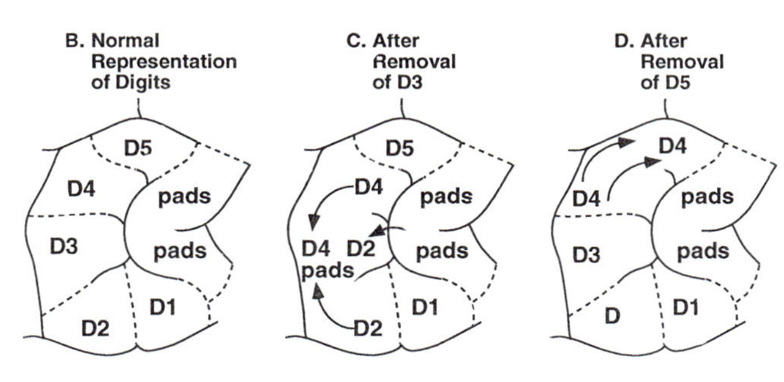

FIGURE 4.2 Reorganization of primary somatosensory cortex (S1) in raccoon after amputation of a digit. (A) Location of S1 on a dorsolateral view of the brain. Digits of the hand are numbered 1–5 from thumb to little finger. (B) Normal representation of the digits in S1. (C) Altered organization of the hand representation after the removal of digit 3. Neurons in the deprived D3 cortex became responsive to stimuli on digits 2 and 4 and the adjoining palmar pad. (D) Activation of the D5 cortex by D4 after removal of digit 5. (Based on Kelahan and Doetsch, 1984; and Rasmusson, 1982.)

tory after sciatic nerve section (Wall and Cusick, 1984; Cusick et al., 1990); and in which the column of cortex activated by a single facial vibrissa enlarged after long-standing lesions of the follicles of other whiskers (Kossut et al., 1988; Levin and Dunn-Meynell, 1991); (2) bats, in which cortex for the climbing digit of the wing became activated by inputs from the adjoining wing after digit amputation (Calford and Tweedale, 1988); and (3) cats, in which some neurons in forepaw cortex acquired receptive fields on the forearm after forepaw denervations (Kalaska and Pomeranz, 1979). Finally, there is recent evidence for the reorganization of primary somatosensory cortex in humans. After the surgical separation of fused fingers, finger representations appeared more distinct when measured with magnetoencephalography (Mogilner et al., 1993). Thus, the somatotopic organization of primary somatosensory cortex is mutable across a range of mammals.

Of course S1 is not the only somatotopic cortical representation. The number of such representations appears to vary across mammalian taxa, and the full number may not have been revealed in any species. In monkeys, seven systematic representations of the body have been identified in cortex, one each in the four architectonic fields of traditional S1 (3a, 3b, 1, and 2) (see Kaas et al., 1979; Kaas, 1983) and additional representations in the second area (S2), the parietal ventral area (PV), and the ventral somatosensory area (VS) (Cusick et al., 1989; Krubitzer and Kaas, 1990). Presumably, all these representations are capable of reorganization, but we presently have direct evidence from only a few. First, in experiments where nerve section in owl and squirrel monkeys produced reorganizations of deprived cortex in area 3b, similar reorganizations occurred in area 1 (Merzenich et al., 1983a,b; see also Garraghty and Kaas, 1991a). Second, although area 3a normally is activated almost exclusively by muscle and other deep receptors, training with tactile stimuli increases the responsiveness of area 3a to tactile stimuli (Jenkins et al., 1990; Recanzone, Merzenich, and Dinse, 1992). Third, the S2 of monkeys normally depends for activation on inputs from anterior parietal cortex (3a, 3b, 1, and 2) (Pons et al., 1987; Garraghty, Pons, and Kaas, 1990). Lesions of the hand representations in anterior parietal cortex partially deprive S2, and the deprived cortex reorganizes to become responsive to the foot and other body parts represented in the intact portions of anterior parietal cortex (figure 4.3) (Pons, Garraghty, and Mishkin, 1988). Thus, areas 3a, 3b, 1, and S2 are all capable of reorganization. However, experiments evaluating the mutability of area 2, PV, and VS have not yet been attempted. Similarly, in other mammals such as cats, where at least five cortical representations of cutaneous receptors exist (Clemo and Stein, 1983; Garraghty et al., 1987), representations outside S1 should be examined for plasticity.

In the visual system, retinotopic reorganizations of primary visual cortex (V1), have been demonstrated after lesions of the retina in cats (Kaas et al., 1990; Chino et al., 1992) and monkeys (Gilbert and Wiesel, 1992; Heinen and Skavenski, 1991). As for lesions of the peripheral nerves in the somatosensory system, lesions of the retina do not damage visual cortex directly but only indirectly produce cortical regions of deprivation. However, unlike the somatosensory system, where cutting a single nerve produces a zone of total deprivation, most of the visual cortex is binocularly driven. Thus,

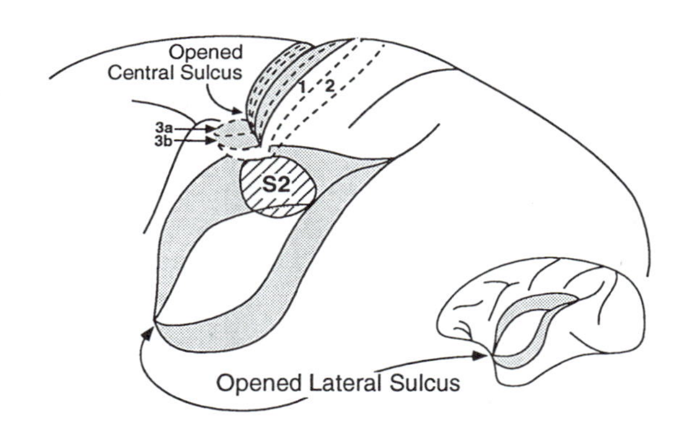

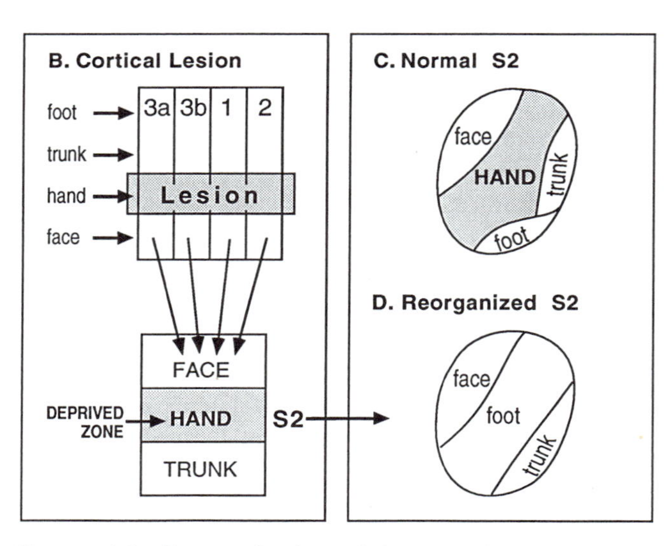

FIGURE 4.3 Reorganization of the second somatosensory area (S2) after lesions of the cortex representing the hand in anterior parietal cortex (areas 3a, 3b, 1, and 2). The immediate effect of the lesion is to silence the hand representation in S2. Some weeks later, this cortex becomes responsive to stimuli on the foot. (Based on Pons et al., 1988.)

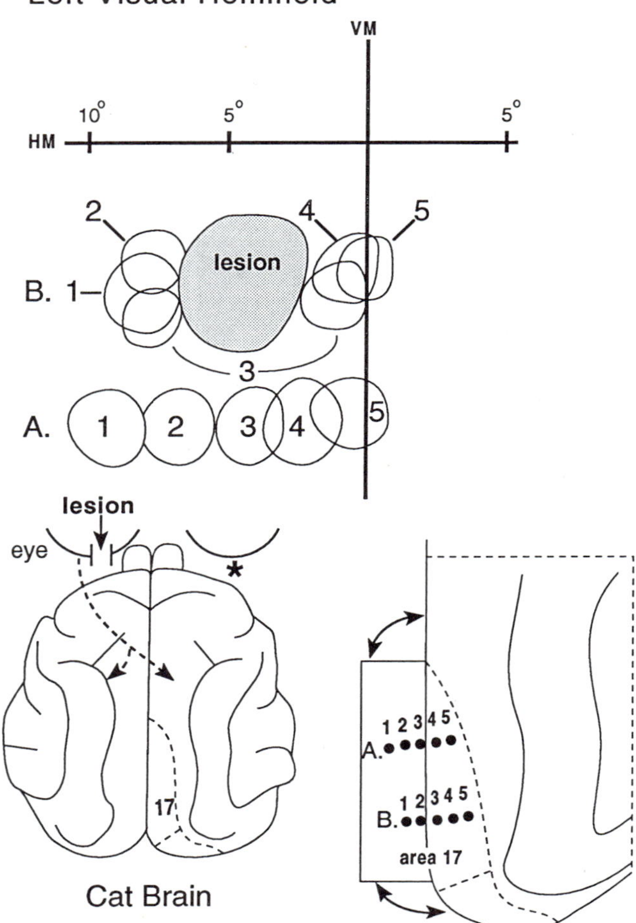

FIGURE 4.4 Reorganization of visual cortex after retinal lesions in cats. A small part of primary visual cortex was deprived of its normal source of activation by totally removing input from the ipsilateral eye and placing a small lesion in the retina of the contralateral eye. Recordings made in rows across area 17 (lower right) normally produce orderly rows of receptive fields in the contralateral visual hemifield (row A, top). In cortex deprived by the lesion, neurons have receptive fields that are displaced from the lesion to parts of the retina surrounding the lesion. Thus, receptive fields are piled up (row B, top), and the representation of the retina surrounding the lesion is expanded. (Based on Kaas et al., 1990.)

total deprivation is produced by placing lesions in matched locations in the two eyes or by removing the complete retina of one eye while placing a restricted lesion in the retina of the other. When retinal lesions are small, on the order of approximately 5° near central vision, a zone of unresponsive cortex is produced in the retinotopic map that becomes reactivated over weeks to months so that neurons acquire new receptive fields involving retinal locations along the margin of the lesion (figure 4.4). Larger lesions may produce a fringe of reorganized cortex around a central core of the cortex that remains unresponsive to visual stimuli.

Recordings in cats with retinal lesions also revealed that area 18 (V2) reorganizes (Kaas et al., 1990). The changes in V2 need not be those simply relayed from V1, as V2 is activated directly via geniculate inputs in cats (see Stone, 1983). The effects of deprivations on the many other visual areas of cats have yet to be determined. Monkeys also have a number of retinotopically organized visual areas (Kaas, 1989), but the potential for reorganization in adults has not been directly studied in extrastriate fields. However, plasticity has been demonstrated in inferior pulvinar. Bender (1983) found that the immediate effect of area 17 lesions was to abolish visual responsiveness totally in the inferior pulvinar but, after 3 weeks, some 15% of neurons recovered visual responsiveness.

Primary auditory cortex (A1) also reorganizes after partial deprivations. A1 normally represents tones from high to low in a progression across one surface dimension of the field, with lines of isorepresentation of the same frequencies extending across the other surface dimension (see Merzenich and Kaas, 1980). We evaluated the potential for adult plasticity in the auditory cortex of adult macaque monkeys by using ototoxic drugs to create a cochlear hearing loss of high frequencies (Schwaber, Garraghty, and Kaas, 1993). By producing a hearing loss for frequencies above 10 kHz, most of the caudal half of A1 was deprived of its normal source of activation. Two to three months after the hearing loss, recordings in A1 showed that regions formerly responsive to high tones were responsive at normal thresholds to tones of 10 kHz or less (figure 4.5).

Similarly, Robertson and Irvine (1989) partially deafened guinea pigs by directly lesioning restricted portions of the organ of Corti. After 1 month or longer, the deprived zone of cortex was reactivated by sound frequencies adjacent to the frequency range damaged by the lesions.

Thus, in both monkeys and rodents, the auditory frequency maps reorganized after partial deafness. Comparable results have been obtained also in cats reared after neonatal, bilateral high-frequency loss (Mount et al., 1991), suggesting that reorganizations may occur in adult cats. More recently, Willott et al., (1993) reported that the high-frequency portion of A1 becomes devoted to middle frequencies in a genetic strain of mice that gradually acquires a loss of high-frequency input as a result of progressive sensorineural pathology of the basal region of the cochlea.

Other auditory areas may demonstrate plasticity as well. For example, besides A1, monkeys have rostral (R) and rostrotemporal (RT) fields that are tonotopi-

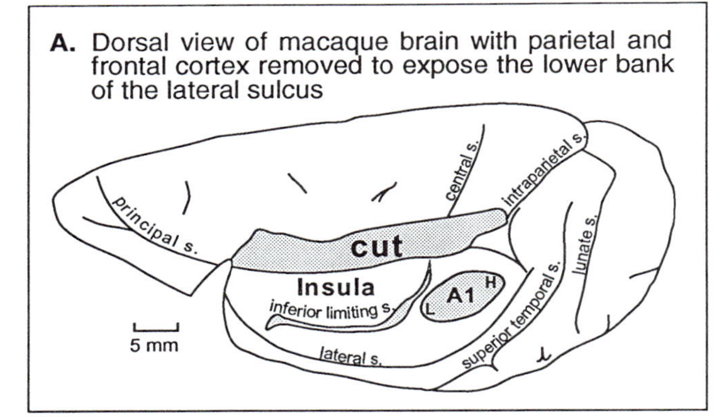

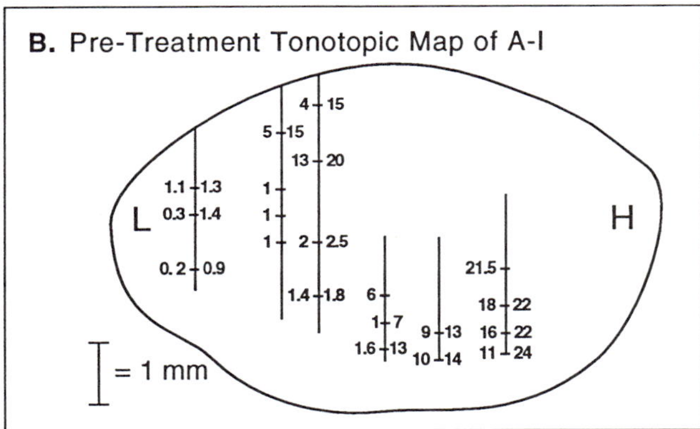

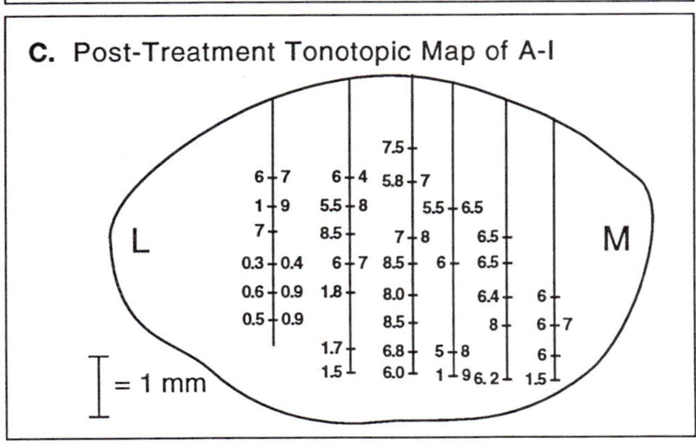

FIGURE 4.5 The reorganization of auditory cortex in macaque monkeys after ototoxic loss of high-frequency hearing. (A) Location of primary auditory cortex (A1) on the lower bank of the lateral sulcus. Overlying portions of parietal and frontal cortex have been cut away to reveal the lower bank and insula. A1 represents tone frequencies from low (L) to high (H). Sulci (S) are named as landmarks. (B) Before treatment, a large part of A1 was mapped with microelectrodes that penetrated lateromedially within A1 (lines). Best frequencies for neurons were determined in kilohertz for sites along the penetrations, as indicated. Low tones were represented rostrally and high tones, caudally. (C) Recordings made months after a high-frequency hearing loss (exceeding 10 kHz) revealed a reorganization of A1, with caudal A1 responsive to tones of less than 10 kHz. Thus, the progression across A1 was from low (L) to middle (M) frequencies. (Based on Schwaber, Garraghty, and Kaas, 1993.)

cally organized (Morel and Kaas, 1992; Morel, Garraghty, and Kaas, 1993). These and other fields need to be examined to determine how they respond to hearing loss.

Recent studies also indicate that motor cortex may reorganize after amputations or section of motor nerves. Several subdivisions of motor cortex exist in monkeys, including primary motor cortex (M1) the supplementary motor area (SMA), and two or more premotor fields (see Stepniewska, Preuss, and Kaas, 1993). At least M1 and probably SMA appear to exist in most other mammals as well (Kaas, 1987). These motor areas contain systematic representations of muscles or movements that can be revealed by electrically stimulating cortex at many sites with microelectrodes. Normal motor maps have somatotopic features that are consistent within a species, and these features appear to be relatively stable over time (Craigs and Rushton, 1976; Donoghue, Suner, and Sanes, 1990). However, removing an effector target by cutting a motor nerve or an amputation changes the motor map in at least M1. This has been shown most clearly in experiments on rats conducted by Sanes, Suner, and Donoghue (1990). Forelimb amputation or facial nerve transection first resulted in a deprived zone in M1, where no movements were evoked by electrical stimulation with microelectrodes. After a recovery period of 1 week or more, representations of other movements expanded in M1 so that stimulation of the deprived cortex evoked movements of intact structures at normal stimulation thresholds (figure 4.6). Similarly, evidence for the reorganization of M1 has been obtained from human patients with long-standing amputations of the arm (Cohen et al., 1991). In these patients, organizations of M1 ipsilateral and contralateral to the amputation were accessed by using local magnetic stimulation through the skull to evoke movements. Muscles of the upper arm in the stump could be activated from a larger area of cortex than could those muscles in the intact arm, suggesting that cortex contralateral to the stump had reorganized to devote more cortex to the remaining muscles.

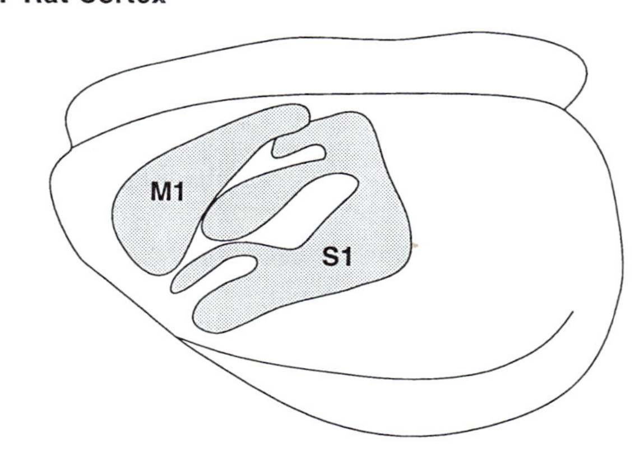

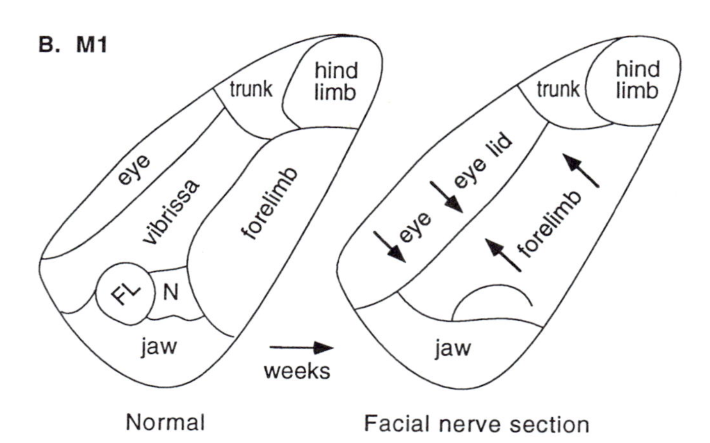

FIGURE 4.6 Reorganization of primary motor cortex of rats after section of the motor nerve to the musculature of the facial vibrissa. (A) Dorsolateral view of a rat brain showing the locations of M1 and primary somatosensory cortex (S1). (B) The somatotopy of M1 (left) becomes transformed (right) by section of the facial nerve. Cortex that formerly evoked vibrissa movements when electrically stimulated evoked eyelid or forelimb movements instead. FL, forelimb; N, neck. (Based on Sanes, Suner, and Donoghue, 1990.)

Where reorganizations occur

Much of the early interest and research on adult plasticity in the somatosensory system stems from the well-known study of Liu and Chambers (1958) purporting to show major axon sprouting of intact peripheral nerve inputs to the dorsal horn of the spinal cortex after synaptic space was made available by selective dorsal root rhizotomies. However, subsequent anatomical studies after such lesions failed to reveal spinal cord sprouting (see Rodin and Kruger, 1984; Rodin, Sampogna, and Kruger, 1983). Likewise, initial electrophysiological studies reported considerable reorganization of the somatotopic map in the spinal cord after such injuries, but later investigations produced more variable results, including no evidence of change (reviewed by Snow and Wilson, 1991). The variable results and lack of agreement about the extent or even existence of somatotopic changes in the spinal cord after injury brought the concept of subcortical plasticity into question. Nevertheless, three general conclusions are supported by the available evidence. First, reorganizations of subcortical structures do occur, and they may occur in all systems and at all subcortical levels. Second, persisting changes in subcortical relay stations of ascending sensory systems are limited in extent; often, they would have little impact on cortical maps. Third, because the anatomical and topographical features of subcortical and cortical maps may differ, there are instances where minor reorganizations at subcortical levels produce major changes in cortical maps.

The dorsal horn of the spinal cord has been rather extensively studied for evidence of plasticity (see Snow and Wilson, 1991, for review). Somatotopic reorganization does appear to occur in the dorsal spinal cord after peripheral nerve or dorsal root injury, but the magnitude of change appears to be close to the level of detectability with microelectrode mapping methods (Wilson and Snow, 1987). Reported changes in somatotopy are spatially limited, and new receptive fields on the skin are very close to original receptive fields. Somatotopic reorganization has also been reported for the dorsal column nuclei after dorsal rhizotomy (Dostrovsky, Millar, and Wall, 1976), but there is no clear evidence of large-scale modifications (see Snow and Wilson, 1991). Furthermore, deactivating parts of the gracile nucleus (McMahan and Wall, 1983) or trigeminal nuclei (Waite, 1984) by nerve section in adult rats produced no clear evidence of plasticity and reactivation. From these studies, it appears that nerve or dorsal root section is followed by no more than limited reactivation of deafferentated portions of the spinal cord and brain stem, yet we have indirect evidence that small changes in the dorsal column nuclei do occur and have large consequences in cortex, as is detailed later.

Reorganization is also expressed at the level of the thalamic relay of somatosensory information, but the evidence is limited. Early evidence that plasticity occurs in the ventroposterior nucleus (VP) of the thalamus comes from a brief report of Wall and Egger (1971). These investigators ablated nucleus gracilis of the medulla in rats, thereby depriving the lateral margin of VP of its normal source of activation from the hind limb. Recordings immediately after the damage revealed no change in somatotopy, but recordings days to weeks later suggested that inputs from the forelimb had expanded their territory to include the deprived hind-limb region of VP. However, the validity of this conclusion has been questioned, in part on the grounds that the magnitude of the reported changes is so small that recordings from distant neurons could explain the apparent expansion (Snow and Wilson, 1991). In another early study, limited recordings from two monkeys 2 weeks after section of fasciculus gracilis suggested an expansion of the forelimb representation into the deactivated hind-limb representation in VP (Pollin and Albe-Fessard, 1979), but the reported results were too sparse to be widely convincing. More recently, Rhoades, Belford, and Killackey (1987) lesioned the principle division of the trigeminal nuclear complex in rats, thereby depriving the face portion of VP (VPM) of its effective source of activation, but leaving the spinal trigeminal inputs intact. Over time, responsiveness to the face returned in VPM, and this return was attributed to the potentiation of the spinal trigeminal input. Finally, there is evidence that major reorganization of the somatotopy in VP occurs after section of the median and ulnar nerves to the glabrous surface of the hand in monkeys (Garraghty and Kaas, 1991b). In monkeys, a large subnucleus devoted to the hand in VP can be distinguished from the rest of the nucleus by partially encapsulating fiber bands. Approximately 90% of this subnucleus is activated by the volar hand (Kaas et al., 1984). After weeks of recovery following nerve section in squirrel monkeys, the complete hand subnucleus was activated by inputs from the dorsal surface of the hand. The reorganization at the level of the thalamus was so extensive that results could not be dismissed as a measurement error. In summary, the experiments of Rhoades, Belford, and Killackey (1987) suggest that alternate inputs can substitute for dominant inputs at the level of the thalamus. The major reorganization demonstrated in the hand subnucleus after peripheral nerve section (Garraghty and Kaas, 1991b) could result from a similar substitution at the thalamic level, but thalamic reorganization in this instance could also reflect changes relayed from the dorsal column nuclei to the thalamus (figure 4.7). Other evidence for reorganization in VP is equivocal.

As for the first relay stations in the somatosensory system, there seems to be only limited reorganization and reactivation in the lateral geniculate nucleus (LGN) of the visual system after retinal lesions (Eysel, Gonzalez-Aquilar, and Mayer, 1981; Eysel, 1982). After photocoagulation of small regions of the retina and weeks of recovery in cats, some neurons of the deprived zone of the LGN recovered responsiveness to visual stimuli directed to regions of the retina just around the

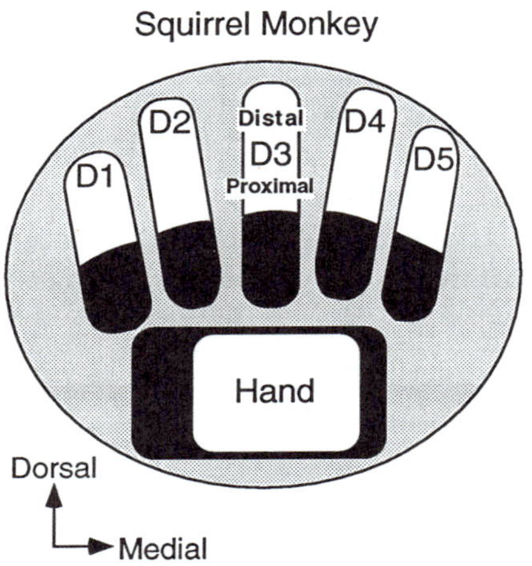

FIGURE 4.7 The somatotopic organization of the cuneate nucleus of the dorsal column–trigeminal complex of the lower brain stem of squirrel monkeys. The glabrous surface of each digit (D1–D5) is represented in a lateromedial sequence just dorsal to the representation of the pads of the palm. Fiber bands separate cell clusters devoted to these body parts. Inputs from the dorsal, hairy skin of each digit and the back of the hand terminate in the same cell clusters as the inputs from the glabrous skin. (Adapted from Florence, Wall, and Kaas, 1991.)

lesion. The results were similar to those obtained from visual cortex (Kaas et al., 1990; Chino et al., 1992), but the magnitude of the filling was much less. Thus, some plasticity does occur at the level of the LGN, but not all the changes observed in cortex are relayed from the thalamus.

We know little yet about the capacity of early stations in the auditory system to reorganize in adult mammals. After hamsters were exposed to tones intense enough to cause stereocilial damage, there was no clear evidence of plastic changes in the tonotopic map in the dorsal cochlear nucleus (Kaltenbach, Gaja, and Kaplan, 1992).

Major reorganizations of cortical sensory maps as a result of minor subcortical reorganizations

Different extents of reorganization imply different mechanisms of change. Small changes could be mediated by the potentiation of ineffective pre-existing connections, whereas major changes would often seem to depend on the growth of new connections. However, maps of the same sensory surface may differ in internal organization from one level to the next, so reorganizations that appear massive at one level may seem less extensive at another level. Topographical differences in maps are most obvious in the somatosensory system. For example, representations of the hand and foot are separated by representations of the limbs and trunk in area 3b of monkeys, whereas they adjoin in S2 (see Krubitzer and Kaas, 1990). These different topographical arrangements may be related to the observations that the foot substitutes for the missing hand input in S2 (Pons, Garraghty, and Mishkin, 1988) but not in area 3b (Florence et al., 1993b). Differences in the somatotopic organizations of maps at different levels may also explain why inputs from the back of the hand so effectively substitute for deactivated inputs from the glabrous hand in area 3b of monkeys (Merzenich et al., 1983a; Garraghty and Kaas, 1991a; Wall, Huerta, and Kaas, 1992). This massive reorganization in cortex largely or completely occurs subcortically, as it is expressed at the level of the VP (Garraghty and Kaas, 1991b). Because the dorsal hand is represented in locations separate from the glabrous hand in VP (Kaas et al., 1984), reorganization at the level of the thalamus would also be extensive. However, the somatotopic arrangement is different at the level of the dorsal column nuclei in the lower brain stem. In the cuneate nucleus of monkeys, the representation of the dorsal hand is distributed in a discontinuous pattern across an array of small clusters of neurons that are isolated from one another by surrounding bands of fibers (see Florence, Wall, and Kaas, 1991). Different cell clusters receive inputs from different digits or pads of the palm, but somewhat segregated parts of the same clusters receive inputs from the volar and hairy surfaces of the same digits (figure 4.8). We hypothesize that the massive substitution of hairy skin inputs for volar skin inputs at thalamic and cortical levels of the somatosensory system of monkeys actually occurs in the cuneate nucleus as a result of short-range substitutions within the cell clusters (see figure 4.8). However, this substitution has not yet been demonstrated in the cuneate nucleus, and a more extensive reorganization at the level of the thalamus remains a viable possibility (figure 4.9) (see Rhoades, Belford, and Killackey, 1987; Garraghty and Kaas, 1991b).

The reversibility of reorganizations

One wonders whether changes in sensory maps produced by partial deprivations or other types of altered sensory activation are reversed if normal patterns of activation return. There is little experimental evidence

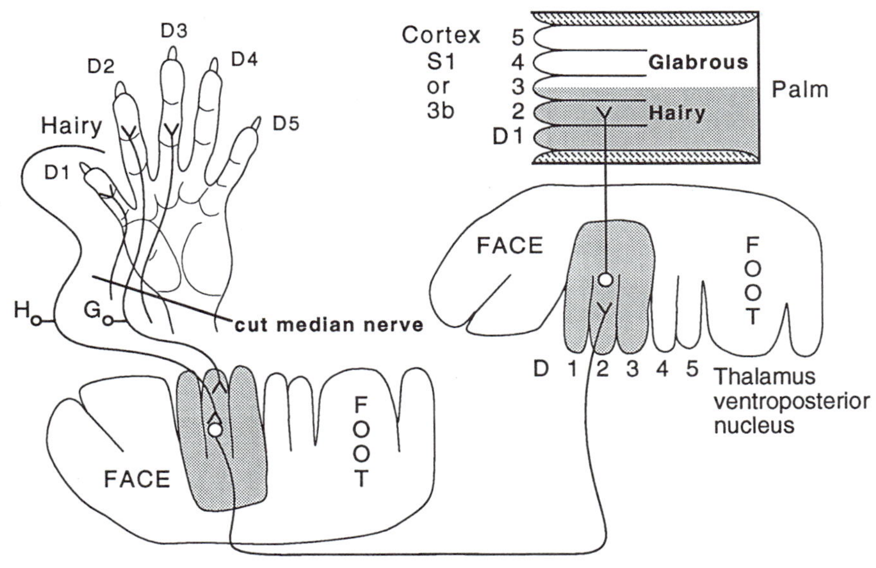

FIGURE 4.8 A substitution of dorsal hairy skin inputs for deactivated glabrous skin inputs in the cuneate nucleus after section of the median nerve may account for the reorganization observed in cortex (Merzenich et al., 1983a) and thalamus (Garraghty and Kaas, 1991b).

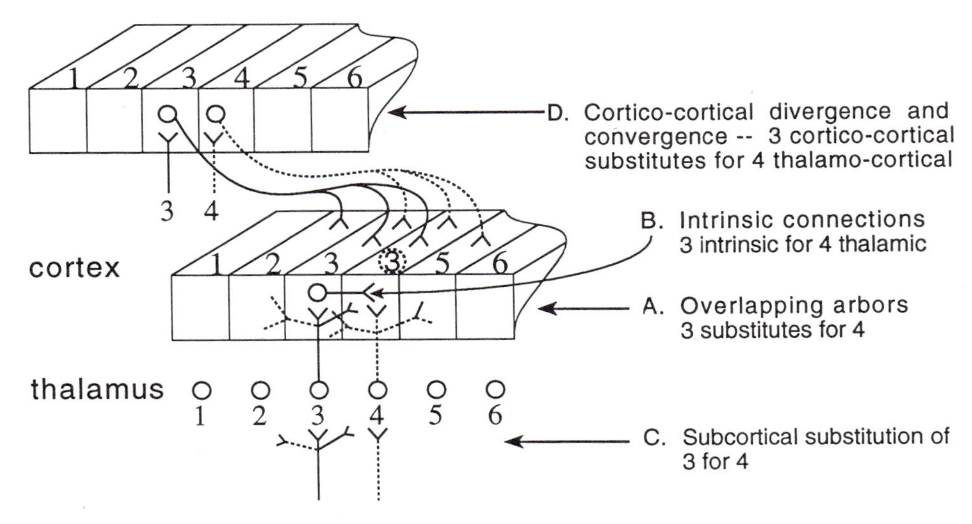

FIGURE 4.9 Most reorganizations probably depend on previously weak inputs substituting for deactivated or less active inputs. Substitutions can take place at all levels of sensory systems.

on the reversibility of the effects of partial deprivation on cortical maps, largely because the methods of producing deprivation are either short-term, such as in anesthetizing a sensory nerve, or irreversible because of damage, such as in an amputation. Even allowing a cut nerve to regenerate does not fully restore a previous pattern of activation of central maps, because the regeneration of cut nerves is generally very disorderly (Wall et al., 1986). Changes in cortical maps produced by anesthetizing a peripheral nerve (e.g., Metzler and Marks, 1979) are reversible after the anesthesia wears off but, of course, this is expected as the period of deprivation seems too limited to produce structural and, hence, persistent changes in the sensory network. In contrast, nerve crush can produce a long period of deprivation. For example, regeneration of the median nerve to the hand crushed at the level of the forearm takes more than 1 month (see Wall, Felleman, and Kaas, 1983). During the course of denervation produced by median nerve crush in monkeys, deprived parts of areas 3b and 1 of somatosensory cortex gradually reorganize, largely over the first 3 weeks, to become fully responsive to new inputs (Merzenich et al., 1983b). If these slowly developing activation patterns were based on major structural changes in the central nervous system, the altered cortical maps could persist after nerve regeneration. However, maps of area 3b and 1 made in individual monkeys before median nerve crush and after regeneration revealed that the normal maps returned completely (Wall, Felleman, and Kaas, 1983). Furthermore, the recovered organizations appeared to match closely the original maps in specific individuals. Thus, the original structural framework must have been maintained during deprivation, and the changes in activity patterns were completely reversible. This reversibility is consistent with the considerable behavioral evidence that normal tactile abilities return after regeneration of crushed nerves in humans (see Wall and Kaas, 1985).

The extent of reorganizations of primary somatosensory areas

The proportion and absolute amount of S1 (3b) that can be reactivated after deprivation varies considerably. At one end of the reorganization spectrum, only approximately 0.3 mm^2 of hind-paw cortex is reactivated by saphenous nerve inputs after sciatic nerve section in rats, and most of the deprived cortex persistently remains unresponsive to tactile stimuli (Wall and Cusick, 1984; Cusick et al., 1990). Similarly, after removal of a single digit, the deprived cortex is reactivated by inputs from adjoining digits in area 3b of monkeys, but digit cortex is not fully reactivated if two or more digits have been lost (Merzenich et al., 1984). In contrast, half to most of hand cortex in area 3b is deprived and then becomes responsive to inputs from the back of the hand after median or median–plus–ulnar nerve section in owl and squirrel monkeys (Merzenich et al., 1983a; Garraghty and Kaas, 1991a; Wall, Huerta, and Kaas, 1992). However, most or all of this reorganization occurs subcortically (Garraghty and Kaas, 1991b), perhaps at the level of the cuneate nucleus, where local modifications in connections could mediate the changes. When the radial nerve to the back of the hand is sectioned in combination with one of the two nerves to the glabrous hand—the median and ulnar nerves—most of the deprived cortex remains unresponsive to cutaneous stimulation (Garraghty et al., 1992). Furthermore, in macaque monkeys, where the median nerve may innervate part of the dorsal, hairy hand in addition to part of the glabrous hand, and the innervation pattern of the cuneate nucleus may differ from that in New World monkeys (see Florence, Wall, and Kaas, 1989), section of the median nerve produces a zone of deprived cortex in area 3b that does not completely recover responsiveness to cutaneous stimuli (unpublished studies). Thus, it appears that S1 or area 3b usually has only a limited capacity for reorganization after partial deprivations. However, there are situations where major changes have been demonstrated.

The most dramatic demonstration of major reorganization in cortex as a result of deprivation comes from the experiments of Pons and colleagues (1991) on monkeys in which all the sensory inputs from the forelimb had been removed by sectioning the dorsal roots of peripheral nerves as they enter the spinal cord. As part of an unrelated study, these monkeys had been deprived of sensory information from the hand, forearm, and upper arm, but not motor outflow, and had received training to use the deafferented limb. Twelve or more years after deafferentation, recordings in these monkeys showed that the complete hand, wrist, forearm, and upper arm regions of areas 3b and 1 were responsive to tactile stimuli on the face. In related experiments, we have studied cortex in monkeys that have had digit or hand amputations as a result of accidental injuries. In one macaque monkey studied 8 years after amputation of an injured hand, much of the deprived cortex in area 3b was responsive to remaining inputs on the wrist, although some of the cortex was unresponsive (Florence et al., 1993b). We do not yet understand how these massive reorganizations occur, but such results imply the growth of new connections (figure 4.10).

It remains uncertain whether primary auditory and visual areas of cortex also differ in the extent of reorgani-

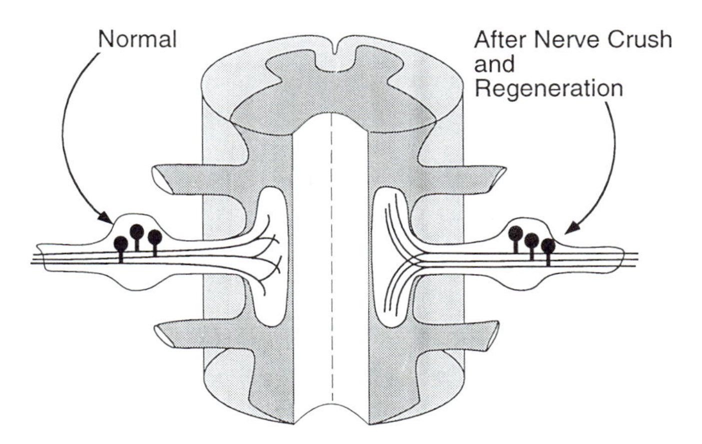

FIGURE 4.10 Some reorganizations may be the result of the growth and formation of new connections. Crushing the median nerve in adult monkeys induces growth in the central arbors of the regenerating afferents so that they overlap more extensively in the dorsal horn of the spinal cord. (Based on Florence et al., 1993.)

zation with variables such as the type or length of deprivation. After larger lesions of the retina in cats, deprived portions of areas 17 and 18 are not completely reactivated, at least within a period of months (Kaas et al., 1990), but longer recovery periods have not been tested. Also, after partial damage to the cochlea, reactivations of auditory cortex have been complete (Robertson and Irvine, 1989; Schwaber, Garraghty, and Kaas, 1993; Willott, Aitkin, and McFadden, 1993). Clearly, the limits of reorganization have not yet been determined.

How reorganizations are mediated

Rapid changes in sensory representations probably are mediated largely by three mechanisms: (1) immediate adjustments in the dynamics of the sensory network by changes in afferent drive (see Nicoletis et al., 1993a,b); (2) the release of neuromodulators that alter the effectiveness of pre-existing synapses (see Dykes, 1990; Juliano et al., 1990; Grasse, Douglas, and Mendelson, 1993); and (3) the potentiation of pre-existing synapses via the NMDA class of glutamate receptors and Hebbian rules (see Bear, Cooper, and Ebner, 1987; Brown, Kairiss, and Keenan, 1990; Diamond, Armstrong-James, and Ebner, 1993). Persisting changes in map structure can result from mechanisms that maintain the rapid alterations and from slowly developing, self-regulatory modifications in neuron structure and function.

Many of the rapidly induced changes in sensory maps that follow sensory pairing and other learning paradigms (e.g., Diamond, Armstrong-James, and Ebner, 1993) probably disappear rapidly. Long-term potentiation is the postulated mechanism of changes in synaptic effectiveness in such situations, and long-term potentiation typically dissipates over minutes to hours (Brown, Kairiss, and Keenan, 1990). Of course, a synaptic pathway, once strengthened above threshold, can be maintained via repeated activation, and there may be other mechanisms of maintaining synaptic strength. In either case, the importance of Hebbian mechanisms and NMDA receptors in allowing more active inputs to substitute for less active inputs in even slowly developing and long-lasting adult plasticity is suggested by experiments wherein the reorganization of area 3b of somatosensory cortex after nerve section in monkeys was prevented by blocking NMDA receptors during the recovery period (Garraghty, Muja, and Hoard, 1993). Similarly, reorganization of the hind-limb portion of S1 of adult cats after selective deafferentation was prevented by the continuous infusion of an NMDA receptor antagonist (Kano, Iino, and Kano, 1991). Modulating neurotransmitters such as acetylcholine and norepinephrine may participate in the stabilization process, as depletion of these substances interferes with aspects of developmental plasticity (Bear and Singer, 1986), and acetylcholine depletion prevents some types of adult plasticity (Juliano, Ma, and Eslin, 1991; Webster et al., 1991).

Changes in activity produced by long-standing deprivations or overstimulations probably induce a host of self-regulatory changes in molecular expression that could change the thresholds for synaptic activation. For example, reduced activity as a result of deprivation reduces the expression of metabolic enzymes such as cytochrome oxidase (e.g., Land and Akhtar, 1987; Wong-Riley and Welt, 1980) and succinic dehydrogenase (e.g., Dawson and Killackey, 1987). Long-standing sensory deprivation in monkeys with major reorganization of area 3b (Pons et al., 1991) also resulted in both decreases and increases of different calcium-binding proteins in the somatosensory thalamus (Rausell et al., 1992). These examples suggest that the expression of many neuroregulatory substances depends on levels of neuronal activity, although changes in expression of any particular substance may or may not have a role in plasticity. Levels of production and release of excitatory and inhibitory transmitters may also be regulated by neuronal activity. Most notably, days of continuous whisker stimulation in rats produced an increase in the expression of the inhibitory neurotransmitter gamma-aminobutyric acid (GABA) in the affected cortex (Welker, Soriano, and van der Loos, 1989). This increase in GABA resulted in a decrease in responsiveness of S1 cortex to normal stimulation of the overstimulated whisker (Welker et al., 1992). Understimulation produced the opposite effect in that the deprived cortex expressed less GABA, and therefore the cortex could be more responsive to weak inputs. There are other examples of activity-related changes in neurotransmitter levels: Median nerve section reduces GABA levels in the deprived cortex of monkeys (Garraghty, Lachica, and Kaas, 1991; see also Rausell et al., 1992), whisker removal has a similar effect on S1 of rats (Land and Akhtar, 1987; Welker, Soriano, and Van der Loos, 1989), and eye removal or eyelid suture depletes GABA in visual cortex of monkeys (Hendry and Jones, 1986). Finally, median and ulnar nerve section in monkeys is followed by a reduction in the neuromodulatory peptide tachykinin in the intrinsic neurons of deprived area 3b (Cusick, 1991).

Reduced activity could also promote several types of neuronal growth that would tend to restore normal activity levels, including the formation of new synapses, extension of dendrites, and the local and even extensive growth of axons and axon arbors. Such structural features of neurons are extremely modifiable during development, but neurons appear to lose morphological plasticity progressively with maturation (Horn, Rose, and Bateson, 1973). Nevertheless, there is evidence that spinal cord damage in adult rats induces increased synaptic turnover in cortex (Ganchrow and Bernstein, 1981), and changes in use of the forelimb correlate with dendrite growth in motor cortex (Greenrough, Larson, and Withers, 1985; Jones and Schallest, 1992). The growth of axons and the formation of new connections can clearly be induced by injury in the developing brain (e.g., Fitzgerald, Woolf, and Shortland, 1990; Rhoades, Belford, and Killackey, 1987), but such growth has been more difficult to demonstrate in adult animals, and there are many situations where there is no apparent sprouting of axons into deprived parts of sensory systems (e.g., Eysel, 1982; Rodin, Sampogna, and Kruger, 1983; Rodin and Kruger, 1984; Rasmusson and Nance, 1986). However, damaged peripheral nerve inputs into the spinal cord can expand into adjacent territories vacated by sectioning dorsal roots or peripheral nerves in adult rats (e.g., Molander, Kinnman, and Aldskogius, 1988; Woolf, Shortland, and Coggeshall, 1992). Similarly, we found that median nerve crush alone induces axonal growth within the deprived zone of spinal cord of adult monkeys (Florence et al., 1993a) (figure 4.10). The nerve crush may induce a growth state, and the lack of evoked activity in the central terminations of the injured axons may allow them to grow into one another's territories. If discorrelations in evoked activity patterns of neurons with different receptive fields have a role in restructuring growth and limiting arbor overlap in adult mammals, as in developing mammals (e.g., Shatz, 1990), then any loss or reduction of evoked activity can be a permissive factor in axon sprouting and growth in the adult nervous system. In adult rats, vibrissae removal is followed by enlargements of the representations of a remaining vibrissa in S1 (Kossut et al., 1988) and by increased expression of growth associated protein (GAP-43) (Dunn-Meynell, Benowitz, and Levin, 1992), suggesting the induction of axonal sprouting.

Presently, there is no clear evidence that any of the reorganizations of sensory representations in adult mammals depend on extensive axon growth; yet there is no obvious alternative explanation for the major reactivations that occur after deafferentation of the complete forelimb (Pons, Garraghty, and Mishkin, 1988) or loss of a hand (Florence et al., in press) in monkeys. In these cases, the long-standing deprivation and reduced activity of central neurons could induce sprouting of nearby axons into deprived tissue, perhaps by the release of some growth factor. Such axon sprouting would seem more likely at the levels of the lower brain stem or thalamus, where distances would be shorter than in cortex. On a smaller scale, a continuous process of neuronal growth and retraction may be important in the maintenance and reshaping of sensory maps throughout life. In an early, innovative study, Rose and coworkers (1960) reported that laminar lesions of visual cortex in adult rabbits, produced by irradiation, resulted in the massive sprouting and growth of axons from normal tissue into the zone of nerve cell loss. The authors suggested that this massive growth was part of the "normal, continuous growth of central neurons."

Perceptual and behavioral consequences of reorganization

Reassignments of neurons in cortical maps from one to another part of a receptor surface could have several different functional consequences. Adding more neurons to the processing circuits for a given part of a receptor surface could enhance the capacity of that circuit. Some evidence is consistent with this view. For example, Recanzone and colleagues (1992) reported that monkeys trained to make a tactile discrimination of the frequency of a vibrating probe on a specific finger improved over time, and the experience increased the size of the cortical representation of that finger. In a similar manner, humans with deafferented zones of skin may express hypersensitivity and increased tactile capacity on skin next to the denervated zone (see Wall and Kaas, 1985), and this skin probably has an increased cortical representation. The fact that, with practice, we are capable of improving performance of most sensorimotor tasks implies that the central processing of afferent inputs is not limited initially by receptor elements and that enlarging the neural pool for given inputs can mediate improvements. Reorganizations that follow partial lesions of cortical areas seem to have clear potential for mediating recovery. Partial lesions of cortical maps in humans and monkeys (e.g. Newsome et al., 1985) commonly are followed by perceptual deficits that rapidly diminish over days to weeks. In these cases, normal performance is impaired by reductions in the numbers of neurons in relevant central circuits, and any reassignment of neurons from other circuits to the damaged circuits could lead to improved performance.

There are also reasons to conclude that adding neurons to a network may not always improve performance. If performance is limited by the density of the receptor array, adding more neurons to a processing circuit may produce no improvement. For example, greatly expanding the normally small cortical representations of the back of the hand in monkeys by sectioning cutaneous nerves to the volar hand may not improve tactile acuity because it is already maximal for the low density of receptors on the dorsal hand.

Instead of improving performance, adding new neurons to central circuits sometimes may degrade capacity. Neurons with reorganized inputs may maintain their original output and effector targets so that errors are made and misperceptions occur. Hand areas of cortex, when activated by the foot or face, may continue to signal stimuli on the hand. Partial deafferentations of skin regions commonly result in mislocalizations of tactile stimuli (see Wall and Kaas, 1985). For example, Ramachandran, Rogers-Ramachandran, and Stewart, (1992) recently reported that patients with forelimb amputations felt tactile stimuli on the face as being both on the face and on the missing forelimb. The fact that most amputees have phantom sensations of missing limbs (Melzack, 1990) suggests that central representations of limb inputs are being activated and continue to mediate the perception of the missing limb. Similarly, phantom auditory perception (tinnitus) distresses a large portion of individuals with inner ear pathological processes; one postulated cause of tinnitus is that decreased input from the periphery causes plastic changes in synaptic weights within the central auditory system, and false sensations result (Jastreboff, 1990).

Conclusions

Over the last 15 years, the plasticity of sensory and motor representations in the brains of adult mammals has been extensively studied, and a number of major conclusions are now supportable.

1. We know that sensory representations, especially at the cortical level, are normally quite dynamic. Receptive field sizes and locations are modified rapidly by natural stimuli outside the classical receptive field, attention and other modifications of the behavioral state, within-field stimuli, learning, and sensory pairing. Most of these modifications are rapidly reversible and thus differ from the more persistent reorganizations of sensory maps in adults that are reminiscent of developmental plasticity. However, if the sensory bias is long-term, the neural change may be enduring.

2. Reorganizations of sensory maps are produced by manipulations that alter patterns of sensory and other neural activity. The most obvious reorganizations follow partial removal or deactivation of afferent drive by nerve section, receptor surface damage, or central lesions. Alterations are also produced by prolonged localized sensory stimulation, sensory experience, and electrical stimulation of sensory maps. More active inputs appear to expand and substitute for less active inputs.

3. Reorganizations have been demonstrated in the somatosensory, auditory, visual, and motor systems at both cortical and subcortical levels, and in a wide range of mammalian species. Thus, the capacity for reorganization seems to be a fundamental characteristic of the adult nervous system, and plasticity may be possible throughout the nervous system. Nevertheless, the most dramatic examples of reorganization are at the cortical level, and changes at early stages of processing may be quite limited.

4. Changes in sensory representations that occur within minutes or hours suggest the rapid potentiation of previously existing synapses. Possible mechanisms include the reduction of afferent drive of off-focus inhibition, the potentiation of synapses via correlated presynaptic and postsynaptic activity, and the involvement of NMDA glutamate receptors, as well as the enhancement of weak activation via the release of neuromodulators. Many of the rapid changes in organization dissipate rapidly as well when premanipulation conditions return.

5. Other modifications in map structure take days to weeks and possibly longer to emerge. These slowly emerging changes are compatible with synaptic and membrane receptor turnover and selection, axon and dendritic growth, and reductions and increases in the production of neurotransmitters and neuromodulators. Most map changes likely depend on a number of rapidly and slowly emerging mechanisms.

6. The reversibility of slowly emerging map changes in somatosensory cortex produced by nerve crush with nerve regeneration suggests that the basic anatomical framework for map organization is relatively stable and that structural organizations that emerge during development tend to persist.

7. Maps at different levels of sensory systems may differ in the way they represent a sensory surface. Thus, local modifications in one map may have widespread consequences in the map at the next level. In addition, maps at higher levels express greater change as a result of accumulating the effects of modifications at earlier levels.

8. There is only limited evidence on the behavioral consequences of map reorganization. Changes in sensory maps could lead to improvements in sensorimotor skills, compensations for sensory losses and impairments, and recoveries of lost abilities following central nervous system damage. Reorganizations could also produce misperceptions and perceptual errors. Modification in sensory systems may mediate all of these behavioral changes.

REFERENCES

Allman, J., F. Miezin, and E. McGuinness, 1985. Stimulus specific responses from beyond the classical receptive field: Neurophysiological mechanisms for local-global comparisons in visual neurons. *Annu. Rev. Neurosci.* 8:407–430.

Alloway, K. D., P. Rosenthal, and H. Burton, 1989. Quantitative measures of receptive field changes during antagonism of GABAergic transmission in primary somatosensory cortex of cats. *Exp. Brain Res.* 78:514–532.

Armstrong-James, M., and K. Fox, 1983. Effects of iontophoresed noradrenaline on the spontaneous activity of neurones in rat primary somatosensory cortex. *J. Physiol. (Lond.)* 335:427–447.

Bear, M. F., L. N. Cooper, and F. F. Ebner, 1987. A physiological basis for a theory of synapse modification. *Science* 237:42–48.

Bear, M. F., A. Kleinschmidt, Q. A. Gu, and W. Singer, 1990. Disruption of experience-dependent synaptic modifications in striate cortex by infusion of an NMDA receptor antagoinst. *J. Neurosci.* 10:909–925.

Bear, M. F., and W. Singer, 1986. Modulation of visual cortical plasticity by acetylcholine and noradrenaline. *Nature* 320:172–176.

Bender, D. B., 1983. Visual activation of neurons in primate pulvinar depends on cortex but not colliculus. *Brain Res.* 279:258–261.

Bonds, A. B., 1991. Temporal dynamics of contrast gain in single cells of the cat striate cortex. *Visual Neurosci.* 6:239–255.

Brown, T. H., P. E. Chapman, E. W. Kairiss, and C. L. Keenan, 1988. Long-term synaptic potentiation. *Science* 242:724–728.

Brown, T. H., E. W. Kairiss, and C. L. Keenan, 1990. Hebbian synapses: Biophysical mechanisms and algorithms. *Am. Rev. Neurosci.* 13:475–511.

Byrne, J. A., and M. B. Calford, 1991. Short-term expansion of receptive fields in rat primary somatosensory cortex after hindpaw digit denervation. *Brain Res.* 565:218–224.

Calford, M. B., and R. Tweedale, 1988. Immediate and chronic changes in responses of somatosensory cortex in adult flying-fox after digit amputation. *Nature* 332:446–448.

Calford, M. B., and R. Tweedale, 1990. Interhemispheric transfer of plasticity in the cerebral cortex. *Science* 249:805–807.

Calford, M. B., and R. Tweedale, 1991a. Acute changes in cutaneous receptive fields in primary somatosensory cortex after digit denervation in adult flying foxes. *J. Neurophysiol.* 65:178–187.

Calford, M. B., and R. Tweedale, 1991b. Immediate expansion of receptive fields of neurons in area 3b of macaque monkeys after digit denervation. *Somatosens. Mot. Res.* 8:249–260.

Chino, Y. M., J. H. Kaas, E. L. Smith III, A. L., Langston, and H. Cheng, 1992. Rapid reorganization of cortical maps in adult cats following restricted deafferentation in retina. *Vision Res.* 32:789–796.

Clemo, H. R., and B. E. Stein, 1983. Organization of a fourth somatosensory area of cortex in cat. *J. Neurophysiol.* 50:910–925.

Code, R. A., D. E. Eslin, and S. L. Juliano, 1992. Expansion of stimulus-evoked metabolic activity in monkey somatosensory cortex after peripheral denervation. *Exp. Brain Res.* 88:341–344.

Cohen, L. G., S. Bandinelli, T. W. Findley, and M.

Hallett, 1991. Motor reorganization after upper limb amputation in man. A study with focal magnetic stimulation. *Brain* 114:615–627.

Craigs, M. D., and D. N. Rushton, 1976. The stability of the electrical stimulation map of the motor cortex of the anesthetized baboon. *Brain* 99:575–600.

Cusick, C. G., 1991. Injury-induced depletion of tachykinin immunoreactivity in the somatosensory cortex of adult squirrel monkeys. *Brain Res.* 568:314–318.

Cusick, C. G., J. T. Wall, D. J. Felleman, and J. H. Kaas, 1989. Somatotopic organization of the lateral sulcus of owl monkeys: Area 3b, S-II, and a ventral somatosensory area. *J. Comp. Neurol.* 282:169–190.

Cusick, C. G., J. T. Wall, J. H. Whiting, and R. G. Wiley, 1990. Temporal progression of cortical reorganization following nerve injury. *Brain Res.* 537:355–358.

Dawson, D. R., and H. P. Killackey, 1987. The organization and mutability of the forepaw and hindpaw representations in the somatosensory cortex of the neonatal rat. *J. Comp. Neurol.* 256:246–256.

Delacour, J., O. Houcine, B. Talbi, 1987. "Learned" changes in the responses of the rat barrel field neurons. *Neuroscience* 23:63–71.

Desimone, R., M. Wassinger, L. Thomas, and W. Schneider, 1990. Attentional control of visual perception: Cortical and subcortical mechanisms. *Cold Spring Harb. Symp. Quant. Biol.* 25:963–971.

Diamond, D. M., and N. M. Weinberger, 1989. Role of context in the expression of learning-induced plasticity of single neurons in auditory cortex. *Behav. Neurosci.* 103:471–494.

Diamond, M. E., M. Armstrong-James, F. F. Ebner, 1993. Experience-dependent plasticity in adult rat barrel cortex. *Proc. Natl. Acad. Sci. U.S.A* 90:2082–2086.

Donoghue, J. P., S. Suner, and J. N. Sanes, 1990. Dynamic organization of primary motor cortex output to target muscles in adult rats: II. Rapid reorganization following motor nerve lesions. *Exp. Brain Res.* 79:492–503.

Dostrovsky, J. O., J. Millar, and P. D. Wall, 1976. The immediate shift of afferent drive of dorsal column nucleus cells following deafferentation: A comparison of acute and chronic deafferentation in gracile nucleus and spinal cord. *Exp. Neurol.* 52:480–495.

Dunn-Meynell, A. A., L. I. Benowitz, and B. E. Levin, 1992. Vibrissectomy induced changes in GAP-43 immunoreactivity in adult rat barrel cortex. *J. Comp. Neurol.* 315: 160–170.

Dykes, R. W., 1990. Acetylcholine and neuronal plasticity in somatosensory cortex. In *Brain Cholinergic Systems*, M. Steriade and D. Biesold, eds. New York: Oxford University Press, pp. 294–313.

Dykes, R. W., P. Landry, R. Metherate, and T. P. Hicks, 1984. Functional role of GABA in cat primary somatosensory cortex: Shaping receptive fields of cortical neurons. *J. Neurophysiol.* 52:1066–1093.

Ebner, F. F., and M. R. Armstrong-James, 1990. Intracortical processes regulating the integration of sensory information. *Prog. Brain Res.* 86:129–141.

Eysel, U. T., 1982. Functional reconnections without new axonal growth in a partially denervated visual relay nucleus. *Nature* 299:442–444.

Eysel, U. T., F. Gonzalez-Aquilar, and U. Mayer, 1981. Time-dependent decrease in the extent of visual deafferentation in the lateral geniculate nucleus of adult cats with small retinal lesions. *Exp. Brain Res.* 41:256–263.

Fiorani, M., M. G. P. Rosa, R. Gattass, and C. E. Rocha-Miranda, 1992. Dynamic surrounds of receptive fields in primate striate cortex: A physiological basis for perceptual completion? *Proc. Natl. Acad. Sci. U.S.A.* 89:8547–8551.

Fitzgerald, M., C. J. Woolf, and P. Shortland, 1990. Collateral sprouting of the central terminals of cutaneous primary afferent neurons in the rat spinal cord: Pattern, morphology, and influence of targets. *J. Comp. Neurol.* 286: 48–70.

Florence, S. L., P. E. Garraghty, M. Carlson, and J. H. Kaas, 1993a. Sprouting of peripheral nerve axons in the spinal cord of monkeys. *Brain Res.* 601:343–348.

Florence, S. L., N. Jain, P. D. Beck, and J. H. Kaas, 1993b. Reorganization of somatosensory cortex and changes in the peripheral nerve innervation patterns in spinal cord and brain stem after amputation of the hand in monkeys. *Soc. Neurosci. Abstr.* 19:1706.

Florence, S. L., J. T. Wall, and J. H. Kaas, 1989. Somatotopic organization of inputs from the hand to the spinal grey and cuneate nucleus of monkeys with observations on the cuneate nucleus of humans. *J. Comp. Neurol.* 286:48–70.

Florence, S. L., J. T. Wall, and J. H. Kaas, 1991. Central projections from the skin of the hand in squirrel monkeys. *J. Comp. Neurol.* 311:563–578.

Foote, S. L., and J. H. Morrison, 1987. Extrathalamic modulation of cortical function. *Annu. Rev. Neurosci.* 10:67–95.

Fox, K., 1992. A critical period of experience-dependent synaptic plasticity in rat barrel cortex. *J. Neurosci.* 12(5): 1826–1838.

Ganchrow, D., and J. H. Bernstein, 1981. Bouton renewal patterns in rat hindlimb cortex after thoracic dorsal funicular lesions. *J. Neurosci. Res.* 6:525–537.

Garraghty, P. E., D. P. Hanes, S. L. Florence, and J. H. Kaas, 1992. The extent of cortical reorganization after nerve injury is limited by the content of the deprivation. *Soc. Neurosci. Abstr.* 18:1548.

Garraghty, P. E., and J. H. Kaas, 1991a. Large-scale functional reorganization in adult monkey cortex after peripheral nerve injury. *Proc. Natl. Acad. Sci. U.S.A* 88: 6976–6980.

Garraghty, P. E., and J. H. Kaas, 1991b. Functional reorganization in adult monkey thalamus after peripheral nerve injury. *Neuroreport* 2:747–750.

Garraghty, P. E., E. A. Lachica, and J. H. Kaas, 1991. Injury-induced reorganization of somatosensory cortex is accompanied by reductions in GABA staining. *Somatosens. Mot. Res.* 8:347–354.

Garraghty, P. E., N. Muja, and R. Hoard, 1993. Role of NMDA receptors in plasticity in adult primary somatosensory cortex. *Soc. Neurosci. Abstr.* 19:1569.

Garraghty, P. E., T. P. Pons, M. F. Hureta, and J. H. Kaas, 1987. Somatosensory organization of S-III in cats. *Somatosens. Res.* 4:333–357.

Garraghty, P. E., T. P. Pons, and J. H. Kaas, 1990. Ablations of areas 3b (SI proper) and 3a of somatosensory cortex in marmosets deactivate the second and parietal ventral somatosensory areas. *Somatosens. Mot. Res.* 7:125–135.

Gilbert, C. D., 1992. Horizontal integration and cortical dynamics. *Neuron* 9:1–13.

Gilbert, C. D., and T. N. Wiesel, 1992. Receptive field dynamics in adult primary visual cortex. *Nature* 356:150–152.

Gonzalez-Lima, F., and J. Aqudo, 1990. Functional reorganization of neural auditory maps by differential learning. *NeuroReport* 1:161–164.

Grasse, K. L., R. M. Douglas, and J. R. Mendelson, 1993. Alterations in visual receptive fields in the superior colliculus induced by amphetamine. *Exp. Brain. Res.* 92:453–466.

Greenrough, W. T., J. R. Larson, and G. S. Withers, 1985. Effects of unilateral and bilateral training in a reaching task on dendritic branching of neurons in the rat motor-sensory forelimb cortex. *Behav. Neural Biol.* 44:301–314.

Haniseh, U-K., T. Rothe, K. Krohn, and R. W. Dykes, 1992. Muscarinic cholinergic receptor binding in rat hindlimb somatosensory cortex following partial deafferentation by sciatic nerve transection. *Neurochem. Int.* 21:313–327.

Hebb, D. O., 1949. *The Organization of Behavior: A Neuropsychological Theory*. New York: Wiley.

Heinen, S. J., and A. A. Skavenski, 1991. Recovery of visual responses in foveal V1 neurons following bilateral foveal lesions in adult monkey. *Exp. Brain Res.* 83:670–674.

Hendry, S. H. C., and E. G. Jones, 1986. Reduction in number of immunostained GABAergic neurons in deprived-eye dominance columns of monkey area 17. *Nature* 320:750–753.

Horn, G., S. P. R. Rose, and P. P. G. Bateson, 1973. Experience and plasticity in the central nervous system. *Science* 181:506–514.

Hubel, D. H., and T. N. Wiesel, 1970. The period of susceptibility to the physiological effects of unilateral eye closure in kittens. *J. Physiol.* 206:419–436.

Jacobs, K. M., and J. P. Donoghue, 1991. Reshaping the cortical motor map by unmasking latent intracortical connections. *Science* 251:944–947.

Jastreboff, P. J., 1990. Phantom auditory perception (tinnitus): Mechanisms of generation and perception. *Neurosci. Res.* 8:221–254.

Jenkins, W. M., M. M. Merzenich, M. T. Ochs, T. Allard, and E. Guic-Robles, 1990. Functional reorganization of primary somatosensory cortex in adult owl monkeys after behaviorally controlled tactile stimulation. *J. Neurophysiol.* 63:82–104.

Jones, T. A., and T. Schallest, 1992. Overgrowth and pruning of dendrites in adult rats recovering from neocortical damage. *Brain Res.* 581:156–160.

Juliano, S. L., W. Ma, M. F. Bear, and D. Eslin, 1990. Cholinergic manipulation alters stimulus-evoked metabolic activity in cat somatosensory cortex. *J. Comp. Neurol.* 297:106–120.

Juliano, S. L., W. Ma, and D. Eslin, 1991. Cholinergic depletion prevents expansion of topographic maps in somatosensory cortex. *Proc. Natl. Acad. Sci. U.S.A.* 88:780–784.

Kaas, J. H., 1983. What, if anything, is S-I? The organization of the "first somatosensory area" of cortex. *Physiol. Rev.* 63:206–231.

Kaas, J. H., 1987. The organization of neocortex in mammals: Implications for theories of brain function. *Annu. Rev. Psychol.* 38:124–151.

Kaas, J. H., 1989. Why does the brain have so many visual areas? *J. Cogn. Neurosci.* 1:121–135.

Kaas, J. H., 1991. Plasticity of sensory and motor maps in adult mammals. *Annu. Rev. Neurosci.* 14:137–167.

Kaas, J. H., L. A. Krubitzer, Y. M. Chino, A. L. Lanston, E. H. Polley, and N. Blair, 1990. Reorganization of retinotopic cortical maps in adult mammals after lesion of the retina. *Science* 248:229–231.

Kaas, J. H., M. M. Merzenich, and H. P. Killackey, 1983. The reorganization of somatosensory cortex following peripheral nerve damage in adult and developing mammals. *Annu. Rev. Neurosci.* 6:325–356.

Kaas, J. H., R. J. Nelson, M. Sur, R. W. Dykes, and M. M. Merzenich, 1984. The organization of the ventroposterior thalamus of the squirrel monkey, *Saimiri sciureus*. *J. Comp. Neurol.* 226:111–140.

Kaas, J. H., R. J. Nelson, M. Sur, C.-S. Lin, and M. M. M. M. Merzenich, 1979. Multiple representations of the body within "S-I" of primates. *Science* 204:521–523.

Kalaska, J., and B. Pomeranz, 1979. Chronic paw denervation causes an age-dependent appearance of novel responses from forearm in "paw cortex" of kittens and adult cats. *J. Neurophysiol.* 42:618–633.

Kaltenbach, J. A., J. M. Czaja, and C. R. Kaplan, 1992. Changes in the tonotopic map of the dorsal cochlear nucleus following induction of cochlear lesions by exposure to intense sound. *Hear. Res.* 59:213–223.

Kano, M., K. Iino, and M. Kano, 1991. Functional reorganization of adult cat somatosensory cortex is dependent on NMDA receptors. *NeuroReport* 2:77–80.

Kasamatsu, T., and J. Pettigrew, 1976. Depletion of brain catecholamines: Failure of ocular dominance shift after monocular occlusion in kittens. *Science* 194:206–209.

Kelahan, A. M., and G. S. Doetsch, 1984. Time-dependent changes in the functional organization of somatosensory cerebral cortex following digit amputation in adult raccoons. *Somatosens. Res.* 2:49–81.

Killackey, H. P., 1989. Static and dynamic aspects of cortical somatotopy: A critical review. *J. Cogn. Neurosci.* 1:3–11.

Kossut, M., P. J. Hand, J. Greenberg, and C. L. Hand, 1988. Single vibrissal cortical column in S-I cortex of rat and its afferations in neonatal and adult vibrissa-deafferented animals: A quantitative 2 DG study. *J. Neurophysiol.* 60:829–852.

Krubitzer, L. A., and J. H. Kaas, 1990. The organization and connections of somatosensory cortex in marmosets. *J. Neurosci.* 10:952–974.

Land, P. W., and N. D. Akhtar, 1987. Chronic sensory denervation affects cytochrome oxidase staining and glutamic acid decarboxylase immunoreactivity in rat ventrobasal thalamus. *Brain Res.* 425:176–181.

Lee, C.-J., and B. L. Whitsel, 1992. Mechanisms underlying somatosensory cortical dynamics: I. In vivo studies. *Cerebral Cortex* 2:81–106.

Lee, S. M., M. G. Weisskopf, and F. F. Ebner, 1991. Horizontal long-term potentiation of responses in rat somatosensory cortex. *Brain Res.* 544:303–310.

Levin, B. E., and A. A. Dunn-Meynell, 1991. Adult barrel cortex plasticity occurs at 1 week but not 1 day after vibrissectomy as demonstrated by the 2-deoxyglucose method. *Exp. Neurol.* 113:237–248.

Liu, C. N., and W. W. Chambers, 1958. Intraspinal sprouting of dorsal root axons. *Arch. Neurol. Psychiatry* 79:46–91.

McMahan, S. B., and P. D. Wall, 1983. Plasticity in the nucleus gracilis of the rat. *Exp. Neurol.* 80:195–207.

Melzack, R., 1990. Phantom limbs and the concept of a neuromatrix. *Trends Neurosci.* 13:88–92.

Merzenich, M. M., and J. H. Kaas, 1980. Principles of organization of sensory-perceptual systems in mammals. In *Progress in Psychobiology and Physiological Psychology*, J. M. Sprague and A. N. Epstein, eds. New York: Academic Press, pp. 1–42.

Merzenich, M. M., and J. H. Kaas, 1982. Organization of mammalian somatosensory cortex following peripheral nerve injury. *Trenos Neurosci.* 5:434–436.

Merzenich, M. M., J. H. Kaas, M. Sur, and C.-S. Lin, 1978. Double representation of the body surface within cytoarchitectonic areas 3b and 1 in "S-I" in the owl monkey (*Aotus trivirgatus.*) *J. Comp. Neurol.* 181:41–74.

Merzenich, M. M., J. H. Kaas, J. Wall, R. J. Nelson, M. Sur, and D. Felleman, 1983a. Topographic reorganization of somatosensory cortical areas 3b and 1 in adult monkeys following restricted deafferentation. *Neuroscience* 8:33–55.

Merzenich, M. M., J. H. Kaas, J. T. Wall, M. Sur, R. J. Nelson, and D. J. Felleman, 1983b. Progression of change following median nerve section in the cortical representation of the hand in areas 3b and 1 in adult owl and squirrel monkeys. *Neuroscience* 10:639–665.

Merzenich, M. M., R. J. Nelson, M. P. Stryker, M. S. Cynader, A. Schoppman, and J. M. Zook, 1984. Somatosensory cortical map changes following digit amputation in adult monkeys. *J. Comp. Neurol.* 224:591–605.

Metzler, J., and P. S. Marks, 1979. Functional changes in cat somatic sensory-motor cortex during short-term reversible epidural blocks. *Brain Res.* 177:379–383.

Mogilner, A., J. A. I. Grossman, U. Ribory, M. Joliet, J. Volkmann, D. Rapaport, R. W. Bensley, and R. R. Llinás, 1993. Somatosensory cortical plasticity in adult humans revealed by magnetoencephalography. *Proc. Natl. Acad. Sci. U.S.A.* 90:3593–3597.

Molander, C., E. Kinnman, and H. Aldskogius, 1988. Expansion of spinal cord primary sensory afferent projection following combined sciatic nerve resection and ruphenous nerve crush: A horseradish peroxidase study in the adult rat. *J. Comp. Neurol.* 276:436–411.

Morel, A., P. E. Garraghty, and J. H. Kaas, 1993. Tonotopic organization, architectonic fields, and connections of auditory cortex in macaque monkeys. *J. Comp. Neurol.* 335:437–459.

Morel, A., and J. H. Kaas, 1992. Subdivisions and connections of auditory cortex in owl monkeys. *J. Comp. Neurol.* 318:27–63.

Mount, R. J., R. V. Harrison, S. G. Stanton, and A. Nagasawa, 1991. Correlation of cochlear pathology with auditory brainstem and cortical responses in cats with high frequency hearing loss. *Scanning Microsc.* 5:1105–1113.

Newsome, W. T., R. H. Wurtz, M. R. Dursteler, and A. Mikami, 1985. Deficits in visual motion processing following icotenic acid lesions of the middle temporal visual area of the macaque monkey. *J. Neurosci.* 5:825–840.

Nicoletis, M. A. L., R. C. S. Lin, D. J. Woodword, and J. K. Chapin, 1993a. Peripheral block of ascending cutaneous information induces immediate spatio-temporal changes in thalamic networks. *Nature* 361:533–536.

Nicoletis, M. A. L., R. C. S. Lin, D. J. Woodword, and J. K. Chopin, 1993b. Dynamic and distributed properties of many-neuron ensembles in the ventral posterior medial thalamus of awake rats. *Proc. Natl. Acad. Sci. U.S.A* 90: 2212–2216.

Nudo, R. J., W. M. Jenkins, and M. M. Merzenich, 1990. Repetitive microstimulation alters the cortical representation of movements in adult rats. *Somatosens. Mot. Res.* 7:463–483.

Pettet, M. W., and C. D. Gilbert, 1992. Dynamic changes in receptive-field size in cat primary visual cortex. *Proc. Natl. Acad. Sci. U.S.A.* 89:8366–8370.

Pollin, B., and P. Albe-Fessard, 1979. Organization of somatic thalamus in monkeys with and without section of dorsal spinal track. *Brain Res.* 173:431–449.

Pons, T. P., P. E. Garraghty, D. P. Friedman, and M. Mishkin, 1987. Physiological evidence for serial processing in somatosensory cortex. *Science* 237:417–420.

Pons, T. P., P. E. Garraghty, and M. Mishkin, 1988. Lesion-induced plasticity in the second somatosensory cortex of adult macaques. *Proc. Natl. Acad. Sci. U.S.A.* 85: 5279–5281.

Pons, T. P., P. E. Garraghty, A. K. Ommaya, J. H. Kaas, E. Taub, and M. Mishkin, 1991. Massive cortical reorganization after sensory deafferentation in adult macaques. *Science* 252:1857–1860.

Ramachandran, V. S., D. Rogers-Ramachandran, and M. Stewart, 1992. Perceptual correlates of massive cortical reorganization. *Science* 258:1159–1160.

Rasmusson, D. D., 1982. Reorganization of raccoon somatosensory cortex following removal of the fifth digit. *J. Comp. Neurol.* 205:313–326.

Rasmusson, D. D., and D. M. Nance, 1986. Non-overlapping thalamocortical projections for separate forepaw digits before and after cortical reorganization in the raccoon. *Brain Res. Bull.* 16:399–406.

Rauschecker, J. P., 1991. Mechanisms of visual plasticity: Hebb synapses, NMDA receptors, and beyond. *Physiol. Rev.* 71:587–615.

Rausell, E., C. G. Cusick, E. Taub, and E. G. Jones, 1992. Chronic deafferentation in monkeys differentially affects nociceptive and nonnociceptive pathways distinguished by specific calcium-binding proteins and down-regulates gamma-aminobutyric acid type A receptors at thalamic levels. *Proc. Natl. Acad. Sci. U.S.A.* 89:2571–2575.

Recanzone, G. H., M. M. Merzenich, and H. R. Dinse, 1992a. Expansion of the cortical representation of a specific skin field in primary somatosensory cortex by intracortical microstimulation. *Cerebral Cortex* 2:181–196.

Recanzone, G. H., M. M. Merzenich, W. M. Jenkins, A. G. Kamil, and H. R. Dinse, 1992. Topographic reorganization of the hand representation in cortical area 3b of owl monkeys trained in a frequency-discrimination task. *J. Neurophysiol.* 67:1031–1056.

Rhoades, R. W., G. R. Belford, and H. P. Killackey, 1987. Receptive-field properties of rat ventral posterior medial neurons before and after selective kainic acid lesions of the trigeminal brain stem complex. *J. Neurophysiol.* 57:1577–1600.

Roberts, W. A., and J. Wells, 1990. Extensive dual innervation and mutual inhibition by forelimb and hindlimb inputs to ventroposterior nucleus projection neurons in the rat. *Somatosens. Mot. Res.* 7:85–95.

Robertson, D., and D. R. F. Irvine, 1989. Plasticity of frequency organization in auditory cortex of guinea pigs with partial unilateral deafness. *J. Comp. Neurol.* 282:456–471.

Rodin, B. E., and L. Kruger, 1984. Absence of intraspinal sprouting in dorsal root axons caudal to a partial spinal hemisection. A horseradish peroxidase transplant study. *Somatosens. Res.* 2:171–192.

Rodin, B. E., S. L. Sampogna, and L. Kruger, 1983. An examination of intraspinal sprouting in dorsal root axons with the tracer horseradish peroxidase. *J. Comp. Neurol.* 215:187–198.

Rose, J. E., L. I. Malis, L. Kruger, and C. P. Baker, 1960. Effects of heavy, ionizing, monoenergetic particles on the cerebral cortex. *J. Comp. Neurol.* 115:243–295.

Sanes, J. N., S. Suner, and J. P. Donoghue, 1990. Dynamic organization of primary motor cortex output to target muscles in adult rats: I. Long-term patterns of reorganization following motor or mixed nerve lesions. *Exp. Brain Res.* 79:479–491.

Scheich, H., C. Simonis, F. Ohl, H. Thomas, J. Tillein, 1992. Learning-related plasticity of gerbil auditory cortex: Feature maps versus meaning maps. In *Advances in Metabolic Mapping Techniques for Brain Imaging of Behavioral and Learning Functions* [NATO ASI Series D], vol. 68, F. Gonzalez-Lima, T. Finkenstadt, and H. Scheich, eds. Boston: Kluwer Academic Publishers, Pp. 447–474.

Schwaber, M. K., P. E. Garraghty, and J. H. Kaas, 1993. Neuroplasticity of the adult primate auditory cortex following cochlear hearing loss. *Am. J. Otol.* 14:252–258.

Shatz, C. J., 1990. Impulse activity and the patterning of connections during CNS development. *Neuron* 5:745–756.

Snow, P. J., R. J. Nudo, W. Rivera, W. M. Jenkins, and M. M. Merzenich, 1988. Somatotopically inappropriate projections from thalamocortical neurons to the SI cortex of the cat demonstrated by the use of intracortical microstimulation. *Somatosens. Res.* 5:349–372.

Snow, P. J., and P. Wilson, 1991. Plasticity in the somatosensory system of developing and mature mammals. *Progress in Sensory Physiology*, vol. II. New York: Springer-Verlag.

Stepniewska, I., T. M. Preuss, and J. H. Kaas, 1993. Architectonic, somatotopic organization, and ipsilateral cortical connections of the primary motor area (M1) of owl monkeys. *J. Comp. Neurol.* 330:238–271.

Stone, J., 1983. *Parallel Processing in the Visual System.* New York: Plenum Press.

Sur, M., R. J. Nelson, and J. H. Kaas, 1982. Representations of the body surface in cortical areas 3b and 1 of squirrel monkeys: Comparisons with other primates. *J. Comp. Neurol.* 211:177–192.

Tromblay, P., E. E. Allen, J. Soyke, C. D. Blaha, R. R. Lane, and B. Gordon, 1986. Doses of 6-hydroxydopamine sufficient to deplete norepinephrine are not sufficient to decrease plasticity in the visual cortex. *J. Neurosci.* 6:266–273.

Turnbull, B. G., and D. D. Rasmusson, 1991. Chronic effects of total or partial digit denervation or raccoon somatosensory cortex. *Somatosens. Mot. Res.* 8:201–213.

Waite, D. M. E., 1984. Rearrangement of neuronal responses in the trigeminal system of the rat following peripheral nerve section. *J. Physiol.* 352:425–445.

Wall, J. T., and C. G. Cusick, 1984. Cutaneous responsiveness in primary somatosensory (S-I) hindpaw cortex before and after partial hindpaw deafferentation in adult rats. *J. Neurosci.* 4:1499–1515.

Wall, J. T., D. J. Felleman, and J. H. Kaas, 1983. Recovery of normal topography in the somatosensory cortex of monkeys after nerve crush and regeneration. *Science* 221:771–773.

Wall, J. T., M. F. Huerta, and J. H. Kaas, 1992. Changes in the cortical map of the hand following postnatal median nerve injury in monkeys: I. Modification of somatotopic aggregates. *J. Neurosci.* 12:3445–3455.

Wall, J. T., and J. H. Kaas, 1985. Cortical reorganization and sensory recovery following nerve damage and regeneration. In *Synaptic Plasticity*, C. W. Cotman, ed. New York: Guilford Press, pp. 231–259.

Wall, J. T., J. H. Kaas, M. Sur, R. J. Nelson, D. J. Felleman, and M. M. Merzenich, 1986. Functional reorganization in somatosensory cortical areas 3b and 1 of adult monkeys after median nerve repair: Possible relationships to sensory recovery in humans. *J. Neurosci.* 6: 218–233.

WALL, P. D., 1977. The presence of ineffective synapses and the circumstances which unmask them. *Philos. Trans. R. Soc. Lond. [Biol.]* 278:361–372.

WALL, P. P., and M. D. EGGER, 1971. Formulation of new connections in adult rat brains after partial deafferentation. *Nature* 232:542–545.

WEBSTER, H. H., U-K. HANISCH, R. W. DYKES, and D. BIESOLD, 1991. Basal forelimb lesions with or without resupine injection inhibit cortical reorganization in rat hindpaw primary somatosensory cortex following sciatic nerve section. *Somatosens. Mot. Res.* 8:327–346.

WEINBERGER, N. M., R. DAVID, and B. LEPUN, 1993. Long-term retention of learning-induced receptive-field plasticity in the auditory cortex. *Proc. Natl. Acad. Sci. U.S.A.* 90:2394–2398.

WELKER, E., S. B. RAU, J. DORFI, P. MELZER, and H. VAN DER LOOS, 1992. Plasticity in the barrel cortex of the adult mouse: Effects of chronic stimulation upon doxyglucose uptake in the behaving animal. *J. Neurosci.* 12: 153–170.

WELKER, E., E. SORIANO, and H. VAN DER LOOS, 1989. Plasticity in the barrel cortex of the adult mouse: Effects of peripheral deprivation on GAD-immunoreactivity. *Exp. Brain Res.* 74:441–452.

WELKER, W. I., 1976. Mapping the brain. Historical trends in functional localization. *Brain Behav. Evol.* 13:327–343.

WELKER, W. I., and S. SEIDENSTEIN, 1959. Somatic sensory representation in the cerebral cortex of the raccoon (*Procyon lotor*). *J. Comp. Neurol.* 111:469–501.

WILLOTT, J. F., L. M. AITKIN, and S. L. MCFADDEN, 1993. Plasticity of auditory cortex associated with sensorineural hearing loss in adult C57BL16J mice. *J. Comp. Neurol.* 329:402–411.

WILSON, P., and P. J. SNOW, 1987. Reorganization of the receptive fields of spinocervical tract neurons following denervation of a single digit in the cat. *J. Neurophysiol.* 57:803–813.

WONG-RILEY, M. T. T., and C. WELT, 1980. Histochemical changes in cytochrome oxidase of cortical barrels after vibrissal removal in neonatal and adult mice. *Proc. Natl. Acad. Sci. U.S.A.* 77:2333–2337.

WOOLF, C. J., P. SHORTLAND, and R. E. COGGESHALL, 1992. Peripheral nerve injury triggers central sprouting of myelinated afferents. *Nature* 355:75–78.

YAMASAKI, D. S., and R. W. WURTZ, 1991. Recovery of function after lesions in the superior temporal sulcus in the monkey. *J. Neurophysiol.* 66:651–673.

5 Dynamic Properties of Adult Visual Cortex

CHARLES D. GILBERT

ABSTRACT Sensory perception requires an ongoing mutability of neuronal response specificity, cortical architecture, and synaptic weights. This is not a requirement of the cortex just during its early development but is a continuing process, extending well into adulthood. Obviously, we need to store and recall percepts, to adapt to changes in the sensory environment, and to normalize and calibrate our discrimination of visual attributes. In the past, it was assumed that these abilities resided in higher-order cortical areas, many stages removed from primary sensory cortex. This view was reinforced by the finding that the balance of input from the two eyes at the cortical level was subject to sensory experience only during a so-called critical period in the first few postnatal months. After the end of this period, the thalamocortical connections become fixed. Recent findings concerning the dependency of receptive field properties and cortical architecture on context and attention, however, have led to a new view of cortical processing. The implications of these findings extend into many aspects of visual perception, including the unification of an object's component contours into a single percept, separation of a figure from its background, perceptual constancies, the storage of percepts by neuronal ensembles, the mechanism of recovery of function following lesions of the central nervous system, and the role of top-down processes (expectation or attention) in perception.

There is a changing view of the receptive field that has bearing on the mutability of adult visual cortex. The characteristics of the receptive field, its size and specificity, are not fixed but are dependent on the stimulus. Originally, receptive fields were characterized by using a simple stimulus, such as a single spot or line. In striate cortex, receptive fields cover a tiny portion of the visual field, with the receptive fields that are located near the center of gaze being a fraction of a degree in diameter and those in the visual periphery being several degrees in diameter. In effect, they constitute a tiny window on a minute part of the visual world, so that the image of an object is broken up into a series of short, oriented line segments (Hubel and Wiesel, 1962). This understanding of the preliminary stage in the analysis of form leads to a problem: Given the atomistic process of breaking up an object into its component contours, how is it ultimately reassembled into a unified percept of a whole object? This process of reassembly is referred to as *segmentation* or *binding*.

The process of visual integration appears to begin at the earliest stages in the visual pathway, as evidenced by the modulatory effects of contextual stimuli. A cell's response to a complex image may not be easily predicted from its response to a simple, single-line stimulus. Although an earlier view may have held that the integration of information across visual space to form a unified percept is a function of high-order cortical areas, it is now evident that, even in primary visual cortex, cells are capable of integrating information across a larger part of visual space than one would have expected from the classical receptive field map. The anatomical substrate for this integration has been found in a plexus of long-range horizontal connections, running parallel to the cortical surface. Cells receiving these connections are capable of integrating information from a relatively large part of cortex and, hence, from a relatively large part of the visual field.

Anatomical substrates for horizontal integration

The visual cortex is retinotopically, or visuotopically, mapped: All cells at any one cortical location, from the pial surface to the white matter, have overlapping receptive fields with a bit of scatter, and there is a regular shift in receptive field position as one moves parallel to the cortical surface. The visual cortex consists of many areas, each representing the visual world on its surface. This concept is known as *visual topography*, which is one aspect of the functional architecture of cortex, or the relationship between the function of cells and their spatial distribution within the cortex. In

CHARLES D. GILBERT The Rockefeller University, New York, N.Y.

striate cortex, there is a minimum distance separating cortical columns having cells with nonoverlapping receptive fields, regardless of position in the cortical map. Whether one considers the foveal representation, where fields are small and there is little scatter in field position, or the visual periphery, where field size and scatter are larger, the cortical distance for nonoverlap is constant at approximately 1.5 mm (figure 5.1) (Hubel and Wiesel, 1974). This value assumes considerable importance when evaluating the functional significance of intrinsic cortical connections.

The ability of cells to be influenced by input from distant visuotopic or somatotopic loci is consistent with the patterns of connectivity observed at many stages

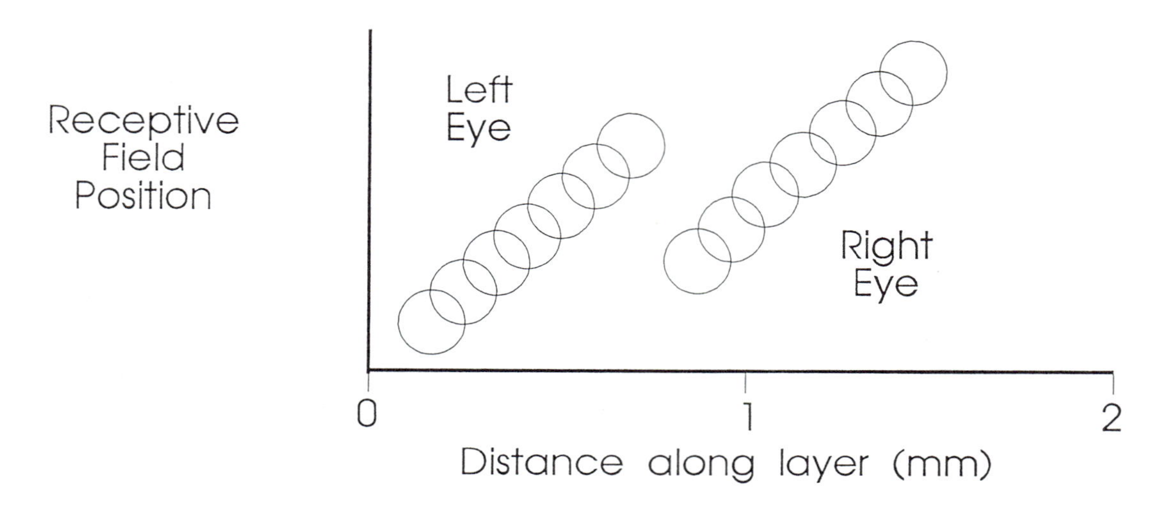

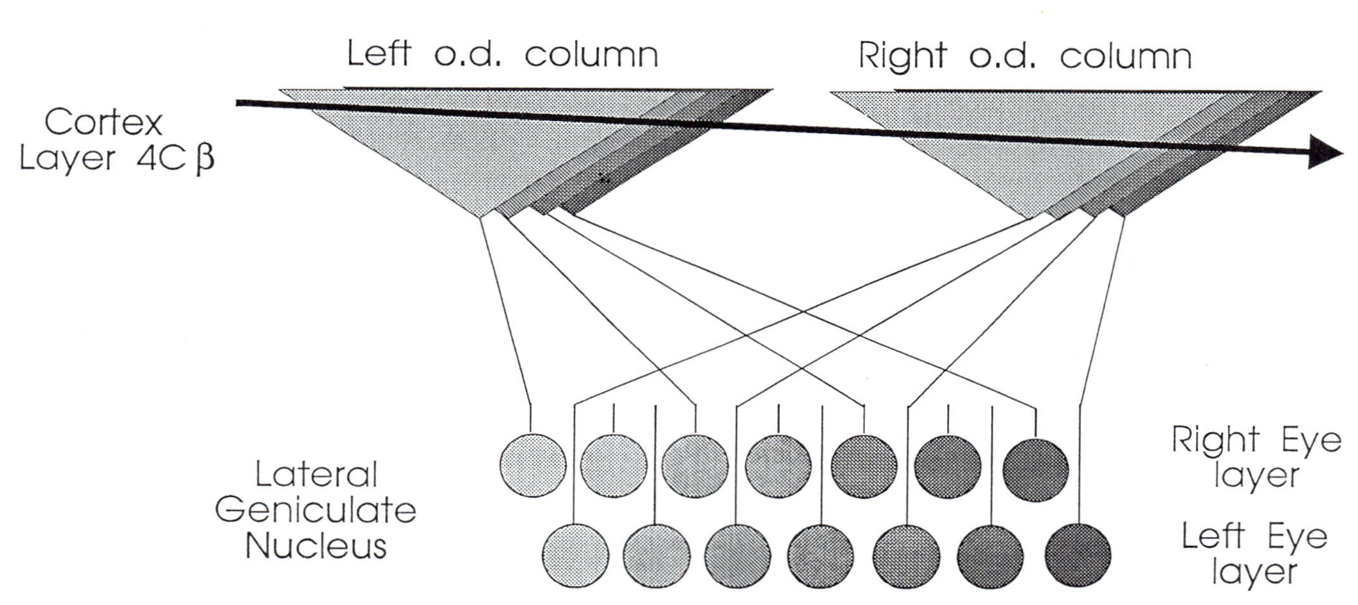

FIGURE 5.1 Receptive fields in layer 4Cβ of macaque monkey striate cortex and spread of thalamic afferents. An individual afferent extends for a substantial portion of an individual ocular dominance column and, taken together with the dendritic spread of the cells in the target layer for the afferents, enables cells potentially to integrate the fields from all the afferent fibers in the column. The lower part of the figure represents the overlap of thalamic projections from an individual layer of the lateral geniculate nucleus. Despite this, the receptive fields show a highly ordered progression and are comparable in diameter to that of a single geniculate afferent. The top part of the figure shows the receptive field maps of cells encountered in a tangential electrode penetration through the layer. (Reprinted from Gilbert, 1992, by permission.)

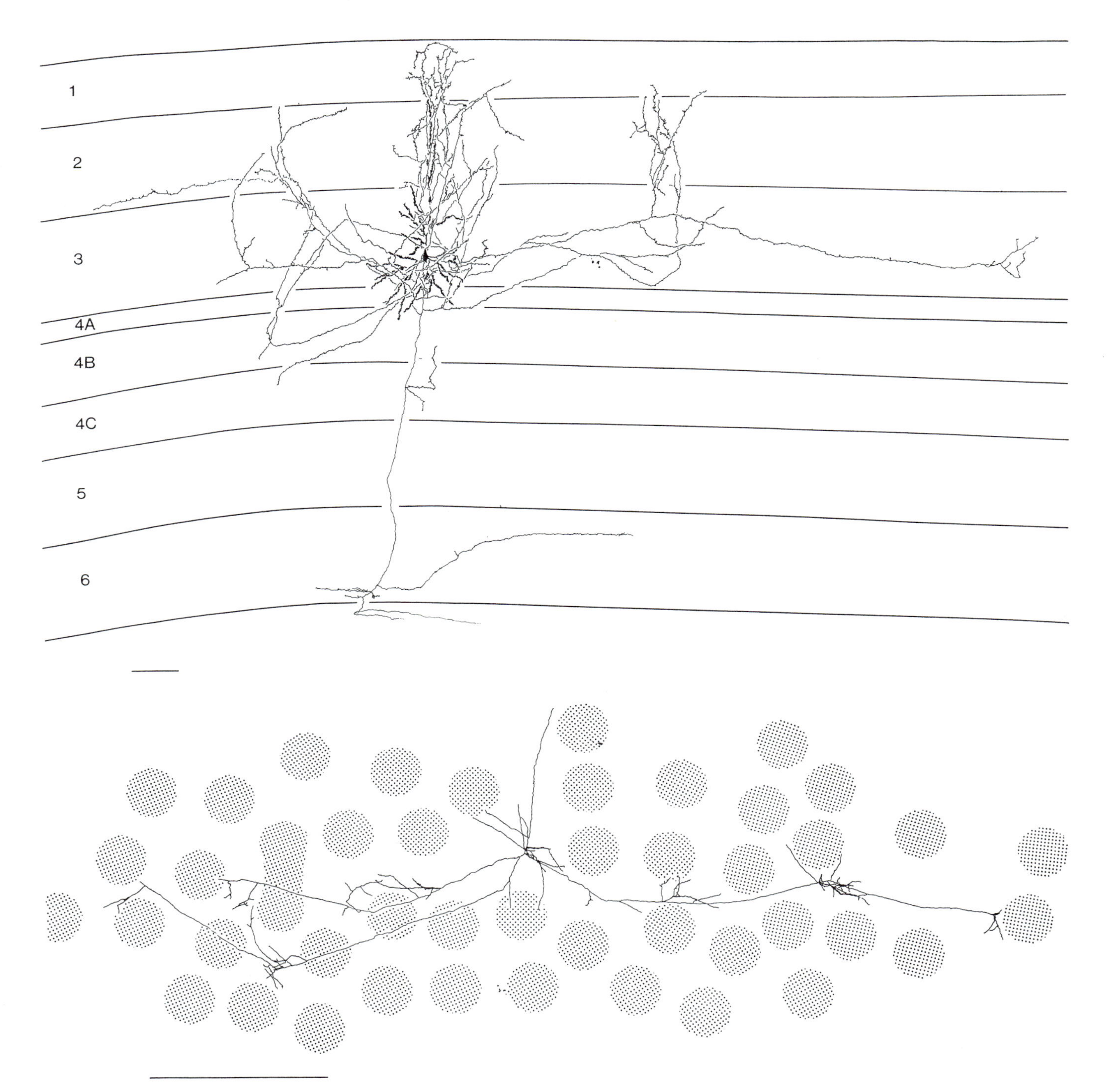

FIGURE 5.2 Examples of the horizontally projecting pyramidal cells in macaque monkey striate cortex. (A) A layer 3 pyramidal cell whose axon extended for more than 4 mm parallel to the cortical surface. The dendrite is seen as the thicker, darker lines in the center and is studded with dendritic spines. The axon gives off several clusters of collaterals within the superficial layers as well as projecting out of the immediate cortical area (McGuire et al, 1991). (B) An axon of another pyramidal cell, seen in surface view. It extends for approximately 6 mm parallel to the cortical surface. The gray patches represent the cytochrome oxidase blobs, a series of darkly staining regions that are aligned in rows along ocular dominance columns (vertically oriented in this rendition). Cells receiving input from axons such as these would be capable of integrating information over a substantial area of cortex representing an area of visual field that would be much larger than the cells' receptive fields. (Reprinted from Gilbert, 1992, by permission.)

along sensory pathways. Connections tend to be highly divergent and convergent. In fact, the observation that ascending and horizontal connections are so widespread has been a common and somewhat puzzling aspect of connectivity in the central nervous system. There are many instances of inconsistencies among the cortical circuitry, receptive field size, and topographical order. A good example of this is the pattern of projection of afferents from the lateral geniculate nucleus to their cortical target layer, 4Cβ (figure 5.2). As originally observed by Hubel and Wiesel (1977), if one records from a succession of cells in layer 4Cβ, moving parallel to the layer across an ocular dominance column, one will encounter cells with tiny, circularly symmetrical receptive fields and a highly ordered shift in receptive field position (see figure 5.2) (Blasdel and Fitzpatrick, 1984). This would seem to conflict with the connectivity of the layer: One would expect an individual cell to have access to all the afferents across the full column width, which measures approximately 350 μm, as the dendritic arbor of a layer 4Cβ cell is roughly 200 to 300 μm in diameter (Lund, 1973; Katz, Gilbert, and Wiesel, 1989) and each afferent covers a large part of the ocular dominance column, often extending 200 μm within a column and sometimes spanning two columns (Blasdel and Lund, 1983; Freund et al., 1989). Yet despite this, the receptive field sizes of 4Cβ cells are comparable to those of single afferents, and there is a strikingly orderly mapping of receptive field position within the column. A comparable example is seen in the somatosensory system with the projection of muscle afferent fibers to the cuneate nucleus. Here, one sees an overlap of approximately 300 afferent fibers at any one point, but the receptive field sizes of cuneate cells are at least an order of magnitude smaller than would be predicted by such convergence (Weinberg, Pierce, and Rustioni, 1990). The only way a cell can have a receptive field of such a restricted diameter is either for the cell to select input from only a few afferents or for local inhibitory processes within the target structure to restrict receptive field size. In any event, the end result is that not all inputs to a cell are equipotent, and only a subset has the capacity to activate it. The nonactivating inputs may be entirely silent, weak, or modulatory. Another intriguing possibility is that different inputs may be functional at different times. This kind of plasticity is suggested by experiments (described later) involving permanent modification of sensory input.

Even more widespread connections are seen in the axon collaterals of cortical pyramidal cells (figure 5.3) (Gilbert and Wiesel, 1979, 1983, 1989; Rockland and Lund, 1982, 1983; Martin and Whitteridge, 1984). These connections allow individual cells to integrate information from a wide area of cortex and, as a consequence of the topographical architecture of cortex, from a large part of the visual field, including loci outside the receptive field. As indicated earlier, horizontal connections spanning 6 to 8 mm would allow communication between cells with widely separated receptive fields. This reveals the puzzling finding that cells integrate information over a larger part of visual space than that covered by their receptive fields and calls into question the very definition of receptive field. The explanation for this seeming contradiction between cortical topography and receptive field structure is that the definition of the receptive field is stimulus-dependent and that a cell's response can be modulated by stimuli lying outside the classical receptive field.

Though the horizontal connections are very widespread, they are fairly specific in terms of the functional properties of the target cells. Rather than contacting all cells within a certain radius, the axon collaterals of the horizontally projecting cells are distributed in discrete clusters. The clustering implies a possible relationship to the functional architecture of the cortex, the tendency for cells with similar functional properties to be grouped into columns of similar functional specificity. In the primary visual cortex, cells with common orientation selectivity and eye preference are distributed in this fashion. Furthermore, as one moves across the cortical surface, there is a systematic clockwise or counterclockwise shift in orientation preference and a progressive shift from left- to right-eye dominance. Several lines of evidence show that the clustering of the horizontal connections allows them to mediate communication between columns of similar orientation preference. The spacing between clusters is roughly the same distance as that required to run through a full cycle of orientation columns (or one hypercolumn for orientation, which is approximately 750 μm wide). A physiological technique, known as *cross-correlation analysis*, demonstrates the functional relationship of cells communicating via the horizontal connections. A cross-correlogram is a histogram of differences in the spike times for a pair of cells. Cells that are connected, or that share a common input, will show a peak in this histogram at a particular delay. Looking across a pop-

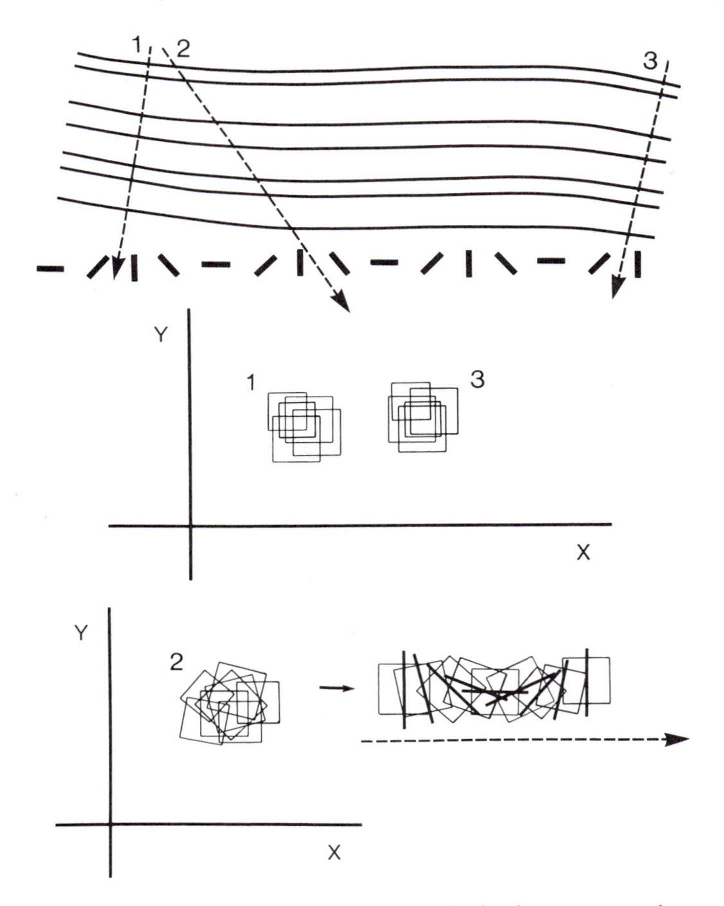

FIGURE 5.3 Relationship among cortical columns, receptive field size and position, and cortical dimensions. The top part of the figure shows a cross-section of cortex with three electrode penetrations: A vertical penetration (1 or 3) encounters cells with overlapping receptive fields having similar orientation specificity (middle). An oblique penetration (2) encounters cells with orientation preference shifting in a systematic clockwise or counterclockwise fashion (bottom), and a full 180° cycle is covered in a distance of approximately 750 μm, a segment of cortex referred to as a *hypercolumn* for orientation. Cells separated by a distance of 1.5 mm, or two hypercolumns, have nonoverlapping receptive fields, taking into account both receptive field size and scatter. This is true of primary visual cortex representing the fovea, where receptive field size and scatter is small, or peripheral retina, where size and scatter are large (Hubel and Wiesel, 1974). Thus, horizontal connections of the magnitude seen in cortex (up to 6 mm) mediate communication between cells separated by as many as eight hypercolumns and, therefore, with nonoverlapping receptive fields. Put another way, cells receiving input from the horizontal fibers integrate information over a much larger part of the visual field than that covered by their receptive fields, as measured by conventional techniques. (Reprinted from Gilbert, 1992, by permission.)

ulation of cells, cross-correlation analysis shows that cells in columns of similar functional specificity showed correlated firing even when separated by distances as great as 2 mm (Ts'o, Gilbert, and Wiesel, 1986; Ts'o and Gilbert, 1988).

Another rule governing the clustering is the division of a cortical area into compartments serving different visual submodalities, such as color, form, and motion. In the superficial layers of striate cortex, for example, there are patches of cells specific for color, with the cells in the intervening areas involved in the analysis of form due to their orientation specificity. This distribution is coincident with the histochemical staining pattern for cytochrome oxidase (CO), with the color-selective cells located within CO-stained blobs (Livingstone and Hubel, 1984a). The horizontal connections are specific for these compartments, with connections running from one blob to another and from interblob to interblob (Livingstone and Hubel, 1984b). On an even more detailed level of specificity, blobs are specific for a given color opponency (red-green or blue-yellow), and the horizontal connections run between blobs of similar color opponency (Ts'o and Gilbert, 1988).

An anatomical technique, combining retrograde tracing with 2-deoxyglucose autoradiography, confirmed the functional specificity of the horizontal connections. Injecting a retrograde tracer, such as rhodamine-filled latex microspheres, reveals the distribution of cells projecting to the injection site. After such an injection, cells are labeled across 8 mm of cortex, which represents an area of visual field roughly an order of magnitude larger than the receptive fields of the recipient cells. Within the area of label, the cells are distributed in clusters. When the retrograde tracing technique is combined with the 2-deoxyglucose technique, one can compare the distribution of labeled cells with some aspect of the columnar functional architecture—the distribution of columns of a particular orientation, for example. These experiments show that the horizontal connections run between columns of similar orientation specificity (Gilbert and Wiesel, 1989).

The widespread connections seen within each cortical area and the highly convergent and divergent connections running between cortical areas allow a progressive integration of information over larger parts of sensory space as one moves upward along sensory pathways. The horizontal plexus is a common feature of cortical connectivity and has been observed in a number of sensory, motor, and associational cortical areas. The connections between different cortical areas have a similar distribution to the intrinsic horizontal connections in that they are widespread and cells give rise to clusters of terminals in the subsequent cortical area; conversely, a site in a given cortical area receives input from clusters of cells in the antecedent cortical area. The clustered intrinsic and corticocortical connections have been seen in other visual areas, including V2, V3, and MT (Gilbert and Kelly, 1975; Zeki, 1976; Gilbert and Wiesel, 1979, 1983; Tigges et al., 1981; Rockland and Lund, 1982; Weller, Wall, and Kaas, in somatosensory and auditory cortex (Imig and Brugge, 1978; Jones, Coulter, and Hendry, 1978; Imig and Reale, 1981; DeFelipe, Hendry, and Jones, 1986) and even in frontal cortex (Goldman and Nauta, 1977). Just as the horizontal connections show columnar specificity in visual cortex, similar specificities are observed in auditory cortex (Imig and Reale, 1981).

Synaptic physiology of horizontal connections

In establishing the functional role of the horizontal connections, one of the first questions one would ask is whether they are excitatory or inhibitory. Because pyramidal cells are the source of the long-range horizontal connections, and the principal target of their axon collaterals are other pyramidal cells, it would be reasonable to believe that the effect of these connections would be excitatory (given that pyramidal cells are excitatory and smooth stellate cells are inhibitory). Even if inhibitory interneurons represent only 20% of the target cells (McGuire et al., 1991), however, the relative influence of inhibition in the horizontal connections could be greater than one would expect from such a small proportion. Because GABAergic cells are more readily activated by injected current than are excitatory cells (Schwartzkroin and Mathers, 1978; McCormick et al., 1985; Giffin, Doyle, and Nerbonne, 1988; Prince and Huguenard, 1988), one could expect a substantial amount of inhibition mediated by the horizontal connections. In fact, as observed in an in vitro cortical slice preparation, the balance between excitation and inhibition generated by activating the horizontal connections varies widely, from cell to cell and also according to the level of recruitment of the horizontal inputs. Consequently, it is consistent with the pattern of connectivity for the horizontal inputs to be responsible for inhibitory receptive field properties as well as facilitatory ones. Moreover, the level of excitation induced by activating the horizontal inputs depends on the level of depolarization of the target cell: The more depolarized the cell is, the larger the excitatory postsynaptic potential (EPSP), owing to voltage-dependent Na^+ conductances. Thus, one can think of the effect of the horizontal connections as being state-dependent, influenced by the level of activation of other inputs converging onto the cell (Hirsch and Gilbert, 1991). This has implications for determining the role of horizontal connections in the dynamic, context-dependent receptive field properties observed in vivo and the modulatory influence of placing stimuli outside the receptive field. The synaptic physiology of the horizontal connections suggests that their influence would depend on the character of the stimuli presented concurrently within the receptive field, which would involve local and interlaminar connections, and outside the receptive field, which would involve the horizontal connections.

Another finding that is important in understanding the functional potential of the horizontal connections is the fact that their synaptic potentials can be altered under the appropriate conditioning regime. The strength of the synaptic connections made by horizontal fibers in primary visual cortex can be strengthened by use (Hirsch and Gilbert, 1992). This is observed when the horizontal inputs are activated while the cell is depolarized with current injection. The ability to strengthen these connections depends on the level of inhibition present when activating the horizontal inputs; with greater inhibition, there is less potentiation. Therefore, any influence that modulates the level of inhibition also will alter the ability to modify the circuit. Though conditioning the horizontal inputs usually potentiates only the same connection, on occasion one also observes a heterosynaptic facilitation of late EPSPs, with the horizontal conditioning influencing the interlaminar inputs. The fact that the horizontal connections can be potentiated in this way suggests that, though under normal circumstances they may play a modulatory role in influencing the response characteristics of a cell, their synaptic potentials can be elevated to an activating level.

Long-term changes in cortical topography

The modifiability of horizontal connections may represent the mechanism underlying dynamic changes in receptive field size and cortical topography. As described previously for visual cortex, sensory cortices are divided into a number of areas, each of which is topographically mapped. Somatosensory cortex is mapped according to position on body surface, and visual cortical areas are mapped according to visual field position. When sensory input is removed, by amputation of a digit or by retinal lesions, an area of cortex is silenced, devoid of functioning input. Over time, the function of the silenced area is altered, and it develops a representation of a different body part, or a different part of the retina, than it represented before the lesion (Merzenich et al., 1984, 1988; Calford and Tweedale, 1988; Sanes et al., 1988; Gilbert and Wiesel, 1990; Kaas et al., 1990; Heinen and Skavenski, 1991). Alterations in cortical topography and receptive field size can also be observed after increased sensory input, such as that induced by sensory stimulation or training on a tactile discrimination task (Jenkins et al., 1990; Recanzone et al., 1992).

In the lesion experiments, the cortical area is remapped effectively, with an enlarged representation of the somatotopic or visuotopic areas immediately surrounding the lesioned sites (figure 5.4). The cells with receptive fields originally located within the lesioned area are initially silenced and then, over a period of 2 months, regain visual input. The renewed receptive fields have shifted to positions outside the lesioned area, effectively producing an expanded representation of the part of retina surrounding the lesion. In cortical terms, the size of the shift is on the order of 3 to 4 mm, similar in extent to the horizontal connections. This degree of mutability of the cortical map is a surprising finding for adult animals, in whom one would have expected that cortical organization would be fully developed and fixed early in life.

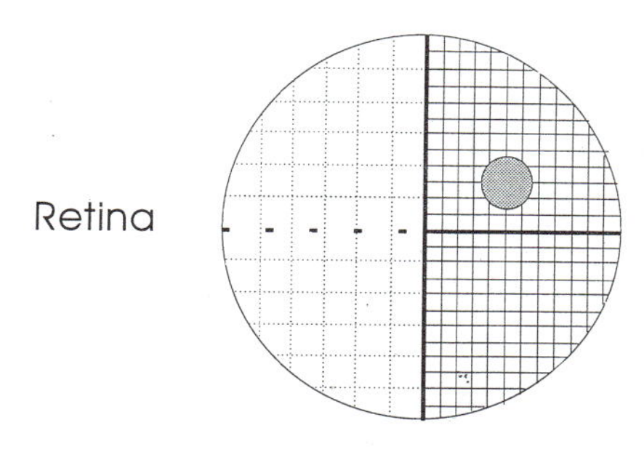

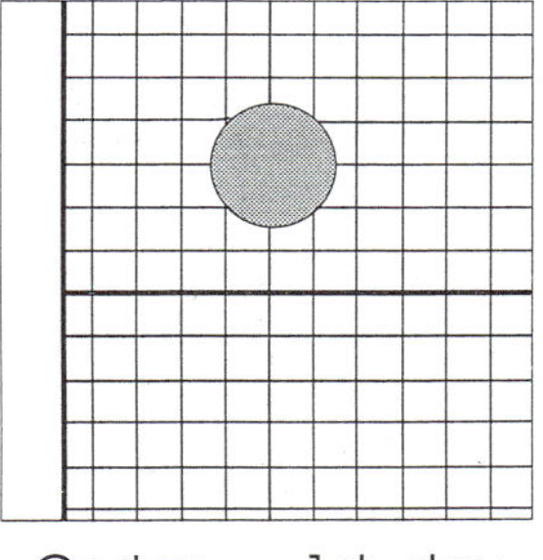

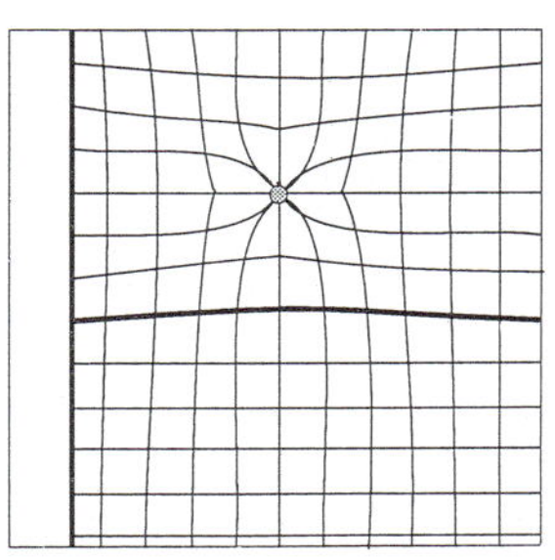

FIGURE 5.4 Effect of making retinal lesions on cortical topography. The visuotopic axes are represented as the Cartesian grid superimposed on the retina, and the corresponding maps on the cortex are displayed in surface view. A lesion, made with a diode laser, destroys the photoreceptor layer in a restricted part of the retina (shaded area, top), effectively removing visual input from the cortical region representing that retinal area (lower left). To get an idea of the dimensions involved, in the primate, a lesion subtending 5° of visual field, centered approximately 4° in the periphery, silences an area of cortex 10 mm in diameter. Over a period of 2 months, the topography of the cortex is reorganized (lower right), with a decreased representation of the silenced input and an increased representation of the perilesional retina (Gilbert and Wiesel, 1990, 1992; Kaas et al., 1990; Heinen and Skavenski, 1991). (Reprinted from Gilbert, 1992, by permission.)

At what stage along the sensory pathway does this reorganization first occur? Some of the reorganization seen in the cortex may be due to changes at earlier levels in the visual pathway, such as the lateral geniculate nucleus of the thalamus. In the somatosensory system, between the periphery and primary somatosensory cortex, sensory information travels over a number of stages, from sensory receptor to afferent fiber to spinal cord to brain stem to thalamus to primary somatosensory cortex. Even in the dorsal horn of the spinal cord, there are widespread connections formed by sensory fibers, and spinal cord sensory maps have been shown to change following peripheral nerve injury (Wall and Werman, 1976; Devor and Wall, 1978). A similar issue pertains to the retinal lesion experiments. The visual pathway runs from retina to the lateral geniculate nucleus in the thalamus to the primary visual cortex. In studies where the lesion involves destruction of retinal ganglion cells, reorganization has been seen in the lateral geniculate nucleus (Eysel, Gonzalez-Aguilar, and Mayer, 1981). Most of the topographical changes reported in the visual cortex, however, must be mediated by intrinsic cortical connections. At a time when the cortical scotoma completely fills in, there remains a visually inactive region in the lateral geniculate nucleus (Gilbert and Wiesel, 1992). Furthermore, the lateral spread of geniculocortical arbors is insufficient to account for the extent of reorganization seen in the cortex (Darian-Smith, Gilbert, and Wiesel, 1992; Gilbert, and Wiesel, 1992). Because the extent of filling in the cortex is large relative to the spread of geniculocortical afferents within the cortex, it is tempting to think of the long-range horizontal connections as the source of visual input to the region of the original cortical scotoma. These connections extend for distances similar in scale to the magnitude of the reorganization seen in the subjects with long-term survival.

Although the reorganization just described results from retinal lesions, there is reason to believe that this degree of cortical plasticity may follow more central lesions of the visual system. One such example of a reorganization of cortical function is that observed after central lesions in cortical areas serving eye movements. Lesions in area MT produce a deficit in smooth pursuit eye movements, those involved with tracking moving objects. After a period of a few days, the animal recovers this ability, presumably indicating that other areas of cortex have taken over the function of the ablated cortex (Newsome et al., 1985a, b).

The complete filling in of a large cortical scotoma (6–8 mm in diameter) takes a few months to occur, but striking changes are observable over a much shorter term. Surprising changes occur within minutes of the lesioning procedure: Cortical receptive fields located near the boundary of the lesion expand in size by an order of magnitude. There is even a measure of remapping of cortical topography within this short time span, with a fill-in spanning a couple of millimeters (Gilbert and Wiesel, 1992). Another way this has been shown is to create a laser lesion in one retina, wait a period of time ranging from minutes to months, and then enucleate the other eye. Immediately following the enucleation, the area of cortex initially silenced to stimulation of the lesioned area recovers input from the surrounding retinal area (Chino et al, 1992). Both studies showed shifts in topographical representation over a cortical distance of approximately 2 mm, as compared to reorganization of 6–10 mm seen in the longer-term recoveries. These findings might reflect the fact that under ordinary circumstances, cells are capable of integrating information over a large part of visual space. Usually the inputs from outside the classical receptive field serve to modulate the response of the cell but, under the appropriate pattern of stimulation, they can be boosted to a suprathreshold level. Removing the input that contributes to the receptive field center, for example, might allow the more peripheral inputs to be potentiated.

Short-term plasticity has also been reported in the somatosensory system, with changes in cortical topography following peripheral lesions. One observation was made in rats after section of the facial motor nerve that innervates the vibrissae (whiskers). Within a period of hours following transection, stimulating the original vibrissal representation of cortex evokes forelimb muscle activation (Donoghue, Suner, and Sanes, 1990). In the flying fox, digit amputation leads to an expansion in the sensory receptive fields of cortical cells originally representing the digit alone to the arm and wing, and this occurs within 15 minutes (Calford and Tweedale, 1991).

Context dependency of receptive field properties

Though much of the earlier work on cortical reorganization has employed various lesioning procedures, it is now evident that altered patterns of sensory stimulation can have profound effects on the functional properties of cortical cells, without requiring actual damage to the periphery.

In addition to long-term changes in a cell's functional properties after permanent alteration of its input, many cells alter their functional specificities with changes in the sensory context within which a stimulus is presented. As illustrated in figure 5.5, a long history of experimentation in visual psychophysics has established the principle that the perception of the attributes of a feature at a given position in the visual field is dependent on context: The perceived position of a point within the focal plane, or of its position in depth, can be shifted by the presence of nearby points (Westheimer, Shimamura, and McKee, 1976; Westheimer and McKee, 1977; Butler and Westheimer, 1978; Badcock and Westheimer, 1985; Westheimer, 1986). The appearance of the orientation of a line depends on nearby lines (Gibson and Radner, 1937; Westheimer, 1990; for review see Howard, 1986), as seen in the tilt illusion (see figure 5.5). One set of lines can have a "repulsive" effect on the perceived orientation of adjoining lines, such that lines oriented clockwise to the surrounding lines are shifted further clockwise and lines oriented counterclockwise are shifted further counterclockwise. Color perception depends on the rel-

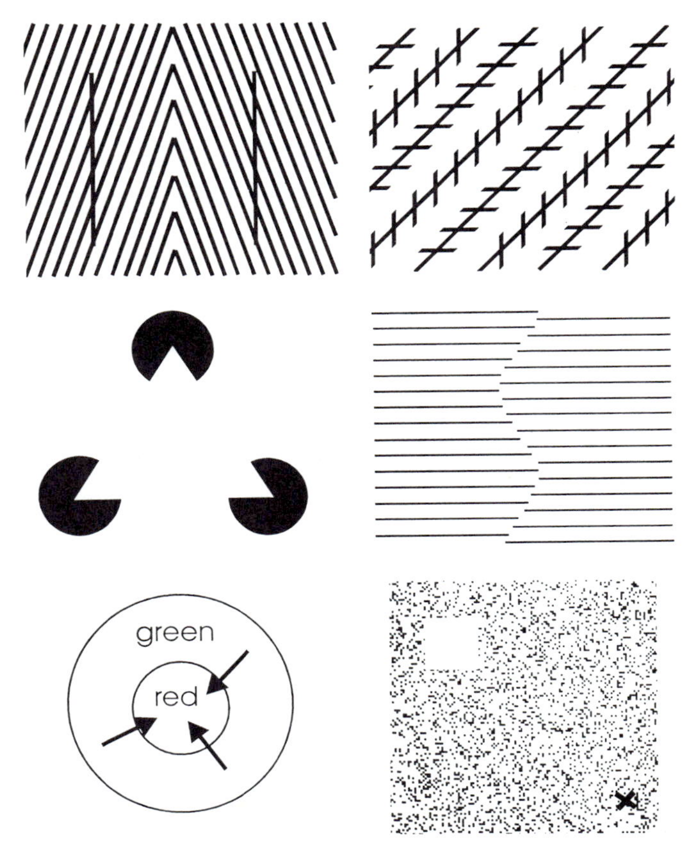

FIGURE 5.5 Illusions demonstrating the contextual sensitivity of various visual attributes. The top two examples show the tilt illusion, where lines of differing orientation cause parallel lines to appear tilted relative to one another. The examples in the center show illusory contours (left, Kanizsa triangle; right, an illusory curve produced by offset parallel lines). The examples at bottom represent perceptual fill-in. Stabilizing on the retina, the circle forming the boundary between the red disc and green annulus would lead to the percept of a single large green disc, with the green surface propagating into the center (cf. Yarbus, 1957; Krauskopf, 1961; Crane and Piantanida, 1983; Paradiso and Nakayama, 1991). On lower right, a dynamic random dot display would fill into the occluded area (or artificial scotoma) near the corner, producing a percept of a uniform field of random dots (fixation point indicated by the *X* in the lower right-hand corner; Ramachandran and Gregory, 1991). (Reprinted from Gilbert, 1992, by permission.)

ative content of light of different wavelengths that is reflected from different surfaces in the visual field (Land, 1974). Illusory contours in one part of the visual field are generated by real contours in another part of the visual field (Kanizsa, 1979). Perceptual fill-in is a process whereby information about color, texture, or movement in one part of the visual field is propagated to parts of the visual field that are occluded. When one masks a portion of the visual field and keeps the boundary of the mask stationary on the retina, the color or pattern of the surrounding area fills into the masked area (Yarbus, 1957; Krauskopf, 1961; Crane and Piantanida, 1983; Paradiso and Nakayama, 1991; Ramachandran and Gregory, 1991).

The physiological bases for such phenomena are increasingly being attributed to input from beyond the classical receptive field. Certain experiments demonstrate that the surrounding influences are capable of modifying the functional specificity of cells.

A hint of the mutability of receptive field structure is seen in the immediate effects of making retinal lesions. The fact that these changes could be induced immediately after the lesioning procedure suggested that an artificial scotoma, produced by masking a portion of the visual field, might be capable of producing similar effects. In fact, when the artificial scotoma is placed over the receptive field of a cell, and the area of visual field surrounding the artificial scotoma is stimulated, the receptive field can expand severalfold in diameter. When stimuli are placed within the receptive field, the field collapses down to its original size and, by alternately putting in and removing the mask from the conditioning stimulus, the field can go through several cycles of expansion and shrinkage (figure 5.6) (Pettet and Gilbert, 1992). Several potential mechanisms may account for the expansion, including a potentiation of the excitatory influence of the horizontal connections or an adaptation of their inhibitory influences. In either event, it seems that contextual stimuli are capable of unmasking influences distant from the receptive field center, and some form of modulation of synaptic connections must be involved.

Receptive field expansion could explain illusory contours or fill-in phenomena in the following way: Any cell represents a labeled line for a visual field locus corresponding to the receptive field center. When the receptive field expands, even stimuli located near the boundary of the expanded field cause the cell to signal the presence of stimuli at the same location, leading to a shifted percept of the location of the stimulus.

Another perceptual domain in which spatial interactions are observed is that of orientation. The fact that the horizontal connections mediate integration of information across visual space and that they relate

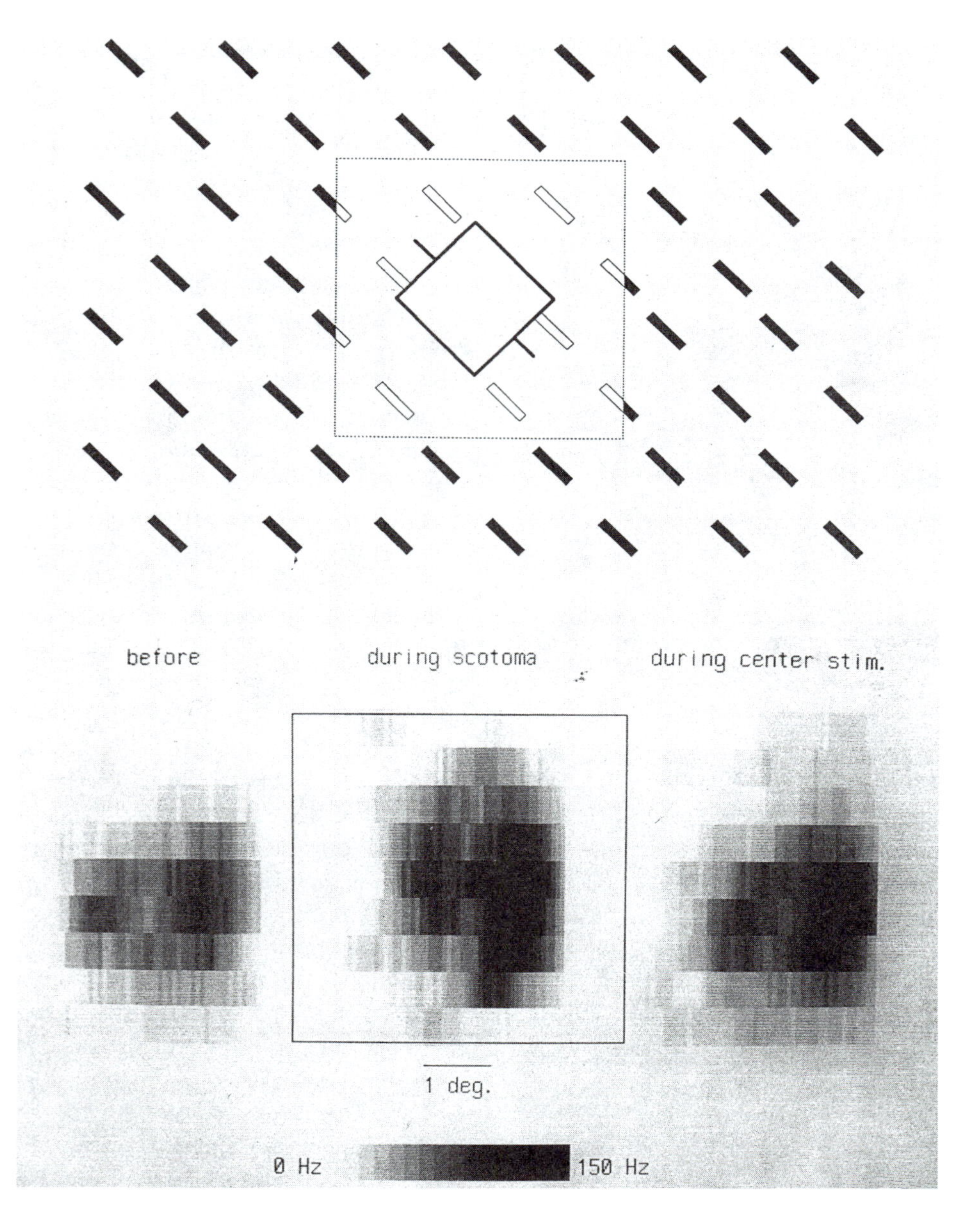

FIGURE 5.6 Effect of an artificial scotoma on receptive field size. Just as the retinal lesions are capable of producing a considerable expansion in receptive field size within minutes after making the lesion (Gilbert and Wiesel, 1992), mimicking the lesion by occluding a small part of the visual field also causes receptive fields located within the occluded area to expand (Petter and Gilbert, 1992). The upper part of the figure shows the conditioning stimulus, a pattern of lines moving outside the receptive field (square with solid outline, the orientation specificity being indicated by the two lines protruding from the square). The lines disappear when they more within the masked area (the mask is indicated by the dotted line but is not explicitly drawn in the stimulating pattern; the stimulating lines are shown as the blackened rectangles, and their disappearance is indicated by the open rectangles). After 10 minutes of conditioning, the receptive field expands. This is illustrated in the bottom part of the figure, which gives a two-dimensional response profile of the cell, the darker portions showing the greater response. The size and position of the occluder is indicated by the outline in the center, and the enlarged receptive field is indicated within the outline. Stimulating the center of the receptive field causes it to collapse in size, as indicated in the lower right. The receptive field can be caused to expand and contract alternately by a sequence of surround followed by center stimulation. (Reprinted from Gilbert, 1992, by permission.)

columns of similar orientation specificity makes it plausible that they would play a role in orientation-dependent contextual phenomena such as the tilt illusion. How is this context dependency represented in the response properties of cells in the visual cortex? One can characterize the specificity of a cell's response to a particular attribute in terms of its response to the range of values of a particular parameter: For example, cells with oriented receptive fields respond to edges over a range of orientations, with an optimal response at a particular orientation. Typically, the orientation tuning curve of cells have a half bandwidth (full width at half height) in the order of 20° to 40°. The peak height and position and tuning bandwidth are referred to as the *filter characteristics* of the cell for that particular attribute. Our ability to perceive differences in orientation of 1° must depend on the relative responses of a number of these orientation-tuned cells. At the level of such a neuronal ensemble, the effect of a contextual pattern can be seen as an alteration of the filter characteristics of a subset of the cells in the ensemble responsible for making the estimate of orientation.

The context dependency of orientation tuning can explain the tilt illusion. There are several kinds of changes that have been observed at the single-cell level that are consistent with the tilt illusion, including iso-orientation inhibition, shifts in the orientation of the optimal response (i.e., shift in the pack of the orientation tuning curve), and changes in orientation-tuning bandwidth. Which, if any, of these changes are most likely to account for the tilt illusion has yet to be established, but the clear message of these and other findings is that to a limited degree, the receptive field properties of a cell are mutable, capable of being influenced by the context within which a feature is presented (figure 5.7) (Gilbert and Wiesel, 1990).

Contextual sensitivity has been demonstrated for other visual attributes. Cells responding to illusory contours have been found in primate area 18 (von der Heydt and Peterhans, 1989). Many cells involved in the analysis of movement are specific for direction of stimulus movement. Their directional specificity can be altered by the presence of a background, moving in a different direction from the central stimulus (Allman, Miezin, and McGuinness, 1985; Tanaka et al., 1986; Gulyas et al., 1987). In the domain of color, the response of cells in visual area V4 does not depend exclusively on the wavelength of light reflected from an object but also on the light reflected from surrounding surfaces (Zeki, 1983). This property is believed to be related to the perceptual phenomenon of color constancy, where an object's color appears to be the same under different conditions of illumination, despite the fact that it reflects different wavelengths of light under different conditions of illumination. Receptive field properties can also be modulated by input from other senses: In area 7 of parietal cortex, cells have visual receptive fields but, in addition, their firing is modulated by the position of gaze of the eye, which can be mediated either by proprioceptive input from the extraocular muscles or by an "efferents copy" of the signal directing eye position to the parietal cortex (Andersen, Essick, and Siegel, 1985). It has been suggested that the visual system may also need to undergo constant error detection and calibration be able to maintain the ability to make precise spatial judgments (Moses, Schechtman, and Ullman, 1990). This calibration or normalization process, in a fashion similar to perceptual constancies, could involve comparisons between disparate locations in visual space, suggesting an additional role for the long-range horizontal connections.

The cortical plasticity that has been observed physiologically may also have a correlate in learning effects (i.e., changes in perceptual performance over time resulting from repeated discrimination trials). Hyperacuity is subject to training, so that after a few hundred trials of attempting to make a judgment of whether two lines are aligned or offset, one's ability to see tiny offsets (as little as 15 seconds of arc) improves markedly, and the improvement seems specific to the orientation of the stimulus pattern (McKee and Westheimer, 1978; Poggio, Fahle, and Edelman, 1992). Stereoacuity also improves with training, and the training is specific for stimulus size and visual field position (Fendick and Westheimer, 1983; Westheimer and Truong, 1988). A third example is training effects for texture segmentation (picking out a pattern of oriented lines from a background of differently oriented lines) (Karni and Sagi, 1991). The specificities of these training routines for stimulus pattern and visual position are suggestive of a process occurring early in the visual pathway.

Although horizontal connections, in many ways, represent an ideal substrate for many of the effects just described, one should also keep in mind that recurrent projections from higher-order cortical areas are also widespread in their distribution and generally arise from cells with larger receptive fields. Thus, it is impor-

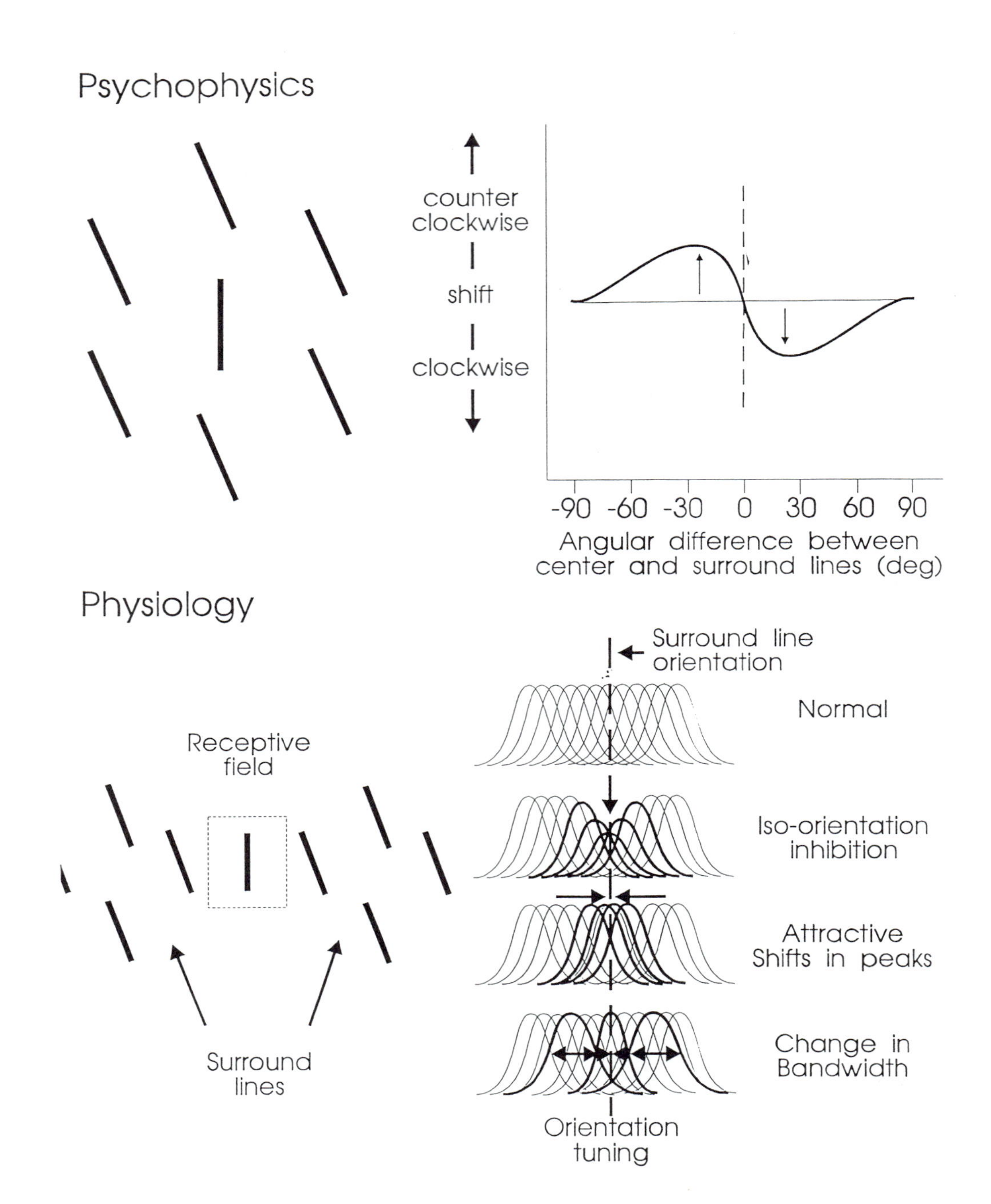

tant to distinguish the roles of intrinsic horizontal connections and feedback projections in modulating receptive field properties.

Under the realm of experience-dependent changes, there have also been numerous observations in other sensory modalities. Use-dependent changes caused by altering tactile experience have been reported in somatosensory cortex. The documented changes are observed after several weeks of training. Though they do not fall under the rubric of the short-term changes described earlier, they do represent an important departure from the experiments involving lesions (digit amputation, sensory nerve transection), in that they raise the possibility of shifting maps with normal sensory experience. For example, animals trained to detect differences in the frequency of a vibrating tactile stimulus develop larger cortical representations of the stimulated digits and larger receptive fields in the expanded areas (Recanzone et al., 1992). This develops further the earlier finding of changes in the representation of passively stimulated areas (Jenkins et al., 1990). Tactile stimulation alone can produce changes in cortical responsivity over tens of presentations, spanning a time period of seconds (both increases and decreases have been observed), with no corresponding effect on the mechanoreceptor afferents (Lee and Whitsel, 1992). A degree of plasticity has also been reported in the auditory system, where one can shift the best frequency of a unit toward a conditioned stimulus frequency (Weinberger, Ashe, and Edeline, in press).

In the motor cortex, changes in the cortical representation of motor activity can be induced by postural changes. These maps are assessed by measuring electromyographic activity while electrically stimulating different cortical sites. Cortical regions that ordinarily are not associated with forearm movement when the elbow is kept flexed can, within 20–30 minutes after holding the elbow in extension, cause forelimb movement. In effect, the forelimb area expanded into an area originally representing the vibrissae (Sanes, Wang, and Donoghue, 1992).

FIGURE 5.7 Relationship between psychophysical demonstrations of the tilt illusion and physiological observations of surround effect on orientation tuning. At the top, a subject is asked to determine whether the central line is tilted to the right or left of vertical in the presence of surrounding lines oriented differently from that of the central line. The graph at top right indicates that there is a repulsive shift in perceived orientation, with lines oriented counterclockwise to the surround lines appearing to shift further counterclockwise and lines oriented clockwise to the surround lines appearing to shift further clockwise. Possible physiological correlates of this phenomenon are illustrated below. The estimate of orientation is obtained by comparing the relative firing levels of cells having different orientation specificities (e.g., as might be represented by the tuning curves of 11 oriented cells, spaced 15° apart). In this model, the presence of lines of a particular orientation surrounding the receptive field would alter the tuning of a subset of these cells, or *orientation filters*, which would alter the estimate of orientation by the ensemble. Changes consistent with the repulsive perceived shift include iso-orientation inhibition (suppression of the firing of cells with the same preferred orientation as that of the surrounding lines), an attractive shift in the peak orientation tuning of some members of the ensemble toward the orientation of the surrounding lines, or changes in the bandwidth of orientation tuning (sharpening the tuning of cells with the same preferred orientation as the surrounding lines, or broadening the tuning of cells with preferred orientations differing from that of the surrounding lines). (Reprinted from Gilbert, 1992, by permission.)

Attentional modification of receptive fields

In considering the dynamic changes in receptive field properties, it is important to make at least brief mention of the role of attention. Attentional influences have been observed in experiments with awake, behaving animals. One dramatic demonstration of this was made in visual area 4 (V4). Here, cells have large receptive fields but, under certain circumstances, the responsive area of the receptive field can be greatly restricted. If the monkey does not attend to a stimulus in a given location, the response of the cell is greatly reduced compared to when the monkey attends to the stimulus (Moran and Desimone, 1985). As the monkey moves its focus of attention to different parts of the receptive field, the site at which one can elicit the most vigorous response moves along with the attentional focus. The specificity of a cell to visual properties other than position can be influenced if the behavioral task requires attention to that property, such as color or orientation. For example, when the animal is required to match a test stimulus to a conditioning stimulus, the firing of a cell during presentation of the test stimulus can depend on the orientation of the previously presented conditioning stimulus. When not attending, the same neuron might not be orientation-specific at all. The firing of this "visual" neuron can depend on nonretinal input—for example when using a grooved plate that the animal manipulates, outside its field of view, as the conditioning stimulus (Haenny, Maunsell, and Schiller, 1988; Spitzer, Desimone, and Moran, 1988; Maunsell et al., 1989; Spitzer and Richmond, 1990). Attentional influences have been observed in parietal as well as in prestriate cortex (Mountcastle et al., 1987; Petersen, Robinson, and Currie, 1989). Attention to selected aspects of a stimulus can determine the relative activation of different cortical areas that specialize in analyzing a given attribute (Corbetta et al., 1990, 1991).

These observations show that a neuron may be affected by "top-down" influences, where higher-order representations may determine a cell's response to inputs coming from earlier areas. The prevalence of recurrent projections in corticocortical sensory pathways suggests not only an anatomical substrate for the attentional influences but also that such influences may descend down to the earliest stages of sensory processing within the central nervous system. A thalamic nucleus, the pulvinar, has also been implicated in the modulation of cortical activity by attention (LaBerge and Buchsbaum, 1990; Robinson et al., 1991; for review see Robinson and Petersen, 1992). One plausible differentiation between the roles of horizontal connections that are intrinsic to a cortical area and the feedback projections is that the horizontal connections mediate context dependency and the feedback projections mediate some of the attentional effects.

There are several missing links that need to be forged in establishing the relationship among horizontal connections, changes in receptive field properties and cortical topography, and the psychophysical observations of contextual influences on perception: To what extent are the changes due to horizontal connections within a particular area, as compared to other sources of lateral integration, such as spread of thalamic afferents and feedback projections from higher cortical areas? For the responsible connections, does the underlying synaptic mechanism involve a potentiation of excitatory connections, adaptation of inhibitory connections, or an enrichment of the terminal arbor of axons? How might the tendency for horizontal connections to run between columns of similar functional specificity relate to their physiological effect? Given the multiplicity of effects of contextual sensitivity on the response properties of cells, which of these phenomena are causative of the perceptual effects?

The functional role of the dynamic changes in receptive field properties depends in part on the temporal window within which the changes take place. Alterations that require a few months to develop may play a role in recovery of function after lesions of the central nervous system. Alterations occurring over minutes may be related to learning and memory and to normalization of the tuning properties of cells to different visual attributes. Changes occurring in milliseconds would be required for processing visual information as one assimilates each view of the world obtained when glancing about one's surroundings.

Theoretical considerations

The contextual and attentional influences on the specificity of neuronal responses is leading to a change in our view of neural processing. At various stages in the description of the receptive field, the visual system has been considered to operate under different mechanisms: Neurons have been characterized as responding to trigger features which, in the extreme, would lead one to postulate the existence of the so-called grandmother cell, a single cell specific for the recognition of a single person (Maturana et al., 1960, Barlow, 1972). Various versions of the nervous system as a passive filter have been suggested, whereby each stage along the visual pathway acts in a stereotypical way, independent of the scene, with each cell representing a filter with fixed characteristics, passing the product of its analysis on to the next stage in the pathway. Included in these schemes are ideas of cells as orientation filters for object boundaries (Hubel and Wiesel, 1962), Fourier analyzers considering objects as the sum of sine waves of different frequencies and phases (Campbell and Robson, 1968; Maffei and Fiorentini, 1973; Schiller, Finlay, and Volman, 1976; Movshon, Thompson, and Tolhurst, 1978; De Valois and De Valois, 1980), and a host of other combinations of *basis sets*, a limited number of rudimentary shapes that, when put together, could represent any arbitrary objects. The developing evidence for the dynamic nature of receptive field properties has led to alternative hypotheses which, in the extreme, assume that neural connections are random and that an individual neuron is capable of assuming any set of filter characteristics. One version of this treats neurons as members of groups that shift their allegiance, along with their specificity, to different groups by altering the strength of connections with other cells in the group (Edelman, 1987), and it contends that one cannot assign a fixed function to a particular area. Other neural network models promote the idea of on-line modification of connections and adjustment of neuron function (Rumelhart, Hinton, and Williams, 1986; Zipser and Andersen, 1988; Lehky and Sejnowski, 1990), though these models do not necessarily entail a continuing modification of the network once it is set up. Although these concepts have a degree of validity, one should keep in mind that, as described previously, there is considerable specificity in the pattern of neural connections and, if a neuron can shift its specificity, its flexibility is limited and it can operate only within a defined range, not the full set of parameters of a given attribute.

Horizontal connections have been implicated in another model of lateral integration, that of temporal correlations of neural activity. It has been reported that neuronal responses tend to oscillate at 40–60 Hz and that neurons in distant columns show synchronization in their oscillations (Eckhorn et al., 1988; Gray et al., 1989; Gray and Singer, 1989). One suggested implication of this phenomenon is that it serves to bind different components of an image together and is a means of performing perceptual grouping or segmentation and of figure-ground segregation (von der Malsburg and Schneider, 1986; Sporns, Tononi, and Edelman, 1991) and even as the substrate of visual awareness (Crick and Koch, 1990). Just as the specificity of the horizontal interactions had been established with cross-correlation studies (Ts'o, Gilbert, and Wiesel, 1986; Ts'o and Gilbert, 1988), a similar rule applied to the synchronization of the oscillations, which also tended to occur between cells with similar orientation preference (Engel et al., 1990; Gray et al., 1990). In all these studies, the horizontal connections were suggested as the anatomical substrate of the correlated activity. It is not yet clear, however, whether the temporal characteristics of the neuronal assembly per se are utilized for binding or whether the contributions to receptive field structure and stimulus specificity that are made by the horizontal connections are more relevant to the process of segmentation. One would expect neurons that are interconnected or that share a common input to have phase-locked firing and, as such, the oscillations could be an epiphenomenon of the cortical circuitry, without necessarily having a perceptual significance. It will ultimately be necessary to show that the oscillations appear and disappear along with the presence or absence of the perceptual linkage between different components of the image. Even then it may not be possible to tease apart the roles of the oscillations and the dynamic changes in receptive field properties, both of which may be a reflection of the mutability in the synaptic physiology of the horizontal connections.

Conclusions

A number of studies discussed in this chapter have led to a view that cells in the visual system are endowed with dynamic properties, influenced by context, expectation, and long-term modifications of the cortical network. These observations will be important for understanding how neuronal ensembles produce a system that perceives, remembers, and adapts to injury. The advantage to being able to observe changes at early stages in a sensory pathway is that one may be able to understand the way in which neuronal ensembles encode and represent images at the level of their receptive field properties, cortical topographies, and the patterns of connections between cells participating in a network.

ACKNOWLEDGMENTS This chapter is adapted from earlier publications (Gilbert, 1992, 1993). The research was supported by a grant from the National Institutes of Health (EY07968) and a McKnight Development Award.

REFERENCES

ALLMAN, J. M., F. MIEZIN, and E. MCGUINNESS, 1985. Direction and velocity specific responses from beyond the classical receptive field in the middle temporal visual area (MT). *Perception* 14:105–126.

ANDERSEN, R. A., G. K. ESSICK, and R. M. SIEGEL, 1985. Encoding of spatial location by posterior parietal neurons. *Science* 230:456–458.

BADCOCK, D. R., and G. WESTHEIMER, 1985. Spatial location and hyperacuity: The centre-surround localization function has two substrates. *Vision Res.* 25:1259–1269.

BARLOW, H. B., 1972. Single units and sensation: A neuron doctrine for perceptual psychology? *Perception* 1:371–394.

BLASDEL, G. G., and D. FITZPATRICK, 1984. Physiological organization of layer 4 in macaque striate cortex. *J. Neurosci.* 4:880–895.

BLASDEL, G. G., and J. S. LUND, 1983. Termination of afferent axons in macaque striate cortex. *J. Neurosci.* 3:1389–1413.

BUTLER, T., and G. WESTHEIMER, 1978. Interference with stereoscopic acuity: Spatial, temporal, and disparity tuning. *Vision Res.* 18:1387–1392.

CALFORD, M. B., and R. TWEEDALE, 1988. Immediate and chronic changes in responses of somatosensory cortex in adult flying-fox after digit amputation. *Nature* 332:446–448.

CALFORD, M. D., and R. TWEEDALE, 1991. Acute changes in cutaneous recpetive fields in primary somatosensory cortex after digit denervation in adult flying fox. *J. Neurophysiol.* 65:178–187.

CAMPBELL, F. W., and J. G. ROBSON, 1968. Application of Fourier analysis to the visibility of gratings. *J. Physiol.* 197:551–566.

CHINO, Y. M., J. H. KAAS, E. L. SMITH III, A. L. LANGSTON, and H. CHENG, 1992. Rapid reorganization of cortical maps in adult cats following restricted deafferentation in retina. *Vision Res.* 32:789–796.

CORBETTA, M., F. M. MIEZIN, S. DOBNEYER, G. L. SHULMAN, and S. E. PETERSEN, 1990. Attentional modulation of neural processing of shape, color and velocity in humans. *Science* 248:1556–1559.

CORBETTA, M., F. M. MIEZIN, S. DOBNEYER, G. L. SHULMAN, and S. E. PETERSEN, 1991. Selective and divided attention during visual discriminations of shape, color and speed: Functional anatomy by positron emission tomography. *J. Neurosci.* 11:2383–2402.

CRANE, H. D., and T. P. PIANTANIDA, 1983. On seeing reddish green and yellowish blue. *Science* 221:1078–1079.

CRICK, F., and C. KOCH, 1990. Some reflections on visual

awareness. *Cold Spring Harb. Symp. Quant. Biol.* 55:953–962.
DARIAN-SMITH, C., C. D. GILBERT, and T. N. WIESEL, 1992. Cortical reorganization following binocular focal retinal lesions in the adult cat and monkey. *Abstr. Soc. Neurosci.* 18:11.
DEFELIPE, J., S. H. C. HENDRY, and E. G. JONES, 1986. A correlative electron microscopic study of basket cells and large GABAergic neurons in the monkey sensory-motor cortex. *Neuroscience* 17:991–1009.
DE VALOIS, R. L., and K. K. DE VALOIS, 1980. Spatial vision. *Annu. Rev. Psychol.* 31:309–341.
DEVOR, M., and P. D. WALL, 1978. Reorganization of spinal cord sensory map after peripheral nerve injury. *Nature* 276:75–76.
DONOGHUE, J. P., S. SUNER, and J. N. SANES, 1990. Dynamic organization of primary motor cortex output to target muscles in adult rats: II. Rapid reorganization following motor nerve lesions. *Exp. Brain Res.* 79:492–503.
ECKHORN, R., R. BAUER, W. JORDAN, M. BROSCH, W. KRUSE, M. MUNK, and H. J. REITBOECK, 1988. Coherent oscillations: A mechanism of feature linking in the visual cortex? *Biol. Cybern.* 60:121–130.
EDELMAN, G. M., 1987. *Neural Darwinism: The Theory of Neuronal Group Selection.* New York: Basic Books.
ENGEL, A. K., P. KONIG, C. M. GRAY, and W. SINGER, 1990. Stimulus-dependent neuronal oscillations in cat visual cortex: Inter-columnar interaction as determined by cross-correlation analysis. *Eur. J. Neurosci.* 2:588–606.
EYSEL, U. T., F. GONZALEZ-AGUILAR, and U. MAYER, 1981. Time-dependent decrease in the extent of visual deafferentation in the lateral geniculate nucleus of adult cats with small retinal lesions. *Exp. Brain Res.* 41:256–263.
FENDICK, M., and G. WESTHEIMER, 1983. Effects of practice and the separation of test targets on foveal and peripheral stereoacuity. *Vision Res.* 23:145–150.
FREUND, T. F., K. A. C. MARTIN, I. SOLTESZ, P. SOMOGYI, and D. WHITTERIDGE, 1989. Arborisation pattern of post-synaptic targets of physiologically identified thalamocortical afferents in straite cortex of the macaque monkey. *J. Comp. Neurol.* 289:315–336.
GIBSON, J. J., and M. RADNER, 1937. Adaptation, after-effect and contrast in the perception of tilted lines. *J. Exp. Psychol.* 20:453–467.
GIFFIN, K., J. P. DOYLE, and J. M. NERBONNE, 1988. Comparison of excitable membrane properties in identified neurons from mammalian visual cortex. *Soc. Neurosci. Abstr.* 14:297.
GILBERT, C. D., 1992. Horizontal integration and cortical dynamics. *Neuron* 9:1–13.
GILBERT, C. D., 1993. Rapid dynamic changes in adult cerebral cortex. *Curr. Opin. Neurobiol.* 3:100–103.
GILBERT, C. D., and J. P. KELLY, 1975. The projections of cells in different layers of the cat's visual cortex. *J. Comp. Neurol.* 163:81–106.
GILBERT, C. D., and T. N. WIESEL, 1979. Morphology and intracortical projections of functionally identified neurons in cat visual cortex. *Nature* 280:120–125.
GILBERT, C. D., and T. N. WIESEL, 1983. Clustered intrinsic connections in cat visual cortex. *J. Neurosci.* 3:1116–1133.
GILBERT, C. D., and T. N. WIESEL, 1989. Columnar specificity of intrinsic horizontal and corticocortical connections in cat visual cortex. *J. Neurosci.* 9:2432–2442.
GILBERT, C. D., and T. N. WIESEL, 1990. The influence of contextual stimuli on the orientation selectivity of cells in primary visual cortex of the cat. *Vision Res.* 30:1689–1701.
GILBERT, C. D., and T. N. WIESEL, 1992. Receptive field dynamics in adult primary visual cortex. *Nalure* 356:150–152.
GOLDMAN, P. S., and W. J. H. NAUTA, 1977. Columnar distribution of corticocortical fibers in the frontal association, limbic and motor cortex of the developing rhesus monkey. *Brain Res.* 122:393–413.
GRAY, C. M., A. K. ENGEL, P. KONIG, and W. SINGER, 1990. Stimulus-dependent neuronal oscillations in cat visual cortex: Receptive field properties and feature dependence. *Eur. J. Neurosci.* 2:607–619.
GRAY, C. M., P. KONIG, A. K. ENGEL, and W. SINGER, 1989. Oscillatory responses in cat visual cortex exhibit inter-columnar synchronization which reflects global stimulus properties. *Nature* 338:334–337.
GRAY, C. M., and W. SINGER, 1989. Stimulus-specific neuronal oscillations in orientation columns of cat visual cortex. *Proc. Natl. Acad. Sci. U.S.A.* 86:1698–1702.
GULYAS, B., G. A. ORBAN, J. DUYSENS, and H. MAES, 1987. The suppressive influence of moving texture background on responses of cat striate neurons to moving bars. *J. Neurophysiol.* 57:1767–1791.
HAENNY, P. E., J. H. R. MAUNSELL, and P. H. SCHILLER, 1988. State dependent activity in monkey visual cortex. *Exp. Brain Res.* 69:245–259.
HEINEN, S. J., and A. A. SKAVENSKI, 1991. Recovery of visual responses in foveal V1 neurons following bilateral foveal lesions in adult monkey. *Exp. Brain Res.* 83:670–674.
HIRSCH, J. A., and C. D. GILBERT, 1991. Synaptic physiology of horizontal connections in the cat's visual cortex. *J. Neurosci.* 11:1800–1809.
HIRSCH, J. A., and C. D. GILBERT, 1992. Long-term changes in synaptic strength along specific intrinsic pathways in the cat visual cortex. *J. Physiol.* 461:247–262.
HOWARD, I. P., 1986. The perception of posture, self-motion and the visual vertical. In *Handbook of Perception and Human Performance*, vol. 1, K. R. Boff, ed. New York: Wiley.
HUBEL, D. H., and T. N. WIESEL, 1962. Receptive fields, binocular interaction and functional architecture in the cat's visual cortex. *J. Physiol.* 160:106–154.
HUBEL, D. H., and T. N. WIESEL, 1974. Uniformity of monkey striate cortex: A parallel relationship between field size, scatter and magnification factor. *J. Comp. Neurol.* 158: 295–306.
HUBEL, D. H., and T. N. WIESEL, 1977. Functional architecture of macaque striate cortex. *Proc. R. Soc. Lond. [Biol.]* 198:1–59.
IMIG, T. J., and J. F. BRUGGE, 1978. Sources and terminations of callosal axons related to binaural and frequency

maps in primary auditory cortex of the cat. *J. Comp. Neurol.* 182: 637–660.

Imig, T. J., and R. A. Reale, 1981. Ipsilateral corticocortical projections related to binaural columns in cat primary auditory cortex. *J. Comp. Neurol.* 203:1–14.

Jenkins, W. M., M. M. Merzenich, M. T. Ochs, T. Allard, and E. Guic-Rocbles, 1990. Functional reorganization of primary somatosensory cortex in adult owl monkeys after behaviorally controlled tactile stimulation. *J. Neurophysiol.* 63:82–104.

Jones, E. G., J. D. Coulter, and S. H. C. Hendry, 1978. Intracortical connectivity of architectonic fields in the somatic sensory, motor and parietal cortex of monkeys. *J. Comp. Neurol.* 181:291–348.

Kaas, J. H., L. A. Krubitzer, Y. M. Chino, A. L. Langston, E. H. Polley, and N. Blair, 1990. Reorganization of retinotopic cortical maps in adult mammals after lesions of the retina. *Science* 248:229–231.

Kanizsa, G., 1979. *Organization in Vision. Essays on Gestalt Perception.* New York: Praeger.

Karni, A., and D. Sagi, 1991. Where practice makes perfect in texture discrimination: Evidence for primary visual cortex plasticity. *Proc. Natl. Acad. Sci. U.S.A.* 88:4966–4970.

Katz, L. C., C. D. Gilbert, and T. N. Wiesel, 1989. Local circuits and ocular dominance in monkey striate cortex. *J. Neurosci.* 9:1389–1399.

Krauskopf, J., 1961. Heterochromatic stabilized images: A classroom demonstration. *Am. J. Psychol.* 80:632–637.

LaBerge, D., and M. S. Buchsbaum, 1990. Positron emission tomographic measurements of pulvinar activity during an attention task. *J. Neurosci.* 10:613–619.

Land, E. H., 1974. The retinex theory of colour vision. *Proc. R. Inst. Gt. Brit.* 47:23–57.

Lee, C. J., and B. L. Whitsel, 1992. Mechanisms underlying somatosensory cortical dynamics: I. In vivo studies. *Cerebral Cortex* 2:81–106.

Lehky, S. R., and T. J. Sejnowski, 1990. Neural network model of visual cortex for determining surface curvature from images of shaded surfaces. *Proc. R. Soc. Lond. [Biol.]* 240:251–278.

Livingstone, M. S., and D. H. Hubel, 1984a. Anatomy and physiology of a color system in the primate visual cortex. *J. Neurosci.* 4:309–356.

Livingstone, M. S., and D. H. Hubel, 1984b. Specificity of intrinsic connections in primate primary visual cortex. *J. Neurosci.* 4:2830–2835.

Lund, J. S., 1973. Organization of neurons in the visual cortex, area 17, of the monkey (*Macaca mulatta*). *J. Comp. Neurol.* 147:455–496.

Maffei, L., and A. Fiorentini, 1973. The visual cortex as a spatial frequency analyzer. *Vision Res.* 13:1255–1267.

Martin, K. A. C., and D. Whitteridge, 1984. Form, function and intracortical projections of spiny neurons in the striate visual cortex of the cat. *J. Physiol.* 353:463–504.

Maturana, H. R., J. Y. Lettvin, W. S. McCulloch, and W. H. Pitts, 1960. Anatomy and physiology of vision in the frog (*Rana pipiens*). *J. Gen. Physiol.* 43:129–176.

Maunsell, J. H. R., T. A. Nealey, G. Sclar, and D. D. DePriest, 1989. Representation of extraretinal information in monkey visual cortex. In *Neural Mechanisms of Visual Perception*, D. M. Lam and C. D. Gilbert, eds. The Woodlands, Texas: Portfolio Publishing, pp. 223–235.

McCormick, D. A., B. W. Connors, J. W. Lighthall, and D. A. Prince, 1985. Comparative electrophysiology of pyramidal and sparsely spiny stellate neurons of the neocortex. *J. Neurophysiol.* 54:782–806.

McGuire, B. A., C. D. Gilbert, P. K. Rivlin, and T. N. Wiesel, 1991. Targets of horizontal connections in macaque primary visual cortex. *J. Comp. Neurol.* 305:370–392.

McKee, S. P., and G. Westheimer, 1978. Stereoscopic acuity for moving retinal images. *J. Opt. Soc. Am.* 68:450–455.

Merzenich, M. M., R. J. Nelson, M. P. Stryker, M. S. Cynader, A. Schoppmann, and J. M. Zook, 1984. Somatosensory cortical map changes following digital amputation in adult monkeys. *J. Comp. Neurol.* 224:591–605.

Merzenich, M. M., G. Recanzone, W. M. Jenkins, T. T. Allard, and R. J. Nudo, 1988. Cortical representational plasticity. In *Neurobiology of Neocortex*, P. Rakic and W. Singer, eds. New York: Wiley, pp. 41–68.

Moran, J., and R. Desimone, 1985. Selective attention gates visual processing in the extrastriate cortex. *Science* 229:782–784.

Moses, Y., G. Schechtman, and S. Ullman, 1990. Self-calibrated collinearity detector. *Biol. Cybern.* 63:463–475.

Mountcastle, V. B., B. C. Motter, M. A. Steinmetz, and A. K. Sestokas, 1987. Common and differential effects of attentive fixation on the excitability of parietal and prestriate (V4) cortical visual neurons in the macaque monkey. *J. Neurosci.* 7:2239–2255.

Movshon, J. A., I. D. Thompson, and D. J. Tolhurst, 1978. Spatial and temporal contrast sensitivity of neurons in areas 17 and 18 of the cat's visual cortex. *J. Physiol.* 283:101–120.

Newsome, W. T., R. H. Wurtz, M. R. Dursteler, and A. Mikami, 1985a. Deficits in visual motion perception following ibotenic acid lesions of the middle temporal visual area of the macaque monkey. *J. Neurosci.* 5:825–840.

Newsome, W. T., R. H. Wurtz, M. R. Dursteler, and A. Mikami, 1985b. Punctate chemical lesions of striate cortex in the macaque monkey: Effect on visually guided saccades. *Exp. Brain. Res.* 58:392–399.

Paradiso, M. A., and K. Nakayama, 1991. Brightness perception and filling-in. *Vision Res.* 31:1221–1236.

Petersen, S. E., D. L. Robinson, and J. N. Currie, 1989. Influences of lesions of parietal cortex on visual spatial attention in humans. *Exp. Brain Res.* 76:267–280.

Pettet, M. W., and C. D. Gilbert, 1992. Dynamic changes in receptive field size in cat primary visual cortex. *Proc. Natl. Acad. Sci. U.S.A.* 89:8366–8370.

Poggio, T., M. Fahle, and S. Edelman, 1992. Fast perceptual learning in visual hyperacuity. *Science* 256:1018–1021.

Prince, D. A., and J. R. Huguenard, 1988. Functional

properties of neocortical neurons. In *Neurobiology of Neocortex*, P. Rakic and W. Singer, eds. New York: Wiley, pp. 153–176.

Ramachandran, V. S., and T. L. Gregory, 1991. Perceptual filling in of artificially induced scotomas in human vision. *Nature* 350:699–702.

Recanzone, G. H., M. M. Merzenich, W. M. Jenkins, A. G. Kamil, and H. R. Dinse, 1992. Topographic reorganization of the hand representation in cortical area 3b of owl monkeys trained in a frequency-discrimination task. *J. Neurophysiol.* 67:1031–1056.

Robinson, D. L., J. W. McClurkin, C. Kertzman, and S. E. Petersen, 1991. Visual responses of pulvinar and collicular neurons during eye movements of awake, trained macaques. *J. Neurophysiol.* 66:485–496.

Robinson, D. L., and S. E. Petersen, 1992. The pulvinar and visual salience. *Trends Neurosci.* 15:127–132.

Rockland, K. S., and J. S. Lund, 1982. Widespread periodic intrinsic connections in the tree shrew visual cortex. *Brain Res.* 169:19–40.

Rockland, K. S., and J. S. Lund, 1983. Intrinsic laminar lattice connections in primate visual cortex. *J. Comp. Neurol.* 216:303–318.

Rumelhart, D. E., G. E. Hinton, and R. J. Williams, 1986. Learning internal representation by error propagation. In *Parallel Distributed Processing: Explorations in the Microstructure of Cognition*, vol. 1, D. E. Rumelhart and J. L. McClelland, eds. Cambridge, Mass.: MIT Press, pp. 318–362.

Sanes, J. N., S. Suner, J. F. Lando, and J. P. Donoghue, 1988. Rapid reorganization of adult rat motor cortex somatic representation patterns after motor nerve injury. *Proc. Natl. Acad. Sci. U.S.A.* 85:2003–2007.

Sanes, J. N., J. Wang, and J. P. Donoghue, 1992. Immediate and delayed changes of rat motor cortical output representation with new forelimb configurations. *Cerebral Cortex* 2:141–152.

Schiller, P. H., B. L. Finlay, and S. F. Volman, 1976. Quantitative studies of single-cell properties in monkey striate cortex: III. Spatial frequency. *J. Neurophysiol.* 39: 1334–1351.

Spitzer, H., R. Desimone, and J. Moran, 1988. Increased attention enhances both behavioral and neuronal performance. *Science* 240:338–340.

Spitzer, H., and B. J. Richmond, 1990. Task difficulty: Ignoring, attending to, and discriminating a visual stimulus yield progressively more activity in inferior temporal neurons. *Exp. Brain Res.* 38:120–131.

Sporns, O., G. Tononi, and G. M. Edelman, 1991. Modeling perceptual grouping and figure-ground segregation by means of active reentrant connections. *Proc. Natl. Acad. Sci. U.S.A.* 88:129–133.

Tanaka, K., K. Hikosaka, H. Saito, M. Yukie, Y. Fukada, and E. Iwai, 1986. Analysis of local and wide-field movements in the superior temporal visual areas of the macaque monkey. *J. Neurosci.* 6:134–144.

Tigges, J., M. Tigges, S. Anschel, N. A. Croos, W. D. Letbetter, and R. L. McBride, 1981. Areal and laminar distribution of neurons interconnecting the central visual cortical areas 17, 18, 19 and MT in squirrel monkey (*Saimiri*). *J. Comp. Neurol.* 202:539–560.

Ts'o, D., and C. D. Gilbert, 1988. The organization of chromatic and spatial interactions in the primate striate cortex. *J. Neurosci.* 8:1712–1727.

Ts'o, D., C. D. Gilbert, and T. N. Wiesel, 1986. Relationships between horizontal and functional architecture in cat striate cortex as revealed by cross-correlation analysis. *J. Neurosci.* 6:1160–1170.

von der Heydt, R., and E. Peterhans, 1989. Mechanisms of contour perception in monkey visual cortex: I. lines of pattern discontinuity. *J. Neurosci.* 9:1731–1748.

von der Malsburg, C., and W. Schneider, 1986. A neural cocktail-party processor. *Biol. Cybern.* 54:29–40.

Wall, P. D., and R. Werman, 1976. The physiology and anatomy of long ranging afferent fibres within the spinal cord. *J. Physiol.* 255:321–334.

Weinberg, R. J., J. P. Pierce, and A. Rustioni, 1990. Single fiber studies of ascending input to the cuneate nucleus of cats: I. Morphometry of primary afferent fibers. *J. Comp. Neurol.* 300:113–133.

Weinberger, N. M., J. Ashe, and J. M. Edeline (in press). Learning-induced receptive field plasticity in the auditory cortex: Specificity of information storage. In *Neural Bases of Learning and Memory*, J. Delacour, ed. Singapore: World Scientific.

Weller, R. E., J. T. Wall, and J. H. Kaas, 1984. Cortical connections of the middle temporal visual area (MT) and the superior temporal cortex in owl monkeys. *J. Comp. Neurol.* 228:81–104.

Westheimer, G., 1986. Spatial interaction in the domain of disparity signals in human stereoscopic vision. *J. Physiol.* 370:619–629.

Westheimer, G., 1990. Simultaneoous orientation contrast for lines in the human fovea. *Vision Res.* 30:1913–1921.

Westheimer, G., and S. P. McKee, 1977. Spatial configurations for visual hyperacuity. *Vision Res.* 17:941–949.

Westheimer, G., K. Shimamura, and S. McKee, 1976. Interference with line orientation sensitivity. *J. Opt. Soc. Am.* 66:332–338.

Westheimer, G., and T. T. Truong, 1988. Target crowding in foveal and peripheral stereoacuity. *Am. J. Optom. Physiol. Optics* 65:395–399.

Yarbus, A. L., 1957. The perception of an image fixed with respect to the retina. *Biophysics* 2:683–690.

Zeki, S. M., 1976. The projections to the superior temporal sulcus from areas 17 and 18 in the rhesus monkey. *Proc. R. Soc. Lond. [Biol.]* 193:199–207.

Zeki, S. M., 1983. Colour coding in the cerebral cortex: The reaction of cells in monkey visual cortex to wavelengths and colours. *Neuroscience* 9:741–765.

Zipser, D., and R. A. Andersen, 1988. A back-propagation programmed network that simulates response properties of a subset of posterior parietal neurons. *Nature* 331:679–684.

6 Time as Coding Space in Neocortical Processing: A Hypothesis

WOLF SINGER

ABSTRACT One of the basic functions in sensory processing consists of the organization of distributed neuronal responses into representations of perceptual objects. This requires selection of responses on the basis of certain *Gestalt criteria* and establishment of unambiguous relations among the subset of selected responses. In this chapter, I propose that response selection and binding together of selected responses is achieved by the temporal synchronization of neuronal discharges. The predictions derived from this hypothesis are examined by combining multielectrode recordings from the mammalian visual cortex with neuroanatomical and behavioral analyses. The results indicate that spatially distributed neurons can synchronize their discharges on a millisecond time scale. The probability of synchronization depends both on the functional architecture of cortical connections and on the configuration of stimuli, reflects some of the *Gestalt criteria* used for perpetual grouping, and correlates well with the animals' perceptual abilities.

Analysis of the activity of single neurons in the central nervous system of awake behaving animals has revealed numerous and fascinating correlations between the responses of individual nerve cells and behavioral performance. This evidence has been construed as supportive of the hypothesis that activation of individual neurons can represent a code for highly complex and integrated functions, a notion that is commonly addressed as the *single neuron doctrine* (Barlow, 1972). However, there have always also been proposals that additional coding principles might be realized in the nervous system of higher vertebrates and mammals. Most of these proposals are extensions of Donald Hebb's postulate that neuronal representations of sensory or motor patterns should consist of assemblies of cooperatively interacting neurons rather than of individual cells. This coding principle implies that information is contained not only in the activation level of individual neurons but also, and actually to a crucial extent, in the relations between the activities of distributed neurons. If true, a complete description of a particular neuronal state would have to take into account not only the rate and the specificity of individual neuronal responses but also the relations between discharges of distributed neurons.

WOLF SINGER Max Planck Institute for Brain Research, Frankfurt, FRG

Evidence of relational codes

Over the last decade, these speculations have received some support from both experimental results and theoretical considerations. Search for individual neurons responding with the required selectivity to individual perceptual objects was only partly successful and has thus far revealed specificity for only faces and a limited set of objects with which the animal had been familiarized extensively before (Gross, Rocha-Miranda, and Bender, 1972; Desimone et al., 1984, 1985; Baylis, Rolls, and Leonard, 1985; Perrett, Mistlin, and Chitty, 1987; Miyashita, 1988; Rolls, 1991; Sakai and Miyashita, 1991). Even in these cases, it is likely that a particular face or object evokes responses in a very large number of neurons. Recordings from motor centers such as the deep layers of the tectum and areas of the motor cortex provided no evidence for command neurons such as exist in simple nervous systems and code for specific motor patterns. Rather, these studies provided strong support for a population code, as the trajectory of a particular movement could be predicted correctly only if the relative contributions of a large number of neurons were considered (Georgopoulos, 1990; Mussa-Ivaldi, Giszter, and Bizzi, 1990; Sparks, Lu, and Rohrer, 1990).

Arguments favoring the possibility of relational codes have also been derived from the concept of coarse coding. There are numerous examples that behaviorally determined discrimination between stimulus features can be superior to the discriminating abilities of individual sensory neurons (for review, see Lehky and Sejnowski, 1990). It has been proposed, therefore, that the information about the precise location of a stimulus and about specific features is not contained solely in the responses of the few neurons that are optimally activated but is encoded in the graded responses of the ensemble of neurons that respond to a particular stimulus. Further indications of the putative significance of relational codes are provided by theoretical studies that attempted to simulate certain aspects of pattern recognition and motor control in artificial neuronal networks. Single-cell codes were found appropriate for the representation of a limited set of well-defined patterns, but the number of required representational elements scaled very unfavorably with the number of representable patterns. Moreover, severe difficulties were encountered with functions such as scene segmentation and figure-ground distinction because single-cell codes turned out to be too rigid and inflexible, again leading to a combinatorial explosion of the required representational units.

By implementing population or relational codes, some of these problems can be alleviated (Hebb, 1949; Braitenberg, 1978; Edelman and Mountcastle, 1978; Grossberg, 1980; Palm, 1982, 1990; Crick, 1984; von der Malsburg, 1985; Singer, 1985, 1990; Edelman, 1987, 1989; Crick and Koch, 1990; Abeles, 1991). The essential feature of assembly coding is that individual cells can participate at different times in the representation of different sensory or motor patterns. The assumption is that just as a particular feature can be present in many different patterns, a neuron coding for this feature can be shared by many different representations. This reduces substantially the number of cells required for the representation of different patterns and allows for considerably more flexibility in the generation of new representations. However, whenever population codes are utilized, problems arise that are commonly addressed as *binding problems*. For the evaluation of population codes, it is necessary to identify in an unambiguous way those responses that participate in the representation of a particular content. In the case of coarse coding of individual stimulus features, responses representing the same feature need to be associated selectively and hence have to be distinguished from responses to other features. Similar distinctions are required if, at higher levels of processing, whole perceptual objects are represented by cell assemblies. Again, responses of cells participating in the representation of the same object need to be identified and segregated from responses to other objects. If only a single feature is present, binding problems do not arise, because in that case all responses can be associated with one another indiscriminately. However, the environment of the organism usually is crowded with different perceptual objects and features. Hence, a large number of neurons will be active simultaneously and, because of broad tuning, many of them will be coactivated by different features that may even belong to different perceptual objects. For a successful association of responses coding for the same feature, for the selective association of features belonging to the same object, and for the segregation of different objects from one another and from background, it is therefore indispensable to have a mechanism that permits, from the many simultaneous responses, selection of those that can be related to one another in a meaningful way and so avoid false conjunctions.

The proposal most pertinent to the experimental data reviewed in this chapter is that the population responses that need to be bound together are distinguished by a temporal code (von der Malsburg, 1985; von der Malsburg and Schneider, 1986). A similar suggestion, although formulated less explicitly, was made previously by Milner (1974). Both hypotheses assume that the responses of neurons participating in the encoding of related contents get organized in time through reciprocal interactions, with the effect that they eventually discharge synchronously. Thus, neurons having joined into an assembly that codes for the same feature or, at higher levels, for the same perceptual object would be identifiable as members of the assembly because their responses would contain episodes during which their discharges were synchronous. If the temporal window for the evaluation of coincident firing is kept in the millisecond range, relations between the activities of spatially distributed neurons can be defined very selectively (von der Malsburg, 1985; Shimizu et al., 1986; Abeles, 1991). Hence, if such temporal patterning is combined with population coding, the number of different features or patterns that can be represented by a given set of neurons is substantially larger than if binding is achieved by convergence.

Predictions

If an assembly of cells coding for a common feature or a common perceptual object or for a particular motor act is distinguished by the temporal coherence of the responses of the constituting neurons, predictions can be derived (as listed here), some of which are accessible to experimental testing.

1. Spatially segregated neurons should exhibit synchronized response episodes if activated by a single stimulus or by stimuli that can be grouped together into a single perceptual object.
2. Synchronization should be frequent among neurons within a particular cortical area, but it should occur also between cells distributed across different cortical areas if these cells respond to a common feature or perceptual object.
3. The probability that neurons synchronize their responses both within a particular area and across areas should reflect some of the Gestaltcriteria used for perceptual grouping.
4. Individual cells must be able to change rapidly the partners with which they synchronize their responses, if stimulus configurations change and require new associations.
5. If more than one object is present in a scene, several distinct assemblies should form. Cells belonging to the same assembly should exhibit synchronous response episodes, whereas no consistent temporal relations should exist between the discharges of neurons belonging to different assemblies.
6. Synchronization should occur as the result of a self-organizing process that is based on mutual and parallel interactions between distributed cortical cells.
7. The connections determining synchronization probabilities should be highly specific, as the criteria according to which distributed responses are bound together reside in the functional architecture of these connections.
8. The synchronizing connections should allow for interactions at levels of processing where responses of neurons already express some feature selectivity in order to permit feature-specific associations. This predicts that corticocortical connections contribute to synchronization.
9. The synchronizing connections should be endowed with adaptive synapses, allowing for use-dependent long-term modifications of synaptic gain to permit the acquisition of new grouping criteria when new object representations are to be installed during perceptual learning.
10. These use-dependent synaptic modifications should follow a correlation rule whereby synaptic connection should strengthen if presynaptic and postsynaptic activity often are correlated, and they should weaken in case there is no correlation. This is required to enhance grouping of cells that code for features

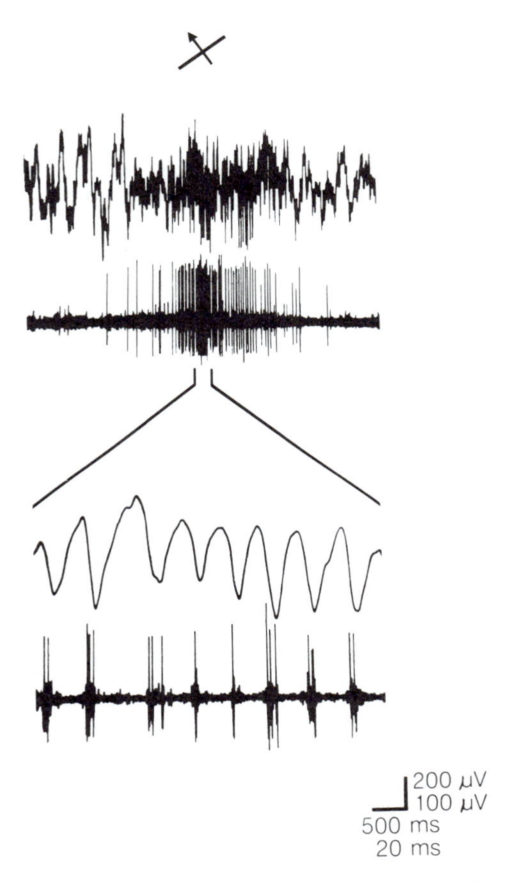

FIGURE 6.1 Multiunit activity (MUA) and local field potential (LFP) responses, recorded from area 17 in an adult cat, to the presentation of an optimally oriented light bar moving across the receptive field. Oscilloscope recordings of a single trial showing the response to the preferred direction of movement. In the upper two traces, on a slow time scale, the onset of the neuronal response is associated with an increase in high-frequency activity in the LFP. The lower two traces display the activity at the peak of the response on an expanded time scale. Note the presence of rhythmical oscillations in the LFP and MUA (35–45 Hz) that are correlated in phase with the peak negativity of the LFP. Upper and lower voltage scales are for the LFP and MUA, respectively. (Adapted from Gray and Singer, 1989.)

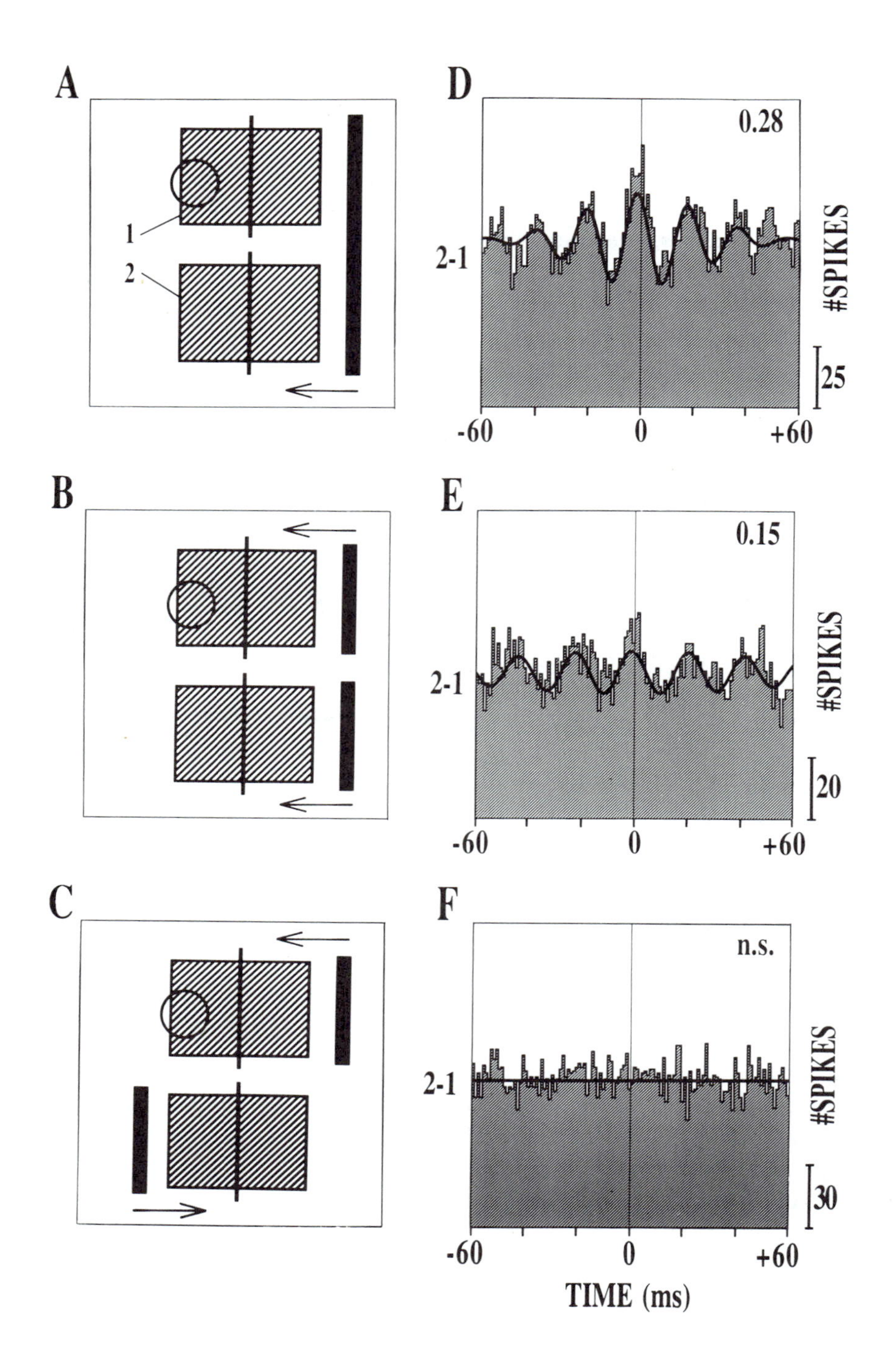

which often occur in consistent relations as is the case for features constituting a particular object.

11. These grouping operations should occur over multiple processing stages because search for meaningful groupings has to be performed at different spatial scales and according to different feature domains. This could be achieved by distributing the grouping operations over different cortical areas as neighborhood relations differ in different areas because of remapping of input connections.

Experimental testing of predictions

Systematic search for the predicted stimulus-dependent synchronization phenomena between spatially distributed cortical neurons had been initiated by the observation that adjacent neurons in the cat visual cortex can engage in highly synchronous discharges when presented with their preferred stimulus (Gray and Singer, 1987). Neurons recorded simultaneously with a single electrode were found to engage transiently in synchronous discharges. In these multiunit recordings, the synchronous discharges appear as clusters of spikes that often follow one another at rather regular intervals of 15 to 30 ms. These sequences of synchronous rhythmical firing usually last no more than a few hundred milliseconds and may occur several times during a single passage of moving stimuli (figure 6.1). Accordingly, autocorrelograms computed from such response epochs often exhibit a periodic modulation (Gray and Singer, 1987, 1989; Eckhorn et al., 1988; Gray et al., 1990; Schwarz and Bolz, 1991; Livingstone, 1991).

This phenomenon of local response synchronization has been observed with multiunit and field potential recordings in several independent studies in different areas of the visual cortex of anesthetized cats (areas 17, 18, 19, and posteromedial lateral suprasylvian) (Eckhorn et al., 1988, 1992; Gray and Singer, 1989; Gray et al., 1990; Engel et al., 1991c); in area 17 of awake cats (Raether, Gray, and Singer, 1989; Gray and Viana di Prisco, 1993) in the optic tectum of awake pigeons (Neuenschwander and Varela, 1993); and in the visual cortex of anesthetized (Livingstone, 1991) and awake behaving monkeys (Kreiter and Singer, 1992; Eckhorn et al., 1993). Similar synchronization phenomena have been observed in a variety of nonvisual structures (for review, see Singer, 1993).

FIGURE 6.2 Long-range synchronization is influenced by stimulus coherence. Multiunit activity was recorded from two sites in area 17 of cat visual cortex that were separated by 7 mm. The two cell groups preferred vertical orientations. (A, B, C) Plots of the receptive fields. The colinear arrangement of the fields allowed the comparison of three different stimulus paradigms: (A) a long continuous light bar moving across both fields, (B) two independent light bars moving in the same direction, and (C) the same two bars moving in opposite directions. The circle represents the center of the visual field, and the thick line drawn across each receptive field indicates the preferred orientation. (D, E, F) The respective cross-correlograms obtained with each stimulus paradigm. Using the long light bar, the two oscillatory responses were synchronized, as indicated by the strong modulation of the cross-correlogram with alternating peaks and troughs (D). If the continuity of the stimulus was interrupted, the synchronization became weaker (E), and it totally disappeared if the motion of the stimuli was incoherent (F). This change of the stimulus configuration affected neither the strength nor the oscillatory nature of the two responses (not shown). The graph superimposed on each of the correlograms represents a Gabor function that was fitted to the data to assess the strength of the modulation (Engel et al., 1990). The number in the upper right corner indicates the relative modulation amplitude, a measure of correlation strength that was determined by computing the ratio of the amplitude of the Gabor function to its offset. ns, not significant. Scale bars indicate the number of spikes. (Reprinted from Engel et al., 1992, by permission.)

Multielectrode recordings have revealed that similar response synchronization can occur between spatially segregated cell groups within the same visual area (Gray et al., 1989, 1992; Engel et al., 1990; König et al., 1993) but also between different cortical areas (Eckhorn et al., 1988; Engel et al., 1991c; Murthy and Fetz, 1992; Nelson et al., 1992) and even across hemispheres (Engel et al., 1991a; Munk et al., 1992). Measurements in awake cats (Raether, Gray, and Singer, 1989; Gray and Viana di Prisco, 1993) and monkeys (Ahissar et al., 1992; Kreiter and Singer, 1992; Murthy and Fetz, 1992) have shown that this phenomenon of response synchronization is not confined to anesthesia but is readily demonstrable in various neocortical areas (visual, auditory, somatosensory and motor) of alert behaving animals. These single-unit data are complemented by a large body of evidence derived from field potential and electroencephalographic recordings, all of which indicate that distributed groups of neurons can engage in synchronous activity (for a review of the extensive literature, see Singer, 1993).

The dependence of response synchronization on stimulus configuration

As outlined previously, the hypothesis of temporally coded assemblies requires that the probabilities with which distributed cells synchronize their responses should reflect some of the *Gestalt criteria* applied in perceptual grouping. Another and related prediction is that individual cells must be able to change the partners with which they synchronize, the selection of partners occurring as a function of the patterns used to activate the cells.

Gray and colleagues (1989) recorded multiunit activity from two locations in cat area 17 separated by 7 mm. The receptive fields of the cells were nonoverlapping, had nearly identical orientation preferences, and were spatially displaced along the axis of preferred orientation. This enabled stimulation of the cells with bars of the same orientation under three different conditions: two bars moving in opposite directions, two bars moving in the same direction, and one long bar moving across both fields coherently. No significant correlation was found when the cells were stimulated by oppositely moving bars, and only a weak correlation was present for the coherently moving bars, but the long-bar stimulus resulted in a robust synchronization of the activity at the two sites (figure 6.2). This effect occurred despite the fact that the overall number of spikes produced by the two cells and the oscillatory patterning of the responses were similar in the three conditions.

In related experiments, Engel and coworkers (1991a, c) demonstrated in the cat that the synchronization of activity between cells in areas 17 and PMLS and between area 17 in each of the two hemispheres exhibits a similar dependence on the properties of the visual stimulus (figure 6.3). These findings indicate that the global properties of visual stimuli can influence the magnitude of synchronization between widely separated cells located within and between different cortical areas. Single contours, but also spatially separate contours that move coherently and therefore appear as parts of a single figure, are more efficient in inducing synchrony among the responding cell groups than incoherently moving contours that appear as parts of independent figures.

These results indicate clearly that synchronization probability depends not only on the spatial segregation of cells and on their feature preferences, the latter being related to the cells' position within the columnar architecture of the cortex, but also—and to a crucial extent—on the configuration of the stimuli. Thus far, synchronization probability appears to reflect rather well some of the *Gestalt criteria* for perceptual grouping. The high synchronization probability of nearby cells corresponds to the binding criterion of vicinity, the dependence on receptive field similarities agrees with the criterion of similarity, the strong synchronization observed in response to continuous stimuli obeys the criterion of continuity, and the lack of synchrony in responses to stimuli moving in opposite directions relates to the criterion of "common fate".

Experiments have also been performed to test the prediction that simultaneously presented but different contours should lead to the organization of two independently synchronized assemblies of cells (Engel, König, and Singer, 1991b; Kreiter, Engel, and Singer, 1992). If groups of cells with overlapping receptive fields but different orientation preferences are activated with a single moving light bar, they synchronize their responses (Engel et al., 1990; Engel, König, and Singer, 1991b). Usually, synchrony is established between all responding neurons, including those that are activated only suboptimally. This agrees with the postulate derived from the hypothesis of coarse coding that all responses of cells participating in the representation of a stimulus ought to be bound together. However, if such a set of groups is stimulated with two independent spatially overlapping stimuli that move in different directions, the activated cells split into two independently synchronized assemblies. Cells whose feature preferences match better with stimulus 1 form one synchronously active assembly, and those matching better with stimulus 2 the other (figure 6.4). Thus, although the two stimuli now evoke graded responses in all the recorded groups, cells representing the same stimulus remain distinguishable because their responses exhibit synchronized response epochs while showing no consistent correlations with responses of cells activated by different stimuli. To extract this information, a readout mechanism is required that is capable of evaluating coincident firing on a millisecond time scale. A recent analysis of the integrative properties of cortical pyramidal cells suggests that in these cells, the window for effective temporal summation may indeed be as short as a few milliseconds (Softky and Koch, 1993).

Another important issue tackled by these experiments is determining whether individual cells can actually change the partners with which they synchronize when stimulus configurations change. Cell groups that engaged in synchronous response episodes when activated with a single stimulus no longer did so when activated with two stimuli but then synchronized with other groups. This agrees with the prediction of the assembly hypothesis that interactions between distributed cell groups should be variable and influenced by the constellation of features in the visual stimulus.

Experience-dependent development of synchronizing connections

The hypothesis requires that synchronization probability depends, to a substantial degree, on interactions between the neurons whose responses actually represent the features that need to be bound together. As

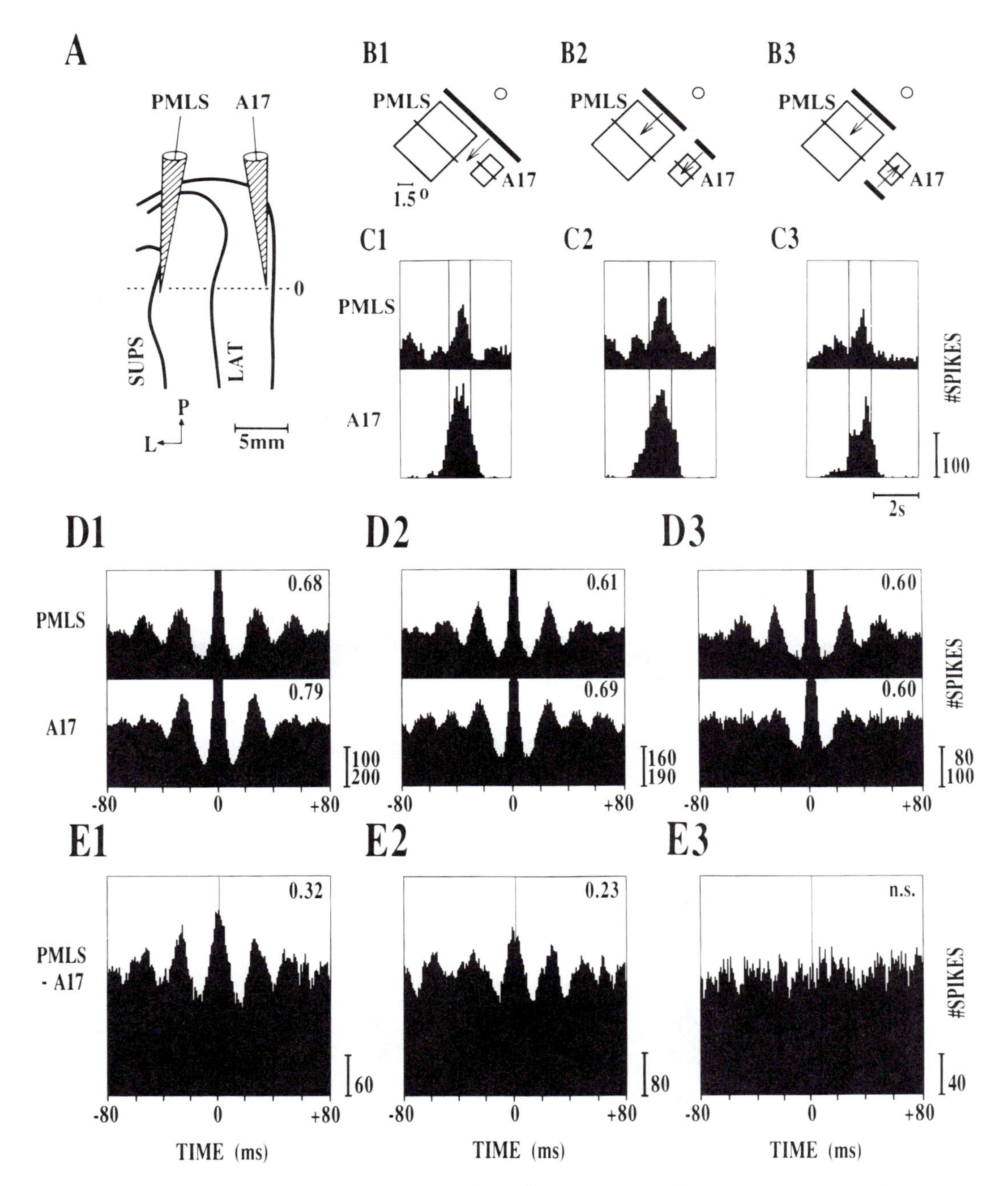

FIGURE 6.3 Interarea synchronization is sensitive to global stimulus features. (A) Position of the recording electrodes. PMLS, A17, area 17; LAT, lateral sulcus; SUPS, suprasylvian sulcus; P, posterior; L, lateral. (B1–B3) Plots of the receptive fields of the PMLS and area 17 recording. Diagrams depict the three stimulus conditions tested. Circle indicates the visual field center. (C1–C3) Peristimulus time histograms for the three stimulus conditions. The vertical lines indicate 1-s windows for which autocorrelograms and cross-correlograms were computed. (D1–D3) The autocorrelograms computed for the three stimulus paradigms are compared. Note that the modulation amplitude of the correlograms is similar in all three cases (indicated by the number in the upper right corner). (E1–E3) Cross-correlograms computed for the three stimulus conditions. The number in the upper right corner represents the relative modulation amplitude of each correlogram. Note that the strongest correlogram modulation is obtained with the continuous stimulus. The cross-correlogram is less regular and has a lower modulation amplitude when two light bars are used as stimuli and there is no significant modulation (n.s.) with two light bars moving in opposite directions. (Reprinted from Engel et al., 1991c, by permission.)

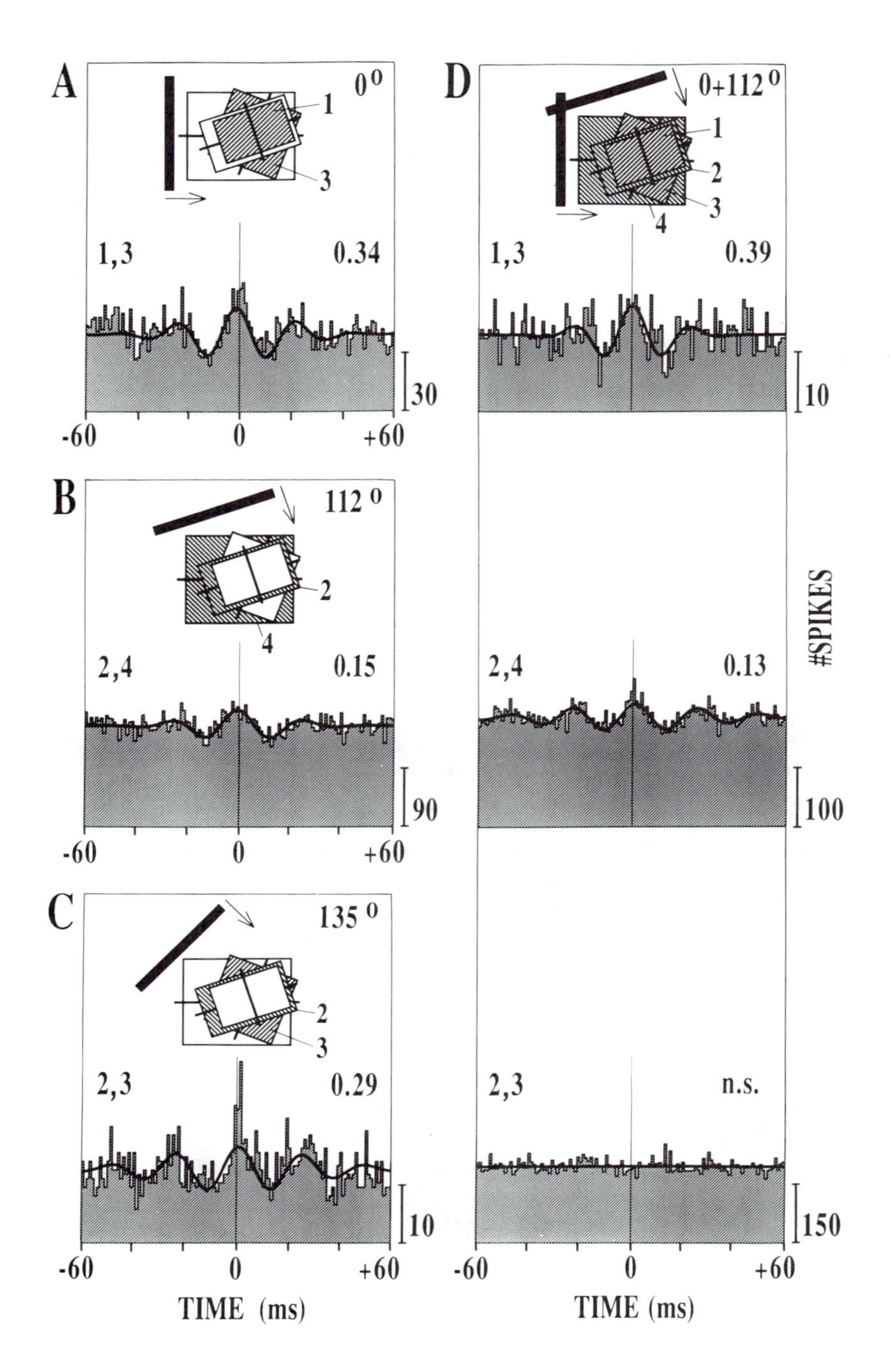

cells in subcortical centers possess only very limited feature selectivity, one is led to postulate that corticocortical connections should also contribute to the synchronization process. This postulate is supported by the finding that synchronization between cells located in different hemispheres is abolished when the corpus callosum is cut (Engel et al., 1991a; Munk et al., 1992), which is direct proof that (1) corticocortical connections contribute to response synchronization and (2) synchronization with zero phase lag can be brought about by reciprocal interactions between spatially distributed neurons despite considerable conduction delays in the coupling connections. Thus, synchrony is not necessarily an indication of common input but may be the result of a dynamic organization process that establishes coherent firing by reciprocal interactions.

The theory of assembly coding implies that the criteria according to which particular features are grouped together, rather than others, reside in the functional architecture of the assembly forming coupling connections. It is of particular interest, therefore, to study the development of the synchronizing connections, to identify the rules according to which they are selected, to establish correlations between their architecture and synchronization probabilities and, if possible, to relate these neuronal properties to perceptual functions.

In mammals, corticocortical connections develop mainly postnatally (Innocenti, 1981; Luhmann, Martinez-Millan, and Singer, 1986; Price and Blakemore, 1985a; Callaway and Katz, 1990) and attain their final specificity through an activity-dependent selection process (Innocenti and Frost, 1979; Price and Blakemore, 1985b; Luhmann, Singer, and Martinez-Millan, 1990; Callaway and Katz, 1991). Recently, it has been found that strabismus, when induced in 3-week-old kittens, leads to a profound rearrangement of corticocortical connections. Normally, these connections link cortical territories irrespective of whether these are dominated by the same or by different eyes. In the strabismics, by contrast, the tangential intracortical connections link with high selectivity only territories served by the same eye. The functional correlate of these changes in the architecture of corticocortical connections is a modification of synchronization probabilities. In strabismics, response synchronization no longer occurs between cell groups connected to different eyes, whereas it is normal between cell groups connected to the same eye (König et al., 1990, 1993).

These results have several implications. First, they are compatible with the notion that tangential intracortical connections contribute to response synchronization (as discussed earlier). However, as strabismus also abolishes convergence of projections from the two eyes onto common cortical target cells, this result is also compatible with the view that synchrony is caused by common input. Second, these results agree with the postulates of the assembly hypothesis that the assembly forming connections should be susceptible to use-dependent modifications and be selected according to a correlations rule. Third, the modifications of intracortical connections and synchronization probabilities add to the list of substrate changes that may be related to the specific perceptual deficits associated with early-onset squint. Strabismic subjects usually develop normal monocular vision in both eyes, but they become unable to fuse signals conveyed by different eyes into coherent percepts even if these signals are made retinotopically contiguous by optical compensation of the squint angle (von Noorden, 1990). Thus, in strabismics, binding mechanisms appear to be abnormal or missing between cells driven from different eyes. The lack of corticocortical connections and the lack of response synchronization could be one of the reasons for this deficit, as is the loss of binocular neurons.

These correlations are, at the least, compatible with the view that the architecture of corticocortical connections, by determining the probability of response synchronization, could set the criteria for perceptual grouping. Because this architecture is shaped by experience, it is possible that some of the binding

FIGURE 6.4 Stimulus dependence of short-range interactions. Multiunit activity was recorded from four different orientation columns of area 17 of cat visual cortex separated by 0.4 mm. The four cell groups had overlapping receptive fields and orientation preferences of 22° (group 1), 112° (group 2), 157° (group 3), and 90° (group 4), as indicated by the thick line drawn across each receptive field in (A)–(D). The figure shows a comparison of responses to stimulation with single moving light bars of varying orientation (left) and responses to the combined presentation of two superimposed light bars (right). For each stimulus condition, the shading of the receptive fields indicates the responding cell groups. Stimulation with a single light bar yielded synchronization between all cells activated by the respective orientation. Thus, groups 1 and 3 responded synchronously to a vertically oriented (0°) light bar (A), groups 2 and 4 to a light bar at an orientation of 112° (B), and cell groups 2 and 3 to a light bar of intermediate orientation (C). Simultaneous presentation of two stimuli with orientations of 0° and 112°, respectively, activated all four groups (D). However, in this case, the groups segregated into two distinct assemblies, depending on which stimulus was closer to the preferred orientation of each group. Thus, responses were synchronized between groups 1 and 3, which preferred the vertical stimulus, and between 2 and 4, which preferred the stimulus oriented at 112°. The two assemblies were desynchronized with respect to each other, and so there was no significant synchronization between groups 2 and 3. The cross-correlograms between groups 1 and 2, 1 and 4, and 3 and 4 were also flat (not shown). Note that the segregation cannot be explained by preferential anatomical wiring of cells with similar orientation preference (Ts'o, Gilbert, and Wiesel, 1986) because cell groups can readily be synchronized in all possible pair combinations in response to a single light bar. The correlograms are shown superimposed with their Gabor function. The number to the upper right of each correlogram indicates the relative modulation amplitude. ns, not significant. Scale bars indicate the number of spikes. (Reprinted from Engel et al., 1991b, by permission.)

and segmentation criteria are acquired or modified by experience.

Impaired response synchronization correlates with perceptual disturbances

Further indications for a relation between experience-dependent modifications of synchronization probabilities and functional deficits come from a recent study of strabismic cats in which amblyopia had developed. Strabismus, when induced early in life, not only abolishes binocular fusion and stereopsis but also may lead to amblyopia of one eye (von Noorden, 1990). This condition develops when the subjects solve the problem of double vision not by alternating use of the two eyes but by constantly suppressing the signals coming from the deviated eye. The amblyopic deficit usually consists of reduced spatial resolution and distorted and blurred perception of patterns. A particularly characteristic phenomenon in amblyopia is crowding, the drastic impairment of the ability to discriminate and recognize figures if these are surrounded with other contours. The identification of neuronal correlates of these deficits in animal models of amblyopia has remained inconclusive because the contrast sensitivity and the spatial resolution capacity of neurons in the retina and the lateral geniculate nucleus were found normal.

In the visual cortex, identification of neurons with reduced spatial resolution or otherwise abnormal receptive field properties remained controversial (for a discussion, see Blakemore and Vital-Durand, 1992; Crewther and Crewther, 1990). However, multielectrode recordings from striate cortex of cats exhibiting behaviorally verified amblyopia have revealed highly significant differences in the synchronization behavior of cells driven by the normal and the amblyopic eye, respectively. The responses to single moving bars that were recorded simultaneously from spatially segregated neurons connected to the amblyopic eye were much less well synchronized with one another than the responses recorded from neuron pairs driven through the normal eye (Roelfsema et al., submitted). This difference was even more pronounced for responses elicited by gratings of different spatial frequency. For responses of cell pairs activated through the normal eye, the strength of synchronization tended to increase with increasing spatial frequency, whereas it tended to decrease further for cell pairs activated through the amblyopic eye (figure 6.5).

Apart from these highly significant differences between the synchronization behavior of cells driven though the normal and the amblyopic eye, no other differences were found in the commonly determined response properties of these cells. Thus, cells connected to the amblyopic eye continued to respond vigorously to gratings whose spatial frequency had been too high to be discriminated with the amblyopic eye in the preceding behavioral tests.

These results suggest that disturbed temporal coordination of responses, such as reduced synchrony, may be one of the neuronal correlates of the amblyopic deficit. Indeed, if synchronization of responses on a millisecond time scale is used by the system to tag and identify the responses of cells that code for the same feature or contour, disturbance of this temporal patterning could be the cause for the crowding phenomenon. If responses evoked by nearby contours can no longer be associated unambiguously with either contour but become confounded, perceptual deficits that closely resemble the crowding phenomenon are expected.

The reduced synchronization among cells driven by the amblyopic eye might be attributable to abnormalities in the network of corticocortical connections linking cell groups dominated by this eye. It is conceivable that the continuous suppression of the signals provided from this eye has impeded the experience-dependent specification of the respective intracortical synchronizing connections. As reduced synchrony also impairs the transmission of responses and hence their saliency, the present results can further account for the fact that amblyopic patients find it difficult to attend to the signals conveyed by the amblyopic eye when both eyes are open. In that case, signals from the amblyopic eye usually are eliminated from further processing and are not perceived.

Conclusions

The experimental results reviewed in this chapter are compatible with predictions derived from the hypothesis that synchronization of neuronal responses on a time scale of milliseconds may be exploited in cortical processing. In a distributed network such as neocortex, where any given cell contacts any other cell with only a few synapses but where individual cells receive converging input from many thousands of different cells, synchronization of discharges is a particularly

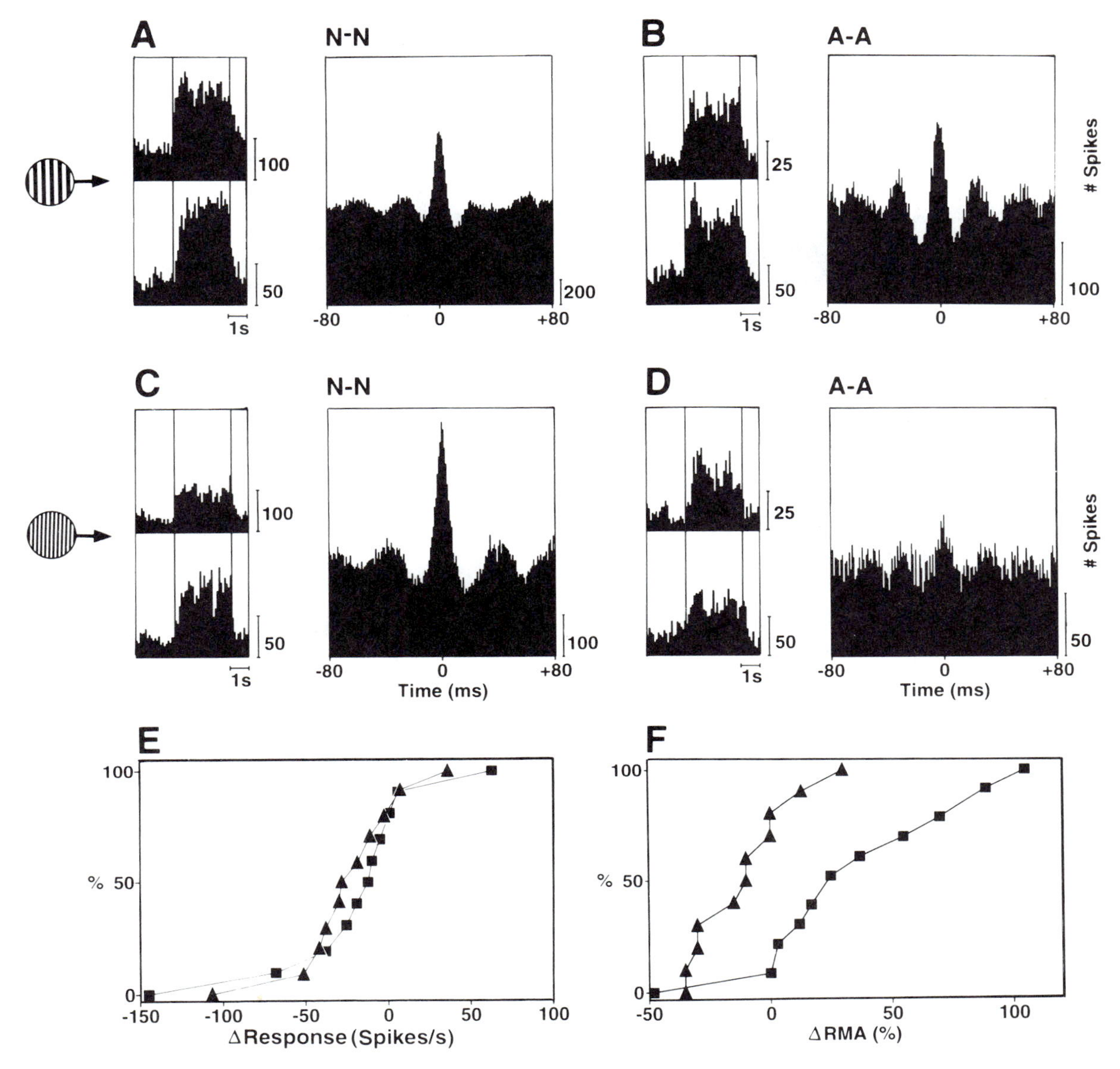

FIGURE 6.5 Amplitudes and synchronization of responses to gratings of different spatial frequencies recorded from cats with strabismic amblyopia. Responses to gratings of low (A, B) and high (C, D) spatial frequencies, recorded simultaneously from two cell groups driven by the normal eye (N-sites) (A, C) and two cell groups driven by the amblyopic eye (A-sites) (B, D), respectively. The left and right panels show the response histograms and the corresponding cross-correlograms. Note that response amplitudes decrease at the higher spatial frequency in both cases, whereas the relative modulation amplitude increases for the N-N pair but decreases for the A-A pair. (E) Cumulative distribution functions of the differences between the amplitude of responses to low- and high-spatial-frequency gratings of optimal orientation. N-sites are indicated by squares (n = 53), A-sites by triangles (n = 35); Abscissa, responses to high-spatial-frequency minus responses to low-spatial-frequency gratings. Note the similarity of the two distributions ($p > .1$). (F) Cumulative distribution functions of the differences between relative modulation amplitudes (ΔRMA of cross-correlograms obtained for responses to high- and low-spatial-frequency gratings of N-N pairs (squares; n = 24) and A-A pairs (triangles; n = 11). ΔRMA values (abscissa) were calculated by subtracting the relative modulation amplitude obtained with the low spatial frequency from that obtained with the high spatial frequency. The difference between the ΔRMA distributions of N-N pairs and A-A pairs is highly significant ($p < .001$). (Used by permission from Roelfsema et al., submitted.)

efficient mechanism to increase the saliency of responses. In addition, synchronization establishes unambiguous relations between responses, as it enhances with great selectivity the saliency of only those response episodes that contain coincident discharges. Hence, synchronization can, in principle, be used to select, with high spatial and temporal precision, those constellations of responses that should be considered for further processing. The proposal is that this selection is achieved in a distributed and highly parallel operation by the system of corticocortical association fibers. Their function would consist essentially of adjusting the *timing* of discharges rather than of modulating discharge rates. Eventually, however, synchronization will also influence firing rates, as synchronous inputs are more effective than asynchronous inputs in driving cells. Hence rate codes and synchronization codes can coexist in the same network. The prediction is that rate coding should prevail at levels of processing that are close to sensory input and to motor output, whereas the intermediate computations should be based predominantly on the shifting of temporal relations.

More direct investigations of this conjecture are now required. As temporal relations such as synchrony can be evaluated only by simultaneously recording the activity of different cells, critical tests require the application of multielectrode recordings, and most likely it will also be necessary to develop new analytical methods to detect reliably transient temporal relations among the activities of widely distributed groups of neurons. For the brain, a synchronous discharge of several thousand cells is likely to be a highly significant event, even if it occurs only once, but for the experimenter, such episodes may pass undetected as long as only the responses of a few neurons can be examined at one time.

REFERENCES

Abeles, M., ed., 1991. *Corticonics*. Cambridge: Cambridge University Press.

Ahissar, E., E. Vaadia, M. Ahissar, H. Bergmann, A. Arieli, and M. Abeles, 1992. Dependence of cortical plasticity on correlated activity of single neurons and on behavioral context. *Science* 257:1412–1415.

Barlow, H. B., 1972. Single units and cognition: A neurone doctrine for perceptual psychology. *Perception* 1:371–394.

Baylis, G. C., E. T. Rolls, and C. M. Leonard, 1985. Selectivity between faces in the responses of a population of neurons in the cortex in the superior temporal sulcus of the monkey. *Brain Res.* 342:91–102.

Blakemore, C., and F. Vital-Durand, 1992. Different neural origins for "blur" amblyopia and strabismic amblyopia. *Ophthalmic Physiol. Opt.* 12.

Braitenberg, V., 1978. Cell assemblies in the cerebral cortex. In *Lecture Notes in Biomathematics*, vol. 21, R. Heimand and G. Palm, eds. Berlin: Springer-Verlag, pp. 171–188.

Callaway, E. M., and L. C. Katz, 1990. Emergence and refinement of clustered horizontal connections in cat striate cortex. *J. Neurosci.* 10:1134–1153.

Callaway, E. M., and L. C. Katrz, 1991. Effects of binocular deprivation on the development of clustered horizontal connections in cat striate cortex. *Proc. Natl. Acad. Sci. U.S.A.* 88:745–749.

Crewther, D. P., and S. G. Crewther, 1990. Neural sites of strabismic amblyopia in cats: Spatial frequency deficit in primary cortical neurons. *Exp. Brain Res.* 79:615–622.

Crick, F., 1984. Function of the thalamic reticular complex: The searchlight hypothesis. *Proc. Natl. Acad. Sci. U.S.A.* 81:4586–4590.

Crick, F., and C. Koch, 1990. Towards a neurobiological theory of consciousness. *Semin. Neurosci.* 2:263–275.

Desimone, R., T. D. Albright, C. G. Gross, and C. Bruce, 1984. Stimulus-selective properties of inferior temporal neurons in the macaque. *J. Neurosci.* 4:2051–2062.

Desimone, R., S. J. Schein, J. Moran, and L. G. Ungerleider, 1985. Contour, color and shape analysis beyond the striate cortex. *Vision Res.* 24:441–452.

Eckhorn, R., R. Bauer, W. Jordan, M. Brosch, W. Kruse, M. Munk, and H. J. Reitboeck, 1988. Coherent oscillations: A mechanism for feature linking in the visual cortex? *Biol. Cybern.* 60:121–130.

Eckhorn, R., A. Frien, R. Bauer, T. Woelbern, and H. Kehr, 1993. High frequency (60–90 Hz) oscillations in primary visual cortex of awake monkey. *NeuroReport* 4: 243–246.

Eckhorn, R., T. Schanze, M. Brosch, W. Salem, and R. Bauer, 1992. Stimulus-specific synchronizations in cat visual cortex: Multiple microelectrode and correlation studies from several cortical areas. In *Induced Rhythms in the Brain*, E. Basar and T. H. Bullock, eds. Boston: Birkhäuser, pp. 47–80.

Edelman, G. M., 1987. *Neural Darwinism: The Theory of Neuronal Group Selection*. New York: Basic Books.

Edelman, G. M., 1989. *The Remembered Present*. New York: Basic Books.

Edelman, G. M., and V. B. Mountcastle, 1978. *The Mindful Brain*. Cambridge, Mass.: MIT Press.

Engel, A. K., P. König, C. M. Gray, and W. Singer, 1990. Stimulus-dependent neuronal oscillations in cat visual cortex: Inter-columnar interaction as determined by cross-correlation analysis. *Eur. J. Neurosci.* 2:588–606.

Engel, A. K., P. Konig, A. K. Kreiter, T. B. Schillen, et al., 1992. Temporal coding in the visual cortex: New vistas on integration in the nervous system. *Trends Neurosci.* 15: 218–226.

Engel, A. K., P. König, A. K. Kreiter, and W. Singer, 1991a. Interhemispheric synchronization of oscillatory neuronal responses in cat visual cortex. *Science* 252:1177–1179.

Engel, A. K., P. König, and W. Singer, 1991b. Direct physiological evidence for scene segmentation by temporal coding. *Proc. Natl. Acad. Sci. U.S.A.* 88:9136–9140.

Engel, A. K., A. K. Kreiter, P. König, and W. Singer, 1991c. Synchronization of oscillatory neuronal responses between striate and extrastriate visual cortical areas of the cat. *Proc. Natl. Acad. Sci. U.S.A.* 88:6048–6052.

Georgopoulos, A. P., 1990. Neural coding of the direction of reaching and a comparison with saccadic eye movements. *Cold Spring Harb. Symp. Quant. Biol.* 55:849–859.

Gray, C. M., A. K. Engel, P. König, and W. Singer, 1990. Stimulus-dependent neuronal oscillations in cat visual cortex: Receptive field properties and feature dependence. *Eur. J. Neurosci.* 2:607–619.

Gray, C. M., A. K. Engel, P. König, and W. Singer, 1992. Synchronization of oscillatory neuronal responses in cat striate cortex: Temporal properties. *Visual Neurosci.* 8:337–347.

Gray, C. M., P. König, A. K. Engel, and W. Singer, 1989. Oscillatory responses in cat visual cortex exhibit intercolumnar synchronization which reflects global stimulus properties. *Nature* 338:334–337.

Gray, C. M., and W. Singer, 1987. Stimulus-specific neuronal oscillations in the cat visual cortex: A cortical functional unit. *Soc. Neurosci. Abstr.* 13:404.3.

Gray, C. M., and W. Singer, 1989. Stimulus-specific neuronal oscillations in orientation columns of cat visual cortex. *Proc. Natl. Acad. Sci. U.S.A.* 86:1698–1702.

Gray, C. M., and G. Viana di Prisco, 1993. Properties of stimulus-dependent rhythmic activity of visual cortical neurons in the alert cat. *Soc. Neurosci. Abstr.* 19:359.8.

Gross, C. G., E. C. Rocha-Miranda, and D. B. Bender, 1972. Visual properties of neurons in inferotemporal cortex of the macaque. *J. Neurophysiol.* 35:96–111.

Grossberg, S., 1980. How does the brain build a cognitive code? *Psychol. Rev.* 87:1–51.

Hebb, D. O., 1949. *The Organization of Behavior*: A Neuropsychological Theory. New York: Wiley.

Innocenti, G. M., 1981. Growth and reshaping of axons in the establishment of visual callosal connections. *Science* 212:824–827.

Innocenti, G. M., and D. O. Frost, 1979. Effects of visual experience on the maturation of the efferent system to the corpus callosum. *Nature* 280:231–234.

König, P., A. K. Engel, S. Löwel, and W. Singer, 1990. Squint affects occurrence and synchronization of oscillatory responses in cat visual cortex. *Soc. Neurosci. Abstr.* 16: 523.2.

König, P., A. K., Engel, S. Löwel, and W. Singer, 1993. Squint affects synchronization of oscillatory responses in cat visual cortex. *Eur. J. Neurosci.* 5:501–508.

Kreiter, A. K., A. K. Engel, and W. Singer, 1992. Stimulus-dependent synchronization of oscillatory neuronal activity in the superior temporal sulcus of the macaque monkey. *Eur. Neurosci. Assoc. Abstr.* 15:1076.

Kreiter, A. K., and W. Singer, 1992. Oscillatory neuronal responses in the visual cortex of the awake macaque monkey. *Eur. J. Neurosci.* 4:369–375.

Lehky, S. R., and T. J. Sejnowski, 1990. Neural model of stereoacuity and depth interpolation based on distributed representation of stereo disparity. *J. Neurosci.* 10:2281–2299.

Livingstone, M. S., 1991. Visually evoked oscillations in monkey striate cortex. *Soc. Neurosci. Abstr.* 17:73.3.

Luhmann, H. J., L. Martinez-Millan, and W. Singer, 1986. Development of horizontal intrinsic connections in cat striate cortex. *Exp. Brain Res.* 63:443–448.

Luhmann, H. J., W. Singer, and L. Martinez-Millan, 1990. Horizontal interactions in cat striate cortex: I. Anatomical substrate and postnatal development. *Eur. J. Neurosci.* 2:344–357.

Milner, P. M., 1974. A model for visual shape recognition. *Psychol. Rev.* 81:521–535.

Miyashita, Y., 1988. Neuronal correlate of visual associative long-term memory in the primate temporal cortex. *Nature* 335:817–820.

Munk, M. H. J., L. G. Nowak, G. Chouvet, J. I. Nelson, and J. Bullier, 1992. The structural basis of cortical synchronization. *Eur. J. Neurosci. [Suppl.]* 5:21.

Murthy, V. N., and E. E. Fetz, 1992. Coherent 25- to 35-Hz oscillations in the sensorimotor cortex of awake behaving monkeys. *Proc. Natl. Acad. Sci. U.S.A.* 89:5670–5674.

Mussa-Ivaldi, F. A., S. F. Giszter, and E. Bizzi, 1990. Motor-space coding in the central nervous system. *Cold Spring Harb. Symp. Quant. Biol.* 55:827–835.

Nelson, J. I., P. A. Salin, M. H. J. Munk, M. Arzi, and J. Bullier, 1992. Spatial and temporal coherence in cortico-cortical connections: A cross-correlation study in areas 17 and 18 in the cat. *Visual Neurosci.* 9:21–38.

Neuenschwander, S., and F. J. Varela, 1993. Visually triggered neuronal oscillations in the pigeon: An autocorrelation study of tectal activity. *Eur. J. Neurosci.* 5:870–881.

Palm, G., 1982. *Neural Assemblies*. Berlin: Springer-Verlag.

Palm, G., 1990. Cell assemblies as a guideline for brain research. *Concepts Neurosci.* 1:133–137.

Perrett, D. I., A. J. Mistlin, and A. J. Chitty, 1987. Visual neurones responsive to faces. *Trends Neurosci.* 10: 358–364.

Price, D. J., and C. Blakemore, 1985a. The postnatal development of the association projection from visual cortical area 17 to area 18 in the cat. *J. Neurosci.* 5:2443–2452.

Price, D. J., and C. Blakemore, 1985b. Regressive events in the postnatal development of association projections in the visual cortex. *Nature* 316:721–724.

Raether, A., C. M. Gray, and W. Singer, 1989. Intercolumnar interactions of oscillatory neuronal responses in the visual cortex of alert cats. *Eur. Neurosci. Assoc. Abstr.* 12:72.5.

Roelfsema, P. R., P. König, A. K. Engel, R. Sireteanu,

and W. Singer, 1993. Reduced neuronal synchrony: A physiological correlate of strabismic amblyopia in cat visual cortex. Manuscript submitted for publication.

Rolls, E. T., 1991. Neural organization of higher visual functions. *Curr. Opin. Neurobiol.* 1:274–278.

Sakai, K., and Y. Miyashita, 1991. Neural organization for the long-term memory of paired associates. *Nature* 354: 152–155.

Schwarz, C., and J. Bolz, 1991. Functional specificity of the long-range horizontal connections in cat visual cortex: A cross-correlation study. *J. Neurosci.* 11:2995–3007.

Shimizu, H., Y. Yamaguchi, I. Tsuda, and M. Yano, 1986. Pattern recognition based on holonic information dynamics: Towards synergetic computers. In *Complex Systems-Operational Approaches*, H. Haken, ed. Berlin: Springer-Verlag, pp. 225–240.

Singer, W., 1985. Activity-dependent self-organization of the mammalian visual cortex. In *Models of the Visual Cortex*, D. Rose and V. G. Dobson, eds. New York: Wiley, pp. 123–136.

Singer, W., 1990. Search for coherence: A basic principle of cortical self-organization. *Concepts Neurosci.* 1:1–26.

Singer, W., 1993. Synchronization of cortical activity and its putative role in information processing and learning. *Annu. Rev. Physiol.* 55:349–374.

Singer, W., in press. Putative functions of temporal correlations in neocortical processing. In *Large Scale Neuronal Theories of the Brain*, C. Koch and J. Davis, eds. Cambridge, MA: MIT Press.

Softky, W. R., and C. Koch, 1993. The highly irregular firing of cortical cells is inconsistent with temporal integration of random EPSPs. *J. Neurosci.* 13:334–350.

Sparks, D. L., C. Lee, and W. H. Rohrer, 1990. Population coding of the direction, amplitude and velocity of saccadic eye movements by neurons in the superior colliculus. *Cold Spring Harb. Symp. Quant. Biol.* 55:805–811.

Ts'o, D., C. Gilbert, and T. N. Wiesel, 1986. Relationship between horizontal interactions and functional architecture in cat striate cortex as revealed by cross-correlation analysis. *J. Neurosci.* 6:1160–1170.

von der Malsburg, C., 1985. Nervous structures with dynamical links. *Berl. Bunsenges. Phys. Chem.* 89:703–710.

von der Malsburg, C., and W. Schneider, 1986. A neural cocktail-party processor. *Biol. Cybern.* 54:29–40.

von Noorden, G. K., 1990. *Binocular Vision and Ocular Motility: Theory and Management of Strabismus*. St. Louis: Mosby.

7 Toward a Molecular Basis for Sensory Perception

GORDON M. SHEPHERD

ABSTRACT Concepts of the neural basis of sensory perception have classically been based on the visual system. However, recent studies at the molecular and cellular level are providing evidence for a set of common principles applicable to all sensory systems that is adapted for the specific tasks in each of the different systems. The olfactory system is providing a simple model for investigating some of these principles and adaptations.

In olfactory receptor cells, information carried in molecules is mapped into neural circuits. This creates molecular images in neural space, which serve as the basis for odor perception and discrimination. Current studies suggest that olfactory cells have molecular receptive fields, analogous to the spatial and color-coded receptive fields in visual cells. Studies of olfactory cortex have identified a basic cortical circuit that appears to be embedded in other types of cerebral cortex. Identification of such a circuit is a critical step toward a consensus on a canonical cortical circuit that might be a basis for more realistic neural network simulations of sensory processing and higher cortical functions.

Cognitive neuroscience has emerged as a new discipline at the confluence of three fields: cognitive studies, experimental neuroscience, and computational modeling. At the Summer Institute for Cognitive Neuroscience, this was symbolized by a logo consisting of a Kanizsa triangle, the corners of which each represented one of these fields. This logo expresses well the unity of the new field, but it also symbolizes the main challenge, because the sides appear only as illusions; the task for cognitive science is to make the connections real!

The Kanizsa triangle also expresses the conviction that each of the three fields is essential if a true cognitive neuroscience is to emerge. Cognitive studies without reference to brain structures or functions are insufficient, as are computational models that do not reflect real neural properties. Similarly, experimental results in neuroscience need to be analyzed for their relevance to behavior and incorporated into more realistic neural and network models. This places a burden on workers in each field to identify the principles in their specialties that are relevant to defining principles governing the combined results of all the fields.

As a step in this direction, each Summer Institute program began with a tutorial on some of the principles of neural organization and function that are relevant to the neural substrate for cognitive function and for more realistic computational models. Those principles have been summarized elsewhere (Shepherd, 1988a; Shepherd and Koch, 1990). Building on the general outline of that tutorial, I will discuss first the importance of the concept of levels of organization for identifying the neural substrates that are relevant to cognitive functions. This will be followed by a discussion of the comparative approach to sensory perception as a way in which progress toward a theoretical basis for cognitive neuroscience can be made. The olfactory system will be introduced as a specific model for sensory perception that is providing insight into neural mechanisms, expressing general principles that cut across most sensory systems. Finally, this chapter addresses the organization of olfactory cortex and the critical importance of identifying basic, or canonical, types of circuits that can provide the basis for realistic simulations of cortical functions.

Levels of neural organization

A fundamental concept in biology is that function arises out of levels of organization. Thus, movement of actin and myosin molecules does not directly move muscles; rather, actin and myosin are organized into sarcomeres, which are organized into muscle fibers, which are organized into muscle groups, and these groups produce coordinated muscle actions around a

GORDON M. SHEPHERD Section of Neurobiology, Yale University School of Medicine, New Haven, Conn.

joint that enable the translation of molecular properties into behavioral functions. A very similar idea applies to nervous organization: Between the gene and gene product on the one hand and a behavior or cognitive function on the other are many levels of nervous organization. These levels begin with functional clusters of gene products at neural sites and synapses and ascend through microcircuits and dendritic subunits to whole neurons, which are in turn organized into local circuits within regions and then distributed circuits between regions, to mediate the range of behaviors of the organism. This concept has been discussed extensively in recent publications (Shepherd, 1988b; Sejnowski et al., 1988; Shepherd and Koch, 1990; Churchland and Sejnowski, 1992). The issue of whether the properties at a given level are inherently derived from, or isomorphic with, the level below (as appears largely to be the case in muscle) or are emergent properties of that level (as is often assumed in the nervous system) is interesting and unresolved (see later).

This concept of levels of organization (figure 7.1) is important, in practice, for determining the adequacy of neuroscientific data for explaining the neural basis of brain function. As an example, among the most interesting recent data relevant to brain functions are the results of brain scans using such methods as positron emission tomography (PET) and magnetic resonance imaging (MRI). As shown in figure 7.1A, these data are at the highest levels of organization; they provide overall maps of the locations of activity but give little insight into the neural mechanisms at lower levels that are responsible for generating the maps. In contrast is the approach that may be summarized as *molecular reductionism* (figure 7.1B), driven by the belief that the cloning of a gene or the identification of a neurotransmitter or neurohormone, by itself, can entirely explain a given behavior. If this is not true of muscle, it is even less likely in the nervous system. Partial success at filling in the levels is being achieved in certain types of diseases, such as parkinsonism, in which loss of a specific neurotransmitter can be correlated with a specific cell type and with the circuits in which it is involved (figure 7.1C). Figure 7.1 thus provides a general scheme for testing the adequacy of our current knowledge for accounting for a given behavior or cognitive function.

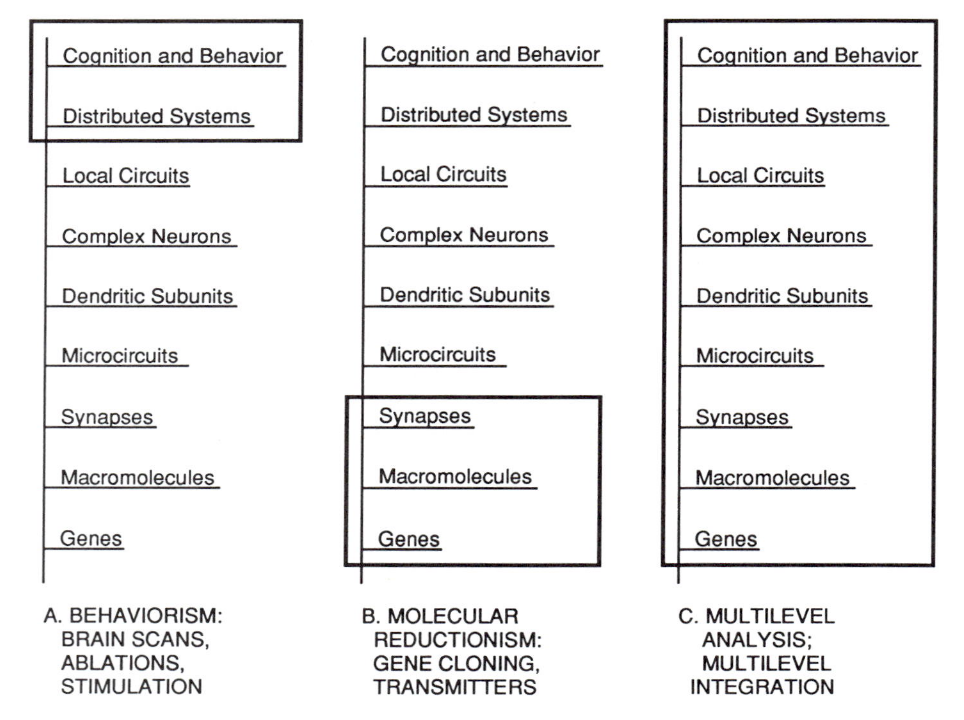

FIGURE 7.1 Illustrations of the analysis of brain functions by different experimental approaches at different levels of organization (see text).

Sensory perception: A comparative approach

Sensory perception has played a central role in our progress toward understanding the neural basis of cognitive functions. In this role, the visual system has been our main model. Given the importance of vision in humans, few would doubt the appropriateness of this model. However, judging from current textbooks, this pre-eminence of vision has carried with it a neglect of other sensory systems in formulating general principles of sensory perception (figure 7.2).

Vision has not always been the near-exclusive focus as a model for sensory perception, as is evident from a brief review of history. Experimental psychology traces its origins to Ernst Weber and his mid-nineteenth century studies of the sense of touch and the sensation of weight. His conclusion that the smallest perceptible difference between two weights can be stated as a constant ratio independent of the magnitudes of the weights (cf. Boring, 1950) was extended to vision and audition and then took mathematical form in the hands of Gustav Fechner in 1860 as the Weber-Fechner law, in which the magnitude of a sensation is proportional to the logarithm of the stimulus (Fechner, 1966). A century later, S. S. Stevens' revision of this law to a power law was based initially on evidence obtained from studies of the intensity of smell (see Stevens, 1975). In the meantime, Adrian and Zotterman (1926) started the analysis of sensation at the single-cell level by recording responses of single fibers from a muscle spindle to different degrees of stretch. Mountcastle (1957) made the next leap forward with the demonstration that cells responsive to touch are organized in the cortex in columns. Thus, the record shows that, until the 1950s, various sensory modalities, including somaesthesia, proprioception, audition, and smell, as well as vision, contributed to the basic principles of psychophysics and the neural basis of perception.

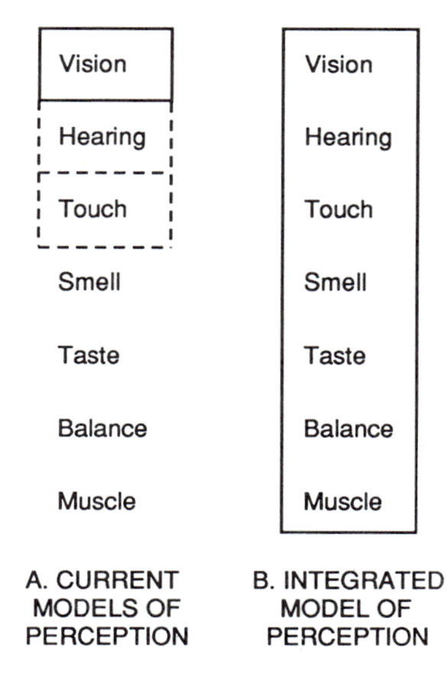

FIGURE 7.2 Analysis of the neural basis of perception, based primarily on the visual system (A) and on a comparative approach utilizing all major systems (B).

The psychophysics of vision has an even longer history, beginning with Newton in the seventeenth century and Thomas Young's theory of color in 1802. Among nineteenth-century investigators, Hermann von Helmholtz stands out for his seminal contributions to vision as well as hearing. In the twentieth century, the attractiveness of the visual system for analysis of spatial aspects of perception came to the fore with the demonstration, by Hartline and his colleagues in the 1930s, of the organization of antagonistic center and surround response in the *Limulus* eye and in the mammal by Kuffler (1953) and Barlow (1953). The pioneering explorations of Maturana and coworkers (1960) and especially Hubel and Wiesel (1962) extended this approach to central pathways and opened the way to analysis of the various submodalities of vision, including shape, directional selectivity, depth perception, and color. These studies provided a plethora of data with which David Marr and his colleagues (Marr, 1981) initiated the analysis of network properties of visual processing from a computational perspective.

In the modern era, many studies of the visual system (such as center-surround receptive fields, cortical columns, and parallel processing streams) have provided models of wide and general interest, but many others (such as motion detection and depth perception) have been more tied to the specific problems of visual processing. Identification of a set of principles that applies across all sensory systems thus has been elusive. Such a set would embrace all the major sensory systems, as illustrated in figure 7.2B. Progress toward this goal therefore requires a comparative systems approach, in which mechanisms in one system are seen as adaptations of a general set or superfamily of mechanisms at the equivalent level of the hierarchy of organization depicted in figure 7.1. It also requires the parallel development of general theoretical concepts, such as the Weber-Fechner-Stevens laws, against which the mechanisms in any given system can be tested and assessed.

Molecular basis of sensory transduction

Such a basic concept applying to all sensory physiology was the idea of "specific nerve energies," introduced by Johannes Muller in 1835. Muller meant that each sensory quality ("energy") is received by nerve fibers or organs preferentially sensitive to that quality only. In more modern form, this idea has meant that a particular sensation depends on transduction mechanisms specific for a given form of stimulus energy in a given set of sensory terminals or end organs. Because sensory stimuli differ so dramatically from one another (photons giving rise to visual sensations share little similarity with sound waves giving rise to hearing or odor molecules giving rise to smell), it might appear that mechanisms of transduction are not the place to start searching for general principles that apply across all systems.

In fact, as happens often in science, different adaptations arise from variations on a common theme of mechanisms. In the past few years, a confluence of studies of anatomy, physiology, and molecular biology in the different systems has revealed surprising similarities among the systems at the molecular level (Shepherd, 1992c) (see figure 7.3). The main point, for present purposes, is that transduction of different sensory stimuli takes place through adaptations of basic membrane-signaling mechanisms. Thus, transduction of stimulus energies as diverse as photons, odor molecules, and sweet-tasting compounds begins with an interaction between the stimulus and a member of the family of seven–transmembrane domain (7TD) re-

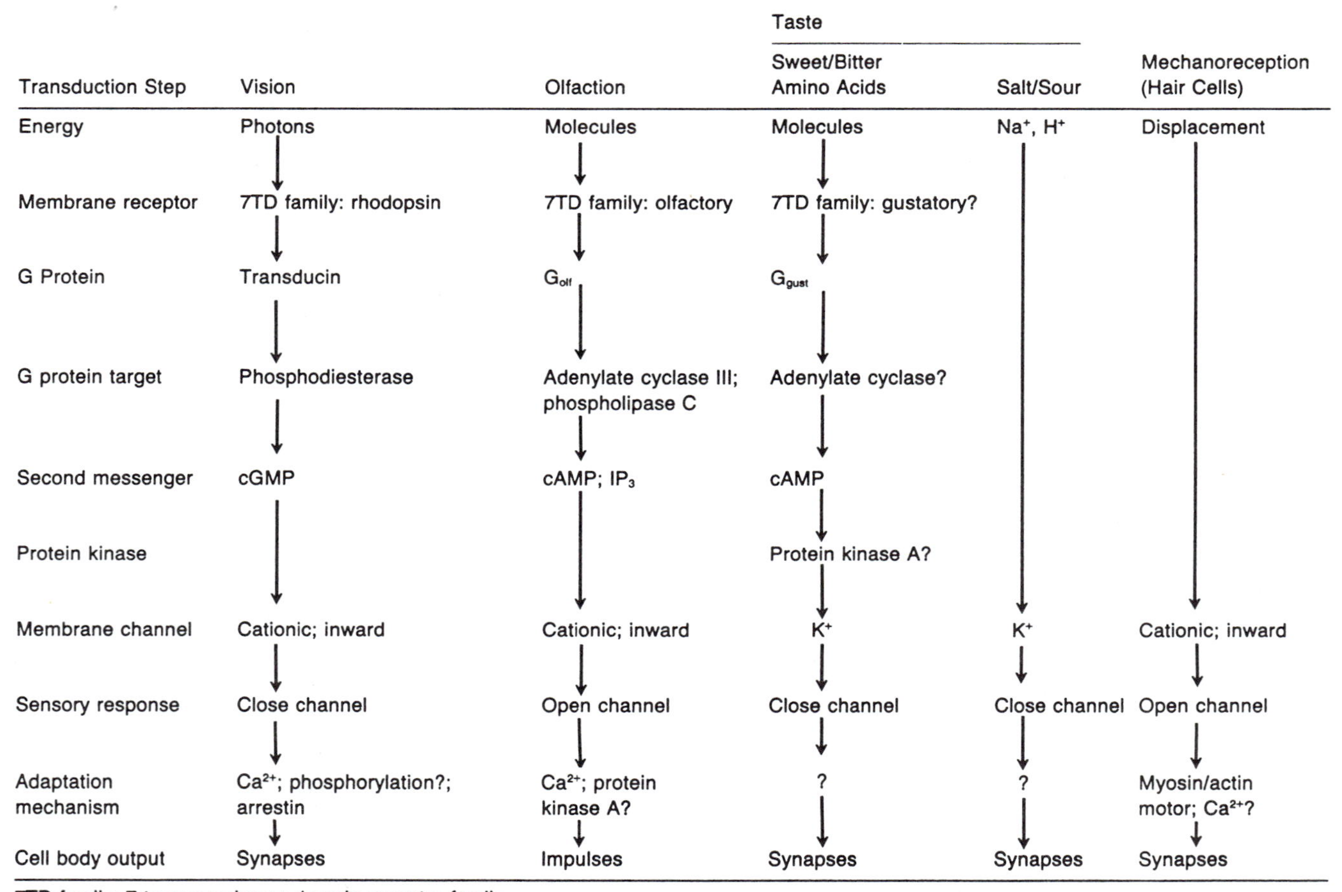

Transduction Step	Vision	Olfaction	Taste: Sweet/Bitter Amino Acids	Taste: Salt/Sour	Mechanoreception (Hair Cells)
Energy	Photons	Molecules	Molecules	Na^+, H^+	Displacement
Membrane receptor	7TD family: rhodopsin	7TD family: olfactory	7TD family: gustatory?		
G Protein	Transducin	G_{olf}	G_{gust}		
G protein target	Phosphodiesterase	Adenylate cyclase III; phospholipase C	Adenylate cyclase?		
Second messenger	cGMP	cAMP; IP_3	cAMP		
Protein kinase			Protein kinase A?		
Membrane channel	Cationic; inward	Cationic; inward	K^+	K^+	Cationic; inward
Sensory response	Close channel	Open channel	Close channel	Close channel	Open channel
Adaptation mechanism	Ca^{2+}; phosphorylation?; arrestin	Ca^{2+}; protein kinase A?	?	?	Myosin/actin motor; Ca^{2+}?
Cell body output	Synapses	Impulses	Synapses	Synapses	Synapses

7TD family: 7 transmembrane domain receptor family.

FIGURE 7.3 Comparison of the main steps in sensory transduction in different sensory receptors cells. 7TD, seven–transmembrane domain receptor molecule; G_{olf}, olfactory-specific GTP-binding protein; G_{gust}, taste-specific guanylate triphosphate–binding protein. (Modified from Shepherd, 1991).

ceptors. The same framework is present for linking receptor activation through G proteins to a second messenger and thence to an action on a membrane ionic channel; the differences are found in various isomers of comparable constituents, different G protein targets, and the presence of alternative second-messenger pathways, as well as in whether the second messenger acts directly on the membrane channel or indirectly through a protein kinase. Even when a second messenger appears not to be involved, as in the case of salt and sour transduction in taste and mechanoreception in hair cells, the mechanisms are variations on well-known mechanisms for modulating ionic membrane channels in other systems.

One suspects that Johannes Muller would be delighted with this new information and would seize the opportunity to recast his doctrine of specific nerve energies into a molecular form. We might assist him by suggesting that each sensory modality is based on an adaptation of a superfamily of membrane-signaling mechanisms that converts the stimulus energy into an allosteric molecular change, leading to the gating of ionic current in a membrane channel.

The generation of a sensory current is only the first step in the transduction sequence that converts a stimulus into a response that can be perceived. Here, too, a common set of operations that applies to all sensory systems is beginning to emerge (Block, 1992; Shepherd, 1991, see also Corey and Roper, 1992). As outlined in figure 7.4, the main operations start with *detection* and depend on the sensitivity of the transduction mechanism in individual cells and the summa-

Transduction Operations	Operations in Single Sensory Cells	Operations in Cell Populations
Detection ↓	Perireceptor mechanisms: filters; carriers; tuning; inactivation Sensitivity Rapidity	Perireceptor mechanisms: filters; carriers; tuning; inactivation Different thresholds
Amplification ↓	Positive feedback Active processes Signal-noise enhancement	Positive feedback Signal-noise enhancement
Encoding, discrimination ↓	Intensity coding Quality coding Temporal differentiation	Different dynamic ranges Quality independent of intensity Center-surround antagonisms Opponent mechanisms Construction of maps
Adaptation and termination ↓	Desensitization Negative feedback Temporal discrimination Repetitive responses	Temporal discrimination
Sensory channel gating ↓	Open or close voltage gating?	
Electrical response ↓	Depolarization or hyperpolarization	
Transmission to brain	Electrotonic spread Active properties Synaptic output or impulse discharges	Spatial patterns: maps and image formation Temporal patterns: directional selectivity, etc.

FIGURE 7.4 Comparison of the main operations in sensory transduction that are common to most sensory receptor cells. (Modified from Shepherd, 1991.)

tion of responses across cell populations. Most systems *amplify* weak signals through positive feedback and active processes that enhance the signal-to-noise ratio. *Discrimination* between signals must then take place; this includes intensity discrimination (as embodied in the Weber-Fechner-Stevens laws) and quality discrimination (which typically involves interactions between populations of receptor cells). In response to prolonged or repeated stimuli, most receptor cells *adapt*, or desensitize, at rates reflecting their specific modality or submodality. All these processes converge on the *gating* of the sensory membrane conductance to generate the *receptor potential*, which ultimately leads to encoding of the response in an *impulse discharge* that is sent to the higher centers in the brain.

In summary, recent work suggests that a basic set of operations, comprising detection, amplification, discrimination, and adaptation, applies across all sensory systems in transducing sensory stimuli. The precise neural mechanism for a given operation varies in the different receptor cells, but each basic operation is essential in the sequence for transducing a local stimulus into a neural response in a single cell and mapping the stimulus field into an ensemble of cells. In the case of the individual receptor cell, the molecular components of each step are now identified and are being analyzed. The synaptic components of the circuits that process the population responses are also being analyzed. It is therefore not unreasonable to anticipate that sensory transduction will soon be understood in terms of a general theory of signal transformation at the molecular and membrane level.

How does the brain know?

A general theory of sensory transduction has the promise of providing a molecular basis of sensory perception. To do this, however, a second question must be answered: How does the brain know what information has been transduced?

One principle is that information is transferred from the transduction step by distributing it in populations of sensory cells. As indicated in figure 7.4, populations of cells are needed to encode *spatial* information, whether it be a visual scene, the surface of the body, or the frequency maps in the cochlea. Similarly, *temporal* information requires cell populations; motion sensitivity in the visual pathway depends on sequential activation of different cells. Finally, *quality* discrimination requires comparisons between cells activated to different extents by different qualities; for example, perception of different colors is based on comparison of activation of different cones. Because these various operations occur in parallel throughout the activated populations, a related principle is that sensory processing is carried out by operations acting in parallel as well as in series.

All these properties are well understood in the various sensory systems. Our present interest is in the transmission of sensory information that can be characterized at the molecular level. Best understood is information about different wavelengths of light that is transduced by three different cone pigments in the retina. The three cone pigments have been cloned and sequenced, such that in current studies single amino acids in the opsin molecules can be correlated with perceptions of different colors (e.g., red and green) (see Baylor, 1992; Nathans et al., 1992). This might imply that the perception of color can be reduced to a molecular basis (as in figure 7.1B), without need of knowledge about other levels of organization in the different parts of the visual pathway. However, analysis of color processing has shown that it is extremely complex, including single- and double-opponent mechanisms in the retina and parallel processing streams that include the parvocellular layers of the lateral geniculate, the cytochrome oxidase blobs of visual area 1 (V1) and stripes of visual area 2 (V2), and the V4 association area of the cerebral cortex (cf. Zeki, 1980; Livingstone and Hubel, 1988).

Molecular image processing in the olfactory system

A problem with color processing is that the color of an object is inextricably bound up with its spatial properties, which require their own complex steps in transmission from transduction to perception. In fact, vision consists of many submodalities related to spatial aspects of the visual world, and the reconstruction of that world by the nervous system draws heavily on neural machinery.

It is therefore instructive to consider the olfactory system, in which spatial attributes of the stimulus field are lacking. As far as is known, we do not map spatial patterns of odor in the external world onto our olfactory sensory epithelium. Neural space in this system is therefore available for processing nonspatial attributes of the stimulus. Because the stimuli are odor molecules, one may postulate that neural space in this system is used to process information carried in odor molecules as the basis for the perceptions we call *smell*. Although we do not know yet which information in an odor molecule is transferred by sensory transduction, it is likely to be carried, as in other ligand-receptor interactions, by critical determinants on the odor molecules: These may be referred to as *odotopes*, being equivalent to the epitopes on an antigen molecule or the pharmacophores of a neurotransmitter or drug ligand. Figure 7.5 depicts a general view of the way that a typical odor molecule may access the pocket of the presumed olfactory receptor belonging to the superfamily of seven transmembrane proteins.

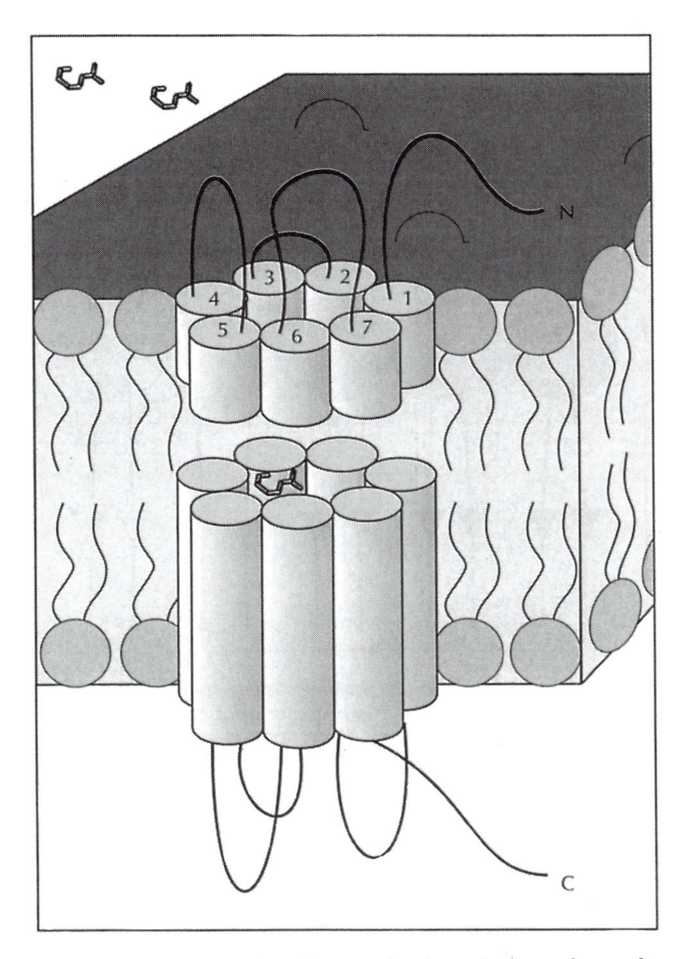

FIGURE 7.5 Model of the interaction between an odor molecule and an olfactory receptor molecule in the membrane of an olfactory sensory cell. The receptor is a member of the family of receptors with seven transmembrane domains. Representative odor ligands are shown diffusing in the mucus and interacting with the receptor in the pocket formed by the transmembrane domains. (Reprinted from Shepherd and Firestein, 1991a, by permission.)

The first indication that odor information might be distributed in neural space was provided by Adrian (1950), who showed that stimulation of an anesthetized rabbit with different odors produces different gradients of activation of mitral cells in the olfactory bulb; one odor might give maximal activation anteriorly and less posteriorly, whereas another odor might give the opposite gradient, and a third might stimulate maximal activation in the middle. Leveteau and MacLeod (1966) then reported all-or-nothing field po-

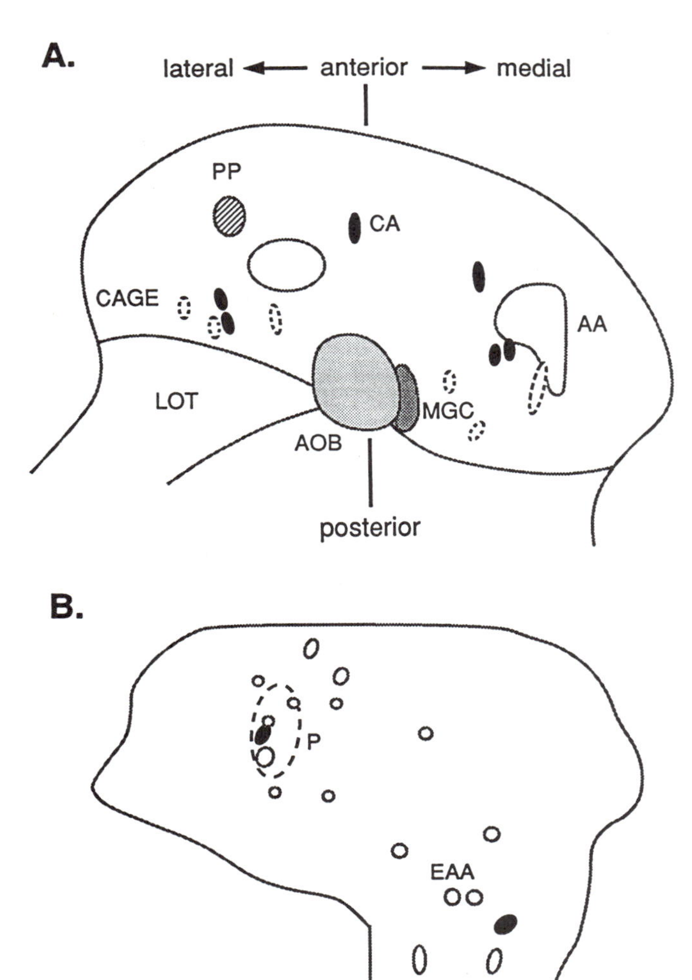

FIGURE 7.6 Maps of the glomerular layer of the olfactory bulb, showing the peak activity patterns of 2-deoxyglucose localization produced by sensory stimulation with different odors. AA, amyl acetate; CA, camphor; PP, peppermint; P, propionic acid; EAA, ethyl ethanoacetate (moderately dense foci shown by open profiles, intense foci shown by filled profiles); AOB, accessory olfactory bulb; MGC, modified glomerular complex; LOT, lateral olfactory tract. (Reprinted from Shepherd and Firestein, 1991b, by permission.)

tentials across the glomerular layer in response to specific odors and suggested that these represented activation of individual glomeruli.

The first clear evidence for glomerular activation by odors was provided by the 2-deoxyglucose (2DG) technique, which showed in the awake behaving rat that individual or small groups of glomeruli are activated by weak odor stimuli and larger domains of glomeruli by stronger stimulation (Sharp, Kauer, and Shepherd, 1975, 1977). For different odors, these domains are overlapping but different. Figure 7.6 summarizes the studies from various laboratories to date (recent studies of c-*fos* expression support the 2DG patterns [Guthrie et al., 1991; Sallaz and Jourdan, 1993]). These patterns are sufficiently distinct that they could serve as part of the basis for odor discrimination. Indeed, simulations with a network model for pattern recognition indicate that the odor patterns are sufficiently distinct to admit of rapid discrimination (cf. Sejnowski, Kienker, and Shepherd, 1985). Based on these studies, we have theorized that the olfactory system constructs in its neural space an odor image representing the salient features (determinants, olfactophores, odotopes) of the odor ligands, just as the visual system constructs an image representing the salient features of the visual scene. This image is first laid down in the receptor sheet and then is processed in the glomerular sheet of the olfactory bulb and the populations of mitral-tufted

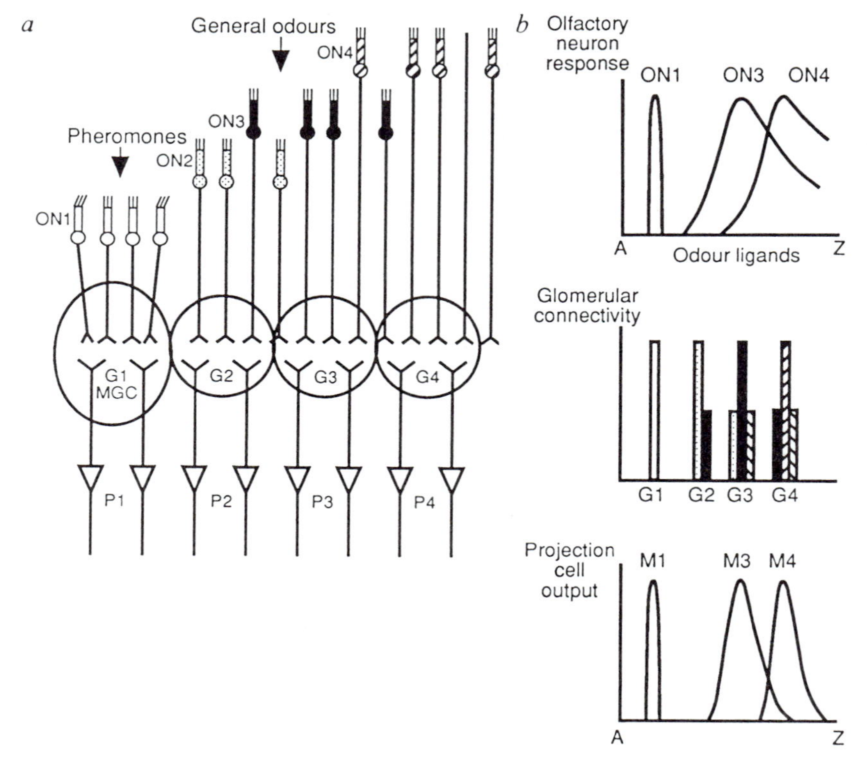

FIGURE 7.7 Principles underlying the initial processing of molecular information in the olfactory system. (a) Anatomical organization of sets of olfactory neurons (ON1–ON4), each neuron expressing only one or a few members of the seven–transmembrane domain receptor family. Specialist cells (ON1) are narrowly responsive to pheromone molecules and project to special glomeruli (G1, macroglomerular complex [MGC], as demonstrated in the insect). Generalist cells (ON 2–4) are broadly responsive to many odor ligands. Each set has primary projections to a given glomerulus (see G2–4), with secondary projections to related glomeruli. Projection neurons (P1–4) send the glomerular output to the olfactory cortex. (b) Functional properties. Top diagram shows the narrow response spectrum of specialist cells (ON1) and broad overlapping response spectra of generalist cells (ON2–4) when tested with different odor ligands (A–Z). Middle diagram shows the relative connectivity of the different sets of cells to the different glomeruli. Bottom diagram shows the similar narrow response spectrum of the output cells from the macroglomerulus (M1) and the changed response spectra of the output cells from other glomeruli, reflecting glomerular processing and actions of other intrabulbar circuits. (Reprinted from Shepherd, 1992b, by permission.)

cell output neurons before being sent to the olfactory cortex.

Our current view of the organization underlying the construction of odor images is summarized in figure 7.7. At one extreme are the specialist receptor cells (ON1 in figure 7.5) of invertebrates, represented by cells in male insects that are specifically tuned to the sex-attractant pheromone emitted by the female. These, as a rule, all project to a special macroglomerular complex (MGC; G1 in figure 7.5) and thus constitute a labeled line for this specific information. At the other extreme are the so-called generalist receptor cells, with broad overlapping response spectra. These represent the main type in vertebrates. The simplest hypothesis is that each subset expresses one type of receptor with a broad affinity, but expression of several types with related affinities is also possible. Three subsets (ON2–4) with overlapping spectra are shown in figure 7.7. The real spectra are not as regular as what is shown here, but the diagram conveys the general concept of a multidimensional overlapping space. Similar to the processing of wavelength information, it is the comparison between the overlapping spectra that enables the different ligands to be distinguished independently of their concentration (cf. Gouras, 1992).

Different odor ligands activate different subsets of olfactory sensory neurons, which project to different subsets of glomeruli in the olfactory bulb. The scheme in figure 7.7 illustrates a simple connectivity rule: The majority of a receptor cell subset projects to a given glomerulus, with smaller projections to other glomeruli. This accounts for the specificity of glomeruli to different odors indicated by the 2DG data. Within a glomerulus, a great deal of processing takes place through densely packed microcircuits involving dendrodendritic synaptic connections between the dendritic terminals of the relay neurons (mitral–tufted cells) and interneurons (periglomerular cells). My colleagues and I postulate that these contribute to the operation of the glomerulus as a functional unit in the processing of odor information. Diffusable intercellular messengers such as NO may contribute to this function (Breer and Shepherd, 1993).

Within the olfactory bulb are two levels of synaptic processing—the glomerular layer and the layer of mitral–tufted cell bodies. We have already indicated that the immediate processing of the input takes place within the glomeruli. When activation of a glomerulus is sufficient, the periglomerular cells generate impulses in their axons that connect to other glomeruli in the vicinity. There are thus intraglomerular and interglomerular stages of processing within the glomerular layer. Possibly the interglomerular connections provide for antagonistic interactions, perhaps the basis for opponent odor processing, analogous to opponent color mechanisms in the retina.

Processing at the level of the mitral–tufted cell bodies provides for control of the output to the olfactory cortex. The output neurons form anatomical subsets in accordance with their connections to a given glomerulus and corresponding functional subsets to the extent that a given glomerulus operates as a functional unit, as described earlier. The mitral–tufted cells are richly interconnected by reciprocal dendrodendritic synapses to a large population of granule cell interneurons (not shown in figure 7.7). These microcircuits provide for extensive recurrent and lateral inhibitory control of the mitral–tufted cell output over a much larger area than occurs within the glomeruli; in other words, mitral–tufted cell bodies have weighted effects from the outputs of many neighboring glomeruli.

One of the predictions of the model is that mitral–tufted cells may have more specific and narrower tuning for odor molecules than the receptor cells. Recent studies by Imamura, Mataga, and Mori (1992) support this prediction. In these studies, carbon series for different types of compounds, such as alcohols, esters, and aldehydes, were systematically tested in lightly anesthetized rabbits while recording from mitral cells. The results showed that mitral cells tend to respond to a narrow range of carbon atoms in a given series and to a similar range in other series (figure 7.8). Mori calls this the *molecular receptive field* for this cell, analogous to the spatial receptive field of a retinal ganglion cell. These studies further indicate that cells tuned to these types of compounds are preferentially localized in certain regions of the olfactory bulb, which is in accord with the suggestions from the 2DG that the glomeruli to which these mitral cells are attached are localized in these regions.

This organization bears several similarities to other systems. The straight-through pathways and the two levels for lateral interactions are reminiscent of the retina, as was pointed out many years ago (Shepherd, 1970) and supported more recently (DeVries and Baylor, 1993). The glomeruli constitute one of the most distinctive types of modules in the nervous

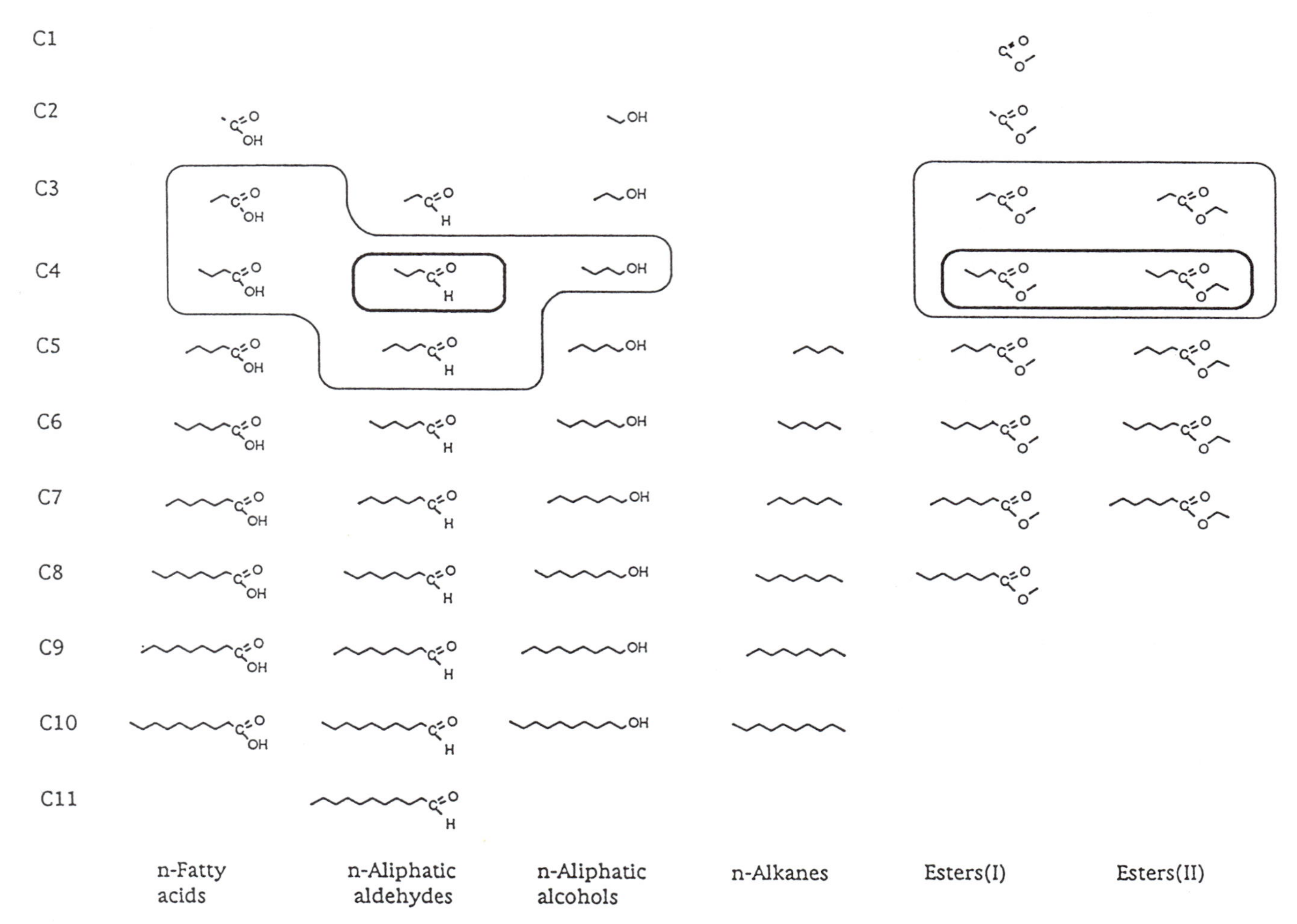

FIGURE 7.8 Illustration of the concept of a molecular receptive field. Systematic tests were carried out to determine the responses of neurons in the olfactory bulb to odor molecules of different types and different numbers of carbon atoms (C1–11). The neuron illustrated here responded most strongly to the compounds enclosed with heavy lines and moderately to the closely related compounds enclosed with lighter lines but showed no responses to the other compounds. The enclosed compounds thus constitute the molecular receptive field of this neuron. (Reprinted from Imamura et al., 1992, by permission.)

system and, as such, are similar to the modular organization of cortical barrels and cortical columns, as well as patches in the corpus striatum. The simple rule of connectivity illustrated in figure 7.7 may be applicable to the projection patterns of some of these other types of modules. Thus far, all the synaptic circuits in the olfactory bulb appear to have their counterparts in other systems. This implies that such basic operations as feedforward, feedback, and lateral inhibition, which are used in the visual system to process spatial information, may be used in the olfactory system to process molecular information.

Cortical operations and canonical circuits

The output of the olfactory bulb is directed to the olfactory cortex, which is composed of several distinct cortical areas: anterior olfactory nucleus, pyriform cortex, olfactory tubercle, parts of the amygdaloid complex, and the lateral entorhinal cortex (Haberly, 1990). The pyriform cortex is generally recognized as the main olfactory cortex involved in sensory processing; its output goes to medial dorsal thalamus and, from there, to specific areas of frontal cortex to mediate olfactory perception.

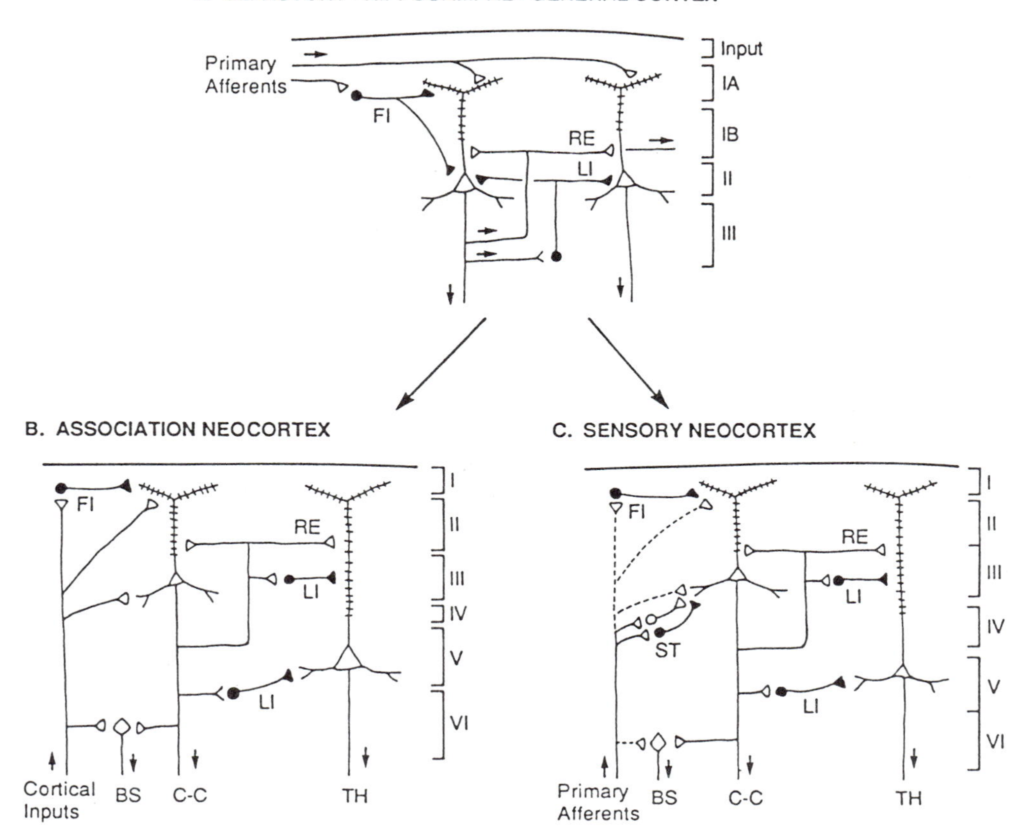

FIGURE 7.9 Basic (canonical) circuits reportedly found in different types of cortex. (A) Basic circuit that is characteristic of olfactory cortex, hippocampus, and dorsal cortex of reptiles. (B, C) Similar circuits embedded in agranular association and granular sensory neocortex. BS, brain stem; C-C, corticocortical; FI, feedforward inhibition; LI, lateral inhibition; RE, recurrent excitation; ST, stellate cell; TH, thalamus. (Modified from Shepherd, 1988.)

Investigation of the olfactory cortex has revealed the basic outlines of its functional organization. In brief, the mitral and tufted cell axons in the lateral olfactory tract (LOT) terminate in the superficial layer of the cortex, making excitatory connections onto the spines of the distal dendrites of pyramidal neurons. The pyramidal neurons consist of superficial and deep types, both characterized by apical and basal dendrites covered with spines. The pyramidal cell axons give off collaterals, which make excitatory connections of two types: Some are recurrent to the superficial layers, where they run horizontally and provide for re-excitatory connections to the pyramidal cells, whereas others excite deep interneurons, the axons of which provide for recurrent and lateral inhibitory connections to the pyramidal cells. There are also feedforward inhibitory connections from LOT terminals through superficial interneurons onto pyramidal cell dendrites.

This organization may be summarized in a basic circuit diagram, as shown in figure 7.9. An important generalization emerging out of work over the past 20 years is that this basic circuit can be identified in a variety of types of cortex. Thus, the hippocampus shares this type of organization, with the exception that the CA1 and CA3 fields are separated in their types of inputs and recurrent excitatory circuits, and the dentate granule cells function as a short-range input population to the CA3 field. Analysis of dorsal forebrain cortex in the turtle has also revealed a close similarity to the olfactory pattern; variations in this case include a prominent population of inhibitory interneurons that is a convergent node for both feedforward and feedback inhibition of the pyramidal cells (Kriegstein and Connors, 1986). In fact, these similarities are not unexpected in view of the commonly believed evolutionary origin of olfactory cortex, hippocampus, and dorsal cortex from primitive three-layered cortex.

A variety of studies reveals that the basic circuit also appears to be embedded in neocortex. The simplest comparison is with agranular cortex, characteristic of motor areas and some association areas. As figure 7.9 indicates, in this cortex the olfactory-hippocampal-turtle organization can be viewed as amplified radially by multiple layers of pyramidal neurons and their associated intrinsic excitatory and inhibitory circuits, and elaborated horizontally into different areas by variations on these cell populations and circuits. More complex is granular cortex, characterized by the additional population of stellate cells in layer 4 (see figure 7.9). These can be interpreted in the same way as the dentate granule cells as well as the granule cells of the cerebellar cortex, in that they provide a short-range feed-forward excitatory pathway, mediating important staging operations for information directed to these respective cortical areas. Figure 7.9 indicates how this can be incorporated into the basic circuit.

It is notable that very similar types of circuits have emerged from the independent studies of Chagnac-Amitai and Connors (1989) and Douglas and colleagues (1989). In addition, the basic circuit has been the foundation for network models of olfactory cortex, which have emphasized its possible relevance as a model for cortical operations (Granger et al., 1989; Haberly, 1985; Wilson and Bower, 1989).

These considerations suggest that there is a basic cortical circuit, laid down early in forebrain evolution, which is elaborated into different specific types of cortical circuits characteristic of the different types of cortex. The adaptability of this basic pattern is exemplified not only by the differences between the main types of cortex illustrated in figure 7.9 but also by the many different cytoarchitectonic areas of the neocortex of higher mammals, reaching its greatest diversity in humans. There is, in this respect, an analogy with the results of molecular cloning of genes for different proteins. A dominant theme in this work is that genes tend to be grouped into families and the families into superfamilies. Each family expresses proteins that are variations on a common pattern (i.e., the same number of transmembrane domains but with different amino acid residues mediating the binding of different signal molecules). At the neuronal level, it has been suggested that there are "canonical" neuron types that are differentiated into subtypes based on details of morphology, membrane properties, and neuroactive substances (Shepherd, 1992a). At the level of neural circuits, we may similarly postulate that there are families and superfamilies of cortical circuits, constituting certain basic or canonical patterns of functional properties and synaptic connections that can mediate fundamental information-processing functions and also be adapted for a variety of related specialized tasks.

This is important first because the identification of principles is crucial to interpreting the explosion of experimental data from work on different regions. For cognitive neuroscience, it is additionally critical if progress is to be made in constructing more realistic computational models for the neural networks that underlie cortical functions (Sejnowski, Koch, and Churchland, 1988; Shepherd, 1990). In constructing these models, theorists can be guided only by what experimentalists tell them must be incorporated into the models to capture the essential properties of the cortex. The concepts of basic or canonical types of neurons and circuits are therefore critical in indicating the minimum essential elements that must be incorporated into network models if they are to operate by the mechanisms actually used by the nervous system to express a given type of behavior or cognitive function. By reaching agreement on these essential elements, experimental neuroscience, cognitive studies, and computer simulations can change their relations within the Kanizsa triangle from illusion to reality.

REFERENCES

Adrian, E. D., 1950. The electrical activity of the mammalian olfactory bulb. *Electroencephalog. Clin. Neurophysiol.* 2: 377–388.

Adrian, E. D., and Y. Zotterman, 1926. The impulses produced by sensory nerve endings: II. The response of a single end-organ. *J. Physiol. (Lond.)* 61:151–171.

Barlow, H. B., 1953. Summation and inhibition in the frog's retina. *J. Physiol. (Lond.)* 119:69–88.

Baylor, D. A., 1992. Transduction in retinal photoreceptor cells. In *Sensory Transduction*, D. P. Corey and S. D. Roper, eds. New York: Rockefeller University Press, pp. 151–174.

Block, S. M., 1992. Biophysical principles of sensory transduction. In *Sensory Transduction*, D. P. Corey and S. D. Roper, eds. New York: Rockefeller University Press, pp. 1–18.

Boring, E. G., 1950. *A History of Experimental Psychology.* New York: Appleton.

Breer, H., and G. M. Shepherd, 1993. Implications of the NOS/cGMP system for olfaction. *Trends Neurosci.* 16:5–9.

Chagnac-Amitai, Y., and B. W. Connors, 1989. Synchronized excitation and inhibition driven by intrinsically bursting neurons in neocortex. *J. Neuropysiol.* 62:1149–1162.

Churchland, P. S., and T. J. Sejnowski, 1992. *The Computational Brain.* Cambridge, Mass.: MIT Press.

Corey, D. P., and S. D. Roper, eds., 1992. *Sensory Transduction.* New York: Rockefeller University Press.

DeVries, S. H., and D. A. Baylor, 1993. Synaptic circuitry of the retina and olfactory bulb. *Cell/Neuron* 72(10):139–149.

Douglas, R. J., K. A. C. Martin, and D. Whitteridge, 1989. A canonical microcircuit for neocortex. *Neural Comput.* 1:480–488.

Fechner, G., 1966. In *Elements of Psychophysics*, vol. 1, D. H. Howe and E. G. Boring, eds. New York: Holt, Rinehart and Winston. (Original work published 1860)

Gouras, P., 1992. Color vision. In *Principles of Neural Science*, E. R. Kandel, J., Schwartz, and T. Jessell, eds. New York: Elsevier, pp. 467–480.

Granger, R., J. Ambros-ungerson, and G. Lynch, 1989. Derivation of encoding characteristics of layer II cerebral cortex. *J. Cogn. Neurosci.* 1:61–97.

Guthrie, K. M., A. J. Anderson, M. Leon, and C. M. Gall, 1991. Spatially distributed increases in c-*fos* mRNA in odor-activated regions of the main olfactory bulb. *Soc. Neurosci.* 17:141.

Haberly, L. B., 1990. Olfactory cortex. In *The Synaptic Organization of the Brain*, G. M. Shepherd, ed. New York: Oxford University Press, pp. 317–345.

Haberly, L. B., 1985. Neuronal circuitry in olfactory cortex: Anatomy and functional implications. *Chem. Senses* 10: 219–238.

Hubel, D. H., and T. N. Wiesel, 1962. Receptive fields, binocular interaction and functional architecture in the cat's visual cortex. *J. Physiol. (Lond.)* 160:106–154.

Imamura, K., N. Mataga, and K. Mori, 1992. Coding of odor molecules by mitral/tufted cells in rabbit olfactory bulb: I. Aliphatic compounds. *J. Neurophysiol.* 68:1986–2002.

Kriegstein, A. R., and B. W. Connors, 1986. Cellular physiology of the turtle visual cortex: Synaptic properties and intrinsic circuitry *J. Neurosci.* 6:178–191.

Kuffler, S. W., 1953. Discharge patterns and functional organization of mammalian retina. *J. Neurophysiol.* 16:37–68.

Leveteau, J., and P. MacLeod, 1966. Olfactory discrimination in the rabbit olfacory glomerulus. *Science* 153: 175–176.

Livingstone, M., and D. Hubel, 1988. Segregation of form, color, movement, and depth: Anatomy, physiology, and perception. *Science* 240:740–749.

Marr, D., 1981. *Vision.* New York: Freeman.

Maturana, H. R., J. Y. Lettvin, W. S. McCulloch, and W. H. Pitts, 1960. Anatomy and physiology of vision in the frog (*Rana pipiens*). *J. Gen. Physiol.* 43:129–175.

Mountcastle, V. B., 1957. Modality and topographic properties of single neurons of cat's somatic sensory cortex. *J. Neurophysiol.* 20:408–434.

Nathans, J., C.-H. Sung, C. J. Weitz, C. M. Davenport, S. L. Merbs, and Y. Wang, 1992. Visual pigments and inherited variation in human vision. In *Sensory Transduction*, D. P. Corey and S. D. Rope, eds. New York: Rockefeller University Press, pp. 109–131.

Sallaz, M., and F. Jourdan, 1993. C-*fos* expression and 2-deoxyglucose uptake in the olfactory bulb of odour stimulated awake rats. *NeuroReport* 4:55–58.

Sejnowski, T. J., P. K. Kienker, and G. M. Shepherd, 1985. Simple pattern recognition models of olfactory discrimination. *Soc. Neurosci. Abstr.* 11:970.

Sejnowski, T. J., C. Koch, and P. S. Churchland, 1988. Computational neuroscience. *Science* 241:1299–1307.

Sharp, F. R., J. S. Kauer, and G. M. Shepherd, 1975. Local sites of activity-related glucose metabolism in rat olfactory bulb during odor stimulation. *Brain Res.* 98:596–600.

Sharp, F. R., J. S. Kauer, and G. M. Shepherd, 1977. Laminar analysis of 2-deoxyglucose uptake in olfactory bulb and olfactory cortex of rabbit and rat. *J. Neurophysiol.* 40:800–813.

Shepherd, G. M., 1970. The olfactory bulb as a simple cortical system: Experimental analysis and functional implications. In *The Neurosciences: Second Study Program*, F. O. Schmitt, ed. New York: Rockefeller University Press, pp. 539–552.

Shepherd, G. M., 1988a. A basic circuit for cortical organization. In *Perspectives on Memory Research*, M. C. Gazzaniga, ed. Cambridge, Mass.: MIT Press, pp. 93–134.

Shepherd, G. M., 1988b. *Neurobiology*, ed. 2. New York: Oxford University.

Shepherd, G. M., 1990. The significance of real neuron architectures for neural network simulations. In *Computational Neuroscience*, E. Schwartz, ed. Cambridge, MA: MIT Press, pp. 82–96.

Shepherd, G. M., 1991. Sensory transduction: Joining the mainstream of membrane signalling. *Cell* 67:845–851.

Shepherd, G. M., 1992a. Canonical neurons and their computational organization. In *Single Neuron Computation*, T. McKenna, J. Davis, and S. F. Zornetzer, eds. Cambridge, MA: MIT Press, pp. 27–59.

Shepherd, G. M., 1992b. Modules for molecules. *Nature* 358: 457–458.

Shepherd, G. M., 1992c. Toward a consensus working model for olfactory transduction. In *Sensory Transduction*, D. P. Corey and S. D. Roper, eds. New York: Rockfeller University, pp. 19–37.

Shepherd, G. M., and S. Firestein, 1991a. Making scents of olfactory transduction. *Curr. Biol.* 1:204–206.

Shepherd, G. M., and S. Firestein, 1991b. Toward a pharmacology of odor receptors and the processing of odor images. *J. Steroid Biochem. Molec. Biol.* 39:583–592.

Shepherd, G. M., and C. Koch, 1990. Introduction to synaptic circuits. In *The Synaptic Organization of the Brain*, G. M. Shepherd, ed. New York: Oxford University, pp. 3–31.

Stevens, S. S., 1975. *Psychophysics: Introduction to Its Perceptual, Neural, and Social Prospects*. New York: Wiley.

Wilson, M. A., and J. M. Bower, 1989. The simulation of large-scale neural networks. In *Methods in Neuronal Modeling: From Synapses to Networks*, C. Koch and I. Segev, eds. Cambridge, Mass.: MIT Press, pp. 291–334.

Young, T., 1802. On the theory of light and colours. *Philos. Trans. R. Soc. Lond.* pp. 12–48.

Zeki, S., 1980. The representation of colours in the cerebral cortex. *Nature* 284:412–418.

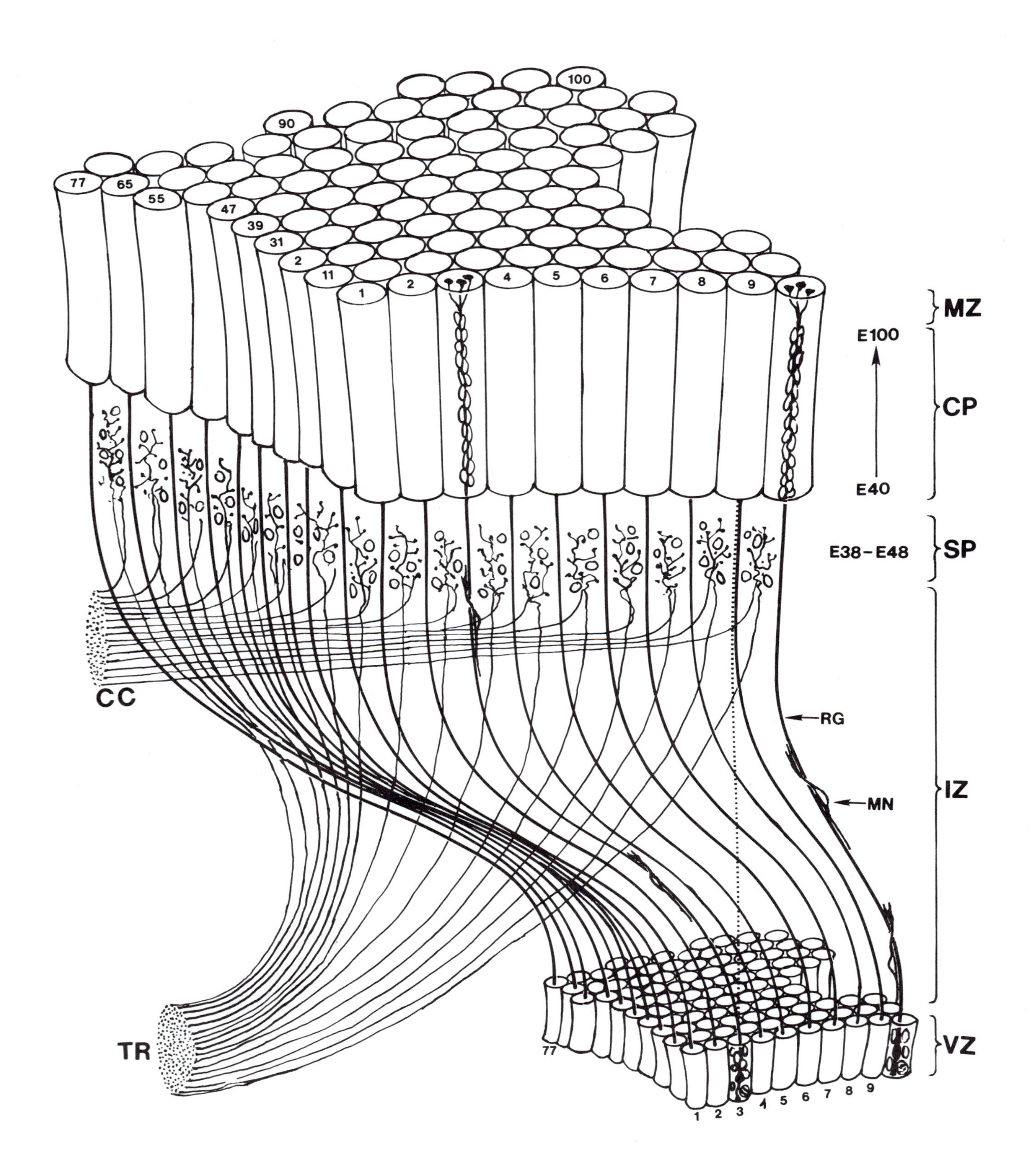

100
90
77
65
55
47
39
31
2
11
1
2
4
5
6
7
8
9
MZ
E100
CP
E40
E38–E48
SP
CC
RG
MN
IZ
TR
77
VZ
1
2
3
4
5
6
7
8
9

II NEURAL AND PSYCHOLOGICAL DEVELOPMENT

A three-dimensional illustration of the basic developmental events and cell-cell interactions during early stages of corticogenesis, before establishment of the final pattern of cortical connections. The figure emphasizes radial mode of neuronal migration, which underlies elaborate columnar organization in primates. The cohorts of neurons generated in the ventricular zone (VZ) traverse the intermediate zone (IZ) and subplate zone (SP), containing "waiting" afferents from several sources, and finally pass through earlier generated deep layers before settling at the interface of the cortical plate (CP) and marginal zone (MZ). The relation between a proliferative mosaic at the VZ and corresponding protomap within the SP and CP is preserved during cortical expansion by transient radial glial scaffolding. (For details see Rakic, 1988b, and chapter 8, this volume.)

Introduction

PASKO RAKIC

In recent years, we have witnessed enormous methodological, factual, and conceptual advances in both developmental neurobiology and neuropsychology. These advances have profoundly changed our understanding of how the brain is organized, how it develops, and how it works. Yet the connections between neuroanatomy, neurochemistry, and neurodevelopment on the one hand and behavioral research in cognition on the other are rather tenuous. Although it is frequently acknowledged that these fields can benefit enormously from closer interaction, examination of reporting practices among researchers in these fields shows little cross-citation. In an attempt to accelerate communication and span the distance between these fields, this section of the text contains contributions from three neurobiologists and three cognitive neuroscientists. Neurobiologists usually base their ideas on data from experiments performed on animals, whereas psychologists use human subjects to reach their conclusions.

In the first of six chapters in this section is a summary of basic cellular events that underlie development of the cerebral cortex: cell proliferation, migration, and aggregation in a form of stratified and radial organization of adult neocortex. Perhaps particularly relevant to the ontogeny of cognition, this chapter provides up-to-date quantitative data on the course of postnatal synaptogenesis in the cerebral cortex. Chapter 9 (Levitt) presents additional and more detailed experimental data about the determination of cell fate at early embryonic ages. Important factors include both innate positional information and cues from extrinsic sources. Levitt demonstrates how advanced methods in molecular neurobiology and transplant technology can be used to study basic questions of cell determination—a major step in building a brain with areal, nuclear, and circuit specializations. Chapters 10 (Spelke, Vishton, and von Hofsten), 11 (Baillargeon), and 12 (Premack and Premack) reveal how children can use their cortical synapses for far more complex logical deductions at earlier stages of their development than hitherto recognized. Methods of examining cognitive functions in children are becoming more precise and the questions asked more subtle and sophisticated. Finally, chapter 13 (Neville) elucidates how maturation of cortical function involves changes from more diffuse to more refined mechanisms. Here, advances in noninvasive methods and computerized imaging have opened new opportunities for viewing cortical maturation and localization in vivo. Just how far from establishing a common ground between developmental neurobiology and developmental psychology are we?

There is no disagreement among the authors of these chapters that development of cognition depends on the functional capacity of the underlying neuronal circuitry. Likewise, there is no doubt that we cannot fully understand the development of this circuitry without taking into consideration the behavior of the organism. Although the fields of neurobiology and neuropsychology still occupy opposite ends of the spectrum of heterogeneous disciplines that comprise modern neuroscience, several points of interaction hold great promise. One link is the necessity for both areas of research to delineate a precise timetable of developmental events. For example, the course of synaptogenesis in the cerebral cortex during postnatal life can be correlated not only with biochemical maturation of neurotransmitters and neurotransmitter receptors but also with the emergence and elaboration of specific cognitive functions. More specifically, the fact that synapses between neurons develop and reach a crest in numbers and densities simultaneously across the entire cortical mantle can be correlated with the acquisition of perceptual and association functions in children across a wide range of domains. In this area of research, the traditional view of hierarchical development needs to be reexamined in view of new findings in both developmental neurobiology and cognitive neuroscience.

The other point of convergence among authors of these chapters is the concept of progressive differentiation, from initially diffuse states and neuronal connections to more refined capacities and more localized circuitry. The concept, as it applies to connectivity of the cerebral cortex, originated from research in the binocular visual system of the macaque monkey, but now the functional development of specific auditory functions, from diffuse to more focused, can be studied in humans by using noninvasive imaging methods. In animal research, the molecular mechanisms that control overproduction and activity-dependent elimination of synaptic contacts is the major focus of developmental neurobiologists. We have learned that the size of the dendritic tree, the number of neurons, and normal synaptic architecture, including details of cortical topography, are achieved only through the precise temporal pattern of neuronal activity. Neuroscientists and developmental psychologists basically agree that, in the cerebral cortex, long-term functional rearrangements in cortical representation occur in response to changes in sensory input.

The third point of convergence between the fields is a common interest in unraveling the relative participation of intrinsic and extrinsic factors in the development of adult cortical circuitry and the establishment of its functional capacity. Intrinsic, genetic mechanisms are currently a subject of renewed interest and empirical support. For example, eye enucleation at embryonic stages that deprives an animal of all input from the retina does not prevent the emergence of certain cellular and molecular features that subserve color vision in the normal visual cortex. Likewise, congenitally deaf subjects, deprived of auditory stimuli, show many electrophysiological features in the cerebrum that are similar to those found in controls with normal hearing, suggesting that, to some extent, the cortical substrate for hearing may be specified initially.

The fourth common goal of this section's authors is the desire to enhance understanding of the causes and pathogenesis of mental disorders of higher brain function. Severe cognitive disorders in children and adolescents are now being characterized by genetic, molecular, or neuroanatomical malfunctions only because we are able to correlate anatomical or biochemical defects and behavioral abnormality. Most intriguing is the possibility that the anomalies have different origins. At each level of analysis, there has been an explosion of new information and synthesis, and more of the same is expected to come from the increased collaboration of scientists of different backgrounds.

There are, therefore, basic similarities in the way neurobiologists and cognitive neuroscientists think about issues and pursue their goals. Stronger correlations, which are needed, are now becoming possible. Also needed is a theoretical framework that can connect neurobiological and cognitive science in a context of specific developmental mechanisms. The challenge to the theorist is to link specific models that address cellular phenomena with global theories that reflect behavioral states. It is hoped that bringing together contributions that concern these developmental issues in animal and human subjects will stimulate further research and strengthen the bonds between the fields.

8 Corticogenesis in Human and Nonhuman Primates

PASKO RAKIC

ABSTRACT The development of the cerebral cortex, with its distinct cellular, molecular, and functional characteristics, is central to our understanding of human cognitive capacity. Recent advances in developmental neurobiology have helped us to gain new insights into the formation of the cerebral cortex and the pathogenesis of disorders of higher brain function. A major event in the prenatal development of the brain is the production of neurons and migration of cells from the place of their origin to their final positions in the cortex. The final positioning of a neuron determines its fate, its shape, synaptic connectivity and, ultimately, its function. Studies of cell migration enable analysis of developmental mechanisms of cell motility and identification of the molecules that help neurons translocate their bodies over long distances and find their proper "addresses" in the brain. Another critical cellular event that begins before birth but is completed primarily during infancy and adolescence is the establishment of the fine wiring arrangement of neural connections. At that stage, environmental stimulation sculpts the final pattern of neural organization from an initial state of excess cells, axons, and synapses. Recent studies show that the number of synapses as well as the density of neurotransmitter receptors in the cerebral cortex are eliminated simultaneously in sensory, motor, and associational areas in close correlation with the emergence of metabolic activity and complex cognitive functions. Data from research on nonhuman primates provide the biological basis for understanding the development of cortical diversity and human congenital brain diseases, including some psychiatric disorders.

It is generally agreed that the cognitive prowess of the human species is due primarily to extraordinary functional capacity of the cerebral cortex. It is, therefore, essential that we directly examine the genetic, molecular, cellular, and physiological development of the cerebral cortex in primates. This chapter is based mostly on normal and experimental studies of neocortical development in the rhesus monkey. The large size of the cerebrum and the presence of visible cytoarchitectonic landmarks in this species allows precise delineation of cortical areas, while the rhesus monkey's gestation of 165 days and protracted sexual maturation of more than 3 years enables accurate timing of developmental events. This may be why some of the basic principles of corticogenesis, including guided neuronal migration, formation of transient subplate zone, segregation of terminals from the initially diffuse state, and the concept of selective elimination of cortical axons and synapses, were described initially in primates. Furthermore, the similarity of neuronal organization of the macaque and human cortex promises that the developmental mechanisms may be comparable in both species. In the past two decades, considerable progress has been made in elucidating the critical cellular events of neuronal proliferation, migration, determination, biochemical differentiation, and synaptogenesis in the cerebral cortex of nonhuman primates that helps our understanding of the emerging cognitive functions and the pathogenesis of various psychiatric disorders in humans.

Traditionally, research in corticogenesis has been confined to the delineation of timing of developmental events, description of cell behavior, and determination of critical stages, mostly based on the analysis of normal material processed by classical neuroanatomical and neurocytological methods. Even more advanced methods of electron microscopy, axonal tracing, immunocytochemistry, and in situ hybridization have relied on the dynamic interpretation of normative data. Only recently have experiments such as prenatal manipulation of cortical development, homotopic and heterotopic transplantation, or creation of chimeric and transgenic animals been conducted to challenge the normal course of development. Collectively, these studies have indicated that development of the cerebral cortex proceeds from early stages, which seem to be specified intrinsically by instructions within cell lineages, to the later phases, which are much more dependent on neuronal activity. The series of cellular events

PASKO RAKIC Section of Neurobiology, Yale University School of Medicine, New Haven, Conn.

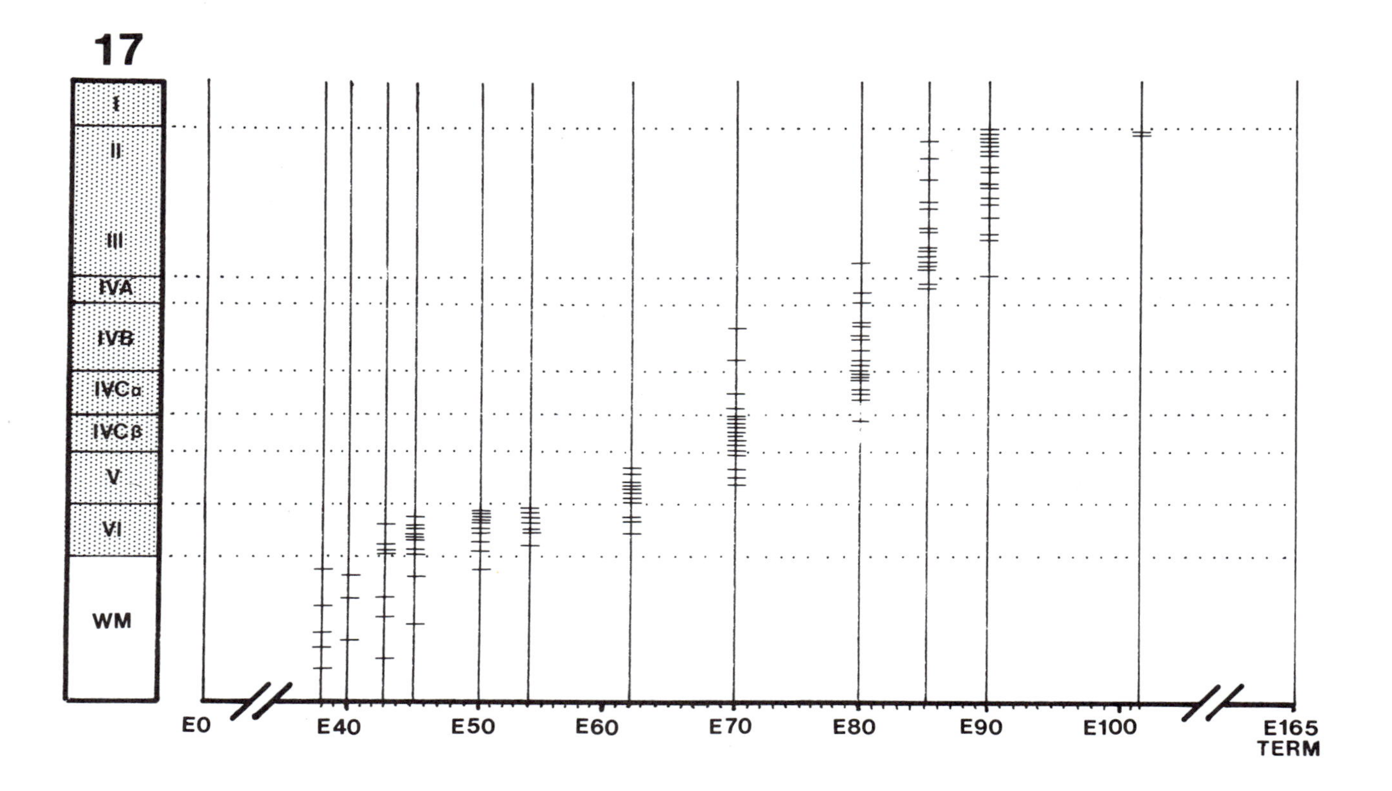
17
I
II
III
IVA
IVB
IVCα
IVCβ
V
VI
WM
E0
E40
E50
E60
E70
E80
E90
E100
E165
TERM

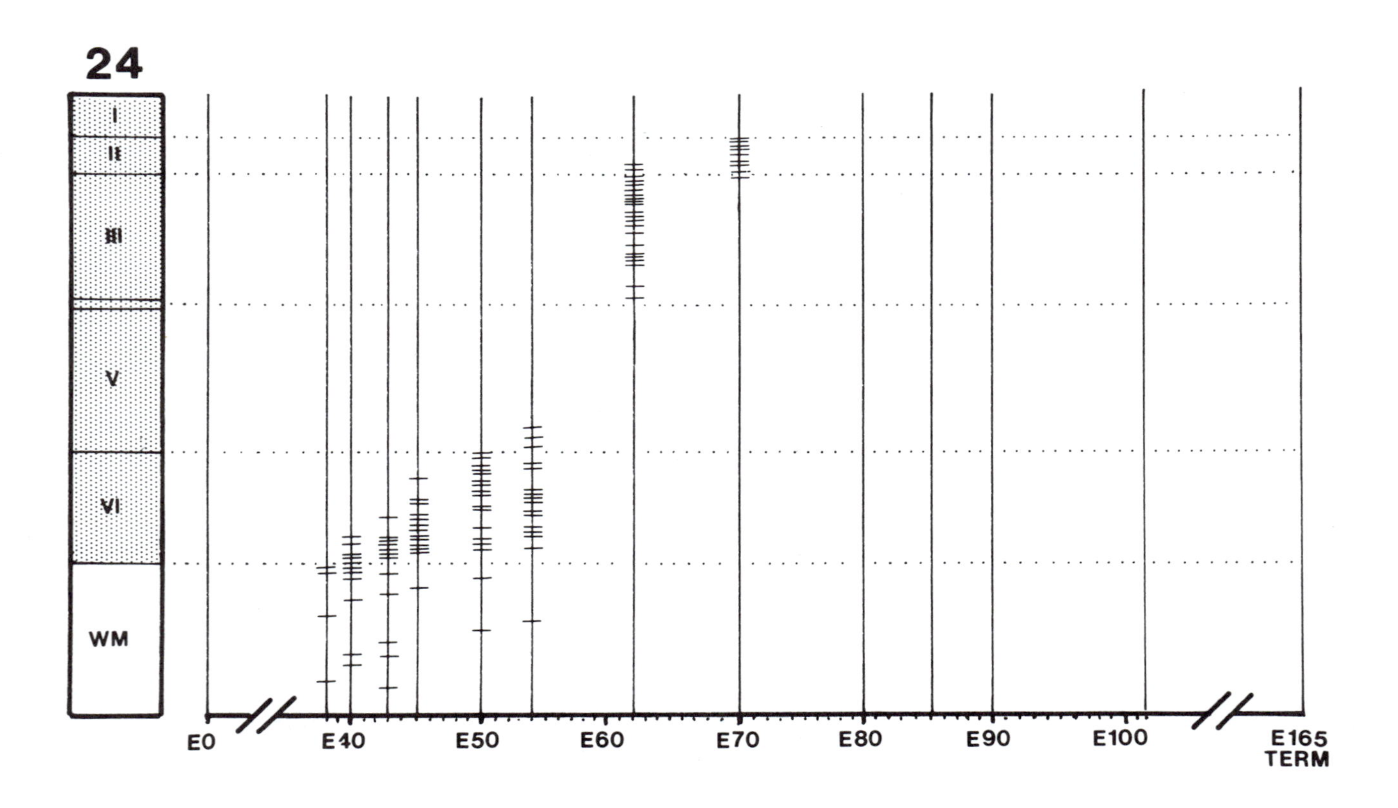
24
I
II
III
V
VI
WM
E0
E40
E50
E60
E70
E80
E90
E100
E165
TERM

involved in formation of the cerebral cortex are described as they emerge in the primate brain during prenatal and postnatal development.

Prenatal generation of cortical neurons

It is well established that the newborn child already has an impressive-looking cerebral cortex with well-delineated individual layers and cytoarchitectonic areas (Conel, 1939). However, with classical histological methods, it could not be determined when the genesis of various classes of cortical neurons actually begins and ends. The labeling of DNA replication with ^{3}H-thymidine provided precise data on both the onset and termination of corticogenesis in the rhesus monkey. Use of this approach revealed that the genesis of cortical neurons in this primate species starts in the first quarter of the 165-day gestational period, approximately the fortieth embryonic day (E40). The active production of cortical neurons lasts between 30 and 60 days, depending on the cytoarchitectonic area (Rakic, 1974, 1982). For example, the genesis of neurons destined for the anterior limbic cortex (Brodmann's area 24) is completed by E70, whereas neurons composing the primary visual cortex (area 17) are being produced until E100 (figure 8.1). Autoradiographic analysis by my colleagues and I shows that no neocortical neuron is generated during the remainder of gestation (Rakic, 1974) or at any time during the nearly 30-year life span of the macaque monkey (Rakic, 1985). Analysis of supravital DNA synthesis in slices of fresh fetal brain (Rakic and Sidman, 1968; Sidman and Rakic, 1982) indicates that corticogenesis in humans begins in the sixth embryonic week, or at approximately E42 (figure 8.2). Use of additional cytological criteria of morphological maturation also indicates that in humans, with a gestational period of approximately 280 days, the first cortical neurons are generated on or near E42, but their production lasts until E125 or so (Sidman and Rakic, 1973, 1982; Rakic, 1988a). Thus, unlike most mammalian species in which corticogenesis continues until—or even after—birth, primates (including humans) acquire their full complement of cortical neurons during the middle third of gestation (see figure 8.2).

FIGURE 8.1 Diagrammatic representation of the positions of heavily labeled neurons in the cortex of juvenile monkeys, each of which had been injected with ^{3}H-thymidine at selected embryonic days: (top) area 17; (bottom) area 24 of Brodmann (1909). On the left side of each diagram is a drawing of the cortex, in which subdivisions into cortical layers are indicated by Roman numerals. Embryonic days (E) are represented on the horizontal line, starting on the left with the first fetal month (E36) and ending on the right at term (E165). Positions of vertical lines indicate the embryonic day on which one animal received a pulse of ^{3}H-TdR. On each vertical line, short horizontal markers indicate positions of all heavily labeled neurons encountered in one 2.5-mm-long strip of cortex. WM, white matter. (Modified from Rakic, 1974, 1982, by permission.)

Origination of cortical neurons near cerebral ventricle

Examination of a series of rhesus monkey embryos, sacrificed shortly after exposure to the ^{3}H-thymidine, revealed that most neurons destined for the neocortex are produced in the proliferative cell layer lining the lumen of the cerebral ventricle (Rakic, 1975). This transient embryonic cell layer, named the *ventricular zone* by the Boulder Committee (1970), is organized as a pseudostratified epithelium in which precursor cells divide asynchronously (figure 8.3). A portion of these cells' precursor body is attached to the ventricular surface with an end-foot while the nucleus moves intermittently, first away from the ventricular surface to synthesize DNA and then back to the surface to undergo mitotic division (reviewed in Sidman and Rakic, 1973). A combination of Golgi, electron microscopic, and immunocytochemical analyses established that precursors of neuronal and glial cell lines in this species coexist in the ventricular zone from the onset of corticogenesis, which in monkey occurs at E40 (Levitt, Cooper, and Rakic, 1983; Cameron and Rakic, 1991). Divergence of glial and neuronal cell lines have also been observed in postmortem human fetal cerebrum of an equally early developmental stage (Choi, 1986).

Autoradiographic analysis of ^{3}H-thymidine-labeled cells showed that after E40, some postmitotic neurons detach their end-feet from the ventricular surface and begin the journey to their final areal and laminar positions in the cortical plate (Rakic, 1974, 1975). Soon afterward, in both human and monkey, one can find another proliferative layer situated between the ventricular and intermediate zones and hence named the *subventricular zone* by the Boulder Committee (1970). Unlike the ventricular zone, this zone consists of dividing cells that are not attached to the ventricular sur-

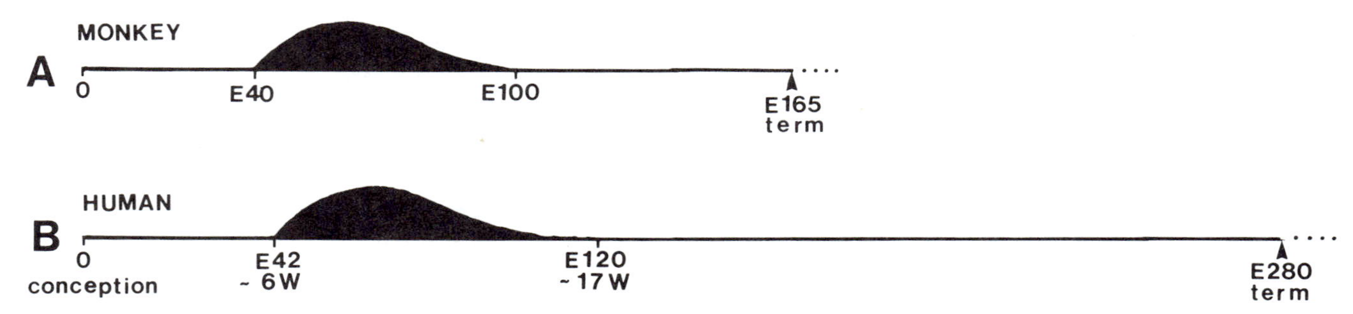

FIGURE 8.2 Diagrammatic representation of the time of neuron origin in rhesus monkey (A) and human (B). The data in monkey are obtained by ^{3}H-thymidine autoradiography (Rakic, 1974), whereas the data in human are based on the number of mitotic figures in the ventricular zone, supravital DNA synthesis, and the presence and density of migrating neurons in the intermediate zone of the fetal cerebrum. (Adapted from Rakic and Sidman, 1968; Rakic, 1978; Sidman and Rakic, 1982.)

face. It increases in width at later stages of corticogenesis, concomitantly with the decline in width of the ventricular zone (Rakic, 1975). It has been proposed that the subventricular zone produces mostly local circuit neurons destined for the supragranular cortical layers and, particularly at later stages, glial cells (Rakic, 1975, 1982).

Development of the transient subplate zone

During embryonic and prenatal development, the telencephalic wall consists of several cellular layers or zones that do not exist in the mature brain. These are, starting from the cavity of cerebral vesicle, the ventricular and subventricular zones just mentioned, the intermediate and subplate zones, the cortical plate, and the marginal zone, situated below the pial surface (see figure 8.3) (Rakic, 1982). Although most of these basic cellular layers were described in the classical literature (e.g., His, 1904; Sidman and Rakic, 1973), the transient subplate zone has only recently been recognized as a separate, developmentally important entity (e.g., Kostovic and Rakic, 1980, 1990). This zone consists of neurons that are generated early and are situated beneath the cortical plate among numerous axons, dendrites, glial fibers, and migrating neurons. In the past two decades, this zone has been implicated in some basic developmental events that lead to cortical development (see the references in Shatz, Chun, and Luskin, 1988; Kostovic and Rakic, 1990).

It was suggested initially, on the basis of findings made in macaque fetuses, that the transient subplate zone may serve as a waiting compartment for afferents generated in the subcortical structures and other areas of the cerebrum that arrive at the cerebral wall ahead of their target neurons in the cortical plate (Rakic, 1977). A subsequent study in both monkey and human fetal brains showed that the axons from the brain stem, basal forebrain, thalamus, and ipsilateral and contralateral cerebral hemisphere in the subplate zone arrive sequentially and, after waiting a variable and partially overlapping period of time, enter the overlying cortical plate (Kostovic and Rakic, 1990). It was suggested that the subplate neurons also play a role in specification of overlying cytoarchitectonic areas (Rakic, 1977) and, more recently, in the formation of ocular dominance columns (Gosh et al., 1990).

Studies of the subplate zone in developing rhesus monkeys and comparison of its size in primate and subprimate species indicates that this transient embryonic zone expanded during evolution of the cerebral cortex, presumably to subserve the increasing number of its associative corticocortical connections and the formation of cerebral convolutions in higher primates (Goldman-Rakic and Rakic, 1984; Kostovic and Rakic, 1990). Many of the subplate neurons eventually degenerate, although some appear to persist in the normal adult cerebrum as a set of interstitial cells scattered within the subcortical white matter (Kostovic and Rakic, 1980; Luskin and Shatz, 1985). An abnor-

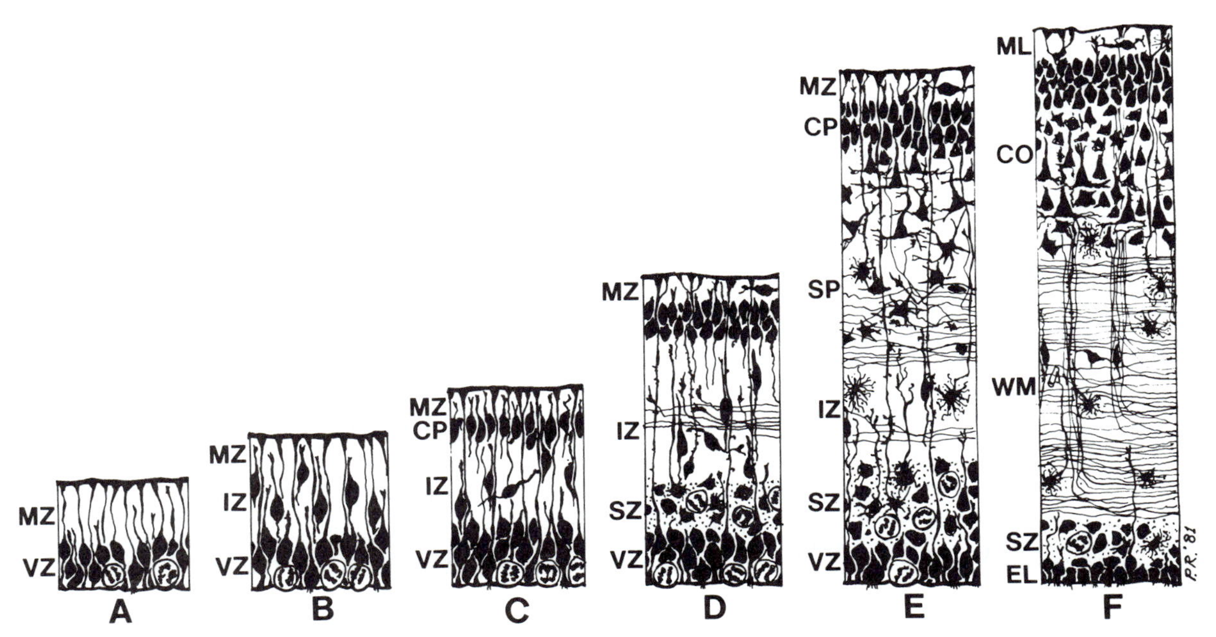

FIGURE 8.3 Schematic illustration of histogenesis of the cerebral neocortex. This modified version of Rakic's drawing used by the Boulder Committee (1970) was updated in 1992 to include new findings on gliogenesis, mode of neuronal migration, and formation of the transient subplate zone. Each column represents the full thickness of the developing cerebral wall so that the ventricular surface is at the bottom and the pial surface at the top. During development of the cerebrum, one can recognize a series of transient developmental zones that are essentially without a direct counterpart in the mature brain. (A) State at which the cerebral wall consists only of proliferative ventricular (VZ) and acellular marginal (MZ) zones. (B) The intermediate zone (IZ) containing displaced postmitotic cells is added. Also, cells of neuronal lineage and radial glial cells that stretch across the cerebral wall become distinguishable at this stage. (C) The cortical plate (CP) is formed by cells that have migrated from the ventricular zone. (D) Another proliferative layer, the subventricular zone (SZ), appears external to the ventricular zone. Horizontally or obliquely disposed axons that originate mostly from the thalamus enter the intermediate zone. (E) Formation of the subplate zone (SP), which consists of horizontally deployed fibers and large, mostly multipolar neurons as well as radially oriented bipolar cells migrating external to the differentiating pyramidal neurons of the deep cortical layers. (F) Final stage of neocortical development into cortex (CO) with the ventricular zone transformed into an ependymal layer (EL) and the remnant of proliferation in the subventricular zone. The subplate zone disappears, whereas the intermediate and marginal zones become transformed into white matter (WM) and molecular layer (ML), respectively. (Reprinted from Rakic, 1982, by permission.)

mally large vestige of subplate neurons may form heterotopic masses in the form of a double cortex, and these are believed to be the source of intractable epileptic discharges in children (Palamini et al., 1991b).

Migration of postmitotic neurons to the cortical plate

The basic tenet of corticogenesis is that neurons which constitute the cortex are not generated in the cortex itself (Sidman and Rakic, 1973). After their last mitotic division near the cerebral ventricle, postmitotic neurons migrate toward the pial surface and form the cortical plate. On the way, they traverse the intermediate and subplate zones, which are already occupied by numerous axons and other cellular elements. The basic mode of neuronal migration across this terrain is similar in all mammalian species examined, except that in large primates the pathway is much longer and the phase of most massive migration coincides with the rapid growth of the cerebral wall and the onset of buckling of its surface into sulci and gyri (figure 8.4). Therefore, in primates, postmitotic cells traverse a longer and more complex pathway in order to attain permanent positions in the cortex (Rakic, 1978). Therefore, the mechanisms that ensure the proper placement of neurons are more prominent in primates and, consequently, defects of migration are more frequent.

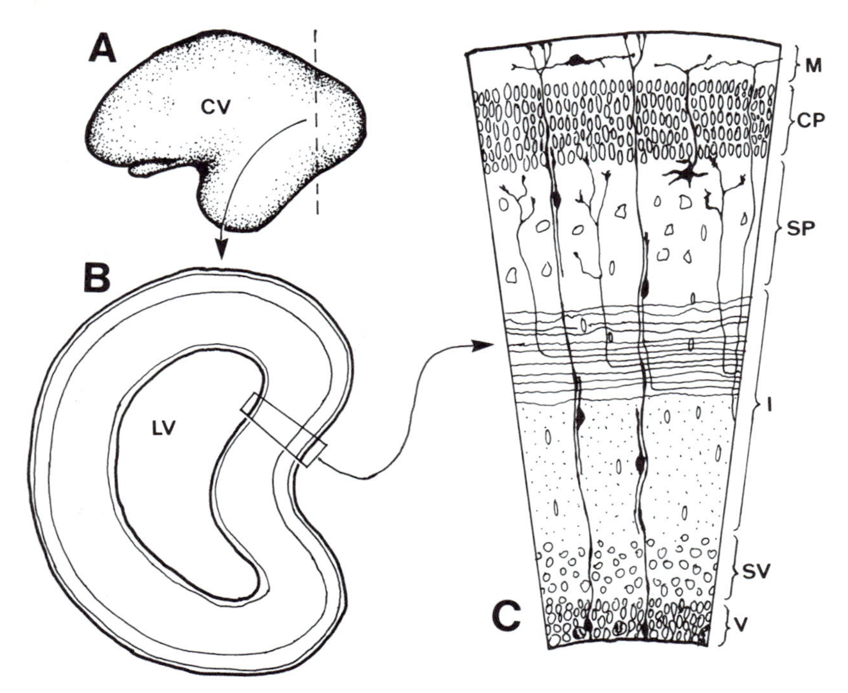

FIGURE 8.4 Cytological organization of the primate cerebral wall during the first half of gestation. (A) Cerebral vesicle (CV) of 60- to 65-day-old monkey fetus is still smooth and lacks the characteristic convolutions that will emerge in the second half of gestation. (B) Coronal section across the occipital lobe at the level indicated by a vertical dashed line in (A). The lateral ventricle (LV) at this age is still relatively large, and only the identification of incipient calcarine fissure marks the position of the prospective visual cortex. (C) A block of the tissue dissected from the upper bank of calarine fissure. At this early stage, one can recognize all embryonic layers: ventricular zone (V), subventricular zone (SV), intermediate zone (I), subplate zone (SP), cortical plate (CP), and marginal zone (M). Note the presence of migrating neuron (dark bipolar profiles) moving along radial glial fibers, which span the full thickness of the cortex. The early afferents from the brain stem and thalamus invade the cerebral wall and accumulate in the subplate zone, where they make transient synapses before entering the cortical plate. (Reprinted from Rakic, 1991, by permission.)

Electron microscopic analysis carried out on serial sections of the rhesus monkey fetal cerebral wall suggested that a majority of postmitotic neurons find their way through the intricate cellular lattice of the intermediate and subplate zones by following the elongated shafts of the radial glial cells (Rakic, 1972). These nonneuronal, spindle-shaped cells stretch their outward-directed process across the fetal cerebral wall from the beginning of corticogenesis and become most prominent during midgestation, when many of them temporarily stop dividing (Schmechel and Rakic, 1979). During the migratory period, cohorts of postmitotic cells follow a radial pathway consisting of single or, more often, multiple glial fibers that form fascicles which stretch between the ventricular and pial surface (see figure 8.4). While moving along the glial fibers, they also come into contact with myriad other axonal and dendritic processes but remain preferentially attached to the glial membrane (Rakic, 1972, 1990).

The neuron-glia relationship leads to the concept of a gliophilic mode of cell migration that is likely to be mediated by the same type of complementary heterotypical adhesion molecules (Rakic, 1981b). However, some migrating neurons were also transferring from one to the other radial glial fascicle (Rakic et al., 1974), whereas others became displaced more tangentionally and moved at an angle or perpendicular to the orientation of the glial palisade. Many of these cells, that do not obey glial contraints, move along axonal fascicles, displaying a neurophilic mode of migration (Rakic, 1985, 1990). Although lateral dispersion of postmitotic neurons has been known to occur (e.g., slanted bipolar cells situated in the intermediate zone in figure 8.3C), it has attracted much more interest since the use of retrovirus-mediated labeling, which showed considerable dispersion of clonally related neurons (Walsh and Cepko, 1992) in addition to those moving radially (Luskin, Perlman, and Sanes, 1988).

The studies in chimeric mice have provided direct evidence that the overwhelming majority of postmitotic neurons move radially to the cortex (Nakatsuji, Kadokowa, and Sumori, 1991; Tan and Breen, 1993). It has been assessed that more than 85% of neurons in the ferret (O'Rourke et al., 1992) and possibly an even higher fraction in primates obey radial constraints imposed by the radial glial palisades (Rakic, 1988b).

Multiple molecular mechanisms of neuronal migration

The application of new methods allows analysis of the molecular mechanisms underlying gliophilic neuronal migration. Initially, it was suggested that at least a pair of complementary heterotypical molecules situated on the neural and glial cell surface is essential for recognition and selection of the migratory pathway (Rakic, 1981b). The spatiotemporal order of neuronal migration from the place of origin to final destination can be achieved even with the single pair of heterotypical molecules because, once movement of the cell is initiated, neurons tend to remain in contiguity with a given radial glial fascicle (Rakic, 1991). The strong bonds are formed presumably by the cell adhesion molecules. This concept is supported also by tissue culture experiments, which have demonstrated that proper adhesion occurs in cultures with cross-matching between neurons and radial glial cells obtained from different brain structures or cortical regions (Hatten, 1990; Hatten and Mason, 1990). Several candidates for recognition and adhesion molecules are currently being tested (see, for example, Edelman, 1983; Schachner et al., 1985; Hatten and Mason, 1986; Hatten, 1990; Cameron and Rakic, in press). It now is generally assumed as likely that more than one class of molecules is involved in neuronal migration (Rakic, Cameron, and Komuro, 1994).

Progress has also been made in understanding the mechanism of the actual physical displacement of the neuronal perikarya. For example, it was shown that ion channels situated on the membrane of the leading process and cell soma of migrating neurons regulate the influx of calcium ions, which may trigger polymerization of cytoskeletal and contractile proteins essential for cell motility and translocation of the nucleus and surrounding cytoplasm (Komuro and Rakic, 1992, 1993). These studies indicate that neuronal migration is a multifaceted developmental event involving cell-cell recognition, differential adhesion, transmembrane signaling, and intracytoplasmic structural changes (Rakic, Cameron, and Komuro, 1994). Understanding of the molecular mechanism of neuronal migration may help explain the pathogenesis of previously inexplicable genetic and acquired conditions, such as childhood epilepsy, mental retardation, and developmental dyslexia (Galaburda et al., 1985; Rakic, 1988a; Volpe, 1987; Aicardi, 1991; Palamini, 1991a).

Laminar settling of cortical neurons

After neurons arrive at the cortical plate, they abruptly stop movement at the interface between layer II and the marginal zone (layer I). As a result, neurons arriving at progressively later stages have to pass between, and come in close contact with, previously generated cells (see figure 8.1). The relationship between neurons that originated at different times, called commonly the *inside-out* gradient of neurogenesis, was suspected by classical neuroembryologists but has been proved only since the introduction of the ^{3}H-thymidine autoradiographic method of labeling permanently postmitotic neurons (reviewed in Sidman and Rakic, 1973). The inside-out gradient of neurogenesis is more prominent in primates, where each injection of ^{3}H-thymidine labels a highly selective sample of cortical neurons (Rakic, 1974), than in rodents, where simultaneously generated neurons are more widespread over adjacent cortical layers (Angevine and Sidman, 1961). Although each generation of migrating neurons reaches the most superficial stratum of the developing cortical plate, the position of these neurons in the adult cortex depends on the number of subsequently produced cells that eventually become situated external to them (Rakic, 1975). In animals injected with ^{3}H-thymidine within the last 2 months of intrauterine life or after birth, only glia were labeled in the adult (Rakic, 1974). Thus, although neurons of the monkey visual cortex are generated comparatively earlier in the gestational period than in most other nonprimate mammals, these neurons follow the same inside-out pattern of development described previously in other species (Angevine and Sidman, 1961). Comparison of the timing of cell origin with the schedule of the cells' morphological, biochemical, and functional differentiation indicates that the time of the last cell division and the onset of differentiation may be independent developmental processes, probably regulated by different genetic mechanisms.

Comparison of the positions of prenatally labeled neurons in the adult monkey cerebrum reveals that simultaneously generated neurons occupy different layers in various cytoarchitectonic areas (see figure 8.1). The failure of such cells to attain their proper position according to a specified time schedule may severely compromise the orderly development of the laminar pattern in the cortex (Rakic, 1988a). As a result, the antimitotic agents—such as cytotoxic drugs or ionizing radiation—have different but highly predictable effects on the developing cortex, depending on the time of initial insult (Yurkewicz et al., 1984; Algan, Goldman-Rakic, and Rakic, 1992).

Determination of neuronal phenotypes

Until approximately E40 in the macaque embryo, all neuronal progenitors produce only new generations of progenitors. This so-called symmetrical mode of cell division, which results in exponential growth, predominates in the early stages of development and has been inferred from the ^{3}H-thymidine analysis of cell proliferation kinetics (Rakic, 1975). After E40, however, some progenitor cells begin to undergo asymmetrical division, which generates one postmitotic cell that is committed to a neuronal line and leaves the proliferative pool and another all that remains in the ventricular zone and continues to divide. Because many cells in this zone divide symmetrically, producing two progenitors, the ventricular zone increases in size and becomes polymorphic epithelium during middle stages of corticogenesis (Rakic, 1975, 1988b). At later stages, however, an increasingly larger number of postmitotic neurons leave the proliferative pool, leading eventually to the exhaustion of the proliferative zones. Recently, it was calculated that nearly half of the germinal cells in the ventricular zone and virtually all those in the subventricular zone in the mouse telencephalon divide symmetrically (Takahashi, Nowakowski, and Caviness, 1994).

Immunocytochemical analyses in fetal monkeys provided the first evidence that proliferative cells in the ventricular zone concomitantly produce neurons and radial glial cells (Levitt, Cooper, and Rakic 1981, 1983). Furthermore, examination of the fate of the ^{3}H-thymidine-labeled cells suggests that the ventricular zone simultaneously produces multiple neural classes, which terminate either in the single or adjacent cortical layers (Rakic, 1988b). In recent years, several lines of evidence have indicated that, the possible fates of postmitotic neurons rapidly become restricted after their last cell division (McConnell, 1990). For example, neurons that remain in ectopic positions near the cerebral ventricle (as a consequence of x-irradiation during embryonic stages) eventually acquire a type of morphology and develop the connections expected from the time of their origin (Jensen and Killackey, 1984). Likewise, ventricular cells transplanted from embryos into the telencephalic wall of a newborn ferret migrate to the host cortex and assume laminar positions, morphological characteristics, and a pattern of connections appropriate for the stage they have achieved in the donor (McConnell, 1988). Furthermore, a subset of migrating neurons sends its axons to the contralateral hemisphere and becomes committed to a callosal neuronal phenotype before entering layer II of the ipsilateral cortical plate (Schwartz, Rakic, and Goldman-Rakic, 1991).

This early commitment of cortical neurons is evident in the reeler mutant mouse, in which the position of cortical laminae is reversed but neurons differentiate into phenotypes expected from the time of their origin rather than from their ectopic location (Caviness and Rakic, 1978). Finally, recent analysis, in which RNA retrovirus-mediated gene transfer was used to mark permanently the progeny of dividing cells with the cytochemically detectable enzyme β-galactosidase, revealed that neurons of different types originate from separate clones (Parnavelas et al., 1991). These findings collectively indicate that the range of morphologies and patterns of synaptic contacts of cortical neurons may be specified, in considerable measure, before the neurons reach their final positions.

Radial unit hypothesis of cortical development

Neurons in the cerebral cortex eventually form a morphologically identifiable stack of radially deployed neurons, variously termed *ontogenetic* or *embryonic columns* (Rakic, 1972, 1978). As described in the preceding section and illustrated schematically in figure 8.5, the radial unit hypothesis predicts that the ventricular zone produces cohorts of postmitotic neurons, most of which are gliophilic and, therefore, constrained in their migratory pathways from randomly mixing while on their way to the cortex (Rakic, 1972, 1978, 1988b). Accord-

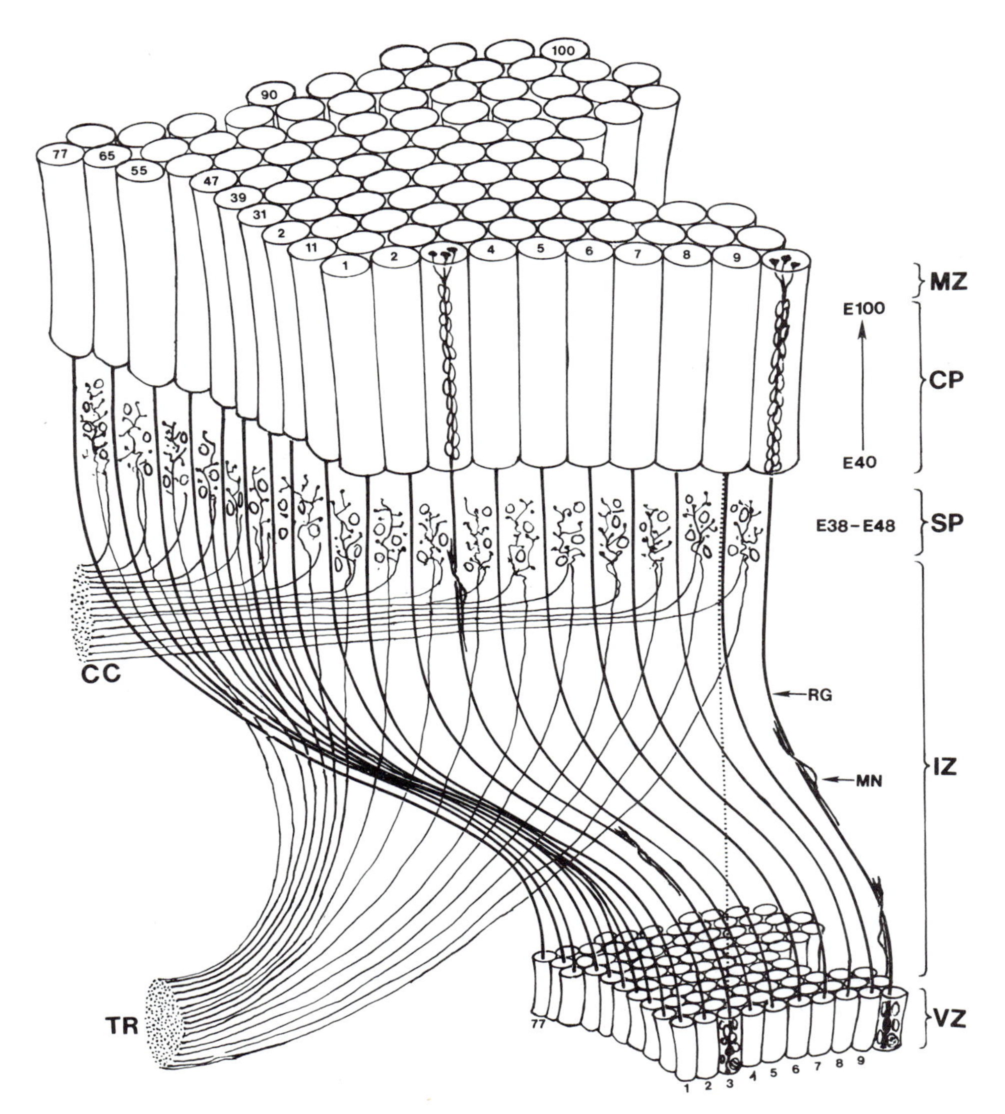

FIGURE 8.5 A cartoon of the relation between a small patch of the proliferative ventricular zone (VZ) and its corresponding area within the cortical plate (CP) in the developing cerebrum. Although the cerebral surface in primates expands and shifts during prenatal development, ontogenetic columns (outlined by cylinders) remain in register with the corresponding proliferative units by the grid of radial glial fibers. Cohorts of neurons produced between E40 and E100 by a given proliferative unit migrate in succession along the radial glial fascicles (RG) and stack up in reverse order of arrival within the same ontogenetic column. Each migrating neuron (MN) first traverses the intermediate zone (IZ) and then the subplate (SP) that contains interstitial cells and "waiting" afferents from the thalamic radiation (TR), and ipsilateral and contralateral corticocortical connections (CC). After entering the cortical plate, neurons bypass earlier-generated neurons and settle at the interface between the CP and marginal zone (MZ). As a result, proliferative units 1 to 100 produce ontogenetic columns to 100 in the same relative position to one another without a lateral mismatch (for example, between proliferative unit 3 and ontogenetic column 9, indicated by a dashed line). Thus, the specification of cytoarchitectonic protomap depends on the spatial distribution of their ancestors in the proliferative units, whereas the laminar position and phenotype of neurons within ontogenetic columns depends on the time of their origin. (Reprinted from Rakic, 1988b, by permission.)

ing to this hypothesis, the number of radial units in each species and individual is specified in the proliferative zones at early embryonic stages, before cortical neurons form synaptic contacts with thalamic afferents.

The radial unit hypothesis of cortical development suggests that an array of columns or radially organized modules of synaptically interrelated neurons in the adult neocortex may be a reflection of their developmental history (Rakic, 1978, 1988b). It was shown initially by Mountcastle (1957) that neurons situated within a single cellular column in the somatosensory cortex become responsive to a specific modality and a narrow receptive field of stimulation at the skin. The radial arrangement of the retinal receptive field has subsequently been shown to exist in the visual cortex (Hubel and Wiesel, 1972). A similar anatomical and functional columnar organization was found in the association cortices (Goldman and Nauta, 1977; Szentagothai, 1978; Goldman-Rakic, 1981; Mountcastle, 1979). Video microscopy, which can identify synchronously activated cells by the fluorescence of calcium-sensitive dyes, reveals radial columns of Ca^{2+}-activated neurons that are continuous throughout the full width of the cortex (Yuste and Katz, 1991). The columnar organization of cortical neurons is reflected in the pattern of thalamocortical as well as ipsilateral and contralateral corticocortical connections. The relatively constant size of the radial terminal field units among species with vastly different sizes of cerebral cortex and modality-specific patterns of its surface supports the idea that, during evolution, the neocortex expands by the addition of new radial units rather than by enlarging them (Bugbee and Goldman-Rakic, 1983).

It has been suggested that the number of proliferative units at the ventricular surface sets a limit of cortical size during individual embryonic development and evolution by regulating the number of columns in the cortex (Rakic, 1978). If this is so, then a small change in the length of the cell cycle or number of symmetrical mitotic divisions at early embryonic stages can produce a large difference in the number of proliferative units that subsequently generate cortical columns (Rakic, 1988b). The mechanisms by which large functional advances are produced as a consequence of relatively small changes in the timing of cellular events is known as *heterochrony* in evolutionary biology (Gould, 1977).

According to the radial unit hypothesis, the scenario in the developing cerebrum may proceed as follows: The ventricular zone initially enlarges predominantly by symmetrical division of germinal cells. Therefore, an additional round of cell divisions at this stage can double the number of proliferative units in the ventricular zone and, consequently, increase the number of radial ontogenetic columns in the cortex by a factor of two (Rakic, 1991). After initiation of corticogenesis, some progenitors in the ventricular zone begin to divide asymmetrically so that an extra round of cell divisions has a much smaller effect on the overall thickness of the cortex. Our working hypothesis is that one set of regulatory genes that operates at early stages determines the type and the number of mitotic divisions in the ventricular zone. It is generally believed that regulatory genes, similar to those found in the fruit fly, operate in the embryonic mammalian telencephalon, and several candidates are currently being tested (see, for example, Lu et al., 1992; Proteus et al., 1991). Subsequently, another set of genes may induce asymmetrical division and control the fate of postmitotic cells. Interaction of migrating cells with incoming afferents may generate variations in the composition of ontogenetic radial columns.

Protomap hypothesis of cytoarchitectonic diversity

Despite the considerable amount of research done in the field of corticogenesis, it is not known how individual and species-specific maps of cytoarchitectonic areas have emerged. Experimental evidence indicates that both intrinsic and extrinsic factors are implicated in this complex process. For example, one influential hypothesis is that all cortical neurons are initially equipotential and that laminar and areal differences in cortical organization are induced exclusively by extrinsic influences exerted via thalamic afferents (Creutzfeldt, 1977). Indeed, there is overwhelming evidence that many features of cortical organization are thalamus-dependent. The most dramatic examples in this category are ocular dominance columns in the visual cortex (Rakic, 1976a, 1977), barrel fields in the somatosensory cortex (Van der Loos and Dorfl, 1978; Killackey, 1989; O'Leary, 1989; Schlaggar and O'Leary, 1992), and topographical maps of peripheral representation in both sensory and motor areas (Merzenich et al., 1988, this volume; Gilbert, this volume).

However, as reviewed later, there is also considerable evidence that cells of the cerebral vesicles may contain some basic programs that foreshadow species-specific cytoarchitectonic organization before they become connected with thalamic input. To reconcile the information obtained from existing experimental and descriptive data on cortical development, I formulated a working model, known as the *protomap hypothesis* (Rakic, 1988b), which postulates that the pattern of cortical cytoarchitectonic areas emerges through synergistic, interdependent interactions between developmental programs intrinsic to cortical neurons and extrinsic signals supplied by specific input from the subcortical structures. According to this hypothesis, neurons of the embryonic cortical plate, indeed the ventricular zone where they originate, set up a primordial species-specific cortical protomap of cortical regions that preferentially attract the appropriate thalamic afferents (reviewed in Rakic, 1994).

The initial indication that cellular events in the ventricular zone foreshadow prospective regional differences in the overlying cerebral mantle came from the ^{3}H-thymidine labeling, which revealed that the ventricular region subjacent to the prospective area 17 produces more neurons per radial unit (Rakic, 1976b) and has a substantially higher mitotic index than the adjacent region situated below area 18 (Kennedy and Dehay, 1993). Because area 17 has more neurons per radial unit of the same width (Rockel, Hirous, and Powell, 1980), region-specific differences can be the result of the production level of neurons at the ventricular zone before postmitotic cells arrive to the cortex and become exposed to the input from the periphery via thalamic afferents. Additional evidence comes from the finding that correct topological connections in the visual cortex are established in congenitally anophthalmic mice and in animals enucleated at early embryonic stages in the absence of any information from the photoreceptors at the periphery (Kaiserman-Abramoff, Graybiel, and Nauta, 1980; Olivaria and Van Sluyters, 1984; Rakic, 1988b). Furthermore, the primary visual cortex in monkeys acquires area-specific cytochrome oxidase patches in the absence of any input from the receptors in the retina (Kuljis and Rakic, 1990). Finally, homotopic, but not heterotopic, fetal transplants can result in functional sparing following neonatal damage to the frontal cortex (Barth and Stainfield, in press).

The role of afferents from the periphery in cortical differentiation is well illustrated in adult animals in which binocular enucleation was performed in the first half of gestation. In such cases, the number of geniculate neurons is reduced to fewer than one-half of the age-matched controls, and the size of area 17 is reduced proportionately (Rakic, 1988b; Rakic, Suner, and Williams, 1991). The laminar and areal distribution of major neurotransmitter receptors in the remaining visual cortex develops a remarkably normal pattern (Rakic and Lidow, 1994). Furthermore, cytochrome oxidase blobs in layers II and III, which are believed to subserve color vision (Livingstone and Hubel, 1988), were segregated and maintained in early binocular enucleates (Kuljis and Rakic, 1990). Finally, synaptic density per unit volume of neuropil, as revealed by quantitative electron microscopy, developed within the normal range in all layers (Bourgeois and Rakic, 1987). These results indicate that the certain cytological, synaptic, and biochemical characteristics of area 17 can develop in the absence of information from the retina. However, the size of the surface devoted to the visual cortex and the details of its cellular and synaptic organization, including formation of ocular dominance columns, depends critically on interaction with thalamic input (Rakic, 1988b; Suner and Rakic, 1994a,b).

More recently, the protomap hypothesis of cortical specification received support from studies that show the emergence of area-specific molecules in the cerebral cortex independently of thalamic input (e.g., Arimatsu et al., 1992; Barbe and Levitt, 1992; Ferri and Levitt, 1993; Cohen-Tannoudji, Babinet, and Wassef, 1994; Levitt, this volume). For example, the use of transgenic mice combined with transareal transplantation revealed that expression of some genes restricted to the region of somatosensory cortex appear in the cells originated from the sector of the ventricular zone subjacent to the somatosensory area, irrespective of the type of afferents they receive from the thalamus (Cohen-Tannoudji, Babinet, and Wassef, 1994). Although the embryonic cerebral wall exhibits area-specific molecular differences, it should be underscored that the *protomap* in the embryonic cerebrum provides only a set of species-specific biological constraints. The position of interareal borders, the overall size of the cytoarchitectonic areas, and the details of cellular, synaptic, and topographical organization of the adult cerebral cortex are achieved only through a cascade of reciprocal interactions between cortical neurons and cues from the specific afferents arriving at the cortical plate from a variety of extracortical sources (Rakic, 1988b, 1994).

Overproduction of cortical neurons and axons

With few exceptions, most brain structures in higher vertebrates have a larger number of neurons and connections during the course of development than they have in adulthood (Cowan, 1973; Easter et al., 1985; Purves, 1988). The primate cerebral cortex is not an exception in this respect. For example, there are 60% more retinal ganglion cells (Rakic and Riley, 1983) and 40% more geniculate neurons (Williams and Rakic, 1988) in the fetal monkey than in the adult. Quantitative studies of the cerebral cortex in this species show considerable overproduction of cortical neurons that are eliminated during well-delineated stages of prenatal development. Judging from the density and distribution of pyknotic (degenerating) cells, there are approximately 35% more neurons in the primary visual cortex during midgestation than in the adult rhesus monkey (Williams, Ryder, and Rakic, 1987). The axons of some cortical cell classes are also more numerous during development, though their production and elimination seems to occur predominantly at later stages. For example, the brain of the newborn monkey has almost 200 million callosal axons compared to the fewer than 50 million recorded in the adult (LaMantia and Rakic, 1990). These axons are eliminated at the rate of 8 million per day, or 50 per second, during the first 3 weeks after birth, and 5 axons per second until the adult value is reached around puberty (figure 8.6). Other classes of interhemispheric connections, such as the anterior and hippocampal commissures, are overproduced and lost at a similar rate during approximately the same period (LaMantia and Rakic, 1994).

The functional significance of this large loss of axons is not fully understood, but it occurs after topographical and columnar organization has been achieved. For example, in adult rhesus monkey, area 17 does not have callosal connections and, unlike the developing cat, this region is never interconnected (Kennedy and Dehay, 1993). Likewise, callosal axons in the prefrontal cortex are relatively precisely connected in the middle of gestation, well in advance of the axonal loss phase that, in this species, occurs mostly postnatally (Schwartz and Goldman-Rakic, 1990). Therefore, axonal elimination in the cerebral cortex probably is involved mainly in synaptic remodeling at the local, rather than global, level (La Mantia and Rakic, 1990). Indeed, as discussed next, loss of interhemispheric axons coincides with the period of rapid synaptic production rather than synaptic elimination.

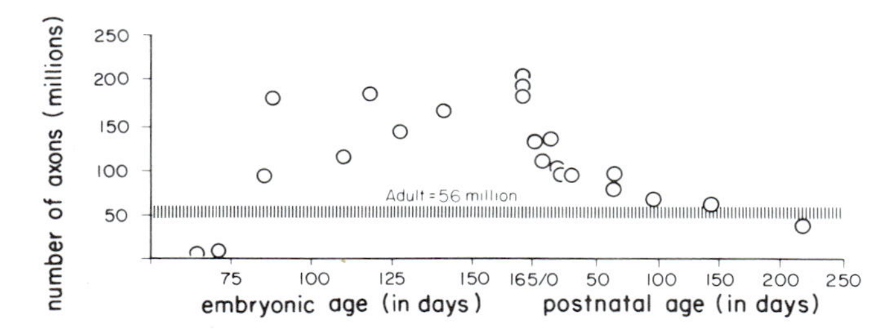

FIGURE 8.6 The estimated total number of axons in corpus callosum in the rhesus monkey as a function of prenatal and postnatal ages in postconceptual days. Each point represents a single animal. Error values (<14%) are smaller than or equal to the size of the dots used to represent each data point. The hatched line indicates the average total number of callosal axons in eight adult monkeys. (Reprinted from LaMantia and Rakic, 1990, by permission.)

Overproduction of cortical synapses

In the cerebral cortex, as in most other brain structures, growing axons eventually have to establish synaptic junctions with a specific neuron class or, even more precisely, with only a part of their body and dendritic tree. This precise pattern of connectivity is achieved through initial overproduction and subsequent elimination of a large number of synapses. For example, in the cerebral cortex of both the human and the rhesus monkey, synapses are more dense and more numerous during infancy than in adulthood (O'Kusky and Colonnier, 1982; Rakic et al., 1986; Huttenlocher and deCourten, 1987). In the newborn rhesus monkey, the density of synapses per unit volume of neuropil and their total number in most areas studied is approximately equal to the adult values. During the first 2–3 months of life, synaptic density continues to increase until it reaches approximately a 30–40% higher level than in the adult (figure 8.7). The phase of high synaptic density lasts throughout adolescence and decreases significantly during sexual maturation, which in this species occurs during the third year of life (Rakic et al., 1986; Zecevic, Bourgeois, and Rakic, 1989; Zecevic and Rakic, 1991; Bourgeois and Rakic, 1993; Bourgeois, Goldman-Rakic, and Rakic, 1994). The decline in number of synapses is due primarily to

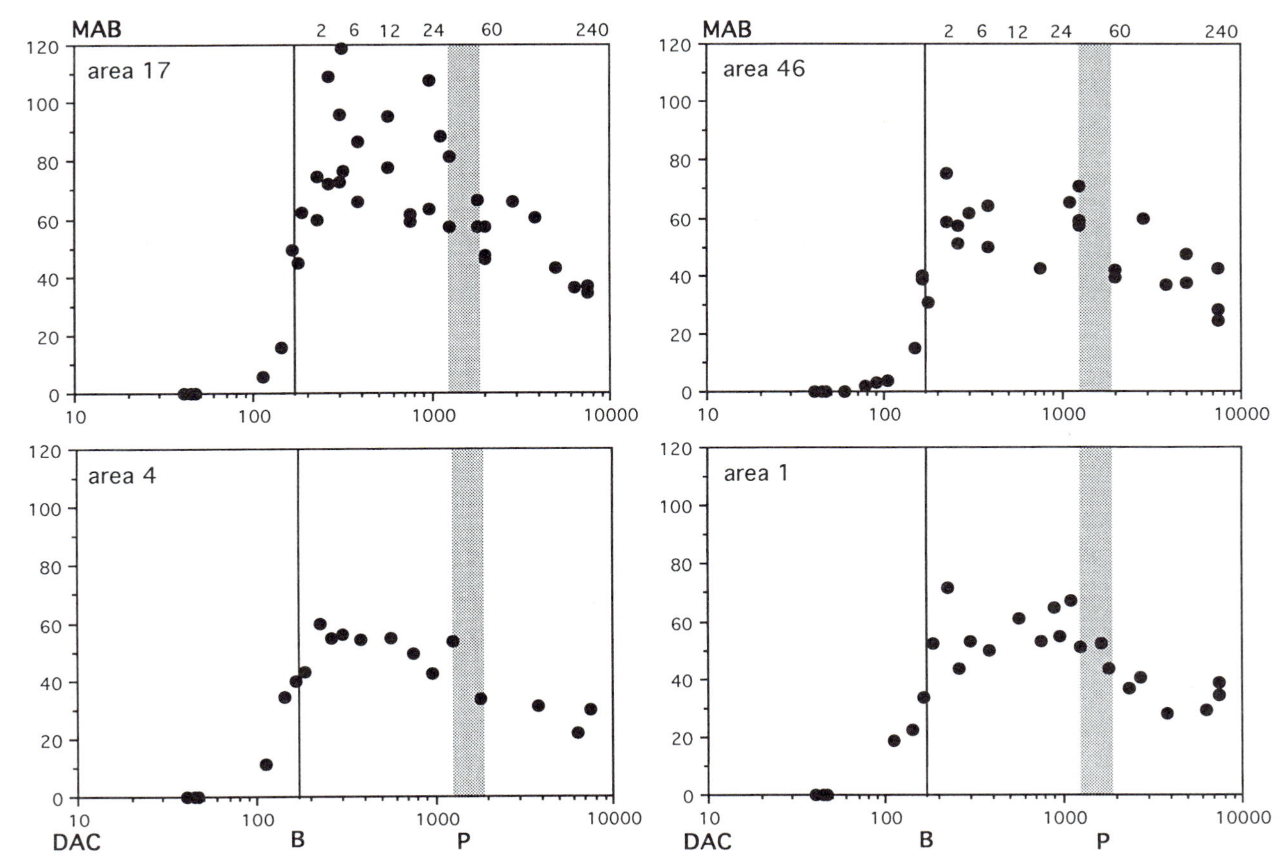

FIGURE 8.7 Histograms of the density of synapses per 100 μm^3 of neuropil in visual (area 17), prefrontal (area 46), motor (area 4), and somatosensory (area 1) cortices at various prenatal and postnatal ages. Each black circle represents the value obtained from a single electron-microscopical probe. Age is presented in conceptional days on a logarithmic scale in order to fit the entire life span of the monkey onto a single graph. Time of birth (B) and puberty (P) are indicated by dotted vertical lines. DAC, day after conception; MAB, month after birth. (For details on synaptogenesis for each cytoarchitectonic area, see the primary articles by Zecevic, Bourgeois, and Rakic, 1989; Zecevic and Rakic, 1991; Bourgeois and Rakic, 1993; Bourgeois, Goldman-Rakic, and Rakic, 1994).

the elimination of asymmetrical junctions located on dendritic spines, whereas synapses on dendritic shafts and cell bodies remain relatively constant.

The course of decline in synaptic density and their absolute number is, perhaps, best documented for macaque monkey primary visual cortex. The number of synapses lost in area 17 of a single cerebral hemisphere approaches 1.8×10^{11} (Bourgeois and Rakic, 1993). The magnitude of this loss is stunning when expressed as a loss of approximately 2500 synapses per second in area 17 of each hemisphere during a period of nearly $2-3\frac{1}{2}$ years. The density of major neurotransmitter receptors in the visual cortex also reaches a maximum level at between 2 and 4 months of age and then declines to the adult level during the period of sexual maturation (Lidow, Goldman-Rakic, and Rakic, 1991; Lidow and Rakic, 1992). This synchronized development of synapses and neurotransmitter receptors, illustrated graphically in figure 8.8, suggests unusual coordination between biochemical and structural differentiation and supports the hypothesis that these cellular events may be related to maturation of function.

An unexpected finding in this series of studies was that the "overshoot" phase of synapses and neurotransmitter receptors occurs simultaneously in visual, somatosensory, motor, and prefrontal cytoarchitectonic areas (see figure 8.8) (Rakic et al., 1986; Lidow, Goldman-Rakic, and Rakic, 1991; Lidow and Rakic, 1992). Isochronic development of anatomically and functionally diverse regions stands in contrast to the

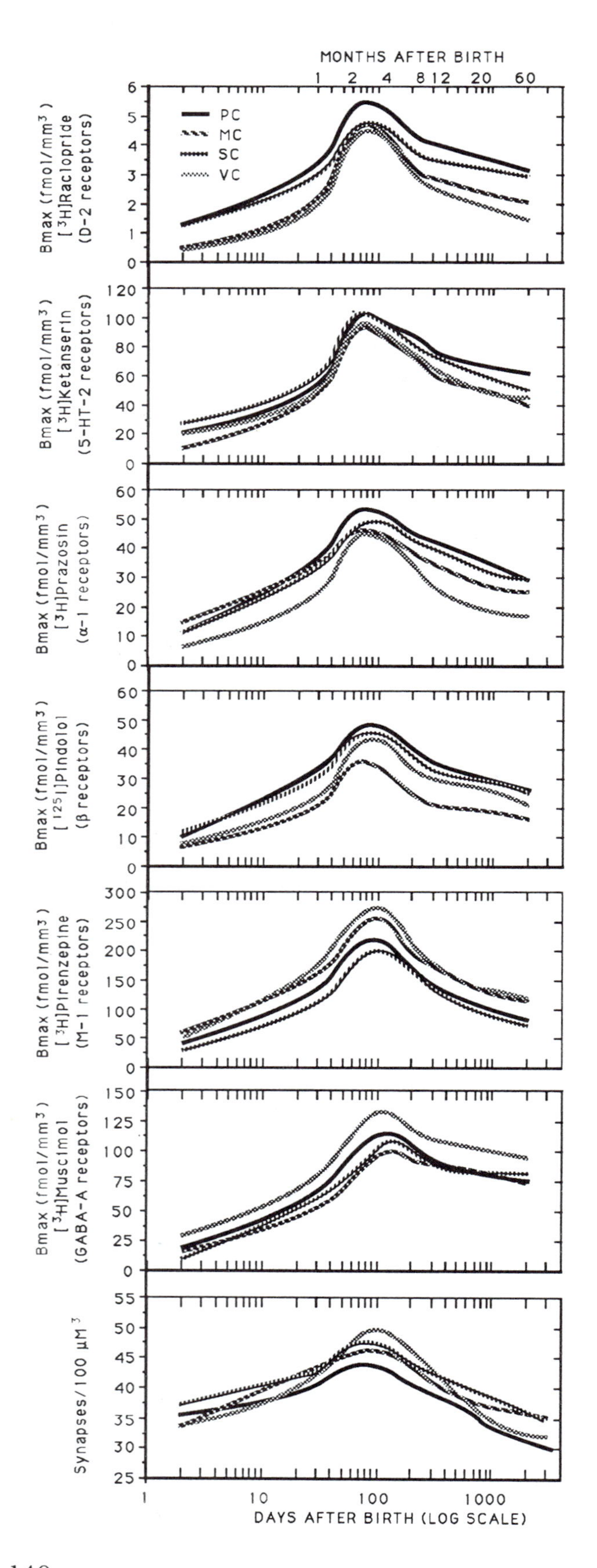

FIGURE 8.8 Developmental modifications in the overall changes (across all layers) in density of the synapses and specific binding of radioligands labeling a representative selection of neurotransmitter receptor subtypes in the prefrontal (PC), primary motor (MC), somatosensory (SC), and primary visual (VC) cortical regions of the developing rhesus monkey. For receptor densities, the lines were obtained by locally weighted least square fit with 50% smoothing based on mean B_{max} values obtained from the measurements of the entire cortical thickness in at least two animals at birth, 1, 2, 4, 8, 12, 36, and 60 months of age. Age is presented in postnatal days on a logarithmic scale. (Primary data are from Rakic et al., 1986, and Lidow, Goldman-Rakic, and Rakic, 1991.)

traditional view of hierarchical development of cortical regions (e.g., Filimonov, 1949; Yakovlev and Lecours, 1967) and indicates that the establishment of cell-to-cell communication in this structure may be orchestrated by a single genetic or humoral signal. This phase of primate life may provide a lengthy unparalleled opportunity for competitive, activity-driven stabilization among various initially overproduced intercortical and intracortical connections, which comprise the largest fraction of cortical synapses (Goldman-Rakic, 1987). Therefore, the period of supernumerary synapses in the cerebral cortex can be considered a stage of maximal opportunity and minimal commitments, providing enormous capacity for the generation of cortical diversity beyond genes.

Competitive synaptic elimination

Studies in a variety of mammalian species support the hypothesis of competitive interactions as a mechanism of attaining point-to-point connectivity during the segregation of initially more numerous and more diffuse projections (Easter et al., 1985; Edelman, 1993). The balance between overproduction and elimination of neurons and axons determines the size and site of territories devoted to a given terminal field. The first, and perhaps best, documented example of the development, which proceeds from diffuse to sharply defined terminal fields, is the primate binocular visual system (Rakic, 1976a, 1977; Hubel, Wiesel, and LeVay, 1977). This basic principle has since been shown in a variety of cortical areas and in mammalian species (Innocenti, 1981; Shatz, 1983; Easter et al., 1985; O'Leary, 1989). In most of these examples, the formation of cortical connections is achieved through dynamic cellular interactions that involve at least two well-defined steps (Rakic, 1976a, 1977). In the first step, one structure projects to its target structure but without regard for the individual nerve cell or its parts. In the second step, axonal terminals sort out and establish connections with selective sets of neurons or their dendrites. Among the cellular mechanisms involved in the sorting process, selective, activity-dependent competition, which leads to selective elimination, is perhaps the most prominent.

The theory of competitive elimination is supported by evidence drawn from the studies of monocular eye deprivation, which results in a decrease in the ocular dominance columns subserving the occluded eye and enlargement of the functional eye (Hubel, Wiesel, and LeVay, 1977). Even more dramatic changes occur when the eye in a monkey fetus is enucleated before geniculocortical fibers have entered the cortical plate (Rakic, 1981a). In such cases, the size of area 17 remains the same, but there is no trace of ocular dominance columns in the cerebral cortex. However, binocular connections fail to withdraw to appropriate territories even after blockade of electrical activity by tetrodotoxin, which prevents the inward currents of sodium ions (Dubin, Start, and Archer, 1986; Stryker and Harris, 1986). At present, it is not known whether, or to what extent, tetrodotoxin arrests the development in the overlapping phase or prevents active growth of some axons.

Competitive elimination of synapses is the focus of intense research in developmental neurobiology. In particular, studies in the mammalian visual system have provided extensive evidence that an activity-dependent competition between inputs for common sites in the target structures determines the final pattern of connections (Hubel, Wiesel, and LeVay, 1977; Shatz, 1990). Manipulation of development that affects separately presynaptic and postsynaptic activity promises to provide some ideas about the role of neurotransmitters and their receptors during development. The mechanism may, in principle, be similar to long-term potentiation that increases synaptic strength by matching presynaptic and postsynaptic activation (e.g., Nicoll, Kauer, and Malenka, 1988). Coordinated participation of cells situated on both sides of the synapse is consistent with the Heb's hypothesis of learning in the adult organism. In this context, developmental neurobiology and cognitive neuroscience are approaching a possible common territory. Most scientists would agree that a qualitative or quantitative change in synaptoarchitecture during development and learning must involve neurotransmitters, second messengers, and differential gene expression. These changes may arise by genomic rearrangement or by several other mechanisms that include third messengers, but the activity-dependent regulation of development seems to occur at the level of gene transcription. Nevertheless, a large gap still remains between understanding the behavioral phenomena and understanding the underlying cellular and molecular events. Our basic goal should be to find out how innately prespecified synaptic connections adjust their structural and functional properties in response to spontaneous and environmentally induced functional activity.

REFERENCES

Aicardi, J., 1991. The agyria-pachigyria complex: A spectrum of cortical malformations. *Brain Dev.* 13:1–8.

Algan, O., P. S. Goldman-Rakic, and P. Rakic, 1992. Selective laminar and areal deletion of cortical neurons by prenatal X-irradiation in the macaque monkey. *Abstr. Soc. Neurosci.* 17:1446.

Angevine, J. B., Jr., and R. L. Sidman, 1961. Autoradiographic study of cell migration during histogenesis of cerebral cortex in the mouse. *Neuron* 192:766–768.

Arimatsu, Y., M. Miyamoto, I. Nihonmatsu, K. Hirata, Y. Urataini, Y. Hatanka, and K. Takiguchi-Hoyashi, 1992. Early regional specification for a molecular neuronal phenotype in the rat neocortex. *Proc. Natl. Acad. Sci. U.S.A* 89:8879–8883.

Barbe, M. F., and P. Levitt, 1992. Attraction of specific thalamic input by cerebral grafts depends on the molecular identity of the implant. *Proc. Natl. Acad. Sci. U.S.A.* 89: 3706–3710.

Barth, T. M., and B. B. Stainfield, 1994. Homotopic, but not heterotopic, fetal cortical transplants can result in functional sparing following neonatal damage to the frontal cortex in rat. *Cerebral Cortex*, in press.

Boulder Committee, 1970. Embryonic vertebrate central nervous system: Revised terminology. *Anat. Rec.* 166:257–261.

Bourgeois, J.-P., P. S. Goldman-Rakic, and P. Rakic (in press). Synaptogenesis in the prefrontal cortex of rhesus monkey. *Cerebral Cortex.* 4:78–96.

Bourgeois, J.-P., and P. Rakic, 1987. Distribution, density and ultrastructure of synapses in the visual cortex in monkeys devoid of retinal input from early embryonic stages. *Abstr. Soc. Neurosci.* 13:1044.

Bourgeois, J.-P., and P. Rakic, 1993. Changing of synaptic density in the primary visual cortex of the rhesus monkey from fetal to adult stage. *J. Neurosci.* 13:2801–2820.

Brodmann, K., 1909. *Vergleichende Lokalisationslehre der Grosshirninde*. Leipzig: Brath.

Bugbee, N. M., and P. S. Goldman-Rakic, 1983. Columnar organization of cortico-cortical projections in squirrel and rhesus monkeys: Similarity of column width in species differing in cortical volume. *J. Comp. Neurol.* 220:355–364.

Cameron, R. S., and P. Rakic, 1991. Glial cell lineage in the cerebral cortex: Review and synthesis. *GLIA* 4:124–137.

Cameron, R. S., and P. Rakic, 1994. Identification of membrane proteins that comprise the plasmalemmal junction between migrating neurons and radial glial cells. *J. Neurosci.* 14:3139–3155.

Caviness, V. S., Jr., and P. Rakic, 1978. Mechanisms of cortical development: A view from mutations in mice. *Annu. Rev. Neurosci.* 1:297–326.

Choi, B. H., 1986. Glial fibrillary acidic protein in radial glial cells of early human fetal cerebrum: A light and electronmicroscopic study. *J. Neuropathol. Exp. Neurol.* 45:408–418, 1986.

Cohen-Tannoudji, M., C. Babinet, and M. Wassef, 1994. *Early intrinsic regional specification of the mouse somatosensory cortex*. Manuscript submitted for publication.

Conel, J. L., 1939. *The Postnatal Development of the Human Cerebral Cortex: I. The Cortex of the Newborn*. Cambridge, Mass.: Harvard University Press.

Cowan, W. M., 1973. Neuronal death as regulative mechanism in the control of cell number in the neuronal system. In *Development and Aging in the Nervous System*, M. Rockstein, ed. New York: Academic Press, pp. 19–41.

Creutzfeldt, O. D., 1977. Generality of the functional structure of the neocortex. *Naturwissenschaften* 64:507–517.

Dubin, M., L. A. Start, and S. M. Archer, 1986. A role for action-potential activity in the development of neuronal connections in the kitten retinogeniculate pathway. *J. Neurosci.* 6:1021–1036.

Easter, S., Jr., D. Purves, P. Rakic, and N. C. Spitzer, 1985. The changing view of neural specificity. *Science* 230: 507–511.

Edelman, G. M., 1983. Cell adhesion molecules. *Science* 219: 450–457.

Edelman, G. M., 1993. Neural Darwinism: Selection and reentrant signaling in higher brain function. *Neuron* 10: 115–125.

Ferri, R. T., and P. Levitt, 1993. Cerebral cortical progenitors are fated to produce region-specific neuronal populations. *Cerebral Cortex* 3:187–198.

Filimonov, I. N., 1949. Cortical cytoarchitecture—general concepts. Classification of the architectonic formations. In *Cytoarchitecture of the Cerebral Cortex in Man,* S. A. Sarkisov, I. N. Filimov, and N. S. Preobrazenskaya, eds. Moscow: Medgiz, pp. 11–32.

Galaburda, A. M., G. F. Sherman, G. D. Rosen, F. Aboitiz, and N. Geschwind, 1985. Developmental dyslexia: Four consecutive patients with cortical abnormalities. *Ann. Neurol.* 18:222–223.

Goldman, P. S., and W. J. H. Nauta, 1977. Columnar organization of association and motor cortex: Autoradiographic evidence for cortico-cortical and commissural columns in the frontal lobe of the newborn rhesus monkey. *Brain Res.* 122:369–385.

Goldman-Rakic, P. S., 1981. Development and plasticity of primate frontal association cortex. In *The Organization of the Cerebral Cortex*, F. O. Schmitt, F. G. Worden, S. G. Dennis, and G. Adelman, eds. Cambridge, Mass.: MIT Press, pp. 69–97.

Goldman-Rakic, P. S., 1987. Development of cortical circuitry and cognitive functions. *Child Dev.* 58:642–691.

Goldman-Rakic, P. S., and P. Rakic, 1984. Experimental modification of gyral patterns. In *Cerebral Dominance: The Biological Foundation*, N. Geschwind and A. M. Galaburda, eds. Cambridge, Mass.: Harvard University Press, pp. 179–192.

Gosh, A., A. Antonini, S. K. McConnell, and C. J. Shatz, 1990. Requirement of subplate neurons in the formation of thalamocortical connections. *Nature* 347:179–181.

Gould, S. J., 1977. *Ontogeny and Phylogeny*. Cambridge, Mass.: Belknap Press.

Hatten, M. E., 1990. Riding the glial monorail: A common mechanism for glial-guided migration in different regions of the developing brain. *Trends Neurosci.* 13:179–184.

Hatten, M. E., and C. A. Mason, 1986. Neuron-astroglia interactions in vitro and in vivo. *Trends Neurosci.* 9:168–174.

Hatten, M. E., and C. A. Mason, 1990. Mechanisms of glial-guided neuronal migration in vitro and in vivo. *Experientia* 46:907–916.

His, W., 1904. *Die Entwicklung des Monschlichen Gehirns Wahrend der Ersten Monate*. Leipzig: Hirzel.

Hubel, D. H., and T. N. Wiesel, 1972. Laminar and columnar distribution of geniculo-cortical fibers in the macaque monkey. *J. Comp. Neurol.* 146:421–450.

Hubel, D. H., T. N. Wiesel, and S. LeVay, 1977. Plasticity of ocular dominance columns in monkey striate cortex. *Philos. Trans. R. Soc. Lond.* 278:377–409.

Huttenlocher, P. R., and C. deCourten, 1987. The development of synapses in striate cortex of man. *Hum. Neurobiol.* 6:1–9.

Innocenti, G. M., 1981. Growth and reshaping of axons in the establishment of visual connections. *Science* 212:824–827.

Jensen, K. F., and H. P. Killackey, 1984. Subcortical projections from ectopic neocortical neurons. *Proc. Natl. Acad. Sci. U.S.A.* 81:964–968.

Kaiserman-Abramoff, I. R., A. M. Graybiel, and W. J. H. Nauta, 1980. The thalamic projection to cortical area 17 in a congenitally anophthalmic mouse strain. *Neuroscience* 5:41–52.

Kennedy, H., and C. Dehay, 1993. Cortical specification of mice and men. *Cerebral Cortex* 3:171–186.

Killackey, H. P., 1989. Neocortical expansion: An attempt toward relating phylogeny and ontogeny. *J. Cogn. Neurosci.* 2:1–17.

Komuro, H., and P. Rakic, 1992. Selective role of N-type calcium channels in neuronal migration. *Science* 257:806–809.

Komuro, H., and P. Rakic, 1993. Modulation of neuronal

migration by NMDA receptors. *Science* 260:95–97.

Kostovic, I., and P. Rakic, 1980. Cytology and time of origin of interstitial neurons in the white matter in infant and adult human and monkey telencephalon. *J. Neurocytol.* 9:219–242.

Kostovic, I., and P. Rakic, 1990. Developmental history of transient subplate zone in the visual and somatosensory cortex of the macaque monkey and human brain. *J. Comp. Neurol.* 297:441–470.

Kuljis, R. O., and P. Rakic, 1990. Hypercolumns in primate visual cortex develop in the absence of cues from photoreceptors. *Proc. Natl. Acad. Sci. U.S.A.* 87:5303–5306.

LaMantia, A. S., and P. Rakic, 1990. Axon overproduction and elimination in the corpus callosum of the developing rhesus monkey. *J. Neurosci.* 291:520–537.

LaMantia, A. S., and P. Rakic, 1994. Axon overproduction and elimination in the anterior commissure of the developing rhesus monkey. *J. Comp. Neurol.* 340:328–336.

Levitt, P., M. L. Cooper, and P. Rakic, 1981. Coexistence of neuronal and glial precursor cells in the cerebral ventricular zone of the fetal monkey: An ultrastructural immunoperoxidase analysis. *J. Neurosci.* 1:27–39.

Levitt, P., M. L. Cooper, and P. Rakic, 1983. Early divergence and changing proportions of neuronal and glial precursor cells in the primate cerebral ventricular zone. *Dev. Biol.* 96:472–484.

Lidow, M., P. S. Goldman-Rakic, and P. Rakic, 1991. Synchronized overproduction of neurotransmitter receptors in diverse regions of the primate cerebral cortex. *Proc. Natl. Acad. Sci. U.S.A.* 88:10218–10221.

Lidow, M. S., and P. Rakic, 1992. Scheduling of monoaminergic neurotransmitter receptor expression in the primate neocortex during postnatal development. *Cerebral Cortex* 2:401–416.

Livingstone, M. S., and D. H. Hubel, 1988. Segregation of form, color, movement, and depth: Anatomy, physiology and perception. *Science* 240:740–749.

Lu, S., L. D. Bogard, M. T. Murtha, and F. Rudle, 1992. The expression pattern of a new murine homeobox gene Dbx displays extreme spatial restriction in embryonic forebrain and spinal cord. *Proc. Natl. Acad. Sci. U.S.A.* 89:8053–8057.

Luskin, M. B., A. L. Pearlman, and J. R. Sanes, 1988. Cell lineage in the cerebral cortex of the mouse studied *in vivo* and *in vitro* with a recombinant retrovirus. *Neuron* 1:635–647.

Luskin, M. B., and C. J. Shatz, 1985. Studies of the earliest generated cells of the cat's visual cortex: Cogeneration of subplate and marginal zones. *J. Neurosci.* 5:1062–1075.

McConnell, S. K., 1988. Development and decision-making in the mammalian cerebral cortex. *Brain Res. Rev.* 13:1–23.

McConnell, S. K., 1990. The specificity of neuronal identity in the mammalian cerebral cortex. *Experientia* 46:892–929.

Merzenich, M. M., G. Recanzone, W. M. Jenkins, T. T. Allard, and R. J. Nudo, 1988. Cortical representational plasticity. In *Neurobiology of the Neocortex*, P. Rakic and W. Singer, eds. New York: Wiley.

Mountcastle, V. B., 1957. Modality and topographic properties of single neurons of cat's somatic sensory cortex. *J. Neurophysiol.* 20:408–434.

Mountcastle, V. B., 1979. An organizing principle for cerebral function: The unit module and the distributed system. In *The Neurosciences: Fourth Study Program*, F. O. Schmitt and F. G. Worden, eds. Cambridge, Mass.: MIT Press, pp. 21–42.

Nakatsuji, M., Y. Kadokowa, and H. Sumori, 1991. Radial columnar patches in the chimeric cerebral cortex visualized by use of mouse embryonic stem cells expressing β-galactooxidase. *Dev. Growth Differentiation* 33:571–578.

Nicoll, R. A., J. A. Kauer, and R. C. Malenka, 1988. The current excitement in long-term potentiation. *Neuron* 1:97–103.

O'Kusky, J., and M. Colonnier, 1982. Postnatal changes in the number of neurons and synapses in the visual cortex (area 17) of the *Macaca* monkey: A stereological analysis in normal and monocularly deprived animals. *J. Comp. Neurol.* 210:291–306.

O'Leary, D. D. M., 1989. Do cortical areas emerge from a protocortex? *Trends Neurosci.* 12:400–406.

Olivaria, J., and R. C. Van Sluyters, 1984. Callosal connections of the posterior neocortex in normal-eyed, congenitally anophthalmic and neonatally enucleated mice. *J. Comp. Neurol.* 230:249–268.

O'Rourke, N. A., M. E. Dailey, S. J. Smith, and S. K. McConnell, 1992. Diverse migratory pathways in the developing cerebral cortex. *Science* 258:299–302.

Palamini, A., F. Anderman, J. Aicardi, O. Dulac, F. Chavis, G. Ponsot, J. M. Pinard, J. Goutieres Livingston, D. Tampieri, E. Anderman, and Y. Robitaille, 1991a. Diffuse cortical dysplasia, or the "double cortex" syndrome. *Neurology* 41:1656–1662.

Palamini, A., F. Anderman, A. Olier, D. Tampieri, and Y. Robitaille, 1991b. Focal neuronal migration disorders and intractable partial epilepsy: Results of surgical treatment. *Ann. Neurol.* 30:750–757.

Parnavelas, J. G., J. A. Barfield, E. Franke, and M. B. Luskin, 1991. Separate progenitor cells give rise to pyramidal and nonpyramidal neurons in the rat telencephalon. *Cerebral Cortex* 1:463–491.

Proteus, M. H., E. J. Brice, A. Buffone, T. B. Usdin, R. D. Ciaranello, and J. R. Rubinstein, 1991. Isolation and characterization of a library of cDNA clones that are preferentially expressed in the embryonic telencephalon. *Neuron* 7:221.

Purves, D., 1988. *Body and Brain. A Trophic Theory of Neural Connections*. Cambridge, Mass.: MIT Press.

Rakic, P., 1972. Mode of cell migration to the superficial layers of fetal monkey neocortex. *J. Comp. Neurol.* 145:61–84.

Rakic, P., 1974. Neurons in the monkey visual cortex: Systematic relation between time of origin and eventual disposition. *Science* 183:425–427.

Rakic, P., 1975. Timing of major ontogenetic events in the

visual cortex of the rhesus monkey. In *Brain Mechanisms in Mental Retardation*, N. A. Buchwald and N. Brazier, eds. New York: Academic Press, pp. 3–40.

Rakic, P., 1976a. Prenatal genesis of connections subserving ocular dominance in the rhesus monkey. *Nature* 261:467–471.

Rakic, P., 1976b. Differences in the time of origin and in eventual distribution of neurons in areas 17 and 18 of the visual cortex in the rhesus monkey. *Exp. Brain Res.* Suppl. 1:244–248.

Rakic, P., 1977. Prenatal development of the visual system in the rhesus monkey. *Philos. Trans. R. Soc. Lond. [Biol.]* 278:245–260.

Rakic, P., 1978. Neuronal migration and contact guidance in primate telencephalon. *Postgrad. Med. J.* 54:25–40.

Rakic, P., 1981a. Development of visual centers in the primate brain depends on binocular competition before birth. *Science* 214:928–931.

Rakic, P., 1981b. Neuron-glial interaction during brain development. *Trends Neurosci.* 4:184–187.

Rakic, P., 1982. Early developmental events: Cell lineages, acquisitions of neuronal positions, and areal and laminar development. *Neurosci. Res. Prog. Bull.* 20:439–445.

Rakic, P., 1985. Limits of neurogenesis in primates. *Science* 227:154–156.

Rakic, P., 1988a. Defects of neuronal migration and pathogenesis of cortical malformations. *Prog. Brain Res.* 73:15–37.

Rakic, P., 1988b. Specification of cerebral cortical areas. *Science* 241:170–176.

Rakic, P., 1990. Principles of neuronal cell migration. *Experientia* 46:882–891.

Rakic, P., 1991. Radial unit hypothesis of cerebral cortical evolution. *Exp. Brain Res.* Suppl. 21:25–43.

Rakic, P., 1994. *Cortical dialectics: Does parcellation of the cortex require a protomap?* Manuscript submitted for publication.

Rakic, P., J.-P. Bourgeois, M. E. Eckenhoff, N. Zecevic, and P. S. Goldman-Rakic, 1986. Concurrent overproduction of synapses in diverse regions of the primate cerebral cortex. *Science* 232:232–235.

Rakic, P., R. S. Cameron, and H. Komuro (in press). Recognition, adhesion and transmembrane signaling in guided neuronal cell migration. *Curr. Opin. Neurobiol.* 4: 63–69.

Rakic, P., and M. Lidow, 1994. *Distribution and density of neurotransmitter receptors in the absence of retinal input from early embryonic stages.* Manuscript submitted for publication.

Rakic, P., and K. P. Riley, 1983. Regulation of axon numbers in the primate optic nerve by prenatal binocular competition. *Nature* 305:135–137.

Rakic, P., and R. L. Sidman, 1968. Supravital DNA synthesis in the developing human and mouse brain. *J. Neuropathol. Exp. Neurol.* 27:246–276.

Rakic, P., L. J. Stensaas, E. P. Sayre, and R. L. Sidman, 1974. Computer-aided three-dimensional reconstruction and quantitative analysis of cells from serial electron-microscopic montages of fetal monkey brain. *Nature* 250: 31–34.

Rakic, P., I. Suner, and R. W. Williams, 1991. A novel cytoarchitectonic area induced experimentally within the primate visual cortex. *Proc. Natl. Acad. Sci. U.S.A.* 88: 2083–2087.

Rockel, A. J., R. W. Hirous, and T. P. S. Powell, 1980. The basic uniformity in structure of the neocortex. *BRAIN* 103:221–244.

Schachner, M., A. Faissner, G. Fischer, G. Keilhauer, J. Kruse, V. Kunemund, J. Lindner, and H. Wernecke, 1985. Functional and structural aspects of the cell surface in mammalian nervous system development. In *The Cell in Contact: Adhesions and Junctions as Morphogenetic Determinants*, G. M. Edelman, W. E. Gall, and J. P. Thiery, eds. New York: Wiley.

Schlaggar, B. L., and D. D. M. O'Leary, 1992. Potential of visual cortex to develop an array of functional units unique to somatosensory cortex. *Science* 252:1556–1560.

Schmechel, D. E., and P. Rakic, 1979. Arrested proliferation of radial glial cells during midgestation in rhesus monkey. *Nature* 227:303–305.

Schwartz, M. L., and P. S. Goldman-Rakic, 1990. Early specification of callosal connections in the fetal monkey prefrontal cortex. *J. Comp. Neurol.* 307:144–162.

Schwartz, M. L., P. Rakic, and P. S. Goldman-Rakic, 1991. Early phenotype expression of cortical neurons: Evidence that a subclass of migrating neurons have callosal axons. *Proc. Natl. Acad. Sci. U.S.A.* 88:1354–1358.

Shatz, C. J., 1983. The prenatal development of the cat's retinogeniculate pathway. *J. Neurosci.* 3:492–499.

Shatz, C. J., 1990. Input activity and the patterning of connections during CNS development. *Neuron* 5:745–756.

Shatz, C. J., J. J. M. Chun, and M. B. Luskin, 1988. The role of the subplate in the development of mammalian telencephalon. In *Cerebral Cortex*, A. Peters and E. G. Jones, eds. New York: Plenum, pp. 35–38.

Sidman, R. L., and P. Rakic, 1973. Neuronal migration with special reference to developing human brain: A review. *Brain Res.* 62:1–35.

Sidman, R. L., and P. Rakic, 1982. Development of the human central nervous system. In *Histology and Histopathology of the Nervous System*, W. Haymaker and R. D. Adams, eds. Springfield, Ill.: Charles C Thomas, pp. 3–145.

Stryker, M. P., and W. A. Harris, 1986. Binocular impulse blockade prevents the formation of ocular dominance columns in cat visual cortex. *J. Neurosci.* 6:2117–2133.

Suner, I., and P. Rakic, 1994a. Correlation of the neuronal numbers in the lateral geniculate nucleus and primary visual cortex. Manuscript submitted for publication.

Suner, I., and P. Rakic, 1994b. Effects of prenatal reduction of geniculocortical input on specification of primary visual cortex. Manuscript submitted for publication.

Szentagothai, J., 1978. The neuronal network of the cerebral cortex: A functional interpretation. *Prog. Brain Res.* 201:219–248.

Takahashi, T., R. S. Nowakowski, and V. S. Caviness, Jr. (in press). Mode of cell proliferation in the developing mouse neocortex. *Proc. Natl. Acad. Sci. U.S.A.*

Tan, S. S., and S. Breen, 1993. Radial mosaicism and tangential cell dispersion both contribute to mouse neocortical development. *Nature* 363:338–640.

Van der Loos, H., and J. Dorfl, 1978. Does the skin tell the somatosensory cortex how to construct a map of the periphery? *Neuroscience Newsletter* 7:23–30.

Volpe, J. J., 1987. *Neurology of the Newborn, ed 2.* Philadelphia: Saunders.

Walsh, C., and C. L. Cepko, 1992. Widespread dispersion of neuronal clones across functional regions of the cerebral cortex. *Science* 255:434–440.

Williams, R. W., and P. Rakic, 1988. Elimination of neurons in the rhesus monkey's lateral geniculate nucleus during development. *J. Comp. Neurol.* 272:424–436.

Williams, R. W., K. Ryder, and P. Rakic, 1987. Emergence of cytoarchitectonic differences between areas 17 and 18 in the developing rhesus monkey. *Abstr. Soc. Neurosci.* 13:1044.

Yakovlev, P. I., and A. R. Lecours, 1967. The myelogenetic cycles of regional maturation of the brain. In *Regional Development of the Brain in Early Life*, A. Minkowski ed. Oxford: Blackwell, pp. 3–70.

Yurkewicz, L., K. L. Valentino, M. K. Floeter, J. W. Fleshman, and E. G. Jones, 1984. Effect of cytotoxic deletions of somatic sensory cortex in fetal rats. *Somatosens. Mot. Res.* 1:303–327.

Yuste, R., and L. Katz, 1991. Control of postsynaptic Ca^{++} influx in developing neocortex by excitatory and inhibitory neurotransmitters. *Neuron* 6:333–344.

Zecevic, N., J.-P. Bourgeois, and P. Rakic, 1989. Synaptic density in motor cortex of rhesus monkey during fetal and postnatal life. *Dev. Brain Res.* 50:11–32.

Zecevic, N., and P. Rakic, 1991. Synaptogenesis in monkey somatosensory cortex. *Cerebral Cortex* 1:510–523.

9 Experimental Approaches that Reveal Principles of Cerebral Cortical Development

PAT LEVITT

ABSTRACT Structure-function relationships in the cerebral cortex have their foundation in the initial formation of specific areas and circuits during development. The events that underlie cortical formation, as in all other brain regions, include cell proliferation, phenotype differentiation, cell migration, and circuit formation. Novel experimental approaches are revealing new concepts of the molecular and cellular basis for each of these ontogenetic events. Immunocytochemistry, retrovirus lineage markers, cell and tissue transplantation, and tissue culture approaches are among the strategies used to determine the timing of cellular commitment to specific phenotypes and the ways in which early decisions in development may affect the final organization of the cerebral cortex. The composite results indicate that early genetic mechanisms play an important role in formation of the differentiated cerebral cortex but rely heavily on environmental interactions to guide each of the specific histogenic events.

The mammalian cerebral cortex may be the most complex of brain areas, particularly in terms of its structural characteristics that reflect the higher-order functions which are subserved by the circuits of this region of the central nervous system (CNS). The cerebral cortex is a layered structure formed by collections of neurons and glial cells. Layers contain characteristic sets of neurons, which produce specific types of efferents and receive specific classes of afferents. These vertical ensembles are organized into domains tangentially. The domains reflect the outcome of a complex developmental process that results in two basic features, distinguishing each of the areas: (1) well-defined laminar patterns that vary between areas, based on the relative number of specific cell classes that accumulate in defined layers, and (2) specific afferents (arising both cortically and subcortically) and efferents that define functional circuits. What is most remarkable about the cerebral cortex, when comparing changes from lissencephalic to gyrencephalic animals, is the dramatic expansion in size (1200 cm^2 per hemisphere in humans versus 4 cm^2 per hemisphere in mice; Kaas, 1993), which is accompanied by increased specificity in parcellation of functional domains. The initial conceptualization of the theory of architectonics as it relates to cortical divisions was detailed by Brodmann (1909) and has withstood the test of time, but a longstanding problem remains: How is such regionalization of the cortical sheet, as simple as 10 areas in the hedgehog to more than 100 areas in humans, generated ontogenetically in any mammalian species (figure 9.1)? The mechanisms underlying areal specialization are of major import in eventually being able to link structural and functional maturation.

PAT LEVITT Department of Neuroscience and Cell Biology, University of Medicine and Dentistry of New Jersey, Robert Wood Johnson Medical School, Piscataway, N.J.

Origins of pattern in the brain

The CNS arises from a pseudostratified neuroepithelium of ectodermal origin that forms the wall of the neural tube. The cells of the neuroepithelium, know as *germinal, precursor*, or *progenitor cells*, give rise to all the neurons and macroglia (astrocytes and oligodendrocytes) of the brain; microglia are monocytic in origin. One hallmark of this epithelium is the consistent finding that cells must leave the cell cycle to differentiate into neurons. This exit is accompanied by extensive cellular movement from the region of cell proliferation lining the lumen of the tube, the ventricular zone, to areas of final residence in the developing brain. There is early regionalization of the neural tube, indicated most profoundly by specific expression of genes en-

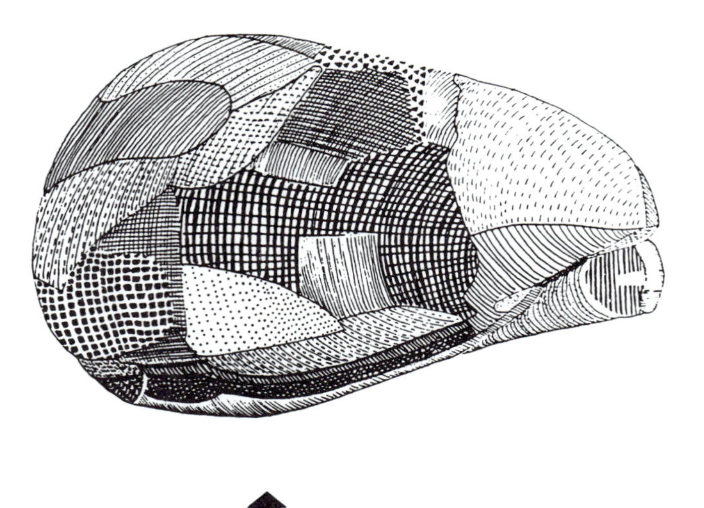

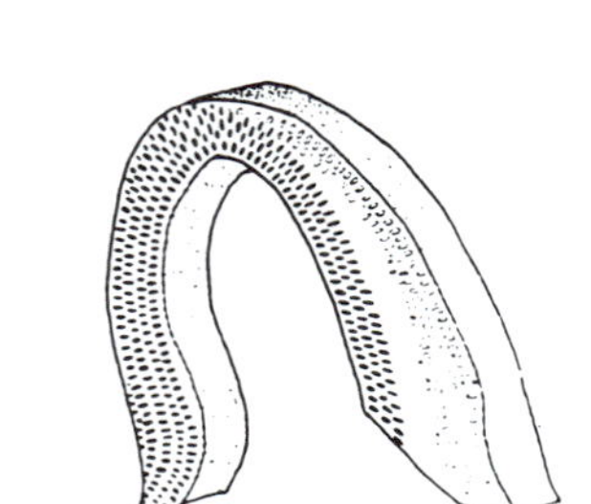

FIGURE 9.1 Schematic drawing depicting the basic developmental problem of producing a highly complex adult brain structure from an apparently simple embryonic structure. The neuroepithelium of the rodent cerebral cortex is depicted on the bottom of this figure, composed of precursor cells (black circles) lining the ventricular surface and, on differentiation, forms the layered cortex (open circles). The precursor cells appear histologically identical, yet they eventually give rise to a highly complex cerebral cortex (top of figure) that exhibits structural and functional diversity in distinct areas. These are depicted by the different drawn patterns on the hemisphere. (This figure combines a modified drawing from Smart and McSherry, 1982, with one from Krieg, 1946.)

coding transcription factors, the homeobox genes. The invertebrate countparts of these genes control characteristics of the segmented body plan. In vertebrates, the patterned expression of the homeobox genes corresponds to early segmental organization of the diencephalon through spinal cord (reviewed by Wilkinson and Krumlauf, 1990; Graham, 1992). There is an early commitment of the precursor cells to give rise to specific CNS segments, shown most convincingly by use of tissue transplants in the chick, in which a piece of neuroepithelium is placed into inappropriate (heterotopic) regions of a host neural tube: Even at extremely early embryonic ages, such experiments result in gene expression and phenotypical differentiation of the transplanted tissue that reflects its origin and not its new location (Lumsden et al., unpublished data).

This early regionalization of the neuraxis underscores the segregation of progenitors into discrete domains that correspond to major areas of the brain, such as the diencephalon, mesencephalon, pons, cerebellum, and medulla. In the brain stem, Lumsden and colleagues (Lumsden and Keynes, 1989; Clarke and Lumsden, 1993) have shown that early segmental restrictions also translate into limited, predictable patterns of neuronal differentiation. The early commitment to produce specific neuronal classes indicates that the anatomical organization of the mature brain stem into particular nuclear patterns may be specified, at least in part, through lineage-based control of cell production during early embryogenesis. This would mean that the mitotically active cells in the middle and caudal areas of the neural tube, almost from the time that the plate closes to form the tube, are predetermined to produce specific types of neurons.

Embryonic origin of the cerebral cortex

The cells giving rise to the cerebral cortex lie in the dorsalmost aspect of the rostral neural tube, arising as a secondary, telencephalic bulge rostral to the lamina terminalis. This region of the neural tube is somewhat unique in that, thus far, there appears to be a striking lack of *patterned* homeobox gene expression.

At the inception of the neural tube, the nuclei of cells stretch across the thickness of the wall to form the ventricular zone (this and subsequent terminology according to the Boulder Committee, 1970). The processes of these cells extend to the pia mater, forming a cell-free zone, the marginal zone. Initially, all cells are active in the cell cycle, proliferating extensively. Each cell undergoes a series of stereotyped cytokinetic movements during the cell cycle, with the position of the cell's nucleus in the wall reflecting the specific phase of the cycle. The nuclei lying closest to the ventricular surface undergo mitosis, whereas those farthest from that surface are in S phase (Sauer, 1935; Sauer and Walker, 1959). It is unclear how cells know when to exit the cycle, but this information is of obvious importance for defining more clearly the mechanisms that underlie the control of neuronal and glial production.

The time of exit of a cell from the cycle has been defined as the *birthdate* or *time of origin* of the neuron, representing the onset of phenotypical differentiation. Both neurons and glia have defined birthdates, although glial cells, by entering phase G_0, maintain an ability to divide under certain physiological conditions. Neurons apparently do not have this capacity.

Neurons produced from different regions of the telencephalic ventricular zone arise, or are born, at distinct times. In gyrencephalic animals, temporal patterns of neurogenesis are readily defined and extend over protracted periods of development (for reviews, see Rakic, this volume; Sidman and Rakic, 1973; Rakic, 1977; Jacobson, 1991). This is not the case in lissencephalic vertebrates, such as rats, where such temporal gradients exist but are less obvious (see Bayer and Altman, 1991, for a complete review). Thus, as a comparison, all the neurons of the rodent cerebral cortex are generated in a 1-week period between embryonic days (E) 13 and 21, representing the last third of gestation, whereas in the rhesus monkey, cortical neurogenesis begins at E40 and extends to E105, representing the middle third of gestation (Rakic, 1974). The time of origin has served as an accurate method for following specific populations of neurons through their development. The ability to mark neurons on their birthday can be accomplished with the radiolabeled base thymidine (^{3}H-thymidine) (for examples, see Sidman, Miale, and Feder, 1959; Sidman, 1970; Rakic, 1974; Bayer and Altman, 1991) or more recently, with the base analog bromodeoxyuridine (Miller and Nowakowski, 1988; Takahaski, Nowakowski, and Caviness, 1993). Cells incorporate the analogs in S phase. Those that subsequently exit the cycle will remain labeled heavily and thus are duly noted as having been generated on the day of the analog administration. Cells continuing to divide will dilute the label to undetectable levels. Permanent incorporation of label allows one to document the site of origin and movement of neurons to their final destinations in the cerebral cortex. This method allowed investigators to determine how the production of neurons in the cerebral cortex related to the unique laminar features of this brain area (Angevine and Sidman, 1961; Hicks and D'Amato, 1968; Rakic, 1974; Smart and McSherry, 1982; Bayer and Altman, 1991). In all mammals, neurons that will take up residence in the deep layers of the cerebral cortex are born before those destined for the more superficial layers. The temporal differences in neurogenesis are maintained during the period of cell migration, with the first-generated neurons settling in their final position prior to subsequent generations of neurons. This means that the later-generated neurons must move past the older neurons to reach their final position, a mechanism that remains a mystery.

As neurons are produced and begin to migrate, the cerebral wall takes on a new appearance (figure 9.2). The first neurons that are produced form a structure called the *preplate*, a layer of cells that eventually will be split by the neurons destined for the cerebral cortex proper, which settle to form the cortical plate. The cortex proper is formed as lamination begins in the cortical plate. The preplate neurons end up below the cortical plate in a zone known as the *subplate* and in the marginal zone apposed to the pial surface, serving as the forerunner of layer I. The preplate may have very special developmental functions (see discussion later).

Cell migration in the CNS is obviously a key step in the production of the vertical specificity that highlights the laminated cerebral cortex. In all species examined, including primates, the movement of postmitotic neurons occurs along well-defined routes, organized by specialized, early developing nonneuronal cells known as *radial glia*. This specialized epithelial cell was first noted by Ramón y Cajal (1890) and was believed to form a support, or scaffold, for the rather fragile developing CNS. It was not until the late 1960s that the radial epithelial, or glial, cell was rediscovered as a potentially significant element in brain formation (Astrom, 1967; Mugnaini and Forstronen, 1967). These and subsequent anatomical analyses using conventional methods such as electron microscopy and silver impregnation revealed a remarkable arcade of radially aligned processes emanating from the glial cell bodies that sit along the ventricular zone and extend through the cerebral wall to terminate as glial end-feet along the pial surface. The breakthrough in understanding this scaffold occurred when Rakic (1972, 1978) was able to document close associations between migrating neurons and the radial glial processes, strongly suggesting that the glial fibers served as a temporary scaffold for cells moving to their final destination. Following the extensive migration of cells along these paths to reach the cortical plate, the radial glial cells in the cerebral hemispheres undergo transforma-

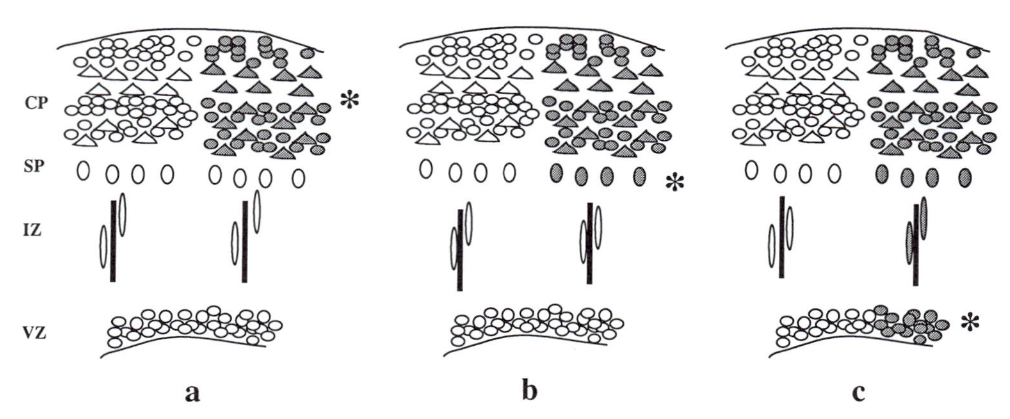

FIGURE 9.2 Schematic diagram of the formation of the cerebral cortex, highlighting three different views of the timing of specification. The cortical plate (CP) neurons form from the precursor cells of the ventricular zone (VZ). Radial migration, through the intermediate zone (IZ), of elongated postmitotic neurons along radial fibers (thick line) is depicted. The filled cells represent any phenotypical destinction that can be detected between two cortical areas. Note the modest cytoarchitectonic differences emerging in the cortical plate between two adjacent areas. Each asterisk (*) indicates the location in the developing cerebral wall where differences between cells composing each area first emerge. In (a) distinctions between neighboring cortical areas are not present until neurons settle in the cortical plate. This would represent tabula rasa, where homogeneous neuronal populations are generated throughout the ventricular zone, migrate to the subplate (SP) and CP, and become specified at some later developmental time in the cortex proper. In (b) a protocortex is generated in which the ventricular zone is homogeneous but produces SP neurons that are distinct between areas. The neurons in the cortical plate, however, express similar properties until some later developmental time. In (c) a protomap exists in the precursor pool that reflects phenotypical differences detected as neurons are generated and migrate to their appropriate location in the overlying cortex.

tions into mature astrocytes (Schmechel and Rakic, 1979; Levitt and Rakic, 1980).

Recently, time-lapse video microscopy has been used in combination with very low light cameras (to avoid tissue damage) to visualize, over several days, fluorescently labeled cells in live slices of the developing cerebral hemisphere. The results indicate that for the vast majority of neurons, the radial glial fiber serves as the sole path by which neurons achieve their final position (O'Rourke et al., 1992). The analysis of a transgenic mouse carrying a reporter gene (see later), which marks half the cells in the brain, also indicates that dispersion of cells in the cortex, once they leave the ventricular zone, is mostly radial (Tan and Breen, 1993). Here, radial columns of cells, containing primarily either the transgene or the host wild-type genotype, extended from layer VI through II in the cerebral cortex. In a third novel technical approach, labeling of cell cohorts with retrovirus lineage markers (see discussion of method in the next seetion) reveals again the dramatic and quite specific cell-cell interactions that guide cell migration (Misson et al., 1991). This is most evident in regions of the developing cerebral hemisphere in which the path of migration must follow a tortuous course due to the changing three-dimensional organization of the wall. In these regions, neurons exhibit changes in shape identical to those seen for the radial glial fibers, and appositions remain close (Rakic, 1978; Smart and McSherry, 1982; Misson et al., 1991). Neurologically mutant mice were identified years ago that highlighted specific defects in migratory behavior of neurons in the cerebral and cerebellar cortices (Caviness and Rakic, 1978). Analysis of these mutant animal models suggested that interactions between migrating neurons and radial glial cells are probably defective (Hatten and Mason, 1990; Gao, Liu, and Hatten, 1992; Hatten, 1993).

Lineage and phenotypical diversity

The cerebral cortex, as all other regions of the brain, is composed of a variety of neurons and glial cells. A basic question that has puzzled developmental biologists and neuroanatomists for more than a century is whether the population of precursor cells in the ventricular zone, which gives rise to all mature neuronal and macroglial phenotypes, is organized into specific, defined populations that account for the cellular diversity in the cerebral cortex (see figure 9.2). On the basis of cellular appearance, two major hypotheses were formulated. Schaper (1897) believed that the neuroepithelium was composed of identical cellular units that showed no obvious preference for generating neurons or glia. He believed that the timing of the cell's exit from the cycle defined cellular diversity. In a contrasting hypothesis, Wilhem His (1889) considered the nuclear stratification in the ventricular zone to represent the segregation of precursor cells giving rise to neurons and glia. The spongioblast, identified by His, was believed to produce nonneuronal cells. His's concept of segregated progenitors was not too far off the mark (as shown later), but the use of histological appearance was not adequate to support his arguments. In fact, Sauer (1935) demonstrated subsequently that the nuclear stratification of cells in the proliferative cell pool was due simply to interkinetic movement of the nucleus during different stages of the cell cycle. Fujita (1963), using ^{3}H-thymidine autoradiography, was able to show that early in histogenesis, all precursors could be labeled with an appropriate dose of the base analog, and thus she concluded that the precursor pool was homogeneous. Unfortunately, Fujita's use of the method was inappropriately applied to the question of precursor segregation and cell lineage.

Immunochemical Approaches to the Lineage Problem What had been missing from all these studies was the ability to mark separate populations of precursor cells. Use of antibodies directed against cell type–specific antigens failed, in the CNS, to reveal any heterogeneity among the progenitor cells. Either there was a total lack of expression of antigens that represented specific phenotypes, such as neurotransmitters (Olson and Seiger, 1972; Pickel et al., 1980), or the antigens, such as certain intermediate filaments, were expressed by all the precursor cells (Tapscott et al., 1981; Traub, 1985; Lendahl, Zimmerman, and McKay, 1990). Antibodies against the specific radial glial and astrocyte marker, glial fibrillary acidic protein (GFAP), were used initially in developing rodent brains and did not label cells until early postnatally. For years, it had been assumed that this marker, along with many others, would not be useful for revealing early differentiation of cells giving rising to astroglia.

While investigating the expression of GFAP in radial glial cells of the developing monkey brain, we noted the rather striking heterogeneous nature of the GFAP immunostaining in the ventricular zone giving rise to the visual cortex (Levitt and Rakic, 1980). At the light microscopic level, it appeared as though some populations of cells in ventricular zone were GFAP-negative, whereas others were GFAP-positive. This was the first evidence that precursor cells, giving rise to astroglia, formed a segregated pool. This finding was confirmed by ultrastructural immunocytochemical analysis, in which cells caught in the act of mitosis were visualized with GFAP immunoreactivity in their cytoplasm (Levitt, Cooper, and Rakic, 1981). Indeed, subsequent quantitative analysis documented that during ontogeny in the primate brain, there is an increasing number of immunolabeled progenitor cells, believed to give rise to radial glial cells and astroglia, and a decreasing number of unlabeled cells, some of which were assumed to generate neurons (Levitt, Cooper, and Rakic, 1983).

Retrovirus Labeling Expands Understanding of Cortical Cell Lineage In our laboratory, my colleagues and I were unable, by immunocytochemical analysis, to address the issue of the identity of the unlabeled cells and, more directly, whether there was an early segregation of precursor cells giving rise to specific neuron populations in the developing cerebral cortex. Development of new analytical methods, allowing one to trace generations of cells derived from a single precursor cell, resolved this question.

Host eukaryotic cells that are infected with retroviruses will incorporate permanently into their own genome the genetic information that is carried by the virus (Sanes, Rubenstein, and Nicolas, 1986; Price, 1987; Cepko, 1988). This requires the host cell to be actively engaged in the cell cycle. To keep the number of infected cells at a minimum, the retroviruses are modified to be replication-incompetent, preventing secondary infections after the initial "hit." In addition, statistical analysis is used to estimate the number of virus particles that should be used to infect a small number of precursor cells. Most important in this technique, the retrovirus is modified, through molecular construction, to carry genetic information encoding an enzyme (reporter) that can be used to visualize, by simple histochemical reaction, all the cells that carry the retrovirus genes. This means that infection of a single progenitor cell in the developing brain results in the permanent incorporation of the retrovirus RNA (through DNA synthesis via a reverse transcriptase) in the host cell DNA. All cells produced from this initially labeled cell would carry the retrovirus construct, including the reporter gene, and would be visualized following a simple histochemical reaction of the tissue. The labeled cells would be related clonally to the parent precursor, thus forming a well-identified cohort. The relationship of the cells as part of a single clone is defined by the expression of the histochemical product.

To date, most retroviral labels have used the bacterial enzyme LacZ (homolog to eukaryotic β-galactosidase). The monoclonality of the cells depends on the assumption that the injection of the retrovirus at a specific concentration results in a minimum (approaching one) number of infections. This method has revealed some interesting and surprising features of histogenesis in the CNS.

Initial studies with the retroviruses revealed that a single infection event resulted in clones of various sizes, depending on the time at which the retrovirus was injected into the brain. In the optic tectum of the chick (Gray et al., 1988), neurons and glial cells were labeled by a single injection, suggesting strongly that the same precursor cell could give rise to neurons and glia. Similarly, clones containing both neurons and glia were obtained in the developing retina (Turner and Cepko, 1987). When a similar application of retrovirus was made into the lateral ventricle of the fetal rat, a different result was obtained (Luskin, Pearlman, and Sanes, 1988; Price and Thurlow, 1988; Walsh and Cepko, 1988; Vaysse and Goldman, 1990). Here, clones of various sizes were seen, but generally each was composed exclusively of neurons or glia and only very rarely was a clone of mixed phenotypes seen. In cell culture experiments, astrocyte and neuron lineages appear to be segregated, although one recent study suggests that the same precursor can give rise to neurons and oligodendrocytes (Williams, Read, and Price, 1991). These mixed clones were not evident from in vivo analysis. In general, it appears that in the production of major cell classes in the cerebral cortex, the precurors generating neurons are segregated by lineage from those giving rise to glia.

Identification of cell types, using the retrovirus-labeling method, had been based solely on morphological appearance of the cells following histochemical reaction. In many instances, the product does not fill the cell, making identification of the phenotype very difficult. In addition, because clonally related cells often ended up in different cortical layers, it was assumed in the initial studies using the retrovirus method that the labeled cells represented different classes of neurons (Walsh and Cepko, 1988; Austin and Cepko, 1990), even though nonpyramidal neurons are present in all layers. Initial conclusions from this type of analysis in the cerebral cortex suggested that while neurons and glia were produced from separate progenitors, a variety of neurons could be generated from a single precursor cell.

This view was reinvestigated recently, using a more definitive method of analysis (Parnavelas et al., 1991). Ultrastructural analysis of the retrovirus-labeled cells enables one to characterize a neuron specifically, based on classical morphology and synaptic organization. Though extraordinarily time-consuming, the efforts revealed an answer far different from that provided by previous light microscopic studies. In all instances, and based on accepted morphological criteria for identifying pyramidal and nonpyramidal neurons, clones almost always were composed of only one of the neuronal classes. This ultrastructural analysis resolved a number of issues. First, and as predicted from the earlier immunocytochemical studies, glial clones and neuronal clones were segregated, indicative of an early commitment of the precursor pool in the cerebral wall to generate one major cell class or the other, but not both. Second, progenitors destined to give rise to long-projecting pyramidal neurons were separated relatively early from those producing nonpyramidal neurons, many of which are interneurons. Timing does not seem to be a factor, because the same precursor cell can give rise to neurons in different cortical layers, which are generated at different embryonic ages. There is no evidence for a precursor cell that gives rise to cells only on a specific day of development. Although independent phenotypical markers for these different neuronal precursor pools have not been identified, the lineage analysis clearly shows that segregated progenitors exist within the ventricular zone giving rise to the cerebral cortex. The experiments do not, however, tell us whether these precursors retain a capacity to give rise to many different cells classes or are committed, perhaps irreversibly, to only one class.

Specification in the cerebral cortex: Laminar and tangential domains

What is most intriguing about the cerebral cortex is its rather remarkable structural organization into well-defined layers and areas. Both anatomical features reflect the dominant organizational concept in the cerebral cortex, that of functional segregation. Indeed, circuitry is established based on the laminar and tangential areal organization, both of which arise developmentally. If we are to understand how specific circuits are formed and what impact the development of anomalous circuits has on cortical function, then the mechanisms that guide layer and area assembly must be defined. Basic, descriptive analysis of the spatiotemporal appearance of these features has been important but has failed to provide the cellular basis for the complex pattern formation that is the hallmark of the mammalian cerebral cortex. Recently, a variety of experimental strategies have been used to investigate these basic developmental issues.

Early Fating of Laminar Organization Laminar organization is generated through temporal control of the production of neuronal populations destined for each layer. As noted earlier, a neuron that will reside in layer VI of a specific area of the cortex will be produced from a precursor cell prior to a neuron that will occupy layer III. Settling patterns—that is, the manner in which migrating cells arrive at and cease movement in the cortical plate—reflect this inside-out motif of histogenesis. When does a cell acquire its laminar identity and how does it know when to cease migration and take up residence in the proper vertical domain? Recently, cell transplants were used to identify more clearly the degree of intrinsic information that guides a neuron's laminar destiny, particularly in comparison to the potential environmental cues that might be used to generate the cortical layers (McConnell, 1988a, b). Retroviral lineage studies had already shown that different precursor cells are not set aside to produce neurons destined for different layers. Thus, the same precursor cell theoretically can produce neurons for all layers.

McConnell (1988 a, b) used a strategy of manipulating the environment through which identified cortical cells must migrate in order to determine how instructive the extrinsic cues might be. In the ferret, neurons destined for layers V and VI are produced early, around E29–E34. At birth (60-day gestation period), these neurons have already completed migration to their final location, but the neurons destined for the more superficial layers (II and III) are in the process of migration. How would a cell clearly destined for deep layers behave when placed in an environment in which only superficial neurons are moving? The transplanted cells were marked by their time of origin by injecting ^{3}H-thymidine into a pregnant dam prior to harvesting the donor cortical cells. The cells were injected deep into the developing cerebral hemisphere, close to the ventricular zone and subcortical white matter of newborn ferrets, and then, weeks later, were visualized autoradiographically to determine their location. In this *heterochronic* transplant (donor and host developmental stage do not match temporally), a relatively small percentage of the original cells migrated. Those that did move, however, exhibited some striking features. Heavily ^{3}H-thymidine-labeled cells, representing neurons that were born at or close to the time of injection, were found in deep layers of the host ferret visual cortex. These neurons normally would migrate to layers V and VI and did just that in the newborn host, even when placed in an environment in which neurons were migrating only to more superficial layers.

For technical reasons, heterochronic transplants into early fetal ferrets cannot be performed. Results from *isochronic* transplants (donor and host cells of the same developmental stage), using cells that were generated much later and thus were destined for layers II and III, showed that these neurons could migrate to their appropriate laminae. The experiments demonstrated a rather early and specific commitment of postmitotic neurons to populate specific layers in the developing cerebral cortex. Moreover, this information appears to be maintained intrinsically, even when the cell is placed in a temporally abnormal environment.

A series of more complex studies have been carried out by McConnell and Kaznowski (1991). Although the initial studies indicated that neurons which had just become postmitotic were committed to a specific laminar fate, the data did not pinpoint the timing of the commitment by a cell to reside in a specific cortical layer. The subsequent experiments revealed a remarkable aspect of cortical development. Cells actively in the cycle were specified to migrate to their appropriate layer (based on time of origin) if they were transplanted just after labeling in S phase and prior to mitosis. Once a cell passes through phase G_2 and divides, the commitment is lost and the phenotype can be driven by environmental cues. The plasticity of this system is lost when the cell permanently exits the cycle. In addition, there clearly is information maintained by the cell in certain phases of the cell cycle that precludes influences by any other information from the environment. The early and specific commitment of cells to reside in certain laminae highlights the existence of fundamental "prepatterning" in the premature cerebral wall.

In some ways, these data support the view that the ventricular zone contains a *protomap* (Rakic, 1988) of the mature cerebral cortex, with populations of cells destined to give rise to certain layers and areas with specific characteristics. The relevance, however, of this precursor map in generating specific tangential domains has been challenged. Perhaps the early laminar fating is not surprising, given the fact that all neocortical areas contain the same basic six layers, with variations in only specific numbers of neurons. A more challenging problem, however, is to understand the mechanisms underlying production of the rather diverse differences in cortical areas, marked by unique cytoarchitectonics and connections. Is the early protomap applicable here, or are later-expressed phenotypes, such as anatomical diversity, regulated by a different set of constraints?

Concepts, Strategies, and Caveats in Investigating Area Specification Arguments over defining the mechanisms underlying cerebral cortical specification revolve around these conflicting hypotheses (see figure 9.2): that the ventricular zone contains a protomap (Rakic, 1988), in which area prepatterns already exist so that committed precursor cells produce neurons destined for specific functional regions versus the absence of any preexisting pattern in the cortex proper (tabula rasa) until late in development, when commitment is driven by environmental interactions between maturing cortical neurons and incoming thalamic fibers. A third view has been postulated (O'Leary, 1989) in which the earliest-generated neurons, forming the preplate (subplate; see above), contain areal information, as a protocortex, that is necessary for initial, specific interactions such as the formation of thalamocortical connections. There is recent experimental support for the role of the subplate in regulating thalamic afferent targeting to appropriate areas of the cortex (Ghosh et al., 1990; Ghosh and Shatz, 1993). Excitotoxic lesions of the subplate neurons in the developing visual or auditory cortex were performed. In these animals, the specific thalamic axons for the appropriate cortical area failed to grow into the neocortex, remaining in the intermediate zone fiber region. In the protocortex concept, the pattern in the cortex proper is driven by subsequent interactions between the cortical plate neurons and their incoming thalamic axons. Tantamount to the latter two hypotheses, if specific environmental interactions are prevented, then areal specificiation, as neuroanatomists define it, should not occur.

Analysis of the descriptive data obtained from studies of the time of neuron origin, initial circuit patterns, and differentiation at the molecular level supports the idea that not all areas of the cerebral cortex are created equal. A number of laboratories have challenged this conclusion based on several different experimental approaches. Recent use of time-lapse video microscopy and the retrovirus-labeling technique to explore this question has raised some doubts concerning the early and specific allocation of ventricular zone cells to generate unique functional areas.

An original tenet of the protomap concept is the ability predictably to relate, spatially, locations in the ventricular zone with generations of neurons produced in the region that subsequently reside in the overlying cortex. This would require unfailing radial migration from the ventricular zone to the cortical plate, a phenomenon that was predicted by Rakic based on radial glial architecture in the primate cerebral wall. Time-lapse analysis in the developing rodent brain reveals two interesting phenomena (O'Rourke et al., 1992). First, approximately 85% of cells followed by video analysis appear to migrate radially, supporting the original contention. There were, however, examples of diverse, nonradial routes taken by fluorescently labeled cells. It was suggested that some postmitotic cells could move to locations in the developing cerebral cortex other than those lying directly above their site of genesis. One potential problem with this interpretation is the inability to identify the phenotype of the cells that move nonradially. Glial cells and brain macrophages probably move in many directions and most likely are not relevant to the architecture of unique cortical areas.

Retrovirus-labeling studies in the rodent indicate that cells originating from the same progenitor can settle in locations at great distances from one another, across several functional domains (figure 9.3). By simultaneously using retroviruses with basic differences in their genetic sequences, Walsh and Cepko (1992) were able to analyze the distribution of clonally related cells that may end up physically separated. In this method, all infected cells are visualized histochemically as usual by the reporter gene, and then specific cells are dissected out of the tissue. Each retroviral sequence is

Multiple Retroviral Labeling and Clonal Dispersion

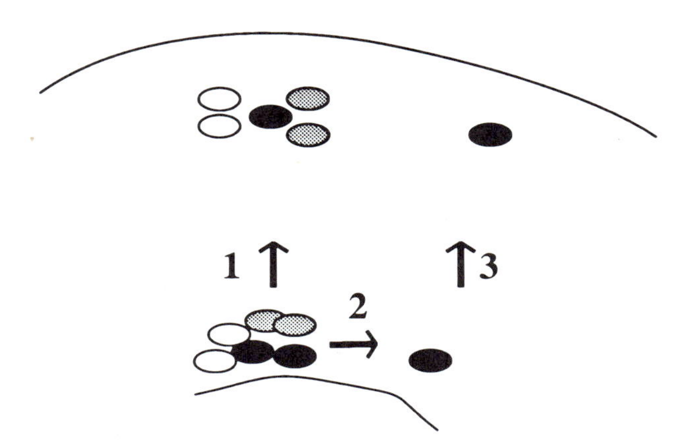

FIGURE 9.3 Multiple retroviral markers have been used to detect clonal dispersion. In this schematic drawing, the white, gray, and black circles represent clonally related cells. The more typical event depicted in (1) shows that a single precursor cell marked in the ventricular zone (lower part of diagram) will produce cells that carry the same retroviral label and will migrate to the overlying cortex when they differentiate into neurons. They maintain close spatial associations after migrating radially. Clonal dispersion can take place if one of the marked daughter cells of the originally infected progenitor moves within the ventricular zone tangentially, even just two to three cell diameters (2). This cell, on differentiating, would then migrate radially to the overlying cortex (3). Its other cohorts, which did not disperse in the ventricular zone from the original site of infection, also migrate radially. In this scheme, the two black cells in the clone will end far apart from each other, in separate cortical domains. The geometry of the cerebral wall, with a narrow base and widening top, lends itself to such a mechanism of dispersion.

detected by Polymerase Chain Reaction (PCR) with defined primer sequences that are specific for each retrovirus used. Thus, two cells that end up several millimeters apart normally are not considered related when the single retrovirus-labeling method is used. In the multiple-labeling method, if two cells carry the same sequence of retroviral RNA, they are considered clonally related. Conversely, two cells are not of the same lineage if they contain different sequences, even if they reside near each other. Quantitative analysis of labeled clones in the neocortex indicates that in 20–40% of the cohorts, cells in a single clonal group can spread up to 1 mm (Walsh and Cepko, 1992; Walsh, 1993). Viewing of labeled clones from the time of injection in the fetal rats indicates that all clones begin as tightly clustered groups of cells but, within 1 week, can spread tangentially within the ventricular and subventricular zones (Walsh and Cepko, 1993), which has been interpreted as strong evidence supporting the view that a rather nonspecified neocortex is generated from a homogeneous group of precursor cells. These data, however, may be consistent with the concept of early specification. Because of the geometry of the developing cerebral wall, with a distinctive cone shape, a small movement of precursor cells in the ventricular zone, representing the base, would result in a very large spread as the cells migrate to the cortical plate (top of the cone). Thus, a cell, once having exited the cycle, might move only radially (supported by the time-lapse data of O'Rourke et al. 1992) but could be related by lineage to a cell that ends up in another cortical domain.

Is there experimental support for such a mechanism? Movement of precursor cells in the ventricular zone has recently been viewed by time-lapse microscopy (Fishell, Mason, and Hatten, 1993), and single cells were followed traversing tangentially an average of 40–60 μm. This represents only four to six cell diameters but suggests that such minor movements could result in major differences in generated area phenotypes. One interpretation of this is that the ventricular zone contains uncommitted, homogeneous precursor cells that can give rise to neurons composing any cortical area. However, an alternate mechanism, which remains consistent with early cortical specification, is one in which the fate of the progenitors would ultimately depend on the region of the cerebral wall in which they reside just prior to exiting the cell cycle. This implies the presence of a dynamic map in the ventricular zone, one that is defined by the influence of specific. environmental cues in each region of the cerebral wall on precursor cells as they leave the cycle. Recent work by my colleagues and I supports this view.

How important is environment in cortical specification? Classical approaches using tissue transplantation have attempted to address this. Changing the environment that is faced by cortical cells at various stages of their differentiation can reveal the potential of developing cells to respond to unusual surroundings. I noted the effective use of this strategy by McConnell and colleagues (1988a, 1991) when assessing mechanisms of cortical lamination. The settling patterns of the transplanted cells were used as the assay to establish the expressed phenotype of the neurons. When one examines phenotypes of different cortical areas, the choice of assay becomes a more complex issue and one that is critically important. Different traits are expressed at distinct developmental times. For example, thalamocortical connections are established in a very specific pattern during fetal development in the rat (Wise and Jones, 1978; Barbe and Levitt, 1992; Crandall and Caviness, 1984; De Carlos and O'Leary, 1992; Erzurumlu and Jhaveri, 1992), but cytoarchitectonic differences in areas are not evident until postnatal ages. Molecular differences between cortical areas have recently been recognized to occur early in fetal development (Horton and Levitt, 1988; Arimatsu et al., 1992). Different traits in the transplants thus can be examined, allowing the effects of environmental influences on specific phenotypical features of the donor tissue to be evaluated. The ability to monitor different traits affords the opportunity to determine whether each phenotypical feature is under identical or different regulatory constraints. Equally important is consideration of the stage of development at which a cell finds itself at the time of transplantation and how this temporal factor influences the cell's response to environmental signals. When transplanting pieces of tissue (usually measuring 1–3 mm^3), this becomes a complex issue, because cell differentiation in the developing cerebral cortex is asynchronous. The grafts, then, will usually be composed of cells that are in different states of differentiation. In the dissociated cell transplants, one assesses the phenotypical expression of single cells. In contrast, investigators usually analyze the behavior of subpopulations of cells in the grafts (O'Leary, 1989; Levitt et al., 1993). Keeping these constraints in mind, several studies have generated revealing aspects of cortical development (O'Leary and Stanfield, 1989; Schlagger and O'Leary, 1991; Barbe and Levitt, 1991, 1992). Some of the data, at first glance, appear contradictory, but they are, in fact, quite consistent with the idea that consideration of developmental timing and the assay of multiple phenotypes are very important for accurately defining mechanisms underlying the generation of a mature, differentiated cortex from the embryonic cerebral wall.

Critical developmental periods for specifying cortex O'Leary's laboratory has employed the transplantation strategy to evaluate the commitment of tissues destined to form somatosensory and visual cortex (Stanfield and O'Leary, 1985; O'Leary and Stanfield, 1989; Schlagger and O'Leary, 1991, 1993). A retrograde double-labeling strategy was used in a neuroanatomical study of the organization of corticospinal projecting neurons in the developing cortex of the newborn rat. O'Leary and Stanfield found that all neocortical areas contained pyramidal neurons in layer V that project to the spinal cord. The visual cortex, however, fails to maintain this projection and loses the axon collaterals postnatally. Sensorimotor regions maintain this projection through maturation. When embryonic tissue from occipital regions of the cerebral hemisphere, containing developing visual cortex, was transplanted to a cavity placed in the newborn somatosensory cortex, the pyramidal neurons that normally lose the spinal projection maintained it. The reciprocal transplant of sensorimotor tissue into newborn occipital cortex resulted in a loss of the spinal projection. These experiments provided some of the first examples that the expression of some cortical traits may be controlled by as yet undefined environmental cues. It was suggested as well that the neocortex may be established as a uniform sheet with few, if any, area-specific properties that are expressed prenatally. A second set of studies was done to evaluate more closely plasticity of area-specific traits.

Because rodent somatosensory cortex exhibits a unique anatomical organization, the layer IV barrels (representing whisker afferents via the trigeminal system), one can readily use this as an assay of an area-specific phenotype (O'Leary, 1989). Barrel fields require specific input from the ventrobasal thalamus in order to form during development (Woolsey and Wann, 1976), which normally occurs during the first postnatal week. The structures can be visualized by a variety of histological methods. In a series of studies, grafts of developing visual and somatosensory cortex were placed homotopically and heterotopically into host newborn rats (Schlagger and O'Leary, 1991). Tissues excised at E17 or E18, when placed into a novel environment, contain cells that express traits identical to their new environment. Thus, visual cortex grafts, when placed in the somatosensory region of the host, contain cells that form anatomically distinctive barrels. Transplants of embryonic somatosensory cortex fail to form such distinctive barrels when placed in visual cortex of host rats.

The data suggest two events. First, because barrel field formation requires thalamic input from a specific nucleus, the grafts of embryonic visual cortex must have cells that differentiate in an appropriate manner to receive ventrobasal input, afferents that they normally do not receive. This, in turn, leads to the formation of typical barrel structures. These results, together with the initial neuroanatomical analysis of corticospinal projections, suggested strongly that environmental factors play a major role in defining neocortical areas. The ability of late prenatal cortical tissue to respond to new environmental cues supported the view that the cortex is generated with all areas expressing identical traits, only to be driven postnatally by undefined environmental signals to express area-specific phenotypes.

This conclusion appears to be in conflict with the recent identification of molecules expressed in the cerebral cortex during fetal development in a region-specific manner (Horton and Levitt, 1988; Arimatsu et al., 1992). In addition, experimental evidence using transplantation and cell culture approaches suggests that cortical neurons may commit to certain area-specific traits rather early in development. Results from the apparently conflicting studies are reconciled, based on our recent studies, in a hypothesis that my colleagues and I recently formulated (Levitt, Ferri, and Barbe, 1993).

A new hypothesis of cortical specification We discovered a novel membrane protein that is expressed by adult cortical and subcortical areas classically defined as limbic (Levitt, 1984; Levitt et al., 1986; Zacco et al., 1990) and in a phylogenetically conserved manner from fish to primate (unpublished observations). The limbic system–associated membrane protein (LAMP) is a recently cloned member of the immunoglobulin superfamily of adhesion molecules (Pimenta, Fischer, and Levitt, 1992), with a phosphatidyl inositol linkage (Zhukareva and Levitt, 1992), and it has the ability to mediate specific limbic cell adhesion and, most important, the formation of connections between limbic regions (Keller et al., 1989; Barbe and Levitt, 1992; Zhukareva and Levitt, 1993). We found early on, by immunocytochemistry in the fetal rat, that LAMP is expressed specifically in developing limbic cortical areas, such as prefrontal, perirhinal, and entorhinal areas, from their inception (Horton and Levitt, 1988). Nonlimbic areas, such as primary sensory and motor regions, are not immunoreactive. Though we have yet to visualize immunoreactive cells in the ventricular zone that give rise to limbic cortex, at E15 cells in the process of migration, clearly prior to settling in the cortical plate, are LAMP-immunoreactive. This analysis suggested an early molecular differentiation of the cerebral cortex and provided a novel approach to examining issues of cortical specification.

We combined tissue transplantation, using donor grafts of different developmental stages, with analysis of molecular differentiation and connectivity (Barbe and Levitt, 1991, 1992) (figure 9.4). Having in hand a specific marker for developing neurons residing in limbic perirhinal cortex enabled us to conduct grafting studies to assay phenotypical plasticity of the fetal cerebral cortex. We focused on perirhinal cortex, which expresses LAMP beginning on E15, and sensorimotor cortex, whose neurons never express LAMP (Horton and Levitt, 1988). The expression of LAMP in 2- to 4-week-old grafts was monitored using tissue from three different embryonic ages (Barbe and Levitt, 1991). At E12, almost all the cells are progenitors. At E14–E15, nearly half of the neurons have been generated, and thus the graft contains mitotically active and postmitotic cells. At E17, approximately 75–80% of the neurons are generated and, of course, LAMP is clearly detectable in perirhinal cortex at this age. The results were striking for several reasons. More than 70% of the neurons in the transplanted perirhinal tissue from E17 were LAMP-immunoreactive, irrespective of the location of the graft. Thus, even when located in the sensorimotor region of the host, the majority of transplanted neurons were LAMP-immunoreactive. In contrast, sensorimotor grafts had few LAMP-positive neurons, no matter where the tissue was placed in the host. Using E14–E15 perirhinal tissue, approximately half of the grafted neurons ex-

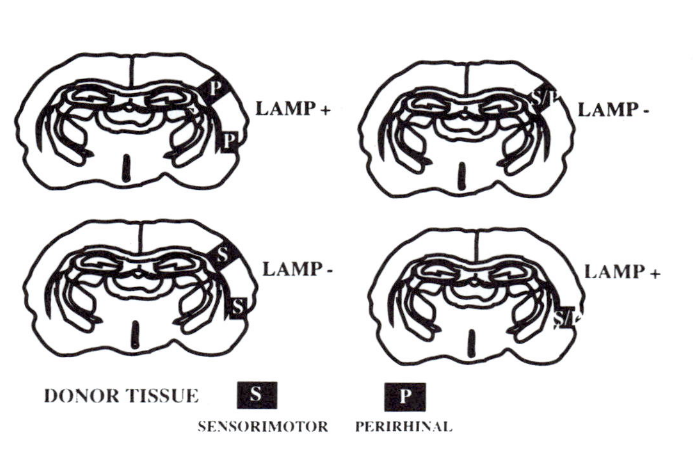

FIGURE 9.4 Summary of transplant results (Barbe and Levitt, 1992) show that both E14 and E17 perirhinal (P) transplants, irrespective of their location in the host, will contain a substantial number (50–70%) of LAMP-positive (LAMP+) neurons. In contrast, the sensorimotor (S) donor tissue will have a much smaller percentage of LAMP-positive cells, and only when placed in the perirhinal region of the host. Using precursor cell transplants (E12 tissue), both sensorimotor and perirhinal grafts will contain a high percentage (>85%) of LAMP-positive neurons when the tissue is placed in the limbic perirhinal location. Neither graft will contain any LAMP-positive cells if they are placed into sensorimotor cortex of the newborn host rat.

pressed LAMP, again irrespective of the location of the transplant. Sensorimotor grafts at this age contained increased numbers of cells that expressed LAMP (up to 15%), but only when the tissue was placed in the host perirhinal cortex. In both instances, the tissue contained a higher proportion of precursor cells than at E17, when neurogenesis in the tissue from both regions is virtually complete.

The experiments using very young fetal grafts provided our first indication that developmental timing is a particularly critical aspect in controlling a cell's commitment to a specific phenotype. Grafts of presumptive sensorimotor or perirhinal cortex from E12 fetuses, which contained only precursor cells, differentiated according to their location in the cortex of the host and not to their origin. We found that either presumptive sensorimotor or perirhinal areas of the cerebral wall contained LAMP-positive neurons when placed into host perirhinal cortex (up to 85%), but the same fetal grafts contained few LAMP-positive cells when placed in sensorimotor cortex of hosts. In these experiments, transplantation of almost exclusively precursor cells evokes a remarkable response; cells never fated to express LAMP can do so within a *limbic* milieu, whereas those progenitors destined to express this area-specific protein do so only when exposed to the appropriate environmental signals. Key to all of these experiments was the use of embryonic tissue harvested at different ages. We were able to demonstrate that the relationship between a precursor cell and its surroundings is lost after a certain developmental stage, presumably when the cell becomes postmitotic and thus, specification of cortex based on this molecular phenotype occurs early.

Our ability to manipulate the molecular phenotype of the grafts may simply highlight a feature that is readily modified in the progenitor cells under certain conditions, unrelated to subsequent differentiation of cortical areas. We next assayed another phenotype unique to different cortical areas, (Barbe and Levitt, 1992) thalamic connectivity. LAMP expression was induced or prevented in perirhinal and sensorimotor grafts according to the paradigms used in the first study, and the connections of the grafts with host thalamic nuclei were subsequently analyzed using DiI labeling. The oldest tissue we used in these experiments was the E14–E15 tissue. In our first studies, perirhinal grafts placed in sensorimotor regions had approximately 50% LAMP-positive cells. The connectivity analysis revealed that this transplant received both limbic and nonlimbic thalamic projections. Based on the number of retrogradely labeled neurons, there was an approximate 1 : 1 ratio of limbic-to-nonlimbic input into the grafts, which is remarkably similar to the mixture of LAMP-positive and LAMP-negative neurons in the grafts. When this experiment was performed with precursor cells (E12), which fail to express LAMP in the sensorimotor environment, only sensorimotor thalamic projections developed. In contrast, E12 precursors from presumptive sensorimotor cortex, when placed into host perirhinal cortex, differentiated into LAMP-positive neurons and also received limbic thalamic projections. Transplants of older sensorimotor tissue, placed into perirhinal cortex of hosts, failed to elicit limbic connections, an expected result given the lack of LAMP expression in these grafts. The data show that cortical areas are committed to express unique patterns of connectivity that reflect their molecular phenotype. The chimeric phenotype of the E14 perirhinal grafts, in which limbic and nonlimbic neurons reside, highlights the tight correlation between LAMP expression and thalamocortical connectivity.

Resolving conflicting hypotheses Can the different results from all the transplant experiments be reconciled? It has been suggested that perhaps there is early parcellation of the cortical sheet between presumptive limbic cortex and neocortex, with the former being phylogenetically older and presumably more conserved between species (Schlagger and O'Leary, 1991, 1993). Given the rather remarkable expansion of association cortex, including inferotemporal and prefrontal lobes, from lissencephalic to gyrencephalic animals, it is difficult to reconcile the concept of early segregation of embryonic cerebral cortex based on this division. We have proposed an alternative view (Levitt, Ferri, and Barbe, 1993), one that suggests a very strong connection between commitment to phenotype and developmental timing. Certain traits—for example, unique molecular properties of neurons in different areas of the cerebral cortex—may be specified very early. This differentiation would drive subsequent connectivity. We demonstrated this with our analysis of LAMP expression, where commitment appears to coincide with cells becoming postmitotic (as shown for laminar commitment; McConnell and Kaznowski, 1991). It is clear, however, that precursor cells are receptive to environmental signals that can regulate cell phenotype. Thus, in the transplants of E17–E18 visual cortex, it is possible that the cells which respond to the new environment, those destined for layer IV, were not postmitotic at the time of transplantation (Miller, 1986, 1988). Because the analysis of barrel formation is qualitative, it is difficult to determine how many of the transplanted cells responded. Even a modest number would be readily visualized by sensitive anatomical methods. It is interesting, in this context, that physiological (Castro et al., 1991) and behavioral (Barth and Stanfield, 1994) analysis of E17–E18 cortical transplants has shown that grafts in heterotopic locations do not subserve area-specific functions. This could be due to lack of a critical mass of cells in the older grafts that respond to the new milieu and would support, as we have suggested, the proposal that many aspects of cortical differentiation are determined early. Alternatively, barrel structures, which are specified by thalamic afferents, may be a feature of cortex that is readily expressed by layer IV cells from any area when placed under certain environmental conditions.

Does the concept of early specification of cortical areas fit with the retrovirus-labeling data, where clonally related cells end up in different cortical domains? This outcome may reflect the rather dynamic composition of the ventricular zone, in which small movements could result in cells derived from the same precursor expressing different area-specific phenotypes. We have shown that precursor cells are susceptible to environmental cues and so we would suggest that, until a cell leaves the cycle, movement in the ventricular zone could greatly influence the expressed phenotype of the cell. Once a cell exits, however, it is committed to follow a specific developmental path, though there are many points along the way at which the neuron can make choices about expressing certain traits based on specific environmental cues to which it is exposed. We have termed this *progressive acquisition of phenotypes* (Levitt, Ferri, and Barbe, 1993) (figure 9.5), a concept introduced previously by developmental biologists (see Greenwald and Rubin, 1992, for review). Clearly, each phenotypical characteristic is almost certain to exhibit developmental plasticity for different periods of time. Perhaps this explains the rather late commitment of cortical areas to maintain a corticospinal projection.

The protomap—possibly existent but not irreversible The graft experiments provide evidence that precursor cells in the cerebral wall are not irreversibly committed to express phenotypes unique to specific cortical areas. Thus, they may be capable of giving rise to any region of cerebral cortex under certain conditions. The failure, in our grafts, of the E12 transplanted cells irreversibly to commit to a limbic phenotype does not address the issue of whether the progenitors, as a population, are fated in certain regions of the ventricular zone to give rise to specific cortical areas. We have recently examined this question using an in vitro approach, by which one can readily expose cells to specific environmental influences (Ferri and Levitt, 1993).

Dorsal (presumptive sensorimotor) and lateral (presumptive perirhinal) areas of the cerebral wall were collected at E12, and the cells were dissociated and grown as monolayers in equivalent, defined culture medium without serum. To our surprise, we observed significant differences in the developmental potential of the precursor cells from each region. By 4 days in culture, more than 80% of the perirhinal precursors that differentiate into neurons express LAMP. In con-

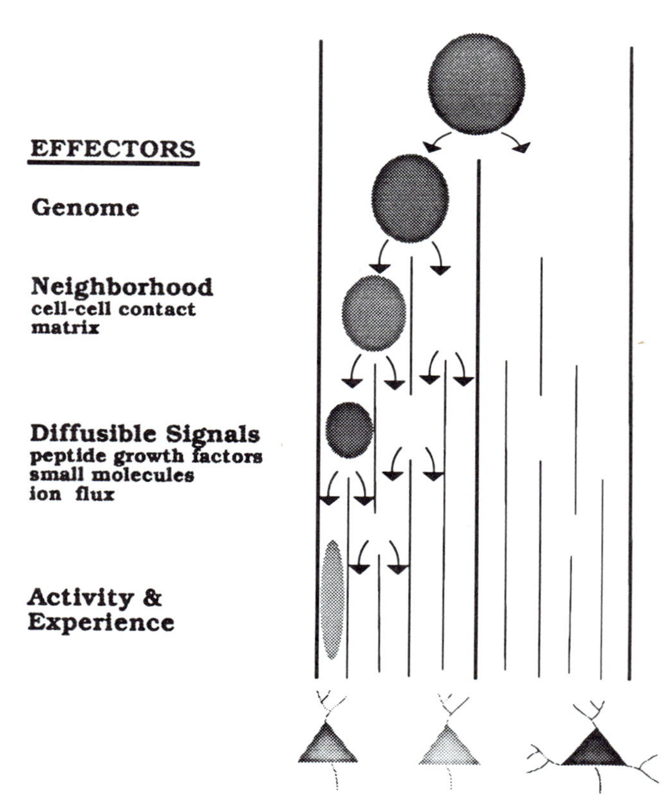

FIGURE 9.5 Summary of progressive acquisition of phenotypes, the proposal that a precursor cell from the cerebral cortex moves through a series of progressive restrictions in its choice to express certain phenotypical traits. The initial choice may be controlled by genetic bias and environmental cues. There are many extrinsic factors along the developmental path that will guide the choices made by the differentiating cortical neuron, but progression through each path ultimately limits the choices that the cell can make. There is continued flexibility of the later-developing traits, but the neuron would still express the features to which it already committed at earlier developmental stages. (Reprinted from Levitt, Ferri, and Barbe, 1993, by permission.)

trast, only 15–20% of the neurons that form from the dorsal wall cells express LAMP. As in our transplant studies, we have taken precursor cells from their normal milieu, but in culture we have failed to provide them with an alternative CNS environment. The fact that the cells from each part of the cerebral wall behaved distinctly indicates that there must be some basic differences between the cells at the time of isolation. We believe that the cerebral wall is composed of *fated*, though uncommitted, pools of precursor cells. Their commitment would depend on specific environmental signals in the local milieu. Recent in vitro experiments using a new marker of lateral cortex have reported similar results (Arimatsu et al., 1992).

Conclusions

Most of the experimental data generated over the last few years are consistent with the notion that the cerebral cortex, like other regions of the CNS, is organized into domains early in development. Specification, either in the vertical or tangential domain, depends on developmental timing. McConnell (1991) has clearly shown reversible commitment of cells in the cycle, but this becomes rapidly irreversible on their exit. Although we have yet to analyze directly the link between cell cycle and area commitment in our system, grafting and culture studies indicate a very strong correlation. Clearly, not all cortical phenotypes are expressed at the same time in development, highlighting the fact that differentiation is not synchronous in the cortical sheet (Levitt, Ferri, and Barbe, 1993). One would predict, then, that different environmental cues will control expression of specific traits (see figure 9.5). Some features will be committed to rather early, whereas other traits (such as connectivity) may preserve exquisite plasticity, even in the mature brain. Why is there such flexibility? The ability of the CNS, including the cerebral cortex, to respond to different environmental challenges—for example, by reorganizing at the anatomical level after injury, and both structurally and at the molecular level during learning—may be a reflection of this developmentally maintained adaptability.

REFERENCES

ANGEVINE, J. B., Jr., and R. L. SIDMAN, 1961. Autoradiographic study of cell migration during histogenesis of cerebral cortex in the mouse. *Nature* 192:766–768.

ARIMATSU, Y., M. MIYAMOTO, I. NIHONMATSU, K. HIRATA, Y. URATANI, Y. HATANAKA, and K. TAKAGUCHI-HAYAHI, 1992. Early regional specification for a molecular neuronal phenotype in the rat neocortex. *Proc. Natl. Acad. Sci. U.S.A.* 89:8879–8883.

ASTROM, K.-E., 1967. On the early development of the isocortex in fetal sheep. *Prog. Brain Res.* 26:1–59.

AUSTIN, C. P., and C. L. CEPKO, 1990. Cellular migration patterns in the developing mouse cerebral cortex. *Development* 110:713–732.

Barbe, M. F., and P. Levitt, 1991. The early commitment of fetal neurons to limbic cortex. *J. Neurosci.* 11:519–533.

Barbe, M. F., and P. Levitt, 1992. Attraction of specific thalamic afferents by cerebral grafts is dependent on the molecular fate of the implant. *Proc. Natl. Acad. Sci. U.S.A.* 89:3706–3710.

Barth, T. M., and B. B. Stanfield, 1994. Homotopic, but not heterotopic, fetal cortical transplants can result in functional sparing following neonatal damage to the frontal cortex. *Cerebral Cortex*, in press.

Bayer, S. A., and J. Altman, 1991. *Neocortical Development.* New York: Raven Press.

Boulder Committee, 1970. Embryonic vertebrate central nervous system: Revised terminology. *Anat. Rec.* 166:257–261.

Brodmann, K., 1909. *Vergleichende Lokalisationslehre der Grosshirnrinde.* Leipzig: Barth.

Castro, A. J., T. P. Hogan, J. C. Sorensen, B. S. Klausen, E. H. Danielson, J. Zimmer, and E. J. Neafsey, 1991. Heterotoptic neocortical transplants. An anatomical and electrophysiological analysis of host projections to occipital cortical grafts placed into sensorimotor cortical lesions made in newborn rats. *Dev. Brain Res.* 58:231–236.

Caviness, V. S., Jr., and P. Rakic, 1978. Mechanisms of cortical development: A view from mutations in mice. *Annu. Rev. Neurosci.* 1:297–326.

Cepko, C. L., 1988. Retrovirus vectors and their applications in neurobiology. *Neuron* 1:343–353.

Clarke, J. D. W., and A. Lumsden, 1993. Segmental repetition of neuronal phenotype sets in the chick embryo hindbrain. *Development* 118:151–162.

Crandall, J. E., and V. S. Caviness, 1984. Thalamocortical connections in newborn mice. *J. Comp. Neurol.* 228: 542–556.

De Carlos, J. A., and D. D. M. O'Leary, 1992.Growth and targetting of subplate axons and establishment of cortical pathways. *J. Neurosci.* 12:1194–1211.

Erzurumlu, R. S., and S. Jhaveri, 1992. Emergence of connectivity in the embryonic rat parietal cortex. *Cerebral Cortex* 2:336–352.

Ferri, R. T., and P. Levitt, 1993. Cerebral corticalprogenitors are fated to produce region-specific neuronal populations. *Cerebral Cortex* 3:187–198.

Fishell, G., C. A. Mason, and M. E. Hatten, 1993. Dispersion of neural progenitors within the germinal zones of the forebrain. *Nature* 362:636–638.

Fujita, S., 1963. The matrix cell and cytogenesis in the developing central nervous system. *J. Comp. Neurol.* 120: 37–42.

Gao, W. -Q., X. -L. Liu, and M. E. Hatten, 1992. The *weaver* gene encodes a nonautonomous signal for CNS neuronal differentiation. *Cell* 68:841–854.

Ghosh, A., A. Antonini, S. K. McConnell, and C. J. Shatz, 1990. Requirement for subplate neurons in the formation of thalamocortical connectiosn. *Nature* 347:179–181.

Ghosh, A., and C. J. Shatz, 1993.A role for subplate neurons in the patterning of connections from thalamus to neocortex. *Development* 117:1031–1047.

Graham, A., 1992. Patterning the rostrocaudal axis of the hindbrain. *Semin. Neurosci.* 4:307–315.

Gray, G. E., J. C. Glover, J. Majors, and J. R. Sanes, 1988. Radial arrangement of clonally related cells in the chicken optic tectum: Lineage analysis with a recombinant virus. *Proc. Natl. Acad. Sci. U.S.A.* 85:7356–7360.

Greenwald, I., and G. M. Rubin, 1992. Making a difference: The role of cell-cell interactiosn in establishing separate identities for equivalent cells. *Cell* 68:271–282.

Hatten, M. E., 1993. The role of migration in central nervous system neuronal development. *Curr. Opin. Neurobiol.* 3:38–44.

Hatten, M. E., and C. A. Mason, 1990. Mechanisms of glial-guided neuronal migration in vitro and in vivo. *Experientia* 46:907–916.

Hicks, S. P., and C. J. D'Amato, 1968. Cell migrations to the isocortex in the rat. *Anat. Rec.* 160:619–634.

His, W., 1889. Die neuroblasten und deren entstehung im embryonal maarke. *Abh. Math. Phys. Cl. Kgl. Sach. Ges. Wiss.* 15:313–372.

Horton, H. L., and P. Levitt, 1988. A unique membrane protein is expressed on early developing limbic system axons and targets. *J. Neurosci.* 8:4653–4661.

Jacobson, M., 1991. *Developmental Neurobiology.* New York: Plenunm Press.

Kaas, J. H., 1993. Evolution of multiple ares and modules within neocortex. *Prespect. Dev. Neurobiol.* 1:101–108.

Keller, F., K. Rimvall, M. F. Barbe, and P. Levitt, 1989. A membrane glycoprotein associated witht he limbic system mediates the formation of the septo-hippocampal pathway in vitro. *Neuron* 3:551–561.

Krieg, W. J. S., 1946. Connections of the cerebral cortex: I. The albino rat. Topography of cortical areas. *J. Comp. Neurol.* 84:221–323.

Lendahl, U., L. B. Zimmerman, and R. D. G. McKay, 1990. CNS stem cells express a new class of intermediate filament. *Cell* 60:585–595.

Levitt, P., 1984. A monoclonal antibody to limbic system neurons. *Science* 223:299–301.

Levitt, P., M. L. Cooper, and P. Rakic, 1981. Coexistence of neuronal and glial precursor cells in the ventricular zone of the fetal monkey cerebrum: An ultrastructural immunoperoxidase analysis. *J. Neurosci.* 1:27–39.

Levitt, P., M. L. Cooper, and P. Rakic, 1983. Early divergence and changing proportions of neruonal and glial precursor cells in the primate cerebral ventricular zone. *Dev. Biol.* 96:472–484.

Levitt, P., R. T. Ferri, and M. F. Barbe, 1993. Progressive acquisition of cortical phenotypes as a mechanism for specifying the developing cerebral cortex. *Perspect. Dev. Neurobiol.* 1:65–74.

Levitt, P., E. Pawlak-Byczkowska, H. L. Horton, and V. Cooper, 1986. Assembly of functional systems in the brain: Molecular and anatomical studies of the limbic system. In *The Neurobiology of Down Syndrome*, C. J. Epstein,

ed. New York: Raven Press, pp. 195–209.
LEVITT, P., and P. RAKIC, 1980. Immunoperoxidase localization of glial fibrillary acidic protein in radial glial cells and astrocytes of the developing rhesus monkey brain. *J. Comp. Neural.* 193:817–848.
LUMSDEN, A., and R. KEYNES, 1989. Segmental patterns of neuronal development in the chick hindbrain. *Nature* 337: 424–428.
LUSKIN, M. B., A. L. PEARLMAN, and J. R. SANES, 1988. Cell lineage in the cerebral cortex of the mouse studied in vivo and in vitro with a recombinant retrovirus. *Neuron* 1:635–647.
MCCONNELL, S. K., 1988a. Fates of visual cortical neurons in the ferret after isochronic and heterochronic transplantation. *J. Neurosci.* 8:945–974.
MCCONNELL, S. K., 1988b. Development and decision making in the mammalian cerebral cortex. *Brain Res. Brain Res. Rev.* 13:1–23.
MCCONNELL, S. K., and C. E. KAZNOWSKI, 1991. Cell cycle dependence of laminar determination in developing neocortex. *Science* 254:282–285.
MILLER, M. W., 1986. The migration and neurochemical differentation of gamma-aminobutyric acid (GABA)–immunoreactive neurons in rat visual cortex as demonstrated by a combined immunocytochemical-autoradiographic technique. *Dev. Brain Res.* 28:41–46.
MILLER, M. W., 1988. Development of projection and local circuit neurons in neocortex. In *Cerebral Cortex*, vol. 7, E. G. Jones and A. Peters, eds. New York: Plenum Press, pp. 133–175.
MILLER, M. W., and R. S. NOWAKOWSKI, 1988. Use of bromodeoxyuridine-immunohistochemistry to examine the proliferation, migration and time of origin of cells in the central nervous system. *Brain Res.* 457:44–52.
MISSON, J. P., C. P. AUSTIN, T. TAKAHASHI, C. L. CEPKO, and V. S. CAVINESS, Jr., 1991. The alignment of migrating neural cells in relation to the murine neopallial radial glial fiber system. *Cerebral Cortex* 1:230–240.
MUGNAINI, E., and P. F. FORSTRONEN, 1967. Ultrastructural studies on the cerebellar histogenesis: I. Differentiation of granule cells and development of glomeruli in the chick embryo. *Z. Zellforsch.* 77:115–143.
O'LEARY, D. D. M., 1989. Do cortical areas emerge from a protocortex? *Trends Neurosci.* 12:400–406.
O'LEARY, D. D. M., and B. B. Stanfield, 1989. Selective elimination of axons extended by developing cortical neurons is dependent on regional locale: Experiments utilizing fetal cortical transplants. *J. Neurosci.* 9:2230–2246.
OLSON, L., and A. SEIGER, 1972. Early prenatal ontogeny of central monoamine neurons in the rat: Fluorescence histochemical observations. *Z. Anat. Entwickl.-Gesch.* 137:301–316.
O'ROURKE, N. A., M. E. DAILEY, S. J. SMITH, and S. K. MCCONNELL, 1992. Diverse migratory pathways in the developing cerebral cortex. *Science* 258:299–302.
PARNAVELAS, J. G., J. A. BARFIELD, E. FRANKE, and M. B. LUSKIN, 1991. Separate progenitor cells give rise to pyramidal and nonpyramidal neurons in the rat telencephalon. *Cerebral Cortex* 1:463–468.
PICKEL, V. M., L. A. SPECHT, K. K. SUMAL, T. H. JOH, D. J. REIS, and A. HERVONEN, 1980. Immunocytochemical localization of tyrosine hydroxylase in the human fetal nervous system. *J. Comp. Neurol.* 194:465–474.
PIMENTA, A. F., I. FISCHER, and P. LEVITT, 1992. The partial nucleotide sequence of a cDNA encoding the limbic system-associated membrane protein (LAMP). *Mol. Cell. Biol.* 3:323a.
PRICE, J., 1987. Retroviruses and the study of cell ineage. *Development* 101:409–419.
PRICE, J., and L. THURLOW, 1988. Cell lineage in the rat cerebral cortex: A study using retroviral mediated gene transfer. *Development* 104:473–482.
PURVES, D., and J. W. LICHTMAN, 1985. *Principles of Neural Development.* Sunderland, Mass.: Sinauer Associates, Inc.
RAKIC, P., 1972. Mode of cell migration to the superficial layers of fetal monkey neocortex. *J. Comp. Neurol.* 145:61–84.
RAKIC, P., 1974. Neurons in rhesus monkey visual cortex: Systematic relation between time of origin and eventual disposition. *Science* 183:425–427.
RAKIC, P., 1977. Prenatal development of the visual system of the rhesus monkey. *Philes. Trans. R. Soc. Lond. [Biol.]* 278:245–260.
RAKIC, P., 1978. Neuronal migration and contact guidance in the primate telencephalon. *Postgrad. Med. J.* 54:25–40.
RAKIÇ, P., 1988. Specification of cerebral cortical areas. *Science* 241:170–176.
RAMÓN Y CAJAL, S., 1890. Sur l'origine et les remifications des fibres nerveuses de la moelle embryonnaires. *Anat. Anz.* 5:85–95.
SANES, J. R., L. R. RUBENSTEIN, and J. F. NICOLAS, 1986. Use of a recombinant retrovirus to study post-implantation cell lineage in mouse embryos. *EMBO J.* 5:3133–3142.
SAUER, F. C., 1935. Mitosis in the neural tube. *J. Comp. Neurol.* 62:377–405.
SAUER, F. C., and B. E. WALKER, 1959. Radioautographic study of interkinetic nuclear migration in the neural tube. *Proc. Soc. Exp. Biol. Med.* 101:557–560.
SCHAPER, A., 1897. The earliest differentiation in the central nervous system of vertebrates. *Science* 5:430–431.
SCHLAGGER, B. L., and D. D. M. O'LEARY, 1991. Potential of visual cortex to develop an array of functional units unique to somatosensory cortex. *Science* 252:1556–1560.
SCHLAGGER, B. L., and D. D. M. O'LEARY, 1993. Patterning of the barrel field in somatosensory cortex with implications for specification of neocortical areas. *Perspect. Dev. Neurobiol.* 1:81–91.
SCHMECHEL, D. E., and P. RAKIC, 1979. A Golgi study of radial glial cells in developing monkey telencephalon: Morphogenesis and transformation into astrocyte. *Anat. Embryol. (Berl.)* 156:115–152.
SIDMAN, R. L., 1970. Autoradiographic methods and principles for study of the nervous system with thymidine-H^3.

In *Contemporary Research Methods in Neuroanatomy*. W. J. H. Nauta and S. O. E. Ebbeson, eds. New York: Springer, pp. 252–274.

Sidman, R. L., I. L. Miale, and N. Feder, 1959. Cell proliferation and migration in the primitive ependymal zone: An autoradiographic study of histogenesis in the nervous system. *Exp. Neurol.* 1:322–333.

Sidman, R. L., and P. Rakic, 1973. Neuronal migration, with special reference to developing human brain: A review. *Brain Res.* 62:1–35.

Smart, I. H. M., and G. M. McSherry, 1982. Growth patterns in the lateral wall of the mouse telencephalon: II. Histological changes during and subsequent to the priod of isocortical neuron production. *J. Anat.* 134:415–442.

Stanfield, B. B., and D. D. M. O'Leary, 1985. Fetal occipital cortical neurons transplanted to the rostral cortex can extend and maintain a pyramidal tract axon. *Nature* 313: 135–137.

Takahashi, T., R. S. Nowakowski, and V. S. Caviness, Jr., 1993. Cell cycle parameters and patterns of nuclear movement in the neocortical proliferative zone of the fetal mouse. *J. Neurosci.* 13:820–833.

Tan, S.-S., and S. Breen, 1993. Radial mosaicism and tangential cell dispersion both contribute to mouse neocortical development. *Nature* 362:638–640.

Tapscott, S. J., G. S. Bennett, Y. Toyama, F. Kleinbart, and H. Holtzer, 1981. Intermediate filament proteins in the developing chick spinal cord. *Dev. Biol.* 86:40–54.

Traub, P., 1985. *Intermediate Filaments*. Berlin: Springer-Verlag.

Turner, D. L., and C. L. Cepko, 1987. A common progenitor for neurons and glia persists in rat retina late development. *Nature* 328:131–136.

Vaysse, P. J.-J., and Goldman, J. E., 1990. A clonal analysis of glial lineages in neonatal forebrain development in vitro. *Neuron* 5:227–235.

Walsh, C., 1993. Cell lineage and regional specification in the mammalian neocortex. *Perspect. Dev. Neurobiol.* 1:75–80.

Walsh, C., and C. L. Cepko, 1988. Clonally related cortical cells show several migration patterns. *Science* 241:1342–1345.

Walsh, C., and C. L. Cepko, 1992. Wide dispersion of neuronal clones across functional regions of the cerebral cortex. *Science* 255:434–440.

Walsh, C., and C. L. Cepko, 1993. Clonal dispersion in proliferative layers of developing cerebral cortex. *Nature* 362:634–636.

Wilkinson, D. G., and R. Krumlauf, 1990. Molecular approaches to the segmentation of the hindbrain. *Trends Neurosci.* 13:335–339.

Williams, B. P., J. Read, and J. Price, 1991. The generation of neurons and oligodendrocytes from a common precursor cell. *Neuron* 685–693.

Wise, S. P., and E. G. Jones, 1978. Developmental studies of the thalamocortical and commissural connections in the rat somatic sensory cortex. *J. Comp. Neurol.* 178:187–208.

Woolsey, T. A., and J. R. Wann, 1976. Areal changes in mouse cortical barrels following vibrissal damage at different postnatal ages. *J. Comp. Neurol.* 170:53–66.

Zacco, A., V. Cooper, S. Hyland-Fisher, P. D. Chantler, H. L. Horton, and P. Levitt, 1989. Isolation, biochemical characteization and ultrastructural localization of the limbic system associated membrane protein (LAMP), a protein expressed on neurons comprising functional neural circuits. *J. Neurosci.* 10:73–90.

Zhukareva V., and P. Levitt, 1992. Homophilic binding and phosphatidylinositol linkage of the limbic system associated membrain protein (LAMP) are consistent with a role in specific recognition events in the developing brain. *Mol. Cell. Biol.* 3:197a.

10 Object Perception, Object-directed Action, and Physical Knowledge in Infancy

ELIZABETH S. SPELKE, PETER VISHTON, AND CLAES VON HOFSTEN

ABSTRACT Human infants' perceptions and actions on objects are sensitive to physical constraints on object motion. Object perception is guided by constraints that objects move as connected and bounded wholes, on continuous and unobstructed paths, on contact with other objects; object-directed reaching and visual tracking are guided by the constraint that objects move smoothly. Infants' sensitivity to these constraints suggests that humans begin at an early age to develop knowledge of objects and their behavior. The contrasting constraints guiding object perception and object-directed actions suggest that separate systems of knowledge underlie these different achievements.

A central question for psychologists and neuroscientists concerns the nature and organization of human knowledge. This question may be approached through studies of perception (the primary processes by which people gain knowledge of their immediate surroundings) and action (the primary processes by which people put their knowledge to practical use). In this chapter, we explore some insights into knowledge, perception, and action that come from studies of early human development. Developmental studies of object perception and object-directed action provide evidence that knowledge of objects begins to emerge in the first months of life and that this knowledge has a particular content and organization. We hope that the findings of such studies will serve as signposts for neuroscientists, guiding the search for the physical basis of knowledge in the developing human brain.

ELIZABETH S. SPELKE and PETER VISHTON Department of Psychology, Cornell University, Ithaca, N.Y. CLAES VON HOFSTEN Department of Psychology, Umeå University, Umeå, Sweden

Object perception

Object perception is a fascinating achievement, because the boundaries, internal unity, and persisting identities of objects are radically underdetermined by the visual information available in natural scenes. Because objects touch and overlap in complex ways, no simple relationships clearly indicate where one object ends and the next begins (see Marr, 1982). Because the back of every opaque object is hidden and the front surfaces of most objects are partly concealed behind nearer objects, the visual information specifying an object is highly incomplete. Because objects and perceivers are movable, finally, objects frequently enter and leave the field of view. Despite these ambiguities and changes, adults immediately and effortlessly perceive objects as stably bounded, as complete and solid, and as persisting and moving on definite paths when they are hidden. An important task for cognitive scientists and neuroscientists is to explain how these perceptions occur; studies of object perception in infancy provide one approach to this task.

FIGURE-GROUND ORGANIZATION According to the Gestalt psychologists (e.g., Koffka, 1935), the ability to perceive an object as a bounded figure in front of an unbounded background is fundamental to all perception and depends on a tendency to confer the simplest organization on visual scenes. Studies of infants' perception of figure-ground relations provide evidence that this ability exists at an early age. For example, Craton (cited in Arterberry, Craton, and Yonas, 1993) investigated 5-month-old infants' perception of a two-dimensional display of moving dots that is seen by adults as a unitary figure moving in front of a background. After familiarization with the display, infants were presented, in alternation, with a stationary object of the shape of the region adults perceive as the figure and a stationary object with the shape of the region adults perceive as the ground. Looking times to these objects were compared to one another and to the looking times of infants in a no-habituation baseline condition, on the well-documented assumption that infants would look longer at the display they perceived to be more novel (see Bornstein, 1985). If infants perceived the figure-ground relations as adults do, then the infants in the experimental condition were expected to look longer at the test display with the shape of the ground, because only the figure should have been perceived as having a definite shape during the familiarization period. This visual preference was observed, providing evidence that infants perceived the figure-ground relation.

Further experiments have explored the stimulus conditions specifying figure-ground relations to infants. Infants appear to perceive a unitary object in front of a background when the figure and its borders move together either laterally (e.g., Craton and Yonas, 1990) or in depth (e.g., Ball and Tronick, 1971). Certain patterns of nonrigid motion also specify figure-ground relations (Bertenthal, 1993). Finally, infants appear to perceive figure-ground relations in stationary displays in which figure and ground are separated in depth (e.g., Termine et al., 1987). In contrast, infants do not appear to perceive figure-ground relations in stationary, two-dimensional displays in which figure and ground differ only in color and texture (Spelke and Born [1982] in Spelke, 1988) or in which the figure-ground organization is specified by configurational properties such as edge alignment and figural simplicity (Craton and Yonas, 1990). These last findings cast doubt on the thesis that figure-ground organization depends on a tendency to group displays into the simplest, most homogeneous units. The basic process of figure-ground organization appears to depend on an analysis of three-dimensional surface motions and arrangements, in accord with the constraint that moving objects maintain their internal connectedness and their external boundaries.

Perception of Objects in Multiple-Object Arrays

Young infants also perceive the boundaries and the unity of objects in more cluttered visual scenes in which multiple objects touch or overlap. For example, an experiment investigated 3-month-old infants' percep-

(a)

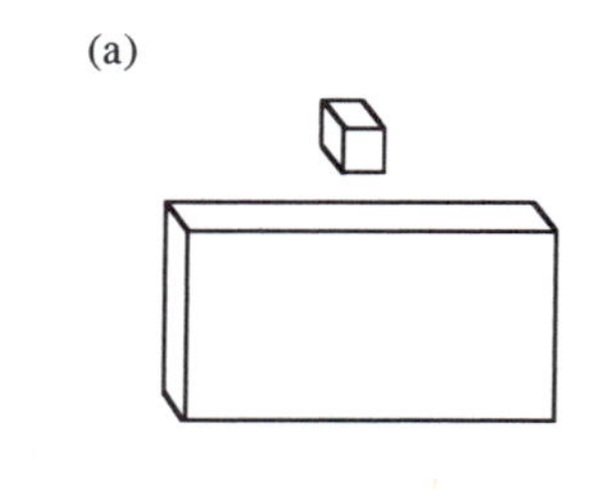

(b)

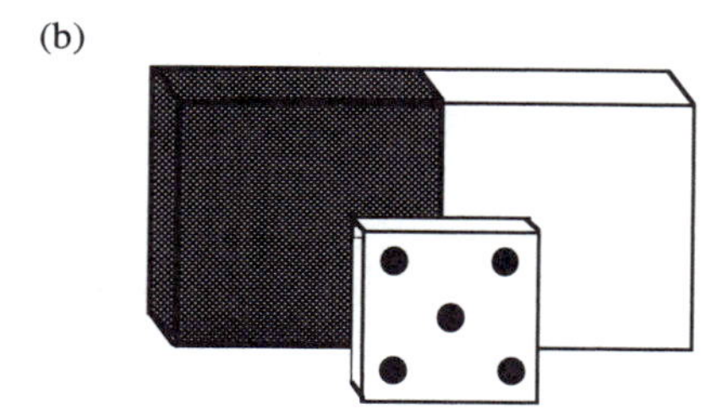

(c)

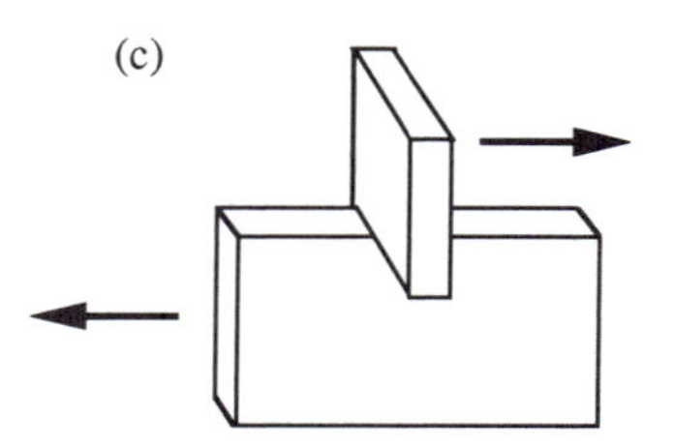

(d)

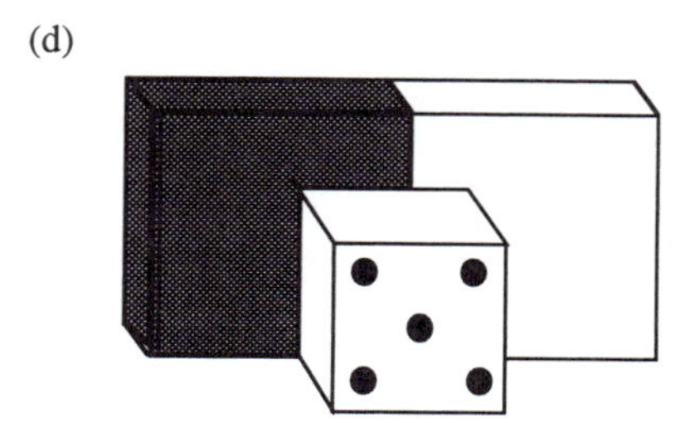

Figure 10.1 Displays for experiments on infants' perception of object boundaries: (a) two stationary, visibly separated objects (after Spelke, Hofsten, and Kestenbaum, 1989); (b) two stationary objects separated in depth; (c) two adjacent objects undergoing relative motion; (d) two stationary objects adjacent in depth (after Spelke, 1990).

tion of the boundaries of two stationary objects that were separated in depth, by familiarizing infants with this display and then presenting test displays in which either one object was displaced relative to the other object or the two objects were displaced together (Kestenbaum, Termine, and Spelke, 1987) (figure 10.1b). If infants perceived a boundary between the two objects, they were expected to look longer at the display in which the two objects moved together. This looking preference was obtained, providing evidence that infants perceived the objects as separate units.

Further research provides evidence that infants also perceive the boundary between two objects that are visibly separated and the boundary between two objects that undergo different rigid motions, but remain adjacent throughout their motion (Spelke, Hofsten, and Kestenbaum, 1989) (figure 10.1a, c). In contrast, young infants evidently do not perceive the boundary between two adjacent objects that are stationary or move together (Kestenbaum et al, 1987; Spelke, Hofsten, and Kestenbaum, 1989) (figure 10.1d). Although young infants perceive object boundaries by analyzing surface arrangements and motions, grouping surfaces into units that are connected and remain connected over motion, they do not appear to perceive object boundaries by analyzing surface colors and textures, dividing the layout into units that are simple and regular.

In visual scenes, objects often are partially occluded by nearer objects, such that different visible surfaces of an object appear in spatially separated regions of the layout (figure 10.2). To investigate whether infants perceive the unity of an object over this pattern of occlusion, Kellman and Spelke (1983) familiarized 4-month-old infants with an object of a simple shape that moved horizontally behind a central occluder (see figure 10.2a). Then the infants were presented with test displays consisting either of the connected, unitary object perceived by adults or of the surfaces that were visible in the original occlusion display, separated by a gap (see figure 10.2d). Infants looked longer at the latter test display, providing evidence that they perceived the center-occluded object as one connected unit. In further studies, infants were found to perceive the unity of this object when its ends underwent common motion in any direction, including motion in depth, but not when its ends were stationary and underwent a common retinal displacement produced by movement of the infant (see Kellman, 1993). These findings suggest that perceived, three-dimensional motion, not retinal displacement, underlies infants' perception of object unity.

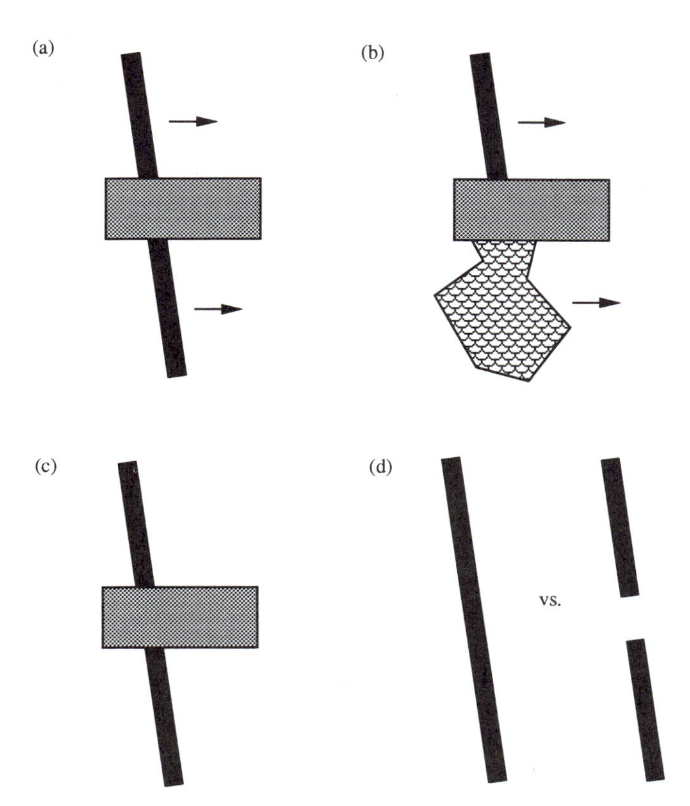

FIGURE 10.2 Displays for experiments on infants' perception of partly occluded objects: (a) Aligned surfaces of the same color, texture, and shape move rigidly behind a central occluder. (b) Misaligned surfaces of different colors, textures, and shapes move rigidly behind a central occluder. (c) Aligned surfaces of the same color, texture, and shape are stationary behind a central occluder. (d) Test displays for an experiment on perception of object unity over occlusion. (After Kellman and Spelke, 1983)

Experiments provide evidence that infants' perception of the unity of a center-occluded object is not affected by the colors, textures, shapes, or alignment relations of the object's visible surfaces. If the ends of a center-occluded object move together, for example, infants' perception of a connected object is equally strong, whether the ends form an object of a homogeneous color and texture and a simple shape (figure 10.2a) or not (figure 10.2b). When a center-occluded object is stationary, moreover, infants' perception of object unity appears to be indeterminate between one connected object and two objects separated by a gap, even when the visible ends of the object share the same color and texture, are aligned, and combine to form an object of a simple shape (see figure 10.2c). Perception of the unity of a partly occluded object therefore appears to depend only on the common motion of the object's visible surfaces. This perception may reflect a sensitivity to a further constraint on object motion: Surfaces normally move together only if they are in contact.

Perception of Objects that Move from View Adults perceive objects to persist when they are fully occluded and to maintain their unity and identity over successive encounters. In experiments using preferential-looking methods similar to those described earlier, infants also have been found to perceive the persistence and the identity of an occluded object under certain conditions. For example, Craton and Yonas (1990) familiarized 4-month-old infants with a disc that moved from a position where it was fully visible to a position where it was fully hidden (figure 10.3a). The object followed an irregular path, such that it was most often seen at a position where it was half in view. Infants subsequently were shown a complete disc and a truncated disc corresponding in shape to the visible surface of the disc when it was half visible. Infants looked longer at the truncated disc, providing evidence that they perceived a persisting object with a constant form over the period of occlusion. In further studies, infants have been found to perceive the persistence of an object during occlusion primarily by analyzing the motion relationships among the object's visible surfaces (Van de Walle and Spelke, 1993) (figure 10.3b). Infants have shown very limited abilities to perceive the form of an object that moves in and out of view by analyzing configural relationships among the object's visible edges (Arterberry, Craton, and Yonas, 1993).

Experiments have investigated infants' perception of object identity over successive encounters by familiarizing infants with events in which objects appear in succession on two sides of a screen (see figure 10.3c–f) and then comparing the looking times of those infants and of infants in a no-familiarization control condition to nonoccluded displays of one versus two objects (Xu and Carey, 1992; Spelke et al., in press, c). These studies provide evidence that both 4- and 10-month-old infants perceive object identity by analyzing the spatiotemporal continuity or discontinuity of object motion: When two object appearances are linked by a connected path of motion (figure 10.3c), infants perceive a single object; when two object appearances are not so linked (figure 10.3d), infants perceive two distinct objects. In contrast, experiments provide no evidence that infants perceive the identity or distinctness of objects by analyzing either changes in objects' speed of motion or changes in objects' colors, textures, and shapes. When objects move into view in alternation on the two sides of a wide screen (figure 10.3e), infants' perception appears to be indeterminate between one and two objects, both when the timing of the appearances specifies uniform, constant speed of motion behind the occluder and when it specifies an abrupt change in the speed of motion. Infants appear to perceive an indeterminate number of objects not only when the objects are visually indistinguishable (figure 10.3e) but also when they differ in color, texture, and shape (figure 10.3f). Perception of object identity evidently accords with the constraint that objects move continuously but not with constraints that objects move smoothly and maintain constant colors, textures, and shapes.

An Interim Summary: Principles of Object Perception A consideration of infants' successes and failures in the preceding reported experiments supports several suggestions about early developing mechanisms for perceiving objects. First, the conditions under which infants perceive figure-ground relations, object unity and boundaries, and object persistence over a period of occlusion are similar. In all three situations, infants perceive objects by analyzing the arrangements and the motions of surfaces but not the colors, textures, or forms of surfaces. These findings suggest that abilities to perceive figure-ground organization, object unity and boundaries, and object identity are related.

Second, infants' perception of objects depends on analyses of the arrangements and motions of surfaces in the three-dimensional visual layout, not on analyses of the arrangements and motions of elements in any two-dimensional retinal projection of the layout. For example, infants perceive the boundaries of objects that are separated in depth but adjacent in the visual field, and they perceive the unity of a center-occluded object when the ends of the object undergo a three-dimensional but not a retinal displacement. These findings and others (see Spelke, 1990, and Kellman, 1993, for discussion) suggest that the processes underlying object perception occur relatively late in visual analysis.

Third, object perception accords with basic physical constraints on object motion. Three principles, each

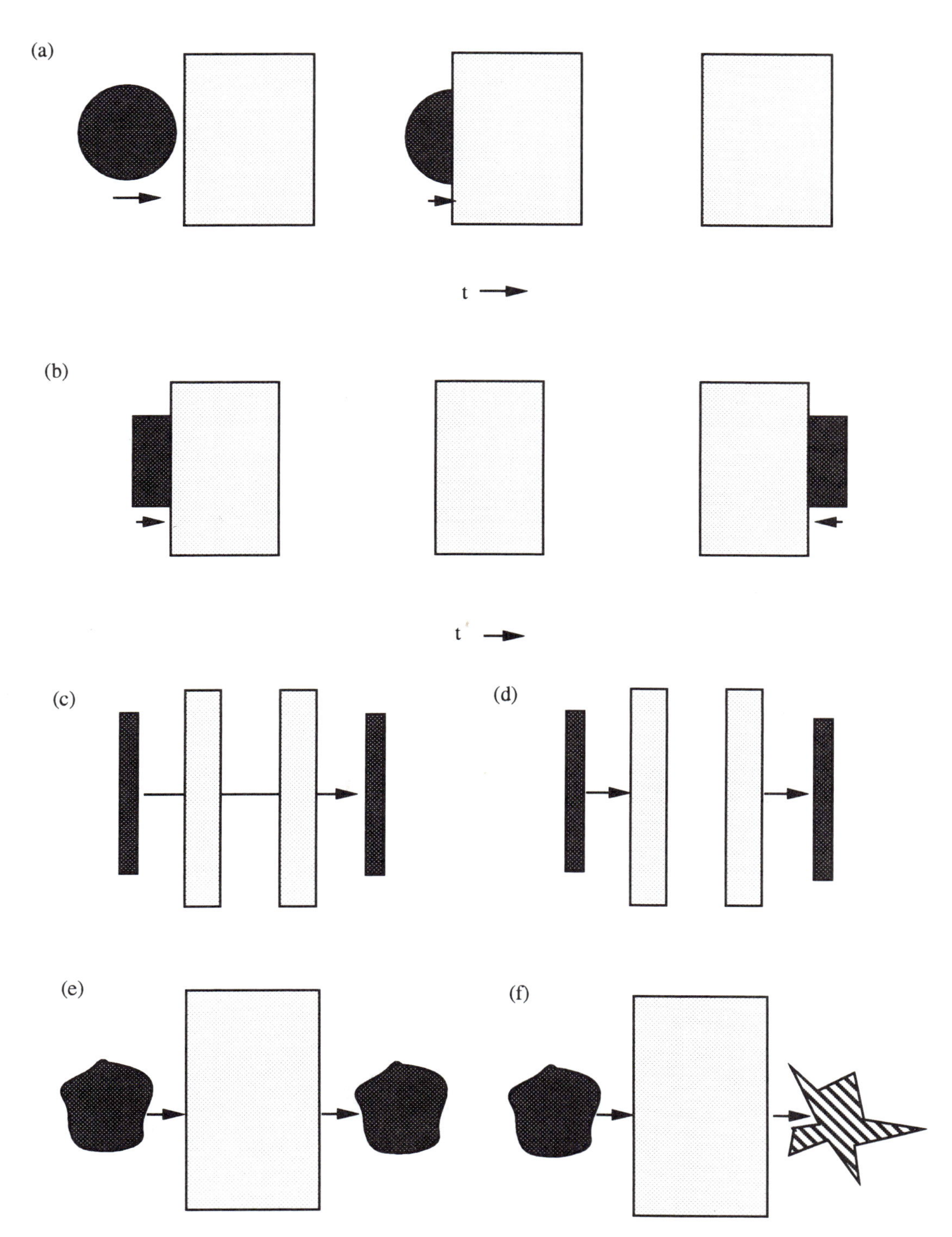

FIGURE 10.3 Displays for experiments on infants' perception of object persistence over occlusion: (a) An object moves from a fully visible to a fully occluded position. (b) An object moves from one partly visible position to a fully hidden position and then to a second partly visible position. (c) An object moves continuously behind two separated occluders. (d) Objects move discontinuously behind two separated occluders. (e) An object moves visibly at a constant speed and is occluded for an appropriate or an inappropriately brief duration. (f) Objects that are different in color, texture, and shape appear in alternation on the two sides of an occluder. (After Craton and Yonas, 1990; Xu and Carey, 1992; Spelke et al., in press, c; Van de Walle and Spelke, 1993) The drawing does not accurately depict the scale or shapes of the objects.

A. The principle of cohesion: A moving object maintains its connectedness and boundaries

Motion in accord with cohesion

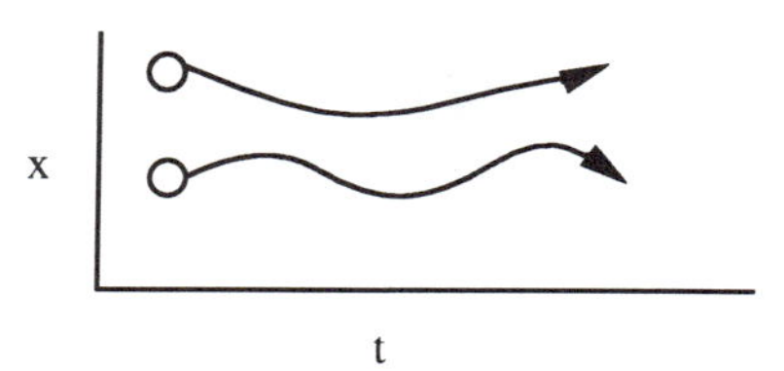

Motion in violation of cohesion

connectedness violation

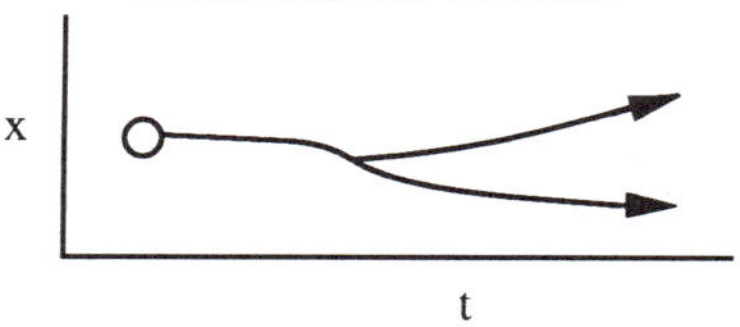

boundedness violation

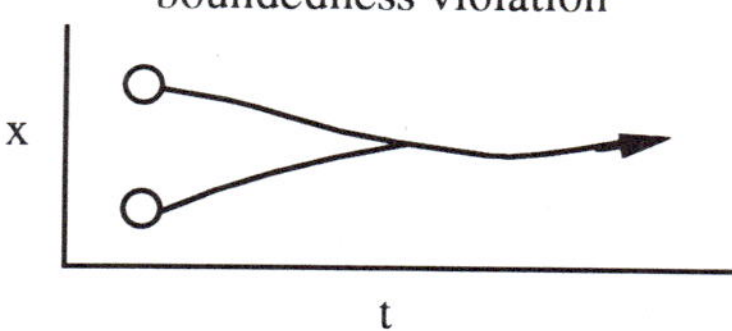

B. The principle of continuity: A moving object traces exactly one connected path over space and time

Motion in accord with continuity

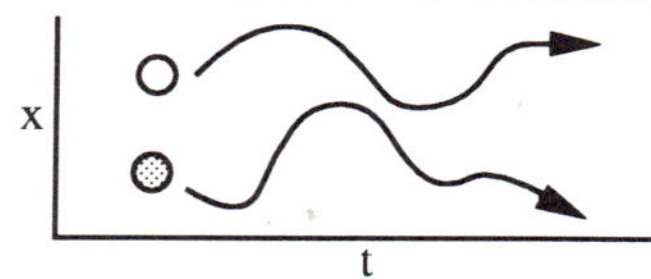

Motion in violation of continuity

continuity violation

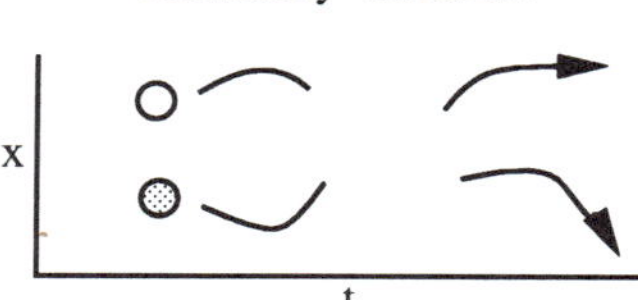

solidity violation

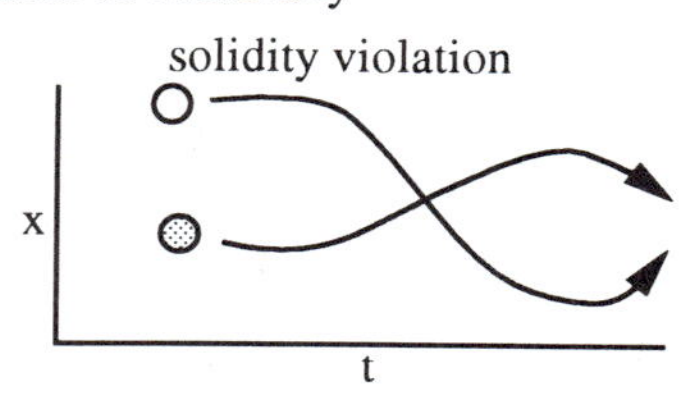

C. The principle of contact: Objects move together if and only if they touch

Motion in accord with contact

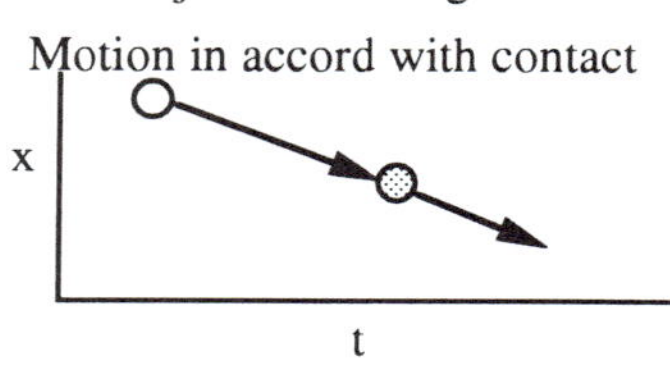

Motion in violation of contact

action on contact violation

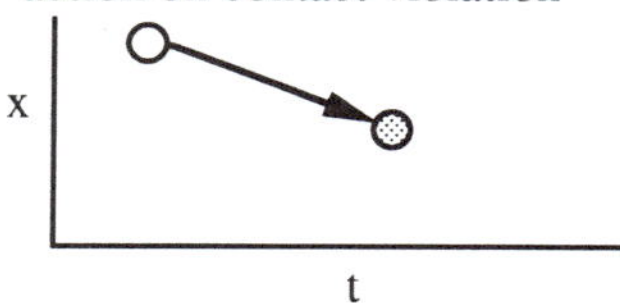

no action at a distance violation

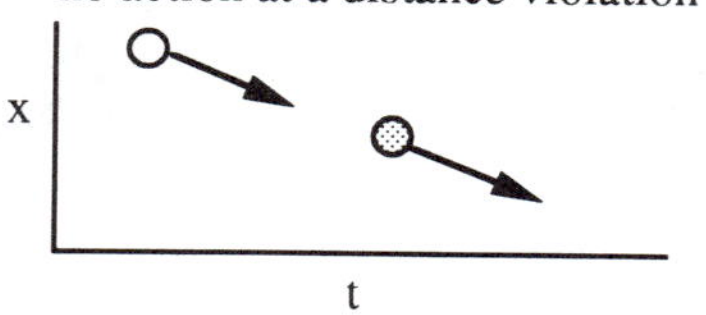

FIGURE 10.4 Principles of object perception and the constraints they encompass.

encompassing two constraints on object motion, suffice to account for the findings of all the studies cited thus far (figure 10.4). According to the *cohesion principle*, objects maintain their connectedness and their boundaries as they move. Objects therefore are distinct from the background, and two objects are distinct from one another, if they are separated in space or undergo separate motions. According to the *contact principle*, distinct objects move together if and only if they touch. The two commonly moving ends of a partly occluded object therefore are perceived to be in contact behind the occluder, whether they appear simultaneously (as in figure 10.2a) or successively (as in figure 10.3b). According to the *continuity principle*, an object traces exactly one connected path over space and time: The path of one object contains no gaps, and the paths of two objects do not intersect. An object therefore is seen as persisting when it moves from view, and two object appearances that are linked by a continuous path of motion are seen as appearances of a single object.

The cited experiments suggest limits to young infants' abilities to perceive objects. In particular, infants do not appear to perceive objects by grouping together surfaces with a common color, texture, or shape, by grouping surfaces into units with a simple form, or by grouping together surfaces that move smoothly. Infants fail to perceive objects in accord with the latter properties, even though infants within the same age range have been shown to be sensitive to surface color, surface texture, configural object properties such as symmetry and alignment, and kinematic relationships such as smoothness of motion (see Spelke, 1990). Gestalt configurational properties are detectable by infants, but they do not appear to provide the basis for infants' perception of objects. These findings cast doubt on the thesis that object perception results from a general tendency to confer the most regular organization on perceptual experience.

Observing Object Motions We turn now to studies investigating in a different way infants' sensitivity to constraints on object motion, by comparing infants' looking times to fully visible events in which an object moves either naturally or unnaturally. This research is based on the assumption that infants will look longer at an unnatural object motion if they are sensitive to the constraints that it violates (see Baillargeon, 1993). A variety of experiments provide evidence that infants infer that objects will move in accord with some, but not all, of the constraints to which adults are sensitive. Specifically, infants appear to be sensitive to the constraints on objects that are captured by the principles of cohesion, continuity, and contact.

For example, experiments by Spelke and colleagues (in press, a) investigated 3-month-old infants' reactions to events in which an object either moved as a whole (consistent) or spontaneously broke apart (inconsistent with the cohesion principle). Infants first were familiarized with a stationary object, and then they were tested with two events in which a hand grasped the top of the object and lifted it into the air (figure 10.5). In one event, the object moved as a whole and came to rest in midair. In the other event, the top half of the object rose into the air while the bottom half of the object remained on the surface. Looking times to the outcomes of these events were recorded, beginning when all or part of the object came to rest in midair, and these looking times were compared to the looking times obtained in a baseline condition, in which the same outcome displays appeared with no preceding events. The infants in the main experiment looked longer at the event outcome in which the object broke in two, providing evidence that infants inferred that the object would move in accord with the cohesion principle.

Further experiments using similar methods provide evidence that 6-month-old infants infer that one visible object will not pass through a second object, in accord

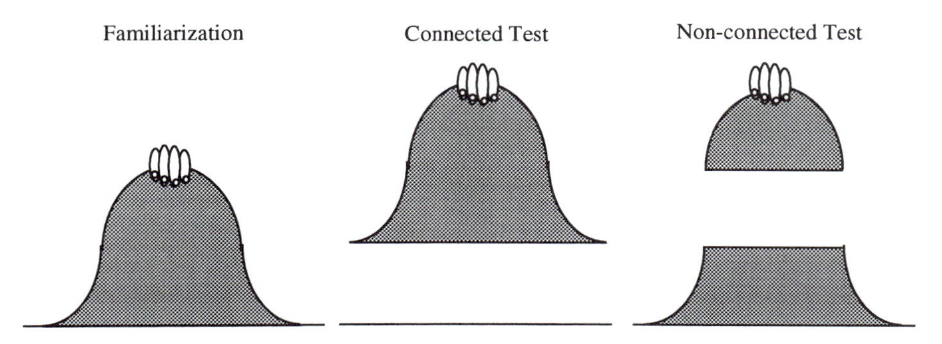

Figure 10.5 Displays for a study of infants' extrapolations of visible object motion in accord with the cohesion principle. (After Spelke et al., in press, a)

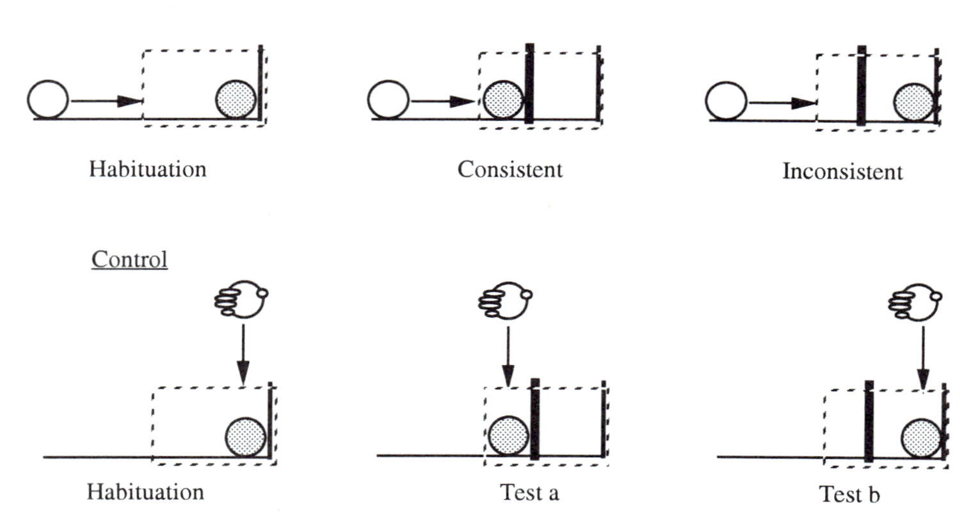

FIGURE 10.6 Displays for a study of infants' extrapolations of hidden object motion in accord with the continuity principle. (After Spelke et al., 1992) Dotted lines indicate the position of the screen; arrows indicate the path of visible object motion; the shaded figure indicates the final position of the object.

with the continuity principle (Sitskoorn and Smitsman, 1993) and that one object will not cause a change in the motion of a second object if the objects do not touch, in accord with the principle of contact (Leslie, 1988). Experiments suggest, however, that young infants are not sensitive to all the constraints on physical objects that are appreciated by adults. In particular, infants do not show consistent reactions to violations of the principle of inertia: They show no visual preference for an event in which a moving object spontaneously stops moving or changes direction (e.g., Leslie, 1988). These experiments suggest that young infants do not appreciate that objects move smoothly in the absence of obstacles.

OBSERVING EVENTS WITH HIDDEN OBJECTS Infants' inferences about the motions of hidden objects have been investigated by means of preferential-looking experiments in which critical object motions occur while the object is out of sight. These experiments provide evidence that infants infer that hidden objects will move in accord with the same principles of cohesion, contact, and continuity that guide their inferences about visible objects.

For example, Spelke and coworkers (1992) familiarized $2\frac{1}{2}$-month-old infants with an event in which a ball rolled behind a screen on an open stage, the screen was raised to reveal the object at rest at the end of the stage (figure 10.6), and looking time to this outcome display was recorded. Then infants were tested with events in which a barrier was introduced at the center of the stage in the path of the ball's motion, the ball was rolled behind the screen as before, and the screen was raised to reveal the ball either at a novel position on the near side of the barrier (consistent) or at its familiar position on the far side of the barrier (inconsistent with the continuity principle). Infants' looking times to the two event outcomes were compared to the looking times of infants in a control condition, who were presented with the same outcome displays preceded by equally consistent events. The infants in the experimental condition looked longer at the inconsistent event outcome. Because the ball could arrive at the inconsistent outcome position only by passing through or by jumping discontinuously over the barrier, the experiment provides evidence that the infants extrapolated the hidden object's motion on a connected and unobstructed path, in accord with the continuity principle (for further evidence, see Baillargeon, 1993).

Further experiments using similar methods provide evidence that infants extrapolate hidden object motion in accord with the principles of cohesion and contact: Infants infer that a hidden object will maintain its connectedness (Carey, Klatt, and Schlaffer, 1992) and that it will set a second object in motion only if the objects touch (Ball, 1973). Experiments nevertheless suggest that young infants do not infer that a hidden object will move in accord with inertia: Presented with events in which a linearly moving object rolled from view behind a screen, infants looked no longer at an event outcome in which the object reappeared at a position that was 90° displaced from the line of its previous motion than at an outcome in which the object reappeared on the line of its previous motion (Spelke et al., in press, b).

Infants' extrapolations of hidden object motions closely resemble their extrapolations of visible object motions, and both kinds of extrapolations accord with the same principles that guide object perception. The existence of common principles underlying perception of visible objects, sensitivity to the naturalness of visible object motion, and inferences about the course of hidden object motion suggests that these abilities depend in part on common mechanisms, attuned to basic constraints on objects' arrangements and motions. This suggestion may appear surprising, given the obvious differences between the seemingly immediate and effortless process of perceiving objects and the often more lengthy and difficult process of reasoning about object motion. If the suggestion is correct, then cognitive scientists and neuroscientists may gain considerable insight into the mechanisms underlying reasoning processes by studying the mechanisms of object perception.

Object-directed action

Thus far, we have considered infants as observers but not as actors. Perceivers of all ages also act on objects, however, and successful action requires knowledge of how objects behave. To lift a cup, for example, one must know where and how to grasp it and where and how to apply force to raise and balance it. Casual observation suggests that much of this knowledge is lacking at early ages. Does any knowledge guide infants' actions on objects? We consider this question by focusing on the early development of object-directed reaching and visual tracking.

REACHING FOR STATIONARY OBJECTS To manipulate their surroundings, infants must direct their reaching and grasping toward the edges of objects. Even newborn infants direct their reaching movements approximately toward an object that moves slowly and irregularly in front of them, but neonates typically do not grasp objects (Hofsten, 1982). At approximately 4 months of age, infants begin to succeed both at reaching for objects and at grasping them (e.g., White, Castle, and Held, 1964). Presented with an object that is suspended in front of a background, 4- to 6-month-old infants' reaching is roughly appropriate to the object's distance (e.g., Yonas and Granrud, 1985), and their grasping is roughly appropriate to the object's orientation (Hofsten and Fazel-Zandy, 1984). Furthermore, a large object is approached using both hands and a small one using only one hand (e.g., Clifton et al., 1991), although the adjustments of hand opening are not systematically related to object size (Hofsten and Ronnquist, 1988).

To reach effectively in scenes containing multiple objects, infants must reach for an object as a unit. Reaching is most successful if the hand closes on the edges of a single object; it is less effective if the hand contacts the center of an object or closes on the edges of distinct objects. Studies of infants' reaching for arrays of multiple objects provide evidence that infants direct their reaches to object edges (Hofsten and Spelke, 1985). For example, 5-month-old infants who were presented with two objects that were arranged in depth reached for the borders of the nearer and smaller object, usually without touching the farther object, provided that the objects were spatially separated or underwent distinct motions. In contrast, when the objects touched and either were stationary or moved together, infants reached for the borders of the two-object assembly, either contacting only the larger, more distant object, or contacting both objects. These studies provide evidence that infants direct their reaches to object boundaries that are specified spatially or kinematically.

Studies of the stimulus information guiding infants' reaching for objects converge with studies of the stimulus information underlying object perception. Five-month-old infants reach for an object that is specified by a pattern of common motion relative to its background (Arterberry, Craton, and Yonas, 1993), even in the absence of any static information for the figure-ground relationship (Craton and Yonas, 1990). Conversely, 5-month-old infants fail to reach appropriately for an object whose boundaries are specified only by discontinuities in surface orientation or edge alignment (Hofsten and Spelke, 1985). These findings suggest an early coordination between object perception and action.

Reaching for Visibly Moving Objects To catch a moving object, one must aim for the position that the object will occupy at the time the reach is completed. Successful reaching for moving objects therefore depends on capacities to extrapolate an object's motion (Hofsten, 1983). Experiments provide evidence that infants reach for moving objects as early as they reach for stationary objects and that they aim for future object positions. Infants appear to reach for objects by extrapolating their motions smoothly.

Evidence that reaching is guided by smooth extrapolations of object motion comes from a series of studies by Hofsten (e.g., 1983). Infants aged 4–8 months were presented with an object that moved at a constant speed on a semicircular trajectory. After observing several cycles of motion, infants began to reach for the object. Measurements of the direction of displacement of the infant's hands at the time a reaching movement began served to assess the infant's initial aiming for the object. Even on the first trial on which a reach occurred, infants aimed for a future object position by extrapolating a smooth, curvilinear path of motion.

More recent experiments obtained converging evidence for smooth extrapolations of object motion by measuring 8-month-old infants' reaching for an object that abruptly halted. When an object moving rapidly on a semicircular arc suddenly stopped while the infant was reaching for it, the reach continued ahead of the object's arrested position toward a position that the object would have attained if it had continued to move smoothly (see color plate 1). These reaching errors provide evidence that infants guide their reaching prospectively, at least 200 ms in advance of an object's currently perceived position (Hofsten and Rosander, 1993).

Because the infants in the studies just cited typically viewed several cycles of object motion before they began reaching, it is possible that the infants learned that an object would move on a smooth path. A final study provides evidence that infants extrapolate smooth object motion even under conditions where such learning cannot occur (Hofsten et al., 1993). Six-month-old infants were presented with an object that moved on four trajectories with equal frequency: two linear trajectories that intersected at the center of the display and two trajectories with a sudden turn at the center (figure 10.7). Because linear and nonlinear trajectories were equally frequent, infants could not learn to predict, at the display's center, whether the object would continue in linear motion or turn. At the time the object reached the center, however, the infants' reaching movements were aimed to a position further along the line of the object's motion. Predictive reaching for moving objects evidently is guided by the principle that a linearly moving object continues in linear motion.

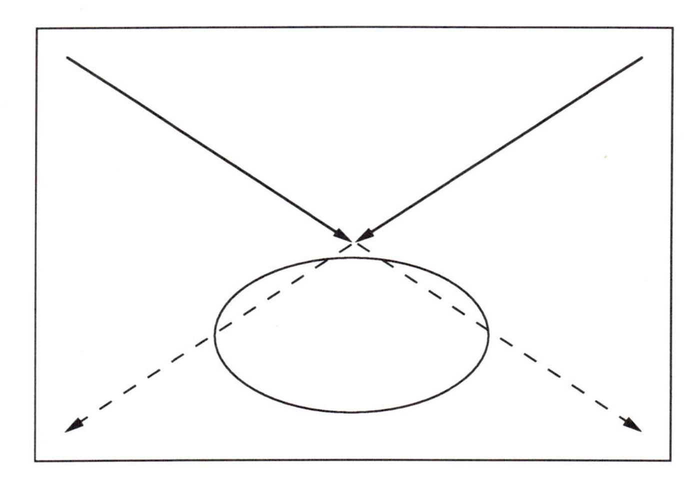

Figure 10.7 Displays for a study of infants' predictive reaching for linearly versus nonlinearly moving objects. (After Hofsten, Spelke, Vishton, and Feng, 1993) Solid arrows indicate the two paths of motion up to the intersection point of the paths, where predictive reaching was measured; dashed arrows indicate the two paths of motion after that point; the circle indicates the optimal area for reaching for the object.

Searching for Hidden Objects The most extensive studies of infants' reaching and visual following have focused on objects that leave the infants' view; the findings of such studies present a complex picture of the development of object-directed actions. In some respects, studies of infants' actions on hidden objects reveal sensitivity to the same constraints on object motion as studies of infants' perception of visible and hidden object motion. In other respects, studies of actions on hidden objects reveal different capacities and developmental patterns than any of the studies described earlier.

We first consider infants' reaching for objects obscured by darkness. If an object appears in a lighted room and then the lights are extinguished, 5-month-old infants reach reliably for the object, aiming for its previously visible borders (Hood and Willats, 1986). If a sound first is paired with a visible object and then is played in the dark, infants reach for the borders of the unseen object with which that sound had been associated (Clifton et al., 1991). These actions provide evidence that infants perceive an object as persisting when it ceases to be visible.

These reaching patterns contrast with infants' reactions to a visible object that moves out of view behind an occluder. Numerous studies, beginning with classical observations by Piaget (1954; see Harris, 1987, for a recent review), provide evidence that infants younger than 8 months do not reach for occluded objects and may not even follow such objects visually. Infants fail to retrieve hidden objects, even after they are trained to retrieve out-of-reach but visible objects (Munakata, 1992). Instead, infants act as if an occluded object is no longer obtainable or even of interest.

If an object moves in and out of view repeatedly on a smooth trajectory, infants as young as 3 months begin to adjust their visual tracking to the object's motion, looking toward its point of reappearance after occlusion (Nelson, 1971). The emergence of this tracking pattern suggests that infants have come to anticipate the object's reappearance. Research by Moore, Borton, and Darby (1978) investigated the principles guiding this anticipation, by presenting 5- and 9-month-old infants with events in which an object's occluded motion either was natural (as in figure 10.3c and 10.3e) or violated one of three constraints on object motion: the continuity constraint (as in figure 10.3d), the constraint that objects move smoothly (the object moved as in figure 10.3e but was occluded for an inappropriately brief duration), and the constraint that objects maintain constant appearances as they move (as in figure 10.3f). Visual tracking was recorded on both the natural and the violation trials, on the assumption that infants would track the event smoothly if the event proceeded as they had anticipated and that they would interrupt their tracking if it did not. At both ages, infants showed reliable disruptions of visual tracking when changes occurred in the speed of object motion or in the object's size, shape, and color (Moore, Borton, and Darby, 1978). In contrast, only the older infants showed a disruption of visual tracking when they viewed an event in which the object motion was spatially discontinuous. These findings are exactly opposite to the findings of preferential-looking experiments (Spelke et al., in press, c; Xu and Carey, 1992): Young infants' visual search appears to be guided by the principle that objects move smoothly and the principle that objects maintain constant visual appearances and not by the principle of continuity.

Finally, numerous experiments have investigated the principles that guide older infants' manual search for hidden objects. (Piaget [1954] and Harris [1987] offer psychological analyses of infants' developing search patterns; Goldman-Rakic [1987], offers a neurobiological analysis.) After infants begin to reach for hidden objects at approximately 8 months of age, their search is subject to striking errors. Infants can be induced to search for an object at places to which the object could not move on any connected, unobstructed path: Object search fails to accord with the continuity principle. When an object moves in succession behind two occluders, infants sometimes look for the object behind the first occluder or behind both occluders at once, as if the object could have left parts of itself in both locations: Object search fails to accord with the cohesion principle. Finally, manual search for hidden objects is not appropriately guided by the contact principle: If a hidden object stands inside a visible object that moves, infants fail to search for the hidden object in a position that is appropriate to the motion of the visible object. In each of these cases, infants' failure to search for hidden objects contrasts with their successful extrapolations of the motions of objects that they observe but on which they do not act (see Baillargeon, 1993).

Studies of infants' search for hidden objects suggest two patterns of dissociation in infants' knowledge. First, there is a dissociation between patterns of manual search for hidden objects and patterns of visual search for hidden objects. At 8 months, for example, visual search accords with the continuity principle but manual search does not (compare Piaget, 1954, to Moore, Borton, and Darby, 1978). Second, there is a dissociation between patterns of manual search for objects and perceptions and inferences about objects that are observed but not manipulated. When infants begin to search for hidden objects at 8 months, their search does not appear to accord with any of the principles that guide their inferences about the motions of hidden objects in events that they observe without overt action (compare Piaget, 1954, to Baillargeon, 1993).

SUMMARY: OBJECT-DIRECTED ACTION Although studies of infants' reaching for stationary, visible objects reveal a close convergence between object-directed reaching and object perception, studies of infants' reaching for visibly moving objects and infants' looking and reaching for hidden objects suggest discrepancies between the knowledge guiding various actions and the knowledge guiding perception. Of the numerous attempts to explain infants' developing action patterns, none appears fully successful. In particular, infants' errors of reaching for and visually tracking hidden objects cannot plausibly be explained by limitations on infants' abilities to represent unseen objects (Clifton et al, 1991), to engage in coordinated action (Munakata, 1992), to remember past events (Baillargeon, 1993), or to understand how objects move (Spelke et al., in press, b). These explanations may fail, because all assume that a single body of knowledge underlies human actions on objects. In contrast, the findings of the studies cited suggest that multiple representational systems guide infants' actions on objects and that these systems are distinct, in part, from the systems by which infants perceive objects and make sense of physical events.

Themes and prospects

A number of themes emerge from the studies we have reviewed. First, human infants appear to share some of the human adult's capacities to perceive, reason, and act; these cognitive capacities trace back to an early point in human life. Like studies of the developing brain (see Rakic, this volume), studies of perceptual and cognitive development cast doubt on the view that psychological capacities and their underlying neural structures develop from the periphery inward, such that infants first sense and respond reflexively to external events and only later come to perceive and reason about the significance of such events. Perception, action, and reasoning all appear to emerge early in infancy and to develop in synchrony thereafter.

Second, young infants' abilities to perceive, act on, and reason about objects appear to be sharply limited, relative to the abilities of adults. Because infants possess only a subset of mature abilities, detailed studies of object perception, object-directed action, and physical reasoning may shed light on the nature of these capacities in their mature state by revealing associations and dissociations among different abilities. Studies of infants already suggest that capacities to perceive and reason about objects are closely related to one another and that both these capacities are surprisingly distinct from capacities to act on objects. These findings may serve as guides for investigations of the neural mechanisms subserving perception, reasoning, and action in the young human brain.

Third, infants' perceptions of objects and reasoning about object motion appear to accord with the constraints on objects that are most fundamental and reliable. Although not all objects are regular in shape and substance, move on smooth paths, or maintain constant shapes, colors, and textures throughout their existence, all objects move cohesively and continuously. Before the rise of modern technology, moreover, all inanimate objects interacted only on contact. Kellman (1993) has suggested that perceptual and cognitive mechanisms that are attuned to the most reliable environmental constraints provide the firmest foundation for learning, because the information they deliver, although incomplete, is rarely in error. Studies of object perception appear to conform to Kellman's principle.

If the earliest-developing knowledge encompasses the most reliable constraints on objects, then knowledge of these constraints is likely to remain central to human perception and reasoning throughout life. Indeed, the principles of cohesion, continuity, and contact appear to be central to mature conceptions of objects: As adults, we hesitate to consider something an object if it fails to move cohesively and continuously on contact with other objects. Object perception and physical reasoning therefore appear to develop by enrichment around constant core principles. Further studies of infants may serve to probe both the detailed nature of core mature knowledge and the physical embodiment of this knowledge in the brain.

Fourth, infants' object-directed actions do not appear to be guided by the same fundamental constraints on objects as object perception. It is not the principles of cohesion, continuity, and contact that guide infants' actions on moving objects but the principle that objects move smoothly. According to the principle of inertia in classical mechanics, a moving object will continue to move in the same direction and at the same speed unless acted on by external forces. In addition, forces applied to a moving object will only gradually change the object's speed or direction. Although the inertia principle does not appear to guide either infants' or adults' reasoning about object motion (diSessa, 1983;

McCloskey, Washburn, and Felch, 1983; Spelke et al., in press, b), smooth changes in object motion are of crucial importance when one tries to coordinate one's action with events in the world. The inertial constraints on the body and the processing constraints on neural activity introduce a time lag between external events and internal adjustments to those events: It takes time to transmit information through the nervous system and further time before the contraction of a muscle has an effect. By smoothly extrapolating object motion, it is possible to overcome these time lags in action systems. The knowledge guiding object-directed actions may diverge from that guiding object perception and physical reasoning, therefore, because the demands on a knowledge system underlying action differ from the demands on a knowledge system underlying perception (Kellman, 1993).

The divergence between the knowledge guiding perception and reasoning about objects and that guiding object-directed tracking and reaching highlights our final theme. It is tempting, in view of the integrated and flexible cognitive performance of human adults, to assume that human cognition rests on a single, complex system of knowledge. Studies of infants, like studies of neurologically impaired adults (e.g., Shallice, 1987), studies of animals (e.g., Gallistel, 1990), and studies of neural development (e.g., Neville, this volume) cast doubt on this assumption: Cognition does not appear to depend on a single, homogeneous knowledge system but rather on a set of distinct systems for representing the world. The studies reviewed in this chapter suggest that the representational system underlying object perception and physical reasoning is distinct from the representational system or systems underlying many object-directed actions. Multiple, largely autonomous systems of knowledge may underlie human cognitive functioning.

In adults, distinct systems of knowledge may work together, such that a wide range of distinct beliefs can jointly influence our thinking and deliberate action (Fodor, 1983; Sperber, in press). In infancy, distinct knowledge systems may be less interconnected: Infants' actions on objects do not appear to be guided by the knowledge guiding infants' perceptions of objects, or the converse. This contrast suggests that cognitive systems become increasingly interactive over the course of human development (see Rozin, 1976; Karmiloff-Smith, 1992; Hermer, 1993). Linking different knowledge systems together may constitute a major cognitive task for the developing child.

If these suggestions are correct, then studies of human development, both in psychology and in neuroscience, may help shed light on the functional organization and the neural basis of human knowledge in two ways. First, studies of infants may serve to probe the nature and organization of a single system of knowledge under conditions that are relatively free from the interactive effects of other knowledge systems. Second, studies of cognitive development may serve to probe the processes by which humans come to relate different cognitive systems to one another. These processes, in turn, may provide a key to understanding the flexibility and productivity of mature human thinking.

ACKNOWLEDGMENTS This work was supported by grants to Elizabeth Spelke from the National Institutes of Health (HD-23103) and the National Science Foundation (INT-9214114), by a fellowship to Elizabeth Spelke from the James McKeen Cattell Foundation, by a fellowship to Peter Vishton from Cornell University, and by grants to Claes von Hofsten from the Swedish Research Council. We thank Jean-Pierre Lecanuet, Henriette Bloch, the Laboratoire de Psychobiologie du Developpement (EPHE-CNRS), and the Laboratoire de Psychologie du Developpement et de l'Education de l'Enfant (Universite Rene Descartes/CNRS) for their assistance in preparing this manuscript.

REFERENCES

ARTERBERRY, M. E., L. G. CRATON, and A. YONAS, 1993. Infants' sensitivity to motion-carried information for depth and object properties. In *Carnegie-Mellon Symposia on Cognition: Vol. 23. Visual Perception and Cognition in Infancy*, C. E. Granrud, ed. Hillsdale, N.J.: Erlbaum.

BAILLARGEON, R., 1993. The object concept revisited: New directions in the investigation of infants' physical knowledge. In *Carnegie-Mellon Symposia on Cognition: Vol. 23. Visual Perception and Cognition in Infancy*, C. E. Granrud, ed. Hillsdale, N.J.: Erlbaum.

BALL, W. A., 1973, April. The perception of causality in the infant. Paper presented at the Society for Research in Child Development, Philadelphia, Penn.

BALL, W. A., and E. TRONICK 1971. Infant responses to impending collision: Optical and real. *Science* 171:818–820.

BERTENTHAL, B. I., 1993. Infants' perception of biomechanical motions: Intrinsic image and knowledge-based constraints. In *Carnegie-Mellon Symposia on Cognition: Vol. 23. Visual Perception and Cognition in Infancy*, C. E. Granrud, ed. Hillsdale, N.J.: Erlbaum.

BORNSTEIN, M., 1985. Habituation of attention as a measure

of visual information processing in human infants: Summary, systematization, and synthesis. In *Measurement of Audition and Vision in the First Year of Life: A Methodological Overview*, G. Gottlieb and N. A. Krasnegor, eds. Norwood, N.J.: Ablex, pp. 253–300.

CAREY, S., L. Klatt, and M. SCHLAFFER, 1992. Infants' representations of objects and nonsolid substances. Unpublished manuscript, MIT, Cambridge, Mass.

CLIFTON, R. K., P. ROCHAT, R. Y. LITOVSKY, and E. E. PERRIS, 1991. Object representation guides infants' reaching in the dark. *J. Exp. Psychol. [Hum. Percept.]* 17:323–329.

CRATON, L. G., and A. YONAS, 1990. The role of motion in infants' perception of occlusion. In *The Development of Attention: 'Research and Theory*, J. T. Enns, ed. Amsterdam: Elsevier–North Holland, pp. 21–46.

DISESSA, A., 1983. Phenomenology and the evolution of intuition. In *Mental Models*, D. Gentner and A. Stevens, eds. Hillsdale, N.J.: Erlbaum, pp. 15–34.

FODOR, J. A., 1983. *The Modularity of Mind*. Cambridge, Mass.: Bradford/MIT Press.

GALLISTEL, C. R., 1990. *The Organization of Learning*. Cambridge, Mass.: Bradford/MIT Press.

GOLDMAN-RAKIC, P. S., 1987. Development of cortical circuitry and cognitive function. *Child Dev.* 58:601–622.

HARRIS, P. L., 1987. The development of search. In *Handbook of Infant Perception*, P. Salapatek and L. B. Cohen, eds. New York: Academic Press.

HERMER, L., 1993. A geometric module for spatial orientation in young children. In *Proceedings of the Fifteenth Annual Meeting of the Cognitive Science Society*. Hillsdale, N.J.: Erlbaum.

HOFSTEN, C. von, 1982. Eye-hand coordination in the newborn. *Dev. Psychol.* 18:450–461.

HOFSTEN, C. von, 1983. Catching skills in infancy. *J. Exp. Psychol. [Hum. Percept.]* 9:75–85.

HOFSTEN, C. von, and S. FAZEL-ZANDY, 1984. Development of visually guided hand orientation in reaching. *J. Exp. Child Psychol.* 38:208–219.

HOFSTEN, C. von, and L. RONNQVIST, 1988. Preparation for grasping an object: A developmental study. *J. Exp. Psychol. [Hum Percept.]* 14:610–621.

HOFSTEN, C. von, and K. ROSANDER, 1993. Perturbation of target motion during catching. Manuscript in preparation.

HOFSTEN, C. von, and E. S. SPELKE, 1985. Object perception and object-directed reaching in infancy. *J. Exp. Psychol. [Gen.]* 114:198–212.

HOFSTEN, C. von, E. S. SPELKE, P. VISHTON, and Q. FENG, 1993. *Principles of predictive catching in 6-month-old infants*. Paper presented at the Thirty-Fourth Annual Meeting of the Psychonomic Society, Washington, D.C.

HOOD, B., and P. WILLATS, 1986. Reaching in the dark to an object's remembered position: Evidence for object permanence in 5-month-old infants. *Br. J. Dev. Psychol.* 4:57–65.

KARMILOFF-SMITH, A., 1992. *Beyond Modularity*. Cambridge, Mass.: Bradford/MIT Press.

KELLMAN, P. J., 1993. Kinematic foundations of infant visual perception. In *Carnegie-Mellon Symposia on Cognition: Vol. 23. Visual Perception and Cognition in Infancy*, C. E. Granrud, ed. Hillsdale, N.J.: Erlbaum.

KELLMAN, P. J., and E. S. SPELKE, 1983. Perception of partly occluded objects in infancy. *Cogn. Psychol.* 15:483–524.

KESTENBAUM, R., N. TERMINE, and E. S. SPELKE, 1987. Perception of objects and object boundaries by three-month-old infants. *Br. J. Dev. Psychol.* 5:367–383.

KOFFKA, K., 1935. *Principles of Gestalt Psychology*. New York: Harcout, Brace and World.

LESLIE, A. M., 1988. The necessity of illusion: Perception and thought in infancy. In *Thought Without Language*, L. Weiskrantz, ed. Oxford, Engl.: Oxford University Press.

MCCLOSKEY, M., A. WASHBURN, and L. FELCH, 1983. Intuitive physics: The straight-down belief and its origin. *J. Exp. Psychol. [Learn. Mem. Cogn.]* 9:636–649.

MARR, D., 1982. *Vision*. San Francisco: Freeman.

MOORE, M. K., R. BORTON, and B. L. DARBY, 1978. Visual tracking in young infants: Evidence for object identity or object permanence? *J. Exp. Child Psychol.* 25:183–198.

MUNAKATA, Y., 1992. Rethinking object permanence: A gradualistic account. Paper presented at the annual joint University of Pittsburgh–Carnegie Mellon University conference.

NELSON, K. E., 1971. Accommodation of visual tracking patterns in human infants to object movement patterns. *J. Exp. Child Psychol.* 12:182–196.

PIAGET, J., 1954. *The Construction of Reality in the Child*. New York: Basic Books.

ROZIN, P., 1976. The evolution of intelligence and access to the cognitive unconscious. *Prog. psychobiol. Physiol. Psychol.* 6:245–279.

SHALLICE, T., 1987. Impairments of semantic processing: Multiple dissociations. In *The Cognitive Neuropsychology of Language*, M. Coltheart, G. Sartori, and R. Job, eds. Hillsdale, N.J.: Erlbaum.

SITSKOORN, M. M., and A. W. SMITSMAN, 1993. Infants' perception of object relations: Passing through or support? Unpublished manuscript.

SPELKE, E. S., 1988. Where perceiving ends and thinking begins: The apprehension of objects in infancy. In *Minnesota Symposium on Child Psychology: Vol. 20. Perceptual Development in Infancy*, A. Yonas, ed. Hillsdale, N.J.: Erlbaum.

SPELKE, E. S., 1990. Principles of object perception. *Cogn. Sci.* 14:29–56.

SPELKE, E. S., K. BREINLINGER, K. JACOBSON, and A. PHILLIPS (in press, a). Gestalt relations and object perception in infancy. *Perception*.

SPELKE, E. S., K. BREINLINGER, J. MACOMBER, and K. JACOBSON, 1992. Origins of knowledge. *Psychol. Rev.* 99: 605–632.

SPELKE, E. S., C. VON HOFSTEN, and R. KESTENBAUM, 1989. Object perception and object-directed reaching in infancy: Interaction of spatial and kinetic information for object boundaries. *Dev. Psychol.* 25:185–196.

SPELKE, E. S., G. KATZ, S. E. PURCELL, S. M. EHRLICH, and K. BREINLINGER, in press, b. Early knowledge of object motion: Continuity and inertia. *Cognition*.

SPELKE, E. S., R. KESTENBAUM, D. SIMONS, and D. WEIN,

(in press, c). Spatiotemporal continuity smoothness of motion, and object identity in infancy. *British Journal of Developmental Psychology*.

SPERBER, D. (in press). The modularity of thought and the epidemiology of representations. In *Domain Specificity in Cognition and Culture*, L. A. Hirschfeld and S. A. Gelman, eds. Cambridge: Cambridge University Press.

TERMINE, N., T. R. HYRNICK, R. KESTENBAUM, H. GLEITMAN, and E. S. SPELKE, 1987. Perceptual completion of surfaces in infancy. *J. Exp. Psychol. [Hum. Percept.]* 13:524–532.

VAN DE WALLE, G., and E. S. SPELKE, 1993, March. Integrating information over time: Infant perception of partly occluded objects. Paper presented at the meeting of the Society for Research in Child Development, New Orleans, La.

WHITE, B. L., P. CASTLE, and R. HELD, 1964. Observations on the development of visually directed reaching. *Child Dev.* 35:349–364.

XU, F., and S. CAREY, 1992. Infants' concept of numerical identity. Paper presented at the Boston University Language Acquisition Conference, Boston, Mass.

YONAS, A., and C. E. GRANRUD, 1985. Reaching as a measure of infants' spatial perception. In *Measurement of Audition and Vision in the First Year of Life*, G. Gottlieb and N. Krasnegor, eds. Norwood, N.J.: Ablex, pp. 301–322.

11 Physical Reasoning in Infancy

RENÉE BAILLARGEON

ABSTRACT How do infants learn about the physical world? Current research on the development of infants' reasoning about various types of physical phenomena (e.g., support and collision phenomena) points to two developmental patterns that recur across ages and phenomena. The first pattern is that, when learning about a new physical phenomenon, infants first form a preliminary, all-or-none concept that captures the essence of the phenomenon but few of its details. With further experience, this initial concept is progressively elaborated. Infants slowly identify discrete and continuous variables that are relevant to the initial concept, study their effects, and incorporate this accrued knowledge into their reasoning, resulting in increasingly accurate predictions over time. The second developmental pattern is that, after identifying a continuous variable as being relevant to an initial concept, infants succeed in reasoning about the variable qualitatively before they are able to do so quantitatively. This chapter reviews some of the evidence for these two developmental patterns. It is argued that the patterns reflect, at least indirectly, the nature and properties of the mechanisms infants bring to the task of learning about the physical world.

A long-standing concern of infancy research has been the description of infants' knowledge about the physical world. Traditionally, this research tended to focus on infants' understanding of occlusion events. When adults see an object occlude another object, they typically assume that the occluded object continues to exist behind the occluder. Piaget (1952, 1954) was the first to examine whether infants hold the same assumption. He concluded that it is not until infants are approximately 9 months old that they begin to appreciate that objects continue to exist when masked by other objects. This conclusion was based mainly on analyses of infants' performance in manual search tasks. Piaget noted that, prior to 9 months or so of age, infants do not search for objects they have observed being hidden. If an attractive toy is covered with a cloth, for example, young infants make no attempt to lift the cloth and grasp the toy, even though they are capable (beginning at approximately 4 months) of performing each of these actions. Piaget took this finding to suggest that young infants do not yet understand occlusion events and incorrectly assume that objects cease to exist when concealed by other objects.

In subsequent years, numerous reports were published confirming Piaget's (1952, 1954) observation that young infants typically fail to search for hidden objects (for reviews of this early research, see Gratch, 1976; Schuberth, 1983; and Harris, 1987). Piaget's *interpretation* of his observation, however, eventually came into question. Researchers came to realize that young infants might perform poorly in search tasks not because of incorrect beliefs about occlusion events but because of difficulties associated with the planning of means-end search sequences (e.g., Bower, 1974; Baillargeon, Spelke, and Wasserman, 1985; Baillargeon et al., 1990; Diamond, 1991). This led investigators to seek alternative methods for exploring infants' beliefs about occluded objects, methods that did not require infants to perform means-end action sequences.

A well-established finding in infancy research, infants' tendency to look longer at novel than at familiar stimuli (for reviews, see Banks, 1983; Olson and Sherman, 1983; Fagan, 1984; Bornstein, 1985; and Spelke, 1985), suggested an alternative method for investigating infants' intuitions about occlusion events. In a typical experiment, infants are presented with two test events: a possible and an impossible event. The possible event is consistent with the belief that objects continue to exist when occluded; the impossible event, in contrast, violates this belief. The rationale is that if infants possess such a belief, they will perceive the impossible event as more novel or surprising than the possible event and will therefore reliably look longer at the impossible than at the possible event.

Using this violation-of-expectation method, investigators have demonstrated that, contrary to traditional claims, even very young infants appreciate that objects continue to exist when occluded (see Harris, 1989;

RENÉE BAILLARGEON Department of Psychology, University of Illinois, Champaign, Ill.

Spelke et al., 1992; and Baillargeon, 1993, for recent reviews). Next, two experiments, conducted with infants aged $3\frac{1}{2}$ and $2\frac{1}{2}$ months, are described that illustrate this conclusion.

In the first experiment (Baillargeon and DeVos, 1991), $3\frac{1}{2}$-month-old infants were habituated to a toy carrot standing on end that slid back and forth along a horizontal track whose center was occluded by a screen; the carrot disappeared at one edge of the screen and reappeared, after an appropriate interval, at the other edge (figure 11.1). On alternate trials, the infants saw a short or a tall carrot slide along the track. Following habituation, the midsection of the screen's upper half was removed, creating a large window. The infants then saw a possible and an impossible test event. In the possible event, the short carrot moved back and forth along the track; this carrot was shorter than the window's lower edge and so did not appear in the window when passing behind the screen. In the impossible event, the tall carrot moved along the track; this carrot was taller than the window's lower edge and hence should have appeared in the window but did not in fact do so. The infants tended to look equally at the short- and the tall-carrot habituation events but looked reliably longer at the impossible than at the possible test event. These results indicated that the infants (1) believed that each carrot continued to exist behind the screen; (2) appreciated that each carrot could not disappear at one end of the screen and reappear at the other end without having traveled the distance behind the screen; (3) were aware that the height of each carrot determined whether it would appear in the screen window; and hence (4) were surprised by the impossible event in which the tall carrot failed to appear in the window.

The results of this experiment provided evidence that, by $3\frac{1}{2}$ months of age, infants believe that objects continue to exist when occluded. The next experiment examined whether $2\frac{1}{2}$-month-old infants possess the same belief (Spelke et al., 1992). The infants were habituated to an event in which a ball rolled from left to right along a platform and disappeared behind a screen (figure 11.2). Next, the screen was removed to reveal the ball resting against a barrier at the end of

Habituation Events

Short Carrot Event

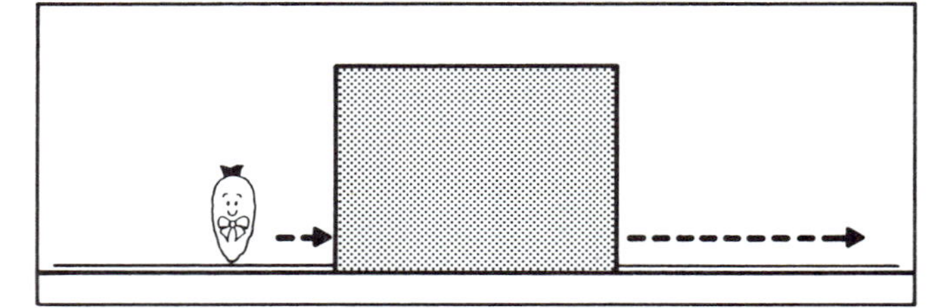

Tall Carrot Event

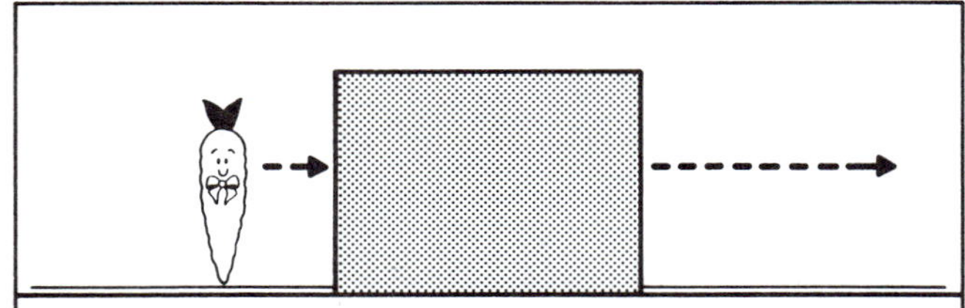

Test Events

Possible Event

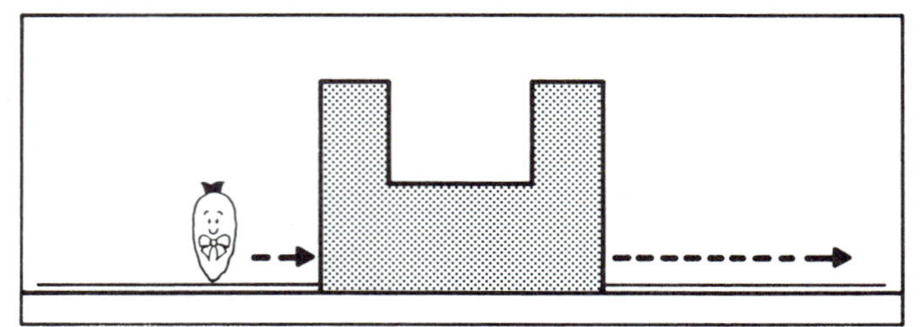

Impossible Event

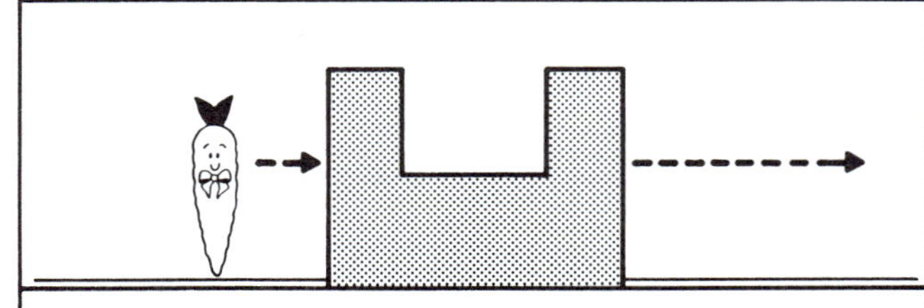

Figure 11.1 Test events used in Baillargeon and DeVos (1991).

Habituation Event

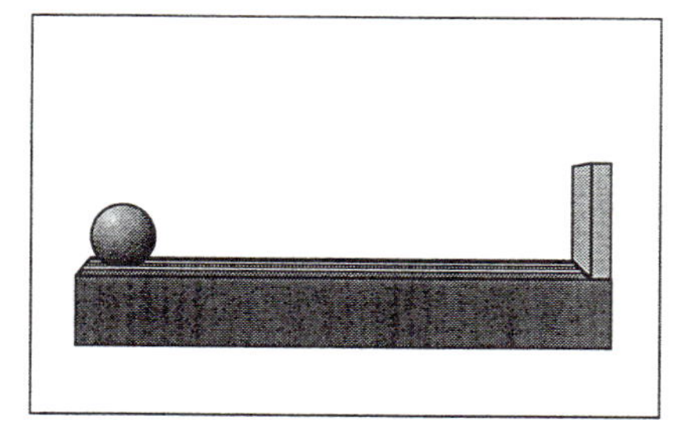
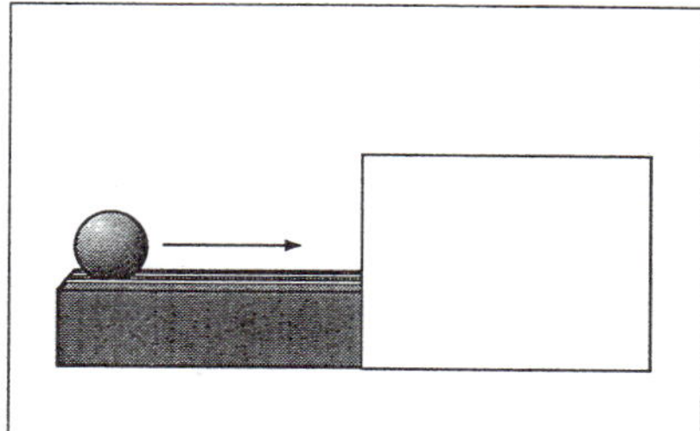
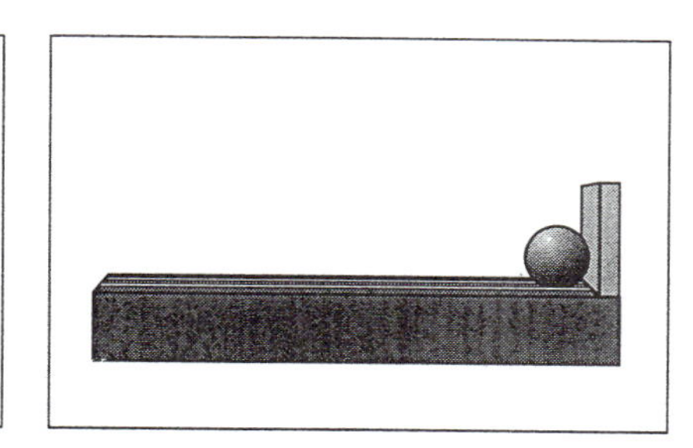

Test Events

Possible Event

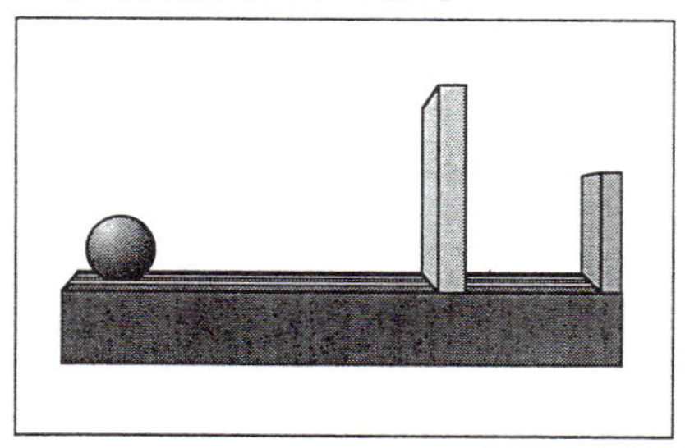
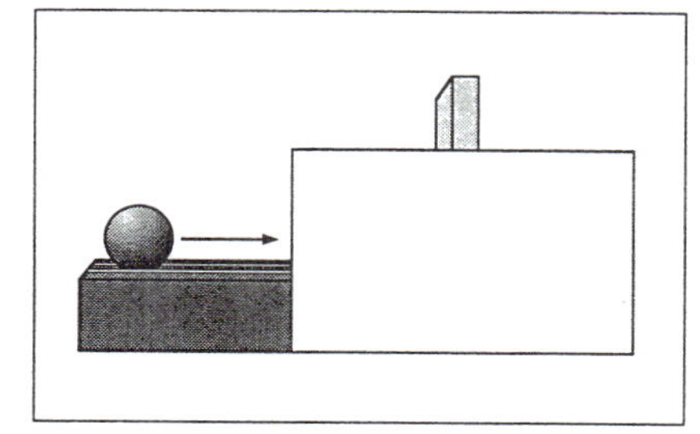
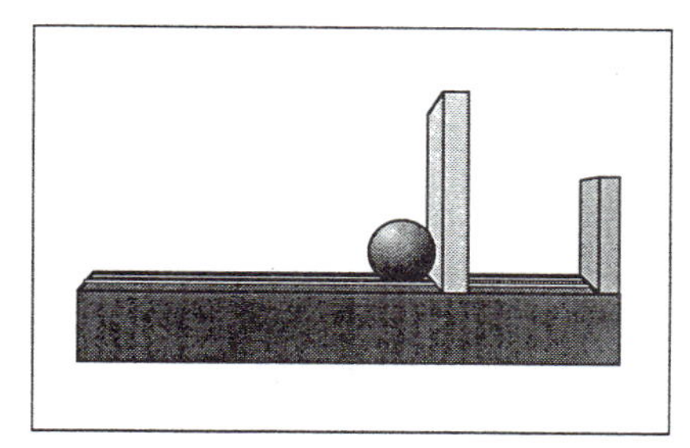

Impossible Event

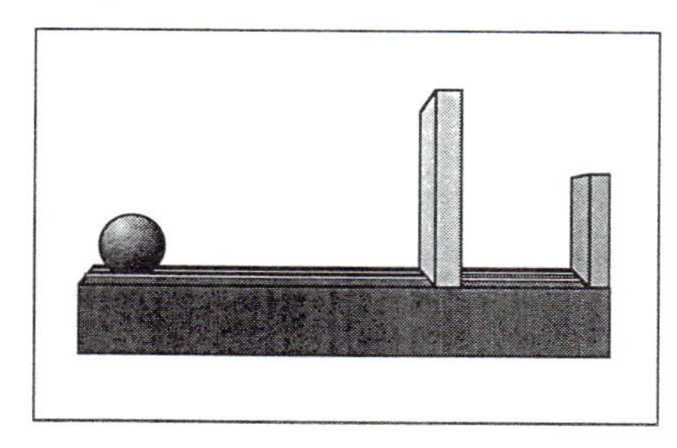
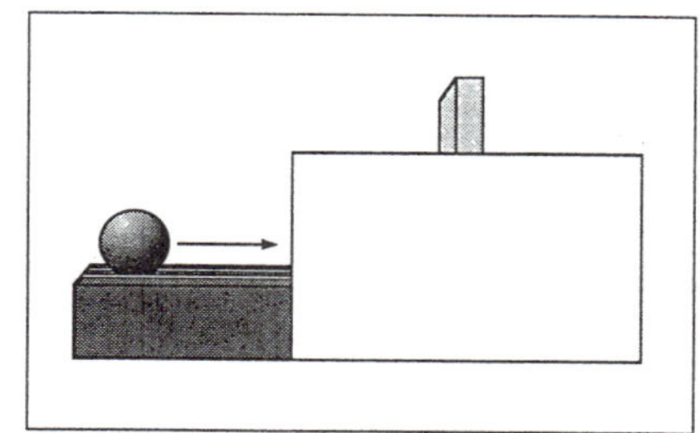
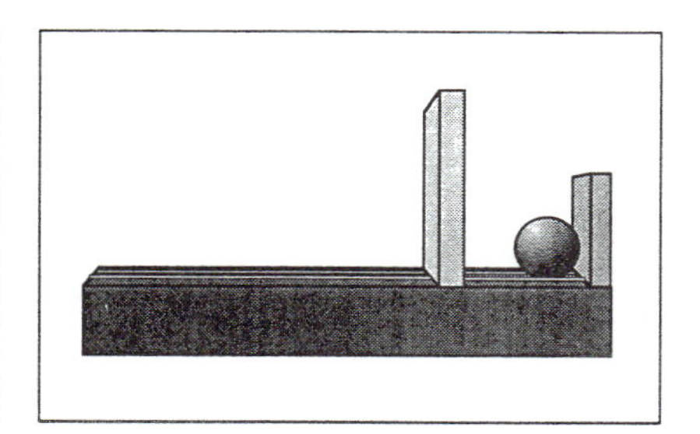

FIGURE 11.2 Test events used in Spelke and colleagues (1992). Schematic drawing based on the authors' description.

the platform. Following habituation, the infants saw two test events that were similar to the habituation event except that a tall, thin box stood behind and protruded above the screen. At the end of the possible event, the screen was removed to reveal the ball resting against the box. At the end of the impossible event, the screen was removed to reveal the ball resting against the barrier, as in the habituation event. The infants looked reliably longer at the impossible than at the possible event, suggesting that they (1) believed that the ball continued to exist behind the screen; (2) understood that the ball could not roll through the space occupied by the box; and hence (3) were surprised by the impossible event in which the ball was revealed on the far side of the box. This interpretation was supported by the results of a control condition in which the ball was lowered to the same final positions as in the possible and the impossible events.

The results of the two experiments just described indicated that, contrary to what had traditionally been claimed, even very young infants believe that objects continue to exist when masked by other objects. By virtue of their designs, the experiments also provided evidence that $2\frac{1}{2}$- to $3\frac{1}{2}$-month-old infants share adults' beliefs that objects cannot appear at two successive points in space without having traveled the distance between them and that objects cannot move through the space occupied by other objects.

How can we explain the presence of such sophisticated physical knowledge at such an early age? Over the past few years, my colleagues and I have begun to build a model of the development of infants' physical reasoning. The model is based on the assumption that infants are born not with substantive beliefs about objects, as researchers such as Spelke (1991; Spelke, Phillips, and Woodward, in press) and Leslie (1988, in press) have proposed, but with highly constrained mechanisms that guide the development of infants' reasoning about objects. The model is derived from findings concerning infants' intuitions about different physical phenomena (e.g., support, collision, and unveiling phenomena). Comparison of these findings points to two developmental patterns that recur across ages and phenomena. We assume that these patterns reflect, at least indirectly, the nature and properties of infants' learning mechanisms. These patterns are described along with some of the evidence supporting them (for further discussion of the model, see Baillargeon, in press, a, b, and Baillargeon, Kotovsky, and Needham, in press).

First pattern: Identification of initial concept and variables

The first developmental pattern is that, when learning about a new physical phenomenon, infants first form a preliminary, all-or-none concept that captures the essence of the phenomenon but few of its details. With further experience, this *initial concept* is progressively elaborated. Infants slowly identify discrete and continuous *variables* that are relevant to the initial concept, study their effects, and incorporate this accrued knowledge into their reasoning, resulting in increasingly accurate predictions over time.

To illustrate the distinction between initial concepts and variables, I will summarize experiments on the development of young infants' knowledge about support phenomena (conducted with Amy Needham, Julie DeVos, and Helen Raschke), collision phenomena (conducted with Laura Kotovsky), and unveiling phenomena (conducted with Julie DeVos).

Knowledge about Support Phenomena Our research on young infants' ability to reason about support phenomena has focused on simple problems involving a box and a platform. Our first experiment asked whether $4\frac{1}{2}$-month-old infants understand that a box can be stable when released *on* but not *off* a platform (Needham and Baillargeon, 1993). The infants again saw a possible and an impossible test event (figure 11.3). In the possible event, a gloved hand deposited a box on a platform and then withdrew a short distance, leaving the box supported by the platform. In the impossible event, the hand deposited the box beyond the platform and then again withdrew, leaving the box suspended in midair with no apparent means

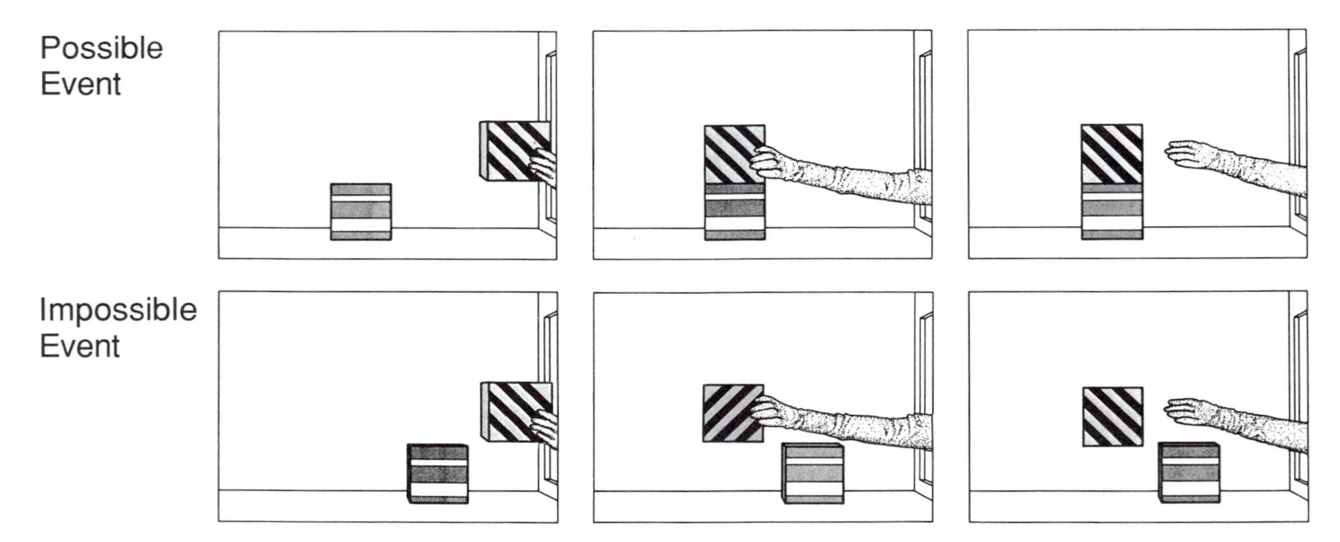

Figure 11.3 Test events used in Needham and Baillargeon (1993).

of support. Additional groups of $4\frac{1}{2}$-month-old infants were tested in two control conditions. In one, the infants saw the same test events as the infants in the experimental condition except that the hand never released the box, which was therefore continually supported. In the other control condition, the infants again saw the same test events as the infants in the experimental condition except that the box fell to the floor of the apparatus when released by the hand beyond the platform.

The infants in the experimental condition looked reliably longer at the impossible than at the possible event, whereas the infants in the two control conditions tended to look equally at the test events they were shown. Together, these results indicated that the infants in the experimental condition realized that the box could not remain stable without support and hence expected the box to fall in the impossible event and were surprised that it did not.

The results of this first experiment suggested that, by $4\frac{1}{2}$ months of age, infants expect a box to be stable if released on but not off a platform. Additional experiments conducted with different procedures yielded similar results with infants aged $5\frac{1}{2}$ months (Leslie, 1984; Kolstad and Baillargeon, 1994) and 3 months (Needham and Baillargeon, 1994).[1] Our next experiment (Baillargeon, Raschke, and Needham, 1994) asked whether $4\frac{1}{2}$-month-old infants not only understand that the box must be *in contact* with the platform in order to be stable but also appreciate what *type* of contact is needed for the box to be stable (figure 11.4). In the possible event, a gloved hand placed a small square box against the side of a large, open platform, on top of a smaller, closed platform. The impossible event was identical to the possible event except that the closed platform was much shorter so that the box now lay well above it.

The results indicated that the female infants looked reliably longer at the impossible than at the possible event, suggesting that they realized the box was inade-

Possible Event

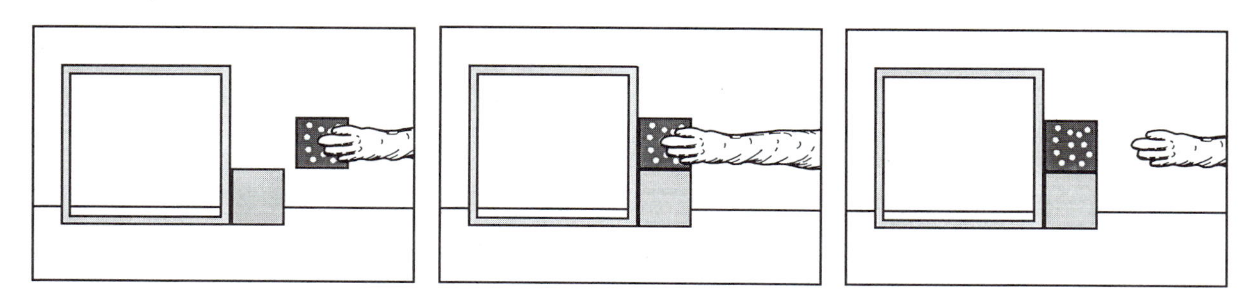

Impossible Event

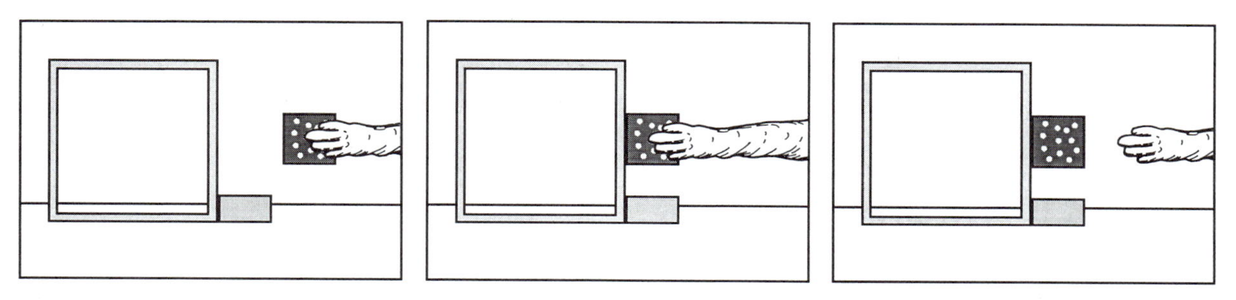

FIGURE 11.4 Test events used in Baillargeon, Raschke, and Needham (1994).

quately supported when it contacted only the side of the open platform and hence expected the box to fall in the impossible event and were surprised that it did not. A control condition in which the hand retained its grasp on the box provided evidence for this interpretation: The infants in this condition tended to look equally at the two test events.

In contrast to the female infants, the male infants in the experimental condition tended to look equally at the impossible and the possible events, as though they believed that the box was adequately supported in both events. Because female infants mature slightly faster than male infants (e.g., Haywood, 1986; Held, in press), gender-related differences such as the one described here are not uncommon in infancy research (e.g., Baillargeon and DeVos, 1991). Given this evidence, it is likely that, when tested with the same experimental procedure, slightly younger female infants (i.e., infants aged $3\frac{1}{2}$ or 4 months) would perform like the $4\frac{1}{2}$-month-old male infants, and slightly older male infants (i.e., infants aged 5 or $5\frac{1}{2}$ months) would perform like the $4\frac{1}{2}$-month-old female infants. An experiment is currently under way to confirm this last prediction.

The results of the last experiment indicated that by $4\frac{1}{2}$ months of age, infants have begun to realize that a box can be stable when placed on but not against a platform. Our next experiment examined whether infants are aware that, in judging the box's stability, one must consider not only the *type* but also the *amount* of contact between the box and the platform (Baillargeon, Needham, and DeVos, 1992). Subjects were $5\frac{1}{2}$- and $6\frac{1}{2}$-month-old infants. The infants watched test events in which a gloved hand pushed a box from left to right along the top of a platform (figure 11.5). In the possible event, the box was pushed until its leading edge reached the end of the platform. In the impossible event, the box was pushed until only the left 15% of its bottom surface remained on the platform. Prior to the test events, the infants saw similar habituation events except that a much longer platform was used so that the box was always fully supported.

The results indicated that the $5\frac{1}{2}$-month-old infants tended to look equally at the two test events, as though they judged that the box was adequately supported in both events. In contrast, the $6\frac{1}{2}$-month-old infants looked reliably longer at the impossible than at the possible event, suggesting that they realized that the box was inadequately supported when only its corner rested on the platform and thus were surprised by the impossible event in which the box did not fall. A control condition in which the hand fully grasped the box provided evidence for this interpretation. In a subsequent experiment (Baillargeon, Needham, and DeVos, 1992), we found that $6\frac{1}{2}$-month-old infants expected the box to be stable when 70%, as opposed to 15%, of its bottom surface rested on the platform.

Together, the results of the experiments reported in this section suggest the following developmental sequence: By 3 months of age, if not before, infants expect the box to fall if it loses contact with the platform and to remain stable otherwise. At this stage, *any* contact between the box and the platform is deemed sufficient to ensure the box's stability. At least two developments take place between 3 and $6\frac{1}{2}$ months of age. First, infants become aware that the locus of contact between the box and the platform must be taken into account when judging the box's stability. Infants initially assume that the box will remain stable if placed either on or against the platform. By $4\frac{1}{2}$ to (presumably) $5\frac{1}{2}$ months of age, however, infants come to distinguish between the two types of contact and recognize that only the former ensures support. The second development is that infants begin to appreciate that the amount of contact between the box and the platform affects the box's stability. Initially, infants believe that the box will be stable even if only a small portion (e.g., the left 15%) of its bottom surface rests on the platform. By $6\frac{1}{2}$ months of age, however, infants expect the box to fall unless a significant portion of its bottom surface (e.g., 70%) lies on the platform.

One way of describing this developmental sequence is that, when learning about the support relation between two objects, infants first form an initial concept centered on a contact/no-contact distinction. With further experience, this initial concept is progressively revised. Infants identify first a discrete (type of contact) and later a continuous (amount of contact) variable and incorporate these variables into their initial concept, resulting in more successful predictions over time.

Knowledge about Collision Phenomena Our research on infants' reasoning about collision phenomena has focused on simple problems involving a moving and a stationary object. Our first experiment (Kotovsky and Baillargeon, 1994b) asked whether $2\frac{1}{2}$-month-old infants expect a stationary object to be displaced when hit by a moving object. The infants in the experiment

Habituation Events

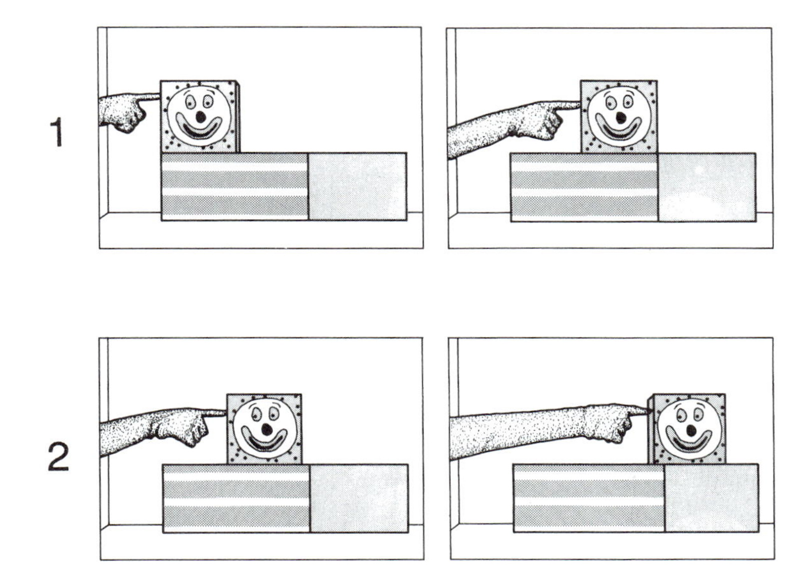

Test Events

Possible Event

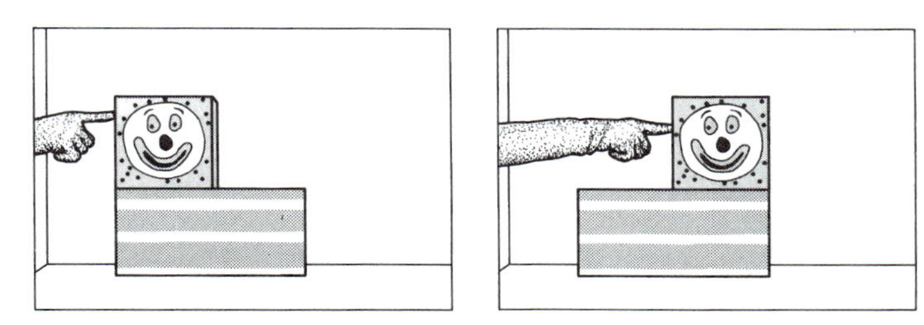

Impossible Event

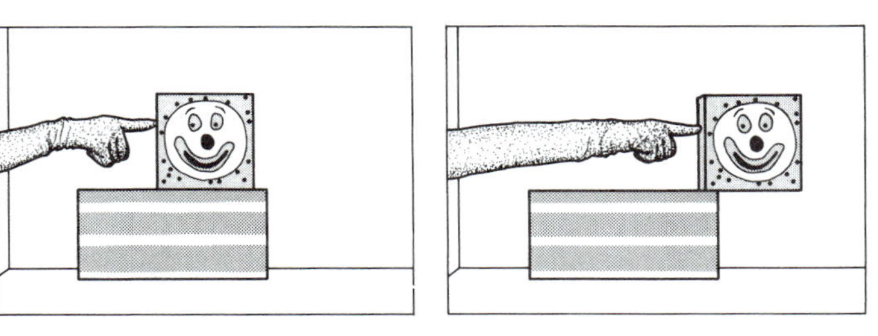

FIGURE 11.5 Test events used in Baillargeon, Needham, and DeVos, (1992).

sat in front of an inclined ramp; to the right of the ramp was a narrow track (figure 11.6). The infants were first habituated to a large cylinder that rolled down the ramp; small stoppers prevented the cylinder from rolling past the ramp. Following habituation, a large wheeled toy bug was placed on the track. In the possible event, the bug was placed 10 cm from the ramp, and it was *not* hit by the cylinder and thus remained stationary after the cylinder rolled down the ramp. In the impossible event, the bug was placed

Habituation Events

Far-Wall Event

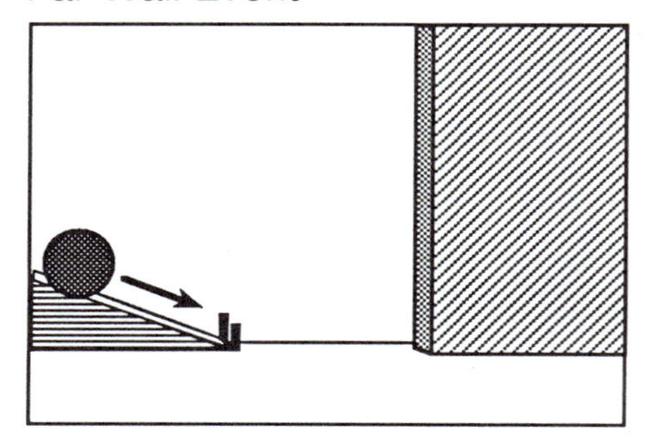

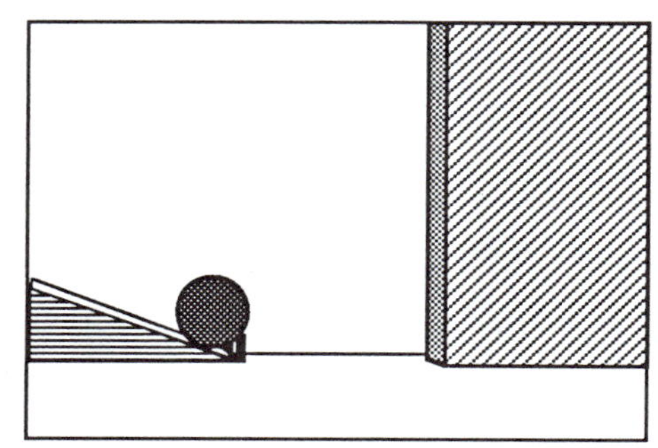

Near-Wall Event

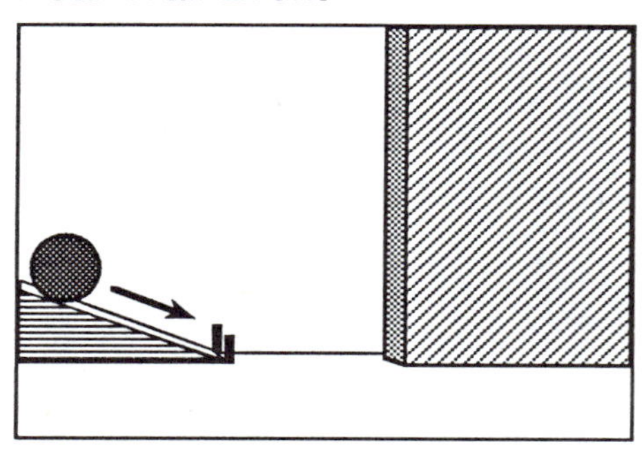

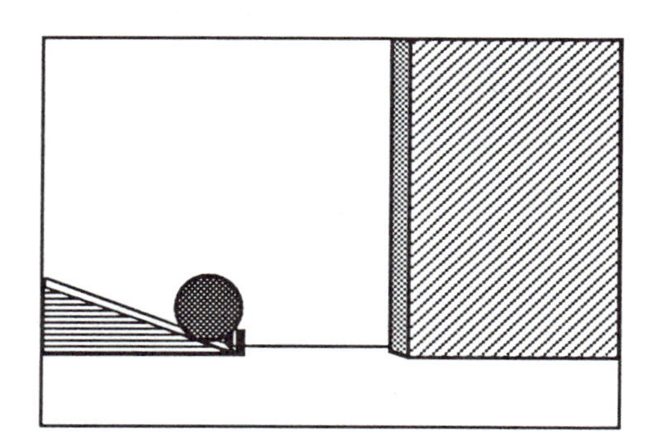

Test Events

Possible Event

Impossible Event

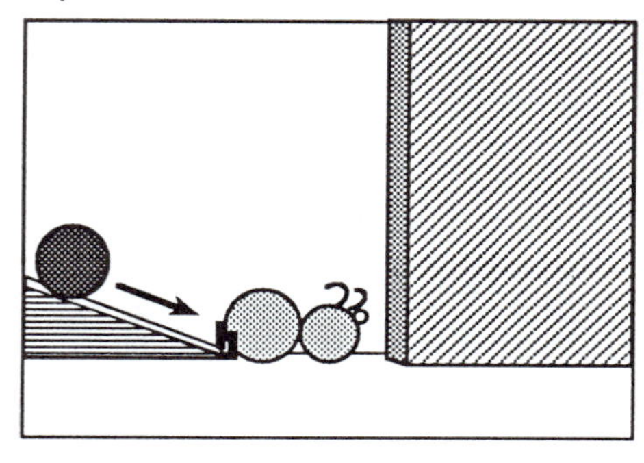

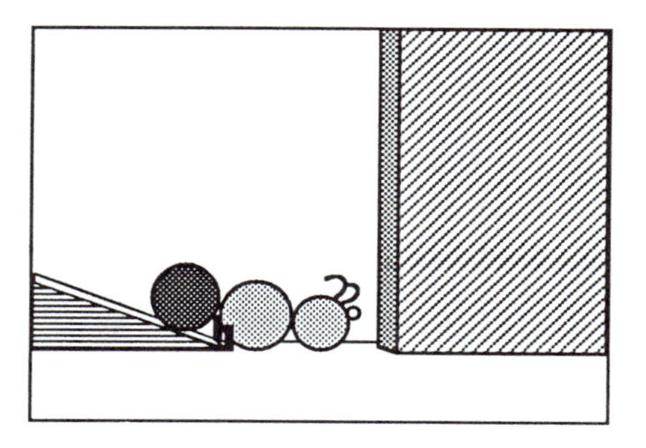

FIGURE 11.6 Test events used in Kotovsky and Baillargeon (1994b).

directly at the bottom of the ramp, and it *was* hit by the cylinder but again remained stationary. Adult subjects typically expect the bug to roll down the track when hit by the cylinder; the experiment thus tested whether $2\frac{1}{2}$-month-old infants would share the same expectation as adults and would be surprised by the impossible event in which the bug remained stationary.

A second group of $2\frac{1}{2}$-month-old infants was tested in a control condition identical to the experimental condition with one exception. In each test event, the right wall of the apparatus was adjusted so that it stood against the front end of the bug, preventing its displacement (recall that, according to the results of Spelke et al., 1992, $2\frac{1}{2}$-month-old infants recognize that an object cannot move through the space occupied by another object).[2]

The infants in the experimental condition looked reliably longer at the impossible than at the possible event, whereas the infants in the control condition tended to look equally at the two events they were shown. Together, these results indicated that the infants in the experimental condition expected the bug to be displaced when hit by the cylinder and hence were surprised by the impossible event in which the bug remained stationary. The results of this first experiment indicated that, by $2\frac{1}{2}$ months of age, infants expect a stationary object to be displaced when hit by a moving object.

Our next experiment asked whether infants could use the size of the moving object to predict how far the stationary object should be displaced (Kotovsky and Baillargeon, 1994a, in press). One group of $6\frac{1}{2}$-month-

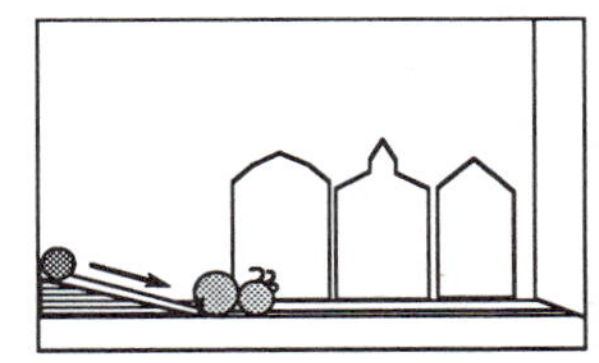

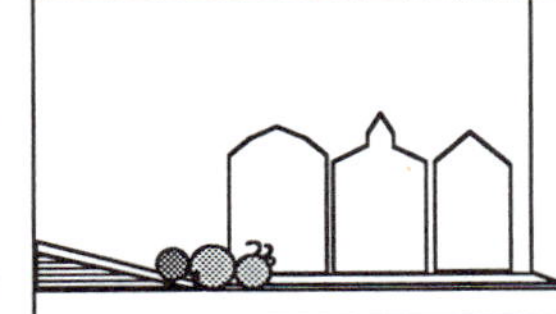

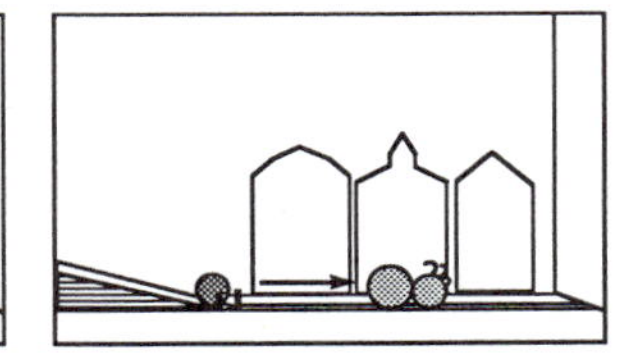

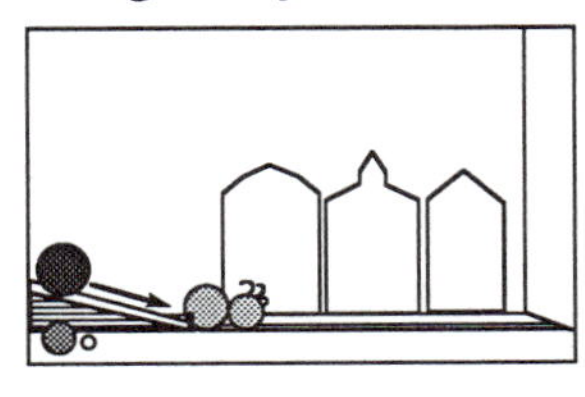

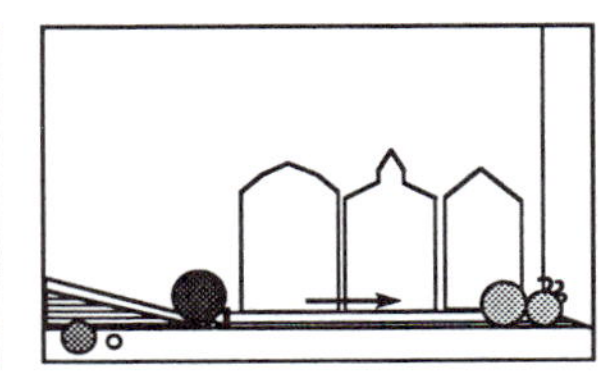

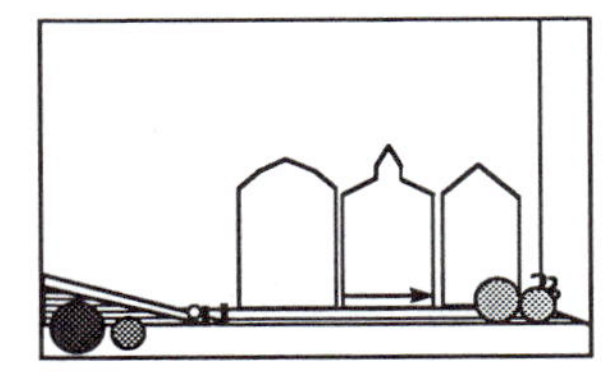

FIGURE 11.7 Test events used in the midpoint condition in Kotovsky and Baillargeon (1994a, experiment 1).

old infants (midpoint condition) was habituated to a blue, medium-size cylinder that rolled down a ramp and hit a toy bug, causing it to roll to the middle of a track (figure 11.7). Two new cylinders were introduced in the test events: a yellow cylinder that was larger than the habituation cylinder, and an orange cylinder that was smaller than the habituation cylinder. Both cylinders caused the bug to travel farther than in the habituation event: The bug stopped only when it reached the end of the track and hit the right wall of the apparatus.

When asked how far the bug would roll when hit by any one cylinder, adult subjects were typically reluctant to hazard a guess: They were aware that the length of the bug's trajectory depended on a host of factors (e.g., the weight of the cylinder and bug, the smoothness of the ramp and track, and so on) about which they had no information. After observing that the bug rolled to the middle of the track when hit by the medium cylinder, however, adult subjects readily predicted that the bug would roll farther with the larger and less far with the smaller cylinder and were surprised when this last prediction was violated.[3] The experiment thus tested whether $6\frac{1}{2}$-month-old infants, like adults, would understand that the size of the cylinder affected the length of the bug's displacement and would be able to use the information conveyed in the habituation event to calibrate their predictions about the test events.

A second group of infants (endpoint condition) was tested in a condition identical to the midpoint condition except that they were given a different calibration point in the habituation event. As shown in figure 11.8, the medium cylinder now caused the bug to roll to the end of the track, just as in the test events.

After seeing that the bug rolled to the end of the track when hit by the medium cylinder, adult subjects expected the bug to do the same with the large cylinder

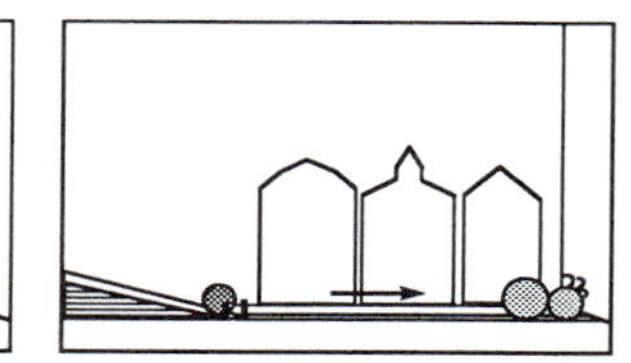

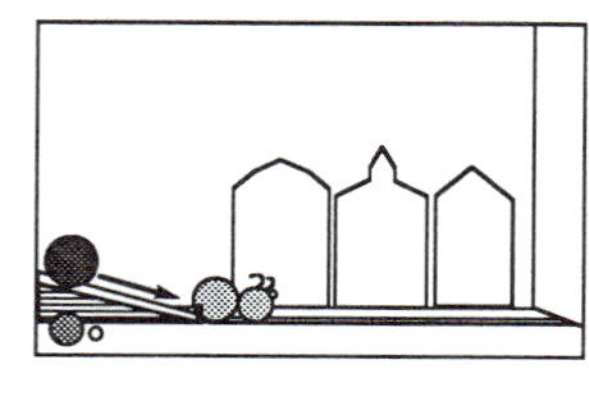

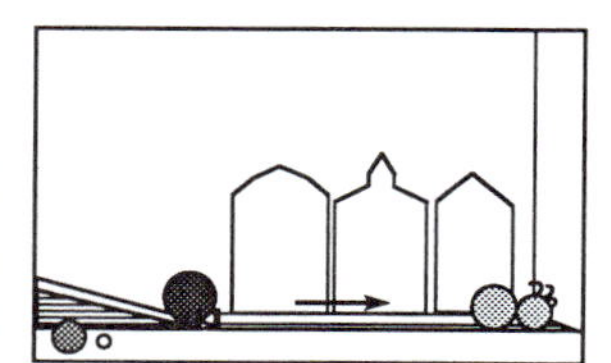

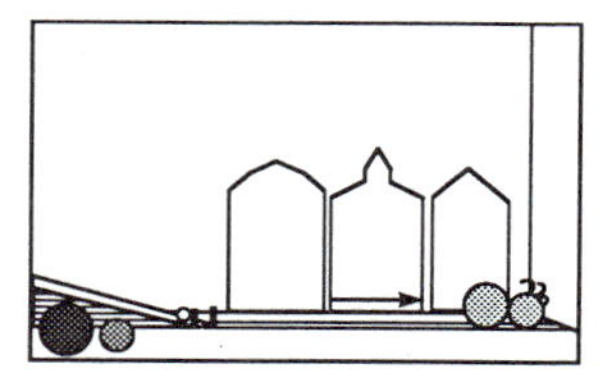

FIGURE 11.8 Test events used in the endpoint condition in Kotovsky and Baillargeon (1994a, experiment 1).

and were not surprised to see the bug do the same with the small cylinder (subjects simply concluded that the track was too short to show effects of cylinder size). The experiment thus tested whether $6\frac{1}{2}$-month-old infants, like adults, would perceive both of the endpoint condition test events as possible.

The results indicated that the infants in the midpoint condition looked reliably longer at the small-cylinder than at the large-cylinder event, whereas the infants in the endpoint condition tended to look equally at the two events. Together, these results indicated that the infants were aware that the size of the cylinder should affect the length of the bug's trajectory and used the habituation event to calibrate their predictions about the test events. After watching the bug travel to the middle of the track when hit by the medium cylinder, the infants were surprised to see the bug travel farther with the smaller but not the larger cylinder. In contrast, after watching the bug travel to the end of the track with the medium cylinder, the infants were not surprised to see the bug do the same with either the small or the large cylinder.

In a subsequent experiment, $5\frac{1}{2}$-month-old infants were tested using the same procedure (Kotovsky and Baillargeon, 1994a). The performance of the female infants was identical to that of the $6\frac{1}{2}$-month-old infants. The male infants, in contrast, tended to look equally at the test events in both the midpoint and the endpoint conditions. At least two interpretations could be advanced for this negative finding. One was that the male infants were still unaware that the size of the cylinder should affect the length of the bug's displacement. The other interpretation was that the male infants had difficulty remembering how far the bug traveled in the habituation event and hence could not make use of this information to predict what should happen in the small-cylinder and large-cylinder events.

Midpoint Condition

Habituation Event

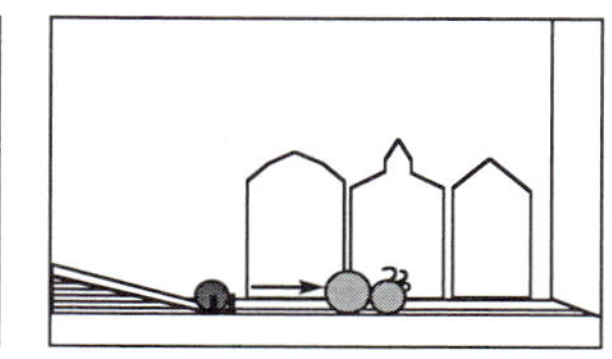

Test Events

Familiar Event

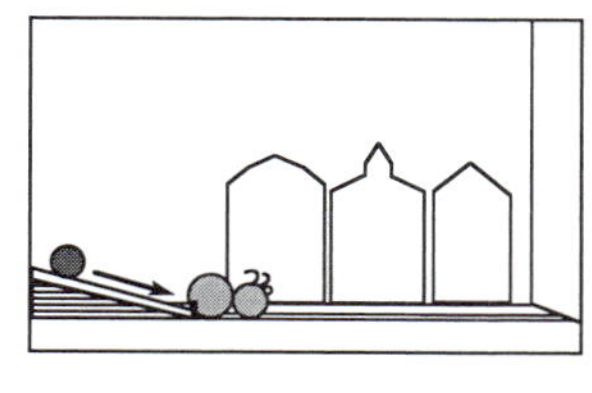

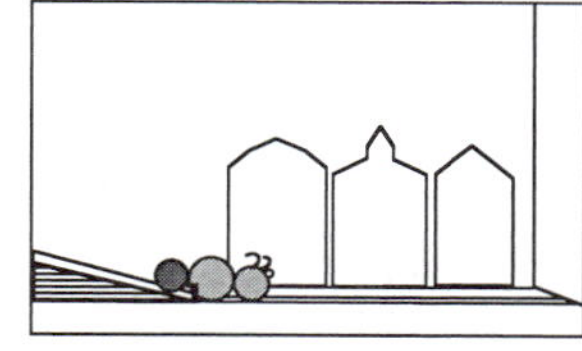

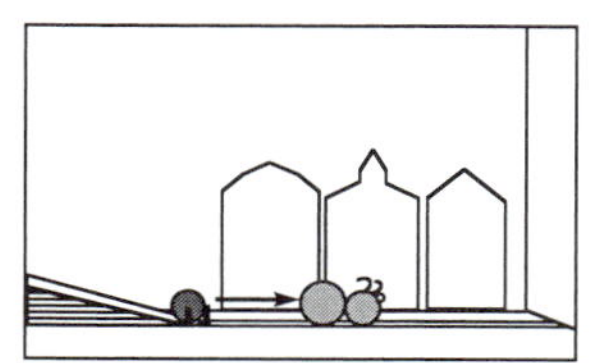

Novel Event

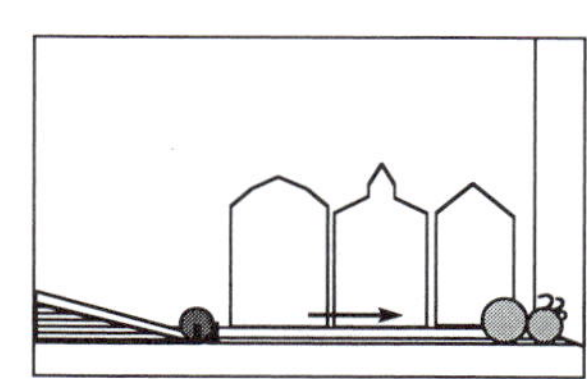

FIGURE 11.9 Test events used in the midpoint condition in Kotovsky and Baillargeon (1994a, experiment 3).

To examine this second interpretation, two groups of $5\frac{1}{2}$-month-old male infants were tested in a simple memory experiment (Kotovsky and Baillargeon, 1994a). The infants in the midpoint condition, as before, were habituated to the medium cylinder rolling down the ramp and hitting the bug, causing it to roll to the middle of the track (figure 11.9). Following habituation, the infants saw two test events. One (familiar test event) was identical to the habituation event. In the other event (novel test event), the medium cylinder now caused the bug to roll to the end of the track. The infants in the endpoint condition saw similar habituation and test events, except that the bug rolled to the end of the track in the habituation event so that which test event was familiar and which was novel were reversed (figure 11.10).

The results revealed a significant overall preference for the novel over the familiar test event, indicating that the infants had no difficulty recalling how far the bug rolled in the habituation event. Such a finding, combined with the negative finding obtained in the last experiment, suggests this conclusion: After observing that the medium cylinder causes the bug to roll to the middle of the track, $5\frac{1}{2}$-month-old male infants expect the bug to do the same when hit by the *same* cylinder but have no expectation as to how far the bug should roll when hit by cylinders of different sizes. Infants seem unaware that they possess information they can use to reason about the novel cylinders.

Together, the results of these collision experiments point to the following developmental sequence: By $2\frac{1}{2}$ months of age, infants expect a stationary object to be displaced when hit by a moving object; however, they are not yet aware that the size of the moving object can be used to predict how far the stationary object will be displaced. If shown that a medium cylinder causes a bug to roll to the middle of a track, for example, infants have no expectation that the bug should travel farther

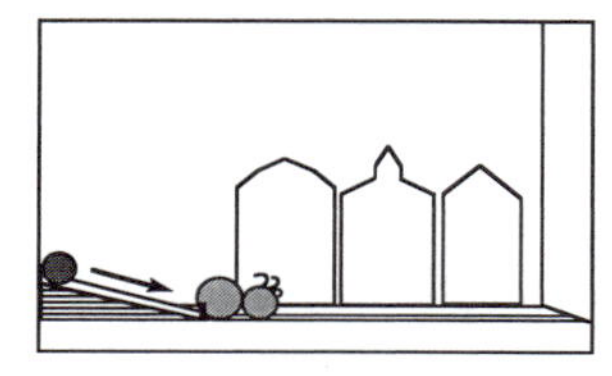

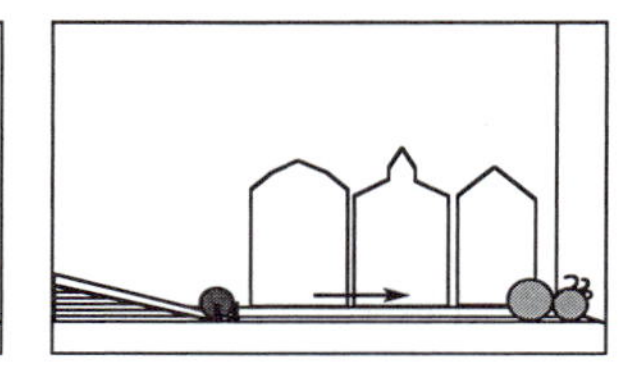
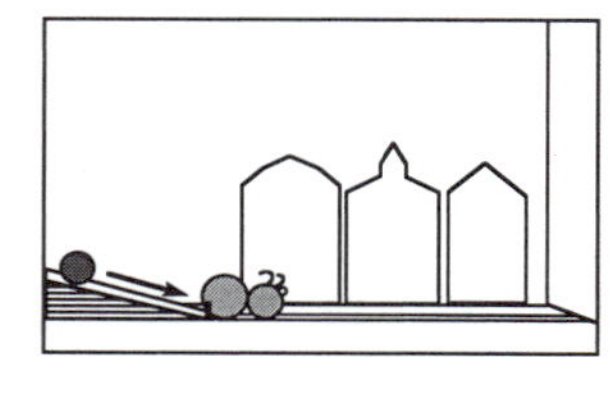

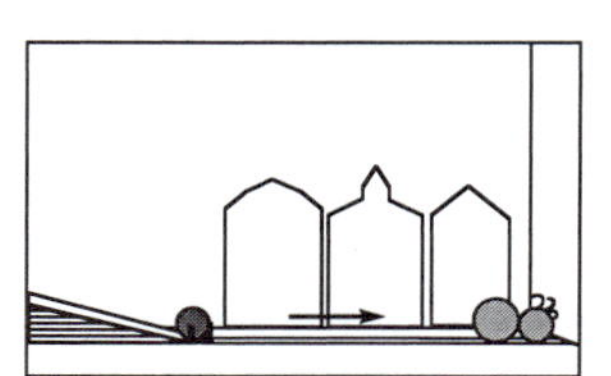
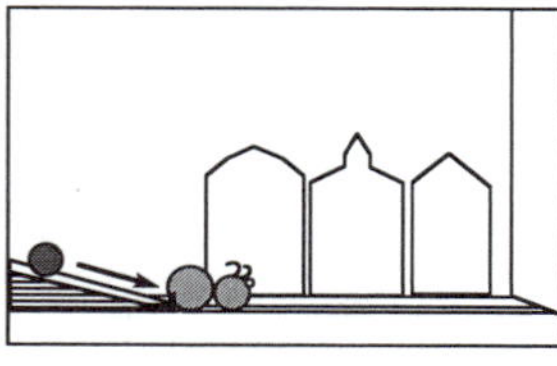

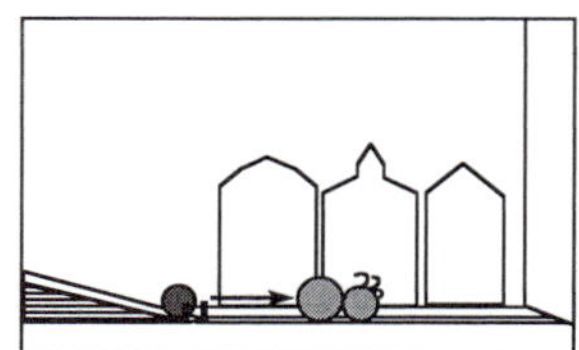

FIGURE 11.10 Test events used in the endpoint condition in Kotovsky and Baillargeon (1994a, experiment 3).

when hit by a larger cylinder and less far when hit by a smaller cylinder. By $5\frac{1}{2}$ to $6\frac{1}{2}$ months of age, however, infants recognize not only that a stationary object should be displaced when hit by a moving object but also that *how far* the stationary object is displaced depends on the size of the moving object.

One interpretation of these findings is that, when learning about collision events between a moving and a stationary object, infants first form an initial concept centered on an impact/no-impact distinction. With further experience, infants begin to identify variables that influence this initial concept. By $5\frac{1}{2}$ to $6\frac{1}{2}$ months of age, infants realize that the size of the moving object can be used to predict how far the stationary object will be displaced.

Knowledge about Unveiling Phenomena Our experiments on unveiling phenomena have involved problems in which a cloth cover is removed to reveal an object. Our first experiment examined whether $9\frac{1}{2}$-month-old infants realize that the presence (absence) of a protuberance in a cover signals the presence (absence) of an object beneath the cover (Baillargeon and DeVos, 1994a). At the start of the possible event, the infants saw two covers made of a soft, fluid fabric; the left cover lay flat on the floor of the apparatus, and the right cover showed a marked protuberance (figure 11.11). Next, two screens were pushed in front of the covers, hiding them from view. A hand then reached behind the right screen and reappeared first with the cover and then with a toy bear of the same height as the protuberance shown earlier. The impossible event was identical except that the location of the two covers at the start of the event was reversed, so that it should have been impossible for the hand to retrieve the bear from behind the right screen.

The infants looked reliably longer at the impossible than at the possible event, suggesting that they understood that the bear could have been hidden under the cover with a protuberance but not under the flat cover. This interpretation was supported by the results of a second condition in which the hand reached behind the left as opposed to the right screen so that the bear's position in the impossible and the possible events was reversed.

The results of this first experiment indicated that, by 9 months of age, infants can use the existence of a protuberance in a cloth cover to infer the existence of an object beneath the cover. Our next experiment (Baillargeon and DeVos, 1994a) investigated whether infants could also use the size of the protuberance to infer the size of the object under the cover (figure 11.12). At the start of the possible event, the infants saw two covers made of a soft fabric: on the left was a small cover with a small protuberance; on the right was a large cover with a large protuberance. (The small protuberance was 10.5 cm high and the large protuberance 22 cm high; the difference between the two was

Possible Event

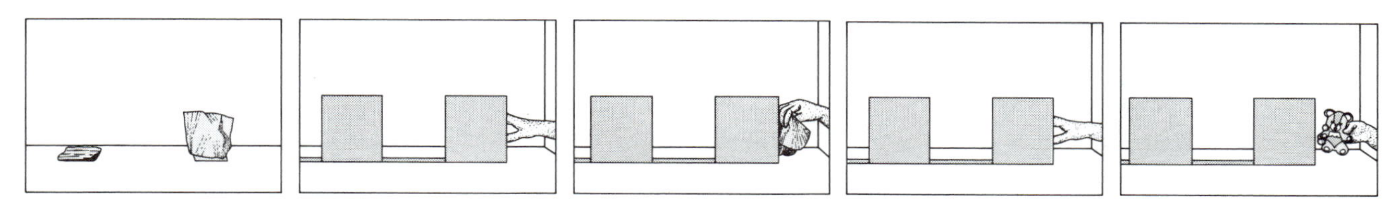

Impossible Event

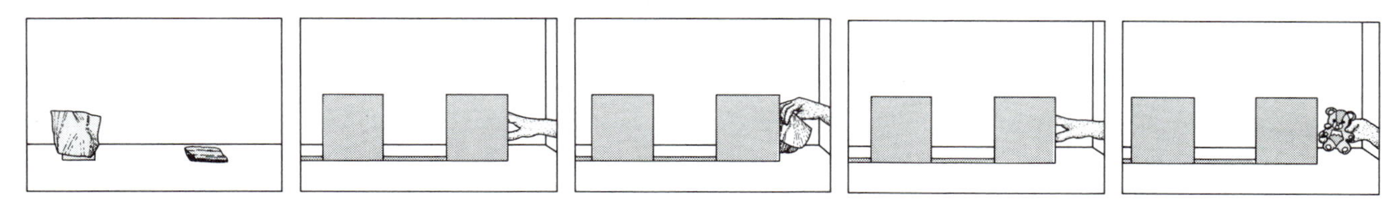

Figure 11.11 Test events used in Baillargeon and DeVos (1994a, experiment 1).

Possible Event

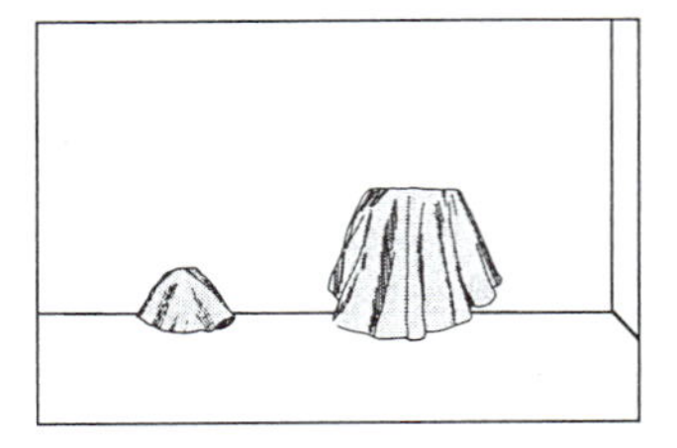
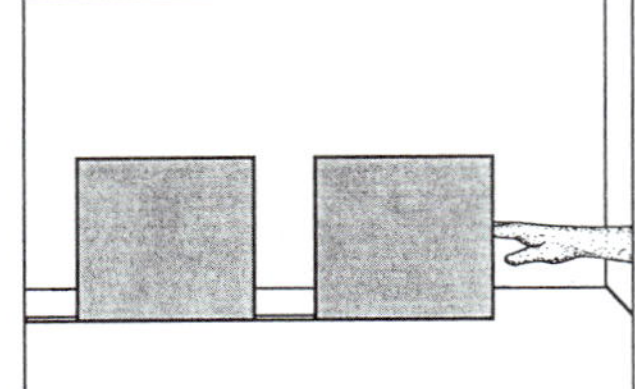

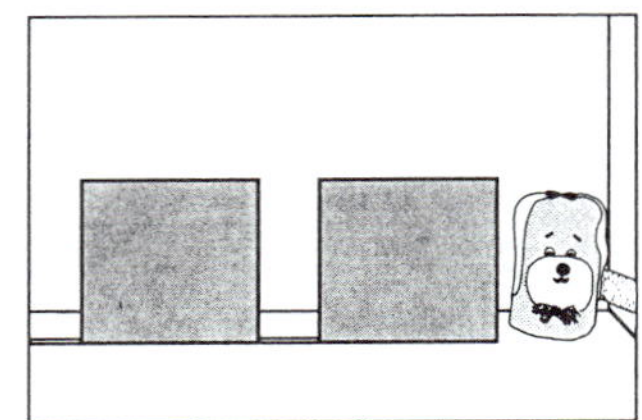

Impossible Event

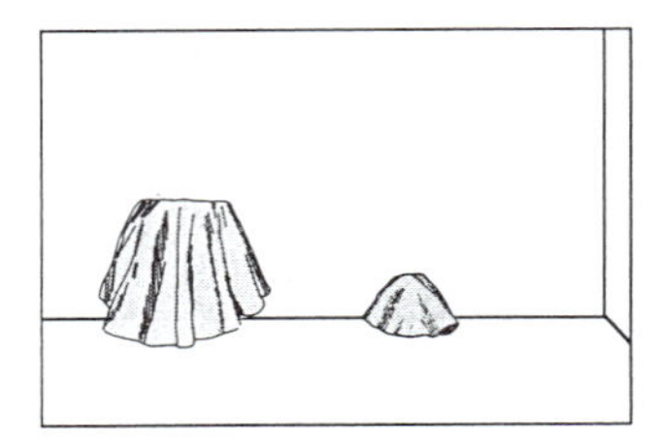
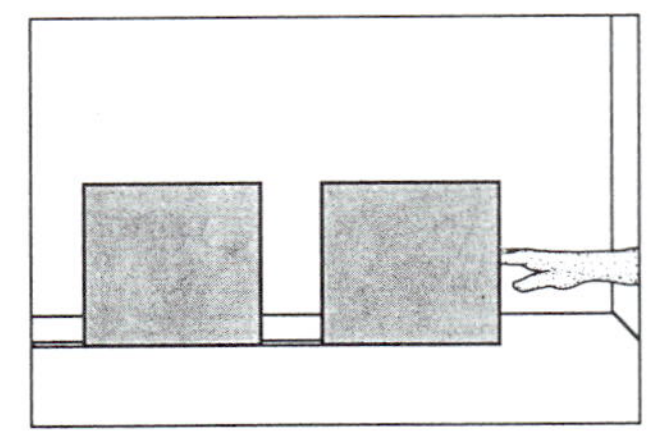
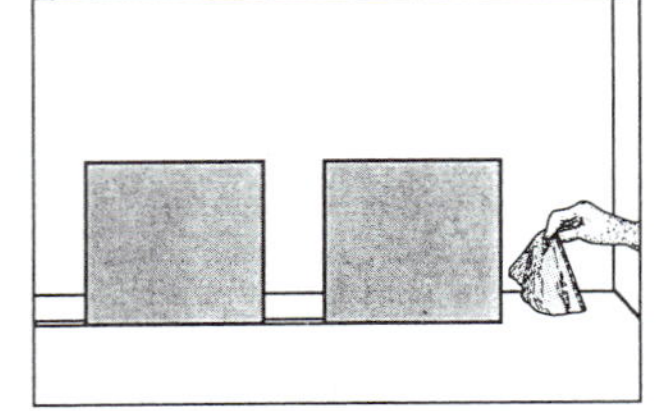
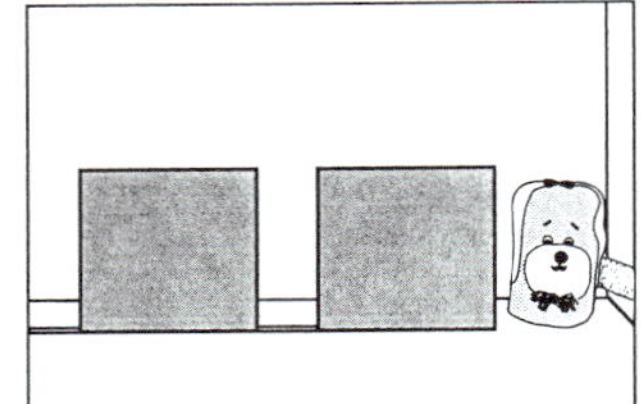

FIGURE 11.12 Test events used in Baillargeon and DeVos (1994a, experiment 3).

thus easily detectable.) Next, screens were pushed in front of the covers, and a gloved hand reached behind the right screen twice in succession, reappearing first with the cover and then with a large toy dog 22 cm tall. The impossible event was identical to the possible event except that the location of the two covers at the start of the event was reversed, so that the hand now appeared to retrieve the large dog from under the cover with the small protuberance.

Unlike the infants in the last experiment, the infants in this experiment tended to look equally at the impossible and at the possible events, suggesting that they believed that the large dog could have been hidden under the cover with either the small or the large protuberance. The same result was obtained in a subsequent experiment that made use of a slightly different procedure (Baillargeon and DeVos, 1994a). How should these negative findings be explained? At least two hypotheses could be proposed. One was that the infants were not yet aware that the size of the protuberance in each cover could be used to infer the size of the object hidden beneath the cover. The other explanation was that the infants recognized the significance of the protuberance's size but had difficulty remembering this information after the cover was hidden from view.

The results of another experiment provided evidence for the first of these two interpretations. The infants in this experiment (Baillargeon and DeVos, 1994b) were given a reminder of the size of the protuberance in the cover behind the screen (figure 11.13). Subjects were $9\frac{1}{2}$- and $12\frac{1}{2}$-month-old infants. At the start of the possible event, the infants saw the cover with the small protuberance; to the right of this cover was a second, identical cover. After a brief pause, the first cover was hidden by the screen; the second cover remained visible to the right of the screen. Next, the hand reached behind the screen's right edge and removed first the cover and then a small toy dog 10.5 cm in height. The hand held the small dog next to the visible cover, allowing the infants to compare their sizes directly. The impossible event was identical to the possible event, except that the hand retrieved the large toy dog from behind the screen.

The $12\frac{1}{2}$-month-old infants looked reliably longer at the impossible than at the possible event, suggesting that they realized that the small but not the large dog could have been hidden under the cover behind the screen. This interpretation was supported by the results of a control condition in which the infants simply saw each dog held next to the visible cover (as in the rightmost panels in figure 11.13); no reliable preference was found for the large-dog over the small-dog display.

In contrast to the $12\frac{1}{2}$-month-old infants, the $9\frac{1}{2}$-month-old infants tended to look equally at the impos-

Possible Event

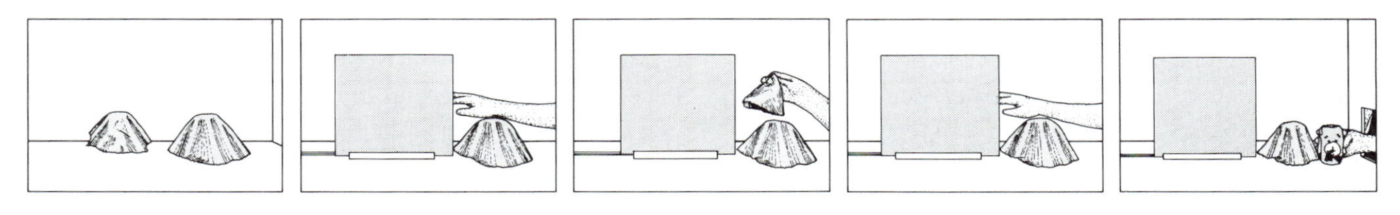

Impossible Event

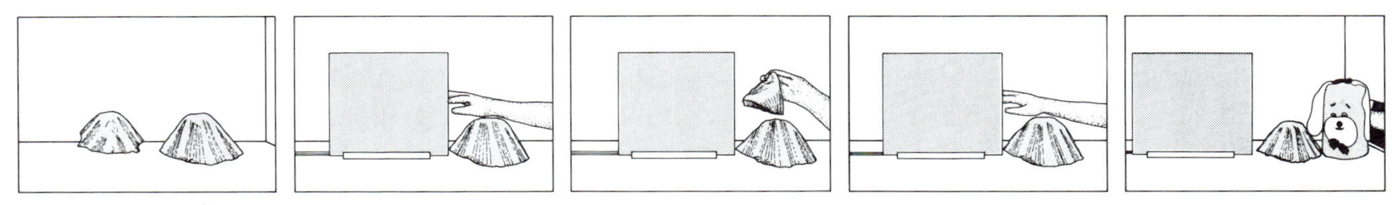

FIGURE 11.13 Test events used in Baillargeon and DeVos (1994b, experiment 2).

sible and the possible events. Thus, despite the fact that the infants had available a reminder—an exact copy—of the cover behind the screen, they still failed to show surprise at the large dog's retrieval. It might be argued that infants younger than $12\frac{1}{2}$ months of age are simply unable, when reasoning about hidden objects, to take advantage of reminders such as the visible cover. As will be seen later, however, even young infants can make use of visual reminders to make predictions concerning hidden objects.

The results summarized in this section suggest the following developmental sequence: By 9 months of age, infants realize that the existence of a protuberance in a cloth cover signals the existence of an object beneath the cover: They are surprised to see an object retrieved from under a flat cover but not from under a cover with a protuberance. However, infants are not yet aware that the size of the protuberance can be used to infer the size of the hidden object. When shown a cover with a small protuberance, they are not surprised to see either a small or a large object retrieved from under that cover. Furthermore, providing a reminder of the protuberance's size has no effect on infants' performance. Under the same conditions, however, $12\frac{1}{2}$-month-old infants show reliable surprise at the large object's retrieval.

One interpretation of these findings is that, when learning about unveiling phenomena, infants first form an initial concept centered on a protuberance/no-protuberance distinction. Later on, infants identify a continuous variable that affects this concept: They begin to appreciate that the size of the protuberance in the cover can be used to predict the size of the object hidden under the cover.

DISCUSSION How can the various developmental sequences described in this section be explained? As was mentioned earlier, our assumption is that these sequences reflect not the gradual unfolding of innate beliefs but the application of highly constrained, innate learning mechanisms to available data. In this approach, the problem of explaining the age at which specific initial concepts and variables are understood is that of determining what data—observations or manipulations—are necessary for learning and when these data become available to infants.

To illustrate, consider the developmental sequence revealed in our support experiments. One might propose that 3-month-old infants have already learned that objects fall when released in midair (Needham and Baillargeon, 1994) because this expectation is consistent with countless observations (e.g., watching their caretakers drop peas in pots, toys in baskets, clothes in hampers) and manipulations (e.g., noticing that their pacifiers fall when they open their mouths) available virtually from birth.

Furthermore, one might speculate that infants do not begin to recognize until $4\frac{1}{2}$ months (Baillargeon,

Raschke, and Needham, 1994) what type of contact is needed between objects and their supports because it is not until this age that infants have available pertinent data from which to abstract this variable. Researchers have found that unilateral, visually guided reaching emerges at approximately 4 months of age (e.g., White, Castle, and Held, 1964). With this newfound ability, infants may have the opportunity to place objects deliberately against other objects and to observe the consequences of these actions. The gender-related difference revealed in our experiment, in this account, would be traceable to female infants engaging in these manipulations slightly ahead of the male infants.

In a similar vein, one could suggest that it is not until $6\frac{1}{2}$ months that infants begin to appreciate how much contact is needed between objects and their supports (Baillargeon, Needham, and DeVos, 1992) because, once again, it is not until this age that infants have available data from which to learn such a variable. Investigators have reported that the ability to sit without support emerges at approximately 6 months of age; infants then become able to sit in front of tables (e.g., on a parent's lap or in a high chair) with their upper limbs and hands relieved from the encumbrance of postural maintenance and thus free to manipulate objects (e.g., Rochat and Bullinger, in press). For the first time, infants may have the opportunity to deposit objects on tables and to note that objects tend to fall unless a significant portion of their bottom surfaces is supported.

In the natural course of events, infants would be unlikely to learn about variables such as type or amount of contact from visual observation alone, because caretakers rarely deposit objects against vertical surfaces or on the edges of horizontal surfaces. There is no a priori reason, however, to assume that infants could not learn such variables if given appropriate observations. We are currently planning "teaching" experiments to explore this possibility.

Second pattern: Use of qualitative and quantitative strategies

In the preceding section we proposed that, when learning about a novel physical phenomenon, infants first develop an all-or-none initial concept and later identify discrete and continuous variables that affect this concept. The second developmental pattern suggested by current evidence concerns the strategies infants use when reasoning about continuous variables. Following the terminology used in computational models of everyday physical reasoning (e.g., Forbus, 1984), a strategy is said to be *quantitative* if it requires infants to encode and use information about absolute quantities (e.g., object A is this large or has traveled this far from object B, where *this* represents some absolute measure of A's size or distance from B). In contrast, a strategy is said to be *qualitative* if it requires infants to encode and use information about relative quantities (e.g., object A is larger than or has traveled farther than object B). After identifying a continuous variable, infants appear to succeed in reasoning about the variable qualitatively before they succeed in doing so quantitatively.

To illustrate the distinction between infants' use of qualitative and quantitative strategies, I will report experiments on the development of infants' ability to reason about collision phenomena (conducted with Laura Kotovsky), unveiling phenomena (conducted with Julie DeVos), and arrested-motion phenomena.

Reasoning about Collision Phenomena Earlier in this chapter (and in Kotovsky and Baillargeon, 1994a), I reported that $6\frac{1}{2}$-month-old infants and $5\frac{1}{2}$-month-old female infants were surprised, after observing that a medium-size cylinder caused a bug to roll to the middle of a track, to see the bug roll farther when hit by a smaller but not a larger cylinder (see figure 11.7). These and other findings indicated that the infants were aware that the size of the cylinder affected the length of the bug's trajectory.

In these experiments, each test event began with a pretrial in which the small, medium, and large cylinders lay side by side at the front of the apparatus. A gloved hand tapped on the cylinder to be used in the event (e.g., the small cylinder in the small-cylinder event). After the computer signaled that the infant had looked at the cylinder for 4 cumulative seconds, the hand grasped the cylinder and deposited it at the top of the ramp to begin the test event. The pretrial was included to enable the infants to compare directly the sizes of the cylinders.

In a subsequent experiment (Kotovsky and Baillargeon, 1994c), $6\frac{1}{2}$- and $7\frac{1}{2}$-month-old infants saw habituation and test events identical to those used in the midpoint condition in our initial experiments, with one exception: Only one cylinder was present in the apparatus in each event. During the pretrial preceding each test event, the gloved hand again tapped on the cylinder, but the other cylinders were absent so that the infants were no longer able to compare the cylinders' sizes visually. Under these conditions, the $6\frac{1}{2}$-month-old infants no longer showed surprise when the small cylinder caused the bug to roll to the end of the track; only the $7\frac{1}{2}$-month-old infants looked reliably longer at the impossible than at the possible event.

Our interpretation of these results is that, at $5\frac{1}{2}$ to $6\frac{1}{2}$ months of age, infants are able to reason about the cylinder's size only qualitatively: They can predict the effect of modifications in the cylinder's size only when they are able to encode such modifications in relative terms (e.g., "This cylinder is smaller than the one next to it, which was used in the last trial"). When forced to encode and compare the absolute sizes of the cylinders, because the cylinders are never shown side by side, the infants fail the task. By $7\frac{1}{2}$ months of age, however, infants have already overcome this initial limitation and succeed in our task even when they must rely on their representation of the absolute size of each cylinder to do so.[4]

Reasoning about Unveiling Phenomena Earlier in this chapter (and in Baillargeon and DeVos, 1994b), I reported that $12\frac{1}{2}$-month-old infants were surprised to see a large but not a small dog retrieved from under a cover with a small protuberance (see figure 11.13). These and control results indicated that the infants were aware that the size of the protuberance in the cover could be used to infer the size of the object hidden under the cover.

In our initial experiment, the infants were tested with a second, identical cover present to the right of the screen. Each dog, after it was retrieved from behind the screen, was held next to the visible cover, allowing the infants to compare in a single glance the size of the dog to that of the cover. In a subsequent experiment (Baillargeon and DeVos, 1994b), $12\frac{1}{2}$- and $13\frac{1}{2}$-month-old infants were tested with the same test events, except that only one cover was present: The infants no longer were provided with a second cover to remind them of the size of the cover behind the screen (figure 11.14). Under these conditions, only the $13\frac{1}{2}$-month-old infants looked reliably longer at the impossible than at the possible event, suggesting that they were surprised to see the large but not the small dog retrieved from the cover behind the screen. This interpretation was supported by a control condition in which a cover with a large rather than a small protuberance stood behind the screen.

The results of this last experiment suggested that the $12\frac{1}{2}$-month-old infants could not succeed at our task without a reminder of the size of the cover behind the screen. In our next experiment, we examined whether infants would remain successful if a second, identical

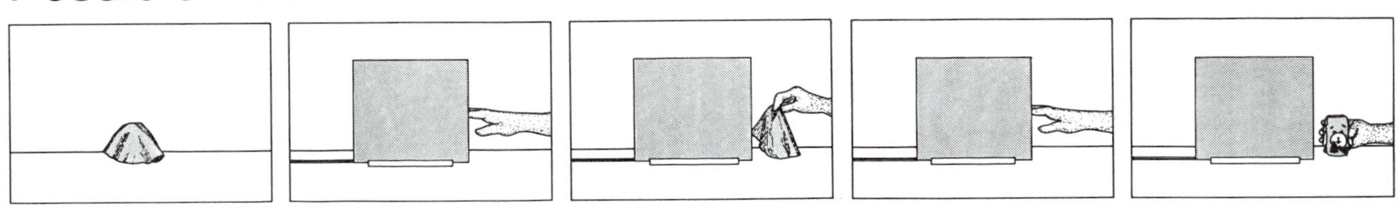

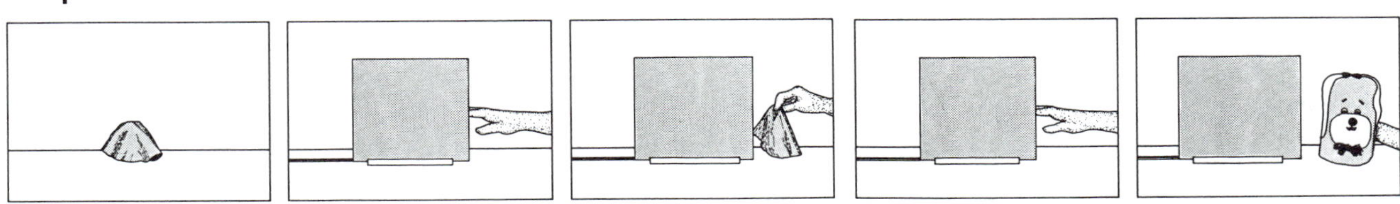

Figure 11.14 Test events used in Baillargeon and DeVos (1994b, experiment 1).

Possible Event

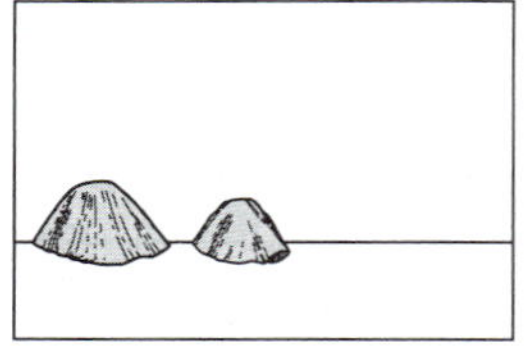
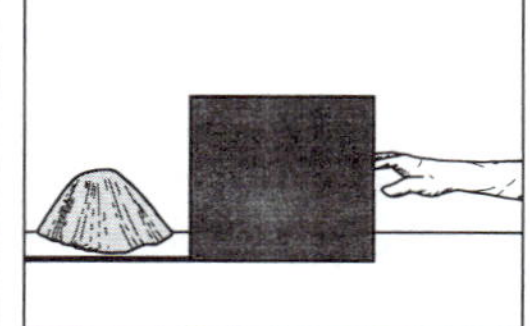
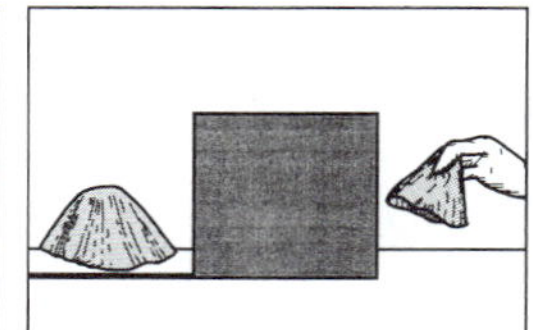
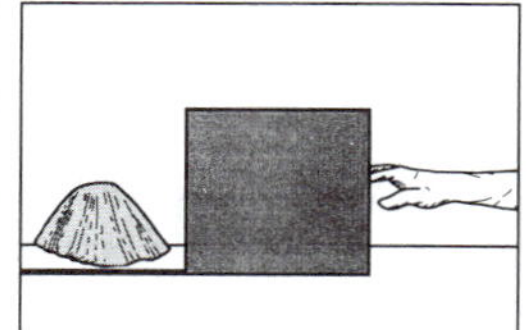

Impossible Event

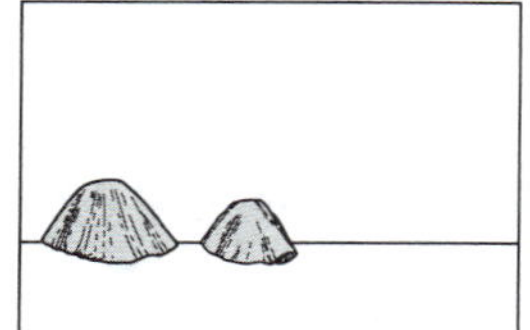
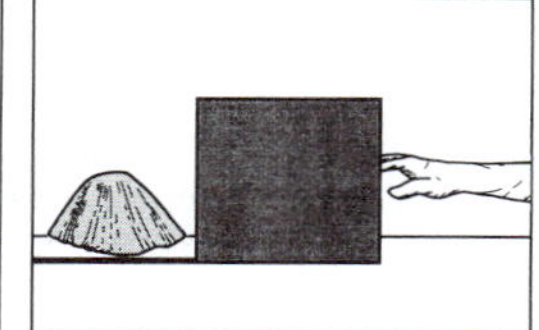
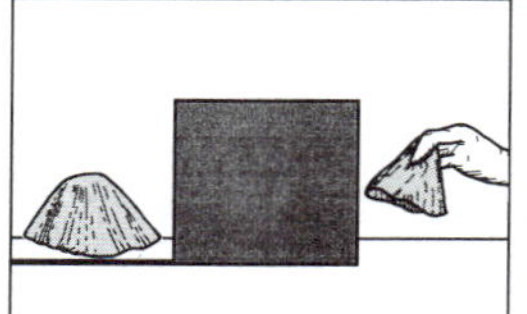
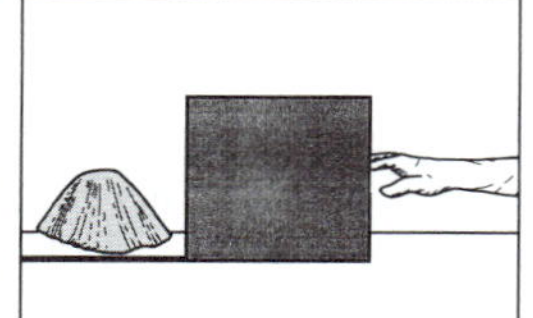
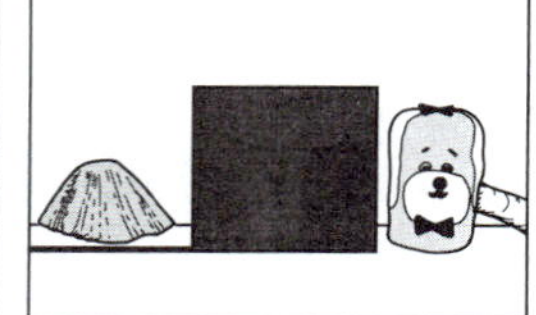

FIGURE 11.15 Test events used in Baillargeon and DeVos (1994b, experiment 3).

cover was again included in the test displays but was placed to the left rather than to the right of the screen (figure 11.15). The infants still had in their visual fields an exact copy of the hidden cover; however, they were no longer able to compare in a single glance the size of each dog to that of the visible cover. The results were once again negative: The infants failed to show surprise at the large dog's retrieval.

Together the results of these experiments suggest that, at $12\frac{1}{2}$ months of age, infants are able to reason only qualitatively about the size of the protuberance in the cover: They can determine which dog could have been hidden under the cover only if they are able to compare, in a single glance, the size of the dog to that of a second, identical cover (e.g., "The dog is bigger than the cover"). When infants are forced to represent the absolute size of the protuberance in the cover, they fail the task. By $13\frac{1}{2}$ months of age, however, infants have already progressed beyond this initial limitation; they no longer have difficulty representing the absolute size of the protuberance and comparing it to that of each dog.

REASONING ABOUT ARRESTED-MOTION PHENOMENA Our research on arrested-motion phenomena has focused on problems involving a large box placed in the path of a rotating screen. One experiment examined $4\frac{1}{2}$-month-old infants' ability to use the height and location of the box to predict at what point the rotating screen would reach the box and stop (Baillargeon, 1991). At the start of each habituation event (figure 11.16), the infants saw a screen that lay flat against the floor of the apparatus, toward them; the screen then rotated 180° about its distant edge until it lay flat against the apparatus floor, toward the back wall. Following habituation, a box was placed behind the screen; this box was progressively occluded as the screen rotated upward. In the possible event, the screen rotated until it reached the occluded box (112° arc). In the impossible event, the screen stopped only after it rotated through the top 80% of the space occupied by the box (157° arc)—to adults, an extreme and easily detectable violation.

A second group of infants (two-box condition) saw the same test events as the infants in the first (one-box) condition, with one exception: A second, identical box was placed to the right of and in the same fronto-parallel plane as the box behind the screen (figure 11.17). The second box stood out of the screen's path and thus remained visible throughout the test events. In the possible event, the screen stopped when aligned with the top of the second box; in the impossible event, the screen rotated past the top of the visible box.

The infants in the two-box condition looked reliably longer at the impossible than at the possible event, suggesting that they realized that the screen's 157°

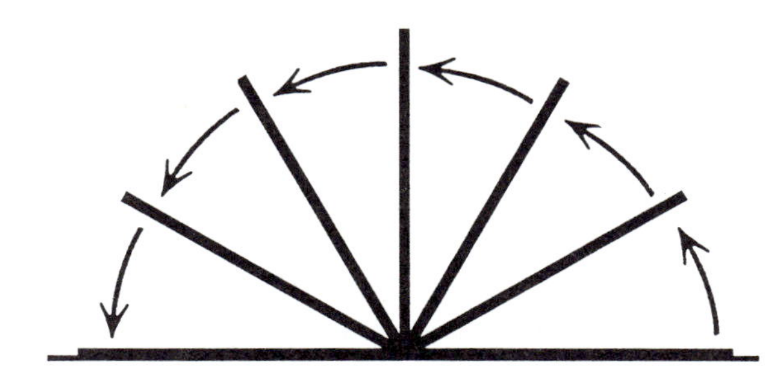

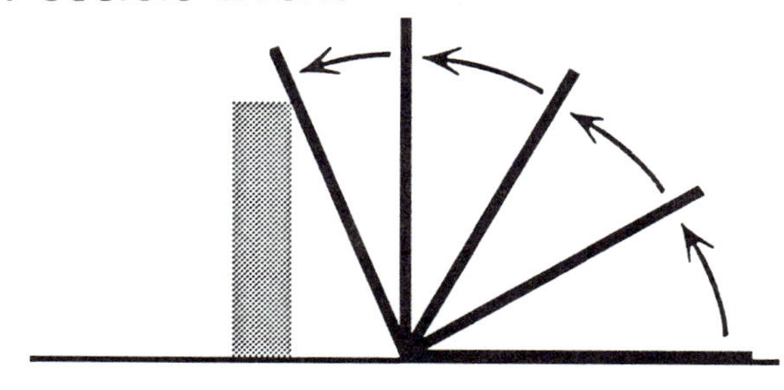

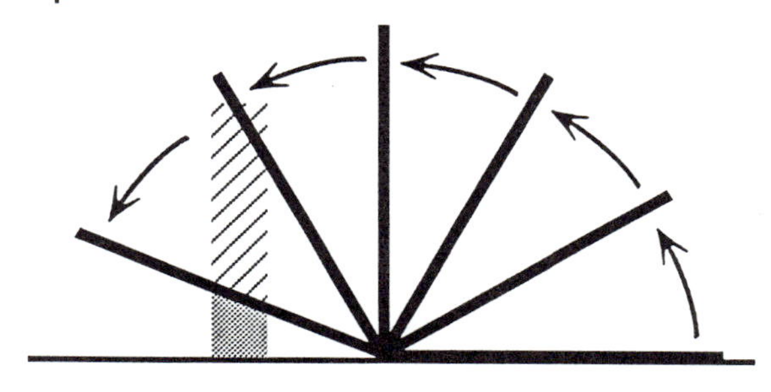

FIGURE 11.16 Test events used in the one-box condition in Baillargeon (1991, experiment 2). Side view.

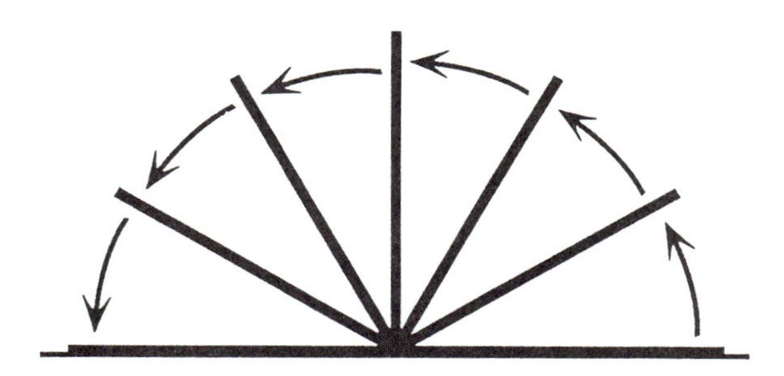

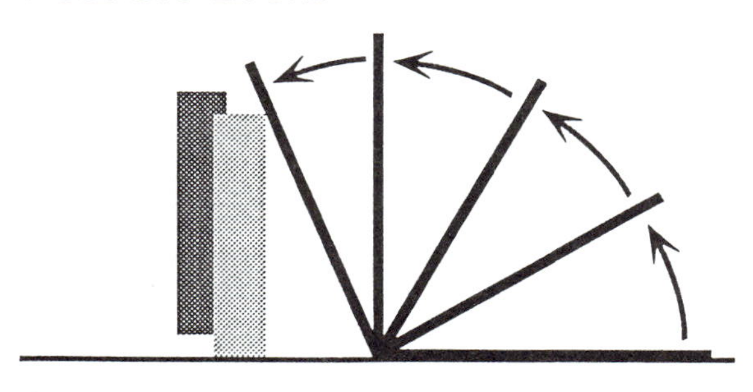

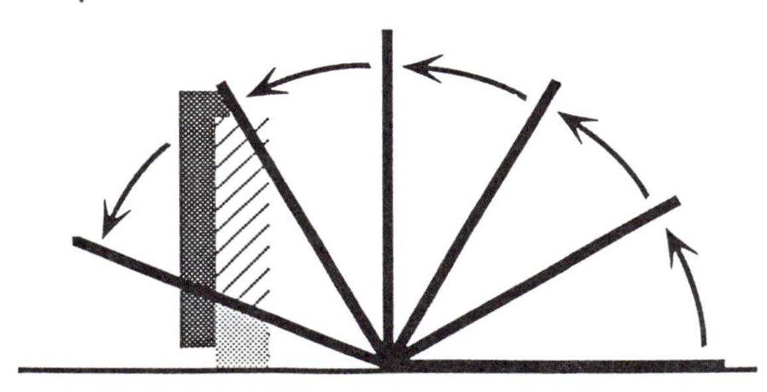

FIGURE 11.17 Test events used in the two-box condition in Baillargeon (1991, experiment 4). Side view.

stopping point was inconsistent with the height and location of the occluded box. This interpretation was supported by a control condition in which the box behind the screen was removed; when only the box to the right of the screen was present, the infants tended to look equally at the events.

In contrast to the infants in the two-box condition, the infants in the one-box condition tended to look equally at the impossible and the possible events, as though they judged both the 112°- and the 157°-screen stopping points to be consistent with the box's height and location. Together, the results of the one- and two-box conditions indicated that the infants were aware that the height and location of the box behind the screen could be used to predict at what point the screen would stop but could detect the 80% violation shown in the impossible event only when provided with a copy of the occluded box.

Habituation Event

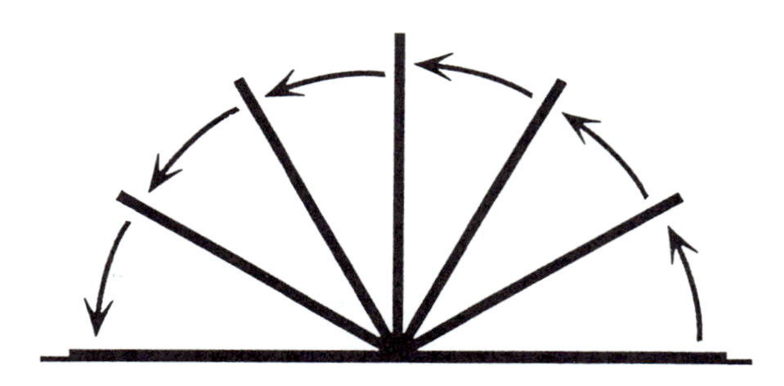

Test Events

Possible Event

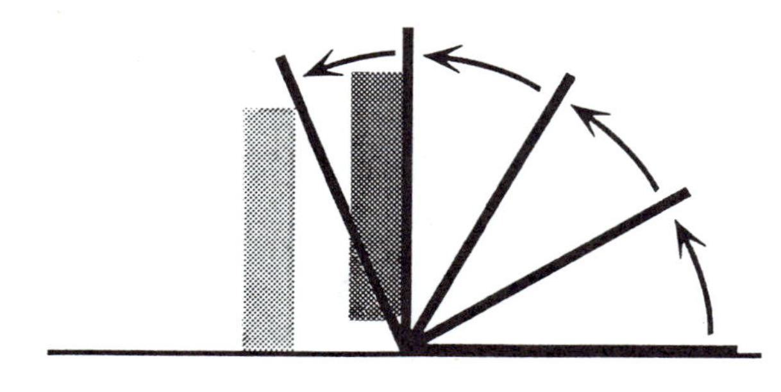

Impossible Event

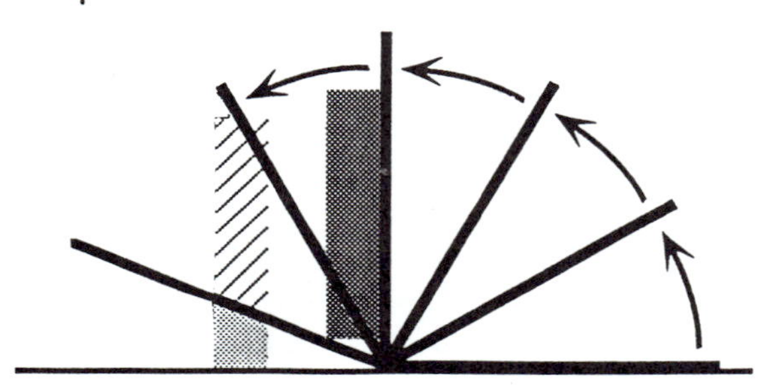

FIGURE 11.18 Test events used in Baillargeon (1991, experiment 4). Side view.

A subsequent experiment revealed that, not only did $4\frac{1}{2}$-month-old infants require the presence of a second box to detect the 80% violation, but this box had to be placed in the same frontoparallel plane as the occluded box (Baillargeon, 1991). When the second box was placed to the right but 10 cm in front of the box behind the screen, the infants no longer showed surprise at the screen's 157° stopping point (figure 11.18). In this experiment, the infants still had a reminder of the occluded box's height; however, they could no longer use a visual comparison strategy to solve the task. When the two boxes were in the same frontoparallel plane, as in the first experiment, all the infants needed to do to solve the task was to compare the height of the screen (at its stopping point) to that of the second box. When the second box was in front of the occluded box, however, this alignment strategy was no longer valid, because the screen rotated past the top of the second box in both the possible and the impossible events.

The results of these experiments thus paralleled those obtained with $12\frac{1}{2}$-month-old infants in the unveiling experiments summarized in the last section (Baillargeon and DeVos, 1994b). Recall that those infants were able to judge which dog could have been hidden under the cover behind the screen only when they could compare, in a single glance, the size of each dog to that of a second, identical cover. The infants failed the task when no second cover was used or the location of the second cover did not allow direct visual comparison with each dog.

In a final experiment (Baillargeon, 1991), $6\frac{1}{2}$-month-old infants were tested in the one-box condition described above. Unlike the $4\frac{1}{2}$-month-old infants, these older infants looked reliably longer at the impossible than at the possible event, suggesting that they (1) represented the height and location of the occluded box; (2) used this information to estimate at what point the screen would reach the occluded box; and therefore (3) were surprised by the impossible event in which the screen continued rotating past this point. A control condition carried out without the box supported this interpretation.

Together the results of the experiments just described suggest that at $4\frac{1}{2}$ months of age, infants realize that, when a box is placed in the path of a rotating screen, the box's height and location affect at what point the screen will stop. However, infants can reason only qualitatively about the screen's stopping point: They succeed at detecting violations only when they are able to compare visually the height of the screen to that of a second, identical box (e.g., "The screen is aligned with the top of the box"). When forced to reason about the absolute height and location of the box behind the screen, infants fail to detect even extreme violations (for further evidence of qualitative reasoning about arrested-motion phenomena in 4-month-old infants, see Spelke et al., 1992). By $6\frac{1}{2}$ months of age, however, infants have progressed beyond this point; they can use their representation of the box's height and location to estimate at what point the screen will stop.

Discussion How should the developmental sequences described in this section be explained? These sequences are unlikely to reflect the gradual maturation of infants' quantitative reasoning or information-processing abilities, because the same pattern recurs at different ages for different physical phenomena. To what other phenomenon-specific changes should the sequences be attributed? One possibility is that, when first reasoning about a continuous variable, infants have difficulty encoding or retaining quantitative information about the variable.

Some evidence for this explanation comes from an experiment that examined $12\frac{1}{2}$-month-old infants' ability to encode and remember the size of a protuberance in a cloth cover (Baillargeon and DeVos, 1994b). At the start of each test event, the infants saw a cover with a small protuberance (figure 11.19); this cover was identical to that used in our previous unveiling experiments (Baillargeon and DeVos, 1994a, 1994b). Next, the cover was hidden by a screen. A gloved hand then reached behind the screen, retrieved the cover *with* its protuberance, and deposited it on the apparatus floor. In the possible event, the cover was identical to that shown at the start of the event. In the impossible event, the cover was more than twice as large as the initial cover. The infants tended to look equally at the two events, suggesting that they had not encoded or could not remember the size of the cover shown at the beginning of each event.

This negative result sheds light on the failure of the $12\frac{1}{2}$-month-old infants in the one-cover experiment to show surprise at the large toy dog's retrieval (Baillargeon and DeVos, 1994b) (see figure 11.14). Clearly, if the infants did not know the size of the hidden cover, they could not judge which size dog could have been hidden under the cover. From this perspective, the finding that $12\frac{1}{2}$-month-old infants were also unsuccessful when a second cover was placed to the left of the screen (Baillargeon and DeVos, 1994b) (see figure 11.13) suggests that they either could not encode infor-

Possible Event

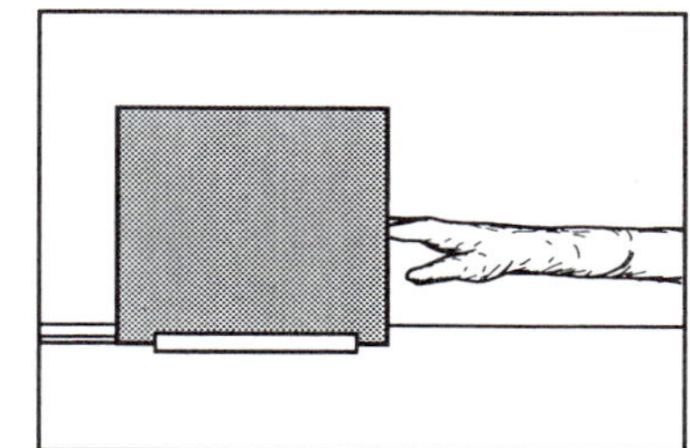

Impossible Event

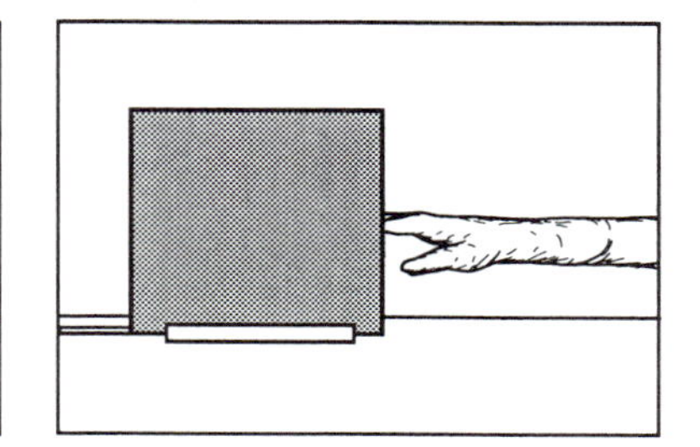
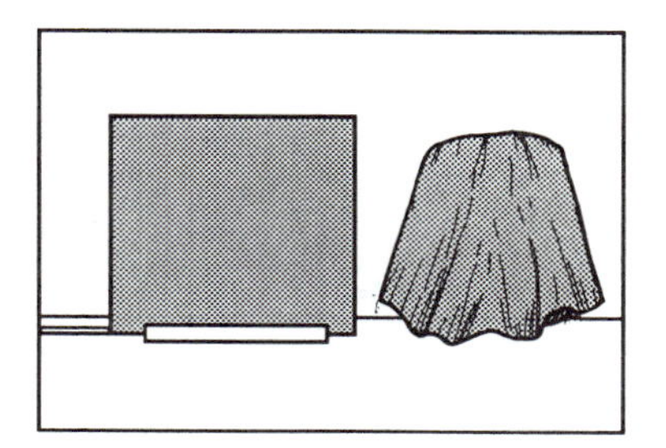

Figure 11.19 Test events used in Baillargeon and DeVos (1994b, experiment 4).

mation about the absolute size of the second cover or could encode this information but could not retain it even for the very brief interval required to shift their gaze from the cover to the dog and compare their representation of each item.

Other explanations could be advanced for the developmental sequences described in this section. For example, it could be that infants *are* able to encode and retain quantitative information about newly identified continuous variables but that these initial quantitative representations are so imprecise that they do not allow infants to detect even the marked violations shown in the present experiments. Further research is needed to evaluate this and other related explanations.

Concluding remarks

The model described in this chapter suggests that in learning to reason about a novel physical phenomenon, infants first form an all-or-none concept and then add to this initial concept discrete and continuous variables that are discovered to affect the phenomenon. Furthermore, after identifying continuous variables, infants succeed in reasoning first qualitatively and only later quantitatively about the variables.

This sketchy description may suggest a rather static view of development in which accomplishments, once attained, are retained in their initial forms. Nothing could be farther from the truth, however. Our data suggest that the variables infants identify, like the qualitative and quantitative strategies they devise, all evolve over time. To illustrate, when judging whether a box resting on a platform is stable, infants initially focus exclusively on the amount of contact between the box's bottom surface and the platform and, as a consequence, treat symmetrical and asymmetrical boxes alike. By the end of the first year, however, infants appear to have revised their definition of this variable to take into account the shape (or weight distribution) of the box (e.g., Baillargeon, in press, a). Similarly, evidence obtained with the rotating screen paradigm suggests that infants' quantitative reasoning continues to improve over time (e.g., $6\frac{1}{2}$-month-old infants can detect 80% but not 50% violations, whereas $8\frac{1}{2}$-month-old infants can detect both), as does their qualitative reasoning (e.g., $6\frac{1}{2}$-month-old infants will make use of a second box to detect a violation even if this second box differs markedly in color from the box behind the screen, whereas $4\frac{1}{2}$-month-old infants will not) (Baillargeon, 1993, in press, a).

The model of the development of infants' physical reasoning proposed here leaves many questions unanswered. In particular, what are the innate constraints that guide infants' identification of initial concepts and variables? Are these constraints purely formal, as we suggested earlier, or will it be necessary, to explain learning, to include substantive information about the nature or properties of objects? Furthermore, what consitutes a physical phenomenon? Should all events that reflect the operation of a same principle (e.g., impenetrability or gravity) be viewed as instances of the same phenomenon, or should phenomena be defined more narrowly, as in the preceding examples, in terms of specific types of interactions between objects?

In an attempt to shed light on these and related questions, we have opted for a dual research strategy. The first is to examine the development of infants' understanding of additional physical phenomena (e.g., arrested-motion, occlusion, and containment phenomena) to determine how easily these developments can be captured in terms of the patterns described in the model. With respect to arrested-motion phenomena, for example, one could ask whether infants younger than $4\frac{1}{2}$ months of age realize that a rotating screen should stop when a box stands in its path but are not yet aware that the height and location of the occluded box can be used to predict at what point the screen will stop. Our second strategy, which was alluded to earlier, is to attempt to teach infants initial concepts and variables to uncover what kinds of observations and how many observations infants require for learning. Would infants younger than $6\frac{1}{2}$ months of age, for example, be able to abstract the variable "amount of contact" in reasoning about support if provided with a set of pertinent visual obversations? We hope that the pursuit of these two strategies eventually will allow us to specify the nature of the mechanisms that infants bring to the task of learning about the physical world.

ACKNOWLEDGMENTS This research was supported by grants from the Guggenheim Foundation, the University of Illinois Center for Advanced Study, and the National Institute of Child Health and Human Development (HD-21104). I would like to thank Laura Kotovsky and Karl Rosengren for their careful reading of the manuscript; Beth Cullum for her help with the data analyses; and Lincoln Craton, Julie

DeVos, Marcia Graber, Myra Gillespie, Valerie Kolstad, Laura Kotovsky, Beth Cullum, Amy Needham, Helen Raschke, and the undergraduate assistants at the Infant Cognition Laboratory at the University of Illinois for their help with the data collection. I thank also the parents who kindly agreed to have their infants participate in the research.

NOTES

1. Spelke and her colleagues have also investigated young infants' intuitions about support relations between objects (e.g., Spelke et al., 1992, 1994a, b). Their results, however, have tended to be negative. See Baillargeon, Kotovsky, and Needham (in press) for a description of these results and possible explanations of the discrepancy between these and the present results.
2. To render the test events shown to the infants in the experimental and the control conditions more comparable, the right wall of the apparatus was also moved in the experimental test events. In each event, the wall was positioned 10 cm from the front end of the bug. In addition, the infants saw the two wall positions on alternate habituation trials (see figure 11.6). Analysis of the habituation data revealed that the infants showed no reliable preference for either wall position.
3. The small, medium, and large cylinders were made of identical material, so their sizes and weights could be expected to covary. Because our data are insufficient to determine whether infants based their predictions on the cylinders' sizes or weights, we will refer only to the sizes of the cylinders.
4. We have discussed at length how infants encode information about the size of the cylinder; but what about the distance traveled by the bug in each event? It seems likely that infants encode this information not in quantitative terms (e.g., "The bug traveled *x* as opposed to *y* distance") but rather in qualitative terms, using as their point of reference the track itself (e.g., "The bug rolled to the middle or the end of the track"), their own spatial position (e.g., "The bug stopped in front of me or rolled past me"), or the brightly decorated back wall of the apparatus (e.g., "The bug stopped in front of this or that section of the back wall").

REFERENCES

Baillargeon, R., 1991. Reasoning about the height and location of a hidden object in 4½- and 6½-month-old infants. *Cognition* 38:13–42.

Baillargeon, R., 1993. The object concept revisited: New directions in the investigation of infants' physical knowledge. In *Carnegie-Mellon Symposia on Cognition: Vol. 23. Visual Perception and Cognition in Infancy*, C. E. Granrud, ed. Hillsdale, N.J.: Erlbaum.

Baillargeon, R. (in press, a). A model of physical reasoning in infancy. In *Advances in Infancy Research*, vol. 9, C. Rovee-Collier and L. Lipsitt, eds. Norwood, N.J.: Ablex.

Baillargeon, R. (in press, b). How do infants learn about the physical world? *Curr. Dir. Psychol. Sci.*

Baillargeon, R., and J. DeVos, 1991. Object permanence in young infants: Further evidence. *Child Dev.* 62:1227–1246.

Baillargeon, R., and J. DeVos, 1994a. The development of infants' intuitions about unveiling events. Manuscript submitted for publication.

Baillargeon, R., and J. DeVos, 1994b. Qualitative and quantitative reasoning about unveiling events in 12½- and 13½-month-old infants. Manuscript in preparation.

Baillargeon, R., M. Graber, J. DeVos, and J. Black, 1990. Why do young infants fail to search for hidden objects? *Cognition* 36:255–284.

Baillargeon, R., L. Kotovsky, and A. Needham (in press). The acquisition of physical knowledge in infancy. In *Causal Understandings in Cognition and Culture*, G. Lewis, D. Premack, and D. Sperber, eds. Oxford, Engl: Oxford University Press.

Baillargeon, R., A. Needham, and J. DeVos, 1992. The development of young infants' intuitions about support. *Early Dev. Parenting* 1:69–78.

Baillargeon, R., H. Raschke, and A. Needham, 1994. Should objects fall when placed on or against other objects? The development of young infants' reasoning about support. Manuscript in preparation.

Baillargeon, R., E. Spelke, and S. Wasserman, 1985. Object permanence in 5-month-old infants. *Cognition* 20: 191–208.

Banks, M. S., 1983. Infant visual perception. In *Handbook of Child Psychology*, vol. 2, P. Mussen, series ed., and M. M. Haith and J. J. Campos, eds. New York: Wiley.

Bornstein, M. H., 1985. Habituation of attention as a measure of visual information processing in human infants. In *Measurement of Audition and Vision in the First Year of Postnatal Life*, G. Gottlieb and N. Krasnegor, eds. Norwood, N.J.: Ablex.

Bower, T. G. R. 1974. *Development in Infancy*. San Francisco: Freeman.

Diamond, A. 1991. Neuropsychological insights into the meaning of object concept development. In *The Epigenesis of Mind: Essays on Biology and Cognition*, S. Carey and R. Gelman, eds. Hillsdale, N.J.: Erlbaum.

Fagan, J. F., III, 1984. Infant memory: History, current trends, relations to cognitive psychology. In *Infant Memory: Its Relation to Normal and Pathological Memory in Humans and Other Animals*, M. Moscovitch, ed. New York: Plenum Press.

Forbus, K. D., 1984. Qualitative process theory. *Artif. Intell.* 24:85–168.

Gratch, G., 1976. A review of Piagetian infancy research: Object concept development. In *Knowledge and Development: Advances in Research and Theory*, W. F. Overton and J. M. Gallagher, eds. New York: Plenum, pp. 59–91.

Harris, P. L., 1987. The development of search. In *Handbook of Infant Perception*, vol. 2, P. Salapetek and L. B. Cohen, eds. New York: Academic Press., pp. 155–207.

Harris, P. L., 1989. Object permanence in infancy. In *Infant Development*, A. Slater and J. G. Bremmer, eds. Hillsdale, N.J.: Erlbaum, pp. 103–121.

Haywood, K. M., 1986. *Lifespan Motor Development*. Champaign, Ill.: Human Kinetics Publishers.

Held, R. (in press). Development of cortically mediated visual processes in human infants. In *Neurobiology of Early Infant Behaviour*, C. von Euler, H. Forssberg, and H. Lagercrantz, eds. London: Macmillan.

Kolstad, V., and R. Baillargeon, 1994. Appearance- and knowledge-based responses to containers in $5\frac{1}{2}$- and $8\frac{1}{2}$-month-old infants. Manuscript in preparation.

Kotovsky, L., and R. Baillargeon (in press). Calibration-based reasoning about collision events in 11-month-old infants. *Cognition*.

Kotovsky, L., and R. Baillargeon 1994a. The development of infants' reasoning about collision events. Manuscript in preparation.

Kotovsky, L., and R. Baillargeon 1994b. Should a stationary object be displaced when hit by a moving object? Reasoning about collision events in $2\frac{1}{2}$-month-old infants. Manuscript submitted for publication.

Kotovsky, L., and R. Baillargeon 1994c. Qualitative and quantitative reasoning about collision events in infants. Manuscript in preparation.

Leslie, A. M., 1984. Infant perception of a manual pick-up event. *Br. J. Dev. Psychol.* 2:19–32.

Leslie, A. M., 1988. The necessity of illusion: Perception and thought in infancy. In *Thought Without Language*, L. Weiskrantz, ed. Oxford, Engl.: Oxford University Press.

Leslie, A. M. (in press). ToMM, ToBy, and Agency: Core architecture and domain specificity. In *Causal Understandings in Cognition and Culture*, G. Lewis, D. Premack, and D. Sperber, eds. Oxford, Engl.: Oxford University Press.

Needham, A., and R. Baillargeon, 1993. Intuitions about support in $4\frac{1}{2}$-month-old infants. *Cognition*. 47:121–148.

Needham, A., and R. Baillargeon, 1994. Reasoning about support in 3-month-old infants. Manuscript submitted for publication.

Olson, G. M., and T. Sherman, 1983. Attention, learning, and memory in infants. In *Handbook of Child Psychology* vol. 2, P. H. Mussen, series ed., and M. M. Haith and J. J. Campos, eds. New York: Wiley, pp. 1001–1080.

Piaget, J., 1952. *The Origins of Intelligence in Children*. New York: International University Press.

Piaget, J., 1954. *The Construction of Reality in the Child*. New York: Basic Books.

Rochat, P., and A. Bullinger (in press). Posture and functional action in infancy. In *Francophone Perspectives on Structure and Process in Mental Development*, A. Vyt, H. Bloch, and M. Bornstein, eds. Hillsdale, N.J.: Erlbaum.

Schuberth, R. E., 1983. The infant's search for objects: Alternatives to Piaget's theory of concept development. In *Advances in Infancy Research*, vol. 2, L. P. Lipsitt and C. K. Rovee-Collier, eds. Norwood, N.J.: Ablex, pp. 137–182.

Spelke, E. S., 1985. Preferential looking methods as tools for the study of cognition in infancy. In *Measurement of Audition and Vision in the First Year of Postnatal Life*, G. Gottlieb and N. Krasnegor, eds. Norwood, N.J.: Ablex.

Spelke, E. S., 1991. Physical knowledge in infancy: Reflections on Piaget's theory. In *The Epigenesis of Mind: Essays on Biology and Cognition*, S. Carey and R. Gelman, eds. Hillsdale, N.J.: Erlbaum.

Spelke, E. S., K. Breinlinger, J. Macomber, and K. Jacobson, 1992. Origins of knowledge. *Psychol. Rev.* 99: 605–632.

Spelke, E. S., A. Phillips, and A. L. Woodward (in press). Physical conceptions in infancy: Objects, people, and shadows. In *Causal Understandings in Cognition and Culture*, G. Lewis, D. Premack, and D. Sperber, eds. Oxford, Engl.: Oxford University Press.

Spelke, E. S., K. Jacobson, M. Keller, and D. Sebba, 1994a. Developing knowledge of gravity: II. Infants' sensitivity to visible object motion. Manuscript submitted for publication.

Spelke, E. S., A. Simmons, K. Breinlinger, and K. Jacobson, 1994b. Developing knowledge of gravity: I. Infants' sensitivity to hidden object motion. Manuscript submitted for publication.

White, B. L., P. Castle, and R. Held, 1964. Observation of the development of visually directed reaching. *Child Dev.* 35:349–364.

12 Origins of Human Social Competence

DAVID PREMACK AND ANN JAMES PREMACK

ABSTRACT We present a theory of human social competence as a domain-specific module. According to this module, infants perceive the spontaneous motion of objects as having an internal cause and interpret this cause as intention. Further, they attribute goals to intentional objects and value to interactions between these objects and expect such objects to reciprocate. When one intentional object coerces another to move, the infant attributes possession to the former. When intentional objects move together freely, the infant interprets them as a group and assigns them special properties. These properties are explained by infants using a theory of mind that attributes mental states—perception, desire, belief—to these intentional objects.

Humans distinguish two classes of objects in terms of motion: animate objects, which start and stop their own motion, and physical objects, which move only when acted on by another object. Although animate objects move in both space (by changing their location) and in place (by movement within their own boundaries, as in respiration or ingestion), only movement in space is interpreted as intentional. *Goal* and *value* are the central properties of intentional objects; *reciprocation*, *possession*, and *group* are secondary properties. An understanding of these five properties of intentional objects forms the basis of human social competence.

Whereas animate objects move in two ways—in space and in place—physical objects move only when launched by contact with another object. It is interesting to note that major divisions in human science—psychology, biology, and physics—coincide with these three kinds of movement.

Fundamental invariances are associated with each of the two classes of objects. A physical object moves as a unit and conserves momentum; two physical objects cannot occupy the same place at the same time. Invariances such as these were present during the Pleistocene and, judging from sensitivities demonstrable in present-day infants (e.g., Spelke, 1982; Baillargeon, 1987), were detected by our ancestors. There are comparable invariances in the animate case. For instance, the movement patterns of respiration and ingestion (motion in place) and of goal-directed action (motion in space) also were present during the Pleistocene and, we will argue here, also were detected by our ancestors.

Presumably, all species developed mechanisms for coping with some or many of these invariances. Individuals that did were more likely to survive and reproduce than those that did not, and so these mechanims evolved, becoming part of the species. A classical example of such a development is found in the frog. On detecting small objects moving in a quick, jerky way, the frog's mouth opens, unleashing its tongue. The frog's remarkable "bug detector" is, of course, only one of many such examples (e.g., Lettvin et al., 1959).

Species might have differed in the number of invariances they accommodated; probably humans detected more and almost certainly read them in a more fine-grained way. This quantitative factor alone, however, does not do justice to the human treatment of invariances. Frogs and most other species react to invariances, but humans seek to explain them. This pursuit of explanation, with its reliance on causal analysis, is a human specialization. The framework for explanation is provided by the domain-specific theories that have evolved in humans, which represent a fundamental emergent. Although precursors can be discerned in other species (e.g., Premack 1988a, b, 1992), the development of such theories is slight.

Social competence theory: A framework for explanation

What is a theory and how does it provide a framework for explanation? A theory has three components, each

DAVID PREMACK and ANN JAMES PREMACK Laboratoire de Psychobiologie du Developpement, CNRS, Paris, France.

linked to a specific content. The first component identifies the class of items to which the theory applies, the second specifies the privileged changes to which members of the class are subject, and the third explains the changes.

Consider the present theory of human social competence, which we offer as an example of a basic human emergent. We label the three components of the theory *intentional*, *social*, and *theory of mind* (TOM), as a means of conveying their content. The first component identifies the class of items to which the theory applies: objects that start and stop their own motion in space (hereafter called *self-propelled objects*). The perception of these objects leads to interpretation of intentional (defined later).

The second component is activated by the output of the first, by the representation of intentional objects. This component specifies the changes that characterize intentional objects. For example, they instantiate goals; they interact in ways to which value is attributed; they reciprocate; they join groups; they have possessions.

The last component, TOM, provides an explanation for those changes that members of the class undergo. According to TOM, these changes are caused by perception, desire, and belief, the three fundamental mental states attributed to humans.

We have said that only humans hold theories, but a common position that contradicts this view is that *all* species hold theories about the world and that they differ only in the nature of the theory they hold. Von Uexkuell (1957), pillar of ethology, has been read (albeit somewhat incorrectly) as contributing to this view. He offers the example of the tick which, rather than perceiving cow or horse, perceives a temperature gradient, causing it either to cling to its perch or to drop and connect with the "temperature" below. This difference is one of perception, however, not of theory.

The bee perceives blue patches, which it uses in expressing the directional component of its dance, but the bee's blue patch is no more the sun than the tick's temperature gradient is a cow or horse. Blue patches and temperature gradients are not objects, nor can they be objects that move in categorically different ways. Invariances such as unitary movement, conservation of motion, and the like do not apply to what the tick or bee perceives. They perceive features, not objects. They lack the concept of object, indeed of conceptual structure in general, which is presupposed by theory. They neither engage in causal analysis nor pursue explanation. The fact is not that bees and ticks have a different theory of the world but that they have no theory.

Much recent work on animal cognition is not an improvement on Von Uexkuell and affords little clarification of human complexity. The work focuses on remarkable adaptations found in nonprimates, nonmammals, and even nonvertebrates. The spider, for instance, to whose back a weight is attached, adjusts the thickness of the thread used in making its web in proportion to the attached weight. When dancing, bees make complex computations that take into account the changing position of the sun. Among vertebrates, we encounter fledglings that record constellations and use them when migrating and rodents that, in returning to the nest by dead reckoning, apparently carry out the functional equivalent of double-integration.

This complexity certainly attests to the "miraculous powers of the blind watchmaker" (Dawkins, 1986), the sagacity of evolution in producing such exquisite mechanisms, but it does little to illuminate human complexity. The contribution of evolution to the complexity of bee and bird bears little relation to its contribution to human complexity.

The bee, for instance, in modifying the directional component in its dance according to changes in the sun's position, has no idea that there is such an object as the sun. The same is true of fledglings that record the composition of constellations and use them in migration. No more than does the bee know there is such an object as a sun does the fledgling know there is such an object as a star. Human complexity consists in knowing that heavenly bodies are physical objects to which all invariances of that domain apply *and* in seeking to explain them. This fundamental aspect is not illuminated by the computations of bees or birds.

The complexity that evolution has conferred on rodents, in giving them a navigational system with the functional equivalent of double-integration, should be compared with the complexity evolution has conferred on humans. In the seventeenth century, Newton and Leibniz independently invented the calculus, evolution having thereby produced a device with a sagacity equivalent to its own. That is, while evolution "invented" the calculus, as witnessed in rodent navigation, it later "invented" a device that itself invented the calculus. This comparison helps place in perspective the nature of human emergents and clarifies why most reports on animal complexity offer little illumination of human complexity.

Social Theory

Our theory of human social competence seeks to explain the theory held by humans. Like all such theories, it has an external (psychophysical) and an internal (conceptual) component. On the psychophysical side, it describes the external conditions that bring about interpretations: for instance, the conditions of self-propelledness, goal-directedness, and positive social interaction. On the conceptual side, it describes the content of the interpretations: the meaning of *intention*, *goal*, *possession*, and so forth.

We have little to say about the psychophysical side except to make occasional suggestions, based largely on computer animations. These animations can be extremely helpful in solving certain psychophysical problems for, in permitting quick manipulation of stimulus conditions, they allow one to observe test cases and confront one's intuitions directly. The serious psychophysical work, however, remains to be done.

For example, while it is clearly change that fosters the impression of self-propelledness, it is not as clear what qualifies as change. A change, such as that of an unchanging object moving from rest to motion, will undermine the impression of self-propelledness. Let us recall the upward and downward deflection of a ball about a midpoint a cycle; the repetition of this cycle of motion defeats the impression of self-propelledness. To retain the impression, the ball must reach different heights and depths, ascending and descending at different angles and speeds. Change must occur both within and between cycles.

In addition, the construction of self-propelled objects may be unique and affect their movement. Typically, they have a pliable outer membrane and a jointed body. When such an object is struck by another object, it will move differently from a rigid object that is struck in the same way. Does an infant take these structural factors into account?

Further, is motion privileged, or will a change in object quality have an equal effect? Suppose an object undergoes spontaneous changes in color or shape or size: Will this lead to the same interpretation as when an object undergoes spontaneous changes in motion? These are questions we intend to pursue in other studies. Here we will concentrate on providing an account of the content of the interpretations that humans ascribe to the critical stimuli.

SELF-PROPELLEDNESS AND ITS INTERPRETATION According to present theory, the infant's interpretation of a self-propelled object is as automatic as the interpretation of causality it places on appropriate collisions between physical objects. What is this interpretation? Of the two classes of objects found in the world, each associated with different kinds of movement, one class moves only when acted on (by another object). These objects are caused to move, *cause* being the interpretation that, as Michotte (1963) has shown, humans place on such an event when there is temporal and spatial contiguity between the colliding objects. Evidently, it is also the interpretation that 6-month-old infants place on the same event (Leslie and Keeble, 1987).

The significant feature of this case is the visible, external presence of the cause. When, however, an object moves by itself, there is no visible, no external, cause. Can we conclude that the movement had no cause? Humans abhor uncaused events. It can mean only that the cause was internal and therefore invisible.

Thus, the first division we find in the world, as depicted in figure 12.1, is that between objects whose movement is externally, (thus visibly) caused and objects whose movement is internally (thus invisibly) caused. This difference distinguishes physical and animate objects. The alternative, that between movements which are and are not caused, is unacceptable to our species.

The contrast between visible and invisible is, of course, a fundamental, profoundly human distinction that makes an early appearance in the present case but has many other applications. It is a distinction engendered by a mind or brain that is utterly committed to explanation but capable of realizing the commitment with no other conceptual tool than that of cause. Hence, when events demand explanation but lack visible causes, causes must be invented. Although psychologists may find it of special interest that invisible makes an early appearance in their domain, mind being a principal exemplar of invisible, psychology has no unique claim on the concept. Invisible events suffuse all sciences—physics and biology no less than psychology.

Internally caused movement is of two kinds, as shown in figure 12.1: movement in space and movement in place. The first of these categories perfectly defines *intentional*, which is, according to the present

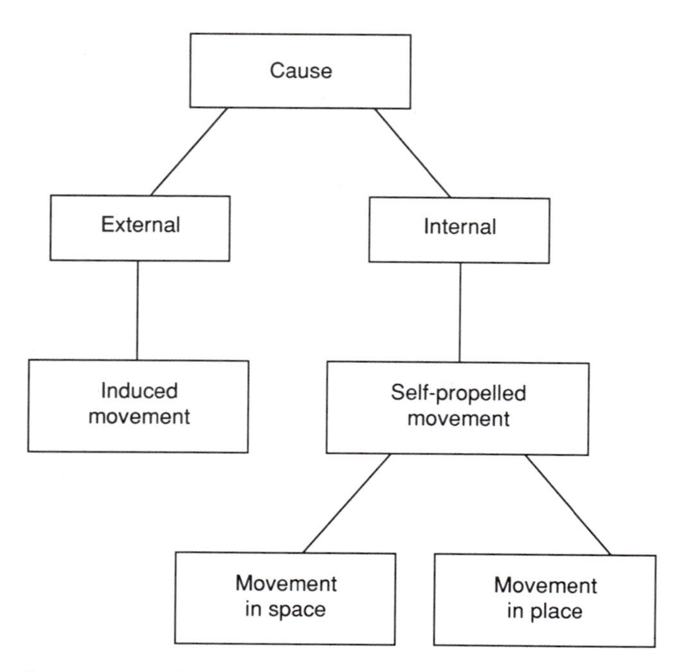

FIGURE 12.1 Schematic outline of division of world into two object classes, their associated types of movement, and external versus internal causal process.

theory, the interpretation humans place on internally caused movement in space.

Movement in place, the other category of internally caused movement, brings about the interpretation of *animate* but not intentional. A classical example is that of respiration, which is easily simulated on the computer by a ball whose sides alternately collapse and recover. Such a "breathing" object appears to be alive, but not until it moves in space does it also appear to be intentional.

Stalking and ingestion illustrate the same distinction. Although both indicate life, only stalking (movement in space) is intentional; ingestion (movement in place) is not. All internally caused movement indicates animacy or life, but only movement that in addition carries the object from one location to another indicates intentionality.

The intentionality judgments of children bear witness to these distinctions. The judgments of children even as old as 4 or 5 years do not conform to those of adults (e.g., King, 1971; Berndt and Berndt, 1975; Smith, 1978). Adults emphatically deny intentionality to certain kinds of acts (e.g., acts that are involuntary, accidental, or have unintended effects), but children do not. When shown videotapes of simple actions, 4- and, to some extent, even 5-year-old children call *intentional* the action of an individual who sneezed, slipped while crossing the floor, or knocked a flower into a basket in the course of reaching for a glass of water, thus accepting as intentional every act that adults rejected. Children denied intention to only one kind of action, that which was *not* self-propelled—for instance, a seated individual who was forcibly moved by the encroachment of a piece of furniture, or a person's arm that was lifted by a second person who, hooking an umbrella handle about the former's wrist, raised the umbrella (Smith, 1978).

For adults, intention is clearly a mental state, one that may even involve planning. For young children, however, it is less a state of mind than a kind of movement, as it is, we claim, for the infant. Though highly compatible with the present theory, the Smith results are nonetheless perplexing.

Children, as the TOM literature (e.g., Leslie, 1987; Wellman, 1990; Wimmer and Perner, 1983) clearly shows, are highly capable of attributing mental states. Why, then, at 4 and 5 years of age do they not map *intentional* onto mental state rather than movement? The answer must be this: Rather than reacting automatically to a task with their full capacity, children respond in keeping with task demands. If a task makes weak demands, as does reading a videotape, the child, rather than use his or her full capacities, reverts to those of infancy and thus maps intention onto movement.

GOAL: EXEMPLARS VERSUS GENERAL THEORY When activated by appropriate movement—that of a self-propelled object—the first component in the infant's social system outputs the interpretation *intentional*. The next interpretation that it outputs is *goal-directed*. Intentional is a necessary, though not sufficient, condition for this interpretation. Goal-directed is attributed only to intentional objects that meet additional criteria.

What is a goal in the eyes of an infant? Goal is not an isolated concept, of course, but one embedded in motivation theory. An individual can be said to understand these concepts implicitly when he or she is able to use the theory to which the concepts belong—for instance, to produce and comprehend certain motivational relations between intentional objects.

How does an infant or child acquire motivational theory? There are two legitimate alternatives. (We can dismiss a third, namely that a child learns the theory piece by piece through general associative mechanisms.) First, all infants acquire a completely general theory, whether of language or of motivation, by maturation. Second, infants are sensitive to certain exemplars that attract their attention. They extract instances of them from the perceptual stream and store mental representations of them. In time, they discern what the exemplars have in common and work toward a theory that will explain their common features.

We do not endow the infant with an abstract concept of goal, or the motivational theory of which this concept is a part, but suggest that the infant can recognize a few exemplars with which it subsequently builds a motivational theory. What are the goals an infant might recognize? We suggest the following three, though we do not claim that this list is exhaustive: (1) escape from confinement, (2) contact with another intentional object, and (3) overcoming gravity. The latter may seem dubious in view of the unclear status of the infant's recognition of gravity or the need objects have for support (cf. Spelke, 1988; Baillargeon, Spelke, and Wasserman, 1985), but we retain it, suggesting that the infant may recognize it later as a goal, once the child masters the concept of the need for support.

In this view, infants attribute a goal to objects that they consider intentional when such objects, for example, seek to climb a hill, contact another intentional object, or escape from confinement. What infants accept as proper instances of these cases will depend in part on the intensity of the action. When an act not only conforms to the exemplar but also exceeds baseline intensity, the infant will be more likely to attribute goal than when the act does not exceed baseline intensity.

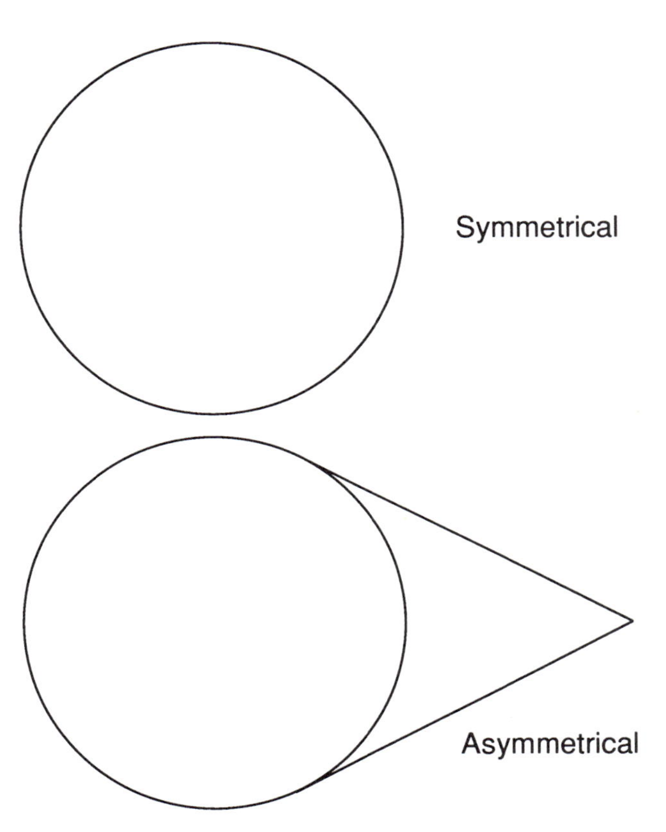

FIGURE 12.2 Example of asymmetrical object, with which infant associates directional movement.

ASYMMETRY AND DIRECTIONAL MOVEMENT The prominence we have given to infants' use of motion should not suggest that they make no use of object quality. Infants, according to the present theory, distinguish symmetrical and asymmetrical self-propelled objects and assign special properties to the latter. Specifically, they expect asymmetrical intentional objects to move directionally. For instance, if we elongate the ball-like objects of our examples or, as shown in figure 12.2, add a nose to them, the infant will expect the object to move in the direction of its asymmetry or nose.

We could test this by setting a group of symmetrical and asymmetrical intentional objects into motion. In the case of asymmetrical objects, an infant should look in the direction of the asymmetry, anticipating that the object will move in that direction, whereas if the objects are symmetrical, the infant should not look in any one direction more than in any other (the child has no anticipations concerning directionality of movement for symmetrical objects). Other tests are possible, of course, but the point is simply to communicate the idea.

Does the putative association of directional movement with asymmetry suggest another glimpse of biology? Is the infant attributing a gradient of sensitivity to the asymmetrical object, treating one end as the head? This strong claim is better replaced by a weaker one that may be more easily proved. The infant may attribute perception at a distance—vision—to the asymmetrical end.

We might show this by placing a baffle in the path of an ongoing asymmetrical object. If moving in the direction of its asymmetry, nose first, for example, the object should be able to "see" and therefore should negotiate the baffle easily, moving through smoothly. However, if moving backward, it should not be able to see and therefore should bump into the baffle and have to grope its way through. An infant who attributes vision, and not only directional movement, to the asymmetrical end should not be surprised by this outcome. Results of this kind would support the claim that infants not only assign directional movement to asymmetrical intentional objects but also attribute vision to the head or asymmetrical end.

Summary of Intentional System The first unit of the three that make up social competence is activated by self-propelled movement in space. Objects having this property are interpreted as intentional, as being the locus of internal causes. The infant expects such objects to engage in goal-directed behavior.

Infants recognize a limited number of goals: We suggest these are escape from confinement, overcoming gravity, and the company of other intentional objects. Infants recognize these cases best when the intensity of action exceeds baseline. They develop motivational theory by discerning what these exemplars have in common and by working out a theory that will explain their common features. Not only motion but also shape contribute to the capacities infants attribute to intentional objects. If the object is asymmetrical, the infant expects it to move directionally and to have a capacity for vision located in the area of the asymmetry.

Social component of social theory

We turn now to the second unit in the system that comprises the infant's social competence. This unit is activated by interactions between intentional objects; hence, we call it the *social system*. Moreover, this unit is internal, activated by mental representations from the first unit, and therefore differs from the first, which is a standard peripheral system activated by perceptual inputs.

The social unit reduces the potentially unlimited interactions between intentional objects to a few types. The major tool used in this analysis is that of *value*, which serves the social unit essentially as goal and intention serve the intentional unit.

Value, like its commonsense counterpart, comes in two forms, positive and negative. *Positive* is that which an intentional object approaches and *negative* that from which it withdraws. At rock bottom, these are the only meanings that *positive* and *negative* have ever had in biology or psychology.

Value Theory The social unit attributes value to the interaction between intentional objects, positive or negative. In distinguishing positive from negative events, the infant uses two criteria. The simpler of these is intensity of motion. Hard actions, such as when one object hits another, are coded negative; soft actions, such as when one object caresses another, are coded positive.

This simple criterion of intensity-based value attribution seems analogous to the intensity-based discrimination of primitive organisms, which approach weak stimuli and withdraw from strong ones. The infant must, of course, have default values for force of movement (just as it must have such values for intensity of movement in order to detect trying). Otherwise, it cannot discriminate hard from soft or either of these from normal.

Intensity is only the first and simplest of the several criteria infants use in the attribution of value. In addition, they make extensive use of a small family of criteria that are functionally equivalent to the distinction between helping and hurting. Once an infant perceives an intentional object as having a goal, it can perceive a second intentional object as either helping or hindering the first object in its realization of the goal. For instance, if one object is perceived as trying to pass through a break in a wall, a second can be perceived as either helping or hindering it in doing so.

Liberty and Aesthetics Although the principal source for the attribution of helping and hurting is goal, alternate sources of helping and hurting can be found in liberty and aesthetics. Each of these gives a special twist to the conditions that lead humans to attribute helping or hurting.

An infant's view of liberty can be seen in the following example: When shown two bouncing objects, one of which becomes trapped in a virtual hole, the infant will interpret the action of a second object that restores the motion of the first as helping and will code it positive.

This criterion is more complex than meets the eye. It requires the infant to judge what the object would have done had it not been trapped. Suppose the infant is shown one object stopping the motion of another: Will the child judge this act as positive or negative? If the object had not been interrupted, would it have stopped or continued its motion? If it would have stopped, the action of the second object was based on an accurate anticipation and, being a form of assistance, would be judged positive. However, if the object would have continued its motion, the action of the second object was an interference and would be judged negative. Although deciding whether an act is helpful or hurtful can be complex, complexity can be circumvented by confining the criterion to cases in which an object's routine is simple enough to be well-known. In such predictable cases, the infant could indeed know what the object would have done if left alone.

Now consider an example of aesthetics. We show an infant two bouncing balls: The one that bounces higher and faster is preferred by the infant. The preferred ball moves into the vicinity of the other and demonstrates its superior bounces several times, as though offering an example. It even assists the other directly, placing itself below, lifting it, helping it to bounce higher. The infant will interpret the actions in this and the previous example as helping, coding them both positive. In this latter case, a positive action consists not in helping an object maintain its liberty but in assisting the object to reach a preferred state.

Direct Test of Value Theory A direct test of the value theory can be made by habituating different groups of infants to each of the four major sources of value—caressing, hitting, helping, and hurting—and then transferring all of them to a common dishabituation condition. In the example depicted in figure 12.3, we have used both a positive and a negative dishabituation condition, that of caressing in one case and that of hitting in another, though any of the other four conditions could equally have been substituted.

In this test, we are interested in how dishabituation will be affected by the habituation that precedes it. Infants who are shifted to a dishabituation condition that is negative (hitting or hurting) after habituation to positive conditions should show greater dishabituation than those so shifted after habituation to negative conditions. Conversely, infants shifted to a dishabituation condition that is positive (caressing or helping) after habituation to negative conditions, should show greater dishabituation than those so shifted after habituation to positive conditions.

In addition, the theory requires that both members of each pair—hitting and hurting, in one case;

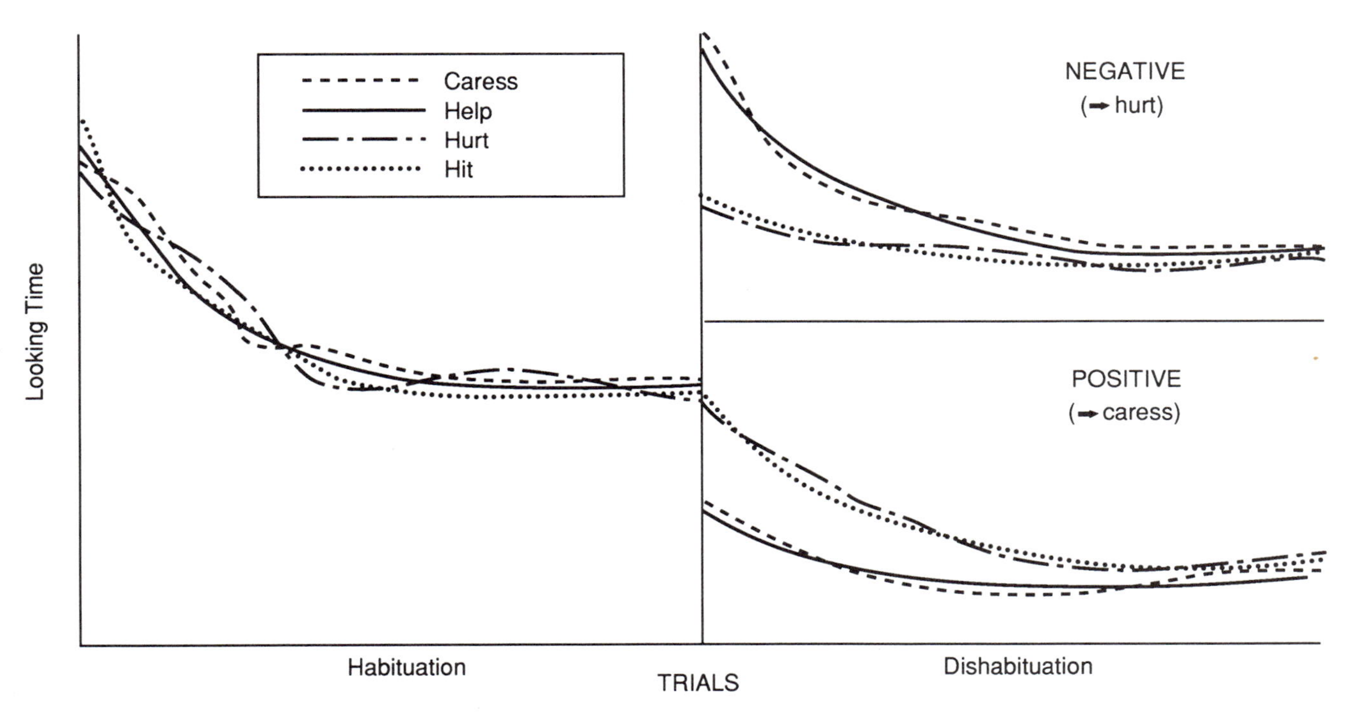

Figure 12.3 Hypothetical habituation-dishabituation functions illustrating basic relations between the two criteria on which either positive or negative value is attributed to interactions between intentional objects. (See text for fuller explication.)

caressing and helping, in the other—produce, ordinally, the same dishabituation outcome. For positives derived from both caressing and helping, the habituation effects should be ordinally equivalent, their physical differences notwithstanding. The theory requires that the same hold for negatives derived from hitting and from hurting; they too should produce ordinally equivalent habituation effects, despite the physical differences of the acts. The paradigm affords a direct test of the value theory, and the hypothetical results of such a test are plotted in figure 12.3.

Consider now a counterintuitive prediction made by the present theory. Note that value, being a social property, is attributed exclusively to the *interaction* between intentional objects and not to the action of an individual. Therefore, if a goal-seeking object either succeeds or fails to obtain its goal (e.g., if it escapes confinement or fails to do so, reaches another intentional object or fails to reach it, regains its liberty or loses it), no value is attached to these acts. Value is not attached to the achievements or failures of individual objects.

We may even suppose that infants vary in their preference for observing failure or success. Further, when old enough, infants may identify with objects they perceive as "trying," vicariously enjoying the success of some, vicariously suffering the failure of others. Yet whatever may prove to be the case with regard to identification and vicarious emotion, the present theory makes no provision for the assignment of value on such a basis. Positive and negative are not indexes of the infant's pleasure or displeasure.

A test of this counterintuitive assumption can be made by using an individual object's success (or failure) as a condition in the habituation-dishabituation paradigm illustrated in figure 12.3. If value attaches to an individual's success or failure as it does to the social interactions—caressing or helping versus hitting or hurting—then the former should have the same effect as the latter in this paradigm. For instance, infants shifted to a dishabituation condition that is, say, negative (hitting or hurting) after habituation to success should show greater dishabituation than those shifted after habituation to failure. Conversely, infants shifted to a dishabituation condition that is positive (caressing or helping) after habituation to failure should show greater dishabituation than those shifted after habituation to success.

However, according to the present theory, this prediction will not be confirmed. Because value should not attach to either success or failure, neither condition should have systematically different effects on dishabituation, as would be predicted for the habituation of conditions to which value does attach. Differences produced by the effects of success or failure should be both minor and explicable in terms of physical similarity. The effects of success and failure, rather than being consistent, should vary from case to case, the action of success being more similar to that of the dishabituation condition on some occasions, the action of failure more similar on others. Results of this kind would confirm that value is not attributed to the success or failure of individual goal seeking but is a social property, confined strictly to the interaction between intentional objects.

Value versus Preference Finally, it is essential to distinguish value from preference. Value is both principled and domain-specific, whereas preference is neither. Value is domain-specific because it applies exclusively to the interaction between intentional objects, and it is principled because the contrast between positive and negative is based on rules, either those that apply to intensity (caressing or hitting), or those that apply to second-order goals (helping or hurting).

Preference, by contrast, is neither principled nor domain-specific. One can compare the preferences of any items—not only one soup to another or one movie to another but also a soup to a movie. Soup, movies, and indefinitely many other items can be placed on the same preference continuum; no property could be less domain-specific than preference.

In addition, preference is unprincipled: There are no known rules that govern its computation. Although one can give elaborate instructions for the measurement of preference (e.g., choice behavior, response probability), one can give no instructions for its computation. Unlike the explicit criteria used to compute value, no criteria are known for the computation of preference.

Note that the fundamental predictions concerning the habituation-dishabituation results depicted in figure 12.3 do not depend in any way on an infant's preferences. Some infants may prefer looking at positive cases (caressing or helping), whereas others prefer looking at negative cases (hitting or hurting) but, whatever their preferences, they should have no effect on the habituation-dishabituation relations predicted in figure 12.3. That is, we do not need two sets of habituation-dishabituation functions, one for infants who prefer looking at positive events and another for those who prefer looking at negative ones.

Infants habituated to a positive case and then shifted to a negative case should show greater dishabituation than infants habituated to a negative case and shifted to another negative. This should hold equally whether infants prefer looking at positive cases or negative ones. Habituation-dishabituation relations are determined by the value an infant attributes to the interactions, not by the infant's preferences for the interactions. Which is to say, if an infant prefers looking at hitting or hurting, the negative cases, this does not mean that for this infant these cases are positive. It means only that he or she would rather look at negative cases than at positive ones. One's preference for a case does not affect its value.

When we say an infant codes hitting as positive and caressing as negative, we make the following claim: If we show an infant that intentional object A hits intentional object B, the infant should expect that B will avoid or withdraw from A on subsequent occasions. Conversely, if we show the infant that A caresses B, the infant should expect that B will approach A on subsequent occasions. The same predictions follow, of course, for helping and hurting and for all other cases that we label positive or negative.

Again, these predictions are independent of the infant's preferences. An infant may, for example, like to look at cases of hitting more than at cases of caressing; nonetheless, it should have the same expectations as those just described. If shown that object A hits object B, it should expect that B will avoid A on subsequent occasions. This should further consolidate the distinction between preference and value that we drew earlier.

RECIPROCATION In reciprocation, the action of one intentional object is followed by a reversal of roles, one that preserves some feature of the original action: its form, value, or both. For instance, if A acted negatively on B and B subsequently acted negatively on A, this would constitute reciprocation with preservation of value.

We can distinguish weak and strong forms: On the strong form, all actions will be reciprocated with preservation of value; on the weak form, *if* an action is reciprocated, it will preserve value. The weak form seems more tenable: All reciprocated actions will preserve value but not all actions will necessarily be reciprocated.

Infants expect, according to the present theory, that if an act is reciprocated, it will preserve value; they do not expect that it will preserve form. If, for instance, object A caressed object B, B need not caress A in order to reciprocate. Instead, object B could help A in any of indefinitely many ways, all of which would preserve value and fulfill the infant's expectations.

POWER: GROUP VERSUS POSSESSION Although value is the major conceptual tool of this module, as an infant develops, the concept of power enters, figuring ever more prominently and finally operating along with value. The older infant will notice, as the younger may not, which of two intentional objects controls the other and, thus, which is more powerful, and it will use this concept as a basis for distinguishing two major social conditions, possession and group.

Possession Intentional objects have the capacity to possess, and they are seen to possess objects to which they are connected in an appropriate way. The concept of possession is thus another fundamental of the present system and, like all its other parts, it is an interpretation made by the infant on an appropriate input. We confront our familiar pair of questions: What constitutes an appropriate input, and what is the content of the interpretation?

Basically, the input that gives rise to the interpretation of possession is a distinctive kind of comovement between two objects, at least one of which is construed as intentional. An early suggestion as to the nature of the comovement can be found in the work of Kummer and Cords (1991). Their interesting data suggest that monkeys treat objects as being possessed when they are connected to and move with another monkey. Only when objects are both connected to and move with a monkey do other monkeys refrain from attempting to take the object from its owner. They treat such objects as though they are part of the monkey's body, in much the same way as they treat the monkey's arm or leg. The monkey data concern only the possession of nonintentional objects, but possession is not restricted to such objects. As the human concept of slavery demonstrates, intentional objects can be possessed no less than physical objects, and the intentional case is important because it demonstrates that possession is not determined merely by comovement and connection.

When two intentional objects are connected and move together, which is the possessor? To answer this question, we must consider a third factor, the power relations between the two objects. Possession requires not only connection and comovement, but also that the possessed object be less powerful than the possessor.

Power of an intentional object can be manifested in many ways (e.g., through size, strength, attractiveness); ultimately, however, it is manifested in its ability to control the movement of another. A decision as to which object is the more powerful must be made only when the possessed object is intentional, for a nonintentional object is incapable of controlling movement and is inescapably the weaker and always the object possessed.

Although monkeys may need physical connection for the interpretation of possession, human infants probably do not. A physical connection between two objects, such as the cord that attached the object to a monkey, may make it easier to perceive that one object is controlled by the other, but it is not essential. What is essential is that the objects differ in power and that one controls the other. It is this relation that an infant will interpret as possession.

Ordinarily, acts on nonintentional objects have no consequences for the infant; one can pound or pet an unintentional object, but in neither case will the infant attribute value to the action. Possession, however, changes the status of nonintentional objects. The infant will code acts on such objects as positive or negative. In addition, the infant will expect acts on these objects to be reciprocated. He or she expects the owner to reciprocate acts on its possessions just as it reciprocates acts on itself.

Possession, for practical reasons, is an important social primitive: It has the consequence of bringing under evaluation objects to which value is not ordinarily attributed. Once value is assigned to actions, expectations concerning reciprocation follow and, in many respects, it is reciprocation that is the crux of morality.

Group The concept of group, like that of possession, concerns the relation among objects but differs in that all the objects in a group are intentional; moreover, they are all of equal power. In group, no object controls the movement of the other. An infant will interpret as a group intentional objects of equal size and strength that move together. Group is the comovement of objects of equal power, a relation among equals therefore, whereas possession is a relation among unequals. Although comovement is found in both cases, power is distributed differently among the comoving elements—equally in the case of group, unequally in the case of possession.

The predilection to bring physically similar objects together, found even in the 10-month-old child (Sugarman, 1983), takes an earlier form in the infant. The infant expects physically alike intentional objects to form groups and physically unlike ones not to. For instance, when shown a set of white intentional objects, an infant will expect them to cohere and move together; it will have the same expectation for a set of black intentional objects, but not for a mixture of white and black objects. However, if the infant is shown that the mixture of black and white objects cohere and move together, it will interpret them as a group. Although an infant expects like objects to flock together, when shown the uncoerced comovement of unlike objects it accepts them as a group in the same degree that it accepts any other. The criterion of uncoerced comovement takes precedence over physical similarity.

The concept of group entails some extremely powerful consequences. First, the infant expects group members to share reciprocation (i.e., it expects each group member to reciprocate acts perpetrated on other group members). For instance, when the infant observes object C act positively (negatively) on object B, it expects not only B but also A, B's co–group member, to act positively (negatively) on C.

Second, the infant expects group members to act alike. This represents an extension of the infant's normal inductive assumptions. Even as the infant expects the same individual to repeat itself in some degree, so the infant expects group members to repeat one another in some degree. For instance, the infant's normal inductive assumptions may take this form: When object A acts positively on B some number of times, the infant expects object A to continue doing so. The infant extends this assumption to members of a group. Having perceived one group member to act positively on B for some number of times, it expects other group members to do the same. In effect, the infant treats different group members as though they were the same individual acting at different times; it equates multiple tokens of a type (group members) with multiple instances of a token (repeated instances of the individual).

Third, the infant expects group members to act positively toward one another. This positive expectation for group members contrasts with an infant's neutral expectations for interactions between unattached intentional objects. As regards the latter, the infant has no expectations, positive or negative.

These are internal consequences; they concern how a group member perceives another—what one is expected to give another member and what one expects to receive in return. The formation of a group also gives rise to external consequences. These concern how group members perceive nonmembers. From the perspective of an observer, the formation of a group does not change the status of those who are not members. Nonmembers do not acquire the special prerogatives that members extend to one another; nonetheless, they do not lose their original status (i.e., they remain intentional objects with respect to all actions judged by standard moral criteria). Thus, an observer of group formation perceives members as gaining prerogatives (relative to one another) but does not regard nonmembers as having lost prerogatives. This is not, however, how a group member perceives a nonmember.

In distinguishing members from nonmembers, a group member both extends prerogatives to members and withdraws them from nonmembers. Thus, becoming a member of a group entails two consequences: the loss of one's neutral predilections toward others and the acquisition of two new attitudes—positive toward members of the group and negative toward nonmembers.

Are external group properties part of the infant's knowledge and therefore properly part of the theory, or are they something a child learns? We propose them as part of the infant's knowledge, with two caveats. First, knowledge of external group properties develops later than knowledge of internal properties of the group. Second, this primitive may be associated with a sex difference. Knowledge of external group properties may be more likely to develop, or to attain a greater strength, in the male infant. If we regard gender as producing overlapping distributions rather than categorical difference, then the mean of the distributions will be greater for the male than for the female infant. Indeed, the distributions may be such that some female infants (but probably no male infants) will escape knowledge of external group properties altogether. Whereas all infants will expect one group member to act positively toward another, and most will expect a group member to act negatively toward a nonmember, some female infants will not expect the latter. If a female infant, lacking negative expectancy, is found to have the expectancy as a child, it will be because of what she has learned.

Intergroup hostility, a staple of human history, may owe more to the devaluation of nongroup members by members than to any other factor. From the momentum of the Crusades and continuing into the present, intergroup hostility appears to have secured a permanent future in human affairs. Frequently, the hostility between human groups bears a stark resemblance to that of chimpanzee groups. Chimpanzees patrol territorial boundaries, seek isolated out-group members, and kill them when they are found (Goodall, 1986). Sex difference is a prominent feature of such intergroup hostility. Only male chimpanzees patrol, seek out, and kill (Goodall, 1986); similarly, primarily male humans carry out warfare; women remain home cooking, cleaning, and caring for children. How might we explain this difference?

We suggested one explanation—namely, the knowledge of external group properties in the male infant. Now we add a second, an explanation that could act in concert with the first. Males appear more prone to form and join groups than females. Male chimpanzees spend much of their time in like sex groups, whereas females, rather than bonding with one another, spend long periods alone with their offspring (Goodall, 1986). But why is the bond mothers form with their children less dangerous than that of group membership? Why does it not lead to the devaluation of other? Mother and child are not comembers of a group, their relation is that of possession, and possession does not entail devaluation of other. Membership in a group, however, involves automatic devaluation of other.

If male humans resemble male chimpanzees in having a penchant for groups, spending more time in them and participating with greater fervor, it will be the male who, as a group member, perceives nonmembers as less than fully qualified intentional objects. Is this a characteristic of male humans? Although the question is relatively straightforward, we can find no relevant data.

Theory of mind component of social theory

Representations of the social module are sent to TOM, the last unit in the system, where interpretation takes a unique form, that of an explanation. The explanations that are the outputs of this unit are states of mind, such as *see*, *want*, and *belief*, or some variation of these basic states (of which humans distinguish literally hundreds).

Children definitely attribute states of mind; 4-year-olds do so at close to an adult level of competence, passing the so-called false belief test, which is seen as a kind of rubicon (Wimmer and Perner, 1983). But do infants attribute states of mind, as the present theory claims? There is, unfortunately, no evidence. Infants have not yet been tested for attribution of mental states.

Moreover, we cannot ask infants to choose between alternatives or answer questions, as we do children. We can test them with implicit procedures only—preference or habituation-dishabituation. There is an advantage in using these procedures, however, for they make weaker test demands than the explicit and may disclose capacities obscured by the greater test demands of explicit procedures. For example, individuals who cannot use a concept instrumentally may nonetheless prove to have the concept when judged by habituation-dishabituation data. This is what we found (Premack, 1988a, b) when comparing young children and chimpanzees on match to sample (explicit procedure) in one case and habituation-dishabituation (implicit procedure) in the other.

There is a twofold advantage in testing the infant on TOM: First, we test the prediction of the present theory that the infant has TOM; second, we test the broader possibility that the infant has implicit precursors for TOM not revealed by the explicit tests given children. Indeed, given the centrality of false belief in theorizing about social attribution, it is essential to determine whether this capacity too may have an implicit precursor in the infant.

We begin with a test designed to determine whether infants attribute simple states of mind such as perception and desire. If the answer is positive, we can proceed to a version of a test for false belief. Using a computer screen, we show the infant an intentional object, a self-propelled ball which, for convenience, we will henceforth call *Bill*. Bill directs his action at an apple perched on top of a hill, trying repeatedly to climb the hill and reach the apple. However, before he can do so, the apple falls and rolls down the hill, disappearing into a black container at the bottom.

This animation provides the infant with information that any individual who attributes states of mind would need to draw the following conclusion: Bill both wants the apple and knows where he can get it (for he saw where it landed when it fell). That Bill wants the apple follows from his goal-directed behavior, his repeated attempts to obtain it; that he knows where the apple is follows from the fact that Bill was properly oriented when the apple fell. He had sufficient light and an unobstructed view, vision-dependent factors that not only young children (Flavel et al., 1981) but even chimpanzees (Premack, 1988a, b) take into account. Finally, as a special inducement to the infant, we gave Bill a nose, providing the asymmetry which, as we posited earlier, infants use as a basis for attributing vision.

An individual who attributed states of mind, given the preceding information, would be quite clear as to what Bill will do next. He will try to retrieve the apple, and he will go to the black container. But what would an infant do? We can answer by showing two groups of infants one or the other of two scenes. In one, Bill descends the hill and heads for the black container, and in the other he descends and heads for the white container. Which group has its expectations disconfirmed, is surprised, and therefore looks longer?

If the groups do not differ, then we cannot claim that infants attribute states of mind. On the other hand, if they not only differ but differ in the appropriate way—the group shown Bill heading for the white container looks longer—we have a legitimate basis for the claim. Infants who pass this test can be said to attribute states of mind and can be advanced to the next test, for false belief.

After the apple has rolled down the hill into the black container, we either remove Bill from the scene temporarily or divert his attention from the black container, during which time the apple is moved from the black to the white container. A 4-year-old child would know that Bill, not having seen the apple moved from one container to the other, still believes the apple is in the black container. Moreover, the 4-year-old would credit Bill with this false belief while at the same time the child knows where the apple actually is. But do infants? We once again show one group of infants that Bill seeks to retrieve the apple by going to the black container, the other group that he does so by going to the white one.

If the groups do not differ in their looking time, we have no basis for claiming false belief in infants, but if they differ and differ in the appropriate way, we do. Infants capable of false belief should expect Bill to go to the black container, for that is where he last saw the apple, and should look longer if shown that he heads instead to the white container. Infants who are incapable of false belief but capable of attributing mental states might expect Bill to go to the white container and therefore look longer if he heads for the black one. These infants fail to separate their knowledge from Bill's and mistakenly assign veridical knowledge to Bill. Although the present theory predicts that the infant has the capacity for TOM only and does not predict false belief, the test for false belief follows that for TOM so easily that it should be carried out.

Conclusion

Though the complexity of most species consists in making automatic computations regarding items they perceive, human complexity consists in having evolved theories that seek to explain perceived items. It treats these items as members of a domain (to which all invariances of the domain apply) and seeks to explain the invariances or privileged changes to which members of the domain are subject.

The domain with which we have dealt is that of objects which change their location in space by starting and stopping their own motion. Such objects contrast with others that also start and stop their own motion but move only in place, and both of these contrast with a third kind of object that has no capacity for self-propulsion but moves only when acted on by another object. These three kinds of motion are automatically interpreted not only by the adult human but also by the infant. All three are interpreted as being *caused*, either externally when one object launches another or internally when there is no second object and the object is self-propelled. Although in principle one could, on the basis of these observations, distinguish between movement that is and is not caused, humans reject this alternative; they distinguish instead between movement that is caused externally and internally. Internally caused motion belongs to the category of invisible events, a category humans find congenial and to which they turn frequently in providing explanations.

Human explanation takes place in the context of natural theories that have three components, one identifying the items to which the theory applies, another specifying the changes to which class members are subject, and a third explaining the changes. We presented an example of such a theory, that of the human theory of self-propelled objects (which change their position in space). Such objects are interpreted as intentional. Intentional objects engage in the following actions: They pursue goals, interact in ways to which value is attributed, reciprocate (with preservation of value), join groups, and take possessions. All these actions are explained by the mental states *perceive*, *want*, and *belief* and their multiple variations. The purpose of human social competence is to understand this set of actions, a goal that is achieved, we suggest, by the theory we have described.

REFERENCES

Baillargeon, R., 1987. Object permanence in $3\frac{1}{2}$ and 4 $\frac{1}{2}$-month-old infants. *Dev. Psychol.* 23:655–664.

Baillargeon, R., E. S. Spelke, and S. Wasserman, 1985. Object permanence in five-month-old infants. *Cognition* 20: 191–208.

Berndt, T. J., and E. G. Berndt, 1975. Children's use of motives and intentionality in person perception and moral judgment. *Child Dev.* 46:905–912.

Dawkins, R., 1986. *The Blind Watchmaker.* New York: Norton.

Flavell, J. H., B. A. Everett, K. Croft, and E.R. Flavell, 1981. Young children's knowledge about visual perception: Further evidence for the Level 1-Level 2 distinction. *Dev. Psychol.* 17:99–103.

Goodall, J., 1986. *The Chimpanzees of Gombe Stream.* Cambridge, Mass.: Harvard University Press.

King, M., 1971. The development of some intention concepts in young children. *Child Dev.* 42:1145–1152.

Kummer, H., and M. Cords, 1991. Cues of ownership in longtailed macaques, *Macaca fascicularis. Anim. Behav.* 42: 529–549.

Leslie, A. M., and S. Keeble, 1987. Do six-month-old infants perceive causality? *Cognition* 25:267–287.

Lettvin, J. Y., R. R. Maturana, W. S. McCulloch, and W. H. Pitts, 1959. What the frog's eye tells the frog's brain. *Proc. Inst. Rad. Eng.* 47:1940–1951.

Michotte, A., 1963. *The Perception of Causality.* London: Methuen.

Premack, D., 1988a. "Does the chimpanzee have a theory of mind?" revisited. In *Machiavellian Intelligence*, R. W. Byrne and A. Whiten, eds. London: Oxford University Press.

Premack, D., 1988b. Minds with and without language.

In *Thought without Language*, L. Weiskrantz, ed. Oxford: Clarendon Press.

PREMACK, D., 1993. Prolegomenon to evolution of cognition. In *Exploring Brain Functions: Models in Neuroscience*, T. A. Poggio and D. A. Glaser, eds. New York: Wiley.

SMITH, M. C., 1978. Cognizing the behavior stream: The recognition of intentional action. *Child Dev.* 49:7360–743.

SPELKE, E. S., 1982. Perceptual knowledge of objects in infancy. In *Perspectives on Mental Representation*, J. Mehler, M. Garrett, and E. Walker, eds. Hillsdale, N.J.: Erlbaum.

SPELKE, E. A., 1988. The origins of physical knowledge. In *Thought without Language*, L. Weiskrantz, ed. Oxford: Clarendon.

SUGARMAN, S., 1983. *Children's Early Thought: Developments in Classification*. Cambridge: Cambridge University Press.

VON UEXKUELL, J., 1957. A walk through the world of animals and man. In *Instinctive Behavior*, C. Schiller, ed. New York: International Universities.

WELLMAN, H., 1990. *The Child's Theory of Mind*. Cambridge, Mass.: MIT Press.

WIMMER, H., and J. PERNER, 1983. Beliefs about beliefs: Representation and constraining function of wrong beliefs in young children's understanding of deception. *Cognition* 13:103–128.

13 Developmental Specificity in Neurocognitive Development in Humans

HELEN J. NEVILLE

ABSTRACT Studies of behavior and event-related brain potentials from normal adults, congenitally deaf adults, and normally developing children during visual attentional and language processing are summarized. The results suggest that different subsystems within vision and within language display different degrees of experience-dependent modification. Within vision, the absence of competition from auditory input has most marked effects on the organization of systems important in processing peripheral information. Within language, delayed exposure to a language has pronounced effects on development of systems important in grammatical processing and many fewer effects on lexical development. Various accounts for these differential effects of early experience are discussed.

Studies of the structural, chemical, and physiological development of the human brain document a protracted and variable maturational time course (which includes both progressive and regressive events) that lasts at least 10 or 15 years (Conel, 1939–1963; Chugani and Phelps, 1986; Huttenlocher, 1990). The end result is a highly differentiated mosaic of subsystems specialized to mediate distinct aspects of sensory and cognitive processing. During this time, sensory and cognitive skills including language are undergoing massive change at highly variable rates. From the perspective of cognitive neuroscience, the great power of studying development is the opportunity to link variability in cognitive functions to concomitant and specific alterations in cerebral function. In the adult, the brain is highly specified and behavior is optimal and so the amount of variability in both systems is typically low. In contrast, the developing organism displays a high degree of change both in different neural systems and in cognition and thus provides an important opportunity to link variability within one trajectory to variability in the other.

HELEN J. NEVILLE The Salk Institute, La Jolla, and the Department of Cognitive Science, University of California, San Diego, Calif.

The research summarized here is directed toward describing the complementary roles of maturational factors and experience in forming the neural systems specialized for different types of cognitive and language processing. Current investigators acknowledge the role of biological constraints imposed by the genotype and that experience is critical in the expression of these constraints. However, a key issue concerns the nature of the role that experience plays. The prevailing view has been that environmental input acts as a trigger to initiate or to facilitate a course of development that is strongly biologically predetermined. According to this view, different neural systems possess intrinsic constraints that make them capable of processing just some and not other kinds of information. On the other hand, a growing body of literature has begun to document a central role for inputs from the environment in setting up and actually defining the functional characteristics of the brain systems they contact (Frost, 1984; Sur, Pallas, and Roe, 1990; Schlaggar and O'Leary, 1991). This general issue is central to a basic understanding of how functionally specialized neural systems are formed in normal development, and it is also central to an understanding of the nature, timing, and extent of modification of neural systems by experience in cases of abnormal development. Additionally, understanding of these issues with respect to cognitive and language development can provide biological evidence pertinent to different conceptions of the inventory and identity of different cognitive systems, how they interact, and the conditions that foster their development.

Role of experience in neurobehavioral development

Much of our current knowledge of the effects of experience on neural development has come from studies of simple sensory processing in animals (Rauschecker and Marler, 1987). These studies clearly demonstrate that specific input at specific times is necessary for normal sensory and related cortical development. Moreover, within a sensory modality or cognitive domain, different neuronal response properties and behavioral capabilities can display different degrees of modifiability by input, and the time periods when they are affected by afferent input can differ (Mitchell, 1981; Harwerth et al., 1986; Curtiss, 1989; Neville, 1990; Neville, Mills, and Lawson, 1992; Maurer and Lewis, 1993). This variability in the degree of experience-dependent modification may depend on the rate of maturation of the neural systems that mediate different functions, with later-developing cortical areas having more opportunity to be affected by incoming input. The results showing that visual processes believed to arise within the retina (e.g., the sensitivity of the scotopic visual system) display relatively short sensitive periods are in accord with this idea. By contrast, binocular functions that rely on later-developing cortical neurons display considerably longer sensitive periods (Harwerth et al., 1986).

Another variable that may contribute to differences in the degree of modifiability of cerebral systems is the extent of overlap, both divergent and convergent, of inputs to a given cortical system. In studies of experience-dependent modification within the somatosensory system of adult monkeys, Merzenich and colleagues have reported that cortical field 3b, which displays the most constrained anatomical distribution of inputs, displays less modifiability than do more convergently or divergently connected cortical areas (e.g., S1 and S11) (Merzenich, 1987; Merzenich et al., 1988).

Yet another factor that may affect the degree of activity-mediated changes within a given system is the degree of precision displayed by the system at maturity. Chalupa and Dreher (1991) review the literature describing the formation of connections within the mammalian visual system and hypothesize that the high-precision system required for high-acuity vision displays less initial redundancy and subsequent pruning of neural projections than do systems characterized by poorer acuity and stereoscopic vision.

Most of the data relevant to these proposals have come from experimental animals. In this chapter, research is summarized that suggests alterations of early experience in humans have marked and specific effects on neurocognitive development and variable effects within both vision and language and that consideration of such developmental differences may help to differentiate functional subsystems within these domains.

The research approach employed herein is to compare different and specific aspects of cerebral organization in normal adults with that in adults who have had specific alterations in early experience (i.e., to look at developmental effects retrospectively) and to further test the hypotheses generated from such research through the study of infants and children at different ages and stages of cognitive development (i.e., to look at development prospectively). The modifiability by experience of the development of both visual and language systems is studied. Normal-hearing adults and congenitally deaf adults are compared to assess the effects of lack of auditory input on visual development and to assess the effects of deaf subjects' abnormal language experience on language-relevant systems of the brain. In a related series of experiments, the development of sensory and language systems in infants and children are compared.

Generally, the picture that emerges from consideration of the retrospective developmental data is that distinct functional subsystems within vision and language display different degrees of modifiability following altered early experience. Additionally, the data from the prospective studies suggest that the different functional subsystems identified in the studies of adults are not uniformly present early in development but instead gradually become differentiated over a variable time course that is characterized by a loss or reduction of redundancy and increasing specificity.

Electrophysiological approaches

It is possible to study aspects of the physiology of sensory and language processing in humans by recording event-related brain potentials (ERPs) to different stimuli presented in paradigms designed to elicit specific aspects of sensory, motor, or cognitive processing (e.g., Hillyard and Picton, 1987). ERPs are voltage fluctuations in the electroencephalogram extracted by signal-averaging techniques. The latencies of different positive and negative components in an ERP reveal the time course of activation (within microseconds) of the underlying neuronal populations. The distribution of ERP activity between and within the hemispheres is determined by the anatomical position and geometry of the contributing neurons. The spatial resolution of the ERP can be enhanced by transforming voltage maps of electrical activity into current source density (CSD) maps that provide a reference-free estimate of the instantaneous electrical currents flowing from the brain perpendicular to the scalp at each location at the specified time point (Nunez and Katznelson, 1981; Perrin et al., 1989). Thus, ERP recordings can provide information about the timing, sequence, and location of sensory and language processes. From recent reviews of the contributions that ERP studies have made to our understanding of sensory, cognitive, and language processes (Hillyard and Picton, 1987; Rugg, 1990; Pritchard, Shappell, and Brandt, 1991; Van Petten and Kutas, 1991; Swick, Kutas, and Neville, in press), it is evident that the utility of this technique is greatest when employed in conjunction with behavioral measures so that the interpretation of behavioral and neurophysiological data constrain each other. This is the approach my colleagues and I have taken in our research.

Visual Spatial Processing In an early study, we observed that ERPs to peripheral and foveal visual stimuli differed in morphology and distribution over the scalp in normal adults in a manner consistent with the hypothesis that they were generated by different cortical systems. Results from congenitally deaf adults in the same paradigm showed that whereas ERPs to foveal stimuli were similar in the two groups, over superior temporal cortical areas ERPs to peripheral stimuli were two to three times larger in deaf than in hearing subjects (Neville, Schmidt, and Kutas, 1983). We hypothesized that the "transient" visual system, proposed to mediate the processing of peripheral, spatial information may, through a process of competitive interactions, take over what would normally be auditory cortical fields either in primary sensory or multimodal cortical areas.

We extended these studies to the area of visual spatial attention to test the generality of the findings and to obtain concurrent measures of behavior. We tested the hypothesis that attention to central and peripheral visual space is mediated by different neural systems in normal-hearing adults and, further, that the systems important in processing peripheral visual information are more altered by auditory deprivation than are those important in processing central visual information. Methodological and statistical details of the experiments are described in Neville and Lawson (1987a, b, c).

Attention to peripheral and central visual space in normal-hearing adults The effects of focused attention to peripherally and centrally located visual stimuli were compared via an analysis of ERPs while normal-hearing subjects focused their eyes straight ahead and attended to a white square located 18° in the left or right visual field or in the center of the visual field. Their task was to detect the direction of motion of the white square in the attended visual field. Effects of attention were assessed by comparing ERPs to the same physical stimulus when subjects were attending it versus when they were attending one of the other locations. Attention to both peripheral and foveal stimuli produced typical enhancements of the early ERP components including a negativity at 150 ms (N150; see Hillyard et al., this volume). However, the distribution over the scalp of the attention-related changes varied according to stimulus location. The attention-related increase in the amplitude of the N150 component to the peripheral stimuli was greater over the parietal region of the hemisphere contralateral to the attended visual field. By contrast, the largest effects of foveally directed attention occurred over the occipital regions, where the increase was bilaterally symmetrical (see left side of figure 13.1). Additionally, for peripheral stimuli only, the effects of attention on the ERPs were significantly larger for moving than for stationary stimuli. A longer-latency positive component (300–600 ms) was larger over the right than the left hemisphere during attention to the lateral visual fields but was symmetrical in amplitude when central stimuli were attended.

These results suggest that different pathways are modulated when attention is deployed to central and peripheral space. Moreover, the pattern of results is compatible with the hypothesis that attention to the center of the visual field modulates activity in the retinogeniculate pathway that projects to striate cortex, whereas attention to the visual periphery is associated with activity in the pathway that includes projections

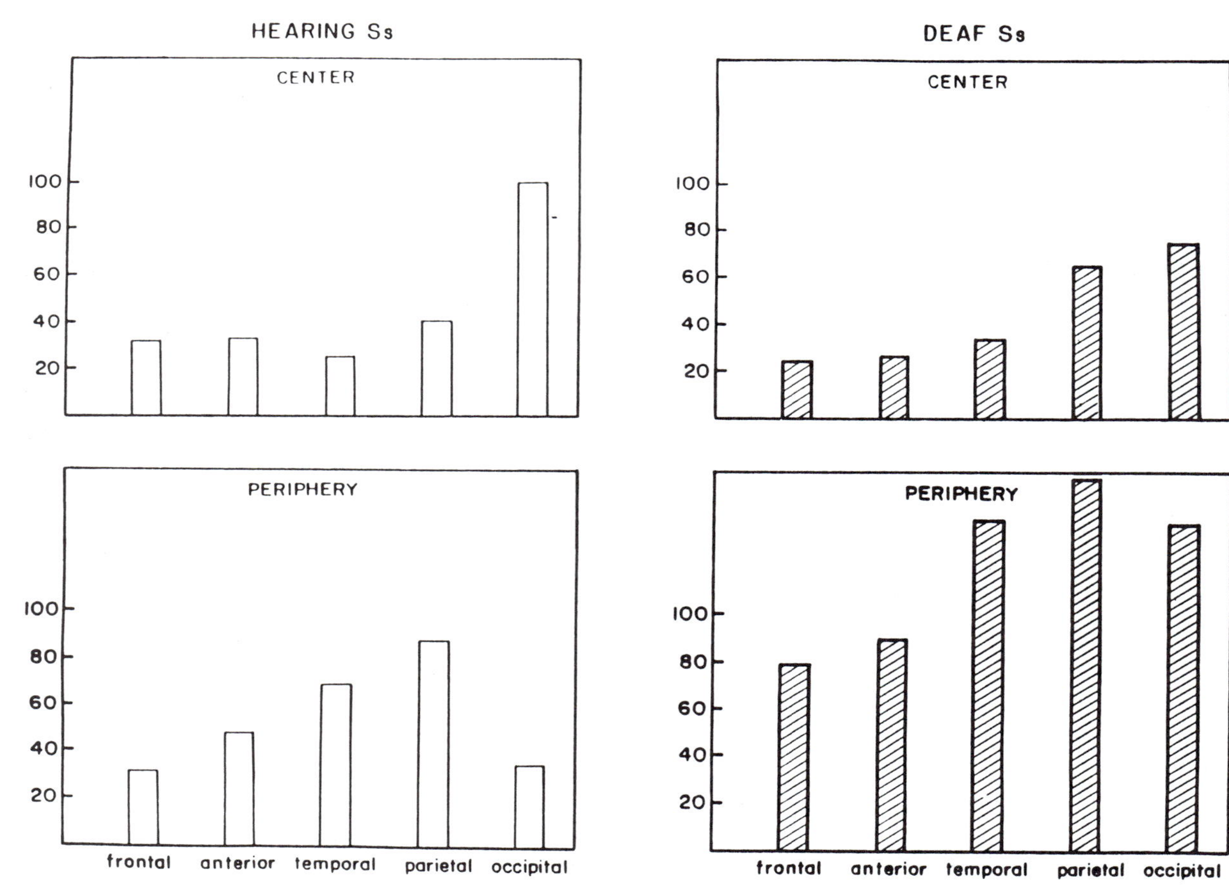

FIGURE 13.1 Percent by which N150 amplitude was increased from nonattend to attend conditions for hearing and deaf subjects (Ss). (Top) Center standards, mean of left and right frontal, anterior temporal, temporal, parietal, and occipital sites. (Bottom) Peripheral standards, mean of contralateral frontal, anterior temporal, temporal, parietal, and occipital sties. (Reprinted from Neville and Lawson, 1987b)

to parietal cortex of the contralateral hemisphere. The fact that the addition of motion selectively increased activity in response to the peripheral stimuli is of interest in view of the sensitivity to motion of neurons along the "dorsal" cortical visual pathway and suggested that the periphery may have an increased representation along that processing stream.

A recent anatomical study in rhesus monkeys is pertinent to this hypothesis. Baizer, Ungerleider, and Desimone (1991) reported segregation of the representations of the peripheral and central visual fields along the dorsal and ventral processing streams, respectively (DeYoe and Van Essen, 1988). Fluorescent dyes injected into posterior parietal cortex were traced to cortical areas implicated in spatial and motion analysis and to peripheral field representations in prestriate cortex. By contrast, dyes injected into temporal cortex were traced along the ventral pathway to areas implicated in form and color analysis and to representations of the central visual field.

Visual attention in congenitally deaf adults To assess experience-dependent changes within these pathways, we employed the same paradigm to study the effects of congenital auditory deprivation on the systems that mediate attention to peripheral and central space. With attention to the centrally located stimuli, ERPs from congenitally deaf and normal-hearing subjects displayed similar attention-related changes and they displayed similar accuracy and speed in detecting the direction of motion of the central stimuli. In contrast, with attention to peripheral visual stimuli, ERPs from the deaf subjects displayed attention-related increases that were several times larger than those of hearing subjects (see figure 13.1, right side). Also, deaf subjects were significantly faster than hearing subjects at detecting the direction of motion of peripheral targets. An important control group, hearing subjects born to deaf parents, who undergo much of the altered language experience of the deaf, including the acquisition of American Sign Language, did not show the enhanced ERPs or behavior. These results suggest that auditory deprivation leads to major alterations in the development of the peripheral visual system.

The specific pattern of group differences included increased activity over occipital areas bilaterally and suggested that the changes may include both compensatory hyperactivity of remaining sensory systems and functional reallocation of auditory and/or multimodal areas. These results suggest the hypothesis that the dorsal visual processing stream, which contains the majority of the peripheral representation, is more modifiable in response to alterations in afferent input than is the ventral processing pathway. This hypothesis agrees broadly with the proposal put forward by Chalupa and Dreher (1991) that components of the visual pathway that are specialized for high-acuity vision exhibit fewer developmental redundancies ("errors"), decreased modifiability, and more specificity than do those displaying less acuity and precision. It may also be that the dorsal visual pathway has a more prolonged maturational time course than the ventral pathway, permitting extrinsic influences to exert an effect over a longer time. Although little evidence bears directly on this hypothesis, recent anatomical data suggest that, in monkey, the peripheral retina is slower to mature (Packer, Hendrickson, and Curcio, 1990). Additionally, considerable data suggest that the development of the Y-cell pathway (which is strongest in the periphery of the retina) is more affected by visual deprivation than is development of the W- and X-cell pathways (Sherman and Spear, 1982). Moreover, studies of humans show that visual deprivation secondary to congenital cataracts has pronounced effects on peripheral vision that markedly reduce visual field extent (Mioche and Perenin, 1986; Bowering et al., 1989).

Critical-period effects and mechanisms We observed that individuals who became deaf after the age of 4 years displayed neither the increased visual ERP over temporal brain regions that we had reported previously nor the enhanced posterior visual attention effects that accompanied congenital auditory deprivation (Neville, Schmidt, and Kutas, 1983; Neville and Lawson, 1987c). We considered several mechanisms that might mediate the effects themselves and the developmental time limits on them (Neville, 1990). One possibility is that the effects are mediated by an early, normally transient redundancy of connections between auditory and visual systems such as that observed in cats and hamsters (Innocenti and Clark, 1984; Frost, 1984, 1989, 1990). With normal sensory experience, these redundant intermodal connections are eliminated or suppressed. However, in the absence of competition from auditory input, visual afferents may become stabilized on what would normally be auditory neurons. The critical-period effects suggest this redundancy may exist between birth and 4 years of age.

One way we tested this hypothesis was to look at the time course of differentiation of sensory responses in normal development. In normal adults, auditory stimuli elicit ERP responses that are large over temporal brain regions but small or absent over occipital regions (figure 13.2, bottom). By contrast, in 6-month-old children, we observed that auditory ERPs are equally large over temporal and occipital brain regions, consistent with the idea that there is less specificity and more redundancy of connections between auditory and visual cortex at this time. This result is in accord with behavioral studies showing considerable intermodal integration during this time period (Spelke, 1987; Meltzoff, 1990; Rose, 1990). Between 6 and 36 months, however, there is a gradual decrease in the amplitude of the auditory ERP over visual areas (see figure 13.2), suggesting that a redundancy of connections between auditory and visual areas may exist early in human development but gradually decreases between birth and 3 years of age. This loss of redundancy may be the boundary condition that determines when auditory deprivation can result in alterations in the organization of the visual system. Ongoing behavioral and ERP studies of deaf and hearing adults and children will contribute to our understanding of the different consequences of early sensory experience on central and peripheral visual processing.

Semantic and Grammatical Processing It seems reasonable to assume that the rules and principles governing development of the sensory systems also

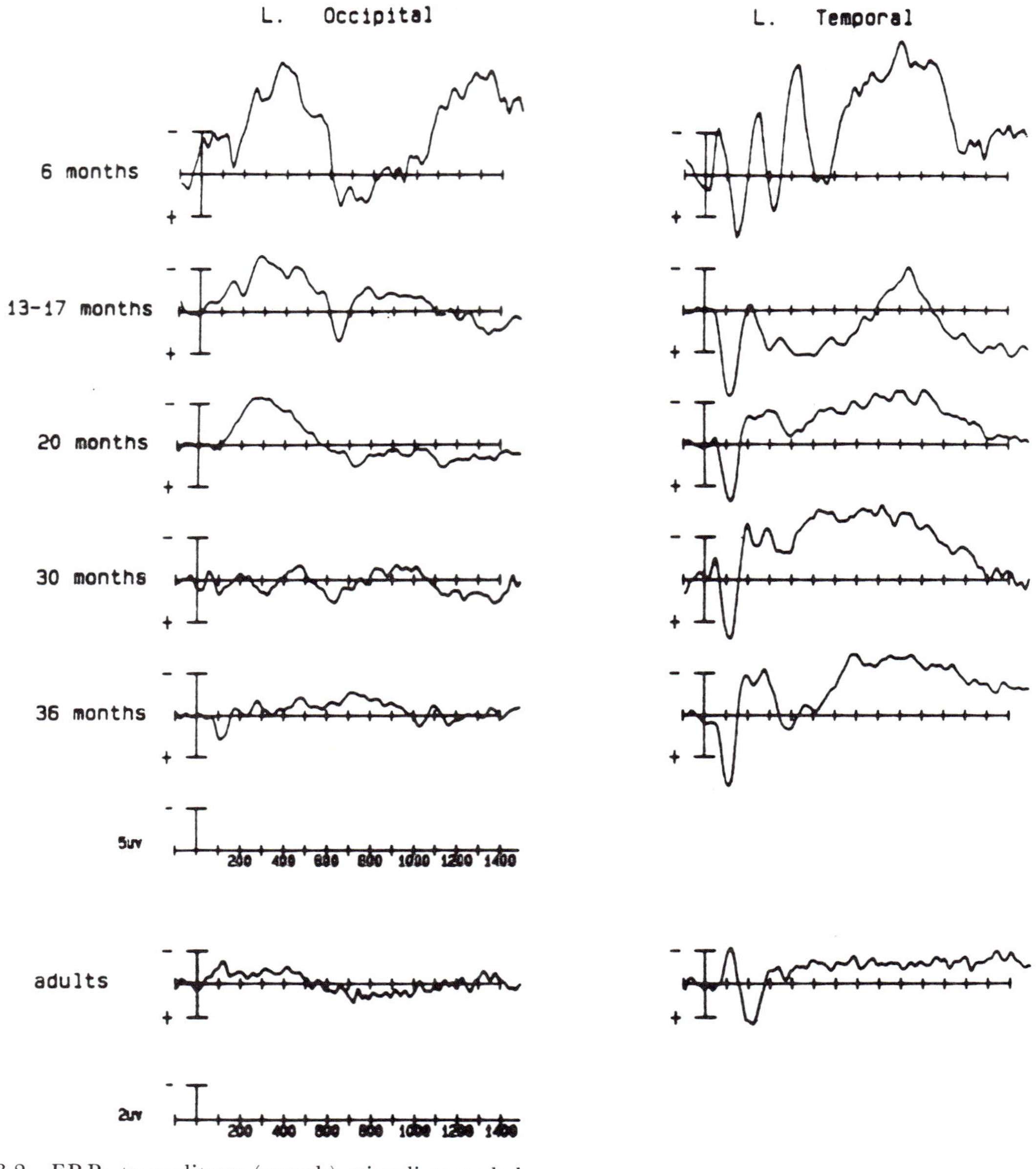

FIGURE 13.2 ERPs to auditory (speech) stimuli recorded over temporal and occipital regions in normal adults (bottom) and in children aged 6–36 months.

guide development of cognitive and language-relevant brain systems. Thus, differences in the rate of differentiation and degree of specification may be apparent within language and may help to identify different functional subsystems. In a series of studies, my colleagues and I have taken this type of developmental approach to address the idea (predominant in linguistic and psycholinguistic research) that language processing is decomposable into separate subsystems including semantic processing (e.g., as occurs in processing the meaning of nouns, verbs, and adjectives that make reference to specific objects and events, labeled *open-class words* because the set openly admits new members) and grammatical processing (e.g., as occurs in processing the structural or relational information provided in English primarily by words such as articles, conjunctions, and auxiliaries, labeled *closed-class words* because they belong to a closed set) (Chomsky, 1965; Garrett, 1980). Empirical studies have provided support for this distinction and have led investigators to propose that open- and closed-class words may be represented in different systems (*lexicons*) that may not be subject to the same processing constraints. However, other investigators point out inconsistencies in the data adduced in favor of the traditional view and instead argue that grammatical phenomena can be accounted for on the basis of the semantic content of closed-class words, which are hypothesized to be represented together with open-class words in a single lexicon. By this view, any behavioral or clinical differences between open- and closed-class words are attributed to their different frequencies in the language (closed-class are more frequent), their different lengths (closed-class words are shorter), or to differences in imageabilities (closed-class words are low in imageability) (Bates and MacWhinney, 1987; Carpenter and Just, 1989).

Normal adults One approach we have taken to this controversy has been to compare the timing and distribution of ERPs elicited by open- and closed-class words in sentences and to assess whether any observed differences can be accounted for in terms of general processing differences (i.e., of frequency, length, and imageability) rather than in terms of the different functions these words perform in language (see Neville, Mills, and Lawson, 1992, for details). In normal-hearing adults, the ERP response to meaning-bearing, open-class words was characterized by a negative component that became maximal approximately 350 ms (N350) after word onset. This component was large over posterior brain regions of both hemispheres. In previous studies, we and others have linked this component to the process of "look-up" or search through the lexicon (Holcomb, 1988; Rugg, Furda, and Lordist, 1988; Holcomb and Neville, 1990; Rugg, 1990). In contrast, ERPs to closed-class words displayed a negative potential largest at 280 ms that was localized to anterior temporal regions of the left hemisphere.

An assessment of the effects of frequency, length, and imagibililty revealed that these variables did not account for the different results for the two word classes. CSD analyses of the N280 peak to closed-class words indicate it is strongly localized to anterior temporal regions of the left hemisphere (see color plate 2, top row), is absent to open-class words in general, and is absent even to very short, highly frequent, open-class words. Thus, these results are consistent with the idea that nonidentical neural mechanisms are active in representing or accessing the open- and closed-class vocabularies important in processing the semantic and grammatical aspects of language, respectively. More generally, these results lend biological support to the distinction.

Deaf adults To further investigate this proposal, we assessed the hypothesis that the systems which mediate semantic and grammatical processing are differentially vulnerable to altered early language experience. We compared the results from normal-hearing subjects to those from a group of deaf subjects whose early language experience was abnormal (i.e., American sign language was their first language, and English was their second language and was learned late).

The pattern of results lends support for this hypothesis. The deaf subjects displayed ERP responses to the open-class words (N350) that were virtually identical to those observed in the normal-hearing subjects. Further comparisons of the responses to semantically anomalous and appropriate sentences revealed similar responses in deaf and hearing subjects. These results suggest that aspects of semantic processing are robust following deaf subjects' altered early language experience.

In contrast to these results, the ERP responses of the deaf and hearing subjects to the closed-class words were markedly different. The ERPs of the deaf subjects lacked the negative potential (N280) over anterior regions of the left hemisphere, and they did not display any other evidence of asymmetry between the hemispheres. Color plate 2 compares the CSD analyses of the N280 peak in hearing subjects (top row) and deaf subjects (bottom row). The patterns of current flow within the right hemisphere are similar in the deaf and hearing subjects. However, over the left hemisphere, only the hearing subjects display the sharply focused N280 sink over anterior temporal regions. These findings suggest that the neural systems mediating grammatical processing are more vulnerable to alterations in the nature and timing of early language experience than are those linked to semantic processing. Thus, these results are in accord with the idea that different brain systems with different developmental time courses and associated vulnerabilities mediate semantic and grammatical aspects of language processing and, more generally, they suggest that it is biologically meaningful to consider these as separate subsystems within language. Recent studies of bilinguals provide support for these hypotheses (Weber-Fox and Neville, 1992, and submitted; Neville and Weber-Fox, in press).

Normal development The data from deaf and hearing adults suggest that the system indexed by the N350 to open-class words and the response to semantic anomalies may develop along a different time course from that indexed by the N280 response and that these differences may underlie their different developmental vulnerabilities. Hence, we tested normal children, aged 4 to 20 years, on both auditory and visual versions of this paradigm, employing a different set of sentences more appropriate for children. As described in Holcomb, Coffey, and Neville (1992), the typical response to semantic anomalies is present in the youngest children tested (i.e., 4-year-olds). By contrast, studies of the development of the N280 response demonstrate that the mature pattern is not evident until the mid-teens (Neville et al., in preparation). Indeed, early in development, ERP responses to both open- and closed-class words are characterized by a similar posterior negativity at approximately 350 ms that in adults is characteristic only of open-class words. By 11 years of age, this response is reduced or absent to closed-class words. Subsequently, an earlier negativity with an anterior left-hemisphere distribution emerges in response to closed-class words. By 15–16 years, a clear N280 response with a typical distribution is evident. The gradual differentiation of the localized response to closed-class words is compatible with a process of decreasing redundancy and increasing specificity within this neural system. Moreover, the time course of this effect suggests that it may be linked to the increasing automaticity that behavioral studies show becomes characteristic of the processing of closed-class words around this age (Friederici, 1983).

Variability with grammatical knowledge The results presented in the preceding section were interpreted as consistent with the hypothesis that the N280 response to closed-class words indexes some aspect of grammatical processing. To provide further evidence on this point, we compared this response in populations that differ in performance on tests of knowledge of English grammar. We reasoned that a more distinct and asymmetrical N280 response would be apparent in individuals displaying greater knowledge of English grammar.

As reported in Neville (1991), in contrast to the majority of deaf subjects, the few deaf subjects who scored high ($>95\%$) on the Saffran and Schwartz grammaticality judgment task displayed an N280 response that was prominent and asymmetrical, just as in normal-hearing subjects. These data argue strongly that the N280 response is linked to grammatical knowledge and, more generally, that the development of left-hemisphere specialization is tied to the acquisition of grammar and not to motor or auditory or phonological skills.

Further evidence along these lines comes from our study of normal children, age 8–13 years (Neville et al., in preparation). We compared ERPs to closed-class words in those who scored well and poorly on a test of knowledge of English grammar (the sentence-structure subtest of the Clinical Evaluation of Language Functions test). The two groups of children were the same mean age ($11\frac{1}{2}$ years). However, they displayed significantly different degrees of asymmetry in the N280 response that positively correlated with the test scores (i.e., lower scores, less asymmetry).

Variability with maturational changes in cortical volume In normal development, the N280 response initially displays a widespread distribution that is gradually reduced over posterior regions of both hemispheres and over anterior regions of the right hemisphere, resulting in an increasing localization of the N280 response to anterior regions of the left hemisphere. This pattern is consistent with anatomical data from humans and other animals showing extensive regressive changes during postnatal development of cerebral cortex, including marked decreases in synaptic density (Rakic et al., 1986; Huttenlocher, 1990). Although the neuroanatomical data from humans is based on relatively few samples, recent studies using magnetic resonance imaging (MRI) to examine brain structure have revealed a similar pattern (Jernigan et al., 1991). In normal humans, marked decreases in cortical gray matter are observed from 8 years until 30 years of age. These may be associated with pruning of the neuropil such as that observed in anatomical studies. The observed reductions in cortical gray matter did not occur uniformly across the cerebral cortex. From 8 years, they were most marked in the superior cortical regions and were not significant in the inferior cortical surfaces. This pattern is consistent with Huttenlocher's findings (1990) of regional differences in the decrease in synaptic density with age in humans. For example, frontal cortex continues to change beyond the time when occipital cortex has matured.

The decreasing redundancy of cortical gray matter and connections may be linked to decreases in plasticity and increases in functional specificity of cortical processes during the peripubertal period. One way my colleagues and I explored this idea was to investigate the hypothesis that increases in the differentiation (including the asymmetry) of the N280 response to closed-class words would be correlated with decreases in superior cortical gray matter on MRI (Jernigan, Coffey, and Neville, unpublished observations). Eighteen normal children, ranging in age from 8 years, 11 months, to 14 years, 7 months, participated in an MRI study and an ERP study, like that described earlier, in which ERPs were recorded as children read sentences containing open- and closed-class words. MRI acquisition and analysis was performed under the protocol described in Jernigan et al. (1991).

As shown in table 13.1, the dorsal anterior cortical volume was significantly negatively correlated with the asymmetry in N280 from each of the electrodes where it occurred (i.e., frontal, anterior temporal, and temporal sites). By contrast, during this time period, age alone was not predictive of volume.

To explore this effect further, we divided the group of children on the basis of a median split on values of superior cortical gray (SCG) volume. Children in the high and low SCG groups (N = 9 in both cases) were of the same mean age (10 years and 2 months). As seen

TABLE 13.1

Multiple regressions on dorsal anterior cortex volume

	Effect of Age		Effect of ERP Asymmetry (left-right)	
Electrode Site	β	p	β	p
Frontal	0.002	0.971	0.184	0.009
Anterior temporal	−0.015	0.817	0.157	0.029
Temporal	−0.048	0.469	0.160	0.028

β, standardized regression coefficient; p, significance level for β.

in figure 13.3, the low SCG group displayed a significant asymmetry to closed-class words, whereas the high SCG group did not. The specificity of this effect for grammatical function words is apparent by comparing this pattern with that observed for open-class words, which yielded equivalent responses in the high and low SCG groups. The individual variability in cortical gray matter and differentiation of the N280 response may be related to differences in peripubertal fluctuation of hormones. Tests of this hypothesis are currently under way.

Effects of primary language acquisition on cerebral organization
The research summarized in the preceding sections implies that language experience determines the development and organization of language-relevant systems of the brain. A strong test of this hypothesis would be to chart the changes in brain organization as children acquire primary language and to separate these from more general maturational changes. We have begun to research this issue (Mills, Coffey, and Neville, 1993, 1994). In one study, we examined patterns of neural activity relevant to language processing in 20-month-old infants to determine whether changes in cerebral organization occur as a function of specific changes in language development when chronological age is held constant (fully described in Mills, Coffey, Neville, 1993, 1994). ERPs were recorded as children listened to a series of words whose meaning was understood by the child, words whose meaning the child did not understand, and backward words. Specific and different ERP components discriminated comprehended words from unknown and from backward words. Distinct lateral and anteroposterior specializations were apparent in ERP responsiveness to the different types of words. Moreover, the results suggested that in children of the same age, language abilities are positively correlated with increased specialization of the neural systems active during language processing.

Conclusions

The results from the language studies taken as a whole point to different developmental time courses and developmental vulnerabilities of aspects of grammatical and semantic or lexical processing. They thus provide support for conceptions of language that distinguish these subprocesses within language. Similarly, within visual processing, peripheral functions were more

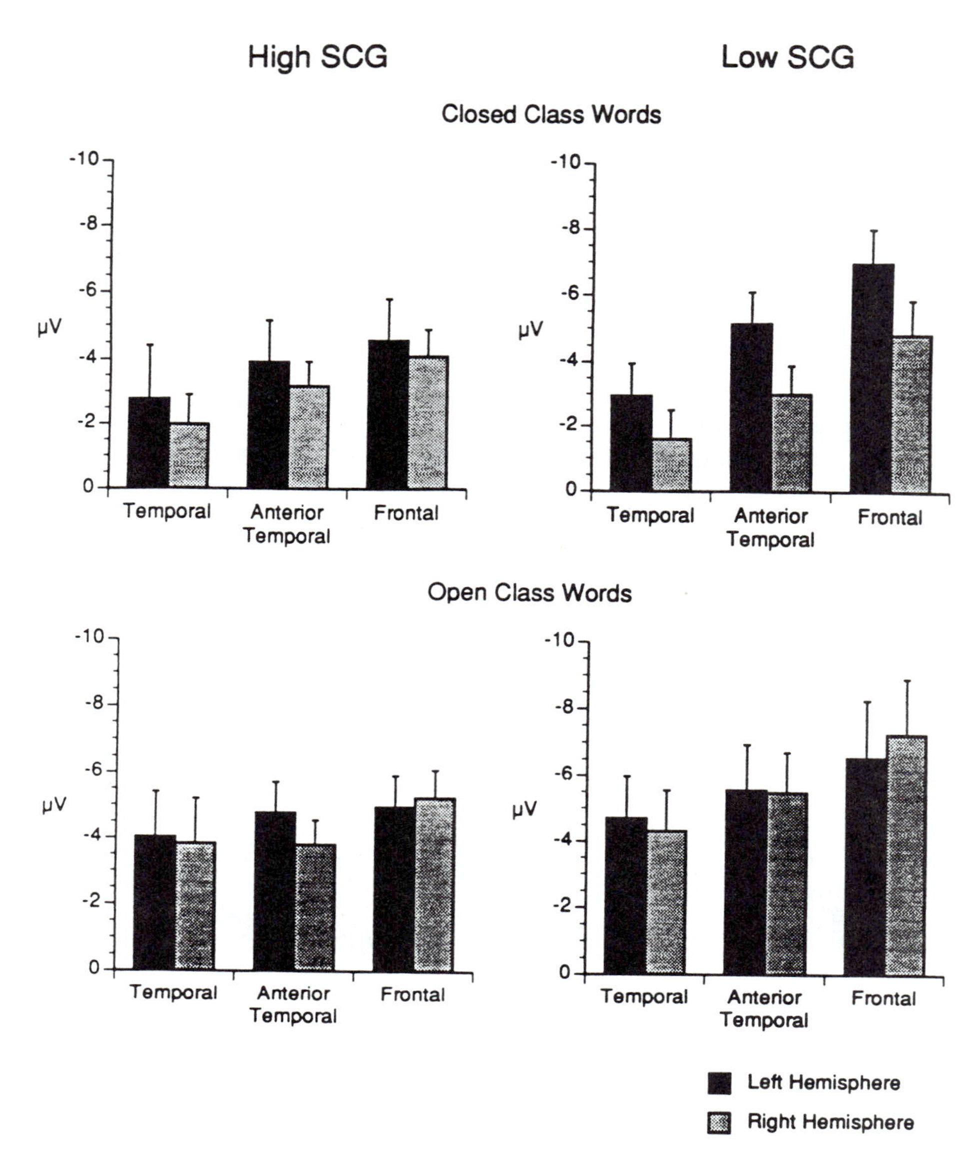

FIGURE 13.3 Amplitude of negative response to closed- and open-class words recorded from temporal, anterior temporal, and frontal regions of the left and right hemispheres of children (mean age, 10 years and two months) with high or low values of superior cortical gray matter (SCG).

altered than were foveal functions after auditory deprivation, providing support for conceptions of visual system organization that distinguish functions along these lines. The relative unmodifiability of responses associated with foveal processing might reflect a greater degree of developmental specificity of high-precision systems in vision, as suggested by Chalupa and Dreher (1991). However, it seems unlikely that the relative developmental stability in the measures of semantic processing used here could be accounted for in a similar fashion, as lexical acquisition is entirely dependent on external input. Indeed, it may be that lexical development (i.e., the establishment of object-word associations) depends on processes similar to those important in the formation of other temporally based coincidence maps. Within somatosensation, vision, and audition, such sensory maps (like lexical acquisition) display considerable plasticity throughout adulthood (Merzenich, 1987; Merzenich et al., 1988; Recanzone, Merzenich, and Jenkins, 1992; Recanzone et al., 1992; Recanzone, Merzenich, and Schreiner, 1992; Recanzone, Schreiner, and Merzenich, 1993). Moreover the general, real-world experiences that determine lexical acquisition are highly similar in the groups of subjects that differed in aspects of linguistical experience. By contrast, functions such as the acquisition of grammar that are dependent on the perception of rule-based invariances appear to be maturationally constrained and display distinct time periods in development when they require specific types of environmental input. The results summarized here suggest that grammatical processing is such a moderately specified system.

The research reported in this chapter demonstrates that there are maturational constraints on developmental plasticity that differ for different aspects of sensory and cognitive function even within a particular domain. This type of developmental evidence can contribute to fundamental descriptions of the architecture of different cognitive systems. Additionally, it may contribute eventually to the design of educational and habilitative programs for both normally and abnormally developing children.

ACKNOWLEDGMENTS The research summarized here has been supported by National Institutes of Health grants DC 00128, DC 00481, NS 29561, NS 22343, and DC 01289.

REFERENCES

Baizer, J. S., L. G. Ungerleider, and R. Desimone, 1991. Organization of visual inputs to the inferior temporal and posterior parietal cortex in macaques. *J. Neurosci.* 11:168–190.

Bates, E., and B. MacWhinney, 1987. Competition, variation and language learning. In *Mechanisms of Language Acquisition*, B. MacWhinney, ed. Hillsdale, N.J.: Erlbaum.

Bowering, E., D. Maurer, T. L. Lewis, P. Brent, and H. P. Brent, 1989. Development of the visual field in normal and binocularly deprived children (abstrad). *Invest. Ophthalmol. Vis. Sci.* 30(suppl.):377.

Carpenter, P. A., and M. A. Just, 1989. The role of working memory in language comprehension. In *Complex Information Processing: The Impact of Herbert A. Simon*, (D. Klahr and K. Kotovsky, eds. Hillsdale, N.J.: Erlbaum, pp. 31–68.

Chalupa, L. M., and B. Dreher, 1991. High precision systems require high precision "blueprints": A new view regarding the formation of connections in the mammalian visual system. *J. Cogn. Neurosci.* 3(3):209–219.

Chomsky, N., 1965. *Aspects of the Theory of Syntax*. Cambridge, Mass: MIT Press.

Chugani, H. T., and M. E. Phelps, 1986. Maturational changes in cerebral function in infants determined by FDG positron emission tomography. *Science* 23:840–843.

Conel, J. L., 1939–1963. *The Postnatal Development of the Human Cerebral Cortex*, vols. 1–6. Cambridge, Mass.: Harvard University Press.

Curtiss, S., 1989. The independence and task-specificity of language. In *Interaction in Human Development*, M. Bornstein and J. Bruner, eds. Hillsdale, N.J.: Erlbaum, pp. 105–137.

Deyoe, E. A., and D. C. Van Essen, 1988. Concurrent processing streams in monkey visual cortex. *Trends Neurosci.* 11(5):219–226.

Friederici, A., 1983 Children's sensitivity to function words during sentence comprehension. *Linguistics* 21:717–739.

Frost, D., 1984. Axonal growth and target selection during development: Retinal projections to the ventrobasal complex and other "nonvisual" structures in neonatal Syrian hamsters. *J. Comp. Neurol.* 230:576–592.

Frost, D., 1989. Transitory neuronal connections in normal development and disease. In *Brain and Reading*, C. Von Eula, ed. London: Macmillan.

Frost, D., 1990. Sensory processing by novel, experimentally induced cross-modal circuits. In *The Development and Neural Bases of Higher Cognitive Function*, A. Diamond, ed. New York: New York Academy of Sciences Press, pp. 92–112.

Garrett, M., 1980. Levels of processing in sentence production. In *Language Production: Vol. 1. Speech and Talk*, B. Butterworth, ed. New York: Academic Press.

Harwerth, R. S., E. L. Smith III, G. C. Duncan, M. L. J. Crawford, and G. K. von Noorden, 1986. Multiple sensitive periods in the development of the primate visual system. *Science* 232:235–238.

Hillyard, S. A., and T. W. Picton, 1987. Electrophysiology of cognition. In *Handbook of Physiology: Sec. 1. The Nervous System: vol. 5 Higher Functions of the Brain (part 2)*, F. Pulm, ed. Bethesda, Md.: American Physiological Society, pp. 519–584.

Holcomb, P. J., 1988. Automatic and attentional processing: An event-related brain potential analysis of semantic priming. *Brain Lang.* 35:66–85.

Holcomb, P. J., S. A. Coffey, and H. J. Neville, 1992. Visual and auditory sentence processing: A developmental analysis using event-related brain potentials. *Dev. Neuropsychol.* 8:203–241.

Holcomb, P. J., and H. J. Neville, 1990. Auditory and visual semantic priming in Lexical decision: A comparison using event-related brain potentials. *Lang. Cognitive Proc.* 5:281–312.

Huttenlocher, P. R., 1990. Morphometric study of human cerebral cortex development. *Neuropsychologia* 28:517–527.

Innocenti, G. M., and S. Clark, 1984. Bilateral transitory projection to visual areas from auditory cortex in kittens. *Dev. Brain Res.* 14:143–148.

Jernigan, T. L., D. A. Trauner, J. R. Hesselink, and P. A. Tallal, 1991. Maturation of human cerebrum observed in vivo during adolescence. *Brain* 114:2037–2049.

Maurer, D., and T. L. Lewis, 1993. Visual outcomes in infant cataract. In *Infant Vision: Basic and Clinical Research*, K. Simons, ed. New York: Oxford University Press.

Meltzoff, A. N., 1990. Towards a developmental cognitive science: The implications of cross-modal matching and imitation for the development of representation and memory in infancy. In *The Development and Neural Bases of Higher Cognitive Functions*, vol. 608, S. Diamond, ed. New York: New York Academy of Sciences Press, pp. 1–37.

Merzenich, M., 1987. Dynamic neocortical processes and the origins of higher brain functions. In *The Neural and Molecular Bases of Learning*, J. Changeux, M. Konishi, and M. Bernhard, eds. Dahlem Konferenzen: Wiley, pp. 337–358.

Merzenich, M., G. Recanzone, W. Jenkins, T. Allard, and R. Nudo, 1988. Cortical representational plasticity. In *Neurobiology of Neocortex*, P. Rakic, W. Singer, and W. Bernhard, eds. Dahlem Konferenzen: Wiley, pp. 41–67.

Mills, D. M., S. A. Coffey, and H. J. Neville, 1993. Language acquisition and cerebral specialization in 20-month-old infants. *J. Cogn. Neurosci.* 5:326–342.

Mills, D. M., S. A. Coffey, and H. J. Neville, 1994. Variability in cerebral organization during primary language acquisition. In *Human Behavior and the Developing Brain*, G. Dawson and K. Fischer, eds. New York: Guilford Publications, pp. 427–455.

Mioche, L., and M. T. Perenin, 1986. Central and peripheral residual vision in humans with bilateral deprivation amblyopia. *Brain Res.* 62:259–272.

Mitchell, D., 1981. Sensitive periods in visual development. In *Development of Perception*, R. Aslin, J. Alberts, and M. Petersen, eds. New York: Academic Press, pp. 3–43.

Neville, H. J., 1990. Intermodal competition and compensation in development: Evidence from studies of the visual system in congenitally deaf adults. In *The Development and Neural Bases of Higher Cognitive Function*, A. Diamond, ed. New York: New York Academy of Sciences Press, pp. 71–91.

Neville, H. J., 1991. Neurobiology of cognitive and language processing: Effects of early experience. In *Brain Maturation and Cognitive Development: Comparative and Cross-Cultural Perspectives*, K. R. Gibson and A. C. Petersen, eds. Hawthorne, N.Y.: Aldine de Gruyter Press, pp. 355–380.

Neville, H. J., and D. Lawson, 1987a. Attention to central and peripheral visual space in a movement detection task: I. Normal hearing adults. *Brain Res.* 405:253–267.

Neville, H. J., and D. Lawson, 1987b. Attention to central and peripheral visual space in a movement detection task: II. Congenitally deaf adults. *Brain Res.* 405:268–283.

Neville, H. J., and D. Lawson, 1987c. Attention to central and peripheral visual space in a movement detection task: III. Separate effects of auditory deprivation and acquisition of a visual language. *Brain Res.* 405:284–294.

Neville, H. J., D. L. Mills, and D. S. Lawson, 1992. Fractionating language: Different neural subsystems With different sensitive periods. *Cerebral Cortex* 2:244–258.

Neville, H. J., A. Schmidt, and M. Kutas, 1983. Altered visual evoked potentials in congenitally deaf adults. *Brain Res.* 266:127–132.

Neville, H. J., and C. Weber-Fox (in press). Cerebral subsystems within language. In *Structure and Functional Organization of the Neocortex*, Christoph Nothdurft, ed. New York: Springer-Verlag.

Nunez, P., and R. Katznelson, 1981. *Electric Fields of the Brain*. New York: Oxford University Press.

Packer, O., A. E., Hendrickson, and C. A. Curcio 1990. Developmental redistribution of photoreceptors across the *Macaca nemestrina* (pigtail macaque) retina. *J. Comp. Neurol.* 298:472–493.

Perrin, D., J. Pernier, O. Bertrand, and J. F. Eschallier, 1989. Spherical splines for scalp potential and current density mapping. *Electroencephalogr. Clin. Neurophysiol.* 72: 184–187.

Pritchard W. S., S. A. Shappell, and M. E. Brandt, 1991. Psychophysiology of N200/N400: A review and classification scheme. In *Advances in Psychophysiology*, vol. 4, P. K. Ackles, J. R. Jennings, and M. G. H. Coles, eds. London: Kingsley Publishers, pp. 43–106.

Rakic, P., J. P. Bourgeois, M. F. Eckenhoff, N. Zecevic, and P. S. Goldman-Rakic, 1986. Concurrent overprotection of synapses in diverse regions of the primate cerebral cortex. *Science* 232:232–235.

Rauschecker, J. P., and P. Marler, 1987. *Imprinting and Cortical Plasticity*. New York: Wiley.

Recanzone, G., M. Merzenich, and W. Jenkins, 1992. Frequency discrimination training engaging a restricted skin surface results in an emergence of a cutaneous response zone in cortical area 3a. *J. Neurophysiol.* 67:1057–1070.

Recanzone, G., M. Merzenich, W. Jenkins, K. Grajski, and H. Dinse, 1992. Topographic reorganization of the hand representation in cortical area 3b of owl monkeys trained in a frequency-discrimination task. *J. Neurophysiol.* 67: 1031–1056.

Recanzone, G., M. Merzenich, and C. Schreiner, 1992. Changes in the distributed temporal response properties of SI cortical neurons reflect improvements in performance on a temporally based tactile discrimination task. *J. Neurophysiol.* 67:1071–1091.

Recanzone, G., C. Schreiner, and M. Merzenich, 1993. Plasticity in the frequency representation of primary auditory cortex following discrimination training in adult owl monkeys. *J. Neurosci.* 12:87–103.

Rose, S. A., 1990. Cross-modal transfer in human infants. In *The Development and Neural Bases of Higher Cognitive Functions*, vol. 608, S. Diamond, ed. New York: New York Academy of Sciences Press, pp. 38–50.

Rugg, M. D., 1990. Event-related brain potentials dissociate repetition effects of high- and low-frequency words. *Memory and Cognition* 18:367–379.

Rugg, M., J. Furda, and M. Lorist, 1988. The effects of task on the modulation of event-related potentials by work repetition. *Psychophysiology* 25:55–63.

Schlaggar, B., and D. O'Leary, 1991. Potential of visual cortex to develop an array of functional units unique to somatosensory cortex. *Science* 252:1556–1560.

Sherman, S. M., and P. D. Spear, 1982. Organization

of visual pathways in normal and visually deprived cats. *Physiol. Rev.* 62:738–855.

Spelke, E. W., 1987. The development of intermodal perception. In *Handbook of Infant Perception: From Perception to Cognition*, vol. 2, P. Salapatek and L. Cohen, eds. New York: Academic Press, pp. 233–273.

Sur, M., S. Pallas, and A. Roe, 1990. Cross-modal plasticity in cortical development: Differentiation and specification of sensory neocortex. *Trends Neurosci.* 13:227–233.

Swick, D., M. Kutas, and H. J. Neville (in press). Localizing the neural generators of event-related brain potentials. In *Localization and Neuroimaging in Neuropsychology*, A. Kertesz, ed. New York: Academic Press.

Van Petten, C., and M. Kutas, 1991. Influences of semantic and syntactic context on open and closed class words. *Memory and Cognition* 19:95–112.

Weber-Fox, C. M., and H. J. Neville, 1992. Maturational constraints on cerebral specializations for language processing: ERP and behavioral evidence in bilingual speakers. *Soc. Neurosci.* 18(1):335.

Weber-Fox, C. M., and H. J. Neville, submitted. Maturational constraints on functional specializations for language processing: ERP and behavioral evidence in bilingual speakers.

HC
ER
36
46
TF
TH
STPa
AITd
AITv
7b
7a
FEF
STPp
CITd
CITv
VIP
LIP
MSTℓ
MSTd
FST
PITd
PITv
DP
VOT
MDP
MIP
PO
MT
V4t
BD
V4
ID
PIP
V3A
V3
VP
M
V2
BD
ID
PULV.
M
V1
BD
ID
M
K
P
SC
M
other
P

III SENSORY SYSTEMS

A hierarchy of visual areas in the macaque, based on laminar patterns of anatomical connections. Approximately 90% of the known pathways are consistent with this hierarchical scheme (Felleman and Van Essen, 1990). Compartments within areas V1, V2, and V4 are discussed in relation to figure 24.6. (Modified from Van Essen, Anderson, and Felleman, 1992, by permission.)

Introduction

COLIN BLAKEMORE AND
J. ANTHONY MOVSHON

PERCEPTION IS the gateway to cognition. The first and most fundamental tasks of cognitive systems are to register, transform, and act on sensory inputs. Furthermore, many cognitive functions are continuously regulated by the sensory consequences of actions. In a very real sense, an understanding of cognition depends on an understanding of the sensory processes that drive it. This section presents a variety of perspectives on the neural foundations of perception, derived from several different methods and several different sensory systems.

In the context of modern cognitive science, it is easy to overlook the formidable tasks faced, and effortlessly mastered, by sensory processing. The task of sensory systems is to provide a faithful representation of biologically relevant events in the external environment. In most cases, the raw signals transduced by receptors seem woefully inadequate to this task and far removed from the complex structures in the world that give rise to them. Yet our sensory systems all contrive, by subtle and complex calculations, to create efficient and informative representations of that world. These representations are at the same time enormously richer and enormously simpler than the simple measurements of light intensity, force, and chemical composition that support them. They are richer because they contain representations of objects, states, and events that are abstracted from the primitive sensory signals; they are simpler because they represent the distillation of the vast quantities of raw measurement information offered to the central nervous system by each sensory surface. To understand fully the richness of sensory processing, we must appreciate both the necessary volume of computation and the sophisticated deductive processes that give rise to our sensory experience.

The most basic questions in this section concern the nature and meaning of the signals carried by single sensory neurons. The issues raised and addressed by such authors as Vallbo, Johnson, Barlow, von der Heydt, and Newsome have roots that extend back six decades to the earliest recordings of sensory unit activity. How are sensory signals encoded by single neurons? How reliable and efficient is this encoding? How are single-neuron signals transformed and elaborated by successive levels of processing? Answers to these questions, and questions about the answers, continue to command the attention of those who study sensory systems. In favorable cases, remarkably sophisticated signals can be discerned in single neurons, and the efficient use of these signals by "higher" processes can, in principle, be used to reveal much of the information the organism uses to control behavior.

It is nonetheless plain that the richness of sensory experience can be understood only in terms of the activity of populations of neurons. The structure and organization of these groups is the concern of chapters by Konishi, Ts'o, Suga, King, Sejnowski, Shapley, and Van Essen. Several mutually orthogonal principles emerge from these analyses. On the one hand, sensory systems seem to partition the tasks that confront them so that semi-independent functional pathways evolve to deal efficiently with submodality-specific sensory processes. On the other hand, within each stream and level of representation, complex and sophisticated interactions occur. These interactions allow information to be shared cooperatively among neurons within a processing level, enhancing the functional possibilities of a simple hierarchical structure. Superimposed on—and dependent on—these interactions are a great variety of sensory maps. Some of these are relatively simple, governed by the topographical nature of the afferent receptor surface. Others achieve elegant representations of critical sensory qualities by successive stages of neural computation. These maps seem to be best understood in terms of the stimulus space in the world rather than the constrained nature of the initial encoding of that world. A complete account of these maps and their formation remains elusive. What factors establish and maintain them throughout life? To what degree are special maps created for special problems, and to what degree are general mapping principles applied in all systems? What kinds of signal processing are occurring at levels where the nature of particular maps is difficult is discern?

Beyond the level of particular maps and representations are questions about the combinations of information that must occur to permit a comprehensive sensory account of the environment. Chapters by Parker, Ullman, Maunsell, Young, and Desimone all raise issues that are fundamentally combinatorial. How are different and separately computed kinds of information about the same sensory quality fused to yield full-formed representations of the world? How are those representations modulated by the behavioral state and needs of the animal, and by memories of other states, needs, and events? These issues occupy the no-man's-land between classical sensory processes and central cognitive ones, and our ability to address them with rigor and precision has made possible a new and more integrated view of the relation between sensory and cognitive neuroscience.

14 Single-afferent Neurons and Somatic Sensation in Humans

ÅKE B. VALLBO

ABSTRACT Afferent systems serving discriminative touch are discussed on the basis of data extracted by the methods of microneurography and microstimulation. With microneurography, the somatosensory input is tapped by recording from a single fiber in a peripheral nerve of an attending human subject who may also participate in psychophysical tests. With microstimulation, identified nerve fibers are stimulated selectively by trains of pulses to launch a controlled sensory input. The findings have modeled in detail the peripheral sensory sheet of the glabrous skin. Relationships between properties and firing of primary afferents and psychophysical data have been explored with regard to detection, intensity, and spatial attributes of percepts. The dual nature of sensory threshold has been demonstrated. Moreover, central subsystems within the human brain have been identified whose functional properties match remarkably well the characteristics of the primary afferents serving them, as evidenced by the attributes of percepts produced when single-unit impulses are injected.

It remains an important issue in cognitive neuroscience to define coincidental and causal relationships between neural activity and mental events. When sensory signals in primary afferents are considered in this context, the hypothesis being addressed is often whether a particular sensation or sensory attribute is dependent on a specific impulse pattern in a specific set of afferent units.

An obvious approach to exploring such hypotheses is to correlate impulse firing in primary afferents with sensory experience as analyzed in psychophysical tests, to assess whether the candidate neural activity appears or grows in parallel with the sensation reported by the subject.

Mountcastle and his coworkers pioneered the study of neural correlates of somatic sensation at the single-unit level (e.g., Werner and Mountcastle, 1965; Mountcastle, Talbot, and Kornhuber, 1966; Mountcastle, 1967, 1984a,b; Talbot et al., 1968). Their approach was to analyze neural activity in anesthetized monkeys and psychophysical responses in humans when identical tactile stimuli were delivered.

A basic assumption in these studies was that touch receptors in human and monkey produce identical responses to skin deformations. When data on single-unit firing of human afferents became accessible by microneurography (Vallbo and Hagbarth, 1968), it was possible to test the validity of this assumption. In addition, cross-species inferences could be avoided altogether because the method permits the assessment of coincidence between features of the impulse response in human afferents and sensations in the human mind elicited by the same stimulus.

The scope of this chapter is confined to sensory signals in peripheral nerves of human subjects. Moreover, the emphasis is on mechanisms related to touch, particularly discriminative touch in the human hand. Areas that are not addressed include temperature and pain mechanisms (Hallin, Torebjörk, and Wiesenfeld, 1982; Konietzny, 1984; Ochoa and Torebjörk, 1989; LaMotte, Lundberg, and Torebjörk, 1992), skin, joint, and muscle receptor activity in relation to position sense and kinesthesia (Gandevia, McCloskey, and Burke, 1992), and the role of cutaneous mechanoreception in motor control (Johansson and Cole, 1992), although very interesting advances have been made in these areas in recent years.

Discriminative touch in relation to cognition

In the context of cognition, discriminative touch in the human hand has an outstanding role compared to other somatosensory submodalities and also compared to tactile sense in most other skin regions. First, the sensory systems for discriminative touch have an exquisite capacity to extract the detailed information that is essential for many cognitive functions, whereas other

ÅKE B. VALLBO Department of Physiology, University of Göteborg, Göteborg, Sweden cutaneous systems are far less capable in this respect. Second, discriminative touch is a crucial element in exploration and manipulation, actions that are vitally important for the construction of internal models of physical objects (Johansson and Westling, 1984, 1987; Gordon et al., 1991). Finally, it is widely accepted that exploratory motor activity producing associated afference is fundamental to the intellectual development of children (Piaget, 1976).

Methods to analyze single afferents in sensation

The information on which this chapter is primarily based was obtained largely from experiments in human subjects using three different techniques—microneurographical recording of neural activity, intraneural microstimulation, and conventional psychophysics.

Microneurography allows the recording of impulses in single nerve fibers of attending human subjects (Vallbo and Hagbarth, 1968; Vallbo et al., 1979). Although the technique is demanding in practice, the procedure is basically straightforward. Insulated tungsten needle electrodes are inserted percutaneously into a peripheral nerve and adjusted manually until the desired type of nerve activity is recorded. Impulse trains from identified sensory units with either myelinated or unmyelinated axons can readily be resolved while the subject focuses his or her attention on a psychophysical task.

It still is not clear which conditions are essential for the discrimination of single-unit action potentials in microneurographical recordings, although the criticism that unit recording is possible only when pressure from the needle electrode blocks most of the fibers in the impaled fascicle has been refuted (Wall and MacMahon, 1985; Torebjörk, Vallbo, and Ochoa, 1987).

Microstimulation is refinement of the microneurographical technique whereby the needle electrode designed for recording nerve impulses is switched for electrical stimulation through the tip in a nerve fascicle (Ochoa and Torebjörk, 1983; Vallbo et al., 1984). The purpose is usually to deliver a train of pulses in order to inject a series of action potentials via an identified sensory unit into the central nervous system (CNS) of the attending human subject.

Microneurography is a powerful tool for assessing coincidence between neural activity and psychophysical events in human subjects, allowing one to draw conclusions about whether a causal relationship is consistent or incongruous with observed data. Microstimulation, on the other hand, has the potential to provide a direct test of a causal relationship between activity in an afferent unit and the construction of a percept within the mind of the subject. The combination of these two methods allows a double and complementary approach to the general question of the relationship between neural activity and sensation.

Tactile afferent units in the human skin

Glabrous Skin Studies over the last two decades have resulted in a detailed description of the peripheral sensory sheet of the glabrous skin of the human hand (Johansson and Vallbo, 1983; Vallbo and Johansson, 1984), which houses four different types of sense organs with myelinated axons responding to light skin deformations (figure 14.1). Two types are characterized by small and well-demarcated receptive fields and are denoted *type I units* (Johansson, 1978; Johansson and Vallbo, 1980). One of these two types reacts exclusively to movements and is called fast-adapting type I (FAI), whereas the other is slowly adapting type I (SAI) because it responds to time-invariant deformations as well. The end organs of these two types are very likely connected to Meissner's corpuscles and Merkel cells, respectively.

The functional characteristics of the other two types are distinctly different because their receptive fields are large and their borders are indistinct (i.e., the sensitivity decreases gently from a central point, often over several centimeters). Again, one of the two types responds to movements exclusively (FAII), whereas the

		RECEPTIVE FIELDS	
		Small, sharp borders	Large, obscure borders
ADAPTATION	Fast, no static response	FAI Meissner	FAII Pacini and Golgi-Mazzoni
	Slow, static response present	SAI Merkel	SAII Ruffini

Figure 14.1 Types of tactile afferent units in the glabrous skin of the human hand. Presumed end organs are indicated.

other is statically sensitive as well (SAII). In fact, the latter type seems to fire incessantly to a sustained deformation. The end organs of the two type II units are very likely pacinian corpuscles and related end organs and Ruffini endings, respectively.

The notation just presented (types FAI and FAII, SAI and SAII) is akin to Iggo's original notation of slowly adapting mechanoreceptors in the hairy skin of subhuman species (Iggo and Muir, 1969; Chambers et al., 1972). It is strongly recommended that the alternative abbreviations—*RA* and *QA* for presumed Meissner (FAI) and *PC* for pacinian corpuscles (FII)—be abandoned as they easily may lead to confusion. In this chapter, the four types of units will be referred to as Meissner, Merkel, Ruffini, and Pacini units, respectively, for the sake of simplicity, although it should be noted that the inference from physiological unit type to morphological structure in human is an extrapolation in some respects.

The human hand is unique among primates in that it has a fair proportion of Ruffini units in the glabrous skin. The evidence that SAII sense organs constitute a distinct entity is based not only on their general response properties but also on their receptive field geography, assessed with a precision that surmounts previous studies in humans as well as animals (Johansson, 1978). It should be noted, however, that the SAII unit type was originally demonstrated in the hairy skin of cat and monkey (Chambers et al., 1972).

HAIRY SKIN Hairy skin, which covers the major part of the human body, has slightly different sensory equipment from the glabrous skin of the hand. Available evidence suggests that Merkel, Ruffini, and Pacini endings are present, whereas Meissner units may partly or largely be replaced by hair follicle units (Järvilehto, Hämäläinen, and Laurinen 1976; Konietzny and Hensel, 1977). However, the picture is not yet clear, in part because studies in humans have largely been limited to the peripheral parts of the extremities where hairs are few. More detailed analyses of unit properties in the hairy skin are needed.

UNMYELINATED TOUCH RECEPTORS In addition to the unit types already listed, which have large myelinated afferents (Johansson and Vallbo, 1983), human hairy skin is innervated by unmyelinated units sensitive to light touch. It was long believed that such a system, originally described by Douglas and Ritchie (1962) in cats and rabbits, was lacking in human skin because low-threshold unmyelinated units had not been encountered in microneurographical recordings. However, it recently was shown that such units are present in the human facial area, suggesting a vestige of an older tactile system in a specialized skin region (Nordin, 1990). Shortly afterward, touch-sensitive units with unmyelinated afferents were found to be abundant in the hairy skin of the forearm (Vallbo et al., 1993). It is therefore likely that humans also are equipped with a general system of unmyelinated fibers for touch, though their role in sensation remains a fascinating enigma.

Sensory contribution of single afferents as revealed by microstimulation

The procedure often pursued in microstimulation experiments is first to insert a microneurographical electrode in a cutaneous fascicle innervating the glabrous skin and then to adjust its position until a stable recording from a mechanoreceptive unit is obtained such that its general response properties can be assessed and its receptive field plotted. The electrode is then reconnected to a stimulator, and a train of brief electrical pulses is delivered (figure 14.2). While the pulse intensity is increased successively between trains from zero up to a few microamperes, the subject is asked to report any sensation from the glabrous skin area (Ochoa and Torebjörk, 1983; Vallbo et al., 1984).

When the stimulus intensity reaches a critical level, often at approximately 1 μA, the subject describes a clear and well-localized sensation. Although the attributes of these sensations may vary in details, a number of notable invariances are apparent in samples collected from many volunteers. The quality of the sensation is regularly in the realm of touch, rather than temperature, pain, or paresthesia: that is, it is a sensation of innocuous skin deformation.

Although the quality of sensation is definitely mechanical, the subject often describes it as slightly odd or exotic in that he or she finds it difficult to specify a real mechanical stimulus that would elicit an identical sensation. Moreover, the spatial properties are characteristic, because the sensation is not pointlike but has a definite, albeit small, extent over the skin surface, typically 2–3 mm in diameter. This size corresponds fairly well to the average size of the receptive field of the most common groups of tactile units in this area (i.e.,

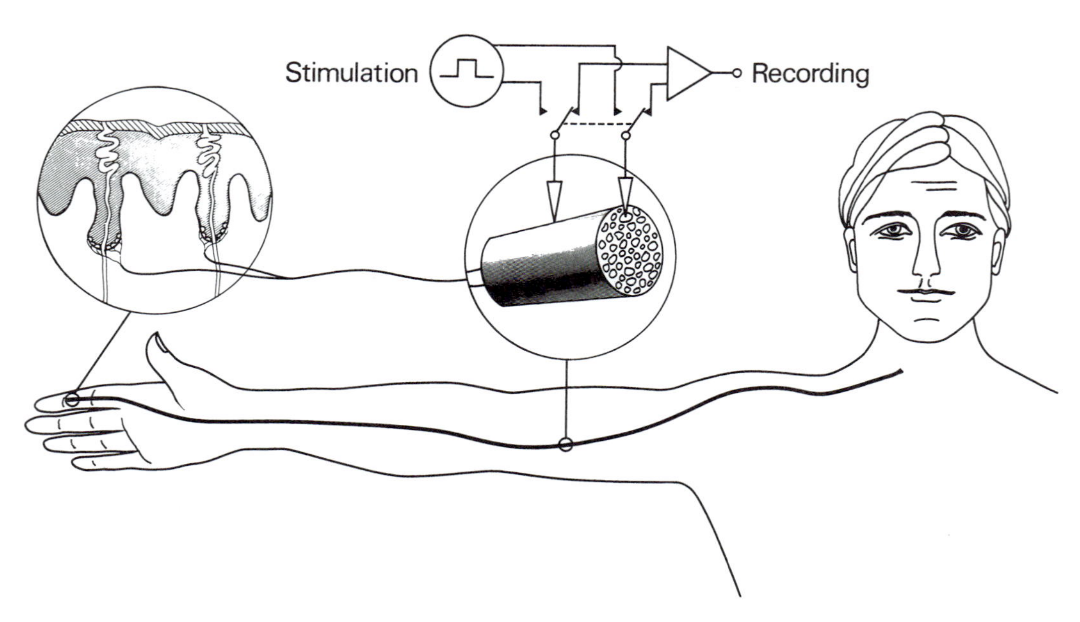

FIGURE 14.2 Microstimulation method. An intraneural needle electrode for recording single-afferent activity is switched to an electrical stimulator. Controlled trains of pulses excite the identified axon to inject a series of nerve impulses into the central nervous system.

Meissner and Merkel units) (Johansson and Vallbo, 1980; Vallbo et al., 1984; Phillips, Johansson, and Johnson, 1992). Hence, the spatial and qualitative attributes of the sensation are consistent with the interpretation that liminal electrical pulses excite one of the large myelinated nerve fibers (i.e., an $A\beta$ fiber), which is connected to mechanoreceptive end organs in the skin. The term *elementary sensation* has been coined to denote this type of sensation evoked by microstimulation (Ochoa and Torebjörk, 1983).

When the stimulus intensity is increased further, an additional sensation of localized skin deformation usually is reported at a different location (figure 14.3), often quite remote from the first one (e.g., on a different finger). The obvious interpretation is that now two different tactile units are excited, and impulse trains in the two constitute the afferent signals for the construction of two separate and readily identifiable elementary sensations.

With higher stimulus intensities, the subject may report that she or he simultaneously perceives three or even four such elementary sensations at different locations. However, at some point as the stimulus intensity continues to be raised, the percept suddenly changes altogether in all subjects tested. The discrete tactile sensations are replaced by a single perceptive entity, often a paresthesialike sensation covering a large skin area. In other cases, a painful sensation localized at a small and distinct point takes over. It seems likely that the paresthesia is due to activation of a larger number of tactile afferents, whereas the painful sensation is due to activation of one or several nociceptive units, which require higher stimulus because their fiber diameters are smaller.

The detailed characteristics of the elementary sensation often match the physiological properties of the particular afferent axon studied in the recording mode. Particularly, the spatial attributes of the sensation often correspond remarkably well to those of the afferent's actual receptive field. Hence, the locations often match almost exactly, as will be described later. Most fields are round, as are the reported spatial extents of the sensations. In a few cases when oval and elongated receptive fields are encountered, the subjects often report similar shape for the sensation and even similar orientation of the long axes (Vallbo et al., 1984).

A spatial match between receptive field of the recorded afferent unit and the first percept reported by the subject at liminal stimulus intensity is not always found, however. Sometimes the location of the sensation is quite remote from the receptive field, strongly suggesting that two different units are involved (i.e.,

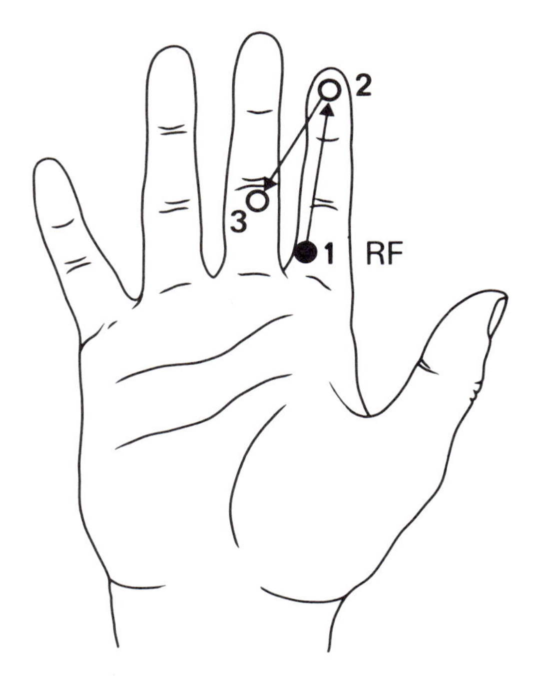

FIGURE 14.3 Recruitment of a number of elementary sensations by microstimulation (as in figure 14.2) with successively higher stimulus intensities. The first sensation to appear was located at the base of the index (1) exactly coinciding with the receptive field (RF) of the afferent unit discriminated in the recording mode. When stimulus intensity was increased, the subject reported an additional sensation at the tip of the index (2), and with still higher intensities the subject reported that he simultaneously felt three separate elementary sensations, the last one on the long finger (3). (Modified from Vallbo et al., 1984, by permission of Oxford University Press.)

one unit is recorded, whereas another fiber is excited by the electrical pulses injected through the same electrode). Although the factors accounting for a discrepancy of this kind have not been explored, control experiments with exposed nerves have demonstrated that it is not an uncommon phenomenon (Calancie and Stein, 1988).

When the spatial correspondence is adequate, the quality of sensation usually matches the properties of the afferent unit, provided that fairly long trains of impulses are used (Ochoa and Torebjörk, 1983). Hence, activation of units that respond exclusively to moving stimuli (i.e., Meissner and Pacini units) is associated with sensations of vibration. Oscillatory movements are the only mechanical stimuli that produce long-lasting trains of impulses in such units. It follows that the brain should interpret this sensory message as an oscillatory skin movement. In contrast, when slowly adapting Merkel units are excited, the subjects report sensations of sustained touch or light pressure (e.g., "as if a small leaf were steadily pressed against the skin").

However, it is not universally true that impulses in a single cutaneous mechanoreceptor evoke a sensation. On the contrary, there is strong evidence that microstimulation of a single Ruffini unit regularly fails to produce a sensation (Ochoa and Torebjörk, 1983; Vallbo et al., 1984). This appears to be true with some of the Meissner, Merkel, and Pacini units as well. It remains an interesting problem to explore systematically whether the Meissner, Merkel, and Pacini units that fail to elicit a sensation have unique physiological properties and whether they are overrepresented in particular skin areas.

The lack of sensations when single Ruffini afferents are activated does not justify the conclusion that these units have no role whatever in sensation. It may mean simply that spatial summation is required from a number of Ruffini units for the brain to construct a conscious sensation. This would fit with the hypothesis that this system has a kinesthetic or proprioceptive role, suggested by the finding that Ruffini units respond strongly to active movements and altered finger positions (Hulliger et al., 1979; Edin and Abbs, 1991; Edin, 1992).

The convincing evidence that an impulse train in a single afferent unit often evokes a sensation whose attributes match the physiological properties of the afferent unit constitutes strong proof for the *labeled line theory*, even down to the level of submodalities within the tactile system. An inference of major physiological significance is that a single afferent may carry sufficient information to produce a sensation distinctly characterized with regard to spatial and qualitative properties. A particular strength of the microstimulation data is that they prove a causal relation between specified somatosensory afference and perception in humans and not merely a coincidence.

The problem of a sensory threshold

Since the dawn of psychophysics and sensory physiology, the concept of a sensory threshold has been a central element in sensory research. Yet it has been an enigma that a distinct threshold cannot be demonstrated in psychophysical tests. A number of theories have been advanced to explain the experimental finding that the probability of psychophysical detection gradually rises with stimulus intensity (Corso, 1963). Emphasis has usually been put on processes within the CNS as, for example, in the signal detection approach (Swets, 1961; Green and Swets, 1966), which assumes that there exists within sensory systems a considerable amount of noise that varies randomly over time and mixes with a true sensory signal. The brain would be able to retrieve the true signal securely only if it had a significant amplitude in relation to the noise. Moreover, a tacit assumption seems to be that repeated stimuli of constant amplitude give rise to invariant neural activity at the level of the primary afferents.

Stimulation Method When the threshold problem was analyzed with tactile stimuli in the glabrous skin of the hand, an initial move was to check the assumption of invariant afferent input using microneurographical recordings from single afferents. The test stimuli had minimal spatial extent and duration, and they were delivered against a neutral background to avoid interference from simultaneous activity in other cutaneous mechanoreceptors. An invariant input signal proved difficult to achieve. Not until a sophisticated stimulation method was designed that had the precision to elicit a single action potential in a single afferent nerve fiber was it possible to study properly the nature of the psychophysical threshold (Westling, Johansson, and Vallbo, 1976).

Differing Psychophysical Threshold Between Skin Regions Measurements of the subjects' detection threshold and the thresholds of primary afferents in the glabrous skin revealed interesting relations even when group data are simply compared (figure 14.4).

First, there are consistent differences between separate skin regions of the hand with regard to subjects' ability to detect weak stimuli. The psychophysical thresholds were low in the major part of the glabrous skin, approximately 10 μm. However, in the center of the palm and over the skin creases at the finger joints, the psychophysical thresholds were considerably higher. Hence, two separate skin regions—a low-threshold region and a high-threshold region—could be discriminated on the basis of psychophysical detection. An immediate question is whether the different psychophysical thresholds are due to different properties of the sensory sheet within the two skin regions or

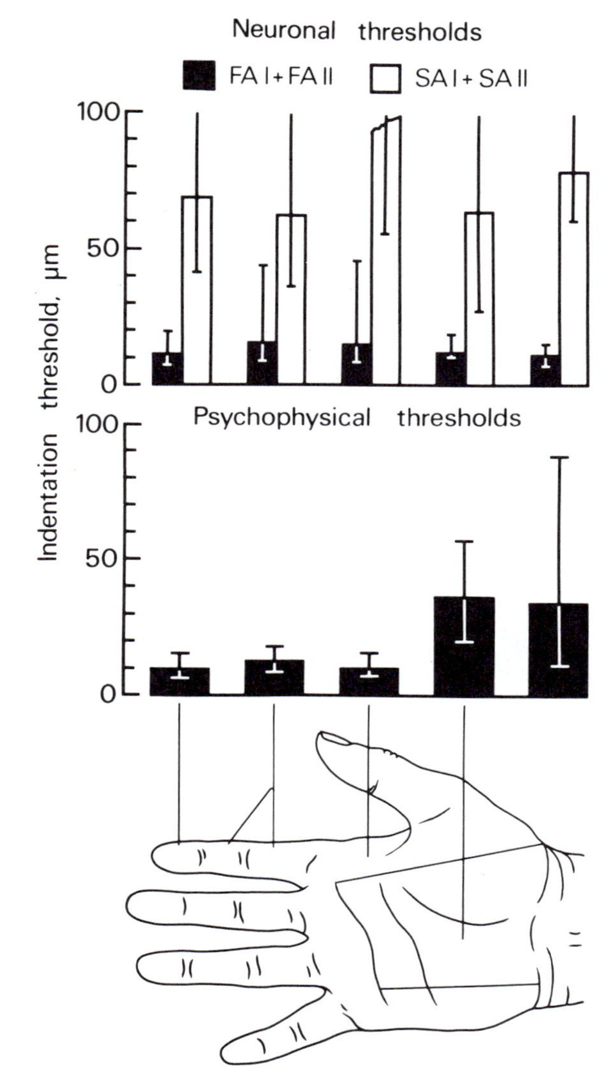

Figure 14.4 Psychophysical detection thresholds and thresholds of tactile units. The upper histogram shows thresholds for evoking a single nerve impulse in afferent units in various regions of the glabrous skin area (total, 128 units). Triangular skin indentations with a slope velocity of 4 mm/s were used. The two types of fast-adapting units (Meissner and Pacini) are pooled because they have about the same threshold distribution for this type of stimulus. The two types of slowly adapting units (Merkel and Ruffini) are pooled as well. The lower histogram shows psychophysical thresholds in corresponding skin regions to identical stimuli (162 test points). From left to right, the columns give data from the terminal phalanx, the rest of the finger, the peripheral part of the palm, and the central part of the palm and, to the extreme right, data from the lateral aspects of the fingers and the regions of the creases taken together. Column heights give medians and bars, the twenty-fifth and seventy-fifth percentiles. (Modified from Johansson and Vallbo, 1979b.)

to different processing of the afferent information by central mechanisms.

Data on neural thresholds of the four types of primary afferents fell into two distinct groups. However, in contrast to the psychophysical thresholds, this grouping was not related to skin area (see figure 14.4). Instead, the two sets of mechanoreceptive thresholds were clearly related to unit type. The fast-adapting units (i.e., Meissner and Pacini units) all had low thresholds (approximately 10 μm), whereas the slowly adapting units exhibited considerably higher thresholds regardless of locations within the glabrous skin area.

A comparison between neural and psychophysical data clearly shows that the thresholds of the slowly adapting units were higher than the main set of psychophysical thresholds. Hence, it can safely be concluded that psychophysical tactile detection is not dependent on Merkel or Ruffini units because they fail to respond to stimuli that subjects readily perceive in most parts of the glabrous skin.

The fast-adapting units, on the other hand, respond to stimuli that human subjects can just detect. Actually, their thresholds match the psychophysical thresholds well. Hence, subjects' ability to detect minimal touch stimuli must be based on a very small afferent input from Meissner or Pacini units (i.e., a single action potential either in a small group of units or even in a single afferent unit). Circumstantial evidence strongly favors Meissner over Pacini as the critical unit type.

Perceptual Detection of Single Action Potential

The findings presented in figure 14.4 are condensed from a larger set of data as the individual observation represents the midpoint of a threshold curve, as illustrated in figure 14.5A, which displays data collected while the subject's detection response and the impulse response from a Meissner unit were simultaneously recorded. Here the neural and psychophysical thresholds were practically identical, as found in many cases in the low-threshold regions of the hand.

The very close agreement between the two threshold curves strongly suggests that the particular Meissner unit recorded provided the critical signal for detection. This interpretation was further supported by correspondence of neural and psychophysical responses on individual trials. When a very sensitive Meissner unit was recorded and a near-threshold stimulus was presented repetitively, a good correspondence was found between unit response and psychophysical detection —that is, when an action potential was elicited, the subject reported detection and vice versa (Vallbo and Johansson, 1976).

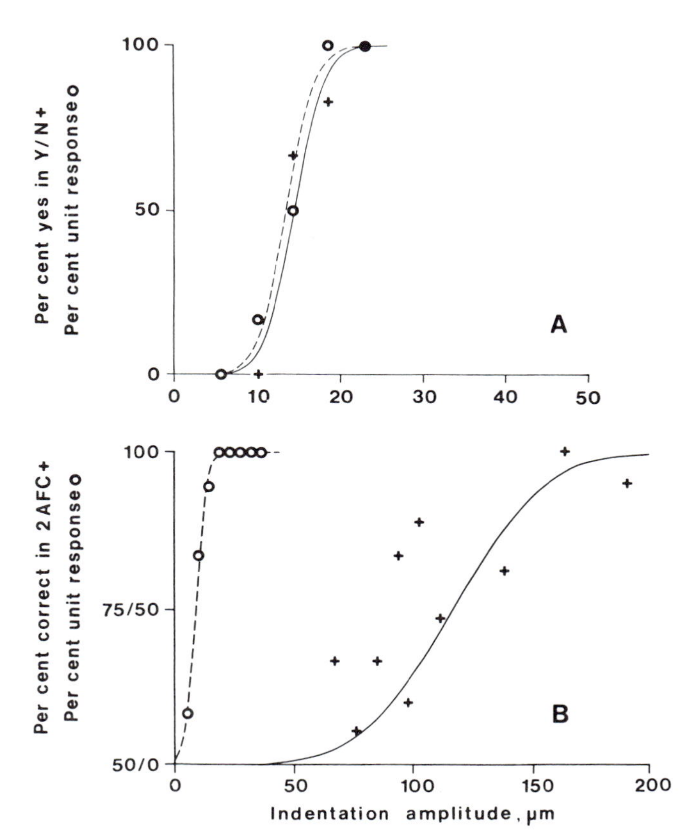

Figure 14.5 Neural and psychophysical threshold curves from sample test points within the low-threshold and high-threshold regions of the glabrous skin (cf. figure 14.4). (A) and (B) show data from a target point at the tip of the thumb and the center of the palm, respectively. Both target points were located within the receptive fields of an identified Meissner unit whose impulse response was the data base for the neural threshold curves. Circles and interrupted lines represent neural response (i.e., the percentage of tests in which the afferent unit responded with one impulse). Crosses and continuous lines represent psychophysical responses. In (A) the yes/no procedure was used, whereas the two-alternative forced choice was used in (B). Chance performance in the latter gives 50% correct responses. Curves are the best-fitting cumulative normal distribution function (maximum likelihood estimation). (Reprinted from Vallbo and Johansson, 1976, by permission.)

The probability that not more than one afferent was activated at psychophysical threshold was explored further using several independent approaches. On the basis of a population model of the tactile sensory sheet of the glabrous skin, it was estimated that a single afferent would be excited at psychophysical detection threshold in most tests (Johansson and Vallbo, 1979b). The model was based on extensive experimental data, including relative unit densities in separate subregions of the glabrous skin; a fiber count of the median nerve at the wrist, which provided a factor to scale the sample data to a population estimate; threshold distribution of the candidate unit types; and the microstructure of the receptive fields (Johansson and Vallbo, 1976, 1979a, 1980; Johansson, 1978).

It may seem paradoxical that it would be possible to stimulate a single afferent in the glabrous skin despite the high degree of receptive field overlap at any point in the hand (Johansson and Vallbo, 1979a, 1980). This paradox is related to the fact that the receptive field of any afferent unit is not invariant. Rather, size as well as geography depend on the characteristics of the stimulus employed. The standard stimulus to define receptive field properties, and hence receptive field overlap, is cruder than the refined method used in the detection study (Westling et al., 1976; Johansson and Vallbo, 1979b).

Additional support for psychophysical detection of a single Meissner action potential was provided by microstimulation; it was found that a single electrical pulse exciting a nerve fiber often was sufficient for psychophysical detection (Ochoa and Torebjörk, 1983; Vallbo et al., 1984). This was demonstrated with several, but not all, Meissner units explored. With the Merkel units, on the other hand, a train of pulses was regularly required, indicating that temporal summation was necessary to produce a sensation in the slowly adapting system.

Hence, several findings support the contention that central processing units in the brain which handle the sensory signals from the Meissner units in the low-threshold region of the hand are essentially noise-free or, to be more exact, any noise that may be present is negligible in relation to the size of the sensory signal evoked by a quantal afferent input when subjects pay maximal attention to the detection task. It is significant that with the signal detection approach the combination of practically 100% hits and 0% false alarms often was encountered. Consequently, nothing could be gained by forcing the subject to modify his or her decision criterion, which is a commonly used procedure in threshold studies that adhere to signal detection theory.

Summation for Detection of Stimuli Figure 14.5B shows data from a point in the center of the palm where the psychophysical threshold is high. The left-hand curve represents the response probability of a Meissner unit whose receptive field center was located at the target point, whereas the right-hand curve is the psychophysical threshold at the same point. Obviously, the subject's detection threshold was much higher than the unit's threshold. This was dramatically illustrated during the recording: A number of impulses were elicited by stimuli that the subject did not notice. Clearly, the failure to detect weak stimuli was not due to a shortage of afferent impulses. Moreover, the population model indicated that several additional afferents were activated (Johansson and Vallbo, 1976, 1979b). Hence, both spatial and temporal summation were required to produce a sensation of touch in the central part of the palm. It is also evident from figure 14.5B that the psychophysical data were much more scattered and the slope of the curve was much shallower compared to data from the low-threshold region illustrated in figure 14.5A. These findings are consistent with the interpretation that more noise is present in this part of the central processing unit, according to signal detection theory (Swets, 1961; Green and Swets, 1966).

Although threshold stimuli are seldom interesting in daily life, the detection data from the separate regions within the glabrous skin of the hand disclose an interesting principle. The contrast between the two skin regions with regard to neural and psychophysical thresholds indicates that central processing of identical afferent information is not uniform but differs between skin regions. It seems likely that this difference reflects the relative importance of the two skin regions as sensory receiving areas. It is obvious that the skin of finger and the palmar pads are more apt to extract sensory information in many manipulatory or exploratory tasks than the center of the palm and the regions of the skin folds at the finger joints.

Nature of the Threshold to Touch in the Glabrous Skin The detection data just presented imply that signal detection theory is irrelevant for most of the glabrous skin area because subjects are capable of detecting a minimal sensory input. Hence, the sensory threshold is a step function when the input is defined in terms of nerve response rather than amplitude of the physical stimulus. When a *test stimulus* was defined as one that elicits an action potential and a *catch trial* as one that does not, test series often produced 100% hits with 0% false alarms. In contrast, findings from the high-threshold region of the glabrous skin area might be consistent with the signal detection approach, although this aspect has not been explored in depth.

Neural code related to intensity of touch sensation

A long-standing problem in psychophysics is the definition of how a sensation grows as stimulus amplitude is increased. Stevens (1957, 1967) demonstrated that very consistent data are obtained if the subject is asked to estimate the intensity of his or her sensation simply by assigning a number to it (magnitude estimation). Physiological functions relating these magnitude estimates to stimulus amplitude all conform to a mathematical power function. Stevens's power law implies that anywhere along an intensive continuum, equal ratios between stimulus magnitudes give rise to equal ratios between perceptive intensities, providing that stimulus amplitudes above threshold stimulus are considered. Hence, modalities and submodalities differ only in the rate of growth, which is expressed as a difference in the exponent of the fitted power equation.

Physiologists have identified neural codes of stimulus amplitudes in terms of impulse discharge rates and successive recruitment of units. On the basis of collateral experiments on humans and monkeys using localized skin indentations (Werner and Mountcastle, 1965; Mountcastle, Talbot, and Kornhuber, 1966; Harrington and Merzenich, 1970), it was suggested that the shape of the psychophysical function is determined at the receptor level, whereas transformations at more central levels in the nervous system are in sum linear (Mountcastle, 1967).

When it became possible to apply data entirely from human subjects, thus avoiding cross-species inferences, the hypothesis was not supported, although an early study seemed to do so (Gybels and van Hees, 1972; Knibestöl and Vallbo, 1980). Interesting discrepancies are obvious between shapes of neural and psychophysical response functions extracted from the same points in the glabrous skin of individual subjects (figure 14.6).

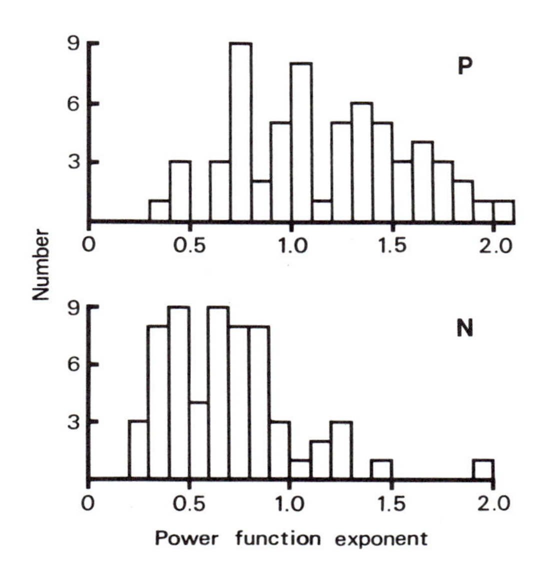

FIGURE 14.6 Characteristics of (A) psychophysical magnitude estimation functions (P) and (B) neural stimulus-response functions (N) of slowly adapting tactile units. Histograms give the exponents of the power equations fitted to experimental data of the type illustrated in figure 14.7. The two sets of data represent pairs of neural and psychophysical responses to rectangular indentations of varying amplitudes and standardized duration at 60 target points in the glabrous skin of 26 subjects. The average exponent of the psychophysical magnitude estimation functions (upper histogram) is 1.0 and range from 0.4 to 2.1. Neural data are total number of impulses in Merkel afferents ($n = 58$) and a few Ruffini afferents ($n = 2$). The average exponent is 0.7, and the range, 0.2–1.4. (Modified from Knibestöl and Vallbo, 1980.)

Psychophysical functions are, on average, linear—that is, the exponent is 1.0 (see figure 14.6, upper histogram) (Knibestöl and Vallbo, 1980)—but the variation is considerable. Neural functions, in contrast, are not linear but mostly decelerating—that is, the exponent is less than 1.0—and there is considerable variation between different units. These findings demonstrate that the sensation of touch grows at a higher rate than the impulse discharge of individual Merkel or Ruffini afferents when stimulus amplitude is increased.

In considering the hypothesis that the average psychophysical function is determined at the receptor level, it is interesting to explore not only whether the neural and psychophysical functions are identical but also whether there exists a weaker dependence or none at all (i.e., whether low exponents in the neural functions are associated with relatively low exponents of the psychophysical function and vice versa). A correlation of this kind would suggest that the transformations in the CNS are, if not linear, at least roughly uniform among subjects. However, the experimental data do not support this presumption because when pairs of neural and psychophysical exponents, derived from identical target points in the subjects' hands, were analyzed, no significant correlation was found between the two ($n = 46$, $r = -0.04$, $p > .6$) (Knibestöl and Vallbo, 1980).

An analysis of the variations revealed interesting relations. It was found that interindividual differences are prominent, whereas individual subjects tend to produce functions of uniform general shape. An illustrative example is presented in figure 14.7, which displays data from three different points in the hand of one subject. This figure demonstrates a considerable

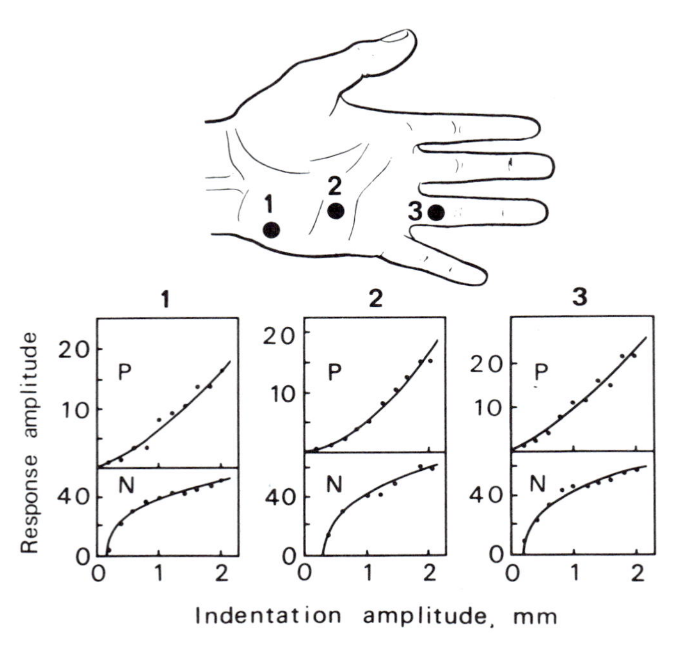

FIGURE 14.7 A case of consistent discrepancy with regard to growth rate between neural stimulus-response functions (N) and psychophysical magnitude-estimation functions (P) in one subject. Data were obtained at three different target points, indicated by black dots (1, 2, 3), located in the center of the receptive fields of three Merkel units. Structure of database identical to that of figure 14.6. Obviously, there is a pronounced discrepancy between the psychophysical functions, which were all accelerating, and the neural functions which were all decelerating. (Reprinted from Knibestöl and Vallbo, 1980, by permission.)

and consistent difference in growth rate, as the subject produced accelerating psychophysical functions at all three points, whereas the stimulus response functions of three Merkel units at these very points were strongly decelerating. The intraindividual consistency and interindividual variation of psychophysical functions provide independent support for the interpretation that central mechanisms play a role in defining how the intensity of the touch sensation grows with stimulus amplitude (cf. Cavonius and Chapman, 1974).

It warrants emphasis that no other reasonable measure of single unit activity in slowly adapting afferents than total impulse response is more favorable to the hypothesis that the average psychophysical function is determined at the receptor level, whereas other codes (e.g., activity in other types of afferents as well as recruitment of units) remain to be explored.

In summary, it seems impossible to maintain the hypothesis that the magnitude of sensation associated with sustained skin indentation is determined simply by the impulse rate in slowly adapting afferents in the glabrous skin. This would also weaken "the general hypothesis that, as regards sensory functions, the brain operates in a linear manner" (Mountcastle, 1967). The human data are fairly consistent with the interpretation that the CNS plays a role in reshaping the intensity function and individual brains seem to differ in terms of their functional properties in this role (cf. Cavonius and Chapman, 1974).

Temporal resolution of tactile stimuli: Detection of vibratory stimuli related to afferent neural input

GLABROUS SKIN Temporal resolution within the sensation of touch has been analyzed extensively using vibratory stimuli. On the basis of psychophysical data, it has been proposed that two receptor systems are involved (Verrillo, 1963, 1965). This interpretation was substantiated considerably by Mountcastle and his coworkers, who presented strong evidence from collateral studies on human subjects and monkeys that Meissner units account for detection in a low-frequency range (approximately 5–40 Hz) and Pacini units for detection in a high-frequency range of vibration (approximately 60–300 Hz) (Talbot et al., 1968; Mountcastle et al., 1969; Mountcastle, LaMotte, and Carli, 1972). Although these analyses strictly refer only to the *detection* of vibratory stimuli (Wall, 1971), there is evidence that an extrapolation to suprathreshold stimuli is justified (Talbot et al., 1968; Mountcastle et al., 1969; Mountcastle, LaMotte, and Carli, 1972).

A reasonable assumption behind these conclusions is that cross-species inferences from monkey to human are fully justified with regard to the dynamic properties of mechanoreceptors. This assumption was supported when studies of afferents from the corresponding skin area in humans demonstrated a similar differential sensitivity to vibratory stimuli for Meissner and Pacini units (Johansson, Landström, and Lundström, 1982). It is interesting, however, that the difference between the two afferent systems deteriorates at higher intensities and therefore "the possibilities to selectively stimulate the different types of units simply by selecting particular stimulus frequencies are limited, particularly when suprathreshold stimuli are applied" (Johansson, Landström, and Lundström, 1982).

Talbot and colleagues (1968) also emphasize that the quality of the sensation elicited from the two receptor systems differs. A vibratory stimulus in the frequency range of 5–40 Hz, which preferentially excites the Meissner units, is reported to give a sensation that is well localized, is felt on the skin surface, and has the character of flutter. Higher frequencies, which mainly excite the Pacini system, elicit a sensation of deep vibrations over a larger area with indistinct borders (Talbot et al., 1968). Obviously these differences match remarkably well the receptive field properties of the afferent units as well as the locations of the end organs, because Meissner units have well-demarcated receptive fields and the end organs are located close to the skin surface, whereas Pacini units have fields with indistinct borders and the corpuscles are located deep in the subcutaneous tissue.

Independent support from human subjects for a dichotomy of vibratory sensibility was provided by microstimulation of single units of either type (Ochoa and Torebjörk, 1983). Although quantitative information has not been reported, qualitative statements maintain that Meissner and Pacini units evoke sensations that differ in several respects. When a Meissner afferent is stimulated at a very low frequency, a sensation of repeated tapping is reported. With slightly higher frequencies, the sensation is described as a flutter on the skin surface, whereas "the oscillatory element progressively fused" at approximately 100 Hz. In contrast, when Pacini units are stimulated, no sensation whatever appears until higher stimulus rates are attained and the vibratory quality of the sensation remains up to 200–300 Hz. It was also found that the perceived oscillatory frequency increased with the stimulus frequency. With regard to the basic quality of sensation, however, the findings were not quite as expected because tickling sensations were more often reported than vibration (Ochoa and Torebjörk, 1983; Schady, Torebjörk, and Ochoa, 1983). Obviously, more data are needed to elucidate these issues.

Physiological as well as psychophysical data from human subjects support the original proposal that two different sensory systems (i.e., the Meissner and the Pacini systems) are responsible for our ability to discriminate temporal aspects of tactile stimuli. The differential properties of the two systems have been demonstrated at the level of the primary afferents in human and monkey, at the thalamic and cortical levels in the monkey, and at the perceptive level in humans. Two complementary methods to activate selectively the two systems in humans (i.e., oscillatory skin movements and electrical microstimulation of identified single afferents) both provide data that are consonant with the dichotomy proposed by the Mountcastle group (Talbot et al., 1968; Mountcastle et al., 1969; Mountcastle, LaMotte, and Carli, 1972).

Hairy Skin Two separate systems for temporal resolution appear to be present also in the hairy skin, because a similar dichotomy has been demonstrated on the dorsal aspect of the hand. Again, this was first described in related experiments in humans and monkeys. Hair follicle units would seem to account for the low-frequency range and Pacini for the high-frequency range (Merzenich and Harrington, 1969). In humans, microneurographical studies indicated that Meissner units rather than hair follicle units account for the low-frequency range at the dorsum of the hand where hairs are scarce. Pacini units were not encountered in these recordings, but it was assumed that they are responsible for human detection of high-frequency vibration (Järvilehto, Hämäläinen, and Laurinen, 1976; Konietzny, and Hensel, 1977).

Obviously, cross-species inferences between human and monkey in this area of study are complicated because the hand dorsum is largely covered by hair in monkeys but not in humans. Nonetheless, the data would be consonant if Meissner and hair follicle systems are equivalent with regard to temporal resolution and perceptive effects.

Lack of Correspondence Between Afferent Signals and Sensory Attributes In relation to the analysis of neural and psychophysical responses to oscillatory tactile movements, it is of considerable theoretical interest that skin movements are not always perceived by human subjects, despite heavy modulations of afferent activity. At low frequencies (approximately 5 Hz or lower), the firing of slowly adapting units may be deeply modulated by sinusoidal indentations with amplitudes well below perceptive threshold in humans (i.e., the changing indentation is felt as stationary [Talbot et al., 1968; Konietzny and Hensel, 1977]). These observations suggest a poor time-resolution of the central somatosensory system connected to the Merkel units. A related characteristic has been demonstrated by microstimulation. When a single Merkel unit is stimulated, the subject does not perceive any sensation whatever until a series of impulses is injected (Ochoa and Torebjörk, 1983; Vallbo et al., 1984). Hence, considerable temporal summation is required to produce a sensation through this system, indicating a striking contrast to the Meissner system, where single impulses may be detected in psychophysical tests. It would be interesting to explore the responses of cortical neurons when liminal discharges for perceptive detection are launched through primary afferents of the two systems, in order to assess whether the activity differs at levels presumably close to where sensations are produced (cf. Libet, 1973).

Spatial resolution of tactile stimuli

Spatial Discrimination There is general agreement that spatial acuity is high in the glabrous skin of the hand, particularly in the fingertips (Weber, 1835; Weinstein, 1968), although this conclusion is based largely on a much-debated method, the two-point discrimination test (Johnson and Phillips, 1981). Several physiological findings suggest that spatial acuity in the hand depends on type I (Meissner and Merkel) rather than type II (Ruffini and Pacini) units. The receptive fields of the former are small and have distinct borders in contrast to the latter, which have large fields with gently changing sensitivity profiles over an area that may cover several square centimeters (Johansson, 1978). Moreover, it has been demonstrated that variations in density of type I units between subregions of the hand correlate fairly well with the two-point limen, whereas the density of type II units is more uniform over all the glabrous skin area (Vallbo and Johansson, 1978). In the fingertips, where spatial resolution is maximal, there are approximately 200 type I units per square centimeter, whereas in the main parts of the finger and in the palm there are no more than approximately 65 and 30 type I units per square centimeter, respectively (Johansson and Vallbo, 1980). Thus, the type I units constitute a fine-grain system of small overlapping fields that are particularly concentrated in the fingertips, whereas the density gradually falls off in the proximal direction, as does the spatial discrimination capacity.

A fair correspondence also exists between two-point threshold at the fingertip and the diameters of the type I receptive fields. The average two-point limen in this region, as reported in independent studies, clusters at approximately 2 mm (Weber, 1835; Weinstein, 1968; Vallbo and Johansson, 1978). This value corresponds to the diameters of a subset of type I receptive fields, whereas the average diameter is slightly larger (i.e., approximately 2.5–4 mm) (Johansson and Vallbo, 1980; Vallbo et al., 1984; Phillips, Johansson, and Johnson, 1992). Hence, it seems likely that one Meissner or Merkel receptive field would often fall between the two points of the caliper at two-point threshold in the fingertips, where the unit density is high.

To what extent and in which respects these relations are truly determinants of spatial discrimination is far from clear. It remains an open question whether the properties of the peripheral sensory sheet set the spatial acuity. Moreover, it is not clear which neural code is utilized. Also, the two-point limen is an equivocal measure of spatial acuity (Johnson and Phillips, 1981). Finally, the receptive field size is a relative measure that depends, to some extent, on the test stimuli.

A clear difference between the type I and type II systems with regard to spatial resolution has been beautifully demonstrated with the ingenious dot-scanning method (Johnson and Lamb, 1981; see also Johnson, this volume), in which an array of embossed dots is scanned over the receptive field of a single unit while its impulse responses are recorded. When this method was applied to the human finger-pad skin, it was found that the Meissner and Merkel systems both are capable of resolving a dot pattern down to a spacing of approximately 1.5 mm, whereas the corresponding value for the type II units is 3.5 mm (Phillips, Johansson, and Johnson, 1992).

LOCALIZATION *Locognosia*, or the ability to localize a touch stimulus on the skin surface, is another aspect of the spatial capacity of the somatosensory system that has been studied for some time (e.g., Weinstein, 1968). In these studies, a pinpoint touch stimulus usually is delivered and the subject then is asked to indicate on the skin or on a map where he or she felt the stimulus. Obviously, the afferent input for such test stimuli involves a number of units with overlapping receptive fields, at least in the glabrous skin of the hand (Johansson and Vallbo, 1980). Schady, Torebjörk, and Ochoa (1983) were interested in exploring whether the precision in such tests is vitally dependent on the combined input from a number of afferents or whether impulses from a single afferent would be sufficient. When intraneural microstimulation was used to activate single afferents from the glabrous skin of the hand, these investigators found that subjects' performance in the locognosia test was similar to that with an actual touch stimulus, suggesting that one axon may carry sufficient information for optimal localization of tactile stimuli (Schady, Torebjörk, and Ochoa, 1983). However, it seems reasonable to question whether the subject's performance is dependent on sensory mechanisms exclusively, because the subject is asked to perform a complex behavioral response requiring the engagement of memory as well as motor processes.

More precise localization with microstimulation input is suggested when another method is used that does not engage the complex processes of the standard locognosia test. By this method, subjects are asked to make a direct and immediate comparison between two sensations and to assess whether they are located on the same spot (Vallbo et al., 1984). One of the two sensations is elicited by a localized skin indentation and the other by microstimulation of a single afferent, the time interval between the two being on the order of 1 second. Subjects usually succeeded in determining whether the two sensations were located on exactly the same spot.

Conclusions

The material surveyed in this chapter has largely been collected by methods that connect single-afferent recordings with sensory reports from the subject. It is then possible to link the precision of single-unit activity with subjects' reports of perceptive effects either evoked by sensory input tapped from a peripheral nerve fiber or injected by microstimulation. It cannot be denied, however, that the neural analysis refers to one sample unit out of a number of afferents which together carry the sensory input in many tests. Hence, it may be pertinent to complement single-unit analysis with estimates of population response, which in turn requires an analysis of the entire sensory sheet within the cutaneous test area.

Single-afferent studies on human subjects have been pursued for a couple of decades, and a wealth of problems have been elucidated, although more systematic studies obviously are needed to elaborate issues that have been merely peripherally reported. However, the data extracted thus far are consonant with Mountcastle's (1984a) original proposal that, to a greater extent than was previously believed, "our quantitative relations to the external world are set at the interface between the relevant receptor sheet and the environment," although his "general hypothesis that, as regards sensory functions, the brain operates in a linear manner" (Mountcastle, 1967) has not been fully supported.

Analyses of the somatosensory system using afferent impulses as the input signal and psychophysical responses as the output signal have enhanced our knowledge of the functional properties of central mechanisms in sensation. The findings have substantiated the interpretation that separate individual central processing units, including the level where percepts are produced, handle the afferent information from separate types of cutaneous mechanoreceptors. Although the idea of specificity in sensory systems is generally accepted, it has not been possible to demonstrate in animal experiments, where sensation is difficult to investigate, the degree of segmentation of the tactile system in labeled lines producing percepts that conform with the functional properties of the afferent units. On the other hand, the findings obviously agree with data from animal experiments, including those on cellular properties in the somatosensory cortex, where segregation in relation to unit properties exists (Kaas et al., 1979; Mountcastle, 1984b).

Although it is obvious that the tactile system is fractionated into separate subsystems, it should be remembered that several subsystems are simultaneously activated in most natural situations. The principles for interaction between subsystems remain to be explored.

It is a fundamental problem, in analyses of neural activity related to cognition (whether in the behavioral or sensory domains), to identify causal relationships as opposed to purely coincidental ones. The method of microstimulation has made it possible to explore directly the causal relationships between afferent activity and percepts of human subjects. It seems worthwhile to explore with more distinct methods the remarkable precision of sensory attributes reported by subjects on activation of single afferent units.

ACKNOWLEDGMENTS This study was supported by grant 3548 from the Swedish Medical Research Council.

REFERENCES

Calancie, B. M., and R. B. Stein, 1988. Microneurography for the recording and selective stimulation of afferents: An assessment. *Muscle Nerve* 11:638–644.

Cavonius, C. R., and R. M. Chapman, 1974. A possible basis for individual differences in magnitude-estimation behaviour. *Br. J. Psychol.* 65:85–91.

Chambers, M. R., K. H. Andres, M. von Duering, and A. Iggo, 1972. The structure and function of the slowly adapting type II mechanoreceptor in hairy skin. *Q. J. Exp. Physiol.* 57:417–445.

Corso, J. F., 1963. A theoretico-historical review of the threshold concept. *Psychol. Bull.* 60:356–370.

Douglas, W. W., and J. M. Ritchie, 1962. Mammalian nonmyelinated nerve fibers. *Physiol. Rev.* 42:297–334.

Edin, B. B., 1992. Quantitative analysis of static strain sensitivity in human mechanoreceptors from hairy skin. *J. Neurophysiol.* 67:1105–1113.

Edin, B. B., and J. H. Abbs, 1991. Finger movement responses of cutaneous mechanoreceptors in the dorsal skin of the human hand. *J. Neurophysiol.* 65:657–670.

Gandevia, S. C., D. I. McCloskey, and D. Burke, 1992. Kinaesthetic signals and muscle contraction. *Trends Neurosci.* 15:62–65.

Gordon, A. M., H. Forssberg, R. S. Johansson, and G. Westling, 1991. The integration of haptically acquired size information in the programming of precision grip. *Exp. Brain Res.* 83:483–488.

Green, D. M., and J. A. Swets, 1966. *Signal Detection Theory and Psychophysics.* New York: Wiley.

Gybels, J., and J. van Hees, 1972. Unit activity from mechanoreceptors in human peripheral nerve during intensity discrimination of touch. *Excerpta Medica* 253:198–206.

Hallin, R. G., H. E. Torebjörk, and Z. Wiesenfeld, 1982. Nociceptors and warm receptors innervated by C fibres in human skin. *J. Neurol. Neurosurg. Psychiatry* 45:313–319.

Harrington, T., and M. M. Merzenich, 1970. Neural coding in the sense of touch: Human sensations of skin indentation compared with the responses of slowly adapting mechanoreceptive afferents innervating the hairy skin of monkeys. *Exp. Brain Res.* 10:251–264.

Hulliger, M., E. Nordh, A.-E. Thelin, and Å. B. Vallbo, 1979. The responses of afferent fibres from the glabrous skin of the hand during voluntary finger movements in man. *J. Physiol. (Lond.)* 291:233–249.

Iggo, A., and A. R. Muir, 1969. The structure and function of a slowly adapting touch corpuscle in hairy skin. *J. Physiol. (Lond.)* 200:763–796.

Järvilehto, T., H. Hämäläinen, and P. Laurinen, 1976. Characteristics of single mechanoreceptive fibres innervating hairy skin of the human hand. *Exp. Brain Res.* 25: 45–61.

Johansson, R. S., 1978. Tactile sensibility in the human hand: Receptive field characteristics of mechanoreceptive units in the glabrous skin area. *J. Physiol. (Lond.)* 281: 101–123.

Johansson, R. S., and K. J. Cole, 1992. Sensory-motor coordination during grasping and manipulative tasks. *Curr. Opin. Neurobiol.* 2:815–823.

Johansson, R. S., U. Landström, and R. Lundström, 1982. Responses of mechanoreceptive units in the glabrous skin of the human hand to sinusoidal skin displacements. *Brain Res.* 244:17–25.

Johansson, R., and Å. B. Vallbo, 1976. Skin mechanoreceptors in the human hand. An inference of some population properties. In *Sensory Functions of the Skin*, Y. Zotterman, ed. Oxford: Pergamon, pp. 171–184.

Johansson, R. S., and Å. B. Vallbo, 1979a. Tactile sensibility in the human hand: Relative and absolute densities of four types of mechanoreceptive units in glabrous skin. *J. Physiol. (Lond.)* 286:283–300.

Johansson, R. S., and Å. B. Vallbo, 1979b. Detection of tactile stimuli. Thresholds of afferent units related to psychophysical thresholds in the human hand. *J. Physiol. (Lond.)* 297:405–422.

Johansson, R. S., and Å. B. Vallbo, 1980. Spatial properties of the population of mechanoreceptive units in the glabrous skin of the human hand. *Brain Res.* 184:353–366.

Johansson, R. S., and Å. B. Vallbo, 1983. Tactile sensory coding in the glabrous skin of the human hand. *Trends Neurosci.* 6:27–32.

Johansson, R. S., and G. Westling, 1984. Roles of glabrous skin receptors and sensorimotor memory in automatic control of precision grip when lifting rougher or more slippery objects. *Exp. Brain Res.* 56:550–564.

Johansson, R. S., and G. Westling, 1987. Signals in tactile afferents from the fingers eliciting adaptive motor responses during precision grip. *Exp. Brain Res.* 66:141–154.

Johnson, K. O., and G. D. Lamb, 1981. Neural mechanisms of spatial tactile discrimination: Neural patterns evoked by Braille-like dot patterns in the monkey. *J. Physiol. (Lond.)* 310:117–144.

Johnson, K. O., and J. R. Phillips, 1981. Tactile spatial resolution: I. Two-point discrimination, gap detection, grating resolution, and letter recognition. *J. Neurophysiol.* 46:1177–1191.

KAAS, J. H., R. J. NELSON, M. SUR, C.-S. LIN, and M. M. MERZENICH, 1979. Multiple representations of the boy within the primary somatosensory cortex of primates. *Science*. 204:251–253.

KNIBESTÖL, M., and Å. B. VALLBO, 1980. Intensity of sensation related to activity of slowly adapting mechanoreceptive units in the human hand. *J. Physiol. (Lond.)* 300:251–267.

KONIETZNY, F., 1984. Peripheral neural correlates of temperature sensations in man. *Hum. Neurobiol.* 3:21–32.

KONIETZNY, F., and H. HENSEL, 1977. The dynamic response of warm units in human skin nerves. *Pflugers Arch.* 370: 111–114.

LAMOTTE, R. H., L. E. R. LUNDBERG, and H. E. TOREBJÖRK, 1992. Pain, hyperalgesia and activity in nociceptive C units in humans after intradermal injection of capsaicin. *J. Physiol. (Lond.)* 448:749–764.

LIBET, B., 1973. Electrical stimulation of cortex in human subjects, and conscious sensory aspects. In *Handbook of Sensory Physiology*, A. Iggo, ed. Berlin: Springer, pp. 743–790.

MERZENICH, M. M., and T. HARRINGTON, 1969. The sense of flutter-vibration evoked by stimulation of the hairy skin of primates: Comparison of human sensory capacity with the responses of mechanoreceptive afferents innervating the hairy skin of monkeys. *Exp. Brain Res.* 9:236–260.

MOUNTCASTLE, V. B., 1967. The problem of sensing and the neural coding of sensory events. In *The Neurosciences*, G. C. Quarton, T. Melnechuck, and F. O. Schmitt, eds. New York: Rockefeller University Press, pp. 393–408.

MOUNTCASTLE, V. B., 1984a. Neural mechanisms in somesthesis: Recent progress and future problems. In *Somatosensory Mechanisms*, C. von Euler, O. Franzén, U. Lindblom, and D. Ottoson, eds. London: Macmillan, pp. 3–16.

MOUNTCASTLE, V. B., 1984b. Central nervous mechanisms in mechanoreceptive sensibility. In *Handbook of Physiology: Sec. 1. The Nervous System: Vol. 3. Sensory Processes*, I. Darian-Smith, ed. Bethesda, Md.: American Physiological Society, pp. 789–878.

MOUNTCASTLE, V. B., R. H. LAMOTTE, and G. CARLI, 1972. Detection thresholds for stimuli in humans and monkeys: Comparison with threshold events in mechanoreceptive afferent nerve fibers innervating the monkey hand. *J. Neurophysiol.* 35:122–136.

MOUNTCASTLE, V. B., W. H. TALBOT, and H. H. KORNHUBER, 1966. The neural transformation of mechanical stimuli delivered to the monkey's hand. In *Touch, Heat and Pain (Ciba Foundation Symposium)*, A. V. S. De Reuck and J. Knight, eds. London: Churchill Livingstone, pp. 325–351.

MOUNTCASTLE, V. B., W. H. TALBOT, H. SAKATA, and J. HYVÄRINEN, 1969. Cortical neuronal mechanisms in flutter-vibration studied in unanaesthetized monkeys. Neuronal periodicity and frequency discrimination. *J. Neurophysiol.* 32:452–484.

NORDIN, M., 1990. Low-threshold mechanoreceptive and nociceptive units with unmyelinated (C) fibres in the human supraorbital nerve. *J. Physiol. (Lond.)* 426:229–240.

OCHOA, J., and E. TOREBJÖRK, 1983. Sensations evoked by intraneural microstimulation of single mechanoreceptor units innervating the human hand. *J. Physiol. (Lond.)* 42: 633–654.

OCHOA, J., and E. TOREBJÖRK, 1989. Sensations evoked by intraneural microstimulation of C nociceptor fibres in human skin nerves. *J. Physiol. (Lond.)* 415:583–599.

PHILLIPS, J. R., R. S. JOHANSSON, and K. O. JOHNSON, 1992. Responses of human mechanoreceptive afferents to embossed dot arrays scanned across fingerpad skin. *J. Neurosci.* 12:827–839.

PIAGET, J., 1976. *The Grasp of Consciousness*. Cambridge: Harvard University Press.

SCHADY, W. J. L., H. E. TOREBJÖRK, and J. L. OCHOA, 1983. Cerebral localisation function from the input of single mechanoreceptive units in man. *Acta Physiol. Scand.* 119:277–285.

STEVENS, S. S., 1957. On the psychophysical law. *Psychol. Rev.* 64:153–181.

STEVENS, S. S., 1967. Intensity functions in sensory systems. *Int. J. Neurol.* 6(2):202–209.

SWETS, J. A., 1961. Is there a sensory threshold? *Science* 134: 168–177.

TALBOT, W. H., I. DARIAN-SMITH, H. H. KORNHUBER, and V. B. MOUNTCASTLE, 1968. The sense of flutter-vibration: Comparison of the human capacity with response patterns of mechanoreceptive afferents from the monkey hand. *J. Neurophysiol.* 31:301–334.

TOREBJÖRK, H. E., Å. B. VALLBO, and J. L. OCHOA, 1987. Intraneural microstimulation in man. Its relation to specificity of tactile sensations. *Brain* 110:1509–1529.

VALLBO, Å. B., and K.-E. HAGBARTH, 1968. Activity from skin mechanoreceptors recorded percutaneously in awake human subjects. *Exp. Neurol.* 21:270–289.

VALLBO, Å. B., K. E. HAGBARTH, H. E. TOREBJÖRK, and B. G. WALLIN, 1979. Somatosensory, proprioceptive, and sympathetic activity in human peripheral nerves. *Physiol. Rev.* 59:919–957.

VALLBO, Å. B., and R. S. JOHANSSON, 1976. Skin mechanoreceptors in the human hand: Neural and psychophysical thresholds. In *Sensory Functions of the Skin*, Y. Zotterman, ed. Oxford: Pergamon, pp. 185–199.

VALLBO, Å. B., and R. S. JOHANSSON, 1978. The tactile sensory innervation of the glabrous skin of the human hand. In *Active Touch: The Mechanisms of Recognition of Objects by Manipulation*, G. Gordon, ed. Oxford: Pergamon, pp. 29–54.

VALLBO, Å. B., and R. S. JOHANSSON, 1984. Properties of cutaneous mechanoreceptors in the human hand related to touch sensation. *Hum. Neurobiol.* 3:3–14.

VALLBO, Å. B., H. OLAUSSON, J. WESSBERG, and U. NORRSELL, 1993. A system of unmyelinated afferents for innocuous mechanoreception in the human skin. *Brain Res.* 628: 301–304.

VALLBO, Å. B., K. Å. OLSSON, K. G. WESTBERG, and F. J. CLARK, 1984. Microstimulation of single tactile afferents from the human hand. Sensory attributes related to unit type and properties of receptive fields. *Brain* 107:727–749.

VERRILLO, R. T., 1963. Effect of contactor area on the

vibrotactile sensitivity. *J. Acoust. Soc. Am.* 35:1962–1966.

Verrillo, R. T., 1965. Temporal summation in vibrotactile sensitivity. *J. Acoust. Soc. Am.* 37:843–846.

Wall, P. D., 1971. Somatosensory mechanisms. In *Handbook of Electroencephalography and Clinical Neurophysiology*, A. Rémond, ed. Amsterdam: Elsevier, pp. 9-1–9-6.

Wall, P. D., and S. B. MacMahon, 1985. Microneurography and its relation to perceived sensation: A critical review. *Pain* 21:209–229.

Weber, E. H., 1835. Ueber den Tastsinn. *Arch. Anat. Physiol. Wissenschaft. Med.* pp. 152–159.

Weinstein, S., 1968. Intensive and extensive aspects of tactile sensitivity as a function of body part, sex, and laterality. In *The Skin Senses*, D. R. Kenshalo, ed. Springfield: Charles C Thomas, pp. 193–218.

Werner, G., and V. B. Mountcastle, 1965. Neural activity in mechanoreceptive cutaneous afferents: Stimulus-response relations, Weber functions, and information transmission. *J. Neurophysiol.* 28:359–397.

Westling, G., R. Johansson, and Å. B. Vallbo, 1976. A method for mechanical stimulation of skin receptors. In *Sensory Functions of the Skin*, Y. Zotterman, ed. Oxford: Pergamon, pp. 151–158.

15 Neural Mechanisms of Tactile Form Recognition

KENNETH O. JOHNSON, STEVEN S. HSIAO, AND I. ALEXANDER TWOMBLY

ABSTRACT Psychophysical and neurophysiological studies have identified the SAI system as the neural system responsible for tactile form recognition and have provided a partial picture of the underlying neural mechanisms. By SAI system, we mean the primary slowly adapting type I (SAI) afferent population and all of the central mechanisms that convey its signals to memory and perception. Human performance in form recognition experiments is near the acuity limit imposed by the spacing of primary SAI receptor units and is affected little by the mode and manner of contact between the skin and the stimulus. The SAI transduction mechanisms produce an isomorphic neural image that is selectively sensitive to discontinuities and curvature in spatial form. The responses of slowly adapting neurons in SI and SII cortex are heterogeneous and nonisomorphic, which implies that a progressive, stepwise transformation to a nonisomorphic representation of form is effected in the central pathways leading to form recognition. The use of linear and nonlinear network models for analyzing the observed cortical neuronal responses to scanned spatial stimuli is discussed.

This chapter concerns our study of tactile form recognition and the theoretical framework that surrounds it. In the somatosensory system, as in vision, spatial form is transduced by a two-dimensional sheet of receptors, which produces an isomorphic neural image of spatial form (Phillips, Johnson, and Hsiao, 1988). Form recognition implies the existence of stored neural representations of previously encountered forms and a mechanism that matches the neural representation of the current form with the stored representations. One possibility is that the matching process is based on isomorphic images such as those produced by the receptor sheet, but the experimental data and practical considerations imply that this is not so. Electrophysiological recordings from single neurons in the central somatosensory pathways show a progressive enlargement of receptive fields, loss of specificity for skin locus, and the emergence of sensitivity to complex stimulus features; that is, the isomorphic neural representation of tactile stimuli found in the periphery is transformed gradually to some different form of representation (Phillips, Johnson, and Hsiao, 1988). Among the practical problems that confound mechanisms based on isomorphic pattern matching are variations in size, position, and orientation between successive presentations of the same and similar objects. It appears that through evolution there has emerged a transformation that segregates the constant features of form from those that vary from one presentation to the next, a transformation that produces a final sensory representation suitable for direct matching with stored representations of form, even if those representations were derived from stimuli at different loci or different modalities (see later). The objective of the research described here is to understand this transformation and the high-level representations on which tactile form recognition is based.

Among the four low-threshold afferent fiber populations that innervate the skin of the hand, only the slowly adapting type I (SAI) afferent fiber population responds with sufficient acuity to account for human performance in form recognition tasks (see Johnson and Hsiao, 1992). SAI afferent fibers provide an acute, high-contrast neural image of tactile stimuli with an innervation density of approximately one fiber per square millimeter of skin. The tactile form-processing system, as we envision it, comprises a single finger pad, the SAI fibers innervating it, and all the central nervous system (CNS) pathways that convey information about form to memory and the mechanisms of form recognition. At subcortical levels, the critical junctions include the nucleus cuneatus of the dorsal column nuclei and the nucleus ventralis posterolateralis of the thalamus.

KENNETH O. JOHNSON, STEVEN S. HSIAO, and I. ALEXANDER TWOMBLY Philip Bard Laboratories of Neurophysiology, Department of Neuroscience, The Johns Hopkins University School of Medicine, Baltimore, Md.

In the cerebral cortex, the path is not known with certainty, but ablation studies and studies of cortical magnification suggest that area 3b and SII cortex are critical cortical junctions (Randolph and Semmes, 1974; Sur, Merzenich, and Kaas, 1980; Burton, 1986; Pons et al., 1987). Both anatomical and neurophysiological data show that each of these synaptic junctions effects a partial transformation that results cumulatively in a major transformation of the primary isomorphic representation of form (Bankman, Johnson, and Hsiao, 1990).

Primary neural representation of form

A wide range of stimuli has been used in psychophysical and neurophysiological studies of tactile form recognition, including gratings, letters, Braille patterns, punctate patterns of various kinds, and textured surfaces (Craig, 1979; Johnson and Phillips, 1981; Loomis, 1985; Heller, 1987; Gardner and Palmer, 1989; Phillips, Johansson, and Johnson, 1990). The responses of the four human primary afferent fiber types to Braille dot patterns scanned across their receptive fields are illustrated in figure 15.1. Only the SAI and RA (cutaneous rapidly adapting) afferent fibers innervate the skin with sufficient density to provide a basis for form recognition. The responses of SAI and RA afferents are homogeneous within their respective classes; thus, plots such as the one illustrated in figure 15.1 are approximations of the spatially distributed SAI and RA neural population responses.

Evidence that the SAI System is Responsible for Form Recognition SAI responses to spatially complex stimuli (see figure 15.1) provide the CNS with a high-resolution isomorphic neural image of stimulus form. The very low spatial acuity exhibited by slowly adapting type II (SAII) and pacinian afferents in figure 15.1 is consistent with the deep location of their receptors compared with the SAI and RA receptors; their

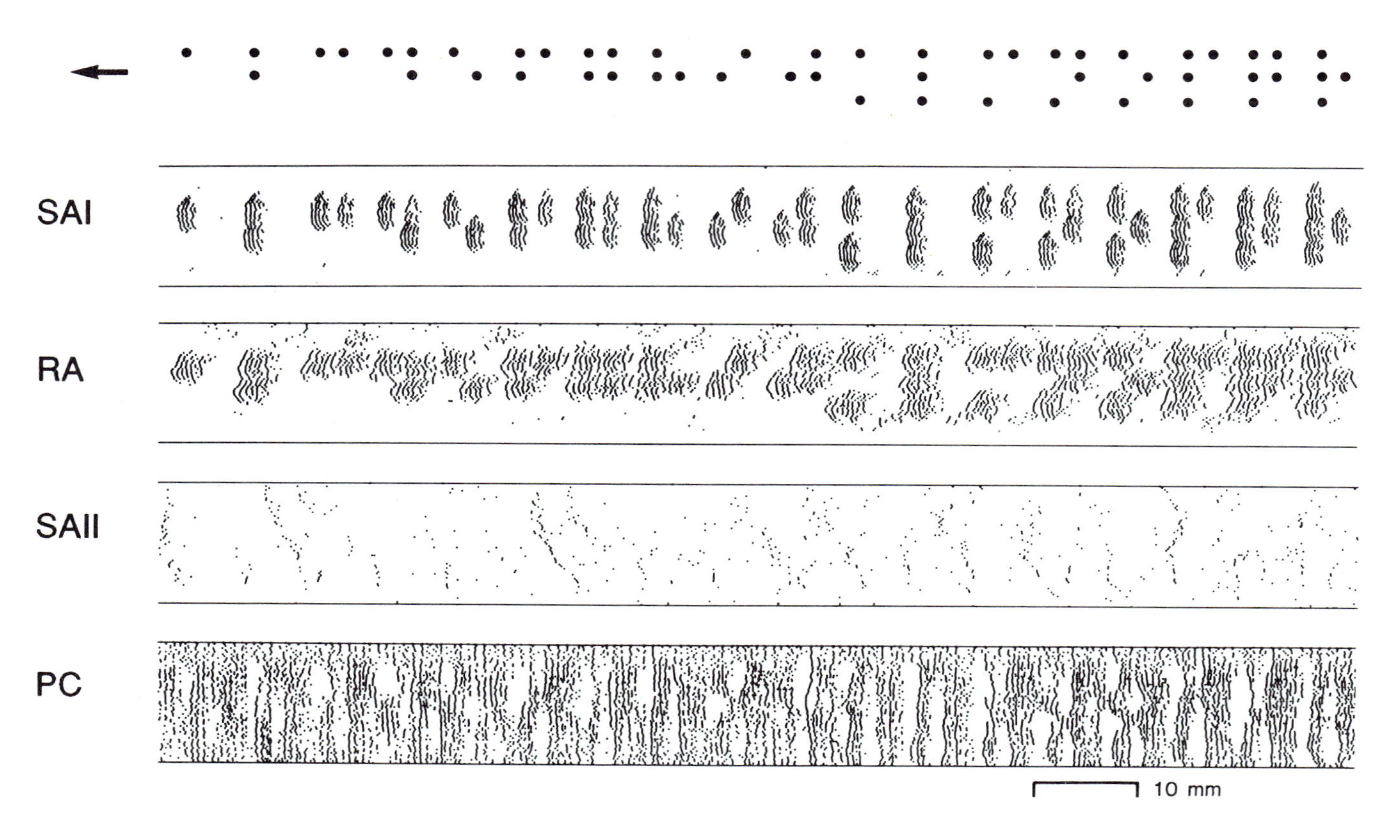

Figure 15.1 Responses of human SAI, RA, SAII, and pacinion (PC) afferent fibers to the Braille patterns *A–R*. The Braille patterns were scanned (60 mm/s) repeatedly over the afferent fibers' receptive fields, which were located on the distal finger pads. After each scan, the patterns were shifted vertically by 200 μm. Each black tick in the bottom four panels represents an action potential evoked by the Braille pattern. (Adapted from Phillips, Johansson, and Johnson, 1990)

insensitivity to spatial form and their sparse innervation densities eliminate them from further consideration in relation to form recognition. The inference that form recognition depends on the neural representation transmitted by the SAI rather than the RA system is based on the close match between SAI responses and human psychophysical performance, on the failure of RA afferent responses to account for human form discrimination behavior, and on the poor spatial resolution exhibited by human subjects when the SAI afferents are not engaged (reviewed in Johnson and Hsiao, 1992). When spatial patterns are pressed into the skin without horizontal movement in a psychophysical task, form discrimination behavior begins to rise above chance levels when the element sizes exceed 0.5 mm and then passes through threshold (performance midway between chance and perfect discrimination) at element sizes of 0.9–1.0 mm (figure 15.2) (Johnson and Phillips, 1981). The responses of SAI and RA afferents to these stationary stimuli are very different. The SAI neural discharge registers surface gaps as small as 0.5 mm (figure 15.3), whereas the RA afferents require gaps of 3.0 mm or more before any modification of firing rate is evident (Phillips and Johnson, 1981a). Thus, there is approximately a 6:1 difference in the spatial acuity of individual SAI and RA afferents. Finally, pattern recognition studies employing the Optacon (a tactile array of 144 vibrating probes), which stimulates RA afferents strongly but not SAI afferents, yield spatial thresholds that are three to four times greater than when SAI afferents are active (Johnson and Hsiao, 1992).

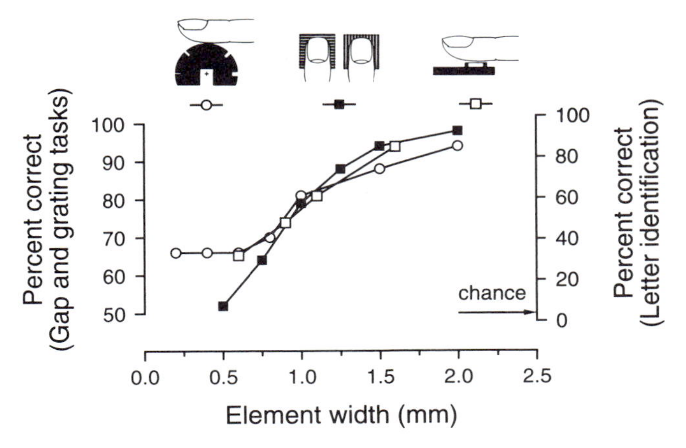

FIGURE 15.2 Human performance in gap detection, grating orientation, and letter recognition tasks. The abscissa represents the fundamental element width for each task, which was gap size for the gap detection task, bar width (half the grating period) for the grating orientation task, and a distance equal to one fifth the letter height for the letter recognition task. (Adapted from Johnson and Phillips, 1981).

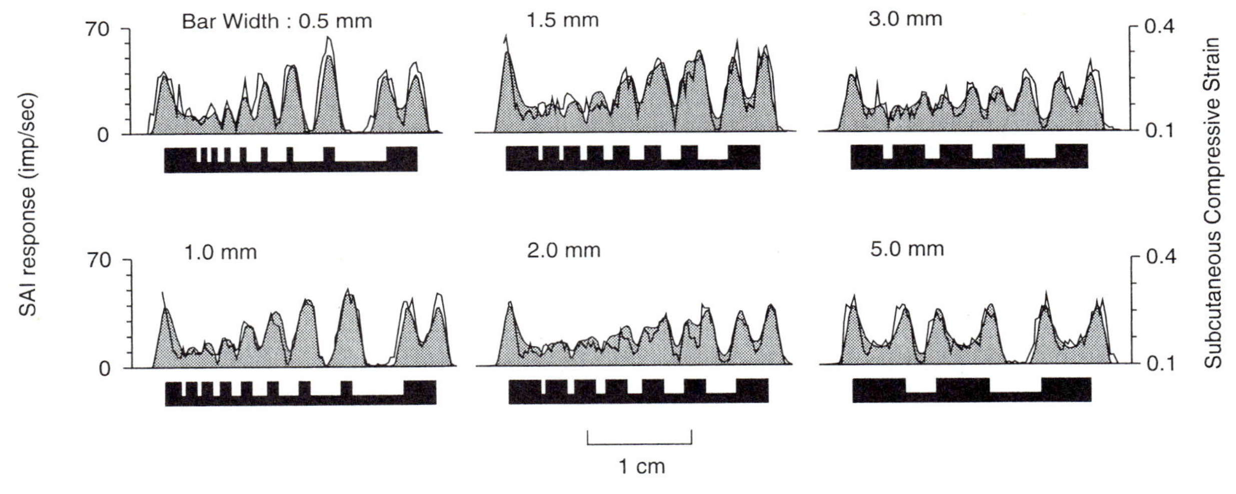

FIGURE 15.3 Responses of a typical SAI afferent fiber to repeated static indentations by six aperiodic mechanical gratings. Below each graph is a cross-sectional outline of the grating used to obtain the neural responses illustrated in the graph. The abscissa of each graph represents the location of the center of the SAI receptive field relative to the grating. The left ordinate and the solid line in each graph represent the mean discharge rate (impulses per second) evoked by the grating. The right ordinate and the shaded graph represent the maximum compressive strain 0.8 mm beneath the surface of a homogeneous, isotropic model of skin. The bar widths are 0.5, 1.5, and 3.0 mm across the top row and 1.0, 2.0, and 5.0 mm across the bottom row. The gap widths vary from 0.5 to 5.0 mm. (Adapted from Phillips and Johnson, 1981b)

Transduction Mechanisms Producing the Primary SAI Representation The SAI transduction mechanisms accomplish more than just translation of the form of the stimulus to an isomorphic neural image. The responses of SAI primary afferent fibers exhibit surround suppression and produce neural images with enhanced representations of discontinuities and gradients in spatial stimuli. Discharge rates evoked by edges, bars, and points are as much as 20 times greater than the responses evoked by uniform surfaces (Johansson, Landstrom, and Lundstrom, 1982; Phillips and Johnson, 1981; Vierck, 1979). The explanation for these responses is found in the complex subcutaneous strain fields produced by even simple mechanical stimuli. Although the cutaneous and subcutaneous tissues are not homogeneous and isotropic, the responses of SAI afferent fibers behave as though they are. Figure 15.3 compares the responses of a typical SAI afferent fiber to six mechanical gratings applied at 870 different locations relative to its receptive field and the maximum compressive strain 800 μm beneath the surface of a continuum mechanical model of skin (Phillips and Johnson, 1981). Srinivasan and Dandekar (1992) have recently suggested that strain energy density, which is closely related to maximum compressive strain, may match the neural data more closely than does maximum compressive strain. Other strain components respond very differently to skin indentation. For example, horizontal tensile strain exhibits little, if any, sensitivity to spatial discontinuities and matches closely the responses of RA afferents to mechanical gratings. Thus, the selective sensitivity to spatial discontinuities appears to be the result of receptor specialization for a particular component of local tissue strain (Phillips and Johnson, 1981a).

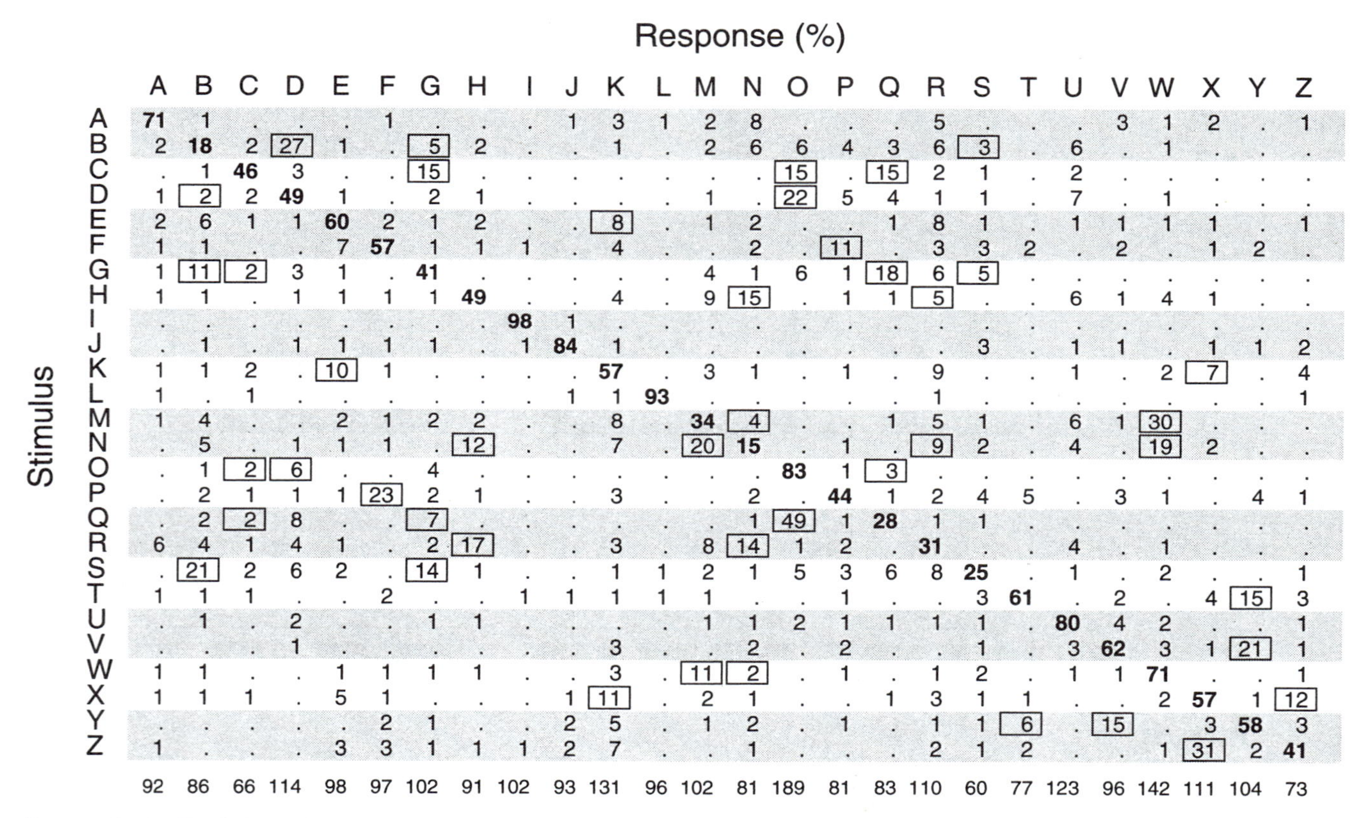

Response (%)

Stimulus	A	B	C	D	E	F	G	H	I	J	K	L	M	N	O	P	Q	R	S	T	U	V	W	X	Y	Z
A	**71**	1	.	.	.	1	.	.	.	1	3	1	2	8	.	.	.	5	.	.	.	3	1	2	.	1
B	2	**18**	2	27	1	.	5	2	.	.	1	.	2	6	6	4	3	6	3	.	6	.	1	.	.	.
C	.	1	**46**	3	.	.	15	.	.	.	.	.	.	.	15	.	15	2	1	.	2	.	.	.	.	.
D	1	2	2	**49**	1	.	2	1	.	.	.	.	1	.	22	5	4	1	1	.	7	.	1	.	.	.
E	2	6	1	1	**60**	2	1	2	.	.	8	.	1	2	.	.	1	8	1	.	1	1	1	1	.	1
F	1	1	.	.	7	**57**	1	1	1	.	4	.	.	2	.	11	.	3	3	2	.	2	.	1	2	.
G	1	11	2	3	1	.	**41**	.	.	.	.	.	4	1	6	1	18	6	5	.	.	.	.	.	.	.
H	1	1	.	1	1	1	1	**49**	.	.	4	.	9	15	.	1	1	5	.	.	6	1	4	1	.	.
I	.	.	.	.	.	.	.	.	**98**	1	.	.	.	.	.	.	.	.	.	.	.	.	.	.	.	.
J	.	1	.	1	1	1	1	.	1	**84**	1	.	.	.	.	.	.	.	3	.	1	1	.	1	1	2
K	1	1	2	.	10	1	.	.	.	.	**57**	.	3	1	.	1	.	9	.	.	1	.	2	7	.	4
L	1	.	1	.	.	.	.	.	.	1	1	**93**	.	.	.	.	.	1	.	.	.	.	.	.	.	1
M	1	4	.	.	2	1	2	2	.	.	8	.	**34**	4	.	.	1	5	1	.	6	1	30	.	.	.
N	.	5	.	1	1	1	.	12	.	.	7	.	20	**15**	.	1	.	9	2	.	4	.	19	2	.	.
O	.	1	2	6	.	.	4	.	.	.	.	.	.	.	**83**	1	3	.	.	.	.	.	.	.	.	.
P	.	2	1	1	1	23	2	1	.	.	3	.	.	2	.	**44**	1	2	4	5	.	3	1	.	4	1
Q	.	2	2	8	.	.	7	.	.	.	.	.	.	1	49	1	**28**	1	1	.	.	.	.	.	.	.
R	6	4	1	4	1	.	2	17	.	.	3	.	8	14	1	2	.	**31**	.	.	4	.	1	.	.	.
S	.	21	2	6	2	.	14	1	.	.	1	1	2	1	5	3	6	8	**25**	.	1	.	2	.	.	1
T	1	1	1	.	.	2	.	.	1	1	1	1	1	.	.	1	.	.	3	**61**	.	2	.	4	15	3
U	.	1	.	2	.	.	1	1	.	.	.	.	1	1	2	1	1	1	1	.	**80**	4	2	.	.	1
V	.	.	.	1	.	.	.	.	.	.	3	.	.	2	.	2	.	.	1	.	3	**62**	3	1	21	1
W	1	1	.	.	1	1	1	1	.	.	3	.	11	2	.	1	.	1	2	.	1	1	**71**	.	.	1
X	1	1	1	.	5	1	.	.	.	1	11	.	2	1	.	.	1	3	1	1	.	.	2	**57**	1	12
Y	.	.	.	.	.	2	1	.	.	2	5	.	1	2	.	1	.	1	1	6	.	15	.	3	**58**	3
Z	1	.	.	.	3	3	1	1	1	2	7	.	.	1	.	.	.	2	1	2	.	.	1	31	2	**41**
	92	86	66	114	98	97	102	91	102	93	131	96	102	81	189	81	83	110	60	77	123	96	142	111	104	73

Figure 15.4 Pooled confusion matrix for 64 subjects performing a letter identification task. Subjects identified Helvetica letters (6.0 mm high, raised 0.5 mm above the background), which are identical in form to the letters along the rows and columns of the matrix. Each entry represents the percentage of trials that a stimulus letter (rows) was identified as the letter within the column (e.g., the letter N was reported on 8% of the trials where the stimulus was the letter A). Boxes identify letter pairs whose combined confusion rates exceeded 16%. (Reprinted from Vega-Bermudez, Johnson, and Hsiao, 1991, by permission.)

Psychophysics of tactile form recognition

The psychophysical task used as the basis for our studies of form processing is letter recognition. Pattern recognition studies with raised letters show that the threshold letter height (height required for 50% correct identifications) is approximately 5.0 mm (see figure 15.2) and that the mode of presentation—stationary or scanned (Phillips, Johnson, and Browne, 1983)—and the subject's experience (Vega-Bermudez, Johnson, and Hsiao, 1991) have a small effect. Subjects' thresholds and even the details of their classification behavior are identical between active and passive contact with the letters. A pooled confusion matrix for 64 subjects identifying scanned letters is illustrated in figure 15.4. Although some details of the confusion matrix are related to the geometrical structure of the letters, many are not. For example, it is predictable that B and D would be confused but not that B would be called D even more frequently than it is called B. That strong bias is not related to some general bias toward the response D because, excluding the responses to B and D themselves, B is reported more frequently than is D. The explanation for this behavior appears to lie in the fact that the primary neural representation of a B looks more like a D than a B (figure 15.5). This example is typical: Half the erroneous responses are confined to 22 of 325 possible confusion pairs (enclosed in boxes in figure 15.4), and all but two of those pairs are strongly biased toward one member of the confusion pair. Quantitative analyses of the confusion matrix suggest that all the subjects' responses are uncontaminated by generalized biases of any kind and are linked instead to the structure of the neural representations (Vega-Bermudez, Johnson, and Hsiao, 1991).

The behavior of naive subjects in the tactile letter recognition task raises intriguing questions about the central mechanisms of tactile pattern recognition. The subjects contributing the data displayed in figure 15.4 had no previous experience in tactile letter identification, were not allowed to touch the letters before testing began, and were told nothing about the validity of their responses during testing. It is difficult to imagine how they performed the task except to suppose that the neural images evoked by the letters were compared to stored neural images based on visual memory. When asked, subjects reported that they were matching the tactile impressions with idealized images based on their visual memory of letters. The explanation offered earlier for the biased confusion between B and D—that B is called D because its neural representation looks like a D—is based on comparison with previously stored visual images of letters. There appear to be only two possible explanations for this behavior. One is that the final sensory representation in the two systems is the same, allowing direct pattern matching between tactile neural images and visual memory. The other is that the two systems effect different transforms and that there is some internal transformation between tactile and visual memory that allows intermodal pattern matching.

Visual and tactile form recognition performance have been compared directly, and these comparisons have provided evidence that the neural mechanisms

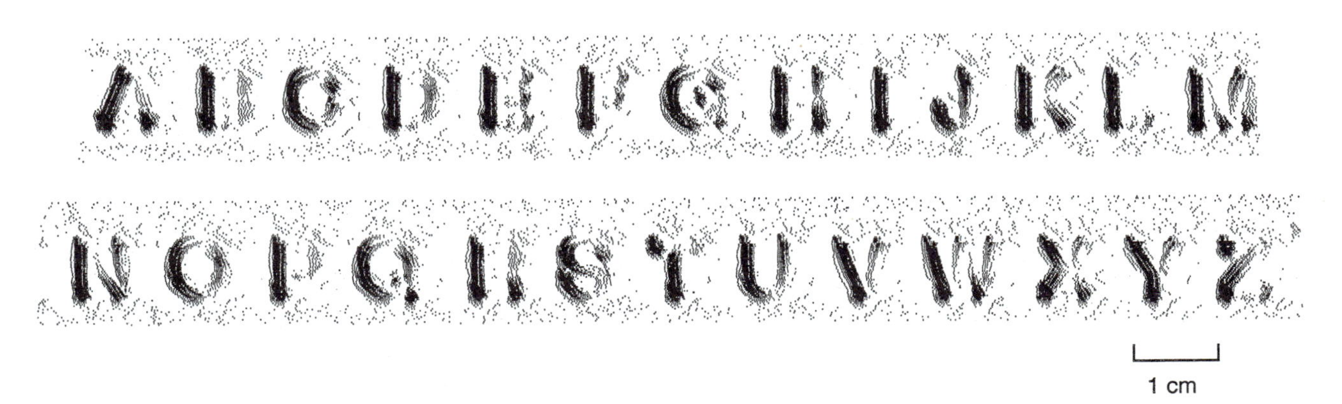

FIGURE 15.5 Response of a typical monkey SAI afferent fiber to the same letters used in the form recognition study illustrated in figure 15.4. Methods are the same as in figure 15.1.

underlying form recognition in the two systems are similar (Phillips, Johnson, and Browne, 1983). When letter heights are expressed as multiples of the primary receptor unit spacings in the two systems (1.0 mm for SAI afferents innervating the finger pad, 0.4 minutes of arc for cones in the fovea) (O'Brien, 1951; Johansson and Vallbo, 1979), subjects' visual and tactile psychometric functions (percent correct versus letter height) are nearly identical. Thus, spatial information at the fovea and at the fingertip appear to be preserved to equal degrees in the two systems. The confusion matrices for letter recognition are similar but not identical.

The correspondence between human and monkey that is assumed in these studies has been tested in a number of ways. Monkeys have been trained in a letter identification task with a subset of the letters used in human studies (Hsiao, O'Shaughnessy, and Johnson, 1993). The monkeys' percent correct for those letters is the same as in humans (Vega-Bermudez, Johnson, and Hsiao, 1991). The neural responses of peripheral afferent fibers to scanned dot patterns (see figure 15.1) have been studied in both species and the results are very similar (Johnson and Lamb, 1981; Connor et al. 1990; Phillips, Johansson and Johnson, 1990; Phillips, Johansson and Johnson, 1992). The glabrous area of human and monkey distal pads are significantly different, being 1000 mm² for humans and 300 mm² for monkeys, but the mean spacing between afferents (approximately 1 mm for both SAI and RA afferents) is similar (Johansson and Vallbo, 1979; Darian-Smith and Kenins, 1980).

Dimensionality of neural representations

We define the dimensionality of a neural representation as the smallest number of variables required to specify all physiologically possible ensembles of neural activity excluding stochastic variation. In this discussion of the neural mechanisms of form recognition, we assume that impulse rate is a sufficient description of the neural activity of individual SAI neurons.

The dimensionality of neural representations is discussed here for two reasons. First, the discussion provides a perspective on the heterogeneity of the cortical neuronal discharge patterns that we observe. Second, dimensionality is a well-defined global measure (like entropy and information) that plays an important role in mathematical theories of transformation and representation. Neural processes operating on a representation can decrease its dimensionality but not increase it. Knowledge of the dimensionalities of the inputs and outputs of a neuronal transformation provides important clues about the transform and its purpose. If the dimensionality of the output is roughly the same as the input, then the transformed representation is most likely an altered view of the original data (i.e., one that reveals relationships obscured in the original view or that allows computations which were prohibitively complex in the original form). This type of transform is complete as no information is lost. If, on the other hand, the dimensionality of the output is greatly reduced, then we are likely to view the transformation as extraction of some kind (e.g., of features, relationships, or regions of the input). The dimensionality of the primary neural representation is measurable. The dimensionality of the final sensory representation is, in principle, measurable as well.

Dimensionality of the Primary Neural Representation The dimensionality of the SAI neural representation is defined as the smallest number of variables required to specify the mean discharge rates evoked by stimuli that indent the skin and remain stationary or move laterally across the skin surface. If it is assumed that the impulse train of a neuron is characterized by its mean rate (one variable), then the dimensionality can, at most, equal the number of neurons in the population. If the spatial acuity of individual neurons is low, the dimensionality may be much less than the number of neurons in the population. The afferent population response can encode spatial form composed of two-dimensional spatial frequencies up to, but not exceeding, the maximum spatial frequency, f, to which individual afferents are sensitive. The number of variables required to specify all such two-dimensional waveforms is $4f^2A$, where A is the area innervated by the afferent population (approximately 1000 and 300 mm² for humans and monkeys, respectively); therefore, the acuity limit on dimensionality is $4f^2A$ (Shannon-Whittaker sampling theorem) (Rosenfeld and Kak, 1982). The dimensionality of the SAI primary representation is not limited by acuity ($f = 1$ cycle per millimeter for SAI afferents). Rather, it is limited by the number of afferent fibers innervating a single finger pad (approximately 700 in humans and 400 in monkeys) (Johansson and Vallbo, 1979; Darian-Smith and Kenins, 1980). Acuity does, however, limit the dimensionality of the RA population response ($f = 0.17$ cycle/mm) to approximately 100 in humans and 40 in monkeys. The number of RA fibers innervating a single finger pad is an order of magnitude greater than the dimensionality of the population response, yielding a redundancy ratio of nearly 10.

DIMENSIONALITY OF THE FINAL SENSORY REPRESENTATION If the transform is linear, and possibly if it is nonlinear, the dimensionality of the final sensory representation is, in principle, measurable through psychophysical experimentation. A result from linear transform theory is that any reduction of dimensionality between the input and output of a transformation is accompanied by the emergence of a null space of equal dimensionality: That is, for every dimension that is lost in transformation, there is introduced in the input space a dimension whose variation, no matter how large, cannot be detected in the output. Conversely, the dimensionality of the final sensory representation is measurable as the dimensionality of the space of all discriminable stimuli.

Regardless of the complications caused by nonlinearities, any substantial loss of dimensionality (e.g., by an order of magnitude) in the representation on which form recognition is based should be demonstrable in psychophysical experiments. The range of spatial forms that have been used in psychophysical experiments yield a consistent result (as discussed earlier), which is that discrimination performance begins to break down when the element density (gaps, bars, points) exceeds one per square millimeter. Those experiments provide no evidence of reduction of dimensionality between the periphery and the final sensory representation on which discrimination is based. However, the test patterns used thus far have spanned only a small fraction of the input space.

Central neural representation of form

The issue of dimensionality bears on the neural responses at each stage of processing leading to form recognition: The number of different neural response functions at each processing stage must be at least as great as the dimensionality of the neural representation at that stage. Furthermore, that dimensionality must be at least as great as the dimensionality of the final neural representation on which recognition is based. When the representation is strictly somatotopic, as in the periphery, the response functions differ only in locus of sensitivity. When the representation is fully distributed relative to a single finger pad, as in SII cortex where the receptive fields are at least as large as a single finger pad (Robinson and Burton, 1980), the response functions must differ in form rather than locus. Thus, if SII cortex is a junction along the pathway to recognition and there is no loss of dimensionality, then there must be at least 700 different response functions representing spatial information from each finger pad. Even if there is an order of magnitude reduction of dimensionality, a large number of different response functions is expected.

NEURONAL RESPONSES IN AREA 3b Among several hundred SI and SII neurons that have been studied with scanned, raised patterns, there is no evidence of response homogeneity or of a small number of response classes. All the cortical neuronal responses illustrated in this chapter were obtained using 8-mm-high letters A–Z followed by a single 0.5-mm dot, all raised 0.5 mm above the background (Johnson and Phillips, 1988). Two examples of responses from slowly adapting neurons in area 3b are illustrated in figure 15.6. The impulse rasters in figure 15.6 without superimposed stimulus patterns illustrate the lack of isomorphism as compared with the peripheral slowly adapting responses (see figure 15.5). The rasters with superimposed stimulus patterns illustrate the precision of the responses in relation to the features of the stimulus patterns. The neuron illustrated in figure 15.6A is unusual among SI neurons in that the responses to the letters appear to be explained by the response to the single raised dot that follows the letter Z. It is evident that the neuron's receptive field contains bands of excitation and inhibition in a 4 × 7-mm region oriented at 30–45° relative to the scanning direction. It is also evident that the receptive field is more extensive than the 4 × 7-mm core region; the vertical lines above and below the response to the raised dot are not due to lines on the stimulus pattern. The neuron illustrated in figure 15.6B responds in a much more complex way for which we have, as yet, no explanation. The segment chosen for illustration shows one of its interesting features—the relationship between the discharge and the occurrence of internal horizontal bars. Peripheral neurons respond weakly to internal horizontal bars, yet that feature seems to be crucial in these responses. The difference in responses evoked by the B and D suggest that the inter-

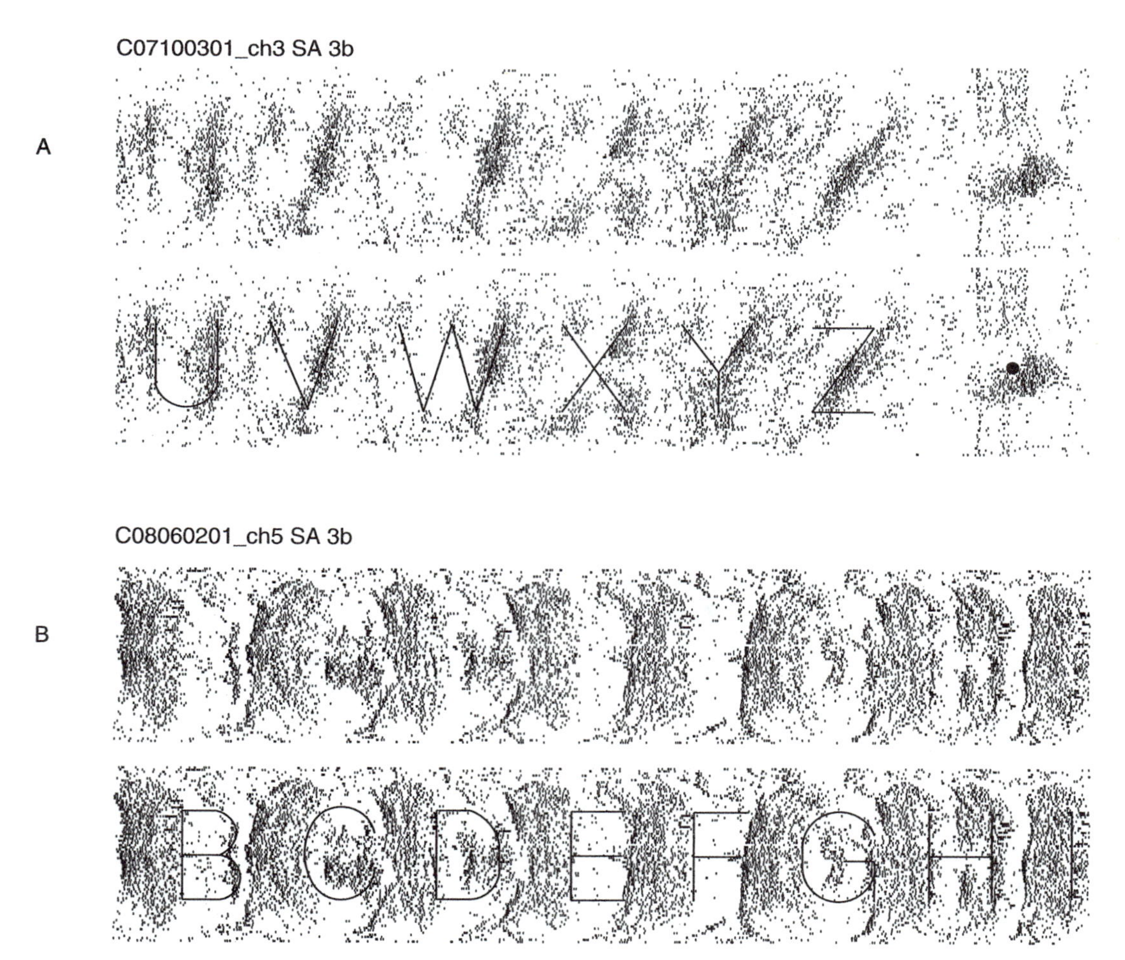

FIGURE 15.6 SI cortical neuronal responses to letters, 8-mm high, scanned at 50 mm/s. Both neurons were recorded from area 3b and were classified as slowly adapting (i.e., they responded to sustained skin indentation with sustained discharges). The bottom set of rasters in panels (A) and (B) illustrates the neuron's response, with the stimulus superimposed, showing how the neuron's activity was related to the spatial features of the stimuli.

nal horizontal bar in the B inhibits the response. That interpretation is corroborated by the responses to the E and F. Likewise, the internal horizontal bar of the G diminishes the response relative to the C. However, the response to the interior of the H is as robust as to the D, showing that the response is not dictated simply by the presence of the internal bar. This neuron yielded no significant response to the scanned dot. More SI neurons are illustrated later in conjunction with attempts to identify the neurons' response functions based on their responses to raised letters.

The two neurons shown in figure 15.6 illustrate the precision that is typical of area 3b slowly adapting neuronal responses as well as two examples of the wide range of response types observed in area 3b (Phillips, Johnson, and Hsiao, 1988; Bankman, Johnson, and Hsiao, 1990). The heterogeneity of observed response properties suggests that the number of response classes in area 3b is large. The nonisomorphic character of the responses and their heterogeneity imply that the isomorphism of the periphery is replaced by a different form of representation in area 3b. The responses observed in area 3b are like those that would be expected at an intermediate processing stage in a chain of stepwise transformations leading from an isomorphic representation, where the response functions differ only in locus of sensitivity, to a fully distributed representation, where the response functions differ in form but not locus (Bankman, Johnson, and Hsiao, 1990).

NEURONAL RESPONSES IN SII CORTEX The responses in SII cortex appear to be sensitive to features over

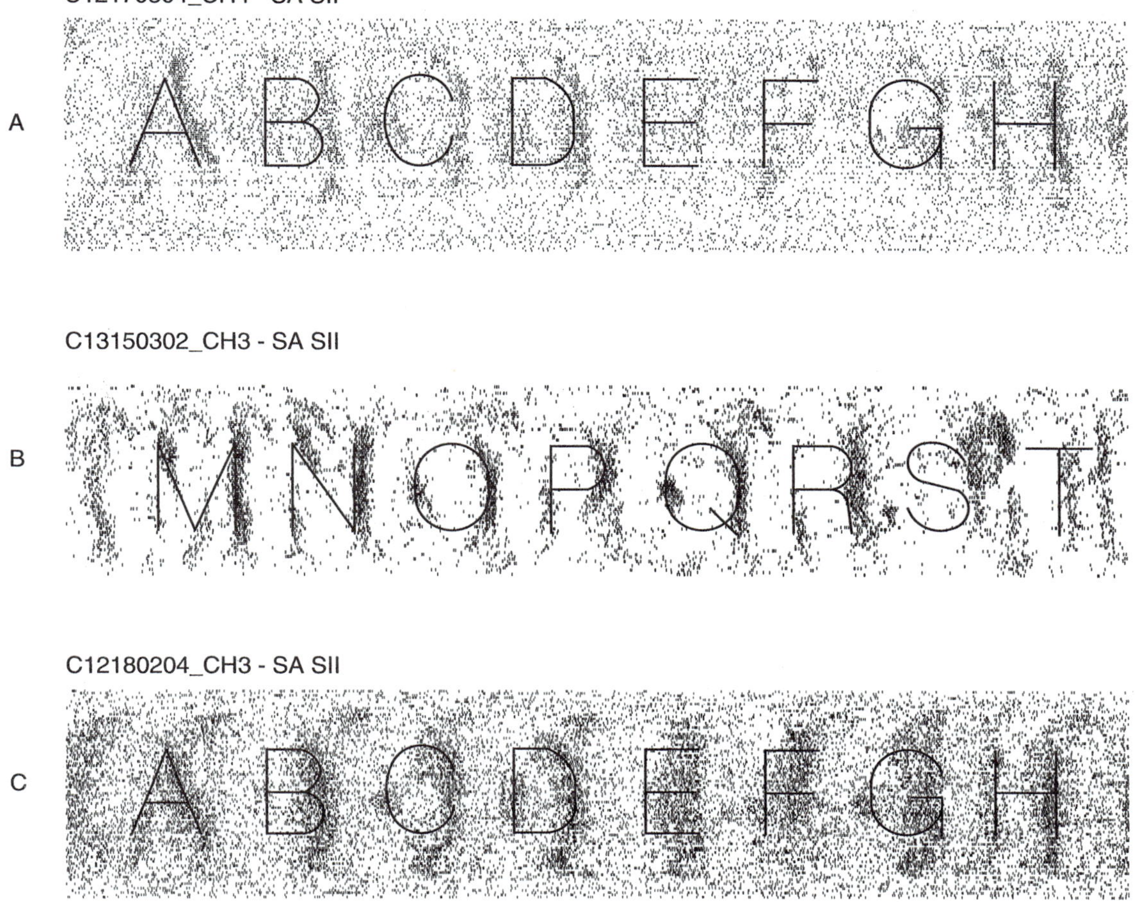

FIGURE 15.7 SII cortical neuronal responses to letters, 8-mm high, scanned at 50 mm/s. All three neurons exhibited slowly adapting type responses.

a larger area than in SI cortex. Three examples are shown in figure 15.7. A feature of all three responses is greater variation in total discharge between letters than is observed in the periphery or SI cortex. The neuron illustrated in figure 15.7A responds vigorously to the trailing edges of the letters A, D, and H, but responds hardly at all to the letter E and minimally to the letter F. Conversely, the neuron illustrated in figure 15.7C responds most vigorously to the letters E, F, and G. Figure 15.7B illustrates a neuron that responds only minimally to the letter T. Another feature of all three responses that is common among SII neurons (though rare among SI neurons) is greater discharge at the trailing than at the leading edges of the letters, even though peripheral afferent fibers all respond more vigorously to leading than to trailing edges (see figure 15.5). Noteworthy also is the intense localized response to leading edges when they are curved (e.g., O and Q) but not when they are straight (see figure 15.7B). All the regions of activity above and below the letters are caused by the letters; the plastic surfaces around the letters are smooth and evoke minimal responses in most peripheral afferents.

Two aspects of the responses illustrated in figures 15.6 and 15.7 warrant further elaboration. First, although the emphasis here is entirely on spatial form, the data illustrate the temporal capacity of the underlying mechanisms and the repeatability of responses between scans, which are separated by 6–16 s depending on the scanning velocity (20–50 mm/s). The spatial detail that is evident, particularly in figure 15.6, would not be possible if the neurons did not respond with very high temporal fidelity and relatively low variability from one sweep to the next. Second, these recordings were all obtained from monkeys performing a visual discrimination task; many of the responses may have been different if the monkey had been performing a tactile pattern recognition task.

Effects of Attention The attention required for letter recognition appears to have a significant effect on the response properties of neurons in SI and particularly SII cortex. Hsiao, O'Shaughnessy, and Johnson (1993) trained a monkey to perform a tactile-visual letter-matching task and a visual dimming-detection task. During neurophysiological recordings from single neurons in SI and SII cortex, raised letters were scanned continuously over the receptive fields of the neurons while the animal was switched periodically between the two tasks. Significant differences between the discharge rates evoked by raised letters in the two tasks were observed in approximately 50% of neurons in SI cortex and 80% of neurons in SII cortex. The effects in SII cortex were divided between increased (58%) and decreased (22%) rates, whereas in SI cortex, only increased rates were produced by attention to the tactile stimuli. The attentional effects were expressed not only as changes in overall discharge rate but, in some cases, as modifications of the form of the responses evoked by the letters, implying that attention was modifying the transformation in some manner (figure 15.8). Whether attentional effects were observed depended on the behavioral relevance of individual letters. During brief periods in the tactile task when a behavioral response could not yield a reward (time-out and reward periods), the neuronal responses were not significantly different from the responses evoked by the same letters during the visual task.

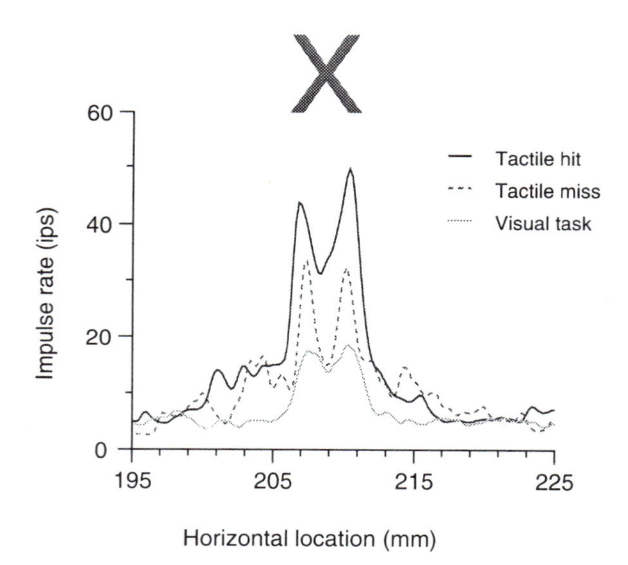

Figure 15.8 Response of a slowly adapting neuron in SII cortex to the letter X (6.0 mm high, 20 mm/s) moving across a monkey's finger while it correctly identified (tactile hit) or failed to identify (tactile miss) the letter or while it was performing a visual task unrelated to the tactile stimulus (visual task). The figure shows that the monkey's focus of attention has a significant effect on the spatial profile of the response. (Adapted from Hsiao, O'Shaughnessy, and Johnson, 1993)

Transformation between the primary and central representations

The responses of SI and SII neurons to scanned letters conform to our expectations of a stepwise transformational process leading from an isomorphic representation to a distributed neural representation. Although it is clear that neurons in SI and SII cortex are responding with high spatiotemporal acuity, the representational rules replacing isomorphism in SI and SII are not inferred easily from responses such as those illustrated in figures 15.6 and 15.7.

The key to understanding any neural representation is the neuronal response functions that are its basis. The methods for inferring the structure of complex response functions from a large sample of input-output data are well established for linear (Ljung, 1987) but not for nonlinear response functions. Among the difficulties presented by the possibility of nonlinearity is that a neuron may respond differently to different classes of stimuli. Because we have chosen letters as the common stimuli in psychophysical and neurophysiological experiments, we have tried to infer the response functions from the discharge evoked by raised letters. Our greatest success has come from the use of artificial, feedforward neural networks trained to match the responses of cortical neurons to scanned stimuli (discussed later). The reasoning is that if one can train a neural network to mimic the responses of a cortical neuron to a wide range of scanned, spatial forms, then that neural network should embody the essence of the neuron's response function (Lehky and Sejnowski, 1988).

The result of our studies, which are ongoing, is that neural networks can be trained to mimic the responses of SI cortical neurons to scanned, raised letters and that provisions for nonlinear spatial interactions are required to match the cortical responses well. However, it has proved difficult to extract the response functions from the networks.

Network Study Methods We have used a variety of feedforward network architectures, ranging from fully connected networks with one level of adjustable weights to locally connected networks with fixed weights and many levels of processing. The first layer in all the networks is a two-dimensional array of nodes into which are fed the firing patterns recorded from actual primary SAI afferent fibers as raised letters are scanned across the skin (Bankman, Johnson, and Hsiao, 1990). In the networks with adjustable connection weights, the final layer contains only a single model neuron whose impulse rate profile is compared to the rate profile of an actual cortical neuron. The internal weights are adjusted by backpropagation to achieve the least mean squared error between the discharge rate of the network's output node and the discharge of a single cortical neuron. The networks are trained on responses to the letters A–M, which typically contain 50,000 discrete data points. The training method is illustrated in figure 15.9. The generality of the result is tested by comparing the network and cortical responses to the letters N–Z.

Figure 15.9 Network training method. The response of a peripheral SAI fiber to the letters A–M [response to the subset A–D shown in bottom row] is fed into a two-dimensional array (6 × 6 mm) representing a subset of peripheral SAI afferents in the finger pad. The activity in the array is propagated through zero, one, or two layers of intermediate nodes (one layer shown) to a single output node. The network is trained using backpropagation to minimize the mean squared error between the model outputs (middle row) and the actual cortical response (top row) to the letters A–M. After training, the neural responses to the letters N–Z are used to test the model's responses to new stimuli (see figure 15.10).

Network Study Results A consistent result from these studies is that networks with more than one level of synaptic weights, which can mimic nonlinear mechanisms, provide better matches to the cortical discharge patterns than do networks with only a single level of connections between the input layer and the output node, which can model only linear spatial integration (i.e., summation of excitatory and inhibitory inputs from different spatial regions). In some cases, networks with a single level of processing cannot match the cortical neuronal responses at all. The results illustrated in figure 15.10 are all based on networks constructed to mimic, in a crude manner, the three levels of synaptic processing that occur at the nucleus cuneatus, the ventrobasal complex of the thalamus, and SI cortex.

Having obtained networks that function like specific SI neurons, at least insofar as scanned letters are concerned, one ought to be able to infer the networks' response functions, but this has proved difficult. These networks contain a very large number of adjustable weights, typically thousands, making it difficult to infer the form of the response functions by simple inspection of the weights. A first-order approximation to the spatial structure of the network's response function is obtained by passing a point stimulus over the network's "skin"; more precisely, the peripheral discharge produced by scanning a point over the skin is fed into the network's input layer. Sample "receptive fields," explored in this way, are shown in figure 15.11. These mixed regions of excitation and inhibition were not observed on manual inspection of the neurons' receptive fields, probably because the neurons were not spontaneously active and the dimensions separating the subregions were small. If the response function from a multilayer network is linear, then the receptive fields shown in figure 15.11D should, if incorporated into a network with one level of connection weights, produce the same neuronal response as the multilevel networks from which they were derived.

The degree to which the linearized network reproduces the cortical responses differs markedly between

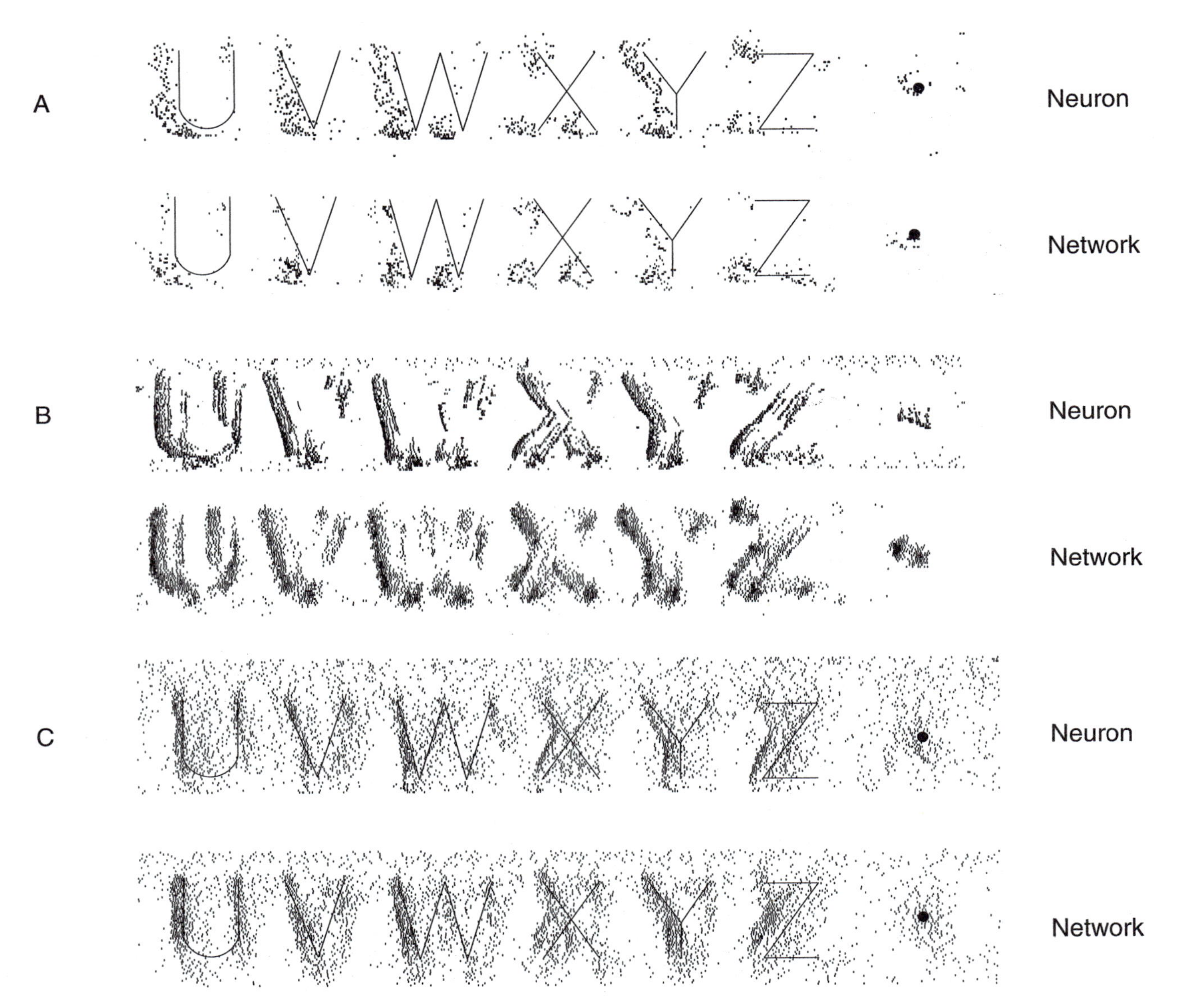

FIGURE 15.10 Responses of three area 3b, slowly adapting neurons to the letters U–Z and a raised dot (top rasters in A, B, and C) compared with the responses predicted from feedforward networks trained to match responses to the letters A–M (bottom rasters in A, B, and C). The networks contained two layers of intermediate nodes between the input array and output node. The fidelity of the networks' responses indicates that they have "captured" the essence of the neurons' response functions.

neurons. The linearized network fails to match the responses of the neuron illustrated in figure 15.11A. Throughout, it inserts neural responses where none exist in the cortical response and fails to respond where the cortical neuron responds vigorously. The backpropagation method was unable to produce a linear network whose responses matched, even roughly, the cortical responses illustrated in figure 15.11A; the best, least-squares fit with a linear network (not shown) consisted of responses only to the bottoms of the letters. Thus, we infer that the mechanisms underlying the responses of this neuron are highly nonlinear. On the other hand, the linearized response illustrated in figure 15.11C is clearly a good first approximation to the cortical response, and thus the receptive field illustrated in figure 15.11D appears to be a good first approximation to the neuron's processing function. The neuron illustrated in figures 15.10B and 15.11B

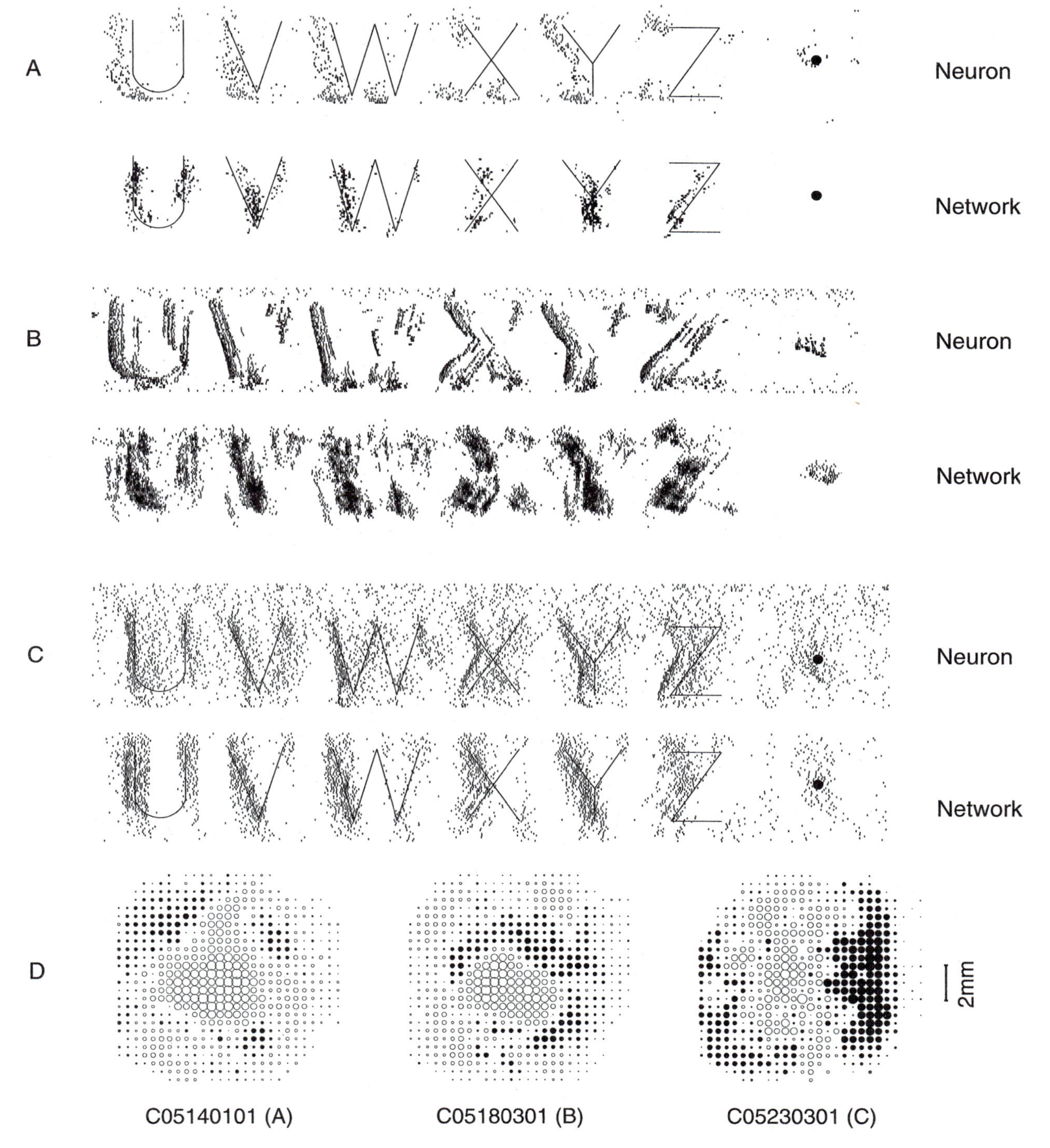

FIGURE 15.11 Responses of the same three area 3b, slowly adapting neurons as in figure 15.10 compared with the responses predicted from linear networks derived from the receptive fields of the nonlinear networks whose responses are illustrated in figure 15.10. The receptive fields illustrated in (D) were derived by probing the three-level networks with the neural responses of an SAI afferent to a raised dot. The circles within the receptive fields represent activation of the network's output unit. Open circles represent positive (excitatory), and closed circles negative (inhibitory), subregions within the receptive field. The linear networks were constructed as single output nodes with firing rates proportional to a weighted sum of primary afferent firing rates. The weights were adjusted to produce the same receptive fields as those for the three-level, nonlinear network.

represents an intermediate case, in which the gross features of the response are retained in the linear approximation.

Value of Network Studies The importance of these studies relates to the importance of the study of cortical neuronal responses to complex stimuli and the paucity of methods for analyzing those responses. These studies suggest that feedforward networks can model even very complex neuronal response functions (see also Lehky, Sejnowski, and Desimone, 1992). If a method can be found for decomposing these networks in a meaningful way, the approach could be valuable in investigating the neural mechanisms underlying form recognition.

Conclusions

Psychophysical and neurophysiological studies have identified the SAI system as the neural system responsible for tactile form recognition, and they have provided a partial picture of the neural mechanisms of form recognition. Psychophysical experiments show that human performance in form recognition experiments is near the limit imposed by the spacing of primary receptor units and is affected little by changes in scanning velocity, contact force, and whether the stimulus is contacted actively or passively. Only the SAI afferents provide sufficient information to account for that psychophysical performance; the RA, pacinian, and SAII systems fail to resolve spatial patterns that are discriminated easily by humans and monkeys. SAI transduction mechanisms produce an isomorphic neural image of spatial form that is selectively sensitive to discontinuities and curvature in spatial form. Neurons in SI and SII cortex, but particularly those with a slowly adapting response, respond to scanned, raised patterns such as those used in psychophysical experiments with high spatial and temporal acuity. The heterogeneity of these responses implies that the peripheral isomorphic representation of form is not maintained in the central pathways and that a progressive, stepwise transformation leads to a nonisomorphic representation of form. Cortical neuronal recordings in monkeys trained to perform a tactile letter recognition task show that the animal's focus of attention has a significant effect on the neural mechanisms underlying this transformation. The key to understanding this transformation is to understand the responses of individual cortical neurons and the role they play in the representation of form. In this chapter, we discuss the use of linear and nonlinear network models to analyze the response functions of single neurons. The most important result of these studies is the capacity of nonlinear networks to model the responses of cortical neurons to complex spatial stimuli. However, to be truly useful, a method for achieving a meaningful decomposition of the networks must be found.

ACKNOWLEDGMENTS The work reported in this chapter was supported by National Institutes of Health grant NS18787 and the W. M. Keck Foundation.

REFERENCES

Bankman, I. N., K. O. Johnson, and S. S. Hsiao, 1990. A neural network model of transformation in the somatosensory system. *Cold Spring Harb. Symp. Quant. Biol.* 55:611–620.

Burton, H., 1986. Second somatosensory cortex and related areas. In *Cerebral Cortex*, vol. 5, E. G. Jones and A. Peters, eds. New York: Plenum, pp. 31–98.

Connor, C. E., S. S. Hsiao, J. R. Phillips, and K. O. Johnson, 1990. Tactile roughness: Neural codes that account for psychophysical magnitude estimates. *J. Neurosci.* 10:3823–3836.

Connor, C. E., and K. O. Johnson, 1992. Neural coding of tactile texture: Comparisons of spatial and temporal mechanisms for roughness perception. *J. Neurosci.* 12: 3414–3426.

Craig, J. C., 1979. A confusion matrix for tactually presented letters. *Percept. Psychophys.* 26:409–411.

Darian-Smith, I., and P. Kenins, 1980. Innervation density of mechanoreceptive fibers supplying glabrous skin of the monkey's index finger. *J. Physiol. (Lond.)* 309:147–155.

Gardner, E. P., and C. I. Palmer, 1989. Simulation of motion on the skin: II. Cutaneous mechanoreceptor coding of the width and texture of bar patterns displaced across the OPTACON. *J. Neurophysiol.* 62:1437–1460.

Heller, M. A., 1987. The effect of orientation on visual and tactual braille recognition. *Perception* 16:291–298.

Hsiao, S. S., D. M. O'Shaughnessy, and K. O. Johnson, 1993. Effects of selective attention of spatial form processing in monkey primary and secondary somatosensory cortex. *J. Neurophysiol.* 70:444–447.

Johansson, R. S., U. Landstrom, and R. Lundstrom, 1982. Sensitivity to edges of mechanoreceptive afferent units innervating the glabrous skin of the human hand. *Brain Res.* 244:27–32.

Johansson, R. S., and A. B. Vallbo, 1979. Tactile sensibility in the human hand: Relative and absolute densities of four types of mechanoreceptive units in glabrous skin. *J. Physiol. (Lond.)* 286:283–300.

Johnson, K. O., and S. S. Hsiao, 1992. Tactual form and texture perception. *Annu. Rev. Neurosci.* 15:227–250.

Johnson, K. O., and G. D. Lamb, 1981. Neural mechanisms of spatial tactile discrimination: Neural patterns evoked by braille-like dot patterns in the monkey. *J. Physiol. (Lond.)* 310:117–144.

Johnson, K. O., and J. R. Phillips, 1981. Tactile spatial resolution: I. Two-point discrimination, gap detection, grating resolution, and letter recognition. *J. Neurophysiol.* 46:1177–1191.

Johnson, K. O., and J. R. Phillips, 1988. A rotating drum stimulator for scanning embossed patterns and textures across the skin. *J. Neurosci. Methods* 22:221–231.

Lehky, S. R., and T. J. Sejnowski, 1988. Network model of shape-from-shading: Neural function arises from both receptive and projective fields. *Nature* 333:452–454.

Lehky, S. R., T. J. Sejnowski, and R. Desimone, 1992. Predicting responses of nonlinear neurons in monkey striate cortex to complex patterns. *J. Neurosci.* 12:3568–3581.

Ljung, L., 1987. *System Identification: Theory for the User.* Englewood Cliffs, N.J.: Prentice Hall.

Loomis, J. M., 1985. Tactile recognition of raised characters: A parametric study. *Bull. Psychon. Soc.* 23:18–20.

O'Brien, B., 1951. Vision and resolution in the central retina. *J. Opt. Soc. Am.* 41:882–894.

Phillips, J. R., R. S. Johansson, and K. O. Johnson, 1990. Representation of Braille characters in human nerve fibers. *Exp. Brain Res.* 81:589–592.

Phillips, J. R., R. S. Johansson, and K. O. Johnson, 1992. Responses of human mechanoreceptive afferents to embossed dot arrays scanned across fingerpad skin. *J. Neurosci.* 12:827–839.

Phillips, J. R., and K. O. Johnson, 1981a. Tactile spatial resolution: II. Neural representation of bars, edges, and gratings in monkey afferents. *J. Neurophysiol.* 46:1192–1203.

Phillips, J. R., and K. O. Johnson, 1981b. Tactile spatial resolution: III. A continuum mechanics model of skin predicting mechanoreceptor responses to bars, edges, and gratings. *J. Neurophysiol.* 46:1204–1225.

Phillips, J. R., K. O. Johnson, and H. M. Browne, 1983. A comparison of visual and two modes of tactual letter resolution. *Percept. Psychophys.* 34:243–249.

Phillips, J. R., K. O. Johnson, and S. S. Hsiao, 1988. Spatial pattern representation and transformation in monkey somatosensory cortex. *Proc. Natl. Acad. Sci. U.S.A.* 85:1317–1321.

Pons, T. P., P. E. Garraghty, D. P. Friedman, and M. Mishkin, 1987. Physiological evidence for serial processing in somatosensory cortex. *Science* 237:417–420.

Randolph, M., and J. Semmes, 1974. Behavioral consequences of selective subtotal ablations in the postcentral gyrus of *Macaca mulatta*. *Brain Res.* 70:55–70.

Robinson, C. J., and H. Burton, 1980. Somatotopographic organization in the second somatosensory area of *M. fascicularis*. *J. Comp. Neurol.* 192:43–67.

Rosenfeld, A., and A. C. Kak, 1982. *Digital Picture Processing*, vol. 1, ed. 2. Orlando: Academic Press.

Srinivasan, M. A., and K. Dandekar, 1992. Role of mechanics in cutaneous mechanoreceptor responses. *Soc. Neurosci. Abstr.* 17:105.

Sur, M., M. M. Merzenich, and J. H. Kaas, 1980. Magnification, receptive-field area, and hypercolumn size in areas 3b and 1 of somatosensory cortex in owl monkeys. *J. Neurophysiol.* 44:295–311.

Vega-Bermudez, F., K. O. Johnson, and S. S. Hsiao, 1991. Human tactile pattern recognition: Active versus passive touch, velocity effects, and patterns of confusion. *J. Neurophysiol.* 65:531–546.

Vierck, C. J., 1979. Comparisons of punctate, edge and surface stimulation of peripheral slowly adapting, cutaneous afferent units of cats. *Brain Res.* 175:155–159.

16 Neural Mechanisms of Auditory Image Formation

MASAKAZU KONISHI

ABSTRACT The goal of central sensory physiology is an explanation of perception. The study of sound localization in the barn owl is used as an example to discuss how systems neurophysiology and perception can be closely integrated. The single-unit approach, which observes the responses of one neuron at a time, relies on the anatomical connections and the stimulus selectivities of neurons for inference about neural processing and encoding. The choice of neuronal selectivities is crucial for the success of any attempts to combine neurophysiology and perception. The top-down approach starts with high-order neurons selective for the same stimulus that releases a particular behavioral response. Study of lower-order neurons can show the flow of information relevant to the selectivity and the algorithm by which the selectivity is created. This strategy has worked in the auditory systems of the barn owl and mustached bat. The results show that their auditory systems process perceptually relevant stimuli in parallel and hierarchically organized networks.

We study sensory systems to explain perception. Conversely, we must refer to specific perceptual tasks to understand the designs of sensory systems. Knowledge of perception is, therefore, indispensable for the study of sensory systems. Few perceptual phenomena can be explained in terms of neuronal activities because few studies integrate perception and neurophysiology. The main problem in shifting from perception to neurophysiology is deciding what behavioral task to choose and what to look for in the brain. The single-unit approach uses neuronal connections and stimulus selectivities as the clues for inference about signal processing and encoding by sensory systems. This approach assumes implicitly that neuronal stimulus selectivities equal the information carried by the neurons.

The concept of processing streams in the visual system of the macaque monkey makes this assumption (Felleman and Van Essen, 1991). Color-sensitive neurons are likely to contribute to the perception of colors. We can also infer how the selectivities for other features such as orientation, direction, and binocular disparity might be involved in the perception of forms, movement, and depth. Such an inference is possible because these features appear to be relevant to the perception of forms, movement, and depth (De Yoe and Van Essen, 1988). This reasoning is, however, merely another implicit assumption.

It often is impossible to deduce the information carried by a neuron. For example, the neurons of the mammalian cochlear nuclei can be classified by their post-stimulus-time histograms such as "chopper" and "primary-like" (Pfeiffer, 1966), but what information is conveyed by these patterns is unknown. The identification of a processing stream requires the study of stimulus selectivities in successive stages of the system, although this procedure does not ensure that a processing stream will be discovered. We may understand how the stimulus selectivities of certain higher-order neurons can be derived from the choppers, but we may not be able to appreciate the relevance of these stimulus selectivities to perception. The concept of processing streams is meaningful only if what is processed refers to particular perceptual tasks.

The best way to avoid pursuing a path of uninterpretable results is to adhere to the fundamental tenet that the brain of an animal is designed for processing stimuli that are important for the animal's survival and reproduction. Bats use echolocation for navigation and prey capture. This behavioral specialization predicted that the auditory system of bats must be designed for the processing of sonar information, a forecast that led investigators to look for cortical cells selective for sonar signals such as neurons sensitive to echo delays (Suga, 1984; see Suga, this volume). The barn owl is another specialized animal that locates prey by listening. This ability suggested the existence of neurons that carry information about sound locations. The owl's inferior

MASAKAZU KONISHI Division of Biology, California Institute of Technology, Pasadena, Calif.

colliculus contains space-specific neurons that respond only to sound coming from a particular direction (Knudsen and Konishi, 1978a,b). In both the bat and the owl, had the investigators not known the perceptual problems the animals must solve, they would not have looked for neurons selective for these natural stimuli. It would seem naive to expect to find single neurons selective for complex biological signals, but such neurons exist, and the study of them has led to an understanding of the computational algorithm and its neural implementation for the perception of the biological signals. In this chapter, the example of the barn owl is used to elaborate the approach just described.

Perception of sound images

Perception of a phantom sound image by listening with two ears is known as *binaural fusion*. This sensation occurs either when the signals for the two ears are correlated or when the combination of the two sounds is meaningful, such as speech sounds (Sayers and Colin, 1957; Liberman and Mattingly, 1985; Stern, Zelberg, and Trahiotis, 1988). The brain uses signals from two sources to create a single image. Binaural fusion is the only human auditory sensation that has been analyzed at both the behavioral and neuronal level in an animal. The barn owl responds to sound delivered through earphones as though it experiences binaural fusion (Moiseff and Konishi, 1981). The owl uses interaural time (ITD) and interaural intensity differences (IID) for localization in azimuth and elevation, respectively (Moiseff, 1989). These binaural cues vary systematically as functions of the direction of sound propagation, such that combinations of ITD and IID define the owl's auditory space. When an owl hears through earphones a binaural signal containing a combination of ITD and IID, it rapidly turns its head as if there were a single sound source in space. ITD varies almost linearly with the azimuthal angle of the sound source. A comparison of ITDs and head-turning angles shows that ITD predicts the azimuthal angle of sound perceived by the owl.

Neural representation of sound images

Combinations of ITD and IID create the images of sound sources. The question for neurophysiological studies is, therefore, how the combinations are represented in the owl's brain. The external nucleus of the inferior colliculus contains neurons that respond only to sound coming from a particular range of direction, which can be expressed as an area on an imaginary globe surrounding the owl's head. These neurons are the space-specific neurons and their preferred spatial areas are called *receptive fields* (Knudsen, Konishi, and Pettigrew, 1977). The space-specific neurons are broadly tuned to frequency, respond only to binaural stimuli, and are selective for combinations of ITD and IID (Moiseff and Konishi, 1981). ITD and IID determine, respectively, the azimuthal and elevational dimensions of a neuron's receptive field (Olsen, Knudsen, and Esterly, 1989). Just as the sound-localizing response requires binaurally correlated signals, so do the space-specific neurons because they do not respond to random noise delivered to the two ears. There is thus a good match between perceptual and neuronal stimulus requirements.

Parallel frequency channels

As in other vertebrate ears, the inner ear of the owl breaks down complex sound into its component frequency bands. The frequency selectivity of a primary auditory fiber is largely determined by the place of its innervation on the basilar membrane. Different frequency channels remain separate throughout the brain stem, and both ITD and IID are initially processed and encoded in each frequency band. The owl uses the rate and timing of impulses to encode stimulus intensity and phase in the frequency band to which the neuron is tuned. Primary auditory fibers fire action potentials at or near a particular phase angle of the stimulus tone. This phenomenon is known as *phase locking* and occurs in all vertebrate ears. The owl's auditory system uses phase-locked spikes for the derivation of ITD.

Simultaneous computations of ITD and IID

The space-specific neurons are separated from the inner ear by several stations, although the number of intervening synapses may be much greater because each station may contain multiple serial connections. The selectivity of space-specific neurons for ITD and IID pairs suggests that the flow of information about ITD and IID can be tracked from lower-order stations to the external nucleus by the use of these binaural cues as stimuli. Such a survey indicates that neurons in all of the lower-order stations respond to monaural stimuli and either to ITD or IID but not to both (Moiseff and Konishi, 1983). It also indicates where neuronal sensitivities to ITD and IID are established for the first time in the ascending pathway.

Anatomical studies show that the ITD-sensitive stations and IID-sensitive stations form two parallel pathways, one originating from the cochlear nucleus magnocellularis and the other from the cochlear nucleus angularis (Takahashi and Konishi, 1988a,b) (figure 16.1). An acid test for an information flow should show that the manipulation of the flow affects the response of the ultimate recipient of the information. Partial inactivation of the cochlear nucleus magnocellularis by an injection of a local anesthetic dramatically changes the tuning of a space-specific neuron to ITD without affecting its response to IID. The converse occurs when the cochlear nucleus angularis is partially inactivated (Takahashi, Moiseff, and Konishi, 1984). Thus, the owl's auditory system processes ITD and IID independently. Recall that space-specific neurons respond only to specific ITD and IID presented simultaneously. The owl must compute both cues simultaneously to form an auditory image of place.

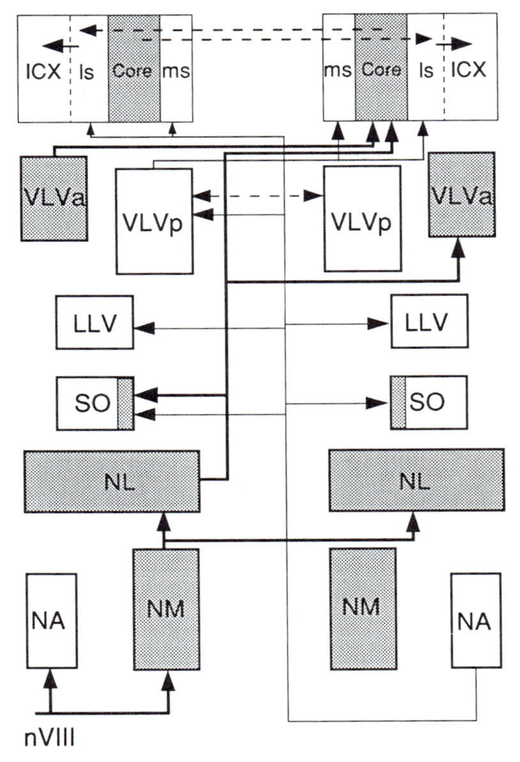

FIGURE 16.1 Parallel auditory pathways. The owl's brainstem auditory system consists of two parallel pathways. The shaded areas indicate the nuclei of the time-processing pathway and the unshaded areas the nuclei of the intensity-processing pathway. The thick and thin arrows indicate anterograde connections in the time and intensity pathways, respectively. The dashed lines with arrowheads on both ends represent reciprocal connections. The connections are symmetrical in two sides of the brainstem, but only one side is shown for simplicity. nVIII, auditory nerve; NA, nucleus angularis; NM, nucleus magnocellularis; NL, nucleus laminaris; SO, superior olivary nucleus; LLV, nucleus ventralis lemnisci lateralis, pars ventralis; VLVa, nucleus ventralis lemnisci lateralis, pars anterior; VLVp, nucleus ventralis lemnisci lateralis, pars posterior; ICX, external nucleus of the inferior colliculus; Is, lateral shell of the central nucleus of the inferior colliculus; Core, core of the central nucleus of the inferior colliculus; ms, medial shell of the central nucleus of the inferior colliculus.

The cochlear nuclei that constitute the starting points of the two pathways differ from each other in both physiology and morphology. Neurons of nucleus magnocellularis have large cell bodies and short or no dendrites. These cell bodies carry large end bulbs of Held where primary auditory fibers terminate. These morphological traits are adaptations for the preservation of temporal information conveyed by the afferent fibers. Phase locking can occur in this nucleus at much higher frequencies (8.5 kHz) than those recorded in other animals. These neurons are, however, relatively insensitive to variations in sound intensity. The other cochlear nucleus, nucleus angularis, contains neurons with normal dendritic fields, and afferent fibers form bouton-type synapses on them. Neurons of nucleus angularis phase lock poorly or not at all, but they are sensitive to variations in sound intensity (Sullivan and Konishi, 1984). Each cochlear nucleus is thus specialized for the transmission of information about time or intensity but not both.

Detection and encoding of ITD

Jeffress (1948) was the first to propose a neural circuit for the detection and encoding of ITD. His model uses afferent axons as delay lines and postsynaptic binaural neurons as coincidence detectors. A coincidence detector fires maximally or only when spikes from the ipsilateral and contralateral sources arrive simultaneously. When a binaural neuron receives ipsilateral and contralateral inputs by axonal paths of different lengths, the impulses from the two sides reach the neuron at different times. Coincidence of impulses occurs only if the arrival times of sound in the two ears are adjusted for the elimination of this temporal disparity. The model can detect different ITDs, because this temporal disparity varies systematically from neuron to neuron. The detection and encoding of ITD should be distinguished from each other. The Jeffress model does not use spike timing or number but the addresses of neurons in the array to distinguish one ITD from another. This concept of place coding of sound location was a revolutionary idea.

The owl's auditory system uses mechanisms similar to those proposed by Jeffress (1948). Neurons of nucleus magnocellularis project bilaterally to nucleus laminaris, which is, therefore, the first binaural station in the time-processing pathway. These two nuclei form circuits for the computation of ITD by the methods of delay lines and coincidence detection (Sullivan and Konishi, 1986; Carr and Konishi, 1990) (figure 16.2). Axons from the ipsilateral nucleus magnocellularis follow somewhat circuitous routes to reach the dorsal surface of nucleus laminaris, whereas those from the contralateral nucleus cross the midline of the medulla to reach the ventral surface of nucleus laminaris. Axons from the two sides course across the body of nucleus laminaris to the surface opposite to the point of entry. As the axons from the two sides interdigitate in a countercurrent fashion within nucleus laminaris, they innervate large laminaris cell bodies that have short or no dendrites. Nucleus laminaris is tonotopically organized, and all three components—ipsilateral and contralateral afferent axons and their recipient laminaris neurons—are narrowly tuned to the same frequency range. The detection and encoding of ITD by the magnocellularis and laminaris circuits are therefore carried out in each frequency band. Recordings of phase-locked impulses in these axons show that conduction delays vary systematically between the dorsal and ventral surfaces of nucleus laminaris. The con-

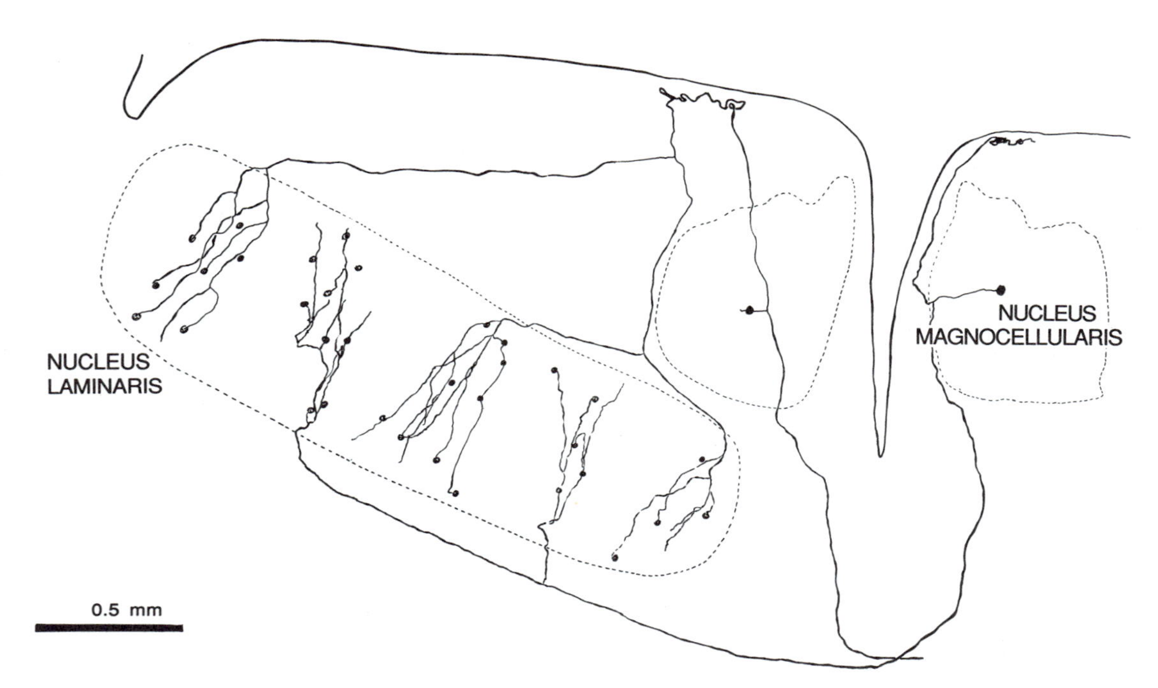

FIGURE 16.2 The innervation of nucleus laminaris. Neurons of nucleus magnocellularis project bilaterally to nucleus laminaris, which is, therefore, the first binaural station in the time-processing pathway. These two nuclei form circuits for the computation of ITD by the methods of delay lines and coincidence detection. Axons (here only one axon and its branches are shown) from the ipsilateral nucleus magnocellularis follow somewhat circuitous routes to reach the dorsal surface of nucleus laminaris, whereas those from the contralateral nucleus cross the midline of the medulla to reach the ventral surface of nucleus laminaris. Axons from the two sides course across the body of nucleus laminaris to the surface opposite to the point of entry. As the axons from the two sides interdigitate in a countercurrent fashion within nucleus laminaris, they innervate large laminaris cell bodies which have short dendrites or none at all.

duction delay at a given distance from either surface appears to be constant and independent of sound frequency.

A comparison of the monaural and binaural responses of laminaris neurons provides evidence for coincidence detection. The period histograms of a laminaris neuron made by stimulation of the ipsilateral and contralateral ears may show a disparity in the arrival times of phase-locked spikes from the two sides. This is due to the difference between the ipsilateral and contralateral axonal paths to the neuron. The disparity equals the neuron's most favorable ITD in magnitude but is opposite in sign. This match is expected if the neuron works as a coincidence detector. Laminaris neurons are, however, not all-or-nothing coincidence detectors, for their responses decrease gradually from maximum to minimum as stimulus timing changes from perfect coincidence to out of phase by 180°.

Ambiguities in encoding ITD

Laminaris neurons respond best to a single interaural time difference that is independent of frequency. Such a time difference is termed a characteristic delay (CD) in mammalian auditory physiology (Goldberg and Brown, 1969; Carr and Konishi, 1990). But the neurons of nucleus laminaris and all lower-order neurons below the level of the external nucleus of the inferior colliculus also respond optimally to other ITDs that vary with frequency. These ITDs consist of CD $\pm nT$, where T is the period of the stimulus tone and n is an integer. This phenomenon, called *phase ambiguity*, is due to the use of phase-locked impulses for the measurement of ITD (figure 16.3). Laminaris neurons fire maximally when they receive binaurally synchronous trains of phase-locked impulses. The coincidence of the two trains recurs every time ITD is changed by integer multiples of the period of the stimulus tone, causing the laminaris neurons to discharge maximally. Thus, the CD corresponds to the ITD that is present in the signal and CD $\pm nT$ are false ITDs. Phase ambiguity occurs at all audible frequencies including those below 1 kHz.

Phase ambiguity is largely an artifact of delivering sound through earphones, although it can occur in free field at the high-frequency end of the owl's audible range. The time taken for sound to travel a distance equal to the separation of the ears is approximately 170 μs in barn owls. ITDs that are larger than this value do not occur in free field. Despite the artifical nature of phase ambiguity, the owl's auditory system solves the problem in a higher-order station by suppressing neuronal responses to all ITDs except the CD.

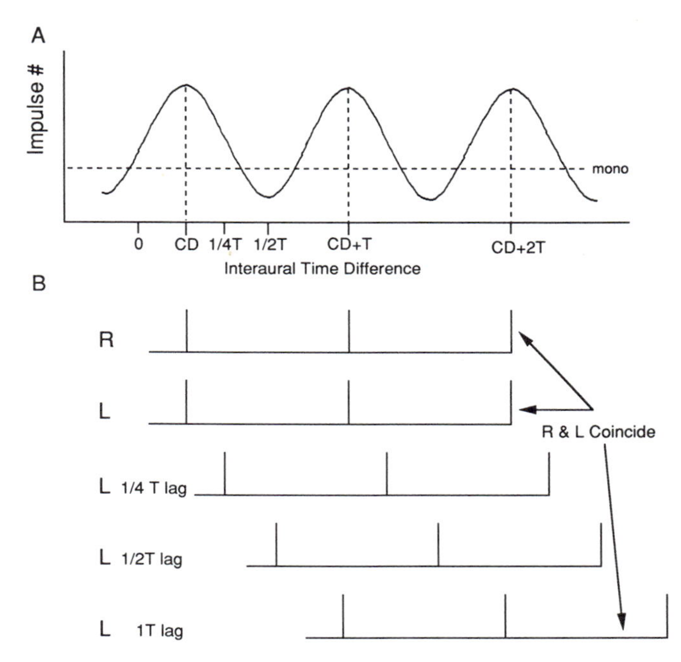

FIGURE 16.3 Responses of laminaris neurons to ITD. (A) Laminaris neurons respond both to monaural and binaural stimuli. They respond best to a particular ITD that causes impulses from the left and right cochlear nuclei to reach them simultaneously. This ITD is independent of frequency and called a characteristic delay (CD). As ITD departs from neurons' CD, neuronal responses do not decline suddenly but gradually, indicating that the neurons are not all or nothing coincidence detectors. When ITD differs from the CD by one half of the period (1/2T) of the stimulus tone, it elicits fewer impulses than that released by monaural stimulation (horizontal dashed line). As ITD departs from this least favorable value, neuronal responses recover gradually until they reach a new peak, indicating that laminaris neurons respond to multiple ITDs. This phenomenon, called phase ambiguity, occurs because phase-locked impulses encode time as explained in (B). Consider trains of phase-locked impulses from the left (L) and right (R) cochlear nuclei arriving at a laminaris neuron. When the two trains are aligned impulse by impulse (i.e., coincidence), the neuron respond best. If the tone in the left ear is delayed relative to the tone in the right ear, the left train of impulses is also delayed from the right one. But when the left train lags by one period of the stimulus tone, the two trains are aligned again, causing the laminaris neuron to respond maximally. The same condition recurs every time one train is advanced or delayed by nT, where T is the period of the stimulus tone and n is an integer.

Initial detection and encoding of IID

The space-specific neurons show a bell-shaped or inverted V-shaped tuning curve for IID. This response property indicates tuning to a narrow range of IID and the lack of excitation or inhibition to monaural stimulation. Neuronal sensitivity for IID emerges in nucleus ventralis lemnisci lateralis, par posterior (VLVp), one of the lemniscal nuclei in the pons. VLVp receives a direct excitatory input from the contralateral nucleus angularis and an inhibitory input from the opposite VLVp. Thus, stimulation of the ipsilateral ear inhibits and the contralateral ear excites VLVp neurons. The responses of these cells vary as a function of IID, because both the degrees of excitation and inhibition depend on sound intensity. Furthermore, different VLVp neurons prefer different ranges of IID. These variations are due to differences in the strength of and sensitivity to inhibition. Both the strength and sensitivity increase from dorsal to ventral in the nucleus. There is, therefore, a map of IID in VLVp (Manley, Koeppl, and Konishi, 1988).

Convergence of parallel pathways

The time and intensity pathways are separate, but they must join together before or in the external nucleus where neurons are selective for combinations of ITD and IID. The site of the convergence is the "lateral shell" which adjoins the external nucleus (Fujita and Konishi, 1989). The lateral shell is also where the bell-shaped response of neurons to IID is created from inputs from the VLVp of both sides. As mentioned earlier, VLVp neurons show excitation and inhibition to stimulation of the contralateral and ipsilateral ears, respectively. This rule is transformed to a new relationship in which neurons with preference for louder sound in the right ear are found in the top of the external nucleus and those with left-ear biases at the bottom, independently of the side of the brain in which the neurons occur. The dorsal and ventral neurons have their receptive fields above and below eye level, respectively. This relationship is consistent with the way the owl perceives the elevation of sound sources, namely, louder sound in the right and left ears means the source being above and below eye level, respectively.

Another property of space-specific neurons is broad frequency tuning, which indicates the convergence of different frequency channels on single neurons. This convergence appears to occur not in one step but gradually between the lateral shell and the external nucleus (Mazer, personal communication). One of the consequences of this frequency convergence is the elimination of phase ambiguity. Neurons tuned to different frequencies but to the same CD project to the same space-specific neuron and endow this cell with the common CD, but the responses of different channels to non-CD ITDs are not expressed in the response of the cell. In this way, the selectivity of space-specific neurons for combinations of ITD and IID is created.

Flow of information and algorithm

The anatomical and physiological findings discussed thus far allow us to construct not only the flow of information about ITD and IID but also the algorithm for the synthesis of the selectivity of space-specific neurons for combinations of ITD and IID (Konishi, 1992). An algorithm is a set of step-by-step processes for the solution of a problem. The algorithm for the computation of ITD-IID pairs is shown in figure 16.4.

A map of auditory space

The space-specific neurons represent the results of all processes that are carried out by lower-order neurons for the creation of sensitivity to ITD-IID pairs. Furthermore, the anatomical sites of space-specific neurons are arranged according to the locations of their receptive fields (Knudsen and Konishi 1978a,b). Consider a two-dimensional space defined by azimuth (or ITD) and elevation (or IID) and group the space-specific neurons by these two variables. The simplest arrangement should contain columns of neurons tuned to the same ITD and layers of the same neurons grouped together according to their IID selectivity. This scheme produces only a sheet of neurons, but the sheet can be considered to have a thickness along which the same ITD and IID selectivities are repeated.

The owl's auditory system uses this scheme to form a map of auditory space in the external nucleus. The distribution of neurons having receptive fields in different coordinates is not uniform but biased for those within a radius of 15° from the center of the face, where the owl localizes sound most accurately. There is thus an auditory "fovea" to which the owl brings the target by turning its head. An auditory space map has also been found in the superior colliculus of the ferret, al-

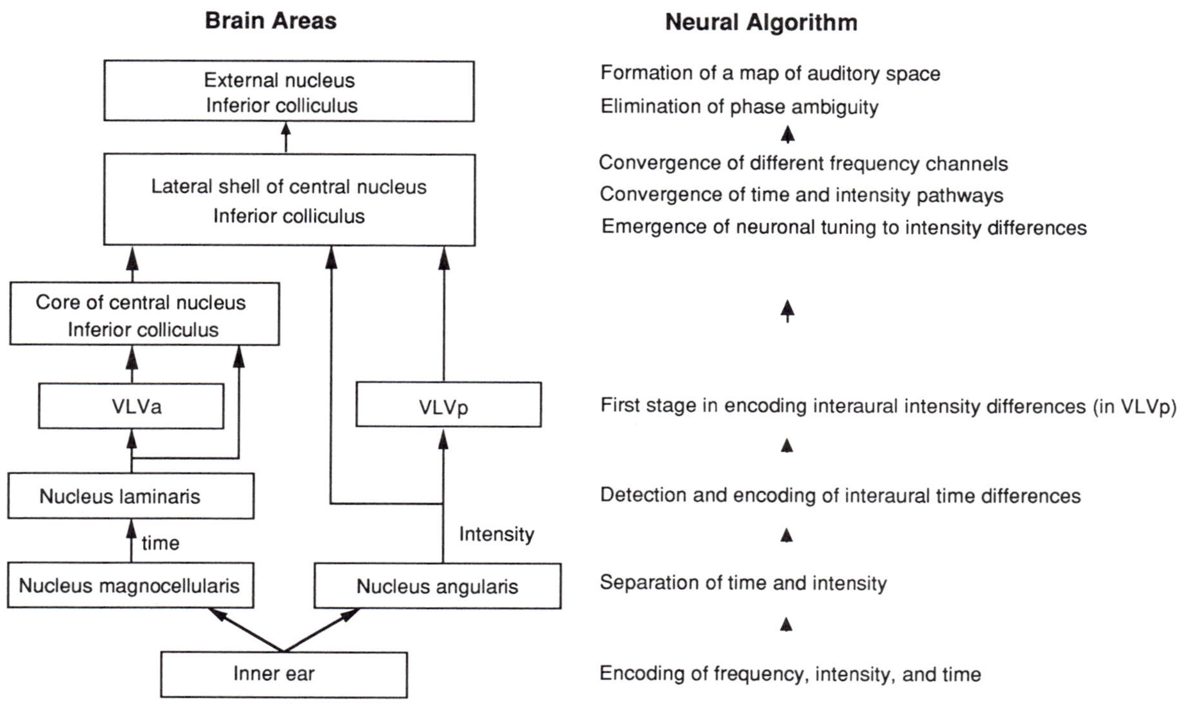

Brain Areas

External nucleus
Inferior colliculus

Lateral shell of central nucleus
Inferior colliculus

Core of central nucleus
Inferior colliculus

VLVa

VLVp

Nucleus laminaris

time

Intensity

Nucleus magnocellularis

Nucleus angularis

Inner ear

Neural Algorithm

Formation of a map of auditory space
Elimination of phase ambiguity

Convergence of different frequency channels
Convergence of time and intensity pathways
Emergence of neuronal tuning to intensity differences

First stage in encoding interaural intensity differences (in VLVp)

Detection and encoding of interaural time differences

Separation of time and intensity

Encoding of frequency, intensity, and time

FIGURE 16.4 Flow of information and neural algorithm. This chart explains how the selectivity of space-specific neurons for combinations of interaural time (ITDs) and interaural intensity differences (IIDs) is created. The processing of these cues occurs in parallel and hierarchical networks in the barn owl's auditory system. Primary auditory fibers convey the codes for frequency, intensity, and time to the cochlear nuclei, which separate the codes for time and intensity. Nucleus laminaris, the first site of binaural convergence in the time-processing pathway, contains circuits for the detection and encoding of ITDs. The nucleus conveys the neuronal selectivity for ITDs to two higher-order stations, VLVa (nucleus ventralis lemnisci lateralis, pars anterior) and the core of the central nucleus of the inferior colliculus. VLVp (nucleus ventralis lemnisci lateralis, pars posterior) carries out the first stage of detecting IIDs and sends the results to the lateral shell of the central nucleus. This station is the site of convergence of the time and intensity pathways and different frequency channels. The lateral shell projects to the external nucleus, which is the highest station, where the space-specific neurons represent the final results of all computations carried out by lower-order neurons. The arrows represent anterograde connections, and the arrowheads on the right indicate steps in signal processing. Note: Neither the side of the hemisphere where projections go nor the relative size of the nuclei is shown.

though the design of this map appears to differ from that of the owl's binaural mechanisms (King and Hutchings, 1987; see King and Carlile, this volume). The owl's auditory space map is not due to a direct projection of the inner ear sensory epithelium but to central mechanisms, yet this map has all the properties of brain maps such as center-excitatory and surround-inhibitory organization among its member neurons and topographical projections to higher centers (Knudsen and Konishi 1978c; Konishi, 1986). It projects to the optic tectum where it forms a joint auditory-visual map (Knudsen, 1984). This map in turn projects to a motor map which controls the speed, direction, and amplitude of head turning responses (Du Lac and Knudsen, 1990). The behavioral significance of the auditory space map has been recently studied. A focal lesion of the auditory space map impairs the accuracy of localization only within the range that is represented by the lesioned part of the map (Wagner, 1993). Thus, the translation of auditory space into the coordinates of the motor map does not require the distribution of neural activities over the entire map but the activity of a local population of space-specific neurons.

Conclusion

The ultimate goal of auditory physiological studies is an understanding of auditory perception. The study of local circuits prevails in auditory physiology. Such research is necessary, but thinking and working at the level of systems are essential to achieve the goal. The systems level means, in the present context, a set of nuclei or networks that carries out the algorithm for a perceptual task. The system of brainstem nuclei involved in sound localization is a case in point. The work on the owl covered essentially all the auditory nuclei below the level of the thalamic auditory station. This exploration was, however, not for the cataloging of neuronal response types as such but for the discovery of processing streams and neural mechanisms underlying sound localization. Search for the flow of information differs from the compilation of neuronal response properties.

In both the owl and mustached bat, the discovery of higher-order neurons that responded selectively to perceptually meaningful stimuli marked the pivotal phase of the neurophysiological study. Furthermore, these neurons occur in orderly arrays to form maps of the stimulus variables to which they are tuned, such as echo delays and space. Such an organization suggested the functional significance of these neurons. If these neurons had been rare and occurred randomly, it would not have been possible to investigate the origin of their selectivity. The philosophical basis for search for such higher-order neurons is the hypothesis that perceptually relevant stimuli are processed by hierarchically organized networks in which the neurons at the top of the hierarchy represent the results of all computations carried out in the pathways leading to them. There is no reason to assume that this type of representation is possible only for relatively simple stimuli. Similar parallel and hierarchical processing networks are likely to give rise to the selectivity of the face neurons in the inferotemporal cortex of the monkey and sheep (Kendrick and Baldwin, 1987; Perrett, Mistlin, and Chitty, 1987; Fujita et al., 1992; Young, 1993).

Aside from the discovery of processing streams and neural mechanisms for perceptual tasks, the systems approach enables us to compare algorithms between different sensory systems and animals. The auditory systems of the owl and mustached bat share some design features with other well-studied sensory sytems such as the electrosensory system of electric fish (Heiligenberg, 1991). All these systems initially separate different sensory information in different parallel pathways, process it in hierarchically organized networks within each pathway, and bring different information together by convergence of the pathways. These findings suggest the existence of rules of signal processing that transcend different sensory systems and animals (Konishi, 1990, 1991).

REFERENCES

Carr, C. E., and M. Konishi, 1990. A circuit for detection of interaural time differences in the brainstem of the barn owl. *J. Neurosci.* 10:3227–3246.

De Yoe, E. A., and D. C. Van Essen, 1988. Concurrent processing streams in monkey visual cortex. *Trends Neurosci.* 11:219–226.

Du Lac, S., and E. I. Knudsen, 1990. Neural maps of head movement vector and speed in the optic tectum of the barn owl. *J. Neurophysiol.* 63:131–146.

Felleman, D. J., and D. C. Van Essen, 1991. Distributed hierarchical processing in the primate cerebral cortex. *Cerebral Cortex* 1:1–47.

Fujita, I., and M. Konishi, 1989. Transition from single to multiple frequency channels in the processing of binaural disparity cues in the owl's midbrain. *Soc. Neurosci. Abstr.* 15:114.

Fujita, I., K. Tanaka, M. Ito, and K. Cheng, 1992. Columns for visual features of objects in monkey inferotemporal cortex. *Nature* 360:343–346.

Goldberg, J. M., and P. B. Brown, 1969. Responses of binaural neurons of dog superior olivary complex to dichotic tonal stimuli: Some physiological mechanisms of sound localization. *J. Neurophysiol.* 32:613–636.

Heiligenberg, H., 1991. *Neural Nets in Electric Fish.* Cambridge, Mass.: MIT Press, p. 179.

Jeffress, L. A., 1948. A place theory of sound localization. *J. Comp. Physiol. Psych.* 41:35–39.

Kendrick, K. M., and B. A. Baldwin, 1987. Cells in temporal cortex of conscious sheep can respond preferentially to the sight of faces. *Science* 236:448–450.

King, A. J., and M. E. Hutchings, 1987. Spatial response properties of acoustically responsive neurons in the superior colliculus of the ferret: a map of auditory space. *J. Neurophysiol.* 57:596–624.

Knudsen, E. I., 1984. Auditory properties of space-tuned units in the owl's optic tectum. *J. Neurophysiol.* 52:709–723.

Knudsen, E. I., and M. Konishi, 1978a. A neural map of auditory space in the owl. *Science* 200:795–797.

Knudsen, E. I., and M. Konishi, 1978b. Space and frequency are represented separately in auditory midbrain of the owl. *J. Neurophysiol.* 41:870–884.

Knudsen, E. I., and M. Konishi, 1978c. Center-surround organization of auditory receptive fields in the owl. *Science* 202:778–780.

Knudsen, E. I., M. Konishi, and J. D. Pettigrew, 1977. Receptive fields of auditory neurons in the owl. *Science* 198:1278–1280.

Konishi, M., 1986. Centrally synthesized maps of sensory space. *Trends Neurosci.* 9:163–168.

Konishi, M., 1990. Similar algorithms in different sensory

systems and animals. *Cold Spring Harb. Symp. Quant. Biol.* 55:575–584.

Konishi, M., 1991. Deciphering the brain's codes. *Neural Computation* 3:1–18.

Konishi, M., 1992. The neural algorithm for sound localization in the owl. *Harvey Lect.* 86:47–64.

Liberman, A., and I. G. Mattingly, 1985. The motor theory of speech perception revisited. *Cognition* 21:1–36.

Manley, G. A., C. Koeppl, and M. Konishi, 1988. A neural map of interaural intensity difference in the brainstem of the barn owl. *J. Neurosci.* 8:2665–2676.

Moiseff, A., 1989. Bi-coordinate sound localization by the barn owl. *J. Comp. Physiol.* 164:637–644.

Moiseff, A., and M. Konishi, 1981. Neuronal and behavioral sensitivity to binaural time difference in the owl. *J. Neurosci.* 1:40–48.

Moiseff, A., and M. Konishi, 1983. Binaural characteristics of units in the owl's brainstem auditory pathway: Precursors of restricted spatial receptive fields. *J. Neurosci.* 3: 2553–2562.

Olsen, J. F., E. I. Knudsen, and S. D. Esterly, 1989. Neural maps of interaural time and intensity differences in the optic tectum of the barn owl. *J. Neurosci.* 9:2591–2605.

Perrett, D. I., A. J. Mistlin, and A. J. Chitty, 1987. Visual neurons responsive to faces. *Trends Neurosci.* 10: 358–364.

Pfeiffer, R. R., 1966. Classification of response patterns of spike discharges for units in the cochlear nucleus: Tone burst stimulation. *Exp. Brain Res.* 1:220–235.

Sayers, B. McA., and E. C. Colin, 1957. Mechanisms of binaural fusion in the hearing of speech. *J. Acoust. Soc. Am.* 29:973–987.

Stern, R. M., A. S. Zelberg, and C. Trahiotis, 1988. Lateralization of complex binaural stimuli: A weighted image model. *J. Acoust. Soc. Am.* 84:156–165.

Suga, N., 1984. The extent to which biosonar information is represented in the bat auditory cortex. In *Dynamic Aspects of Neocortical Function*, G. M. Edelman, W. E. Gall, and W. M. Cowan, eds. New York: Wiley, pp. 315–373.

Sullivan, W. E., and M. Konishi, 1984. Segregation of stimulus phase and intensity in the cochlear nuclei of the barn owl. *J. Neurosci.* 4:1787–1799.

Sullivan, W. E., and M. Konishi, 1986. Neural map of interaural phase difference in the owl's brainstem. *Proc. Natl. Acad. Sci. U.S.A.* 83:8400–8404.

Takahashi, T. T., and M. Konishi, 1988a. Projections of nucleus angularis and nucleus laminaris to the lateral lemniscal nuclear complex of the barn owl. *J. Comp. Neurol.* 274:212–238.

Takahashi, T. T., and M. Konishi, 1988b. Projections of the cochlear nuclei and nucleus laminaris to the inferior colliculus of the barn owl. *J. Comp. Neurol.* 274:190–211.

Takahashi, T., A. Moiseff, and M. Konishi, 1984. Time and intensity cues are processed independently in the auditory system of the owl. *J. Neurosci.* 4:1781–1786.

Wagner, H., 1993. Sound-localization deficits induced by lesions in the barn owl's auditory space map. *J. Neurosci.* 13:371–386.

Young, M. P., 1993. Modules for pattern recognition. *Curr. Biol.* 3:44–46.

17 Neural Coding for Auditory Space

ANDREW J. KING AND SIMON CARLILE

ABSTRACT A neural representation of auditory space is computed centrally using spatial cues that are generated by the acoustical properties of the ears and head. The functional organization of neurons that utilize these cues varies at different levels of the mammalian brain. In the superior colliculus, a map of auditory space is found that is derived from a combination of interaural differences in sound level and spectral cues provided by the outer ears. Growth-related changes occur in the value of these localization cues, which are mirrored in the gradual emergence of the auditory space map during early postnatal life. This spatial representation can undergo substantial refinement during development in response to both auditory and visual experience and is also continually updated by changes in the orientation of the sense organs.

Localization of sound sources

The ability of animals to determine the location of a sound source represents one of the most remarkable attributes of the auditory system. Locations in visual space or on the body surface are encoded directly by the distribution of activity within the receptor cells of the retina and skin. As a result of spatially ordered afferent projections, this information is represented centrally in the form of topographical maps. In contrast, the peripheral auditory system extracts the frequency content, rather than the spatial coordinates, of a sound source. Therefore, sound source location must be derived computationally within the brain using cues arising at each ear. For a sound source located off the midline, the difference in the path lengths to each ear will give rise to a difference in time of arrival of the sound (interaural time difference [ITD]), whereas the acoustical shadow cast by the head may produce an interaural pressure level (or intensity) difference (IID). Lord Rayleigh (1907) observed that the head is a most effective acoustical obstacle for short-wavelength, high-frequency sounds, and therefore suggested that IIDs are used to determine sound location at these frequencies. On the other hand, because the auditory system can encode phase information at relatively low frequencies only, interaural phase difference cues are restricted to low-frequency sounds. There is considerable psychophysical (see Middlebrooks and Green, 1991) and physiological evidence (reviewed by Irvine, 1986, 1992) consistent with this so-called duplex theory of sound localization. In addition, other possible binaural cues include the ongoing time differences in the envelopes of high-frequency, spectrally complex stimuli as well as onset time differences.

Most previous studies of sound localization have examined the sensitivity of human subjects or single neurons recorded in animals to IIDs and ITDs using headphones to vary the intensity and the onset time or phase of the stimulus in each ear. Neurons that are selective for specific ITDs or IIDs have been recorded at different levels of the central auditory pathway. Indeed, there are instances, such as in the barn owl, where the binaural interactions exhibited by auditory neurons appear to be sufficient to explain their selectivity for sound location (see Konishi, this volume; Olsen, Knudsen, and Esterly, 1989). For individual frequencies, both ITDs and IIDs are inherently ambiguous because a specific interaural disparity can arise from any location on the surface of an imaginary cone extending out from the interaural axis, the so-called cone of confusion (see Irvine, 1986). Consequently, the general applicability to localization under more natural free-field conditions of dichotic studies employing spectrally simple stimuli has been questioned (Wightman, Kistler, and Perkins, 1987). The duplex theory cannot explain the capacity of the auditory system to determine the location of a sound source on the midsagittal plane or to distinguish between locations in front of and behind the interaural axis or, indeed, to localize sounds under monaural listening conditions.

ANDREW J. KING and SIMON CARLILE University Laboratory of Physiology, Oxford, England

The mammalian outer ear filters the sound in a location-dependent fashion (e.g., Middlebrooks, Makous, and Green, 1989; Carlile, 1990b; Musicant, Chan, and Hind, 1990; Rice et al., 1992; Carlile and King, 1994). These spectral transformations are unique to each location in space and therefore provide cues that are sufficient to explain sound localization by monaural listeners (e.g., Fisher and Freedman, 1968; Oldfield and Parker, 1986; Butler, Humanski, and Musicant, 1990). Pinnal spectral cues are also believed to be utilized when judging the elevation of a sound source (Middlebrooks, 1992). As well as providing monaural cues, the directional filtering of both pinnae results in a complex pattern of interaural spectral differences, which may provide an unambiguous cue to sounds located off the midline (Middlebrooks, Makous, and Green, 1989; Carlile, 1990a; Musicant, Chan, and Hind, 1990; Rice et al., 1992; Carlile and King, 1994).

It is clear that multiple binaural and monaural cues are involved in auditory localization. The relative importance of different cues varies not only with the nature of the sound source but also with the spatial location of the source (Middlebrooks and Green, 1991) and the species of the listener (Lewis, 1983). Physiological investigations into the manner in which a neural image of auditory space is synthesized under natural conditions should ideally consider all the cues that are potentially available.

Role of the auditory cortex in sound localization

Convergence of information from the two ears first occurs in the nuclei of the brain stem, where the initial processing of ITD and IID cues takes place (Irvine, 1986, 1992). Neurons sensitive to variations in the values of these binaural disparity cues are also found at all subsequent levels of the auditory pathway. Despite the considerable amount of subcortical processing, it is clear that the auditory cortex of carnivores and primates plays a pivotal role in mediating sound localization behavior (reviewed by Clarey, Barone, and Imig, 1992). Thus, unilateral auditory cortical lesions produce localization deficits in the contralateral hemifield (Jenkins and Masterton, 1982). Moreover, lesions made in discrete isofrequency regions of the primary auditory cortex (A1) result in localization deficits that are restricted to frequencies deprived of their cortical representation (Jenkins and Merzenich, 1984). These findings were initially believed to reflect an impairment in the ability of animals with cortical lesions to make certain types of motor response toward the location of a sound source (e.g., Heffner, 1978). More recently, Heffner and Heffner (1990) have argued that, following lesions of the auditory cortex, animals do exhibit sensory deficits but, more importantly, lose their ability to associate sounds with positions in space.

Thus, at this level of the brain, contralateral sound locations appear to be encoded in frequency-specific channels. In contrast, there have been no reports that small auditory cortical lesions lead to behavioral deficits limited to a circumscribed region within one hemifield, although spatially restricted impairments do result from small lesions in the space-mapped regions of the barn owl's midbrain (Wagner, 1993). Furthermore, decorticate animals are capable of indicating the location of a sound source by, for example, making reflexive head movements in the direction of novel sounds (Thompson and Masterton, 1978).

Representation of auditory space in the auditory cortex

The behavioral results raise important questions concerning the manner in which auditory space is represented in A1 and other cortical areas. The binaural properties of many A1 neurons resemble those found at lower levels of the brain and, in most cases, the ITDs and IIDs to which they are most sensitive correspond to locations in the contralateral hemifield (Clarey, Barone, and Imig, 1992). Free-field studies have revealed that some A1 neurons exhibit a degree of spatial tuning, whereas others are insensitive to the location of broadband or tonal stimuli (Middlebrooks and Pettigrew, 1981; Imig, Irons, and Samson, 1990; Rajan, Aitkin, and Irvine, 1990). Most azimuth-sensitive neurons respond maximally to locations in the contralateral hemifield, although others prefer ipsilateral or midline locations. Within an isofrequency region, neurons with similar spatial response properties tend to be clustered together, although occasional shifts in spatial tuning among sequentially encountered units have also been reported. However, in terms of the distribution of spatially tuned neurons, these findings are not consistent with a topographical representation of sound source location in A1. Rather, the neural code for auditory space may be contained in more complex neuronal response patterns or in the functional interconnections between groups of cortical neurons.

Representation of auditory space in the superior colliculus

The position of a sound in space is encoded in the deeper layers of the mammalian superior colliculus (SC) in a manner that is qualitatively different from that found in A1. The auditory spatial receptive fields of SC neurons tend to be large, although the great majority are tuned to the extent that they respond maximally to a specific region of space, which is usually in the contralateral hemifield. Consequently, the topography of this representation can be expressed in terms of the distribution of preferred sound directions or best positions. The auditory best positions of SC neurons vary in a continuous and systematic fashion across the nucleus (King and Palmer, 1983; Middlebrooks and Knudsen, 1984; Wong, 1984; King and Hutchings, 1987). Thus, anterior space is represented rostrally and posterior

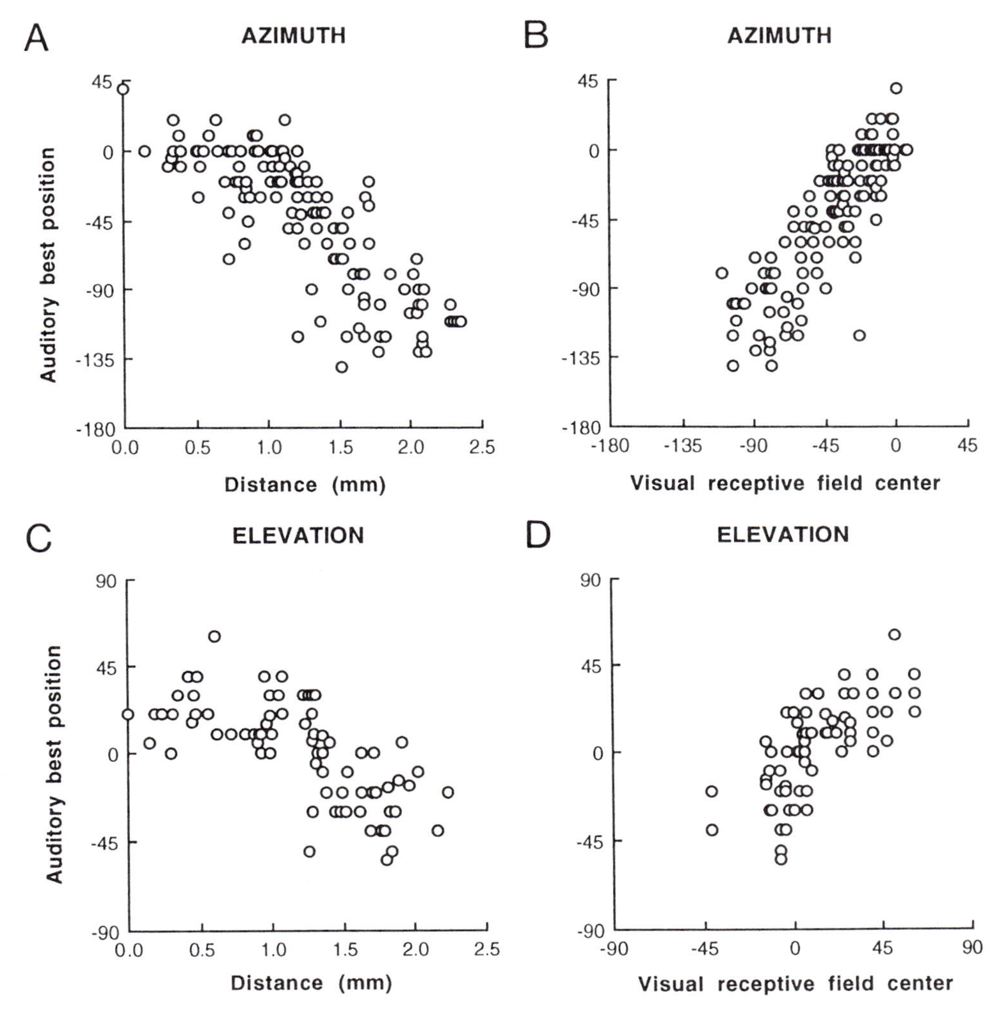

FIGURE 17.1 The representation of auditory space in the superior colliculus (SC) of the ferret. The loudspeaker azimuth where the maximum response was obtained (auditory best position) is plotted against (A) the distance of each unit from the rostrolateral margin of the SC and (B) the location of the center of the visual receptive field of superficial layer units recorded in the same electrode penetration. The best positions in elevation are plotted in (C) against the distance of the units from the rostromedial margin of the SC and in (D) against the corresponding coordinates of the visual receptive fields.

space in caudal SC, whereas superior sound locations are represented medially and inferior locations on the lateral side of the nucleus (figure 17.1A, C). This two-dimensional map of auditory space is similar to that in the mesencephalicus lateralis pars dorsalis (Knudsen and Konishi, 1978) and the optic tectum (Knudsen, 1982) of the barn owl. At the same time, it should be stressed that, as with other sensory maps, this is a point-to-area representation, so that a single sound source may evoke some activity throughout much, or even all, of the deeper layers of the SC (Middlebrooks and Knudsen, 1984). Although the maximum activity will be focused at a restricted site, it is likely that a spatially distributed code provides the basis by which stimulus location is represented precisely within the SC (McIlwain, 1991).

The significance of the map of auditory space can be considered in terms of the sensorimotor function of the SC. The deeper layers of this nucleus also receive visual and tactile inputs, whereas the overlying superficial layers are generally regarded as exclusively visual (reviewed by Sparks, 1988). The representations of all three modalities are topographically organized and are closely aligned with one another. The spatial relationship between the visual map in the superficial layers and the auditory map in the deeper layers of the ferret SC is shown in figure 17.1B and D. The azimuth and elevation dimensions of both maps cover approximately the same region of space. Neurons in the deeper layers often receive inputs from more than one modality, and several studies have noted that the visual and auditory receptive fields of these cells tend to be superimposed (e.g., Gordon, 1973; Harris, Blakemore, and Donaghy, 1980; Middlebrooks and Knudsen, 1984).

The auditory space map in the SC is fundamentally different from the visual and somatosensory maps in that it is a computational representation rather than simply reflecting the topographical projections from the sensory epithelia. However, all three sensory maps appear to share common coordinates in the SC and are in register with the motor output maps that are also found in this structure (Sparks, 1988). This is clearly an efficient way of allowing different modality cues arising from the same region in space to initiate orientation movements that redirect the eyes, pinnae, and head toward that location. The presence of topographically aligned sensory maps also provides a means by which multiple sensory cues can interact both within (e.g., King, Carlile, and Chevassut, 1990) and between different modalities (Stein and Meredith, 1990). The strength and nature of these interactions, which depend on the spatiotemporal relationships of the different stimuli, may determine both the responses of SC neurons and the accuracy of orienting behavior (Stein and Meredith, 1990).

The presence of a two-dimensional map of auditory space in the SC makes this a very useful structure in which to study the neural coding of sound localization cues. The functional organization of the SC also indicates that auditory space should be coded in a dynamic fashion (Knudsen, 1991) to meet two very different requirements, those imposed by the mobility of the eyes and ears and those resulting from developmental changes in the relative geometry of different sense organs. These issues will be addressed in the remainder of this chapter.

Acoustical basis of the auditory space map

Responses to Dichotic Stimulation In contrast to A1, auditory neurons in the SC are broadly tuned for sound frequency and are not tonotopically organized (King and Palmer, 1983; Wise and Irvine, 1983; Hirsch, Chan, and Yin, 1985; Middlebrooks, 1987). This is a common and necessary feature of nuclei that contain unambiguous codes for auditory space. Because SC neurons tend to be biased toward the upper end of the audible frequency range, dichotic studies have concentrated on their sensitivity to IIDs (Wise and Irvine, 1983, 1985; Hirsch, Chan, and Yin, 1985). Most auditory neurons in the SC of the cat are binaurally driven and are sensitive to variations in IID. The majority of cells exhibit sigmoidal IID response functions, which are produced by an inhibitory influence of the ipsilateral ear on the contralateral excitatory response. Others show peaked functions, based either on binaural facilitatory interactions or a mixture of facilitation and inhibition. In terms of the auditory space representation, the most interesting findings are that these binaural classes are segregated within different regions of the SC (figure 17.2A) and that there is a topographical variation in the sensitivity of neurons with sigmoidal response functions (figure 17.2B). This produces a gradient in sensitivity to IIDs from values of, or close to, 0 dB in rostral SC to progressively larger values favoring the contralateral ear at more caudal locations. Although this is clearly consistent with the rostrocaudal representation of sound azimuth, these

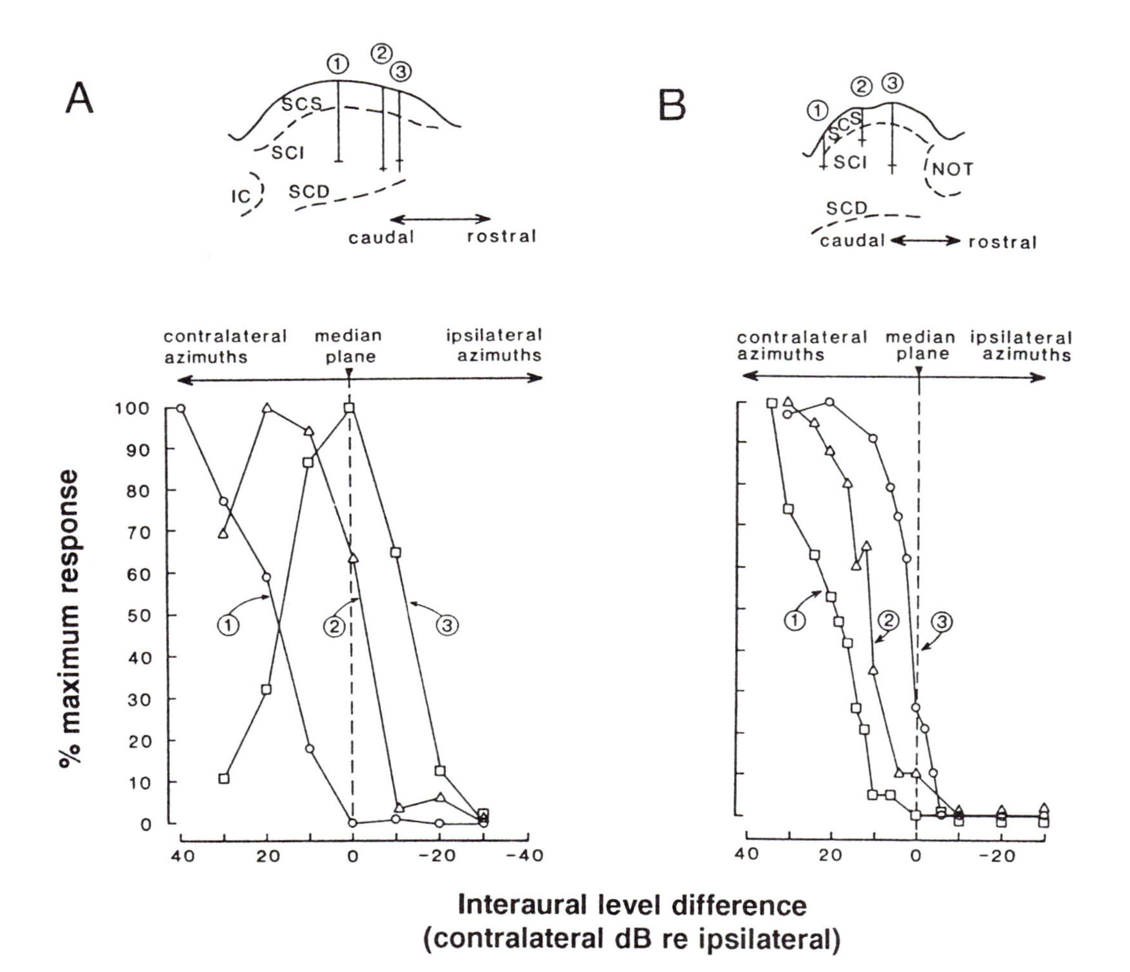

FIGURE 17.2 Topographical organization of binaural sensitivity along the rostrocaudal axis of the superior colliculus (SC) in the cat. In the upper panels, the locations of single units recorded in two separate experiments (A, B) are indicated in line drawings of sagittal sections of the SC. The normalized IID functions for each unit are shown by the corresponding numbers in the lower panels. In these dichotic experiments, the interaural sound level of a broadband noise was varied while the average binaural stimulus level was held constant. The IID values are indicated on the abscissa, with the corresponding horizontal locations indicated at the top of each plot. A topography of IID sensitivity is indicated by the systematic shift from peaked to sigmoidal functions, favoring progressively greater IIDs, as the electrode location was moved from rostral to caudal SC. IC, inferior colliculus; SCS, superficial layers of the SC; SCI, intermediate layers of the SC; SCD, deep layers of the SC; NOT, nucleus of the optic tract. (Modified from Wise and Irvine, 1985, by permission.)

binaural responses correspond to locations in the posterior as well as the anterior hemifield and therefore do not, by themselves, explain the capacity of SC neurons to distinguish between front and back.

RESPONSES TO FREE-FIELD STIMULATION The contribution of different localization cues to the synthesis of the map of auditory space can also be inferred from free-field studies, by appropriate manipulation of the acoustical information available. Passive displacement of the pinnae in anesthetized cats alters the directional responses of SC cells in a manner that is predictable from the change in the spatial distribution of IID values (Middlebrooks and Knudsen, 1987). Similarly, plugging one ear produces level-dependent changes in spatial tuning that can be explained in terms of the binaural response properties of these neurons (Palmer and King, 1985; Middlebrooks, 1987). Thus, following occlusion of the ear ipsilateral to the recording site, most units become progressively more broadly tuned at increasing sound levels, which is consistent with a loss of inhibitory input from that ear. At sound levels near threshold, these cells retain their normal spatial tuning and, at least in the guinea pig (Palmer and King, 1985), their azimuthal best positions are topographically organized. A near-threshold map of auditory space has also been reported after ablation of the ipsilateral cochlea in adult guinea pigs (Palmer and King, 1985) and neonatal ferrets (King, Moore, and Hutchings, 1994). Under these conditions, the two-dimensional map of auditory space, which closely resembles that found in normal animals, must be based on monaural pinnal cues alone (figure 17.3). However, even if the auditory input is restricted to one ear throughout postnatal life, no topographical order is present in the representation of either azimuth or elevation at the higher sound levels that would normally stimulate both ears. Whereas this monaural map provides a neural correlate for the capacity of human observers to localize sounds using one ear alone, it would appear that binaural inputs are required to maintain the integrity of the representation over a range of sound levels.

The monaural map of auditory space is presumably derived from spectral cues generated by the outer ear, which vary systematically and unambiguously with both the horizontal and vertical location of the sound source (Middlebrooks, Makous, and Green, 1989; Carlile, 1990b; Musicant, Chan, and Hind, 1990; Rice et al., 1992; Carlile and King, 1994). For example, in the ferret (Carlile, 1990b; Carlile and King, 1994), the transmission of a broadband stimulus shows a maximum gain in anterior space at 7–20 kHz but is reduced at posterior locations for a frequency band centered on 14 kHz (see color plate 3A). A pronounced spectral notch also is apparent that shifts in frequency from 24 to 30 kHz as the location of the sound source is changed from the anterior midline to the interaural axis. In the cat, cells in the dorsal cochlear nucleus

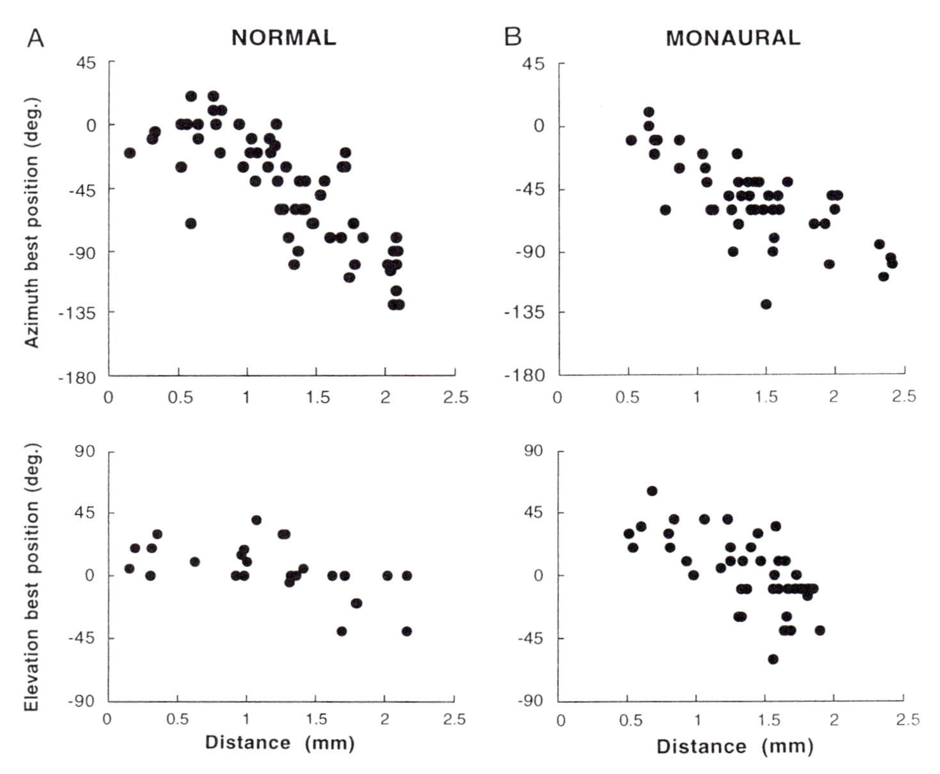

FIGURE 17.3 The representation of auditory space in the superior colliculus (SC) of (A) normal and (B) monaurally deafened ferrets for near-threshold, broadband stimuli. The azimuth and elevation components of the auditory best positions are plotted against the location of each unit measured in terms of the distance from the rostrolateral and rostromedial margins of the SC, respectively. At stimulus levels of approximately 10 dB above threshold, the azimuthal representations in the intact and monaural animals are statistically indistinguishable. There are, however, differences in the representation of elevation under these two conditions. (Data from King, Moore, and Hutchings, 1994.)

are sensitive to the frequency of similar notches in the spectral transfer function of the outer ear, suggesting that they may be tuned to this monaural localization cue (Young et al., 1992). Location-dependent variation in sensitivity to monaural spectral cues has yet to be demonstrated for individual neurons in the SC. However, it has been reported that the frequency tuning of sound-evoked field potentials changes systematically along the rostro-caudal axis of the nucleus in a manner that matches the azimuthal variation in spectral filtering by the auditory periphery (Carlile and Pettigrew, 1987).

As well as underlying the monaural spatial tuning, the outer ear plays a vital role in the generation of an unambiguous map of auditory space at higher sound levels, which stimulate both ears. Interaural spectral differences arise from the difference in filtering of the sound by each outer ear and may provide unambiguous localization cues for locations away from the midline (see color plate 3B). Bilateral removal of the outer ear (pinnectomy) disrupts both monaural and binaural spectral cues, so that the clear differences that normally exist between anterior and posterior locations are no longer apparent (see color plate 3). An equivalent change is observed in the suprathreshold representation of auditory space in the SC (figure 17.4). Whereas

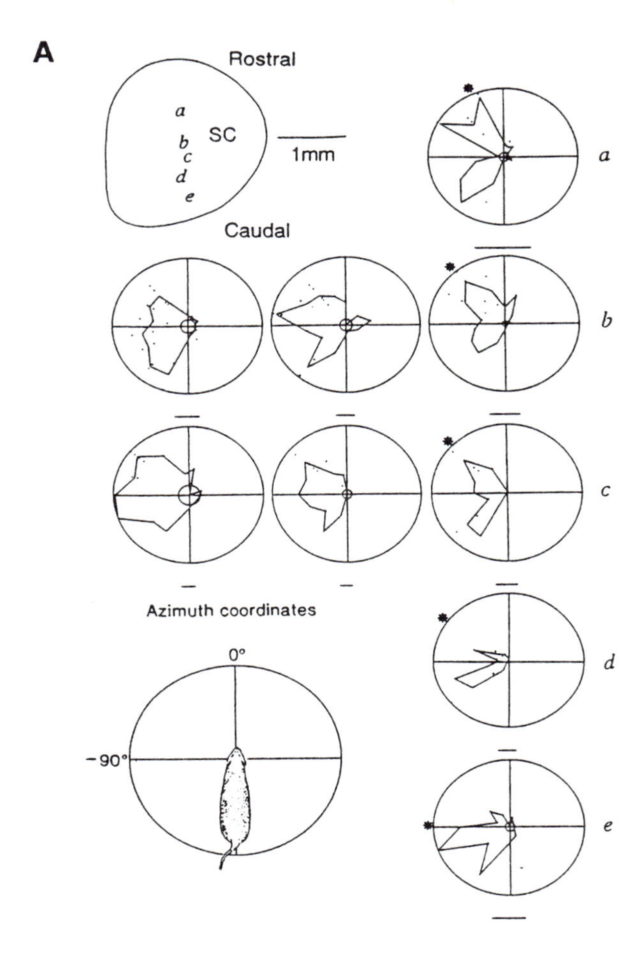

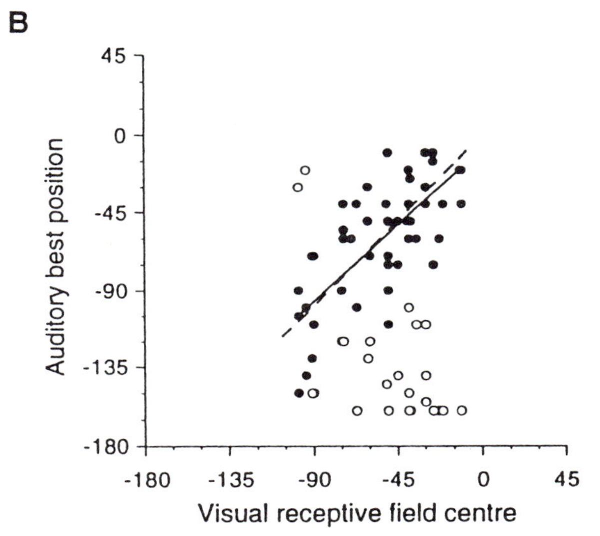

FIGURE 17.4 Representation of the azimuthal dimension of auditory and visual space in the right superior colliculus (SC). (A) Spatial response profiles are shown for nine auditory neurons recorded from five different rostrocaudal electrode penetrations (*a–e*) in the deeper layers of the SC of a bilaterally pinnectomized ferret. Recording locations are indicated on the plan view of the SC. The number of spikes elicited by a suprathreshold (25–30 dB above threshold), broadband stimulus presented at different locations on the audiovisual horizon is plotted as radial distance from the origin of each polar plot. The calibration line below each plot indicates a response of one spike per stimulus. The location of the center of the visual receptive fields of units recorded in the superficial layers of the same electrode tracks is indicated by the asterisks. More than half of the auditory response profiles exhibited two directional lobes separated by a sharp null in the response, indicating that these units were tuned ambiguously for locations about the interaural axis. (B) The auditory best position of each of the directional lobes of the auditory responses is plotted against the location of the center of the visual receptive field for 47 auditory single units from five bilaterally pinnectomized ferrets. The best positions of the directional lobes nearest to the associated visual response are indicated by the filled circles, whereas the second lobes of the bilobed responses are shown by the open circles. The solid line indicates the linear regression for the filled circles, and the broken line indicates the equivalent regression for the intact animals. The associations indicated by the two lines of best regression are statistically indistinguishable. (Data from Carlile and King, 1994)

the great majority of auditory units in the SC are normally tuned to a single location in space, more than half have bilobed azimuthal response profiles following pinnectomy, indicating that they are now tuned to two distinct locations. One lobe is tuned appropriately according to the location of each unit within the nucleus, whereas the second lobe is tuned to sound locations not normally represented at these recording sites (see figure 17.4). The best positions of the appropriately tuned lobes are topographically organized, forming a map of sound azimuth that is not significantly different from that seen in normal animals. However, the second lobe in many of the response profiles indicates that this representation of space is ambiguous. The bilobed responses of these cells are equivalent to the front-back localization confusions demonstrated by human observers when the cavities of their pinnae have been filled (Musicant and Butler, 1984; Oldfield and Parker, 1984).

Given the marked changes in the spatial pattern of monaural and binaural spectral cues after pinnectomy, it seems most unlikely that the residual spectra are responsible for the virtually normal selectivity of SC cells for anterior sound locations. In contrast to the barn owl's optic tectum, in which tuning for ITDs is primarily responsible for the representation of sound azimuth (Olsen, Knudsen, and Esterly, 1989; see Konishi, this volume), few neurons in the mammalian SC appear to be sensitive to physiological variations in ITD values (Hirsch, Chan, and Yin, 1985). The topographical variation in the position of the anterior lobe of the bilobed response profiles is therefore presumably based on the gradient in sensitivity to residual IIDs. The spectral cues generated by the outer ears appear to be used by the auditory system to distinguish between locations on either side of the interaural axis. As a consequence, the SC contains an unambiguous representation of a specific region of auditory space, which matches the other sensory maps.

Dynamic coding of auditory space

If auditory space is to be reliably coordinated with the motor pathways that control orientation behavior, a mechanism must exist by which information concerning the position of mobile sense organs is integrated with the signals relating to target location. For example, the gaze shifts required to fixate an auditory target depend not only on the location of the sound source with respect to the head and ears but also on the starting position of the eyes. In head-restrained cats, the accuracy of saccadic eye movements to auditory targets is compensated to a large degree for variations in

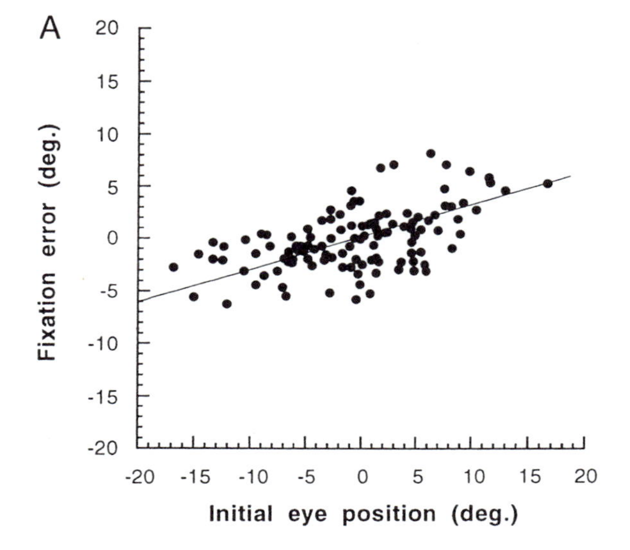

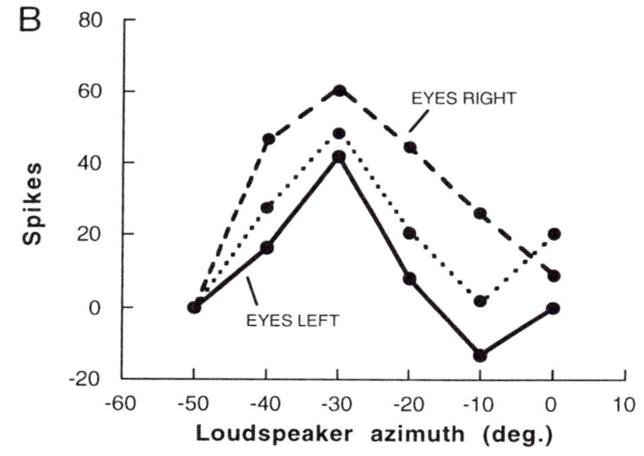

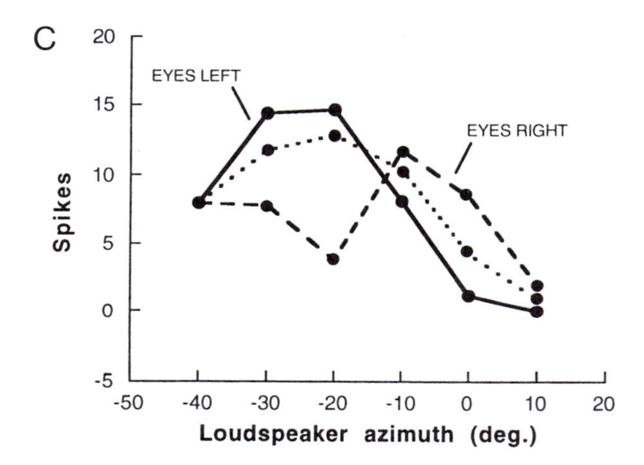

initial eye position, as few additional localization errors are made when the eyes are initially deviated from the central orbital position (figure 17.5A).

Because the coordinates of visual space are centered on the retina, whereas those of auditory space are centered on the ears and head, the registration of the different maps in the SC should also, unless compensated, depend on the position of the eyes and pinnae relative to the head (Sparks, 1988). Recordings from awake primates (Jay and Sparks, 1987) and cats (Harris, Blakemore, and Donaghy, 1980; Hartline et al., 1989, 1994; Peck and Wartman, 1989) have shown that the responses of many, but not all, auditory neurons in the SC are altered as the direction of gaze changes. This may take the form of a modulation of activity throughout the receptive field (figure 17.5B) or a shift in the borders of the spatial response profile in the direction of the change in eye position (figure 17.5C). These changes presumably cause a compensatory shift in the distribution of sound-evoked activity in the SC and have the effect of partially transforming the auditory space representation into retinocentric coordinates. Similar effects of eye position on the auditory spatial tuning of units in the primate frontal cortex have also been reported (Russo and Bruce, 1989).

The need for a mechanism to calibrate the auditory space representation according to changes in eye position clearly depends on the oculomotor range of the animal in question. Because of the dependence of the map of auditory space in the mammalian SC on IIDs and spectral cues, animals with mobile pinnae will encounter additional problems in representing auditory space and in maintaining the alignment of different sensory maps. However, little change is seen in the accuracy of auditory localization in head-restrained cats over a wide range of initial pinnal positions, and signals about pinnal position may also be incorporated into the auditory receptive fields of SC neurons (Hartline, King, and Northmore, 1989). Thus, auditory and possibly other sensory signals in the SC (Knudsen, 1991) appear to be modulated as instant-to-instant changes in the spatial orientation of different sense organs alter the response required for object localization.

FIGURE 17.5 (A) Influence of initial eye position on the accuracy of saccadic eye movements toward noise bursts presented from different loudspeaker locations. The horizontal eye position (measured with a scleral search coil) at the time of sound onset is plotted, relative to the cat's primary orbital position, on the abscissa. The angular deviation of the sound-evoked fixations from the average fixation position for each of the speakers is plotted on the ordinate. For clarity, data for the four noise targets have been pooled. The linear regression has a slope of 0.24, indicating that the saccadic localization of auditory targets was compensated for in this animal by 76% of the eyes' initial deviation from primary position. (B) Auditory spatial response profile of a unit recorded from the deeper layers of the SC in an alert cat, showing the modulating effect of a change in azimuthal eye position. Solid lines, eyes 14° left; dotted lines, eyes 5° left; dashed lines, eyes 10° right. (C) Another unit recorded from the same animal showing a lateral shift in its auditory spatial response profile in the same direction as the change in eye position. Solid lines, 10° left; dotted lines, eyes center; dashed lines, eyes 15° right. These responses were calculated from multiple regression equations in which the pinnal and vertical eye positions were held constant. (Adapted from Hartline et al., 1994)

Developmental plasticity in the coding of auditory space

POSTNATAL MATURATION OF SPECTRAL CUES AND AUDITORY SPATIAL TUNING Both binaural and monaural localization cue values depend on the size and shape of the head and outer ears, and might therefore be expected to vary between individuals. Moreover, growth of the head and ears will alter the relationship between the values of these localization cues and positions in space (Moore and Irvine, 1979; Carlile, 1991a,b). In the ferret, the spatial pattern of monaural and binaural spectral cues changes markedly over a period of several weeks during postnatal development (see color plate 4). The principal features of the monaural spectral cues become adultlike almost 4 weeks after the onset of hearing, which occurs at approximately 4 weeks after birth (Morey and Carlile, 1990), although the interaural spectral differences associated with each position in space continue to change beyond this age.

The presence of an adultlike map of auditory space during the period when the cues are changing in value would require continuous adjustment of the mechanisms responsible for the spatial tuning of SC neurons. However, soon after the onset of hearing, the spatial response profiles are mostly omnidirectional, and there is no sign of any topographical order in the auditory representation (Withington-Wray, Binns, and Keating, 1990; King and Carlile, 1991). Over the next few weeks, the neurons become more selective for sound location, and the topographical variation in best posi-

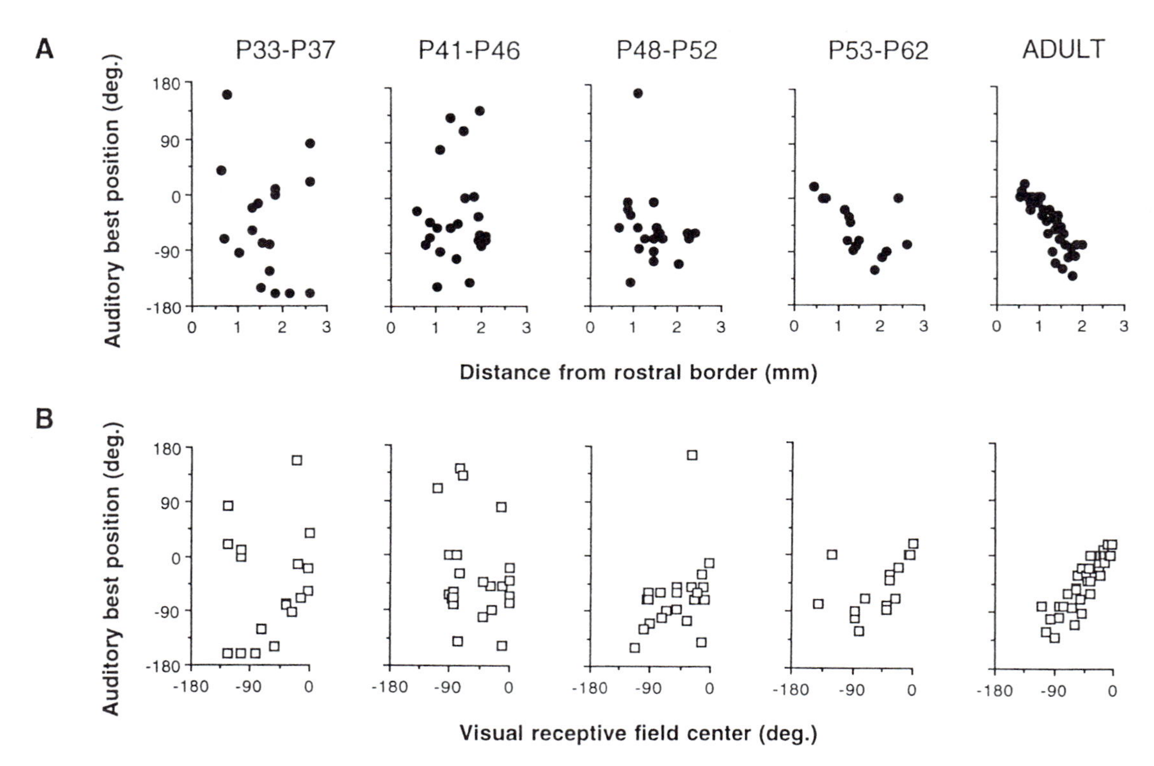

FIGURE 17.6 Development of the representation of auditory space in the superior colliculus (SC). The auditory and visual receptive fields of multiunit responses were mapped in ferrets at different postnatal ages. The neonatal responses were grouped by age as indicated at the top of the figure. (A) The auditory best positions plotted against the rostrocaudal location of the recording electrode. P, postnatal day. (B) Registration between the auditory and visual representations is shown by plotting the auditory best positions against the location of the center of the visual field recorded in the same electrode penetration. There is a gradual emergence of the topography of the auditory representation and a concomitant improvement in registration between the visual and auditory representations in the SC, which, by the eighth to ninth postnatal week, is approaching that in the adult.

tions gradually emerges (figure 17.6A). An adultlike representation of sound azimuth is apparent in the ferret SC at approximately 60 days after birth, which corresponds approximately with the age by which the spectral cues have matured.

In contrast to the gradual appearance of the auditory space map, an adultlike visual map is in place in the superficial layers of the SC in the youngest animals examined after eye opening. Consequently, the registration of the two maps gradually improves over the same time course as the development of the auditory map (figure 17.6B). This is compatible with a hierarchical sequence of development in which primary topographical maps mature before computational representations.

The maturation of responses in the central auditory system is determined in part by the development of the auditory periphery (Rubel, 1984; King and Moore, 1991). The changing appearance of the auditory space map may therefore simply reflect the acoustical cues achieved by depriving animals of all spatial cues early auditory cortex of the kitten possess adult-like sensitivity to ITDs and IIDs (Brugge, 1988). However, this does not mean that those neurons, particularly in a space-mapped region such as the SC, are tuned to the adult cue values. Binaural response properties of auditory cortical neurons may be continually adjusted throughout life to accommodate potential changes in localization cue values (Merzenich, Jenkins, and Middlebrooks, 1984), although this susceptibility to experiential factors is likely to be greatest when the head and ears are growing. A capacity for refinement appears to be particularly important for the auditory space map in the SC, which becomes aligned with the retinotopic map of visual space despite developmental changes in the relative geometry of the eyes and ears.

The Role of Sensory Experience in Development of the Auditory Space Map Rearing ferrets (King et al., 1988) and barn owls (Knudsen, 1985) with abnormal binaural cues leads to a compensatory adjustment in the auditory space map in the SC. Despite the presence of a plug in one ear, the auditory responses of SC neurons are spatially tuned at all sound levels, and their best positions are topographically organized and in register with the visual map (figure 17.7A, B). That this process involves an adjustment in selectivity to binaural cue values has been confirmed in barn owls (Mogdans and Knudsen, 1992). However, a similar period of monaural occlusion in adult animals (Knudsen, 1985; King, Hutchings, and Moore, 1992) does not appear to lead to a compensatory change in the neural mechanisms responsible for spatial selectivity (figure 17.7C).

Spectral localization cues appear to play a critical role in development of the auditory space map. If these are made ambiguous by early removal of the outer ears, spatially tuned responses fail to develop (King and Carlile, 1989). A similarly disruptive effect is achieved by depriving animals of all spatial cues early in life, as was accomplished by rearing guinea pigs in an environment of omnidirectional white noise (Withington-Wray et al., 1990).

Although these findings suggest that auditory experience plays an important role in shaping the auditory responses of SC neurons according to the acoustical cues available to individual animals, it appears to be the visual map that is responsible for aligning the sensory and motor maps during development (reviewed by King and Moore, 1991; Knudsen, 1991). Thus, altering or eliminating the visual cues available also changes auditory spatial tuning, suggesting that vision may play an instructive role in the development of the auditory space map. For example, surgically induced lateral deviation of one eye results in a corresponding change in the auditory spatial tuning of SC neurons, so that the registration of the auditory and visual maps is preserved (figure 17.8A). On the other hand, rotation of the eye appears to exceed the capacity of the mechanism that effects this adaptive realignment, because the auditory representation is no longer topographically

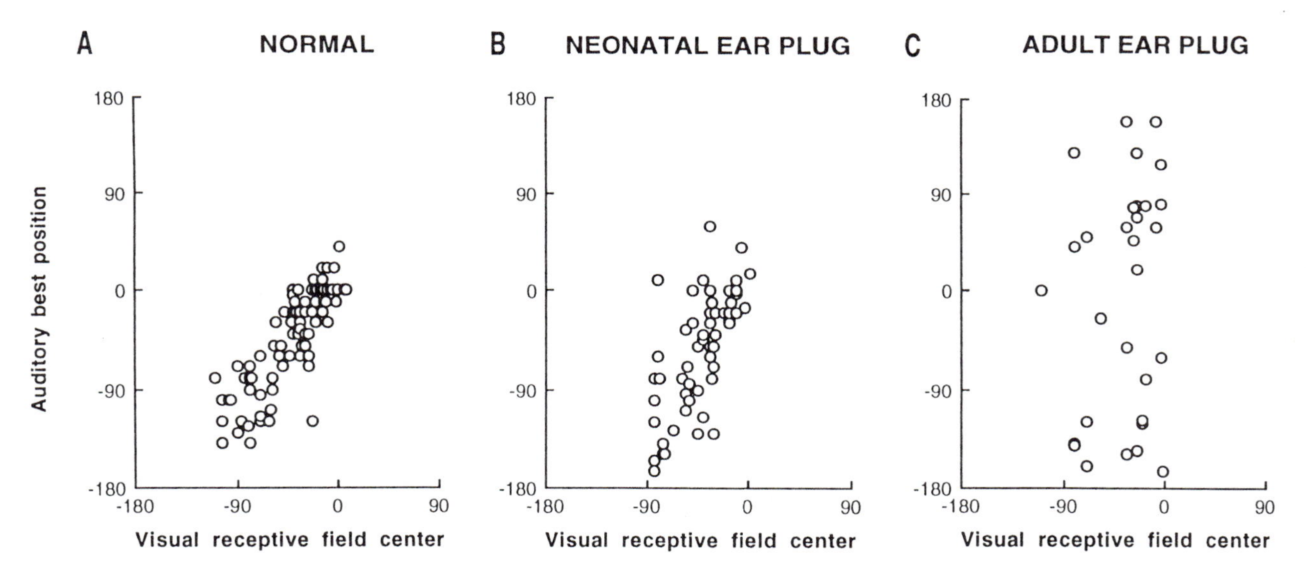

Figure 17.7 Effects of chronic monaural occlusion on the map of sound azimuth in the superior colliculus (SC). (A) Relationship between suprathreshold auditory best position and visual receptive field center (in degrees) for units recorded in the same electrode penetrations in normal, adult ferrets. (B) Corresponding data from adult ferrets that had been reared from before the onset of hearing with the ear ipsilateral to the recording site occluded by a plug that caused a frequency-dependent attenuation of 10–40 dB. Despite the presence of the earplug, the auditory responses were spatially tuned and topographically organized. (C) Data from ferrets that had been subjected to a similar period of monaural occlusion, this time commencing when they were at least 15 months old. The lack of correspondence between the visual and auditory coordinates in the latter group indicates that the topography of the auditory representation has been disrupted by the presence of the earplug. This suggests that auditory neurons in the SC have a much greater capacity to adjust to altered binaural cues during infancy than in adulthood. (Based on King et al., 1988)

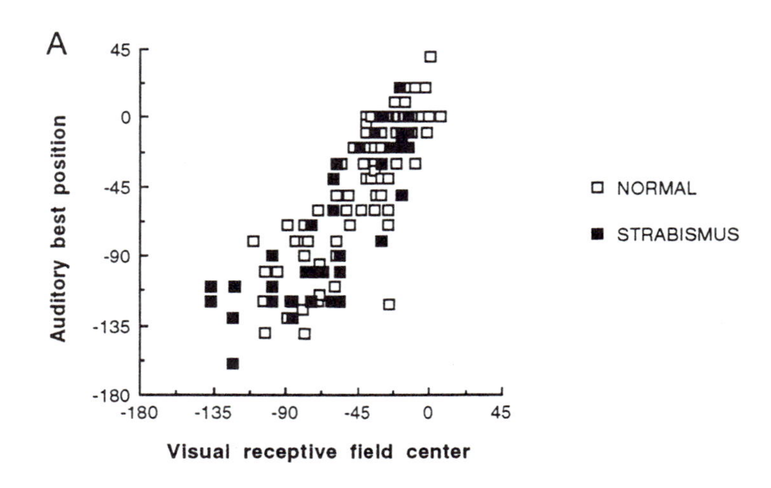

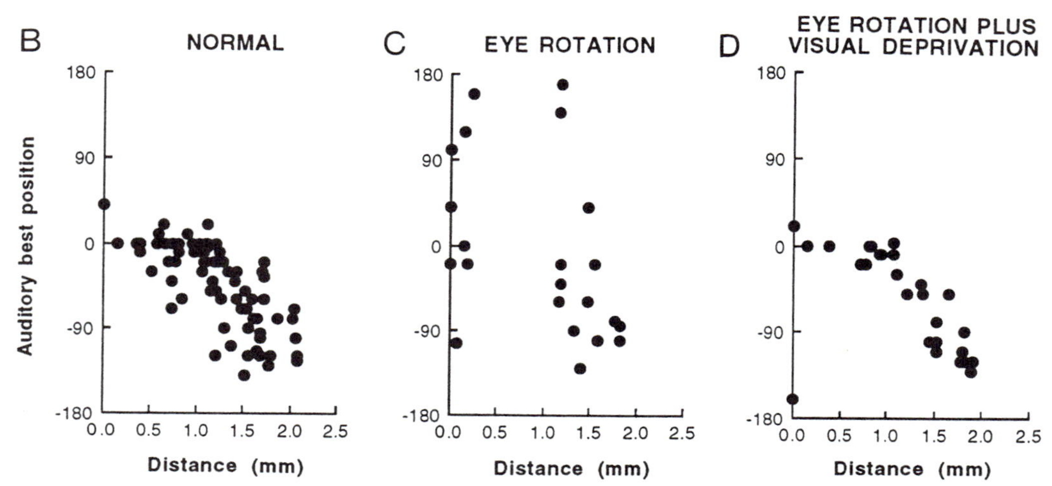

FIGURE 17.8 The effect of neonatal changes in eye position on the representation of sound azimuth in the superior colliculus (SC). (A) Relationship between suprathreshold auditory best position and visual receptive field center (in degrees) for units recorded in the same electrode penetration. The open squares represent data from normal, adult ferrets. The filled squares are from adult ferrets in which an outward deviation of the contralateral eye had been induced surgically before eye opening. The change in eye position shifted the region of both visual and auditory space represented in the SC by an amount corresponding to the 15–20° change in eye position. Consequently, the alignment of the visual and auditory maps is very similar in both sets of animals. (B–D) Variation in best azimuth at suprathreshold sound levels as a function of rostrocaudal distance in the SC. (B) Data from normal, adult ferrets. (C, D) Data from two groups of adults in which the contralateral eye had been rotated by 110–210° around its anteroposterior axis (after section of all six extraocular muscles). This surgery was also carried out before normal eye opening. Eye rotation leads to a disruption of the auditory map (B) but not if the animals are visually deprived by eyelid suture at the same time (C). (Based on King et al., 1988)

organized. However, the change in auditory spatial tuning following this procedure occurs only if the animals are allowed to see the displaced visual cues (figure 17.8B–D). The auditory responses therefore appear to be influenced by the visual signals reaching the SC rather than by the change in eye position per se. The basis by which vision refines development of the map of auditory space is unknown but is likely to involve the detection of correlated activity across different sensory maps.

Future directions: Virtual auditory environments

Studies employing free-field stimuli are important for demonstrating how auditory space is represented in the brain. On the other hand, the dichotic presentation of stimuli over headphones allows an analysis of the contributions of individual cues to the generation of the computational representations of auditory space. However, when sounds are presented over headphones the filtering characteristics of the outer ear are eliminated and the percept generated in a human listener is of a sound located inside the head. Changing the interaural cue values simply moves the internal location of the apparent sound source nearer to one ear or the other. If a complex stimulus is first filtered in a way that resembles the filtering of the outer ear and is then presented through headphones, the illusion of sounds located in the external world can be generated (Wightman, Kistler, and Perkins, 1987; Middlebrooks and Green, 1991). This impression is sometimes referred to as virtual auditory space (VAS), and indicates the special role played by the spectral transformations of the outer ear in generating our perception of external auditory space.

VAS stimulation allows the generation of biologically realistic signals that are equivalent to free-field stimuli but, in addition, provides complete control over all aspects of the localization cues presented to each ear (Carlile and King, 1993). This powerful technique is being employed in psychophysical (Wightman and Kistler, 1992) and physiological (Brugge et al., 1992) studies of the relative potency of different cues for localizing single, static sound sources. Moreover, VAS stimulation will facilitate the investigation of a number of complex auditory processes, such as the analysis of moving auditory targets, the segregation and localization of concurrent sounds and of sounds against background noise, and the role of reverberant sound fields in auditory distance processing.

ACKNOWLEDGMENTS We are grateful for the support provided by a Wellcome Senior Research Fellowship (A. J. K.) and a Beit Memorial Fellowship (S. C.). We also thank the Lister Institute of Preventive Medicine, Medical Research Council, and McDonnell-Pew Centre for Cognitive Neuroscience for financial support, and Alan Palmer, David Moore, and Colin Blakemore for their comments on an earlier version of the manuscript. Simon Carlile's present address is Department of Physiology, University of Sydney, New South Wales, Australia.

REFERENCES

BRUGGE, J. F., 1988. Stimulus coding in the developing auditory system. In *Auditory Function: Neurobiological Bases of Hearing*, G. M. Edelman, W. E. Gall, and W. M. Cowan, eds. New York: Wiley, pp. 113–136.

BRUGGE J. F., J. C. K. CHAN, J. E. HIND, A. D. MUSICANT, P. W. F. POON, and R. A. REALE, 1992. Neural encoding of virtual acoustic space. *J. Acoust. Soc. Am.* 92:2333–2334.

BUTLER, R. A., R. A. HUMANSKI, and A. D. MUSICANT, 1990. Binaural and monaural localization of sound in two-dimensional space. *Perception* 19:241–256.

CARLILE, S., 1990a. The auditory periphery of the ferret: I. Directional response properties and the pattern of interaural level differences. *J. Acoust. Soc. Am.* 88:2180–2195.

CARLILE, S., 1990b. The auditory periphery of the ferret: II. The spectral transformations of the external ear and their implications for sound localization. *J. Acoust. Soc. Am.* 88: 2196–2204.

CARLILE, S., 1991a. The auditory periphery of the ferret: Postnatal development of acoustic properties. *Hear. Res.* 51:265–278.

CARLILE, S., 1991b. Postnatal development of the spectral transfer functions and interaural level differences of the auditory periphery of the ferret. *Abstr. Soc. Neurosci.* 17: 232.

CARLILE, S., and A. J. KING, 1993. Auditory neuroscience: From outer ear to virtual space. *Curr. Biol.* 3:446–448.

CARLILE, S., and A. J. KING, 1994. Monaural and binaural spectrum level cues in the ferret: Acoustics and the neural representation of auditory space. *J. Neurophysiol.* 71:785–801.

CARLILE, S., and A. G. PETTIGREW, 1987. Distribution of frequency sensitivity in the superior colliculus of the guinea pig. *Hear. Res.* 31:123–136.

CLAREY, J. C., P. BARONE, and T. J. IMIG, 1992. Physiology of thalamus and cortex. In *The Mammalian Auditory Pathway: Neurophysiology*, A. N. Popper and R. R. Fay, eds. New York: Springer-Verlag, pp. 232–334.

FISHER, H. G., and S. J. FREEDMAN, 1968. Localization of sound during simulated unilateral conductive hearing loss. *Acta Otolaryngol. (Stockh.)* 66:213–220.

GORDON, B., 1973. Receptive fields in deep layers of cat superior colliculus. *J. Neurophysiol.* 36:157–178.

HARRIS, L. R., C. BLAKEMORE, and M. DONAGHY, 1980. Integration of visual and auditory space in the mammalian superior colliculus. *Nature* 288:56–59.

HARTLINE, P. H., A. J. KING, D. D. KURYLO, D. P. M. NORTHMORE, and R. L. P. VIMAL, 1989. Effects of eye position on auditory localization and auditory spatial representation in cat superior colliculus. *Invest. Ophthalmol. Vis. Sci.* Suppl. 30:181.

HARTLINE, P. H., A. J. KING, and D. P. M. NORTHMORE, 1989. How do pinna orientation and movement affect auditory localization and superior collicular receptive fields of alert cats? In *Neural Mechanisms of Behavior*, J. Erber, R. Menzel, H.-J. Pflüger, and D. Todt, eds. Stuttgart: Verlag, p. 242a.

HARTLINE, P. H., R. L. P. VIMAL, A. J. KING, D. D. KURYLO, and D. P. M. NORTHMORE, 1994. Effects of eye position on auditory localization and neural representation of space in superior colliculus of cats. Manuscript submitted for publication.

Heffner, H., 1978. Effect of auditory cortex ablation on localization and discrimination of brief sounds. *J. Neurophysiol.* 41:963–976.

Heffner, H. E., and R. S. Heffner, 1990. Effect of bilateral auditory cortex lesions on sound localization in Japanese macaques. *J. Neurophysiol.* 64:915–931.

Hirsch, J. A., J. C. K. Chan, and T. C. T. Yin, 1985. Responses of neurons in the cat's superior colliculus to acoustic stimuli: I. Monaural and binaural response properties. *J. Neurophysiol.* 53:726–745.

Imig, T. J., W. A. Irons, and F. R. Samson, 1990. Single-unit selectivity to azimuthal direction and sound pressure level of noise bursts in cat high-frequency primary auditory cortex. *J. Neurophysiol.* 63:1448–1466.

Irvine, D. R. F., 1986. The auditory brainstem. *Prog. Sensory Physiol.* 7:1–279.

Irvine, D. R. F., 1992. Physiology of the auditory brainstem. In *The Mammalian Auditory Pathway: Neurophysiology*, A. N. Popper and R. R. Fay, eds. New York: Springer-Verlag, pp. 153–231.

Jay, M. F., and D. L. Sparks, 1987. Sensorimotor integration in the primate superior colliculus: II. Coordinates of auditory signals. *J. Neurophysiol.* 57:35–55.

Jenkins, W. M., and R. B. Masterton, 1982. Sound localization: Effects of unilateral lesions in central auditory system. *J. Neurophysiol.* 47:987–1016.

Jenkins, W. M., and M. M. Merzenich, 1984. Role of cat primary auditory cortex for sound-localization behavior. *J. Neurophysiol.* 52:819–847.

King, A. J., and S. Carlile, 1989. Generation of an auditory space map in the ferret superior colliculus requires the presence of monaural localization cues. *Abstr. Soc. Neurosci.* 15:746.

King, A. J., and S. Carlile, 1991. Maturation of the map of auditory space in the superior colliculus of the ferret. *Abstr. Soc. Neurosci.* 17:231.

King, A. J., S. Carlile, and T. J. T. Chevassut, 1990. Spatial organization of auditory receptive fields in the ferret superior colliculus. *Br. J. Audiol.* 24:198.

King, A. J., and M. E. Hutchings, 1987. Spatial response properties of acoustically responsive neurons in the superior colliculus of the ferret: A map of auditory space. *J. Neurophysiol.* 57:596–624.

King, A. J., M. E. Hutchings, and D. R. Moore, 1992. Neurophysiological consequences of chronic, unilateral hearing loss in adult ferrets. *Br. J. Audiol.* 26:199–200.

King, A. J., M. E. Hutchings, D. R. Moore, and C. Blakemore, 1988. Developmental plasticity in the visual and auditory representations in the mammalian superior colliculus. *Nature* 332:73–76.

King, A. J., and D. R. Moore, 1991. Plasticity of auditory maps in the brain. *Trends Neurosci.* 14:31–37.

King, A. J., D. R. Moore, and M. E. Hutchings, 1994. Topographic representation of auditory space in the superior colliculus of adult ferrets after monaural deafening in infancy. *J. Neurophysiol.* 71:182–194

King, A. J., and A. R. Palmer, 1983. Cells responsive to free-field auditory stimuli in guinea-pig superior colliculus: Distribution and response properties. *J. Physiol.* 342:361–381.

Knudsen, E. I., 1982. Auditory and visual maps of space in the optic tectum of the owl. *J. Neurosci.* 2:1177–1194.

Knudsen, E. I., 1985. Experience alters the spatial tuning of auditory units in the optic tectum during a sensitive period in the barn owl. *J. Neurosci.* 5:3094–3109.

Knudsen, E. I., 1991. Dynamic space codes in the superior colliculus. *Curr. Opin. Neurobiol.* 1:628–632.

Knudsen, E. I., and M. Konishi, 1978. A neural map of auditory space in the owl. *Science* 200:795–797.

Lewis, B., 1983. Directional cues for auditory localization. In *Bioacoustics: A Comparative Approach*, B. Lewis, ed. London: Academic Press, pp. 233–257.

McIlwain, J. T., 1991. Distributed spatial coding in the superior colliculus: A review. *Visual Neurosci.* 6:3–13.

Merzenich, M. M., W. M. Jenkins, and J. C. Middlebrooks, 1984. Observations and hypotheses on special organizational features of the central auditory nervous system. In *Dynamic Aspects of Neocortical Function*, G. M. Edelman, W. E. Gall, and W. M. Cowan, eds. New York: Wiley, pp. 397–424.

Middlebrooks, J. C., 1987. Binaural mechanisms of spatial tuning in the cat's superior colliculus distinguished using monaural occlusion. *J. Neurophysiol.* 57:688–701.

Middlebrooks, J. C., 1992. Narrow-band sound localization related to external ear acoustics. *J. Acoust. Soc. Am.* 92: 2607–2624.

Middlebrooks, J. C., and D. M. Green, 1991. Sound localization by human listeners. *Annu. Rev. Psychol.* 42:135–159.

Middlebrooks, J. C., and E. I. Knudsen, 1984. A neural code for auditory space in the cat's superior colliculus. *J. Neurosci.* 4:2621–2634.

Middlebrooks, J. C., and E. I. Knudsen, 1987. Changes in external ear position modify the spatial tuning of auditory units in the cat's superior colliculus. *J. Neurophysiol.* 57: 672–687.

Middlebrooks, J. C., J. C. Makous, and D. M. Green, 1989. Directional sensitivity of sound-pressure levels in the human ear canal. *J. Acoust. Soc. Am.* 86:89–108.

Middlebrooks, J. C., and J. D. Pettigrew, 1981. Functional classes of neurons in primary auditory cortex of the cat distinguished by sensitivity to sound location. *J. Neurosci.* 1:107–120.

Mogdans, J., and E. I. Knudsen, 1992. Adaptive adjustment of unit tuning to sound localization cues in response to monaural occlusion in developing owl optic tectum. *J. Neurosci.* 12:3473–3484.

Moore, D. R., and D. R. F. Irvine, 1979. A developmental study of the sound pressure transformation by the head of the cat. *Acta Otolaryngol. (Stockh.)* 87:434–440.

Morey, A. L., and S. Carlile, 1990. Auditory brainstem of the ferret: Maturation of the brainstem auditory evoked response. *Dev. Brain Res.* 52:279–288.

Musicant, A. D., and R. A. Butler, 1984. The influence of pinnae-based spectral cues on sound localization. *J. Acoust. Soc. Am.* 75:1195–1200.

Musicant, A. D., J. C. K. Chan, and J. E. Hind, 1990. Direction-dependent spectral properties of cat external ear: New data and cross-species comparisons. *J. Acoust. Soc. Am.* 87:757–781.

Oldfield, S. R., and S. P. A. Parker, 1984. Acuity of sound localisation: A topography of auditory space: II. Pinna cues absent. *Perception* 13:601–617.

Oldfield, S. R., and S. P. A. Parker, 1986. Acuity of sound localisation: A topography of auditory space: III. Monaural hearing conditions. *Perception* 15:67–81.

Olsen, J. F., E. I. Knudsen, and S. D. Esterly, 1989. Neural maps of interaural time and intensity differences in the optic tectum of the barn owl. *J. Neurosci.* 9:2591–2605.

Palmer, A. R., and A. J. King, 1985. A monaural space map in the guinea-pig superior colliculus. *Hear. Res.* 17: 267–280.

Peck, C. K., and F. S. Wartman, 1989. Effects of eye position on auditory responses in cat superior colliculus. *Invest. Ophthalmol. Vis. Sci.* Suppl. 30:181.

Rajan, R., L. M. Aitkin, and D. R. F. Irvine, 1990. Azimuthal sensitivity of neurons in primary auditory cortex of cats: II. Organization along frequency-band strips. *J. Neurophysiol.* 64:888–902.

Rayleigh (Lord), 1907. On our perception of sound direction. *Philos. Mag.* 13:214–232.

Rice, J. J., B. J. May, G. A. Spirou, and E. D. Young, 1992. Pinna-based spectral cues for sound localization in cat. *Hear. Res.* 58:132–152.

Rubel, E. W., 1984. Ontogeny of auditory system function. *Annu. Rev. Physiol.* 46:213–229.

Russo, G. S., and C. J. Bruce, 1989. Auditory receptive fields of neurons in frontal cortex of rhesus monkey shift with direction of gaze. *Soc. Neurosci. Abstr.* 15:1204.

Sparks, D. L., 1988. Neural cartography: Sensory and motor maps in the superior colliculus. *Brain Behav. Evol.* 31:49–56.

Stein, B. E., and M. A. Meredith, 1990. Multisensory integration: Neural and behavioral solutions for dealing with stimuli from different sensory modalities. *Ann. N.Y. Acad. Sci.* 608:51–65.

Thompson, G. C., and R. B. Masterton, 1978. Brain stem auditory pathways involved in reflexive head orientation to sound. *J. Neurophysiol.* 41:1183–1202.

Wagner, H., 1993. Sound-localization deficits induced by lesions in the barn owl's auditory space map. *J. Neurosci.* 13:371–386.

Wightman, F. L., and D. J. Kistler, 1992. The dominant role of low-frequency interaural time differences in sound localization. *J. Acoust. Soc. Am.* 91:1648–1661.

Wightman, F. L., D. J. Kistler, and M. E. Perkins, 1987. A new approach to the study of human sound localization. In *Directional Hearing*, W. A. Yost and G. Gourevitch, eds. New York: Springer-Verlag, pp. 26–48.

Wise, L. Z., and D. R. F. Irvine, 1983. Auditory response properties of neurons in deep layers of cat superior colliculus. *J. Neurophysiol.* 49:674–685.

Wise, L. Z., and D. R. F. Irvine, 1985. Topographic organization of interaural intensity difference sensitivity in deep layers of cat superior colliculus: Implications for auditory spatial representation. *J. Neurophysiol.* 54:185–211.

Withington-Wray, D. J., K. E. Binns, S. S. Dhanjal, S. G. Brickley, and M. J. Keating, 1990. The maturation of the superior collicular map of auditory space in the guinea pig is disrupted by developmental auditory deprivation. *Eur. J. Neurosci.* 2:693–703.

Withington-Wray, D. J., K. E. Binns, and M. J. Keating, 1990. The developmental emergence of a map of auditory space in the superior colliculus of the guinea pig. *Dev. Brain Res.* 51:225–236.

Wong, D., 1984. Spatial tuning of auditory neurons in the superior colliculus of the echolocating bat, *Myotis lucifugus*. *Hear. Res.* 16:261–270.

Young, E. D., G. A. Spirou, J. J. Rice, and H. F. Voigt, 1992. Neural organization and responses to complex stimuli in the dorsal cochlear nucleus. *Philos. Trans. R. Soc. Lond. [Biol.]* 336:407–413.

18 Processing of Auditory Information Carried by Species-specific Complex Sounds

NOBUO SUGA

ABSTRACT Human speech and animal sounds share three major types of information-bearing elements (IBEs). The IBEs and combinations of the IBEs are characterized by information-bearing parameters (IBPs). For the representation of auditory information at a hypothetical auditory center, five hypotheses are conceivable: frequency axis, amplitude-frequency coordinate, detector, IBP filter, and synchronization hypotheses. The auditory periphery of the mustached bat shows specializations for detection and frequency analysis of species-specific complex biosonar signals. It contains neurons that are utilized for fine level-tolerant frequency analysis and others that are utilized for fine time analysis. The bat's central auditory system processes different types of auditory information in a parallel-hierarchical way and represents these in different areas of the auditory cortex. These areas are physiologically distinct and exhibit functional organizations that support four of the five hypotheses.

Information-bearing elements and parameters in species-specific sounds

In higher vertebrates, communication sounds are usually complex. The amplitude spectra of these sounds commonly change with time. The sonagrams of animal sounds, including human speech sounds, exhibit three basic patterns or components: constant-frequency (CF), noise-burst (NB), and frequency-modulated (FM) components (figure 18.1A). For instance, the sonagram of a consonant consists of a vertical bar or band indicating the scatter of sound energy over many frequencies. This is called a *fill* or noise burst. The fricative consonants /*s*/ and /*sh*/, for example, can be recognized by their high-energy NBs concentrated at approximately 2–3 kHz and above 4.4 kHz, respectively. The sonagram of a vowel consists of several horizontal bars called *formants*, which are spectral peaks characterizing a vowel. The lowest is the first formant (F_1), the second lowest is the second formant (F_2), and so on. These may be considered CF components. Vowels are identified by the combinations of F_1, F_2, and F_3, so that they are expressed by loci in frequency-versus-frequency coordinates (figure 18.1C) (Peterson and Barney, 1952). When two phonemes are combined to form a monosyllabic word, new components called *transitions* appear. These are FM components (see figure 18.1A) and are very important for the identification of consonants within words. For example, plosive consonants /k/, /t/, /p/, /g/, /d/, and /b/ are identified by the transitions before F_1 and F_2 of a vowel (figure 18.1D) (Liberman et al., 1956). Human speech sounds thus consist of three types of *information-bearing elements* (IBEs).

These three types of IBEs are also found in sounds produced by many species of animals. For instance, the mustached bat, *Pteronotus parnellii*, emits complex biosonar signals consisting of long CF and short FM components. It also emits at least 19 syllable types of nonbiosonar sounds, which are presumably communication calls. These syllables consist of CF, FM, or NB components (Kanwal et al., 1993).

Auditory information is carried not only by the acoustic parameters characterizing each of these three types of IBEs but also by parameters representing relationships among these elements in the frequency,

NOBUO SUGA Department of Biology, Washington University, St. Louis, Mo.

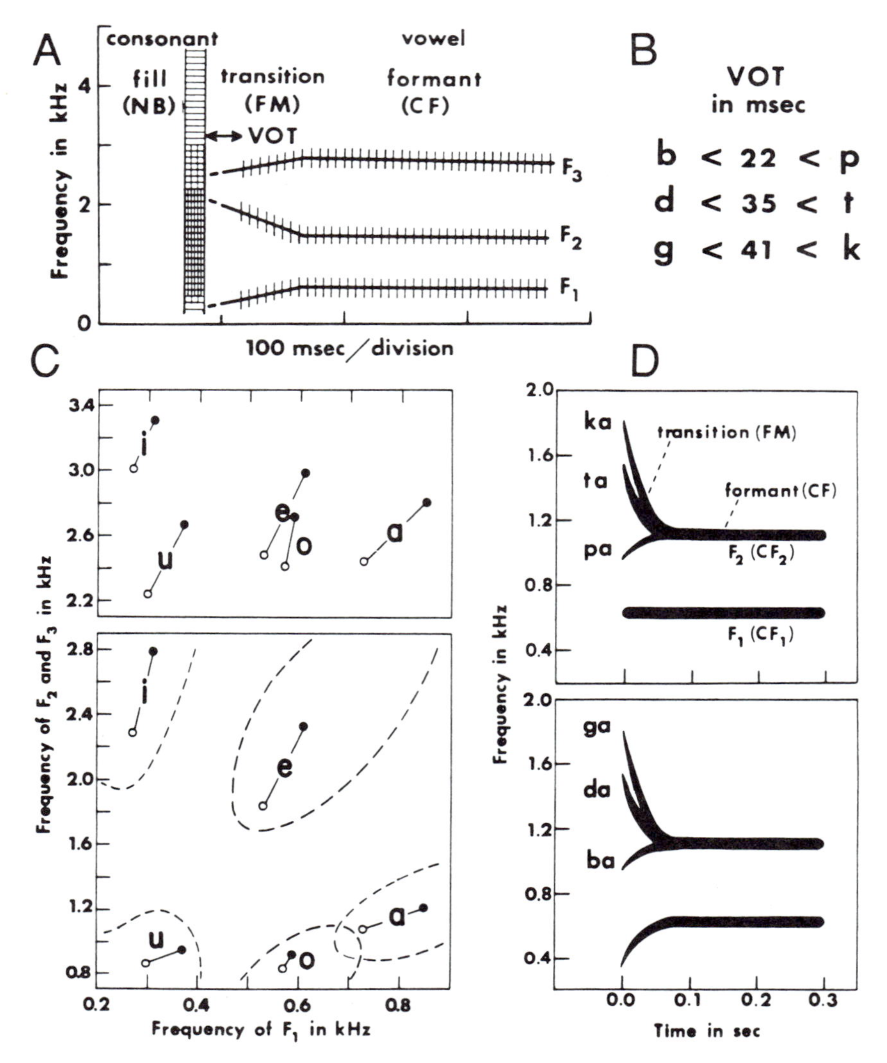

FIGURE 18.1 Information-bearing elements (IBEs) in human speech sounds. (A) Schematized sonagram of a monosyllabic sound shows four types of IBEs: fill or noise-burst component (NB), transition or frequency-modulated component (FM), formant or constant-frequency component (CF), and voice onset time (VOT) or time interval between two acoustical events. F_1, F_2, and F_3 are the first, second, and third formants of a vowel, respectively. (B) The phonetic boundary of VOT is 22 ms for the consonants /*b*/ and /*p*/, 35 ms for /*d*/ and /*t*/, and 41 ms for /*g*/ and /*k*/. For example, when the VOT is shorter than 22 ms, a monosyllabic sound is recognized as /*ba*/. When it is longer than 22 ms, however, it is recognized as /*pa*/. (Based on Lisker and Abramson, 1964.) (C) The relationship between the frequencies of the F_1 and F_2 or F_3 of five vowels (/*a*/, /*e*/, /*i*/, /*o*/, and /*u*/). The average frequencies of the formants differ between male (open circles) and female (filled circles) speakers. The areas surrounded by dashed lines in the lower figure represent formant frequencies for unanimous classification of vowels. (Based on Peterson and Barney, 1952.) (D) Schematized sonograms of monosyllabic sounds: /*a*/ changes into /*pa*/, /*ta*/, or /*ka*/ through the addition of different FM components of F_2. The sounds /*pa*/, /*ta*/, and /*ka*/ become /*ba*/, /*da*/, and /*ga*/, respectively, through the addition of an FM component to F_1. (Based on Liberman et al., 1956.) (Reprinted from Suga, 1984, by permission.)

amplitude, or time domains. For instance, voice-onset time is the time interval between two acoustical events and is an important cue for speech recognition (see figure 18.1A, B) (Lisker and Abramson, 1964). Target-range information important for a bat is carried not by an echo alone but by the delay of the echo from an emitted biosonar pulse. A parameter is a continuum and can have any value. However, only a limited part of the continuum is important for each species. This limited part of the continuum has been called the *infor-*

mation-bearing parameter (IBP) (Suga, 1982, 1984). An identical IBP may have different biological significance for different species. Conversely, different IBPs may have an identical biological significance for different species. It must be noted that IBPs for sound localization (i.e., binaural cues) are created by a receiver.

Representation of auditory information in the auditory cortex

In the mammalian cochlea, sensory cells may be viewed as filters arranged along a frequency axis on the basilar membrane for frequency analysis. Their outputs are coded by primary auditory neurons that discharge action potentials (impulses) at higher rates for larger outputs. Therefore, at the periphery, the frequency of an acoustical signal is expressed by the location of activated neurons and its amplitude by their discharge rates. The duration of the signal and the time interval between signals are expressed by the temporal pattern of neural activity. The peripheral auditory system has an anatomical axis for frequency only. The activity of individual peripheral neurons cannot uniquely express the properties of an acoustic signal. For instance, a peripheral neuron tuned to 40 kHz responds not only to a pure tone of 40 kHz but also to an FM sound sweeping across 40 kHz, regardless of sweep direction, and to a noise burst containing 40 kHz, regardless of bandwidth. Therefore, the neuron cannot code the type of acoustic signal stimulating the ear. The properties of an acoustic signal are appropriately expressed only by the spatiotemporal pattern of the activity of all peripheral neurons.

Action potentials sent into the brain by peripheral neurons are transmitted to many auditory nuclei and finally to the cerebral cortex. In the ascending auditory system, each level consists of a few sudivisions, and each subdivision has a frequency axis. This means that there are multiple representations of the cochlea at each level as a result of divergent projections of ascending nerve fibers (multiple cochleotopic or tonotopic representations). The divergent projections are incorporated with convergent interactions, so that response properties of neurons are created that are very different from those of peripheral neurons. For example, neurons that selectively respond to one of the three types of IBEs are created in the central auditory system. They are called *CF-*, *FM-*, or *NB-specialized neurons* (Suga, 1969; Casseday and Covey, 1992). For processing complex species-specific sounds, the central auditory system must extract different types of auditory information from various combinations of IBEs, so that different types of combination-sensitive neurons are also created in the central auditory system (see Suga's review article, 1990).

The auditory cortex of mammals consists of many areas or subdivisions, and the multiple cochleotopic representations are prominent there. This suggests that separate auditory areas are concerned with representing (processing) different types of auditory information. What kind of information is represented in each area? How is each area functionally organized? Neurophysiological data obtained suggest at least five possible functional organizations within the auditory cortex. Therefore, five working hypotheses of the functional organization of an auditory center may be proposed to explain the neural representation of auditory information. These hypotheses are not mutually exclusive but valid depending on the type of auditory information and species (Suga, 1982).

THE FREQUENCY AXIS HYPOTHESIS The frequency axis hypothesis suggests that the hypothetical auditory center has no anatomical axis representing an acoustic parameter other than frequency and that a complex acoustic signal is represented by the spatiotemporal pattern of activity of primary-like neurons arranged along the frequency axis. A frequency axis (tonotopic representation) has been demonstrated in the primary auditory cortex of many species of mammals, including bats (Suga, 1965b; Suga and Jen, 1976), cats (e.g., Merzenich et al., 1975), and monkeys (e.g., Brugge and Merzenich, 1973). Therefore, this hypothesis cannot be ruled out. However, it has been considered that it is unrealistically simple, because most cortical neurons show nonmonotonic impulse-count functions and excitatory and inhibitory frequency-tuning curves. In the cat and the ferret, response properties of neurons in the primary auditory cortex vary along the axis orthogonal to the frequency axis (Imig and Adrian, 1977; Schreiner et al., 1992; Schreiner and Sutter, 1992; Shamma et al., 1993).

THE FREQUENCY-AMPLITUDE COORDINATE HYPOTHESIS The frequency-amplitude coordinate hypothesis suggests that a complex acoustical signal is represented by the spatiotemporal pattern of activity of neurons that are tuned to particular combinations of frequency and

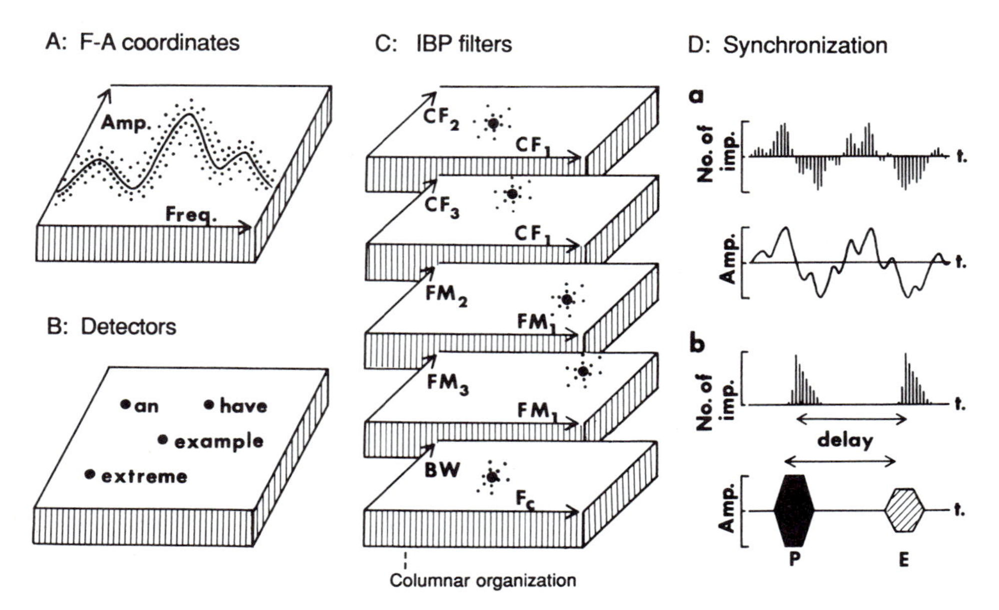

FIGURE 18.2 Four hypotheses (A–D) for the representation of auditory information by neural activity in a hypothetical center. In all hypotheses, auditory information is represented by the spatiotemporal pattern of neural activity. The response properties of individual neurons and the interpretation of their functions are, however, different according to the hypotheses. (A) The frequency-amplitude coordinate hypothesis. (B) The detector hypothesis. (C) The information-bearing parameter (IBP) filter hypothesis. Here coordinates are formed by neurons tuned to IBPs that characterize biologically important signal elements or combinations of the elements. CF_1, CF_2, CF_3, FM_1, FM_2, and FM_3 are different signal elements. F_c, center frequency of noise bursts; BW, bandwidth of noise bursts. (D) The synchronization hypothesis, which is an extended version of Wever's volley theory. In D_a, the lower trace represents a sound wave; the upper trace is the compound period histogram displaying the response of a single neuron to it. In D_b, the lower trace represents an oscillogram of a biosonar pulse (P) and its echo (E); the upper trace is a peristimulus-time (PST) histogram displaying the response of a single neuron to the P-E pair. (Reprinted from Suga, 1982, by permission.)

amplitude and that are arranged along the coordinates of frequency versus amplitude at a hypothetical auditory center (figure 18.2A). Each of these neurons has a small excitatory area, is tuned to a particular combination of frequency and amplitude, and responds to different types of sounds containing a component that stimulates this area.

As described earlier, a frequency axis has been demonstrated in the auditory cortex of many species of mammals. Neurons tuned to particular amplitudes as well as frequencies are common in the particular subdivisions of the auditory cortex of bats (Suga, 1977; Suga and Manabe, 1982), cats (Phillips and Orman, 1984), and monkeys (Brugge and Merzenich, 1973). Therefore, the central auditory systems of these animals have neurons necessary for forming the frequency-amplitude coordinates. However, an amplitude axis (amplitopic representation) has been found only in the Doppler-shifted CF (DSCF) processing area of the auditory cortex of the mustached bat. Recently, Pantev and colleague (1989) have claimed that the human auditory cortex shows amplitopic representation, and Schreiner and coworkers (1992) have obtained data suggesting the presence of an amplitude axis in the cat's primary auditory cortex.

An amplitude spectrum can be expressed in the coordinates of frequency versus threshold, as in the auditory periphery (Liberman, 1978, 1982). Because best amplitudes of cortical neurons of the mustached bat are linearly related to their minimum thresholds, the amplitude axis is related to the threshold axis. However, threshold representation by neurons with a monotonic impulse-count function is different from the amplitopic representation and has two disadvantages compared to the latter: First, for threshold representation, more neurons are involved in representing the amplitude spectrum of a stronger acoustical stimulus, and second, the boundary between excited and nonexcited neurons is unclear and variable.

A threshold representation found in the dog by Tunturi (1952) is intriguing, but it has not been accepted by auditory physiologists, probably because the threshold representation was found only for the stimulation of the ipsilateral ear, which does not supply the main excitatory input to the auditory cortex. The main excitatory input comes from the contralateral ear and is uniformly lower in threshold than that from the ipsilateral ear. Also, the functional significance of the threshold representation is not clear, because most of the cortical auditory neurons show relatively phasic ON responses to tone bursts or nonmonotonic impulse-count functions.

THE DETECTOR HYPOTHESIS The detector hypothesis suggests that a biologically important acoustical signal is represented by the excitation of detector neurons that selectively respond to that particular signal (figure 18.2B). Different types of detector neurons are, of course, arranged in a particular spatial pattern in the hypothetical auditory center. The spatiotemporal pattern of neural activity in the auditory center thus will change according to the sequence of biologically important sounds. The distinguishing feature of the detector hypothesis is a one-to-one correspondence between a particular signal and the excitation of a detector neuron or neurons in a single cortical column. Neurons specialized to respond selectively to certain types of acoustical stimuli have been found in several species of animals, such as bats (reviewed by Suga, 1973, 1984), monkeys (Symmes, 1981), song birds (Margoliash, 1983), and frogs (Fuzessery and Feng, 1982, 1983), but it remains to be ascertained whether such specialized neurons are appropriately called *detectors*.

THE IBP FILTER HYPOTHESIS The IBP filter hypothesis falls between the previous two hypotheses. In auditory neurophysiology, the statement that neurons respond to sound *x* but not to others is justified only when the filter properties of the neurons (i.e., tuning curves of the neurons to individual parameters characterizing sound *x*) are studied. The neurons act as filters that correlate acoustical signals with their filter properties—that is, stored information. The degree of correlation is expressed by the magnitude of the output of the filters. The neurons are maximally excited only when the properties of acoustical signals perfectly match their filter properties. All neurons in the auditory system, including peripheral ones, act as filters. Specialized neurons expressing the outputs of neural circuits tuned to particular IBPs or particular combinations of IBPs are called *IBP filters* (Suga, 1978).

Communication sounds are commonly characterized by many different parameters: frequency, FM rate, FM depth, amplitude, amplitude modulation (AM) rate, AM depth, harmonic structure, duration, interval, and so on. Some parameters are IBPs. The IBP filter hypothesis states that the auditory center represents IBPs or combinations of IBPs by spatiotemporal patterns of activities of specialized neurons (IBP filters) acting as a kind of cross-correlator (figure 18.2C). It also states that different types of IBP filters are aggregated separately in identifiable areas of the auditory center. This hypothesis is strongly supported by the data obtained from the auditory cortex of the mustached bat, which are described later.

THE SYNCHRONIZATION HYPOTHESIS The synchronization hypothesis is an expanded version of the volley theory (Wever, 1949). When IBEs are lower than 5 kHz, peripheral neurons of mammals produce discharges synchronized with the sound waves. The envelope of a compound period histogram of a neural response thus reproduces the stimulus waveform (figure 18.2D, a). The synchronization hypothesis states that neurons in the hypothetical auditory center represent acoustical signals by synchronous discharges. In small mammals, many of the predominant components of their communication sounds are higher than 5 kHz and hence cannot be coded by synchronous discharges, but the rate of sound emission or amplitude modulation usually is less than 1000 per second. For example, echolocating bats emit ultrasonic signals (biosonar pulses) and listen to echoes. The rate of pulse emission usually is less than 200 per second. The delay of an echo from the emitted pulse is the primary cue for target ranging. Peripheral neurons show discharges synchronized with each emitted pulse and its echo. Therefore, range information is coded by the time interval between a pair of grouped discharges (figure 18.2D, b). The synchronization hypothesis states that range information is represented in the auditory cortex by paired grouped discharges but not by an excitation of neurons specialized for responding to a particular echo delay.

In the cat's medial geniculate body, only 3% of neurons that are activated by tone bursts up to 1.0 kHz show clear phase-locked responses, and phase-locked responses to sounds higher than 1.0 kHz rarely are found (Rouiller et al., 1979). It is expected that both the population of phase-locking neurons and the degree of phase locking are smaller in the auditory cortex than in the medial geniculate body. Therefore, the synchronization hypothesis can be true only for the representation of some of the properties of acoustical signals, such as a low-frequency fundamental and AM or FM occurring periodically at low rates.

In the FM-FM area of the auditory cortex of the mustached bat, neurons are tuned to particular time intervals between the pulse and its echo. The time intervals are systematically represented by loci of activated neurons in this area (Suga and O'Neill, 1979; O'Neill and Suga, 1982). Therefore, the *temporal code* at the periphery is changed into a *place code* at the auditory center (Suga, 1990).

Synchronous (phase-locked) discharges play an important role in information processing for sound localization (Rose et al., 1966; Yin and Kuwada, 1984; Sullivan and Konishi, 1984). However, the location of

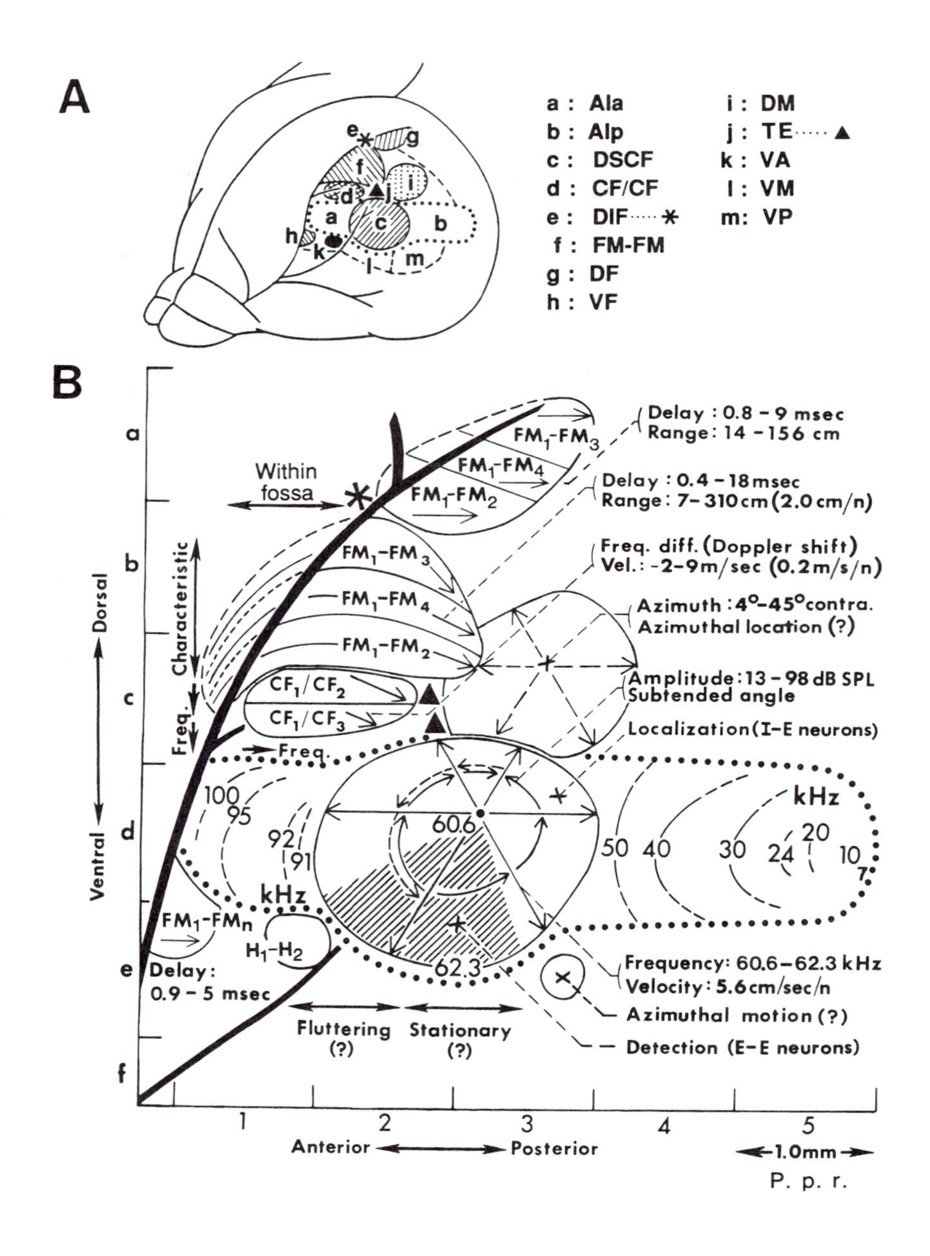

a sound source is represented in the midbrain not by the magnitude of phase-locked discharges but by the locus of activated neurons that are tuned to particular interaural time or phase differences (Knudsen and Konishi, 1978; Konishi, this volume).

Neural processing of biosonar information

The auditory cortex of the mustached bat consists of many areas or subdivisions, each of which is electrophysiologically distinct. The functional organizations of these cortical areas are very different from one another. Some of these areas are clearly specialized for processing certain types of biosonar information (figure 18.3). The DSCF area is part of the primary auditory cortex, and its organization fits the frequency-amplitude coordinate hypothesis. The FM-FM, dorsal fringe (DF), ventral fringe (VF), and CF/CF areas are organized in a way that fits the IBP filter hypothesis. The anterior (AIa) and posterior (AIp) portions of the primary auditory cortex have a frequency axis, but it is not yet known what is expressed along the axis orthogonal to the frequency axis. All neurons in the areas just listed as well as others show stimulus-locked responses when an acoustical stimulus is delivered at low repetition rates. Therefore, the synchronization hypothesis is valid to some extent.

Among several species of animals studied, the functional organization of the auditory cortex has been most extensively explored in the mustached bat, and its auditory cortex shows the functional organizations hypothesized in figure 18.2. Therefore, the neural processing of biosonar information in this species is summarized here.

FIGURE 18.3 Functional organization of the auditory cortex of the mustached bat. (A) Dorsolateral view of the left cerebral hemisphere. The auditory cortex consists of a number of areas (a–m). The DSCF, CF/CF, DIF, FM-FM, DF, VF, and DM areas (c, d, e, f, g, h, and i, respectively) are specialized for processing biosonar information. The branches of the median cerebral artery are shown by the branching lines. The longest branch is on the fossa. (B) Graphical summary of the functional organization of the auditory cortex. The tonotopic representation of the primary auditory cortex (the area surrounded by a dotted line in zone d) and the functional organization of the DSCF, CF/CF, FM-FM, DF, VF, and DM areas are indicated by lines and arrows. The DSCF area has axes representing either target velocity (echo frequency, 60.6–62.3 kHz) or subtended target angle (echo amplitude, 13–98 dB SPL) and is divided into two subdivisions suitable for either target detection (shaded) or target localization (unshaded). These subdivisions are occupied mainly by excitatory-excitatory (E-E) or inhibitory-excitatory (I-E) binaural neurons, respectively. The FM-FM, DF, and VF areas each consist of three major types of delay-tuned, combination-sensitive neurons: FM_1-FM_2, FM_1-FM_3, and FM_1-FM_4. These three types of delay-tuned neurons form separate clusters, each of which has an axis representing target ranges (e.g., from 7 to 310 cm in the FM-FM area. The axis orthogonal to the range axis probably represents fine target characteristics. The CF/CF area consists of two major types of velocity-tuned combination-sensitive neurons: CF_1/CF_2 and CF_1/CF_3. These two types of velocity-tuned neurons are independently clustered, and each cluster has two frequency axes and represents target velocities from -2 to $+9$ m/s. Velocity-tuned neurons are also clustered in the DIF area. The VA area contains only H_1-H_2 combination-sensitive neurons. The DM area appears to have an azimuthal axis representing contralateral 4°–45°. (Revised from Suga, 1984.)

BIOSONAR SIGNAL For capture of prey (flying insects) and orientation, the mustached bat emits biosonar pulses (orientation sounds), each of which consists of a long CF component followed by a short FM component (figure 18.4). Because each pulse contains four harmonics (H_{1-4}), there are eight components, defined as CF_{1-4} and FM_{1-4} (figure 18.4B). In the emitted pulse, the second harmonic (H_2) is always predominant, and the frequency of CF_2 is approximately 61 kHz. The frequency of the CF component is slightly different among individual bats and between the two sexes. Accordingly, tonotopic representation in the auditory cortex is sexually dimorphic (Suga et al., 1987). In FM_2, frequency sweeps down from 61 kHz to approximately 49 kHz. The CF components are suited for target detection and the measurement of target velocity (the relative motion of a target evokes a steady Doppler shift of echo frequency, and wing beats of an insect evoke periodic Doppler shifts of echo frequency). On the other hand, the short FM components are suited for ranging, localizing, and characterizing a target. Thus, different parameters of echoes received by the bat carry different types of information about a target. As described in the following sections, the central auditory system of the mustached bat consists of subsystems for processing different types of auditory information in parallel, and each subsystem processes a particular type of auditory information in a parallel-hierarchical way (Suga, 1988, 1990).

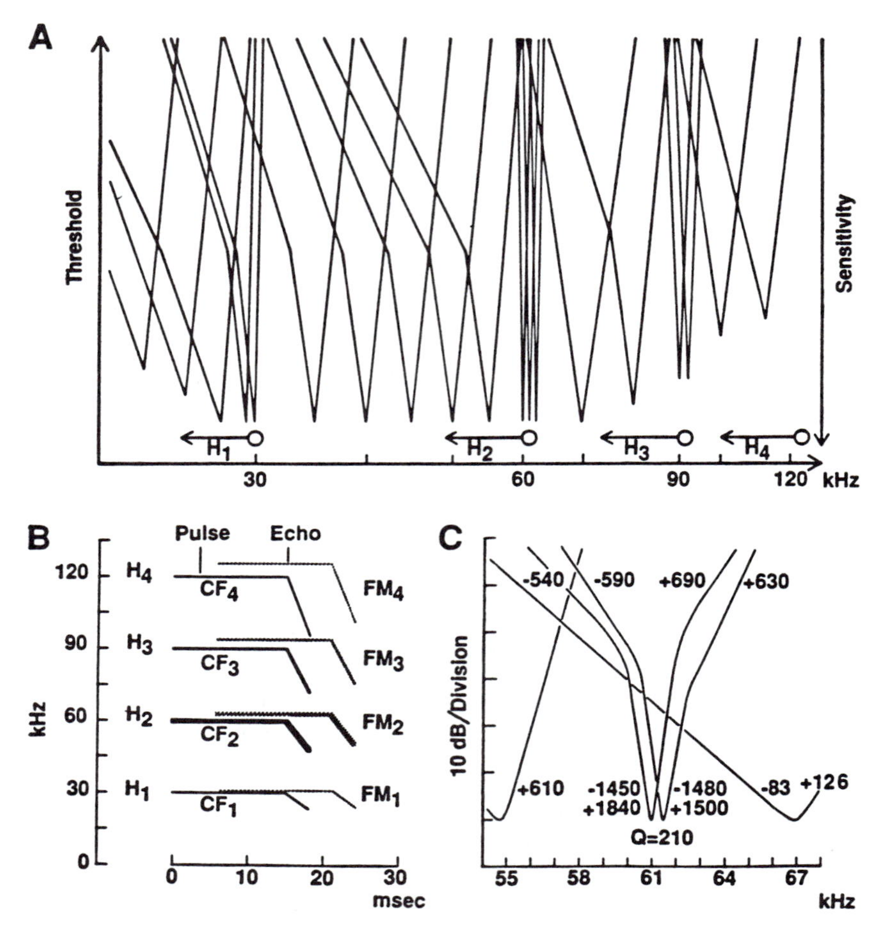

FIGURE 18.4 Biosonar signals and frequency-tuning curves of peripheral neurons of the mustached bat. (A) Schematized frequency-tuning curves of 19 peripheral neurons. (B) Schematized sonograms of a biosonar pulse and its Doppler-shifted echo. H_{1-4}, first through fourth harmonics; CF_{1-4}, constant-frequency components in the four harmonics; FM_{1-4}, frequency-modulated components in the four harmonics. (C) Averaged frequency-tuning curves of peripheral neurons tuned to 54.8, 61.0, 61.5, or 67.0 kHz. The numbers in C indicate the slopes of the curves in dB per octave. Q is a quality factor indicating the sharpness of a frequency-tuning curve at approximately 61 kHz. (Based on Suga and Jen, 1977.)

FREQUENCY ANALYSIS AT THE PERIPHERY The eight components (CF_{1-4} and FM_{1-4}) of biosonar signals all differ from one another in frequency and are therefore analyzed simultaneously in different regions of the basilar membrane (figure 18.5, bottom). For simplicity, we may consider that there are eight channels for the processing of these signal elements: CF_1 channel, CF_2 channel, and so on. Peripheral neurons are more sharply tuned in frequency to the CF components than to the FM components (figure 18.4A). In other words, the neurons utilized for velocity measurement are specialized for fine frequency analysis. Among the CF channels, CF_2 is unique because it is associated with an extraordinarily sharply tuned local resonator in the cochlea (Pollak et al., 1972; Suga and Jen., 1977; Suga et al., 1975). The slopes of the frequency-tuning curves of single peripheral neurons tuned to 61 kHz are as steep as −1450 and 1850 dB per octave (figure 18.4C). This is remarkable, considering that the slopes of variable analog filters commercially available are only 24 or 48 dB per octave. The CF_2 channel is also unique in its size. It occupies approximately one-third of the ascending auditory system, through the auditory cortex (Pollak and Casseday, 1989; Suga and Jen, 1976). The sharp frequency tuning and the large population of neurons for coding CF_2 are adaptations for the analysis of relative target velocity and wing beats of insects (Johnson et al., 1974). To facilitate such analysis in

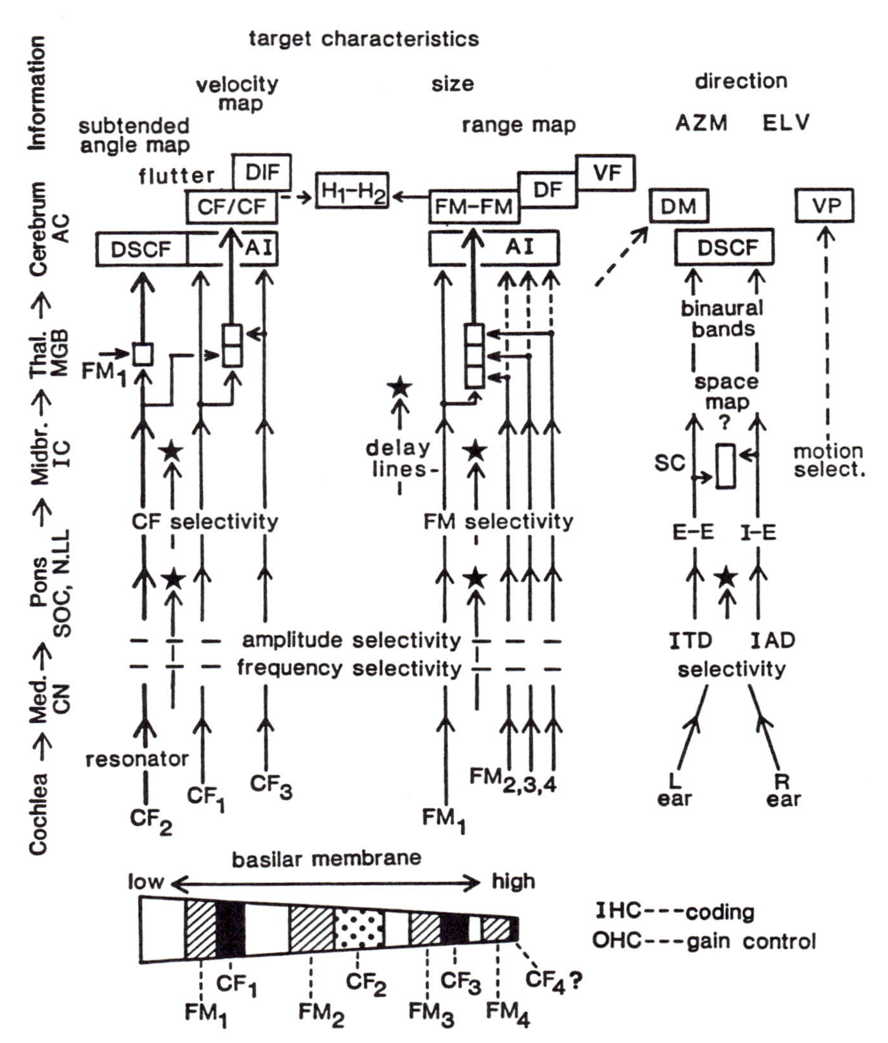

FIGURE 18.5 Parallel-hierarchical processing of different types of biosonar information carried by complex biosonar signals. The eight components (CF_{1-4} and FM_{1-4}) of the biosonar pulse and its echo are analyzed at different locations on the basilar membrane in the cochlea (bottom). The signal elements coded by auditory nerve fibers are sent up to the auditory cortex (AC) through a number of auditory nuclei (left margin): cochlear nucleus (CN), superior olivary complex (SOC), nucleus of lateral lemniscus (NLL), inferior colliculus (IC), and medial geniculate body (MGB). During the ascent of the signals, frequency selectivity is increased in some neurons, and amplitude, constant-frequency (CF), and frequency-modulation (FM) selectivities are added to some neurons (arrows with a star). Each star indicates that the addition of selectivity takes place in the auditory nuclei and cortex as well as in the nucleus where the arrow starts. The CF_2 channel is associated with a sharply tuned local resonator and is disproportionately large. It projects to the DSCF area of the auditory cortex. In certain portions of the MGB, two channels processing different signal elements (e.g., CF_1 and CF_2 or FM_1 and FM_2) converge to produce two major types of combination-sensitive neurons: velocity-tuned (CF/CF) neurons and delay-tuned (FM-FM) neurons. Velocity-tuned neurons project to the CF/CF area of the auditory cortex, where target velocity information is systematically represented (left top). Velocity-tuned neurons also cluster in the DIF area. Delay-tuned neurons project to the FM-FM, DF, and VF areas of the cortex where target-range information is systematically represented (center top). The delay lines utilized by them are mostly created in the inferior colliculus. The DSCF area has frequency-versus-amplitude coordinates for systematic representation of the properties of an echo CF_2 and consists of two subdivisions, mainly containing inhibitory-excitatory (I-E) or excitatory-excitatory (E-E) binaural neurons (right column). Signal processing in the auditory system is parallel-hierarchical. (Revised from Suga, 1988.)

frequency, the mustached bat in flight adjusts the frequency of its biosonar pulse, thereby stabilizing the frequencies of Doppler-shifted echoes from background at the best frequency to stimulate the CF_2 channel (Schnitzler, 1970).

However, even for peripheral neurons tuned to 61 kHz (figure 18.4A, C), frequency-tuning curves resemble inverted triangles, so that there is considerable ambiguity in frequency coding by single neurons at high stimulus amplitudes. All peripheral neurons respond not only to tone bursts (CF tones) but also to FM sounds and noise bursts (figure 18.6B, top) (e.g., Suga, 1973, 1978). They are not at all selective in responding to a particular type of sound because they are insensitive to amplitude spectrum.

Processing of Frequency and Amplitude in the Central Auditory System Lateral inhibition that sharpens frequency tuning of neurons occurs in the ascending auditory system from the cochlear nucleus through the auditory cortex (e.g., Suga, 1964; Suga et al., 1975; Yang et al., 1992). Most sharpening occurs in subthalamic nuclei (see figure 18.5) (Kuwabara and Suga, 1988). Therefore, sharply frequency-tuned neurons are more frequently found in the supracollicular nuclei than the cochlear nucleus. It is particularly important to know that lateral inhibition does not sharpen the tip 10-dB portion of frequency-tuning curves but rather the curves' tail portion, so that the ambiguity in frequency coding remains small even at high stimulus amplitudes (Suga and Manabe, 1982). Neurons with such level-tolerant frequency tuning curves sandwiched between inhibitory tuning curves respond well to tone bursts at particular frequencies but poorly to FM sounds or noise bursts when these stimuli stimulate both the excitatory and inhibitory

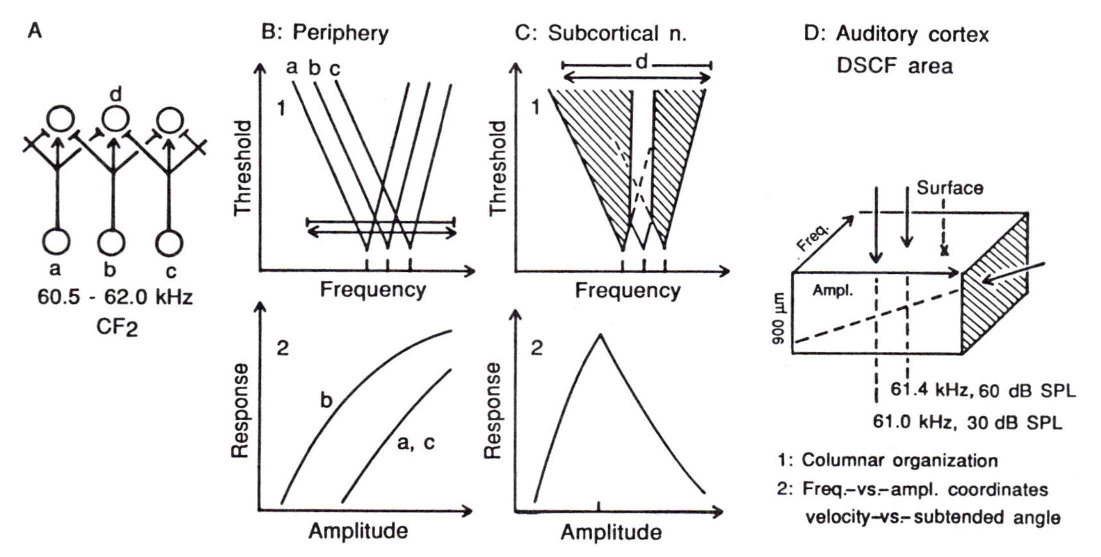

Figure 18.6 Sharpening of frequency-tuning curves and creation of amplitude tuning by inhibition. (A) Neural circuit for lateral inhibition. The arrow and bar heads represent excitatory and inhibitory synpases, respectively. (B) Frequency-tuning curves (B1) and impulse-count functions (B2) of three peripheral neurons: a, b and c in A. The impulse-count functions are measured with a tone burst at a best frequency of neuron b. The double-headed arrow in B1 indicates that the neurons respond to upward as well as downward sweeping FM sounds and that the threshold of the response is slightly higher than that to a CF tone at a neuron's best frequency. The horizontal bar in B1 indicates that these neurons respond to a noise burst and that the threshold of the response is slightly higher than those to the FM sounds. (C) A subcortical neuron (d in A) has a sharp level-tolerant frequency-tuning curve sandwiched between inhibitory tuning curves (shaded) as a result of lateral inhibition (C1). This subcortical neuron is tuned to a weak stimulus because of the inhibition. The impulse-count function measured with a tone burst at a best frequency of neuron d is highly nonmonotonic (C2). The double-headed arrow and horizontal bar in C1 indicate that neuron d does not respond to FM sounds and noise bursts (Based on Suga, 1965a,b, 1969, 1973.) (D) In the DSCF area in the auditory cortex, each cortical column is characterized by a particular combination of best frequency and best amplitude. It has the frequency-versus-amplitude coordinates. (Based on Suga, 1965b, 1977, and Suga and Manabe, 1982.)

tuning curves (figure 18.6C, top) (Suga 1969, 1973). The extent of neural sharpening in frequency tuning is most dramatic in the CF_2 channel so that many neurons in this channel selectively respond to CF tones only in a very narrow frequency range, regardless of amplitude (based on Suga and Tsuzuki, 1985).

At the periphery, auditory neurons show monotonic impulse-count functions: the stronger the stimulus, the larger is the number of impulses discharged (figure 18.6B, bottom) (e.g., Suga and Manabe, 1982). They are not tuned in amplitude. Many neurons in the ascending auditory system, however, are tuned in amplitude as well as frequency and thus have a *best amplitude* (figure 18.6C, bottom). Such amplitude tuning or selectivity is created by inhibition occurring in different auditory nuclei, including the cochlear nucleus (see figure 18.5). Therefore, amplitude-selective neurons are more frequently found in the supracollicular nuclei than in the cochlear nucleus. The creation of amplitude selectivity is prominent in the CF_2 channel compared with other channels, so that the great majority of neurons in the CF_2 channel are tuned both in frequency and amplitude (Suga, 1977; Suga and Manabe, 1982).

As described earlier, the CF_2 channel is unique. It projects to the DSCF processing area in the primary auditory cortex through the ventral division of the medial geniculate body (MGBv) (see figure 18.5). The DSCF area is very large (see figure 18.3). It is 900 μm thick and shows the columnar organization. Each functional column contains 45–50 neurons tuned to a specific combination of best frequency and best amplitude. Along the cortical surface of the DSCF area, both best frequency and best amplitude are systematically mapped, forming the frequency-versus-amplitude coordinates (see figures 18.3B and 18.6D). The frequency axis is radial, representing 60.6–62.3 kHz, i.e., the frequency range of the CF_2 components of potential echoes (Suga and Jen, 1976; Suga and Manabe, 1982; Suga et al., 1987). The best frequencies of single neurons often change at a rate of approximately 20 Hz per column at a ventral portion of the DSCF area. The amplitude axis is circular, representing 13–98 dB SPL. This amplitopic representation is for the systematic representation of subtended target angles (Suga, 1977; Suga and Manabe, 1982). The echoes from flying insects are to be represented by periodic changes in spatiotemporal pattern of neural activity in the frequency-versus-amplitude coordinates. Inactivation of the DSCF area with muscimol (an agonist of gamma-aminobutyric acid) indicates that this area is necessary for fine frequency analysis at approximately 61 kHz (Riquimaroux et al., 1991).

DSCF neurons show a facilitative response to the combination of FM_1 and CF_2. This facilitation is broadly tuned to an approximately 21-ms echo delay and may assist the bat in remaining within 4 m of vegetation so that it can effectively find flying insects (Fitzpatrick et al., 1993). Nearly 50% of DSCF neurons are sensitive to periodic frequency modulation that would be found in echoes from flying insects, but the remaining 50% are not (Suga et al., 1983).

In addition to the frequency-versus-amplitude coordinates, the DSCF area has binaural bands or subdivisions that are composed mainly of either excitatory-excitatory (E-E) or inhibitory-excitatory (I-E) neurons (see figure 18.3B and 18.5, right column) (Manabe et al., 1978). Therefore, the DSCF area has an interdigitated organization.

The CF_1 and CF_3 channels, respectively, project to the primary auditory cortex posterior and anterior to the DSCF area. In these areas, tonotopic representation is systematic, as found in the cat's primary auditory cortex, and amplitopic representation has not been found, unlike in the DSCF area (Asanuma et al., 1983). Therefore, the DSCF area is unique both in its very fine tonotopic (echo frequency) representation and in its amplitopic (subtended target angle) representation.

Extraction of Velocity Information in the Central Auditory System Doppler shift is a primary cue for obtaining velocity information. A 1.0-kHz Doppler shift will be introduced to a 61-kHz sound by a relative target velocity of 2.84 m/s. For the measurement of Doppler shift, the bat's auditory system must compute a difference in frequency between pulse and echo CF components. The ascending auditory system creates sharply frequency-tuned, level-tolerant neurons by lateral inhibition occurring at subthalamic nuclei, as shown in figures 18.6 and 18.7 (in *Pteronotus*: Suga et al., 1975; in *Myotis*: Suga, 1964, 1965a,b, 1968, 1969; in *Eptesicus*: Casseday and Covey, 1992; in cats: Evans and Nelson, 1973; Young and Brownell, 1976). In the dorsal division of the medial geniculate body (MGBd) of the thalamus, a part of the CF_1 channel and a part of the CF_2 or CF_3 channel are integrated, so that neurons in the MGBd respond poorly to the CF_1,

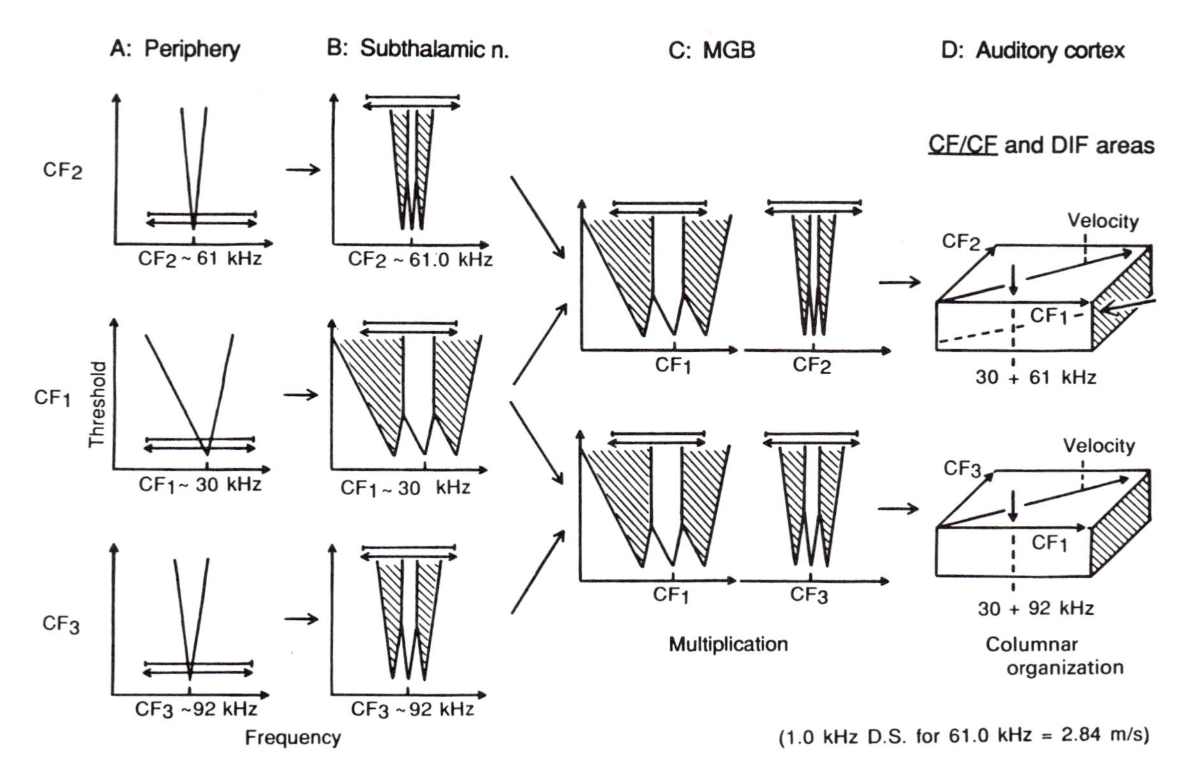

FIGURE 18.7 Neural mechanisms for creating frequency-versus-frequency coordinates. Triangular frequency-tuning curves at the periphery (A) are changed into level-tolerant frequency-tuning curves in the subthalamic auditory nuclei (B) by lateral inhibition. Then, the CF_1 and CF_2 or CF_3 channels are integrated in the medial geniculate body to create two types of velocity-tuned CF/CF combination-sensitive neurons (C). These neurons project to the CF/CF area in the auditory cortex (D) and form the CF_1/CF_2 and CF_1/CF_3 subdivisions. In each subdivision, a cortical column is characterized by a particular combination of two frequencies. Each subdivision has frequency-versus-frequency coordinates, in which relative velocities of targets are systematically mapped. (Composite of Suga, 1965b, and Suga et al., 1983a.)

CF_2, and CF_3 tones when delivered alone, but respond strongly when the CF_1 tone is delivered together with either the CF_2 or the CF_3 tone. Therefore, there are two major types of CF/CF combination-sensitive neurons: CF_1/CF_2 and CF_1/CF_3. These CF/CF neurons project separately to the CF/CF area in the auditory cortex and form the CF_1/CF_2 and CF_1/CF_3 subdivisions (see figures 18.3, 18.5, 18.7) (Suga et al., 1981, 1983; Olsen and Suga, 1991a). The synaptic mechanism for facilitative response involves *N*-methyl-D-aspartate (NMDA) receptors (Butman, 1992). A small number of CF/CF neurons are tuned to a combination of three tones: CF_1, CF_2, and CF_3 (Suga et al., 1979, 1983). The response properties of these multi-combination-sensitive $CF_1/CF_{2,3}$ neurons are basically the same as those of single-combination-sensitive CF_1/CF_2 and CF_1/CF_3 neurons.

Each $CF_1/CF_{2\text{ or }3}$ neuron has two frequency-tuning curves for facilitation: One is tuned to approximately 29 kHz CF_1, and the other is tuned to approximately 61 kHz CF_2 or 92 kHz CF_3 (see figure 18.7). A critical parameter for the facilitation of almost all CF/CF neurons is the deviation of the CF_2 or CF_3 frequency from the exact harmonic relationship with the CF_1 frequency, which represents the amount of Doppler shift in terms of CF_2 or CF_3. Some CF/CF neurons, however, are tuned to a combination of two tones that are exact harmonics. The facilitative frequency-tuning curves (facilitative areas) of CF/CF neurons are very sharp, level-tolerant, and sandwiched between inhibitory frequency-tuning curves (inhibitory areas). Therefore, these CF/CF neurons are specialized to respond to particular combinations of two frequencies, regardless of stimulus amplitude. The width of the facilitative frequency-tuning curves of some CF/CF neurons for CF_2 or CF_3 is only $\pm 1.0\%$ of their best frequencies regardless of stimulus amplitude. They do not respond to combinations of FM sounds in the pulse-echo pairs

(Suga et al., 1979, 1983a; Suga and Tsuzuki, 1985) and do not respond well to noise bursts, because the noise bursts stimulate the inhibitory areas sandwiching the facilitative or excitatory areas (Suga, 1969).

In the CF/CF area, neurons tuned to a particular combination of two CF tones are arranged orthogonally to the cortical surface, so that each location along the cortical surface represents a particular combination of two frequencies. The combination of two frequencies systematically varies along the cortical surface, so that there are frequency-versus-frequency coordinates. Along these coordinates, amounts of Doppler shifts in terms of CF_2 or CF_3 (i.e., relative target velocities) are systematically represented. Therefore, there is a velocity axis along which relative velocities from 8.6 to -1.2 m/s are represented (see figures 18.3B and 18.7D) (Suga et al., 1981, 1983). Velocities between 0 and 6 m/s are overrepresented, perhaps because these are the relative velocities encountered by the bat during the approach and terminal phases of echolocation.

CF/CF neurons show sharp, level-tolerant frequency tuning and are remarkably specialized to respond to a particular frequency relationship between the two CF tones. Their response properties are apparently a result of the parallel-hierarchical processing of complex sounds: Amplitude tuning and level-tolerant frequency tuning in the CF_1, CF_2, and CF_3 channels are created in parallel, whereas the creation of CF/CF neurons, by utilizing the response properties created in these CF channels, is hierarchical. (The reason why velocity information is processed by CF_1/CF_2 and CF_1/CF_3 neurons is discussed later.)

In the CF/CF area, one-third of neurons show synchronous discharges to periodic frequency modulations of CF_2 or CF_3 when delivered together with CF_1. The remaining two-thirds do not (Suga et al., 1983b). The CF/CF area represents relative target velocities by locations of optimally excited neurons and represents the wing beats of insects by synchronous discharges.

EXTRACTION OF DISTANCE INFORMATION IN THE CENTRAL AUDITORY SYSTEM For echolocation, the bat emits a biosonar pulse and listens to its echo. The echo delay is a primary cue for measuring target range. A 1.0-ms echo delay corresponds to a target distance of 17.3 cm at 25°C. At the periphery, neurons respond to both the pulse and its echo, because they are tonic ON responders with a short recovery period (e.g., Suga, 1964). The time interval between the two grouped discharges is directly related to the echo delay. In the central auditory system, however, echo delay is represented by neurons tuned to it, as described later (in *Pteronotus*: Suga et al., 1978, 1983; O'Neill and Suga, 1979, 1982; Suga and O'Neill, 1979; Suga and Horikawa, 1986; Edamatsu et al., 1989; Edamatsu and Suga, 1993; in *Eptesicus*: Feng et al., 1978; in *Myotis*: Sullivan, 1982a,b; Berkowitz and Suga, 1989; Wong et al., 1992).

In the FM_{1-4} channels, inhibition increases frequency selectivity of some neurons and adds amplitude selectivity to many neurons at higher levels (see figure 18.5). The extent of sharpening of frequency tuning is much less in the FM channels than in the CF channels (based on Olsen and Suga, 1991b; Suga et al., 1983). Interestingly, FM selectivity is also added to some neurons through inhibition or facilitation, so that these FM-specialized neurons selectively respond to FM sounds (in *Pteronotus*: O'Neill, 1985; in *Myotis*: Suga, 1965a,b, 1969, 1978). Phasic ON responders (onset detectors) are also created in the ascending auditory system. For example, the nucleus of the lateral lemniscus contains such neurons (Covey and Casseday, 1991; O'Neill et al., 1992; Suga and Schlegel, 1973). Delay lines that spread neural responses in time are mostly created along each iso-best frequency slab of the inferior colliculus (Hattori and Suga, 1989) (figure 18.8). Delay lines longer than 4 ms are created by inhibition (Olsen and Suga, 1991b) occurring in the inferior colliculus (Saitoh and Suga, 1992). In the MGBd, a part of the FM_1 channel (which is associated with delay lines) and a part of either the FM_2, FM_3, or FM_4 channel (none of which is associated with delay lines) are integrated to create three major types of delay-tuned, FM-FM combination-sensitive neurons: FM_1-FM_2, FM_1-FM_3, and FM_1-FM_4. These neurons respond poorly to single FM sounds but respond strongly to FM_1 combined with either FM_2, FM_3, or FM_4. The delay of the FM_2, FM_3, or FM_4 from the FM_1 (hereafter, echo delay) is the critical parameter for the facilitative responses of FM-FM neurons. These delay-tuned neurons act as delay-dependent amplifiers for processing target range information (Olsen and Suga, 1991b).

Suga and Schlegel (1973) studied the lateral lemniscus and the inferior colliculus of the little brown bat and first proposed a neural network model for ranging that was composed of onset detectors, recovery periods

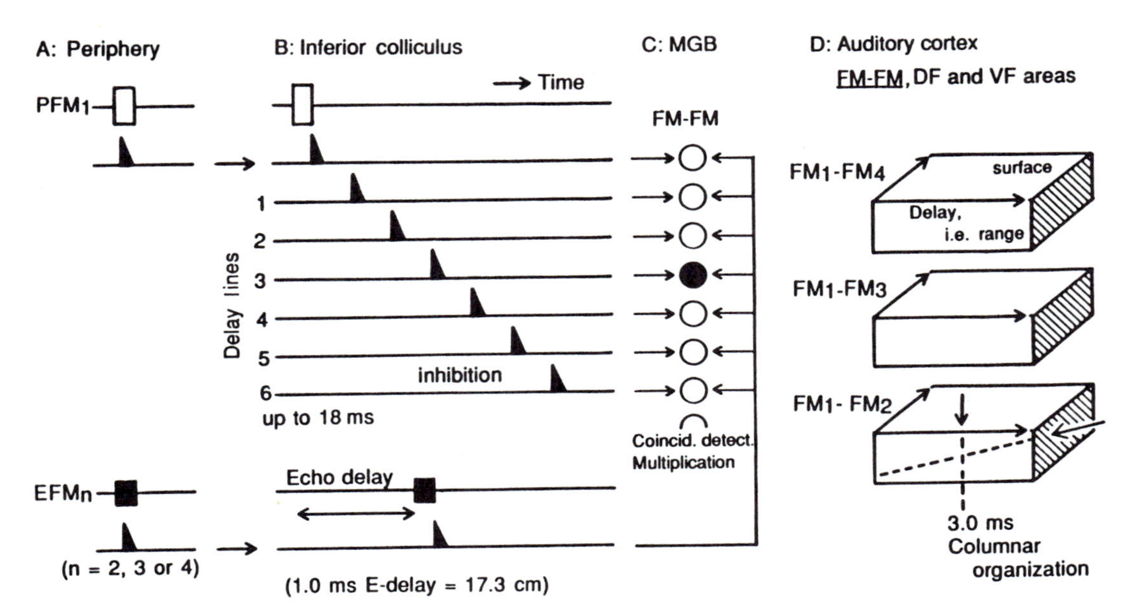

FIGURE 18.8 Neural mechanisms for creating delay-tuned neurons and a delay (range) axis. Delay-tuned neurons utilize delay lines (0.4–18 ms long) that are mostly created by inferior collicular neurons tuned to frequencies swept by the FM_1 channel (top portion of A and B). However, the delay-tuned neurons do not use delay lines created by neurons tuned to frequencies swept by FM_n ($n = 2$, 3, or 4; bottom portion of A and B). An array of delay-tuned neurons in the medial geniculate body receive signals from both the FM_1 and FM_n channels (C). In a delay-tuned neuron (filled circle) where an echo (FM_n) delay from the pulse (FM_1) is equal to the delay line associated with it, both the signals arrive at the same time. The amount of coincidence is amplified by NMDA receptors at the synapse of the delay-tuned neuron. Three types of thalamic delay-tuned neurons separately project to the FM-FM area of the auditory cortex, forming three subdivisions within the FM-FM area (D). Each subdivision of the FM-FM area shows columnar organization characterized with a particular value of echo delay. It has an echo delay axis (i.e., target-range axis). Delay-tuned neurons also cluster in the DF and VF areas. For simplicity, phasic ON response (onset detector), amplitude tuning, and FM sensitivity are eliminated from the model. (Adapted from Suga, 1990.)

acting as delay lines, and temporary recovery (facilitation) for coincidence detection. The neural mechanisms for ranging in the mustached bat also consist of these elements. However, there are certain differences in mechanisms between these two species (Sullivan, 1982a,b; Berkowitz and Suga, 1989). The following is a discussion of delay lines, coincidence detection, and amplification (facilitation) in the mustached bat. As described earlier, thalamic delay-tuned neurons receive one input from the FM_1 channel through delay lines and the other input from the FM_n channel ($n = 2$, 3, or 4) without delay lines. When an echo FM_n returns, say, with a 3.0-ms delay from the pulse FM_1, a delay-tuned neuron associated with a 3.0-ms delay line receives the inputs from both the FM_1 and FM_n channels at the same time. The extent of the coincidence is expressed by the degree to which the delay-tuned neuron is facilitated. The facilitation is mediated by NMDA receptors where the FM_1 and FM_n channels synapse on the delay-tuned neuron (Suga et al., 1990; Butman, 1992). Because the shortest delay line utilized by delay-tuned neurons is 0.4 ms, no delay-tuned neurons strongly respond to the combination of FM_1 and FM_n within a single pulse.

The three types of delay-tuned neurons in the MGBd separately project to the FM-FM area of the auditory cortex, forming FM_1-FM_2, FM_1-FM_3, and FM_1-FM_4 subdivisions. In a 100-μm-wide zone along the boundary between these subdivisions, multi-combination-sensitive neurons such as FM_1-$FM_{3,4}$, FM_1-$FM_{2,4}$, FM_1-$FM_{2,3}$, and FM_1-$FM_{2,3,4}$ are located. The response properties of these multi-combination-sensitive neurons are basically the same as those of single-combination-sensitive FM-FM neurons (Misawa and Suga, 1990).

In each subdivision of the FM-FM area, neurons tuned to a particular echo delay are arranged orthogonally to a cortical surface, so that each location along the cortical surface represents a particular echo delay. The echo delay represented by neurons (best delay) systematically increases along the rostrocaudal axis of the FM-FM area. Thus, FM-FM neurons form an echo-delay axis in each subdivision for the systematic representation of target range information (see figure 18.8). The FM-FM area represents target ranges from 7 to 310 cm (see figure 18.3B) (Suga and O'Neill, 1979; O'Neill and Suga, 1982). The response properties of delay-tuned neurons are a result of the parallel-hierarchical processing: Amplitude selectivity, FM selectivity, and delay lines in the four FM channels are created in parallel, whereas the response properties of delay-tuned neurons are created by combining the FM channels in a hierarchical manner. (It is explained later why distance information is processed by FM_1-FM_n neurons.)

Thalamic delay-tuned neurons project not only to the FM-FM area but also to the DF and VF areas of the auditory cortex. These three areas are mutually connected (Fitzpatrick, Olsen, and Suga, in preparation). The DF and VF areas each consist of three subdivisions containing one of the three types of delay-tuned neurons. The DF area represents target ranges of up to 160 cm (see figure 18.3) (Suga and Horikawa, 1986), whereas the VF area represents target ranges of up to 80 cm (Edamatsu et al., 1989). These three range-tuned areas are probably related to biosonar behavior in different situations. The FM-FM area is very likely to be involved in ranging when the bat is in a target-directed flight. On the other hand, the VF area appears to be suited for ranging when the bat is at rest (Edamatsu and Suga, 1993). The functional significance of these multiple-range (delay) axes remains to be further studied.

PROTECTION OF NEURAL COMPUTATION FROM MASKING CF/CF and FM-FM neurons extract velocity or distance information by comparing the higher harmonics of the Doppler-shifted echo with the first harmonic of an emitted pulse. The first harmonic (CF_1 and FM_1) is the weakest component. It has less than 1% of the total energy of the emitted pulse. However, it is used as the reference to measure relative velocity and distance of a target. Why does the auditory system create such heteroharmonic-sensitive neurons instead of homoharmonic-sensitive neurons? This heteroharmonic sensitivity is one of the mechanisms used to protect the neural processing of biosonar information from masking by the sounds produced by conspecifics (Suga and O'Neill, 1979). Let us consider the situation in which hundreds of bats are flying in a cave and the first harmonic of their pulses is completely suppressed by the antiresonance of the vocal tract. Then, in the air, there are many second, third, and fourth harmonics produced by conspecifics. The combinations of these harmonics in the sounds produced by the conspecifics cannot excite the heteroharmonic-sensitive FM-FM and CF/CF neurons, because the first harmonic is absent. When the animal itself emits pulses, the first harmonic is present at the larynx and stimulates both the ears by bone conduction. The FM-FM and CF/CF neurons are then turned on by the first harmonic to process target distance or velocity information. In this way, the neural computation of biosonar information in the auditory cortex is protected from the masking that would otherwise be caused by the sounds emitted by many conspecifics flying nearby.

Important principles for processing species-specific sounds

A comparison of the data obtained from the mustached bat with those obtained from other species illustrates the specialized neural mechanisms for processing biosonar signals and also the general neural mechanisms that are probably shared by many different species for processing other types of species-specific sounds (Suga, 1989). Some of these mechanisms are listed here:

1. The peripheral auditory system has evolved for both reception and frequency analysis of biologically important sounds. The sharpness of frequency tuning, sensitivity, or population can be greater for peripheral neurons tuned to frequencies of sounds that are most important to the species.

2. The frequency tuning of some central neurons is sharpened by lateral inhibition, which eliminates the "skirt" of a frequency-tuning curve. The more important the frequency analysis of particular components of sounds, the more pronounced is the neural sharpening for neurons tuned to these components.

3. The cochlea, or part of it, projects in parallel to different subdivisions of a nucleus or nuclear complex at each level of the ascending auditory system. These multiple cochleotopic (tonotopic) representations result from divergence of axons, which is usually asso-

ciated with convergence of axons for sorting out different types of auditory information. This combined divergence-convergence occurs repeatedly in the central auditory system and is the anatomical basis of parallel-hierarchical processing of information for both acoustical pattern recognition and sound localization.

4. Through divergence and convergence, neural filters are created that are tuned to various IBPs other than frequency. These IBP-tuned neurons (hereafter, IBP neurons or filters) act as cross-correlators that correlate incoming signals with their filter properties (i.e., with neurally stored information).

5. Species-specific complex sounds are processed by combination-sensitive neurons (i.e., IBP filters tuned to different combinations of signal elements).

6. Different types of IBP filters are clustered separately at particular locations of the central auditory system. In other words, the system contains functional subdivisions or areas specialized for processing particular types of auditory information important to a species.

7. In each subdivision or area, IBP filters are systematically arranged so that they form an axis or map representing the IBP. With the exception of frequency, there is no peripheral anatomical basis for the IBP axis. The IBP axis or map is a result of neural interactions, so it is a computational axis or map.

8. The axis or population of neurons representing an IBP is apportioned according to the species-specific importance of the IBP.

9. The bandwidth of IBP filters is not so narrow as to express a particular value of an IBP by the excitation of only a few neurons located at a single location along the IBP axis. Even after the sharpening of the tuning of IBP filters by lateral inhibition, an IBP value is expressed by a spatiotemporal pattern of excitation of many neurons distributed along the IBP axis.

10. The auditory cortex consists of specialized areas mainly excited by species-specific sounds and an unspecialized area (primary auditory cortex) excited by more general sounds as well as by the species-specific sounds. The primary auditory cortex is tonotopically organized and contains neurons somewhat similar to peripheral neurons, probably, for processing general sounds with a simple spatiotemporal pattern of neural activity.

ACKNOWLEDGMENT The research on the mustached bat has been supported by grants from NIDCD (DC00175), ONR (N00014-90-J-1068), and The McKnight Foundation. I thank D. C. Fitzpatrick, K. K. Ohlemiller, and J. F. Olsen for their comments on this chapter.

REFERENCES

Asanuma, A., D. Wong, and N. Suga, 1983. Frequency and amplitude representations in anterior primary auditory cortex of the mustached bat. *J. Neurophysiol.* 50:1182–1196.

Berkowitz, A., and N. Suga, 1989. Neural mechanisms of ranging are different in two species of bats. *Hear. Res.* 41:255–264.

Brugge, J. F., and M. M. Merzenich, 1973. Patterns of activity of single neurons of the auditory cortex in monkey. In *Basic Mechanisms of Hearing*, A. R. Moller, ed. New York: Academic Press, pp. 745–766.

Butman, J. A., 1992. Synaptic mechanisms for target ranging in the mustached bat. Washington University Ph.D. thesis.

Casseday, J. H., and E. Covey, 1992. Frequency tuning properties of neurons in the inferior colliculus of an FM bat. *J. Comp. Physiol.* 319:34–50.

Covey, E., and J. H. Casseday, 1991. The monaural nuclei of the lateral lemniscus in an echolocating bat: Parallel pathways for analyzing temporal features of sound. *J. Neurosci.* 11:3456–3470.

Edamatsu, H., M. Kawasaki, and N. Suga, 1989. Distribution of combination-sensitive neurons in the ventral fringe area of the auditory cortex of the mustached bat. *J. Neurophysiol.* 61:202–207.

Edamatsu, H., and N. Suga, 1993. Differences in response properties of neurons between two delay-tuned areas in the auditory cortex of the mustached bat. *J. Neurophysiol.* 69:1700–1712.

Evans, E. F., and P. E. Nelson, 1973. The responses of single neurons in the cochlear nucleus of the cat as a function of their location and the anesthetic state. *Exp. Brain Res.* 17: 402–427.

Feng, A. S., J. A. Simmons, and S. A. Kick, 1978. Echo detection and target-ranging neurons in the auditory system of the bat *Eptesicus fuscus*. *Science* 202:645–648.

Fitzpatrick, D. C., J. S. Kanwal, J. A. Butman, and N. Suga, 1993. Combination-sensitive neurons in the primary auditory cortex of the mustached bat. *J. Neurosci.* 13:931–940.

Fuzessery, Z. M., and A. S. Feng, 1982. Frequency selectivity in the anuran auditory midbrain: Single unit responses to single and multiple tone stimulation. *J. Comp. Physiol. A.* 146:471–484.

Fuzessery, Z. M., and A. S. Feng, 1983. Mating call selectivity in the thalamus and midbrain of the leopard frog (*Rana p. pipiens*): Single and multiunit analysis. *J. Comp. Physiol. A.* 150:333–344.

Hattori, T., and N. Suga, 1989. Latency map in the inferior colliculus of the mustached bat. *Assoc. Res. Otolaryngol. Abstr.* 94.

Imig, T. J., and H. O. Adrian, 1977. Binaural columns in the primary field (AI) of cat auditory cortex. *Brain Res.* 138:241–257.

Johnson, R. A., O. W. Henson, Jr., and L. R. Goldman, 1974. Detection of insect wing beats by the bat, *Pteronotus parnellii. J. Acoust. Soc. Am.* 55:S53.

Kanwal, J. S., K. K. Ohlemiller, and N. Suga, 1993. Communication sounds of the mustached bat: Classification and multidimentional analyses of cell structure. *Assoc. Res. Otolaryngol. Abstr.* 111, 442.

Knudsen, E. I., and M. Konishi, 1978a. A neural map of auditory space in the owl. *Science* 200:795–797.

Knudsen, E. I., and M. Konishi, 1978b. Space and frequency are represented separately in auditory midbrain of the owl. *J. Neurophysiol.* 41:870–884.

Kuwabara, N., and N. Suga, 1988. Mechanisms for production of "range-tuned" neurons in the mustached bat: Delay lines and amplitude selectivity are created by the midbrain auditory nuclei. *Assoc. Res. Otolaryngol. Abstr.* 200.

Liberman, A. M., P. C. Delattre, L. J. Gerstman, and F. S. Cooper, 1956. Tempo of frequency change as a cue for distinguishing classes of speech sounds. *J. Exp. Psychol.* 52:127–137.

Liberman, M. C., 1978. Auditory-nerve response from cats raised in a low-noise chamber. *J. Acoust. Soc. Am.* 63:442–455.

Liberman, M. C., 1982. Single-neuron labeling in the cat auditory nerve. *Science* 216:1239–1241.

Lisker, L., and A. S. Abramson, 1964. A cross-language study of voicing in initial stops: Acoustical measurements. *Word* 20:384–422.

Manabe, T., N. Suga, and J. Ostwald, 1978. Aural representation in the Doppler-shifted-CF processing area of the primary auditory cortex of the mustache bat. *Science* 200: 339–342.

Margoliash, D., 1983. Acoustic parameters underlying the responses of song-specific neurons in the white-crowned sparrow. *J. Neurosci.* 3:1039–1057.

Merzenich, M. M., P. L. Knight, and G. L. Roth, 1975. Representation of cochlea within primary auditory cortex in the cat. *J. Neurophysiol.* 38:231–249.

Misawa, H., and N. Suga, 1990. Multi-combination-sensitive neurons in the FM-FM area of the auditory cortex of the mustached bat. *Assoc. Res. Otolaryngol. Abstr.* 274.

Olsen, J. F., and N. Suga, 1991a. Combination-sensitive neurons in the medial geniculate body of the mustached bat: Encoding of relative velocity information. *J. Neurophysiol.* 65:1254–1274.

Olsen, J. F., and N. Suga, 1991b. Combination-sensitive neurons in the medial geniculate body of the mustached bat: Encoding of target range information. *J. Neurophysiol.* 65:1275–1296.

O'Neill, W. E., 1985. Responses to pure tones and linear FM components of the CF-FM biosonar signal by single units in the inferior colliculus of the mustached bat. *J. Comp. Physiol.* 157:797–815.

O'Neill, W. E., J. R. Holt, and M. Gordon, 1992. Responses of neurons in the intermediate and ventral nuclei of the lateral lemniscus of the mustached bat to sinusoidal and pseudorandom amplitude modulations. *Assoc. Res. Otolaryngol. Abstr.* 140, 418.

O'Neill, W. E., and N. Suga, 1979. Target range-sensitive neurons in the auditory cortex of the mustache bat. *Science* 203:69–73.

O'Neill, W. E., and N. Suga, 1982. Encoding of target-range information and its representation in the auditory cortex of the mustached bat. *J. Neurosci.* 47:225–255.

Pantev, C., M. Hoke, K. Lehnertz, and B. Lutkenhoner, 1989. Neuromagnetic evidence of an amplitopic organization of the human auditory cortex. *Electroencephalog. Clin. Neurophysiol.* 72:225–231.

Peterson, G. N., and H. L. Barney, 1952. Control methods used in a study of the vowels. *J. Acoust. Soc. Am.* 24:175–184.

Phillips, D. P., and S. S. Orman, 1984. Responses of single neurons in posterior field of cat auditory cortex to tonal stimuli. *J. Neurophysiol.* 51:147–163.

Pollak, G. D., and J. H. Casseday, 1989. *The Neural Basis of Echolocation in Bats.* New York: Springer-Verlag.

Pollak, G., O. W. Henson, Jr., and A. Novick, 1972. Cochlear microphonic audiograms in the "pure tone" bat *Chilonycteris parnellii parnellii. Science* 176:66–68.

Riquimaroux, H., S. J. Gaioni, and N. Suga, 1991. Cortical computational maps control auditory perception. *Science* 251:565–568.

Rose, J. E., N. Gross, C. D. Geisler, and J. E. Hind, 1966. Some neural mechanisms in the inferior colliculus of the cat which may be relevant to localization of a sound source. *J. Neurophysiol.* 29:288–314.

Rouiller, E., Y. Ribaupierre, de, and F. Ribaupierre, de, 1979. Phase-locked responses to low frequency tones in the medial geniculate body. *Hear. Res.* 1:213–226.

Saitoh, I., and N. Suga, 1992. Effects of inhibitory amino acid antagonists on delay tuning of FM-FM neurons of the mustached bat. *Assoc. Res. Otolaryngol.* 141, 421.

Schnitzler, H.-U., 1970. Echoortung bei der Fledermaus *Chilonycteris rubiginosa. Zool. Vergl. Physiol.* 68:25–38.

Schreiner, C. E., J. R. Mendelson, and M. L. Sutter, 1992. Functional topography of cat primary auditory cortex: Representation of tone intensity. *Exp. Brain Res.* 92: 105–122.

Schreiner, C. E., and M. L. Sutter, 1992. Topography of excitatory bandwidth in cat primary auditory cortex: Single-neuron versus multiple-neuron recordings. *J. Neurophysiol.* 68:1487–1502.

Shamma, S. A., J. W. Fleshman, P. R. Wiser, and H. Versnel, 1993. Organization of response areas in ferret primary auditory cortex. *J. Neurophysiol.* 69:367–383.

Suga, N., 1964. Recovery cycles and responses to frequency modulated tone pulses in auditory neurons of echolocating bats. *J. Physiol. (Lond.)* 175:50–80.

Suga, N., 1965a. Analysis of frequency modulated sounds by neurones of echolocating bats. *J. Physiol. (Lond.)* 179:26–53.

Suga, N., 1965b. Responses of cortical auditory neurones to

frequency modulated sounds in echolocating bats. *Nature* 206:890–891.

Suga, N., 1968. Analysis of frequency-modulated and complex sounds by single auditory neurons of bats. *J. Physiol. (Lond.)* 198:51–80.

Suga, N., 1969. Classification of inferior collicular neurones of bats in terms of responses to pure tones, FM sounds and noise bursts. *J. Physiol. (Lond.)* 200:555–574.

Suga, N., 1973. Feature extraction in the auditory system of bats. In *Basic Mechanisms in Hearing*, A. R. Moller, ed. New York: Academic Press, pp. 675–744.

Suga, N., 1977. Amplitude-spectrum representation in the Doppler-shifted-CF processing area of the auditory cortex of the mustache bat. *Science* 196:64–67.

Suga, N., 1978. Specialization of the auditory system for reception and processing of species-specific sounds. *FASEB J.* 37:2342–2354.

Suga, N., 1982. Functional organization of the auditory cortex: Representation beyond tomotopy in the bat. In *Cortical Sensory Organization*, Vol. 3, C. N. Woolsey, ed. Clifton, N.J.: Humana Press, pp. 157–218.

Suga, N., 1984. The extent to which biosonar information is represented in the bat auditory cortex. In *Dynamic Aspects of Neocortical Function*, G. M. Edelman, W. E. Gall, and W. M. Cowan, eds. New York: Wiley, pp. 315–373.

Suga, N., 1988. Auditory neuroethology and speech processing: Complex sound processing by combination-sensitive neurons. In *Functions of the Auditory System*, G. M. Edelman, W. E. Gall, and W. M. Cowan, eds. New York: Wiley, pp. 679–720.

Suga, N., 1989. Principles of auditory-information processing derived from neuroethology. In *Principles of Sensory Coding and Processing. J. Exp. Biol.* 146:277–286.

Suga, N., 1990. Cortical computational maps for auditory imaging. *Neural Networks* 3:3–21.

Suga, N., and J. Horikawa, 1986. Multiple time axes for representation of echo delays in the auditory cortex of the mustached bat. *J. Neurophysiol.* 55:776–805.

Suga, N., and P. H.-S. Jen, 1976. Disproportionate tonotopic representation for processing species-specific CF-FM sonar signals in the mustache bat auditory cortex. *Science* 194:542–544.

Suga, N., and P. H.-S. Jen, 1977. Further studies on the peripheral auditory system of "CF-FM" bats specialized for fine frequency analysis of Doppler-shifted echoes. *J. Exp. Biol.* 69:207–232.

Suga, N., K. Kujirai, and W. E. O'Neill, 1981. How biosonar information is represented in the bat cerebral cortex. In *Neuronal Mechanisms of Hearing*, J. Syka, and L. Aitkin, eds. New York: Plenum Press, pp. 197–219.

Suga, N., and T. Manabe, 1982. Neural basis of amplitude-spectrum representation in auditory cortex of the mustached bat. *J. Neurophysiol.* 47:225–255.

Suga, N., H. Niwa, and I. Taniguchi, 1983. Representation of biosonar information in the auditory cortex of the mustached bat, with emphasis on representation of target velocity information. In *Advances in Vertebrate Neuroethology*, P. Ewert, R. R. Capranica, and D. J. Ingle, eds. New York: Plenum Press, pp. 829–867.

Suga, N., H. Niwa, I. Taniguchi, and D. Margoliash, 1987. The personalized auditory cortex of the mustached bat: Adaptation for echolocation. *J. Neurophysiol.* 58:643–654.

Suga, N., J. F. Olsen, and J. A. Butman, 1990. Specialized subsystems for processing biologically important complex sounds: Cross-correlation analysis for ranging in bat's brain. *Cold. Spring Harbor Symp. Quant. Biol.* 55:585–597.

Suga, N., and W. E. O'Neill, 1979. Neural axis representing target range in the auditory cortex of the mustached bat. *Science* 206:351–353.

Suga, N., W. E. O'Neill, K. Kujirai, and T. Manabe, 1983. Specificity of combination-sensitive neurons for processing of complex biosonar signals in the auditory cortex of the mustached bat. *J. Neurophysiol.* 49:1573–1626.

Suga, N., W. E. O'Neill, and T. Manabe, 1978. Cortical neurons sensitive to combininations of information-bearing elements of biosonar signals in the mustache bat. *Science* 200:778–781.

Suga, N., W. E. O'Neill, and T. Manabe, 1979. Harmonic-sensitive neurons in the auditory cortex of the mustached bat. *Science* 203:270–274.

Suga, N., and P. Schlegel, 1973. Coding and processing in the auditory systems of FM-signal-producing bats. *J. Acoust. Soc. Am.* 54:174–190.

Suga, N., J. A. Simmons, and P. H.-S Jen, 1975. Peripheral specialization for fine analysis of Doppler-shifted echoes in the auditory system of the "CF-FM" bat *Pteronotus parnellii*. *J. Exp. Biol.* 63:161–192.

Suga, N., and K. Tsuzuki, 1985. Inhibition and level-tolerant frequency-tuning curves in the auditory cortex of the mustached bat. *J. Neurophysiol.* 53:1109–1145.

Sullivan, W. E., III, 1982a. Neural representation of target distance in auditory cortex of the echolocating bat *Myotis lucifugus*. *J. Neurophysiol.* 48:1011–1032.

Sullivan, W. E., III, 1982b. Possible neural mechanisms of target distance coding in auditory system of the echolocating bat *Myotis lucifugus*. *J. Neurophysiol.* 48:1033–1047.

Sullivan, W. E., and M. Konishi, 1984. Segregation of stimulus phase and intensity coding in the cochlear nucleus of the barn owl. *J. Neurosci.* 4:1787–1799.

Symmes, D., 1981. On the use of natural stimuli in neurophysiological studies of audition (review). *Hear. Res.* 4: 203–214.

Tunturi, A. R., 1952. A difference in the representation of auditory signals for the left and right ears in the iso-frequency contours of the right middle ectosylvian auditory cortex of the dog. *Am. J. Physiol.* 168:712–727.

Wever, E. G., 1949. *Theory of Hearing*. New York: Wiley.

Wong, D., M. Maekawa, and H. Tanaka, 1992. The effect of pulse repetition rate on the delay sensitivity of neurons in the auditory cortex of the FM bat *Myotis lucifugus*. *J. Comp. Physiol.* 170:393–402.

Yang, L., G. D. Pollak, and C. Resler, 1992. GABAergic

circuits sharpen tuning curves and modify response properties in the mustache bat inferior colliculus. *J. Neurophysiol.* 68:1760–1774.
YIN, T. C. T., and S. KUWANDA, 1984. Neural mechanisms of binaural interaction. In *Dynamic Aspects of Neocortical Function*, G. M. Edelman, W. E. Gall, and W. M. Cowan, eds. New York: Wiley, pp. 263–313.
YOUNG, E. D., and W. E. BROWNELL, 1976. Responses to tones and noise of single cells in dorsal cochlear nucleus of unanesthetized cats. *J. Neurophysiol.* 39:282–300.

19 Parallel Neural Pathways and Visual Function

ROBERT SHAPLEY

ABSTRACT The two main channels from the primate retina through the lateral geniculate nucleus (LGN) to visual cortex are usually called the *P* and *M* pathways—after the parvocellular (P) and magnocellular (M) layers of the LGN through which they travel. The two pathways appear to begin to diverge at the earliest possible point, in the retina at the photoreceptor–bipolar cell synapse. They remain segregated from each other through the retina, LGN, and the primary afferent layers of the primary visual cortex, V1. Neurons in the P pathway are much more sensitive for color than they are for black-white modulation; M cells have much higher sensitivity for black-white than for colored patterns. There are several other functional differences: receptor inputs, sampling density, dynamics of response, and functional connections with the circuitry of the visual cortex. Here we show that these functional differences are caused by different patterns of cone photoreceptor connections to P and M pathways in the retina, where functional segregation begins. The different patterns of cone inputs to P and M neurons mean that M cells can be silenced at equiluminance if only the receptive field center mechanism is activated, whereas P cells respond vigorously to equiluminant (red-green, chromatically modulated) stimuli. These neurophysiological findings carry implications for understanding P and M contributions to motion perception. New results indicate that there are different motion pathways and that P and M signals are connected differentially to these motion mechanisms.

Parvocellular and magnocellular pathways in retina and LGN

The concept of parallel neural pathways in the primate visual system emerged from the study of the layering of nerve cells in the lateral geniculate nucleus (LGN) in monkeys and humans. In the main body of an Old World primate's LGN, there are six clearly segregated layers of cells. The four more dorsal layers are composed of small cells and are named the parvocellular layers. The two more ventral layers, composed of larger neurons, are the magnocellular layers.

Parvocellular (P) neurons are almost without exception color-opponent. *Color opponency* means that they receive signals of opposite sign from different types of cone photoreceptor. The property of color opponency is conferred on them by their ganglion cell inputs from the class of ganglion cells called *P cells* by Shapley and Perry (1986). Many magnocellular neurons give the same sign of response to all wavelengths of light; this property is referred to as *broad-band* spectral sensitivity. However, only some (approximately half) of the magnocellular (M) neurons are truly broad-band; the others are color-opponent, according to the preceding definition. Some of these are the cells Wiesel and Hubel (1966) called *type IV cells*, which have an excitatory receptive field center (ON-center) mechanism that is spectrally broad-band and an antagonistic inhibitory surround mechanism that is selectively sensitive to red light. However, we have also seen OFF-center cells with type IV properties—that is, with a broad-band inhibitory receptive field center and an excitatory surround that is selectively sensitive to red. The properties of the magnocellular neurons, both broad-band and type IV, are determined almost completely by their retinal ganglion cell inputs (Kaplan and Shapley, 1986). Retrograde-labeling experiments of Leventhal, Rodieck, and Dreher (1981) and Perry, Oehler, and Cowey (1984) showed that magnocellular cells receive input from a class of retinal ganglion cells that are somewhat larger in cell body size and dendritic extent than P cells. This group of ganglion cells was called *M cells* by Shapley and Perry (1986).

Functional roles of P and M pathways

The obvious anatomical segregation and functional specialization of the P and M pathways has led to hypotheses about their possible roles in specific tasks for

ROBERT SHAPLEY Center for Neural Science, New York University, New York, N.Y.

visual perception—for example, selective response to motion (Livingstone and Hubel, 1987, 1988). Evidence has been offered to refute Livingstone and Hubel's idea that signals that have to do with object color, transmitted via the P pathway, do not contribute to motion, stereoscopic vision, or figure-ground segregation. However, much of this controversy depends on an overly simplified view of the visual system in which there is a single "place" where motion direction is computed, or a single site for segregation of figure from ground. It seems more conservative and reasonable at this point, when we have rather limited knowledge of visual computations in cerebral cortex, to conclude that P and M signals are segregated for some functional purpose. Furthermore, we can infer some of the functional purpose of this segregation by examining the relative *strength* of color and luminance in their contributions to particular visual tasks such as motion or stereoscopic vision. Controversial points concerning neurophysiological bases of visual performance persist, but I believe these can be resolved by a re-examination of the primary data about visual properties of P and M pathways—the main goal of this chapter.

Contrast gain in M and P pathways

Aside from their spectral sensitivities, another visual property that distinguishes parvocellular from magnocellular neurons is contrast gain. *Contrast* means the variation in the amount of light in a stimulus, normalized by the mean amount of light. For example, in a periodic grating pattern in which the peak light intensity is P and the least intensity is T (for trough), then contrast (C) is defined as,

$$C = (P - T)/(P + T)$$

Contrast gain is defined as the change in response of the neuron per unit change in contrast, in the limit as the contrast goes to zero. The different contrast gains of parvocellular and magnocellular LGN neurons are illustrated in figure 19.1. The response as a function of contrast grows much more steeply for the magnocellular than for the parvocellular neuron. The ratio of the average contrast gains of the population of magnocellular neurons to the average of parvocellular neurons is approximately 8 under midphotopic conditions (Kaplan and Shapley, 1982; Hicks, Lee, and Vidyasagar, 1983; Derrington and Lennie, 1984). Ehud Kaplan and I have shown that this contrast gain

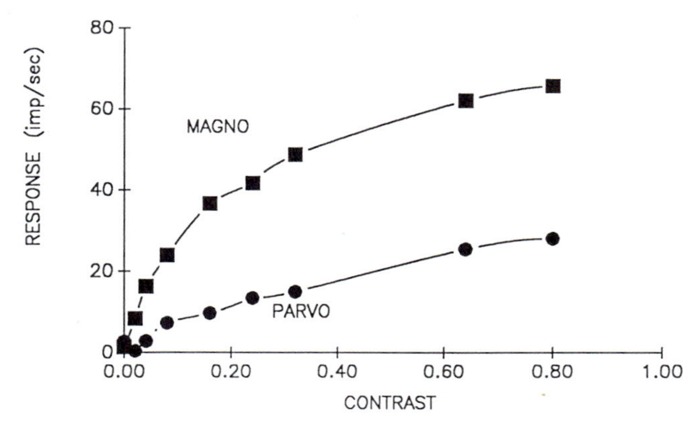

FIGURE 19.1 Response amplitude as a function of achromatic contrast for magnocellular (MAGNO) and parvocellular (PARVO) neurons. Drifting gratings were used, of optimal spatial frequency and 4-Hz drift rate for each neuron. The vertical coordinate is the amplitude of the best-fitting first harmonic in the neural response at the drift rate of 4 Hz.

difference in LGN neurons is already set up in the retina. We found that the M retinal ganglion cells had approximately eight times the contrast gain of P ganglion cells (Kaplan and Shapley, 1986).

Primate photoreceptors and spectral sensitivity of P and M cells

There are three cone photoreceptor types in human and macaque retinas. The spectral sensitivities of these photoreceptors have been determined for macaque retina by Baylor, Nunn, and Schnapf (1987) and for human retina by Schnapf, Kraft, and Baylor (1987), using suction electrodes to measure cone photocurrent directly. The photocurrent measurements agree with estimates of cone spectral sensitivity based on human psychophysics (Smith and Pokorny, 1975). The Smith and Pokorny cone action spectra are three smooth functions of wavelength peaking at 440 nm (S cones), 530 nm (M cones), and 560 nm (L cones).

Color exchange, or silent substitution (Estevez and Spekreijse, 1974), is a technique for identifying contributions from particular photoreceptors or spectral response mechanisms. For example, if one chooses two monochromatic lights with wavelengths and quantum fluxes such that they are equally effective at stimulating the L cone, then temporal alternation between these two lights should cause no variation in the response of the L cone. The same argument works for the V_λ luminance mechanism, which presumably is the spectral sensitivity of a neural mechanism that receives additive inputs from L and M cones. Two lights that, when exchanged, produce no response from the V_λ luminance mechanism are called *equiluminant*.

The results of a simulated color exchange experiment on cones and a broad-band cell with a V_λ spectral sensitivity are illustrated in figure 19.2. The spectral distributions of the light sources were those of the red and green phosphors on standard color television sets, designated *P22* phosphors. The experiment that is simulated here is color exchange between the red (denoted *R*) and green (denoted *G*) phosphors. There are three gedanken experiments illustrated in figure 19.2; one on a pure L-cone response mechanism, one on a pure M-cone response mechanism, and one on a Luminance mechanism (*LUM* in the figure). In each of these simulated experiments, the R phosphor modulation is fixed at a high modulation depth, say 0.8, and then the G modulation depth is varied from 0 to 0.8. The *x* axis is the ratio of the G modulation depth to the R modulation depth in units of luminance modulation, so that when the G/R ratio is 1.0, the G phosphor is equiluminant with the R phosphor at all times during the stimulus cycle. When the luminance modulation of the green phosphor is approximately 0.5 that of the red (G/R ratio 0.5), the response of the M cones is nulled. When the G/R ratio is approximately 1.3, the L-cone response is nulled. Notice that the shape of each predicted response curve relative to the G/R ratio for each of these spectral mechanisms is similar; near the null, the response relative to G/R ratio forms a V shape. This prediction is based on the assumption of small-signal linearity, a good assumption in the case of macaque P and M pathways.

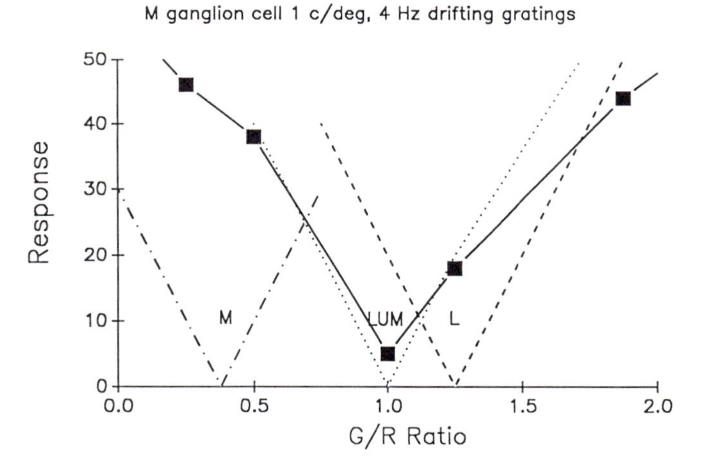

FIGURE 19.2 Response versus green-red (G/R) contrast ratio in a color scan experiment. Red gun contrast was held fixed at 0.8, and green gun contrast varied from 0 to 0.8. The modulation depth of red and green gun stimuli were equal in luminance units (G/R ratio of 1.0) when green gun contrast reached 0.32. Hypothetical response functions for L and M cones are drawn, as well as the predicted response function for a luminance mechanism (LUM). The data are from a typical magnocellular neuron driven by drifting G/R gratings at 4 Hz. The spatial frequency was 1 c/deg in this experiment.

One can prove that a spectral mechanism that sums the responses of M and L cones will have a null in a color exchange experiment at a G/R ratio *between* the nulls of the two cones. If the spectral sensitivity of the summing mechanism is $S(\lambda) = K^*\mathrm{L}(\lambda) + \mathrm{M}(\lambda)$, where K is a number between zero and infinity, then when K approaches zero, the color-exchange null approaches the M-cone null from above along the G/R axis. When K goes to infinity, the color-exchange null approaches the L-cone null, from below on the G/R axis. The null of the luminosity curve, V_λ, between the cone nulls in figure 19.2, is a case in point. For that curve, K, the ratio of L-cone strength to M-cone strength is approximately 2; that is, $V_\lambda = 2\mathrm{L}(\lambda) + \mathrm{M}(\lambda)$. Real data are also shown in figure 19.2. The points plotted there are from the amplitudes of response of an M ganglion cell to drifting red-green, heterochromatic gratings. The response plummeted near a G/R ratio of unity, the equiluminant point. This is the usual result: M cells respond weakly or not at all to equiluminant color exchange (see later).

Next, we consider what happens in a color-exchange experiment on a color-opponent neuron. In such a cell, L- and M-cone signals are not summed but subtracted. The results shown in figure 19.3 would occur. The response of the simplest type of color-opponent cell is indicated in this figure by the sloping dashed line: This is the response of a cell in which the strength of L- and M-cone signals are equal but the sign is opposite (M − L). What is striking about this simple calculation is that such an opponent neuron will have no null response between the cone nulls along the G/R axis (also indicated in figure 19.3). This result is general for any neural mechanism with a spectral sensitivity $K^*\mathrm{L}(\lambda) - \mathrm{M}(\lambda)$, where K (again the relative strength of L cone to M cone) is a number greater than zero and less than infinity. As K approaches zero, the null of the mecha-

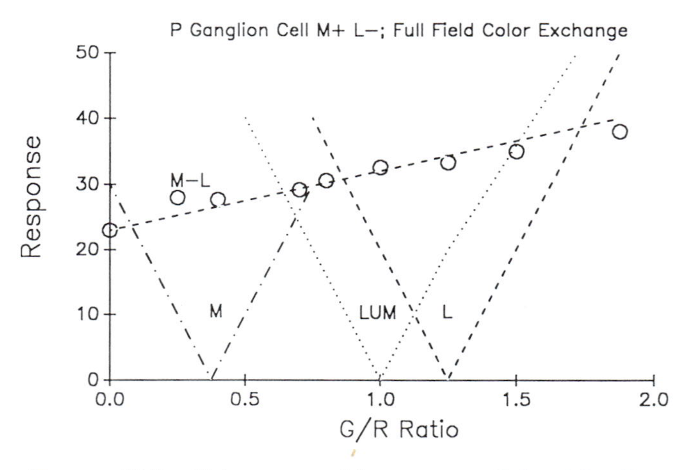

FIGURE 19.3 Color scan with a parvocellular M + L − neuron. Here the stimulus was full-field (7° square) color exchange at 4 Hz. The data are well fitted by a prediction based on pure subtractive cone interaction with equal weights for the L and M cones. LUM, luminance mechanism; G/R, green-red ratio.

nism approaches the M-cone null from below; as K goes toward infinity, the null of the mechanism approaches the L-cone null from above. As an example of data from P cells, data from an M + L − neuron are plotted in figure 19.3. The stimulus in this experiment was full-field, red-green color exchange at 4 Hz. Notice that the response amplitudes were relatively constant at all G/R ratios, meaning that the cell responded to all color exchanges. Such a neuron must have approximately equal input from M and L cones, but the cone inputs must be antagonistic.

Responses of M and P neurons to equiluminant stimuli

The human sensitivity to light across the visible spectrum under photopic, daylight conditions is called the *photopic luminosity function*, denoted V_λ. The most straightforward way to determine V_λ would be to measure psychophysically the sensitivity for increments of light of different wavelength on a photopic background. However, the photopic luminosity function is not measured in this way, mainly because such measurements are variable between and within observers (Sperling and Harwerth, 1971; King-Smith and Carden, 1976). Rather, the procedure known as heterochromatic flicker photometry has been employed. Monochromatic light of a given wavelength is flickered against a white light at a frequency of 20 Hz or more, and the radiance of the monochromatic light is adjusted until the perception of flicker disappears or is minimized (Coblentz and Emerson, 1917).

The *luminance* of a light source is its effectiveness in stimulating the visual neural mechanism that has as its spectral sensitivity V_λ, the photopic luminosity function. Thus, the luminance of any light may be computed by multiplying its spectral radiance distribution, wavelength by wavelength, by V_λ.

One particular color-exchange experiment has become crucial—namely, measuring responses of P and M neurons to equiluminant color exchange. The interesting result is that for M cells and magnocellular neurons, studied with stimuli that produce responses from the receptive field center mechanism, the position of the null on the color-exchange axis is close to that predicted from the human photopic luminosity function, V_λ (Lee, Martin, and Valberg, 1988; Shapley and Kaplan, 1989; Kaplan, Lee, and Shapley, 1990). There is rather little variance in the G/R ratio at the M-magnocellular equiluminant point.

There are other experiments which indicate that, under stimulus conditions wherein the center of the receptive field is not the only response mechanism contributing to the response, M cells and magnocellular neurons do not have a color-exchange null at equiluminance. Lee, Martin, and Valberg (1988) reported that large discs that stimulate center and surround have nulls away from equiluminance. Shapley and Kaplan (1989) reported that heterochromatic sine gratings of low spatial frequency may not produce a color null in magnocellular neurons. Derrington, Krauskopf, and Lennie (1984), using the technique of modulation in color space, found that many magnocellular units exhibited properties expected of color-opponent cells. It is likely that all these results are related to the earlier finding of Wiesel and Hubel (1966) that many magnocellular neurons had a receptive field surround that was more red-sensitive than the receptive field center. Such neurons could behave as color-opponent cells to stimuli that covered both center and surround if the spectral sensitivities of center and surround were different enough. Similar M ganglion cells were reported by DeMonasterio and Schein (1980). Thus, in psychophysical experiments, if the stimulus is designed to tap the receptive field center of cells in the M pathway, it will elicit a spectral sensitivity function like V_λ. Such a stimulus will be nulled in a color-exchange experiment at equiluminance. However, stimuli covering a large area or of a very low spatial frequency, which are detected by the M-magnocellular pathway but which do not isolate the M central receptive field mechanism, may appear to be detected by a color-opponent pathway in color-exchange experiments, because the type IV M-magnocellular cells constitute a color-opponent pathway when stimulated with stimuli of this nature.

There is another result that indicates a failure of nulling at equiluminance in magnocellular neurons. This is the second-harmonic distortion discovered by Schiller and Colby (1983). In color exchange experiments with large-area stimuli, these investigators often found strong frequency-doubled responses. Such results were not reported by Derrington, Krauskopf, and Lennie (1984), who rarely found frequency doubling (20% of the time) in their experiments. Shapley and Kaplan (1989) reported that frequency doubling was dependent on spatial frequency of the pattern used for color-exchange. Center-isolating stimuli elicited no frequency doubling, but doubling could be observed when spatial frequency was so low (less than 0.5 c/deg) that the receptive field surround could contribute to the M cell's response.

There is a conflict in the data between the report of Logothetis and colleagues (1990) and our own work and that of others regarding the responses of M cells and P cells at equiluminance. First, Logothetis's group reported that a large fraction of magnocellular neurons in their LGN sample were not silenced at equiluminance under a variety of stimulus conditions. They infer from this result that magnocellular neurons as a group are not silenced at equiluminance. This may simply be a difference of opinion in interpretation of results. Logothetis and coworkers (1990) state that they used a variety of spatiotemporal stimuli to evaluate equiluminant responses and, as I have noted previously, the spatial pattern of the stimulus has a crucial effect on whether magnocellular neurons are silenced at equiluminance. Once again, our conclusion is that the magnocellular neuron population *can* be silenced with equiluminant stimuli, (e.g., midspatial frequency grating patterns) that stimulate only the magnocellular receptive field centers. Using a variety of spatiotemporal stimuli, some of which isolate center responses and some which do not, is irrelevant for this conclusion. Therefore, Logothetis's group's conclusion that magnocellular cells as a group cannot be silenced at equiluminance is false.

Another result of Logothetis and colleagues' study (1990) is more problematic because there seems to be a conflict between data. The issue is the behavior of parvocellular neurons at equiluminance. Logothetis and coworkers state that a significant fraction of parvocellular neurons were silenced at equiluminance, but this finding has not been duplicated by others (Derrington, Krauskopf, and Lennie, 1984; Lee, Martin, and Valberg, 1988; Shapley, Reid, and Kaplan, 1991). Our work (Reid and Shapley, 1992) on the cone connections to parvocellular neurons shows that P cells are designed specifically to respond at equiluminance because of spatially overlapping inputs from different cone types of opposite sign. Hence, it is difficult to understand how Logothetis's group obtained contrary results. This is not a trivial point as from this result these investigators concluded that the parvocellular population response went through a dip at equiluminance too and that therefore a failure of a central response at equiluminance means nothing about P and M input to that response. I should point out that even if Logothetis and colleagues' data were firm, their conclusion would not follow from their data. Approximately one-fourth of their sample of parvocellular neurons were silenced at equiluminance, but three-fourths were responsive. There is no evidence from their experiments that the population response of parvocellular neurons would experience a dip in response at equiluminance; at other color balances, other fractions of parvo neurons would (from their data) be silenced and there would be no special drop in population response at the color balance of equiluminance. Therefore, when equiluminant stimuli are presented in psychophysical experiments, the large majority of the parvocellular population will be excited (similar to the fraction of parvocellular cells that will be excited by other color balances).

Spatial pattern of cone inputs to P and M cells

The precise mapping of cone types to receptive field mechanisms is a crucial bit of information for understanding achromatic and chromatic contrast gain in the P and M pathways. Wiesel and Hubel (1966) postulated that color-opponent cells in LGN received excitatory (or inhibitory) input from one cone type in the receptive field center and antagonistic inputs from a complementary cone type in the receptive field surround. An alternative hypothesis is that there is mixed receptor input to the receptive field surround and only (or predominantly) one cone input to the center of the receptive field (see Shapley and Perry, 1986; Lennie, Haake, and Williams, 1990). The detailed quantitative evidence needed to decide between these hypotheses was obtained by Clay Reid and me in experiments on spatial mapping with cone-isolating, color-exchange stimuli (Reid and Shapley, 1992). We found that both receptive field center and surround were cone-specific.

To understand the photoreceptor inputs to receptive fields, we measured the two-dimensional spatiotemporal weighting functions of parvocellular and magnocellular neurons (Shapley, Reid, and Kaplan, 1991; Reid and Shapley, 1992). We adapted an approach to spatiotemporal functional analysis pioneered by Erich Sutter (1987), called *maximal length shift register sequences*, or *m-sequences*. In our experiments, each picture element or pixel in a 16 × 16 array is modulated in time by a sixteenth order m-sequence. The sequences are derived by a computation to be described elsewhere (Reid, Victor, and Shapley, in preparation), with a technique similar to that described by Sutter (1987). One important property of the m-sequence is that its power spectrum is white. Such a signal can be used to recover the temporal weighting function of a linear system by cross-correlation. Following Sutter (1987), we have exploited the whiteness of the m-sequences by making each of the 256 pixels, in the 16 × 16 spatial stimulus array, a time-shifted version of the same basic m-sequence. Therefore, the temporal stimulus at each spatial position is uncorrelated with all the others for all times less than the time-shift between pixels (usually 4 seconds). Then the temporal weighting function at each of the positions in the array can be recovered by computation of a single cross-correlation between neuron output and m-sequence input.

We combined the method of 2-D mapping and m-sequences with the technique of cone isolation described previously, to measure the 2-D spatial map of cone inputs to neurons in the P and M pathways. Thus, the two binary states of the m-sequence pixels were two colors that would stimulate only a single cone class but would be identical stimuli for the other two cone types. Then the spatiotemporal response measured would be for one cone class only. We also used black-white stimuli to characterize the conventional achromatic distribution of sensitivity across the receptive field.

Our most important results on parvocellular neurons with this technique are illustrated in figure 19.4. There

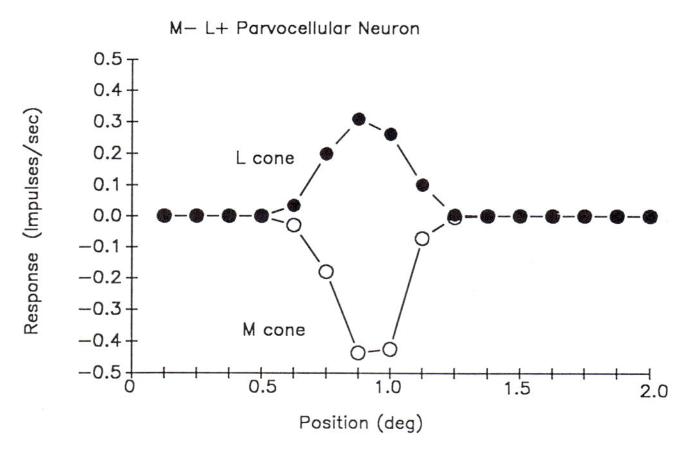

FIGURE 19.4 L- and M-cone line-weighting functions derived from first-order kernels (Reid and Shapley, 1992): parvocellular neuron. The data are from an M − L+ neuron. Each data point is the integral along the vertical axis of the stimulus screen of the first-order kernel derived from the m-sequence measurements of the spatial distribution of sensitivity. Cone-isolating picture elements were used, so these maps are of the cone line-weighting functions. The data are taken from the spatial weighting function measured at the time of peak response, 48 ms after stimulus onset. Three features of the data are noteworthy: first, the spatial overlap of L- and M-cone inputs; second, the opposite sign of inputs from L and M cones; and third, the absence of center-surround organization in the cone inputs.

are shown one-dimensional slices through the excitatory (solid) and inhibitory (dashed) spatial kernels, taken at the time of peak response (48 ms) for the two different cone types (L and M) that converge onto a single P cell, an M − L+ opponent neuron. This cell was classified as an M-cone center neuron because of the stronger M-cone input. The important point is the substantial spatial overlap of the cone inputs. Another crucial point is the absence of any sign change with position in either cone kernel. In other words, there was no center-surround interaction within one cone type in this representative P cell. The implication is that the standard center and surround mechanisms were cone-specific.

M ganglion cell fields were different in significant ways. For one thing, M- and L-cone inputs to the *center of the receptive field* had the same response signature and were therefore synergistic rather than antagonistic, and their relative strengths were approximately in the 2:1 ratio of cone inputs to the photopic luminosity function, V_λ. Equiluminant G/R m-sequences elicited spatial kernels with small or absent central responses in

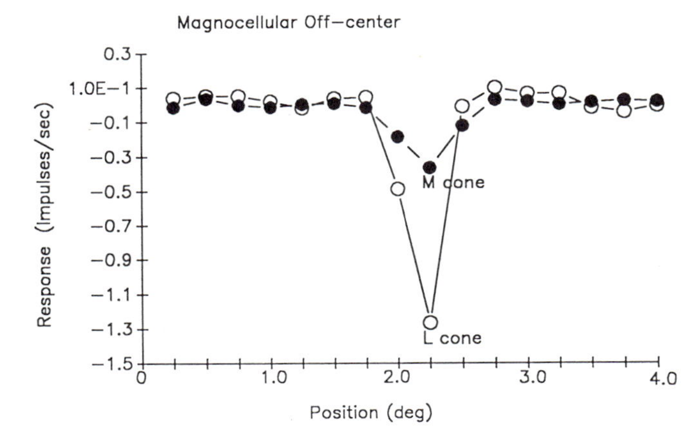

FIGURE 19.5 L- and M-cone line-weighting functions derived from first-order kernels (Reid and Shapley, unpublished): magnocellular neuron. Here the same sort of line-weighting function as for the parvocellular neuron in figure 19.4 is shown for a magnocellular neuron. There are several differences. First, the vertical scale is much compressed, indicating the much stronger input from L cones to this neuron. Second, the sign of the inputs to this OFF-center neuron is the same from L and M cones, which is completely different from the situation in parvocellular neurons. Third, there is some hint of center-surround organization in both cone inputs, more so in the L-cone input. These data are again from 48 ms after stimulus onset.

M cells. Examples of first-order cone responses from a typical magnocellular neuron are given in figure 19.5. Here again I have drawn profiles cut through the 2-D spatial distributions of sensitivity for each cone photoreceptor. In many M ganglion cells and magnocellular cells, the receptive field surround received much stronger relative input from L cones than did the center. Such neurons were type IV–like, though some had OFF centers.

Parvocellular and magnocellular signals: contribution to motion perception

I conclude with some speculative remarks about how parvocellular and magnocellular signals might or might not be segregated in the neural pathways that analyze motion of objects. The experience of motion perception is altered when equiluminant patterns are used as stimuli. This is particularly true at low speeds and at low and intermediate color contrast, at which equiluminant patterns appear to stand still although they are moving physically (Cavanagh, Tyler, and Favreau, 1984; Lindsey and Teller, 1990; Fiorentini, Burr, and Morrone, 1991). These results have been interpreted as evidence that parvocellular chromatic signals do not serve as input to the motion pathway (Livingstone and Hubel, 1987). Subsequently, there have been several different experiments demonstrating that color does support motion perception (Gorea and Papathomas, 1989; Cavanagh and Anstis, 1991; Fiorentini, Burr, and Morrone, 1991) and that lesions of magnocellular LGN do not affect some motion tasks (Schiller, Logothetis, and Charles, 1990; Merigan and Maunsell, 1990). The conclusion of this work seems to be that parvocellular and magnocellular signals both support motion perception but that there is something strange about how the brain computes speed when equiluminant stimuli are used.

However, two articles about color and motion (Lindsey and Teller, 1990; Fiorentini et al., 1991) and two about spatial effects in coherent motion perception (Lorenceau and Shiffrar, 1992; Rubin and Hochstein, 1993), taken together, seem to suggest a different hypothesis, namely, that there are multiple motion pathways operating in parallel, and that P and M pathways may provide differential inputs to the different motion pathways. The strongest psychophysical evidence favoring dual contributions of P and M signals to motion perception was provided by Cavanagh and Anstis (1991). These investigators used their technique of motion nulling to derive an *equivalent contrast* for moving chromatic patterns that were used to null out the motion of luminance patterns moving in the opposite direction. If one accepts the logic that motion signals from chromatic and achromatic patterns are simply summed at a common motion integrator, then their experiments imply an equivalency of chromatic and luminance signals for motion.

However, DeValois and Switkes (1983) and Switkes, Bradley, and DeValois (1988) have established that there are nonlinear masking effects between static chromatic and achromatic grating patterns of the type used by Cavanagh and Anstis (1991). It is not known yet whether similar chromatic-achromatic masking occurs in a motion detection paradigm such as the one employed by Cavanagh and Anstis, but it is a possibility. Therefore, some part of the inferred contribution of color signals to motion might instead be interpreted as chromatic masking of achromatic signals prior to the motion computation.

The results of Cavanagh and Anstis (1991) for a normal subject are redrawn in figure 19.6, where one

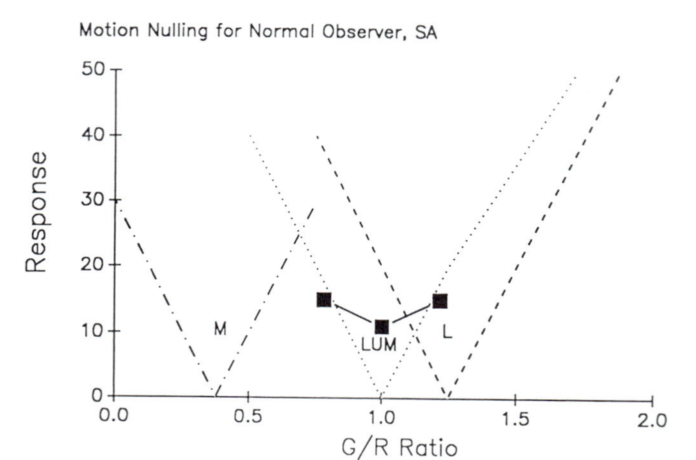

FIGURE 19.6 Cavanagh and Anstis's (1991) motion-nulling input strength compared to null points for the cones and for luminance. What is plotted is the effective contrast for a colored grating (in the motion-nulling paradigm of Cavanagh and Astis) versus its green-red (G/R) ratio (which is equivalent to effective luminance contrast; for instance, a G/R ratio of 1.22 is equivalent to 0.05 luminance contrast, whereas a G/R ratio of 0.78 is equivalent to a luminance contrast of −0.05). It is worth nothing that the data show a mild dip at equiluminance, which means that a luminance mechanism (Lum), presumably wired to magnocellular inputs, must be involved in the nulling of motion in these experiments. The low spatial frequency used by these authors (1 c/deg) guarantees that all parvocellular neurons must have been working in a color-opponent mode and would not show any local minimum in response at equiluminance.

can see that there is a mild dip in effectiveness of heterochromatic gratings near equiluminance, for motion perception. As I have discussed previously, any local minimum of response or performance near equiluminance is a strong indicator of magnocellular inputs, especially in this case in which the investigators used moderately low spatial frequency patterns for which parvocellular neurons would be almost completely color-opponent. This alters somewhat the interpretation of the Cavanagh and Anstis results and allows one to consider other psychophysical evidence that suggests that parvocellular-borne color signals might contribute to a *different kind* of motion perception.

Two studies comparing motion discrimination to simple detection indicate that color does not contribute to motion perception in the same way as does luminance. The first of these is the article by Lindsey and Teller (1990) which indicates that, under the conditions of their experiments, the ratio of color contrast needed for motion direction discrimination is approximately three or four times higher than needed for color detection. They also showed that motion was special in this regard: Form discriminations, for instance, were achievable at as low a color contrast as would permit the equiluminant pattern still to be detected. Luminance discrimination-detection ratios for motion were around unity also. The work of Fiorentini, Burr, and Morrone (1991) complements the results of Lindsey and Teller (1990) by showing that the chromatic discrimination-detection threshold ratio is high for low temporal frequencies (and, therefore, low speeds) but approaches unity at high speed. Luminance motion has a discrimination-detection threshold ratio of unity for all temporal frequencies. The recent work of Hawken and colleagues (Hawken, Gegenfurtner, and Tang, 1994) indicates that *speed matching* shows a similar pattern of differences: At low speeds, color and luminance behave very differently, but at high speeds, they are very similar.

Recent work now reveals two main motion pathways for the perception of the direction of motion of spatial patterns: a contour (1-D) pathway and a terminator pathway (0-D) (Lorenceau and Shiffrar, 1992; Rubin and Hochstein, 1993). What this means is that the motion direction of any single contour is computed as some sort of weighted sum of the motion of its terminators (0-D motion) and the motion orthogonal to the contour (1-D motion). The weighting depends upon contrast, speed or temporal frequency, contour length, and location in the visual field. Both of these pairs of investigators noted that low luminance contrast tends to favor orthogonal (1-D) motion.

What I suggest here, admittedly speculatively, is that these two different motion pathways could receive differential inputs from parvocellular and magnocellular sources that drive directionally selective neurons in primary visual cortex. Specifically, I suggest that the 1-D pathway arises from directionally selective simple and complex cells in layer IVcα, whereas the 0-D pathway begins with the directionally selective end-stopped neurons that reside in layers II–III of striate cortex. The 1-D pathway might receive predominantly magnocellular signals, whereas the 0-D pathway could receive more balanced inputs from both P and M channels. Although this suggestion does not resolve all the complexities of color and motion interactions, it provides a specific hypothesis one can use to analyze and explain the data in hand. For instance, the characteristics of perception of low-speed motion seem to fit with the terminator pathway. High-speed motion seems to be dominated by the magnocellularly driven 1-D, or contour, motion pathway.

The situation in motion perception may be a paradigm for other visual submodalities such as form and stereo perception. For instance, Tyler (1990) has proposed that there may be several parallel mechanisms for extracting depth from stereopsis in the cortex and that these may receive differentially weighted inputs from parvocellular and magnocellular signals. Clearly, Tyler's proposal is very close to my own view about what is going on in motion perception, and it may be a general principle for interpreting how parvocellular and magnocellular channels are integrated in the visual cortex.

ACKNOWLEDGMENTS I thank my New York colleagues, Ehud Kaplan, Clay Reid, Michael Hawken, Karl Gegenfurtner, and John Krauskopf, and my Jerusalem colleagues, Nava Rubin and Shaul Hochstein, who have contributed many ideas to this chapter. This work has been supported by National Institutes of Health grant EY 01472 and by a grant from the US-Israeli Binational Science Foundation.

REFERENCES

Baylor, D. A., B. J. Nunn, and J. L. Schnapf, 1987. Spectral sensitivity of cones of the monkey *Macaca fascicularis*. *J. Physiol.* 390:145–160.

Cavanagh, P., and S. M. Anstis, 1991. The contribution of color to motion in normal and color-deficient observers. *Vision Res.* 31:2109–2148.

Cavanagh, P., C. W. Tyler, and O. E. Favreau, 1984. Perceived velocity of moving chromatic gratings. *J. Opt. Soc. Am. [A]* 1:893–899.

Coblentz, W. W., and W. B. Emerson, 1917. Relative sensibility of the average eye to light of different colors and some practical applications to radiation problems. *Bull. Bur. Stand.* 14:167–236.

DeMonasterio, F. M., and S. J. Schein, 1980. Protan-like spectral sensitivity of foveal Y ganglion cells of the retina of macaque monkeys. *J. Physiol.* 299:385–396.

Derrington, A. M., J. Krauskopf, and P. Lennie, 1984. Chromatic mechanisms in lateral geniculate nucleus of macaque. *J. Physiol.* 357:241–265.

Derrington, A. M., and P. Lennie, 1984. Spatial and temporal contrast sensitivities of neurones in the lateral geniculate nucleus of macaque. *J. Physiol.* 357:219–240.

DeValois, K. K., and E. Switkes, 1983. Simultaneous masking interactions between chromatic and luminance gratings. *J. Opt. Soc. Am.* 73:11–18.

Estevez, O., and H. Spekreijse, 1974. A spectral compensation method for determining the flicker characteristics of the human colour mechanism. *Vision Res.* 14:823–830.

Fiorentini, A., D. Burr, and C. Morrone, 1991. Temporal characteristics of colour vision: VEP and psychophysical evidence. In *From Pigments to Perception*, A. Valberg and B. B. Lee, eds. New York: Plenum, pp. 139–150.

Gorea, A., and T. V. Papathomas, 1989. Motion processing by chromatic and achromatic pathways. *J. Opt. Soc. Am. [A]* 6:590–602.

Hawken, M. J., K. R. Gegenfurtner, and C. Tang, 1994. Contrast dependence of colour and kuminance motion mechanisms in human vision. *Nature*, 367:268–270.

Hicks, T. P., B. B. Lee, and T. R. Vidyasagar, 1983. The responses of cells in macaque lateral geniculate nucleus to sinusoidal gratings. *J. Physiol.* 337:183–200.

Kaplan, E., B. B. Lee, and R. Shapley, 1990. New views of primate retinal function. In *Progress in Retinal Research*, vol. 9, N. Osborne and G. Chader, eds. Oxford: Pergamon, pp. 273–336.

Kaplan, E., and R. Shapley, 1982. X and Y cells in the lateral geniculate nucleus of macaque monkeys. *J. Physiol.* 330:125–143.

Kaplan, E., and R. Shapley, 1986. The primate retina contains two types of ganglion cells, with high and low contrast sensitivity. *Proc. Natl. Acad. Sci. U.S.A.* 83:2755–2757.

King-Smith, P. E., and D. Carden, 1976. Luminance and opponent-color contributions to visual detection and adaptation and to temporal and spatial integration. *J. Opt. Soc. Amer.* 66:709–717.

Lee, B. B., P. R. Martin, and A. Valberg, 1988. The physiological basis of heterochromatic flicker photometry demonstrated in the ganglion cells of the macaque retina. *J. Physiol.* 404:323–347.

Lennie, P., P. W. Haake, and D. R. Williams, 1991. The design of chromatically opponent receptive fields. In *Computational Models of Visual Processing*, M. S. Landy and J. A. Movshon, eds. Cambridge, Mass.: MIT Press, pp. 71–82.

Leventhal, A. G., R. W. Rodieck, and B. Dreher, 1981. Retinal ganglion cell classes in the old-world monkey: Morphology and central projections. *Science* 213:1139–1142.

Lindsey, D., and D. Teller, 1990. Motion at isoluminance: Discrimination/detection ratios for moving isoluminant gratings. *Vision Res.* 30:1751–1762.

Livingstone, M. S., and D. H. Hubel, 1987. Psychophysical evidence for separate channels for the perception of form, color, motion, and depth. *J. Neurosci.* 7:3416–3468.

Livingstone, M. S., and D. H. Hubel, 1988. Segregation of form, color, movement, and depth: Anatomy, physiology, and perception. *Science* 240:740–749.

Logothetis, N. K., P. H. Schiller, E. R. Charles, and A. C. Hurlbert, 1990. Perceptual deficits and the role of color opponent and broad band channels in vision. *Science* 247:214–217.

Lorenceau, J., and M. Shiffrar, 1992. The influence of terminators on motion integration across space. *Vision Res.* 32:263–273.

Merigan, W. M., and J. H. R. Maunsell, 1990. Macaque

vision after magnocellular lateral geniculate lesions. *Visual Neurosci.* 5:347–352.

Perry, V. H., R. Oehler, and A. Cowey, 1984. Retinal ganglion cells that project to the dorsal lateral geniculate nucleus in the macaque monkey. *Neuroscience* 12:1101–1123.

Reid, R. C., and R. Shapley, 1992. Spatial structure of cone inputs to receptive fields in primate lateral geniculate nucleus. *Nature* 356:716–718.

Rubin, N., and S. Hochstein, 1993. Isolating the effect of one-dimensional motion signals on the perceived direction of moving two-dimensional objects. *Vision Res.* 33:1385–1396.

Schiller, P. H., and C. L. Colby, 1983. The responses of single cells in the lateral geniculate nucleus of the rhesus monkey to color and luminance contrast. *Vision Res.* 23: 1631–1641.

Schiller, P. H., N. K. Logothetis, and E. R. Charles, 1990. The role of color-opponent and broad-band channels in vision. *Visual Neurosci.* 5:321–346.

Schnapf, J., T. Kraft, and D. A. Baylor, 1987. Spectral sensitivity of human cone photoreceptors. *Nature* 325:439–441.

Shapley, R., and E. Kaplan, 1989. Responses of magnocellular LGN neurons and M retinal ganglion cells to drifting heterochromatic gratings. *Invest. Ophthalmol. Vis. Sci.* Suppl. 30:323.

Shapley, R., and V. H. Perry, 1986. Cat and monkey retinal ganglion cells and their visual functional roles. *Trends Neurosci.* 9:229–235.

Shapley, R., R. C. Reid, and E. Kaplan, 1991. Receptive field structure of P and M cells in the monkey retina. In *From Pigments to Perception*, A. Valberg and B. B. Lee, eds. New York: Plenum, pp. 95–104.

Smith, V. C., and J. Pokorny, 1975. Spectral sensitivity of the foveal cone photopigments between 400 and 500 nm. *Vision Res.* 15:161–172.

Sperling, H. G., and R. S. Harwerth, 1971. Red-green cone interactions in the increment-threshold spectral sensitivity of primates. *Science* 172:180–184.

Sutter, E., 1987. A practical non-stochastic approach to nonlinear time-domain analysis. In *Advances in Methods of Physiological Systems Modelling*, vol. 1. Los Angeles: University of Southern California.

Switkes, E., A. Bradley, and K. K. DeValois, 1988. Contrast dependence and mechanisms of masking interactions among chromatic and luminance gratings. *J. Opt. Soc. Am. [A]* 7:1149–1162.

Tyler, C. W., 1990. A stereoscopic view of visual processing streams. *Vision Res.* 30:1877–1896.

Wiesel, T. N., and D. H. Hubel, 1966. Spatial and chromatic interactions in the lateral geniculate body of the rhesus monkey. *J. Neurophysiol.* 29:1115–1156.

20 Functional Compartments in Visual Cortex: Segregation and Interaction

DANIEL Y. TS'O AND ANNA W. ROE

ABSTRACT Our understanding of the neural mechanisms underlying visual processing has been greatly advanced by the study of how neurons are organized in the visual cortex according to functional properties. The functional organizations of visual areas V1 and V2 are very distinct and highly suggestive of a visual processing architecture. Cells tuned for particular contour orientations, wavelength, retinal disparities, and other properties are clustered in different subcompartments within V1 and V2. These subcompartments are not entirely segregated from one another but interact, producing more elaborate receptive field types. We will examine how these functional domains cooperate to analyze the visual world in terms of a variety of parameters.

One of the most striking features of the primate visual cortex is the high degree of organization based on functional properties. Cells found in the visual cortex are selective for specific aspects of visual stimuli, including orientation, movement, depth, and color. Our understanding of visual processing has been greatly advanced by the study of the organization of these properties. In particular, the classical studies of Hubel and Wiesel (1968, 1974, 1977) established the columnar organization of such properties as ocular dominance and orientation selectivity and helped formulate a modular description of the organization of visual cortex. One fundamental demonstration arising from these studies is the interlacing of several functional maps (e.g., ocular dominance and orientation) within the same cortical structure. This arrangement neatly facilitates the processing of multiple dimensions for a given region in visual space.

More recent studies have further elaborated on the functional architecture of primary visual cortex (V1) as well as of extrastriate visual areas. An important discovery was the patchy staining of visual cortex for the mitochrondrial enzyme cytochrome oxidase (Wong-Riley, 1979). In V1, this staining procedure, in tangential section, reveals a lattice of oval patches or blobs that physiological studies suggest are involved in color processing (figure 20.1). In other visual areas, cytochrome oxidase histology shows very different staining patterns. For example, in V2, instead of blobs, one sees a series of darkly stained bands or stripes (see figure 20.1).Electrophysiological studies show that these stripes contain clusters of cells selective for color or disparity. These and other lines of evidence indicate that although each visual area is highly organized, the specific properties that are organized and the geometry of the organization are different from area to area. Furthermore, response properties that are well organized (e.g., retinotopy, ocular dominance, and orientation in V1) in earlier cortical stages often become less well or not at all organized in higher areas (e.g., V4, MT). In contrast, other properties, though present at early stages (e.g., disparity or directionality in V1) are not well organized until higher areas (table 20.1). This observation suggests that there may be some theoretical or developmental limit on the number of properties that can be organized in the cortical structure. Most importantly, this shifting of the organizing factors from area to area further reinforces the notion that the properties that are well organized in an area indicate or even dictate the role of that area in visual processing.

In this chapter, we will discuss the functional properties of cells in visual areas V1 and V2, as well as the organizations that they form. In addition, we will examine the segregation between the different functional compartments and the degree to which these compartments interact.

DANIEL Y. TS'O and ANNA W. ROE Division of Neuroscience, Baylor College of Medicine, Houston, Tex.

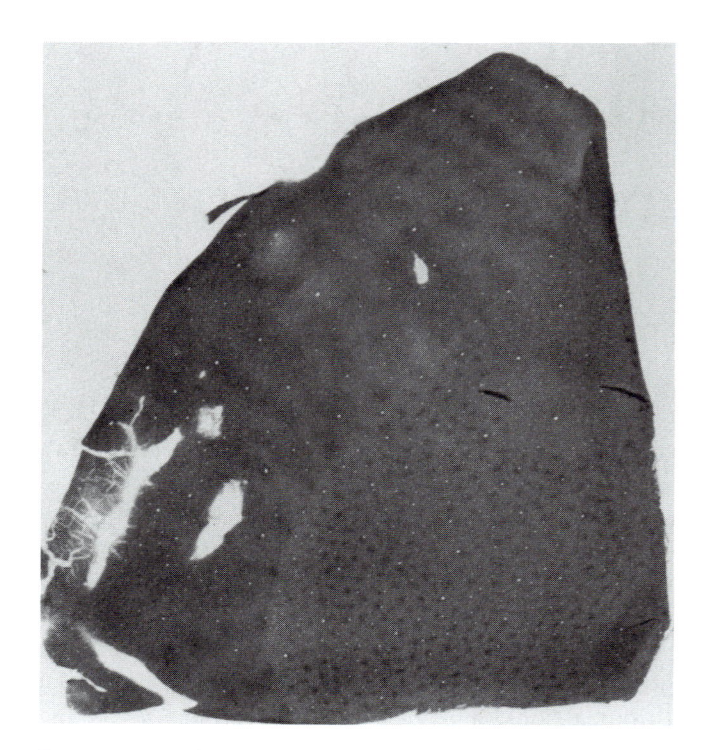

FIGURE 20.1 Cytochrome oxidase histology of V1 and V2. A tangential section from visual cortex of the squirrel monkey, showing the cytochrome oxidase–rich blobs of V1 (the polka-dotted pattern). The lightly staining regions between neighboring blobs are known as the *interblobs*. In V2 are the thick and thin and pale stripes.

Functional properties and organization in V1

Physiological studies in V1 have found a wide range of response properties, several of which have been demonstrated to be systematically represented. In addition to a precise retinotopy, V1 was found to contain a pronounced columnar organization for ocular dominance and orientation (see color plate 5). Also introduced by the classical work of Hubel and Wiesel was the notion of a hypercolumn, a complete set of columns of one type, representing, for example, all orientations, arranged in an orderly manner. Interwoven with the columns of one functional property are the columns of other functional properties, such that a complete set of hypercolumns for all properties form the basic module of cortical organization required to process a given region in visual space. Based on electrophysiological studies, the original "icecube model" of Hubel and Wiesel (1977) depicted an orthogonal arrangement of ocular dominance and orientation hypercolumns (figure 20.2).

This model of cortical organization has been examined more recently with optical imaging methods. Two such studies (Bartfeld and Grinvald, 1992; Blasdel, 1992a, b) found that the orientation columns do run perpendicularly to the ocular dominance columns at the ocular dominance borders. Orientation selectivity, then, changes smoothly and regularly at the ocular dominance borders, as in the original icecube model. However, the orientation changes can become more

TABLE 20.1
Functional properties organized in several visual areas

Functional Property	V1	V2	V4	MT
Retinotopy	Yes	Yes, multiple maps	Degraded	Degraded
Ocular dominance	Yes	No	No	No
Orientation	Yes	Yes, in some regions		
Color	Yes, blobs and bridges	Yes, color stripes	Yes?	No
Directionality	No?	No?	No	Yes
Disparity	No	Yes, disparity stripes	No?	Yes?
End-inhibition	No	Yes?		

Note: As can be seen, though cells with a given property may exist within an area, those cells may or may not be organized. What properties are organized within a given area changes from one to the next: Properties that in previous areas (areas lower in the hierarchy) are well organized may become less well organized, whereas other properties previously not organized, though present, become organized.

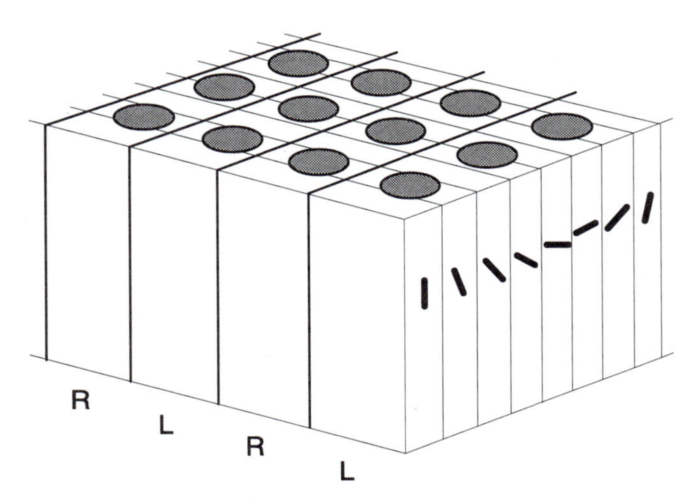

FIGURE 20.2 Modified icecube model, depicting the orthogonal arrangement of the ocular dominance columns of V1 (left- and right-eye columns) and the orientation columns. In addition, the blobs are shown, lying in the centers of the ocular dominance columns. (Based on Horton, 1984.)

rapid and discontinuous away from the borders and toward the centers of ocular dominance. Thus, these studies indicate that some modification of the icecube model is required, particularly in light of our current understanding of color processing in V1.

FUNCTIONAL ORGANIZATION OF COLOR More recently, an additional important feature of the modular organization of the monkey striate cortex was revealed by the finding of a patchy labeling of striate cortex to a stain developed for the mitochrondrial enzyme cytochrome oxidase (Wong-Riley, 1979; Horton and Hubel, 1981; Humphrey and Hendrickson, 1983; Horton, 1984). High concentrations of this enzyme are believed to indicate long-term elevated metabolic activity. The regions of dense cytochrome oxidase reactivity, in tangential sections of macaque striate cortex, are centered on ocular dominance columns and laid out as a matrix of oval patches or blobs, approximately 150 × 200 μm each. Optical imaging studies have further shown that the blobs coincide with centers of monocularity (Ts'o et al., 1990).

Livingstone and Hubel (1984a) found that the blobs contained cells with receptive fields that were monocular, unoriented, and often color-selective. Many of these cells could be categorized into groups or types that were similar to types found in the lateral geniculate nucleus (LGN) (Wiesel and Hubel, 1966). However, several blob color cell types did not correspond to LGN types. One prominent example were cells described as double-color-opponent, having spatially opponent fields with color opponency in both center and surround. Subsequent studies (Ts'o and Gilbert, 1988) showed that the majority of these cells did not have true double-color-opponent properties (Daw, 1968), as their surround was always suppressive and the response to stationary isoluminant color contrast was poor. Accordingly, this class of color cell was named *modified type II*, suggesting that the suppressive surround represents a modification of the LGN type II field. Color, then, is yet another functional dimension that is integrated into the cortical matrix. These data have prompted a revision of the icecube model that incorporates the blobs into the centers of each ocular dominance column (Livingstone and Hubel, 1984a) (see figure 20.2).

Color opponency–specific blobs The view of the organization of the blob color system has been further refined by the finding that blobs are color opponency–specific. In addition to confirming (at least in the superficial cortical layers) a patchy or columnar organization of color in V1 (see also Gouras, 1974; Michael, 1981), Ts'o and Gilbert (1988) reported that single vertical electrode penetrations within blobs encounter color cells of only one color opponency, either red versus green or blue versus yellow (see also Dow and Vautin, 1987). Furthermore, cells recorded in neighboring vertical penetrations within a single blob share the same color opponency. These findings suggest that individual blobs are dedicated to the processing of one color opponency system. Some blobs were found that did not contain any color cells but contained only broad-band center-surround (type III) cells. This finding may be relevant in interpreting the role of the blobs in nocturnal primates, such as the owl monkey, that have very poor color vision.

The color-specific blobs across cortex do not share equal representation in V1. Based on multiple electrode penetrations into the perifoveal cortical representation of V1, Ts'o and Gilbert (1988) found a 3:1 ratio of red/green to blue/yellow blobs, paralleling biases found in the retina (5:2) (see Gouras, 1968; DeMonasterio and Gouras, 1975; Schiller and Malpeli, 1978) and the LGN (12:1) (see Wiesel and Hubel, 1966; Dreher, Fukada, and Rodieck, 1976; Kruger,

1977). Furthermore, blue-yellow blobs seem to cluster together, suggesting a nonuniform or patchy distribution of blue cone inputs to cortex.

Bridges An intriguing observation first made by Horton (1984) is that the ocular dominance columns often seem paired such that the blobs of a given left-eye dominance band are connected by bridges to the blobs of only one neighboring right-eye dominance band and not the other (figure 20.3). This pattern of pairing gives the appearance, in cytochrome oxidase histology, of railroad tracks or ladders, with the bridges appearing as ties or rungs crossing two ocular dominance bands (the rails of the track). Horton speculated that the origins of this arrangement may be tied to a gradual separation of the ocular dominance columns during development.

Recordings in the bridges suggest that they, like the blobs, also contain unoriented cells that are often color-selective (Ts'o and Gilbert, 1988; Landisman and Ts'o, 1992). Thus, the bridges are apparently an additional component of the V1 color system. Binocular, unoriented, color-selective cells were found in bridges between blobs of opposite ocular dominance. This view has been confirmed by optical imaging studies designed to reveal the color-selective regions of cortex (Landisman, Grinvald, and Ts'o, 1991; Landisman and Ts'o, 1992) that directly show that the bridge regions are color-selective. Bridges have also been found spanning neighboring blobs of different color opponency. Color response properties in such bridges are often found to be neither red/green nor blue/yellow but mixed in spectral selectivity.

Interposed between the interblob and the blob regions are the so-called periblob regions, in which color-oriented cells are often found (Dow and Vautin, 1987; Ts'o and Gilbert, 1988). Some initial studies using cross-correlation analysis indicate that color-oriented cells receive input from color blob cells having matching color specificity (Ts'o and Gilbert, 1988). The distribution and receptive field properties of color-oriented cells and preliminary cross-correlation data all suggest that these cells represent an interplay between the blob and interblob regions or, alternatively, between the form and color pathways.

It should be noted that other investigators using different methods of color stimulation and classification have reported little or no correlation between the presence of color-selective cells and the blobs (Lennie, Krauskopf, and Sclar, 1990; Leventhal, Thompson, and Liu, 1993). One possible difference may be the means for localizing the blobs and the consideration of the presence of the bridge regions that also contain color-selective cells. The reconciliation of these disparate views must await future studies.

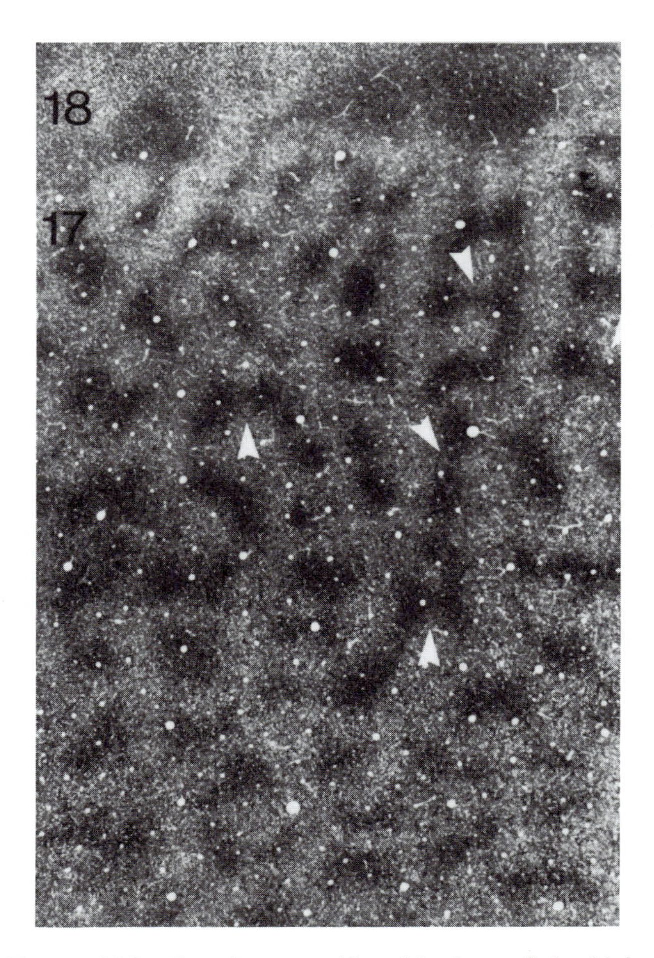

FIGURE 20.3 Cytochrome oxidase histology of the blobs and bridges of V1 (area 17) of the macaque monkey. The blobs are located in the center of the ocular dominance columns, which run perpendicular (roughly bottom to top in this figure) to the V1/V2 (area 17/18) border. The bridges (white arrowheads) span neighboring blobs, both across opposite ocular dominance columns and along the same column. Also apparent is the pairing of ocular dominance columns such that the blobs of a given column are joined via bridges with the blobs of only one of its two neighboring ocular dominance columns. This pattern produces an appearance of ladders or railroad tracks, in which the bridges appear as the rungs or ties.

Other receptive field types Many other receptive field properties have been described in physiological studies of V1. However, of these, few have been demonstrated to be systematically organized within V1. Several properties clearly have a laminar rather than columnar distribution, such as the directional cells of layer IVB (Hawken, Parker, and Lund, 1988) or the distribution of simple versus complex cells (Gilbert, 1977). No clear organization in V1 for properties such as disparity or end inhibition has yet been described.

HORIZONTAL INTERACTIONS BETWEEN COLUMNAR COMPARTMENTS The vertical or interlaminar cortical connections are believed to underlie the basic columnar organization of visual cortex. The early Golgi studies that demonstrated these vertical connections (Lorento de No, 1949) did not hint at the long lateral extent of axonal arbors present in the cortex. Evidence for the long-range horizontal connections in the cortex was first seen in degeneration studies and later in extracellular injections of tracers into primary visual cortex. Patchy patterns of label were seen to extend up to several millimeters from the injection site (Rockland and Lund, 1983; Livingstone and Hubel, 1984b; Gilbert and Wiesel, 1989). The clustered or patchy distribution of these connections, as well as their strong asymmetries, perhaps most clearly shown in intracellular HRP studies (Gilbert and Wiesel, 1979, 1983; Martin and Whitteridge, 1984), suggested a relationship with the underlying functional architecture. This relationship was studied physiologically using the cross-correlation technique (Ts'o, Gilbert, and Wiesel, 1986), discussed in the next section. The combination of extracellular tracer injections with stains for other functional markers (such as cytochrome oxidase or 2-deoxyglucose) provided further indications that these lateral connections were specific to the functional organization. For example, a given orientation column characterized by single-unit recordings was found to receive input in a latticelike arrangement from regions of similar orientation specificity (as revealed by 2-deoxyglucose labeling) and avoid regions of orthogonal orientation preference (Gilbert and Wiesel, 1989). In the V1 blob system, focal injections into single cytochrome oxidase blobs in layers II and III label preferentially nearby blobs up to a millimeter away, whereas injections into interblob regions label interblob regions (Livingstone and Hubel, 1984b). In contrast to the axon arborization, the functional specificity of dendritic arborizations (at least of blob cells) may be less strict (Hubener and Bolz, 1992; Malach, 1992).

Cross-correlation studies of V1 intrinsic connectivity The degree of functional specificity in horizontal dimension also has been investigated with physiological methods. The method of cross-correlation analysis provides a statistical measure of the temporal relationship between the firing of two cells and therefore an indication of the connectivity or interactions between the two cells. Cross-correlation studies reveal that horizontal interactions in both cat (Ts'o, Gilbert, and Wiesel, 1986; Hata et al., 1991) and monkey (Ts'o and Gilbert, 1988; Schwarz and Bolz, 1991) occur preferentially between cells of similar orientation selectivity (figure 20.4), even those with receptive fields separated by as much as 3°. Although some excitatory monosynaptic interactions were observed, common input interactions were most prevalent (40–50%). This prevalence of common input might be expected given the small synaptic efficacies involved and the high degree of interconnectivity in the cortical network.

Connectivity of color processing Because the blob regions represent a major component of the functional and cytoarchitectonic organization of the monkey striate cortex, it is natural to ask how the blobs might interact with the nonblob regions. Livingstone and Hubel (1984b), using focal extracellular injections of HRP, demonstrated connections between adjacent blobs and also between neighboring nonblob (interblob) regions. No connections between blob and interblob regions were found. This segregation of connections was confirmed and extended with results using cross-correlation analysis, which showed that blob-blob connections existed between cells with matching color opponency (figure 20.5) and that interblob-interblob connections existed between cells with matching orientation preference (Ts'o and Gilbert, 1988). Preliminary anatomical results, using focal injections at blob sites whose color opponency had been previously determined, seem to support this arrangement (Ts'o, 1989). Cross-correlation studies have also shown that lateral connections contribute to the construction of specific color-receptive field properties. For example, type II cells (cells with color-opponent centers but no surrounds) provide monosynaptic input to modified type II cells (cells with color-opponent centers and broadband surrounds) and type I cells (basic center-surround color cells) and modified type II cells contribute directly to oriented color cells in V1 (Ts'o and Gilbert, 1988). These findings suggest that lateral connections within

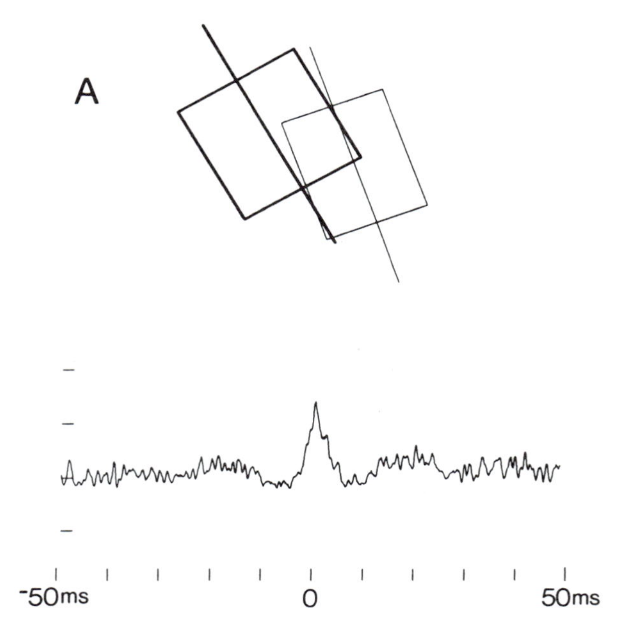

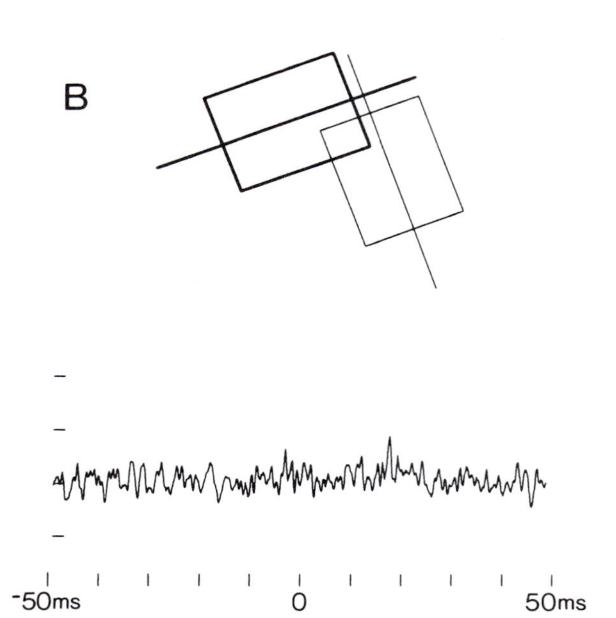

Figure 20.4 Correlograms obtained from two cell pairs. (A) The cell pair had similar receptive properties: The first cell had an orientation preference of 120°, directional preference to the right, and an ocular dominance group of 2; the second cell had identical orientation and direction preference and an ocular dominance group of 3. (B) The first cell here was the same as the first cell in (A). The second cell had different receptive field properties: an orientation preference of 20°, upward directionality, and an ocular dominance of 5.

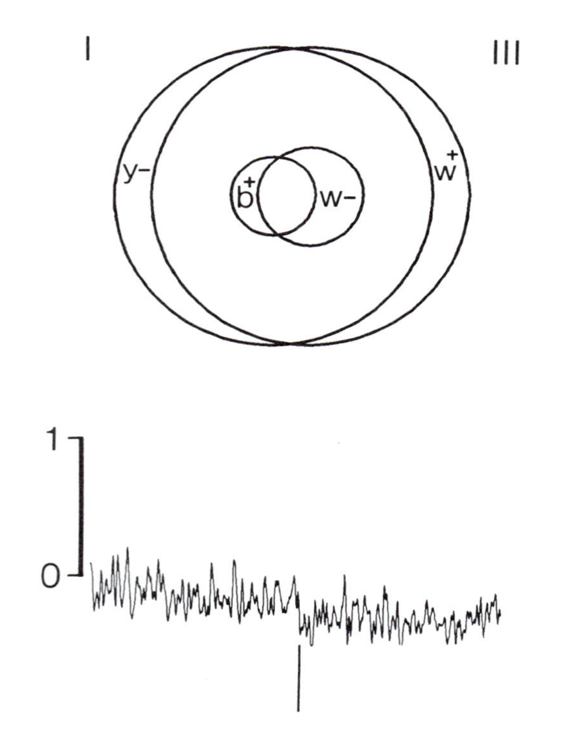

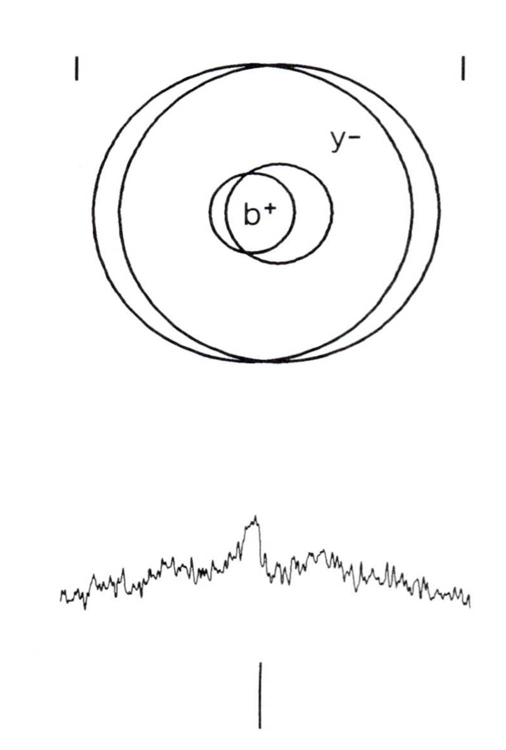

Figure 20.5 (Left) Blue-ON yellow-OFF type I to type III OFF-center correlogram. (Right) Blue-ON yellow-OFF type I to Blue-ON yellow-OFF type I cell. Correlograms were normalized by the baseline correlation. The shape and pattern of the peak indicates an excitatory monosynaptic connection from one type I to the other. $N_a = 4378$; $N_b = 19{,}370$; contribution, 13.2%; effectiveness, 3.0%. W (white) indicates broadband spectral properties.

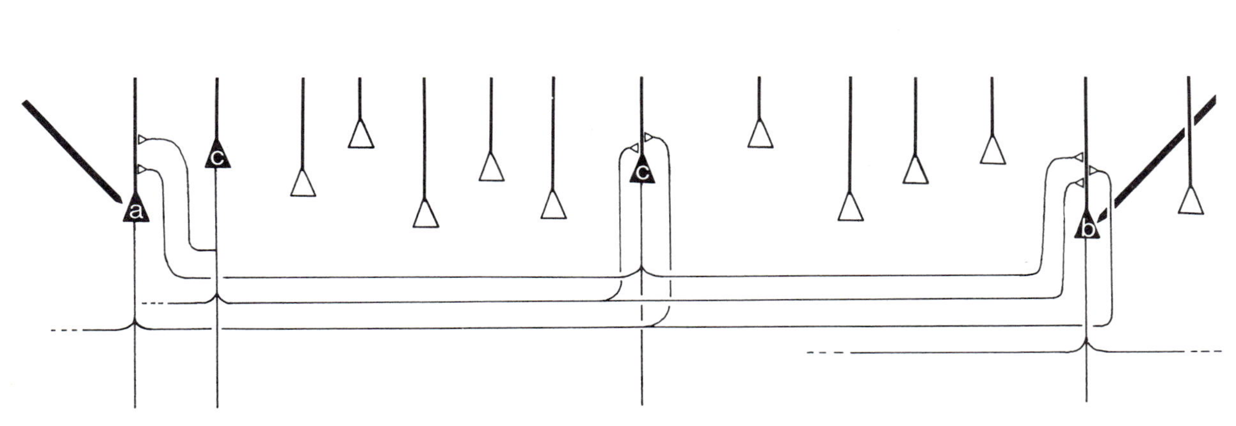

FIGURE 20.6 Schematic representing results on long-range horizontal connections between columns of cells with similar receptive field properties. Recording from the two cells a and b, having similar orientation preferences (shown at the top), cell a is seen to provide a direct input to cell b. In addition, the pair receives common input from several c cells in cortical columns with like orientation specificity with intervening columns of differing specificity. The common input connections outnumber the direct inputs from a to b.

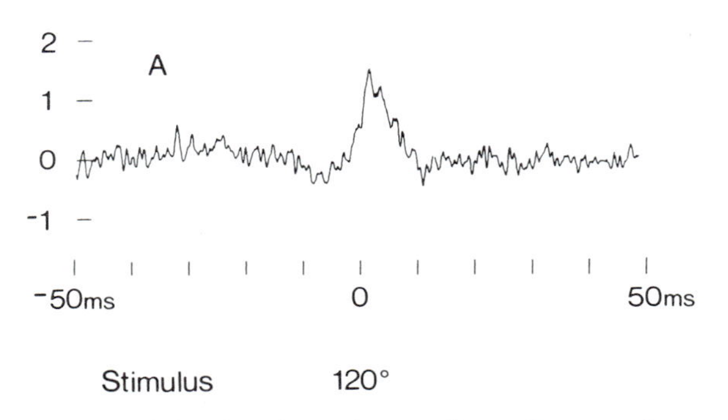

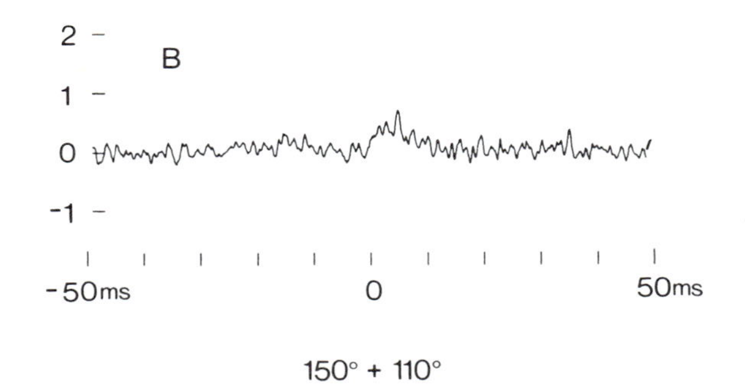

FIGURE 20.7 Stimulus dependency of correlograms from a cell pair: changes with different orientations of visual stimulation. One cell of the pair had an orientation preference of 100° and the second cell had an orientation preference of 130°. The two cells had matched directionality and ocular dominance and were separated by 350 μm. The pair was first stimulated with a single light slit oriented at 120° (A), and then with two slits simultaneously at 110° and 150° (B). Note the substantial change in the strength of the correlogram with the change in stimulation, though both conditions provided good stimulation to both cells.

the cortex have a like-to-like pattern, within both the orientation domain (figure 20.6) and the color domain. Undoubtedly, similar rules of connectivity may be found within other functional groupings.

Initial studies have shown that many of these cortical patterns of cross-correlation are not static but rather change their strength and signature with changes in state or stimulus context (figure 20.7). This type of finding, which also has been observed in the cortico-cortical interactions between V1 and V2 (Ts'o, Roe, and Shey, 1993) as well as in other cortical areas (Aersten et al., 1991; Ahissar et al., 1992), emphasizes the dynamic nature of cortical processing. However, attaining an understanding of the origins of such changes remains challenging.

Functional properties and functional organization in V2

INPUTS TO V2 Unlike the visual system in the cat, the second visual area (V2) in the primate receives its visual inputs primarily from V1. Direct geniculate input to V2 is sparse and scattered and arises from the interlaminar zones and the S layers of the LGN (Bullier and Kennedy, 1983). Reversible inactivation of V1 by cortical cooling renders cells in V2 visually unresponsive (Schiller and Malpeli, 1977; Girard and Bullier, 1989), suggesting that V1 is the primary source of visual drive for V2.

STRIPES IN V2 In contrast to V1, cytochrome oxidase staining in the second visual area, V2, reveals a pattern of bands or stripes of dense labeling, separated by pale interstripes (Livingstone and Hubel, 1982, 1984a; Humphrey and Hendrickson, 1983; Tootell et al., 1983; Horton, 1984; DeYoe and Van Essen, 1985; Shipp and Zeki, 1985). The V2 stripes are functionally distinct and contain different populations of functional cell types. In contrast to V1, there is no apparent organization to ocular dominance in V2 and most V2 cells are binocularly responsive. Cells in the thin stripes are color-selective, some of which are similar to the modified type II cells seen in V1. Interstripe regions are characterized by oriented, non-color-selective fields that commonly exhibit end-stopping. Thick-stripe cells usually are oriented, lack end-stopping, and often are disparity-sensitive (DeYoe and Van Essen, 1985; Hubel and Livingstone, 1987; Ts'o, Gilbert, and Wiesel, 1990). In addition to these physiological findings, the pattern of V2 corticocortical connections seen in anatomical tracer studies suggests that thin, pale, and thick stripes are the V2 components of the color, form, and motion-disparity pathways, respectively (see Van Essen and DeYoe, this volume). Some reports, however, suggest a less clear-cut relationship between V2 cell properties and the V2 stripe organization. Studies using achromatic and isoluminant chromatic gratings to characterize V2 cells quantitatively have found a less striking segregation of cell types (Levitt and Movshon, 1990). These differences in findings may be related to differences in the methods of cell classification or approaches to correlating physiology with anatomy.

Stripe substructure Within a single stripe in V2, there is some evidence to indicate clustering of cell properties such as color, disparity, and orientation (DeYoe and Van Essen, 1985; Tootell and Hamilton, 1989; Ts'o et al., 1989). Other functional methods used to image stripe organization, such as 2-deoxyglucose labeling (Tootell et al., 1983; Tootell and Hamilton, 1989) and optical imaging of intrinsic signals (Ts'o et al., 1990; Ts'o, Gilbert, and Wiesel, 1990; Ts'o, Roe, and Shey, 1993) clearly indicate clustering of functional activity along the length of the stripes. In thin stripes, for example, physiological recordings encounter patches of red-green opponent–dominated patches as well as blue-yellow-dominated patches. Disparity-tuned thick-stripe cells, as mentioned previously, tend to fall into near cell–dominated, far cell–dominated, tuned excitatory, and tuned inhibitory clusters (Ts'o, Gilbert, and Wiesel, 1990, 1991). In pale stripes, physiological recordings frequently encounter a cluster of similarly oriented cells adjacent to another cluster of oriented cells selective for a very different orientation. It is clear that V2 stripes are, in fact, not homogeneous structures but rather collections of distinct functional modules (see color plate 6).

Higher-order receptive field types in V2 Several types of higher-order receptive fields have been described in V2. Two types of color cells are the spot cell and the oriented color cell. Spot cells (Baizer, Robinson, and Dow, 1977; Hubel and Livingstone, 1987) are unoriented color-selective cells that respond optimally to a small spot. However, unlike a standard center-surround cell, the spot cell has a spatial independence such that a small spot (e.g., 0.25°) is effective over a relatively large portion (e.g., 4 × 4°) of the visual field (Hubel and Livingstone, 1985, 1987). These spot cells are not found in V1 and may represent a further elaboration in V2 of the modified type II cells of V1. One possible circuit would involve convergence from modified type II cells with matching color opponency and both eye dominances and with receptive field centers scattered over the large area corresponding to the spot cell's field (typically four to eight times larger than a modified type II cell at the same eccentricity). Such convergence would require input either intrinsically from the modified type II cells within V2 or via corticocortical connections from V1-modified type II cells over a relatively wide area of cortex.

Although it is still unclear how oriented color cells are organized in either V1 or V2, preliminary evidence suggests that in V1 they are localized in the border regions between the color-selective blobs and the orientation-selective interblobs (Ts'o and Gilbert, 1988). In V2, they are similarly located at such border zones (i.e., thin, pale borders). It is possible that oriented color cells in V2 receive input directly from oriented color cells in V1 (and thus comprise a separate oriented color cell pathway) or, alternatively, they may be constructed de novo via convergent input from unoriented color cells in V1 or V2 (Roe and Ts'o, 1992).

An organization apparently absent in V1 that is prominent in V2 is the organization for disparity. Disparity cells, defined by their selectivity for stimuli presented at a specific retinal disparity, are common to the thick stripes of V2. Most remarkable is the predominance of obligatory binocular cells, disparity cells that respond vigorously to stimulation of both eyes when stimuli are present at the optimal disparity but are absolutely silent to monocular stimulation of any type. Studies describing near cells (crossed disparity) and far cells (uncrossed disparity) in V2 suggest a possible clustered organization of near cells and far cells in the V2 thick stripes (Ts'o, Gilbert, and Wiesel, 1990, 1991). Another example of a mixed-property cell occurring at a stripe border region is the color-disparity cell. Seen in both cytochrome oxidase–stained sections and in optical images of functional organization in V2 are the regions where thick and thin stripes appear to merge, resulting in the absence of an intervening pale stripe. These color-disparity borders often are characterized by patches of color-disparity cells and may result from the convergence of the two functional compartments (Ts'o, Gilbert, and Wiesel, 1990, 1991).

Form processing also gains in complexity in V2. Unlike standard orientation-selective cells, some V2 cells, called *subjective contour* cells (von der Heydt and Peterhans, 1989; Peterhans and von der Heydt, 1989), respond to stimulus patterns that are perceived to contain contours not actually present in the stimulus. These investigators proposed a model for the construction of the V2 subjective contour cell from the convergence of appropriately positioned end-inhibited cells in V1 and V2.

Topography in V2 Yet another distinguishing feature of V2 is its retinotopic organization. It was previously known that receptive fields in V2 are larger than those in V1 and retinotopy in V2 is less precise than that in V1 (Gatass, Gross, and Sandell, 1981). With the discovery of cytochrome oxidase and the stripe organization in V2, a new issue arose concerning visuotopic mapping in V2. In V1, there is a point-to-point mapping of the visual field onto the cortex such that roughly a square millimeter of cortex, which spans a left-eye and a right-eye ocular dominance column, represents a specific locus in visual space. In V2, each thick-pale–thin-pale cycle spans approximately 4 mm of cortex. Therefore, roughly speaking, at any single isoeccentricity, each cycle of thin-pale–thick-pale stripes potentially receives input from a cortical area spanning four to five ocular dominance hypercolumns, or approximately the width of ten blobs. How does the region of visual space that is represented across several hypercolumns map onto a set of three functionally distinct stripes? Evidence from receptive fields recorded in long tangential penetrations across the stripes in V2 suggest that there is a multiple and discontinuous mapping of the visual field in V2 (Roe and Ts'o, 1993). In other words, any region of visual space is represented at least in triplicate across the thin, pale, and thick stripes in V2. An electrode traveling across any single stripe will encounter a continuous progression of receptive fields, followed by a jump back in receptive field progression at the stripe border, followed by another continuous progression representing the same region of visual space in a different functional domain. Therefore, several ocular dominance columns of V1 must map in a topographical manner onto each of the color, form, and disparity stripes across the V1/V2 border. Results from anatomical studies are consistent with this prediction (Livingstone and Hubel, 1984a).

Connectivity, interactions, and the relationship between V1 and V2

The previous section has described the difference in the functional organizations of V1 and V2 as well as differences in the types of receptive fields they contain. In V2, the emphasis in organization has shifted from one dominated by lower-level features (such as ocular dominance, orientation, color, and point-to-point mapping) to one concerned with more integrated features (such as disparity, contour recognition on a more global level, spatial invariance of color features, and independent topography within selected feature domains). What corticocortical connectivities underlie these transformations in functional properties and functional organization? To approach this question, we first examine the anatomical basis of V1/V2 connectivity.

V1/V2 Anatomical Connectivity Anatomical studies suggest the functional segregation of inputs from V1

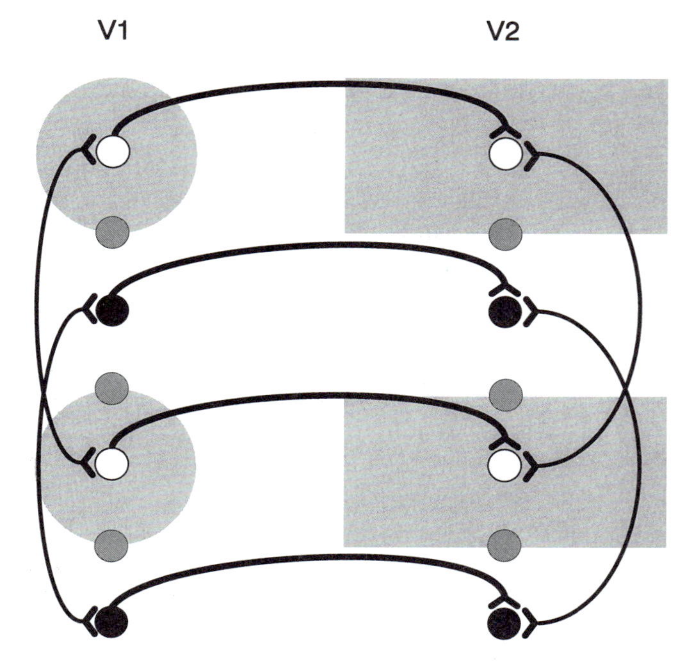

FIGURE 20.8 Schematic of V1/V2 connections. The diagram depicts intrinsic connections among blobs (shaded circles) and among interblobs (black dots) within V1, as well as intrinsic connections within V2, among thin stripes (shaded bands) and among pale-stripe zones (thick-stripe regions omitted for simplicity). In addition, the V1/V2 corticocortical connections between blob and thin stripes and between interblob and pale stripes are shown. This simplistic view of V1/V2 interactions seems inadequate in light of the cross-correlation results demonstrating a high degree of divergence and convergence, the prevalence of common input interactions, and the rarity of clear consistent temporal delay in the response of V2 relative to V1.

to V2 (figure 20.8), as well as segregation of outputs from V2 into higher cortical areas. Focal tracer injections into thin stripes in V2 preferentially label blob regions in V1, and those into pale stripes of V2 label interblob regions (Hubel and Livingstone, 1984a); thick stripes of V2 are reciprocally connected to layer IVB of V1 (Livingstone and Hubel, 1987). The terminal arbors of single axons projecting from V1 to V2 appear to extend along the length of V2 stripes, again suggesting that direct V1 input is channeled into single stripes in V2 rather than distributed divergently across stripes (Rockland and Virga, 1990). Outputs from the thin and interstripes of V2 continue on to area V4, and those from the thick stripes to cortical area MT (DeYoe and Van Essen, 1985; Shipp and Zeki, 1985; DeYoe et al., 1988; Felleman et al., 1988). These data and the accompanying physiological characterizations of the blob and interblob regions in V1 (Livingstone and Hubel, 1984a) and stripes in V2 (Hubel and Livingstone, 1987) suggested a segregation of form, color, and motion-stereo processing streams. This view of visual processing has been both supported and criticized (for reviews, see Livingstone and Hubel, 1988; Schiller and Logothetis, 1990; Merigan and Maunsell, 1993).

Anatomical studies to date have demonstrated the general patterns of connectivity between the V2 stripes and other compartments in other cortical areas (see Van Essen and DeYoe, this volume). However, we know that the stripes are actually composed of a series of functionally distinct patches. It is then expected that V2 connectivity at a finer level bears some relationship to the subcompartments within a single stripe. Furthermore, it is likely the corticocortical connections play a significant role in the generation of particular higher-order receptive field properties in V2 and elsewhere. Investigation of these issues will require examining the connectivity at a finer scale, both anatomically and physiologically.

V1/V2 FUNCTIONAL CONNECTIVITY Cross-correlation analysis has been used to examine the patterns of V1/V2 corticocortical connectivity (Bullier, Munk, and Nowak, 1992; Roe and Ts'o, 1992), and two notable characteristics have emerged. The first concerns the degree of convergence and divergence between V1 and V2. As stated previously, each stripe in V2 is likely to receive input from as many as four to five ocular dominance hypercolumns. Thus, V1/V2 connectivity is likely to be highly convergent and divergent. Indeed, cross-correlation studies indicate a high degree of convergence and divergence in V1/V2 connectivity in comparison to V1/V1 cell pairs (Bullier, Munk, and Nowak, 1992; Roe and Ts'o, 1992). V1/V2 correlations are typically 50–200 ms wide compared to V1/V1 or V2/ V2 correlation peaks, which are typically 5–20 ms wide. This amount of temporal dispersion in the correlograms found between V1 and V2 suggests a higher degree of synaptic jitter, perhaps due to a larger number of weaker synapses participating in the interaction. V1/V2 cross-correlation peaks are also often centered on zero, indicating the presence of common input. Similarly, broad and centered cross-correlations have been reported between cell pairs in cat area 17/18 (Nelson et al., 1992). Such peaks could result from common input arising in the pulvinar (Livingstone and Hubel, 1982; Bullier and Kennedy, 1983; Horton, 1984); however, a more likely source may be a V1 cell or assembly of cells that provide inputs to both the V1 cell and the V2 cell under study. These findings also suggest that individual V1 and V2 cells participate in large corticocortical networks.

The second important aspect of V1/V2 connectivity relates to functional specificity. In V1, cross-correlation studies (Ts'o, Gilbert, and Wiesel, 1986; Hata et al., 1991) have revealed that strong correlation peaks occur between cells of similar functional specificity (e.g., similar color preference, orientation) but not between those of different functional specificity (see color plate 7). In general, V1/V2 cell pairs also exhibit this pattern of behavior (Roe and Ts'o, 1992). For example, in the color domain, color cell pairs with matched color specificities exhibit peaked correlations, whereas those with different color specificities exhibit flat correlations. Similarly, only broadband oriented cell pairs with similar orientation selectivity have peaked correlations. Receptive field overlap is an important determinant for functional interaction in some domains but not in others. Thus, interactions between V1 and V2 cells are specific not only in terms of the types and number of cells with which they interact but also in terms of the extent of cortex across which they interact.

These findings contribute to a view of corticocortical processing which extends beyond that of simply channeling information from one cortical level to another. Functional specificity, which was initially suggested by anatomical studies, actually occurs at a much more specialized level, at the level of specific functional properties such as color, orientation, and spatial location. However, this specificity is maintained despite the high degree of convergence and divergence present in V1/V2 projections. This convergent-divergent yet specific type of connectivity is consistent with the pattern of multiple and segregated topographical representation in V2. It is also likely to be important for the generation of specific cell types in V2. For example, the spot cell may summate inputs from color cells arising from a large extent of V1, all of which have the same color specificity. Similarly, the subjective contour cell may arise due to converging inputs from similarly oriented broadband cells arising from many locations in V1. Thus, corticocortical projections participate in specific functional transformations in sensory representation. These transformations determine both the mode of topographical representation in a target cortical area and the generation of specific functional properties there.

Conclusion

Visual areas V1 and V2 are segregated into functional compartments specialized for such properties as orientation, color, and disparity. However, the segregation is far from absolute, and much interaction occurs between neighboring compartments. Whether the purpose of compartmentalization is simply for developmental or anatomical convenience or for facilitating specific computations within a given functional domain or both, it is likely that the organization of multiple feature domains within a cortical area reflects the processing strategies employed and is an important clue to the role of that cortical area in visual processing.

REFERENCES

Aersten, A., E. Vaadia, M. Abeles, E. Ahissar, H. Bergman, B. Karmon, Y. Lavner, E. Margalit, I. Nelken, and S. Rotter, 1991. Neural interactions in the frontal cortex of a behaving monkey: Signs of dependence on stimulus context and behavioral state. *J. Hirnforsch.* 32:735–743.

Ahissar, E., E. Vaadia, M. Ahissar, H. Bergman, A. Arieli, and M. Abeles, 1992. Dependence of cortical plasticity on correlated activity of single neurons and on behavioral context. *Science* 257:1412–1415.

Baizer, J. S., D. L. Robinson, and B. M. Dow, 1977. Visual responses of area 18 neurons in awake, behaving monkey. *J. Neurophysiol.* 40:1024–1037.

Bartfeld, E., and A. Grinvald, 1992. Relationships between orientation-preference pinwheels, cytochrome oxidase blobs, and ocular-dominance columns in primate striate cortex. *Proc. Natl. Acad. Sci.* 89:11905–11909.

Blasdel, G. G., 1992a. Differential imaging of ocular dominance and orientation selectivity in monkey striate cortex. *J. Neurosci.* 12:3115–3138.

Blasdel, G. G., 1992b. Orientation selectivity, preference, and continuity in monkey striate cortex. *J. Neurosci.* 12: 3139–3161.

Bullier, J., and H. Kennedy, 1983. Projection of the lateral geniculate nucleus onto cortical area V2 in the macaque monkey. *Exp. Brain Res.* 53:168–172.

Bullier, J., and M. H. J. Munk, and L. G. Nowak, 1992. Synchronization of neuronal firing in areas V1 and V2 of the monkey. *Soc. Neurosci. Abstr.* 18:11.

Daw, N. W., 1968. Color-coded ganglion cells in the goldfish retina: Extension of their receptive fields by means of new stimuli. *J. Physiol. (Lond.)* 197:567–592.

DeMonasterio, F. M., and P. Gouras, 1975. Functional properties of ganglion cells of the rhesus monkey retina. *J. Physiol. (Lond.)* 251:167–196.

DeYoe, E. A., D. J. Felleman, J. J. Knierim, J. Olavarria, and D. C. Van Essen, 1988. Heterogeneous subregions of macaque visual area V4 receive selective projections from V2 thin-stripe and interstripe subregions. *Invest. Ophthalmol. Vis. Sci.* Suppl. 29:115.

DeYoe, E. A., and D. C. Van Essen, 1985. Segregation of efferent connections and receptive field properties in visual area V2 of the macaque. *Nature* 317:58–61.

Dow, B. M., and R. G. Vautin, 1987. Horizontal segregation of color information in the middle layers of foveal striate cortex. *J. Neurophysiol.* 57:712–739.

Dreher, B., Y. Fukada, and R. Rodieck, 1976. Identification, classification and anatomical segregation of cells with X-like and Y-like properties in the lateral geniculate nucleus of old-world primates. *J. Physiol. (Lond.)* 258:433–452.

Felleman, D. J., E. A. DeYoe, J. J. Knierim, J. Olavarria, and D. C. Van Essen, 1988. Compartmental organization of projections from V2 to extrastriate areas V3, V3A, and V4t in macaque monkeys. *Invest. Ophthalmol. Vis. Sci.* Suppl. 29:115.

Gatass, R., C. G. Gross, and J. H. Sandell, 1981. Visual topography of V2 in the macaque. *J. Comp. Neurol.* 201: 519–539.

Gilbert, C. D., 1977. Laminar differences in receptive field properties of cells in cat primary visual cortex. *J. Physiol.* 268:391–421.

Gilbert, C. D., and T. N. Wiesel, 1979. Morphology and intracortical projections of functionally characterised neurones in the cat visual cortex. *Nature* 280:120–125.

Gilbert, C. D., and T. N. Wiesel, 1983. Clustered intrinsic connections in cat visual cortex. *J. Neurosci.* 3:1116–1133.

Gilbert, C. D., and T. N. Wiesel, 1989. Columnar specificity of intrinsic horizontal and corticocortical connections in cat visual cortex. *J. Neurosci.* 9:2432–2442.

Girard, P., and J. Bullier, 1989. Visual activity in area V2 during reversible inactivation of area 17 in the macaque monkey. *J. Neurophysiol.* 62:1287–1302.

Gouras, P., 1968. Identification of cone mechanisms in monkey ganglion cells. *J. Physiol. (Lond.)* 199:533–547.

Gouras, P., 1974. Opponent-color cells in different layers of foveal striate layers. *J. Physiol. (Lond.)* 238:583–602.

Hata, Y., T. Tsumoto, H. Sato, and H. Tamura, 1991. Horizontal interactions between visual cortical neurons studied by cross correlation analysis in the cat. *J. Physiol.* 441:593–614.

Hawken, M. J., A. J. Parker, and J. S. Lund, 1988. Laminar organization and contrast sensitivity of direction-selective cells in the striate cortex of the old world monkey. *J. Neurosci.* 8:3541–3548.

Horton, J. C., 1984. Cytochrome oxidase patches: A new cytoarchitectonic feature of monkey visual cortex. *Philos. Trans. R. Soc. Lond.* 304:199–253.

Horton, J. C., and D. H. Hubel, 1981. Regular patchy distribution of cytochrome oxidase staining in primary visual cortex of macaque monkey. *Nature* 292:762–764.

Hubel, D. H., and M. S. Livingstone, 1985. Complex-unoriented cells in a subregion of primate area 18. *Nature* 315:325–327.

Hubel, D. H., and M. S. Livingstone, 1987. Segregation of form, color, and stereopsis in primate area 18. *J. Neurosci.* 7:3378–3415.

Hubel, D. H., and T. N. Wiesel, 1968. Receptive fields and functional architecture of monkey striate cortex. *J. Physiol. (Lond.)* 195:215–243.

Hubel, D. H., and T. N. Wiesel, 1974. Sequence regularity and geometry of orientation columns in the monkey striate cortex. *J. Comp. Neurol.* 158:267–293.

Hubel, D. H., and T. N. Wiesel, 1977. Functional architecture of macaque monkey visual cortex. *Proc. R. Soc. Lond.* 198:1–59.

Hubener, M., and J. Bolz, 1992. Relationships between dendritic morphology and cytochrome oxidase compartments in monkey striate cortex. *J. Comp. Neurol.* 324:67–80.

Humphrey, A. L., and A. E. Hendrickson, 1983. Background and stimulus-induced patterns of high metabolic activity in the visual cortex (area 17) of the squirrel and macaque monkey. *J. Neurosci.* 3:345–358.

Kruger, J., 1977. Stimulus dependent color specificity of monkey lateral geniculate neurones. *Exp. Brain Res.* 30: 297–311.

Landisman, C. E., A. Grinvald, and D. Y. Ts'o, 1991. Optical imaging reveals preferential labeling of cytochrome oxidase-rich regions in reponse to color stimuli in areas V1 and V2 of macaque monkey. *Soc. Neurosci. Abstr.* 17: 1089.

Landisman, C. E., and D. Y. Ts'o, 1992. Color processing in the cytochrome oxidase-rich blobs and bridges of macaque striate cortex. *Soc. Neurosci. Abstr.* 18:592.

Lennie, P., J. Krauskopf, and G. Sclar, 1990. Chromatic mechanisms in striate cortex of macaque. *J. Neurosci.* 10: 649–669.

Leventhal, A. G., K. G. Thompson, D. Liu, L. M. Neuman, and S. J. Ault, 1993. Form and color are not segregated in monkey striate cortex. *Invest. Opthalmol. Vis. Sci.* Suppl. 34:813.

Levitt, J. B., and J. A. Movshon, 1990. Receptive fields and functional architecture of macaque V2. *Invest. Opthalmol. Vis. Sci.* Suppl. 31:89.

Livingstone, M. S., and D. H. Hubel, 1982. Thalamic inputs to cytochrome oxidase–rich regions in monkey visual cortex. *Proc. Natl. Acad. Sci. U.S.A.* 79:6098–6101.

Livingstone, M. S., and D. H. Hubel, 1984a. Anatomy and physiology of a color system in the primate visual cortex. *J. Neurosci.* 4:309–356.

Livingstone, M. S., and D. H. Hubel, 1984b. Specificity of intrinsic connections in primate primary visual cortex. *J. Neurosci.* 4:2830–2835.

Livingstone, M. S., and D. G. Hubel, 1987. Connections between layer 4B of area 17 and the thick cytochrome oxidase stripes of area 18 in the squirrel monkey. *J. Neurosci.* 7:3371–3377.

LIVINGSTONE, M. S., and D. H. HUBEL, 1988. Segregation of form, color, movement, and depth: Anatomy, physiology, and perception. *Science* 240:740–749.

LORENTE DE NO, R., 1949. Cerebral cortex: Architecture, intracortical connections, motor projections. In *Physiology of the Nervous System*, ed. 3, J. F. Fulton, ed. London: Oxford University Press, pp. 288–313.

MALACH, R., 1992. Dendritic sampling across processing streams in monkey striate cortex. *J. Comp. Neurol.* 315: 303–312.

MARTIN, K. A. C., and D. WHITTERIDGE, 1984. Form, function and intracortical projections of spiny neurones in the striate visual cortex of the cat. *J. Physiol.* 353:463–504.

MERIGAN, W. H., and J. H. R. MAUNSELL, 1993. How parallel are the primate visual pathways? *Annu. Rev. Neurosci.* 16:369–402.

MICHAEL, C. R., 1981. Columnar organization of color cells in layer 4Cb of the monkey striate cortex. *J. Neurophysiol.* 46:587–604.

NELSON, J. I., P. A. SALIN, M. L.-J. MUNK, M. ARZI, and J. BULLIER, 1992. Spatial and temporal coherence in cortico-cortical connections: A cross-correlation study in areas 17 and 18 in the cat. *Visual Neurosci.* 9:21–37.

PETERHANS, E., and R. VON DER HEYDT, 1989. Mechanisms of contour perception in monkey visual cortex: II. Contours bridging gaps. *J. Neurosci.* 9:1749–1763.

ROCKLAND, K. S., and J. S. LUND, 1983. Intrinsic laminar lattice connections in primate visual cortex. *J. Comp. Neurol.* 216: 303–318.

ROCKLAND, K. S., and A. VIRGA, 1990. Organization of individual cortical axons projecting from area V1 (area 17) to V2 (area 18) in the macaque monkey. *Visual Neurosci.* 4:11–28.

ROE, A. W., and D. Y. TS'O, 1992. Functional connectivity between V1 and V2 in the primate. *Soc. Neurosci. Abstr.* 18:11.

ROE, A. W., and D. Y. TS'O, 1993. Visual field representation within primate V2 and its relationship to the functional distinct stripes. *Invest. Ophthalmol. Vis. Sci.* Suppl. 34:812.

SCHILLER, P. H., and N. K. LOGOTHETIS, 1990. The color-opponent and broad-band channels of the primate visual system. *Trends Neurosci.* 13:392–398.

SCHILLER, P. H., and J. G. MALPELI, 1977. The effect of striate cortex cooling on area 18 cells in the monkey. *Brain Res.* 126:366–369.

SCHILLER, P. H., and J. G. MALPELI, 1978. Functional specificity of lateral geniculate nucleus laminae of the rhesus monkey. *J. Neurophysiol.* 41:788–797.

SCHWARZ, C., and J. BOLZ, 1991. Functional specificity of a long-range horizontal connection in cat visual cortex: A cross-correlation study. *J. Neurosci.* 11:2995–3007.

SHIPP, S., and S. ZEKI, 1985. Segregation of pathways leading from area V2 to areas V4 and V5 of macaque monkey. *Nature* 315:322–325.

TOOTELL, R. B. H., and S. L. HAMILTON, 1989. Functional anatomy of the second visual area (V2) in the macaque. *J. Neurosci.* 9:2620–2644.

TOOTELL, R. B. H., M. S. SILVERMAN, R. L. DE VALOIS, and G. H. JACOBS, 1983. Functional organization of the second cortical visual area of primates. *Science* 220:737–739.

TS'O, D. Y., 1989. The functional organization and connectivity of color processing. In *Neural Mechanisms of Visual Perception*, D. M. Lam and C. D. Gilbert, eds. Woodlands, Tex.: Gulf, pp. 87–115.

TS'O, D. Y., R. D. FROSTIG, E. E. LIEKE, and A. GRINVALD, 1990. Functional organization of primate visual cortex revealed by high-resolution optical imaging. *Science* 249: 417–420.

TS'O, D. Y., and C. D. GILBERT, 1988. The organization of chromatic and spatial interactions in the primate striate cortex. *J. Neurosci.* 8:1712–1727.

TS'O, D. Y., and C. D. GILBERT, R. D. FROSTIG, A. GRINVALD, and T. N. WIESEL, 1989. Functional architecture of visual area 18 of macaque monkey. *Soc. Neurosci. Abstr.* 15:161.

TS'O, D. Y., C. D. GILBERT, and T. N. WIESEL, 1986. Relationships between horizontal interactions and functional architecture in the cat striate cortex as revealed by cross-correlation analysis. *J. Neurosci.* 6:1160–1170.

TS'O, D. Y., C. D. GILBERT, and T. N. WIESEL, 1990. Functional architecture of color and disparity in visual area 2 of macaque monkey. *Soc. Neurosci. Abstr.* 16:203.

TS'O, D. Y., C. D. GILBERT, and T. N. WIESEL, 1991. Orientation selectivity of and interactions between color and disparity subcompartments in area V2 of macaque monkey. *Soc. Neurosci. Abstr.* 17:1089.

TS'O, D. Y., A. W. ROE, and J. SHEY, 1993. Functional connectivity within V1 and V2: Patterns and dynamics. *Soc. Neurosci. Abstr.* 19:1499.

VON DER HEYDT, R., and E. PETERHANS, 1989. Mechanisms of contour perception in monkey visual cortex: I. Lines of pattern discontinuity. *J. Neurosci.* 9:1731–1748.

WIESEL, T. N., and D. H. HUBEL, 1966. Spatial and chromatic interactions in the lateral geniculate body of the rhesus monkey. *J. Neurophysiol.* 29:1115–1156.

WONG-RILEY, M. T. T., 1979. Changes in the visual system of monocularly sutured or enucleated cats demonstrable with cytochrome oxidase histochemistry. *Brain Res.* 171: 11–28.

21 The Visual Analysis of Shape and Form

SHIMON ULLMAN

ABSTRACT Aspects of visual shape analysis are approached in this chapter in the context of two major visual tasks that rely on the analysis of shape and spatial relations: object recognition and spatial cognition. During visual recognition, objects' shapes are processed and compared with shape representations stored in memory; this task therefore provides a framework for studying aspects of shape representation and shape processing by the visual system. A number of alternative approaches to shape representation and shape comparison for the purpose of recognition are reviewed and compared. The second task, spatial cognition, has to do with a general capacity of the visual system to analyze shape properties and spatial relations in the field of view. It is proposed that this capacity is obtained by the application of so-called visual routines to the visual representations. These are sequences of basic operations that are wired into the visual system and that can be assembled to extract a large variety of shape properties and spatial relations in response to either externally imposed or internally generated visual tasks.

The visual analysis of shape and form plays a primary role in human visual perception. We often use nonshape aspects of visual appearance, such as the color, texture, or motion of objects, but many visual tasks rely primarily on the analysis of shape. I will not attempt to define precisely here the terms *shape* and *form*, though their meanings will become clearer in the pages that follow. The word *form* is used to include both the shape of objects and spatial relations among them. Roughly, the common use of shape in vision relates to visual tasks that rely on the perception of extended configurations (unlike color, for instance, that is assigned to a point) and that depend on the exact configuration, including orientation, position, and distances between elements (unlike texture, for example, that relies more on statistical properties [Julesz, 1981]).

It often is difficult to discuss issues of shape and form in vision in the abstract. Questions such as what is shape, how is shape represented, and how is shape processed do not have a general answer and are thus better examined within the framework of particular visual tasks. Hence, I will discuss shape and form analysis within the framework of selected visual tasks, though because the scope of the subject is so broad, not all aspects of shape analysis will be covered. There are various visual tasks that rely on shape analysis but will not be considered. There are also general factors affecting the perception of shape, such as figural adaptation (Over, 1971), contextual effects and frames of reference (Rock, 1974), and figural illusions (Coren and Girgus, 1978) that will not be discussed. Instead, I will focus on key issues in the visual analysis of shape and form in the context of two major tasks that rely heavily on shape, form, and spatial relations. The first is the task of visual object recognition. In the course of recognition, the visual system analyzes objects' shapes and compares them with shape representations stored in memory. Recognition is therefore a natural framework to discuss a number of key issues in shape representation and processing. In addition to recognition, we have a more general capacity to analyze shape and spatial relations: We can rely on our visual sense to obtain answers to questions such as, Is x round or elongated, is x longer than y, or can x fit inside y? Problems of this type can often be answered by merely looking at the objects in the scene. These aspects of perception do not require object recognition, but they do require efficient analysis of shape properties and spatial relations. This facet of form analysis by the visual system I refer to as *visual cognition.* Finally, I discuss briefly physiological aspects of shape analysis by the primate visual system.

SHIMON ULLMAN Department of Applied Mathematics, The Weizmann Institute of Science, Rehovot, Israel, and the Department of Brain and Cognitive Sciences, Massachusetts Institute of Technology, Cambridge, Mass.

Shape-based visual recognition

In the course of object recognition, the projected shape of a three-dimensional (3-D) object can be used to identify the object and to access information associated with similar objects we have seen in the past. We can sometimes use for recognition visual, but nonshape, cues such as color, texture, and motion. For example, various material types and different scene classes (e.g., lake scenery, mountainous terrain) can be recognized visually without relying on precise shape. Certain objects (such as a tiger or a giraffe) can sometimes be identified on the basis of texture and color pattern, and some objects can be recognized visually on the basis of their characteristic motion rather than their specific shape (Johansson, 1973; Cutting and Kozlowski, 1977). Most common objects are recognized, however, mainly by their shape properties, with color, texture, or motion playing only a secondary role. We will be able to recognize readily, say, a pink elephant, despite its unusual color, based on its shape properties. The process of visual object recognition can therefore serve as a useful framework for analyzing a number of key issues related to visual shape analysis, such as: How is the shape of of an object described internally in the visual system? What aspects of the shape are extracted and stored in memory, and how are stored shapes compared with new images? How is form invariance achieved for various shape transformations, such as scale, position, and viewing direction in 3-D?

These and other questions can be grouped under two main headings: those related to *shape encoding* (or representation) and those related to *shape processing*. The encoding issue has to do with the way the shape is represented internally. For example, the input shape may be represented internally in terms of a set of basic features, such as lines, edges, corners, intersections, and the like, or using elementary shape primitives, such as shape "codons" (Hoffman and Richards, 1986) or 3-D shape primitives such as the "geons" proposed by Biederman (1985). The processing issue has to do with internal processes that can manipulate shape descriptions (e.g., scale or rotate them, extract their major axes, compare two different shapes). Such processes may depend on the particular visual task for which the shape is being used. As we will see, different approaches to shape-based recognition make different proposals regarding shape encoding and processing by the visual system.

SHAPE EQUIVALENCE One extreme example of the use of shape in recognition is to store a large number of different views associated with each object and then compare the image of the currently viewed object with all the shapes stored in memory (Abu-Mostafa and Psaltis, 1987). This approach avoids any elaborate shape processing, such as feature extraction, decomposing the shape into constituent parts, or extracting invariant shape properties (as discussed further later). Several mechanisms, known as *associative memories*, have been proposed for implementing this direct shape comparison. These mechanisms, usually embodied in neuronlike networks, can store a large set of patterns (P_1, P_2 ... P_n), and then, given an input shape Q, they can retrieve the pattern P_i that is most similar to Q (Willshaw, Buneman, and Longuet-Higgins, 1969; Kohonen, 1978; Hopfield, 1982). In terms of the encoding and processing issues listed earlier the recoding in this approach is minimal and, in terms of processing, the main operation is limited to simple, direct shape comparison.

The problem with using such direct shape comparison, without more elaborate encoding and processing, is that the notion of shape similarity employed by such schemes is too restricted. The typical similarity measure used is the so-called Hamming distance, which is defined for two binary vectors: If u and v are two binary vectors (i.e., strings of 1s and 0s) then the Hamming distance between u and v is simply the number of coordinates in which they differ. Such a simple shape similarity measure is insufficient for two reasons. First, the space of all possible views of all the objects to be recognized is likely to be prohibitively large. Second, the shape to be recognized often will not be sufficiently similar to any image seen in the past. This variability in appearance is of fundamental importance to shape perception. It implies, in particular, that certain shapes are seen as equivalent in the sense that they are perceived spontaneously as different images of the same object. Two figures that are separated by a large Hamming-like distance can be spontaneously perceived as having similar shapes; for example, one may be simply a scaled version of the other. Many of the shape-processing mechanisms in vision are directly related to different types of shape variability, and it is therefore worth considering briefly the main sources of natural variability of shapes and their implications for shape processing.

The first source is the effect of viewing position,

including changes in the shape's scale, position, and orientation in the image, as well as distortions caused by changes of orientation in 3-D space. The problem of separating, at least to some degree, external shape changes from shape changes caused by viewing conditions is a major challenge in visual shape analysis and will be discussed further below.

The second source of shape variability has to do with photometrical effects (i.e., the positions and distribution of light sources in the scene, shadows, highlights, etc.). Illumination effects can change drastically the light-intensity distribution falling on the retina, but they usually have minor effects on shape perception. The ability to achieve this form of invariance can be explained in part by the extraction of shape features that are relatively insensitive to illumination, such as intensity edges.

The third source is the effect of object setting. Objects are rarely seen in isolation: They are usually seen against some background, next to, or partially occluded by other objects. This raises problems such as figure-ground separation, grouping, figure completion, and subjective contours (Kanisza, 1979; Nakayama, Shimojo, and Silverman, 1989).

The fourth and final variation is the effect of changing shape. Many objects, such as the human body, can maintain their identity while changing their 3-D shape. To identify such objects correctly, the visual system must be able to deal, therefore, with the effects induced by some systematic changes of objects' shape. A related problem is raised by object classification (as opposed to the identification of an individual object), because classification introduces allowable shape changes (either 2-D or 3-D) within the class in question. For example, we can recognize a shape as a triangle, irrespective of specific geometrical shape (as well as position and size). To recognize a triangle, it clearly is not necessary to store in memory a large number of representative shapes: All the shapes in this class have certain properties in common, and these regularities can be exploited by the shape analysis and recognition mechanisms. This can be achieved by recoding of the shape (e.g., generating a more abstract description in terms of vertices and line segments). Alternatively, it can be achieved by appropriate shape processing (e.g., by shifting and scaling the input shape in the course of matching it to a stored model). Similarly, shape encoding and shape processing can be used to compensate for other sources of shape variations. We may expect shape analysis in the visual system to employ strategies for both shape encoding and shape processing that will enable robust and efficient use of the shape for recognition and other tasks.

It still is unknown what types of representations and processing the visual system actually uses. The main approaches to shape-based recognition that have been suggested, and their implications to the general issues of shape encoding and processing, are discussed and compared next.

Shape Invariances A common approach to shape-based recognition has been to assume that objects have certain invariant shape properties that are common to all their views (Pitts and McCulloch, 1947) and that these invariances are extracted by the visual system. For example, in pattern recognition systems, a *compactness measure*, defined as the ratio between the shape's apparent area and its perimeter length squared, has been employed. Shapes that tend to be round and compact will have a high score on this measure, whereas long and narrow objects will have a low score. Furthermore, the measure will be unaffected by rotation, translation, and scaling in the image plane. Certain Fourier descriptors (coefficient in the Fourier transform) and object moments are additional examples of invariant measures that have been proposed. Other types of invariant measures are discussed in Mundy and Zisserman (1992).

Formally, a property of this type can be defined as a function from shapes to the real numbers. The use of an invariant shape property assumes that this function can be computed from the viewed shape alone (without reference to stored models) and that the resulting value will be common to the different views of the same object. It is also important that such a function be relatively simple to compute. Otherwise, in recognizing, for example, different instances of the letter A, one may define a function whose value is 1 if the viewed object is the letter A, and 0 otherwise. This function would be an invariant of the letter A, but the problem of computing this invariance would be, of course, equivalent to the original problem of recognizing the letter.

In some approaches, a shape property defined for a given object (or class of objects) is not expected to remain entirely invariant, only to lie within a restricted range. Properties of different objects may have partially overlapping ranges, but the hope is that by defining a number of different properties, it will become possible to classify the shapes of objects and classes. This leads naturally to the concept of feature spaces, which have been used extensively in pattern recognition (Duda and Hart, 1973). If n different properties are measured, shape is characterized by a vector of n real numbers. It then becomes possible to represent a given shape by a point in an n-dimensional space, R^n. The set of all the views induced by a given object define, in this manner, a subspace of R^n (Tou and Gonzalez, 1974). This representation could become useful for identifying and classifying objects, provided that the subspaces have a simple structure. They may be described, for instance, by a collection of spheres (Reilly, Cooper, and Elbaum, 1982) or be separated by linear hyperplanes (Minsky and Papert, 1969).

A well-known psychological theory that belongs to the general category of invariant properties theories (but does not use feature spaces) is Gibson's theory of high-order invariances (Gibson, 1950, 1979). Gibson postulated that invariant properties of objects may be reflected in so-called higher-order invariances in the optical array. Such invariances may be based, for example, on spatial and temporal gradients of texture density. A set of invariances may be picked up, according to this theory, by the visual system and may be used to characterize objects and object classes.

How useful have such shape invariance methods been for explaining recognition and shape analysis by the visual system? The invariant properties approach, including the construction of feature spaces and their separation into subspaces, have probably been studied more extensively than any other method for shape-based recognition. It has met with some success within certain limited domains: A number of artificial vision systems perform simple recognition of industrial parts based on the measurement of global properties such as area, elongation, perimeter length, and different moments (see a review in Bolles and Cain, 1982). In such domains, simple shape invariances may indeed be common to all the members of a given class. In other cases, such invariances may not exist; there is no particular reason to assume the existence of relatively simple shape properties (particularly global ones) that are preserved across the transformations that an object may undergo. It is not surprising, therefore, that despite considerable effort, generally applicable invariant properties for visual object recognition proved difficult to find. Shape invariances are likely to be useful but insufficient; additional mechanisms are required for shape analysis and recognition.

Shape Ecomposition and Structural Descriptions

A second general approach to shape analysis and its use in recognition relies on the decomposition of objects' shapes into constituent parts. This approach clearly has some intuitive appeal. Many objects seem to contain natural parts: A face, for example, contains the eyes, nose, and mouth as distinct parts that can often be recognized on their own.

Decomposition into constituent parts The shape decomposition approach assumes that each object can be decomposed into a small set of generic shapes. The basic shapes are generic in the sense that all objects can be described as different combinations of the same basic components. The decomposition must also be stable—that is, preserved across views. Shape-based recognition proceeds in this approach by locating the parts, classifying them into the different types of generic shape primitives, and then describing the objects in terms of their constituent parts.

One approach to shape decomposition has been to apply the decomposition process repeatedly: The shape is first partitioned into a small number of major parts, and each part is then broken down again into more elementary primitives. An example of a shape hierarchy of this kind is to detect straight-line segments as the most basic parts and then to detect higher-level shapes such as corners and vertices, based on the already detected line segments. These parts can be combined, in turn, into higher-level structures (e.g., certain configurations of lines and vertices can be combined to define triangles). Such approaches are known as *feature hierarchies*. The simple basic-level parts are *features* (a term also used in many other contexts), and higher-level structures are constructed hierarchically (Selfridge, 1959; Sutherland, 1959; Barlow, 1972; Milner, 1974). This approach has been motivated in part by physiological findings (Hubel and Wiesel, 1962, 1968) in the cat and monkey, that can be interpreted as the extraction by the visual cortex of elementary features such as oriented edge fragments and line segments.

A close relative of the feature hierarchy approach is the syntactical pattern recognition method (Fu, 1974; Leeuwenberg, 1971). Here, too, the first stage consists of identifying simple parts in the input image, followed by the grouping of elementary parts into higher-order ones. The emphasis in the syntactical approach is on the construction of higher-order shape descriptions using methods borrowed from the syntactical analysis of formal languages. (See also the hierarchical-generative process of Leyton [1986].)

Structural descriptions Another approach to shape decomposition can be viewed as a mixture of part decomposition and the invariant properties approach. The invariances in this approach are defined using parts and their spatial relations. For example, the total number of parts of a given type may be an invariant of the shape. A triangle, for instance, always contains three lines, three vertices, and no free line terminators. This is, in fact, how *perceptrons*, which are simple parallel pattern recognition devices, have been used to recognize triangles independent of shape, location, and size (Minsky and Papert, 1969). Such schemes combine feature extraction with simple invariants (the existence and lack of certain features), without attempting to describe interpart relationships. Similarly, in the pandemonium scheme (Selfridge, 1959) a shape is classified based on the existence of certain parts, or features, without describing spatial relations among different parts.

In other schemes, relations between constituent parts are used to capture invariants that are common to a family of shape variations. In the capital letter *A*, for example, two of the line segments meet at a vertex, and this property holds for most variations of the letter (figure 21.1). Here, again, part decomposition is obtained first and, in the next stage, simple invariances are defined in terms of the constituent parts. The invariances are expressed in terms of relations between two or more parts, such as "above," "touching," "to the left of," "longer than," "containing" (Barlow, Narasimhan, and Rosenfeld, 1972). For 2-D applications, in which objects are restricted to move parallel to the image plane, simple relations such as distances and angles measured in the image would remain invariant (Bolles and Cain, 1982). In the more general 3-D case, part decomposition schemes often try to employ relations that would remain invariant over a wide range of different viewing positions (Marr and Nishihara, 1978; Biederman, 1985; Lowe, 1985).

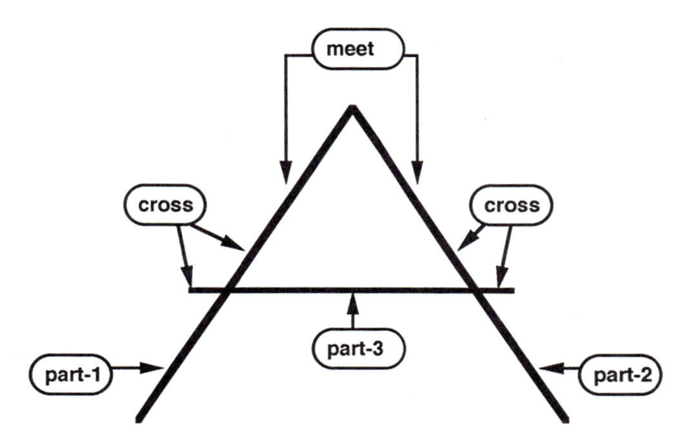

FIGURE 21.1 A simplified structural description of the capital letter *A*. The shape is described in terms of its parts, and the spatial relations between parts.

When augmented with descriptions of relations among parts, the shape decomposition approach leads to the notion of structural descriptions. The use of such structural descriptions in vision has become, in recent years, a popular approach to shape analysis and recognition (Pinker, 1984; Farah, 1991). Early computational examples of this approach include Grimsdale and colleagues (1959), Clowes (1967), and Winston (1970/1975). As psychological models, early examples of structural descriptions theories applied to human vision are Sutherland's (1968) theory and Milner's (1974) model of visual shape recognition. The main basic-level features used in these two theories are edges and line segments. In a second level, invariant properties and relations are defined using, for example, the total number of parts (such as the number of line segments of a given orientation) and length ratios of line pairs. A number of psychological experiments, involving primarily the recognition of 2-D characters at different orientations (Corbalis et al., 1978; White, 1980) have been interpreted as supporting this general approach to recognition.

A recent example of a structural description recognition scheme applied to 3-D objects is Biederman's (1985) theory of recognition by components. According to this scheme, objects are described in terms of a small set of primitive parts called *geons*. These primitives are similar to the generalized cylinders used by Marr and Nishihara (1978). They include simple 3-D shapes such as boxes, cylinders, and wedges. More complex objects are described by decomposing them into their constituent geons, together with a description of the spatial relations between components. The number of primitive geons is assumed to be small (approximately 50), and objects are typically composed of a small number of parts (fewer than 10).

An important aspect of most structural description approaches is an attempt to produce shape descriptions that are so-called object-centered rather than viewer-centered (Marr and Nishihara, 1978). This means that the descriptions are invariant, or at least insensitive, to viewing parameters such as scale and orientation. Invariance is achieved primarily by basing the description on intrinsic shape properties, such as the shape's axes of elongation or symmetry.

In any scheme that relies on decomposition into parts, it is crucial to have a reliable and stable procedure for identifying part boundaries. Otherwise, the same object may give, under slightly different viewing conditions, different descriptions in terms of its constituent parts. In Biederman's (1985) scheme, certain "nonaccidental" relationships between contours in the image are used to determine the part decomposition. These relations include, for example, the colinearity of points or lines, symmetry and skewed symmetry, and parallelism of curve segments. Robust shape decomposition remains, however, one of the main difficulties in the structural description approach.

For a variety of 3-D shapes, the notion of part decomposition appears to be natural. A table, for instance, is often composed of a flat surface supported from below by four legs. Such a description appears much more natural than trying to characterize 2-D table shapes in terms of simple properties such as total area, perimeter length, and so on, as used in the invariant properties approach. It is also true that, as argued by Palmer (1977) and Biederman (1985), human observers sometimes find it easy to identify the parts of an object even when the object is unfamiliar.

At the same time, it appears that the use of structural descriptions has at least two limitations. The first problem is that the decomposition into generic parts often falls considerably short of characterizing the shape in question. For example, a dog, a fox, and a cat (as well as several other animals) probably have similar and perhaps identical decompositions into main parts. This may be useful at some level—for instance, as an intermediate level in the recognition process or perhaps for basic-level classification (Rosch et al., 1976; Biederman, 1985). Eventually, however, these animals are distinguishable not because each one has a different arrangement of parts but because of differences in the detailed shape at particular locations.

A second limitation of the structural description approach is that many objects are difficult to decompose into standard shapes that are sufficient to characterize the objects. Structural descriptions therefore are unlikely to be sufficient for shape-based recognition. It remains unclear, however, to what extent abstract structural descriptions are indeed used by the visual system for shape analysis and recognition.

Shape Alignment In the use of both shape invariances and shape decomposition, the emphasis was on elaborate, bottom-up shape encoding. The input shape was assumed to be processed through a series of stages to produce a new and more useful description. The shape description could be in terms of shape invariances or a structural description, and it is supposed to capture the essential characteristics of the shape and thereby make subsequent recognition relatively straightforward.

In shape alignment schemes, the emphasis is different: Instead of sophisticated shape encoding, recognition is achieved by alignment processes that compensate for the differences between an observed and a stored shape. The basic component of alignment is the application of shape transformations to compensate for possible differences between a viewed shape and previously stored patterns. Such compensating transformations can be applied either to internal 3-D shape models or to 2-D shapes and their combinations.

Explicit 3-D alignment In 3-D alignment, a 3-D model is brought into alignment with a viewed shape. The process bears some similarity to mental rotation (Shepard and Cooper, 1982) but must take place on a faster time scale (Taar and Pinker, 1989). The compensating transformation that will bring the stored model into the closest agreement with the input shape can be extracted using a small number of corresponding features in the image and stored model (Ullman, 1989). The full alignment process then proceeds as follows. A small number of corresponding features is identified in the stored 3-D model and the input shape. From this, the compensating transformation required to bring the model into registration with the image is deduced. The transformation is applied internally, and the transformed model is compared with the viewed shaped.

The details of this process will not be reviewed here (Lowe, 1985; Huttenlocher and Ullman, 1990). The important aspects of this approach to shape analysis are the assumption that the visual system stores explicit models of 3-D object shapes and the use of processes that manipulate 3-D shape descriptions, including rotation in 3-D space, to compensate for the differences between a viewed shape and a stored model.

The combination of 2-D images In the 3-D alignment scheme, as well as in other approaches, it has been assumed that the visual system somehow stores and manipulates 3-D shapes. This seems almost inevitable, as the visual system must use 2-D shapes to analyze 3-D objects.

An interesting alternative developed recently is to dispense with explicit 3-D models and use instead combinations of 2-D shapes. Mathematically, this proposal is based in part on the fact that it is possible to represent the viewed shape of a 3-D object, from any viewing direction, using the linear combination of a small number of views (Ullman and Basri, 1991).

Figure 21.2 illustrates the linear combination property. Figure 21.2a shows three different views of a car (Volkswagen Beetle). The figure shows only those edges that were extracted in all three views; as a result, some of the edges are missing. Figure 21.2b shows two new views of the car. These new images were not obtained from novel views of the car but were generated instead by using linear combinations of the first three views. Figure 21.2c shows two new views of the car obtained from new viewing positions. In figure 21.2d, these new views and the linear combinations obtained in figure 21.2c are superimposed. It can be seen that the novel views are matched well by linear combinations of the three original views.

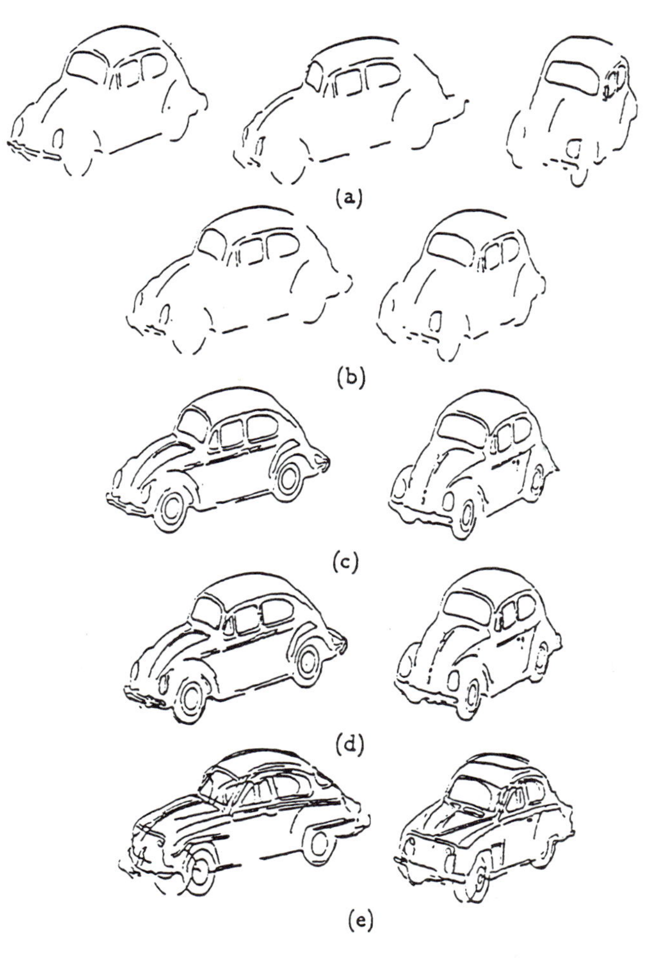

FIGURE 21.2 The linear combination of 2-D views. (a) Three views of a car (Volkswagen Beetle). Only some of the edges have been extracted. (b) Two new views of the car, obtained by linear combinations of the views in (a). (c) Two novel views of the car, obtained from new viewing positions. (d) Superposition of the images in (c) and the linear combinations in (b). The new views are matched well by linear combinations of the original views. (e) The best matching linear combination to a different car (a Saab), with a closely similar 3-D shape. The match is less precise than (d), illustrating that the linear combination of 2-D views can replace explicit 3-D models to make fine distinctions between similar objects seen from novel viewing positions. (Reprinted from Ullman and Basri, 1991, by permission. © 1991 IEEE)

Based on the properties of image combination, a number of schemes have studied the possibility of using combinations of 2-D images rather than manipulating 3-D shapes internally (Poggio and Edelman, 1990; Ullman and Basri, 1991). One way of performing image combination is to use image interpolation (Poggio and Edelman, 1990), by which the visual system can use not only previously stored shapes but also shapes constructed internally by interpolation between stored patterns. Computationally, such an approach can be used instead of explicitly manipulating 3-D shapes internally. Psychologically, experiments have begun to explore the use of view combination by the visual system for shape-based recognition (Bülthoff and Edelman, tem 1992), but more research is required to study and compare the use of 2-D and 3-D shape representations by the human visual system.

CONCLUDING REMARKS In concluding this section, it may be useful to summarize briefly the differences in shape encoding and shape processing proposed by the different approaches. It should be noted, however, that despite the differences, the approaches are not mutually exclusive but may complement one another in useful ways (Ullman, 1989).

In the invariant properties approach, a shape is encoded by extracting a set of invariances. A popular possibility within this general approach is to perform a set of shape measurements and then represent the shape as a vector of the results, or as a point in a multidimensional space. The main processing envisioned within this framework centers around clustering and classification in multidimensional spaces.

In the shape decomposition approach, a shape is encoded in terms of basic shape primitives, which may be 2-D or 3-D in nature, and their spatial relationships. Shape-related processing in this approach consists primarily of constructing and comparing such structural descriptions.

In the alignment, or transformational, approach, though some shape abstractions are performed (Ullman, 1989), the shape descriptions are less abstract and more pictorial in nature compared with the alternative schemes. Shape processing includes primarily the application of alignment transformations and possibly forms of image combination or interpolation.

Visual cognition: the visual analysis of shape properties and spatial relations

The visual analysis of shape, form, and spatial relations is not limited to the task of shape-based recognition. We can use vision to establish a wide range of shape properties and spatial relations in response to a variety of either externally imposed or internally generated visual tasks. A few examples are shown in figure 21.3. In figure 21.3a, the task is to determine visually whether the *x* lies inside or outside the figure. Observers can establish this relation effortlessly, unless the boundary becomes highly convoluted. The answer appears to simply pop out, and we cannot give a full account of how the decision was reached. It is interesting to note that this capacity appears to be associated with relatively advanced visual systems. The pigeon, for example, which shows an impressive capacity for figure classification and recognition (Herrnstein, 1984), is essentially unable to perform this task in a general manner. It can attain only very limited success and appears to base its decision on simple local cue (Herrnstein et al., 1989).

In figure 21.3b, the task involves elongation judgments, in this case, of ellipse figures. The visual system usually is precise at detecting such elongations: Reliable judgments of elongation can be obtained when the major axis of the ellipse is 4–5% longer than the minor axes (Cave, 1983). Interestingly, the judgments are more difficult when the axes themselves, without the ellipses, are present. This suggests that the judgment is not based on the extraction and comparison of the main axes, and it is as yet unclear what mechanisms, in fact, subserve this and related shape judgments. In figure 21.3c, the task is to determine whether two dots lie on a common contour. Again, a solution is obtained by merely looking at the figure. In figure 21.3d, the task is more complex—to determine whether the black disc can be moved to the location of the *x* without colliding with any of the other shapes.

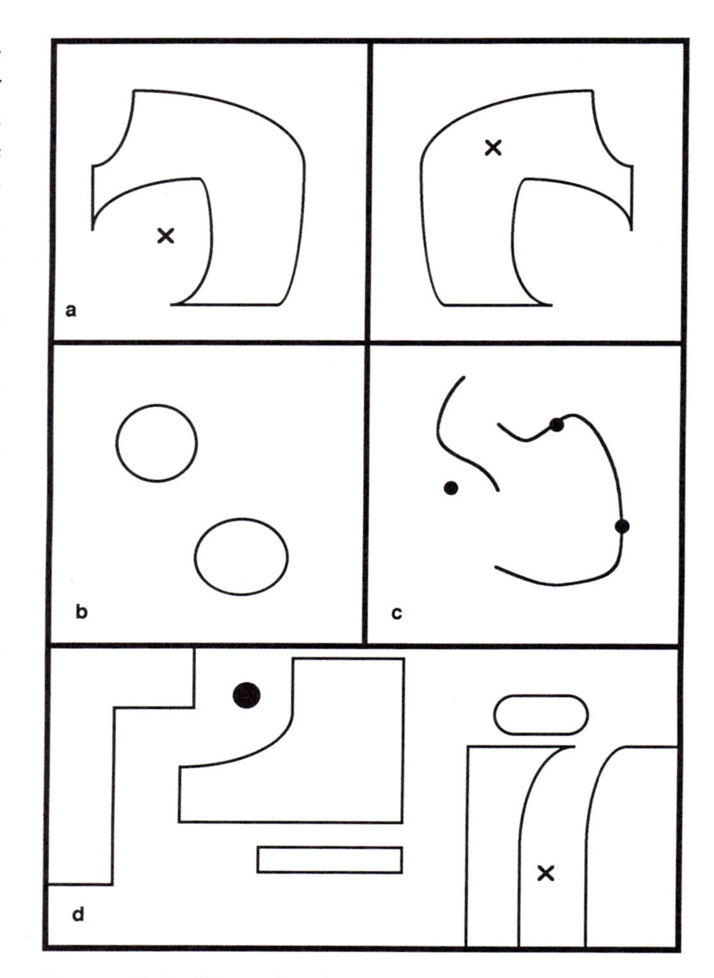

FIGURE 21.3 Examples of general visual cognition tasks involving the analysis of shape properties and spatial relations. (a) Inside-outside relation: It is easy to determine whether the *x* lies inside or outside the closed figure. (b) Elongation judgment. (c) The task is to determine whether two black dots lie on a common contour. (d) The task is to determine whether the black disc can be moved to the location of the *x* without colliding with nearby shapes.

These figures and tasks are artificial, but similar visual cognition problems also occur in natural settings, in the course of object manipulation, navigation, and so on, and are solved routinely by the visual system. In addition, we make use of visual aids such as diagrams, charts, sketches, and maps, because they draw on the system's natural capacity to manipulate and analyze spatial information, and this ability is used to help our reasoning and decision-making processes.

Spatial analysis of this kind does not require object recognition—that is, it does not depend on object naming or on whether we have seen the objects in the past. It does require, however, the analysis of shape and spatial relations among shapes. The visual system can perform a wide variety of such visual cognition tasks with remarkable adeptness that cannot be mimicked at present by artificial computer vision systems.

The mechanisms underlying such visual cognition are currently poorly understood. It remains unclear what internal shape processes are used, even in simple configurations such as comparing the length of two line segments. Can the system somehow apply an internal yardstick to the segments, shift them internally to test their overlap, or use some other method? These questions remain unanswered at present. One possible approach, discussed in more detail in Ullman (1984) warrants mention. This approach assumes that the perception of shape properties and spatial relations is achieved by the application of so-called visual routines to the early visual representations. These visual routines are highly efficient sequences of basic operations that are wired into the visual systems. Routines for different properties and relations are then composed from the same set of basic operations, using different sequences. Using a fixed set of basic operations, the visual system can assemble different routines and in this manner extract an essentially unbounded variety of shape properties and spatial relations. To understand visual cognition in general, it will therefore be required to identify the set of basic operations used by the visual system. An explanation of how we determine a particular relation such as "above," "inside," or "touching" would require a specification of the visual routine used to extract the shape or property in question.

Possible basic visual operations are discussed in Ullman (1984). Examples include the operations of indexing, coloring, and boundary tracing. Indexing is the operation of shifting the processing focus to a new location (usually to a salient location). Area coloring is the filling in of a closed region, labeling it as distinct from the surrounding background. Boundary tracing is the operation of moving along a boundary or a contour, integrating information from different parts of the contour (Jolicoeur, Ullman, and MacKay, 1986; Pylyshyn, 1988).

Operations of this type can be used as part of visual routines for extracting different properties and relation. For example, to establish whether an *x* lies inside or outside a closed curve *C*, an indexing operation will allow the analysis to start at the location of the *x* without scanning the entire image exhaustively. Areas coloring, or filling in, the region from the *x* outward may then play a useful role in establishing the inside-outside relation. Boundary tracing may be used in closing boundary gaps in the case of a discontinuous contour. Similarly, indexing and boundary tracing can be used to establish whether the two dots in figure 21.3c lie on a common contour and, more generally, to integrate information that belongs to a single contour and separate it from irrelevant information associated with other contours nearby (Jolicoeur, Ullman, and Mackay, 1991).

More complex visual routines have been used in computational studies by Chapman (1991, 1992) in the context of a sophisticated computer system that interprets its environment visually in the course of playing an interactive video game. Although the tasks were varied and sometimes complex, different combinations of a small and fixed set of basic operations proved sufficient to perform all the visual analysis regarding shapes and their relations required for playing the game successfully.

From a psychological standpoint, there has been some evidence (Jolicoeur, Ullman, and MacKay, 1986, 1991; Pringle and Egeth, 1988) supporting the notion that visual routines are, in fact, applied by the visual system in the course of extracting shape properties and spatial relations. The area of general visual cognition still is relatively unexplored, however, from both the computational and psychological perspectives. We do not yet understand in any detail how the visual system can establish, with remarkable flexibility and efficiency, a variety of spatial properties and relations, and we still cannot mimic this performance in artificial systems.

Finally, it should be noted that the representations and processes used for general visual cognition may be very different from those used by the system for shape-based recognition. As noted in the introduction, different visual tasks pose different requirements, and these requirements may give rise to representations and processes that are specialized, at least to some degree, to the task at hand.

Biological aspects of shape analysis

From single-cell recordings in the primary visual cortex, we know something about how the process of shape analysis in the visual system begins. Studies starting with the work of Hubel and Wiesel (1962, 1968) have shown that at the early processing stages, in the primary visual cortex (V1), the image is represented in terms of simple local features such as oriented edges and lines. V1 also shows some degree of internal separation and specialization for different aspects of the stimulus. For example, color information appears to be processed by only a subset of the cells in this area. Separation of function exists to some degree in subsequent processing stages and, in particular, it seems possible that some visual areas are more directly concerned with shape-related processing than others. A rough division has been suggested between the dorsal stream of processing leading from V1 to parietal cortex and the more ventral occipitotemporal stream going to inferotemporal cortex, where shape processing—leading, for instance, to object recognition—seems to take place (Ungerleider and Mishkin, 1982). It also appears that within the ventral stream, some populations of cells may be more involved in shape analysis than others. For example, some cells (e.g., in area IT) respond to specific shapes irrespective of color and maintain shape-specific response across large changes of color variations (Gross, 1992), whereas other populations may be more directly involved in color processing.

The notion that IT cortex is involved with shape processing and recognition is also supported by brain lesion studies. Damage to IT can cause deficits in object recognition and particularly the precise identification of individual objects (as compared with broad classification) (Damasio, Damasio, and Tranel, 1990).

Single-cell recordings in IT revealed some cells responding to complex shapes. A particular population of such cells, in the STS region of IT, responds selectively to face images (Perret, Rolls, and Caan, 1982; Perret et al., 1985), whereas other cells in a nearby region have been reported to respond to hand images. Some of the face-selective cells appear to respond to complete face images, others to face parts. Many show considerable invariance along several dimensions, such as size, position, orientation in the image plane, and color. In terms of a 3-D viewing angle, some cells show a selective response to an individual face over considerable changes in viewing direction. At the same time, the response is not entirely object-centered—that is, the selectivity does not cover the full range from profile to full-face.

In addition to cells responding to complex and meaningful stimuli, other cells in IT have been reported to respond to more elementary shapes and shapes that do not correspond to familiar objects (Fujita et al., 1992).

Shape-selective cells in IT respond in a graded fashion and will respond not only to a single shape but to similar shapes as well. The shape of a specific face, for instance, may still be represented by such cells in the population response of a number of cells. This may be a general principle of shape encoding, that complex shapes are represented by the combined activity of cells either broadly tuned to different shapes or tuned to different parts of the entire shape.

Little is known about shape processing along the way from V1 to anterior IT. The initial studies in V1 regarding simple, complex, and end-stopped cells, have raised the possibility of a processing hierarchy, in which higher processing stages extract increasingly elaborate shape features. There is, at present, little evidence from single-cell recordings to support a systematic hierarchy of shape features, although the notion cannot be ruled out. Area V2, which is placed one step higher than V1 in the cortical hierarchy, appears to be involved in some aspects of form processing: Some of the units in this area are sensitive to subjective contours; others show sensitivity to a colinear arrangement of dots moving coherently with respect to the background (Peterhans and von der Heydt, 1991), but these are not necessarily higher-order elements in a feature hierarchy compared with the processing in V1.

Some evidence appears to be broadly compatible with a transformational, or alignment, view of shape processing in some of the higher visual areas. For example, damage to area V4 and posterior IT (Schiller and Lee, 1991; Weizkrantz, 1990), appears to affect more the ability to compensate for transformations such as size, orientation, or illumination changes than the ability to recognize nontransformed shapes. According to this view, shape analysis in intermediate areas may be concerned primarily not with the creation of increasingly elaborate shape description but with alignment and compensatory processes.

It seems likely that shape analysis by the visual system involves internal transformations and other processes applied to shapes, as well as different levels of shape encoding and abstraction. Within the visual system, shape representation and processing involves cell populations in multiple visual areas. This makes the physiological study of shape processing difficult to observe and interpret at the single-unit level. Additional techniques (such as multiunit recording, optical imaging, positron emission tomography, and functional magnetic resonance imaging) and the combination of several research disciplines (neuroscience, psychophysics, and computational studies) are therefore likely to play an important role in elucidating the analysis of shape and form by the visual system.

ACKNOWLEDGEMENT This work was supported in part by National Science Foundation grant IRI-8900207.

REFERENCES

ABU-MOSTAFA, Y. S., and D. PSALTIS, 1987. Optical neural computing. *Sci. Am.* 256:66–73.

BARLOW, H. B., 1972. Single units and sensation: A neuron doctrine for perceptual psychology? *Perception* 1:371–394.

BARLOW, H. B., R. NARASIMHAN, and A. ROSENFELD, 1972. Visual pattern analysis in machines and animals. *Science* 177:567–575.

BIEDERMAN, I., 1985. Human image understanding: Recent research and a theory. *Comput. Vis. Graphics Image Proc.* 32:29–73.

BOLLES, R. C., and R. A. CAIN, 1982. Recognizing and locating partially visible objects: The local-feature-focus method. *Int. J. Robotics Res.* 1(3):57–82.

BÜLTHOFF, H. H., and S. EDELMAN, 1992. Psychophysical support for 2-D view interpolation theory of object recognition. *Proc. Natl. Acad. Sci. U.S.A.* 89:60–64.

CAVE, K., 1983. The importance of axes and curvature in simple shape recognition. Unpublished undergraduate thesis, Harvard University, Cambridge, Mass.

CHAPMAN, D., 1991. *Vision, Instruction, and Action.* Cambridge, Mass.: MIT Press.

CHAPMAN, D., 1992. Intermediate vision: Architecture, implementation, and use. *Cogn. Sci.* 16(4):491–537.

CLOWES, M. B., 1967. Perception, picture processing, and computers. In *Machine Intelligence*, vol. 1, N. L. Collins and D. Michie, eds. Edinburgh: Oliver & Boyd, pp. 181–197.

CORBALLIS, M. C., N. J. ZBRODOFF, L. I. SCHETZER, and P. B. BUTLER, 1978. Decisions about identity and orientation of rotated letters and digits. *Memory and Cognition* 6:98–107.

COREN, S., and J. S. GIRGUS, 1978. *Seeing Is Deceiving: The Psychology of Visual Illusions.* Hillsdale, N.J.: Erlbaum.

CUTTING, J. E., and L. T. KOZLOWSKI, 1977. Recognizing friends by their walk: Gait perception without familiarity cues. *Bull. Psychon. Soc.* 9(5):353–356.

DAMASIO, A. R., H. DAMASIO, and D. TRANEL, 1990. Impairment of visual recognition as clues to the processing of memory. In *Signal and Sense: Local and Global Order in Perceptual Maps*, G. M. Edelman, W. E. Gall, and W. M. Cowan, eds. New York: Wiley.

DUDA, R. O., and P. E. HART, 1973. *Pattern Classification and Scene Analysis.* New York: Wiley.

FARAH, M. J., 1991. Patterns of co-occurrence among associative agnosias: Implications for visual object representation. *Cogn. Neuropsychol.* 8(1):1–19.

FU, K. S., 1974. *Syntactic Methods in Pattern Recognition.* New York: Academic Press.

FUJITA, I., K. TANAKA, M. ITO, and K. CHENG, 1992. Columns for visual features of objects in monkey inferotemporal cortex. *Nature* 360:343–346.

GIBSON, J. J., 1950. *The Perception of the Visual World.* Boston: Houghton Mifflin.

GIBSON, J. J., 1979. *The Ecological Approach to Visual Perception.* Boston: Houghton Mifflin.

GRIMSDALE, R. L., F. H. SUMNER, C. J. TUNIS, and T. KILBURN, 1959. A system for the automatic recognition of patterns. *Proc. Inst. Elec. Eng.* 106(26):210–221.

GROSS, C. G., 1992. Representation of visual stimuli in inferior temporal cortex. *Philos. Trans. R. Soc. Lond. [Biol.]* 335: 3–10.

HERRNSTEIN, R. J., 1984. Objects, categories, and discriminative stimuli. In *Animal Cognition*, H. L. Roitblat, T. G. Bever, and H. S. Terrace, eds. Hillsdale, N.J.: Erlbaum.

HERRNSTEIN, R. J., W. VAUGHAN, D. B. MUMFORD, and S. M. KOSSLYN, 1989. Teaching pigeons an abstract relational rule. *Percept. Psychophys.* 46:56–64.

HOFFMAN, D., and W. RICHARDS, 1986. Parts of recognition. In *From Pixels to Predicates*, A. P. Pentland, ed. Norwood, N.J.: Ablex.

HOPFIELD, J. J., 1982. Neural networks and physical systems with emergent collective computational abilities. *Proc. Natl. Acad. Sci. U.S.A.* 79:2554–2558.

HUBEL, D. H., and T. N. WIESEL, 1962. Receptive fields, binocular interaction, and functional architecture in the cat's visual cortex. *J. Physiol. (Lond.)* 160:106–154.

HUBEL, D. H., and T. N. WIESEL, 1968. Receptive fields and functional architecture of monkey striate cortex. *J. Physiol. (Lond.)* 195:215–243.

HUTTENLOCHER, D. P., and S. ULLMAN, 1990. Recognizing solid objects by alignment with an image. *Int. J. Comput. Vis.* 5(2):195–212.

JOLICOEUR, P., S. ULLMAN, and M. MACKAY, 1986. Curve tracing: A possible elementary operation in the perception of spatial relations. *Memory and Cognition* 14:129–140.

JOLICOEUR, P., S. ULLMAN, and M. MACKAY, 1991. Visual curve tracing properties. *J. Exp. Psychol.* 17(4):997–1022.

JOHANSSON, G., 1973. Visual perception of biological motion and a model for its analysis. *Percept. Psychophys.* 14(2):201–211.

JULESZ, B., 1981. Textons, the elements of texture perception, and their interactions. *Nature* 290:91–97.

KANISZA, G., 1979. *Organization in Vision: Essays in Gestalt Perception.* New York: Praeger.

KOHONEN, T., 1978. *Associative Memories: A System Theoretic Approach.* Berlin: Springer-Verlag.

LEEUWENBERG, E., 1971. A perceptual coding language for visual and auditory patterns. *Am. J. Psychol.* 84(3):307–349.

LEYTON, M., 1986. Principles of information structure common to six levels of the human cognitive system. *Inform. Sci.* 38(1):1–120.

LOWE, D. G., 1985. *Perceptual Organization and Visual Recognition.* Boston: Kluwer Academic.

MARR, D., and H. K. NISHIHARA, 1978. Representation and recognition of the spatial organization of three-dimensional shapes. *Proc. R. Soc. [Biol.]* 200:269–291.

MILNER, P. M., 1974. A model for visual shape recognition. *Psychol. Rev.* 81(6):521–535.

MINSKY, M., and S. PAPERT, 1969. *Perceptrons.* Cambridge, Mass.: MIT Press.

MUNDY, J. L., and A. ZISSERMAN, 1992. *Geometric Invariance in Computer Vision.* Cambridge, Mass.: MIT Press.

NAKAYAMA, K., S. SHIMOJO, and G. H. SILVERMAN, 1989. Stereoscopic depth: Its relation to image segmentation, grouping, and the recognition of occluded objects. *Perception* 18:55–68.

OVER, R., 1971. Comparison of normalization theory and neural enhancement explanation of negative aftereffects. *Psychol. Bull.* 75(4):225–243.

PALMER, S. E., 1977. Hierarchical structure in perceptual representation. *Cogn. Psychol.* 9:441–474.

PERRET, D. I., E. T. ROLLS, and W. CAAN, 1982. Visual neurons responsive to faces in the monkey temporal cortex. *Exp. Brain Res.* 47:329–342.

PERRET, D. I., P. A. J. SMITH, D. D. POTTER, A. J. MISTLIN, A. S. HEAD, A. D. MILNER, and M. A. REEVES, 1985. Visual cells in the temporal cortex sensitive to face view and gaze direction. *Proc. R. Soc. [Biol.]* 223:293–317.

PETERHANS, E., and R. VON DER HEYDT, 1991. Elements of form perception in monkey prestriate cortex. In *Representations of Vision*, A. Gorea, ed. Cambridge, Engl.: Cambridge University Press.

PINKER, S., 1984. Visual cognition: An introduction. *Cognition* 18:1–63.

PITTS, W., and W. S. MCCULLOCH, 1947. How we know universals: The perception of auditory and visual forms. *Bull. Math. Biophys.* 9:127–147.

POGGIO, T., and S. EDELMAN, 1990. A network that learns to recognize three-dimensional objects. *Nature* 343:263–266.

PRINGLE, R., and H. E. EGETH, 1988. Mental curve tracing with elementary stimuli. *J. Exp. Psychol.* 14(4):716–728.

PYLYSHYN, Z., 1988. Here and there in the visual field. In *Computational Processing in Human Vision*, Z. Pylyshyn, ed. Norwood, N.J.: Ablex.

REILLY, D. L., L. N. COOPER, and C. ELBAUM, 1982. A neural model for category learning. *Biol. Cybern.* 45:35–41.

ROCK, I., 1974. *Orientation and Form.* New York: Academic Press.

ROSCH, E., C. B. MERVIS, W. D. GRAY, D. M. JOHNSON, and BOYES-BRAEM, 1976. Basic objects in natural categories. *Cogn. Psychol.* 8:382–439.

SCHILLER, P. H., and K. LEE, 1991. The role of the primate extrastriate area V4 in vision. *Science* 251:1251–1253.

SELFRIDGE, O. G., 1959. Pandemonium: A paradigm for learning. In *Proceedings of a Symposium on the Mechanization of Thought Processes.* London: H. M. Stationary Office.

SHEPARD, R. N., and L. A. COOPER, 1982. *Mental Images and Their Transformations.* Cambridge, Mass.: MIT Press/Bradford Books.

SUTHERLAND, N. S., 1959. Stimulus analyzing mechanisms. In *Proceedings of a Symposium on the Mechanization of Thought Processes.* London: H. M. Stationary Office.

SUTHERLAND, N. S., 1968. Outline of a theory of visual pattern recognition in animal and man. *Proc. R. Soc. [Biol.]* 171:297–317.

TAAR, M., and S. PINKER, 1989. Mental rotation and orientation dependence in shape recognition. *Cogn. Psychol.* 21:233–282.

TOU, J. T., and R. C. GONZALES, 1974. *Pattern Recognition Principles.* Reading, Mass.: Addison-Wesley.

ULLMAN, S., 1984. Visual routines. *Cognition* 18:97–159.

ULLMAN, S., 1989. Aligning pictorial descriptions: An approach to object recognition. *Cognition* 32(3):193–254.

ULLMAN, S., and R. BASRI, 1991. Recognition by linear combination of models. *IEEE Trans. PAMI* 13(10):992–1006.

UNGERLEIDER, L. G., and M. MISHKIN, 1982. Two cortical visual systems. In *Analysis of Visual Behavior*, D. J. Ingle, M. A. Goodale, and R. J. W. Mansfield, eds. Cambridge, Mass.: MIT Press.

WEISKRANTZ, L., 1990. Visual prototypes, memory, and the inferotemporal lobe. In *Vision, Memory and the Temporal Lobe*, E. Iwai and M. Mishkin, eds. New York: Elsevier, pp. 13–28.

WHITE, M. J., 1980. Naming and categorization of tilted alphanumeric characters do not require mental rotation. *Bull. Psychon. Soc.* 15(3):153–156.

WILLSHAW, D. J., O. P. BUNEMAN, and H. C. LONGUET-HIGGINS, 1969. Non-holographic associative memory. *Nature* 222:960–962.

WINSTON, P. H., 1970. Learning structural descriptions from examples. Ph.D. Thesis, MIT, Cambridge, Mass. In *The Psychology of Computer Vision*, 1975, P. H. Winston ed. New York: McGraw-Hill.

22 Multiple Cues for Three-dimensional Shape

ANDREW J. PARKER, BRUCE G. CUMMING, ELIZABETH B. JOHNSTON, AND ANYA C. HURLBERT

ABSTRACT The visual information available from scenes of the natural world is encapsulated, to a large degree, by the geometrical relationships between three-dimensional (3-D) solid objects and their projections into the two retinal images. Various perceptual cues for 3-D shape are used by human observers. These cues are created by transformations of the geometrical information by spatial or temporal displacement of the vantage point or the viewed object in the cases of motion and stereo, and by the link between surface properties and geometry in the cases of shading, texture, and specularities. This chapter discusses the interactions between these various cues, when they are presented in a controlled manner in naturalistic images synthesized by computer graphics techniques. It is argued that biological visual systems may monitor the statistics of cues in the visual environment as a means of discovering the appropriate rules for combining information.

Shape, geometry, and cues

Human vision acquires information about three-dimensional (3-D) shape from a number of different sources. Binocular vision, motion parallax, texture gradients, outline contour, and shading all contribute information about 3-D shape, although often the kind of information acquired from each source is rather different. For example, the point-by-point distribution of stereo disparities across a scene can yield estimates of the depth of one point relative to another, whereas the pattern of shading or texture across a surface can signal the local orientation of the surface (Horn, 1977; Witkin, 1981). The understanding that different sources of information may be exploited in different ways and the steady development of experimental techniques for manipulating these sources of information independently has led to a considerable corpus of knowledge about how individual cues signal 3-D shape. More recently, interest has turned to the question of how these cues might actually be used in a consistent way when the visual system is inspecting natural scenes, in which multiple cues are available. In this chapter, we review some of these developments and relate them to earlier work that studied individual cues in isolation.

It is essential to make more precise our intuitive notion of shape and how shape might be indicated by visual cues. Wheatstone's papers (Wheatstone, 1838, 1852) contain a surprising number of the necessary elements. Wheatstone first identifies a source of information used by human vision to obtain 3-D shape which, in his example, is binocular disparity—the spatial differences in the relative position of features observed from the vantage points of the right and left eyes. He proceeds to demonstrate that different 3-D shapes give rise to characteristic patterns of binocular disparity, offering examples of the interocular differences in the foreshortening of a flat, tilted plane and the interocular differences in line curvature that can be obtained from viewing curved 3-D objects. He then points the way to placing on a more secure mathematical basis the relationship between the shape itself, particularly its geometry, and the cues by which it is signaled to the human observer. His suggestion is to exploit the tools of differential geometry (he mentions particularly the work of Monge). This proposal shows a remarkable degree of prescience as this approach has been adopted by several researchers of machine and human vision in the last 15 years (Brady et al., 1985; Besl and Jain, 1986; Koenderink, 1990; Rogers and

ANDREW J. PARKER and BRUCE G. CUMMING University Laboratory of Physiology, Oxford, England.
ELIZABETH B. JOHNSTON Sara Lawrence College, Bronxville, N.Y.
ANYA C. HURLBERT Department of Physiology, Medical School, University of Newcastle, Newcastle, U.K.

Bradshaw, 1993) and has been considerably extended outside the case of binocular disparity.

These considerations focus thinking on the link between geometry and shape. A fundamental problem is to understand which geometrical features are needed by human vision to support particular forms of behavior. For example, to reach forward and close a hand on a target requires matching the closure of the hand to the absolute dimensions of the shape. This could be done by continuous visual monitoring of the hand as it approaches and closes on the shape or, perhaps less likely and certainly less accurately, it could be done by obtaining an absolute estimate of the shape and its distance and then directly matching the closure of the hand to it. Presumably, tactile cues are also involved, and so a further possibility is that the task is solved with high-level knowledge of the nature of the 3-D shape, perhaps by a learned association between grasp and familiar shapes.

For other tasks, different considerations apply. For identification or recognition of an object as the same shape, regardless of viewing distance or position in the visual field, determination of the absolute size is unimportant (unless, of course, some features of the object become invisible in the retinal image by a loss of acuity for spatial detail at large viewing distances or large eccentricities). Indeed, the idea that there might be some constancy for 3-D shape subsumes, for certain types of tasks, the separate ideas of size constancy and depth constancy. Koenderink (1990; Koenderink and van Doorn, 1992) has expressed this notion of an object having the same shape by introducing a scale-independent measure of shape, the *shape index*. Still other considerations may apply for understanding the relationships between component parts of an object, for example a cup and its handle (Hoffman and Richards, 1984). In this kind of case, a great deal can be established about component parts without a full metrical 3-D representation of the object. Often much can be gained from a simple qualitative assignment of surface regions with labels such as *convex*, *concave*, *cylindrical*, or *saddle-shaped* (see Weinshall, 1990).

Hence, for this view of visual processing the important components are first to find out what aspects of 3-D geometry can be determined by various potential cues to 3-D shape, which will depend in part on theo-

FIGURE 22.1 Two views of a 3-D scene containing some simple geometrical objects. The scenes contain several of the classically identified cues to 3-D organization: stereo (if the two views are combined binocularly); motion (if the two views are presented sequentially in time); texture, formed with regular and irregular elements; shading, including mutual interreflections, outline contour, and occlusions; and linear perspective. The images were created using a purpose-built software package, which implements a novel version of the radiosity method from computer graphics that has been developed in our laboratory in collaboration with the Oxford Engineering Department. (We would like to acknowledge the contributions of Mr. Chris Christou and Dr. Andrew Zisserman to this development.)

retical analysis of the information content of various cues to shape and on an experimental determination of their functional effectiveness in biological visual systems; and second to consider, again both theoretically and experimentally, what forms of interactions may occur between cues of different types when they are independently manipulated but are all contributing to the determination of a single 3-D shape. Later, these criteria are employed to develop a critique of experimental methods for investigating the way in which multiple cues are used by human observers. Before that, we need to consider the cues that are potentially available and how they might be manipulated.

Manipulating the visual cues in images

NATURAL AND SYNTHETIC IMAGES Many of the cues that can be used by human observers are illustrated in the pair of images in figure 22.1 which depicts a simple scene with some geometrical objects within it. The two images are representations of the viewpoints of the left and right eyes. The differences in the contours and details between the two images indicate the potential contribution of binocular disparity. It is equally possible to consider the two images as views of the scene obtained at two different times by a monocular observer translating relative to the scene. The same geometrical differences can then be considered as motion parallax information (Rogers and Graham, 1979; Rogers and Koenderink, 1986), but it should be remembered that the geometry of motion parallax is inevitably more complicated than that of stereo. First, the distance and direction of the translation of the moving observer are potentially different from the fixed interocular shift offered by the lateral separation of the eyes. Second, objects may move in the scene relative to one another so that patterns of motion may often be created on the retina by means other than self-motion of the observer. Among the other cues to 3-D shape shown within the images of the scene in figure 22.1 are shading, texture, perspective, and outline contour.

Because this scene is evidently a synthetic image rather than a natural image, it is appropriate to consider the arguments for and against the use of synthetic images in the study of 3-D shape. The most obvious uncertainty that arises from the use of synthetic images is the possibility that the images lack some of the cues exploited by human observers when inspecting natural scenes. A slightly more complex version of this argument would suggest that there is some special way in which the combination of cues is presented in natural images.

A purist (or pessimist) might argue that it is impossible to carry out meaningful experiments unless natural images are used. Even photographs or video sequences are potentially suspect. On the other hand, this contention creates some difficulty for setting up the kinds of experiments that need to be undertaken. If it is understood that an experiment on cues to 3-D shape will involve the parametrical variation of just one or two of the multiple cues an observer might use, then either the experimenter must be able to obtain unambiguous measurements of the cues available in natural images or a synthetic image must be used. In fact, there is a close link between the two alternatives because both demand the adoption of a physically based model of image formation. In the first case, the model is needed to interpret how visible features of the image have arisen from the interactions of light with the physical objects within the scene. In the second, the model is used to construct a synthetic image based on the same physical principles. In both cases, the physical aspects of image formation need to be addressed and the applicability of any particular proposed model needs to be established.

ALTERING THE LINKS BETWEEN DIFFERENT CUES IN IMAGES Early attempts to manipulate the relationships between cues often exploited ingenious optomechanical devices. Notably, Helmholtz (1962) mentions a variety of binocular devices, such as the telestereoscope and the pseudoscope, for altering the normal geometry of binocular viewing and thus disrupting the natural relationships among stereo disparity, vergence, and other cues to depth. Wallach's study of the kinetic depth effect used stereoscopic projection to examine the interrelationship between motion cues and binocular disparities in the specification of rigid 3-D shapes (Wallach and O'Connell, 1953). More recently, Rogers and Graham (1979) examined adaptation to depth defined by motion parallax and its relationship to depth from binocular disparities. Taken as a whole, these studies represent the extreme possibilities for systematically altering the relationship between multiple cues within an image. Devices such as the telestereoscope rely on natural images for their input and then apply certain well-defined transformations to those images.

The Rogers and Graham (1979) study made use of stimulus patterns that were specifically designed to isolate one cue at a time and then studied the interactions among those cues.

Computer Graphics Approaches The range of options for those working on these problems has been greatly enhanced by advances in computer graphics. The components of a modern synthetic approach can be identified as follows:

The process of image formation in natural scenes is studied to abstract the general physical principles involved in the interactions of light with surfaces and objects within the scene;

These processes are simulated within a computer environment that allows the parametric variation of the physical processes to yield images containing different cues to 3-D shape.

Although the scheme is potentially very general, it must be acknowledged that there are some significant limitations. First, there are some aspects of natural scenes for which the physical process of image formation is poorly understood. For example, some types of glossiness, especially on metallic surfaces, are hard to capture. Equally, surfaces made from spatially complex and slightly irregular elements are hard to model adequately: Examples are details of human skin, animal fur, or bird feathers. Difficulty also arises with haphazardly folded, soft-fabric materials, which are especially problematic to specify dynamically. A second type of limitation is that there are some features that can be modeled but that cannot then be adequately displayed with present-day technology. An obvious example of this kind is the limited range of luminances that can be presented on a conventional cathode ray tube (CRT) display, which significantly restricts the ability to display the kinds of light distributions that are actually generated in natural scenes. Nonetheless, video sequences of natural scenes acquired with a television camera and replayed on a standard CRT monitor have a striking degree of realism. This suggests that our main limitation in graphics is in the area of image understanding rather than display technology.

Models of interactions between cues

Current thinking about the interactions between cues for 3-D shape has been greatly influenced by work on computer vision, particularly from the perspective of artificial intelligence. This work identified a goal of abstracting high-level descriptions of shape (the sort of descriptions on which geometrical reasoning and robot pathplanning could be based) from low-level descriptions of images (for example, in the form of the locations and orientations of edges). Moreover, it was quickly realized that it would be difficult to move directly from descriptions at a low level, such as edges in the image, to high-level representations of 3-D shape. Thus, an intermediate level of description was proposed, which had various names and styles, most notably *intrinsic images* (Barrow and Tenenbaum, 1978), *the* $2\frac{1}{2}$*-D sketch* (Marr and Ullman, 1981; Marr, 1982), and *dense depth maps* (Brady et al., 1985; Besl and Jain, 1986).

This intermediate level of description was intended to provide a means by which information about the properties of surfaces in the visual scene could be collected suitably for the extraction of geometrical information about 3-D shape. Unfortunately, the processing of visual images rarely yields information about the depth of surfaces in a scene in the form of a continuous map of depths at every possible location in the visual field. For example, algorithms for recovering depth through binocular stereopsis typically give only an estimate of depth at points in the visual image where there are robust features in the luminance domain. This is because the algorithm requires such features to establish a match between the right and left eyes' images in order to extract depth from the binocular information. One response to this problem was to propose a stage of interpolation to convert the sparse samples of depth into a continuous representation. Various schemes have been advanced (Grimson, 1982; Blake and Zisserman, 1987).

The other response to this problem appeals to the existence of other cues to depth and surface properties within the original visual images. There are a number of ways in which these other cues might be used. It could be that some cues are used merely to fill in the gaps created by the absence of others. Alternatively, different cues could mutually support one another. The whole issue of how to combine information acquired from different sources has been the subject of considerable study in the machine vision community (Clark and Yuille, 1990). In that field, the treatments often deal with issues outside the scope of this chapter, such as the combination of information from visual and infrared sources or the use of sonar or laser-based range finders. This work has produced a considerable array of statistical techniques for data combination. Closer to our present interests, Poggio, Gamble, and Little (1988) have investigated the fusion of information from purely visual sources in a general statistical framework.

Three possibilities for cue interaction have been identified (Bülthoff and Mallot, 1988; Maloney and Landy, 1989; Clark and Yuille, 1990; Johnston, Cumming, and Parker, 1993). The first is *vetoing*, where by one cue completely overrides another. For example, if the right and left images of a stereo pair are swapped in the stereoscope, then the depth from stereo is reversed but other cues are unaffected. With complex scenes, such as those that arise from natural images, the perceived depth relationships within the scene as a whole are retained (Schriever, 1925). In this case, the information from one of the depth cues appears to be completely discarded (although the reversed stereo occasionally is visible as a local intrusion into the depth profile of the scene) and the observer's perception is dominated by the other cues present.

A second possibility is that the depth signals from two or more cues are pooled in a simple *linear estimation* procedure. This would mean that the depth signaled by the combination of these cues can be calculated simply from a knowledge of the depth signaled by each cue when it is presented by itself, together with a simple weighting factor applied to each cue before they are combined. This is what Clark and Yuille (1990) term *weak fusion* between depth cues. Figure 22.2 depicts an idealized data set that would be obtained if the interaction between two cues were to take on a linear form.

Clark and Yuille (1990) also identify a third form of interaction called *strong fusion*, in which the depth cues interact prior to the estimation of depth from either cue. In a particular example, called *promotion* by Maloney and Landy (1989), one depth cue provides additional information that assists the interpretation of another cue. For example, horizontal binocular disparities can yield only relative depth between points in the visual scene until they are combined with an estimate of the absolute distance to some particular point. This estimate could come from vergence angle or from visual cues such as the horizontal gradient of vertical disparity (Mayhew and Longuet-Higgins, 1982; Cumming, Johnston, and Parker, 1991). In either case, if such information is combined with binocular stereo to yield

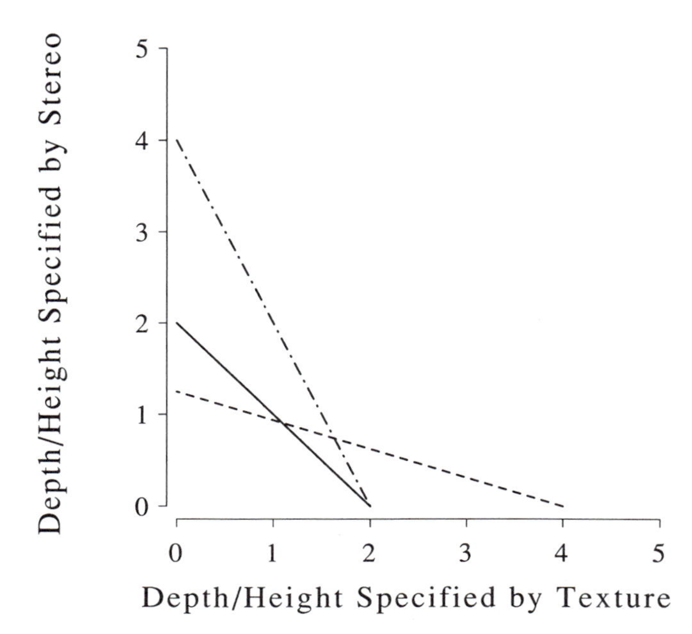

FIGURE 22.2 The expected form of experimental results if cues from two different sources interact linearly in a quantitative shape judgment task. The depth of the figure is expressed relative to its height, which remained constant througout all manipulations. The solid line indicates a simple linear combination of stereo and texture with equal weights for each. This is a prediction of the simple model that the perceived depth d is given by:

$$d = \alpha_S S + \alpha_T T$$

where S and T indicate the depth signaled by stereo and texture alone, α_S and α_T are the weights assigned to stereo and texture, and $\alpha_S + \alpha_T = 1$. The dashed line shows the effect of changing the relative weighting of stereo and texture so that stereo has three times the weighting of texture. The dot-dashed line explores the effect of another possible contributing factor—namely, that the two cues might be equally weighted but the depth signaled by one cue acting alone might not be geometrically correct. In this case, the slope of the interaction is altered even without an alteration in the weighting of the two cues. (Reprinted from *Vision Research*, 33 (5/6), E. B. Johnston, B. G. Cumming, and A. J. Parker, Integration of depth modules: Stereopsis and texture, pages 813–826, copyright 1993, with kind permission from Pergamon Press Ltd., Headington Hill Hall, Oxford OX3 DBW, U.K.)

a complete set of points whose *absolute depths* are known before there is any further combination with other depth cues, this would be an example of promotion.

Methods of studying interactions between cues

Attention should now be turned to the question of how to set up an experimental investigation of cue interactions in human vision. There are a number of psychophysical approaches that could be adopted. However, these differ somewhat in the nature of the information they will provide.

Discrimination Judgments One possibility is to examine discrimination judgments on the grounds that if multiple independent sources of information are available, then discrimination performance should improve. A difficulty with a simple discrimination task is that observers are likely to use any cue available to solve the task with which they have been presented. Thus, if two images differ in the way in which texture (for example) signals 3-D shape, it is possible that the observer will solve a discrimination task successfully without recourse to 3-D shape information. The task could become, in effect, a simple 2-D texture discrimination judgment and thus of little relevance. Nonetheless, there is evidence that discrimination improves when multiple cues are present (Parker et al., 1991). Blake, Bülthoff, and Sheinberg (1993) have shown this rather clearly for separate components of texture cues to 3-D shape. A second problem with interpreting the outcome of discrimination experiments is that the predicted outcome is affected greatly by the assumption of *independence* between different cues. If this assumption is violated, such as in the scheme of promotion (Maloney and Landy, 1989), then deviations from simple statistical combinations may be expected. Indeed, if there is some form of strong fusion of cues to depth, then it is hard to make a straightforward prediction about discrimination performance. On the other hand, a deviation from simple statistical combinations based on the independence assumption is evidence that something other than weak fusion is at work.

Matching Tasks A second psychophysical approach is to ask the experimental subject to match the depth between two figures or to set a probe target to match the depth or surface structure of a figure. By varying the cues to depth in the primary test target, it is argued that some measure of the depth signaled by combinations of cues can be estimated. These techniques approach the requisite question more directly than the discrimination tasks, but there are three particular weaknesses. First, if a probe target, such as a spot presented binocularly at a particular stereoscopic depth, is flashed so that it is superimposed on the test figure, then the depth signal from the probe itself could interfere with the perception of the test figure. Second, any matching task of this kind can yield information only about the *relative* depth sensations evoked by the primary test target and the figure used to match its depth. Third, it is necessary to conduct some control experiments to provide some assurance that the matching task is not being fulfilled by the subject's treating the task as a trivial 2-D judgment and thus using only a limited set of cues, much in the way that was just discussed for discrimination tasks. Within these limitations, some valuable information has been gathered by examining human performance on tasks of this kind (Bülthoff and Mallot, 1988; Koenderink, van Doorn, and Kappers, 1992)

Interactions at Threshold Another way to test for interactions between cues to 3-D shape involves examining the interactions at threshold. If the sensitivity for detecting a 3-D figure that is depicted by a pair of cues can be predicted simply on the basis of statistically independent detection of each separate cue, then this may be construed as evidence in favor of the weak fusion model. Two groups recently investigated this approach. Bradshaw and Rogers (1992) measured the thresholds for detecting a 3-D surface defined by motion parallax and binocular stereopsis. They found that the threshold for both cues presented together exceeded the threshold calculated for statistical independence of the two cues, which argues for an early interaction between stereo and motion cues to 3-D shape, as do the results of Johnston, Cumming, and Landy (1994), discussed further below.

Cumming and Parker (1994) measured the thresholds for the detection of motion in depth in dynamic random-dot stereograms. In these figures, the motion in depth is defined only by the change in binocular disparity over time and not by other cues that would be available naturally, such as interocular velocity differences. In the second part of the experiment, controlled levels of interocular velocity difference were added to determine whether this would enhance the detectability of the pure disparity cue. There was no enhancement, which suggests that, at least under the conditions examined, interocular velocity differences are not exploited by human observers for the recovery of motion in depth. Thus, the nature of the interaction between stereo and motion is not a simple binocular combination of local monocular position and local monocular velocities. Further work is needed to understand this issue.

3-D Shape Judgments Two additional possibilities that have been the focus of work in our laboratory have both involved judgments about the 3-D shape of single figures. The first approach asks subjects to adjust a 3-D shape until it conforms to a simple canonical figure such as a sphere or a cylinder of circular cross-section. The measure of performance is then how much physical depth is required to produce a sensation of 3-D shape corresponding to the required form. Two major advantages of this approach are that it is a measure requiring some form of absolute judgment and it can be turned into the generally desirable form of psychophysical judgment, in which observers are required to make a simple forced-choice decision about the figure. A weakness is that it requires the subject to have a clear notion about the canonical form of the figure. However, in practice, performance is highly consistent: Subjects adopt reproducible criteria and, for cases such as stereo, they exhibit discrimination thresholds close to those that would be predicted on the basis of the monocular changes in the 2-D image cues.

The other approach we have examined tests a more qualitative aspect of shape perception. This exploits the fact that some targets are perceptually ambiguous, in the sense that they have two alternative interpretations for their organization in depth. (One of the classical examples of this is the Necker cube.) In the figures studied for this test, the surface in depth has the form of a stiff sheet bent into a sinusoidal profile. Such figures are ambiguous in that either the upper portion or the lower portion (or indeed both portions) may appear to be convex rather than concave. Typically, even though there may be enough information in the stimulus to resolve the ambiguity, human observers make mistakes when the number of cues to depth is limited. One measure of the power of additional cues is the extent to which they stabilize the percept into a consistent and veridical interpretation.

Some experimental results

Volumetric Texture Rendering An observation that highlights the issue of cue combination is to consider the texture cue available in a conventional Julesz-type random-dot stereogram. Although the binocular disparities could indicate that the stereo pair depicts a smoothly curved surface that varies in depth, the texture information would continue to indicate that the surface is flat and frontoparallel, even though the texture alone cannot specify the particular depth plane in which the figure is located. Thus, there is conflict between the texture and the stereo cues. To examine the interaction of texture and stereo cues, a different method of generating textured figures for the production of pairs of images for binocular inspection is needed. In order for the stereo information and the texture information to be in accord with each other in a way that could arise in a natural image, stereograms were created by ray tracing from a solid 3-D representation of an object.

It is possible to conceive of a number of different ways in which texture might be included in the representation of a solid object. For example, texture might arise from an initially uniform pattern over a figure that becomes deformed and altered as the shape of the figure changes: This route to texture formation seems common in biological systems, such as the uneven pattern of stripes on the skin of the zebra (Bard, 1977). A second alternative is that material might be deposited on a smooth surface, such as pebbles on a beach. A third way in which surfaces may become textured is that for a solid material textured all the way through, the exposed surface simply reveals a pattern of textured surface markings, as happens when wood or stone is carved to shape. Other possibilities exist, but all these cited have obvious links with the formation of texture on natural objects.

The third of these options offers a simple method for making stereo pairs that include texture variation but are similar in form to those made by the Julesz (1971) method of generating stereograms. A 3-D block of texture was made up from individual cubic volume elements (voxels) whose gray-level values were assigned on a random basis. Visually, this block appears as a uniformly textured figure, rather like granite. (A small amount of smoothing was introduced by applying a 3-D Gaussian filter to the block; this prevented aliasing effects when the blocks were finally portrayed by means of the pixels of the computer display.) The required 3-D shape was then created by cutting away the undesired portions to leave a smoothly curved surface, typically a cylinder or a sphere. The resulting shape was then portrayed as a stereo pair by producing two images viewed from vantage points separated by the interocular separation of the psychophysical observer.

Variations of the texture cue were induced by stretching or compressing the rectangular block in the depth dimension (see figures 22.3, 22.4). A compression of the

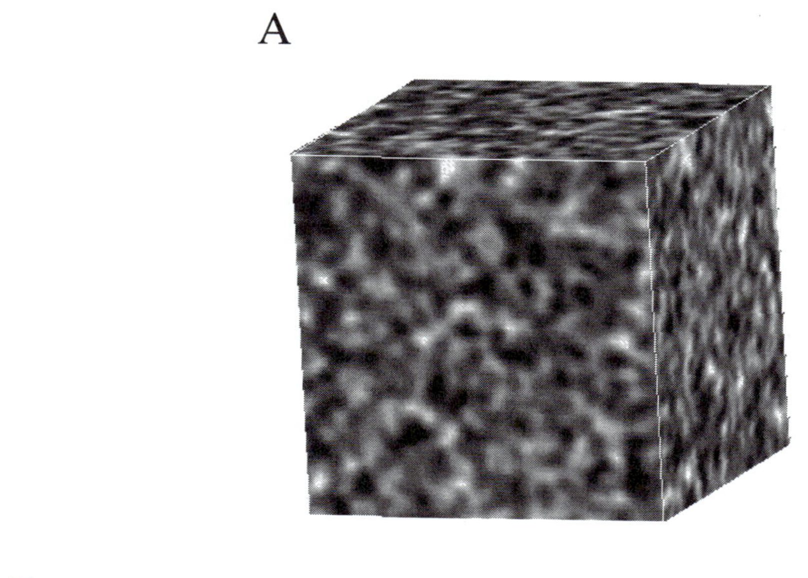

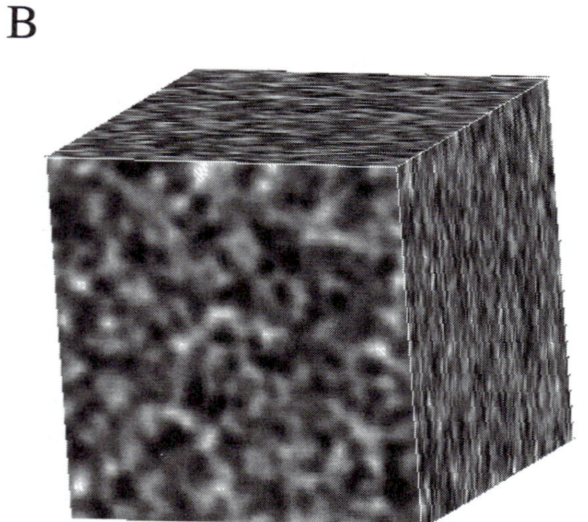

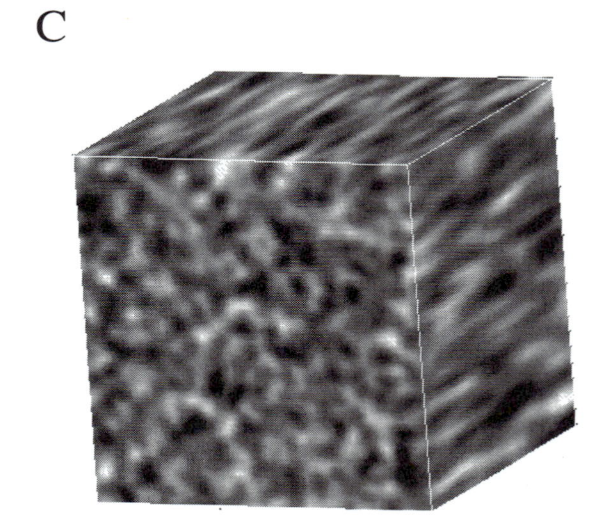

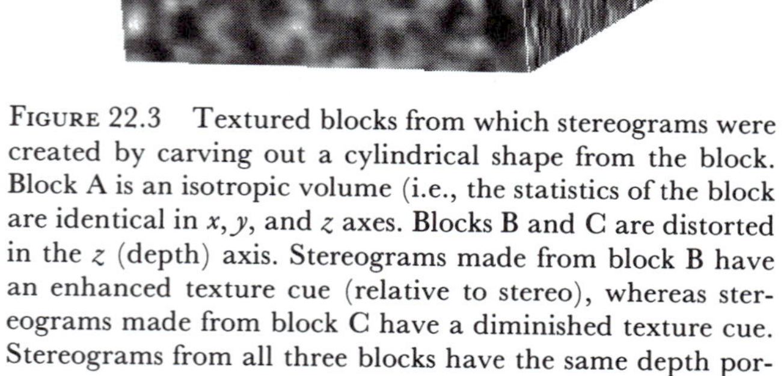

FIGURE 22.3 Textured blocks from which stereograms were created by carving out a cylindrical shape from the block. Block A is an isotropic volume (i.e., the statistics of the block are identical in x, y, and z axes. Blocks B and C are distorted in the z (depth) axis. Stereograms made from block B have an enhanced texture cue (relative to stereo), whereas stereograms made from block C have a diminished texture cue. Stereograms from all three blocks have the same depth portrayed by binocular disparity because the geometry of the projection is identical for all three cases. (Reprinted from *Vision Research*, 33 (5/6), E. B. Johnston, B. G. Cumming, and A. J. Parker, Integration of depth modules: Stereopsis and texture, pages 813–826, copyright 1993, with kind permission from Pergamon Press Ltd., Headington Hill Hall, Oxford OX3 DBW, U.K.)

texture means that a given depth range of the curved surface must then pass through more texture elements than if it had been created from the standard uniformly textured block. However, because the stereo cue is determined by the geometry of the interocular separation and the vantage points of the two eyes, no change in the stereo cues is introduced by this maneuver. It is therefore possible to introduce into these figures independent manipulations of the texture and stereo cues.

STEREO AND TEXTURE The task of each subject was the quantitative shape judgment task mentioned previously. Subjects were required to determine whether a

A

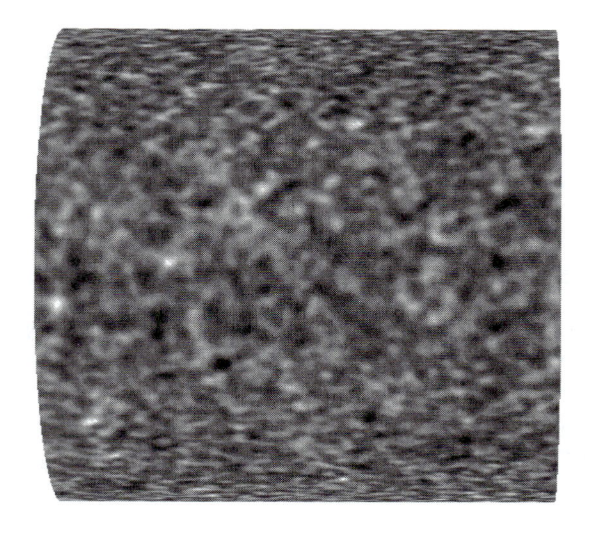

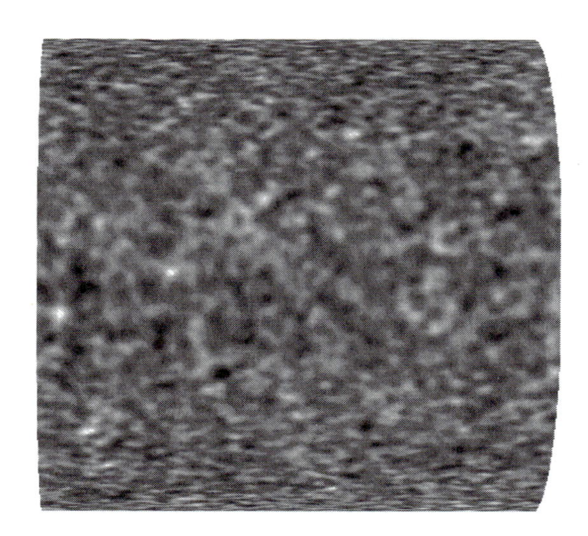

B

FIGURE 22.4 Two stereo pairs made from the blocks in figure 22.3. (A) This stereo pair was made from block A in figure 22.3 and has concordant texture and stereo depth cues. (B) This stereogram has a texture cue corresponding to a flat surface and would be created by using a block in which all volume elements are identical along the z-axis and vary only in x and y. (Reprinted from *Vision Research*, 33 (5/6), E. B. Johnston, B. G. Cumming, and A. J. Parker, Integration of depth modules: Stereopsis and texture, pages 813–826, copyright 1993, with kind permission from Pergamon Press Ltd., Headington Hill Hall, Oxford OX3 DBW, U.K.)

cylindrical surface was more or less flattened in depth than a cylinder of circular cross-section: Thus, if subject's depth sensation was diminished with respect to the physically ideal solution, he or she would require a surface of elliptical cross-section (with the major axis aligned in the depth direction) to perceive that the surface was circular. Performance on this task was measured for a range of stereo and texture combinations.

The results are shown in figure 22.5, which shows that as the strength of the texture cue is increased, the amount of depth from stereo that needs to be intro-

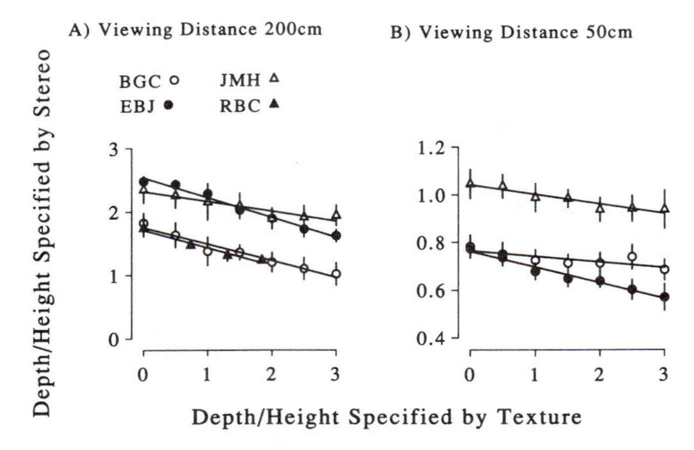

FIGURE 22.5 Experimental results for the interaction between texture and stereo cues gathered using the quantitative shape judgment task described in the text. The results are plotted in the same format as the schematic in figure 22.2. Data are shown for four subjects at two different viewing distances: (A) 200 cm; (B) 50 cm. As the texture cue is increased, subjects required less binocular disparity to make the perceived depth of the figure correspond to a cylinder of circular cross-section. The results are well described by a simple linear model. The viewing distance–dependent distortions of stereopsis (Johnston, 1991) are also evident here, in that with zero texture (i.e., texture corresponding to a flat plane), subjects require more depth at large viewing distances than geometrically expected. At least two subjects at the shorter viewing distance (B) need less depth than the geometrical expectation. (Reprinted from *Vision Research*, 33 (5/6), E. B. Johnston, B. G. Cumming, and A. J. Parker, Integration of depth modules: Stereopsis and texture, pages 813–826, copyright 1993, with kind permission from Pergamon Press Ltd., Headington Hill Hall, Oxford OX3 DBW, U.K.)

duced to give the perception of a circular cross-section decreases. In fact, the interrelation between stereo and texture in these results is well described by a linear interaction. Although this means that the interaction of stereo and texture can be summarized by quoting a simple relative weighting of the two cues, the absolute values of these weights are more difficult to determine. One problem, which is common to a number of methods for estimating the interactions between cues, is that it is invalid to assume that each cue alone will give rise to a veridical perception of depth, even if the cue is introduced in its geometrically correct form. Indeed, it is already known that this is not true for stereo (Johnston, 1991). Despite this caveat, it is evident that texture is an extremely weak cue compared with binocular stereopsis, which may reflect the unreliability of texture relative to binocular stereo within the natural environment. Decisions about texture can be based only on a statistical estimation that local variations in texture actually reflect local variations in surface orientation and not a local change in the properties of the material.

STEREO AND MOTION The combination of stereo and motion provides an interesting case with respect to the 3-D shape judgment tasks introduced earlier. Motion cues alone are, in fact, capable of signaling the true physical shape of a 3-D object, provided that at least three views are seen (Ullman, 1979). (Ullman also pointed out that two views might be enough if there is sufficient perspective information but noted that recovery of the information might be prone to noise.) It is important to appreciate that this claim applies only to the shape of the object and not to its size. In fact, because the velocity of the object in the 3-D scene cannot be recovered unless the viewing distance is known, the absolute size of the object cannot be recovered solely from measurements of motion made at the retina. Ullman's (1979) result does not apply to this ambiguity, but it does demonstrate that, with adequate motion cues, observers should see 3-D shape correctly, even if the actual size of the figure is misperceived. The speed-distance ambiguity is compatible only with uniform linear scaling of the object in question.

It has been pointed out that combinations of stereo and motion could have a special status for recovering 3-D shape owing to the close links in the geometry of the two sources of information (Regan and Beverly, 1979; Richards, 1985); stereo and motion together could recover absolute size as well as shape. Experimentally, the most obvious example of a strong interaction between depth cues comes from a comparison of stereo and motion, where by the observer is given only two frames of the motion target, which are presented in an apparent motion sequence. Not surprisingly, such two-frame motion figures are subject to distortions of shape (Johnston, Cumming, and Landy, 1994), already noted for binocular stereopsis (Johnston, 1991). However, with binocular two-frame motion sequences, observers recover 3-D shape correctly, just as they do with monocular multiframe motion sequences. In one sense, this is not surprising as the binocular two-frame motion sequence gives the observer *four* independent views of the shape, which is more than sufficient information to fulfill the three-frame requirement (Ullman, 1979). However, the fact that binocular stereopsis and motion flow are combined in this way is clear evidence of a strong interaction between the two cues. Exactly how well this effect generalizes to a range of viewing conditions is as yet unclear (Tittle et al., 1993).

An approach very similar to that described for texture-stereo interactions has been used to study the interactions between motion and stereo (Johnston, Cumming, and Landy, 1994). Using stimuli very similar in form to those depicted in figures 22.3 and 22.4, the investigators systematically altered the interrelations between the binocular stereo cues and the motion parallax cues. When both cues were present (i.e., neither zero motion nor zero stereo), the interaction could again be described by a linear weighting. Thus although, in a sense, motion flow can calibrate stereo, both theoretically and in practice, stimuli with discordant combinations of stereo and motion information give rise to the intermediate sensations of depth that are predicted by the linear pooling model. The simplicity of this result is somewhat surprising, given the previous evidence for a more sophisticated geometrical combination of cues between motion flow and binocular stereo.

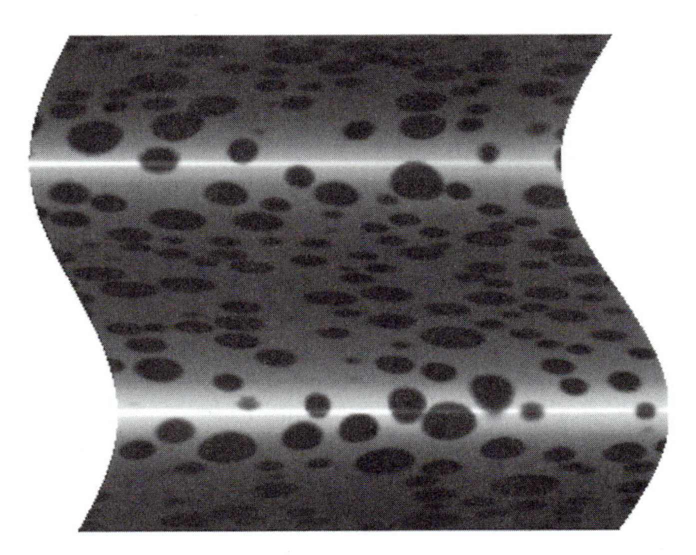

FIGURE 22.6 A figure of sinusoidal sheet that was rotated back and forth about a vertical axis. When in motion, the figure conveys a number of cues to 3-D shape: motion disparities, shading and specular highlights (both of which change consistently under motion), outline contour, and texture. With all cues present, the surface is generally seen as a rigid, rotating form with convex and concave regions correctly assigned. However, with some cues omitted, the shape is often seen with incorrect concave-convex assignments or even as a nonrigid form (almost always as two convex regions twisting relative to each other—only very occasionally do some subjects report seeing two concave regions). The ability of cues to stabilize the physically correct interpretation of motion can be used as a measure of their effectiveness.

MULTIPLE CUES FOR QUALITATIVE SHAPE JUDGMENTS Qualitative shape judgments have been used to study the interactions between mutiple cues. Figure 22.6 depicts a single view from a motion sequence of a shape that could be portrayed with shading, specular highlights, texture, and binocular stereopsis as well as motion parallax. The motion sequence was constructed by taking a succession of perspective views of the 3-D sinusoidal surface as it oscillated in angular position by rotating about a vertical axis through the middle of the figure. These were replayed as a "cine-loop" of pictures on the computer, to give a smooth motion sequence. An additional important feature is that the motion of the specular highlights, if present, was geometrically correct for a single light source illuminating the shape (Blake and Bülthoff, 1991). Notably, the specular highlight does not move as a surface feature but according to the laws of reflection at mirrorlike surfaces. Thus, for concave surfaces, the highlight is a real inverted image of the light source, whereas for convex regions, it is a virtual upright image. In all the stimuli used in these motion experiments, the relative motions of the surface and the specularities are determined by their geometrically correct depth relationships.

The value of this stimulus is best understood by considering a highly reduced form of it, in which the motion cue is signaled by dots alone. In fact, if the positions of the dots are calculated by a parallel projection, then it is geometrically impossible to disambiguate the curvature of the surface. Even with a perspective projection to the observer's monocular vantage point, the perceptual decision is almost impossible to assign correctly. Thus, if observers are asked to select whether the upper or the lower surface region is convex (or concave), then they do so almost at chance levels for motion parallax depicted with dot stimuli. This is the classical motion parallax ambiguity (Wallach and O'Connell, 1953). In fact, a somewhat surprising third perceptual interpretation often appeared with dot patterns. In this case, *two* convex regions were seen, joined by a nonrigid segment across the middle of the stimulus. The stimulus rarely was perceived as having two concave regions. Nonrigid interpretations of the perceptually reversed Necker cube (Dosher, Sperling, and Wurst, 1986) and other figures have been noted.

In general, as more and more cues were added to the figure, the percept was stabilized toward the geometrically correct, rigid interpretation. A particularly strong monocular cue was the perspective cue given by the outline contour of the figure. Indeed, if the outline contour was arranged so that the 3-D surface actually got wider as it receded in depth, then the dominating effect of perspective signaled by the outline contour could be easily appreciated, even in the presence of otherwise robust cues from motion parallax, texture, and shading. Binocular stereo was, however, completely effective in sorting out the ambiguity of concave versus convex regions of the image.

This last result makes for an interesting contrast with motion parallax information. Motion parallax, at least in multiple-frame motion sequences, gives a good quantitative estimate of the true shape of convex surfaces (Johnston, Cumming, and Landy, 1994). By contrast, the quantitative estimates of the shape of convex surfaces provided by stereo are often awry, although the deviations seem to have a systematic basis (Johnston, 1991). If regions of the image have plenty of monocular features or dense texture on which binocular stereopsis can be established, then binocular stereo can be highly effective in revealing the sign of curvature (i.e., concave or convex). Clearly, if binocular stereo and motion flow cues are combined by strong fusion, then these potential ambiguities should be less apparent in natural viewing.

Synopsis: modules, statistics, and learning

The finding that a number of interactions between cues can be described by simple linear pooling has some significant implications. It would be easy to dismiss this result as unexceptional, yet the simplicity of the linear model is perhaps the most surprising feature of the results. Cues arise from different sources and are very likely processed in different ways in the visual nervous system, but the final perceptual interactions can often be characterized by a model of almost trivial simplicity. Why should this occur?

One argument that can be advanced is that the success of the linear model reflects a structural organization for the processing of multiple cues for 3-D shape. The obvious hypothesis is that the linear model indicates that distinct and separate modules are responsible for processing different cues. In making this argument, it is important to recall one key feature of recent results: The addition of cues from multiple sources does not always improve perception. There are circumstances under which adding information actually causes the perceptual judgment to depart from the geometrically correct solution. The results *cannot* be summarized by the dictum that adding more information makes things better. Blake, Bülthoff, and Sheinberg (1993) have pointed out that linear combination rules are predicted by models based on pooling of statistical evidence with maximum likelihood estimators. In this context, it would seem that the weights assigned to cues are perhaps derived only from estimates of the reliability of each cue assessed independently. Otherwise, there might be more extensive evidence for strong interactions between cues. In this sense, each module for recovering shape from a particular cue could be considered an independent entity able to carry out its own self-calibration.

The exceptions to the rule of linear interactions are clearly present but are difficult to characterize simply. The case of stereo and motion may be particularly indicative that such interactions will occur; computational analysis suggests that it is especially valuable to consider the interactions between cues *prior* to the extraction of depth from either. A further possible deviation from linear interactions might conceivably occur when three cues are available, two of which are concordant and the third not. In this case, the third cue may be discarded because it is inconsistent, even though the same cue might contribute significantly under other conditions. This idea has not been tested experimentally for human vision.

The whole problem of 3-D shape and the cues that sustain human performance gives rise to a number of issues concerning the calibration of sensor information. How do humans acquire and store the parameters of interocular separation and distance (or vergence angle) that would be necessary for the geometrically correct interpretation of stereo? What are the statistics of the 3-D structure of natural scenes for parameters such as the stereo disparity gradient or the more general pattern of surface gradients and curvatures?

How do humans monitor the statistics of textures, so as to allow them to infer distortions of surface shape from surface texture? In the case of motion parallax created by self-motion of the observer, what information is used to integrate visual and somatosensory cues? There are numerous other examples. It is hard to escape the idea that somehow these parameters are acquired through mechanisms of learning over a time period that is longer than those used for the perceptual experiments studied here.

Many of the biological advantages of humans as a species arise from skill in movement through the visual environment and, above all, through skill in the manipulation of objects by hand. The way in which the visual system provides a geometrical engine for calculating the information necessary to support these skills is an essential component of our biological success.

ACKNOWLEDGMENTS Much of the work from our laboratory mentioned in this chapter has been made possible by grants from the McDonnell-Pew Program in Cognitive Neuroscience, The Wellcome Trust, Science and Engineering Research Council, and the Medical Research Council.

REFERENCES

BARD, J., 1977. A unity underlying the different zebra striping patterns. *J. Zool. Lond.* 183:527–539.

BARROW, H. G., and J. M. TENENBAUM, 1978. Recovering instrinsic scene characteristics from images. In *Computer Vision Systems*, A. R. Hanson and E. M. Riseman, eds. Orlando, Fla.: Academic Press, pp 3–26.

BESL, P. J., and A. K. JAIN, 1986. Invariant surface characteristics for 3D object recognition in range images. *Comput. Vis. Graphics Image Proc.* 33:33–80.

BLAKE, A., and H. H. BÜLTHOFF, 1991. Shape from specularities: Computation and psychophysics. *Philos. Trans. R. Soc. Lond. [Biol.]* 331:237–252.

BLAKE, A., H. H. BÜLTHOFF, and D. SHEINBERG, 1993. Shape from texture: Ideal observers and human psychophysics. *Vision Res.* 33(12):1723–1737.

BLAKE, A., and A. ZISSERMAN, 1987. *Visual Reconstruction*. Cambridge, Mass.: MIT Press.

BRADSHAW, M. F., and B. J. ROGERS, 1992. Subthreshold interactions between binocular disparity and motion parallax. *Invest. Ophthalmol. Vis. Sci.* 33(4):1332.

BRADY, J. M., J. PONCE, A. YUILLE, and H. ASADA, 1985. Describing surfaces. In *Proceedings of the second International Symposium on Robotics Research*, H. Hanafusa and H. Inoue, eds. Cambridge, Mass.: MIT Press, pp. 5–16.

BÜLTHOFF, H. H., and H. A. MALLOT, 1988. Integration of depth modules: Stereo and shading. *J. Opt. Soc. Am. [A]* 5(10):1749–1758.

CLARK, J. J., and A. L. YUILLE, 1990. *Data Fusion for Sensory Information Processing Systems*. Boston: Kluwer Academic.

CUMMING, B. G., E. B. JOHNSTON, and A. J. PARKER, 1991. Vertical disparities and perception of three-dimensional shape. *Nature* 349(6308):411–413.

CUMMING, B. G., and A. J. PARKER, 1994. Binocular mechanisms for detecting motion-in-depth. *Vision Res.* 34:483–496.

DOSHER, B. A., G. SPERLING, and S. A. WURST, 1986. Tradeoffs between stereopsis and proximity luminance covariance as determinants of perceived 3-D structure. *Vision Res.* 26(6):973–990.

GRIMSON, W. E. L., 1982. A computational theory of visual surface interpolation. *Philos. Trans. R. Soc. Lond. [Biol.]* 298:395–427.

HELMHOLTZ, H. VON, 1962. *Treatise on Physiological Optics*, ed. 3, J. P. C. Southall, trans. New York: Dover Publications.

HOFFMAN, D. D., and W. A. RICHARDS, 1984. Parts of recognition. *Cognition* 18:65–96.

HORN, B. K. P., 1977. Understanding image intensities. *Artif. Intell.* 8(2):201–231.

JOHNSTON, E. B., 1991. Systematic distortions of shape from stereopsis. *Vision Res.* 31:1351–1360.

JOHNSTON, E. B., B. G. CUMMING, and M. S. LANDY, 1994. Integration of stereopsis and motion shape cues. *Vision Res.* in press.

JOHNSTON, E. B., B. G. CUMMING, and A. J. PARKER, 1993. Integration of depth modules: stereopsis and texture. *Vision Res.* 33:813–826.

JULESZ, B., 1971. *Foundations of Cyclopean Perception*. Chicago: University of Chicago Press.

KOENDERINK, J. J., 1990. *Solid Shape*. Cambridge, Mass.: MIT Press.

KOENDERINK, J. J., and A. J. VAN DOORN, 1992. Surface shape and curvature scales. *Image Vis. Comput.* 10(8):557–565.

KOENDERINK, J. J., A. J. VAN DOORN, and A. M. L. KAPPERS, 1992. Surface perception in pictures. *Percept. Psychophys.* 55(5):487–496.

MALONEY, L. T., and M. S. LANDY, 1989. A statistical framework for robust fusion of depth information. In *Visual Communications and Image Processing: IV. Proceedings of the SPIE*, W. A. Pearlman, ed. Society for Photometric and Illumination Engineering, pp. 1154–1163.

MARR, D., 1982. *Vision*. San Francisco: Freeman.

MARR, D., and S. ULLMAN, 1981. Directional selectivity and its use in early visual processing. *Proc. R. Soc. [Biol.]* 211: 151–180.

MAYHEW, J. E. W., and H. C. LONGUET-HIGGINS, 1982. A computational model of binocular depth perception. *Nature* 297:376–379.

PARKER, A. J., J. S. MANSFIELD, E. B. JOHNSTON, and Y. YANG, 1991. Stereo, surfaces and shape. In *Computational Models of Visual Processing*, M. S. Landy and J. A. Movshon, eds. Cambridge, Mass.: MIT Press.

POGGIO, T., E. B. GAMBLE, and J. J. LITTLE, 1988. Parallel integration of visual modules. *Science* 242:436–439.

Regan, D., and K. I. Beverly, 1979. Binocular and monocular stimuli for motion in depth: Changing-disparity and changing-size feed the same motion-in-depth stage. *Vision Res.* 19:1331–1342.

Richards, W. A., 1985. Structure from stereo and motion. *J. Opt. Soc. Am. [A]* 2:343–349.

Rogers, B. J., and M. F. Bradshaw, 1993. Vertical disparities, differential perspective and binocular stereopsis. *Nature* 361:253–255.

Rogers, B. J., and M. E. Graham, 1979. Motion parallax as an independent cue for depth perception. *Perception* 8:125–134.

Rogers, B. J., and J. J. Koenderink, 1986. Monocular aniseikonia: A motion parallax analogue of the induced effect. *Nature* 322:62–63.

Schriever, W., 1925. Experimentalle studien über stereoskopisches sehen. *Z. Psychol.* 96:113–170.

Tittle, J., V. Perotti, J. Todd, and J. F. Norman, 1993. The perception of relative surface orientation from binocular disparity and motion. *Invest. Ophthalmol. Vis. Sci.* 34(4): 1132.

Ullman, S., 1979. *The Interpretation of Visual Motion.* Cambridge, Mass.: MIT Press.

Wallach, H., and D. N. O'Connell, 1953. The kinetic depth effect. *J. Exp. Psychol.* 45:207–217.

Weinshall, D., 1990. Qualitative depth from stereo, with applications. *Comput. Vis. Graphics Image Proc.* 49:222–241.

Wheatstone, C., 1838. Contributions to the physiology of vision: I. On some remarkable, and hitherto unobserved, phenomena of binocular vision. *Philos. Trans. R. Soc. Lond.* 128:371–394.

Wheatstone, C., 1852. Contributions to the physiology of vision: II. On some remarkable, and hitherto unobserved, phenomena of binocular vision (continued). *Philos. Trans. R. Soc. Lond.* 142:1–17.

Witkin, A. P., 1981. Recovering surface shape and orientation from texture. *Artif. Intell.* 17:17–47.

23 Form Analysis in Visual Cortex

RÜDIGER VON DER HEYDT

ABSTRACT The functional properties of various types of orientation-selective neurons of areas V1 and V2 of the visual cortex are reviewed, and their role in visual form perception is discussed, with special consideration of contour perception and the phenomenon of illusory contours. The stages of processing are illustrated by a computational model. Complex cells show a variety of functionally different subtypes. Whereas some are selective for pattern (e.g., responding to a narrow dark bar but not to light bars or edges), others are pattern-invariant. These can be conceived as constituting a "local oriented energy" operator, generating a representation of the contrast borders that is independent of border profile (edge and line) and contrast polarity. Common types of end-stopped cells can be considered directional derivatives of this representation. These may be important in highlighting occlusion features such as T-junctions, corners, and line ends. Contour cells of V2 are believed to combine the activity of complex cells of a given orientation with signals from end-stopped cells of orthogonal orientation, thereby inferring occluding contours by combining evidence from contrast borders with evidence from occlusion features. This mechanism detects occluding contours on images of natural scenes better than mechanisms using just contrast border detection, but it does generate illusory contours.

This chapter deals with some aspects of visual cortical processing that seem to be related to the perception of form. I will confine my discussion to mechanisms found in cortical areas V1 and V2 of the monkey and their presumed analogs in the human visual cortex. The term *form analysis* might lead one to expect too much; what these areas seem to provide are the elements for an analysis of form, such as representations of certain patterns of local intensity variation and perhaps pieces of contour with some attributes such as figure-ground direction. I will focus particularly on the question of how the system infers contours that are not well defined in the image and how it attempts to identify real object contours. Certainly, one would like to understand the nature of the final encoding of form in the brain for the purposes of memorizing, recognizing, and handling objects, but apparently little is known about this. On the other hand, many recent studies of biological and technical vision systems have pointed out that intelligent low-level mechanisms are more important than we had previously believed and the category *intermediate-level processing* has been created to characterize these stages (Fischler and Firschein, 1987). The large size of the primary and secondary visual areas—V1, V2, and V4—that prepare information for further processing in the inferotemporal cortex also emphasizes this point (cf. Felleman and Van Essen, 1991). The processing of complex forms is discussed further in chapters 24, 29, and 30.

RÜDIGER VON DER HEYDT Philip Bard Laboratories of Neurophysiology, Department of Neuroscience, Johns Hopkins University School of Medicine, Baltimore, Md.

Can we distinguish cortical mechanisms of form analysis from the processing of the other aspects of vision such as color, motion, or location and orientation in space? Looking at objects, it seems rather obvious that these aspects are independent. We certainly believe that our perception of the shape of an object is independent of its color; likewise, if we painted a table and a chair with the same color paint, we would be surprised if they did not appear in the same color (although this may happen). Form, color, motion, position in space, and so forth appear to be orthogonal dimensions of perception. However, a closer look reveals that this is not true. There are many examples to the contrary: Colors change when an object is viewed under large and small visual angles (contrast versus assimilation). Perception of form is much more difficult if an object has a color of exactly the same brightness as the background of a different color (equiluminance); the object's three-dimensional shape may not be perceived correctly at all. Simple rotation of a figure by 45° can change its perceived shape (e.g., a tipped square becomes a diamond).

Whatever the degree of independence among different perceptual qualities, it is important to realize that the perceptual dimensions are abstractions, obviously conceived in analogy to the physical properties of objects. In general, they are not qualities of the visual stimulus that the system could easily extract and separate at an early stage. The system strives to identify objects in the chaos of intensity values sensed by its receptors and to assign attributes to them that remain constant so that we can recognize something (i.e., attributes that correlate with certain physical properties), and it does this with varying success. Although perception appears to be immediate and effortless, the internal generation of the attributes may in fact involve huge amounts of computation, occupying billions of cells at every moment we look at the world.

Therefore, one can hardly expect to find subsystems for the analysis of form, color, motion, and such very early in the visual cortex, because those computations are generally complex and not likely to be completed in a few synaptic steps. This point is a frequent source of confusion and controversy when, for example, orientation selectivity and end-stopping of cortical neurons are believed to imply form analysis or when direction selectivity is linked to movement perception or chromatic sensitivity to color, for instance. Also color processing may require orientation-selective cells at some stage, motion contributes to form perception (e.g., by creating dynamic contours and 3-D perception), chromatic contrast contributes to form and motion perception, and so on.

We do not know how the higher centers of form representation in the inferotemporal cortex use the basic image representations of areas V1 and V2. There is no general agreement even on the question of what is achieved by the processing of V1 regarding the goal of form perception. Is it a decomposition of the image into linear elements and corners and some degree of position invariance, as implied by Livingstone and Hubel (1987), or a filtering and representation of spatial frequency components, as emphasized by Wilson and colleagues (1990)? Many perceptual phenomena demonstrate that vision involves an active organization of visual information (Kanizsa, 1979). These phenomena can provide tools for determining cortical function; in physiological experiments, they may help to identify the meaning of a neuronal response (Baumgartner, von der Heydt, and Peterhans, 1984). A conspicuous example is the illusory contours that human observers perceive in certain figures (figure 23.1). Contours indeed seem to be fundamental to the process of encoding and analyzing forms; we can recognize many objects just by their contours, and the illusory contours, generated by the system in the absence of direct sensory information, seem to reveal its built-in assumptions about the shapes of objects. If neural signals can be found that represent these contours, they are likely to be involved in form analysis. Later, evidence for illusory contour representation is reviewed.

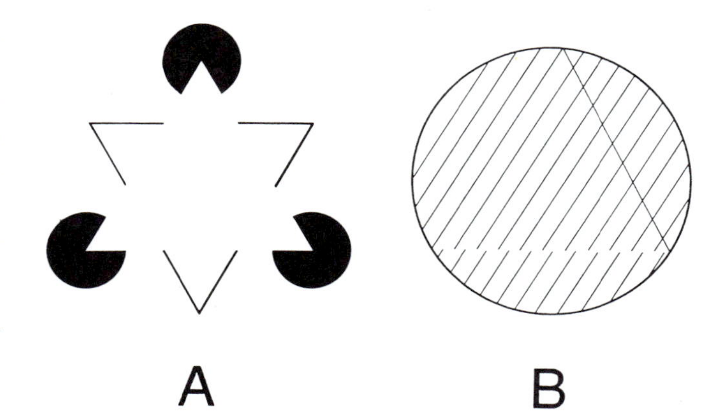

FIGURE 23.1 Examples of illusory contours and contour masking. (Left) The Kanizsa triangle. (After Kanizsa, 1979.) Where is the triangle in the right panel? Solution: Its right flank is perceived as drawn, and its base is perceived although not drawn, whereas its left flank is drawn but not perceived! (Modified from Galli and Zama, 1931.)

The next section briefly summaries some basic properties of receptive fields in areas V1 and V2, concentrating on the results in the monkey. This survey is a jumping-off point to introduce a computational model of cortical contour processing, which begins with a simulation of simple cells. Then we will look in more detail at the neural mechanism of contour perception, specifically the phenomenon of illusory contours. These studies are an example of an attempt to relate neurophysiology to perception and to understand visual processing at the three levels of theory, algorithm, and implementation, outlined by David Marr (1982).

Receptive fields of cells in the visual cortex

The primary visual cortex (also called *area 17*, or *V1*) is the area of the cortex that receives, via the relay stage of the lateral geniculate nucleus, the input from the eyes. In primates, it is the main visual entrance to the cortex.

The neurons in this area have been classified according to their responses to visual stimuli. These are described in terms of receptive fields, as each cell is sensitive to light changes only within a small patch of the visual field of one or both eyes. In the simplest case, a receptive field has a characteristic two-dimensional sensitivity distribution. Depending on the position in the receptive field, light either increases or decreases the cell's activity (frequency of action potentials)—that is, the cell is either excited or inhibited. The excitatory and inhibitory inputs tend to be balanced so that most cortical cells hardly respond to uniform illumination. The impulse frequency code implies an output nonlinearity because the maintained firing rate is often low or zero (rectification nonlinearity).

CONCENTRIC RECEPTIVE FIELDS Four major classes of receptive fields—*concentric*, *simple*, *complex*, and *end-stopped*—have been found in V1 (Hubel and Wiesel, 1977). The concentric, or center-surround, receptive field can be described by a Mexican-hat function, the difference of a narrow and a broad circular Gaussian function. They come in two main types, one that is excited by light in the center and inhibited by light in the surround and another exhibiting the converse behavior. Cells of the lateral geniculate nucleus have concentric receptive fields, as do their target cells in the input layer 4C of V1. In addition, some of the cells in the upper (output) layers of V1 show concentric receptive field organization (see chapter 20).

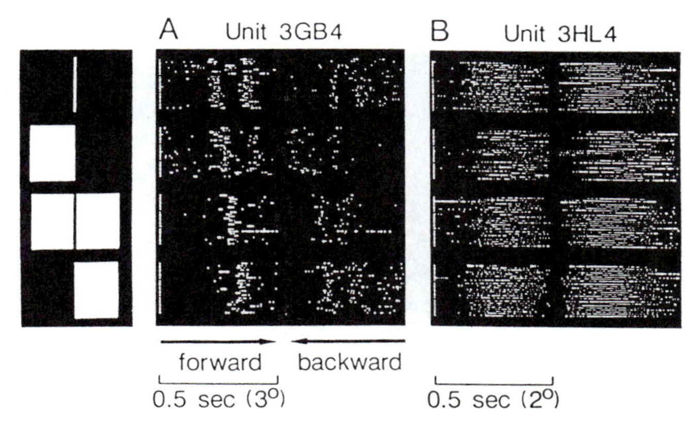

FIGURE 23.2 Responses of a simple and a complex cell to light and dark bars and edges. Each pattern was moved back and forth 24 times in random sequence. The cells responded with volleys of action potentials as represented by the rows of dots in (A) and (B). It can be seen that the receptive field of the simple cell (A) had three excitatory regions; at two positions, it was activated by the light bar and at an intermediate position by the dark bar (this is obvious for the forward sweeps of the stimulus and less clear for the backward sweeps). The edge responses were similarly interlaced. The complex cell (B) responded similarly to all four patterns with no evidence for spatially separate subregions. Both cells were recorded in area V1 of an alert monkey during periods of active fixation of gaze. The occasional, irregular displacements of the responses are due to small eye movements. (Reprinted from von der Heydt and Peterhans, 1988, by permission.)

SIMPLE RECEPTIVE FIELDS Most of the cells in V1, however, have oriented receptive fields and respond preferentially or exclusively to oriented light patterns, such as light-dark boundaries, lines, or gratings. Each cell has a preferred orientation to which it is tuned. The tuning width is typically approximately 40° at half-amplitude, but widths between 10° and 90° occur (De Valois, Yund, and Hepler, 1982). The preferred orientations are distributed quasi-continuously, with some emphasis on vertical and horizontal. As for concentric fields, simple receptive fields have separate regions of excitatory and inhibitory influence, but these are organized as parallel bands. There can be two, three, or more such bands of alternating polarity (see the example in figure 23.2A). According to this structure, a grating stimulus can have excitatory or inhibitory effects, or no effect, depending on the position of the grating (its spatial phase), and moving gratings cause strongly modulated responses (Maffei and Fiorentini, 1973; Schiller, Finlay, and Volman, 1976; De Valois, Albrecht, and Thorell, 1982).

Most simple cells can be conceived as linear spatial filters (disregarding the output nonlinearity, as mentioned previously) (Movshon, Thompson, and Tolhurst, 1978a; Maffei et al., 1979; see von der Heydt, 1987, for a review). The spatial summation can be described again by differences of Gaussians, now with displaced centers (Heggelund, 1981, 1986a,b; Parker and Hawken, 1988), or by Gabor functions (sine and cosine functions multiplied by Gaussians) (Marcelja, 1980; Daugman, 1984). The Gabor-function model expresses in mathematical form the dual function of these receptive fields and their trade-off between feature localization (Hubel and Wiesel, 1962, 1968) and spatial frequency selectivity (Maffei and Fiorentini, 1973; De Valois, Albrecht, and Thorell, 1982; De Valois and De Valois, 1988). When stimulated with sinusoidal gratings of appropriate orientation, they respond best when the grating periodicity matches that of the alternating receptive field subregions. Spatial frequency tuning curves of simple cells resemble Gaussian functions, as predicted from this model. The peak frequency and the width of tuning vary from cell to cell, the width averaging 1.4 octaves at half-amplitude (range, 0.5–3),

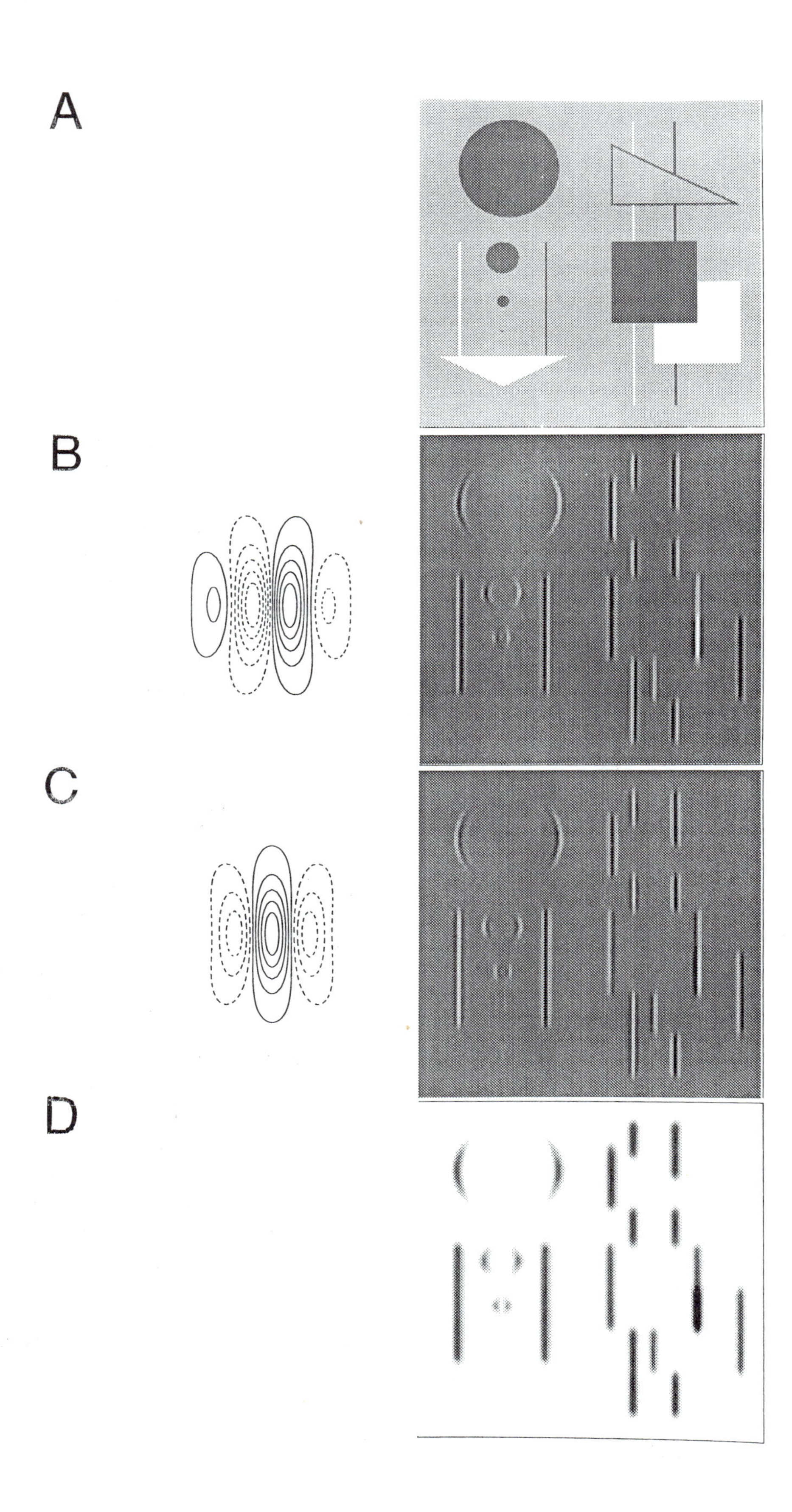
A
B
C
D

whereas the preferred frequencies of different cells, at a fixed position in visual field, cover a range as large as 5 octaves—for example, from 0.5 to 20 cycles per degree or so for foveal receptive fields (De Valois, Albrecht, and Thorell, 1982). Toward the periphery, the range shifts to lower frequencies (De Valois, Albrecht, and Thorell, 1982), perhaps simply related to the increase in receptive field size (Hubel and Wiesel, 1974).

Model receptive fields of even and odd symmetry (bar and edge detectors) corresponding in orientation and spatial frequency tuning to the average cortical simple cell are shown in figure 23.3, which illustrates also the effect of applying these filters on a test image. These filters correspond to the differential response between pairs of simple cells of opposite polarity, thus eliminating the rectification nonlinearity.

Complex Receptive Fields Complex cells have a similar preference for oriented patterns but do not show distinct bands of excitatory and inhibitory effects (see figure 23.2B). In contrast to simple cells, they show little or no effect of the phase of gratings and respond to moving gratings with unmodulated activity (Maffei and Fiorentini, 1973; Schiller, Finlay, and Volman, 1976; De Valois, Albrecht, and Thorell, 1982). Light and dark lines produce excitation at any position, with the strength of response varying gradually over the receptive field, approximating a Gaussian function. However, simultaneous stimulation with two lines (Movshon, Thompson, and Tolhurst, 1978b) and spatial frequency tuning (De Valois, Albrecht, and Thorell, 1982) reveal subunits with a receptive field structure similar to that of simple cells, as if the complex cell summed input from several simple cells with displaced receptive fields (Hubel and Wiesel, 1962; see von der Heydt, 1987, for a discussion of other models). The modulus (square root of the sum of the squares) of the responses of even and odd pairs of simple field filters, first used in computational vision (Granlund, 1978), has proved to be a good basis for modeling human motion detection and feature localization (Adelson and Bergen, 1985; Morrone and Owens, 1987; Morrone and Burr, 1988). It captures the essential features of typical complex cells (Pollen and Ronner, 1982; Heitger et al., 1992). Figure 23.3D shows the result of applying this operation to our test image. The traces of edges and lines are similar and invariant with change of contrast polarity in this representation. The same is true for line-edge combinations, which are common in natural images. Thus, this type of complex cell seems to account for some of the invariances of form perception. The interesting point is that edges and lines can be localized accurately by the maxima of the complex cell activity, despite the complicated oscillatory response patterns produced by simple cells.

Figure 23.3 The image representations obtained by operations that model the function of simple and complex cells. (A) Test image. (B, C) Results of convolving the test image with even and odd simple fields, as shown to the left. (D) Modulus of (B) and (C) (complex representation). (Modified from Heitger et al., 1992.)

End-Stopped Receptive Fields The term *end-stopped cell* refers to neurons that are similar to simple or complex cells in their preferences for edges, lines, or gratings but in which their responses decrease when the stimulating pattern is extended over a certain length. They are activated best by short lines or short pieces of edges and respond weakly or not at all to long stimuli extending across the receptive field. However, in general, the ends of lines or corners are also effective (see figure 23.4). In the monkey, nearly all end-stopped cells of V1 and V2 respond well to line ends or corners, and approximately half of them are asymmetrical, responding more than twice as much to stimuli extending in one direction than in the other (Peterhans and von der Heydt, 1990). A model of symmetrical and asymmetrical end-stopped cells appears in figure 23.5 (Heitger et al., 1992).

Discussion It is important to note that these models are by no means exhaustive descriptions of the variety of complex and end-stopped cells that can be found in areas V1 and V2 of the monkey. The binocular and stereoscopic properties are ignored here. Many complex cells also show obvious pattern selectivity. For example, discrimination between bar or edge patterns and the contrast polarities is frequently found both in V1 and V2. Examples recorded in V2 are shown in figure 23.6. Another conspicuous subtype of complex cells is the grating cell (von der Heydt, Peterhans, and Dürsteler, 1992). These are selectively activated by periodic patterns such as gratings or checkerboards but fail altogether to respond to bars and edges (figure 23.7). They may be involved in texture coding. It is surprising to find such rather specialized functions in

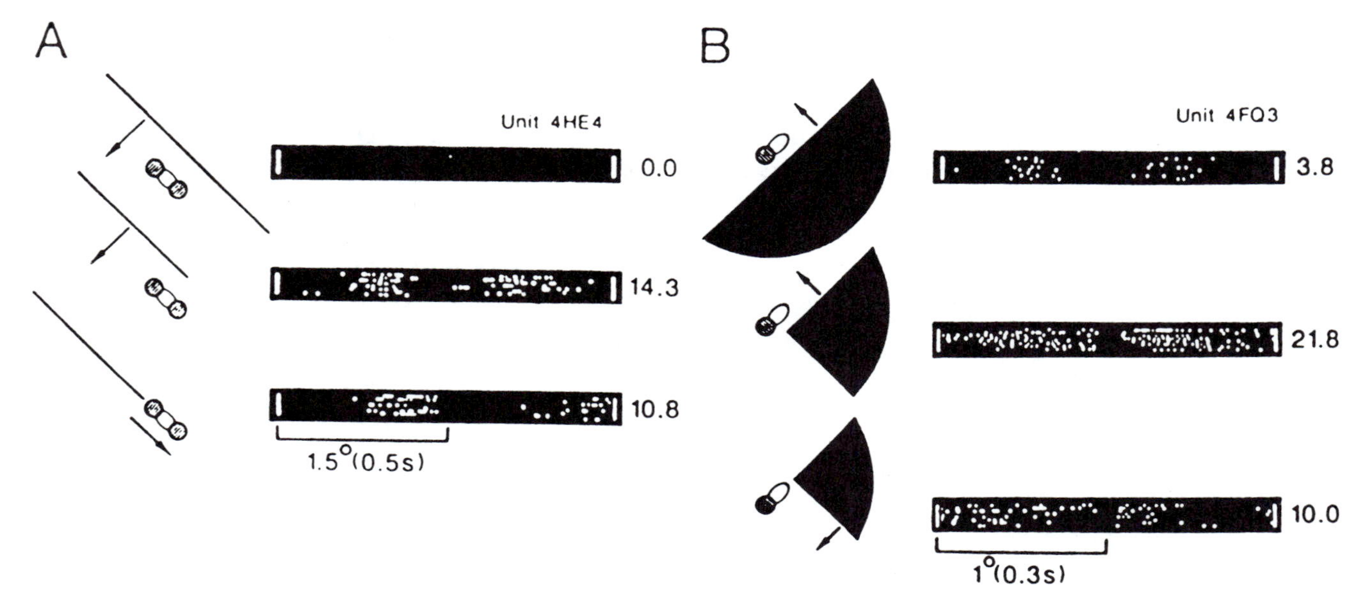

FIGURE 23.4 Two examples of end-stopped cells. Long lines or edges at the preferred orientation did not activate cell (A) and produced only weak responses in cell (B) (top row), but the end of a line or a corner evoked strong responses (middle and bottom rows). It did not make much difference which way the stimulus was moved. The receptive fields are symbolized by ellipses for the excitatory portion and hatched circles for the inhibitory end-zones. (Reprinted from von der Heydt and Peterhans, 1988, by permission.)

V1, where some 4% of all cells are grating cells. These are concentrated in the superficial layers.

The receptive field types in V2—the area adjacent to V1 that receives its afferent input from V1—are similar to those of V1 at first glance, but the receptive fields are larger. Again, there are unoriented and oriented fields and fields with and without end-stopping, but simple cells are rare. Sensitivity to disparity, movement, or wavelength is found in many cells. There is a close correspondence between the patterns of anatomical segregation of function in the two areas (Hubel and Livingstone, 1987; see also chapter 20). A remarkable new feature is that many neurons in V2 respond to figures with illusory contours as if these contours were defind by edges or lines (see the following section). These cells occur in the pale and the thick dark stripes but not in the thin dark stripes of the cytochrome-oxidase pattern (Peterhans and von der Heydt, 1993; also see chapter 20).

Cortical mechanisms of contour perception

Figure 23.8 shows the responses of a cell recorded in area V2 to a light bar and to an illusory contour stimulus presented at various orientations (the stimuli are depicted only for two orientations). It can be seen that the cell responded to the illusory contour stimulus and that the responses occurred at the same orientations for the illusory contour as for the bar. Because there was no other feature of this orientation in the stimulus, and a line grating without discontinuity produced no responses (bottom of figure 23.8B), we conclude that the cell signaled an illusory contour as well as a real contour! (Note also the absence of responses to the illusory contour stimulus when presented with the gratings at the preferred orientation. I will return to this observation later.)

Figure 23.9 shows typical results obtained in V2 with the two kinds of stimuli, plotted as orientation tuning curves. Responses to bars or edges are plotted with continuous lines, those to the contour created by abutting gratings with stippled lines. For the latter, the orientation axis refers to the orientation of the anomalous contour.[1] Each of the four cells showed a peak in the orientation tuning corresponding to the anomalous contour. More than 40% of the orientation-selective cells in V2 responded in this manner. The anomalous contour could be moving (A–C) or stationary (D). In general, the tunings obtained with light bar and anomalous contour were similar in shape and usually

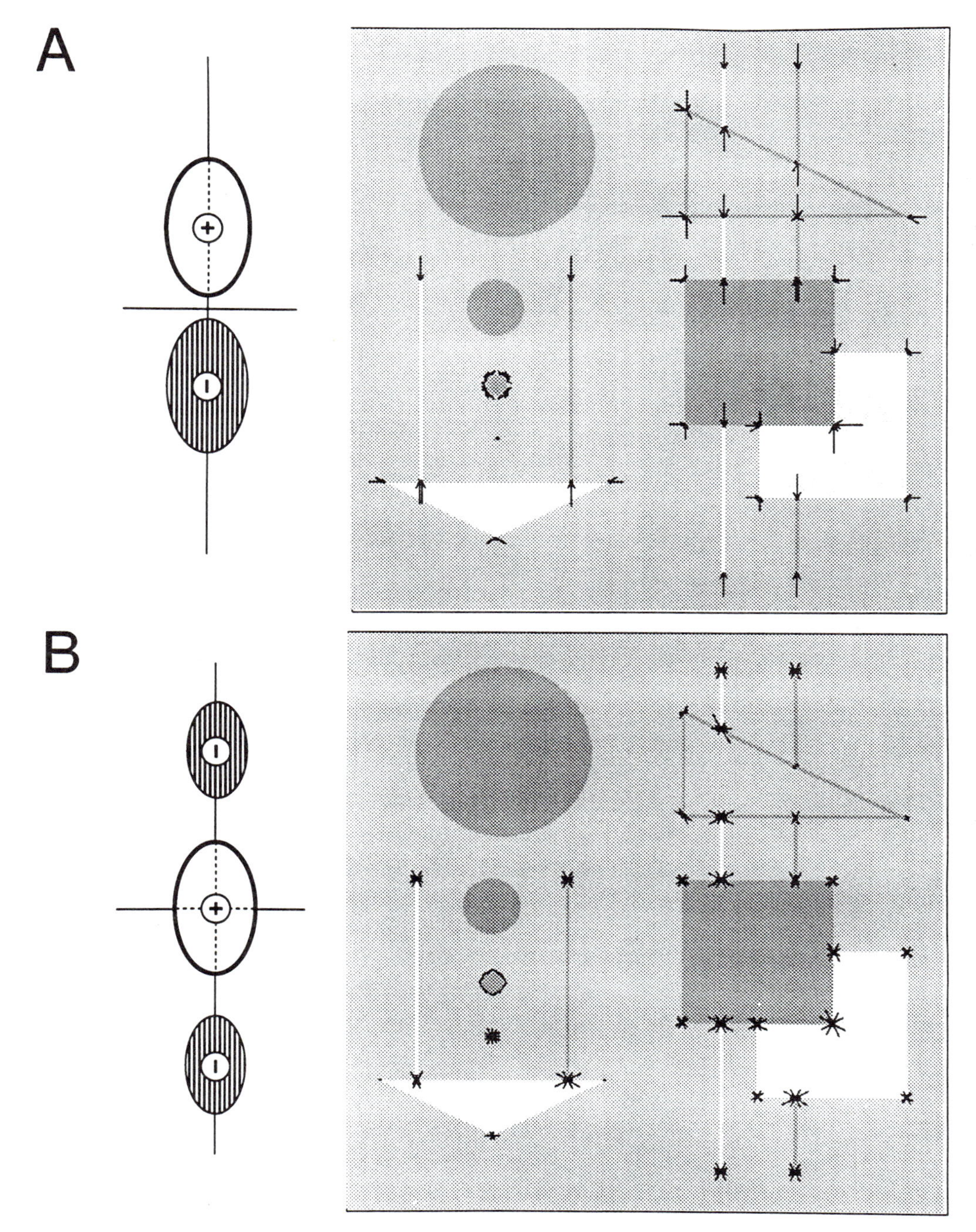

FIGURE 23.5 The representations of the test image obtained by operations that model end-stopped cells. (A) Asymmetrical (single-stopped) operator. (B) Symmetrical (double-stopped) operator. The responses are represented at the key points, which are defined by the local maxima of the combined activity of single- and double-stopped operators. The rays at the key points indicate the strength of signal in the various directions (30° angular spacing was used). The single-stopped operator indicates terminations. (Modified from Heitger et al., 1992.)

peaked at nearly the same orientation (figure 23.9A–C). However, a few cells clearly preferred slightly different orientations: Figure 23.9D shows the case with the largest difference. The strengths of light-bar and anomalous contour responses often were very different, and their ratio varied from cell to cell. The neuron of figure 23.9A responded better to the anomalous contour stimulus than to bars and edges of any size and

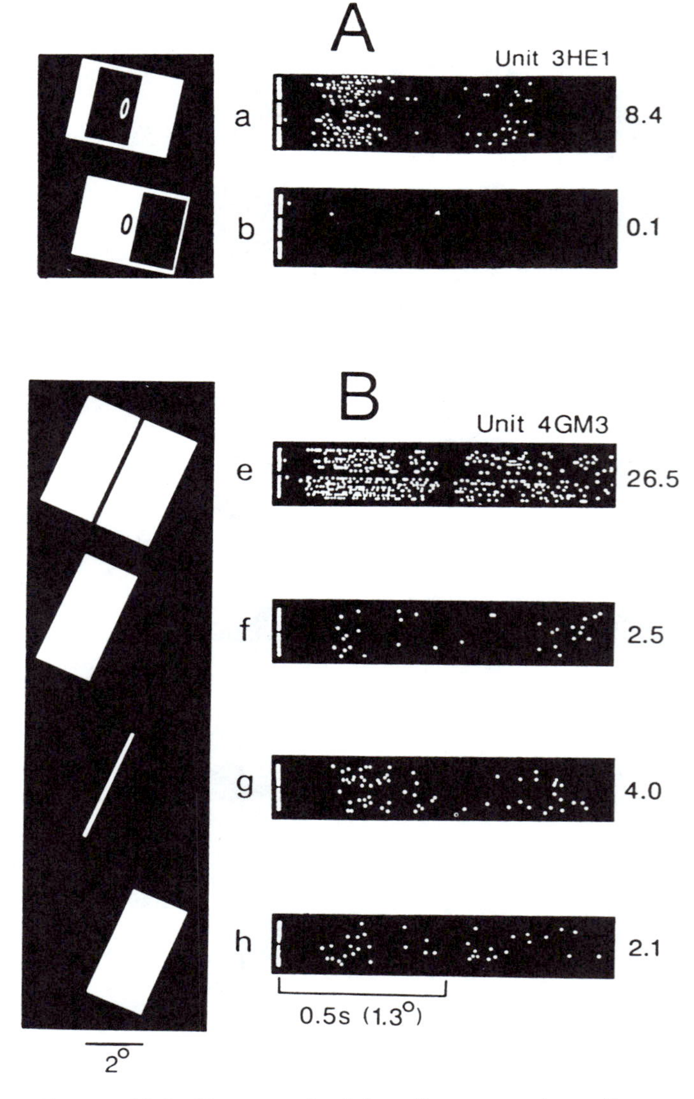

FIGURE 23.6 Pattern selectivity of two complex cells recorded in area V2. The cell in (A) responded best to edges and was selective for the edge polarity. The cell in (B) responded selectively to narrow dark bars. Each stimulus was presented 24 times (A) or 16 times (B). Mean numbers of spikes per presentation are shown on the right. (Reprinted from Peterhans and von der Heydt, 1989, by permission.)

contrast, whereas the neuron of figure 23.9B responded much better to a thin light bar. These variations are important as they indicate that responses and orientation selectivity for bars or edges and anomalous contours are generated by different mechanisms.

The cell of figure 23.9A shows a second peak corresponding to the orientation of the lines of the illusory contour stimulus, as one would expect. The cell of figure 23.9B also responded to the lines of the gratings (not illustrated). However, some cells failed to show this response (e.g., the cell of figure 23.8 and those of figure 23.9C and D). Half the cells that signaled the illusory contour displayed this behavior; they seemed to ignore the gratings. Poor responses or unresponsiveness to gratings in cells that responded well to isolated bars or edges has also been observed in V1 (Born and Tootell, 1991; von der Heydt, Peterhans, and Dürsteler, 1992). This feature seems complementary to the exclusive preference for periodic patterns in grating cells (von der Heydt, Peterhans, and Dürsteler, 1992), shown in figure 23.7. The mechanism might be side inhibition (De Valois, Thorell, and Albrecht, 1985; Born and Tootell, 1991), and its function is probably to prevent contour cells from responding to textures. The perceptual effect can be observed in figure 23.1B.

The responses of cells in V1, in contrast to V2, did not reflect the presence of the anomalous contour. Also, varying the line spacing and positioning failed to produce such responses. There was only a single exception in 60 cells tested. Simple, complex, and end-stopped cells were not different in this respect. Most cells responded to the stimulus according to the orientation of the gratings, but they seemed to be blind for the anomalous contour.

The perception of an illusory contour in a stimulus such as that of figure 23.8B depends on the number of lines that terminate at the contour. A contour is not perceived at the end of just one line; it appears perhaps with two or three lines and increases further in strength or distinctness when more lines are added. We found a similar increase in the strength of the neuronal responses to these stimuli. Figure 23.10 shows the relative strength of contour cell responses when the number of inducing lines was varied. The responses of 10 cells were normalized, each to its maximum bar or edge response, and averaged. One can see that this mean response increased with the number of lines and leveled off at approximately 9–12 lines.

In the case of the Kanizsa triangle (figure 23.1A) and similar figures in which illusory contours are induced by solid shapes, do cells of V2 also represent these contours? In such figures, the illusory contours appear as interpolations between pairs of edges. Therefore, one has to be cautious in relating neuronal responses to perception if one wants to exclude the trivial case in which a neuron is excited because those edges stimulate part of its response field. In fact, perceptual

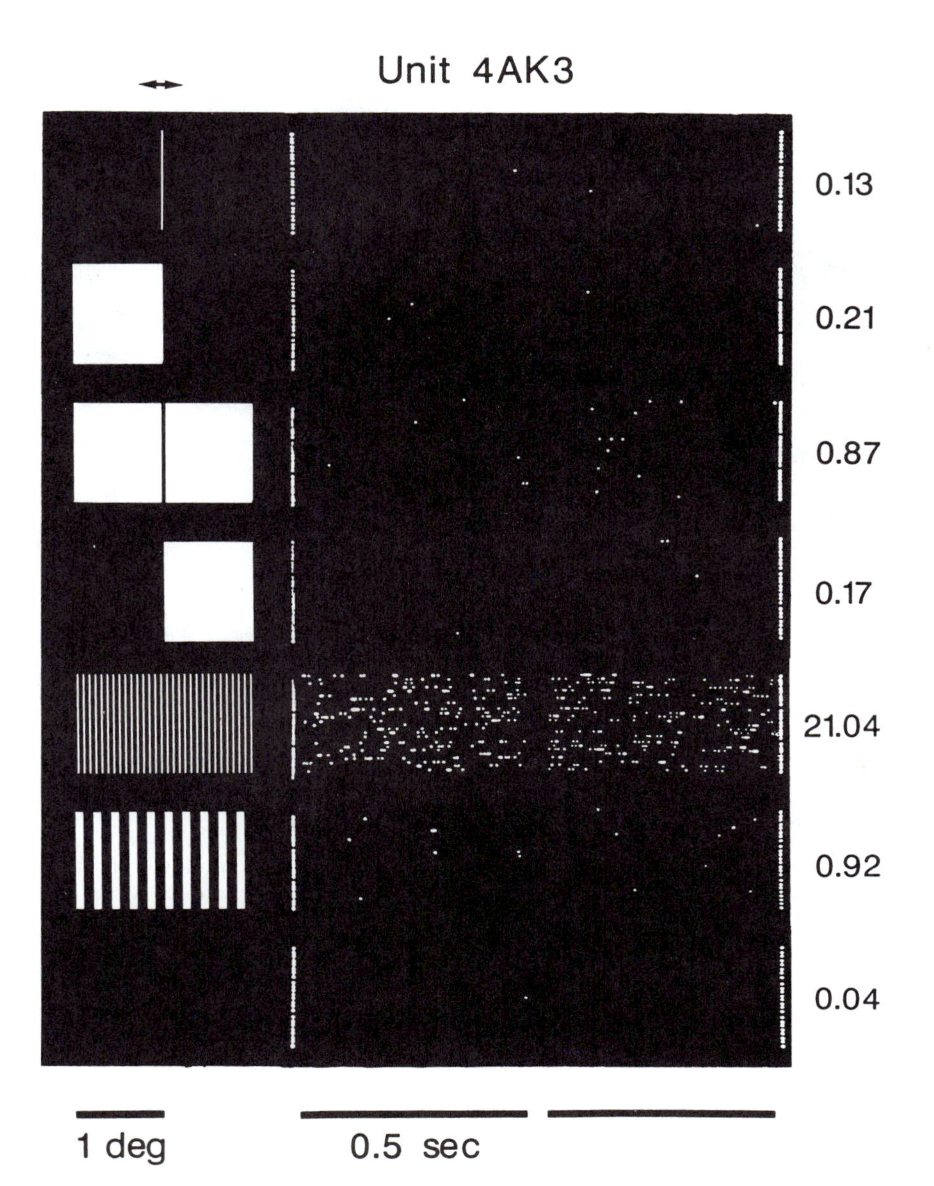

FIGURE 23.7 Pattern selectivity of a grating cell recorded in layers II–III of area V1. The cell responded selectively to gratings of 15 cs/deg ± 0.5 octaves. Conventions as for figure 23.6. The bottom panel shows spontaneous activity. (Reprinted from von der Heydt, Peterhans, and Dürsteler, 1992, by permission.)

observations argue against such an explanation. The illusory contours disappear when one of the "bridge heads" (defining end feature) is taken away or when the configuration is modified slightly—for example, by the addition of thin closing lines to the inducers (figure 23.11, inset). A filter response would only be halved in the first case and would remain practically unaffected in the latter, where the addition is negligible in terms of luminous flux.

Many cells of V2 did indeed respond as if they were signaling these illusory contours and showed exactly those effects that characterize perception. This is depicted in figure 23.11, which demonstrates the phenomenon of nonlinear summation (the whole is more than the sum of its parts) as well as the effect of the closing lines. (For perception, see the inset at the bottom of figure 23.11.) In the experiments, moving illusory figures were used.

As with the line-induced contours, a fraction of cells in V2 were found to signal illusory contours induced by solid shapes. Altogether, 55 of 150 cells tested in V2 (37%) signaled anomalous contours, compared to only

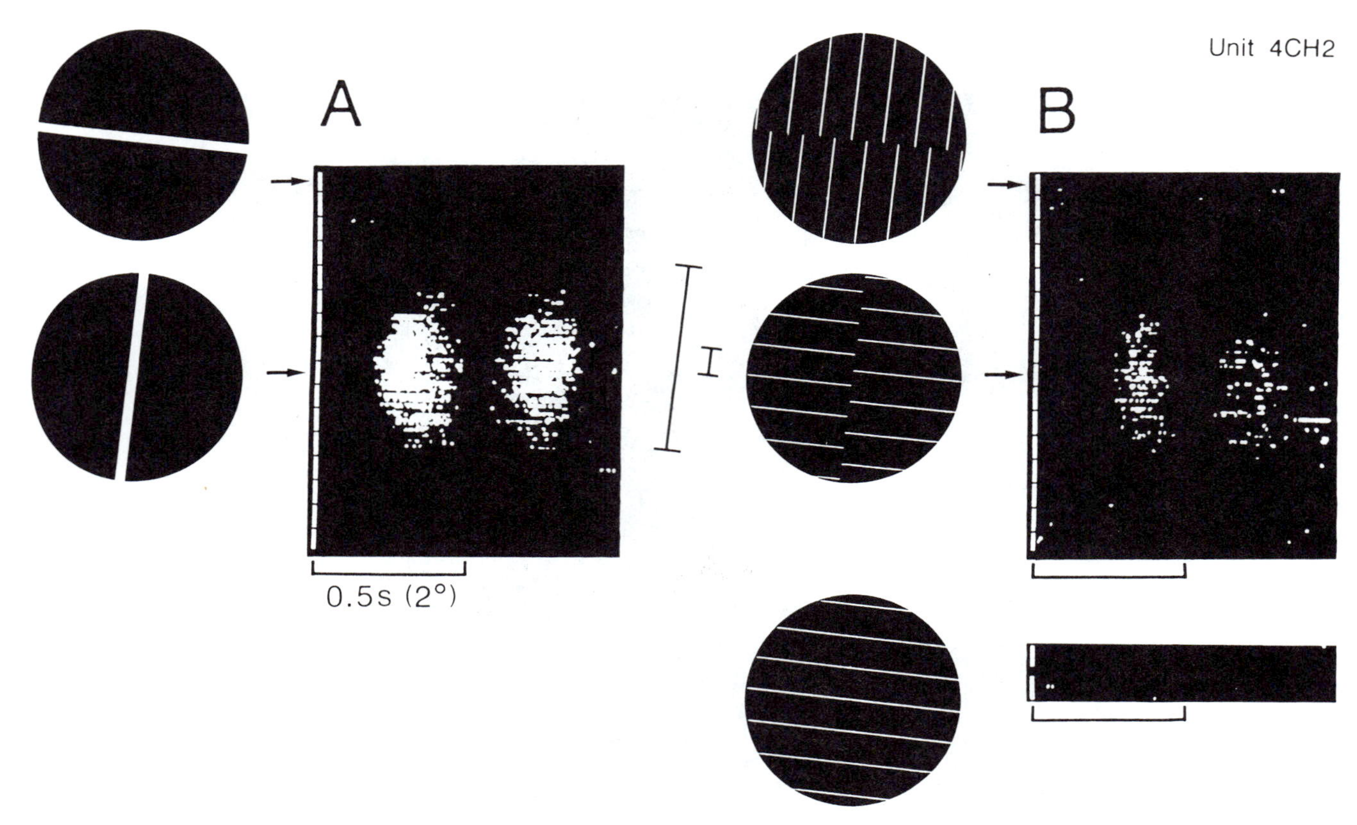

FIGURE 23.8 Responses of a neuron recorded in V2 to a light bar (A) and an illusory contour stimulus (B) presented at 16 orientations spanning 180°. (Bottom display) A grating without discontinuity produced no responses. Each orientation was presented eight times. The rows of dots represent the responses. The two clusters in the left and right halves of the display correspond to forward and backward movements. The illusory contour stimulus evoked responses at the same orientations at which the bar was effective. Neurons that show such a response are called *contour cells*. (Reprinted from von der Heydt and Peterhans, 1989a, by permission.)

1 of 77 cells tested in V1. Often, the same individual cell was found to respond to both kinds of stimuli. Of 15 cells that signaled the line-induced contour, 9 also-signaled the illusory bar contour, and of 23 cells that were unresponsive to the former, 22 were unresponsive also to the latter.

THE THEORETICAL LEVEL Physiological data per se are not a satisfactory explanation for a perceptual phenomenon. What might be the significance of illusory contours? V2 is a large area, and 37% of its orientation-selective cells is certainly an enormous number of cells. Why does the system put so much effort into representing these contours? We must consider that the visual cortex has been designed for interpreting images of 3-D scenes and not for analyzing pictures. In such images, information from near and far objects is jumbled. Some objects are partially occluded by others, and those in front appear embedded in a background. If the system is to recognize something, it must detect the *occluding contours* in the image (i.e., the lines of discontinuity of depth). If we look around, most objects appear to be bounded by clear contours but, in the array of intensity values that make up an image, the contours usually are not clear at all. This is demonstrated in figure 23.12A: It is not easy for us to see the problem by looking at these photographs because our visual system is so efficient. However, if one tries, by computer, to find the contours on the basis of the contrast borders in such an image, one realizes they are incomplete and faulty, whatever method of edge detection is used (Granlund, 1978; Marr and Hildreth, 1980; Canny, 1986; Morrone and Owens, 1987). Figure 23.12B shows the edge map generated by the complex cell operator introduced in figure 23.3. One can see that there are many gaps in these contours and that foreground and

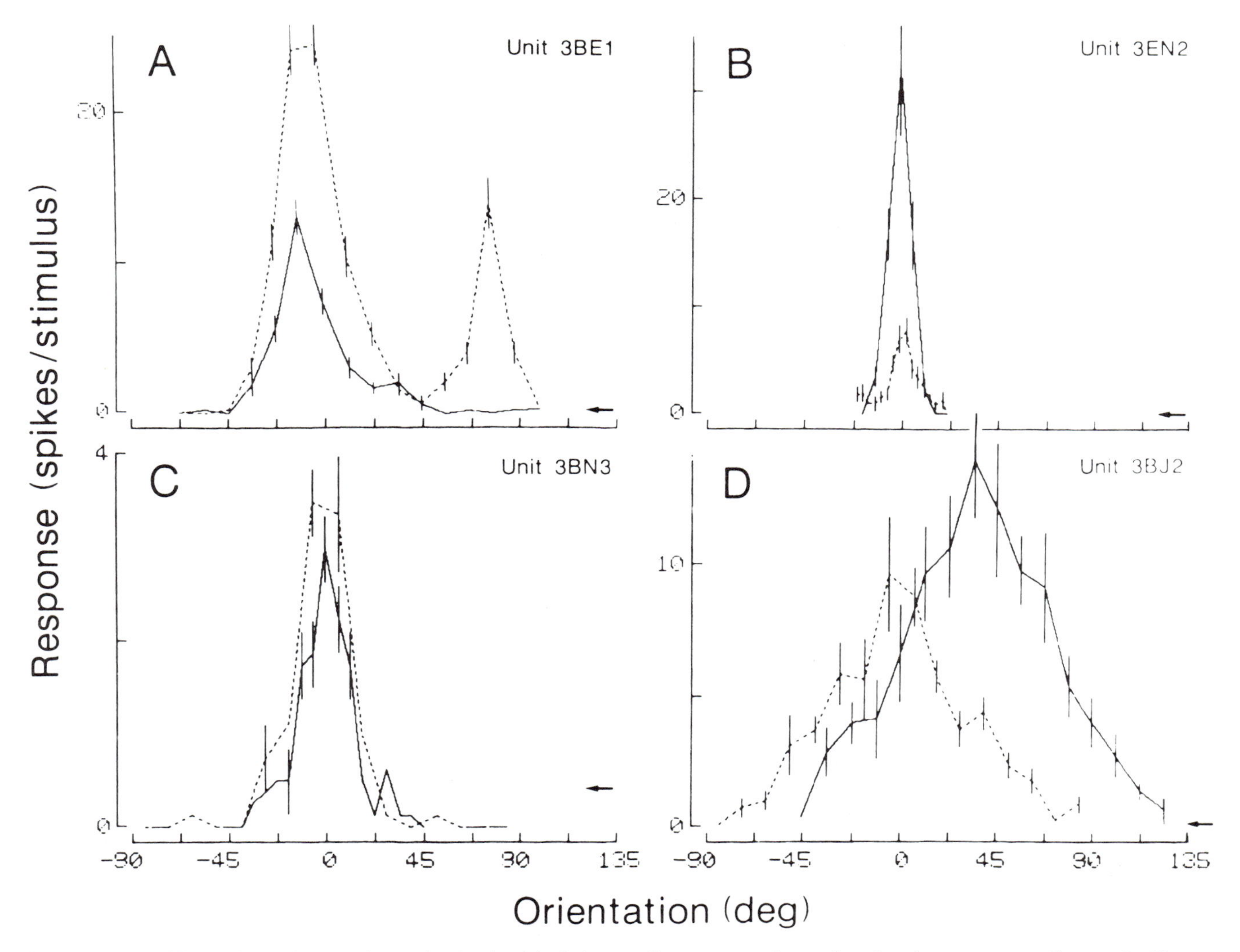

FIGURE 23.9 The orientation tunings obtained with light bar and illusory contour in four contour cells of V2. Continuous lines stand for bars, broken lines for illusory contours, horizontal arrows indicate the levels of spontaneous activity. One can see the peaks of tuning corresponding to the illusory contour. (Reprinted from von der Heydt and Peterhans, 1989a, by permission.)

background often merge, forming spurious corners. Obviously, it would be difficult to identify the leaves, wires, or stones from such a representation.

Theoretically, there are many different ways to detect occluding contours. Disparity difference, motion parallax, and other depth cues can be used. Even in static, monocular images, several strategies are available. Edge detection exploits only the luminance or chromatic difference at the contour, but in general there is also a discontinuity of pattern. Abrupt terminations of lines and edges are indicators of occlusion; if the background is rich in these structures, which is certainly the case for natural scenes, their terminations are significant events. Of course, a single terminating edge or line has little significance, but the occurrence of several terminations along a straight or curved line is evidence for occlusion. One can see that this strategy is particularly useful for complex scenes with many instances of occlusion—that is, just when edge detection is least reliable. Thus, edge detection and detection of pattern discontinuity complement each other.

None of these cues (i.e., computational strategies) can guarantee success all the time; only a combination of several cues may lead to reliable contour detection. Therefore, it is plausible to assume that the cortical contour mechanism employs modules, each for an ori-

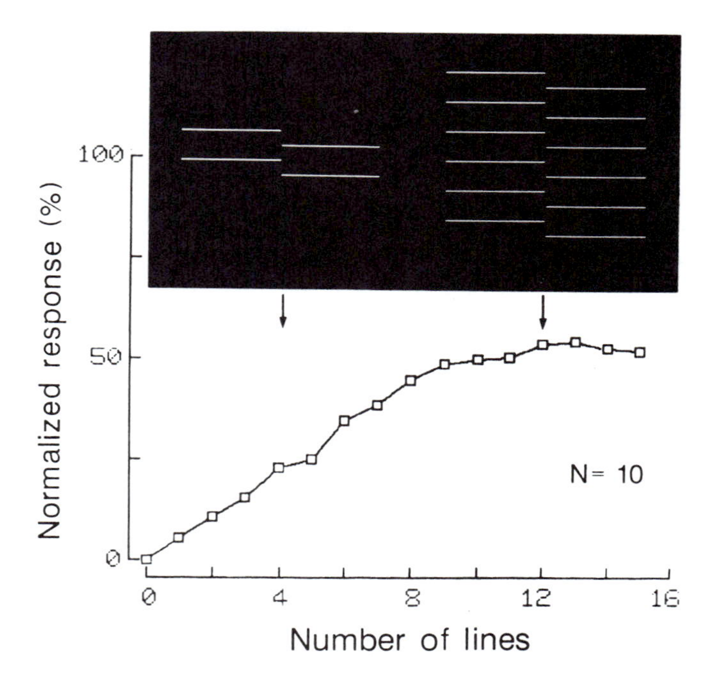

FIGURE 23.10 The mean relative strength of the illusory contour signals as a function of the number of lines in the stimulus. Average data from 10 contour cells. The responses reach a maximum at approximately 12 lines, corresponding to a length of nearly 2.4°. The stimuli of half-maximum and maximum responses are shown above the curve, to demonstrate the perceptual difference. (Reprinted from von der Heydt and Peterhans, 1989a, with permission.)

entation and position in the visual field, that combine a variety of elementary circuits, each of which is designed to exploit one particular cue. Specifically, the illusory contour responses recorded thus far in V2 seem to result from a combination of circuits for edge detection and for detection of pattern discontinuity.

A MODEL OF CONTOUR PERCEPTION The receptive fields of end-stopped cells are ideally suited for the detection of terminations of edges and lines, so we believe that the contour cells of V2 pool signals of end-stopped cells with receptive fields scattered along an axis given by the cell's preferred orientation (figure 23.13). By combining these signals (2) with the edge signal (1), the contour cell estimates the probability that an occluding contour of that orientation is present in its receptive field. We call the contribution from (2) the *grouping signal*. Mostly end-stopped fields orthogonal to this orientation should contribute, because occlusion is more likely to occur for background structures orthogonal rather than parallel to the occluding contour. A multiplicative connection between pairs of end-stopped cells straddling the receptive field center is assumed in order to explain the observation of nonlinear summation shown in figure 23.11 and the corresponding effect in perception. The gradual increase of response strength with the number of inducing lines in the stimulus of figure 23.10 results from summation of these signals at Σ, and the effect of the closing lines (figure 23.11) is simply a consequence of stimulating the inhibitory end-zones of the relevant end-stopped cells.

Also the brightness illusions and depth effects that are often associated with illusory figures find a natural explanation in this theory (von der Heydt and Peterhans, 1989b). Surprisingly, the model also predicts geometrical illusions of the Zöllner type. The corresponding orientation distortion could, in fact, be demonstrated in the contour cells of V2 (von der Heydt and Peterhans, 1989c).

We have developed a computational model according to the scheme of figure 23.13 (Heitger and von der Heydt, 1993). It is built on the simple, complex, and end-stopped cell models explained earlier. Figure 23.14 shows the representations produced at various levels of the model, for the Kanizsa triangle and for the four-armed Ehrenstein figure in which an illusory disc is perceived. The complex cell representation (see figure 23.14B) serves as the edge map and also provides the input for the end-stopped representation. For illustration, the orientation channels have been pooled. Figure 23.14C shows the signals of the asymmetrical (single-stopped) end-stopped cells. These are illustrated only at certain key points, defined by the local maxima of the pooled end-stopped cell activity. The rays indicate the strength of the signals in the various directions (a total of 12 were used). These signals are used for computing the grouping signal, which is shown in figure 23.14D, again with orientation channels pooled. This stage consists of convolution with orientation-selective kernels, corresponding to the scatter of end-stopped fields converging at Σ in figure 23.13, and a nonlinear pairing operation (multiplication). The size of the kernels (i.e., the extent of grouping.) was modeled according to the range of summation found in V2 contour cells (see figure 23.10). Details of the grouping operation are explained in the legend of figure 23.14. The grouping signal is then added, channel by channel, to the edge map B, and the result, which corresponds to the activity at the level of the contour cells in V2, is illustrated in figure 23.14E. Figure 23.14F shows

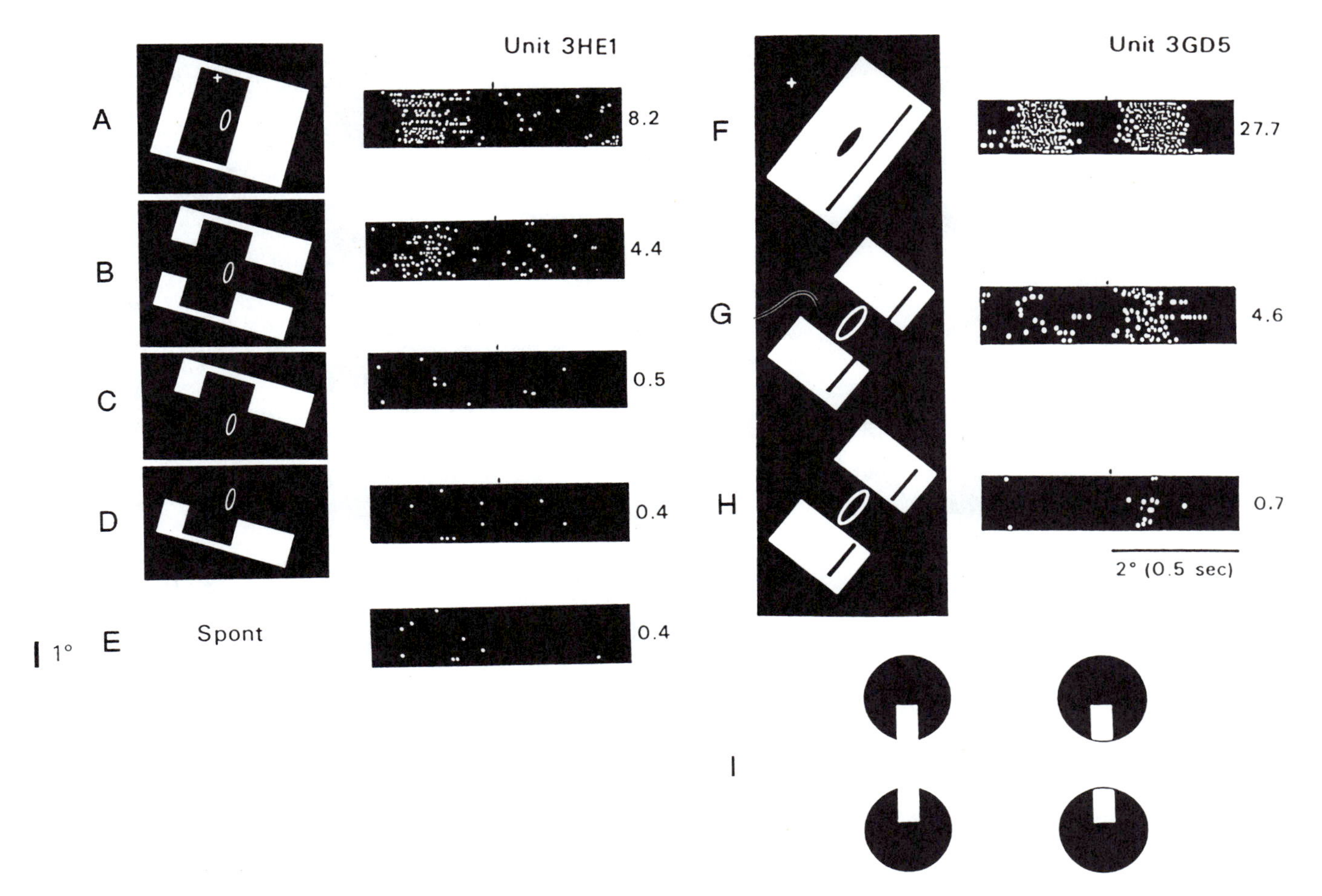

FIGURE 23.11 Responses of contour cells of V2 to illusory contour figures akin to the Kanizsa triangle. The occluding (black) figures where moved back and forth, whereas the occluded (white) rectangles were stationary. The ellipses represent the minimum response fields of the cells (i.e., the minimum region outside which a bar or edge stimulus did not evoke a response), which were determined beforehand. (A) and (F) show the responses to solid figures (rectangle or bar). The illusory contour figures produced responses (B, G), but either half of such a figure did not evoke a response (C, D); neither did a figure in which the perception of an illusory bar was abolished by adding small intersecting lines (H)—see inset below (H). (E) shows the spontaneous activity. (Modified from Peterhans and von der Heydt, 1989)

the local maxima of E, which can be compared with perception.

The Kanizsa triangle and the Ehrenstein figure are crucial tests for a model. The Kanizsa triangle seems to require interpolation between given pieces of lines. However, interpolation would not produce the circle in the Ehrenstein figure but rather would complete the lines to a cross. The model generated illusory contours in many other figures that agreed well with perception. The shapes were nearly invariant against scaling over two to three octaves.

Figure 23.12C shows the result obtained using the photographs of figure 23.12A. The circles mark three regions where improvements in contour representation are obvious. One can see that many gaps in the edge map (B) are closed here, thereby converting spurious corners into T-junctions. Therefore, this contour representation should be much easier to read by subsequent processing stages. There are apparent ambiguities in this representation; the model tends to produce "transparent" (foreground and background) contours. Heitger and von der Heydt (1993) point out that these ambiguities might be resolved by combining this contour map with the map of key points, thereby linking pieces of smooth contours with those key points that turned out to be object features (true corners). This seems to call for another stage of form description in which local elements are linked to figures. Unfortu-

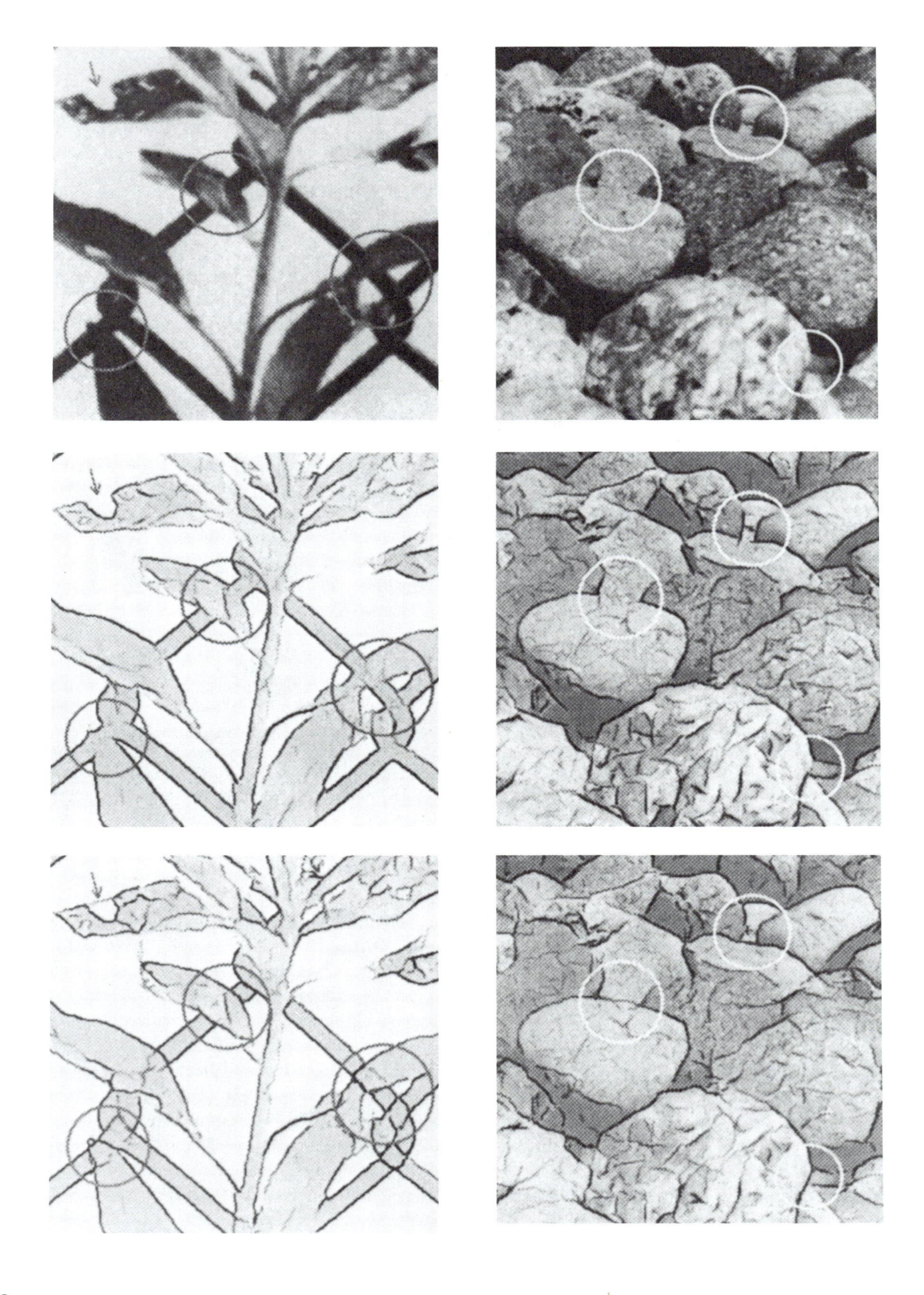

FIGURE 23.12 Demonstration of the general problem of contour detection. (Top) Photographs of natural scenes. Often, the contours are not defined by contrast borders (circles). (Middle) Map of the contrast borders extracted by the "complex cell" model of figure 23.3. The output of the orientation channels of the model have been pooled for illustration, and overlaid on a low-contrast copy of the original. Shades of gray of contour lines represent strength of contour signal. Note the missing contour segments and false connections between foreground and background (circles). (Bottom) Map of the contours generated by the complete model as explained in figures 23.13 and 23.14. The illusory contour mechanism fills in missing segments and resolves false connections. (Reprinted from Heitger and von der Heydt, 1993, by permission)

nately, we have only foggy ideas about this stage in the human visual system, although recent neurophysiological experiments in the monkey have shed some light on the subject (Tanaka et al., 1991).

Cognitive and low-level theories of visual perception

Thus far, I have presented a picture of a form perception system in which the processing is entirely bottom-up. Indeed, the main thrust of the neurophysiological experiments on illusory contour responses and of the modeling is that a relatively complex perceptual phenomenon, which seems to involve intelligent processes, can be explained by low-level visual mechanisms. This is in contrast to the view espoused by Gregory (1972) and Rock and Anson (1979), who suggested that much of perception is to be considered as a high-level cognitive inference: The system tries to infer the real-world situation that would most likely account for the given sensory data (perceptual hypothesis). In the case of the illusory triangle (see figure 23.1A), the system might consider the possibilities that three sectored black discs and three angles happen to be arranged in that particular way or, alternatively, that there is a white triangle overlying three complete discs and an outline triangle, accepting the latter as the more likely explanation.

In contrast, many perceptual findings argue for a low-level explanation—for example, the tilt aftereffects produced by illusory and real contour stimuli (Paradiso, Shimojo, and Nakayama, 1989), the preattentive discrimination of illusory contour figures (Gurnsey, Humphrey, and Kapitan, 1992), and the summation effects (Soriano, Spillmann, and Bach, 1993). These studies showed close parallels to the neurophysiological

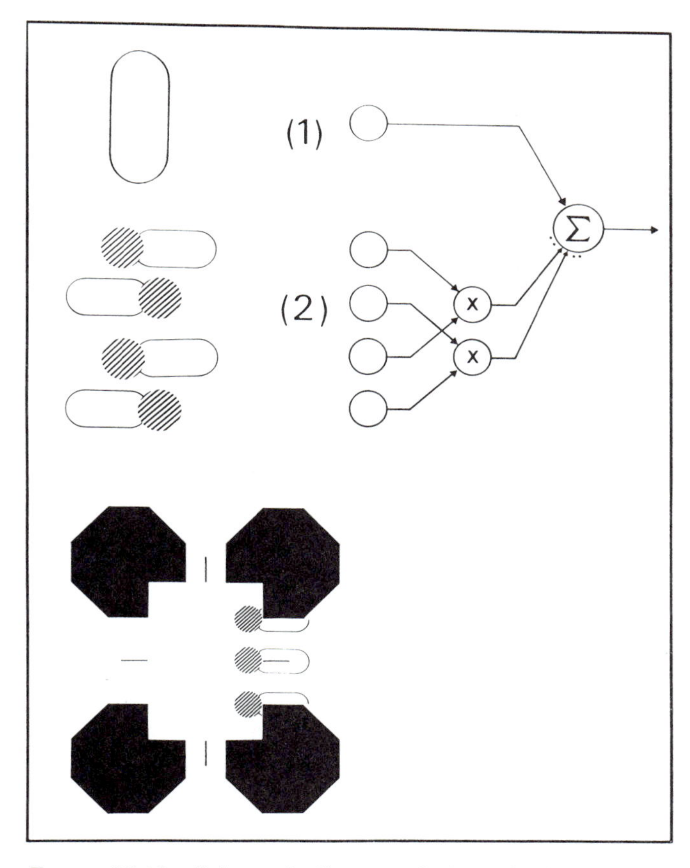

FIGURE 23.13 Schematic diagram of a hypothetical contour mechanism. The contour cell ($\sum$) is assumed to sum the signals of two parallel paths, (1) an edge detection path, with input from simple or complex cells, and (2) a grouping path from orthogonal end-stopped cells. It is assumed that the end-stopped receptive fields cover the receptive field of the contour cell densely, and that multipliers connect any pair of end-stopped cells in the upper and lower halves, since the experiments showed no obvious effects of position or spacing of the lines. (Reprinted from von der Heydt and Peterhans, 1989c, by permission)

results described previously. As Paradiso, Shimojo, and Nakayama (1989) point out, the term *inference* need not be restricted to the highest levels of analysis. Inferences could be made also at a relatively low level, by a set of autonomous simple computations with limited information (see also Nakayama and Shimojo, 1992). The cooperative feedback network of Grossberg and Mingolla (1985a,b) is an example of such a model of contour perception. The contour model presented in this chapter is even simpler because it does not involve feedback loops and generates the contours in a one-shot process. The probability reasoning implied by the cognitive theory can be part of the low-level process—

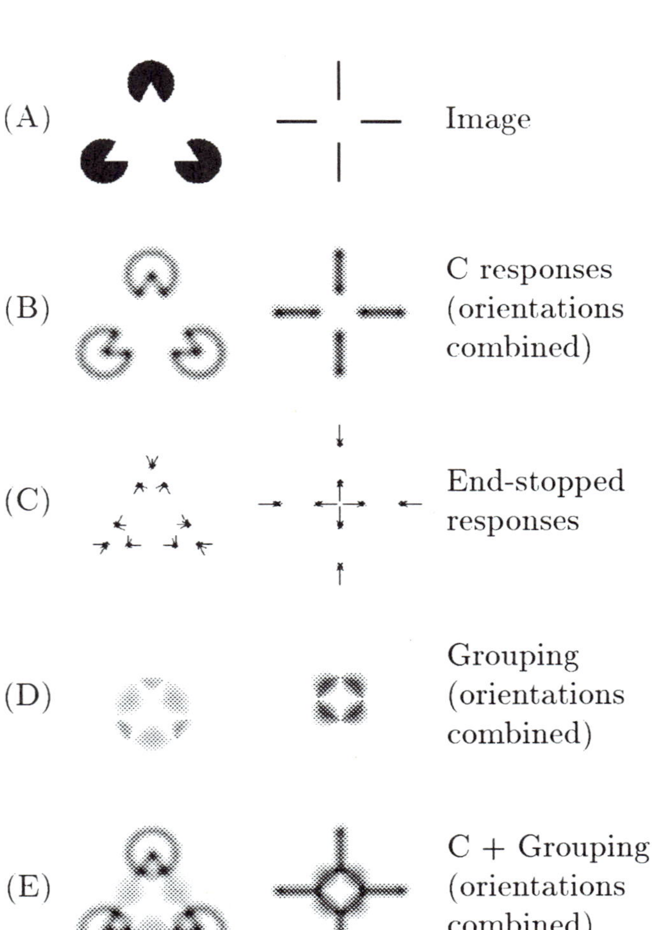

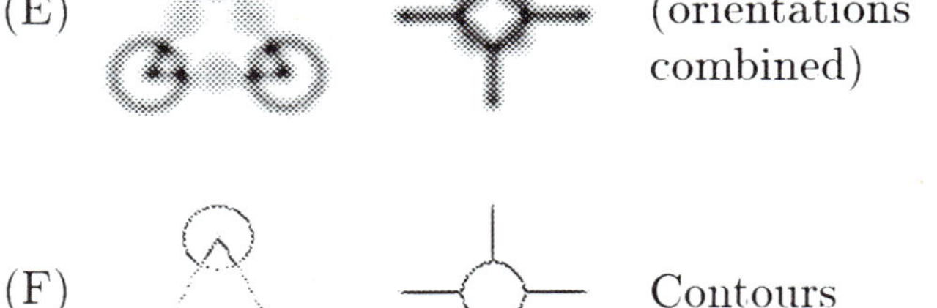

FIGURE 23.14 Illustration of the image representations produced by a computer implementation of the model. From the original images (A), complex and end-stopped cell representations (B, C) are derived, as explained in figures 23.3 and 23.5. The summation of end-stopped responses was implemented by convolution with an elongated grouping field. For the computation, the sum over the products of pairs of cells of figure 23.13 has been replaced by the product of the sums over the two half fields. To improve the flexibility of the contours, straight opposite as well as angled (180° ± 60°) half-fields were used, and the results were summed. The grouping operation also involved a gradual differentiation of key points according to their type (corner vs. line end) and selectivity for direction of termination. The size of the grouping field was fixed and four times larger than the simple and complex fields. The result of grouping is shown in (D). (E) The sum of (D) and (B). (F) Local maxima of (E). The orientation channels are pooled for illustration. Shades of gray represent strength of contour signal (B, D, E, F). (Modified from Heitger and von der Heydt, 1993)

that is, the knowledge about image statistics may be "wired in."

As a result, cognitive and low-level explanations may become hard to distinguish. Nonetheless, I believe it is important to retain the distinction on at least the conceptual level, because the representations of image information and the algorithms implied are very different. Discussion about the definition of *depth cues* may illustrate this. Coren (1972) argued that illusory contours result from the presence of depth cues in the stimulus. Most investigators today probably agree with his point that seeking for a 3-D interpretation is the pivot. In the discussion, he suggests that "a 'depth cue' be defined as some aspect of a configuration which can be read as consistent with a given spatial arrangement of objects at different relative distances" (Coren, 1972). In the cognitive theory, this means that the system uses internal representations of objects and their possible layout in space and checks whether the given image is consistent with one or the other arrangement. In the low-level theory, the depth cues are image features that excite end-stopped cells in such a way that the contour mechanism is activated. In either case, the term *cue* represents an algorithm, and the algorithms are obviously very different.

NOTE

1. Because *illusory contour* refers to a perception, and thus to a specific observer, I use the term *anomalous* for contours that are defined by the configuration but not given by a contrast border. In the case of the abutting gratings, for example, the anomalous contour is defined as the line connecting the endpoints of lines; however, an illusory contour may or may not be perceived, or could be perceived at different orientation.

REFERENCES

ADELSON, E. H., and J. R. BERGEN, 1985. Spatiotemporal energy models for the perception of motion. *J. Opt. Soc. Am. [A]* 2:284–299.

BAUMGARTNER, G., R. VON DER HEYDT, and E. PETERHANS,

1984. Anomalous contours: A tool in studying the neurophysiology of vision. *Exp. Brain Res.* Suppl. 9:413–419.

BORN, R. T., and R. B. H. TOOTELL, 1991. Single-unit and 2-deoxyglucose studies of side inhibition in macaque striate cortex. *Proc. Natl. Acad. Sci. U.S.A.* 88:7071–7075.

CANNY, J., 1986. A computational approach to edge detection. *IEEE Trans. PAMI* 8:679–698.

COREN, S., 1972. Subjective contours and apparent depth. *Psychol. Rev.* 79:359–367.

DAUGMAN, J. G., 1984. Spatial visual channels in the Fourier plane. *Vision Res.* 24:891–910.

DE VALOIS, R. L., D. G. ALBRECHT, and L. G. THORELL, 1982. Spatial frequency selectivity of cells in macaque visual cortex. *Vision Res.* 22:545–559.

DE VALOIS, R. L., and K. K., DE VALOIS, 1988. *Spatial Vision.* New York: Oxford University Press.

DE VALOIS, R. L., L. G. THORELL, and D. G. ALBRECHT, 1985. Periodicity of striate-cortex-cell receptive fields. *J. Opt. Soc. Am. [A]* 2:1115–1123.

DE VALOIS, R. L., E. W. YUND, and N. HEPLER, 1982. The orientation and direction selectivity of cells in macaque visual cortex. *Vision Res.* 22:531–544.

FELLEMAN, D. J., and D. C. VAN ESSEN, 1991. Distributed hierarchical processing in the primate cerebral cortex. *Cerebral Cortex* 1:1–47.

FISCHLER, M. A., and O. FIRSCHEIN, 1987. *Readings in Computer Vision: Issues, Problems, Principles, and Paradigms.* Los Altos, Calif.: Morgan Kaufmann.

GALLI, A., and A. ZAMA, 1931. Beiträge zur Theorie der Wahrnehmung. *Z. Psychol.* 123:308–348.

GRANLUND, G. H., 1978. In search of a general picture processing operator. *Comput. Vis. Graphics Image Proc.* 8:155–178.

GREGORY, R. L., 1972. Cognitive contours. *Nature* 238:51–52.

GROSSBERG, S., and E. MINGOLLA, 1985a. Neural dynamics of perceptual grouping: textures, boundaries, and emergent segmentations. *Percept. Psychophys.* 38:141–171.

GROSSBERG, S., and E. MINGOLLA, 1985b. Neural dynamics of form perception: Boundary completion, illusory figures, and neon color spreading. *Psychol. Rev.* 92:173–211.

GURNSEY, R., G. K. HUMPHREY, and P. KAPITAN, 1992. Parallel discrimination of subjective contours defined by offset gratings. *Percept. Psychophys.* 52:263–276.

HEGGELUND, P., 1981. Receptive field organisation of simple cells in cat striate cortex. *Exp. Brain Res.* 42:89–98.

HEGGELUND, P., 1986a. Quantitative studies of the discharge fields of single cells in cat striate cortex. *J. Physiol. (Lond.)* 373:277–292.

HEGGELUND, P., 1986b. Quantitative studies of enhancement and suppression zones in the receptive field of simple cells in cat striate cortex. *J. Physiol. (Lond.)* 373:293–310.

HEITGER, F., L. ROSENTHALER, R. VON DER HEYDT, E. PETERHANS, and O. KÜBLER, 1992. Simulation of neuronal contour mechanisms: From simple to endstopped cells. *Vision Res.* 32:963–981.

HEITGER, F., and R. VON DER HEYDT, 1993. A computational model of neural contour processing: Figure-ground segregation and illusory contours. In *Proceedings of the Fourth International Conference on Computer Vision.* Washington D.C.: IEEE Computer Society Press, pp. 32–40.

HUBEL, D. H., and M. S. LIVINGSTONE, 1987. Segregation of form, color, and stereopsis in primate area 18. *J. Neurosci.* 7:3378–3415.

HUBEL, D. H., and T. N. WIESEL, 1962. Receptive fields, binocular interaction and functional architecture in the cat's visual cortex. *J. Physiol. (Lond.)* 160:106–154.

HUBEL, D. H., and T. N. WIESEL, 1968. Receptive fields and functional architecture of monkey striate cortex. *J. Physiol. (Lond.)* 195:215–243.

HUBEL, D. H., and T. N. WIESEL, 1974. Sequence regularity and geometry of orientation columns in the monkey striate cortex. *J. Comp. Neurol.* 158:267–294.

HUBEL, D. H., and T. N. WIESEL, 1977. Ferrier lecture: Functional architecture of macaque monkey visual cortex. *Proc. R. Soc. Lond. [Biol.]* 198:1–59.

KANIZSA, G., 1979. *Organization in Vision. Essays on Gestalt Perception.* New York: Praeger.

LIVINGSTONE, M. S., and D. H. HUBEL, 1987. Psychophysical evidence for separate channels for the perception of form, color, movement, and depth. *J. Neurosci.* 7:3416–3468.

MAFFEI, L., and A. FIORENTINI, 1973. The visual cortex as a spatial frequency analyser. *Vision Res.* 13:1255–1267.

MAFFEI, L., C. MORRONE, M. PIRCHIO, and G. SANDINI, 1979. Responses of visual cortical cells to periodic and non-periodic stimuli. *J. Physiol. (Lond.)* 296:27–47.

MARČELJA, S., 1980. Mathematical description of the responses of simple cortical cells. *J. Opt. Soc. Am.* 70:1297–1300.

MARR, D., 1982. *Vision. A Computational Investigation into the Human Representation and Processing of Visual Information.* San Francisco: Freeman.

MARR, D., and E. HILDRETH, 1980. Theory of edge detection. *Proc. R. Soc. Lond. [Biol.]* 207:187–217.

MORRONE, M. C., and D. C. BURR, 1988. Feature detection in human vision: A phase-dependent energy model. *Proc. R. Soc. Lond. [Biol.]* 235:221–245.

MORRONE, C., and R. OWENS, 1987. Feature detection from local energy. *Pattern Recogn. Lett.* 6:303–313.

MOVSHON, J. A., I. D. THOMPSON, and D. J. TOLHURST, 1978a. Spatial summation in the receptive fields of simple cells in the cat's striate cortex. *J. Physiol. (Lond.)* 283:53–77.

MOVSHON, J. A., I. D. THOMPSON, and D. J. TOLHURST, 1978b. Receptive field organization of complex cells in the cat's striate cortex. *J. Physiol. (Lond.)* 283:79–99.

NAKAYAMA, K., and S. SHIMOJO, 1992. Experiencing and perceiving visual surfaces. *Science* 257:1357–1363.

PARADISO, M. A., S. SHIMOJO, and K. NAKAYAMA, 1989. Subjective contours, tilt aftereffects, and visual cortical organization. *Vision Res.* 29:1205–1213.

PARKER, A. J., and M. J. HAWKEN, 1988. Two-dimensional spatial structure of receptive fields in monkey striate cortex. *J. Opt. Soc. Am. [A]* 5:598–605.

Peterhans, E., and R. von der Heydt, 1989. Mechanisms of contour perception in monkey visual cortex: II. Contours bridging gaps. *J. Neurosci.* 9:1749–1763.

Peterhans, E., and R. von der Heydt, 1990. Neurons with end-stopped receptive fields detect occlusion cues. *Soc. Neurosci. Abstr.* 16(1):293.

Peterhans, E., and R. von der Heydt, 1993. Functional organization of area V2 in the alert macaque. *Eur. J. Neurosci.* 5:509–524.

Pollen, D. A., and S. F. Ronner, 1982. Spatial computation performed by simple and complex cells in the visual cortex of the cat. *Vision Res.* 22:101–118.

Rock, I., and R. Anson, 1979. Illusory contours as the solution to a problem. *Perception* 8:665–681.

Schiller, P. H., B. L. Finlay, and S. F. Volman, 1976. Quantitative studies of single-cell properties in monkey striate cortex: III. Spatial frequency. *J. Neurophysiol.* 39: 1334–1351.

Soriano M., L. Spillmann, and M. Bach, 1993. The phase-shifted grating illusion: A comparison between magnitude estimates and neuronal responses. Submitted for publication.

Tanaka, K., H. Saito, Y. Fukada, and M. Moriya, 1991. Coding visual images of objects in the inferotemporal cortex of the macaque monkey. *J. Neurophysiol.* 66:170–189.

von der Heydt, R., 1987. Approaches to visual cortical function. *Rev. Physiol. Biochem. Pharmacol.* 108:69–150.

von der Heydt, R., and E. Peterhans, 1988. Contour processing in primate visual cortex. In *Mustererkennung 1988*, H. Bunke, O. Kübler, and P. Stucki, eds. Berlin: Springer, pp. 111–127.

von der Heydt, R., and E. Peterhans, 1989a. Mechanisms of contour perception in monkey visual cortex: I. Lines of pattern discontinuity. *J. Neurosci.* 9:1731–1748.

von der Heydt, R., and E. Peterhans, 1989b. Ehrenstein and Zöllner illusions in a neuronal theory of contour processing. In *Seeing Contour and Colour*, J. J. Kulikowski, C. M. Dickinson, and I. J. Murray, eds. Oxford: Pergamon, pp. 729–734.

von der Heydt, R., and E. Peterhans, 1989c. Cortical contour mechanisms and geometrical illusions. In *Neural Mechanisms of Visual Perception*, D. M. K. Lam and C. D. Gilbert, eds. The Woodlands, Tex.: Portfolio Publishing, pp. 157–170.

von der Heydt, R., E. Peterhans, and M. R. Dürsteler, 1992. Periodic-pattern-selective cells in monkey visual cortex. *J. Neurosci.* 12:1416–1434.

Wilson, H. R., D. Levi, L. Maffei, J. Rovamo, and R. L. De Valois, 1990. The perception of form: Retina to striate cortex. In *Visual Perception: The Neurophysiological Foundations*, L. Spillmann and J. S. Werner, eds. New York: Academic Press, pp. 231–272.

24 Concurrent Processing in the Primate Visual Cortex

DAVID C. VAN ESSEN AND EDGAR A. DEYOE

ABSTRACT Visual processing in primates involves several major subcortical centers plus a mosaic of dozens of distinct areas in the cerebral cortex. Anatomical and physiological evidence indicates that these structures are arranged in a hierarchy that includes at least 10 stages of cortical processing as well as several subcortical stages. At each stage of the hierarchy there is evidence for concurrent processing streams, including the magnocellular, parvocellular, and koniocellular streams at subcortical levels and the magno-dominated, blob-dominated, and interblob-dominated streams at cortical levels. Substantial cross-talk occurs between streams at many stages of the hierarchy. This convergence and divergence may provide the flexibility needed for the visual system to carry out a wide range of tasks using information derived from a variety of different low-level cues.

Overview

The images impinging on our retinas provide a steady barrage of information in the domains of space, time, and spectral composition, which we use for a wide variety of visually mediated tasks. Some tasks are related to conscious perception, and these can be further subdivided into tasks related to object identification (what it is) and tasks related to the location of different objects (where it is). Other tasks are related to visuomotor function (e.g., making saccades, tracking a moving object, and navigating through the environment). To carry out these diverse functions, the visual system has evolved a highly complex organization that includes dozens of distinct visual areas in the cerebral cortex plus a large number of subcortical visual centers. In addition, some of these visual centers are themselves divisible into repeating domains (compartments) that differ in their connectivity and physiological characteristics. These and related findings have led naturally to the proposition that the visual system contains distinct processing streams to carry out different aspects of visual function more or less in parallel (Stone, Dreher, and Leventhal, 1979; DeYoe and Van Essen, 1988; Livingstone and Hubel, 1988; Zeki and Shipp, 1988; Desimone and Ungerleider, 1989; Merigan and Maunsell, 1993). Although there is widespread support for this hypothesis in broad terms, it has proved difficult to decipher what exactly constitutes a visual processing stream and to ascertain the key functional differences between streams. In this chapter, we review the progress that has been made on this journey, illustrate why it has been more arduous than some may have expected, and indicate some of the unresolved issues for future investigation.

CONCURRENT PROCESSING The term *concurrent processing* can be applied broadly to situations in which information is processed simultaneously along several distinct routes, or streams. We will use this term, rather than the commonly used alternative *parallel processing*, because it more readily encompasses the spectrum of possible interactions that can occur between streams. At one end of the spectrum are situations in which each stream is totally separate and independent of the other (figure 24.1A). A second class includes situations in which the different streams remain distinct and parallel to one another but with interactions between streams occurring at one or more levels (figure 24.1B). A third class includes situations in which the convergence and divergence between levels leads to complete merging or splitting of streams (figure 24.1C).

A useful analogy in discussing these issues is the organization of an industrial factory. A typical factory takes many types of raw materials for its inputs and produces a range of outputs—not just a single item—as its product line. The intervening production steps often involve combining intermediate components in

DAVID C. VAN ESSEN Department of Anatomy and Neurobiology, Washington University School of Medicine, St. Louis, Mo.
EDGAR A. DEYOE Department of Cell Biology and Anatomy, Medical College of Wisconsin, Milwaukee, Wis.

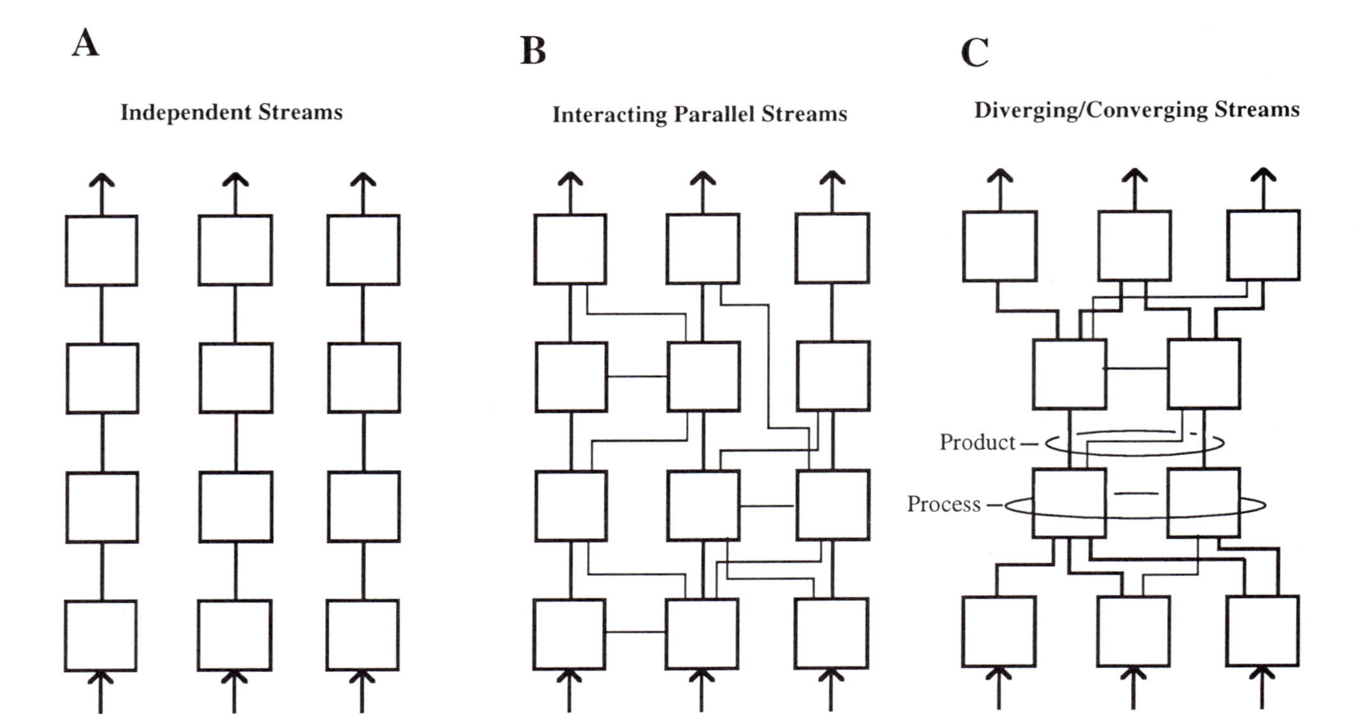

FIGURE 24.1 Three conceptually distinct types of concurrent processing, each of which involves a set of processing centers (boxes) interconnected by lines that represent their outputs (products). (A) The simplest arrangement is one in which each processing stream is completely independent of the others. (B) An alternative pattern involves streams that have significant cross talk with one another but nonetheless are dominated by "main-line" pathways and thus remain parallel to one another. (C) A third pattern shows a greater degree of convergence and divergence, such that some processing centers are not dominated by any single input. Note that these distinctions are based on an analysis of the ascending flow of signals (products), but each scheme is compatible with descending signals at some or all levels.

any of several possible ways. In this respect, the flow of intermediate products involves the type of merging and splitting of processing streams illustrated in figure 24.1C. The main disadvantage of this strategy is the need for complex communication links between modules. In exchange, there are two obvious advantages: It provides for efficient compartmentalization of function, insofar as each step of the manufacturing process can be assigned to a module well suited for that particular task; and it is inherently flexible, because each module can access whatever inputs it needs and can distribute its outputs as necessary to achieve the desired set of final products. A major thrust of this chapter will be to argue that similar principles apply to the overall design of the primate visual system.

The nature of concurrent processing will be discussed at three distinct levels in this chapter. First, we will discuss briefly at a conceptual level how certain types of information need to be represented and transformed to mediate specific aspects of visual perception. Next, we will explore in some detail the underlying anatomical infrastructure, as revealed by analyzing the pathways that interlink different visual areas and compartments within areas. Finally, we will consider neurophysiological and behavioral findings that suggest how the types of abstract information discussed in the first section might be mapped onto the physical structures outlined in the second section. Only by pursuing the analysis at all three levels is there hope of deciphering how the visual system actually works.

FROM CUES TO PERCEPTIONS Our perceptions of objects in the world are derived from various low-level cues, including orientation, velocity, spectral composition, and binocular disparity. These cues represent information that can be extracted from retinal images at relatively early stages of processing. Each of these cues makes an obvious contribution to the specific aspects of perception indicated in figure 24.2: Velocity cues contribute to motion perception, disparity cues to depth

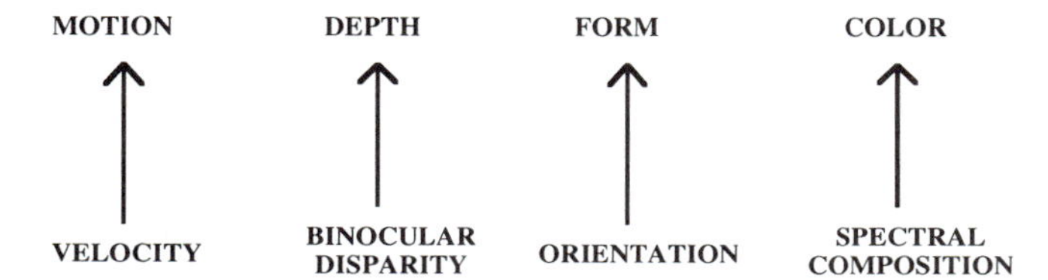

FIGURE 24.2 A simple one-to-one mapping between low-level cues that can be readily extracted from visual images (bottom row) and specific attributes of visual perception (top row).

perception, orientation cues to form perception, and spectral cues to color perception. If this 1 : 1 mapping were to constitute the sole route for information flow in the visual system, it would suggest that processing occurs independently in each stream. This, in turn, would motivate anatomists and physiologists to search for functional streams that are each concerned with the processing of a single cue and lead to a single aspect of perception. For example, one might hope to find a color-processing stream completely dedicated to extracting spectral information and using this information exclusively to generate percepts of color.

An alternative perspective emphasizes that each of the low-level cues just listed contributes to multiple aspects of perception (cf. DeYoe and Van Essen, 1988; Stoner and Albright, 1993). This mapping from single cues to multiple attributes is illustrated in figure 24.3. Velocity cues (figure 24.3A) contribute not only to motion perception but also to form perception (using structure from motion processing) and to depth perception (using motion parallax processing). Likewise, disparity cues contribute to form and motion perception as well as to depth perception (figure 24.3B), and orientation cues contribute directly to depth perception as well as to form (figure 24.3C). In addition to these direct contributions, there are several instances in which cues can have a concealed (indirect) contribution to a particular aspect of perception; these are indicated by dashed lines in figure 24.3. This is particularly notable for spectral cues (figure 24.3D), which can contribute to form, depth, and motion perception,

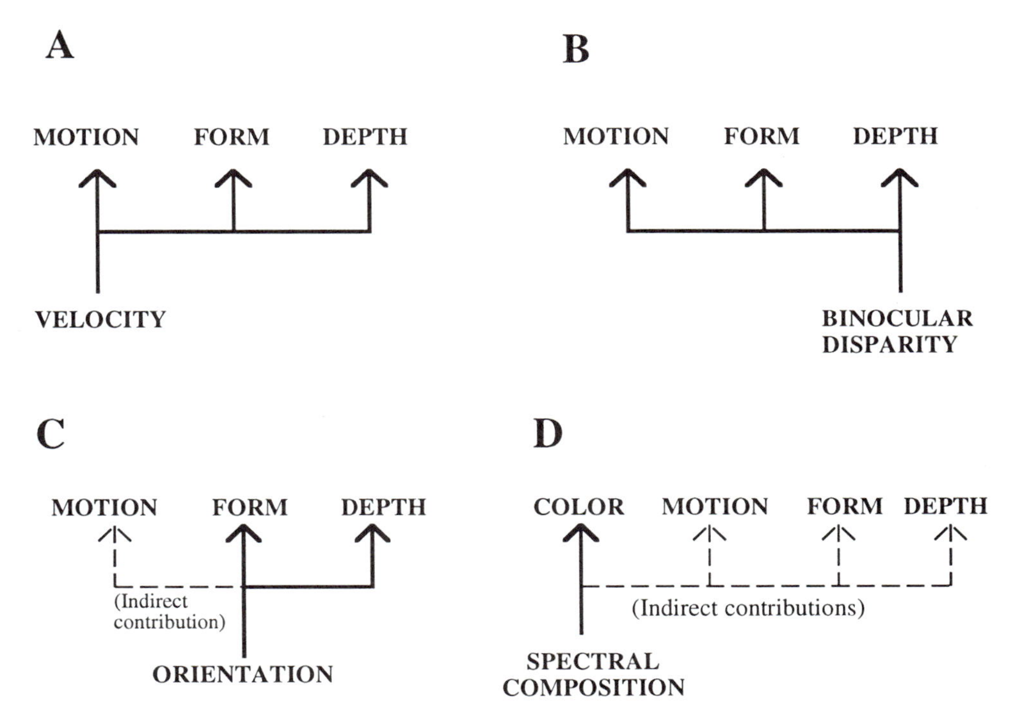

FIGURE 24.3 A schematic illustration of how individual cues can contribute to multiple aspects of perception. Solid lines indicate direct contributions, in which the values of velocity (A), disparity (B), orientation (C), and wavelength (D) are intimately involved in the computation of motion, form, depth, or color perception. Dashed lines indicate indirect (concealed) contributions of cues to percepts (DeYoe and Van Essen, 1988). For example, a discontinuity in texture orientation or in spectral composition can signify a border whose geometrical characteristics are independent of the values of the cues that delineate the border.

albeit not always to the same degree as luminance-based cues (Papathomas, Gorea, and Julesz, 1991; Dobkins and Albright, 1993).

These arguments indicate that both divergence and convergence of information about different low-level cues are important in the function of the visual system and, hence, that there are extensive interactions between streams at the abstract level illustrated in figure 24.3. It is logical to expect a corresponding degree of interaction in whatever functional streams exist at the structural level. Support for this notion comes from the anatomical and behavioral studies reviewed next.

Anatomical organization of the macaque visual system

AREAS, HIERARCHIES, AND THE DORSAL-VENTRAL STREAM The remainder of this chapter will concentrate on the visual system of the macaque monkey, the most intensively studied and best-understood laboratory animal. Color plate 8 shows the layout of different visual areas in the macaque cerebral cortex as they are currently understood. These areas are shown on a two-dimensional unfolded map of the entire cerebral cortex of the right hemisphere; the major subcortical visual centers are also included on the lower left. This display format allows ready visualization of all areas, even those that are deeply buried within one or another of the numerous cortical sulci present in the intact brain (see insets, upper left). The representation of surface area is preserved for subcortical as well as cortical structures, so that their relative sizes are all evident at a glance. Among the subcortical structures, the nuclei within the pulvinar complex are collectively much larger than the lateral geniculate nucleus (LGN), even though the latter is the primary relay of visual information from the retina to the cortex. At the cortical level, there are two very large areas, V1 and V2, each occupying approximately 10% of the entire cortical sheet. The remaining extrastriate areas are all individually much smaller, but collectively they occupy the remainder of the occipital lobe, much of the parietal and temporal lobes, and even part of the frontal lobe.

The 32 distinct visual areas shown in this scheme have been identified on the basis of one or more of the following criteria: (1) a distinctive pattern of inputs and outputs; (2) a distinctive architecture; (3) a topographically organized representation of part or all of the contralateral visual hemifield; and (4) a characteristic functional signature revealed by physiological recordings or restricted cortical lesions. There are uncertainties about the exact location of many of the borders and, in some cases, even the identities of some of the areas indicated in this scheme. Alternative partitioning schemes have been suggested for some regions, especially in the temporal and parietal lobes (see Desimone and Ungerleider, 1989; Felleman and Van Essen, 1991). Thus, this scheme should be regarded as a progress report that will assuredly be subject to continued revision and refinement. Also, not all these areas are exclusively visual in function, as some have inputs from other sensory modalities or are closely linked to visuomotor function.

Accompanying the great expansion in extent of visual cortex relative to that of the LGN is a corresponding increase in the number of neurons involved in visual processing. Each LGN contains approximately 1 million neurons projecting to V1, whereas V1 in each hemisphere contains nearly 250 million neurons (1200-mm^2 area, 2×10^5 neurons/mm^2), a density that is notably high because of the increased thickness of layer 4 (O'Kusky and Colonnier, 1982). All the extrastriate areas combined contain 400 million or so neurons per hemisphere (4000-mm^2 area, $\times 10^5$ neurons/mm^2), based on surface area estimates of Felleman and Van Essen (1991) and the cell density estimates of Rockel, Hiorns, and Powell (1980). Altogether, there are approximately 1.3 billion visual cortical neurons (counting both hemispheres), which corresponds to nearly 600 cells for each LGN input.

Within the visual cortex, information flow is mediated by an impressively rich set of cortico-cortical connections that has been revealed by the systematic application of sensitive pathway–tracing techniques. On average, each visual area has approximately 10 distinct inputs and 10 outputs (Felleman and Van Essen, 1991). The total number of reported corticocortical connections is 305, which is nearly a third of the total possible number of pathways interconnecting 32 areas, with many more presumably yet to be discovered. These pathways differ greatly in their robustness, though. Some (such as those between V1 and V2 and between V2 and V4) involve a large fraction of the output cells in the area of origin and have very dense terminations in the target area, whereas others are only faintly detectable using the most sensitive available technique. Despite these large differences, there has been only limited success to date in developing methodology that yields systematic quantification of the absolute or relative strengths of different pathways.

Given the degree of complexity implied by the sheer number of identified pathways, a pessimist might legitimately despair of understanding how information flows and is processed within this intricate network. Fortunately, several organizing principles have been recognized that indicate an important degree of order within the system. The first such principle is that connections tend to be arranged in reciprocal pairs (e.g., V1 projects to V2, and V2 projects back to V1); only a few convincing exceptions to this rule have been found. This observation is significant because it rules out simple schemes in which information flow occurs in a strictly bottom-up fashion. It also has led to another important principle, because the majority of reciprocal pathways are asymmetrical with regard to the cortical layers in which connections originate and terminate. The resulting hypothesis is that some pathways are of the forward (ascending) type, whereas their reciprocal counterparts are of the feedback (descending) type (Rockland and Pandya, 1979; Maunsell and Van Essen, 1983). (A third type, the lateral pathway, is symmetrical in the two directions.) Using these objective criteria, each area can be specified as higher, lower, or at the same level as the areas with which it is connected. Systematic application of this process to all sets of interconnected areas leads to the overall hierarchical scheme depicted in figure 24.4 (Felleman and Van Essen, 1991; Van Essen, Anderson, and Felleman, 1992). It contains 10 hierarchical levels within the visual cortex plus two subcortical levels at the bottom (the retina and LGN) and linkages to areas in the limbic system at the top.

It is instructive to compare this hierarchy to an alternative scheme for representing cortical connectivity, described by Young (1992; see also chapter 29). His topological connectivity scheme for macaque visual cortex ignores information about laminar patterns and instead relies only on the presence or absence of connections between areas, using a multidimensional scaling method to place areas close together if they share many targets in common. The two schemes are strikingly similar in the relative positions of visual areas, given the different criteria used in their generation. Nonetheless, there are numerous minor differences and also a few more substantial ones. The broad similarities reflect the fact that connections occur more frequently between areas at nearby levels of the cortical hierarchy than between areas separated by many levels (Felleman and Van Essen, 1991). The most significant difference is that Young's scheme clearly shows the segregation into dorsal and ventral processing streams that have been described by Desimone and Ungerleider (1989). Altogether, these different schemes provide complementary rather than conflicting ways of analyzing cortical connectivity. Both represent a principled, objective, and compact representation of a vast amount of anatomical data, and both provide a framework for making testable hypotheses about cortical function.

Compartments and Streams Concurrent processing originates in the retina, where distinct classes of retinal ganglion cell project to separate target layers in the LGN. Many studies have concentrated on two major processing streams, parvocellular (P) and magnocellular (M) streams, which are the most prominent components of the retinogeniculostriate projection. Evidence obtained mainly in the 1980s suggested that within V1 and V2, the P stream splits into two largely separate streams, with the projections to parietal cortex arising mainly from the M stream and the projections to inferotemporal cortex arising mainly from the P stream (Maunsell and Newsome, 1987; DeYoe and Van Essen, 1988; Livingstone and Hubel, 1988; Desimone and Ungerleider, 1989). Although major aspects of this scheme remain intact, recent evidence suggests that a number of significant revisions are in order. The most important of these relate to (1) the number of subcortical streams, (2) the degree to which these streams interact and merge within the cortex, and (3) the existence of compartments within individual areas at higher levels of the cortical hierarchy, not just areas V1 and V2. Figure 24.5 provides an updated anatomical framework that reflects these findings (see also chapters 19 and 20 for additional discussion of the different processing streams at subcortical and early cortical levels).

At the subcortical level, three distinct streams have now been identified in primates, including the macaque (Casagrande and Norton, 1991). The two best studied and most prominent remain the aforementioned P and M streams. P cells constitute the great majority of retinal ganglion cells and LGN neurons (~80%) and thus are presumably the most important in terms of overall amount of information carried through the optic nerve. M cells constitute only 10% or so of ganglion cells and LGN neurons, but they have received con-

HC
ER
36
46
TF
TH
STPa
AITd
AITv
7b
7a
FEF
STPp
CITd
CITv
VIP
LIP
MSTℓ
MSTd
FST
PITd
PITv
DP
VOT
MDP
MIP
PO
MT
V4t
BD
V4
ID
PIP
V3A
V3
VP
M
V2
BD
ID
PULV.
M
V1
BD
ID
M
K
P
SC
M
other
P

siderable scrutiny for various reasons, including the fact that they are much larger than P cells in soma size, dendritic extent, and axonal diameter. The third (and least understood) stream in the retinogeniculostriate pathway is the koniocellular (K) stream. This includes the small intercalated cells that lie between the individual P and M layers as well as the small-called (S) layers that lie ventral to the M layers of the LGN. Interestingly, K cells are as numerous as M cells (Benson et al., 1991; Hendry, personal communication). They have direct projections to the cytochrome oxidase blobs of V1 (Casagrande and Lachica, 1992; Lachica and Casagrande, 1992), and they receive selective inputs from the superior colliculus (Harting et al., 1991) as well as from small-diameter axons presumed to arise from the retina (Conley and Fitzpatrick, 1989). The small size of K cells, the sparseness of their cortical terminations, and their sluggish visual responsiveness (at least in galagos; see Norton et al., 1988) suggests that the K stream may play a relatively minor role in the transmission of ascending sensory information.

The magno-dominated stream At the cortical level, the three major processing streams have differing patterns of convergence from the three subcortical streams. The one having the most modest convergence was previously identified as the M stream (DeYoe and Van Essen, 1988) but will now be referred to as the magno-dominated, or MD, stream to reflect the fact that it is driven predominantly, but not exclusively, by the manocellular system. In V1, the MD stream includes layers 4Cα, to which the magnocellular layers of the LGN project, and layer 4B, to which layer 4Cα strongly projects (Fitzpatrick, Lund, and Blasdel, 1985). In V2, it includes the cytochrome oxidase (CO) thick stripes, which receive their main ascending inputs from layer 4B of V1 (Livingstone and Hubel, 1987). At higher levels, the MD stream includes areas V3 and MT, which receive strong inputs from both layer 4B of V1 and the thick stripes of V2 as well as three areas to which area MT strongly projects: the dorsal and lateral subdivisions of the medial superior temporal area MSTd and MSTl, respectively) in the superior temporal sulcus and the vertral intraparietal area (VIP) in the intraparietal sulcus. Anatomical evidence suggestive of cross talk with other streams is evident even within the geniculorecipient layers of V1, where both dendritic and local axonal projections cross the 4Cα/4Cβ boundary (Lund, 1988; Lund and Yoshioka, 1991). In V2, cross talk occurs between the CO thick stripes and the other stripe compartments, especially thin stripes (Rockland, 1985; Livingtone and Hubel, 1987). At higher levels, there are numerous additional sources of cross talk, some of which are shown in figure 24.6 and others of which are discernible in figure 24.4.

Another intriguing characteristic of the MD stream is a relatively high incidence of cells that are immunoreactive for the monoclonal antibody Cat-301 (DeYoe et al., 1990). The domains of high Cat-301 immunoreactivity include the M layers of the LGN, portions of layers 4Cα and 4B of V1 (and also layer 6, which is not shown in figure 24.5), the V2 thick stripes, and areas V3, MT, MST, and VIP. Within these extrastriate areas, immunoreactivity is concentrated in a subset of cells in layers 3, 5, and 6. It is striking that the correlation of Cat-301 immunoreactivity with a particular processing stream persists through five distinct levels of the visual hierarchy, even though the functional or developmental significance of the pattern remains unclear.

It is difficult, from anatomical information alone, to assess accurately the relative influence of the different subcortical streams on any given area or compartment in the MD stream. Besides the inherent limitations to quantifying the strengths of different projections, there is also the uncertainty of whether a given projection has, say, a predominantly direct excitatory drive, versus a net suppressive or modulatory influence. A valuable approach to transcending these anatomical limitations has come from selective inactivation techniques, which involve recording from target cortical structures before, during, and after transient blockage of specific LGN layers. Using this approach while recording from MT reveals that MT responses are dramatically reduced after inactivation of M layers of the LGN, whereas after P-layer inactivation the reduction in responses is much smaller (Maunsell, Nealy, and DePriest, 1990). A potential complication is that M-layer inactivation also affects the majority of neurons in the K stream but, for reasons cited earlier, it seems unlikely

FIGURE 24.4 A hierarchy of visual areas in the macaque, based on laminar patterns of anatomical connections. Approximately 90% of the known pathways are consistent with this hierarchical scheme (Felleman and Van Essen, 1990). Compartments within areas V1, V2, and V4 are discussed in relation to figure 24.5. (Modified from Van Essen, Anderson, and Felleman, 1992, by permission.)

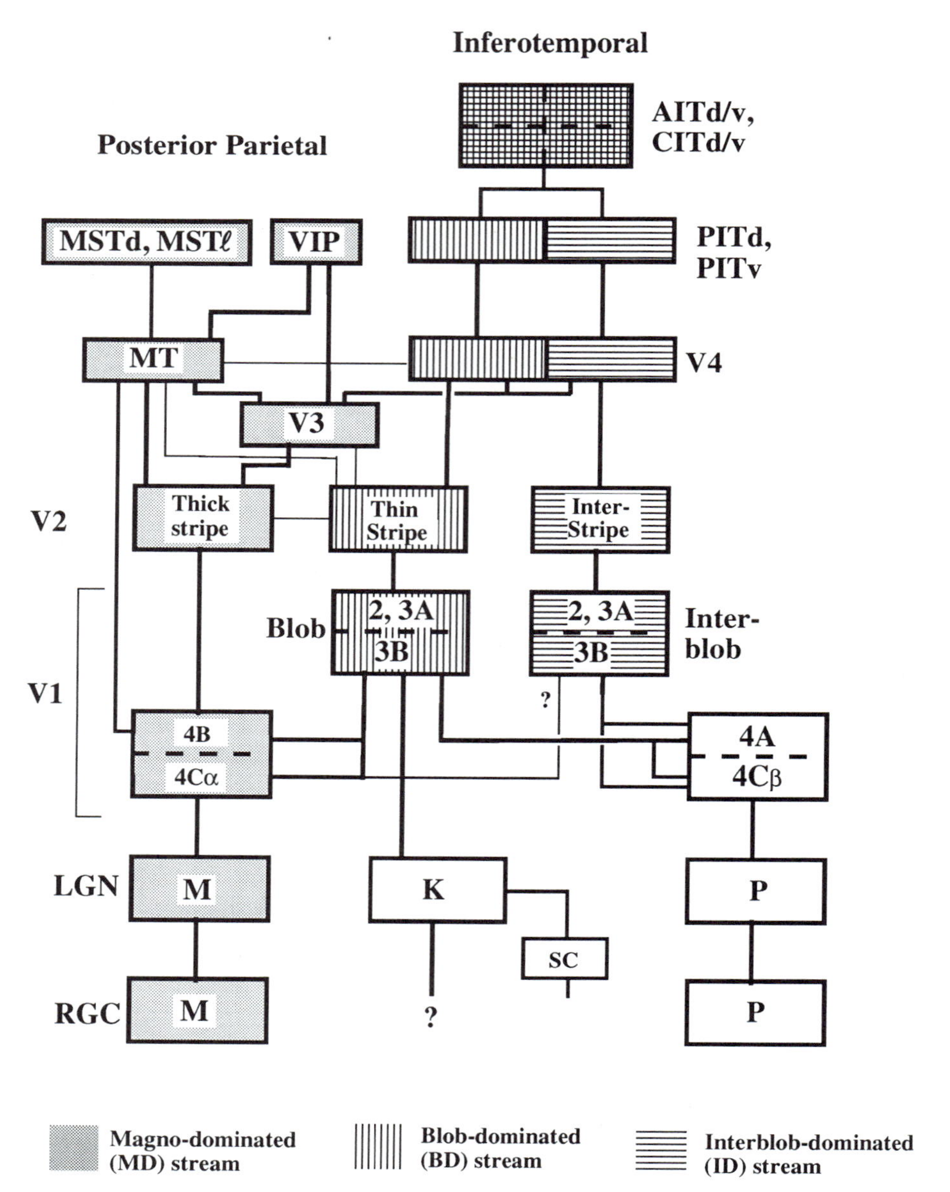

FIGURE 24.5 Subcortical and cortical processing streams in the macaque. Three subcortical streams (P, M, and K, in order of prominence) have been identified anatomically and physiologically. In the cortex, there are also three major streams: the magno-dominated (MD), blob-dominated (BD), and interblob-dominated (ID) streams. The MD stream, as the name implies, is dominated by a single subcortical source, but the BD and probably also the ID streams receive strong convergence from multiple subcortical streams. At the top of the cortical hierarchy, the MD stream leads primarily to posterior parietal areas implicated in spatial localization and the BD and ID streams to inferotemporal areas implicated in pattern recognition. Abbreviations in this and following figures: RGC, retinal ganglion cell; LGN, lateral geniculate nucleus; MT, middle temporal area; MSTd, l, dorsal and lateral subdivisions of the medial superior temporal area; VI, ventral intraparietal area; PITd, v, dorsal and ventral subdivisions of the posterior inferotemporal area; CITd, v, central inferotemporal area; AITd, v, anterior inferotemporal area.

that this is a major confounding factor. These physiological observations are thus in rough accord with the preponderant but not exclusive M-stream input to MT suggested by anatomical studies.

The blob-dominated and interblob-dominated streams The cortical stream having the greatest degree of convergence is that associated with the CO blobs of V1. At one time, the blobs were believed to receive their major ascending input via P cells of the LGN relaying through layers 4A and 4Cβ of V1 (Fitzpatrick, Lund, and Blasdel, 1985; Livingstone and Hubel, 1988). They were therefore considered part of the so-called P-B (parvo-blob) stream (DeYoe and Van Essen, 1988). It is now evident that the blobs instead receive strong inputs from all three geniculostriate streams. Input from the K stream arises via a direct projection to the blobs, but that from the M stream is notably robust by virtue of strong projections arising from layers 4Cα and 4B (Fitzpatrick, Lund, and Blasdel, 1985; Casagrande and Lachica, 1992; Lachica, Beck, and Casagrande, 1992; Yoshioka, Levitt, and Lund, 1993). LGN inactivation experiments provide physiological confirmation that many cells in the blobs can be driven after P-layer inactivation, presumably by a combination of M and K inputs (Nealey, Ferrera, and Maunsell, 1991). Because the blobs are a locus of major integration from all three subcortical streams and appear not to be dominated by any one in particular, we will refer to them and to their major targets collectively as the *blob-dominated* (BD) stream. These are indicated by vertical hatching in figure 24.5. The major output of the blobs is a selective projection to the CO thin stripes of V2 (Livingstone and Hubel, 1984). Beyond V2, the BD stream includes distinct subregions of V4 and probably also of posterior inferotemporal cortex (Van Essen et al., 1990; DeYoe, Glickman, and Wieser, 1992; Felleman, McClendon, and Lin, 1992; DeYoe et al., 1993).

The third cortical stream is associated with the CO interblobs of V1, and we will refer to it as the *interblob-dominated* (ID) stream. The interblobs receive major ascending inputs from the parvocellular LGN layers by way of layers 4A and 4Cβ (Lachica, Beck, and Casagrande, 1992; Yoshioka, Levitt, and Lund, 1993), but they may also receive MD inputs by way of layer 4Cα (Yoshioka, Levitt, and Lund, 1993). LGN inactivation experiments suggest that interblob neurons in V1 can be driven through the M or K geniculostriate streams (Nealey, Ferrera, and Maunsell, 1991). The major output from the interblobs is to the CO interstripes of V2. Beyond V2, the ID stream appears to include subregions in V4 and posterior inferotemporal cortex that complement the subregions belonging to the BD stream.

V4 also receives strong projections from a number of other visual areas (V3, V3A, V4t, and MT), some of which are associated with the MD stream (see above) and some of which show indications of compartmental organization (Born and Tootell, 1992; DeYoe et al., 1993). LGN inactivation experiments reveal comparably strong effects on V4 neurons after either M-layer or P-layer inactivation (Ferrera, Nealey, and Maunsell, 1992), suggesting that V4 receives roughly balanced M and P inputs, along with an indeterminate degree of influence from the K stream.

Beyond the occipital lobe, the MD stream projects most heavily to areas in posterior parietal cortex, whereas the BD and ID streams project most heavily to areas in inferotemporal cortex. This has been illustrated most graphically in dual-tracer experiments involving injections of one tracer into parietal cortex and a separate tracer into inferotemporal cortex (Morel and Bullier, 1990; Baizer, Ungerleider, and Desimone, 1991). However, the segregation is by no means perfect, as is evident from the degree of cross talk discernible in figures 24.4 and 24.5. An even higher degree of convergence is apparent in the projections to the frontal lobe and to polysensory cortex in the superior temporal sulcus.

Compartmental dimensions A notable characteristic of the cortical compartments described in the preceding section is that their dimensions change dramatically at successively higher levels of the hierarchy, both in absolute extent and in relation to the overall size of each area. Figure 24.6 provides a semiquantitative illustration of these relationships between compartmental size and overall area size. V1 has a surface area of nearly 1200 mm^2 per hemisphere (Van Essen, Newsome, and Maunsell, 1984), and it contains approximately 7000 CO blobs, based on a density of 6 blobs/mm^2 (Wong-Riley and Carroll, 1984; Livingstone and Hubel, 1984). The overall shape of the unfolded V1 is nearly 5 cm long (along the fovea-to-periphery axis) and 3 cm wide (along the axis of constant eccentricity), so this corresponds to an array of blobs approximately 125 × 75 in extent. In terms of the numbers of neurons involved, each V1 module (one blob, the surrounding interblob, and the associated cortical domain) contains nearly 30,000 neurons for every 130 or so LGN afferents.

V2 is comparable in overall size to V1, but it is very different in shape because it has a split representation of the horizontal meridian that leads to separate regions for lower fields (V2d) and upper fields (V2v); its dimensions when unfolded are approximately 100 mm long and 10 mm wide (Van Essen et al., 1986). The stripes in V2 run roughly orthogonal to its long dimension,

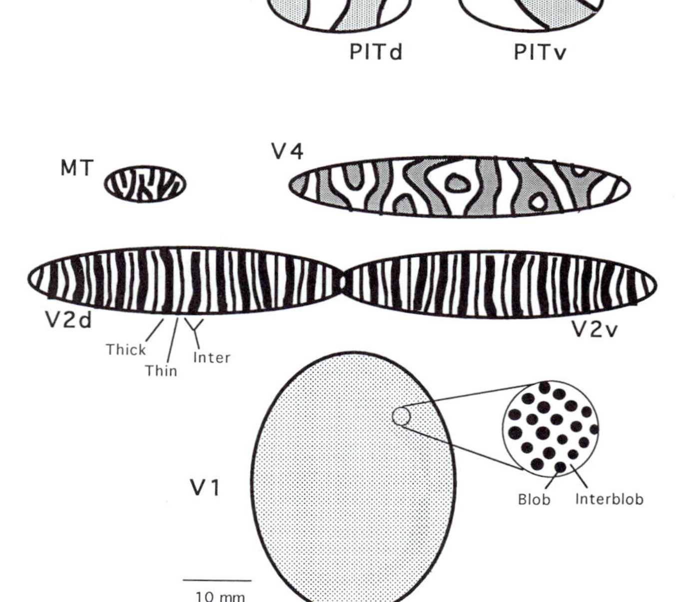

FIGURE 24.6 A schematic illustration of the relative sizes of different visual areas and of the compartments therein. At successive stages of the cortical hierarchy, the areas become progressively smaller, whereas the compartments become progressively larger.

and the mean periodicity of the stripes is 4 mm or so, for a total of nearly 28 complete sets of stripes (Van Essen et al., 1990). V4 approaches only half the size of V1 and V2, and its degree of elongation after unfolding is intermediate between them ($\sim$60 mm × $\sim$10 mm). The geometry of the compartments in V4 is not yet known because no histochemical or immunocytochemical markers have yet been discovered, but indirect evidence from the studies outlined previously suggests that the periodicity is on the order of 5 mm or more, indicating that the number of such compartments in V4 is perhaps only one or two dozen. MT, at the same hierarchical level as V4, also appears, on physiological grounds, to be compartmentally organized (Born and Tootell, 1992). Interestingly, the size of compartments is much smaller than for the other extrastriate visual areas, but this may be related to the much smaller overall size of MT. Finally, the posterior inferotemporal areas are each only approximately 200 mm^2 in extent, and their periodicity appears to be even coarser, suggesting they contain at most only a few compartments. Indeed, it becomes difficult to distinguish unequivocally between areas and compartments at this level.

Mapping function onto structure

Given the high degree of anatomical complexity described in the preceding section, it should not be surprising to find a corresponding richness in neurophysiological aspects of visual processing. Numerous single-unit studies have been directed at characterizing receptive field properties of neurons in different areas and compartments and in ascertaining what properties are common to cells in each processing stream. These studies have been complemented by experiments dealing with the behavioral effects of selective lesions or inactivation. Some of the relevant experimental data are reviewed in other chapters dealing with specific aspects of visual function (see chapters 19, 20, 23, 25, and 29). Here we will review briefly the key experimental findings as they relate to different cortical streams and will also touch on several broader issues relating to the nature of neural representations and transformations.

SUBCORTICAL REPRESENTATIONS Within the retina, the signals encoded by photoreceptors are transformed into a compact and efficient representation that conveys a large amount of perceptually useful information along the limited number of fibers contained within the optic nerve. Contributing to this transformation are several distinct strategies that are relevant to our subsequent discussion.

The first of these strategies is *stereotyped filtering characteristics*. Retinal ganglion cells have concentric receptive fields with antagonistic center and surround mechanisms. In the spatial frequency domain, each cell has an optimal spatial frequency that is matched to its receptive field center size, but it also conveys some information about low spatial frequencies because the center mechanism generally is stronger than the surround. The spatial scale of receptive fields increases steeply with eccentricity, in parallel with the well-known differences in visual acuity. Finally, P cells and most M cells show linear spatial summation characteristics, but a subset of M cells have strong spatial nonlinearities.

The second strategy is *division of labor*. P cells and M cells differ in their temporal characteristics, with P cells being maximally sensitive to lower temporal frequencies and M cells to higher temporal frequencies (Marrocco, McClurkin, and Young, 1982; Hicks, Lee, and Vidyasagar, 1983; cf. Van Essen and Anderson, 1990). In the spatial domain, M cells have threefold larger dendritic fields than P cells (up to 10-fold larger in the human fovea [Dacey and Peterson, 1992]), and they have a much lower sampling density (Perry, Oehler, and Cowey, 1984; Grünert et al., 1993). This suggests that high spatial frequencies are subserved primarily by P cells, although the reported physiological differences are smaller than predicted by the anatomy (Marrocco, McClurkin, and Young, 1982; Blakemore and Vital-Durand, 1986; Kaplan, Lee, and Shapley, 1990). The division of labor, illustrated schematically in the bottom level of figure 24.7, allows the combined M and P population to provide full coverage of a broader range of spatial and temporal frequencies than could be handled by either type alone.

The third strategy is *multiplexing*. The cone inputs to the center and surround mechanisms of P cells differ, which allows these cells to convey spectral as well as spatial information. On the other hand, spectral information is only a small fraction (arguably just a few percent) of the total information content in natural images (Buchsbaum, 1987; Van Essen and Anderson, 1990). Thus, P cells primarily carry spatial (and temporal) information, with a modest amount of spectral information multiplexed alongside (but see Reid and Shapley, 1992, for an alternate view). A different form of multiplexing occurs in M cells, some of which have spatial or spectral nonlinearities that can convey information about texture, dynamic change, and spectral contrast.

Confirmation of this physiologically based characterization has come from recent behavioral observations after selective lesions of P or M layers of the LGN (Schiller, Logothetis, and Charles, 1990; Merigan, Katz, and Maunsell, 1991; Merigan and Maunsell, 1993). In brief, P-layer lesions lead to a large deficit in color discrimination and to moderate deficits in form and depth discrimination, especially at low temporal frequencies and high spatial frequencies. In contrast, M-layer lesions lead to deficits in motion, depth, and form perception, especially at high temporal frequencies. These results make a strong case for functional convergence, in which P and M streams contribute jointly to many aspects of perception but make largely separate contributions to the perception of color and movement. Thus, the one stream–one function notion of completely independent streams is not supported by these critical data.

Neuronal Representations at the Cortical Level

An overarching objective is to attain a full description and understanding of the representations and transformations that occur at each stage of cortical processing. Returning momentarily to the factory analogy raised earlier in this chapter, the goal is to identify at each level the intermediate products that emerge from the different streams and to infer what processing has led to each of these products (cf. figure 24.1C). For several reasons, this problem is much more formidable than at the retinal level. There are numerous visual areas with which to deal and numerous layers and compartments within each area. Although the overall information space is no larger than at the retinal level (because information can only be destroyed, not created, at successive stages of processing), it is carved into pieces that are much smaller (because cortical cells are more selective) and more diverse (because tuning characteristics are more variable). It is even difficult to determine which are the most appropriate dimensions for characterizing cells. Hence, before discussing specific receptive field characteristics, several general comments are in order.

A key initial question is what are the most appropriate measures for characterizing neuronal receptive field properties. A standard, time-honored approach is to present a series of stimuli that are varied systematically along one or more dimensions and to analyze the neural responses by generating tuning curves (for the one-dimensional case) or tuning surfaces (for two or more dimensions). The tuning profile can then be assessed according to the degree of selectivity (e.g., using an index that compares the most effective vs. the least effective stimulus) and the breadth of tuning (e.g., using measures of the bandwidth of the response). However, there are other important characteristics of cortical cells that are not readily captured by these measures. One problem is that neurons often have irregular tuning profiles (e.g., multiple peaks) along any given dimension. Another problem is that conventional selectivity indices do not reflect directly how much noise and uncertainty lurk in the underlying neural responses. This has often been addressed by computing

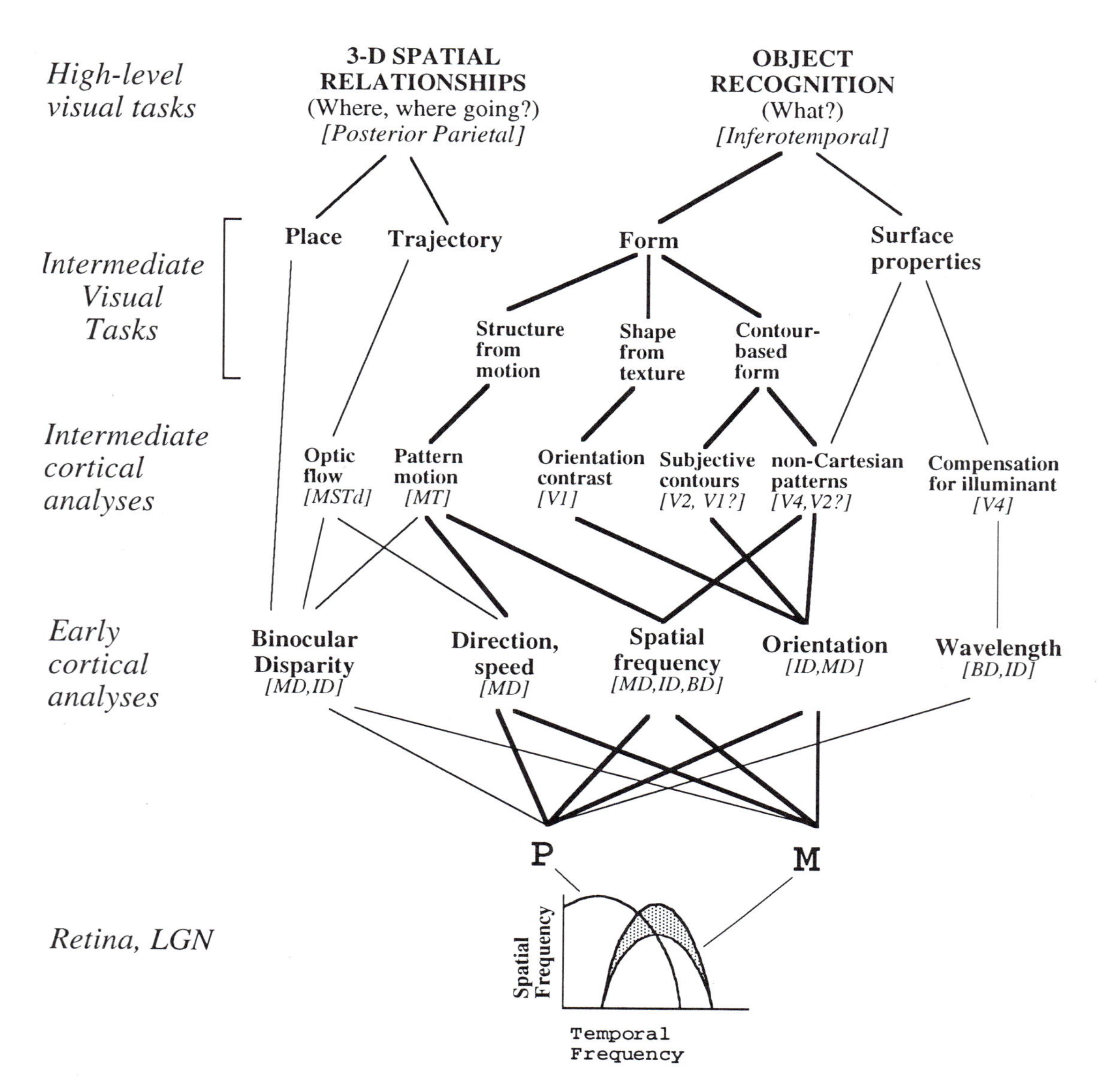

error bars for individual data points but, for technical reasons, it is difficult to determine statistical confidence limits rigorously using conventional standard errors.

These problems are compounded when comparing the physiological properties of different areas and streams to infer their function. Typically, this involves a comparison of the incidence (percent of sampled cells) of a particular type of selectivity in different areas or streams. In this respect, it is of considerable importance that cortical neurons generally do not fall into a simple dichotomy of being either completely selective or completely nonselective along any given dimension. When tested quantitatively, the population of neurons in any particular region is typically distributed along a continuum with regard to the degree of selectivity and breadth of tuning (e.g., Burkhalter and Van Essen, 1986; Felleman and Van Essen, 1987; Peterhans and von der Heydt, 1993). Consequently, to estimate the incidence of selectivity of any particular type, it often is necessary to settle on a somewhat arbitrary cutoff to delineate between selective and nonselective cells. The criteria that have been applied vary widely across investigators; as a consequence, there are large discrepancies in the reported incidence of different types of selectivity (cf. Felleman and Van Essen, 1987).

For these and other reasons, there has been increasing interest in developing alternative approaches that can provide quantitative measures of how much information is actually conveyed by the responses of a given neuron. Given a standardized stimulus set for comparing the responses of different cells, classical information theory (Shannon, 1948) provides a basis for quantitatively determining how well a cell's response differentiates among the test stimuli, regardless of the particular shape of its tuning profile but taking into account the response variability (Optican and Richmond, 1987; Optican et al., 1991). Such analyses resemble those based on signal detection theory (Britten et al., 1992) in that they are founded on an estimation of the probability that a particular response is associated with each stimulus condition. The greater the likelihood that different response values (e.g., spike rates) are evoked by different stimuli, the greater is the information conveyed. In this approach, noise (i.e., response variability) plays a key role: The lower the noise, the greater the confidence that a particular response was elicited by one stimulus and not another; consequently, the greater the information. The significance of these considerations has been demonstrated in a recent study of V4 in which some cells were shown to have high selectivity indices for color or spatial frequency but nevertheless conveyed relatively little reliable information about those parameters (DeYoe, Glickman, and Wieser, 1992). In the long run, our understanding of cortical function will benefit from systematic application of both of these complementary approaches, but most of the currently available experimental data derive from the measurement of tuning curves and selectivities, using many different stimulus dimensions relating to form, motion, and spectral processing.

FIGURE 24.7 Convergence and divergence of information at different stages of visual processing. Lines represent some (but by no means all) of the major routes of information flow from subcortical parvocellular (P) and magnocellular (M) streams (bottom level), through the selectivities made explicit in the firing patterns of cells at early and intermediate levels of cortical processing (second and third levels). At the upper levels of the diagram, the connecting lines show how these streams may contribute to several intermediate visual tasks, which in turn converge onto the two most general tasks of vision represented at the level of inferotemporal and posterior parietal cortex. The specific relationships between the intermediate visual tasks and the pathways and areas shown in figure 24.4 are yet to be determined, though the general flow of information is largely determined by the types of analyses needed to support the ultimate tasks of object recognition and space perception. Outlined in bold print is a group of processing streams that begin with contributions from both M and P cells but then diverge through several streams before converging again to contribute to the perception of object form. Note how the flow of information more closely resembles the architecture of figure 24.1C rather than figure 24.1A or 24.1B. Types of information flow that have been omitted from the diagram (to avoid further clutter) include the contribution of disparity cues to the representation of shape and the contribution of spectral cues to the analysis of both shape and motion. At the subcortical level, the two tuning curves show how the representation of spatiotemporal information space is distributed across the M- and P-cell populations; the shaded zone represents the region where M cells provide incomplete coverage (nonlinear in some cells) because of their lower sampling density. A similar principle of distinct yet overlapping representation of information space is also true for anatomical pathways at later stages, though this is difficult to represent in a diagram of this type. The processing streams and visual areas shown in the second and third layers are assigned based on the incidence of selectivities encountered physiologically. Ideally, these assignments will eventually be based on quantitative measures of the information conveyed by cells in each area and by the effects of selective removal or inactivation. MD, magno-dominated; ID, interblob-dominated; BD, blob-dominated.

DISTRIBUTION OF RESPONSE SELECTIVITIES Physiological studies of single-unit properties in different areas and compartments have led to insights along several lines. These are summarized schematically in figure 24.7, which is intended to be illustrative rather than exhaustive in its portrayal of the types of analysis occurring at different stages of processing.

1. Selectivity occurs along many low-level dimensions, either as a characteristic initially established in the cortex or as a property that becomes refined over that existing at subcortical levels. These properties are all encountered at the level of V1, but they also occur at a coarser spatial scale in extrastriate areas, which indicates that they must be independently computed more than once. There is not a 1 : 1 matching between each type of tuning and a particular processing stream. Instead, each stream has a characteristic profile of selectivities among its constituent neurons, and many cells show tuning along multiple dimensions (DeYoe and Van Essen, 1988; Peterhans and von der Heydt, 1993). In this respect, the representation of information by the different streams is analogous to the shared representation of spatial and temporal frequencies by the M and P streams subcortically. Most notably, the MD stream has a high incidence of selectivity for binocular disparity, direction and speed, spatial frequency, and orientation (see figure 24.7, second row from bottom).

The BD stream has a high incidence of wavelength selectivity as well as spatial frequency tuning, whereas the incidence of orientation selectivity is high in the ID stream, along with selectivity for disparity, spatial frequency, and wavelength. Thus, there is significant multiplexing of information in all three streams. In V4, color-selective and orientation-selective cells have been reported to occur in separate clusters in V4 (Zeki, 1983a), but this is unlikely to be an all-or-none segregation as there are many cells that are highly selective for both color and form (Schein and Desimone, 1990). On the other hand, a recent study of V4 based on information theoretical measures suggests that there may indeed be a considerable degree of segregation between cells that convey reliable spectral versus spatial frequency information (DeYoe, Glickman, and Wieser, 1992). Color-selective cells have also been reported in inferotemporal cortex (Fuster, 1990; Komatsu et al., 1992), but their distribution within BD versus ID compartments there has not been established.

2. More complex types of tuning have also been identified (see figure 24.8, middle row). In the motion domain, several studies have reported emergent motion-related properties that are absent or less robust in V1. These include selectivity in area MST for motion characteristics related to optic flow, including rotation, expansion, and contraction (Tanaka and Saito, 1989; Duffy and Wurtz, 1991), as well as pattern-selective cells in MT that are selective for combined grating patterns rather than just for their component gratings (Movshon et al., 1985). In the domain of form vision, some higher-order characteristics, such as selectivity for orientation contrast, are encountered even at the level of V1 (Knierim and Van Essen, 1992). Despite its appearance at such an early hierarchical level, orientation contrast selectivity appears only with a temporal delay relative to standard orientation selectivity, which fits with the intuitive notion that it is a more advanced type of analysis. Other types of higher-order characteristics related to form vision include responsiveness to subjective contours in the thick stripes and interstripes of V2 (Peterhans and von der Heydt , 1993) and perhaps also in V1 (Shapley, this volume) and to noncartesian patterns (e.g., concentric, radial, or hyperbolic gratings) in V4 and perhaps also V2 (Gallant, Braun, and Van Essen, 1993). Finally, selectivity for spectral contrast and for surface color independent of the spectral composition of the illuminant have been reported for V4 (Zeki, 1983b; Schein and Desimone, 1990).

3. At higher levels of the hierarchy, a key issue is how neural response properties are related to the distinction between posterior parietal areas implicated in the analysis of three-dimensional spatial relationships and inferotemporal areas implicated in pattern recognition (Desimone and Ungerleider, 1989). Support for this dichotomy comes from reports that neurons in inferotemporal cortex respond well to complex shapes, including natural stimuli such as faces and hands (Desimone et al., 1984; Tanaka et al., 1991; Miyashita, 1993). To support object recognition across a broad range of objects and to permit generalizations from one object to another, there must be convergence of inputs from a variety of intermediate analyses (see figure 24.7, second and third rows from top). In the case of form and pattern recognition, there should be converging input from analyses of shape based on texture (and shading) gradients and motion patterns as well as on contour information (bold lines in the figure). How these intermediate tasks are parceled out among the various visual areas and streams remains unclear. This is true also, though perhaps to a lesser extent, for posterior parietal cortex, where response properties fit with the suggested involvement in the analysis of spatial relationships, eye movements, and target selection for visual attention (Andersen, 1987; Colby, 1991). However, a simple dichotomy between temporal and parietal cortices leaves uncertain the assignment of intermediate visual tasks such as structure from motion, which is based on the analysis retinal motion patterns but should be particularly useful in shape recognition.

Summary of Physiological Findings The MD stream provides information about optic flow and other aspects of motion that is used, especially by posterior parietal cortex, for the analysis of spatial relationships and visuomotor control. The BD and ID streams provide information about color and form for use in pattern and shape recognition. A key unresolved issue is the functional difference between the BD and ID streams. The notion that the BD stream is exclusively involved in color perception and the ID stream is the sole mediator of form perception, while attractive (Livingstone and Hubel, 1988), now seems unlikely on the basis of the evidence just reviewed. An intriguing alternative that seems computationally better grounded is based on the distinction between the geometrical shape of an object (its three-dimensional configuration) and its intrinsic surface characteristics (its color and also textural composition, including both spatial frequency and orientation, or grain). These distinctions are closely related to the processing associated with the "boundary contour system" (implicated in shape determination) and the "feature contour system" (implicated in inferring surface characteristics) articulated in Grossberg's (1987a,b) computational theory of early vision. An interesting physiological prediction of this hypothesis is that the BD stream might be enriched in cells responsive to and selective for textural characteristics (as distinct from simple contours) as well as in color-selective cells, whereas the ID stream might be enriched in cells selective for the contours that form object outlines. Similarly, there might also be perceptual distinctions related to shape versus surface characteristics that could be revealed by selective lesions of the ID versus BD streams.

Conclusions: Filters, feature detectors, and factories revisited

The observations discussed in the preceding section are relevant to the question of whether cortical neurons should be regarded as feature detectors that explicitly encode uniquely discriminable features or objects, be they edges, corners, or bananas (Barlow, 1972; also chapter 26). The feature detector school of thinking presumes that the firing rate of a cell reflects the confidence that the cell's preferred stimulus is present within its receptive field. This, in turn, motivates neurophysiologists to place primary emphasis on determining what the absolutely best stimulus is for any given cell and to pay less attention to the detailed shape of the tuning curve. We instead align ourselves with the alternative view of considering neurons as filters (either linear or nonlinear) that are relatively broadly tuned along multiple dimensions, so that any particular characteristic of a visual stimulus can be inferred only from comparisons across a population of cells. This line of thinking emphasizes that the information conveyed by visual neurons lies mainly on the slopes of their multidimensional tuning curves. In this regard, the primary objectives of receptive field characterization are to identify the set of stimulus dimensions along which any given cell conveys reliable information and to characterize the shapes of these tuning profiles, be they simple or complex. Even if the process fails to reveal the ultimate stimulus that drives a cell better than any other conceivable stimulus, this would not be a matter of great concern, because such a stimulus might never actually appear in the receptive field of that neuron over the life of the animal.

Finally, we return briefly to the factory analogy, raised earlier, to address two remaining points, one relating to feedback and the other to explicit control systems. As is evident in figures 24.4, 24.5, and 24.7, the overall architecture of visual cortex consists of a variety of processing streams (assembly lines) that run in parallel at any given level but converge and diverge in order to accommodate the analysis of a wide variety of different types of visual information. In a factory, feedback is essential for proper functioning, both for short-range fine-tuning operations and for a variety of long-range interactions. Likewise, in the visual system, there are powerful feedback connections at every level beyond the optic nerve. The function of these feedback pathways is not well understood, but they will surely turn out to be important in a variety of ways.

For a factory to run smoothly, it is desirable to have an explicit management (control) structure whose functions are distinct from those of the assembly lines. Likewise, for the visual system, there are good reasons for postulating the existence of explicit control systems related to phenomena such as directed visual attention (cf. Olshausen, Anderson, and Van Essen, 1993; Van Essen, Anderson, and Olshausen, in press). This general type of organization allows multiple processes (computations) to operate concurrently and interactively, as long as they contribute to the efficient utilization of resources for the specific constellation of products (tasks) that the factory was designed to produce.

ACKNOWLEDGMENTS We thank W. Press for comments on the manuscript, S. Danker for help in manuscript preparation, and many colleagues for valuable discussions. Work from our laboratories was supported by National Institutes of Health grants EY02091, EY08406, and EY10244.

REFERENCES

Allman, J. M., F. Meizin, and E. McGuinness, 1985. Stimulus specific responses from beyond the receptive field: Neurophysiological mechanisms for local-global comparisons in visual neurons. *Annu. Rev. Neurosci.* 8:407–429.

Andersen, R. A., 1987. Inferior parietal lobule function in spatial perception and visuomotor integration. In *Handbook of Physiology: Sec. 1. The Nervous System*, V. B. Mountcastle, F. Plum, and S. R. Geiger, eds. Bethesda, Md.: American Physiological Society, pp. 483–518.

Baizer, J. S., L. G. Ungerleider, and R. Desimone, 1991. Organization of visual inputs to the inferior temporal and posterior parietal cortex in macaques. *J. Neurosci.* 11:168–190.

Barlow, H. B., 1972. Single units and sensation: A neuron doctrine for perceptual psychology? *Perception* 1:371–394.

Benson, D. L., P. J. Isackson, S. H. Hendry, and E. G. Jones, 1991. Differential gene expression for glutamic acid decarboxylase and type II calcium-calmodulin–dependent protein kinase in basal ganglia, thalamus, and hypothalamus of the monkey. *J. Neurosci.* 11:1540–1564.

Blakemore, C., and F. Vital-Durand, 1986. Organization and postnatal development of the monkey's lateral geniculate nucleus. *J. Physiol.* 380:453–491.

Born, R. T., and R. B. H. Tootell, 1992. Segregation of global and local motion processing in primate middle temporal visual area. *Nature* 357:497–499.

Britten, K. H., M. N. Shadlen, W. T. Newsome, and J. A. Movshon, 1992. The analysis of visual motion: A comparison of neuronal and psychophysical performance. *J. Neurosci.* 12:4745–4765.

Buchsbaum, G., 1987. Color signal coding: Color vision and color television. *Color Res. Appl.* 12:266–269.

Burkhalter, A., and D. C. Van Essen, 1986. Processing of color, form, and disparity in visual areas VP and V2 of ventral extrastriate cortex in the macaque monkey. *J. Neurosci.* 6:2327–2351.

Casagrande, V. A., and E. A. Lachica, 1992. What are the cytochrome oxidase (CO) blobs and interblobs really segregating? *Invest. Ophthalmol. Vis. Sci.* 33:900.

Casagrande, V. A., and T. T. Norton, 1991. Lateral geniculate nucleus: A review of its physiology and function. In *Vision and Visual Dysfunction: Vol. 4. The Neural Basis of Visual Function*, A. G. Leventhal, ed. New York: Macmillan.

Colby, C., 1991. The neuroanatomy and neurophysiology of attention. *J. Child Neurol. Suppl.* 6:S90–S118.

Conley, M., and D, Fitzpatrick, 1989. Morphology of retinogeniculate axons in the macaque. *Visual Neurosci.* 2: 287–296.

Dacey, D. M., and M. R. Petersen, 1992. Dendritic field size and morphology of midget and parasol ganglion cells of the human retina. *Proc. Natl. Acad. Sci. U.S.A.* 89:9666–9670.

Desimone, R., T. D. Albright, C. G. Gross, and C. Bruce, 1984. Stimulus-selective properties of inferior temporal neurons in the macaque. *J. Neurosci.* 4:2051–2062.

Desimone, R., and L. Ungerleider, 1989. Neural mechanisms of visual processing in monkeys. In *Handbook of Neuropsychology*, vol. 2, F. Boller and J. Grafman, eds. pp. 267–299, Amsterdam: Elsevier.

DeYoe, E. A., D. J. Felleman, D. C. Van Essen and E. McClendon, 1994. Multiple processing streams in occipitotemporal visual cortex. Manuscript submitted for publication.

DeYoe, E. A., S. Glickman, and J. Wieser, 1992. Clustering of visual response properties in cortical area V4 of macaque monkeys. *Soc. Neurosci. Abstr.* 18:592.

DeYoe, E. A., S. Hockfield, H. Garren, and D. C. Van Essen, 1990. Antibody labeling of functional subdivisions in visual cortex: CAT-301 immunoreactivity in striate and extrastriate cortex of the macaque monkey. *Visual Neurosci.* 5:67–81.

DeYoe, E. A., and D. C. Van Essen, 1988. Concurrent processing streams in monkey visual cortex. *Trends Neurosci.*, 11:219–226.

Dobkins, K. R., and T. D. Albright, 1993. What happens if it changes color when it moves? Psychophysical experiments on the nature of chromatic input to motion detectors. *Vision Res.* 33:1019–1036.

Duffy, C. J., and R. H. Wurtz, 1991. Sensitivity of MST neurons to optic flow stimuli: I. A continuum of response selectivity to large-field stimuli. *J. Neurophysiol.* 65:1329–1345.

Felleman, D. J., E. McClendon, and K. Lin, 1992. Modular segregation of visual pathways in occipital and temporal lobe visual areas in the macaque monkey. *Soc. Neurosci. Abstr.* 18:390.

Felleman, D. J., and D. C. Van Essen, 1987. Receptive field properties of neurons in area V3 of macaque monkey extrastriate cortex. *J. Neurophysiol.* 57:889–920.

Felleman, D. J., and D. C. Van Essen, 1991. Distributed hierarchical processing in primate visual cortex. *Cerebral Cortex*, 1:1–47.

Ferrera, V. P., T. A. Nealey, and J. H. R. Maunsell, 1992. Mixed parvocellular and magnocellular geniculate signals in visual area V4. *Nature* 358:756–758.

Fitzpatrick, D., J. S. Lund, and G. G. Blasdel, 1985. Intrinsic connections of macaque striate cortex: Afferent and efferent connections of lamina 4C. *J. Neurosci.* 5: 3329–3349.

Fuster, J. M., 1990. Inferotemporal units in selective visual attention and short-term memory. *J. Neurophysiol.* 64:681–697.

Gallant, J. L., J. Braun, and D. C. Van Essen, 1993. Selectivity for polar, hyperbolic, and cartesian gratings in macaque visual cortex. *Science* 259:100–103.

Grossberg, S., 1987a. Cortical dynamics of three-dimensional form, color, and brightness perception: I. Monocular theory. *Percept. Psychophys.* 41:87–116.

Grossberg, S., 1987b. Cortical dynamics of three-dimensional form, color, and brightness perception: II. Binocular theory. *Percept. Psychophys.* 41:117–158.

Grünert, U., U. Greferath, B. B. Boycott, and H. Wässle, 1993. Parasol (Pα) ganglion-cells of the primate fovea: Immunocytochemical staining with antibodies against $GABA_A$-receptors. *Vision Res.* 33:1–14.

Harting, J. K., M. F. Huerta, T. Hashikawa, and D. P. Van Leishout, 1991. Projection of the mammalian superior colliculus upon the dorsal lateral geniiculate nucleus:

Organization of tectogeniculate pathways in nineteen species. *J. Comp. Neurol.* 304:275–306.
HICKS, T. P., B. B. LEE, and T. R. VIDYASAGAR, 1983. The responses of cells in macaque lateral geniculate nucleus to sinusoidal gratings. *J. Physiol.* 337:183–200.
KAPLAN, E., B. B. LEE, and R. SHAPLEY, 1990. New views of primate retinal function. In *Progress in Retinal Research*, vol. 9, N. Osborne and G. Cader, eds. Oxford: Pergamon, pp. 273.
KNIERIM, J. J., and D. C. VAN ESSEN, 1992. Visual cortex: Cartography, connections, and concurrent processing. *Curr. Opin. Neurobiol.* 2:150–155.
KOMATSU, H., Y. IDEURA, H. KAJI, and S. YAMANE, 1992. Color selectivity of neurons in the inferior temporal cortex of the awake macaque monkey. *J. Neurosci.* 12:408–424.
LACHICA, E. A., P. D. BECK, and V. A. CASAGRANDE, 1992. Parallel pathways in macaque monkey striate cortex: Anatomically defined columns in layer III. *Proc. Natl. Acad. Sci.* 89:3566–3570.
LACHICA, E. A., and V. A. CASAGRANDE, 1992. Direct W-like geniculate projections to the cytochrome oxidase (CO) blobs in primate visual cortex: Axon morphology. *J. Comp. Neurol.* 319:141–158.
LIVINGSTONE, M. S., and D. H. HUBEL, 1984. Anatomy of physiology of a color system in the primate visual cortex. *J. Neurosci.* 4:309–356.
LIVINGSTONE, M. S., and D. H. HUBEL, 1987. Connections between layer 4B of area 17 and the thick cytochrome oxidase stripes of area 18 in the squirrel monkey. *J. Neurosci.* 7:3371–3377.
LIVINGSTONE, M., and D. HUBEL, 1988. Segregation of form, color, movement, and depth: Anatomy, physiology, and perception. *Science* 240:740–749.
LUND, J. S., 1988. Excitatory and inhibitory circuitry and laminar mapping strategies in primary visual cortex of the monkey. In *Signal and Sense: Local and Global Order in Perceptual Maps*, G. M. Edelman, W. E. Gall, and W. M. Cowan, eds. New York: Wiley.
LUND, J. S., and T. YOSHIOKA, 1991. Local circuit neurons of macaque monkey striate cortex: III. Neurons of laminae 4B, 4A, and 3B. *J. Comp. Neurol.* 311:234–258.
MARROCCO, R. T., L. W. MCCLURKIN, and R. A. YOUNG, 1982. Spatial action and conduction latency classification of cells of the lateral geniculate nucleus of macaques. *J. Neurosci.* 2:1275–1291.
MAUNSELL, J. H. R., T. P. NEALEY, and D. D. DEPRIEST, 1990. Magnocellular and parvocellular contributions to responses in the middle temporal visual area (MT) of the macaque monkey. *J. Neurosci.* 10:3323–3334.
MAUNSELL, J. H. R., and W. T. NEWSOME, 1987. Visual processing in monkey extrasriate cortex. *Annu. Rev. Neurosci.* 10:363–402.
MAUNSELL, J. H. R., and D. C. VAN ESSEN, 1983. The connections of the middle temporal visual area (MT) and their relationship to a cortical hierarchy in the macaque monkey. *J. Neurosci.* 3:2563–2586.
MERIGAN, W. H., L. M. KATZ, and J. H. R. MAUNSELL, 1991. The effects of parvocellular lateral geniculate lesions on the acuity and contrast sensitivity of macaque monkeys. *J. Neurosci.* 11:994–1001.
MERIGAN, W. H., and J. H. R. MAUNSELL, 1993. How parallel are the primate visual pathways? *Annu. Rev. Neurosci.* 16:369–402.
MIYASHITA, M., 1993. Inferior temporal cortex: Where visual perception meets memory. *Annu. Rev. Neurosci.* 16: 245–264.
MOREL, A., and J. BULLIER, 1990. Anatomical segregation of two cortical visual pathways in the macaque monkey. *Visual Neurosci.* 4:555–578.
MOVSHON, J. A., E. A. ADELSON, M. S. GIZZI, and W. T. NEWSOME, 1985. The analysis of moving visual patterns. In *Pattern Recognition Mechanisms*, C. Chagas, R. Gattass, and C. Gross, eds. Rome: Vatican Press.
NEALEY, T. A., V. P. FERRERA, and J. H. R. MAUNSELL, 1991. Magnocellular and parvocellular contributions to the ventral extrastriate cortical processing stream. *Soc. Neurosci. Abstr.* 17:525.
NORTON, T. T., V. A. CASAGRANDE, G. E. IRVIN, M. A. SESMA, and H. M. PETRY, 1988. Contrast-sensitivity functions of W-, X-, and Y-like relay cells in the lateral geniculate nucleus of bush baby, *Galago crassicaudatus*. *J. Neurophysiol.* 59:1639–1656.
O'KUSKY, J., and M. COLONNIER, 1982. A laminar analysis of the number of neurons, glia, and synapses in the visual cortex (area 17) of adult macaque monkeys. *J. Comp. Neurol.* 210:178–290.
OLSHAUSEN, B. A., C. H. ANDERSON, and D. C. VAN ESSEN, 1993. A neurobiological model of visual attention and invariant pattern recognition based on dynamic routing of information. *J. Neurosci.*, 13:4700–4719.
OPTICAN, L. M., T. J. GAWNE, B. J. RICHMOND, and P. J. JOSEPH, 1991. Unbound measures of information and channel capacity from multivariate neuronal data. *Biol. Cybernt.* 65:305–310.
OPTICAN, L. M., and B. J. RICHMOND, 1987. Temporal encoding of two-dimensional patterns by single units in primate inferior temporal cortex: III. Information theoretic anaysis. *J. Neurophysiol.* 57:162–178.
PAPATHOMAS, T. V., A. GOREA, and B. JULESZ, 1991. Two carriers for motion correspondence: Color and luminance. *Vision Res.* 31:1883–1891.
PERRY, V. H., R. OEHLER, and A. COWEY, 1984. Retinal ganglion cells that project to the dorsal lateral geniculate nucleus in the macaque monkey. *Neuroscience* 12:1101–1123.
PETERHANS, E., and R. VON DER HEYDT, 1993. Functional organization of area V2 in the alert macaque. *Eur. J. Neurosci.* 5:509–524.
REID, R. C., and R. M. SHAPLEY, 1992. Spatial structure of cone inputs to receptive fields in primate lateral geniculate nucleus. *Nature* 356:716–718.
ROCKEL, A. J., R. W. HIORNS, and T. P. S. POWELL, 1980. The basic uniformity in structure of the neocortex. *Brain* 103:221–244.

ROCKLAND, K. S., 1985. A reticular pattern of intrinsic connections in primate area V2 (area 18). *J. Comp. Neurol.* 235:467–478.

ROCKLAND, K. S., and D. N. PANDYA, 1979. Laminar origins and terminations of cortical connections of the occipital lobe in the rhesus monkey. *Brain Res.* 179:3–20.

SCHEIN, S. J., and R. DESIMONE, 1990. Spectral properties of V4 neurons in the macaque. *J. Neurosci.* 10:3369–3389.

SCHILLER, P. H., N. K. LOGOTHETIS, and E. R. CHARLES, 1990. Role of the color-opponent and broad-band channels in vision. *Visual Neurosci.* 5:321–346.

SHANNON, C. E., 1948. A mathematical theory of communication. *Bell Systems Tech. J.* 27:379–423.

STONE, J., B. DREHER, and A. LEVENTHAL, 1979. Hierarchical and parallel mechanisms in the organization of visual cortex. *Brain Res. Rev.* 1:345–394.

STONER, G. R., and T. D. ALBRIGHT, 1993. Image segmentation cues in motion processing: Implications for modularity in vision. *J. Cogn. Neurosci.* 5:129–149.

TANAKA, K., and H.-A. SAITO, 1989. Analysis of motion of the visual field by direction, expansion/contraction, and rotation cells clustered in the dorsal part of the medial superior temporal area of the macaque monkey. *J. Neurophysiol.* 62:626–641.

TANAKA, K., H.-A. SAITO, Y. FUKADA, and M. MORIYA, 1991. Coding visual images of objects in the inferotemporal cortex of the macaque monkey. *J. Neurophysiol.* 66: 170–189.

VAN ESSEN, D. C., and C. H. ANDERSON, 1990. Information processing strategies and pathways in the primate retina and visual cortex. In *Introduction to Neural and Electronic Networks*, S. F. Zornetzer, J. L. Davis, and C. Lau, eds. Orlando, Fla.: Academic Press, pp. 43–72.

VAN ESSEN, D. C., C. H. ANDERSON, and D. J. FELLEMAN, 1992. Information processing in the primate visual system: An integrated systems perspective. *Science* 255:419–423.

VAN ESSEN, D. C., C. H. ANDERSON, and B. A. OLSHAUSEN, (in press). Dynamic routing strategies in sensory, motor, and cognitive processing. In *Large Scale Neuronal Theories of the Brain*, C. Koch and J. Davis, eds. Cambridge, Mass.: MIT Press.

VAN ESSEN, D. C, D. J. FELLEMAN, E. A. DEYOE, J. OLAVARRIA, and J. J. KNIERIM, 1990. Modular and hierarchical organization of extrastriate visual cortex in the macaque monkey. *Cold Spring Harb. Symp. Quant. Biol.* 55:679–696.

VAN ESSEN, D. C, W. T. NEWSOME, and J. H. R. MAUNSELL, 1984. The visual field representation in striate cortex of the macaque monkey: Asymmetries, anisotropies and individual variability. *Vision Res.* 24:429–448.

VAN ESSEN, D. C, W. T. NEWSOME, J. H. R. MAUNSELL, and J. L. BIXBY, 1986. The projections from striate cortex (V1) to areas V2 and V3 in the macaque monkey: Asymmetries, areal boundaries, and patchy connections. *J. Comp. Neurol.* 244:451–480.

WONG-RILEY, M., and CARROLL, E. W., 1984. Effect of impulse blockage on cytochrome oxidase activity in monkey visual system. *Nature* 307:262–264.

YOSHIOKA, T., J. B. LEVITT, and J. S. LUND, 1993. Anatomical basis for channel interactions in macaque cortical visual area V1. *Invest. Ophthalmol. Vis. Sci.* 34:1173.

YOUNG, M. P., 1992. Objective analysis of the topological organization of the primate cortical visual system. *Nature* 358:152–155.

ZEKI, S., 1983a. The distribution of wavelength and orientation selective cells in different areas of monkey visual cortex. *Proc. R. Soc. Lond. [Biol.]* 217:449–470.

ZEKI, S., 1983b. Colour coding in the cerebral cortex: The responses of wavelength-selective and colour-coded cells in monkey visual cortex to changes in wavelength composition. *Neuroscience* 9:767–781.

ZEKI, S. M., and SHIPP, S., 1988. The functional logic of cortical connections. *Nature* 335:311–317.

25 Visual Motion: Linking Neuronal Activity to Psychophysical Performance

WILLIAM T. NEWSOME, MICHAEL N. SHADLEN, EHUD ZOHARY, KENNETH H. BRITTEN, AND J. ANTHONY MOVSHON

ABSTRACT We describe an experimental and theoretical analysis of the neural systems underlying perceptual discriminations of motion direction. Single-neuron recordings obtained while alert monkeys performed the discrimination task revealed two surprising relationships between physiological responses and psychophysical performance: First, on average, psychophysical sensitivity was comparable to the sensitivity of single cortical neurons. Second, decisions made near psychophysical threshold were partially correlated with the trial-to-trial response fluctuations of *single* cortical neurons. Both these results suggested that psychophysical decisions may be based on the responses of a small number of cortical neurons, a conclusion that seems implausible given the large population of directionally selective neurons available within the visual cortex.

We developed a biologically constrained model that permits us to simulate actual experiments and to identify conditions that could give rise to the unexpectedly close relationships between physiology and psychophysical performance. Both relationships can be reproduced in the simulations, even for large numbers of input neurons, if the activity of the input neurons is weakly intercorrelated and if additional noise sources are present at the processing stage where the responses are pooled. Given realistic values for pooling noise, the amount of correlated activity necessary to bring simulated and actual data into agreement corresponds well with that observed between adjacent neurons during multiunit recording experiments in the visual cortex. In sum, our efforts lead to a reasonably straightforward account of the sensory processes underlying performance on our task and suggest future experiments that may shed light on the neural mechanisms that form and implement decisions.

WILLIAM T. NEWSOME, MICHAEL N. SHADLEN, EHUD ZOHARY, and KENNETH H. BRITTEN Department of Neurobiology, Stanford University School of Medicine, Stanford, Calif.
J. ANTHONY MOVSHON Howard Hughes Medical Institute, Center for Neural Science and Department of Psychology, New York University, New York, N.Y.

How does cognitive function emerge from the coordinated activity of neural systems within the brain? This deceptively simple question encompasses a host of vexing problems, both practical and theoretical, in fields as diverse as neurophysiology, computational modeling, decision theory, and the quantitative assessment of behavior. Novel insights into the neural basis of cognition frequently arise when ideas and techniques from several of these disciplines are directed in concert toward a single, well-defined problem.

In this chapter, we describe our recent efforts to address one such problem in cognitive neuroscience: the perception of visual motion. Our investigation involved neurophysiological recording from directionally selective neurons in the visual cortex of alert monkeys while the animals performed a rigorously controlled direction discrimination task. Having recorded both the neural signals present in the cortex and the animals' perceptual judgments, we sought to link neural activity to psychophysical performance by constructing a biologically constrained model that receives as input neural signals such as those we recorded and produces as output judgments of the same reliability as those produced by our monkeys.

Our basic strategy in this investigation (and some of our problems as well) can be appreciated by considering figure 25.1, a drawing taken from the introduction to Norma Graham's recent book, *Visual Pattern Analyzers* (1989). A subject views a visual pattern that varies in luminance as a function of two spatial dimensions and time [$L(x, y, t)$] with the goal of performing a visual discrimination. The pattern is filtered and encoded by low-level analyzers within the visual system that provide inputs to higher-level mechanisms includ-

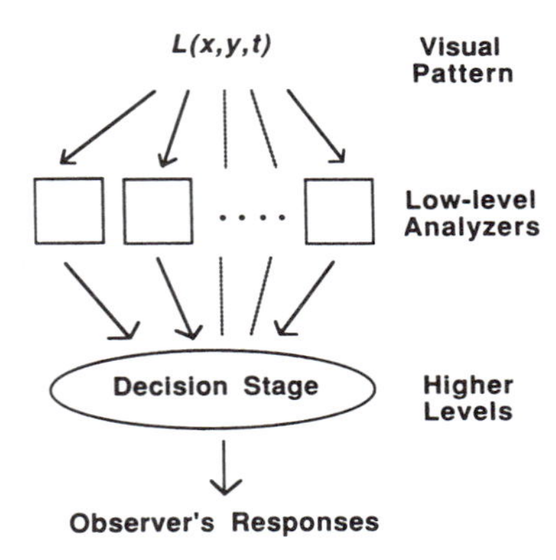

FIGURE 25.1 A general framework for thinking about the neural basis of psychophysical performance. $L(x, y, t)$, luminance, two spatial dimensions, and time. (Reprinted from Graham, 1989, by permission.)

ing a decision stage. The result of the decision is revealed by the observer's behavioral responses.

From the perspective of cognitive neuroscience, our ultimate goal is to identify and understand the actual neural systems that produce the observer's behavioral responses to a particular stimulus set. At present, physiological analysis of sensory systems (including the present study) is restricted almost entirely to the study of low-level analyzers or intermediate mechanisms that still precede the decision stage. As Graham (1989, pages 3–12) points out, however, it is important to remember that even in the most carefully designed paradigm, the observer's responses are always the product of both the low-level analyzers and a decision process. As figure 25.1 suggests, then, our models for linking neurophysiological data to psychophysics in a quantitative manner should include plausible decision rules that translate neuronal activity into performance. Explicit declaration and modeling of decision rules causes the physiologist to think carefully about the neural processes that transform sensory signals into measurable behavior.

With this general framework in mind, we can state three tactics that were essential to the design of this study:

1. We employed a set of visual stimuli that selectively modulates directionally selective neurons within the brain; the stimulus set contains no spurious features that might permit successful discrimination on the basis of nonmotion cues such as form or position. By designing the stimulus set to match the properties of directionally selective neurons, we ensured that these cells would serve as low-level analyzers for our psychophysical paradigm.

2. We used a very simple psychophysical paradigm so as to reduce the necessary processing steps between low-level analyzers and the decision stage. In essence, we tried to render any intermediate processing stages "transparent" so that the monkeys' psychophysical responses would reflect as directly as possible the outputs of the low-level analyzers.

3. All experiments were carried out with weak motion stimuli near psychophysical threshold. It is near threshold, where the monkey performs imperfectly, that the relationship between neural activity and psychophysical decision can be gauged most revealingly.

Thus, we attempted to create an experimental paradigm that offers a reasonable prospect of understanding the neural processes underlying a simple form of cognitive behavior.

The paradigm

VISUAL STIMULI We trained monkeys to discriminate opposed directions of coherent motion (up vs. down,

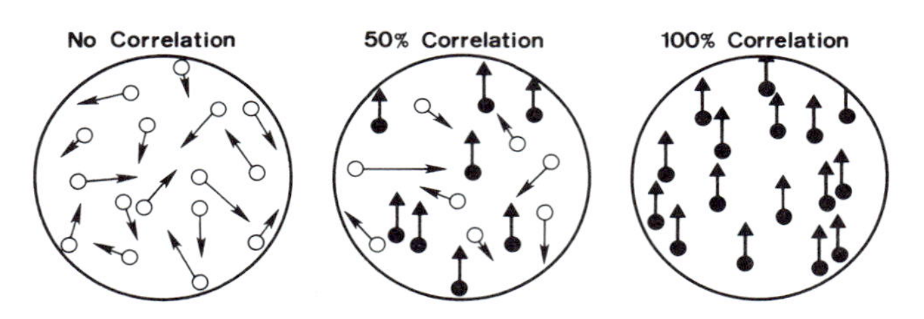

FIGURE 25.2 A schematic representation of the visual stimulus set employed in these experiments. The stimuli were patterns of dynamic random dots plotted sequentially on the face of an oscilloscope. Each dot was visible briefly before being replaced 45 ms later by a dot at another location on the screen. In the form of the stimulus illustrated in the left-hand panel, the locations of the replacement dots were selected completely at random (0% correlation). This display consisted of dynamic noise with no net motion in any single direction. In the form of the stimulus illustrated at the right, each dot was replaced by a partner dot with a fixed offset in space and time (100% correlation). In this display, each dot moved uniformly in the specified direction. The center panel illustrates a mixed stimulus in which 50% of the dots carried the uniform motion signal and the remaining dots provided a masking motion noise. (Reprinted from Newsome and Paré, 1988, by permission.)

right vs. left, etc.) in the set of visual stimuli schematized in figure 25.2. The stimuli were dynamic random dots plotted sequentially on the face of a cathode ray tube (CRT) at a very high rate (6.67 kHz). After 45 ms, a dot is either displaced in a specified direction (correlated motion) or replaced by another dot at a random location on the screen (noise). In one extreme form of the display, illustrated in the left-hand panel of figure 25.2, all dots were positioned randomly so that the display was pure noise. In this form, the display contained many local motion events (due to fortuitous pairings of dots in space and time) but contained no *net* motion in any single direction. At the other extreme, illustrated in the right-hand panel of figure 25.2, all dots were displaced uniformly so that the display contained noise-free motion in a specified direction. Our software permitted us to create any stimulus intermediate between these two extremes by specifying the percentage of dots that carried the correlated motion signal. In the center panel of the figure, for example, 50% of the dots are in correlated motion while the remaining 50% provide a masking motion noise. The percentage of dots in correlated motion governs the strength of the motion signal without affecting the overall luminance, contrast, or average spatial and temporal structure of the stimulus. (See Britten et al., 1992, for a detailed description of this stimulus set.)

As suggested in the preceding section, this set of stimuli gives us several experimental advantages. First, it allows us to manipulate the direction and strength of the motion signal while keeping other properties of the stimulus constant. Second, the observer must extract the motion signal from a spatially dispersed and constantly changing field of random dots; there are no familiar form or position cues in the display that might provide a spurious means of identifying the direction of motion. Finally, we are able to conduct experiments near psychophysical threshold simply by reducing the strength of the motion signal to an appropriate level. Thus, the stimulus set permits manipulation of directionally sensitive mechanisms within the brain and ensures that psychophysical decisions will be based on the output of these mechanisms.

Electrophysiological Recording While the monkeys performed the direction discrimination task, we recorded the activity of single directionally selective neurons in visual area MT (or V5) of the cerebral cortex. More than 90% of the neurons in MT are directionally selective (Zeki, 1974; Maunsell and Van Essen, 1983b; Albright, 1984), and these neurons are known to play a prominent role in processing the motion signals contained in our visual stimuli (Newsome and Paré, 1988; Salzman, Britten, and Newsome, 1990; Salzman et al., 1992).

The essential characteristic of directionally selective neurons is that they respond with a stream of action potentials when motion is presented within their receptive fields in a particular "preferred" direction. The activity of these cells is minimized, and frequently inhibited, by motion in the opposite, or "null," direction. Most directionally selective neurons also respond to a range of directions centered around the preferred, but the deepest modulation in the activity of such cells is caused by motion along the preferred-null axis. For each neuron studied, therefore, we adjusted the psychophysical task so that the monkey discriminated opposed directions of motion aligned along the cell's preferred-null axis. In addition, the random-dot stimuli were positioned on the CRT screen so that they covered, but did not extend beyond, the boundaries of the cell's receptive field. To keep the random-dot pattern positioned on the receptive field during a trial, the monkey was required to hold its eyes steady on a specified fixation point during each stimulus presentation. (The monkey's eye position was measured continuously using the scleral search coil technique, and trials were aborted if the monkey failed to fixate properly; for details see Britten et al., 1992). For each neuron, therefore, we created a psychophysical situation in which decisions were most likely to be influenced by the activity of the neuron under study.

Action potentials from single neurons were recorded using standard physiological techniques, and the time of occurrence of each action potential was recorded for subsequent analysis. For all of the analyses in this chapter, the response of the neuron on a particular trial was considered to be the total number of action potentials recorded during presentation of the visual stimulus.

Psychophysical Procedures Each trial began with the onset of a fixation point. After the monkey fixated, the random-dot stimulus appeared for 2 seconds within the neuron's receptive field. Correlated motion occurred in either the preferred or null direction of the cell under study; the direction was chosen randomly for each trial. At the end of the 2-second viewing period, the monkey indicated its judgment of the direction of correlated motion by making an eye movement to one of two light-emitting diodes corresponding to the two possible directions of motion. Correct choices were rewarded with water or juice, whereas incorrect choices were punished with a brief time-out period between trials.

Each experiment consisted of 15 or more trials in each of the preferred and null directions at each of several correlation levels spanning psychophysical threshold. In addition, 30 trials were included at 0% correlation. On the latter trials, rewards were given randomly with a probability of 0.5, without regard for the monkey's choice, as there was no net motion in either the preferred or null direction. For each experiment, the monkey's choices were compiled into a psychometric function in which the percentage of correct responses was plotted against the strength of the motion signal. Psychophysical threshold, considered to be the motion strength for which the monkey chose correctly on 82% of the trials, was computed from a sigmoidal curve fitted to the psychophysical data. Under optimal conditions, psychophysical thresholds for monkey and human observers can be as low as 2–4% correlation. Thresholds varied widely in our experiments, however, because the psychophysical conditions were adjusted to match the receptive field location, receptive field size, the preferred-null axis, and the preferred speed for each neuron studied. The geometrical mean psychophysical threshold for all experiments in this study was approximately 14% correlation.

Three experimental observations

Our experiments thus far have yielded three basic observations that must inform and constrain models linking the activity of directionally selective neurons to performance on our task. Two of these observations concern the relationship of neuronal responses to psychophysical decisions, and the third concerns the relationship between the responses of neighboring neurons in MT. We will summarize each of these results in turn.

A Comparison of Neuronal and Psychophysical Sensitivity In the first analysis, we compared the sensitivities of single MT neurons to the psychophysical sensitivity of the monkeys (Newsome et al., 1989; Britten et al., 1992). The gist of this comparison can be quickly grasped by considering our stimulus set, as illustrated in figure 25.2. Psychophysical sensitivity to the motion signals is assessed by computing the stimulus strength at which the monkey discriminates the direction of the moving dots with 82% accuracy, as described in the preceding section. It is easy to see that directionally selective neurons will exhibit a similar threshold phenomenon in their responses to this stimulus set. For strong motion signals, a directionally selective neuron consistently generates larger responses to preferred direction motion than to null direction motion. The neuron thus signals the direction of motion reliably, even though its responses to any given stimulus are somewhat variable. For very weak motion signals, however, a directionally selective neuron provides no information about the direction of motion as it responds roughly equally to both directions. Thus, the sensitivity of a single neuron can be expressed as a threshold, below which the neuron fails to signal the direction of motion in a reliable manner. This neuronal threshold may then be compared to the monkey's psychophysical threshold measured on the same set of trials.

Neuronal thresholds can be computed from our physiological data using methods based on the theory of signal detection (Green and Swets, 1966; see Britten et al., 1992, for a complete exposition). The relationship between neuronal and psychophysical sensitivity can then be expressed as the ratio of neuronal to psychophysical threshold. Figure 25.3 illustrates the distribution of such threshold ratios for 216 experiments conducted in three rhesus monkeys. Values less than unity indicate neurons that were more sensitive (i.e., lower thresholds) than was the monkey psychophysically, whereas ratios greater than unity indicate cells that were less sensitive than the monkey. The outcome of the comparison is unequivocal: The sensitivity of MT neurons is, on average, very similar to the psychophysical sensitivity of the monkeys (geometrical mean ratio, 1.19).

This is a most puzzling finding. Given the massive numbers and complex interconnections of neurons within the visual cortex, one would expect even simple psychophysical decisions to be based on the responses of a large pool of neurons. Thus, noise in the responses of individual neurons could be averaged out in the pooling process, leading to psychophysical sensitivity that is substantially better than that exhibited by

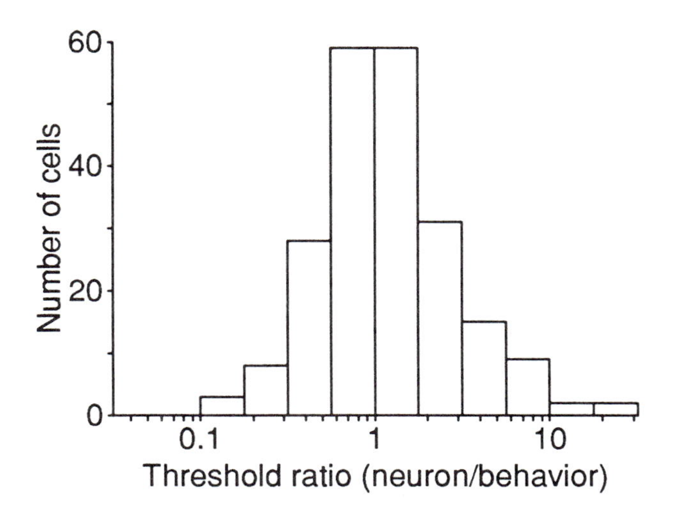

FIGURE 25.3 The distribution of threshold ratios obtained in 216 single-neuron recording experiments in three rhesus monkeys. The distribution is centered near unity, showing that MT neurons were, on average, as sensitive to the motion signals as were the monkeys psychophysically. (Reprinted from Britten et al., 1992, by permission.)

individual neurons (e.g., Pirenne, 1943; Tolhurst, Movshon, and Dean, 1983). In the context of our psychophysical paradigm, however, the data in figure 25.3 clearly show that any model linking neuronal activity to psychophysical performance must account for *roughly equal* neuronal and psychophysical sensitivities.

A TRIAL-TO-TRIAL COVARIATION BETWEEN NEURONAL RESPONSE AND PSYCHOPHYSICAL DECISION The second analysis we performed can be appreciated by considering a monkey's decisions on the direction discrimination task for motion conditions near or below psychophysical threshold. Over the course of a single experiment, the monkey sees each stimulus condition (motion strength and direction) 15 or more times in a randomized sequence. For a given near-threshold motion condition, the monkey will decide in favor of the preferred direction of motion on some proportion of the trials and in favor of the null direction on the remaining trials. This noisy pattern of decisions occurs even when the visual stimulus itself is identical on each trial for a given motion condition.

Consider also that the responses of an individual neuron to successive presentations of a particular stimulus are very noisy. Such responses can be characterized by a *mean* number of action potentials discharged, but there is substantial trial-to-trial variance

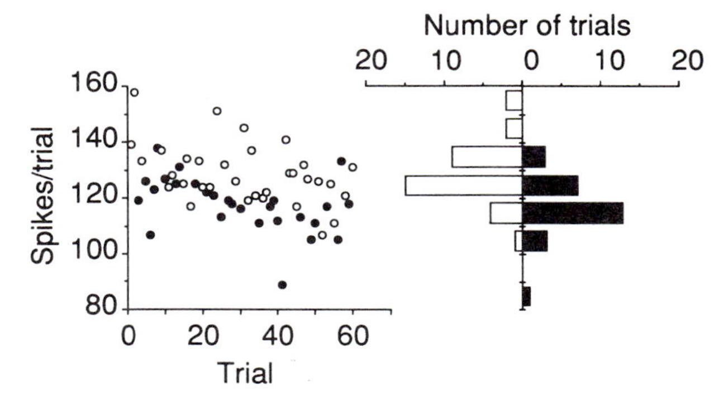

FIGURE 25.4 Covariation of neuronal response and psychophysical decision. The responses of a single neuron are plotted for 60 successive presentations of a 0% correlation stimulus. The monkey's psychophysical decision on each trial is represented by the symbol used for that data point. An open circle indicates that the monkey chose the neuron's preferred direction on a particular trial, whereas a closed circle indicates a choice of the null direction. On average, the monkey tended to select the preferred direction on trials in which the neuron generated a larger response, as indicated by the fact that the open circles tend to fall toward the upper half of the scatterplot. The histogram to the right is a frequency distribution that shows the cumulative response distribution for all trials in which the monkey decided in favor of the preferred direction (open bars) and for all trials in which the monkey decided in favor of the null direction (solid bars). The response of the neuron is plotted on the vertical axis of the histogram, and the number of times the response was observed is plotted on the horizontal axis. The frequency histograms were formed by counting the number of data points in horizontally oriented bins extending across the full width of the scatterplot. The distribution of responses for preferred direction choices is offset upward (toward larger responses) relative to the distribution for null direction choices.

about the mean (e.g., Dean, 1981; Britten et al., 1993). We therefore asked whether the monkey's decisions were correlated with the neuron's responses on successive presentations of a given motion condition. In particular, we wondered whether the monkey would be more likely to choose the neuron's preferred direction on trials in which that neuron's response was larger than average.

Figure 25.4 illustrates such an effect for one neuron. The left-hand panel shows responses for 60 presentations of a 0% correlation stimulus (these presentations were randomly interleaved among the usual range-of-motion conditions for a single experiment). The monkey's psychophysical decision on each trial is indicated by the symbol for each data point; open circles indicate decisions favoring the preferred direction, whereas closed symbols represent decisions favoring the null direction. Because the 0% correlated stimulus contained no net motion in either direction, the monkey's decisions varied randomly between the preferred and null directions. However, the monkey tended to choose the preferred direction on trials in which the neuron responded with a larger number of action potentials (i.e., the open circles are, on average, displaced upward relative to the closed circles). The effect is seen most clearly in the frequency histograms in the right-hand panel of figure 25.4, which were formed by summing the data points in the scatterplot within horizontally oriented bins. Thus, the histogram formed by the open bars shows the cumulative distribution of neuronal responses on trials in which the monkey chose the preferred direction, whereas the closed bars show the distribution of responses on trials in which the monkey chose the null direction. The preferred decision distribution (open bars) is offset in the direction of larger neuronal responses (upward) with respect to the null decision distribution, confirming our qualitative impression of the data in the scatterplot.

The offset between the two histograms in figure 25.4 reflects the degree of association between neuronal response and psychophysical decision, and this offset can be quantified using a metric based on signal detection theory (Newsome et al., 1989). This metric, which we refer to as a *sender operating characteristic*, or SOC value, ranges from a value of 0.5 if the two distributions are identical to a value of 1.0 if the two distributions are completely nonoverlapping. In the former case, trial-to-trial fluctuations in neuronal response convey no information about the monkey's eventual decision; in the latter case, trial-to-trial fluctuations in responsiveness would predict the decision perfectly. The metric can assume values below 0.5 if neuronal response and psychophysical decision are *anticorrelated*—that is, if the monkey tends to choose the preferred direction on trials in which the neuron yields *weak* responses.

The distribution of SOC values for all 216 neurons in our study is shown in figure 25.5. The mean SOC value was 0.54 for this sample of neurons, which is significantly greater than 0.5 (t test, $p < .0001$). On the whole, therefore, trial-to-trial fluctuations in the responses of MT neurons were weakly correlated with fluctuations in the monkeys' choices of motion direction; larger neuronal responses lead to an increased probability of a preferred direction choice. In other words, some portion of the variance in the monkey's decision process can be accounted for by the variance in responsiveness of *single* MT neurons.

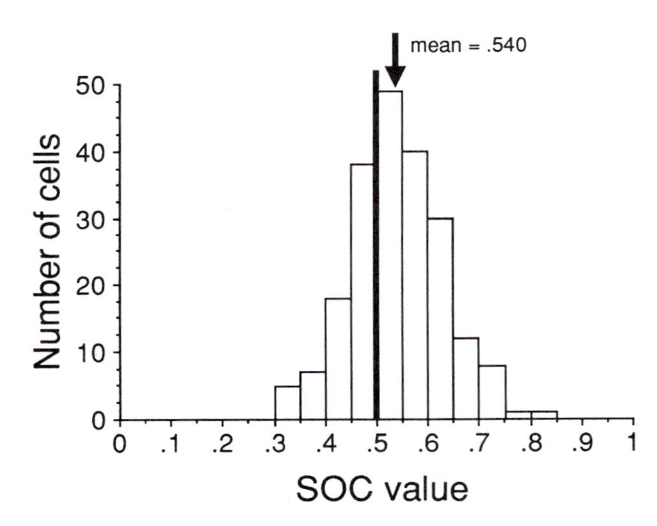

FIGURE 25.5 The distribution of SOC values for all 216 neurons in our data set. The distribution is significantly offset to the right of 0.5 (t test, $p < .0001$), indicating a significant covariation of neuronal response and psychophysical decision across the entire sample of MT neurons.

Like the first, this finding was completely unexpected. If decisions are based on the pooled responses of hundreds or thousands of neurons, no single neuron should account for a measurable portion of the variance in a monkey's decisions. The relationship could, of course, arise artifactually if an independent factor is responsible for the variation in *both* neuronal response and perceptual decision (e.g., eye position, attention, small of amounts of stimulus variance). We have conducted several experiments to control for such factors, and we are reasonably confident that the relationship cannot be explained in this way (Britten, et al., in preparation). We believe, therefore, that the covariation between neuronal response and perceptual decision arises because trial-to-trial fluctuations in the activity of MT neurons actually *influence* the outcome of near-threshold decisions. Again, any model of the neural processes underlying performance on our task should reproduce this effect.

CORRELATED RESPONSES OF ADJACENT MT NEURONS
Both results presented thus far were surprising precisely because they violated common conceptions of the effects of pooling signals from independent neurons. Pooling should lead to increased psychophysical sensitivity re-

lative to that of individual neurons, and pooling should dilute the effects on the decision process of random fluctuations in the responses of any single neuron. An alternative possibility, however, is that neurons contributing to the pool are not independent but are instead *correlated* because of common inputs or because of synaptic interactions within the pool. In an extreme case, neurons contributing to the pool might be perfectly correlated, such that no benefits could accrue from pooling because each neuron would carry identical patterns of signal and noise. High levels of correlated firing would also give rise to the observed covariation between neuronal response and perceptual decision as all neurons within the pool would essentially act as one. To deduce the neural processes underlying performance, therefore, it is critical to determine the amount of correlated firing among MT neurons.

We recorded simultaneously from 35 pairs of MT neurons while a monkey performed the direction discrimination task (Zohary, Shadlen, and Newsome, 1992). The recordings were made from a single microelectrode in conjunction with a spike-sorting device that assigned action potentials to individual neurons by a template-matching algorithm (Worgotter, Daunicht, and Eckmiller, 1986). Because the recordings were made with a single electrode, the data were obtained from nearby neurons having similar receptive field properties. Presumably, then, our methods maximized the probability of observing correlated firing.

We recorded the responses of each pair of neurons to successive presentations of a given visual stimulus and then computed a correlation coefficient to determine the amount of correlated firing between the two neurons. After finding that this correlation coefficient did not vary systematically with the direction or strength of the motion signal, we averaged the condition-specific values to obtain a single correlation coefficient for each pair of neurons. Figure 25.6 shows the distribution of correlation coefficients obtained for the 35 pairs in our sample. The mean correlation coefficient for all 35 pairs was 0.143, and the distribution in figure 25.6 is offset significantly to the right of 0 (t test, $p < .0001$). Thus, trial-to-trial fluctuations in responsiveness are indeed correlated between adjacent MT neurons, and the assumption of independence within a pool appears to be wrong. The average magnitude of the correlation was nevertheless small. A major challenge for our modeling efforts is to determine whether this modest level of correlation among MT neurons can account for the

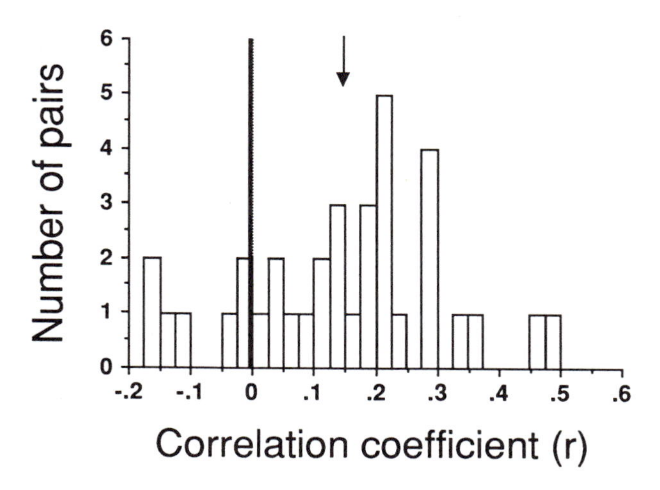

FIGURE 25.6 The distribution of correlation coefficients measured for 35 pairs of simultaneously recorded MT neurons. The distribution is significantly offset to the right of 0.0, indicating that the responses of adjacent MT neurons are not independent. Across the sample of 35 pairs, the mean correlation coefficient was 0.143.

surprising relationships between neuronal activity and psychophysical performance described in the preceding two sections.

Monte Carlo simulations

We have performed an extensive series of Monte Carlo simulations with the goal of developing a biologically constrained model of the neural processes underlying performance on our task. An adequate model should:

1. Accept as input neuronal responses such as those we recorded from MT and produce as output decisions regarding direction of motion.
2. Incorporate modest amounts of correlation within the pool of neurons contributing to the simulated decisions.
3. Produce simulated psychophysical sensitivity that is comparable to the observed psychophysical sensitivity of our monkeys and thus to the mean sensitivity of the contributing neurons.
4. Produce a trial-to-trial covariation between neuronal response and simulated decision (SOC effect) of a magnitude comparable to that observed in our experiments.
5. Conform to constraints 1–4 for large pool sizes.

The basis of constraints 1–4 should be self-evident by now, but constraint 5 warrants further comment. Reasonable agreement between simulated and observed psychophysical sensitivity can be obtained by incorporating a very small number of neurons (<10) into the pool that influences the decision stage. Under these circumstances, psychophysical sensitivity is not substantially improved by pooling because of the limited number of neurons available. In addition, small numbers of neurons will produce an SOC effect because the response fluctuations of any individual neuron within the pool will account for a measurable portion of the decision variance near threshold. However, the anatomy and physiology of the cortical motion system would seem to preclude such a simple solution. For any given location in the visual field, MT may contain several columns of neurons preferring each direction of motion. Each column, in turn, comprises on the order of 10^3 neurons (assuming columnar dimensions of $100 \times 100\ \mu m$ and cell packing densities similar to those in striate cortex; see Albright, Desimone, and Gross, 1984, and O'Kusky and Colonnier, 1982). Even if we consider only efferent neurons from these columns, it would seem that hundreds (if not thousands) of neurons carrying appropriate directional signals must be available from MT itself, and directional neurons are present in other cortical areas as well. Thus, we suspect that a realistic model must incorporate a large number of neurons in the pools that influence the decision. Simulations were therefore performed with varying numbers of input neurons to determine how the results are influenced by pool size.

SIMULATION PROCEDURES Figure 25.7 is a block diagram that shows how a single trial is simulated, beginning with the stimulus-driven responses of the input neurons and ending with a decision concerning motion direction. We assume that decisions in our task are based on a comparison of the pooled responses of two groups of directionally selective neurons, each group responding preferentially to one of the two directions being discriminated. If, for example, up is being discriminated from down, responses are combined from a population of neurons responding best to upward motion (U_1, U_2, ... U_n in figure 25.7) to generate a pooled up signal, and responses are combined from a separate population preferring downward motion (D_1, D_2, ... D_n) to generate a pooled down signal. The pooled signals are then compared and the decision is cast in favor of the pool generating the largest signal. Note that this decision rule is consistent with psychophysical data which suggest that judgments of motion direction are based on a comparison of directional mechanisms tuned to opposite directions of motion (Sekuler, Pantle, and Levinson, 1978; Anstis, 1986). Note also that the rule is simple so that the decisions reflect the output of the low-level analyzers (directional neurons) in a fairly direct manner.

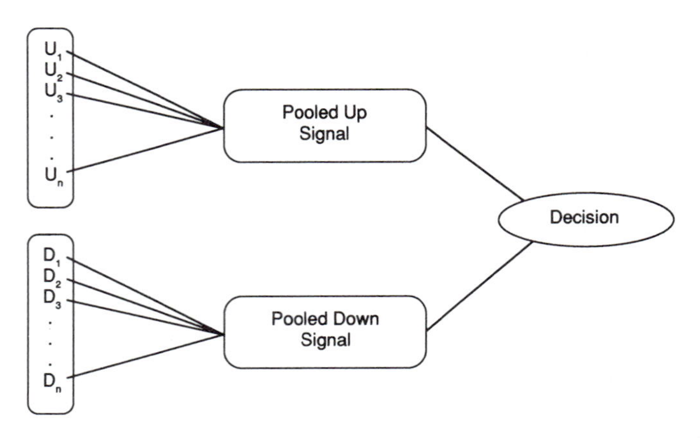

FIGURE 25.7 A block diagram of the pooling model. The diagram illustrates the procedure used to simulate a decision following one presentation of a visual stimulus in which upward motion is discriminated from downward motion.

All simulations presented in this chapter are based on physiological and psychophysical data obtained from one of our three monkeys because this is the only animal for whom we have measured the actual amount of intercorrelated activity among MT neurons (multiunit recordings described previously). Thus, our simulations seek to produce the observed behavior of this animal based on the measured responses of its MT neurons.

The responses at the input stage to the model are selected from actual recordings made from 51 MT neurons in this animal. Although our recordings were made from MT neurons with many different preferred directions and speeds, we assume that the data are representative of the variation that exists among neurons preferring any *single* direction and speed of motion. Thus, data from our 51 neurons can be used to simulate both the upward and the downward inputs to the model with appropriate adjustment of each neuron's preferred direction. For each of the 51 neurons, we have previously computed analytical expressions describing the mean response and response variance as a function of the direction and strength of the motion signal (Britten et al., 1993). To simulate the response of one neuron to a single stimulus presentation, we draw randomly from the response distribution appropriate for that neuron and that visual stimulus. In this manner, the simulations are faithful both to the mean response and to the response variance of any single neuron contributing to the pool, and to the pronounced differences between neurons as well.

The basic strategy of the simulations was to present a sequence of visual stimuli to a population of N neurons (selected randomly with replacement from the set of 51) whose responses provide the inputs to the two pools. The sequence of stimuli was chosen to mimic an actual measurement of psychophysical threshold. Thus, several near-threshold motion conditions were presented for 50 repetitions each. For each stimulus presentation, the response of each neuron was selected as outlined previously, and the individual responses were linearly summed within each pool (up and down, see figure 25.7). A decision was simulated for each stimulus presentation according to the decision rule outlined earlier (largest pooled signal wins), and the resulting data were compiled into psychometric functions expressing simulated performance as a function of motion strength. Thresholds were computed from the simulated psychometric functions using the same techniques we applied to real psychophysical data. To estimate the expected performance for a pool size of N neurons, we repeated the simulation 1000 times, each time beginning with a new selection of N neurons from our data set of 51, and calculated the geometrical mean threshold. This entire process was then repeated for different values of N to determine how the results varied with pool size.

With these results in hand, simulated psychophysical performance could be compared to the simulated neuronal inputs and to the actual performance of the animal to determine whether constraints 2–5 were met for each set of initial conditions tested (pool size, correlated responses, etc.). Note that constraint 1 is incorporated into the design of the model. (The appendix to Britten et al., 1992, contains a detailed description of our simulation techniques.)

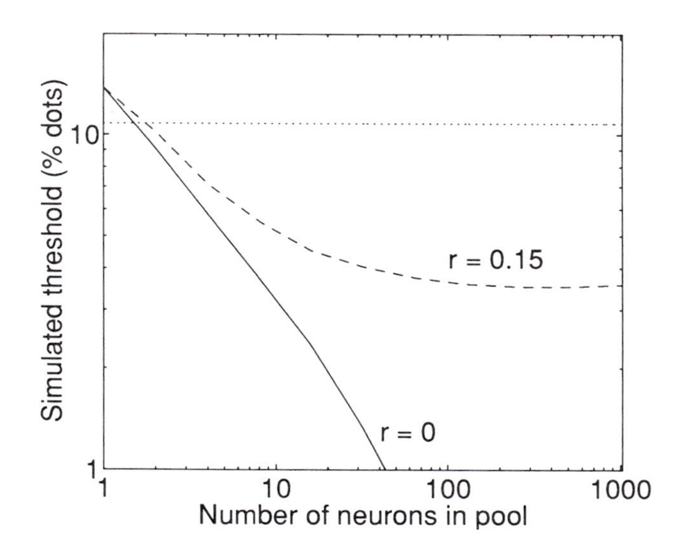

Figure 25.8 The effect of correlation within the input pools on simulated psychophysical thresholds. Simulated thresholds are plotted as a function of the number of neurons in the input pools. The solid line shows the results when the responses of the input neurons are independent ($r = 0$), and the dashed line indicates the results when the input neurons are weakly intercorrelated ($r = 0.15$). The horizontal dotted line depicts the mean psychophysical threshold observed during the actual experiments.

Simulation Results Figure 25.8 shows simulated thresholds as a function of the number of neurons providing input to the two pools. The horizontal dotted line indicates the mean psychophysical threshold obtained from the monkey during the 51 experiments. The solid line indicates simulated performance when the responses of each input neuron are completely independent of all other neurons in the pool (correlation coefficient, r, between the pooled neurons equals 0). The dashed line illustrates simulated performance when correlated activity such as we observed in MT was incorporated into the pool. In the latter simulations, the responses of individual neurons were not completely independent but were partially correlated with the responses of every other neuron in the pool ($r = 0.15$ on average).

Several interesting results are apparent in figure 25.8. Not surprisingly, simulated psychophysical threshold falls rapidly with pool size if the responses of the pooled neurons are independent; pooling averages out the noise carried by individual neurons, thus permitting more accurate discrimination of weak signals. On the other hand, a modest amount of correlated activity within the pool sharply limits the beneficial effects of pooling. Simulated performance becomes asymptotic for pools of roughly 100 neurons, and further large increases in pool size improve performance only marginally. To appreciate this result intuitively, realize that every correlated neuron carries a small amount of common noise that can never be purged by averaging within the pool. This common noise imposes a fundamental limit on performance. Finally, note that simulated performance matches observed performance best for a pool size of 1–2 neurons. This is a simple confirmation of the result illustrated in figure 25.3; the average psychophysical sensitivity of our monkeys was nearly equal to the average sensitivity of *single* MT neurons. For large pool sizes, therefore, a discrepancy remains between simulated and observed performance, even when correlated activity is introduced into the pool. We will show how this discrepancy can be resolved shortly, but first we shall consider whether our simulations produce a trial-to-trial covariation of neuronal response and simulated decision (constraint 4).

Because the simulations incorporated the response variance of the input neurons, and because each of our simulated experiments was based on multiple presentations of each stimulus condition, the SOC analysis illustrated in figure 25.4 for a real MT neuron can also be performed for each neuron in our simulated pools. Figure 25.9 depicts average SOC values for the simulated neurons as a function of pool size. Again, the solid line shows the outcome when the pooled neurons were completely independent, and the dashed line indicates the effect of modest correlation within the pools ($r =$ 0.15 on average). The dotted line shows the average SOC value actually observed in our recordings from MT neurons.

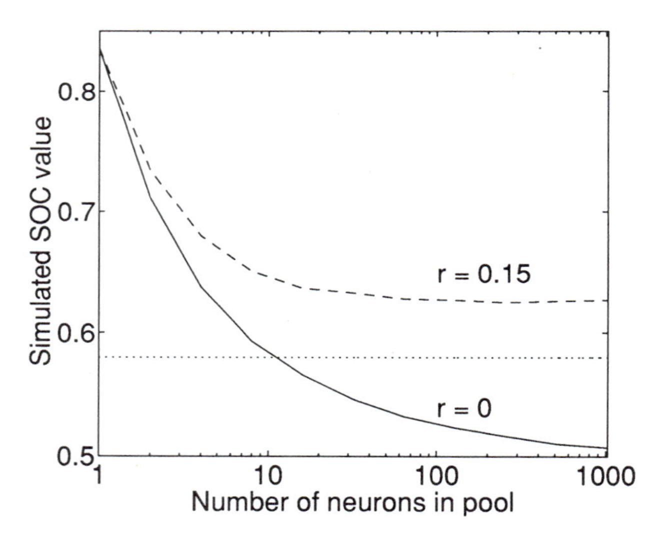

FIGURE 25.9 The effect of correlation within the input pools on simulated SOC values. Simulated SOC value is plotted against the number of neurons in the input pools. The solid line shows the results when the responses of the input neurons are independent ($r = 0$), and the dashed line indicates the results when the input neurons are weakly intercorrelated ($r = 0.15$). The horizontal dotted line depicts the mean SOC value observed during the actual experiments.

When the pooled responses were completely independent, the SOC value fell rapidly toward 0.5 (no relationship between neuronal response and decision) as pool size increased. This result is expected as increased pool size reduces the impact of any single neuron on the decision stage. When correlation was introduced to the pool, however, the simulated SOC values approached an asymptote near 0.63 as pool size increased. Again, an intuitive understanding of this effect is that a small fraction of the trial-to-trial response variance is nearly identical for all correlated neurons and cannot be diluted by adding more correlated neurons to the pool. This correlated variance, though small, has a measurable influence on the outcome of near-threshold decisions. We note, however, that the asymptotic SOC value of 0.63 in our simulations differs from the average SOC value of 0.58 observed in the MT recordings (dotted line, figure 25.9). Thus, introducing correlated activity within the pools causes the SOC effect to appear in our simulations, even for arbitrarily large pool sizes, but the simulated effect is larger than that observed experimentally.

Interestingly, correction of a single flawed assumption in the original model can resolve the quantitative discrepancies between experiment and simulation *for both the thresholds and the SOC values.* In all our simulations thus far, the inputs to the model were pooled by perfect linear addition of individual responses. Realistically, however, the pooling stage itself must consist of neurons whose noise will degrade the summed signal. To simulate this effect, we added noise to the summed signals at the pooling stage of the model (figure 25.7); the decision on any given trial thus reflected both the summed input signals and the added pooling noise. The amount of noise added on each trial was selected randomly from a normal distribution whose width could be varied systematically to explore the effects of different amounts of pooling noise.

Figure 25.10 is a three-dimensional plot that illustrates the effect of pooling noise on simulated psychophysical thresholds and on simulated SOC values. The amount of pooling noise added in each simulation is indicated on the horizontal axis by its coefficient of variation (cv, the ratio of standard deviation to the sum of the input signals). Simulations were carried out with three different levels of correlated activity incor-

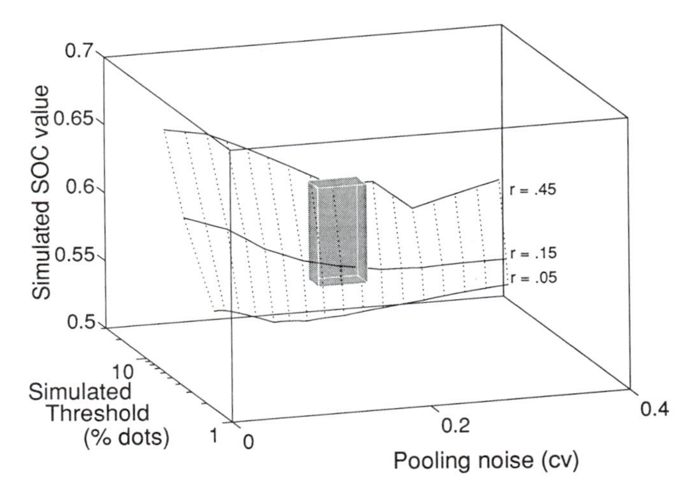

FIGURE 25.10 The effect of pooling noise on simulated thresholds and SOC values. Simulated results are shown for three different levels of correlation within the input pools (solid lines). Pooling noise is quantified by a coefficient of variation (cv) of the summed signals at the pooling stage. The shaded box shows the region of the 3-D space in which thresholds and SOC values for the $r = 0.15$ simulation are in joint agreement with the observed data. For both threshold and SOC values, the region of reasonable agreement is considered to be the experimentally observed mean plus or minus two standard errors of the mean. The curves for $r = 0.05$ and $r = 0.45$ do not intersect the shaded box.

porated within the pools. The middle curve in the figure shows the data for a mean correlation coefficient of 0.15, as observed experimentally in MT, and the other two curves illustrate results for correlation values of 0.05 and 0.45, bracketing those observed in MT. All thresholds and SOC values shown in the figure are asymptotic values for arbitrarily large pool sizes. Not surprisingly, pooling noise causes the simulated thresholds to rise; noise introduced at this stage of the pathway renders less reliable performance overall. At the same time, pooling noise causes simulated SOC values to *decrease* because individual decisions are less securely coupled to the fluctuating responses of the input neurons. Thus, the curves in figure 25.10 sweep through trajectories from relatively high points in the nearest corner of the cube to relatively low points in the far corner of the cube.

The interesting new finding in figure 25.10 is that a limited range of pooling noise values can bring both the simulated thresholds and the simulated SOC values into quantitative agreement with the observed data. The shaded box indicates this space of joint solutions for an intrapool correlation coefficient of 0.15. The depth dimension of the box represents the mean of the observed psychophysical thresholds, plus or minus two standard errors of the mean (SEM), and the vertical dimensions are determined by the mean of the observed SOC values, plus or minus two SEM. The horizontal width of the box indicates the range of pooling noise values that cause the simulated data to enter this solution space (for $r = 0.15$). (Our preliminary analyses indicate that this range of pooling noise values is similar to that expected if real neurons implement the pooling operation.)

For an intrapool correlation of 0.05 (lower curve), no amount of pooling noise can cause the simulated data to satisfy jointly the experimental constraints for threshold and SOC value. Either constraint can be met singly, but the two cannot be met simultaneously. This result can be appreciated by realizing that the 3-D path of the lower curve in figure 25.10 passes beneath all possible points representing joint agreement with experimentally observed threshold and SOC values. Similarly, the curve for an intrapool correlation of 0.45 passes over and behind the solution space. Thus, the range of intra-pool correlation coefficients that makes sense of our experimental data is rather constrained and includes the correlation value actually measured from MT neurons.

Importantly, adding noise to either the input stage or the decision stage of the model is much less effective at satisfying the experimental constraints enumerated at the outset of this section. Noise can be added at the input stage by assuming that the input pool contains neurons whose sensitivity to the motion signals in our stimuli is substantially less than that of the neurons we actually recorded. These neurons might, for example, have preferred directions or speeds that are imperfectly matched to the motion signal being discriminated. Adding noise to the input pool in the form of "insensitive" neurons degrades performance and can therefore bring simulated thresholds into correspondence with observed thresholds. Adding insensitive neurons to the input pool has no effect on SOC values, however, as long as the insensitive neurons share the same level of correlated activity as the rest of the neurons in the pool. The amount of correlated activity alone determines the magnitude of the SOC values for asymptotically large pool sizes. Although noise at the pooling stage appears necessary for reconciling simulated and observed SOC values, the ranges of acceptable agreement between the simulated and observed data in figure 25.10 are sufficiently broad that some mixture of input and pooling noise might satisfy the experimental constraints adequately.

Finally, we suggest that adding noise to the last stage of the model, after the decision computation has been performed, is inconsistent with our experimental data. Noise added at this stage would degrade performance equally for all motion strengths, resulting in performance levels less than 100% correct for even strong motion signals. In practice, our monkeys usually performed perfectly for strong motion signals and rarely performed below 95% correct.

Conclusions

The final set of simulations (see figure 25.10) reflects a model that satisfies all the experimental constraints laid out at the beginning of the preceding section. The model accepts as input realistic neuronal responses whose variances are weakly correlated (like the MT neurons studied in our multiunit recordings) and produces judgments of motion direction as output. Complete psychophysical experiments can be simulated to analyze the relationships between input responses and performance. The model produces simulated thresholds equal to the observed psychophysical thresholds and thus to the average thresholds of the input neurons. In addition, the model yields a trial-to-trial covariation between neuronal response and psychophysical decision that is comparable in magnitude to that observed in MT neurons. Both of these effects are present for arbitrarily large pools of input neurons.

Thus, the paradoxes raised by our original observations can now be resolved. We suggest that psychophysical sensitivity is no better than average neuronal sensitivity owing to both the presence of weakly correlated activity within the pools of neurons contributing to the decision and the presence of additional noise sources at the pooling stage. The trial-to-trial covariation of decision and neuronal response is accounted for completely by correlated activity within the pools of input neurons. In fact, correlation values such as those observed in MT would yield higher SOC values were it not for the presence of additional noise sources at the pooling stage.

Several intriguing questions remain about the nature of signal pooling within the motion pathway. Directionally selective neurons are present at several levels of the pathway, but successive levels are characterized by extensive convergence of their afferent projections and by progressively larger receptive field sizes (see Maunsell and Van Essen, 1983a, and Ungerleider and Desimone, 1986, among many others). Thus, pooling of directionally selective inputs probably occurs several times as signals are transmitted through the pathway.

We have recently recorded from one possible pooling stage beyond MT, the medial superior temporal area (MST), a neighboring area within the superior temporal sulcus that also contains a large proportion of directionally selective neurons. Interestingly, the absolute sensitivity of MST neurons is similar to that of MT neurons (and, thus, to psychophysical threshold) despite the large receptive fields of these neurons and the extensive convergent inputs they receive from MT (Celebrini and Newsome, 1992). Thus, the expected gain in sensitivity from convergent MT afferents may be offset rather precisely by additional noise sources within MST itself. In the context of our psychophysical task, therefore, convergent projections may be necessary for noise-free transmission of motion signals from one stage of the pathway to the next. Successive processing stages along the motion pathway are important for complex analyses of visual motion (Movshon et al., 1985; Tanaka et al., 1986; Andersen et al., 1990; Duffy and Wurtz, 1991), but for simple directional signals such as those used in our task, some stages of the pathway may merely represent obstacles to faithful transmission of information to the decision stage.

From the perspective of cognitive neuroscience, a particularly interesting question finally arises: Where within the nervous system is the decision implemented? In general, it seems likely that many decisions are implemented within structures that link sensory systems to high-level motor planning centers. In our paradigm, the operant response that reveals the monkey's decision is a saccadic eye movement to a specified visual target. Thus, the logical places to search for neural correlates of the decision process are the pathways and structures that link MT to oculomotor planning centers such as the lateral intraparietal area (Gnadt and Andersen, 1988; Barash et al., 1991a; Barash et al., 1991b; Duhamel, Golby, and Goldberg, 1992), the frontal eye fields (Goldberg and Bruce, 1990), and the superior colliculus (Mays and Sparks, 1980; Glimcher and Sparks, 1992). The anatomy of these pathways is well-known, and single-unit recordings in the relevant structures during performance on our task should enable us to identify the site or sites where this transformation occurs.

Future experiments must also address the fascinating possibility that the neural structures underlying the decision process may be different in monkeys trained to indicate their decisions with arm movements rather than eye movements. For arm movement responses, candidate integrative structures might include area 5 of the parietal lobe and the arm representations within the premotor and supplementary motor cortices. Appropriate modifications of our paradigm should allow us to explore whether decision circuits are specific to particular motor responses or whether the circuits should be regarded as a supramotor decisional network that is independent of specific operant responses.

ACKNOWLEDGMENTS We thank Judy Stein for excellent technical assistance during the course of the experiments reported herein. Ongoing financial support for this research has been provided by the National Eye Institute (EY05603 and EY02017). Michael Shadlen is supported by a Postdoctoral Research Fellowship for Physicians from the Howard Hughes Medical Institute, and Ehud Zohary is supported by a training grant from the McDonnell-Pew Program in Cognitive Neuroscience.

REFERENCES

ALBRIGHT, T. D., 1984. Direction and orientation selectivity of neurons in visual area MT of the macaque. *J. Neurophysiol.* 52:1106–1130.

ALBRIGHT, T. D., R. DESIMONE, and C. G. GROSS, 1984. Columnar organization of directionally selective cells in visual area MT of macaques. *J. Neurophysiol.* 51:16–31.

ANDERSEN, R., R. SNOWDEN, S. TREUE, and M. GRAZIANO, 1990. Hierarchical processing of motion in the visual cortex of monkey. *Cold Spring Harb. Symp. Quant. Biol.* 55:741–748.

ANSTIS, S., 1986. Motion perception in the frontal plane: Sensory aspects. In *Handbook of Perception and Human Performance*, vol. 1, K. Boff, L. Kaufman, and J. Thomas, eds. New York: Wiley, pp. 16-1–16-27.

BAIR, W., C. KOCH, W. NEWSOME, and K. BRITTEN, 1992. Power spectrum analysis of MT neurons from awake monkey. *Soc. Neurosci. Abstr.* 18:12.

BARASH, S., R.M. BRACEWELL, L. FOGASSI, J. W. GNADT, and R. A. ANDERSEN, 1991a. Saccade-related activity in the lateral intraparietal area: I. Temporal properties; comparison with area 7a. *J. Neurophysiol.* 66:1095–1108.

BARASH, S., R. M. BRACEWELL, L. FOGASSI, J. W. GNADT, and R. A. ANDERSEN, 1991b. Saccade-related activity in the lateral intraparietal area: II. Spatial properties. *J. Neurophysiol.* 66:1109–1124.

BRITTEN, K. H., M. N. SHADLEN, W. T. NEWSOME, and J. A. MOVSHON, 1992. The analysis of visual motion: A comparison of neuronal and psychophysical performance. *J. Neurosci.* 12:4745–4765.

BRITTEN, K. H., M. N. SHADLEN, W. T. NEWSOME, and J. A. MOVSHON, 1993. Responses of neurons in macaque MT to stochastic motion signals. *Visual Neurosci.* 10:1157–1169.

CELEBRINI, S., and W. T. NEWSOME, 1992. Responses of neurons in area MST during direction discrimination performance: A comparison of neuronal and psychophysical sensitivity. *Soc. Neurosci. Abstr.* 18:1101.

DEAN, A. F., 1981. The variability of discharge of simple cells in cat striate cortex. *Exp. Brain Res.* 44:437–440.

DUFFY, C. J., and R. H. WURTZ, 1991. Sensitivity of MST neurons to optic flow stimuli: I. A continuum of response selectivity of large-field stimuli. *J. Neurophysiol.* 65:1329–1345.

DUHAMEL, J.-R., C. L. COLBY, and M. E. GOLDBERG, 1992. The updating of the representation of visual space in parietal cortex by intended eye movements. *Science* 255:90–92.

GLIMCHER, P. W., and D. L. SPARKS, 1992. Movement selection in advance of action in the superior colliculus. *Nature* 355:542–545.

GNADT, J. W., and R. A. ANDERSEN, 1988. Memory related motor planning activity in posterior parietal cortex of monkey. *Exp. Brain Res.* 70:216–220.

GOLDBERG, M. E., and C. J. BRUCE, 1990. Primate frontal eye fields: III. Maintenance of a spatially accurate saccade signal. *J. Neurophysiol.* 64:489–508.

GRAHAM, N. V. S., 1989. *Visual Pattern Analyzers.* Oxford: Oxford University Press.

GREEN, D. M., and J. A., SWETS, 1966. *Signal Detection Theory and Psychophysics.* New York: Wiley.

MAUNSELL, J. H. R., and D. C. VAN ESSEN, 1983a. The connections of the middle temporal visual area (MT) and their relationship to a cortical hierarchy in the macaque monkey. *J Neurosci.* 3:2563–2586.

MAUNSELL, J. H. R., and D. C. VAN ESSEN, 1983b. Functional properties of neurons in the middle temporal visual area (MT) of the macaque monkey: I. Selectivity for stimulus direction, speed and orientation. *J. Neurophysiol.* 49: 1127–1147.

MAYS, L. E., and D. L. SPARKS, 1980. Dissociation of visual and saccade-related responses in superior colliculus neurons. *J. Neurophysiol.* 43:207–232.

MOVSHON, J. A., E. H. ADELSON, M. S. GIZZI, and W. T. NEWSOME, 1985. The analysis of moving visual patterns. In *Pattern Recognition Mechanisms*, C. Chagas, R. Gattass, and C. Gross, eds. New York: Springer-Verlag, pp. 117–151.

NEWSOME, W. T., K. H. BRITTEN, J. A. MOVSHON, and M. SHADLEN, 1989. Single neurons and the perception of visual motion. In *Neural Mechanisms of Visual Perception. Proceedings of the Retina Research Foundation*, D. M.-K. Lam and C. D. Gilbert, eds. The Woodlands, Tex.: Portfolio Publishing, pp. 171–198.

NEWSOME, W. T., and E. B. PARÉ, 1988. A selective impairment of motion perception following lesions of the middle temporal visual area (MT). *J. Neurosci.* 8:2201–2211.

O'Kusky, J., and M. Colonnier, 1982. A laminar analysis of the number of neurons, glia, and synapses in the visual cortex (area 17) of adult macaque monkeys. *J. Comp. Neurol.* 210:278–290.

Pirenne, M. H., 1943. Binocular and uniocular threshold of vision. *Nature* 152:698–699.

Salzman, C. D., K. H. Britten, and W. T. Newsome, 1990. Cortical microstimulation influences perceptual judgements of motion direction. *Nature* 346:174–177.

Salzman, C. D., C. M. Murasugi, K. H. Britten, and W. T. Newsome, 1992. Microstimulation in visual area MT: Effects on direction discrimination performance. *J. Neurosci.* 12:2331–2355.

Sekuler, R., A. Pantle, and E. Levinson, 1978. Physiological basis of motion perception. In *Handbook of Sensory Perception*, vol. 8, H. L. Teuber, ed. Berlin: Springer-Verlag, pp. 67–96.

Tanaka, K., H. Hikosaka, H. Saito, Y. Yukie, Y. Fukada, and E. Iwai, 1986. Analysis of local and wide-field movements in the superior temporal visual areas of the macaque monkey. *J. Neurosci.* 6:134–144.

Tolhurst, D. J., J. A. Movshon, and A. F. Dean, 1983. The statistical reliability of signals in single neurons in cat and monkey visual cortex. *Vision Res.* 23:775–785.

Ungerleider, L. G., and R. Desimone, 1986. Cortical connections of visual area MT in the macaque. *J. Comp. Neurol.* 248:190–222.

Worgotter, F., W. Daunicht, and R. Eckmiller, 1986. An on-line spike form discriminator for extracellular recordings based on an analog correlation technique. *J. Neurosci. Methods* 17:141–151.

Zeki, S. M., 1974. Functional organization of a visual area in the posterior bank of the superior temporal sulcus of the rhesus monkey. *J. Physiol.* 236:549–573.

Zohary, E., M. N. Shadlen, and W. T. Newsome, 1992. Correlated activity of neurons in area MT. *Soc. Neurosci. Abstr.* 18:1101.

26 The Neuron Doctrine in Perception

HORACE BARLOW

ABSTRACT A brief history of the 100-year-old neuron doctrine is given, and it is argued that the neuron remains the important unit of function for developing a rational account of how behavior is generated. We now know that neurons are much more reliable and diverse in their properties than was once believed, and in recent years it has been shown that a very small collection of cortical neurons can have a measurable influence on the response made by the whole animal. Important gaps in our understanding of the properties of single neurons remain and are pointed out. The distinction between representation by *cardinal cells* and *ensemble encoding* is emphasized. The former employs a smaller number of active elements, and these represent significant and meaningful subsets of sensory stimuli, whereas the latter typically employs larger numbers of active units, and these represent arbitrary subsets of sensory stimuli, as the bits in the ASCII code represent nearly arbitrary subsets of keyboard characters. It is argued that neurons are the *only* elements in the brain that can combine evidence to form well-based decisions and that a single one of them may be sufficient for such a decision.

Origin and relevance of the neuron doctrine

The neuron doctrine summarizes the anatomical truth that brains are composed of cells and structures formed by cells, as is true of all other tissues. Schwann formulated this cell theory in 1839, nearly two centuries after the invention of the microscope, but it took dozens of European scholars more than 50 years to agree that it also applied to the brain, and the neuron doctrine was not stated in the form shown in table 26.1 until 1891. A fascinating account of the history of its development has been given by Shepherd (1991). The key evidence was produced by Cajal in a very few years starting in 1887, when he adopted the silver impregnation method that had been introduced by Golgi in 1875, but much background work was completed before then, including important contributions from unexpected figures such as Helmholtz, Freud, and Nansen, best known in quite different spheres.

HORACE BARLOW NEC Research Institute, Princeton N.J., Center for Neuroscience and Department of Psychology, New York, N.Y., and Physiological Laboratory, Cambridge, England

Three main reasons accounted for the fact that the neuron doctrine was so long in being widely accepted: First, methods of preparing tissues for optical microscopy, and the instruments themselves, were only slowly improved to the levels required to see that axons, the axon terminals, and dendrites were all connected to cell bodies and arose from them. Second, the diversity of cell types in the brain, and their unusual shapes and properties, made it a genuinely difficult problem. Third, mistaken theoretical assumptions impeded progress. Shepherd (1991) quotes Deiters, who in 1865 described the first preparations showing axons originating from nerve cells in the central nervous system (CNS), to explain the support enjoyed by the rival reticular theory, which postulated protoplasmic continuity between the processes in the brain. Deiters was convinced that the dendrites of these cells did not anastomose with one another, but he left open the possibility of continuity between axon terminals because, he said, "Theory demanded it."

This reasoning was based on the false theoretical assumption that protoplasmic continuity was required to mediate the interactions between nerve cells that, as everyone agreed, must occur in the brain. The crucial step in discarding this assumption was Cajal's demonstration that axon terminals made systematic, obviously nonaccidental, patterns of contact with other cells, and that there was no protoplasmic continuity at these points. The existence of these orderly, apparently purposeful patterns suggested to him that cells could influence one another at points of contact (i.e., synapses) and, once this was accepted, the protoplasmic continuity postulated by the reticular theory became unnecessary, although the evidence denying its existence was incomplete (Shepherd, 1991). Even today it remains

Table 26.1

Outline history of the neuron doctrine

1839	Schwann formulated the cell theory. Nerve cell bodies had been recognized but not their connection with either axons or neuropil (*"punktsubstanz"*)
1865	Deiters showed the origin of axis cylinders from ventral horn cells in the spinal cord.
1875	Golgi discovered the silver impregnation method, which stains all the parts of a single neuron.
1887	Ramón y Cajal adopted the Golgi method and showed nonaccidental patterns of contact between axon terminals and other nerve cells, with no evidence of cell continuity.
1891	Waldeyer formulated and promulgated the neuron doctrine in a widely read German periodical: **"The nerve cell is the anatomical, physiological, metabolic, and developmental unit of the nervous system"**
Barriers to this doctrine's formulation and acceptance:	
Technical: The methods of histological preparation and the optical quality of microscopes had to be improved.	
The genuine difficulty of the problem: Nerve cells have extraordinarily complex and variable structure.	
Theoretical: Some clung to the belief that cell continuity was necessary for cells to interact.	

worthwhile looking out for universally held false assumptions that impede progress.

The neuron doctrine is deeply embedded in modern neuroanatomical thinking and is also accepted without much question in modern developmental and biochemical studies. In contrast, many neurophysiologists and psychologists, especially those interested in the field of perception, balk at the implications of the doctrine. Two main reasons for this can be identified. The first is an intuitive revolt against *any* scientific explanation for the working of our minds, about which our direct personal experience seems to tell us so much. We perceive the world with seamless unity, we perform astonishingly well-integrated motor actions ourselves and observe them in other living things and, above all, we have an intelligent awareness of the material and social worlds with which we interact. Wonder and astonishment at what the brain does are fully justified, but the reductionist attempt to explain its actions by the organized activity of individual nerve cells is not thereby doomed to failure, and the conclusion this chapter tends toward is that, although it has far to go, this theory is actually making steady progress.

The second reason for disputing that neurons should be regarded as the units of function is a good one. The brain and behavior can be studied at many different levels, and it is not initially clear that the cellular level has greater importance than other ones, for which the natural unit of function is usually *not* the nerve cell. For example, to understand drug actions receptor molecules in cell membranes are nowadays the relevant unit. At the other extreme, one has to consider human societies and individual people to understand how a politician or spiritual leader influences his or her followers, and this is unlikely to be much illuminated by considering the individual cells that make up the brains in question.

Table 26.2 attempts to separate the levels of analysis for which the neuron doctrine is important from those for which it is not. The left column lists various possible levels of analyzing behavior, from the most reductionist causes at the top to their final consequences at the bottom. In the right column, four of these levels are identified as bottlenecks, because upper causal levels in the table influence lower consequential levels through these bottlenecks; everything has to flow through them and, furthermore, what flows through them provides a good summary of the results of the upper mechanisms. If one knew all the genes acting in the brain and when they were expressed, this would provide a fairly complete conceptual input for understanding how their organized activity produces the many types of working neuron. Similarly, the properties of single neurons provide the raw material for explaining how their organized activity produces behavior, and the behavior of individuals provides the raw material for explaining social phenomena, including many of the selective factors that have molded the gene pool of the species. The top level is, of course, a subset of the bottom level, so the relationship is circular. The boldface central area shows that the neuron doctrine is all-important in determining how behavior is produced by single

TABLE 26.2
Levels in the analysis of behavior

Genes acting in the brain	First bottleneck
Cell biochemistry Cytoplasmic structures Receptor molecules Transmitter substances Membrane biophysics	Phenomena at these levels can explain how single neurons work, ultimately as a result of the organized activity of genes.
Single neurons	Second bottleneck
Small groups of neurons Anatomical parts of the brain Sensory psychophysics Perception Learning	Analysis at these levels can explain how individuals behave in terms of the organized activity of single neurons.
Individual behavior	Third bottleneck
Individual Family Social group Tribe Species	Analysis at these levels can explain how an individual's behavior affects his or her inclusive fitness and thus helps to show how the gene pool has been formed by selection.
Gene pool of species	Fourth bottleneck

neurons, and the table also shows the relation between this particular problem and others for which other levels of analysis are more appropriate.

Progress in the elucidation of single-neuron function

Table 26.3 lists a number of experimental studies that have led to the demonstration that single neurons in mammalian cortex are sensitive and powerful pattern recognition devices. The studies chosen are somewhat arbitrary, and no mention is made of many others that are perhaps equally important, but steady advances are being made, and I believe a more thorough survey would refute the suggestion that progress has slowed since the breakthroughs of the fifties and sixties.

Over the past 50 years, there has been an astonishing change in how we regard cells in the CNS, and especially in the cortex. At the beginning of this period, it was believed that there was such an incredibly large number of such cells ($10^5/mm^3$ of cortex, and more than 10^{10} altogether) that it would be absurd and meaningless to consider the role of a single one, and therefore averaging the activity of large numbers of them was the only sensible approach. Now it is possible to record from a single neuron in the cortex of an awake, behaving monkey, determine how well it performs its task of pattern recognition, and compare this performance with that revealed by the behavioral responses of the same animal. The fact that thresholds are comparable (Britten et al., 1992) would have astounded the cortical neurophysiologist of 50 years ago, but it is not unexpected from the material I reviewed 20 years ago (Barlow, 1972). At that time, it was already clear that individual neurons are not noisy and unreliable, as many had supposed, but are individually efficient and capable of quite complex pattern recognition tasks. Furthermore, they are extremely diverse in their properties so that using the average of their electrical outputs for analysis is as inappropriate as averaging the sounds of words would be when attempting to analyze language (though electroencephalograms and evoked potentials are, of course, useful tools for some problems). How the brain actually combines the outputs of many neurons remains an unresolved issue, but the lower-envelope principle bears reiteration.

THE LOWER-ENVELOPE PRINCIPLE De Valois, Abramov, and Mead (1967) related wavelength discrimination to the responses of the different classes of color-opponent lateral geniculate nucleus (LGN) neurons, and Talbot and coworkers (1968) related the sense of tactile vibration to the responses of different cutaneous touch receptors. Both sets of results provided evidence for an important *lower-envelope principle*, which states

TABLE 26.3

Steps in understanding the behavior of single sensory neurons

1938–40	Hartline (1940) and Granit (1947) record from retinal ganglion cells. They show there is not simple transduction, but rather coding.
1950–53	Barlow (1950, 1953) and Kuffler (1953) discover lateral inhibition in frog and cat retina. "Fly-detectors" link retinal neurons to behavior.
1959–62	Lettvin and colleagues (1959) in frog retina and Hubel and Wiesel (1962) in cat cortex establish the idea of trigger features and hierarchies.
1967–8	De Valois and colleagues (1967) and Talbot and colleagues (1968) show evidence of the lower-envelope principle in wavelength and tactile discriminations.
1963–70	Hubel and Wiesel (1970) link deficits of cortical neurons following early visual deprivation to amblyopic defects.
1971	Barlow and coworkers show that retinal ganglion cell thresholds in cat are comparable with behavioral thresholds.
1972	Gross, Rocha-Miranda, and Bender (1972) describe units in the inferotemporal area responding selectively to hands and faces.
1985	Movshon and associates (1985) show how neural responses in MT correlate with coherence and slippage in moving plaids.
1989	Vallbo shows that a single impulse in a single touch fiber gives rise to a tactile sensation in humans.
1989	Newsome, Britten, and Movshon (1989) show that single neurons in MT have thresholds close to the behavioral threshold for detecting coherent motion of random-dot patterns (see also Britten et al., 1992).

that sensory thresholds are set by the class of sensory unit that has the lowest threshold for the particular stimulus used and is little influenced by the presence or absence of responses in the enormous number of other neurons that are less sensitive to that stimulus. This result is compatible with the notion that outputs are either combined optimally, giving more weight to the responses with greater signal-to-noise ratio, or are combined by a process of "probability summation," in which a positive response is given when any neuron in a group gives a significant response, and a negative response only when all neurons in that group fail to give a significant response. Whether this principle will fit the new results to be described next remains to be seen (see also Newsome et al, this volume).

SENSITIVITY OF SINGLE NEURONS Absolute measurements have shown that single units can be extremely sensitive. In the cat, for instance, there is a reliably detectable change in the maintained discharge of some retinal ganglion cells when only a few quanta are absorbed in the photoreceptors, a number comparable with that required at the behavioral absolute threshold (Barlow, Levick, and Yoon, 1971). Similar results have been obtained for the skin-indentation threshold in monkeys' fingers (Talbot et al., 1968), as well as for auditory thresholds (Kiang et al., 1965; Evans, 1975).

These studies suffer from the fact that they were performed on anesthetized animals, but the technique of recording from single neurons in the brains of awake behaving monkeys, pioneered by Evarts and championed for sensory work by Mountcastle and his school, resolves this problem. It is now possible to compare the messages occurring in the neurons of the cerebral cortex with the behavioral performance of an animal measured at the same time as the records were being obtained.

Newsome, Britten, and Movshon (1989) and Britten and colleagues (1992) recorded from single neurons in the middle temporal area (MT, or V5) of awake behaving monkeys (see also chapter 25). In this area, the majority of neurons respond selectively to specific types of movement in the visual image. Their sensitivity was tested by varying the proportion of dots that moved in a coherent direction on a display, the rest of the dots being formed and replaced in the same manner as the coherent ones but being displaced randomly. Monkeys can be trained to signal the apparent direction of motion of these patterns and can detect coherent motion of only a few percent of the dots, a figure comparable to the threshold for human subjects. When a suitable neuron in MT was found, the parameters of the stimulus were first adjusted to be optimal for that cell, which required adjustments to the position and size of the display and to the direction and velocity of the coherently moving dots. Then, while the responses of the cell were being recorded, the trained monkey also signaled in which direction it judged that the pattern was moving. The experimenters calculated how often the monkey *would* have been able to get the correct answer if it had had at its disposal the information provided by the single isolated cell, together with one tuned to the reverse direction of motion but having properties identical in every other way to the one recorded. The results showed that the average performance calculated from single neurons in this way matched very well the monkey's actual performance, both the mean threshold and the mean slope of the psychometric functions being almost identical.

Note that this is a task that, by its nature, requires the collection of information from a large patch of the visual field, and it is one that cannot be done reliably until such information has been collected. The neurons in MT characteristically have large receptive fields covering an area in the visual field 10 to 1000 times that of the neurons in primary visual cortex. Each neuron collects information from a large set of primary visual cortex neurons, and these must all be ones that respond optimally to motion in the same, or nearly the same, direction and velocity. Hence, it appears that MT has neurons tuned to pick up coherent motion in different directions and velocities over patches of the visual field varying in position and size, but all large. Clearly, a very large number of neurons are required to cover the whole visual field with such matched filters, which have a wide range of positions, sizes, and perhaps disparities, and which are tuned to a wide range of directions and velocities of movement. MT contains a number of cells that is probably sufficient for this task, though it has not been shown that there is a large excess.

The performance of MT cells is a dramatic demonstration of the fact that single neurons can be sensitive and reliable detectors of complex spatiotemporal patterns of excitation. It also very clearly demonstrates how and why single neurons are capable of this: They are the final common path for information converging from the appropriately tuned subset of units in just the region containing the information and, equally important, they exclude irrelevant information that would only contribute unwanted noise to the discrimination task.

There are other examples of cells in other areas of the cortex collecting information from appropriately tuned subsets of units (Zeki, 1980; Desimone et al., 1985), being thus enabled to make discriminations that would otherwise be impossible, and this may be a general paradigm for the task performed by cortical neurons. The tricky part is to discover what is collected and what discriminations this enables.

It is possible that information may be combined in more subtle ways than that required for detecting coherent motion. Movshon and colleagues (1985) found neurons in MT that responded as if they were directionally tuned to the motion of the plaid formed by a crossed pair of gratings rather than to either of the components. This is interesting because it demonstrates a new form of selectivity being generated in a peristriate area, and it may require more than simple pooling of inputs from cells projecting from V1 and V2. The properties of color constancy that Zeki (1980) claims for cells in V4 also require a more subtle form of combining afferent information than simple pooling (Desimone et al., 1985).

Sensing Single Impulses Finally, some experiments on the tactile system of humans must be mentioned because they demonstrate that a single impulse in a single sensory fiber can, under some conditions, be a perceptible event. These experiments were conducted by inserting a fine metal microelectrode into the nerve from the hand (Vallbo, 1989; see also chapter 14). With fortunate placement of the electrode, it is possible to isolate the activity of a single fiber and, in such a case, a region can be found where light touch causes a succession of brief, equal-sized action potentials. One can reduce the intensity of the mechanical stimulus until it causes, on average, only a single action potential: Will such a minimal stimulus ever be felt? The answer, which may initially seem surprising, is yes: It is felt just as a brief and very light touch. This is not universally true; for some types of receptor and at some positions on the skin, far more than one impulse is needed. Nonetheless, single action potentials are not insignificant events in the representation of our sensations; we are accustomed to atoms and molecules in the world around us being orders of magnitude smaller than anything we can perceive directly, but this does not appear to be true of nerve impulses, which are the atomic events that build up our sensations.

Face Cells in Inferotemporal Cortex The experiments reviewed in the preceding sections support and strengthen the view that single neurons are reliable and selective, but there have been many other experimental studies of single-unit or multiunit activity in the cortex that do not concur. The early results from Gross's laboratory (Gross, Rocha-Miranda, and Bender, 1972) seemed to provide convincing evidence for cells with highly specific response characteristics in inferotemporal cortex of the macaque, but some later results complicate the picture. First, many of those working on single neurons in this area, including Gross and his colleagues, now say that faces are not distinguished from each other and other objects by the firing of individual cells but by population or ensemble encoding (Gross, 1992; Rolls, 1992; Young and Yamane, 1992; Gross et al., 1993). There are problems with this interpretation, which are considered later in this chapter, but other results paint a much more complicated picture that is an extension rather than a retraction of Gross's original findings.

It is generally agreed that the face cells themselves are a very inhomogeneous group, some of them being for generic faces of any type, orientation, expression, and the like, whereas others are specific for the viewpoint or even the face of a particular individual. Perrett's group (1990) finds families of cells in neighboring regions that, by their responses, classify body movements and actions seen by the monkey, and these include cells that would be classified as face cells, but which are sensitive to changes in the direction of gaze. In one case, these investigators were able to show that such a cell responded to downward movement of the head at a distance but, when close enough, it would respond to downward movement of the eyes alone (Perrett et al., 1992). This work points in a different direction from ensemble encoding.

It is critical to understand that the analysis of responses in higher areas is a genuinely difficult problem, because literally anything might turn out to be a specific and effective stimulus. The discovery that some cells are highly responsive to polar gratings (Gallant, Braun, and Van Essen, 1993) demonstrates this, for many of these cells responded poorly, and therefore equivocally and unreliably, to the normal range of stimuli. However, if a cell is claimed to represent a face, then it is necessary to show that it fires at a certain rate nearly every time a face is presented and only very rarely reaches that rate at other times. Instead, the generally accepted practice is to show that the mean of a number of responses differs from the mean maintained level at the significance level $p = 0.05$. This is the test required to justify an experimenter's claim about influences on the mean firing rate, but it does not justify the claim that a particular cell codes for a particular stimulus. For this, the error rate on single trials must be low, and a t test on the means is not the appropriate test, especially when precise information about how many responses were averaged is omitted, as is often the case. If the claims of Perrett's group (1990, 1992) can be confirmed, however, they are very much more important than the qualifications and retractions of the other groups, for they suggest these areas are doing new and qualitatively different types of analysis.

Other Experimental Findings Promising results from recordings in inferotemporal cortex (Fujita et al., 1992) that may reveal an "alphabet" of characteristics used in object identification should be mentioned, especially as these may link with the "geons" suggested by Biederman (1987). Some progress has been made in finding the neuronal basis for visual search (Chelazzi, Miller, and Duncan, 1993) and memory (Desimone, 1992), but these results do not seem to bear very directly on the issues important for the neuron doctrine. There is, however, one major discrepancy with my 1972 view that emerges from many of these experiments.

The facts now indicate that high-level neurons have higher maintained activity, and are easier to activate, than seemed to be the case in 1972. The change is probably due to improvements in the general condition of the preparations, the increased use of monkeys (which seem to have a more active cortex than cats), and use of awake rather than anesthetized preparations. It is still possible that the cortex contains many cells that have both very low maintained discharge and high selectivity, as these properties would cause them to be overlooked, but this view is less plausible than it was 20 years ago. The theoretical significance of this change will be considered in the next section.

Origins of concepts relating single neurons to perception and behavior

The rise in the status of single neurons naturally led to several new terms and concepts, which are summarized

Table 26.4
Origin of some concepts relating neurons and behavior

Pontifical neurons	Sherrington (1941) declared the brain is rather a "million-fold democracy whose each unit is a cell."
Nerve cell assembly	Hebb (1949) wanted to relate psychology to single neurons but thought reverberatory effects in groups of interconnected cells were necessary to explain the persistence of excitation.
Redundancy reduction	Attneave (1954) pointed out that many aspects of perception indicate that redundancy is being exploited. Barlow (1959) showed the relevance of this for sensory mechanisms and intelligence.
Grandmother cells	Lettvin (see appendix to this chapter), to illustrate that cells knew the kantian "*Ding an sich*," imagined the consequences of surgical removal of "mother" cells.
Sparse coding in associative nets	Willshaw, Buneman, and Longuet-Higgins (1969) showed the merit of sparse coding for increasing the memory of associative nets.
Economy of impulses, and Cardinal cells	Barlow (1969, 1972) argued for representation by neurons that are only rarely active and have carefully chosen selective properties.
Yellow Volkswagen cells	Harris (1980) saw the impossibility of having cells selective for every contingency that causes pattern-selective adaptation.
Coarse coding	Hinton (1981) suggested that if features rarely occur together, it is better to have overlapping tuning curves.
Parallel Distributed Processing	Rumelhart and McClelland (1986) edited a massive three-volume book that showed how groups of simplified artificial neurons could perform many brainlike tasks.
Projective zones	Lehky and Sejnowski (1988) showed the importance of the pattern of outgoing connnections from hidden units (or cells).

in table 26.4. Four of them—pontifical neurons, grandmother cells, cardinal cells, and yellow Volkswagen cells—are closely related. Sherrington (1941) was aware of the extensive convergence of sensory pathways and asked the natural question: Do all the pathways that are active for a given sensory scene converge and produce activity in a single pontifical cell, whose role is to represent that scene? He was appalled by the idea and retreated without further discussion to his "million-fold democracy."

Grandmother Cells The term *grandmother cell* is widely used to describe the notion that there are cells in the brain that become active when and only when a grandmother, or some other arbitrary but specific feature, is present to the senses. It is briefly mentioned (and the concept rejected) in my 1972 paper, but Jerry Lettvin invented the term, and because no account of its origin is available elsewhere, he has kindly allowed a letter explaining it to be reproduced as an appendix to this article. Note that his story originally postulated mother (not grandmother) cells, which is a significant difference, for most would agree that mothers have such central significance for behavior that one would rather expect special provisions to be made for their perceptual representation, whereas grandmothers, except in the human species, are without much behavioral significance. Note also that there were 18,000 such cells rather than a single one, as is often assumed.

Yellow Volkswagen Cells Those who object to the notion that complex and specific perceptions can be represented by a single species of cell usually point to the immense number that would be required for all possible perceptions. This issue was most specifically addressed by Harris (1980), who was worried by the fact that contingent adaptation effects were being discovered wherever anybody looked for them. The popular explanation for such effects attributed them to the fatigue of the responding neurons, and selective adaptation was being proclaimed a noninvasive psychophysical method capable of revealing the classes of neural units in the brain. On this interpretation, the results implied that there were preexisting cells tuned to all the joint modes of excitation for which adaptation occurred, but as Harris (1980) pointed out, although it may be reasonable to postulate "yellowness cells" and even "Volkswagen cells," one surely cannot have cells for every possible adjective-noun combination such as "yellow Volkswagens." This provided a motive for suggesting that these adaptational effects are caused by changes in the interactions between cells (Barlow, 1990) resulting from changed associative structure, and that they do not indicate the changed responsiveness of units selective to particular conjunctions.

Redundancy Reduction Cardinal cells (Barlow, 1972) are like pontifical, grandmother, and yellow Volkswagen cells, but involve an additional concept, redundancy reduction. This idea was introduced in modern form by Attneave (1954), who pointed out that borders lie at the edge of relatively constant regions of the image and thus contradict expectations based on these uniform regions. Shannon-type information increases with unexpectedness, so these were the informative, nonredundant, parts of the image. Attneave (1954) therefore suggested that perception involves selection *for* the informative and *against* the redundant or non-informative parts of sensory messages.

This idea can explain many aspects of sensory coding and may give some insight into the basic nature of intelligence (Barlow, 1959, 1961). There is reason to believe that distinguishing information and redundancy is always necessary if sensory messages are to be used for any but the simplest task. Redundancy is *not* an unnecessary addition to the message, as the term might suggest, nor does it consist of repetition or reduplication that simply protects the message from degradation by noise, though it does have that effect. The redundancy in a message is any regularity, repetition, or symmetry that allows a prediction to be made about other parts of the message. Even more broadly, any measurable statistic of the message that differs from what the channel would allow specifies a form of redundancy, because it defines an additional constraint on the message. Now the statistical structure in a sensory message is imposed on it by important features in the environment, so redundancy actually conveys knowledge of the environment, and the term *knowledge* can be substituted for *redundancy* in many contexts. Thus, it should be clear that redundancy is not reduced in these codes simply for economy or because it is unimportant, but because it has a different status from the information and has to be handled in a different way. Ideally, a redundancy-reducing code passes on information only, but stores knowledge of the environment in order to recognize and *not* repeat what is already known.

Cardinal Cells and Economy of Impulses If redundancy is reduced, the capacity of the channels carrying the messages must be reduced, even if the information they carry is preserved. In the brain, this can be done either by decreasing the number of nerve fibers carrying the messages, or by decreasing the mean frequency of impulses in the fibers. The number of neurons decreases from photoreceptors to optic nerve but, when the visual pathway enters the cortex, the number of neurons increases enormously, so it is attractive and plausible to regard the cortical encoding as achieving redundancy reduction by the "economy of impulses" (Barlow, 1969). The idea behind cardinal cells is that they achieve as complete representation of the sensory scene as possible with the minimum number of active neurons. They differ from pontifical or grandmother cells in that they do not represent arbitrary or capricious features in the environment, but features useful for their representative role; they are more like Lettvin's original mother cells than the more popular grandmother cells. The idea that object recognition is based on a limited alphabet of forms (geons) is clearly related (Biederman, 1987; Fujita et al., 1992). Note that cardinal cells can be active in combinations and thus have something of the descriptive power of words.

The notion of sparse representation by cardinal cells is initially made more plausible by the fact that there is a considerable decrease in the firing rate of neurons between optic nerve and V1, although this decrease may not continue at higher levels. Also, unless there are many cells with a very much lower activity ratio than those recently described, the decrease in mean rate is not nearly great enough even to keep the capacity of the system constant: Figures given by Hawken and Parker (1991) show that there are 3000 to 10,000 cortical neurons per retinal sample point in the fovea of macaque.

Willshaw, Buneman, and Longuet-Higgins (1969) demonstrated that the memory capacity of associative nets is increased when the input is sparsely coded—that is, the input vectors contain a low proportion of active bits. Gardner-Medwin and Recce (1987) point out other advantages of sparseness. Field (1993) has recently argued that sparse codes are well adapted to represent the type of statistical structure that is found in natural images, and Webber (1993) has developed a theory of the symmetries present in natural images and decribes an algorithm for finding them. However, the important question remains, How sparse? How many units are typically required to represent a scene? Originally, I suggested the answer to this was 1000 but, for reasons already given, this estimate probably is too low.

Coarse Coding *Coarse coding* is representation by a small number of variables used in combination, rather than by a large number of variables most of which will be zero on any particular occasion (Hinton, 1981). For instance, to represent the position and orientation of an object in the six-dimensional space of three translations and three orientations, one could divide this hyperspace into resolvable cells and then state which one of these cells was occupied. Alternatively, one could block out the hyperspace into many incompletely overlapping regions and, by choosing their number and size appropriately, the position and orientation of a single object could be accurately deduced from the combination of these larger regions that were occupied. This method has the advantage of using far fewer units, though at the cost of possible ambiguity if more than a single object is present.

It is evident that coarse coding is a tactic to change from a sparse coding to a more densely distributed representation, and it thus draws attention to an important process that was given insufficient emphasis in my previous article (Barlow, 1972)—namely, the need to combine the outputs of many selective units to form an element that generalizes in an appropriate way. There are physiological examples of this: For instance, a complex cell in the hierarchical model of Hubel and Wiesel (1962) collects the outputs of simple cells with the same orientation but different positions in the visual field and thus provides a cell that is still selective for orientation but is invariant for position over a limited range. The direction-selective ganglion cells in the rabbit retina do the same (Barlow and Levick, 1965). This example is instructive because the selectivity and generalization seem to be performed by different parts of the same cell; the selectivity depends on the pattern of termination on the dendritic branches, whereas the cell body combines inputs from these branches to generalize. The problem with grandmother cells, and perhaps cardinal cells as originally specified, is that they are overspecialized; a hierarchy that combines selective and generalizing steps in alternation (Barlow, 1993) is very much more attractive and has been developed as a pattern recognition device by Fukushima (1975, 1993).

Cell Assemblies The concept of an assembly of nerve cells was introduced by Hebb (1949) in the seminal book in which he also described hebbian synapses and phase sequences. He can be regarded as the prophet of connectionism, for his main aim was to develop a conceptual system "... which relates the individual nerve cell to psychological phenomena." The idea of cell assemblies is slightly out of line with this aim, so one wonders why he introduced it.

In the 1940s, knowledge of nerve cells was confined mainly to the millisecond time scale, and slower neuronal phenomena, such as the actions of internal transmitters and neuromodulators or the long-lasting changes of internal ion concentration that result from activity, were not understood (Magleby, 1986; Zucker et al., 1991; Regehr and Tank, 1992). Hebb saw that slow processes, of the order of a second in duration, were crucial in understanding psychological phenomena, so he seized on the idea of reverberating circuits (Lorente de No, 1938) to provide short-term persistence of a neuronal excitatory effect; his cell assemblies were collections of neurons sustaining a sort of buzz of impulses by means of these reverberating chains of activity passing through hebbian synapses that had been strengthened by previous activity. We still do not know which of the slow consequences of activity are important but, on rereading Hebb's book, it is easy to see that the mere existence of long-lasting effects in single cells makes his concept of neuronal assemblies superfluous, just as the mere existence of synapses made protoplasmic continuity and the reticular theory unnecessary. The concept played an important part in getting across Hebb's main message—that mechanisms at a cellular level could explain behavioral phenomena—but self-re-exciting reverberatory circuits do not seem to be nearly as widespread as Hebb thought. The concept of cell assemblies in that form is no longer viable, but can it be revived in any other form?

Abeles (1982) has developed the radically new concept that the unit of transmission of information around the cortex is a synchronously firing group of neurons he calls a *synfire group*, with one such group firing the next group in a chain. Different groups can overlap in membership, and any cell can belong to more than one such group. This is a challenging idea that is compatible with some of the physiology (Abeles et al., 1993) and many of the remarkable facts coming to light about the detailed connectivity of cortical neurons (Braitenberg and Shüz, 1991). Despite the fact that correlations with behavioral and psychophysical facts have not been established, as they have for single neurons, this is an idea worth keeping one's eye on.

Associative Nets and Parallel Distributed Processing The remaining items in table 26.4 derive from the connectionist program (summarized in the three-volume synopsis edited by Rumelhart and McClelland, 1986), which showed that networks of model neurons are capable of many of the things Hebb claimed for real ones, without having recourse to reverberating cell assemblies, by the way. The backpropagation algorithm (Rumelhart, Hinton, and Williams, 1986) was the most powerful tool, but associative nets— which are simple devices in which an input array makes synapses with an output array—are particularly important in relation to the neuron doctrine. It can be shown that the appropriate synaptic modification rule enables such a net to learn to give a particular output vector in response to a particular input vector, and that a single net is capable of holding many such input-output vector pairs (Longuet-Higgins, Willshaw, and Buneman, 1970; Palm, 1980; Kohonen, 1980, 1984; Baum, Moody, and Wikzek, 1988). They are also capable of generalization, in the sense that an incomplete input vector can give the output vector appropriate for the complete vector of which it is a part.

Those who regard associative nets as promising models of the nervous system point out that they avoid the bottleneck of a single units (see table 26.2) by transforming an input vector (representing a pattern on the input) directly to an output vector (representing a pattern on the output). Though this may prove to be a useful way of understanding how sensory input patterns are transformed into motor output patterns, it is not clear how noise-resistant such transformations are (see figure 26.1), and it is certainly not the only valid way of looking at associative nets. Each bit of the output is activated by a particular subset of the possible input vectors, and these constitute the bottlenecks of table 26.2. One can then regard the specific input vector that corresponds to a specific output vector as the intersection of the subsets of input vectors for each of the bits active in the output. This is a more analytical approach, and whether it is better or worse than considering the output vector as a whole depends partly on whether the output components are meaningful and significant when considered in isolation, a point taken up later.

Projective Zones Projective zones are an example of a concept easily neglected physiologically but brought to prominence by the connectionist program. When one records from a neuron one easily forgets that its role in the brain is not only to respond in specific conditions and not at other times, but also to signal to other neurons when these conditions arise. In other words, one forgets to ask, where does the axon of this neuron go to? In addition, one can easily fail to realize that the arrangement of projective zones is itself an important computational operation. An example of this has been described already: The organization of the projective zones of individual neurons in V1 and V2 creates a neuron in MT that is a sensitive detector of coherent motion. This is a functional example of what nontopographical mapping (Barlow, 1981) can achieve, and it seems likely to be a mechanism of very wide importance in the cortex.

Artificial Neural Nets as Statistical Estimators When the results of network simulations of cognitive tasks were first presented, the approach was criticized on the grounds that it gave no insight about *how* the task had been done. It has recently become clear that neural networks perform efficient statistical estimations (White, 1989; Geman, Bienenstock, and Doursat, 1992; Ripley, 1992), which explains their success. Probably the early simulations were entered into without a clear appreciation that it was the statistical aspects of the cognitive tasks selected for simulation that made them difficult; if this had been appreciated, traditional statistical methods would presumably have been deployed initially and, if professional statisticians are to be believed, these would have given results at least as good as the network methods!

This tale has an important moral: If artificial neural nets, designed to imitate cognitive functions of the brain, are truly performing tasks that are best formulated in statistical terms, then is this not likely also to be true of cognitive functions in general? The idea that the brain is an accomplished statistical decision-making organ agrees well with the notions to be sketched in the last section of this article.

Questions and problems

Table 26.5 brings together some questions and problems about the role of single neurons in perception. The first three are primarily points that need to be resolved experimentally; the remainder raise theoretical issues.

Measures of Neural Responses Some measure of average impulse activity over a short period is a practical necessity, but the question which is the best measure to use has not received much attention. Mean rate in spikes per second is the measure generally used and, for N impulses separated by intervals i in a period T, this is nearly $N(\Sigma i)^{-1}$, since $\Sigma i \approx T$. Note that this is not the mean of the instantaneous rates given by successive impulse intervals, which would be $N^{-1}\Sigma i^{-1}$. Because interval values are extremely variable, different mathematical means may give significantly different results, particularly for dynamic range. Ideally, one should use a measure that reflects the influence of the cell on its postsynaptic contacts, but it would also be worth determining whether any measure gives a better signal-to-noise ratio or extends the dynamic range.

The approach followed by de Ruyter van Steveninck and Bialek (1988) is more radically different. These investigators, rather than asking what neural response a particular physical stimulus gives, ask what impulses in a neuron signify about the physical stimulus being applied, and their approach is surely correct because it is relevant to the actual use made of sensory messages. It leads to much higher estimates of transmission rates (Bialek et al., 1991); Rieke, Warland, and Bialek (1991) obtained information transfer rates up to nearly 300 bits/per second in single-nerve fibers from cricket cercus. This is two orders of magnitude greater than estimates of bit rates using the methods that have become established in vertebrate neurophysiology; note that these bit rates bear on the rate at which the experimenter gets information from the preparation as well as the rate at which the brain gets information from the sense organ. The main reason for the higher rates is that the random stimuli chosen to drive the sense organs were much richer than the narrow selection of often-repeated stimuli used traditionally, but another factor is that correlations were sought for each individual impulse rather than simply for the so-called mean rate. This work is leading to new generalizations about the encoding of neural messages (Bialek et al., in press). The use of reverse correlation methods in cortical studies is a step in the same direction and is proving fruitful (Ohzawa, DeAngelis, and Freeman, 1990; DeAngelis, Ohzawa, and Freeman, 1991).

The Problem of Dynamic Range It has been shown that cortical neurons are sensitive enough to account for behavioral thresholds, but they appear to have a very much narrower dynamic range than the intact animal. Expressing this in terms of the signal detection theory measure d′, or signal-to-noise ratio, a value of approximately 2 units was found for single cells in macaque V1, whereas integrating just-noticeable differences for intensity discrimination in humans gives a figure of 20 or more under equivalent circumstances (Barlow et al., 1987). A similar problem has been iden-

Table 26.5

Unresolved questions about neurons in perception

What is the best measure of the short-term average activity of a single neuron?
Why is the dynamic range of a single neuron so small compared with that obtained from psychophysical just-noticeable difference experiments?
Can measures of oscillations, synchronization, or correlation among neurons improve the relation between psychophysical and neural performance?
When comparing single-unit and behavioral thresholds, how does one take into account the problem of false-positive responses resulting from the very large numbers of neurons in the brain?
How sparse is the distributed representation in the brain?
Is the relation between unit activity and input pattern tight enough to support population or ensemble coding?
Can a single neuron mediate a perceptual discrimination?

tified in hearing (Evans, 1978). The experiments need to be repeated on unanesthetized animals using a variety of measures of mean response for, as it stands, this narrow dynamic range appears to be a very serious limitation on single neurons as computational elements.

OSCILLATIONS, SYNCHRONIZATION, AND CORRELATIONS There has recently been much effort devoted to multiunit recording (Palm, Aertsen, and Gerstein, 1988; Gerstein and Gochin, 1992; Vaadia and Aertsen 1992) and to the fine temporal structure in the firing of neurons in the cortex (von der Malsburg and Schneider, 1986; Eckhorn et al., 1988; Gray et al., 1989; Singer, 1990). These are two comparatively unexplored features of neural responses that may reveal many new facets of neural coding, but thus far the yield has been somewhat disappointing. The important experimental question is whether, by using measures based on these results, one can explain behavioral performance better than one can using normal measures from single neurons. This has not been accomplished so far. Vaadia and Aertsen (1992) give an example of direction sensitivity in the auditory cortex that could be seen only in the interaction between two neurons, not in the responses of either alone, but it required nearly 300 stimuli to make the correlation evident, whereas it should be evident on nearly every trial if it is to be of much use to the animal itself. Furthermore, there are surely single neurons that show directional selectivity with greater sensitivity than this, and the natural assumption must be that it is these, not the correlated firing, that signals direction to the animal. In the case of oscillations, there is the serious possibility that they may arise entirely in the LGN and have none of the significance that has been attributed to them (Ghose and Freeman, 1992).

FALSE-POSITIVE RESPONSES False-positive responses are a problem that arises when comparing single-unit and behavioral thresholds. A threshold response consists of a significant change in the maintained discharge, which is high in most retinal ganglion cells (10–60/per second) and present at a lower rate in many cortical neurons. While the experimenter has to contend with the noisy maintained discharge of the one cell from which he or she is recording, the whole animal must contend with noise from all the cells that might have been stimulated: How is the cat or monkey brain to know which particular ganglion cell or neuron to attend to or, for that matter, in which period of time to attend? Obviously, the enormous numbers of neurons carrying the sensory representation make the possibility of a false-positive response much greater for the whole brain than it is for the experimenter looking at only one neuron. Pelli (1991) gives a very clear analysis of the problem from a signal detection theory point of view.

Perhaps the rate of firing of a neuron should be read as an indication of the certainty that its trigger feature is present (Barlow, 1972), for then setting a high threshold for the recipient neuron would eliminate all but the most strongly firing neurons, and these would be the ones signaling their trigger feature with greatest certainty. Such an arrangement might also explain the lower-envelope principle (Devalois, Abramov, and Mead, 1967; Talbot et al., 1968). If the right measure of certainty is used, this also offers the prospect of combining information from different neurons in an optimal or near-optimal way, imitating the action of the "ideal homunculus" (see Földiák, 1992).

FORMS OF REPRESENTATION We consider next what has come to be called *ensemble* or *population encoding* (Gross, 1992; Rolls, 1992; Young and Yamane, 1992; Gross et al., 1993), in which information is carried by the pattern of impulses in a group of neurons. Let us first see how this view differs from others.

Mutually exclusive representation Strictly speaking, the only alternative to a distributed representation, in which several units can be simultaneously active, is one that is mutually exclusive—that is, one in which only a single unit can be active at any one time. Possibly, Sherrington's pontifical cells were believed to form a mutually exclusive set, but grandmother cells, cardinal cells, and yellow Volkswagen cells certainly were not. Hence, there is no dispute about whether the representation is distributed; rather, differences of opinion revolve around the kind of distributed representation the brain uses.

Sparseness One parameter for characterizing the representation is its sparseness—that is the average proportion of units that are active at any one time. As mentioned previously, the evidence now indicates that this proportion is higher in the cortex than appeared to be the case 20 years ago. In the theoretical framework presented here, this signifies that generalization plays a major role in successive stages of transformation (Barlow, 1993), whereas previously selectivity had been emphasized (Barlow, 1972). Little information is available about the actual activity ratios, even though these should be easily determinable experimentally by observing activity in sensory pathways during normal sensory experience. The matter is of considerable current interest because of the suggestion that "good" features will show strongly kurtotic distributions of activity (Barlow and Tolhurst, 1992; Field, 1993). Note that if impulse frequency signals certainty, sparseness can be adjusted to any required level by adjusting the threshold of a postsynaptic unit, because this determines the rates of firing that produce significant effects.

Direct representation In any distributed representation, individual elements are active for particular subsets of the input. The question is whether this directly represented subset is significant and meaningful or whether it is just an arbitrary collection of the possible inputs. The latter is nearly true for the ASCII code, where a single bit in the seven-bit word gives singularly little useful information about the character represented. However, this is not a necessary feature of distributed representations; with an extra bit or two, it would be possible to devise a code in which a unique bit signals any numeral, another one any letter, and other ones any vowel or punctuation mark. It is convenient to have a name for this kind of distributed representation; we call it a *direct* representation (Gardner-Medwin and Barlow, 1993), in which important categories of the input are in one-to-one correspondence with the activity of particular representational elements. This type of representation would be much more useful than other types of ensemble encoding when it comes to learning, as the following argument illustrates.

Learning with direct and distributed representations Learning that there is a significant association with a feature requires estimating four numbers, those needed to fill the 2×2 contingency table for the occurrences of the feature itself and for the occurrence of the reinforcing event. We can assume that information about the occurrence of the reinforcing event is available everywhere, but a potentially serious problem arises when the other feature is represented only in a distributed manner, because the representational elements whose numbers one can count are active not only for the feature of interest but also for other features. This introduces errors, and to compensate for them requires a larger sample of information. The extent of this efficiency loss can be estimated and varies greatly with details (Gardner-Medwin and Barlow, 1993), but this factor implies that learning can always occur with fewer trials or errors for directly represented features (i.e., for the subsets of the input states coded by single units).

Tightness of coding The tightness of the relation between input stimulus and the firing of a neuron is another problem area. If a neuron x is claimed to code for stimulus y, then it must nearly always fire above a critical rate when y is present and almost never fire above that rate when it is absent. Advocates of ensemble encoding do not seem to realize that their schemes demand an even closer and more reliable relation between stimulus and firing of a neuron when this is part of a distributed code, or ensemble encoding, than when it directly represents a feature of interest. Figure 26.1 shows what happens with a 5% error rate in the ASCII code, which is often taken as an example of ensemble encoding. The errors in representation become horrendous when *all* the neurons taking part in it have to be correct, for the error rate is then $1 - (1 - P)^N$ for an error rate P in each of the N neurons. Note that for features that are directly represented as described in the preceding section, the lower error rate P applies; the higher rate is a consequence of using *combinations* of such directly represented features. Clearly, in any distributed representation, the directly represented features are more reliable than features represented

Random alteration of 5.2% of the bits in the ASCII code for

"TIGHT CODING IS ESSENTIAL"

gave

"TIDHF'CODINg KS ESSEOTAAL"

The character is wrong if any of its bits are altered, and this has expectation $1 - 0.948^7 = 0.312$.
The observed rate in this example is 28%.

FIGURE 26.1 Effects of errors in ensemble encoding.

by patterns, and a sensible system will, as far as is possible, represent important features directly, not by combinations.

Much of the work on single neurons in higher visual areas is bedevilled by insufficient attention to the tightness of the coding. No one can be blamed for not finding what makes a cell fire reliably, because the possibilities are endless. Faced with units responding poorly, it is natural to analyze exhaustively the weak and unreliable responses that the available stimuli do produce, but units that fail to show a tight relation between the physical stimulus and the neuronal response on nearly every trial are uninformative for the animal, for the experimenter, and for the reader of the experimenter's publications; it is no solution to say that such units belong to an ensemble or population code, because this aggravates the problem of unreliability. Of course, error-correcting procedures might be proposed but have not been to date.

A possible psychophysical linking principle In 1972, I suggested that ". . . active high-level neurons directly and simply cause the elements of our perception," and I still think this simple idea has some merit, even though I now believe that interactions with other individuals and society have to be taken into account when considering the conscious aspects of perception (Barlow, 1987). In a well-designed psychophysical experiment, these aspects of consciousness are not important, so the simple idea that single units signal the elements of perception can be preserved by formulating the following psychophysical linking principle:

> Whenever two stimuli can be distinguished reliably, then some analysis of the physiological messages they cause *in some single neuron* would enable them to be distinguished with equal or greater reliability.

This principle is intended to pose the problem of the relation between single neurons and perception in a testable form. Without the italicized phrase, it must be true, unless one postulates supernatural, nonphysical processes occurring in the brain. With the phrase intact, the principle asserts that it is not *necessary* to have more than a single neuron, but it allows probability summation, so if there are several neurons in parallel in a group, many of them may be active at significant levels for many of the positive responses. However, a single neuron may be enough, as will become clearer in the next section.

The role of single units in perception

In this section, the aim is to sketch a plausible hypothesis about how the brain might compute the representation underlying perception using single neurons as its computational elements. First, let us review the properties of single neurons that are relevant to this task.

Neurons as Computational Elements From a computational point of view, a neuron has three parts: a receptive part, usually the dendrites and cell body, that receives inputs from a set of presynaptic neurons; a conducting part, its axon; and a transmitting part, its axon terminals, which influence the receptive parts of other neurons. We can regard the receptive part as applying a test to the pattern of activity in the neurons that impinge on it, and generating a response if the condition is met but not if it is not met. For the integrate-and-fire model dating back to McCulloch and Pitts (1943) and used in many neural net models, the test is simply whether the summed effects of the excitatory and inhibitory presynaptic terminals cause sufficient depolarization to exceed threshold, but neurons seem to be capable of much more complex tests (see Sensitivity of single neurons, above).

The number of presynaptic neurons impinging on a given cell in the cortex is approximately 10,000 (Braitenberg and Shüz, 1991). A full quantitative account of this receptive part would require the tuning curves that describe its responsiveness in all the dimensions, plus specifications of any interactions of one parameter on the tuning for another; qualitatively, it may be sufficient to specify its trigger feature—the class of stimuli in this multidimensional space that will excite the unit under normal circumstances. For completeness, one should also consider the past history of the neuron, for the receptive characteristics can change when the input is changed by adaptation or deprivation, but for the moment let us confine ourselves to steady-state conditions of normal use.

The conducting axon simply transmits impulses from the receptive part to the transmitting part at rates up to approximately 100 per second in the cortex. The transmitting parts of a neuron, its axon terminals, end on a group of other neurons, again numbering nearly 10,000 in the cortex. The ability of axons to find their correct projective zones is one of the most remarkable characteristics of nerve cells, and it is responsible for the creation of the topographical maps that are such a notable feature of the cortex. There is some selection of which branches of axons survive following a stage of exuberant growth during development (Innocenti, 1991) but, because the course of an axon cannot be changed rapidly, in adults the projective zone must be a relatively fixed property. The excitatory effects of the terminals can be modified by presynaptic inhibition, and there is evidence of retrograde influences from the postsynaptic neuron on the terminals but, over a short time scale in vertebrates, the neuron can probably be considered as transmiting to a selected but relatively fixed group of neurons.

Neurons are capable of generating rhythmical activity both individually and by virtue of their connections with other neurons (Selverston 1988; Getting, 1989; Harris-Warrick and Marder, 1991). Both the properties of the neurons involved and their functional role are very well understood in some instances in invertebrates. Interestingly, neuromodulators have been shown to have profound effects in modifying these properties so that they cause entirely new patterns of rhythmical output to be generated. Cells in mammalian cortex and thalamus also generate rhythms (Eckhorn et al 1988; Grey et al., 1989), which seem especially prominent in the olfactory system of vertebrates (Adrian, 1942; Freeman, 1975) as well as invertebrates (Gelperin and Tank, 1990), but the reasons for this are unknown. Although these rhythmical propensities should be borne in mind, it is difficult to include them in our account until their role is better understood.

The simple summary of a neuron as a computing element in perception is, therefore, that it tests an input group of neurons for the presence of its trigger feature and transmits information about the outcome of this test to another group of neurons in its projective zone. Both the input conditions and the exact site of termination can be modified by experience, the former more rapidly and more easily than the latter.

Guessing Right The ability to make reasonably good decisions in complex and uncertain circumstances is surely the behavioral capacity that should puzzle us most, not only in animals but especially in ourselves. After all, it is "guessing right" (or the reputation for being able to do so) that earns business executives their high salaries, and the ability is almost equally well rewarded (and tested objectively at much more frequent intervals) in sport, where outguessing your opponent is at least as important as strength, speed, or dexterity. In academic life, knowledge and reasoning power are the supposed criteria of excellence, but these abstract qualities reveal themselves in practice by good guesswork about revealing experiments to perform or fertile topics to write about. Guessing right when others guess wrong is the most prized achievement in all professions and occupations.

The Role of Guesswork in Perception The usual way to analyze behavior is to split what the brain does into two halves: Sensory messages are used to make a representation of the external world, we say, and the brain then uses this representation to decide a course of action that will accomplish its goals. This seems natural because we fool ourselves that we perceive a straightforward representation of the world and decide simply according to what that representation shows, but even the most elementary knowledge of the psychology of perception tells us that it is very far from being a straightforward representation. Ask people what they see before them, and they will list objects and events that certainly require complex decision-making processes to identify. Thus, by the time the perceptual representation has been generated, a large proportion of the brain's guesswork has already been done, whether or not we are consciously aware of this.

How to Guess Right A paradigm for the ability to guess right arose from the task of radar operators in World War II who had to decide whether a bump on an already bumpy oscilloscope trace was an enemy aircraft or just a random disturbance (Woodward, 1953). Signal detection theory applied these insights to the problem of sensory discrimination by human observers (Swets, 1964). Three separable steps are needed to make the most reliable and sensitive guesses: First, relevant and irrelevant sources of information must be identified and a measure of the relevant information made by combining it appropriately while excluding the irrelevant. Second, a criterion must be set according to the consequences of different types of error. Third, a decision must be made and communicated when the measure meets this criterion.

Neurons as Decision Makers Table 26.6 compares the three requirements for good decision making with the summarized role of neurons as computational elements. Note that there is a surprising homology between them. Neurons are thus well qualified to be the

TABLE 26.6
Neurons as decision makers

Requirements for Good Decision Making	*What Neurons Do*
Combine relevant evidence to form a decision variable	Collect information from their presynaptic terminals
Set a criterion for positive decision	Test for the presence of their trigger features
Communicate when it is exceeded	Transmit impulse if it is present

decision makers in forming our perception; and if we return to the psychophysical linking principle, we can strengthen the suggestion that the brain may rely on single neurons for discriminations based on perception.

Suppose for a moment that the linking principle is wrong and that one would have to analyze the messages in a group of two or more neurons to perform a discrimination with the reliability and sensitivity shown by a psychophysical experiment. To account for this performance, the brain itself would also have to combine information from two or more neurons, but how? Neurons are the only elements available to do the computation, but if one or more of them were used, the information would be collected at that higher level, and the experimenter should be able to record from a neuron there and more closely match the performance of the intact brain. If two or more neuronal records were still required to match up to psychophysical performance, then this same argument applies again. The point is that if nerve cells are the only mechanisms in the brain that can combine evidence and make a decision, then it seems that the psychophysical linking principle *must* be true. Even if it is, though, there is a large gap to bridge: The problem is to determine how neurons as micro–decision makers can be combined in a network to achieve the macro-decision-making capacities that underlie perception and behavior.

I do not believe there is very much room for sloppiness and inefficiency either in the performance of the individual components of this network or in its organization. Because our ability to outguess our competitors is a major factor determining our survival, there must be powerful selection pressure to eliminate inefficiency; some of the examples of a single neuron's performance reviewed above confirm this for the components, and the difficulty of emulating even our most elementary perceptual achievements argues for the effectiveness of the whole network. Shortcuts and simplifications are needed to help us describe the operation of this nearly miraculous net, but I cannot see any alternative to regarding single neurons as the basic functional unit, just as the neuron doctrine proclaimed.

Conclusions

The arguments of this chapter can be summarized as follows:

1. The neuron doctrine is all-important for achieving Donald Hebb's goal of relating "the individual nerve cell to behavior," but the single neuron is not the most appropriate unit of function for many other aspects of neuroscience (see table 26.2).

2. Many slow and persistent effects are now known to occur both presynaptically and postsynaptically following excitation, and these render Hebb's concept of reverberating cell assemblies unnecessary and obsolete. However, we do need useful concepts about the functional role of small groups of neurons.

3. During the past 20 years, important ideas about neural organization have come from work on artificial nets using much oversimplified models of nerve cells (see table 26.4). Groups of interconnected units can simulate cognitive performance by doing sophisticated statistical computations with reasonably high efficiency, which suggests that statistical problems lie at the core of cognitive function.

4. Neurons are much more pattern-selective and reliable than was formerly believed, but there are important unresolved questions, summarized in table 26.5.

5. Advocates of grandmother cells, cardinal cells, ensemble encoding, and dense distributed representations all agree that the elements of perception are used in combination, as are words in language: Perception certainly uses a distributed, not a mutually exclusive, representation. Disagreements center around the representation's sparseness (i.e., what proportion of units is typically active), whether the elements directly

represent meaningful and significant features of the sensory input or arbitrary and meaningless ones (as in the ASCII code), and how tight the coding is (i.e., how reliably the firing of a unit is correlated with the sensory stimulus). It is argued here that the tightness of the coding demands more experimental attention, especially as it is even more important for ensemble encoding than for the direct representation of significant and meaningful features (see figure 26.1).

6. Neurons collect and combine information impinging on their synapses, generate a signal when a criterion is exceeded, and transmit this to other neurons elsewhere in the brain. These are the operations required for making good decisions in uncertain circumstances (see table 26.6). The problem is to learn how these micro–decision makers are combined in series and parallel to mediate the macrodecisions of the whole animal.

7. A psychophysical linking principle is proposed which states that a single neuron can provide a sufficient basis for a perceptual discrimination. This appears to be necessarily true if neurons are the only means available to the brain for making a decision based on evidence combined from other neurons.

REFERENCES

ABELES, M., 1982. *Local Cortical Circuits: An Electrophysiological Study*. Berlin: Springer-Verlag.

ABELES, M., Y. PRUT, H. BERGMAN, E. VAADIA, and A. AERTSEN, 1993. Integration, synchronicity and periodicity. In *Spatio-Temporal Aspects of Brain Function, A. Aertsen and W. von Seelen, eds. Amsterdam: Elsevier.*

ADRIAN, E. D., 1942. Olfactory reactions in the brain of the hedgehog. *J. Physiol. (Lond.)* 100:459–473.

ATTNEAVE, F., 1954. Informational aspects of visual perception. *Psychol. Rev.* 61:183–193.

BARLOW, H. B., 1950. The receptive fields of ganglion cells in the frog retina. In *Proceedings of the Eighteenth International Physiological Congress*. Copenhagen: Bianco Lunos Bogtrykkeri, pp. 88–89.

BARLOW, H. B., 1953. Summation and inhibition in the frog's retina. *J. Physiol. (Lond.)* 119:69–88.

BARLOW, H. B., 1959. Sensory mechanisms, the reduction of redundancy, and intelligence. In *The Mechanisation of Thought Processes*, London: Her Majesty's Staionery Office, pp. 535–539.

BARLOW, H. B., 1961. Possible principles underlying the transformations of sensory messages. In *Sensory Communication*, W. Rosenblith, Ed. Cambridge, Mass.: MIT Press, pp. 217–234.

BARLOW, H. B., 1969. Trigger features, adaptation, and economy of impulses. In *Information Processing in the Nervous System*, K. N. Leibovic, ed. New York: Springer-Verlag, pp. 209–226.

BARLOW, H. B., 1972. Single units and sensation: A neuron doctrine for perceptual psychology? *Perception* 1:371–394.

BARLOW, H. B., 1981. Critical limiting factors in the design of the eye and visual cortex. The Ferrier lecture, 1980. *Proc. R. Soc. Lond. [Biol.]* 212:1–34.

BARLOW, H. B., 1987. The biological role of consciousness. In *Mindwaves*, C. Blakemore and S. Greenfield, eds. Oxford: Basil Blackwell.

BARLOW, H. B., 1990. A theory about the functional role and synaptic mechanism of visual after-effects. In *Vision: Coding and efficiency*, C. B. Blakemore, ed. Cambridge, Engl.: Cambridge University Press.

BARLOW, H. B., 1993. Object identification and cortical organisation. In *The Functional Organisation of the Human Visual Cortex*, E. Ottoson, B. Gulyas, and P. Roland, eds. Oxford: Pergamon.

BARLOW, H. B., T. P. KAUSHAL, M. HAWKEN, and A. J. PARKER, 1987. Human contrast discrimination and the contrast discrimination of cortical neurons. *J. Opt. Soc. Am. [A]* 4:2366–2371.

BARLOW, H. B., and W. R. LEVICK, 1965. The mechanism of directionally selective units in the rabbit's retina. *J. Physiol.* 178:477–504.

BARLOW, H. B., W. R. LEVICK, and M. YOON, 1971. Responses to single quanta of light in retinal ganglion cells of the cat. *Vision Res.* 11 (suppl. 3): 87–101.

BARLOW, H. B., and D. J. TOLHURST, 1992. Why do you have edge detectors? In *1992 Optical Society of America Annual Meeting*, 1992, Albuquerque, N. Mex. Technical Digest series, Vol. 23. Washington D. C.: Optical Society of America, p. 172.

BAUM, E. B., J. MOODY, and F. WILCZEK, 1988. Internal representations for associative memory. *Biol. Cybern.* 59(4): 217–228.

BIALEK, W., M. DEWEESE, F. RIEKE, and D. WARLAND, (in press). Bits and brains: information flow in the nervous system. *Physica [A]*.

BIALEK, W., F. RIEKE, R. DE RUYTER VAN STEVENINCK, and D. WARLAND, 1991. Reading a neural code. *Science* 252: 1854–1857.

BIEDERMAN, I., 1987. Recognition by components: A theory of human image understanding. *Psychol. Rev.* 94(2):115–147.

BRAITENBERG, V., and A. SCHÜZ, 1991. *Anatomy of the Cortex: Statistics and Geometry*. Berlin: Springer-Verlag.

BRITTEN, K. H., M. N. SHADLEN, W. T. NEWSOME, and J. A. MOVSHON, 1992. The analysis of visual motion: A comparison of neuronal and psychophysical performance. *J. Neurosci.* 12:4745–4765.

CHELAZZI, L., E. K. MILLER, and J. DUNCAN, 1993. A neural basis for visual search in inferior temporal cortex. *Nature* 363:345–347.

DEANGELIS, G. C., I. OHZAWA, and R. D. FREEMAN, 1991. Depth is encoded in the visual cortex by a specialised receptive field structure. *Nature* 352:156–159.

DE RUYTER VAN STEVENINCK, R., and W. BIALEK, 1988. Real-time performance of a movement-sensitive neuron in blowfly visual system: Coding and information transfer in short spike sequences. *Proc. R. Soc. Lond. [Biol.]* 234:379–414.

DESIMONE, R., S. J. SCHEIN, J. MORAN, and J. UNGERLEIDER, 1985. Contour, color and shape analysis beyond the striate cortex. *Vision Res.* 25:441–452.

DESIMONE, R., 1992. The physiology of memory—recordings of things past. *Science* 258:245–246.

DE VALOIS, R. L., I. ABRAMOV, and W. R. MEAD, 1967. Single cell analysis of wavelength discrimination at the lateral geniculate nucleus in the macaque. *J. Neurophysiol.* 30:415–433.

ECKHORN, R., R. BAUER, W. JORDAN, M. BROSCH, W. KRUSE, M. MUNK, and H. J. REITBOECK, 1988. Coherent oscillations: A mechanism of feature linking in the cerebral cortex? Multiple electrodes and correlation analysis in the cat. *Biol. Cybern.* 60:121–130.

EVANS, E. F., 1975. Cochlear nerve and nucleus. In *Handbook of Sensory Physiology*, Vol. 5, part 2, W. D. Keidel and W. D. Neff, eds. Berlin: Springer-Verlag.

EVANS, E. F., 1978. Place and time coding of intensity in the peripheral auditory system: Some physiological pros and cons. *Audiology* 17:369–420.

FIELD, D. J., 1993. What is the goal of sensory coding? Manuscript submitted for publication.

FÖLDIÁK, P., 1992. The "ideal homunculus": Statistical inference from neural population responses. In *Computation and Neural Systems*, F. Eeckman and J. Bower, eds. Norwell, Mass.: Kluwer Academic.

FREEMAN, W. F., 1975. *Mass Action in the Nervous System*. New York: Academic Press.

FUJITA, I., K. TANAKA, M. ITO, and K. CHENG, 1992. Columns for visual features of objects in monkey inferotemporal cortex. *Nature* 360:343–346.

FUKUSHIMA, K., 1975. Cognitron: A self-organising multilayered neural network. *Biol. Cybern.* 20:121–136.

FUKUSHIMA, K., 1993. An improved neocognitron architecture and learning algorithm. In *Computational Learning and Cognition*, (Proceedings of the third N.E.C. Research Sympossium, Princeton, N.J.), E. Baum, ed. Philadelphia: Society for Industrial and Applied Mathematics, pp. 29–43.

GALLANT, J. L., J. BRAUN, and D. C. VAN ESSEN, 1993. Selectivity for polar, hyperbolic and cartesian gratings in macaque visual cortex. *Science* 259:100–103.

GARDNER-MEDWIN, A. R., and H. B. BARLOW, 1993. Sparse coding and the efficiency of associative learning. Manuscript in preparation.

GARDNER-MEDWIN, A. R., and M. L. RECCE, 1987. The benefits of a low activity ratio in auto-associative recall. In a report of RSRE conference held at Malvern, U.K.: Her Majesty's Stationery Office.

GELPERIN, A., and D. W. TANK, 1990. Odour-modulated collective network oscillations of olfactory interneurons in a terrestial mollusc. *Nature* 345:437–440.

GEMAN, S., E. BIENENSTOCK, and R. DOURSAT, 1992. Neural networks and the bias/variance dilemma. *Neural Computation* 4(1):1–58.

GERSTEIN, G. L., and P. M. GOCHIN, 1992. Neuronal population coding and the elephant. In *Information Processing in the Cortex*, A. Aertsen and V. Braitenberg, eds. Berlin: Springer-Verlag, pp. 139–160.

GETTING, P. A., 1989. Emerging principles governing the operation of neural networks. *Annu. Rev. Neurosci.* 12:185–204.

GHOSE, G. M., and R. D. FREEMAN, 1992. Oscillatory discharge in the visual system: Does it have a functional role? *J. Neurosci.* 68(5):1558–1574.

GRANIT, R., 1947. *Sensory Mechanisms of the Retina*. London: Oxford University Press.

GRAY, C. M., P. KOENIG, A. K. ENGEL, and W. SINGER, 1989. Oscillatory responses in the cat visual cortex exhibit inter-columnar synchronization which reflects global stimulus properties. *Nature* 338:334–337.

GROSS, C. G., 1992. Representation of visual stimuli in inferior temporal cortex. *Proc. R. Soc. Lond. [Biol.]* 335:3–10.

GROSS, C. G., C. E. ROCHA-MIRANDA, and D. B. BENDER, 1972. Visual properties of neurons in infero-temporal cortex of macaque. *J. Neurophysiol.* 35:96–111.

GROSS, C. G., H. R. RODMAN, P. M. GOCHIN, and M. W. COLOMBO, 1993. Inferior temporal cortex as a pattern recognition device. In Computational Learning and Cognition (*Proceedings of the Third NEC Research Symposium, Philadelphia*) E. B. Baum, Ed. Princeton, N.J.: Society for Industrial and Applied Mathematics.

HARRIS, C. S., 1980. Insight or out of sight? Two examples of perceptual plasticity in the human adult. In *Visual Coding and Adaptability*, C. S. Harris, ed. Hillsdale, N.J.: Erlbaum, pp. 95–149.

HARRIS-WARRICK, R. M., and E. MARDER, 1991. Modulation of neural networks for behavior. *Annu. Rev. Neurosci.* 14:39–57.

HARTLINE, H. K., 1940. The receptive fields of optic nerve fibres. *Am. J. Physiol.* 130:690–699.

HAWKEN, M. J., and A. J. PARKER, 1991. Spatial receptive field organisation in monkey V1 and its relationship to the cone mosaic. In *Computational Models of Visual Processing*, M. S. Landy and J. A. Movshon, eds. Cambridge, Mass.: MIT Press.

HEBB, D. O., 1949. *The Organization of Behavior*. New York: Wiley.

HINTON, G. F., 1981. Shape representation in parallel systems. In Proceedings of the *Seventh International Joint Conference on Artificial Intelligence*. Vancouver, B.C., Canada, pp. 1088–1096.

HUBEL, D. H., and T. N. WIESEL, 1962. Receptive fields, binocular interaction, and functional architecture in the cat's visual cortex. *J. Physiol.* 195:215–243.

HUBEL, D. H., and T. N. WIESEL, 1970. The period of susceptibility to the physiological effects of unilateral eye closure in kittens. *J. Physiol. (Lond.)* 206:419–436.

INNOCENTI, G. M., 1991. The development of projections from cerebral cortex. *Prog. Sens. Physiol.* 12:65–114.

KIANG, N. Y.-S., T. WATANABE, E. C. THOMAS, and L. F.

CLARKE, 1965. *Discharge Patterns of Single Fibers in the Cat's Auditory Nerve*. Cambridge, Mass.: MIT Press.

KOHONEN, T., 1980. *Content Addressable Memories*. Berlin: Springer-Verlag.

KOHONEN, T., 1984. *Self-Organization and Associative Memory*. Berlin: Springer.

KUFFLER, S. W., 1953. Discharge patterns and functional organisation of mammalian retina. *J. Neurophysiol.* 16:37–68.

LEHKY, S. R., and T. J. SEJNOWSKI, 1988. Network model of shape-from-shading: Neural function arises from receptive and projective fields. *Nature* 333:452–454.

LETTVIN, J. Y., H. R. MATURANA, W. S. MCCULLOCH, and W. H. PITTS, 1959. What the frog's eye tells the frog's brain. *Proceedings of the Institute of Radio Engineers* 47:1940–1951.

LONGUET-HIGGINS, H. C., D. J. WILLSHAW, and O. P. BUNEMAN, 1970. Theories of associative recall. *Q. Rev. Biophys.* 3:223–244.

LORENTE DE NO, R., 1938. Analysis of the activity of chains of internuncial neurons. *J. Neurophysiol.* 1:207–244.

MAGLEBY, K. L., 1987. Short term changes in synaptic efficacy. In *Synaptic Function*, G. M. Edelman, W. E. Gall, and W. M. Cowan, eds. New York: Wiley.

MCCULLOCH, W. S., and W. H. PITTS, 1943. A logical calculus of the ideas immanent in nervous activity. *Bull. Math. Biophys.* 5:115–133.

MOVSHON, J. A., E. H. ADELSON, M. S. GIZZI, and W. T. NEWSOME, 1985. The analysis of moving visual patterns. In *Pattern Recognition Mechanisms*, C. Chagas, R. Gattass, and C. Gross, eds. Rome: Vatican Press, pp. 117–151. (Reprinted in *Experimental Brain Research*, 1986, *Suppl.* 11: 117–151.)

NEWSOME, W. T., K. H. BRITTEN, and J. A. MOVSHON, 1989. Neuronal correlates of a perceptual decision. *Nature* 341: 52–54.

OHZAWA, I., G. C. DEANGELIS, and R. D. FREEMAN, 1990. Stereoscopic depth discrimination in the visual cortex: Neurons ideally suited as disparity detectors. *Science* 249: 1037–1041.

PALM, G., 1980. On associative memory. *Biol. Cybern.* 36:19–31.

PALM, G., A. M. H. J. AERTSEN, and G. L. GERSTEIN, 1988. On the significance of correlations among neuronal spike trains. *Biol. Cybern.* 59:1–11.

PELLI, D. G., 1991. Noise in the visual system may be early. In *Computational Models of Visual Processing*, M. S. Landy and J. A. Movshon, eds. Cambridge, Mass.: MIT Press, pp. 147–151.

PERRETT, D., M. HARRIES, A. J. MISTLIN, and A. J. CHITTY, 1990. Three stages in the classification of body movement by visual neurons. In *Images and Understanding* H. B. Barlow, C. Blakemore, and E. M. Weston-Smith, eds. Cambridge, Engl.: Cambridge University Press, pp. 94–107.

PERRETT, D. I., J. K. HEITANEN, M. W. ORAM, and P. J. BENSON, 1992. Organisation and functions of cells responsive to faces in the temporal cortex. *Philos. Trans. R. Soc. Lond. [Biol.]* 335:31–38.

REGEHR, W. D., and D. W. TANK, 1992. Calcium concentration dynamics produced by synaptic activation of CA1 Hippocampal pyramidal cells. *J. Neurosci.* 12(11):4202–4223.

RIEKE, F., D. WARLAND, and W. BIALEK, (1991, Oct.). *Coding efficiency and information capacity in sensory neurons*. Princeton, N.J.: NEC Research Institute.

RIPLEY, B. D., 1992. Statistical aspects of neural networks. In *Seminaire Europeen de Statistique*, Sandbjerg. Denmark: Chapman and Hall.

ROLLS, E. T., 1992. Neurophysiological mechanisms underlying face processing within and beyond the temporal cortical visual areas. *Philos. Trans. R. Soc. Lond. [Biol.]* 335: 11–22.

RUMELHART, D. E., G. E. HINTON, and R. J. WILLIAMS, 1986. Learning internal representations by error propagation. In *Parallel Distributed Processing*, D. E. Rumelhart and J. McClelland, eds. Cambridge, Mass.: MIT Press, pp. 318–362.

RUMELHART, D. E., and J. MCCLELLAND, 1986. *Parallel Distributed Processing* (Vols. 1–3). Cambridge, Mass.: MIT Press.

SELVERSTON, A. I., 1988. A consideration of invertebrate central pattern generators as computational data bases. *Neural Networks* 1:109–117.

SHEPHERD, G. M., 1991. *Foundations of the Neuron Doctrine*. Oxford: Oxford University Press.

SHERRINGTON, C. S., 1941. *Man on His Nature*. Cambridge, Engl.: Cambridge University Press.

SINGER, W., 1990. Search for coherence: A basic principle of cortical self-organization. *Concepts Neurosci.* 1(1):1–26.

SWETS, J. A., ed., 1964. *Signal Detection Theory by Human Observers*. New York: Wiley.

TALBOT, W. P., I. DARIAN-SMITH, H. H. KORNHUBER, and V. B. MOUNTCASTLE, 1968. The sense of flutter-vibration: Comparison of human capacity with response patterns of mechano-receptive afferents from the monkey hand. *J. Neurophysiol.* 31:301–334.

VAADIA, E., and A. AERTSEN, 1992. Coding and computation in the cortex: Single-neuron activity and cooperative phenomena. In *Information Processing in the Cerebral Cortex*, A. Aertsen and V. Braitenberg, eds. Berlin: Springer-Verlag.

VALLBO, Å. B., 1989. Single fibre microneurography and sensation. In *Hierarchies in Neurology: A Reappraisal of a Jacksonian Concept*, C. Kennard and M. Swash, eds. London: Springer, pp. 93–109.

VON DER MALSBURG, C., and W. SCHNEIDER, 1986. A neural cocktail-party processor. *Biol. Cybern.* 54(1):29–40.

WEBBER, C. J. S., 1993. Symmetric self-organisation: A geometrical theory of invariant response and pattern-constituent discovery. Manuscript submitted for publication.

WHITE, H., 1989. Learning in artificial neural networks: A statistical perspective. *Neural Computation* 1:425–464.

WILLSHAW, D. J., O. P. BUNEMAN, and H. C. LONGUET-HIGGINS, 1969. Nonholographic associative memory. *Nature* 22:960–962.

Woodward, P. M., 1953. *Probability and Information Theory with Applications to Radar*. Oxford: Pergamon.

Young, M. P., and S. Yamane, 1992. Sparse population coding of faces in the infero-temporal cortex. *Science* 256: 1327–1331.

Zeki, S., 1980. The representation of colours in the cerebral cortex. *Nature* 284:412–418.

Zucker, R. W., K. R. Delaney, R. Mulkey, and D. W. Tank, 1991. Presynaptic calcium in transmitter release and posttetanic potentiation. *Ann. N.Y. Acad. Sci.* 635:191–207.

APPENDIX
J. Y. LETTVIN ON GRANDMOTHER CELLS

Rutgers University
March 8, 1991

Dear Horace,

It is with a sense of proud shame that I accept fathership of the "grandmother" cell. A paternity knocks but once. How the seminal idea came to term may amuse you.

In the fall of '69, Professor Giorgio de Santillana took ill and I assumed his undergraduate course, switching the topic to "Biological Foundations for Perception and Knowledge." As I remember, the class was huge, several hundred.

Around Christmas, coming to the problem of how neurons can somehow represent individual substances, I prepared this story:

In the distant Ural mountains lives my second cousin, Akakhi Akakhievitch, a great if unknown neurosurgeon. Convinced that ideas are contained in specific cells, he had decided to find those concerned with a most primitive and ubiquitous substance—mother. Starting with geese and bears, he gradually progressed over the years to Trotskyites under death sentence in Siberia. And he located some 18,000 neurons clustered in von Seelendonck's area (rarely recognized among the degenerate materialist Sherringtonian exponents in the West) that responded uniquely only to the animal's mother, however displayed, whether animate or stuffed, seen from before or behind, upside down or on a diagonal, or offered by caricature, photograph, or abstraction.

He had put the mass of data together and was preparing his paper, anticipating a Nobel prize, when into his office staggered Portnoy, world-reknowned for his Complaint. (Roth had just published the novel). On hearing Portnoy's story, he rubbed his hands with delight and led Portnoy to the operating table almost hid under microstereotactic apparatus, assuring the mother-ridden schlep that shortly he would be rid of his problem.

With great precision he ablated every one of the several thousand separate neurons and waited for Portnoy to recover. We must now conceive the interview in the recovery room.

"Portnoy?"
"Yeah."
"You remember your mother?"
"Huh?"
(Akakhi Akakhievitch can scarcely restrain himself. Dare he take Portnoy with him to Stockholm?)
"You remember your father?"
"Oh, sure."
"Who was your father married to?"
(Portnoy looks blank.)
"You remember a red dress that walked around the house with slippers under it?"
"Certainly."
"So who wore it?"
(Blank)
"You remember the blintzes you loved to eat every Thursday night?"
"They were wonderful."
"So who cooked them?"
(Blank)
"You remember being screamed at for dallying with shikses?"
"God, that was awful."
"So who did the screaming?"
(Blank)

And so it went. Every attribute had by his mother was clear—for no attribute is the substance. To whatever detail Akakhi Akakhievitch pressed, the detail was remembered. Even though the details were compresent, the attributes bundled, somehow or another Portnoy looked blank when asked if they suggested his mother. That there was a simple person to whom all the attributes belonged was freely conceded. All the relations had by this person were remembered, including mothering (which is not a mother). It made no difference—Portnoy had no mother. "Mother" he could conceive—it was generic. "My mother" he could not—it was specific.

Akakhi Akakhievitch went into panic. It occurred to him that if there were but one logician or philosopher on the Nobel committee he was doomed. He had abolished the cells for knowing mother as *ding an sich*. He had let the Kantians be his guide and was now trapped.

If he had only gone after "grandmother cells" he would have been safe simply because grandmothers are notoriously ambiguous and often formless. A patient man, he went back to the geese and bears and "grandmother cells."

Since over the next seven years I gave this course in tandem with Teuber's wonderful introduction to psychology, which had an equally large audience, there was much overlap and he heard of my paradigm. Hence the enclosed comment by him.

Certainly I can't claim priority on these grounds, if for no other reason than that Akakhi Akakhievitch had precedence—and his colleagues in the Irkutsk Academy of Speculative Neurosurgery certainly must have known of it before me. It may even be that some ethologist had speculated on the matter, or that some British analytic philosopher chose the "grandmother cell" to illustrate the folly of nervous science.

I can only tell you how I came to the idea but can't say if I had heard of these cells from other sources and then used them but forgot their provenance.

There is a meeting soon that we both are to attend—I think it is at Cornell. I am looking forward to it if only to see you again. You probably do not know that had it not been for you and your kindness in bringing some people to that demonstration in my laboratory, my work would probably still be discredited. I have been profoundly grateful ever since.

Affectionately,
Jerome Lettvin
(N.J. Professor of Neuroengineering)

27 A Model of Visual Motion Processing in Area MT of Primates

TERRENCE J. SEJNOWSKI AND STEVEN J. NOWLAN

ABSTRACT Motion perception requires the visual system to satisfy two conflicting demands: first, spatial integration of signals from neighboring regions of the visual field to overcome noisy signals, and second, sensitivity to small velocity differences to segment regions corresponding to different objects (Braddick, 1993). We have developed a computational model for the visual processing of motion in area MT that accounts for these conflicting demands. The model has two types of units, similar to those found in area MT. One type of unit in the model integrates information about the direction motion to estimate the local velocity; these local velocity units compete among themselves to determine the most likely local velocity. A second type of unit selects regions of the visual field where the velocity estimates are most reliable; these selection units have nonclassical receptive field surrounds by virtue of competition with pools of similar units across the visual field. The output of the model is a distributed segmentation of the image into patches that support distinct objects moving with a common velocity.

The processing of motion in the primate's visual cortex begins in area V1, where cells with reliable selectivity for direction of motion are found (Maunsell and Newsome, 1987); however, these cells do not detect true velocity but instead are tuned to a limited range of spatiotemporal frequencies and exhibit spatially restricted receptive fields so that they can report only the perpendicular component of the velocity for straight edges. This so-called aperture problem is illustrated in figure 27.1. To overcome these limitations and compute true local velocity measurements, it is necessary to integrate motion responses from cells with a variety of directions and spatiotemporal frequency tunings over a wider area of the visual field (Heeger, 1987; Grzywacz and Yuille, 1990).

TERRENCE J. SEJNOWSKI and STEVEN J. NOWLAN Howard Hughes Medical Institute, The Salk Institute, and University of California, San Diego, Calif.

Neurons that respond selectively to velocity over a wide range of spatial frequencies are found in visual area MT, which receives a direct projection from area V1 (Albright, 1992; Maunsell and Newsome, 1987; Rodman and Albright, 1989). A class of cells in MT, the "pattern cells" of Movshon and colleagues (1985), respond to the direction of overall motion of plaid patterns composed of two differently oriented gratings rather than to the direction of the individual components. Psychophysical studies suggest that the perceived velocity of such patterns generally is close to the velocity that is uniquely consistent with the constraints imposed by the individual component's motions (Adelson and Movshon, 1982), although other possibilities have been suggested (Wilson et al., 1992; Rubin and Hochstein, 1993). In this chapter, we are concerned with the properties of such pattern motion cells and their responses to motion stimuli that have previously been used in psychophysical and physiological experiments.

We have proposed a computational model for the formation of local velocity responses in MT and for the combination of these local velocity estimates into a representation of the velocity of objects in the visual scene (Nowlan and Sejnowski, 1993, 1994a,b). The model departs from previous suggestions of how local velocity is estimated from visual scenes (Horn and Schunk 1981; Heeger 1987, 1992; Nagel, 1987; Grzywacz and Yuille 1990). Our model does not compute an accurate estimate of local velocity at all image locations, called the *optical flow field*, but instead evaluates the reliability of local velocity estimates and forms a coarse representation of the motion of objects in the

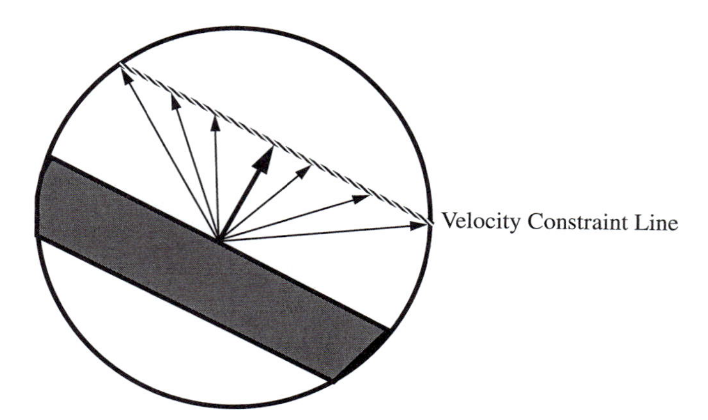

FIGURE 27.1 The aperture problem. The receptive field of a single cortical neuron is restricted to a fraction of the visual field. When a bar is introduced into the receptive field of the cell, only the velocity component orthogonal to the edge of the bar, indicated by the heavy arrow, can be measured by any local velocity mechanism. Any of the velocities indicated by the other arrows, all of which terminate on the velocity constraint line, would produce the same motion within this aperture. The aperture problem cannot be solved without appeal to information outside the classical receptive field.

visual scene by combining only the most valid subsets of local velocity measurements. The same general strategy may be used for computing other properties of objects in other areas of cortex. Selection can be used whenever several objects must be represented at the same time within the same population of neurons.

The model

DESIGN PRINCIPLES The model provides a framework for studying how the signals carried by cortical neurons could be used to estimate the velocity of moving objects in the presence of occlusion. Our goal is threefold: First, the model should provide a computationally robust algorithm for computing the velocities of moving objects in visual scenes. Second, it should be consistent with the known physiology and anatomy of visual cortex. Third, its performance should agree with psychophysical studies from primates. The processing units in the network model are meant to capture the responses that are observed at the level of the average firing rates of neurons. We have not attempted to account for how these responses are actually synthesized by cortical neurons, nor have we replicated in detail the interactions that occur between neurons. However, the computational operations in the model are relatively simple ones, such as summation and normalization, which can be implemented by real neurons in a variety of ways (see the discussion later in this chapter).

Soft-maximization is an example of a simple computational operation that occurs at several stages of processing in our model. The purpose of soft-maximization is to enhance the firing rate of the unit that has the highest firing rate in a population and to normalize all other responses so that the total activity is a constant, regardless of the initial firing rates. This can be accomplished mathematically by the following function:

$$R_k = e^{\alpha R'_k} / \sum_i e^{\alpha R'_i} \tag{1}$$

where R'_k are the initial firing rates and R_k are the normalized firing rates in a population of units indexed by k, and α is a constant that determines the degree of separation of the highest firing rate from all the others. The summation occurs here over all units in the population, but the operation can be performed within any subpopulation. In our model, it was applied in three separate subpopulations of units. As the amplification constant α is increased, the differences are further enhanced until, in the limit as α becomes very large, one unit fires at its maximum rate and all the rest are reduced to zero, a limit called *winner-take-all*. The value of α can be different for each use in the model, but once it has been chosen it is fixed for all stimuli. In a more sophisticated model, α could adapt dynamically. (See the discussion later for ways that this soft-maximization operation could be implemented with neural mechanisms found in the cerebral cortex.)

The general design of the model is a cascade of stages, each consisting of an array of processing units that are locally connected and arranged in a roughly retinotopically organized map (figure 27.2). Within each stage, there are a number of layers or channels, each covering the visual field. For example, for each of several directions of movement there will be an array of neurons that together provide an overlapping map of the visual field, in the same way that orientation is represented in area V1. Processing is divided into four main stages, which are summarized here and described in greater detail in Nowlan and Sejnowski (1994a,b). The retinal and motion-energy stages were accomplished by fixed filters, each of which could be computed by a simple feedforward network of converging and diverging connections. The subsequent stages of processing were adaptive filters in the sense that the connection strengths in the model were not predeter-

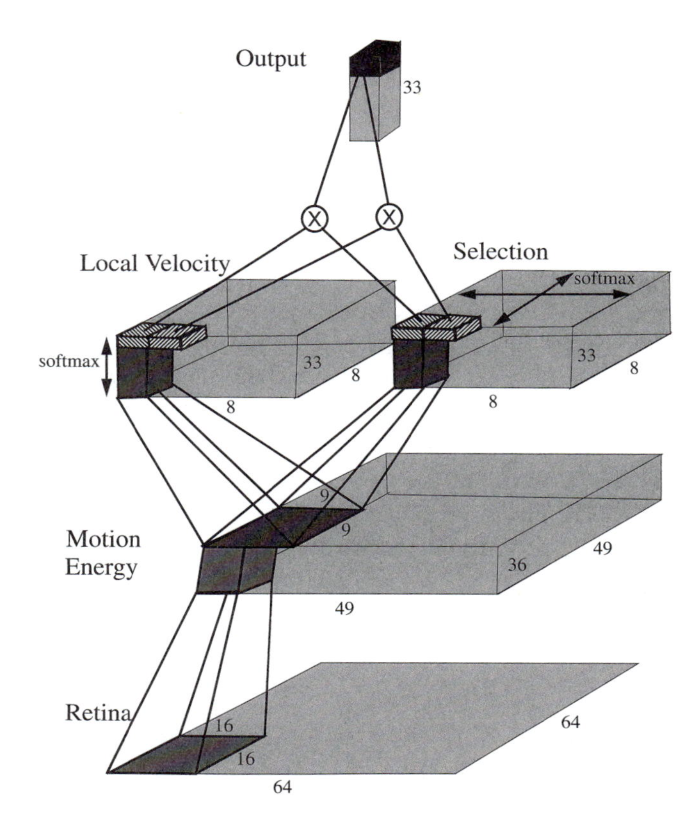

FIGURE 27.2 Architecture of a network model for a small region of area MT. The inputs to the model are "movies" containing 64 frames of 64 × 64–pixel arrays, with each pixel having a range of 256 gray levels. The movies are presented to the retina and processed sequentially by each layer. The first layer of processing, which corresponds to the primary visual cortex area V1, is composed of motion-energy units. Each motion-energy column contains 36 units that receive inputs from a 16 × 16 region of the retina for 16 successive frames of the movie. Each motion-energy unit is tuned to a different combination from four directions of motion, three spatial frequencies, and three temporal frequencies. The next layer of processing contains two independent networks of units, each receiving inputs from a 9 × 9 spatial array of motion-energy columns. Each local velocity and selection column contains 33 units that represent eight different directions and four different speeds as well as one unit for null motion. For each local velocity unit, there is a corresponding selection unit that receives inputs from the same 9 × 9 array of motion-energy columns. Because adjacent motion-energy columns have only partially overlapping receptive fields, the effective receptive field areas of the local velocity and selection columns are 81 times larger than the receptive field area of the motion-energy units. Competition occurs within each local velocity column, as indicated by the double-pointed arrow: Units that are most strongly activation are enhanced, and the weakest are suppressed by soft-maximization. For the selection units, there is a competition across space. Soft-maximization is applied between selection units having the same preferred velocity (double-pointed arrows). The information within the local velocity and selection networks is then combined to form the final estimate of velocities on the output layer, which is a single column of 33 units. The value of each output unit is computed by multiplying the outputs of corresponding pairs of local velocity and selection units and summing across a slab of these units. This output is the final estimate of the velocities of all the objects moving with the 64 × 64–pixel array. The velocity of more than one object may be represented in the array at the same time, as along as the velocities are not too similar. There were a total of 8.8 million weights in the network but, because of translational symmetries, only 138,600 of them were independent.

mined but were instead determined by an optimization procedure. Once the adaptive weights were found, they were fixed, and all the results presented here were obtained from one network with fixed connection strengths.

RETINAL PROCESSING The 64 × 64–pixel input array was roughly equivalent to an array of photoreceptors that represents the intensity at each pixel location by one of 256 gray levels. A motion stimulus consisted of a sequence of images that were processed first through the retinal stage and subsequently through spatiotemporal filters, as described in the next section. The retinal stage of processing contrast-enhanced each image with a difference-of-gaussian filter at each location of the array, which removed the constant component of intensity across the image, smoothed the noise, and enhanced the edges.

MOTION-ENERGY FILTERS In the first stage of motion processing, a distributed representation of motion was extracted that served as a model for the inputs to MT. We used the motion-energy model of Adelson and Bergen (1985), which consisted of arrays of spatiotemporal filters, each tuned to a particular direction of motion and sensitive to a particular combination of spatial and temporal frequency. Altogether, there were 36 channels of motion-energy filters tuned to four directions and nine combinations of spatial and temporal frequencies. Each filter received input from a 16 × 16 patch of the retinal output. These filters were broad and overlapping in their selectivities, and their velocity tuning depended on the spatial characteristics of the moving pattern. Soft-maximization was applied to the pool of 36 motion-energy channels to normalize their outputs and make them report relative rather than absolute values for contrast to the next stage of processing (Albrecht and Geisler, 1991; Heeger, 1992). The properties of motion-energy filters resembled those of directionally tuned complex cells in the visual cortex of cats and monkeys (Nakayama, 1985; Maunsell and Newsome, 1987; Emerson, Bergen, and Adelson, 1992).

LOCAL VELOCITY NETWORK For a single, rigidly moving object in the visual field with no occluding or transparent objects, relatively simple averaging schemes can be used to estimate the local velocity (Heeger 1987, in press; Grzywacz and Yuille 1990). A linear weighted summation can be performed for each direction of motion and speed and the maximum taken across these channels. In our model, there were eight different best directions corresponding to equally spaced compass points and four different best velocities, for a subtotal of 32 different combinations, plus one more unit that represented zero velocity, giving a total of 33 velocity-tuned units. There was an array of 8×8 locations, each containing 33 velocity units, and each of these velocity units received inputs from 9×9 motion-energy units (see figure 27.2).

These velocity units were broadly tuned around their best direction and velocity so that a given motion stimulus produced a pattern of activity in these 33 units. The unit with the largest response, representing the most likely local velocity in that patch of the visual field, was enhanced by soft-maximization, which also reduced the weaker responses (equation 1). The output of each unit can be considered the evidence for a particular velocity in a particular region of the image. The constraint that the sum of activity in the pool of 33 neurons must equal one can be interpreted as the constraint that the total evidence across all velocities must sum to one. Using soft-maximization rather than a winner-take-all limit means that the population can represent more than one velocity in each pool.

SELECTION NETWORK For multiple moving objects and visual scenes that include occlusion and transparency, the local velocity estimates may not be accurate, and it is not at all obvious which features of the image are relevant for determining whether a local region contains reliable information. If information from several objects is within the receptive field of a pool of local velocity units, the output will be ambiguous. These locations may provide little or no unambiguous information and should be ignored as long as other parts of the object contain reliable velocity estimates. The purpose of the selection network was to identify the regions of the image that contain the most reliable estimates. The inputs to the selection network were the same motion-energy array used for the local velocity network. There was a selection unit corresponding to each local velocity unit.

Before the selection units were combined pairwise with the local velocity units, the outputs of the directionally selective units in the selection network competed spatially across the visual field. The soft-maximization operation (equation 1) was applied separately to each of the 33 selection channels. The purpose of this comparison was to identify the spatial locations containing the most reliable information and to suppress those locations containing the least reliable information. Finally, the outputs of each channel in the 8×8 local velocity network and the 8×8 selection network were multiplied, point by point, then summed to produce a final estimate of the velocity:

$$v_k(t) = \sum_{x,y} I_k(x,y,t) S_k(x,y,t) \quad (2)$$

where $v_k(t)$ is global evidence for a visual target at time t moving at a particular velocity k, the $I_k(x,y,t)$ is the local evidence for velocity k computed by the local velocity pathway from region (x,y) at time t, and $S_k(x,y,t)$ is the weight assigned by the selection pathway to that region.

CREATING THE NETWORK The properties of the local velocity and selection units were not set a priori but were determined by an optimization procedure. We specified the input representation of motion to the model, the problem that we believe the system was designed to solve, and a network architecture that reflected some of the important constraints imposed by cortical physiology and anatomy (Churchland and Sejnowski, 1992). We then used an optimization procedure called *mixtures of experts* to adjust the model parameters to solve a velocity estimation problem for prototypical input patterns, as described elsewhere (Nowlan, 1990; Nowlan and Sejnowski, 1994b). The input patterns used for optimizing the performance of the network were 500 "movies" of moving objects, such as the example shown in figure 27.3. The known veloc-

a) 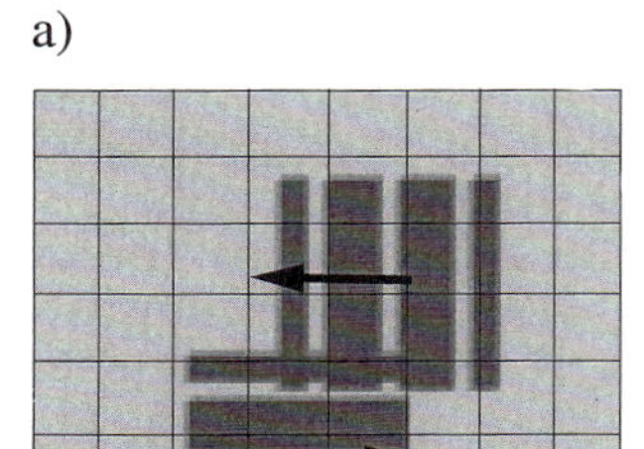

b) 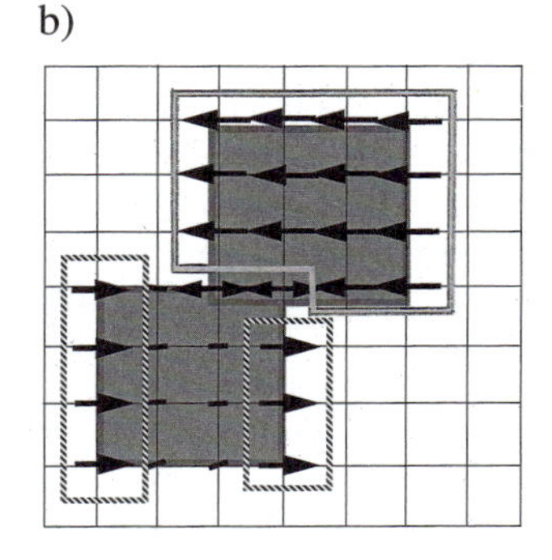

FIGURE 27.3 Responses of local velocity and selection units to partially occluding objects. (a) Input to the model consists of two blobs moving in opposite directions (one to the right at 0.25 pixels per frame and the second to the left at 0.5 pixels per frame). The arrows indicate the direction and speed of motion. (b) Hashed lines indicate the region of motion selected for the rightward-moving blob, whereas the solid lines indicate a separate region of support for the motion of the leftward blob. The local velocity estimates in the region of overlap are ambiguous, showing two directions of motion simultaneously.

ity of each object was used to modify the weights in the appropriate local velocity network, and the weights in the selection networks were changed to identify the most reliable regions.

Although the velocity and selection networks appear to be symmetrical, the optimization procedure treated the selection network differently because its role was to gate the velocity network rather than compute the velocity itself. The optimization procedure modified the input weights to the local velocity and selection units to minimize the error of the output units, but only the local velocity units were given feedback about the correct output velocity; the selection units were instead given information about which local velocity units carried the most reliable estimates. There was also a difference in the way these two networks were normalized: soft-maximization was applied to the local velocity units only within a column but, for the selection network, the soft-maximization was spatially applied separately for each velocity channel. The optimization procedure converged to sets of weights that gave robust estimates of object velocities despite occlusion; more importantly, the model performed well on psychophysical stimuli, such as those presented in the next section, which were not used during optimization (Nowlan and Sejnowski, 1993, 1994a,b).

To illustrate how the selection network handles the problem of segmentation, consider two objects moving in opposite directions (see figure 27.3a). One object is striped horizontally and moves to the right with a speed that is half that of the second object, which is striped vertically and moves to the left. The local velocity estimates and selected regions are shown in figure 27.3b. Selected regions are denoted with a dashed line if they correspond to the rightward motion of the first object or with a stippled line if they correspond to the leftward motion of the second object. In regions where the two objects overlap, the second object occludes the first object totally, and intensity in these regions is the intensity of the second object.

In figure 27.3, the local velocity estimates in the region of overlap of the two objects are marked with double arrowheads. In these regions, the activity in the local velocity pool is not concentrated in a single unit but is distributed bimodally with activity peaks corresponding to two opposing directions of motion. This ambiguous region is not included in the region of support for the motion of either object. Note that for the rightward-moving object, the selected regions correspond to only the leading and trailing edges. For the leftward-moving object, the contrast stripes are perpendicular to the direction of motion, providing strong motion signals over most of the region covered by the object. As a result, the region of support for this object is correspondingly much larger.

The properties of the selection units were determined indirectly by optimizing the network to produce the correct velocity in the final output layer of units. To a first approximation, the selection units in our network detected discontinuities in the distribution of motion-energy inputs but, because they were optimized to respond primarily to the patterns of discontinuities that characterize regions of reliable support for object motion, the algorithm that they implement is more restrictive. Thus, not all discontinuous patterns of velocity activate the selection units equally well (figure 27.4). For example, some units preferred motion end-stopping within their receptive fields. For a given input, the spatial regions selected tended to form disjointed subsets over which to integrate the local velocity field. This allowed the model to account for interpenetrating motion fields for which the assumption of spatial continuity of the velocity field is invalid. We have empirically determined the properties of the local velocity and se-

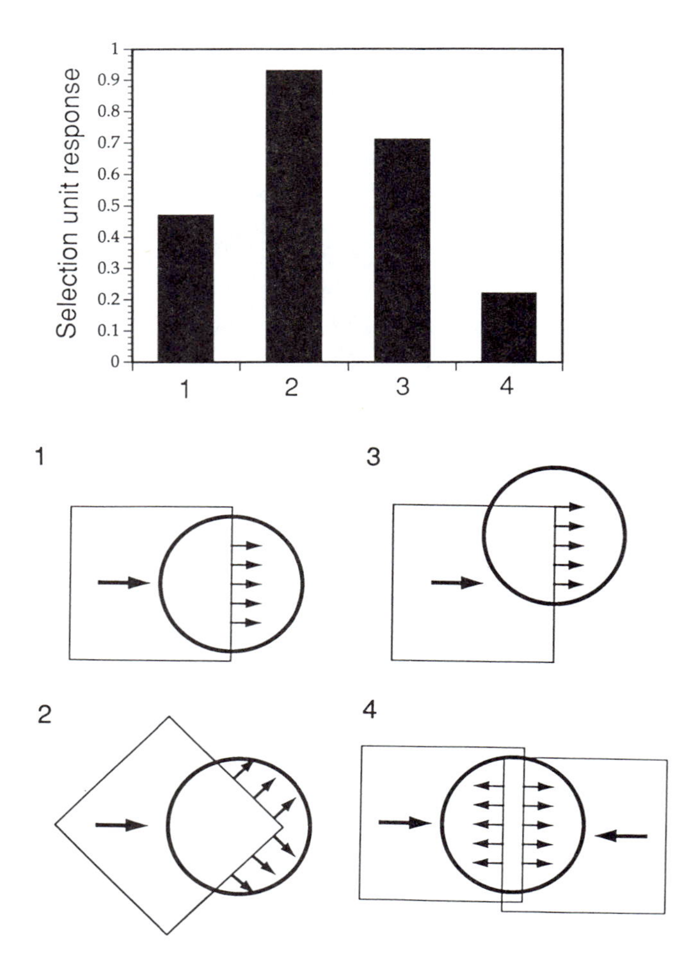

FIGURE 27.4 Responses of a selection unit to motion discontinuities. (a) The strength of the response of a selection unit tuned to rightward motion to four different distributions of motion-energy responses within the selection unit's receptive field. The four motion-energy distributions are represented schematically in (b). In the first three examples, the square is moving in the direction indicated by the large arrow against a stationary background. In the last example, two squares, one semitransparent, move against a stationary background. In each example, the circle indicates the receptive field of the selection unit, and the small arrows indicate the direction of local motion reported by the motion energy units within this receptive field. In example 1, the receptive field is centered over an edge moving to the right and sees a uniform distribution of rightward motion-energy responses producing a moderate response (0.48) from the unit. In contrast, the response to a similar motion in example 3 is much stronger (0.74) because, in this case, the receptive field covers a corner region that contains a discontinuity between a region of rightward motion-energy response and a region containing no motion-energy response. The strongest response from this selection unit (0.93) is seen in example 2, where the receptive field encloses a discontinuity between two orthogonal sets of local motion-energy measurements. Finally, example 4 shows that the selection unit is directionally tuned: Local motion-energy responses corresponding to two opposed motions (generated, in this case, by one transparent object moving in front of a second moving object) will suppress the response of a selection unit. In this figure, the responses of a single isolated selection unit are shown. The overall response of a selection unit in the network is determined by its own local response and the responses of similarly tuned selection units in other regions of the image with which a selection unit competes.

lection units by presenting the model with a variety of standard motion stimuli, such as oriented gratings.

Unit response properties

The first stage of our model is intended to correspond approximately to primary visual areas V1 and V2. The choice of normalized motion-energy responses as the representation of image motion in this first stage of the model reflects the measured response properties of simple and complex cells in mammalian primary visual cortex (Tolhurst and Movshon, 1975; Holub and Morton-Gibson, 1981; Emerson et al., 1987; McLean and Palmer, 1989). The spatial and temporal response characteristics of these filters and the broader tuning in temporal frequency compared to spatial frequency were chosen based on measured responses in primary visual cortex (Adelson and Bergen, 1985; Heeger, in press). The inverse relationship between motion-energy filter center frequencies and receptive field sizes also matches the relationship found in visual cortex (Hochstein and Shapley, 1976; Maffei and Fiorentini, 1977; Andrews and Pollen, 1979). The normalization of the motion-energy responses is suggested by the saturating contrast response curves of cells in primary visual areas (Ohzawa, Sclar, and Freeman, 1985; Heeger, 1992).

The local velocity and selection stages of the model we associate primarily with the middle temporal cortical area (MT) of visual contex, although some of the functions in the model may be occurring in the medial superior temporal area (MST) and possibly parietal cortex. Although the response properties of units in these stages were not specified in advance, a number of architectural decisions constrained the model. The localized receptive fields and roughly retinotopic organization of the units in both the first and second stages of the model are matched by the organization of visual cortical areas in the early stages of visual processing

(Maunsell and Newsome, 1987). In particular, the area of the receptive fields for units in the selection and local velocity layers of the model were 81 times larger than receptive fields in the motion-energy layer. The receptive fields in MT are, on average, 100 times larger than those found in V1 (Gattass and Gross, 1981), and the spatial scale of directional responses is proportionally larger (Mikami, Newsome, and Wartz, 1986). We have also constrained the feedforward "weights" of units in both the local velocity and selection pathways to be purely excitatory. Thus, the properties of the units in these networks are due primarily to excitatory inputs, with inhibitory effects being expressed only in the competitive renormalization used in all parts of the model.

The renormalization used in our model has been suggested previously to account for some aspects of neural responses in both V1 (Heeger, 1992) and MT (Snowden et al., 1991). The architecture proposed in our model makes two strong assumptions about the nature of this normalization in areas such as MT. The local velocity pathway requires competitive interactions among units with similar receptive field locations but different tuning properties (this is similar to the interactions proposed by Heeger, 1992, and Snowden et al., 1991). The selection pathway requires interactions among cells covering most of the visual field, but these long-range interactions occur only among cells with similar tuning properties. Neuronal mechanisms that might be used to implement the competitive soft-maximization operation are examined later in this chapter.

We explored the response characteristics of selection and local velocity units in the optimized model using drifting sinusoidal gratings. The local velocity units all exhibited very similar tuning curves, which tended to be symmetrical about the optimal direction of response and narrowly tuned. The average tuning bandwidth for these units was 53°. The tuning curves for selection units showed considerably more variation. Selection units tend, on average, to have much broader tuning curves (average tuning bandwidth, 84°). A small number of selection units showed a bimodal directional tuning curve. In addition, selection units tended to have maximal orientation responses for bars oriented close to their preferred direction of motion, whereas local velocity units showed maximal responses for bars nearly orthogonal to the preferred direction of motion. The broader directional tuning and similarity of orientation and direction tuning suggests that the selection units resemble the pattern (Movshon et al., 1985) or type II cells (Albright, 1984) found in monkeys, whereas the local velocity units are more similar to the component or type I cells. The responses of the local velocity and selection units to plaid patterns also supports this identification (Nowlan and Sejnowski, 1994a,b). The directional tunings of both local velocity and selection units are sharper than the tunings found by Albright (1984) but similar to those reported by Maunsell and Van Essen (1983).

We tested the units in the model for velocity tuning with gratings oriented optimally for each unit and spanning a range of temporal and spatial frequencies. The velocity units were tuned to a fairly narrow range of velocities over a broad range of spatial and temporal frequencies, and some cells with this type of spatiotemporal frequency response have been found in MT (Newsome, Gizzi, and Movshon, 1983). In general, the units in our model were tuned to velocity over a broader range of spatial and temporal frequencies than typically found in MT, perhaps because there were many fewer units than MT cells to represent the same range. Thus, a single local velocity unit may represent a population of cells in MT, each of which is sensitive to velocity over a narrower range of spatial and temporal frequencies.

The spatial frequency and temporal frequency sensitivities were more separable for selection units than for the local velocity units because they could be better approximated by products of purely spatial and purely temporal functions. In addition, the response of the selection unit showed a slight dip in the midrange frequencies that is not seen in velocity unit responses. Nearly all the selection units showed some degree of sensitivity of their velocity tuning to variation in spatial frequency. Many MT neurons also show some sensitivity of their velocity tuning to spatial frequency (Maunsell and Newsome, 1987), but this variation has not been systematically correlated with other types of variation (such as tuning width).

There is another important qualitative difference in the responses of local velocity and selection units that was apparent when gratings were either restricted to the receptive field region or presented across the entire visual field. The responses of selection units were strongly suppressed when the grating was presented to the entire visual field rather than just to the receptive field of a unit (figure 27.5a). The responses of local

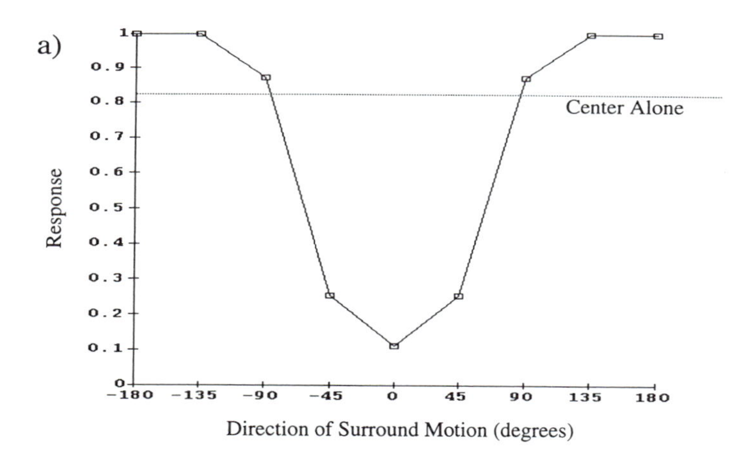

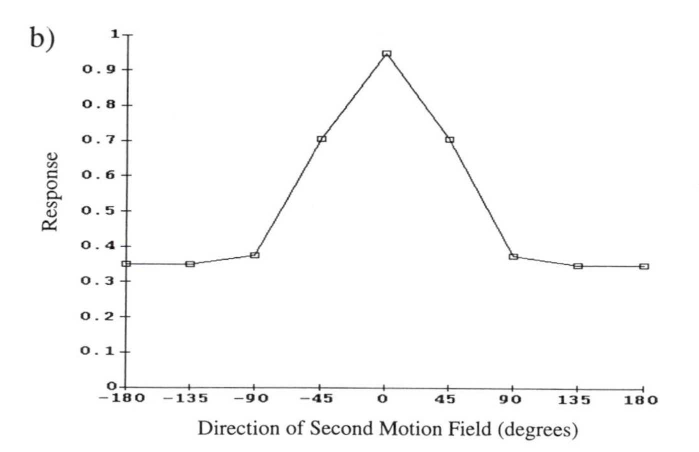

FIGURE 27.5 Interactions between units in the selection and velocity networks. (a) Response of selection unit to an optimal sinusoidal grating when a second grating is presented outside the receptive field of the selection unit. The dashed line shows the response to the grating within the receptive field alone. The response is displayed as a function of the orientation of the surround grating, measured relative to the orientation of the grating within the receptive field. When the surround motion matches the receptive field motion, there is strong inhibition of responses. This inhibition decreases as the surround and receptive field motions differ more in direction, becoming weak facilitation when the difference in orientation is greater than 90°. (b) Response of local velocity unit to a dynamic transparent random-dot stimulus. The stimulus consisted of two fields of random dots moving within the receptive field. The first field was chosen so its motion produced a maximal response from the unit. The direction of the second field relative to the first was adjusted, and the response of the unit was plotted as a function of the direction of this second field of dots. The presence of a second motion always suppressed the response of the unit, with this suppression saturating for motions orthogonal to the preferred direction of motion of the unit (at which point the local velocity pool has a bimodal activity distribution).

velocity units, in comparison, were weakly facilitated by whole-field presentation. The suppression of response due to motion in regions surrounding the selection unit's receptive field was strongest when the surrounding motion matched the optimal motion for eliciting a response in the selection unit's receptive field. The degree of suppression decreased as the surround motion differed more from the optimal motion for the unit (see figure 27.5a). In fact, motion in a direction more than 90° away from optimal for the selection unit tended to facilitate the response of the selection unit. This is a consequence of the shift in balance that occurs in the soft-maximization operation at different spatial locations. This type of modulation of the receptive field response by the nonclassical surround in MT was first reported by Allman, Miezin, and McGuinnes, 1985 (see the discussion later).

The substructure of the receptive fields in MT were studied by Snowden and colleagues (1991), whose stimuli consisted of random dots moving within the classical receptive field. One field of dots always moved in the preferred direction of the cell, whereas a second field, if present, moved in a different direction, producing two distinct motions simultaneously within the cell's receptive field. We presented similar stimuli to the local velocity and selection units in our model. The local velocity units exhibited responses that are qualitatively very similar to the responses observed by Snowden's group (figure 27.5b). The presence of a second motion in the receptive field always suppressed the response of a local velocity unit, with the degree of suppression increasing until the second motion was orthogonal to the preferred direction of the unit. These response characteristics are explained by the local competition within each pool of velocity units.

Transparent plaids

Moving plaid patterns have been used to study how the visual system integrates multiple motion signals into a coherent motion percept (Adelson and Movshon, 1982; Ramachandran and Cavanaugh, 1987; Welch, 1989; Stoner, Albright, and Ramachandran, 1990; Wilson, Ferrera, and Yo, 1992). These stimuli consist of two independently moving gratings that are superimposed (figure 27.6a). When human observers are presented with either grating alone, they always reliably report the motion of the grating. However, if the two gratings have similar properties and are superim-

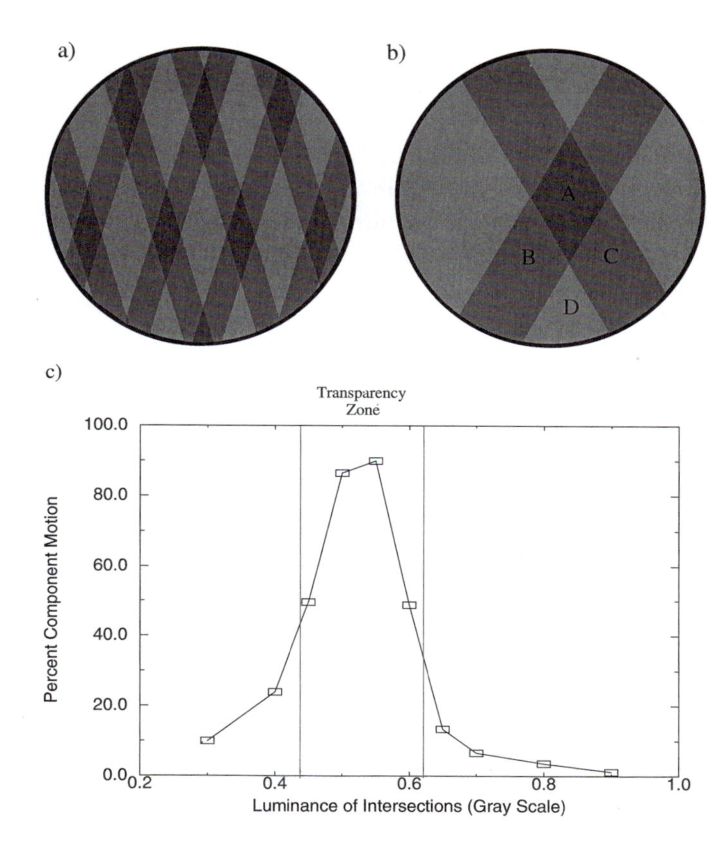

FIGURE 27.6 Plaid pattern stimuli used to test the model on transparency. (a) The plaid patterns were composed of two square-wave gratings of dark bars on a lighter background (Stoner, Albright, and Ramachandran, 1990). (b) Detail of pattern in (a). In the experiments, the intensity of region A was varied, whereas the intensity of regions B, C, and D were held constant. Depending on the luminance of A, the observer would see either two drifting gratings or a single coherent pattern motion. (c) Responses of model to similar plaid patterns. There was a transparency zone, in which the model responses were consistent with two separate component motions. Outside this transparency zone, the model reported only a single pattern motion, consistent with human perception.

posed, the two independent motions of the gratings seem do disappear; most observers see the two gratings cohere and form a single pattern moving in a direction different from either of the gratings alone. Cells that respond in the direction of pattern rather than component motion are found in MT, which suggests the importance of MT in producing the percept of coherent motion (Movshon et al., 1985; Rodman and Albright, 1989).

Stoner, Albright, and Ramachandran (1990) have found that the percept of coherent pattern motion can be affected by altering the luminance of the region of intersection of the two gratings. Some cells in MT also tend to respond to either the direction of pattern or component motion, depending on the luminance of this intersection region (Albright, 1992). The stimuli used by Stoner, Albright, and Ramachandran consisted of two square-wave gratings with thin bars on a lighter background (figure 27.6a). The luminance of the intersection regions was varied, whereas the luminance of the bars and the background was left the same. Stoner and colleagues (1990) manipulated the luminance of the intersection regions and showed that there is a range of luminances for which humans see one transparent grating lying on top of the other and reliably report the presence of the two grating motions rather than the coherent pattern motion.

We presented the model with a series of plaid patterns consisting of two square-wave gratings in which the luminance of the intersection region of the plaid pattern was varied systematically through a series of values that spanned the transparency zone. The response of the output units in the model varied nonlinearly and nonmonotonically with the luminance of the intersection region. As shown in figure 27.6c, the performance of the model closely matched the psychophysical results reported by Stoner, Albright, and Ramachandran (1990). We analyzed the network to determine how it made its decision and found that when it reported component motion, the total support for motion was concentrated on the portions of the gratings that were outside the intersection regions but, during pattern motion, support for motion was concentrated on the regions surrounding the intersections of the two gratings (Nowlan and Sejnowski, 1993). This is another example of how the selection network responds to patterns of motion-energy discontinuities (see figure 27.4).

Discussion

In our model of motion processing in MT, the conflicting demands for spatial averaging and spatial segmentation were satisfied by two separate networks, one that computes the local velocity estimate and a second that selects regions where reliable velocity estimates are possible. The outputs of these two networks were combined multiplicatively to produce reliable estimates of the velocities of objects without assuming spatial continuity. In our model, there was only a single population of output units representing the velocities in a small patch of the visual field. In MT, there is an array of such units that can, in turn, serve as the input to other areas that represent nonuniform flow fields, such as expansion and rotation, which are preferred by neurons in MST (Saito et al., 1986; Duffy and Wurtz, 1991).

Robust Estimation Many models of motion processing perform spatial averaging by assuming the spatial continuity of the velocity field (Marr and Ullman, 1981; Adelson and Movshon, 1982; Hildreth, 1984; Heeger, 1987; Grzywacz and Yuille, 1990). The problems for this approach posed by occlusion and transparency in motion stimuli were overcome in our model by the selection network, which estimated not the local velocity but rather the confidence with which the local velocity could be measured. Because each spatial region is considered to be independent, parts of different objects that interpenetrate are not averaged. The local regions of support for each velocity are then combined by a global competitive mechanism. Our selection pathway can be regarded as a feedforward mechanism for computing regions of support for robust velocity estimation (Li, 1985).

The selection units respond primarily to motion-energy gradients. Velocity gradients have been used by Koch, Wang, and Mathur (1989) to determine the boundaries between regions with different uniform velocities and by Smith and Grzywacz (1993) to determine where to apply a winner-take-all operation. In our model, the selection units form a pattern recognition network that weights motion-energy patterns in a graded fashion according to their degree of robustness; the spatial competition imposed by soft maximization ensures that the most reliable regions gain the strongest support (see figure 27.4).

Selective Attention Selection may represent a fundamental aspect of cortical processing that occurs with many preattentive phenomena (Bergen and Julesz, 1983; Treisman, 1988). The same mechanisms that are used to implement the covert, preattentive form of selection in our model could also be used for overt attentional processing. Top-down influences could enhance the probability that a selection is made to a particular property of the input. Attentional modulation of single-unit responses has not been reported in MT, but other motion areas such as MST may be better candidates. Moran and Desimone (1985) have observed in area V4 that the response of a neuron to its preferred stimulus is reduced if the monkey attends to a nonpreferred stimulus but only if the nonpreferred stimulus is presented within the receptive field for the neuron. This is evidence for the type of local competition that is required within our local velocity network.

The inhibitory effects of conflicting motion signals within the receptive field of local velocity units in our model are similar to effects found by Snowden and colleagues (1991) for some MT cells. The suppression reached a maximum when the second field of dots was roughly orthogonal to the preferred direction of the cell and was relatively constant after that. The soft-maximization operation used in the model produced precisely this type of suppression; increasing the number of dots in the nonpreferred direction causes the slope of the response as a function of dot density in the preferred direction to decrease, but the response always saturates at the same level. The competition among the local velocity units in the model makes the prediction that a second random-dot pattern moving in a cell's preferred direction, but at a speed significantly different from the optimal speed for the cell, will also have a suppressive effect on the cell's response to an optimal stimuli.

Nonclassical Surrounds The inhibitory effects of surround motion on selection units in our model are very similar to inhibitory surround effects that have been observed in many cells in MT (Allman, Miezin, and McGuinnes, 1985; Tanaka et al., 1986). The effect of whole-field motion on the responses of selection units in our model derives from the competition among the selection units representing the same candidate velocity across all regions of the visual field. When evidence for a particular velocity is present in all regions of the image equally, very little support needs to be assigned to any one region. Thus, the presence of similar motion in surrounding regions tended to suppress the selection units. On the other hand, if support for the velocity were concentrated in a small region of the image, the selection units in these regions had much stronger responses.

Recently, Born and Tootell (1992) have used the 2-deoxyglucose technique to identify the spatial organization of two broad classes of cell in MT based on how these cells responded to whole-field motion. *Band* cells responded to some directions of whole-field motion, whereas *interband* cells showed no strong response to any direction of whole-field motion. If we identify the interband cells of Born and Tootell with our selection units, the model makes the strong prediction that long-range intrinsic connections will occur primarily between groups of interband cells. The long-range horizontal connections between pools of selection units may modulate local inhibitory circuits.

Soft-Maximization A soft-maximization operation can be implemented by local mutual inhibition between the units in the population (Feldman and Ballard, 1982; Heeger, 1992), but other circuits that are faster and more reliable also can perform the task (Grzywacz and Yuille, 1990). An effective way to control the gain of the soft-maximization operation, the value of α required in equation 1, is to control local amplification. Recurrent excitation within networks of cortical pyramidal neurons appears to amplify inputs and the degree of amplification is controlled by inhibitory circuits (Douglas, Martin, and Whitteridge, 1989). We have assumed that the average firing rate of a neuron carries the output signal and that the neurons fire asynchronously. Synchronization of the spike firing among a pool of cells would also enhance their impact on mutual postsynaptic targets and, in principle, could be used to implement a fast soft-maximization operation (Steriade, McCormick, and Sejnowski, 1993).

Random Dots We have also applied our model of motion processing to dynamic random-dot displays (Nowlan and Sejnowski, 1994a,b), where there are no regions of consistent motion and the velocity of neighboring dots can be very different (Braddick, 1974; Morgan and Ward, 1980; Nakayama and Tyler, 1981; Williams and Sekuler, 1984). Newsome, Britten, and Movshon, (1989) computed psychometric functions for the ability of a monkey to identify correctly the direction of motion as a function of the percentage of dots moving coherently and showed that these psychometric functions were closely matched by "neurometric" functions computed from the responses of single MT neurons. Some neurons carried signals that were as reliable as the behavior of the monkey. When we used similar dynamic random-dot stimuli in our model, the percent of correct responses plotted as a function of coherence level for the model was qualitatively similar to the psychometric response curve from neurons in MT. The threshold for the model depended on the size of the coherent region. Nonetheless, the model was able to process this type of motion display properly even though it was not used to optimize the original model.

Predictions

In this chapter, we have introduced a novel strategy for computing the reliability of local velocity estimates. The purpose of the selection units, although they are velocity-tuned, is not to represent the local velocity. Likewise, the fact that a neuron is velocity-tuned is not sufficient evidence to conclude that the neuron's function is to represent local velocity, which suggests that some neurons in the visual cortex may be more concerned with grouping information than with representing the information itself. A similar algorithm may be used in other regions of sensory cortex to assess the importance of information processing within local neighborhoods. For example, selection networks could be used in binocular vision to assess the reliability of stereoscopic correspondences and also to group nearby regions of the visual field that contain parts of the same object based on similar binocular disparities, even when there are transparencies.

Selection in our model occurred in two stages. First, the selection network at each spatial location rapidly computed a selection value for each broadly tuned velocity unit. This feedforward operation was followed by soft-maximization normalization across spatial locations for each velocity. The intrinsic horizontal axonal system within neocortex is the most likely substrate for this process. We predict that one function of these intrinsic collaterals is to compare the relative activity levels within different columns. The physiological effects of these collaterals should, in some circumstances, be the suppression (rather than enhancement) of activity in neighboring columns.

The selection network provides a partial solution to the problem of image segmentation. Previous attempts to segregate figure from ground have implicitly assumed that objects were spatially continuous and that the first step was to find a bounding contour. Our approach to segmentation does not make this assumption; the selection network may group information that is spatially separated by intervening ground or by other objects. This leaves open the problem of how motion is integrated with other properties of the object. Integration could be achieved by referencing each selected estimate back to a high-resolution spatial map, such as area V1, which predicts that the cortical feedback projections carry information about segmentation and that neural correlates of segmentation should be observed in single neurons in area V1. Visual stimuli to test this hypothesis would allow comparison of the same image over the receptive field of a neuron while it is part of either a figure or the background.

In contrast to previous approaches that have attempted to construct a motion flow field throughout space or for all parts of an object (Marr, 1982), we attempt only to represent explicitly those selected parts that are particularly salient and unambiguous. Separate regions that share the same local velocity will be grouped and assigned to one object. Such reduced representations may also have advantages for indexing object representations in the ventral processing stream. With fewer salient features to match, the combinatorial problem of finding the correct match is greatly simplified. Reduced representations may also be helpful in learning the causal relationships between representations as salience is already a part of the representation (Ballard and Whitehead, 1990; Churchland, Ramachandran, and Sejnowski, 1994).

REFERENCES

Adelson, E. H., and J. R. Bergen, 1985. Spatiotemporal energy models for the perception of motion. *J. Opt. Soc. Am. [A]* 2:284–299.

Adelson, M., and J. A. Movshon, 1982. Phenomenal coherence of moving visual patterns. *Nature* 300:523–525.

Albrecht, D. G., and W. S. Geisler, 1991. Motion sensitivity and the contrast-response function of simple cells in the visual cortex. *Visual Neurosci.* 7:531–546.

Albright, T. D., 1984. Direction and orientation selectivity of neurons in visual area MT of the macaque. *J. Neurophysiol.* 52:1106–1130.

Albright, T. D., 1992. Form-cue invariant motion processing in primate visual cortex. *Science.* 255:1141–1143.

Allman, J., F. Miezin, and E. McGuinnes, 1985. Stimulus-specific responses from beyond the classical receptive field: Neurophysiological mechanisms for local-global comparisons in visual neurons. *Annu. Rev. of Neurosci.* 8:407–430.

Andrews, B. W., and T. A. Pollen, 1979. Relationship between spatial frequency selectivity and receptive field profile of simple cells. *J. Neurophysiol. (Lond.)* 287:163–176.

Ballard, D. H., and S. D. Whitehead, 1990. Active perception and reinforcement learning. *Neural Computation* 2: 409–419.

Bergen, J. R., and B. Julesz, 1983. Rapid discrimination of visual patterns. *IEEE Trans. Systems Man Cybern.* 13:857.

Born, R. T., and R. B. H. Tootell, 1992. Segregation of global and local motion processing in primate middle temporal visual area. *Nature* 357:497–500.

Braddick, O. J., 1974. A short-range process in apparent motion. *Vision Res.* 14:519–527.

Braddick, O. J., 1993. Segmentation versus integration in visual motion processing. *Trends Neurosci.* 16:263–268.

Churchland, P. S., V. S. Ramachandran, and T. J. Sejnowski, 1994. A critique of pure vision. In *Large-Scale Neuronal Theories of the Brain*, C. Koch and J. Davis, eds. Cambridge, Mass.: MIT Press.

Churchland, P. S., and T. J. Sejnowski, 1992. *The Computational Brain.* Cambridge, Mass.: MIT Press.

Douglas, R. J., K. A. C. Martin, and D. Whitteridge, 1989. A canonical microcircuit for neocortex. *Neural Computation* 1:480–488.

Duffy, C. J., and R. H. Wurtz, 1991. Sensitivity of MST neurons to optic flow stimuli: II. Mechanisms of response selectivity revealed by small-field stimuli, *J. Neurophysiol.* 65:1346–1359.

Emerson, R. C., J. R. Bergen, and E. H. Adelson, 1992. Directionally selective complex cells and the computation of motion energy in cat visual cortex. *Vision Res.* 32:203–218.

Emerson, R. C., M. C. Citron, W. J. Vaughn, and S. A. Klein, 1987. Nonlinear directionally selective subunits in complex cells of cat striate cortex. *J. Neurophysiol.* 58:33–65.

Feldman, J., and D. Ballard, 1982. Connectionist models and their properties. *Cogn. Sci.* 6:205–254.

Gattass, R., and C. G. Gross, 1981. Visual topography of striate projection zone (MT) in the posterior superior temporal sulcus of the macaque. *J. Neurophysiol.* 46:621–638.

Grzywacz, N. M., and A. L. Yuille, 1990. A model for the estimation of local image velocity by cells in the visual cortex. *Proc. R. Soc. Lond. [Biol.]* 239:129–161.

Heeger, D. J., 1987. Model for the extraction of image flow. *J. Opt. Soc. Am. [A]* 4:1455–1471.

Heeger, D. J., 1992. Normalization of cell responses in cat striate cortex. *Visual Neurosci.* 9:181–198.

Hildreth, E. C., 1984. *The Measurement of Visual Motion.* Cambridge, Mass.: MIT Press.

Hochstein, S., and R. M. Shapley, 1976. Quantitative analysis of retinal ganglion cell classifications. *J. Physiol. (Lond.)* 262:237–264.

Holub, R. A., and M. Morton-Gibson, 1981. Response of visual cortical neurons of the cat to moving sinusoidal gratings: Response-contrast functions and spatiotemporal integration. *J. Neurophysiol.* 46:1244–1259.

Horn, B. K. P., and B. G. Schunk, 1981. Determining optical flow. *Artif. Intell.* 17:185–203.

Koch, C., H. T. Wang, and B. Mathur, 1989. Computing motion in the primate's visual system. *J. Exp. Biol.* 146: 115–139.

Li, G., 1985. Robust regression. In *Exploring Data, Tables,*

Trends and Shapes, D. C. Hoaglin, F. Mosteller, and J. W. Tukey, eds. New York: Wiley.

MAFFEI, L., and A. FIORENTINI, 1977. Spatial frequency rows in the striate visual cortex. *Vision Res.* 17:257–264.

MARR, D., 1982. *Vision.* New York: W. H. Freeman.

MARR, D., and S. ULLMAN, 1981. Directional selectivity and its use in early visual processing. *Proc. R. Soc. Lond. [Biol.]* 211:151–180.

MAUNSELL, J. H. R., and W. T. NEWSOME, 1987. Visual processing in monkey extrastriate cortex. *Annu. Rev. Neurosci.* 10:363–401.

MAUNSELL, J. H. R., and D. C. VAN ESSEN, 1983. Functional properties of neurons in the middle temporal visual area (MT) of the macaque monkey: I. Selectivity for stimulus direction, speed and orientation. *J. Neurophysiol.* 49:1127–1147.

McLEAN, J., and L. A. PALMER, 1989. Contribution of linear spatiotemporal receptive field structure to velocity selectivity of simple cells in area 17 of cat. *Vision Res.* 29:675–679.

MIKAMI, A., W. T. NEWSOME, and R. H. WURTZ, 1986. Motion selectivity in macaque visual cortex: II. Spatiotemporal range of directional interactions in MT and V1. *J. Neurophysiol.* 55:1328–1339.

MORAN, J., and R. DESIMONE, 1985. Selective attention gates visual processing in the extrastriate cortex. *Science* 229: 782–784.

MORGAN, M. J., and R. WARD, 1980. Conditions for motion flow in dynamic visual noise. *Vision Res.* 20:431–435.

MOVSHON, J. A., E. H. ADELSON, M. S. GIZZI, and W. T. NEWSOME, 1985. The analysis of moving visual patterns. In *Pattern Recognition Mechanisms*, C. Chagas, R. Gattass, and C. Gross, eds. New York: Springer-Verlag, pp. 117–151.

NAGEL, H. H., 1987. On the estimation of optical flow: Relations between different approaches and some new results. *Artif. Intell.* 33:299–324.

NAKAYAMA, K., 1985. Biological image motion processing: A review. *Vision Res.* 25:625–660.

NAKAYAMA, K., and C. W. TYLER, 1981. Psychophysical isolation of movement sensitivity by removal of familiar position cues. *Vision Res.* 21:427–433.

NEWSOME, W. T., K. H. BRITTEN, and J. A. MOVSHON, 1989. Neuronal correlates of a perceptual decision. *Nature* 341: 52–54.

NEWSOME, W. T., M. S. GIZZI, and J. A. MOVSHON, 1983. Spatial and temporal properties of neurons in macaque MT. *Invest. Ophthalmol. Vis. Sci.* 24:106.

NOWLAN, S. J., 1990. *Competing experts: An experimental investigation of associative mixture models.* (Tech. Rep. No. CRG-TR-90-5). University of Toronto, Toronto, Canada: Department of Computer Science.

NOWLAN, S. J., and T. J. SEJNOWSKI, 1993. Filter selection model for generating visual motion signals. In *Advances in Neural Information Processing Systems*, Vol. 5, S. J. Hanson, J. D. Cowan, and C. L. Giles, eds. San Mateo, Calif.: Morgan Kaufmann, pp. 369–376.

NOWLAN, S. J., and T. J. SEJNOWSKI, 1994a. Model of motion processing in area MT of primates. *J. Neurosci.* in press.

NOWLAN, S. J., and T. J. SEJNOWSKI, 1994b. Filter selection model for motion segmentation and velocity integration. *J. Opt. Soc. Am.* in press

OHZAWA, I., G. SCLAR, and R. D. FREEMAN, 1985. Contrast gain control in the cat's visual system. *J. Neurophysiol.* 54: 651–667.

RAMACHANDRAN, V. S., and P. CAVANAUGH, 1987. Motion capture anisotropy. *Vision Res.* 27:97–106.

RODMAN, H. R., and T. D. ALBRIGHT, 1989. Single-unit analysis of pattern-motion selective properties in the middle temporal visual area (MT). *Exp. Brain. Res.* 75:53–64.

RUBIN, N., and S. HOCHSTEIN, 1993. Isolating the effect of one-dimensional motion signals on the perceived direction of moving two-dimensional objects. *Vision Res.* 33:1385–1396.

SAITO, H., M. YUKIE, K. TANAKA, K. HIKOSAKA, Y. FUKADA, and E. IWAI, 1986. Integration of direction signals of image motion in the superior temporal sulcus of the macaque monkey. *J. Neurosci.* 6:145–157.

SMITH, J. A., and N. M. GRZYWACZ, 1993. A local model for transparent motions based on spatio-temporal filters. In *Computation and Neural Systems*, J. Bower and F. Eeckman, eds. Norwell, Mass.: Kluwer Academic.

SNOWDEN, R. J., S. TREUE, R. G. ERICKSON, and R. A. ANDERSEN, 1991. The response of area MT and V1 neurons to transparent motion. *J. Neurosci.* 11:2768–2785.

STERIADE, M., D. McCORMICK, and T. J. SEJNOWSKI, 1993. Thalamocortical oscillations in the sleeping and aroused brain. *Science.* 262:679–685.

STONER, G. R., T. D. ALBRIGHT, and V. S. RAMACHANDRAN, 1990. Transparency and coherence in human motion perception. *Nature* 344:153–155.

TANAKA, K., H. HIKOSAKA, H. SAITO, Y. YUKIE, Y. FUKADA, and E. IWAI, 1986. Analysis of local and wide-field movements in the superior temporal visual areas of the macaque monkey. *J. Neurosci.* 6:134–144.

TOLHURST, D. J., and J. A. MOVSHON, 1975. Spatial and temporal contrast sensitivity of striate cortical neurons. *Nature* 257:674–675.

TREISMAN, A., 1988. Features and objects: The fourteenth Bartlett memorial lecture. *Q. J. Exp. Psychol. [A]* 40:201.

WELCH, L., 1989. The perception of moving plaids reveals two motion-processing stages. *Nature* 337:734–736.

WILLIAMS, D. W., and R. SEKULER, 1984. Coherent global motion percepts from stochastic local motions. *Vision Res.* 24:55–62.

WILSON, H. R., V. P. FERRERA, and C. YO, 1992. A psychophysically motivated model for two-dimensional motion perception. *Visual Neurosci.* 9:79–97.

28 Attentional Mechanisms in Visual Cortex

JOHN H. R. MAUNSELL AND VINCENT P. FERRERA

ABSTRACT Many studies have described changes in neuronal activity related to behavioral state or attention. In visual cerebral cortex, a variety of extraretinal effects have been reported. To date, however, relatively little effort has been directed at understanding the degree of regional specialization for extraretinal signals. Given the high degree of specialization that has been found for the processing of different classes of visual sensory information, it seems likely that corresponding specialization exists for extraretinal representations in visual cortex. We have recorded from individual neurons in several different cortical areas in monkeys performing a match-to-sample task. The results of these studies suggest that different visual areas are specialized for the types of extraretinal information they represent.

Most neurophysiological studies of visual cortex have been directed at understanding how neurons represent information in the retinal image. Recently, however, more attention has been directed at recording from behaving animals in order to study extraretinal representations in visual cortex. A wide variety of extraretinal signals have been described, including changes in neuronal activity related to eye position or eye movements or changes related to attention to a particular stimulus dimension or spatial location.

Although diverse extraretinal signals have been described in visual cortex, relatively little effort has been directed at understanding the distribution of these signals across different regions of visual cortex. There are many reasons to believe that different regions of visual cortex may be specialized for various types of extraretinal signals. Chief among these is the high degree of specialization seen in the processing of signals that arise from the retina. The cortex of the macaque monkey contains more than 30 distinct visual areas (Felleman and Van Essen, 1991). Each of these areas has its own, sometimes limited, representation of the visual field and is believed to play a distinct role in processing visual information. Cortical visual areas appear to be organized in a hierarchical fashion, with areas on higher levels responsible for representing increasingly complex types of visual information. For example, while neurons in the primary visual area, V1, respond strongly to the appearance of edges in particular orientations, many neurons in higher stages of processing in the temporal lobe respond only to complex patterns or forms, including faces or hands (Desimone, 1991).

In addition to this hierarchical arrangement, there is also parallel segregation in visual cortex. Large regions of visual cortex appear to be segregated into two streams of processing, the temporal and parietal pathways (Ungerleider and Mishkin, 1982). Each of these pathways includes different visual areas and is believed to be involved in different types of visual processing. The parietal pathway, which includes many areas containing neurons with a high degree of direction selectivity, is considered important for the analysis of motion and spatial relationships. The temporal pathway includes other areas whose neurons appear more selective for pattern or form. Elements of these pathways are illustrated in figure 28.1. On the basis of the neurophysiological distinctions and other lines of evidence, Ungerleider and Mishkin (1982) proposed that the temporal pathway might be important for determining *what* objects are, while the parietal pathway might be more involved in visual assessment of *where* objects are. This dichotomy has proved to be a valuable framework in which to evaluate the function of visual cortex, although it should be recognized that the functional and anatomical segregation of these pathways is not complete (see Merigan and Maunsell, 1993). Recordings from neurons in behaving monkeys have found extraretinal signals in the two pathways that appear to be consistent with their putative functional roles. For example, neurons in the parietal path-

---

JOHN H. R. MAUNSELL Division of Neuroscience, Baylor College of Medicine, Houston, Tex.
VINCENT P. FERRERA Department of Physiology, University of California at San Francisco, San Francisco. Calif.

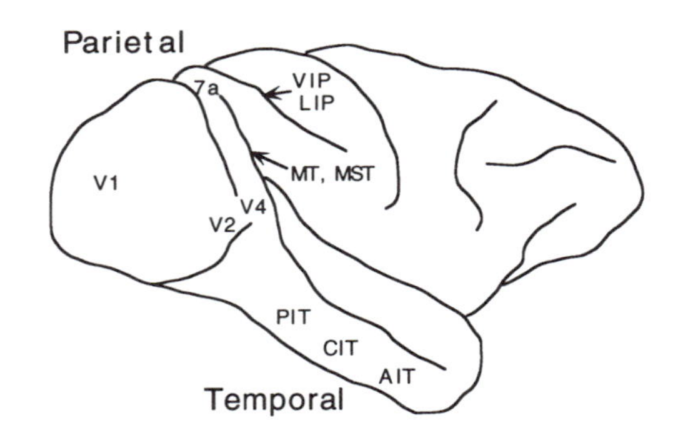

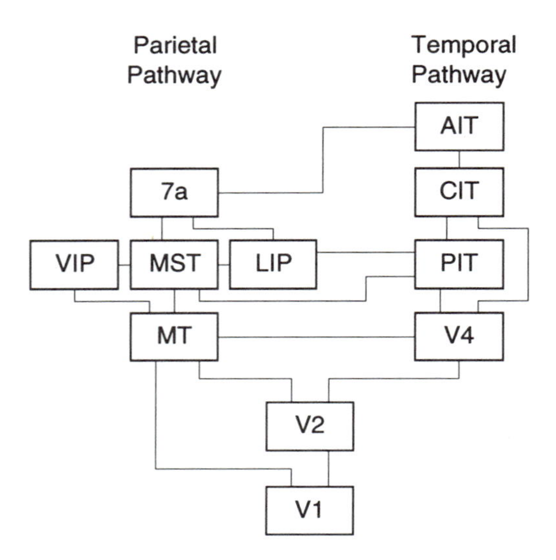

FIGURE 28.1 The organization of monkey visual cortex. Depicted is a lateral view of cerebral hemisphere from a macaque monkey, with the approximate locations of selected visual areas labeled. Some of these areas lie entirely buried within the cortical folds and are not visible on this surface view. The block diagram shows the functional relationships among these cortical areas. The areas are assigned to levels in a hierarchy based on defining anatomical characteristics of their axonal connections (Maunsell and Van Essen, 1983b). Areas at higher levels represent later stages of cortical processing. The areas split into two streams of processing, the parietal and temporal pathways, which are believed to mediate different classes of visual functions. AIT, anterior inferotemporal area; CIT, central inferotemporal area; LIP, lateral intraparietal area; MT, middle temporal area; MST, medial superior temporal area; PIT, posterior inferotemporal area; VIP, ventral intraparietal area; V1, visual area one; V2, visual area two; V4, visual area four; 7a, Brodmann's area 7a.

way respond differently depending on the orientation of the eyes in their orbits (Andersen and Mountcastle, 1983; Andersen, 1987), whereas some neurons in the temporal pathway are selectively activated when an animal is required to remember a particular pattern (Miyashita, 1988).

Most investigations of extraretinal signals have used tasks adapted to the likely function of the area being studied. There have been few direct comparisons between the pathways using the same visual stimuli or behavioral tasks. As a result, it is unclear whether extraretinal representations are specialized for particular regions of cortex, either between the temporal and parietal pathways or between different levels of processing within each pathway. We have examined extraretinal signals in different regions of cortex with the goal of learning more about the distribution and strength of particular extraretinal representations. Throughout these studies, we have used match-to-sample tasks and examined the extent to which the activity of neurons in various areas is affected by requiring the animal to remember different sample stimuli. The results from these experiments have shown considerable differences in the prevalence and strength of extraretinal signals in different areas and suggest that the temporal and parietal pathways may be specialized for extraretinal representations to the same extent as they are specialized for sensory representations.

V4 neurons during match-to-sample orientation task

One of the principal questions that we wished to address was how the effects of attention were distributed across different levels of processing in visual cortex. Because most investigations had focused on areas that were among the highest levels, we chose to examine area V4. V4 lies at an intermediate level of the temporal pathway in the hierarchy illustrated in figure 28.1. One advantage to studying V4 is that most of its neurons respond to relatively simple stimuli. Neurons in V4 are frequently selective for color and orientation (Desimone and Schein, 1987; Schein and Desimone, 1990).

We examined the effects of attention in V4 using a match-to-sample orientation task, which is illustrated in figure 28.2. At the start of each trial, a sample stimulus appeared on a video monitor that was positioned in front of the monkey. The sample was a black and white or colored grating that was drawn in a randomly selected orientation. The animal signaled that it was ready by fixating a small spot in the center of the sample stimulus and pressing a lever. The sample stimulus then disappeared, leaving only the fixation spot.

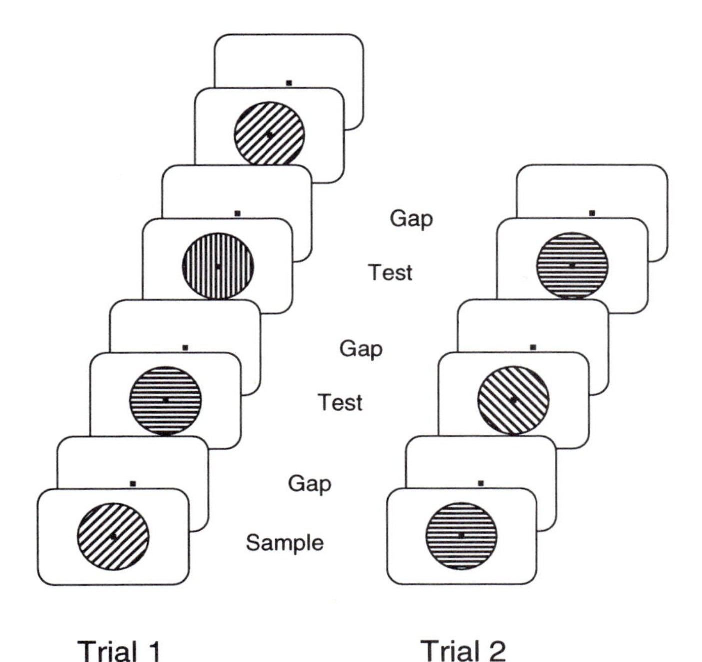

FIGURE 28.2 Match-to-sample orientation task. This matching task required the monkey to remember the orientation of a sample grating that appeared at the start of a trial and to release a lever at the appearance of a grating with the same orientation during a test sequence that followed. Two representative trials are shown, with time increasing up and to the right. The orientation of the sample stimulus and the sequence of the test stimuli were both randomly selected so that the matching stimulus might come anywhere from first to fourth in the test sequence. The sample stimulus and the test stimuli all were separated by brief gaps during which only a fixation spot was present on the display. These gaps were usually 600–800 ms, and each test stimulus was presented for approximately 250–400 ms. The animal was required to release the lever within 600 ms or so after the appearance of the matching stimulus. Only four orientations were used to ensure that each combination of sample and test orientations occurred many times during recording from each neuron.

Shortly thereafter, a sequence of test stimuli appeared, each centered on the fixation spot and separated from other stimuli by gaps during which only the fixation spot was present. The animal's task was to release the lever when it saw a test stimulus that had an orientation that was the same as the sample stimulus. Four orientations were used, any one of which might be selected as the sample on a given trial, and the test sequences were randomly selected such that the matching stimulus might come anywhere from first to fourth in the test sequence.

The animal was required to keep its gaze on the fixation spot throughout each trial, so that the stimuli were all centered on the fovea. We recorded from a region of V4 that contained neurons with central receptive fields, which were covered by the stimuli. In this way, we could measure how individual neurons responded to the appearance of each stimulus. Using this matching task, it was possible to see how well neurons responded to different stimulus orientations and to construct orientation tuning curves. Of greater interest was the question of how the responses to different test stimulus orientations depended on the orientation for which the animal was searching.

Many of the neurons recorded in V4 discriminated stimulus orientation and were not affected by the orientation that the animal was seeking at the time the stimuli appeared. The data in figure 28.3A were collected from one such neuron. This neuron preferred the right oblique stimulus. The upper two plots in figure 28.3A are response histograms for two particular test stimulus sequences. In the upper sequence, the animal had been instructed to search for a left oblique orientation, and the matching stimulus appeared in the fourth position. In the lower sequence, the animal was instructed to search for horizontal, and the match also appeared fourth in the test sequence. The neuron responded strongly when the right oblique stimulus was presented, and it did not matter which orientation had been given as the sample. This was true across all the trials the animal performed while data were collected from this neuron. The plot labeled *stimulus orientation* in the figure shows the orientation selectivity for this neuron. It includes all the responses to each of the four orientations that appeared in the test sequences and shows the strong preference for the right oblique orientation. The plot labeled *sample orientation* replots exactly the same data but as a function of the orientation that the animal was seeking. Each point includes responses to each of the four stimuli. They are sorted according to the sample that was presented at the start of the trial in which they appeared. (Note that these are responses to the test stimuli, not the sample stimuli.) The responses of this neuron were unaffected by which sample orientation the animal was remembering.

Other neurons behaved differently. Data from one such neuron are plotted in figure 28.3B. The format is exactly the same as that for figure 28.3A. The response plots at the top of the figure show that this neuron did not respond consistently to particular stimulus orienta-

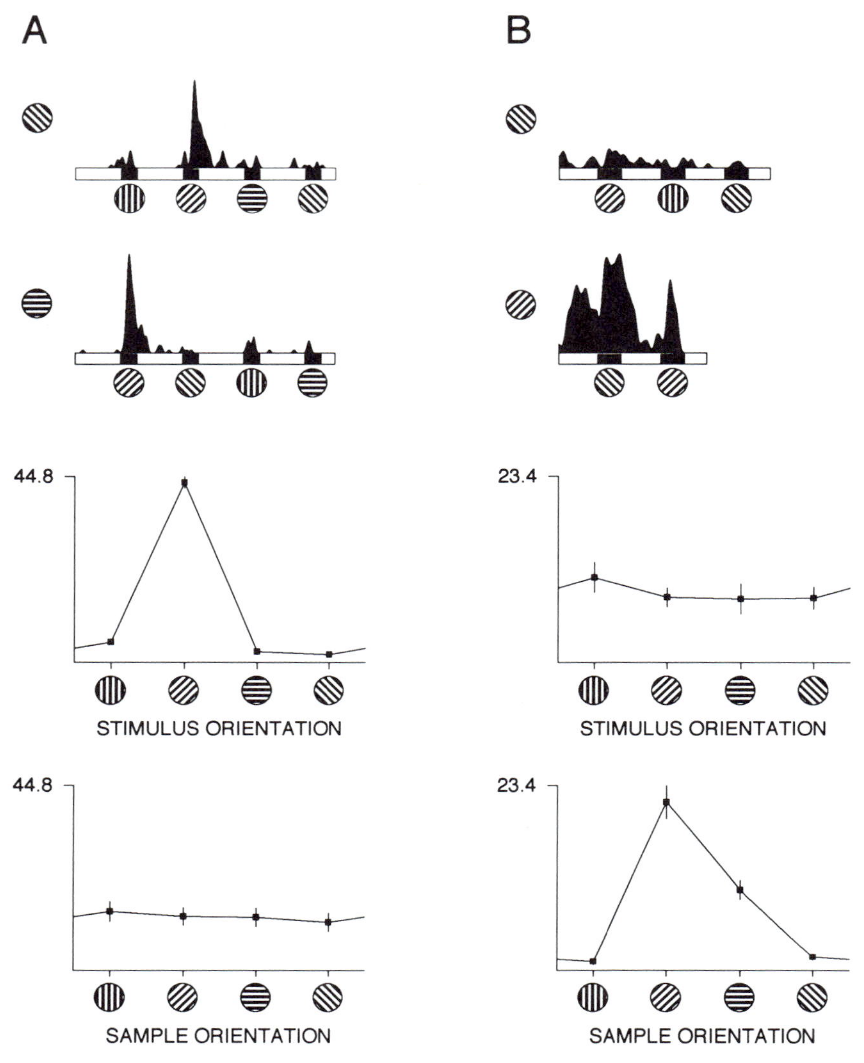

FIGURE 28.3 Responses of V4 neurons recorded during the orientation match-to-sample task. (A) and (B) plot data from two different V4 neurons. In each panel, the upper two plots show responses to selected trial sequences. The grating to the left of each plot shows the sample orientation that had been presented, and gratings below the plots show the orientations of the test stimuli that appeared. Each test stimulus was on during the period of time marked by the black bar above the grating. Only the fixation spot was present during the intervening periods. The amplitude of the plot is the rate of firing. Each plot is the average of approximately eight repetitions of the indicated sequence. Each presentation was randomly interleaved with many other sequences that are not shown. The neuron in (A) responded consistently to the appearance of the right oblique stimulus, whereas that in (B) responded strongly to any stimuli when they appeared during trials in which the animal was searching for a right oblique stimulus. It failed to respond to the same stimuli when they appeared in other trials. Plots labeled *stimulus orientation* compile responses from all trial sequences (regardless of sample orientation). The vertical axes in these and other plots are scaled in units of impulses per second. Plots labeled *sample orientation* replot the same data but, in this case, the data are sorted according to which sample orientation had been presented. Thus, each point includes responses to all test stimulus orientations.

tions. Responses to the oblique stimuli were much stronger during the trial in which the animal was searching for right oblique than they were during the trial in which the animal was searching for left oblique. Examining responses from all the trials made clear that this neuron was more responsive to any orientation during any trial in which the right oblique orientation had been the sample and was less responsive during other trials. This is documented in the plot labeled *sample orientation*, which shows that the average response to any stimulus during trials with the right oblique sample were much stronger than the others. The other plot, stimulus orientation, shows that this neuron did not discriminate among the visual stimuli that appeared in the test sequences: It did not have conventional orientation tuning.

The behavior of this neuron cannot be explained in terms of stimulus selectivity. The sample stimuli that caused the differences in response were removed from the visual display well before the data in figure 28.3B were collected. It also cannot be explained as a consequence of the animal attending to some stimuli and not others; to perform the task, the animal had to attend to every test stimulus. It is likewise improbable that the change in neuronal activity arises from the animal finding one stimulus more interesting or more difficult, because different neurons were more or less responsive for different sample orientations. We believe that this change in activity is related to the animal remembering the sample orientation. These neurons convey information about the orientation the animal is remembering in the same way that neurons with conventional orientation tuning convey information about the orientation of contours on the retina. The animal needs to maintain a neuronal representation of the sample orientation throughout each trial, and we believe that neurons such as the one illustrated in figure 28.3B may contribute to such a representation.

In a population of 429 isolated neurons recorded in V4 (Maunsell et al., 1991), 25% (107) had statistically significant differences in their responses as a function of the sample orientation (two-way analysis of variance, $p < .05$). The strength of behavioral effects differed greatly among the neurons that we recorded in V4 and, for most cells, the effects were weaker than those shown in figure 28.3B. To evaluate the range of effects, we computed for each neuron an index that compared the average level of response during trials with the sample that produced the strongest rate of firing to the average level of response during trials with the sample associated with the weakest rate of firing. We used the index $(P - L)/(P + L)$, which varies from 0 when the preferred and least preferred are exactly the same (no effect) to 1 when there is no response in trials with the least preferred sample orientation. The distribution for all the V4 neurons is plotted in the upper panel of figure 28.4. The upper x axis shows corresponding ratios of average preferred to least-preferred responses. Values that were statistically significant are plotted in solid black. The median index value among the fraction of the population with significant effects was 0.31, which equals a ratio of 1.9 : 1.0. The data in figure 28.3 lie near the ends of this distribution, with the index for the responses in figure 28.3A being 0.10 and that for figure 28.3B being 0.89.

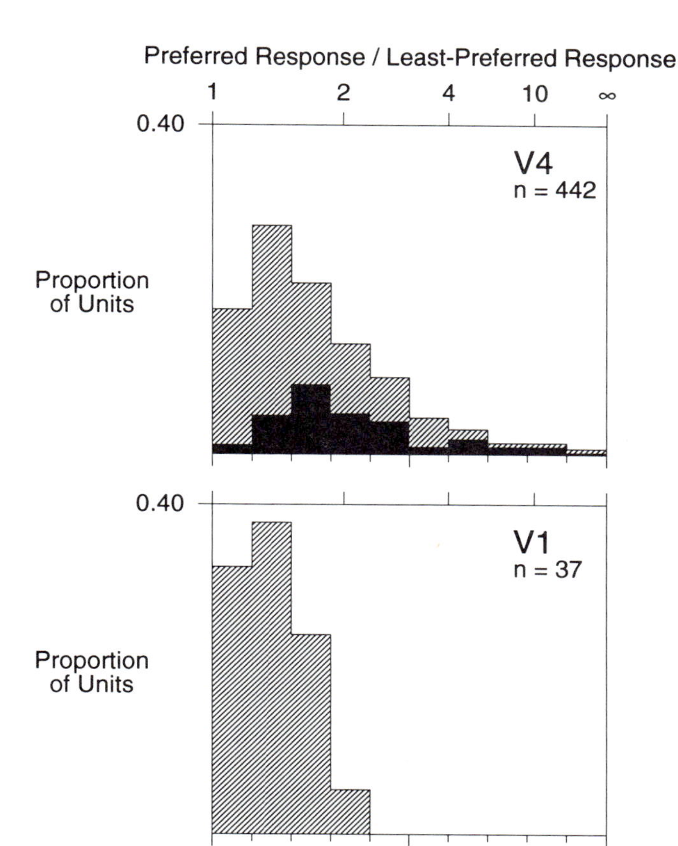

FIGURE 28.4 Strength of behavioral effects in V4 and V1 during orientation matching. The strength of behavioral effects was quantified using the index $(P - L)/(P + L)$, where P represents the average rate of firing during trials that had the sample orientation that led to the strongest rate of firing, and L represents the corresponding rate during trials that had the sample orientation that led to the weakest rate of firing. This index varies from 0 (no difference) to 1 (no response on trials with the least preferred sample orientation). The upper x axis shows corresponding ratios of preferred to least preferred responses. Statistically significant values ($p < .05$) are drawn in black; the remaining values are hatched. No significant behavioral effects were seen in V1, and the V4 distribution, unlike the V1 distribution, had a long tail including values from a subpopulation that showed substantial modulations as a function of the orientation that the animal was seeking.

For comparison, we also collected data from a small sample of neurons in V1 while the animal performed the same orientation-matching task. The distribution of the strength of behavioral effects for V1 are plotted in the lower panel of figure 28.4. None of the 37 neurons recorded in V1 had significant changes in the rate of firing as a function of sample orientation. We found no neurons in V1 in which responses were greatly affected by the orientation that the animal was seeking.

The motion pathway

The results in V4 lead to the question of whether similar properties might be found in other relatively early stages of extrastriate cortex. As described previously, two streams of processing have been identified in primate visual cortex. V4 lies squarely in the temporal pathway, with the bulk of its outputs going directly or indirectly to areas in inferotemporal cortex. We wanted to determine whether areas in the parietal pathway might show similar behavior during a match-to-sample task.

One of the characteristics of neurons in the parietal pathway is that they respond well to moving stimuli and frequently have a clear direction preference. Responses to stationary stimuli are relatively weak. We therefore used different stimuli in testing areas in the parietal pathway. The stationary gratings that were used in recording from V4 were replaced with fields of dynamic random dots, which moved coherently in one of four directions. The fields of dots were also offset from the center of gaze so that they fell on the receptive field of the neuron being recorded. Otherwise, this match-to-sample direction task was the same as the match-to-sample orientation task, with a randomly selected direction of motion being presented as a sample stimulus at the start of each trial and the animal required to release a lever when the same direction appeared in a subsequent test sequence.

The middle temporal visual area (MT) is a key element in the parietal pathway. We recorded from 66

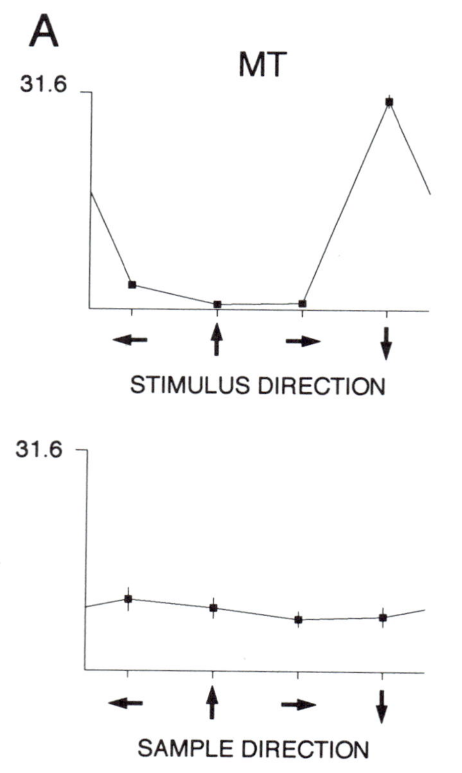

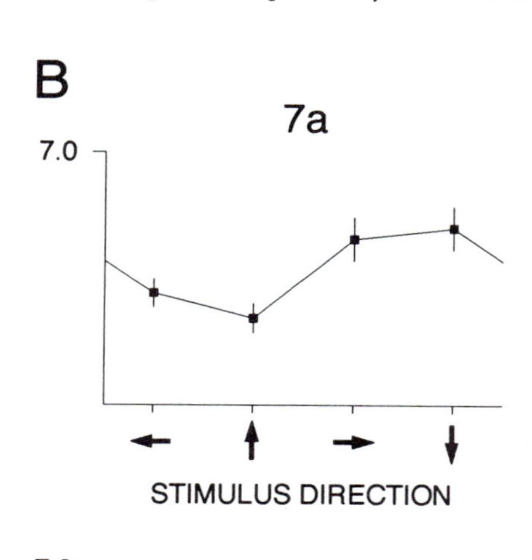

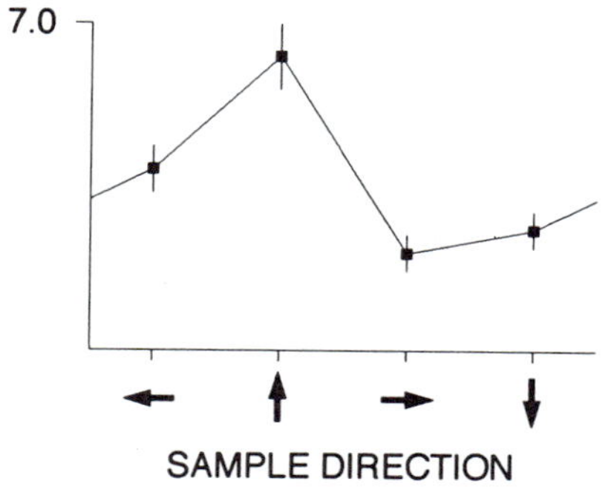

FIGURE 28.5 Responses of an MT neuron (A) and a neuron in area 7A (B) during the match-to-sample direction task. The format of these plots is the same as those in figure 28.3. The MT neuron was strongly direction-selective but showed no significant behavioral effects. This was typical of MT neurons. The neuron in area 7A showed appreciable behavioral effects, but these were among the strongest seen in any of the three areas in the parietal pathway.

MT neurons while the animal performed the direction-matching task. Most of these neurons responded strongly to the presentations of the moving random-dot stimuli. The data in figure 28.5A were recorded from one MT neuron. The plots in this figure are analogous to those that appear in figure 28.3. The upper plot in figure 28.5A shows that the MT neuron was very direction-selective, responding well only when the dots moved down. The lower plot shows that there were no behavioral effects for this neuron: Responses to the moving dot patterns were not affected by the direction for which the animal was searching. Other MT units had statistically significant behavioral effects, but overall the behavioral effects in MT were weak relative to those encountered in V4. The distribution of indices for the strength of behavioral effects in MT is shown in the uppermost histogram in figure 28.6. This distribution is much more concentrated near 0 (no effect) than was the distribution of values for V4 (see figure 28.4).

Because only weak behavioral effects were found in MT, we examined later stages in the parietal pathway, the medial superior temporal area (MST) and area 7a. Stronger behavioral effects were found here, but even these areas did not attain the level seen in V4 during the orientation-matching task. Figure 28.5B plots data recorded from a unit in area 7a. This unit had one of the stronger behavioral modulations encountered in the parietal pathway. The distributions of index values for MST and 7a appear in figure 28.6. It is clear that effects were stronger in these areas than in MT, albeit not overwhelming.

The relatively weak behavioral effects found at even the later stages of the parietal pathway led us to wonder about neuronal activity in V4 during the direction-matching task. Superficial consideration suggested that the properties of V4 were not particularly well suited to the direction-matching task. Estimates of the prevalence of direction selectivity have suggested that only 5–10% of V4 neurons are strongly direction-selective (Van Essen and Zeki, 1978; Zeki, 1978; Desimone and Schein, 1987) compared to more than 90% in MT (Zeki, 1974, 1978; Maunsell and Van Essen, 1983a; Albright, 1984; Albright, Desimone, and Gross, 1984). We were therefore surprised to find that an appreciable proportion of V4 neurons showed behavioral effects during the direction-matching task. The distribution of indices are plotted in figure 28.6. Comparison with the data from the other areas shows that behavioral effects during the direction-matching task were as strong or stronger in V4 than those at all levels of the parietal pathway. Thus, it appears that the contributions of visual areas to particular behavioral tasks may not always be predictable based on the response properties of their neurons.

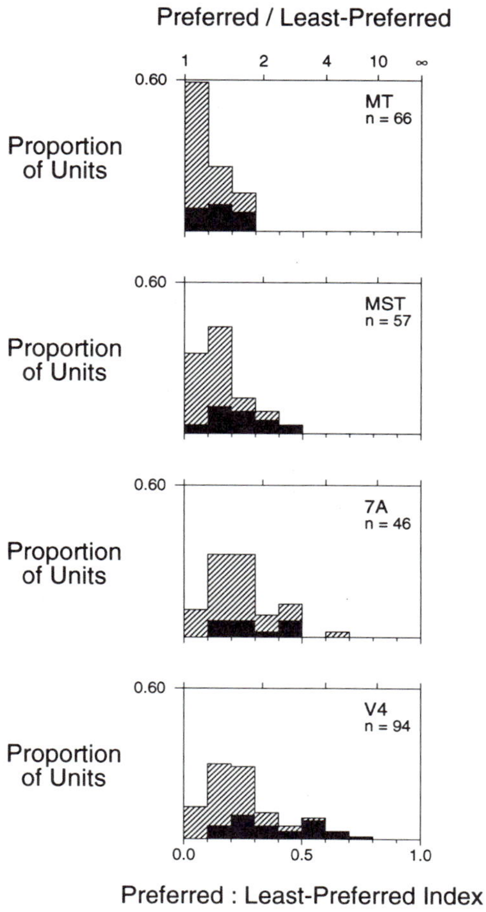

FIGURE 28.6 Strength of behavioral effects during direction matching. The strength of behavioral effects was quantified using the same index as in figure 28.4. Values that were statistically significant are drawn in black. Stronger behavioral effects were seen in ascending the parietal pathway from MT to MST and 7A. Nevertheless, effects in V4 during direction matching (bottom panel) were stronger, although V4 contains relatively few neurons with conventional direction selectivity.

The distribution of behavioral effects between cortical pathways

The predominance of behavioral effects in V4 during the match-to-sample direction task suggests that the parietal pathway may not be the primary contributor to all types of motion analysis, despite the preponderance of direction-selective neurons within it. Although the what-where distinction laid out by Ungerleider and Mishkin (1982) has proved useful for describing the distinctions between the parietal and temporal pathways, it would be surprising if such simple terms completely described their functional differences. Dichotomous classes of visual functions have been suggested many times, using different terms that can be equally well defended: Among these are *evaluating-orienting* (Ingle, 1967), *what-where* (Schneider, 1967), *focal-ambient* (Trevarthen, 1968), *examining-noticing* (Weiskrantz, 1972), *figural-spatial* (Breitmeyer and Ganz, 1976), *foveal-ambient* (Stone, Dreher, and Leventhal, 1979), and *object-spatial* (Mishkin, Ungerleider, and Macko, 1983).

What are the functional differences between the temporal and parietal pathways? Recently there have been specific attempts to refine the what-where categorization. Previc (1990) has suggested that the temporal and parietal pathways may be more closely associated with near and far vision. More relevant to the current findings is the suggestion by Goodale and Milner (1992) that whereas the temporal pathway is concerned with identifying objects (what), the essence of parietal functions may be visual guidances (how) rather than spatial relationships per se (where). These investigators proposed that neural mechanisms in the parietal pathway are adapted for motion analysis specifically related to guiding visual behaviors. In their view, the process of identifying a direction of motion would not lie in the domain of the parietal pathway because it is an evaluation involving no body movements. Instead, as a process of recognition, this would be mediated by the temporal pathway. Thus, the parietal pathway would subserve motion analysis only for that subset of motions directly involved in guiding body movements.

Much of what is known about the neurophysiology of the cortical areas in the parietal pathway is consistent with this view. Neurons in the dorsomedial portion of MST (MSTd) respond best to the motion of large-field stimuli, and some prefer stimulus rotation or expansion (Saito et al., 1986; Tanaka, Fukada, and Saito, 1989; Duffy and Wurtz, 1991). These properties have led to suggestions that these neurons would respond strongly to optic flow resulting from locomotion and that they might be important for navigation. MSTd has also been shown to contain disparity-sensitive neurons whose preferred direction of motion reverses in going from near to far disparities (Roy, Komatsu, and Wurtz, 1992). Such neurons could similarly be stimulated optimally by optic flow patterns during locomotion. Extraretinal signals have been described that are related to pursuit eye movements (Komatsu and Wurtz, 1988; Newsome, Wurtz, and Komatsu, 1988) or to saccadic eye movements (Mountcastle et al., 1984; Gnadt and Andersen, 1988; Andersen et al., 1990; Goldberg, Colby, and Duhamel, 1990). Some parietal neurons are also active during hand-projection movements in the dark (Mountcastle et al., 1975; Robinson, Goldberg, and Stanton, 1978).

There are some observations that appear to contradict the idea that the parietal pathway is specifically adapted for visual guidance. Some extraretinal signals are seen in the absence of any movement or obvious intention to make a movement. The responses of many parietal neurons are modulated in a spatially selective way depending on whether the subject is required to attend to that stimulus (Bushnell, Goldberg, and Robinson, 1981). Neurons have also been shown to be more or less responsive depending on the angle of gaze (Andersen and Mountcastle, 1983). It can be argued that these extraretinal signals are related to planned movements, but the activity appears to be much more closely related to maintaining a representation of spatial orientation rather than immediate visual guidance. Nevertheless, it can be argued that such signals encode information about body orientation that is essential for many types of movement. Overall, the available data are consistent with the view that the parietal pathway may be particularly important for visually guided behaviors, whereas passive evaluation or recognition of extrapersonal movement would be more in the domain of the temporal pathway.

Outstanding issues related to extraretinal signals in visual cortex

There are many issues about extraretinal signals in visual cortex that remain unresolved. One indication from the current results is that we need to know more about classes of extraretinal signals. Most studies of behavioral effects have selected a task that appeared in some way suited to a particular cortical region and then examined neuronal properties in that region while the task was performed. The current results suggest that it is likely to be informative to differentiate the involvement of various cortical areas in different behavioral tasks. For example, it may prove revealing to assay extraretinal signals in a wide variety of cortical areas using a single task. Showing that various areas have a greater or lesser relationship to a particular task could be far more valuable than demonstrating that a particular area is involved in that task. It is possible that effects of "attention" are represented in as many diverse forms as are aspects of the retinal image and that the specialization of cortical areas extends to extraretinal representations as well as retinal.

Closely related to this is the question of how extraretinal signals vary in prevalence across levels of cortical processing. Many studies have failed to find appreciable extraretinal signals in V1, and there are few reports of failures to find extraretinal signals in the highest levels of visual cortex. The current data and those from other studies that have examined successive levels of processing (Mountcastle et al., 1987; Haenny and Schiller, 1988) are consistent with the idea that the prevalence of extraretinal signals increases smoothly and systematically in ascending the cortical hierarchy, but more direct tests are needed.

Finally, and perhaps most important for the near future, is the question of the overall strength of extraretinal signals in visual cortex. Remarkably few studies have provided data on the strength of behavioral effects in visual cortex. Too many have simply reported the number of neurons that reached a particular criterion. Our data show effects that are weak. Although occasional neurons have dramatic effects (e.g., figure 28.3B), a 2 : 1 change is more typical among the neurons with significant behavioral effects. This is a modest modulation compared to the changes in activity that can result from visual stimulation. It is notable that other studies of extraretinal signals in higher visual cortex have described effects that are not markedly stronger (Bushnell, Goldberg, and Robinson, 1981; Mountcastle, Andersen, and Motter, 1981; Richmond, Wurtz, and Sato, 1983; Richmond and Sato, 1987).

Understanding how much of the modulation of neuronal activity in visual cortex depends on retinal stimulation and how much arises from extraretinal sources relates to a basic question about the function of visual cortex. On the one hand, visual cortex might be a fundamentally sensory structure, with extraretinal modulations playing a relatively small role. In that case, the visual cortex might play the role of rendering an optimized representation of the visual environment to be used by other structures in making decisions and generating behaviors. At the other extreme, one could imagine that cortical processing was highly dynamic, with the representations in different cortical areas radically altered by descending influences to suit the immediate needs of the organism. These extremes might be combined such that early cortical levels were basically sensory, with representations in successive levels becoming increasingly dynamic until little fixed sensory response remained at the final stages of visual cortex. Other intermediate situations can be imagined as well. Although the evidence in hand is crude, it suggests that most of the activity in early and intermediate levels of visual cortex depends primarily on retinal stimulation and that extraretinal signals provide only a modest modulation of the sensory representation. Establishing the relative strength of extraretinal influences in higher visual cortex remains an important goal. If neuronal activity in those regions is dominated by extraretinal inputs, it would suggest that visual cortex performs a systematic transformation from sensory coordinates onto what might be considered behavioral coordinates. Alternatively, if extraretinal effects remain modest throughout visual cortex, it would instead suggest that relatively clear lines might be drawn between sensory structures and those that are immediately involved in generating and executing decisions and behavioral acts.

ACKNOWLEDGMENTS We thank John Assad, Jay Gibson, and Anne Sereno for comments on an earlier version of this chapter. The research described and preparation of the manuscript were supported by Office of Naval Research N00014 90 1070, a McKnight Neuroscience Development Award, and National Institutes of Health grant F32 NS08658.

REFERENCES

ALBRIGHT, T. D., 1984. Direction and orientation selectivity of neurons in visual area MT of the macaque. *J. Neurophysiol.* 52:1106–1130.

ALBRIGHT, T. D., R. DESIMONE, and C. GROSS, 1984. Columnar organization of directionally selective cells in visual area MT of the macaque. *J. Neurophysiol.* 51:16–31.

ANDERSEN, R. A., 1987. Inferior parietal lobule function in spatial perception and visuomotor integration. In *Handbook of Physiology: Sec. 1. The Nervous System*, V. B. Mountcastle, F. Plum, and S. R. Geiger, eds. Bethesda, Md.: American Physiological Society, pp. 483–518.

ANDERSEN, R. A., R. M. BRACEWELL, S. BARASH, J. W. GNADT, and L. FOGASSI, 1990. Eye position effects on visual, memory, and saccade-related activity in areas LIP and 7a of macaque. *J. Neurosci.* 10:1176–1196.

ANDERSEN, R. A., and V. B. MOUNTCASTLE, 1983. The influence of the angle of gaze upon the excitability of the light-sensitive neurons of the posterior parietal cortex. *J. Neurosci.* 3:532–548.

BREITMEYER, B. G., and L. GANZ, 1976. Implications of sustained and transient channels for theories of visual pattern masking, saccadic suppression and information processing. *Psychol. Rev.* 83:1–36.

BUSHNELL, M. C., M. E. GOLDBERG, and D. L. ROBINSON, 1981. Behavioral enhancement of visual responses in monkey cerebral cortex: I. Modulation in posterior parietal cortex related to selective visual attention. *J. Neurophysiol.* 46:755–771.

DESIMONE, R., 1991. Face-selective cells in the temporal cortex of monkeys. *J. Cogn. Neurosci.* 3:1–8.

DESIMONE, R., and S. J. SCHEIN, 1987. Visual properties of neurons in area V4 of the macaque: Sensitivity to stimulus form. *J. Neurophysiol.* 57:935–868.

DUFFY, C. J., and R. H. WURTZ, 1991. Sensitivity of MST neurons to optic flow stimuli: I. A continuum of response selectivity to large-field stimuli. *J. Neurophysiol.* 65:1329–1345.

FELLEMAN, D. J., and D. C. VAN ESSEN, 1991. Distributed hierarchical processing in the primate cerebral cortex. *Cerebral Cortex* 1:1–47.

GNADT, J. W., and R. A. ANDERSEN, 1988. Memory related motor planning activity in posterior parietal cortex of macaque. *Exp. Brain Res.* 70:216–220.

GOLDBERG, M. E., C. L. COLBY, and J.-R. DUHAMEL, 1990. Representation of visuomotor space in the parietal lobe of the monkey. *Cold Spring Harb. Symp. Quant. Biol.* 55:729–740.

GOODALE, M. A., and A. D. MILNER, 1992. Separate visual pathways for perception and action. *Trends Neurosci.* 15: 20–25.

HAENNY, P. E., and P. H. SCHILLER, 1988. State dependent activity in monkey visual cortex: I. Single cell activity in V1 and V4 on visual tasks. *Exp. Brain Res.* 69:225–244.

INGLE, D., 1967. Two visual mechanisms underlying the behavior of fish. *Psychol. Forsch.* 31:44–51.

KOMATSU, H., and R. H. WURTZ, 1988. Relation of cortical areas MT and MST to pursuit eye movements: I. Localization and visual properties of neurons. *J. Neurophysiol.* 60: 580–603.

MAUNSELL, J. H. R., G. SCLAR, T. A. NEALEY, and D. D. DEPRIEST, 1991. Extraretinal representations in area V4 in the macaque monkey. *Visual Neurosci.* 7:561–573.

MAUNSELL, J. H. R., and D. C. VAN ESSEN, 1983a. Functional properties of neurons in the middle temporal visual area of the macaque monkey: I. Selectivity for stimulus direction, speed and orientations. *J. Neurophysiol.* 49:1148–1167.

MAUNSELL, J. H. R., and D. C. VAN ESSEN, 1983b. Anatomical connections of the middle temporal visual area in the macaque monkey and their relationship to a hierarchy of cortical areas. *J. Neurosci.* 3:2563–2586.

MERIGAN, W. H., and J. H. R. MAUNSELL, 1993. How parallel are the primate visual pathways? *Annu. Rev. Neurosci.* 16:369–402.

MISHKIN, M., L. G. UNGERLEIDER, and K. A. MACKO, 1983. Object vision and spatial vision: Two cortical pathways. *Trends Neurosci.* 6:414–417.

MIYASHITA, Y., 1988. Neural correlate of visual associative long-term memory in the primate visual cortex. *Nature* 335:817–820.

MOUNTCASTLE, V. B., R. A. ANDERSEN, and B. C. MOTTER, 1981. The influence of attentive fixation upon the excitability of the light-sensitive neurons of the posterior parietal cortex. *J. Neurosci.* 1:1218–1235.

MOUNTCASTLE, V. B., J. C. LYNCH, A. P. GEORGOPOULOS, S. SAKATA, and C. ACUNA, 1975. Posterior parietal association cortex of the monkey: Command functions for operations within extrapersonal space. *J. Neurophysiol.* 38:871–908.

MOUNTCASTLE, V. B., B. C. MOTTER, M. A. STEINMETZ, and C. J. DUFFY, 1984. Looking and seeing: The visual functions of the parietal lobe. In *Dynamic Aspects of Neocortical Function*, G. M. Edelman, W. E. Gall, and W. M. Cowan, eds. New York: Wiley, pp. 159–194.

MOUNTCASTLE, V. B., B. C. MOTTER, M. A. STEINMETZ, and A. K. SESTOKAS, 1987. Common and differential effects of attentive fixation on the excitability of parietal and prestriate (V4) cortical visual neurons in the macaque monkey. *J. Neurosci.* 7:2239–2255.

NEWSOME, W. T., R. H. WURTZ, and H. KOMATSU, 1988. Relation of cortical areas MT and MST to pursuit eye movements: II. Differentiation of retinal from extraretinal inputs. *J. Neurophysiol.* 60:604–619.

PREVIC, F. H., 1990. Functional specialization in the lower and upper visual fields in humans: Its ecological origins and neurophysiological implications. *Behav. Brain Sci.* 13: 519–575.

RICHMOND, B. J., and T. SATO, 1987. Enhancement of inferior temporal neurons during visual discrimination. *J. Neurophysiol.* 58:1292–1306.

RICHMOND, B. J., R. H. WURTZ, and T. SATO, 1983. Visual responses of inferior temporal neurons in awake rhesus monkey. *J. Neurophysiol.* 50:1415–1432.

ROBINSON, D. L., M. E. GOLDBERG, and G. B. STANTON, 1978. Parietal association cortex in the primate: Sensory mechanisms and behavioral modulations. *J. Neurophysiol.* 41:910–932.

ROY, J.-P., H. KOMATSU, and R. H. WURTZ, 1992. Disparity sensitivity of neurons in monkey extrastriate area MST. *J. Neurosci.* 12:2478–2492.

SAITO, H.-A., M. YUKIE, K. TANAKA, K. HIKOSAKA, Y. FUKADA, and E. IWAI, 1986. Integration of direction signals of image motion in the superior temporal sulcus of the macaque monkey. *J. Neurosci.* 6:145–157.

SCHEIN, S., and R. DESIMONE, 1990. Spectral properties of V4 neurons in the macaque. *J. Neurosci.* 10:3369–3389.

SCHNEIDER, G. E., 1967. Contrasting visuomotor functions of tectum and cortex in the golden hamster. *Psychol. Forsch.* 31:52–62.

STONE, J., B. DREHER, and A. LEVENTHAL, 1979. Hierarchical and parallel mechanisms in the organization of visual cortex. *Brain Res. Brain Res. Rev.* 1:345–394.

Tanaka, K., Y. Fukada, and H. Saito, 1989. Underlying mechanisms of the response specificity of expansion/contraction and rotation cells in the dorsal part of the medial superior temporal area of the macaque monkey. *J. Neurophysiol.* 62:642–656.

Trevarthen, C. B., 1968. Two mechanisms of vision in primates. *Psychol. Forsch.* 31:299–337.

Ungerleider, L. G., and M. Mishkin, 1982. Two cortical visual systems. In *The Analysis of Visual Behavior*, D. J. Ingle, R. J. W. Mansfield, and M. S. Goodale, eds. Cambridge, Mass.: MIT Press, pp. 549–586.

Van Essen, D. C., and S. M. Zeki, 1978. The topographic organization of rhesus monkey prestriate cortex. *J. Physiol.* 277:193–226.

Weiskrantz, L., 1972. Behavioral analysis of the monkey's visual system. *Proc. R. Soc. Lond. [Biol.]* 182:427–455.

Zeki, S. M., 1974. Functional organization of a visual area in the posterior bank of the superior temporal sulcus of the rhesus monkey. *J. Physiol.* 236:549–573.

Zeki, S. M., 1978. Uniformity and diversity of structure and function in rhesus monkey prestriate visual cortex. *J. Physiol.* 277:273–290.

29 Open Questions about the Neural Mechanisms of Visual Pattern Recognition

MALCOLM P. YOUNG

ABSTRACT This chapter examines two open questions that concern the neural mechanisms of visual recognition—namely, how these mechanisms are organized and how cells in the anterior inferotemporal cortex (IT) participate in recognition. Although it is fairly certain that the occipitotemporal areas of primate visual cortex form a discriminable, hierarchically organized subsystem that is involved in visual recognition, there remain many uncertainties concerning the patterns of connectivity within and between these areas and concerning possible functional relations between these structures and those outside the occipitotemporal region. It is uncertain how IT cells participate in visual recognition in the general case, although the responses of small populations of IT cells have been shown to be sufficient to account for the recognition of some faces.

The cortical visual system occupies approximately half the monkey's cerebral cortex (Felleman and Van Essen, 1991). It is composed of perhaps a billion neurons, which exhibit complex patterns of visual feature preferences and are distributed in more than 30 discriminable visual processing regions (Zeki and Shipp, 1988; Felleman and Van Essen, 1991). These cortical areas are interconnected by hundreds of ipsilateral and interhemispheric corticocortical connections, as well as by a very rich subcortical network (Kaas and Huerta, 1988; Young, 1992a, 1993a). Given the evident anatomical and physiological complexity of the visual system, it is encouraging that so much progess, as detailed in the chapters of this book, has been made in tying neural structure and function to visual performance.

On the other hand, there are many important remaining uncertainties in our present knowledge of the neural mechanisms of vision. This chapter explores some of these open questions, particularly those that concern the neural mechanisms of visual pattern recognition. Because the domain of the unknown is more extensive than that of the known, I have further restricted myself to just two sets of uncertainties. These are, first, how the neural systems believed to support visual object recognition are organized, and second, how cells in inferotemporal cortex (IT) represent visual objects.

MALCOLM P. YOUNG University Laboratory of Physiology, Oxford University, Oxford, England

Brain structures involved in visual pattern recognition

For two reasons, the focus of discussions of visual pattern recognition has fallen mainly on a number of cortical regions that are situated in the occipital and temporal cortex. First, bilateral ablation of the most rostral of these areas, the anterior part of IT, results in severe and enduring deficits in tasks requiring visual discrimination or recognition (Gross, 1972; Ungerleider and Mishkin 1982). Lesions of other occipitotemporal cortical stations, such as V4, which lie on the presumed pathway between V1 and IT, can also result in recognition deficits (Heywood, Gadotti, and Cowey, 1992; Schiller and Lee, 1991). These results imply that areas between the occipital and temporal poles are essential for intact visual recognition.

Second, the feature preferences of cells in the occipitotemporal pathway appear similar to those expected in the higher stages of a pattern recognition network. Populations of cells in the anterior inferotemporal cortex (AIT), for example, respond more enthusiastically to moderately complex pattern stimuli than to simple bars or spots (Tanaka et al., 1990; Fujita et al., 1992), and a large minority of cells show a preference for handlike stimuli (Gross, Rocha-Miranda, and Bender, 1972), faces (Bruce, Desimone, and Gross, 1981; Perrett,

Rolls, and Caan, 1982; Desimone et al., 1984; Yamane, Kaji, and Kawano, 1988; Young and Yamane, 1992), or other complex patterns that monkeys can learn to discriminate (Miyashita, 1988; Miyashita and Chang, 1988). Many of these cells exhibit remarkable specificity for particular stimuli, such as faces, while not appearing sensitive to the position, size, characteristics of illumination, color, or spatial frequency of those stimuli (Bruce, Desimone, and Gross, 1981; Perrett, Rolls, and Caan, 1982; Desimone et al., 1984; Rolls, Baylis, and Leonard, 1985; Perrett, Mistlin, and Chitty, 1987).

The specificity of cells in AIT for particular categories of complex stimuli, and their apparently invariant responses given differences in presentation context, have suggested for many that these cells have essentially "solved" the difficult problem of invariant pattern recognition and represent the final stage of an efficient recognition mechanism (e.g., Perrett, Mistlin, and Chitty, 1987; Rolls, 1992). Taking into account these results and those from testing the behavioral effects of occipitotemporal lesions, there is little doubt that the areas of the occipitotemporal stream are involved in visual object recognition.

There remains, however, considerable doubt concerning *how* these areas are involved. What are the details of relations between the cells in these areas, and what might their relations with structures outside the classical object pathway contribute to visual function? What are the mechanisms by which the cell properties in IT are generated? How do cells in IT contribute to visual object recognition?

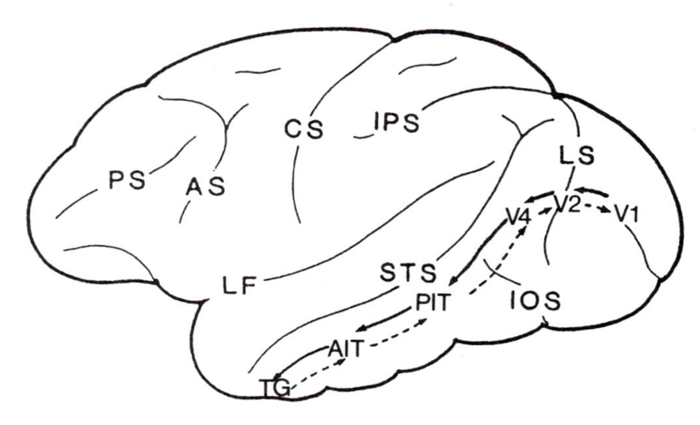

FIGURE 29.1 Lateral view of the macaque cortex, showing some of the stations of a presumed pathway from V1 to the temporal polar cortex (TG), which may be invloved in mediating visual pattern recognition.

How are the neural mechanisms of visual object recognition organized?

It has often been suggested that the mechanisms of visual pattern recognition are distributed in a sequence of visual areas running from V1 through V2, V4, and posterior inferotemporal cortex (PIT), to AIT (see, for example, (Perrett, Mistlin, and Chitty, 1987; Tanaka et al., 1991; Rolls, 1992) (figure 29.1). In this scheme, the goal of this serial arrangement is to provide output that consists of economical representations of visual objects that are invariant to changes in illumination, size, and retinal location. This goal is achieved by developing receptive field properties at successive cortical stations by stacking the areas, which are often thought to contain competitive learning networks (e.g., Rolls, 1992). In this framework, either the computations performed by the occipitotemporal areas are largely independent of the dorsal stream of processing, which contributes little more than foveation of visual targets, or the dependence of occipitotemporal computations on other structures is not made explicit.

A contrasting view is that mechanisms of pattern processing are not segregated from mechanisms of other types of visual processing, and there is no parallel processing in the cortical visual system except in the sense that maps of visual space are reproduced in multiple visual areas (e.g., Martin, 1992). In this model, the occipitotemporal areas are so closely integrated with the other areas of the visual system that it makes little sense to treat them as a discriminable subsystem. This conclusion is reached by considering the connections that exist between the supposedly separate streams as, for instance, in the argument that there cannot possibly be two discriminable visual streams because the middle temporal area (MT), a prototypical dorsal stream area, and V4, a prototypical ventral stream area, are reciprocally interconnected.

It is interesting to juxtapose the simplicity of these ideas with the sophistication of the visual system. At its simplest, characterization of the organization of the ventral stream of processing as a simple hierarchy between V1 and AIT asserts an organizational feature of the cortical visual system while taking into account only eight connections among five cortical areas. Similarly, the contrary argument can rest, at its simplest, on two connections, one from MT to V4, the other from V4 to MT. However, the cortical visual system consists of at least 32 visual areas connected by more than 300 corticocortical connections (Felleman and Van Essen, 1991). It is hardly plausible that a statement about the organization of a system can be supported by considering fewer than 3% (8 of 300) or fewer than 1% (2 of 300) of the connections that define the system's architecture. Clearly, slightly more sophisticated versions of this type of reasoning suffer from the same difficulty as, for example, in an argument against the existence of parallel processing based on the fact that the magnocellular (M) and parvocellular (P) streams mix in V1 and V4, and that there is a nonstriate pathway into MT (Martin, 1992).

This problem in reasoning from connectional data to the organization of a system arises because there are far too many connections to consider them all at once. It is easy to focus attention on a few connections that seem to support a case while forgetting about all the less convenient ones. Without an objective reference, how can disagreements of this type be resolved? Are the areas of the occipitotemporal pathway segregated from parietal areas, or are they not?

ANALYSIS OF CONNECTIONAL DATA When faced with a large amount of data—in this case, connectional data—rather than attempting simply to intuit the information that the data embody, the natural approach is to analyze it. There are two methods available at present to analyze connectional information to show the relations between brain structures. Hierarchical analysis (Rockland and Pandya, 1979; Maunsell and Van Essen, 1983; Felleman and Van Essen, 1991) considers the cortical laminae in which connections originate and terminate. By assuming that terminations in cell-rich layers are ascending and that terminations in cell-sparse layers are descending, it is possible to arrange the cortical areas into a largely consistent unidimensional hierarchy. This type of analysis is important because it indicates both that the cortical visual system, including its occipitotemporal component, is hierarchically organized, and the probable direction of the flow of signals in the system. It depends, however, on detailed data about the laminar origin and termination of connections, which are not available for many connections. Also, it is less applicable to structures that do not have clear laminar organization and does not give any insight into organizational features that are not hierarchical. The latter limitation means that this type of analysis is silent on the issue of whether the occipitotemporal areas are wholly integrated with, or discriminable from, the dorsal stream areas: The left-to-right positions of areas in the familiar hierarchical diagram of Felleman and Van Essen (1991) could be shuffled at random without doing violence to the analytical rules and the data that constrain the diagram.

The other method available for the analysis of connectional data is topological analysis (Scannell, Young, and Blakemore, 1992; Young, 1992a,b, 1993a; Scannell and Young, 1993; Young and Scannell, 1993). This approach uses optimization to produce multidimensional representations of the organization of a brain system that can respect almost any connection pattern (Scannell and Young, 1993; Young, 1993a). It indicates organizational features that are complementary to those shown by hierarchical analysis. It can use the most widely available neuroanatomical data, such as the presence or absence of connections, but gives direct insight into the likely direction of flow of signals only where there are nonreciprocal connections (Young, 1993a).

The analysis proceeds by examining a matrix of connections. The values in a connection matrix are *proximities* that define spatial relations between points representing the brain structures in a space. The defined spatial relations can be perfectly reflected in the configuration of points in this space, so that connected points are close together and unconnected ones far apart, but only when the space has a large number of dimensions. The connectional organization can be made understandable by reducing the dimensionality of the space to three or fewer dimensions, while preserving as much as possible the proximities between the points of the configuration. The low-dimensional configuration of points produced by the analysis optimally fits the connection matrix so that the proximities of the points of the structure are as close as possible to the rank order of the proximities of areas in the connection matrix. This dimensional reduction is achieved by nonmetric multidimensional scaling (MDS) (Shepard, 1962; Young and Harris, 1990; Young, 1992a).

Analysis of the connections among the areas of the primate cortical visual system by this method yields a structure (see color plate 9) that is consistent with, and statistically significantly related to, the hierarchical relations among the areas revealed by the hierarchical analysis of Felleman and Van Essen (1991) (Young, 1992a,b). In addition, the analysis shows a clear structural dichotomy between dorsal areas and ventral areas that is very similar, and statistically significantly related, to the visual dichotomy suggested by Ungerleider and Mishkin (1982) (Young, 1992a). These results suggest that the occipitotemporal areas may be both organized hierarchically and somewhat segregated from the dorsal stream areas.

The topological analysis of the visual cortex, however, provides a compelling result only if it is robust. The solution must be robust in the face of two situations. First, it must be robust against the differences in density or strength of the different projections: Perhaps when data relating to strong, moderate, or weak connection densities are included, the solution will be markedly different. In fact, in no analysis of any sensory system of either the cat or monkey (Scannell, Young, and Blakemore, 1992; Young, unpublished data) has the inclusion of this information given rise to a solution that explains less than 90% of the corresponding "binary" solution. Even if the data are treated as metric (which they are not) and relatively large differences in density are introduced, the solutions remain similar, probably for two reasons. Connectivity is sparse, and so even a weak connection is a rare attractive constraint, and structures that are topologically close (i.e., those that have a very similar pattern of connectivity) tend to exchange strong connections.

Second, the solution must be robust against changes in status of some of the possible connections that have not thus far been reported: Some of these connections will exist. In fact, the most violent perturbation of the connection data, in which all unreported connections (as opposed to connections that have been sought and reported absent) are assumed to exist, an assumption that would turn cortical neuroanatomy on its head, results in a solution that is 76% similar to that in figure 29.1 (Young, 1992a) (color plate 10). The solutions are constrained to be similar because a sufficiently large number of connections have been sought and were found to be absent (see Felleman and Van Essen, 1991). Poorly studied areas, such as VOT and V4t, have their positions shifted by a large number of hypothetical new connections, but the organizational features of the solution are very similar. Thus, the grossest possible perturbation of the data does not radically disturb the conclusions, and they are unlikely to be overturned by growth in our information about visual cortical connectivity. Inevitably the solutions will evolve, as the hierarchical diagrams have done, but the results from this type of analysis currently appear to be robust.

Hence, results from topological analysis of connection matrices (Young, 1992a), hierarchical analysis of laminar origin and termination patterns (Felleman and Van Essen, 1991), and analysis of the behavioral effects of cortical lesions (Ungerleider and Mishkin, 1982) converge in concluding that the occipitotemporal areas are hierarchically organized and form a subsystem that can be discriminated structurally from other components of the cortical visual system. Disputation of these conclusions should account for why analyses of completely different data by completely different methods come to such similar conclusions. Hence, the evidence from lesion studies and from analysis of connectivity is that the occipitotemporal areas do indeed form a discriminable, hierarchically organized subsystem of the cortical visual system.

Spatial Organization of Forward and Backward Projections Nonetheless, there are many remaining uncertainties concerning the organization of processing within occipitotemporal areas. Many accounts of visual recognition, for example, suppose that there is a steady convergence of information, brought about by wide "fan-out" (in which the axonal arbors of cells at the lower level might contact a large number of cells at the higher) from one level of the hierarchy to the next (e.g., Barlow, 1985; Perrett, Mistlin, and Chitty, 1987; Rolls, 1992; Perrett and Oram, 1993). Some of these accounts further suppose that short-range excitation and longer-range inhibition obtain at all stations of the hierarchy, so that the system possesses a competitive learning architecture (Rolls, 1992). These ideas are summarized in figure 29.2.

Little is known currently of the spatial organization of forward projections in the hierarchy, such as whether cells at a lower station actually do contact a large number of cells at a higher station through an extensive axonal arbor (although interesting anatomical work has begun in several laboratories). Similarly, there is little information on whether short-range excitation and longer-range inhibition obtain at all stations of the hierarchy, as required for a competitive learning architecture (but see Lund, Levitt, and Yoshioka, in press). Nonetheless, it is chastening that the lone cross-correlation study in extrastriate cortex that has targeted these issues, the investigation of IT by Gochin and coworkers (1991), showed that IT may not be organized as suggested. This study found that the do-

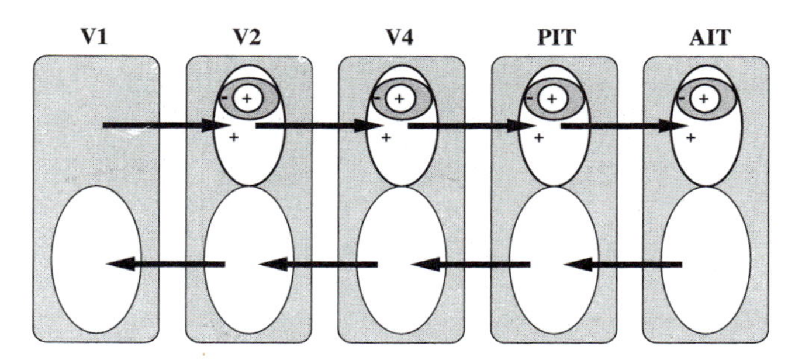

FIGURE 29.2 Schematic of some accounts of the organization of the occipitotemporal areas (see, for example, Rolls, 1992), illustrating the idea that the arbors of projection neurons passing from lower to higher stations are hypothesized to distribute output over large regions of the higher area (i.e., wide fan-out). This is conceived to allow convergence onto cells at successive cortical stages, so that information may be smoothly concentrated as one ascends the network. Local interactions at every level are considered to form a competitive learning architecture in which cells' stimulus selectivities are refined by lateral inhibition extending beyond the domain of local excitation, so that cell groups compete for activity. The backprojections distribute output over a wide area of the lower station to which they connect.

main of common input, the area of spread of input projections, is usually restricted to less than 500 μm. Hence, there does not appear to be the expected wide homogeneous "fan-out" of projections into IT. Further, Gochin and coworkers (1991) showed that direct intracortical excitation and inhibition in IT occur over roughly the same spatial scales and, hence, that the range of excitation may not be smaller than that of inhibition. This result is strikingly different from that in cat cortical area 17 (e.g., Toyama, Kimura, and Tanaka, 1981), which was influential in developing theories on competitive learning networks. Because there does not appear to be a domain of local inhibition extending beyond that of local excitation by which to implement competition, the result suggests that IT may *not* exhibit a competitive learning architecture.

The distribution of receptive fields and stimulus selectivity along the occipitotemporal pathway may also be inconsistent with steady convergence of information in the hierarchy. There is, for instance, a clear discontinuity at the border between PIT and AIT (Tanaka et al., 1991). Posterior to this discontinuity, cells in the dorsal part of PIT have contralateral receptive fields near the fovea, which generally subtend less than 5°. In the ventral part of PIT as well, cells have contralateral receptive fields (Boussaoud, Desimone, and Ungerleider, 1991). A large proportion of cells in PIT require only relatively simple trigger features. Anterior to the discontinuity, cells typically have very large bilateral receptive fields that almost always include the fovea, and there is a much larger proportion of cells that require trigger features more complex than simple bars, gratings, and spots (Tanaka et al., 1991). This discontinuity is not anticipated by models that suggest a simple serial convergence. In addition, the discontinuity in receptive field position and size implies some elaborate callosal interactions in this region, the features of which have not yet been explored.

It is interesting that almost nothing is known concerning the function of, and spatial organization of interactions mediated by, backward projections, which constitute a substantial proportion of the connections of the system. Backprojections appear to terminate in wider cortical domains than those of forward projections, but it is not known what function, if any, these connections serve. To date, there has been no unequivocal report of a physiological communication arising from the action of a corticocortical backprojection: Effects on V1 of cooling prestriate areas (e.g., Mignard and Malpeli, 1991) could be mediated by connections other than corticocortical backprojections, such as thalamic relays. It is not even known with certainty whether the effect of corticocortical backprojections is excitatory or inhibitory, although these backprojections are more likely to be excitatory as they arise from pyramidal cells and generally terminate in zones that contain few inhibitory interneurons. Figure 29.3 summarizes some of the uncertainties about ipsilateral and callosal interactions among the occipitotemporal areas.

RELATIONS AMONG OCCIPITOTEMPORAL AND EXTRA-OCCIPITOTEMPORAL AREAS There are additional uncertainties concerning the relations among the occipitotemporal areas and areas outside the occipitotemporal subsystem. Although the areas of the ventral stream are structurally discriminable from those of the dorsal stream, there are many opportunities for crosstalk (though not at every station) between the two subsystems (see color plate 9). These cross-talk connections provide the opportunity for the areas of the parietal cortex and caudal superior temporal sulcus to influence directly the computations of the temporal areas. The role of the dorsal areas in recognition,

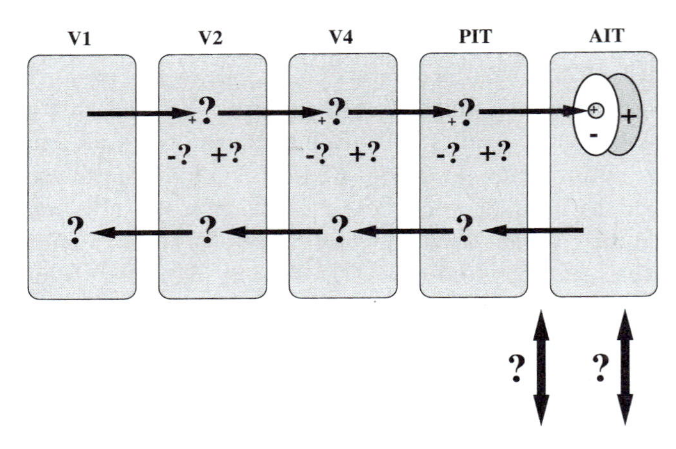

FIGURE 29.3 Schematic of some missing data that would bear on the organization of occipitotemporal processing illustrated in figure 29.2. Little is currently known about the arbors of the forward projections, the arbors of the backward projections, the organization of local connectivity, or the arrangement of interhemispheric connections that mediate the discontinuity in receptive field position at the PIT/AIT border. The single cross-correlation study outside V1 and V2 that has targeted these issues (Gochin et al., 1991) suggests that forward projections into AIT may not exhibit wide fan-out, as has been hypothesized, because the domain of common input seems restricted to 500-μm patches. In contrast to results from cat area 17, which gave impetus to the development of competitive learning models, Gochin and colleagues (1991) found that the domain of intracortical excitation extended over much the same area as intracortical inhibition.

therefore, may go much beyond mere foveation. These possible interactions are an understudied feature of cortical processing, though an interesting theoretical lead has been provided by De Yoe and Van Essen (1988), and a recent investigation of cells in AIT showed that they could respond differentially to patterns defined by motion cues (Sagy, Vogels, and Orban, 1993).

Such interconnections are examples of the wider opportunities for signaling into IT along multiple routes that are not limited to permutations of the classical occipitotemporal hierarchy. Some of these multiple routes can be illustrated by examining structures for the entire cerebral cortex derived by topological analysis (e.g., Young 1993a) (see color plate 11). These alternative routes might involve parahippocampal, limbic, dorsal stream, frontal, pulvinar and dorsal thalamic, striatal, and tectal structures. Given this rich connectivity, it is possible that some properties of AIT cells may not be constructed and maintained only by a feedforward network involving the classical object route. Limbic inputs to cells that appear sensitive to more than one view of an object (Perrett, Mistlin, and Chitty, 1987) are possible indications of the importance of extrinsic inputs, as are the recognition and discrimination deficits that can attend limbic lesions and lesions to the basal ganglia (Mishkin and Appenzeller, 1987; Ridley, Baker, and Murray, 1988), which structures are not components of the classical occipitotemporal pathway. I think it presently unclear to what extent the ventral component of the visual system can perform without its interactions with these discriminably different systems.

How do IT cells participate in the recognition of visual objects?

The preferences of many cells in AIT are difficult to account for by reference to simple stimulus features, such as orientation, motion, position, or colour, and appear to lie in the domain of shape (Gross, Rocha-Miranda, and Bender, 1972; Perrett, Rolls, and Caan, 1982; Desimone et al., 1984; Tanaka et al., 1991). Selectivity for visual patterns and objects, such as faces, hands, arbitrary geographical shapes, and fractal patterns, may suggest that cells in this area contribute an elaborated recognition code for the identification of objects. However, the details of the participation of IT cells in recognition cannot be addressed without careful consideration of what features, or combinations of features, cells with pattern selectivity are really interested in. Cells with apparent selectivity for faces, for instance, might be triggered by the presence of anything from two roughly collinear bars (most faces have eyebrows), or a colored ovoid, to the full configural and textural information present in a typical face stimulus. There are currently two systematic approaches to the issue of what feature constellations IT cells code.

THE SIMPLIFICATION APPROACH One approach, which has been widely employed (e.g., Gross, Rocha-Miranda, and Bender, 1972; Perrett, Rolls, and Caan, 1982; Desimone et al., 1984) but which has been brought to a particularly fine focus by Tanaka's group (Fujita et al., 1992; Tanaka et al., 1990, 1991), has been to try to determine the sufficient feature or features of a cell "on line," by simplifying the stimuli that excite it. This method begins by presenting a large number of patterns or objects while recording from a neuron, to find objects that excite the cell. Then the component features of the effective stimulus, as judged to be present by the experimenters, are segregated and presented again singly or in combination. By assessing the enthusiasm of the cell for each of the simplified stimuli, an informal descent in feature space takes place, with the aim of finding the simplest combination of stimulus features that maximally excites the cell.

This process requires very painstaking work. Even the simplest real-world object contains a rich set of possible elementary features, not all of which will be obvious to the experimenters. A humble carrot, for example, has elements of color, shape, orientation, depth, curvature, and texture, and may show specular reflections and shading. Elements in any one of these domains, and any combination of the elements, may be the triggering stimulus for a cell excited by an image of the carrot, so there is a vast number of possible combinations of the elementary features of any real-world object. It is therefore impracticable to present all possible feature combinations systematically, and the simplified stimuli that are actually presented are typically a subset of the possible combinations, guided by the intuitions of the experimenters. Hence, it is not possible to conclude that the best simplified stimulus is optimal for the cell, only that it was the best of those presented. It cannot even be assumed that the cell codes only one minimum in feature space (i.e., that there is only one optimal set of features); it is possible that the cell could exhibit two or more minima, corresponding to very different feature combinations (Young, 1993b). It follows from these considerations that identification in this way of the preference of an IT cell may not reveal the stimulus conditions necessary for the cell to participate in recognition.

It is a rather unreasonable requirement, however, that a study in IT should identify the features necessary for a cell's response: All the possible visual patterns to which a cell could respond cannot be presented. Indeed, even without identifying necessary features, this approach has recently provided insights into the topographical organization of pattern-selective cells in this part of the cortex. By repeating the simplification protocol at many cortical sites, Fujita and colleagues (1992) determined that closely adjacent cells (recorded simultaneously through one electrode) usually responded to very similar feature combinations. In vertical penetrations, they consistently recorded cells that responded to the same "optimal" stimulus as for the first cell tested, indicating that cells with similar stimulus preferences extend through most cortical layers. In tangential penetrations, cells with similar preferences were found in patches of approximately 500 μm. Fujita and colleagues (1992) found, in most of these tangential tracks, that cells immediately outside the cluster showed no response to the stimulus category preferred by that cluster but that, after intervals varying up to 1 mm, further clusters of cells could be found that were again activated by the same stimuli.

These findings imply that cells in IT, as elsewhere, are organized into functional columns or modules (Mountcastle, 1978), as has long been suspected, and the results have naturally been interpreted as revealing aspects of the organization of IT that bear on its role in recognition (Fujita et al., 1992; Stryker, 1992; Perrett and Oram 1993; Young, 1993b). According to the results of this method, cells in IT show a preference for iconic patterns that are simpler, in general, than the visual objects that confront an animal. One interpretation of these results, then, is that each cell or column in IT contributes the signaling of an iconic pattern "partial," whenever it is present. Thus, IT might consist of a large number of detectors of pattern partials, each of which provides a basis function for recognition (Fujita et al., 1992). Together the detectors constitute an "al-

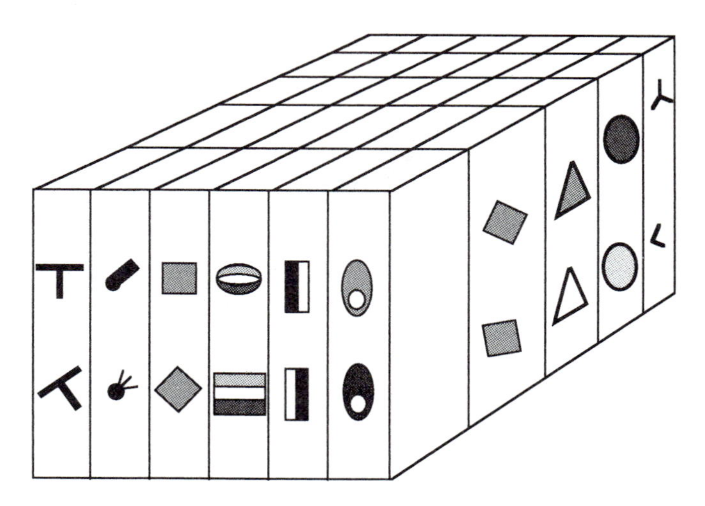

FIGURE 29.4 The visual alphabet conception of IT cortex. This schematic illustrates the division of IT into modules. The cells in each module respond to similar pattern partials and, in this way, are conceived to code a set of basis functions, or pattern primitives, that in combination might represent the appearance of a particular object (see, for example, Stryker, 1992), assuming that response to the presence of a pattern partial is not disturbed by the presence of other features in the visual field.

phabet in which our visual memories are written" (Stryker, 1992), so that the representation of any complex object would be a distributed array of cells each coding different partials (Perrett and Oram, 1993) (figure 29.4). Dividing down the areal extent of IT by the observed size of the modules reveals that the number of signaled partials in the alphabet is on the order of 1000 (Fujita et al., 1992; Stryker, 1992; Perrett and Oram, 1993). The idea is that, although, the number of partials is very small by comparison to the number of possible visual patterns that must be recognized, the number of possible combinations offered by this alphabet is very large. These combinations of partials could signal a practically unbounded set of visual patterns, in the same way that the number of words that can be constructed from an alphabet or syllabary is very large indeed. The representation of a common or garden-variety visual object would thus be conceived to be a coarsely coded population response across most or all of IT.

This idea has the virtues of simplicity and elegance, but it assumes that an IT cell will reliably signal the presence of the particular pattern partial that excites it regardless of whatever else is present in the visual field. Unfortunately, it is already clear from published work that this sometimes is not the case; an example is shown in figure 29.5 (Tanaka et al., 1991). In this example, the simplification protocol converged on an inverted T shape as the preferred pattern partial for this cell (see figure 29.5A). Using the model just outlined, any more complex object that contains this pattern partial should evoke a strong response from the cell as the cell contributes the information that the partial is present. Just such a more complex object is shown in figure 29.5C, together with the cell's response to it. The cell did not respond well to a plus sign, in which the preferred partial is still present, in concert with a bar below its center. Hence, the presence of other visual features can disrupt the response of a cell to its partial, a result that is the opposite of that assumed in the visual alphabet conception of IT.

In the case illustrated in figure 29.5, the cell may be signaling a combination of the presence of its partial *and* the absence of something else, perhaps in an exclu-

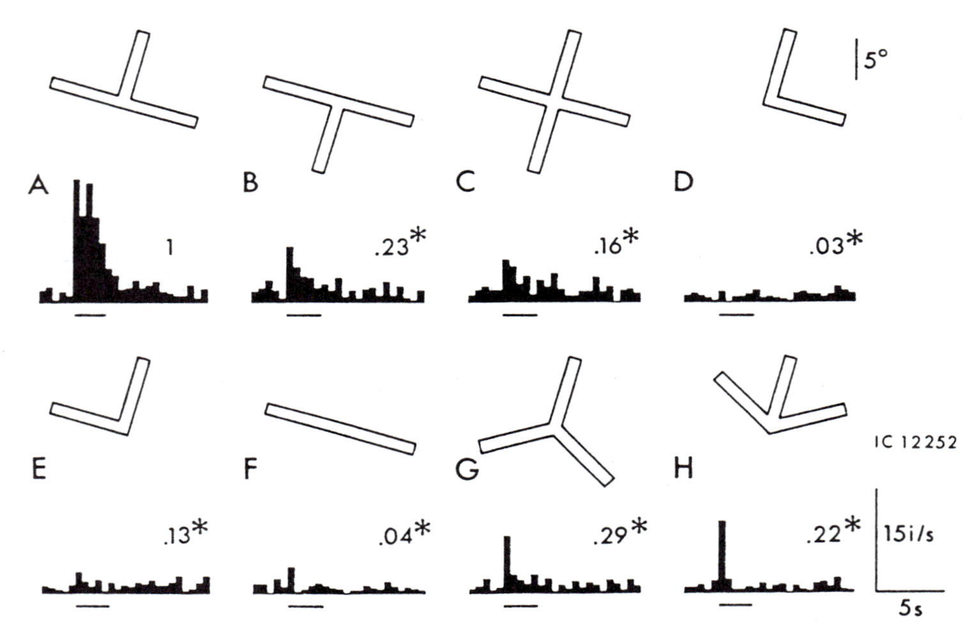

FIGURE 29.5 An experimental test of the assumption underlying the visual alphabet conception of IT. In this case, the simplification protocol converged on an inverted T shape (A) as the preferred pattern partial for this cell. The assumption underlying the visual alphabet conception of IT is that the cell will respond strongly to any more complex object that contains this partial. A more complex object containing this partial is shown at (C). The response of the cell to this only slightly more complex object is very small, despite the fact that the preferred pattern partial is still present. Clearly, this is in disagreement with the assumption underlying the visual alphabet conception of IT. (Reprinted with permission from K. Tanaka et al., Coding visual images of objects in the inferotemporal cortex of the macaque monkey. *J. Neurosci.* 6:134–144, 1991.)

sive-or computation. The characteristics of what must be absent for the cell to respond have not been defined, except to show that a bar below the center of the inverted T is a member of this set, and so the simplification protocol has gone only half-way toward defining the conditions sufficient for the cell's response. If other cells behave in a similar way, the characterization by this method of the pattern partials preferred by IT cells cannot be sufficient to account for the performance of the cells in the recognition of even slightly more complex objects.

These considerations suggest that, at present, the iconic characterizations of the simplification approach capture neither necessary nor sufficient descriptions of the behavior of IT cells in recognition. More work is needed to show in what way the characterization of cell preferences by the simplification method contributes to understanding the role IT cells play in object recognition.

THE NUMERICAL MODELING APPROACH A second systematic approach to the issue of how IT cells participate in recognition has been to quantify cells' responses to a set of stimuli to make a numerical model. This numerical model can then be compared by regression-like methods to other numerical models that capture the various ways in which the stimuli differ, to find which dimensions of the stimuli are good predictors of the physiological responses. This statistical modeling idea (Young and Tanaka, 1990) has been applied to both single-cell responses (Yamane, Kaji, and Kawano, 1988) and the responses of populations of cells (Young and Yamane, 1992, 1993) in IT. In essence, this approach involves giving up any attempt to identify

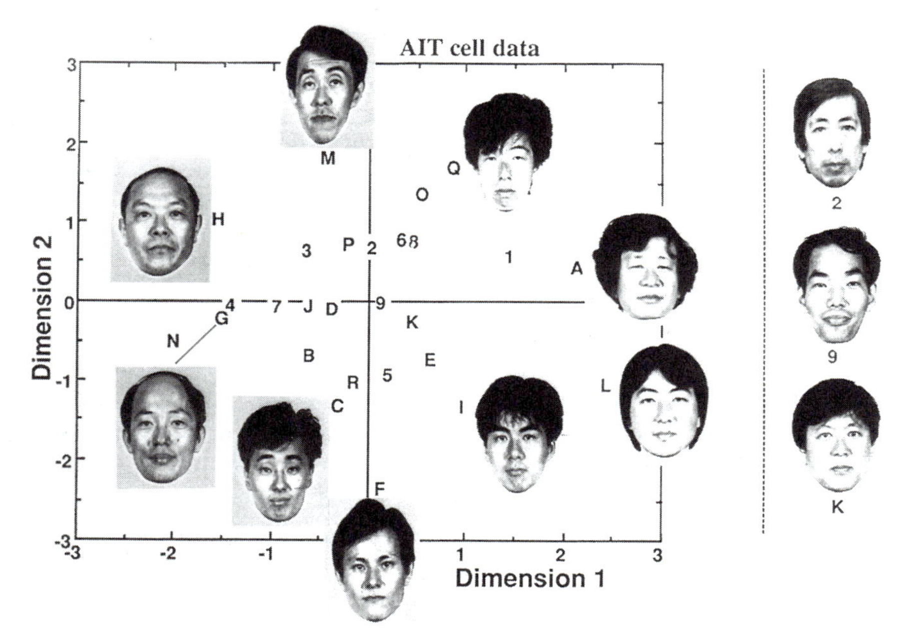

FIGURE 29.6 Diagram of the numerical model of the AIT population responses to faces. Single-unit activity was recorded during exposure to disembodied heads of Japanese men in full face. These faces were presented on a television screen for 600 ms, and responses were quantified as the difference between the mean firing rate in the prestimulus baseline and that in the 500-ms period beginning 100 ms after stimulus onset. Proximity data were derived from each matrix, in which cells were represented by columns and the response to each face by rows, by computing euclidean distances between the face responses. Nonmetric multidimensional scaling was applied to the proximity data, and configurations in one to six dimensions were produced so that analyses in different dimensions could be compared in a scree test. The scree test revealed that a two-dimensional solution was satisfactory, as it explains 70% of the variability in the data. In the figure, faces that evoked a similar pattern of response across the population are close together, whereas faces that evoked a very different population response are far apart. Example faces are shown at the position of the point corresponding to the population response to that face in the diagram or, for faces whose population response was represented near the center, at the side of the diagram. (Reprinted with permission from M. P. Young and S. Yamane, Sparse population coding of faces in the inferotemporal cortex. *Science* 256:1327–1331, 1992. © AAAS.)

the necessary conditions for cells' responses and, by examining the internal relations between a set of stimuli and the responses to them, determining whether the cells' activities are sufficient to carry information about the stimuli.

In the populational analysis, the responses of AIT cells to a set of 27 faces, whose physical characteristics had been extensively quantified (Yamane, Kaji, and Kawano, 1988), were examined. Unit recordings were made while the monkeys performed a face discrimination task, in which the monkeys responded at a greater than 90% correct performance level. A quantitative numerical model of the population responses to the faces was made by applying MDS to the data, an approach used previously for qualitative analysis of population encoding of complex stimuli (Baylis, Rolls, and Leonard, 1985). This analysis produced a two-dimensional configuration that accounted for 70% of the variance in the data. The configuration of points representing the similarity of the population responses to the face stimuli is shown in figure 29.6. Faces plotted close together evoked a similar pattern of response across the population, whereas faces plotted far apart evoked very different population responses.

The population responses were quantitatively compared with numerical models that encoded the properties of the faces by Procrustes rotation, a regression-like procedure (Gower, 1971). The general physical similarity model, computed from measurements of the faces, was statistically significantly related to the population response model, so the responses of these cells, en masse, were more similar the more physically similar were the faces. In addition, measurement variables encoding the relations between the eyes and the hairline were significantly related to the population model. No other models were significantly related. Hence, these analyses suggest that the AIT population may have been coding the general physical properties of the presented faces, with a particular emphasis on the upper part of the face (Young and Yamane, 1992; 1993).

The results suggest that neurons responsive to faces in AIT share dimensions of specificity and that these shared dimensions correspond to physical properties of the faces. It was also possible to show that these cells formed a readable population code that could sometimes identify particular faces. This was done by applying the population vector technique developed by Georgopoulos and colleagues (1982). The responses of each cell were plotted against the angle of each face stimulus in the derived space, and a sine function was fitted to the data. Single-cell vectors for each stimulus were derived from the peak angle of the sine function for each cell (for direction) and from the response of the cell to each stimulus (for length). A population vector for each stimulus was then computed by summation. Figure 29.7 shows plots of stimulus vectors, single-cell vectors, and population vectors for arbitrarily chosen stimuli. The direction of the population vectors was, in general, close to the direction of the stimulus vectors.

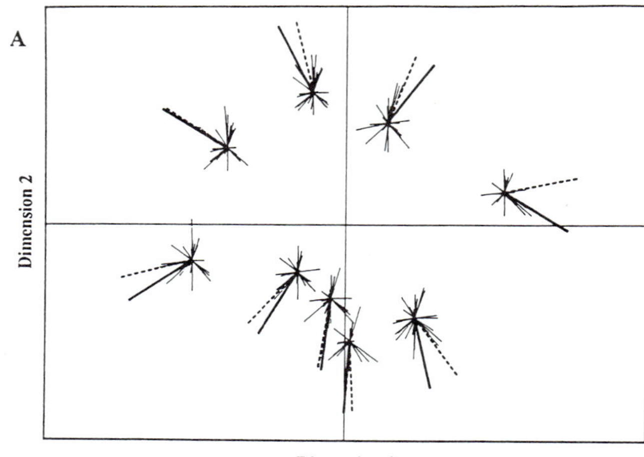

FIGURE 29.7 Population vector plots for the AIT data, showing single-cell vectors (light lines), population vectors (heavy lines), and stimulus vectors (dashed lines). Each set of vectors is plotted at the position of its corresponding face stimulus in the space. Both population vectors and stimulus vectors have been arbitrarily elongated to show more clearly the angular discrepancies between them. The cross hairs pass through the centroid of the space. Despite the small samples of cells, the population vectors typically matched their corresponding stimulus vectors closely, suggesting both that population encoding is present and that the population code is sparse. (Reprinted with permission from M. P. Young and S. Yamane, Sparse population coding of faces in the inferotemporal cortex. *Science* 256:1327–1331, 1992. © AAAS.)

The results of this quantitative approach suggest that the activity of cells responsive to faces in IT is best predicted by fairly complex combinations of cues related to the configuration of facial elements rather than to simple stimulus features and that cells participate in the recognition of faces by signaling in the form of a sparse population code. This approach, however, suffers from a number of specific limitations. For example, it may be applied to investigate only stimuli, such as faces, for which there is available a good deal of physical and psychological information about the ways they differ. For most of the objects that primates can recognize, information about the dimensions in which these objects differ is lacking, and so this approach cannot currently bear directly on the mechanisms of recognition in the general case. This is especially unfortunate because the representation of faces in IT may be different from the representation of other stimuli (Tanaka et al., 1991). In addition, the method is very laborious, requiring a numerical characterization of all the candidate models, and therefore, as in other approaches to investigating object coding in IT, the method requires a priori assumptions about what elements or primitives the cells might be interested in. The populational version of this approach is poor at examining the dependence of cell responses on combinations of features or dimensions, because there is no extension of multiple regression that can accommodate more than one dependent variable.

Conclusions

It is fairly certain that the classical occipitotemporal areas form a discriminable, hierarchically organized subsystem, but the functional details of ascending, descending, local, and callosal interactions in this pathway are presently patchy. Similarly, the functional relations between ventral stream areas and other structures, some of which appear necessary for intact recognition and discrimination performance, are uncertain. As for how cells in these areas participate in object recognition, information on processing stimuli other than faces is difficult to interpret. Depending on whether one sees the proverbial glass as half-empty or half-full, these considerations imply either that rather little progress has been made in understanding the neural mechanisms of visual recognition or that there exist many opportunities for exciting and fundamental work in this field.

ACKNOWLEDGMENTS This work was supported by a Royal Society University Research Fellowship, the Medical Research Council, and the Oxford McDonnell-Pew Centre for Cognitive Neuroscience. I am very grateful to Colin Blakemore, Ichiro Fujita, Anya Hurlbert, Andrew Parker, and Keiji Tanaka for their comments on earlier versions of this chapter.

REFERENCES

BARLOW, H. B., 1985. The role of single neurons in the psychology of perception. *Q. J. Exp. Psychol. [A]* 37:121–145.

BAYLIS, G. C., E. T. ROLLS, and C. M. LEONARD, 1985. Selectivity between faces in the responses of a population of neurons in the cortex of the superior temporal sulcus of the monkey. *Brain Res.* 342:91–102.

BOUSSAOUD, D., R. DESIMONE, and L. G. UNGERLEIDER, 1991. Visual topography of area TEO in the macaque. *J. Comp. Neurol.* 306:554–575.

BRUCE, C., R. DESIMONE, and C. G. GROSS, 1981. Visual properties of neurons in a polysensory area in superior temporal sulcus of the macaque. *J. Neurophysiol.* 46:369–384.

DESIMONE, R., T. D. ALBRIGHT, C. G. GROSS, and C. J. BRUCE, 1984. Stimulus-selective properties of inferior temporal neurons in the macaque. *J. Neurosci.* 8:2051–2068.

DE YOE, E. A., and D. C. VAN ESSEN, 1988. Concurrent processing streams in monkey visual cortex. *Trends Neurosci.* 11:219–223.

FELLEMAN, D. J., and D. C. VAN ESSEN, 1991. Distributed hierarchical processing in the primate cerebral cortex. *Cerebral Cortex* 1:1–47.

FUJITA, I., K. TANAKA, M. ITO, and K. CHENG, 1992. Columns for visual features of objects in monkey inferotemporal cortex. *Nature* 360:343–346.

GEORGOPOULOS, A. P., J. F. KALASKA, R. CAMINITI, and J. T. MASSEY, 1982. On the relations between the direction of two-dimensional arm movements and cell discharge in primate motor cortex. *J. Neurosci.* 2:1527–1537.

GOCHIN, P. M., E. K. MILLER, C. G. GROSS, and G. L. GERSTEIN, 1991. Functional interactions among neurons in macaque inferior temporal cortex. *Exp. Brain Res.* 84: 505–516.

GOWER, J. C., 1971. Statistical methods of comparing different multivariate analyses of the same data. In *Mathematics in the Archeological and Historical Sciences*, D. G. Kendall, and P. Tavtu, eds. Edinburgh: Edinburgh University Press, pp. 138–149.

GROSS, C. G., 1972. Visual functions of the inferotemporal cortex. In *Handbook of Sensory Physiology*, Vol. 7, part B, R. Jung, ed. Berlin: Springer, pp. 451–482.

GROSS, C. G., C. E. ROCHA-MIRANDA, and D. B. BENDER, 1972. Visual properties of neurons in the inferotemporal cortex of the macaque. *J. Neurophysiol.* 35:96–111.

HEYWOOD, C. A., A. GADOTTI, and A. COWEY, 1992. Cortical area V4 and its role in the perception of colour. *J. Neurosci.* 12:4056–4065.

KAAS, J. H., and M. F. HUERTA, 1988. The subcortical visual system of primates. *Comp. Primate Biol.* 4:327–391.

LUND, J. S., J. B. LEVITT, and T. YOSHIOKA, (in press). Topography of excitatory and inhibitory connectional anatomy in visual cortex as a crucial determinant of functional organization: An issue urgently in need of modelers' attention. In *Computational Vision Based on Neurobiology* (SPIE technical proceedings).

Martin, K. A. C., 1992. Visual cortex: Parallel pathways converge. *Curr. Biol.* 2:555–557.

Maunsell, J. H., and D. C. Van Essen, 1983. The connections of the middle temporal visual area (MT) and their relationship to a cortical hierarchy in the macaque monkey. *J. Neurosci.* 3:2563–2586.

Mignard, M., and J. G. Malpeli, 1991. Paths of information flow through visual cortex. *Science* 251:1249–1251.

Mishkin, M., and T. Appenzeller, 1987. The anatomy of memory. *Sci. Am.* 256:62–72.

Miyashita, Y., 1988. Neuronal correlate of visual associative long-term memory in the primate temporal cortex. *Nature* 335:817–820.

Miyashita, Y., and H. S. Chang, 1988. Neuronal correlate of pictorial short-term memory in the primate temporal cortex. *Nature* 331:68–70.

Mountcastle, V. B., 1978. An organizing principle for cerebral function: The unit module and the distributed system. In *The Mindful Brain*, V. B. Mountcastle and G. M. Edelman, eds. Cambridge, Mass.: MIT Press, pp. 7–50.

Perrett, D. I., A. J. Mistlin, and A. J. Chitty, 1987. Visual cells responsive to faces. *Trends Neurosci.* 10:358–364.

Perrett, D. I., and M. W. Oram, 1993. Neurophysiology of shape processing. *Image Vis. Comp.* 11:317–333.

Perrett, D. I., E. T. Rolls, and W. Caan, 1982. Visual neurones responsive to faces in the monkey temporal cortex. *Exp. Brain Res.* 47:329–342.

Ridley, R. M., H. F. Baker, and T. K. Murray, 1988. Basal nucleus lesions in monkeys: Recognition memory impairment or visual agnosia? *Psychopharmacology* 95:289–290.

Rockland, K. S., and D. N. Pandya, (1979) Laminar origins and terminations of cortical connections of the occipital lobe in the rhesus monkey. *Brain Res.* 179:3–20.

Rolls, E. T., 1992. Neurophysiological mechanisms underlying face processing within and beyond the temporal cortical visual areas. *Philos. Trans. R. Soc. Lond.* 335:11–21.

Rolls, E. T., G. C. Baylis, and C. M. Leonard, 1985. Role of low and high spatial frequencies in the face-selective responses of neurons in the cortex in the superior temporal sulcus in the monkey. *Vision Res.* 25:1021–1035.

Sagy, G., R. Vogels, and G. A. Orban, 1993. Cue invariant shape selectivity of macaque inferior temporal neurons. *Science* 260:995–997.

Scannell, J. W., and M. P. Young, 1993. The connectional organization of neural systems in the cat cerebral cortex. *Curr. Biol.* 3:191–200.

Scannell, J. W., M. P. Young, and C. Blakemore, 1992. Optimization analysis of the connections between areas of cat cerebral cortex. *Soc. Neurosci. Abstr.* 18:313.10

Schiller, P. H., and K. Lee, 1991. The role of the primate extrastriate area V4 in vision. *Science* 251:1251–1253.

Shepard, R. N., 1962. Multidimensional scaling with an unknown distance function. *Psychometrika* 27:125–140.

Stryker, M. P., 1992. Elements of visual perception. *Nature* 360:301–302.

Tanaka, K., H. Saito, Y. Fukada, and M. Moriya, 1990. Integration of form, texture and color information in the inferotemporal cortex of the macaque. In *Vision Memory and the Temporal Lobe*, E. Iwai, and M. Misitkin, eds. New York: Elsevier, pp. 101–109.

Tanaka, K., H. Saito, Y. Fukada, and M. Moriya, 1991. Coding visual images of objects in the inferotemporal cortex of the macaque monkey. *J. Neurosci.* 6:134–144.

Toyama, K., M. Kimura, and K. Tanaka, 1981. Organization of cat visual cortex as investigated by cross-correlation technique. *J. Neurophysiol.* 46:202–214.

Ungerleider, L. G., and M. Mishkin, 1982. Two cortical visual systems. In *Analysis of Visual Behavior*, D. G. Ingle, M. A. Goodale, and R. J. Q. Mansfield, eds. Cambridge, Mass.: MIT Press, pp. 549–586.

Yamane, S., S. Kaji, and K. Kawano, 1988. What facial features activate face neurons in the inferotemporal cortex of the monkey? *Exp. Brain Res.* 73:209–214.

Young, F. W., and D. F. Harris, 1990. Multidimensional scaling. In *SPSS Base System User's Guide*, M. J. Norusis, ed. Chicago: SPSS Inc.

Young, M. P., 1992a. Objective analysis of the topological organization of the primate cortical visual system. *Nature* 358:152–155.

Young, M. P., 1992b. Optimization analysis of the organization of cortico-cortical connections between areas of monkey cerebral cortex. *Soc. Neurosci. Abstr.* 18:313.9.

Young, M. P., 1993a. The organization of neural systems in the primate cerebral cortex. *Proc. R. Soc. Biol. Sci.* 252:13–18.

Young, M. P., 1993b. Visual cortex: Modules for pattern recognition. *Curr. Biol.* 3:44–46.

Young, M. P., and J. W. Scannell, 1993. Analysis and modelling of the organization of the mammalian cerebral cortex. In *Experimental and Theoretical Advances in Biological Pattern Formation*, P. Maini, ed. New York: Springer-Verlag.

Young, M. P., and K. Tanaka, 1990. Derivation of population vectors and population distributions from the response specificities of visual cortical cells. *Soc. Neurosci. Abstr.* 16: 502.16.

Young, M. P., and S. Yamane, 1992. Sparse population coding of faces in the inferotemporal cortex. *Science* 256: 1327–1331.

Young, M. P., and S. Yamane, 1993. An analysis at the population level of the processing of faces in the inferotemporal cortex. In *Brain Mechanisms of Perception and Memory: From Neuron to Behavior*, T. Ono, L. R. Squire, D. I. Perrett, and M. Fukuda, eds. Oxford: Oxford University Press.

Zeki, S., and S. Shipp, 1988. The functional logic of cortical connections. *Nature* 335:311–317.

30 Multiple Memory Systems in the Visual Cortex

ROBERT DESIMONE, EARL K. MILLER, LEONARDO CHELAZZI, AND ANDREAS LUESCHOW

ABSTRACT Attention gates access to memory, but the converse is also true: The contents of memory guide attention. Memory for objects is reflected both in the maintained activity of inferotemporal (IT) cells in the absence of any stimuli and in their response to current stimuli. New or not recently seen stimuli preferentially activate adaptive memory cells in IT cortex, and the output of these cells may drive attentional systems, causing the organism to orient to the new stimuli in a scene. Many times, though, we need to search for a particular familiar object and suppress orienting to novel ones. In this case, top-down mechanisms are able to bias those cells that code the searched-for item, resulting in enhanced activation when the stimulus occurs. Such top-down feedback may derive from prefrontal cortex. An even more pronounced case of memory-guided attention occurs in visual search, where the representation of a target item in memory is used to guide the search of an array containing several stimuli. The interaction between the neural representation of the array and the memory trace of the target results in the target's "capturing" the responses of IT cells. Thus, interactions between memory and attention in IT cortex result in the selection of objects that are foveated and acted upon.

Generally speaking, we remember those things to which we attend and, conversely, we often attend to those things that stir our memory. Indeed, attention and memory are so closely intertwined that their mechanisms may be nearly indistinguishable in visual cortex. To understand memory, then, it is useful to consider some basic constraints on visual processing and the role of attention.

Despite the massively parallel architecture of the visual cortex, its ability to process fully, discriminate, and store in memory independent objects at the same time is surprisingly limited. These late limited-capacity stages of vision require focal attention. Fortunately, though, only a small portion of the crowded visual environment usually is relevant for behavior at a given moment in time. Objects in a scene must *compete* for focal attention as well as behavioral action, and this competition appears to be carried out at many levels of the visual system.

This competition among stimuli involves both automatic and cognitive components, which can be loosely regarded as *bottom-up*, or preattentive, processes and *top-down*, or attentive, ones. Preattentive processes underlie much of figure-ground segregation and depend largely on the intrinsic properties of stimuli in the scene itself. The competition is weighted toward stimuli that differ from their background, which will tend to attract focal attention. A red apple on a green field, for example, will automatically stand out in the scene. This type of figure-ground segregation commonly is believed to involve neurons whose response to an otherwise effective stimulus in their receptive field may be suppressed by the presence of similar stimuli in the receptive field surround (Allman, Miezin, and McGuinness, 1985; Desimone et al., 1985).

In addition to preattentive scene segmentation, attentive processes are also engaged in selecting relevant objects. Attentive selection processes—at least the ones we are considering—are driven by the task at hand rather than by the stimuli themselves. One can attentively bias the competition among stimuli in favor of a particular item of food, for example, that does not pop out from its background preattentively based on a difference in luminance, color, motion, and so on. In fact, attentive processes can override preattentive ones. A red apple on a green field will be extracted preattentively and compete strongly for focal attention, but a gardener may choose to ignore the apple and attend to the field.

ROBERT DESIMONE, EARL K. MILLER, LEONARDO CHELAZZI, and ANDREAS LUESCHOW Laboratory of Neuropsychology, National Institute of Mental Health, Bethesda, Md.

We will describe recent neurophysiological evidence suggesting analogous automatic and cognitive processes within memory. An adaptive memory mechanism sensitive to stimulus repetition automatically biases visual processing toward novel or infrequent stimuli. Cognitive, working memory mechanisms come into play when one searches for a particular item in a temporal sequence. The former mechanism seems to operate preattentively, whereas the latter mechanism may be equivalent to attentive selection. Together, these two types of mnemonic processes influence which object in a crowded scene will capture attention.

Novelty and familiarity

All our experiments described in this chapter are based on variations of the delayed matching-to-sample (DMS) task. In the standard form of DMS, a sample stimulus is followed, after a delay, by a test stimulus, and the animal is rewarded for indicating whether it matches the sample. Some of the important variations include whether the test stimuli are novel or familiar, whether the delay periods are blank or filled with intervening items, whether the test stimuli are presented one at a time or in a spatial array, and whether the delay is long or short. In all the studies to be described, animals were required to maintain fixation throughout the trial, unless an eye movement was specifically used as a behavioral response. Furthermore, all the inferotemporal (IT) recordings were in the most anteroventral portion, a region known to be important for visual memory (Horel et al., 1987; Zola-Morgan et al., 1989; Gaffan and Murray, 1992; Suzuki et al., 1993; Meunier et al., 1993).

To test whether IT neurons might play a role in recognition memory, we recorded their responses to novel and familiar stimuli (Miller, Li, and Desimone, 1991; Li, Miller, and Desimone, 1993). Although we used the DMS task, the main purpose of the task was to engage the monkeys' attention to the stimuli. Each cell was tested with a set of stimuli that the animal had never seen before, and our interest was in the incidental memories that would accrue as the animals gradually became familiar with these new stimuli during the session. We therefore concentrated on across-trial changes in response to the sample stimuli rather than on within-trial effects.

The sample stimuli were a set of 20 digitized complex objects (shapes, faces, complex patterns, etc.) presented on a computer display. A new set was chosen randomly for each cell. Each trial began with a sample stimulus, followed by a sequence of one to four test stimuli, the final one of which was always a match. On average, a single trial lasted approximately 4 seconds, and a trial with a given sample stimulus was repeated in the session after 3 or 35 intervening trials with other stimuli. The nonmatching stimuli in each trial were chosen from a fixed set of highly familiar stimuli, so that the novel stimuli occurred only as samples and matching test items.

IT cells code object features such as shape, color, and texture (Gross, Rocha-Miranda, and Bender, 1972; Schwartz et al., 1983; Desimone et al., 1984; Tanaka et al., 1991) and, correspondingly, virtually every cell studied gave selective responses to the different stimuli used in the task, responding better to some stimuli than to others. In addition to these sensory effects on responses, there were strong memory effects. For approximately one third of the cells in anterior IT cortex, responses to the initially novel stimuli systematically declined over the session, as the stimuli became familiar.

On average, responses reached a minimum (at 40% of the peak response) after about six to eight trials of a given sample stimulus, showing few further changes in subsequent trials. The memory of a given stimulus was long-lasting; responses to a familiar stimulus remained suppressed even after 35 intervening trials, or 150 intervening stimuli, which was the maximum tested. Furthermore, these effects were highly stimulus-specific. When we introduced new stimuli after responses had reached a floor with the initial set, responses to the new set were at full strength. An example of this is shown in figure 30.1. We have termed this stimulus-specific decrement in response with familiarity *adaptive mnemonic filtering*. It is worth pointing out, though, that the cells were not novelty detectors, in the sense of cells that respond best to any novel stimulus. Rather, they gave their best responses to stimuli that were both novel and had the appropriate sensory features. Most of the remaining cells showed no change in response over the session. Rolls and colleagues (1989) have described similar systematic changes in the responses of face-selective cells in lateral IT cortex as monkeys have become familiar with new faces, and Riches, Wilson, and Brown (1991) have described response decrements with familiarity in IT cortex of monkeys passively viewing stimuli.

Figure 30.2 shows how the adaptive mnemonic filter

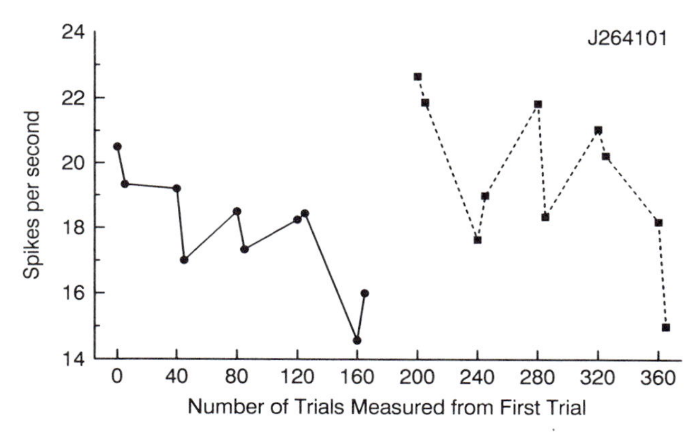

FIGURE 30.1 Average responses of a single IT neuron to repeated presentations of two sets of 20 initially novel stimuli. The solid line shows the average response to the first 10 presentations of the stimuli from the first set, and the dashed line shows the average response to the first 10 presentations of stimuli from the second set. After the response to the first set had reached a floor, the second set was substituted for the first. The response rebounded and then declined as the new stimuli became familiar. The staircase appearance of the lines is caused, in part, by the different number of intervening trials (3 or 35) between repetitions of a given stimulus. Response decrements were typically larger with the smaller number of intervening trials. (Adapted from Li, Miller, and Desimone, 1993.)

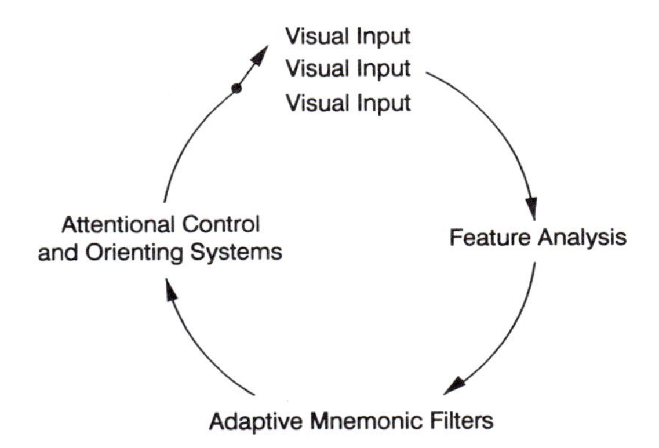

FIGURE 30.2 Interaction between systems for memory and attention. Out of the multiple visual inputs entering extrastriate cortex in parallel, a novel stimulus will maximally activate IT adaptive mnemonic filter cells, which will, in turn, drive attentional and orienting systems. This leads to increased attention and, typically, foveation of the novel stimulus. As the stimulus becomes familiar, activation of adaptive mnemonic filter cells decreases, which reduces the drive on the attentional and orienting systems. Once the drive has sufficiently diminished, the system is free to process other, competing stimuli.

mechanism in IT cortex may drive attentional and orienting systems, biasing the organism toward novel stimuli. As the organism learns about the new stimulus, activity among IT cells will decline, reducing the drive on the orienting systems and freeing the organism to orient to another new stimulus. Thus, it is behavior that completes the loop, resulting in an organism driven to seek out new stimuli and incorporate knowledge about them into the structure of the cortex. Such a scheme has some similarities to the adaptive resonance theory (ART) of Carpenter and Grossberg (1987). In ART also, novel stimuli cause an activation of attentional systems, which allows learning in the system to take place. The threshold for activating attentional systems by novelty in ART is internally adjustable, to keep already learned categories of stimuli from being modified by stimuli that differ only slightly from those that have been previously seen.

How could response decrements among IT cells result in memory formation? Ironically, it may be the cells whose responses do not change with familiarity that actually carry the memory of the learned stimulus. The first experience with a stimulus may activate some cells that do not code its most critical features. As these critical features are learned with experience, responses of the inappropriate cells largely drop out, which may be the cells whose responses decline in our recordings. This would be analogous to the situation in neural development in which initial prolific anatomical connections are pruned back to the set that carries the best information.

A related question is the relationship of these response changes to the different types of memory defined at the psychological and neuropsychological level. A broad distinction usually is made between explicit memory, or memory for specific facts and events, and implicit memory, or memory underlying procedural learning, perceptual learning, habit formation, priming, habituation, and so on (Mishkin, 1982; Zola-Morgan and Squire, 1984, 1985; Squire, Knowlton, and Musen, 1993). Neural processes that change with familiarity might contribute to any or all of the behaviorally defined memory systems. To pinpoint this contribution will require not only further experimental work but also better theoretical models of how the different types of memory are established. Neurophysiology is sometimes like the game-show "Jeopardy": It provides answers and the trick is to figure out the right questions.

Short-term memory

Most studies of memory in IT cortex have been directed at short-term memory—that is, memory lasting, at most, a few seconds. In animals performing the DMS task, responses of IT neurons to test stimuli vary depending on whether or not they match the sample (Gross, Bender, and Gerstein, 1979; Mikami and Kubota, 1980; Baylis and Rolls, 1987; Miller, Li, and Desimone, 1991; Riches, Wilson, and Brown, 1991; Eskandar, Richmond, and Optican, 1992; Miller, Li, and Desimone, 1993). Furthermore, some studies have found high maintained activity in IT cortex during the retention interval, as though the cells are actively maintaining a memory of the sample (Fuster and Jervey, 1981; Miyashita and Chang, 1988; Fuster, 1990). In nearly all the studies showing such effects, the animals viewed a blank screen during the delay periods. However, outside the laboratory, our retention intervals seldom are empty but rather are filled with new stimuli entering the visual system and presumably activating the same neurons that are involved in storing, retrieving, and recalling memories. The problem is particularly perplexing because Baylis and Rolls (1987) found that intervening stimuli in the retention interval eliminated the differential responses of neurons in lateral IT cortex to matching and nonmatching stimuli.

Delayed Matching to Sample with Intervening Stimuli We re-examined this question in anteroventral IT cortex using a DMS paradigm with intervening stimuli similar to that used in our study of familiarity but with a few necessary differences (Miller, Li, and Desimone, 1991, 1993). Unlike in the familiarity study, the stimuli were already familiar to the animal and, presumably, any change in response with familiarity had already taken place before the recordings. Here we were interested in the maintenance of the sample memory for the few seconds within each trial.

The monkey viewed a sequence of one to six test stimuli following the sample and was rewarded for responding to a matching test stimulus. For example, the monkey might see a sample stimulus *A* followed by a sequence of test stimuli—*B*, *C*, *D*, *E*, *A*—and would be rewarded for choosing *A* as the match. Unlike in the study of familiarity, the same stimuli that appeared as sample and matching stimuli on one trial appeared as nonmatching stimuli on others. The stimuli used for each neuron were a set of six digitized complex objects randomly chosen from a larger set of familiar stimuli. Each stimulus was on for 500 ms, and the blank delays between pairs of stimuli in the sequence lasted 700 ms.

Consistent with previous reports (Fuster and Jervey, 1981; Miyashita and Chang, 1988; Fuster, 1990), many cells had stimulus-specific maintained activity in the delay interval following the sample. For example, if a cell responded better to sample *A* than to sample *B*, it often had higher maintained activity in the delay following *A* than in the delay following *B*. However, this maintained activity was abolished after the first intervening item in the sequence. Thus, maintained activity in IT cortex is unlikely to be the sole mechanism for the maintenance of short-term memories, at least under these conditions. What, then, is its function? One clue is that the activity is under the monkey's voluntary control; it disappears when the monkey does not need to remember the stimulus (Chelazzi and Desimone, unpublished data). There is also some preliminary evidence that the activity may be more prevalent when the animal must compare stimuli across different retinal locations and that it may be predictive of whether the animal performs the trial correctly, particularly with long delays (Colombo and Gross, personal communication; Lueschow and Desimone, unpublished data). Maintained activity in IT cortex may reflect a type of visual rehearsal, which is easily disrupted (just as rehearsal of a telephone number is easily disrupted by hearing new numbers) but is nonetheless an important aid to short-term memory. As we will describe later, it seems that memories of past stimuli are more commonly expressed in IT responses to current stimuli.

Nearly all the cells gave stimulus-selective responses to the different stimuli used in the task. For half the cells, responses to the test stimuli did not vary significantly depending on whether they were matching or nonmatching. These cells appeared to communicate only sensory information. However, for the other half, responses to test stimuli were a joint function of the sensory features of that stimulus and stored memory traces. For the overwhelming majority of cells showing such effects, responses to matching stimuli were suppressed compared to nonmatching stimuli (figure 30.3). (Only a few cells showed effects in the opposite direction.) Furthermore, this suppression was maintained even when up to five other stimuli intervened between the sample and the final match, which was the maximum tested. This is the same direction of response

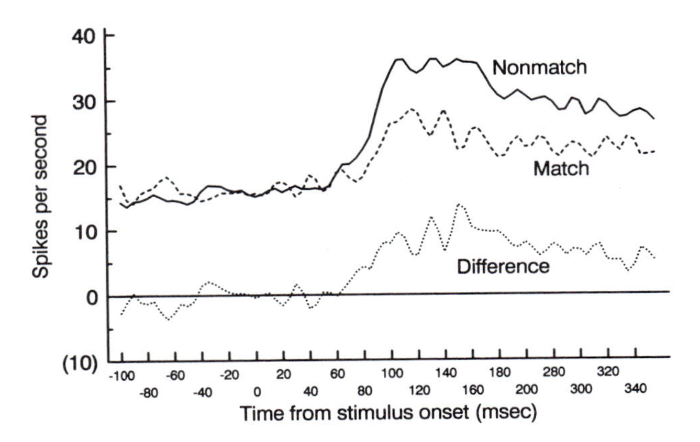

FIGURE 30.3 Average responses of a population of adaptive mnemonic filter cells in IT cortex to matching and nonmatching stimuli. The difference line plots the difference between the two histograms. The suppression of response to matching stimuli begins almost immediately with the start of the sensory-evoked response. (Adapted from Miller, Li, and Desimone, 1993.)

change with memory we found in the study of familiarity, but in this case the suppression was caused by the recent presentation of an already familiar stimulus. The fact that the suppression survived intervening stimuli in our study but not in the one by Baylis and Rolls (1987) is presumably due to the difference in recording sites—anteroventral IT cortex in our case versus lateral IT cortex in the Baylis and Rolls study.

The response suppression to matching stimuli is not based on a simple, or low level, comparison to the sample memory. Preliminary evidence shows that responses to matching test items are suppressed even when they are presented at a different retinal location from the sample or in a different size (Lueschow, Miller, and Desimone, 1993).

The comparison of sample and test stimulus may be high-level, but it also is very fast. The suppressive effects begin extremely rapidly in IT cortex—virtually at the onset of the visual response, which starts as soon as 80 ms after stimulus onset (see figure 30.3). This rapidity argues against the idea that the suppressive effects are due to on-line feedback to IT cortex from other structures during the test stimulus response. Sensitivity to repetition may be an intrinsic property of the visual cortex.

We argued earlier that the decrement of IT responses with familiarity may bias the organism to attend to novel stimuli. Much the same can be said of the decrements that are found with repetition of familiar items. Together, these properties may function as a type of temporal figure-ground mechanism that causes new or unexpected items to stand out in the visual scene, just as the spatial receptive field–surround structure of cortical neurons causes stimuli with contrasting color or motion to stand out. In a sense, the past may function as a type of surround that is compared with the present stimulus in the receptive field. This type of memory could serve as a preattentive stimulus filter operating in parallel over the visual field. Stand-out based on stimulus novelty has been shown in psychophysical studies in people (Wang, Cavanagh, and Green, 1992). Likewise, several lines of physiological evidence suggest that the matching effects in IT cortex are not the result of a cognitive memory mechanism but rather an automatic storage and retrieval process sensitive to stimulus repetition.

First, the effect of the sample memory on subsequent responses is not limited to matching stimuli: All responses, both matching and nonmatching, are affected by the sample memory trace, a result that has also been reported recently by Eskandar, Richmond, and Optican (1992). There is suggestive evidence that the more similar the test stimulus to the sample (whether matching or nonmatching), the greater the suppression, as though the memory trace acts like a filter through which all stimuli must pass. The matching test stimulus may show the most suppression simply because it has the greatest similarity to the previously seen sample.

Second, suppressive effects of stimulus repetition are found in IT cortex in animals passively viewing stimuli and even in animals under anesthesia (Miller, Gochin, and Gross, 1991; Riches, Wilson, and Brown, 1991). Thus, a volitional mechanism does not seem to be required for memory storage, though there may be a voluntary "reset" of the repetition suppression between trials. (We did not find suppressive effects spanning trials when the stimuli were already familiar to the monkey.)

Third, suppressive effects of repetition begin early in the visual system. Nelson (1991) has reported orientation-specific suppression lasting a few hundred milliseconds in V1 of the cat. Maunsell and coworkers (1991) and Haenny and Schiller (1988) have also reported stimulus repetition effects on V4 responses, although these are not always suppressive. We speculate that both the temporal range over which comparisons are made and the complexity of the features compared increases as one moves anteriorly through the visual system. Our results, together with the results of Baylis and Rolls (1987) argue that the temporal range of these effects is greatest in anteroventral IT cortex. More direct evidence that the adaptive memory mechanism in IT cortex works automatically is presented in the next section.

Dual Mechanisms of Short-term Memory As described previously, responses to matching test stimuli are suppressed compared to nonmatching ones. If this match suppression were due to a cognitive, or voluntary, working memory mechanism, we would expect the suppression to be confined to stimuli that matched the sample and not be present for stimuli that matched other, irrelevant stimuli in the trial sequence. If, on the other hand, suppression is due to a temporal figure-ground mechanism sensitive to any stimulus repetition, then the responses to even nonmatching test stimuli might be suppressed if they matched other test stimuli within the trial.

This was tested in IT recordings in two short-term memory paradigms (Miller and Desimone, 1994). The first was the same used in the original study described earlier. That is, a sample stimulus, *A*, was followed by several test stimuli—*B*, *C*, *D*, *E*, *A*—and the animal indicated which test stimulus matched the sample. We will refer to these as *standard trials*. The second paradigm was a variation of the first, in which the intervening test stimuli sometimes matched each other. That is, sample *A* might be followed by *B*, *C*, *B*, *D*, *A*, and the animal would have to respond to the final matching *A*, ignoring the repeated *B*s. We will refer to these as *ABBA trials*. Standard and ABBA trials were intermixed within the session.

We introduced the ABBA trials in two animals that had been trained previously with only the standard trials. Surprisingly, we found that both very often responded to the repeated *B*s in the ABBA trials. That is, the animals seemed to have learned the standard task by following a simple repetition rule—namely, respond to any repeated stimulus within the sequence. It fact, it may be easier for the animal to perform the task using an automatic repetition mechanism than to use a mechanism that requires active maintenance of the sample memory. This raises the question of what strategy has been used by animals in all previous behavioral or neurophysiological studies using the DMS task.

After the animals were given additional training and learned to perform the ABBA trials correctly, we recorded from cells in the same anteroventral portion of IT cortex studied in our earlier short-term memory study. As in that earlier study, many cells showed stimulus-specific activity in the delay following the sample, but this activity was abolished by intervening stimuli. Examining the test responses, half the cells showed significant effects of the sample on subsequent test stimulus responses, the same proportion as in the earlier study. Furthermore, within the class showing significant memory effects, the majority showed suppressed responses to the matching test stimuli, as found previously. However, the ABBA trials demonstrated that the responses of these cells were equally suppressed by the repeated *B*s (i.e., the repeated nonmatching stimuli), even though the monkeys had learned to ignore the repeated nonmatches and respond to only the one item that matched the sample memory. This result argues that mnemonic suppression is caused by simple stimulus repetition.

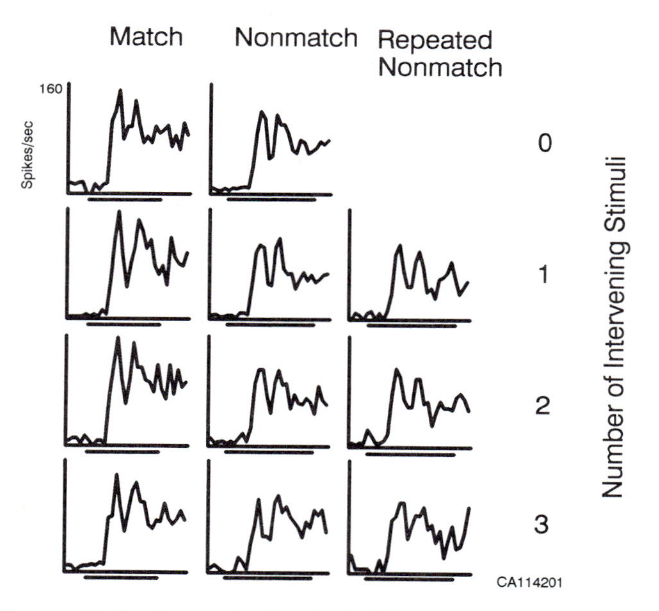

Figure 30.4 Example of a single IT neuron showing mnemonic enhancement in the ABBA task. Histograms show the responses to an individual stimulus appearing as a match, nonmatch, or repeated nonmatch following zero to three intervening stimuli. The bar beneath each histogram indicates when the stimulus was on. This neuron gave enhanced responses to matching stimuli and weaker responses to both nonmatches and repeated nonmatches.

What, then, mediates the animals' ability to perform the ABBA trials? As we found in our previous study of short-term memory, a subgroup of cells with significant memory effects gave enhanced responses to the matching test stimulus. Although these cells were relatively uncommon (fewer than 9% of the cells with memory effects) in our earlier study using standard trials only, in this new study with ABBA trials they comprised 35% of the cells with significant memory effects. Furthermore, the responses of these cells were enhanced only for the one test stimulus that matched the sample held in memory. There was no enhancement for repeated nonmatches (i.e., the repeated *B*s in the sequence). These cells were biased, or primed, to detect just the one item for which the monkey was looking (i.e., the stimulus that matched the sample memory) (figure 30.4).

Two findings indicate that these enhanced responses to the match stimulus are not directly linked to the behavioral response. First, responses to the intervening nonmatch stimuli are not enhanced on error trials when the animal mistakenly responds to them. Second, the enhancement occurs at the very onset of the visual response (i.e., at approximately 80 ms after stimulus onset), which seems far too early for the stimulus to have been discriminated and the behavioral response chosen.

In conclusion, there are at least two, and perhaps three, short-term memory mechanisms in IT cortex (figure 30.5). One is an automatic suppressive mecha-

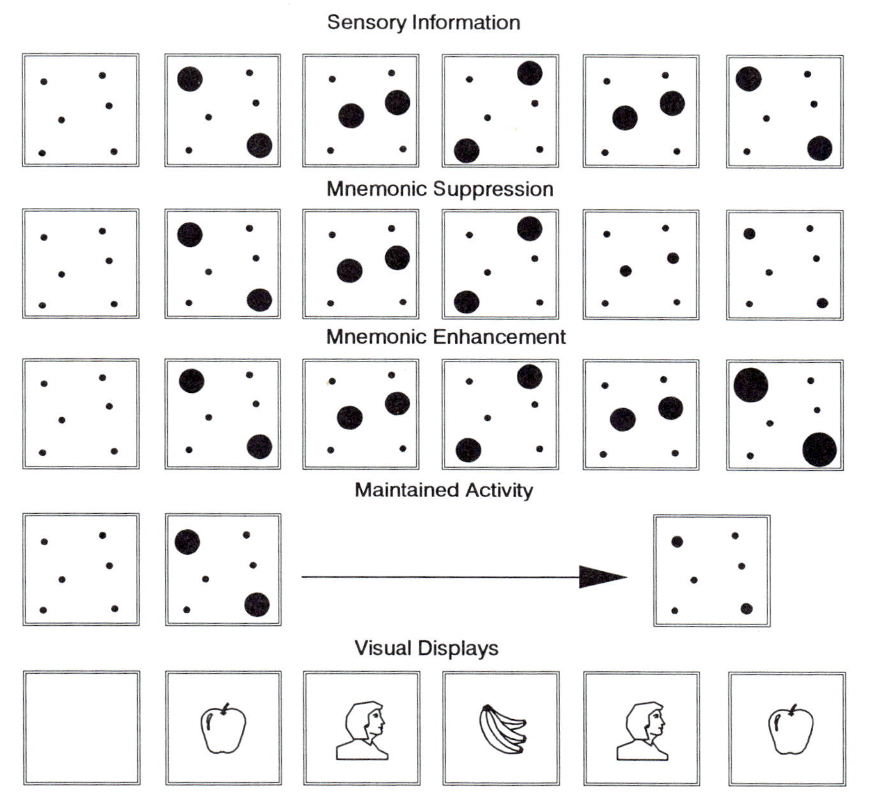

FIGURE 30.5 The top four rows are schematic representations of activity in a theoretical population of neurons during performance of a DMS task. Stimulus sequences are shown in the bottom row. Each of the circles represents a neuron, and the size of the circle represents its level of activation. Sensory neurons (top row) are activated by stimuli that contain features to which the neurons are tuned. Their activity level is constant regardless of whether the stimuli have been seen before. Neurons showing mnemonic suppression or enhancement are also activated by stimulus features. Suppression occurs for some cells when a stimulus has recently been seen, whereas enhancement occurs for other cells when the current stimulus matches the memory trace of a specific stimulus the animal actively holds in memory (i.e., not to any repeated stimulus). Neurons showing maintained activity may initially be activated by a visual stimulus and show sustained activation throughout the retention interval(s) as if they are actively holding the stimulus on-line.

nism based on simple stimulus repetition, a mechanism that may be intrinsic to the visual cortex. Another is a voluntary enhancement mechanism, linked to stimuli actively maintained in memory. This type of short-term memory seems to correspond to working memory, described in psychological studies (Baddeley, 1986). Evidence presented in the next section suggests that this mechanism may depend, in part, on top-down feedback from prefrontal cortex to IT cortex. Maintained activity in the absence of any stimulus may be a third memory mechanism, also linked to working memory (see the next section). There is no reason why all three mechanisms might not operate at the same time, supporting different aspects of memory-guided behavior. For example, in addition to detecting the matching *A* in the ABBA design, human observers are well aware of the repeated *B*s.

Prefrontal priming of IT neurons

What primes IT neurons to detect the matching stimulus in the ABBA task? One obvious candidate is ventral prefrontal cortex, which is heavily interconnected with IT cortex. Behavioral studies have shown that lesions in this region severely disrupt performance of DMS tasks (Mishkin and Manning, 1978; Goldman-Rakic, 1987; Fuster, 1989). Furthermore, Fuster (1973) and Funahashi, Bruce, and Goldman-Rakic (1989) have shown that in dorsal prefrontal cortex there are neurons whose activity is related to spatial working memory, and Wilson, O'Scalaidhe, and Goldman-Rakic (1993) have recently found evidence for an equivalent role of ventral prefrontal neurons in working memory for objects. In both the spatial and object cases, animals were required to maintain the memory of an item for a number of seconds before making some type of discriminative response. Many prefrontal neurons showed high, stimulus-specific, maintained activity during the retention interval in such tasks, and it has been proposed that this maintained activity helps maintain the stimulus trace in working memory. Fuster, Bauer, and Jervey (1985) have suggested that prefrontal cortex is involved in the mnemonic functions of IT cortex.

As described earlier in this chapter, we found stimulus-specific maintained activity in the delay following the sample in IT cortex, but this activity was abolished by intervening stimuli, at least for the overwhelming majority of neurons. To test whether delay activity is maintained after intervening stimuli in ventral prefrontal cortex, we recorded from prefrontal neurons in an animal performing the ABBA task (Chelazzi et al., 1993b). Our preliminary results show that some prefrontal neurons do have stimulus-specific activity in the delay following the sample and, unlike in IT cortex, this activity is present throughout the trial. Intervening stimuli may disrupt the ongoing maintained activity, but it quickly recovers. If this maintained information about the sample is somehow fed back from prefrontal cortex to the appropriate circuits in IT cortex, starting at the time of sample presentation or immediately afterward, this might explain the early onset of the enhanced IT responses to the matching stimulus (the sought item in the ABBA sequence) (figure 30.6).

The relationship between the delay activity in prefrontal and IT cortex is unclear. A connection between the two is reported by Fuster, Bauer, and Jervey, (1985), who found that cooling of prefrontal cortex reduces the incidence of cells showing delay activity in IT cortex. On the other hand, it is curious that whereas the sample-selective maintained activity is typically

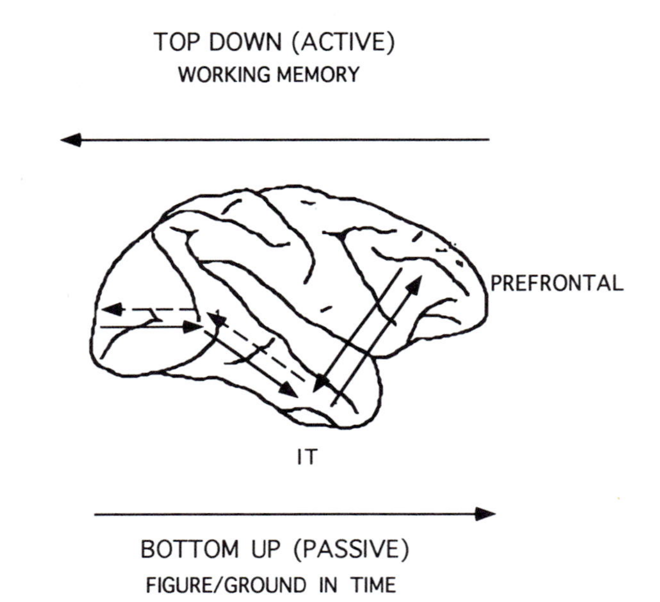

FIGURE 30.6 Dual mechanisms of short-term memory. Simple stimulus repetition engages passive, or bottom-up, mechanisms in IT cortex and possibly earlier visual areas. These mechanisms mediate a type of temporal figure-ground or stand-out of stimuli that are new or not recently seen. By contrast, working memory is believed to involve an active, or top-down mechanism, in which neurons in IT cortex (and possibly earlier visual areas) are primed to respond to specific items held in short-term memory. This priming of IT neurons may require feedback from prefrontal cortex.

abolished after intervening stimuli in IT cortex, it often is larger following intervening stimuli in prefrontal cortex, suggesting that the two phenomena may be complementary.

Visual search

We have seen how IT memory mechanisms are engaged by temporal sequences of stimuli, with neuronal responses biased in favor of stimuli that are novel or not recently seen or the sought item in the sequence. We can now extend these analyses to stimuli distributed in space rather than time, in the type of visual search that underlies finding a face in a crowd. The mechanism for this type of search requires a means both to hold the object of search in memory and to use this stored object to resolve competition among the array elements. Although there are many similarities between visual search and DMS, the terminology of the paradigms is different. The object of the search, usually presented at the start of the trial or trial sequence, is commonly termed the *cue* (rather than the *sample*) the sought stimulus in the array is the *target* (rather than the *matching item*) and the other items in the array are *nontargets* or *distractors* (rather than *nonmatching items*).

In our version of the search task (Chelazzi et al., 1993a), the stimuli were complex digitized patterns, such as those used in the study of memory. For each cell, we selected one stimulus that was a good one for the cell (i.e., activated the cell when presented alone) and at least one poor stimulus that elicited little or no response on its own. At the start of each trial, either a good or poor stimulus was presented as a cue, which the animal was required to fixate. After a 1500- to 3000-ms delay, a choice array containing a good and a poor stimulus was presented 4°–5° extrafoveally, and the animal was rewarded for making a saccadic eye movement to the target stimulus that matched the cue (figure 30.7). Target-cue identity (good or poor stimulus) as well as target location was varied, so that the animal had to find the target based on its remembered features. IT receptive fields are large and, typically, both the cue location at the fovea and both extrafoveal array locations are included within the contralateral field. Responses to stimuli in the ipsilateral field tended to be weak, however, so first we will describe the results when the choice array was located in the contralateral visual field.

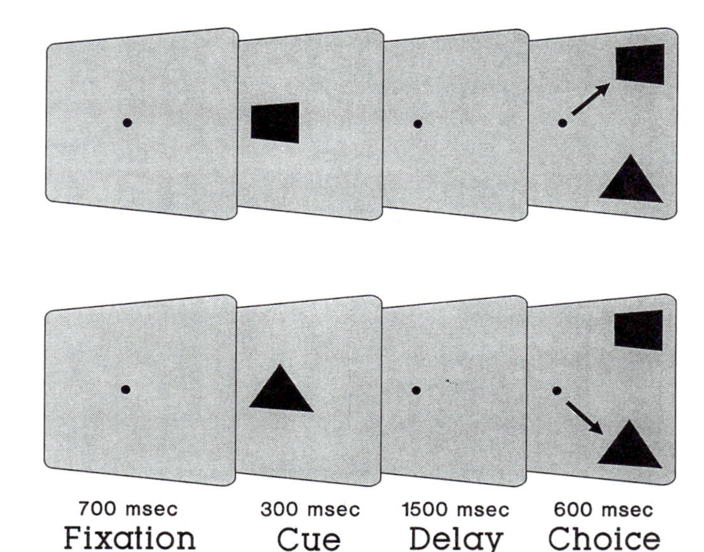

FIGURE 30.7 Sequence of stimulus events in the visual search task. Stimuli are represented, for simplicity, by a square and triangle. Each trial began with a cue followed, after a 1,500- to 3000-ms delay, by two choice stimuli. To receive a reward, the animal was required to saccade to the choice stimulus that matched the cue. (Adapted from Chelazzi et al., 1993a)

We compared responses to identical choice arrays but with either the good or poor stimulus as the target (i.e., preceded by either the good or poor stimulus as the cue). Because the sensory displays were identical under the two conditions (and positions randomized), any difference in response to the arrays could be attributed to the effects of target selection.

As might be expected from the earlier memory experiments in IT cortex, the cue often elicited stimulus-specific activity during the delay interval of the task. This activity was greater when the cue was a good stimulus for the cell than when it was a poor stimulus (7.9 versus 5.6 spikes per second). Furthermore, this delay activity sometimes persisted into the initial response to the choice array itself.

Except for this cue-selective activity persisting beyond the delay interval, the initial response to the choice array was, on average, about the same regardless of which stimulus was the target, that is, the two stimuli in the choice array tended initially to activate cells in IT cortex tuned to either of their features. However, within 200 ms after stimulus onset, or approximately 100 ms before the onset of the eye movement, responses diverged dramatically depending on which stimulus

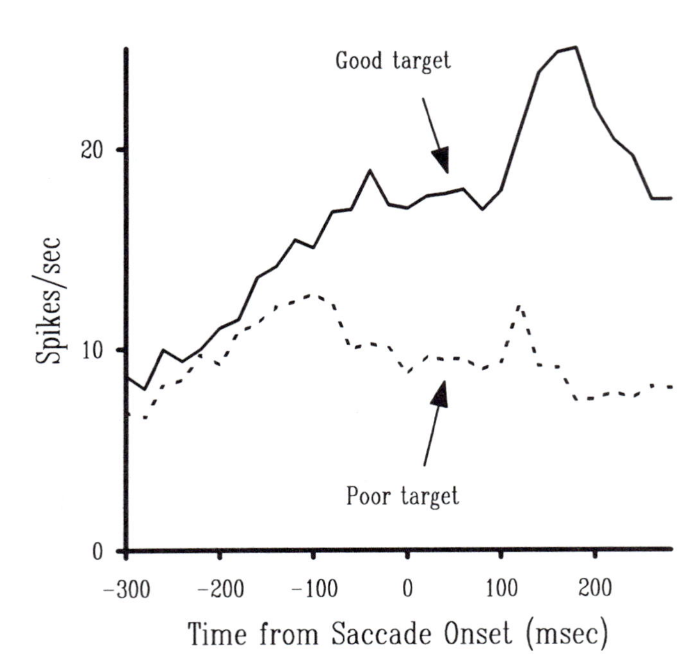

FIGURE 30.8 Average responses to identical choice arrays in which either the good (solid lines) or poor (dashed lines) stimulus was the target. The animals' saccade to the target occurred at time 0. Beginning approximately 120 ms before the saccade, the response is determined by whether the target is the good or poor stimulus for the recorded cell: That is, well before the saccade, the target has captured the responses of IT cells, so that neuronal responses are determined solely by the target stimulus. (Adapted from Chelazzi et al., 1993a)

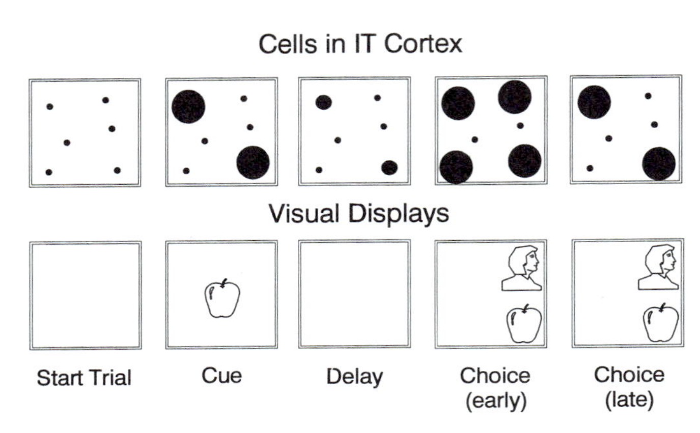

FIGURE 30.9 Representation of neuronal activity in IT cortex during visual search. For conventions, see figure 30.5. The cue activates those IT neurons tuned to any of its features, and these cells remain active (at a reduced firing rate) during the delay period. When the choice array first appears, IT neurons are activated by both the target and distractor. Later, but well before the eye movement to the target, neurons tuned to the target remain active, whereas neurons tuned to the distractor are suppressed. (Based on Chelazzi et al., 1993a)

was the target (figure 30.8). The target appeared to capture the responses of the cells, so that only cells selective for the target stimulus remained active. If the target happened to be a good stimulus for the recorded cell, responses remained high, but if the target happened to be a poor stimulus, responses were suppressed, practically down to the baseline level. We found the same results when the choice array was increased to five stimuli, the maximum tested.

We previously found that when monkeys attend to a stimulus at one location and ignore a stimulus at another, IT responses to the ignored stimulus are suppressed (Moran and Desimone, 1985). These results show that IT responses to ignored stimuli are suppressed even when the target location is not known in advance and the target features themselves guide selection.

Because target selection in IT cortex happens well in advance of the eye movement, this information is presumably available to oculomotor systems in order to direct the eyes to the target. Could IT cells provide the necessary target location information? Although large, IT receptive fields are not uniform. Many fields have a "hot spot" at the fovea, which then extends asymmetrically into the upper or lower contralateral visual field. Retinal location might be just another coarsely coded stimulus feature for IT cells.

We have narrowed the critical interactions underlying target selection to a 200-ms interval following the onset of the choice array. Cue-selective activity in the delay often persists into this initial response period, which may give the IT cells selective for the target stimulus a competitive advantage (figure 30.9). The working memory mechanism we found in our previous memory experiments (Miller and Desimone, 1994) seems to bias the competition even further toward these cells, as many IT cells give enhanced initial responses (over and above the delay activity) when their good stimulus is the sought target. *Working memory* and *attentive selection* in this case are equivalent. In the population average histogram (see figure 30.8), this enhancement tends to be masked by the responses of other cells showing suppression, due to the action of the adaptive memory mechanism engaged by the repetition of cue-target stimuli.

The idea that competition among IT cells mediates target selection is supported by the fact that three fourths of the cells with significant target effects in the 90 ms preceding the saccade also show cue-selective activity during the delay. Furthermore, the target effects are very much reduced when target and distractors are located in opposite hemifields. In this configuration, the IT cells activated most strongly by the different stimuli are located in opposite hemispheres where, presumably, they are less able to interact with one another.

If we are correct about the equivalence of the working memory and attentive selection mechanisms in IT cortex, then understanding the neuronal interactions underlying target selection takes on added importance. Biased competition within networks of cells could mediate many selection phenomena in attention, including spatial selection. Biased competition is an alternative to models of attention based on gating of specific inputs into individual cells (Olshausen, Anderson, and Van Essen, 1993; also see Desimone, 1992).

ACKNOWLEDGMENTS We gratefully acknowledge the advice and support of Mortimer Mishkin in all phases of the experiments. We also thank Lin Li, who collaborated on the memory studies; John Duncan, who collaborated on the visual search study; Rebecca Hoag, who trained the monkeys; and Tom White, who wrote all the computer programs. This work was supported in part by the Human Frontiers Science Program Organization.

REFERENCES

ALLMAN, J., F. MIEZIN, and E. MCGUINNESS, 1985. Direction- and velocity-specific responses from beyond the classical receptive field in the middle temporal area (MT). *Perception* 14:105–126.

BADDELEY, A., 1986. *Working Memory*. Oxford: Clarendon Press.

BAYLIS, G. C., and E. T. ROLLS, 1987. Responses of neurons in the inferior temporal cortex in short term and serial recognition memory tasks. *Exp. Brain Res.* 65:614–622.

CARPENTER, G. A., and S. GROSSBERG, 1987. A massively parallel architecture for a self-organizing neural pattern recognition machine. *Comput. Vis. Graph. Image Process.* 37: 54–115.

CHELAZZI, L., E. K. MILLER, J. DUNCAN, and R. DESIMONE, 1993a. A neural basis for visual search in inferior temporal cortex. *Nature* 363:345–347.

CHELAZZI, L., E. K. MILLER, A. LUESCHOW, and R. DESIMONE, 1993b. Dual mechanisms of short-term memory: Ventral prefrontal cortex. *Soc. Neurosci. Abstr.* 23:975.

DESIMONE, R., 1992. Neural circuits for visual attention in the primate brain. In *Neural Networks for Vision and Image Processing*, G. Carpenter and S. Grossberg, eds. Cambridge, Mass.: MIT Press, pp. 343–364.

DESIMONE, R., T. D. ALBRIGHT, C. G. GROSS, and C. BRUCE, 1984. Stimulus-selective properties of inferior temporal neurons in the macaque. *J. Neurosci.* 4:2051–2062.

DESIMONE, R., S. J. SCHEIN, J. MORAN, and L. G. UNGERLEIDER, 1985. Contour, color and shape analysis beyond the striate cortex. *Vision Res.* 25:441–452.

ESKANDAR, E. N., B. J. RICHMOND, and L. M. OPTICAN, 1992. Role of inferior temporal neurons in visual memory: I. Temporal encoding of information about visual images, recalled images, and behavioral context. *J. Neurophysiol.* 68:1277–1295.

FUNAHASHI, S., C. J. BRUCE, and P. S. GOLDMAN-RAKIC, 1989. Mnemonic coding of visual space in the monkey's dorsolateral prefrontal cortex. *J. Neurophysiol.* 61:331–349.

FUSTER, J. M., 1973. Unit activity in prefrontal cortex during delayed-response performance: Neuronal correlates of transient memory. *J. Neurophysiol.* 36:61–78.

FUSTER, J. M., 1989. *The Prefrontal Cortex*. New York: Raven Press.

FUSTER, J. M., 1990. Inferotemporal units in selective visual attention and short-term memory. *J. Neurophysiol.* 64:681–697.

FUSTER, J. M., R. H. BAUER, and J. P. JERVEY, 1985. Functional interactions between inferotemporal and prefrontal cortex in a cognitive task. *Brain Res.* 330:299–307.

FUSTER, J. M., and J. P. JERVEY, 1981. Inferotemporal neurons distinguish and retain behaviorally relevant features of visual stimuli. *Science* 212:952–955.

GAFFAN, D., and E. A. MURRAY, 1992. Monkeys (*Macaca fascicularis*) with rhinal cortex ablations succeed in object discrimination learning despite 24-hr intertrial intervals and fail at matching to sample despite double sample presentations. *Behav. Neurosci.* 106:30–38.

GOLDMAN-RAKIC, P. S., 1987. Circuitry of primate prefrontal cortex and regulation of behavior by representational memory. In *Handbook of Physiology: Sec. 1. The Nervous System*, F. Plum, ed. Bethesda, Md.: American Physiological Society, pp. 373–417.

GROSS, C. G., D. B. BENDER, and G. L. GERSTEIN, 1979. Activity of inferior temporal neurons in behaving monkeys. *Neuropsychologia* 17:215–229.

GROSS, C. G., C. E. ROCHA-MIRANDA, and D. B. BENDER, 1972. Visual properties of neurons in inferotemporal cortex of the macaque. *J. Neurophysiol.* 35:96–111.

HAENNY, P. E., and P. H. SCHILLER, 1988. State dependent activity in monkey visual cortex: I. Single cell activity in V1 and V4 on visual tasks. *Exp. Brain Res.* 69:225–244.

HOREL, J. A., D. E. PYTKO-JOINER, M. L. VOYTKO, and K. SALSBURY, 1987. The performance of visual tasks while segments of the inferotemporal cortex are suppressed by cold. *Behav. Brain Res.* 23:29–42.

LI, L., E. K. MILLER, and R. DESIMONE, 1994. The representation of stimulus familiarity in anterior inferior temporal cortex. *J. Neurophysiol.* 69:1918–1929.

LUESCHOW, A., E. K. MILLER, and R. DESIMONE, 1993. Effect of stimulus transformations on short-term memory mechanisms in inferior temporal cortex. *Soc. Neurosci. Abstr.* 23: 975.

MAUNSELL, J. H. R., G. SCLAR, T. A. NEALEY, and D. D.

DePriest, 1991. Extraretinal representations in area V4 in the macaque monkey. *Visual Neurosci.* 7:561–573.

Meunier, M., J. Bachevalier, M. Mishkin, and E. A. Murray, 1993. Effects on visual recognition of combined and separate ablations of the entorhinal and perirhinal cortex in rhesus monkeys. *J. Neurosci.* 13:5418–5432.

Mikami, A., and K. Kubota, 1980. Inferotemporal neuron activities and color discrimination with delay. *Brain Res.* 182:65–78.

Miller, E. K., and R. Desimone, 1994. Parallel neuronal mechanisms for short-term memory. *Science* 263:520–522.

Miller, E. K., P. M. Gochin, and C. G. Gross, 1991. Habituation-like decrease in the responses of neurons in inferior temporal cortex of the macaque. *Visual Neurosci.* 7: 357–362.

Miller, E. K., L. Li, and R. Desimone, 1991. A neural mechanism for working and recognition memory in inferior temporal cortex. *Science* 254:1377–1379.

Miller, E. K., L. Li, and R. Desimone, 1993. Activity of neurons in anterior inferior temporal cortex during a short-term memory task. *J. Neurosci.* 13:1460–1478.

Mishkin, M., 1982. A memory system in the monkey. *Philos. Trans. R. Soc. Lond. [Biol.]* 298:83–95.

Mishkin, M., and F. J. Manning, 1978. Non-spatial memory after selective prefrontal lesions in monkeys. *Brain Res.* 143:313–323.

Miyashita, Y., and H. S. Chang, 1988. Neuronal correlate of pictorial short-term memory in the primate temporal cortex. *Nature* 331:68–70.

Moran, J., and R. Desimone, 1985. Selective attention gates visual processing in extrastriate cortex. *Science* 229:782–784.

Nelson, S. B., 1991. Temporal interactions in the cat visual system: I. Orientation-selective suppression in the visual cortex. *J. Neurosci.* 11:344–356.

Olshausen, B. A., C. H. Anderson, and D. C. Van Essen, 1993. A neurobiological model of visual attention and invariant pattern recognition based on dynamic routing of information. *J. Neurosci.* 13:4700–4719.

Riches, I. P., F. A. Wilson, and M. W. Brown, 1991. The effects of visual stimulation and memory on neurons of the hippocampal formation and the neighboring parahippocampal gyrus and inferior temporal cortex of the primate. *J. Neurosci.* 11:1763–1779.

Rolls, E. T., G. C. Baylis, M. E. Hasselmo, and V. Nalwa, 1989. The effect of learning on the face selective responses of neurons in the cortex in the superior temporal sulcus of the monkey. *Exp. Brain Res.* 76:153–164.

Schwartz, E. L., R. Desimone, T. D. Albright, and C. G. Gross, 1983. Shape recognition and inferior temporal neurons. *Proc. Natl. Acad. Sci. U.S.A.* 80:5776–5778.

Squire, L. R., B. Knowlton, and G. Musen, 1993. The structure and organization of memory. *Annu. Rev. Psychol.* 44:453–495.

Suzuki, W. A., S. Zola-Morgan, L. R. Squire, and D. G. Amaral, 1993. Lesions of the perirhinal and parahippocampal cortices in the monkey produce a modality-general and long-lasting memory impairment. *J. Neurosci.* 13:2430–2451.

Tanaka, K., H. Saito, Y. Fukada, and M. Moriya, 1991. Coding visual images of objects in the inferotemporal cortex of the macaque monkey. *J. Neurophysiol.* 66:170–189.

Wang, Q., P. Cavanagh, and M. Green, 1992. Familiarity and pop-out in visual search. *Invest. Opthalmol. Vis. Sci.* 33:1252.

Wilson, F. A. W., S. P. O'Scalaidhe, and P. S. Goldman-Rakic, 1993. Dissociation of object and spatial processing domains in primate prefrontal cortex. *Science* 260:1955–1958.

Zola-Morgan, S., and L. R. Squire, 1984. Preserved learning in monkeys wuth medial temporal lesions: Sparing of motor and cognitive skills. *J. Neurosci.* 4:1072–1085.

Zola-Morgan, S., and L. R. Squire, 1985. Medial temporal lesions in monkeys impair memory on a variety of tasks sensitive to human amnesia. *Behav. Neurosci.* 99:22–34.

Zola-Morgan, S., L. R. Squire, D.G. Amaral, and W. A. Suzuki, 1989. Lesions of perirhinal and parahippocampal cortex that spare the amygdala and hippocampal formation produce severe memory impairment. *J. Neurosci.* 9: 4355–4370.

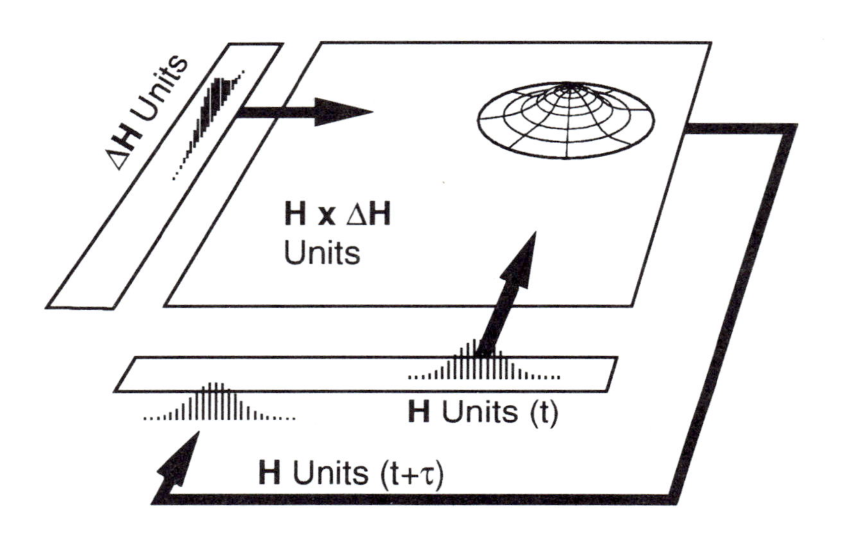
ΔH Units
H x ΔH
Units
H Units (t)
H Units (t+τ)

IV STRATEGIES AND PLANNING: MOTOR SYSTEMS

Head-direction (H) cells can be observed in several parts of the brain, including dorsal presubiculum, posterior neocortex, and lateral dorsal thalamus. Directional selectivity could be maintained solely on the basis of inertial (vestibular) cues using an associative mapping function relating the head direction (H) at time $t + \tau$ *to that at time* t *by means of the angular deviation* (ΔH). *Angular deviation is equivalent to angular velocity over a fixed time interval* (τ).

Introduction

EMILIO BIZZI

THE PLANNING and execution of limb and eye movements results from the activation of discrete modules distributed in a number of cortical and subcortical areas. The chapters in this part of the text highlight the specific functional properties of these modules.

To plan an arm trajectory, the central nervous system (CNS) first must locate the position of an object with respect to the body and represent the initial position of the arm. Recordings from single neurons in the parietal cortex and superior colliculus in awake monkeys have significantly contributed to our understanding of how space is represented. As discussed in Andersen's chapter, the main conclusion of his studies is that in the parietal cortical areas there are retinotopic neurons whose activity is tuned by signals derived from somatosensory sources. These neurons represent an object in the coordinate of the head. Recently, Brotchie and Andersen (1991) and Andersen and colleagues (1993), recording from parietal area 7a and the lateral interparietal area (LIP), have shown that the visual receptive field of parietal neurons is modified by signals representing both eye and head position. This result suggests that parietal area 7a contains a representation of space in body-centered space.

Neurons representing object location in body-independent (allocentric) coordinates have also been found. Andersen and collaborators have shown that there is a population of visually responsive cells in LIP whose activity is influenced by eye position and by the vestibular system. Presumably, these cells encode the locations of visual stimuli in world coordinates. In addition to these parietal cells, the neurons in the hippocampus also elaborate signals representing space in world-centered coordinates (see the chapter by McNaughton, Knierem, and Wilson).

To specify the limb's trajectory toward a target, the CNS must locate not only the position of an object with respect to the body but also the initial position of the arm. The conventional wisdom is that proprioception provides information about arm configuration to be used in the programming of the arm's trajectory. However, Ghez and collaborators demonstrated that directional errors are present during reaching when subjects rely on proprioceptive cues alone. It is of interest that these errors disappear when static vision of the hand or of the target is allowed. On the basis of these results, Ghez's group concluded that information about the initial position of the limb derives from a number of sources but especially from the visual afferences. With respect to the oculomotor system, the relationship of the target to the initial position of the eyes is critical for planning the amplitude of saccades. As discussed by Wurtz and Munoz, the saccade is driven by a motor error between the current position of the visual axis and the representation of the target in the colliculus.

The current view on the formation of arm trajectories is that the CNS formulates the appropriate command for the desired trajectory on the basis of knowledge about the initial arm position and the target's location. Recent psychophysical evidence reviewed by Bizzi and Mussa-Ivaldi as well as Jordan supports the hypothesis that the planning of limbs' movements constitute an early and separate stage of information processing. According to this view, during planning the brain is mainly concerned with establishing movement kinematics, a sequence of positions that the hand is expected to occupy at different times within the extrapersonal space. Later, during execution, the dynamics of the musculoskeletal system are controlled in such a way as to enforce the plan of movement within different environmental conditions.

Jordan, Bizzi and Mussa-Ivaldi, and Georgopoulos present evidence indicating that the planning of arm trajectories is specified by the CNS in extrinsic coordinates. The analysis of arm movements has revealed kinematic invariances. Remarkably, these simple and invariant features were detected only when the hand motion was described with respect to a fixed cartesian reference frame, a fact suggesting that CNS planning takes place in terms of the hand's motion in space. Even more complex curved movements performed by human subjects in an obstacle-avoidance task displayed invariances in the hand's motion and not in joint motion. The data derived from straight and curved movements were utilized by Flash and Hogan (1985), who showed that the kinematic invariances could be derived from a single organizing principle based on optimizing endpoint smoothness. As discussed by Jordan, their mathematical model is compatible with a hypothesis based on planning in space but not with one based on joint coordinates. It follows that if actions are planned in spatial or extrinsic coordinates, then for the execution of movement, the CNS must convert the desired direction and velocity of the limb into signals that control muscles.

Investigators of motor control have been well aware of the computational complexities involved in the production of muscle forces. A variety of proposals have been made to explain these complexities. Bizzi and Mussa-Ivaldi address the notion that the CNS may transform the desired hand motion into a series of equilibrium positions. The forces needed to track the equilibrium trajectory result from the intrinsic elastic properties of the muscles.

Recently, a set of experiments performed in frogs with spinal cords that were surgically disconnected from the brain stem has provided neurophysiological support for the equilibrium-point hypothesis. By microstimulation of the spinal cord, Bizzi, Mussa-Ivaldi, and Giszter (1991) have shown that this region is organized to produce the neural synergies necessary for the expression of equilibrium points. Microstimulation of the lumbar gray matter resulted in a limited number of force patterns. The limited force pattern may be viewed as representing an elementary alphabet from which, through superimposition, a vast number of movements could be fashioned by impulses conveyed by suprospinal pathways.

In summary, the chapters presented in this part indicate that the planning and execution of arm movements are implemented by a series of processes, each one involving a number of discrete circuits distributed in the cortical and subcortical areas. While aspects of this processing, particularly the early representation of space, are well understood, the complex issue of translating the planning of arm movements into muscle forces remains, to a certain extent, a matter of conjecture. However, the recent discovery of structures in the brain stem and spinal cord that are specialized for the production of muscle synergies may provide a way to understand how descending and segmental impulses impinging on these areas may produce dynamic events.

REFERENCES

Andersen, R. A., L. H. Snyder, L. Chiang-Shan, and B. Stricanne, 1993. Coordinate transformations in the representation of spatial information. *Curr. Opin. Neurobiol.* 3:171–176.

Bizzi, E., F. A. Mussa-Ivaldi, and S. Giszter, 1991. Computations underlying the execution of movement: A biological perspective. *Science* 253:287–291.

Brotchie, P. R., and R. A. Andersen, 1991. A body centered coordinate system in posterior parietal cortex. *Abstr. Soc. Neurosci.* 17:1281.

Flash, T., and N. Hogan, 1985. The coordination of arm movements: An experimentally confirmed mathematical model. *J. Neurosci.* 5:1688–1703.

31 Toward a Neurobiology of Coordinate Transformations

EMILIO BIZZI AND FERDINANDO A. MUSSA-IVALDI

ABSTRACT We discuss a perspective in motor control that is based on the distinction between planning and execution of motor behaviors. This distinction has emerged from experimental observations which suggest that the central nervous system plans movements in terms of spatial or extrinsic coordinates rather than body-centered coordinates. What, then, are the processes that transform a motor plan into the signals that activate the muscles? We review a number of theoretical and experimental findings suggesting that the coordinate transformations leading to the execution of a motor plan are implemented by a small number of control modules. In particular, some recent experiments in the spinalized frog have suggested that these control modules may be organized by the spinal cord as simple synergies of springlike muscles. Subsequent theoretical work has revealed that a wide variety of motor control policies can be obtained from a simple linear combination of these few control modules.

In this chapter, we address a conundrum that has long faced investigators in motor control: if movements are specified by the central nervous system (CNS) in terms of goals, as psychophysical evidence seems to imply, what are the processes in the CNS that transform the neural representation of these goals into signals that activate the muscles? In the last few years, psychophysical and electrophysiological observations made by a number of investigators have indicated that motor goals as simple as reaching and pointing are planned by the CNS in terms of extrinsic coordinates representing the motion of the hand in space. In the first part of this chapter, we review the evidence that motor behavior is represented in the higher centers of the brain in terms of extrinsic coordinates. In the second part, we propose a novel hypothesis describing one possible way in which motor plans could be transformed into motor commands.

EMILIO BIZZI and FERDINANDO A. MUSSA-IVALDI Department of Brain and Cognitive Sciences, Massachusetts Institute of Technology, Cambridge, Mass.

Evidence for motor planning in spatial (extrinsic) coordinates

In what coordinate frames does the CNS represent movement? The observations by Morasso (1981) suggest that the planning of arm movements is carried out in extrinsic coordinates that represent the motion of the hand in space. Morasso instructed human subjects to point with one hand to different visual targets that were randomly activated. His analysis of the movements showed two kinematic invariances: (1) The hand trajectories were approximately straight segments, and (2) the tangential hand velocity for different movements always appeared to have a bell-shaped configuration: The time needed to accelerate the hand was approximately equal to the time needed to bring it back to rest. Because these simple and invariant features were detected in the coordinates of the hand only, these results suggest that CNS planning takes place in terms of hand motion in space. Morasso's observations were extended to more complex curved movements performed by human subjects in an obstacle avoidance task (Abend, Bizzi, and Morasso, 1982). Again, kinematic invariances were present in the hand and not in the joint motion. Later, Flash and Hogan (1985) showed that the kinematic behavior described by Morasso (1981) and Abend's group (1982) could be derived from a single organizing principle based on optimizing endpoint smoothness. It is interesting that their mathematical model is compatible with a hypothesis based on planning in space but not with that based on joint coordinates.

The idea that movements of the hand are planned by the CNS in terms of extrinsic coordinates is related to the hypothesis that the planning and execution of movements constitute two separate stages of information processing (Bernstein, 1967; Hogan et al., 1987). According to this view, during planning, the brain is concerned with establishing movement kinematics—that is, a sequence of positions that the hand is expected to occupy at different times within the extrapersonal space. During execution, the dynamics of the musculoskeletal system are controlled in such a way as to enforce the planning of movements within different environmental conditions. This separation between planning and execution (or between kinematics and dynamics) has been challenged by Uno, Kawato, and Suzuki (1989), who have suggested that the kinematic features observed by Morasso may be accounted for by a model in which the rate of change of joint torque, rather than endpoint smoothness, is minimized. In essence, these authors proposed that the observed kinematics may be a "side effect" of the computational processes underlying the control of dynamics.

However, in contrast with this view, recent experimental findings by Flash and Gurevich (1992) and by Shadmehr and Mussa-Ivaldi (in press) have demonstrated that the kinematics of movement are indeed planned independently of the dynamic conditions in which movement occurs. Shadmehr and Mussa-Ivaldi (1993) asked human subjects to execute movements of the hand in different directions while holding the handle of a two-joint planar robot. At the beginning of the experiment, the robot acted as a passive manipulandum, and the subject produced straight-line hand movements (as described in Morasso's experiment). In a subsequent stage of the experiment, the robot was programed to generate a field of forces that depended linearly on the measured velocity of the hand. The effect of the field was rather complex, consisting of a combination of destabilizing forces and a viscous drag in two orthogonal directions. At first, subjects reacted to the unexpected field with a visibly perturbed trajectory of the hand (figure 31.1). Subjects were then asked to perform several target acquisition movements within the field of forces. They were not given any instruction regarding hand trajectory.

As shown in figure 31.1, the kinematics of the hand movements returned gradually to the original straight line with a bell-shaped tangential velocity profile, as observed before the onset of the perturbing field. A similar adaptive behavior was observed by Flash and Gurevich (1992), who used a springlike load instead of a velocity-dependent field as a perturbation. In both cases, at the end of the adaptation period, the basic kinematics observed by Morasso (1981) were restored in a totally changed dynamic environment. Clearly, this finding is not compatible with the idea that the movement kinematics are derived from a single optimization principle applied to the dynamics of the motor system. On the contrary, the experimental findings on the adaptation to external fields strongly suggest that movement kinematics are programed as a desired trajectory to be enforced in different dynamic conditions.

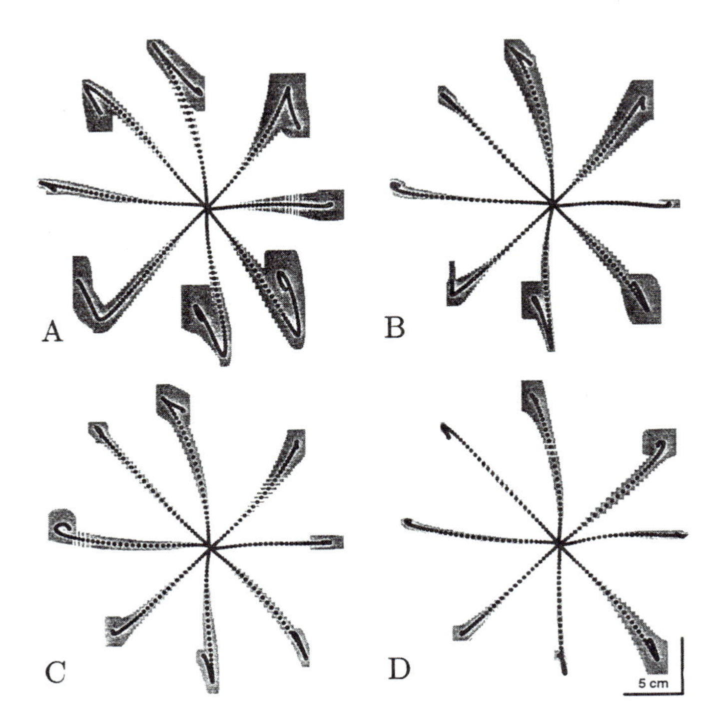

FIGURE 31.1 Average plus standard deviation of hand trajectories during exposure to a force field. Performance during (A) the first 250 targets, (B) the second 250 targets, (C) the third 250 targets, and (D) the final 250 targets. All trajectories shown are under no-visual-feedback condition. (From Shadmehr and Mussa-Ivaldi, in press)

In addition to the psychophysical evidence, recordings from cells in cortical and subcortical areas of monkeys have shown a correlation between the cells' firing pattern and the direction of the hand's motion (Georgopoulos, Kittner, and Schwartz, 1988a, b; Kalaska et al., 1989; Caminiti, Johnson, and Urbano, 1990). It appears that the activity of certain classes of cortical cells is represented in spatial coordinates without any specification about how muscles are to be engaged to produce the forces necessary for the movement.

Taken together, the psychophysical and electrophysical evidence points toward a hierarchical organization of the motor system, with the higher centers of the CNS representing motor goals in spatial coordinates. The next step, of course, is the transformation of the motor goals into impulses for the large number of muscles that are simultaneously activated during the execution of even the simplest kind of limb movement.

Execution of planned actions: The production of muscle's forces

Investigators of motor control have been well aware of the computational complexities involved in the production of muscles' forces (Saltzman, 1979). A variety of proposals have been made to explain these complexities. For instance, Hollerbach and Atkeson (1987), inspired by work in robotics, have suggested that the CNS first derives the motion of the joints from the planned path of the limb's endpoint (inverse kinematics). Next, the CNS computes the necessary joint torques (inverse dynamics) and then distributes the torques to a number of muscles. This hypothesis rests on the assumption that the CNS can estimate accurately limb inertias, center of mass, and the moment arm of muscles about the joints. This line of thinking seems to us improbable because the implied feedforward computations would lead to large motor instabilities every time the small errors in evaluating the various parameters occur. Alternative proposals have been made that do not depend on the solution of the complicated inverse dynamics problem. Specifically, it has been proposed that the CNS may transform the desired hand motion into a series of equilibrium positions (Bizzi et al., 1984). The forces needed to track the equilibrium trajectory result from the intrinsic elastic properties of muscles and from local feedback loops (Feldman, 1974, 1986; Bizzi, Polit, and Morasso, 1976; Bizzi et al., 1992; Hogan, 1984, 1985).

Support for the equilibrium point hypothesis derives from data from psychophysical and behavioral experiments. Recently, Bizzi, Mussa-Ivaldi, and Giszter (1991) and Giszter, Mussa-Ivaldi, and Bizzi (1993) provided neurophysiological underpinnings for the equilibrium point hypothesis. To this end, they investigated the spinal circuitry of the spinalized frog.

Force Fields Evoked by Stimulation of the Spinal Cord Bizzi, Mussa-Ivaldi, and Giszter (1991; Giszter, Mussa-Ivaldi, and Bizzi, 1993) elicited motor responses in the leg of spinalized frogs by microstimulating the spinal gray matter. They induced contractions in the leg muscles of the frog by placing a microelectrode in the intermediate neuropil zone of the gray matter. Very small amounts of current were used in these experiments, enough only to elicit a measurable mechanical response in the leg. The key feature of this experiment was the way in which the muscle's mechanical response was measured. As shown in figure 31.2E, a six-axis force transducer was attached at the level of the ankle. The transducer, which was connected to an *x-y* manipulator, allowed the experimenters to position the leg at a number of locations in the frog's work space. At each location, two types of forces were observed: a resting force caused by the elastic and reflex properties of muscles prior to microstimuation and then, after a latency (ranging from 30 to 150 ms), a force evoked by the electrical activation of the spinal gray matter. Usually, the forces elicited by stimulation rose to a plateau and then declined to the baseline level after a variable period (300 ms–1 s) following the termination of the stimulus. It is important to note that the stimulation of each site in the spinal cord produced a repeatable set of forces at a specific location of the leg in space. As shown in figure 31.2B, the forces varied greatly according to the placement of the limb in the work space. These variations, of course, are due to changes in muscle length, in moment arm of muscles, and in afferent feedback. The ensemble of forces recorded at different locations of the work space constitutes a force field.

Circuitry Subserving the Different Force Fields With surgical techniques, Giszter, Mussa-Ivaldi, and Bizzi (1993) and Loeb et al. (1992) ruled out the possibility that the force fields resulted from stimulation of the intraspinal component of sensory afferents. Microstimulation performed in chronically deafferented and acutely spinalized frogs showed that both the strength and the convergent tendency of the force field were preserved. Giszter, Mussa-Ivaldi, and Bizzi (1993) also found that force fields were essentially unchanged in frogs transected 6 weeks before microstimulation. These studies indicate that long ascending and descending systems, as well as reflex pathways, are not a necessary part of the circuitry subserving the pattern of force fields.

In a separate series of experiments, Giszter, Mussa-Ivaldi, and Bizzi (1993) also ruled out the possibility that the convergent force field resulted from directly activating motoneurons. Three sets of findings oppose this possibility: First, in most instances, microstimulation delivered to the ventral horn resulted in parallel or

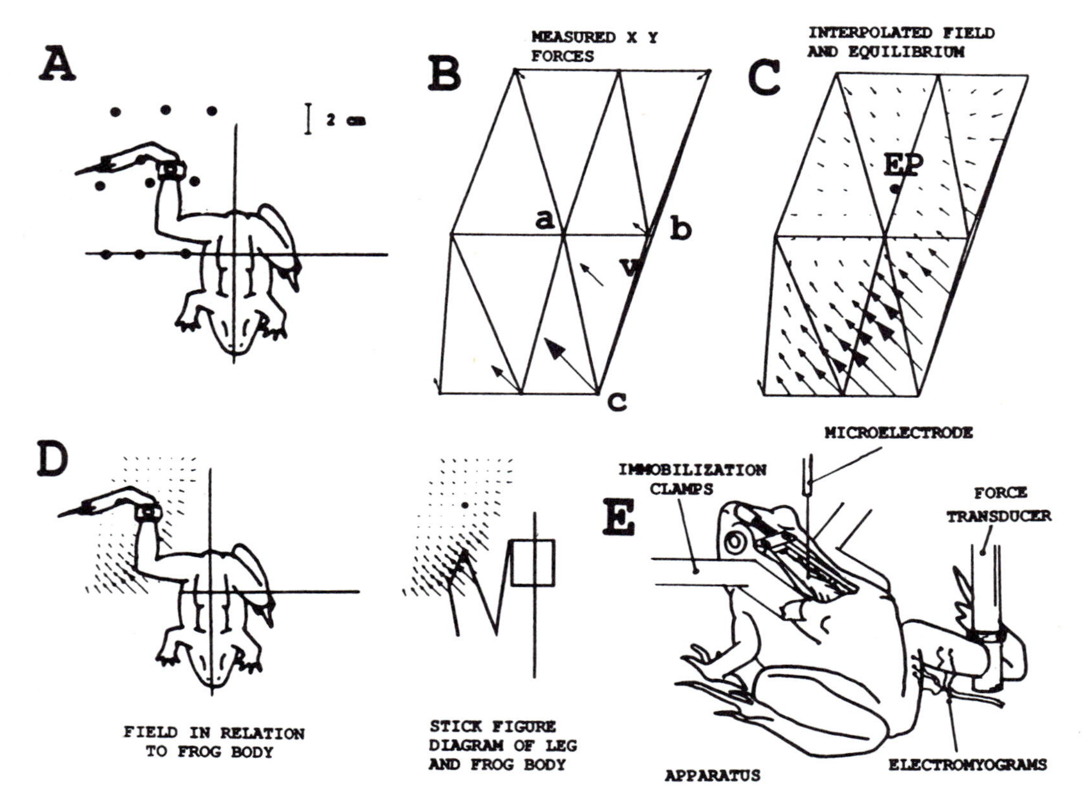

FIGURE 31.2 The apparatus and method of construction of a force field. (A) A collection of forces are recorded at several different spatial locations (black dots). (B) A minimum perimeter (Delaunay) tesselation of the nine points is constructed. Within each triangle, vectors are estimated using an exact linear interpolation based on the three corner vectors. Thus, vector v is calculated using vectors at vertices a, b, and c. (C) The interpolated field is used to find any equilibrium point (EP). (D) On left, the interpolated force field is shown in relation to the frog in the apparatus. On right, this representation is reduced to a stylized construction that is used to express the relation of frog body axis and leg to the interpolated force field in the remaining figures. (E) The apparatus. The spine is clamped, and the pelvis is held clamped by restraints (not shown). With an electrode in the spinal gray matter, the mechanical response to stimulation is recorded at the force sensor, which is attached to the limb at the ankle. The limb configuration is fully constrained by the pelvic restraint and the force sensor. (From Giszter, Mussa-Ivaldi, and Bizzi, 1993)

divergent, rather than convergent, force fields (figure 31.3C). Second, we investigated whether force fields may result from the random activation of motoneurons by combining the isometric responses of individual muscles in a simulation. To this end, we stimulated individual leg muscles through a pair of implanted electrodes and obtained muscle force fields. Next, we simulated random combinations of the measured muscle fields and obtained sets of combined fields by adding together all these modulated fields. We found that, in a set of 20,000 simulated combinations, the number of fields having an equilibrium point within the tested work space was only 8.4%. Thus, random recruitment of motoneurons could not account for the convergent force fields observed in the majority of our experiments. Third, experiments utilizing sulphorhodamine (a substance that is actively absorbed by active neurons) in conjunction with heptanol (a substance that blocks gap junctions) demonstrated an extensive uptake of rhodamine by motoneurons during microstimulation. The uptake was extensive—two orders of magnitude larger than the uptake that would have been expected from direct electrical activation of motoneurons. Therefore, we concluded that activation of motoneurons must have resulted through a synaptic mechanism involving neurons with long branches.

On the basis of the cell types known to exist in the spinal cord, it is likely that activation of the motoneurons after microstimulation of the spinal gray matter is mediated by propriospinal interneurons. These inter-

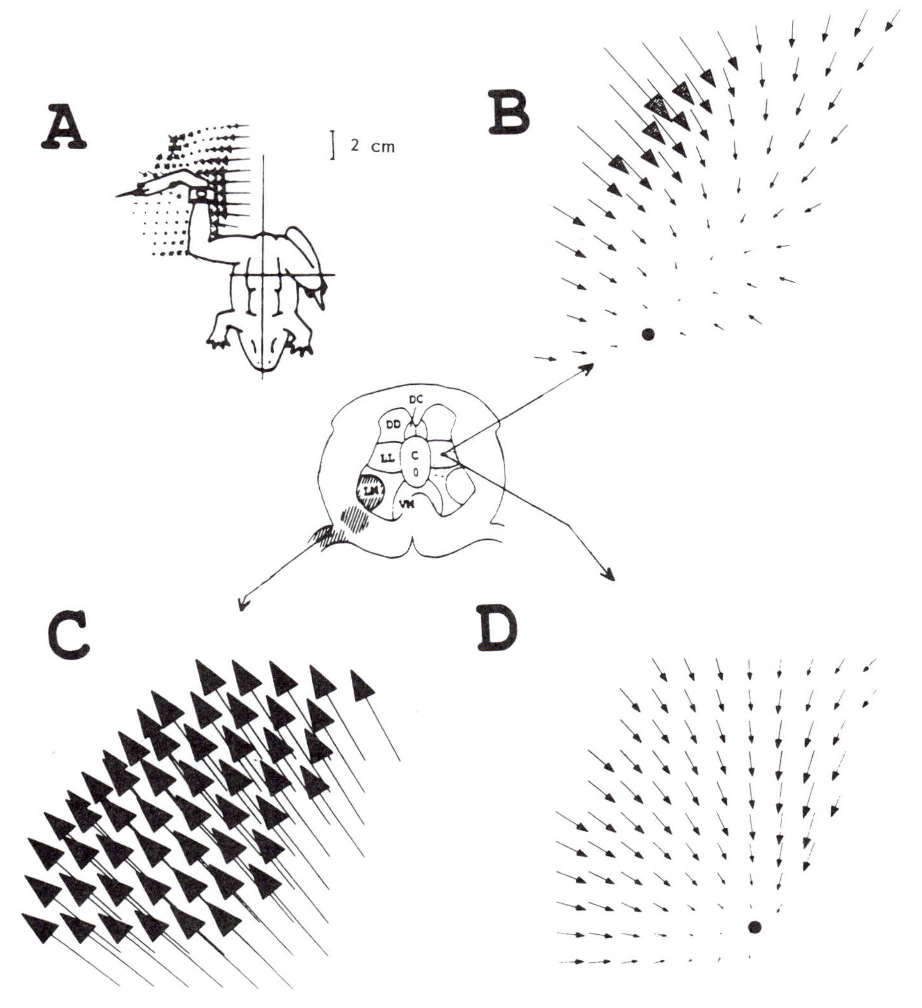

FIGURE 31.3 Examples of force fields elicited by microstimulation. In the center is a transverse section of the spinal cord. DC, dorsal column nucleus; DD, dorsal neuropil region; LL, lateral neuropil region; C, central neuropil region; VM, ventromedial neuropil region; LM, lateral motoneuronal region. (A) Convergence force field (CFF) shown in reference to the frog's leg. (B) CFF recorded from a chronically deafferented frog. The animal was prepared for microstimulation 3 weeks after bilateral section of dorsal roots 7–10. The equilibrium point is indicated by a filled circle. Stimulation parameters are similar to those in (D). (C) Parallel force field recorded from the region of the motoneurons. Parameters of stimulation as in (B) and (D). (D) The CFF elicited by microstimulating a small area within the stippled region of the spinal cord with constant current pulses of 1–6 mA applied at 40 Hz for 300 ms. The spread of current was ~100 mm in radius. Force vectors were recorded at each location of the 4 × 4 grid. At a given latency (t) from the onset of stimulus, we applied a piecewise linear interpolation to the set of measured forces. The equilibrium point is indicated by a filled circle. (From Bizzi, Mussa-Ivaldi, and Giszter, 1991)

neurons, which are known to have long ascending and descending branches, are ideally suited to spread the excitation from the central gray matter (Alstermark and Sasaki, 1986).

TEMPORAL EVOLUTION OF THE FORCE FIELDS Force fields evoked by microstimulation of the spinal gray matter displayed a temporal evolution in the sense that individual forces reached a peak magnitude and then reverted to their resting position. In most instances (70%), the force vectors resulted in a field that was convergent and that was characterized, at any given time, by a single equilibrium point. Analysis of the temporal evolution of the forces showed a smooth movement of the equilibrium point from its position at rest to a position reached when the forces achieved

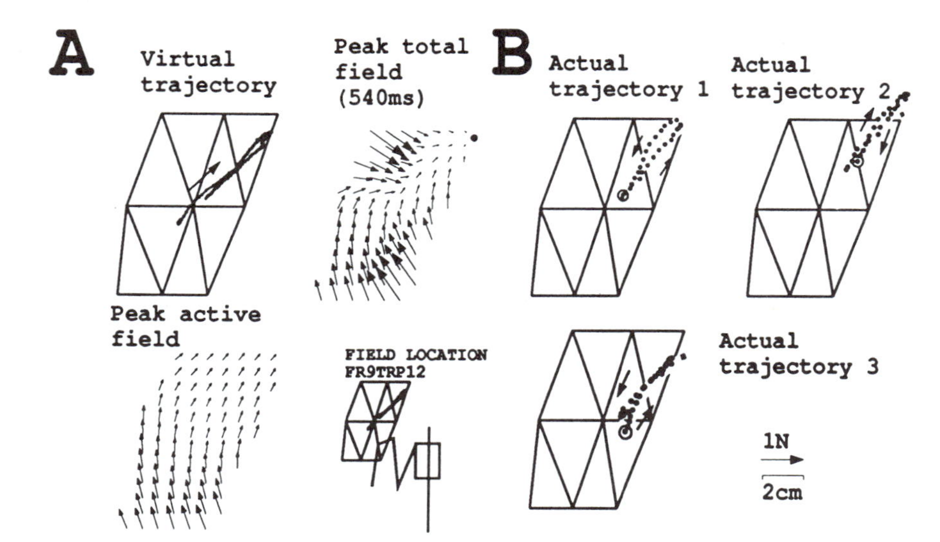

FIGURE 31.4 Comparison of actual and virtual trajectory. (A) Force field measurements. Shown is the virtual trajectory of equilibrium points (upper left), which derives from the time course of the total convergent force field shown at peak amplitude. The active field (lower left) had a pattern of forces that would specify a flow moving caudally and medially. The relation of the trajectory to the frog's body is shown schematically at lower right. (B) Trajectory measurements. Three sample limb trajectories from the suspended limb are shown. These were recorded at 30 Hz using a video system. We related the kinematic data to the grid of points used to collect the forces sampled for the field. The three trajectories begin at different locations in the grid, due to different initial resting fields as a result of the leg's suspension (see text). The actual pattern of movement is well predicted by the total field. This was especially evident in the first trajectory where resting equilibria almost coincide. The direction of the movement was well predicted by the active pattern in those trajectories where initial resting postures differed. (From Giszter, Mussa-Ivaldi, and Bizzi, 1993)

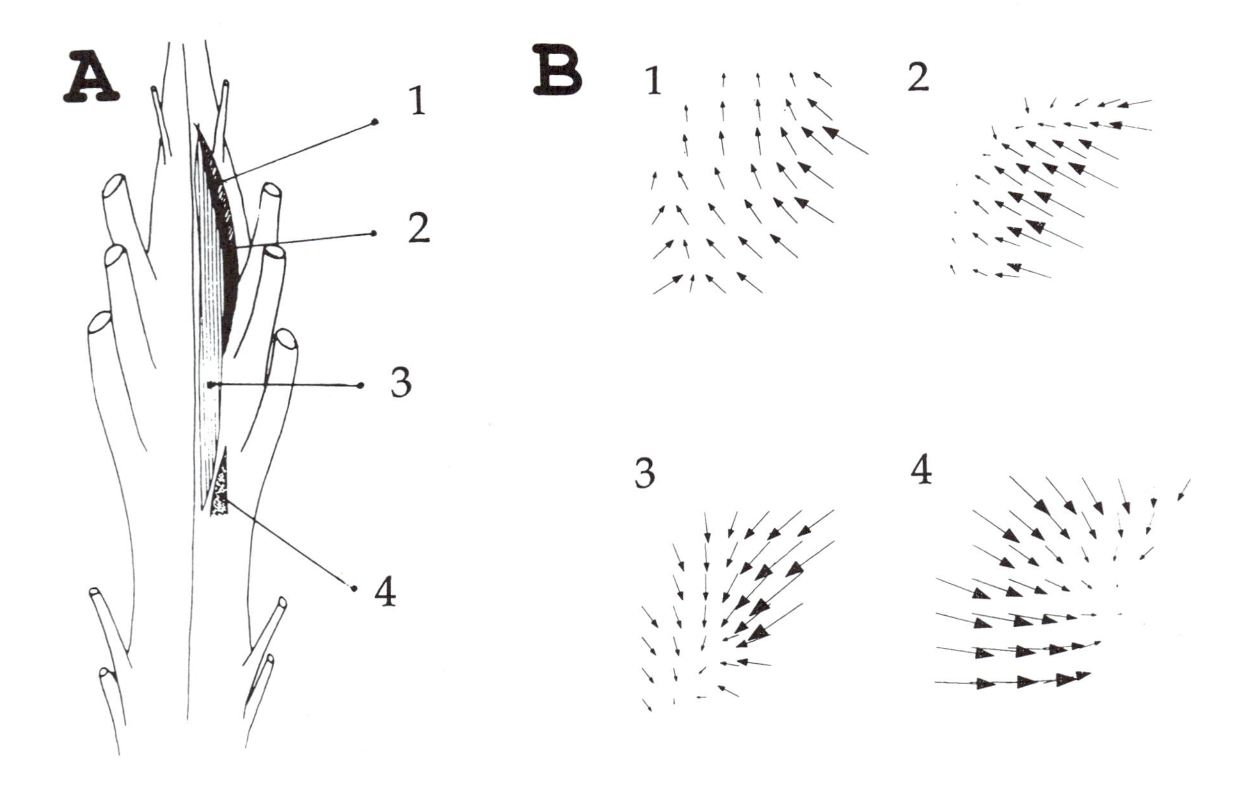

peak amplitude. Bizzi, Mussa-Ivaldi, and Giszter (1991; Giszter, Mussa-Ivaldi, and Bizzi, 1993) called the motion of the equilibrium point a *virtual trajectory*, and it corresponds to the position toward which the limb's endpoint would be attracted if it were free to move. In a separate group of experiments, Giszter's group (1993) showed that when the frog's leg was free, actual and virtual trajectories were very similar (figure 31.4).

A Map of Force Fields Bizzi, Mussa-Ivaldi, and Giszter (1991; Giszter, Mussa-Ivaldi, and Bizzi, 1993) showed that microstimulation of the lumbar gray matter resulted in a limited number of force patterns. In a preliminary series of experiments, a number of regions of the spinal gray matter from which the same force pattern was elicited were identified. These regions formed stripes that are oriented rostrocaudally (figure 31.5). They also extend dorsoventrally over a distance of 300 mm in depth (figure 31.6). It should be emphasized that the different force patterns were most clearly observed in the data when the baseline forces, representing the resting fields, were subtracted from the total field. The active force fields were identified by means of a point-by-point vector subtraction of the prestimulus resting forces from the forces resulting from microstimulation. As shown by Giszter, Mussa-Ivaldi, and Bizzi (1993), focusing on the active field simplifies the description and the interpretation of microstimulation data in the sense that a few clear patterns emerge. Another remarkable feature of the active fields is that they remain constant through time. Thus, the orientation of the forces and, hence, the equilibrium point of the active field remain constant even for a range of variation in the stimulus parameters. The fields elicited by different stimulus strengths were amplitude-scaled versions of one another.

The invariant characteristics of the active force fields and the limited number of distinct force patterns represent new properties of the functional organization of the spinal cord. It is tempting to speculate that the limited number of active force fields might produce movements and postures through a combination of the active fields. The connectivity and the branching of descending as well as reflex pathways might provide the substrate for combining force fields. Although experimental verification of this hypothesis is not yet available, combinations of active fields by way of simultaneous microstimulation of two spinal cord sites lend plausibility to this hypothesis.

Simultaneous Microstimulation Bizzi, Mussa-Ivaldi, and Giszter (1991) have shown that simultaneous stimulation of two sites, each generating a force field, results in a force field proportional to the vector sum of the two fields. It is important to note that the superimposition applies to the vectors of the fields and not to the equilibrium points (figure 31.7). Vector summation of force fields implies that the complex nonlinearities that characterize the interactions both among neurons and between neurons and muscles are in some way eliminated. More importantly, this result has led to a novel hypothesis for explaining movement and posture based on combinations of a few elementary elements. The limited force pattern may be viewed as representing an elementary alphabet from which, through superimposition, a vast number of movements could be fashioned. Through a computational analysis, Mussa-Ivaldi (1992), along with Giszter (1992) has verified that this novel view of the generation of movement and posture is plausible.

Physiologically, it also is possible to relate natural movements to the microstimulation results. Supraspinal initiation of movements results from patterns of neural activity distributed by the branches of descending fibers throughout fairly wide regions of the spinal cord. Conceivably, these branches may stimulate local clusters of cells that, in turn, generate force fields. If we make the assumption that these fields sum like the force fields generated by microstimulation, then a variety of movements and postures may result as a consequence of the pattern of terminal arborization of descending pathways.

Figure 31.5 (Left) Regions of the lumbar spinal cord containing the neural circuitries that specify the force fields, A through D. Within each region, similar sets of convergence force fields (CFFs) are produced. The diagram is based on 40 CFFs elicited by microstimulation of premotor regions in three spinalized frogs. (Right) Four types of CFFs. To facilitate comparison among CFFs recorded in different animals, we subtracted the passive force field from the force field obtained at steady-state. Passive field represents the mechanical behavior generated by the frog's leg (recorded at the ankle) in the absence of any stimulation. The force field is at steady-state when the forces induced by the stimulation of the spinal cord have reached their maximal amplitude. (From Bizzi, Mussa-Ivaldi, and Giszter, 1991)

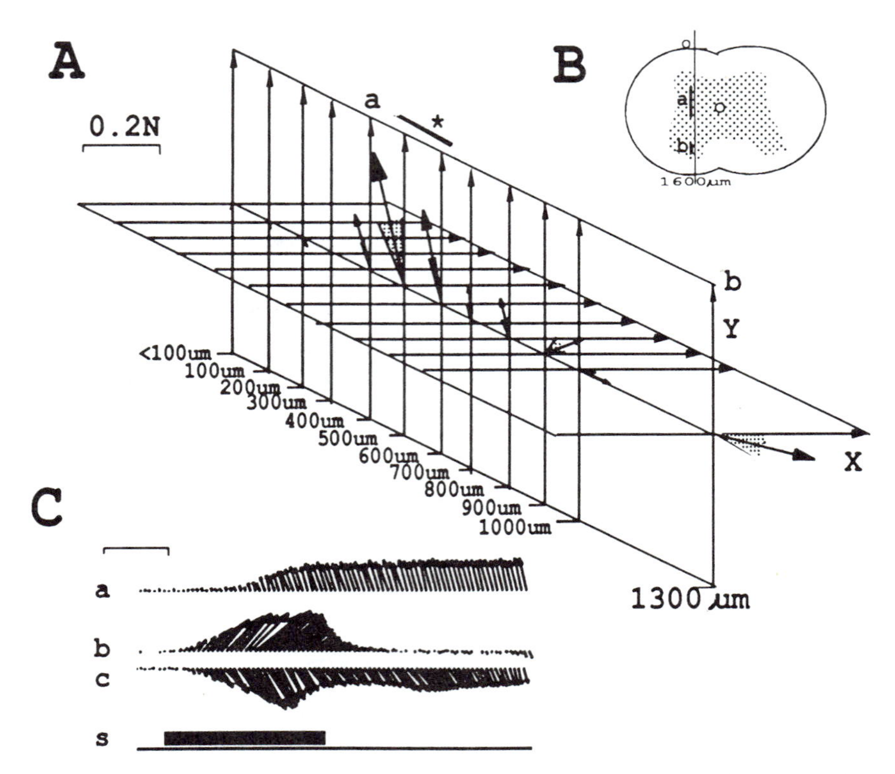

FIGURE 31.6 The orientation of average force with depth. (A) The orientation of the average force vector at successive depths in the spinal cord. The force vector displayed is composed of components *Fx* and *Fy*. The baseline forces were removed. These force vectors were elicited by a 2.7-µA, 60-Hz, 300-ms train. The shaded triangles are standard deviations of the individual force vectors. Note the constant orientation of the vectors for the 500-µm depth starting at *a*. When the electrode penetrated deeper (to *b*) among the motoneuron somas, a different orientation was found. (B) Schematic cross-section of cord showing electrode track locations. (C) Growth over time of the vectors at spinal cord sites (scale bar [s], 0.2 N). The force vector typically showed one of three behaviors displayed here. The top trace is the force actually recorded at point *a* in (A). *a*, the most common response is a smooth rise to a force plateau; *b*, a typical transient response; *c*, a typical transient/plateau response. (From Giszter, Mussa-Ivaldi, and Bizzi, 1993)

The spinal cord areas that we have identified through microstimulation may not only be sites for integrating descending signals but may also be utilized by the reflex pathways. Recently, Giszter, Mussa-Ivaldi, and Bizzi (1993) verified that force fields almost identical to those induced by microstimulation are elicited following cutaneous stimulation of the frog's leg. Furthermore, costimulation of different cutaneous sites results in a force field that is a simple vector summation, just as with microstimulation of two spinal cord sites. Although it is tempting to speculate that reflex and descending systems converge on the zones we have described in the spinal gray matter, more studies are needed to assess this convergence.

The hypothesis that the premotor zones in spinal gray matter may be one of the structures underlying the transformation from extrinsic to intrinsic coordinates is consistent with the results obtained by other groups of investigators. For instance, Masino and Knudsen (1990) have demonstrated the existence of a few separate circuits for controlling horizontal and vertical head movements in the owl. These structures, which are located in the brain stem, receive inputs from the tectum and transform the tectal movement vectors into neck motoneural activation.

Generation of motor behaviors through the summation of force fields

The results of microstimulation of the frog's spinal cord led us to the hypothesis that the spinal premotor circuits are organized so as to implement a relatively

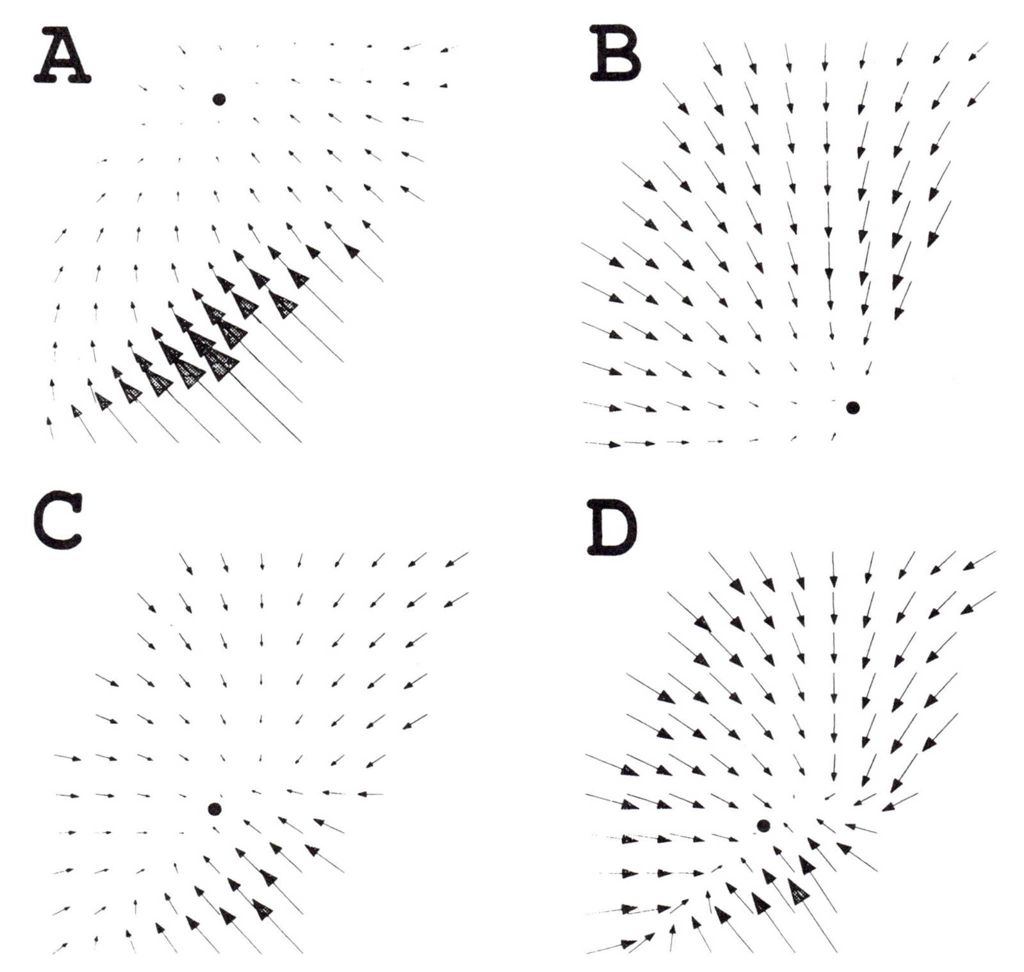

FIGURE 31.7 Combinations of multiple stimuli. (A) and (B) show the individual fields resulting from stimulation at two different sites in the premotor areas of the lumbar spinal cord. The equilibrium of field (A) is in extension and (B) is in flexion. (C) is the computed field ⟨AB⟩ predicted by a simple vectorial summation of fields (A) and (B). (D) is the actual field evoked by stimulation of (A) and (B) together. The equilibrium point is indicated by a filled circle. (From Bizzi, Mussa-Ivaldi, and Giszter, 1991)

small number of control modules. From an architectural point of view, each module establishes—presumably via a pattern of propriospinal connections—a common drive to a set of agonist and antagonist muscles. From a functional point of view, each module generates a field of forces. A limb's posture is implicitly encoded as a stable equilibrium point within this field. Equilibrium is defined as the point at which the force vanishes. Stability is ensured by the convergence of the surrounding force vectors. In addition, we found that the force fields generated by spinal microstimulations summed vectorially when two stimuli were applied simulteneously at two different spinal sites. Having established these facts, we have yet to understand how such a crude limb-positioning system may be the basis for the rich motor repertoire that the spinal cord is capable of generating.

What variety of movements can be generated by adding a few convergent vector fields? This question was addressed by Mussa-Ivaldi and Giszter (1992) in a theoretical study based on the observation that any generic movement may be formally represented as a vector field—that is, as a mapping that associates an output force to any value of a system's state (defined by position and velocity). Let us consider, for example, two hypothetical alternatives in the motor control

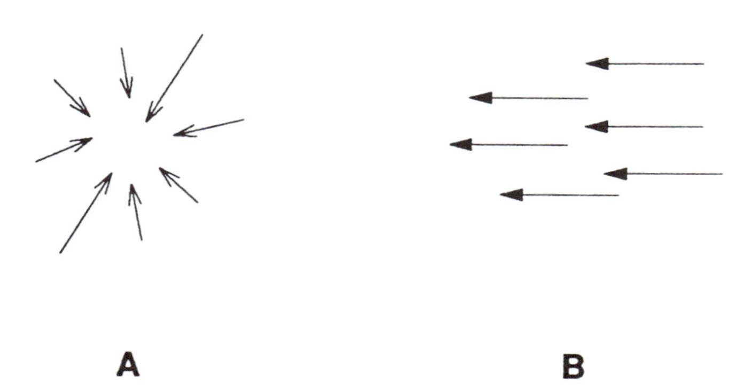

FIGURE 31.8 Force fields and control strategies. (A) A convergent field of forces corresponding to position control. (B) A uniform and parallel field of forces corresponding to force control.

spectrum: the so-called position and force control strategies. Position control may be enforced by responding to an externally imposed displacement with a "restoring force." Thus, position control is represented by a field of forces that converge toward a point in the work space (figure 31.8A). In contrast, an ideal force controller generates a field of parallel forces, such as the one shown in figure 31.8B. This parallel pattern expresses the requirement that the output force should not be influenced by errors in the predicted system's state.

The idea of representing a desired motor output as a field of forces has been exploited in robotics as a way to plan trajectories within obstacle-ridden environments. Kathib (1986), Hogan (1984), and Rimon and Koditschek (1989) have shown that an obstacle can be represented as a field of repulsive forces, pushing the arm away from the region of space to be avoided. In contrast, the goal of a reaching movement is represented as a field of attractive forces pulling the arm toward the goal position. It follows that a successful reaching trajectory is derived simply by letting the arm go within the net field determined by the combination of attractive and repulsive forces. According to this view, any desired behavior of a limb may be described, in principle, as a field of forces.

In the robotics approach just, described, once a field has been appropriately planned, then the execution is a straightforward process in which the state variables of the robotic arm are measured and converted into the appropriate forces. In other words, there are no a priori constraints affecting the force field that can be implemented by the controllers. In contrast, our investigations of the spinalized frog, as well as our measure of multijoint arm stiffness in humans (Mussa-Ivaldi, Hogan, and Bizzi, 1985), have suggested that the biological control system is highly constrained by the viscoelastic muscle properties, by the geometrical properties of muscles, and by the organization of the spinal modules. Given these neural and biomechanical constraints, can a desired and arbitrary force field be approximated by a vectorial combination of the fields generated by each module? The question of what repertoire can be obtained by adding the outputs of a set of modules is mathematically equivalent to a problem of functional approximation.

To address this problem, Mussa-Ivaldi (1992) has extended the paradigm of basis function networks (Poggio and Girosi, 1990) to the approximation of vector fields. According to this approach, a pattern of vectors sampled at a set of locations in the work space is approximated by a linear combination of elementary fields called *basis fields*. The problem of reproducing an arbitrary field is then equivalent to the problem of finding an appropriate set of combination coefficients, given a number of vectors expressing the desired feature. For example, to express a desired force control policy, it is sufficient to choose a number of sampled vectors, parallel to one another and of equal amplitude. Of course, the quality of the approximation will depend on both the functional form and the number of basis fields.

Mussa-Ivaldi and Giszter (1992) suggested that the force fields generated by the premotor circuits in a frog's spinal cord may be regarded as a particular set of basis fields. Following this approach, they found that the vectorial combination of as few as four convergent force fields is an adequate mechanism for generating a broad variety of movements in the horizontal plane. In particular, a local field of parallel forces—corresponding to a force control policy—may be obtained by scaling and adding together a set of convergent fields. The converse, of course, is not true. Note that if the basis fields were a set of ideal force generators implementing uniform patterns of forces, there would be no way to obtain anything other than another uniform force pattern from a simple linear summation of the basis fields. Thus, the apparent burden implied by the viscoelastic properties of muscles and the structure of the force fields generated by the spinal cord turns out to be a crucial feature not only for the inherent stability of springlike systems (Hogan, 1984) but also for the possibility of constructing a repertoire of behavior with a simple summation mechanism.

Note that it is straightforward to incorporate time in this paradigm by assuming that the output of each spinal module is not a static but a time-varying field of forces. In this case, the activation of the *i-th* module induces an output force that is a function of position (x) and time (t): $\phi_i(x, t)$. This is a time-varying vector field that corresponds to the representation of an elementary *pattern generator* in the form of a mechanical wave. According to this view and taking into account the finding that the fields generated by spinal stimulation add vectorially, we suggest that a broad class of motor behaviors may be obtained as a weighted superposition of elementary waves:

$$\sum_i c_i \phi_i(x, t)$$

In this expression, the coefficients c_i constitute a set of modulators by means of which the descending commands may implement different planned behaviors. If one such behavior is a particular time-varying field, $F(x, t)$, then the task of implementing it would be equivalent to finding modulators that provide a good approximation for $F(x, t)$:

$$F(x,t) \sim \sum_i c_i \phi_i(x, t)$$

ACKNOWLEDGMENTS This work was supported by National Institutes of Health grants NS09343 and AR26710 and Office of Naval Research grant N00014/90/J/1946.

REFERENCES

ABEND, W., E. BIZZI, and P. MORASSO, 1982. Human arm trajectory formation. *Brain* 105:331–348.

ALSTERMARK, B., and S. SASAKI, 1986. Integration in descending motor pathways controlling the forelimb in the cat: 15. Comparison of the projection from excitatory C3-C4 propriospinal neurones to different species of forelimb motoneurones. *Exp. Brain Res.* 63:543–566.

BERNSTEIN, N. A., 1967. *The Coordination and Regulation of Movements.* New York: Pergamon.

BIZZI, E., N. ACCORNERO, W. CHAPPLE, and N. HOGAN, 1984. Posture control and trajectory formation during arm movement. *J. Neurosci.* 4:2738–2744.

BIZZI, E., N. HOGAN, F. A. MUSSA-IVALDI, and S. GISZTER, 1992. Does the nervous system use equilibrium-point control to guide single and multiple joint movements. *Behav. Brain Sci.* 15:603–613.

BIZZI, E., F. A. MUSSA-IVALDI, and S. GISZTER, 1991. Computations underlying the execution of movement: A biological perspective. *Science* 253:287–291.

BIZZI, E., A. POLIT, and P. MORASSO, 1976. Mechanisms underlying achievement of final head position. *J. Neurophysiol.* 39:435–444.

CAMINITI, R., P. B. JOHNSON, and A. URBANO, 1990. Making arm movements within different parts of space: Dynamic aspects in the primate motor cortex. *J. Neurosci.* 10:2039–2058.

FELDMAN, A. G., 1974. Change of muscle length due to shift of the equilibrium point of the muscle-load system. *Biofizika* 19:534–538.

FELDMAN, A. G., 1986. Once more on the equilibrium-point hypothesis (λ model) for motor control. *J. Mot. Behav.* 18:17–54.

FLASH, T., and I. GUREVICH, 1992. Arm movement and stiffness adaptation to external loads. In *Proceedings of the Annual Conference of the IEEE Engineering in Medicine and Biology Society*, Orlando, Fla. IEEE Publishing Services, New York, pp. 885–886.

FLASH, T., and N. HOGAN, 1985. The coordination of arm movements: An experimentally confirmed mathematical model. *J. Neurosci.* 5:1688–1703.

GEORGOPOULOS, A. P., R. E. KITTNER, and A. B. SCHWARTZ, 1988a. Primate motor cortex and free arm movements to visual targets in three-dimensional space: I. Relations between single cell discharge and direction of movement. *J. Neurosci.* 8:2913–2927.

GEORGOPOULOS, A. P., R. E. KITTNER, and A. B. SCHWARTZ, 1988b. Primate motor cortex and free arm movements to visual targets in three-dimensional space: II. Coding of the direction of movement by a neuronal population. *J. Neurosci.* 8:2928–2937.

GISZTER, S. F., F. A. MUSSA-IVALDI, and E. BIZZI, 1993. Convergent force fields organized in the frog's spinal cord. *J. Neurosci.* 13:467–491.

HOGAN, N., 1984. An organizing principle for a class of voluntary movements. *J. Neurosci.* 4:2745–2754.

HOGAN, N., 1985. The mechanics of multi-joint posture and movement control. *Biol. Cybern.* 52:315–331.

HOGAN, N., E. BIZZI, F. A. MUSSA-IVALDI, and T. FLASH, 1987. Controlling multijoint motor behavior. In *Exercise and Sport Sciences Reviews 15*, K. B. Pandolf, ed. New York: MacMillan, pp. 153–190.

HOLLERBACH, J. M., and C. G. ATKESON, 1987. Deducing planning variables from experimental arm trajectories: Pitfalls and possibilities. *Biol. Cybern.* 56:279–292.

KALASKA, J. F., D. A. D. COHEN, M. L. HYDE, and M. PRUD'HOMME, 1989. A comparison of movement direction-related versus load direction-related activity in primate motor cortex, using a two-dimensional reaching task. *J. Neurosci.* 9:2080–2102.

KATHIB, O., 1986. Real-time obstacle avoidance for manipulators and mobile robots. *Int. J. Robot Res.* 5:90–99.

LOEB, E. P., S. F. GISZTER, P. BORGHESANI, and E. BIZZI, 1992. Effects of dorsal root cut on the forces evoked by spinal microstimulation in the frog. *Somatosens. Mot. Res.* 10:81–95.

Masino, T., and E. I. Knudsen, 1990. Horizontal and vertical components of head movement are controlled by distinct neural circuits in the barn owl. *Nature* 345:434–437.

Morasso, P., 1981. Spatial control of arm movements. *Exp. Brain Res.* 42:223–227.

Mussa-Ivaldi, F. A., 1992. From basis functions to basis fields: Vector field approximation from sparse data. *Biol. Cybern.* 67:479–489.

Mussa-Ivaldi, F. A., and S. F. Giszter, 1992. Vector field approximation: A computational paradigm for motor control and learning. *Biol. Cybern.* 67:491–500.

Mussa-Ivaldi, F. A., N. Hogan, and E. Bizzi, 1985. Neural, mechanical and geometrical factors subserving arm posture in humans. *J. Neurosci.* 5:2732–2743.

Poggio, T., and F. Girosi, 1990. Networks for approximation and learning. *Proc. IEEE* 78:1481–1497.

Rimon, E., and D. E. Koditschek, 1989. The construction of analytic diffeomorphisms for exact robot navigation on star worlds. In *Proceedings of the 1989 IEEE International Conference on Robotics and Automation, Scottsdale, Ariz.*, IEEE Publishing Service, New York, pp. 21–26.

Saltzman, E., 1979. Levels of sensorimotor representation. *J. Math. Psychol.* 20:96–1063.

Shadmehr, R., and F. A. Mussa-Ivaldi, in press. Adaptive representation of dynamics during learning of a motor task. *J. Neurosci.*

Uno, Y., M. Kawato, and R. Suzuki, 1989. Formation and control of optimal trajectory in human multijoint arm movement: Minimum torque-change model. *Biol. Cybern.* 61:89–101.

32 Motor Cortex and Cognitive Processing

APOSTOLOS P. GEORGOPOULOS

ABSTRACT This chapter summarizes key observations and concepts concerning the involvement of the motor cortex in motor cognitive processes operating on the direction of movement in space. Large populations of neurons in motor cortex are engaged with reaching movements. Single cells are directionally broadly tuned, but the neuronal population carries an unambiguous directional signal. The outcome of this population code can be visualized as a vector that points in the direction of the upcoming movement during the reaction time, during an instructed delay period, and during a memorized delay period. Moreover, when a mental transformation is required for the generation of a reaching movement in a direction different from a reference direction, the population vector provides a direct insight into the nature of the cognitive process by which the required transformation is achieved.

Most of the brain deals with motor function. A number of areas of the cerebral cortex and a number of subcortical structures, including the cerebellum and large portions of the basal ganglia, brain stem and spinal cord, are concerned with the specification, control, and ongoing modification of self-initiated or stimulus-elicited movements. The variety and complexity of the structures involved and the intricacy of the aspects of motor function controlled can be appreciated by considering the large variety of motor syndromes produced by disease processes affecting motor structures. These motor deficits range, in two extremes, from paralysis, which is loss of voluntary movement, to apraxia, which is loss of particular motor skills in the absence of paralysis.

A crucial node in brain control of motor function is occupied by the motor cortex, a strip of cerebral cortex located just in front of the central sulcus, in the precentral gyrus. The motor cortex provides major outputs to the spinal cord and brain stem and is heavily interconnected with other cortical areas and with the major subcortical structures, the cerebellum and the basal ganglia. The question of how the motor cortex performs its function has been intensely investigated in the past century, but it is only during the last 25 years that major advances have been made as a result of the availability of a direct technique by which the ongoing activity of cells in the motor cortex could be recorded in awake, behaving animals performing various motor tasks (Lemon, 1984). Several studies showed that changes in motor cortical cell activity precede the development of the motor output and relate quantitatively to its intensity and spatial characteristics (for reviews, see Evarts, 1981; Georgopoulos, 1990). Specifically, when reaching in space, cell activity relates primarily to the direction of the movement and less to its extent (see Georgopoulos, 1990). Under isometric conditions, when a static force is exerted, cell activity relates to the magnitude (see Evarts, 1981) and direction (Kalaska et al., 1989) of force, and when a force pulse is developed, cell activity relates to the incremental dynamic force exerted but not to the total force output (Georgopoulos et al., 1992).

Coding of motor direction by single cells and neuronal populations

A major aspect of the spatial characteristics of the motor output, be it movement or isometric force, is its direction in space. Cells in the motor cortex (Georgopoulos et al., 1982; Georgopoulos, Schwartz, and Kettner, 1986; Schwartz, Kettner, and Georgopoulos, 1988; Kalaska et al., 1989; Caminiti, Johnson, and Urbano, 1990; Schwartz, 1993), as well as in other structures (Kalaska, Caminiti, and Georgopoulos, 1983; Fortier, Kalaska, and Smith, 1989; Caminiti et al., 1991), are broadly tuned to the direction of movement. This means that the cell activity is highest

APOSTOLOS P. GEORGOPOULOS Brain Sciences Center, Veterans Affairs Medical Center, and Departments of Physiology and Neurology, University of Minnesota Medical School, Minneapolis, Minn.

A

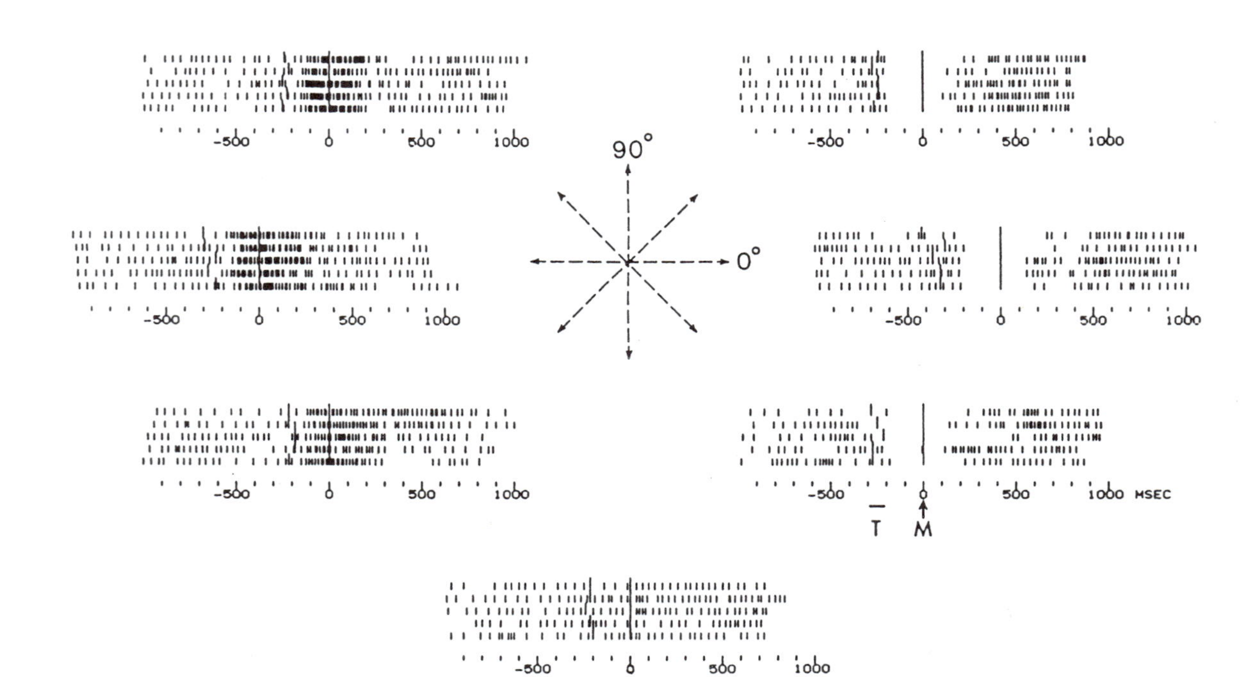

B

IMPULSES/SEC
60
40
20
0
45
135
225
315°
DIRECTION OF MOVEMENT

for a movement in a particular direction (the cell's preferred direction) and decreases progressively with movements farther away from this direction. The changes in cell activity relate to the direction and not the endpoint of the reaching movement (Georgopoulos, Kalaska, and Caminiti, 1985). Quantitatively, the crucial variable on which cell activity depends is the angle formed between the direction of the movement and the cell's preferred direction: The intensity of cell activity is a linear function of the cosine of this angle (Georgopoulos et al., 1982; Schwartz, Kettner, and Georgopoulos, 1988). An example is shown in figure 32.1. The preferred directions of single cells range throughout the three-dimensional directional continuum (Schwartz, Kettner, and Georgopoulos, 1988) (see color plate 12, top panel).

The broad directional tuning indicates that a given cell participates in movements of various directions and that a movement in a particular direction will involve activation of a whole population of cells. Given that single cells are directionally tuned, we proposed a vectorial neural code for the direction of reaching by the neuronal ensemble (Georgopoulos et al., 1983; Georgopoulos, Schwartz, and Kettner, 1986; Georgopoulos, Kettner, and Schwartz, 1988): (1) A particular vector represents the contribution of a directionally tuned cell and points in the cell's preferred direction; (2) cell vectors are weighted by the change in cell activity during a particular movement; and (3) the sum of these vectors (i.e., the population vector) provides the unique outcome of the ensemble coding operation. We found that the population vector points in the direction of the movement (Georgopoulos et al., 1983; Georgopoulos, Schwartz, and Kettner, 1986; Georgopoulos, Kettner, and Schwartz, 1988) (figure 32.2 and color plate 12, middle panel). Ninety-five percent confidence intervals on the direction of the population vector can be generated using statistical bootstrapping techniques (Georgopoulos, Kettner, and Schwartz, 1988) (color plate 12, lower panel). The population vector approach has proved useful not only in studies of motor cortex (Georgopoulos et al., 1983; Georgopoulos, Schwartz, and Kettner, 1986; Georgopoulos, Kettner, and Schwartz, 1988; Kalaska et al., 1989; Caminiti, Johnson, and Urbano, 1990) but also in studies of other motor areas, including the cerebellum (Fortier, Kalaska, and Smith, 1989), the premotor cortex (Caminiti et al., 1991), and areas 5 (Kalaska, Caminiti, and Georgopoulos, 1983; Kalaska, 1988) and 7 (Steinmetz et al., 1987) of the parietal cortex.

FIGURE 32.1 Directional tuning of a cell recorded in the arm area of the motor cortex during two-dimensional reaching. (Top) Impulse activity during five trials of reaching in the directions indicated in the drawing at the center. Short vertical bars indicate the occurrence of an action potential. Rasters are aligned to the onset of movement (M). Longer vertical bars preceding the onset of movement indicate the onset of the target (T); those following the movement indicate, successively, the entrance to the target zone and the delivery of reward. (Bottom) Average frequency of discharge ($\pm$standard error of the mean) from the onset of the stimulus until the entry to the target zone is plotted against the direction of movement. Continuous curve is a cosine function fitted to the data using multiple regression analysis. (Reprinted from Georgopoulos et al., 1982, by permission. Copyright by Society for Neuroscience.)

SOME GENERAL PROPERTIES OF THE NEURONAL POPULATION VECTOR *The neuronal population vector predicts the direction of reaching for movements of different origin.* In these experiments, monkeys made movements that were begun from different points and were in the same direction but described parallel trajectories in three-dimensional space. Under these conditions, the population vector in the motor cortex predicted well the direction of the reaching movement (Kettner, Schwartz, and Georgopoulos, 1988; Caminiti et al., 1991), even if the preferred directions of individual cells shifted systematically in the horizonal plane with different movement origins (Caminiti, Johnson, and Urbano, 1990).

The direction of reaching is predicted well by neuronal population vectors in different cortical layers. The average absolute angle between the population vector calculated from cells in the upper layers (II and III) and the direction of movement was $4.31° \pm 2.98°$ (mean $\pm$ standard deviation, $n = 8$ movement directions), compared to $2.32° \pm 2.06°$ for the lower layers (V and VI) (Georgopoulos, 1990). This finding suggests that the ensemble operation of the population vector can be realized separately in the upper and lower layers, which is important, for that information can then be distributed to different structures according to the differential projections from the upper and lower layers (Jones and Wise, 1977).

The neuronal population coding of the direction of reaching is resistant to loss of cells. The population coding just described is a distributed code and, as such, does not

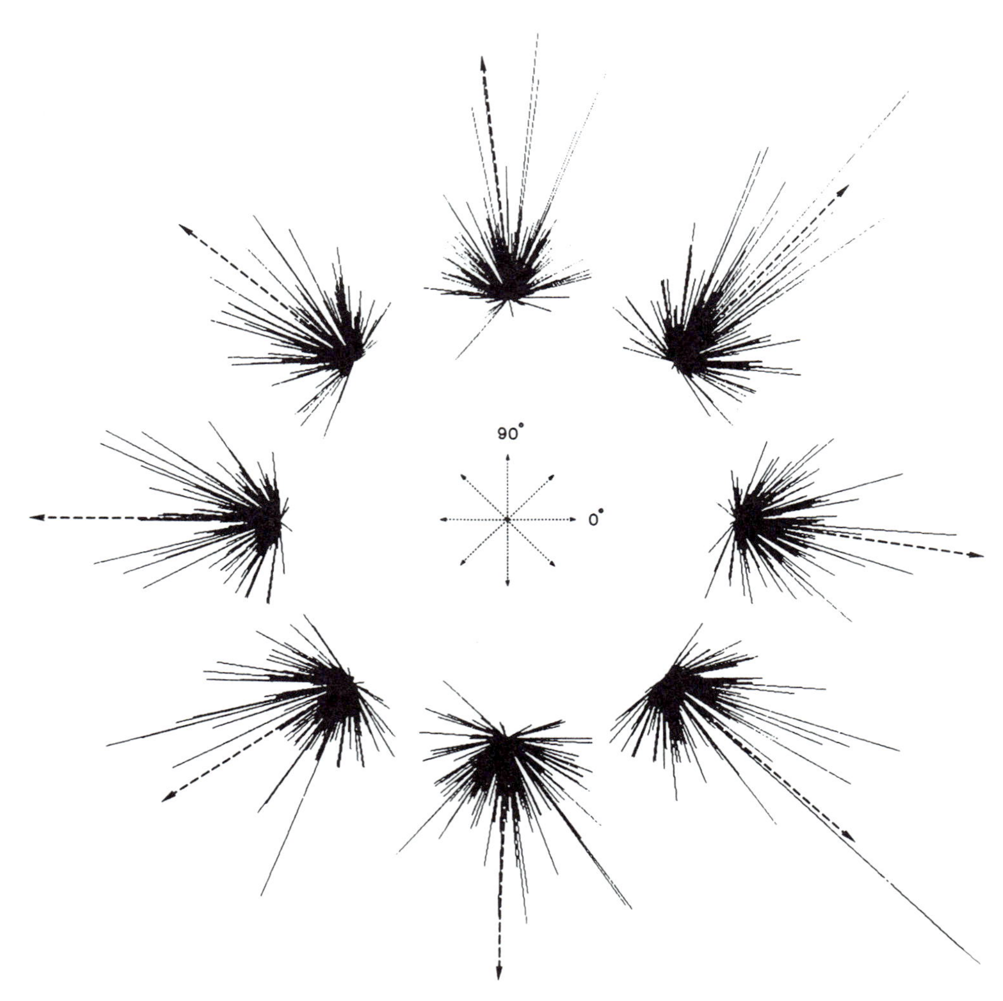

FIGURE 32.2 Population vector analysis applied to eight movement directions in two-dimensional space. Each cluster represents the same population; the movement directions are shown in the diagram at the center. The population vector (interrupted arrow) points in or near the direction of movement. (Reprinted from Georgopoulos et al., 1983, by permission of the publisher.)

depend exclusively on any particular cell. This robustness was evaluated by calculating the population vector from progressively smaller samples of cells randomly selected from the original population (Georgopoulos, Kettner, and Schwartz, 1988). Indeed, the direction of the population vector can be reliably estimated from as few as 100–150 cells.

The neuronal population vector transmits directional information comparable to that transmitted by the direction of movement. In the standard task used in our studies, monkeys (Georgopoulos et al., 1982) or human subjects (Georgopoulos and Massey, 1988) moved a manipulandum from the center of a planar working surface to a target on a circle. In this case, the direction of the target is the ideal direction: If the subject's movements were straight lines from the center to the target, the subject's performance would be perfect and we could say that the movement transmitted the maximum possible information. However, movements rarely end dead on target and, therefore, the information transmitted is rarely maximal: The more the dispersion of the movement endpoint around the target, the less information is transmitted. Now, this dispersion can be parcellated into errors in the amplitude of movement and errors in the direction of movement; accordingly, the information transmitted by the amplitude and the direction of movement can be studied separately. Because my colleagues and I were interested in the control of the direction of movement, we asked subjects to "move in the direction of the target" without imposing restrictions on the amplitude or the endpoint of the movement: This provided a purely directional task (Georgopoulos and Massey, 1988). The calculation of the information transmitted by the direction of movement involves the construction of a performance matrix, in which the ideal and actual directions are tabulated and from which the information transmitted can be computed (Georgopoulos and Massey, 1988). Essentially, the same technique can be used to calculate the information transmitted by the direction of the population vector. Because the population vector is the vectorial sum of weighted contributions of individual cells, and because these weights can vary from trial to trial due to intertrial variability in neuronal discharge, then the direction of the population vector can vary somewhat from trial to trial: This variation can be treated in exactly the same way as the direction of movement, and the information transmitted can be calculated.

Indeed, we calculated the information transmitted by the direction of movement and the direction of the population vector (Georgopoulos and Massey, 1988) and found the following: First, the information transmitted by both of these measures increases with input information but more slowly than the maximum possible, and it tends to saturate at high levels of input information. This loss of information is probably due to noise generated during the initial (perceptual) and successive (perceptual-motor) processing stages. Second, the information transmitted by the population vector was consistently higher than that transmitted by the movement by approximately 0.5 bits. This means that an additional loss of information is incurred between the motor cortex and the movement. However, this loss differs from the previous one, for it does not increase with increasing stimulus information but remains constant at approximately 0.5 bits at all levels of input information. Such a loss could be suffered during processing in other motor structures or at the stage of biomechanical implementation of the movement.

The neuronal population vector predicts the direction of dynamic isometric force (Georgopoulos et al., 1992). This finding established the fact that the coding of directional information applies to the motor output in general, even in the absence of joint motion. Moreover, it showed that the direction specified by the motor cortex is not that of the total force exerted by the subject but that of the dynamic component of the force—that is, the component of the force remaining after a constant, static force is subtracted.

TIME-VARYING PROPERTIES OF THE NEURONAL POPULATION VECTOR *The neuronal population vector predicts the movement trajectory in continuous, tracing movements* (Schwartz, 1993). Monkeys were trained to trace smoothly with their index finger sinusoids displayed on a screen, from one end to the other. The direction of the population vectors, calculated successively in time along the trajectory, changed throughout the sinusoidal movement, closely matching the smoothly changing direction of the finger path. Moreover, a neural "image" of the sinusoidal trajectory of the movement was obtained by connecting successive population vectors tip to tail. This finding suggests that the length of the population vector carries information concerning the instantaneous velocity of the movement.

The neuronal population vector predicts the direction of reaching during the reaction time. This is the simplest case of predicting the direction of an upcoming movement. Given that the changes in cell activity in the motor cortex precede the onset of movement by approximately 160–180 ms, on average (Georgopoulos et al., 1982), it is an important finding that the population vector predicts the direction of the upcoming movement during that period during which the movement is being planned (Georgopoulos et al., 1984; Georgopoulos, Kettner, and Schwartz, 1988). An example is shown in figure 32.3.

The neuronal population vector predicts the direction of reaching during an instructed delay period. Experiments were conducted in which monkeys were trained to withhold the movement for a period of time after the onset of a visual cue signal and to move later in response to a go signal. During this instructed delay period, the population vector in the motor cortex, computed every 20 ms, gave a reliable signal concerning the direction of the movement that was triggered later for execution (Georgopoulos, Crutcher, and Schwartz, 1989).

Neural mechanisms of cognitive processing: neuronal populations as keys for understanding

The results summarized in the preceding sections underscore the operational usefulness of the neuronal population vector for monitoring in time the directional tendency of the neuronal ensemble. We took advantage of this property and used the population vector as a probe to decipher the neural processing of directional information during two cognitive opera-

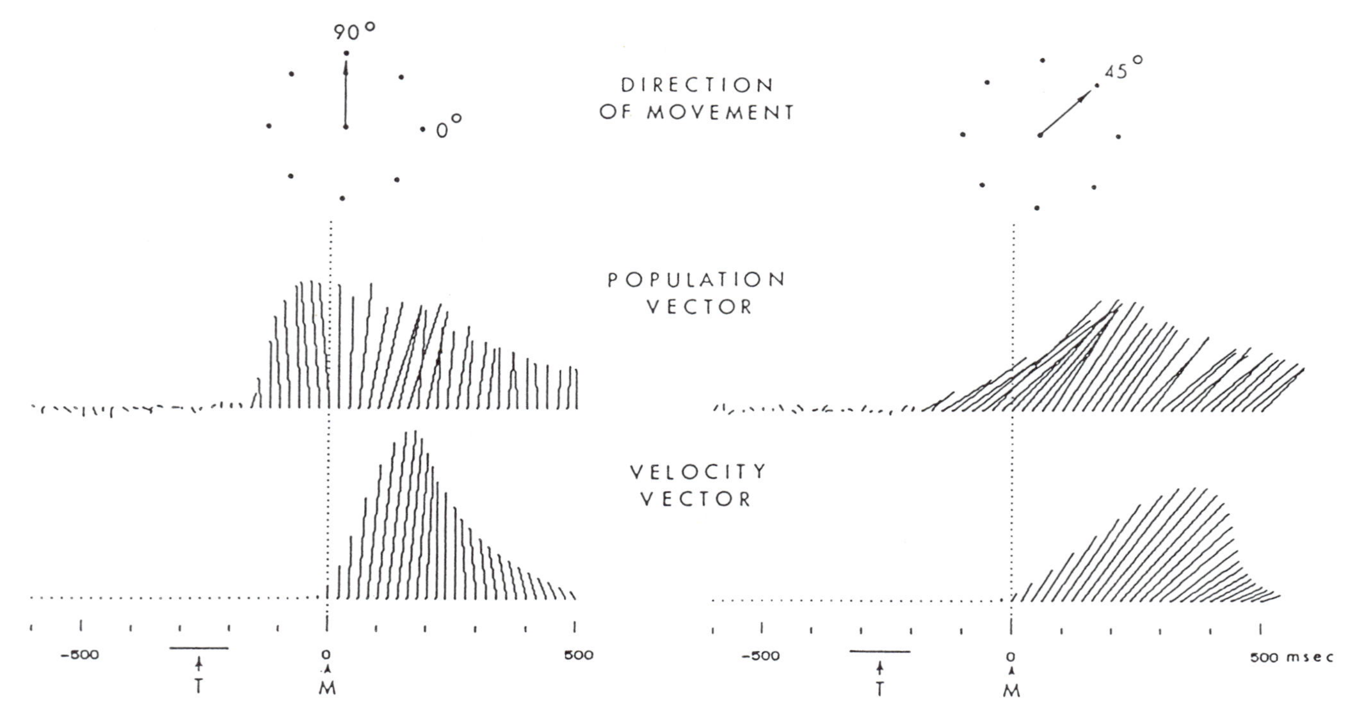

FIGURE 32.3 The population vector points in the direction of movement well before the movement begins. The results for two movement directions in two-dimensional space are illustrated (top). The population vector was calculated every 20 ms (middle). The average instantaneous (20-ms bin) velocity of the movement also is shown (bottom). Before the target onset (T), the population vector is very short and its direction varies from moment to moment. Well before the onset of movement (M), it increases in length and its direction points to the direction of the upcoming movement. This finding suggests that even the earliest inputs to the motor cortex are relevant to the direction of the upcoming movement. (Reprinted from Georgopoulos et al., 1984, by permission of the publisher.)

tions, one involving memory holding and the other mental rotation.

MEMORY HOLDING In these experiments (Smyrnis et al., 1992), two rhesus monkeys were trained to move a handle on a two-dimensional working surface in directions specified by a light on the plane. They first captured with the handle a light on the center of the plane and then moved the handle in the direction indicated by a peripheral light (*cue* signal). The signal to move (*go* signal) was given by turning off the center light. The following tasks were used: In the *nondelay* task, the peripheral light was turned on at the same time as the center light went off. In the *memorized delay* task, the peripheral light stayed on for 300 ms (cue period) and the center light was turned off 450–750 ms later (delay period). Finally, in the *nonmemorized-delay* task, the peripheral light stayed on continuously, whereas the center light went off 750–1050 ms after the peripheral light came on. Recordings in the arm area of the motor cortex ($N = 171$ cells) showed changes in single-cell activity in all tasks. The population vector was calculated every 20 ms following the onset of the peripheral light. We were interested in two aspects of the information carried by the population vector. One concerns its direction, which can be interpreted as the directional information carried by the directional signal. The other concerns the length of the population vector, which can be regarded as the strength of the directional signal carried. The direction of the population vector during the memorized delay period was close to the direction of the target (figure 32.4). It is interesting that the population vector length was similar in the cue period, but it was longer during the memorized versus the nonmemorized part of the delay. This is demonstrated in figure 32.5, which illustrates the time course of the length of the population vector in the two delay tasks. Three phases can be distinguished in this time course. First, there is an initial increase of vector length during the 300 ms of the delay period; this increase is

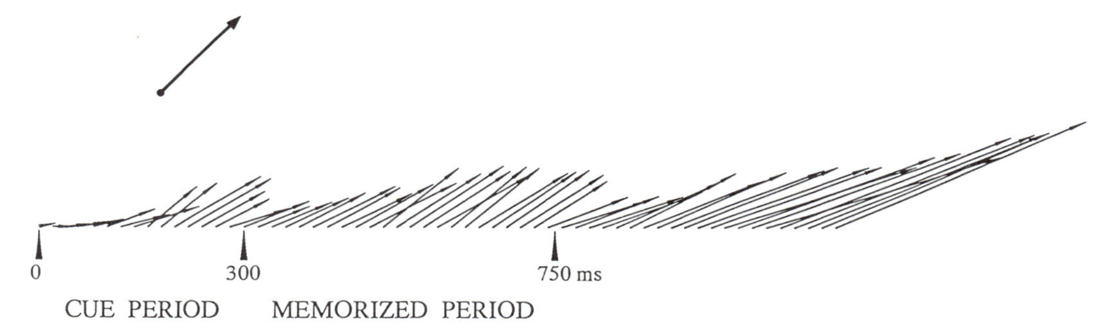

FIGURE 32.4 Population vectors in the memorized delay task for the direction indicated are plotted every 20 ms. The arrow on top indicates the direction of the cue signal present during the first 300 ms of the delay period. (Reprinted from Smyrnis et al., 1992, by permission.)

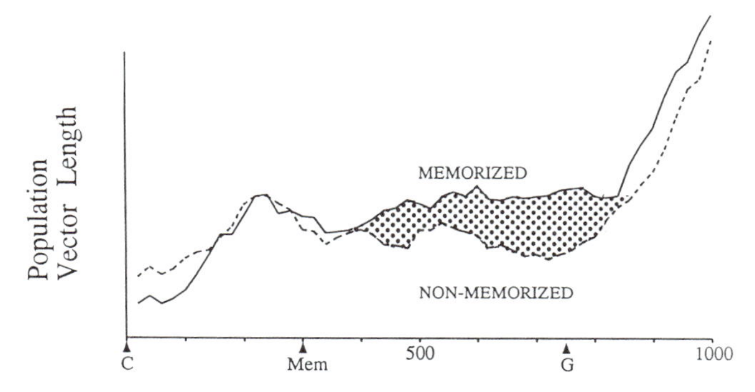

FIGURE 32.5 Length of mean result of the population vector is plotted against time for the two delay tasks. G, minimum time of onset of the go signal. (Reprinted from Smyrnis et al., 1992, by permission.)

similar for both tasks. Second, this increase subsides during the rest of the nonmemorized delay period but continues at a somewhat higher level during the memorized delay period; the latter difference is indicated by stippling in figure 32.5. Finally, there is a steep increase in the population vector length following the go signal at the end of the delay period. Thus, the memorized task is distinguished from the nonmemorized one by the higher population signal in that part of the delay period during which the instructed direction had to be held in memory.

The directional information carried by the population vector in the memorized task identified the memorized information in a direct fashion. Moreover, this analysis provided an insight concerning the time course of encoding and holding directional information. For that purpose, we used the length of the population vector, which can be regarded as reflecting the strength of the directional signal in the neuronal ensemble. The population vector length showed an initial increase that started approximately 100 ms after the cue onset and peaked at 250 ms. This increase was very similar in both the memorized and the nonmemorized delay tasks (see figure 32.5). We interpret this initial peak as reflecting an *encoding* process. A second phase ensued that differed in the memorized and nonmemorized tasks in that a higher, sustained signal was present during the memorized delay period but not during the nonmemorized delay (see stippled area in figure 32.5). We interpret this as reflecting a *holding-in-memory* process. After onset of the go signal, the population vector length increased similarly in all tasks used. These findings are interesting because the increase in the signal during the memorized delay period was observed *in the absence* of the target, though one would expect that the signal would be stronger in the presence rather than the absence of the visual stimulus. This finding strengthens our interpretation of this increase as a memory signal, in contrast to a sensory one, and raises the more general possibility that the motor cortex may be particularly involved when only part of the visual information about an upcoming movement is provided.

MENTAL ROTATION The second cognitive process my colleagues and I chose to study involved a transformation of an intended movement direction. In these studies, we first carried out psychological experiments in human subjects. Then we trained monkeys to perform the same task and recorded the activity of single cells in the brain of each of these animals during performance of the task. Finally, we tried to connect the neural results with those of the human studies and interpret the latter on the basis of the former, the objective being to get as close as possible in relating neurophysiology and cognitive psychology. The following is a description of these steps as they were applied to a particular problem of a mental transformation of movement direction.

The task required subjects to move a handle at an angle from a reference direction defined by a visual stimulus on a plane. Because the reference direction changed from trial to trial, the task required that in a given trial, the direction of movement be specified according to this reference direction. In the *psychological studies* (Georgopoulos and Massey, 1987) human subjects performed blocks of 20 trials in which the angle above and its departure (counterclockwise or clockwise) were fixed, although the reference direction varied. Seven angles (5°–140°) were used. The basic finding was that the time to initiate a movement (reaction time) increased in a linear fashion with the angle. The most parsimonious hypothesis to explain these results is that subjects arrive at the correct direction of movement by shifting their motor intention from the reference direction to the movement direction, traveling through the intermediate angular space. This idea is very similar to the mental rotation hypothesis proposed by Shepard and Cooper (1982) to explain the monotonic increase of the reaction time with orientation angle when a judgment must be made regarding whether a visual image is normal or mirror-image: In both cases, a mental rotation is postulated. In fact, the mean rates of rotation and their range among subjects were very similar in the perceptual (Shepard and Cooper, 1982) and motor (Georgopoulos and Massey, 1987) studies. Moreover, when the same subjects performed both perceptual and motor rotation tasks, their processing rates were positively correlated (Pellizzer and Georgopoulos, 1993), which indicates similar processing constraints for both tasks.

In the *neurophysiological studies* (Georgopoulos et al., 1989; Lurito, Georgakopoulos, and Georgopoulos, 1991), two rhesus monkeys were trained to move the handle 90° and counterclockwise from the reference direction; these trials were intermixed with others in which the animals moved in the direction of the target. When the time-varying neuronal population vector was calculated during the reaction time, it was found that it rotated from the stimulus (reference) to the movement direction through the counterclockwise angle when the animal had to move away from the stimulus, or in the direction of the stimulus when the animal had to move toward it (figure 32.6). It is remarkable that the population vector rotated at all, but especially that it rotated through the smaller, 90° counterclockwise angle. These results showed clearly that the cognitive process in this task involved rotation of an analog signal. The occurrence of a true rotation was further documented by showing that there was a transient increase during the middle of the reaction time in the recruitment of cells with preferred directions between the stimulus and movement directions: This indicated that rotation of the population vector was not the result of varying activation of just two cell groups, one with preferred directions centered on the stimulus and another in the movement direction. Therefore, this rotation process, sweeping through the directionally tuned ensemble, provided for the first time a direct visualization of a dynamic cognitive process. In this respect, it is noteworthy that the population vector is, in essence, a measure that can take continuous values in direction.

In summary, the results of these studies provide the neural correlates of a dynamic cognitive representation (Freyd, 1987). The essential contribution of this work is in the neural identification and visualization of the time-varying, dynamic representation of direction in the motor cortex when a transformation of this direction is required and achieved. Interestingly, the mean rotation rate and the range of rates observed for different reference directions were very similar to those obtained in the human studies.

Functional imaging of the human motor cortex

Although the single-cell recording technique has contributed significantly to our knowledge of cortical function, it is limited in that it can usually be applied only to a restricted brain area at one time. Other techniques, including positron emission tomography, can provide the greater picture of areas of activation in the brain during performance of a task. A new major tool is the blood oxygenation level–dependent (BOLD) contrast imaging of the brain using nuclear magnetic resonance (Ogawa et al., 1990). The technique is noninvasive and sensitive, does not require averaging of data from more than one subject, and possesses adequate resolution. It has already produced scans with both high spatial resolution and functional, task-dependent activation of the human motor cortex (Ban-

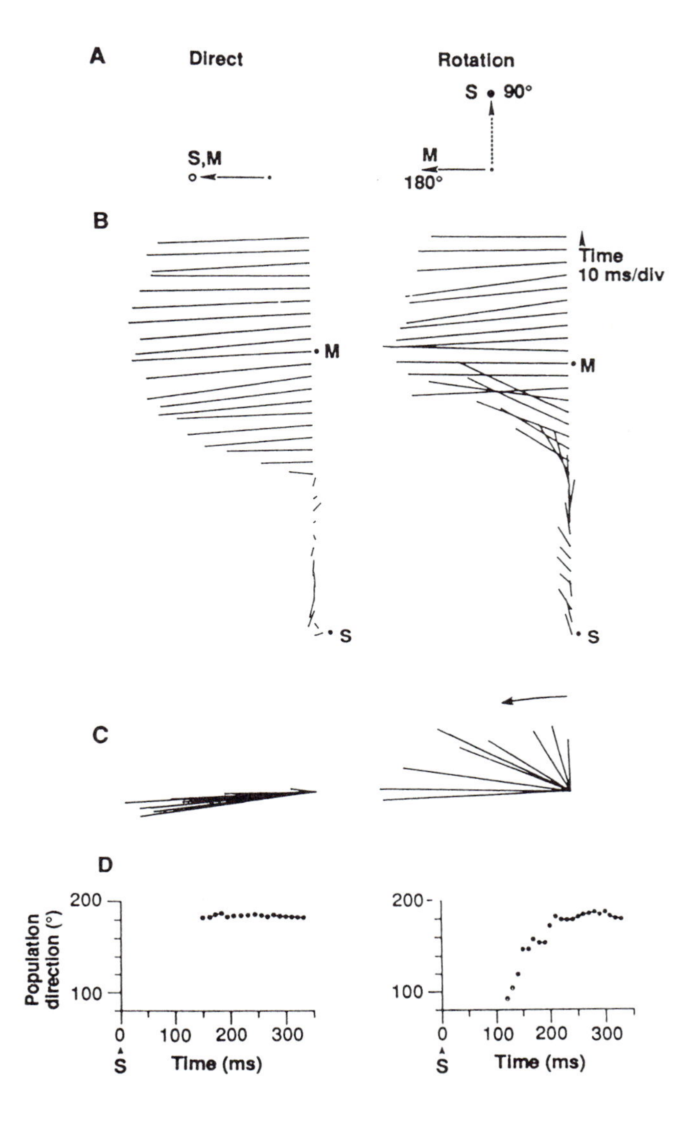

FIGURE 32.6 Results from a direct (left) and rotation (right) movement. (A) Task. Unfilled and filled circles indicate dim and bright light, respectively. Interrupted and continuous lines with arrows indicate stimulus (S) and movement (M) direction, respectively. (B) Neuronal population vectors calculated every 10 ms from the onset of S, at positions shown in (A), until after the onset of M. When the population vector lengthens, for the direct case (left) it points in the direction of the movement, whereas for the rotation case it points initially in the direction of the stimulus and then rotates counterclockwise (from 12 o'clock to 9 o'clock) and points in the direction of the movement. (C) Ten successive population vectors from (B) are shown in a spatial plot, starting from the first population vector that increased significantly in length. Note the counterclockwise rotation of the population vector (right panel). (D) Scatter plots of the direction of the population vector as a function of time, starting from the first population vector that increased significantly in length following stimulus onset. For the direct case (left panel), the direction of the population vector is in the direction of the movement (~180°); for the rotation case (right panel), the direction of the population vector rotates counterclockwise from the direction of the stimulus (~90°) to the direction of the movement (~180°).

dettini et al., 1992; Kim et al., 1993). Especially at high magnetic fields (4 tesla), functional images of high contrast and spatial resolution (e.g., less than 0.7 mm in image plane) depicting activated gray-matter areas can be obtained. An example of such imaging of the human motor cortex during a finger task is shown in color plate 13. This method holds great promise for understanding the cooperation among various brain regions in motor processing.

Neural modeling of the motor cortex

The motor cortex can be essentially regarded as a neural network. The elements of the network are directionally tuned, and one possible operation of the network is the computation of the neuronal population vector. The broad directional tuning of single cells seems to be a general aspect of the population operation, as this property appeared in the hidden layer of a three-layer network trained to calculate the population vector (Lukashin, 1990). Another question concerns the interactions among cells. It is known that there are extensive local interconnections (Huntley and Jones, 1991) as well as functional neuronal interactions (Stefanis and Jasper, 1964; Asanuma and Brooks, 1965; for a review see Fetz, Toyama, and Smith, 1991). A three-part objective then is to (1) identify these interactions among the directionally tuned cells in the motor cortex and discover the rules that govern their presence, (2) study an artificial neural network made of directionally tuned elements with massive interconnections, and (3) compare the results obtained in the real (i.e., motor cortical) and artificial networks.

We found the following (Georgopoulos, Taira, and Lukashin, in press): First, interactions between cells in the motor cortex relate to the directional tuning of the cells in a pair: Interactions are more than twice as frequent when the cells are tuned than when they are not and, for directionally tuned cells, the interaction ranges from strongly positive (i.e., excitatory) to strongly negative (i.e., inhibitory) as the angle between the preferred directions of the cells in a pair varies from

0° (i.e., same preferred direction) to 180° (i.e., opposite preferred directions). Second, the same trend was observed between the directionally tuned elements in a massively interconnected, dynamic artificial network during a stable period of computation of the neuronal population vector. Third, it was found that for the network to be stable, the strength of the synaptic interactions must be low: In the best (i.e., most stable) case, the mean synaptic strength tends to $2/N$, where N is the number of elements in the network. This is in keeping with the fact that cortical cells in an area are extensively but weakly interconnected (Martin, 1988). These findings validate the correspondence between the motor cortical and the artificial neural network and open the possibility of using this network in the temporal domain to explore the mechanisms of the cognitive operations described in the preceding section.

Conclusion

Single-cell recording, imaging, and dynamic computer modeling techniques are powerful tools that will continue to advance our knowledge. There is already a good database from the most painstaking and time-consuming studies of single-cell recordings in behaving monkeys, as reviewed previously; the imaging and modeling studies should proceed at a faster pace. Then, we should arrive at a much better understanding of the mechanisms by which the motor cortex and other brain areas process motor and cognitive information.

ACKNOWLEDGMENT This work was supported by U.S. Public Health Service grant NS 17413.

REFERENCES

ASANUMA, H., and V. B. BROOKS, 1965. Recurrent cortical effects following stimulation of internal capsule. *Arch. Ital. Biol.* 103:220–246.

BANDETTINI, P. A., E. C. WONG, R. S. HINKS, R. S. TIKOFSKY, and J. S. HYDE, 1992. Time course EPI of human brain function during task activation. *Magn. Reson. Med.* 25:390–397.

CAMINITI, R., P. B. JOHNSON, C. GALLI, S. FERRAINA, Y. BURNOD, and A. URBANO, 1991. Making arm movements within different parts of space: The premotor and motocortical representation of a coordinate system for reaching at visual targets. *J. Neurosci.* 11:1182–1197.

CAMINITI, R., P. B. JOHNSON, and A. URBANO, 1990. Making arm movements within different parts of space: Dynamic aspects in the primate motor cortex. *J. Neurosci.* 10:2039–2058.

EVARTS, E. V., 1981. Motor cortex and voluntary movement. In *Handbook of Physiology*, Sec. 1, vol. 2 (part 2), J. M. Brookhart, V. B. Mountcastle, V. B. Brooks, and S. R. Geiger, eds. Bethesda, Md.: American Physiological Society pp. 1083–1120.

FETZ, E. E., K. TOYAMA, and W. SMITH, 1991. Synaptic interactions between cortical neurons. In *Cerebral Cortex*, A. Peters and E. G. Jones, eds. New York: Plenum, pp. 1–47.

FORTIER, P. A., J. F. KALASKA, and A. M. SMITH, 1989. Cerebellar neuronal activity related to whole-arm reaching movements in the monkey. *J. Neurophysiol.* 62:198–211.

FREYD, J. J., 1987. Dynamic mental representations. *Psychol. Rev.* 94:427–438.

GEORGOPOULOS, A. P., 1990. Neurophysiology and reaching. In *Attention and Performance*, vol. 13, M. Jeannerod, ed. Hillsdale, N.J.: Erlbaum, pp. 227–263.

GEORGOPOULOS, A. P., J. ASHE, N. SMYRNIS, and M. TAIRA, 1992. Motor cortex and the coding of force. *Science* 256: 1692–1695.

GEORGOPOULOS, A. P., R. CAMINITI, J. F. KALASKA, and J. T. MASSEY, 1983. Spatial coding of movement: A hypothesis concerning the coding of movement direction by motor cortical populations. *Exp. Brain Res.* Suppl. 7: 327–336.

GEORGOPOULOS, A. P., M. D. CRUTCHER, and A. B. SCHWARTZ, 1989. Cognitive spatial motor processes: 3. Motor cortical prediction of movement direction during an instructed delay period. *Exp. Brain Res.* 75:183–194.

GEORGOPOULOS, A. P., J. F. KALASKA, and R. CAMINITI, 1985. Relations between two-dimensional arm movements and single cell discharge in motor cortex and area 5: Movement direction versus movement endpoint. *Exp. Brain Res.* Suppl. 10:176–183.

GEORGOPOULOS, A. P., J. F. KALASKA, R. CAMINITI, and J. T. MASSEY, 1982. On the relations between the direction of two-dimensional arm movements and cell discharge in primate motor cortex. *J. Neurosci.* 2:1527–1537.

GEORGOPOULOS, A. P., J. F. KALASKA, M. D. CRUTCHER, R. CAMINITI, and J. T. MASSEY, 1984. The representation of movement direction in the motor cortex: Single cell and population studies. In *Dynamic Aspects of Neocortical Function*, G. M. Edelman, W. M. Cowan, and W. E. Gall, eds. New York: Wiley, pp. 501–524.

GEORGOPOULOS, A. P., R. E. KETTNER, and A. B. SCHWARTZ, 1988. Primate motor cortex and free arm movements to visual targets in three-dimensional space: II. Coding of the direction of movement by a neuronal population. *J. Neurosci.* 8:2928–2937.

GEORGOPOULOS, A. P., J. LURITO, M. PETRIDES, A. B. SCHWARTZ, and J. T. MASSEY, 1989. Mental rotation of the neuronal population vector. *Science* 243:234–236.

GEORGOPOULOS, A. P., and J. T. MASSEY, 1987. Cognitive spatial-motor processes: 1. The making of movements at various angles from a stimulus direction. *Exp. Brain Res.* 65:361–370.

GEORGOPOULOS, A. P., and J. T. MASSEY, 1988. Cognitive

spatial motor processes: 2. Information transmitted by the direction of two-dimensional arm movements and by neuronal populations in primate motor cortex and area 5. *Exp. Brain Res.* 69:315–326.

GEORGOPOULOS, A. P., A. B. SCHWARTZ, and R. E. KETTNER, 1986. Neuronal population coding of movement direction. *Science* 233:1416–1419.

GEORGOPOULOS, A. P., M. TAIRA, and A. V. LUKASHIN (in press). Cognitive neurophysiology of the motor cortex. *Science.*

HUNTLEY, G. W., and E. G. JONES, 1991. Relationship of intrinsic connections to forelimb movement representations in monkey motor cortex: A correlative anatomic and physiological study. *J. Neurophysiol.* 66:390–413.

JONES, E. G., and S. P. WISE, 1977. Size, laminar and columnar distribution of efferent cells in the sensory-motor cortex of monkeys. *J. Comp. Neurol.* 175:391–438.

KALASKA, J. F., 1988. The representation of arm movements in postcentral and parietal cortex. *Can. J. Physiol. Pharmacol.* 66:455–463.

KALASKA, J. F., R. CAMINITI, and A. P. GEORGOPOULOS, 1983. Cortical mechanisms related to the direction of two-dimensional arm movements: Relations in parietal area 5 and comparison with motor cortex. *Exp. Brain Res.* 51: 247–260.

KALASKA, J. F., D. A. D. COHEN, M. L. HYDE, and M. PRUD'HOMME, 1989. A comparison of movement direction-related versus load direction-related activity in primate motor cortex, using a two-dimensional reaching task. *J. Neurosci.* 9:2080–2102

KETTNER, R. E., A. B. SCHWARTZ, and A. P. GEORGOPOULOS, 1988. Primate motor cortex and free arm movements to visual targets in three-dimensional space: III. Positional gradients and population coding of movement direction from various movement origins. *J. Neurosci.* 8:2938–2947.

KIM, S.-G., J. ASHE, A. P. GEORGOPOULOS, H. MERKLE, J. M. ELLERMANN, R. S. MENON, S. OGAWA, and K. UĞURBIL, 1993. Functional imaging of human motor cortex at high magnetic field. *J. Neurophysiol.* 69:297–302.

LEMON, R. N., 1984. *Methods for Neuronal Recording in Conscious Animals.* Chichester, Engl.: Wiley.

LUKASHIN, A. V., 1990. A learned neural network that simulates properties of the neuronal population vector. *Biol. Cybern.* 63:377–382.

LURITO, J. T., T. GEORGAKOPOULOS, and A. P. GEORGOPOULOS, 1991. Cognitive spatial-motor processes: 7. The making of movements at an angle from a stimulus direction: Studies of motor cortical activity at the single cell and population levels. *Exp. Brain Res.* 87:562–580.

MARTIN, K. A. C., 1988. From single cells to simple circuits in the cerebral cortex. *Q. J. Exp. Physiol.* 73:637–702.

OGAWA, S., T-M. LEE, A. R. KAY, and D. W. TANK, 1990. Brain magnetic resonance imaging with contrast dependent blood oxygenation. *Proc. Natl. Acad. Sci. U.S.A.* 87: 9868–9872.

PELLIZZER, G., and A. P. GEORGOPOULOS, 1993. Common processing constraints for visuomotor and visual mental rotations. *Exp. Brain Res.* 93:165–172.

SCHWARTZ, A. B., 1992. Motor cortical activity during drawing movements: Single-unit activity during sinusoid tracing. *J. Neurophysiol.* 68:528–541.

SCHWARTZ, A. B. (in press). Motor cortical activity during drawing movements: Population representation during sinusoid tracing. *J. Neurophysiol.* 70:28–36.

SCHWARTZ, A. B., R. E. KETTNER, and A. P. GEORGOPOULOS, 1988. Primate motor cortex and free arm movements to visual targets in three-dimensional space: I. Relations between single cell discharge and direction of movement. *J. Neurosci.* 8:2913–2927.

SHEPARD, R. N., and L. COOPER, 1982. *Mental Images and Their Transformations*, Cambridge, Mass.: MIT Press.

SMYRNIS, N., M. TAIRA, J. ASHE, and A. P. GEORGOPOULOS, 1992. Motor cortical activity in a memorized delay task. *Exp. Brain Res.* 92:139–151.

STEFANIS, C., and H. JASPER, 1964. Recurrent inhibition in pyramidal tract neurons. *J. Neurophysiol.* 27:855–877.

STEINMETZ, M. A., B. C. MOTTER, C. J. DUFFY, and V. B. MOUNTCASTLE, 1987. Functional properties of parietal visual neurons: Radial organization of directionalities within the visual field. *J. Neurosci.* 7:177–191.

33 Coordinate Transformations and Motor Planning in Posterior Parietal Cortex

RICHARD A. ANDERSEN

ABSTRACT The posterior parietal cortex is neither a strictly visual nor a strictly motor structure; rather it performs visuomotor integration functions including coordinate transformations for the determination of spatial locations and the formation of plans for movement. Coordinate transformations are an essential aspect of visually guided behavior and are required because sensory information is derived in the coordinates of the retina and must be transformed to the coordinates of muscles for movement. These transformations produce in the posterior parietal cortex a representation of space that uses a population code and is formed by a specific operation that systematically combines visual and eye position signals to form *planar gain fields*. Activity related to the planning of eye movements has been found in the lateral intraparietal area, a recently discovered cortical area that appears to be specialized for saccades. This planning-related activity appears to encode the movements that the animal intends to make.

Recent neurophysiological experiments suggest there exist intermediate and abstract representations of space interposed between sensory input and motor output. These intermediate representations are formed by combining information from various modalities. A *head-centered* representation refers to a coordinate system framed with respect to the head and is formed by combining information about eye position and the location of a visual stimulus imaged on the retinas (figure 33.1) A *body-centered* coordinate representation likewise is achieved by combining head, eye, and retinal position information (see figure 33.1). An even more complicated representation is one in *world-centered* coordinates (see figure 33.1), which can be achieved by combining vestibular signals with eye position and retinal position signals. There is reason to believe that the brain contains and uses all these representations. This chapter focuses on head-centered representations of space formed by the combining of eye and head position signals in the posterior parietal cortex. The studies outlined here provide a glimpse into the internal operations of the brain that are the basis of our spatial perceptions and actions.

A second major aspect of sensorimotor integration is the planning of movements. At some point in this integration process, sensory signals give way to signals related to what the animal intends to do. In the second part of this chapter, I will discuss evidence that this step from sensory representation to decisions to make movements utilizes the neural circuitry within the posterior parietal cortex. Such studies have been aided by the recent discovery of a small area within the posterior parietal cortex, the lateral intraparietal area (LIP), which appears to be specialized in the processing of saccadic eye movements. This area carries not only sensory information related to the targets for eye movements but also signals related to the planning of eye movements.

Representation of space in area 7a of posterior parietal cortex

Lesions to the posterior parietal cortex produce profound spatial deficits in both humans and nonhuman primates. My colleagues and I thus chose to examine how space is represented in the posterior parietal cortex by examining the spatial receptive field properties of neurons in behaving monkeys. One might have imagined that locations in head-centered coordinates are encoded using receptive fields similar to retinal receptive fields but anchored in head-centered, rather than

RICHARD A. ANDERSEN Division of Biology, California Institute of Technology, Pasadena, Calif.

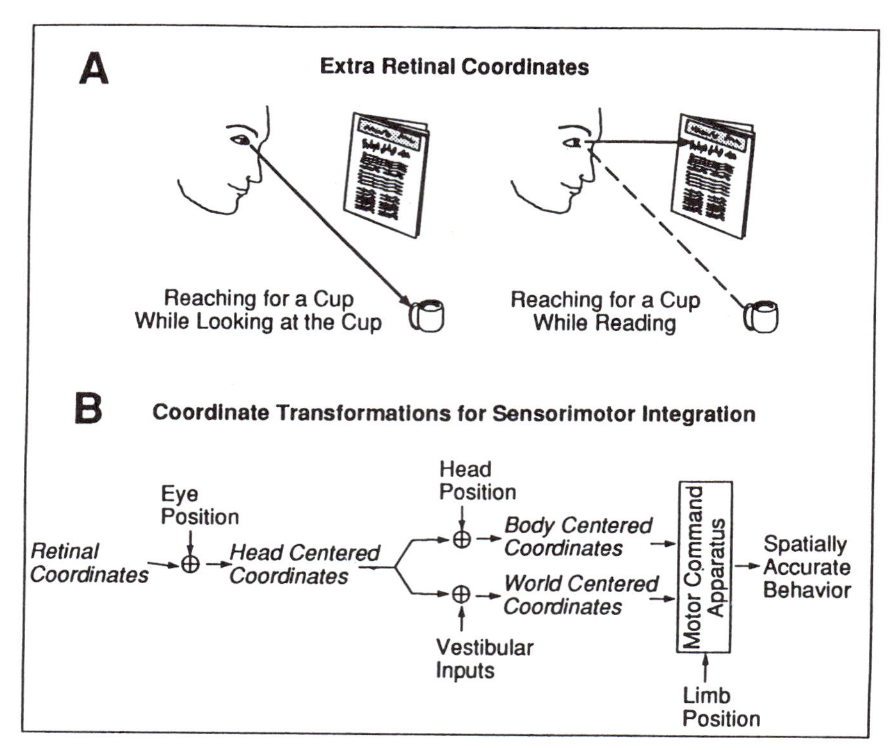

FIGURE 33.1 (A) Demonstration of why representations of space in extraretinal coordinates are required for accurate motor behaviors. The term *extraretinal* refers to the encoding of visual stimuli in higher-level coordinate frames than simple retinal coordinates. In the sketch on the left, a person is fixating the cup and it is imaged on the foveas, whereas on the right the subject is fixating the newspaper and the cup is imaged on a peripheral part of the retinas. In both cases, the individual is able to localize the cup with a reaching movement. Because different parts of the retinas are stimulated in the two conditions, information about eye position must also be available to determine accurately that the cup was at the same location in space. (B) Schematic showing how extraretinal coordinate frames can be computed from retinal coordinates. Visual stimuli are imaged on the retinas and are input to the brain in retinal coordinates. Eye position signals can be added to form representations in head-centered coordinates, and body-centered coordinates can be formed by adding head position information. One way of forming world coordinates is to add vestibular signals, which code the location of the head in the world, to a head-centered coordinate frame. For illustrative purposes, the figure shows these signals being added sequentially. It is not known yet whether there is a hierarchical organization of extraretinal coordinate frames in the brain or whether several of these signals come together at once to form body- and world-coordinate frames, combined with information about limb position derived from proprioceptive inputs, to encode accurate reaching movements. (From Andersen et al., in press, with permission)

retinal, coordinates. If this were the case, each time the eyes would move the receptive field would change the location on the retina from which it derives its input, in order to code the same location in space.

PLANAR GAIN FIELDS Early investigations of area 7a of the posterior parietal cortex showed, however, that locations in head-centered coordinates could be coded in an entirely different format (Andersen and Mountcastle, 1983; Andersen, Essick, and Siegel, 1985). The receptive fields of the neurons did not change their retinal locations when eye position changed. Rather the visual and eye position signals interacted to form *planar gain fields*, in which the amplitude of the visual response was modulated by eye position (Andersen, Essick, and Siegel, 1985) (figure 33.2). The gain fields were said to be planar because the amplitude of the response to stimulation of the same patch of retina varied linearly with horizontal and vertical eye position (Andersen, Essick, and Siegel, 1985).

These results indicated that spatial locations are not represented explicitly at a single-cell level using receptive fields in space. However, the location of a target in head-centered coordinates could still be easily deter-

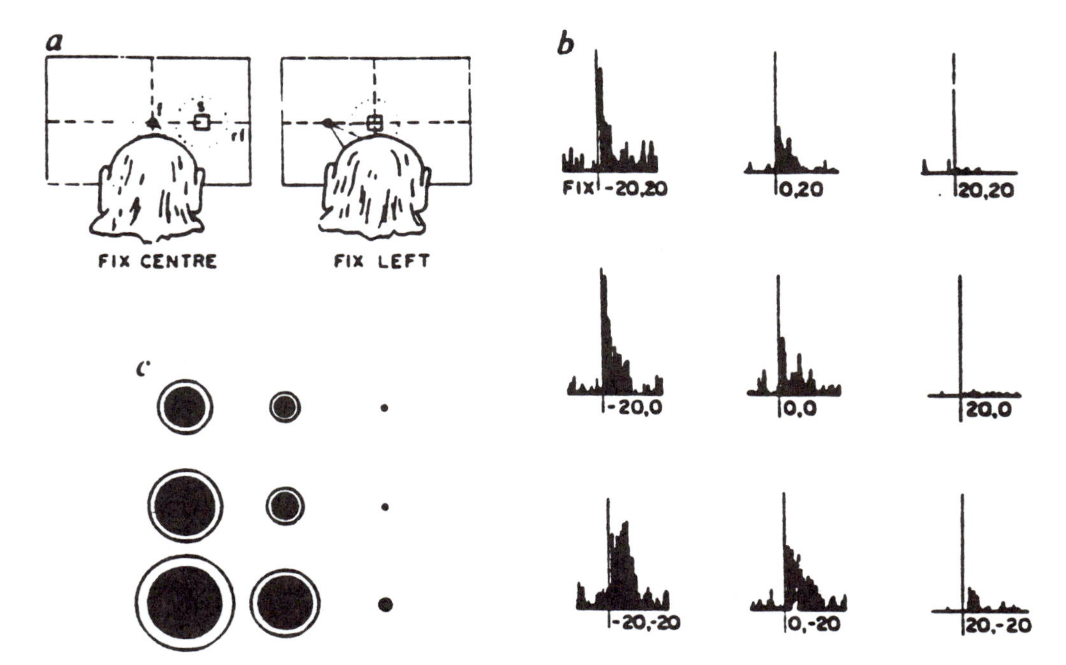

FIGURE 33.2 (a) Experimental protocol for determining spatial gain fields, with the projection screen viewed from behind the monkey's head. To determine the effect of eye position, the monkey, whose head is fixed, fixates on a point (f) at one of nine symmetrically placed locations on the projection screen. The stimulus (S) is always presented at the same retinal location, chosen as the maximum response zone of the retinal receptive field. The stimulus consists of 1°- or 6°-diameter spots flashed for 500 ms. Each measurement is repeated eight times. (b) Peristimulus histograms of a typical gain field determination. The nine histograms are located in the same relative positions as the fixations that produced them. The vertical line indicates the time of visual stimulus onset. (c) A graphical method for illustrating these data, in which the diameter of the darkened inner circle, representing the visually evoked gain fields, is calculated by subtracting the background activity recorded 500 ms before the stimulus onset from the total activity during the stimulus. The annulus diameter corresponds to the background activity that is due to an eye position signal alone, recorded during the 500 ms before the stimulus presentation. (Reprinted with permission from *Nature* 331:679–684. Copyright 1988 Macmillan Magazines Ltd.)

mined if the activity of several area 7a neurons were examined together; in other words, the representation of space is *distributed* in this area. Figure 33.3 demonstrates why this representation is distributed. The contour plot of activity is made for the variables of location in head-centered space and eye position. When examined in this fashion, it can be seen that area 7a neurons are tuned to a particular location in head-centered space but only for a limited range of eye positions. The location of maximum response in head-centered coordinates is a conjunction of the preferred eye position of the cell and the most responsive part of its retinal receptive field. To derive a signal for location in head-centered space independent of eye position requires the activity of a subset of parietal neurons, and thus the code is a distributed one. A distributed code of this sort is, of course, not unique to the posterior parietal cortex. Middle temporal (MT) neurons, for instance, are tuned to a limited range of temporal and spatial frequencies. For stimuli of different shapes to be perceived as moving at the same speed would require a population of cells tuned to different temporal and spatial frequencies. Thus, the perception of speed, independent of the exact texture or shape of a moving stimulus, appears to use a distributed code not unlike the one for spatial location.

Looking at the behavior of single cells as merely components of a much larger, distributed network has been critical in advancing our understanding of how the brain computes locations in space. Neural networks trained to convert inputs of eye position and retinal position into an output of locations in head coordinates develop a distributed representation in the hidden layer interposed between the input and output layers (Zipser and Andersen, 1988). This distributed representation appears to be the same as that found in area

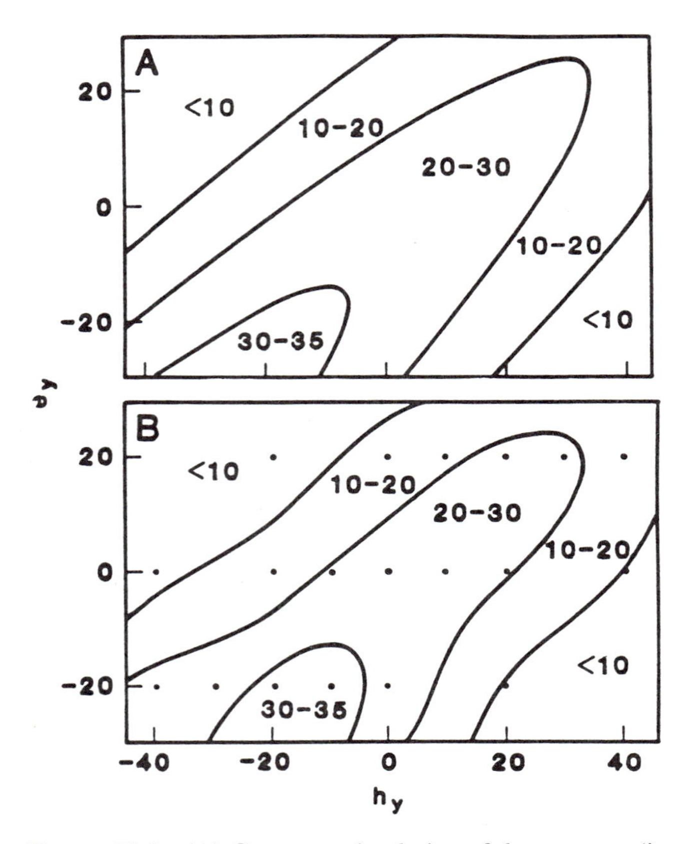

FIGURE 33.3 (A) Computer simulation of the response (in spikes per second) of an area 7a neuron predicted by multiplying the vertical axis of a planar gain field by the vertical axis of a gaussian receptive field. The results are represented on the contour plot with the stimulus head-centered coordinates (h_y) plotted along the abscissa and eye position (e_y) along the ordinate. (B) Contour plot of actual recording data for a cell with the same gain field and receptive field characteristics as the model neuron plotted in (A). Each data point represents the mean evoked response to eight repetitions of the stimulus. The average standard error for these data points was two spikes per second. (Reprinted with permission from Andersen et al., 1985. Copyright 1985 by the AAAS)

7a, with the hidden units exhibiting planar gain fields. A mathematical analysis of this network indicates that the planar gain fields are the basis of an algorithm for adding eye and retinal position vectors in a distributed network (Goodman and Andersen, 1990; Brotchie, Andersen, and Goodman, 1993). Thus, the method of integrating these two signals is not random but is systematic and requires that the gain fields be planar.

One of our neural network models for area 7a was trained to produce output units with receptive fields in head-centered coordinates (Zipser and Andersen, 1988). The middle layer of this model produced gain fields similar to those found in area 7a, suggesting that gain fields are an intermediate stage between retinal and spatial receptive fields. A possible objection to this model is that cells resembling its output (receptive fields in space) are not routinely found. However, we also trained a second network with an output representation similar to the activity found in oculomotor structures and motor centers in general. In this format, activity varies monotonically as a function of location with respect to the head. We have shown that such a network can be trained to make eye movements and have argued that receptive fields in space are an unnecessary encoding of spatial location (Goodman and Andersen, 1989; Andersen et al., 1990). Instead, we believe that cells with planar gain fields are an intermediate step in the transformation from visual to motor coordinates.

OTHER AREAS WITH GAIN FIELDS Recently, gain fields have been found in several areas besides 7a. In monkeys, these areas include cortical area LIP (Andersen et al., 1990), cortical area V3a (Galletti and Battaglini, 1989), the inferior and lateral pulvinar (Robinson, McClurkin, and Kertzmann, 1990) and premotor and prefrontal cortex (Boussaoud, Barth, and Wise, 1993), and in cats in the superior colliculus (Peck et al., 1992). In the cases where data were collected for a sufficient number of eye positions, the gain fields usually were linear for horizontal and vertical eye positions. These results suggest that gain fields are a typical format for representing spatial information in many areas of the brain.

It is interesting that the newer data just cited show that planar gain fields appear to be the predominant method of representing space and performing coordinate transformations. A clue to the predominance of this form of representation comes from Mazzoni, Andersen, and Jordan (1991a). We found that networks with multiple hidden layers trained to make coordinate transformations have gain fields in all the hidden layers. The planar gain field is an economical method for compressing spatial information (Goodman and Andersen, 1990).

It has been suggested that receptive fields in space also may exist in some cortical areas (Battaglini et al., 1990; Fogassi et al., 1992; MacKay and Riehle, 1992). These reports are preliminary, and further work is needed to substantiate such claims.

Distance, body-centered coordinates, and world-centered coordinates

The data in the preceding section indicate that there are representations with respect to the head in the two dimensions of elevation and azimuth. Recent recording experiments suggest that the third dimension of distance from the head is also contained within these representations, and the method of encoding distance is in the form of gain fields. Gnadt and Mays (1991; Gnadt, 1992) found LIP neurons in which the vergence angle modulated the magnitude of the visually evoked responses but not the disparity tuning of the cells. These types of gain fields are also predicted by neural network models similar to the Zipser-Andersen model but are trained to localize in depth (Lehky, Pouget, and Sejnowski, 1990).

Our earlier experiments tested the interaction of eye position and retinal position signals for animals with their heads mechanically immobilized. As a result, head-centered representations could not be distinguished from body-centered representations. With this in mind, Brotchie, Andersen, and Goodman (1993) have examined the effect of head position on the visual response of cells in the posterior parietal cortex. Neural network simulations performed prior to the experiments suggested that posterior parietal neurons should have gain fields for head position as well as eye position if they are representing space in body-centered coordinates. Furthermore, the eye and head gain fields of individual parietal neurons should have the same gradients (two-dimensional slopes), even though the gradients of different cells may vary considerably. The recording experiments from areas 7a and LIP bore out these predictions. Approximately half the cells with eye position gain fields were found to have similar head position gain fields. These results suggest that there may be two representations of space in the posterior parietal cortex, one in head-centered coordinates (units with gain fields for eye position) and the other in body-centered coordinates (units with gain fields for eye and head position).

Finally, recent recordings from our laboratory have shown that vestibular signals are integrated with the various other signals (Snyder, Brotchie, and Andersen, in press). When monkeys are rotated in the chair in the dark, many cells that show tonic activity related to eye position exhibit similar changes in activity for movement of the head relative to the room. Presumably, these cells are receiving an integrated vestibular signal. Because these cells code both location of the eye in the head and location of the head in the world, they are coding the direction of gaze in the world. Another subset of cells has retinal receptive fields with gain fields for eye position and chair rotation. The modulation of the visual response by a vestibular signal suggests that this population of cells may code locations of visual stimuli in world coordinates.

Biologically plausible learning rule

One criticism of neural network models has been that the learning rule used for training the networks is unlikely to be used by the nervous system. Mazzoni, Andersen, and Jordan, (1991a, b) trained a neural network to perform the transformation from retinal to head-centered coordinates using a reinforcement learning rule developed by Barto and Jordan (1987) that is more biologically plausible than backpropagation (figure 33.4). They found the reinforcement-trained networks produced the same gain fields that are produced by the backpropagation-trained networks and are found in the brain, which suggests that the algorithm discovered for computing the coordinate transformation is largely independent of the exact learning rule used to generate it. Likewise, it suggests that posterior parietal neurons could learn or adjust spatial representations using a learning paradigm that is more reasonable in terms of what is currently known about learning mechanisms in the central nervous system.

Microstimulation experiments

Goodman and Andersen (1989) examined the effects of microstimulating the Zipser-Andersen model for eye movements. Their model was connected to a simplified set of oculomotor muscles (four instead of six), and individual hidden units were maximally activated to simulate microstimulation. The most typical result of stimulating individual hidden units was the change in amplitude pattern seen with stimulation of area LIP (figure 33.5b). This model suggests that the change in amplitude pattern is indicative of a distributed representation of space in LIP. Because the representation of head-centered space is distributed, a single cell does not drive the eyes to a goal in space; rather, such behavior requires the activity of many LIP neurons. To demonstrate directly how these changes in amplitude patterns could code single locations among a group of

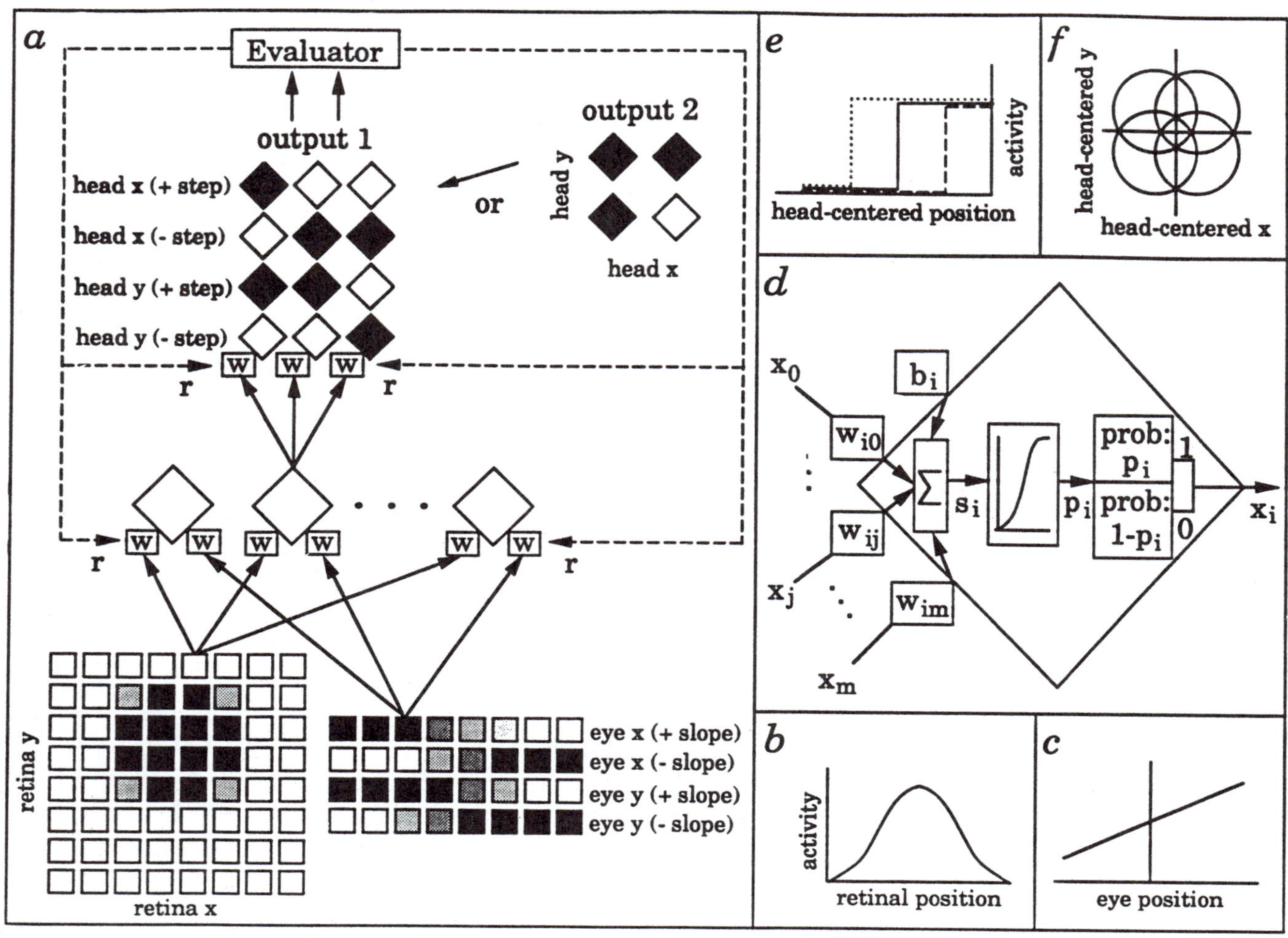

FIGURE 33.4 (a) Network structure. (b) Retinal input is encoded by 64 units with gaussian receptive fields, whereas eye position (c) is represented by 32 units with linear activation functions. In the retinal input, each unit has an output between 0 and 1, a $1/e$ width of 15°, and a receptive field peak 10° apart from that of its horizontal and vertical neighbors. In the eye position input, the output of each unit (between 0 and 1) is a linear function of horizontal or vertical orbital angle, with random slope and intercept. These input formats reproduce properties of certain area 7a neurons that respond only to visual stimuli or to changes in eye position. The shading of each unit is proportional to its activity, with black representing maximum activity. The hidden and output layers are composed of binary stochastic elements (d), which produce an output of 1 with probability (prob) p equal to the logistical function of the sum of the weighted inputs ($s_i = \sum_{j=0}^{m} w_{ij}x_j$), and 0 with probability $1 - p$. The jth unit in the network provides input x_j to ith unit via the connection w_{ij}; m is the number of inputs to the units, and b is a bias. The network used from two to eight hidden units. The output units encode head-centered locations according to one of two output formats. (e) In the binary-monotonic format, each unit produces an output of 1 or 0, depending on whether the encoded locations are to the right or to the left (or, for some units, above or below) a certain reference point. For example, a typical output layer consisted of four sets of three units, giving an output of 1 when the x (or y) craniotopic coordinate is greater than (or less than) -40, 0, or $+40$ degrees. This format is analogous to the eye position input format, in that four groups of units encode an increase in horizontal or vertical position angle by increasing or decreasing their activation monotonically. (f) Another format used is the binary-gaussian one, in which four units give an output of 1 when the spatial position is within 100° of their receptive field centers, which are located at (±60°, ±60°). This format is analogous to that of the retinal input, in that a position angle is encoded topographically by units with overlapping receptive fields.

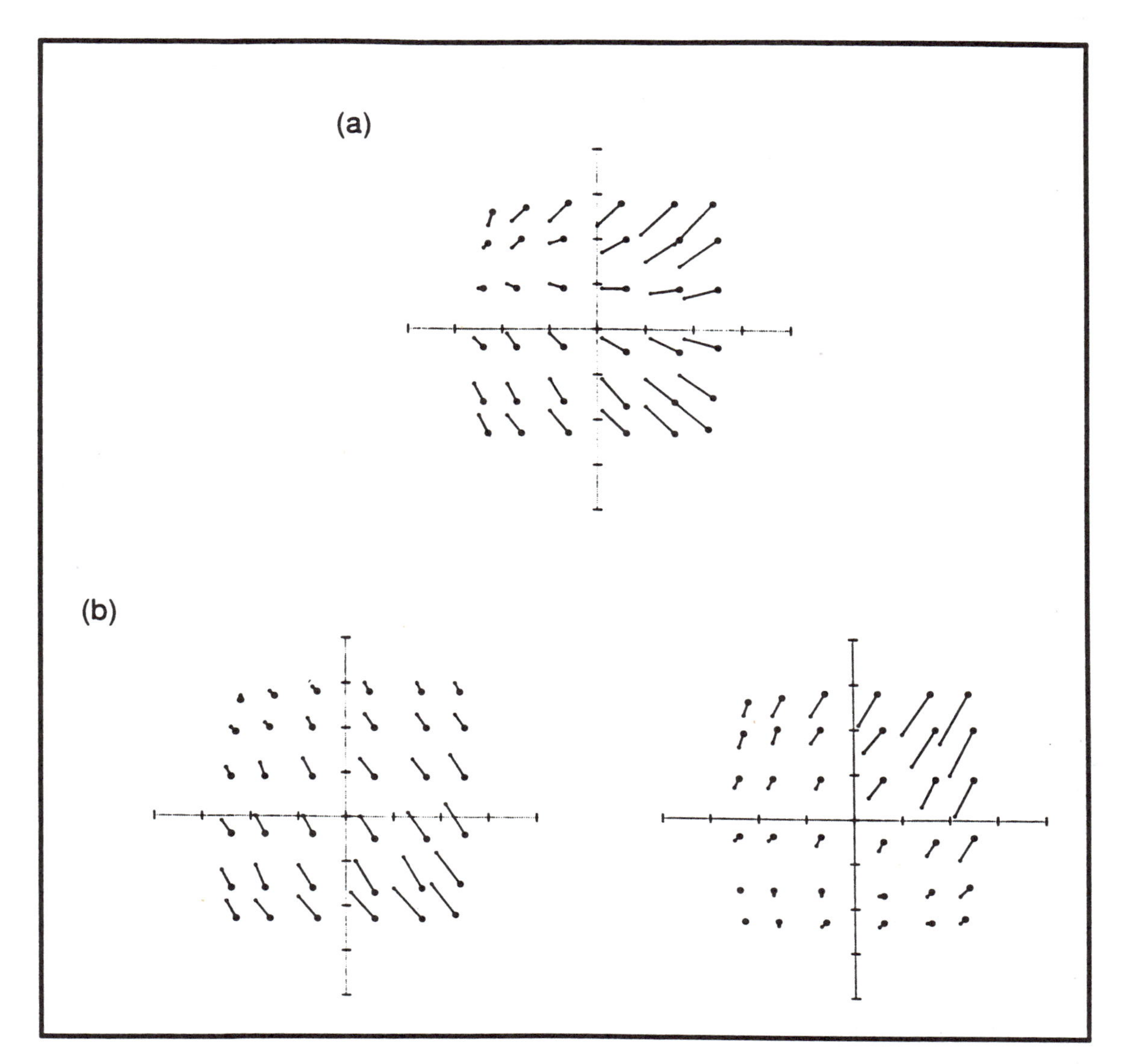

FIGURE 33.5 (a) Eye movements recorded when two hidden units of a monotonic output network are stimulated simultaneously. (b) Eye movements recorded when each of the two hidden units is stimulated alone. Note that the result of simultaneous stimulation, illustrated in (A), is more or less the vector addition of the two saccade fields that results from stimulation of the individual units. (Reprinted with permission from *J. Cogn. Neurosci.* 1:317–326. Copyright MIT Press 1989)

neurons, Goodman and Andersen (1989) showed that stimulation of two or more hidden units produced a pattern of eye movements that converge toward a single goal in head-centered space (figure 33.5a).

Area LIP

It has been appreciated for some time that the posterior parietal cortex is involved in the processing of eye movements. Balint (1909) described bilateral lesions to the posterior parietal cortex in human patients that resulted in the inability to will saccades, although spontaneous saccades were unaffected. In monkeys, electrical stimulation of the posterior parietal area produces saccadic eye movements (Fleming and Crosby, 1955; Wagman, 1964), and lesions to the parietal cortex also produce deficits in saccades (Keating and Gooley, 1988; Lynch and McLaren, 1989).

In the mid-1970s, Mountcastle and his colleagues embarked on cell-recording experiments within the inferior parietal lobule (which encompasses approximately the posterior half of the posterior parietal cortex) and reported cells selective for saccades as well as neurons selective for smooth pursuit, reach, and fixation. Mountcastle's group (1975) reported that in electrode penetrations perpendicular to the cortex, all cells tended to have the same functional properties, an observation consistent with a columnar organization. Lynch and coworkers (1977) later reported that these columns were not segregated by functional types into particular parts of the posterior parietal cortex, suggesting that repeating columns of each functional class

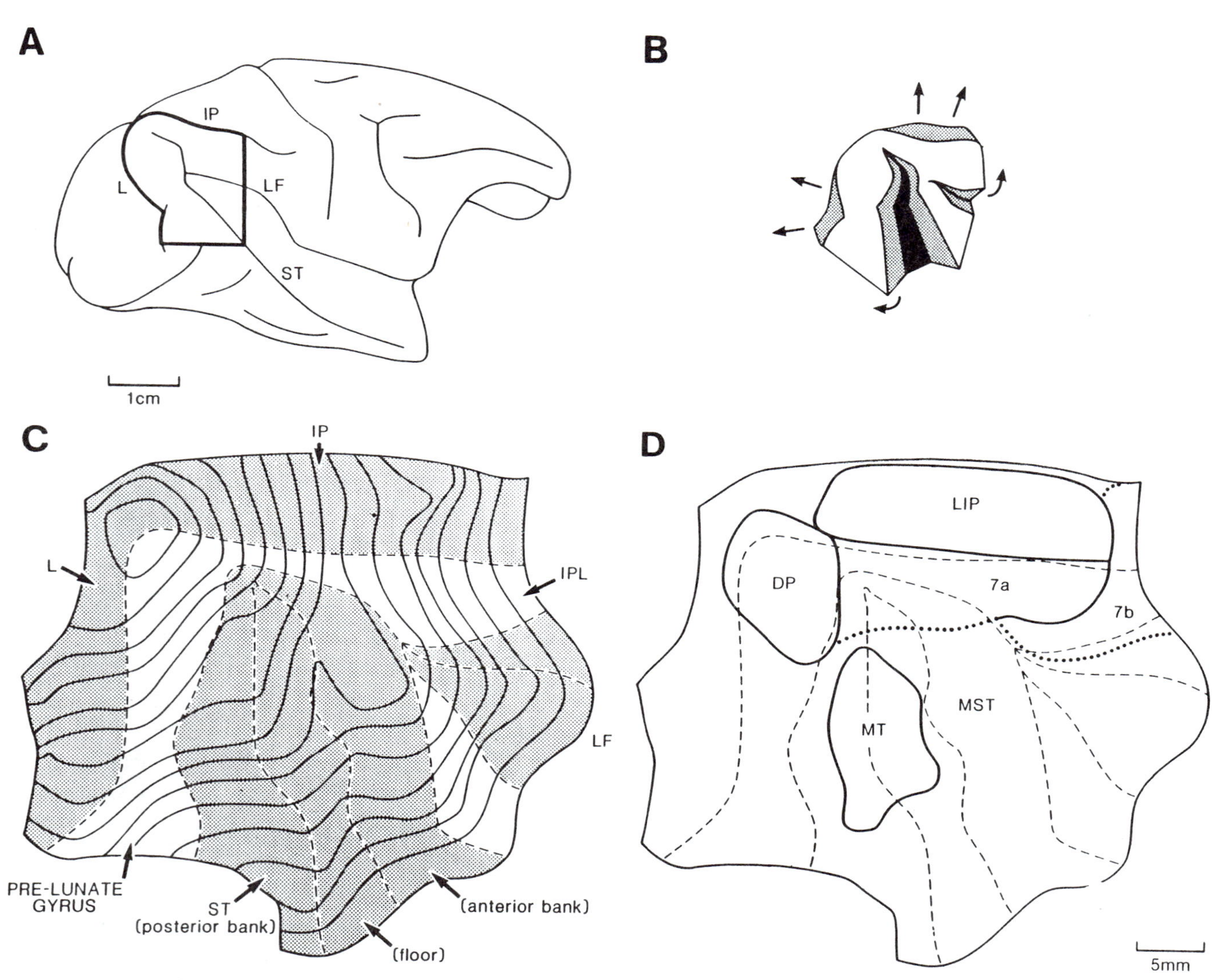

FIGURE 33.6 Parcellation of inferior parietal lobule and adjoining dorsal aspect of the prelunate gyrus used in this study. The cortical areas are represented on flattened reconstructions of the cortex. (A) Lateral view of monkey cortical hemisphere. The darker line indicates the area to be flattened. Cortical areas: L, lateral; IP, intraparietal; LF, lateral fissure; ST, superior temporal. (B) The same cortex isolated from the rest of the brain. Stippled areas are cortex buried in sulci, and the blackened area is the floor of the superior temporal sulcus. The arrows indicate movement of local cortical regions resulting from the mechanical flattening. (C) The completely flattened representation of the same area. The stippled areas represent cortical regions buried in sulci, and the contourlike lines are tracings of layer IV taken from frontal sections through this area. Cortical areas as in (A); IPL, inferior parietal lobule. (D) Locations of several of the cortical areas. The dotted lines indicate borders of cortical fields that are not precisely determinable. Cortical areas: LIP, lateral intraparietal; DP, dorsoparietal; MT, middle temporal; MST, medical superior temporal. (Reprinted with permission from *J. Comp. Neurol.* 296:65–113. Copyright 1990 Wiley-Liss)

are rather evenly distributed across the inferior parietal lobule. These results had to be interpreted with some caution, however, as results were pooled from several different brains and referenced to sulcal patterns that vary considerably from animal to animal. Following up on these observations, Andersen, Asanuma, and Cowan (1985) reasoned that these different types of columns could be selectively labeled with anatomical tracers by assuming that they had connections with different brain structures. Thus, for instance, columns of cells with saccade-related activity would likely project to the frontal lobe in the region of the frontal eye fields . However, when retrograde tracers were injected into the frontal eye fields and adjoining dorsolateral prefrontal cortex, label was found predominantly within the lateral bank of the intraparietal cortex. Andersen, Asanuma, and Cowan (1985) named this area the *lateral intraparietal area* because it was located on the lateral bank of the intraparietal sulcus, lateral to the ventral intraparietal area (VIP) described earlier by Maunsell and Van Essen (1983) (figure 33.6). Subsequent recording experiments showed that most LIP cells had activity related to eye movements, and a majority of these responded prior to saccades (Gnadt and Andersen, 1988; Andersen et al., 1990; Barash et al., 1991a, b). Other studies showed that reach activity was confined largely to area 7b (Hyvärinen and Shelepin, 1979; Robinson and Burton, 1980a, b; Hyvärinen, 1981) and smooth-pursuit activity to the medial superior temporal area (MST) (Newsome, Wurtz, and Komatsu, 1988) (see figure 33.6). Fixation activity typically varies with direction of gaze (Lynch et al., 1977; Sakata, Shibutani, and Kawano, 1980; Andersen, Essick, and Siegel, 1987) and appears primarily to convey information about eye position (Andersen, Essick, Siegel, 1987; Andersen, 1989). These eye position–related activities are typically found in areas LIP and 7a (see figure 33.6). Thus, many of the functional types discovered by Mountcastle and colleagues (1975) are actually segregated into small cortical fields, of which area LIP is one, and not into interdigitated cortical columns.

Physiology

Visual and Saccade-related Responses There was, briefly, some controversy about whether parietal neurons had saccade-related activity. When Mountcastle and colleagues (1975) first observed saccade responses, they proposed that the area issued general commands to make saccadic eye movements. Soon thereafter, Robinson, Goldberg, and Stanton (1978) observed visual responses from parietal neurons and they challenged Mountcastle's command hypothesis, arguing that the cells were responding in a sensory fashion to the saccade targets as visual stimuli rather than in a motor fashion related to the eye movement. Using a memory saccade task that separated sensory from motor responses, Andersen, Essick, and Siegel (1987) showed that posterior parietal neurons had both visual and saccade-related activity. These results suggested that it was more appropriate to consider posterior parietal cortex as being involved in sensorimotor integration rather than as strictly a sensory or a motor structure (Andersen, 1987).

The visual and saccade activity in LIP has recently been studied in great detail by our laboratory (Barash et al., 1991a, b) and compared to visual and saccade activity in area 7a. Saccade-related responses in LIP generally begin prior to eye movements, whereas most area 7a saccade responses are postsaccadic, beginning after initiation of eye movements. This observation, and the reduced activity related to planning eye movements in 7a compared to LIP (see later), led these investigators to propose that LIP participates in the planning of eye movements, whereas area 7a appears to subserve other functions. These studies also showed that visual responses to saccade targets are generally weaker in 7a and have a longer latency and that the spontaneous activity in LIP is greater than in 7a. The visual receptive fields and motor fields of LIP neurons generally were found to overlay one another.

Memory Activity Andersen and colleagues (Gnadt and Andersen, 1988; Andersen et al., 1990; Barash et al., 1991a, b) described memory-related activity in LIP using a task that required monkeys to make saccades to remembered locations in the dark. The cells remained active during the period in which the animal withheld its response while remembering the location of an extinguished saccade target (figure 33.7). Using a double-saccade task similar to the one developed by Mays and Sparks (1980), Gnadt and Andersen were able to distinguish between whether the cells were coding the location of the sensory stimulus or whether they were coding the intention to make a saccade of a particular amplitude and direction. These investigators found that the activity could be evoked

A) 200 msec Delay

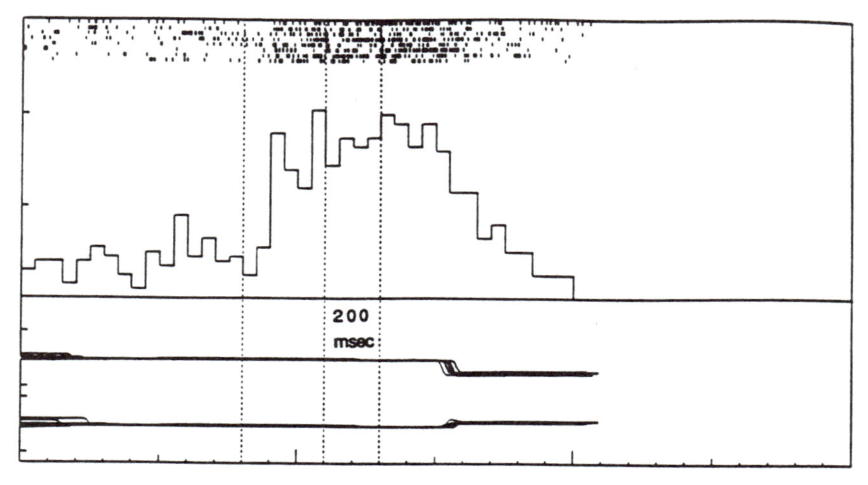

B) 1000 msec Delay

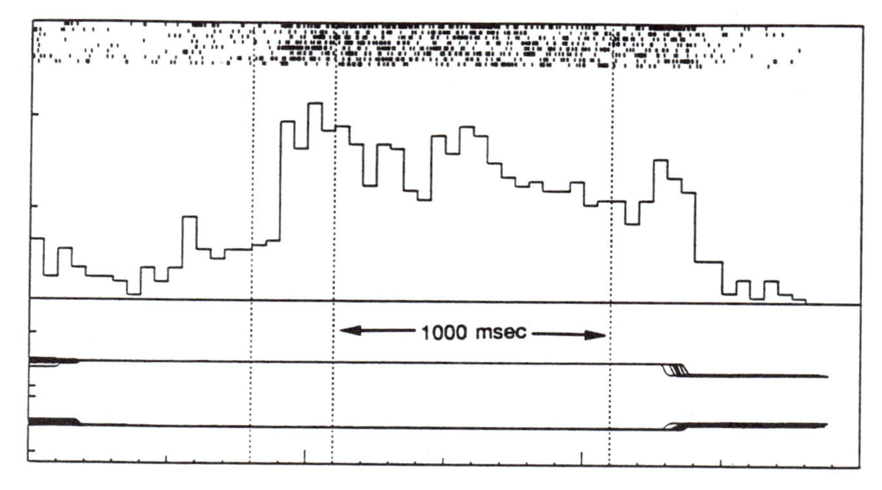

C) 1300 msec Delay

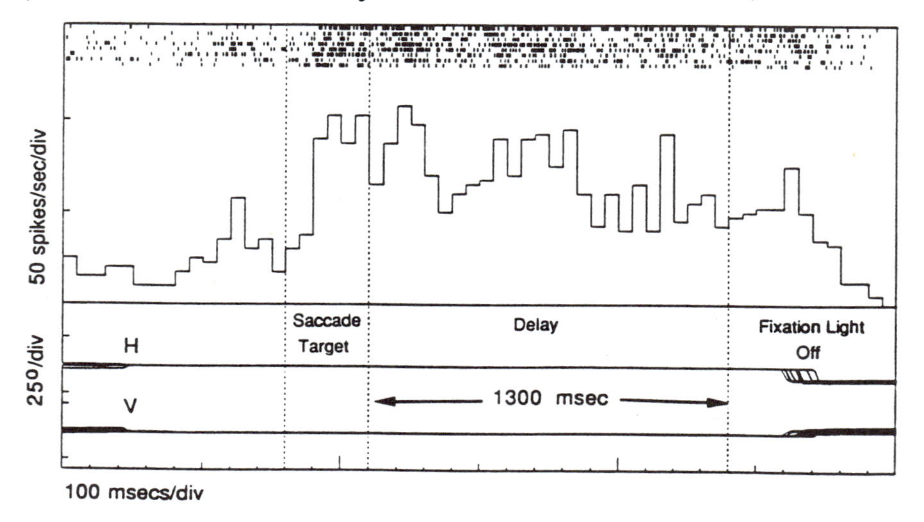

even when the sensory stimulus did not fall in the receptive field but the eye movement was made into the cell's motor field, arguing for the latter alternative (figure 33.8). They interpreted these activities as being part of a motor plan that has been triggered and remains active during the waiting period. Goldberg, Colby, and Duhamel (1990) and Barash and associates (1991a, b) confirmed these double-saccade results. We (Barash et al., 1991a, b) showed that the memory activity of LIP cells is directionally tuned, and these memory fields coincide with the visual and saccade fields. These results suggested the hypothesis that the memory activity reflects the intention of the monkey to make the next saccade.

To test the idea that this activity is related to intention, we (Bracewell et al., 1991) trained monkeys in a change-in-plan task, for which an animal was required to make an eye movement to a remembered target after a delay. However, during the delay period of some trials, the target would flash on at a new location, requiring a change in the direction of the planned saccade. It was found that the activity of the cells would turn on and off in a manner consistent with the motor plan that the animal had to formulate, as required by the task. The remarkable aspect of this result was that the plan to make an eye movement in a particular direction could be determined by examining the activity of the LIP cells without the animal emitting any behavior. To test this intention idea further, we (Bracewell et al., 1991) trained monkeys to make saccades to the remembered location of auditory targets. We found that many of the cells exhibited intended movement activity for both visual and auditory stimuli, consistent with the idea that the activity was more related to the plan to make an eye movement than to the modality of the sensory stimulus. Finally, in a memory double-saccade task, we (Bracewell et al., 1991; Mazzoni et al., 1992) showed that the memory activity was present only for the next intended movement. In this task, two targets were flashed briefly, and the animal had to remember the location of both targets. If the second target fell in the visual receptive field of a cell but both saccades were of a different amplitude and direction from the motor field of the cell, then the cells usually showed no response. Even when the task was configured so that the second saccade target fell in the visual receptive fields and the second saccade was made into the motor fields, the cells did not become active until after the first saccade.

In general, it is difficult to determine whether neural activity is related to attention or intention; for instance, Goldberg, Colby, and Duhamel (1990) have argued that the memory activity is related to the animal's *visual attention*, and not intention, to make movements. A visual attention interpretation would be consistent with the change-in-plan results but not with the auditory memory results, and so the interpretation must be broadened to posit that the attentional activity in this area is multimodal. Moreover, the memory double-saccade results do not appear to be consistent with any simple attention hypothesis, because the animal must attend and memorize both visual targets but most cells will have little or no memory activity for the visual targets in their receptive fields if the task does not require eye movements into their motor fields. A correlation of the memory activity with intention seems the most straightforward interpretation for these data.

FIGURE 33.7 Memory saccade task with different delays demonstrating the memory character of the activity during the delay. Delays are (A) 200 ms, (B) 1000 ms, and (C) 1300 ms. The rasters show the actual neural activity used to make the histograms. The period between the first two dotted vertical lines represents the time the saccade target is present, and the period between the second and third lines is the delay period. The fixation light goes off coincident with the third dotted vertical line. Both horizontal (H) and vertical (V) eye position traces are shown. In this experiment, the saccade target appeared 15° to the left. There is a vertical component in the leftward eye movement; this upward component for horizontal eye movements is common for saccades to remembered locations made in the dark. (Reprinted with permission from Andersen et al., 1990. Copyright 1990 Society for Neuroscience)

Conclusions

Recent experiments, reviewed in this chapter, are shedding light on the nature of abstract representations of space. Spatial representations are derived by integrating visual signals with information about eye position, vergence angle, and head position. These signals are brought together in the posterior parietal cortex to form a specific, distributed representation of space that is typified by linear gain fields.

One issue for further research is whether the different representations of space, outlined previously, share

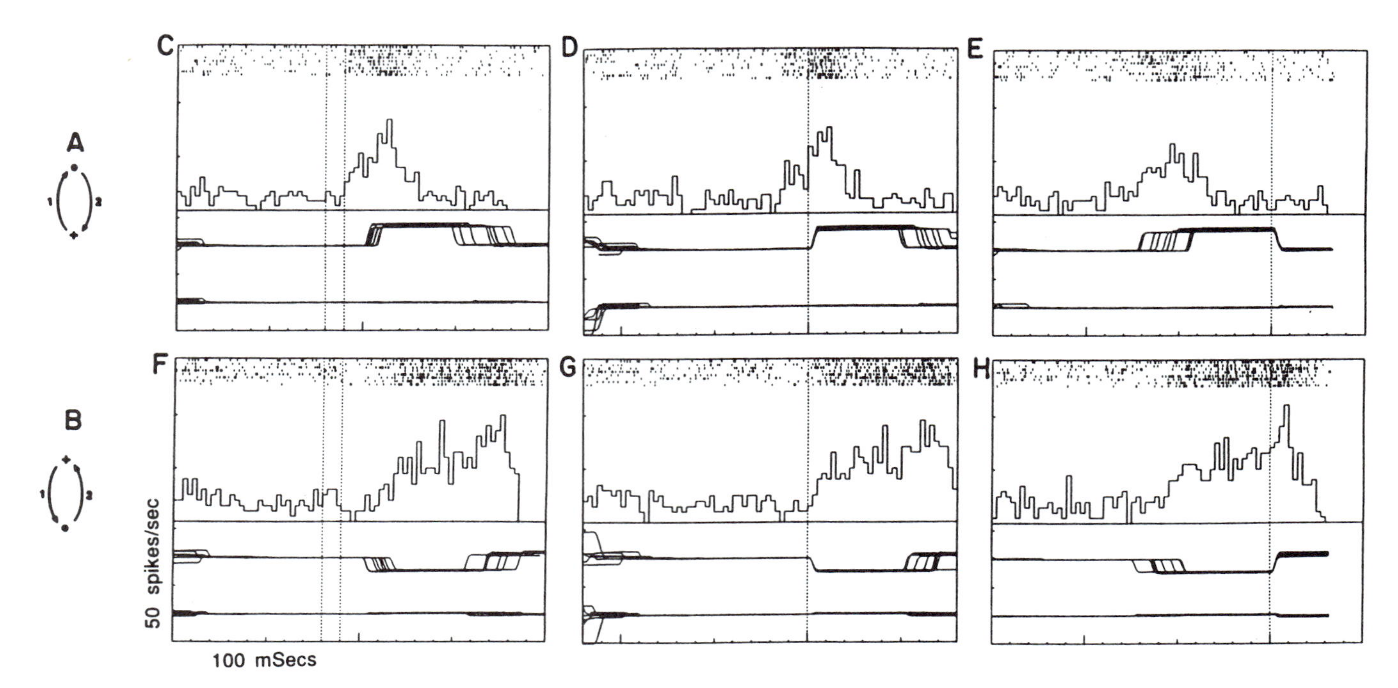

FIGURE 33.8 Back-saccade paradigm. (A, B) Scheme of the two saccades in the task. The first saccade is to the (single) target, whereas the second saccade is made in the dark back to the location of the original fixation point. (C–H) Activity in the back-saccade task of an LIP neuron. The preferred direction of this neuron, for the light-sensitive (LS), memory (M), and saccade (S) phases, is upward. Hence, in the top row, the visual stimulation and the first movement are in the preferred direction, and the second movement is in the opposite, nonpreferred direction. In the bottom row, the visual stimulation and the first saccade are in the nonpreferred direction, but the second saccade is in the preferred direction. (C) and (F) are aligned on the sensory stimuli. The first dotted vertical line denotes the offset of the fixation spot and the simultaneous onset of the target. The second dotted line represents the target offset. (D) and (G) are aligned on the beginning of the first saccade, and the dotted line denotes the time the first saccade begins. (E) and (H) are aligned on the beginning of the second saccade, and the dotted line denotes the time the second saccade begins. Shown in each panel, from the top, are the spike rasters, where each horizontal trace represents a trial and each tick within a line marks the time of occurrence of a spike; the resulting histogram; and the horizontal and vertical eye position traces of the various trials, superimposed. (Reprinted with permission from Barash et al., 1991a)

the same neural circuits. LIP is fascinating in that it appears to be an example of such an area. Cells in LIP integrate eye and head position with retinal signals to code space in head and body coordinates. Many cells here also carry vergence and disparity signals, enabling the representation of distance with respect to the body.

A related issue is whether the coordinate transformations proceed in a hierarchical fashion. For instance, are the body-centered cells of the posterior parietal cortex constructed by adding head position signals to the head-centered representation? Alternatively, the entire representation could be body-centered, with some cells exhibiting only retinal and eye position signals within this highly distributed representation (training networks to code in body-centered coordinates often generate some hidden units that carry only eye and retinal signals). Is information about shoulder position added to the body-centered representation of space in areas 7a and LIP to generate arm-referenced representations? Are there additional representations of visual targets that code with respect to the hand? These and many other questions make this a rich area for future research.

In the past few years, substantial progress has been made in researching the role of LIP in processing saccades. This area appears to make interesting, high-level contributions to the processing of eye movements. Cells in this area integrate information on eye position, head position, and vergence angle as well as the usual retinal location information and appear to represent targets in head- and body-centered spatial coordinates. Recent studies also reveal memory-related activities in

LIP that may function in the formation of motor plans. These results point to a central role for LIP in directing gaze.

REFERENCES

Andersen, R. A., 1987. The role of the inferior parietal lobule in spatial perception and visual-motor integration. In *Handbook of Physiology: Sec. 1. The Nervous System:* Vol. 5. *Higher Functions of the Brain* (part 2), F. Plum, V. B. Mountcastle, and S. R. Geiger, eds. Bethesda, Md.: American Physiological Society, pp. 483–518.

Andersen, R. A., 1989. Visual and eye-movement functions of the posterior parietal cortex. *Annu. Rev. Neurosci.* 12: 377–403.

Andersen, R. A., C. Asanuma, and W. M. Cowan, 1985. Callosal and prefrontal associational projecting cell populations in area 7a of the macaque monkey: A study using retrogradely transported fluorescent dyes. *J. Comp. Neurol.* 232:443–455.

Andersen, R. A., R. M. Bracewell, S. Barash, J. W. Gnadt, and L. Fogassi, 1990. Eye position effects on visual, memory and saccade-related activity in areas LIP and 7A of macaque. *J. Neurosci.* 10:1176–1196.

Andersen, R. A., G. K. Essick, and R. M. Siegel, 1985. Encoding of spatial location by posterior parietal neurons. *Science* 230:456–458.

Andersen, R. A., G. K. Essick, and R. M. Siegel, 1987. Neurons of area 7 activated by both visual stimuli and oculomotor behavior. *Exp. Brain Res.* 67:316–322.

Andersen, R. A., and V. B. Mountcastle, 1983. The influence of the angle of gaze upon the excitability of the light sensitive neurons of the posterior parietal cortex. *J. Neurosci.* 3:532–548.

Andersen, R. A., L. Snyder, C.-S. Li, and B. Stricanne, in press. Coordinate transformations is the representation of spatial information. *Curr. Opinion Neurobiol.*

Balint, R., 1909. Seelenlahmung des "Schauens," Optische Ataxie, Raumliche Storung der Aufmerksamkeit. *Psychiatr. Neurol.* 25:51–81.

Barash, S., R. M. Bracewell, L. Fogassi, J. W. Gnadt, and R. A. Andersen, 1991a. Saccade-related activity in the lateral intraparietal area I. Temporal properties. *J. Neurophysiol.* 66:1095–1108.

Barash, S., R. M. Bracewell, L. Fogassi, J. W. Gnadt, and R. A. Andersen, 1991b. Saccade-related activity in the lateral intraparietal area II. Spatial properties. *J. Neurophysiol.* 66:1109–1124.

Barto, A. G., and M. I. Jordan, 1987. Gradient following without backpropagation in layered networks. In *Proceedings of the IEEE International Conference on Neural Networks* 2:629–636.

Battaglini, P. P., P. Fattori, C. Galletti, and S. Zeki, 1990. The physiology of area V6 in the awake, behaving monkey. *J. Physiol.* 423:100P.

Boussaoud, D., T. M. Barth, and S. P. Wise, (in press). Effects of gaze on apparent visual responses of frontal cortex neurons. *Exp. Brain Res.*

Bracewell, R. M., S. Barash, P. Mazzoni, and R. A. Andersen, 1991. Neurons in the macaque lateral intraparietal cortex (LIP) appear to encode the next intended saccade. *Soc. Neurosci. Abstr.* 17:1282.

Brotchie, P. R., R. A. Andersen, and S. Goodman, 1993. The influence of head position on the representation of space in the posterior parietal cortex. Manuscript submitted for publication.

Fleming, J. F. R., and E. C. Crosby, 1955. The parietal lobe as an additional motor area: The motor effects of electrical stimulation and ablation of cortical areas 5 and 7 in monkeys. *J. Comp. Neurol.* 103:485–512.

Fogassi, L., V. Gallese, G. di Pellegrino, L. Fadiga, M. Gentilucci, G. Luppino, M. Matelli, A. Pedotti, and G. Rizzolatte, 1992. Space coding by premotor cortex. *Exp. Brain Res.* 89:686–690.

Galletti, C., and P. P. Battaglini, 1989. Gaze-dependent visual neurons in area V3A of monkey prestriate cortex. *J. Neurosci.* 9:1112–1125.

Gnadt, J. W., 1992. Area LIP: Three-dimensional space and visual to oculomotor transformation. *Behav. Brain Sci.* 15:745–746.

Gnadt, J. W., and R. A. Andersen, 1988. Memory related motor planning activity in posterior parietal cortex of macaque. *Exp. Brain Res.* 70:216–220.

Gnadt, J. W., and L. E. Mays, 1991. Depth-tuning in area LIP by disparity and accommodative cues. *Soc. Neurosci. Abstr.* 17:1113.

Goldberg, M. E., C. L. Colby, and J.-R. Duhamel, 1990. The representation of visuomotor space in the parietal lobe of the monkey. *Cold Spring Harb. Symp. Quant. Biol.* 55:729–739.

Goodman, S., and R. A. Andersen, 1989. Microstimulation of a neural-network model for visually guided saccades. *J. Cogn. Neurosci.* 1:317–326.

Goodman, S. J., and R. A. Andersen, 1990. Algorithm programmed by a neural network model for coordinate transformation. In *Proceedings of the International Joint Conference on Neural Networks, San Diego*, Vol. 2. Ann Arbor, Mich.: IEEE Neural Networks Council, pp. 381–386.

Hyvärinen, J., 1981. Regional distribution of functions in parietal association area 7 of the monkey. *Brain Res.* 206: 287–303.

Hyvärinen, J., and Y. Shelepin, 1979. Distribution of visual and somatic functions in the parietal associative area 7 of the monkey. *Brain Res.* 169:561–564.

Keating, E. G., and S. G. Gooley, 1988. Disconnection of parietal and occipital access to the saccadic oculomotor system. *Exp. Brain Res.* 70:385–398.

Lehky, S. R., A. Pouget, and T. J. Sejnowski, 1990. Neural models of binocular depth perception. *Cold Spring Harb. Symp. Quant. Biol.* 55:765–777.

Lynch, J. C., and J. W. McLaren, 1989. Deficits of visual attention and saccadic eye movements after lesions of parieto-occipital cortex in monkeys. *J. Neurophysiol.* 61: 74–90.

Lynch, J. C., V. B. Mountcastle, W. H. Talbot, and T. C. T. Yin, 1977. Parietal lobe mechanisms for directed visual attention. *J. Neurophysiol.* 40:362–389.

MacKay, W. A., and A. Riehle, 1992. Planning a reach: Spatial analysis by area 7a neurons. In *Tutorials in Motor Behavior 2*, G. Stelmach and J. Requin, eds. New York: Elsevier.

Maunsell, J. H. R., and D. C. Van Essen, 1983. The connections of the middle temporal visual area (MT) and their relationship to a cortical hierarchy in the macaque monkey. *J. Neurosci.* 3:2563–2586.

Mays, L. E., and D. L. Sparks, 1980. Dissociation of visual and saccade-related responses in superior colliculus neurons. *Science* 43:207–232.

Mazzoni, P., R. A. Andersen, and M. I. Jordan, 1991a. A more biologically plausible learning rule for neural networks. *Proc. Natl. Acad. Sci.* 88:4433–4437.

Mazzoni, P., R. A. Andersen, and M. I. Jordan, 1991b. A more biologically plausible learning rule than backpropagation applied to a network model of cortical area 7a. *Cerebral Cortex* 1:293–307.

Mazzoni, P., R. M. Bracewell, S. Barash, and R. Andersen, 1992. *Soc. Neurosci. Abstr.* 18:148.

Mountcastle, V. B., J. C. Lynch, A. Georgopoulos, H. Sakata, and C. Acuna, 1975. Posterior parietal association cortex of the monkey: Command function for operations within extrapersonal space. *J. Neurophysiol.* 38:871–908.

Newsome, W. T., R. H. Wurtz, and H. Komatsu, 1988. Relation of cortical areas MT and MST to pursuit eye movements: I. Differentiation of retinal from extraretinal inputs. *J. Neurophysiol.* 60:604–620.

Peck, C. K., J. A. Baro, and S. M. Warder, 1992. Eye position effects on visual responses in the superior colliculus. *Invest. Ophthalmol. Vis. Sci. (Suppl.)* 33:1357.

Robinson, C. J., and H. Burton, 1980a. Organization of somatosensory receptive fields in cortical areas 7b, retroinsular postauditory and granular insula of *M. fascicularis*. *J. Comp. Neurol.* 192:69–92.

Robinson, C. J., and H. Burton, 1980b. Somatic submodality distribution within the second somatosensory (SII), 7b, retroinsular, postauditory and granular insular cortical areas of *M. fascicularis*. *J. Comp. Neurol.* 192:93–108.

Robinson, D. L., M. E. Goldberg, and G. B. Stanton, 1978. Parietal association cortex in the primate: Sensory mechanisms and behavioral modulations. *J. Neurophysiol.* 41:910–932.

Robinson, D. L., J. W. McClurkin, and C. Kertzmann, 1990. Orbital position and eye movement influences on visual responses in the pulvinar nuclei of the behaving macaque. *Exp. Brain Res.* 82:235–246.

Sakata, H., H. Shibutani, and K. Kawano, 1980. Spatial properties of visual fixation neurons in posterior parietal association cortex of the monkey. *J. Neurophysiol.* 43:1654–1672.

Snyder, L. H., P. R. Brotchie, and R. A. Andersen, in press. World-centered encoding of location in posterior parietal cortex of monkey. In *Proceedings of Neural Control of Movement Third Annual Meeting*.

Wagman, I. H., 1964. Eye movements induced by electric stimulation of cerebrum in monkeys and their relationship to bodily movements. In *The Oculomotor System*, M. B. Bender, ed. New York: Harper & Rows, pp. 18–39.

Zipser, D., and R. A. Andersen, 1988. A back-propagation programmed network that simulates response properties of a subset of posterior parietal neurons. *Nature* 331:679–684.

34 Role of Monkey Superior Colliculus in Control of Saccades and Fixation

ROBERT H. WURTZ AND DOUGLAS P. MUNOZ

ABSTRACT Saccadic eye movements and the intervening periods of visual fixation represent one of the simplest behavioral systems that have been studied in the primate. The superior colliculus (SC) is a key structure in both systems, although it is only part of a system extending from cerebral cortex to the pons. Recent experiments on the control of saccades have identified two types of collicular cells that discharge in relation to saccades. *Burst cells* have a discrete burst of activity before onset of saccades. Many have a discharge whose end coincides closely with the end of the saccade, a clipped discharge. A modified feedback system controlling the position of the eyes is consistent with key aspects of the discharge of these cells and the initiation of saccades by the SC. Other cells show a slow buildup of activity before the onset of a saccade and, during a saccade, show a moving wave front of activity across the SC. These *buildup cells* are proposed also to be inside a feedback loop controlling the amplitude of the saccade. The control of visual fixation between saccades depends on the activity of *fixation cells* within the anterior colliculus whose activity increases during such active fixations. Initial studies of the interaction between these systems of saccade and fixation control are consistent with the idea that one inhibits the other.

Rapid eye movements and the intervening periods of visual fixation represent a simple behavioral system in primates. Saccades move the eyes rapidly from one point of interest to another. Little visual information is available during these movements, which usually last no more than 40 ms. It is in the periods of fixation following saccades, in which the fine-grained foveas of the eyes are directed at the objects of interest, that we obtain nearly all visual information. In this chapter, we consider some of the underlying neuronal mechanisms that control these saccadic eye movements and visual fixation.

Monkeys, like humans, also make saccadic eye movements separated by periods of visual fixation, and this behavioral similarity between the two species of primates has made the monkey a superb model for analysis of the neuronal systems underlying the control of visual fixation and saccadic eye movements. The parts of the brain related to the generation of saccades have been studied extensively, and a series of areas extending from cerebral cortex to the pons have been identified as part of the system that controls the generation of saccades (Wurtz and Goldberg, 1989). These areas include the posterior parietal cortex, the frontal and supplementary eye fields in the frontal lobe, the caudate nucleus and substantia nigra pars reticulata in the basal ganglia, the superior colliculus (SC) on the roof of the midbrain, and regions of the mesencephalic and pontine reticular formation that are closely related to the motor nuclei driving the extraocular muscles. In this chapter, we concentrate on the role of the monkey SC in the control of visual fixation and saccadic eye movements.[1]

Organization of superior colliculus

The SC, a layered structure on the roof of the midbrain, receives important input from the cortical and deep telencephalic areas and, in turn, projects to the pontine and mesencephalic gaze centers. The alternating fiber and cell body layers clearly have different functions, as indicated by the differential relation of neuronal activity in these layers to visual stimulation

ROBERT WURTZ Laboratory of Sensorimotor Research, National Eye Institute, Bethesda, Md.
DOUGLAS MUNOZ MRC Group in Sensory-Motor Physiology, Department of Physiology, Queen's University, Kingston, Canada

and the generation of saccadic eye movements. A cell in the superficial layers increases its discharge rate following the onset of a visual stimulus in a particular part of the visual field, the visual receptive field of the cell. The receptive fields of these cells form a retinotopically coded map of the contralateral visual field.

The discharge of the cells in the intermediate layers of the SC is tightly linked to the occurrence of saccades (Sparks, 1978). These saccade-related cells also frequently begin to discharge at a fixed latency after visual stimulation but discharge more vigorously before the onset of saccadic eye movements. The increased discharge rate of saccade-related cells accompanies saccades made to points within the area of the visual field that define the movement field of the cell (Wurtz and Goldberg, 1972). The movement field of each saccade cell has a central region with a maximal discharge and a gradient of response that fades away as the saccade vector deviates from this central area (Wurtz and Goldberg, 1972; Sparks and Mays, 1980).

Cells in different parts of the colliculus have movement fields that represent different amplitudes and directions of saccades. If we accept that the central point of each cell's movement field gives the maximum discharge of the cell, we can look at this point for many cells throughout the intermediate layers and see a map of preferred saccade directions and amplitudes—a movement map (figure 34.1). By electrically stimulating the SC, Robinson (1972) demonstrated that the amplitude of the saccade increases along a line running from the rostrolateral pole to the caudomedial pole of the SC, as shown on the map in figure 34.1, with the smallest saccades represented in the rostral SC and the largest in the caudal SC. Upward and downward directions are represented medially and laterally, respectively.

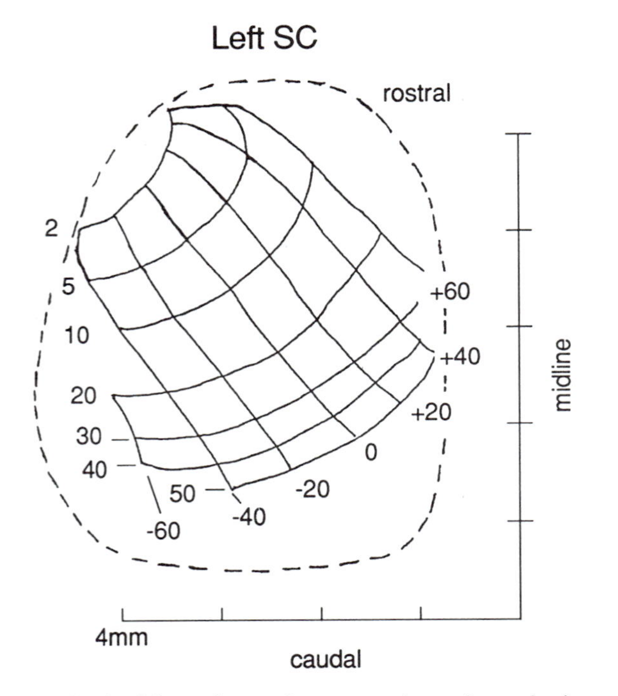

FIGURE 34.1 Map of saccade vectors throughout the intermediate layers of the SC. (After Robinson, 1972.)

An impending saccade is preceded by neural activity in one area of the SC movement map. Saccades to different parts of the field are accompanied by activity in different areas of the map. The activity of one cell does not determine the amplitude and direction of a saccade, but the population of all active cells does (Lee, Rohrer, and Sparks, 1988; Sparks and Mays, 1990).

Two types of saccade cells

Whereas neurons concentrated in the intermediate layers of the monkey SC discharge in relation to saccadic eye movements, two types of cells with different activity patterns recently have been described (Munoz and Wurtz, 1992). One population of neurons, *burst cells*, discharged a high-frequency burst of action potentials immediately prior to saccade onset. A second population of saccade-related cells had a slow buildup of activity in addition to movement-related activity; we refer to these neurons as *buildup cells*.

We distinguish the two classes of saccade cells by using several different saccade tasks, including the visually guided saccade task in which the monkey makes a saccade to a visual target that appears as the fixation point disappears (figure 34.2). The left column of the figure shows the cell discharge aligned on the onset of the visual target, and the right column shows the same discharge aligned on the onset of the saccade. Both the burst cell (figure 34.2A) and the buildup cell (figure 34.2B) began to discharge approximately 60–70 ms after target onset (figure 34.2, left) and then generated a more vigorous discharge in association with saccade onset (figure 34.2, right). The burst cell paused briefly between its presumed visual- and movement-related responses, whereas the buildup cell demonstrated no clear pause.

To determine whether the discharge of the buildup cell was related to the preparation for the saccade or to

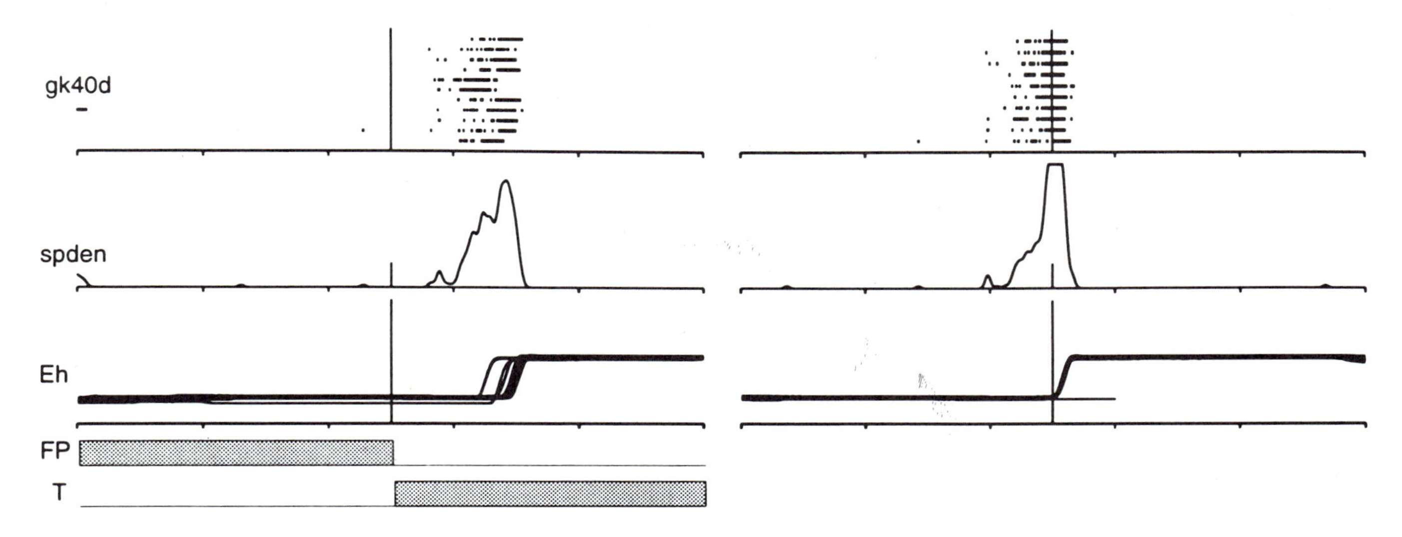

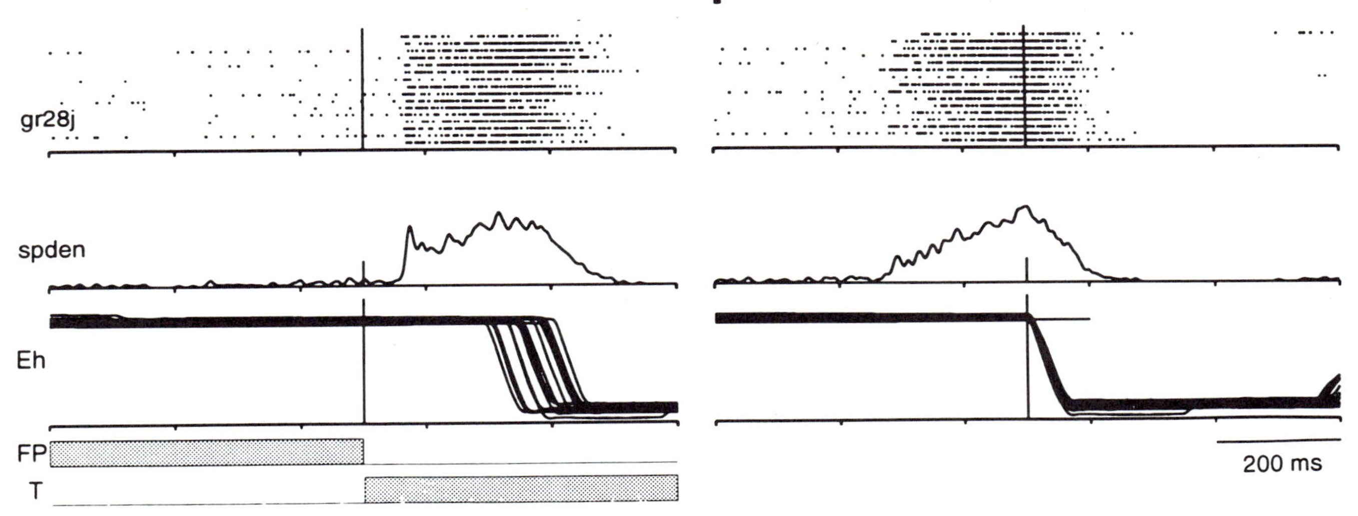

FIGURE 34.2 Discharge of burst and buildup cells. Shown in each panel are the individual rasters, the spike density profile (spden), and the horizontal eye position traces (Eh) for 8 to 10 trials. The traces in the left column are aligned on target onset, and the same data are aligned on saccade onset in the right column. Saccade direction and amplitude were selected for the strongest movement-related response. (A) Burst cell. The discharge shown for this cell is for the optimal saccadic amplitude and direction of 10° to the right. This cell was classified as a burst cell rather than a preparatory cell because the presaccadic activity consisted only of the burst before the saccade. (B) Buildup cell. Discharge is for saccades 25° to the left. The buildup of activity is independent of the visual stimulus. FP, fixation point; T, target. (From Munoz and Wurtz, 1994)

the presence of the visual stimulus, we used a memory-guided saccade paradigm in which the target was flashed briefly and the monkey had to make a saccade to the spatial location of the target after the offset of the fixation point. In this task, burst cells gave only weak responses to the target flash and then exhibited a burst of activity in association with the saccade to the remembered location of the target flash. In contrast, the buildup cells discharged in a sustained manner from the time of the flash until the saccade was made, even though the target was no longer visible.

This sustained discharge could also be elicited in the absence of any retinal stimulation as the discharge occurred before the visual target had been presented in a gap saccade paradigm. In this latter task, the fixation point went off, leaving a period of darkness preceding target onset; the buildup cell was silent until after fixation point offset and then began to discharge in the period of darkness, even though the target had not yet appeared. The intensity of this anticipatory activity was similar to that seen in the memory-guided task when the target had appeared but the saccade was delayed. The buildup cell then increased its discharge after onset of the target and continued to discharge until the monkey made the saccade to the visible target. In contrast, the burst cell remained silent during the period of darkness between fixation point offset and target onset and then discharged a brief phasic burst approximately 60–70 ms after target onset, followed by a second, more robust burst of spikes synchronized with saccade initiation.

Although there is a continuum of cell types extending from those with burst cell characteristics to those demonstrating a buildup of activity, we have divided cells into these two groups. In general, we classified neurons as burst cells if they lacked significant sustained activity in the interval between the visual- and movement-related bursts seen in the visually guided task and during the instructed delay period in the memory-guided tasks. Neurons that had, in addition to a movement-related response, a sustained response related to preparation to make the saccade were classified as buildup cells.

Another major difference between burst and buildup neurons was in their movement fields. Figure 34.3 compares the discharge of a burst cell and a buildup cell associated with various amplitude saccades whose directions matched the optimal direction of each cell. The optimal saccade amplitude for the cells was approximately 8°. The burst cell (figure 34.3A) discharged maximally for saccades that were close to the optimal amplitude and, when saccade amplitude was greater or less than optimal, the discharge of the cell diminished. The saccade-related responses of the buildup cell (figure 34.3B) diminished if either the amplitude of the saccade was smaller than optimal or the direction deviated from optimal. However, buildup cells continued to discharge for saccades of optimal direction whose amplitudes were greater than optimal. In net, the burst cell had a movement field that was closed—that is, the response field had a distal border because the cell did not discharge for saccades significantly larger than optimal. In contrast, the buildup cell had an open-ended movement field: The cell discharged for all saccades of optimal direction that were equal to or greater than the optimal amplitude.

The latency between the onset and termination of the saccade to the peak of the cell discharge also differed for the burst and buildup cells. The timing of the peak discharge of burst cells relative to saccade onset did not vary with saccade amplitude; it always occurred around the time of saccade onset. For buildup cells, however, the occurrence of peak discharge relative to saccade onset depended on saccade amplitude; as saccade amplitude increased, the time from saccade onset to peak discharge also increased.

Several other characteristics of these cells should be noted. The saccade-related discharge patterns just described for both burst and buildup cells occurred regardless of whether the saccade was made to a visual target (as in figure 34.2) or to a remembered target. The saccade-related portion of both cell types' discharge was similar whether or not the target was visible during the movement. There were also no differences in the shapes of the movement fields for saccades made under these two conditions. The activity recorded from buildup cells during saccades that were larger than the optimal amplitude was not related to the programming of subsequent corrective saccades. With large saccades, cells with open-ended movement fields were active, regardless of whether there was a subsequent corrective saccade.

We determined the relative location of burst and buildup cells below the collicular surface. The depth of each cell was determined relative to the depth of the first multicell visual responses that we encountered on that penetration as the electrode entered the superficialmost layers of the SC. The visual cells, lacking saccade-related activity, were located in approximately the first millimeter of the SC; burst cells were found immediately beneath the visual cells, approximately 1–2 mm below the dorsal surface; the buildup cells were located ventral to and somewhat intermingled with the burst cells, approximately 1.5–2.5 mm below the dorsal surface of the SC. Both types of saccade-related cells were in the intermediate lay-

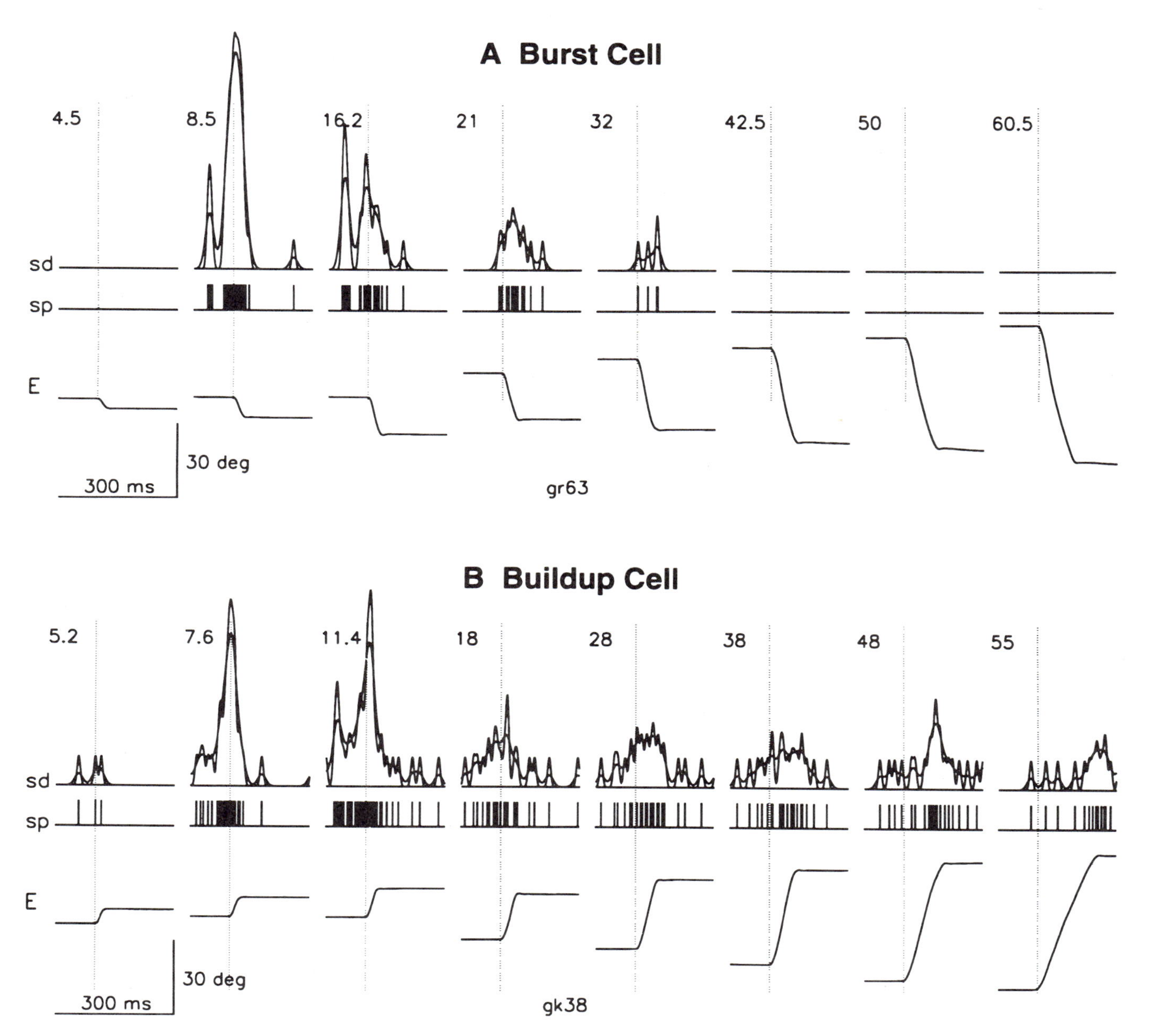

FIGURE 34.3 Movement fields of SC burst cell (A) and buildup cell (B). Saccade amplitude was systematically varied along the optimal direction across the movement field of the cell. The optimal saccade amplitude for both cells was approximately 8°. Shown in each panel is a single trial with two spike density profiles (sd), the individual action potentials (sp), and the radial eye position (E). The burst cell had a discrete movement field, whereas the buildup cell had an open-ended movement field. (From Munoz and Wurtz, 1994)

ers of the SC, with the buildup cells lying deeper than the burst cells.

Fixation cells in rostral superior colliculus

On the saccade map of the SC (see figure 34.1), large-amplitude saccades are represented caudally and small saccades are represented rostrally. Some cells in the rostral pole, however, do not increase their discharge rate before saccades but instead do so during periods of active fixation. This observation was made in the SC of the cat (Munoz and Guitton, 1989, 1991) and, more recently, in the monkey (Munoz and Wurtz, 1993a, b, c). Figure 34.4 shows the discharge of such a

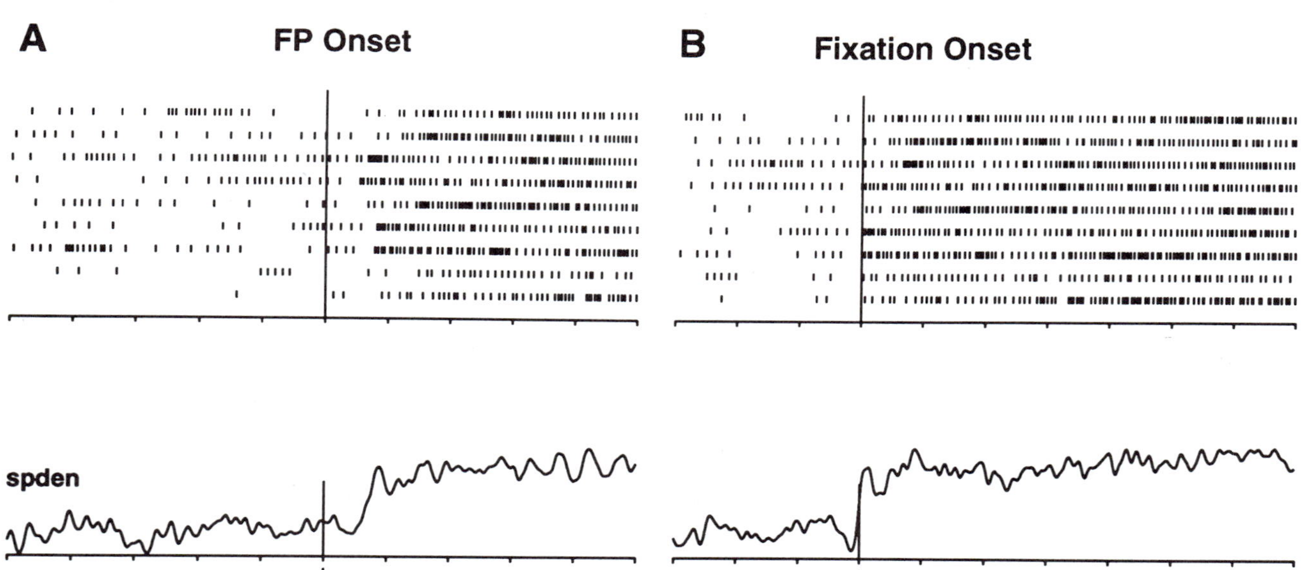
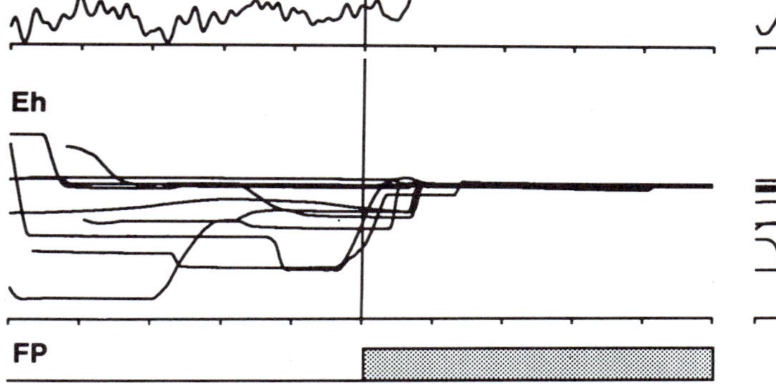
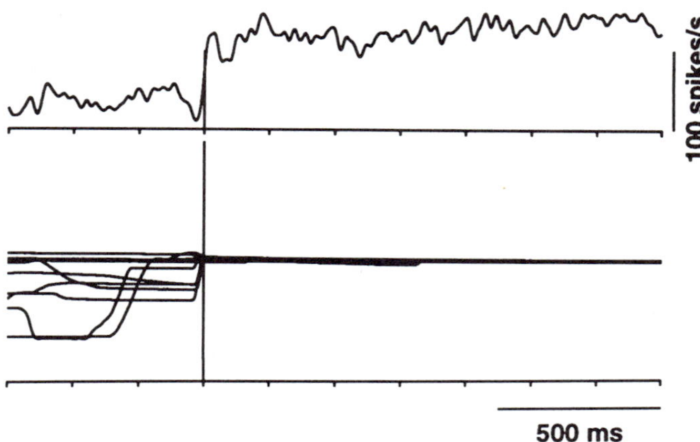

FIGURE 34.4 Example of the discharge of a fixation cell during active fixation. Rasters are aligned on (A) the onset of the fixation target (fixation point [FP] onset) and (B) the time when the eye entered the computer-controlled fixation window (fixation onset). The traces shown from top to bottom are the individual rasters, the spike density function (spden), and the horizontal eye position traces (Eh). (From Munoz and Wurtz, 1993a)

fixation cell while the monkey was looking about in the experimental room and then after it made a saccade to the visual target. In figure 34.4A, the raster and spike density display are aligned on this onset of the target, whereas in figure 34.4B they are aligned on the time when the monkey achieved fixation of the target. The discharge rate of the cell went up with acquisition of the target, not with target onset. In addition, the discharge was low while the monkey fixated a point on the blank screen but increased with active fixation of the target.

These fixation cells in the monkey have a number of other characteristics. The discharge was not simply the result of the visual stimulus falling on the foveal receptive field of a visually sensitive neuron. When we blinked the target off briefly, but the monkey continued to fixate, the cell continued to discharge. We have used this continued discharge as a criterion for the identification of fixation cells. In contrast, the discharge of other cells lying in the anterior colliculus more dorsal to fixation cells does pause with removal of the fixation point, indicating that the response of these cells is a visual one.

Another salient characteristic of collicular fixation cells is that they pause during saccades between actively fixated targets. Just as the duration of saccades increases with saccade amplitude, the duration of the pause also increases with larger saccadic amplitudes. Thus, the fixation cells and the saccade cells have patterns of discharge that are reciprocal: Fixation cells (figure 34.5, left) are active during fixation and silent during saccades, whereas saccade cells (figure 34.5, right) are silent during fixation but burst at the time of the saccades.

Fixation cells were located in the rostral pole of the SC at a depth similar to that of the buildup cells. We

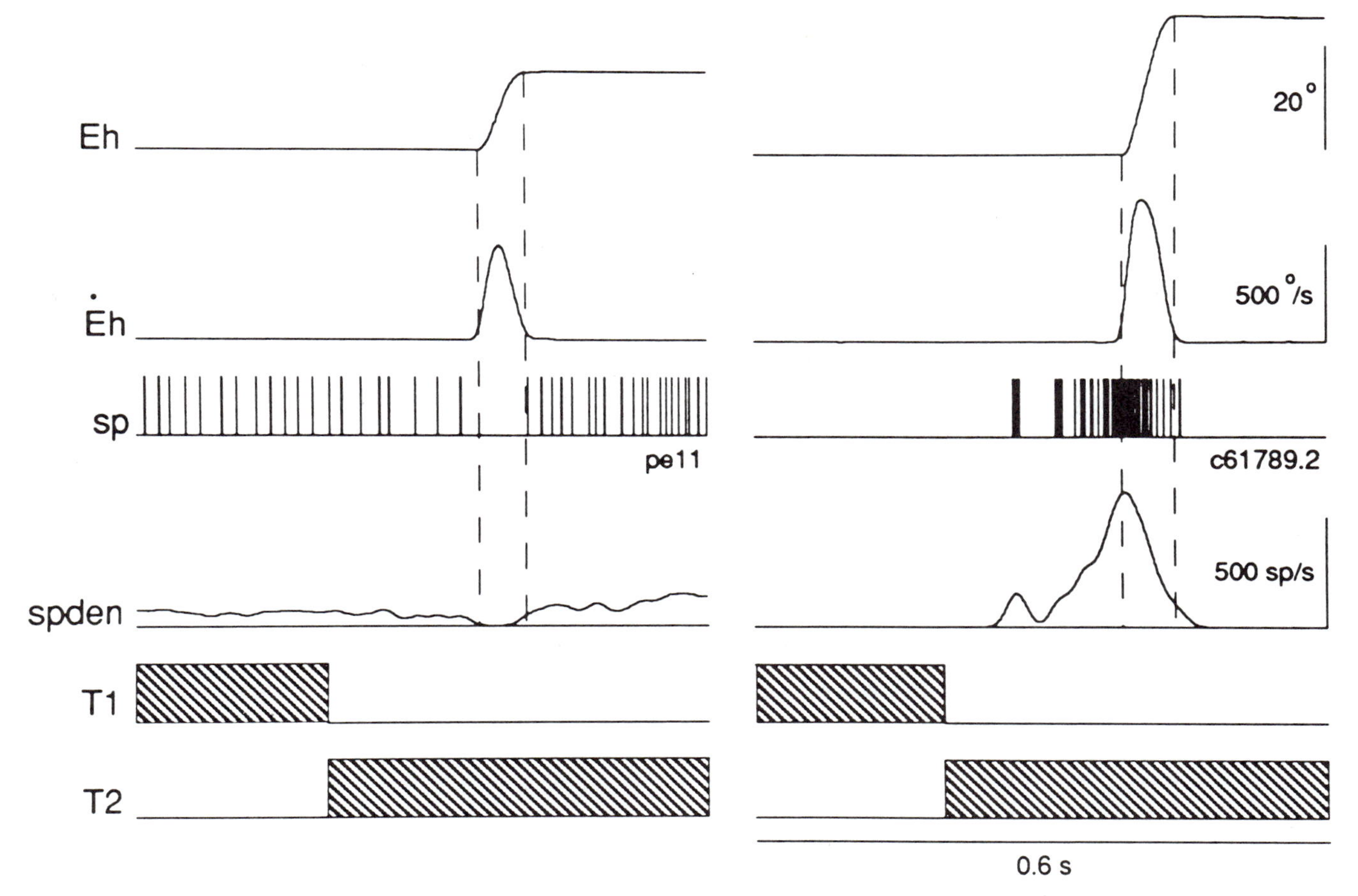

FIGURE 34.5 Comparison of the discharge of a fixation cell located in the anterior pole of the SC (left) and a saccade cell located in the caudal left SC (right). Same conventions as in figure 34.2. (From Munoz and Wurtz, 1993c)

consider fixation cells as the extension of the buildup cell layer into the rostral pole.

Interaction between fixation and saccades

The reciprocal relationship of activity between saccade and fixation cells within the SC suggests that they might be mutually inhibitory. This mutual inhibition, in turn, suggests that if the activity of the fixation cells were artificially altered, the frequency of the saccades might be changed. We both increased and decreased the activity of the fixation cells to test the effect of this on the generation of saccades.

Figure 34.6 shows the logic of our experiments. Our hypothesis, like that developed for the cat (Munoz and Guitton, 1991), is that fixation cells in the rostral pole of the SC exert control over the saccadic system by inhibiting the saccade cells in the SC as well as by activating the brainstem omnipause neurons that gate the burst neurons in the paramedian pontine reticular formation. Though we show only the interaction of the fixation cells with the rest of the SC in figure 34.6, the effect of the fixation cells on these other midbrain and pontine areas should be regarded as being represented on the schematic drawing by the rest of the SC. Our strategy was to alter the activity of fixation cells in the rostral SC while leaving the rest of the SC undisturbed and then to test subsequent saccade generation. We first increased the activity in the fixation zone by applying low-frequency electrical stimulation (figure 34.6B) to increase activity of cells adjacent to the site of stimulation. According to our hypothesis, this manipulation should lead to increased activity in the fixation zone of the SC and decreased activity in the rest of the SC related to saccades.

Figure 34.7 illustrates the effect of stimulating both fixation zones simultaneously while the monkey made saccades in the visually guided saccade paradigm.

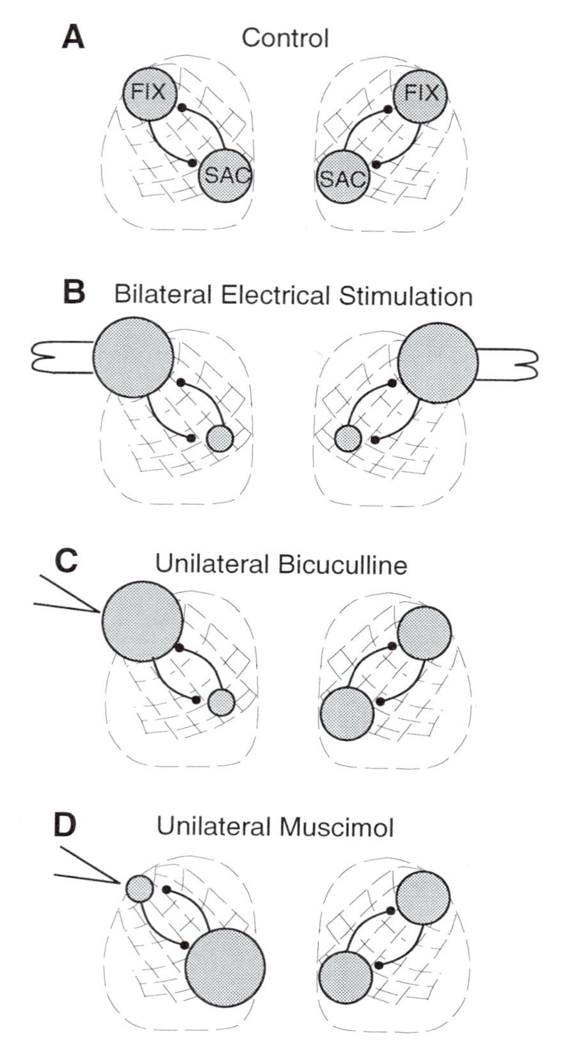

FIGURE 34.6 (A) Schematic motor map of the deeper layers of the monkey SC showing the putative connections between the collicular fixation zone (FIX) and the saccade zone (SAC). See text for details. (B) Activation of FIX cells in both colliculi with bilateral electrical stimulation would lead to increased inhibition of SAC cells. (C) Artificial activation of FIX cells with a unilateral injection of GABA antagonist (bicuculline) would lead to increased activity in the FIX zone and then to increased inhibition of SAC cells. (D) Artificial inhibition of FIX cells with a unilateral injection of a GABA agonist (muscimol) would lead to decreased activity in the FIX zone and then disinhibition of SAC cells. (From Munoz and Wurtz, 1993b)

Stimulation trials (solid traces) were interleaved with control trials during which no stimulation occurred (dotted traces). The small vertical tick on the eye position and velocity traces indicates the cue for the monkey to initiate the saccade (simultaneous offset of the fixation point and onset of the peripheral target). The monkey made saccades approximately 200 ms after target onset in the control condition to targets 20° to the left or right. A long-duration, low-frequency train of stimulation (500 ms, 150 Hz, 30 μA, marked by the horizontal bar under the eye position traces) delayed saccade initiation. The monkey could generate the saccade to the new target only after the stimulation ceased. All centrifugal saccades were affected as were centripetal saccades. Even with the delay in initiation, the saccades reached the target, as indicated by the equal amplitude of the normal and delayed saccades in figure 34.7. This accuracy endured even if the target was no longer present after the stimulation ended, as was the case when the monkey made a saccade to the remembered location of the target. When we applied stimulation to the fixation zone during the saccade, saccades were interrupted in midflight.

When we positioned the electrode outside the location on the motor map where fixation cells were recorded, the effect was not evident. Also, when we stimulated at locations above, below, or rostral to the location of fixation cells with similar parameters, no effect was seen on saccade generation.

Whereas electrical stimulation allowed us to increase activity within the fixation zone, injection of GABAergic drugs allowed us either to increase the activity with a GABA antagonist (bicuculline, figure 34.6C) or to decrease activity with a GABA agonist (muscimol, figure 34.6D). We found that bicuculline and muscimol injections had a profound effect on the monkey's ability to fixate a target and make saccades to a new target. Application of bicuculline increased saccade latencies, whereas muscimol reduced latencies and led to instability of fixation.

The effect of reducing fixation activity with muscimol is illustrated in figure 34.8. We used a memory-guided saccade task to maximize the requirements for fixation. In this saccade task (figure 34.8A), we flashed a spot of light (T2) for 80 ms at a point in the contralateral visual field, but we required the monkey to continue fixating until after the fixation point (T1) went off several hundred (400–800) milliseconds later. If the monkey delayed the saccade until after offset of the fixation point (T1), it was rewarded as a correct response, but if the saccade occurred earlier, it was not rewarded for this incorrect response. The histograms at the top of figure 34.8B show that the normal monkey easily made almost entirely correct saccades. Most of

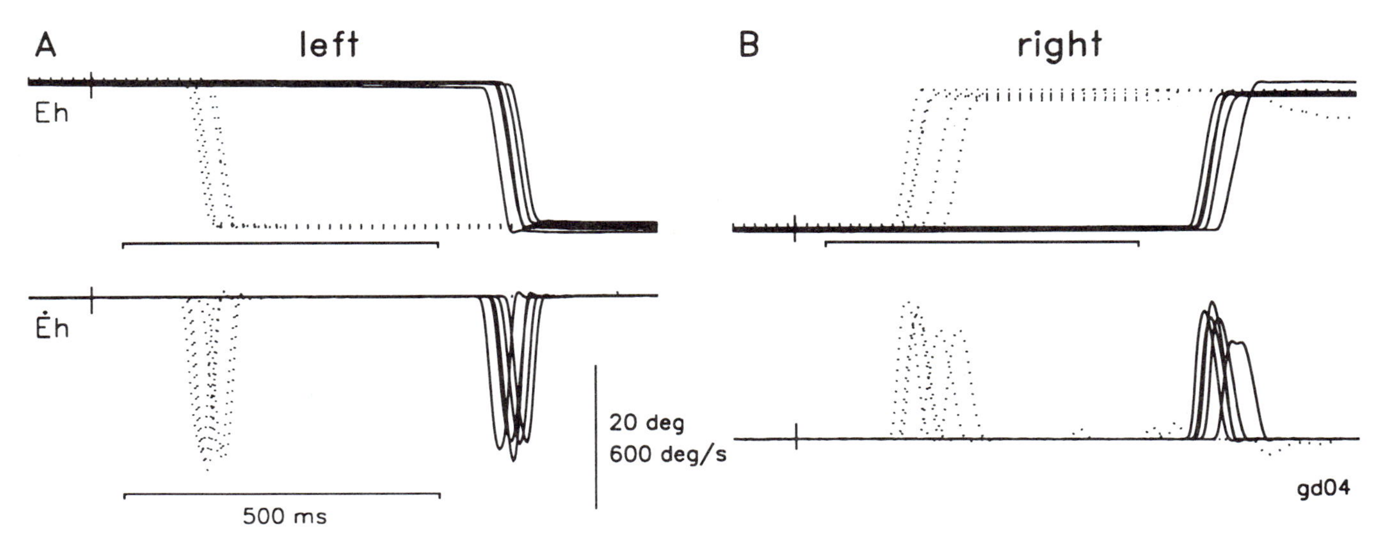

FIGURE 34.7 Suppression of saccades by bilateral stimulation of both fixation zones simultaneously. Five control trials (dotted traces) and five stimulation trials (solid traces) are superimposed in each panel as the monkey made visually guided saccades. The vertical tick on the eye position traces indicates the time of target onset, and the horizontal bar under the eye position traces indicates the time of stimulation. Low-frequency, long-duration stimulation (500 ms, 150 Hz, 30 μA) of both fixation zones prevented the initiation of centrifugal saccades. (Eh, horizontal eye position traces.) (After Munoz and Wurtz, 1993b)

these saccades were initiated approximately 200 ms after T1 offset. After injection of muscimol into the rostral colliculus (bottom of figure 34.8B), the monkey had difficulty delaying initiation of the saccade after the flash of the target, and many saccades occurred just after the flash. Note that these saccades on the incorrect trials were to the right target but at the wrong time. Thus, the execution of the saccade was not disrupted—it was simply delayed—exactly as we would expect if we had removed an inhibition on saccade generation.

The incorrect saccades shown on the frequency histograms in the lower half of figure 34.8B not only occurred before the fixation point went off, but many occurred within a latency of 80–100 ms after the target light flashed. Such a short regular latency is characteristic of express saccades previously observed in the monkey (Fischer and Boch, 1983). Of particular relevance to the fixation cells in the SC is the proposal (Fischer, 1987) that express saccades occur most frequently when fixation has already been broken, which is exactly what we propose is the consequence of the functional removal of the rostral SC.

The fixation cells that we have identified are almost certainly part of a larger system within the brain, as cells whose discharge is modified by fixation under some conditions have been observed in parietal cortex (Mountcastle, Andersen, and Motter, 1981) and substantia nigra pars reticulata (Hikosaka and Wurtz, 1983). Extrastriate cortical lesions also alter the frequency of express saccades, which is consistent with the role of cortex in the control of visual fixation (Weber and Fischer, 1990).

Burst cells temporally related to saccades

The most intensively studied saccade-related cells have been the burst cells (see figures 34.2A, 34.3A). The work on these cells has been summarized recently by Sparks and Hartwich-Young (1989). Occurrence of the burst is highly correlated with occurrence of the saccade (Sparks, 1978), and each saccade is accompanied by activity at a specific location on the SC movement map that specifies the amplitude and direction of the saccade. The sum of activity from all cells presumably is responsible for bringing the saccade on target.

In fact, bringing the eye on target has been a vexing problem for models of the saccadic system, and fitting the SC into these models has been even more difficult.

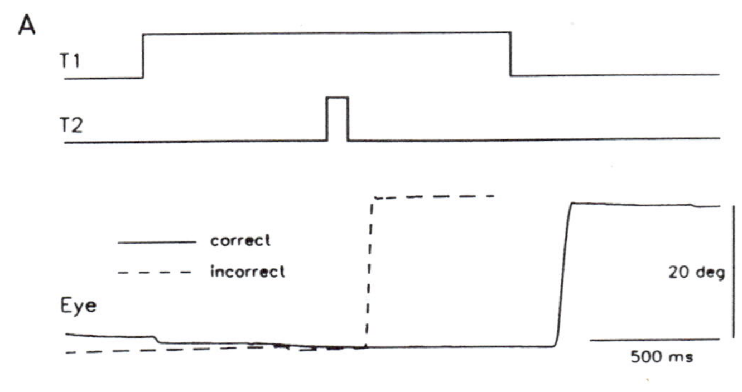

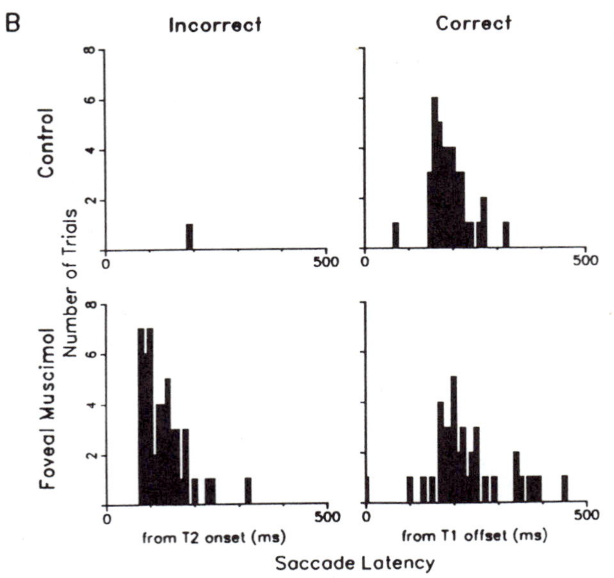

FIGURE 34.8 Shortening of saccade latency as a result of inactivation of fixation cells. (A) Examples of saccades occurring after the fixation point goes off (T1; correct) and shortly after the flash in the peripheral field (T2; incorrect). (B) Frequency of occurrence of these two types of saccades before (top) and after (bottom) the injection of muscimol. (From Munoz and Wurtz, 1993c)

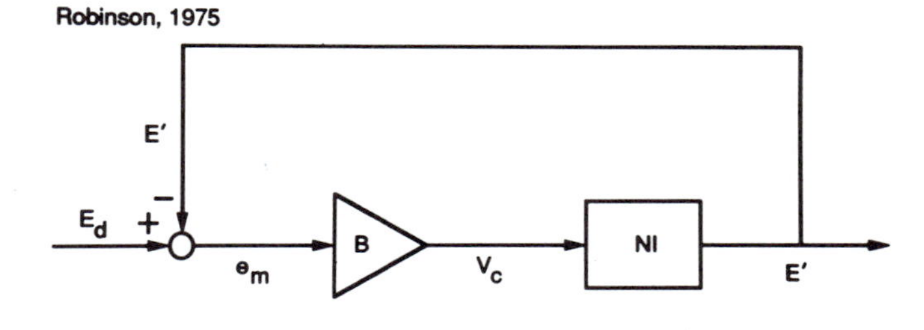

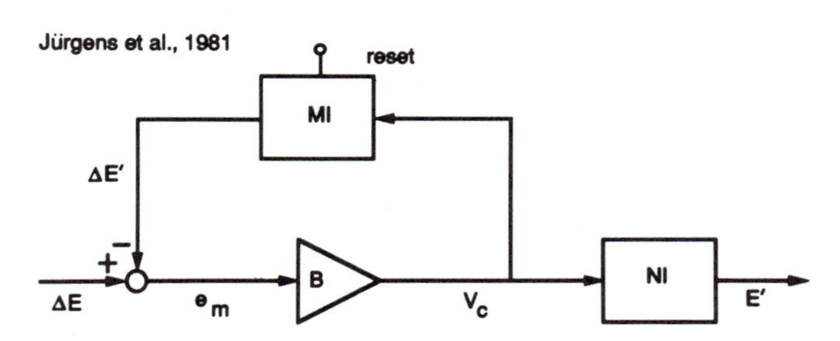

FIGURE 34.9 Two models of feedback control of saccadic amplitude. (Top) Robinson model. (Bottom) Jürgens, Becker, and Kornhuber model. Both are simplified to show only essential elements related to the negative feedback hypothesis (see text for details). E′, new eye position; ΔE′, change in E′; E_d, desired eye position; e_m, error signal; B, burst cells; MI, model integrator; V_c, velocity command; NI, neural integrator. (From Wurtz and Munoz, in press.)

Robinson (1975) proposed a local feedback model to solve this problem. The essential features of his model are shown in the upper half of figure 34.9. A burst of activity (a velocity command, V_c) from pontine burst cells (B) drives the saccade, and the same V_c signal is integrated by a neural integrator (NI) to hold the eye at the new position. At the same time, an internal representation of this position signal (E′) is fed back for comparison with the desired eye position (E_d). As long as a difference persists between these values, an error signal (e_m) is generated that continues to drive the burst generator (V_c). When the difference (e_m) reaches zero, the activity driving the eye stops, and the eye reaches the target.

One problem with this model, however, is that the E_d signal is in cranial or head-centered coordinates rather than the eye-movement (retinotopic) coordinates of the map of movement fields layed out on the SC (see figure 34.1). The saccade cells discharge the same burst regardless of whether the monkey began the saccade close to the center of the visual field or from one side or the other. If this map were in spatial coordinates, where the eye starts should substantially alter the discharge of the cell (Jürgens, Becker, and Kornhuber, 1981).

A subsequent model by Jürgens, Becker, and Kornhuber (1981) offered a resolution between the feedback model of Robinson (1975) and the retinotopic coordinate system seen in the SC (figure 34.9, bottom). These investigators added a second integrator (a model integrator [MI]) that was reset after each saccade so that the feedback was an internal feedback of *change* in eye position (ΔE′) rather than absolute eye position (E′). This internal representation of change in position ΔE′ was compared with the change in eye position required

(ΔE), and any difference is the e_m that would drive the eye to the new position, as in the Robinson model.

This placement of the system in retinotopic coordinates made the signals at the summing junction consistent with the retinotopic coordinate system of the SC and inspired a reinvestigation of the SC to determine whether these signals could be found. The change in eye position required (ΔE) should be present and remain throughout the saccade, whereas the feedback signal ($\Delta E'$) should be present during the saccade and should be reset after it. The difference signal (e_m) is maximal at saccade onset and decreases during the saccade to approach zero as the saccade reaches the target.

Waitzman and coworkers (1988, 1991) found that many saccade-related burst cells in the SC demonstrated clipped responses—that is, the intense burst portion of the discharge came to an end at the time the eye stopped moving. For example, the discharge of the burst cell in figure 34.2A ended close to the time the saccade ended (its response was clipped off by the saccade). Furthermore, the dynamics of the change in discharge in many clipped cells revealed a nearly linear decline over the duration of the saccade and a reduction in motor error (Waitzman et al., 1991).

With this previously unappreciated observation that many burst cells ended as the saccade ended, Waitzman's group (1988, 1991) constructed a modified Becker and Jürgens model of the saccadic system that incorporated the SC specifically into the model (figure 34.10). The retinotopic map of burst or clipped cells in the SC is now envisioned as conveying the e_m signal so that the SC is *inside* the feedback loop. The ΔE signal may be the increased activity in one region of the visual map in the superficial layers of the SC or in the inputs from cortex to SC. No correlate of $\Delta E'$ is evident; the signal may be conveyed by the terminals ending on the clipped cells. The e_m signal in the model would be conveyed to the brain stem by the clipped cells in the SC.

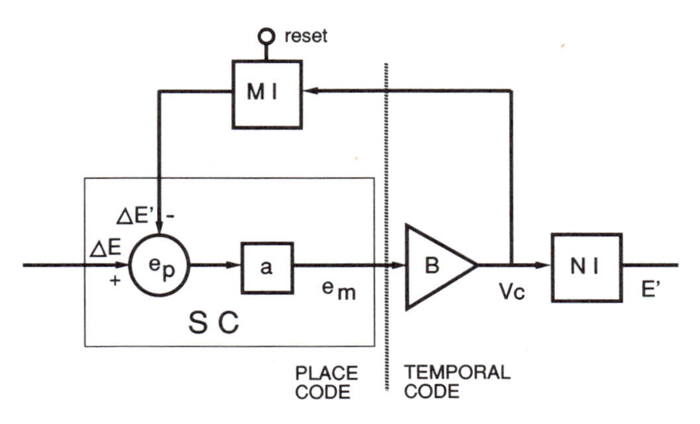

FIGURE 34.10 Model of negative feedback system for control of saccadic amplitude with SC in the feedback loop. a, scale factor; e_p, collicular locus; other abbreviations as in figure 34.9. (From Waitzman et al., 1991.)

A consequence of this model is worth noting. A change in eye position is controlled by the feedback loop, and only errors in position after the output of the summing junction are controlled. Any changes of cell activity within the colliculus would be expected to be associated with changes of eye velocity because they would influence the strength of the burst but not the final eye position controlled by the loop. Thus, a relation between eye velocity and cell discharge (Rohrer, White, and Sparks, 1987) and the effect on eye velocity of chemical lesions of SC (Hikosaka and Wurtz, 1985) both would be consistent with the model.

Finally, note that the SC is in retinotopic coordinates and the preferred change in eye position is conveyed by activity within this map, a place code. The projections from the SC, however, go from this place code to the temporal code that is used eventually to drive the eye muscles. Although we do not know how this conversion occurs, a reasonable hypothesis is that areas within the SC where a hill of neuronal activity is related to large saccades have a stronger innervation onto the pontine bursters than do those associated with small saccades (Edwards and Henkel, 1978; Wurtz and Albano, 1980). The temporal code also must be translated back into a spatial code in the feedback loop, but whether this translation starts from a position signal (Waitzman et al., 1991) or a velocity signal (Lefevre and Galiana, 1990; Droulez and Berthoz, 1991) is not known.

Buildup cells spatially related to saccades

A salient distinction between burst cells and buildup cells in the SC is a difference in the pattern of activity in the cells lying at different positions on the SC motor map. Figures 34.11 and 34.12 illustrate this by showing the normalized spike density profiles of cells in each of the burst and buildup cell layers. The cells have different optimal saccade amplitudes and are located in dif-

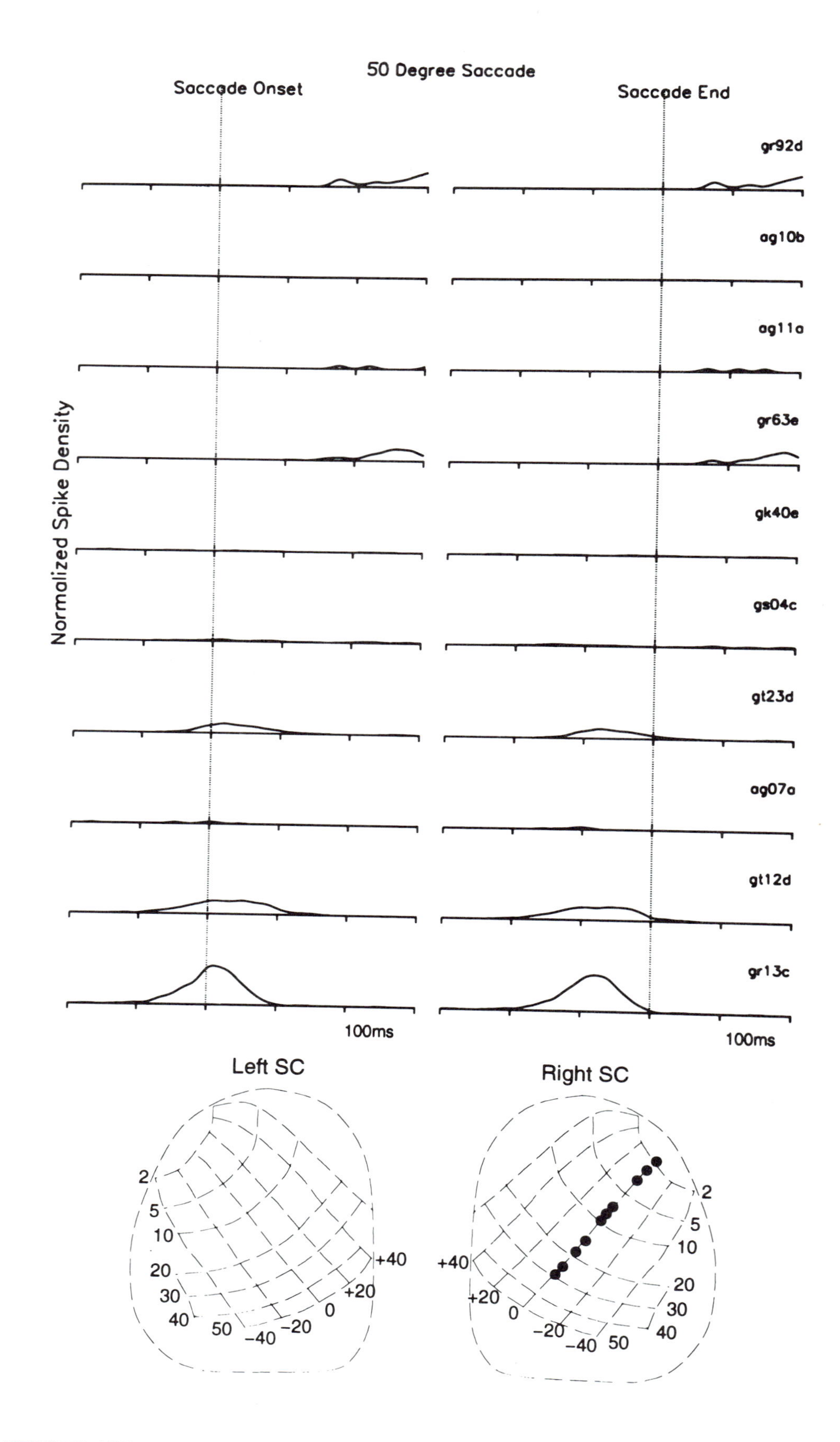
50 Degree Saccade
Saccade Onset
Saccade End
gr92d
ag10b
ag11a
gr63e
gk40e
gs04c
gt23d
ag07a
gt12d
gr13c
Normalized Spike Density
100ms
100ms
Left SC
Right SC
2
5
10
20
30
40
50
-40
-20
0
+20
+40

ferent locations within the SC, which are shown schematically by filled circles on the SC motor map. The discharge for each cell is that accompanying a 50° saccade. The peak discharge of the caudalmost cells in both layers occurred near the time of saccade onset. In the burst cell layer (see figure 34.11), only cells in the caudal SC discharged with the 50° saccade, and cells lying more rostral to the initially active zone remained silent. The level of discharge of burst cells in the initially active zone simply diminished so that by saccade termination, these neurons were almost silent. However, the buildup cells lying rostral to the initially active cells were activated sequentially at some point during the 50° saccade (see figure 34.12). Looking at the left column in figure 34.12, where cell responses are aligned on saccade onset, the peak discharge began before the 50° saccade for the caudalmost cell and gradually moved later for more rostrally located cells. A clear-moving front of activity is therefore visible beginning in the caudal SC and moving to the rostral SC. Again, looking at the burst cell layer, no such movement is evident.

In the rostral SC (see figure 34.12, top), activity was confined to the fixation cells 200 ms before the onset of the 50° saccade. As activity began in the buildup cell layer and then later in the burst cell layer, fixation-related activity in the rostral pole simultaneously diminished. At saccade onset, fixation-related activity had ceased and cells in both layers of the caudal SC were maximally active. The fixation cells began to discharge again at the end of the saccade.

Thus, the buildup cells seem to differ from the burst cells in the activity contained in various parts of the movement map. In the burst cell layer, the neural activity hill reaches peak height at saccade onset and diminishes in size during the saccade. This characteristic contributed to the formation of the model for saccade generation that includes the SC in a feedback loop controlling saccade amplitude, as described previously. In the buildup cell layer, the activity in the buildup cells seems to change as if a front of activity were moving across the SC during the course of the saccade. At the start of a large-amplitude saccade, neural activity is centered in the caudal SC but, as the vector error between the current position of the visual axis and the target decreases during the saccade, cells located progressively more rostral in the SC (i.e., those preferring smaller and smaller amplitude gaze shifts) begin to discharge.

This observation of a shift in activity across the SC during a saccade was first made for movement-related cells in the cat SC (Munoz, Guitton, and Pélisson, 1991) and also led to the conclusion that the SC is within the feedback loop controlling the amplitude of saccades but for reasons quite different from those described earlier. Munoz, Guitton, and Pélisson (1991) argued that when this shifting activity reached the fixation cells located in the rostral SC, the saccade was terminated, thus closing a loop. The location of the SC in relation to a feedback control system for the amplitude of saccades had remained a puzzle for almost two decades, but these two sets of experiments both reached the conclusion that the SC was in the feedback loop.

FIGURE 34.11 Independence of burst cells from sequential activation. Each curve is for one cell that is located progressively further caudally in the SC, as indicated by the dots on the SC map at bottom of figure. Activity was obtained when the monkey made 50° saccades. Same discharge is aligned on saccade onset (left column) and on saccade end (right column). Only cells in the caudalmost SC were activated during the 50° saccade. (From Munoz and Wurtz, 1994.)

Conclusions

Our understanding of a system within the brain that controls the generation of saccadic eye movements has developed substantially over the past 20 years, but the SC has remained a central structure in this system. The recognition that burst cells within the colliculus discharge in a manner consistent with their location within a feedback loop that governs the amplitude of the saccade has allowed the spatial map within the colliculus to be more readily understood in the *spatial-to-temporal* transition that is necessary to activate the temporally driven eye muscles. The study of a second set of saccade-related cells, the buildup cells, that appear to have a moving front of activity across the SC during a saccade, has contributed a second map of movement-related activity. Although there many facets have yet to be worked out, these observations further constrain the position of the SC in any model of the control of saccades.

In contrast to the long-standing recognition of a saccadic system, the recognition of a system for active fixation is relatively recent (Munoz and Guitton,

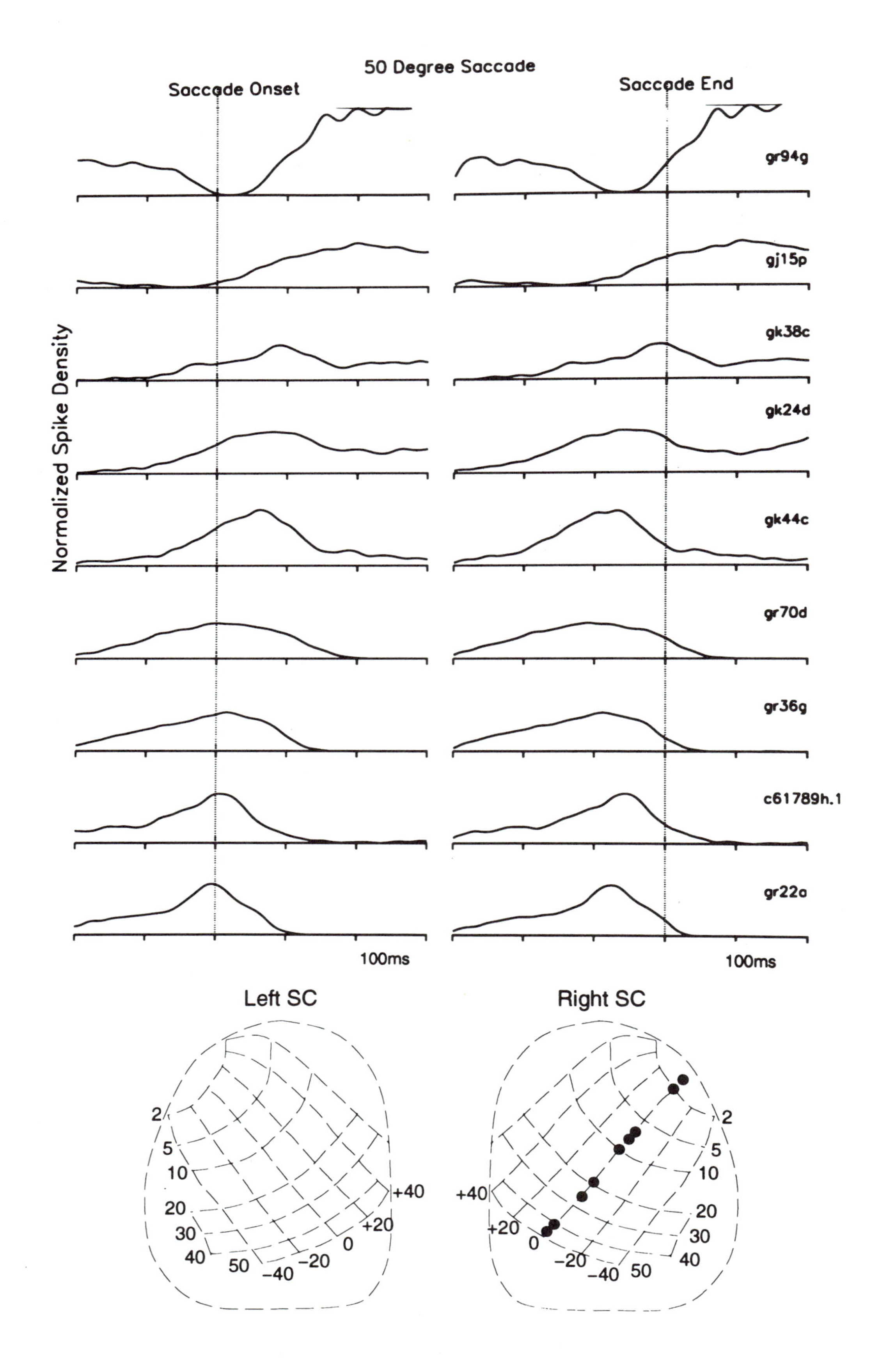
50 Degree Saccade
Saccade Onset
Saccade End
Normalized Spike Density
gr94g
gj15p
gk38c
gk24d
gk44c
gr70d
gr36g
c61789h.1
gr22a
100ms
100ms
Left SC
Right SC
2
5
10
20
30
40
50
+40
+20
0
-20
-40

FIGURE 34.12 Successive activation of buildup cells whose fields are located progressively closer to the anterior pole of the SC. Same organization as in figure 34.11. (From Munoz and Wurtz, 1994.)

1989). The role of the rostral colliculus in this function is now becoming established in both the cat and the monkey. Interaction between this fixation system and the saccadic system at the level of the SC allows us to study the integration of these two systems in a relatively simple environment. The factors involved in such an interaction may aid in our understanding of such integration in other more complex sensorimotor systems.

NOTE

1. The experiments reported in this chapter are based largely on the original research reports of Waitzman and colleagues (1991) and Munoz and Wurtz (1993a, b, c). The chapter is also derived from a recent summary of the SC (Wurtz and Munoz, in press).

REFERENCES

DROULEZ, J., and A. BERTHOZ, 1991. The concept of dynamic memory in sensorimotor control. In *Motor Control: Concepts and Issues*, D. R. Humphrey and H.-J. Freund, eds. Chichester, Engl.: Wiley, pp. 137–161.

EDWARDS, S. B., and C. K. HENKEL, 1978. Superior colliculus connections with the extraocular motor nuclei in the cat. *J. Comp. Neurol.* 179:451–468.

FISCHER, B., 1987. The preparation of visually guided saccades. *Rev. Physiol. Biochem. Pharmacol.* 106:1–35.

FISCHER, B., and R. BOCH, 1983. Saccadic eye movements after extremely short reaction times in the monkey. *Brain Res.* 260:21–26.

HIKOSAKA, O., and R. H. WURTZ, 1983. Visual and oculomotor functions of monkey substantia nigra pars reticulata: II. Visual responses related to fixation of gaze. *J. Neurophysiol.* 49:1254–1267.

HIKOSAKA, O., and R. H. WURTZ, 1985. Modification of saccadic eye movements by GABA-related substances: I. Effect of muscimol and bicuculline in monkey superior colliculus. *J. Neurophysiol.* 53:266–291.

JÜRGENS, R., W. BECKER, and H. H. KORNHUBER, 1981. Natural and drug-induced variations of velocity and duration of human saccadic eye movements: Evidence for a control of the neural pulse generator by local feedback. *Biol. Cybern.* 39:87–96.

LEE, C., W. H. ROHRER, and D. L. SPARKS, 1988. Population coding of saccadic eye movements by neurons in the superior colliculus. *Nature* 332:357–360.

LEFEVRE, PH., and H. L. GALIANA, 1990. Velocity versus position feedback to the superior colliculus in gaze control modelling. *Soc. Neurosci. Abstr.* 16:1084.

MOUNTCASTLE, V. B., R. A. ANDERSEN, and B. C. MOTTER, 1981. The influence of attentive fixation upon the excitability of the light-sensitive neurons of the posterior parietal cortex. *J. Neurosci.* 1:1218–1235.

MUNOZ, D. P., and D. GUITTON, 1989. Fixation and orientation control by the tecto-reticulo-spinal system in the cat whose head is unrestrained. *Rev. Neurol. (Paris)* 145:567–579.

MUNOZ, D. P., and D. GUITTON, 1991. Control of orienting gaze shifts by the tectoreticulospinal system in the head-free cat: II. Sustained discharges during motor preparation and fixation. *J. Neurophysiol.* 66:1624–1641.

MUNOZ, D. P., D. GUITTON, and D. PÉLISSON, 1991. Control of orienting gaze shifts by the tectoreticulospinal system in the head-free cat: III. Spatiotemporal characteristics of phasic motor discharges. *J. Neurophysiol.* 66:1642–1666.

MUNOZ, D. P., and R. H. WURTZ, 1992. Two classes of cells with saccade related activity in the monkey superior colliculus. *Soc. Neurosci. Abstr.* 18:699.

MUNOZ, D. P., and R. H. WURTZ, 1994. Saccade related activity in monkey superior colliculus. Manuscript in preparation.

MUNOZ, D. P., and R. H. WURTZ, 1993a. Fixation cells in monkey superior colliculus: I. Characteristics of cell discharge. *J. Neurophysiol.* 70:559–575.

MUNOZ, D. P., and R. H. WURTZ, 1993b. Fixation cells in monkey superior colliculus: II. Reversible activation and deactivation. *J. Neurophysiol.* 70:576–589.

MUNOZ, D. P., and R. H. WURTZ, 1993c. Superior colliculus and visual fixation. *Biomedical Res.* 14 Suppl 1:75–79.

ROBINSON, D. A., 1972. Eye movements evoked by collicular stimulation in the alert monkey. *Vision Res.* 12:1795–1808.

ROBINSON, D. A., 1975. Oculomotor control signals. In *Basic Mechanisms of Ocular Motility and Their Clinical Implications*, G. Lennerstrand and P. Bach-y-Rita, eds. Oxford: Pergamon, pp. 337–374.

ROHRER, W. H., J. M. WHITE, and D. L. SPARKS, 1987. Saccade-related burst cells in the superior colliculus: Relationship of activity with saccadic velocity. *Soc. Neurosci. Abstr.* 13:1092.

SPARKS, D. L., 1978. Functional properties of neurons in the monkey superior colliculus: Coupling of neuronal activity and saccade onset. *Brain Res.* 156:1–16.

SPARKS, D. L., and R. HARTWICH-YOUNG, 1989. The deep layers of the superior colliculus. In *The Neurobiology of Saccadic Eye Movements: Reviews of Oculomotor Research*, vol. 3, R. H. Wurtz and M. E. Goldberg, eds. Amsterdam: Elsevier, pp. 213–256.

SPARKS, D. L., and L. E. MAYS, 1980. Movement fields of saccade-related burst neurons in the monkey superior colliculus. *Brain Res.* 190:39–50.

SPARKS, D. L., and L. E. MAYS, 1990. Signal transformations required for the generation of saccadic eye movements. *Annu. Rev. Neurosci.* 13:309–336.

WAITZMAN, D. M., T. P. MA, L. M. OPTICAN, and R. H.

Wurtz, 1988. Superior colliculus neurons provide the saccadic motor error signal. *Exp. Brain Res.* 72:649–652.

Waitzman, D. M., T. P. Ma, L. M. Optican, and R. H. Wurtz, 1991. Superior colliculus neurons mediate the dynamic characteristics of saccades. *J. Neurophysiol.* 66: 1716–1737.

Weber, H., and B. Fischer, 1990. Effect of a local ibotenic acid lesion in the visual association area on the prelunate gyrus (area V4) on saccadic reaction times in trained rhesus monkeys. *Exp. Brain Res.* 81:134–139.

Wurtz, R. H., and J. E. Albano, 1980. Visual-motor function of the primate superior colliculus. *Annu. Rev. Neurosci.* 3:189–226.

Wurtz, R. H., and M. E. Goldberg, 1972. Activity of superior colliculus in behaving monkey: III. Cells discharging before eye movements. *J. Neurophysiol.* 35:575–586.

Wurtz, R. H., and M. E. Goldberg, 1989. *The Neurobiology of Saccadic Eye Movements: Reviews of Oculomotor Research*, vol. 3. Amsterdam: Elsevier.

Wurtz, R. H., and D. P. Munoz (in press). Saccadic and fixation systems of oculomotor control in monkey superior colliculus. In *Neuroscience: From Neural Networks to Artificial Intelligence. Proceedings of a U.S.-Mexico Seminar Held in the City of Xalapa in the State of Veracruz on December 9–11, 1991*, P. Rudomin, M. A. Arbib, F. Cervantes-Perez, and R. Romo, eds. Berlin: Springer-Verlag.

35 Contributions of Vision and Proprioception to Accuracy in Limb Movements

CLAUDE GHEZ, JAMES GORDON, MARIA FELICE GHILARDI, AND ROBERT SAINBURG

ABSTRACT We have studied movement errors in normal human subjects and in patients deafferented by large-fiber sensory neuropathy. In normals, movement extent and direction were subject to different sources of variable and systematic errors, suggesting that these parameters are programed independent. Moreover, vision of the hand and the target were necesary to program direction accurately. These data suggest that the planning of reaching movements takes place in an extrinsic, hand-centered coordinate system.

In deafferented patients, simple movements aimed to visual targets showed large errors in direction and extent because of failure to compensate for directional variations in limb inertia. In movements with direction reversals, distinctive errors appeared because of failure to program elbow muscle contractions in accord with interaction torques produced at the elbow by variations in acceleration of the upper arm. Both inertial and reversal errors were substantially reduced when patients had recently had the opportunity to monitor movements of their arm visually. We conclude that the programing of accurate trajectories requires a frequently updated internal model of the state and properties of the limb by proprioceptive input. It is proposed that such internal models are critical for the transformation from extrinsic to intrinsic coordinates used to plan the joint angle changes and torques needed to execute the movement.

It is generally understood that the accuracy of limb movements depends largely on precisely calibrated feedforward commands that direct the hand to the target (Georgopoulos, 1986). Although vision and proprioception are both essential if movements are to be accurate, the nature of the information provided by these two modalities is not fully understood. For example, it normally is taken for granted that vision simply provides information about the location of the target. Whether vision is needed also to determine the initial position of the hand is not known. Some investigators hypothesize that the relationship of the target to the limb is critical (Burnod et al., 1992; Flanders, Helms Tillery, and Soechting, 1992). For these authors, movement trajectories are driven by a motor error representing the difference between the intended final limb configuration and its initial configuration, determined proprioceptively. Whether the extent and direction of movement can, in fact, be programed accurately by the comparison of visual information obtained from a target and information about arm configuration obtained proprioceptively has not been examined in any detail.

Significant insights into the role of proprioception in trajectory formation have been obtained by studying the motor deficits of patients with large-fiber sensory neuropathy (Rothwell et al., 1982; Sanes et al., 1985; Forget and Lamarre, 1987; Forget and Lamarre, 1990; Ghez et al., 1990). In these patients, the selective degeneration of large-diameter afferent fibers may abolish completely all sense of joint position as well as stretch reflexes. Studies of such patients have documented the importance of proprioceptive input for the regulation of steady-state force and for detecting and correcting trajectory errors due to mechanical perturbations (Rothwell et al., 1982; Sanes et al., 1985). Evidence from such studies suggests that loss of proprioception does not alter or impair the strategies that subjects use to make single-joint movements: Like intact controls, deafferented patients produce move-

CLAUDE GHEZ, MARIA FELICE GHILARDI, and ROBERT SAINBURG Center for Neurobiology and Behavior, New York State Psychiatric Institute, College of Physicians and Surgeons, Columbia University, New York, N.Y.

JAMES GORDON Program in Physical Therapy, College of Physicians and Surgeons, Columbia University, New York, N.Y.

ments or forces of different magnitudes by scaling a stereotyped pattern of contraction in agonist and antagonist muscles (Rothwell et al., 1982; Sanes et al., 1985; Forget and Lamarre, 1987). Nevertheless, it is generally agreed that proprioceptive information is essential for the correct calibration of motor commands (Paillard and Brouchon, 1974) and that motor learning might be impaired in deafferentation (Rothwell et al., 1982). However, the aspects of trajectory control that require such learning remain undefined.

The studies described in this chapter examine how visual and proprioceptive information about the limb contributes to accuracy of planar hand movements. We first address the question of whether extent and direction are explicitly planned features of reaching movements. We approach this question by analyzing the sources of variability of movement endpoints and by defining how vision of the limb alters systematic trajectory errors in movements made without visual feedback. We then examine the trajectory deficits exhibited by deafferented patients to determine how proprioceptive information contributes to accuracy. Patients are found to demonstrate extent and direction errors that vary with movement direction, as do normals, but in exaggerated form. Finally, we consider the role of proprioceptive input in more complex movements requiring precise coordination of elbow and shoulder joint motions. Our findings indicate that proprioceptive input plays a critical role in motor planning by updating an internal model of biomechanical characteristics of the limb.

Methods

Subjects were 11 neurologically intact individuals between the ages of 28 and 42 years and 3 patients with large-fiber sensory neuropathy. All three patients (MA, age 42; GL, age 54; and CF, age 60) had virtually complete loss of position, vibration, and discriminative touch sensation in both upper extremities, including elbow and shoulder, and tendon reflexes were absent. On the other hand, pain, temperature, and coarse touch sensations were preserved. Somatosensory evoked potentials from upper as well as lower extremities were absent. Sensory nerve conductions were slowed, a finding consistent with a loss of large-diameter afferent fibers. Muscle strength and electromyography were, however, normal. Although the degree of sensory loss in the upper extremities was roughly similar for the three patients, lower-extremity involvement was most severe in patients GL and CF, who were wheelchair-bound. Patient MA, on the other hand, could walk, albeit with a wide base.

The results reported here were obtained using two tasks in which subjects faced a vertical computer screen and moved a hand-held cursor on a horizontal digitizing tablet. In general, the tablet was at shoulder level and the subject's arm was supported in the horizontal plane by a sling suspended from the ceiling. This was done to counter the effects of gravity and to simplify biomechanical analyses. For selected experiments, however, the tablet was at waist level. The first task was a simple reaching task in which the computer screen was used to display the position of the cursor on the tablet along with two circles, indicating a starting and a target location. At the beginning of a trial, subjects were to position the cursor in the start circle. Then, after an unpredictable time, a go tone was presented, and subjects were to move the cursor to the target with a single, uncorrected movement. Knowledge of results (KR) generally was provided by displaying the hand path on the screen after the movement although, in some experiments, the errors made with and without KR were compared. Targets in different directions and at different distances from the starting position were presented in a pseudorandom order. To prevent the correction of errors detected visually and to identify errors related to the planning of movement, the screen cursor was blanked after presentation of the tone. Each subject's hand and arm were hidden from view by a drape and a two-way mirror. The influence of information about the location and properties of the limb was analyzed in certain experiments by allowing each subject to see his or her limb at rest or during movements. Data consisted of hand positions sampled by the computer at 200 Hz.

In the second task, subjects viewed the computer screen that now displayed one of six possible straight-line segments, in pseudorandom order, along with the cursor. They were instructed first to position the cursor at one end of the line. On presentation of an auditory cue, they were to move the cursor straight to the end of the line and return to the origin in a single uninterrupted movement. The cursor was again blanked during the movement to prevent visual corrections. The instructions stressed that the outward and return motion should overlap and that movement reversals should be sharp and without discernible pause between the outward and inward segments. Movements were to be carried out at a comfortable speed.

During these reversal movements, the arm and forearm were supported in the horizontal plane by a low-inertia brace equipped with ball-bearing joints under the shoulder and elbow joints. Joints distal to the elbow were immobilized with a thermoplastic splint attached to the brace, and the scapula was immobilized with straps to restrict movement to the shoulder and elbow. Precision, single-turn, linear potentiometers (Beckman Instruments) were used to monitor the elbow and shoulder joint angles. The tip of a magnetic pen, controlling the screen cursor, was attached to the end of the hand splint, 1 cm above the digitizing tablet. Surface electromyographic (EMG) activity from biceps, brachioradialis, and triceps was recorded with active electrodes (Liberty Mutual, Inc.). EMG and potentiometer signals were sampled at 1000 Hz and were acquired by a Macintosh computer equipped with external A/D converters (MP-100 Biopacq).

Results

We begin our analysis by presenting results obtained in normal subjects. We then analyze the changes in trajectory formation in deafferented patients.

Independently Planned Extent and Direction

Despite the absence of visual feedback, movements made by intact subjects were reasonably accurate, their endpoints clustering around the appropriate targets. As in studies of planar hand movements performed with vision, hand paths were nearly straight and trajectory profiles were bell-shaped (figure 35.1A). In general, endpoint distributions had elliptical shapes whose major axes were oriented in the directions of the movements; however, the eccentricity of these ellipses decreased with distance (figure 35.1B). Thus, while extent variability was fairly large for small movements, relative extent variability decreased progressively with distance. In contrast, directional variability was essentially constant and unaffected by distance. The fact that extent and directional variability were differently influenced by target distance suggests that these two features of the response were specified by distinct mechanisms.

The straight hand paths and the scaling of peak accelerations implies that both the directions and extents of the movements were largely specified by the time of movement initiation. There were, however, systematic differences in the scaling of velocities (figure 35.1C) and accelerations for movements in different directions. To explore this directional dependency, we presented subjects with targets in 24 directions at a constant distance from a common starting position. As illustrated for one subject in figure 35.2B, peak accelerations varied markedly with movement direction, and acceleration vectors formed an ellipsoidal shape. Accelerations in the 60° and 240° directions were more than twice as large as for movements in the 150° and 330° directions.

Why should accelerations vary with the direction of the movement? In our convention and with the hand's initial position in the midline, the 150° direction corresponded approximately to the axis of the forearm, whereas the 60° direction was perpendicular to it. Because the inertial load at the hand is greatest in the direction of the forearm and lowest at right angles to it, we hypothesized that the systematic differences in acceleration resulted from directional variations in inertia. To test this idea, we used the standard equations of motion of a two-link manipulator and morphometric data from each subject to compute the directional variations in inertia at the hand (Hogan, 1985). As has been demonstrated by Hogan (1985), inertia at the hand varies with direction, forming an elliptical contour whose major axis is oriented in the direction of the forearm (see figure 35.2B). This means that if a constant force were to be applied at the hand, the resulting initial accelerations would show a corresponding directional variation but rotated 90°. The resulting contour is referred to as a *mobility ellipse*. It can be seen in figure 35.2B that the directional variations in acceleration closely match the changes in initial acceleration due to the inertial anisotropy (compare solid lines fitted to points and dotted lines representing the computed mobility ellipse). Although movements in the directions of lowest inertia are slightly hypermetric, movement extents showed much less directional variation than the accelerations. This is because differences in hand acceleration were largely compensated by directional variations in movement time. It should be noted that these variations in movement time were substantial and ranged, across subjects, from 150 to 300 ms for average movement times of approximately 400 ms.

These findings indicate that subjects do not program the magnitude of the force they use to accelerate the hand at the onset of movement to take into account

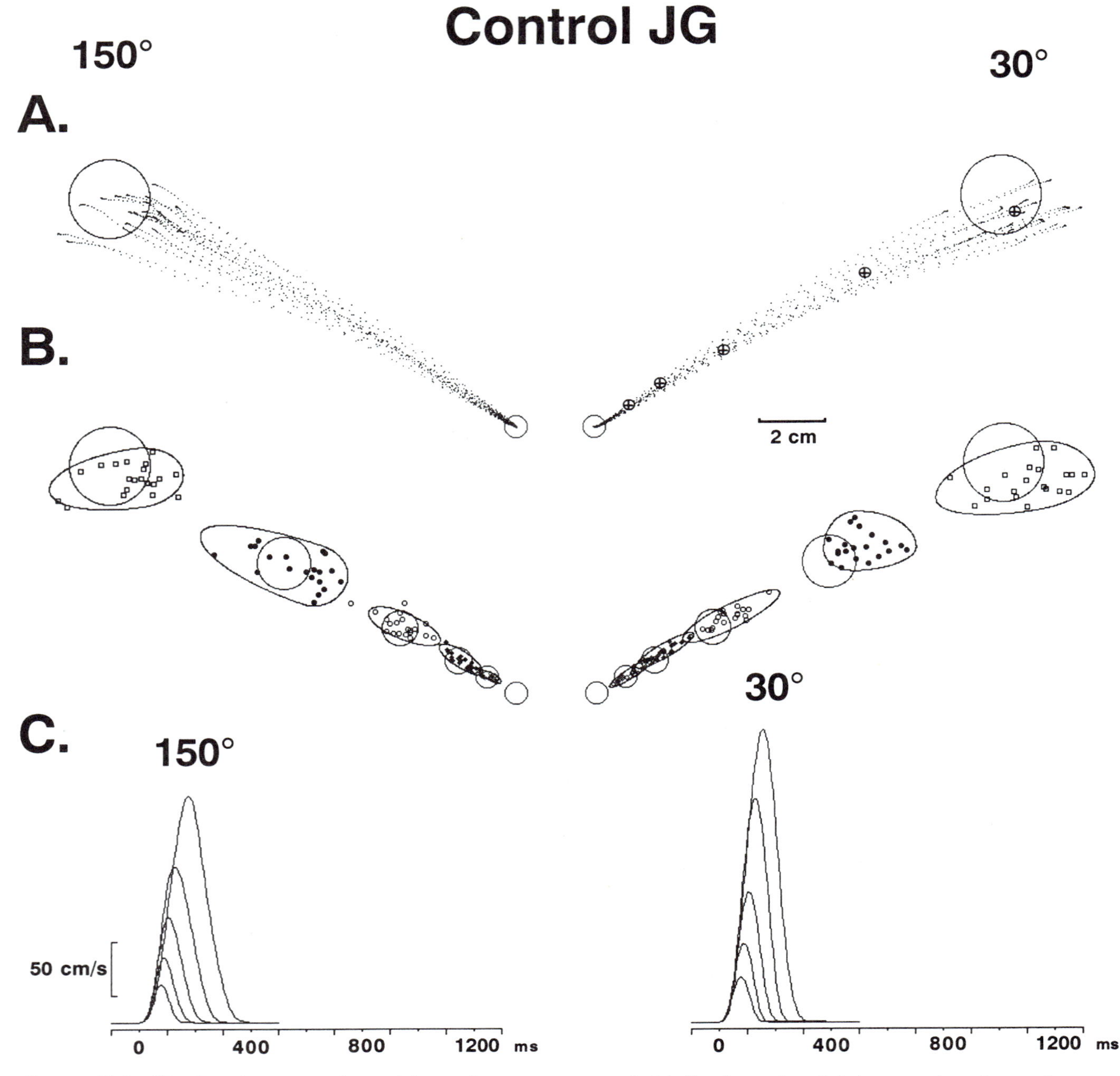

FIGURE 35.1 Hand paths are nearly straight, and tangential velocities of movements scale with target distance. Control subject made 20 movements to targets at five distances in two directions presented in pseudorandom order. Tablet was at waist level. (A) Hand paths to the most distant targets (19.6 cm) in the two directions (150° and 30°) plotted at 20-ms intervals. Note that the centers of the endpoint distributions of movements to nearer targets (small circles with crosshair) lie along the slightly curved paths to the most distant target at 30°. (B) Endpoint distributions. Each distribution is surrounded by a contour whose orientation was computed by the method of principal components and whose size and shape is based on the interquartile range in each of the major axes. Small circles show target locations. (C) Tangential velocities of movements.

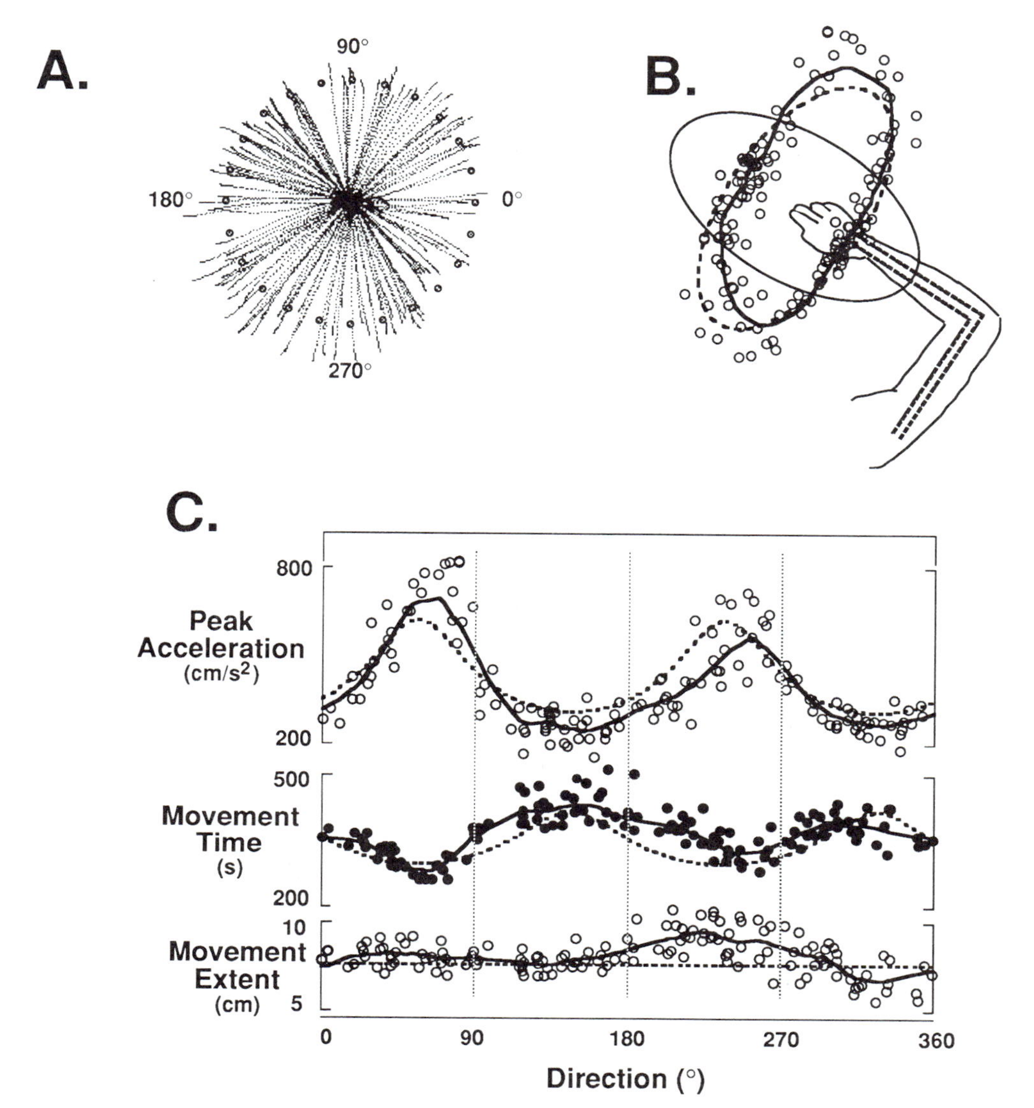

FIGURE 35.2 Subjects compensate for variations in inertial resistance by modulating movement time. (A) Hand paths of movements (six per target) aimed to targets in 24 directions at 7.5 cm from a central starting position (plotted at 5-ms intervals). (B) Solid ellipse represents inertial resistance of the tip of the hand to force applied in different directions. Dashed ellipse represents mobility of the hand. The small circles show the directions and magnitudes of peak accelerations for movements in (A) in a polar format, with the origin at the index finger of the hand. The solid curve is a LOWESS fit to these points (Cleveland, 1979). (C) Scatter plots and fitted solid lines show peak acceleration, movement time, and movement extent as a function of direction for the movements in (A). Dashed line on peak acceleration is predicted peak acceleration for a constant force applied to the hand in all directions. Dashed line on movement time represents predicted movement time under the assumption that trajectories have invariantly shaped velocity profiles. Dashed line on movement extent indicates target distance.

directional differences in limb inertia.[1] Instead, they adapt movements to these differences by adjustments in movement time. Although it is not inconceivable that neural mechanisms could vary movement time to compensate for expected variations in load, it does not seem likely that such control could be as precise as observed here. It is more plausible that the variation in movement time for different directions occurs because, as suggested by several groups (Bizzi et al., 1984; Feldman et al., 1990), the nervous system does not specify directly the precise kinematic features of the movement. Rather it specifies a virtual trajectory whose implementation lags behind the actual trajectory by a delay that reflects the biomechanical properties of the physical plant. This lag would be expected to be greater for directions in which resistive forces are greater. Whether the resulting compensation for differences in inertia occurs purely because of the biomechanical characteristics of the muscles and joints (Bizzi et al., 1984) or because of feedback mechanisms —as envisaged, for example, in the lambda model of Feldman and coworkers (1990)—remains to be determined.

Planned movement in a Hand-centered Coordinate System Although directional errors were gen-

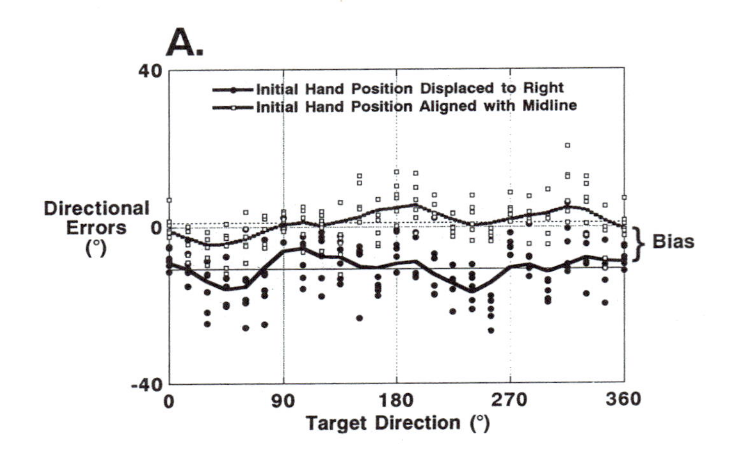

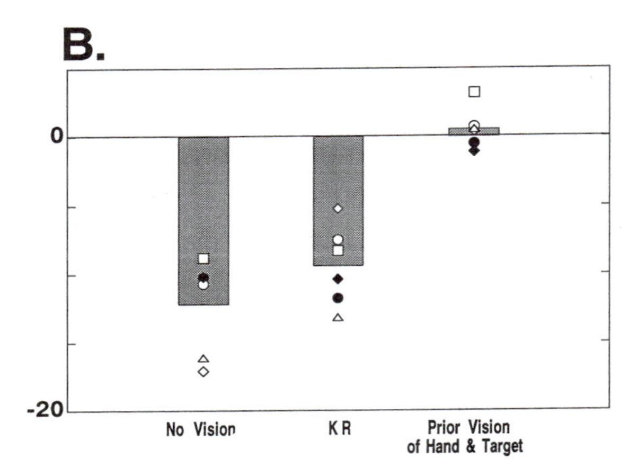

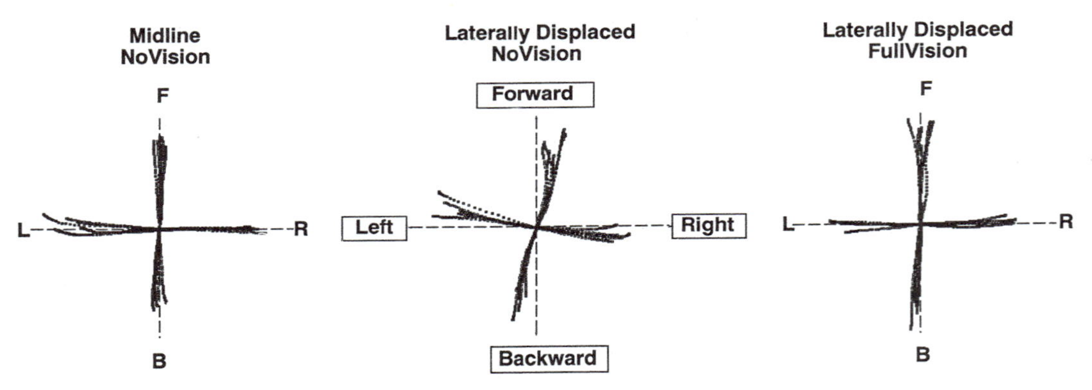

Figure 35.3 Deviations of the initial position of the hand from the midline produce directional biases that can be corrected only by vision of the hand relative to the target. (A) Directional errors for movements to 24 radially arranged targets with starting positions of the hand in the midline (open circles) and displaced 44 cm to the right (filled circles). Clockwise errors are negative; counterclockwise errors are positive. Fitted lines are computed, as previously, by LOWESS. In both initial positions, the elbow angle was approximately 90°. With the hand in the midline, the forearm segment angle was approximately 140° and, with the hand in the lateral position, the forearm segment angle was nearly 90° relative to the mediolateral axis. (B) Mean directional errors for movements from the displaced initial position under three conditions: No vision, without vision of the hand or knowledge of the results (KR); KR, without vision of the hand but with KR; and prior vision, in which subjects were allowed to see their hand and target before movement but there was no KR. Bars are mean directional errors of six subjects whose individual means are shown by different symbols. (C) Hand paths for movements in four directions and two initial positions when subject was given the verbal instruction to move the hand in the directions indicated. NoVision indicates subject was allowed no vision of the hand throughout block; in FullVision, subject was allowed to see the hand during movement.

erally small, even when movements were performed without visual feedback, this was true only when the initial position of the hand was near the subject's midline. Figure 35.3A shows the relationships of directional errors to target direction for two initial hand positions. For both positions, there are small systematic variations in directional errors for different target directions.[2] In addition, however, lateral displacement of the initial hand position (in this case, by extending the initial shoulder angle and maintaining the same elbow angle) produced a directional bias that shifted the directions of all responses clockwise by an average of 11°.

Comparisons of movements from a wide variety of initial positions showed that the magnitude of this directional bias varied systematically with the mediolateral distance of the hand from a plane through the body midline. It was, however, relatively unaffected by variations in anteroposterior location of the initial position of the hand. Similarly, directional bias was identical for movements of varying extents.

These directional biases were remarkably robust, present in all subjects, and surprisingly resistant to learning. Thus, they were only minimally reduced in movements made with KR (i.e., display of hand path after each movement) in comparison to movements without KR (figure 35.3B). In contrast, the bias always disappeared when subjects were allowed to visualize the location of their hand relative to targets placed directly in the workspace (see figure 35.3B).

The directional biases did not, however, depend on the presence of a visual display. Indeed, the same clockwise bias was evident when subjects were verbally instructed to move the hand directly forward, directly backward, to the left, or to the right from different initial positions (figure 35.3C). As in the experiments with targets presented on a computer monitor, this bias disappeared when each subject was allowed to view his or her hand during the movement. These directional biases therefore represent transformational errors related, perhaps, to distortions in the subjects' representation of the location of their hand in peripersonal space. Whatever the precise geometry of subjects' representation of peripersonal space, the finding of directional biases that are independent of movement extent lends further support to the hypothesis that extent and direction are planned independently. The fact that these errors are corrected by viewing the hand in relation to the target indicates that this information cannot be derived correctly from proprioceptive input alone.

Roles of Proprioceptive Information in the Planning and Execution of Multijoint Movements

Prevention of inertial errors by vision of the limb's response to a prior motor command The directional biases just described emphasize that proprioception does not provide sufficient information about limb configuration for neural controllers to compute hand direction accurately. What, then, is the function of proprioceptive information in programing movements? If it plays a role in "calibrating" motor commands or updating an internal representation of the limb, what aspects of the limb are represented? To answer these questions, we studied movements performed by three patients deprived of proprioceptive sensation because of large-fiber sensory neuropathy but with intact motor function (see under Methods). We examined the effects of visual feedback by comparing movements with and without vision of the screen cursor. If proprioceptive sensation functioned solely by correcting movements through negative feedback, performance would be expected to improve only when movements were performed when the cursor was displayed on the screen. Vision of the limb by itself would be expected to have little, if any, effect because targets were not displayed in relation to the limb and thus error information, needed for feedback, was not readily available. However, if vision served to improve programing (e.g., by updating a subject's internal model of his or her arm), vision of the arm should improve performance, and this improvement should persist for some time when vision was no longer available.

In contrast to the straight and accurate movements made by controls, movements of deafferented patients were highly curved, and endpoint errors were greatly increased (compare figure 35.4A, B with figure 35.1A and B). Nevertheless, the normal scaling strategy for producing movements of different extents was preserved (figure 35.4C). The hand paths of movements in different directions showed a striking dependence of both extent and direction errors on the direction of the movement (figure 35.5A, no vision; compare to control paths in figure 35.2A). The variations in movement extent corresponded closely to directional variations in

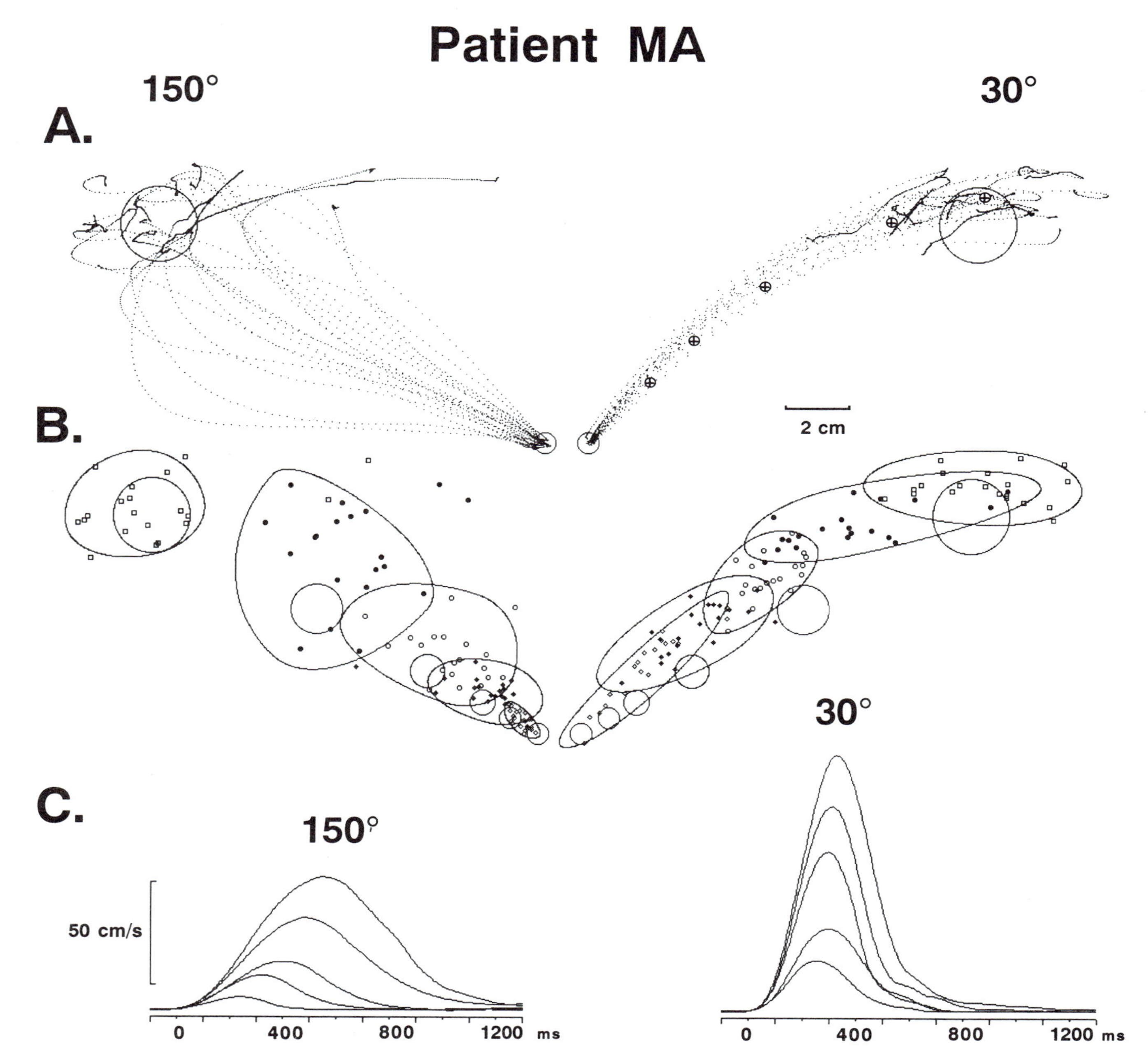

FIGURE 35.4 (A) Hand paths, (B) endpoint distributions, and (C) tangential velocities of movements to targets in two directions in patient MA (as per figure 35.1). Note the consistent curvature of movements to 30°, which were performed principally by external rotation of the shoulder, and the fact that the endpoint distributions of movements to nearer targets, shown by small circles with a crosshair, all lie along the paths to the most distant target. See text for further comments.

inertia. Thus, whereas in controls variations in movement extent show little or no dependence on inertial variations in peak acceleration (figure 35.5B, left), in the patients the two parameters were highly correlated (figure 35.5B, right, no-vision condition). Correspondingly, variations in movement time were negatively correlated with inertial variations in acceleration in controls (figure 35.5C, left) but not in the patients (figure 35.5C, right, no-vision condition). Thus, a significant cause of the patients' errors in movements per-

A. Successive Trials

No Vision of hand

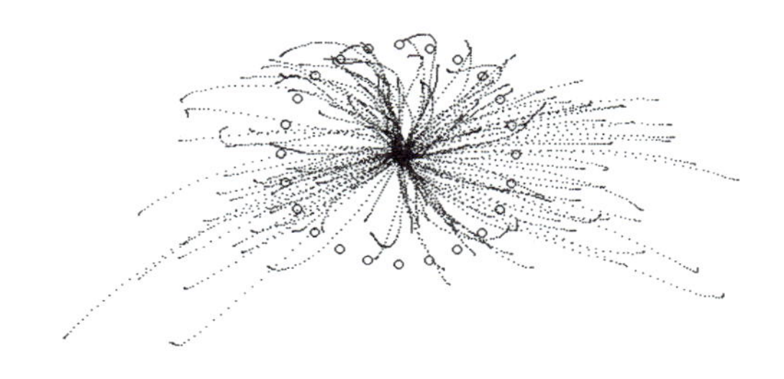

Alternate Trials

No Vision of hand

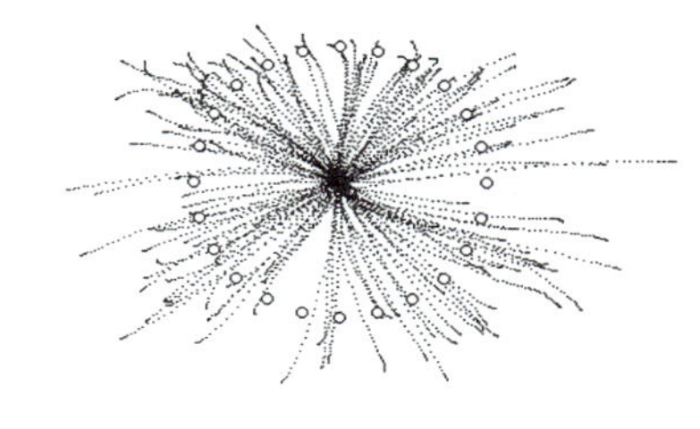

Vision of hand

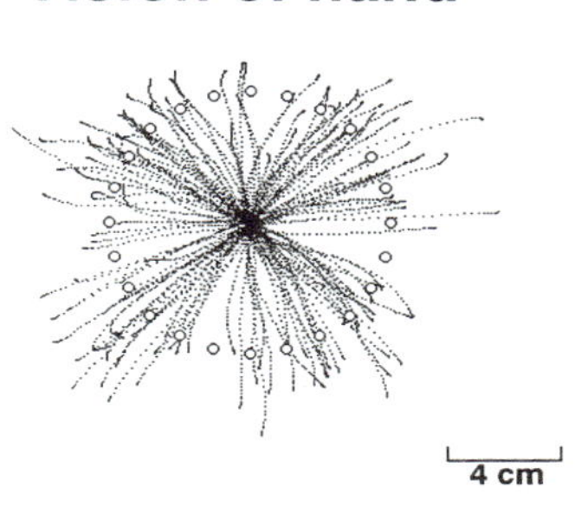

B. Extent Variance Accounted for by Peak Acceleration

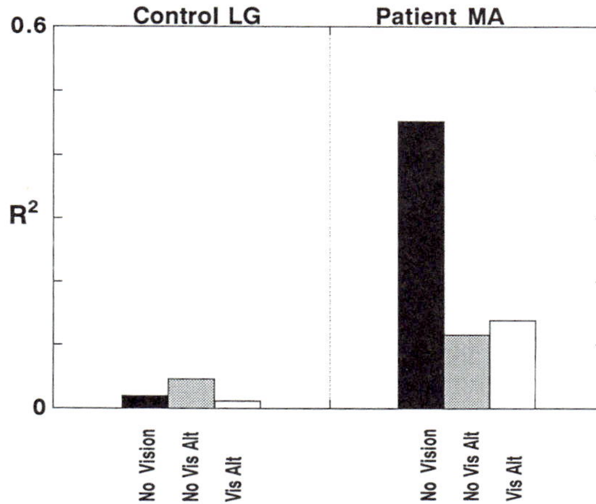

C. Movement Time Variance Accounted for by Peak Acceleration

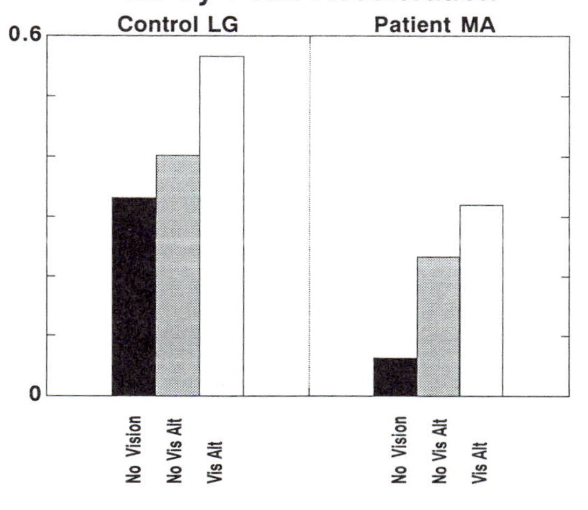

FIGURE 35.5 Vision of the hand prior to or during movement reduces inertial errors. (A) Hand paths for movements made in 24 directions by patient MA under three conditions. The left plot shows hand paths when the subject could not see her hand at all. The middle and right plots show movement during a session in which trials were presented alternately with no vision of the hand and with vision of the hand. Paths plotted every 20 ms. (B) Squared correlation coefficient (r^2) between peak acceleration and movement extent for movements made by a control subject (LG) and deafferented subject (MA) in the three conditions described in (A). (C) Squared correlation coefficient (r^2) between peak acceleration and movement time for movements made by a control subject (LG) and deafferented subject (MA) in the three conditions described in (A).

formed without vision was the failure to compensate for variations in limb inertia by modifying movement time according to movement direction.

Whereas vision of the limb produced little change in the errors made by controls, it produced a remarkable improvement in the patients, and this effect outlasted the presence of vision information. This was ascertained by allowing patients to see their limb on alternate trials with randomized directions. We then compared movements performed with and without vision in these alternating blocks, with movements performed in blocks of trials in which all movements were made without vision. Hand paths of responses made by patient MA with and without vision of the limb in the alternating block are plotted separately in figure 35.5A (center and right), and errors are analyzed separately in parts B–D. The paths are straighter and less hypermetric in low-inertia directions in both alternate vision blocks compared with the block performed without vision (see figure 35.5A, left). In both sets of alternate vision responses, movement extent becomes substantially less dependent on acceleration (see figure 35.5B), whereas movement time develops a significant dependence on initial acceleration, as in the control (see figure 35.5C). Thus, vision of the limb either during or before movement improves accuracy by allowing patients to adapt their motor commands to inertial and other mechanical properties of the limb.

Errors were also decreased when patients simply viewed their limb between trials during a period of immobility. However, in all patients, this improvement was less than when they were allowed to see their limb in motion during a preceding response. Viewing the limb in motion provides information about the limb's dynamic response to a preceding neural command, information that cannot be obtained simply from a static view of the limb. It is likely that this information allows subjects to recalibrate dynamic internal models of their limb and that these models are critical for the programing process (Atkeson, 1989; Jordan and Rumelhart, 1990; Morasso and Sanguineti, 1992). Interestingly, however, the improvement provided by vision was short-lived and lasted only 5 to 10 trials (approximately 1 minute). This suggests that in control subjects, the huge amounts of information provided by proprioception are critical to recalibrating continuously the internal limb representations.

Proprioceptive information is needed to control interaction torques The prominent and variable curvature of their hand movements (see figures 35.4, 35.5) suggests that patients also had difficulty coordinating elbow and shoulder motions. This was confirmed by comparing the time of directional changes in joint motion in three-dimensional movements performed by patients and controls (Sainburg, Poizner, and Ghez, in press).

Incoordination of elbow and shoulder movements was particularly striking when patients attempted to reverse the direction of their hand movements. Because direction reversals are associated with high angular accelerations that, in turn, produce large interaction torques, we suspected that failure to synchronize joint motions might reflect a failure to control this aspect of limb biomechanics (Hollerbach and Flash, 1982; Schneider et al., 1989).

To characterize the effects of interaction torques on the coordination of elbow and shoulder, we used a task that allowed these torques to be varied systematically (Sainburg et al., 1992). Each subject was to move his or her hand along a straight target line and, at the end of the line, to reverse direction and return to the point of origin. Directions were selected so that the shoulder angular excursion would increase progressively over six target directions (figure 35.6). The lengths of the target paths were adjusted to maintain the elbow angular excursion nearly constant; the arm was supported in the horizontal plane to eliminate the effects of gravity.

Whereas controls always reversed direction sharply, patients could do so only for directions in which shoulder excursions were minimal. The sharp reversals in controls corresponded to nearly synchronous direction reversals of the shoulder and elbow joints. In patients, as shoulder excursions increased, hand paths devel-

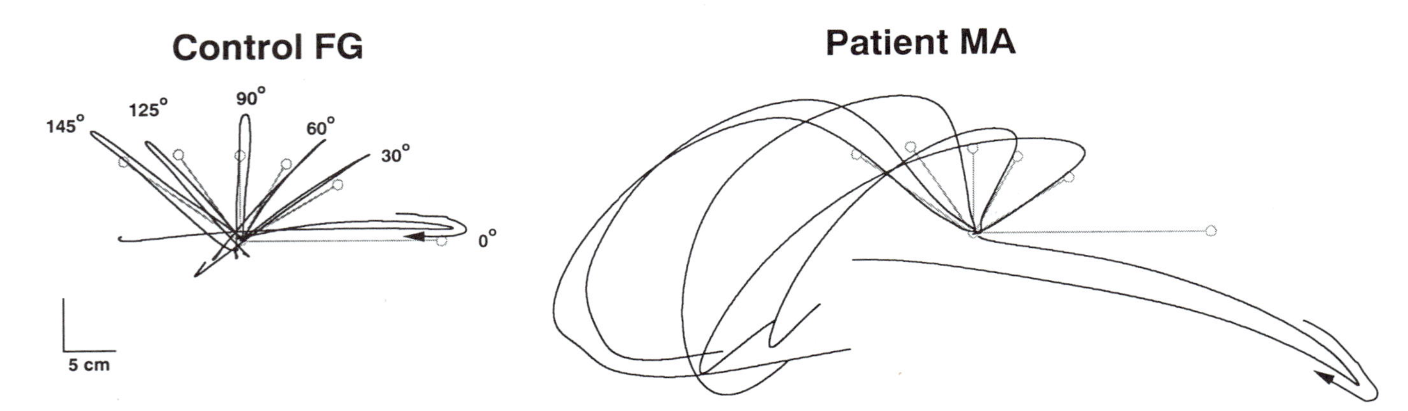

FIGURE 35.6 Deafferented patients are unable to reverse direction abruptly. Sample hand paths of movements performed by control subject (left) and patient (right) are drawn over the corresponding target lines (gray). Circles at the end of each line are shown for clarity and did not exist in the actual visual presentation to the subjects.

oped a marked medial deviation because the elbow reversed direction prematurely.

To determine whether the premature elbow flexion could have resulted from a failure to counter an interaction torque produced by shoulder deceleration, we next computed the torques acting on the elbow (Winter, 1990). Using a modification of the method introduced by Hoy, Zernicke, and Smith (1985), we sub-

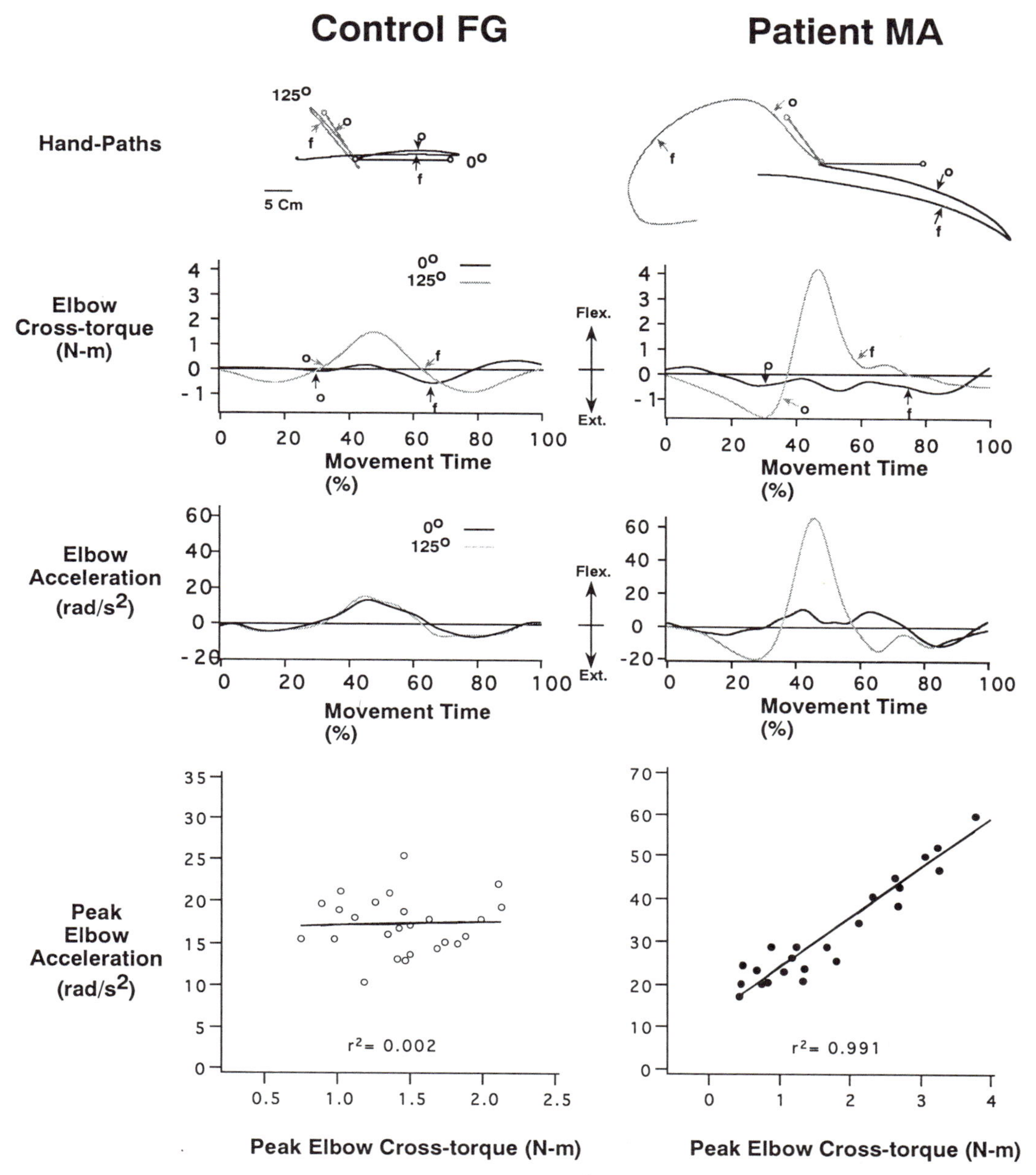

FIGURE 35.7 Reversal errors are produced by cross-torques acting at the elbow. (Top) As in figure 35.6, the hand paths (solid lines) are drawn over the targets (gray lines) for movements in the 0° and 125° directions performed by a control subject (left) and a patient (right). Elbow joint cross-torque profiles are shown for the movements illustrated at top. Elbow joint angular acceleration profiles also are shown. Initiation (o) and final (f) points of the flexor acceleration phase of elbow joint motion are marked by arrows in all graphs in plots of cross-torques and elbow accelerations. At bottom, peak elbow flexor acceleration is plotted against peak flexor cross-torque for all movements performed by control subject (left) and by patient (right). Note the absence of correlation in the movements performed by control subject ($r^2 = .002$), whereas correlation is high in the movements performed by patient ($r^2 = .911$).

divided the torques acting at the elbow into three components, which we termed *self-torque*, *cross-torque*, and *residual torque* (Cooper, Martin, and Ghez, in press).[3] The self-torque at the elbow represents the torque required to overcome the inertial resistance of the forearm and, as such, varies with elbow angular acceleration alone. The cross-torque is the interaction torque produced at the elbow by motions at other joints. The residual torque represents the inverse of the summed cross- and self-torques; thus, it includes the effects of active muscle contraction as well as the elasticity and viscosity of muscles and connective tissue elements (Hoy, Zernicke, and Smith, 1985).

This partitioning of joint torques showed that for hand directions with significant shoulder excursions, the cross-torque acting on the elbow during movement reversals was indeed in the flexor direction. It was small or negligible for movements at 0° and 30° but increased progressively for movements at 60°–145° for both controls and patients. In controls, elbow angular accelerations remained independent of cross-torque across directions, whereas in the patients, elbow accelerations were closely correlated with cross-torques (figure 35.7).

Controls were able to control elbow acceleration in spite of varying cross-torques by varying the timing and degree of activation of elbow flexors and extensors. Thus, when cross-torque was low, as in the 0° movements, biceps and brachialis EMG recordings became active prior to the onset of flexor acceleration, while the triceps that had been extending the forearm became silent. In contrast, at 125° and 145°, biceps and brachialis were silent. Flexor acceleration was initiated by the cross-torque from the shoulder, which was then modulated by triceps activation. In contrast, patients were unable to modulate substantially the timing of elbow muscle activation across directions. Instead EMG recordings of elbow muscles showed that they attempted to regulate movement trajectory by increaseing joint stiffness through cocontraction of flexors and extensors. This was not a successful strategy, however, perhaps because the loss of stretch reflexes may substantially alter the functional stiffness of muscles (Sanes and Shadmehr, in press).

These results confirm and extend the observations of Smith and Zernicke and their colleagues and demonstrate a critical role of proprioceptive input in compensating for interaction torques (Smith and Zernicke, 1987; Koshland and Smith, 1989a, b). Interestingly, as was the case for inertial errors, vision of the limb enabled the patients to improve interjoint coordination significantly. Pilot data in two patients indicates that when they could see their limb in motion, the patients were better able to synchronize elbow and shoulder reversals. Similarly, the timing and patterns of biceps and triceps activation were better adapted to the direction of movement. The compensation provided by vision of the limb was, however, far from complete, and interaction torques still produced significant distortions in the hand paths. This suggests that vision cannot fully compensate for the loss of proprioception in ensuring interjoint coordination.

Several factors may account for this. First, it is possible that muscle proprioceptors contribute to interjoint coordination in part through feedback mechanisms operating directly through spinal connections (Smith and Zernicke, 1987; Koshland and Smith, 1989a, b; Nichols, 1989; Soechting and Lacquaniti, 1989; Lacquaniti, Borghese, and Carrozzo, 1991). Indeed, while classical physiological studies have emphasized their monosynaptic connections, recent experiments suggest that both Ia and Ib receptors have extensive connections with interneurons acting on a variety of motoneuron groups (Baldissera, Hultborn, and Illert, 1981). Second, somatosensory receptors may provide more high-frequency information concerning limb dynamics than is available through vision. Third, the amount of information that can be processed in parallel by somatosensory channels may be much greater than through vision (Gordon and Ghez, 1992). It remains for future experiments to decide among these alternatives.

Conclusions

The finding that extent and direction variability have different determinants supports the idea that extent and direction represent explicitly planned dimensions of reaching movements (Rosenbaum, 1980; Bonnet, Requin, and Stelmach, 1982; Bock et al., 1990; Gordon et al., in press; Gordon, Ghilardi, and Ghez, in press B). This is equivalent to stating that reaching movements are specified vectorially as suggested by the results of unit-recording studies in motor cortex (Georgopoulos, 1986). Because different factors independently influence errors in extent and direction, our data suggest that these two features of the hand path are programed by channels operating relatively independently. This principle, which we found to govern the specification (or programing) of single-joint movements (Favilla, Hening, and Ghez, 1989; Ghez, Hening, and Favilla, 1990), therefore appears to extend to multijoint movements as well (see also Favilla et al., 1990).

Our finding that the initial planning of extent and direction does not take into account limb geometry and inertia implies that essential aspects of trajectory planning take place in an extrinsic coordinate system. The nature of the systematic directional errors made by normal subjects further suggests that this system is centered at the inital position of the hand. Whether a shoulder-centered system, suggested by the work of Soechting and coworkers (see Flanders et al., 1992, for discussion), represents a task-specific alternative or whether it represents a later stage in processing is not clear at present. It should be noted that this representation of target location in extrinsic space eventually will have to be transformed into its equivalent in intrinsic coordinates, where muscles or joint torques are represented explicitly.

We envisage that processes related to the extent channel adaptively set a temporal profile of descending activation (Ghez, 1979; Ghez et al., 1983; see also Gottlieb et al., 1992) according to the type of load (viscous, elastic, etc.) that the subject expects. The extent channel appears to use visual information about target distance to scale the activation profile according to a calibration rule. The direction channel would specify the relative changes in joint angle at the elbow and shoulder corresponding to movements in a particular direction (Mel, 1991). Movements planned in this way would not be perfectly straight. Moreover, one would expect to find similar errors related to incorrect specification of relative joint angle changes for movements at different distances. Such consistent curves should result in movement endpoints to nearer targets distributed along the curvature of the paths to the most distant target. Such corresponding patterns occurred frequently in deafferented patients (e.g., figure 35.4, 30° movements) but were noted also in normal subjects (e.g., figure 35.1, 30° movements) (see also Gordon, Ghilardi, and Ghez, in press).

Our findings indicate that, although trajectories are adjusted to reach targets at different distances by a simple scaling rule, significant aspects of movement kinematics are emergent properties of the system and are not controlled explicitly. Thus, our results provide support for the view that descending mechanisms specify a virtual trajectory that is distinct and leads the actual trajectory of the limb (Kelso and Holt, 1980; Bizzi et al., 1984; Flash, 1987; Feldman et al., 1990). Thus, the detailed shape of the acceleration profile and movement duration appears to arise from the interplay of descending control signals, segmental mechanisms, and muscle properties with inertial and other biomechanical characteristics of the limb. However, the existence of inertial extent errors is difficult to explain if one assumes a simple equilibrium control system. A clue may lie in the large lags that appear to exist between virtual and actual trajectories (which must be at least as great as the difference in movement time between movements in low- and high-inertia directions). This suggests that central mechanisms need to provide a terminal control signal cocontracting agonist and antagonist muscles to ensure that the final position be achieved without oscillations. Such a clamping system at the end of movement has been proposed by others in the context of single-joint movements (Ghez, 1979; Ghez et al., 1983; Feldman, 1986; Feldman et al., 1990; Gottlieb, 1992). Errors in setting the parameters or the timing of this terminal clamping command may be responsible for the inertial errors in normals and in patients.

The occurrence of direction-dependent biases that are corrected by vision of the hand and target demonstrates the importance of knowledge of the initial position of the hand in the planning of movement direction. Our findings further suggest that proprioceptive information does not provide the static cues necessary to specify correctly a direction of movement that will reach an arbitrary target in space. The fact that directional biases (and the corresponding underestimate of the distance of the hand from the midline) are similar in intact and deafferented subjects strongly suggests that this aspect of initial state information arises from other sources, especially vision of the hand in relation to the target.

The remarkable trajectory errors made by patients with large-fiber sensory neuropathy indicate that proprioceptive information is critical if accuracy is to be achieved. The function of such information, however, is not limited to the correction of errors through feedback; in addition, it operates by generating and recalibrating internal models of the mechanical properties of the limbs (Atkeson, 1989; Jordan and Rumelhart, 1990; Ghez, Hening, and Gordon, 1991; Morasso and Sanguineti, 1992). These internal models appear to be essential for the transformation of direction and extent information into an intrinsic coordinate system of muscles or joint torques. Vision of the limb in motion allowed the patients to reduce substantially inertial errors in movement extent as well as the curvature and directional errors that we presume are due to uncontrolled interaction torques. It seems plausible that these dynamic properties of the limb are computed from the information about the limb's response to centrally monitored voluntary commands.

Our results suggest that representations of the dynamic properties of the limb are especially crucial for achieving accurate control over interaction torques that develop during multijoint movements. The dramatic breakdown of this control during movement reversals in deafferented patients attests to this. Because of delays inherent in transmission and in excitation-contraction coupling, it is difficult to imagine that the normal control of such interaction forces could be accomplished through feedback mechanisms alone. Instead, failure of feedforward control mechanisms that depend on a proprioceptively updated internal model of the limb are more likely to account for the reversal errors seen in patients. Hence, it appears that the motor systems predict interaction torques and control their effects so as to achieve the kinematic results required by the behavioral task. In such a system, internal models would provide the means for predicting the unfolding scenario of goal-directed movements. These internal models may also be critical for interpreting corollary discharge information and errors that arise in the course of such movements.

ACKNOWLEDGMENTS This work was supported by NIH grant NS 22713 and by the McKnight Foundation.

NOTES

1. It should be noted that the similarity between the distribution of acceleration vectors and the mobility ellipse indicates only that the major source of variability in peak acceleration is the variation in limb inertia. This suggests that the forces at the hand were actually constant or independent of direction. However, in a different task situation, direct measurements of forces at the onset of movements in different directions (Shadmehr, Mussa-Ivaldi, and Bizzi, in press) indicated that force at the hand showed directional variations that were *not* matched to limb inertia but appeared to be explained by the stiffness fields of the arm. Similarly, the patterns of activation of elbow and shoulder muscles initiating targeted movements in different directions from different initial positions cannot be accounted for simply by assuming that subjects direct either the force or the acceleration precisely in the direction of the target (Karst and Hasan, 1991a). It is difficult to compare our data with either of those studies because the tasks were somewhat different and neither errors nor response trajectories were reported in the published material.
2. The occurrence of systematic directional errors that varied with movement direction resulted in a distribution of movement directions with four distinct peaks. Although similar from subjects to subject, neither the errors nor the peaks in the distributions can be completely explained on the basis of inertial anisotropy. One hypothesis that appears attractive and that we currently are testing is that these systematic directional errors reflect errors in the selection of elbow and shoulder muscle activation patterns. Indeed, recent work by Karst and Hasan (1991a,b) indicates that for horizontal plane movements such as the ones studied here, movements are initiated with one of four stereotypical patterns. Like the peaks in direction distributions for our subjects, those patterns depend on the forearm segment angle.
3. The inertial moment (I_f), center of mass (r_f), and mass (m_f) of the forearm segment were computed from regression equations that include body weight and segment length (Winter, 1990).

Torque Component	*Formula*
Residual torque	$(I_f + m_f r_f^2 + m_f r_{se} r_f \cos\theta_e)\ddot{\theta}_s + (I_f + m_f r_f^2)\ddot{\theta}_e + (m_f r_{se} r_f \sin\theta_e)\dot{\theta}_s^2$
Self-torque	$-(I_f + m_f r_f^2)\ddot{\theta}_e$
Cross-torque	$-(I_f + m_f r_f^2 + m_f r_{se} r_f \cos\theta_e)\ddot{\theta}_s - (m_f r_{se} r_f \sin\theta_e)\dot{\theta}_s^2$

where θ_e = elbow angle
θ_s = shoulder angle
r_{se} = upper arm length

REFERENCES

Atkeson, C. G., 1989. Learning arm kinematics and dynamics. *Annu. Rev. Neurosci.* 12:157–183.

Baldissera, F., H. Hultborn, and M. Illert, 1981. Integration in spinal neuronal systems. In *Handbook of Physiology: Sec. 1. The Nervous System:* Vol. 2. *Motor Control* (part 1). V. B. Brooks, ed. Bethesda, Md.: American Physiological Society, pp. 509–595.

Bizzi, E., N. Accornero, W. Chapple, and N. Hogan, 1984. Posture control and trajectory formation during arm movement. *J. Neurosci.* 4:2738–2744.

Bock, O., M. Dose, D. Ott, and R. Eckmiller, 1990. Control of arm movements in a 2-dimensional pointing task. *Behav. Brain Res.* 40:247–250.

Bonnet, M., J. Requin, and G. E. Stelmach, 1982. Specification of direction and extent in motor programming. *Bull. Psychon. Soc.* 19:31–34.

BURNOD, Y., P. GRANDGUILLAUME, I. OTTO, S. FERRAINA, P. B. JOHNSON, and R. CAMINITI, 1992. Visuomotor transformations underlying arm movements toward visual targets: A neural network model of cerebral cortical operations. *J. Neurosci.* 12:1435–1453.

CLEVELAND, W. S., 1979. Robust locally weighted regression and smoothing scatterplots. *J. Am. Stat. Assoc.* 74:829–836.

COOPER, S. E., J. H. MARTIN, and C. GHEZ, 1993. Differential effects of localized inactivation of deep cerebellar nuclei on reaching in the cat. *Soc. Neurosci. Abstr.* 19:1278.

FAVILLA, M., J. GORDON, M. F. GHILARDI, and C. GHEZ, 1990. Discrete and continuous processes in the programming of extent and direction in multijoint arm movements. *Soc. Neurosci. Abstr.* 16:1089.

FAVILLA, M., W. HENING, and C. GHEZ, 1989. Trajectory control in targeted force impulses: VI. Independent specification of response amplitude and direction. *Exp. Brain Res.* 75:280–294.

FELDMAN, A. G., 1986. Once more on the equilibrium-point hypothesis (λ model) for motor control. *J. Mot. Behav.* 18:17–54.

FELDMAN, A. G., S. V. ADAMOVICH, D. J. OSTRY, and J. R. FLANAGAN, 1990. The origin of electromyograms—explanations based on the equilibrium point hypothesis. In *Multiple Muscle Systems: Biomechanics and Movement Organization*, J. M. Winters and S. L.-Y. Woo, eds. New York: Springer-Verlag, pp. 195–213.

FLANDERS, M., S. I. HELMS TILLERY, and J. F. SOECHTING, 1992. Early stages in a sensorimotor transformation. *Behav. Brain Sci.* 15:309–362.

FLASH, T., 1987. The control of hand equilibrium trajectories in multi-joint arm movements. *Biol. Cybern.* 57:257–274.

FORGET, R., and Y. LAMARRE, 1987. Rapid elbow flexion in the absence of proprioceptive and cutaneous feedback. *Hum. Neurobiol.* 6:27–37.

FORGET, R., and Y. LAMARRE, 1990. Anticipatory postural adjustment in the absence of normal peripheral feedback. *Brain Res.* 508:176–179.

GEORGOPOULOS, A. P., 1986. On reaching. *Annu. Rev. Neurosci.* 9:147–170.

GHEZ, C., 1979. Contributions of central programs to rapid limb movements in the cat. In *Integration in the Nervous System*, H. Asanuma and V. J. Wilson, eds. Tokyo: Igaku-Shoin, pp. 305–320.

GHEZ, C., J. GORDON, M. F. GHILARDI, C. N. CHRISTAKOS, and S. E. COOPER, 1990. Roles of proprioceptive input in the programming of arm trajectories. *Cold Spring Harb. Symp. Quant. Biol.* 55:837–847.

GHEZ, C., W. HENING, and M. FAVILLA, 1990. Parallel interacting channels in the initiation and specification of motor response features. In *Attention and Performance: XIII. Motor Representation and Control*, M. Jeannerod, ed. Hillsdale, N.J.: Erlbaum, pp. 265–293.

GHEZ, C., W. HENING, and J. GORDON, 1991. Organization of voluntary movement. *Curr. Opin. Neurobiol.* 1:664–671.

GHEZ, C., D. VICARIO, J. H. MARTIN, and H. YUMIYA, 1983. Sensory motor processing of targeted movements in motor cortex. In *Motor Control Mechanisms in Health and Disease*, J. E. Desmedt, ed. New York: Raven, pp. 61–92.

GORDON, J., and C. GHEZ, 1992. Roles of proprioceptive input in control of reaching movements. In *Children with Movement Disorders, Medicine and Sport Science: 36*. H. Forssberg and H. Hirschfeld, eds. Basel: Karger, pp. 124–129.

GORDON, J., M. F. GHILARDI, S. E. COOPER, and C. GHEZ (in press). Accuracy of planar reaching movements: II. Systematic extent errors resulting from inertial anisotropy. *Exp. Brain Res.*

GORDON, J., M. F. GHILARDI, and C. GHEZ (in press). Accuracy of planar reaching movements: I. Independence of direction and extent variability. *Exp. Brain Res.*

GOTTLIEB, G. L., 1992. Kinematics is only a (good) start. *Behav. Brain Sci.* 15:749.

GOTTLIEB, G. L., M. L. LATASH, D. M. CORCOS, T. J. LIUBINSKAS, and G. C. AGARWAL, 1992. Organizing principles for single joint movements: V. Agonist-antagonist interactions. *J. Neurophysiol.* 67:1417–1427.

HOGAN, N., 1985. The mechanics of multi-joint posture and movement control. *Biol. Cybern.* 52:315–331.

HOGAN, N., 1988. Planning and execution of multijoint movements. *Can. J. Physiol. Pharmacol.* 66:508–517.

HOLLERBACH, J. M., and T. FLASH, 1982. Dynamic interactions between limb segments during planar arm movement. *Biol. Cybern.* 44:67–77.

HOY, M. G., R. F. ZERNICKE, and J. L. SMITH, 1985. Contrasting roles of inertial and muscle moments at knee and ankle during paw-shake response. *J. Neurophysiol.* 54:1282–1294.

JORDAN, M. I., and D. E. RUMELHART, 1990. *Forward Models: Supervised Learning with a Distal Teacher (Occasional Paper No. 40)*. Cambridge, Mass.: Center for Cognitive Science.

KARST, G. M., and Z. HASAN, 1991a. Initiation rules for planar, two-joint arm movements: Agonist selection for movements throughout the work space. *J. Neurophysiol.* 66:1579–1593.

KARST, G. M., and Z. HASAN, 1991b. Timing and magnitude of electromyographic activity for two-joint arm movements in different directions. *J. Neurophysiol.* 66:1594–1604.

KELSO, J. A. S., and K. G. HOLT, 1980. Exploring a vibratory system analysis of human movement production. *J. Neurophysiol.* 43:1183–1196.

KOSHLAND, G. F., and J. L. SMITH, 1989a. Mutable and immutable features of paw-shake responses after hindlimb deafferentation in the cat. *J. Neurophysiol.* 62:162–173.

KOSHLAND, G. F., and J. L. SMITH, 1989b. Paw-shake response with joint immobilization: EMG changes with atypical feedback. *Exp. Brain Res.* 77:361–373.

LACQUANITI, F., N. A. BORGHESE, and M. CARROZZO, 1991. Transient reversal of the stretch reflex in human arm muscles. *J. Neurophysiol.* 66(3):939–954.

MEL, B. W., 1991. A connectionist model may shed light on neural mechanisms for visually guided reaching. *J. Cogn. Neurosci.* 3(3):231.

Morasso, P., 1981. Spatial control of arm movements. *Exp. Brain Res.* 42:223–227.

Morasso, P., and V. Sanguineti, 1992. Equilibrium point and self-organization. *Behav. Brain Sci.* 15:781–782.

Nichols, T. R., 1989. The organization of heterogenic reflexes among muscles crossing the ankle joint in the decerebrate cat. *J. Physiol. (Lond.)* 410:463–477.

Paillard, J., and M. Brouchon, 1974. A proprioceptive contribution to the spatial encoding of position cues for ballistic movements. *Brain Res.* 71:273–284.

Rosenbaum, D. A., 1980. Human movement initiation: Specification of arm, direction, and extent. *J. Exp. Psychol. [Gen.]* 109:444–474.

Rothwell, J. L., M. M. Traub, B. L. Day, J. A. Obeso, P. K. Thomas, and C. D. Marsden, 1982. Manual motor performance in a deafferented man. *Brain* 105:515–542.

Sainburg, R. L., M. F. Ghilardi, F. Ferracci, H. Poizner, and C. Ghez, 1992. Deafferented subjects fail to compensate for interaction torques during multi-joint movements. *Soc. Neurosci. Abstr.* 18:647.

Sainburg, R. L., H. Poizner, and C. Ghez, 1993. Loss of proprioception produces deficits in interjoint coordination. *J. Neurophysiol.* 70:2136–2147.

Sanes, J., and R. Shadmehr (in press). Organization of motor control and postural stiffness of deafferented humans. *Can. J. Physiol. Pharmacol.*

Sanes, J. N., K.-H. Mauritz, M. C. Dalakas, and E. V. Evarts, 1985. Motor control in humans with large-fiber sensory neuropathy. *Hum. Neurobiol.* 4:101–114.

Schneider, K., R. F. Zernicke, R. A. Schmidt, and T. J. Hart, 1989. Changes in limb dynamics during the practice of rapid arm movements. *J. Biomech.* 22:805–817.

Shadmehr, R., F. A. Mussa-Ivaldi, and E. Bizzi (in press). Postural force fields of the human arm and their role in generating multi-joint movements. *J. Neurosci.*

Smith, J. L., and R. F. Zernicke, 1987. Predictions for neural control based on limb dynamics. *Trends Neurosci.* 10: 123–128.

Soechting, J. F., and F. Lacquaniti, 1989. An assessment of the existence of muscle synergies during load perturbations and intentional movements of the human arm. *Exp. Brain Res.* 74(3):535–548.

Winter, D. A., 1990. *Biomechanics and Motor Control of Human Movement.* New York: Wiley.

36 The Superior Colliculus: A Window for Viewing Issues in Integrative Neuroscience

DAVID L. SPARKS AND JENNIFER M. GROH

ABSTRACT The superior colliculus (SC) is a fruitful site for investigating a variety of interesting problems in integrative neuroscience. Situated at the interface between sensory and motor processing, the SC contains computational maps of cells responsive to visual, auditory, and somatosensory stimuli as well as cells that fire before saccadic eye and head movements. In this chapter, the computational issues of the coordination of different motor programs, population coding of saccades, target selection, derivation of eye position signals from corollary discharge, and coordinate transformations necessary for sensorimotor integration will be discussed with respect to the SC.

The intermediate and deeper layers of the superior colliculus (SC) are sites where auditory, visual, and somatosensory signals converge and an area where motor commands for orienting movements of the eyes and head are generated (for recent reviews, see Sparks, 1986; Sparks and Mays, 1990). Studies of the SC concerned with the translation of sensory signals into motor commands have been forced to address some of the most important and perplexing problems of integrative neuroscience. In the process, significant advances have been made in understanding (1) the role of computational maps in processing sensory and motor signals; (2) the coordination of different motor programs; (3) the extraction of information from the spatial and temporal pattern of activity in populations of neurons; (4) the role of corollary discharge in guiding movements; and (5) mechanisms for selecting a single response for execution from a large repertoire of potential responses. The experimental strategies developed for studying the SC have also been fruitful when applied to studies of other brain areas. In this chapter, we review the advances just listed, emphasizing studies related to the question of sensorimotor integration, the translation of sensory signals into motor commands.

DAVID L. SPARKS and JENNIFER M. GROH Department of Psychology, Institute of Neurological Sciences, University of Pennsylvania Philadelphia, Pa.

Computational maps

Many neural computations are performed by computational maps in the brain (see Knudsen, du Lac, and Esterly, 1987, for a review). For example, sensory neurons may be organized in two- or three-dimensional arrays with systematic variations in the value of a computed parameter (e.g., the angle of orientation of a line segment, the direction or velocity of stimulus movement) across each dimension of the array. Neurons, as elements of the map, may be thought of as an array of preset processors or filters, each tuned slightly differently, and operating in parallel on incoming signals. Consequently, signals are transformed, almost instantaneously, into a distribution of neural activity within the computational map. The values of the stimulus parameters are represented as locations of peaks of activity within the map (Knudsen, du Lac, and Esterly, 1987).

Computational maps may also be used for programing movements. Systematic variations in the direction, amplitude, or velocity of a movement can be represented topographically across a neural array. The most thoroughly studied motor map is the map of saccadic eye movements found in the SC, first described in detail using microstimulation techniques (Robinson, 1972). As illustrated in figure 36.1, the amplitude and direction of stimulation-induced saccades are a function of the site of stimulation in the SC. The neuronal bases of this map and a discussion of how information about the metrics (direction and amplitude) of sac-

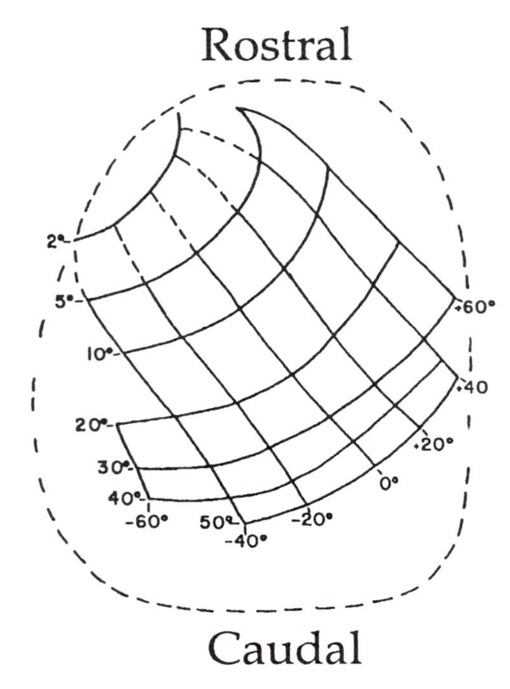

FIGURE 36.1 Motor map of the SC. Isoamplitude lines (2–50) run from medial to lateral, and isodirection lines (−60 to +60) run from anterior to posterior. (From Robinson's [1972] microstimulation data.)

cades is extracted from the computational map are described in later sections.

Coordination of motor programs

Combined movements of the eyes and head, made in response to external stimuli or produced by stimulation of the SC, have been studied extensively in attempts to understand how different motor programs are coordinated. Note that the motor map illustrated in figure 36.1 extends only to approximately 45°, but visual stimuli appearing at an eccentricity of 80° in the visual field can be detected. What happens when a sudden stimulus appears at 80°? Can an orienting movement of only 45° occur because the collicular motor map extends only to 45°? Note that the experiments on which the map shown in figure 36.1 are based were performed in monkeys with restrained heads. That the collicular motor map extended beyond the oculomotor range was first demonstrated in experiments in which cats were the subjects (Roucoux, Crommelinck, and Guitton, 1980). Roucoux, Crommelinck, and Guitton (1980) found that stimulation in some regions of the SC produced large gaze shifts involving coordinated rotations of the eyes and head. Both the eyes and head began to move at nearly the same time and *in the same direction*. In many cases, the amplitude of the stimulation-induced head movement was larger than the amplitude of the eye movement, but when gaze—the sum of eye and head positions—reached the angle coded by the site being stimulated in the motor map, the head continued to move whereas the vestibulo-ocular reflex (VOR) became active, such that the eye changed direction and moved in the opposite direction by an amount that compensated for the continued movement of the head. Gaze angle remained constant.

Results of early experiments (Stryker and Schiller, 1975) indicated that stimulation of the monkey SC did not produce coordinated movements of the eyes and head such as those described in the cat. Also, in cats (Munoz, Guitton, and Pélisson, 1991) and barn owls (du Lac and Knudsen, 1990), the amplitude and velocity of gaze shifts depends on the duration, frequency, and intensity of stimulation, whereas the metrics of movements produced by stimulation in the monkey SC were believed to be independent of stimulation parameters.

What is the current status of these apparent differences in the functional organization of cat and monkey SC? Recent experiments conducted by Freedman, Stanford, and Sparks (1993) used a broader range of stimulation parameters than did early experiments studying the effects of microstimulation in the monkey. These newer studies indicate that there are no fundamental differences in the functional organization of the cat and monkey SC: Stimulation of the caudal SC in monkeys produces large gaze shifts that involve coordinated movements of the eyes and head (figure 36.2). Moreover, the parameters of stimulation do influence the metrics of the movements in monkeys when the head is free to move (figure 36.3).

The microstimulation experiments in the cat and monkey and behavioral studies using cats, monkeys, and humans as subjects (Bizzi, Kalil, and Morasso, 1972; Morasso, Bizzi, and Dichgans, 1973; Harris, 1980; Roucoux, Crommelinck, and Guitton, 1980; Laurutis and Robinson, 1986; Tomlinson and Bahra, 1986a, 1986b, Guitton and Volle, 1987; Munoz, Guitton, and Pélisson, 1991) indicate that eye-head coordination is not achieved by a single-motor program. At least two major modes of operation are observed. When the desired gaze shift is small, this may be accomplished by movements of the eyes alone; vol-

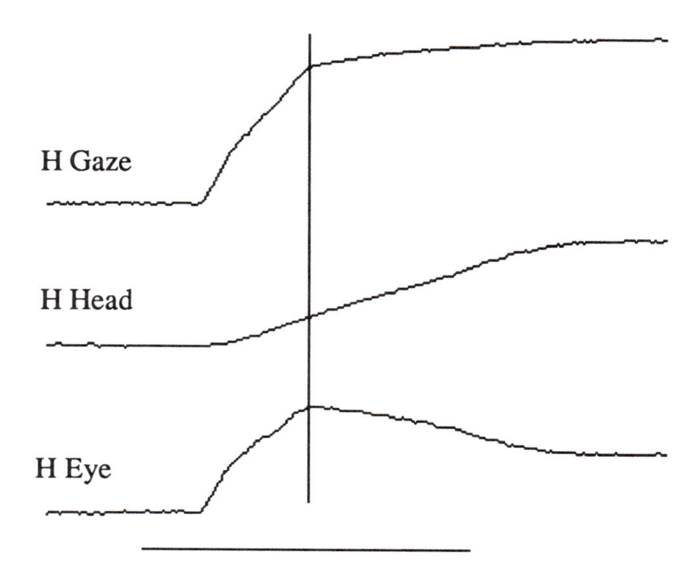

Figure 36.2 Coordinated movement of eyes and head produced by stimulation of the monkey SC. A suprathreshold stimulation train delivered for 200 ms (horizontal calibration mark) produced a predominantly horizontal (H) gaze shift of 17°. The vestibulo-ocular reflex is apparent at the end of the high velocity gaze shift (vertical line) as the head continued moving. (Freedman, Stanford, and Sparks, unpublished)

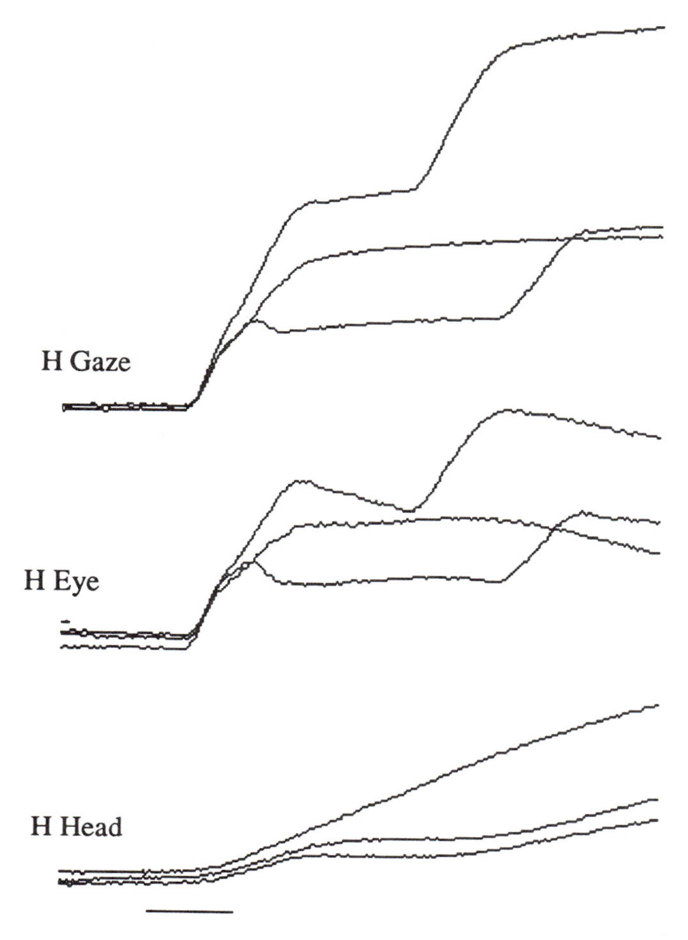

Figure 36.3 Effects of varying train duration on gaze shifts produced by microstimulation of the monkey SC. Three traces of horizontal (H) gaze, horizontal eye, and horizontal head position are superimposed. The changes in eye, head, and gaze position were produced by suprathreshold stimulation trains 75, 100, or 200 ms in duration. The horizontal calibration line marks the time and duration of the 75-ms stimulation train. Note that the amplitude of the gaze shift increases as the duration of the stimulation train is increased. The 200-ms stimulation train produced a staircase of changes in gaze position: an initial change of 34° followed, after approximately 100 ms, by a second, smaller horizontal gaze shift. (Freedman, Stanford, and Sparks, unpublished)

untary or involuntary movements of the head are countered by the VOR. Very large gaze shifts (e.g., 80°–90°) are accomplished by moving both the eyes and head in the same direction toward the target. During the initial phase of these movements, the VOR is switched off but later, after the line of sight is near the target, the VOR is reactivated. Strategies for achieving intermediate-amplitude gaze shifts are not as clear, but the VOR may still be active with reduced gain.

In summary, the motor representation in the SC involves the coordination of at least two motor programs, eye and head. The motor command signals observed in the SC are abstract and very different from those needed by motor neurons sending signals to individual muscles.

Population coding

To this point, only the motor representations revealed by electrical stimulation of different regions of the SC have been considered. How are these higher-order motor commands represented at the level of the activity of single neurons?

Many neurons in the SC discharge before saccadic eye movements. These cells have movement fields; that is, each neuron discharges maximally prior to saccades having a particular direction and a particular amplitude, regardless of the initial position of the eye in the orbit (Wurtz and Goldberg, 1972; Sparks, Holland, and Guthrie, 1976; Sparks and Mays, 1980). A gradient of response amplitude is observed across the movement field. Movements to the center of the field are preceded by a vigorous discharge, but movements deviating from this optimal direction and amplitude are accompanied by less vigorous activity. The size of the movement field is a function of the amplitude of the optimal movement. Neurons discharging prior to small saccades have small and sharply tuned fields, whereas neurons discharging prior to large saccades have large movement fields and relatively coarse tuning. The movement fields of SC neurons are also characterized by a temporal gradient. Cells begin firing earlier for saccades near the center of the movement field than they do for movements to the periphery of the field (Sparks and Mays, 1980).

The topographical organization of movement fields within the SC forms the neuronal substrate of the motor map revealed by microstimulation (see figure 36.1).

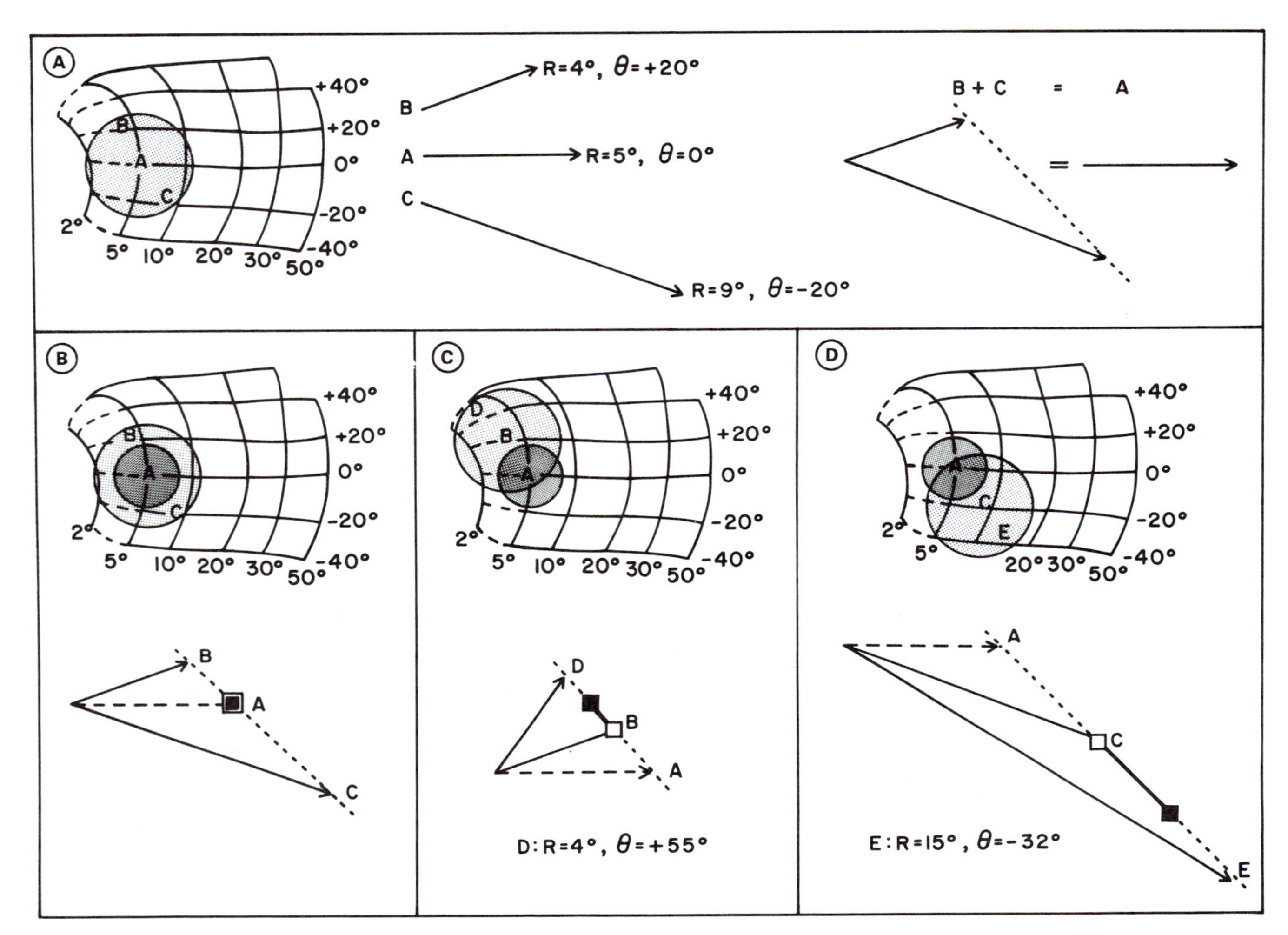

FIGURE 36.4 (A) The population-averaging scheme of Sparks, Holland, and Guthrie (1976). (Left) Motor map of the left SC. The stippled area represents the hypothetical extent of cells active before saccades to a target located 5° to the right of the fixation stimulus. The active population is assumed to be symmetrical in shape. (Middle) Cells at locations A, B, and C fire most vigorously for the movements shown. (Right) Weighted averaging of activity at points B and C yields the same movement as activity at the center of the active population (A). (B–D) The predicted effect of deactivating a subset of cells in the active population. The site of deactivation (darkly stippled circle) remains the same in each panel, but the location of the active population (lightly stippled area) is different in each panel because saccades to three different targets are required. Beneath each map are the saccade vectors associated with neural activity at each of the locations illustrated. The open square represents the vector of the intended, or programed, saccade associated with activity in the lightly stippled area. The dashed line represents the vector of the movement tendency produced by neurons at the deactivated site. These neurons will not contribute to the metrics of the saccade, and a saccade to the approximate location of the filled square is predicted. (Reprinted from Lee, Rohrer, and Sparks, 1988, by permission.)

Neurons discharging prior to small saccades are located anteriorly, and neurons firing before large saccades are found posteriorly. Cells near the midline discharge prior to movements with up components, and cells laterally discharge maximally before movements with down components.

The issue of population coding arises because the movement fields of collicular neurons are large and coarsely tuned (Sparks, Holland, and Guthrie, 1976; Sparks and Mays, 1980). Because each neuron fires before a broad range of saccades, a large population of neurons is active before each saccade. It is not known definitively how the signals that are needed to control precisely the direction and amplitude of a saccade are extracted from the activity of this large population of coarsely tuned cells. One possibility is that the location of the most intense activity within the population is determined at a subsequent stage of neural processing (Knudsen, du Lac, and Esterly, 1987). Another possibility is that each member of the active population contributes to the movement, and the exact trajectory of a saccade is determined by the average or sum of the population response (McIlwain, 1976; Sparks, Holland, and Guthrie, 1976; Tweed and Vilis, 1985; van Gisbergen, van Opstal, and Tax, 1987; van Opstal and van Gisbergen, 1989). Results of recent experiments (Lee, Rohrer, and Sparks, 1988; Sparks, Lee, and Rohrer, 1990), in which small subsets of the active population were reversibly deactivated, support the population-averaging hypothesis.

The population-averaging model (figure 36.4) assumes that the region of collicular neurons active before a given saccade occupies a symmetrical area within the motor map. Only the neurons in the center of the active population discharge maximally before the programed movement but, for each subset of active neurons (B in figure 36.4) producing a movement tendency with a direction and amplitude other than the programed movement, there will be a second subset of active neurons (C in figure 36.4) producing an opposing movement tendency such that the resultant of the two movements will have the programed direction and amplitude. According to this hypothesis, each member of the active population contributes to the ensuing saccade; there is no need to sharpen the population response to generate an accurate saccade. Thus, as illustrated in figure 36.4, with the same region of colliculus inactivated, an animal required to make different saccades should demonstrate a predictable pattern of errors.

Rhesus monkeys were trained on a saccadic eye movement task. A glass pipette (Malpeli and Schiller, 1979) was used for recording extracellular unit activity, microstimulation, and pressure injection of various agents into the SC that affect neuronal activity. The location of the pipette tip in the collicular motor map was determined from plots of movement fields and measurements of the direction and amplitude of stimulation-induced saccades. The behavioral effects of the injections were assessed by comparing saccades to a selected set of visual targets before and after the injections.

As predicted by the population-averaging hypothesis, a systematic pattern of errors in direction and amplitude was observed after small injections of lidocaine (figure 36.5). Saccades to targets requiring more of an upward component than the so-called best saccade had too much of an upward component, and saccades to targets requiring movements with more of a downward

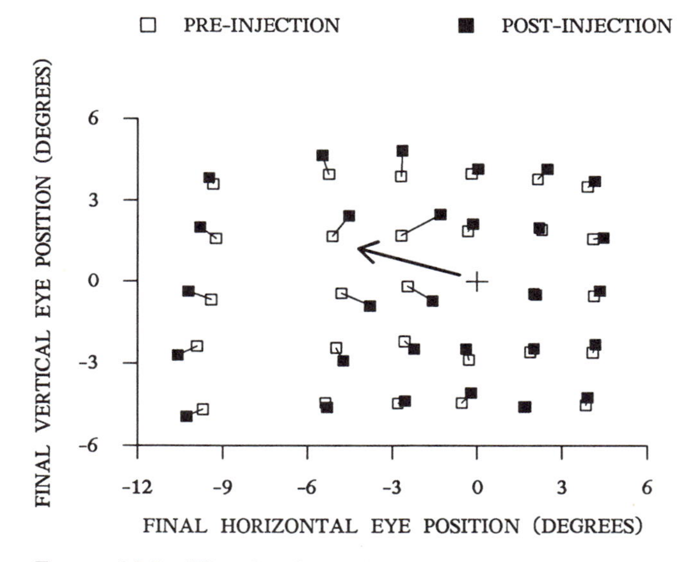

FIGURE 36.5 The plot shows the position of the initial fixation (+) and the endpoint (arrowhead) of the best saccade. Each symbol represents the average endpoint of three to five visually guided saccades. Unfilled symbols represent movements occurring before the injections; filled symbols represent postinjection trials for matching targets. Lines connecting the squares represent the average error introduced by modifying the spatial and temporal pattern of neuronal activity within the SC. Effects of a single 200-nl injection of lidocaine. (From Lee, Rohrer, and Sparks, 1988.)

component than the best saccade had too much of a downward component. Also as predicted, movements to targets requiring a saccade smaller in amplitude than the best saccade were hypometric, whereas saccades to targets requiring a movement larger than the best saccade were hypermetric.

Results of these experiments support the major predictions of the vector-averaging model. According to this model, saccadic accuracy results from the averaging of the movement tendencies produced by the entire active population rather than the discharge of a small number of finely tuned cells. Small changes in the direction and amplitude of saccades are produced by slight shifts in the location of the large population of active cells. Moreover, because the contribution of each neuron to the direction and amplitude of the movement is relatively small, the effects of variability (or noise) in the discharge frequency of a particular neuron are minimized. Thus, the large movement fields (resulting in a large population of neurons being active during a specific movement) may contribute to, rather than detract from, saccadic accuracy.

Our findings may generalize to other sensory and motor systems. Georgopoulos, Schwartz, and Kettner (1986) described a distributed population code in the motor cortex for the control of reaching movements of the arm. Vogels (1990) developed a population-coding model of visual orientation discrimination in which the orientation of the stimulus is represented by an ensemble of broadly tuned units in a distributed manner. Heiligenberg (1991) has shown that the jamming avoidance response of weakly electric fish is driven by a distributed system of contributions resulting from the evaluation of inputs from pairs of points within a somatotopic map.

Spatial attention: Target and response selection

Do the sensory and motor maps in the SC represent all potential saccade targets and all potential movements or only those targets that have been selected as a saccade goal and only those responses that are chosen for execution? Neural signals related to the metrics of a movement, or to the onset of potential targets, are readily observed. Isolation of neural signals related to the process of target or movement selection is difficult because selection is a covert process. Glimcher and Sparks (1992) designed experiments to test whether neural activity in the SC is related to saccade selection. Their tasks used a cue to specify which of two physically identical visual stimuli was the goal of an impending saccade. This cue was spatially and temporally isolated from the potential targets as well as from visual cues signaling movement initiation. Response selection output elements must be differentially sensitive to target and distractor stimuli, and changes in their activity must be linked to the time at which sufficient information is available for response selection, as is the case for prelude bursters. They studied the low-frequency activity (prelude) that precedes the vigorous saccade-related burst in one class of collicular neurons (prelude burster). The low-frequency activity begins shortly after information becomes available for correct saccade selection, and this activity was predictive of saccade choice.

One of the trial types used to separate the effects of movement selection and visual stimulation on neural activity is illustrated in figure 36.6. Trials were initiated when a fixation stimulus was illuminated yellow. After a variable interval, two potential targets appeared. Which was the target and which was the distractor was unknown until the fixation LED later changed color to red or green. Red indicated that the upper stimulus would serve as the target and the lower stimulus as the distractor. Green indicated the opposite. The informative change in color of the fixation stimulus was separated by an additional delay from the cue that signaled saccade initiation (offset of the fixation stimulus).

By using relatively long delay intervals in the behavioral task, we noted that the onset of prelude activity can precede burst onset by up to 7 seconds. This early responding occurred only after sufficient information became available to permit accurate response selection. We found that prelude bursters responded differentially to distractors and targets presented at the same location in their response field (figure 36.7). These data are consistent with the hypothesis that the low-frequency activity of the prelude burst cells reflects the output of the covert process of response selection. For these cells, responses to distractor stimuli are heavily filtered but not completely eliminated. Further studies will be required to determine whether the SC actively participates in the selection process or whether collicular neurons are merely passively reflecting the results of processing occurring at other levels.

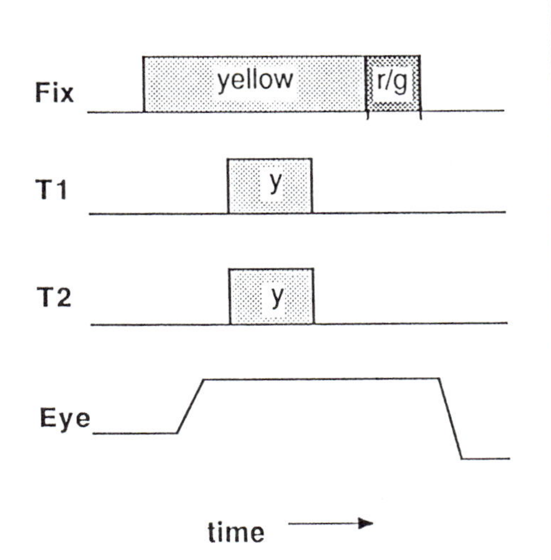

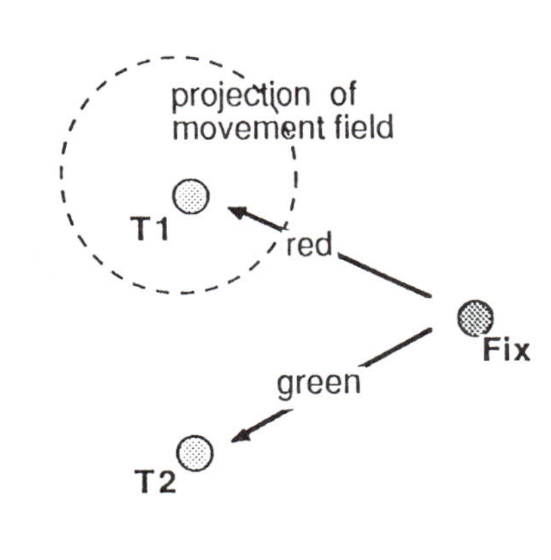

FIGURE 36.6 Schematic diagram of one trial type used to separate the effects of movement selection and visual stimulation on prelude burster activity. Animals first fixated a central yellow LED (Fix). After a variable delay, two yellow LEDs were illuminated briefly at eccentric locations (T1, T2). After another variable delay, the color of the fixation LED changed to red or green and signaled which of the two eccentric stimuli would be the endpoint of a rewarded saccade (the target) and which served as a distractor. After another variable interval, the fixation LED was extinguished and the animal made a saccade to the target's location. The last delay temporally isolated target specification from response initiation.

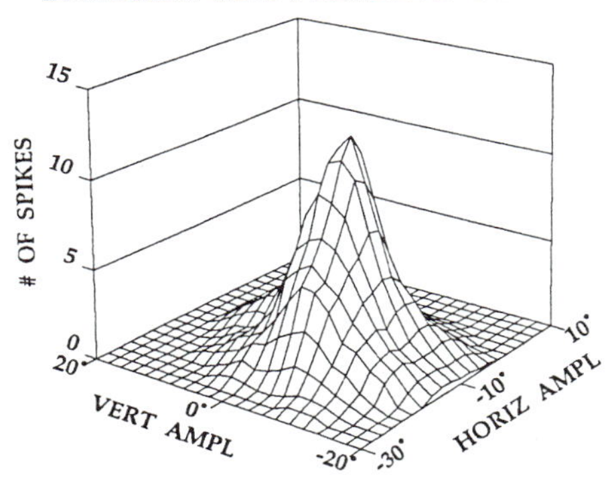

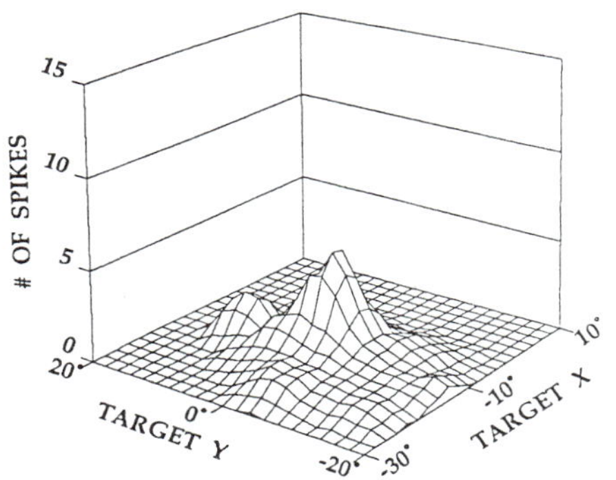

FIGURE 36.7 Four hundred movement trials were used to construct these plots of the sensitivity of one neuron to targets and distractors. (Left) Target-movement prelude field. For each of the 400 trials, the number of spikes in the 200 ms before the movement initiation cue was plotted against the horizontal and vertical amplitude of the movement. Prelude activity occurring well in advance of the movement is spatially tuned. (Right) Distractor-nonmovement prelude field. Here the same unit activity for each of the 400 trials is plotted against the *X*, *Y* coordinates of the distractor. (From Glimcher and Sparks, 1992.)

Corollary discharge

Precise information about the position of the eyes in the orbits is required for the spatial localization of visual targets (see next section). The question of whether eye position signals originate from a central copy of the oculomotor command or arise peripherally from extraocular muscle proprioceptors has been a central issue in oculomotor physiology.

Porter, Guthrie, and Sparks (1983) localized cell bodies innervating extraocular muscle sensory receptors using the technique of retrograde transport of

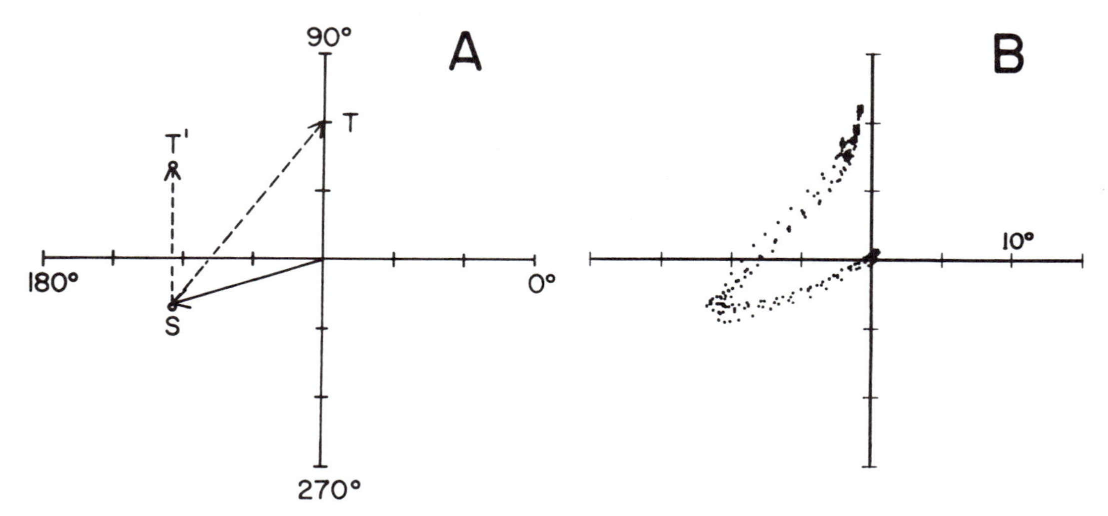

FIGURE 36.8 Compensation for stimulation-induced perturbations in eye position. (A) The experimental paradigm. In complete darkness, while monkeys were fixating on a center target represented by the intersection of the axes, an eccentric target (T) was flashed for 60–100 ms. After the offset of T but before the animal initiated a saccade, electrical stimulation of the SC drove the eyes to another position in the orbit (S). Retinocentric models predict that the animal will look to position T′; spatial models predict that the animal will look to T. (B) Typical results. The trajectories of the eye movements occurring on four stimulation trials are superimposed. Each dot represents the horizontal and vertical position of the eye sampled at 2-ms intervals. The target was flashed 20° above the fixation point. Stimulation drove the eyes downward and leftward. After a brief delay, the animal compensated for the stimulation-induced movement by making a saccade that directed gaze to the approximate position of the target in space. See text and Sparks and Mays (1983) for more detail.

horseradish peroxidase (HRP) after its injection into the muscles. The oculomotor nerves proper, trigeminal ganglia, and the brain stem of three species of macaque were examine, but labeled sensory cells were found only within the ipsilateral trigeminal ganglia. Section of the ophthalmic nerve at its junction with the trigeminal ganglion prior to injection of HRP into ipsilateral extraocular muscles eliminated all labeling of trigeminal sensory neurons. Thus, in rhesus monkeys, extraocular muscle proprioceptive fibers are anatomically isolated along their peripheral course from axons carrying motor commands. At this site, they can be transected without damaging the motor fibers.

Mays and Sparks (1980a) and Sparks and Mays (1983) developed a behavioral task requiring eye position information, a stimulate-and-compensate behavioral task. While subjects maintained central fixation, an eccentric target was flashed briefly. Randomly, on 30% of the trials, after the offset of the target but before a visually guided saccade began, microstimulation of the SC drove the eyes to a different orbital position. Monkeys compensated for this stimulation-induced perturbation in eye position by generating a saccade to the approximate position of the now-absent target (figure 36.8). Note that the compensatory movement was not a passive rotation to a preset orbital position (Sparks and Mays, 1983; Sparks and Porter, 1983). Saccades to the position of the target on stimulation trials also could not be based on a visual update of target location; the eccentric target was extinguished before the onset of the stimulation-induced saccade. Because the occurrence of stimulation was unpredictable, compensation for stimulation-induced saccades could not be programed in advance. Because trials were run in total darkness, targets could not be localized with respect to an external visual frame of reference. Therefore, saccades to the targets must have been directed by signals combining information about the retinal locus of stimulation and precise information about the change in eye position produced by stimulation of the SC.

Guthrie, Porter, and Sparks (1983) tested the performance of animals with bilateral ophthalmic nerve transections on the stimulate-and-compensate behavioral task. Animals with bilateral nerve sections did compensate for stimulation-induced perturbations of eye position (Guthrie, Porter, and Sparks, 1983). As this compensation is possible only if information about the stimulation-induced eye movement is still available, and transection of ophthalmic nerves eliminated extraocular muscle proprioception, eye position information must have been provided by a centrally generated copy of the motor command.

Sensorimotor integration

How do sensory signals elicit orienting movements of the eyes? To produce a saccadic eye movement, the brain must determine the target's location in a sensory coordinate framework, compute the position of the target with respect to the eyes, and execute a saccade to bring the target onto the fovea. Saccades can be generated to visual, auditory, and somatosensory targets, but the sensory coordinate frameworks for each of these modalities are different. The signals must be transformed into a common eye-centered motor coordinate framework before the motor circuitry in the SC for generating the movement can be accessed. The issues of spatial localization of sensory targets and coordinate transformations for generating motor commands to targets of different modalities will be discussed next.

LOCALIZATION OF VISUAL TARGETS Visual targets excite a particular region of the retina. Neural representations of these targets presumably are encoded in retinal coordinates at least through the early stages of visual processing. If the eyes have not moved since presentation of a brief visual stimulus, then the retinal error vector corresponds to the saccade vector necessary to bring the target's position onto the fovea. However, saccades can be made to the remembered location of a briefly flashed target despite the occurrence of an intervening eye movement, either one induced by stimulation (as described earlier) or one directed to a visual target (Mays and Sparks, 1980b). Retinal information is insufficient for generating eye movements to brief visual targets under these conditions: Eye position information is necessary as well.

Evidence for a neural map of visual space in nonretinal coordinates in the SC came from the double-saccade experiment of Mays and Sparks (1980b) (figure 36.9). In this experiment, monkeys were presented with two targets flashed briefly in sequence. Monkeys made saccades to first one and then the other target. The saccade to the second target was made accurately, despite the fact that its vector differed from the original retinal error vector. The intermediate layers of the SC were found to contain visually responsive cells that would fire if the remembered location of the target lay in its receptive field, even though that region of the retina had never been stimulated. Similar quasi-visual (QV) cells have since been found in the frontal eye fields (Bruce and Goldberg, 1990) and parietal cortex (Gnadt and Andersen, 1988; Duhamel, Colby, and Goldberg, 1992).

The existence of these QV cells encoding the position of the target relative to the current position of the eyes indicates that a coordinate transformation has taken place: Information about the position of the eyes in the orbit has been combined with the retinal image of the target. Mays and Sparks called the resulting coordinate framework *motor error*, referring to the difference or error between the current position of the eyes and the target position. Other authors have used the term *updated retinal coordinates* to describe similar kinds of cells (Duhamel, Colby, and Goldberg, 1992).

This coordinate transformation may occur through any of several different mechanisms. Absolute eye position information (the position of the eyes in the orbit) could be used to transform the visual signals into a head-centered coordinate frame. Mathematically, this could be done by adding the retinal error vector and the eye position vector. After an eye movement, subtraction of the new eye position vector would yield the current motor error of the remembered target. If movements of the head in space are included, then a similar addition and subtraction of head position (either with respect to the body or with respect to inertial space) would be required as well. For the sake of simplicity, we will assume the head to be fixed in space.

Such a mechanism suggests the existence of a head-centered (or, if head movements are included, body-centered or spatial) visual map. Such a map is not necessary, however. The subtraction of a vector representing change in eye position from the retinal signal is mathematically equivalent to the addition and subtraction of absolute eye position vectors. This mechanism would produce the eye-centered motor error coordinate framework directly.

No head-centered, body-centered, or spatial map of visual space has ever been found, perhaps due to the paucity of experiments that might have discovered such a representation. Although such a map of visual space might facilitate integration of information across saccades or the perception of visual stability as the eyes

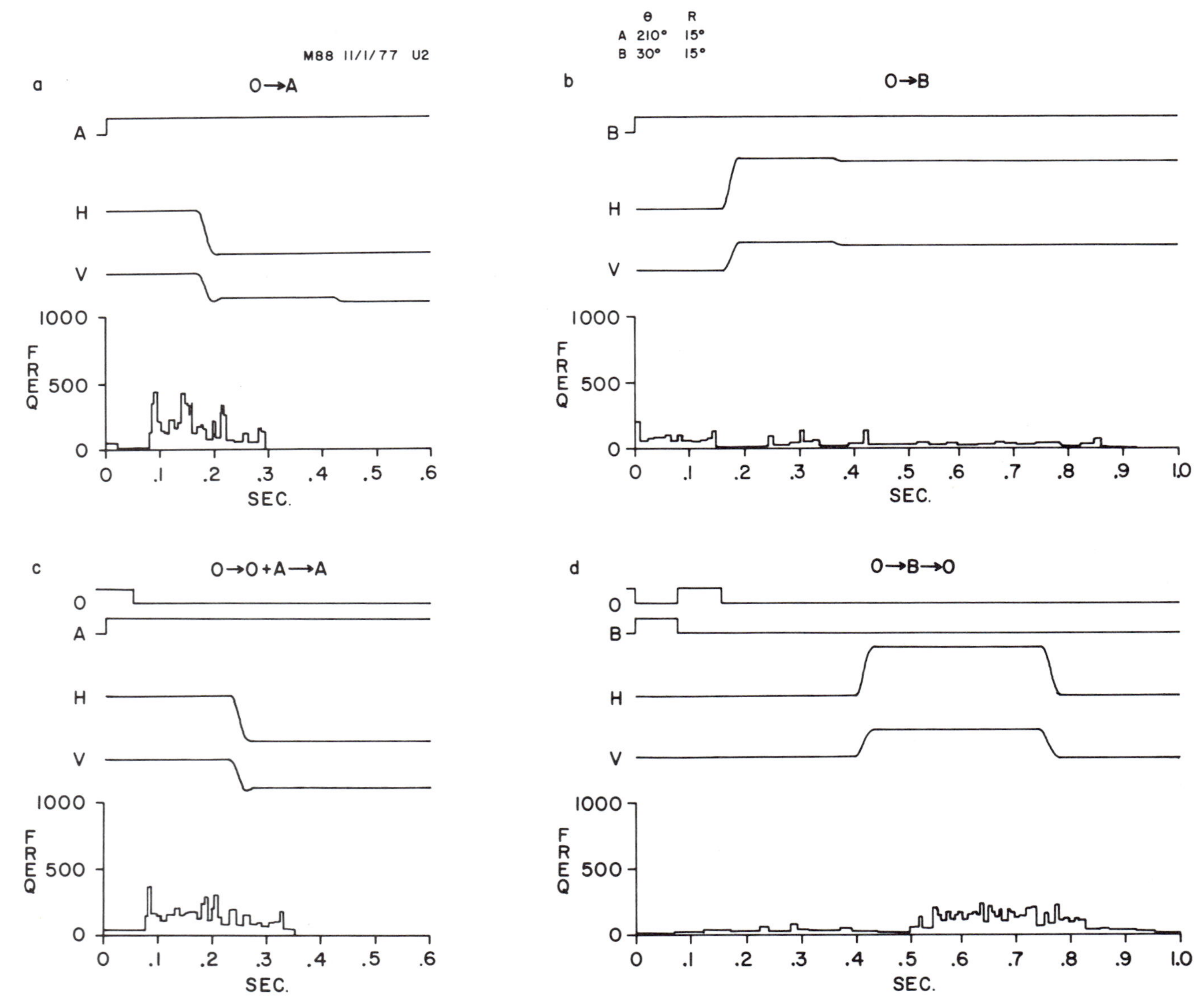

FIGURE 36.9 Discharge pattern of a quasi-visual cell. (a) Response of the cell to a target (A) presented in the receptive field of the cell. (b) Response of the cell to a target (B) presented in the opposite direction. (c) Response of the cell to the target A in the receptive field on a delayed-saccade trial, in which target onset and saccade onset were separated by a longer interval. The cell's activity was temporally coupled with the onset of the target, and no saccade-related burst occurred. (d) Activity of the cell on a double-saccade trial, in which saccades to B and then back to the fixation 0 were required in sequence. The B-to-0 saccade was the same as the 0-to-A saccade. The cell began firing after the saccade to B and before the saccade back to 0, though target 0 was no longer illuminated. H, horizontal eye position; V, vertical eye position. (From Mays and Sparks, 1980b.)

move, it is worth noting that, except for saccades and pursuit movements, the eyes move less with respect to the visual scene than do either the head or body, thanks to the vestibular and optokinetic mechanisms for stabilizing retinal images. That the location of visual targets may be stored in an updated eye-centered frame of reference is therefore also plausible.

Models for interactions between retinal and eye position signals fall into two classes: those that produce head-centered visual signals and those that produce updated eye-centered visual signals without employing explicit head-centered coordinates. The models also fall into two classes in terms of the coding format of the visual information. Whereas eye position signals gener-

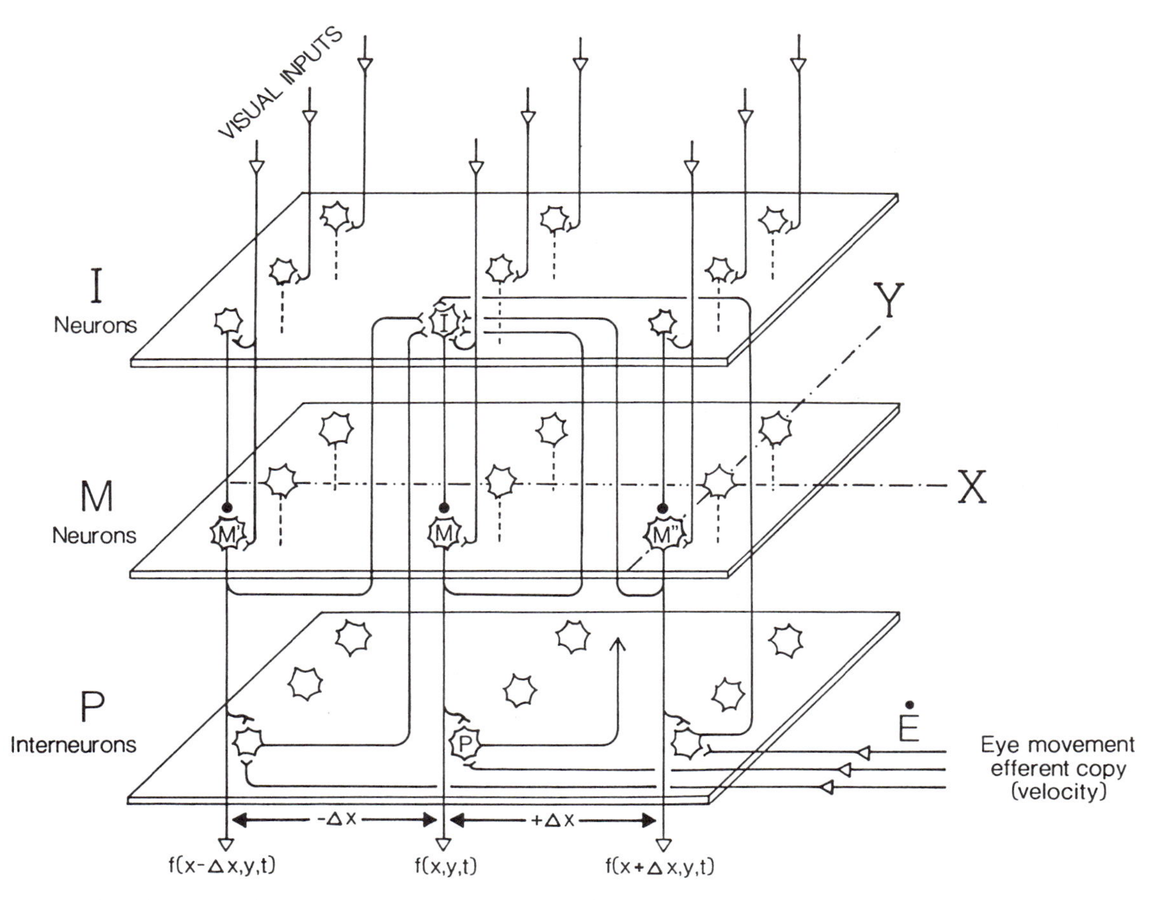

FIGURE 36.10 Architecture of the Droulez and Berthoz (1991) neural network model. The model has three layers of units: input (I) neurons, main (M) neurons, and interneurons (P). The input neurons receive weighted inputs from the main and interneurons of neighboring modules as well as direct excitatory retinal visual inputs. The main output neurons also receive visual input. The interneurons receive a velocity signal (E). When an eye movement occurs, feedback from these interneurons shifts activity across the map.

ally are agreed to exist in the form of a firing rate code, visual signals are place-coded topographical maps in some models (Zipser and Andersen, 1988; Droulez and Berthoz, 1991; Groh and Sparks, 1992a) and firing rate codes in others (Zipser and Andersen, 1988; Mittelstaedt, 1990).

Droulez and Berthoz (1991) proposed a model (figure 36.10) that used eye velocity information to shift visual activity on an eye-centered map analogous to the QV map in the SC or frontal eye fields (FEF). The visual signals were encoded topographically, whereas eye velocity signals were encoded in firing rates. The site of activity for a remembered visual target was shifted across the map when a nonzero eye velocity signal was received. The site of activity was stationary when the eyes were stationary. No head-centered visual signals were present in the model.

Mittelstaedt (1990) proposed a variety of classes of models that would transform visual signals from retinal to head-centered coordinates. These basic solutions differ in the source of the eye position information (proprioception or efference copy) and in the circuitry for combining eye position and retinal signals. They employ channels representing single values of a parameter (similar to rate coding) for both retinal and head-centered visual signals as well as for the eye position signals.

Zipser and Andersen (1988) proposed two variations for a retinal-to-head coordinates model (figure 36.11). The basic model consists of a three-layer neural network. In one version of the model, the head-centered output was encoded in a topographical map; in the other, the output was encoded in the firing rate. Both versions received place-coded retinal input and rate-

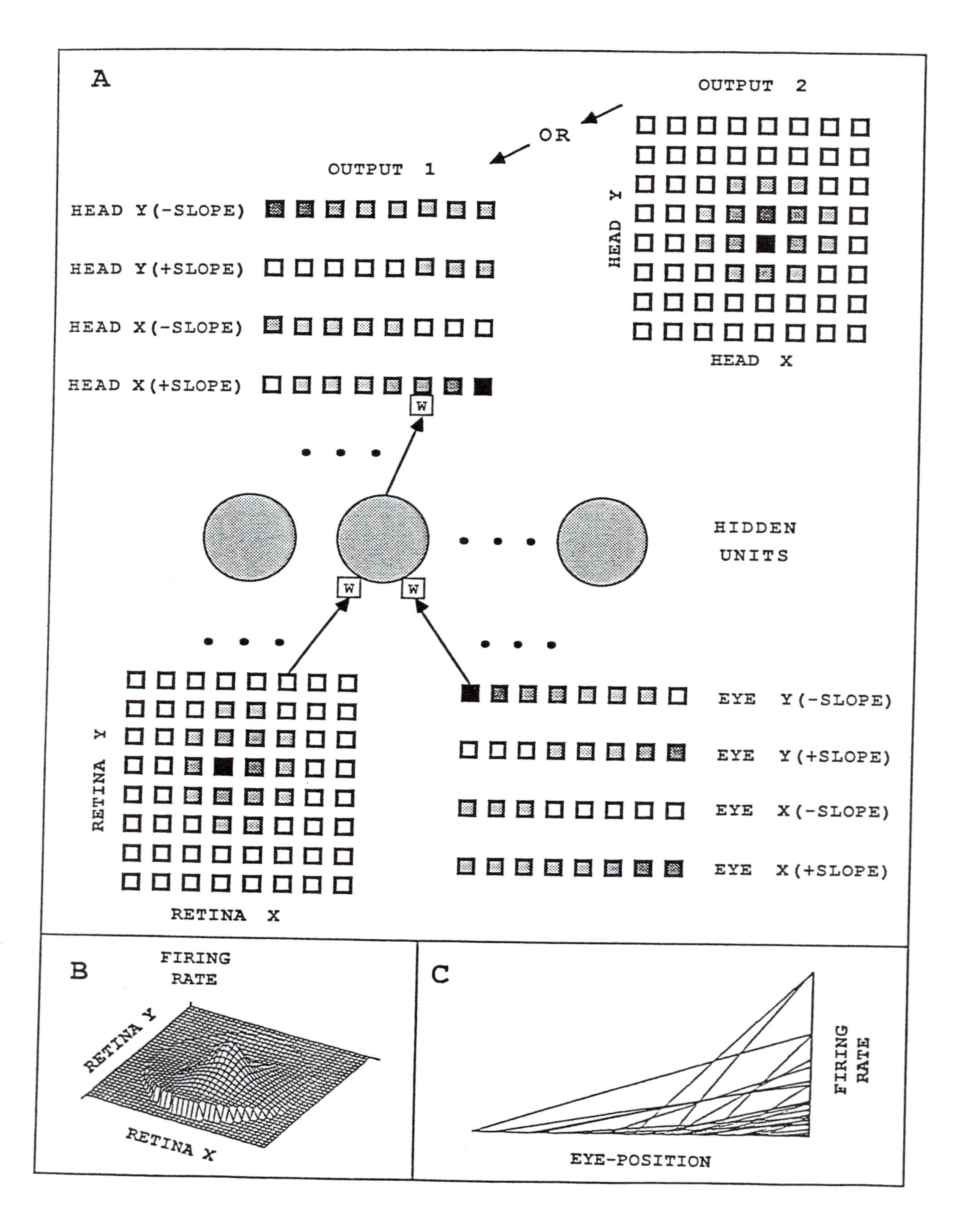
A
OUTPUT 2
OR
OUTPUT 1
HEAD Y(-SLOPE)
HEAD Y(+SLOPE)
HEAD X(-SLOPE)
HEAD X(+SLOPE)
HEAD Y
HEAD X
W
HIDDEN UNITS
W
W
RETINA Y
RETINA X
EYE Y(-SLOPE)
EYE Y(+SLOPE)
EYE X(-SLOPE)
EYE X(+SLOPE)
B
FIRING RATE
RETINA Y
RETINA X
C
FIRING RATE
EYE-POSITION

coded eye position signals. Backpropagation was used to modify the synaptic weights of the intermediate layer. Cells in parietal cortex that exhibited combinations of retinal visual and eye position–dependent activity served as the inspiration for this model.

Recent models developed by Groh and Sparks (1992a) to address auditory coordinate transformations are also applicable to the visual system. These models can be used to transform retinal signals into head-centered coordinates. In the vector addition model (figure 36.12), the place-coded retinal signal was decomposed into rate-coded horizontal and vertical components. Rate-coded horizontal and vertical eye position signals were added through excitatory synaptic connections. The resulting rate-coded, head-centered visual signal was then converted into a place code using a graded synaptic weighting pattern and inhibitory interneurons.

A second model proposed by Groh and Sparks employed local dendritic circuitry to execute a different neural algorithm (figure 36.13). A pattern of local connections at each dendrite involving excitation from a retinal visual input and inhibition from eye position units and interneurons resulted in dendrites that would selectively excite the soma if, and only if, the target was in that unit's head-centered receptive field. This model produced a place code of head-centered space without using a rate-coded intermediate stage.

The use of rate-coded visual signals in some of the models just described limits their ability to encode multiple targets. A rate-coded visual signal can encode only one target at a time unless the targets are multiplexed. The Droulez and Berthoz (1991) model does not employ rate-coded visual signals and was shown to be capable of shifting multiple visual targets simultaneously. The dendrite model of Groh and Sparks (1992a) also can transform the coordinates of multiple visual targets. The version of the Zipser and Andersen (1988) model producing place-coded output was not tested with multiple targets. Nonetheless, the limitation imposed by rate-coding does not eliminate the other models from contention. Eye movements occur serially, so single stimuli may be chosen as targets of eye movements before they are transformed into the appropriate coordinate framework for generating an eye movement. Models that can represent only single visual targets cannot account for another of the proposed uses for a head-centered representation of visual space—namely, the perception of visual stability when the eyes move—as this percept presumably requires the encoding of the entire visual scene.

Figure 36.11 The Zipser and Andersen (1988) backpropagation network. (A) Inputs to the model consist of retinal visual signals and rate-coded eye position signals (horizontal and vertical, positive and negative slopes). These units project to a layer of hidden units, which in turn project to the output layer of head-centered visual signals. Two versions of the model were implemented, one using rate-coded output units, the other using an array of place-coded output units with gaussian receptive fields. (B, C) Activity patterns of the input units.

Localization of Auditory Targets Unlike the visual system, the receptotopic organization of the auditory system does not yield spatial information about the stimulus. The cochlea produces a neural representation of the component frequencies of a sound. The sound's location must be derived from the differences in sound arrival time at the two ears, the differences in sound intensity at the two ears, and the spectral cues produced by the folds of the pinnae. These cues provide a measure of the sound's position with respect to the ears and head.

Many of the data on neural maps of sound location have come from barn owls (see Konishi et al., 1988, for review). The brain stem of the barn owl contains a variety of signals related to sound position, culminating in a map of auditory space in the optic tectum, the avian homolog of the SC. This map is in register with the visual map in the same structure. However, the barn owl differs from cats and primates in that its eyes do not move with respect to the head. Therefore, the head-centered coordinates of the barn owl's auditory map can maintain a constant registry with the retinal coordinates of its visual map.

Early work in the SC of the cat revealed both auditory and visual spatial receptive fields (Wickelgren, 1971). However, these experiments were conducted in the anesthetized preparation, so no dissociation between head- and eye-centered coordinate frames occurred. Jay and Sparks (1984, 1987b) trained monkeys to make saccades to auditory targets. They found that the auditory receptive fields in the primate SC shifted with changes in eye position (figure 36.14). The auditory map was encoded in an eye-centered coordinate framework that remained in register with the visual map encoded in the same coordinates.

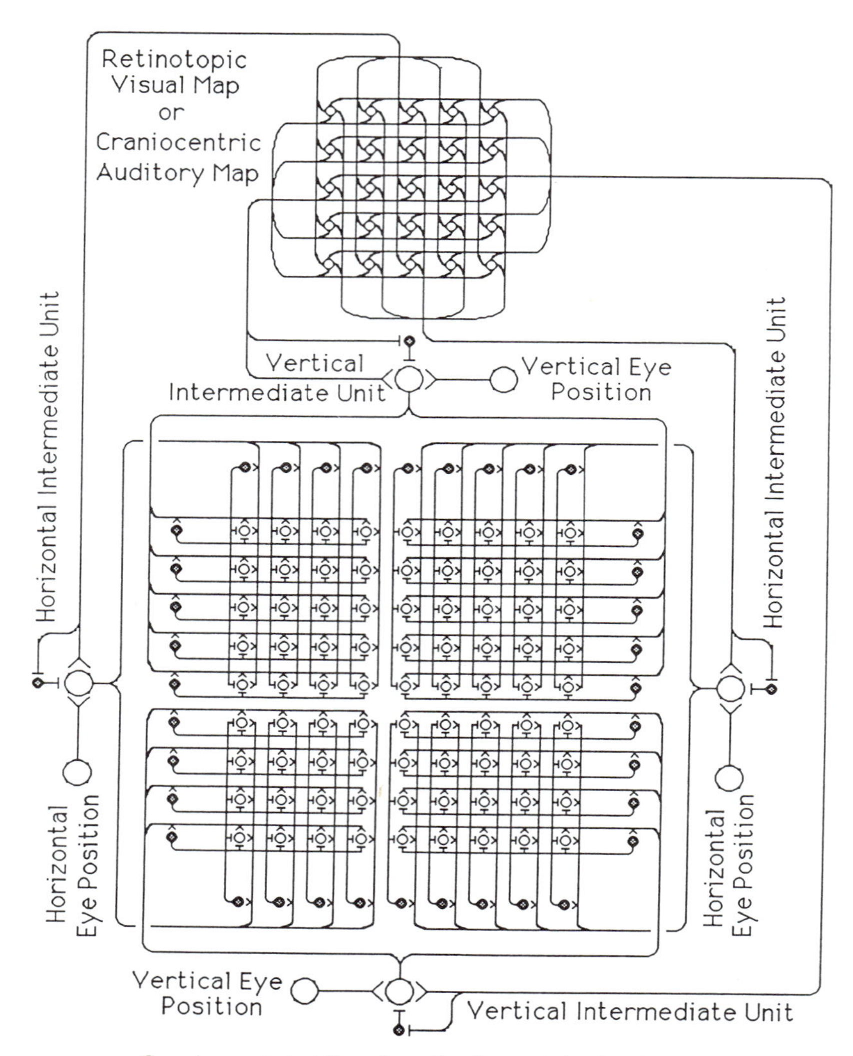

FIGURE 36.12 The vector addition model of Groh and Sparks (1992a). Inputs consist of a retinotopic visual map and rate-coded eye position units. Units in the map project to intermediate units with a weighting proportional to their position in the map. The rate code of eye position is added, yielding a rate code of head-centered target position. This rate code is converted into a place code in the output map through a combination of graded synaptic weights, thresholds, and inhibition. Inhibitory interneurons prevent the intermediate units from being active in the absence of a visual target. For the auditory system, the input map encodes head-centered auditory target position, and the eye position signals are subtracted instead of added. The output encodes eye-centered auditory target position.

Thus, a coordinate transformation for auditory signals also is known to occur. This transform is from head-centered to eye-centered coordinates. Unlike the visual system, the coding format of the input to this transform is unknown. No head-centered auditory map has been found in the primate, though very few studies have been done. Some spatial tuning was found in the auditory cortex of awake monkeys by Ahissar and colleagues (1992), but the coordinate framework of these responses is unknown because eye position was not monitored.

The models proposed by Groh and Sparks (1992a) were originally designed to execute this coordinate transform. Mathematically, the transform consists of the subtraction of an eye position vector from a vector representing target position with respect to the head. The vector subtraction model (like the vector addition model for visual signals) decomposes a place-coded

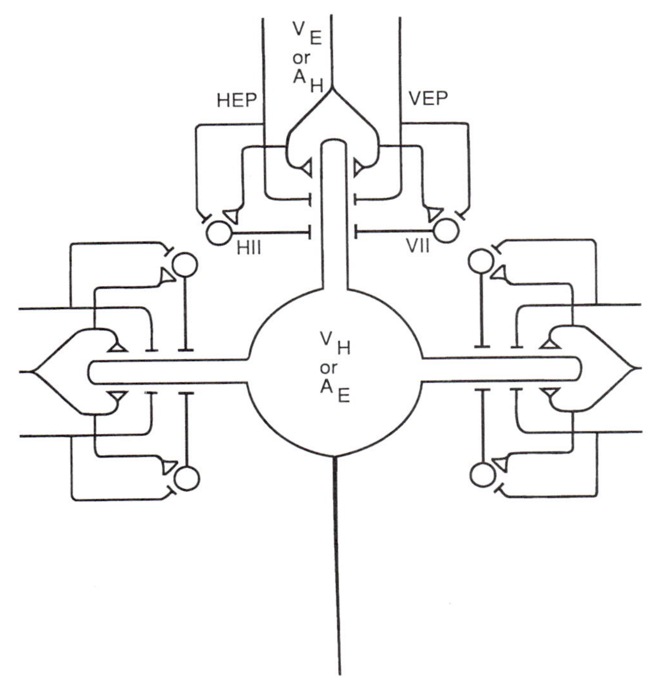

FIGURE 36.13 The dendrite model of Groh and Sparks (1992a). This unit is part of a map of units encoding head-centered visual target position and has a receptive field in head-centered coordinates. Each dendrite receives input from one particular retinal location, as well as eye position signals. The pattern of synaptic weights permits the dendrite to compare retinal location and eye position, exciting the soma if the right combination is detected. Different dendrites monitor different retinal locations and require different eye positions to activate the soma but produce selectivity for the same head-centered location as the other dendrites on a given unit. For the auditory system, the dendrites receive inputs from head-centered auditory units. Combining with eye position signals produces eye-centered receptive fields. A_H, auditory input, head coordinates; V_E, vertical eye position; HEP, horizontal eye position; VEP, vertical eye position; HII, horizontal inhibitory interneuron; VII, vertical inhibitory interneuron; V_H, visual signal, head-centered coordinates; A_E, auditory signal, eye-centered coordinates.

head-centered signal into rate-coded horizontal and vertical components and subtracts a rate-coded eye position signal. The resulting rate-coded eye-centered signal then is converted back into a place code, analogous to the auditory map in the SC. The essence of the algorithm employed by the vector subtraction-addition models is the conversion from a place to a rate code and back again. The synapses at the intermediate stage must produce a monotonic function of the position of the auditory target with respect to the eyes (or visual target with respect to the head), but it need not be a linear one.

The rate-coded intermediate stage imposes the same limitation on number of targets for the auditory system as it does for the visual system. However, it is possible that the head-centered auditory signals serving as input to the coordinate transformation are encoded in a firing rate rather than a map. If this is so, the system may already be limited to single targets, and the most parsimonious algorithm is the simple subtraction of firing rates through inhibitory synapses.

The dendrite model for the auditory transformation is like that described for the visual transformation, with local dendritic circuitry selecting appropriate target–eye position combinations. Multiple auditory targets can be represented, at a cost of a much larger network of units.

Which of these various models for visual and auditory coordinate transformations most closely approximates the brain itself depends on the format of head-centered auditory signals. In the cat, visual and auditory signals both are present in the region around the anterior ectosylvian sulcus, which projects heavily to the SC (Clarey and Irvine, 1986; Meredith and Clemo, 1989). The spatial selectivity and coordinate framework of these auditory signals is as yet unknown. The SC probably is not the only place in the brain where sensory signals are encoded in a coordinate framework different from that in which they arise, and it may well receive its sensory input from other dynamic maps.

Coding of sensory signals in a common coordinate framework is believed to be necessary not only for coordination of motor programs but also for sensory and perceptual processes such as binding of signals from the same source. Sounds tend to be localized in space to visual objects that are deemed likely to have emitted the sound. This allows us to watch movies and perceive the dialog as coming from the images of characters on the screen, despite the fact that the sounds usually are emitted by speakers on the sides of the theater. Ventriloquists perform a similar trick.

Sensory modalities interact with one another for the purpose of calibration as well. The optic tectum of the barn owl has served as a model system for studying the calibration of spatial audition by visual signals. In blind-reared owls, maps of auditory space in the tectum are distorted, stretched, and even upside-down (Knudsen, Esterly, and du Lac, 1991). Ocular prisms

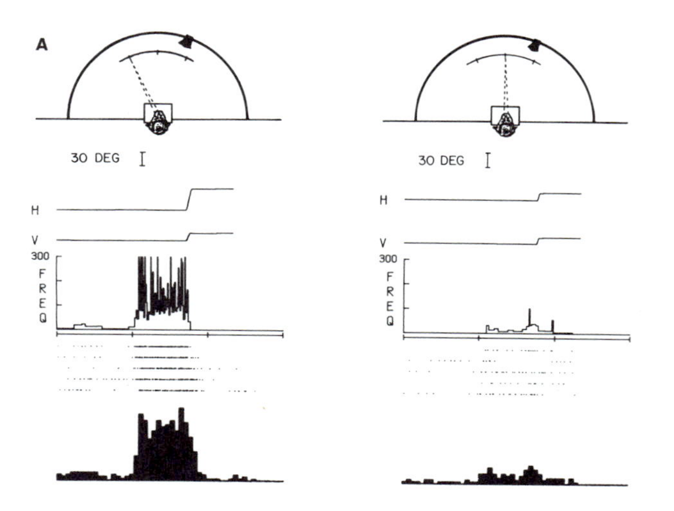

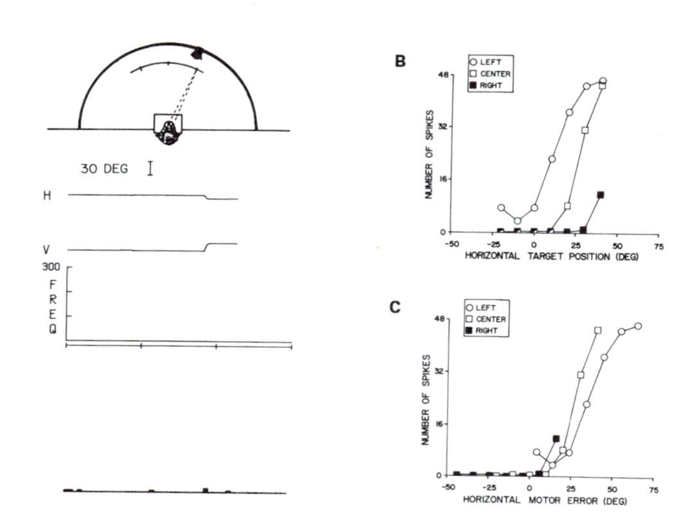

FIGURE 36.14 Auditory receptive fields shift with eye position. The monkey was trained to perform saccades to auditory targets. (A) Top row shows a top view of the monkey and apparatus. The auditory stimulus was delivered by a movable speaker mounted on a hoop. Visual fixation lights could be illuminated at various locations. The lower panels show the activity of an auditory cell in the SC. When the animal fixated to the left, the cell fired vigorously in response to an auditory stimulus delivered by the speaker positioned to the right. When the animal fixated straight ahead, the discharge was much weaker and, when the animal fixated to the right of the speaker, the cell was silent. (B) Summary of the response of the cell as a function of the horizontal position of the target for the three different fixation positions. (C) When the same data are plotted as a function of the horizontal component of the saccade to the target, the curves are more closely aligned. H, horizontal eye position; V, vertical eye position. (From Jay and Sparks, 1984.)

affect the owl's orienting behavior to sound as well (Knudsen and Knudsen, 1989).

LOCALIZATION OF SOMATOSENSORY TARGETS A tactile stimulus is detected by cutaneous receptors on the body. Like the visual system, the receptotopic organization of the somatosensory system encodes spatial information regarding the position of the stimulus on the body surface. However, the spatial information from the receptors directly activated by the stimulus is incomplete. Joint receptors, Golgi tendon organs, and muscle spindles also provide information regarding the position of the limb receiving the tactile stimulus. The brain must compute the spatial position of the stimulus with respect to some body axis using the somatotopic information from the cutaneous receptors and the kinesthetic information regarding joint position.

The process of localizing somatosensory targets, then, is a combination of body position sense and stimulus location on the body surface. The resulting coordinate framework is body-centered. To generate a saccade to a tactile stimulus, the position of the stimulus with respect to the eyes must be determined, requiring yet another kind of coordinate transformation. How this might be accomplished by the brain is not known.

The SC of the anesthetized cat (Stein, Magalhaes-Castro, and Kruger, 1975) and mouse (Drager and Hubel, 1975a, b, 1976) have been found to contain sensory responses to somatosensory stimuli. These responses show a rough topography and registry with visual responses but, in the anesthetized preparation, the visual and body axes do not move with respect to one another, so the coordinate framework of the responses cannot be determined. The coordinate framework of somatosensory responses in the SC of the awake primate is as yet unknown.

Just as vision affects the localization of auditory targets, so too does it affect the localization of the body in space. Visual input strongly influences the perceived position of the limbs (e.g., Shimojo, 1987). The converse is also true: Position sense can affect perception of the location of auditory and visual targets in contact with the body. Lackner and Shenker (1985) used vibration of arm muscles to induce the illusion of limb motion and limb displacement in human subjects sitting in the dark. This phenomenon is known as the *vibratory myesthetic illusion.* Subjects could make smooth pursuit movements tracking the illusory motion of the hand. If subjects held a single light source such as a fiber-optic strand in their hands, they perceived concurrent motion and displacement of the visual stimulus corresponding to the motion and displacement of the hand. The subjects perceived that they were tracking the visual target so that there was no retinal slip. However, eye position measurements revealed that the eyes were stationary. A similar phenomenon occurred when subjects held an auditory speaker in their hands.

MOTOR CONVERGENCE Once targets of different sensory modalities have been translated into a common coordinate framework, they can access a common motor circuitry for generating a saccade. All eye movements are generated by the same extraocular muscles. Above this level lies brainstem circuitry specific to all saccadic eye movements. Input to this saccade generator is provided by the SC and FEF, which represent saccade vectors in topographical motor maps. Two questions regarding the activity of these motor maps with respect to the generation of saccades are of interest: First, are the same cells active for saccades of all types? Second, if so, what differences in the activity profile of these cells might account for differences in the dynamics of saccades to targets of different modalities?

Motor activity in the SC is found in two classes of cells: saccade-related burst cells, which fire a high-frequency burst shortly before a movement into the movement field of the cell, and visuomotor cells, which have a response coupled with the onset of the visual stimulus (Wurtz and Goldberg, 1972; Sparks, 1978; Mays and Sparks, 1980b). The majority of burst cells discharge before saccades to auditory stimuli (Jay and Sparks, 1987a). Most visuomotor cells exhibit motor but not sensory activity for auditory saccades as well. Recent work by Groh and Sparks (1993) suggests that somatosensory saccades are represented by the same population of motor cells.

Saccades to auditory (Jay and Sparks, 1990) and somatosensory (Groh and Sparks, 1992b) stimuli have a lower peak velocity than saccades of comparable amplitude to visual targets. This velocity difference suggests a difference in the manner in which sensory signals access common motor circuitry. Jay and Sparks (1987a) proposed that the lower velocity of auditory saccades was due to the slightly smaller population of motor cells active for such saccades. An alternative is that the discharge of motor cells is less vigorous for auditory and somatosensory saccades than for visual saccades. That activity patterns in the SC can influence saccade velocity is supported by several experiments. In the cat, the activity of premotor cells was correlated with the velocity of the movement (Berthoz, Grantyn, and Droulez, 1986; Munoz and Guitton, 1987; Munoz, Guitton, Pélisson, 1991). Lee, Rohrer, and Sparks (1988) reported that saccade velocity was reduced following lidocaine injections. Frequency of stimulation in the SC was correlated with saccade velocity (Stanford et al., 1993). Whether differences in collicular activity produce the sensory modality–dependent velocity profile differences has yet to be determined. Because discharge rate is spatially tuned for the movement vector, and because the spatial tuning of motor cells can vary under different conditions (Stanford and Sparks, 1994), this possibility can be tested only by a thorough sampling of the movement field for both visual and auditory or somatosensory saccades, which has not yet been done.

Conclusions

Studies of the SC have considerably enhanced our understanding of a number of important and perplexing problems in integrative neuroscience. The experimental strategies developed for studying neurons in the SC have been used to advantage in other areas. The tasks developed for studying coordinate frameworks of visual and auditory signals have been extended to parietal (Gnadt and Andersen, 1988), frontal (Russo and Bruce, 1989; Bruce and Goldberg, 1990), and prefrontal (Bruce, 1990) cortical areas. The idea that sensory and motor signals are encoded by the spatial and temporal pattern of activity within large populations of neurons has a long history. The detailed map of saccade direction and amplitude found in the SC and the recent advent of methods for reversibly deactivating small populations of neurons permit explicit tests of hypotheses about the extraction of information from population responses. These methods could be extended to other candidate population codes. Tasks developed for studying neuronal correlates of target and response selection in the SC may allow a more direct approach to studies of this important historical problem. Coordination of the eye and head components of gaze saccades may serve as a useful model for more complex cases such as eye-hand coordination. Recent investigations (Schlag, Schlag-Rey, and Dassonville, 1989) into the timing and nature of corollary discharge signals of eye position in the oculomotor system continue to add to the rich history of psychophysical experiments on such signals.

ACKNOWLEDGMENTS We acknowledge invaluable programing assistance from Kathy Pearson. This work was supported by National Institutes of Health grant EY01189 to D.L.S. and by National Science Foundation and National Defense Science and Engineering Graduate Fellowships to J.M.G.

REFERENCES

Ahissar, M., E. Ahissar, H. Bergman, and E. Vaadia, 1992. Encoding of sound-source location and movement: Activity of single neurons and interactions between adjacent neurons in the monkey auditory cortex. *J. Neurophysiol.* 67:203–215.

Berthoz, A., A. Grantyn, and J. Droulez, 1986. Some collicular efferent neurons code saccadic eye velocity. *Neurosci. Lett.* 72:289–294.

Bizzi, E., R. Kalil, and P. Morasso, 1972. Two modes of active eye-head coordination in monkeys. *Brain Res.* 40: 45–48.

Bruce, C. J., 1990. Integration of sensory and motor signals for saccadic eye movements in the primate frontal eye fields. In *Signal and Sense: Local and Global Order* in Perceptual Maps. New York: Wiley-Liss, pp. 261–314.

Bruce, C. J., and M. E. Goldberg, 1990. Primate frontal eye fields: III. Maintenance of a spatially accurate saccade signal. *J. Neurophysiol.* 64:489–508.

Clarey, J. C., and D. R. F. Irvine, 1986. Auditory response properties of neurons in the anterior ectosylvian sulcus of the cat. *Brain Res.* 386:12–19.

Drager, U. C., and D. H. Hubel, 1975a. Physiology of visual cells in mouse superior colliculus and correlation with somatosensory and auditory input. *Nature* 253:203–204.

Drager, U. C., and D. H. Hubel, 1975b. Responses to visual stimulation and relationship between visual, auditory and somatosensory inputs in mouse superior colliculus. *J. Neurophysiol.* 38:690–713.

Drager, U. C., and D. H. Hubel, 1976. Topography of visual and somatosensory projections to mouse superior colliculus. *J. Neurophysiol.* 39:91–101.

Droulez, J., and A. Berthoz, 1991. A neural network model of sensoritopic maps with predictive short-term memory properties. *Proc. Natl. Acad. Sci. U.S.A.* 88:9653–9657.

Duhamel, J.-R., C. L. Colby, and M. E. Goldberg, 1992. The updating of the representation of visual space in parietal cortex by intended eye movements. *Science* 255: 90–92.

du Lac, S., and E. I. Knudsen, 1990. Maps of head movement vector and speed in the optic tectum of the barn owl. *J. Neurophysiol.* 63:131–146.

Freedman, E. G., T. R. Stanford, and D. L. Sparks, 1993. An analysis of the metrics and dynamics of visually guided and collicular stimulation-induced gaze shifts in the monkey. *Soc. Neurosci. Abstr.* 19:786.

Georgopoulos, A. P., A. B. Schwartz, and R. E. Kettner, 1986. Neuronal population coding of movement direction. *Science* 233:1416–1419.

Glimcher, P., and D. L. Sparks, 1992. Movement selection in advance of action: Saccade-related bursters of the superior colliculus. *Nature* 355:542–545.

Gnadt, J. W., and R. A. Andersen, 1988. Memory related motor planning activity in posterior parietal cortex of macaque. *Exp. Brain Res.* 70:216–220.

Groh, J. M., and D. L. Sparks, 1992a. Two models for transforming auditory signals from head-centered to eye-centered coordinates. *Biol. Cybern.* 67:291–302.

Groh, J. M., and D. L. Sparks, 1992b. Characteristics of saccades to somatosensory targets. *Soc. Neurosci. Abstr.* 18: 701.

Groh, J. M., and D. L. Sparks, 1993. Motor activity in the primate superior colliculus (SC) during saccades to somatosensory and visual targets. *Invest. Ophthalmol. Vis. Sci.* 34:1137.

Guitton, D., and M. Volle, 1987. Gaze control in humans: Eye-head coordination during orienting movements to targets within and beyond the oculomotor range. *J. Neurophysiol.* 58:427–459.

Guthrie, B. L., J. D. Porter, and D. L. Sparks, 1983. Corollary discharge provides accurate eye position information to the oculomotor system. *Science* 221:1193–1195.

Harris, L. R., 1980. The superior colliculus and movements of the head and eyes in cat. *J. Physiol. (Lond.)* 300:367–391.

Heiligenberg, W., 1991. The jamming avoidance response of the electric fish *Eigenmannia*: Computational rules and their neuronal implementation. *Semin. Neurosci.* 3:3–18.

Jay, M. F., and D. L. Sparks, 1984. Auditory receptive fields in primate superior colliculus shift with changes in eye position. *Nature* 309:345–347.

Jay, M. F., and D. L. Sparks, 1987a. Sensorimotor integration in the primate superior colliculus: I. Motor convergence. *J. Neurophysiol.* 57:22–34.

Jay, M. F., and D. L. Sparks, 1987b. Sensorimotor integration in the primate superior colliculus: II. Coordinates of auditory signals. *J. Neurophysiol.* 57:35–55.

Jay, M. F., and D. L. Sparks, 1990. Localization of auditory and visual targets for the initiation of saccadic eye movements. In *Comparative Perception: I. Basic Mechanisms*, M. Berkley and W. Stebbins, eds. New York: Wiley, pp. 351–374.

Knudsen, E. I., S. du Lac, and S. D. Esterly, 1987. Computational maps in the brain. *Annu. Rev. Neurosci.* 10:41.

Knudsen, E. I., S. D. Esterly, and S. du Lac, 1991. Stretched and upside-down maps of auditory space in the optic tectum of blind-reared owls; acoustic basis and behavioral correlates. *J. Neurosci.* 11:1727–1747.

Knudsen, E. I., and P. F. Knudsen, 1989. Vision calibrates sound localization in developing barn owls. *J. Neurosci.* 9:3306–3313.

Konishi, M., T. Takahashi, H. Wagner, W. E. Sullivan, and C. E. Carr, 1988. Neurophysiological and anatomical substrates of sound localization in the owl. In *Auditory Function*, G. M. Edelman, W. E. Gall, and W. M. Cowan, eds. New York: Wiley.

Lackner, J. R., and B. Shenker, 1985. Proprioceptive influences on auditory and visual spatial localization. *J. Neurosci.* 5:579–583.

Laurutis, V. P., and D. A. Robinson, 1986. The vestibulo-ocular reflex during human saccadic eye movements. *J. Physiol. (Lond.)* 373:209–234.

Lee, C., W. H. Rohrer, and D. L. Sparks, 1988. Population coding of saccadic eye movements by neurons in the superior colliculus. *Nature* 332:357–360.

Malpeli, J. G., and P. H. Schiller, 1979. A method of reversible inactivation of small regions of brain tissue. *J. Neurosci. Methods* 1:143–151.

Mays, L. D., and D. L. Sparks, 1980a. Saccades are spatially, not retinocentrically, coded. *Science* 208:1163–1165.

Mays, L. E., and D. L. Spark, 1980b. Dissociation of visual and saccade-related responses in superior colliculus. *J. Neurophysiol.* 43:207–232.

McIlwain, J. T., 1976. Large receptive fields and spatial transformations in the visual system. *Int. Rev. Physiol.* 10: 223.

Meredith, M. A., and H. R. Clemo, 1989. Auditory cortical projection from the anterior ectosylvian sulcus (field AES) to the superior colliculus in the cat: An anatomical and electrophysiological study. *J. Comp. Neurol.* 289:687–707.

Mittelstaedt, H., 1990. Basic solutions to the problem of head-centric visual localization. In *Perception and Control of Self-Motion*, R. Warren and A. H. Wertheim, eds. Hillsdale, N.J.: Erlbaum, pp. 267–287.

Morasso, P., E. Bizzi, and J. Dichgans, 1973. Adjustment of saccade characteristics during head movements. *Exp. Brain Res.* 16:492–500.

Munoz, D. P., and D. Guitton, 1987. Tecto-reticulo-spinal neurons have discharges coding the velocity profiles of eye and head orienting movements. *Soc. Neurosci. Abstr.* 13: 393.

Munoz, D. P., D. Guitton, and D. Pélisson, 1991. Control of orienting gaze shifts by the tectoreticulospinal system in the head-free cat: III. Spatiotemporal characteristics of phasic motor discharges. *J. Neurophysiol.* 66:1642–1666.

Porter, J. D., B. L. Guthrie, and D. L. Sparks, 1983. Innervation of monkey extraocular muscles: Localization of sensory and motor neurons by retrograde transport of horseradish peroxidase. *J. Comp. Neurol.* 218:208–219.

Robinson, D. A., 1972. Eye movements evoked by collicular stimulation in the alert monkey. *Vision Res.* 12:1795–1808.

Roucoux, A., M. Crommelinck, and D. Guitton, 1980. Stimulation of the superior colliculus in the alert cat: II. Eye and head movements evoked when the head is unrestrained. *Exp. Brain Res.* 39:75–85.

Russo, G. S., and C. J. Bruce, 1989. Auditory receptive fields of neurons in frontal cortex of rhesus monkey shift with direction of gaze. *Soc. Neurosci. Abstr.* 15:1204.

Schlag, J., M. Schlag-Rey, and P. Dassonville, 1989. Interactions between natural and electrically evoked saccades: II. At what time is eye position sampled as a reference for the localization of a target? *Exp. Brain Res.* 76:548–558.

Shimojo, S., 1987. Attention-dependent visual capture in double vision. *Perception* 16:445–447.

Sparks, D. L., 1978. Functional properties of neurons in the monkey superior colliculus: Coupling of neuronal activity and saccade onset. *Brain Res.* 156:1–16.

Sparks, D. L., 1986. Translation of sensory signals into commands for the control of saccadic eye movements: Role of the primate superior colliculus. *Physiol. Rev.* 66:118–171.

Sparks, D. L., R. Holland, and B. L. Guthrie, 1976. Size and distribution of movement fields in the monkey superior colliculus. *Brain Res.* 113:21–34.

Sparks, D. L., C. Lee, and W. H. Rohrer, 1990. Population coding of the direction, amplitude, and velocity of saccadic eye movements by neurons in the superior colliculus. *Cold Spring Harb. Symp. Quant. Biol.* 55:805–811.

Sparks, D. L., and L. E. Mays, 1980. Movement fields of saccade-related burst neurons in the monkey superior colliculus. *Brain Res.* 190:39–50.

Sparks, D. L., and L. E. Mays, 1983. The spatial localization of saccade targets: I. Compensation for stimulation-induced perturbations in eye position. *J. Neurophysiol.* 49: 45–63.

Sparks, D. L., and L. E. Mays, 1990. Signal transformations required for the generation of saccadic eye movements. *Annu. Rev. Neurosci.* 13:309–336.

Sparks, D. L., and J. D. Porter, 1983. The spatial localization of saccade targets: II. Activity of superior colliculus neurons preceding compensatory saccades. *J. Neurophysiol.* 49:64–74.

Stanford, T. R., E. G. Freedman, J. M. Levine, and D. L. Sparks, 1993. The effects of stimulation parameters on the metrics and dynamics of saccades evoked by electrical stimulation of the primate superior colliculus. *Soc. Neurosci. Abstr.* 19:786.

Stanford, T. R., and D. L. Sparks, 1994. Systematic errors for saccades to remembered targets: Evidence for a dissociation between saccade metrics and activity in the superior colliculus. *Vision Res.* 34:93–106.

Stein, B. E., B. Magalhaes-Castro, and L. Kruger, 1975. Superior colliculus: Visuotopic-somatotopic overlap. *Science* 189:224–225.

Stryker, M. P., and P. H. Schiller, 1975. Eye and head movements evoked by electrical stimulation of monkey superior colliculus. *Exp. Brain Res.* 23:103–112.

Tomlinson, R. D., and P. S. Bahra, 1986a. Combined eye-head gaze shifts in the primate: I. Metrics. *J. Neurophysiol.* 56:1542–1557.

Tomlinson, R. D., and P. S. Bahra, 1986b. Combined eye-head gaze shift in the primate: II. Interactions between saccades and the vestibuloocular reflex. *J. Neurophysiol.* 56:1558–1570.

Tweed, D., and T. Vilis, 1985. A two dimensional model for saccade generation. *Biol. Cybern.* 52:219.

Van Gisbergen, J. A. M., A. J. Van Opstal, and A. A. M. Tax, 1987. Collicular ensemble coding of saccades based on vector summation. *Neuroscience* 21:541.

Van Opstal, A. J., and J. A. M. Van Gisbergen, 1989. A nonlinear model for collicular spatial interactions underlying the metrical properties of electrically elicited saccades. *Biol. Cybern.* 60:171.

Vogels, R., 1990. Population coding of stimulus orientation by striate cortical cells. *Biol. Cybern.* 64:25–31.

Wickelgren, B. G., 1971. Superior colliculus: Some receptive field properties of bimodally responsive cells. *Science* 173:69–73.

Wurtz, R. H., and M. E. Goldberg, 1972. Activity of superior colliculus in behaving monkey: III. Cells discharging before eye movements. *J. Neurophysiol.* 35:575.

Zipser, D., and R. A. Andersen, 1988. A back-propagation programmed network that simulates response properties of a subset of posterior parietal neurons. *Nature* 331:679–684.

37 Vector Encoding and the Vestibular Foundations of Spatial Cognition: Neurophysiological and Computational Mechanisms

BRUCE L. MCNAUGHTON, JAMES J. KNIERIM, AND MATTHEW A. WILSON

ABSTRACT Recent neurophysiological and behavioral studies suggest that the influence of the vestibular system on spatial orientation runs far deeper than its well-known role in the perception of attitude and motion and in the stabilization of visual images. The vestibular system also appears to play a central role in establishing the fundamental directional reference framework that is used to construct cognitive representations of the environment and to compute optimal trajectories to remembered locations on the basis of visual landmarks. This so-called sense of direction appears to involve the integration of angular velocity signals that arise primarily in the vestibular system. The integration process seems to be implemented without an appreciable time constant (leak) and is automatically reset every 360°, which leads to the suggestion that the integration may actually be implemented by linear associative mapping, an operation that is, at least theoretically, easy to implement with real or artificial neural networks. Evidence from behavioral studies suggests that such allocentric directional information, in conjunction with distance information obtained from other sources, leads to a vector-based internal representation of spatial relationships. In this view, optimal trajectories are computed by subtracting vectors to landmarks, remembered at the goal location, from the perceived vectors at the current location. Vector subtraction can also be performed by linear associative mapping. Consideration of the kind of internal representation necessary for such mapping leads to some new perspectives on the nature of spatial representations in both parietal cortex and hippocampus, as well as to some experimentally testable predictions.

BRUCE L. MCNAUGHTON, JAMES J. KNIERIM, and MATTHEW A. WILSON Department of Psychology and ARL Division of Neural Systems, Memory and Aging, University of Arizona, Tucson, Ariz.

The indigenous people of Puluwat Atoll in the Caroline Islands make their lives on the sea. They often take their small outrigger canoes on voyages that may cover hundreds of miles, and they may be out of sight of land for many days. During the voyage, they may change course frequently as they tack upwind or put in at some intermediate island in the event of bad weather. They seldom miss their targets. Young would-be navigators receive many years of training from an experienced teacher in the system of *etak*, which is basically a system of cognitively facilitated dead reckoning, in which the navigator's self-motion through the water is tracked in relation to the *imagined* positions of "reference islands" that lie alongside but out of sight of the desired course (Gladwin, 1970). In regions of ocean where no convenient reference island exists, the cognitive map of the navigator contains mythical islands, which serve the identical function. The judgment of self-motion is apparently more accurate within a strong cognitive spatial reference framework.

In total darkness, rodents can conduct a quasi-random journey for several meters from their nest sites in search of food or missing young and then return on a direct bearing. They continue to reach their target successfully after a slow rotation (below vestibular threshold) of the entire environment, including the animal and nest; however, if the environment is rotated above vestibular threshold, the rodent misses the target by the corresponding angle. Mittelstaedt and Mittelstaedt (1980) concluded that rodents are capable of path integration (i.e., keeping track of their own self-motion relative to the nest) over considerable distances. Rats also can maintain their bearings when rotated in an enclosed box at approximately 30 rpm for up to nearly 10 revolutions (Matthews, Campbell, and Deadwyler, 1988). Both of these capabilities are destroyed by lesions to the hippocampal formation. Why should a capability that appears to depend on low-level vestibular sensations be disrupted by damage to the hippocampus, which itself receives no direct vestibular input?

The hippocampus is the highest level of association cortex in the mammalian brain (Swanson, Köhler, and Bjorklund, 1987; Felleman and Van Essen, 1991) and is an essential component in the initial registration of episodic or declarative memories. Hippocampal damage leads to severe deficits in acquiring cognitive representations of space (O'Keefe and Nadel, 1978). Hippocampal neurons are selectively active in specific locations within a particular environment (O'Keefe and Dostrovsky, 1971). Groups of hippocampal cells transmit an ensemble code for space sufficiently accurate to predict a rat's position within a few centimeters, by simultaneously recording from as few as 70–150 hippocampal pyramidal cells (Wilson and McNaughton, 1993) (see color plate 14). In a familiar environment, the firing of hippocampal "place cells" can be controlled entirely by the locations of the visual landmarks. If the animal is introduced into the environment after the landmark array has been rotated, "place fields" rotate accordingly; yet, if either the landmarks are removed (O'Keefe and Speakman, 1987) or the lights are extinguished (McNaughton, Leonard, and Chen, 1989; Quirk, Muller, and Kubie, 1990) after the animals have been given their initial bearings relative to the landmark array, hippocampal neurons continue to fire in normal relation to the now-invisible landmarks, at least for several minutes. Interestingly, hippocampal neuronal activity and spatial selectivity cease almost completely if active self-motion is prevented by restraint (Foster, Castro, and McNaughton, 1989). The hippocampus is thus a crucial part of a system that is capable, in principle, of representing the relationships among landmarks. After more than two decades of research, however, it is still a matter of conjecture why a given hippocampal neuron fires where it does and what is actually represented. One of the unresolved problems involves the dependency of hippocampal firing on directional information. Under some conditions, such as random foraging for food, firing of hippocampal place cells does not depend at all on the direction the animal is facing in the place field (Muller, Kubie, and Ranck, 1987). In other situations, such as running on linear tracks toward specific goals, most of the variance in firing within the place fields is accounted for by the direction the animal is facing (McNaughton, O'Keefe, and Barnes, 1983).

Central representation of a landmark array

The history of thought concerning the reference framework in which objects (landmarks) are represented in the cognitive system has been characterized by the idea that stimuli, which are initially defined in egocentric coordinates (i.e., retinocentric, craniocentric, or somatocentric), are somehow remapped into an allocentric coordinate system (i.e., one that is independent of the orientation of the body in space) (Tolman, 1948; O'Keefe and Nadel, 1978). Most of the debate in this field has concerned the nature of allocentric representation and how it is implemented in the brain.

Let us reduce the problem to its essence. With reference to figure 37.1A, assume that a rat has a nest at location B in an environment in which the only distinguishable feature is a radially symmetrical landmark at location A. Later, the animal is placed or arrives at some different location C. There is insufficient information for it to compute the required trajectory back to its nest, because the set of egocentric angles and distances that the animal may have learned at B does not uniquely determine B but applies equally well to all locations B' on the circle centered at A with radius AB. One of two additional items of information is required. The classically assumed basis of allocentric representation is to make use of a second, *discriminable* landmark, visible from B. The alternative solution is to give the animal some sort of compass (i.e., an allocentric directional reference). These two solutions are neither computationally nor adaptively equivalent.

The classical solution is based on coding and storing relationships among multiple landmarks. For example, given two landmarks A and D (figure 37.1B), the animal might store the given distances BD and BA as well as either the computed distance DA or the given egocentric angle α. Some time later, finding itself at C and

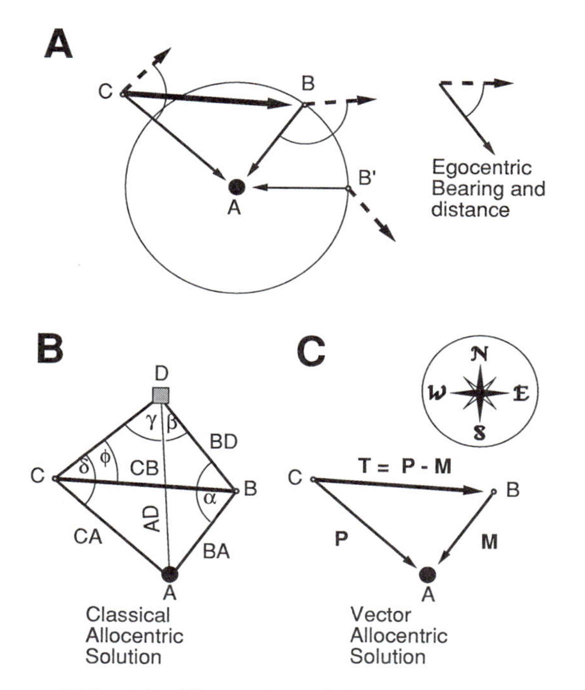

FIGURE 37.1 (A) Illustration of the essential problem of generating an optimal trajectory from the present location C to a remembered location B on the basis of a landmark located at A. This problem is underdetermined. In the absence of additional information, all points at distance AB from A are equivalent. (B) The classical allocentric encoding scheme makes use of stored relationships among landmarks. (C) The vector subtraction hypothesis relies on the presence of an allocentric directional reference. Vectors: **T**, trajectory; **P**, perceptual; **M**, memory.

wishing to return to B, the animal must compute the angle β from the stored information available, add this to the angle γ, which is computed from the current given input, and then compute the required distance CB and angle ϕ from a combination of the stored and current inputs. This exercise is not mathematically difficult but does require that neurons compute the trigonometric and square root functions necessary for its solution. The main problem, however, is that one cannot predict in advance which (if any) of the landmarks available at B will also be available at C; hence, if multiple landmarks exist, their mutual relationships must be stored. The storage space required increases approximately as the square of the number of landmarks.

Next, consider the compass solution (*vector encoding*). Using a compass and the distance AB, the animal need store only the vector quantity **M** (the memory vector). Later, when presented with the vector **P** (the perceptual vector), the desired trajectory vector (**T**) is obtained by subtracting the memory vector from the perceptual vector (figure 37.1C):

$$\mathbf{T} = \mathbf{P} - \mathbf{M}$$

Vector subtraction is an example of a simple associative mapping because, for each ordered vector pair, there is a unique solution (output). This is the sort of operation that neural networks (real or artificial) are ideally suited to implement (figure 37.2). Moreover, the amount of information that must be stored increases only linearly with the number of landmarks, and the computation can be performed using any single landmark.

What might we expect the receptive fields of vector encoding neurons to look like? A simple idea is that they should constitute a set of two-dimensional radial basis functions (gaussians), with landmark distance represented on one axis and bearing represented on the other axis. Vector-encoding cells thus should fire only when the animal is located at a certain distance and bearing from a landmark (i.e., when the animal is at a particular location). *This is exactly the defining characteristic of hippocampal place cells!*

Collett, Cartwright, and Smith (1986) investigated the cognitive representation of landmarks in a series of simple behavioral experiments. Rodents were trained to search for a single buried food item in relation to two cylindrical landmarks in an otherwise featureless environment. After learning, the cognitive representation was probed by doubling the distance between the landmarks. The classical allocentric model predicts that the animals would search in the one location that was the closest approximation to the original, as though the environment had been somehow stretched on a rubber sheet. What happened was that the animals spent nearly equal time searching at two locations, each at the correct bearing and distance from one of the landmarks (figure 37.3).

In a similar experiment, an equilateral triangle of identical cylinders was used, with the target located at the center. When the distance of one cylinder along the perpendicular bisector of the other two was doubled, the animals searched in the correct location, not in the topologically stretched location predicted by the classical allocentric model. When two of the cylinders were removed, the animals searched at three locations around the remaining cylinder, each at the correct distance and bearing from one of the original cylinders.

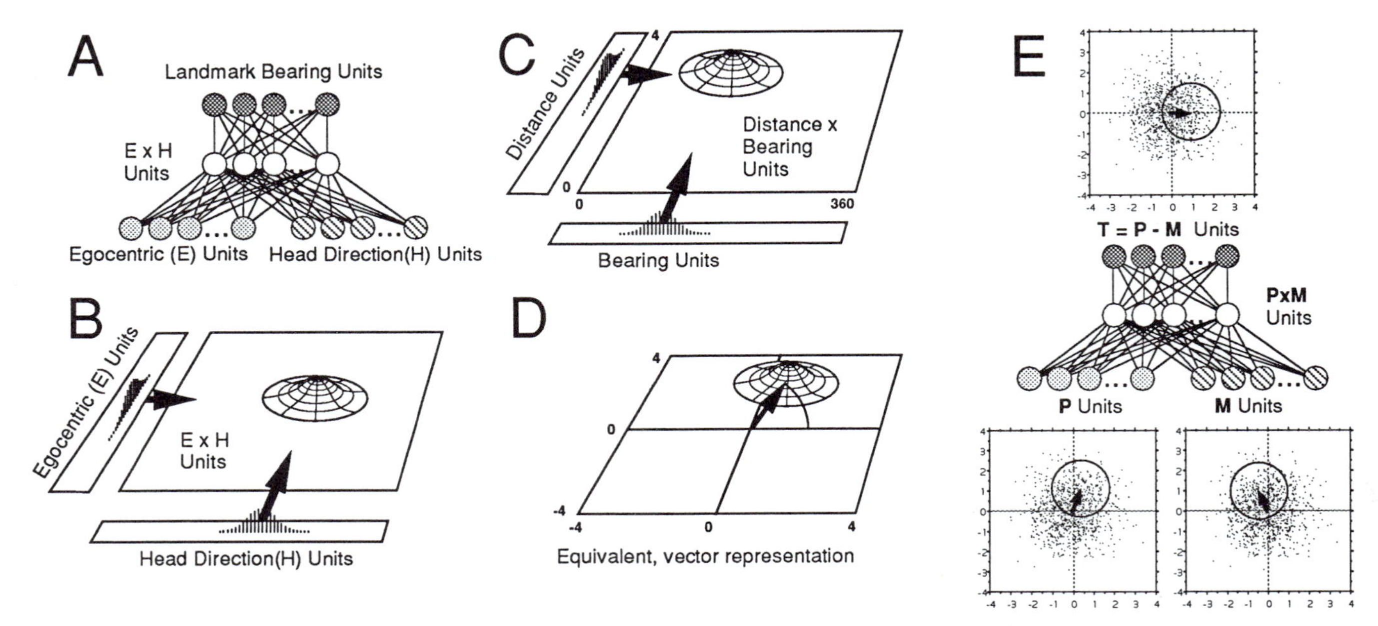

FIGURE 37.2 (A) Cells encoding the allocentric bearing of a landmark currently being attended to (landmark bearing units) can be imagined to arise through associative mapping using a standard three-layer network. (B) The necessary internal representation is a cross-feature code in which different units are activated by different combinations (i.e., two-dimensional radial basis functions) of egocentric (E) bearing (e.g., retinal location) and head (H) bearing relative to the allocentric direction reference. (C) Vector cells can be thought of as representing the cross-feature code for distance and allocentric bearing of the landmark currently being attended to. (D) The cartesian equivalent of vector cells illustrating that vector cells are place cells, although the converse may or may not be true. (E) Hypothetical scheme for vector subtraction using associative mapping in a standard three-layered neural network. The input representation is assumed to be two sets of vector-encoding (place) cells. The output is the desired trajectory (**T**). The sufficient internal representation is a set of units reflecting the interaction (cross-feature code) of the currently perceived vector (**P**) and the remembered vector (**M**). Such cells would have place fields that are modulated by the location of the upcoming movement.

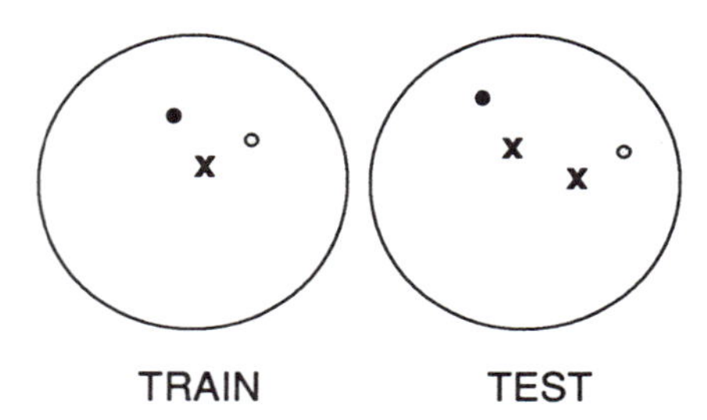

FIGURE 37.3 Illustration of the experiment by Collett, Cartwright, and Smith (1986), in which gerbils were trained to dig for food at a fixed location in relation to two different cylindrical landmarks in an otherwise essentially featureless environment. After successful training, the separation of the landmarks was increased. Rather than searching at the topologically correct single location, the animals searched at two different sites, each at the correct distance and bearing from one of the landmarks. The authors concluded that landmark vectors from the goal are stored separately for each landmark and are used subsequently to compute trajectories from different locations to the goal by vector subtraction.

The classical prediction would have been a circular search pattern. The conclusion offered by Collett, Cartwright, and Smith (1986) was that animals encode vectors to each landmark independently. When faced with a conflict, rather than taking the vector sum of the possible solutions, each vector "votes" for its target, and the location with the most votes wins. When there is no clear winner, the system is unstable and undergoes transitions (attentional shifts) among the different solutions. How such voting may occur is suggested later.

Head-direction cells and the directional reference system

The mammalian brain contains a tightly coupled system of neurons that produces an accurate representation of bearing within a spatial reference framework

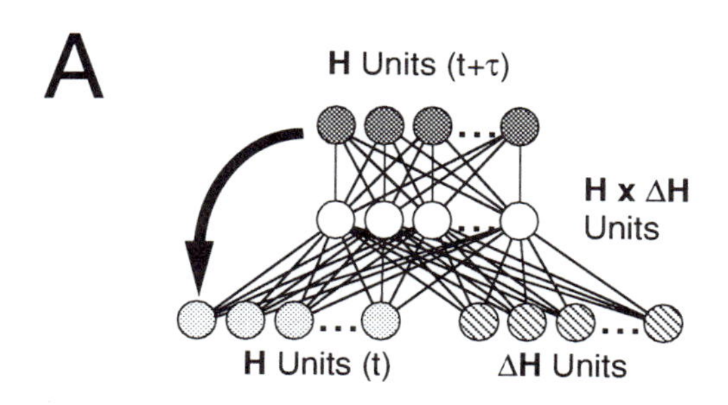

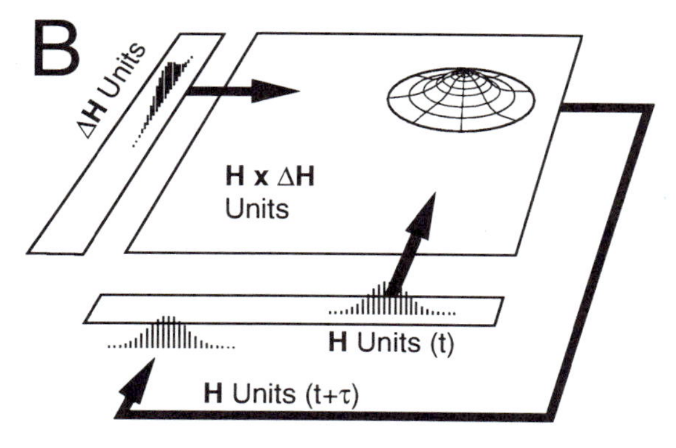

FIGURE 37.4 Head-direction (H) cells can be observed in several parts of the brain, including dorsal presubiculum, posterior neocortex, and lateral dorsal thalamus. Directional selectivity could be maintained solely on the basis of inertial (vestibular) cues using an associative mapping function relating the head direction (H) at time $t + \tau$ to that at time t by means of the angular deviation (ΔH). Angular deviation is equivalent to angular velocity over a fixed time interval (τ).

(figure 37.4). The complete anatomy of this system is still unclear, but it includes cells in both the subicular complex (Taube et al., 1990a, b) and parietal and posterior cingulate (retrosplenial) cortex (Chen et al., 1990) as well as the lateral dorsal (LD) thalamic nuclei (Mizumori and Williams, 1993), which relay information from the pretectum. Cells within this system are tuned (some sharply, some coarsely) to the orientation of the rat's head in space and hence are referred to as *head-direction* (H) *cells*. If the tuning curves are computed at different locations within the environment, the preferred orientation vectors of H cells are everywhere parallel. In a familiar environment, the direction tuning is controlled by the prominent visual landmarks; however, if the landmark array is rotated slowly in the animal's presence, the direction tuning sometimes exhibits a clear lag relative to the visual cues. The tuning persists in darkness but may drift. If the animal is moved passively, from a familiar environment into an unfamiliar one, directional tuning is maintained with respect to the inertial reference framework in the unfamiliar environment, so long as the movement is smooth and continuous. If the animal is intentionally disoriented during transit, H cells adopt an arbitrary preferred orientation. Tuning curves are equally distributed around 360°. In all cases in which two or more H cells with different preferred orientations have been simultaneously recorded, a shift in the preferred direction of one cell is accompanied by precisely equivalent shifts in the others. H-cell tuning is a function of azimuth only; pitch and roll are irrelevant.

McNaughton, Chen, and Markus (1991) hypothesized that the primary source of information for H cells is the vestibular apparatus and suggested a computational model for how bearing is updated in this system. The model depends on the fact that there is a one-to-one relationship between initial bearing and final bearing after a rotation of a fixed number of degrees. This can be expressed as a linear associative mapping and solved by a simple neural network with an internal layer in which two-dimensional response functions (gaussians) are used to represent all possible combinations of initial bearing (H) and angular displacement (ΔH) (i.e., angular velocity) over some small fixed interval τ. With the appropriate synaptic weightings, such H × ΔH cells can be used to elicit the representation of the correct target orientation from the output cells. In this case, the output units either are coupled to the H input or are, in fact, the same units. Such a network is capable of behaving as an inertial compass. It acts as an integrator of angular velocity signals without the necessity of resetting each 360° and without the leakage problem that is often associated with physical integration devices. Cells with H × ΔH response profiles were subsequently identified in the posterior parietal cortex of freely behaving rats. These cells had preferred allocentric orientations but showed higher activity during rotation in one direction through the preferred bearing than during rotation in the other direction.

Like any inertial compass, the head-direction system is subject to cumulative error unless periodically corrected with visual sightings. Presumably, this is why H cells tend to drift out of calibration in total darkness, in the absence of local tactile cues. McNaughton, Chen, and Markus (1991) suggested that this correction is accomplished by the formation of associations between active H cells and coactive cells sensitive to visual stimuli in egocentric coordinates (i.e., local view cells). Such associations, once established, would enable visual stimuli to override the current state of an H system that had drifted out of calibration in the absence of visual input. There was a clear prediction from this hypothesis: In a familiar enclosed environment, under conditions of illumination, the vestibular input during transient rotation of the environment (including the floor) should be ignored. The H-cell tuning should follow the rotation because it has become coupled to the visual stimuli. In an unfamiliar environment, however, there will have been insufficient associative strength accumulated for this coupling to occur. The vestibular cues hence will not be overridden, and the H-cell tuning will be maintained with respect to the inertial reference framework. In three separate experiments in which the opportunity has arisen to test this hypothesis, the results have been consistent with the prediction (see color plate 15).

Landmark stability and spatial learning

The perceived stability of objects within the overall spatial reference framework is essential for their use as landmarks. For example, rats are able to learn to search at some bearing and distance from an object if the object is maintained in constant relation to a white curtain located on one wall of the enclosure, several meters away. If the object is moved from trial to trial, the rats cannot learn this problem unless the food is located at the object itself (Biegler and Morris, 1993). In contrast, we recently have found that when several landmarks define the food location, as in Collett, Cartwright, and Smith's experiments (1986), but the whole array is rotated from trial to trial, the animals quickly learn to ignore the fixed remote visual stimuli and orient on the basis of the landmarks (Gothard and McNaughton, unpublished). This appears to be a classical case of figure-ground ambiguity. We presume that, under these conditions, the H system becomes aligned on the basis of the landmark array.

The H system as fundamental determinant of the hippocampal spatial code

Muller, Kubie, and Ranck (1987) studied hippocampal place cells under conditions in which the visuospatial information was intentionally reduced to a single landmark, a white panel covering 90° of the wall in an otherwise homogeneous, gray, cylindrical environment. Perfectly normal place fields were observed under this situation. Once the animal has been familiarized with such an environment, the place cells are under control of the cue card; thus, if the card is rotated prior to introducing the animal, the firing fields of hippocampal cells rotate accordingly. In an attempt to replicate these findings, we made an interesting serendipitous discovery. If, from the first experience, the animals are disoriented before introduction into the cylinder from a sound attenuating box, place fields sometimes exhibit rotational instability in the cylinder, even when the cue card is held in a fixed location relative to the laboratory. The cells fire at the correct distance from the walls but often at a random bearing (see color plate 16). The relationships among fields of simultaneously recorded cells is preserved, although the bearing reference of the whole system can sometimes drift slowly during a trial. The cells behave as if the animal were in the dark. Indeed, when H cells are recorded simultaneously with place cells, the place field bearings and the H-cell orientation preference are consistent with one another (see color plate 17). The most likely explanation seems to be that, when the animal is placed into the cylinder after disorientation, its H system is oriented randomly. Consequently, the animal perceives that the cue card, the only polarizing visual stimulus, is located at different positions from trial to trial. The cue card thus is rejected from the spatial reference framework. Preliminary confirmation of this has been obtained by first training animals without disorientation. They were moved smoothly, on an open platform, into the cylinder on each trial. Place fields were consistent from trial to trial, even if the disorientation procedure was subsequently initiated. Moreover, rotation of the cue card showed that it had gained control of the spatial reference framework under these conditions, presumably because of the prior repeated consistent association it had developed with the H system.

Vector representation

To compute a trajectory using vector subtraction, a minimum requirement is the existence of vector-encoding cells. Both hippocampus and parietal cortex (McNaughton, Leonard, and Chen, 1989) have place cells (whether they represent the same, if any, stimulus invariants is an open question). The proposition that a cell is location-specific, irrespective of egocentric orientation, is equivalent (at least until tested) to the proposition that it encodes the location and bearing of at least one fixed landmark. With a limited number of ad hoc assumptions, this proposition actually can account for several key observations on hippocampal place cells. Let us formulate the specific hypothesis as follows:

1. Each place cell represents a vector (i.e., a distance and bearing) (see figure 37.2). The bearing is set by the H system. Distance is presumably derived from various sources, including stereopsis, vergence, motion parallax, size constancy, texture gradients, and so on, and is scaled in some asymptotic fashion (i.e., nearer distances are more densely represented than far ones, and the breadth of tuning for distance increases with distance).

2. In any particular environment, place-cell vectors may become bound to one or more landmarks (typically one, occasionally two, rarely more than two). Obviously, it cannot be bound to two landmarks in the same place as it is assumed that each cell encodes only one distance-bearing pair. The binding is determined by the particular set of features to which a given cell is responsive, however, and sometimes there will be more than one landmark with the correct feature set to activate the cell when the animal is in the appropriate position.

Vector subtraction

The proposition that the trajectory vector (**T**) is the difference between the current landmark vector (**P**) and the stored landmark vector (**M**) requires a means for vector subtraction. There have been several proposals regarding how this might be accomplished. For example, Touretzky, Redish, and Wan (1993) suggested that vectors are coded using populations of cells that are sinusoidally tuned to different bearings and whose firing rate is modulated by distance. The overall population firing rate thus signals distance, whereas bearing is determined by which cells are firing fastest. This constitutes a distributed neuronal representation of vectors in phasor form. Subtraction is performed by first rotating **M** by 180° and then adding it to **P**.

We propose an alternative solution, based on the associative mapping principles that are the bread and butter of neural networks. A mapping is a function of one or more variables that assigns to each element in the domain (input) a unique value in the range (output). For vector subtraction, the trajectory vector **T** equals $f(\mathbf{P}, \mathbf{M})$, where the function f is vector subtraction. This maps each possible combination of **P** and **M** onto one and only one value of **T**. The mapping is not linearly separable. The internal representation (so-called hidden layer) necessary to solve it must contain units that encode $\mathbf{P} \times \mathbf{M}$ interactions. This is the same principle as the gain fields of the posterior parietal cortex (Andersen, Essick, and Siegel, 1985), which encode interactions between retinal location (**R**) and eye position (**E**) and which have been shown (Zipser and Andersen, 1988) to be a sufficient representation for transforming retinal coordinates into head (**H**) coordinates. If this coordinate transformation is written in vector notation ($\mathbf{H} = \mathbf{R} - \mathbf{E}$), it becomes clear that vector subtraction is not something unknown to parietal cortex. Although Andersen and his colleagues have used various learning algorithms to generate the properties of the internal representation, the necessary connection strengths can be "hand-wired" once one knows what is required to perform the mapping. In other words, a simple developmental rule could generate the necessary parietal cortical circuitry, without backpropagation or other learning mechanisms, although plasticity of these connections is by no means excluded.

To compute $\mathbf{P} - \mathbf{M}$, both vectors must be available as input to the associative mapping network performing the subtraction. In other words, a representation of the stored vector must be activated from memory for at least a brief time during the planning of the movement. It might be the case that the **P** and **M** vectors need to be represented in separate populations of neurons; however, there is no a priori reason for this. The problem can be solved by sequential activation with appropriate delay units. Although not interpreted in the framework of vector subtraction, precisely this kind of predictive activity was recorded in some parietal neurons of monkeys before intentional eye movements during active fixation tasks (Duhamel, Colby, and Goldberg, 1992). Just prior to saccades, the receptive fields of such cells shift transiently to the location on the retina at which the image of the object of attention will fall after the eye movement.

At any given instant, which place (vector) cells are active at a given location is determined by which landmarks are currently attended to (attention, for example, is known to be a powerful determinant of parietal cortical neuronal activity). Two somewhat different versions of the hypothesis emerge from this point. One might assume that cells bound to different landmarks at a given location must not be coactive, as this would mean that there is more than one **P** vector active. This would lead to the classical binding problem for the subtraction operation because there would be uncertainty regarding which vector to subtract from which. It turns out, however, that this problem might not be insurmountable. If n different **P** and **M** vectors are coactive, there are n^2 different ways they can be subtracted; however, n of these solutions will be identical (i.e., when each **M** vector is subtracted from the correct **P** vector). The other $n(n-1)$ solutions will differ from the correct solution as well as differing, in general, from one another. Simple thresholding could thus ensure that only the correct **T** cells will be activated.

Construction of vector cells

The associative mapping idea provides a simple hypothesis as to how place cells could come to encode landmark bearings in allocentric coordinates (i.e., independent of which direction the animal faces). The basis of the coordinate transformation from egocentric to allocentric is the same principle that we have applied to the other transformations described earlier. Rats do not make tracking saccades. Thus, the egocentric orientation (E) of the attended landmark is specified essentially by its bearing from the head axis, although it is actually derived from retinal position. The orientation of the head axis in the inertial reference frame (H) is given by head-direction cells. The allocentric bearing is thus a simple linear combination of E and H, and the required internal representation is a set of cells encoding E $\times$ H interactions (see figure 37.2). In a sense, these are identical to the cells already found by Andersen, Essick, and Siegel (1985) in monkey parietal cortex, which encode retinal location modulated by head orientation (i.e., gain fields). The distance component of landmark vectors (place cells) may come from one of several sources of information, including stereopsis, vergence, image size, texture gradients, motion parallax, and path integration. Such cells should respond to the attended landmark when it is at a certain distance, irrespective of bearing, and hence should have circular firing fields around the attended landmark.

Vector-based account of some existing data

In the following we discuss how the foregoing conceptual framework can provide a consistent account for certain key experimental results from single neuron recording experiments.

1. When the H system is disoriented in a landmark-poor environment (simple cylinder), place cells shift their locations according to the current H reference. This must happen if the H system provides the landmark bearing.

2. Reversible inactivation of the LD thalamus, which seems to be a principal relay in the H system, disrupts hippocampal place fields and choice accuracy on a spatial memory problem (Mizumori and Williams, 1992).

3. Removal of any subset of landmarks does not disrupt place fields. From single-neuron recording, one might simply assert that the correct landmarks have not been removed; however, the same result is obtained if the complete array of controlled landmarks is removed, after the rat has first been allowed to get its bearings with the landmarks present (O'Keefe and Speakman, 1987). A landmark may include something as basic as the slope, in allocentric coordinates, of the tangent at a patch of the circular curtain that surrounded the apparatus. Thus, one cannot conclude that the complete set of landmarks has been removed. The controlled (asymmetrical) landmarks are, however, necessary to establish the initial reference direction. For example, this would explain the data illustrated in color plates 15–17. On the other hand, failure of a landmark vector cell to shut off when its demonstrated landmark is removed would require the assumption of autoassociative memory for particular landmark vector configurations. Vector subtraction is, of course, invertible, and the representation of current position relative to a landmark can, in principle, be computed from the difference between the given vector at the starting location (**P**) and the movement (**T**) just made ($\mathbf{M} = \mathbf{P} - \mathbf{T}$). This could be a basis for path integration and would also explain the maintenance of correct spatial firing by hippocampal cells in darkness, under conditions in which the initial position is known. In this sense, a "landmark" may, on occasion, be nothing more than a start location, with no particular sensory feature associated with it.

4. Permutation of the spatial arrangement of landmarks, however, usually causes radical rearrangement

of place fields. This is a straightforward prediction of the vector hypothesis.

5. When the partition between the familiar and unfamiliar halves of an enclosed rectangular (120 × 60–cm) arena is removed, the cells that become active in the novel half are mostly drawn from the pool of cells that were inactive in the familiar half. This could be because the cells that become active in the novel half represent vectors whose distance components are greater than the dimensions of the familiar half, and they become preferentially bound to the landmarks in the familiar half through attentional mechanisms.

6. When rats run on linear tracks to and from specific goal locations (e.g., the eight-arm maze), place fields are more robust than normal but also depend strongly on the direction in which the rat is moving on the track. This is easily accounted for if it is assumed that the rat's attention is fixated strongly on different landmarks on the inward and outward journeys. When rats forage randomly for food, attention presumably shifts among landmarks, making the activity of any one cell less robust but omnidirectional.

7. In a cylindrical environment, placing barriers in the middle of a place field appears to make the field less robust (Muller, Kubie, and Ranck, 1987). Presumably, attention is focused on the new landmark. Other cells become active, whereas the previous local ones become less active.

8. In a cylindrical environment with two identical white cards located symmetrically (i.e., 180° apart), symmetrical place fields appear only very rarely (Sharp, Kubie, and Muller, 1990). The landmarks are presumably concave wall segments and contrast boundaries. The symmetry is presumably broken by the H-system input, which makes the vectors for equivalent features different, even if the feature is merely the slope of the tangent to any small wall segment. Interestingly, however, which of the two symmetrical locations is adopted depends on the history of how the animal was introduced into the arena. Presumably, the direction reference is set at the first moment of entry, partly on the basis of the pre-existing setting.

Predictions for new experiments

The vector encoding hypothesis also leads to several testable predictions, some of which would be very difficult to account for under other current hypotheses.

1. In an experiment involving two explicit landmarks (e.g., Collett, Cartwright, and Smith, 1986; see figure 37.3), the place fields will be larger with distance from the landmarks and elongated toward their respective landmarks (due to nonlinear distance scaling but linear angle scaling).

2. In the experiment by Collett's group, when the landmark separation was increased, the animals searched alternately at two locations, each at the correct distance and bearing from one of the landmarks (rather than at the one best-compromise location). The hypothesis that place cells are landmark vectors predicts either that different sets of cells will be active when the animal is searching near one or the other landmark or that the same cells will be active but their respective place fields will shift in two coherent sets, one bound to each landmark (figure 37.5). The former is the stronger prediction. The alternative (nonvector) hypothesis predicts that the entire spatial reference framework (i.e., the positions of all place fields) shifts alternately between the two landmarks after their separation is increased. Failure of this critical experiment would cast the vector-landmark binding hypothesis into serious doubt.

3. The vector-landmark binding hypothesis for place cells leads to the following interesting, if somewhat bizarre, prediction: If information about bearing is removed, leaving only landmark and distance information, and if place cells continue to fire, they will fire in a circle of fixed radius about their bound landmarks. As discussed previously, disruption of the LD thalamic relay for H information leads to disruption of place fields (Mizumori and Williams, 1992); however, those studies were conducted on an eight-arm maze, which

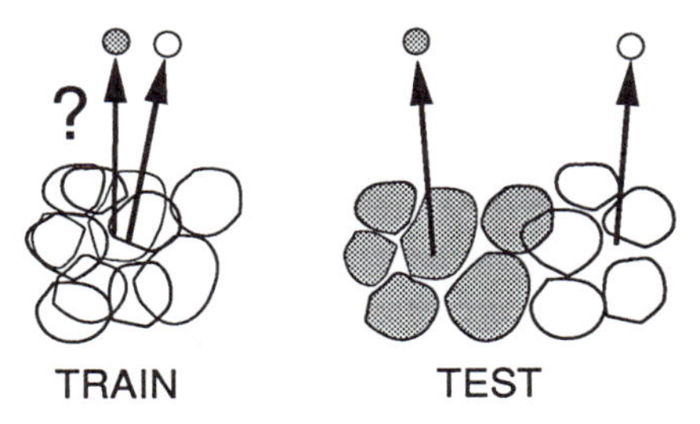

FIGURE 37.5 The vector hypothesis makes a strong prediction about what should happen if place cells are recorded during a repetition of the Collett, Cartwright, and Smith (1986) experiment. When the separation between landmarks is increased, the place fields should segregate according to the landmarks to which they are bound.

would have precluded observing such an effect, because of both the spatial constraint on sampling as well as the unusual firing characteristics of place cells on linear tracks.

4. During the time that the trajectory is being planned, cells with place fields near the goal (**M** vectors) will become active. This should occur in parietal cortex and may also occur in the hippocampus. **M**-vector activity might be simultaneous with the **P**-vector activity, or it might temporarily override it. In a delayed, two-tone, conditional spatial response task, we expect to observe **M**-vector activity (i.e., activity of cells with place fields at the target location) at some point during the tone-go-response sequence, most likely only for a brief period after the tone but possibly throughout the interval or immediately following the go signal. **M**-vector activity does not need to persist once the trajectory has been computed, although it might do so; however, on the basis of the delayed response data in primates (e.g., Goldman-Rakic, Funahashi, and Bruce, 1990), it seems more likely that the actual **T** vectors will persist. A result that can be interpreted as **M** vector activation in parietal visual cells was described above (Duhamel, Colby, and Goldberg, 1992).

5. Parietal cortex and possibly hippocampus will contain neurons that encode landmark vectors but that apparently are modulated by the goal of the impending movement. This is equivalent to saying that there is **P** $\times$ **M** interaction. In a task involving two different goals, these cells will fire more when the animal is in the place field and one of the goals is the impending target. This is precisely equivalent to the parietal cortical case of visual cells with retinal receptive fields that are modulated by the goal of the impending foveation movement (i.e., they are modulated by eye position).

Conclusions

A directional reference signal that is maintained internally on the basis of inertial signals generated by self-motion appears to be a fundamental determinant of the cognitive representation of space. These signals provide, in principle, a basis for simple computation of appropriate optimum trajectories to goals that is consistent with behavioral observations. Although much work remains to be done, the principles of associative mapping and cross-feature coding appear adequate to enable the appropriate internal representations needed for these computations to be carried out. There appears to be no need to invoke special-purpose circuitry for these calculations.

ACKNOWLEDGMENTS We thank W. E. Skaggs, K. Gothard, and C. A. Barnes for useful discussions related to this chapter. This work was supported by grants MH46823 and NS09052 from the Public Health Service and by the Office of Naval Research.

REFERENCES

Andersen, R. A., G. K. Essick, and R. M. Siegel, 1985. Encoding of spatial location by posterior parietal neurons. *Science* 230:456–458.

Biegler, R., and R. G. M. Morris, 1993. Landmark stability is a prerequisite for spatial but not discrimination learning. *Nature* 361:631–633.

Chen, L. L., B. L. McNaughton, C. A. Barnes, and E. R. Ortiz, 1990. Head-directional and behavioral correlates of posterior cingulate and medial prestriate cortex neurons in freely moving rats. *Soc. Neurosci. Abstr.* 16:441.

Collett, T. S., B. A. Cartwright, and B. A. Smith, 1986. Landmark learning and visuospatial memories in gerbils. *J. Comp. Physiol. [A]* 158:835–851.

Duhamel, J. R., C. L. Colby, and M. E. Goldberg, 1992. The updating of the representation of visual space in parietal cortex by intended eye movements. *Science* 255:90–92.

Felleman, D. J., and D. C. Van Essen, 1991. Distributed hierarchical processing in the primate cerebral cortex. *Cerebral Cortex* 1:1–47.

Foster, T. C., C. A. Castro, and B. L. McNaughton, 1989. Spatial selectivity of rat hippocampal neurons: Dependence on preparedness for movement. *Science* 244:1580–1582.

Gladwin, T., 1970. *East Is a Big Bird: Navigation and Logic on Puluwat Atoll.* Cambridge, Mass.: Harvard University Press.

Goldman-Rakic, P. S., S. Funahashi, and C. J. Bruce, 1990. Neocortical memory circuits. *Cold Spring Harb. Symp. Quant. Biol.* 55:1025–1038.

Matthews, B. L., K. A. Campbell, and S. A. Deadwyler, 1988. Rotational stimulation disrupts spatial learning in fornix-lesioned rats. *Behav. Neurosci.* 102:35–42.

McNaughton, B. L., L. L. Chen, and E. J. Markus, 1991. "Dead reckoning," landmark learning, and the sense of direction: A neurophysiological and computational hypothesis. *J. Cogn. Neurosci.* 3(2):190–202.

McNaughton, B. L., B. Leonard, and L. Chen, 1989. Cortical-hippocampal interactions and cognitive mapping: A hypothesis based on reintegration of the parietal and inferotemporal pathways for visual processing. *Psychobiology* 17(3):230–235.

McNaughton, B. L., J. O'Keefe, and C. A. Barnes, 1983. The stereotrode: A new technique for simultaneous isolation of several single units in the central nervous system from multiple unit records. *J. Neurosci. Methods* 8:391–397.

MITTELSTAEDT, M. L., and H. MITTELSTAEDT, 1980. Homing by path integration in a mammal. *Naturwissenschaften* 67:566.

MIZUMORI, S. J. Y., and J. D. WILLIAMS, 1992. Interdependence of hippocampal and lateral dorsal thalamic representations of space. *Soc. Neurosci. Abstr.* 18:708.

MIZUMORI, S. J. Y., and J. D. WILLIAMS, 1993. Directionally selective mnemonic properties of neurons in the lateral dorsal nucleus of the thalamus of rats. *J. Neurosci.* 13: 4015–4028.

MULLER, R. U., J. L. KUBIE, and J. B. RANCK, JR., 1987. Spatial firing patterns of hippocampal complex-spike cells in a fixed environment. *J. Neurosci.* 77:1935–1950.

O'KEEFE, J., and J. DOSTROVSKY, 1971. The hippocampus as a spatial map. Preliminary evidence from unit activity in the freely moving rat. *Brain Res.* 34:171–175.

O'KEEFE, J., and L. NADEL, 1978. *The Hippocampus as a Cognitive Map.* Oxford: Clarendon Press.

O'KEEFE, J., and A. SPEAKMAN, 1987. Single unit activity in the rat hippocampus during a spatial memory task. *Exp. Brain Res.* 68:1–27.

QUIRK, G. J., R. U. MULLER, and J. L. KUBIE, 1990. The firing of hippocampal place cells in the dark depends on the rat's recent experience. *J. Neurosci.* 10:2008–2017.

SHARP, P. E., J. L. KUBIE, and R. U. MULLER, 1990. Firing properties of hippocampal neurons in a visually symmetrical environment: Contributions of multiple sensory cues and mnemonic properties. *J. Neurosci.* 10:3093–3105.

SWANSON, L. W., C. KÖHLER, and A. BJORKLUND, 1987. The limbic region: I. The septohippocampal system. In *Handbook of Chemical Neuroanatomy: 5. Integrated Systems of the CNS* (part 1), T. H. A. Bjorklund and L. W. Swanson, eds. Amsterdam: Elsevier.

TAUBE, J. S., R. U. MULLER, and J. B. RANCK, JR. 1990a. Head direction cells recorded from the postsubiculum in freely moving rats: I. Description and quantitative analysis. *J. Neurosci.* 10:420–435.

TAUBE, J. S., R. U. MULLER, and J. B. RANCK, JR. 1990b. Head direction cells recorded from the postsubiculum in freely moving rats: II. Effects of environmental manipulations. *J. Neurosci.* 10:436–447.

TOLMAN, E. C., 1948. Cognitive maps in rats and men. *Psychol. Rev.* 55:189–208.

TOURETZKY, D. S., A. D. REDISH, and H. S. WAN, 1993. Neural representation of space using sinusoidal arrays. *Neural Computation* 5:869–884.

WILSON, M. A., and B. L. MCNAUGHTON, 1993. Dynamics of the hippocampal ensemble code for space. *Science* 261: 1055–1058.

ZIPSER, D., and R. A. ANDERSEN, 1988. A back-propagation programmed network that simulates response properties of a subset of posterior parietal neurons. *Nature* 331:679–684.

38 Computational Motor Control

MICHAEL I. JORDAN

ABSTRACT Some of the computational approaches that have been developed in the area of motor control are reviewed in this chapter. The focus is on problems relating to sensorimotor transformations and the adaptive control of dynamic systems. Discussion is organized around three related topics—optimization principles, internal models, and motor learning—each of which has broad relevance to issues in motor control.

The study of motor control is fundamentally the study of sensorimotor transformations. For the motor control system to move its effectors to apply forces to objects in the world or to position its sensors with respect to objects in the world, it must coordinate a variety of forms of sensory and motor data. These data are generally in different formats and may refer to the same entities but in different coordinate systems. Transformations between these coordinate systems allow motor and sensory data to be related, closing the sensorimotor loop. Equally fundamental is the fact that the motor control system operates with dynamic systems whose behavior depends on the way energy is stored and transformed. The study of motor control is therefore also the study of dynamics.

These two interrelated issues—sensorimotor transformations and dynamics—underlie much of the research in the area of motor control. Many of the questions asked in the empirical study of motor control revolve around issues such as the nature of the sensorimotor transformations involved in particular tasks, the way in which these transformations adapt to reflect changes in the environment, the way in which motor plans are converted into time-varying behavior, and the problem of coordinating a distributed control system in the face of sensorimotor delays. In this chapter, I discuss what is at stake from a theoretical point of view. Some of the computational tools that have been developed to explore these issues are addressed, and models of specific phenomena are described. The discussion focuses on three related topics—optimization principles, internal models, and motor learning. Let us begin with optimization theory, which provides a general set of tools for dealing with controlled dynamic systems.

MICHAEL I. JORDAN Department of Brain and Cognitive Sciences, Massachusetts Institute of Technology, Cambridge, Mass.

Optimization

Let us consider the act of reaching to grasp an object. How should we describe the trajectories that the hand follows during reaching movements? One theoretical approach to this problem has been to utilize the tools of optimization theory. Rather than describing the kinematics of a movement directly, in terms of the time-varying values of positions or angles, the movement is described more abstractly, in terms of a global measure such as total efficiency, smoothness, accuracy, or duration. For example, we might postulate that the trajectory followed by the hand is the one that minimizes the total energy expended during the movement. The theoretical goal is to formulate a single such postulate, or a small set of postulates, that account for a wide variety of the data on reaching.

KINEMATICS OF REACHING Experimental observations of unconstrained point-to-point reaching movements have noted that several aspects of movements tend to remain invariant, despite variations in movement direction, movement speed, and movement location (Morasso, 1981; Flash and Hogan, 1985). First, as shown in figure 38.1, the motion of the hand tends to follow roughly a straight line in space. This observation is not uniformly true; significant curvature is observed for certain movements, particularly vertical movements and movements near the boundaries of the workspace (Soechting and Lacquaniti, 1981; Atkeson and Hollerbach, 1985; Uno, Kawato, and Suzuki, 1989). The tendency to make straight-line movements, however, characterizes a reasonably large class of

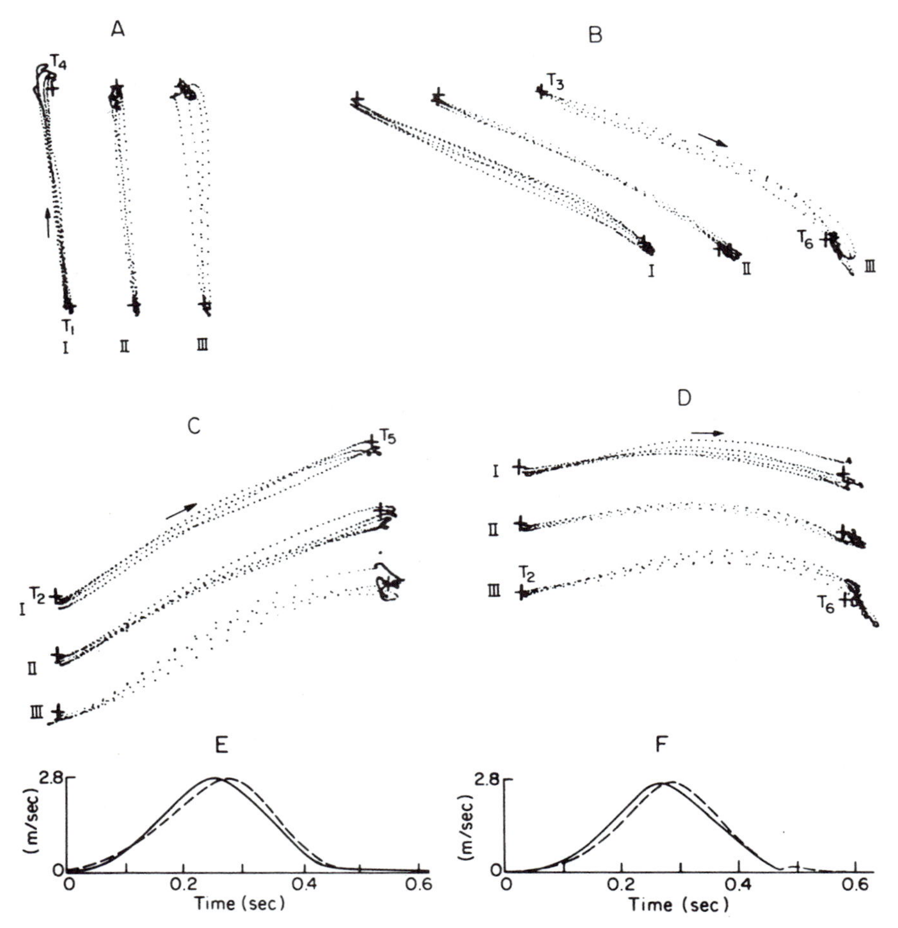

FIGURE 38.1 Paths and trajectories of movements in the plane. (A–D) Paths for slow (I), medium (II), and fast (III) movements. (E, F) hand speed from the fast (solid) and slow (dashed) movements in (C) and (D), respectively. (Reprinted from Flash, 1987, by permission.)

movements and is somewhat surprising given that the muscles act in rotating coordinate systems attached to the joints.

Second, the movement of the hand is smooth. Higher derivatives of the hand motion, such as the velocity and acceleration, tend to vary smoothly in time. Consider the plot of tangential speed shown in figure 38.1E and F. Scrutiny of the curve in the early phase of motion reveals that the slope of the plot of speed against time is initially zero and increases smoothly. This is striking given that the positional error is maximal at the beginning of the movement.

Third, the shape of the plot of hand speed is unimodal and roughly symmetrical (bell-shaped). There are exceptions to this observation as well, particularly for movements in which feedback plays an important role (Beggs and Howarth, 1972; MacKenzie et al., 1987; Milner and Ijaz, 1990) but, again, this observation characterizes a reasonably large class of movements.

DYNAMIC OPTIMIZATION Consider the mathematical machinery that is needed to describe optimal movements. Letting T denote the duration of a movement and $x(t)$ denote the value of the degree of freedom x at time t, a movement is a function $x(t)$, $t \in [0, T]$. There are an infinite number of such functions. An optimization approach proposes to choose between these functions by comparing them on the basis of a measure of cost. The cost function is a *functional*—that is, a function which maps functions into real numbers. To every movement corresponds a number that provides a basis of comparison between movements. The cost function in dynamic optimization is generally taken to be of the following form:

$$\mathcal{J} = \int_0^T g(x(t), t)\, dt \qquad (1)$$

where $g(x(t), t)$ is a function that measures the instantaneous cost of the movement. The instantaneous cost (g) typically quantifies aspects of the movement that are considered undesirable, such as jerkiness or error or high-energy expenditure. Integrating the instantaneous cost provides the total measure of cost ($\mathcal{J}$).

The mathematical techniques that have been developed for optimizing expressions such as equation 1 fall into two broad classes: those based on *calculus of variations* and those based on *dynamic programming* (Kirk, 1970). Both classes of techniques provide mathematical conditions that characterize optimal trajectories. For certain classes of problems, these conditions provide equations that can be solved once and for all, yielding closed-form expressions for optimal trajectories. For other classes of problems, the equations must be solved numerically.

Optimization Principles One of the most influential models of reaching kinematics is the minimum jerk model proposed by Hogan (1984) and Flash and Hogan (1985). Jerk is the third derivative of position, or the rate of change of acceleration. Letting $x(t)$ denote the position at time t, the minimum jerk model is based on the following cost function:

$$\mathcal{J} = \int_0^T \left[\frac{d^3x}{dt^3}\right]^2 dt \qquad (2)$$

where T is the duration of the movement. Using the calculus of variations, Hogan showed that the trajectory that minimizes this cost function is of the following form:

$$x(t) = x_0 + (x_f - x_0)\,[10(t/T)^3 - 15(t/T)^4 + 6(t/T)^5] \qquad (3)$$

where x_0 and x_f are the initial and final positions, respectively. The minimum jerk profile is shown in figure 38.2.

Why jerk, rather than acceleration or velocity? Flash and Hogan (1985) argue that the trajectories obtained by penalizing these lower-order derivatives are not sufficiently smooth. In particular, the minimum squared acceleration profile has a nonzero acceleration at the initial time (see figure 38.2). As we noted earlier, the behavioral data show that movements have zero initial acceleration.

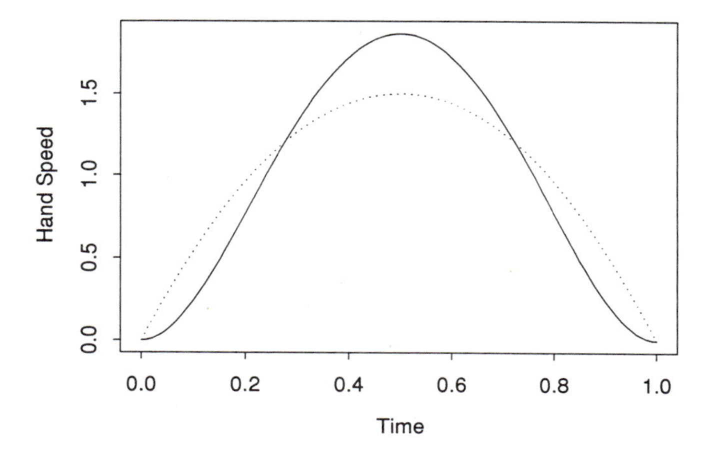

Figure 38.2 Hand speed as a function of time for a minimum squared jerk trajectory (solid line) and a minimum squared acceleration trajectory (dotted line).

The minimum jerk model takes the smoothness of motion as the basic primitive. How does the model fare with respect to the other data on reaching described earlier? First, as seen in figure 38.2, the trajectories predicted by the minimum jerk principle are bell-shaped. Second, as argued by Flash and Hogan (1985), the natural generalization of equation 2 to three spatial dimensions involves taking the sum of the squares of the jerk along each dimension, which is equivalent to three independent minimizations. Thus, the resulting trajectories are all of the form of equation 3, differing only in the values of the initial and final positions. Because these differences simply scale the time-varying part of equation 3, the result is straight-line motion in space.

One aspect of the minimum jerk model that is unsatisfying is the need to prespecify the duration T. Hoff (1992) has extended the minimum jerk model by allowing the duration to be a free parameter. Because longer movements can always be made smoother than short movements, Hoff created a trade-off between duration and smoothness by penalizing duration. This is accomplished via the following cost function:

$$\mathcal{J} = \int_0^T \left\{\gamma\left[\frac{d^3x}{dt^3}\right]^2 + 1\right\} dt \qquad (4)$$

where T is free. The term of unity in the integrand increases the cost of movements as a function of their duration and trades off against the cost due to smoothness. The parameter γ quantifies the relative magnitudes of these two costs. Hoff (1992) has shown that this model can reproduce the results from experiments in which the locations of targets are switched just before the onset of movement (Pélisson et al., 1986; Georgopoulos, Kalaska, and Massey, 1981). The model successfully predicts both movement trajectories and movement durations. (We discuss target-switching experiments further in the following section.)

Uno, Kawato, and Suzuki (1989) have presented data that are problematical for the minimum jerk model. First, they studied trajectories when an external force (a spring) acted on the hand. They found that subjects made curvilinear movements in this case, which is not predicted by the minimum jerk model.[1] Second, they studied movements with via points and observed that symmetrically placed via points did not necessarily lead to symmetrical paths of the hand in space. Finally, they studied large-range movements and observed significant curvilinearity in the paths of motion. These observations led Uno et al. to suggest an alternative optimization principle in which forces play a role. They proposed penalizing the rate of change of torque, a quantity that is locally proportional to jerk (under static conditions). This principle is captured by the following cost function:

$$\mathcal{J} = \int_0^T \sum_{i=1}^{n} \left[\frac{d\tau_i}{dt}\right]^2 dt \qquad (5)$$

where $d\tau_i/dt$ is the rate of change of torque at the ith joint. Uno, Kawato, and Suzuki (1989) showed that this *minimum torque change* cost function predicts trajectories that correspond to the trajectories they observed empirically. Flash (1990) has criticized these results, however, and Dornay and colleagues (1989) have developed a further model that is based on optimizing the change in muscle tension.

Control, Planning, and Learning Note that a description of movement in terms of an optimization principle does not necessarily imply that there is a computational process of optimization underlying the actual control of movement. Optimization theory simply stipulates that the system operates at the minimum of the cost function (the measure that is optimized) but does not commit itself to any particular computational process that puts the system at the minimum. Nonetheless, it is worth asking how optimal behavior might be achieved and maintained. This leads to a perspective in which motor planning or motor learning is identified as the *process* of optimization. The theoretical goal, in this case, is to identify a mechanism that moves the system toward the minimum of the cost function, such that the steps taken by such a mechanism correspond to the adjustments observed in experiments on motor planning or motor learning.

One of the virtues of the minimum jerk model is that the optimal trajectories can be solved for analytically (cf. equation 3). The simplicity of this computation suggests that the motor control system could actually compute the minimum jerk trajectory, essentially enshrining the quintic polynomial as the basic primitive for movement. Flash (1987) has proposed a model along these lines by combining the minimum jerk model with the equilibrium point model of control (cf. Polit and Bizzi, 1979; Feldman, 1986).[2] She proposes that the equilibrium point follows a minimum jerk trajectory and that the hand lags behind. This model has been the subject of controversy because it has been observed (Katayama and Kawato, 1992) that the stiffness values used by Flash were rather large and that unrealistic movement trajectories are obtained if more reasonable stiffness values are used.

If minimum jerk trajectories are the primitives of movement, it is interesting to ask how these primitives might be combined to yield more complex motion. Flash and Henis (1992) have proposed that minimum jerk trajectories can be superimposed. They modeled target-switching experiments by proposing that when a second target appears, the subject initiates a movement that begins at the first target and ends at the second target: This movement is superimposed on the ongoing movement toward the first target.

Hoff (1992) has taken the minimum jerk model in a somewhat different direction. Recall that the minimum jerk model yields a parameterized quintic polynomial, in which the boundary conditions of the movement provide the parameter values. Hoff shows that these boundary conditions can be updated incrementally, yielding a *feedback control law* that produces minimum jerk trajectories on-line. The advantage of the feedback formulation is that the system responds naturally to perturbation, updating the minimum jerk trajectory to reflect the new boundary conditions. Hoff (1992) shows that this model provides an alternative account of the target-switching experiments and can account for a variety of interactions between the transport and grasp components of perturbed reaching movements.

The minimum torque change model and Hoff's online duration model do not yield analytical solutions to the optimization problem, which implies that solutions must be found numerically. There are essentially two approaches to this problem, one based on the calculus of variations and another stemming from dynamic programming. Kawato (1990) and Jordan (1990) have proposed neural network models that solve the equations obtained from the calculus of variations. Solving these equations requires that dependencies backward and forward in time be satisfied. Kawato has proposed a cascade of networks to implement these dependencies, whereas Jordan has proposed a recurrent network that essentially satisfies the same set of dependencies. Hirayama, Kawato, and Jordan (1993) have shown that this approach provides an account of speed-accuracy trade-offs in fast movements—in particular, that it can account for Fitts' law (for alternative accounts of Fitts' law, see Bullock and Grossberg, 1988; Meyer et al., 1990). Hoff (1992) has also provided an account of Fitts' law and has proposed a dynamic programming approach to solving for optimal trajectories.

The direct numerical approaches to solving for optimal trajectories require a complex, time-consuming optimization process. Although models based on such approaches have shown promise in accounting for psychophysical data, their biological plausibility is questionable. Jordan, Flash, and Arnon (in press) have explored an alternative approach to trajectory formation that requires substantially less computation. The model is a *spatial deviation* model, based not on a temporal smoothness constraint such as minimum jerk or minimum torque change but rather on the spatial error from a straight-line path in space. The solution procedure successively approximates a straight-line path by using an error-correcting learning algorithm (see under "Motor learning"). Note that this approach assumes that the system is constructed so as to prefer straight-line paths. Jordan, Flash, and Arnon (in press) show that this assumption, along with the implicit smoothing due to the low-pass characteristics of the muscles and the arm dynamics, suffice to yield smooth, bell-shaped hand-speed profiles. As in Uno, Kawato, and Suzuki's (1989) minimum torque change model, the dynamics of the plant influence the shape of the speed profiles.

Finally, it also is possible to consider models of trajectory formation that are not based on optimization principles. One noteworthy example is the VITE model of Bullock and Grossberg (1988). The VITE model is based on an internal feedback loop that integrates the error between the desired final position of an effector and the estimated actual position of the effector. An unadorned feedback loop would produce a large acceleration in the early phases of the movement, at odds with the empirical data. Thus, Bullock and Grossberg invoke a time-varying feedback gain, which they refer to as a GO *signal*. The GO signal is initially of small amplitude, which effectively attenuates the large acceleration that would otherwise result. Bullock and Grossberg (1988) show that their model captures the basic data on the kinematics of reaching and can also account for a number of subtleties, such as the form of the asymmetries observed in speed profiles as a function of movement duration (cf. Beggs and Howarth, 1972).

Sensorimotor transformations and internal models

The basic task of a control system is to manage the relationships between sensory variables and motor variables. There are two basic kinds of transformations that can be considered: sensory-to-motor transformations and motor-to-sensory transformations. The transformation from motor variables to sensory variables is accomplished by the environment and the musculoskeletal system; these physical systems transform efferent motor actions into reafferent sensory feedback. It is also possible, however, to consider internal transformations, implemented by neural circuitry, that mimic the external motor-to-sensory transformation. Such internal transformations, or *internal forward models*, allow the motor control system to predict the consequences of particular choices of motor actions.[3] We can also consider internal models that perform a transformation in the opposite direction, from sensory variables to motor variables. Such transformations are known as *internal inverse models*, and they allow the motor control system to transform desired sensory consequences into the motor actions that yield these consequences. Internal inverse models are the basic module in open-loop control systems.

INTERNAL FORWARD MODELS Let us consider some examples of forward models and inverse models. Duhamel, Colby, and Goldberg (1992) have recently presented data that imply a role for an internal forward model in the saccadic system. These investigators proposed that one of the roles of the lateral intraparietal area (LIP) of cortex is to maintain a retinal representation of potential saccadic targets. Such a representation simplifies the task of saccade generation because the transformation from a retinal representation to a motor representation is relatively simple. The retinal representation must be updated, however, whenever the eyes move in the head. This updating process requires an internal forward model that embodies knowledge of the retinal effects of eye movements. In particular, for a given eye movement (a motor action), the brain must predict the motion of objects on the retina (the sensory consequences). This predicted motion is added to the current retinal representation to yield an updated retinal representation.

Another role for forward models is to provide a fast internal loop that helps stabilize feedback control systems. Feedback control in biological systems is subject to potential difficulties with stability, because the sensory feedback through the periphery is delayed by a significant amount (Miall et al., in press). If an internal forward model of the motor plant is available, the control system can use the forward model rather than the actual physical plant to predict its errors. In effect, the control system controls the forward model rather than the actual system. Because the loop through the forward model is not subject to peripheral delays, the difficulties with stability are lessened. The control signals obtained within this inner loop are sent to the periphery, and the physical system moves along in tandem. Of course, there will be inevitable disturbances acting on the physical system that are not modeled by the internal model; consequently, the feedback from the actual system cannot be neglected entirely. However, the *predictable* feedback can be canceled by delaying the output from the forward model. Only the unpredictable components of the feedback, which are likely to be small, are used in correcting errors within the feedback loop through the periphery. This kind of feedback control system, which uses a forward model both for mimicking the plant and for canceling predictable feedback, is known in the engineering literature as a *Smith predictor*. Miall and associates (in press) have proposed that the cerebellum acts as a Smith predictor.

Another interesting example of a forward model is found in the literature on speech production. Lindblom, Lubker, and Gay (1979) studied an experimental task in which subjects produced vowel sounds while their jaw was held open by a bite block. Lindblom's group observed that the vowels produced by the subjects had formant frequencies in the normal range, despite the fact that unusual articulatory postures were required to produce these sounds. Moreover, the formant frequencies were in the normal range during the first pitch period, before any possible influence of acoustical feedback. Lindblom, Lubker, and Gay (1979) proposed a model of the control system for speech production that involved placing a forward model of the vocal tract in an internal feedback pathway.

To summarize, an internal forward model can be used to predict errors, to compute fast internal corrections, and to cancel delayed feedback. As pointed out in the classical literature on efference copy (cf. Gallistel, 1980), forward models can also be used to cancel the effects of egomotion on sensory reafference, leading to a stable percept. Finally, forward models can play a role in motor learning, as will be discussed later.

Internal Inverse Models Internal inverse models also play an important role in motor control. A particularly clear example of an inverse model arises in the vestibulo-ocular reflex (VOR), which couples the movement of the eyes to the motion of the head, thereby allowing an organism to keep its gaze fixed in space. This is achieved by causing the motion of the eyes to be equal and opposite to the motion of the head (Robinson, 1981). In effect, the VOR control system must compute the motor command that is predicted to yield a particular eye velocity. This computation is an internal inverse model of the physical relationship between muscle contraction and eye motion.

There are many other such examples of inverse models. Indeed, inverse models are a fundamental module in open-loop control systems: They allow the control system to compute an appropriate control signal without relying on error-correcting feedback.[4]

An important issue to stress in our discussion of internal models is that internal models are not required to be detailed or accurate models of the external world. Often an internal model need provide only a rough approximation of some external transformation to play a useful role. For example, an inaccurate inverse model can provide an initial open-loop push that is corrected by a feedback controller. Similarly, an inaccurate forward model can be used inside an internal feedback loop because the feedback loop corrects the errors. This issue will be addressed again in the following section on motor learning, where we will see that an inaccurate forward model can be used to learn an accurate controller.

Motor learning

In the previous section, we have seen several ways in which internal models can be used in a control system. Inverse models are the basic building block of open-loop control. Forward models can also be used in open-loop control and have additional roles in state estimation and compensation for delays. It is important to emphasize that an internal model is a form of knowledge about the environment (cf. Ghez et al., 1990; Lacquaniti, Borghese, and Carrozzo, 1992). Many motor control problems involve interacting with objects in the external world, and these objects generally have unknown mechanical properties. In addition, there are changes in the musculoskeletal system due to growth or injury. These considerations suggest an important role for adaptive processes. Through adaptation, the motor control system is able to maintain and update its internal knowledge of external dynamics.

In this section, five different approaches to the problem of motor learning will be discussed: direct inverse modeling, feedback error learning, distal supervised learning, reinforcement learning, and unsupervised bootstrap learning. All of these approaches provide mechanisms for learning general sensorimotor transformations. They differ principally in the kinds of data and the kinds of auxiliary supporting structure that they require.

The first three schemes that will be discussed are instances of a general approach to learning known as *supervised learning*. A generic supervised learning system is shown in figure 38.3. A supervised learner requires a *target output* corresponding to each input. The error between the target output and the actual output is computed and is used to drive the changes to the parameters inside the learning system. This process generally is formulated as an optimization problem in which the cost function is one half the squared error:

$$J = \frac{1}{2} \|\mathbf{y}^* - \mathbf{y}\|^2 \tag{6}$$

The learning algorithm adjusts the parameters of the system so as to minimize this cost function. For details on particular supervised learning algorithms, see Hertz, Krogh, and Palmer (1991).

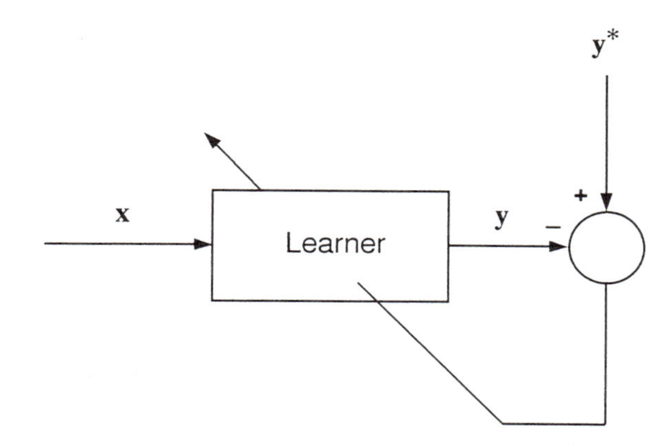

Figure 38.3 A generic supervised learning system. The vector y is the actual output and the vector y* is the target output. The error between the target and the actual output is used to adjust the parameters of the learner.

In the following sections, we assume that the controlled system or *plant* (the musculoskeletal system and any relevant external dynamic systems) is described by a set of state variables $\mathbf{x}[n]$, an input $\mathbf{u}[n]$, and an output $\mathbf{y}[n]$. These variables are related by the following dynamic equation:

$$\mathbf{x}[n+1] = f(\mathbf{x}[n], \mathbf{u}[n]) \tag{7}$$

where n is the time step and f is the *next-state equation*. We also require an *output equation* that specifies how the output $\mathbf{y}[n]$ is obtained from the current state:

$$\mathbf{y}[n] = g(\mathbf{x}[n]) \tag{8}$$

We use the notation $\mathbf{y}^*[n]$ to refer to a desired value (a target value) for the output variable and the notation $\hat{\mathbf{x}}[n]$ to refer to an internal estimate of the state of the controlled system.

Direct Inverse Modeling How might a system acquire an inverse model of the plant? One approach is to present various test inputs to the plant, observe the outputs, and provide these input-output pairs as training data to a supervised learning algorithm by reversing the role of the inputs and outputs. That is, the plant output is provided as an input to the learning controller, and the controller is required to produce as output the corresponding plant input. This approach, shown diagrammatically in figure 38.4, is known as *direct inverse modeling* (Widrow and Stearns, 1985; Miller, 1987; Atkeson and Reinkensmeyer, 1988; Kuperstein, 1988). Note that we treat the plant output as being observed at time n. The input to the learning

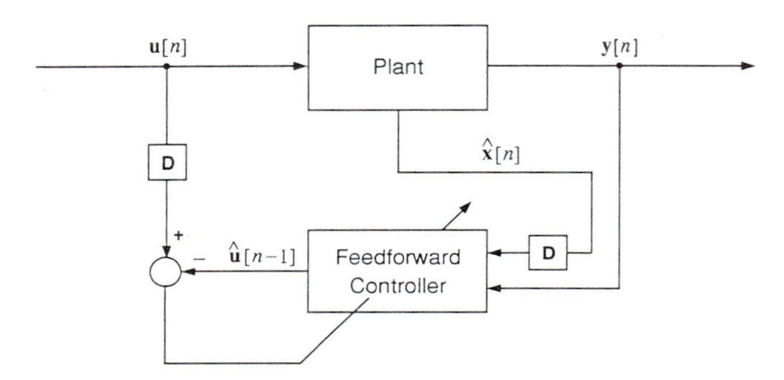

FIGURE 38.4 The direct inverse modeling approach to learning a controller. The state estimate $\hat{x}[n]$ is assumed to be provided by an observer (not shown). See text for details.

controller is the current plant output $\mathbf{y}[n]$ and the delayed state estimate $\hat{\mathbf{x}}[n-1]$. The controller is required to produce the plant input that gave rise to the current output, in the context of the delayed estimated state. This is a supervised learning problem in which the plant input $\mathbf{u}[n]$ serves as the target in the following cost function:

$$\mathcal{J} = \tfrac{1}{2} \| \mathbf{u}[n] - \hat{\mathbf{u}}[n] \|^2 \qquad (9)$$

where $\hat{\mathbf{u}}[n]$ denotes the controller output at time n.

An example of a problem for which the direct inverse modeling approach is applicable is the *inverse kinematics* problem, which requires learning a sensorimotor transformation between the desired position of the hand in spatial coordinates and a corresponding set of joint angles for the arm to achieve that position. To learn such a transformation, the system tries a random joint angle configuration (a vector $\mathbf{u}[n]$) and observes the resulting hand position (the vector $\mathbf{y}[n]$). The system gathers a number of such pairs and uses a supervised learning algorithm to learn a mapping from $\mathbf{y}[n]$ to $\mathbf{u}[n]$.

The direct inverse modeling approach is well-behaved for linear systems and indeed can be shown to converge to correct parameter estimates for such systems under certain conditions (Goodwin and Sin, 1984). For nonlinear systems, however, a difficulty arises that is related to the general degrees-of-freedom problem in motor control (Bernstein, 1967). The problem is due to a particular form of redundancy in nonlinear systems (Jordan, 1992). In such systems, the optimal parameter estimates (i.e., those that minimize the cost function in equation 9) in fact may yield an incorrect controller.

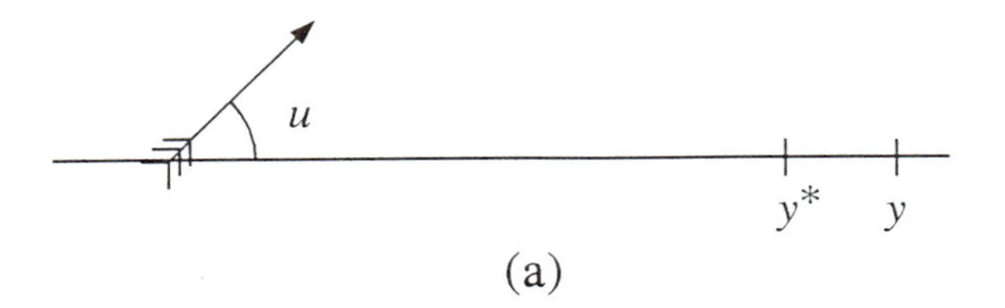

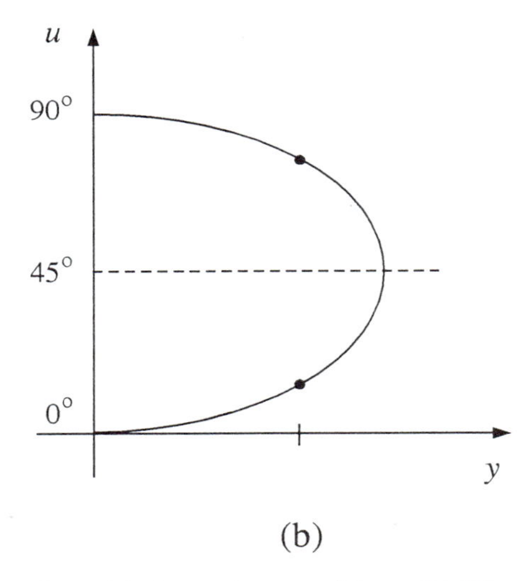

FIGURE 38.5 (a) An archery problem. (b) The parabolic relationship between distance traveled (y) and angle (u) for a projectile. For each value of y, there are two corresponding values of u, symmetrically placed around 45°.

Consider the following simple example. Figure 38.5 shows a one degree-of-freedom archery problem: A controller chooses an angle u and an arrow is projected at that angle. Figure 38.5(b) shows the parabolic relationship between distance traveled and angle. Note that for each distance, there are two angles which yield that distance. This implies that a learning system using direct inverse modeling sees two different targets paired with any given input. If a least-squares cost function is used (cf. equation 9), then the system produces an output that is the average of the two targets which, by the symmetry of the problem, is 45°. Thus, the system converges to an incorrect controller that maps each target distance to the same 45° control signal.

FEEDBACK ERROR LEARNING Kawato, Furukawa, and Suzuki (1987) have developed a direct approach to motor learning known as *feedback error learning*. Feedback error learning makes use of a feedback controller

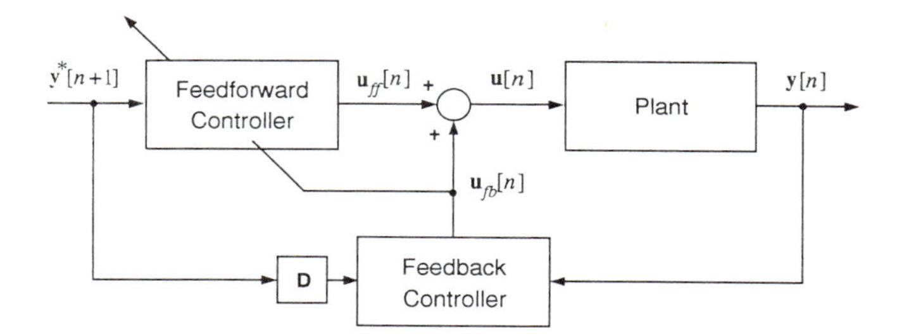

FIGURE 38.6 The feedback error learning approach to learning a feedforward controller. The feedback control signal is the error term for learning the feedforward controller. See text for details.

to guide the learning of the feedforward controller.[5] Consider a composite feedback-feedforward control system in which the total control signal is the sum of the feedforward component and the feedback component:

$$\mathbf{u}[n] = \mathbf{u}_{ff}[n] + \mathbf{u}_{fb}[n] \qquad (10)$$

In the context of a direct approach to motor learning, the signal $\mathbf{u}[n]$ is the target for learning the feedforward controller (cf. figure 38.4). The error between the target and the feedforward control signal is $(\mathbf{u}[n] - \mathbf{u}_{ff}[n])$ which, in the current case is simply $\mathbf{u}_{fb}[n]$. Thus, an error for learning the feedforward controller can be provided by the feedback control signal (figure 38.6).

An important difference between feedback error learning and direct inverse modeling regards the signal used as the controller input. In direct inverse modeling, the controller is trained off-line; that is, the input to the controller for the purposes of training is the actual plant output, not the desired plant output. For the controller actually to participate in the control process, it must receive the desired plant output as its input. The direct inverse modeling approach therefore requires a switching process: The desired plant output must be switched in for the purposes of control, and the actual plant output must be switched in for the purposes of training. The feedback error learning approach provides a more elegant solution to this problem. In feedback error learning, the desired plant output is used for both control and training. The feedforward controller is trained on-line; that is, it is used as a controller while it is being trained. Although the training data that it receives—pairs of actual plant inputs and desired plant outputs—are not samples of the inverse dynamics of the plant, the system nonetheless converges to an inverse model of the plant because of the error-correcting properties of the feedback controller.

By utilizing a feedback controller, the feedback error learning approach also solves another problem associated with direct inverse modeling. Direct inverse modeling is not *goal-directed*; it is not sensitive to particular output goals (Jordan and Rosenbaum, 1989). This is seen by simply observing that the goal signal ($\mathbf{y}^*[n]$) does not appear in figure 38.4. The learning process samples randomly in the control space, which may or may not yield a plant output near any particular goal. Even if a particular goal is specified before the learning begins, the direct inverse modeling procedure must search throughout the control space until an acceptable solution is found. In the feedback error learning approach, however, the feedback controller serves to guide the system to the correct region of the control space. By using a feedback controller, the system makes essential use of the error between the desired plant output and the actual plant output to guide the learning. This fact links the feedback error learning approach to the indirect approach to motor learning (discussion next). In the indirect approach, the learning algorithm is based directly on the output error.

DISTAL SUPERVISED LEARNING In this section, an indirect approach to motor learning known as distal supervised learning is described. Distal supervised learning avoids the nonconvexity problem (cf. figure 38.5) and certain other problems associated with direct approaches to motor learning (Jordan, 1990; Jordan and Rumelhart, 1992). In distal supervised learning, the controller is learned indirectly through the intermediary of a forward model of the plant. The forward model must itself be learned from observations of the inputs and outputs of the plant. This approach is therefore composed of two interacting processes, one process in which the forward model is learned and another process in which the forward model is used in the learning of the controller (figure 38.7).

The forward model of the plant is a mapping from states and inputs to predicted plant outputs, and it is trained using the *prediction error* ($\mathbf{y}[n] - \hat{\mathbf{y}}[n]$), where $\hat{\mathbf{y}}[n]$) is the output of the forward model. The second process, training the controller, accomplished in the following manner. The controller and the forward

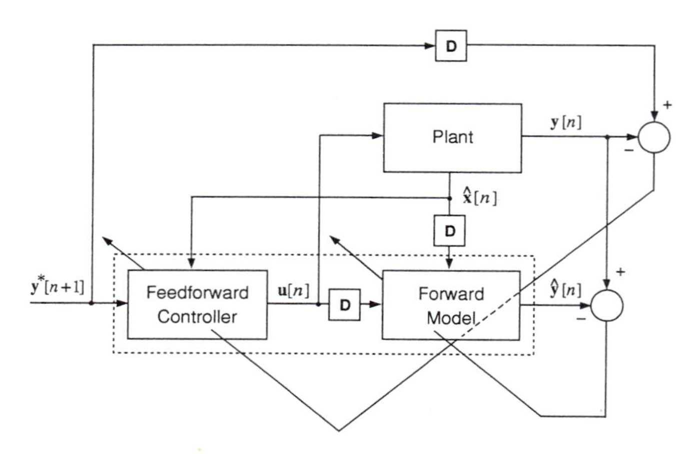

FIGURE 38.7 The distal supervised learning approach. The forward model is trained using the prediction error $(y[n] - \hat{y}[n])$. The subsystems in the dashed box constitute the composite learning system. This system is trained by using the performance error $(y^*[n] - y[n])$ and holding the forward model fixed. The state estimate $\hat{x}[n]$ is assumed to be provided by an observer (not shown).

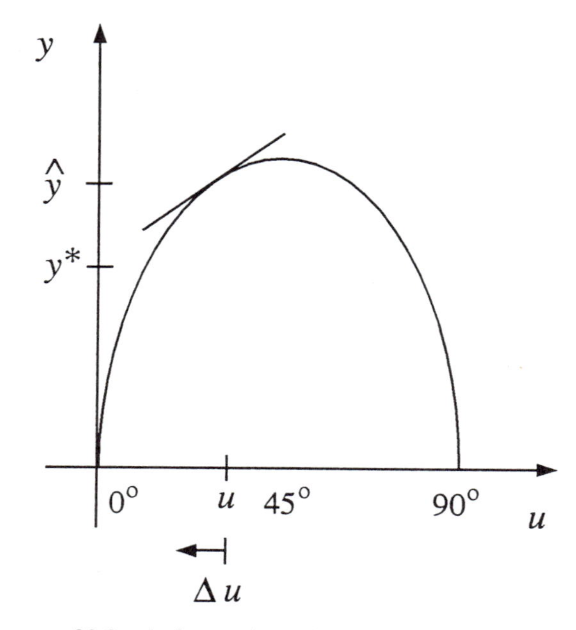

FIGURE 38.8 A forward model for the archery problem. Given a value of u, the forward model allows the output y to be predicted, the error $y^* - y$ to be estimated, and the slope to be estimated at u. The product of the latter two quantities provides information about how to adjust u to make the error smaller.

model are joined together and are treated as a single composite learning system. If the controller is to be an inverse model, then the composite learning system should be an identity transformation (i.e., a transformation whose output is the same as its input). This suggests that the controller can be trained indirectly by training the composite learning system to be an identity transformation. This is a supervised learning problem in which the entire composite learning system (the system inside the dashed box in figure 38.7) corresponds to the box labeled "learner" in figure 38.3. During this training process, the parameters in the forward model are held fixed. Thus, the composite learning system is trained to be an identity transformation by a constrained learning process in which some of the parameters inside the system are held constant. By allowing only the controller parameters to be altered, this process trains the controller indirectly.

Training a system to be an identity transformation means that its supervised error signal is the difference between the input and the output. This error signal is just the *performance error* $(\mathbf{y}^*[n] - \mathbf{y}[n])$ (cf. figure 38.7), the observed error in motor performance: That is, the learning algorithm trains the controller by correcting the error between the desired plant output and the actual plant output.

Let us return to the archery problem presented earlier. Assume that the system has already acquired a perfect forward model of function relating u to y, as shown in figure 38.8. The system can now utilize the forward model to recover a solution u for a given target y^*. This can be achieved in different ways depending on the particular supervised learning technique that is adopted. One approach involves using the local slope of the forward model to provide a correction to the current best guess for the control signal. (This corresponds to using gradient descent as the algorithm for training the composite learning system.) As seen in the figure, the slope provides information about the direction in which to adjust the control signal in order to reduce the performance error $(y^* - y)$. The adjustments to the control signal are converted into adjustments to the parameters of the controller using the chain rule.

An advantage of working with a forward model is that the nonconvexity of the problem does not prevent the system from converging to a unique solution. The system simply heads downhill toward one solution or the other (see figure 38.8). Moreover, if particular kinds of solutions are preferred (e.g., the left branch versus the right branch of the parabola), then additional constraints can be added to the cost function to force the system to search in one branch or the other (Jordan, 1990).

Suppose, finally, that the forward model is imperfect. In this case, the error between the desired output and the predicted output is the quantity ($\mathbf{y}^*[n] - \hat{\mathbf{y}}[n]$), the *predicted performance error*. Using this error, the best the system can do is to acquire a controller that is an inverse of the forward model. Because the forward model is inaccurate, the controller is inaccurate. However, the predicted performance error is not the only error available for training the composite learning system. Because the actual plant output ($\mathbf{y}[n]$) can still be measured after a learning trial, the true performance error ($\mathbf{y}^*[n] - \mathbf{y}[n]$) remains available for training the controller. This implies that the output of the forward model can be discarded; the forward model is needed only for the structure that it provides as part of the composite learning system (e.g., the slope that it provides in figure 38.8). Moreover, for this purpose, an exact forward model is not required. Roughly speaking, the forward model need provide only coarse information about how to improve the control signal based on the current performance error, not precise information about how to make the optimal correction. If the performance error is decreased to zero, then an accurate controller has been found, regardless of the path taken to find that controller.

Reinforcement Learning Reinforcement learning algorithms differ from supervised learning algorithms by requiring significantly less information to be available to the learner. Rather than requiring a vector performance error or a vector target in the control space, reinforcement learning algorithms require only a scalar evaluation of performance. A scalar evaluation signal simply tells a system whether it is performing well; it does not provide information about how to correct an error. A variety of reinforcement learning algorithms have been developed recently (Sutton, 1988; Watkins, 1989), and ties have been made to optimal control theory (Barto, Sutton, and Watkins, 1990). A strength of reinforcement learning algorithms is their ability to learn in the face of delayed evaluation.

In the simplest reinforcement learning paradigm, there is a set of possible responses at each point in the state, goal space. Associated with the ith response is a probability p_i of selecting that response as an output on a given trial. Once a response is selected and transmitted to the environment, a scalar evaluation or reinforcement signal is computed as a function of the response and the state of environment. The reinforcement signal then is used in changing the selection probabilities for future trials: If reinforcement is high, the probability of selecting a response is increased; otherwise, the probability is decreased. Typically, the probabilities associated with the remaining (unselected) responses are also adjusted in some manner, so that the total probability sums to one.

Reinforcement learning algorithms are able to learn in situations in which very little instructional information is available from the environment. In particular, such algorithms need make no comparison between the input goal (the vector $\mathbf{y}^*$) and the result obtained (the vector $\mathbf{y}$) to find a control signal that achieves the goal. When such a comparison can be made, however, reinforcement learning still is applicable but may be slower than other algorithms that make use of such information. Although in many cases in motor control it would appear that a comparison with a goal vector is feasible, the question is empirical and as yet unresolved (Adams, 1984): Does feedback during learning serve to strengthen or weaken the action just emitted or to provide structural information about how to change the action just emitted into a more suitable action?

Bootstrap Learning The techniques that we have discussed until now all require that either an error signal or an evaluation signal be available for adjusting the controller. It is also possible to consider a form of motor learning that needs no such corrective information. This form of learning improves the controller by building on earlier learning. Such an algorithm is referred to in the adaptive signal processing literature as *bootstrap learning* (Widrow and Stearns, 1985; Nowlan, 1991).

How might a system learn without being corrected or evaluated? Let us work within the framework discussed previously for reinforcement learning, in which one of a set of responses is chosen with probability p_i. Suppose that, due to prior learning, the system performs a task correctly a certain fraction of the time but still makes errors. The learning algorithm is as follows: The system selects actions according to the probabilities that it has already learned and rewards its actions indiscriminately: That is, it rewards both correct and incorrect actions. We argue that the system can still converge to a control law in which the correct action is chosen with probability one. The reason for this is that the responses of nonselected actions are effectively weakened, because the sum of the p_i must be one: That is, if p_i is strengthened, p_j must be decreased for all j not equal to i. Given that the system starts with the correct action having a larger probability of being selected than the other actions, that action has a larger probability of being strengthened and thus an even larger probability of being selected. Thus, if the initial balance in favor of the correct action is strong enough, the system can improve.

We know of no evidence that such a learning process is utilized by the motor control system, but the possibility appears never to have been investigated directly. The algorithm has intuitive appeal because there are situations in which it seems that mere repetition of a task can lead to improvement. The simplicity of the algorithm is certainly appealing from the point of view of neural implementation.

ACKNOWLEDGMENTS Preparation of this chapter was supported in part by grants from the Human Frontier Science Program, the McDonnell-Pew Foundation, and the Office of Naval Research (N00014-90-J-1942). Michael Jordan is a National Science Foundation Presidential Young Investigator.

NOTES

1. The minimum jerk model is entirely kinematic (forces do not enter into the optimization); thus, it predicts straight-line motion regardless of external forces. This argument assumes that subjects are able, in principle, to compensate for the external force.
2. The equilibrium point model proposes that movement is achieved by altering the relative stiffnesses of the agonist and antagonist muscles across a joint; this changes the equilibrium position of the joint and causes the arm to move.
3. Forward models are known by a variety of names in the motor control literature, including *efference copy* and *corollary discharge* (cf. Gallistel, 1980). In this section and the next, we describe some new proposals for possible roles for forward models in motor control and motor learning.
4. Note, however, that inverse models are not the *only* way to implement open-loop control schemes. As we have seen, an open-loop controller can also be implemented by placing a forward model in an internal feedback loop. See Jordan (in press) for further discussion.
5. A *feedforward controller* is an internal inverse model of the plant that is used in an open-loop control configuration. See Jordan (in press) for further discussion of feedforward and feedback control.

REFERENCES

Adams, J. A., 1984. Learning of movement sequences. *Psychol. Bull.* 96:3–28.

Atkeson, C. G., and J. M. Hollerbach, 1985. Kinematic features of unrestrained vertical arm movements. *J. Neurosci.* 5:2318–2330.

Atkeson, C. G., and D. J. Reinkensmeyer, 1988. Using associative content-addressable memories to control robots. Paper presented at the IEEE Conference on Decision and Control, San Francisco.

Barto, A. G., R. S. Sutton, and C. J. C. H. Watkins, 1990. Sequential decision problems and neural networks. In *Advances in Neural Information Processing Systems*, vol. 2, D. Touretzky, ed. San Mateo, Calif.: Morgan Kaufmann, pp. 686–693.

Beggs, W. D. A., and C. I. Howarth, 1972. The movement of the hand towards a target. *Q. J. Exp. Psychol.* 24:448–453.

Bernstein, N., 1967. *The Coordination and Regulation of Movements.* London: Pergamon.

Bullock, D., and S. Grossberg, 1988. Neural dynamics of planned arm movements: Emergent invariants and speed-accuracy properties during trajectory formation. *Psychol. Rev.* 95:49–90.

Dornay, M., Y. Uno, M. Kawato, and R. Suzuki, 1989. Simulation of optimal movements using the minimum-muscle-tension-change model. In *Advances in Neural Information Processing Systems*, vol. 4, J. Moody, S. Hanson, and R. Lippmann, eds. San Mateo, Calif.: Morgan Kaufmann, pp. 627–634.

Duhamel, J. R., C. L. Colby, and M. E. Goldberg, 1992. The updating of the representation of visual space in parietal cortex by intended eye movements. *Science* 255:90–91.

Feldman, A. G., 1986. Once more on the equilibrium-point hypothesis (λ model) for motor control. *J. Mot. Behav.* 18:17–54.

Flash, T., 1987. The control of hand equilibrium trajectories in multi-joint arm movements. *Biological Cybernetics* 57:257–274.

Flash, T., 1990. The organization of human arm trajectory control. In *Multiple Muscle Systems: Biomechanics and Movement Organization*, J Winters and S. Woo, eds. Berlin: Springer-Verlag, pp. 282–301.

Flash, T., and E. Henis, 1992. Arm trajectory modification during reaching towards visual targets. *J. Cogn. Neurosci.* 3:220–230.

Flash, T., and N. Hogan, 1985. The coordination of arm movements: An experimentally confirmed mathematical model. *J. Neurosci.* 5:1688–1703.

Gallistel, C. R., 1980. *The Organization of Action: A New Synthesis.* Hillsdale, NJ: Lawrence Erlbaum.

Georgopoulos, A. P., J. F. Kalaska, and J. T. Massey, 1981. Spatial trajectories and reaction times of aimed movements: Effects of practice, uncertainty and change in target location. *J. Neurophysiol.* 46:725–743.

Ghez, C., J. Gordon, M. F. Ghilardi, C. N. Christakos, and S. E. Cooper, 1990. Roles of proprioceptive input in the programming of arm trajectories. *Cold Spring Harb. Symp. Quant. Biol.* 55:837–847.

Goodwin, G. C., and K. S. Sin, 1984. *Adaptive Filtering Prediction and Control*. Englewood Cliffs, N.J.: Prentice Hall.

Hertz, J., A. Krogh, and R. G. Palmer, 1991. *Introduction to the Theory of Neural Computation*. Redwood City, CA: Addison-Wesley.

Hirayama, M., M. Kawato, and M. I. Jordan, 1993. The cascade neural network model and a speed-accuracy tradeoff of arm movement. *Journal of Motor Behavior* 25: 162–175.

Hoff, B. R., 1992. *A computational description of the organization of human reaching and prehension* (Computer Science Tech. Rep. No. USC-CS-92-523). Los Angeles: University of Southern California.

Hogan, N., 1984. An organising principle for a class of voluntary movements. *J. Neurosci.* 4:2745–2754.

Jordan, M. I., 1990. Motor learning and the degrees of freedom problem. In *Attention and Performance*, vol. 13, M. Jeannerod, ed. Hillsdale, N.J.: Erlbaum.

Jordan, M. I., 1992. Constrained supervised learning. *J. Math. Psychol.* 36:396–425.

Jordan, M. I. (in press). Computational aspects of motor control and motor learning. In *Handbook of Perception and Action: Motor Skills*. H. Heuer and S. Keele, eds. New York: Academic Press.

Jordan, M. I., T. Flash, and Y. Arnon (in press). A model of the learning of arm trajectories from spatial targets. *J. Cogn. Neurosci.*

Jordan, M. I., and D. A. Rosenbaum, 1989. Action. In *Foundations of Cognitive Science*, M. I. Posner, ed. Cambridge, Mass.: MIT Press.

Jordan, M. I., and D. E. Rumelhart, 1992. Forward models: Supervised learning with a distal teacher. *Cogn. Sci.* 16:307–354.

Katayama, M., and M. Kawato, 1992. *Virtual trajectory and stiffness ellipse during multi-joint movements predicted by neural inverse models*. Tech. Rep. TR-A-0144. ATR Human Information Processing Laboratories, Nara, Japan.

Kawato, M., 1990. Computational schemes and neural network models for formation and control of multijoint arm trajectory. In *Neural Networks for Control*, W. T. Miller III, R. S. Sutton, and P. J. Werbos, eds. Cambridge, Mass.: MIT Press.

Kawato, M., K. Furukawa, and R. Suzuki, 1987. A hierarchical neural-network model for control and learning of voluntary movement. *Biol. Cybern.* 57:169–185.

Kirk, D. E., 1970. *Optimal Control Theory*. Englewood Cliffs, N.J.: Prentice Hall.

Kuperstein, M., 1988. Neural model of adaptive hand-eye coordination for single postures. *Science* 239:1308–1311.

Lacquaniti, F., N. A. Borghese, and M. Carrozzo, 1992. Internal models of limb geometry in the control of hand compliance. *J. Neurosci.* 12:1750–1762.

Lindblom, B., J. Lubker, and T. Gay, 1979. Formant frequencies of some fixed-mandible vowels and a model of speech motor programming by predictive simulation. *J. Phonetics* 7:147–161.

MacKenzie, C. L., R. G. Marteniuk, C. Dugas, D. Liske, and B. Eickmeier, 1987. Three dimensional movement trajectories in Fitts' task: Implications for control. *Q. J. Exp. Psychol.* 39:629–647.

Meyer, D. E., J. E. K. Smith, S. Kornblum, R. A. Abrams, and C. E. Wright, 1990. Speed-accuracy tradeoffs in aimed movements: Toward a theory of rapid voluntary action. In *Attention and Performance*, vol. 13, M. Jeannerod, ed. Hillsdale, N.J.: Erlbaum.

Miall, R. C., D. J. Weir, D. M. Wolpert, and J. F. Stein (in press). Is the cerebellum a Smith predictor? *J. Mot. Behav.*

Miller, W. T., 1987. Sensor-based control of robotic manipulators using a general learning algorithm. *IEEE Journal of Robotics and Automation*, 3:157–165.

Milner, T. E., and M. M. Ijaz, 1990. The effect of accuracy constraints on three-dimensional movement kinematics. *Neuroscience* 35:365–374.

Morasso, P., 1981. Spatial control of arm movements. *Experimental Brain Research*, 42:223–227.

Nowlan, S. J., 1991. *Soft competitive adaptation: Neural network learning algorithms based on fitting statistical mixtures* (Tech. Rep. No. CMU-CS-91-126). Pittsburgh: Carnegie Mellon University.

Pélisson, D., C. Prablanc, M. A. Goodale, and M. Jeannerod, 1986. Visual control of reaching movements without vision of the limb: II. Evidence of fast unconscious processes correcting the trajectory of the hand to the final position of a double-step stimulus. *Exp. Brain Res.* 6:303–311.

Polit, A., and E. Bizzi, 1979. Processes controlling arm movements in monkeys. *Science* 201:1235–1237.

Robinson, D. A., 1981. The use of control system analysis in the neurophysiology of eye movements. *Annu. Rev. Neurosci.* 4:463–503.

Soechting, J. F., and F. Lacquaniti, 1981. Invariant characteristics of a pointing movement in man. *J. Neurosci.* 1:710–720.

Sutton, R. S., 1988. Learning to predict by the methods of temporal differences. *Machine Learning* 3:9–44.

Uno, Y., M. Kawato, and R. Suzuki, 1989. Formation and control of optimal trajectory in human multijoint arm movement: A minimum torque change model. *Biol. Cybern.* 61:89–101.

Watkins, C. J. C. H., 1989. Learning from delayed rewards. Unpublished doctoral thesis, University of Cambridge, England.

Widrow, B., and S. D. Stearns, 1985. *Adaptive Signal Processing*. Englewood Cliffs, N.J.: Prentice Hall.

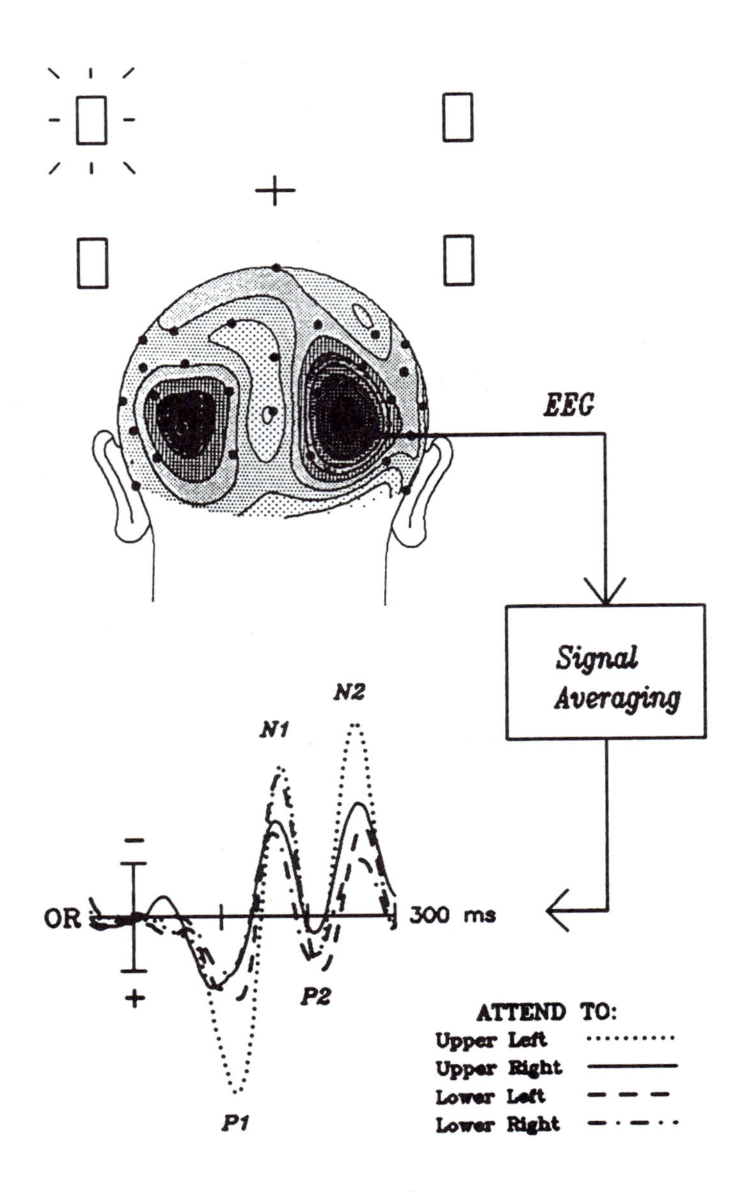

EEG
Signal Averaging
N2
N1
OR
300 ms
P2
P1
ATTEND TO:
Upper Left
Upper Right
Lower Left
Lower Right

V ATTENTION

Amplitude changes in visual ERP components in the spatial attention task of Mangun, Hillyard, and Luck (1993). (Reprinted from Hillyard, 1993, by permission.)

Introduction

MICHAEL I. POSNER

OUR TREATMENT of attention centers around the issue of orienting to sensory stimuli, an issue that is central to current progress in cognitive neuroscience. It represents a major theme of my chapter and the chapters by Hillyard and coworkers, Johnson, LaBerge, Rafal and Robertson, and Stein and colleagues. Just as the structure of sensory systems has led the way in the study of neuroscience, orienting to sensory systems has been a major vehicle for bringing the computational and experimental approach of cognitive science together with methods for imaging brain activity. In the chapters on sensory orienting, we introduce methods of cellular recording, positron emission tomography, study of brain lesions, use of event-related electrical and magnetic potentials, and brain development. These methods are applied to the visual and auditory modalities and to their integration within a few brain structures.

Integration of these methods of analysis and experimentation is the common frontier for all cognitive neuroscience. For example, we know from many cognitive experiments that behavioral measures of detection such as reaction time and d' are often enhanced by attention, but enhanced compared to what? The data are clearest when the attended signal is compared to an unattended signal presented at the same time. However, when one goes beyond the advantage of attended signals over stimultaneous unattended signals, the problem of an appropriate baseline becomes apparent. Evidence from studies using positron emission tomography show increases in blood flow in the region being attended, but there are problems in interpreting blood flow. While an increase in blood flow means increased neuronal activity, if that neuronal activity is largely due to inhibitory neurons, it could well be that the computation performed by that brain area might not be enhanced by more activity. When we examine scalp electrical activity, we see an amplification of the early components of the attended message with respect to the unattended, but this could be due to either an increased gain at the attended location or a reduced signal at the unattended location or some combination. We know that, at least in some locations and conditions, inhibition of unattended activity seems dominant at the cellular level. The chapters in our section provide snapshots of the overall process as viewed through various methods of experimentation. They show a general convergence among these methods in concluding, for example, that attention produces a relative increase in neural activity to the attended stimulus as compared to the unattended one. However, extending beyond that statement to claim that the attended message is amplified or that the unattended is filtered is to reach the frontier of our current knowledge, a frontier that we may come to understand with the aid of the methods and results summarized in this section.

Orienting is not all of attention but is one specific example. It would not have been possible to do for other areas of attention the thorough examination that we have provided for orienting, both because the attention section is limited in space and because there is less evidence to provide the basis for comparing and integrating methods. Instead, we devote a single chapter to each of two other major topics in attention. Robbins and Everitt summarize and organize work on arousal that stems from the early reticular concept and now involves an understanding of how the different extracortical transmitter systems influence cortical computations. This area has been skilfully explored with animal models and is being increasingly related to higher-level attentional control.

Duncan provides a glimpse of current studies that involve what are sometimes called *executive control systems*. The issues raised by executive control are among the most complex and basic in the whole field. Some of these are summarized in my overview chapter, and they take concrete form in the figure on p. 610. The placement of executive function with the anterior cingulate is based, in part, on the many recent studies of blood flow and metabolism that have shown activity in this area when subjects attend (see also chapter 39). Duncan raises the issue of whether frontal function might be related to the current goals of the person and what this indicates for the nature of intelligence.

The study of attention is in an exciting arena in which mental and physical concepts interact with each other. We hope that these chapters will be sufficiently illuminating to provide a beacon to the field of cognitive neuroscience as a whole.

39 Attention in Cognitive Neuroscience: An Overview

MICHAEL I. POSNER

ABSTRACT First this chapter examines briefly the history of the concept of attention within psychology and neuroscience in relation to the chapters that appear in this section. Then various methods used in the chapters to link cognitive approaches to attention to underlying brain systems are reviewed. Three anatomical networks that subserve the processes of orienting, higher-level attention, and alertness are discussed. The overview ends with a summary of possible future development and application of this knowledge to attentional pathologies and other attentional states.

History

The problem of selective attention is one of the oldest in psychology. William James (1907) wrote, at the turn of the century, "Everyone knows what attention is. It is the taking possession by the mind in clear and vivid form of one out of what seem several simultaneous objects or trains of thought."

The dominance of behavioral psychology postponed research into the internal mechanisms of selective attention in the first half of this century. The finding that integrity of the brainstem reticular formation was necessary to maintain the alert state provided some anatomical reality to the study of the arousal mechanisms underlying one aspect of attention (Moruzzi and Magoun, 1949). This approach has led to many new findings relating individual subcortical transmitter systems of arousal (e.g., dopamine, norepinephrine) to the computations underlying selection of information (see chapter 44).

The quest for information-processing mechanisms to support the more selective aspects of attention began with studies of listening following World War II. A filter was proposed that was limited for information (in the formal sense of information theory) and located between highly parallel sensory systems and a limited-capacity perceptual system (Broadbent, 1958).

Selective listening experiments supported a view of attention that suggested early selection of a relevant message, with nonselective information being lost to conscious processing. Physiological studies (Skinner and Yingling, 1977) suggested that selection of the relevant channels might involve a thalamic gating mechanism using the nucleus reticularis thalami and controlled from prefrontal sites. Peripheral gating mechanisms still represent a potential source of selection that might be especially important in lower mammals. However, in human information-processing studies, it was clear that unattended information often was processed to a high level as evidenced by the fact that an important message on the unattended channel might interfere with the selected channel (Posner, 1978). This suggested selection at higher cortical levels must be involved. More recent monkey studies suggested that thalamic mechanisms might work in conjunction with extrastriate areas to gate information at cortical levels (Moran and Desimone, 1985; see also chapter 41). The ability to select input channels and the levels at which selection occurs has remained an active feature of current studies of attention (see chapters 40–43).

In the 1970s, psychologists began to distinguish between automatic and controlled processes (Posner, 1978). It was found that visual words could activate other words similar in meaning (their semantic associates), even when the person had no awareness of the words' presence. These studies indicated that the parallel organization found for sensory information extended to semantic processing. Thus, selecting a word meaning for active attention appeared to suppress the availability of other word meanings. Attention was viewed less as an early sensory bottleneck and more as a system for providing priority for motor acts, consciousness, and memory (Allport, 1980). These higher-

MICHAEL I. POSNER Department of Psychology, Institute of Cognitive and Decision Sciences, University of Oregon, Eugene, Oreg.

level mechanisms of attention involve frontal areas and provide important means of coordination among cognitive activities (see chapter 45).

Another approach to problems of selectivity arose in work on the orienting reflex (Sokolov, 1963; Kahneman, 1973). The use of slow autonomic systems (e.g., skin conductance as measures of orienting) made it difficult to analyze the cognitive components and neural systems underlying orienting. In the mid-1970s, neurobiologists began to study information processing in alert monkeys (Wurtz, Goldberg, and Robinson, 1980). Because the visual system had been relatively well explored using microelectrodes, much of this work involved the visual system. During the last 15 years, there has been a steady advancement in our understanding of the neural systems related to visual orienting from studies using single-cell recording in alert monkeys. This work showed a relatively restricted number of areas in which the firing rates of neurons were enhanced selectively when monkeys were trained to attend to a location. At the level of the superior colliculus, selective enhancement could be obtained only when eye movement was involved but, in the posterior parietal lobe, selective enhancement occurred even when the animal maintained fixation. An area of the thalamus, the lateral pulvinar, was similar to the parietal lobe in containing cells with the property of selective enhancement (Colby, 1991). Indeed the thalamic areas relate to the earlier rat models (Skinner and Yingling, 1977) but are believed to perform selection at higher levels of analysis (see chapter 41). Vision remains the central system for the integration of cognitive and neuroscience approaches to selectivity (see chapters 40–42, 45, and 46).

Until recently, there has been a separation between human information-processing and neuroscientific approaches to attention using nonhuman animals. The former tended to describe attention either in terms of a bottleneck that protected limited-capacity central systems from overload or as a resource that could be allocated to various processing systems in a way analogous to the use of the term in economics. On the other hand, neuroscientific views emphasized several separate neural mechanisms that might be involved in orienting and maintaining alertness. Currently, there is an attempt to integrate these two within a cognitive neuroscience of attention (Posner and Petersen, 1990; Näätänen, 1992). The chapters in part V of this book all take this viewpoint.

Methods

The effort to link attention to specific brain systems depends on having methods available to secure these links (Sejnowski and Churchland, 1989). The basic dimensions for classifying methods are based on their spatial localization and their temporal precision. Major methods used for spatial localization in the studies reported in this section include depth recordings, positron emission tomography (PET), and functional magnetic resonance imaging (MRI). The poststimulus latency histograms from single or small numbers of cells can provide both temporal and spatial precision, but their invasive nature requires a protocol based on either surgical interventions in humans or the use of animal models. Temporal precision with normal human subjects is a feature of various cognitive methods involving reaction time or speed accuracy trade-off measures (Bower, 1989) and is also studied by the use of event-related electrical or magnetic potentials (Näätänen, 1992).

An important aspect of understanding the current developments in this field is to track the convergence of evidence from various methods of study. The chapters in part V of this book have been selected to provide a background in various methods. These include performance studies using reaction time or interference during multiple tasks (chapters 40 and 45), study of changes of attention with development (chapter 46), recording from scalp electrodes (chapter 42), lesions in humans and animals (chapters 40, 43, and 44), and various methods for imaging and recording from restricted brain areas (chapter 41), including individual cells (chapter 43).

Current progress in the anatomy of the attention system rests most heavily on two important methodological developments. First, the use of microelectrodes with alert animals showed that attention altered the activity of individual cells (Colby, 1991). Second, anatomical (e.g., computerized tomography or MRI) and physiological (e.g., PET, functional MRI) methods of studying parts of the brain allowed more meaningful investigations of localization of cognitive functions in normal people (Raichle, 1987).

We can distinguish two different types of anatomy related to attention. By the *source* of attention, we mean those anatomical areas that seem to be specific to attention rather than being primarily involved in other forms of processing. (The three attentional networks discussed next are attentional in this sense.) However, when attention operates during task performance, it will operate at the site where the computation involved in the task is usually performed. Thus, when subjects attend to the color, form, or motion of a visual object, they amplify blood flow in various extrastriate areas (Corbetta et al., 1991). These areas are known to be involved in the passive registration of the same information. We will expect that most brain areas, especially cortical areas, will show attention effects in this sense, although they are not part of the brain's attention system.

To move beyond the specification of anatomical areas, it is useful to use methods sensitive to the time dynamics of information processing, which usually requires analysis in the millisecond range. The temporal precision of various methods is changing rapidly, but the imaging methods based on blood flow or volume require changes in blood vessels that limit their temporal precision to hundreds of milliseconds at best. Currently, combined studies using anatomical methods and those sensitive to time-dynamic change provide a convenient way to trace the rapid time-dynamic changes that occur in the course of human information processing (see chapter 42).

It would be ideal if imaging methods were developed that provided the desired combination of high temporal and spatial resolution. However, all the current technologies have their own limitations: For example, cellular recording is limited to animals and is associated with a host of sampling problems; magnetoencephalography is expensive and, unless many channels are used, must be repositioned for examination of each area; scalp electrical recording suffers from difficulty in localizing the generators directly from scalp distributions; and functional MRI is limited by the time for blood vessels to reflect brain activity. Each of the extant methods will lead to modifications that may address some or all of these problems, and entirely new methods may become available. In the following section, the focus is on issues that will need to be addressed with whatever methods prove most useful.

Networks

Three fundamental working hypotheses characterize the current state of efforts to develop a combined cognitive neuroscience of attention. First, there exists an attentional system of the brain that is at least somewhat anatomically separate from various data-processing systems. By *data-processing systems*, we mean those that can be activated passively by input or output. Second, attention is carried out by networks of anatomical areas. It is neither the property of a single brain area nor is it a collective function of the brain working as a whole (Mesulam, 1981, 1990; Posner and Petersen, 1990). Third, the brain areas involved in attention do not carry out the same function, but specific computations are assigned to different areas (Posner et al., 1988; Mesulam, 1990).

It is not possible to specify the complete attentional system of the brain, but something is known about the areas that carry on three major attentional functions: orienting to sensory stimuli, particularly locations in visual space; detecting target events, whether sensory or from memory; and maintaining the alert state. The authors of the chapters in this section address these functions, but they do so using differing methods, theories, and assumptions. Hillyard and colleagues, Johnson, LaBerge, Rafal and Robertson, and Stein and associates deal with sensory orienting primarily, Robbins and Everitt with arousal and the alert state, and Duncan with detection and higher-level functions. The concentration on sensory orienting reflects the fact that this function has been the one selected for much of the effort to link cognitive and neural functions.

Orienting Network Much of the work in orienting has involved orienting to visual locations because of its close connection to shifts in eye position, which can be observed so easily from outside. Usually, we define visual orienting in terms of eye movements that place the stimulus on the fovea. Foveal viewing improves the efficiency of processing targets in terms of acuity, but it also is possible to change the priority, given a stimulus, by attending to its location covertly, without any change in eye or head position (Posner, 1988). When a person or a monkey is cued to attend to a location, events that occur at that location are responded to more rapidly, give rise to enhanced scalp electrical activity, and can be reported at lower threshold. This improvement in efficiency is found within the first 50–150 ms after a cue occurs at the target location. Similarly, if people are asked to move their eyes to a target, an improvement in efficiency at the target location begins well before the eyes move. This covert shift of attention appears to function as a way of guiding the eyes to appropriate areas of the visual field. In fact, there is evidence that rapid saccades require a shift of attention to the location before they will occur.

Three areas of the monkey brain have shown selective enhancement when monkeys attend to eccentric visual stimuli (Wurtz, Goldberg, and Robinson, 1980; Colby, 1991). These are the posterior parietal lobe, the superior colliculus, and the pulvinar. Brain injury to any of these three areas that have been found to show selective enhancement of neuronal firing rates also causes a reduction in one's ability to shift attention covertly (Posner, 1988). However, each area seems to produce a somewhat different deficit. This underlies the general principle that individual areas carry out separate operations, even though they serve together to carry out the network's function.

Damage to the posterior parietal lobe has its greatest effect on the ability to disengage from attentional focus to a target located in a direction opposite the side of the lesion. Apparently, this effect is mediated both by hyperattention to the ipsilesional cue and difficulty in the attractive ability of the contralesional targets. The effects of the parietal lobes of the two cerebral hemispheres are not identical (DeRenzi, 1982): Damage to the right parietal lobe has a greater overall effect than does damage to the left parietal lobe. There is dispute about the reasons for the asymmetries. One proposal is that the right parietal lobe is dominant for spatial attention and controls attention to both sides of space, whereas the left parietal lobe plays a subsidiary role. This view has been supported by studies using PET, which show that the right parietal lobe is affected by attention shifts in both visual fields, whereas the left parietal lobe is influenced only by right visual field shifts of attention (Corbetta et al., 1993). According to another account, the right parietal lobe is influenced more by the global aspects of figure, whereas the left parietal lobe is influenced more by local aspect (Robertson, Lamb, and Knight, 1988). A third view attributes the asymmetry to differences in arousal in the two cerebral hemispheres (Posner and Petersen, 1990). These positions are not mutually exclusive and are discussed in more detail by Rafal and Robertson in chapter 40.

Lesions of the superior colliculus and the surrounding midbrain areas also affect the ability to shift attention. However, in this case, the shift is slowed whether or not attention is first engaged elsewhere. This finding suggests that a computation involved in moving to the target is impaired. In addition, patients with damage in this midbrain area also return to former target locations as readily as to fresh locations to which they have never attended (see chapter 40). Normal subjects and patients with parietal and other cortical lesions show a reduced probability of returning attention to an already examined location.

Patients with lesions of the thalamus and monkeys with chemical lesions of one thalamic nucleus (the pulvinar) also show difficulty in covert orienting. This difficulty appears to be in selective attention to a target on the side opposite the lesion, so as to avoid responding in error to distracting events that occur at other locations. A study of patients with unilateral lesions of this thalamic area showed a slowing of responses to a cued target on the side opposite the lesion, even when the subject had plenty of time to orient there (Posner, 1988). This contrasted with the results found with parietal and midbrain lesions, in which responses are nearly normal on both sides once attention has been cued to the location. Alert monkeys with chemical lesions of this area made faster-than-normal responses when cued to the side opposite the lesion, irrespective of the side of the cue (Petersen, Robinson, and Morris, 1987). Data from normal human subjects, required to filter out irrelevant visual stimuli, showed selective metabolic increases in the pulvinar opposite the stimulus being attended (LaBerge and Buchsbaum, 1990). The role of this area has been subject to detailed anatomical and computational analysis in chapter 41.

These findings make two important points. First, they confirm the idea of anatomical areas carrying out individual cognitive operations. Second, they suggest a particular hypothesis of the circuitry involved in covert attention shifts. The parietal lobe first disengages attention from its present focus; then the midbrain is active to move the index of attention to the area of the target, and the pulvinar is involved in restricting input to the indexed area.

Although the circuitry just described remains speculative, it is clear that patients with parietal lesions have difficulties in pattern recognition, implying that somehow the parietal lobe damage comes to affect the processing of patterns (DeRenzi, 1982). The dorsal pathway extending from the primary visual cortex to the parietal lobe appears to mediate selective visual attention. Considerable anatomical data suggest that a second ventral cortical pathway, leading from the striate cortex to the infratemporal cortex, is involved in processing color and form during pattern recognition (Ungerleider and Mishkin, 1982). There is evidence from single-cell recording in alert monkeys that visuospatial attention affects this pattern recognition system (Moran and Desimone, 1985). Attention to a visual location affects the processing of stimuli within the receptive fields of neurons of the V4 area. Although it is not known how attention gains access to V4, one likely candidate is via the pulvinar, which had close connections to both the parietal system and V4 (see chapter 41).

Cognitive studies of normal humans have been important in exploring how attention influences pattern recognition processes. A major distinction is between the processing of simple features (e.g., line orientation and color) and that of items defined by a combination of features (e.g., a red vertical line). Simple features appear to be processed in parallel: That is, the search time is not affected by the number of nontarget items in the display (Treisman and Schmidt, 1982). When targets are defined by a combination of attributes (e.g., the red vertical line) that are located within displays of highly similar nontargets (e.g., red horizontal lines and green vertical lines), the search appears to be a serial process and takes longer as the number of distractors increases (Duncan and Humphries, 1989). There is evidence that this visual orienting system is also involved in visual search.

One theory of how attention affects pattern recognition is that it works to combine separate features into unitary percepts. According to this view, simple features are not combined until one orients attention to them. It is for this reason that attention is necessary to search for a conjunction of features. When a target is made of features that are also present in distractors, there can be illusory conjunctions, due to an improper conjunction of elements from different locations. It is to avoid such illusory combinations that one attends selectively to each item present in the array (Treisman and Schmidt, 1982). It is also clear that attentional searches can be guided by nonlocation features (e.g., search among only the red objects in a multicolored field for a T-shaped figure) (Wolfe, Cave, and Franzel, 1989). The evidence from PET studies (Corbetta et al., 1991) suggests that the source of attention input for these nonlocation features is from a network different from that which mediates orienting, which may explain why two different attentional operations can be executed in parallel.

There is another aspect of the visual orienting attention system. Just as we can attend to a spatial location, we can also attend to a small or large object. If one views a large letter composed of small ones, it is possible to attend either to the overall form or to its constituents. The size of feature selected relates to the type of sine wave to which the system will be most sensitive (spatial frequency). When attending to local objects, people are relatively good at detecting high-frequency probes but, when attending to global objects, they do relatively better for low-frequency stimuli (see chapter 40).

There is evidence from both normal people and patients that the right hemisphere is biased toward global processing and the left toward local processing. When given a large letter made of small letters, patients with right-sided parietal lesions copy the local letter but miss the global organization, whereas patients with left-hemisphere lesions copy the global orientation while missing the local constituents (Robertson, Lamb, and Knight, 1988).

We have concentrated on visual orienting because that has been the area in which integration between cognitive and neuroscientific studies has been most advanced. Selection of spatial position and temporal scale appear to be the most important functions performed by the visual orienting network. However, the earliest studies of selective attention used the ears or both the eyes and the ears as channels for the presentation of sensory information (see chapter 42). There is good evidence that one can bias processing toward one ear or one particular frequency. When this is done, the electrical signal from the selected channel is amplified with respect to information on unselected channels (Näätänen, 1992; see also chapter 42). If required to do so, subjects do well in attending to several channels at once. However, an exception to this generally good parallel processing arises when targets occur on more than one channel. The interference between targets can happen between as well as within sensory channels (Duncan, 1980). The reasons for this form of sensory interference are discussed in the next section.

Executive Network There are limits to how much we can attend at one time (Kahneman, 1973; Näätänen, 1992; Bourke, Duncan, and Nimmo-Smith, in press). In perception, people are more successful at attending to different aspects of the same object than attending to those same aspects when present in different objects (Duncan, 1980). Some of the limitations that arise in attending to different tasks simultaneously are related to the similarity of the attended information. It is more difficult to attend to sources of information when both are presented to the same modality than when they are presented to separate modalities. In a similar way, tasks that must be transformed into similar codes or that deal with similar semantic content are more difficult to attend to simultaneously than is true when contents or codes are different.

Beyond this, there is a more general limitation on how much one can attend at one time. This general limitation can be demonstrated most clearly when all the specific sources of interference are removed (Bourke, Duncan, and Nimmo-Smith, in press). One illustration of this limitation is when people must detect two targets that occur simultaneously. In this situation, there is a great deal of interference. This effect is apparent even when targets occur in separate modalities, and it can be seen when the only task is to decide whether one or two targets have been presented (Duncan, 1980). This finding underscores the idea that some common system is involved whenever a signal (sensory or memorial) is to be consciously noted. There is also a good deal of evidence that the storage of recently presented information, the generation of ideas from long-term memory, and the development of complex schema all interfere with the detection of new signals (Duncan, 1980).

Perhaps because of these limitations, much of perceptual input goes unattended while some aspects become the focus of attention. Attending, in this sense, is jointly determined by environmental events and current goals and concerns. When appropriately balanced, these two kinds of inputs will lead to the selection of information relevant to the achievement of goals and lends coherence to behavior. The system must, however, remain sufficiently flexible to allow goals and concerns to be reprioritized on the basis of changing environmental events. This balance appears to be adversely affected by major damage to the frontal lobes (see chapter 45).

There is evidence that areas of the midfrontal lobe, including the anterior cingulate gyrus, are involved in executive attention (Vogt, Finch, and Olson, 1992). The anterior cingulate gyrus has been found to be active during both language tasks and tasks involving selection of visual targets in studies of blood flow and metabolism. Moreover, experimental studies show that the degree of blood flow in the anterior cingulate gyrus increases as the number of targets to be detected increases. Thus, this area appears to be sensitive to the mental operations of target detection.

The internal organization of the anterior cingulate gyrus shows alternating bands of cells with close connections to the dorsolateral frontal cortex and the posterior parietal lobe (Goldman-Rakic, 1988) (figure 39.1). This organization suggests an integrative role because studies have implicated the lateral frontal cortex in semantic processing while, as we have seen, the posterior parietal lobe is important for spatial attention. The anterior cingulate gyrus might provide an important connection between widely different aspects of attention (e.g., attention to semantic content and visual location).

A persistent issue in cognitive psychology is whether we should consider an executive network as exercising voluntary control, for it raises the issue of a homunculus and the possibility of infinite regress. Nonetheless, there is little doubt that there is some central control over our behavior and thought patterns. In particular, the study of human expertise in problem solving and other behavior has always considered a central executive system that can describe at least a significant portion of the mental operations involved in problem solving. At issue is how close the central system exercising control of voluntary behavior is to the system involved in target detection as discussed earlier (Posner and Rothbart, 1992). There is no question that one can dissociate attention in the sense of awareness from voluntary control. During dreams, for example, we are well aware of events going on in a dream, but we seem unable to exercise voluntary control over them. In contrast, lesions of the anterior cingulate gyrus can produce evidence for a lack of control over our own behavior (Goldberg, 1981). Patients may think that someone else controls the activity of their arms or that someone else controls their thought processes. A major area for the neuropsychology of attention will be understanding the complex relationship between awareness, limited capacity, and control from higher levels of cognition.

Just as the visual orienting system involves a vertical network of areas, so we might expect a similar network to be associated with detection. The anterior cingulate

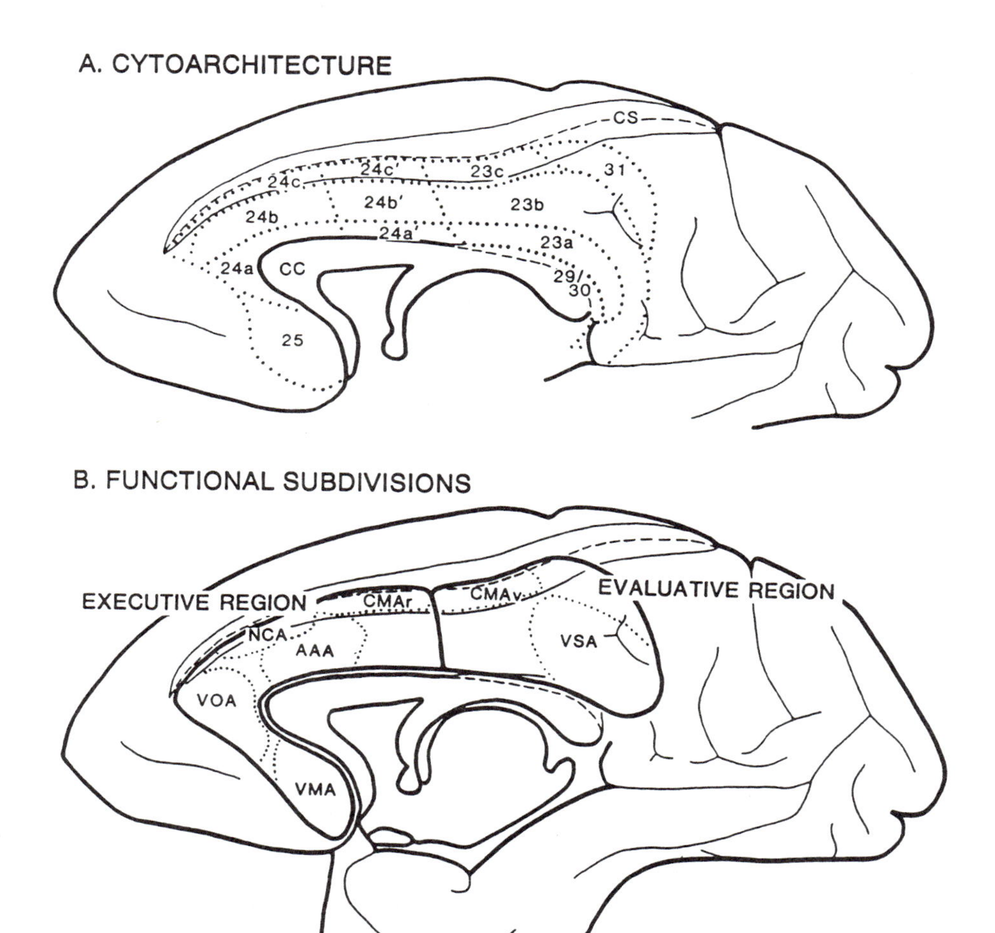

FIGURE 39.1 Two perspectives on the regional organization of the cingulate gyrus. Each view represents the medial surface of the rhesus monkey brain. (A) Cytoarchitectural areas are delineated with dotted lines. The depths of the cingulate sulcus (CS) and callosal sulcus are identified with dashed lines. The corpus callosum is indicated (CC). This map is based on a recent review and refinement of primate cytoarchitecture (Vogt et al., 1992). (B) The two major functional division of cingulate cortex are the executive and evaluative regions (both outlined with thick lines). The executive region has further subdivisions including the visceromotor (VMA), vocalization (VOA), nociceptive (NCA), rostral cingulate motor (CMAr), and attention to action (AAA) areas. The evaluative region includes the ventral cingulate motor (CMAv) and visuospatial (VSA) areas. (Reprinted from Vogt, Finch, and Olson, 1992 by permission)

gyrus receives major dopaminergic input from the ventral tegmental area of the basal ganglia, and it is known that disorders such as Parkinson's disease and schizophrenia, known to influence aspects of the of the basal ganglia dopamine system, have been shown to have general effects on attention.

ALERTING NETWORK The earliest anatomy of attention involved maintenance of the alert state. Cognitive psychologists have studied changes in alerting, both by using long boring tasks with low target probability, such as is required by the military when monitoring radar screens for possible enemy planes or missiles, and by the use of warning signals such as used in footraces to get the runners to prepare to move quickly from the start position (Parasuraman and Davies, 1984). In both of these situations, there is evidence that an increase in alertness improves the speed of target detection. The trade-off between improved speed and reduced accuracy with warning signals has led to a view that alerting does not act to improve the buildup of information concerning the nature of the target but instead acts on the attentional system to enhance the speed of actions taken toward the target (Posner, 1978).

Our understanding of the neural systems related to alerting has improved over the last few years. Patients with lesions of the right frontal area have difficulty maintaining the alert state (Posner and Petersen, 1990). In addition, experimental studies of blood flow in normal people during tasks that demand sustained vigilance show right frontal lobe and parietal activation (Pardo et al., 1990).

The neurotransmitter norepinephrine appears to be involved in maintaining the alert state. This norepinephrine pathway arises in the midbrain, but the right frontal area appears to have a special role in its cortical distribution (Posner and Petersen, 1990). Among posterior visual areas in the monkey, norepinephrine pathways are selective for areas involved in visuospatial attention (Morrison and Foote, 1986). At least one study shows that during the maintenance of vigilance, the metabolic activity of the anterior cingulate gyrus is reduced over a resting baseline value (Cohen et al., 1988). These anatomical findings would support the subjective observation that, while waiting for infrequent visual signals, one has to be prepared to orient but also has to empty one's head of any ideas that might interfere with detection (Posner and Petersen, 1990).

There are clear effects of blocking norepinephrine and maintenance of the alert state. The best example comes from studies of alert monkeys in which clonidine and guanfacine were used to reduce the level of norepinephrine output (Witte, Gordon-Lickey, and Marrocco, 1992). Both of these drugs eliminated the normal effect of warning signals in reducing reaction time. It appears likely that an understanding of the site of action of these drugs would greatly enhance our understanding of the way in which the alerting system affects the selective aspect of attention (see chapter 44).

Applications

Much remains unknown concerning the functional anatomy of attention, particularly the anterior portions of the system. Studies of blood flow and metabolism in normal people should be adequate to provide candidate areas involved in aspects of attention. It will then be possible to test further the general proposal that these constitute a unified system and that constituent computations are localized.

We have begun to understand the circuitry that underlies the visual orienting network. However, more detailed cellular studies in monkeys are necessary to test these hypotheses and to understand more completely the time course and the control structures involved in covert shifts of attention. Even more fascinating is the possibility that the microstructure of areas involved in attention will be somehow differently organized than those areas carrying out passive data processing. Such differences could provide a clue as to the way in which brain tissue might relate to subjective experience. Even with our current knowledge, ideas about attention have proved useful in integrating aspects of social developmental psychology with psychopathology.

The idea of attention as a network of anatomical areas makes relevant study of both the comparative anatomy of these areas and their development in infancy. In the first few months of life, infants develop nearly adult abilities to orient to external events (see chapter 46), but the cognitive control produced by the system involving detection and control requires many months or years of development. Studies of orienting and motor control are beginning to lead to an understanding of this developmental process. As more about the maturational processes of brain and transmitter system is understood, it may be possible to match developing attentional abilities with changing biological mechanisms. The neural mechanisms of attention must support not only common development among infants in their regulatory abilities but also the obvious differences among infants in their rates and success of attentional control (Posner and Rothbart, 1992).

There are many disorders that often are supposed to involve attention, including neglect, closed-head injury, schizophrenia, and attention deficit disorder. The specification of attention in terms of anatomy and function should be useful in clarifying the underlying bases for these disorders. In particular, many of these disorders involve a cortical-level computation and its influence from subcortical transmitter networks.

The development of theories of deficits might also foster the integration of psychiatric and higher-level neurological disorders, both of which might affect the brain's attentional system. In the case of Parkinson's disease, for example, we know that dopamine pathways from the basal ganglia influence cortical motor and cognitive computations. As the disease progresses, there are obvious effects on many cognitive computations that involve attention, such as memory search and shifts of covert attention. Attentional problems are also important in the early stages of schizophrenia, in which deficits in covert attention shifts have been shown to be related to difficulties in generating coherent verbalizations and sequences of thought (Early et al., 1989a, b). However, progression of the disease appears to lead to withdrawal and depression. Although detailed discussion of the disorders of attention is beyond the scope of this part of the text, the references listed by the chapter authors may be useful for readers interested in following these issues.

REFERENCES

Allport, A., 1980. Attention and performance. In *Cognitive Psychology: New Directions*, G. Claxton, ed. London: Routledge and Kegan Paul.

Bourke, P. A., J. Duncan, and I. Nimmo-Smith (in press). A general factor involved in dual task performance decrement. *Q. J. Exp. Psychol.*

Bower, G. H., and J. P. Clapper, 1989. Experimental methods in cognitive science. In *Foundations of Cognitive Science*, M. I. Posner, ed. Cambridge, Mass.: MIT Press, pp. 245–300.

Broadbent, D. E., 1958. *Perception and Communication*. London: Pergamon.

Cohen, R. M., W. E. Semple, M. Gross, H. J. Holcomb, S. M. Dowling, and T. E. Nordahl, 1988. Functional localization of sustained attention. *Neuropsychiatr. Neuropsychol. Behav. Neurol.* 1:3–20.

Colby, C. L., 1991. The neuroanatomy and neurophysiology of attention. *J. Child Neurol.* 6:S90–S118.

Corbetta, M., F. M. Miezin, S. Dobmeyer, G. L. Shulman, and S. E. Petersen, 1991. Selective and divided attention during visual discrimination of color, shape and speed: Functional anatomy by positron emission tomography. *J. Neurosci.* 11:2383–2402.

Corbetta, M., F. M. Miezin, G. L. Shulman, and S. E. Petersen, 1993. A PET study of visuospatial attention. *J. Neurosci.* 13:1202–1226.

DeRenzi, E., 1982. *Disorders of Space Exploration and Cognition.* New York: Wiley.

Duncan, J., 1980. The locus of interference in the perception of simultaneous stimuli. *Psychol. Rev.* 87:272–300.

Duncan, J., and G. W. Humphries, 1989. Visual search and stimulus similarity. *Psychol. Rev.* 96:433–458.

Early, T. S., M. I. Posner, E. M. Reiman, and M. E. Raichle, 1989. Hyperactivity of the left striato-pallidal projection: I: Lower level theory. *Psychiatr. Dev.* 2:85–108.

Early, T. S., M. I. Posner, E. M. Reiman, and M. E. Raichle, 1989b. Hyperactivity of the left striato-pallidal projection: II: Phenomenology and thought disorder. *Psychiatr. Dev.* 2:109–121.

Goldberg, G., 1981. Medial frontal cortex infraction and the alien hand sign. *Neurology* 38:683–686.

Goldman-Rakic, P. S., 1988. Topography of cognition: Parallel distributed networks in primate association cortex. *Annu. Rev. Neurosci.* 11:137–156.

James, W., 1907. *Psychology*. New York: Holt, Rinehart and Winston.

Kahneman, D., 1973. *Attention and Effort.* Englewood Cliffs, N.J.: Prentice Hall.

LaBerge, D., and M. S. Buchsbaum, 1990. Positron emission tomographic measurements of pulvinar activity during an attention task. *J. Neurosci.* 10:613–619.

Mesulam, M. M., 1981. A cortical network for directed attention and unilateral neglect. *Ann. Neurol.* 10:309–315.

Mesulam, M. M., 1990. Large scale neurocognitive networks and distributed processing for attention, language and memory. *Ann. Neurol.* 28:587–613.

Moran, J., and R. Desimone, 1985. Selective attention gates visual processing in extrastriate cortex. *Science* 229:782–784.

Morrison, J. H., and S. L. Foote, 1986. Noradrenergic and seretoninergic innervation of cortical, thalamic and tectal visual structures in old and new world monkeys. *J. Comp. Neurol.* 243:117–128.

Moruzzi, G., and H. V. Magoun, 1949. Brainstem reticular activation of the EEG. *Electroencephalogr. Clin. Neurophysiol.* 1:445–473.

Näätänen, R., 1992. *Attention and Brain Function.* Hillsdale, N.J.: Erlbaum.

Parasuraman, R., and D. R. Davies, 1984. *Varieties of Attention.* New York: Academic Press.

Pardo, J. V., P. Pardo, K. Janer, and M. E. Raichle, 1990. The anterior cingulate cortex mediates processing selection in the Stroop attention conflict paradigm. *Proc. Nat. Acad. Sci. U.S.A.* 87:256–259.

Petersen, S. E., D. L. Robinson, and J. D. Morris, 1987. Contributions of the pulvinar to visual spatial attention. *Neuropsychologia* 25(1A):97–106.

Posner, M. I., 1978. *Chronometric Explorations of Mind.* Hillsdale, N.J.: Erlbaum.

Posner, M. I., 1988. Structures and functions of selective attention. In *Master Lectures in Clinical Neuropsychology and Brain Function: Research, Measurement, and Practice*, T. Boll and B. Bryant, eds. Washington, D.C.: American Psychological Association, pp. 171–202.

Posner, M. I., and S. E. Petersen, 1990. The attention system of the human brain. *Annu. Rev. Neurosci.* 13:25–42.

Posner, M. I., S. E. Petersen, P. T. Fox, and M. E. Raichle, 1988. Localization of cognitive functions in the human brain. *Science* 240:1627–1631.

Posner, M. I., and M. K. Rothbart, 1992. Attention and conscious experience. In *The Neuropsychology of Conscious-*

ness, A. D. Milner and M. D. Rugg, eds. London: Academic Press, pp. 91–112.

RAICHLE, M., 1987. Circulatory and metabolic correlates of brain function in normal humans. In *Handbook of Physiology* (sec. 1–5, part S), V. B. Mountcastle and F. Plum, eds. Bethesda, Md.: American Physiological Society.

ROBERTSON, L. C., M. R. LAMB, and R. T. KNIGHT, 1988. Effects of temporal-parietal junction on perceptual and attentional processing in humans. *J. Neurosci.* 8:3757–3769.

SEJNOWSKI, T. J., and P. S. CHURCHLAND, 1989. Brain and cognition. In *Foundations of Cognitive Science*, M. I. Posner, ed. Cambridge, Mass.: MIT Press, pp. 301–356.

SKINNER, J. E., and C. YINGLING, 1977. Central gating mechanisms that regulate event-related potentials and behavior. *Prog. Clin. Neurophysiol.* 1:30–69.

SOKOLOV, E. N., 1963. Higher nervous functions: The orienting reflex. *Annu. Rev. Physiol.* 25:545–580.

TREISMAN, A., and H. SCHMIDT, 1982. Illusory conjunctions in the perception of objects. *Cogn. Psychol.* 14:107–141.

UNGERLEIDER, L. G., and M. MISHKIN, 1982. Two cortical visual systems. In *Analysis of Visual Behavior*, D. J. Ingle, M. A. Goodale, and R. J. W. Mansfield, eds. Cambridge, Mass.: MIT Press, pp. 540–580.

VOGT, B. A., D. M. FINCH, and C. R. OLSON, 1992. Overview: Functional heterogeneity in cingulate cortex: The anterior executive and posterior evaluative regions. *Cerebral Cortex* 2(6):435–443.

WITTE, E. A., M. E. GORDON-LICKEY, and R. T. MARROCCO, 1992. Pharmacological depletion of catecholamines modifies covert orienting in rhesus monkey. *Soc. Neurosci. Abstr.* 18:537.

WOLFE, J. K. M., K. R. CAVE, and S. L. FRANZEL, 1989. Guided search: An alternative to the feature integration model for visual search. *J. Exp. Psychol. [Hum. Percept.]* 15:419–433.

WURTZ, R. H., M. E. GOLDBERG, and D. L. ROBINSON, 1980. Behavioral modulation of visual responses in the monkey: Stimulus selection for attention and movement. *Prog. Psychobiol. Physiol. Psychol.* 9:43–83.

40 The Neurology of Visual Attention

ROBERT RAFAL AND LYNN ROBERTSON

ABSTRACT Advances in cognitive science and in anatomical and functional neuroimaging are leading to a clearer understanding of the psychobiology underlying clinical disorders of visual attention. We have three golas: to understand how attentional processes may operate in regulating normal perception and action; to identify the neural basis of visual attention; and to understand the psychophysiological basis for the disabilities of these patients in everyday life. We review what has been learned about the role of specific brain structures in orienting attention to locations in the visual field and the role of attention in object recognition. Both syndrome-based observations and anatomically based studies are considered to illustrate how each approach contributes to the understanding of the functions of visual attention and the neural substrates serving these functions.

The gentleman in figure 40.1 had recently suffered a large stroke affecting the right cerebral cortex, including frontal, parietal, and temporal lobes. He had severe left hemispatial neglect and did not orient to people who approached him from his left side. When his attention was engaged by someone to his right, he did not respond to people on his left who spoke to him; even a $10 bill held a few inches from the left side of his face went unnoticed. This striking deficiency in lateralized orienting, known as *unilateral neglect*, is a common clinical problem which, if persistent, is extremely disabling.

It might appear that this patient is blind on his left side. Testing for the clinical sign called *extinction* shows, however, that he does not have a visual sensory defect but a disorder in visually attending. Figure 40.1 shows one conventional way to examine for extinction. The patient is asked to keep his or her eyes fixed on the examiner's face and to report objects presented in one or both visual fields. When presented a single object, he would orient to either the left or right field, so he is not blind in the left field. Of course, he also has no trouble reporting the key when it is presented in his right visual field. In figure 40.1A, two coins are presented simultaneously, one in each field, but the patient orients only to the one on the right and does not notice the one on the left. When asked, he is quite certain that nothing happened on the left side and that he saw only one coin.

Now contrast visual neglect with hemianopia. Patients with unilateral lesions that destroy all the primary visual (striate) cortex do have a visual sensory defect obliterating the entire contralesional visual field. When presented objects or light flashes anywhere in the visual field opposite the lesion (often right up to the midline), they will be unaware of them and deny seeing them. This occurs even when there is no other stimulus present, as in a darkened room, and even when they are told to expect signals there. The margins of scotoma are sharp and mapped decisively in retinotopic coordinates.

It might seem that this is a more severe deficit than extinction. Patients with extinction can, after all, see and report signals in their contralesional field (i.e., opposite the lesion), at least under some circumstances. Yet neglect is unambiguously a more serious disability. Patients with neglect are compromised even in simple activities of daily life. They often fail to eat food on one side of their plate or to shave or apply makeup to one side of their face. Fortunately, it is rare for such severe neglect to persist. After a time, the attention defect may not be obvious at all and may be elicited only by testing for extinction. Eventually, clinical extinction (as tested at the bedside) may become inconsistent, and the problem may be manifest only with more sensitive experimental testing using tachistoscopic stimulus presentation.

In this chapter, we review some of what is known about the psychobiology underlying clinical disorders

ROBERT RAFAL and LYNN ROBERTSON Department of Neurology and Center for Neuroscience, University of California, Davis; VA Medical Center, Martinez, Calif.

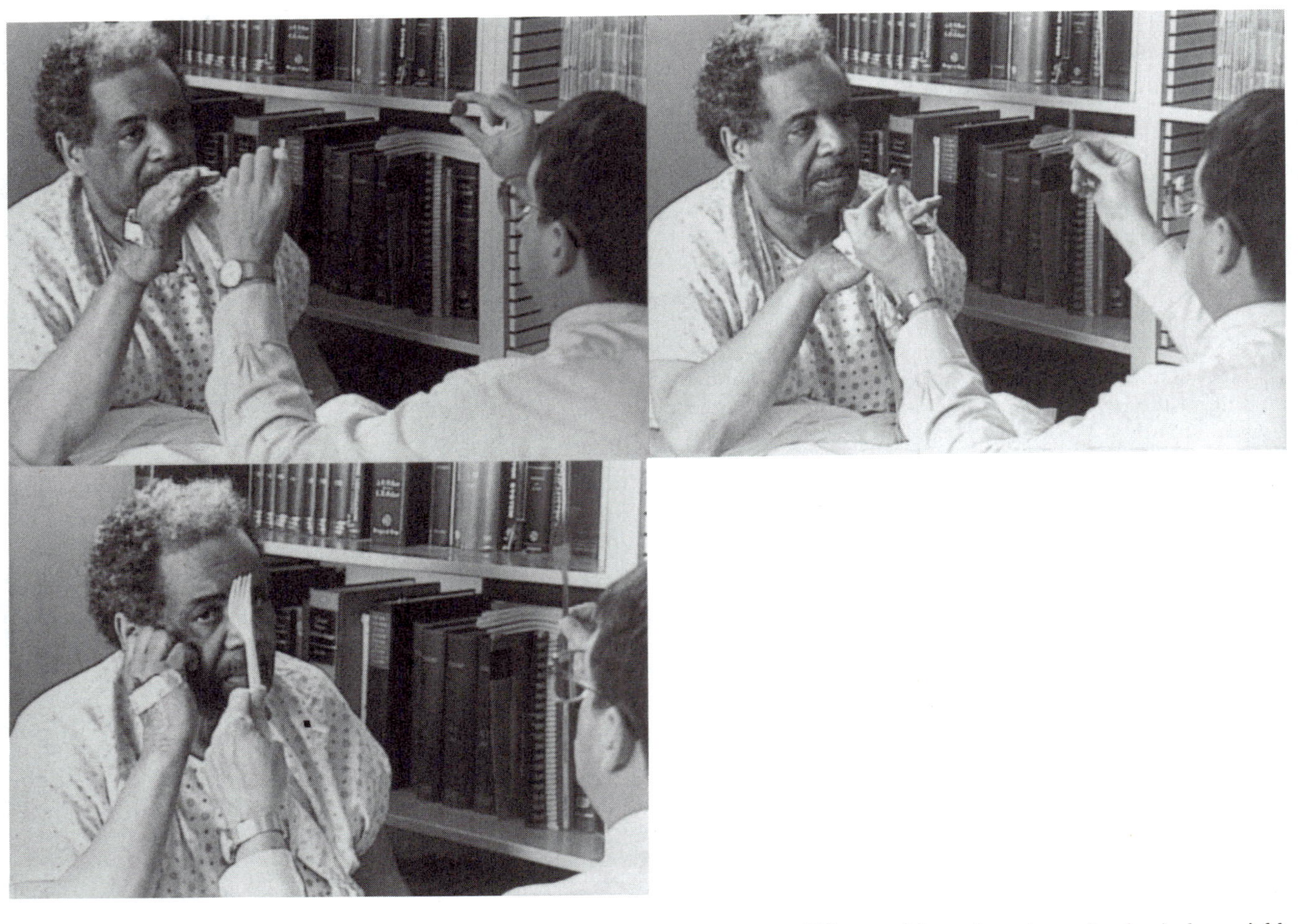

Figure 40.1 Extinction in a patient with recent right frontoparietal infarction. Extinction is influenced by what the competing objects are. (Top left) When shown two identical objects (coins), the patient does not notice and fails to report the one in his left (contralesional) field. (Top right) When shown two different objects (a coin and a key), he quickly reports both. (Bottom) When shown two different kinds of forks (a white, plastic picnic fork and a silver, metal dinner fork), there is extinction of the fork in his contralesional field.

of neglect, hemianopia, and rarer disorders. Our goals are three: to understand how attentional processes may operate in regulating normal perception and action; to identify the neural basis of visual attention; and to understand the psychophysiological basis for the disabilities of these patients in everyday life. The methodological approaches to these issues may be very different. Each focuses on different questions and requires investigation at a different level of analysis.

By studying patients with certain syndromes, we can learn much about the function of attention and what kinds of information processing can proceed without attention. For example, the study of patients with visual neglect can contribute much to the study of early vision: what kind of visual processing occurs at a preattentive level, how the information outside conscious awareness influences behavior, and the role attention plays in integrating perceptual features and selecting objects and actions in guiding goal-directed behavior.

To understand the neural substrates of and the circuitry involved in visual attention, a different approach may be necessary. Patients with such striking clinical syndromes as neglect typically have large lesions that affect several brain areas. Moreover, neglect, defined symptomatically as a defect in spatial orienting behavior, can occur with lesions in very different brain regions: parietal lobe, frontal eye fields, cingulate cortex, basal ganglia, thalamus, or midbrain. The reason for defective orienting may be very different in each case. The neural mechanisms for visual attentional orienting constitute a distributed network that may include all of these regions (Mesulam, 1981).

To learn what role the parietal lobe, midbrain, thalamus, or frontal cortex plays in this distributed network requires another approach—not the study of patients with florid syndromes of attentional dysfunction, but the study of groups of patients with well-circumscribed lesions affecting one of the structures in the network but sparing others. This approach generally requires the selection of patients with relatively small lesions who have no clinically obvious attentional disorder in the chronic stage of illness, at a time when the remote (diaschisis) effects on other parts of the network have resolved.

This kind of analysis has become possible only in the past decade or so with the development of high-resolution neuroimaging techniques, such as computerized tomography (CT) and magnetic resonance imaging (MRI). During this same period that the lesion method has matured in the study of human cognition, other methods of dynamic brain imaging such as event-related potential recordings (from scalp or implanted electrodes), positron emission tomography (PET), magnetoencephalography, and fast-scan MRI have provided converging findings both in brain-injured and normal human subjects. By putting together these various strands of research, a more complete picture is beginning to emerge.

We will first review what has been learned about the role of specific brain structures in orienting attention to locations in the visual field. Then we will consider the role of attention in object recognition. Both syndrome-based observations and anatomically based studies will be considered to illustrate how each approach contributes to the understanding of visual attention.

Neural mechanisms for orienting to locations in the visual field

We will begin by summarizing findings in neurological patients using the paradigm described in Posner's chapter (this volume), designed for measuring the orienting of attention in space. *Covert orienting* refers to allocating attention to a point in space independent of eye position. Although we usually move attention (covert) and the eyes (overt) in concert, it is possible to show that overt and covert orienting can occur separately.

In a typical spatial attention experiment using the precuing method, the subject is asked to respond, by pressing a key, to the appearance of a target at a peripheral location. The target is preceded by a precue, which may summon attention to the target location (valid cue) or to the wrong location (invalid cue) or may have only an alerting value and provide no spatial information (neutral cue). The cue is typically a flash of light at one of the possible target locations or an arrow, usually in the center of the display, instructing the subject where to expect the forthcoming target signal. In most of the studies that have been done in neurological patients, the task is simply to detect the appearance of a large, bright visual signal and make a simple reaction-time (RT) key-press response.

Disengaging Attention After Parietal Lesions

This paradigm for measuring covert shifts of attention has now been widely employed using simple detection RT in patients with both cortical and subcortical brain lesions. Figure 40.2 shows the results of a recent experiment in patients with chronic lesions of the tempero-parietal junction (TPJ) (Senechal, dissertation data, University of California at Davis). Patients in this study did not have visual neglect or clinical extinction. Each had a single, chronic (at least 2 years' postictus) cortical lesion restricted to posterior association cortex and including the TPJ. In this experiment, a peripheral cue was used to summon attention to either the contralesional or ipsilesional field. The target could then appear either at the cued location (valid cue) or in the opposite visual field (invalid cue). Because the goal of this experiment was to examine reflexive orienting, the precue did not predict the location of the forthcoming target, which was as likely to appear at the cued or uncued location. Figure 40.2 shows the detection RT difference between the contralesional and ipsilesional fields for the two cuing conditions. Detection RT in the contralesional field was not slower than in the ipsilesional field if attention had first been summoned there by a valid cue. Therefore, the patients were able to move attention to the contralesional field in response to the cue. When, however, a cue summoned attention toward the ispsilesional field and the target subsequently occurred in the opposite, contralesional field (invalid cue), detection RT was slowed.

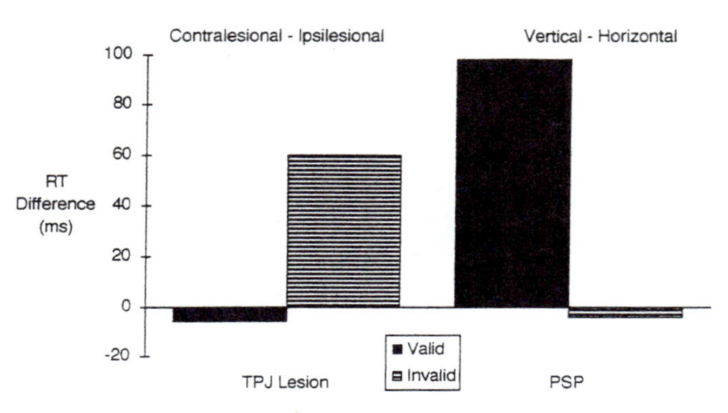

FIGURE 40.2 Covert orienting to peripheral visual signals in eight patients with lesions of the midbrain (progressive supanuclear palsy [PSP]) and in eight patients with lesions of the temperal-parietal junction (TPJ). The subjects' task was to press a response key as soon as they detected a target. A precue (brightening of one of the peripheral boxes) was presented 150 ms prior to onset of the target to summon attention there. This precue did not predict the location of the subsequent target because the experiment was designed to examine purely reflexive orienting. The effectiveness of the cue in summoning attention is inferred from the difference in reaction time (RT) when the target appears at the cued location (valid cue condition) in comparison with an uncued location (invalid cue condition). The results are depicted in terms of difference between the more or less affected field—contralesional and ipsilesional in the case of lateralized cerebral lesions, vertical and horizontal in the case of patients with PSP. Detection RT differences between fields are shown for the valid and invalid cue conditions 150 ms after cue onset for both PSP and TPJ lesion groups. Note that field differences for the PSP patients occurred only in the valid cue condition. This difference occurred because valid cues were not effective in summoning attention in the vertical plane. The opposite pattern of results is found with TPJ lesions. These patients are able to orient contralesionally to benefit from a valid cue. In contrast, they are impaired only in the invalid cue condition. When attention is summoned by an invalid cue in the ipsilesional field, detection RTs for contralesional targets are longer.

These findings suggest that lesions of the TPJ do not affect the ability to move attention toward the contralesional field; rather, they impair the ability to disengage attention to detect a contralesional target signal.

This extinctionlike RT pattern has now been found in several studies (Posner et al., 1984, 1987a, b; Baynes, Holtzman, and Volpe, 1986), and it has been shown that the degree of this disengage deficit correlates with the severity of clinical neglect (Morrow and Ratcliff, 1988). Later we consider some of the properties of this disengage deficit, and try to specify how it relates to clinical neglect and extinction.

MIDBRAIN MECHANISMS FOR REFLEXIVE ORIENTING The encephalization of visual function in cerebral cortex is a relatively new development in phylogeny. The geniculostriate pathway is fully developed only in mammals. The demands of increasingly complex visual cognition presumably generated the evolutionary pressures leading to the development of a completely new, parallel visual pathway in mammals. In lower vertebrates, vision is mediated by input through the retinotectal pathway to the superior colliculus of the midbrain. What function does the phylogenetically older midbrain pathway serve in humans?

There is converging evidence for a special role of the midbrain pathway in reflexive orienting in humans from four sources: First, patients with lesions of the midbrain have been studied to define what visually guided behaviors are impaired. Second, patients with hemianopia due to lesions of the visual cortex are examined to determine what visually guided behaviors are preserved when only the midbrain visual pathway is available. Third, extrageniculate vision is studied in normal subjects by comparing the orienting of attention into the temporal and nasal hemifields. This approach exploits a difference in the normal anatomy of cortical and subcortical visual systems: The phylogenetically older midbrain pathway is asymmetrically represented, with the temporal hemifield receiving more visual information. Fourth, the development of visual orienting in neonates has been studied to identify what kinds of visual processing and orienting are mediated by the midbrain pathway before the geniculostriate pathway matures.

ORIENTING OF VISUAL ATTENTION IN PROGRESSIVE SUPRANUCLEAR PALSY Progressive supranuclear palsy (PSP) is a progressive degenerative disorder affecting subcortical nucleii of the diencephalon, midbrain, cerebellum, and brain stem. Because the basal ganglia and the substantia nigra are involved, the clinical picture shares many features with Parkinson's disease. In addition, however, there is degeneration, unique to this disease, involving the superior colliculus and adjacent peritectal region. This pathological process results in the distinctive paralysis of voluntary eye movements, especially pronounced in the vertical plane. The study of these patients affords a special opportunity to understand the function of the midbrain extrageniculate pathways in regulating human visually guided behavior.

The midbrain pathology of this disease not only produces a compromise of eye movements but also results in a striking and distinctive global derangement of visually guided behavior (Rafal, 1992). Although visual acuity is not affected, patients with PSP behave as if they were blind, even at a stage in the disease when their eyes are not totally paralyzed. They do not orient to establish eye contact with persons who approach them or engage them in conversation; they also do not look down at their plate while eating but rather grope for their food without looking.

Using the same paradigm described earlier for studying patients with TPJ lesions, Rafal and colleagues (1988) showed that PSP patients are slow not only in moving their eyes but also in moving covert visual attention. Because PSP causes greater impairment in vertical than in horizontal saccades, a comparison of the speed of orienting covert attention in the vertical and horizontal planes was made. Parkinson's disease patients, who served as controls in this study, showed comparable attention shifts in the vertical and horizontal plane. The results for the PSP patients are shown in figure 40.2. The difference in detection RT for vertically and horizontally placed targets is shown for valid and invalid cues. In the invalid cue condition, detection RTs for vertical and horizontal targets were comparable. However, in the vertical plane, valid cues were not effective in summoning attention, and detection RT did not differ between valid and invalid cues. Hence, RTs differed between vertical and horizontal targets in the valid cue condition but not in the invalid cue condition.

Converging evidence that the retinotectal pathway functions in normal reflexive orienting was obtained from a study of normal adults (Rafal, Henik, and Smith, 1991). This study exploited a lateralized neuroanatomical arrangement of retinotectal pathways that distinguishes them from those of the geniculostriate system—namely, more direct projections to the colliculus from the temporal hemifield. Benefits and costs in orienting to an exogenous cue were greater when attention was summoned by signals in the temporal hemifield.

Blindsight: Saccade Inhibition by Signals in the Hemianopic Field What visuomotor function is preserved when only the retinotectal pathway is competent to process visual input? Humans who become hemianopic due to destruction of the primary visual (striate) cortex are rendered clinically blind in the entire hemifield contralateral to the lesion and cannot see even salient signals, such as a waving hand, within the scotoma (the blind area). They are unable to report such events and deny any awareness of them. In no other species, including monkeys, do striate cortex lesions produce such profound and lasting blindness. It is clear that the geniculostriate pathway in humans is dominant over the phylogenetically older retinotectal pathways. It is perhaps not surprising that the neuroscientific community came to view the older pathway as being vestigial in humans, providing little service to normal vision.

Yet it is striking how well these patients compensate for their visual loss. With time, function in everyday life seems almost normal, with little hint that such patients have lost half their visual field. This remarkable compensation may be mediated, in part, by preserved retinotectal visual pathways, which process information that, while not accessible to conscious awareness, can nevertheless trigger orienting responses toward the hemianopic field. This so-called blindsight has been demonstrated by requiring hemianopic subjects to move their eyes or reach toward signals that they cannot see and by employing forced choice discrimination tasks (Weiskrantz, 1986). The physiological mechanisms responsible for blindsight remain uncertain, and the role of the retinotectal pathway is controversial. In some patients, there is residual vision, which could be mediated by spared geniculostriate fibers and which could reflect degraded cortical vision near the perceptual threshold (Fendrich, Wessinger, and Gazzaniga, 1992). Other investigators propose that blindsight reflects processing of visual input from retinotectal afferents to the superior colliculus (Weiskrantz, 1986).

Rafal and coworkers (1990) showed that a visual signal in the hemianopic field, for which the patient had no conscious awareness, reflexively activates the colliculus to produce an inhibition of orienting to the opposite field. This experiment studied patients who had suffered an occipital stroke that destroyed geniculostriate pathways; they were blind in the visual field opposite the lesion and could not report the presence or absence of stimuli presented there. They maintained fixation on the middle of a video display and on each trial made a response to the appearance of a target in the intact hemifield. Blindsight was inferred from the effect of an unseen visual distractor in the hemianopic field on response to a target in the intact field.

Each eye was tested separately so that the effects of the unseen distractors presented in the temporal and nasal hemifields could be compared. The results showed that unseen distractor signals presented to the blind temporal (but not nasal) hemifield of hemianopic patients increased the latency of saccades directed to targets presented in the intact visual field. Braddick and associates (1992) have recently shown that babies who are hemianopic after hemispherectomy do make eye movements toward signals in the temporal blind hemifield. These results provide direct evidence that there is a reflexive activation of retinotectal pathways to prime the oculomotor system and that this activation inhibits saccades to the opposite field.

Effects of Cortical Lesions on Reflexive and Voluntary Orienting To define the role of specific brain areas such as the temporoparietal junction, the inferior parietal lobule, or the frontal eye fields in visual orienting, it is necessary to study groups of patients with lesions in these areas and to compare them to patients with lesions in adjacent cortex but sparing the region of interest. Moreover, it is important to examine patients with chronic lesions who are clinically stable. Acute lesions in one area have distributed physiological effects affecting functionally and anatomically related structures. These effects, referred to as *diaschesis*, are revealed through dynamic metabolic brain-imaging techniques such as PET and SPECT scanning.

This approach, which has been most effectively developed by Robert T. Knight at UC Davis, proceeds from the careful selection of patients with chronic unilateral lesions. Groups of patients with a single lesion involving only the dorsolateral prefrontal cortex or only the posterior association cortex are tested *after* they have recovered from the acute stage of the illness (at least 6 months after the stroke). Patients with more than one lesion, dementia, mental illness, or other debilitating medical problems are excluded from these group studies. Most of these patients are functioning independently at a rather high level.

Figure 40.3 shows the neuroimaging reconstructions of the patient groups in whom we have measured voluntary and reflexive saccades. One study (Henik et al., in press) examined the effects of lesions of the dorsolateral prefrontal cortex (DLPFC), which extended into the frontal eye fields (FEF); a group of patients with DLPFC lesions sparing the FEF served as controls. Another experiment, conducted with K. Yamashita, focused on the effects of lesions of posterior association cortex that involved the TPJ; a group of patients with lesions of parietal lobe that spared the TPJ served as controls.

In these studies, reflexive saccades were made to a peripheral visual target appearing in either the ipsilesional or contralesional field. Voluntary saccades were made in response to an arrowhead in the center of the display that pointed toward either the ipsilesional or contralesional field. For each type of eye movement, saccadic latencies toward the ipsilesional field were compared to the saccadic latencies toward the contralesional field. To determine whether any effects found were specific for eye movement and not to shifts of covert attention, each patient also was tested in a detection control condition in which they viewed the same display but kept the eyes fixated and made key-press responses to the target.

As shown in figure 40.4, FEF lesions had opposite effects on endogenously activated and visually guided saccades to external signals. Voluntary saccades were slower to the contralesional field, whereas exogenously triggered saccades were faster to the contralesional field. These results indicate that FEF lesions have two separate effects on eye movements: First, the FEFs are involved in generating endogenous saccades, and lesions in this region therefore increase their latency. Second, the FEFs have inhibitory connections to the midbrain, and lesions in this area result in disinhibition and a consequent decrease in latency for reflexive saccades to exogenous signals. In contrast, lesions of the TPJ *increased* the latency for reflexive saccades to contralesional targets but did affect contralesional voluntary saccades. Thus, lesions of the FEF and the TPJ have opposite effects on the latency of reflexive saccades.

Figure 40.3 Neuroimaging reconstruction of representative patient groups with chronic unilateral lesions of dorsolateral prefrontal cortex (DLPFC) or posterior association cortex tested in the saccade experiments. DLPFC group subdivided into those in whom the lesions extend into the frontal eye fields (N = 9) (top left), and those in whom the FEF is spared (N = 7) (top right). The posterior association cortex lesion group is subdivided into those in whom the lesion involves the TPJ (N = 5) (bottom left), and those in whom the lesion spares the TPJ (N = 5) (bottom right).

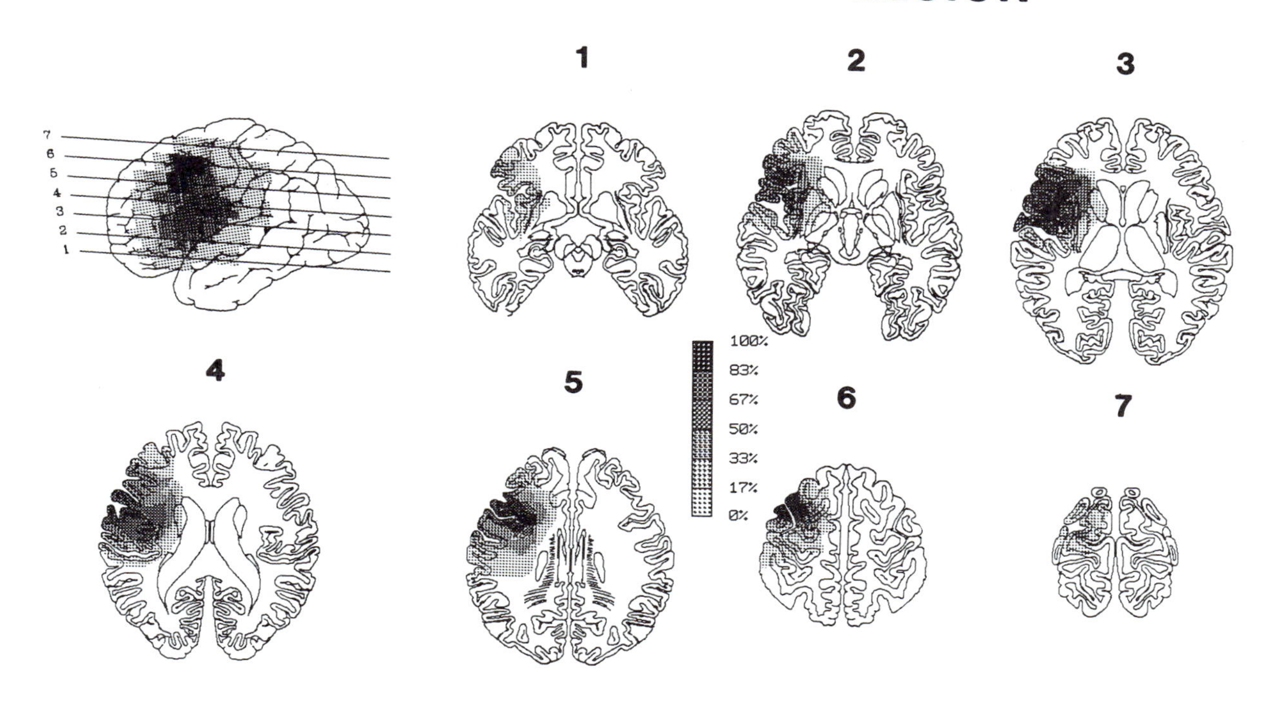
A
FRONTAL EYE FIELD LESION
1
2
3
4
5
6
7
100%
83%
67%
50%
33%
17%
0%

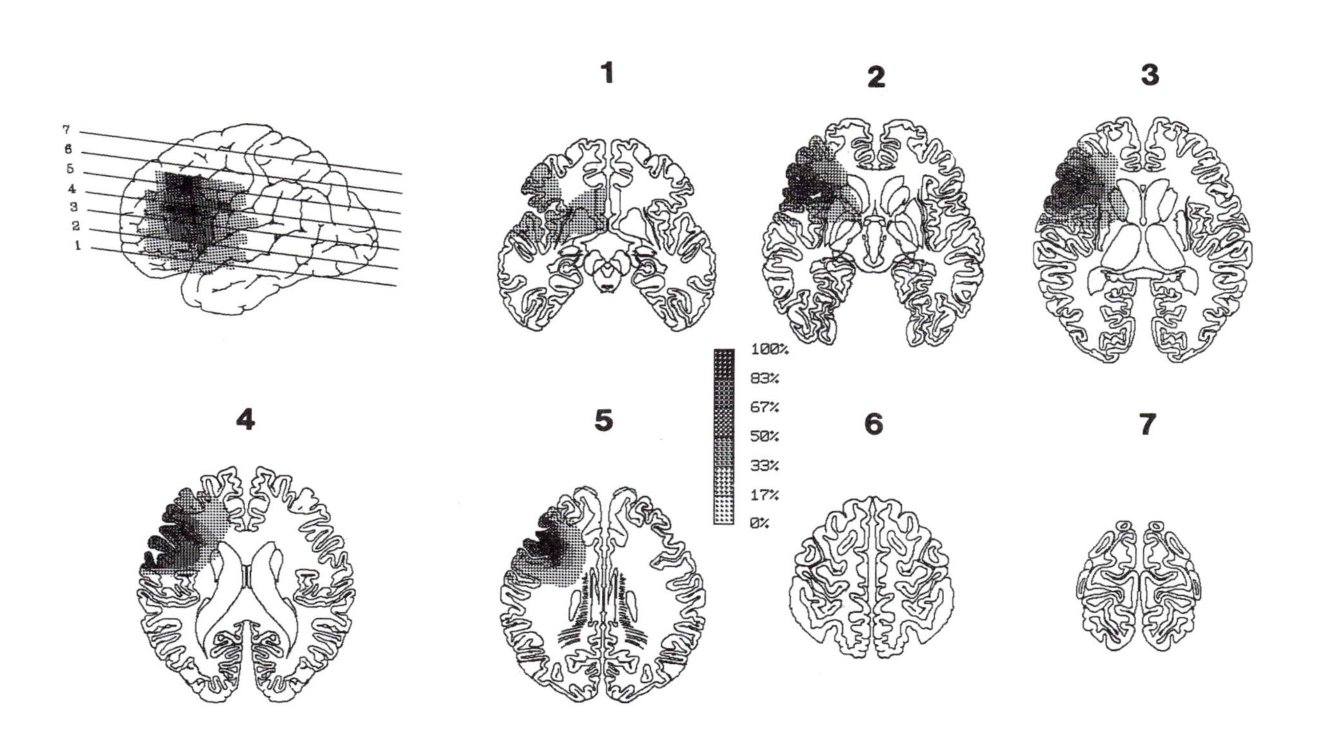
B
NO FRONTAL EYE FIELD LESION
1
2
3
4
5
6
7
100%
83%
67%
50%
33%
17%
0%

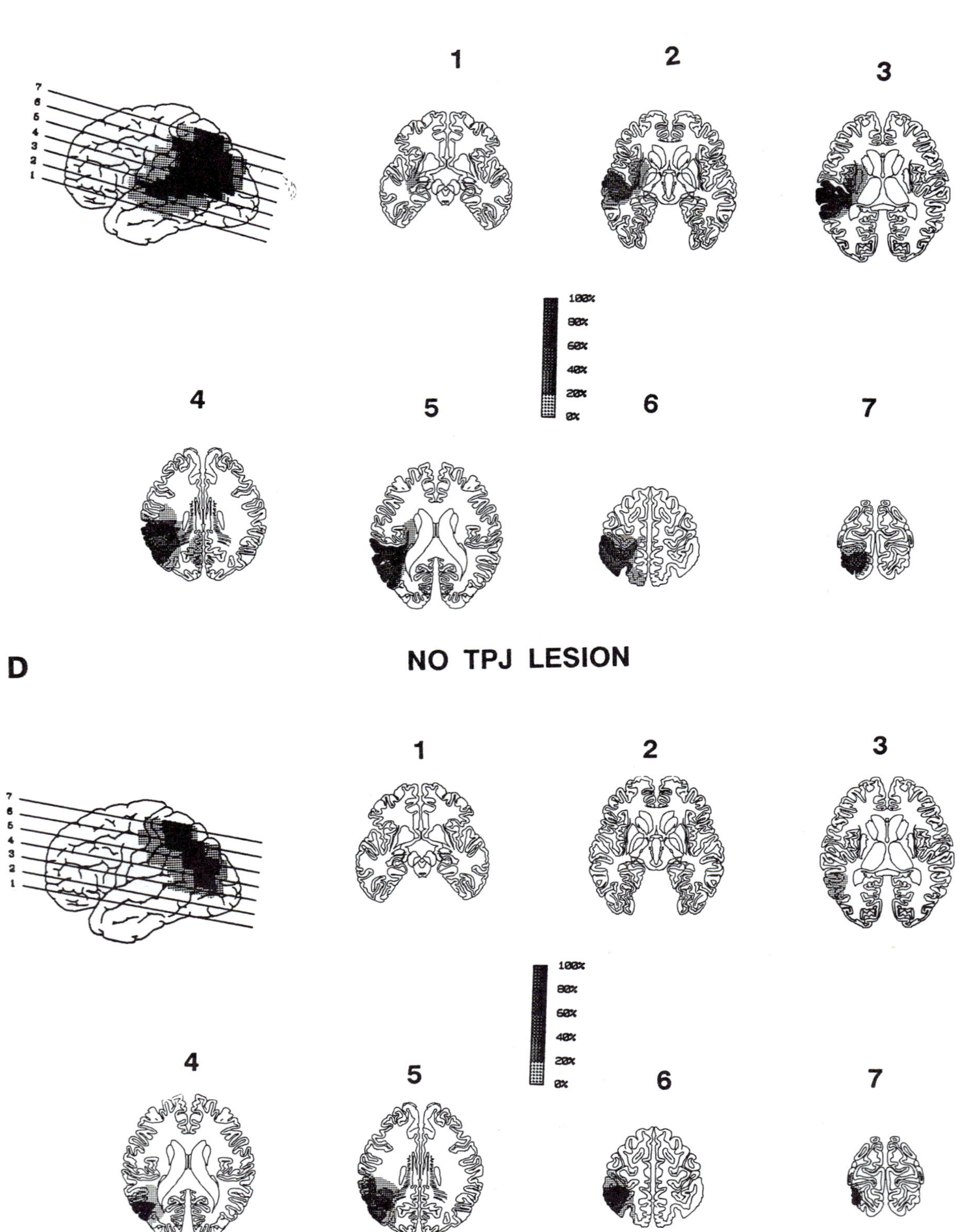

FIGURE 40.3 (cont.)

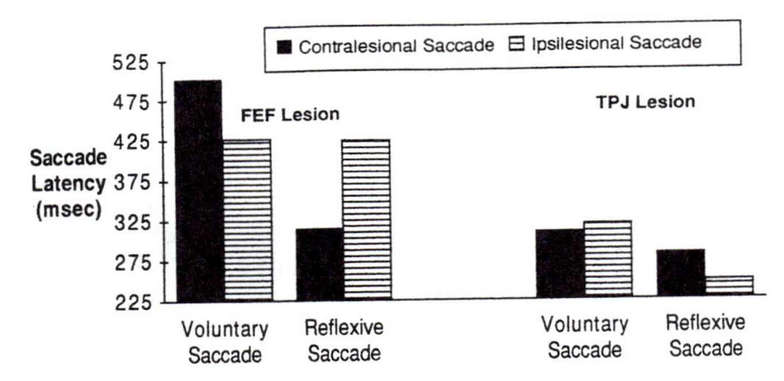

FIGURE 40.4 Latencies (in milliseconds) for reflexive and voluntary saccades toward ipsilesional and contralesional fields in frontal eye field and temporal-parietal junction lesion groups. In patients with DLPFC lesions, only those involving the FEF (left) had lateralized effects on saccade latencies. Latencies for contralesional reflexive saccades were less than for those to the ipsilesional field, whereas latencies for voluntary saccades were longer toward the contralesional field. In the patients with posterior association cortex lesions, those with TPJ lesions (right) showed an increased latency for contralesional saccades without having a lateralized effect on voluntary saccades. (Note that the experiments done in the two patient groups shown in this figure were not identical, and the absolute saccade latencies cannot be compared directly. The cue-target intervals were shorter and, hence, the pace of the task was quicker in the posterior lesion group experiment. This difference between absolute saccade latencies between the experiments was also true for the normal and neurological controls tested in those experiments.)

The dissociations observed in these studies provide evidence that there are separate brain mechanisms for covert orienting and for eye movements: Midbrain systems are critical for reflexive orienting, whereas cortical structures—the FEF and TPJ—are involved in voluntary orienting. Having dissected the neurology of exogenously and endogenously triggered attention shifts and eye movements, the job now is to determine how they work together in regulating normal orienting behavior. We are beginning to get a picture of what these neural structures are and how each, with its own evolutionary history and place in ontogeny, contributes to adaptive orienting. How these subsystems are integrated and coordinated is a key question for understanding both normal and deranged orienting behavior.

The nervous system routinely goes about its business through an orchestration of facilitatory and inhibitory processes. Even the simplest mental operation is likely to be associated with activation of some structures and inhibition of others. Similarly, a lesion is likely to produce not only loss of some functions but disinhibition of others. The data from the patients with FEF lesions shown in figure 40.4 is a rather striking example of this. Note that the latencies for reflexive and voluntary saccades toward the ipsilesional (i.e., the "good") field are not different. Reflexive saccades to visual signals normally are triggered with shorter latencies than voluntary saccades. These patients appear not to make reflexive saccades toward their ipsilesional field; rather, when a signal occurs there, they are obliged to use the slower voluntary system to trigger the eye movement. We might consider the possible cascade of effects from FEF lesions as follows: The FEF lesion not only impairs the ability to make voluntary saccades to signals in the contralesional field; it also results in disinhibition of the superior colliculus on the side of the lesion. This disinhibition results in relatively quicker reflexive saccades toward the contralesional field and may account for the difficulty these patients have in inhibiting reflexive glances in an antisaccade task (Guitton, Buchtel, and Douglas, 1985). The complexity of these interactions needs to be kept in mind when attempting to understand not only normal brain function but also neurological disabilities after brain damage. As we shall see later, cortical-subcortical interactions may play an important role in the symptomatology of visual neglect and may be relevant to developing rehabilitation strategies.

Objects and parts of objects in the visual scene

Spatial relationships between objects in the world come in various forms. Objects are not only located next to one another with spatial gaps between them, but they also occur in more complex spatial relationships such as within one another or as a part of one another. Windows and doors are parts of houses, but they are special types of parts. They are local objects embedded within more global objects. Experimental data as well as our own experience tell us that the visual system not only perceives the hierarchical structure of such objects but can selectively attend to different levels of objects. Attention can be directed to a house or it can be directed to its windows and doors or to a wall or the material from which it is made. One look around the world will convince the reader that this type of spatial relationship between objects is extremely common.

Briefly, the picture that is beginning to emerge is one in which mechanisms associated with the left hemisphere, specifically the TPJ, are biased to process local information in normals, whereas mechanisms associated with the right are biased to process global information. This is the case even when patterns vary in visual angle. Other evidence suggests that these biases emerge only when forms at different levels must be identified (presumably reaching conscious awareness). Still other mechanisms preattentively respond to the structure of the objects at the two levels to produce interactive effects between levels. Finally, attentional mechanisms can change the asymmetrical biases either by selectively altering the size of an attentional window or by modulating spatial frequencies associated with the two levels. These attentional mechanisms can be disrupted independently by damage to different cortical areas. The evidence for these distinct mechanisms will be the subject of the following sections.

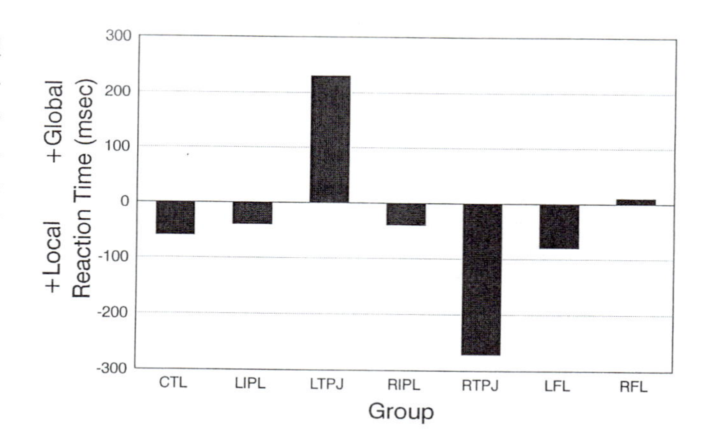

FIGURE 40.5 Mean global advantage (positive) or local advantage (negative) for normal controls and different patient groups. CTL, controls; LIPL, left inferior parietal lobe; LTPJ, left temporal-parietal junction; RIPL, right inferior parietal lobe; RTPJ, right temporal-parietal junction; LFL, left dorsolateral prefrontal cortex; RFL, right dorsolateral prefrontal cortex.

HEMISPHERIC DIFFERENCES AND THE TPJ In a series of studies, we presented hierarchical patterns with local and global forms to groups of patients with chronic, single unilateral lesions centered in different areas of the cortex. The stimuli in these studies were large letters constructed from small letters. The goal was to use these hierarchical stimuli to determine which neural regions could be responsible for so-called Gestalt processing deficits, long associated with right hemisphere damage, and the processing of details, long associated with left hemisphere damage. Across studies, we found the expected asymmetrical deficits only in patients with lesions involving the TPJ.

In one study, a single hierarchical pattern appeared in the middle of the screen for 100 ms, and subjects were instructed to indicate, by pressing one of two switches, which of two predesignated targets appeared. As shown in figure 40.5, abnormal global or local biases occurred only for TPJ patient groups. Patients with left-hemisphere damage in this region were slower to respond to local targets, and patients with right-hemisphere damage in this region were slower to respond to global targets. Patients with parietal damage that did not extend into this region did not produce this asymmetry nor did patients with dorsolateral frontal lobe damage.

Patients with more anterior temporal lobe lesions due to temporal lobectomy have also been reported to have difficulty with global information with right- but not left-hemisphere resection (Doyon and Milner, 1991). Left-hemisphere lobectomy did not affect local processing. These findings may have occurred because the amount of tissue removed was greater on the right than the left. Overall, the right temporal lobectomy resections extended further posteriorly, which would be closer in location to the TPJ. Recent studies using evoked potentials have also revealed asymmetries to hierarchical patterns in normal subjects over posterior temporal scalp electrodes, consistent with those predicted by the patient data (Heinze and Munte, 1993).

The global/local asymmetry can be observed in TPJ groups whether the stimuli are shown in central or peripheral vision, and asymmetries occur whether the stimuli are presented in the contralateral or ipsilateral visual field to the lesion. Visual resolution falls off rapidly in sensory and primary vision as one moves out from the fovea, and local patterns suffer more than global patterns when stimuli are presented in the peripheral visual field. A mean global advantage with peripheral presentation in normals changes to a mean local advantage with central presentation. More importantly, the relationship between the advantage shown by normals and TPJ lesion groups does not change; the patients diverged from normals for both central and peripheral presentation. This result demonstrated that the asymmetry was superimposed on primary visual processes that favor global or local properties in sensory vision. Further evidence to support this conclusion was found in a study where we varied the visual angle of centrally presented hierarchical patterns (Lamb, Robertson, and Knight, 1990). Although the overall local advantage changed across visual angles for normals, the divergence from normals for the two TPJ lesion groups remained relatively constant.

Stimulus features that might account for this divergence over visual angle and peripheral versus central presentation have not been tested directly in the TPJ lesion groups. However, studies in normals suggest that the differences are due to deficits in responding to various spatial frequencies that differentiate levels, a hypothesis originally proposed by Sergent (1982) using hierarchical patterns and half-field presentation in normals. Subsequent investigators have supported this conclusion by presenting pure sine wave gratings in the right or left visual field (Kitterle, Christman, and Hellige, 1990; Christman, Kitterle, and Hellige, 1991), and several studies have demonstrated links between spatial frequency analysis and identifying global or local forms (Shulman and Wilson, 1987; Hughes, Fendrich, and Reuter-Lorenz, 1990; Badcock et al., 1991; LaGasse, 1993).

In a hierarchical pattern, the global letter invariably will contain lower frequencies than the local letters, independent of visual angle, whereas absolute frequencies will vary with visual angle or stimulus size. Central versus peripheral presentation would also change the frequency spectrum represented by the visual system in terms of absolute frequencies, yet the local letters would still contain relatively higher frequencies than the global letters, independent of presented location. The data supporting the conclusion that the global-local asymmetry over right and left TPJ lesion groups is superimposed on primary visual information is consistent with the idea that the globally and locally biased mechanisms respond to *relative* spatial frequencies at the two levels. These mechanisms, associated with right and left TPJ regions respectively, could accomplish this analysis by performing a type of high- or low-pass filter on initial frequency registration. Interposed between these higher-order filters and the sensory representation would have to be some mechanism that selects the frequency range to be piped through these filters to produce the relative effects.

Attention to Levels and Left Inferior Parietal Lobe In one of our earlier studies, we investigated whether attentional manipulations could alter asymmetrical biases (Robertson, Lamb, and Knight, 1988). We varied the probability, and thus the expectation, that a target would appear at a given level in a divided attention paradigm. Normal subjects respond to global targets more rapidly when global targets are expected and to local targets when they expected. This was true also for TPJ lesion groups, who showed the asymmetrical biases revealed in figure 40.5. In other words, attentional manipulations also influenced these subjects. Conversely, these manipulations did not produce expectancy effects in a group of patients with lesions centered in inferior parietal lobe (IPL) only a few centimeters away.

Hierarchical patterns were presented in the center of the screen to matched controls and patient groups with lesions in different areas. The subject's task was to identify targets that appeared at one level or the other. Subjects were not required to report at which level the target appeared, only what the target was on each trial. On every trial, a target appeared either at the global or the local level but never at both. There were two targets designated before the session began, and the subject's task was to press one switch if one target appeared and another if the other target appeared. The remaining level always contained a nontarget letter. One second after the response, a new stimulus was presented, and again the subject's task was to identify which target was present by pressing the appropriate switch.

Attention was investigated by varying the probability that the target would appear at the global or local level between blocks of trials. In one block of trials, the target was more likely to be global and in another it was more likely to be local. Normal subjects, whether younger or older, produced reliable crossover effects in RT and errors consistent with a change in attentional distribution between blocks (figure 40.6). When global targets were more likely over a block of trials, RT was faster to report the identity of the global than local form and, when local targets were more likely, RT was faster to report the identity of the local than the global form. The top portion of figure 40.6 shows the data from a group of controls and different patient groups. The only group that showed a reduced crossover effect was the one with lesions centered in area 39 of the left IPL (LIPL). The bottom half of figure 40.6 shows a replication of this finding; the only procedural change between the two experiments was that, in the second, subjects were informed of the probability manipulations and had to report which condition they were in

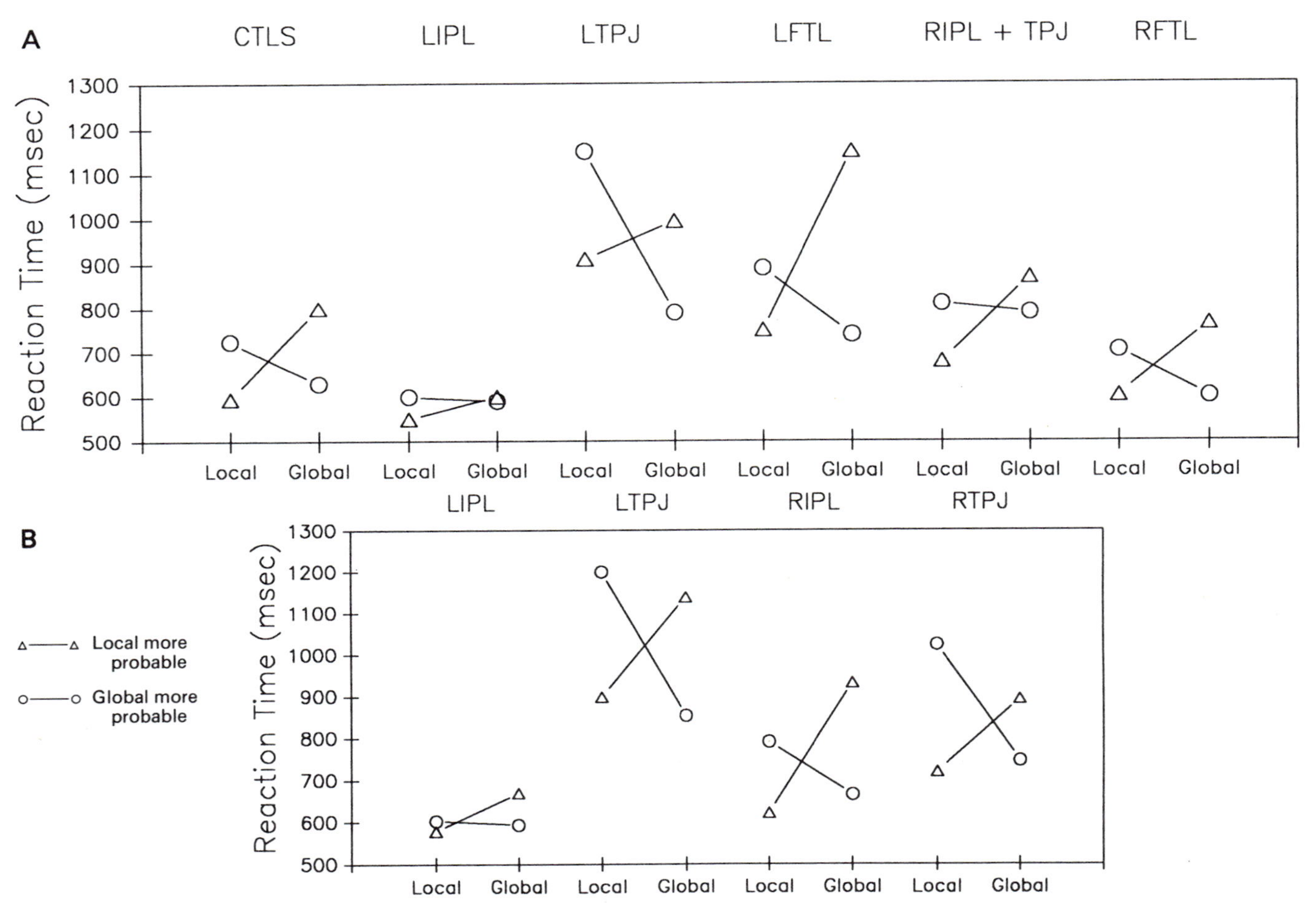

FIGURE 40.6 Mean reaction time for a group of normal controls and groups of patients as a function of varying probability of a target level. CTLS, controls; LIPL, left inferior parietal lobe; LTPJ, left temporal-parietal junction, LFTL, left frontal lobe; RIPL, right inferior parietal lobe; RFTL, right frontal lobe; RTPJ, right temporal-parietal junction. (A) Data collected without informing the subjects of the probability manipulations. (B) Data from a new group of subjects when they were informed of the probability manipulations.

before and after each block of trials. Again, only the LIPL group showed reduced crossovers.

We originally interpreted the results in LIPL groups as indicating a deficit in controlled attentional allocation between levels. However, we recently reanalyzed the RT data on a trial-by-trial basis to examine possible sequential or level "priming" effects across trials. The question was whether the crossovers exhibited in figure 40.6 where due to tonic effects of the overall expectancy within a block of trials or phasic effects that occurred on a trial-by-trial basis. This analysis revealed that the crossover could be attributed to a mechanism that kept track of what *level* had been selected for identification on the previous trial (i.e., what level had been selectively attended.) There were priming effects specific to a level that left an attentional print from one trial to the next, independent of whether the targets were the same or different, and they accounted for much of the crossover effects in RT to global and local targets over blocks of trials (figure 40.7).

These level priming effects could not be attributed to response or target repetition. A global *H* preceded by another global *H* was responded to nearly as rapidly as a global *H* preceded by a global *S*. Furthermore, a global *H* preceded by a local *H* was responded to more slowly. The same relationship held for local targets. A local *H* preceded by another local *H* was responded to nearly as rapidly as when preceded by a local *S*, whereas a local *H* preceded by a global *H* was responded to

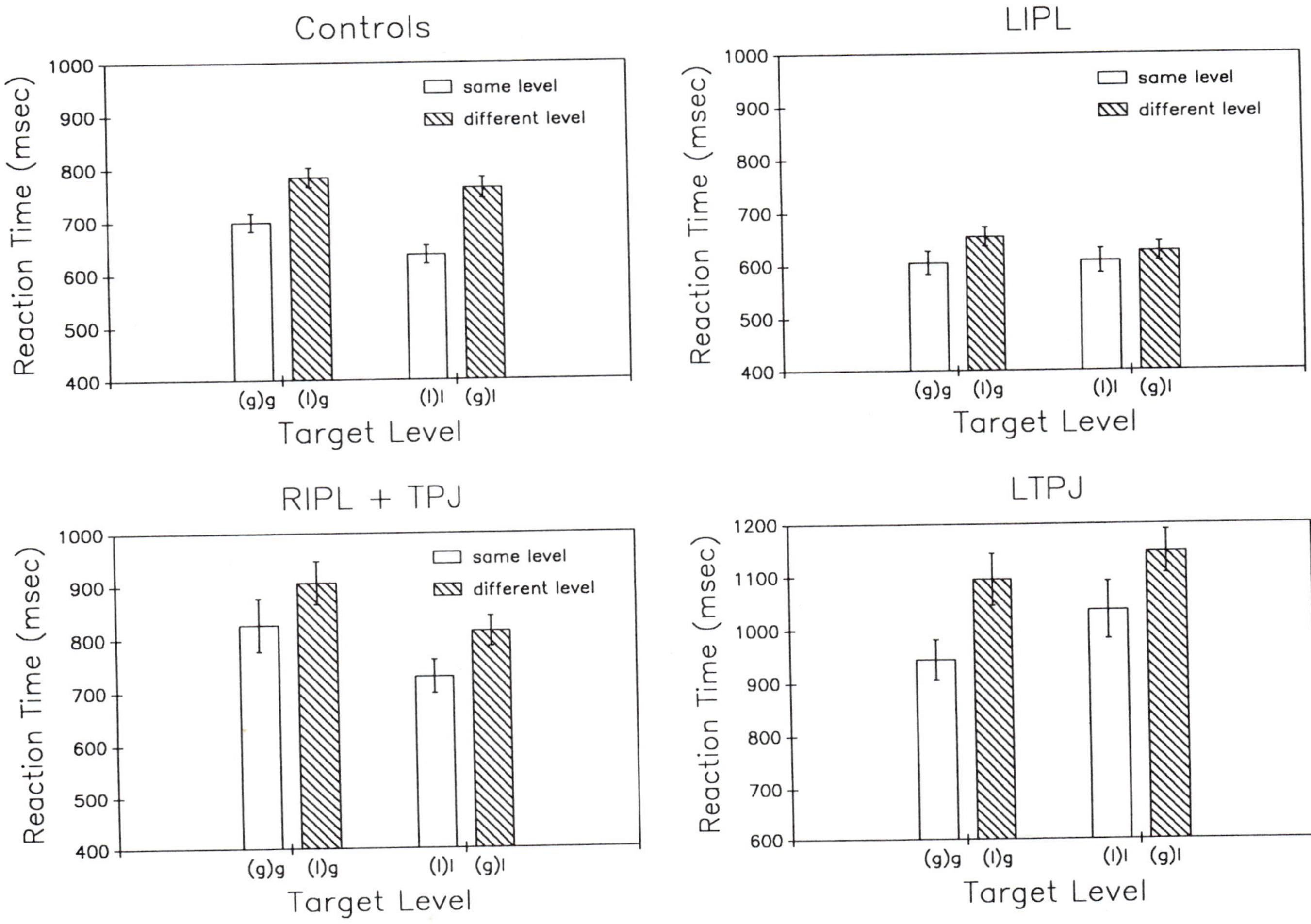

FIGURE 40.7 Mean reaction time for trial $N + 1$ targets at the global [g] or local [l] level, with the preceding trial N target level represented in parentheses for normal controls, a left inferior parietal lobe group (LIPL + TPJ), a right inferior parietal lobe group (RIPL), and a left temperal-parietal junction lesion (LTPJ) group.

more slowly. These findings demonstrate that it is not the internal representation of the attended target per se that is carried over from trial N to trial $N + 1$. Rather, it is the level of structure of the attended target on the previous trial. Data from subsequent studies show that the sequential effects are independent of stimulus location on trial N and trial $N + 1$: They are level-specific and not location-specific.

What is the feature or dimension that is sustained over time, independent of the representation of the form itself? One promising candidate is spatial frequency. Shulman and Wilson (1987) directed normal subjects to respond to the global level of hierarchical stimuli, followed by a detection response to sinusoidal gratings that differed in spatial frequency. These investigators found that high-frequency gratings were better detected when subjects had just responded to a local letter, whereas low-frequency gratings were better detected when subjects had just responded to a global letter. In a recent study, we found that our sequential effects also could be attributed to a sustaining effect of spatial frequency (Robertson, submitted). We do not yet know what the effects of these manipulations are in patient groups with lesions in different areas, but we do know that other features that can be used to differentiate objects (color and location) will produce similar sequential effects in normals. Together these effects suggest that the mechanism disrupted with LIPL damage is one that, under normal conditions, maintains a feature representation of the level that was attended on the previous trial. It is likely that this representation contains other features as well. We assume that attention was allocated to the target level in order to identify the pattern. When the next stimulus appears, attention was still "located" on the same feature or value that was used to discriminate forms. Under such conditions, there would be a cost associated with moving attention to a different value, somewhat analogous to the costs observed when attention must move from a cued location to an unexpected location.

Integration and Segregation of Global and Local Structure The speed of identifying a global or local target in a hierarchical pattern is not sufficient to explain how global and local forms are perceived as separate objects that are also integrated into a single whole. When we look at a hierarchically structured scene, we do not integrate local objects in a way that they disappear into the more global pattern. Although we see the local objects as parts of the global form, they do not merge with the global form as does a part such as the edge of a table or the texture of the wall. Rather, we comprehend the fact that different objects exist at different levels but are also integral with one another.

Evidence from patient groups is consistent with the hypothesis presented by Robertson and Lamb (1991) that interference effects between levels are supported by interhemispheric communication between posterior regions of the corpus callosum. Interference effects between levels occur when forms at the two levels are dissimilar to one another. For instance, in a large letter *H* made of small *H*s, the global *H* and local *H* are consistent and are as similar as any global and local pattern can be. In a large letter *S* made of small *H*s, the global *S* and local *H* are inconsistent and less similar. Likewise, a global *A* is more similar to a local *H* than it is to a local *E*. Response time is typically faster when the two levels are similar than when they are dissimilar (e.g., Lamb and Robertson, 1989). This interaction is completely absent in the same TPJ lesion groups that show asymmetrical performance in target identification speed, and this absence occurs whether the lesion is on the left or right. In other words, whether there is an overall global or local advantage, the interaction effects between levels can be eliminated (Lamb, Robertson, and Knight, 1989, 1990). Subsequent studies with three commissurotomized subjects showed that normal interaction effects were also eliminated in these patients with full commissurotomy, whereas determinations of visual field by speed of global-local identification were consistent with more efficient processing of global forms by the right hemisphere and more efficient processing of local forms by the left hemisphere (Robertson, Lamb, and Zaidel, 1993).

Finally, in a patient with bilateral inferior parietal-occipital damage, normal interaction effects were found with a nearly complete inability to identify global forms (see section below on Bálint's syndrome). These data together demonstrate that the interaction effects between levels, and mechanisms that identify forms at different levels, use different pathways and can be affected independently.

Attention to objects and locations

Disengagement of Attention and Neglect in Visual Search Earlier, we discussed how the phenomenon of extinction, in which one visual stimulus competes with another, might be attributable to a deficit in disengaging attention from a location. We now turn to studies of attentional search in a cluttered field in which many objects are competing for attention. This is a more typical situation in the world and the context in which neglect is manifest: a defect in exploratory behavior toward the contralesional side that is dependent on the number of objects in the field.

Eglin, Robertson, and Knight (1989) used a visual search task of the type developed by Treisman and Gelade (1980). They varied the side of a predesignated conjunction target (one requiring the conjuction of more than one feature to identify) among a variable number of distractors. Normally, subjects are asked to report the presence or absence of a target. To overcome the bias patients with neglect would have to say no whenever the target appeared in their contralesional field, Eglin et al. (1989) changed the method to a localization task and timed subjects' ability to point to a target. Under these conditions, even patients with relatively severe neglect continued to search into the neglected space because they knew that a target must be present. The target and number of distractors varied and were crossed with right and left sides of the display.

As long as no distractors appeared on the ipsilesional side of the display, no differences were found in locating a target on the neglected and intact sides. In other words, in displays that were limited to the ipsilesional side of a page, there were no objects to attract attention to the intact side and so nothing from which to disengage attention. Under these circumstances, these patients searched the display on the left as readily as they searched displays on the right. For bilateral displays, all patients eventually moved attention to the neglected side if the target was not found on the intact side, and this contralateral delay increased as a function of the number of distractors or objects on the intact side. Each distractor on the intact side added approximately three times the base search rate to the contralateral delay. Disengaging attention from the ipsilesional field of distractors to move attention to the contralesional field was evident and depended on the number of items in the display.

Mark, Kooistra, and Heilman (1988) describe an elegantly simple demonstration that patients with neglect have difficulty in disengaging attention only when ipsilesional items are present. They used a line cancellation task, a conventional bedside method for demonstrating and measuring neglect. The patient is shown a page filled with lines and is asked to cross them all out. Typically, a patient with left hemineglect fails to cross out many of the items on the left side of the page. Mark, Kooistra, and Heilman (1988) compared this conventional cancellation task with another condition in which the patient was asked to erase all the lines. As each line was erased, and thus no longer present, the patient no longer had to disengage from it before moving on. Performance was strikingly better in this erasure task than in the conventional line cancellation task.

PREATTENTIVE VISUAL PROCESSING IN THE NEGLECTED FIELD What is extinguished in visual neglect? Are extinguished stimuli processed and, if so, in what way does this information affect behavior? Certainly, awareness of the signal is compromised. Is it the case that all perceptual processing of the extinguished object is below the threshold of awareness? If not, to what level is information about the extinguished object processed?

Figure 40.1 shows that "extinguished" objects can be processed to an extent that information about their identity is available, when certain conditions are present, even though the subject reports being unaware of the stimulus. When shown two identical objects—two coins (see figure 40.1A), the subject reports seeing only the one on his right (the ipsilesional field). When he is simultaneously presented two different objects—a coin and a key—he promptly identifies not only the one on his right but also the one in his left (contralesional) field (see figure 40.1B). Information in the unattended field clearly is processed sufficiently for the visual system to know that it is different from its counterpart in the ipsilesional field. This difference signal evidently triggers an orienting response leading to detection and, ultimately, identification.

What happens when two objects of the same type appear that differ in physical features? Figure 40.1C demonstrates that extinction returned when the patient was shown two different types of fork, a silver, metal dinner fork and a white, plastic picnic fork: He reported only the fork on his right. Even though these objects differ visually, the fact that they are classified the same seemed to determine whether the patient could report both.

This may indicate that an extinguished object can be fully processed up to a level of semantic classification or response selection, even when the patient is unaware of its presence. To test this possibility experimentally, Baylis, Driver, and Rafal (1993) tested five patients with clinical evidence of extinction. Colored letters were presented either unilaterally or bilaterally, and subjects were asked to report what they saw on each side. The critical trials were those in which bilateral targets were presented and in which the patient reported seeing "nothing" in the contralesional field. In one condition, they were asked to report only the letters (X or E) and, in another, to report only the colors (red or green).

This study confirmed that extinction occurred much more frequently when the bilateral stimuli were identical *in the attribute to be reported*. On blocks in which the task was to report the name of the letter, extinction was not ameliorated if the stimuli were of different colors; and vice versa for the color report blocks.

It is becoming clear that a great deal of visual information can be processed without visual attention. Preattentive visual processes apparently are based on elementary visual features and Gestalt grouping principles, which parse the visual scene into candidate objects. The phenomena depicted in figure 40.1 demonstrate that extinction of an object in the contralesional field can vary depending on the similarity of the two objects in the dimension to be reported.

Other evidence shows that even irrelevant information can be processed preattentively to guide attentional search over the visual field. In a study involving neglect patients, Grabowecky, Robertson, and Treisman (1993) placed irrelevant distractors around a section in the center of a display where attentional search was required to find a target. This central target area always contained 1 target and 15 distractors. To induce a serial search through the 16 central items, only conjunction search was included. In each display, irrelevant distractors appeared on the contralesional or ipsilesional side of the central search area or on neither or on both sides.

As in Eglin, Robertson, and Knight (1989), there was a contralateral delay into the neglected field. However, the most important finding was that in six of the seven patients, irrelevant information on the ipsilesional side increased the delay significantly but, when irrelevant information was on both sides, the contralateral delay approached that for conditions where no irrelevant distractors were present. In other words, information in the ipsilesional field produced less extinction when balanced by information in the contralesional field.

These data demonstrate that patients with neglect have preattentive information about the spatial extent or spatial structure of the field. The spatial extent biased attention toward the ipsilesional side of the display when only ipsilesional distractors were present. When irrelevant distractors then were added to the contralesional side, the mass of items extended across the field, and attentional search began closer to the center (Grabowecky, Robertson, and Treisman, 1993). This pattern emerged whether or not the irrelevant distractors had features in common with the target.

Other preattentive mechanisms also are evident in patients with unilateral neglect, such as those that respond to symmetry. The effect of symmetry on perceiving objects as figure as opposed to ground was first demonstrated by the Gestalt psychologists. Observation in a patient with visual neglect (Driver, Baylis, and Rafal, 1993) demonstrated that symmetry can be processed at a preattentive level. This patient showed the same effect of symmetry as normals in distinguishing figure from ground, even though he performed at chance when asked to determine whether the shapes were symmetrical. That is, even though his deficits prevented him from reporting whether shapes were symmetrical, he nevertheless perceived symmetrical shapes as the objects in the visual scene.

Behrman and Tipper (in press) showed dumbbell-shaped stimuli to patients with left-sided neglect and asked them to detect targets appearing in one rounded end or the other. Under static conditions, detection was slower for targets appearing in the left end of the dumbbell. When the dumbbell rotated, however, such that the end on the left moved into the right field, and the right end moved into the left field, detection was slower for the part of the dumbbell that had moved from the neglected field. That is, it took its neglect with it.

In summary, visual attention can be allocated to locations or objects. The visual array is preattentively represented at a rather high level. It is segregated into figure-ground, the extent of the stimulus display is registered, orientation and symmetry are processed, and even semantic relationships are present to the extent that they will affect performance. Attentional orienting operates on candidate objects parsed preattentively to select for recognition and purposeful action.

BÁLINT'S SYNDROME AND OBJECT-BASED ATTENTION Perhaps the most dramatic evidence for object-based attentional selection is seen in patients with bilateral lesions of posterior parietal lobes or the parieto-occipital junction who manifest simultaneous agnosia. In this classical syndrome of Bálint (Husain and Stein, 1988), the patient can see only one object at a time. In a case report in 1919 of a 30-year-old World War I veteran who had a gunshot wound through the parieto-occipital regions, Holmes and Horax (1919, p. 402) detailed this syndrome and provided an analysis that stands today: "[T]he essential feature was his inability to direct attention to, and to take cognizance of, two or more objects ..." Because of this constriction of visual attention (what Bálint referred to as the *psychic field of gaze*), the patient can attend to only one object at a time, regardless of the size of the object.

Patients with Bálint's syndrome have great difficulty making judgments comparing two objects or parts of objects. They cannot say which of two objects is smaller or closer, nor can they say which of two lines is longer or whether two angles are the same or different. Holmes and Horax (1919) noted in their patient:

> Though he failed to distinguish any difference in the length of lines, even if it was as great as 50 per cent, he could always recognize whether a quadrilateral rectangular figure was a square or not ... He explained that in order to decide whether a figure was or was not a square, he did not compare the length of its sides but, 'on a first glance I see the whole figure and know whether it is a square or not ... His power of

recognizing promptly the shape of a simple geometrical figure demonstrates that in this we do not naturally depend on the comparison of lines and angles, but that we apprehend shapes as a whole and accept them as unities.

Bálint first observed the dramatic "constriction of visual attention" (Holmes and Horax, 1919) in a patient whom he examined in 1903. His patient did "not take notice of things lying to either side of the object . . ."

Patients with Bálint's syndrome might be viewed as having difficulty disengaging attention to move in any direction—essentially a bilateral disengage deficit (Farah, 1990). However, these patients cannot attend to more than one object even if the objects spatially overlap, suggesting that the restriction is object-based rather than space-based. For example, one of our patients was unable to look at the investigator and tell him whether he was wearing glasses, although she had no difficulty seeing the investigator's face or the glasses if viewed individually. Shown a cross filling a circle, the patient with Bálint's syndrome will report one or the other. Shown a six-pointed star constructed of two triangles of different colors, the patient may see only one triangle (Luria, 1964). No strictly space-based model of attention can accommodate such a striking phenomenon: These patients are unable to disengage from one object to attend to another even at the same location.

Humphreys and Riddoch (1992) have provided elegant experimental evidence for object-based attentional restriction in Balint's syndrome. Two patients with Bálint's syndrome were shown displays with 32 circles, each of which was all red, all green, or half-red and half-green. The task was to report whether each display contained one or two colors, and the critical test was when the displays contained two colors. In one condition, the spaces between the circles contained randomly placed black lines. In two further conditions, the lines connected either pairs of same-colored circles or pairs of different-colored circles. Both patients were better at correctly reporting the presence of two colors when the lines connected different-colored pairs of circles. Circles connected by a line are perceived as a single object (e.g., as a dumbbell). When each object contained both red and green, the patients could report the presence of the two colors better than when each object contained both red or both green. If each object contained only a single color, the patients had great difficulty reporting two colors as their attention tended to lock onto a single object.

The degree to which small and, at times, barely perceptible features in the visual scene can grab the attention of the patient and blind him or her to salient features is indeed extraordinary. One of our patients was shown a simple drawing on a sheet of stationary. After looking for several seconds, she shook her head and, in great frustration, reported: "I'm sorry doctor, but I can't make it out. The watermark on the paper is so distracting."

When shown a hierarchical stimulus of a large letter constructed of small letters, one of our patients with Bálint's syndrome (R.M.) could not see both the local and global letter. This 56-year-old man had suffered bilateral parieto-occipital strokes. Although he was lucid and had good visual acuity and intact visual fields, he was functionally blind. He had spatial disorientation and had to be escorted in the hospital (although, by this time, he was able to get about in his home all right). Although eye movement to command was intact and optokinetic nystagmus preserved, he had great difficulty following moving objects with his eyes. He had great difficulty seeing more than one object at a time and could not tell which of two objects was closer to him. Unlike some patients with dense Bálint's syndrome, however, he could judge whether an object was moving toward or away from him. He did not blink to a visual threat and could not reach accurately toward objects. Given a pencil and asked to place a dot in the center of a circle drawn on a piece of paper, he usually did not even get the point in the circle. He had no difficulty recognizing faces, objects, shapes, words, or colors. Visual contrast sensitivity was intact. Examination by Dr. V. Ramachandran revealed that he was able to perceive apparent motion, to see at least some illusory contours, experienced illusory motion, and could perceive shape from shading and shape from motion. Over a period of several months, this patient was shown many hierarchical figures and, in every instance, he reported the local element first. On only two occasions during this entire period did he also report the global element. His attention was consistently captured by the local element.

Although his missing of the global form was reminiscent of that seen in patients with lesions of the right TPJ, this area was spared in this man. Also, unlike patients with left TPJ lesions, his local bias was not due to a failure to process information at the global level.

In RT studies with figures in which the global and local levels could be either incongruent or congruent with the local level, he consistently reported the local letter. Unlike patients with TPJ lesions and patients with full commissurotomy who exhibit no interference from global information, R.M. consistently showed global interference. Thus, his failure to identify the global element was not due to a failure to process the information at the global level but to an inability to attend explicitly to it.

As with patients with unilateral extinction, he also showed large interference effects from spatially separated stimuli of which he was not aware. In one experiment, this same patient was shown colored squares in the center of the display, which could be either red or green, and his RT to name the color was measured. The target colored square in the center remained visible until he responded. On each trial, another colored square was flashed briefly (for 16 ms) 8° above, below, or to the right of the color patch. This distractor was either the same color as the target or incongruent. He had very little difficulty discriminating this distractor when it was shown alone without the center target, yet when presented with the target, he was unable to report it, even when encouraged to try to ignore the color in the center. The patient also was unable to say whether one or two stimuli or one or two colors were present. Although he did not demonstrate any awareness of the distractor patch when presented with the center target, this patch nevertheless strongly affected his RT to name the center target color. It is apparent that information from unattended objects is processed in this patient.

Coslett and Saffran (1991) showed that this unattended information can be processed to a semantic level in a patient with Bálint's syndrome. With brief, simultaneous presentation of two words or drawings, their patient identified both stimuli significantly more frequently when the stimuli were semantically related than when they were unrelated.

We found in R.M. that the summoning of his attention to a given object and his locking onto it were not simply random. A distinctive feature in his visual field captured his attention and strongly biased what he was likely to see. For instance, he was more likely to report the *Q* in a background of *O*s than an *O* in a background of *Q*s. He was also able more accurately to report a red *O* among green *X*s in a visual display than a red *O* among red *X*s and green *O*s. When the display required serial attentional search, his performance was impaired.

It is clear that preattentive processing has a large effect on normal attentional orienting. The study of patients with neglect or extinction, as well as those with simultaneous agnosia in Bálint's syndrome, can tell us what features of objects and properties of displays of objects are preattentively processed and which require attention. Visual attention does not work like a simple spotlight that facilitates processing of information at the place at which it is aimed. It may be summoned to a location reflexively by the midbrain extrageniculate system, by information in the visual scene that preattentive processes indicate is the likely location of a potential object of interest, or by voluntary processes that guide strategic search. Once attention is deployed to a location, it is captured by objects there, and it is the object—not only the location—from which attention must be disengaged in order to move elsewhere. Still, while attention is allocated to an object, preattentive processes that guide subsequent orienting behavior continue to operate outside the ken of attention.

Hemispheric Asymmetry for Disengaging from Locations and Objects The plight of the patient with Bálint's syndrome suggests that bilateral parietal lobe lesions disrupt attending to both location-based and object-based representations. A recent study in patients with chronic, unilateral parietal lesions suggests that the right and left parietal lobes may make different contributions to location- and object-based processing. Egly, Driver, and Rafal (in press) used a modification of the covert orienting paradigm to measure both space- and object-based attentional shifts in normal subjects and patients with chronic lesions of the posterior association cortex. Part of an object was cued, and then the target was presented either within the same object or in a different object (figure 40.8A). In normal subjects, cuing effects (or costs) were greater for shifting attention between than within objects; so the effects of moving between locations in space and the additional effect of shifting attention between objects were identified. Several findings in the patient groups are noteworthy. First, both patient groups (those with a left- or right-hemisphere lesion) produced larger costs to respond to targets in their contralesional than their ipsilesional fields, again replicating the deficit in disengaging attention. The important new finding

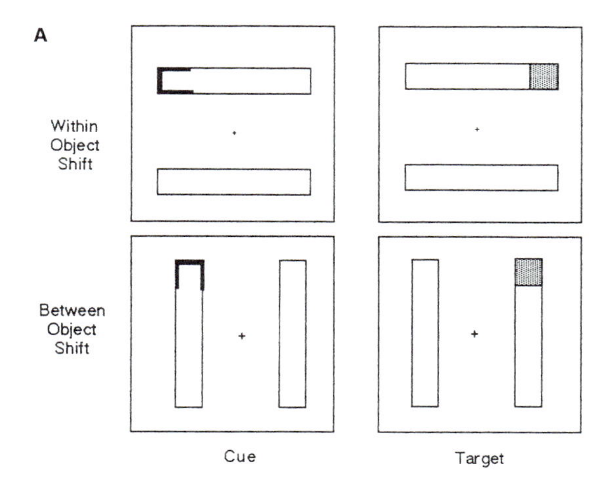

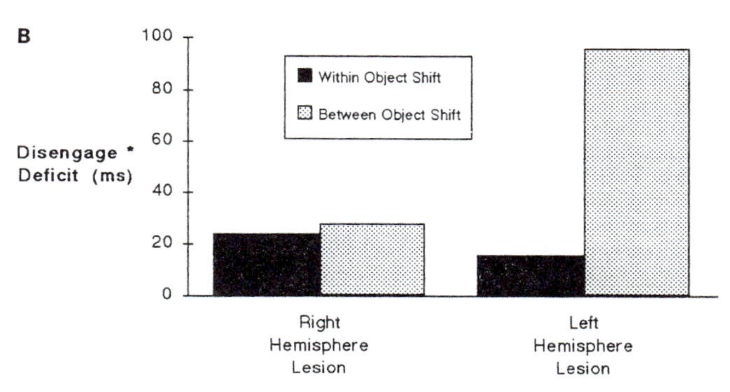

FIGURE 40.8 (Top) Comparison of the time to disengage covert attention to move within an object or between objects. One end of a rectangle is cued to indicate the location where a detection target is likely to occur. Occasionally, this cue is invalid and the target appears at a different location. The uncued location where this target appears either may be within the same object or may require an attention shift to a different object. (Bottom) Results from the experiment in patients with left (N = 5) and right (N = 8) parietal lesions. The size of the disengage deficit (the greater costs in reaction time to reallocate attention from ipsilesional to contralesional field compared to reallocating attention from contralesional to ipsilesional field) is shown in each patient group in the within-object and between-object shift conditions. In the right parietal lesion group, the disengage deficit is present in both between- and within-object shift conditions and is not larger in the between-object shift condition. In the left parietal lesion group, the disengage deficit is manifest only in the between-object shifting condition.

was a difference between the two patient groups in shifting attention between objects (figure 40.8B). Although both patient groups showed problems in disengaging attention from their intact field, the disengage deficit in patients with left-hemisphere lesions occurred only for between-object shift of attention from the ipsilesional to the contralesional field. This finding suggests that the right parietal lobe may be critical for shifting attention between locations, whereas the left parietal lobe is critical for shifting attention between objects. We have recently confirmed this conclusion in a commissurotomized patient with disconnected neocortices. In his right visual field, normal costs for shifting between objects were observed; no object-based costs were observed for attention shifts in the left visual field (Egly et al., in press).

Symptom complex of the neglect syndrome—a synthesis

We conclude this chapter with a review of what has been learned about the roles of various cortical and subcortical structures in visual attention and with an attempt to synthesize this information in a way that considers how disruptions in this distributed system contribute to the individual component symptoms of visual neglect. Extinction, one of the component symptoms of the neglect syndrome, is due to a defect in orienting to contralesional objects when attention is engaged on an ipsilateral location or object. Other components of the neglect syndrome can be related to deficits in endogenously generated exploratory and manipulative behavior within contralesional space (as demonstrated clinically by line bisection, cancellation, or drawing tasks). Patients with lesions in different regions may manifest different aspects of the neglect syndrome. The common denominator is asymmetrical spatial orienting, but this may occur in different patients for different reasons (Kinsbourne, 1987). The attentional biases of neglect may reflect an abnormality of reflexive orienting or of voluntary orienting or a failure of integration of these two orienting modes (Butter, 1987).

A number of pathophysiological mechanisms have been advanced to account for these difficulties (table 40.1). Some of them will be reviewed here, and we will consider which might contribute to another salient characteristic of the clinical syndrome of neglect—namely, that it tends to be more severe and more persistent with of right-hemisphere lesions.

MOTOR NEGLECT AND DEFECTS IN INTENTION Neglect behavior can result from a motor bias, which is separable from defects in spatial attention. By using pulleys or mirrors (Coslett et al., 1990; Mijovic, 1991; Tegner and Levander, 1991) in cancellation tasks, it has been

Table 40.1

Pathophysiological mechanisms contributing to the neglect syndrome

Deficit	Putative Neural Substrate
Disengaging covert attention	Parietal lobe (right—from locations; left—from objects)
Voluntary shifts of attention	Superior parietal lobule
Hyperengaging to local elements	Right temporal-parietal junction
Arousal	Mesencephalic reticular formation; intralaminar thalamic nucleii; right hemisphere > left hemisphere
Representation of space	Parietal lobes
Generating endogenous saccades	Frontal eye fields
Hyperreflexive orienting to ipsilesional field	Superior colliculus
Motor intention	Premotor cortex
Motivation for exploratory behavior	Cingulate gyrus
Spatial working memory	Dorsolateral prefrontal cortex

possible to dissociate perceptual (attentional) and motor (intentional) components of neglect behavior. With these manipulations, ipsilesional hand movements produce perceptual orienting in the contralesional direction. In patients whose problem is a contralesional motor bias, this manipulation facilitates task performance. In patients whose problem is attentional, the same manipulation will not help. These studies have demonstrated that one or both kinds of deficit can contribute to neglect behavior. The degree to which each contributes varies from patient to patient.

As expected, these studies suggest that frontal lesions are more likely to produce more motor bias (what has been called a *defect of intention* or *hemihypokinesia*), whereas posterior lesions are more likely to cause a deficit in attentional orienting. Our studies of frontal and posterior lesions on reflexive and voluntary saccades are consistent with this fractionation of neglect symptomatology into defect in intention and attention. Figure 40.4 showed FEF lesions caused a defect in directing voluntary saccades toward the contralesional field, whereas posterior lesions did not but did cause deficit in detection of signals in the contralesional field. The cortical anatomy relating to defects in motor bias remains to be worked out more precisely. Lesions of the dorsolateral prefrontal cortex and basal ganglia seem likely to cause hemihypokinesia. Lesions of the cingulate gyrus may also cause hemihypokinesia, perhaps due to a failure of limbic structures to motivate voluntary action effectively (Mesulam, 1981)

Spatial Representation Neglect may cause not only a failure to attend to contralesional signals but also loss of conscious access to the contralesional side of internally generated visual images. Bisiach and Luzzatti (1978) asked patients with left hemineglect to imagine themselves in the Piazza del Duomo in Milan. In one condition, they asked the patients to imagine themselves at one end of the square, looking toward the cathedral dominating the other end of the square, and to describe what they would be able to see. In another condition, the patients were asked to imagine themselves standing on the cathedral steps facing the opposite way. In both circumstances, the patients reported fewer landmarks on the contralesional side of their mental image. These kinds of observations have engendered an account of neglect in which cortical lesions produce a degradation of the contralesional representation of space. The critical lesion for producing a defect in spatial representation has yet to be determined. Although the parietal lobe is a candidate (Andersen, 1987; Duhamel, Colby, and Goldberg, 1992), areas of lateral prefrontal cortex involved in working memory may also be implicated (Goldman-Rakic, 1987).

Hemispherical Asymmetry for the Control of Attention or Spatial Representation One explanation that has been advanced to account for the fact that neglect is more common, more severe, and more protracted after right-hemisphere lesions assumes that the right hemisphere is dominant for spatial attention. These accounts propose that the right hemisphere has neural mechanisms for attending to both visual fields, whereas the left hemisphere attends predominantly to the right visual field (Heilman, Valenstein, and Watson, 1985; Heilman et al., 1987; Weintraub and Mesulam, 1987). PET studies in normals have supported this hypothesis, finding metabolic changes in superior parietal lobes in a spatial orienting task (Corbetta et al., 1993).

Hemispherical Asymmetry for the Control of Arousal Selective orienting within the visual field may also be affected by the level of arousal. One putative mechanism contributing to clinical neglect is hypoarousal. Unilateral lesions of ascending activating pathways, especially those originating in the mesencephalic reticular formation, can produce neglect (Heilman et al., 1985). The greater prevalence of neglect after right-hemisphere lesions has also been ascribed to a right-hemisphere dominance for arousal (Heilman and Valenstein, 1979; Heilman and Van Den Abell, 1980; Posner et al., 1987).

Contribution to Neglect of the Local Bias Caused by Right TPJ Lesions We saw earlier that lesions of the right TPJ caused a strong local bias in visual processing. These patients can get locked onto local details or parts of objects and fail entirely to perceive the critical big picture. This piecemeal perception carries over into everyday life in failures to interpret adequately social situations or critical elements necessary for everyday problem solving. The local bias produced by right TPJ lesions is an important factor contributing to neglect and may be a major reason that neglect is more severe after right- than left-hemisphere lesions (Robertson and Eglin, 1993). The interaction of the local bias with difficulty in disengaging attention is an important factor in producing the classical constructional signs of neglect in paper-and-pencil tasks. In performing a cancellation task, for example, a patient with a lesion in the right TPJ is prone to get locked onto the local level (i.e., the item being crossed out). Once locked on to the local item, he or she will have more difficulty disengaging from that item than might a patient with a comparable left-hemisphere lesion who is able to maintain the big picture of the global display.

Hyperorienting to the Ipsilesional Field One model of the neurobiological basis of spatial attention postulates that each hemisphere, when activated, mediates an orienting response in the contralateral direction (Kinsbourne, 1987). According to this account, neglect results from a unilateral lesion because of a breakdown in the balance of hemispherical rivalry such that the nonlesioned hemisphere generates an unopposed orienting response to the side of the lesion (Ladavas, Del, and Provinciali, 1989; D'Erme et al., 1992).

One explanation for this hyperreflexive orienting postulates a loss of excitation of the superior colliculus on the side of the hemispheric lesion. As a result, the opposite superior colliculus (that is, the contralesional side) becomes disinhibited, which results in exaggerated reflexive orienting to signals in the ipsilesional field. This hyperorienting may, in turn, account for the difficulties in disengaging attention to move in a contralesional direction.

Experiments in the cat have confirmed that this kind of cortical-subcortical interaction is important in regulating visually guided orienting behavior. Sprague (1966) rendered cats blind in one visual field by removing occipital and parietal cortex. He then showed that vision in this field improved if the opposite superior colliculus were removed or if the inhibitory connections between the two colliculi were disconnected. This Sprague effect is believed to work in the following way: Parieto-occipital projections to the ipsilateral superior colliculus normally exert a tonic facilitation. After parietal lesions, the colliculus loses this tonic activation. Because there are inhibitory connections between the two colliculi, the opposite superior colliculus becomes disinhibited, and this, in turn, produces disinhibited reflexive orienting to ipsilesional signals.

The Sprague effect suggests that one aspect of neglect can be aggravated by disinhibition of subcortical visual pathways on the side opposite a cortical lesion and that prevention of visual input to this colliculus can alleviate neglect. It obviously is not an option to remove surgically the contralesional superior colliculus in humans who have suffered parietal lobe strokes. Nevertheless, we have seen that it is possible to decrease collicular activation, and reflexive orienting, by occluding one eye with a patch. Posner and Rafal (1987) suggested that patching the eye on the side of the lesion might help reduce symptoms of neglect. Studies in both monkeys and patients with neglect indicate that this maneuver may have some benefit (Butter, Kirsch, and Reeves, 1990).

Conclusions

The orienting of attention to a point in space usually is accompanied by overt movements of the head, eyes, or body. This is so whether attention is summoned exogenously, as in turning toward a movement seen out of the corner of the eye, or is deployed endogenously, as when we decide to look both ways before crossing the street. In everyday life, there are constant competing demands on attention by the outside world as well as from internally generated goals. A distributed network of neural structures orchestrates the orienting of attention and reconciles these competing demands.

When these neural processes are working as they should, we seem to handle the competing demands of the outside world and those needed for planned activity so seamlessly that these subsystems appear to be one. However, damage to the brain reveals how dependent we are on the efficient coordination of attention for coherent perception, thought, and action. Advances in cognitive science and anatomical and functional neuroimaging have helped us to begin to understand the specific contributions of each of the brain areas. By helping us to integrate cognitive science and neuroscience, this understanding should lead to more rational treatment approaches for these syndromes.

ACKNOWLEDGMENTS This work was supported by a Veterans Administration Medical Research Council grant and by U.S. Public Health Service grants RO1 AA006637 to L. R. and RO1 MH45414 to R. R. The order of authorship was determined by coin toss.

REFERENCES

ANDERSEN, R. A., 1987. Inferior parietal lobule function in spatial perception and visuomotor integration. In *Handbook of Physiology: 5. Higher Functions of the Brain* (part 2), F. Plum, ed. Bethesda, Md.: American Physiological Society, pp. 483–518.

BADCOCK, J. C., F. A. WHITWORTH, D. R. BADCOCK, and W. J. LOVEGROVE, 1991. Low-frequency filtering and the processing of local-global stimuli. *Perception* 19:617–629.

BAYLIS, G., R. RAFAL, and J. DRIVER, 1993. Extinction and stimulus repetition. *J. Cogn. Neurosci.* 5:453–466.

BAYNES, K., H. D. HOLTZMAN, and B. T. VOLPE, 1986. Components of visual attention: Alterations in response pattern to visual stimuli following parietal lobe infarction. *Brain* 109:99–114.

BEHRMAN, M., and S. P. TIPPER (in press). Object-based visual attention: Evidence from unilateral neglect. In *Attention and Performance: XIV. Conscious and Nonconscious Processing and Cognitive Functioning*, C. Umilta and M. Moscovitch, eds. Hillsdale, N. J.: Erlbaum.

BISIACH, E., and C. LUZZATTI, 1978. Unilateral neglect of representational space. *Cortex* 14:129–133.

BRADDICK, O., J. ATKINSON, B. HOOD, W. HARKNESS, G. JACKSON, and K. F. VARGHA, 1992. Possible blindsight in infants lacking one cerebral hemisphere. *Nature* 360(6403):461–463.

BUTTER, C. M., 1987. Varieties of attention and disturbances of attention: A neuropsychological analysis. In *Neurophysiological and Neuropsychological Aspects of Spatial Neglect*, M. Jeannerod, ed. Amsterdam: North-Holland, pp. 1–24.

BUTTER, C. M., N. L. KIRSCH, and G. REEVES, 1990. The effect of lateralized dynamic stimuli on unilateral spatial neglect following right hemisphere lesions. *Restorative Neurol. Neurosci.* 2:39–46.

CHRISTMAN, S., F. L. KITTERLE, and J. HELLIGE, 1991. Hemispheric asymmetry in the processing of absolute versus relative spatial frequency. *Brain Cogn.* 16:62–73.

CORBETTA, M., F. M. MIEZIN, G. L. SHULMAN, and S. E. PETERSEN, 1993. A PET study of visuospatial attention. *J. Cogn. Neurosci.* 13:1202–1226.

COSLETT, H. B., D. BOWERS, E. FITZPATRICK, B. HAWS, and K. M. HEILMAN, 1990. Directional hypokinesia and hemispatial inattention in neglect. *Brain* 113:475–486.

COSLETT, H. B., and E. SAFFRAN, 1991. Simultanagnosia. To see but not two see. *Brain* 113:1523–1545.

D'ERME, P., I. ROBERTSON, P. BARTOLOMEO, A. DANIELE, and G. GAINOTTI, 1992. Early rightwards orienting of attention on simple reaction time performance in patients with left-sided neglect. *Neuropsychologia* 30(11):989–1000.

DOYON, J., and B. MILNER, 1991. Right temporal-lobe contribution to global visual processing. *Neuropsychologia* 29: 343–360.

DRIVER, J., G. BAYLIS, and R. RAFAL, 1993. Preserved figure-ground segmentation and symmetry perception in a patient with neglect. *Nature* 360:73–75.

DUHAMEL, J. R., C. L. COLBY, and M. E. GOLDBERG, 1992. The updating of the representation of visual space in parietal cortex by intended eye movements. *Science* 255:90–92.

EGLIN, M., L. C. ROBERTSON, and R. T. KNIGHT, 1989. Visual search performance in the neglect syndrome. *J. Cogn. Neurosci.* 1:372–385.

EGLY, R., J. DRIVER, and R. RAFAL (in press). Shifting visual attention between objects and locations: Evidence from normal and parietal lesion subjects. *J. Exp. Psychol. [Gen.]*

EGLY, R., R. RAFAL, J. DRIVER, and Y. STARREVELD (in press). Covert orienting in the split-brain reveals hemispheric specialization for object-based attention. *Psychol. Sci.*

FARAH, M. J., 1990. *Visual Agnosia*. Cambridge, Mass.: MIT Press.

FENDRICH, R., C. M. WESSINGER, and M. S. GAZZANIGA, 1992. Residual vision in a scotoma: Implications for blindsight. *Science* 25:1489–1491.

GOLDMAN-RAKIC, P. A., 1987. Circuitry of primate prefrontal cortex and regulation of behavior by representational memory. In *Handbook of Physiology: Sec. 1. The Nervous System*: Vol. 5, J. Mills and M. V. B., eds. Bethesda, Md.: American Physiological Society, pp. 373–417.

GRABOWECKY, M., L. C. ROBERTSON, and A. TREISMAN, 1993. Preattentive processes guide visual search: Evidence from patients with unilateral visual neglect. *J. Cogn. Neurosci.* 5:288–302.

GUITTON, D., H. A. BUCHTEL, and R. M. DOUGLAS, 1985. Frontal lobe lesions in man cause difficulties in suppressing reflexive glances and in generating goal directed saccades. *Exp. Brain Res.* 58:455–472.

HEILMAN, K. M., D. BOWERS, E. VALENSTEIN, and R.

WATSON, 1987. Hemispace and hemispatial neglect. In *Neurophysiological and Neuropsychological Aspects of Spatial Neglect*, M. Jeannerod, ed. Amsterdam: North-Holland, pp. 115–150.

HEILMAN, K. M., and E. VALENSTEIN, 1979. Mechanisms underlying hemispatial neglect. *Ann. Neurol.* 5:166–170.

HEILMAN, K. M., and T. VAN DEN ABELL, 1980. Right hemisphere dominance for attention: The mechanisms underlying hemispheric asymmetries of inattention (neglect). *Neurology* 30:327–330.

HENIK, A., R. RAFAL, and D. RHODES, in press. Endogenously generated and visually guided saccades after lesions of the human frontal eye fields. *J. Cogn. Neurosci.*

HOLMES, G., and G. HORAX, 1919. Disturbances of spatial orientation and visual attention, with loss of stereoscopic vision. *Arch. Neurol. Psychiatry* 1:385–407.

HUMPHREYS, G. W., and M. J. RIDDOCH, 1992. Interactions between objects and space-vision revealed through neuropsychology. In *Attention and Performance XIV*, D. E. Meyers and S. Kornblum, eds. Hillsdale, N.J.: Erlbaum.

HUMPHREYS, G. W., and M. J. RIDDOCH (in press). Interactive attentional systems and unilateral neglect. In *Unilateral Neglect: Clinical and Experimental Studies*, I. H. Robertson and J. C. Marshall, eds. Hillsdale, N. J.: Erlbaum.

HUSAIN, M., and J. STEIN, 1988. Rezso Balint and his most celebrated case. *Arch. Neurol.* 45:89–93.

KINSBOURNE, M., 1987. Mechanisms of unilateral neglect. In *Neurophysiological and Neuropsychological Aspects of Spatial Attention*, M. Jeannerod, ed. Amsterdam: Elsevier.

KITTERLE, F. L., S. CHRISTMAN, and J. HELLIGE, 1990. Hemispheric differences are found in the identification, but not detection, of low versus high spatial frequencies. *Percept. Psychophys.* 48:297–306.

LADAVAS, E., P. M. DEL, and L. PROVINCIALI, 1989. Unilateral attention deficits and hemispheric asymmetries in the control. *Neuropsychologia* 27:353–366.

LAMB, M. R., and L. C. ROBERTSON, 1989. Do response time advantage and interference reflect the order of processing of global and local level information? *Percept. Psychophys.* 46:254–258.

LAMB, M. R., L. C. ROBERTSON, and R. T. KNIGHT, 1990. Component mechanisms underlying the processing of hierarchically organized patterns: Inferences from patients with unilateral cortical lesions. *J. Exp. Psychol. [Learn. Mem. Cogn.]* 16:471–483.

LAMB, M. R., L. C. ROBERTSON, and R. T. KNIGHT, 1989. Attention and interference in the processing of global and local information: Effects of unilateral temporal-parietal junction lesions. *Neuropsychologia* 4:471–483.

LURIA, A. R., 1964. Disorders of "simultaneous perception" in a case of bilateral occipito-parietal brain injury. *Brain* 82:437–449.

MARK, V. W., C. A. KOOISTRA, and K. M. HEILMAN, 1988. Hemispatial neglect affected by nonneglected stimuli. *Neurology* 38:1207–1211.

MCGILNCHEY-BERROTH, R., W. P. MILBERG, M. VERFAELLIE, M. ALEXANDER, and P. T. KILDUFF, 1993. Semantic processing in the neglected visual field: Evidence from a lexical decision task. *Cogn. Neuropsychol.* 10:79–108.

MESULAM, M. M., 1981. A cortical network for directed attention and unilateral neglect. *Ann. Neurol.* 4:309–325.

MIJOVIC, D., 1991. Mechanisms of visual spatial neglect. Absence of directional hypokinesia in spatial exploration. *Brain* 1575–1593.

MORROW, L. A., and G. C. P. RATCLIFF, 1988. The disengagement of covert attention and the neglect syndrome. *Psychobiology* 16:261–269.

POSNER, M. I., A. W. INHOFF, F. J. FRIEDRICH, and A. COHEN, 1987a. Isolating attentional systems: A cognitive-anatomical analysis. *Psychobiology*

POSNER, M. I., and R. D. RAFAL, 1987. Cognitive theories of attention and the rehabilitation of attentional deficits. In *Neuropsychological Rehabilitation*, R. J. Meir, L. Diller, and A. L. Benton, eds. London: Churchill Livingstone.

POSNER, M. I., J. A. WALKER, F. J. FRIEDRICH, and R. RAFAL, 1984. Effects of parietal injury on covert orienting of visual attention. *J. Neurosci.* 4:1863–1874.

POSNER, M. I., J. A. WALKER, F. J. FRIEDRICH, and R. D. RAFAL, 1987b. How do the parietal lobes direct covert attention? *Neuropsychologia* 25:135–146.

RAFAL, R., A. HENIK, and J. SMITH, 1991. Extrageniculate contributions to reflexive visual orienting in normal humans: A temporal hemifield advantage. *J. Cogn. Neurosci.* 3:323–329.

RAFAL, R., J. SMITH, J. KRANTZ, A. COHEN, and C. BRENNAN, 1990. Extrageniculate vision in hemianopic humans: Saccade inhibition by signals in the blind field. *Science* 250:118–121.

RAFAL, R. D., M. I. POSNER, J. H. FRIEDMAN, A. W. INHOFF, and E. BERNSTEIN, 1988. Orienting of visual attention in progressive supranuclear palsy. *Brain* 111:267–280.

RAFAL, R. P., 1992. Visually guided behavior in progressive supranuclear palsy. In *Progressive Supranuclear Palsy: Clinical and Research Approaches*, I. Litvan and Y. Agid, eds. Oxford: Oxford University Press.

ROBERTSON, L. C., submitted. Attentional prints: A source for feature priming effects.

ROBERTSON, L. C., and M. EGLIN, 1993. Attention search in unilateral visual neglect. In *Unilateral Neglect: Clinical and Experimental Studies*, I. Robertson and J. Marshall, eds. London: Taylor and Francis.

ROBERTSON, L. C., and M. R. LAMB, 1991. Neuropsychological contributions to theories of part/whole organization. *Cogn. Psychol.* 23:299–330.

ROBERTSON, L. C., M. R. LAMB, and R. T. KNIGHT, 1988. Effects of lesions of the temporal-parietal junction on perceptual and attentional processing in humans. *J. Neurosci.* 8:3757–3769.

ROBERTSON, L. C., M. R. LAMB, and E. ZAIDEL, 1993. Callosal transfer and hemisphere laterality in response to hierarchical patterns: Evidence from normal and commissurotomized subjects. *Neuropsychology* 7:325–342.

SERGENT, J., 1982. The cerebral balance of power: Confrontation or cooperation? *J. Exp. Psychol. [Hum. Percept.]* 8:253–272.

Shulman, G. L., and J. Wilson, 1987. Spatial frequency and selective attention to local and global information. *Perception* 16:89–101.

Sprague, J. M., 1966. Interaction of cortex and superior colliculus in mediation of peripherally summoned behavior in the cat. *Science* 153:1544–1547.

Tegner, R., and M. Levander, 1991. Through a looking glass. A new technique to demonstrate directional hypokinesia in unilateral neglect. *Brain* 113:1943–1951.

Treisman, A., and G. Gelade, 1980. A feature integration theory of attention. *Cogn. Psychol.* 12:97–136.

Weintraub, S., and M. M. Mesulam, 1987. Right cerebral hemisphere dominance in spatial attention. *Arch. Neurol.* 44:621–625.

Weiskrantz, L., 1986. *Blindsight: A Case Study and Implications*. Oxford: Oxford University Press.

41 Computational and Anatomical Models of Selective Attention in Object Identification

DAVID LABERGE

ABSTRACT Selective attention is characterized computationally in this chapter as the expression of a positive difference between information flow in cell clusters representing a target location and information flow in cell clusters representing surrounding locations—in particular, posterior cortical circuits, such as the V1-to-inferotemporal cortical pathway. A major problem is to specify what mechanism or mechanisms produce an enhancement of the conjectured target-surround differences in cortical pathways. One mechanism that has been proposed by several investigators is the circuitry of the thalamus. Evidence from simulated operations of thalamic circuitry models suggests that thalamic nuclei, and the pulvinar in particular, can take as an input a small difference in firing rates between a target cluster of cells and surrounding cells and produce as an output a manyfold increase in the input difference. This simulated magnification of a target-surround input difference is produced mainly by enhancements in target-cell firings, which contrasts with magnification of a target-surround input difference mainly by attenuation of surround-cell firings. The pulvinar mechanism apparently is directly controlled by axon fibers from the posterior parietal cortex, whose cells are driven, in turn, both by bottom-up inputs originating in the eye and top-down inputs arising from the prefrontal cortex.

In this chapter, selective attention is considered from the combined points of view of computational and behavioral analyses, cognitive systems, brain architecture, brain circuitry, and network simulations of neural circuitry. This multifaceted approach seems entirely appropriate when one considers the daunting mystery of the mind-brain relationship involved in attention.

We begin by asking the computational question: What does attention do for a person at the cognitive level? Then we use the provisional answer to this question to direct the design of behavioral experiments that, it is hoped, will epitomize selective attention and modulate its intensity. The selective attention process is made more amenable to measurement by neurobiological techniques by the construction of behavioral tasks that induce the process to operate at as high an intensity level as is manageable and for an appreciable period of time.

The goals of these measurement experiments are to determine brain areas that are crucial to the selective attention process and to determine properties of circuit and cell activity within these areas during selective attention. A basic assumption of the present cognitive neuroscientific approach to attention research is that if we could discover how neurons and circuits function during selective attention, we would, in effect, know the algorithm used by the system. In short, we would know how the brain performs the task of selective attention.

However, currently there are very few data from measurements of neural circuit activity during selective attentional processing, although there are considerable data from measurements of individual neurons (same of which will be reviewed later). Moreover, one should be wary of the possibility, however remote, that the algorithm used by the neural circuitry is, in principle, not penetrable by measuring the separate activities of the component cells, in the same way that the activity of individual hidden units of distributed networks may not be informative about how the network computes its output (Robinson, 1992). In view of both of these considerations, one can turn to simulating the operations of neural circuits providing that sufficient structure of the circuit has been revealed by neurotracing methods. Later in this chapter, we describe a simulation of the operation of the thalamocortical circuit, whose input-output relationship appears to conform to the input-

DAVID LABERGE Department of Cognitive Sciences, University of California, Irvine, Calif.

output pattern defined computationally for selective attention in many behavioral tasks.

Characterizing selective attention computationally

Information in the environment is useful to the organism in achieving particular goals. To be effective in navigating the physical world and dealing with objects, properties of the visual scene must be perceived and judged. Frequently, the judgment must be made quickly, as in the cases of monitoring a radar screen, driving a fast-moving vehicle, and grasping kitchen utensils while cooking. Chief among the sensory properties leading to a judgment (and consequent action) are the location of objects, the identity of objects, and attributes of objects such as their color, size, direction, and velocity of movement. However, to discover the location and identity of an object requires more than simply delivering the information to a judgment module in the observer's system. Before the appropriate information can be given to a specialized processing module (such as a color detector, motion detector, or identity detector), several problems need to be solved. Among these are the binding problem (for combining more than one attribute into a single object) and the selection problem. Both have been regarded by investigators as problems that are solved by the system's attentional mechanisms. In particular, Treisman (Treisman and Gelade, 1980; Treisman and Gormican, 1988) made the binding (conjoining) of object attributes the core notion of her feature-integration theory of attention, and Broadbent (1958) gave new life to the notion of selection, which dates back at least as far as the time of William James (1890), by representing it as a mechanism that filters information flow in a communication system.

The process of selective attention can be computationally characterized as taking as an input the array of information arising from a cluttered visual field and delivering as an output the information flowing from a confined area of that field. The research goal is to describe the algorithm (i.e., the set of operations) that accomplishes this result.

Considerations of the input and output in a computational description can be clarified as one refines the designs of behavioral tasks. The flanker task adapted by my colleagues and I from Eriksen and Eriksen (1974) and developed extensively before we used it with human subjects in a positron emission tomography (PET) scan experiment provides an illustration. We assume that visual selective attention to spatial location takes as its input the information arising from a cluttered array of objects and delivers as its output the information arising only from the location of the target. One of the simplest ways to embody these notions in a stimulus display is to present an object in a target location and surround it with other objects and ask the subject to identify the object in the target location. As a simple response indicator of the processing, the subject presses a button when the target location contains a specific object (e.g., an *R*) but withholds the button press when a different object (e.g., a *P* or *Q*) appears there instead. Two variables that strongly affect the time to identify the *R* in the target location are the distance between the locations of the target and the nearest distracting (flanking) object (Eriksen and Eriksen, 1974) and the similarity of the nearest distracting object (LaBerge and Brown, 1989). Examples of the first case are obvious. Figure 41.1 displays examples of the second case in descending order of similarity and difficulty: *BRK*, *VRY*, |*R*|, and *R* alone. The difference in response times to the *BRK* and the *R* is approxi-

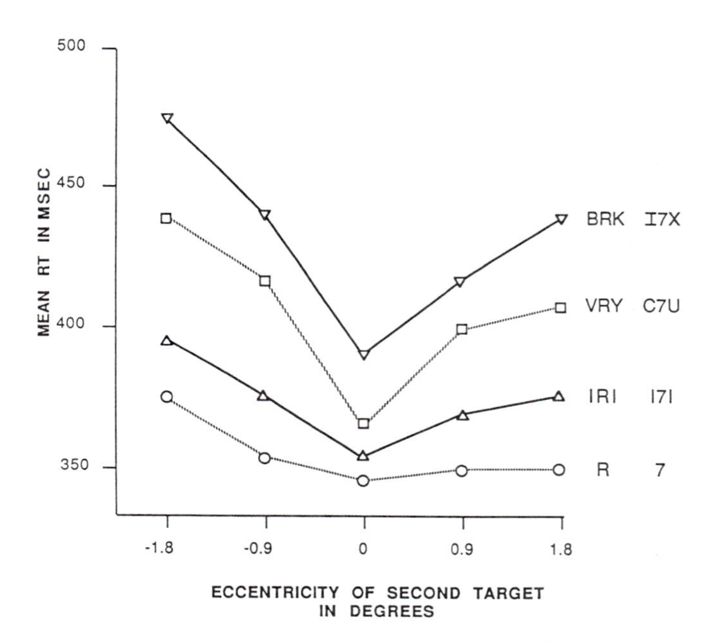

FIGURE 41.1 Mean response times to a target *R* (or *7*) as a function of its location and type of flanker when a single letter *S* was identified in the center location just prior to the *R* target. The *R* was discriminated against *P* or *Q*, and *7* was discriminated against *T* or *Z*. RT, response time. (Reprinted from LaBerge and Brown, 1989, by permission. Copyright 1989 by the American Psychological Association.)

mately 45 ms when the triplet appears at the cued location (where attention is concentrated prior to the display), but this difference increases to 95 ms when the triplet appears a degree or so to the left or right of the cued location.

The physiological measurement of selective attention that we had in mind was a PET scan and, because the purpose of our PET experiment task was to induce a strong degree of attentional processing, we chose objects with the highest degree of target-flanker similarity (i.e., *GOQ*) and placed eight distractors (instead of only two) around the target object (making a nine-item ensemble) and spaced the distractors as close to the target as correct responding would allow. Then, noting from figure 41.1 that attentional processing was increased when the target was presented away from the center of attentional concentration, we presented the displays to the right or left of center rather than in the center. To induce less attentional processing (e.g., as a control condition), one can use one of the other target-flanker combinations, with or without increasing spacing between the target and flanker. In our PET experiment (LaBerge and Buchsbaum, 1990), we chose the isolated *O* as the control condition.

Expression of selective attention in the brain: Levels of description

Assuming that the selective attention process is expressed within the flow of information arising from the visual field, as this information influences modules that identify an object, what specific modification in the stream of information results in the delivery of the target information to the identifier module? Before a hypothesis can be framed to answer this question, one needs to decide the level of information processing appropriate for dealing with the modification of information flow between early visual cortical areas (beginning with V1) and the object identification areas in the inferotemporal cortex (IT). Based on Shepherd (1990) and Churchland, Koch, and Sejnowski (1990) one listing of the main levels of organization of the nervous system is cognitive-behavioral systems, brain areas, neural circuits, neurons, microcircuits, synapses, and molecules and ions.

Given these alternative levels of description, my colleagues and I chose to adopt the circuit level as the appropriate level of description for the expression of attention, but this assumption does not imply that a circuit in only one brain area expresses attention (e.g., in the circuits entering a particular part of the IT, as during identification of an object). It seems likely that the expression of selective attention may occur simultaneously in many regions of the visual cortex (e.g., an object that possesses several attributes could involve selection by location in several maps simultaneously) (see Koch and Ullman, 1985; Treisman and Gormican, 1988).

Cortical areas essential to the operation of selective attention in shape identification

In this section, we examine areas in the cortex that appear to be specialized for the performance of operations necessary to identify a visual shape when a distracting object is present. These cortical areas are IT, the V4 area, the posterior parietal cortex (PPC), and the dorsolateral prefrontal cortex (DLPFC) (figure 41.2).

Whereas identification of a object involves processing of featural information in IT, selective attention to an object in a field of distracting objects (e.g., identifying as *O* the center object in the stimulus *GOQ*) is presumed to involve processing of spatial information in the PPC, where locations of objects are presum-

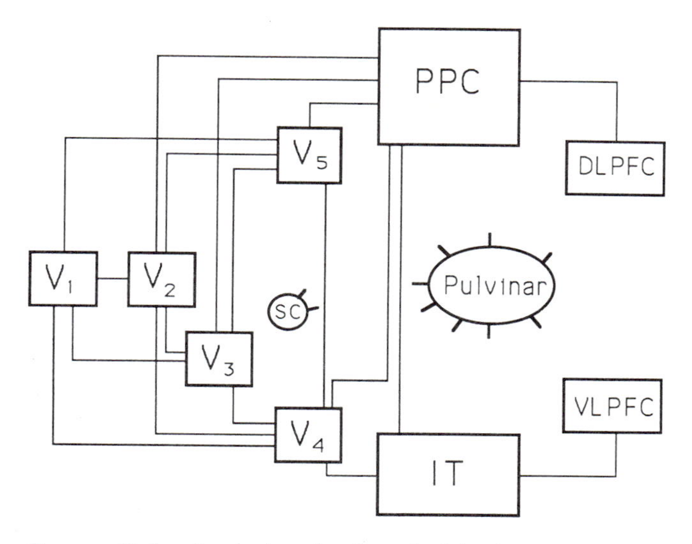

FIGURE 41.2 Cortical and subcortical brain areas serving visual selective attention. PPC, posterior parietal cortex; IT, inferotemporal area; DLPFC, dorsolateral prefrontal cortex; VLPFC, ventrolateral prefrontal cortex; SC, superior colliculus. (Adapted from Andersen et al., 1990; Boussaoud, Ungerleider, and Desimone, 1990; and Felleman and Van Essen, 1991.)

ably indexed. During selective attentional processing, activities in these two brain regions must communicate so that just that featural information which leads to the identification of the object is passed to the shape identification module. The participation of both the spatial and featural streams of information processing in the course of object identification is assumed to be necessary not only when the location of a target object is known (i.e., its location is stored in working memory, as in the present task) but also when the location of the target object is not known (as in search tasks).

The Inferotemporal Cortex The kinds of objects to which cells in IT preferentially respond vary widely from simple to complex, including objects made up of various combinations of color, texture, and shape (e.g., Gross, Rocha-Miranda, and Bender, 1972; Desimone et al., 1984), faces and hands (Bruce, Desimone, and Gross, 1981), and toy animals, vegetables, and other natural objects (Tanaka et al., 1991). Cells in IT also respond selectively to parts of an object, such as facial features (Perrett, Rolls, and Caan, 1982). For reviews of the face-cell literature, the reader is referred to Perrett, Mistlin, and Chitty (1987) and Desimone (1991). Variations in the location and size of objects do not appreciably change the selectivities of these IT cells (Schwartz et al., 1983).

The finding that some IT cells respond to a whole object and others to its parts is relevant to the way we conceptualize the optional identification of the letter *O* and the word *TON* that are embedded in the whole stimulus word *STONE*. While one part of an object is being identified, other parts are regarded as distractors and, to block the information arising from the distracting parts, a selective attentional mechanism would seem to be required to operate on the information flow to the object identification region of IT.

Information about object identity appears to be vector-coded by a population or ensemble of cells and not uniquely coded or locally coded by one cell (for a comparison of vector and local coding schemes, see Churchland and Sejnowski, 1992). For example, face stimuli appear to evoke responses in a distributed population of codes in IT (Perrett, Mistlin, and Chitty, 1987) but, although each cell in the ensemble responds somewhat to every face stimulus, particular faces appear to be coded locally in clumps—that is, by a sparse ensemble code (Baylis, Rolls, and Leonard, 1985; Young and Yamane, 1992), in which a relatively few cells may be sufficient to identify a particular object or face . Within anterior IT, cells with similar selectivities of features (e.g., combinations of intersecting bar features that produce *L*-like, *T*-like, *W*-like, and inverted *Y*-like patterns) appear to cluster together in groups whose size is estimated to correspond to approximately 2000 columns (Fujita et al., 1992). Analysis of the firing patterns of adjacent cells in IT indicate that they carry separate information instead of firing redundantly or interactively (Gawne, Eskandar, and Richmond, 1992).

In view of the foregoing considerations of cell behavior in IT, we conjecture that the identification of an object such as the letter *O* involves the firing of a particular small ensemble of cells that are tuned to fire at a somewhat high rate, whereas a much larger population of cells fires at a relatively low rate, owing to those cells tunings to the many other stimuli that possess features similar to the letter *O*. Identifying the familiar shapes *STONE* and *TON*, in which the object *O* is embedded, would seem to follow the same principle of cell groupings in IT as in the case of the single letter *O*, particularly when the word *TON* is highly familiar and is apparently attended to as a whole.

Area V4 Visual information that flows into IT apparently enters chiefly through the gateway of area V4 (Desimone, Fleming, and Gross, 1980), and this cortical area has yielded the strongest evidence for the modulation effects of attention in the identification of objects (for a review, see Desimone and Ungerleider, 1989). It is not surprising, therefore, that cells in V4 have been found to be responsive to component features of visual form.

The major inputs to V4 are from the early visual areas of V1, V2, and V3, but there are also inputs from a wide variety of other areas, including the temporal, frontal, and parietal areas (for a review, see Felleman and Van Essen, 1991). Of special importance to selective spatial attention in V4 are the direct connections (shown in monkey) with posterior parietal area LIP, the lateral intraparietal area (Andersen et al., 1990; Blatt, Andersen, and Stoner, 1990; Seltzer and Pandya, 1984), where cells are sensitive to the spatial location of a stimulus.

Neurons in V4 are of particular interest in this chapter because many of them respond as if they are expressing attention: That is, they appear to respond to a stimulus display by modulating information flow at particular locations (Moran and Desimone, 1985) and also by modulating information flow corresponding to particular attributes, such as orientation or color (Haenny, Maunsell, and Schiller, 1988; Haenny and Schiller, 1988; Spitzer, Desimone, and Moran, 1988). Using a matching-to-sample task, Moran and Desimone (1983) trained monkeys to attend to a colored bar shape in one location and to ignore a different one in the other location. When both stimuli were within the receptive field of a cell, the cell responded well if the stimulus to which the cell was responsive (e.g., a red bar) appeared at the attended location but suffered considerable attenuation if the stimulus appeared at the ignored location. This effect was not found for cells in either V1 or V2.

Moran and Desimone (1985) also examined the case in which the attended location occurred within the receptive field of a cell and the ignored location occurred outside the receptive field of that cell. In this case, the cell's response to the onset of a target stimulus within its receptive field was not contingent on the location at which the monkey directed attention. One could reason that attention to a location outside the receptive field of a cell does not spread inhibitory effects into the field of that cell and hence there is no contraction of its receptive field nor any attenuation of its response to its preferred stimulus.

In a subsequent study, Desimone and his associates (Luck et al., 1992) presented stimuli one a time, instead of two at a time, and in an alternating fashion between two locations. Their results confirmed the finding that effects of attention at the time of target stimulus onset are manifest only when both the attended and ignored locations are within the cell's receptive field.

Taken together, these two experiments imply that the expression of attention to a location in this task involves the attenuation of information from the uncued location when it lies within a cell's receptive field and not the enhancement of information flow at the attended location. In other words, the expression of attention here involves only a gating (filtering) of information at uncued locations and not either a gain alone at the cued location or a gate-gain combination of the two.

Other single-cell studies have shown a gain effect of attentional manipulations in the responding of V4 cells. Spitzer, Desimone, and Moran (1988) varied the difficulty of orientation discrimination and color discrimination in a matching-to-sample task and found that the sample stimulus of the more difficult discrimination produced not only a relative gain (enhancement) in a cell's response but also a narrowing in the cell's tuning curve, suggesting an effect similar to constriction of the receptive field. Haenny, Maunsell, and Schiller (1988) displayed oriented gratings to monkeys in a match-to-sample task and found that the responses of more than half of the 192 cells observed in V4 responded strongly when the monkey had been cued to expect a particular orientation and responded weakly when the monkey had been cued to expect a different orientation. In a related study, Haenny and Schiller (1988) demonstrated that repetition of a particular oriented grating increased the firing rate of 72% of the 154 V4 cells examined and 31% of the V1 cells examined, but only the V4 cells showed an additional narrowing of orientation tuning. Also, in spatial cuing studies with event-related potentials (ERPs) (Mangun et al., 1987) and PET (Corbetta et al., 1993), electrical and blood flow activity are shown to be enhanced at the brain areas corresponding to the location of a spatial cue.

The foregoing V4 single-cell data indicate that cells show a relative gain in firing rate when a single object is presented and attention is directed to its orientation but, when two objects are presented and attention is directed to the location of one of them, cells showed no change in firing rate if the stimulus was a target and showed attenuation in firing rate if the stimulus was a distractor (providing that the distractor was in the receptive field of the cell). Apparently, identifying an object in a field cluttered with a distractor here involves the expression of attention as a reduction in activity at the location of the distractor, with no apparent change in activity at the location of the target. The two locations in these tasks were always separated by at least 1°, so that when objects appeared in both locations simultaneously, the location of each object was distinct enough to compete for an eye movement at the onset of the stimulus. Although overt eye movements were inhibited by training, the information for the two potential eye movements may provide a means for attending to one location and ignoring the other. This information could conceivably be in a form such that inhibitory information would be directed to the location of the ignored stimulus whereas no excitatory information would be directed to the location of the attended stimulus, thereby producing selection by gating alone. A similar view of selective attention is given by Treisman and Sato (1990) for detecting targets in search tasks in which distractor locations are inhibited rather than target locations being activated.

When distractors are placed sufficiently close to the target so that the ensemble is initially perceived as one object, then there is no basis for competition of eye movements and, therefore, perhaps less need for strong inhibitory processing in V4. The additional processing that presumably is required to align attention around the target area of these kinds of displays may be predominantly facilitatory. Examples of stimulus displays that are presumed initially to elicit only one eye movement signal because they fit the category of "ensemble perceived as one object" are shown in figure 41.1, where target-flanker spacings were on the order of 0.15°. Subjects did not know in advance at which of five locations an ensemble triplet (e.g., *VRY*) would appear and, when it did, eye movement information presumably specified the location of the whole object. At this target-flanker spacing, increases in target-flanker similarity produce robust increases in human response times but, as the target-flanker spacing increases beyond 1°, the target-flanker effects on response time disappear (Eriksen and Eriksen, 1974). Bringing locations of targets and distractors closer may evoke the same kind of attentional gain expressed when highly similar orientations produce confusions in discrimination tasks. When locations of targets and distractors are close (as they almost always are when one part of an object is singled out from adjacent parts), the featural information in these locations is more likely to interact and produce confusions. Consequently, to be successful, selective attention may require more than simply an attenuation of the surround and may call on the additional sharpening properties that lie in enhancements of the target area.

Thus, the mechanism that produces selective attentional expression mainly by variations in enhancements (gains) at the target location may be different from the mechanism that produces selective attentional expression mainly by variations in attenuations (filtering, gating) at the surrounding locations. However, if one assumes that the expression of selective visual attention by spatial location takes place at or near V4 in the V1-IT pathway, then there must exist appropriate anatomical connections between each mechanism and this area.

Posterior Parietal and Dorsolateral Prefrontal Cortices One common way that selection of an object takes place is by means of its location in the visual field or its position within a larger object or group of objects. A major brain area directly connected to V4 that specializes in the processing of spatial information necessary for locational and positional selection is LIP (e.g., Blatt, Andersen, and Stoner, 1990), a subarea of the PPC. The PPC is influenced, in turn, by other brain areas that specialize in spatial information (Selemon and Goldman-Rakic, 1988), notably the DLPFC, and the major subcortical areas specializing in visuospatial functions—the superior colliculus, which computes information that is relatively directly concerned with eye movements, and the thalamus, whose functioning in this context will be described later in this chapter. Treating these cortical and subcortical areas in relative isolation individually or even as a closed group can be misleading, because these structures are interconnected extensively with other brain areas (Goldman-Rakic, 1988).

The evidence supporting the role of the PPC in spatial processing, and particularly spatial attentional processing, comes from a variety of experimental methodologies and measures. These include human lesion studies (Critchley, 1953; Posner et al., 1984; Heilman et al., 1985) and monkey lesion studies (Stein, 1978; Lynch and McClaren, 1989) that show an inability to redirect attention to the contralateral field; single recordings in monkeys that show enhancement of firings to attended visual locations (e.g., Mountcastle et al., 1975; Bushnell, Goldberg, and Robinson, 1981; Motter and Mountcastle, 1981; Hyvarinen, 1982; Andersen, Essick, and Siegel, 1985; Goldberg, Colby, and Duhamel, 1990); ERP studies in humans carried out using magnetic resonance imaging (MRI) scans of individual subject's brains that show increased negativity over the PPC 150–190 ms after a visual event has occurred (Mangun, Hillyard, and Luck, 1992); and cerebral blood flow studies that indicate increased activation in the superior parietal cortex during visuospatial tasks (Haxby et al., 1991) and during a spatial cuing task (Corbetta et al., 1993).

The prefrontal areas, important to the internal voluntary control of spatial attention, project to the PPC areas in a descending fashion from the prefrontal areas. Goldman-Rakic, Chafee, and Friedman (in press) measured activity of cells in both the prefrontal and posterior parietal areas of the same animal who was engaged in performing an oculomotor delayed-response (ODR) task. This task required the animal to move its eyes to the location of a visual target that had been briefly flashed several seconds earlier. Single-cell recordings were taken from the prefrontal and posterior cortices in the same animal. Raster displays and histograms of the activity of cell pairs, one in the prefrontal cortex and the other in the PPC, showed highly similar profiles: Some pairs of cells responded to the location of the cue while it was present, other pairs responded to the direction of the upcoming eye movement during a 3-second delay period, and still other pairs responded to the direction of the eye movement when it occurred.

These results strongly suggest that storage of the location of an object over a short period of time (e.g., between the time of an instructive locational cue and the signal to respond, or even over a block of trials) occurs in prefrontal areas and that cells in spatial maps here can project this coded spatial information to corresponding spatial maps in the PPC in two important ways: in a tonic (sustained) mode as, for example, is required to maintain preparatory attention to a location or to maintain a bodily posture in readiness to make a sensorimotor response toward an object (e.g., reaching or grasping), or in a phasic (brief) mode as, for example, is required to narrow spatial attention to select one object in a cluttered field or to trigger a sensorimotor response to an object.

A mechanism of selective attention: the pulvinar

EVIDENCE FOR PULVINAR INVOLVEMENT IN VISUAL SELECTIVE ATTENTION It is assumed that the expression of selective attention in the occipitotemporal pathway leading to object identification is produced by a mechanism responsive to spatial information, particularly spatial information represented in the PPC and superior colliculus (SC). A structure that is connected to the occipitotemporal areas and to both the PPC and the SC is the thalamus. Moreover, the thalamus has been proposed by several investigators as embodying a mechanism of selective attention within its circuitry (Chalupa, 1977; Yingling and Skinner, 1977; Scheibel, 1981; Crick, 1984; Sherman and Koch, 1986; LaBerge and Brown, 1989).

The thalamus is connected reciprocally with virtually every area of cortex (Jones, 1985), and therefore thalamic nuclei have the capability of influencing activity in any cortical area. Of particular interest in this chapter are the thalamic nuclei that project reciprocally to visual cortical areas in the occipital, temporal, parietal, and frontal lobes. These nuclei are clustered in the posterior part of the thalamus in the division called the *pulvinar*. In the human, the pulvinar is the largest nucleus of the thalamus, occupying approximately two fifths of the thalamic volume. The proportional size of the pulvinar to the thalamus decreases from human to monkey and, in the cat, the pulvinar is so small that it escapes labeling in many brain atlases. Within the primate pulvinar are four major divisions termed the *medial* (PuM), *lateral* (PuL), *inferior* (PuI), and *anterior* (PuA) pulvinar (Jones, 1985).

The pathways of early visual processing in both cortex and SC, as well as pathways in IT, interconnect with the pulvinar in a manner that preserves topographical relations, particularly with the inferior and lateral nuclei of the pulvinar (Allman et al., 1972; Benevento and Rezak, 1976; Burton and Jones, 1976; Bender, 1981; Lund et al., 1981; Ungerleider, Galkin, and Mishkin, 1983; Dick, Kaske, and Creutzfeldt, 1991; Kaske, Dick, and Creutzfeldt, 1991). In the posterior parietal areas, neurolabeling studies have shown connections from LIP and 7a to PuL and PuM and vice versa (Asanuma, Andersen, and Cowan, 1985; Schmahmann and Pandya, 1990). The PuM also connects with lateral prefrontal areas (Goldman-Rakic and Porrino, 1985). The regions within PuM containing cells that connect to area 7a apparently overlap areas in PuL containing cells that connect to the lateral prefrontal area (Asanuma, Andersen, and Cowan, 1985). Therefore, there is not only a clear connectivity between PPC areas and PuL and PuM but also between lateral prefrontal areas and these pulvinar areas.

Several different physiological measures suggest that the pulvinar is responsive to tasks that involve attentional operations. Single-cell recordings show an enhancement of cell firings in the pulvinar to a visual stimulus when that stimulus is a target of an impending eye movement or when the animal attends to it without moving the eyes to it (Peterson, Robinson, and Keys, 1985). When microinjections of mucimol, a GABA agonist, were injected into the dorsal region of PuM of monkeys, shifts of attention to the contralateral visual field were impaired in the spatial orienting task; injections of the GABA antagonist bicucilline into the same area facilitated shifts of attention to the contralateral visual field (Petersen, Robinson, and Morris, 1987).

Human patients with lesions of the posterior thalamus on one side were slower to respond to visual stimuli (cued as well as uncued) in the contralateral field while showing no signs of contralateral neglect (Rafal and Posner, 1987). These authors interpreted the results as indicating an impairment in engaging attention at a new location, in contrast to the impairment in disengaging attention that is characteristic of the neglect syndrome typically produced by lesions in the PPC (e.g., Critchley, 1953; Posner et al., 1984). In monkeys, lesions in the pulvinar produce impairments in the attentional scanning of a visual display (Underleider and Christensen, 1979).

PET studies of brain activity during attention have also suggested involvement of the pulvinar in attention tasks (for a review, see LaBerge, 1990). Corbetta and colleagues (1991) used a task that focused attention in preparation for discriminating among of an array of shapes with respect to their shape, size, color, or velocity of movement, and found that in the right thalamus there was an increase in blood flow during the velocity and shape conditions. Another PET study (LaBerge and Buchsbaum, 1990) involved a task designed to intensify the activity in the brain mechanisms that implement operation of selectively attending to a shape, by asking subjects to identify a target shape *O* (vs. a *C* or a zero) when the target was surrounded on all sides by eight similar flankers (*G* and *Q*). The results showed that, on average, the pulvinar contralateral to the side of the eight-flanker task display showed a significantly greater amount of glucose uptake than the pulvinar contralateral to the side receiving the no-flanker task. These findings appear to support the hypothesis that the pulvinar provides operations essential to the selection of a visual shape when other shapes are positioned nearby.

Another study questions whether the pulvinar is indeed necessary for the selective processing of a target object when another object is presented at the same time, either in the same or in the opposite hemifield (Desimone et al., 1989). The lateral pulvinar on one side was deactivated by injections of muscimol, and a colored bar and a distracting bar (apparently positioned approximately 2° apart) were presented in the visual field contralateral to the affected pulvinar. The results showed no effect of pulvinar deactivation over the case in which no distractor was present, but when a similar unilateral deactivation was induced in the SC, performance was impaired when a distractor appeared in either the same or the opposite visual field (all in the absence of eye movements). Other investigators have observed that lesions in the SC increase the time to shift attention (Posner and Cohen, 1984; Kertzman and Robinson, 1988).

If one assumes that attention to location makes use of eye movement information, then disruptions in the eye movement computations in the SC would be expected to feed into the pulvinar and cause disruptions there as well, as cells in the superficial layers of the SC project via pulvinar to cortical areas (Hartung et al., 1980; Abramson and Chalupa, 1988). The experiment by Desimone's group (1989) apparently indicates that with their displays, in which distractors are positioned at least 2° from the target and discriminations are based on a single feature (e.g., color), pulvinar processing may be bypassed, perhaps by the direct route that exists between areas LIP and V4 (Seltzer and Pandya, 1984; Andersen et al., 1990). In tasks where the spacing between target and distractor is large enough potentially to evoke different eye movements, the mechanism that produces the expression of attention in V4 may be associated with computations in the SC. Additional research is needed to determine clearly the relationships between target-flanker separations and the involvement of selection mechanisms associated with the SC and pulvinar.

One way to determine the suitability of a proposed mechanism for producing selective attention in the occipitotemporal pathways is to examine the underlying circuitry and evaluate its potential for producing output firing patterns appropriate for attentional expression. In the case of the SC, a great deal is known about its connectivity with other brain structures and about the firing behavior of its cells in a variety of visual tasks (e.g., Wurtz, and Goldberg, 1972a, b; Sparks and Mays, 1980), but apparently little is known about its circuit structure (Sparks and Nelson, 1987). In contrast, a good deal is known about the circuit structure of the pulvinar.

Structure of Pulvinar Circuitry Figure 41.3 is a diagram of the thalamic cell types and their interconnections, in which the only significant uncertainty

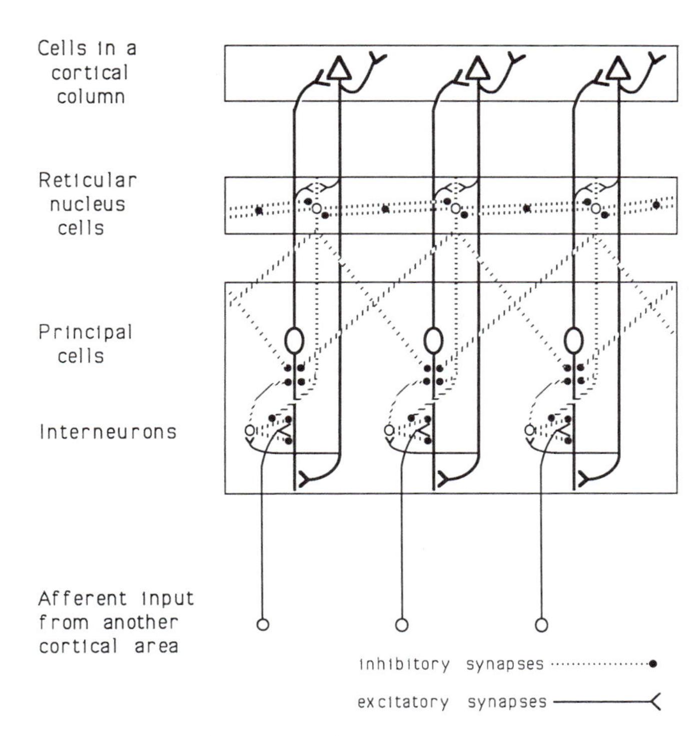

FIGURE 41.3 The standard thalamocortical circuit. Afferent inputs (e.g., from posterior parietal cortex) synapse with principal (relay) cell dendrites in glomeruli that contain interneuron synapses. Outputs of principal cells are mainly to layer 3 of a cortical area (e.g., V4), and returning projections arise mainly from layer 6.

about the connectivity concerns whether the reticular nucleus (RN) projections to principal (P) cells occurs only within a thalamic column (e.g., Crick, 1984; Steriade, Domich, and Oakson, 1986) or only to neighboring columns (e.g., Sherman and Koch, 1986) or to both (LaBerge, Carter, and Brown, 1992). In the simulation of thalamic circuit operations (described in the next section), circuits embodying each of these three versions of the RN–to–P cell connections were run separately, and the patterns of output firing rates to cortical cells were compared. Although almost all the cellular and connectivity knowledge we have of the thalamus is based on the monkey, cat, and rat, it appears relatively safe to extrapolate from data from the monkey (and, in some cases, from the cat and rat) to the human because of identical histochemical staining patterns in the thalami of these two primate species (Hirai and Jones, 1989).

A thalamic nucleus is organized in a columnar fashion, in the sense that information flow from an afferent input through a thalamic P cell is segregated from the information flowing through other thalamic P cells (Jones, 1985; Steriade, Jones, and Llinas, 1990). The only known interactions between channels within the thalamus occurs by means of the lateral inhibitory links between cells of the reticular nucleus. Therefore, it would seem that the separate channels of information flow from thalamic inputs to outputs follow a local coding scheme that produces a precise topological transformation (Jones, 1985). Furthermore, the lateral inhibitory computations of the RN cells could be expected to sharpen local gradients of information flow across the cortical columns (in somewhat the same way that horizontal cells function in the retina), so that information flow that is routed through the thalamus may serve to focus information spatially rather than to spread or distribute it. This principle may hold generally for a cluster of thalamic input fibers carrying a vector-coded representation, even considering that a typical thalamocortical axon spreads within a 2- to 2.5-mm diameter area (Kaske, Dick, and Creutzfeldt, 1991) and that thalamocortical axons terminating in primary visual, auditory, and somatosensory areas may be restricted to a 1- to 2-mm (Ferster and Levay, 1978; Gilbert and Wiesel, 1983). In contrast, the columnar organization in the cortex is characterized by extensive interconnections between columns, both adjacent and remote (for relevant reviews, see Toyama, 1988; Felleman and Van Essen, 1991), which enables heightened activity in one column to spread or distribute activity into adjacent and remote columns.

The remaining part of the circuit diagram shown in figure 41.3 concerns the RN cells and the interneurons, both of which are inhibitory (Jones, 1985; Steriade, Jones, and Llinas, 1990), the RN cells providing recurrent inhibition and the interneurons providing feedforward inhibition. In the case of the RN cells, the main axon fiber enters the (dorsal) thalamus, spreads its terminals widely (Yen et al., 1985), and synapses not only with P cells but also with interneurons (Montero and Singer, 1985), thus enabling RN cells to reduce the inhibitory influence of interneurons on thalamic afferent input (Steriade, Domich, and Oakson, 1986).

Often overlooked in circuit diagrams of cortical and subcortical areas is the ubiquitous presence of axon terminals arising from nuclei in the brain stem and basal forebrain (Jones, 1985; Steriade, Jones, and Llinas, 1990). Neurotransmitter substances secreted by these axon terminals modulate in a diffuse manner both the synaptic connections between thalamic cells and the intrinsic membrane properties of a cell that affect spontaneous firing and thereby influence states of sleep and arousal (McCormick, 1992; see also chapter 44).

SIMULATIONS OF PULVINAR CIRCUIT OPERATIONS Given the relatively well-known anatomical structure of the pulvinar circuit (and thalamic circuit in general), my colleagues and I (LaBerge, Carter, and Brown, 1992) attempted to determine by simulations whether the pulvinar circuit is particularly suitable for computing the output characteristic of selective attentional operations. The results of the simulations suggest that the thalamic circuit selects by an enhancement of the target column inputs relative to surround (figure 41.4). Characterizing attention as an enhancement effect is in accord with the positions taken by Wurtz and Goldberg (1972a), and by Posner and Hillyard in their chapters of this book. In contrast, other neurorelated views of attention have assumed that attention operates mainly to gate (attenuate) distracting sensory input in sensory pathways of the ear (Hernandez-Peon, 1960), in the visual pathways at the level of the lateral geniculate nucleus (LGN) (Crick, 1984; Sherman and Koch, 1986), in the reticular nucleus (Yingling and Skinner, 1977), and in area V4 (Desimone et al., 1989).

In evaluating the results of the present simulations, it should be noted that the finding that the thalamic network can greatly magnify a target-surround input does not prove that the thalamic network in fact operates in this manner. Rather, the present demonstration has the status of an existence proof which shows that the thalamic circuit *could* instantiate a magnification of the target-surround contrast. Thus, for the present displays, in which the target and flankers are closely spaced (within 1°), the mechanism that produces the expression in the V1-IT pathway is assumed to be pulvinar circuitry, and the control of this mechanism is assumed to originate in the lateral prefrontal areas,

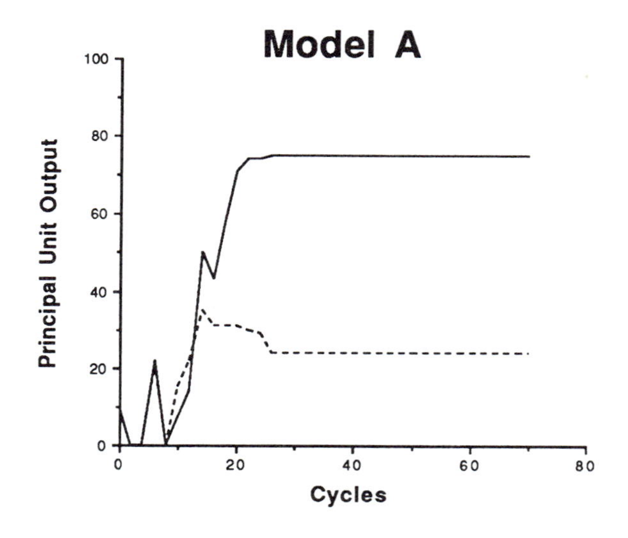

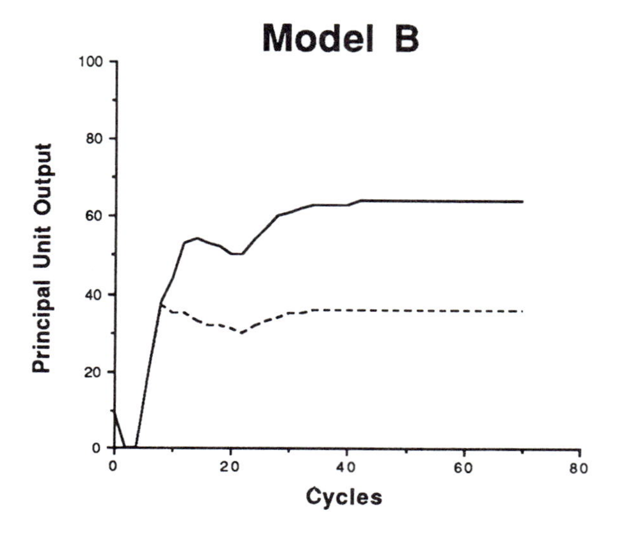

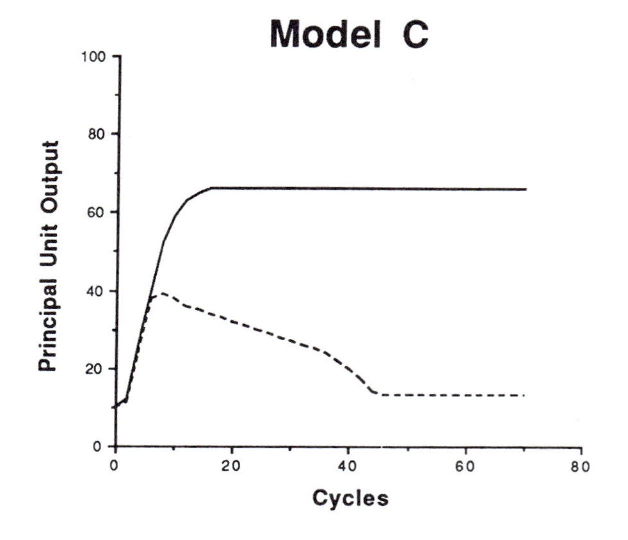

FIGURE 41.4 Simulated trajectories of target (solid lines) and flanker (dotted lines) principal cell firing rates, given an afferent input of 38 for the target, 37 for the flanker, and 5 for the spaces between them. Simulation based on three versions of the network shown in figure 41.3. (Reprinted from LaBerge, Carter, and Brown, 1992, by permission.)

project from there to the PPC, and then project from the PPC to the pulvinar.

Controls on selective attention via pulvinar afferents in triangular circuits linking brain areas

In the foregoing sections, we examined the structure and functions of a typical pulvinar circuit and defined the expression of selective attention with respect to the pattern of output across the group of cortical cells to which the circuit projects. In this section, we describe evidence that indicates the sources of the main afferent inputs to the pulvinar circuits. We regard the afferent inputs as the attentional controls on the pulvinar circuit mechanism that, in turn, produces the expression of attention in a localized cortical area.

Axon inputs from external sources to the pulvinar (and to thalamic nuclei in general) synapse on two structurally and functionally identifiable classes of receptor sites. (We disregard here the relatively diffuse influence of neuromodulatory afferents from the brain stem and basal forebrain.) Both classes of receptor sites are on the principal thalamic cell: One class of sites is located proximal to the soma where axon terminals synapse (along with an inhibitory interneuron) within glomeruli, and the other class of sites is located on the distal dendrites (Sherman and Koch, 1990). A distinguishing feature of the two types of synapses is that the axon terminals at the distal dendrite locations contain round vesicles that are smaller than those at the proximal locations.

The axon terminals found on distal dendrites of P cells are on fibers arising from cortical cells in layer 6, and the diameters of these descending fibers are typically much smaller than the diameters of ascending P-cell axon fibers that synapse on cortical cells (Steriade, Jones, and Llinas, 1990). The descending fibers presumably provide the feedback to the ascending P-cell fibers, forming an excitatory corticothalamocortical loop that is largely responsible for the enhancement property attributed to the thalamic (and pulvinar) circuit in our simulations.

In the pulvinar, the axon terminals that synapse proximal to the soma apparently originate from cells in brain areas other than those cortical areas projecting to the distal dendrites. Neurolabeling of V1 cortical cells identifies axon terminals in both the LGN and the pulvinar (Conley and Raczkowski, 1990); the cortical cells that project back to the LGN are located in the superficial part of layer 6, and the cortical cells that project to the pulvinar are located in layer 5 and in the deeper part of layer 6. The layer 6 cells that project to the LGN presumably terminate on distal dendrites of principal (relay) LGN cells, and the layer 5 and deep 6 cells that project to the pulvinar presumably terminate near the somas of principal cells. These pulvinar P cells presumably project reciprocally to a cortical area upstream from V1. (For evidence of this dual mode of corticothalamic termination in the mediodorsal nucleus of the thalamus, see Schwartz, Dekker, and Goldman-Rakic, 1991).

Injections of two different labels into subareas TEO and TE of the IT region show that the terminal label from TEO cells overlaps cells labeled from TE injections, and the same crossover pattern holds for the terminal label from TE cells (Ungerleider, personal communication). Stated differently, cells in TEO project not only to pulvinar cells in areas where cells return those projections (in the manner of the standard corticothalamic loop) but also to pulvinar areas where cells receive and return projections to cells in TE (in the standard manner). Thus, it appears that TEO projects to TE over two routes, the indirect (bisynaptic) route through pulvinar and the direct (monosynaptic) route, constituting a triangular circuit.

It seems reasonable to generalize the existence of triangular circuits to pairs of cortical areas all along the V1-IT pathway, as diagrammed in figure 41.5, since many labeling studies of visual cortical areas show axon labeling in more than one pulvinar area

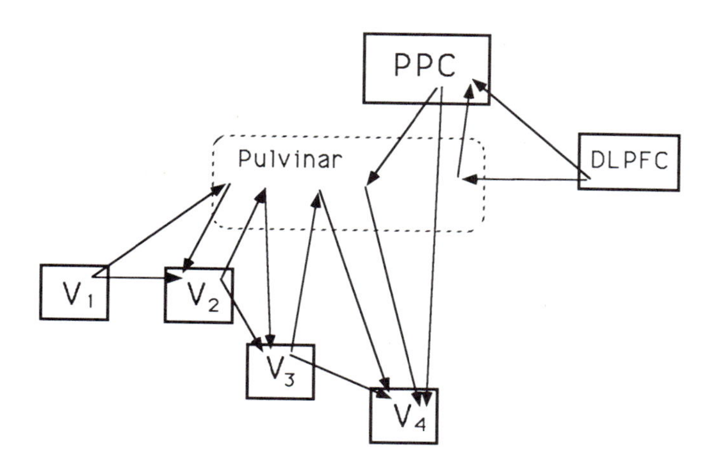

FIGURE 41.5 Diagram of conjectured triangular circuits that connect pairs of cortical areas. PPC, posterior parietal cortex; DLPFC, dorsolateral prefrontal cortex.

(e.g., Dick, Kaske, and Creutzfeldt, 1991; Kaske, Dick, and Creutzfeldt, 1991). Of particular interest to top-down control of attention is the evidence that injections of DLPFC cells label terminals in the medial pulvinar at the same time that they label terminals in PPC and the mediodorsal nucleus of the thalamus (Goldman-Rakic and Porrino, 1985). Presumably, some cells in DLPFC reciprocally project to cells in the mediodorsal (in the manner of the standard corticothalamic loop), whereas other DLPFC cells project to pulvinar as part of a triangular circuit whose two other links are the corticocortical connections between DLPFC and PPC and the pulvinar and PPC (see figure 41.5).

Of particular importance to the present areal circuitry scheme of selective attention is a triangular circuit assumed to exist between some area or areas of PPC and some area or areas within the V1-IT pathway (e.g., a triangular circuit joining PPC and V4 in the monkey, and its homolog in the human). The direct (backward) corticocortical route from PPC to V4 is not expected to magnify small target-surround differences in PPC. The indirect connection between PPC and V4 that takes the longer route of the triangular circuit through the pulvinar presumably produces enhancement and sharpening by the thalamocortical circuitry. Thus, the indirect projection from PPC through the pulvinar and thence to V4 would presumably provide the target-surround enhancement effect that this chapter assumes is the expression of attention in V4.

ACKNOWLEDGMENT The writing of this chapter was supported in part by a grant from the Office of Naval Research.

REFERENCES

ABRAMSON, B. P., and L. M. CHALUPA, 1988. Multiple pathways from the superior colliculus to the extrageniculate visual thalamus of the cat. *J. Comp. Neurol.* 271:397–418.

ALLMAN, J. M., J. H. KAAS, R. H. LANE, and F. M. MIEZIN, 1972. A representation of the visual field in the inferior nucleus of the pulvinar in the owl monkey (*Aotus trivirgatus*). *Brain Res.* 40:291–302.

ANDERSEN, R. A., C. ASANUMA, G. ESSICK, and R. M. SIEGEL, 1990. Corticocortical connections of anatomically and physiologically defined subdivisions within the inferior parietal lobule. *J. Comp. Neurol.* 296:65–113.

ANDERSEN, R. A., G. K. ESSICK, and R. M. SIEGEL, 1985. Encoding of spatial location by posterior parietal neurons. *Science* 230:456–458.

ASANUMA, C., R. A. ANDERSEN, and W. M. COWAN, 1985. The thalamic relations of the caudal inferior parietal lobule and lateral prefrontal cortex in monkeys: Divergent cortical projections from cell clusters in the medial pulvinar nucleus. *J. Comp. Neurol.* 241:357–381.

BAYLIS, G. C., E. T. ROLLS, and C. M. LEONARD, 1985. Selectivity between faces in the responses of a population of neurons in the cortex in the superior temporal sulcus of the monkey. *Brain Res.* 342:91–102.

BENDER, D. B., 1981. Retinotopic organization of macaque pulvinar. *J. Neurophysiol.* 46:672–693.

BENEVENTO, L. A., and M. REZAK, 1976. The cortical projections of the inferior pulvinar and adjacent lateral pulvinar in the rhesus monkey: An autoradiographic study. *Brain Res.* 108:1–24.

BLATT, G. J., R. A. ANDERSEN, and G. R. STONER, 1990. Visual receptive field organization and cortico-cortical connections of the lateral intraparietal area (area LIP) in the macaque. *J. Comp. Neurol.* 299:421–445.

BOUSSAOUD, D., L. G. UNGERLEIDER, and R. DESIMONE, 1990. Pathways for motion analysis: Cortical connections of the medial superior temporal and fundus of the superior temporal visual areas in the macaque. *J. Comp. Neurol.* 296:462–495.

BROADBENT, D. A., 1958. *Perception and Communication.* London: Pergamon.

BRUCE, C. J., R. DESIMONE, and C. G. GROSS, 1981. Visual properties of neurons in a polysensory area in superior temporal sulcus of the macaque. *J. Neurophysiol.* 46:369–384.

BURTON, H., and E. G. JONES, 1976. The posterior thalamic region and its cortical projection in New World and Old World monkeys. *J. Comp. Neurol.* 168:249–301.

BUSHNELL, M. C., M. E. GOLDBERG, and D. L. ROBINSON, 1981. Behavioral enhancement of visual responses in monkey cerebral cortex: I. Modulation in posterior parietal cortex related to selective attention. *J. Neurophysiol.* 46:755–772.

CHALUPA, L. M., 1977. A review of cat and monkey studies implicating the pulvinar in visual function. *Behav. Biol.* 20:149–167.

CHURCHLAND, P. S., C. KOCH, and T. J. SEJNOWSKI, 1990. What is computational neuroscience? In *Computational Neuroscience*, E. L. Schwartz, ed. Cambridge, Mass.: MIT Press, pp. 46–55.

CHURCHLAND, P. S., and T. J. SEJNOWSKI, 1992. *The Computational Brain.* Cambridge, Mass.: MIT Press.

CONLEY, M., and D. RACZKOWSKI, 1990. Sublaminar organization within layer VI of the striate cortex in Galago. *J. Comp. Neurol.* 302:425–436.

CORBETTA, M., F. M. MIEZIN, S. DOBMEYER, G. L. SCHULMAN, and S. E. PETERSEN, 1991. Selective and divided attention during visual discrimination of shape, color, and speed: Functional anatomy by positron emission tomography. *J. Neurosci.* 11:2383–2402.

CORBETTA, M., F. M. MIEZIN, G. L. SHULMAN, and S. E. PETERSEN, 1993. A PET study of visuospatial attention. *J. Neurosci.* 13:1202–1226.

CRICK, F., 1984. The function of the thalamic reticular complex: The searchlight hypothesis. *Proc. Natl. Acad. Sci. U.S.A.* 81:4586–4590.

CRITCHLEY, M., 1953. *The Parietal Lobes*. London: Edward Arnold.

DESIMONE, R., 1991. Face-selective cells in the temporal cortex of monkeys. *J. Cogn. Neurosci.* 3:1–8.

DESIMONE, R., T. D. ALBRIGHT, C. G. GROSS, and C. BRUCE, 1984. Stimulus selective properties of inferior temporal neurons in the macaque. *J. Neurosci.* 4:2051–2062.

DESIMONE, R., J. FLEMING, and C. G. GROSS, 1980. Prestriate afferents to inferior temporal cortex: An HRP study. *Brain Res.* 184:41–55.

DESIMONE, R., and L. G. UNGERLEIDER, 1989. Neural mechanisms of visual processing in monkeys. In *Handbook of Neuropsychology*, vol. 2, F. Boller and J. Grafman, eds. Amsterdam: Elsevier, pp. 267–299.

DESIMONE, R., M. WESSINGER, L. THOMAS, and W. SCHNEIDER, 1989. Effects of deactivation of lateral pulvinar or superior colliculus on the ability to selectively attend to a visual stimulus. *Soc. Neurosci. Abstr.* 15:162.

DICK, A., A. KASKE, and O. D. CREUTZFELDT, 1991. Topographical and topological organization of the thalamocortical projection to the striate and prestriate cortex in the marmoset. *Exp. Brain Res.* 84:233–253.

ERIKSEN, B. A., and C. W. ERIKSEN, 1974. Effects of noise letters upon the identification of a target letter in a nonsearch task. *Percept. Psychophys.* 16:143–149.

FELLEMAN, D. J., and D. C. VAN ESSEN, 1991. Distributed hierarchical processing in the primate cerebral cortex. *Cerebral Cortex* 1:1–47.

FERSTER, D., and S. LEVAY, 1978. The axonal arborization of lateral geniculate neurons in the striate cortex of the cat. *J. Comp. Neurol.* 182:923–944.

FUJITA, I., K. TANAKA, M. ITO, and C. KANG, 1992. Columns for visual features of objects in monkey inferotemporal cortex. *Nature* 360:343–346.

GAWNE, T. J., E. N. ESKANDAR, and B. J. RICHMOND, 1992. The heterogeneity of adjacent neurons in inferior temporal cortex. *Soc. Neurosci. Abstr.* 18:147.

GILBERT, C. D., and T. N. WIESEL, 1983. Clustered intrinsic connections in cat visual cortex. *J. Neurosci.* 3:1116–1133.

GOLDBERG, M. E., C. L. COLBY, and J.-R. DUHAMEL, 1990. The representation of visuomotor space in the parietal lobe of the monkey. *Cold Spring Harb. Symp. Quant. Biol.* 55:729–739.

GOLDMAN-RAKIC, P. S., 1988. Topography of cognition: Parallel distributed networks in primate association cortex. *Annu. Rev. Neurosci.* 11:137–156.

GOLDMAN-RAKIC, P. S., M. CHAFEE, and H. FRIEDMAN, 1993. Allocation of function in distributed circuits. In *Brain Mechanisms of Perception and Memory: From Neuron to Behavior*, T. Ono, L. R. Squire, M. E. Raichle, D. I. Perrett, and M. Fukuda, eds. New York: Oxford University Press, pp. 445–456.

GOLDMAN-RAKIC, P. S., and L. J. PORRINO, 1985. The primate mediodorsal (MD) nucleus and its projections to the frontal lobe. *J. Comp. Neurol.* 242:535–560.

GROSS, C. G., C. E. ROCHA-MIRANDA, and D. E. BENDER, 1972. Visual properties of neurons in inferotemporal cortex of the macaque. *J. Neurophysiol.* 35:96–111.

HAENNY, P. E., J. H. R. MAUNSELL, and P. H. SCHILLER, 1988. State dependent activity in monkey visual cortex: II. Retinal and extraretinal factors in V4. *Exp. Brain Res.* 69:245–259.

HAENNY, P. E., and P. H. SCHILLER, 1988. State dependent activity in monkey visual cortex: I. Single cell activity in V1 and V4 on visual tasks. *Exp. Brain Res.* 69:225–244.

HARTUNG, J. K., M. F. HUERTA, A. J. FRANKFURTER, N. L. STROMINGER, and G. J. ROYCE, 1980. Ascending pathways from the monkey superior colliculus: An autoradiographic analysis. *J. Comp. Neurol.* 192:853–882.

HAXBY, J. V., C. L. GRADY, B. HORWITZ, L. G. UNGERLEIDER, M. MISHKIN, R. E. CARSON, P. HERSCOVITCH, M. B. SCHAPIRO, and S. I. RAPPOPORT, 1991. Dissociation of object and spatial visual processing pathways in human extrastriate cortex. *Proc. Natl. Acad. Sci. U.S.A.* 88:1621–1625.

HEILMAN, K. M., D. BOWERS, H. B. COSLETT, H. WHELAN, and R. T. WATSON, 1985. Directional hypokinesia: Prolonged reaction times for leftward movements in patients with right hemisphere lesions and neglect. *Neurology* 35:855–859.

HERNANDEZ-PEON, R., 1960. Neurophysiological correlates of habituation and other manifestations of plastic inhibition (internal inhibition). *Electroenchephalogr. Clin. Neurophysiol.* Suppl. 13:101–114.

HIRAI, T., and E. G. JONES, 1989. A new parcellation of the human thalamus on the basis of histochemical staining *Brain Res. Rev.* 14:1–34.

HYVARINEN, J., 1982. Parietal association cortex: Posterior parietal lobe of the primate brain. *Physiol. Rev.* 62:1060–1129.

JAMES, W., 1890. *Principles of Psychology*. New York: Holt, Rinehart and Winston.

JONES, E. G., 1985. *The Thalamus*. New York: Plenum.

KASKE, A., A. DICK, and O. D. CREUTZFELDT, 1991. The local domain for divergence of subcortical afferents to the striate and extrastriate visual cortex in the common marmoset: A multiple labelling study. *Exp. Brain Res.* 84:254–265.

KERTZMAN, C., and D. L. ROBINSON, 1988. Contributions of the superior colliculus of the monkey to visual spatial attention. *Soc. Neurosci. Abstr.* 14:831.

KOCH, C., and S. ULLMAN, 1985. Shifts in selective visual attention: Towards the underlying neural circuitry. *Hum. Neurobiol.* 4:219–227.

LABERGE, D., 1990. Thalamic and cortical mechanisms of attention suggested by recent positron emission tomographic experiments. *J. Cogn. Neurosci.* 2:358–372.

LABERGE, D., and V. BROWN, 1989. Theory of attentional operations in shape identification. *Psychol. Rev.* 96:101–124.

LABERGE, D., and M. S. BUCHSBAUM, 1990. Positron emission tomographic measurements of pulvinar activity during an attention task. *J. Neurosci.* 10:613–619.

LABERGE, D., M. CARTER, and V. BROWN, 1992. A network simulation of thalamic circuit operations in selective attention. *Neural Computation* 4:318–331.

Luck, S. J., L. Chelazzi, S. A. Hillyard, and R. Desimone, 1992. Attentional modulation of responses in area V4 of the macaque. *Soc. Neurosci. Abstr.* 19:147.

Lund, J. S., A. E. Hendrickson, M. P. Ogren, and E. A. Tobin, 1981. Anatomical organization of primate visual cortex area VII. *J. Comp. Neurol.* 202:19–45.

Lynch, J. C., and J. W. McClaren, 1989. Deficits of visual attention and saccadic eye movements after lesions of parieto-occipital cortex in monkeys. *J. Neurophysiol.* 61: 74–90.

Mangun, G. R., J. C. Hansen, and S. A. Hillyard, 1987. The spatial orienting of attention: Sensory facilitation or response bias? In *Current Trends in Event-Related Potential Research*, R. Johnson, Jr., J. W. Rohrbaugh, and R. Parasuraman, eds. Amsterdam: Elsevier, pp. 118–124.

Mangun, G. R., S. A. Hillyard, and S. J. Luck, 1992. Electrocortical substrates of visual selective attention. In *Attention and Performance*, vol. 14, D. Meyer and S. Kornblum, eds. Hillsdale, N.J.: Erlbaum.

McCormick, D. A., 1992. Neurotransmitter actions in the thalamus and cerebral cortex and their role in neuromodulation of thalamocortical activity. *Prog. Neurobiol.* 39: 337–388.

Montero, V. M., and W. Singer, 1985. Ultrastructural identification of somata and neural processes immunoreactive to antibodies against glutamic acid decarboxylase (GAD) in the dorsal lateral geniculate nucleus of the cat. *Exp. Brain Res.* 59:151–165.

Moran, J., and R. Desimone, 1985. Selective attention gates visual processing in the extrastriate cortex. *Science* 229: 782–784.

Motter, B. C., and V. B. Mountcastle, 1981. The functional properties of the light-sensitive neurons of the posterior parietal cortex studied in waking monkeys: Foveal sparing and opponent vector organization. *J. Neurosci.* 1:3–26.

Mountcastle, V. B., J. C. Lynch, A. Georgopoulos, H. Sakata, and C. Acuna, 1975. Posterior parietal association cortex of the monkey: Command functions for operations within extrapersonal space. *J. Neurophysiol.* 38:871–907.

Perrett, D. I., A. J. Mistlin, and A. J. Chitty, 1987. Visual cells responsive to faces. *Trends Neurosci.* 10:358–364.

Perrett, D. I., E. T. Rolls, and W. Caan, 1982. Visual neurons responsive to faces in the monkey temporal cortex. *Exp. Brain Res.* 47:329–342.

Petersen, S. E., D. L. Robinson, and W. Keys, 1985. Pulvinar nuclei of the behaving rhesus monkey: Visual responses and their modulation. *J. Neurophysiol.* 54:867–886.

Petersen, S. E., D. L. Robinson, and J. D. Morris, 1987. Contributions of the pulvinar to visual spatial attention. *Neuropsychologia* 25:97–105.

Posner, M. I., and Y. Cohen, 1984. Components of performance. In *Attention and Performance*, vol. 10, H. Bouma and D. Bowhuis, eds. Hillsdale, N. J.: Erlbaum, pp. 531–556.

Posner, M. I., J. A. Walker, F. J. Friedrich, and R. D. Rafal, 1984. Effects of parietal injury on covert orienting of visual attention. *J. Neurosci.* 4:1863–1874.

Rafal, R. D., and M. I. Posner, 1987. Deficits in human visual spatial attention following thalamic lesions. *Proc. Natl. Acad. Sci. U.S.A.* 84:7349–7353.

Robinson, D. A., 1992. Implications of neural networks for how to think about brain function. *Behav. Brain Sci.* 15: 644–655.

Robinson, D. L., and S. E. Petersen, 1992. The pulvinar and visual salience. *Trends Neurosci.* 15:127–132.

Scheibel, A. B., 1981. The problem of selective attention: A possible structural substrate. In *Brain Mechanisms and Perceptual Awareness*, O. Pompeiano and C. Marsen, eds. New York: Raven.

Schmahmann, J. D., and D. N. Pandya, 1990. Anatomical investigation of projections from thalamus to posterior parietal cortex in the rhesus monkey: A WGA-HRP and fluorescent tracer study. *J. Comp. Neurol.* 295:299–326.

Schwartz, E. L., R. Desimone, T. D. Albright, and C. G. Gross, 1983. Shape recognition and inferior temporal neurons. *Proc. Natl. Acad. Sci. U.S.A.* 80:5776–5778.

Schwartz, M. L., J. J. Dekker, and P. S. Goldman-Rakic, 1991. Dual mode of corticothalamic synaptic termlnation in the mediodorsal nucleus of the rhesus monkey. *J. Comp. Neurol.* 309:289–304.

Selemon, L. D., and P. S. Goldman-Rakic, 1988. Common cortical and subcortical target areas of the dorsolateral prefrontal and posterior parietal cortices in the rhesus monkey: Evidence for a distributed neural network subserving spatially guided behavior. *J. Neurosci.* 8:4049–4068.

Seltzer, B., and D. N. Pandya, 1984. Further observations on parieto-temporal connections in the rhesus monkey. *Exp. Brain Res.* 55:301–312.

Shepherd, G. M., ed., 1990. *The Synaptic Organization of the Brain.* New York: Oxford University Press.

Sherman, S. M., and C. Koch, 1986. The control of retinogeniculate transmission in the mammalian lateral geniculate nucleus. *Exp. Brain Res.* 63:1–20.

Sherman, S. M., and C. Koch, 1990. Thalamus. In *The Synaptic Organization of the Brain*, G. M. Shepherd, ed. New York: Oxford University Press.

Sparks, D., and L. Mays, 1980. Movement fields of saccade-related burst neurons in the monkey superior colliculus. *Brain Res.* 190:39–50.

Sparks, D. L., and J. S. Nelson, 1987. Sensory and motor maps in the mammalian superior colliculus. *Trends Neurosci.* 10:312–317.

Spitzer, H., R. Desimone, and J. Moran, 1988. Increased attention enhances both behavioral and neuronal performance. *Science* 240:338–340.

Stein, J. F., 1978. Effects of parietal lobe cooling on manipulation in the monkey. In *Active Touch*, G. Gordon, ed. New York: Pergamon.

Steriade, M., L. Domich, and G. Oakson, 1986. Reticularis thalami neurons revisited: Activity changes during shifts in states of vigilance. *J. Neurosci.* 6:68–81.

Steriade, M., E. G. Jones, and R. R. Llinas, 1990. *Thalamic Oscillations and Signaling*. New York: Wiley.

Tanaka, K., H. Saito, Y. Fukada, and M. Moriya, 1991. *J. Neurophysiol.* 66:170–189.

Toyama, K., 1988. Functional connections of the visual cortex studied by cross-correlation techniques. In *Neurobiology of Neocortex*, P. Rakic and W. Singer, eds. New York: Wiley, pp. 203–217.

Treisman, A., and G. Gelade, 1980. A feature integration theory of attention. *Cogn. Psychol.* 12:97–136.

Treisman, A., and S. Gormican, 1988. Feature analysis in early vision: Evidence from search asymmetries. *Psychol. Rev.* 95:15–48.

Treisman, A., and S. Sato, 1990. Conjunction search revisited. *J. Exp. Psychol. [Hum. Percept.]* 16:459–478.

Ungerleider, L. G., and C. A. Christensen, 1979. Pulvinar lesions in monkeys produce abnormal scanning of a complex visual array. *Neuropsychologia* 17:493–501.

Ungerleider, L. G., T. W. Galkin, and M. Mishkin, 1983. Visuotopic organization of projections from striate cortex to inferior and lateral pulvinar in rhesus monkey. *J. Comp. Neurol.* 217:137–157.

Wurtz, R. H., and M. E. Goldberg, 1972a. Activity of superior colliculus in behaving monkey: III. Cells discharging before eye movements. *J. Neurophysiol.* 35:575–586.

Wurtz, R. H., and M. E. Goldberg, 1972b. Activity of superior colliculus in behaving monkey: IV. Effects of lesions on eye movements. *J. Neurophysiol.* 35:587–596.

Yen, C. G., M. Conley, S. H. C. Hendry, and E. G. Jones, 1985. The morphology of physiologically identified GABAergic neurons in the somatic sensory part of the thalamic reticular nucleus in the cat. *J. Neurosci.* 5:2254–2268.

Yingling, C. D., and J. E. Skinner, 1977. Gating of thalamic input to cerebral cortex by nucleus reticularis thalami. In *Attention, Voluntary Contraction and Event-related cerebral Potentials. Prog. Clin. Neurophysiol.* vol. 1, F. F. Desmedt, ed. Basel: Karper, pp. 70–96.

Young, M. P., and S. Yamane, 1992. Sparse population coding of faces in the inferotemporal cortex. *Science* 256: 1327–1331.

42 Neural Systems Mediating Selective Attention

STEVEN A. HILLYARD, GEORGE R. MANGUN, MARTY G. WOLDORFF, AND STEVEN J. LUCK

ABSTRACT Electrical and magnetic recordings of human brain activity can trace the flow of sensory-perceptual information through the afferent pathways and cortical sensory areas with high temporal resolution. Studies that employed these recordings during both auditory and visual selective attention tasks have demonstrated that attended stimuli evoke enlarged sensory responses in modality-specific cortex. The short latency of these evoked responses, as well as their anatomical and physiological characteristics, provide strong support for early selection theories of attention. These and other techniques for imaging the spatial and temporal properties of human brain activity are beginning to provide an outline of the brain systems that mediate selective attention.

Human observers can readily focus attention on high-priority stimuli in the environment and analyze their properties in considerable detail, often at the expense of less relevant inputs. This selection of a particular subset of the available stimuli for preferential processing is an essential function of selective attention (Kinchla, 1992). In recent years, a number of conceptual and technological advances have enabled systematic investigation of the neural systems of the human brain that mediate attentional processes (see chapter 39). To develop a full picture of the brain circuitry that underlies attention, it is necessary not only to identify the critical participating brain regions but also to determine the precise timing of neural communication among them. Although new methods for measuring regional cerebral blood flow and metabolism are providing a detailed anatomical picture of the brain regions that are active during attention (Petersen, Fiez, and Corbetta, 1992), these methods do not reveal the time course of information transmission along neural pathways.

The time course of brain activity patterns related to attention can be measured at the millisecond level of resolution through recordings of the electrical and magnetic fields that are generated by active populations of nerve cells. These fields can be recorded noninvasively from the surface of the head while subjects engage in attention-demanding tasks. Recent advances in techniques for localizing the intracranial generators of these surface fields (e.g., Dale and Sereno, 1993) have permitted more accurate determinations of the anatomical origins of precisely timed brain activity related to cognitive processes. Together, these approaches for imaging spatial and temporal aspects of human brain activity are providing new insights into the operations of the attention systems of the brain.

Electrical and magnetic fields of the brain

As stimulus information traverses the sensory pathways, the activation of successive relay nuclei and cortical areas gives rise to a sequence of precisely timed evoked potentials and evoked magnetic fields that may be recorded from the surface of the head with appropriate instrumentation. For example, the characteristic sequence of voltage deflections triggered at the scalp by an auditory stimulus is shown in figure 42.1. These positive and negative voltage peaks or components represent the field potentials generated by synchronous nerve cell activity at different sites in the auditory pathways of the brain stem and cerebral cortex. Volt-

STEVEN A. HILLYARD and STEVEN J. LUCK Department of Neurosciences, University of California at San Diego, La Jolla, Calif.

GEORGE R. MANGUN Department of Psychology and Center for Neuroscience, University of California at Davis, Davis, Calif.

MARTY G. WOLDORFF University of Texas Health Sciences Center at San Antonio, Research Imaging Center, San Antonio, Tex.

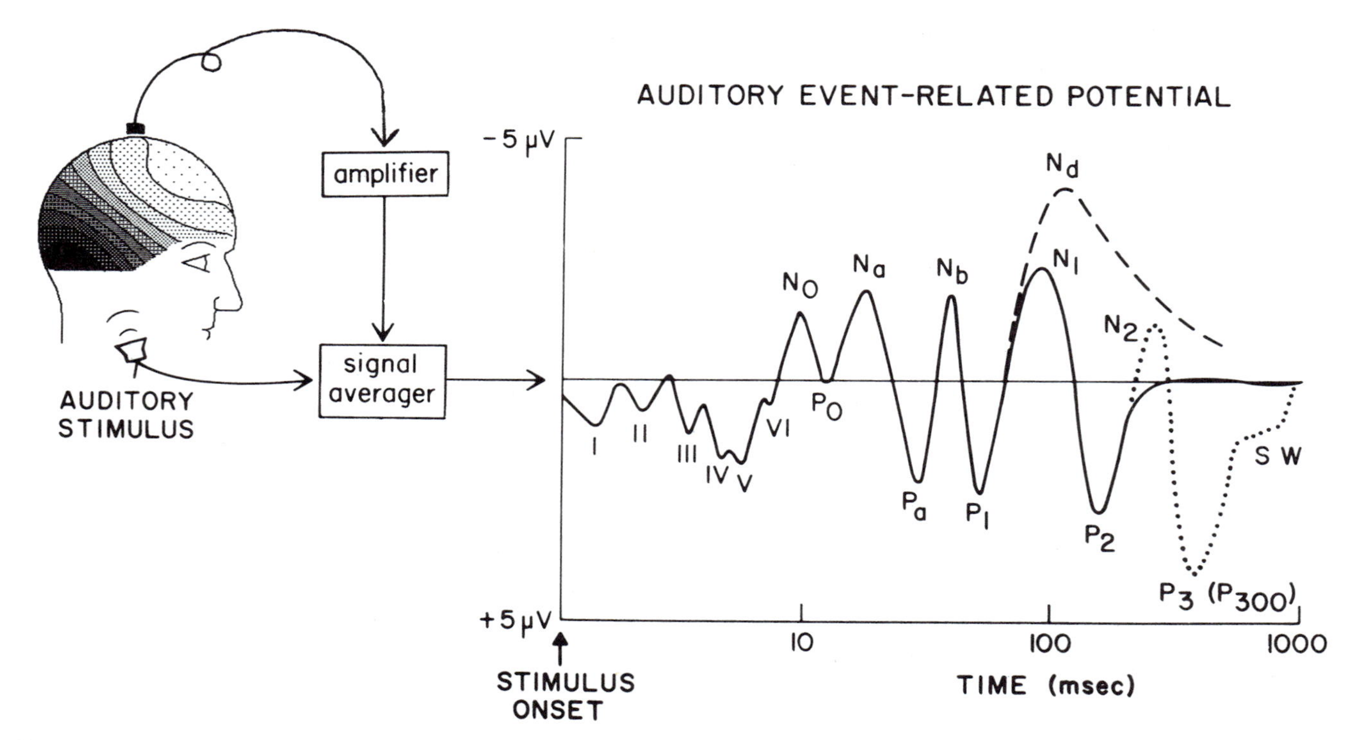

FIGURE 42.1 Component structure of the auditory ERP recorded from the scalp in response to a brief stimulus such as a click or tone. The electrical responses of the auditory pathways are amplified and signal-averaged over many stimulus presentations to produce an averaged waveform with an adequate signal/noise ratio. The logarithmic time base used here allows visualization of the major components (voltage peaks) of the auditory ERP, which occur at specific latencies after stimulus onset. These components include early waves evoked from the auditory brainstem pathways (I–VI), early positive (P) and negative (N) components evoked from the cortex (N_a, P_a, N_b, and P_1), and late cortical components (N_1 and P_2). Additional components that may vary as a function of the attentional and cognitive processing of the stimulus are shown in dotted and dashed lines (N_d, N_2, P_3, and SW). (Reprinted from Hillyard, 1993, by permission.)

age deflections such as these, which may be time-locked to sensory, cognitive, or motor events, are termed *event-related potentials* (ERPs), and their magnetic field counterparts are designated *event-related fields* (ERFs) (e.g., Kaufman and Williamson, 1985; Hillyard, 1993).

At the cellular level, both ERPs and ERFs are generated as a consequence of the flow of ionic currents across nerve cell membranes during synaptic activity. The passage of transmembrane currents into the extracellular fluids produces field potentials that may spread some distance from the active cells, depending on the electrical conductance of the intervening fluids and tissues. In contrast, the concentrated intracellular flow of ionic currents along elongated neuronal processes such as dendrites gives rise to magnetic fields that surround the active cells and pass freely through biological tissues (figure 42.2). When a large population of elongated neurons having a similar orientation are activated concurrently, their summated fields may be of sufficient strength to be recordable as ERPs or ERFs at the surface of the scalp.

Localization of active sources in the brain

Algorithms based on the physics of electrical volume conduction or magnetic field propagation have been developed that allow the intracranial locations of ERP or ERF generators to be estimated on the basis of their surface field distributions (Williamson et al., 1991; Scherg, 1992). This calculation of neural generator position from surface-recorded data is known as the *inverse problem*. Current approaches to solving the inverse problem require that simplifying assumptions be made about the geometry of the underlying cell populations, the number of discrete populations that may be active

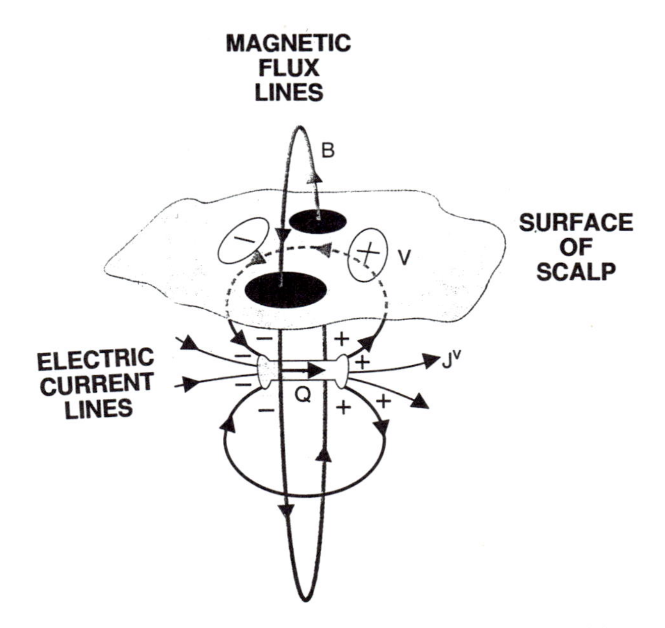

FIGURE 42.2 Theoretical voltage and magnetic field patterns produced at the surface of the scalp by an idealized "current dipole" (Q) in the brain that is oriented parallel to the surface. Such a current dipole would be produced, for example, by a population of elongated neurons of similar orientation that were generating synchronous synaptic potentials. Electrical volume currents (J^v) pass through the conducting media to produce the surface voltage field. In contrast, the magnetic field lines (B) produced by the current dipole emerge from the scalp and may be recorded by superficial sensors. (Adapted from Kaufman and Williamson, 1985.)

and, in the case of ERPs, the conductive properties of the brain, skull, and scalp.

The localization of active neural sources is more straightforward in the case of ERF recordings, because magnetic fields pass through the brain, skull, and scalp without the distortions that occur for electrical potentials (Williamson and Kaufman, 1987). Also, surface magnetoencephalographic (MEG) recordings are selectively sensitive to the magnetic fields arising from current flows in neurons oriented tangentially (parallel) to the surface of the scalp (as in figure 42.2), which means that ERFs primarily reflect the activity of neurons situated in the cortical sulci but not the activity of radially oriented neurons in the cortical gyri. The specific sensitivity of MEG recordings to sulcal sources simplifies localization of the MEG activity. This contrasts with ERP recordings, which are sensitive to current flow from neurons of any orientation with respect to the surface.

Attention and brain physiology

Psychological theories of attention have traditionally been divided on the question of whether selective attention acts at early versus late levels of processing. Early selection theorists have postulated a relative suppression of unattended sensory inputs at an early stage of feature analysis, whereas proponents of late selection have argued that selection takes place only after the identity and meaning of the stimuli have been thoroughly analyzed (reviewed in Näätänen, 1992). Because ERPs and ERFs can trace the flow of sensory information through the afferent pathways with a high degree of temporal resolution, they can provide critical data to distinguish between early and late selection mechanisms. By measuring the timing of attention-induced modifications in stimulus-evoked ERPs and ERFs and specifying the brain areas in which they occur, strong inferences can be made about the level of processing at which selectivity is imposed.

Further information about attentional mechanisms can be derived by examining the waveform of the ERP and ERF changes. If attention acts as a simple gain control (amplification) process that increases the strength of a selected sensory input, this should produce an increase in the amplitude of the corresponding sensory-evoked ERP or ERF component with little or no change in its wave shape or generator source localization. In other words, if the same neural representation is activated by attended and unattended stimuli, but to different degrees, the associated evoked response should differ only in amplitude. On the other hand, the analysis of attended inputs via separate, specialized neural circuits would lead to ERPs and ERFs that differ qualitatively in waveform and localization.

Auditory selective attention

One of the best known and most dramatic examples of auditory selective attention is the so-called cocktail party effect, in which a listener in a noisy environment focuses attention on one conversation while suppressing awareness of competing conversations. A protracted controversy has revolved around the fate of the unattended verbal messages—whether they are attenuated at an early level, prior to analysis of their semantic content, or whether they are fully analyzed for content and subsequently suppressed from awareness at a later stage (reviewed in Näätänen, 1992).

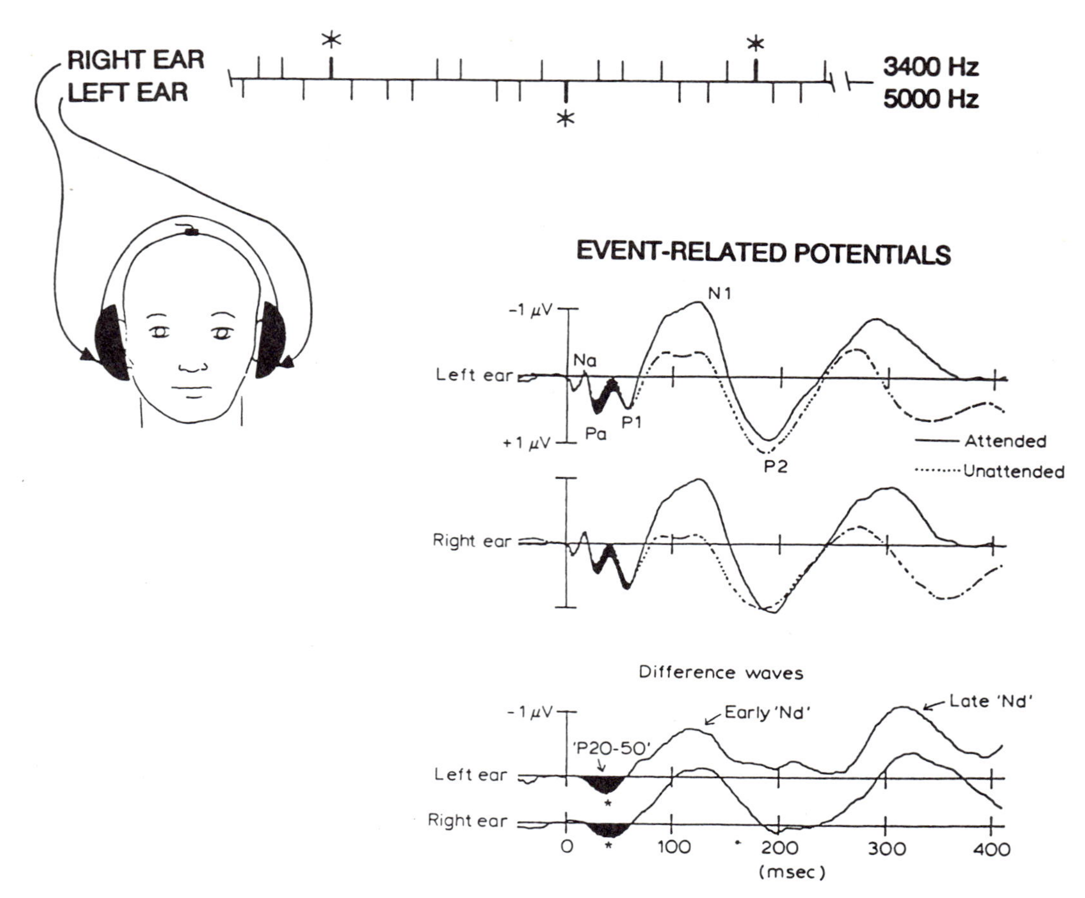

FIGURE 42.3 Dichotic listening task in which early auditory ERP components were modulated during channel-selective attention. Randomized sequences of tones were delivered to left (3400 Hz) and right (5000 Hz) ears at short interstimulus intervals (120–320 ms). Asterisks indicate slightly deviant target tones that the subject attempted to detect in one ear at a time. Averaged ERPs to tones in each ear are shown as a function of whether that ear was attended or unattended. The attention effect consists of an enhanced early positive (P) wave (P20–50, shaded area) and subsequent enlarged negativity (N) at 60–150 ms (N1 or N100). These effects may also be seen in the difference waves formed by subtracting the unattended from the attended ERP (lower tracings). (Data from Woldorff, Hansen, and Hillyard, 1987.)

This level-of-selection issue has been studied extensively by recordings of ERPs in a dichotic listening task that serves as a simplified model of the cocktail party situation (e.g., Woldorff and Hillyard, 1991). In this task, randomized sequences of tones are presented to the left and right ears at a rapid rate of 3–6 per second; the tones in the two ears also differ in pitch, and a fraction of the tones (the targets) differ slightly in some parameter, such as duration or intensity, from the more common standard tones. The subject's task is to pay attention to the tones in one ear at a time and to detect occurrences of the difficult-to-detect targets in that ear.

In several versions of this high-load dichotic listening paradigm, it has been observed that tones in the attended ear elicit an enlarged negative ERP deflection that begins as early as 60–70 ms poststimulus and overlaps the major sensory-evoked negative component (N1 or N100), which peaks at approximately 100 ms (figures 42.1, 42.3). In this discussion, the enlarged negativity elicited by attended stimuli will be termed the *Nd* (negative difference) *wave*, which generally consists of an early phase that overlaps the N100 and a later phase that may persist for several hundred milliseconds beyond the N100 peak.

A protracted controversy has arisen over the question of whether the early portion of the Nd attention effect includes an actual amplification of the evoked N100 component (Hillyard et al., 1973) or whether it arises exclusively from a separate, endogenous neural source that is activated only by attended stimuli (Näätänen and Michie, 1979; Näätänen, 1990). The former effect would be in line with a sensory gain control mechanism, whereas the latter would indicate specialized processing of attended-ear stimuli. This question has proved difficult to resolve because both the evoked N100 and the Nd attention effect are composed of multiple subcomponents that arise from concurrently active neural generator sources in the auditory cortex (Näätänen and Picton, 1987; Woldorff and Hillyard, 1991).

Evidence suggesting that the early Nd may be endogenous has come from a number of studies showing that the scalp topography and amplitude variations of the early Nd can be dissociated from those of the unattended N100 wave (Näätänen and Michie, 1979; Näätänen et al., 1992; Teder et al., 1993; Woods, Alho, and Algazi, 1994). Based on such findings, Näätänen (1990, 1992) has proposed that the early Nd reflects the activation of an endogenous, attention-specific component, termed the *processing negativity*, that differs in cortical origin from the sensory-evoked N100. The processing negativity was hypothesized to index the matching of incoming stimuli with the short-term memory trace of the cues that define the attended channel (e.g., pitch or ear of entry), which was considered a basic mechanism of auditory attention. Näätänen and associates (1992, p. 493) concluded that the evoked N100 itself was unaffected by attention, and consequently they argued that "attention does not modulate initial stimulus representations in audition."

The fact that the early Nd can differ in scalp distribution from the sensory-evoked N100, however, does not prove that this attention-related negativity is exclusively (or even primarily) endogenous in nature. Because the evoked N100 is known to consist of multiple, overlapping subcomponents, it is quite possible that attention could enhance the amplitude of only one (or a subset) of the evoked N100 generators while leaving the others unchanged. Such a selective enhancement would very likely result in an Nd that differs in scalp distribution from the overall unattended N100.

EVIDENCE FOR EARLY GAIN CONTROL IN AUDITORY CORTEX Evidence supporting the view that auditory attention does include an early gain control mechanism that increases N100 amplitude has come from studies showing that the early phase of the Nd coincides precisely in time with the evoked N100 wave (see figure 42.3) and that the early Nd and N100 waves show similar patterns of lateral asymmetry related to ear of stimulation (reviewed in Woldorff and Hillyard, 1991). Even stronger support for this hypothesis was obtained in recent neuromagnetic studies that utilized the dichotic listening task (Rif et al., 1991; Woldorff et al., 1993). Woldorff and colleagues (1993) found that the unattended M100 (the magnetic counterpart of the N100) and its enhancement by attention not only had virtually identical waveforms but were both localized to the same region of auditory cortex. Figure 42.4 shows that the M100 component elicited by right ear tones in the dichotic listening paradigm was increased in amplitude and did not change in latency when right ear tones were attended. The magnetic field of this M100 component was distributed in accordance with a neural generator source in the supratemporal auditory cortex (figure 42.5). In the present context, the key finding was that the neural generators of the M100 to unattended tones, the enlarged M100 to attended tones, and the differential attention effect itself were all localized together in the auditory cortex of the supratemporal plane (figure 42.5B). This pattern accords with the hypothesis that selective listening amplifies early evoked activity in the auditory cortex (relative to unattended inputs) and does, in fact, modulate early stimulus representations.

Further support for an early selection and gain control mechanism comes from the recent finding that attended-ear stimuli during dichotic listening elicit an enlarged positive ERP component as early as 20–50 ms after stimulus onset (see figure 42.3, shaded area). Moreover, this P20–50 attention effect (Woldorff, Hansen, and Hillyard, 1987; Woldorff and Hillyard, 1991) has a magnetic counterpart, the M20–50 effect (see figure 42.4), and its neural generators were also localized to auditory cortex (Woldorff et al., 1993). The timing of this enhanced ERP and ERF activity in humans at 20–50 ms corresponds to the latencies of the earliest attention effects observed for neurons in primary auditory cortex of monkeys (Goldstein, Benson, and Hienz, 1982).

In contrast to these attention effects at the cortical level, recordings of the very-short-latency brainstem evoked potentials (1–10 ms; see figure 42.1) have yet to reveal any consistent attention-related changes in either dichotic listening or cross-modal attention tasks (e.g., Woldorff, Hansen, and Hillyard, 1987; Hackley, Woldorff, and Hillyard, 1990). These negative results

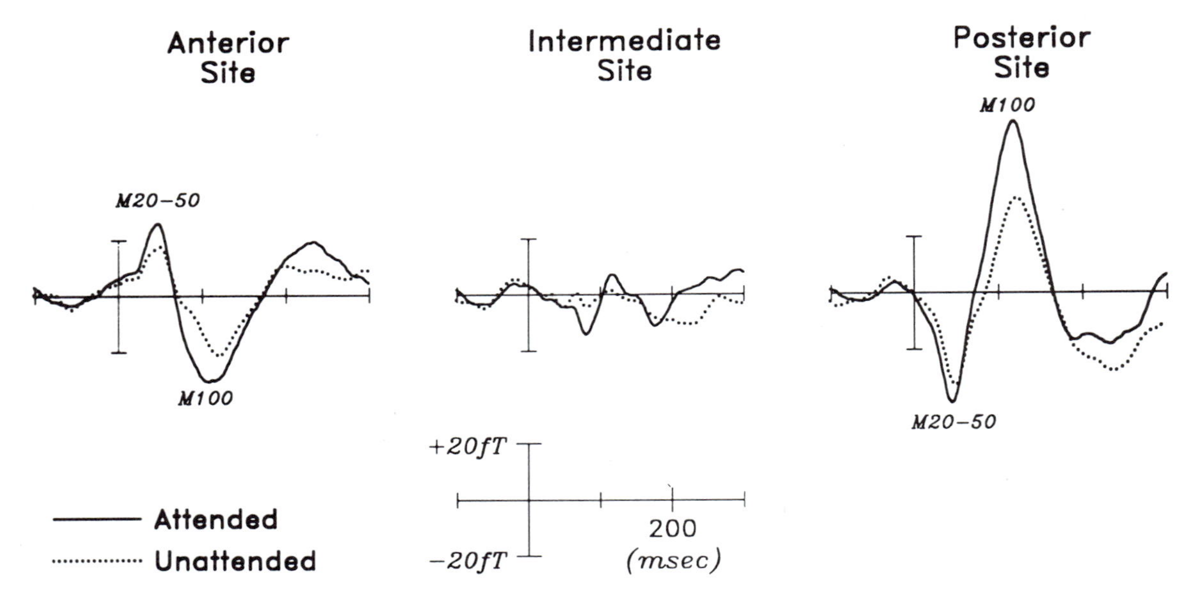

FIGURE 42.4 Event-related magnetic field (ERF) waveforms recorded during the fast-rate dichotic listening paradigm of Woldorff and colleagues (1993). Grand-average ERFs from seven subjects in response to right ear standard tones are shown for anterior, posterior, and intermediate recording sites over the left hemisphere. Note the amplitude enhancement of the sensory-evoked M20–50 and M100 waves in attended (solid) versus unattended (dotted) conditions. Both these waves invert in polarity at the anterior relative to the posterior site, consistent with a vertically oriented current dipole source located in the auditory cortex of the supratemporal plane (see figure 42.5).

imply that subcortical modulation of afferent transmission in the brainstem relay nuclei does not play a role in human auditory attention. Rather, both ERP and ERF recordings indicate that selective listening exerts its control over auditory processing either at the level of the thalamic relay or at a very early cortical stage, thereby producing an altered response in auditory cortex as early as 20 ms after stimulus onset. A thalamic locus of attentional control over sensory transmission to primary cortex would be in accordance with the thalamic gating hypothesis of Skinner and Yingling (1977), who proposed (on the basis of neurophysiological studies in cats) that information flow to the cortex via thalamic relay nuclei could be modulated by attention. Whether information is modulated at thalamic or early cortical levels (or both) during human auditory attention, it is clear that ERP and ERF data provide support for early selection of inputs before perceptual processing is complete.

EARLY AUDITORY SELECTION AND MISMATCH NEGATIVITY If selective listening does involve an early gain control over inputs to auditory cortex, one might expect to find that the early registration and analysis of stimulus features would be suppressed in the unattended channel (relative to the attended channel). Evidence concerning this issue comes from studies of an ERP component known as *mismatch negativity* (MMN), which is specifically elicited (at a latency of 150–200 ms) by any type of physically deviant stimulus in a repetitive sequence of sounds. Both electrical and magnetic recordings indicate that the MMN is generated primarily in auditory cortex (Scherg, Vajsar, and Picton, 1989; Sams et al., 1991). Because elicitation of the MMN requires a comparison of the deviant tone's properties with the sensory memory trace of the preceding repetitive tone, the MMN has been interpreted as an index of early feature analysis in the auditory cortex (Näätänen, 1990). Näätänen and colleagues (reviewed in Alho, 1992; Näätänen, 1990) have reported that the MMN was unaffected by various attentional manipulations, which led them to conclude that all the physical features of auditory stimuli are automatically analyzed and stored in short-term memory whether or not they are attended.

In several recent experiments, however, ERPs were recorded in dichotic listening tasks that were optimized for the selective focusing of attention, and it was found that the MMN was, in fact, suppressed for deviant tones in the unattended ear (Woldorff, Hackley, and

Attended Unattended Attention Effect

A

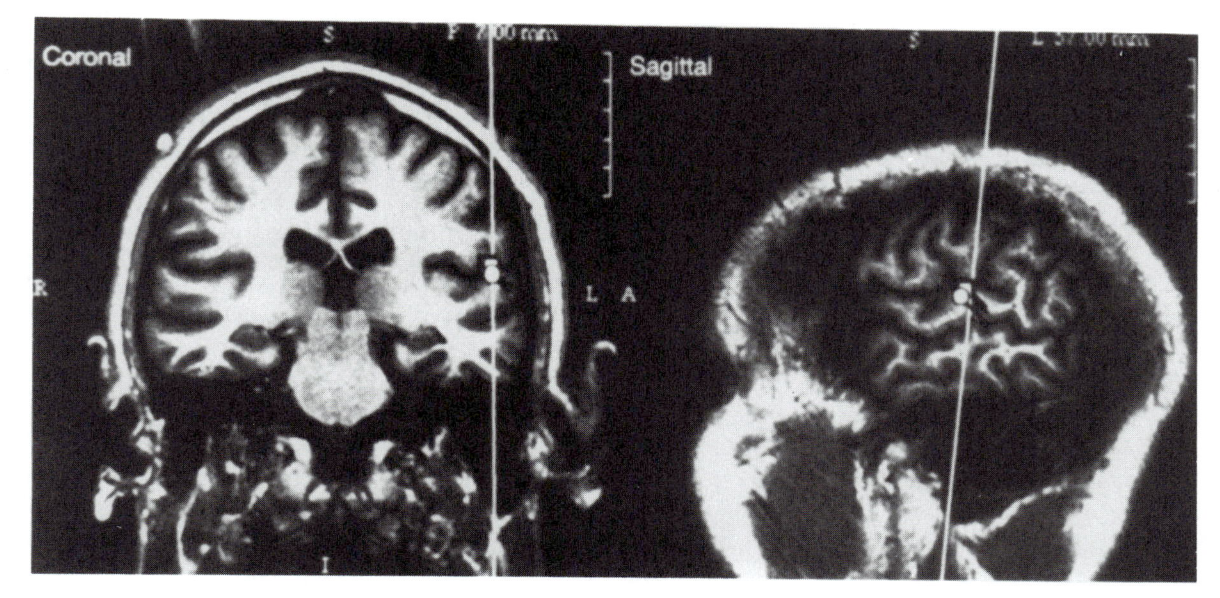

FIGURE 42.5 (A) Magnetic field distributions for the M100 component elicited by right ear tones in a dichotic listening experiment in one subject. Separate mappings are shown for the M100 elicited by attended tones and unattended tones, and for the differential attention effect obtained by subtracting the unattended from the attended responses. These mappings show a dipolar field with magnetic field lines emerging from the head superiorly and entering inferiorly. Arrows indicate the orientations and schematic locations of the single equivalent dipole sources that were calculated to best fit these surface ERF distributions. (B) Magnetic resonance imaging scans showing the calculated locations of the best-fit dipole sources for the attended M100 (white square), the unattended M100 (dark square), and the differential M100 attention effect (white circle), which were all situated within millimeters of one another in the auditory cortex of the supratemporal plane. (Data from Woldorff et al., 1993.)

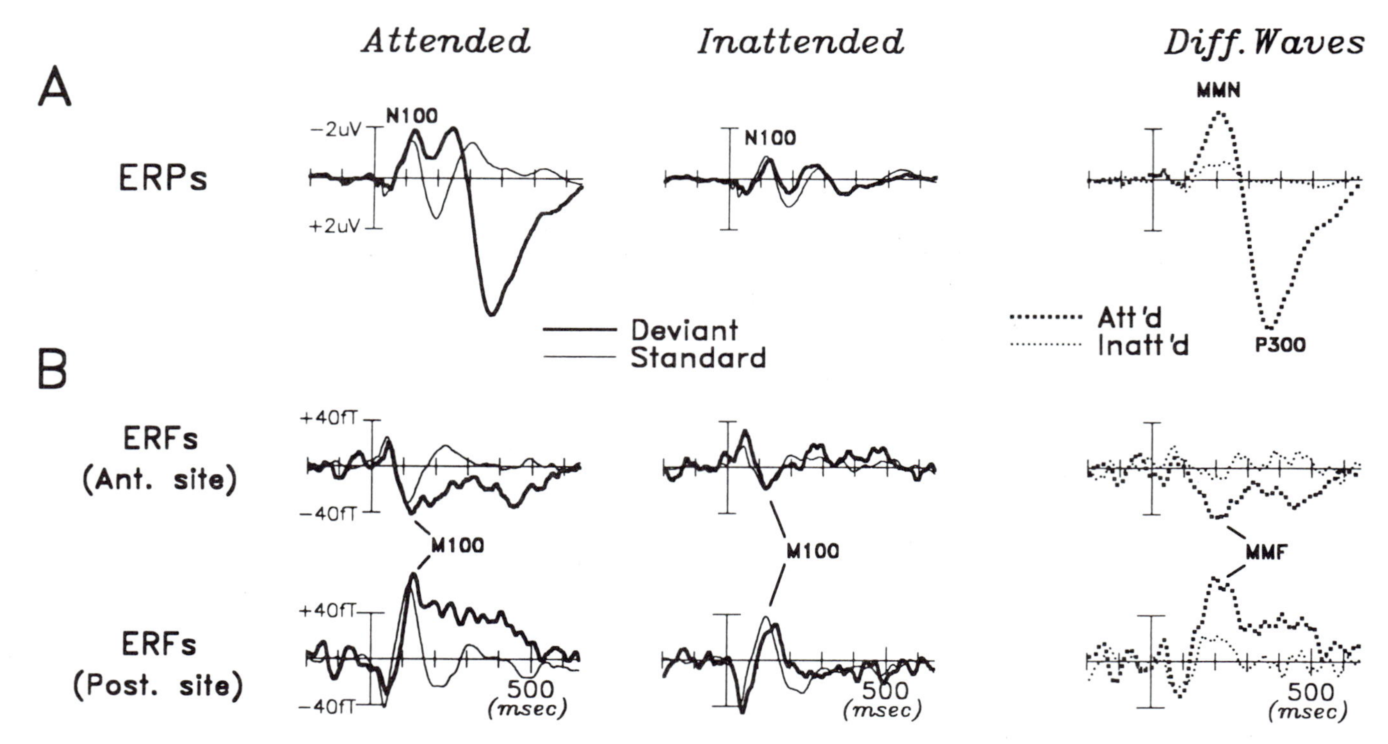

FIGURE 42.6 Suppression of mismatch negativity (MMN) and mismatch field (MMF) during focused auditory attention in fast-rate dichotic listening experiments. (A) ERPs from study by Woldorff and coworkers (1991). Superimposed are the ERPs (scalp site Cz) elicited by standard and deviant (lower-intensity) tones in the attended ear (left column) and in the unattended ear (middle column). In the right column are the corresponding difference waves, which were derived for each attention condition by subtracting the standard-tone ERP from the deviant-tone ERP. Note the strong suppression of the MMN elicited by unattended-ear deviants. (B) ERFs (to right ear tones only) recorded from anterior and posterior sites over the left hemisphere during fast-rate dichotic listening. The attended-tone ERFs (left column), unattended-tone ERFs (middle column), and difference waves (right column) are displayed in parallel to the ERPs shown in (A). Note that the attended deviant tones elicited substantial MMF activity at 200 ms that inverted in polarity at the anterior site relative to the posterior site. This activity was highly suppressed for the unattended tones. (Data from Woldorff et al., 1992.)

Hillyard, 1991; Woldorff et al., 1992). This suppression was observed both for the MMN and its magnetic counterpart, the mismatch field (MMF), as shown in figure 42.6. The magnetic field distributions (figure 42.7) indicated that the MMF that was suppressed by attention represented a modality-specific neural response of the auditory cortex rather than a nonspecific response associated with detection of a task-relevant stimulus (cf. Näätänen, 1991). These data strongly support the view that auditory feature analysis and mismatch comparisons are suppressed in the unattended channel when attention is strongly focused on the other channel.

TIMING OF AUDITORY FEATURE SELECTIONS The studies reviewed earlier indicate that highly focused auditory attention exerts a gain control over early evoked activity in the auditory cortex, reflected in amplitude modulations of the P20–50 and N100 components. In addition to these evoked potential changes, attended tones typically elicit a more prolonged negative ERP (corresponding to Näätänen's processing negativity) that appears to reflect activation of separate, endogenous neural populations specialized for processing attended inputs. Accordingly, the overall negativity elicited by attended-channel sounds (i.e., the Nd) is most likely composed of an enlarged N100 wave together with an overlapping processing negativity that may extend for several hundred milliseconds and may reflect a more detailed analysis of attended-channel features in specialized auditory fields (Näätänen, 1992; Woods, Alho, and Algazi, 1994).

Whatever its exact composition, the waveform of the Nd attention effect can provide precise information

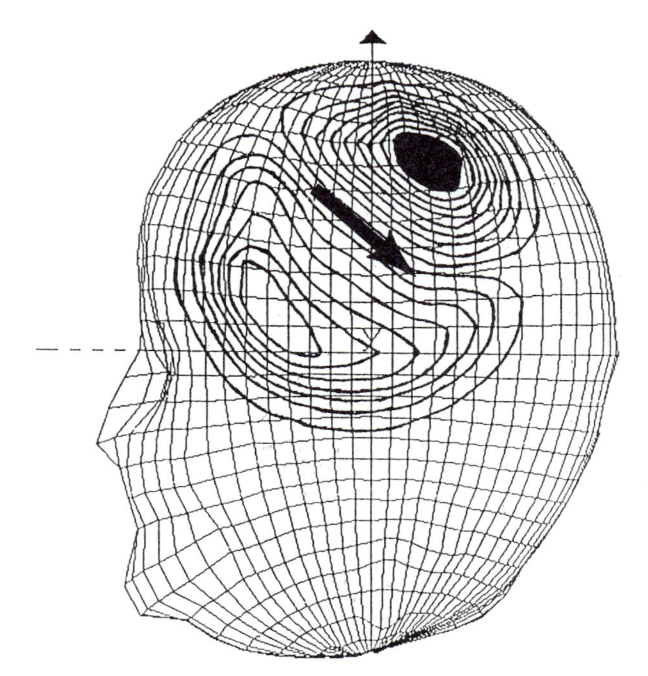

FIGURE 42.7 Topographical distribution of the magnetically recorded mismatch field (MMF) that was supressed in response to unattended deviant tones. This mapping shows the distribution of the *differential* MMF to attended versus unattended deviant tones, obtained by subtracting the unattended from attended difference waves shown in the right column of figure 42.6B. This fleld distribution is highly similar to that of the M100 (figure 42.5) and shows a similar source localization to the auditory cortex. (Data from Woldorff et al., 1992.)

regarding the time points at which different classes of auditory features are selected or rejected. In one set of studies, for example, high- and low-pitched tones were presented to the left and right ears in random order, and subjects were required to focus attention on one of the pitch-ear "feature conjunction" stimuli (Woods and Alain, 1993; Woods, Alho, and Algazi, 1994). Topographically distinct Nd waves were found to be associated with selections of the individual features and the feature conjunction. The timing of these Nd waves indicated that the pitch and location features were selected and processed independently and in parallel for the first 70–100 ms, while evidence for conjunction selection became evident after 110–120 ms. Woods, Alho, and Algazi (1994) pointed out that this delay of 30–50 ms for selection of the conjunction with respect to feature selection is in accordance with estimates of the time required to conjoin features in visual search tasks.

Visual selective attention

The ability of human observers to shift their attention rapidly to selected positions of a visual scene has been well documented in numerous behavioral studies, which have shown that stimuli situated at or near a covertly attended location are detected and discriminated more accurately than are stimuli outside the spotlight of spatial attention (for review, see Van Der Heijden, 1992). As with auditory attention, there has been a long-standing debate about the level of processing at which visual attention operates. Some investigators have argued that visuospatial attention acts at an early stage of visual encoding or feature processing and thus affects early perceptual representations (e.g., Prinzmetal, Presti, and Posner, 1986; Reinitz, 1990). An alternative view, however, is that spatial attention acts at postperceptual levels by biasing decision or response stages to favor specific inputs or by rapidly allocating decision processes to attended information (e.g., Sperling and Dosher, 1986; Mueller and Humphreys, 1991). Because the distinction between perceptual and decision stages has proved difficult to pin down in behavioral studies, the issue of early versus late selection has remained unresolved.

ERP recordings have been used to address the level-of-selection issue for visual attention in a manner analogous to the auditory attention studies described previously. Studies employing several different types of spatial attention paradigms have shown that stimuli presented within the focus of attention produce enhanced sensory-evoked P1 (80–120 ms) and N1 (140–190 ms) components relative to unattended stimuli. This characteristic ERP pattern has been observed in tasks involving sustained attention, spatial cuing, and visual search (reviewed in Luck, Fan, and Hillyard, 1993). These early P1 and N1 changes appear to be unique to spatial attention and do not occur during attention to nonspatial features such as color or spatial frequency, which accords with psychological theories in which location is given special status relative to other features (e.g., Treisman and Gelade, 1980).

NEURAL SYSTEMS MEDIATING VISUOSPATIAL ATTENTION

A recent study by Mangun, Hillyard, and Luck (1993) attempted to localize the cortical areas that give rise to the attention-sensitive P1 and N1 components in a task that involved sustained attention to one of four locations in the visual field (figure 42.8). Rectangles were

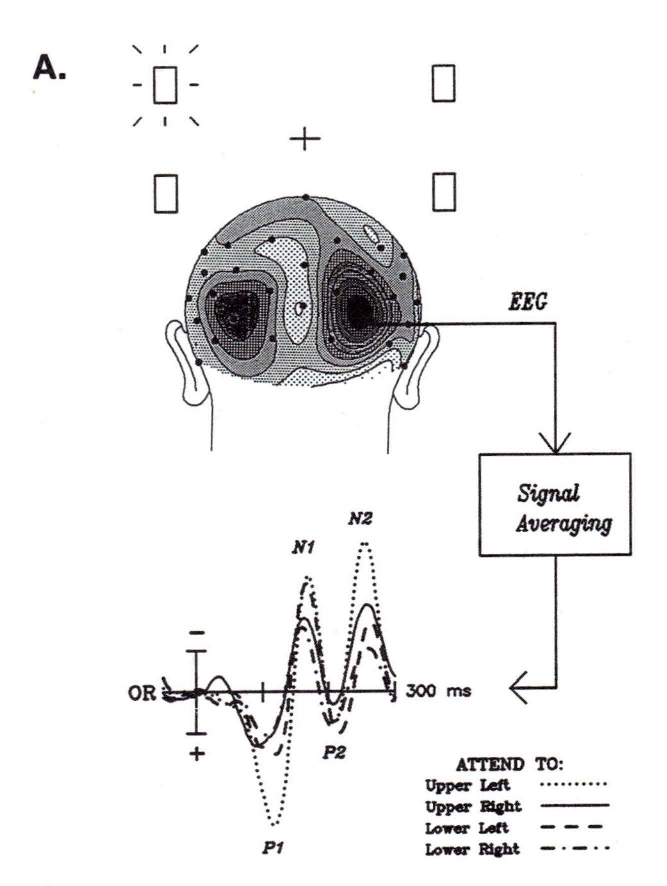

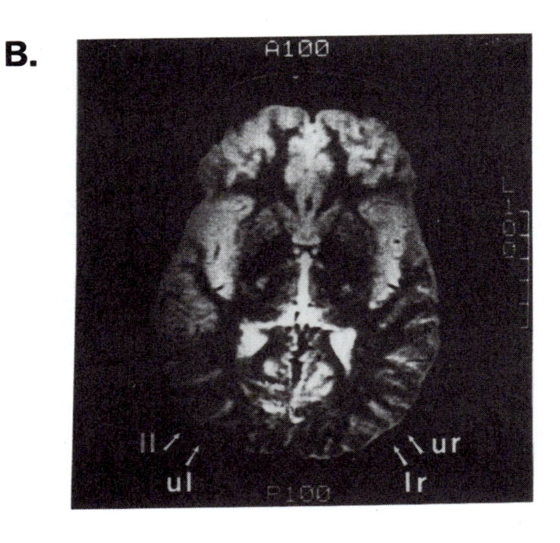

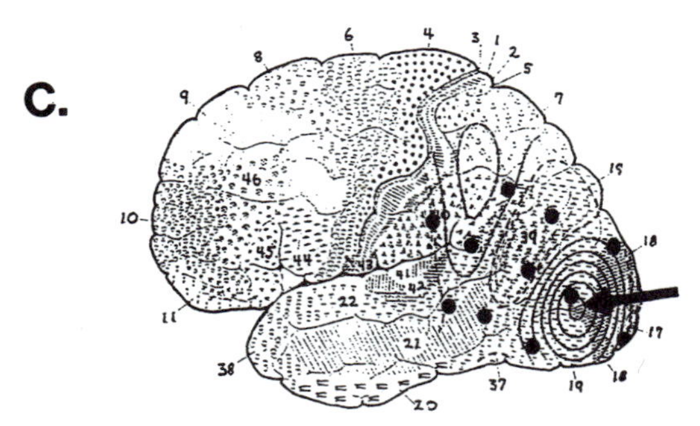

FIGURE 42.8 (A) Amplitude changes in visual ERP components in the spatial attention task of Mangun, Hillyard, and Luck (1993). Stimuli (flashed rectangles) were presented to the four quadrants of the visual field in random order at intervals of 250–550 ms. Subjects were instructed to attend to flashes in only one of the quadrants at a time during each run. ERPs were averaged separately for each type of stimulus and attention condition for each of 30 scalp sites. Shown are grand-average ERPs to the upper left quadrant flashes under each of the four attention conditions. Note the amplification of the P1, N1, and N2 waves when the flashes were attended. The head map shows the scalp current density distribution of the P1 component (at 108 ms), which was maximal over the contralateral occipital scalp. (B) Magnetic resonance imaging (MRI) scan of one subject's brain in a plane that passes through the lateral extrastriate cortex. Arrows show the locations of the current density maxima of the P1 attention effect for stimuli in the four visual quadrants: lower left (ll), upper left (ul), lower right (lr), and upper right (ur). Note that the MRI scan is left-right reversed. (C) Superposition of the average current density contours for the P1 component on a classic map of Brodmann's cortical areas. The maximum source of the P1 attention effect overlies lateral prestriate visual cortex. (Reprinted from Hillyard, 1993, by permission.)

flashed in random order to the four positions, and subjects attended to flashes at only one of the locations during each 50-s run. Figure 42.8A shows that flashes at an attended location evoked markedly larger P1 and N1 waves (and later waves as well) than did flashes outside the focus of attention. The scalp distribution of the enhanced P1 component showed a sharp focus of current density over the ventrolateral extrastriate cortex (figure 42.8B, C), whereas the N1 attention affect was localized somewhat more dorsally, over occipitoparietal cortex. Significantly, both the P1 and N1 waves to attended stimuli were simply increased in amplitude, with no change in either waveform or scalp distribution. This implies that spatial attention acts as a sensory gain control mechanism that modulates the flow of information differentially between attended and unattended zones of the visual fields.

An earlier ERP component that began at approximately 50–60 ms displayed a scalp topography consistent with a generator source in the primary visual (striate) cortex. This component reversed in polarity for upper- versus lower-field stimuli, as would be expected from the retinotopic mapping of the upper and lower half-fields onto striate cortex regions situated on opposing banks of the calcarine fissure. Significantly, this polarity-reversing component did not change in amplitude when evoked by attended versus unattended stimuli. This result is consistent with single-cell recording studies in monkeys, which have found no evidence that selective attention influences activity in primary visual cortex (Colby, 1991). Thus, both human and monkey studies have failed to find evidence for the existence of an attentional mechanism that acts at the thalamic level to modulate sensory transmission from the lateral geniculate nucleus to primary visual cortex (cf. Crick, 1984).

Recent positron emission tomography studies have also indicated that visual selective attention acts to modify neural processing in multiple extrastriate visual areas but not in the striate cortex or lateral geniculate nucleus. Increased cerebral blood flow in discrete zones of the extrastriate cortex has been observed during sustained attention to several types of stimulus features and patterns (Corbetta et al., 1991; Haxby et al., 1991). One such cortical area appears to correspond closely to the anatomical position of the P1 generator source identified in our ERP studies (figure 42.9). The timing of the stimulus selection reflected in the P1 wave (onset at 70–90 ms poststimulus) also corresponds well with the latencies of attention-sensitive single-unit discharge in several areas of macaque extrastriate cortex (Desimone et al., 1990; Colby, 1991).

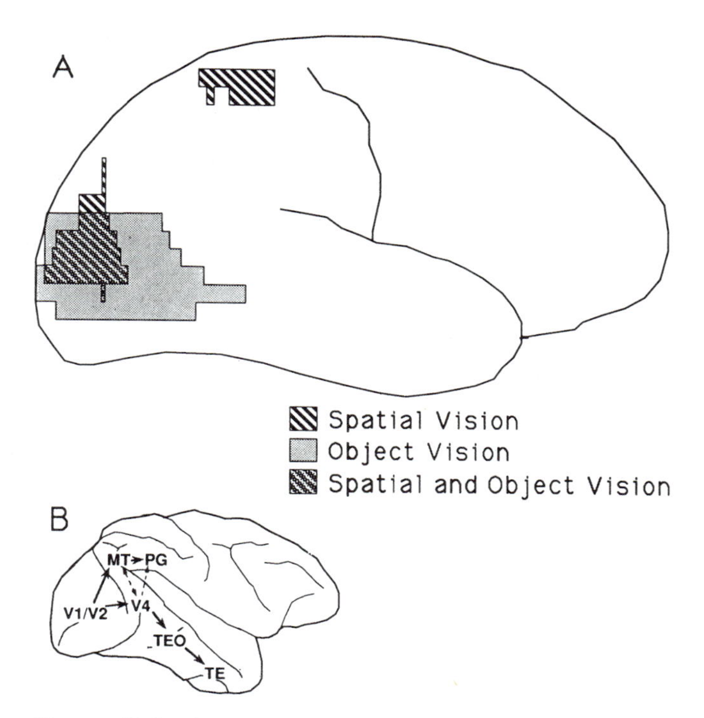

FIGURE 42.9 Comparison of visual cortical areas in the macaque and extrastriate visual areas in humans. (A) Composite of posterior right and left cortical regions in which three or more subjects demonstrated increased normalized cerebral blood flow during face matching (object vision) or dot-location matching (spatial vision) tasks, relative to a control task. (B) A map of the major visual areas in macaque visual cortex and their interconnections. The results this PET study in humans suggest homologies between the lateral occipital area activated by both tasks and lateral portions of area V4 in the macaque. (Reprinted from Haxby et al., 1991, by permission.)

LaBerge (chapter 41) has proposed a computational and anatomical model of visuospatial attention that incorporates current knowledge of visual system neurophysiology with evidence from relevant human PET and ERP studies. In this model, spatial attention does not influence visual processing at the level of the striate cortex but rather acts to modulate the flow of information from striate to extrastriate areas. The pulvinar nucleus of the thalamus was ascribed the key role of selectively biasing information flow in extrastriate visual pathways by means of its reciprocal anatomical projections to those cortical areas. LaBerge has shown how the neural circuitry of the pulvinar may intensify a preferential bias signal for an attended visual field location that originates in the prefrontal cortex and is relayed to the pulvinar via the posterior parietal cortex. In this framework, the enlarged P1 and N1 components observed for attended stimuli may correspond to the enhanced information flow in extrastriate cortex originating from attended zones of the visual field.

SELECTION OF FEATURES AND OBJECTS When stimuli are selected on the basis of features such as color, size, or shape, the associated ERP patterns are very different from the early P1/N1 amplitude modulations produced during attention to location. Attention to such nonspatial features is indexed instead by longer-latency ERP components in the 150- to 350-ms range that are largely endogenous (i.e., components triggered specifically by attended stimuli but not otherwise). Among these components are the posterior selection negativities elicited by stimuli having an attended feature, the N200 (N2) waves associated with discrimination and identification of relevant stimulus dimensions, and the P300 (P3) component that is triggered when task-relevant events have been recognized and classified (Harter and Aine, 1984; Wijers et al., 1989; Kenemans, Kok, and Smulders, 1993). The selection negativity and N200 waves show differences in scalp topography that depend on the specific features being selected, and an important area for future research will be to relate these ERPs to patterns of cerebral blood flow during feature-selective processing (e.g., Corbetta et al., 1991).

As was described for the auditory Nd component, the temporal properties of the selection negativity, N200, and P300 waves yield precise information about the timing and order with which different stimulus features and feature combinations are selected. When selection depends on multiple features of a stimulus, different ERPs may be associated with the processing of each one, and inferences can be made about whether the features are selected and processed in a parallel or interactive fashion. In several studies, the ERP patterns indicated that an initial stage of independent feature selection preceded the identification of feature conjunctions or objects (e.g., Harter and Aine, 1984), consistent with Treisman and Gelade's (1980, page 98) view that "features come first in perception." In other tasks, the selection of one feature of a stimulus was found to be heirarchically dependent on the prior selection of another, more discriminable feature (Hillyard and Münte, 1984; Wijers et al., 1989; Kenemans, Kok, and Smulders, 1993). These studies illustrate the use of physiological measures to elucidate the timing and organization of information-processing stages.

Visual Orienting Recordings of ERPs have also helped to clarify some of the mechanisms underlying the orienting of visual attention. In the well-known task developed by Posner and associates (see Posner, 1980, for review), for example, an advance cue (such as an arrow) indicates the most probable location of a subsequent target stimulus, to which a speeded motor response is required. Posner and colleagues interpreted the speeding of response times (RTs) to validly cued targets as a consequence of improved perceptual processing brought about by the orienting of attention to the precued location (Posner, 1980). This early selection hypothesis was countered, however, by the proposal that the RT changes might instead be caused by postperceptual factors such as decision or response bias (e.g., Sperling and Dosher, 1986).

Support for Posner's early selection view has come from several ERP studies showing that the early sensory-evoked P1 and N1 waves are enlarged in response to valid, relative to invalid, targets (Harter et al., 1989; Mangun and Hillyard, 1991; Anllo-Vento, in press). These ERP findings imply that spatial orienting, like sustained focused attention, involves a facilitation of sensory processing in the extrastriate visual cortex.

Consistent with this hypothesis, several recent studies have found that the detectability (d') of faint luminance targets is enhanced at precued locations (e.g., Downing, 1988; Hawkins et al., 1990). Several authors have noted, however, that attention-induced changes in postsensory, decision-level processes might also affect the detectability of faint targets (Sperling and Dosher, 1986; Mueller and Humphreys, 1991). To evaluate the level of processing at which attention improves signal detectability, we recorded ERPs from subjects performing a threshold-level luminance detection task (Hillyard, Luck, and Mangun, 1994). Subjects in this study were required to detect faint masked targets presented at one of four possible locations (figure 42.10). A central arrow cue designated the most probable location at which the target information would be presented. On neutral trials, arrows pointed to all four locations, indicating an equal probability of target occurrence at each location. Target detectability was found to be highest on valid trials, lowest on invalid trials, and intermediate on neutral trials. In association with these signal detectability effects, the amplitudes of the early, sensory-evoked P1 and N1 components over the lateral occipital scalp were larger on valid than on invalid trials (figure 42.11). This result implies that the improvement in detection sensitivity to precued locations involves facilitated sensory processing and is not solely a consequence of higher decision processes.

The neutral cue trials provided a baseline condition that allowed the facilitation of processing at validly cued locations to be differentiated from the suppression of processing at invalidly cued locations. As shown in figure 42.11, the N1 component was enhanced on valid trials compared to neutral trials but was not suppressed on invalid trials. In contrast, the P1 component was reduced on invalid trials compared to neutral trials but was not enhanced on valid trials. This difference suggests that the P1 and N1 attention effects reflect two qualitatively distinct mechanisms of attention, one that operates by enhancing the processing of information within the focus of attention and another that operates by suppressing information falling outside the attended zone.

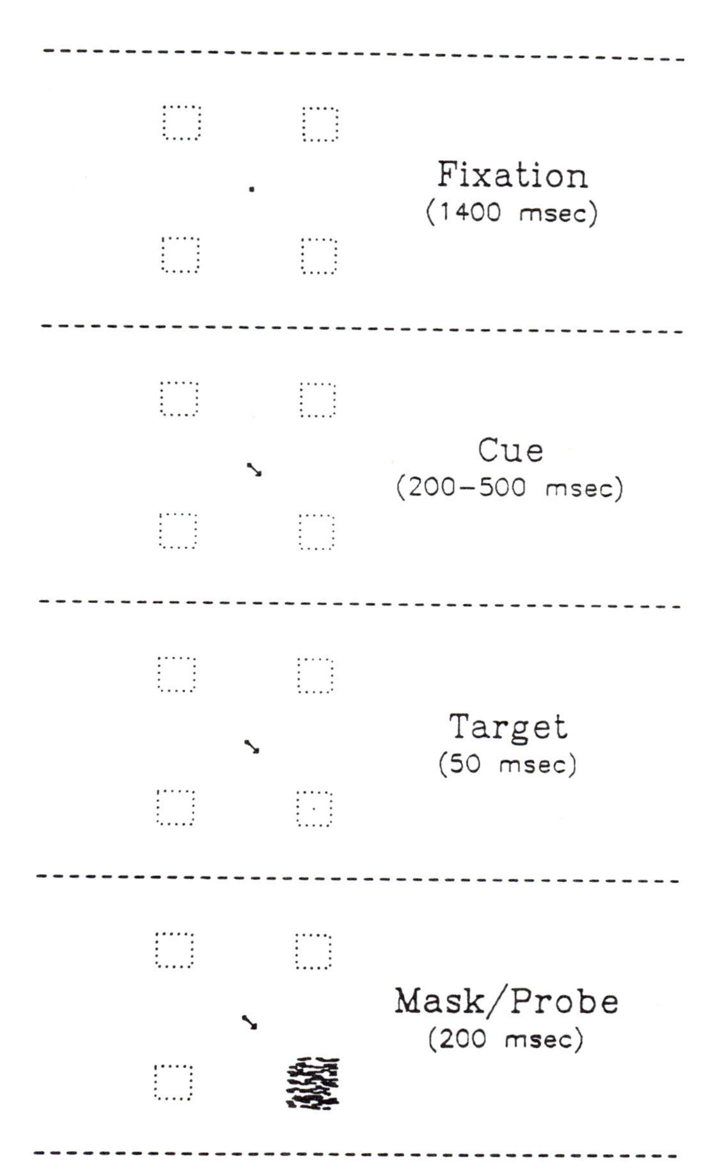

FIGURE 42.10 Sequence of stimulus events in the luminance detection experiment of Hillyard, Luck, and Mangun (1994). Each trial began with a fixation interval, followed by a central arrow cue, a brief luminance target (or absence of a target), and an immediate mask. Subjects reported the presence or absence of the target at the masked location. A single location was cued on 80% of trials, and all four locations were cued (neutral condition) on 20% of trials. When a single location was cued, the target-mask complex was presented at the cued location on 75% of trials or at one of the uncued locations on the remaining 25%. The neutral cue was not predictive of target location.

VISUAL SEARCH Visual search experiments typically have found that a target embedded in a multielement array can be detected rapidly and in parallel when it is distinguished by a simple feature from the surrounding distractor items. In contrast, the detection of a feature conjunction target appears to require the serial application of focal attention to the array items (Treisman and Gelade, 1980). Although visual search and spatial cuing paradigms both appear to involve visuospatial attention, there have been only a few investigations of whether the attentional processes engaged during cued orienting are the same as those deployed during visual search (e.g., Briand and Klein, 1987).

This issue was examined using ERP recordings in a recent study by Luck, Fan, and Hillyard (1993; see also Luck and Hillyard, in press). To obtain ERP indices of the selective processing of specific items within the search array, a probe stimulus technique was used. Subjects were required to discriminate the shape of a uniquely colored item that was present within each array, which was followed, after a 250-ms delay, by an irrelevant probe stimulus (a small square) flashed either at the location of this target item or at the location of a distractor item on the opposite side of the display. Previous studies had indicated that subjects would be able to focus attention on the target item by the time of probe delivery, and we wanted to learn whether this focusing of attention would influence the early components of the probe-evoked potential. It was found that the evoked P1 component at contralateral electrode sites was significantly enlarged in response to probes presented at the location of the target item as compared to probes presented at the location of a distractor item, indicating that the attentional mechanism reflected by changes in P1 amplitude is employed during visual search as well as spatial cuing and sustained attention paradigms.

This increase in the visually evoked P1 component (at 90–130 ms) to probes at the target location provides strong evidence that early sensory transmission is modulated when focal attention is directed to an item in a search array in order to discriminate its properties. Because the probes were not task-relevant and differed

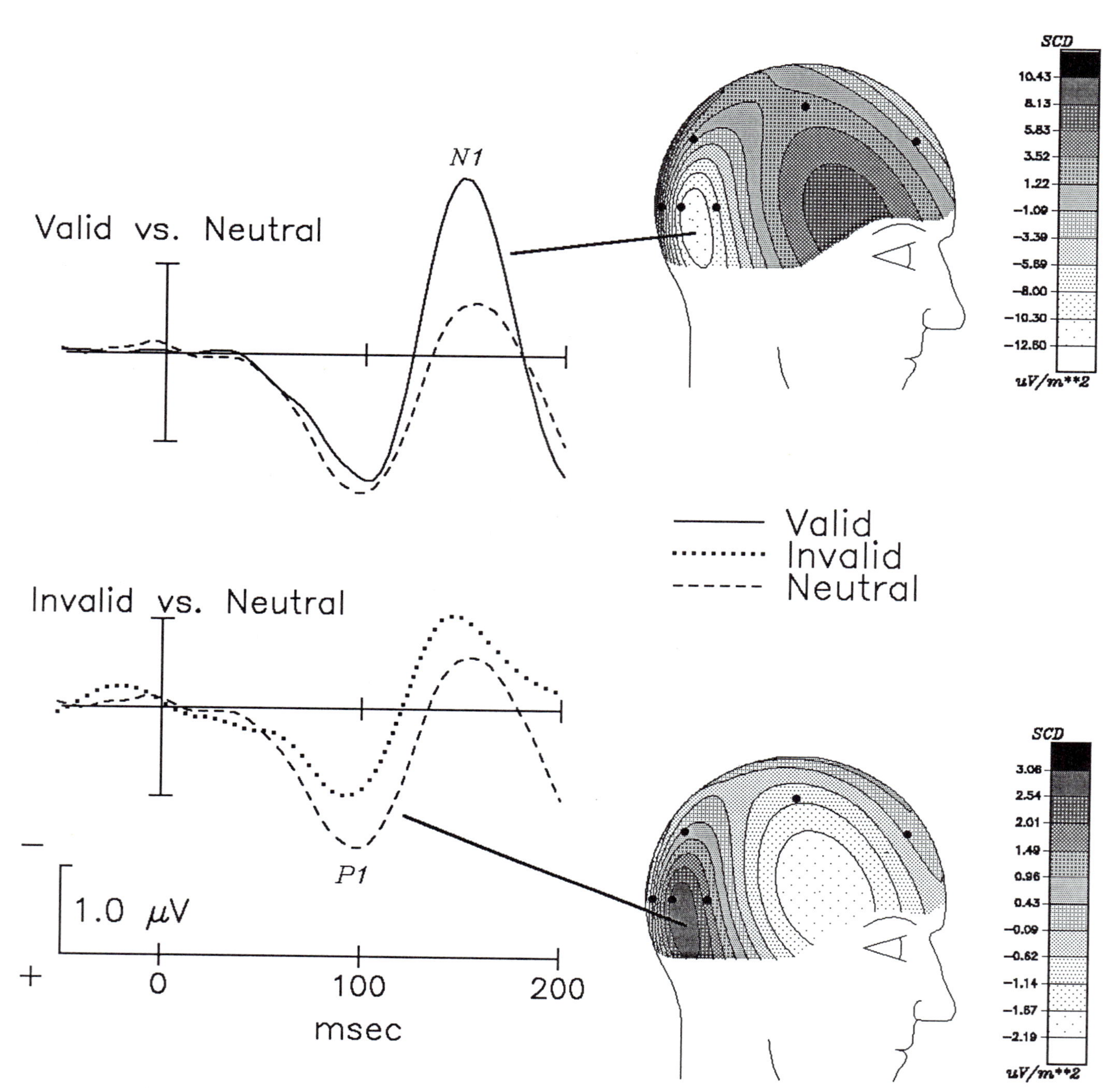

FIGURE 42.11 ERP waveforms and maps of scalp current density (SCD) from the spatial cuing experiment diagrammed in figure 42.10 (Hillyard, Luck, and Mangun, 1994). The near-threshold targets were too dim to evoke a measurable ERP, and therefore the ERPs elicited by the target-mask complex were used to assess sensory processing at cued and uncued locations. When these stimuli were presented at a validly cued location, they elicited an enlarged N1 component compared to stimuli presented on neutral trials (top waveforms), but no enhancement was observed for the P1 component. Conversely, stimuli presented at an invalidly cued location elicited a reduced P1 component compared to stimuli presented on neutral trials (bottom waveforms), but no suppression was observed for the N1 component. These ERP waveforms were recorded at lateral occipital scalp sites contralateral to the position of the eliciting target-mask stimulus.

markedly in color and shape from the target items, these ERP data further indicate that spatial attention acts to select stimuli at an early level solely on the basis of their location (cf. Heinze et al., 1990). Moreover, the P1 attention effect in this study appeared topographically equivalent to that observed previously in spatial cuing and sustained attention paradigms, which supports the idea that a common sensory gain control mechanism operates in all these situations.

Conclusions

The studies of electrical and magnetic brain recordings reviewed here provide unequivocal support for early selection theories of attention. In both auditory and visual tasks that were optimized for the selective focusing of attention, it was found that short-latency evoked responses in modality-specific cortex were modulated in a manner consistent with an early gain control over sensory information flow. The timing of the earliest evoked potential changes (20–50 ms for auditory and 70–90 ms for visual) indicates that attention acts to modify processing in cortical areas that encode elementary stimulus features rather than fully analyzed pattern or object representations. Longer-latency ERP components were found to provide precise information regarding the time points at which multiple stimulus features and feature combinations are selectively processed. As the anatomical generators of these attention-sensitive ERPs and ERFs become better defined by means of source localization and other neuroimaging techniques, a more complete picture will unfold of the brain systems that mediate selective attention.

ACKNOWLEDGMENTS This work was supported by grants from the National Institute of Mental Health (MH-25594 and MH-00930) and the National Institutes of Health (NS-17778), by the Office of Naval Research contract N00014-89-J-1806, and by awards from the Human Frontier Science Project and the Davis and San Diego McDonnell-Pew Centers for Cognitive Neuroscience.

REFERENCES

ALHO, K., 1992. Selective attention in auditory processing as reflected by event-related brain potentials. *Psychophysiology* 29:247–263.

ANLLO-VENTO, L. (in press). Shifting attention in visual space: The effects of peripheral cueing on brain cortical potentials. *Int. J. Neurosci.*

BRIAND, K. A., and R. M. KLEIN, 1987. Is Posner's "beam" the same as Treisman's "glue"? On the relation between visual orienting and feature integration theory. *J. Exp. Psychol. [Hum. Percept.]* 13:228–241.

COLBY, C. L., 1991. The neuroanatomy and neurophysiology of attention. *J. Child Neurol.* 6:90–118.

CORBETTA, M., F. M. MIEZIN, S. DOBMEYER, G. L. SHULMAN, and S. E. PETERSEN, 1991. Selective and divided attention during visual discriminations of shape, color, and speed: Functional anatomy by positron emission tomography. *J. Neurosci.* 11:2383–2402.

CRICK, F., 1984. Function of the thalamic reticular complex: The searchlight hypothesis. *Proc. Natl. Acad. Sci.* 81:4586–4590.

DALE, A. M., and M. I. SERENO, 1993. Improved localization of cortical activity by combining EEG and MEG with MRI cortical surface reconstruction: A linear approach. *J. Cogn. Neurosci.* 5:162–176.

DESIMONE, R., M. WESSINGER, L. THOMAS, and W. SCHNEIDER, 1990. Attentional control of visual perception: Cortical and subcortical mechanisms. *Cold Spring Harb. Symp. Quant. Biol.* 55:963–971.

DOWNING, C. J., 1988. Expectancy and visual-spatial attention: Effects on perceptual quality. *J. Exp. Psychol. [Hum. Percept.]* 44:188–202.

GOLDSTEIN, M. H., JR., D. A. BENSON, and R. D. HIENZ, 1982. Studies of auditory cortex in behaviorally trained monkeys. In *Conditioning: Representation of Involved Neural Functions*, C. D. Woody, ed. New York: Plenum, pp. 307–317.

HACKLEY, S. A., M. WOLDORFF, and S. A. HILLYARD, 1990. Cross-modal selective attention effects on retinal, myogenic, brainstem and cerebral evoked potentials. *Psychophysiology* 27:195–208.

HARTER, M. R., and C. J. AINE, 1984. Brain mechanisms of visual selective attention. In *Varieties of Attention*, R. Parasuraman and D. R. Davies, eds. London: Academic Press, pp. 293–321.

HARTER, M. R., S. L. MILLER, N. J. PRICE, M. E. LALONDE, and A. L. KEYES, 1989. Neural processes involved in directing attention. *J. Cogn. Neurosci.* 1:223–237.

HAWKINS, H. L., S. A. HILLYARD, S. J. LUCK, M. MOULOUA, C. J. DOWNING, and D. P. WOODWARD, 1990. Visual attention modulates signal detectability. *J. Exp. Psychol. [Hum. Percept.]* 16:802–811.

HAXBY, J. V., C. L. GRADY, B. HORWITZ, L. G. UNGERLEIDER, M. MISHKIN, R. E. CARSON, P. HERSCOVITCH, M. B. SCHAPIRO, and S. I. RAPPOPORT, 1991. Dissociation of object and spatial visual processing pathways in human extrastriate cortex. *Proc. Natl. Acad. Sci. U.S.A.* 88:1621–1625.

HEINZE, H. J., S. J. LUCK, G. R. MANGUN, and S. A. HILLYARD, 1990. Visual event-related potentials index focused attention within bilateral stimulus arrays: I. Evidence for early selection. *Electroencephalogr. Clin. Neurophysiol.* 75: 511–527.

HILLYARD, S. A., 1993. Electrical and magnetic brain recordings: Contributions to cognitive neuroscience. *Curr. Opin. Neurobiol.* 3:217–224.

Hillyard, S. A., R. F. Hink, V. L. Schwent, and T. W. Picton, 1973. Electrical signs of selective attention in the human brain. *Science* 182:177–180.

Hillyard, S. A., S. J. Luck, and G. R. Mangun, 1994. The cueing of attention to visual field locations: Analysis with ERP recordings. In *Cognitive Electrophysiology: Event-Related Brain Potentials in Basic and Clinical Research*, H. J. Heinze, T. F. Münte and G. R. Mangun, eds. Boston: Birkhäuser, pp. 1–25.

Hillyard, S. A., and T. F. Münte, 1984. Selective attention to color and locational cues: An analysis with event-related brain potentials. *Percept. Psychophys.* 36:185–198.

Kaufman, L., and S. J. Williamson, 1985. The neuromagnetic field. In *Evoked Potentials. Frontiers in Clinical Neuroscience.* Vol. 3. I. Bodis-Wolner and R. Q. Cracco, eds. New York: Alan Liss, pp. 85–98.

Kenemans, J. L., A. Kok, and F. T. Y. Smulders, 1993. Event-related potentials to conjunctions of spatial frequency and orientation as a function of stimulus parameters and response requirements. *Electroencephalogr. Clin. Neurophysiol.* 88:51–63.

Kinchla, R. A., 1992. Attention. *Annu. Rev. Psychol.* 43:711–742.

Luck, S. J., S. Fan, and S. A. Hillyard, 1993. Attention-related modulation of sensory-evoked brain activity in a visual search task. *J. Cogn. Neurosci.* 5:188–195.

Luck, S. J., and S. A. Hillyard (in press). The role of attention in feature detection and conjunction discrimination: An electrophysiological analysis. *Int. J. Neurosci.*

Mangun, G. R., and S. A. Hillyard, 1991. Modulations of sensory-evoked brain potentials provide evidence for changes in perceptual processing during visual-spatial priming. *J. Exp. Psychol. [Hum. Percept.]* 17:1057–1074.

Mangun, G. R., S. A. Hillyard, and S. J. Luck, 1993. Electrocortical substrates of visual selective attention. In *Attention and Performance*, vol. 14, D. Meyer and S. Kornblum, eds. Cambridge, Mass.: MIT Press, pp. 219–243.

Mueller, H. J., and G. W. Humphreys, 1991. Luminance increment detection: Capacity-limited or not? *J. Exp. Psychol. [Hum. Percept.]* 17:107–124.

Näätänen, R., 1990. The role of attention in auditory information processing as revealed by event-related potentials and other brain measures of cognitive function. *Behav. Brain Sci.* 13:201–288.

Näätänen, R., 1991. Mismatch negativity outside strong attentional focus: A commentary on Woldorff et al. *Psychophysiology* 28:478–484.

Näätänen, R., 1992. *Attention and Brain Function.* Hillsdale, N. J.: Erlbaum.

Näätänen, R., and P. T. Michie, 1979. Early selective attention effects on the evoked potential: A critical review and reinterpretation. *Biol. Psychol.* 8:81–136.

Näätänen, R., and T. Picton, 1987. The N1 wave of the human electric and magnetic response to sound: A review and an analysis of the component structure. *Psychophysiology* 24:375–425.

Näätänen, R., W. Teder, K. Alho, and J. Lavikainen, 1992. Auditory attention and selective input modulation: A topographical ERP study. *NeuroReport* 3:493–496.

Petersen, S. E., J. A. Fiez, and M. Corbetta, 1992. Neuroimaging. *Curr. Opin. Neurobiol.* 2:217–222.

Posner, M. I., 1980. Orienting of attention. *Q. J. Exp. Psychol.* 32:3–25.

Prinzmetal, W., D. E. Presti, and M. I. Posner, 1986. Does attention affect visual feature integration? *J. Exp. Psychol. [Hum. Percept. Perf.]* 12:361–369.

Reinitz, M. T., 1990. Effects of spatially directed attention on visual encoding. *Percept. Psychophys.* 47:497–505.

Rif, J., R. Hari, M. S. Hamalainen, and M. Sams, 1991. Auditory attention affects two different areas in the human supratemporal cortex. *Electroencephalogr. Clin. Neurophysiol.* 79:464–472.

Sams, M., E. Kaukoranta, M. Hamalainen, and R. Näätänen, 1991. Cortical activity elicited by changes in auditory stimuli: Different sources for the magnetic N100m and mismatch responses. *Psychophysiology* 28:21–29.

Scherg, M., 1992. Functional imaging and localization of electromagnetic brain activity. *Brain Topography* 5:103–111.

Scherg, M., J. Vajsar, and T. Picton, 1989. A source analysis of the human auditory evoked potentials. *J. Cogn. Neurosci.* 1:336–355.

Skinner, J. E., and C. D. Yingling, 1977. Central gating mechanisms that regulate event-related potentials and behavior. *Prog. Clin. Neurophysiol.* 1:30–69.

Sperling, G., and B. A. Dosher, 1986. Strategy and optimization in human information processing. In *Handbook of Perception and Human Performance*, vol. 1, K. R. Boff, L. Kaufman and J. P. Thomas, eds. New York: Wiley, pp. 2–65.

Teder, W., K. Alho, K. Reinikainen, and R. Näätänen, 1993. Interstimulus interval and the selective-attention effect on auditory ERPs: "N1 enhancement" versus processing negativity. *Psychophysiology* 30:71–81.

Treisman, A. M., and G. Gelade, 1980. A feature-integration theory of attention. *Cogn. Psychol.* 12:97–136.

Van Der Heijden, A. H. C., 1992. *Selective Attention in Vision.* London: Routledge.

Wijers, A. A., G. Mulder, T. Okita, and L. J. M. Mulder, 1989. Event-related potentials during memory search and selective attention to letter size and conjunctions of letter size and color. *Psychophysiology* 26:529–547.

Williamson, S. J., and L. Kaufman, 1987. Analysis of neuromagnetic signals. In *Methods of Analysis of Brain Electrical and Magnetic Siqnals: EEG Handbook*, vol. 2, A. S. Gevins and A. Redmond, eds. Amsterdam: Elsevier, pp. 405–448.

Williamson, S. J., Z. L. Lu, D. Karron, and L. Kaufman, 1991. Advantages and limitations of magnetic source imaging. *Brain Topography* 4:169–180.

Woldorff, M. G., C. C. Gallen, S. R. Hampson, S. A. Hillyard, C. Pantev, and F. E. Bloom, 1992. Suppression of unattended-channel mismatch-related activity in human auditory cortex during auditory selective attention. *Soc. Neurosci. Abstr.* 18:584.

Woldorff, M. G., C. C. Gallen, S. A. Hampson, S. R. Hillyard, C. Pantev, D. Sobel, and F. E. Bloom, 1993. Modulation of early sensory processing in human auditory cortex during auditory selective attention. *Proc. Natl. Acad. Sci. U.S.A.* 90:8722–8726.

Woldorff, M. G., S. A. Hackley, and S. A. Hillyard, 1991. The effects of channel-selective listening on the mismatch negativity wave elicited by deviant tones. *Psychophysiology* 28:30–42.

Woldorff, M., J. C. Hansen, and S. A. Hillyard, 1987. Evidence for effects of selective attention to the midlatency range of the human auditory event related potential. *Electroencephalogr. Clin. Neurophysiol.* Suppl. 40:146–154.

Woldorff, M., and S. A. Hillyard, 1991. Modulation of early auditory processing during selective listening to rapidly presented tones. *Electroencephalogr. Clin. Neurophysiol.* 79:170–191.

Woods, D. L., and C. Alain, 1993. Feature processing during high-rate auditory selective attention. *Percept. Psychophys.* 53:391–402.

Woods, D. L., K. Alho, and A. Algazi, 1994. Stages of auditory feature conjunction: An event-related brain potential study. *J. Exp. Psychol. [Hum. Percep. Perf.]* 20:81–94.

43 Neural Mechanisms Mediating Attention and Orientation to Multisensory Cues

BARRY E. STEIN, MARK T. WALLACE, AND M. ALEX MEREDITH

ABSTRACT A number of sensory systems, each adapted to deal with a different form of environmental energy, have developed and been elaborated over the course of vertebrate phylogeny. Although sensory information is processed by the nervous system in a modality-specific manner, there is also a great deal of convergence and interaction among sensory modalities. Because such multisensory interactions generally represent a synergy of information processing, they provide numerous advantages for an organism in dealing with minimal or ambiguous environmental cues. As a midbrain site essential for mediating attentive and orientation behaviors to stimuli from different senses, the superior colliculus has provided an effective model for understanding the neural dynamics of multisensory integration and their effects on overt behavior. Many of the fundamental determinants of multisensory integration revealed by this model appear to be operative at many sites in the central nervous system and may govern such higher-order functions as perception and cognition as well as the more immediate behavioral responses involving the midbrain. Furthermore, the neural principles that guide multisensory integration appear to apply equally well to such widely divergent species as hamsters, rats, cats, monkeys, and humans.

Evolution has provided higher organisms with an array of different senses, each of which provides a unique "view" of the environment. Because these sensory modalities can work simultaneously (or substitute for one another in different circumstances), the ability to detect external events is enhanced significantly, with a parallel increase in the circumstances under which an organism can survive.

BARRY E. STEIN and MARK T. WALLACE Department of Neurobiology and Anatomy, Bowman Gray School of Medicine, Wake Forest University, Winston-Salem, N.C.
M. ALEX MEREDITH Department of Anatomy, Medical College of Virginia, Virginia Commonwealth University, Richmond, Va.

Although some of the sensory modalities may have had similar evolutionary origins (e.g., see Gregory, 1967; Northcutt, 1986), each has become associated with unique subjective attributes. Such sensory impressions as hue, pitch, tickle, and itch are inextricably linked to their own sensory systems and have no equivalents in other modalities. The subjective independence of these experiences also gives one the very strong impression that the different sensory systems are functionally independent. However, under many circumstances, the independence of these sensory channels is more illusory than real. There is a host of perceptual studies demonstrating that stimuli in one modality can significantly alter experiences in another. One broad class of these perceptual effects is called the *ventriloquism effect* (Howard and Templeton, 1966), a phenomenon of intersensory bias in which a stimulus from one sensory modality influences judgments about the location of a stimulus in another modality. An example is when a visual cue (the dummy's moving mouth) alters one's perception of the location of an auditory cue—hence the name. We commonly experience the ventriloquism effect while watching a movie or television program. All the sounds originate from the same loudspeakers, which are relatively far from the visual images on the screen. Nevertheless, we perceive each voice as originating from the appropriate character.

Other examples of powerful cross-modality integration are evident when we use visual and vestibular cues to make judgments about the orientation of objects, or of our own bodies, in space. Gravitational and proprioceptive cues can easily bias visual judgments (Clark and Graybiel, 1966; Wade and Day, 1968; Roll, Velay, and Roll, 1991), and pilots must be taught to depend on their instruments in many situations in which their inclination is to trust the accuracy of their own sensory experiences. Even much of what we describe as taste represents a synthesis of gustatory, olfactory, and tactile sensations.

In each of these instances, and in myriad others (for a more complete description of various intersensory perceptual phenomena, see Welch and Warren, 1986; Stein and Meredith, 1993), the nervous system integrates the information it receives from the different sensory modalities to produce its perceptual products. Because the sensory environment is always changing, this multisensory integration is an ongoing and dynamic process, yet it is a process of which we are completely unaware. What neural apparatus is responsible for the synthesis of information from senses that are so clearly separated from one another in the periphery?

Converging sensory inputs

Presumably, whenever different sensory inputs converge onto individual neurons in the central nervous system, there is the potential for cross-modal, or multisensory, integration. There are many sites in the brain where such convergence takes place, most of which are outside the primary projection pathways. Evoked potential and single-neuron recording studies have demonstrated that multisensory convergence takes place in all species examined thus far. It is present in unicellular organisms, comparatively simple metazoans, in the higher primates, and at all intervening levels of complexity. It would, in fact, be unprecedented to find an animal in which the different senses are organized so that they could not interact with one another. It seems likely that during the evolution of the various sensory systems, some mechanisms were preserved and others were introduced so that their combined action could provide information that would be unavailable from their individual operation. Thus, for example, simultaneous cues from two (or more) sensory modalities can enhance the salience of a stimulus and eliminate ambiguities about its identification that might occur when cues from only one modality are available.

The superior colliculus

One of the densest concentrations of multisensory neurons is found in the deep layers of the superior colliculus (figure 43.1) (e.g., see Stein and Arigbede, 1972; Gordon, 1973; Stein, Magalhaes-Castro, and Kruger, 1976; Meredith and Stein, 1983, 1986a; Stein, 1984b; King and Palmer, 1985), and this structure has proved to be an excellent model in which to examine the neural bases of multisensory integration. The superior colliculus receives visual, auditory, and somatosensory inputs (Stein, 1984b) and plays an important role in attending and orienting to these sensory stimuli (e.g., Sprague and Meikle, 1965; Schneider, 1969; Casagrande et al., 1972; Goodale and Murison, 1975; Stein, 1984a; Sparks, 1986; Stein and Meredith, 1990; Meredith, Wallace, and Stein, 1992).

A number of years ago, we began to examine the integrative responses of neurons in the cat superior colliculus to combinations of sensory stimuli. However, before describing the results of these experiments, it is important to review briefly the organization of the different sensory representations in this midbrain structure.

Alignment Among Sensory Maps in the Superior Colliculus Each sensory representation in the superior colliculus is organized in maplike fashion, with the different maps overlapping and aligned with one another. This is a general organizational plan that is independent of species. Points in frontal sensory space are represented toward the front of the superior colliculus, whereas those progressively further caudally are represented progressively further caudally in the structure. Points in superior sensory space are located medially, and those progressively more inferior are represented progressively more laterally (e.g., Siminoff, Schwassmann, and Kruger, 1966; Feldon, Feldon, and Kruger, 1970; Schaefer, 1970; Cynader and Berman, 1972; Drager and Hubel, 1975; Stein, Magalhaes-Castro, and Kruger, 1976; Chalupa and Rhoades, 1977; Finlay et al., 1978; Stein and Dixon, 1979; Graham et al., 1981; Meredith and Stein, 1990).

Together, these aligned maps generate a comprehensive multisensory map (figure 43.2). As a result, stimuli of any modality (i.e., visual, auditory, or somatosensory) that originate from the same event excite neurons in the same region of the superior colliculus. This represents an efficient way of combining the various sources of sensory information to produce a focus of neural activity. The sensory maps are also in register with motor maps (see Stein, Goldberg, and Clamann, 1976, and Harris, 1980, for a discussion in cat, and Sparks, 1986, for a discussion in primate). For exam-

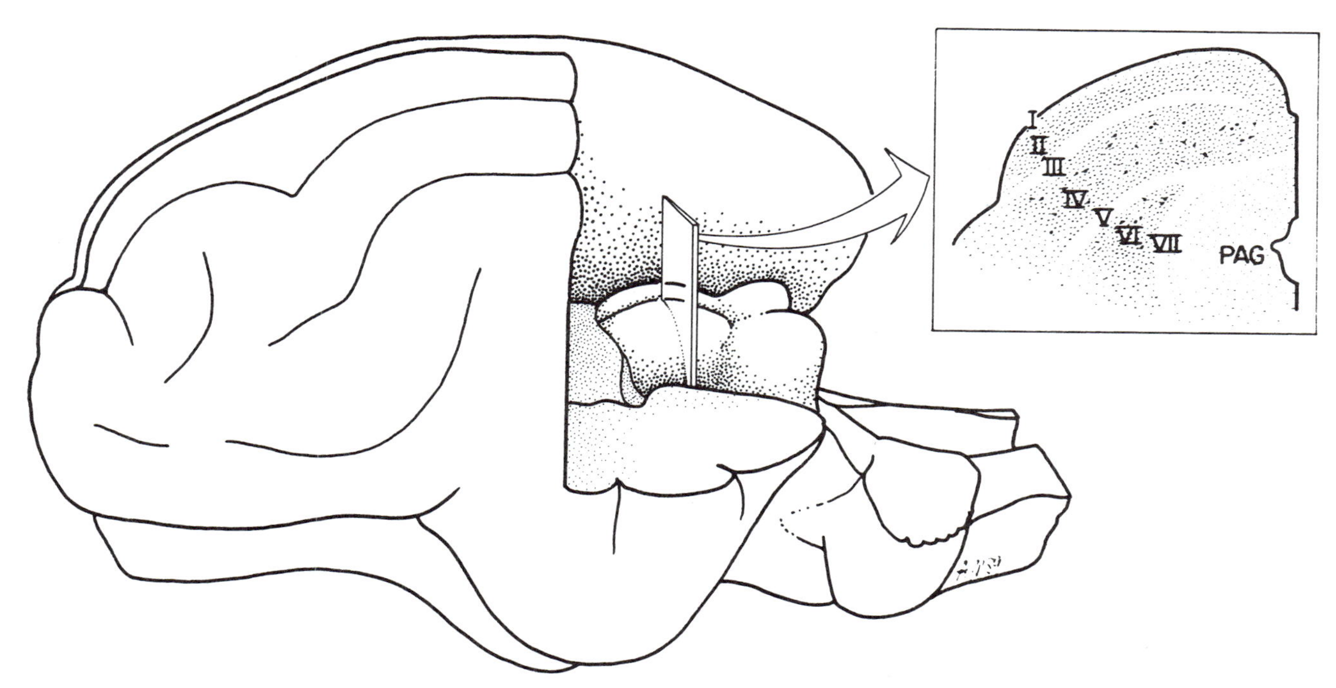

Figure 43.1 Location and laminar pattern of the superior colliculus. In this schematic diagram of the cat brain, the posterior region of the cerebral cortex is removed to reveal the superior colliculus. A coronal section through the superior colliculus (upper right) shows its laminar organization. Superficial layers include I, stratum zonale; II, stratum griseum superficiale; and III, stratum opticum. Deep layers include IV, stratum griseum intermediale; V, stratum album intermediale; V, stratum griseum profundum; and VII, stratum album profundum. PAG, periaqueductal gray. (Reprinted from Meredith and Stein, 1990, by permission.)

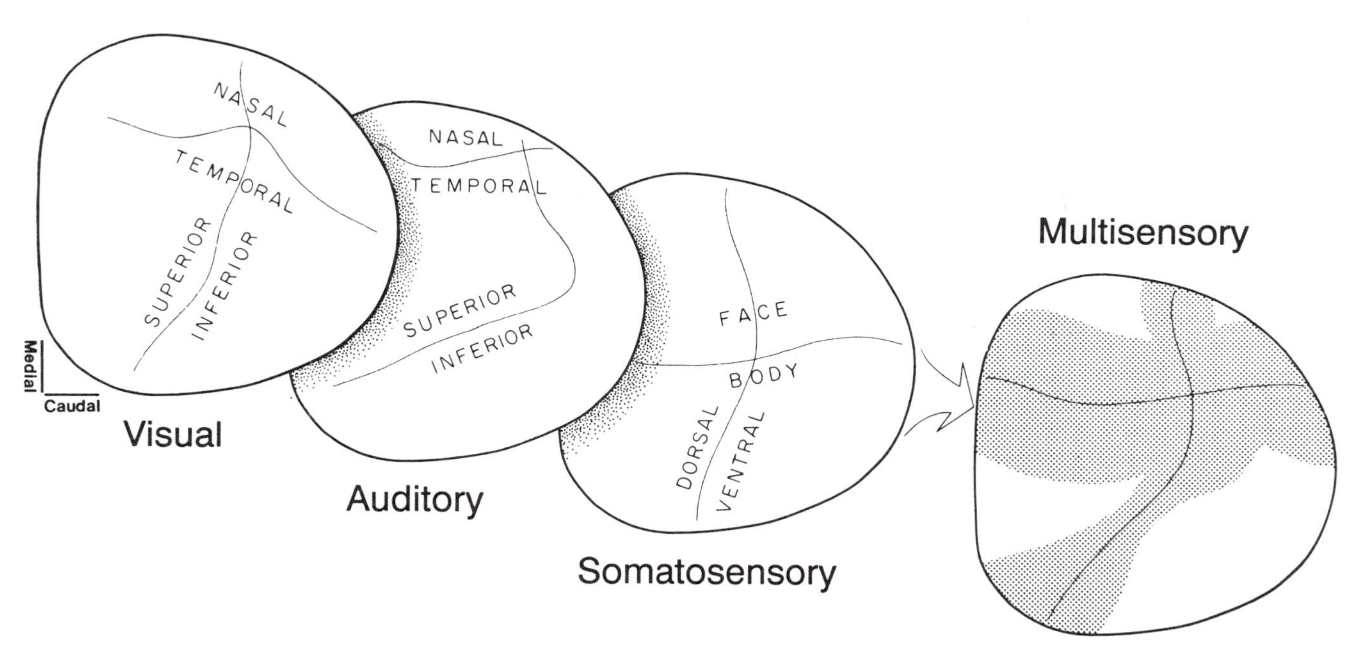

Figure 43.2 Correspondence of visual, auditory, and somatosensory maps in the superior colliculus. The representation of the horizontal and vertical meridians of the different sensory systems in the superior colliculus are very similar. This common coordinate system suggests a representation of multisensory space. (Reprinted from Stein and Meredith, 1993, by permission.)

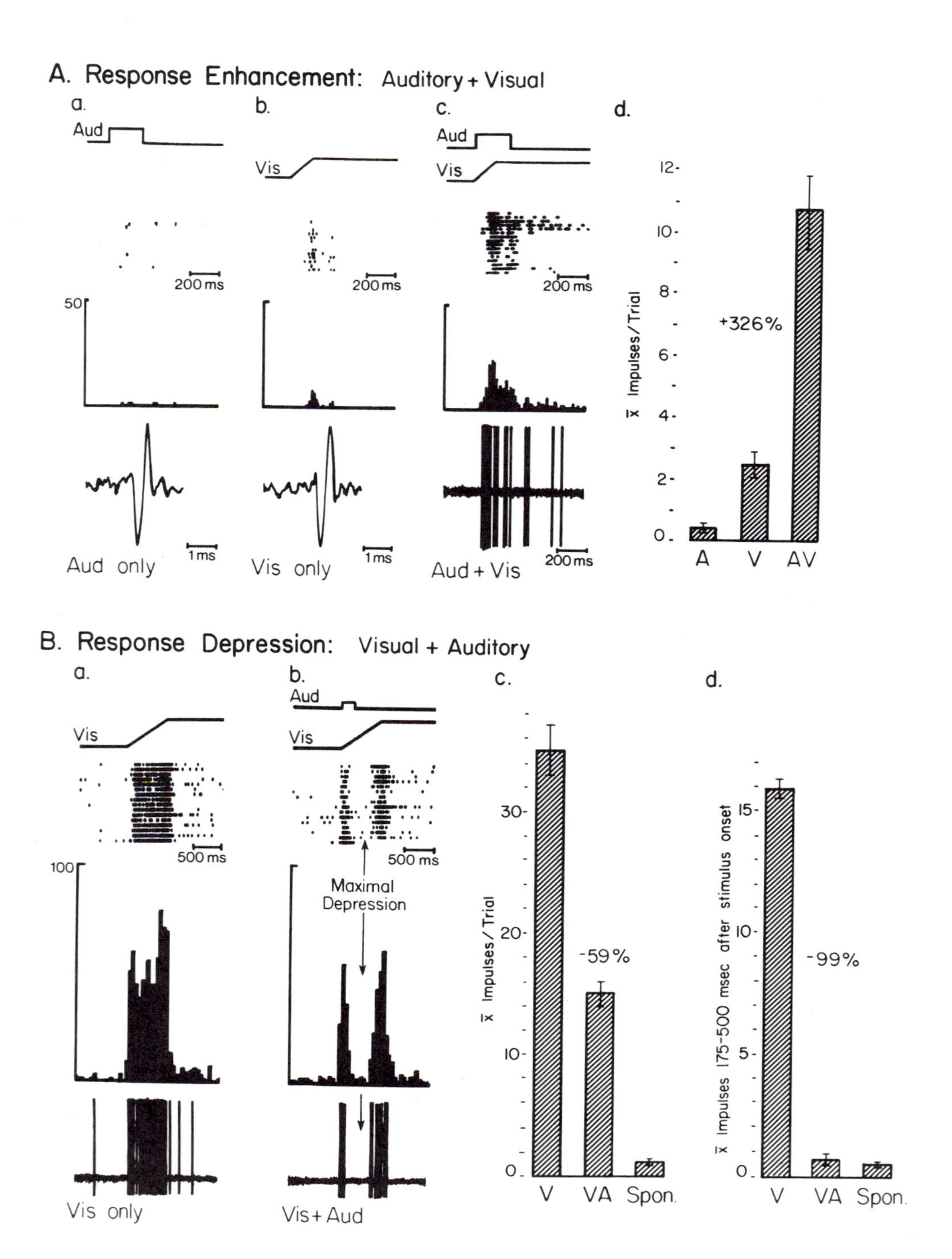

ple, stimuli in inferior sensory space produce a high level of activity in the lateral superior colliculus, a locus from which downward orientations of the eyes, ears, and head are initiated. This organization allows the different sensory inputs to produce identical behaviors via common motor circuits, thereby centering all the sensory organs on a stimulus (Stein, Magalhaes-Castro, and Kruger, 1976).

This alignment of the different maps has significant adaptive value, for it is the spatial relationships among sensory maps that are critical for determining how multisensory inputs are synthesized. The advantages should become clear from the following discussion of the rules of multisensory integration.

Rules Governing Multisensory Integration in Superior Colliculus Neurons That combining stimuli from different modalities could have a dramatic effect on the activity of a neuron was evident in the first multisensory superior colliculus neuron we examined systematically (Meredith and Stein, 1983). Responses to an effective unimodal stimulus (e.g., visual) could be markedly enhanced by combining it with a stimulus from another modality (e.g., auditory) (figure 43.3). After examining additional neurons, we found that some required certain stimulus combinations to respond at all. In other cases, inhibitory interactions were evident, so that a neuron might respond vigorously to one sensory stimulus, not at all to another, and poorly to their combination (see figure 43.3).

To explore how these opposing tendencies might relate to the properties of a particular multisensory neuron, or to the physical nature of the stimuli with which it was confronted, we began systematically manipulating the parameters of the stimuli (Meredith and Stein, 1983, 1986a,b; Meredith, Nemitz, and Stein, 1987; Stein, Meredith, and Wallace, 1993). The results of these studies produced a set of very reliable guidelines for these neurons that are called, for descriptive convenience, *integrative rules*.

Figure 43.3 Response enhancement (A) and response depression (B) in a multisensory superior colliculus neuron. (A) This neuron responded to auditory (a, represented by square wave) and visual (b, represented by ramp) stimuli. The raster below each stimulus trace displays the neuron's responses to 16 successive stimulus presentations. Each dot represents one neuronal impulse. Peristimulus time histograms (bin width = 10 ms) sum the activity shown in the rasters. Representative oscillograms in a and b show this to be a single neuron. When presented alone, the stimuli were weakly effective. However, when the stimuli were combined (c), vigorous responses were elicited. The increase in the number of impulses was significant (paired t-test, $p < .001$) and is summarized in the bar graph shown in (d). (B) Another neuron was vigorously activated by a visual stimulus (a) but was not responsive to auditory or somatosensory stimuli presented alone (not shown). However, when the visual stimulus was combined with an auditory stimulus (b), the visual response was dramatically reduced, as summarized in (c). The depressive effect of the auditory stimulus was virtually complete during the period from 175 to 500 ms after the onset of the auditory stimulus (d), reducing the neuron's activity to its spontaneous level. Peristimulus time histogram bin width = 50 ms. (Reprinted from Meredith and Stein, 1983, by permission.)

Spatial rule The overall spatial registry found among the different sensory maps in the superior colliculus extends to the different sensory receptive fields of an individual multisensory neuron. Thus, a visual and an auditory stimulus originating from the same location in space will fall within the excitatory receptive fields of a visual-auditory multisensory neuron. This is generally the result when the two stimuli originate from the same event. The physiological result of this pairing is an enhancement of the neuron's response (figure 43.4). Many more impulses are evoked by the stimulus combination than by either stimulus alone. In fact, the response is much more than simply the sum of the responses to the individual stimuli; the increase is multiplicative. In contrast, when one of the stimuli (e.g., the auditory stimulus) is moved away from the other to a location outside its excitatory receptive field (as if it originates from a separate event), it is no longer capable of enhancing the neuron's activity. Moreover, if the stimulus now falls within an inhibitory region of the auditory receptive field, it will depress the neuron's responses to the visual stimulus (see figure 43.4) (Meredith and Stein, 1986b). The same effects are seen in multisensory neurons responsive to other combinations of stimuli. These observations indicate that enhancement and depression are dynamic properties that depend on the relative spatial relationships of stimuli and the receptive fields of the neuron.

Temporal rule The multisensory interactions that are so common in superior colliculus neurons might not have been predicted based on the substantial latency differences among the inputs from the different modalities. Responses to auditory stimuli are in the range of 6–25 ms, somatosensory stimuli 12–30 ms, and visual stimuli 40–120 ms. These differences would seem to make it difficult for the inputs to be effective during the same time period so they could interact with one another. However, most stimuli initiate excitatory or inhibitory events that last far longer than these intersensory latency differences. This provides a long interactive "window" in superior colliculus neurons,

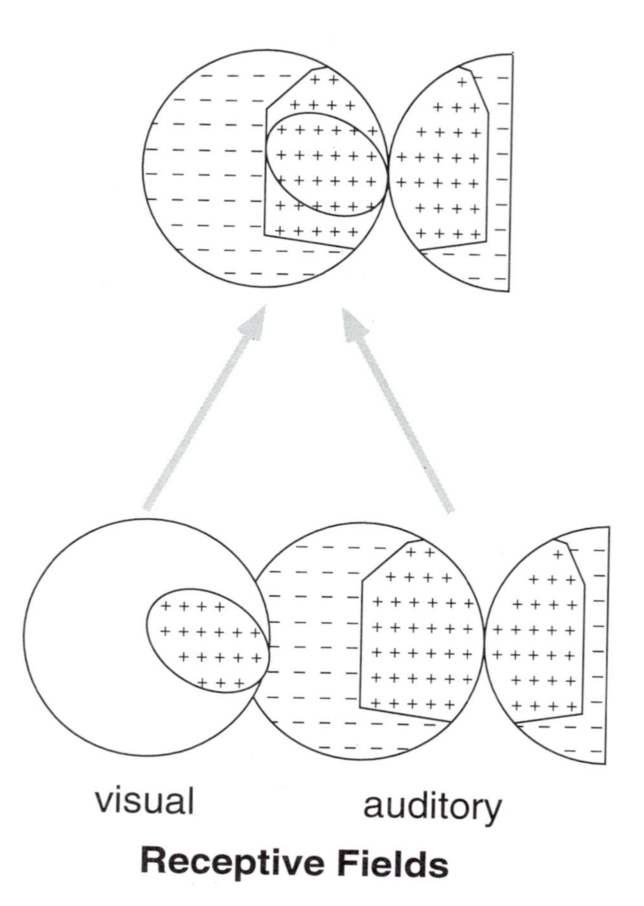

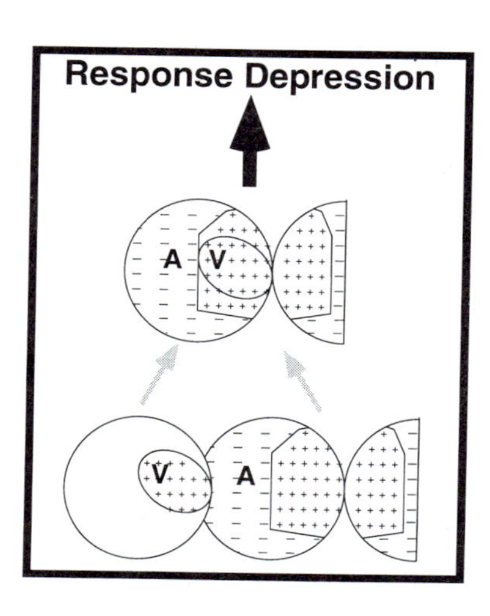

FIGURE 43.4 The spatial relationship between multisensory stimuli and a neuron's receptive field dictates response. On the left is a multisensory receptive field resulting from the overlap of two unimodal receptive fields (+, excitatory regions of the receptive field; −, inhibitory regions). Note that the auditory receptive field has medial and lateral inhibitory regions and that there is no inhibitory region of the visual receptive field. When two stimuli are presented at the same location (V and A represent locations of the visual and auditory stimuli), each falls within the excitatory region of its receptive field, and response enhancement results (top right). When the two stimuli are presented at different locations, one (A) falls within the inhibitory region of its receptive field, and response depression results. Note that the auditory receptive field is represented by a circle, which represents the frontal 180° of auditory space, and a semicircle, which represents the caudal 90° of left auditory space. (Right caudal space is not depicted for reasons of clarity.)

enabling the inputs to interact despite their significant differences in arrival time (Meredith, Nemitz, and Stein, 1987). Maximal interactions have less to do with matching the different input latencies to a given neuron than with overlapping the excitatory or inhibitory events evoked by each stimulus. The key to maximizing multisensory interactions is to overlap the periods of peak activity of the unimodal discharge trains. In many instances, this is best achieved when the stimuli are presented simultaneously, and this is probably not serendipitous (figure 43.5). In other cases, peak overlap occurs only when one stimulus precedes the other by 50–100 ms. The wide temporal window for multisensory interactions can avoid a potentially significant problem. If, for example, integration depended on visual and auditory inputs reaching a neuron simultaneously, these cues could influence one another from only a limited range of distances (see figure 43.5).

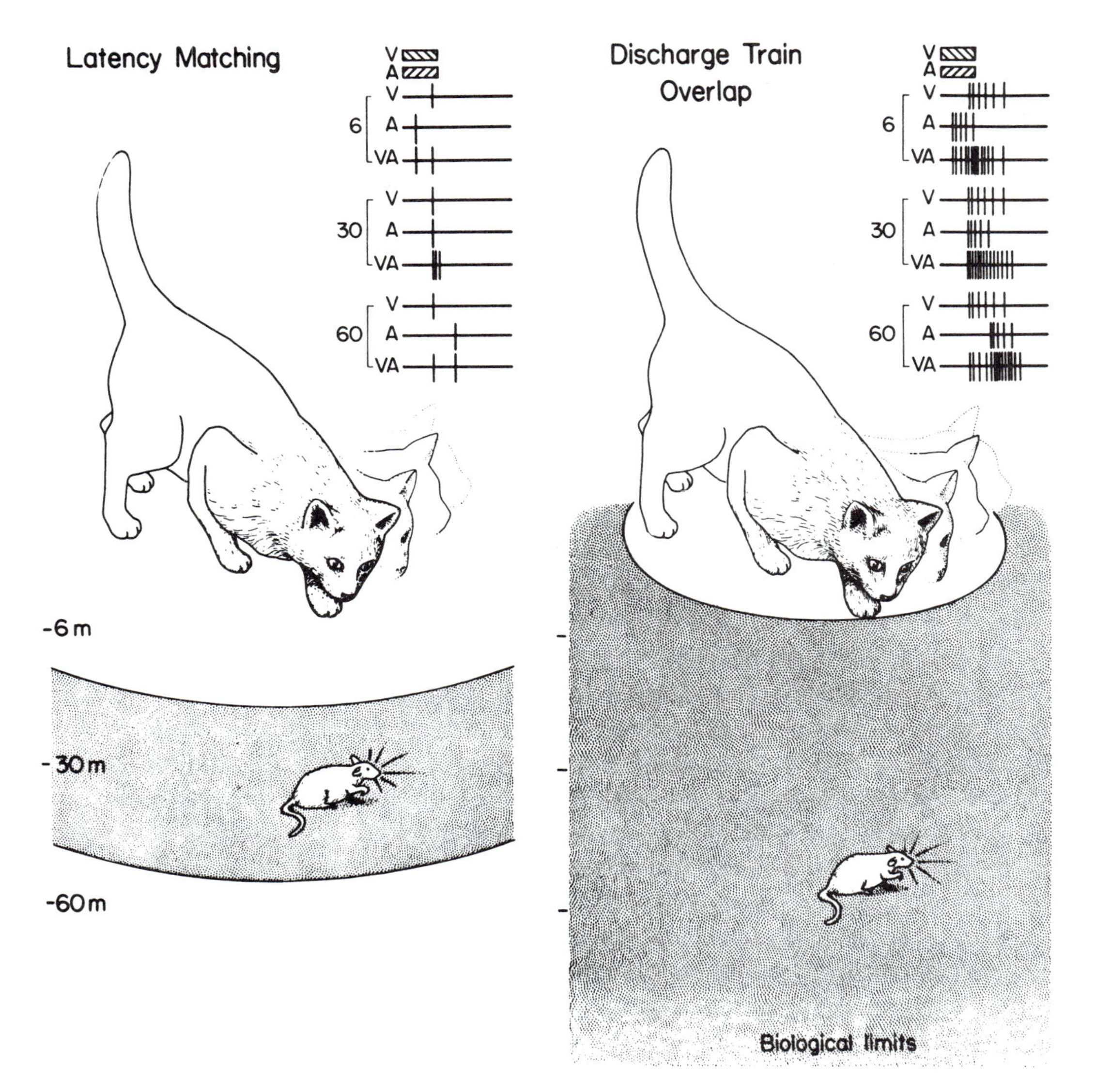

FIGURE 43.5 Overlapping discharge trains, rather than matching input latencies, results in maximal multisensory interactions. There are significant differences among the modalities in the speed with which inputs reach the receptors and the conduction time from receptors to central neurons. If interactions could occur only when input latencies are matched (left), integration in visual-auditory neurons would take place only when the cues originate from a narrow strip 7–39 m from the cat (shaded area). Note that in this case (see the schematized oscillograms above the cat's head), when the stimuli are at 30 m, their inputs produce more than the simple addition of impulses that occurs at 6 and 60 m. However, in actuality, interactions can occur as long as any portion of the two unimodal discharge trains overlap (right). This extends the interactive window, and integration of visual-auditory stimuli can take place at a much wider range of distances. (Reprinted from Meredith, Nemitz, and Stein, 1987, by permission.)

Magnitude, or inverse effectiveness, rule The magnitude of a multisensory interaction, referred to as the *percent interaction*, was measured according to the following formula:

$$CM - SM_{max}/SM_{max} \times 100$$

where CM is the combined modality response and SM_{max} is the best single modality response.

As noted in the discussion of the spatial rule, multisensory enhancements are more than simply additive; they are multiplicative. For example, a response to a combined stimulus might be 10 or more impulses, whereas each response to an individual stimulus might be 2–3 impulses or fewer (e.g., see figure 43.3). In some instances, the individual stimuli evoke no response; only their combination is effective and may produce a surprisingly vigorous response (Meredith and Stein, 1986a). As with most sensory responses, manipulating the physical properties of the individual stimuli alters their effectiveness. These manipulations also affect multisensory integration: the more effective the unimodal stimuli, the lower the magnitude of the

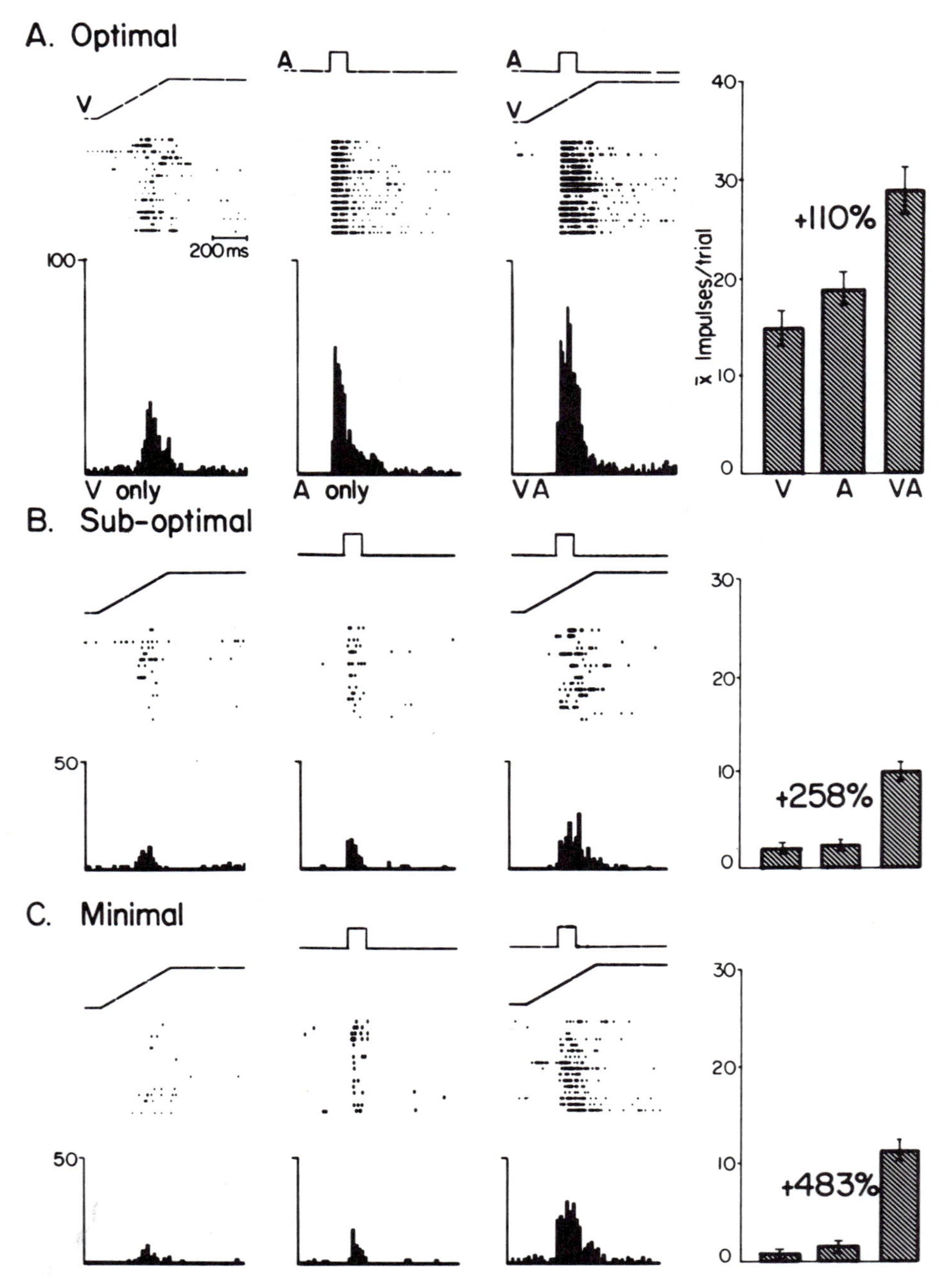

FIGURE 43.6 Multisensory enhancement increases as unimodal stimulus effectiveness decreases. As the physical parameters of the single-modality stimuli are systematically changed (e.g., size of the visual stimulus, intensity of the auditory stimulus), so that progressively fewer discharges are evoked, the percentage of response enhancement resulting from their combination increases. (Reprinted from Meredith and Stein, 1986a, by permission.)

enhancement they generate in combination, a phenomenon referred to as *inverse effectiveness* (see figure 43.6). In behavioral situations, this makes intuitive sense: Potent unimodal stimuli need no enhancement to be effective. However, if during hunting or predator avoidance, combinations of weak unimodal stimuli enhance one another's influence, the adaptive significance of the effect becomes obvious.

Receptive field preservation rule The unimodal receptive field properties of multisensory superior colliculus neurons (e.g., receptive field borders, direction selectivity, velocity selectivity) remain unaltered in the presence of a stimulus from another modality (Stein, Meredith, and Wallace, 1993), and they are essential determinants of which stimuli have access to the circuitry of the superior colliculus. Thus, unimodal selectivity remains constant despite a very changeable sensory world and determines which stimuli can initiate integration in these neurons.

RULES GOVERNING MULTISENSORY ATTENTIVE AND ORIENTATION BEHAVIORS Axons of multisensory neurons form a major component of the output pathways by which the superior colliculus influences attentive and orientation behaviors (Meredith and Stein, 1985; Meredith, Wallace, and Stein, 1992; Wallace, Meredith, and Stein, 1993). Therefore, it seemed likely that the rules of multisensory integration that were evident at the cellular level in the superior colliculus would also govern attentive and orientation behaviors. To test this possibility, experiments were conducted with cats trained to orient to sensory stimuli using a paradigm that paralleled the one used in previous physiological studies (Stein, Huneycutt, and Meredith, 1988; Stein et al., 1989).

The animals were trained to fixate directly ahead and then orient toward, and directly approach, a very dimly illuminated light-emitting diode (LED). The task was a demanding one because the stimuli were low-intensity and thus very difficult to see. During testing, brief low-intensity noise bursts could accompany the visual stimulus, originating either from the same spatial location (spatial coincidence) or from a different location (spatial disparity) (figure 43.7). The ability to detect and orient to the visual stimulus was substantially improved when the auditory stimulus was presented simultaneously and in spatial coincidence, and was significantly degraded when the auditory cue was presented simultaneously but in spatial disparity with the LED (see figure 43.7).

The behavioral product of the multisensory enhancements, like the physiological product, was not a simple addition of unimodal responses but was multiplicative. Thus, there was a very close parallel between the physiology of individual neurons in the superior colliculus and the attentive and orientation behaviors of the animal. As it turned out, both these processes are dependent on descending projections from association cortex, a result that was unexpected when we began to explore the influence of cortex on multisensory responses in the superior colliculus.

The superior colliculus depends on cortex for multisensory integration

PROJECTIONS FROM UNIMODAL SENSORY CORTICES CONVERGE ON INDIVIDUAL SUPERIOR COLLICULUS NEURONS As might be apparent from the preceding discussion, a great deal of effort had been directed to understanding the manner in which the different sensory inputs are integrated in superior colliculus neurons. However, far less attention had been directed toward determining the origin of their multisensory nature. There were two obvious possibilities: Superior colliculus neurons become multisensory because of the direct convergence of unimodal afferents, or superior colliculus neurons become multisensory because they receive inputs from other multisensory neurons.

Sensory inputs to the superior colliculus originate from far too many structures (e.g., see Edwards et al., 1979; Huerta and Harting, 1984; Stein and Meredith, 1991) to explore their influences simultaneously. However, because inputs from the cerebral cortex have been found to be important for many of the complex properties of superior colliculus neurons, an examination of the patterns of corticotectal input seemed a reasonable place to begin. Two broad regions of cortex were most promising because their projections to the superior colliculus are particularly robust (figure 43.8): These are the lateral suprasylvian area or LS (Heath and Jones, 1971), an extraprimary visual area surrounding the suprasylvian sulcus, and the anterior ectosylvian sulcus or AES, a polysensory region that has been subdivided into three modality-specific zones—the fourth somatosensory cortical area, SIV; (Clemo and Stein, 1982), an auditory area, field AES (Clarey

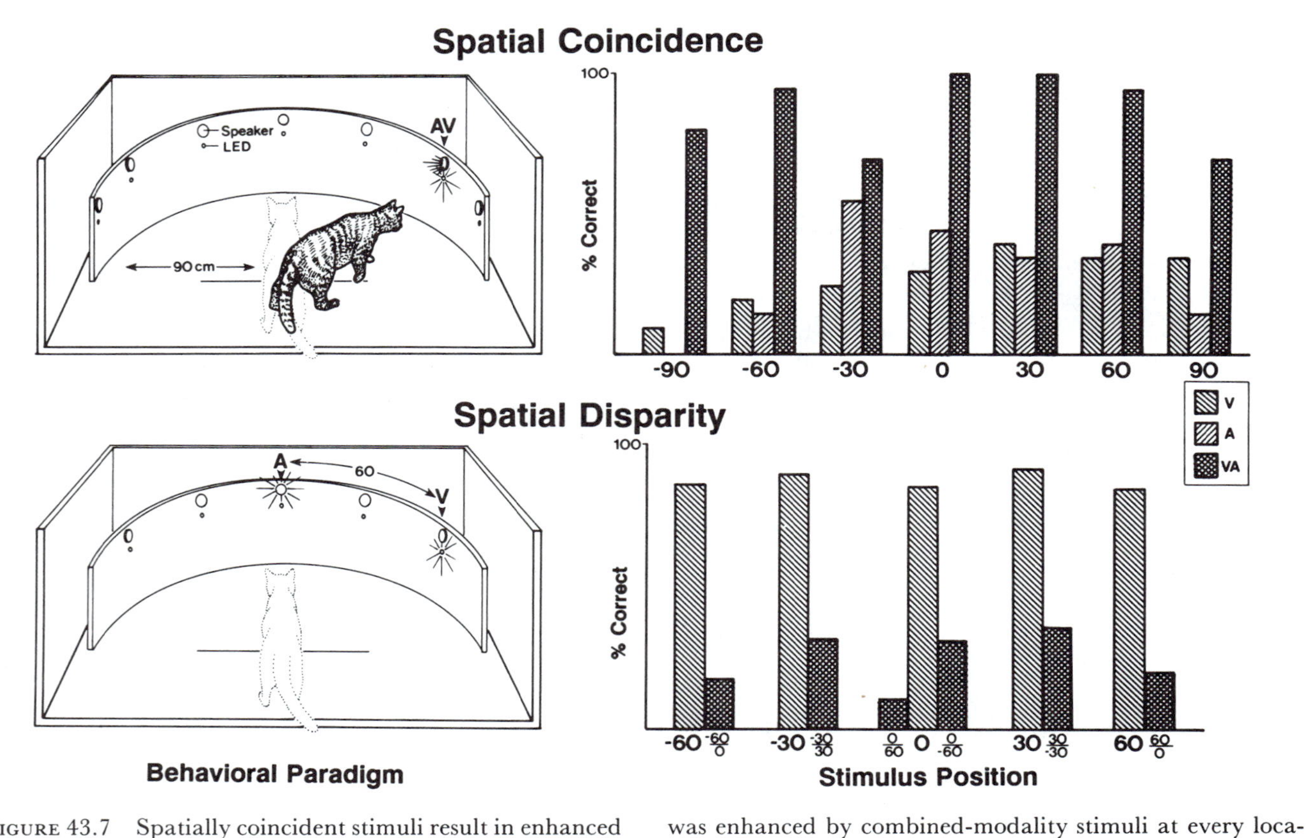

FIGURE 43.7 Spatially coincident stimuli result in enhanced multisensory orientation, whereas spatially disparate stimuli result in depressed multisensory orientation. An array of speakers (larger circle) and LEDs (smaller circle) are located in vertical pairs above a food tray at each of seven regularly spaced (30°) intervals. In the spatially coincident paradigm (top), during training an animal was required to orient to and move directly toward a visual or auditory stimulus to receive a food reward. During testing, low-intensity stimuli were presented individually and then in combination (AV) at the same location at each of the seven eccentricities. The animal's ability to detect and approach the correct position was enhanced by combined-modality stimuli at every location (top right). In the spatially disparate paradigm (bottom), animals were trained to approach a visual stimulus (V) but to ignore an auditory stimulus (A). During testing, the visual stimulus was presented alone or in combination with an auditory stimulus that was 60° out of register with it (e.g., A at 0°, V at 60°). The intensity of the visual stimulus was such that a high percentage of correct responses were elicited to it alone. When a visual stimulus was combined with a spatially disparate auditory stimulus, orientation to the visual stimulus was depressed. (Reprinted from Stein et al., 1989, by permission.)

and Irvine, 1986), and a visual area, AEV (Mucke et al., 1982).

These corticotectal regions are known to play important roles in determining the unimodal sensory responsiveness of deep-layer superior colliculus neurons. Deactivation of LS has been shown to degrade visual responsiveness substantially, eliminating it altogether in some neurons (Stein, 1978; Ogasawara, McHaffie, and Stein, 1984). Similarly, deactivation of the modality-specific portions of AES has been shown to degrade the responses of somatosensory neurons in the superior colliculus and raise the response threshold of auditory neurons (Clemo and Stein, 1986; Meredith and Clemo, 1989).

Although these observations made it clear that influences from cortex are important for the normal sensory responsiveness of deep-layer neurons, it was still not known whether the different sensory cortices send inputs that converge onto individual superior colliculus neurons. This would represent one way by which to create multisensory neurons (as suggested earlier). However, another scheme seemed even more probable. Multisensory neurons have been found intermixed with unimodal neurons in transitional regions of LS and AES (see figure 43.8). Multisensory neurons in LS are located at its rostral aspect, where the visual area abuts somatosensory and auditory cortices, and multisensory neurons in AES are located along the borders

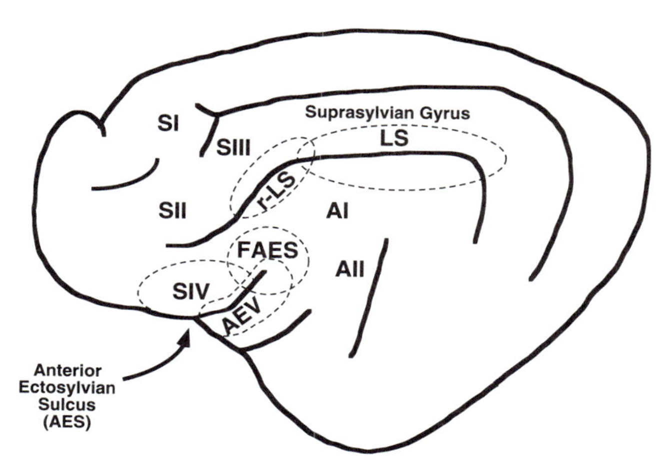

FIGURE 43.8 Location of major corticotectal areas in the cat. The lateral suprasylvian area (LS) lies along both banks of the suprasylvian sulcus. The anterior ectosylvian sulcal cortex (AES) is subdivided into three modality-specific zones —somatosensory (SIV), auditory (FAES), and visual (AEV)—with some overlap at their borders (where most multisensory neurons are found). Similarly, situated at the borders of visual (LS), auditory (AI and FAES), and somatosensory (SII and SIII) cortices is the polysensory rostral lateral suprasylvian (r-LS) zone.

of the different unimodal representations (Clemo et al., 1991; Wallace, Meredith, and Stein, 1992; Stein, Meredith, and Wallace, 1993). These cortical multisensory neurons may project to and determine the multisensory nature of superior colliculus neurons (as suggested earlier). The presence of a single circuit consisting of cortical and superior colliculus multisensory neurons would be a parsimonious organizational plan. However, as is often the case with armchair logic in biology, it does not work that way.

Orthodromic stimulation techniques provided data that were consistent with the first, rather than the second, possibility. Inputs from unimodal regions of cortex (i.e., LS and each of the three unimodal divisions of AES) were shown to converge onto multisensory superior colliculus neurons in various combinations that were appropriate for the modalities of the activated neurons (Wallace, Meredith, and Stein, 1993). For example, visual and auditory cortical inputs, but not somatosensory inputs, were found to converge onto visual-auditory neurons. Combining single-unit cortical recording with antidromic stimulation of the superior colliculus confirmed that the corticotectal neurons were indeed unimodal and that the vast majority of these inputs were monosynaptic.

Most of the neurons in the superior colliculus that were the targets of these converging unimodal cortical projections were efferents, projecting out of the superior colliculus via the crossed tectoreticulospinal (TRS) system (Wallace, Meredith, and Stein, 1993). Because the TRS tract projects to brainstem and spinal cord loci that are involved in orienting movements of the eyes, ears, and head (Grantyn and Grantyn, 1982; Moschovakis and Karabelas, 1985; Guitton and Munoz, 1991; Munoz and Guitton, 1991; Munoz, Guitton, and Pélisson, 1991), AES and LS cortices can exert a potent influence on midbrain-mediated reactions to external stimuli. Consequently, the cortex appears to use the superior colliculus as a means of bringing together information from different sensory modalities in order to influence attentive and orientation behaviors. (This is an issue that will be further explored later.)

MULTISENSORY CORTICAL NEURONS DO NOT PROJECT TO THE SUPERIOR COLLICULUS Orthodromic and antidromic corticotectal studies demonstrated that, in striking contrast to the convergence of unimodal cortical inputs onto superior colliculus neurons, multisensory neurons in AES and LS were not corticotectal (figure 43.9). This finding was surprising given that these multisensory neurons were found in areas rich in corticotectal projections. The fact that their multisensory properties and integrative capabilities closely parallel those in the superior colliculus (see Wallace, Meredith, and Stein, 1992; Stein, Meredith, and Wallace, 1993) made this finding even more surprising.

CORTICAL NEURONS INTEGRATE MULTISENSORY INFORMATION IN THE SAME WAY AS DO SUPERIOR COLLICULUS NEURONS Each of the rules that govern multisensory integration in the superior colliculus was also examined in individual neurons in the transitional (i.e., multisensory) regions of AES and LS. An example of multisensory enhancement in an AES neuron is shown in figure 43.10. Several of the rules of multisensory integration could be demonstrated in this neuron and, ultimately, all the known rules were demonstrated among the samples of AES and LS neurons studied:

1. *Receptive field alignment:* The somatosensory and auditory receptive fields of this neuron exhibited the spatial overlap that is characteristic of multisensory neurons, regardless of their location in the brain or

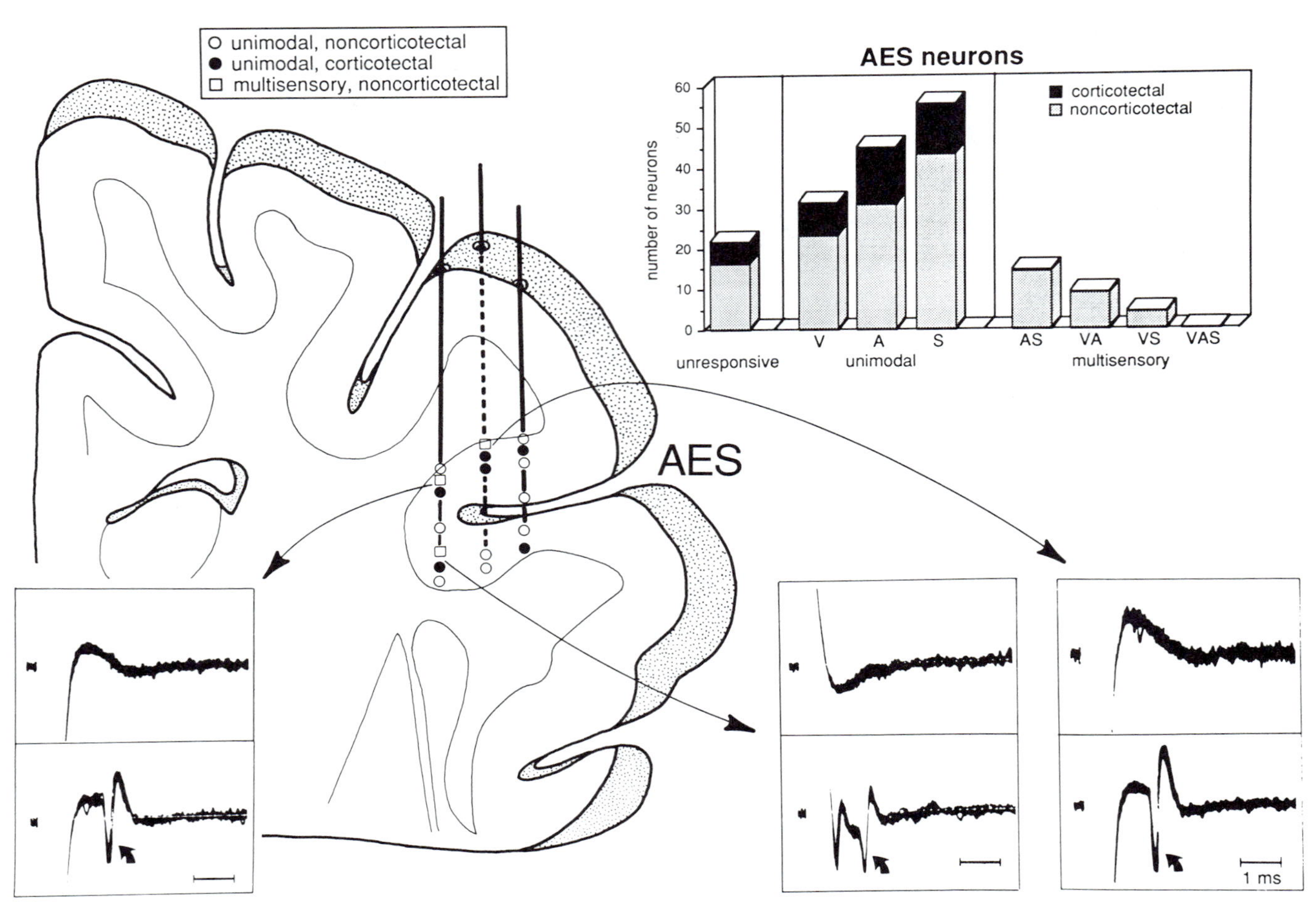

FIGURE 43.9 Multisensory AES neurons are not corticotectal. At top right is a stacked bar graph showing the modality distribution of AES neurons and their corticotectal projection patterns. Whereas many unresponsive and unimodal neurons were found to be corticotectal (black shading), no multisensory neurons were found to project to the superior colliculus. At left is shown how such data were obtained. Depicted are three electrode penetrations (solid and dashed lines) through the AES, shown in coronal section. In two of the three penetrations, multisensory noncorticotectal neurons were found in close apposition to unimodal corticotectal neurons. Arrows point to results seen after stimulation of the superior colliculus for each neuron pair. In each instance, stimulation resulted in antidromic activation of the unimodal neuron (bottom) but no activation of the multisensory neuron (top). (Reprinted from Wallace, Meredith, and Stein, 1993, by permission.)

the species in which they are found (e.g., see Stein, Meredith, and Wallace, 1993).

2. *Spatial rule:* The neuron responded to each unimodal stimulus presented individually but, when the stimuli were presented simultaneously and in spatial register (i.e., each within its excitatory receptive field), there was a significant enhancement of the response (see figure 43.10). When the same two stimuli were separated so that they were spatially disparate, with one falling outside its receptive field, the response was depressed.

3. *Temporal rule:* Varying the temporal interval between the stimuli demonstrated that the magnitude of response enhancement was greatest when the peak periods of unimodal responses were overlapped.

4. *Inverse effectiveness:* In general, combinations of two weak unimodal stimuli resulted in the largest interaction.

5. *Preservation of receptive field properties:* In no case were the receptive field borders or selectivities of a cortical neuron altered during the presentation of multisensory stimuli.

Table 43.1

*Integrative properties of multisensory neurons in different species and different areas of the brain**

Principle	Species and Location					
	Cat SC[1]	Cat AES[2]	Cat r-LS[3]	Rat Parl[4]	Monkey STS[5]	Human EP[6]
Receptive field correspondence	Yes	Yes	Yes	Yes	Yes	?
Spatial	Yes	Yes	Yes	Yes	Yes	Yes
Temporal	Yes	Yes	Yes	Yes	?	Yes
Inverse effectiveness	Yes	Yes	?	?	?	?
Multiplicative	Yes	Yes	Yes	Yes	Yes	?
Receptive field immutability	Yes	Yes	?	?	?	?

SC, superior colliculus; AES, anterior ectosylvian sulcal cortex; r-LS, rostral lateral suprasylvian cortex; Parl, parietal cortex 1; STS, superior temporal sulcal cortex; EP, evoked potential studies from human cortex.

* The various principles of multisensory neurons appear to be universal properties that apply to a number of different species and areas.

Data from [1]Meredith and Stein, 1986a,b; Meredith, Nemitz, and Stein, 1987; and Stein, Meredith, and Wallace, 1993; [2]Wallace, Meredith, and Stein, 1992; and Stein, Meredith, and Wallace, 1993; [3]Stein, Meredith, and Wallace, 1993; [4]Ramachandran, Wallace, and Stein, 1993; [5]Stein, Meredith, and Wallace, 1993; and [6]Costin et al., 1991.

That these integrative principles may be universal properties of multisensory neurons, regardless of location or species, is further reinforced by observations made in rat cortex, in primate cortex, and in evoked potential studies of human subjects (table 43.1).

The fact that multisensory cortical neurons in AES and LS do not project to the superior colliculus raises the intriguing possibility that multisensory integration is carried out in parallel in several independent systems according to the same set of integrative rules (figure 43.11). One multisensory system is composed of superior colliculus neurons that receive converging unimodal cortical inputs and that project into the TRS pathway. A second system is made up of multisensory cortical neurons that do not project to the superior colliculus and that may be involved in higher-order functions more closely linked to perception and cognition. Perhaps the constancies in the rules that govern multisensory integration in superior colliculus and cortical neurons provide coherence among the processes that underlie both the behaviors mediated by the midbrain and the evaluative processes that involve the neocortex.

There is a third system involving multisensory superior colliculus neurons that do not receive converging cortical inputs and do not project into the TRS pathway. Their inputs, outputs, rules of multisensory integration, and functional roles remain to be determined.

Unimodal Cortices Use the Superior Colliculus Neuron as a Substrate for Multisensory Association The findings discussed in the preceding sections led us to wonder just how important cortical inputs are for superior colliculus neurons to perform their multisensory integrative functions. Certainly, if these were the only sources of multisensory input, they would be critical. However, as noted earlier, the superior colliculus gets sensory inputs from a variety of sources, and it is possible that in some neurons, ascending inputs will show convergence patterns that parallel those of the descending inputs. If so, is multisensory integration dependent on one or the other series of converging inputs, or both?

We began to examine this question by evaluating the ability of superior colliculus neurons to integrate multisensory information in the presence and absence of influences from AES or LS or both (Wallace and Stein, 1994). This was done in the following manner: After determining that a given superior colliculus neuron was multisensory, the neuron was presented with each unimodal stimulus individually as well as in combination. This established baseline responses for multisensory integration. The tests were performed a second time after deactivating appropriate regions of cortex by cooling (see Stein 1978; Ogasawara et al., 1984; Clemo and Stein, 1986). For example, for a visual-somatosensory neuron, visual (LS or AEV) cortices

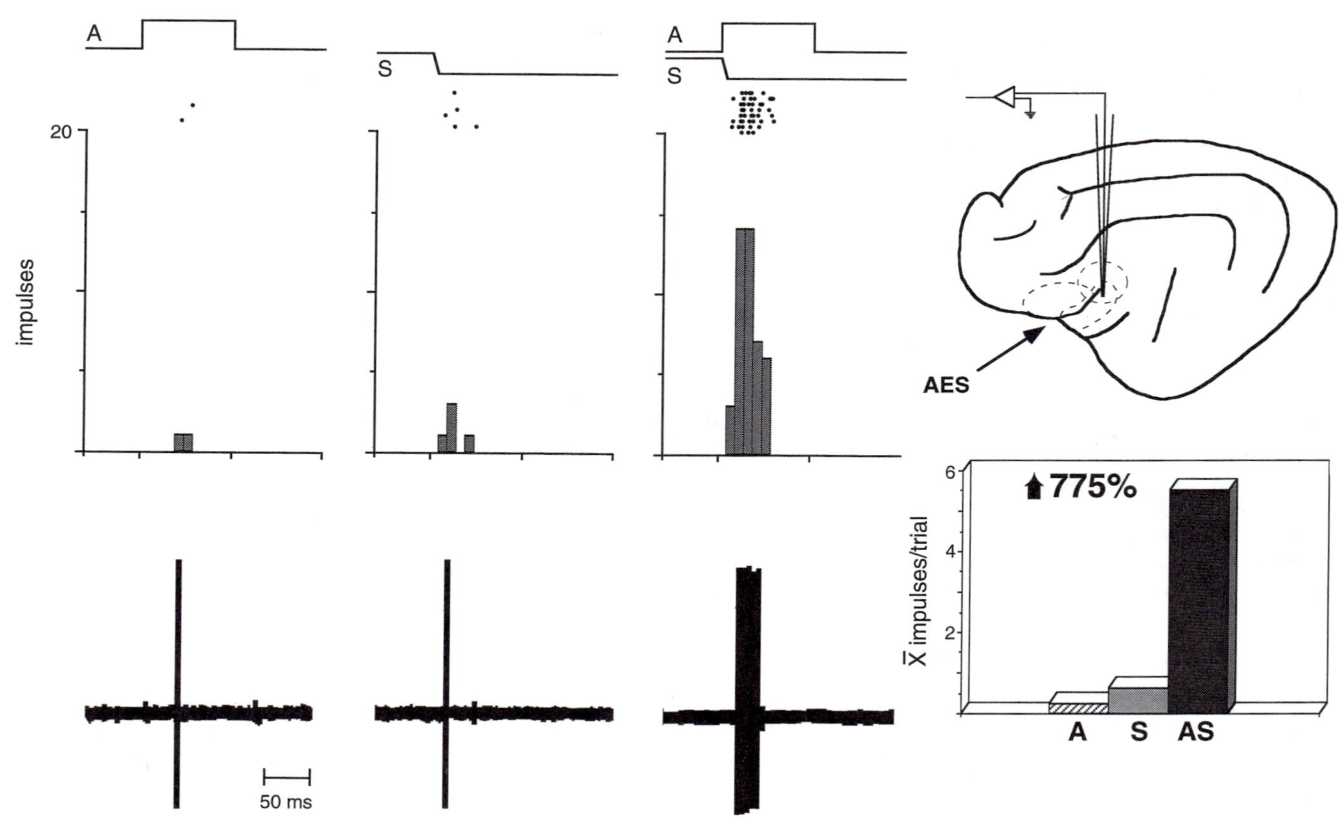

FIGURE 43.10 Response enhancement in a cortical (AES) auditory-somatosensory neuron. When presented alone, the auditory (A; square wave) and somatosensory (S; ramp) stimuli were weakly effective. However, their combination (AS) elicited a 775% increase over the response to the best (i.e., somatosensory) unimodal stimulus. Each dot represents a single neuronal impulse, and the rasters show the results for eight stimulus presentations for each condition (i.e., auditory alone, somatosensory alone, and combined). Peristimulus time histograms (bin width = 20 ms) show the summed responses for the eight trials, and the bottom panels show a representative oscillographic trace. Top right shows the location of this neuron in the AES (dashed lines indicate subdivisions depicted in figure 43.8), and box on bottom right depicts the averaged (mean) responses for each condition.

alone were deactivated, somatosensory (SIV) cortex alone was deactivated, and two or all three cortical areas were then deactivated simultaneously. Cortex was then rewarmed, and a third series of tests was performed.

As might be expected, some superior colliculus neurons received all their excitatory input from cortex, for when each relevant modality-specific cortex was deactivated, the neuron's responses to that modality were eliminated (Wallace and Stein, 1994). Deactivating all the cortices simultaneously rendered the neuron unresponsive to all sensory stimuli. Obviously, in these cases, cortical inputs were critical for the neuron's sensory responses.

In most instances, sensory inputs were provided from sources in addition to AES and LS. Consequently, the sensory responses of these neurons survived cortical deactivation. However, their capacity to integrate multisensory inputs did not. A characteristic example of this loss of integration after cortical deactivation is presented in figure 43.12. This auditory-somatosensory neuron exhibited a fourfold response enhancement to a multisensory stimulus. Deactivation of either field AES or SIV produced fewer impulses in response to the respective unimodal stimuli but resulted in a dramatic reduction in the level of multisensory enhancement. The strength of this effect is underscored by the observation that, typically, weaker responses to unimodal stimuli are associated with greater levels of multisensory enhancement (i.e., inverse effectiveness). Simultaneous deactivation of auditory and somatosensory cortices eliminated multi-

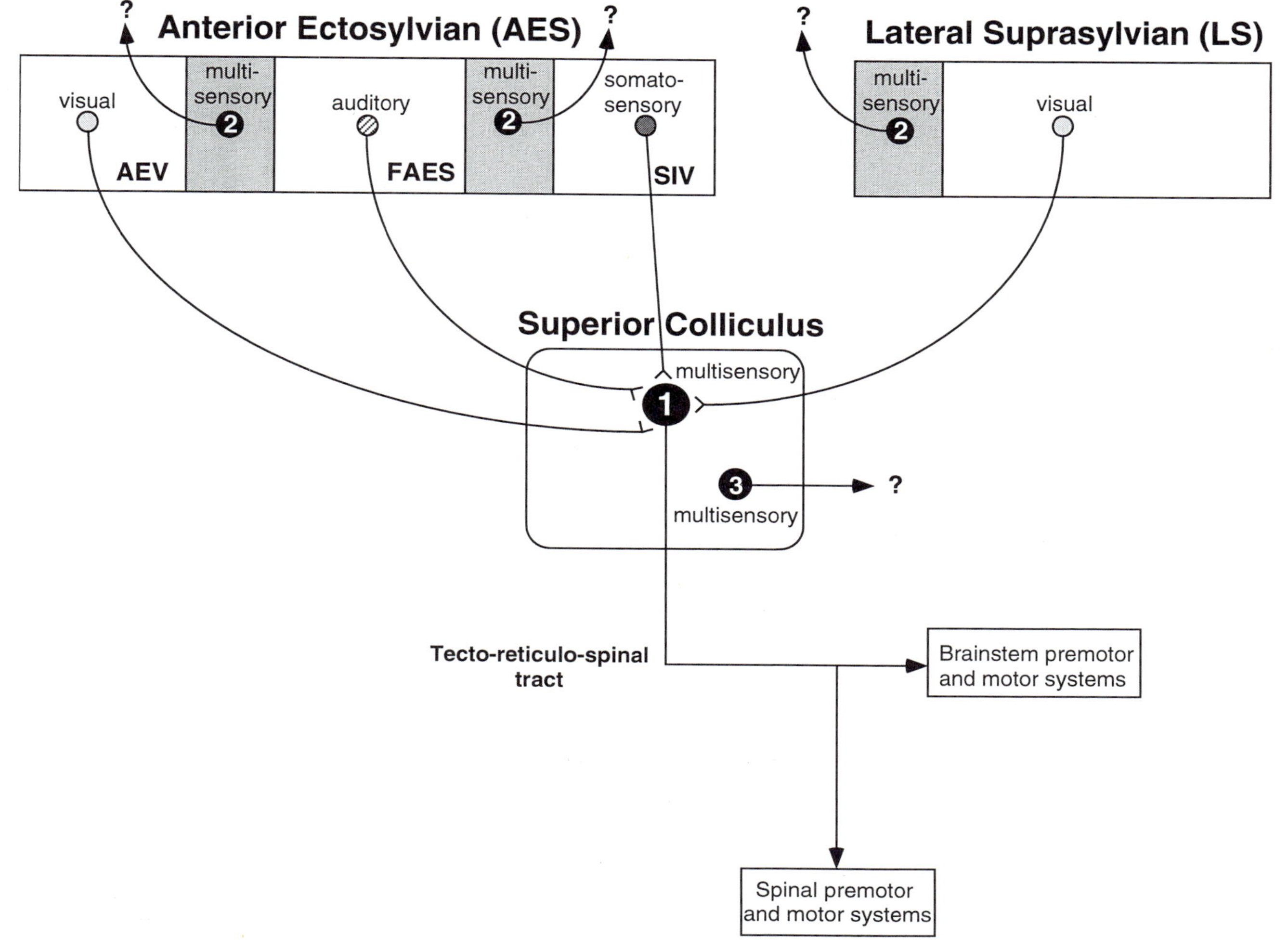

FIGURE 43.11 Schematic diagram of cortical and superior colliculus multisensory circuits. Population 1 is superior colliculus neurons that receive convergent inputs from unimodal cortices and that project to the brain stem and spinal cord via the tectoreticulospinal (TRS) pathway. Population 2 is an independent set of multisensory neurons in AES and rostral LS. Population 3 is multisensory superior colliculus neurons that neither receive convergent cortical input nor project via the TRS tract.

sensory integration in this neuron, despite the fact that it continued to respond to each unimodal cue.

The dependence of multisensory integration on inputs from the AES was apparent in examples of each multisensory neuron type examined (auditory-somatosensory, visual-auditory, and visual-somatosensory). In contrast, there were no examples in which deactivation of LS produced equivalent results. Although there were many instances in which LS deactivation decreased or eliminated responses to visual stimuli, it never showed the specificity for multisensory integration seen after deactivation of AES. In addition, deactivation of primary sensory cortices (e.g., striate cortex, AI–II, SI–III) had no effect on the responses of deep-layer superior colliculus neurons (also see Stein, 1978; Ogasawara, McHaffie, and Stein, 1984; Clemo and Stein, 1986).

These data indicate that many superior colliculus neurons derive multiple sensory inputs from sources other than association cortex (i.e., AES), but their capacity to integrate these inputs depends on influences from association cortex (figure 43.13). (A few exceptions to this scheme were noted, but these may have been the result of incomplete cortical deactivation.) Further confirmation of this conclusion comes from the observation that multisensory superior colliculus neu-

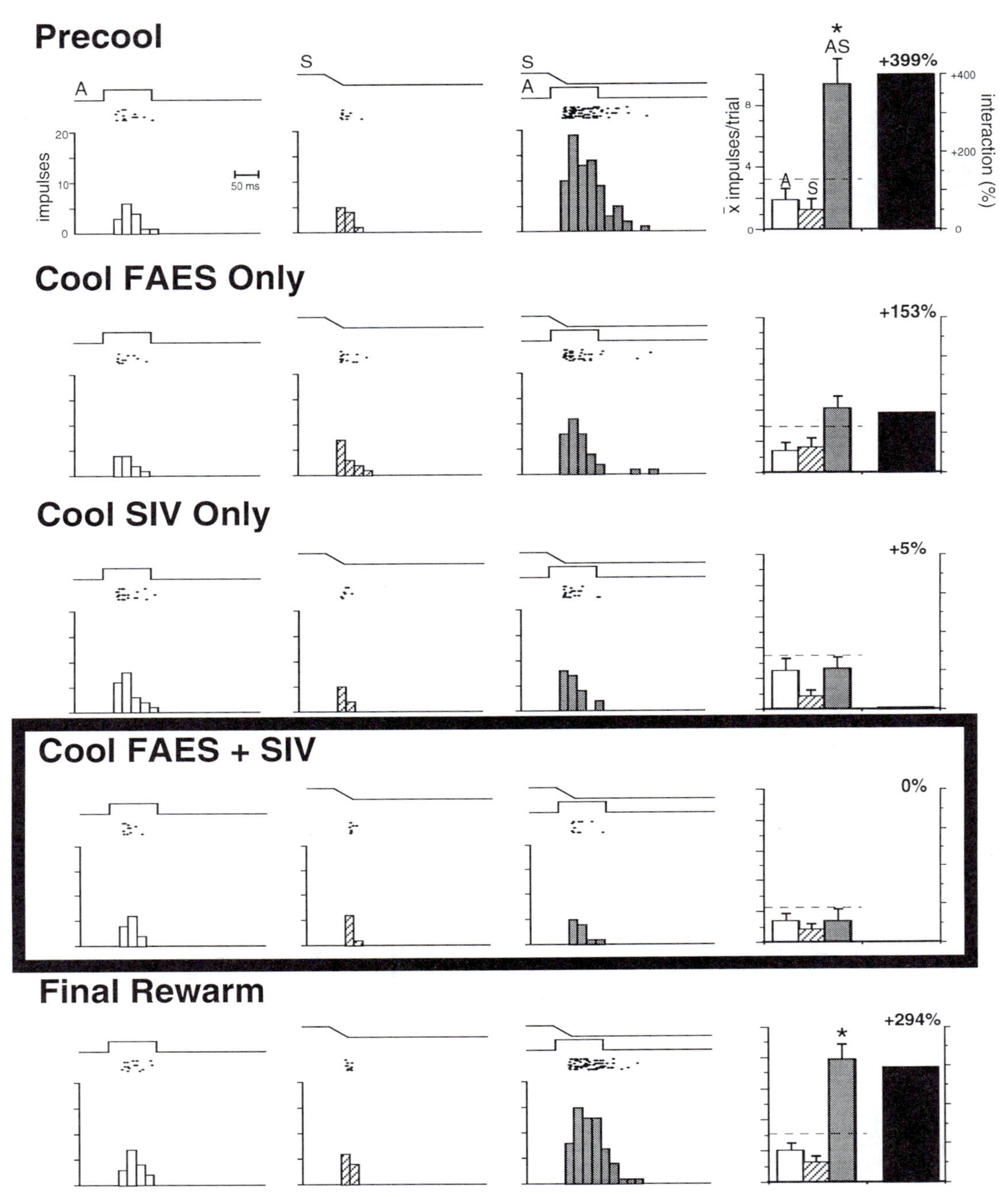

FIGURE 43.12 AES deactivation eliminates multisensory enhancement in the superior colliculus. Prior to cortical deactivation (precool), this auditory-somatosensory neuron showed a modest response to either an auditory (A) or somatosensory (S) stimulus. However, the neuron's response to their combination (AS) was enhanced 399%. Deactivating FAES or SIV significantly diminished this enhancement, and to a degree substantially greater than that predicted by the slight change in the unimodal response. Combined deactivation of FAES and SIV (boxed condition) eliminated multisensory enhancement altogether but did not preclude unimodal responses. Dashed lines depict the expected result to multisensory combinations if the unimodal responses were simply summed. Asterisks denote significant ($p < .001$) multisensory enhancements. Although cortex was rewarmed after each cooling condition (and responses returned to precool levels), only the final rewarm condition is shown. Conventions for this figure are the same as those used in figure 43.11.

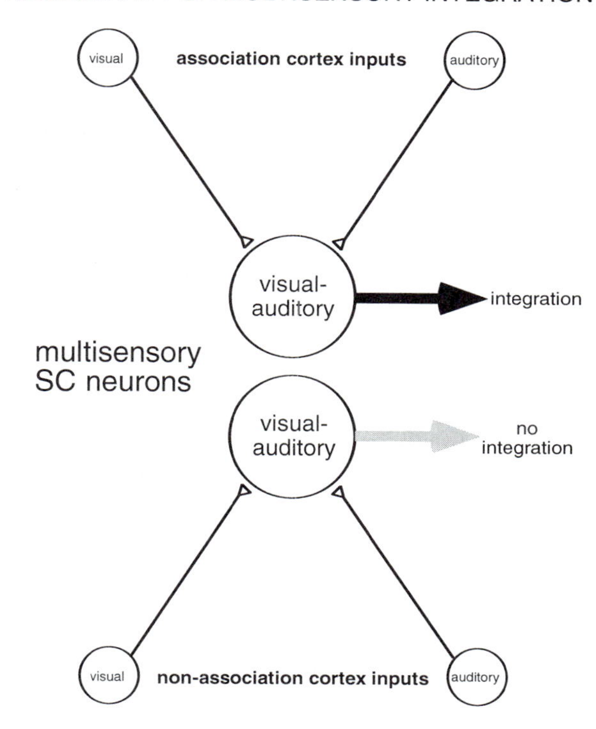

FIGURE 43.13 Input from association cortex is necessary for multisensory integration in superior colliculus neurons. Whereas converging unimodal inputs from nonassociation areas (e.g., ascending projections) may be sufficient to render a superior colliculus (SC) neuron multisensory (bottom), it is only in the presence of inputs from association cortex (i.e., AES) that the neuron is capable of multisensory integration (top).

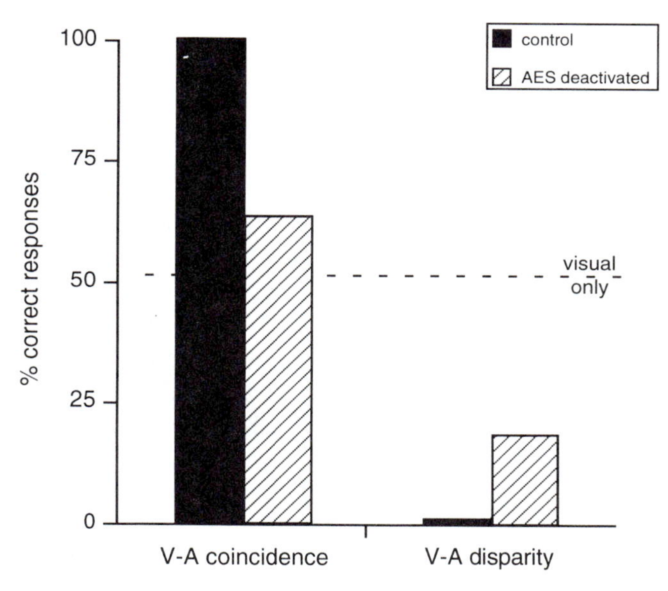

FIGURE 43.14 Deactivation of AES interferes with multisensory orientation behavior. In normal-behaving animals, the probability of correctly responding to a very dim visual stimulus (dashed line) is significantly enhanced when a weak auditory stimulus is presented at the same location (V-A coincidence) and is significantly depressed when the auditory stimulus is presented 45° contralateral to the visual stimulus (V-A disparity). Deactivation of AES with lidocaine degrades behavioral enhancement and depression. These behavioral observations parallel the physiological observations shown in figure 43.12 and are consistent with the schematic shown in figure 43.13.

rons that failed to exhibit cross-modal integration were unaffected by AES deactivation. These neurons are almost certainly the third population of multisensory neurons referred to earlier (see figure 43.11).

CONSEQUENCES OF DEACTIVATION OF CORTEX ON ATTENTIVE AND ORIENTATION BEHAVIORS Because the vast majority of neurons in the superior colliculus that receive convergent cortical inputs project into the TRS tract, the principal pathway mediating attentive and orientation behaviors, one would expect a loss of neural enhancement after AES deactivation to be paralleled by a loss of behavioral enhancement. This expectation was confirmed in experiments with behaving animals. In fact, the effects were very dramatic. The training and testing paradigm used for these tests involved the same stimuli (i.e., LEDs and noise bursts) and the same paradigm as described earlier (see figure 43.7). Animals were trained to orient toward and approach a visual cue. That cue could be presented alone, accompanied by a spatially coincident noise burst, or accompanied by a spatially disparate noise burst. However, in this experimental series, AES could be reversibly deactivated by injecting lidocaine through an indwelling cannula. The deactivation of AES had very little effect on the animal's ability to orient to the visual cue alone, but there was a substantial loss of the multisensory integration that normally occurs when the visual and auditory stimuli are paired. Now, as shown in figure 43.14, the capability of the auditory stimulus to enhance or depress the effectiveness of the visual cue was significantly degraded (Wilkinson, Meredith, and Stein, 1992). This effect was not seen after deactivation of primary auditory cortex, primary visual cortex, or LS, nor was it seen after injections of saline into AES.

Conclusions

The data generated thus far show that the nervous system provides mechanisms of attentive and orientation behaviors that supersede the individual sensory modalities. These systems integrate information to enhance responses to stimuli that originate from the same event and to inhibit responses to unrelated stimuli. These are dynamic processes, and, while the physical properties of the stimuli as well as the genetic and experiential history of the organism will certainly have their effects, they are likely to do so within the fundamental rules that have been described here. These rules appear to be independent of structure and thus independent of the functions mediated by different structures. This could lend coherence to superior colliculus–mediated behavioral responses and cortically mediated perceptual and cognitive processes. In this way, the brain can simultaneously raise or lower the salience of a given complex of stimuli at all levels of information processing.

Integral to this system are the close functional ties between cortex and the superior colliculus. Of primary importance in this scheme in the cat is the AES, a small area of association cortex at the junction of temporal, parietal, and frontal cortices. This area performs at least some of its association functions via its projections to the superior colliculus. In this case, the superior colliculus neuron provides a mechanism by which the influences of different unimodal regions of the AES can be synthesized and a circuit through which this synthesized information can control attentive and orientation behavior. Whether this turns out to be a general feature of association cortex in all mammals remains to be determined. However, preliminary work with monkeys (Stein, Meredith, and Stein, 1993) and human subjects (Costin et al., 1991) suggests that the same principles that govern the physiological and behavioral responses in cats are operative in primates, thereby lending credence to the view that similar cortical and subcortical circuits are involved.

REFERENCES

Casagrande, V. A., J. K. Harting, W. C. Hall, and I. T. Diamond, 1972. Superior colliculus of the tree shrew: A structural and functional subdivision into superficial and deep layers. *Science* 177:444–447.

Chalupa, L. M., and R. W. Rhoades, 1977. Responses of visual, somatosensory, and auditory neurones in the golden hamster's superior colliculus. *J. Physiol. (Lond.)* 207: 595–626.

Clarey, J. C., and D. R. F. Irvine, 1986. Auditory response properties of neurons in the anterior ectosylvian sulcus of the cat. *Brain Res.* 386:12–19.

Clark, B., and A. Graybiel, 1966. Contributing factors in the perception of the oculogravic illusion. *Am. J. Psychol.* 79:377–388.

Clemo, H. R., M. A. Meredith, M. T. Wallace, and B. E. Stein, 1991. Is the cortex of cat anterior ectosylvian sulcus a polysensory area? *Soc. Neurosci. Abstr.* 17:1585.

Clemo, H. R., and B. E. Stein, 1982. Somatosensory cortex: A "new" somatotopic representation. *Brain Res.* 235:162–168.

Clemo, H. R., and B. E. Stein, 1986. Effects of cooling somatosensory cortex on response properties of tactile cells in the superior colliculus. *J. Neurophysiol.* 55:1352–1368.

Costin, D., H. J. Neville, M. A. Meredith, and B. E. Stein, 1991. Rules of multisensory integration and attention: ERP and behavioral evidence in humans. *Soc. Neurosci. Abstr.* 17:656.

Cynader, M., and N. Berman, 1972. Receptive-field organization of monkey superior colliculus. *J. Neurophysiol.* 35: 187–201.

Drager, U. C., and D. H. Hubel, 1975. Responses to visual stimulation and relationship between visual, auditory and somatosensory inputs in mouse superior colliculus. *J. Neurophysiol.* 38:690–713.

Edwards, S. B., C. L. Ginsburgh, C. K. Henkel, and B. E. Stein, 1979. Sources of subcortical projections to the superior colliculus in the cat. *J. Comp. Neurol.* 184:309–330.

Feldon, S., P. Feldon, and L. Kruger, 1970. Topography of the retinal projection upon the superior colliculus of the cat. *Vision Res.* 10:135–143.

Finlay, B. L., S. E. Schneps, K. G. Wilson, and G. E. Schneider, 1978. Topography of visual and somatosensory projections to the superior colliculus of the golden hamster. *Brain Res.* 142:223–235.

Goodale, M. A., and R. C. C. Murison, 1975. The effects of lesions of the superior colliculus on locomotor orientation and the orienting reflex in the rat. *Brain Res.* 88:243–261.

Gordon, B. G., 1973. Receptive fields in the deep layers of the cat superior colliculus. *J. Neurophysiol.* 36:157–178.

Graham, J., H. E. Pearson, N. Berman, and H. E. Murphy, 1981. Laminar organization of superior colliculus in the rabbit: A study of receptive-field properties of single units. *J. Neurophysiol.* 45:915–932.

Grantyn, A., and R. Grantyn, 1982. Axonal patterns and sites of termination of cat superior colliculus neurons projecting in the tecto-bulbo-spinal tract. *Exp. Brain Res.* 46: 243–265.

Gregory, R. L., 1967. Origin of eyes and brains. *Nature* 213:369–372.

Guitton, D., and D. P. Munoz, 1991. Control of orienting

gaze shifts by the tectoreticulospinal system in the head-free cat: I. Identification, localization, and effects of behavior on sensory responses. *J. Neurophysiol.* 66:1605–1623.

HARRIS, L. R., 1980. The superior colliculus and movements of the head and eyes in cats. *J. Physiol.* 300:367–391.

HEATH, C. J., and E. G. JONES, 1971. The anatomical organization of the suprasylvian gyrus of the cat. *Ergeb. Anat. Entwickl. Gesch.* 45:1–64.

HOWARD, I. P., and W. B. TEMPLETON, 1966. *Human Spatial Orientation.* London: Wiley.

HUERTA, M. F., and J. K. HARTING, 1984. The mammalian superior colliculus: Studies of its morphology and connections. In *Comparative Neurology of the Optic Tectum*, H. Vanegas, ed. New York: Plenum, pp. 687–773.

KING, A. J., and A. R. PALMER, 1985. Integration of visual and auditory information in bimodal neurones in the guinea-pig superior colliculus. *Exp. Brain Res.* 60:492–500.

MEREDITH, M. A., and H. R. CLEMO, 1989. Auditory cortical projection from the anterior ectosylvian sulcus (field AES) to the superior colliculus in the cat: An anatomical and electrophysiological study. *J. Comp. Neurol.* 289:687–707.

MEREDITH, M. A., J. W. NEMITZ, and B. E. STEIN, 1987. Determinants of multisensory integration in superior colliculus neurons: I. Temporal factors. *J. Neurosci.* 10:3215–3229.

MEREDITH, M. A., and B. E. STEIN, 1983. Interactions among converging sensory inputs in the superior colliculus. *Science* 221:389–391.

MEREDITH, M. A., and B. E. STEIN, 1985. Descending efferents from the superior colliculus relay integrated multisensory information. *Science* 227:657–659.

MEREDITH, M. A., and B. E. STEIN, 1986a. Visual, auditory, and somatosensory convergence on cells in superior colliculus results in multisensory integration. *J. Neurophysiol.* 56:640–662.

MEREDITH, M. A., and B. E. STEIN, 1986b. Spatial factors determine the activity of multisensory neurons in cat superior colliculus. *Brain Res.* 365:350–354.

MEREDITH, M. A., and B. E. STEIN, 1990. The visuotopic component of the multisensory map in the deep laminae of the cat superior colliculus. *J. Neurosci.* 10:3727–3742.

MEREDITH, M. A., M. T. WALLACE, and B. E. STEIN, 1992. Visual, auditory and somatosensory convergence in output neurons of the cat superior colliculus: Multisensory properties of the tecto-reticulo-spinal projection. *Exp. Brain Res.* 88:181–186.

MOSCHOVAKIS, A. K., and A. B. KARABELAS, 1985. Observations on the somatodendritic morphology and axonal trajectory of intracellularly HRP-labeled efferent neurons located in the deeper layers of the superior colliculus of the cat. *J. Comp. Neurol.* 239:276–308.

MUCKE, L., M. NORITA, G. BENEDEK, and O. CREUTZFELDT, 1982. Physiologic and anatomic investigation of a visual cortical area situated in the ventral bank of the anterior ectosylvian sulcus of the cat. *Exp. Brain Res.* 46:1–11.

MUNOZ, D. P., and D. GUITTON, 1991. Control of orienting gaze shifts by the tectoreticulospinal system in the head-free cat: II. Sustained discharges during motor preparation and fixation. *J. Neurophysiol.* 66:1624–1641.

MUNOZ, D. P., D. GUITTON, and D. PÉLISSON, 1991. Control of orienting gaze shifts by the tectoreticulospinal system in the head-free cat: III. Spatiotemporal characteristics of phasic motor discharges. *J. Neurophysiol.* 66:1642–1666.

NORTHCUTT, R. G., 1986. Evolution of the octavolateralis system: Evaluation and heuristic value of phylogenetic hypotheses. In *The Biology of Change in Otolaryngology*, R. Vanderwater and E. Rubel, eds. New York: Excerpta Medica, pp. 3–14.

OGASAWARA, K., J. G. MCHAFFIE, and B. E. STEIN, 1984. Two visual systems in cat. *J. Neurophysiol.* 52:1226–1245.

RAMACHANDRAN, R., M. T. WALLACE, and B. E. STEIN, 1993. Distribution and properties of multisensory neurons in rat cerebral cortex. *Soc. Neurosci. Abstr.* 19:1447.

ROLL, R., J. L. VELAY, and J. P. ROLL, 1991. Eye and neck proprioceptive messages contribute to the spatial coding of retinal input in visually oriented activities. *Exp. Brain Res.* 85:423–431.

SCHAEFER, K. P., 1970. Unit analysis and electrical stimulation in the optic tectum of rabbits and cats. *Brain Behav. Evol.* 3:222–240.

SCHNEIDER, G. E., 1969. Two visual systems: Brain mechanisms for localization and discrimination are dissociated by tectal and cortical lesions. *Science* 163:895–902.

SIMINOFF, R., O. SCHWASSMANN, and L. KRUGER, 1966. An electrophysiological study of the visual projection to the superior colliculus of the rat. *J. Comp. Neurol.* 127:435–444.

SPARKS, D. L., 1986. Translation of sensory signals into commands for control of saccadic eye movements: Role of primate superior colliculus. *Physiol. Rev.* 66:116–177.

SPRAGUE, J. M., and T. H. MEIKLE, JR., 1965. The role of the superior colliculus in visually guided behavior. *Exp. Neurol.* 11:115–146.

STEIN, B. E., 1978. Nonequivalent visual, auditory and somatic corticotectal influences in cat. *J. Neurophysiol.* 41:55–64.

STEIN, B. E., 1984a. Development of the superior colliculus. *Annu. Rev. Neurosci.* 7:95–125.

STEIN, B. E., 1984b. Multimodal representation in the superior colliculus and optic tectum. In *Comparative Neurology of the Optic Tectum*, H. Vanegas, ed. New York: Plenum, pp. 819–841.

STEIN, B. E., and M. O. ARIGBEDE, 1972. Unimodal and multimodal response properties of neurons in the cat's superior colliculus. *Exp. Neurol.* 36:179–196.

STEIN, B. E., and J. P. DIXON, 1979. Properties of superior colliculus neurons in the golden hamster. *J. Comp. Neurol.* 183:269–284.

STEIN, B. E., and S. J. GOLDBERG, and H. P. CLAMANN, 1976. The control of eye movements by the superior colliculus in the alert cat. *Brain Res.* 118:469–474.

STEIN, B. E., W. S. HUNEYCUTT, and M. A. MEREDITH, 1988.

Neurons and behavior: The same rules of multisensory integration apply. *Brain Res.* 448:335–358.

Stein, B. E., B. Magalhaes-Castro, and L. Kruger, 1976. Relationship between visual and tactile representation in cat superior colliculus. *J. Neurophysiol.* 39:401–419.

Stein, B. E., and M. A. Meredith, 1990. Multisensory integration: Neural and behavioral solutions for dealing with stimuli from different sensory modalities. *Ann. N.Y. Acad. Sci.* 608:51–70.

Stein, B. E., and M. A. Meredith, 1991. Functional organization of the superior colliculus. In: *The Neural Basis of Visual Function*, A. G. Leventhal, ed. Hampshire, U.K.: Macmillan, pp. 85–110.

Stein, B. E., and M. A. Meredith, 1993. *The Merging of the Senses*. Cambridge, Mass.: MIT Press.

Stein, B. E., M. A. Meredith, W. S. Huneycutt, and L. McDade, 1989. Behavioral indices of multisensory integration: Orientation to visual cues is affected by auditory stimuli. *J. Cogn. Neurosci.* 1:12–24.

Stein, B. E., M. A. Meredith, and M. T. Wallace, 1993. Nonvisual responses of visually responsive neurons. *Prog. Brain Res.* 95:79–90

Wade, N. J., and R. H. Day, 1968. Apparent head position as a basis for a visual aftereffect of prolonged head tilt. *Percept. Psychophys.* 3:324–326.

Wallace, M. T., M. A. Meredith, and B. E. Stein, 1992. Integration of multiple sensory modalities in cat cortex. *Exp. Brain Res.* 91:484–488.

Wallace, M. T., M. A. Meredith, and B. E. Stein, 1993. Converging influences from visual, auditory and somatosensory cortices onto output neurons of the superior colliculus. *J. Neurophysiol.* 69:1797–1809.

Wallace, M. T., and B. E. Stein, 1994. Cross-modal synthesis in the midbrain depends on input from cortex. *J. Neurophysiol.* 71:429–432.

Welch, R. B., and D. H. Warren, 1986. Intersensory interactions. In *Handbook of Perception and Human Performance: I. Sensory Processes and Perception*, K. R. Boff, L. Kaufman, and J. P. Thomas, eds. New York: Wiley, pp. 1–36.

Wilkinson, L. K., M. A. Meredith, and B. E. Stein, 1992. Cortical deactivation disrupts multisensory integration. *Soc. Neurosci. Abstr.* 18:1031.

44 Arousal Systems and Attention

TREVOR W. ROBBINS AND BARRY J. EVERITT

ABSTRACT Unitary concepts of arousal have outlived their usefulness, and their psychological fractionation corresponds to a similar chemical differentiation of the reticular formation. Neurobiological characteristics of the monoaminergic and cholinergic systems are described in terms of their anatomical, electrophysiological, and neurochemical properties. Functional studies suggest that the ceruleocortical noradrenergic system, under certain circumstances, is implicated in processes of selective attention, that the mesolimbic and mesostriatal dopaminergic systems contribute to different forms of behavioral activation, and that the cortical cholinergic projections have fundamental roles in the cortical processing of signals, affecting attentional and mnemonic processes. It is impossible to ascribe a unitary function to the ascending serotoninergic systems, which contribute to behavioral inhibition and appear to oppose the functions of the other systems at several levels. Implications of the interactions among these chemically defined arousal systems occurring in stressful or arousing conditions are considered.

Evolving concepts of arousal

The concept of arousal as a general state of central nervous system activity has appeared in several guises and in varied experimental contexts in cognitive and behavioral neuroscience. Classical electrophysiological studies related the concept to the different stages of sleep and wakefulness (Lindsley, 1958; Moruzzi and Magoun, 1949); Hebb (1955) developed the notion of a state that optimized the processing of sensory stimuli in the cerebral neocortex, a form of attentional function; and other theorists have noted the relationship of arousal to other constructs energizing performance, such as drive and motivation (Hull,1949; Yerkes and Dodson, 1908). It is also apparent that any theory of conscious awareness must take into account the alerting consequences of fluctuations in arousal level.

The arousal construct is subject to enormous embarrassment from a number of empirical sources. Various indices of arousal do not intercorrelate to a high degree, as would be expected of a unitary construct (Eysenck, 1982), and putative manipulations of arousal, whether pharmacological or psychological, do not interact in a manner suggestive of an underlying unidimensional continuum (see Robbins and Everitt, 1987). Furthermore, in the context of human performance, there are complicated interactions between arousal and the nature of the task requirements, which have led to alternative formulations based, for example, on the concept of cognitive effort (Kahneman, 1973). This is a particularly important idea and one challenging for theories of underlying neural mechanisms, because it suggests that arousal is not simply a passive process but can be regulated, to some extent, by environmental and task demands.

These doubts in the utility of arousal-like constructs have been paralleled by developments in neuroscience. Fluctuations in arousal level formerly were believed to be mediated by a diffuse network of neurons in the brain stem called the *ascending reticular activating system* (or *reticular formation*). The forebrain projections of this system, in conjunction with the nonspecific thalamic inputs to the neocortex, provided the neural substrate for Hebb's hypothesized modulation of stimulus processing as a function of arousal level. Nonspecific aspects of stimulus input, such as intensity and salience, would interact with the central manifestation of internal states (e.g. of autonomic activity) to affect the detailed analysis of the input in terms of its various attributes at the cortical level. Naturally, such analysis could also provide feedback to the reticular core to modulate arousal level as a function of salience. However, in the decades following Hebb's hypothesis, it has become apparent that there is much more specificity in the reticular formation, in terms of the chemical identification of discrete neural systems as well as the distribution of their projections to the forebrain, than was initially realized. In this chapter, we consider the possibility that this differentiation corresponds to the fractionation of the arousal construct enforced by empiri-

TREVOR W. ROBBINS and BARRY J. EVERITT Departments of Experimental Psychology and Anatomy, University of Cambridge, Cambridge, England cal studies and debated formally by theorists such as Broadbent (1971).

Chemical differentiation of the reticular core

Technical advances in visualizing the nervous system led to demonstrations that several components of the reticular formation were, in fact, sites of monoaminergic or cholinergic cell bodies (see figure 44.1). Much of the early anatomical work was done on the rat, and the chemical pathways are now precisely delineated. The main principles of organization found in the rat apply to the primate brain (see schematics of monoamine innervations in Heimer, 1983), but there have been some important evolutionary developments that will be discussed later. The first main principle to note is that the various systems arise at several levels of the neural axis, notably from the brain stem. For example, the pedunculopontine nucleus of the brain stem contains cholinergic neurons that project diffusely to a variety of structures, including notably the prefrontal cortex, thalamus, and substantia nigra (Satoh and Fibiger, 1986; Garcia-Rill, 1991). In fact, this primi-

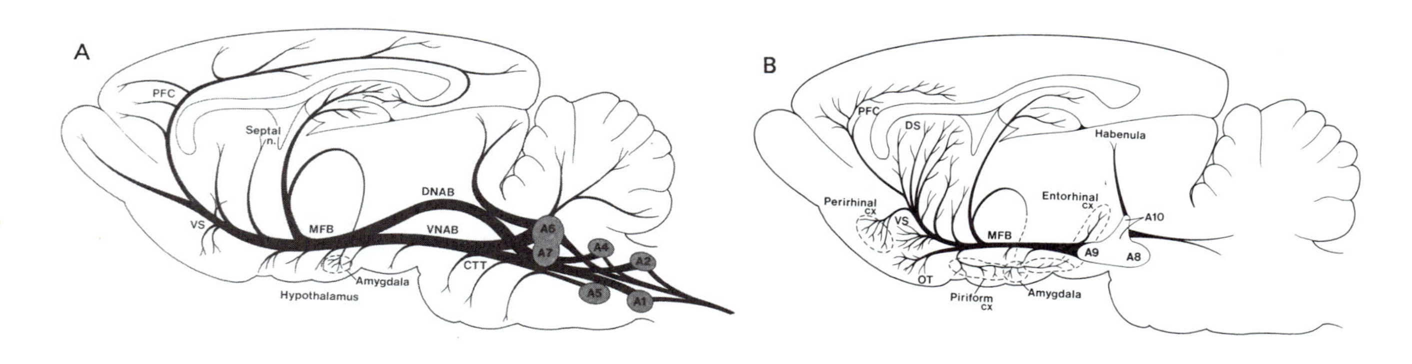

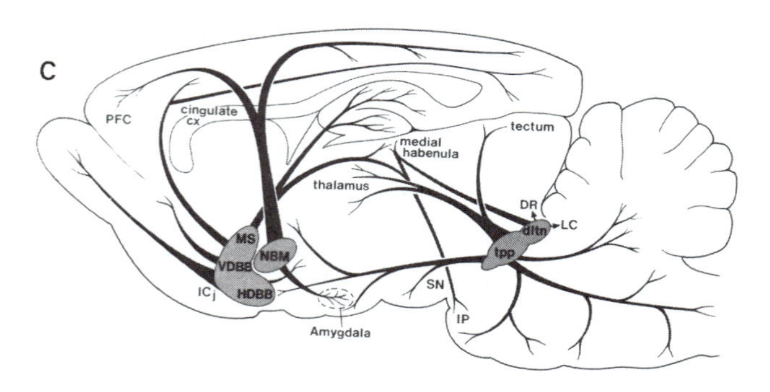

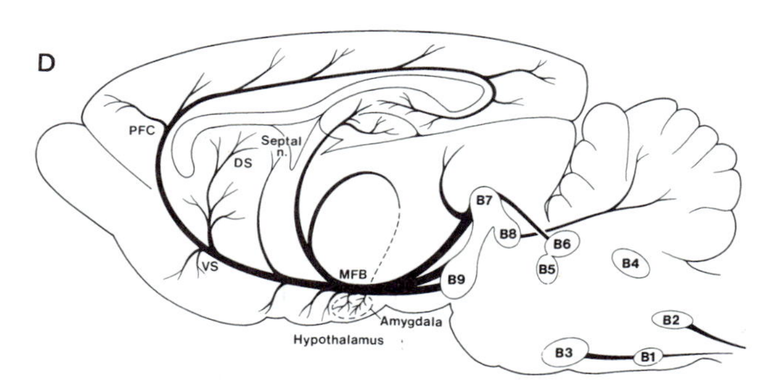

FIGURE 44.1 (Top left) Origin and distribution of the central noradenergic pathways in the rat brain. Note noradrenergic cell groups A1–A7, including the locus ceruleus (A6). DNAB, dorsal noradrenergic ascending bundle; VNAB, ventral noradrenergic ascending bundle. (Top right) Origin and distribution of the central dopamine pathways. Note dopaminergic cell groups A8–A10. (Bottom left) Origin and distribution of the central cholinergic pathways. Note rostral cell groups, NBM, nucleus basalis magnocellularis (Meynert in primates); MS, medial septum; VDBB, vertical limb nucleus of the diagonal band of Broca; HDBB, horizontal limb nucleus. (Bottom right) Origin and distribution of the central serotoninergic pathways. Note cell groups in the raphe nucleus, B4–B9. MFB, medial forebrain bundle; CTT, central tegmental tract; PFC, prefrontal cortex; VS, ventral striatum; DS, dorsal striatum; OT, olfactory tubercle; cx, cortex; ICj, islands of Calleja; SN, substantia nigra; IP, interpeduncular nucleus; dltn, dorsolateral tegmental nucleus; tpp, tegmental pedunculopontine nucleus; DR, dorsal raphe; LC, locus ceruleus.

tive projection is the main caudal portion of an ascending cholinergic system, which also includes the important rostral cell groups of the basal forebrain that innervate the entire neocortical mantle (Wainer and Mesulam, 1990). The locus ceruleus in the pons is now known to consist entirely of noradrenergic (NA) cell bodies, which often send projections that ramify to different parts of the telencephalon, contrasting with the largely diencephalic destination of NA neurons originating in the medulla oblongata (Lindvall and Björklund, 1983). The raphe nuclei of the midbrain are the major source of serotonin (5-hydroxytryptamine [5-HT]) neurons, which innervate, in a somewhat complementary manner, the same range of forebrain destinations as the ceruleocortical NA system. There are two major divisions of the 5-HT neurons, those originating in median and dorsal portions of the raphe nuclei, which may also be distinguished in terms of their terminal domains and their associated receptor subtypes (Jacobs and Azmitia, 1992). Also situated in the midbrain are the dopamine (DA) cell bodies that provide important inputs to the dorsal and ventral striatum and anterior neocortex via, respectively, the mesolimbic, mesostriatal, and mesocortical DA projections (Lindvall and Björklund, 1983). It is noteworthy that this complex, the substantia nigra (pars compacta), is not present in primitive vertebrates and is believed to have evolved from a pedunculopontine nucleus that formerly contained both cholinergic and dopaminergic cells (see Schultz, 1989).

This chapter will illustrate the functional implications of these neurochemical systems for novel conceptualizations of arousal-like processes, but it should be pointed out that these systems probably account for only a fraction of the nonspecific modulation of cortical function. For example, an ascending histaminergic projection originating in the hypothalamus has been identified (Schwartz et al., 1991). Space limitations prevent us from elaborating on the role of nonspecific thalamic afferents to the neocortex, which provide excitatory glutamatergic inputs to the cortex from the nonspecific thalamic nuclei to the superficial layers of neocortex. The nucleus reticularis of the thalamus receives inputs from both the caudal and rostral ascending cholinergic projections and, with other thalamic nuclei such as the pulvinar, has been suggested to have attentional functions—for example, in solving the so-called binding problem (see Crick and Koch, 1990; see also chapter 41).

Principles of organization: Specificity in nonspecificity

Although the anatomical organization of the monoaminergic systems suggests diffuse and nonspecific functions, there is also considerable evidence for specificity. For example, some NA cells in the locus ceruleus send axons that branch to innervate distinct forebrain locations, which means that activity in the cell will simultaneously affect activity in, say, both the hippocampus and the neocortex. However, other cells may favor different sets of locations, and there is considerable topographical organization within the locus ceruleus itself (Foote, Bloom, and Aston-Jones, 1983), which makes it imperative to know how activity in the cells of this nucleus is normally generated. As widespread as the projections of the locus ceruleus are to the forebrain, its innervation of the basal ganglia relative, for example, to that of the 5-HT or DA systems is remarkably sparse. Another example of a restricted projection field is that of the mesocortical DA system, which mainly innervates the frontal cortex in the rat, although there is evidence of greater posterior cortical innervation in the primate brain (Berger, Gaspar, and Verney, 1991).

There is no viable explanation for this heterogeneity of innervation; whether it arises because of different requirements for certain forms of modulation at a neuronal level or whether it is a product of interactions at the level of neural systems or psychological process is not known. Presumably, the very existence of different, chemically defined elements of the reticular core implies that these elements have different functions, even when projecting to common destinations. The factors affecting the innervation of the neocortex by these chemically defined systems include the cytoarchitectonic characteristics of the cortical area, the laminae and functions of the cortical region, the age of the animal, and the species under study. Thus, postcentral sulcus somatosensory regions of the neocortex have much higher tissue concentrations of norepinephrine than other cortical regions in monkeys, especially when compared with occipital cortex. In contrast, 5-HT concentrations are more uniform, with greater variation in the 5-HT metabolite 5-HIAA (increasing levels in more posterior regions) (Levitt, Rakic, and Goldman-Rakic, 1984). Moreover, Old World monkeys appear to receive a more highly developed 5-HT innervation of the visual cortex than New World primates.

A further dimension of importance may prove to be laterality; there is evidence for asymmetrical distributions of neurotransmitter to the two halves of the brain in the case of the mesostriatal DA projections (Glick and Ross, 1981) and the thalamic NA projection (Oke et al., 1978; Oke, Lewis, and Adams, 1980). The latter asymmetry is particularly noteworthy in terms of posterior cortical attentional mechanisms (see chapter 39), with the left pulvinar containing more NA than the right but the opposite being true for the right ventral posterior thalamic nuclei, which carries somatosensory afferents to the cortex.

The diffuse innervation of different cortical laminae by the monoaminergic and cholinergic neurons is also to be contrasted with the highly specific input of the sensory thalamus to layer 4. Nevertheless, there is evidence in monkeys that the NA innervation of layers 5 and 6 in the visual cortex is more dense than that of 5-HT, which projects more strongly to layer 4 (Morrison, Foote, and Bloom, 1984). These authors have also emphasized that the noradrenergic innervation is considerably less dense in the squirrel monkey geniculostriate compared with the pulvinar extrastriate thalamocortical systems, consistent with a role in orienting and visual selective attention rather than sensory aspects of visual processing per se. In contrast, Ishikawa and Tanaka (1977) reported dense innervation of the dorsolateral geniculate nucleus and only a sparse innervation of the pulvinar by NA terminals in the rhesus monkey.

Another form of complementarity between different monoaminergic innervations of the neocortex concerns the different inputs of the DA-ergic and NA-ergic projections to the anterior cingulate system. Whereas the mesocortical DA projection in the rat is mainly distributed in layers II and III and the more superficial part of layer I, the NA-ergic input is to the deeper part of layer I and to layers V and VI (Lindvall and Björklund, 1983). The functional significance of this division is far from clear, although it is again obvious that the NA-ergic innervation can affect cortical outflow via the pyramidal cells of the deep layers. Clearly, both amine systems can influence processing in the anterior attentional system that is associated with executive function (see chapter 39).

Such detailed information may help in the formulation of how these different monoaminergic projections make distinct contributions to information processing within a single cortical column in a given neocortical region. However, it is evident that mere anatomical characterization of these systems only begins to set an agenda for considering their functions. As we shall see, electrophysiological and neurochemical measures have also provided considerable clues to functions manifested in behavioral or cognitive terms.

Neurobiological clues to function

Electrophysiogical and Neurochemical Indications of Presynaptic Activity According to Environmental Circumstance Studies of single-unit recordings in the anesthetized or behaving animal provide important information about the conditions under which a chemically defined arousal system becomes active. Normally, it is informative to measure spontaneous activity of a system in relation to stages of the sleep-waking cycle and the phasic effects, superimposed on this tonic background activity, that are produced by the presentation of environmental stimuli or stressors with impact on the autonomic and endocrine systems. Compared to the precise sensory trigger feature requirements of neurons in primary sensory cortex, it is evident that each of the major systems to be described has far less stringent eliciting factors. For example, NA cells of the rat, cat, or monkey locus ceruleus fire monotonically in relation to stages of sleep and waking, with the highest rates occurring during wakefulness (Foote, Bloom, and Aston-Jones, 1983; Jacobs, 1987). The effects of fluctuations in locus ceruleus activity can apparently be discerned in the cortical electroencephalogram (EEG) (Foote, Bloom, and Aston-Jones, 1983) and in P3 event-related potentials (Pineda, Foote, and Neville, 1989). In addition, locus ceruleus cells respond phasically to novel or noxious stimuli, including conditioned stimuli, but may also be mildly responsive to appetitive events (Aston-Jones and Bloom, 1981; Jacobs, 1987). However, these cells are unresponsive to orientational properties of stimuli, while responding to their overall physical intensity (Watabe, Nakai, and Kasamatsu, 1982). A particularly important observation is the waning responsiveness or habituation of firing in response to repeated presentations of an initially novel stimulus (Aston-Jones and Bloom, 1981: Jacobs, 1987).

These electrophysiological observations generally are endorsed by studies of locus ceruleus function that depend on the ex vivo (Anisman et al., 1987) or in vivo (Abercrombie, Keller, and Zigmond, 1988) monitoring of presynaptic neurochemical changes in indices of NA turnover in various conditions, including exposure to stressors. Although we generally assume that there is a good correlation between firing rate in the cell body and release and utilization of neurotransmitter at the terminal, it should, of course, be realized that this is not always the case. There is some evidence for local regulation of release of both dopamine (Grace, 1991) and norepinephrine (Marrocco et al., 1987) in their terminal domains. For example, Marrocco's group (1987) showed that norepinephrine release in the V2 region of cats in response to a visual stimulus was jointly dependent on which eye was stimulated, and therefore on the gating input from the lateral geniculate nucleus.

Overall, the properties of the locus ceruleus neurons are consistent with a postulated arousal function of the ceruleocortical projection. These properties, however, contrast somewhat with results found for the other monoaminergic and cholinergic systems. For example, Jacobs (1987) finds subtly different relationships between dorsal raphe 5-HT cell firing and the sleep-waking stage in the cat but, more obviously, a lack of response to chronic stressors and an absence of habituation to external sensory events. In contrast, neurochemical measures indicate that the activity of central 5-HT neurons can be sensitive to stressors, and possible discrepancy in electrophysiological and neurochemical measures requires some resolution, perhaps depending on which cell groups (e.g. dorsal versus median raphe) are implicated.

In further contrast, spontaneous activity in the mesencephalic DA neurons of cats is apparently not tied to the sleep-waking stage and hence shows no obvious circadian variation (Jacobs, 1987). In monkeys, Schultz (1992) has defined with considerable care the contingencies that normally activate the mesencephalic DA cells. These include novel stimuli but also primary reinforcers such as food and conditioned stimuli that predict their presentation. However, such conditioned stimuli have much stronger effects if they consist, for example, of light onset rather than light offset. He emphasizes the overall importance of stimulus salience in the activation of these DA neurons. Neurochemical measurements, conducted mainly in rats, show a rather different picture of effective stimuli or states for the induction of changes in midbrain DA cell activity. It is apparent that rewarding or incentive cues can increase the DA signal in the ventral striatum, using the evolving technique of in vivo voltammetry (Phillips, Pfaus, and Blaha, 1991), but other in vivo evidence suggests that DA turnover is increased by stressors, perhaps predominantly in the mesocortical projection but, in some circumstances, also in the mesostriatal and mesolimbic projections (Blanc et al., 1980; Abercrombie et al., 1989; Deutch and Roth, 1990). It seems likely that the consistent discrepancy between electrophysiological and neurochemical indices in analyzing the properties of these systems may reflect the different time bases over which these observations are collected—milliseconds in the case of single-unit activity and seconds to several minutes in the case of neurochemical measures.

The ascending cholinergic neurons have long been implicated in cortical arousal (Steriade and Buzsaki, 1990). In terms of single-unit properties, the basal forebrain cholinergic cells have been less intensively studied than the monoamine neurons, but there are suggestions that such cells are particularly influenced by conditioned visual stimuli and by reinforcers, though also by aversive air puffs (Richardson and DeLong, 1990; Wilson and Rolls, 1990), perhaps processed initially by the limbic afferents, which form the most prominent source of inputs for these neurons (Mesulam and Mufson, 1984).

Activity of the Systems at the Postsynaptic Receptor and Beyond The impact of activity in the monoaminergic and cholinergic systems is, of course, to be found in their terminal domains, at different widespread sites in the forebrain. One important concept that has arisen from the microiontophoretic administration of the monoamines or acetylcholine has been that these neurotransmitters can act to alter the balance of activity between a sensorily evoked response and the background level of neuronal firing. For example, application of norepinephrine to each of the main sensory regions of the neocortex leads to a general reduction in spontaneous firing rate which, in turn, produces a more favorable signal-to-noise (S/N) ratio for the evocation of a neuronal response to a sensory stimulus such as a vocalization, a visually oriented line, or a touch (Foote, Friedman, and Oliver, 1975; Waterhouse and Woodward, 1980; Kasamatsu and Heggelund, 1982). In certain instances, norepinephrine appears to act conditionally; that is, its effect depends on the ongoing activity in the target cell (Segal and Bloom, 1976). Such conditional effects are especially apparent when the activity of cells of the hippocampal dentate gyrus is recorded in response to auditory tones in the behaving rat under different conditions. When the tone is predictive of food, the predominant response is excitatory and is increased by coincident ceruleal stimulation; when the tone is uncorrelated with food and the evoked response is inhibitory, ceruleal stimulation increases the magnitude of the inhibitory response.

Analysis of comparable responses for other systems has perhaps been less systematic. However, there is considerable evidence that iontophoretic application of acetylcholine to the cat visual cortex increases the cells' responsivity to stimulus orientation directly via its excitatory action (Sillito, 1987). Such evidence is consistent with other views that acetylcholine serves to boost S/N ratios at the cortical level (Drachman and Sahakian, 1979). Thus, cortical norepinephrine and acetylcholine both seem to promote increased S/N ratios but via different mechanisms, and perhaps in different contexts. In other studies, iontophoretic application of 5-HT has been shown to blunt evoked responses (Waterhouse, Moises, and Woodward, 1986), and this reduction of S/N ratio may be consistent with the complementary roles of the NA and 5-HT systems. Finally, Rolls and colleagues (1984) have adduced some evidence for alterations in S/N ratios following microiontophoretic application of dopamine to the monkey striatum. Nonetheless, as is the case for the other systems, interpreting the implications of these observations may depend on unraveling the complex sequence of excitatory and inhibitory effects that this system exerts on striatal function.

Another important dimension of the action of these chemically defined arousal systems is their actions beyond the receptor at the level of second messenger systems. The monoaminergic and cholinergic projections activate a variety of receptors, some of which are associated with ion-gated channels but others of which depend on activating biochemical cascades associated with adenylyl cyclase or inositol phosphate (see Bloom, 1988). Clearly, such interactions may underlie the changes in neuronal plasticity that occur consequent to changes in monoaminergic or cholinergic activity. In this way, activity in these pathways may exert long term effects that outlast their immediate (though relatively slow) time courses of action (Rauschecker, 1991).

Behavioral and cognitive functions of the central ascending arousal systems

METHODOLOGICAL APPROACHES Although the neurobiological data provide an essential backdrop for understanding the functions of the monoaminergic and cholinergic systems, this information eventually must be integrated with the behavioral and cognitive effects of manipulation of these neurotransmitter systems using selective drugs. We will review several examples of this in studies employing human volunteers as well as experimental animals.

The major problem for human psychopharmacology is that administration of a drug is necessarily systemic, which means that it is almost impossible to define or limit its initial site of action. Thus, even though the sophisticated experimental paradigms of cognitive psychology can be employed, possible effects of interest are likely to be masked or confounded by other actions of the drug. In contrast, in experimental animals, not only is it possible to define the site of action of a compound with considerable confidence, but powerful neurotoxins can also be used to produce long-lasting and profound depletions in specific neurotransmitter projections. Furthermore, in animals, it is likely to prove increasingly feasible to monitor the activity of different systems, perhaps simultaneously, using in vivo methods in the freely moving animal exposed to defined behavioral situations. Whereas this in vivo approach is theoretically possible for testing cognitive functions in human subjects within the physically constrained environment of the positron emission tomography (PET) scanner, the restricted availability of suitable ligands means that progress using this approach may be slow, although the method of evaluating effects of systemically administered drugs on cerebral blood flow has recently yielded data of interest (Grasby et al., 1992). Specifically, distinct regional patterns of attenuation and augmentation of memory-induced increases in regional cerebral blood flow were found after administration of apomorphine, a dopamine agonist, or buspirone, a partial 5-HT1A agonist. Sites of interaction common to both drugs included an augmentation effect in the posterior cingulate cortex and an attenuation effect in the right prefrontal cortex. However, both drugs also produced specific attenuations—in the left dorsolateral prefrontal cortex for apomorphine and in the retrosplenial area for buspirone—that correlated with the maximal impairment of memory function of each drug. Thus, PET provides a method for correlating the effects of monoaminergic drugs on cognitive function with regional cerebral activity modulated by the effects of such drugs, though questions of causality still remain.

Consequently, the main techniques employed involve the use of specific neurotoxins to destroy monoaminergic neurotransmitter systems, generally at source but occasionally in specific terminal regions. In the case of the cholinergic system, no truly specific neurotoxin exists, but it has proved possible to produce reasonably specific loss of such neurons in the basal forebrain by the serendipitous use of certain excitotoxins. These techniques all generally cause a chronic depletion of neurotransmitter that lasts virtually permanently but are subject to complications posed by recovery of function resulting, for example, from receptor up-regulation and by all the standard problems of interpretation posed by lesion studies. In slightly different contexts, both Ungerstedt (1971) and Abercombie, Keller, and Zigmond (1988) have both emphasized the importance of producing almost total depletion of the neurotransmitter system under investigation. The lesion approach can be counterpointed by the use of acute intracerebral infusions of drugs, although there are many potential problems that must be surmounted with controls for anatomical as well as pharmacological specificity.

The behavioral and cognitive approach demands careful attention. One of the major problems in the field is lack of systematic comparison of effects across similar test situations, which is as true for the behavioral methods as for the neurobiological investigations described earlier. Consequently, our own approach has used common tests across a variety of neuropharmacological interventions so that the separate roles of each neurotransmitter sytem can be assessed. A second problem has been that the range of test situations employed in the same study has typically been low. A priori, it appears likely that the effects of manipulating, say, the widely ramifying ceruleocortical NA system will be exerted on several distinct forms of processing mediated by different neural regions. The challenge then is to identify what is the common contribution of the NA projections to these various processes.

Functions of the Ascending NA Systems The ceruleocortical NA projection appears to fulfill many of the properties of the cortical arousal system envisaged by Hebb (1955). Clearly, according to the neurobiological evidence, activity in this projection should enhance processing in several terminal regions, notably the neocortex and hippocampus, perhaps by effects on attentional mechanisms. But how well does the behavioral evidence support this hypothesis?

Profound depletion of cortical norepinephrine to less than 10% of control values can be effected in the rat brain by infusions of the neurotoxin 6-hydroxydopamine (6-OHDA) into the trajectory of the ceruleocortical pathway (also called the *dorsal noradrenergic bundle*, or DNAB). This depletion occurs in the absence of any discernible effect on other neurotransmitter systems or gross nonspecific damage at the site of the infusion (Mason and Iversen, 1979). There is some loss of hypothalamic norepinephrine as some of the lateral tegmental NA projections run in the dorsal, rather than the ventral, NA bundle (VNAB). Possible contributions of this damage can be assessed (and generally excluded) by more caudal brainstem infusions of the neurotoxin aimed specifically at this projection.

Gross behavioral deficits are not apparent; rats with DNAB lesions appear to eat and drink normally and show no obvious alterations in spontaneous locomotor activity. They are also unimpaired in the acquisition of certain simple associative tasks such as conditioned taste aversion (Dunn and Everitt, 1987). However, we have found consistent deficits in the acquisition of conditional discriminations, in which the rat has to learn a rule to guide response choice (of the form "if stimulus *x*, do *y*; if stimulus *a*, do *b*"), which are not apparent if the rat is pretrained on the task (Everitt et al., 1983). This apparent selectivity for the acquisition phase of the task is confirmed in aversively motivated tasks, in which pavlovian conditioned suppression is impaired in rats with ceruleocortical NA depletion in certain circumstances but is unaffected when the conditioning occurs prior to surgery (Cole and Robbins, 1987a). These results appear to indicate that cortical norepinephrine is more important for learning than performance, a conclusion consistent with early views of Kety (1970) and Crow (1968). However, it is far from clear that the effects are specific to the associative aspects of learning; recent experiments have revealed that rats with DNAB lesions are able to learn certain associations in aversive situations. For example, Selden, Robbins, and Everitt (1990) confirmed that a DNAB lesion impairs aversive stimulus (CS) conditioning but found also that conditioning to the context was enhanced relative to control values. This result suggests that DNAB lesions do not produce their effects on conditioning by modulating anxiety, nor do they generally impair aversive learning. Rather, it appears that the lesion broadens the attentional span of the animal with the result that distal cues are preferentially utilized over proximal cues, even when such cues are less predictive of shock. Thus, conditioning to explicit stimuli is impaired relative to the predominantly distal contextual cues in the conditioning apparatus. Although these effects on aversive conditioning are paralleled by appropriate changes in plasma corticosterone, it is evident that such neuroendocrine changes are secondary to the effects on conditioning. This contrasts with the effects of hypothalamic NA depletion, produced by VNAB lesions, which do not affect aversive conditioning but can affect the plasma corticosterone response to the unconditioned stimulus (Selden, Robbins, and Everitt, 1993).

The shift in balance toward contextual conditioning produced by DNAB lesions is found in several other situations. For example, ceruleocortical norepinephrine depletion appears to increase the environmental suppression over eating in rats in an open-field situation, yet such rats readily eat novel food and so appear to be less discriminating of such explicit stimuli (Cole, Robbins, and Everitt, 1988). Furthermore, the lesion actually enhances spatial learning in the Morris water maze under certain circumstances but, again, impairs acquisition of simultaneous visual discrimination in the same apparatus (Selden et al., 1990). The Morris water maze generally is solved by the rat on the basis of distal, spatial cues external to the maze as it navigates toward a hidden platform, and so the enhanced acquisition resulting from NA depletion can be interpreted as an effect parallel to the enhanced contextual aversive conditioning described previously. In contrast, simultaneous visual discrimination is conducted among visual stimuli present in the water tank and so more immediately proximal to the animal as it swims toward them.

An additional important feature of the water-maze experiments was that they were conducted in cold (approximately 12°C) water; when the more usual warmer temperature (26°C) was used, the deficits were not present. The cold water is clearly a more stressful or arousing environment, as confirmed by the striking effects on core body temperature and plasma corticosterone values (which were, however, equivalent between the lesioned and control groups). Thus, it appears that the effects of DNAB lesions are especially apparent in stressful circumstances, and this has led to the hypothesis that activity in the ceruleocortical projection normally functions to preserve attentional selectivity under high levels of arousal (Everitt, Robbins, and Selden, 1990). The hypothesis is close to one originally propounded by Amaral and Sinnamon (1977) but clearly needs to be tested further using a variety of other ways to manipulate central NA function. For example, it is possible that recent observations of facilitated shifting to attend to visual cues introduced into a spatial-maze situation, after treatment with the $alpha_2$ receptor antagonist idazoxan (Devauges and Sara, 1990), is consistent with this view, as this drug has been shown to increase central norepinephrine turnover through its action at presynaptic receptors.

One important additional principle to have emerged from these studies is a confirmation of the view that DNAB lesions will lead to a wide range of cognitive deficits because of the depletion of different areas with discrete functions. Thus, aversive CS conditioning is also impaired by 6-OHDA lesions of the amygdala itself (Selden et al., 1991), acquisition of the Morris swim maze is known to depend on the integrity of the hippocampus (Morris et al., 1982), and acquisition of the conditional discrimination is retarded by 6-OHDA lesions of the medial prefrontal cortex (Ryan, Everitt, and Robbins, unpublished data).

It should be made clear that the effects of NA manipulations on S/N ratio and apparent attentional mechanisms may also affect the consolidation of memory traces, perhaps via the longer-term changes in postsynaptic function described earlier. Thus, two lines of evidence support the notion that NA-dependent mechanisms of the amygdala contribute to the consolidation of aversive memory traces: Post-trial treatment with intra-amygdaloid NA at certain doses facilitates retention of passive avoidance (Liang, McGaugh, and Yao, 1990), and posttrial intra-amygdaloid propanolol, a postsynaptic beta blocker, produces a retrograde amnesia on the same type of task (Gallagher et al., 1977).

Evidence for Attentional Dysfunction Following Ceruleocortical NA Depletion An adaptation of a simple continuous performance test of attention for human subjects has proved useful in the investigation of attentional impairment in the rat. Using a specially designed apparatus, the rat is trained to detect brief visual stimuli presented unpredictably at one of five locations. Five seconds after the rat pushes a panel at the rear of the apparatus, the next stimulus occurs, and 100 such presentations constitute a single session. Rats with 6-OHDA-induced DNAB lesions are unimpaired in the normal version of the task or when the stimuli are systematically degraded by dimming them (Carli et al., 1983). However, these lesioned animals are impaired in three well-defined sets of circumstances: first, when brief bursts of loud white noise are interpolated just prior to the presentation of each stimulus (Carli et al., 1983) (figure 44.2a); second, when the stimuli are presented unpredictably in time (Cole and Robbins, 1992); and third, when the rats receive treatment with *d*-amphetamine, injected either peripherally or directly into the nucleus accumbens, where it causes a large increase in locomotor activity and also impulsive and premature responding in the setting of the five-choice task (Cole and Robbins, 1987b). These observations are consistent with the notion that the ceruleocortical NA depletion affects controlled rather than automatic responding, perhaps via effects on attentional function.

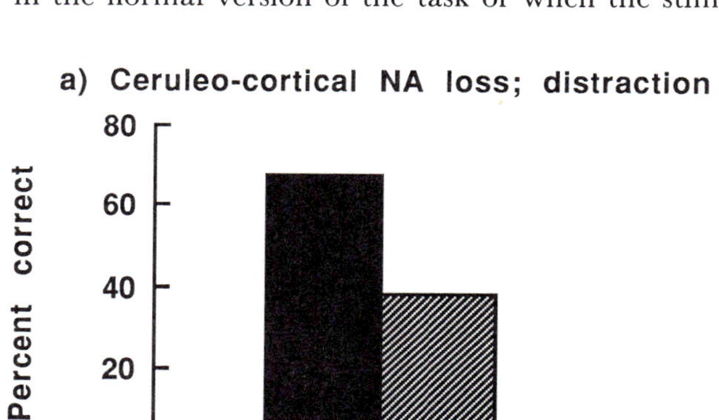

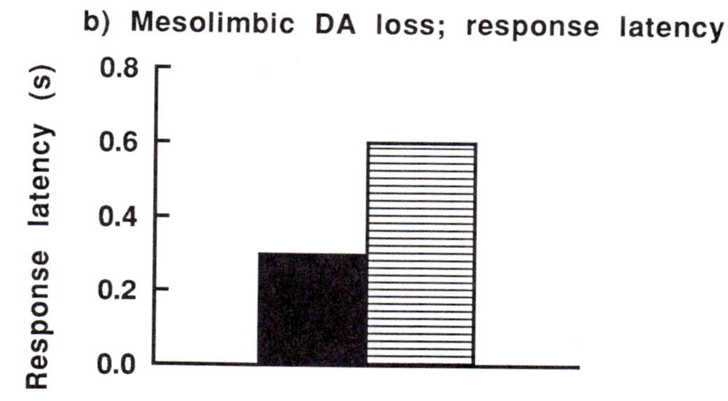

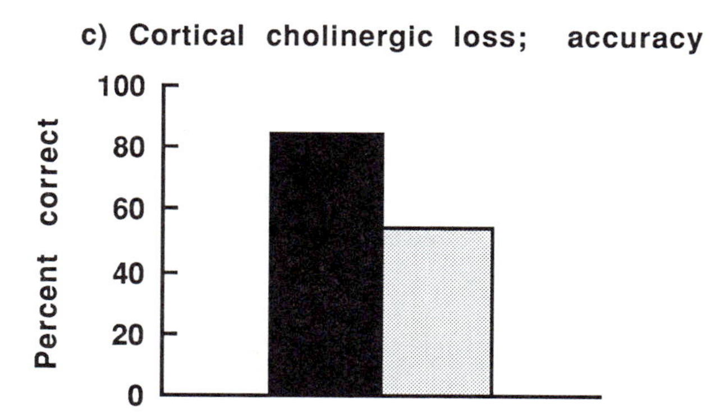

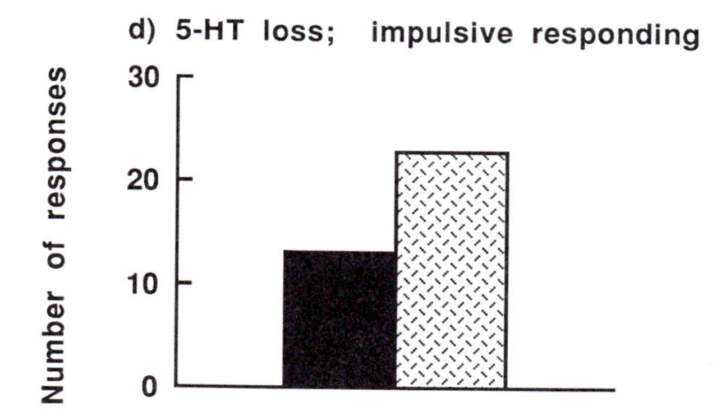

Figure 44.2 Summary diagram illustrating contrasting effects on the five-choice task of selective damage to the noradrenergic (NA), dopaminergic (DA), cholinergic, and serotoninergic (5-HT) systems. The diagram highlights optimal conditions for exposing deficits in each condition. (a) Ceruleocortical NA system; no deficit on baseline but accuracy reduced after distraction with white noise. (b) Mesolimbic DA depletion; primarily, baseline speed and overall probability of responding are affected. (c) Cortical cholinergic system; baseline accuracy is impaired. (d) Serotoninergic depletion; no effects on accuracy, but impulsive responding is increased. See text for further details and references. The control groups are shown indicated by black histograms.

Notably, under conditions when the rat is aroused (by distracting white noise, temporal unpredictability, or increased behavioral output), impairments in discrimination performance become apparent. These arousing conditions correspond in part to what Posner (1978) terms *alerting* (see also chapter 39).

These observations are consistent with two other experiments using animal subjects and possibly with some in human volunteers. Roberts, Price, and Fibiger (1976) found that similar DNAB lesions in rats greatly increase the rats' distractibility in a maze setting. A rather different approach has also found recently some evidence that aged rhesus monkeys treated with clonidine or another alpha$_2$ agonist are prevented from attentional distraction during the delay period of a delayed-response procedure (Arnsten and Contant, 1992). The usual interpretation of this result is that alpha$_2$ agonists are acting predominantly postsynaptically in the prefrontal cortex of the aged animal and thus are boosting diminished NA function.

Effects of Manipulating Central NA Function in Humans Treatment of normal human volunteers with intravenous clonidine normally impairs performance, an effect that is attributed to the drug's sedative properties (i.e., the presynaptic action of the compound in down-regulating central NA function). Thus, clonidine impairs learning of difficult paired verbal associates (Frith et al., 1985) and reduces *d'* indices of attentional performance in a continuous performance task in which the subject has to detect target sequences of numbers (Coull et al., 1992). Moreover, the drug degrades the efficiency of performance on a test of executive function, the Tower of London task (Coull et al., 1992). Of particular interest are the effects of intravenous clonidine on the covert orienting task that is a key paradigm for characterizing the orienting components of visuospatial attention and is sensitive to parietal lobe damage (Posner et al., 1984). Clark, Geffen, and Geffen (1989) have shown that intravenous clonidine apparently reduces the cost of switching attention in the invalid cuing condition in normal human volunteers. There are several difficulties in relating these results to studies with animals, not the least of which is that it is difficult, even in monkeys, to obtain reliable effects of validity cuing in the covert orienting paradigm. However, preliminary results from a study of a single rhesus monkey have shown that clonidine does not modify the validity effect but rather reduces the alerting effect of the warning cue to reduce reaction times (Witte, Gordon-Lickey, and Marrocco, 1992). It is possible that the discrepancies arise from the use of peripherally (Witte, Gordon-Lickey, and Marrocco, 1992) and centrally (Clark, Geffen, and Geffen, 1989) cued versions of the task, which might engage different attentional systems.

In summary, there are some parallels in the effects of manipulations of the cortical noradrenergic system in humans and experimental animals over a range of experimental situations, but important issues remain to be resolved about the specificity of the role of central noradrenergic mechanisms in attentional function.

Functions of the Ascending DA Systems Studies of the functional consequences of manipulations that affect central DA function make it clear that this system is involved in radically different forms of arousal from the ceruleocortical NA projection. The mesolimbic DA system is implicated in incentive motivational processes by which evaluative processing is translated into action, based on evidence from several sources: For example, the reinforcing effects of psychomotor stimulant drugs such as amphetamine and cocaine are mediated in part via this system (see Koob, 1992, for a review); mesolimbic dopamine depletion in the ventral striatum effected by 6-OHDA reduces locomotor activity in novel environments or in the presence of food (Koob et al., 1978); neurochemical measures of dopamine function are consistent with the role of the system in controlling approach behavior to incentive stimuli (Phillips, Pfaus, and Blaha, 1991); and responding for stimuli that predict reward is enhanced by treatments boosting DA function in the ventral striatum (Robbins et al., 1989). Thus, it appears that one of the main functions of this system is to achieve the activation of behavior in response to cues that signal the availability of incentives or reinforcers. Consistent with this analysis is the finding that accuracy of detecting visual target stimuli in the five-choice attentional task is not much affected by mesolimbic DA depletion or low doses of *d*-amphetamine; these treatments do, however, affect the latency (see figure 44.2b) and overall probability of responding—for example, by affecting the level of impulsive or premature responses. The disruptive effects of both *d*-amphetamine and white noise on premature responding are blocked by mesolimbic DA depletion (Cole and Robbins, 1989). Thus, it has been suggested that any apparent effects on attentional function produced by elevating DA activity in the nucleus accumbens are secondary to the behaviorally activating effects of the drug.

The role of the mesostriatal DA system in arousal processes is perhaps best understood in terms of the activation or energization of behavior. This applies as much to well-established responses, such as eating and drinking, as to recently learned behavior (see the review by Dunnett and Robbins, 1992). It applies particularly to the performance of skilled responses in reaction time tasks. For example, rats with mesostriatal DA depletion are retarded in their initiation of pretrained head-turn responses in contralateral space cued by stimuli presented unpredictably to either side of the head (Carli, Evenden, and Robbins, 1985). This retardation of responding is unlikely to arise simply from a lack of arousal or alerting, (Posner, 1978; see also chapter 39), as the lesioned rats benefit to the same extent as controls in the facilitation of reaction time produced by adding loud white noise to the visual cue (data of Carli and Robbins; see Robbins and Brown, 1990). These effects are consistent with some of the symptoms of Parkinson's disease, which is associated with striatal DA loss and has been suggested to depend on a loss of internal cues in control of behavior (Brown and Marsden, 1988). A striking example of this loss of internal activation is provided by a recent study by Brown and Robbins (1991) in which rats were trained in a simple (i.e., precued) and choice reaction time procedure requiring a response to a visual stimulus presented on both sides of space, after a variable foreperiod. The dopamine depletion did not alter the beneficial effect of precuing the response. In fact, as seen in figure 44.3, the main effect of striatal DA depletion was to retard reaction time as a function of the foreperiod delay, the normal speeding of responding being abolished by striatal DA depletion (see figure 44.3). This experiment thus suggests a role for striatal DA in processes of response preparation, such as motor readiness. The findings are consistent with a role for this system in the functioning of a corticostriatal motor loop that outputs to the supplementary motor area.

Although striatal dopamine may have a role in motor response activation, it is clear that the DA system also innervates structures such as the caudate nucleus that have cognitive functions, including the organization of sequences of behavior (e.g., in planning and in switching response sets). L-Dopa withdrawal in patients with Parkinson's disease has been shown to exacerbate such

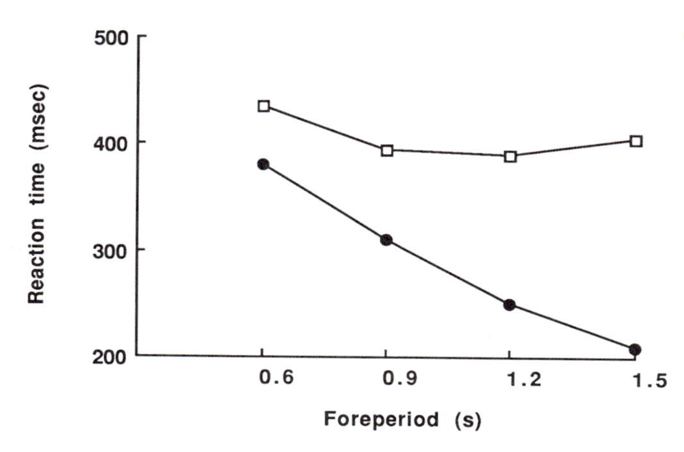

FIGURE 44.3 Effect of striatal dopamine depletion on reaction time as a function of foreperiod. Note the progressive speeding of response with increasing foreperiod and its abolition by striatal dopamine depletion. Data are shown for simple reaction time condition only, in which the rat received advance information about which response was correct after the foreperiod; the same effect was found in a choice reaction time condition. Filled symbols, control; open symbols, striatal dopamine depletion. (Data redrawn from Brown and Robbins, 1991.)

deficits as lengthened thinking latencies in a planning task and the production of efficient sequences of responding in a spatial working memory task (Lange et al., 1992). It is possible that such effects are mediated by the mesostriatal DA system at the level of the caudate nucleus. However, possible roles of the mesocortical DA projection to prefrontal cortex and anterior cingulate gyrus and the modulation of an anterior executive network (see chapter 39) certainly cannot be ruled out, particularly in view of the disruptive effects of apomorphine in the prefrontal cortex in a functional activation study employing PET (Grasby et al., 1992; see above).

FUNCTIONS OF THE ASCENDING 5-HT SYSTEMS Analysis of the raphe 5-HT systems has not reached the same stage as that for the complementary ceruleocortical NA pathways, due in part, until recently, to the lack of availability of suitably specific compounds with which to manipulate this system and also to the fact that manipulations of 5-HT in animals have been found to alter basic functions such as appetite, sleep, locomotor activity, and responses to pain (see Jacobs and Azmitia, 1992), which inevitably complicates the assessment of cognitive performance. The clues to function from neurobiological evidence are also less helpful than in the case of central norepinephrine. Nevertheless, there is evidence that, in certain circumstances, manipulations of 5-HT affect processes concerned with behavioral inhibition (cf. Gray, 1982), particularly in aversive contexts, so that massive depletion of central 5-HT using the neurotoxin 5,7-dihydroxytryptamine (5,7-DHT), infused intraventricularly, greatly elevates rates of punished responding in the rat (see Soubrié, 1986). Similarly, behavioral disinhibition can extend to appetitive situations when, for example, the rat is required to inhibit responding to provide food reinforcement (Wogar, Bradshaw, and Szabadi, 1992). There is limited evidence that such effects of behavioral disinhibition are due to alterations in target structures of the dorsal raphe nuclei (see Soubrié, 1986, and commentary), which include the striatum and amygdala; at a neurochemical level, this might be mediated by interactions with mesostriatal and mesolimbic DA systems.

In terms of effects on learning and memory, a plausible case has recently been advanced that 5-HT depletion or down-regulation can occasionally improve performance (McEntee and Crook, 1991), as in the case of both aversive memory and spatial appetitive learning, whereas treatments boosting 5-HT function can impair performance. This generalization extends to recent human experimentation, buspirone, the 5-HT1A receptor agonist, impairs supra–list learning and interacts with the task in producing changes in cerebral blood flow in posterior regions of the cortex (Grasby et al., 1992; see above). We have some evidence for the hypothesis that 5-HT depletion can improve performance, using the same conditional discrimination that is sensitive to depletions of central norepinephrine (Everitt et al., 1983) or striatal DA (Robbins et al, 1990) or acetylcholine from the cingulate cortex (Marston, Everitt, and Robbins, 1994). Acquisition and performance of this task are facilitated significantly by intracerebroventricular administration of the neurotoxin 5,7-DHT, effects that are not so readily ascribed to behavioral disinhibition (Ward et al., 1991). Perhaps the facilitation arises from removal of possible de-arousing effects of reduced S/N ratios produced by cortical 5-HT activity (Waterhouse, Moises, and Woodward, 1986). In contrast, we have observed no significant improvements on the five-choice attentional task, even under the various test conditions that reveal deficits following central norepinephrine or acetylcholine loss. In fact, the only significant effect of central 5-HT loss is a significant increase in premature or impulsive responding (Harrison et al., 1992) (see figure 44.2d).

Functions of the Ascending Cholinergic Projections Much interest has focused on the functions of the central cholinergic system, based largely on the possible importance of its distribution for various forms of dementia, especially of the Alzheimer type (DAT) (Perry et al., 1978) and the obvious amnesic effects of the anticholinergic drug scopolamine (see Sahakian, 1987). However, it is also clear that the central cholinergic system has important roles in attentional function, which is apparent, in part, from the effects, in patients with DAT, of cholinergic drugs such as the anticholinesterases physostigmine, or THA, as well as nicotine, on tests requiring continuous attention (Sahakian et al., 1989, 1993), and also from the unusual propensity of the anticholinesterases to alleviate attentional neglect in single-case studies of such patients (Muramoto, Sugishita, and Ando, 1984; Sahakian, Joyce, and Lishman, 1987).

In the animal literature, there has been abundant evidence of impaired visual or auditory signal detection performance following peripheral doses of scopolamine, caused selectively by changes in discriminative sensitivity (Warburton, 1977). However, attempts to study the role of specific systems by destroying cholinergic neurons in the basal forebrain have concentrated on more conventional tests of learning and memory. Unfortunately, however, much of the evidence is invalid with respect to interpretations of central cholinergic function because of lack of caution in interpreting the effects of excitotoxic lesions (see Dunnett, Everitt, and Robbins, 1991).

Using the excitotoxic amino acids AMPA or quisqualic acid, which produce large reductions in cortical cholinergic markers in the frontal cortex, several studies have now shown weaker effects on certain tests of learning and memory than after use of other excitotoxins, which produce considerably less effective cortical cholinergic lesions (e.g., see the review in Dunnett, Everitt, and Robbins, 1991). This makes more impressive the discovery of profound deficits on the five-choice task that we had already used to study the effects of central monoaminergic depletion (see Carli et al., 1983; Cole and Robbins, 1987b). Rats with approximately 70% reductions in cortical choline acetyltransferase activity are impaired even in the baseline version of detecting the location of randomly presented light flashes (Robbins et al., 1989) (see figure 44.2c). The deficit is not a trivial consequence of motivational, sensory, or motor loss but is exacerbated when the duration of the visual events is shortened still further from 500 ms. It is mimicked by several other manipulations that reduce cholinergic function, including intracerebrovascular hemicholinium or intrabasalis infusions of muscimol, a GABA-ergic agonist that reduces cortical cholinergic function. Moreover, all three methods of reducing cortical cholinergic function are reversed by systemic administration of the anticholinesterase physostigmine, and so they are almost certainly primary cholinergic deficits (Muir et al., 1992; Muir, Robbins, and Everitt, 1992). It is of further interest that patients with probable DAT show impairments also in a parallel task, which are likewise reversed by an anticholinesterase, THA (Sahakian et al., 1993).

Although apparently attentional in nature, the impairment in continuous performance raises several interesting problems of interpretation. First, recent unpublished data (Muir, Robbins, and Everitt) show that it is a frontal rather than a parietal deficit and clearly has some affinities with the delayed response deficit in primates with lesions of the prefrontal cortex, as the visual stimulus is not present at the time of the response. Second, it is clear that damage to other ascending components of the cortical cholinergic projection—for example, damage to the projection from the vertical limb of the diagonal band of Broca to primarily cingulate cortex—does not produce the same effect but leads instead to impairments on other tasks related to the normal functioning of this area of the cortex, including the acquisition of response rules in a conditional discrimination (Marston et al., 1994). One possibility is that the cholinergic projection aids the processing of stimuli at a cortical level by enhancing the impact of salient information, perhaps via a mechanism leading to increases in S/N ratio, as described previously.

Interactions among the arousal systems

From the evidence surveyed thus far, it is apparent that each of the neurotransmitter systems we have defined has rather different, sometimes context-dependent, functions. The ceruleocortical NA system seems to have a protective function of maintaining discriminability in stressful or arousing circumstances; the mesolimbic and mesostriatal DA systems play a role in the activation of output, whether cognitive or motor in nature; the cholinergic systems appear to enhance stimulus processing at the cortical level; and the 5-HT systems may serve to dampen the actions of each of the others—for example, by promoting behavioral inhibition and cortical de-arousal. These conclusions have been made possible by experiments that have compared the functions of these projections using the same experimental paradigms. The implication is that in many situations, these reticular systems are simultaneously active in varying degrees, to optimize processing capacity and facilitate response output. Accurately simulating the distinct influences of these nonspecific systems clearly will provide a challenge for future attempts to model cognitive function with neural networks.

To some extent, of course, the differential functions of these arousal systems depend on their terminal domains, but an important issue for future studies will be to define the separable functions within a common cortical region, such as the prefrontal cortex; current results suggest that this will be feasible. For example, manipulations of prefrontal cortical acetylcholine and dopamine have contrasting effects on attentional set-shifting and reversal-learning performance in monkeys (Roberts et al., 1992, 1994). Moreover, drugs affecting prefrontal NA and DA function appear to affect delayed response performance in different ways (Arnsten and Contant, 1992; Goldman-Rakic, 1992).

Regulation of the different systems also requires careful consideration. For example, it appears that there are interactions between these projections at several levels. Some of the most striking are the opposed influences of the 5-HT systems on ceruleocortical NA function (see Jacobs and Azmitia, 1992), on mesolimbic DA function (see Soubrié, 1986), and probably on the central cholinergic systems (e.g., in the entorhinal cortex) (Barnes et al., 1989). However, largely on anatomical or pharmacological grounds, other important functional interactions between the cortical DA and NA systems (Tassin et al., 1991) and between basal forebrain cholinergic and NA projections (Zaborszky, Cullinan, and Braun, 1991) or medial septum cholinergic and 5-HT projections (Nilsson et al., 1988) have also been proposed. The true significance of these will become apparent in future investigations.

Two other aspects of this regulation of the central arousal systems are critically important. The first is the nature of the afferents that normally provide input to the cell groups at the origin of the ascending arousal systems. It would be expected that systems able to alter their activity on the basis of the salience of sensory input, dependent on such instructed factors as novelty and conditioning, would be sensitive to descending influences of the forebrain. In this context, it has been pointed out that the prefrontal cortex is perhaps the only cortical structure that sends feedback connections to the monoaminergic and cholinergic cell groups of the brain stem and midbrain; thus, this may be another route by which the cortex can regulate its own input (Goldman-Rakic, 1987). However, the very restricted nature of the afferents to the locus ceruleus has also been carefully documented (n. prepositus hypoglossi and n. paragigantocellularis) (Aston-Jones et al., 1986), and this poses somewhat of a paradox for functional interpretation. Finally, activity of a chemically defined arousal system can, in some circumstances, be self-regulatory; for example, there is evidence that activity in the mesocortical DA projection sometimes is associated with down-regulation of the subcortical DA systems (Blanc et al., 1980; Pycock, Carter, and Kerwin, 1980). In this case, activation of the mesocortical DA projection potentially allows an adjustment of the level of activation of the striatal output systems in the production of response output. These concepts, based largely on neurobiological data and behavioral experiments with animals, may provide important clues to understanding both normal and pathological aspects of cognitive function in human.

ACKNOWLEDGMENTS We acknowledge the financial support of the Wellcome Trust and Medical Research Council (U.K.), as well as the contributions of our colleagues.

REFERENCES

ABERCROMBIE, E. B., K. A. KEEFE, D. S. DIFRISCHIA, and M. J. ZIGMOND, 1989. Differential effects of stress on in vivo dopamine release in striatum, nucleus accumbens, and medial frontal cortex. *J. Neurochem.* 52:1655–1658.

ABERCROMBIE, E. B., R. W. KELLER, and M. J. ZIGMOND, 1988. Characterization of hippocampal norepinephrine release as measured by microdialysis perfusion: Pharmacological and behavioral studies. *Neuroscience* 27:897–904.

AMARAL, D. G., and H. M. SINNAMON, 1977. The locus coeruleus: Neurobiology of a central noradrenergic nucleus. *Prog. Neurobiol.* 9:147–196.

ANISMAN, H., J. IRWIN, W. BOWERS, P. AHLUWALIA, and R. M. ZACHARKO, 1987. Variations of norepinephrine concentrations following chronic stressor application. *Pharmacol. Biochem. Behav.* 26:653–659.

ARNSTEN, A. F. T., and T. A. CONTANT, 1992. Alpha-2 adrenergic agonists decrease distraction in aged monkeys performing the delayed response task. *Psychopharmacology* 108:159–169.

ASTON-JONES, G., and F. E. BLOOM, 1981. Norepinephrine-containing locus coeruleus neurons in behaving rats exhibit pronounced responses to non-noxious environmental stimuli. *J. Neurosci.* 1:897–900.

ASTON-JONES, G., M. ENNIS, V. A. PETTIBONE, W. T. NICKELL, and M. T. SHIPLEY, 1986. The brain locus coeruleus: Restricted afferent control of a broad efferent network. *Science* 234:734–737.

BARNES, J. M., N. M. BARNES, B. COSTALL, R. J. NAYLOR, and M. B. TYERS, 1989. 5-HT receptors mediate inhibition of acetyl choline release in cortical tissue. *Nature* 338: 762–763.

BERGER, B., P. GASPAR, and C. VERNEY, 1991. Dopaminergic innervation of the cerebral cortex: Unexplained differences between rodents and primates. *Trends Neurosci.* 14:21–27.

BLANC, G., D. HERVE, H. SIMON, A. LISOPRAWSKI, J. GLOWINSKI, and J. P. TASSIN, 1980. Response to stress of mesocortical frontal neurons in rats after long-term isolation. *Nature* 284:265–267.

BLOOM, F. E., 1988. Neurotransmitters: Past, present and future directions. *FASEB J.* 2:32–41.

BROADBENT, D. E., 1971. *Decision and Stress.* London: Academic Press.

BROWN, R. G., and C. D. MARSDEN, 1988. Internal versus external control cues and the control of attention in Parkinson's disease. *Brain* 111:323–345.

BROWN, V. J., and T. W. ROBBINS, 1991. Simple and choice reaction time performance following unilateral striatal dopamine depletion in the rat. *Brain* 114:513–525.

CARLI, M., J. L. EVENDEN, and T. W. ROBBINS, 1985. Depletion of unilateral striatal dopamine impairs initiation of contralateral actions and not sensory attention. *Nature* 313:679–682.

CARLI, M., T. W. ROBBINS, J. L. EVENDEN, and B. J. EVERITT, 1983. Effects of lesions to ascending noradrenergic neurones on performance of a 5-choice serial reaction time task; implications for theories of dorsal noradrenergic bundle function based on selective attention and arousal. *Behav. Brain Res.* 9:361–380.

CLARK, C. R., G. M. GEFFEN, and L. B. GEFFEN, 1989. Catecholamines and the covert orienting of attention. *Neuropsychologia* 27:131–140.

COLE, B. J., and T. W. ROBBINS, 1987a. Dissociable effects of lesions to the dorsal or ventral noradrenergic bundle on the acquisition, performance and extinction of aversive conditioning. *Behav. Neurosci.* 101:476–488.

Cole, B. J., and T. W. Robbins, 1987b. Amphetamine impairs the discriminative performance of rats with dorsal noradrenergic bundle lesions on a 5-choice serial reaction time task: New evidence for central dopaminergic-noradrenergic interactions. *Psychopharmacology* 91:458–466.

Cole, B. J., and T. W. Robbins, 1989. Effects of 6-hydroxydopamine lesions of the nucleus accumbens septi on performance of a 5-choice serial reaction time task in rats: Implications for theories of selective attention and arousal. *Behav. Brain Res.* 33:165–179.

Cole, B. J., and T. W. Robbins, 1992. Forebrain norepinephrine: Role in controlled information processing in the rat. *Neuropsychopharmacology* 7:129–141.

Cole, B. J., T. W. Robbins, and B. J. Everitt, 1988. Lesions of the dorsal noradrenergic bundle simultaneously enhance and reduce responsivity to novelty in a food-preference test. *Brain Res. Rev.* 13:325–349.

Coull, J. T., H. C. Middleton, B. J. Sahakian, and T. W. Robbins, 1992. Noradrenaline in the frontal cortex—attentional and central executive function (abstract). *J. Psychopharm.* 95:A24.

Crick, F., and C. Koch, 1990. Towards a neurobiological theory of consciousness. *Semin. Neurosci.* 2:263–275.

Crow, T. J., 1968. Cortical synapses and reinforcement: A hypothesis. *Nature* 219:736–737.

Deutch, A., and R. H. Roth, 1990. The determinants of stress-induced activation of the prefrontal cortical dopamine system. *Prog. Brain Res.* 85:367–403.

Devauges, V., and S. J. Sara, 1990. Activation of the noradrenergic system facilitates an attentional shift in the rat. *Behav. Brain Res.* 39:19–28.

Drachman, D. R., and B. J. Sahakian, 1979. The effects of cholinergic agents on human learning and memory. In *Nutrition in the Brain*, A. Barbeau, J. H. Growdon, and R. J. Wurtman, eds. New York: Raven, pp. 351–366.

Dunn, L. T., and B. J. Everitt, 1987. The effects of lesions to the noradrenergic projections from the locus coeruleus and lateral tegmental cell groups on conditioned taste aversion in the rat. *Behav. Neurosci.* 101:409–422.

Dunnett, S. B., B. J. Everitt, and T. W. Robbins, 1991. The basal forebrain cortical cholinergic system: Interpreting the effects of excitotoxic lesions. *Trends Neurosci.* 14: 494–500.

Dunnett, S. B., and T. W. Robbins, 1992. The functional role of the mesotelencephalic dopamine systems. *Biol. Rev.* 67:491–518.

Everitt, B. J., T. W. Robbins, M. Gaskin, and P. J. Fray, 1983. The effects of lesions to ascending noradrenergic neurons on discrimination learning and performance in the rat. *Neuroscience* 10:397–410.

Everitt, B. J., T. W. Robbins, and N. R. W. Selden, 1990. Functions of the locus coeruleus noradrenergic system: A neurobiological and behavioural synthesis. In *Pharmacology of Noradrenaline*, D. J. Heal and C. A. Marsden, eds. Oxford: Oxford University Press, pp. 349–378.

Eysenck, M. W., 1982. *Attention and Arousal.* Berlin: Springer-Verlag.

Foote, S. L., F. E. Bloom, and G. Aston-Jones, 1983. Nucleus locus coeruleus: New evidence of anatomical and physiological specificity. *Physiol. Rev.* 63:844–914.

Foote, S. L., R. Friedman, and A. P. Oliver, 1975. Effects of putative neurotransmitters on neuronal activity in monkey cerebral cortex. *Brain Res.* 86:229–242.

Frith, C. D., J. Dowdy, N. Ferrier, and T. J. Crow, 1985. Selective impairment of paired associate learning after administration of a centrally acting adrenergic agonist (clonidine). *Psychopharmacology* 87:490–493.

Gallagher, M., B. S. Kapp, R. E. Musty, and P. A. Driscoll, 1977. Memory formation: Evidence for a specific neurochemical system in the amygdala. *Science* 198:423–425.

Garcia-Rill, E., 1991. The pedunculopontine nucleus. *Prog. Neurobiol.* 36:363–389.

Glick, S. D., and J. F. Ross, 1981. Right-sided population bias and lateralization of activity in normal rats. *Brain Res.* 205:222–225.

Goldman-Rakic, P. S., 1987. Circuitry of primate prefrontal cortex and regulation of behavior by representational memory. In *Handbook of Physiology: Sec. 1. The Nervous System*: Vol. 5. *Higher Functions of the Brain* (part 1), F. Plum, ed. Bethesda, Md.: American Physiological Society, pp. 373–417.

Goldman-Rakic, P. S., 1992. Dopamine-mediated mechanisms of the prefrontal cortex. *Semin. Neurosci.* 4:149–159.

Grace, A. A., 1991. Phasic versus tonic dopamine release and the modulation of dopamine system responsivity: A hypothesis for the aetiology of schizophrenia. *Neuroscience* 41:1–24.

Grasby, P. M., K. J. Friston, C. J. Bench, Ç. D. Frith, E. Paulesu, P. J. Cowen, P. F. Liddle, R. S. J. Frackowiak, and R. Dolan, 1992. The effect of apomorphine and buspirone on regional cerebral blood flow during the performance of a cognitive task measuring neuromodulatory effects of psychotropic drugs in man. *Eur. J. Neurosci.* 4:1203–1212.

Gray, J., 1982. *The Neuropsychology of Anxiety.* Oxford: Clarendon Press.

Harrison, A. A., J. L. Muir, T. W. Robbins, and B. J. Everitt, 1992. The effects of forebrain 5-HT depletion on visual attentional performance in the rat (abstract). *J. Psychopharm.* 236:A59.

Hebb, D. O., 1955. Drives and the CNS (conceptual nervous system). *Psychol. Rev.* 62:243–254.

Heimer, L., 1983. *The Human Brain and Spinal Cord.* New York: Springer-Verlag.

Hull, C., 1949. Stimulus intensity dynamism (V) and stimulus generalization. *Psychol. Rev.* 56:67–76.

Ishikawa, M., and C. Tanaka, 1977. Morphological organization of catecholamine terminals in the diencephalon of the rhesus monkey. *Brain Res.* 119:45–55.

Jacobs, B. L., 1987. Brain monoaminergic activity in behaving animals. *Prog. Psychobiol. Physiol. Psychol.* 12:171–206.

Jacobs, B. L., and E. C. Azmitia, 1992. Structure and function of the brain serotonin system. *Physiol. Rev.* 72:165–229.

Kahneman, D., 1973. *Attention and Effort*. Englewood Cliffs, N.J.: Prentice Hall.

Kasamatsu, T., and P. Heggelund, 1982. Single cell responses in cat visual cortex to visual stimulation during iontophoresis of noradrenaline. *Exp. Brain Res.* 45:317–324.

Kety, S. S., 1970. The biogenic amines in the central nervous system: Their possible role in arousal, emotion and learning. In *The Neurosciences: Second Study Program*, F. O. Schmidt, ed. New York: Rockefeller University Press, pp. 324–336.

Koob, G. F., 1992. Dopamine, addiction and reward. *Semin. Neurosci.* 4:139–148.

Koob, G. F., S. J. Riley, S. C. Smith, and T. W. Robbins, 1978. Effects of 6-hydroxydopamine lesions of the nucleus accumbens septi and olfactory tubercle on feeding, locomotor activity and amphetamine anorexia in the rat. *J. Comp. Physiol. Psychol.* 92:917–927.

Lange, K. W., T. W. Robbins, C. D. Marsden, M. James, A. M. Owen, and G. M. Paul, 1992. L-Dopa withdrawal in Parkinson's disease selectively impairs cognitive performance in tests sensitive to frontal lobe dysfunction. *Psychopharmacology* 107:394–404.

Levitt, P., P. Rakic, and P. S. Goldman-Rakic, 1984. Comparative assessment of monoamine afferents in mammalian cerebral cortex. In *Monoamine Innervation of the Cerebral Cortex*, L. Descarrieres, T. R. Reader, and H. H. Jasper, eds. New York: Liss, pp. 41–59.

Liang, K. C., J. L. McGaugh, and H.-Y. Yao, 1990. Involvement of amygdala pathways in the influence of post-training amygdala norepinephrine and peripheral epinephrine on memory storage. *Brain Res.* 508:225–233.

Lindsley, D. B., 1958. The reticular system and perceptual discrimination. In *Reticular Formation of the Brain*, H. H. Jasper, ed. Boston: Little, Brown, pp. 513–534.

Lindvall, O., and A. Björklund, 1983. Dopamine- and norepinephrine-containing neuron systems: Their anatomy in the rat brain. In *Chemical Neuroanatomy*, P. C. Emson, ed. New York: Raven, pp. 229–255.

Marrocco, R. T., R. F. Lane, J. W. McClurkin, C. D. Blaha, and M. F. Askine, 1987. Release of cortical catecholamines by visual stimulation requires activity in thalamocortical afferents of monkey and cat. *J. Neurosci.* 7: 2756–2767.

Marston, H. M., H. L. West, L. S. Wilkinson, B. J. Everitt, and T. W. Robbins, 1994. Effects of excitotoxic lesions of the septum and vertical limb nucleus of the diagonal band on Broca on conditional visual discrimination: Relationship between performance and choline acetyltransferase activity in the cingulate cortex. *J. Neurosci.*, in press.

Mason, S. T., and S. D. Iversen, 1979. Theories of the dorsal bundle extinction effect. *Brain Res.* 1:107–137.

McEntee, W., and T. Crook, 1991. Serotonin, memory and aging. *Psychopharmacology* 103:143–149.

Mesulam, M., and E. J. Mufson, 1984. Neural inputs into the nucleus basalis of the substantia innominata (Ch4) in the rhesus monkey. *Brain* 107:253–274.

Morris, R. G. M., P. Garrud, J. N. P. Rawlins, and J. O'Keefe, 1982. Place navigation impaired in rats with hippocampal lesions. *Nature* 297:681–683.

Morrison, J. H., S. L. Foote, and F. E. Bloom, 1984. Regional, laminar, developmental and functional characteristics of noradrenaline and serotonin innervation patterns in monkey cortex. In *Monoamine Innervation of Cerebral Cortex*, L. Descarries, T. R. Reader, and H. H. Jasper, eds. New York: Liss, pp. 61–75.

Moruzzi, G., and H. W. Magoun, 1949. Brain stem reticular formation and activation of the EEG. *Electroencephalogr. Clin. Neurophysiol.* 1:455–473.

Muir, J. L., S. B. Dunnett, T. W. Robbins, and B. J. Everitt, 1992. Attentional functions of the forebrain cholinergic system: Effects of intraventricular hemicholinium, physostigmine, basal forebrain lesions and intra-cortical grafts. *Exp. Brain Res.* 89:611–622.

Muir, J. L., T. W. Robbins, and B. J. Everitt, 1992. Disruptive effects of muscimol infused into the basal forebrain: Differential interaction with cholinergic mechanisms. *Psychopharmacology* 107:541–550.

Muramoto, O., M. Sugashita, and K. Ando, 1984. Cholinergic system and constructional praxis. *J. Neurol. Neurosurg. Psychiatr.* 47:485–491.

Nilsson, O. G., R. E. Strecker, A. Daszuta, and A. Bjorklund, 1988. Combined serotoninergic and cholinergic denervation of the forebrain produces severe deficits in a spatial learning task in the rat. *Brain Res.* 453:235–246.

Oke, A., R. Keller, I. Mefford, and R. N. Adams, 1978. Lateralization of norepinephrine in human thalamus. *Science* 200:1411–1413.

Oke, A., R. Lewis, and R. N. Adams, 1980. Hemispheric asymmetries of norepinephrine distribution in rat thalamus. *Brain Res.* 188:269–272.

Perry, E. K., B. E. Tomlinson, G. Blessed, K. Bergmann, P. H. Gibson, and R. H. Perry, 1978. Correlations of cholinergic abnormalities with senile plaques and mental test scores in senile dementia. *Br. Med. J.* 2:1457–1459.

Phillips, A. G., J. G. Pfaus, and C. D. Blaha, 1991. Dopamine and motivated behavior. In *The Mesolimbic Dopamine System: From Motivation to Action*, P. Willner and J. Scheel-Kruger, eds. Chichester, Engl.: Wiley, pp. 199–224.

Pineda, J. A., S. L. Foote, and H. J. Neville, 1989. Effects of locus coeruleus lesions on auditory, long-latency, event-related potentials in monkey. *J. Neurosci.* 9:81–93.

Posner, M., 1978. *Chronometric Explorations of Mind*. Hillsdale, N.J.: Erlbaum.

Posner, M., J. A. Walker, F. J. Friedrich, and R. D. Rafal, 1984. Effects of parietal injury on the covert orienting of visual attention. *J. Neurosci.* 4:1863–1874.

Pycock, C. J., C. J. Carter, and R. W. Kerwin, 1980. Effect of 6-hydroxydopamine lesions of the medial prefrontal cortex on neurotransmitter systems in subcortical sites in the rat. *J. Neurochem.* 34:91–99.

Rauschecker, J. P., 1991. Mechanisms of visual plasticity: Hebb synapses, NMDA receptors and beyond. *Physiol. Rev.* 71:587–615.

Richardson, R. T., and M. R. DeLong, 1990. Responses of primate nucleus basalis neurons to water rewards and related stimuli. In *Brain Cholinergic Systems*, M. Steriade and D. Biesold, eds. Oxford: Oxford University Press, pp. 282–293.

Robbins, T. W., and V. J. Brown, 1990. The role of the striatum in the mental chronometry of action: A theoretical review. *Rev. Neurosci.* 2:181–213.

Robbins, T. W., and B. J. Everitt, 1987. Psychopharmacological studies of arousal and attention. In *Cognitive Neurochemistry*, S. M. Stahl, E. C. Goodman, and S. D. Iversen, eds. Oxford: Oxford University Press, pp. 21–36.

Robbins, T. W., B. J. Everitt, H. M. Marston, J. Wilkinson, G. H. Jones, and K. J. Page, 1989. Comparative effects of ibotenic acid and quisqualic acid induced lesions of the substantia innominata on attentional functions in the rat: Further implications for the role of the cholinergic system of the nucleus basalis in cognitive processes. *Behav. Brain Res.* 35:221–240.

Robbins, T. W., J. R. Taylor, M. Cador, and B. J. Everitt, 1989. Limbic-striatal interactions and reward-related processes. *Neurosci. Biobehav. Rev.* 13:155–162.

Roberts, A. C., M. De Salvia, L. S. Wilkinson, P. Collins, J. L. Muir, B. J. Everitt, and T. W. Robbins, 1994. Effects of prefrontal dopamine depletion on an analogue of the Wisconsin Card Sort Text in monkeys: Possible interaction with subcortical dopamine systems. *J. Neurosci.*, in press.

Roberts, A. C., T. W. Robbins, B. J. Everitt, and J. L. Muir, 1992. A specific form of cognitive rigidity following excitotoxic lesions of the basal forebrain in monkeys. *Neuroscience* 47:251–264.

Roberts, D. C. S., M. T. C. Price, and H. C. Fibiger, 1976. The dorsal tegmental noradrenergic projection: An analysis of its role in maze learning. *J. Comp. Physiol. Psychol.* 90:363–372.

Rolls, E. T., S. J. Thorpe, M. Boytim, I. Szabo, and D. I. Perrett, 1984. Responses of striatal neurons in the behaving monkey: 3. Effects of iontophoretically applied dopamine on normal responsiveness. *Neuroscience* 12:1201–1212.

Sahakian, B. J., 1987. Cholinergic drugs and their effects on human cognitive function. In *Handbook of Psychopharmacology*, vol. 20, L. L. Iversen, S. D. Iversen, and S. H. Snyder, eds. New York: Plenum, pp. 393–424.

Sahakian, B. J., G. H. Jones, R. Levy, J. Gray, and D. Warburton, 1989. The effects of nicotine on attention, information processing and short-term memory in patients with dementia of the Alzheimer type. *Br. J. Psychiatr.* 154: 797–800.

Sahakian, B. J., E. M. Joyce, and W. A. Lishman, 1987. Cholinergic effects on constructional abilities and mnemonic processes: A case report. *Psychol. Med.* 17:329–333.

Sahakian, B. J., A. M. Owen, N. J. Morant, S. A. Eagger, S. Boddington, L. Crayton, H. A. Crockford, M. Crooks, K. Hill, and R. Levy, 1993. Further analysis of the cognitive effects of tetrahydroaminoacridine (THA) in Alzheimer's disease: Assessment of attentional and mnemonic function using CANTAB. *Psychopharmacology* 110: 395–401.

Satoh, K., and H. C. Fibiger, 1986. Cholinergic neurons of the lateral tegmental nucleus: Efferent and afferent connections. *J. Comp. Neurol.* 253:277–302.

Schultz, W., 1989. Neurophysiology of basal ganglia. In *Handbook of Experimental Pharmacology*, vol. 88, D. B. Calne, ed. Heidelberg: Springer-Verlag, pp. 1–45.

Schultz, W., 1992. Activity of dopamine neurons in the behaving primate. *Semin. Neurosci.* 4:129–138.

Schwartz, J.-C., J.-M. Arrang, M. Garbarg, H. Pollard, and M. Ruat, 1991. Histaminergic transmission in the mammalian brain. *Physiol. Rev.* 71:1–51.

Segal, M., and F. E. Bloom, 1976. The action of norepinephrine in the rat hippocampus: IV. The effects of locus coeruleus stimulation on evoked hippocampal activity. *Brain Res.* 107:513–525.

Selden, N. R. W., B. J. Cole, B. J. Everitt, and T. W. Robbins, 1990. Damage to ceruleo-cortical noradrenergic projections impairs locally cued but enhances spatially cued water maze acquisition. *Behav. Brain Res.* 39:29–51.

Selden, N. R. W., B. J. Everitt, L. E. Jarrard, and T. W. Robbins, 1991. Complementary roles for the amygdala and hippocampus in aversive conditioning to explicit and contextual cues. *Neuroscience* 42:335–350.

Selden, N. R. W., T. W. Robbins, and B. J. Everitt, 1990. Enhanced behavioral conditioning to context and impaired behavioral and neuroendocrine responses to conditioned stimuli following ceruleocortical noradrenergic lesions: Support for the attentional hypothesis of central noradrenergic function. *J. Neurosci.* 10:531–539.

Selden, N. R. W., T. W. Robbins, and B. J. Everitt, 1993. Diencephalic noradrenaline depletion impairs the corticosterone response to footshock but does not affect conditioned fear. *J. Neuroendocrinol.* 4:773–779.

Sillito, A. M., 1987. Synaptic processes and neurotransmitters operating on the central visual system: A systems approach. In *Synaptic Function*, G. M. Edelman, W. E. Einar, and W. M. Cowan, eds. New York: Wiley, pp. 329–371.

Soubrié, P., 1986. Reconciling the role of the central serotonin neurons in human and animal behavior. *Behav. Brain. Sci.* 9:319–364.

Steriade, M., and G. Buzsaki, 1990. Parallel activation of thalamic and cortical neurons by brainstem and basal forebrain cholinergic neurons. In *Brain Cholinergic Systems*, M. Steriade and D. Biesold, eds. Oxford: Oxford University Press, pp. 3–62.

Tassin, J.-P., D. Herve, P. Vezina, F. Trovero, G. Blanc,

and J. GLOWINSKI, 1991. Relationships between mesocortical and mesolimbic dopamine neurons: Functional correlates of D1 receptor heteroregulation. In *The Mesolimbic Dopamine System: From Motivation to Action*, P. Willner and J. Scheel-Kruger, eds. Chichester, Engl.: Wiley, pp. 175–196.

UNGERSTEDT, U., 1971. Adipsia and aphagia after 6-hydroxydopamine-induced degeneration of the nigrostriatal dopamine system. *Acta Physiol. Scand. Suppl.* 367: 69–93.

WAINER, B. H., and M.-M. MESULAM, 1990. Ascending cholinergic pathways in the rat brain. In *Brain Cholinergic Systems*, M. Steriade and D. Biesold, eds. Oxford: Oxford University Press, pp. 65–119.

WARBURTON, D. M., 1977. Stimulus selection and behavioral inhibition. In *Handbook of Psychopharmacology*, vol. 8, L. L. Iversen, S. D. Iversen, and S. H. Snyder, eds. New York: Plenum, pp. 385–431.

WARD, B., B. J. EVERITT, T. W. ROBBINS, and L. S. WILKINSON, 1991. Facilitated acquisition of a conditional visual discrimination, following 5,7 dihydroxytryptamine-induced lesions of forebrain serotoninergic projections. *Soc. Neurosci. Abstr.* 17:148.

WATABE, K., K. NAKAI, and T. KASAMATSU, 1982. Visual afferents to norepinephrine-containing neurones in cat locus coeruleus. *Exp. Brain Res.* 48:66–80.

WATERHOUSE, B. D., H. C. MOISES, and D. J. WOODWARD, 1986. Interaction of serotonin with somatosensory cortical neuronal responses to afferent synaptic inputs and putative neurotransmitters. *Brain Res. Bull.* 17:507–518.

WATERHOUSE, B. D., and D. J. WOODWARD, 1980. Interaction of norepinephrine with ceruleo-cortical activity evoked by stimulation of somatosensory afferent pathways. *Exp. Neurol.* 67:11–34.

WILSON, F. A. W., and E. T. ROLLS, 1990. Learning and memory is reflected in the responses of reinforcment-related neurons in the primate basal forebrain. *J. Neurosci.* 10:1254–1267.

WITTE, E. A., M. E. GORDON-LICKEY, and R. T. MARROCCO, 1992. Pharmacological depletion of catecholamines modifies covert orienting in rhesus monkeys. *Soc. Neurosci. Abstr.* 226:11.

WOGAR, M. A., C. M. BRADSHAW, and E. SZABADI, 1992. Impaired acquisition of temporal differentiation performance following lesions of the ascending 5-hydroxytryptamine pathways. *Psychopharmacology* 107:373–378.

YERKES, R. M., and J. D. DODSON, 1908. The relation of strength of stimulus to the rapidity of habit formation. *J. Comp. Neurol. Psychol.* 18:459–482.

ZABORSZKY, L., W. E. CULLINAN, and A. BRAUN, 1991. Afferents to basal forebrain cholinergic projections: An update. In *The Basal Forebrain: Anatomy and Function*, T. C. Napier, P. W. Kalivas, and I. Hanin, eds. New York: Plenum.

45 Attention, Intelligence, and the Frontal Lobes

JOHN DUNCAN

ABSTRACT A general distinction between active, voluntary, or controlled and passive, stimulus-driven, or automatic behavior has been used to explain the consequences of frontal lobe lesions, normal differences in general intelligence (or Spearman's *g*), and the effects of dividing attention between (dissimilar) concurrent tasks. It is suggested that these three problems are indeed closely linked, all concerning the process of goal or abstract action selection under conditions of novelty or weak environmental prompts to behavior. One source of evidence concerns goal neglect, or disregard of a task requirement even though it has been understood. Characteristic of frontal lobe patients, such neglect also occurs in normal people from the lower part of the *g* distribution, especially under dual-task conditions. A second source of evidence concerns poor performance of frontal lobe patients on standard tests of fluid rather than crystallized intelligence. The general prediction is that, across a set of tasks or performance measures, profiles of frontal lobe impairment, *g* correlation, and dual-task decrement should agree. It remains an open question whether *g* and dual-task decrements arise as the integrated result of separable frontal functions or reflect the contribution to behavior of specific frontal areas such as dorsolateral prefrontal cortex or the anterior cingulate gyrus.

After damage to the human frontal lobes, there can be a widespread disorganization of behavior, reflected in many different types of error occurring in many different types of task. Disturbances may include disinhibition, impulsivity and distractability, rigidity and perseveration, apathy, and lack of response (Luria, 1966). Such difficulties may be revealed in tasks involving perceptual analysis or classification (Milner, 1963; Luria, 1966), memory (Milner, 1971), simple response selections (Drewe, 1975), spatial or verbal problem solving (Milner, 1965; Luria, 1966), and many others; in daily life, there can be widespread difficulties of planning, self-control, and regard for social conventions (e.g. Eslinger and Damasio, 1985). In general, the normal structure of goal-directed behavior is disturbed, producing activity that seems fragmented, irrelevant, or bizarre (Bianchi, 1922).

JOHN DUNCAN Medical Research Council Applied Psychology Unit, Cambridge, England

In this chapter, I relate these effects of frontal damage to two other phenomena. The first concerns individual differences in the normal population. If a disparate set of tests is administered to a broad sample of people, between-test correlations will be almost universally positive: To some extent at least, a person who performs well in one test will tend also to perform well in others (Spearman, 1927). One possible explanation is that some common factor—conventionally termed general intelligence or Spearman's *g*—makes a contribution to success in all manner of tasks. For example, *g* might reflect the efficiency of some particular information-processing system involved in the organization of many different activities. Accepting this interpretation, it is easy to show which tests are best correlated with *g*, and this is the basis for design of standard intelligence tests such as the Wechsler Adult Intelligence Scale or WAIS (Wechsler, 1955) and Raven's Progressive Matrices (Raven, Court, and Raven, 1988).

The remaining set of findings to be considered concerns the problem of divided attention, or interference between concurrent tasks. In part, such interference depends on task *similarity*: Two visual or two manual tasks, for example, will show stronger interference than tasks with different input and output modalities, suggesting conflicts within special-purpose or modality-specific processing systems (Treisman and Davies, 1973; McLeod, 1978). Even when concurrent tasks have nothing obvious in common, however, some interference between them generally remains (Bourke, Duncan, and Nimmo-Smith, 1993; McLeod and Posner, 1984). Again, this might be taken to reflect conflicts within processing systems whose role in the control of behavior is rather broad.

In this chapter, I present the hypothesis that these three sets of phenomena are closely related. All three concern a process of abstract action or goal selection, important especially under conditions of novelty or weak environmental prompts to behavior. Variations in the efficiency of this process are reflected in Spearman's *g*, and conflicts within it are responsible for many cases of interference between dissimilar, concurrent tasks. Though certainly other brain structures are likely to be involved, the general goal selection function is strongly dependent on the integrity of the frontal lobe and therefore is impaired by frontal lesions. In a final section, I consider the importance of specific frontal regions, in particular dorsolateral prefrontal cortex and anterior cingulate cortex.

Attention, practice, and environmental cues to action

By *attention*, we refer loosely to selectivity in cognitive operations. Such selectivity has many components—at any given time, certain goals are in control of behavior, certain stimulus information is thereby relevant, and so on—and there have been many demonstrations of dissociation among different aspects of attention (e.g., Treisman, 1969; Broadbent, 1971; Allport, 1980; Posner and Petersen, 1990). An obvious link among the three sets of phenomena just described, however, is suggested by attention in the sense of a common distinction between controlled, active, or voluntary, and automatic, passive, or stimulus-driven behavior, applied separately to frontal lobe functions (Bianchi, 1922; Luria, 1966; Fuster, 1980; Norman and Shallice, 1980; Frith et al., 1991), Spearman's *g* (Snow, 1981; Ackerman, 1988), and dual-task interference (James, 1890; Fitts and Posner, 1967; Schneider and Shiffrin, 1977).

This distinction is most directly motivated by the contrast between novel and practiced or habitual behavior. Novelty in two different senses has often been implicated in frontal dysfunction. The first sense concerns *perseveration*, or difficulty changing some line of behavior in the short term. Many examples have been reported in frontal lobe patients and animals, from simple perseveration of inappropriate motor activity (Bianchi, 1922; Luria, 1966; Diamond and Goldman-Rakic, 1989) to perseveration of higher-order mental sets, classification rules, and so forth (Milner, 1963; Owen et al., 1993). The second sense of novelty concerns longer-term *familiarity*. Familiar or regularly practiced lines of activity may be relatively immune to frontal lesions (Luria and Tsvetkova, 1964; Walsh, 1978); complementarily, novel behavior may be interrupted by familiar but inappropriate intrusions or stereotypes (Luria, 1966). Similarly, novelty has been implicated in *g* correlations. Practicing simple, consistent perceptual-motor tasks may reduce correlations with *g* (Fleishman and Hempel, 1954; Ackerman, 1988). Finally, the role of novelty in dual-task interference has been noted since the nineteenth century. Novel behavior is hard to combine with concurrent activities and is experienced as requiring active attention. Habitual behavior, in contrast, may interfere rather little with other activities; it may appear automatic and even involuntary, leaving attention free for other concerns (James, 1890; Bryan and Harter, 1899).

Another relevant variable is the strength of the environmental prompt to action, assumed to distinguish voluntary from stimulus-driven behavior (e.g., Frith et al., 1991). One form of prompt is a direct verbal command or suggestion. It has often been noted that frontal lobe patients may need explicit verbal prompts either to continue with a task or to satisfy some aspect of its requirements (Luria and Tsvetkova, 1964; Hecaen and Albert, 1978). For example, the patient may fail to make progress on some complex problem such as preparing a meal (Penfield and Evans, 1935) or solving a spatial puzzle (Luria and Tsvetkova, 1964), yet may perform perfectly adequately when the problem is broken down into a set of separately specified components. Similarly, explicit verbal instructions detailing a precise strategy or manner of procedure may reduce *g* correlations in a variety of complex tasks, including classroom learning (e.g., Snow, 1981).

For reasons such as these, both frontal lobe functions (Norman and Shallice, 1980) and Spearman's *g* (Ackerman, 1988) have previously been related to the problem of dual-task interference and attentional versus automatic control of behavior. Interestingly, though, it has been generally accepted, at least since the work of Hebb and Penfield (1940), that frontal lobe functions are rather unrelated to conventional intelligence (Teuber, 1972). It is my contention, however, that conventional wisdom is incorrect and that frontal lobe functions, Spearman's *g*, and dual-task interference are indeed closely related through the problem of control of behavior under conditions of novelty and weak environmental prompts.

Goal selection

As an extension to more posterior motor systems, prefrontal cortex is generally considered to be involved in the high-level control of action. Accordingly, it may be useful to relate the effects of novelty and prompts to standard accounts of action control.

For this purpose, I shall adopt the common view that action is represented and controlled as a hierarchy of goals and subgoals (Miller, Galanter, and Pribram, 1960). As an example, consider the action of traveling from Cambridge to Squaw Creek (figure 45.1). In the initial plan, motivated by the invitation to make a presentation, this action is specified abstractly (figure 45.1, top level); it might be realized by flying, sailing, hitchhiking, and the like. Suppose, however, that the decision is to fly; now we may fill in the details of acquiring a passport and tickets, traveling to the airport, and so on (figure 45.1, second level). Each of these components is again specified at some level of abstraction: Traveling to the airport might be satisfied by driving, taking a train, or via some other form of transportation but, if the decision is to drive, further details may again be filled in, as shown at the successive levels of figure 45.1, until a level of actual motor commands is reached (Sacerdoti, 1974). Each entry in figure 45.1 specifies a particular abstract requirement on behavior which is then realized by the details beneath it; in other words, each entry functions as a *goal* to which lower-level actions are directed until it has been achieved.

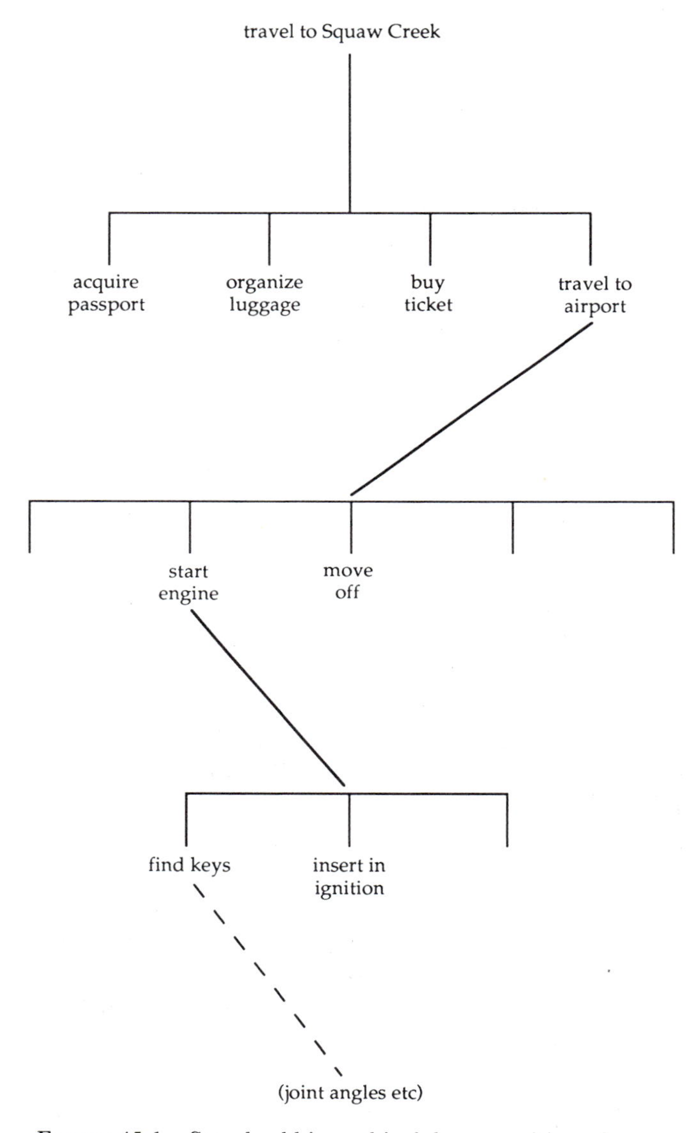

FIGURE 45.1 Standard hierarchical decomposition of an action into a goal-subgoal tree.

The advantages of this hierarchical scheme (i.e., of working at successive levels of abstraction) are well-known. Typically, only some abstract description of an action will be directly required or cued at any given stage of a plan: For example, the action of acquiring tickets makes sense providing only that the airport can be reached *somehow*, no matter how; or, as another example, the invitation to make a presentation prompts only the abstract goal of travel to Squaw Creek, not the detailed working out of this requirement (e.g., which hand to raise in opening the car door). Selecting the relevant abstraction then serves as a framework that guides the more detailed planning beneath it (Sacerdoti, 1974).

This general idea that actions are planned in a hierarchy of abstractions already raises interesting parallels with the frontal lobe literature. As pointed out by Sacerdoti (1974), planning at a detailed level can be chaotic without guidance from more abstract levels. The problem-solving program is lost in the multitude of detailed response options to consider. Given an adequate abstract framework, however, the lower level functions "just as if (the program) were given (several) small problems to solve consecutively" (Sacerdoti, 1974, page 129). In just the same way, as already noted, frontal lobe patients may make progress on some complex problem such as preparing a meal only when it is broken down into separately prompted components.

Given this hierarchical view of action selection, one may ask how particular goals are selected for control

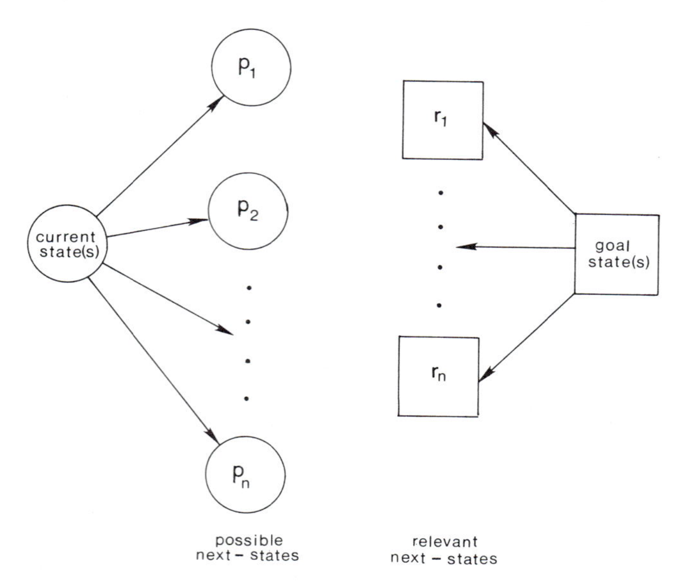

FIGURE 45.2 Activation of candidate goals. The current state activates possible next-states (p_1, p_2, p_n), whereas the goal state activates relevant next-states (r_1, r_n). (Reprinted from Duncan, 1990, by permission.)

of behavior at any particular time. It has often been noted that two opposite influences must be at work (Duncker, 1945; Reitman, 1965; Anderson, 1983), as shown in figure 45.2. First, new candidate goals arise through working backward from active superordinate goals. For example, the goal of acquiring a passport arises from the supergoal of traveling to Squaw Creek. In figure 45.2, these are called *relevant* next-states; they arise through relevance to an already active goal that the system is working to bring about (Duncan, 1990). Second, new candidate goals arise through working forward from the current world state and, in particular, environmental events. For example, the goal of driving to the airport may be temporarily suspended by the sight of an old friend beckoning in the street. In figure 45.2, these are called *possible* next-states. The set of relevant and possible next-states must somehow be weighted in terms of net importance, leading to a final selection of which goals to pursue.

In the context of such a competition among momentary candidate goals, it is easy to see practice and strong environmental prompts as providing one major source of bias (cf. Norman and Shallice, 1980). Frequent experience with selecting a particular goal or action in some particular environmental or behavioral context, just like a direct verbal command, may make this goal very easy to select in any given instance. In this case, the efficiency of goal-weighting procedures may be relatively unimportant, and behavior may seem automatic or stimulus-driven. Without such a strong bias, however, reductions in the efficiency of goal weighting produce poor choices of behavior. It is under these circumstances that behavior is especially sensitive to frontal lobe impairment, Spearman's *g*, and dual-task interference.

Experimental evidence

GOAL NEGLECT The first experiments, carried out in collaboration with several colleagues, directly concern the effects of verbal instructions and prompts in novel behavior. Of course, verbal commands always prompt behavior at some particular level of abstraction; in line with our previous discussion, they prompt goals or task requirements to be satisfied, leaving many details unspecified. Even a command to carry out some simple action (e.g., Tie your shoelaces) is very far from a complete description of the motor activity required. This is why it is appropriate to see verbal commands as one form of environmental input to the general process of goal selection.

As described earlier, one characteristic of frontal lobe patients is that single verbal commands may prove insufficient to prompt appropriate behavior. The command must be repeated or additional prompts given (e.g., Hecaen and Albert, 1978). At the same time, the patient may show that the original command was understood and remembered; there is a mismatch between what is *known* of task requirements and what is actually *attempted* in behavior. For example, the patient may have been told that onset of a light is a cue to squeeze with the hand. When the light is seen, the patient may state, "I should squeeze," yet make no attempt to do so (Luria, 1966). Mismatches may also occur between behavior and the patient's own verbal account of requirements. In the Wisconsin card-sorting task, for example, stimuli must be sorted first according to one attribute (e.g., color), then according to another. Again, incorrect sorting strategies may be continued even though the patient states verbally that they should be abandoned (Milner, 1963). Such behavior is perhaps a special case of the wider clinical picture of *lack of concern* in frontal lobe patients (e.g., Walsh, 1978); though difficulties may be appreciated, they seem not to become the focus of appropriate emotional or cognitive activity.

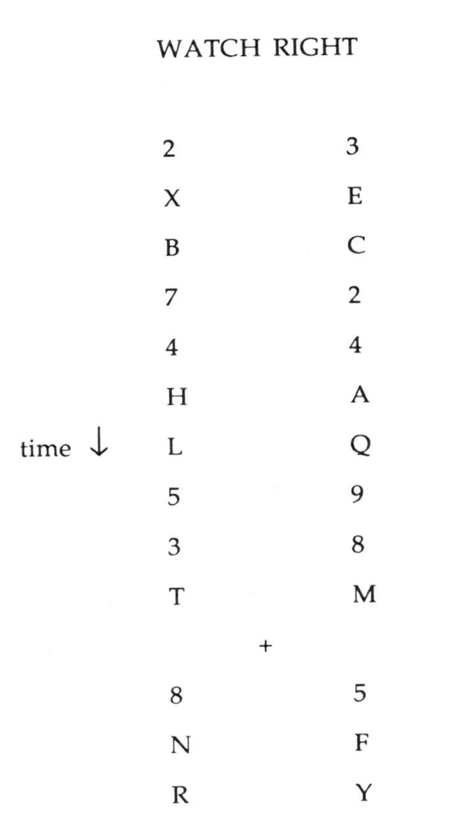

FIGURE 45.3 Sample stimulus sequence for a letter-monitoring trial. Time runs from top to bottom. Each pair of characters is shown for 200 ms and is separated from the next pair by a 200-ms blank interval.

We use the term *goal neglect* to describe such disregard of a task requirement, even though it has been understood. In our research, we observe neglect even in some members of the normal population, and we consider the circumstances under which this occurs and the effects of frontal lobe lesions.

The stimulus sequence from a typical trial of our task is shown in figure 45.3, with time running from top to bottom. The sequence consists of a series of frames, each presented for 200 ms and separated from the next by an additional 200 ms. Each frame consists of a pair of letters or digits, presented side by side in the middle of a computer screen. There are three basic task requirements:

1. Letters should be repeated aloud as soon as they are seen. Digits are to be ignored.

2. The subject must watch for letters on only one side at a time, left or right. The trial begins with an instruction, "WATCH LEFT" or "WATCH RIGHT," written in the center of the screen. In figure 45.3, the trial begins with WATCH RIGHT, so the subject repeats *E*, *C* . . . while ignoring *X*, *B* . . .

3. Finally, near the end of each trial, the subject sees a further cue, which sometimes calls for a switch of sides. The cue is a plus or minus symbol presented in the center of the screen. Again, it lasts for 200 ms and is separated from preceding and following character pairs by intervals of 200 ms. A plus sign means that, for the remainder of the trial, the subject should watch the right, whereas a minus sign means he or she should watch the left. In the example, the subject is to continue watching right and repeat *F* and *Y*. However, if the cue had been a minus sign, a switch to the left would have been required.

The first two of these task requirements are almost always satisfied correctly. The third, however, is sometimes neglected. Most commonly, the subject simply continues to report letters from the initially attended side until the end of the trial. Explicit questions, however, almost always reveal that the rule is actually remembered. As in cases of frontal goal neglect, the task requirement has been understood but exerts no apparent influence over behavior.

Two characteristics of this neglect are important. First, it is confined entirely to novel behavior and, specifically, behavior before the very first correct trial. For any given subject, one sees a series of 0 . . . *n* trials on which the plus or minus cue is disregarded, followed by immediate resolution to almost perfect performance. Once the task requirement gains control over behavior, such control is retained for the remainder of the experiment.

Second, neglect is extremely sensitive to verbal and other prompts, drawing attention to the neglected task requirement. Most effective is explicit verbal feedback from the experimenter, pointing out any errors that have been made after each trial. Such trial-by-trial feedback almost always produces resolution to good performance within a few trials. Obviously, the limitation producing goal neglect does not lie in any absolute inability to perform the task (e.g., because stimuli are presented too quickly). When explicitly asked, neglecting subjects may report either that the plus and minus cues passed unnoticed or that they were disregarded. A typical comment might be, "I realize now that the cues have been going over my head . . ."

On the face of it, then, this task produces behavior that has much in common with the goal neglect of frontal lobe patients. A goal or task requirement sometimes is disregarded even though it has been understood. Such disregard is confined to novel behavior and is very sensitive to repeated verbal prompts. It is as if this aspect of task requirements somehow slips the subject's mind. The next step is to consider its relationship to Spearman's *g* and the effects of frontal lobe lesions.

Data from three groups of normal subjects are shown in figure 45.4. Data come from an initial block of 12 trials, grouped into 3 successive subblocks of 4 trials each. Each subblock is scored as passed if appropriate responses are made to the plus and minus cues and otherwise scored as failed, and each subject receives a score between 0 and 3, indicating the number of failed subblocks. Each subject also receives a score on a standard test of fluid intelligence, Cattell's Culture Fair (Institute for Personality and Ability Testing, 1973). Large-scale factor analyses have shown that this test correlates above 0.8 with Spearman's *g*. Using the published norms for this test, each subject's score is transformed to a *z* score, indicating where that subject lies with respect to the standard or reference population. A score of +1.0, for example, indicates that the subject lies one standard deviation above the mean. In figure 45.4, subjects have been sorted into bins based on these Culture Fair scores, and each panel shows the mean number of failed subblocks in the goal neglect task as a function of Culture Fair score. The left panel (A) comes from a group of 90 young to middle-aged subjects; the middle panel (B) comes from 41 elderly subjects, whose Culture Fair scores, as expected, are lower (Cattell, 1971); and the right panel (C) comes from 38 young to middle-aged subjects performing under somewhat altered task conditions (see later). The results of all three experiments suggest a tight relationship between goal neglect and *g*. Neglect is hardly ever seen among people whose Culture Fair scores are above the population mean but is almost universal at more than one standard deviation below the mean.

In addition to these data from normal subjects, we have investigated the effects of major frontal lobe lesions arising from a variety of causes, including closed head injuries and surgically removed tumors. In this group, again, the first two task requirements—report

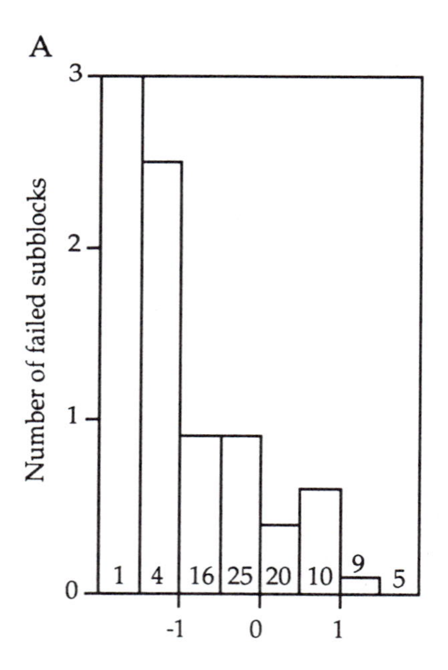

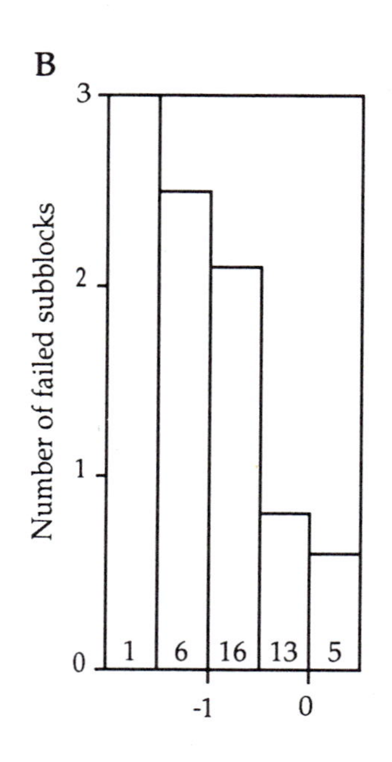

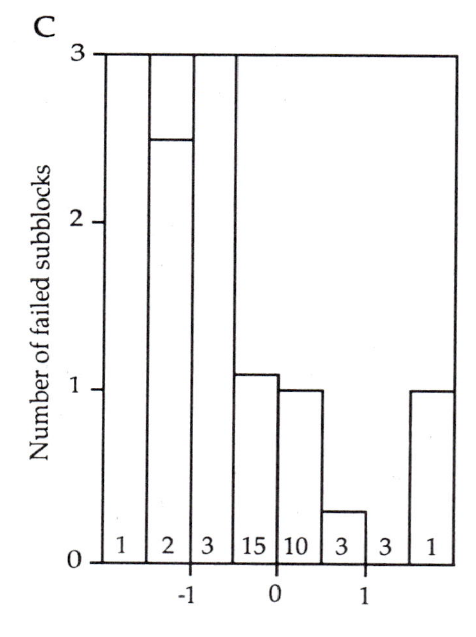

FIGURE 45.4 Relation between goal neglect (number of failed subblocks out of three) and *g* in three experiments. Subjects have been sorted into bins based on performance on the Culture Fair test of *g*. The figure shows the mean number of failed subblocks for subjects in each bin, with the number of subjects falling into the bin displayed at the bottom of each column. Note that, except for the middle panel in which no subjects had Culture Fair scores exceeding +0.5, extreme bins include all subjects beyond a *z* score of ±1.5. (Left) Ninety subjects aged 29–57 years. (Middle) Forty-one subjects aged 60–70 years. (Right) Thirty-eight subjects aged 39–49 years under dual-task conditions.

of letters but not digits and attention to the correct side at the start of the trial—are almost always satisfied. In contrast, the plus or minus cue is almost always disregarded without additional verbal prompting.

A final set of findings both generalizes neglect to a different task requirement and provides some information regarding interference between concurrent tasks or requirements. To the basic letter-monitoring task, we added another: Occasionally, during the course of each trial, a brief dot would be flashed either above or below the stream of alphanumerical characters, and the subject was to respond by pressing one of two alternative keys depending on the dot's position (above or below). Instructions for both dot- and letter-monitoring tasks were given before the task was actually attempted; in one group of subjects, however, the dot task was described first, whereas in a second group it was described last (i.e., when the subject was already bearing in mind the requirements of letter monitoring). The letter-monitoring performance of these subjects is shown in figure 45.4C. In this experiment, however, we were more interested in the possibility of neglect in the dot task. For the first group of subjects—the group for whom the dot task was described first—omissions in this task were rare and unrelated to *g*. For the second group, however, we again observed a pattern of 0 . . . *n* trials on which no response was made to the dots, followed by resolution within a few trials to almost perfect performance. Again, neglect resolved with the introduction of trial-by-trial feedback indicating that dots had been ignored and, again, there was a strong relationship between the number of neglected trials and score on the Culture Fair test of *g* ($r = .66$, $N = 18$). The results show that goal neglect is very sensitive to a particular form of dual-task interference. A task requirement is likely to be disregarded only when several others have already been described, presumably implying a set of already selected or active goals. It seems likely that, under these circumstances, neglect can be produced in low-*g* subjects for task requirements of almost any description.

In summary, these experiments reveal a difficulty that is common to frontal lobe patients and normal people from the lower part of the *g* distribution. This difficulty concerns neglect of a goal or task requirement even though it has been understood and remembered. It is confined to novel behavior and is very sensitive to direct verbal prompting of the neglected requirement. There is also a kind of dual-task interference: A task requirement is likely to be neglected only if it is specified after several others.

Conventional Tests of Intelligence As we have said, though both frontal lobe functions and Spearman's *g* have, in the past, been related to voluntary, active, or attentional control of behavior, they are conventionally believed to be rather unrelated to one another. The next set of results deals with the basis for this apparent paradox.

Although certainly frontal lobe lesions can produce deficits on standard tests of intelligence such as the WAIS (e.g., Hamlin, 1970), there is little to suggest that such deficits are especially severe, compared either with deficits on other tests (Mettler, 1949) or with the effects of other lesions (Weinstein and Teuber, 1957; Warrington, James, and Maciejewski, 1986). On the contrary, it has often been supposed that tests of this sort are especially unsuited to revealing the effects of frontal lobe lesions (Teuber, 1972). Particularly striking are patients who, despite major frontal lobe lesions and evident behavioral difficulties, attain excellent WAIS scores (Eslinger and Damasio, 1985; Shallice and Burgess, 1991).

In considering such results, however, it may be useful to distinguish between two different kinds of intelligence test, or two different ways to obtain an estimate of *g*. Clinically popular tests such as the WAIS depend on averaging across a diverse set of subtests, raising two points. First, many of these subtests emphasize what is often termed *crystallized intelligence*, or previously acquired knowledge (vocabulary, general knowledge, etc.). Such measures may well reflect *g at the time knowledge was originally acquired* rather than current *g* (Cattell, 1971); once learned, knowledge may be rather insensitive to subsequent changes. Second, individual subtests may, in themselves, have only modest *g* correlations (Marshalek, Lohman, and Snow, 1983), suggesting only modest demands on any *g* factor. As Spearman (1927) showed, average performance on any set of tests will give much the same *g* estimate, providing this set is sufficiently large and diverse.

Typical tests of the alternative type are the Progressive Matrices and Culture Fair. In these cases, an attempt is made to avoid heavy dependence on knowledge or education; instead, the tests emphasize novel problem solving, using spatial, verbal, or other materials. Interestingly, such tests of problem solving have high *g* correlations taken alone (i.e., without averaging over very diverse subtests). Such a result suggests especially strong demands on those information-processing functions most involved in *g*. These are generally known as tests of *fluid intelligence* (Cattell, 1971). With their relative independence from prior knowledge and strong *g* correlations, such tests may be especially suitable for investigating functions related to *g* after brain lesions. Specifically, such tests may reveal a substantial decrement following frontal lobe lesions, even when tests such as WAIS do not.

Recently, my colleagues and I have been testing this prediction. We selected patients who, despite major frontal lobe lesions and manifest behavioral impairments, nevertheless achieved high WAIS scores—exactly those patients, in other words, who provide the strongest conventional evidence *against* any link between frontal dysfunctions and *g*. We were able to find three suitable patients, all with WAIS intelligence quotients (IQs) between 125 and 130. (Following normal convention, IQs are scaled to give a mean of 100 and a standard deviation of 15 in the normal population.) All patients had localized frontal lobe lesions, one from an infarction affecting primarily the white matter of the left frontal lobe, one from a bifrontal open head injury, and one from a left frontal lobectomy for removal of tumor. Though all performed well on at least some conventional tests of frontal lobe function, all were poor on the real-world multiple errands task of Shallice and Burgess (1991) and showed the typical, everyday difficulties of such patients (cf. Eslinger and Damasio, 1985; Shallice and Burgess, 1991).

To test fluid intelligence in these patients, we again used Cattell's Culture Fair. Exactly as predicted, the patients' Culture Fair IQs were 22–38 points lower than their WAIS IQs. Because such a comparison could be contaminated by various factors, including differential norms and regression to the mean, we also tested control subjects matched individually to the patients on WAIS IQ, age, sex, and socioeconomic group. The patients' Culture Fair IQs were 23–60 points lower than the Culture Fair IQs of the matched controls. Such results help substantially to clear up the paradox of preserved intelligence in frontal lobe patients. When *g* is measured by a test of current or fluid intelligence, it is strongly dependent on the integrity of the frontal lobe.

An item similar to some of those in the Culture Fair test is shown in figure 45.5. The task is to decide which of the five numbered boxes on the right would correctly complete the matrix on the left. Though we do not know what it is about such a task that makes it such a good measure of *g*, its close relationship to goal neglect and their joint dependence on frontal lobe functions raise the question of what the two might share. The materials in figure 45.5 again imply multiple task requirements. Variations in shape, size, and orientation of line must be taken into account to determine that item 4 is the correct choice (see Carpenter, Just, and Shell, 1990). Novelty and the strength of environmental prompts to action may both be important: Successive items in the test involve very different operations, and task requirements must largely be worked out from the visual materials, without explicit verbal prompts. It seems possible that these general characteristics of standard problem-solving tests contribute both to their high *g* correlations and to their dependence on frontal lobe functions.

Comparison of Profiles The previous two lines of research revealed similarities between the results of frontal lobe damage and the behavior of normal people

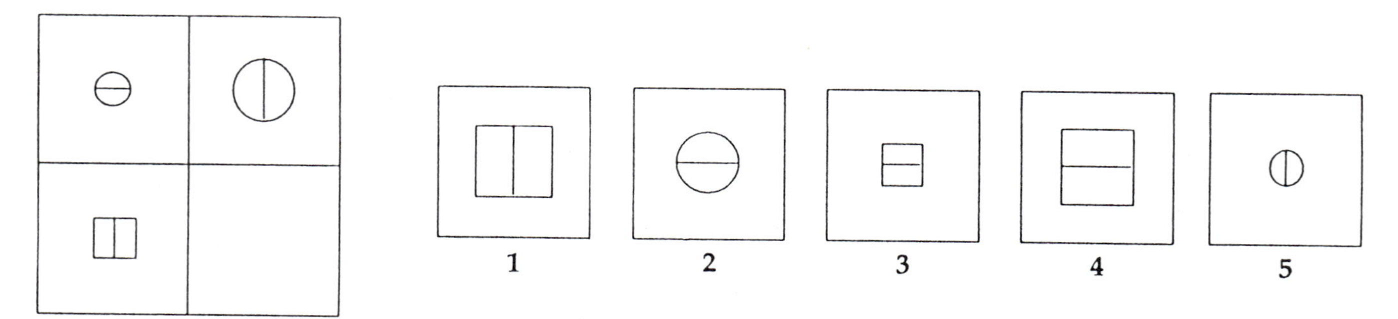

Figure 45.5 Example problem based on the Culture Fair test.

from the lower part of the g distribution. Let us now consider the general basis for research of this sort.

The hypothesis is that frontal lobe impairments, g correlations, and dual-task decrements all reflect a general goal-weighting process. In any set of tasks or performance measures, scores will depend, to a variable degree, on such a process. Tasks very dependent on the process should show large frontal impairments, g correlations, and dual-task decrements, whereas tasks less dependent on the process should show small impairments, correlations, and decrements. In general, across any set of tasks or measures, profiles of frontal impairment, g correlation, and dual-task decrement should be similar.

To put the prediction more formally, consider that the correlation coefficient r between any variable x and g is the slope of the best-fitting straight line relating x to g, both expressed as standard or z scores. In other words, r indicates the expected change in x per unit change in g, both expressed as z scores. If the effects of frontal lobe damage or dual-task interference can be understood as a simple reduction in g, then across any set of measures, performance impairments or decrements, expressed as z scores, should be linearly related to g correlations (Duncan et al., 1993).

In principle, this prediction applies to any set of tasks or performance measures and, in recent work, we have begun to compare profiles of g correlation and dual-task decrement in a range of different tasks, from simple perceptual-motor skills to complex problem solving. An example from the domain of perceptual-motor skill is shown in figure 45.6. In this study, the 12 measures used were components of driving skill, measured on the road in an instrumented car (Duncan et al., 1993). They included measures of scanning pattern (1 to 6), car control (7 to 11), and safety margin (12). To measure g correlations, we administered a standard driving test and the Culture Fair to a group of 90 drivers. To measure dual-task decrements, we had an additional group of 24 subjects drive the test route twice, once with a secondary task (generating letters in random order as they drove, in time with a metronome; see Baddeley, 1986) and once without. The figure shows profiles of g correlation and dual-task decrement, the latter transformed to z scores using the between-subjects standard deviation from our main 90-driver sample. As we might expect of simple components of a highly familiar skill, both g correlations and dual-task decrements were extremely modest. Across the 12 driving measures, however, the two profiles were in fairly good agreement, with a correlation of .67 between them.

Although profiles of frontal lobe impairment and g correlation have yet to be compared, the method offers one rational solution to the problem of evaluating the relative sensitivity of different tasks to frontal lobe lesions. The standard approach in neuropsychology has been to search for tasks that either are or are not sensitive to frontal lobe lesions, but it is difficult to be confident that any task is truly insensitive to the general disorganization of behavior that frontal lobe lesions can produce. In just the same way, almost any task has some sort of positive g correlation, no matter how modest. There is little to be learned from simply establishing that a task has some nonzero g correlation; and similarly, there may be little to learn from simply showing a statistically significant frontal lobe deficit. Instead, we need to compare relative frontal lobe deficits in different tasks. The hypothesis that deficits expressed as z scores should be linearly related to g correlations offers one way to do this.

In summary, frontal lobe patients and normal people from the lower part of the g distribution share a common tendency toward goal neglect (i.e., disregard of a task requirement even though it has been understood) in novel behavior. Weak environmental prompts and multiple concurrent requirements are some of the conditions under which such neglect occurs. Frontal lobe patients also perform poorly on at least one test of novel problem solving of the sort most closely related to (current) g; such tests tend similarly to combine novelty, weak prompts, and multiple concurrent requirements. Spearman's g is related as well to dual-task interference: across tasks or measures, profiles of g correlation and dual-task decrement tend to agree. These findings are consistent with the hypothesis that frontal lobe impairments, Spearman's g, and interference between dissimilar concurrent tasks are closely linked through a general process of goal weighting, most likely to fail when behavior is novel and environmental cues to action are weak.

Functional anatomy

A final issue should be considered. Some degree of functional specialization within the frontal cortex is well established. Lesion studies in animals, for example, show predominantly spatial deficits from dorso-

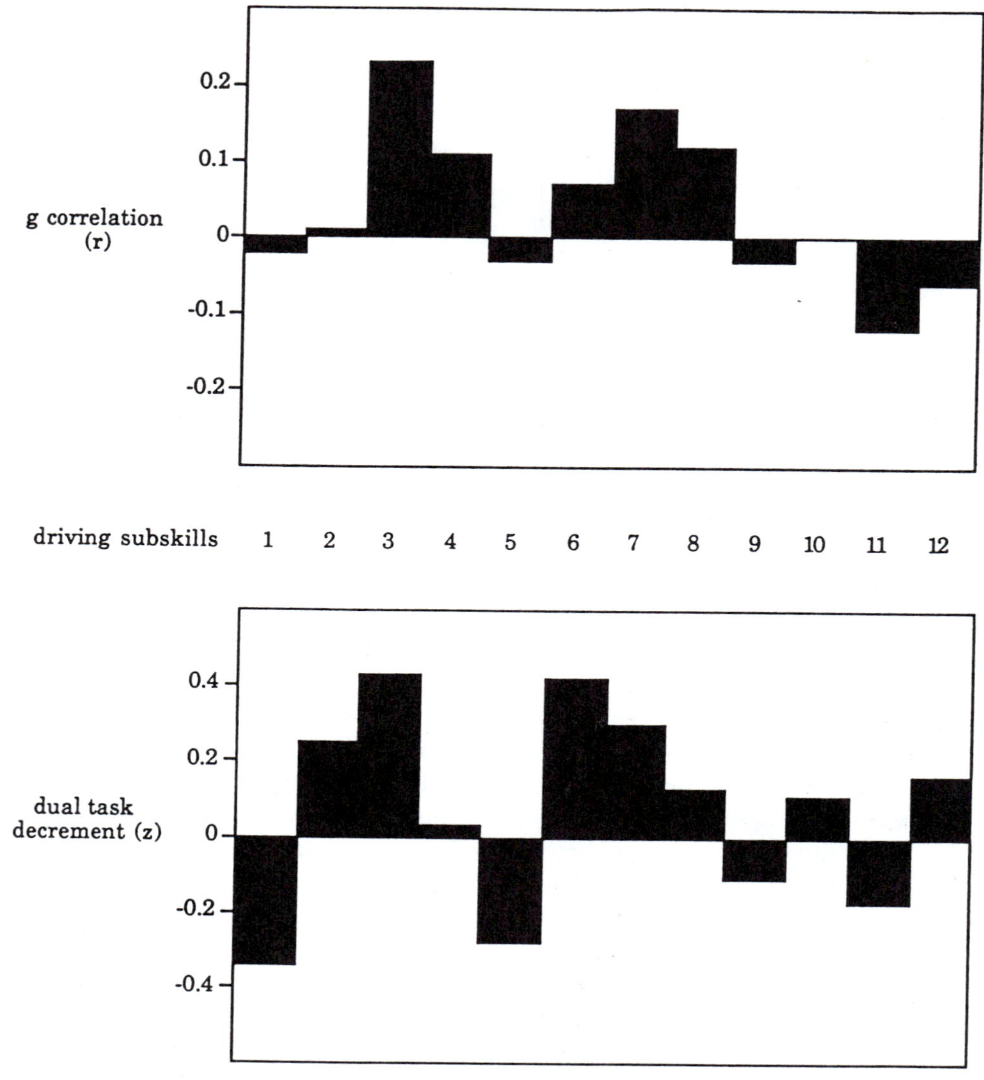

FIGURE 45.6 Profiles of g correlation (r) and dual-task decrement (z) across 12 driving subskills. (Reprinted from Duncan et al., 1993, by permission.)

lateral prefrontal lesions (Mishkin and Manning, 1978; Goldman-Rakic, 1988), recognition memory impairments associated specifically with ventromedial lesions (Kowalska, Bachevalier, and Mishkin, 1991), and so on. Although human lesion data are less clear, it is true that several higher-order motor impairments can be specifically associated with lesions of posterior frontal regions including the frontal eye fields and supplementary motor area (e.g., Paus et al., 1991; Braun et al., 1992). How does such a picture of modular function within frontal cortex relate to the hypothesis that the effects of frontal lobe lesions might resemble either a general reduction in g or the general effects of dual-task interference?

Two possibilities are worth considering. The first is that a general goal-weighting function, reflected in both Spearman's g and generalized dual-task decrements, arises through the integrated activity of frontal systems that ultimately will turn out to be separable. Several authors (e.g., Shallice and Burgess, 1991) have suggested views of this sort (e.g., that general control functions such as inhibition of inappropriate action stereotypes, or shifting of cognitive set, might be localized to different frontal regions). One version of this hypothesis is currently being tested by my colleagues and I. An extensive battery of neuropsychological tests and other clinical measures was given to a series of 90 head-injured subjects. Tests were chosen to include some conventionally sensitive to frontal lobe lesions, but others conventionally less so. Several measures were also taken from each test, indicating errors of perseveration, impulsivity, underactivity, and so forth. By testing head-injured subjects, we aimed for a group with diverse brain lesions, often having a substantial frontal lobe component. Although this study is still in progress, one preliminary finding is lack of evidence for any interesting form of modularity. For example, correlations between different measures of perseveration or impulsivity are certainly no higher than correlations between total task scores. It remains possible that separate action control functions may be independently undertaken by different frontal lobe regions, but we have seen no evidence for this to date.

The alternative possibility is that particular frontal cortical regions have major responsibility for the general goal-weighting function. Despite the evidence for some modularity of function within the frontal lobe, it is easy to find evidence also for fairly general-purpose action selection systems. In the monkey, for example, lesions of the inferior frontal convexity produce deficits in a wide range of tasks, involving different materials, input modalities, and so on (Passingham, 1975; Mishkin and Manning, 1978).

The most direct evidence for such a hypothesis comes from experiments using positron emission tomography to measure task-related changes in activity in the human brain. Two regions of the frontal lobe—dorsolateral prefrontal cortex and anterior cingulate cortex—have now been found to be active during a variety of different complex tasks, including generating verbs in response to nouns (Petersen et al., 1988), generating random finger movements (Frith et al., 1991), making complex visual comparisons between successive stimuli (Corbetta et al., 1991), and recalling verbal material under various conditions (Grasby et al., 1993). Activity in both regions has been found to decrease during learning of a verb generation task (Raichle et al., 1991), associated with the development of stereotyped responses. It has also been pointed out that, in several of the tasks activating these regions, the correct action is not prompted by any feature of the immediate stimulus (Frith et al., 1991). As reflected in Posner's hypothesis of an anterior attention system (Posner and Petersen, 1990), these may be regions of the human frontal lobe that are especially involved in the attentional or voluntary control of many different novel or weakly prompted activities (Corbetta et al., 1993).

As already mentioned, of course, we should be cautious about concluding that such general functions as Spearman's g or generalized dual-task interference *reside* in particular frontal structures. The evidence supports the hypothesis that both g and dual-task interference concern a general goal-weighting function that is heavily dependent on the frontal lobe and therefore disturbed by frontal lesions, but this is not to say that the frontal lobe or specific frontal regions are entirely responsible for it. It remains an open question how such general functions might arise from the combined activity of separate frontal regions and the many cortical and subcortical structures with which they interact.

ACKNOWLEDGMENTS This work was supported by research contract 9652/32 from the Transport and Road Research Laboratory and by grant AFOSR-90-0343 from the Air Force Office of Scientific Research, Air Force Systems Command, U.S.A.F. The U.S. government is authorized to reproduce and distribute reprints for governmental purposes notwithstanding any copyright notation thereon.

REFERENCES

ACKERMAN, P. L., 1988. Determinants of individual differences during skill acquisition: Cognitive abilities and information processing. *J. Exp. Psychol. [Gen.]* 117:288–318.

ALLPORT, D. A., 1980. Attention and performance. In *Cognitive Psychology: New Directions*, G. Claxton, ed. London: Routledge and Kegan Paul, pp. 112–153.

ANDERSON, J. R., 1983. *The Architecture of Cognition*. Cambridge, Mass.: Harvard University Press.

BADDELEY, A. D., 1986. *Working Memory*. Oxford: Oxford University Press.

BIANCHI, L., 1922. *The Mechanism of the Brain and the Function of the Frontal Lobes*. Edinburgh: Livingstone.

BOURKE, P. A., J. DUNCAN, and I. NIMMO-SMITH, 1993. A general factor involved in dual task performance decrement. Manuscript submitted for publication.

BRAUN, D., H. WEBER, T. MERGNER, and J. SCHULTE-MONTING, 1992. Saccadic reaction times in patients with frontal and parietal lesions. *Brain* 115:1359–1386.

BROADBENT, D. E., 1971. *Decision and Stress*. London: Academic Press.

BRYAN, W. L., and N. HARTER, 1899. Studies on the telegraphic language. The acquisition of a hierarchy of habits. *Psychol. Rev.* 6:345–375.

CARPENTER, P. A., M. A. JUST, and P. SHELL, 1990. What

one intelligence test measures: A theoretical account of the processing in the Raven Progressive Matrices Test. *Psychol. Rev.* 97:404–431.

CATTELL, R. B., 1971. *Abilities: Their Structure, Growth and Action* Boston: Houghton Mifflin.

CORBETTA, M., F. M. MIEZIN, S. DOBMEYER, G. L. SHULMAN, and S. E. PETERSEN, 1991. Selective and divided attention during visual discriminations of shape, color, and speed: Functional anatomy by positron emission tomography. *J. Neurosci.* 11:2383–2402.

CORBETTA, M., F. M. MIEZIN, G. L. SHULMAN, and S. E. PETERSEN, 1993. A PET study of visuospatial attention. *J. Neurosci.* 13:1202–1226.

DIAMOND, A., and P. S. GOLDMAN-RAKIC, 1989. Comparison of human infants and rhesus monkeys on Piaget's AB task: Evidence for dependence on dorsolateral prefrontal cortex. *Exp. Brain Res.* 74:24–40.

DREWE, E. A., 1975. Go–no go learning after frontal lobe lesions in humans. *Cortex* 11:8–16.

DUNCAN, J., 1990. Goal weighting and the choice of behaviour in a complex world. *Ergonomics* 33:1265–1279.

DUNCAN, J., P. WILLIAMS, M. I. NIMMO-SMITH, and I. BROWN, 1993. The control of skilled behavior: Learning, intelligence, and distraction. In *Attention and Performance*, vol. 14, D. Meyer and S. Kornblum, eds. Cambridge, Mass.: MIT Press, pp. 323–341.

DUNCKER, K., 1945. On problem solving. *Psychol. Monogr.* 58 (Whole No. 270):1–113.

ESLINGER, P. J., and A. R. DAMASIO, 1985. Severe disturbance of higher cognition after bilateral frontal lobe ablation: Patient EVR. *Neurology* 35:1731–1741.

FITTS, P. M., and M. I. POSNER, 1967. *Human Performance.* Belmont, Calif.: Brooks/Cole.

FLEISHMAN, E. A., and W. E. HEMPEL, 1954. Changes in factor structure of a complex psychomotor test as a function of practice. *Psychometrika* 19:239–252.

FRITH, C. D., K. FRISTON, P. F. LIDDLE, and R. S. J. FRACKOWIAK, 1991. Willed action and the prefrontal cortex in man: A study with PET. *Proc. R. Soc. Lond. [Biol.]* 244:241–246.

FUSTER, J. M., 1980. *The Prefrontal Cortex: Anatomy, Physiology, and Neuropsychology of the Frontal Lobe.* New York: Raven.

GOLDMAN-RAKIC, P., 1988. Topography of cognition: Parallel distributed networks in primate association cortex. *Annu. Rev. Neurosci.* 11:137–156.

GRASBY, P. M., C. D. FRITH, K. I. FRISTON, C. BENCH, R. S. J. FRACKOWIAK, and R. J. DOLAN, 1993. Functional mapping of brain areas implicated in auditory-verbal memory function. *Brain* 116:1–20.

HAMLIN, R. M., 1970. Intellectual function 14 years after frontal lobe surgery. *Cortex* 6:299–307.

HEBB, D. O., and W. PENFIELD, 1940. Human behavior after extensive removal from the frontal lobes. *Arch. Neurol. Psychiatr.* 44:421–438.

HECAEN, H., and M. L. ALBERT, 1978. *Human Neuropsychology.* New York: Wiley.

Institute for Personality and Ability Testing, 1973. *Measuring Intelligence with the Culture Fair Tests.* Champaign, Ill.: The Institute for Personality and Ability Testing.

JAMES, W., 1890. *The Principles of Psychology.* New York: Holt, Rinehart and Winston.

KOWALSKA, D. M., J. BACHEVALIER, and M. MISHKIN, 1991. The role of the inferior prefrontal convexity in performance of delayed nonmatching-to-sample. *Neuropsychologia* 29:583–600.

LURIA, A. R., 1966. *Higher Cortical Functions in Man.* London: Tavistock.

LURIA, A. R., and L. D. TSVETKOVA, 1964. The programming of constructive ability in local brain injuries. *Neuropsychologia* 2:95–108.

MARSHALEK, B., D. F. LOHMAN, and R. E. SNOW, 1983. The complexity continuum in the radex and hierarchical models of intelligence. *Intelligence* 7:107–127.

MCLEOD, P., 1978. Does probe RT measure central processing demand? *Q. J. Exp. Psychol.* 30:83–89.

MCLEOD, P., and M. I. POSNER, 1984. Privileged loops from percept to act. In *Attention and Performance*, vol. 10, H. Bouma and D. G. Bouwhuis, eds. Hillsdale, N.J.: Erlbaum, pp. 55–66.

METTLER, F. A., 1949. *Selective Partial Ablation of the Frontal Cortex: A Correlative Study of Its Effects on Human Psychotic Subjects.* New York: Hoeber.

MILLER, G. A., E. GALANTER, and K. H. PRIBRAM, 1960. *Plans and the Structure of Behavior.* New York: Holt, Rinehart and Winston.

MILNER, B., 1963. Effects of different brain lesions on card sorting. *Arch. Neurol.* 9:90–100.

MILNER, B., 1965. Visually guided maze learning in man: Effects of bilateral hippocampal, bilateral frontal and unilateral cerebral lesions. *Neuropsychologia* 3:317–338.

MILNER, B., 1971. Interhemispheric differences in the localization of psychological processes in man. *Br. Med. Bull.* 27:272–277.

MISHKIN, M., and F. J. MANNING, 1978. Nonspatial memory after selective prefrontal lesions in monkeys. *Brain Res.* 143: 313–323.

NORMAN, D. A., and T. SHALLICE, 1980. *Attention to Action: Willed and Automatic Control of Behavior* (Rep. No. 8006). San Diego: University of California, Center for Human Information Processing.

OWEN, A. M., A. C. ROBERTS, J. R. HODGES, B. A. SUMMERS, C. E. POLKEY, and T. W. ROBBINS, 1993. Contrasting mechanisms of impaired attentional set-shifting in patients with frontal lobe damage or Parkinson's disease. *Brain.* 116:1159–1175.

PASSINGHAM, R., 1975. Delayed matching after selective prefrontal lesions in monkeys (*Macaca mulatta*). *Brain Res.* 92: 89–102.

PAUS, T., M. KALINA, L. PATOCKOVA, Y. ANGEROVA, R. CERNY, P. MECIR, J. BAUER, and P. KRABEC, 1991. Medial vs. lateral frontal lobe lesions and differential impairment of central-gaze fixation maintenance in man. *Brain* 114:2051–2068.

PENFIELD, W., and J. EVANS, 1935. The frontal lobe in man: A clinical study of maximum removals. *Brain* 58:115–133.

Petersen, S. E., P. T. Fox, M. I. Posner, M. Mintun, and M. E. Raichle, 1988. Positron emission tomographic studies of the cortical anatomy of single word processing. *Nature* 331:585–589.

Posner, M. I., and S. E. Petersen, 1990. The attention system of the human brain. *Annu. Rev. Neurosci.* 13:25–42.

Raichle, M. E., J. Fiez, T. O. Videen, P. T. Fox, J. V. Pardo, and S. E. Petersen, 1991. Practice-related changes in human brain functional anatomy. *Soc. Neurosci. Abstr.* 17:21.

Raven, J. C., J. H. Court, and J. Raven, 1988. *Manual for Raven's Progressive Matrices and Vocabulary Scales.* London: H. K. Lewis.

Reitman, W. R., 1965. *Cognition and Thought.* New York: Wiley.

Sacerdoti, E. D., 1974. Planning in a hierarchy of abstraction spaces. *Artif. Intell.* 5:115–135.

Schneider, W., and R. M. Shiffrin, 1977. Controlled and automatic human information processing: I. Detection, search, and attention. *Psychol. Rev.* 84:1–66.

Shallice, T., and P. W. Burgess, 1991. Deficits in strategy application following frontal lobe damage in man. *Brain* 114:727–741.

Snow, R. E., 1981. Toward a theory of aptitude for learning: I. Fluid and crystallized abilities and their correlates. In *Intelligence and Learning*, M. P. Friedman, J. P. Das, and N. O'Connor, eds. New York: Plenum, pp. 345–362.

Spearman, C., 1927. *The Abilities of Man.* New York: Macmillan.

Teuber, H.-L., 1972. Unity and diversity of frontal lobe functions. *Acta Neurobiol. Exp. (Warsz.)* 32:615–656.

Treisman, A. M., 1969. Strategies and models of selective attention. *Psychol. Rev.* 76:282–299.

Treisman, A. M., and A. Davies, 1973. Divided attention to ear and eye. In *Attention and Performance*, vol. 4, S. Kornblum, ed. London: Academic Press, pp. 101–117.

Walsh, K. W., 1978. *Neuropsychology: A Clinical Approach.* New York: Churchill Livingstone.

Warrington, E. K., M. James, and C. Maciejewski, 1986. The WAIS as a lateralizing and localizing diagnostic instrument: A study of 656 patients with unilateral cerebral lesions. *Neuropsychologia* 24:223–239.

Wechsler, D., 1955. *Wechsler Adult Intelligence Scale.* New York: Psychological Corporation.

Weinstein, S., and H.-L., Teuber, 1957. Effects of penetrating brain injury on intelligence test scores. *Science* 125: 1036–1037.

46 The Development of Visual Attention: A Cognitive Neuroscience Perspective

MARK H. JOHNSON

ABSTRACT The relation between the postnatal development of brain systems and the ontogeny of visual attention in the human infant is described. Transitions in overt orienting (involving head and eye movements) are accounted for in terms of the sequential development of particular cortical pathways. Experiments designed to measure covert (internal) shifts of visual attention are summarized, and results are presented which indicate that infants are capable of covert shifts of attention by 4 months of age. The onset of this ability is attributed to the development of neural circuitry involving the frontal eye fields and parietal cortex. Finally, aspects of saccadic and attentional control attributable to development of the prefrontal cortex are outlined.

Neuropsychological studies with adult patients examine the effects on cognition of the loss of certain neural systems or structures, while work with other species allows us to analyze brain-cognitive relations in simpler, and more tractable, neural systems. Development can be used as a method for studying the relation between neural systems and cognition in a way that complements the use of patient or animal data, for in developmental studies we can examine the effects of the addition of new neural systems to the simpler form of the human brain found in the newborn infant.

In recent years, it has become evident that there are multiple pathways involved in the control of eye movements and visual attention in adults (e.g., Schiller 1985; Posner and Petersen, 1990) (figure 46.1). Given the obvious difficulty in analyzing the complex combinations of hierarchical and parallel systems found in the adult, investigating the sequential development of these pathways and the construction of interacting brain systems during ontogeny may be informative (see Johnson, 1990, 1994a). Further, properly directed attention during infancy plays a vital role in ensuring the normal development of other cognitive functions, such as face recognition (Johnson and Morton, 1991; Johnson, 1992). In this chapter, I review studies on the ontogeny of both overt and covert aspects of visual orienting and suggest some ways that this infant data can shed light on the neural basis of visual attention.

Although our understanding of visual attention and orienting in adults is far from complete, a number of distinctions have been proposed that will be helpful in reviewing the ontogeny of attention (figure 46.2). Eye movements that shift gaze from one location to another may be referred to as *overt* orienting. In contrast, shifts of visual attention between spatial locations or objects that occur independently of eye and head movements are referred to as *covert* (Posner, 1980). Only in the past few years have studies designed to detect covert shifts of attention in infancy been performed, and the extent to which these covert processes in infants resemble those in adults requires further research.

A further distinction in the adult literature is that between *endogenous* and *exogenous* cuing of visual orienting (Klein, Kingstone, and Pontefract, 1992). Exogenously driven saccades are often short-latency and reflexive and are triggered by stimuli that appear within the visual field. In contrast, endogenously driven saccades are commonly of longer latency and are described as *intentional* or *volitional*. A common experimental procedure for studying exogenous cuing involves a brief cue stimulus being presented in a particular spatial location. This cue serves to draw attention to that location, with the result that subjects are (initially)

MARK H. JOHNSON Department of Psychology, Carnegie Mellon University, Pittsburgh, Pa.; and MRC Cognitive Development Unit, University College London, London, England

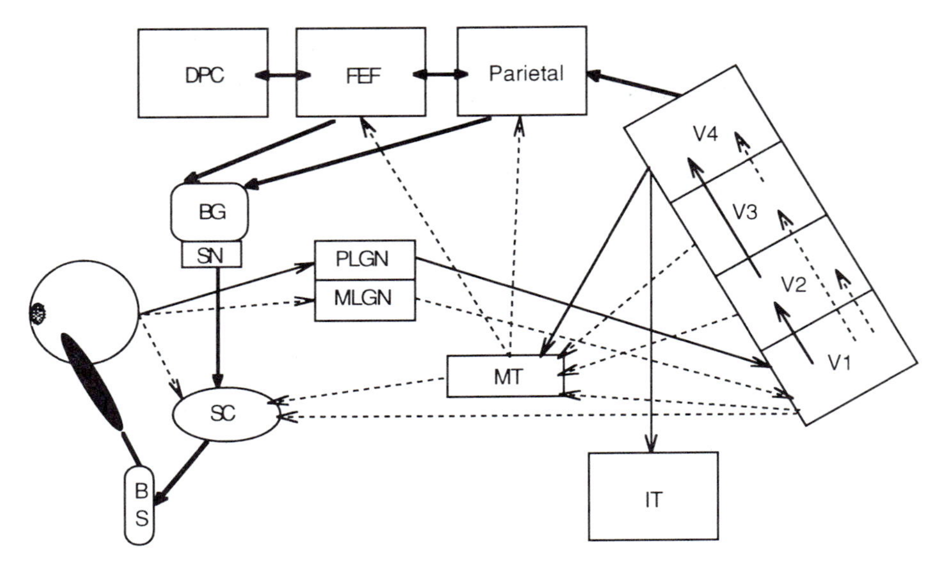

FIGURE 46.1 A diagram representing some of the main neural pathways and structures involved in visual orienting and attention. Solid lines indicate primarily parvocellular input, whereas the dashed arrows represent magnocellular input. V1–V4, visual cortex; FEF, frontal eye fields; DPC, dorsolateral prefrontal cortex; BG, basal ganglia; SN, substantia nigra; MLGN, PLGN, lateral geniculate, magno and parvo portions; MT, middle temporal area; IT, inferotemporal cortex; SC, superior colliculus; BS, brain stem. Bold lines indicate mixed input. (Courtesy of Rick Gilmore).

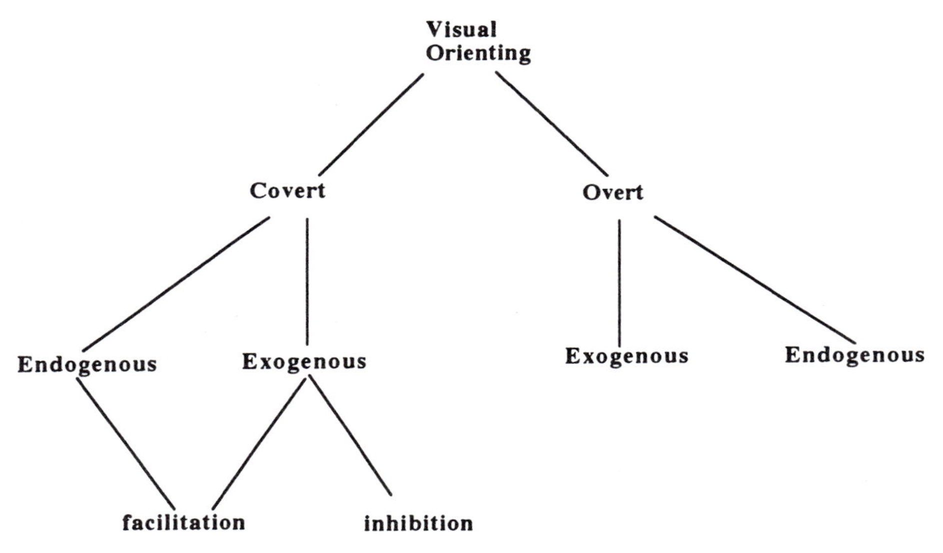

FIGURE 46.2 Diagram representing some dissociations between components of visual orienting and attention. See text for further explanation.

faster to respond to targets that appear in the same place. In an endogenous cuing experiment, a centrally presented arrow pointing to the right or left, an auditory cue, or a verbal instruction to look in a certain direction is used to direct attention. In this chapter, we will see how the ontogeny of visually guided behavior over the first few months of life can be partly viewed as a transition from exogenous to endogenous control of orienting.

A challenge facing all those interested in the neural basis of visual orienting and attention is to understand the mapping between the various neural pathways implicated in these processes (see figure 46.1) and the distinctions revealed by cognitive studies (see figure 46.2). As we shall see, the study of development offers a unique perspective on this mapping.

Overt visual orienting in early infancy

Bronson (1974, 1982) argued that visually guided behavior in the newborn human infant is controlled primarily by means of the subcortical retinocollicular pathway and that it is only by approximately 2 or 3 months of age that the locus of control switches to cortical pathways. The development of visually guided behavior can thus be viewed as a shift from subcortical to cortical processing. The claim that the primary visual pathway is not fully functioning until 2 or 3 months of postnatal age has been supported by a variety of electrophysiological, neuroanatomical, and behavioral studies (for reviews, see Atkinson, 1984, and Johnson, 1990). However, two factors have led to criticism of Bronson's original proposals. First, more recent neurophysiological knowledge about the visual pathways of the mammalian brain has led to reconsideration of the original two-pathways account. For example, it is now known that there are several comparatively independent cortical streams of processing held to have differing information-processing functions (e.g., see chapter 24). Furthermore, our understanding of the integrative abilities of the superior colliculus has greatly increased in recent years (see chapter 43). Thus, both Atkinson (1984) and Johnson (1990) have argued that the original cortical-subcortical dichotomy for the ontogeny of visual processing may inadequately capture the complexity of the transition. Second, the striking perceptual abilities of the very young infant in some tasks have led many psychologists to question the notion of the "decorticate" newborn (e.g., Slater, Morison, and Somers, 1988; Bushnell, Sai, and Mullin, 1989).

As a result of these considerations, Johnson (1990) suggested a specific hypothesis about the partial cortical functioning of newborns and its expansion over the first 6 months of life. This hypothesis accounts for changes in the development of overt visual orienting over the first few months of life, in terms of the maturation of several cortical pathways. These pathways were derived from proposals initially introduced to account for adult primate electrophysiological and lesion data related to oculomotor control (Schiller, 1985) and are as follows:

1. A direct pathway from the retina to the superior colliculus, involved in rapid input-driven (exogenous) eye movements toward simple, easily discriminable stimuli and fed mainly by the temporal visual field.
2. A cortical pathway that goes to the superior colliculus both directly from the primary visual cortex and also via the middle temporal area (MT). This pathway is largely driven by the broadband or magnocellular stream.
3. A cortical pathway that combines both broadband and color-opponent streams of processing in the frontal eye fields (FEFs) and that is involved in the detailed and complex analysis of visual stimuli, such as the temporal sequencing of eye movements within complex arrays and anticipatory saccades.
4. A final pathway for the control of eye movements involving tonic inhibition of the colliculus via the substantia nigra and basal ganglia.

Schiller (1985) proposed that the fourth pathway ensures that the activity of the colliculus can be regulated. More recent findings suggest that this oculomotor pathway forms an integrated system with the FEF and parietal lobes (e.g., Alexander, DeLong, and Strick, 1986) and that it plays some role in regulating subcortical processing by these cortical structures.

Johnson (1990) proposed that the characteristics of visually guided behavior of the infant at particular ages is determined by which of these pathways is functional, and which of these pathways is functional is influenced by the developmental state of the primary visual cortex. The basis of this claim at the neuroanatomical level lies in three sets of observations: (i) The primary visual cortex is the major (though not exclusive) gateway for input to the three cortical pathways (2, 3, 4) (Schiller, 1985); (ii) the primary visual cortex shows a postnatal continuation of the prenatal inside-out pattern of growth of the cortex, with the deeper layers (5 and 6) showing greater dendritic branching, length, and extent of myelinization than more superficial layers (2 and 3) near the time of birth (e.g., Purpura, 1975; Rabinowicz, 1979; Becker et al., 1984; Huttenlocher, 1990); and (iii) there is a restricted pattern of inputs and outputs from the primary visual cortex (e.g., the efferents to V2 depart from the upper layers; see, for example, Rockland and Pandya, 1979, and Burkhalter and Bernardo, 1989).

Johnson (1990) reasoned that if more superficial layers of the primary visual cortex are comparatively less developed than deeper layers in terms of the extent of dendritic growth, then projections from the superficial layers may be weaker, or even nonfunctional, at a stage when projections from the deeper layers are strong. This hypothesized relative restriction on projection patterns yielded the prediction that output from V1 to pathway 2 should be stronger than the projection to pathway 3 at early stages of development. Such a prediction from cellular development has implications for information processing: The functions subserved by pathway 2 involving structure MT should appear earlier in development than those of the pathway 3 involving the FEF. Employing this logic, Johnson (1990) attempted to account for characteristics of the visually guided behavior of the infant in terms of the sequential development of pathways underlying visual orienting. This sequential development is a dynamic process, but a number of phases may be characterized.

It should be stressed that the phases described in the following sections are not meant to be rigid or sudden-onset stages of development. Rather they are to be viewed as snapshots along a dynamic and constantly changing path. Similarly, the strong predictions from Johnson's (1990) account concern the *sequence* of development within an individual infant rather than the exact ages of group effects. In general, the transitions described next can be characterized as a shift from exogenous or automatic eye movement control to a more predictive system influenced by endogenous factors (see Johnson, 1994b).

THE NEWBORN As mentioned previously, evidence from measures of the extent of dendritic arborization and myelinization suggest the hypothesis that only the deeper layers of the primary visual cortex are capable of supporting organized information-processing activity in the human newborn. Because the majority of feedforward intracortical projections depart from outside the deeper layers (5 and 6), some of the cortical orienting pathways (those involving MT and the FEFs) may be receiving only weak or disorganized input at this stage. Nonetheless, evidence from various sources, such as visually evoked potentials, indicate that information from the eye is entering the primary visual cortex in the newborn. Thus, while agreeing with Bronson's (1974) proposal that *most* of the newborn's visual behavior can be accounted for in terms of processing in the subcortical pathway, Johnson (1990) argued that there is *some* information processing occurring in the deeper cortical layers at birth.

There are a number of characteristics of newborn visually guided behavior consistent with predominantly subcortical control. Among these are saccadic pursuit tracking of moving stimuli, preferential orienting to the temporal field, and the externality effect.

Saccadic pursuit tracking of moving stimuli Aslin (1981) reports that tracking in very early infancy has two characteristics. The first is that the eye movements follow the stimulus in a saccadic or steplike manner, as opposed to the smooth pursuit found in adults and older infants. The second characteristic is that the eye movements always lag behind the movement of the stimulus rather than predicting its trajectory. Therefore, when a newborn infant visually tracks a moving stimulus, it could be described as performing a series of separate reorientations. Such behavior is consistent with collicular control of orienting (see Johnson, 1990, for details).

Preferential orienting to temporal field Newborns much more readily orient toward stimuli in the temporal, as opposed to the nasal, visual field (e.g., Lewis, Maurer, and Milewski, 1979). Posner and Rothbart (1980) suggest that midbrain structures such as the colliculus can be driven most readily by temporal field input. This proposal has been confirmed in studies of adult blindsight patients by Rafal and coworkers (1990; see also chapter 40), who established that distractor stimuli placed in the temporal "blind field" had an effect on orienting into the good field, whereas distractor stimuli in the nasal blind field did not. Recent evidence from studies of infants with complete hemispherectomy indicate that the subcortical (collicular) pathway is capable of supporting saccades toward a peripheral stimulus in the cortically blind field (Braddick et al., 1992).

The externality effect Infants in the first few months of life do not attend to stationary pattern elements within a larger frame or pattern (e.g., Maurer and Young, 1983) unless these elements are moving (Bushnell, 1979). Although a variety of explanations have been proposed to account for this phenomenon (e.g., Aslin and Smith, 1988), Johnson (1990) proposed that part of the explanation could involve a collicular mechanism attempting to shift the retinal image of the largest frame or pattern elements into the foveal field.

Bearing in mind that the majority of visuomotor functions in the newborn are likely to be primarily subcortical, we should note that evidence for pattern recognition (Slater, Morison, and Rose, 1982) and orientation discrimination (Atkinson et al., 1988; Slater, Morison, and Somers, 1988) are indicative of at least some cortical functioning.

ONE MONTH OF AGE Between 1 and 3 months of age, infants show obligatory attention (Stechler and Latz, 1966): That is, they have great difficulty in disengaging their gaze from a stimulus in order to saccade elsewhere. Although this phenomenon remains poorly understood, Johnson (1990) suggests that it is due to the development of tonic inhibition of the colliculus via the substantia nigra (pathway 4). This as yet unregulated tonic inhibition of the colliculus has the consequence that stimuli impinging on the peripheral visual field no longer elicit an automatic exogenous saccade as readily as in newborns. An alternative account of obligatory attention was proposed by Posner and Rothbart (1980), who suggest that it is only after development of the parietal lobes that the infant has the ability to disengage from stimuli. Thus, the end of obligatory attention marks the onset of covert attention abilities in the infant (see next section). Whichever of these accounts of obligatory attention proves to be correct, the failure to disengage may be object-centered rather than dependent on spatial location (see Johnson, 1994a).

TWO MONTHS OF AGE Although their eye movements still lag behind the movement of the stimulus, infants at approximately 2 months of age begin to demonstrate periods of smooth visual tracking. Furthermore, they become more sensitive to stimuli placed in the nasal visual field (Aslin, 1981) and become sensitive to coherent motion (R. Spitz, personal communication, 1992). Johnson (1990) proposed that the onset of these behaviors coincides with the functioning of the pathway involving structure MT. The enabling of this route of eye movement control may provide the cortical magnocellular stream with the ability to regulate activity in the superior colliculus.

MORE THAN 3 MONTHS OF AGE After approximately 3 months of age, the pathway involving the FEFs may gain in strength due to further dendritic growth and myelinization within the upper layers of the primary visual cortex. These cellular level developments may greatly increase the infant's ability to make anticipatory eye movements and to learn sequences of looking patterns. Further, with regard to the visual tracking of a moving object, not only do infants now show periods of smooth tracking, but their eye movements often predict the movement of the stimulus in an anticipatory manner. A number of experiments by Haith and colleagues (1988) have demonstrated that anticipatory eye movements can be readily elicited from infants by this age. Haith exposed $3\frac{1}{2}$-month-old infants to a series of picture slides that appeared either on the right or on the left side of the infant. These stimuli were presented either in an alternating sequence with a fixed interstimulus interval (ISI) or with an irregular alternation pattern and ISI. The regular alternation pattern produced more stimulus anticipations, and reaction times to make an eye movement were reliably faster than in the irregular series. Haith (1988) concluded from these results that infants of this age are able to develop expectancies for noncontrollable spatiotemporal events. Canfield and Haith (1991) tested 2- and 3-month-old infants in a similar experiment that included more complex sequences (such as left-left-right, left-left-right). They failed to find significant effects with 2-month-olds, but 3-month-olds appeared to able to acquire at least some of the more complex sequences. Furthermore, if the acquired sequence was changed to another sequence, infants of 3 months made errors consistent with having acquired the first sequence (Arehart and Haith, 1991).

Evidence for covert shifts of visual attention in early infancy

In adults, covert attention may be directed to a spatial location by a very briefly presented visual stimulus. Although subjects do not make a saccade to this stimulus, they are faster to report (often by means of a button press) the appearance of a target stimulus in the cued location than in another location. We may therefore infer that a covert shift of attention to the cued location facilitates subsequent detection at the location.

With infants, we are limited to indirect methods of studying covert shifts of attention. Furthermore, we have the problem that infants do not accept verbal instruction and are poor at motor responses readily elicited from adults (e.g., a button press). One motor response that can be readily elicited even from very young infants is eye movement (overt orienting). Thus, in most of the experiments reported here, measures of overt orienting are used to study covert shifts of attention. This can be done by examining the influence of a cue stimulus, to which infants do not make an eye movement, on their subsequent saccades toward target stimuli. This approach has also been taken in some adult studies purporting to measure shifts of covert attention (e.g., Maylor, 1985; Shephard, Findlay, and Hockey, 1986). Of course, this does not mean that covert shifts of attention are not involved in tasks in which the infant makes a saccade toward the cue, merely that it is more difficult to establish that this is so.

Whereas processes of covert attention develop throughout the school-age years (e.g., Enns and Brodeur, 1989; Tipper et al., 1989), I will focus on developments in the first 6–8 months of life, as these can more readily be associated with the development of neural pathways and structures and are thus more relevant for a cognitive neuroscientific perspective.

Exogenously Cued Covert Shifts of Attention One way in which evidence for covert attention has been provided in adults is by studying the effect on detection of cuing saccades to a particular spatial location. A briefly presented cue serves to draw covert attention to the location, resulting in the subsequent facilitation of detection of targets at that location (Posner and Cohen, 1980; Maylor, 1985). Whereas *facilitation* of detection and responses to a covertly attended location occurs if the target stimulus appears very shortly after the cue offset, *inhibition* of saccades toward that location occurs with longer latencies between cue and target. This latter phenomenon, referred to as *inhibition of return* (Posner et al., 1985), may reflect an evolutionarily important mechanism for preventing attention from returning to a spatial location that has very recently been processed. In adults, facilitation is observed reliably when targets appeared at the cued location within approximately 150 ms of the cue, whereas targets that appear between 300 and 1300 ms after a peripheral (exogenous) cue result in longer detection latencies (e.g., Posner and Cohen, 1980, 1984; Maylor, 1985).

After incurring lesions to the posterior parietal lobe, adults show severe neglect of the contralateral visual field. According to Posner and colleagues (Posner 1988; Posner and Petersen, 1990), this neglect is due to damage to the posterior attention network, a brain circuit that includes not only the posterior parietal lobe but also the pulvinar and superior colliculus (see figure 46.1 for all but the pulvinar). Damage to this circuit is postulated to impair subjects' ability to shift covert attention to a cued spatial location. The involvement of these regions in shifts of visual attention has been confirmed by positron emission tomography (PET) studies (see chapter 41). Both neuroanatomical (Conel, 1939–1967) and PET (Chugani, Phelps, and Mazziotta, 1987) evidence from the human infant indicate that the parietal lobe is undergoing substantive and rapid development between 3 and 6 months after birth. The question arises, therefore, as to whether infants become capable of covert shifts of attention during this time.

Hood and Atkinson (1991; see also Hood, in press) reported that 6-month-old infants displayed faster reaction times (RTs) to make a saccade to a target when it appeared immediately after a brief (100-ms) cue stimulus, whereas a group of 3-month-olds did not. Johnson (1993a, 1994b) employed a similar procedure in which a brief (100-ms) cue was presented on one of two side screens before bilateral targets were presented either 100 or 600 ms later. He reasoned that the 200-ms stimulus onset asynchrony (SOA) may be short enough to produce facilitation, whereas the long SOA trials should result in preferential orienting toward the opposite side (inhibition of return). From figure 46.3, it can be seen that there was no significant difference in the RT to make a saccade to the cued or opposite target between the long and the short ISI trials for a group of 2-month-old infants. However, in a 4-month-old group of infants, a significant facilitation of RTs was noted when making a saccade toward the cued target: At the short ISI, these infants showed a faster RT toward the cued target, whereas at the long ISI, they were slower to respond to the cued target (inhibition). A similar result was found with the direction of saccade (cued or opposite target) data (Johnson, 1993a).

Similar experiments run with slightly older infants indicate that covert shifts of attention may get faster between 4 and 8 months of age. Whereas 4-month-olds

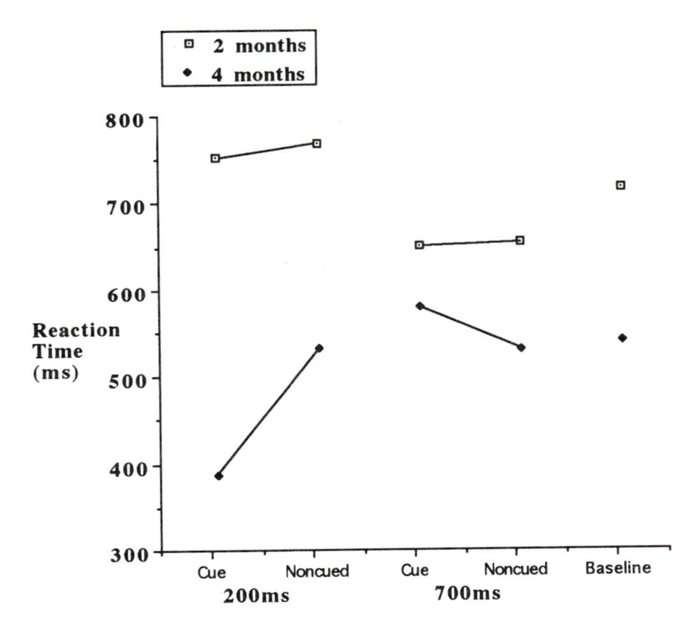

FIGURE 46.3 The mean median reaction time of groups of 2- and 4-month-old infants to respond to cued targets, opposite targets, and no-cue baseline trials, following either a short (200 ms) or long (700 ms) stimulus onset asynchrony. (Data from Johnson, 1993a.)

show clear facilitation when a target is presented 200 ms after the onset of a cue, 6- and 8-month-olds, like adults, show equivalent facilitation only if the target is presented less than 150 ms after cue onset (Johnson, 1993a, b). At present, it remains an open question whether the facilitation produced by the cue in these experiments is the result of direct priming of the eye movement system or whether the eye movements are following an independent covert shift of attention. However, the finding that 4-month-old infants show facilitation to a cue even when it predicts the presentation of a target in the opposite direction (normally also necessitating an eye movement in the opposite direction) is more consistent with the theory that facilitation effects are due to drawing covert attention to the location (Johnson, Posner, and Rothbart, 1994).

While facilitation of responding to a cued spatial location may depend on the functioning of the parietal lobe, inhibition of return (IOR) has been more closely associated with the functioning of the superior colliculus (see chapter 40). For example, supranuclear palsy, a progressive disease that affects the colliculus, selectively reduces IOR (Posner et al., 1985). Because midbrain structures such as the superior colliculus are believed to be well developed at the time of birth (Bronson, 1974; Atkinson, 1984; Chugani, Phelps, and Mazziotta, 1987; Johnson, 1990), we would expect that IOR should be present from around the time of birth in the human infant.

IOR can be studied in infants following either overt or covert shifts of attention, according to whether or not the infant is allowed to make a saccade toward the cue stimulus. Clohessy and colleagues (1991) studied IOR following overt orienting. Infants sat in front of three monitor screens on which colorful dynamic stimuli were presented. At the start of each trial, an attractive fixation stimulus appeared on the central screen. Once the infant had fixated on this stimulus, a cue stimulus was presented on one of the two side monitor screens. When the infant had made a saccade toward the cue, it was turned off, following which the infant returned its gaze to the center screen before an identical target stimulus was presented bilaterally on both side screens. Infants of 3 months of age showed no significant preferential orienting to the bilateral targets as a result of the cue, but infants of 6 months of age made saccades more frequently toward the side opposite that on which the cue had appeared. The authors argued that this preferential orienting toward the opposite side from the cue is indicative of IOR and its development between 3 and 6 months of age.

More recent experiments suggest that the age when IOR can first be demonstrated may depend on the size of the visual angle between the central fixation point and the cue or target. Harman, Posner, and Rothbart (in press) reasoned that if infants have to make several saccades to a target at 30° eccentricity, then they will not show IOR at the target destination. This is because in adult studies, IOR has been linked to saccade planning (Rafal et al., 1989). A target at only 10° eccentricity, however, can easily be reached in one saccade in the very young infant, and thus IOR should be observed. In accordance with their prediction, Harman, Posner, and Rothbart (1992) found evidence of IOR in 3-month-old infants at 10° but not at 30°. Although it is possible that IOR might be found at still younger ages (Valenza et al., 1992), Harman and colleagues suggest that its developmental onset probably is linked to the maturation of cortical structures involved in the development of programed eye movements, namely the FEFs.

To date, evidence for IOR following covert shifts of attention has been detected only in infants of 4 months and older. Hood and Atkinson (1991) used a short cue duration (100 ms) to ensure that the infants did not make a saccade toward this stimulus. Thus, any effects of the cue presentation on subsequent saccades to the target could be attributed to a covert shift of attention during the cue presentation. Six-month-old infants showed IOR under these conditions, but a group of 3-month-old infants did not. Hood (in press) reports an experiment similar to that of Hood and Atkinson (1991) but with an improved method that allows, among other things, more accurate assessment of RTs to make a saccade toward the target. In this experiment, a group of 6-month-olds were exposed to a longer-duration cue (180 ms) before immediately being presented with a single target on either the ipsilateral or the contralateral side. This procedure resulted in a clear difference in mean RT to orient toward the target, depending on whether it appeared in the same spatial location as the cue. Using methods similar to those of Hood and colleagues, Johnson (1993b, 1994b) studied IOR after covert orienting in 2-, 4-, 6-, and 8-month-old infants and confirmed that the phenomenon was present only in the older age groups (for 2- and 4-month-old RT data, see figure 46.3). Because all these studies used peripheral cues and targets that were more than 10° from fixation, the conclusion that infants younger than 4 months do not show IOR under these circumstances must remain preliminary.

The pattern of results obtained with infants after covert and overt orienting suggests one of two possibilities about the neural basis of IOR: (1) that, in addition to the colliculus, IOR involves cortical structures such as the FEF, which become sufficiently mature 3–4 months after birth, or (2) that covert shifts of attention require cortical structures that are not functional until 3–4 months of age, and thus IOR after covert shifts of attention cannot be demonstrated until this age. The second of these possibilities predicts that IOR after overt orienting will be present in the first few months of life and perhaps even from birth. A recent report claims to find IOR in newborns after overt orienting (Valenza et al., 1992). If this preliminary evidence from newborns can be replicated, then the infant data suggest that IOR after overt orienting is exclusively dependent on subcortical structures and that cortical structures are involved only after covert shifts of attention.

Endogenously Cued Covert Shifts of Attention

The direction of attention in response to verbal, symbolic, or memorized cues is referred to as *endogenously driven orienting*. Whereas the endogenous direction of attention undoubtedly involves many of the same structures and circuits as play a role in orienting to exogeneous cues, recent PET evidence indicates additional involvement of the prefrontal cortex when endogenous cues are used (Deiber et al., 1991; Frith et al., 1991; Corbetta et al., 1993).

Most adult studies of endogenously cued shifts of attention have involved the use of verbal or symbolic cues (such as an arrow), but with infants we have to rely on memory as the source of endogenous cuing. This has been done in several laboratories by rewarding infants for making a saccade in a particular direction after presentation of an auditory or visual cue (e.g., de Schonen and Bry, 1987; Colombo et al., 1990; Johnson, Posner, and Rothbart, 1991).

One example of such a study was conducted by Johnson, Posner, and Rothbart (1991), who attempted to train infants to use a stimulus presented in a central location as a cue to predict the peripheral location (right or left of center) at which an attractive target stimulus would subsequently appear. This experiment is similar, in some respects, to studies in adults in which attention is cued to a peripheral location by means of a central (endogenous) cue such as an arrow. Groups of 2-, 3-, and 4-month-old infants were exposed to a number of training trials in which there was a contingent relation between which of two cue stimuli were presented on the central monitor and the location (right or left of center) at which an attractive target stimulus was subsequently presented. After a number of such training trials, we occasionally presented test trials in which the target appeared on both of the side monitors, regardless of which central stimulus preceded it. In these test trials, we measured whether the infants looked more toward the cued location than toward the opposite location. Whereas 2- and 3-month-old infants looked equally frequently to the cued and opposite sides, the group of 4-month-olds looked significantly more often toward the cued location. This result, taken together with a similar finding in an earlier study (de Schonen and Bry, 1987), is consistent with endogenously cued shifts of attention being present in 4-month-old infants.

Several other tasks recently conducted with infants are consistent with an increasing prefrontal cortex endogenous control over shifts of attention and saccades at approximately 4 months of age. For example, Funahashi, Bruce, and Goldman-Rakic (1989, 1990) have devised an oculomotor delayed response task to study the properties of neurons in the dorsolateral prefrontal cortex of macaque monkeys. In this task, the monkey plans a saccade toward a particular spatial location but has to wait for a period (usually between 2 and 5 seconds) before actually executing the saccade. Single-unit recording in the macaque indicates that some cells in the dorsolateral prefrontal cortex code for the direction of the saccade during the delay. Furthermore, reversible microlesions to the area result in selective amnesia for saccades to a localized part of the visual field. A recent PET study on human subjects has confirmed the involvement of prefrontal cortex (and parietal cortex) in this task (Jonides et al., 1993).

Gilmore and Johnson (1994) devised an infant version of the oculomotor delayed response task. The results obtained to date indicate that 6-month-old human infants can perform delayed saccades successfully with delays of up to at least 5 seconds, suggesting some influence of the prefrontal cortex on eye movement control by this age. As an aside, these findings with infants are relevant to claims that the dorsolateral prefrontal cortex is not sufficiently developed to support tasks that require both inhibition and working memory until approximately 10 months (e.g., Diamond, 1991) and may be consistent with a more graded account of the development of these abilities (see Munakata et al., 1994).

Another way in which the development of endogenous control can be studied is to investigate the extent to which endogenous factors can overide or inhibit the influence of exogenous cues. One task that opposes the influence of exogenous and endogenous cues is the so-called antisaccade task (Hallett, 1978). In an antisaccade task, subjects are instructed not to make a saccade toward a stimulus but rather to saccade in the opposite direction where a second stimulus is subsequently presented: That is, subjects have to inhibit a spontaneous, automatic, eye movement toward a stimulus and direct their saccade in the opposite direction. Guitton, Buchtel, and Douglas (1985) reported that while normal subjects and patients with temporal lobe damage could do this easily, patients with frontal lobe damage, and especially those with damage around the FEF, were severely impaired. That is, the frontal lobe patients had difficulty suppressing the automatic saccades toward the first stimulus.

Clearly, one cannot give verbal instructions to a young infant to make a saccade in a direction opposite from the location in which the first stimulus appears. However, we can motivate the infant to want to look at a second (target) stimulus more than at the first (cue) stimulus. This can be achieved by making the second stimulus more dynamic and colorful than the first. Thus, over a series of trials, an infant may learn to withhold a saccade to the first stimulus in order to anticipate the appearance of the second stimulus on the opposite side of center.

Preliminary evidence has been obtained for a steady decrease in the extent of orienting toward the first (cue) stimulus over a number of training trials in 4-month-old infants (from an initial level of more than 80% to a level of less than 50%) (see Johnson, 1994b). Although this preliminary finding needs to be replicated and extended with a larger sample, the large number of trials in which infants made a saccade straight from the fixation point to the second (target) stimulus in the later stages of the experiment indicates that endogenous control can influence responses to exogenous cues.

Another manifestation of the endogenous control of attention concerns so-called sustained attention. *Sustained attention* is the ability of subjects to maintain the direction of their attention toward a stimulus even in the presence of distractors. Richards (1989a, b) has developed a heart rate marker for sustained attention in infants. The heart rate–defined period of sustained attention usually lasts for between 5 and 15 seconds after the onset of a complex stimulus. Figure 46.4 illustrates the heart rate–defined periods of sustained attention delineated by Richards and Casey (1991).

To investigate the effect of sustained attention on the response to exogenous cues, Richards (1989a, b) used an interrupted stimulus method in which a peripheral stimulus (a flashing light) is presented while the infant is gazing at a central stimulus (a television screen with a complex visual pattern). By varying the length of time between the onset of the television image and the onset of the peripheral stimulus, he was able to present

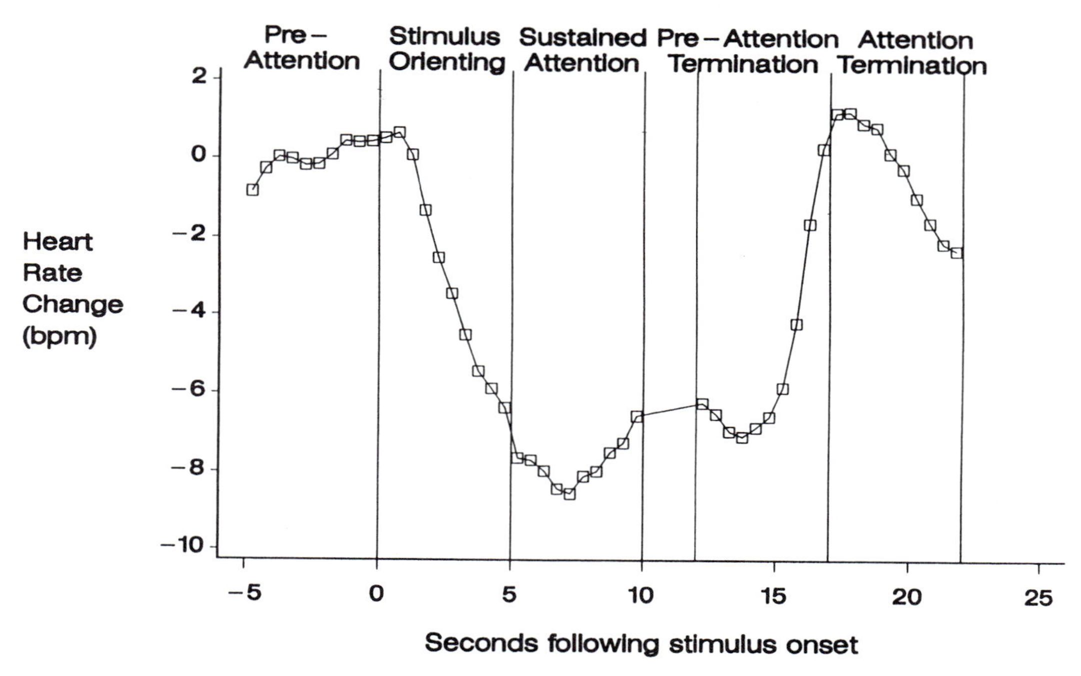

FIGURE 46.4 The heart rate–defined phases of sustained attention. (Reprinted from Richards and Casey, 1991, by permission.)

the peripheral stimulus either within the period of sustained covert attention or outside it. Richards found that during the periods when heart rate was decreased (sustained endogenous attention), it took twice as long for the infant to shift its gaze toward the peripheral stimulus as when heart rate had returned to prestimulus levels (attention termination). Furthermore, those saccades that are made to a peripheral stimulus during sustained attention are less accurate than normal and involve multiple hypometric saccades, characteristics of collicular-generated saccades (Richards, 1991). Thus, the lack of distractibility during periods of sustained attention is likely to be due to cortically mediated pathways inhibiting collicular mechanisms.

If we assume that the cortically mediated pathways responsible for the regulation of collicular saccades are in the frontal cortex, then we should expect there to be no clearly definable period of sustained attention in infants younger than approximately 4 months. Concordant with this suggestion, Richards (1989a) reports that the clearly defined differences in distractibility in relation to heart rate were not found, or were at least very slight, in a group of infants 8 weeks of age.

CORTICAL CONTROL OF ATTENTIONAL SHIFTS AND THE OCULOMOTOR CIRCUIT Three regions of cortex have been identified as being important for shifts of attention: The FEFs are linked to the endogenous control of eye movements, the parietal cortex to covert shifts of attention, and the prefrontal cortex to endogenous control involving delays. Clearly, it would be simple-minded to believe that the functions of these three structures are completely independent of one another. It would also be misleading to describe the onset of functioning in these structures as an all-or-none phenomenon. It is much more likely that their functioning develops in a more collaborative and graded manner. This being likely, it is interesting to note that there are a number of closed-loop circuits that project down from the cortex to the basal ganglia before returning to one of the cortical regions from which it originated (for review, see Alexander, DeLong, and Strick, 1986). One of these pathways is referred to commonly as the *oculomotor circuit*, due to evidence from neurophysiological studies of its involvement in eye movements. It has been proposed that this circuit is crucial for voluntary saccades (Alexander, DeLong, and Strick, 1986). It receives projections from the FEF, parietal cortex, and dorsolateral prefrontal cortex. After passing through a number of subcortical structures such as the caudate and portions of the substantia nigra (which is crucial for tonic inhibition of the colliculus), it returns to the FEF. An attractive possibility is that the oculomotor circuit develops as an integrated whole, giving rise to many of the transitions observed in both overt and covert orienting at approximately 4 months of age.

Conclusions and future directions

In this chapter, I have traced the hierarchical development of neural pathways underlying components of visual orienting and attention. These neural developments were reflected in a general trend at the behavioral level: The control of saccades passed from being driven mainly by exogenous stimuli to being largely under the control of endogenous factors such as memory. This trend is seen in a variety of tasks, including those in which the infant learns to predict the spatial location at which a target will appear. We have also seen how developmental data can complement data from other sources and even help to resolve controversies about the neural basis of attention in adults.

At least three directions for future research are evident at this point: preattentive parsing of the visual field, the study of infants with brain damage, and neuroimaging studies. Rovee-Collier, Hankins, and Bhatt (1992) have begun work on the first of these issues, preattentive processing, by demonstrating that 3-month-olds show some similar "pop-out" effects to those observed in adults. Interestingly, while their 3-month-old subjects appeared to parse the visual field in a way similar to adults in the presence of one target, they did not show evidence of focused attention when presented with multiple target stimuli within an array. From the viewpoint developed in this chapter, we would expect such focused attention to be present from 4 months of age. Furthermore, it is possible that phenomena such as the externality effect observed over the first month or so of life may reflect the inability of infants to segment the visual field in the same way as do adults.

A number of laboratories have begun to apply visual attention tasks such as those discussed in this chapter to infants with focal or diffuse brain damage (e.g., Braddick et al., 1992; Ross, Tesman, and Nass, 1992). Such investigations not only will allow us to test theories about the relation between the developing brain and attention but may also prove to be of clinical and diagnostic value. Finally, noninvasive neuroimaging methods such as high-density event-related potentials are likely to be powerful tools for unraveling the relation between the development of the brain and this aspect of cognition.

ACKNOWLEDGMENTS I wish to thank Mike Posner, Rick Gilmore, and Jeff Shrager for comments on drafts of this chapter, and Leslie Tucker and Kathy Sutton for their assistance with some experiments described as well as with preparation of the manuscript. Financial assistance was provided by National Science Foundation grant DBS-9120433 and the Carnegie Mellon faculty development fund.

REFERENCES

ALEXANDER, G. E., M. R. DELONG, and P. L. STRICK, 1986. Parallel organization of functionally segregated circuits linking basal ganglia and cortex. *Annu. Rev. Neurosci.* 9: 357–382.

AREHART, D. M., and M. M. HAITH, 1991. Evidence for visual expectation violations in 13-week old infants. *Abstracts of the Society for Research in Child Development volume 8.*

ASLIN, R. N., 1981. Development of smooth pursuit in human infants. In *Eye Movements: Cognition and Visual Perception*, D. F. Fisher, R. A. Monty, and J. W. Senders, eds. Hillsdale, N.J.: Erlbaum, pp. 31–51.

ASLIN, R. B., and L. B. SMITH, 1988. Perceptual development. *Annu. Rev. Psychol.* 39:435–473.

ATKINSON, J., 1984. Human visual development over the first six months of life: A review and a hypothesis. *Hum. Neurobiol.* 3:61–74.

ATKINSON, J., B. HOOD, J. WATTAM-BELL, S. ANKER, and J. TRICKLEBANK, 1988. Development of orientation discrimation in infants. *Perception* 17:587–595.

BECKER, L. E., D. L. ARMSTRONG, F. CHAN, and M. M. WOOD, 1984. Dendritic development on human occipital cortex neurones. *Brain Res.* 315:117–124.

BRADDICK, O. J., J. ATKINSON, B. HOOD, W. HARKNESS, G. JACKSON, and F. VARGHA-KHADEM, 1992. Possible blindsight in infants lacking one cerebral hemisphere. *Nature* 360:461–463.

BRONSON, G. W., 1974. The postnatal growth of visual capacity. *Child Dev.* 45:873–890.

BRONSON, G. W., 1982. Structure, status and characteristics of the nervous system at birth. In *Psychobiology of the Human Newborn*, P. Stratton, ed. Chichester, Engl.: Wiley.

BURKHALTER, A., and K. L. BERNARDO, 1989. Organization of corticocortical connections in human visual cortex. *Proc. Natl. Acad. Sci.* 86:1071–1075.

BUSHNELL, I. W. R., 1979. Modification of the externality effect in young infants. *J. Exp. Child Psychol.* 28:111–229.

Bushnell, I. W. R., F. Sai, and J. Mullin, 1989. Neonatal recognition of the mother's face. *Br. J. Dev. Psychol.* 7:3–15.

Canfield, R. L., and M. M. Haith, 1991. Young infants' visual expectations for symmetric and asymmetric stimulus sequences. *Dev. Psychol.* 27:198–208.

Chugani, H. T., M. E. Phelps, and J. C. Mazziotta, 1987. Positron emission tomography study of human brain functional development. *Ann. Neurol.* 22:487–497.

Clohessy, A. B., M. I. Posner, M. K. Rothbart, and S. Vecera, 1991. The development of inhibition of return in early infancy. *J. Cogn. Neurosci.* 3:346–357.

Colombo, J., D. W. Mitchell, J. T. Coldren, and J. D. Atwater, 1990. Discrimination learning during the first year: Stimulus and positional cues. *J. Exp. Psychol. [Learn. Mem. Cogn.]* 16:98–109.

Conel, J. L., 1939–1967. *The Postnatal Development of the Human Cerebral Cortex*, vols. 1–8. Cambridge, Mass.: Harvard University Press.

Corbetta, M., F. M. Miezin, G. L. Shulman, and S. E. Peterson, 1993. A PET study of visuospatial attention. *J. Neurosci.* 13:1202–1226.

Deiber, M. P., R. E. Passingham, J. G. Colebatch, K. J. Friston, P. D. Nixon, and R. S. J. Frackowiak, 1991. Cortical areas and the selection of movement: A study with positron emission tomograpy. *Exp. Brain Res.* 84:393–402.

de Schonen, S., and I. Bry, 1987. Interhemispheric communication of visual learning: A developmental study in 3–6 month old infants. *Neuropsychologia* 25:73–83.

Diamond, A., 1991. Neuropsychological insights into the meaning of object concept development. In *The Epigenesis of Mind: Essays on Biology and Cognition*, S. Carey and R. Gelman, eds. Hillsdale, N.J.: Erlbaum.

Enns, J. T., and D. A. Brodeur, 1989. A developmental study of covert orienting to peripheral visual cues. *J. Exp. Child Psychol.* 48:171–189.

Frith, C. D., K. Friston, P. F. Liddle, R. S. J. Frackowiak, 1991. Willed action and the prefrontal cortex in man: A study with PET. *Proc. R. Soc. Lond.* 244:241–246.

Funahashi, S., C. J. Bruce, and P. S. Goldman-Rakic, 1989. Mnemonic coding of visual space in the monkey's dorsolateral prefrontal cortex. *J. Neurophysiol.* 61:331–349.

Funahashi, S., C. J. Bruce, and P. S. Goldman-Rakic, 1990. Visuospatial coding in primate prefrontal neurons revealed by oculomotor paradigms. *J. Neurophysiol.* 63(4): 814–831.

Gilmore, R., and M. H. Johnson, 1994. Six month olds performance in versions of the oculomotor delayed response task. *Infant Behavior and Development* (ICIS Abstracts).

Guitton, H. A., H. A. Buchtel, and R. M. Douglas, 1985. Frontal lobe lesions in man cause difficulties in suppressing reflexive glances and in generating goal-directed saccades. *Exp. Brain Res.* 58:455–472.

Haith, M. M., C. Hazan, and G. S. Goodman, 1983. Expectation and anticipation of dynamic visual events by 3.5-month-old babies. *Child Dev.* 59:467–479.

Hallett, P. E., 1978. Primary and secondary saccades to goals defined by instructions. *Vision Res.* 18:1279–1296.

Harman, C., M. I. Posner, and M. K. Rothbart (in press). Spatial attention in 3-month olds: Inhibition of return at 10 degree and 30 degree target eccentricities. *Can. J. Psychol.*

Hood, B. M., 1993. Inhibition of return produced by covert shifts of visual attention in 6-month old infants. *Infant Behav. Dev.* 16:245–254.

Hood, B. M. (in press). Visual selective attention in the human infant: A neuroscientific approach. *Adv. Infancy Res.*

Hood, B., and J. Atkinson, 1991. Shifting covert attention in infants. *Abstracts of the Society for Research in Child Development*, Vol. 8.

Huttenlocher, P., 1990. Morphometric study of human cerebral cortex development. *Neuropsychologia* 28:517–527.

Johnson, M. H., 1990. Cortical maturation and the development of visual attention in early infancy. *J. Cogn. Neurosci.* 2:81–95.

Johnson, M. H., 1992. Cognition and development: Four contentions about the role of visual attention. In *Cognitive Science and Clinical Disorders*, D. J. Stein and J. E. Young, eds. San Diego: Academic Press, pp. 43–60.

Johnson, M. H., 1993a. Evidence for covert shifts of attention in early infancy. Manuscript submitted for publication.

Johnson, M. H., 1993b. The temporal dynamics of shifting visual attention during early infancy. Manuscript submitted for publication.

Johnson, M. H., 1994a. Dissociating components of visual attention: A neurodevelopmental approach. In *The Neural Basis of High-Level Vision*, M. J. Farah and G. Radcliffe, eds. Hillsdale, N.J.: Erlbaum, pp. 241–268.

Johnson, M. H., 1994b. Visual attention and the control of eye movements in early infancy. In *Attention and Performance: XV. Conscious and Nonconscious Information Processing*, C. Umilta and M. Moscovitch, eds. Cambridge, Mass.: MIT Press, pp. 291–310.

Johnson, M. H., and J. Morton, 1991. *Biology and Cognitive Development: The Case of Face Recogniton.* Oxford: Blackwell.

Johnson, M. H., M. I. Posner, and M. K. Rothbart, 1991. Components of visual orienting in early infancy: Contingency learning, anticipatory looking, and disengaging. *J. Cogn. Neurosci.* 3:335–344.

Johnson, M. H., M. I. Posner, and M. K. Rothbart, 1994. Facilitation of saccades toward a covertly attended location in early infancy. *Psychol. Sci.* 5:90–93.

Jonides, J., E. E. Smith, R. A. Koeppe, E. Awh, S. Minoshima, and M. A. Mintun, 1993. Spatial working memory in humans as revealed by PET. *Nature* 363:623–625.

Klein, R. M., A. Kingstone, and A. Pontefract, 1992. Orienting of visual attention. In *Eye Movements and Visual Cognition: Scene Perception and Reading*, K. Rayner, ed. New York: Springer-Verlag.

Lewis, T. L., D. Maurer, and A. Milewski, 1979. The development of nasal field detection in young infants. *Diss. Abstr. Int.* 41B:1547.

Maurer, D., and R. E. Young, 1983. The scanning of compound figures by young infants. *J. Exp. Child Psychol.* 35: 437–448.

Maylor, E. A., 1985. Facilitatory and inhibitory components of orienting in visual space. In *Attention and Performance*, vol. 11, M. I. Posner and O. M. Marin, eds. Hillsdale, N.J.: Erlbaum.

Munakata, Y., J. L. McClelland, M. H. Johnson, and R. S. Siegler, 1994. Rethinking object permanence: Do the ends justify the means-ends? *Infant Behavior and Development* (ICIS Abstracts).

Posner, M. I., 1980. Orienting of attention. *Q. J. Exp. Psychol.* 32:3–25.

Posner, M. I., 1988. Localization of cognitive functions in the human brain. *Science* 240:1627–1631.

Posner, M. I., and Y. Cohen, 1980. Attention and the control of movements. In *Tutorials in Motor Behavior*, G. E. Stelmach and J. Requin, eds. Amsterdam: North Holland, pp. 243–258.

Posner, M. I., and Y. Cohen, 1984. Components of visual orienting. In *Attention and Performance*, H. Bouma and D. G. Bouwhis, eds. Hillsdale, N.J.: Erlbaum.

Posner, M. I., and S. E. Petersen, 1990. The attention system of the human brain. *Annu. Rev. Neurosci.* 13:25–42.

Posner, M. I., R. D. Rafal, L. S. Choate, and J. Vaughan, 1985. Inhibition of return: Neural basis and function. *Cogn. Neuropsychol.* 2:211–228.

Posner, M. I., and M. K. Rothbart, 1980. The development of attentional mechanisms. In *Nebraska Symposium on Motivation*, J. H. Flower, ed. Lincoln, Neb.: University of Nebraska Press.

Purpura, D. P., 1975. Normal and aberrant neuronal development in the cerebral cortex of human fetus and young infant. In *Brain Mechanisms of Mental Retardation*, N. A. Buchwald and M. A. B. Brazier, eds. New York: Academic Press.

Rabinowicz, T., 1979. Normal and aberrant neuronal development in the cerebral cortex of human fetus and young infant. In *Human Growth: 3. Neurobiology and Nutrition*, F. Falkner and J. M. Tanner, eds. New York: Plenum.

Rafal, R. D., P. A. Calabresi, C. W. Brennan, and T. K. Sciolto, 1989. Saccade preparation inhibits reorienting to recently attended locations. *J. Exp. Psychol. [Hum. Percept. Perform.]* 15:673–685.

Rafal, R., J. Smith, J. Krantz, A. Cohen, and C. Brennan, 1990. Extrageniculate vision in hemianopic humans: Saccade inhibition by signals in the blind field. *Science* 250:1507–1518.

Richards, J. E., 1989a. Sustained visual attention in 8-week old infants. *Infant Behav. Dev.* 12:425–436.

Richards, J. E., 1989b. Development and stability of HR-defined, visual sustained attention in 14, 20, and 26 week old infants. *Psychophysiology* 26:422–430.

Richards, J. E., 1991. Infant eye movements during peripheral visual stimulus localization as a function of central stimulus attention status. *Psychophysiology* 28:S4.

Richards, J. E., and B. J. Casey, 1991. Heart rate variability during attention phases in young infants. *Psychophysiology* 28:43–53.

Rockland and Pandya, 1979. Laminar origins and terminations of cortical connections of the occipital lobe in the rhesus monkey. *Brain Res.* 179:3–20.

Ross, G., P. Auld, J. Tesman, and R. Nass, 1992. Effects of subependymal and mild intraventricular lesions on visual attention and memory in premature infants. *Dev. Psychol.* 28(6):1067–1074.

Rovee-Collier, C., E. Hankins, and R. Bhatt, 1992. Textons, visual pop-out effects, and object recognition infancy. *J. Exp. Psychol. [Gen.]* 121:435–445.

Schiller, P. H., 1985. A model for the generation of visually guided saccadic eye movements. In *Models of the Visual Cortex*, D. Rose and V. G. Dobson, eds. Chicester, Engl.: Wiley.

Shephard, M., J. M. Findlay, and R. J. Hockey, 1986. The relationship between eye movements and spatial attention. *Q. J. Exp. Psychol.* 38A:475–491.

Slater, A. M., V. Morison, and D. Rose, 1982. Perception of shape by the new-born baby. *Br. J. Dev. Psychol.* 1:135–142.

Slater, A. M., V. Morison, and M. Somers, 1988. Orientation discrimination and cortical function in the human newborn. *Perception* 17:597–602.

Stechler, G., and E. Latz, 1966. Some observations on attention and arousal in the human infant. *J. Am. Acad. Child Psychiatry* 5:517–525.

Tipper, S., T. Bourque, S. H. Anderson, and J. C. Brehaut, 1989. Mechanisms of attention: A developmental study. *J. Exp. Child Psychol.* 48:353–378.

Valenza, E., F. Simion, C. Umilta, and E. Paiusco, 1992. Inhibition of return in newborn infants. Unpublished manuscript, University of Padua, Italy.

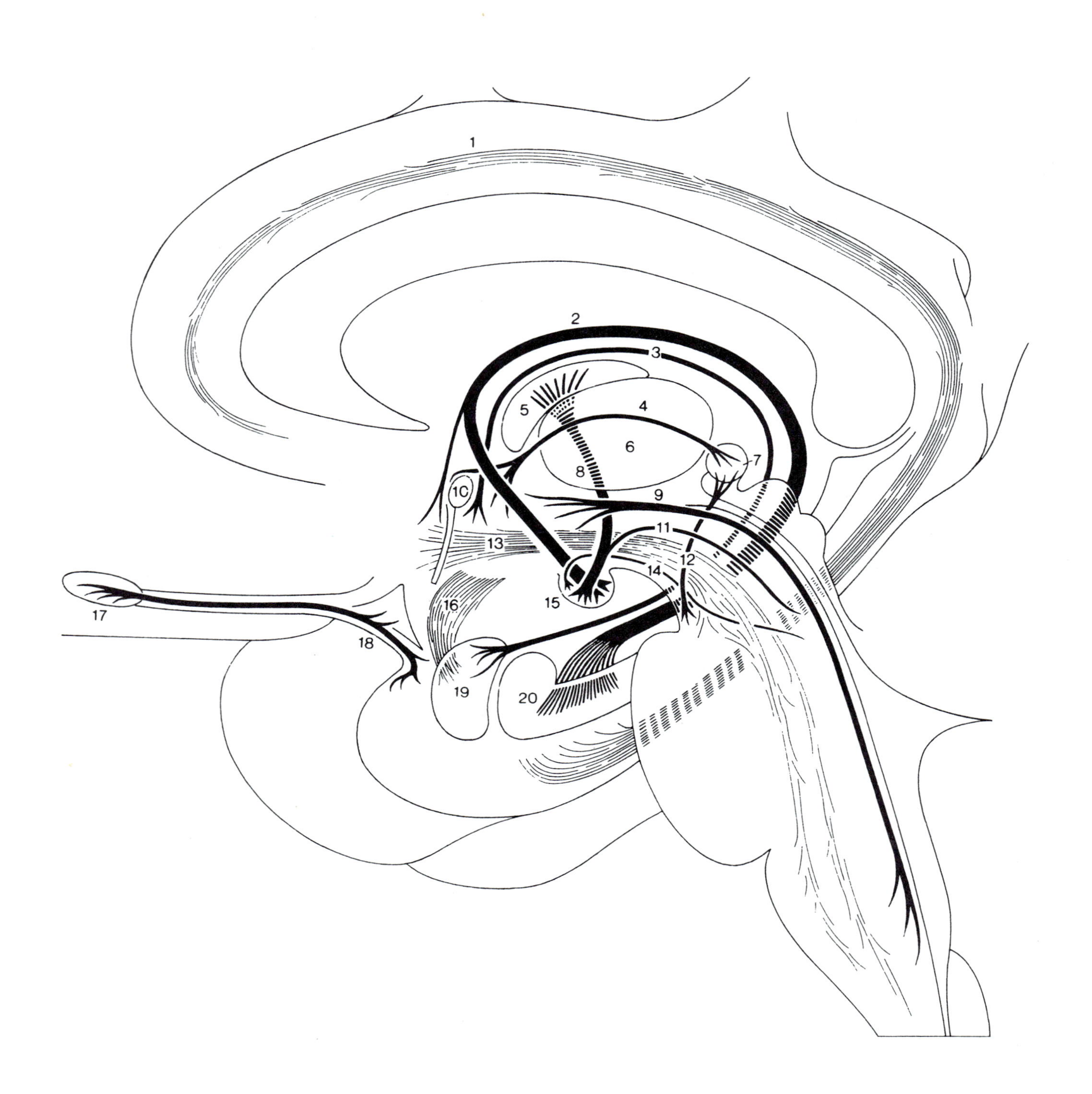

1
2
3
4
5
6
7
8
9
10
11
12
13
14
15
16
17
18
19
20

VI MEMORY

The major pathways of the limbic system and the rhinencephalon. 1, cingulum; 2, fornix; 3, stria terminalis; 4, stria medullaris thalami; 5, nucleus anterior thalami; 6, nucleus medialis thalami; 7, nuclei habenulae; 8, tractus mammillothalamicus; 9, fasciculus longitudinalis dorsalis; 10, commissura anterior; 11, tractus mammillotegmentalis; 12, tractus habenulointerpeduncularis; 13, fasciculus telencephalicus medialis; 14, pedunculus corporis mammillaris; 15, corpus mammillare; 16, ansa peduncularis; 17, bulbus olfactorius; 18, stria olfactoria lateralis; 19, corpus amygdaloideum; 20, hippocampus. (Reproduced from figure 191 of Nieuwenhuys, Voogd, and van Huijzen, 1988, with permission.)

Introduction

ENDEL TULVING

MEMORY IS many things, even if not everything that has been labeled memory corresponds to what cognitive neuroscientists think of as memory. Memory is a gift of nature, the ability of living organisms to retain and to utilize acquired information or knowledge. The term is closely related to *learning*, in that memory in biological systems always entails learning (the acquisition of information) and in that learning implies retention (memory) of such information.

Memory is a trick that evolution has invented to allow its creatures to compress physical time. Owners of biological memory systems are capable of behaving more appropriately at a later time because of their experiences at an earlier time, a feat not possible for organisms without memory.

Memory is a biological abstraction. There is no place in the brain that one could point at and say, Here is memory. There is no single activity, or class of activities, of the organism that could be identified with the concept that the term denotes. There is no known molecular change that corresponds to memory, no known cellular activity that represents memory, no behavioral response of a living organism that *is* memory. Yet the term *memory* encompasses all these changes and activities.

Memory is a convenient chapter heading designating certain kinds of problems that scientists study. Methods of science have been brought to bear on the problems of memory for over a hundred years, in many different organisms, and at many different levels of analysis, extending from molecular mechanisms to phenomena of conscious awareness.

The eight chapters in the Memory section of *The Cognitive Neurosciences* provide summaries of and glimpses into contemporary memory research. The overarching concern has been the search for the identity of the neural substrates of memory and for understanding of the correlation between neural mechanisms and memory processes.

Two recent conceptual developments have played an especially significant role in shaping the temper of today's research. One represents a consequence of the analytical and empirical separation of encoding, storage, and retrieval processes in human memory. As a result, it is no longer adequate to talk about, say, variables affecting "memory performance," or conditions responsible for "memory impairment." *Performance* and *impairment* now have to be specified in terms of specific memory processes and their interaction. The other development concerns the discovery, or the emergence of the concept, of different dissociable forms of memory, or multiple memory systems. Against the backdrop of traditional thought, this is a radical idea whose eventual implications and ramifications may well exceed all current expectations.

Although one finds suggestions in the earlier literature that learning and memory may assume different forms, the prevalent even if unarticulated view used to be that the underlying mechanisms of all learning and memory are basically the same. This unitarian view, rooted in the desire to adhere to the principle of parsimony, succeeded in escaping inimical data for a long time. Yet it overlooked the fact that nature itself is seldom parsimonious, and that, as has been noted, evolution is a tinkerer, not an engineer. Facts that are difficult to fit into the unitarian framework have been appearing on the scene, under the rubric of task dissociations, with accelerated frequency.

The new concepts of memory processes and memory systems have considerably changed the way cognitive neuroscientists pursue their mission. The quest for understanding of the identity and localization of the neural substrates of what we call "memory" has metamorphosed into the search for the neural correlates of encoding, consolidation, storage, and retrieval, separately for the different, dissociable forms of memory. Thus, for instance, it is now reasonably clear that the hippocampal structures are necessary for the encoding and consolidation of some but not other kinds of input, but that they do not play a significant role in the retrieval of any kind of stored information. All eight chapters that constitute the Memory section of *The Cognitive Neurosciences* are concerned with some aspect of this basic issue of the identity and localization of different processes of different forms of memory.

Historically, the basic questions regarding the neural substrates of memory were two. The first one was: Are memories localized in the brain? The "localizationists," who included Broca, Fritsch, Hitzig, and Ferrier, said yes; their detractors, the "integrationists," whose roster included Flourens, Franz, and Lashley, said no. As in all difficult-to-decide issues, the fortunes of the two sides have waxed and waned indecisively over time, and are likely to continue to do so in the foreseeable future. In our Memory section, Squire and Knowlton represent the localizationist camp, whereas Markowitsch holds aloft the integrationist banner. The second question was raised by those who adopted the affirmative position regarding localization: Where are memories localized? During much of the history of the science of memory, the relevant evidence on this question was derived from brain-damaged patients whose lesions could be identified. After the famous case of H. M. appeared on the scene, with well-documented bilateral surgical lesions in the medial temporal lobes, including the hippocampus, the hippocampal structures that were already known to be implicated in memory disorders were rapidly elevated to the position of prime candidate for the "seat of memory."

Today's consensus holds that the limbic system, including the hippocampal structures, plays a critical role in certain forms of memory, even if that role and the exact identities of the "certain forms" are not yet clear. In addition the consensus holds that some other brain regions are also involved in memory processes. One of the most thoroughly studied of these other regions is the frontal lobes, although their role in memory has been somewhat controversial. Shimamura's chapter provides a contemporary review of some of the issues and some of the evidence.

Lesion analysis has yielded a great deal of valuable intelligence about the brain-mind interaction. But it is not without problems, as its critics have been fond of pointing out. This is why recent technological developments that allow objective observation of the activity of living brains engaged in cognitive tasks have been eagerly welcomed by students of memory. Two chapters in the present collection cover these developments.

Rugg discusses the promise of, and problems with, the use of evoked (event-related) potentials to track the happenings at the neural level that accompany memory processes, while Roland et al. present an overview of the recent achievements of, and remaining difficulties with, the use of positron emission tomography (PET) in identifying brain regions that are involved in various operations of memory.

The other chapters bring to the reader the latest word on the "kinds" of memory that localizationists, integrationists, electrophysiological recorders, and brain imagers have to keep apart in their probes into memory. Baddeley presents an analysis of working memory, the human brain/mind's executive functions that integrate cognitive information from a variety of sources, including recent inputs, for use in meeting the needs of the present. Schacter provides an overview of some of the latest discoveries in the domain of implicit memory, which is currently one of the "hottest" areas of memory research. Tulving presents a classificatory scheme of five major human memory systems and considers their interrelation in terms of the seriality, parallelism, and independence of the processes of information acquisition, storage, and retrieval.

The advances that have been made over the last one hundred years in the understanding of the brain/mind of memory have been steady and substantial, even if at times they have appeared to be frustratingly slow. Over the last 20 years or so, however, the pace of discovery has quickened. The painstaking early studies have now laid a solid foundation on which to build; the new techniques and methods will deepen our understanding of the workings of the brain/mind; and, perhaps most important, new insights into the basic character of memory, and the character of the scientific problem of memory, will help to guide research in even more rewarding directions.

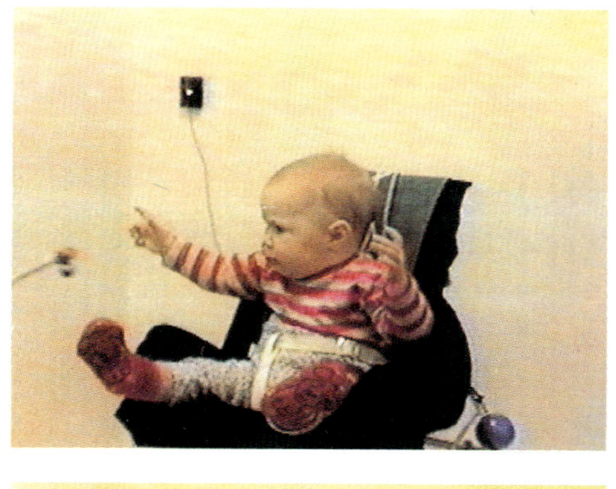
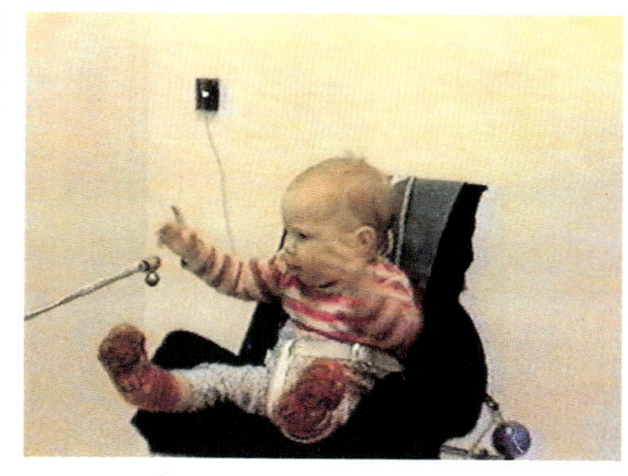
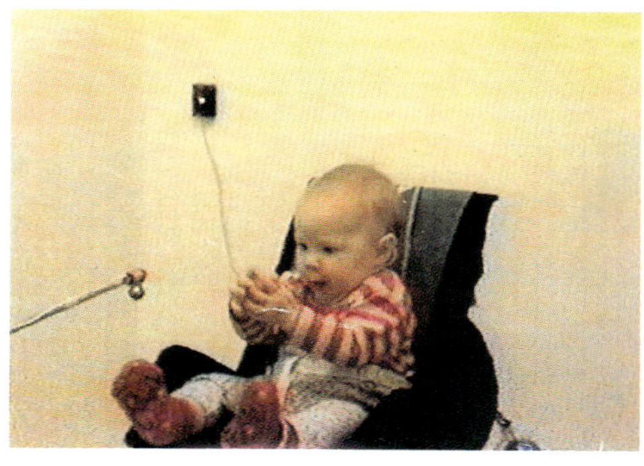
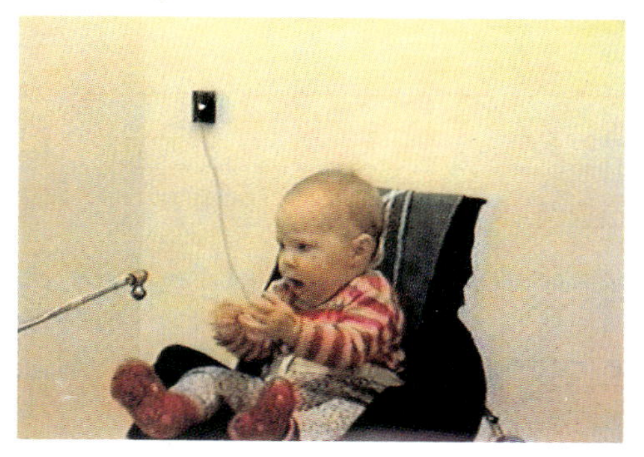

PLATE 1 A reach of an 8-month-old infant for an object that stops 200 ms before the end of the reach. (Upper left) 300 ms into the reach; the object is moving at 60 cm/s. (Upper right) 500 ms into the reach; the object stops. (Lower left) 700 ms into the reach; the infant grasps with both hands at the position the object would have occupied if it had continued to move. (Lower right) 1000 ms into the reach; the infant redirects attention toward the object's true position. (After Hofsten and Rosander, 1993.) (See chapter 10).

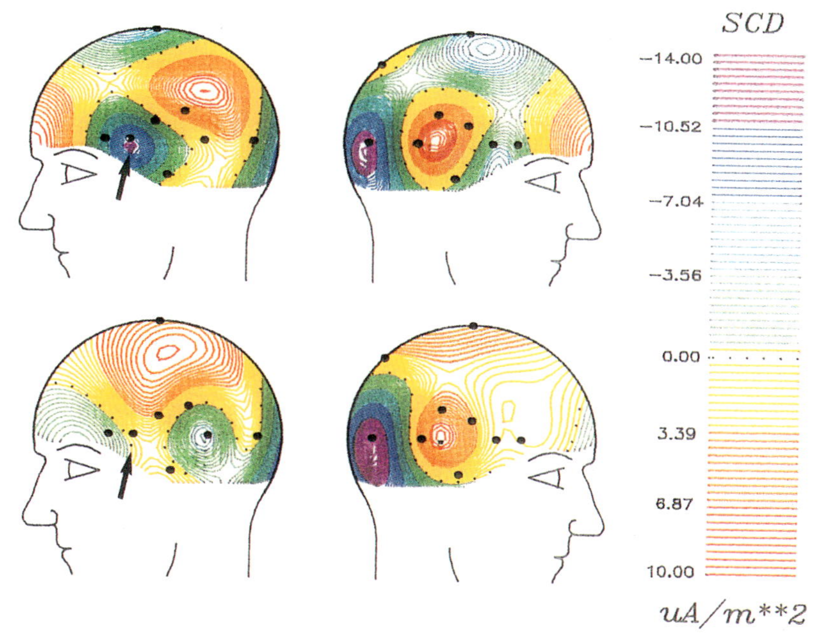

PLATE 2 Scalp topography of current source densities calculated for the N280 (most negative peak in the window 235–400 ms) in response to closed-class words. Maps are averages across 17 hearing subjects (top row) and from 10 congenitally deaf subjects (bottom row). The longer wavelengths represent current sources (current flowing out of the head); the shorter wavelengths represent current sinks. Each map utilizes the same scale. The maps of the hearing and deaf subjects were similar over the right hemisphere. Over the left hemisphere the maps of hearing subjects display a prominent sink over the anterior temporal region that is absent in the maps from the deaf subjects (arrows). (From Neville, Mills, and Lawson, 1992.) (See chapter 13.)

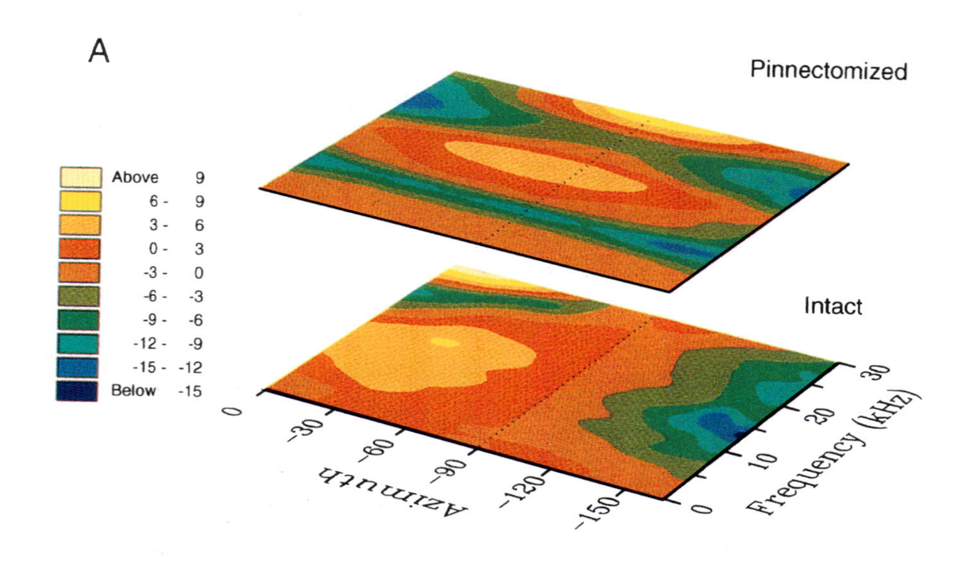

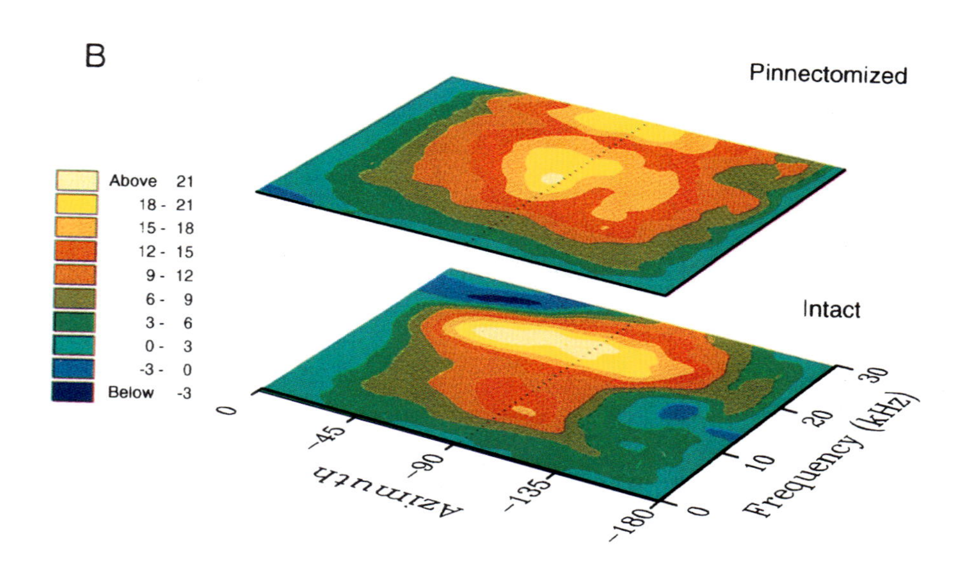

PLATE 3 The mean monaural (A) and binaural (B) spectral cues for ipsilateral sound locations are shown for five ferrets before and after bilateral pinnectomy. A probe microphone implanted across the wall of the auditory canal was used to measure the spectral transfer functions every 10° along the audiovisual horizon. (A) The directionally dependent component of the response, the location dependency function, was determined by subtracting from each transfer function the reference transfer function recorded on the interaural axis (−90°) in the intact animal (see Carlile, 1990b). The gains in transmission are indicated by the contour colors. The anterior midline is at 0° and the ipsilateral interaural axis is at −90° azimuth. A number of spectral features can be identified in normal, intact ferrets, which could be used to discriminate between anterior and posterior locations (see text, chapter 17). All these features were eliminated by pinnectomy. (B) The interaural spectrum level differences (ISDs) were calculated by subtracting the response obtained at each position in the contralateral hemisphere from that obtained at the corresponding (mirror image) angle on the ipsilateral side of the midline. The ISDs are shown for the intact (lower) and pinnectomized (upper) animals, as in (A). As with the monaural spectral cues, pinnectomy clearly results in the abolition of those features of the ISDs that could encode front-back location. (Adapted from Carlile and King, 1994, by permission.)

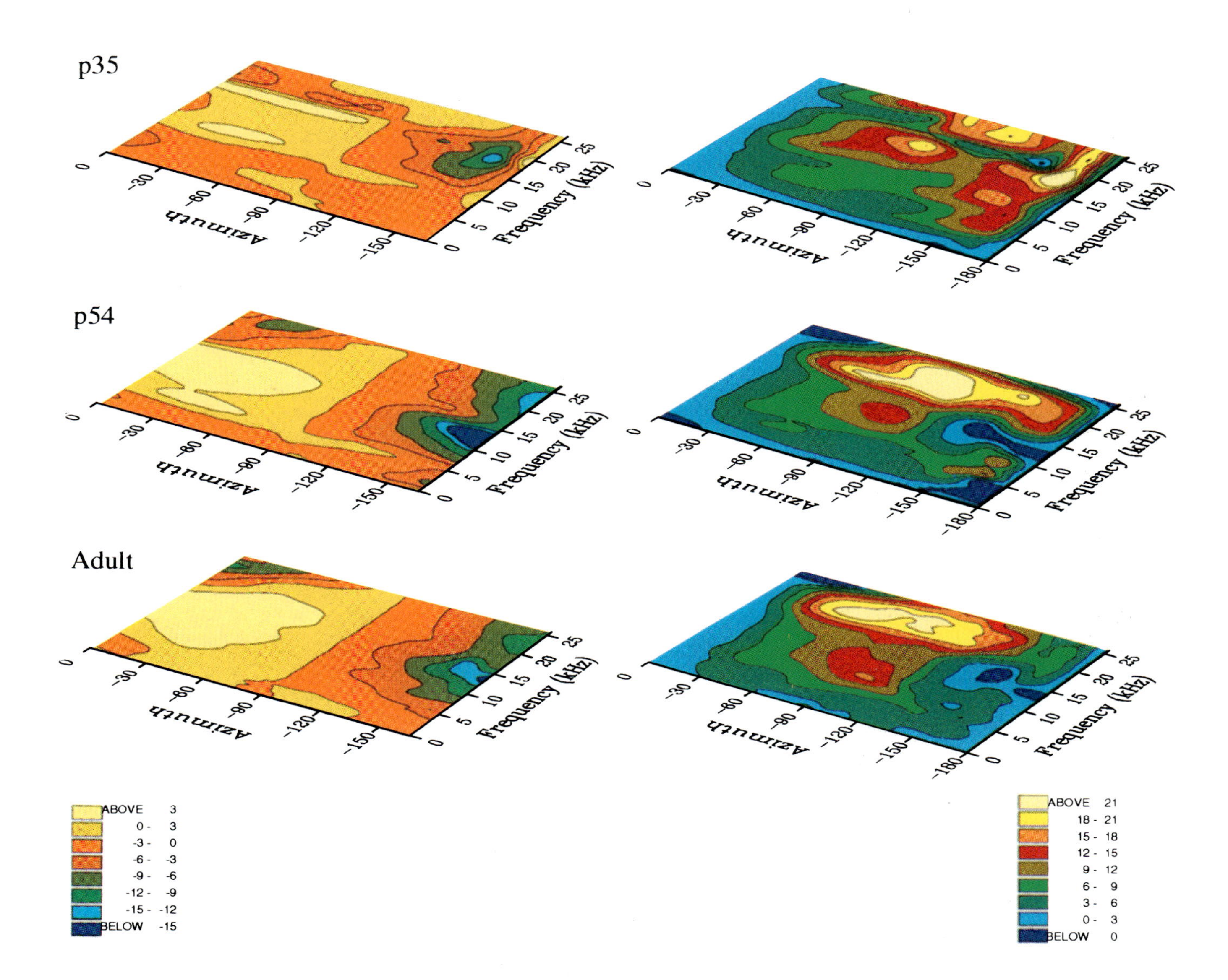

Plate 4 The time course of development of the monaural and binaural spectral cues in the ferret. The location dependency functions (LDFs; left panels) and the interaural spectral differences (ISDs; right panels) recorded from two neonatal ferrets at postnatal day 35 (p35) and p54 are compared with mean data from five adult animals (see also plate 3). By the end of the eighth postnatal week, the horizon LDFs (azimuthal monaural spectral cues) are very similar to those in the adult. However, there are still a number of significant differences in the ISDs between the young and adult ferrets, particularly at the higher frequencies. All other details are as for plate 3. (See chapter 17.)

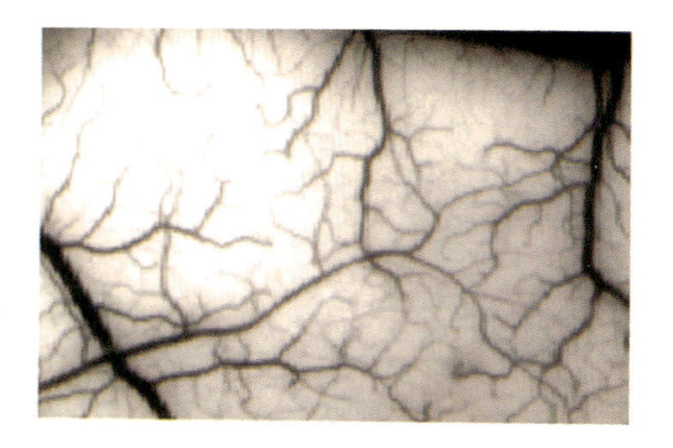

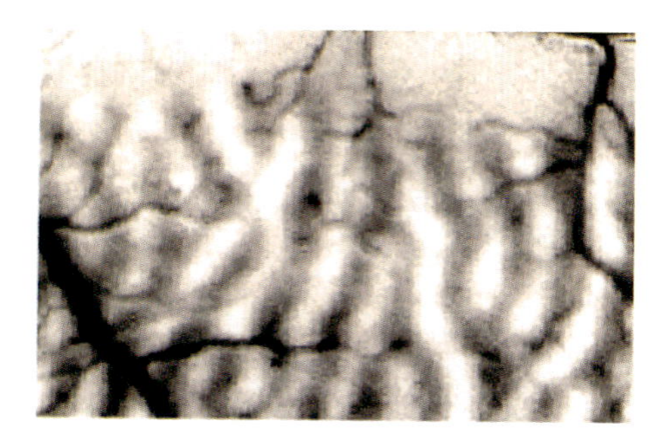

PLATE 5 Optical imaging of ocular dominance and orientation in V1 and V2. (Top) Image of the cortical surface and vasculature, a 9 × 6-mm field of view. V2 can be seen as the topmost quarter of the image, whereas the remaining region is V1. (Middle) Ocular dominance columns in this same portion of V1/V2, derived by subtracting images obtained during right-eye stimulation from those obtained during left-eye stimulation. No ocular dominance columns are present in V2. (Bottom) Pseudocolor-coded map of orientation tuning obtained by the vector addition of images acquired during presentation of visual stimuli of several different orientations. These images demonstrate the coexistence of multiple functional maps (e.g., ocular dominance and orientation) within the same cortical area. (See chapter 20.)

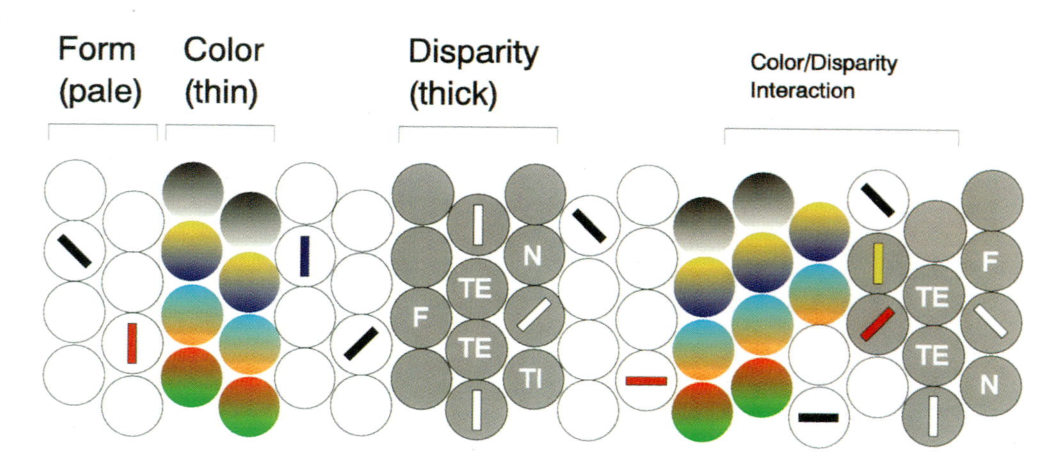

PLATE 6 Model of V2 functional organization. At a coarse level, V2 consists of alternating thick, pale, and thin stripes. Each stripe is actually composed of multiple patches or subcompartments with a particular functional specialization. V2 patches in the color stripes are selective for different color properties, whereas patches in the disparity stripes are responsive to a range of disparities, from near to far. Among oriented regions in pale and disparity stripes, there are regions with highly uniform orientation selectivity and columnar organization, as well as regions in which orientation is poorly organized and not columnar. There are regions in V2 where color and disparity stripe have joined together (depicted on the right). At the border of these regions, cells are tuned for both color and disparity. (See chapter 20.)

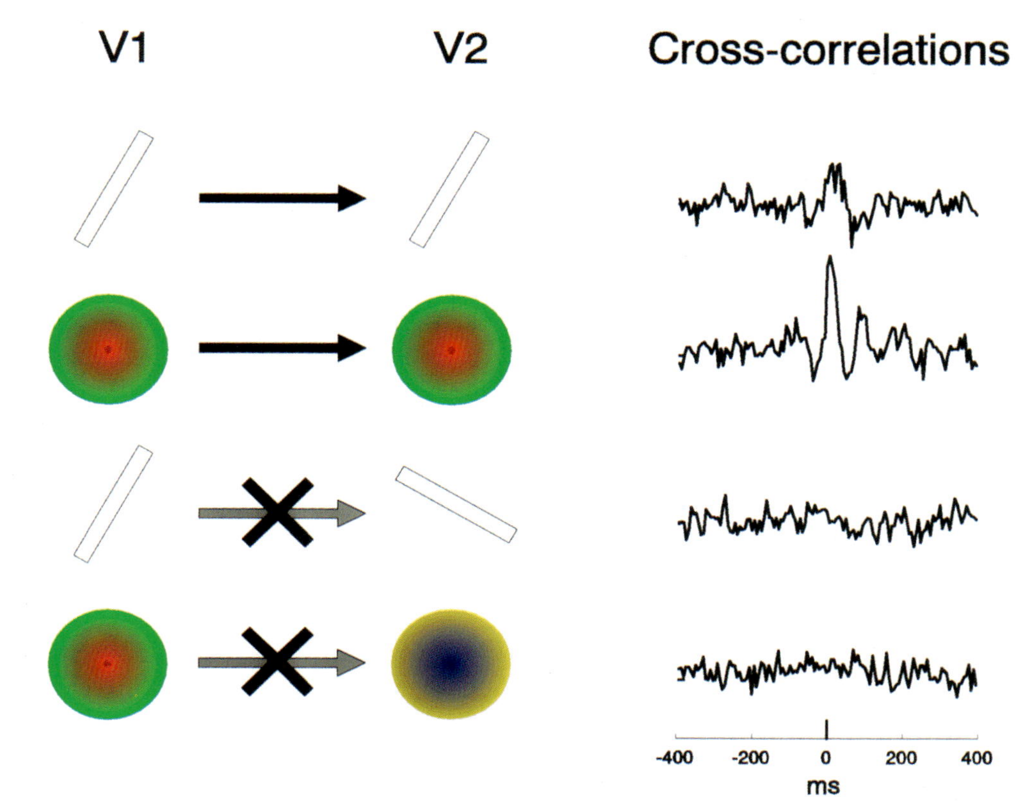

PLATE 7 Examples of V1/V2 correlograms demonstrating functional specificity. In general, peaked correlograms were found only when the receptive field properties of the two cells matched (e.g., similar orientation, similar color preference). Total correlogram width is ±200 ms. (See chapter 20.)

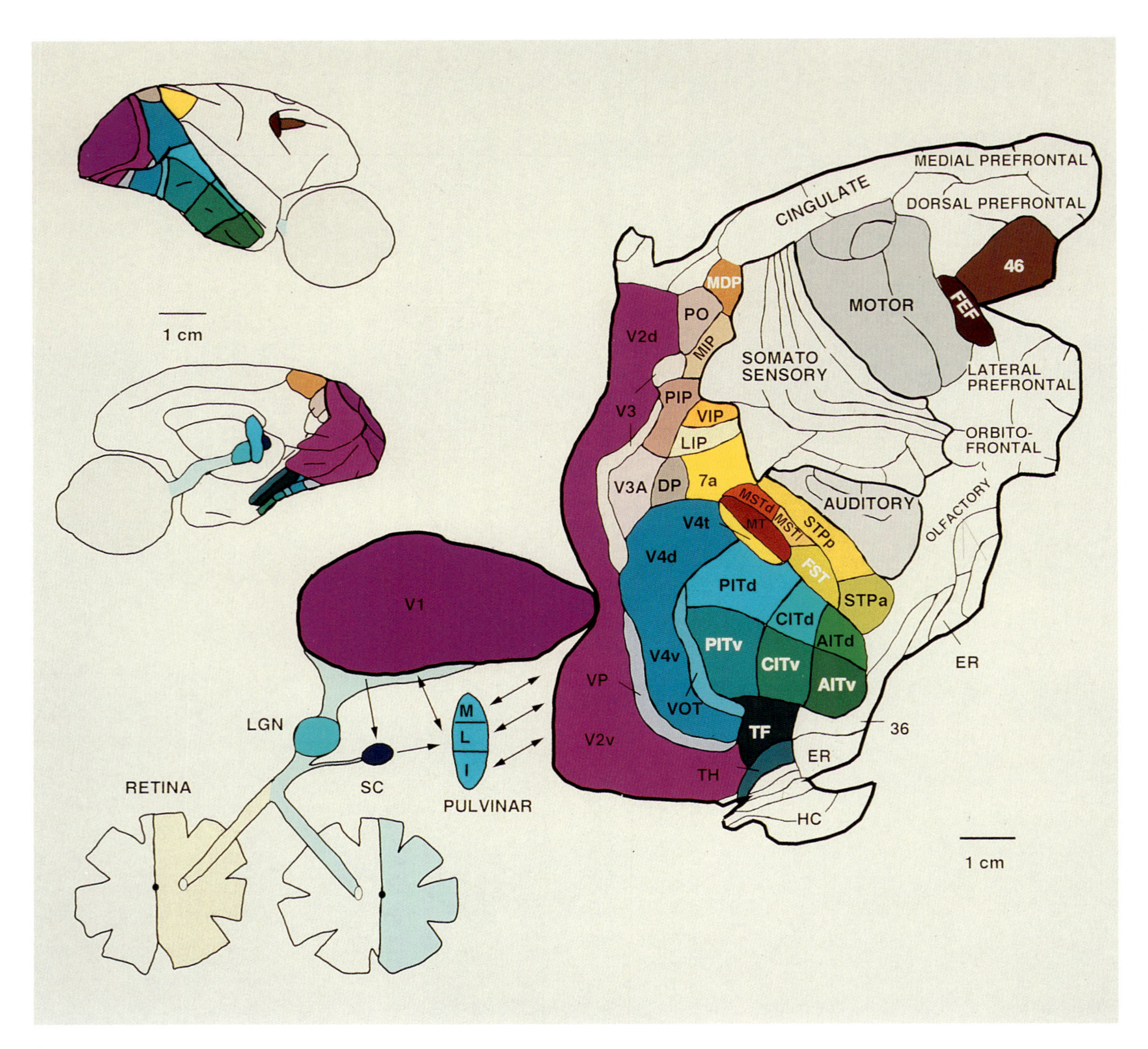

PLATE 8 An overview of the macaque visual system, including lateral and medial views of the right hemisphere (insets, upper left) and an unfolded representation of the entire cerebral cortex and major subcortical visual centers. Visual areas of the cerebral cortex, along with major subcortical visual centers, are shown in various colors. (Reprinted from Van Essen, Anderson, and Felleman, 1992.) (See chapter 24.)

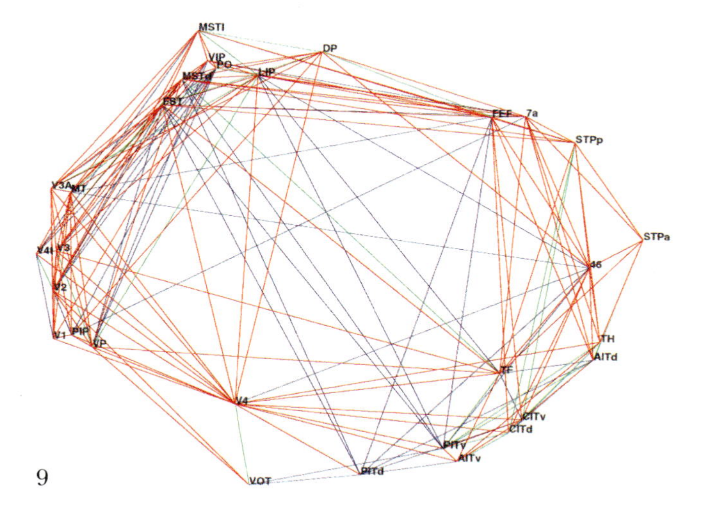

9

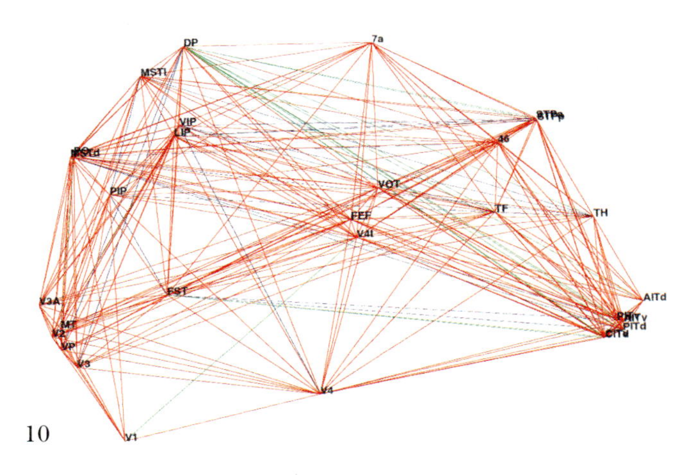

10

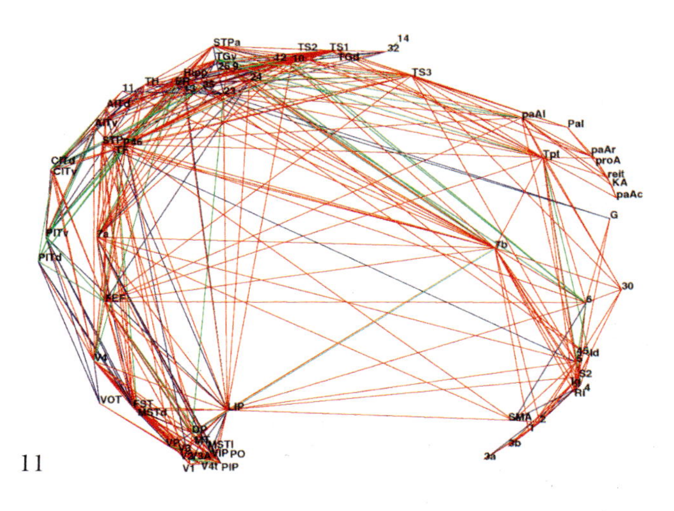

11

Plate 9 The topological organization of the macaque cortical visual system. Reciprocal connections are colored red, one-way projections going from left to right are blue, and one-way projection going from right to left are green. A total of 301 connections are represented, of which 62 are one-way. This nonarbitrary structure is a best-fit representation in two dimensions of the connectional topology of this system, in which the positions of areas are specified as ones that minimize the distance between connected areas and maximize the distance between area that are not connected. The analysis represents, in a spatial framework the organization structure of the network of corticocortical connections between elements of the visual cortex. In detail, the structure was derived by submitting the connection matrix between visual areas to nonmetric multidimensional scaling using ALSCAL (Young and Harris, 1990). Solutions with the level of measurement specified as nominal and ordinal were derived to assess whether a least-squares categorical transformation was required (see Young and Harris, 1990), but there was no perceptible difference between them. Ordinal solutions in one to five dimensions were derived so that solutions with different numbers of dimensions could be compared in a scree test, which showed diminishing returns in numbers of dimensions greater than two. The three-dimensional solution was visualized by animating the structure to obtain structure-from-motion. No new qualitative features were apparent in three dimensions, and the features of the two-dimensional structure were readily apparent. This structure was derived with an ordinal level of measurement in two dimensions, and the configuration of points (60 parameters) accounted for 40% of the variability in the corresponding connection matrix (435 parameters). (Reprinted with permission from *Nature* 358:153–155, 1992. © 1992 Macmillan Magazine Ltd.) (See chapter 29.)

Plate 10 The topological organization of the macaque cortical visual systems as it would be if all possible connections that have not been explicitly sought and discovered to be absent are assumed to exist. The structure was derived by means identical to those that derived plate 9. The solution still exhibits a structural dichotomy between the occipitotemporal areas on the one hand and the posterior parietal and caudal superior temporal sulcal areas on the other. It is still clearly hierarchical. Hence, the qualitative features of this solution, which represents the grosses possible perturbation of the data for this list of visual areas, are identical to those of plate 9. Quantitatively, this solution explains 76% of the variability in plate 9. It is constrained to do so because a sufficiently large number of possible connections have been sought and reported absent. The differences between this figure and plate 9 are mainly in the positions of poorly studied areas, such as V4t and VOT, which have been shifted to the center of the structure by a large number of hypothetical new connections. (See chapter 29.)

Plate 11 Topological organization of the entire macaque cortical processing system as presently known. A total of 758 connections between 72 areas are represented, of which 136 (18%) are one-way. This connectivity represents 15% of the possible connections among these areas. This nonarbitrary format represents, in a spatial framework, the organizational structure of the network of corticocortical connections of this animal. The structure illustrates the relations among areas of the occipitotemporal cortex with structures outside this visual subsystem. AIT, for example, is closely associated with limbic, parahippocampal, and prefrontal structures. (Reprinted with permission from *Proc. R. Soc. Biol. Sci.* 252:13–18, 1993.) (See chapter 29.)

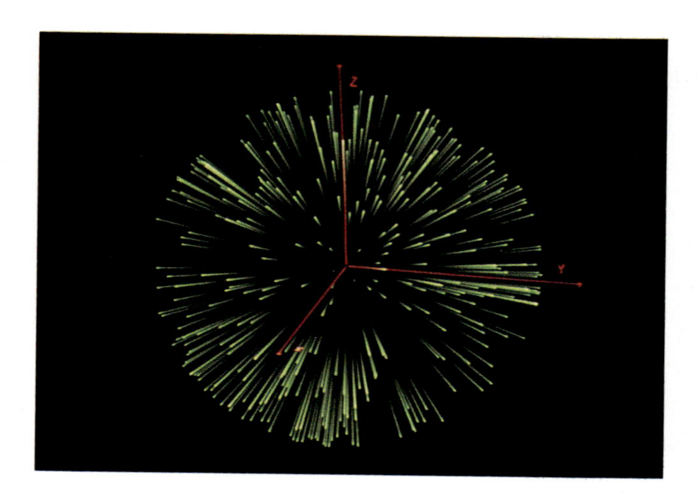

PLATE 12 (Top left) Preferred directions of 475 directionally tuned cells recorded during a three-dimensional reaching task. Lines are vectors of unit length. (Top right) The same populations of cells (light blue lines) shaped for a movement in direction indicated by the yellow line. The preferred directions are the same as in left panel, but their length is proportional to the changes in cell activity associated with the particular movement direction illustrated. The direction of the population vector (orange) is close to that of the movement. (Bottom) Ninety-five percent confidence cone for the direction of the population vector (line in the center of the cone). The movement direction (yellow line) is within the cone. (Reprinted from Georgopoulos, Kettner, and Schwartz, 1988, by permission. Copyright by the Society for Neuroscience.) (See chapter 32.)

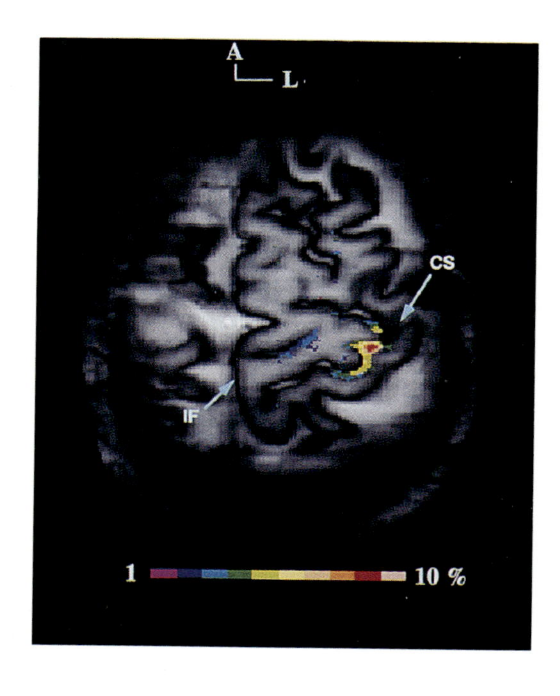

PLATE 13 Functional activation of the motor cortex during contralateral finger movements in one subject. Each color in the functional map represents a 1% increment going from left to right. The color-coded functional map is superimposed on a T_1-weighted anatomical image in gray scale. There is increased signal activity in the caudal lateral precentral gyrus, within the depths of the sulcus at the same level. CS, central sulcus; IF, interhemispheric fissure; A, anterior; L, lateral. (Reprinted from Kim et al., 1993, by permission.) (See chapter 32.)

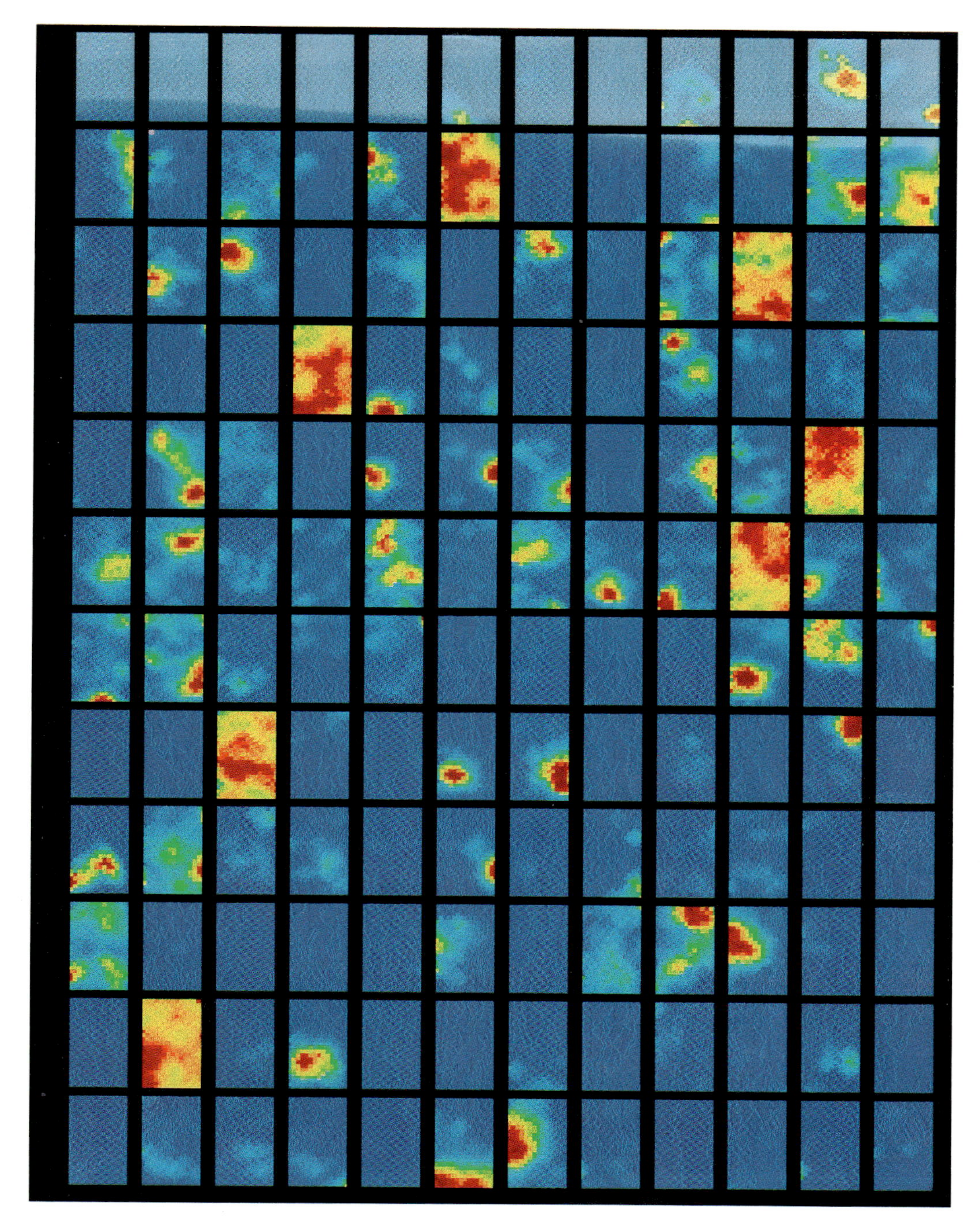

PLATE 14 Spatial firing distributions of 144 hippocampal neurons recorded simultaneously while the rat explored in a 60 × 45-cm apparatus. Each panel represents the firing distribution of one cell. Red indicates high firing; dark blue indicates no firing. The seven high-rate cells are inhibitory interneurons; the rest are pyramidal cells. Spatial selectivity of this sort is what would be predicted if each cell encoded a vector to a specific landmark. The spatial encoding by hippocampal cells is sufficiently robust that the animal's location in the apparatus can be predicted to within a few centimeters solely on the basis of the mean firing rate distributions of such a small sample of simultaneously recorded cells. (Reprinted from Wilson and McNaughton, in press, by permission.) (See chapter 37.)

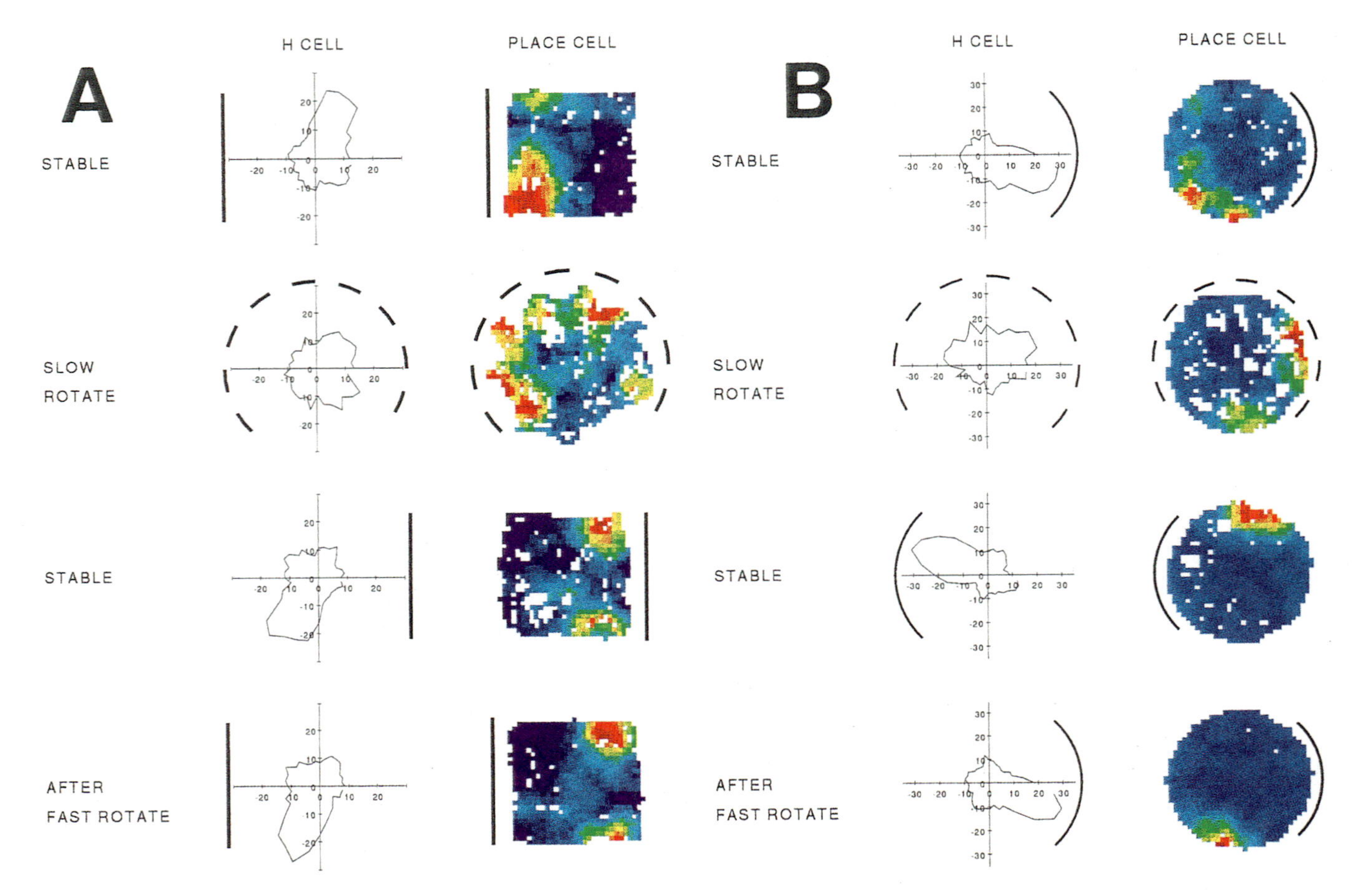

Plate 15 To overcome the effects of cumulative error, an inertially driven H system would need to have modifiable connections from local view cells (McNaughton, Chen, and Markus, 1991). Associations between local view and H cells acquired during initial exploration of an environment would be capable of resetting the H system in cases of subsequent disagreement between the visual input and the allocentric (inertial) directional reference. Simultaneous recordings from a hippocampal place cell and a head-direction (H) cell in lateral dorsal thalamus are illustrated. The experiment was carried out first in an unfamiliar apparatus (A) and then in a familiar apparatus (B). The two apparatuses were approximately the same size, both had high walls painted uniform gray, and both had a single polarizing visual cue, a white card covering one fourth of the total wall surface (indicated by vertical line in A and curved line in B). H-cell firing rate is shown as a function of head orientation relative to laboratory north. Place-cell firing is illustrated as a function of the animal's location in the apparatus (red, high rate; dark blue, no firing). Slow rotation of the apparatus (i.e., below vestibular threshold) in either case caused the firing functions of both cells to rotate accordingly. As predicted, fast rotation caused both cells to remain fixed to the inertial (laboratory) reference frame in the unfamiliar apparatus (box) but to rotate with the visual cue in the familiar apparatus. (See chapter 37.)

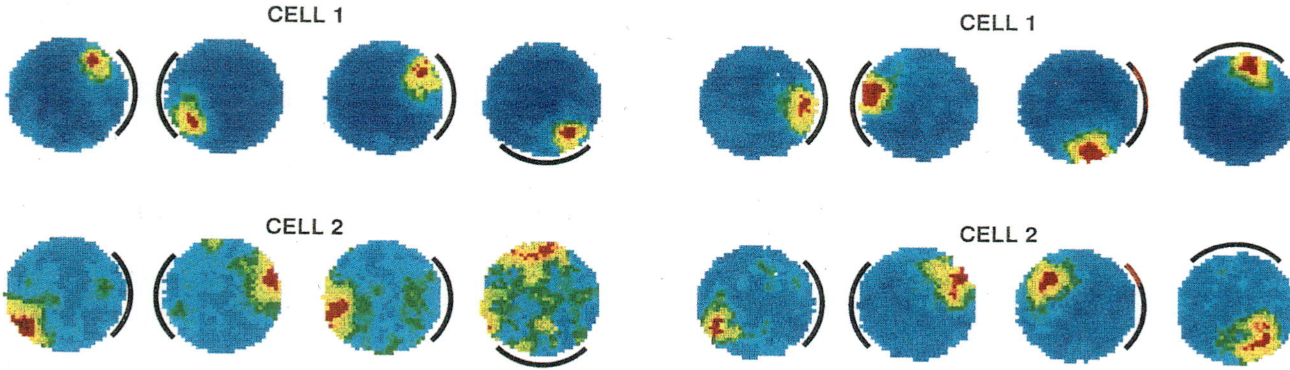

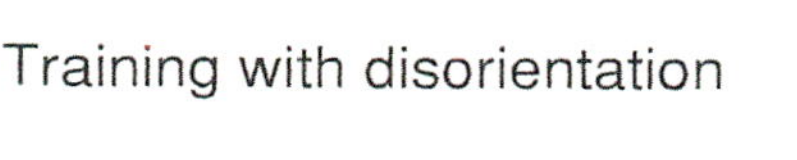

Plate 16 If a rat is trained to forage for small chocolate pellets tossed randomly into a 60-cm-diameter cylinder, painted uniformly gray except for a white card covering one fourth of the circle (dark bar), the binding of hippocampal place cells to the card depends on how the animal has been trained. If it is completely disoriented before each session, place fields exhibit rotational instability. Rats often shift their bearing relative to the card. If trained without disorientation (see text, chapter 37), the card exerts complete control of the place field bearing. One explanation is that if the animal is disoriented on each trial, the card appears at a different bearing relative to the animal's inertial reference on each trial. No consistent association is formed and the card consequently is unable to reset the directional reference, as it does when training is without prior disorientation.

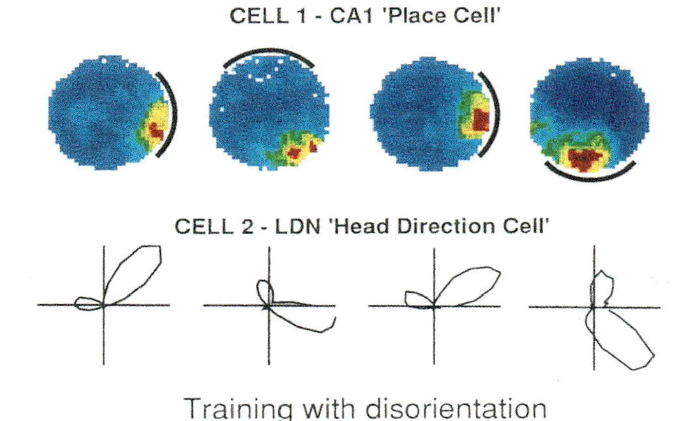

Plate 17 Simultaneous recording of an H cell in lateral dorsal thalamus and a hippocampal place cell in a rat trained with the disorientation procedure. The place cell is bound to the H system, not to the cue card. The H-cell recording in this case actually suffered from some contamination from a second H cell that was not separately resolved. The latter gave rise to the small secondary component of directional firing. The constant angular difference between the two cells' preferred orientations illustrates the fact that the H-cell system is tightly coupled. In the fourth session, the slight broadening of the directional tuning function of the H system and the blurring of the place field were due to drift of the directional reference during the recording session. (See chapter 37.)

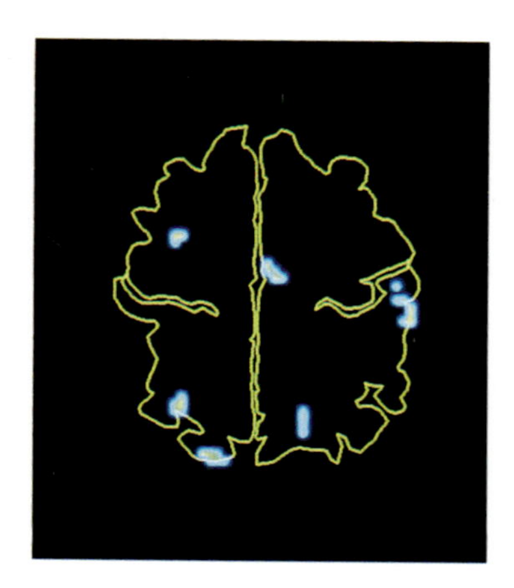

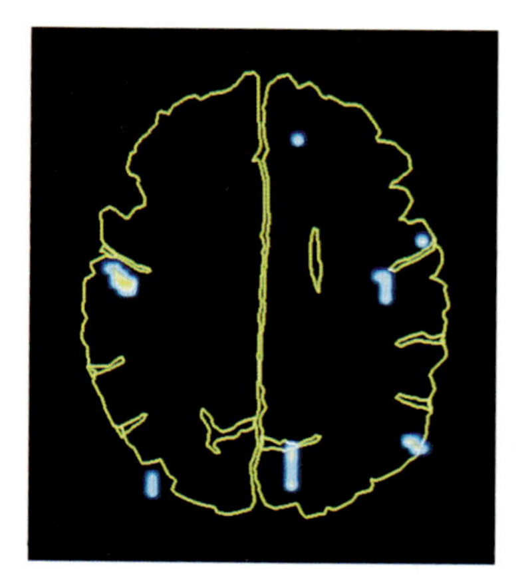

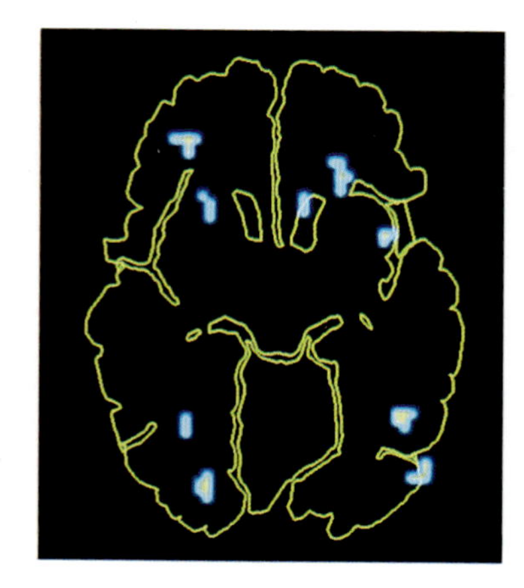

PLATE 18 Atlas anatomically standardized pictures of 10 subjects showing statistically significant clusters (fields) appearing by subtracting the rCBF of pointing at an early phase of learning from the rCBF image of pointing guided by long-term memory of the targets. (From left to right) Section through the superior parietal lobules; the central sulcus is also shown. Section through the frontal eye field and the parietal lobe. Section through the superior part of cuneus; the parieto-occipital sulcus is shown. (See chapter 49.)

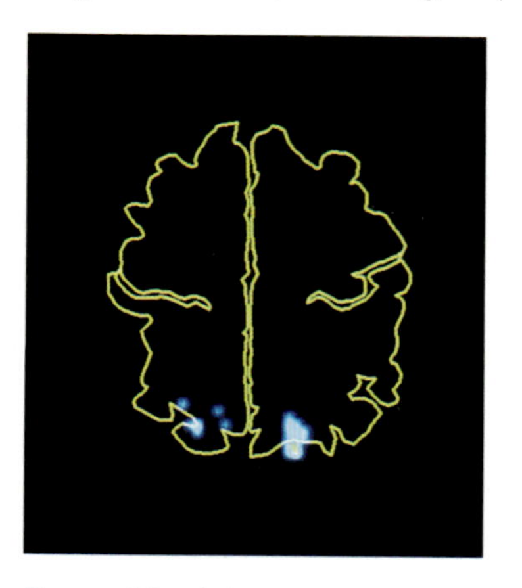

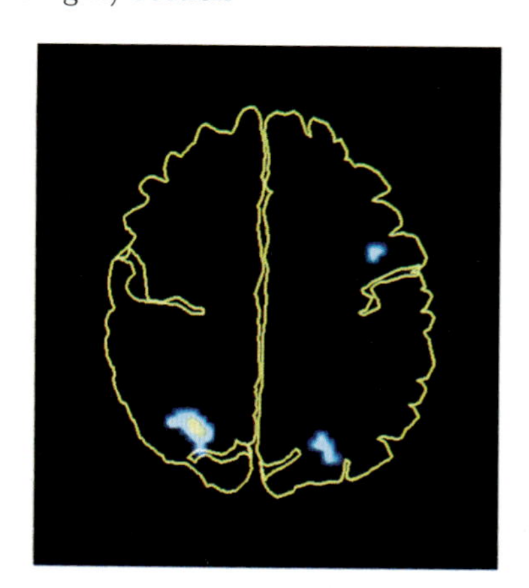

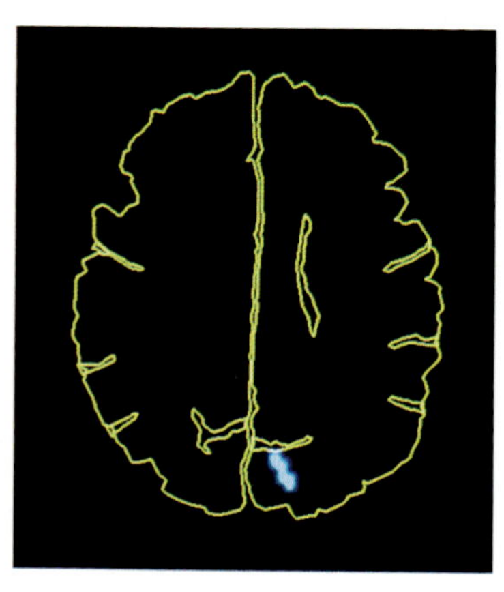

PLATE 19 Atlas anatomically standardized pictures of 10 subjects showing statistically significant and consistently activated clusters (fields) appearing by multiplication of the cluster image of visual learning of large patterns with the cluster image of pointing guided by long-term memory of the targets. (From left to right) Section through the superior parietal lobules. Section through the superior cuneus. Section through the fusiform gyri; the lateral fissure and the inferior temporal sulcus and the caudate nucleus are also shown. (See chapter 49.)

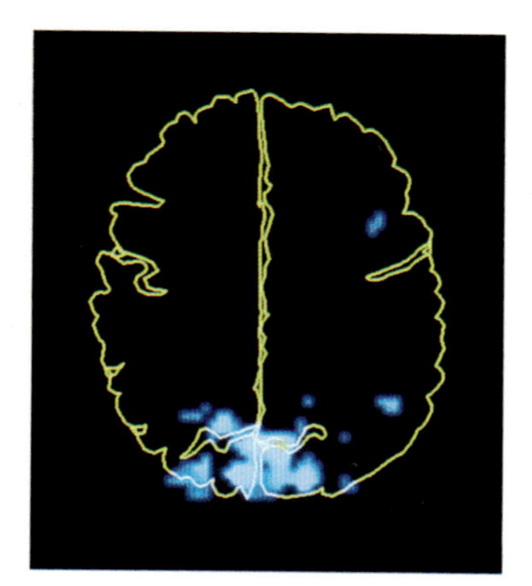

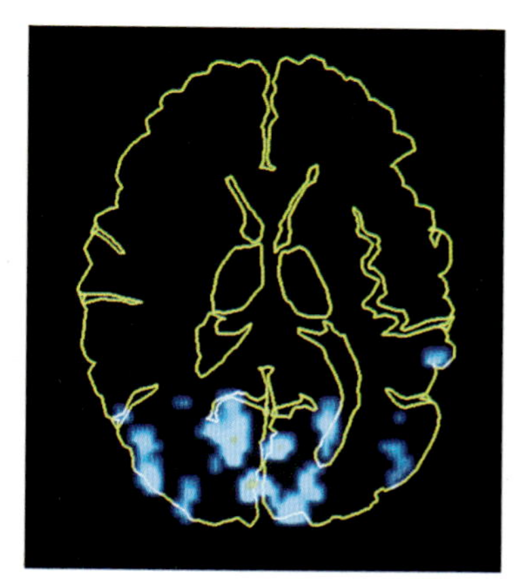

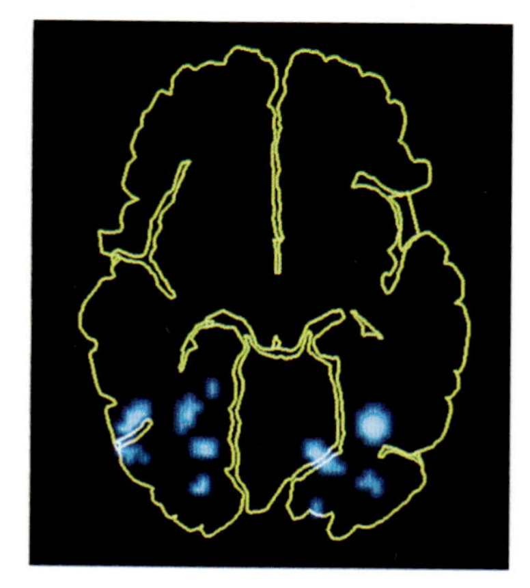

PLATE 20 Atlas anatomically standardized pictures of 11 subjects showing statistically significant and consistently activated clusters (fields) appearing by multiplication of the cluster image of visual learning with that of visual recognition of the same patterns. (From left to right) Section through the superior cuneus. Section through the calcarine cortex. Section through the fusiform gyri. (See chapter 49.)

Left Right

Story in Tamil

AC

VAC

List of French words

Sentences with pseudo-words

Semantically anomalous sentences

Story in French

GIN, CEA-SHFJ, Orsay
LPSC, CNRS, INSERM & EHESS, Paris

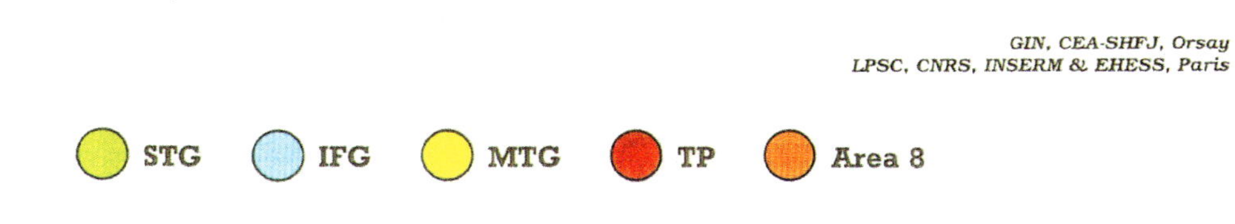

Plate 21 Active brain regions in the five experimental conditions: listening to a story in Tamil ($N = 5$), to a list of French words ($N = 5$), to sentences with pseudo-words ($N = 6$), to semantically anomalous sentences ($N = 6$), and to a story in French ($N = 10$). Observed average regional activations were mapped on a reconstruction of the external brain surface of subject 10 based on high-resolution magnetic resonance images. The anterior commissure vertical plant (VAC) and bicommissural plane (AC-PC) were used to limit the projection of the temporal pole region (TP). The inferior frontal gyrus region (IFG) includes the pars opercularis, triangularis, and orbitaris of the third frontal gyrus. A superior prefrontal area corresponding to Brodmann's area 8 (Area 8) was defined on individual MRI using a stereotactic atlas. STG, superior temporal gyrus; MTG, middle temporal gyrus region. (Reprinted from Mazoyer et al., 1993, with permission.) (See chapter 61.)

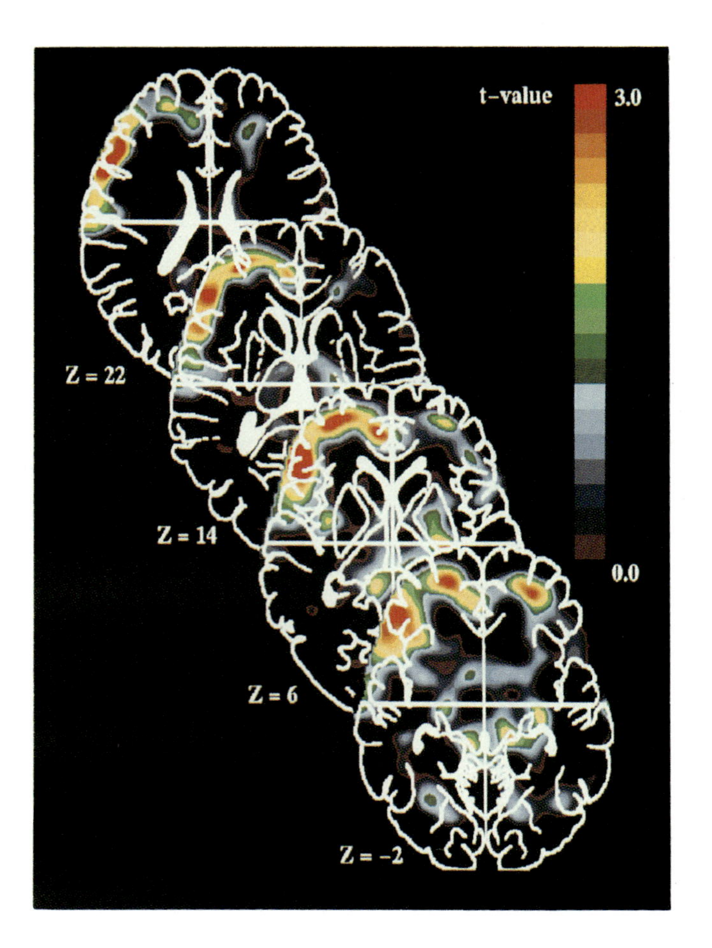

PLATE 22 Extent of increased blood flow in the left prefrontal cortex (PFC) of the depressed subjects with familial pure depressive disease (FPDD). The horizontal image slices shown are from an image of *t*-values, produced by a voxel-by-voxel computation of the *t*-statistic (Drevets et al., 1992). This *t*-image was used to generate hypotheses regarding the regions that might contain differences between the depressed and control subjects. The presence of an abnormality in this region was confirmed in a separate group of subjects (Drevets et al., 1992). The *t*-image slices shown correspond to areas where blood flow is increased in the depressives relative to the controls. The number to the left of each slice locates the image planes in millimeters from the bicommissural line (positive = superior). Left is on the reader's left and anterior is toward the page top; x and y axes mark the center of the bicommissural line. The area of increased blood flow in the left PFC involved much of the ventrolateral PFC and extended through the tissue immediately caudal to the frontal pole to include part of the left medial PFC (which includes the pregenual portion of the anterior cingulate gyrus). (See chapter 76.)

PLATE 23 *T*-image corresponding to increased blood flow in depressed subjects with FPDD at the sagittal plane through the amygdala ($x = 21$ mm left of midline). Anterior is toward the left, z and y axes locate the midpoint of the bicommissural line. The most rostral part the the left prefrontal cortical (PFC) region is also demonstrated (see plate 22). (From Drevets et al., 1992, with permission.) (See chapter 76.)

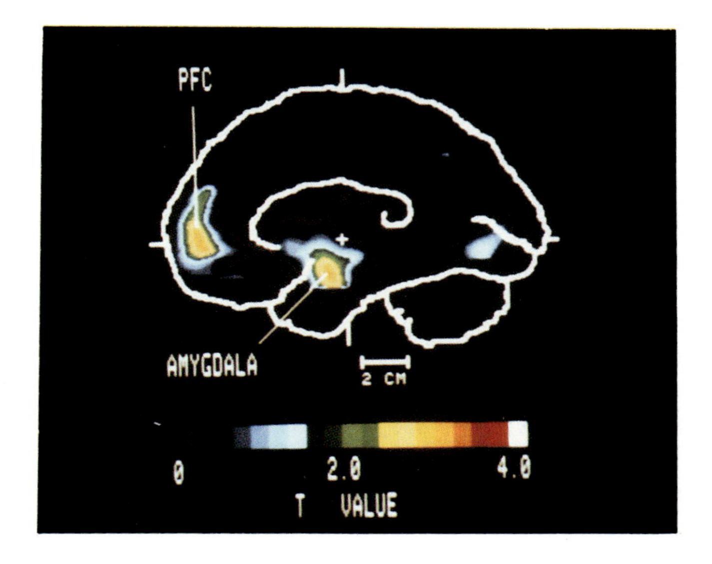

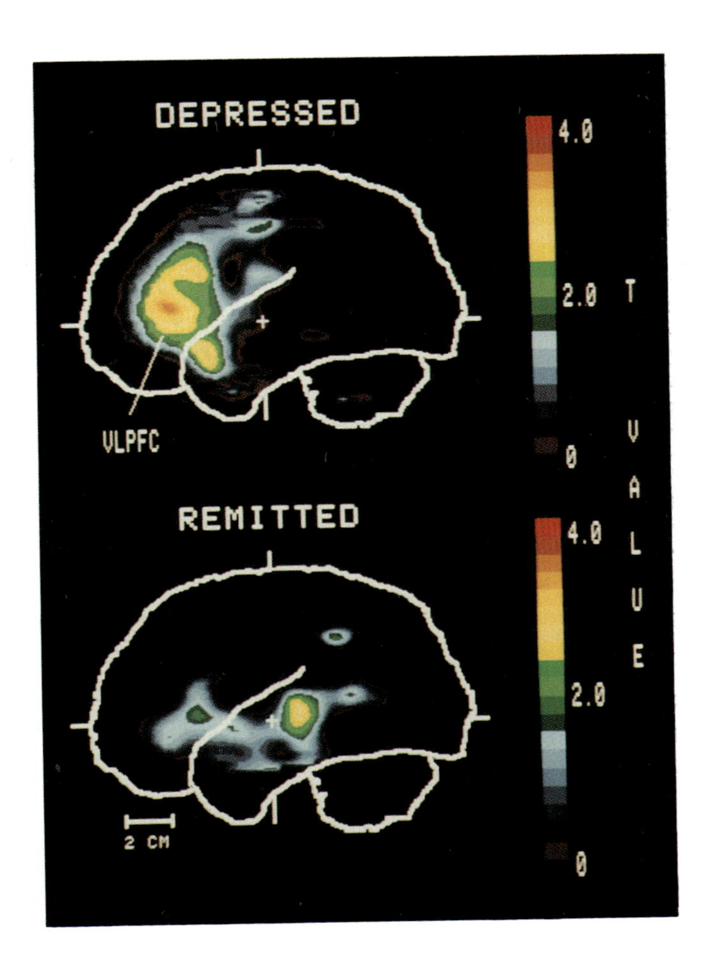

PLATE 24 Increased activity in the left PFC was found in the depressed subjects but not in the remitted subjects. Each *t*-image indicates areas where subjects in the depressed (top) and remitted (bottom) phases of FPDD have increased blood flow relative to the controls. Mean voxel *t*-values for the sagittal planes from 47–57 mm left of the bicommissural line are demonstrated. VLPFC, ventrolateral prefrontal cortex. (From Drevets et al., 1992, with permission.) (See chapter 76.)

PLATE 25 Increased blood flow was observed in the left medial thalamus of the depressed subjects with FPDD. The *t*-image demonstrates areas of increased blood flow in the depressives relative to controls in a coronal plane 7 mm caudal to the midline. Although limitations in spatial resolution preclude localization to discrete thalamic nuclei, the coordinates for the center of mass of the mean difference in blood flow are located in the mediodorsal nucleus (Drevets et al., 1992). Left is to the reader's left. (See chapter 76.)

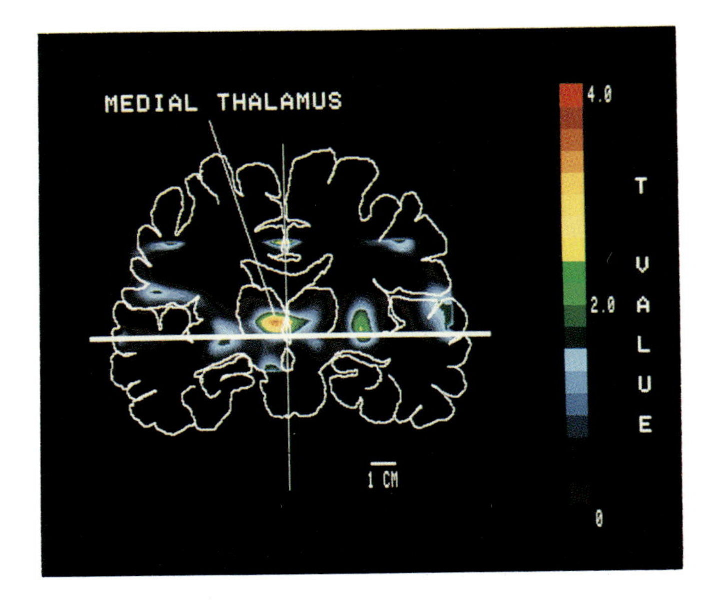

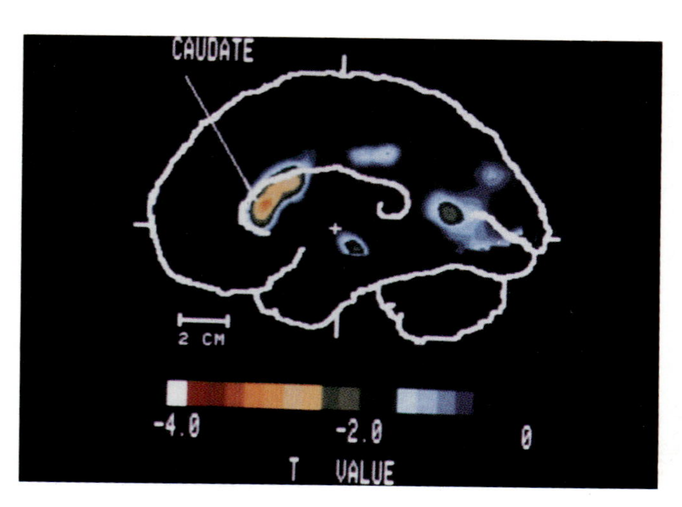

PLATE 26 Decreased blood flow was observed in the left caudate of the depressed subjects with FPDD. The *t*-image demonstrates areas of decreased blood flow in the depressives relative to the controls in a sagittal projection of the greatest voxel *t*-values in all planes between the midline and 10 mm left of midline. Anterior is to the reader's left. The color bar is reversed in this figure relative to the previous figures to indicate negative *t*-values. (From Drevets et al., 1992, with permission.) (See chapter 76.)

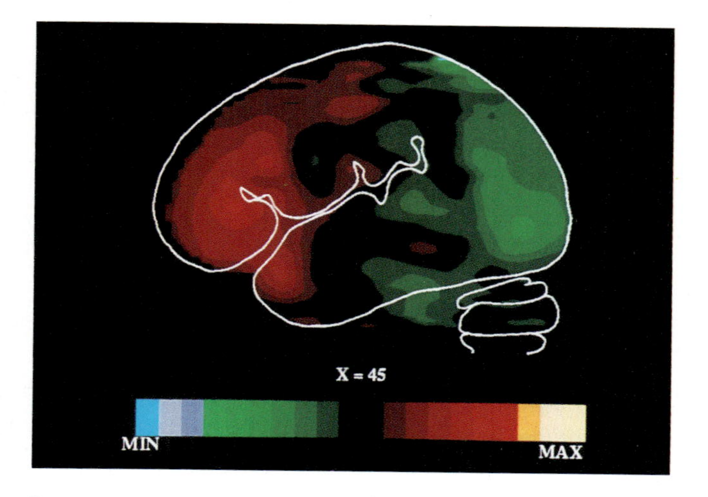

PLATE 27 Sagittal view demonstrating areas of increased and decreased blood flow in depressed subjects with FPDD relative to controls. The peak voxel blood flow difference through the entire brain is displayed. This image grossly illustrates the general pattern of increased blood flow in limbic prefrontal areas and decreased blood flow in posterior systems in the depressed phase of FPDD. (See chapter 76.)

47 Working Memory

ALAN BADDELEY

ABSTRACT Working memory refers to the temporary storage of information in connection with performing other, more complex tasks. A brief account of the application of the concept to animal learning, artificial intelligence, and developmental psychology is followed by a more detailed description of the related concept of short-term memory. A unitary short-term store proved too simple to function as an adequate working memory and later evolved into the concept of a multicomponent working memory. It is proposed that this comprises an attentional control system, the *central executive*, aided by slave systems responsible for the temporary storage and manipulation of either visual material (the *visuospatial sketchpad*), or verbal material (the *phonological loop*). Finally, the role of the phonological loop in language acquisition is briefly discussed.

The term *working memory* (WM) refers to the assumption that some form of temporary storage of information is necessary for performing a wide range of cognitive skills including comprehension, learning, and reasoning. It is important, however, to bear in mind that the term is used with rather different meanings in different research areas. Some of the more prominent uses of the concept will be discussed before the chapter goes on to describe in more detail one approach that has been particularly concerned with attempting to specify the characteristics of WM.

Concepts of working memory

WORKING MEMORY AND ANIMAL LEARNING A number of researchers in the area of animal learning have used the term *working memory* to refer to the capacity to retain information across trials within a given test session (Olton, Becker, and Handelmann, 1980). The classic WM task in animal learning is Olton's radial maze, which comprises a number of runways radiating out from a common start point. Each arm is baited with food, and the animal is given a series of trials. Typically the animal will avoid revisiting an arm from which the food has already been taken, indicating some form of retention of earlier trials. The measure is highly sensitive to certain drugs and brain lesions, and has proved a very popular paradigm. More recently the term *working memory* has been applied to a wider range of tasks, but unfortunately it is not clear how this concept of WM maps onto the term as used in studies of human subjects; it is almost certainly not equivalent, and is probably more closely akin to an aspect of long-term memory than to the models that will be described subsequently.

WORKING MEMORY AND PRODUCTION SYSTEM ARCHITECTURES In their ambitious program to simulate human cognition using computer models, Newell and Simon (1972) developed an approach based on production systems. This involves a rule-governed system in which quite complex behavior can be simulated in terms of a number of relatively simple "productions," whereby when prespecified conditions are met, the production fires. The productions are assumed to be held in working memory, which hence plays a crucial role in such models. However, while models such as Anderson's ACT* (1983), and more recently Newell's SOAR have been quite successful in simulating a wide range of complex cognitive skills, they assume a WM of unlimited capacity, an assumption that Newell (1992) admitted is not consistent with what is known of human working memory.

Other approaches used the production system architecture but made specific assumptions about the limitations of working memory; Kintsch and van Dijk (1978) for example, regard this as one of the major limits on language comprehension. Carpenter and Just, whose work on individual differences in WM will be discussed later, also use a production system architecture to build their models. However, it could be argued that the main impact of their work currently stems from their empirical results rather than from their production system model.

ALAN BADDELEY Medical Research Council Applied Psychology Unit, Cambridge, England

Neo-Piagetian Concepts of Working Memory When Piaget's all-embracing model of cognitive development began to lose ground as a result of extensive criticism, some of his followers produced alternative ways of conceptualizing the development of cognitive processing in children. Pascual-Leone (1970) has proposed a model based on the assumption of a limited-capacity WM system that develops through childhood, an approach that has, however, suffered from its complexity and lack of a single clear specification. A somewhat simpler account was given by Case, who argued that WM capacity remains constant, with improved performance coming from the more efficient use of the limited capacity as a result of faster information processing and better strategies in older children (Case, Kurland, and Goldberg, 1982). However, Hitch, Halliday, and Littler (1984) have subsequently presented evidence suggesting that a better account of the Case et al. results can be given by the Baddeley and Hitch model that will be described later.

Short-term Memory All the approaches to working memory discussed so far have made use of the concept but done relatively little to investigate its detailed characteristics. The approach that has been most concerned with the analysis of WM is that which has evolved from the earlier concept of short-term memory (STM). In the late 1950s, Brown (1958) and Peterson and Peterson (1959) demonstrated that even small amounts of information would be rapidly forgotten if active rehearsal was prevented. The evidence seemed to suggest that forgetting was attributable to the fading of a memory trace, in contrast to forgetting from long-term memory (LTM), which was generally assumed to be the result of interference from later learning. This led to the suggestion of separate long- and short-term memory systems.

In a classic article, Melton (1963) challenged this interpretation, arguing strongly that all the data could be encompassed by a single model that assumed forgetting to be based on interference. Melton's paper stimulated a flurry of new evidence, and led to an important conceptual distinction, between STM as a hypothetical memory *system*, and STM as a label for a particular type of *task*—one that may well prove to be a very impure measure of the underlying system. Waugh and Norman (1965) suggested using STM to refer to a class of task, and proposed William James's earlier term *primary memory* to refer to the hypothetical limited-capacity memory system while calling the system assumed to underlie LTM tasks *secondary memory*. Atkinson and Shiffrin (1968) make a similar distinction between STM tasks and a hypothetical short-term store (STS), which they differentiate from the long-term store (LTS).

In the mid-1960s, new experimental evidence for a dichotomy between LTS and STS began to emerge from a wide range of sources, of which the following three were among the most influential:

Two-component tasks. In the *free recall* task, subjects are given a list of unrelated words, and then asked to recall as many as possible in any order; they tend to be particularly good at recalling the last few items presented, the so-called *recency effect*. The first few items may also be favored (the *primacy effect*), with the middle items showing a rather lower level of recall. However, if recall is delayed by a few seconds, filled by a distracting task such as counting, then the recency effect disappears, while earlier items are comparatively unaffected (see figure 47.1). It is suggested that the recency items are retained in a temporary primary memory, while the earlier items are retained in long-term memory. Such a view is reinforced by the observation that variables that are known to influence LTS also influence earlier items, but do not influence the recency effect (Glanzer, 1972).

Acoustic and semantic coding. Conrad (1964) observed that when subjects attempted to recall strings of visually presented consonants, their errors were acoustically similar to the target item, hence *B* was more likely to be misremembered as *V* than as the visually more similar *R*. It was suggested that these effects indicated that the STS was acoustically based. A later series of experiments (Baddeley, 1966a) used words, confirming that the similar-sounding items (*man*, *mat*, *cap*, *map*, *can*) lead to poorer immediate serial recall than phonologically dissimilar words (*pit*, *day*, *cow*, *pen*, *rig*), whereas similarity of meaning (*huge*, *big*, *large*, *great*, *tall*) caused few problems. When long-term learning was required, however, the pattern reversed and meaning became the dominant factor (Baddeley, 1966b), a finding that was extended by Kintsch and Buschke (1969). These results therefore seemed to suggest that STS was acoustically based, while LTS favored semantic coding.

Neuropsychological evidence. It had been known for many years that patients with dense amnesia might nonetheless have good immediate memory as measured

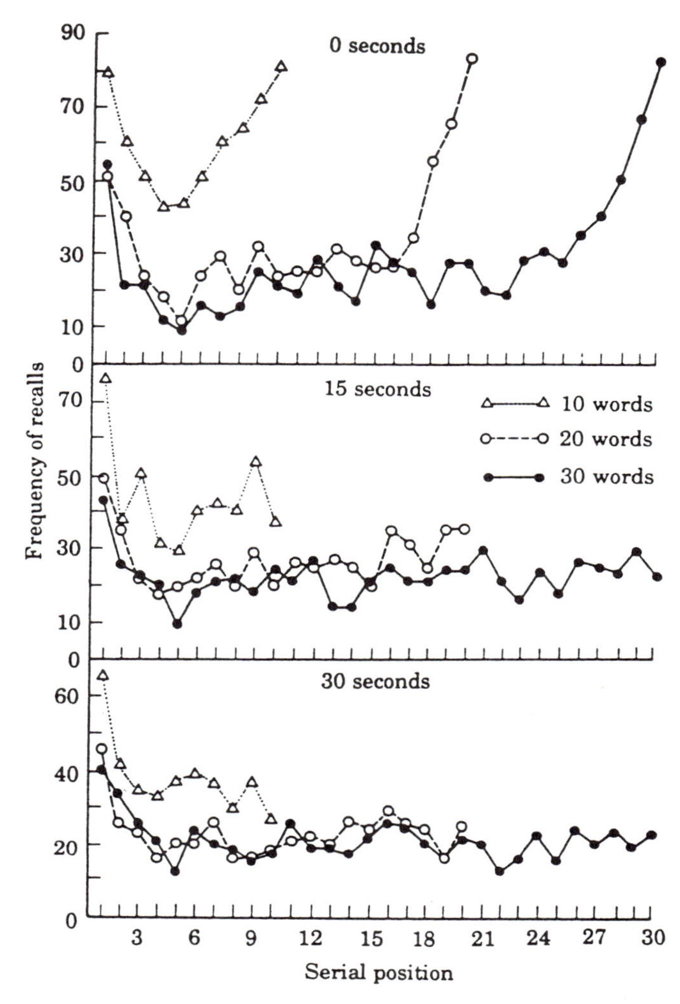

FIGURE 47.1 Serial position curve for lists of 10, 20, or 30 words recalled immediately or after a 15- or 30-s delay. Note that for each list length, the last few items presented are very well recalled on immediate test (the recency effect) but not after a delay. (Data from Postman and Phillips, 1965)

by digit span, the capacity to repeat back sequences of digits such as a telephone number. This suggested that STS might be intact in amnesic patients with grossly impaired long-term learning capacity. Evidence for this was provided by Milner (1966), and amplified using a range of two-component tasks by Baddeley and Warrington (1970). At the same time, Shallice and Warrington (1970) produced evidence of the opposite syndrome, a patient with a memory span of two digits, grossly impaired Peterson short-term forgetting performance, and virtually no recency effect in free recall, who nevertheless appeared to show a normal capacity for long-term learning. This double dissociation appeared to provide the strongest possible evidence for separate storage systems underlying long- and short-term memory. A great many models were developed at about this time, all more or less closely resembling the model developed by Atkinson and Shiffrin (1968), which has for that reason been dubbed the *modal model.*

THE MODAL MODEL As figure 47.2 shows, this model assumes three types of memory store. Information first of all enters a series of brief sensory registers, of which the visual system that is sometimes termed *iconic memory* and an auditory equivalent, *echoic memory*, are the most salient. Such systems are perhaps best regarded as representing the storage processes involved in perception. Information flows on to an STS, which was assumed by Atkinson and Shiffrin to act as a working memory that not only provided temporary storage but also selected and operated a whole range of encoding and retrieval strategies. It was assumed to be essential for retrieval of information from LTS, and for new learning, with the probability of learning being a direct function of the amount of time an item was held in the STS.

For a brief moment there was a feeling that the problems had been solved, then, all too rapidly, difficulties began to appear. A number of studies cast severe doubt on the assumption that simply holding information in STS guaranteed long-term learning (e.g., Craik and Watkins, 1973). Problems with this assumption led Craik and Lockhart (1972) to propose their *levels of processing* framework. Rather than assuming separate stores based on acoustic and semantic information, Craik and Lockhart suggested that the durability of a memory trace was a function of the depth of processing. Hence a superficial judgment such as whether the word *dog* is printed in capital or lowercase letters leads to poorer learning than a slightly deeper judgment such as "Does it rhyme with *log*?," which in turn is poorer than that resulting from a semantic judgment such as "Is it an animal?." This approach led to extensive research, and continues to offer a good rule of thumb that accounts for a wide range of data, although the extent to which the concept offers scope for further theoretical development has been questioned (Baddeley, 1978). Furthermore, the levels of processing approach is perhaps best seen as concerned with the role of coding in long-term memory; Craik and Lockhart continued to assume the need for a primary memory system, but did not regard its investigation as their principal concern.

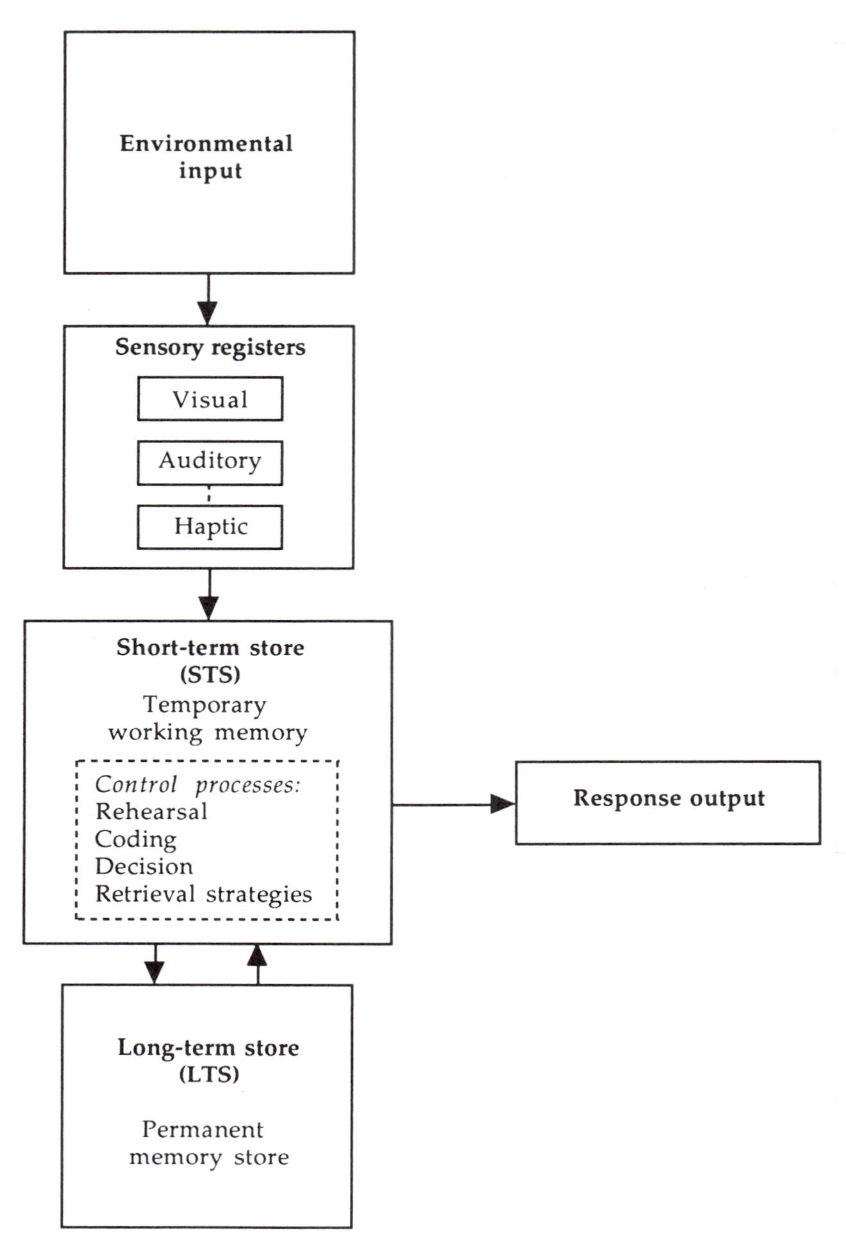

FIGURE 47.2 The flow of information through the memory system as conceived by Atkinson and Shiffrin (1968).

The second major problem with the modal model stemmed from the neuropsychological evidence; if STS is a crucial working memory system that is necessary for learning, then why should patients with massive STS impairment nonetheless show normal long-term learning? This was the problem tackled by Baddeley and Hitch (1974).

A LIMITED-CAPACITY WORKING MEMORY? If the STS acts as a limited-capacity WM, then restricting its capacity should impair such cognitive tasks as learning, comprehending, and reasoning. Why then, should STS patients appear to learn normally, and indeed to have few problems in coping with everyday life? Since such patients were rare, Hitch and I decided to attempt to

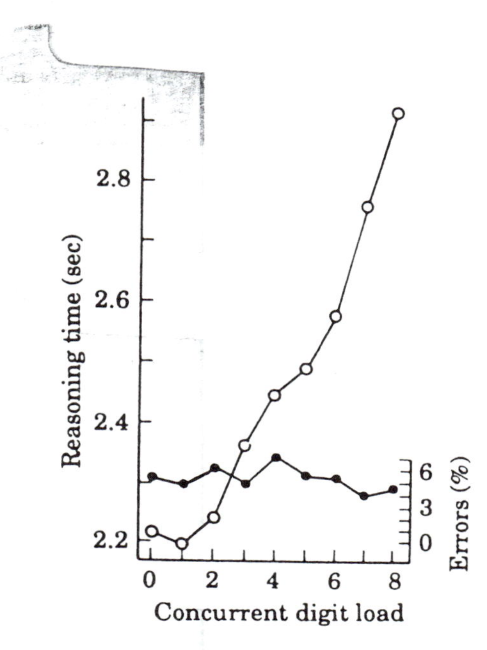

FIGURE 47.3 Speed and accuracy of grammatical reasoning as a function of number of digits concurrently being rehearsed. (From Baddeley, 1986)

simulate an STS deficit using normal subjects by differentially disrupting STS during the performance of learning, comprehension, and reasoning tests. In one study, subjects were required to perform a verbal reasoning task while continuing to repeat from zero to eight digits. The assumption was that the greater the digit load, the less the WM capacity for reasoning, with performance being wiped out when load approached span. As figure 47.3 shows, there was certainly an impact of concurrent task on reasoning speed, although an increase of 50% for a digit load of eight is hardly dramatic. Furthermore, errors remained constant at about 5%, regardless of digit load. A broadly similar pattern of results was obtained for tests of comprehension and of free recall learning, where a concurrent digit load impaired the LTS component but left the recency effect undisturbed.

These results raised two major problems for the modal model; a load of eight digits should have completely filled the STS, so how could the subjects continue to reason so accurately? And why was the recency effect in free recall unaffected by a concurrent digit load? We therefore abandoned the modal model and discarded the assumption that digit span offers a simple measure of a unitary working memory system. On the other hand, the clear evidence that concurrent span does interfere with a wide range of tasks encouraged us to keep the broad hypothesis of a limited-capacity WM that is capable of simultaneously storing and manipulating information. This concept of a limited-capacity general working memory will be discussed, before the chapter goes on to our further proposals to fractionate working memory into subsystems.

INDIVIDUAL DIFFERENCES IN WORKING MEMORY CAPACITY The concept of a limited-capacity WM has been applied very fruitfully to the study of individual differences. In an influential study, Daneman and Carpenter (1980) developed a measure of WM capacity based on a task that required the subject to read a series of sentences and subsequently recall the last word of each. The need simultaneously to comprehend and remember was what made this different from a simple word span task. They found a robust correlation between the subjects' WM span measure and their performance on standard tests of reading comprehension. In a subsequent study, they divided their subjects into those with high or low working memory span, and then demonstrated that low-span subjects were substantially more likely to be misled by inappropriate context, particularly when the correct interpretation involved carrying information across two sentences (Daneman and Carpenter, 1983). A parallel series of studies by Oakhill and her colleagues has been concerned with children who are good readers in the sense of being able to pronounce printed words, but poor comprehenders. She shows that such children perform poorly on WM span tasks, and in addition are bad at drawing inferences from the text (Oakhill, Yuill, and Parkin, 1988). An elegant series of studies on language comprehension in the elderly by Kemper and her colleagues also suggests that a deterioration in WM capacity results in difficulties in producing or comprehending certain types of syntactic structure (Kemper, 1992).

Some of the most exciting recent developments have been concerned with attempting to link the concept of WM to the concept of intelligence. Kyllonen and Christal (1990) tested a large number of U.S. Air Force recruits on a range of working memory tests, together with a battery of standard IQ measures, typically based on tests of reasoning. They found a very high correlation between two sets of measures, but the standard IQ tests tended to be more sensitive to prior knowledge, while WM was somewhat more influenced by processing speed. In a subsequent study, Christal (1991) at-

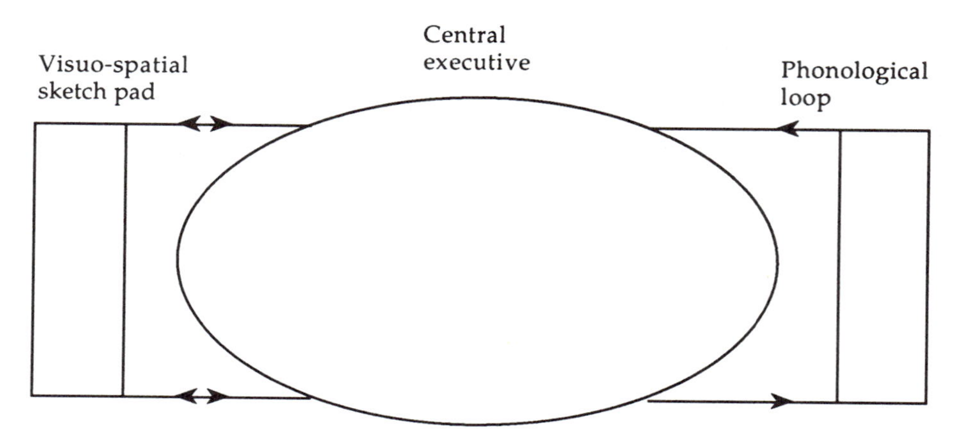

FIGURE 47.4 A simplified representation of the Baddeley and Hitch (1974) working memory model.

tempted to predict performance on a course concerned with learning about logic gates on the basis of both WM measures and the standard U.S. Air Force battery. He found that the WM measures provided the better prediction. The concept of a limited-capacity WM system therefore seems to have considerable potential for application in the area of psychometrics, where it has the additional attraction of offering measures that are less influenced by educational and cultural factors than the more traditional reasoning tests.

Components of working memory

The studies described above have used an operational definition of WM to provide a measure, but have been less concerned with understanding the constituent components of the hypothetical system. Baddeley and Hitch (1974) argued that it was necessary to abandon the concept of a unitary system, and proposed instead the tripartite model illustrated in figure 47.4. This assumes an attentional controller, the *central executive*, that is aided by two slave systems, the *visuospatial sketchpad*, which is able to hold and manipulate visual and spatial images, and the *phonological loop*, which involves a speech-based system. The system has been described in detail elsewhere (Baddeley, 1992a, 1992b), and hence will simply be outlined here.

THE CENTRAL EXECUTIVE The central executive is a limited-capacity system that is responsible for providing the link between the slave systems and LTM, and which is responsible for strategy selection and planning. Initially the executive was neglected in favor of tackling the more tractable slave systems. More recently, however, it has begun to be specified in more detail, initially by incorporating the Norman and Shallice (1980) model of the attentional control of action (Baddeley, 1986). We are now systematically attempting to tease apart the subcomponents of executive control, developing measures of the capacity to coordinate information and applying them with encouraging results to the analysis of the cognitive deficit associated with Alzheimer's disease (Baddeley et al., 1991). Other approaches to the analysis of the executive have involved applying tasks such as random generation to the analysis of complex skills such as chess (for a discussion of recent developments in this area see Baddeley, 1993). There is clearly a danger that a concept such as the central executive may reflect nothing more than a convenient homunculus; by attempting to split off and understand in turn the various executive functions, we hope in due course to make the homunculus redundant, and pension him off.

THE VISUOSPATIAL SKETCHPAD Initial evidence for a separate subsystem concerned with visuospatial information came from studies in which the subject was induced to remember material using either a verbal or a visuospatial code. It was then shown that concurrent articulatory or verbal activity would differentially disrupt one system, whereas visual or spatial activity would disrupt the other. For example, subjects taught to use an imagery mnemonic based on spatial location to remember lists of words performed consistently better than those using a rote rehearsal procedure under control conditions, whereas the difference disappeared when subjects were concurrently required to track a moving spot of light with a stylus (Baddeley and Lieberman, 1980). It seems likely that the system has both a visual component, concerned with factors such as color and shape, and a spatial component concerned with location. Further evidence for this comes from neuropsychological studies that suggest that a dissociation exists between visual and spatial codes. The system appears to be anatomically somewhat dispersed, with bilateral occipital components involved in the more visual aspects of coding, while spatial representation appears to depend more upon parietal regions (Farah, 1988). Single-unit recording work using awake monkeys suggests that there may also be a frontal lobe component (Goldman-Rakic, 1988), while a recent PET-scanning study by Jonides et al. (1993) provides confirmation for the involvement of each of these locations in a visual WM task.

THE PHONOLOGICAL LOOP Evidence for a separate subsystem concerned with storing and manipulating memory for sounds comes from a number of sources. This system appears to comprise two components, a memory store capable of holding phonological information for a period of one or two seconds, coupled with an articulatory control process (Baddeley, 1986, 1992a), a conclusion that is also supported by a recent PET-scanning study (Paulesu, Frith, and Frackowiak, 1993). Memory traces may be refreshed by subvocal articulation, a process that can also be used to feed the store, when the subject names overtly or covertly some item presented visually. The system is assumed to underlie digit span, with the number of items retained being a joint function of the rate at which the memory trace fades and the rate at which it can be refreshed by subvocal rehearsal.

Evidence for this simple model comes from the previously described phonological similarity effect. The system relies on phonological coding, with the result that items that are similar in sound have fewer distinguishing features and hence are more vulnerable to forgetting. The memory trace may also be corrupted if the subject is presented with irrelevant spoken material, which gains obligatory access to the store, corrupting the trace (Colle and Welsh, 1976; Salamé and Baddeley, 1982). The magnitude of the irrelevant speech effect does not depend upon the meaning of the material, since the phonological store is insensitive to semantic factors. The effect is not due to simple distraction, since noise does not impair performance, although nonspeech sounds may be disruptive under some circumstances (Jones, 1993).

The principal source of evidence for the importance of articulation comes from the *word length effect*, a tendency for memory span to decline as the constituent words increase in length. This is assumed to occur because rehearsal occurs in real time, so that long words take longer to rehearse, increasing the opportunity for the memory trace to decay before or during recall (Baddeley, Thomson, and Buchanan, 1975; Cowan, 1984).

Subvocal rehearsal can be prevented if the subject is required to utter repeatedly an irrelevant speech sound (*articulatory suppression*). Suppression eliminates the word length effect, presumably because it prevents rehearsal. It also interferes with the subvocal naming process whereby visual information can be registered in the phonological store. Hence, articulatory suppression will eliminate the phonological similarity effect for visually presented material, but not when presentation is auditory, since this guarantees entry to the store without the need for subvocal naming.

Anarthric patients who have lost the capacity to control their peripheral speech musculature may still show evidence of normal utilization of the phonological loop, suggesting that the system depends upon the central speech programming system rather than its overt operation (Baddeley and Wilson, 1985); and indeed Bishop and Robson (1989) have found comparatively normal articulatory loop functioning even in children who were congenitally dysarthric, suggesting that the system does not even require overt speech for its normal development. On the other hand, disruption of motor programming, such as presumably occurs in dyspraxic patients, does appear to impair performance (Caplan, Rochon, and Waters, 1992).

The previously described STM deficit patients appear to have a deficit in the phonological loop system (Vallar and Shallice, 1990) they presumably avoid the general cognitive disturbance that would be predicted by the modal model because they have preserved central executive and visuospatial sketchpad functioning. On the other hand, the fact that such patients appear to have few problems in coping with everyday life raises the question of what function is served by the phonological loop.

A clue to this question was given by a study in which a patient with a very pure phonological memory deficit was required to learn either pairs of words in her native language (e.g., *house–dog*), or the Russian equivalent of familiar words (e.g., *rose–svieti*). She proved normal at standard paired associate learning but was appallingly bad at new phonological learning (Baddeley, Papagno, and Vallar, 1988). Later studies have attempted to simulate this using articulatory suppression with normal subjects, and have shown that whereas paired associate learning is unaffected by suppression, foreign language vocabulary learning is clearly impaired (Papagno, Valentine, and Baddeley, 1991).

A parallel series of experiments investigated the learning by young children of the vocabulary of their native language. A task involving hearing and repeating back nonwords of gradually increasing length proved to be markedly impaired in children who had been selected as having impaired language development together with normal nonverbal intelligence (Gathercole and Baddeley, 1990). This nonword repetition test correlates well with vocabulary in four-year-olds. Cross-lagged correlational measures suggest that nonword repetition at age four predicts vocabulary a year later significantly better than the reverse of vocabulary predicting nonword repetition, thus suggesting that at this point at least, nonword repetition is the driving force, although the pattern of correlations changes as the children mature (Gathercole and Baddeley, 1989). Broadly speaking, then, the evidence is consistent with the hypothesis that the phonological loop has evolved, probably from more basic auditory perception and verbal production mechanisms, as a device for language acquisition. Impaired functioning of the loop is therefore likely to be much more troublesome for a child, who is just learning language and related skills such as reading, than it is for an adult. Broadly speaking, the association between problems of dyslexia and impaired WM performance suggests that this may well be the case (Gathercole and Baddeley, 1993).

Conclusions

The concept of a limited-capacity working memory has proved useful across a wide range of domains. Paradigms based on individual differences have demonstrated the importance of temporary storage in a wide range of tasks, and have illustrated the practical usefulness of psychometric measures based on the concept. It seems likely that such measures are typically an amalgam of two or more of the subcomponents suggested in the preceding section. It also seems likely that future developments will blend the breadth of studies using individual differences with the analytic approach underlying the studies concerned with fractionation, to provide a more detailed understanding of this important system that lies at the interface between perceiving, remembering, and planning.

REFERENCES

Anderson, J. R., 1983. *The Architecture of Cognition.* Cambridge: Harvard University Press.

Atkinson, R. C., and R. M. Shiffrin, 1968. Human memory: A proposed system and its control processes. In *The Psychology of Learning and Motivation: Advances in Research and Theory*, vol. 2, K. W. Spence, ed. New York: Academic Press, pp. 89–195.

Baddeley, A. D., 1966a. Short-term memory for word sequences as a function of acoustic, semantic and formal similarity. *Q. J. Exp. Psychol.* 18:362–365.

Baddeley, A. D., 1966b. The influence of acoustic and semantic similarity on long-term memory for word sequences. *Q. J. Exp. Psychol.* 18:302–309.

Baddeley, A. D., 1978. The trouble with levels: A re-examination of Craik and Lockhart's framework for memory research. *Psychol. Rev.* 85:139–152.

Baddeley, A. D., 1986. *Working Memory.* Oxford: Oxford University Press.

Baddeley, A. D., 1992a. Working memory. *Science* 255:556–559.

Baddeley, A. D., 1992b. Is working memory working? The Fifteenth Bartlett Lecture. *Q. J. Exp. Psychol.* 44A:1–31.

Baddeley, A. D., 1993. Working memory or working attention? In *Attention: Selection, Awareness and Control: A Tribute to Donald Broadbent*, A. Baddeley and L. Weiskrantz, eds. Oxford: Oxford University Press.

Baddeley, A. D., S. Bressi, S. Della Sala, R. Logie, and H. Spinnler, 1991. The decline of working memory in Alzheimer's disease: A longitudinal study. *Brain* 114:2521–2542.

Baddeley, A. D., and G. Hitch, 1974. Working memory. In *The Psychology of Learning and Motivation*, vol. 8, G. A. Bower, ed. New York: Academic Press, pp. 47–89.

Baddeley, A. D., and K. Lieberman, 1980. Spatial working memory. In *Attention and Performance VIII*, R. S. Nickerson, ed. Hillsdale, N.J.: Erlbaum, pp. 521–539.

Baddeley, A. D., R. Logie, S. Bressi, S. Della Sala, and H. Spinnler, 1986. Dementia and working memory. *Q. J. Exp. Psychol.* 38A:603–618.

Baddeley, A. D., C. Papagno, and G. Vallar, 1988. When long-term learning depends on short-term storage. *J. Mem. Lang.* 27:586–595.

Baddeley, A. D., N. Thomson, and M. Buchanan, 1975. Word length and the structure of short-term memory. *J. Verb. Learn. Verb. Be.* 14:575–589.

Baddeley, A. D., and E. K. Warrington, 1970. Amnesia and the distinction between long- and short-term memory. *J. Verb. Learn. Verb. Be.* 9:176–189.

Baddeley, A. D., and B. Wilson, 1985. Phonological coding and short-term memory in patients without speech. *J. Mem. Lang.* 24:490–502.

Bishop, D. V. M., and J. Robson, 1989. Unimpaired short-term memory and rhyme judgement in congenitally speechless individuals: Implications for the notion of "articulatory coding." *Q. J. Exp. Psychol.* 41A:123–141.

Brown, J., 1958. Some tests of the decay theory of immediate memory. *Q. J. Exp. Psychol.* 10:12–21.

Caplan, D., E. Rochon, and G. S. Waters, 1992. Articulatory and phonological determinants of word-length effects in span tasks. *Q. J. Exp. Psychol.* 45:177–192.

Case, R. D., D. M. Kurland, and J. Goldberg, 1982. Operational efficiency and the growth of short-term memory span. *J. Exper. Child Psychol.* 33:386–404.

Christal, R. E., 1991. Armstrong Laboratory Human Resources Directorate Technical Report AL-TP-1991-0031. Brooks Air Force Base, Texas 78235-5000.

Colle, H. A., and A. Welsh, 1976. Acoustic masking in primary memory. *J. Verb. Learn. Verb. Be.* 15:17–32.

Conrad, R., 1964. Acoustic confusion in immediate memory. *Br. J. Psychol.* 55:75–84.

Cowan, N., 1984. On short and long auditory stores. *Psychol. Bull.* 96:341–370.

Craik, F. I. M., and R. S. Lockhart, 1972. Levels of processing: A framework for memory research. *J. Verb. Learn. Verb. Be.* 11:671–684.

Craik, F. I. M., and M. J. Watkins, 1973. The role of rehearsal in short-term memory. *J. Verb. Learn. Verb. Be.* 12:599–607.

Daneman, M., and P. A. Carpenter, 1980. Individual differences in working memory and reading. *J. Verb. Learn. Verb. Be.* 19:450–466.

Daneman, M., and P. A. Carpenter, 1983. Individual differences in integrating information between and within sentences. *J. Exp. Psychol. [Learn. Mem. Cogn.]* 9:561–584.

Farah, M. J., 1988. Is visual memory really visual? Overlooked evidence from neuropsychology. *Psychol. Rev.* 95: 307–317.

Gathercole, S., and A. D. Baddeley, 1989. Evaluation of the role of phonological STM in the development of vocabulary in children: A longitudinal study. *J. Mem. Lang.* 28:200–213.

Gathercole, S., and A. D. Baddeley, 1990. Phonological memory deficits in language-disordered children: Is there a causal connection? *J. Mem. Lang.* 29:336–360.

Gathercole, S., and A. D. Baddeley, 1993. *Working Memory and Language.* Hove, Sussex: Erlbaum.

Glanzer, M., 1972. Storage mechanisms in recall. In *The Psychology of Learning and Motivation: Advances in Research and Theory*, vol. 5, G. H. Bower, ed. New York: Academic Press.

Goldman-Rakic, P. W., 1988. Topography of cognition: Parallel distributed networks in primate association cortex. *Annu. Rev. Neurosci.* 11:137–156.

Hitch, G. J., M. S. Halliday, and J. Littler, 1984. Memory span and the speed of mental operations. Paper presented at the joint Experimental Psychology Society/ Netherlands Psychonomic Foundation Meeting, Amsterdam.

Jones, D., 1993. Objects, streams and threads of auditory attention. In *Attention: Selection, Awareness and Control. A Tribute to Donald Broadbent*, A. Baddeley and L. Weiskrantz, eds. Oxford: Oxford University Press, pp. 87–104.

Jonides, J., E. E. Smith, R. A. Koeppe, E. Awh, S. Minoshima, and M. A. Mintun, 1993. Spatial working memory in humans as revealed by PET. *Nature* 363:623–625.

Kemper, S., 1992. Adults' sentence fragments—who, what, when, where, and why. *Commun. Res.* 19:444–458.

Kintsch, W., and H. Buschke, 1969. Homophones and synonyms in short-term memory. *J. Exp. Psychol.* 80:403–407.

Kintsch, W., and T. A. van Dijk, 1978. Toward a model of text comprehension and production. *Psychol. Rev.* 85:363–394.

Kyllonen, P. C., and R. E. Christal, 1990. Reasoning ability is (little more than) working-memory capacity. *Intelligence* 14:389–433.

Melton, A. W., 1963. Implications of short-term memory for a general theory of memory. *J. Verb. Learn. Verb. Be.* 2:1–21.

Milner, B., 1966. Amnesia following operation on the temporal lobes. In *Amnesia*, C. W. M. Whitty and O. L. Zangwill, eds. London: Butterworths, pp. 109–133.

Newell, A., 1992. SOAR as a unified theory of cognition: Issues and explanations. *Behav. Brain Sci.* 15:464–492.

Newell, A., and H. A. Simon, 1972. *Human Problem Solving.* Englewood Cliffs, N.J.: Prentice-Hall.

Norman, D. A., and T. Shallice, 1980. Attention to action: Willed and automatic control of behavior. University of California San Diego CHIP Report 99.

Oakhill, J. V., N. Yuill, and A. J. Parkin, 1988. Memory and inference in skilled and less-skilled comprehenders. In *Practical Aspects of Memory: Current Research and Issues*, vol. 2, *Clinical and Educational Implications*, M. M. Gruneberg, P. E. Morris and R. N. Sykes, eds. Chichester: Wiley, pp. 315–320.

Olton, D. S., J. T. Becker, and G. E. Handelmann, 1980. Hippocampal function: Working memory or cognitive mapping. *Physiol. Psychol.* 8:239–246.

Papagno, C., T. Valentine, and A. D. Baddeley, 1991. Phonological short-term memory and foreign-language vocabulary learning. *J. Mem. Lang.* 30:331–347.

Pascual-Leone, J. A., 1970. A mathematical model for the transition rule in Piaget's developmental stages. *Acta Psychol.* 32:301–345.

Paulesu, E., C. D. Frith, and R. S. J. Frackowiak, 1993. The neural correlates of the verbal component of working memory. *Nature* 362:342–345.

Peterson, L. R., and M. J. Peterson, 1959. Short-term retention of individual verbal items. *J. Exp. Psychol.* 58: 193–198.

Postman, L., and L. W. Phillips, 1965. Short-term temporal changes in free recall. *Q. J. Exp. Psychol.* 17:132–138.

Salamé, P., and A. D. Baddeley, 1982. Disruption of short-term memory by unattended speech: Implications for the structure of working memory. *J. Verb. Learn. Verb. Be.* 21:150–164.

Shallice, T., and E. K. Warrington, 1970. Independent functioning of verbal memory stores: A neuropsychological study. *Q. J. Exp. Psychol.* 22:261–273.

Vallar, G., and T. Shallice, eds., 1990. *Neuropsychological Impairments of Short-term Memory.* Cambridge: Cambridge University Press.

Waugh, N. C., and D. A. Norman, 1965. Primary memory. *Psychol. Rev.* 72:89–104.

48 Anatomical Basis of Memory Disorders

HANS J. MARKOWITSCH

ABSTRACT Brain structures that are of critical importance for the transmission of information for long-term storage or for retrieval are reviewed; it is emphasized that in addition to the partition of memory along the dimension time, a contents-based division is necessary to describe the processing of information within brain circuits. Episodic, semantic, priming, and procedural forms of memory are recognized as principal subdivisions. The term *bottleneck structures* is introduced to characterize those regions of the human brain through which (episodic) information usually has to pass to become long-term stored. These regions may be treated individually, but here preference is placed on considering them as embedded in networks of which the Papez circuit and the basolateral circuit are the most prominent examples. Nevertheless, it is recognized that the structures enclosed within these circuits, or intimately connected to them, may not be functionally equivalent, that different forms of pooling are possible, and that the partition into diencephalic, medial temporal, and basal forebrain areas may be the most conventional one. The anatomical locus or loci for the retrieval of old episodic—or individual-specific—information is seen as differing from that of the bottleneck structures and as depending on the (possibly combined) action of lateral temporopolar and (latero-)inferior prefrontal regions. The principal storage sites of episodic information are seen in the cerebral cortex with a dominance in its integrative areas. The loci for storage and processing of semantic memory are also taken to be found there. Finally, this review emphasizes the importance of the major telencephalic nuclei, probably together with the cerebellum, for the processing of priming and procedural memory. Here, the basal ganglia have a central role for skill memories, while certain forms of priming or affective-autonomous information processing may involve nonprimary cortical regions and the amygdala as well.

We are able to learn things as divergent as riding a bike, citing a Latin poem, remembering telephone numbers and mathematical rules, or recognizing and identifying a face or a smell confronted long ago. How the brain and its individual structures organize the processes of information encoding, consolidating, storing, and retrieving is the principal topic discussed in the following. As our nervous system—and that of many other animals—is characterized anatomically by the multitude and diversity of its interconnections, and functionally by its strong dependence on biochemical and bioelectrical processes, an integrated study of transmitter, hormonal, and electrophysiological activity changes in various regions and pathways would be most appropriate for an unraveling of the anatomical basis of memory. Traditional concepts and methodological constraints have, however, resulted in the use of one prevalent strategy—the study of brain-damaged individuals. Methods applicable to non-brain-damaged individuals—such as electrophysiological recordings (see Rugg, this volume) or neuroimaging techniques (see Roland et al., this volume)—help to complement our picture of information processing by the brain, as do various kinds of methods that can also be used in animals (neuroanatomical tracing techniques, biochemical interventions, etc.).

The "lesioning method," as it may be termed somewhat condensed and generalized, allows us—for humans—to analyze the consequences of nature's products (namely of brain damages or malformations), and—for animals—to analyze the consequences of induced, circumscribed, and well-positioned neuronal changes via direct interventions: mechanical, electrical, or chemical destruction of brain tissue, or the temporary functional interruption of brain regions by cooling or substance application. To study "the anatomical basis of memory disorders" consequently (frequently) means to study impairments both as dependent and independent variables and to infer from an abnormal situation to the normal state or condition of the organism in information transmission, or to generalize from one situation (e.g., an object discrimination task) or species (e.g., rat, monkey) to another one (e.g., episodic mem-

HANS J. MARKOWITSCH Physiological Psychology, University of Bielefeld, Bielefeld, Germany ory; human beings). Such inferences are common; they are, however, not straightforward and not without hazards, as I will illustrate below. Nevertheless, as especially in the neurosciences knowledge is dependent on available and used methodology, a good many of our present conclusions stem from evidence obtained in the way just described.

The conceptual basis of neuroanatomy of memory disorders

Amnesia—the failure or lack of memory—may have organic or psychic causes, and may be temporary or permanent (table 48.1). Originally, amnesia was regarded as global, that is, as involving all aspects of memory performance. However, this view was questioned by clinicians who further qualified the term as "simple," "retrograde," "anterograde," "isolated anterograde," "retarded," "catathymous," "temporary," "periodical," "progressive," "systematic," or "persistent," by proposals on memory subdivisions (Tulving, 1972), and by the revelation of spared mnestic functions in amnesics (Warrington and Weiskrantz, 1968). During recent decades the investigation of patients with memory disorders has become increasingly sophisticated, though studies from the old literature already had in fact pointed to the existence of, for example, spared memory functions in severely amnesic patients, or had referred to a temporal gradient in the loss of memories after brain damage ("Ribot's law"; see the overview in Markowitsch, 1992a). How diversified types of memory can be classified nowadays has been listed by Maurer (1992) in his Table 1. He divided memory by the process used in its retrieval ("retrieval through thought," "retrieval through action or through perception"), by the modality of cognitive memory lost in amnesia ("all modalities," "some modalities," "specific 'metamodalities'"), and by the subprocess of cognitive memory used in storage of the memory ("ultrashort-term storage," "short-term storage," "long-term storage").

TABLE 48.1

Overview of patient groups in whom memory disorders may be prominent features

Trauma-based cases with cerebral concussions or compressions
Patients with cerebral infarctions
Patients with intracranial tumors
Patients with bacterial or viral infections
Patients with deficiency diseases or avitaminoses
Patients with intoxications, chronic alcohol abuse, Korsakoff syndrome
Epileptics
Patients with degenerative diseases of the CNS (e.g., Alzheimer's patients)
Patients with organic insufficiencies (e.g., of liver, heart, kidneys)
Patients with a status after anoxia or hypoxia (e.g., after heart attack or drowning)
Psychiatric patients (e.g., schizophrenics)
Patients after drug addiction or as a consequence of drug usage (e.g., after taking anticholinergic or anticonvulsive substances, benzodiazepines, neuroleptics)
Patients after electroconvulsive therapy
Patients with transient global amnesia
Patients with psychogenic amnesia

As on the behavioral side, the anatomical study of memory disorders has experienced continual refinement over the last decades. While the hippocampal region had already been considered as inevitably involved in amnesia (e.g., Scoville and Milner, 1957), other foci have since been discussed: The diencephalon-related Korsakoff's syndrome is but one example (for review see Markowitsch, 1992a, 1992b). The major progress in relating anatomical and functional networks of the human brain came only recently after efficient brain imaging apparatuses were available and is still inmidst its golden season (see, e.g., Steinmetz et al., 1992).

Experiments on animals have supported us considerably in understanding the role of individual brain structures or of their interaction in information processing. First of all, work on animals has elucidated the wiring diagram of the brain. Proposals of brain circuits (e.g., "Papez circuit"; see Irle and Markowitsch, 1982) or of fiber-neuron networks ("ascending reticular activating system," "medial forebrain bundle") were based on animal results. Basic neuronal (and biochemical) mechanisms of information transmission depended entirely on the outcome of experimental manipulations in nonhuman brain systems (e.g., "long-term potentiation," synaptic and receptor mechanisms). And even the use of mammals for modeling human memory disorders after circumscribed brain damage brought fruitful outcome (e.g., McDonald and White, 1993).

On the other hand, no one denies that there are basic differences between the behavior of human and nonhuman species and that there is linearity neither in evolutionary trends nor in the intellectual capacities of man and beast (see Bingham, 1992; Hodos and Campbell, 1969; Shettleworth, 1993). These differences lead some authors to conclude "that there is not yet a sound empirical basis for making cross-species generalizations about the neural structures that mediate performance on the tasks used to assess memory" (Howe and Courage, 1993, p. 310).

Human mediators of knowledge are dominantly of a symbolic character (language), and we may have developed strategies of information processing that are absent in many animals or at best exist in rudimentary forms (prospective memory, metamemory, and the like). Furthermore, there is a shift in functions mediated by a given structure, and brain structures do differ between species with respect to their relative size in the system and with respect to their diversification (Stephan, Baron, and Frahm, 1988). The hippocampal formation is a good example (Stephan, 1975; Bingham, 1992). Some brain structures—especially cortical areas—even are nonexistent in nonhuman species. Not only Gestalt theory but also results from functional neuroimaging techniques or electrophysiological recordings tell us that the brain principally acts as a whole (Baars, 1993); a brain without some structures, with other structures differently developed, and with its possessor acting in a different ecological and social environment, will function differently from another one with other background variables (not to mention individual-specific differences existing within a given species) (Markowitsch, 1988a). Consequently, inferences from animal experimentation will provide only hints to the respective situation in humans, especially when the subject under study is as complex as is memory.

Coming back to human cases while still using clues from animal work, I will introduce Chow's (1967) caveats with respect to the interpretation of results based on brain lesions: "First, if a brain lesion fails to affect a learning task, it cannot be stated that this part of the brain is unimportant in normal animals. Second, if the lesion does influence performance of the task, it does not necessarily mean that it is the only neural structure involved. Third, the aim of the ablation methods is in a way never attainable, for it throws away the object (a region of the brain) one wishes to study" (p. 708).

It is obvious that we usually only discuss the bottleneck structures that are crucial for the *passage* of information but are not the actual loci of engrams themselves (Markowitsch, 1988a). As memory furthermore is not of a static nature, we have to deal with the possibilities that only certain stages of the memory process may be altered and that information retrieval differs when required at different times or within different contexts.

The process of memorizing is dynamically influenced by organismic and environmental variables. The subject's arousal level, intelligence, and the motivational and emotional status are just a small selection of the personality factors involved, while physical stimulus conditions, interference effects, and familiarity with the presented material are examples of external influences. When memory processing is analyzed in brain-damaged individuals—as is principally the case in this review—the etiology and time course of the brain damage, in addition to its locus and extent, are variables that have a considerable influence on the subject's reactibility to the encoding of events (Damasio and Geschwind, 1985). (Pre- and perinatal brain changes may even lead to totally altered forms of information processing, which, together with other facts, indicate that considerable interindividual variation in neuronal information processing is quite likely; see Calabrese et al., 1994.) As a last introductory point, our abilities to describe and analyze memory are only as good as the measures, designs, and theories we employ, and the inability to gain access to a piece of information does not prove its nonexistence. (See, e.g., Rapp and Caramazza's [1993] recent discussion of the access versus storage distinction, as an example of such theories).

Memory in neuropsychology

Memory nowadays is subdivided with respect to both time and contents (see chapters 52 and 54). The time-related organization of memory has, of course, a long tradition, and models such as that of Hebb (1949) are still considered valid. Hebb subdivided memory into short-term and long-term memory, and assumed the existence of a reverberatory circuit in which the information is temporarily held before being transmitted for long-term storage or before being lost.

Thus, attention must be given not only to cases in which learning can be studied from the moment a stimulus is presented, but also to retrospective (or old) memory. At the beginning of this century authors already spoke of the "Janushead" of the mneme, which could look back from the present into the past and at

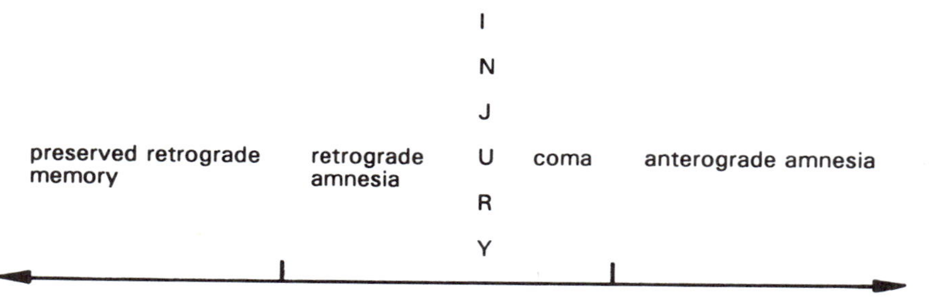

FIGURE 48.1 Sequence of possible consequences of (traumatic) brain injury on memory. (Modified from figure 4-1 of Levin, Benton, and Grossman, 1982, and figure 24.1 of Albert and Moss, 1984.)

the same time witness the passing of time from the present to the future (see Markowitsch, 1992a). This bidirectional time axis is of special importance in cases with abrupt brain damage (figure 48.1), in whom *retrograde amnesia* reflects the usually time-gradient-related loss of past information, and *anterograde amnesia* refers to the inability of storing new information long term. But this time line is also of use in determining dynamic aspects of memory such as the possible process of recovery of temporarily lost retrograde memories.

However, in addition to the dimension of time, memory is contentually subdivided and even hierarchically arranged (e.g., Tulving, 1991) into such divisions as procedural memory, perceptual representation ("sensory" priming), semantic memory, and episodic memory (see chapter 52). In short, *episodic memory* refers to the conscious recollection of the personal past (remembrance of very personal, individual-specific items that are framed in time and space). *Semantic memory* contains the general knowledge of the world (e.g., grammar, arithmetic). *Priming* refers to the facilitated, usually cue-based and unreflected reproduction of information one has been confronted with in the past, and *procedural memory* means the ability to perform skill-based operations (see chapters 52 and 54, and Tulving, 1991).

These divisions have turned out to be of major importance for an understanding of information processing by the brain, and even the results of pharmaceutical interventions support the view of biochemically (and therefore anatomically) dissociable memory systems (Nissen, Knopman, and Schacter, 1987).

In principle, patients with the traditional version of so-called global amnesia are only impaired in the field of episodic memory, and patients with widespread cerebral damage, such as in more advanced stages of Alzheimer's disease, may be impaired in the semantic memory field in addition, while the reverse is true for patients with basal ganglia damage, who may be largely normal in semantic and episodic memory, but manifest major problems in the domains of procedural or skill memory. Short-term memory usually is intact in so-called pure amnesic cases (e.g., Cave and Squire, 1992), but may be impaired to different degrees in patients with additional (e.g., neocortical) damage and consequently also in patients with dementias.

I will discuss these relations in the following, starting with the most commonly investigated domain of episodic memory and the brain structures related to it.

Episodic memory and the limbic system

The concept of the limbic system (Broca, 1878) (figure 48.2) was repeatedly abandoned (Brodal, 1982) and then revived (Isaacson, 1982) during the last decades. LeDoux (1989) criticized the steadily increasing number of structures included in this concept and suggested using the term "limbic forebrain" instead. Nevertheless, since Papez's (1937) "proposed mechanism of emotions," the so-called Papez circuit includes the core structures of the limbic system and those structures that also belong to the so-called bottleneck structures for memory—or, more exactly, for the transmission of information from short- into long-term storage. A conception of this transfer was illustrated by Kornhuber (1988; see figure 48.3). While his ideas cannot reflect the latest inferences on the anatomicofunctional relations of episodic memory, they do provide an intriguingly clear model which can be used as a basis for further elaboration and speculation.

THE PAPEZ AND THE BASOLATERAL LIMBIC CIRCUITS: AMNESIA AS A DISCONNECTION SYNDROME Using Kornhuber's basic scheme as a background, I will now introduce those regions of the human brain that are pre-

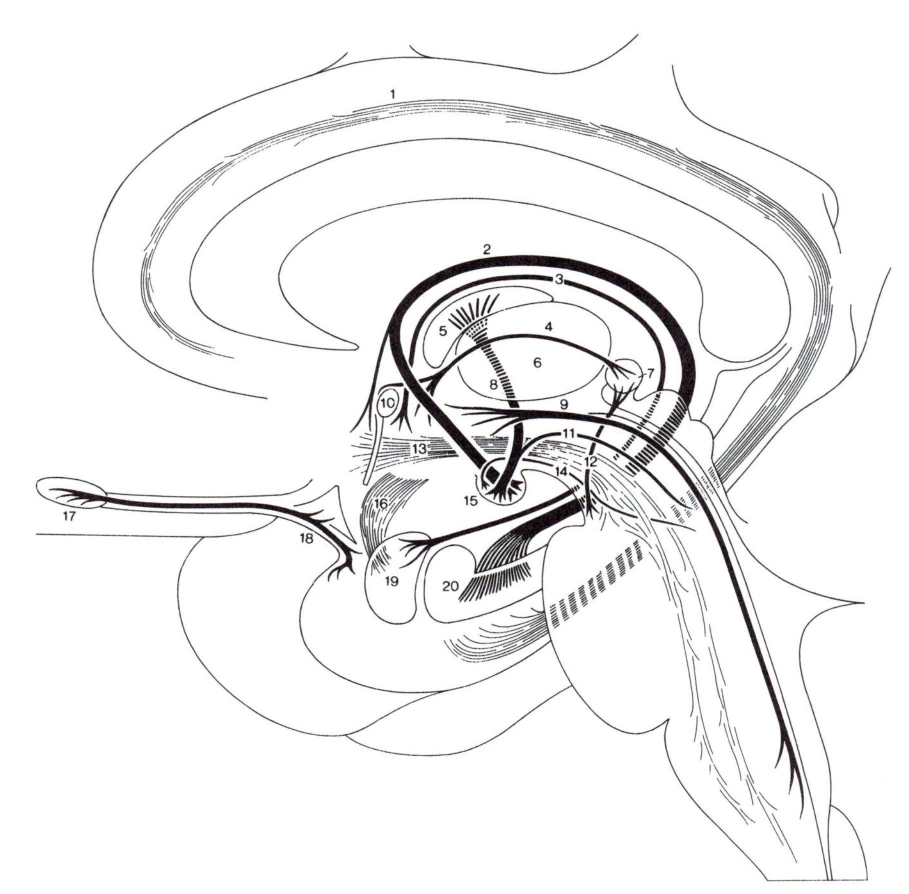

FIGURE 48.2 The major pathways of the limbic system and the rhinencephalon. 1, cingulum; 2, fornix; 3, stria terminalis; 4, stria medullaris thalami; 5, nucleus anterior thalami; 6, nucleus medialis thalami; 7, nuclei habenulae; 8, tractus mammillothalamicus; 9, fasciculus longitudinalis dorsalis; 10, commissura anterior; 11, tractus mammillotegmentalis; 12, tractus habenulointerpeduncularis; 13, fasciculus telencephalicus medialis; 14, pedunculus corporis mammillaris; 15, corpus mammillare; 16, ansa peduncularis; 17, bulbus olfactorius; 18, stria olfactoria lateralis; 19, corpus amygdaloideum; 20, hippocampus. (Reproduced from figure 191 of Nieuwenhuys, Voogd, and van Huijzen, 1988, with permission.)

sently considered to be of central importance for the transmission of episodic information for long-term storage. In discussing these structures, I want to emphasize that my view of information processing in the brain is principally a network-based one which has close affinities to that proposed by Mesulam (1990), but includes as well features of the models given by Damasio (1989) and Deacon (1989). The likelihood that such a widespread structural network exists in memory processing has been underlined by the results of recent functional neuroimaging studies (Fazio et al., 1992; Heiss et al., 1992; Pepin and Auray-Pepin, 1993; Sandson et al., 1991), and its existence has even been suggested for human procedural learning (Grafton et al., 1992).

While Kornhuber combined the Papez circuit with the thalamic mediodorsal nucleus in his circuitry (figure 48.3), some evidence points to the possibility of a rather parallel or complementary role of the medial diencephalic and the medial temporal lobe structures (see Goldman-Rakic, 1988; Markowitsch and Pritzel, 1985; McClelland and Rumelhart, 1986; Mishkin, 1982; and Squire, Knowlton, and Musen, 1993). Furthermore, the individual contribution of single regions and fiber pathways within these two major target systems has been the subject of much investigation and debate.

Before considering the possible role of individual structures (of these two systems and of the so-called basal forebrain memory system), I wish to introduce the assumption of a second limbic circuit which—again—may act in parallel or complementarily in information processing: the basolateral limbic circuit

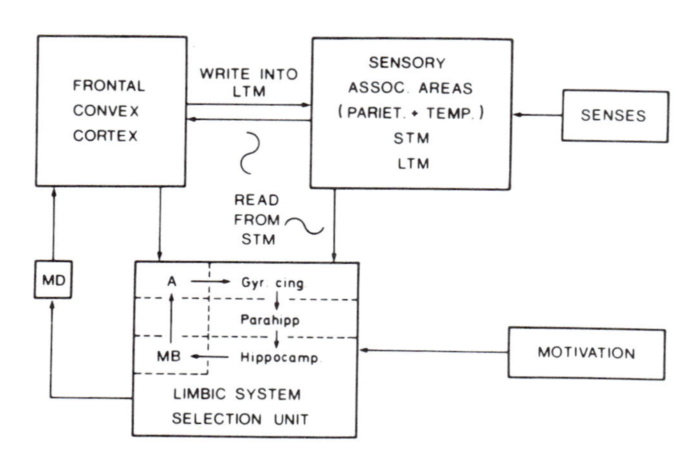

FIGURE 48.3 Kornhuber's (1988) schema of memory processing emphasizing the cooperation of cortical and subcortical mechanism involved in the selection of information between short-term memory (STM) and long-term memory (LTM). MB, mammillary body; A, anterior thalamic nucleus; MD, mediodorsal thalamic nucleus. (Reproduced from figure 4 of Kornhuber, 1988, with permission.)

(Sarter and Markowitsch, 1985a). See figure 48.4. The components of both the basolateral and the Papez circuit or its modifications (see Irle and Markowitsch, 1982; Braak and Braak, 1992) are embedded in the more extended circuitry of the limbic system, as emphasized in figure 48.5, which provides a state-of-the-art representation of the principal neuronal connectivity important for long-term information processing. It is therefore only a matter of course to consider amnesia as a disconnection syndrome as was done by von Cramon, Hebel, and Schuri (1986), von Cramon and Markowitsch (1992), Malamut et al. (1992), Warrington and Weiskrantz (1982), and many others.

Bottleneck Structures for Memory Processing

Especially within the sector of episodic memory (but also for other forms of memory; see Tulving and Schacter, 1992), a material-specific processing of information does occur. Examples for this include the verbal versus nonverbal distinction, which can be well distinguished between the two halves of the cerebrum, or the neutral versus emotional-affective continuum (which may involve, for example, amygdala and septum). As different forms of information involve different brain structures (or partly overlapping brain structures to a quantitatively different extent), it is logical to assume that differing forms of stimulus material also stress different neuronal assemblies within the as-

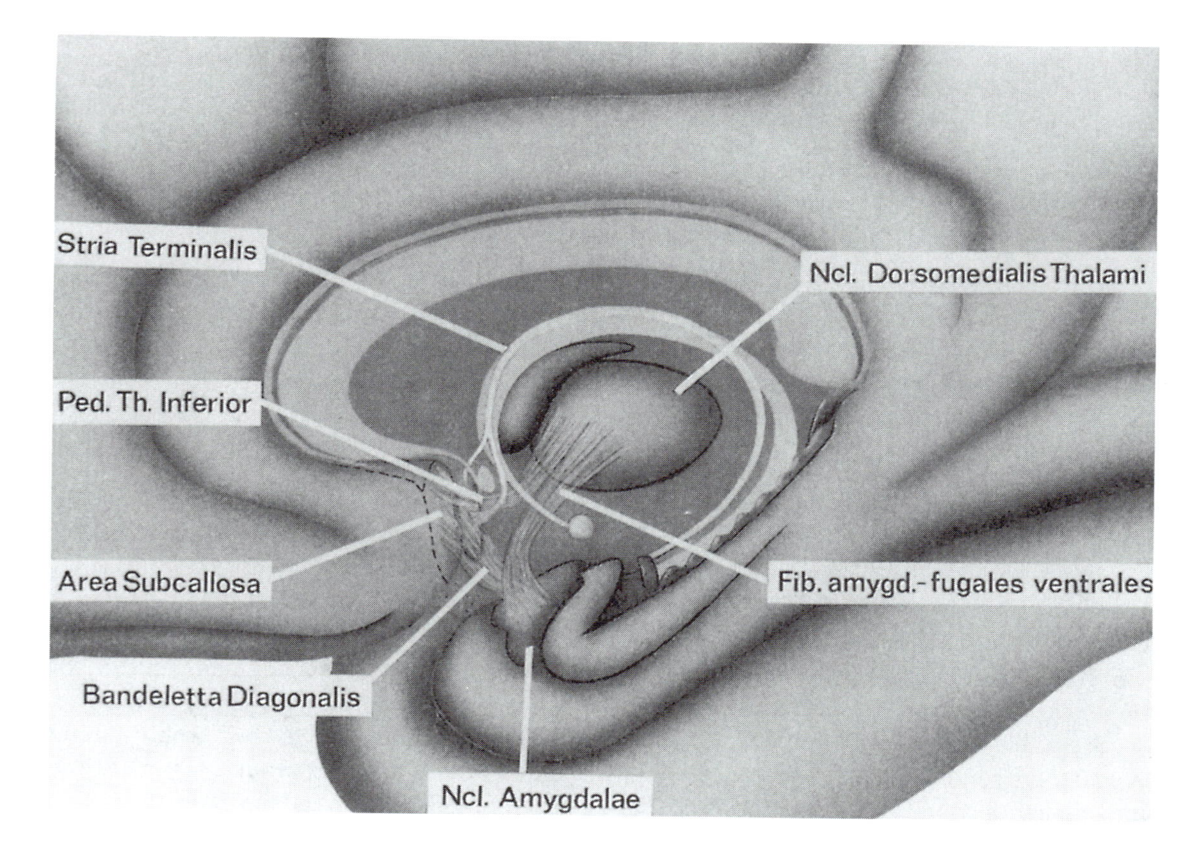

sumed limbic networks. Consequently, the importance of one or the other structure within such a circuit may turn out to be minor for certain kinds of memory processing, while central for the memorizing of other forms of material. The spared remembrance of stimuli with an extraordinary affective content in global amnesics is evidence for differential memory processing (Claparede, 1911; Markowitsch et al., 1984; Markowitsch, Kessler, and Denzler, 1986) and may indicate a specific role of, for example, the amygdala in the emotional charging of information (Sarter and Markowitsch, 1985a, 1985b; cf. the Klüver-Bucy syndrome, Klüver and Bucy, 1939).

The amygdala, one of the three major components of the basolateral limbic circuit, and one of the major atrophying nuclear complexes in Alzheimer's disease (Scott et al., 1992; Unger et al., 1991), is a good example of a region with a selective role in memory. Its contribution to memory processing remained unspecified in older studies on medial temporal amnesia (e.g., Scoville and Milner, 1957), while Mishkin (1978) later considered that its combined destruction together with major portions of the hippocampal formation was necessary for severe and lasting amnesia, a finding that was later confirmed in human patients (Duyckaerts et al., 1985; Heit, Smith, and Halgren, 1988) and monkeys (Zola-Morgan, Squire, and Mishkin, 1982; Saunders, Murray, and Mishkin, 1984). However, counterevidence for its participation in memory processing has been published on the basis of a restricted number of tasks and animals (Zola-Morgan, Squire, and Amaral, 1989b; see also Squire and Zola-Morgan, 1991). On the other hand, evidence from human case reports exists that stresses the role of this structure in memory (Newton et al., 1971; Tranel and Hyman, 1990), a role that has also come to be considered likely on the basis of theoretical considerations (Sarter and Markowitsch, 1985a, 1985b). Tranel and Hyman (1990) found in a 23-year-old patient after bilateral amygdala damage (mineralization due to Urbach-Wiethe disease) "a significant defect in visual, nonverbal memory" (p. 354) and concluded that while empirical evidence for a memory-related role of the amygdala remains equivocal, their "case is consistent with the position that the amygdala *is* a crucial component of the neural substrate of memory in humans" (p. 355). In another study, the interruption of major portions of the basolateral limbic circuit (of which the amygdala is a core structure) was suggested as contributing substantially to a lasting and severe verbal memory impairment (Markowitsch et al., 1990).

FIGURE 48.4 The principal, memory-processing-relevant, connectivity of the basolateral limbic circuit. The figure shows a "subdivision" of the basolateral limbic circuit that may be part of a subcortical memory operating system. The mediodorsal thalamic nucleus, the subcallosal area, and the amygdala are linked which each other by distinct fiber projections. The subcallosal area and amygdala are connected to the mediodorsal thalamic nucleus by fibers traveling through the inferior thalamic peduncle. Fibers arising from the amygdala are labeled ventral amygdalofugal pathways. The subcallosal area and amygdala are interconnected via the diagonal band (bandeletta diagonalis). Additionally mentioned is the stria terminalis. (Reproduced from figure 5.3. of von Cramon, 1992, with permission.)

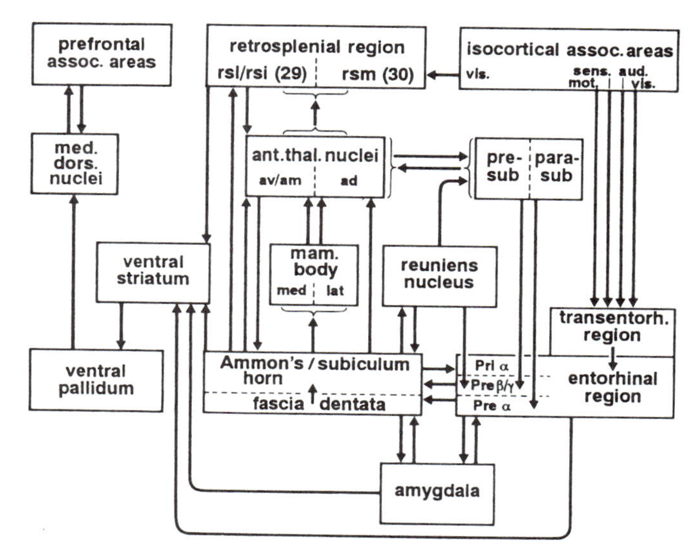

FIGURE 48.5 This scheme illustrates the components and the connectivity of the limbic and associated brain structures that are most intimately involved in memory processing. Delineated are isocortical and limbic input to the entorhinal region, connections between entorhinal region and hippocampal formation, and output to ventral striatum and motor system. ant.thal.nuclei (av/am, ad), anterior thalamic nuclei (anteroventral, anteromedial, anterodorsal); isocortical assoc. areas (mot.sens.aud.vis.), isocortical association areas (somatomotor, somatosensory, auditory, visual); mam.body (med.lat.), mammillary body (medial, lateral); med.dors.nuclei, mediodorsal thalamic nuclei; prefrontal assoc. areas, prefrontal association areas; para-sub, parasubiculum; pre-sub, presubiculum; retrosplenial region (rs/rsi, rsm), lateral, intermediate, and medial field of the retrosplenial region (Brodmann areas 29 and 30). (Reproduced from figure 2 of Braak and Braak, 1992, with permission.)

Part of the discrepancy in obtained results and conclusions may be due to differences in measurements or test procedures applied and to the underlying view one has about the brain's functioning in information processing. Furthermore, it might be of help to distinguish between "a neuronal system which is involved in memory processing and a system which affects memory processes" (Gold, Edwards, and McGaugh, 1975, 104).

What I have here described for the amygdala can be found as well for several other structures of the limbic system such as the mammillary bodies or the fornix. Victor, Adams, and Collins (1989, from screening brains of Korsakoff patients) and Zola-Morgan, Squire, and Amaral (1989a, on the basis of results from two monkeys with bilateral mammillary bodies lesions) considered the role of the mammillary nuclei unimportant for memory, but the case given in Dusoir et al. (1990) together with at least ten other published studies in humans (reviewed in Markowitsch, 1992c) emphasize that it is their destruction that most likely causes persistent amnesia. Counterevidence against the role of the fornix in memory was given in the study of Zola-Morgan, Squire, and Amaral (1989a), while positive evidence was described in the review of Gaffan and Gaffan (1991).

These remarks on the pros and cons of an involvement of specific structures in mnemonic information processing are given to demonstrate that we still are far away from a coherent view of a memory network of the human brain—not to mention the problems possibly inherent in the transfer of results from nonhuman species (see, e.g., Bingham, 1992; Shettleworth, 1993).

Though the preceding discussion stressed a rather "localizationistic" or "isolated" view of memory processing by the brain, I will briefly point to some other relevant bottleneck structures. The hippocampal formation and its surrounding allocortical masses (ento- and perirhinal cortex) are usually attributed a crucial role in the transfer of episodic information for long-term storage (see Squire and Knowlton's chapter in this volume). An exception is Salzmann's (1992) recent review in which he came to the conclusion "that hippocampus and parahippocampus play a significant part in the regulation of selective attention and expectation" (p. 163), but probably not in memory. While the circuitry of this medial temporal system has largely been unraveled, the discussion on memory-relevant structures within the medial diencephalon continues (Markowitsch, 1988b; Markowitsch, von Cramon, and Schuri, 1993). Similarly, the question of whether both systems act in parallel or have distinct functions in memory processing remains open (e.g., Markowitsch, 1992c; McKee and Squire, 1992; Markowitsch, von Cramon, and Schuri, 1993). That circumscribed bilateral (infarction-caused) damage of the dorsal diencephalon can result in very dense anterograde and time-limited retrograde amnesia (without conscious realization of these defects) was demonstrated in case A. B. of Markowitsch, von Cramon, and Schuri (1993). Figure 48.6 exemplifies the profoundness of this defect.

The contribution of individual structures of the "third" memory system—the basal forebrain—is still unresolved as well. Von Cramon, Markowitsch, and Schuri (1993) emphasized a role of the septum; von Cramon and Schuri (1992) a role of the fiber tracts interconnecting the basal forebrain with the hippocampal formation; Irle et al. (1992) surprisingly pointed to the basal ganglia; and Morris et al. (1992) had an amnesic case with damage principally restricted to the diagonal band of the Broca region. It is, however, likely that extensive damage to this system resembles the amnesic effects found in patients with medial diencephalic or medial temporal lobe damage, while less complete destruction may affect certain facets of information processing, for example by causing an emotional flattening or by reducing attention. (On the other hand, less complete damage has also been shown to be much less devastating in its consequences in the other two regional complexes: see, e.g., Markowitsch and Pritzel, 1985; Kritchevsky, Graff-Radford, and Damasio, 1987; Markowitsch, 1992c.)

SPECT-based investigations on the phenomenon of transient global amnesia (Markowitsch, 1990) have likewise failed to reveal indications in favor of a specific memory system, as both cases with transient metabolic undersupply of the medial temporal (Stillhard et al., 1990) and the medial diencephalic system (Goldenberg et al., 1991) were found to exist.

While the associative areas of the cerebral cortex may be the major fields of (episodic) information storage, the nuclei and fiber systems of the forebrain and some of the allocortical areas can be termed bottleneck structures that are crucial for the antecedent transmission of information. Material- and modality-specific memory deficits may occur after selective neocortical damage (e.g., Ellis, Young, and Critchley, 1989;

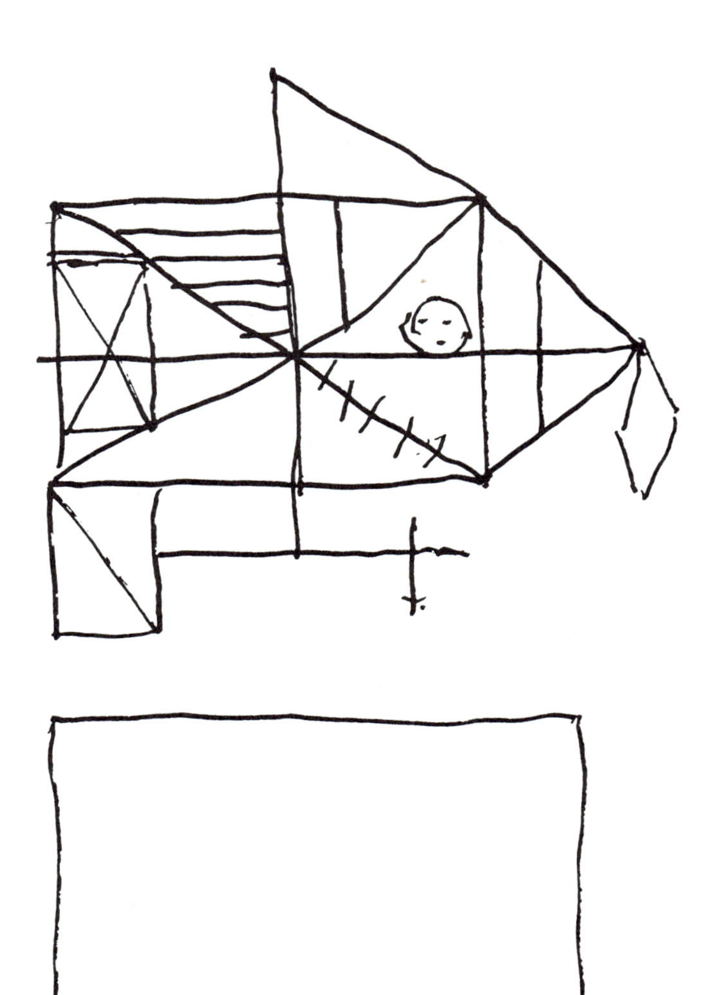

FIGURE 48.6 The performance of a bilateral diencephalic damaged patient with dense amnesia in copying the Rey-Osterrieth figure (top) and his attempt at redrawing it by heart immediately after having seen and copied it. (Results on case A. B. of Markowitsch, von Cramon, and Schuri, 1993.)

Caramazza and Hillis, 1991). Taken together, those structures that have long been implicated in the processing of emotionally laden and mnemonic information still receive the major attention, though the contribution of fiber systems, interconnecting the nuclear and cortical areal masses, has to be considered in addition.

Retrograde amnesia for episodic material

Isolated retrograde amnesia is a rare and usually transient condition. While in the first reports there was either no evidence on the affected brain structures or the damage was widespread and diffuse, more recently described cases suggested the existence of defined cortical foci that may be essential for the retrieval of stored episodic (for reviews see Kopelman, 1993; Markowitsch, Calabrese, Haupts et al., 1993) or semantic information (De Renzi, Liotti, and Nichelli, 1987; Grossi et al., 1988).

A carefully analyzed recent case had bilateral damage in the temporopolar region and some further frontal damage (Kapur et al., 1992). A patient with quite similar brain damage (figure 48.7) and the same etiology (a fall from a horse) had similarly extensive retrograde amnesia for personal events, but maintained the ability to learn information anew and retained solid knowledge on the level of semantic memory (Markowitsch, Calabrese, Haupts et al., 1993; Markowitsch, Calabrese, Liess et al., 1993). From these two reports and a few related cases (reviewed in Markowitsch, Calabrese, Haupts et al., 1993) we conclude that the region of the lateral temporopolar cortex (probably in conjunction with the reciprocally interconnected inferior prefrontal cortex) is crucial for the retrieval of episodic information. Possibly the amygdala participates in this process.

On the semantic memory level, Grossi and coworkers (1988) reported the case of an 18-year-old patient with traumatic brain damage in the region of the left parietal lobe and a rather selective disturbance of old memories from culture. The authors assumed that their results might "contradict the 'embeddedness' hypothesis described by Tulving (1984)" (p. 463). Even more extensively documented was another case (De Renzi, Liotti, and Nichelli, 1987). A schoolteacher was found to have a damaged left anteromedial inferior temporal lobe as a consequence of herpes encephalitis. Her autobiographical memory was normal, and on the semantic level grammatical-syntactic rules were preserved, but her knowledge of the world was drastically impaired. Even cooking created problems as she had lost knowledge about ingredients. On the other hand mere skill memory was well preserved (e.g. working the dough to make ravioli, driving, typewriting, operating the washing machine).

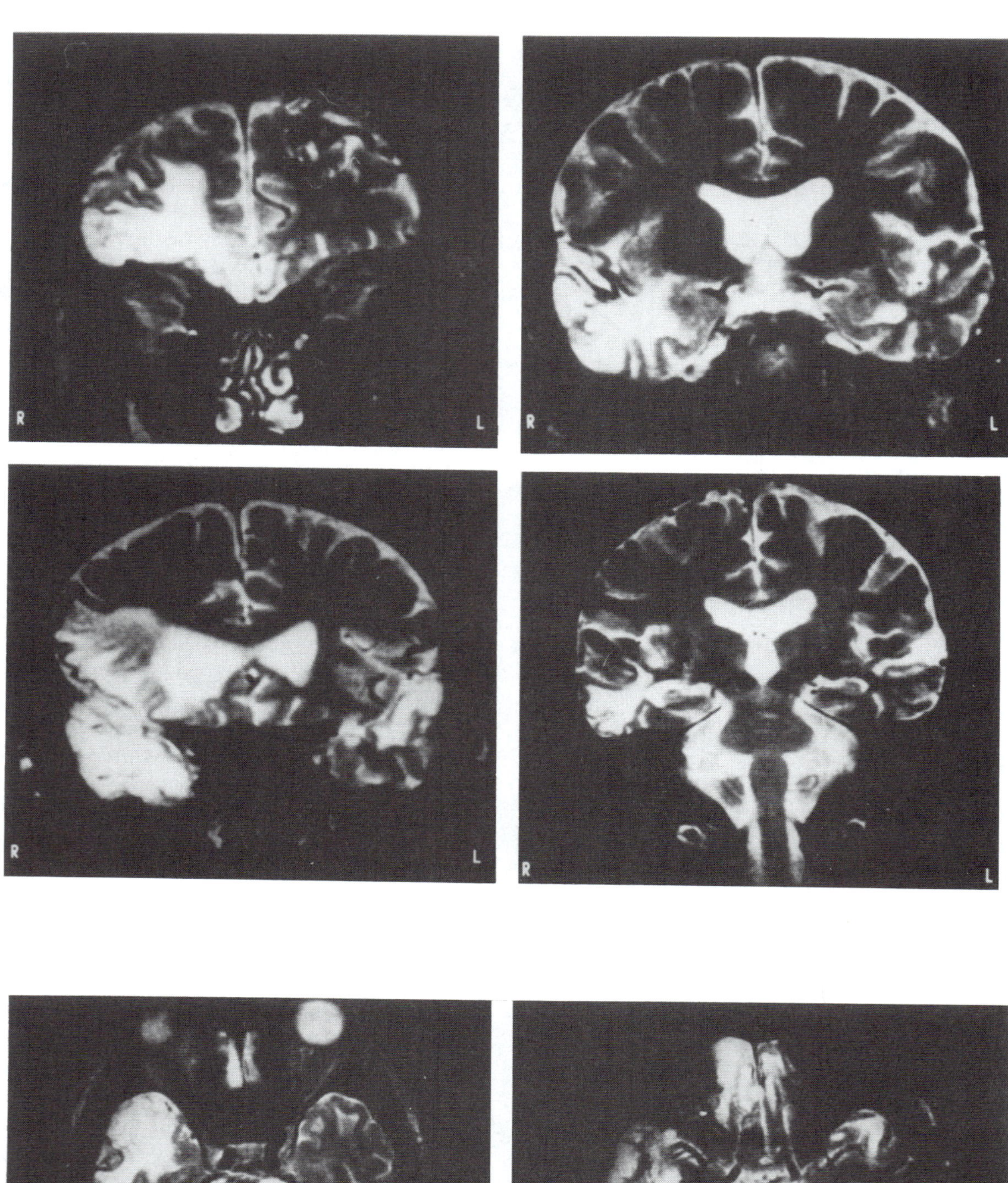
R
L
R
L
R
L
R
L

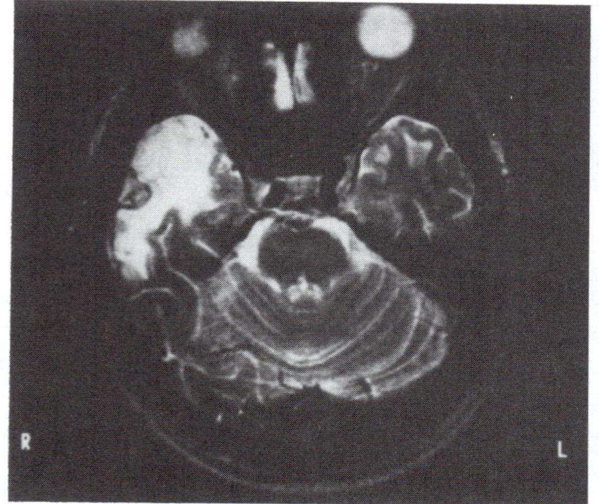
R
L

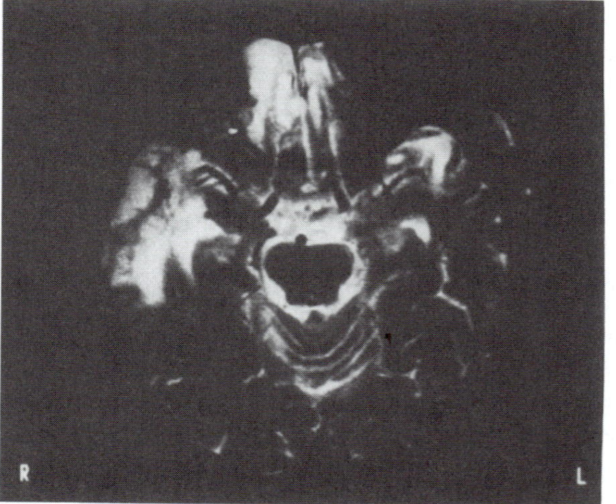
R
L

The anatomical foci of other memory systems

In particular, research on patients with Huntington's and Parkinson's disease has adduced evidence for an involvement of the basal ganglia in various forms of procedural learning and memory (e.g., Brandt and Butters, 1986; Damasio and Tranel, 1991). Heindel et al. (1989) established a performance dissociation between verbal priming and motor learning between Alzheimer's patients on the one side and cases with Huntington's chorea or Parkinson's disease on the other. And Heindel, Salmon, and Butters (1991) concluded from their results on patients with Alzheimer's or Huntington's disease that the acquisition of motor skills is dependent on the corticostriatal system, while neocortical neurons mediate the storage of semantic information. A further structure whose contribution to conditioning and procedural memory, but also to other forms of information processing, is increasingly acknowledged is the cerebellum (Akshoomoff et al., 1992; Fiez et al., 1992). With the probable exception of skill learning, the results on the anatomical loci of nonepisodic memory systems are, however, much less firm than those on the episodic memory level. Results from patients with Alzheimer's (Grosse, Gilley, and Wilson, 1991) or Pick disease (Hodges et al., 1992) and data obtained with functional neuroimaging techniques (Roland et al., 1990) suggest, however, that the undisturbed functioning of major portions of the cerebral cortex is essential for semantic memory processing. Priming may have affinities to the right hemisphere (Henke, Landis, and Markowitsch, 1993) and here perhaps to modality-specific regions of the nonprimary cortex (see Schacter, Chiu, and Ochsner, 1993).

Memory and the brain

While we understand now that the undisturbed functioning of certain anatomical stations is essential for the final storage of mnemonic information, we cannot say in what way individual structures complement each other functionally, and how far the functional equipotentiality goes within a given brain region. It still is only speculative that certain brain regions are particularly apt for extracting (retrieving) information. Also, possible modes of interhemispheric interactions of a given structure and possible differences between initial and later (for example, exercise-related) stages of information acquisition and processing are rarely unraveled. It is not unlikely that at least for some kinds of information the originally and the finally involved memory systems differ, or that in the course of learning and with time there is an interaction between memory systems that involves at least a quantitative change in the contribution and participation of individual systems. Possibly the refined methodologies of positron emission tomography and functional magnetic resonance imaging will lead to a more profound understanding of the brain's code for controlling its environment. These techniques may also be of help in elucidating differences in the processing of implicit and explicit memories and in elucidating proposed interactions between the mere processing of information and its conscious reflection (see Schacter, 1991; Smith, 1991).

FIGURE 48.7 Coronal and horizontal T2-weighted MRI scans showing the main portions of the damage in the anterior and lateral temporal lobes of a patient with severe and selective retrograde amnesia (case E. D. of Markowitsch, Calabrese, Haupts et al., 1993, and Markowitsch, Calabrese, Liess et al., 1993). The top four axial levels increase from top left to bottom right. *R* and *L* denote the left and right halves of the brain.

REFERENCES

AKSHOOMOFF, N. A., E. COURCHESNE, G. A. PRESS, and V. IRAGUI, 1992. Contribution of the cerebellum to neuropsychological functioning: Evidence from a case of cerebellar degenerative disorder. *Neuropsychologia* 30:315–328.

ALBERT, M. S., and M. MOSS, 1984. The assessment of memory disorders in patients with Alzheimer's disease. In *Neuropsychology of Memory*, L. R. Squire and N. Butters, eds. New York: Guilford, pp. 236–265.

BAARS, B. J., 1993. How does a serial, integrated and very limited stream of consciousness emerge from a nervous system that is mostly unconscious, distributed, parallel and of enormous capacity? In *Experimental and Theoretical Studies of Consciousness*, Ciba Foundation Symposium 174. Ciba, pp. 282–303.

BINGHAM, V. P., 1992. The importance of comparative studies and ecological validity for understanding hippocampal structure and cognitive function. *Hippocampus* 2: 213–220.

BRAAK, H., and E. BRAAK, 1992. The human entorhinal cortex: Normal morphology and lamina-specific pathology in various diseases. *Neurosc. Res.* 15:6–31.

BRANDT, J., and N. BUTTERS, 1986. The neuropsychology of Huntington's disease. *TINS* 9:118–120.

BROCA, P., 1878. Anatomie comparée des circonvolutions cérébrales. Le grand lobe limbique et la scissure limbique dans le série des mammifères. *Rev. Anthropol.* 2:385–498.

Brodal, P., 1982. *Neurological Anatomy in Relation to Clinical Medicine*, 2d ed. Oxford: Oxford University Press.

Calabrese, P., G. R. Fink, H. J. Markowitsch, J. Kessler, H. Durwen, J. Liess, M. Haupts, and W. Gehlen, 1994. Left hemispheric neuronal heterotopia: A PET, MRI, EEG, and neuropsychological investigation of a university student. *Neurology*. 44:302–305.

Caramazza, A., and A. E. Hillis, 1991. Lexical organization of nouns and verbs in the brain. *Nature* 349:788–790.

Cave, C. B., and L. R. Squire, 1992. Intact verbal and nonverbal short-term memory following damage to the human hippocampus. *Hippocampus* 2:151–164.

Chow, K. L., 1967. Effects of ablation. In *The Neurosciences*, G. C. Quarton, T. Melnechuk, and F. O. Schmitt, eds. New York: Rockefeller University Press, pp. 705–713.

Claparede, E., 1911. Récognition et moiité. *Arch. Psychol.* 11:79–90.

Damasio, A. R., 1989. Time-locked multiregional retroactivation: A systems level proposal for the neural substrates of recall and recognition. *Cognition* 33:25–62.

Damasio, A. R., and N. Geschwind, 1985. Anatomical localization in clinical neuropsychology. In *Handbook of Clinical Neurology*, vol. 1 (45), J. A. M. Frederiks, ed. Amsterdam: Elsevier, pp. 7–22.

Damasio, A. R., and D. Tranel, 1991. Disorders of higher brain function. In *Comprehensive Neurology*, R. N. Rosenberg, ed. New York: Raven, pp. 639–657.

Deacon, T. W., 1989. Holism and associationism in neuropsychology: An anatomical synthesis. In *Integrating Theory and Practice in Clinical Neuropsychology*, E. Perecman, ed. Hillsdale, N.J.: Erlbaum, pp. 1–47.

De Renzi, E., M. Liotti, and P. Nichelli, 1987. Semantic amnesia with preservation of autobiographic memory: A case report. *Cortex* 23:575–597.

Dusoir, H., N. Kapur, D. P. Byrnes, S. McKinstry, and R. D. Hoare, 1990. The role of diencephalic pathology in human memory disorder. *Brain* 113:1695–1706.

Duyckaerts, C., C. Derouesne, J. L. Signoret, F. Gray, R. Escourolle, and P. Castaigne, 1985. Bilateral and limited amygdalohippocampal lesions causing a pure amnesic syndrome. *Ann. Neurol.* 18:314–319.

Ellis, A. W., A. W. Young, and M. R. Critchley, 1989. Loss of memory for people following temporal lobe damage. *Brain* 112:1469–1483.

Fazio, F., D. Perani, M. C. Gilardi, F. Colombo, S. F. Cappa, G. Vallar, V. Bettinardi, E. Paulesu, M. Alberoni, S. Bressi, M. Franceschi, and G. L. Lenzi, 1992. Metabolic impairment in human amnesia: A PET study of memory networks. *J. Cerebr. Blood Flow Metab.* 12: 353–358.

Fiez, J. A., S. E. Petersen, M. K. Cheney, and M. E. Raichle, 1992. Impaired non-motor learning and error detection associated with cerebellar damage. *Brain* 115: 155–178.

Gaffan, D., and E. A. Gaffan, 1991. Amnesia in man following transection of the fornix: A review. *Brain* 114:2611–2618.

Gold, P. E., R. M. Edwards, and J. L. McGaugh, 1975. Amnesia produced by unilateral, subseizure, electrical stimulation of the amygdala in rats. *Behav. Biol.* 15:95–105.

Goldenberg, G., I. Podreka, N. Pfaffelmeyer, P. Wessely, and L. Deecke, 1991. Thalamic ischemia in transient global amnesia: A SPECT Study. *Neurology* 41:1748–1752.

Goldman-Rakic, P. S., 1988. Topography of cognition: Parallel distributed networks in primate association cortex. *Annu. Rev. Neurosci.* 11:137–156.

Grafton, S. T., J. C. Mazziotta, S. Presty, K. J. Friston, R. S. J. Frackowiak, and M. E. Phelps, 1992. Functional anatomy of human procedural learning determined with regional cerebral blood flow and PET. *J. Neurosci.* 12:2542–2548.

Grosse, D. A., D. W. Gilley, and R. S. Wilson, 1991. Episodic and semantic memory in early versus late onset Alzheimer's disease. *Brain Lang.* 41:531–537.

Grossi, D., L. Trojano, A. Grasso, and A. Orsini, 1988. Selective "semantic amnesia" after closed-head injury: A case report. *Cortex* 24:457–464.

Hebb, D. O., 1949. *The Organization of Behavior*. New York: Wiley.

Heindel, W. C., D. P. Salmon, C. W. Shults, P. A. Walicke, and N. Butters, 1989. Neuropsychological evidence for multiple implicit memory systems: A comparison of Alzheimer's, Huntington's, and Parkinson's disease patients. *J. Neurosci.* 9:582–587.

Heindel, W. C., D. P. Salmon, and N. Butters, 1991. The biasing of weight judgments in Alzheimer's and Huntington's disease: A priming or programming phenomenon? *J. Clin. Exp. Neuropsychol.* 13:189–203.

Heiss, W.-D., G. Pawlik, V. Holthoff, J. Kessler, and B. Szelies, 1992. PET correlates of normal and impaired memory functions. *Cerebrovasc. Brain Metab. Rev.* 4:1–27.

Heit, G., M. E. Smith, and E. Halgren, 1988. Neural encoding of individual words and faces by the human hippocampus and amygdala. *Nature* 333:773–775.

Henke, K., T. Landis, and H. J. Markowitsch, 1993. Subliminal perception of pictures in the right hemisphere. *Consciousness and Cognition* 2:225–236.

Hodges, J. R., K. Patterson, S. Oxbury, and E. Funnell, 1992. Semantic dementia. *Brain* 115:1783–1806.

Hodos, W., and C. B. G. Campbell, 1969. *Scala naturae*: Why there is no theory in comparative psychology. *Psychol. Rev.* 76:337–350.

Howe, M. L., and M. L. Courage, 1993. On resolving the enigma of infantile amnesia. *Psychol. Bull.* 113, 305–326.

Irle, E., and H. J. Markowitsch, 1982. Connections of the hippocampal formation, mamillary bodies, anterior thalamus and cingulate cortex: A retrograde study using horseradish peroxidase in the cat. *Exp. Brain Res.* 47:79–94.

Irle, E., B. Wowra, H. J. Kunert, J. Hampl, and S. Kunze, 1992. Memory disturbances following anterior communicating artery rupture. *Ann. Neurol.* 31:473–480.

Isaacson, R. L., 1982. *The Limbic System*, 2d ed. New York: Plenum.

Kapur, N., D. Ellison, M. P. Smith, D. L. McLellan, and E. H. Burrows, 1992. Focal retrograde amnesia following bilateral temporal lobe pathology. *Brain* 115:73–85.

Klüver, H., and P. C. Bucy, 1939. Preliminary analysis of functions of the temporal lobes. *AMA Arch. Neurol. Psychiatry* 42:979–1000.

Kopelman, M. D., 1993. The neuropsychology of remote memory. In *Handbook of Neuropsychology*, vol. 8, F. Boller and H. Spinnler, eds. Amsterdam: Elsevier, pp. 215–238.

Kornhuber, H. H., 1988. The human brain: From dream and cognition to fantasy, will, conscience, and freedom. In *Information Processing by the Brain*, H. J. Markowitsch, ed. Toronto: Huber, pp. 241–258.

Kritchevsky, M., N. F. Graff-Radford, and A. R. Damasio, 1987. Normal memory after damage to medial thalamus. *Arch. Neurol.* 44:959–962.

LeDoux, J. E., 1989. Cognitive-emotional interactions in the brain. *Cognition Emotion* 3:267–289.

Levin, H. S., A. L. Benton, and R. G. Grossman, 1982. *Neurobehavioral Consequences of Closed Head Injury*. New York: Oxford University Press.

Malamut, B. L., N. Graff-Radford, J. Chawluk, R. I. Grossman, and R. C. Gur, 1992. Memory in a case of bilateral thalamic infarction. *Neurology* 42:163–169.

Markowitsch, H. J., 1988a. Individual differences in memory performance and the brain. In *Information Processing by the Brain*, H. J. Markowitsch, ed. Toronto: Huber, pp. 125–148.

Markowitsch, H. J., 1988b. Diencephalic amnesia: A reorientation towards tracts? *Brain Res. Brain Res. Rev.* 13:351–370.

Markowitsch, H. J., ed., 1990. *Transient Global Amnesia and Related Disorders*. Toronto: Hogrefe and Huber.

Markowitsch, H. J., 1992a. *Intellectual Functions and the Brain*. Toronto: Hogrefe and Huber.

Markowitsch, H. J., 1992b. Diencephalic amnesia. In *Trastornos de la Memoria*, D. Barcia Salorio, ed. Barcelona: Editorial MCR, pp. 269–336.

Markowitsch, H. J., 1992c. *Neuropsychologie des Gedächtnisses*. Göttingen: Hogrefe.

Markowitsch, H. J., P. Calabrese, M. Haupts, H. F. Durwen, J. Liess, and W. Gehlen, 1993. Searching for the anatomical basis of retrograde amnesia. *J. Clin. Exp. Neuropsychol.* 15:947–967.

Markowitsch, H. J., P. Calabrese, J. Liess, M. Haupts, H. F. Durwen, and W. Gehlen, 1993. Retrograde amnesia after traumatic injury of the temporo-frontal cortex. *J. Neurol. Neurosurg. Psychiatry* 56:988–992.

Markowitsch, H. J., J. Kessler, C. Bast-Kessler, and R. Riess, 1984. Different emotional tones significantly affect recognition performance in patients with Korsakoff psychosis. *Int. J. Neurosci.* 25:145–159.

Markowitsch, H. J., J. Kessler, and P. Denzler, 1986. Recognition memory and psychophysiological responses towards stimuli with neutral or emotional content. A study of Korsakoff patients and recently detoxified and long-term abstinent alcoholics. *Int. J. Neurosci.* 29:1–35.

Markowitsch, H. J., and M. Pritzel, 1985. The neuropathology of amnesia. *Prog. Neurobiol.* 25:189–287.

Markowitsch, H. J., D. Y. von Cramon, E. Hofmann, C.-D. Sick, and P. Kinzler, 1990. Verbal memory deterioration after unilateral infarct of the internal capsule in an adolescent. *Cortex* 26:597–609.

Markowitsch, H. J., D. Y. von Cramon, and U. Schuri, 1993. The mnestic performance profile of a bilateral diencephalic infarct patient with preserved intelligence and severe amnesic disturbances. *J. Clin. Exp. Neuropsychol.* 15: 627–652.

Maurer, R. G., 1992. Disorders of memory and learning. In *Handbook of Neuropsychology*. Vol. 7, *Child Neuropsychology*, S. J. Segalowitz and I. Rapin, eds. Amsterdam: Elsevier, pp. 241–260.

McClelland, J. L., and D. E. Rumelhart, 1986. A distributed model of human learning and memory. In *Parallel Distributed Processing: Explorations in the Microstructure of Cognition*. Vol. 2, *Psychological and Biological Models*, J. L. McClelland, D. E. Rumelhart, and the PDP Research Group, eds. Cambridge, Mass.: The MIT Press, pp. 171–215.

McDonald, R. J., and N. M. White, 1993. A triple dissociation of memory systems: Hippocampus, amygdala, and dorsal striatum. *Behav. Neurosci.* 107:3–22.

McKee, R. D., and L. R. Squire, 1992. Equivalent forgetting rates in long-term memory for diencephalic and medial temporal lobe amnesia. *J. Neurosci.* 12:3765–3772.

Mesulam, M.-M., 1990. Large-scale neurocognitive networks and distributed processing for attention, language, and memory. *Ann. Neurol.* 28:597–613.

Mishkin, M., 1978. Memory in monkeys severely impaired by combined, but not by separate removal of amygdala and hippocampus. *Nature* 273:297–298.

Mishkin, M., 1982. A memory system in the monkey. *Philos. Trans. R. Soc. Lond. [Biol.]* 298:85–95.

Morris, M. K., D. Bowers, A. Chatterjee, and K. M. Heilman, 1992. Amnesia following a discrete basal forebrain lesion. *Brain* 115:1827–1847.

Newton, F. H., R. N. Rosenberg, P. W. Lampert, and J. S. O'Brien, 1971. Neurologic involvement in Urbach-Wiethe's disease (lipoid proteinosis). *Neurology* 21:1205–1213.

Nieuwenhuys, R., J. Voogd, and C. van Huijzen, 1988. *The Human Central Nervous System: A Synopsis and Atlas*, 3d ed. Berlin: Springer.

Nissen, M. J., D. S. Knopman, and D. L. Schacter, 1987. Neurochemical dissociation of memory systems. *Neurology* 37:789–794.

Papez, J. W., 1937. A proposed mechanism of emotion. *Arch. Neurol. Psychiatry* 38:725–743.

Pepin, E. P., and L. Auray-Pepin, 1993. Selective dorsolateral frontal lobe dysfunction associated with diencephalic amnesia. *Neurology* 43:733–741.

Rapp, B., and A. Caramazza, 1993. On the distinction between deficits of access and deficits of storage: A question of theory. *Cognitive Neuropsych.* 10:113–141.

Roland, P. E., B. Gulyás, R. J. Seitz, C. Bohm, and S. Stone-Elander, 1990. Functional anatomy of storage, recall, and pattern in man. *Neuroreport* 1:53–56.

Salzmann, E., 1992. Zur Bedeutung von Hippocampus und Parahippocampus hinsichtlich normaler und gestörter Gedächtnisfunktionen. *Fortschr. Neurol. Psychiatr.* 60:163–176.

Sandson, T. A., K. R. Daffner, P. A. Carvalho, and M.-M. Mesulam, 1991. Frontal lobe dysfunction following infarction of the left-sided medial thalamus. *Arch. Neurol.* 48:1300–1303.

Sarter, M., and H. J. Markowitsch, 1985a. The amygdala's role in human mnemonic processing. *Cortex* 21:7–24.

Sarter, M., and H. J. Markowitsch, 1985b. The involvement of the amygdala in learning and memory: A critical review with emphasis on anatomical relations. *Behav. Neurosci.* 99:342–380.

Saunders, R., E. A. Murray, and M. Mishkin, 1984. Further evidence that amygdala and hippocampus contribute equally to recognition memory. *Neuropsychologia* 22:785–796.

Schacter, D. L., 1991. Unawareness of deficit and unawareness of knowledge in patients with memory disorders. In *Awareness of Deficit after Brain Injury: Clinical and Theoretical Issues*, G. P. Prigatano and D. L. Schacter, eds. New York: Oxford University Press, pp. 127–151.

Schacter, D. L., C.-Y. P. Chiu, and K. N. Ochsner, 1993. Implicit memory: A selective review. *Annu. Rev. Neurosci.* 16:159–182.

Scott, S. A., S. T. DeKosky, D. L. Sparks, C. A. Knox, and S. W. Scheff, 1992. Amygdala cell loss and atrophy in Alzheimer's disease. *Ann. Neurol.* 32:555–563.

Scoville, W. B., and B. Milner, 1957. Loss of recent memory after bilateral hippocampal lesions. *J. Neurol. Neurosurg. Psychiatry* 20:11–21.

Shettleworth, S. J., 1993. Varieties of learning and memory in animals. *J. Exp. Psychol. [Anim. Behav.]* 19:5–14.

Smith, M. E., 1991. Making up the brain's mind. *Behav. Brain Sci.* 14:454–455.

Squire, L. R., B. Knowlton, and G. Musen, 1993. The structure and organization of memory. *Annu. Rev. Psychol.* 44:453–495.

Squire, L. R., and S. Zola-Morgan, 1991. The medial temporal lobe memory system. *Science* 253:1380–1386.

Steinmetz, H., Y. Huang, R. Seitz, U. Knorr, G. Schlaug, H. Herzog, T. Hackländer, and H.-J. Freund, 1992. Individual integration of positron emission tomography and high-resolution magnetic resonance imaging. *J. Cerebr. Blood Flow Metab.* 12:919–926.

Stephan, H., 1975. *Allocortex: Handbuch der mikroskopischen Anatomie des Menschen,* vol. 4, part 9. Berlin: Springer.

Stephan, H., G. Baron, and H. Frahm, 1988. Comparative size of brains and brain components. In *Comparative Primate Biology*. Vol. 4, *Neurosciences*, H. D. Steklis and J. Erwin, eds. New York: A. R. Liss, pp. 1–38.

Stillhard, G., T. Landis, R. Schiess, M. Regard, and G. Sialer, 1990. Bitemporal hypoperfusion in transient global amnesia: 99m-HM-PAO SPECT and neuropsychological findings during and after an attack. *J. Neurol. Neurosurg. Psychiatry* 53:339–342.

Tranel, D., and B. T. Hyman, 1990. Neuropsychological correlates of bilateral amygdala damage. *Arch. Neurol.* 47:349–355.

Tulving, E., 1972. Episodic and semantic memory. In *Organization of Memory*, E. Tulving and W. Donaldson, eds. New York: Academic Press, pp. 381–403.

Tulving, E., 1984. Relations among components and processes of memory. *Behav. Brain Sci.* 7:257–263.

Tulving, E., 1991. Concepts of human memory. In *Memory: Organization and Locus of Change*, L. R. Squire, N. M. Weinberger, G. Lynch, and J. L. McGaugh, eds. New York: Oxford University Press, pp. 3–32.

Tulving, E., and D. L. Schacter, 1992. Priming and memory systems. In *Neuroscience Year: Supplement to the Encyclopedia of Neuroscience*, G. Adelman and B. H. Smith, eds. Cambridge, Mass.: Birkhäuser, pp. 130–133.

Unger, J. W., L. W. Lapham, T. H. McNeill, T. A. Eskin, and R. W. Hamill, 1991. The amygdala in Alzheimer's disease: Neuropathology and Alz 50 immunoreactivity. *Neurobiol. Aging* 12:389–399.

Victor, M., R. D. Adams, and G. H. Collins, eds., 1989. *The Wernicke-Korsakoff Syndrome and Related Neurologic Disorders due to Alcoholism and Malnutrition* 2d ed. Philadelphia: F. A. Davis.

Von Cramon, D. Y., 1992. Focal cerebral lesions damaging (subcortical) fibre projections related to memory and learning functions in man. In *Neuropsychological Disorders Associated with Subcortical Lesions*, G. Vallar, S. F. Cappa, and C.-W. Wallesch, eds. New York: Oxford University Press, pp. 132–142.

Von Cramon, D. Y., N. Hebel, and U. Schuri, 1986. Is vascular thalamic amnesia a disconnection syndrome? In *Neurology*, K. Poeck, H. J. Freund, and H. Gänshirt, eds. Berlin: Springer, pp. 195–203.

Von Cramon, D. Y., and Markowitsch, H. J., 1992. The problem of "localizing" memory in focal cerebro-vascular lesions. In *Neuropsychology of Memory*, 2d ed., L. R. Squire and N. Butters, eds. New York: Guilford, pp. 95–105.

Von Cramon, D. Y., H. J. Markowitsch, and U. Schuri, 1993. The possible contribution of the septal region to memory. *Neuropsychologia* 31:1159–1180.

Von Cramon, D. Y., and U. Schuri, 1992. The septo-hippocampal pathways and their relevance to human memory: A case report. *Cortex* 28:411–422.

Warrington, E. K., and L. Weiskrantz, 1968. New method of testing long-term retention with special reference to amnesic states. *Nature* 217:972–974.

Warrington, E. K., and L. Weiskrantz, 1982. Amnesia: A disconnection syndrome? *Neuropsychologia* 20:233–248.

Zola-Morgan, S., L. R. Squire, and D. G. Amaral, 1989a. Lesions of the hippocampal formation but not lesions of the fornix or the mammillary nuclei produce long-lasting memory impairment in monkeys. *J. Neurosci.* 9:898–913.

Zola-Morgan, S., L. R. Squire, and D. G. Amaral, 1989b. Lesions of the amygdala that spare adjacent cortical regions do not impair memory or exacerbate the impairment following lesions of the hippocampal formation. *J. Neurosci.* 9:1922–1936.

Zola-Morgan, S., L. R. Squire, and M. Mishkin, 1982. The neuroanatomy of amnesia: Amygdala-hippocampus versus temporal stem. *Science* 218:1337–1339.

49 Positron Emission Tomography in Cognitive Neuroscience: Methodological Constraints, Strategies, and Examples from Learning and Memory

PER E. ROLAND, RYUTA KAWASHIMA,
BALÁZS GULYÁS, AND BRENDAN O'SULLIVAN

ABSTRACT Positron emission tomographic (PET) measurements of brain activations are carried out by many different approaches, each carrying limitations. In addition, each carries specific limitations in interpretations. Measurements of regional cerebral metabolic rate (rCMR) or regional cerebral blood flow (rCBF) mark the areas of the brain showing rapid changes in metabolism associated with synaptic activity. Atlas-based anatomical standardization, signal averaging together with intertask subtractions, or repeat measurements in single subjects reveal a mosaic of activated fields for which cognitive components can be associated with certain combinations of active fields. Consistency analyses across experimental groups show that cortical functional fields are reproducible entities, and common denominators can be found for the conditions that activate a small subset of fields. Learning and recognition of large, complex visual patterns activate an almost identical set of cortical functional fields in visual areas. A subset of these are activated when the patterns are recalled.

The main objective of using positron emission tomography (PET) in cognitive neuroscience is to localize brain activations. Brain activations are fast biochemical and biophysical activity increases in neurons and glia, of which transmembranic ion fluxes constitute an important part. The bulk of these fast changes occur in the synaptic regions (Sokoloff et al., 1977; Mata et al., 1980; Roland, 1993 for a review). The energy demands are therefore the highest in synaptic regions when the brain activates. The extra demands on glucose and oxygen also result in increases of the regional cerebral blood flow, rCBF. Consequently one can measure brain activation by measuring the regional glucose or oxygen consumption. In practice, however, the rCBF has been the physiological parameter cognitive neurobiologists have chosen, because it is by far the easiest to measure. In normal brains the rCBF is monotonically related to the regional metabolism (Roland, 1993). This relation is spatially very accurate and can be seen at the columnar level (Greenberg et al., 1979; Ginsberg, Dietrich, and Busto, 1987).

So, no matter whether one chooses to measure the rCMR or the rCBF, what will be revealed are the sites in the brain where the synapses change their metabolic activity. Because the methods for the measurement of rCMR and rCBF require a steady state for shorter or longer time intervals, these methods show the average change in rCMR or rCBF for the time of measurement with the PET. This implies that it is not possible, with these methods, to follow time differences in synaptic activation at different sites in the brain.

Measurements of rCBF as a parameter of regional synaptic metabolism

Measurements of the rCBF have now replaced direct measurements of the rCMR, both because of the ease

PER E. ROLAND, RYUTA KAWASHIMA, BALÁZS GULYÁS, and BRENDAN O'SULLIVAN Division of Human Brain Research, Department of Neuroscience, Karolinska Institute, Stockholm, Sweden with which rCBF measurements are done and because measurements of the regional cerebral phosphorylation rate of glucose and glucose oxidation require long times and measurements of the regional cerebral oxygen consumption require repetitive measurements. It is not appropriate here in detail to discuss the different methods of rCBF measurements in combination with PET, but only to give a superficial overview of the disadvantages (see Phelps, Mazziotta, and Shelbert, 1986; Raichle, 1987; and Roland, 1993 for a review of methods).

A PET scanner measures the concentration of a radioactive tracer molecule in the brain at a certain time. Absolute measurements of rCBF in ml/100g/min or ml/g/min require freely diffusible tracers and steady state of the rCBF during the measurement. Therefore the measurement time has to be as short as possible, which is also desirable for physiological reasons. Freely diffusible tracers are gases such as ^{77}Kr, ^{11}C- or ^{18}F-fluoromethane, ^{13}N-N_2O, and ^{15}O-butanol. Except for ^{15}O-butanol, the other tracers are gases that must be inhaled by the test subjects. ^{15}O-butanol has the advantage that it can be injected intravenously, and thus does not require any voluntary effort from the test subjects. Because $H_2{}^{15}O$ is easy to synthesize, most PET centers use this tracer for rCBF determination. It has the disadvantage that it is not freely diffusible (Eichling, et al., 1974). Consequently the behavior of the tracer cannot be described by a single compartmental model, and the mathematics used to describe the behavior of $H_2{}^{15}O$ becomes very complicated. In practice a single compartmental model is often used anyway, sometimes with some correction factors for the increasing diffusion limitation at high rCBFs.

Measurement of the rCBF requires an arterial line to measure also the concentration of tracer in the arterial blood. The rCBF then can be calculated by a dynamic method in which a rapid sequence of pictures are taken with the PET camera (Koeppe, Holden, and Ip, 1985; Roland et al., 1987). Another approach is the so-called autoradiographic method, in which the radioactivity is integrated over typically 40 s and the rCBF is calculated by inserting a brain-blood partition coefficient for water of .95 or .76 for butanol (Herscovitch, Markham, and Raichle, 1983; Roland et al., 1993). Other variants are slow infusion of $H_2{}^{15}O$ and inhalation of ^{15}O-labeled CO_2, which is converted to $H_2{}^{15}O$ in the lungs (Lammertsma et al., 1990). The autoradiographic method is sensitive to the integration time: short integration times favor high flows, but the images become more noisy; long integration times favor slow flows, but they give better statistics. The dynamic method also shows similar but more moderate dependency on the time interval used for rCBF determination.

Many investigators feel that arterial cannulation is more invasive than venous cannulation. Therefore a technique has been developed in which, simply put, the regional radioactive counts are integrated with the PET camera over a short time period following the injection of $H_2{}^{15}O$. This value is nonlinearly related to the true rCBF. Since the rCBF cannot be calculated, there are conversion schemes to relate the integrated counts to the rCBF. Since the counts depend on, among other things, the amount injected, the obtained three-dimensional pictures of the brain have to be normalized. Consequently the monotone relation between the rCBF and the synaptic metabolism does not apply. This entails problems of interpretation.

Whatever method is used to measure or estimate rCBF with PET, the pictures are usually quite noisy. This can be overcome by what is called signal averaging. New PET scanners can detect radioactive coincident events between all detector rings. This makes the measurements of radioactivity much more sensitive. This again permits the same subjects to be examined during several repetitions of a control state and a test state. From this one can add the difference pictures of test minus control, $\Delta rCBF_i$, and calculate a mean difference picture, $\Delta rCBF_i$, for one subject. In such mean difference pictures, one can get a reasonable signal-to-noise ratio and distinguish cortical activations (Watson et al., 1993). Still the most often-used alternative is to create pictures of mean differences in rCBF between test and control state from all individuals in a group doing the same task: $\Delta rCBF = 1/n \, \Sigma \, \Delta rCBF_i$. However, because individual brains are of different sizes and shapes, by such a procedure one will not add rCBF from anatomically homologous regions.

This problem is solved in different ways. Because the PET pictures do not have sufficient anatomical information, detailed pictures of the anatomy of the individual brains are required. These are usually obtained as magnetic resonance tomograms. These magnetic resonance tomograms will have to be brought into register with the PET scans. This is done by having the subjects' heads fixed in identical fashion in the magnetic resonance tomograph and the PET image. In the further establishment of anatomical standardization, some adhere to stereotaxical schemes with anatomical landmarks by proportional reductions and enlargements of individual brains, making them compatible (Talairach et al., 1967; Fox et al., 1988). Others use atlas-based linear and nonlinear transformations of the three-dimensional pictures of the individual brains (Bohm et al., 1986; Evans et al., 1988). The accuracy and precision of these procedures determine the signal-to-noise ratio in the final mean pictures and the accuracy of localizing the activations. For example, in proportional stereotaxical transformations the standard deviation of the position of the superficial part of the central sulcus is more than one cm (Talairach et al., 1967; Steinmetz, Fürst, and Freund, 1990). In atlas-based, non-landmark-based linear and nonlinear transformations it can be less than 2 mm (Seitz et al., 1990).

In some instances authors feel that the data even after anatomical standardization are too noisy. The usual procedure then is to use a spatial filter of the individual pictures to reduce this noise. As a consequence, neither the extent of individual activations nor the extent of mean activations can be determined. However, it is possible to determine a center of each activation or a point where the change is maximal. These points are then referred to in a three-dimensional stereotaxical system as the Talairach coordinates.

The subtraction method

In other primate research in which single-unit recordings are performed, one first measures the action potentials in a non-task state, then imposes the experimental stimulus and monitors the change in action potentials. Subsequently one observes the neurones return to the non-task-related activity. In functional imaging with PET, each measurement of rCBF reflects the average rCBF for a certain state. One is therefore forced to make a comparison between the rCBF picture of a reference state and the rCBF picture of a test state. This is done by subtracting the rCBF picture of the reference state from that of the test state such that one gets a $\Delta rCBF_i$ picture. Originally, such individual subtraction pictures were made just to visualize the region's changing rCBF and show the percent change (Roland and Larsen, 1976). Subsequently, subtractions between different task states also were made, first with the purpose of looking at differences in the loci of activation occurring as a consequence of change in direction of attention (Roland, 1982). In these test states the inputs to the brain were identical, as was the response the subjects had to make. Also, paradigms were developed in which different cognitive components were added from test to test (Petersen et al., 1988). These types of paradigms have recently become predominant in cognitive functional mapping studies. The basic assumption is that if test B contains one extra cognitive component than test A, then a subtraction of $rCBF_{test\ B} - rCBF_{test\ A}$ will yield a picture in which the activations associated with the extra cognitive component are revealed, since all other activations will cancel out.

If the rCBF pictures of the two tests are not normalized, and a field of activations cancel out by subtraction, it means that the rCBF in test A was similar to the rCBF in test B for that field. This also means that the synaptic metabolism for that field was of similar magnitude in test A and in test B. However, it does not imply that the processing in the field was identical in the two conditions. Since every cortical field is probably anatomically connected to between 10 and 20 other cortical areas (Felleman and van Essen, 1991), one cannot from the mere localization and intensity of an activated field deduct what type of information transformation underlies the activation. Conversely, a field presumably may also participate in identical information transformation but show different intensity of activation depending on effects of attention (see above). Another assumption underlying the idea that one can isolate cognitive components by such subtractions is that only task-related activations occur in the averaged rCBF pictures. Thus one could say that a kind of economy principle prevails in the brain: Only fields related to the task are activated (Roland, 1993). In the case that the rCBF pictures have been normalized (for example, to a global value of 50 ml/100 g/min) and the actual rCBF is unknown, one has to rely on another hypothesis, that is, that changes in neuronal and synaptic activity are related to the relative rCBF pattern rather than to an absolute rCBF pattern.

Precise anatomical standardization procedures permit another type of subtraction analysis—intergroup subtraction—because the anatomically standardized brains are nearly identical anatomically. In this case, the test image of the mean rCBF of test A in one group is subtracted from the mean rCBF of test B in another group (Roland and Seitz, 1991; Seitz et al., 1991). This has the advantage that the analysis of the conditions associated with activations of multiple cortical fields can be extended to several experimental groups.

Field activation

Analysis of largely unfiltered subtraction pictures and regions of high rCBF during tests has led to the hypothesis that the cerebral cortex in man activates in fields of a certain size (Roland, 1985). The experimental, anatomical, and physiological bases of fields of activation have been reviewed recently (Roland, 1993). Functional fields of activation appear in individuals as well as in mean ΔrCBF and ΔrCMR images. Their size ranges from some 800 mm^3 to 3000 possibly 4000 mm^3. Fields of activation appear in sensory cortices, motor cortices, homotypical isocortex, mesocortex, hippocampus, and even in the cerebellar cortex. Extended fields of synaptic activity also appear subcortically in conjunction with the cortical activations. All types of brain work—learning, recall, recognition, motor planning, motor execution, perception, and thinking—are associated with activations of multiple fields in cortex.

Similar but smaller fields of activation exist in autoradiographic studies of other primates in sensory and motor tasks. The anatomical and physiological cause for the appearance of field activations seems to be the spread of axon terminals of corticothalamic and corticocortical fibers in combination with the spread of intracortical collaterals migrating from pyramidal neurons. That activations macroscopically appear as fields can also be in accordance with columnar activations in some parts of the cortex, in the sense that the active columns together macroscopically constitute a localized field. Thus, according to the field activation hypothesis, "the cerebral cortex participates in brain work in awake human subjects by activating multiple cortical fields for each task. Activation means that the synapses and neurons in a field increase their biochemical activity. This leads to increases in transmembraneous ion transport and to increases of the regional cerebral metabolism (rCMR) or regional cerebral blood flow (rCBF)" (Roland, 1993).

Learning and recognition of visual patterns

Storage of information in the brain is assumed to be dependent on specific synaptic activity. Presumably the mechanism of increased synaptic transmission efficacy is associated with biochemical processes that in turn increase the synaptic energy metabolism. This in turn should increase the rCBF, and measuring rCBF thus should be one way to monitor changes in brain activity associated with learning. Motor learning has been studied by Seitz et al. (1990), Seitz and Roland (1992), and Kawashima, O'Sullivan, and Roland (1994). Learning of somatosensory information has been studied by Roland et al. (1989) and by Roland, Gulyás, and Seitz (1991), and learning of auditory information by Mazziotta et al. (1982) and Pawlik et al. (1987).

In a recent study of visual learning (Roland and Gulyás, 1994) the rCBF was measured with the freely diffusible tracer $^{11}C{-}CH_3F$ in (1) one control state, a rest state defined by Roland and Larsen (1976), (2) during learning of colored geometrical patterns, and (3) during recognition of the learned patterns. All subjects also had magnetic resonance tomograms of their brains. The rCBF from the control state was subtracted from that of the test state to give individual subtraction images. These were reformatted by the adjustable computerized atlas of Bohm et al. (1986) into standard size and shape and subsequently averaged to give mean ΔrCBF images. In addition, mean rCBF images for each test were created from the individual rCBF images. The anatomically standardized pictures were analyzed for local field activations occurring in the brain as clusters of voxels having high signal-to-noise ratio (Roland et al., 1993). The statistical analysis of the cluster detection was described extensively in a recent report (Roland et al., 1993). Briefly, it was as follows: For each subject, subtraction pictures were made by voxel by subtraction of the rCBF picture of the control from the test picture. For the whole group of experimental subjects these individual subtraction pictures, $\Delta rCBF_i$, were analyzed voxel by voxel for normal distribution of the values. Only voxels for which the ΔrCBF values could be considered normally distributed were included in the further analysis, and the remaining were set to zero. Mean and variance pictures as well descriptive Student's *t*-pictures were calculated: $\Delta rCBF = \Sigma \Delta rCBF$/no. of subjects; $t = \Delta rCBF/\sqrt{\text{variance}}/\sqrt{}$ no. of subjects. Voxels with *t*-values greater than 2.26 were considered clustered if they were attached by side, edge, or corner. Based on an analysis of randomly occurring clusters of voxels having $t > 2.26$ in test minus test pictures, a distribution of the occurrence false positive clusters having 2, 3, 4, or more voxels was made. With reference to this distribution it was decided to reject the hypothesis that all clusters of size 11 and above (having more than 11 voxels) belonged to the distribution of false positives (Roland et al., 1993). The average probability of finding *one* false positive cluster of size 11 and above in the three-dimensional space representing the brain was .07. The descriptive t-image was thresholded by only accepting voxels having $t > 2.26$ and occurring in clusters of size 11 and above; all other voxels were set to zero. This image is called a cluster image. In this image, only clusters of size 11 and above having $t > 2.26$ are shown and considered regions of changed rCBF. In table 49.1, the cluster sizes of the activated fields are shown in mm^3. One voxel had a volume of 44.03 mm^3. The larger the cluster, the smaller the probability that it is a false positive.

During the learning sessions the subjects looked at ten colored complex geometrical patterns with the purpose of learning them. Each pattern was exposed for 10 s covering 33° × 33° field of view. The patterns were always presented in the same order. The PET measurement during learning was made during the second learning session, which took place 3–4 min after the start of the experiment. After 20 learning sessions extending over 55 min, another 55 min elapsed before the subjects in a recognition experiment (during which old patterns were mixed with new similar patterns in ratios between 0.15 and 0.35) recognized the learned patterns. In this recognition session each pattern was exposed for only 100 ms, and a new pattern appeared every second. It was subsequently determined that subjects could recognize the learned patterns with an average probability of 0.97 (Roland and Gulyás, 1994).

TABLE 49.1

Mean volume and localization of visual cortical fields in visual areas during learning and recognition of complex, colored, large visual patterns (control rest) (in mm³)

Area	Right Hemisphere		Left Hemisphere	
	Learn.	Rec.	Learn.	Rec.
V1	4400	1650	3740	3130
Pericalcarine	9890	2370	11710	4730
Lingual g.	1350	2110	3510	2070
Fusiform g.*	3300	7690	2660	2580
G. temp. inf. post.			850	3380
S. temp. inf. ant.	2450	3080		
S. temp. inf. post.	2915	4180	460	
Occ. lat.	5410	2110	4775	3970
S. par. occ. B	1730	630	2155	1180
S. par-occ. post.	1480	2620	1140	720
S. par-occ. ant.	1140	1010	460	460
Cuneus sup.	7230	2870	8410	3510
Occ. sup.	3890	3890	3800	2620
Precuneus post.	5110	1010	3210	1010
STS-GANG post.	2660	1820	2915	3380
G. angularis ant.	2110	2450	510	
S. interparietalis post.	2495	3510	1480	1100
Lob. parietal sup. post.	1350	2030	3210	590

*Can be divided into an anterior and posterior field.
STS, sulcus temporalis superior; GANG, gyrus angularis.

In the early learning phase, functional fields appeared in the following anatomical structures: (1) the primary visual area in and around the calcarine sulcus; (2) visual association areas: rCBF increases covered the rest of the cuneus, the posterior part of precuneus, the lingual gyrus, the fusiform gyrus, the occipital gyri, the angular gyrus, and the posterior part of the superior parietal lobule; (3) prefrontal cortical regions, especially the cortex lining the superior frontal sulcus and the frontal eye field; (4) limbic and paralimbic structures: the anterior hippocampal formation (but not the posterior), the anterior cingular cortex, temporal pole, and anterior sector of insula; (5) the anterior midpart of the neostriatum. The sizes of these fields is shown in table 49.1.

To a very large extent, recognition activated identical visual areas. When the individual $rCBF_i$ from learning was subtracted from $rCBF_i$ from recognition, the only differences were a field in the anterior part of the right fusiform gyrus, which was more activated in recognition, and the primary visual area, which was more strongly activated in visual learning. There were minor differences in the activations of the prefrontal cortex, but no statistically significant differences in hippocampus or the parahippocampal gyrus (Roland and Gulyás, 1994). In these experiments one cannot distinguish the effect of perceiving the colored patterns from the effects of storing them and recognizing them.

In a very recent experiment, Kawashima, O'Sullivan, and Roland (1994) let subjects see for 5 s a colored target covering a 40° × 50° visual field with seven white circular targets having different diameters. Then the targets were turned off. After a delay and on the command, "point," the subjects were to point to the targets in order of size. Repetitive exposures of the targets were made, but each time the subjects saw the targets for only 5 s, and some time elapsed before they were asked to point. The control state in this experimental series was rest in which the subjects had their eyes closed (Roland and Larsen, 1976). A PET measurement was taken at the early phase of learning, when subjects had performed the task several times but had not achieved their final precision. After 30 min of learning the subjects achieved their final and stable precision. After a further 20-min pause, during which the subjects were distracted from internal rehearsal, a subsequent measurement was taken with the subjects pointing repetitively. The rCBF was measured with ^{15}O-butanol as the tracer. The data treatment was as described above.

In these experiments the subjects did not get any visual feedback. They had many learning sessions in which, after a while, they learned the positions of the targets. During the actual pointing sessions they must have kept an internal representation of the target positions in their minds . In the beginning the subjects could have kept the target representations in a working memory. However, after learning had taken place, they must have addressed the long-term memory storage sites for the sensory representations they would need to perform the pointing. To see the effects of addressing these storage sites in particular, the rCBF picture from early learning was subtracted from that during late learning (see color plate 18).

Color plate 18 demonstrates significantly more intense activation of the fields in visual association areas and somatosensory areas. The activated fields that were significantly more activated when the pointings were guided by the representations from long-term memory were located in visual association areas in the superior parts of cuneus and the middle and superior occipital gyri. However, fields where activation was specifically related to long-term memory were also located in the adjacent posterior parts of the parietal lobules, in the angular gyrus, and in the posterior parts of the superior parietal lobule. These regions presumably also contain visual association areas (Roland, 1993; Roland and Gulyás, 1994). The specific activations of fields in somatosensory cortex may indicate that the subjects, during the learning, also accumulated somatosensory representations (Kawashima, O'Sullivan, and Roland, 1993). The specific activations of the visual association areas indicate that the retrieval of the targets from long-term memory requires synaptic activation of these areas, which are also potential storage sites.

Consistency analysis

The activated fields in an activation study of the brain occupy a very small fraction of the entire brain's volume. Accordingly, the number of three-dimensional picture elements, or voxels, that represent the fields of activations in test-minus-control mean subtraction pictures only constitutes a small fraction of the total number of voxels representing the brain. In the case of the pointing study, 309 voxels were representing the activated fields of the 50,000 voxels representing the brain. The 309 voxels occurred as 25 clusters of voxels, in which each cluster corresponded to a field of activation (Kawashima, Roland, and O'Sullivan, 1994). These clusters occur statistically independently of each other (Roland et al., 1993).

Suppose now that one is interested in the question whether some of the fields activated in the visual learning and in pointing toward the targets under the guidance of long-term memory are identical? First, the reason for asking this question might seem obscure, since the two tasks do not seem to have much in common. In the visual learning task, subjects were perceiving complex colored patterns occupying a large part of the visual field. In the pointing task, the subjects did not receive any visual information while the PET measurement was taken. However, they had stored some representation of the targets, and they were using this representation during the actual PET measurement of pointing. In the visual learning task the subjects had to store the representation of the patterns; otherwise they would be unable to recognize the patterns. The target stimulus and the complex geometrical patterns were similar in the sense that both were colored and occupied a large part of the visual field. Secondly, the subjects needed to explore both types of stimuli with eye movements, bringing all parts of the stimuli into foveal vision. In fact the frequency of eye movements in the visual learning (eyes open) and during long-term-memory-guided reaching (eyes closed) was similar. Apart from this, the two tasks differed in almost every aspect.

This is exactly the purpose of consistency analysis. At best, tasks should only have one common component. The principle is that the cluster image of task A minus control A is multiplied with the cluster image of task B minus control B. Since all voxels in the cluster images, except for the clusters representing the active fields, have zero value, only the field that is activated consistently in both tasks appears in the resultant image. If the cluster images that are multiplied stem from two different experimental groups of subjects, the two set of experiments represented in cluster images A and B are independent. Furthermore, since the probability of getting exactly one voxel of overlap in such independent cluster images is very small, and since the probability of getting several voxels of overlap is extremely small, consistency analysis is statistically strong.

Color plate 19 shows the actual multiplication image applied to the present question about fields consistently activated in visual learning and in pointing guided by long-term memory. One can see that fields of activation are common for the two tasks in the superior parts of cuneus, the superior occipital gyri, and the adjacent parts of the posterior superior parietal lobule. First, this demonstrates that fields are reproducible functional elements of the cerebral cortex. Second, it demonstrates that even though only one target screen was shown in the pointing task, it is presumably represented in many fields. Which of these fields are related to eye movements and which are related to representation of large, colored visual field patterns cannot be determined from this consistency analysis. In a study of recall of the learned visual patterns, the posterior part of the superior parietal lobules was activated in exactly the same place (Roland and Gulyás, 1994). Furthermore, color plate 19 demonstrates that the posterior superior parietal lobules also contained fields of activation in this place. Here the effect of eye movements is ruled out by the subtraction. Therefore these sites are strong candidates for visual storage sites of large colored patterns (Kawashima, O'Sullivan, and Roland, 1994).

Consistency analysis of task activations within the same group of subjects is more descriptive, since all subjects have the same control state and so the consistency multiplication is made from cluster images based on the same control state. If the probability of false positives in the cluster images is kept low, as in the example of visual learning and visual recognition, the probability of getting fields consistently activated in two consecutive tasks is even lower. A pertinent question is whether it is identical visual areas that are active in visual learning of patterns and in recognition of the learned patterns. This was investigated by multiplying the cluster images of visual learning minus control with those of recognition minus control. From color plate 20 one can see that the functional fields within the visual area domain of the occipital lobe and adjacent parietal lobe to a large extent were identical in learning and recognition.

ACKNOWLEDGMENTS The research described in the chapter has been sponsored by the Human Frontier Science Program Organization, the Swedish Medical Research Council, the Wenner-Gren Foundation, and a research grant from the New South Wales Institute of Psychiatry to Dr. O'Sullivan.

REFERENCES

BOHM, C., T. GREITZ, G. BLOMQVIST, L. FARDE, P. O. FORSGREN, and D. KINGSLEY, 1986. Applications of a computerized adjustable brain atlas in positron emission tomography. *Acta Radiol. Suppl.* 369:449–452.

EICHLING, J. O., M. E. RAICHLE, R. L. GRUBB, JR., and M. M. TER-POGOSSIAN, 1974. Evidence of the limitations of water as a freely diffusible tracer in brain of the rhesus monkey. *Circ. Res.* 35:358–364.

EVANS, A. C., C. BEIL, S. MARRETT, C. J. THOMPSON, and A. HAKIM, 1988. Anatomical functional correlation using an adjustable MRI-based region of interest atlas with positron emission tomography. *J. Cereb. Blood Flow Metab.* 11:513–530.

FELLEMAN, D. J., and D. C. VAN ESSEN, 1991. Distributed hierarchical processing in the primate cerebral cortex. *Cerebral Cortex* 1:1–47.

FOX, P. T., M. A. MINTUN, E. M. RIEMANN, and M. E. RAICHLE, 1988. Enhanced detection of focal brain responses using intersubject averaging and change-distribution analysis of subtracted PET images. *J. Cereb. Blood Flow Metab.* 8:642–653.

GINSBERG, M. D., D. DIETRICH W. D., and R. BUSTO, 1987. Coupled forebrain increases of local cerebral glucose utilization and blood flow during physiologic stimulation of a somatosensory pathway in the rat: Demonstration by double-label autoradiography. *Neurology* 37:11–19.

GREENBERG, J., P. HAND, A. SYLVESTRO, and M. REIVICH, 1979. Localized metabolic-flow couple during functional activity. *Acta Neurol. Scand.* 60 (Suppl. 72): 12–13.

HERSCOVITCH, P., J. MARKHAM, and M. E. RAICHLE, 1983. Brain blood flow measured with intravenous H_2 ^{15}O. I. Theory and Error Analysis. *J. Nucl. Med.* 24:782–789.

KAWASHIMA, R., B. T. O'SULLIVAN, and P. E. ROLAND, 1994. Functional anatomy of reaching and visuomotor learning in man: A positron emission tomographic study. *Cereb. Cortex.*

KOEPPE, R. A., J. E. HOLDEN, and W. R. IP, 1985. Performance of parameter estimation techniques for the quantification of local cerebral blood flow by dynamic positron computed tomography. *J. Cereb. Blood Flow Metab.* 5:224–234.

LAMMERTSMA, A. A., V. J. CUNNINGHAM, and M.-P. DEIBER, 1990. Combination of dynamic and integral methods for generating reproducible functional images. *J. Cereb. Blood Flow Metab.* 10:675–686.

MATA, M., D. J. FINK, H. GAINER, C. B. SMITH, L. DAVIDSEN, H. SAVAKI, W. J. SCHWARTZ, and L. SOKOLOFF, 1980. Activity-dependent energy metabolism in rat posterior pituitary primarily reflects sodium pump activity. *J. Neurochem.* 34:213–215.

MAZZIOTTA, J. C., M. E. PHELPS, R. E. CARSON, and D. E. KUHL, 1982. Tomographic mapping of human cerebral metabolism: Auditory stimulation. *Neurology* 32:921–937.

PAWLIK, G., W.-D. HEISS, C. BEIL, G. GRÜNEWALD, K. HERHOLZ, K. WIENHARD, and R. WAGNER, 1987. Three-dimensional patterns of speech-induced cerebral and cerebellar activation in healthy volunteers and in aphasic stroke patients studied by positron emission tomography of 2(^{18}F)-fluorodeoxyglucose. In *Cerebral Vascular Disease*, vol. 6, J. S. Meyer, H. Lechner, M. Reivich, and E. O. Ott., eds. Amsterdam, New York, Oxford: Excerpta Medica, pp. 207–210.

PETERSEN, S. E., P. T. FOX, M. I. POSNER, M. MINTUN, M. E. RAICHLE, 1988. Positron emission tomographic studies of the cortical anatomy of single-word processing. *Nature* 331:585–589.

PHELPS, M. E., J. C. MAZZIOTTA, and H. R. SCHELBERT, 1986. *Positron Emission Tomography and Autoradiography.* New York: Raven.

RAICHLE, M. E., 1987. Circulatory and metabolic correlates of brain function in normal humans. In *Handbook of Physiology*, section 1, *Nervous System*, vol. 5, *Higher Functions of the Brain*, F. Plum, ed. New York: Oxford University Press, pp. 643–674.

ROLAND, P. E., 1982. Cortical regulation of selective attention in man: A regional cerebral blood flow study. *J. Neurophysiol.* 48:1059–1078.

ROLAND, P. E., 1985. Application of imaging of brain blood flow to behavorial neurophysiology: The cortical field activation hypothesis. In *Brain Imaging and Brain Function*, L. Sokoloff, ed. New York: Raven, pp. 87–106.

ROLAND, P. E., 1993. *Brain Activation.* Wiley, p. 589.

ROLAND, P. E., L. ERIKSSON, S. STONE-ELANDER, and L. WIDÉN, 1987. Does mental activity change the oxidative metabolism of the brain? *J. Neurosci.* 7:2373–2389.

ROLAND, P. E., L. ERIKSSON, L. WIDÉN, and S. STONE-ELANDER, 1989. Changes in regional cerebral oxidation metabolism induced by tactile learning and recognition in man. *Eur. J. Neurosci.* 1:3–18.

ROLAND, P. E., and B. GULYÁS, 1994. Visual memory, visual imagery and visual recognition of large field patterns: Functional anatomy by positron emission tomography. *Cereb. Cortex.*

ROLAND, P. E., B. GULYÁS, and R. J. SEITZ, 1991. Structures in the human brain participating in visual learning, tactile learning, and motor learning. In *Memory and Locus of Change*, L. Squire, ed. Oxford: Oxford University Press, pp. 95–113.

ROLAND, P. E., and B. LARSEN, 1976. Focal increase of cerebral blood flow during stereognostic testing in man. *Arch. Neurol.* 33:551–558.

ROLAND, P. E., B. LEVIN, R. KAWASHIMA, and S. ÅKERMAN, 1993. Three-dimensional analysis of clustered voxels in ^{15}O-butanol brain activation images. *Human Brain Mapping* 1:3–19.

ROLAND, P. E., and R. J. SEITZ, 1992. Positron emission studies of the somatosensory system in man. *Ciba Found. Symp.* 163:113–124.

SEITZ, R. J., C. BOHM, T. GREITZ, P. E. ROLAND, L. ERIKSSON, G. BLOMQVIST, G. ROSENQVIST, and B. NORDELL, 1990. Accuracy and precision of the computerized brain atlas programme for localization and quantification in positron emission tomography. *J. Cereb. Blood Flow Metab.* 10:443–457.

SEITZ, R. J., and P. E. ROLAND, 1992. Learning of sequential finger movements in man: A combined kinematic and positron emission tomography (PET) study. *Eur. J. Neurosci.* 4:154–165.

SEITZ, R. J., P. E. ROLAND, C. BOHM, T. GREITZ, and S. STONE-ELANDER, 1991. Somatosensory discrimination of shape: tactile exploration and cerebral activation. *Eur. J. Neurosci.* 3:481–492.

SOKOLOFF, L., M. REIVICH, C. KENNEDY, M. H. DES ROSIERS, C. S. PATLAK, K. D. PETTIGREW, O. SAKURADA, and M. SHINOHARA, 1977. The [^{14}C] deoxyglucose method for the measurement of local cerebral glucose utilization: Theory, procedure, and normal values in the conscious and anesthetized albino rat. *J. Neurochem.* 28:897–916.

STEINMETZ, H., G. FÜRST, and H.-J. FREUND, 1990. Variation of perisylvian and calcarine anatomical landmarks within stereotaxic proportional coordinates. *Am. J. Neuroradiol.* 11:1123–1130.

TALAIRACH, J., G. SZIKLA, P. TOURNOUX, A. PROSSALENTIS, M. BORDAS-FERRER, L. COVELLO, M. IACOB, and E. MEMPEL, 1967. *Atlas d'Anatomie Stéréotaxique du Télencéphale*, 1st ed. Paris: Masson, p. 326.

WATSON, J. G., R. MYERS, R. S. FRACKOWIAK, F. V. HAJNAL, R. P. WOODS, J. C. MAZZIOTTA, S. SHIPP, and Z. ZEKI, 1993. Area V5 of the human brain: Evidence from a combined study using positron emission tomography and magnetic resonance imaging. *Cereb. Cortex* 3:79–94.

50 Event-Related Potential Studies of Human Memory

MICHAEL D. RUGG

ABSTRACT Event-related potentials (ERPs) evoked by words at the time of learning differ according to whether the words are successfully retrieved in a subsequent memory test. These differences emerge as early as 250 ms after stimulus onset, indicating that processes that make an event memorable can be active from very early in the course of its processing. ERPs evoked by "old" and "new" items in tests of recognition memory also show robust differences. While the functional significance of these differences is unclear, it is probable that they are found only when old items are recognized explicitly. If so, "old/new" ERP effects may provide a covert method for determining whether previously presented items are processed with or without conscious recognition. ERPs have yet to be convincingly shown to reflect implicit memory. The difficulties of demonstrating such a relationship, and ways of overcoming them, are discussed.

This chapter focuses on event-related potential (ERP) studies relevant to contemporary cognitive and neuropsychological work on long-term memory. The chapter adopts the perspective that normal memory depends upon the operation of dissociable processes and/or systems. The distinction is made between *explicit* ("declarative" or "conscious") and *implicit* ("nondeclarative" or "unconscious") memory, and is drawn on the basis of whether a prior experience affects behavior with or without awareness of memory for the experience (see Schacter, this volume). A second distinction is drawn between two different ways of assessing memory (Richardson-Klavehn and Bjork, 1988): *direct* memory tests, such as free recall, which make specific reference to a previous learning episode and encourage subjects to reflect on the content of their memories, and *indirect* tests, in which task performance can be influenced by, but is not dependent upon, memory for the study items. Direct tests are held to be sensitive predominantly to explicit memory, and indirect tests to implicit memory. It has, however, been pointed out (e.g., Jacoby and Kelley, 1992) that it should not be assumed that these two types of tests each tap only one form of memory.

MICHAEL D. RUGG Wellcome Brain Research Group, School of Psychology, University of St. Andrews, St. Andrews, UK

Event-related potentials

ERPs are small fluctuations in the spontaneous electrical activity of the brain (the electroencephalogram or EEG) time-locked to an event such as the onset of a stimulus. Their small size in relation to the EEG means that ERP waveforms are nearly always obtained by averaging a number of EEG samples. The averaging process largely eliminates EEG fluctuations that have a random temporal relationship with the events to which sampling is time-locked, and leaves a waveform that represents the average of the event-related activity embedded in each sample.

ERP waveforms are fractionated into constituent components by combining the criteria of scalp distribution (waveform features with different distributions are unlikely to share the same sources) and sensitivity to experimental manipulations (features that differ in their sensitivity to a manipulation are likely to reflect functionally distinct processes). Using these criteria, a number of ERP components sensitive to cognitive variables have been distinguished (Hillyard and Picton, 1987). Two components important here are the P300 and the N400. The P300 (also known as the P3, or the *late positive component*) is a positive-going wave, which typically is of maximum amplitude over central-parietal scalp regions. It is evoked in a wide range of circumstances (Pritchard, 1981; Johnson, 1993) and, belying its label, can have a peak latency anywhere between 300 and 600 or more ms, depending on experimental conditions. The N400 is a negative wave that peaks around 400 ms poststimulus. When evoked by visual or auditory words, its amplitude is inversely proportional to the extent that the word has been "primed" by its prior context (Kutas and Van Petten, 1988).

ERPs and memory

ERPs are useful in memory research for several reasons. First, they have high temporal resolution, and hence are well suited to addressing questions about the time-course of brain activity associated with cognitive processing. Second, ERPs can be used as "covert" measures of processing, allowing differences in processing to be detected in the absence of behavioral responding. Third, comparison of the scalp distribution of ERP effects can be used to assess whether stimuli from different experimental conditions evoke different patterns of neural activity, and hence whether they might engage functionally distinct processes. Finally, as knowledge of the functional significance and intracerebral origins of ERPs accumulates, they will provide a method for the "on-line" monitoring of the neural activity underlying specific cognitive processes.

Until the mid-1980s, the majority of ERP studies of memory were conducted within the "memory-scanning" paradigm of Sternberg (1966; see Kutas, 1988, for a review), and had little to say about long-term memory. The past decade however, has seen a proliferation of studies in this area. These have become increasingly well integrated with the empirical and theoretical concerns of workers in the general area of memory research.

Memory encoding

ERP studies of encoding have involved variations on a single paradigm. A series of study items is presented, and EEG epochs time-locked to the presentation of each item are sampled and stored. Memory for these items is subsequently tested. The EEG sampled at study is then used to form two classes of ERP, associated respectively with successfully and unsuccessfully remembered items. Differences between these two ERP classes will be referred to below as *subsequent memory effects*.

A number of studies have described subsequent memory effects (Sanquist et al., 1980; Karis, Fabiani, and Donchin, 1984; Fabiani, Karis, and Donchin, 1986, 1990; Neville et al., 1986; Paller et al., 1987b; Paller, Kutas, and Mayes, 1987a; Paller, McCarthy, and Wood, 1988; Friedman, 1990b; Paller, 1990; Rugg, 1990; Smith, 1993). In every case, the finding has been that subsequently remembered items evoke more positive-going ERPs than items that fail to be retrieved. The most important of these studies are briefly described below.

Selected Studies Karis, Fabiani, and Donchin (1984) and Fabiani, Karis, and Donchin (1990) investigated subsequent memory effects using a procedure in which an "isolation effect" was created by presenting a word in letters of a different size from the remainder of the items on a list. In Karis, Fabiani, and Donchin (1984), subjects were segregated post hoc on the basis of whether they showed large or small recall advantages for isolated compared to nonisolated words. Subjects with large advantages reported the use of "rote" mnemonic strategies, while those with small advantages reported using "elaborative" strategies. As shown in figure 50.1, ERPs to the isolates showed subsequent memory effects in both groups of subjects. Karis, Fa-

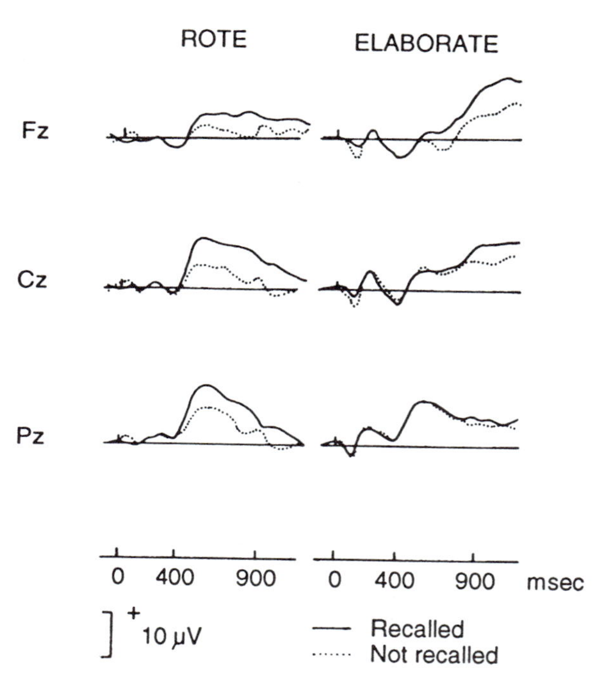

Figure 50.1 Grand-average waveforms from Karis, Fabiani, and Donchin (1984), illustrating ERPs from midline electrodes evoked by physically distinctive study words that were subsequently recalled or not recalled. Rote: ERPs from the 3 subjects using a "rote" strategy to remember the study items. Elaborate: ERPs from the 3 subjects using elaborative mnemonic strategies at study. (Redrawn with permission from Donchin and Fabiani, 1991. Reprinted by permission of J. Wiley and Sons, Ltd.)

biani, and Donchin argued that in the rote subjects, the effects resulted from the modulation of the P300 component, whereas in the "elaborators" the subsequent memory effect arose from changes in the amplitude of a different component, a frontal-maximum "slow positive wave." Fabiani, Karis, and Donchin (1990) had subjects rehearse study items by rote or by elaboration. In the elaborative condition, subsequent memory effects were confined largely to the frontal electrode, whereas in the rote condition the effect was distributed more posteriorly, involving what Fabiani, Karis, and Donchin took to be the P300.

Donchin and Fabiani (1991) argued that whereas nonelaborative mnemonic strategies lead to a subsequent memory effect arising from modulation of the P300, elaborative strategies are associated with ERP differences in another component, a frontal-maximum positive wave. They suggested that this latter effect may be an index of "extended processing" that overrides the mnemonic benefits of the encoding processes reflected by the P300. Donchin and Fabiani hypothesized that the subsequent memory effect on the P300 reflects variations in the "distinctiveness" of study items. According to Donchin and Fabiani, physical distinctiveness is an important determinant of both the efficacy of memory encoding (in the absence of elaborative processing) and the amplitude of P300. Thus, to the extent that performance at test reflects variations in the subjective distinctiveness of items at study, there will be a relationship between study P300 and subsequent memory performance. Donchin and Fabiani's (1991) hypothesis seems predicated on the assumption that variations in recall *within* a sample of physically distinctive items is largely attributable to variation along the same dimension that makes these items, as a class, easier to remember. But unless distinctiveness is to be operationalized in terms of recall probability, with all the problems this entails (Schmidt, 1991), the basis for this assumption is unclear. There seems no reason to reject a priori the possibility that within a sample of physically distinctive items, differences in distinctiveness are not the most important determinant of recall probability.

In contrast to Donchin and Fabiani (1991), other investigators have been more circumspect in associating subsequent memory effects with specific ERP components. It has been argued, however, that these effects do not involve modulation of the N400 (Neville et al., 1986). In Neville et al. (1986) subjects were presented with words that either "fit" or did not fit the sense of a preceding phrase. Figure 50.2 shows that ERPs to subsequently recognized words that fit their context

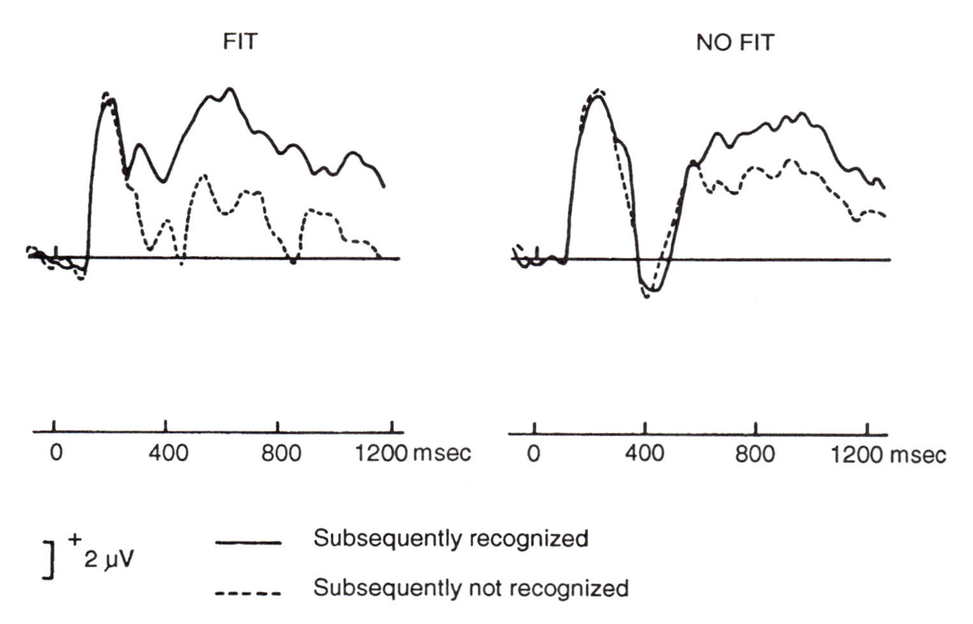

FIGURE 50.2 Grand-average waveforms from a midcentral electrode from Neville et al. (1986), evoked by words during a semantic judgment task and averaged according to whether the words were subsequently recognized. Note the prominent N400 wave in the waveforms evoked by words that did not fit their semantic context (no fit), and the relatively late onset of the subsequent memory effect. (Redrawn with permission.)

were more positive-going than those to unrecognized "fit" words. Figure 50.2 further shows that study ERPs to recognized and unrecognized "no-fit" words also showed subsequent memory effects, but that these commence some 200 ms later than those evoked by fit words. As can be seen in the figure, the ERPs to no-fit words contained a prominent N400 component that did not vary with subsequent memory. Similar findings were reported by Kutas (1988) using a cued recall test. Neville et al. (1986) interpreted their subsequent memory effects as reflecting differences in the "elaborative/consolidation procedure" undergone by words as they are encoded into memory. They suggested that the processes engaged by semantic incongruity (and indexed by N400) were incompatible with those responsible for effective encoding, and that the later onset of subsequent memory effects in the ERPs to the no-fit words may underlie the poorer memory typically seen for such items.

Paller (1990) addressed the question of whether ERP subsequent memory effects were specific to explicit memory. Some subjects received a cued recall test, in which the three-letter stems of the study words served as cues (a direct test). The remainder were given a stem completion test, in which stems of study words were presented with the instruction to complete each stem with the first word that came to mind (an indirect test). Subsequent memory effects were reliable for the cued recall test (and for a later test of free recall), but not for stem completion (figure 50.3). Paller took these findings to mean that explicit and implicit memory rely on qualitatively different encoding processes. But in at least one other study (Paller et al., 1987b), ERP subsequent memory effects *were* found for stem completion. It is noteworthy that in the experiment by Paller and colleagues (1987b), almost 60% of word stems were completed to form study words, against a baseline (guessing) level of around 12%. By contrast, in the later study by Paller (1990), only around 20% of stems were correctly completed, against a baseline rate of about 10%. Thus in the later study subsequent memory effects may have been "diluted" by the 50% or so trials on which stems were correctly completed by chance.

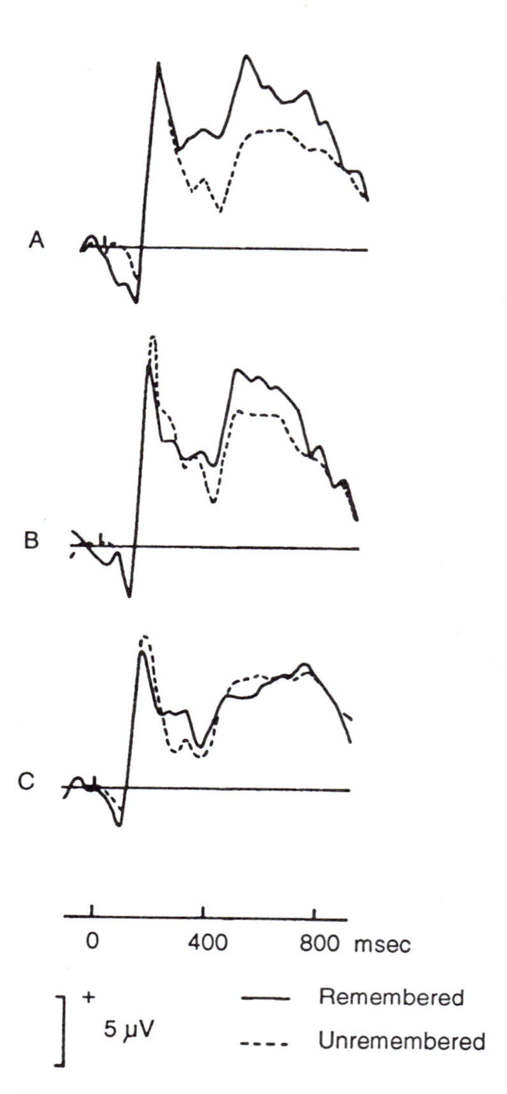

FIGURE 50.3 Grand-average waveforms evoked by study words from a midparietal electrode in Paller (1990). (A) Averaged according to whether the words were subsequently free-recalled; (B) averaged according to whether the words were successfully recalled in a cued recall task; (C) averaged according to whether the words were subsequently employed to complete word stems in a stem completion task. (Copyright 1990 by the American Psychological Association. Adapted by permission.)

SIGNIFICANCE OF SUBSEQUENT MEMORY EFFECTS It is clear that ERPs recorded at study can discriminate words on the basis of their probability of subsequent retrieval. In some experiments, this effect has taken the form of an enhancement in ERPs to remembered words of a frontal-maximum positive wave; in others, the effect has shown an even distribution over the midline, or a posterior maximum. It therefore seems likely that multiple ERP components, presumably reflecting multiple cognitive processes, contribute to subsequent memory effects; the number of these components and the circumstances under which each makes its contribution are unclear. Despite their inconsistencies, the present findings offer some insights into memory encoding. They suggest that processes that make a study item memorable are active as early as 250–300 ms after presentation, and exert an influence even when subjects engage in elaborative processing extending well beyond the recording epoch. The data of Neville et al. (1986) and Kutas (1988) further suggest that the processes reflected by ERP subsequent memory effects are incompatible with certain other kinds of cognitive operation. And investigations of the neural origins of these effects should shed light on the structures active during the initial stages of memory encoding, the time course of their activity, and the nature of their interactions.

Studies of recognition memory

The experiments discussed in this section all compared ERPs evoked by "old" and "new" words in tests of recognition memory. Most studies have employed a study–test paradigm, in which items are first presented in a study task, and then re-presented in a test task along with new ones. A few studies have employed a continuous recognition task, in which items are presented in a single series and must be discriminated on the basis of whether they are being presented for the first or the second time. Differences between ERPs evoked by correctly classified old and new items will be referred to below as *old/new effects*.

Old/new effects have been described in several studies (Sanquist et al., 1980; Karis, Fabiani, and Donchin, 1984; Johnson, Pfeferbaum, and Kopell, 1985; Neville et al., 1986; Rugg and Nagy, 1989; Smith and Halgren 1989; Friedman 1990a, 1990b, 1992; Rugg et al., 1991; Paller and Kutas, 1992; Potter et al., 1992; Rugg and Doyle, 1992, 1994; Smith, 1993). The uniform finding is that correctly classified old items evoke more positive-going ERPs than new items do. This finding was initially interpreted in terms of putative functional properties of the P300 component, rather than with respect to the light it might shed on recognition memory (e.g., Karis, Fabiani, and Donchin, 1984; Neville et al., 1986). Given this background, the study of Smith and Halgren (1989) is doubly important—first, because it suggested that old/new ERP effects result from the modulation of multiple ERP components, and second, because the effects were interpreted in the context of a functional model of recognition memory.

Smith and Halgren (1989) recorded from patients who had undergone left- or right-sided anterior temporal lobectomy, and from normal controls. Following the presentation of a block of words for study, a series of test blocks were presented in rapid succession, each containing the same 10 study words and an equal number of new words. ERPs from controls and right-sided patients exhibited reliable old/new effects, whereas those from the left-sided patients did not (figure 50.4). On the basis of the differing scalp distributions of the early and late parts of these effects, Smith and Halgren (1989) took them to reflect the modulation of two components: N400, the amplitude of which was attenuated in ERPs to old items, and a subsequent late positive component, the amplitude of which was enhanced. These two putative components of the old/new effect will be referred to below as *early* and *late* effects respectively.

Smith and Halgren (1989) suggested that both early and late old/new effects reflect processes that play a functional role in the encoding and retrieval of context-bound "episodic" memories. By their view (see also Halgren and Smith, 1987) the N400 is generated as the semantic attributes of the evoking item are integrated with the "cognitive context" pertaining at the time, leading to the formation of an "episodic trace" of the item. Repetition of the item within the same context triggers the retrieval of the episodic trace formed on its first presentation. This forestalls the generation of N400, and leads instead to an enhanced late positive component. Damage to the left medial temporal lobe prevents these processes from occurring for verbal material, leading to poor episodic memory for such material and to the absence of old/new ERP effects.

Smith and Halgren (1989) noted that although their left-sided patients failed to show old/new ERP effects, they had only a mild deficit in task performance. They accounted for this dissociation by recourse to a "dual-process" account of recognition memory. According to such accounts (e.g., Mandler, 1980, 1991; Jacoby and Dallas, 1981; Jacoby and Kelley, 1992), recognition memory has two bases: an item can be recognized as old following retrieval of a specific prior episode (*recollection*), or on the basis of its *familiarity*. According to Jacoby and colleagues (e.g., Jacoby and Dallas, 1981; Jacoby, 1983; Jacoby and Kelley, 1992), the familiarity of an item (more precisely, familiarity relative to a preexperimental baseline; e.g., Mandler, 1980) is proportional to the ease, or "fluency," with which it is

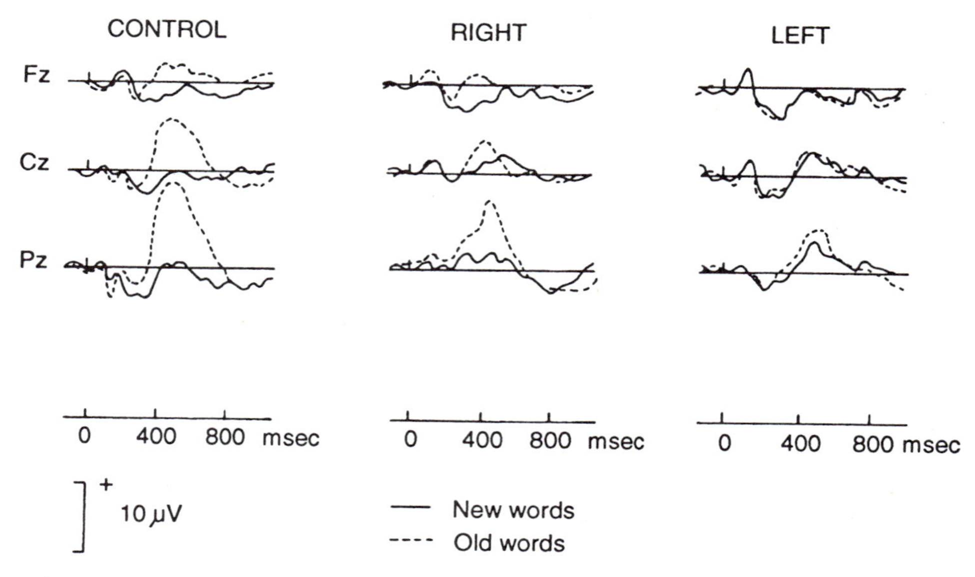

FIGURE 50.4 Grand-average waveforms from midline electrodes evoked by old and new words in the recognition memory test of Smith and Halgren (1989). ERPs are depicted separately for unoperated control subjects, and for subjects who had undergone right- or left-sided anterior temporal lobectomy. (Copyright 1989 by the American Psychological Association. Adapted by permission.)

identified. Fluency is in turn a function of the number and recency of prior experiences with an item (the same variables held to be responsible for implicit memory phenomena such as repetition priming; e.g., Jacoby, 1983).

Smith and Halgren (1989) argued that their left lobectomy patients had poor recollection but were unimpaired at making familiarity-based recognition judgments. They proposed that old/new ERP effects reflect processes specific to recollection, and that differences in item familiarity are not reflected in ERPs, even when such differences are sufficient to permit accurate recognition judgments. Two studies provide evidence that conflicts with these views of the functional significance of old/new effects. Potter et al. (1992) investigated the effects of the anticholinergic drug scopolamine in a continuous recognition memory task. Scopolamine impairs performance on direct memory tests, but appears to spare implicit memory (e.g., Kopelman and Corn, 1988). In view of the presumed overlap between the processes underlying implicit memory and the familiarity component of recognition memory, Potter and colleagues reasoned that the amnestic effects of scopolamine should result mainly from disruption of recollection. Thus, they argued, the ERPs of subjects performing a recognition memory task under scopolamine should resemble those of Smith and Halgren's (1989) left temporal lobectomy patients: recognition performance should be mediated largely by familiarity, and old/new ERP effects should be small. As expected, scopolamine impaired recognition memory. But as figure 50.5 shows, early old/new effects were unaffected by the drug, while late effects were slightly enhanced. Potter et al. concluded that, contrary to Smith and Halgren's (1989) hypothesis, impairment of recollection is not necessarily associated with reduced old/new ERP effects.

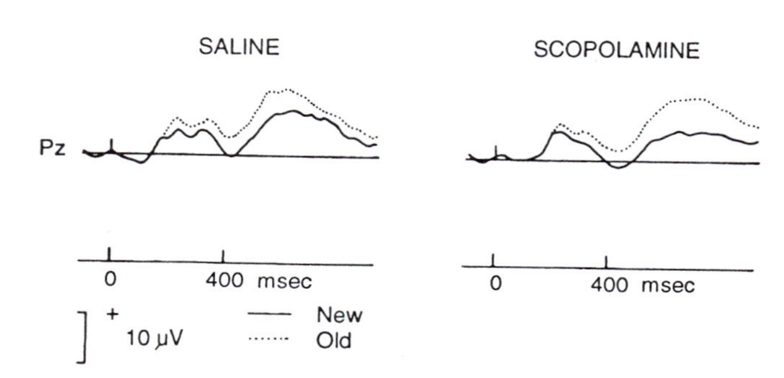

FIGURE 50.5 Grand-average midparietal ERPs evoked by correctly detected new and old words in the continuous recognition memory procedure of Potter et al. (1992), overlaid according to new/old status and whether subjects had received saline or scopolamine before performing the task.

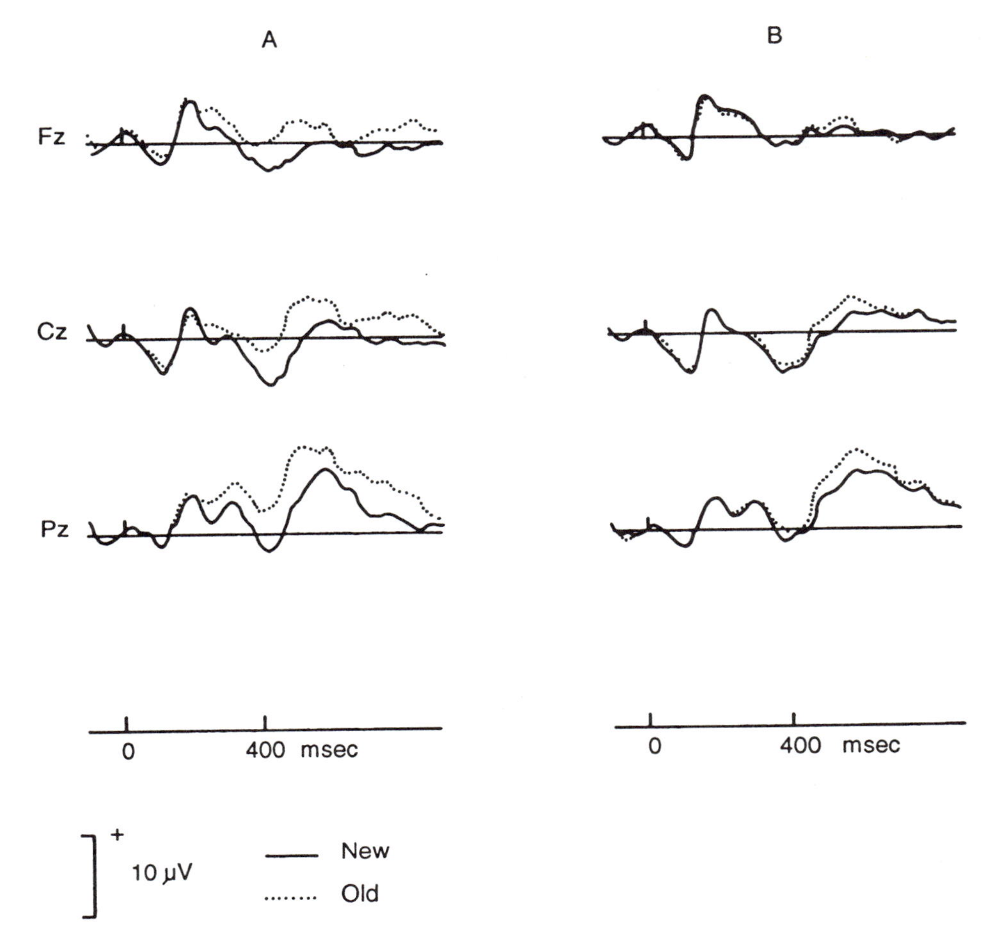

FIGURE 50.6 Midline grand-average ERPs from the two recognition memory tests employed by Rugg and Nagy (1989). A, ERPs evoked by correctly detected new and old words in a continuous recognition memory task in which 19 items intervened between first and second presentations; B, ERPs evoked by correctly detected new and old words after a study–test interval of approximately 45 min. Adapted with permission.

Rugg and Nagy's (1989) findings from healthy subjects also point to the absence of a strong relationship between early old/new effects and memory retrieval. When the interval between first and second presentations of each word was about a minute, both early and late old/new effects were present (figure 50.6A). But when memory was tested over a study–test interval of about 45 minutes, early effects were no longer apparent, although recognition remained well above chance (figure 50.6B). Rugg and Nagy argued that early effects were related more to study–test interval than to any process responsible for discriminating old from new items. This argument is strengthened by the sparse evidence for early old/new effects in other studies in which the study–test interval is more than a few minutes (e.g., Neville et al., 1986; Paller and Kutas, 1992; Rugg and Doyle, 1992; Smith, 1993).

SIGNIFICANCE OF EARLY OLD/NEW EFFECTS Early old/new effects—and thus the sensitivity of the N400 component to a word's old/new status—seem unlikely to reflect processes contributing to recognition memory over intervals of more than a few minutes. It is relevant to note that a substantial literature now exists on the modulation of the N400 by the repetition of items in indirect memory tests (see "ERPs and implicit memory" later in this chapter; Rugg, 1987, 1990; Rugg, Furda, and Lorist, 1988; Karayandis et al., 1991; Van Petten et al., 1991; Besson, Kutas, and Van Petten, 1992; Hamberger and Friedman, 1992), and different views have developed about the functional significance of this component in both memory and other tasks (c.f. Rugg and Doyle, 1994, and Van Petten et al., 1991). The question of whether N400 reflects processes that play a functional role in memory will have to await the resolution of this debate.

Late Old/New Effects In contrast to early effects, late old/new effects remain robust over lengthy study-test intervals, raising the possibility that they reflect processes of functional significance to recognition memory over these intervals. Rugg and Doyle (1992, 1994) addressed the question of whether late effects dissociate the recollective and familiarity components of recognition memory. These experiments were motivated by the findings of Rugg (1990), who employed an indirect memory test (nonword detection) in which words of low and high frequencies of occurrence were repeated. Rugg found that whereas the repetition of low frequency words enhanced the amplitude of a late positive wave, the same region of the ERP was insensitive to the repetition of high frequency words (see also Young and Rugg, 1992). Rugg accounted for this finding with an explanation similar to that advanced within the dual-process framework to account for the "word frequency effect" in recognition memory (the fact that low frequency words are recognized more accurately than those of high frequency). According to this explanation (e.g., Mandler, Goodman, and Wilkes-Gibbs, 1982), old, low frequency words engender higher levels of relative familiarity than old, high frequency items, and are therefore easier to recognize. Rugg (1990) suggested that the frequency by repetition interaction he found for the amplitude of the late positive component showed that this component was sensitive to the relative familiarity of the item evoking it: the greater the relative familiarity, the more positive-going the component.

If Rugg's (1990) hypothesis is correct, and if the word frequency effect does indeed reflect the influence of relative familiarity, late old/new effects should be larger for low than for high frequency words in any recognition memory test in which low frequency words are more accurately recognized. In two experiments in which memory was tested over an interval of about 20 minutes, Rugg and Doyle (1992, 1994) found that late old/new effects were, as predicted, larger for low than for high frequency words (figure 50.7). They argued that these findings were consistent with the idea that late old/new effects do indeed reflect the differing relative familiarities of new and old words.

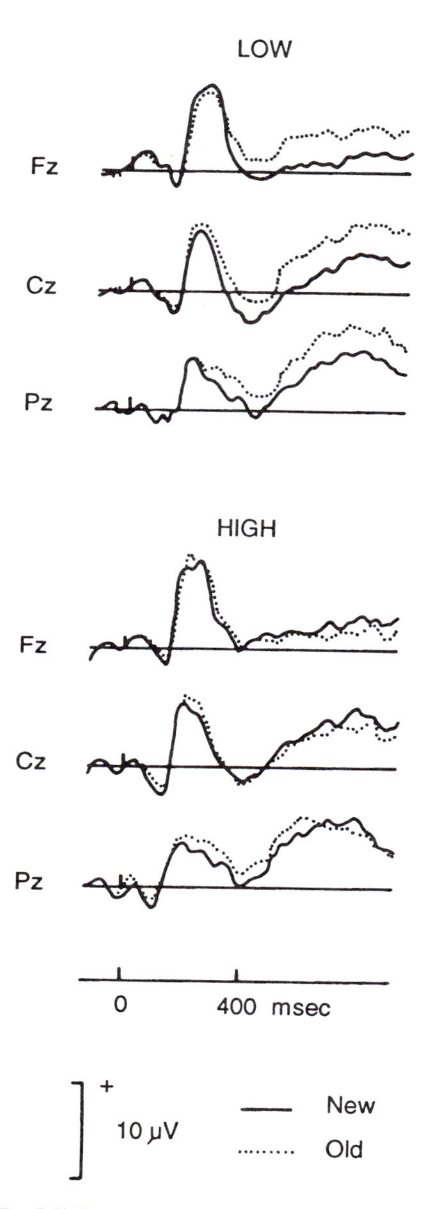

Figure 50.7 Midline grand-average ERPs evoked by correctly and confidently detected low and high frequency old and new words in the recognition test of Rugg and Doyle (1992). (Adapted with permission.)

Paller and Kutas (1992) investigated old/new effects in the absence of overt "old" and "new" responses. Subjects processed words in two study tasks, one requiring orthographic and the other imaginal processing. The test task was indirect, requiring identification of words presented too briefly to allow error-free performance. Test items consisted of old words from the two study tasks, along with words new to the experiment. In line with previous work (see Richardson-Klavehn and Bjork, 1988, for a review), study task had little effect on identification, which was better for words presented in either task than for new items. However, the ERPs evoked at test by correctly identified items differed according to study condition. As shown in figure 50.8, words from the imagery task evoked the more positive ERPs. This difference resembled previously described late old/new effects, including the tendency to be larger over the left than the right hemisphere (Neville et al., 1986; Rugg and Doyle, 1992, 1994). Paller and Kutas assumed that although the words from the two study tasks were equally well primed (as indexed by identification performance), words from the imagery task were more likely to be

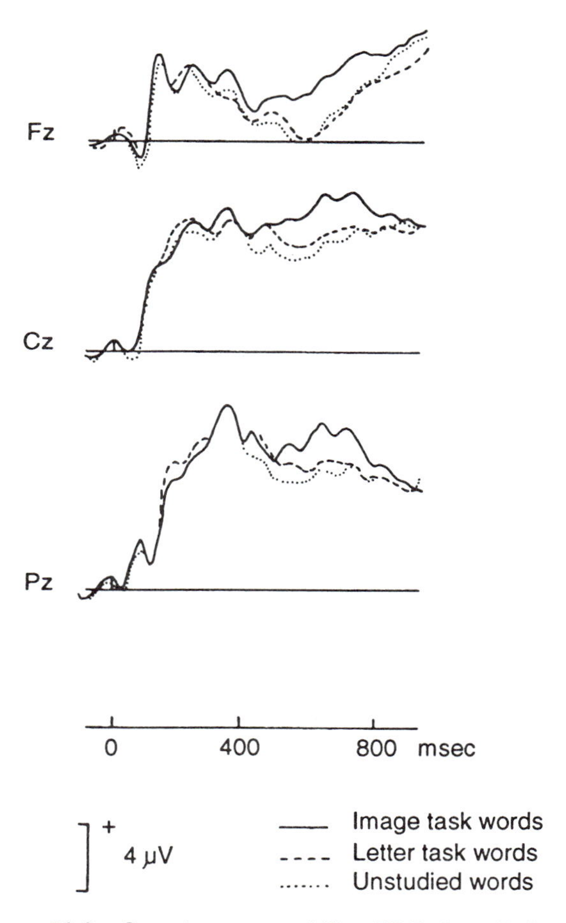

FIGURE 50.8 Grand-average midline ERPs from Paller and Kutas (1992), evoked by correctly identified test words new to the experiment (unstudied), or which had previously been studied in an imagery or a letter-detection task. (Redrawn with permission.)

recognized as having been experienced recently. They argued that the difference between the ERPs evoked at test by items from the two study tasks could therefore be regarded as an ERP signature (or "template") for what they called "conscious recollection."

Paller and Kutas's findings and interpretation are consistent with previous observations (Neville et al., 1986; Rugg and Doyle, 1992, 1994) that late old/new effects are found only with old words correctly identified as such. The findings also show that these effects are a consequence neither of the requirement to categorize test items as "old" or "new", nor of the need to respond differentially to the items (see also Rugg, Brovedani, and Doyle, 1992). The data do not, however, speak to the question of which of the putative components of recognition memory are most closely associated with late old/new effects. While a necessary condition for these effects may be awareness that the evoking items have recently been experienced, it is unclear whether this awareness need include details of where and when the experience occurred (recollection), as opposed to arising from a feeling of familiarity devoid of such contextual detail.

Gardiner and his associates (Gardiner, 1988; Gardiner and Java, 1990, 1991; Gardiner and Parkin, 1990) have developed a method that is purported to dissociate recollection and familiarity. Following a positive recognition judgment, subjects indicate whether the item evoked an explicit memory of its initial presentation (a "remember" or "R" response), or whether instead it was judged old merely on the basis of a feeling of familiarity, with no accompanying "recollective experience" (a "know" or "K" response). Employing the R/K procedure, Smith (1993) contrasted ERPs evoked by old items attracting each type of judgment. As shown in figure 50.9, old/new effects were larger for "R" responses, leading Smith (1993) to conclude that the late old/new ERP effect reflects the operation (or the outcome) of the recollective rather than the familiarity component of recognition memory. But as is clear from figure 50.9, *both* classes of recognition judgment were associated with old/new effects, the scalp distributions of which were indistinguishable. Smith's (1993) data therefore imply that while recollection (as operationalized by the R/K procedure) may be associated with the enhancement of late old/new effects, the effects are not specific to this form of recognition.

Rugg and Doyle's (1992) hypothesis that late old/new effects reflect differences in relative familiarity

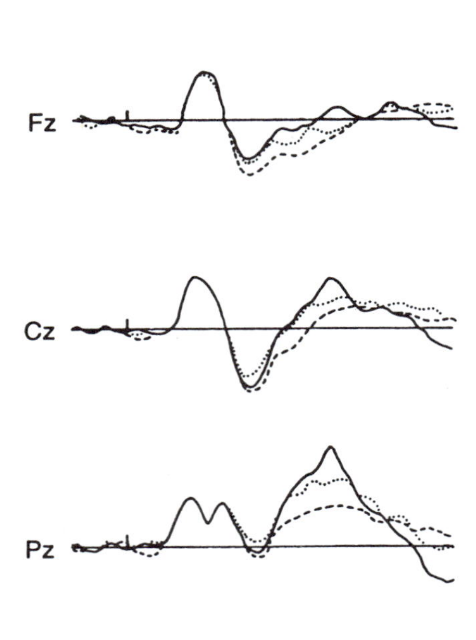

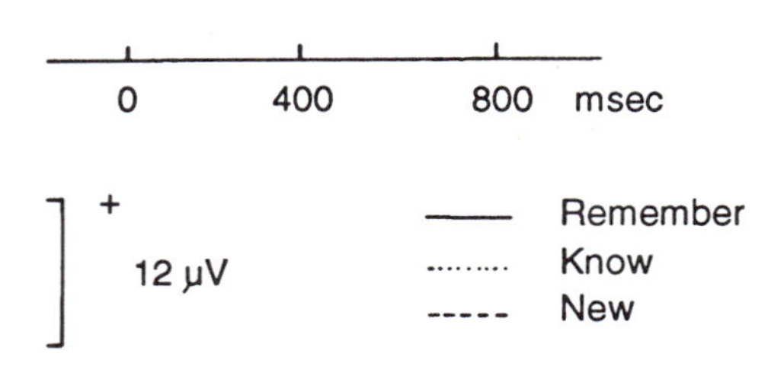

FIGURE 50.9 Grand-average midline ERPs from the recognition memory test of Smith (1993), comparing waveforms evoked by correctly judged new items with those evoked by correctly detected old words associated with "remember" or "know" judgments. (Redrawn with permission.)

rests on the assumption that it is correct to attribute the word frequency effect in recognition to relative familiarity. In contradiction to this assumption, Gardiner and Java (1990) found that the recognition advantage for low frequency words was confined to R responses. Thus they argued that low frequency words are better recognized because they are better recollected, not because they engender higher levels of relative familiarity. Coupled with Smith's (1993) finding that R items evoke larger old/new effects than K items, Gardiner and Java's (1990) data suggest an alternative account of the word frequency by old/new interaction described by Rugg and Doyle (1992, 1994). The interaction could have arisen because ERPs evoked by old words are sensitive not to word frequency but to the proportion of such words that engender a recollective experience.

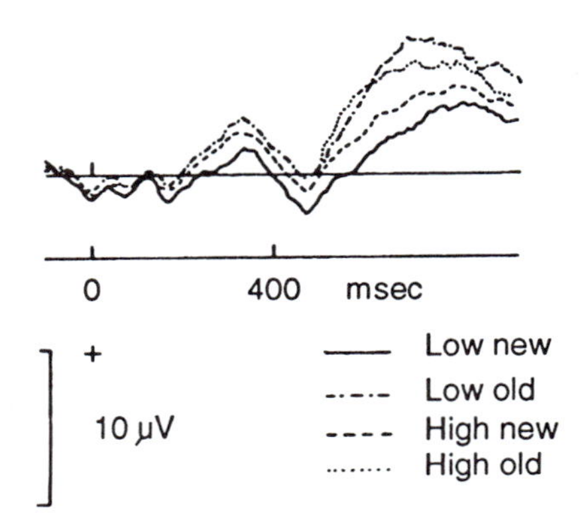

FIGURE 50.10 Grand average ERPs from a left parietal electrode evoked by correctly detected new and old test items in the experiment of Rugg, Wells, and Doyle (1993). ERPs are overlaid according to old/new status and word frequency; only items attracting a "remember" response were used to form ERPs to old items. Note the larger late old/new effects for low frequency words.

To evaluate this possibility, Rugg, Wells, and Doyle (1993) repeated the experiment of Rugg and Doyle (1992), adding the requirement to make an R/K judgment on each word judged old. They too found that the recognition advantage for low frequency words was restricted to items associated with R judgments. ERPs to these items are shown in figure 50.10, where it can be seen that the old/new effect is larger for low than for high frequency words. Thus even among "recollected" items, word frequency exerts an influence on late old/new ERP effects. This may point to the inadequacy of the remember/know procedure, and indicate that R and K judgments do not cleanly dissociate the two bases of recognition memory. Alternatively, it could be that late old/new effects are indeed sensitive to recollection. If this second alternative is correct, Rugg, Wells, and Doyle's (1993) findings imply that recollection may be graded rather than "all-or-none," such that some items engage or evoke it more than others (e.g., low versus high frequency words).

The studies reviewed above leave it unclear whether late old/new effects are more closely associated with recognition based on recollection or familiarity. The findings do, however, raise the possibility that a necessary condition for the emergence of late old/new effects on ERPs is awareness that an item has recently been experienced. Whether the effects reflect processes that contribute to this awareness, or are contingent upon it, is uncertain. Nonetheless, ERPs could provide a covert method for determining whether items in indirect memory tests are processed with or without conscious recognition.

ERPs and implicit memory

The question of whether ERPs are sensitive to implicit memory presents formidable methodological difficulties, since a convincing demonstration that an ERP effect is a reflection of implicit memory will require evidence that the effect occurs without awareness of prior experience with the test item. Thus, to establish an ERP 'correlate' of implicit memory, it is insufficient merely to record ERPs in indirect memory tests resembling those used to study the effects of implicit memory on behavior.

As noted earlier, there is a sizable literature on the modulation of ERPs by the repetition of words and other stimuli in indirect memory tests. Despite the indirect nature of the tasks in which they have been studied, there are two reasons why so-called ERP repetition effects presently shed little light on implicit memory. First, it is not known whether any aspect of these effects is independent of subjects' awareness that they are processing repeated rather than unrepeated items. Second, when evoked by isolated words, the effects are rather short-lived, lasting less than 15 minutes (Rugg, 1990). This stands in marked contrast to repetition effects on behavior, when a single exposure of a word can be sufficient to facilitate its subsequent processing hours later (e.g. Scarborough, Cortese, and Scarborough, 1977; Jacoby, 1983). So even if ERP repetition effects should turn out to be independent of explicit memory, their short-lived nature suggests that they reflect only a subset of the processes mediating implicit memory.

Bentin, Moscovitch, and Heth (1992) claimed that differences between ERPs evoked by repeated and unrepeated words do indeed occur without explicit memory. These claims, however, are based largely on ERP data from an indirect test (lexical decision), in which explicit memory for critical items was not assessed. One finding from Bentin, Moscovitch, and Heth (1992) is not susceptible to this criticism. ERPs evoked by old items misclassified as new in a recognition memory test were reported to differ from those evoked by genuinely new words. This finding is consistent with the idea that ERPs can reflect implicit memory, but is open to an alternative explanation. Subjects may have adopted a criterion for responding "yes" that was sufficiently strict to dissociate a continuous measure of recognition—that is, ERPs—from the discontinuous measure, yes/no responding. A replication adopting a more fine-grained behavioral measure (e.g., confidence ratings) would help resolve this issue.

In conclusion, there is presently little evidence that ERPs reflect implicit memory. In this respect, ERP research lags behind that employing electrodermal activity as a physiological measure of memory (Verfaille, Bauer, and Bowers, 1991; Bauer and Verfaille, 1992). The paucity of relevant ERP data can be rectified by two developments: first, more sophisticated experimental designs allowing the study of variables known to dissociate implicit and explicit memory in normal subjects; and second, studies of severely amnesic patients, in whom ERP signs of implicit memory can be sought without the problem of "contamination" by explicit memory. Following the logic applied in many behavioral studies (see Schacter, this volume), old/new ERP effects common to normal and amnesic subjects would be strong candidates as ERP signs of implicit memory.

ACKNOWLEDGMENTS The author's research is supported by a Major Award from the Wellcome Trust.

REFERENCES

Bauer, R.M., and M. Verfaille, 1992. Memory dissociations: A cognitive psychophysiology perspective. In *Neuropsychology of Memory*, ed. 2, L. R. Squire and N. Butters, eds. New York: Guilford, pp. 58–71.

Bentin, S., M. Moscovitch, and I. Heth, 1992. Memory with and without awareness: Performance and electrophysiological evidence of savings. *J. Exp. Psychol. [Learn. Mem. Cogn.]* 18:1270–1283.

Bentin, S., and B. S. Peled, 1990. The contribution of stimulus encoding strategies and decision-related factors to the repetition effect for words: electrophysiological evidence. *Mem. Cognition* 18:359–366.

Besson, M., M. Kutas, and C. Van Petten, 1992. An event-related potential (ERP) analysis of semantic congruity and repetition effects in sentences. *J. Cognitive Neurosci.* 4:132–149.

Donchin, E., and M. Fabiani, 1991. The use of event-related brain potentials in the study of memory: Is P300 a measure of event distinctiveness? In *Handbook of Cognitive Psychophysiology: Central and Autonomic System Approaches*. J. R. Jennings and M. G. H. Coles, eds. Chichester: Wiley, pp. 471–498.

Fabiani, M., G. Gratton, G. A. Chiarenza, and E. Donchin, 1990. A psychophysiological investigation of the von Restorff paradigm in children. *J. Psychophysiolology* 4:15–24.

Fabiani, M., D. Karis, and E. Donchin, 1986. P300 and recall in an incidental memory paradigm. *Psychophysiology* 23:298–308.

Fabiani, M., D. Karis, and E. Donchin, 1990. Effects of mnemonic strategy manipulation in a Von Restorff paradigm. *Electroencephalogr. Clin. Neurophysiol.* 75:22–35.

Friedman, D., 1990a. Cognitive event-related potential components during continuous recognition memory for pictures. *Psychophysiology* 27:136–148.

Friedman, D., 1990b. ERPs during continuous recognition memory for words. *Biol. Psychol.* 30:61–88.

Friedman, D., 1992. Event-related potential investigations of cognitive development and aging. *Ann. N.Y. Acad. Sci.* 658:33–64.

Gardiner, J. M., 1988. Functional aspects of recollective experience. *Mem. Cognition* 16:309–313.

Gardiner, J. M., and R. I. Java, 1990. Recollective experience in word and non-word recognition. *Mem. Cognition* 18:23–30.

Gardiner, J. M., and R. I. Java, 1991. Forgetting in recognition memory with and without recollective experience. *Mem. Cognition* 19:617–623.

Gardiner, J. M., and A. J. Parkin, 1990. Attention and recollective experience in recognition memory. *Mem. Cognition* 18:579–583.

Halgren, E., and M. E. Smith, 1987. Cognitive evoked-potentials as modulatory processes in human-memory formation and retrieval. *Hum. Neurobiol.* 6:129–139.

Hamberger, M., and D. Friedman, 1992. Event-related potential correlates of repetition priming and stimulus classification in young, middle-aged, and older adults. *J. Gerontol.* 47:395–405.

Hillyard, S. A., and T. W. Picton, 1987. Electrophysiology of cognition. In *Handbook of Physiology, Section 1: Neurophysiology*, F. Plum, ed. New York: American Physiological Society, pp. 519–584.

Jacoby, L. L., 1983. Perceptual enhancement: Persistent effects of an experience. *J. Exp. Psychol. [Learn. Mem. Cogn.]* 9:21–38.

Jacoby, L. L., and M. Dallas, 1981. On the relationship between autobiographical memory and perceptual learning. *J. Exp. Psychol. [Gen.]* 3:306–340.

Jacoby, L. L., and C. Kelley, 1992. Unconscious influences of memory: Dissociations and automaticity. In *The Neuropsychology of Consciousness*, A. D. Milner and M. D. Rugg, eds. London: Academic Press, pp. 201–234.

Johnson, R., 1993. On the neural generators of the P300 component of the event-related potential. *Psychophysiology* 30:94–101.

Johnson, R., A. Pfefferbaum, and B. S. Kopell, 1985. P300 and long-term memory: Latency predicts recognition performance. *Psychophysiology* 22:497–507.

Karayandis, F., S. Andrews, P. B. Ward, and N. McConaghy, 1991. Effects of inter-item lag on word repetition: An event-related potential study. *Psychophysiology* 28:307–318.

Karis, D., M. Fabiani, and E. Donchin, 1984. P300 and memory: Individual differences in the Von Restorff effect. *Cognitive Psycholology* 16:177–216.

Kopelman, M. D., and T. H. Corn, 1988. Cholinergic "blockade" as a model for cholinergic depletion. A comparison of the memory deficits with those of Alzheimer-type dementia and the alcoholic Korsakoff syndrome. *Brain* 111:1079–1110.

Kutas, M., 1988. Review of event-related potential studies of memory. In *Perspectives in Memory Research*, M. S. Gazzaniga, ed. Cambridge, Mass.: MIT Press, pp. 182–217.

Kutas, M., and C. Van Petten, 1988. ERP studies of language. In *Advances in Psychophysiology*, P. K. Ackles, J. R. Jennings and M. G. H. Coles, eds. Greenwich, Conn.: JAI Press, pp. 139–188.

Mandler, G., 1980. Recognising: the judgment of previous occurrence. *Psychol. Rev.* 87:252–271.

Mandler, G., 1991. Your face looks familiar but I can't remember your name: A review of dual process theory. In *Relating Theory and Data: Essays on Human Memory in Honor of Bennet B. Murdock*, W. E. Hockley and S. Lewandowsky, eds. Hillsdale, N.J.: Erlbaum, pp. 207–225.

Mandler, G., G. O. Goodman, and D. L. Wilkes-Gibbs, 1982. The word frequency paradox in recognition. *Mem. Cognition* 10:33–42.

Neville, H. J., M. Kutas, G. Chesney, and A. L. Schmidt, 1986. Event-related brain potentials during initial encoding and recognition memory of congruous and incongruous words. *J. Mem. Lang.* 25:75–92.

Paller, K. A., 1990. Recall and stem-completion have different electrophysiological correlates and are modified differentially by directed forgetting. *J. Exp. Psychol. [Learn. Mem. Cogn.]* 16:1021–1032.

Paller, K. A., and M. Kutas, 1992. Brain potentials during retrieval provide neurophysiological support for the distinction between conscious recollection and priming. *J. Cognitive Neurosci.* 4:375–391.

Paller, K. A., M. Kutas, and A. R. Mayes, 1987a. Neural correlates of encoding in an incidental learning paradigm. *Electroencephalogr. Clin. Neuropsysiol.* 67:360–371.

Paller, K. A., M. Kutas, A. P. Shimamura, and L. R. Squire, 1987b. Brain responses to concrete and abstract words reflect processes that correlate with later performance on a test of stem-completion priming. *Electroencephalogr. Clin. Neurophysiol. Suppl.* 40:360–371.

Paller, K. A., G. McCarthy, and C. C. Wood, 1988. ERPs predictive of subsequent recall and recognition performance. *Biol. Psychol.* 26:269–276.

Potter, D. D., C. D. Pickles, R. C. Roberts, and M. D. Rugg, 1992. The effects of scopolamine on event-related potentials in a continuous recognition memory task. *Psychophysiology* 29:29–37.

Pritchard, W., 1981. Psychophysiology of P300. *Psychol. Bull.* 89:506–540.

Richardson-Klavehn, A., and R. A. Bjork, 1988. Measures of memory. *Annu. Rev. Psychol.* 39:75–543.

Rugg, M. D., 1987. Dissociation of semantic priming, word

and nonword repetition by event-related potentials. *Q. J. Exp. Psychol. [A]* 39:123–148.

Rugg, M. D., 1990. Event-related brain potentials dissociate repetition effects of high and low frequency words. *Mem. Cognition* 18:367–379.

Rugg, M. D., P. Brovedani, and M. C. Doyle, 1992. Modulation of event-related potentials by word repetition in a task with inconsistent mapping between repetition and response. *Electroencephalogr. Clin. Neurophysiol.* 84:521–531.

Rugg, M. D., and M. C. Doyle, 1992. Event-related potentials and recognition memory for low-frequency and high-frequency words. *J. Cognitive Neurosci.* 4:69–79.

Rugg, M. D., and M. C. Doyle, 1994. Event related potentials and stimulus repetition in direct and indirect tests of memory. In *Cognitive Electrophysiology*, H. Heinze, T. Munte, and G. R. Mangun, eds. Boston: Birkhauser, pp. 124–148.

Rugg, M. D., J. Furda, and M. Lorist, 1988. The effects of task on the modulation of event-related potentials by word repetition. *Psychophysiology* 25:55–63.

Rugg, M. D., and M. E. Nagy, 1989. Event-related potentials and recognition memory for words. *Electroencephalogr. Clin. Neurophysiol.* 72:395–406.

Rugg, M. D., R. C. Roberts, D. D. Potter, C. D. Pickles, and M. E. Nagy, 1991. Event-related potentials related to recognition memory. Effects of unilateral temporal lobectomy and temporal lobe epilepsy. *Brain* 114:2313–2332.

Rugg, M. D., T. Wells, and M. C. Doyle, 1993. Event-related potentials, word frequency, and recollection-based recognition. Unpublished.

Sanquist, T. F., J. W. Rohrbaugh, K. Syndulko, and D. B. Lindsley, 1980. Electrocortical signs of levels of processing: Perceptual analysis and recognition memory. *Psychophysiology* 17:568–576.

Scarborough, D. L., C. Cortese, and H. J. Scarborough, 1977. Frequency and repetition effects in lexical memory. *J. Exp. Psychol. [Hum. Percept.]* 3:1–17.

Schmidt, S. R., 1991. Can we have a distinctive theory of memory? *Mem. Cognition* 19:523–542.

Smith, M. E., 1993. Neurophysiological manifestations of recollective experience during recognition memory judgments. *J. Cognitive Neurosci.* 5:1–13.

Smith, M. E., and E. Halgren, 1989. Dissociation of recognition memory components following temporal lobe lesions. *J. Exp. Psychol. [Learn. Mem. Cogn.]* 15:50–60.

Sternberg, S., 1966. High-speed scanning in human memory. *Science* 153:652–654.

Van Petten, C., M. Kutas, R. Kluender, M. Mitchiner, and H. McIsaac, 1991. Fractionating the word repetition effect with event-related potentials. *J. Cognitive Neurosci.* 3:129–150.

Verfaille, M., R. M. Bauer, and D. Bowers, 1991. Autonomic and behavioral evidence of "implicit" memory in amnesia. *Brain Cogn.* 15:10–25.

Young, M. P., and M. D. Rugg, 1992. Word frequency and multiple repetition as determinants of the modulation of event-related potentials in a semantic classification task. *Psychophysiology* 6:664–676.

51 Memory and Frontal Lobe Function

ARTHUR P. SHIMAMURA

ABSTRACT Patients with frontal lobe lesions exhibit a wide array of cognitive deficits, including impairment in planning, problem solving, short-term memory, metamemory, and temporal memory. Some have viewed the memory impairment associated with frontal lobe dysfunction as secondary to other cognitive disorders, such as deficits in attention, inferential reasoning, and cognitive mediation. Others have viewed memory impairment as a primary deficit in frontal lobe mechanisms (e.g., working memory, temporal organization of memory). This chapter reviews aspects of impaired memory performance in patients with frontal lobe lesions. A theoretical framework is proposed that suggests that the cognitive and memory deficits observed in patients with frontal lobe lesions occur as a result of a disruption in inhibitory control of extraneous activity.

About how long is the average man's necktie? How many words can you report that begin with the letter *A*? Which public event occurred more recently—the volcanic eruption of Mount St. Helens or the explosion of the space shuttle *Challenger*? Answers to these rather disparate questions require evaluation and retrieval of information in memory. Moreover, in most individuals the answers are not readily available, and thus some plan or search strategy must be initiated. Interestingly, the ability to answer such questions depends critically on the integrity of the frontal lobes. That is, patients with frontal lobe lesions exhibit significant impairment when memory is queried in these ways. Such findings offer clues to the underlying mechanism associated with frontal lobe function. That is, they provide important links between cognitive and biological function, and thus offer a cognitive neuroscience analysis of brain mechanisms.

The frontal lobes encompass roughly a third of the cerebral cortex in humans (Goldman-Rakic, 1987; Fuster, 1989). Each frontal lobe can be divided into three major subareas—the primary motor cortex, the premotor cortex, and the prefrontal cortex. The primary motor cortex provides the main cortical output for voluntary movements. Damage to this area produces paralysis on the side of the body opposite the brain lesion. The premotor cortex is closely linked to the primary motor cortex and appears to be important for the integration and programming of sequential movements (Luria, 1973). The prefrontal cortex covers a large area anterior to the primary motor and premotor cortices, and its function is the central topic of this chapter. The prefrontal cortex is often divided into an orbital frontal region (anterior tip of the frontal lobes) and a dorsolateral region. There are intricate connection between regions in the prefrontal cortex and posterior cortical regions, such as primary sensory areas, association areas, and limbic structures (Goldman-Rakic, 1987; Fuster, 1989). The prefrontal cortex also has primary subcortical projections to and from the mediodorsal nucleus of the thalamus.

As a result of the extensive connections between the prefrontal cortex and other brain regions, it is not too surprising that damage to this area has been associated with a variety of mental dysfunctions in humans—including disorders of personality, affect, motor control, language, problem solving, and memory (for review, see Milner, Petrides, and Smith, 1985; Benton, 1991; Shimamura, Janowsky, and Squire, 1991). Studies of brain-injured patients and lesion studies using animal models have suggested that different regions within the prefrontal cortex mediate different mental functions. For example, personality and affective disorders have been associated with orbital prefrontal lesions, disorders of language production have been associated with the prefrontal region known as Broca's area, and certain cognitive and memory disorders have been associated with dorsolateral prefrontal lesions.

ARTHUR P. SHIMAMURA Department of Psychology, University of California, Berkeley, Calif.

Apart from these rather crude guidelines, however, the characterization of frontal lobe function has been rather elusive. Some researchers have portrayed frontal lobe function rather mystically, suggesting it is what makes humans unique. In particular, it has been considered to be the basis for high-level, intellectual reasoning and abstraction (see Goldstein, 1939; Halstead, 1947). Yet, Hebb (1945) argued strongly against this view and demonstrated that frontal lobe lesions in humans did not grossly affect intellectual functioning. Others have suggested that the primary deficit is one of severe perseveration of actions, due primarily to problems of response inhibition (Brutkowski, Mishkin, and Rosvold, 1963; Luria, 1973). More recently, the function of the dorsolateral prefrontal cortex has been associated with planning and the temporal organization of behavior (Milner, Petrides, and Smith, 1985; Goldman-Rakic, 1987; Fuster, 1989). Finally, Baddeley (1986) has referred to frontal lobe dysfunction as a "dysexecutive" syndrome, which involves a disruption of working memory. Working memory is related to *short-term memory* and refers to the cognitive system concerned with the on-line supervision of information to be encoded, stored, and retrieved.

This chapter briefly reviews the contribution of the frontal lobes to memory performance. Memory dysfunction associated with frontal lobe lesions is contrasted with memory dysfunction associated with damage to other brain areas, such as the medial temporal area. Finally, a theoretical framework is developed that attempts to characterize the physiological bases for the behavioral anomalies observed in patients with frontal lobe dysfunction.

New learning ability

Patients with frontal lobe lesions do not exhibit significant impairment on standard tests of new learning ability. In these tests, subjects are shown stimuli—such as words, pictures, or abstract designs— and are later asked to recollect the material that they have learned. For example, Janowsky, et al. (1989) demonstrated that patients with frontal lobe lesions perform within the normal range on the Wechsler Memory Scale-Revised (WMS-R), a clinical test battery that assesses verbal and nonverbal memory on a variety of measures, including recognition memory and paired-associate learning. Patients with frontal lobe lesions did, however, exhibit poor performance on the attention-concentration index of the WMS-R. This index assesses short-term memory (e.g., digit span) and corroborates the suggestion by Baddeley (1986) that frontal lobe lesions disrupt working memory. Interestingly, this disruption does not significantly impair the capacity to learn new information or what psychologists describe as the acquisition of long-term memory.

The finding of preserved new learning ability in patients with frontal lobe lesions can be contrasted with the severe learning impairment associated with lesions involving the medial temporal lobe (e.g., hippocampus) or diencephalic midline (e.g., thalamic nuclei). These lesions produce organic amnesia, in which patients have difficulty remembering information and events that occur after the onset of amnesia (for review, see Shimamura, 1989). This severe and debilitating impairment of new learning ability is called *anterograde amnesia*. Various neurological disorders—such as anoxia (i.e., experiencing oxygen loss), ischemia (i.e., loss of blood flow), viral encephalitis, and Alzheimer's disease —can damage the medial temporal lobe and thereby produce anterograde amnesia. Also, damage to diencephalic structures has been implicated in the amnesic disorder associated with Korsakoff's syndrome. Figure 51.1 displays performance on three tests of new learning ability by patients with Korsakoff's syndrome (KOR), other amnesic patients with medial temporal or diencephalic lesions (AMN), and patients with frontal lobe lesions (F). Shown also are data from control groups for each of the patient groups (data from Janowsky et al., 1989).

Impairment of Free Recall Patients with frontal lobe lesions exhibit impairment on one standard test of new learning capacity—free recall (DellaRocchetta, 1986; Jetter, et al., 1986; Janowsky et al., 1989). Figure 51.2 illustrates the disproportionate impairment on tests of free recall compared to performance on tests of recognition memory. In this test, a list of 15 words was presented for study, and free recall was tested immediately after the last word was presented. To assess learning, four additional study–test trials were administered using the same list of words. Patients with frontal lobe lesions exhibited significant impairment on this test, despite good performance on a comparable test of recognition memory. Although near-perfect performance on the immediate tests of recognition may have concealed a difference between patient and control groups, recognition memory memory was still quite good in the

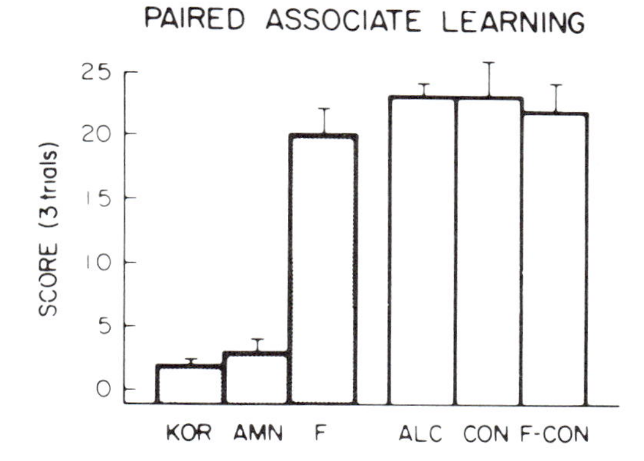

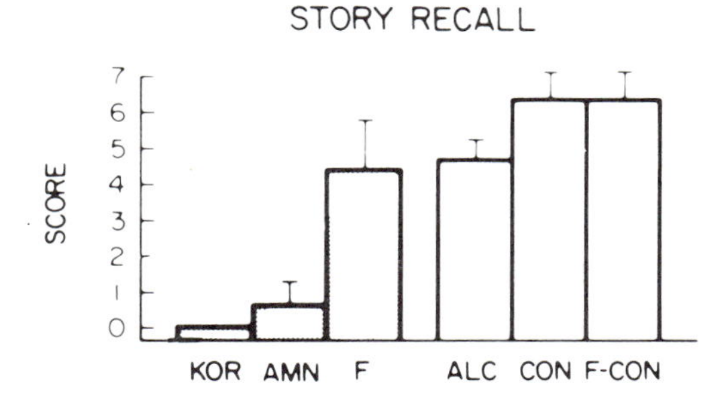

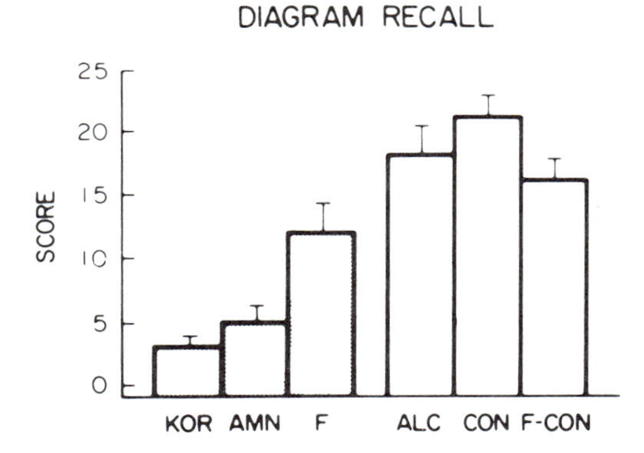

FIGURE 51.1 Performance of patients with frontal lobe lesions, amnesic patients, and control subjects on three tests sensitive to anterograde amnesia. KOR, patients with Korsakoff's syndrome; AMN, 5 other patients with amnesia; F, 7 patients with frontal lobe lesions; ALC, alcoholic control subjects matched to the Korsakoff patients; CON, healthy control subjects matched to the 5 other amnesic patients; F-CON, control subjects matched to the patients with frontal lobe lesions. Brackets show standard error of the mean. From Janowsky et al., 1989)

patients even when a test was administered after a one-week retention interval.

The impairment in free recall stands in contrast to performance on other standard measures of new learning capacity. One possibility is that tests of free recall put heavy demands on internally generated memory strategies and thus require extensive use of search and retrieval processes. In one study (Gershberg and Shimamura, 1991), impared free recall performance was associated with decreased use of memory organizational strategies. Specifically, on a test of free recall for unrelated lists, patients with frontal lobe lesions exhibited low scores on a measure of subjective organization, which is an index of the degree to which subjects cluster the same words together across multiple study-test trials. Also, on a test of free recall for related lists, patients with frontal lobe lesions did not cluster semantically related words as much as did control subjects.

Encoding semantic information and proactive interference

The impairment in free recall in patients with frontal lobe lesions suggests a disruption in the use of memory strategies. Previous findings have suggested that patients with frontal lobe lesions have problems when irrelevant information must be ignored. For example, Perret (1974) showed that patients with frontal lobe lesions are more susceptible to interference effects on the Stroop test, a test of attention in which the naming of ink color is disrupted by the presence of color words (e.g., the word *red* written in green ink). In terms of memory performance, experimental psychologists have used various test procedures to describe the positive and negative consequences of prior learning in the learning of new information (see Postman, 1961; Postman and Underwood, 1973). The term *proactive interference* has been used to describe the negative effects that prior learning has on new learning.

Preliminary findings from patients with frontal lobe lesions suggest that heightened susceptibility to proactive interference can be observed on tests of memory in which subjects must inhibit previously learned responses (Shimamura et al., submitted). We presented three study–test learning trials of a list of 12 related paired associates (e.g., *thief–crime*; *lion–hunter*). Proactive interference was increased by having subjects learn a second list in which each cue word used in the first list was paired with a new target word (e.g., *thief–bandit*; *lion–circus*). This memory test paradigm is called AB–AC paired-associate learning. Although patients appeared to learn the first list nearly as a well as control subjects, they exhibited disproportionate impairment

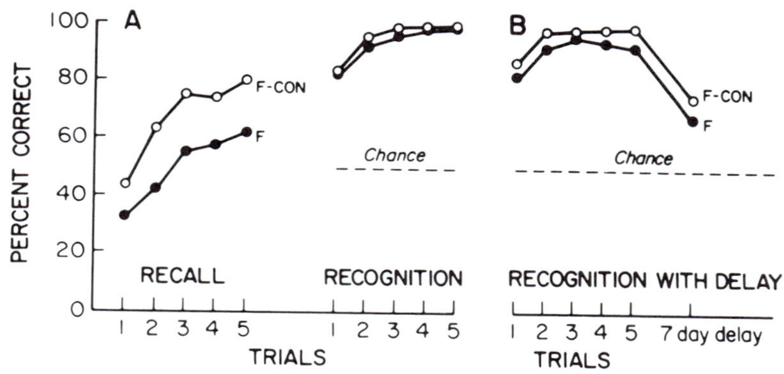

FIGURE 51.2 Performance by patients with frontal lobe lesions, amnesic patients, and control subjects on the Rey Auditory Verbal Learning Test. (A) Recall performance for 15 words during 5 study–test trials and yes/no recognition performance for 15 study words and 15 distractor words during 5 study–test trials. F, patients with frontal lobe lesions; F-CON, control subjects. (B) A second test of yes/no recognition performance by patients with frontal lobe lesions and control subjects for 15 study words and 15 distractor words. The delayed test trial was given 7 days after the 5 study–test trials. (Data from Janowsky et al., 1989)

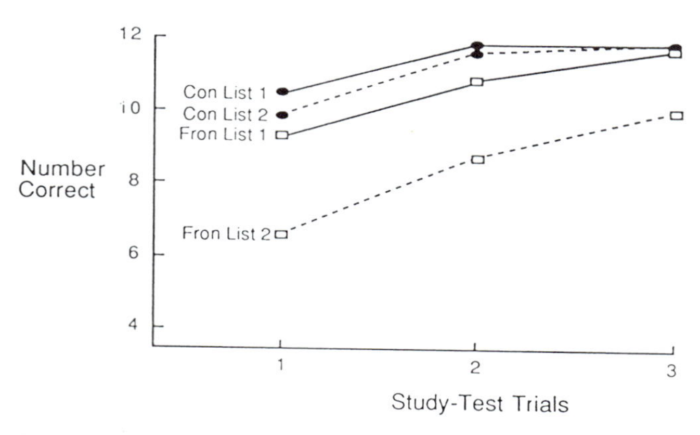

FIGURE 51.3 Performance by patients with frontal lobe lesions and control subjects on a test of AB–AC paired-associate learning. Subjects were given 3 study-test learning trials of a list of 12 related paired associates (LION–HUNTER) and then another 3 study–test learning trials using another list in which the cue word was the same but the target word was different (LION–CIRCUS). Performance on the second list by patients with frontal lobe lesions was particularly affected, presumably because of the increased interference from the first list.

when they were required to ignore the first associations and learn new ones (see Figure 51.3).

RELEASE FROM PROACTIVE INTERFERENCE *Release from proactive interference* is another deficit in memory that has been associated with frontal lobe dysfunction (Moscovitch, 1982). To test this phenomenon, subjects are presented three words and then asked to recall the words following a 15–20 s distraction task (see Wickens, 1970). For the first four trials, words from the same semantic category are presented (e.g., clothing articles: SHOE, SHIRT, PANTS). Subjects typically exhibit a decline in recall performance across the four trials as a result of semantic interference from previous trials (i.e., proactive interference). That is, after several trials, subjects have difficulty discriminating between words presented in the current trial and words presented on previous trials. When words from a different semantic category are presented for study (e.g., bird names: ROBIN, SPARROW, CANARY), normal subjects exhibit improved memory performance. Indeed, they perform as well as they did on the very first trial (Wickens, 1970). This "release" from proactive interference indicates that normal subjects can improve their performance by encoding the shift in the meaning of the word stimuli.

Neuropsychological findings have shown that patients with Korsakoff's syndrome do not exhibit release from proactive interference (Cermak et al., 1976; Squire, 1982). Presumably, these patients do not adequately encode information in a semantic manner, and thus fail to benefit from a shift in semantic categories. Yet, normal release from proactive interference can be observed in other amnesic patients, such as patients with amnesia due to an anoxic or ischemic episode, patients with viral encephalitis, and psychiatric patients prescribed electroconvulsive therapy (Cermak, 1976;

Squire, 1982; Janowsky, et al., 1989). Moscovitch (1982) suggested that failure to release from proactive interference, as observed in patients with Korsakoff's syndrome, was mediated by frontal lobe damage. However, Freedman and Cermak (1986) showed that failure to release from proactive interference in patients with frontal lobe lesions occurred only when new learning impairment (e.g., anterograde amnesia) was also present. Recent findings have corroborated the findings of Freedman and Cermak (1986) by showing that patients with circumscribed dorsolateral prefrontal lesion—who do not exhibit impairment on standard tests of new learning—do not exhibit deficits in the release from proactive interference (Janowsky et al., 1989; see figure 51.4).

Problem Solving and Proactive Interference Problems in the ability to shift between strategies have been observed in tests of problem solving. For example, Milner (see Milner, Petrides, and Smith, 1985) identified severe impairment in patients with frontal lobe lesions on the Wisconsin Card Sorting Test. This problem-solving test measures the ability to identify the relevant dimension or category of a series of stimulus cards that vary in three dimensions (shape, color, number). Patients with frontal lobe lesions often fail to inhibit erroneous responses. In another sorting test (Delis et al., 1992), six cards are used and each card has multiple features (e.g., a card with a horizontal-striped background with the word *tiger* printed above a black triangle). On the basis of these stimulus features, the cards can be sorted into two piles of three cards. Several sorts are based on the physical features of the cards—for example, card background (horizontal versus oblique stripes) or triangle color (white versus black). Other sorts are based on the semantic nature of he words that appeared on each card—for example, land versus water animals or dangerous versus domestic animals. Patients with frontal lobe lesions exhibit impairment in the ability to sort the cards on the basis of the stimulus features. These findings suggest that patients with frontal lobe lesions are particularly disrupted by irrelevant or extraneous features.

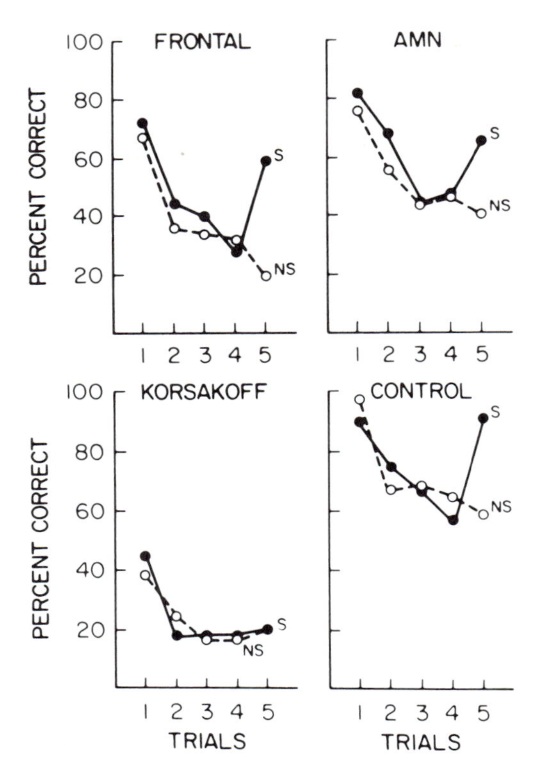

Figure 51.4 Release from proactive interference by 7 patients with frontal lobe lesions, 5 patients with (non-Korsakoff) amnesia (AMN), 7 patients with Korsakoff's syndrome, and 8 healthy control subjects. S, shift of category on fifth trial, NS, no shift of category on fifth trial. Note that only the patients with Korsakoff's syndrome exhibited a failure to release from proactive interference. (From Janowsky et al., 1989)

Retrieving semantic information

The findings from the previous section suggest that patients with frontal lobe lesions exhibit impairment in the ability to encode or register semantic information. This impairment is particularly apparent when extraneous or interfering information is presented prior to new learning, as in the case of proactive interference effects, or when extraneous information is presented during problem-solving tasks. In this section, it is evident that patients with frontal lobe lesions not only exhibit impairment in the encoding of semantic information, they are also impaired in the *retrieval* of semantic information.

Word Finding Impairment in memory retrieval is typically observed in problems of word finding. These "anomic" deficits are often observed in patients with left hemisphere lesions. In tests of "word fluency" (Benton and Hamsher, 1978), subjects are given one minute to produce words beginning with the letter *F*. The same task is then repeated for the letters *A* and *S*, and performance is based on the total number of words produced for the three letters (*F, A,* and *S*). Patients with left frontal lobe lesions exhibit impairment on this test. For example, Janowsky et al. (1989) found that patients with bilateral or left unilateral lesions of the frontal lobes produced significantly fewer words on this task (21.5 words produced) than control subjects (37.5 words produced). Patients with right unilateral frontal lesions, however, were unimpaired on the verbal fluency test (40.7 words produced). In another study, patients with right frontal lobe lesions were found to be impaired on a nonverbal version of the fluency test in which subjects were asked to create as many designs as possible (Jones-Gotman and Milner, 1977).

Word finding deficits can also be observed when patients with frontal lobe lesions are asked to retrieve words from semantic categories. In the supermarket fluency test on the Dementia Rating Scale (Mattis, 1976), subjects are asked to name as many items as they can think of that could be purchased at a supermarket. Patients with frontal lobe lesions exhibited fewer retrievals than control subjects (Janowsky et al., 1989). Impairment on tests of word retrieval may be related to difficulties in organizing and searching information in semantic memory. As in the case of encoding deficits, it may be that retrieval deficits are related to the failure to inhibit irrelevant or extraneous information in memory. For example, when a patient retrieves the word FRUIT in the word fluency or supermarket test, the retrieval of that word may interfere with subsequent searches if the patient cannot inhibit or disregard the word.

METAMEMORY AND COGNITIVE ESTIMATION Interference in retrieval can produce "memory blocking" effects, which can be commonly observed when information is on the "tip of the tongue." Indeed, patients with anomic deficits experience inordinate occasions of this annoying lapse of memory. The "tip-of-the-tongue" phenomenon is a part of the domain of research called *metamemory*. Metamemory refers to knowledge about one's memory capabilities and knowledge about strategies that can aid memory (see Metcalfe and Shimamura, 1994). Patients with frontal lobe lesions exhibited metamemory deficits when they were asked to evaluate what they know. In one test, subjects were given 24 sentences to learn (e.g., *Patty's garden was full of marigolds*). After a delay, cued recall was assessed for the last word in each sentence (e.g., *Patty's garden was full of ______*). If the correct answer to a question could not be recalled, then subjects rated their feeling of knowing—that is, they rated on a four-point scale how likely they would be able to recognize the answer if some choices were given (e.g., very likely to recognize the answer, not very likely to recognize the answer). To verify the accuracy of these "feeling-of-knowing" judgments, the ratings were correlated with performance on an subsequent recognition test. Patients with frontal lobe lesions exhibited deficits in feeling-of-knowing accuracy when memory was tested after a 1- to 3-day delay, despite the fact that they performed normally in the ability to recall the words when cued with the sentence context (Janowsky, Shimanuera, and Squire, 1989a). That is, memory was not significantly impaired, yet the patients were unable to assess what they knew or did not know.

The feeling-of-knowing impairment exhibited by patients with frontal lobe lesions may be related to deficits in other tasks that involve memory retrieval. One such example is the finding that patients with frontal lobe lesions have difficulty making estimates or inferences from everyday experiences (e.g., How tall is the average English woman? Shallice and Evans, 1978). Such answers are not readily available and typically require search and retrieval strategies. Similarly, patients with frontal lobe lesions have difficulty estimating the price of objects (Smith and Milner, 1984). All these deficits—that is, deficits in cognitive estimation as well as in feeling-of-knowing judgments—might be construed as deficits in metamemory (i.e., knowing about what is stored in memory). These findings of impaired metamemory, in conjunction with the findings of preserved new learning, suggest that the frontal lobes may be critically involved in the manipulation and organization of information but not as involved in the storage of new information in memory.

Memory for temporal context

Milner (1971) was the first to report that patients with frontal lobe lesions exhibit an impairment in the temporal organization of memory. She reported a study of recency judgments, in which subjects are shown a series of stimuli (e.g., words, pictures) and on certain occasions are asked to judge which one of two stimuli was presented more recently (see Milner, Corsi, and Leonard, 1991). Another test procedure demonstrated

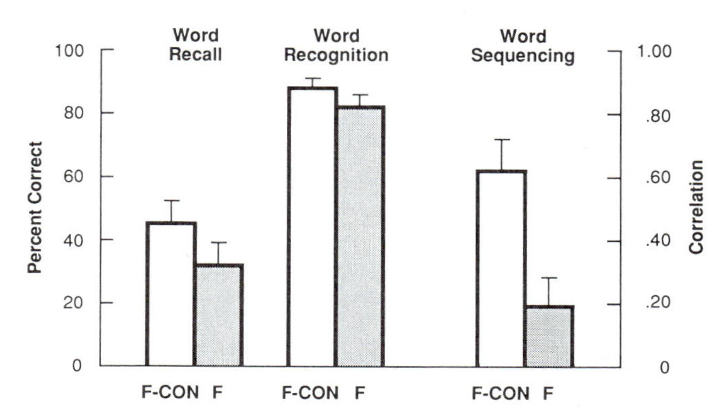

FIGURE 51.5 Following the presentation of 15 words, subjects were asked to recall and recognize the words on the list or, on a different occasion, to reconstruct the order in which the words were presented. Word sequencing performance was based on the correlation of the judged order of the words with the actual list order. F, patients with frontal lobe lesions; F-CON, control subjects for patients with frontal lobe lesions. (From Shimamura, Janowsky, and Squire, 1990)

impaired memory for temporal order memory in patients with frontal lobe lesions (Shimamura, Janowsky, and Squire, 1990). Subjects were presented a list of 15 words, one at a time, and then asked to reconstruct the list order from a random display of the words. Memory for temporal order was assessed by correlating a subject's judged order with the actual presentation order of the words. Patients with frontal lobe lesions exhibited impaired memory for temporal order on this test (see figure 51.5). Neither word recognition nor recall was significantly impaired, though free recall appeared to be mildly affected. Patients with Korsakoff's syndrome and non-Korsakoff's amnesic patients exhibited both impaired word memory and impaired memory for temporal order. However, patients with Korsakoff's syndrome exhibited somewhat greater impairment on the word sequencing test, though this difference was only marginally significant.

In a second study, patients with frontal lobe lesions, patients with Korsakoff's syndrome, and non-Korsakoff's amnesic patients were asked to reproduce the chronological order of 15 factual events that occurred between 1940 and 1985 (e.g., *Jonas Salk discovered the first Polio vaccine*; *The name of the Polish Labor movement led by Lech Walesa was Solidarity*) (Shimamura, Janowsky, and Squire, 1990). The findings from this test corroborated the findings obtained from the word sequencing test. Specifically, patients with frontal lobe lesions exhibited significant impairment on the fact sequencing test but were not significantly impaired on recall or recognition tests about the facts themselves. This study demonstrated that impaired memory for temporal order can occur in patients with frontal lobe lesions, even when the memory for the items is good and the items are distributed across a 45-year period.

SOURCE MEMORY In everyday experiences, one often remembers some factual information but forgets the context in which the information was encountered. For example, one might remember the name of a good restaurant or movie but forget who told you the information or when and where the information was encountered. Such instances represent a loss of source or contextual memory. Findings of specific source amnesia suggest a distinction between memory for factual or semantic information and memory for contextual or episodic memory (Tulving 1972, 1983; Hirst, 1982; Mayes, Meudell, and Pickering, 1985). A common feature of these accounts is that factual (semantic) and context-specific (episodic) memory are proposed to be dissociable components of memory function. Neuropsychological analyses of *source amnesia* suggest that impaired memory for spatial-temporal context may be a disorder related to frontal lobe dysfunction and not a disorder associated with damage to the medial temporal or diencephalic areas that cause amnesia (Schacter, Harbluk, and MacLachlan, 1984; Janowsky et al., 1989b).

To test whether patients with frontal lobe lesions exhibit source memory impairment, Janowsky et al. (1989b) asked patients with frontal lobe lesions, age-matched control subjects, and younger control subjects to learn a set of 20 trivia facts that could not be previously recalled (e.g., *The name of the dog on the Cracker Jacks box is Bingo*). After a 6–8 day retention interval, fact recall was tested for the 20 learned facts (e.g., *What is the name of the dog on the Cracker Jacks box?*) and for 20 new facts. When subjects correctly answered a fact question, they were asked to recollect the source of the information ("Can you tell me where you learned the answer?"; "When was the most recent time you heard that information?"). Two kinds of source errors were scored: errors in which subjects correctly recalled a learned fact but falsely reported that the fact was most recently encountered at some time prior to the learning

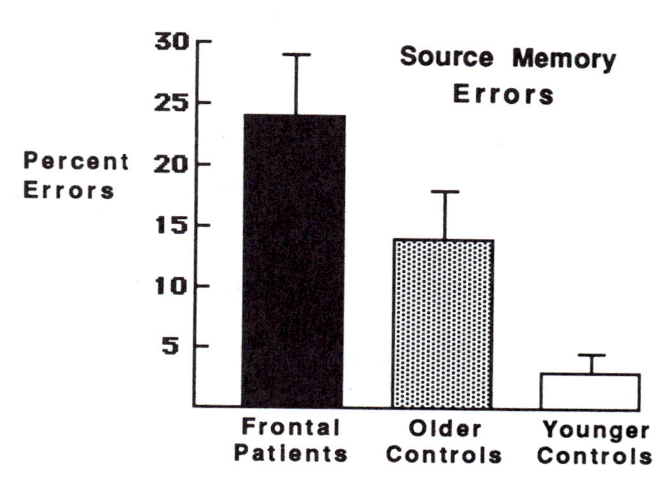

FIGURE 51.6 Source memory performance by patients with frontal lobe lesions (mean age, 64 years), age-matched control subjects (mean age, 62 years), and younger control subjects (mean age, 49 years). Source errors represent instances when a subject failed to recall the time and place a fact was presented. (Data from Janowsky, Shimamura, and Squire, 1989, experiment 2)

session; and errors in which subjects correctly recalled a new fact but falsely reported that the fact was encountered during the learning session.

Source memory ability was significantly impaired in patients with frontal lobe lesions, even though memory for the facts themselves was normal (see figure 51.6). This rather selective memory impairment suggests that the frontal lobes contribute specifically to contextual memory. Interestingly, source memory impairment can also occur with normal aging, as indicated by the finding that older control subjects (mean age 63.9 years) made more source errors than younger control subjects (mean age 49.4 years) (see also McIntyre and Craik, 1987). Neuronal cell loss associated with normal aging does occur prominently in the frontal lobes (see Haug, et al., 1983), which suggests that source memory impairment in normal aging may also be related to subtle frontal lobe dysfunction.

A theoretical interpretation of frontal lobe function

The manifold array of memory and cognitive deficits associated with frontal lobe lesions might suggest that there is no single, underlying mechanism associated with the prefrontal cortex. That is, this brain region may not embody a unitary mechanism with a specific function. Indeed, patients with frontal lobe lesions exhibit numerous disorders that affect memory performance, including impairment of free recall, word finding, metamemory, memory for temporal order, and source memory. These deficits may or may not be related to other cognitive disorders associated with frontal lobe dysfunction, such as deficits in planning, initiation, selective attention, problem solving, and concept shifting. Can all of these disorders be related in any way?

A general deficit that appears to be indicative of many aspects of frontal lobe dysfunction is the failure of these patients to disregard or inhibit irrelevant information. These patients do not seem to be able to control inadvertent information processing. This problem affects memory performance as a result of interference from previously activated memories (i.e., proactive interference). It may also be a major factor in attentional disorders associated with frontal lobe lesions in which on-line cognitive processes are disrupted, such as disorders on the Stroop test or card sorting tests. One possibility is that the prefrontal cortex controls information processing by enabling inhibitory control—specifically inhibitory control of activity in posterior cortical regions. That is, the prefrontal cortex modulates other activity by way of a dynamic filtering or gating mechanism that inhibits extraneous activity. Effective filtering thus increases the availability of relevant sensory and cognitive activity by filtering signals from irrelevant noise. Based on this view, the manifold behavioral disorders that result from damage to different regions in the prefrontal cortex do not occur because each region is performing different computations or doing something different. Instead, each region is performing the same computation, only the result is different because specific prefrontal regions are connected to different areas in posterior cortex, and thus are filtering different aspects of cognitive function.

When frontal lobe function is viewed as a dynamic filtering mechanism, it becomes evident that memory for temporal order (e.g., recency judgments, source memory) may not be directly impaired in patients with frontal lobe lesions. Instead, the nature of the task may place great demands on inhibitory gating. For example, remembering where and when a fact was learned may require extensive retrieval strategies that require several search paths, some of which may not be productive. Inability to inhibit irrelevant search strategies may interfere on tasks requiring extensive search and retrieval. This explanation not only accounts for disorders of spatial-temporal context, but may also explain related cognitive disorders associated with frontal lobe lesions, such as disorders of problem solving, inference making, metamemory, and cognitive estimation. In fact, recent findings indicate that patients with frontal lobe lesions do not always exhibit impairment in recency judgments. They perform well when the experimenters make sure the items are encoded distinctively by having subjects manipulate each item (e.g., lift the cup) (McAndrews and Milner, 1991).

Some physiological evidence also suggests that the frontal lobes are involved in the inhibitory control of posterior cortical activity. Knight and colleagues (Knight, Scabini, and Woods, 1989; Yamaguchi and Knight, 1990) studied scalp evoked potentials in patients with frontal lobe lesions. In one study (Knight, Scabini, and Woods, 1989), the amplitude of middle latency auditory evoked potentials, which are presumed to be generated in primary auditory cortex, was *potentiated* as a result of frontal lobe lesions. Thus, there appeared to be a *disinhibition* of posterior cortical activity as a result of frontal lobe lesions. In a recent PET study of normal individuals (Frith et al., 1991), *increases* in activity in the dorsolateral prefrontal cortex were related to decreases in activity in posterior cortical regions. Thus, as a result of the extensive reciprocal connections between the prefrontal cortex and other cortical areas, it may be that inhibitory control of many aspects of mental function is provided by this proposed gating mechanism. Based on this view, a multitude of cognitive disorders would occur as a result of frontal lobe lesions, not because different areas of the frontal lobes are serving different functions, but because different areas of the frontal lobes are inhibiting different posterior cortical regions (which themselves serve different cognitive functions). This view extends an earlier view of *response* disinhibition as the mediator of frontal lobe dysfunction (see Brutkowski, Mishkin, and Rosvold, 1963; Luria, 1973), a view that was limited and did not adequately account for many findings from human and animal lesion studies.

The theory of inhibitory gating as a primary function of the frontal lobes offers a unified explanation of the neuropsychological findings. Thus, impairment in the domains of attention, memory, and problem solving may be explained by a particular deficit in inhibitory gating. In attention, the deficit is due to a failure to inhibit irrelevant stimulus information. In memory, the deficit is due to a failure to inhibit previous memory associations. In problem solving, the deficit is due to a failure to inhibit irrelevant or erroneous search or decision paths. This theory may also be useful in explaining the kinds of social disinhibitions observed in patients with orbital prefrontal lesions (Lhermitte, 1986). Such patients appear to exhibit a failure to inhibit inappropriate social behaviors (e.g., telling crude jokes, emotional rages). Viewed in this manner, the notion of inhibitory gating can be useful as a way to describe many, if not most, aspects of frontal lobe function. Moreover, a variety of experimental studies—both behavioral and physiological—can be proposed to test predictions based on the theoretical framework.

ACKNOWLEDGMENTS Preparation of this chapter was supported by a grant from the National Institute on Aging (grant no. AG09055).

REFERENCES

BADDELEY, A., 1986. *Working memory*. Oxford: Oxford University Press.

BENTON, A. L., 1991. The prefrontal region: Its early history. In *Frontal Lobe Function and Dysfunction*, H. S. Levin, H. M. Eisenberg, and A. L. Benton, eds. New York: Oxford University Press.

BENTON, A. L., and K. HAMSHER, 1978. *The Multilingual Aphasia Examination*. Iowa City: University of Iowa Press.

BRUTKOWSKI, S., M. MISHKIN, and H. E. ROSVOLD, 1963. Positive and inhibitory motor conditioned reflexes in monkeys after ablation of orbital or dorso-lateral surface of the frontal cortex. In *Central and Peripheral Mechanisms of Motor Function*, E. Gutmann and P. Hnik, eds. Prague: Czechoslovak Academy of Sciences, pp. 133–141.

CERMAK, L. S., 1976. The encoding capacity of a patient with amnesia due to encephalitis. *Neuropsychologia* 14:311–326.

DELIS, D. C., L. R. SQUIRE, A. BIHRLE, and P. MASSMAN, 1992. Componential analysis of problem-solving ability: Performance of patients with frontal lobe damage and amnesic patients on a new sorting test. *Neuropsychologia* 30:683–697.

DELLAROCCHETTA, A. I., 1986. Classification and recall of pictures after unilateral frontal or temporal lobectomy. *Cortex* 22:189–211.

FREEDMAN, M., and L. S. CERMAK, 1986. Semantic encoding deficits in frontal lobe disease and amnesia. *Brain Cogn.* 5:108–114.

FRITH, C. D., K. J. FRISTON, P. F. LIDDLE, and R. S. J. FRACKOWIAK, 1991. A PET study of word finding. *Neuropsychologia* 29:1137–1148.

FUSTER, J. M. 1989. *The Prefrontal Cortex: Anatomy, Physiology, and Neuropsychology of the Frontal Lobe*, ed. 2. New York: Raven.

GERSHBERG, F. B., and A. P. SHIMAMURA, 1991. The role of the frontal lobes in free recall: Interference, organization, and serial position effects. *Society for Neuroscience Abstracts* 17:136.

Goldman-Rakic, P. S., 1987. Circuitry of primate prefrontal cortex and regulation of behavior by representational memory. In *Handbook of Physiology: The Nervous System*, Vol. 5, F. Plum, ed. Bethesda, Md.: American Physiological Society, pp. 373–417.

Goldstein, K., 1939. Clinical and theoretical aspects of lesions of the frontal lobes. *Arch. Neurol. Psychiatry* 41:865–867.

Halstead, W. C., 1947. *Brain and Intelligence*. Chicago: University of Chicago Press.

Haug, H., U. Barmwater, R. Eggers, D. Fischer, S. Kuhl, and N. L. Sass, 1983. Anatomical changes in aging brain: Morphometric analysis of the human prosencephalon. In *Brain Aging: Neuropathology and Neuropharmocology*, J. Cervos-Navarro and H. I. Sarkander, eds. New York: Raven, pp. 1–12.

Hebb, D. O., 1945. Man's frontal lobes: A critical review. *Arch. Neurol. Psychiatry* 54:10–24.

Hirst, W., 1982. The amnesic syndrome: Descriptions and explanations. *Psychol. Bull.* 91:435–460.

Janowsky, J. S., A. P. Shimamura, M. Kritchevsky, and L. R. Squire, 1989. Cognitive impairment following frontal lobe damage and its relevance to human amnesia. *Behav. Neurosci.* 103:548–560.

Janowsky, J. S., A. P. Shimamura, and L. R. Squire, 1989a Memory and metamemory: Comparisons between patients with frontal lobe lesions and amnesic patients. *Psychobiology* 17:3–11.

Janowsky, J. S., A. P. Shimamura, and L. R. Squire, 1989b Source memory impairment in patients with frontal lobe lesions. *Neuropsychologia* 27:1043–1056.

Jetter, W., U. Poser, R. B. Freeman, and J. H. Markowitsch, 1986. A verbal long term memory deficit in frontal lobe damaged patients. *Cortex* 22:229–242.

Jones-Gotman, M., and B. Milner, 1977. Design fluency: The invention of nonsense drawings after focal cortical lesions. *Neuropsychologia* 15:653–674.

Knight, R. T., D. Scabini, and D. L. Woods, 1989. Prefrontal gating of auditory transmission in humans. *Brain Res.* 504:338–342.

Lhermitte, F., 1986. Human autonomy and the frontal lobes. Part 2: Patient behavior in complex and social situations: The "environmental dependency syndrome." *Ann. Neurol.* 19:335–343.

Luria, A. R., 1973. *The Working Brain*. New York: Basic Books.

Mattis, S., 1976. Dementia rating scale. In *Geriatric Psychiatry 1*, R. Bellack and B. Karasu, eds. New York: Grune and Stratton, pp. 77–121.

Mayes, A. R., P. R. Meudell, and A. Pickering, 1985. Is organic amnesia caused by a selective deficit in remembering contextual information? *Cortex* 21:167–202.

McAndrews, M. P., and B. Milner, 1991. The frontal cortex and memory for temporal order. *Neuropsychologia* 29: 859–860.

McIntyre, J. S., and F. I. M. Craik, 1987. Age differences in memory for item and source information. *Can. J. Psychol.* 42:175–192.

Metcalfe, J. and Shimamura, A. P., eds., 1994. *Metacognition: Knowing about Knowing*. Cambridge, Mass.: MIT Press.

Milner, B., 1971. Interhemispheric differences in the localization of psychological processes in man. *Br. Med. Bull.* 127:272–277.

Milner, B., 1982. Some cognitive effects of frontal lobe lesions in man. In *The Neuropsychology of cognitive function*, D. E. Broadbent and L. Weiskrantz, eds. London: The Royal Society, pp. 211–226.

Milner, B., P. Corsi, and G. Leonard, 1991. Frontal-lobe contribution to recency judgements. *Neuropsychology* 29: 601–618.

Milner, B., M. Petrides, and M. L. Smith, 1985. Frontal lobes and the temporal organization of memory. *Hum. Neurobiol.* 4:137–142.

Moscovitch, M., 1982. Multiple dissociations of function in amnesia. In *Human Memory and Amnesia*, L. Cermak, ed. Hillsdale, N.J.: Erlbaum, pp. 337–370.

Perret, E., 1974. The left frontal lobe of man and the suppression of habitual responses in verbal categorical behavior. *Neuropsychologia* 12:323–330.

Postman, L., 1961. The present status of interference theory. In *Verbal Learning and Verbal Behavior*, C. N. Cofer, ed. New York: McGraw-Hill.

Postman, L., and B. J. Underwood, 1973. Critical issues in interference theory. *Memory and Cognition* 1:19–40.

Schacter, D. L., J. Harbluk, and D. McLachlin, 1984. Retrieval without recollection: An experimental analysis of source amnesia. *J. Ver. Learn Ver. Behav.* 23:593–611.

Shallice, T., and M. E. Evans, 1978. The involvement of the frontal lobes in cognitive estimation. *Cortex* 14:294–303.

Shimamura, A. P., 1986. Priming in amnesia: Evidence for a dissociable memory function. *Q. J. Exp. Psychol.* 38:619–644.

Shimamura, A. P., 1989. Disorders of memory: The cognitive science perspective. In *Handbook of Neuropsychology*, F. Boller and Jordan Grafman, eds. Amsterdam: Elsevier, pp. 35–73.

Shimamura, A. P., J. S. Janowsky, and L. R. Squire, 1990. Memory for the temporal order of events in patients with frontal lobe lesions and amnesic patients. *Neuropsychologia* 28:803–813.

Shimamura, A. P., J. S. Janowsky, and L. R. Squire, 1991. What is the role of frontal lobe damage in amnesic disorders? In *Frontal Lobe Function and Dysfunction*. H. S. Levin, H. M. Eisenberg, and A. L. Benton, eds. New York: Oxford University Press, pp. 173–195.

Shimamura, A. P., P. J. Jurica, J. A. Mangels, F. B. Gershberg, and R. T. Knight, submitted. Memory interference effects in patients with frontal lobe lesions: Evidence from tests of paired-associate learning.

Shimamura, A. P., and L. R. Squire, 1986. Memory and

metamemory: A study of the feeling-of-knowing phenomenon in amnesic patients. *J. Exp. Psychol. [Learn. Mem. Cogn.]* 12:452–460.
SMITH, M. L., and B. MILNER, 1984. Differential effects of frontal-lobe lesions on cognitive estimation and spatial memory. *Neuropsychologia* 22:697–705.
SQUIRE, L. R., 1982. Comparisons between forms of amnesia: Some deficits are unique to Korsakoff's syndrome. *J. Exp. Psychol. [Learn Mem Cogn.]* 8:560–571.
TULVING, E., 1972. Episodic and semantic memory. In *Organization of Memory*, E. Tulving, and W. Donaldson, eds. New York: Academic Press, pp. 381–403.
TULVING, E. 1983. *Elements of Episodic Memory*. Oxford: Clarendon.
WICKENS, D. D., 1970. Encoding strategies of words: An empirical approach to meaning. *Psychol. Rev.* 22:1–15.
YAMAGUCHI, S., and R. T. KNIGHT, 1990. Gating of somatosensory inputs by human prefrontal cortex. *Brain Res.* 521: 281–288.

52 Implicit Memory: A New Frontier for Cognitive Neuroscience

DANIEL L. SCHACTER

ABSTRACT Implicit memory refers to nonconscious effects of previous experiences on performance of memory tests that do not require explicit recollection. A growing number of studies during the past decade have shown that implicit and explicit forms of memory can be dissociated experimentally. This chapter provides an overview of implicit memory research from several perspectives that are central to cognitive neuroscience: cognitive studies of normal subjects, neuropsychological investigations of memory-impaired populations, and electrophysiological and neuroimaging studies. Taken together, evidence from these different perspectives supports the hypothesis that implicit and explicit forms of memory depend on different memory systems that are associated with distinct regions of the brain.

The topic of this chapter—implicit memory—is a relative newcomer to the landscape of memory research. In fact, the term *implicit memory* was first introduced to the field less than a decade ago (Graf and Schacter, 1985; Schacter, 1987). As stated in Schacter (1987, 501), "Implicit memory is revealed when previous experiences facilitate performance on a task that does not require conscious or intentional recollection of those experiences." By contrast, explicit memory "is revealed when performance on a task requires conscious recollection of previous experiences." The relatively recent emergence of the terms *implicit memory* and *explicit memory* is largely attributable to the fact that implicit memory constitutes a novel, if not entirely unprecedented, focus for memory research. Most psychological studies of memory have used tasks that involve intentional recollection of previously studied materials, and theoretical accounts have typically focused on data concerning explicit remembering. However, beginning in the 1960s and 1970s, and especially in the early 1980s, evidence began to accumulate indicating that effects of prior experiences could be expressed without, and dissociated from, intentional or conscious recollection. The terms *implicit* and *explicit* memory were put forward in an attempt to capture and describe essential features of the observed dissociations. Related distinctions include declarative versus nondeclarative memory (Squire, 1992), direct versus indirect memory (Johnson and Hasher, 1987), and memory with awareness versus memory without awareness (Jacoby and Witherspoon, 1982).

Despite the recent vintage of the concept, studies of implicit memory have had a profound impact on contemporary research and theorizing. As early as 1988, Richardson-Klavehn and Bjork were able to assert that research on implicit memory constitutes "a revolution in the way that we measure and interpret the influence of past events on current experience and behavior" (1988, 467–477). Since that time interest in the issue has continued and intensified, as studies concerning implicit memory have appeared with astonishing frequency in cognitive, neuropsychological, and even psychiatric journals.

The main purpose of this chapter is to provide an overview of implicit memory research with respect to the concerns of cognitive neuroscience. No attempt is made to provide exhaustive coverage of the area (for recent reviews, see Roediger and McDermott, 1993; Schacter, Chiu, and Ochsner, 1993). Rather, the goal is to acquaint the reader with the major methodological and theoretical issues in contemporary research, and to summarize experimental studies that have examined implicit memory at both the cognitive and neuropsychological levels of analysis. To accomplish this objective, the chapter is divided into four main sections. The first summarizes the historical back-

DANIEL L. SCHACTER Department of Psychology, Harvard University, Cambridge, Mass.

ground of contemporary research, and the second considers some basic terminological and methodological issues. The third and major section reviews cognitive and neuropsychological evidence that illuminates the nature and characteristics of implicit memory. The fourth section summarizes contemporary theoretical approaches to relevant phenomena.

Historical background

Although sustained interest in implicit memory has arisen only recently, a variety of clinical, anecdotal, and experimental observations concerning pertinent phenomena have been made during the past several centuries. I have offered a relatively systematic treatment of historical developments elsewhere (Schacter, 1987), and will here only summarize briefly the immediate precursors to current research.

Contemporary concern with implicit and explicit memory can be traced to two unrelated lines of research that developed during the 1960s and 1970s. First, neuropsychological investigations revealed that densely amnesic patients could exhibit relatively intact learning abilities on certain kinds of memory tasks, such as motor skill learning (e.g., Milner, Corkin, and Teuber, 1968), and fragment-cued recall (e.g., Warrington and Weiskrantz, 1974). Second, cognitive psychologists interested in word recognition and lexical access initiated investigations of the phenomenon known as repetition or direct priming, that is, facilitation in the processing or identification of a stimulus as a consequence of prior exposure to it on tests that do not require explicit remembering. For example, several investigators found that performance on a lexical decision test, where subjects judge whether letter strings constitute real words or nonwords, is facilitated significantly by prior exposure to a target word (for historical review, see Schacter, 1987).

By 1980, then, two independent lines of research indicated that effects of past experience could be demonstrated in the absence of, or without the requirement for, conscious recollection. But the possible links between them were not apparent, or at least were not discussed in scientific publications. The situation changed radically during the next few years. Cognitive studies of normal subjects revealed that priming effects on such tasks as word identification and word completion could be dissociated from recall and recognition (Jacoby and Dallas, 1981; Graf, Mandler, and Haden, 1982; Tulving, Schacter, and Stark, 1982; Graf and Mandler, 1984), and neuropsychological studies of amnesic patients with severe explicit memory deficits demonstrated entirely normal levels of skill learning (Cohen and Squire, 1980; Moscovitch, 1982) and priming (Jacoby and Witherspoon, 1982; Graf, Squire, and Mandler, 1984; Shimamura and Squire, 1984; Schacter, 1985). This convergence of cognitive and neuropsychological evidence provided a basis for the distinction between implicit and explicit memory (Graf and Schacter, 1985; Schacter, 1987).

These developments opened the floodgates for a virtual tidal wave of research. The range of phenomena subsumed under the general label of implicit memory has expanded, theoretical discussion is intense, and the rapid pace of investigation shows no signs of slowing down. Looking back to 1980 from the vantage point of the present, it is no exaggeration to say that we have witnessed the birth and development of a new subfield of memory research. While the vast scope of implicit memory research is in some sense exhilarating, it also means that one cannot hope to cover all of it in a relatively brief chapter. Accordingly, I will focus primarily on studies that have examined phenomena of direct priming, both because more is known about priming than any other implicit memory phenomenon, and because it has played a central role in theoretical discussion and debate. However, an exclusive focus on priming can lead to an overly narrow conception of implicit memory, so I will consider priming in relation to other forms of implicit memory where appropriate.

Methodological issues

The terms *implicit* and *explicit* memory were put forward in an attempt to capture salient features of the phenomena described in the preceding section, without implying commitment to a particular theoretical view of the mechanisms underlying the two forms of memory. Thus, Schacter (1987, 501) noted specifically that "the concepts of implicit and explicit memory neither refer to, nor imply the existence of, two independent or separate memory systems." Rather, these concepts "are primarily concerned with a person's psychological experience at the time of retrieval." The terms *implicit memory test* and *explicit memory test* have been used to characterize tasks on which memory performance can be characterized as either implicit (i.e., unintentional, nonconscious) or explicit (i.e., intentional, conscious).

One difficulty that arises when attempting to operationalize and experimentally examine implicit memory is that tasks that are characterized as implicit memory tests can be influenced by explicit memory. Thus, nominally implicit tests are not always functional measures of implicit memory. For example, when a severely amnesic patient exhibits a priming effect on a stem completion test, we can be relatively confident that the observed effect reflects the exclusive influence of implicit memory. However, when a college student or any other subject with intact explicit memory exhibits a priming effect, it is always possible that he or she has "caught on" to the fact that test stems can be completed with study list items, and has converted the nominally implicit test into a functionally explicit one (Bowers and Schacter, 1990).

This issue is fundamental to all research on implicit versus explicit memory, and procedures have been developed for confronting the problem. Consider, for example, the *retrieval intentionality criterion* suggested by Schacter, Bowers, and Booker (1989). The criterion consists of two key components: (1) The physical cues on implicit and explicit tests are held constant and only retrieval instructions (implicit or explicit) vary; and (2) an experimental manipulation is identified that affects performance on the two tests differently. The basic argument is that when these conditions are met, we can rule out the possibility that implicit test performance is contaminated by intentional retrieval strategies. The logic here is straightforward: If subjects are engaging in explicit retrieval on a nominally implicit test, then their performance on implicit and explicit tests that use identical cues should be affected similarly by a given experimental manipulation; thus, dissociations produced under these conditions indicate that the implicit test is not contaminated. And, indeed, a number of studies have produced dissociations that satisfy the retrieval intentionality criterion (e.g., Graf and Mandler, 1984; Hayman and Tulving, 1989; Roediger, Weldon, Stadler, and Riegler, 1992; Schacter and Church, 1992; for a different approach to the "contamination" problem, see Jacoby, 1991).

Characteristics of implicit memory: Cognitive and neuropsychological research

Cognitive Studies When contemplating the recent surge of research that constitutes the basis of this chapter, a question that naturally arises concerns the reasons for the intensive scrutiny: Why is implicit memory worth knowing about? One compelling answer to this question is that many situations exist in which implicit memory behaves quite differently from, and independently of, explicit memory. Scientists are naturally curious about surprising phenomena that violate their expectations, and implicit memory is surely one of them. Research with normal subjects has produced two main kinds of evidence for dissociation between implicit and explicit memory: *stochastic independence* and *functional independence*.

Stochastic independence Stochastic independence refers to a lack of correlation between two measures of memory at the level of the individual item. To illustrate the concept, consider an early study of priming by Tulving, Schacter, and Stark (1982). Subjects studied a long list of low frequency words (e.g., ASSASSIN), and were later given two successive memory tests: an explicit test of recognition memory in which they indicated via "yes" or "no" responses whether they recollected that a test item had appeared previously on the study list; and an implicit test of fragment completion in which they attempted to complete graphemic fragments of words (e.g., A--A--IN). Priming was observed on the fragment completion test: there was a significantly higher completion rate for fragments that represented previously studied words than for fragments that represented nonstudied words (e.g., -E-S--X for BEESWAX). More importantly, however, a contingency analysis of recognition and fragment completion performance revealed that the probability of producing a studied item on the fragment completion test was uncorrelated with—independent of—the probability of recognizing the same item. This finding of stochastic independence was striking and unexpected, because previous research had indicated that performance on explicit memory tests, such as cued recall, is correlated with, or dependent on, recognition memory (see Tulving, 1985).

Stochastic independence between priming and recognition memory has since been observed in a variety of experiments using different kinds of implicit memory tests (see, e.g., Jacoby and Witherspoon, 1982; Hayman and Tulving, 1989; Witherspoon and Moscovitch, 1989; Schacter, Cooper, and Delaney, 1990), and it has been suggested that such evidence is of great theoretical import (Tulving, 1985). But some have contended that findings of stochastic independence are artifacts of either the experimental procedures or the contingency analyses that are used to assess independence (cf. Shimamura, 1985; Hintzman and Hartry, 1990; Ostergaard, 1992). Many of these criticisms, however, have been answered convincingly (see, for example, Hayman and Tulving, 1989; Schacter, Cooper, and Delaney, 1990; Tulving and Flexser, 1992).

Functional independence Functional independence between implicit and explicit memory occurs when experimental manipulations affect performance on implicit and explicit tasks in different and even opposite ways. A key source of evidence for functional independence is provided by experiments that manipulate the conditions under which subjects study or encode target items. For instance, a seminal finding from the early 1980s involved experiments that varied the level or depth of encoding during a study task. Research in the levels of processing tradition (e.g., Craik and Tulving, 1975) had already established that explicit recall and recognition performance are much more accurate following "deep" encoding tasks that require semantic analysis of target words (e.g., judging the category to which a word belongs) than following "shallow" encoding tasks that only require analysis of an item's surface features (e.g., judging whether a word has more vowels or consonants). In striking contrast, several studies revealed that the magnitude of priming effects on the word identification task (Jacoby and Dallas, 1981) and stem completion task (Graf, Mandler, and Haden, 1982; Graf and Mandler, 1984) are not significantly influenced by levels of processing manipulations. More recent studies have confirmed and extended this general pattern of results in a variety of experimental paradigms (e.g., Bowers and Schacter, 1990; Graf and Ryan, 1990; Roediger et al., 1992).

While the foregoing studies used familiar words as target materials, and visual presentation and testing procedures, recent work indicates that the critical dissociation observed in these experiments can be produced with nonverbal figures (e.g., Schacter, Cooper, and Delaney, 1990) and in the auditory modality (Schacter and Church, 1992). Various other ways of manipulating encoding processes have also produced dissociative effects on implicit and explicit tests (e.g., Jacoby, 1983; Roediger and Challis, 1992).

Additional evidence is provided by studies that have altered the surface features of target items between study and test. For example, it is well established that priming on identification and completion tests is reduced and sometimes eliminated by study-to-test changes in modality of presentation (e.g., Jacoby and Dallas, 1981; Roediger and Blaxton, 1987), even though modality change typically has less impact on explicit memory. In a compelling demonstration, Weldon and Roediger (1987) showed that priming on the word fragment completion test could be eliminated by presenting a picture of a word's referent, rather than the word itself, at the time of study. By contrast, explicit memory was considerably higher following study of the picture than of the word.

Other experiments have investigated the extent to which priming is sensitive to within-modality changes of perceptual information between study and test. For example, a number of studies have assessed whether visual word priming is affected by study-to-test changes in letter case (i.e., upper or lower), typeface, or other perceptual features. The studies have yielded a mixed and complex picture, with some experiments providing evidence of perceptually specific priming (e.g., Jacoby and Hayman, 1987; Roediger and Blaxton, 1987; Hayman and Tulving, 1989) and others revealing no such effects (e.g., Clarke and Morton, 1983; Carr, Brown, and Charalambolous, 1989). Some attempts to resolve the discrepancies have been made (cf. Graf and Ryan, 1990; Marsolek, Kosslyn, and Squire, 1992), but simple answers are not yet available. Rather more consistent evidence of within-modality perceptual specificity has been reported in studies of auditory word priming, where study-to-test changes in speaker's voice can affect priming significantly (Schacter and Church, 1992; Church and Schacter, in press). Finally, several studies of visual object priming have revealed significant effects of changing an object's picture plane orientation between study and test (e.g., Biederman and Cooper, 1991; Cooper, Schacter, and Moore, 1991), although study-to-test changes in object size appear to have no effect on priming (Biederman and Cooper, 1992; Cooper, Schacter, Ballesteros, and Moore, 1992).

The view of priming that emerges from cognitive research, then, depicts a form of memory that is little affected by semantic or conceptual factors, strongly dependent on modality-level information, and sometimes dependent on highly specific, within-modality perceptual information. Note, however, that this characterization applies to priming effects that have been observed on so-called *data-driven* implicit tests, in which subjects' attention is focused primarily on the physical properties of test cues (e.g., Jacoby, 1983; Roediger and Blaxton, 1987). Priming has also been examined on *conceptually driven* implicit tests, which focus subjects' attention on semantic properties of test cues (cf. Blaxton, 1989; Hamman, 1990). Still other implicit tests appear to involve a mixture of data-driven and conceptually driven processes, such as the cued stem completion task developed by Graf and Schacter (1985) to study priming of newly acquired associations. Here, priming depends on some semantic study elaboration (Schacter and Graf, 1986b) but also exhibits modality specificity (Schacter and Graf, 1989).

NEUROPSYCHOLOGICAL STUDIES While cognitive research provides insights into the psychological and behavioral properties of implicit memory, it does not illuminate the neural bases of the critical phenomena. We now consider several kinds of evidence that provide pertinent information. The bulk of this section is devoted to considering studies of patients with memory disorders. However, we will also touch briefly on evidence from electrophysiological, neuroimaging, and pharmacological studies.

Memory-impaired patients The amnesic syndrome has played an important role in the development of implicit memory research, as noted in the introduction. Because it has been reasonably well established that human amnesia is produced by lesions to limbic and diencephalic structures (cf. Weiskrantz, 1985; Squire, 1992), findings of spared skill learning and priming are frequently taken as evidence that these structures are not necessary for, or involved in, these expressions of implicit memory.

The claim that priming can be fully intact in amnesia was not established firmly until the 1980s. In the earlier studies of Warrington and Weiskrantz (1974) that used word fragments as test cues, there was some ambiguity concerning whether subjects were given implicit or explicit test instructions. Moreover, in these studies amnesic patients sometimes exhibited normal performance and sometimes exhibited impaired performance. These issues were clarified by data indicating that amnesic patients show normal performance on fragmented word tests and similar tasks when given implicit memory instructions, and show impaired performance when given explicit memory instructions (cf. Graf, Squire, and Mandler, 1984; Shimamura and Squire, 1984; Cermak et al., 1985; Schacter, 1985).

During the past decade, research has focused on exploring the boundary conditions of preserved priming in amnesic patients. One issue that has assumed center stage during the past decade concerns whether amnesic patients, like normal subjects, exhibit normal priming for novel or unfamiliar materials that do not have preexisting representations in memory. Early evidence indicated that priming of nonwords (e.g., *numby*) is either absent or impaired in amnesic patients (Diamond and Rozin, 1984; Cermak et al., 1985), but methodological and conceptual considerations limit the force of these conclusions (for discussion, see Bowers and Schacter, 1993). Recent experiments have delineated conditions under which amnesic patients can exhibit intact priming for nonwords (e.g., Haist, Musen, and Squire, 1991; but see also Cermak et al., 1991).

A number of investigators have examined whether amnesic patients show priming for newly acquired associations between unrelated words, using variants of the paradigm introduced by Graf and Schacter (1985). The general outcome of these studies has pointed to impaired or absent priming of novel associates (Schacter and Graf, 1986b; Cermak, Bleich, and Blackford, 1988; Shimamura and Squire, 1989), although positive results have been reported in patients with mild memory impairments (e.g., Schacter and Graf, 1986b). In contrast to the inconsistent pattern of results observed with nonwords and unrelated paired associates, more convincing evidence for priming of novel information has been reported in studies using nonverbal materials, including novel objects (Schacter et al., 1991; Schacter, Cooper, and Treadwell, 1993) and dot patterns (Gabrieli et al., 1990; Musen and Squire, 1992). These data indicate that amnesic patients can form some sort of novel representation for unfamiliar objects and patterns.

A related issue concerns whether amnesic patients exhibit implicit memory for specific perceptual features of target materials, as has been observed in some studies of normal subjects that were noted earlier. Kinoshita and Wayland (1993) reported that on a visual fragment completion test, Korsakoff amnesics failed to show more priming when surface features of target words (handwritten vs. typed) were the same at study and test than when they differed. In an as-yet-unpublished study of auditory priming on a low-pass filter identification test, Barbara Church and I found that normal control subjects, but not amnesic patients, showed more priming in a same-voice condition than in a different-voice condition; amnesic patients do, however, show normal auditory priming when voice change effects are not involved (Schacter, Church, and Treadwell, 1994). Although these observations are preliminary, they have potentially important theoretical implications because they suggest that not all aspects of priming are fully observed in amnesic patients (for discussion, see Schacter, in press).

It was noted earlier that an exclusive focus on priming could lead to an overly narrow conception of implicit memory, and this is certainly true when considering amnesic patients. For example, while the evidence for priming of newly acquired associations in amnesic patients is weak, recent work indicates that amnesic patients can show robust implicit learning of new associations under conditions in which learning develops gradually across multiple trials (Musen and Squire, 1993). Similarly, research concerning the learning of skills and procedures, where implicit knowledge is acquired gradually across multiple trials, provides evidence that amnesics can show robust and even normal learning of novel spatiotemporal associations (Nissen and Bullemer, 1987), grammatical rules (Knowlton, Ramus, and Squire, 1992), and procedures for performing computer-related tasks (e.g., Glisky and Schacter, 1989). These kinds of observations suggest that priming of novel associations from a single study exposure depends on different mechanisms than does gradual learning of skills and contingencies.

Although the neuropsychological investigation of implicit memory has been dominated by experiments with amnesic patients, during the past several years the scope of investigation has broadened, and studies of various other patient populations have begun to make important empirical and theoretical contributions. Research with patients suffering from different kinds of dementia has proven particularly revealing. For example, studies of patients with Alzheimer's disease, who are typically characterized by extensive damage to cortical association areas as well as limbic structures, have consistently revealed impairment of priming on the word stem completion task together with spared procedural learning on motor skill tasks (e.g., Butters, Heindel, and Salmon, 1990). By contrast, patients with Huntington's disease, who are typically characterized by damage to basal ganglia, exhibit normal stem completion priming together with impaired acquisition of motor skills (Butters, Heindel, and Salmon, 1990). The double dissociation between priming and skill learning indicates clearly that different mechanisms underlie the two forms of implicit memory. Another example is provided by recent evidence indicating normal auditory priming in a patient with cortical (left hemisphere) damage and a severe auditory comprehension deficit (Schacter et al., 1993). This finding supports the idea discussed earlier that perceptual priming is a presemantic phenomenon that does not depend on the integrity or involvement of conceptual processes.

Electrophysiological, neuroimaging, and pharmacological studies Whereas studies of patient populations are potentially valuable sources of information about the brain processes and systems that subserve implicit and explicit memory, it also would be desirable to obtain relevant evidence from research with intact brains. Although relatively little work has been carried out along these lines, a few beginning steps have been taken.

A number of studies have examined priming effects by recording event-related potentials (ERPs), or electrophysiological changes in the brain that are linked to specific stimulus events, measured at the scalp, and quantified through signal averaging techniques. To take just one example, Paller (1990) measured ERPs to target words during an encoding task in which subjects were instructed to try to remember some words and to forget others (i.e., directed forgetting). The directed forgetting manipulation influenced explicit recall but not stem completion priming. More importantly, Paller found that ERP responses during encoding differed reliably for words that subsequently were or were not recalled, whereas these same ERP responses were unrelated to whether or not an item exhibited priming on the stem completion test. These findings thus provide converging electrophysiological evidence for the dissociative effect of different encoding processes on priming and explicit memory (see also Rugg and Doyle, in press).

Neuroimaging techniques such as positron emission tomography (PET) provide a promising new tool for investigating the neural bases of implicit memory. Little evidence is yet available, but one study by Squire and colleagues (1992) indicates that primed visual stem completion performance is associated with decreased activity in right extrastriate occipital cortex relative to unprimed completion performance. In addition, however, there were significant changes in right hippocampus. One difficulty in interpreting these results is that the priming data were likely contaminated by explicit memory: The study lists were short, a semantic encoding task was used, there were multiple study–test trials, and the completion rate for primed items was extremely high. In a more recent PET study that eliminated explicit contamination, we found that visual priming on the stem completion test was associated with decreased activity in right extrastriate occipital cortex, but failed to observe any evidence of hippocampal activation (Schacter, Albert, Alpert, Rafferty, and Rauch, unpublished data, 1994).

A few studies have revealed that various pharmacological agents differentially affect implicit and explicit memory (Nissen, Knopman, and Schacter, 1987; Danion et al., 1989), and some evidence from anesthetized patients indicates implicit memory for information presented during anesthesia (e.g., Kihlstrom et al., 1990). Although the basis of drug effects on implicit memory is not well understood, further psychopharmacological investigations could elucidate the neurochemical basis of implicit memory.

Theoretical issues: Lessons of implicit memory research

I noted earlier that when we look back to 1980 from the perspective of the present, the explosion of studies concerning implicit memory can be seen to constitute a new subfield of memory research. What lessons have been learned from this work? What do we know about memory now that we did not know then, and that is worth knowing?

At a rather general level, many researchers would agree that a principal lesson centers on the idea that memory is not a unitary or monolithic entity: The effects of past events on current experience and performance can be expressed not only via explicit remembering, but also by subtle changes in our ability to identify, act on, and make judgments about words, objects, and other kinds of stimuli—changes that are frequently independent of the ability to engage in conscious recollection of a prior experience. At a more specific level, a number of researchers have argued that dissociations between implicit and explicit forms of memory are mediated by, and reflect the existence of, distinct and dissociable underlying memory systems (e.g., Cohen and Squire, 1980; Tulving, 1985; Hayman and Tulving, 1989; Gabrieli et al., 1990; Tulving and Schacter, 1990; Schacter, 1990, 1992; Squire, 1992).

What kinds of systems are involved in implicit memory? Because implicit memory is a rather general descriptive term that covers a number of distinct phenomena, formulating an answer to this question requires that one first specify the particular kind of implicit memory that one wishes to explain. Consider, for example, the priming effects on completion, identification, and similar data-driven implicit memory tests that have been studied so extensively. We (Tulving and Schacter, 1990; Schacter, 1990, 1992, in press) have argued that such effects depend to a large extent on a perceptual representation system (PRS) that is in turn composed of several domain-specific subsystems (e.g., visual word form, auditory word form, structural description). The various PRS subsystems are based in posterior regions of cortex and operate at a modality-specific, presemantic level—that is, they represent information about the form and structure, but not the meaning and associative properties, of words and objects (for details and background, see Schacter, 1990, 1992; cf. Moscovitch, 1992). This basic idea accommodates many of the known facts about priming on data-driven tests—that it does not depend on semantic or conceptual processing, does depend on modality-specific perceptual information, and is typically preserved in amnesic patients. The PRS is not, however, involved in all forms of implicit memory. To take just one example, motor skill learning likely depends on a habit or procedural learning system that appears to depend critically on the integrity of corticostriatal circuits (see, e.g., Mishkin, Malamut, and Bachevalier, 1984; Butters, Heindel, and Salmon, 1990).

The hypothesis that dissociations between implicit and explicit memory reflect the operation of multiple memory systems is not universally accepted. Some students, for example, have preferred to retain the notion of a single memory system, and have attempted to account for implicit–explicit dissociations by appealing to different processes operating within the system, using theoretical ideas that have been invoked previously to account for dissociations among explicit memory tests (see, e.g., Jacoby, 1983; Roediger and Blaxton, 1987). However, recent discussions indicate that so-called processing accounts are often complementary to, rather than competitive with, multiple systems views (Hayman and Tulving, 1989; Roediger, 1990; Schacter, 1992). The critical task for future work will be to develop more specific and neurobiologically plausible accounts of the processes and systems that are responsible for implicit and explicit forms of memory.

ACKNOWLEDGMENTS Preparation of this chapter was supported by grant PO1 NS27950-01A1 from the National Institute of Neurological Disorders and Stroke and by grant RO1 MH45398-01A3 from the National Institute of Mental Health. I thank Wilma Koutstaal for comments on an earlier draft of the chapter, and thank Dana Osowiecki for help with preparation of the manuscript.

REFERENCES

BIEDERMAN, I., and E. E. COOPER, 1991. Evidence for complete translational and reflectional invariance in visual object priming. *Perception* 20:585–593.

BIEDERMAN, I., and E. E. COOPER, 1992. Size invariance in visual object priming. *J. Exp. Psychol. [Hum. Percept.]* 18: 121–133.

BLAXTON, T. A., 1989. Investigating dissociations among memory measures: Support for a transfer appropriate processing framework. *J. Exp. Psychol. [Learn. Mem. Cogn.]* 15:657–668.

BOWERS, J. S., and D. L. SCHACTER, 1990. Implicit memory and test awareness. *J. Exp. Psychol. [Learn. Mem. Cogn.]* 16:404–416.

BOWERS, J. S., and D. L. SCHACTER, 1993. Priming of novel information in amnesia: Issues and data. In *Implicit Memory: New Directions in Cognition, Neuropsychology, and Development*, P. Graf and M. E. J. Masson, eds. New York: Academic Press, pp. 303–326.

BUTTERS, N., W. C. HEINDEL, and D. P. SALMON, 1990. Dissociation of implicit memory in dementia: Neurological implications. *Bull. Psychonomic Soc.* 28:359–366.

CARR, T. H., J. S. BROWN, and A. CHARALAMBOUS, 1989. Repetition and reading: Perceptual encoding mechanisms are very abstract but not very interactive. *J. Exp. Psychol. [Learn. Mem. Cogn.]* 15:763–778.

CERMAK, L. S., R. P. BLEICH, and M. BLACKFORD, 1988. Deficits in the implicit retention of new associations by alcoholic Korsakoff patients. *Brain. Cogn.* 7:145–156.

CERMAK, L. S., N. TALBOT, K. CHANDLER, and L. R. WOLBARST, 1985. The perceptual priming phenomenon in amnesia. *Neuropsychologia* 23:615–622.

CERMAK, L. S., M. VERFAELLIE, W. MILBERG, L. LETOURNEAU, and S. BLACKFORD, 1991. A further analysis of perceptual identification priming in alcoholic Korsakoff patients. *Neuropsychologia* 29:725–736.

CHURCH, B. A., and D. L. SCHACTER, in press. Perceptual specificity of auditory priming: Implicit memory for voice intonation and fundamental frequency. *J. Exp. Psychol. [Learn. Mem. Cogn.]*

CLARKE, R., and J. MORTON, 1983. Cross modality facilitation in tachistoscopic word recognition. *Q. J. Exp. Psychol.* 35A:79–96.

COHEN, N. J., and L. R. SQUIRE, 1980. Preserved learning and retention of pattern analyzing skill in amnesics: Dissociation of knowing how and knowing that. *Science* 210: 207–210.

COOPER, L. A., D. L. SCHACTER, S. BALLESTEROS, and C. MOORE, 1992. Priming and recognition of transformed three-dimensional objects: Effects of size and reflection. *J. Exp. Psychol. [Learn. Mem. Cogn.]* 18:43–57.

COOPER, L. A., D. L. SCHACTER, and C. MOORE, 1991. Orientation affects both structural and episodic representations of 3-D objects. *Paper presented to the Annual Meeting of the Psychonomic Society, San Francisco.*

CRAIK, F. I. M., and E. TULVING, 1975. Depth of processing and the retention of words in episodic memory. *J. Exp. Psychol. [Gen.]* 104:268–294.

DANION, J. M., M. A. ZIMMERMAN, D. WILLARD-SCHROEDER, D. GRANGE, and L. SINGER, 1989. Diazepam induces a dissociation between explicit and implicit memory. *Psychopharmacology* 99:238–243.

DIAMOND, R., and P. ROZIN, 1984. Activation of existing memories in anterograde amnesia. *J. Abnorm. Psychol.* 93: 98–105.

GABRIELI, J. D. E., W. MILBERG, M. M. KEANE, and S. CORKIN, 1990. Intact priming of patterns despite impaired memory. *Neuropsychologia* 28:417–428.

GLISKY, E. L., and D. L. SCHACTER, 1989. Extending the limits of complex learning in organic amnesia: Computer training in a vocational domain. *Neuropsychologia* 27:107–120.

GRAF, P., and G. MANDLER, 1984. Activation makes words more accessible, but not necessarily more retrievable. *J. Verb. Learn. Verb. Be.* 23:553–568.

GRAF, P., G. MANDLER, and P. HADEN, 1982. Simulating amnesic symptoms in normal subjects. *Science* 218:1243–1244.

GRAF, P., and L. RYAN, 1990. Transfer-appropriate processing for implicit and explicit memory. *J. Exp. Psychol. [Learn. Mem. Cogn.]* 16:978–992.

GRAF, P., and D. L. SCHACTER, 1985. Implicit and explicit memory for new associations in normal subjects and amnesic patients. *J. Exp. Psychol. [Learn. Mem. Cogn.]* 11:501–518.

GRAF, P., L. R. SQUIRE, and G. MANDLER, 1984. The information that amnesic patients do not forget. *J. Exp. Psychol. [Learn. Mem. Cogn.]* 10:164–178.

HAIST, F., G. MUSEN, and L. R. SQUIRE, 1991. Intact priming of words and nonwords in amnesia. *Psychobiology* 19:275–285.

HAMANN, S. B., 1990. Level-of-processing effects in conceptually driven implicit tasks. *J. Exp. Psychol. [Learn. Mem. Cogn.]* 16:970–977.

HAYMAN, C. A. G., and E. TULVING, 1989. Is priming in fragment completion based on "traceless" memory system? *J. Exp. Psychol. [Learn. Mem. Cogn.]* 14:941–956.

HINTZMAN, D. L., and A. L. HARTRY, 1990. Item effects in recognition and fragment completion: Contingency relations vary for different subsets of words. *J. Exp. Psychol. [Learn. Mem. Cogn.]* 16:955–969.

JACOBY, L. L., 1983. Remembering the data: Analyzing interactive processes in reading. *J. Verb. Learn. Verb. Be.* 22:485–508.

JACOBY, L. L., 1991. A process dissociation framework:

Separating automatic from intentional uses of memory. *J. Mem. Lang.* 30:513–541.

Jacoby, L. L., and M. Dallas, 1981. On the relationship between autobiographical memory and perceptual learning. *J. Exp. Psychol. [Gen.]* 110:306–340.

Jacoby, L. L., and C. A. G. Hayman, 1987. Specific visual transfer in word identification. *J. Exp. Psychol. [Learn. Mem. Cogn.]* 13:456–463.

Jacoby, L. L., and D. Witherspoon, 1982. Remembering without awareness. *Can. J. Psychol.* 36:300–324.

Johnson, M. K., and L. Hasher, 1987. Human learning and memory. *Annu. Rev. Psychol.* 38:631–668.

Kihlstrom, J. F., D. L. Schacter, R. C. Cork, C. A. Hurt, and S. E. Behr, 1990. Implicit and explicit memory following surgical anesthesia. *Psychol. Sci.* 1:303–306.

Kinoshita, S., and S. V. Wayland, 1993. Effects of surface features on word-fragment completion in amnesic subjects. *Am. J. Psychol.* 106:67–80.

Knowlton, B. J., S. J. Ramus, and L. R. Squire, 1992. Intact artificial grammar learning in amnesia: Dissociation of classification learning and explicit memory for specific instances. *Psychol. Sci.* 3:172–179.

Marsolek, C. J., S. M. Kosslyn, and L. R. Squire, 1991. Form specific visual priming in the right cerebral hemisphere. *J. Exp. Psychol. [Learn. Mem. Cogn.]* 18:492–508.

Milner, B., S. Corkin, and H. L. Teuber, 1968. Further analysis of the hippocampal amnesic syndrome: Fourteen year follow-up study of H.M. *Neuropsychologia* 6:215–234.

Mishkin, M., B. Malamut, and J. Bachevalier, 1984. Memories and habits: Two neural systems. In *Neurobiology of Learning and Memory*, G. Lynch, J. L. McGaugh, and N. M. Weinberger, eds. New York, N.Y.: Guilford, pp. 65–77.

Moscovitch, M., 1982. Multiple dissociations of function in amnesia. In *Human Memory and Amnesia*, L. S. Cermak, ed. Hillsdale, N.J.: Erlbaum, pp. 337–370.

Moscovitch, M., 1992. Memory and working-with-memory: A component process model based on modules and central systems. *J. Cognitive Neurosci.* 4:257–267.

Musen, G., and L. R. Squire, 1992. Nonverbal priming in amnesia. *Mem. Cognition* 20:441–448.

Musen, G., and L. R. Squire, 1993. On the implicit learning of novel associations in amnesic patients and normal subjects. *Neuropsychology* 7:119–135.

Nissen, M. J., and P. Bullemer, 1987. Attentional requirements of learning: Evidence from performance measures. *Cognitive Psychology* 19:1–32.

Nissen, M. J., D. S. Knopman, and D. L. Schacter, 1987. Neurochemical dissociation of memory systems. *Neurology* 37:789–794.

Ostergaard, A. L., 1992. A method for judging measures of stochastic dependence: Further comments on the current controversy. *J. Exp. Psychol. [Learn. Mem. Cogn.]* 18:413–420.

Paller, K. A., 1990. Recall and stem-completion priming have different electrophysiological correlates and are modified differently by direct forgetting. *J. Exp. Psychol. [Learn. Mem. Cogn.]* 16:1021–1032.

Richardson-Klavehn, A., and R. A. Bjork, 1988. Measures of memory. *Annu. Rev. Psychol.* 36:475–543.

Roediger, H. L., 1990. Implicit memory: Retention without remembering. *Am. Psychol.* 45:1043–1056.

Roediger, H. L., and T. A. Blaxton, 1987. Retrieval modes produce dissociations in memory for surface information. In *The Ebbinghaus Centennial Conference*, D. S. Gorfein and R. R. Hoffman, eds. Hillsdale, N.J.: Erlbaum, pp. 349–379.

Roediger, H. L., and B. H. Challis, 1992. Effects of identity repetition and conceptual repetition on free recall and word fragment completion. *J. Exp. Psychol. [Learn. Mem. Cogn.]* 18:3–14.

Roediger, H. L., and K. B. McDermott, 1993. Implicit memory in normal human subjects. In *Handbook of Neuropsychology*, H. Spinnler and F. Boller, eds. Amsterdam: Elsevier, pp. 63–131.

Roediger, H. L. I., M. S. Weldon, M. L. Stadler, and G. L. Riegler, 1992. Direct comparison of two implicit memory tests: Word fragment and word stem completion. *J. Exp. Psychol. [Learn. Mem. Cogn.]* 18:1251–1269.

Rugg, M. D., and M. C. Doyle, in press. Event related potentials and stimulus repetition in direct and indirect tests of memory. In *Cognitive Electrophysiology*, T. Munke and G. R. Mangun, eds. Cambridge, Mass.: Birkhauser.

Schacter, D. L., 1985. Priming of old and new knowledge in amnesic patients and normal subjects. *Ann. N.Y. Acad. Sci.* 444:44–53.

Schacter, D. L., 1987. Implicit memory: History and current status. *J. Exp. Psychol. [Learn. Mem. Cogn.]* 13:501–518.

Schacter, D. L., 1990. Perceptual representation systems and implicit memory: Toward a resolution of the multiple memory systems debate. In *Development and Neural Bases of Higher Cognitive Functions*, A. Diamond, ed. New York: New York Academy of Sciences, pp. 543–571.

Schacter, D. L., 1992. Understanding implicit memory: A cognitive neuroscience approach. *Am. Psychol.* 47:559–569.

Schacter, D.L., in press. Priming and multiple memory systems: Perceptual mechanisms of implicit memory. In *Memory systems 1994*, D. L. Schacter and E. Tulving, eds. Cambridge, Mass.: MIT Press.

Schacter, D. L., J. Bowers, and J. Booker, 1989. Intention, awareness and implicit memory: The retrieval intentionality criterion. In *Implicit memory: Theoretical issues*, S. Lewandowsky, J. C. Dunn, and K. Kirsner, eds. Hillsdale, N.J.: Erlbaum, pp. 47–65.

Schacter, D. L., C. Y. P. Chiu, and K. N. Ochsner, 1993. Implicit memory: A selective review. *Annu. Rev. Neurosci.* 16:159–182.

Schacter, D.L., and B. Church, 1992. Auditory priming: Implicit and explicit memory for words and voices. *J. Exp. Psychol. [Learn. Mem. Cogn.]* 18:915–930.

Schacter, D. L., B. Church, and J. Treadwell, 1994. Implicit memory in amnesic patients. Evidence for spared auditory priming. *Psychol. Sci.* 5:20–25.

Schacter, D. L., L. A. Cooper, and S. M. Delaney, 1990. Implicit memory for unfamiliar objects depends on access to structural descriptions. *J. Exp. Psychol. [Gen.]* 119:5–24.

Schacter, D. L., L. A. Cooper, M. Tharan, and A. B. Rubens, 1991. Preserved priming of novel objects in patients with memory disorders. *J. Cognitive Neurosci.* 3:118–131.

Schacter, D. L., L. A. Cooper, and J. Treadwell, 1993. Preserved priming of novel objects across size transformation in amnesic patients. *Psychol. Sci.* 4:331–335.

Schacter, D. L., and P. Graf, 1986a. Effects of elaborative processing on implicit and explicit memory for new associations. *J. Exp. Psychol. [Learn. Mem. Cogn.]* 12:432–444.

Schacter, D. L., and P. Graf, 1986b. Preserved learning in amnesic patients: Perspectives on research from direct priming. *J. Clin. Exp. Neuropsychol.* 8:727–743.

Schacter, D. L., and P. Graf, 1989. Modality specificity of implicit memory for new associations. *J. Exp. Psychol. [Learn. Mem. Cogn.]* 15:3–12.

Schacter, D. L., S. M. McGlynn, W. P. Milberg, and B. A. Church, 1993. Spared priming despite impaired comprehension: Implicit memory in a case of word meaning deafness. *Neuropsychology* 7:107–118.

Shimamura, A. P., 1985. Problems with the finding of stochastic independence as evidence for multiple memory systems. *Bull. Psychonom. Soc.* 23:506–508.

Shimamura, A. P., and L. R. Squire, 1984. Paired-associate learning and priming effects in amnesia: A neuropsychological approach. *J. Exp. Psychol. [Gen.]* 113:556–570.

Shimamura, A. P., and L. R. Squire, 1989. Impaired priming of new associations in amnesia. *J. Exp. Psychol. [Learn. Mem. Cogn.]* 15:721–728.

Sloman, S. A., C. A. G. Hayman, N. Ohta, J. Law, and E. Tulving, 1988. Forgetting in primed fragment completion. *J. Exp. Psychol. [Learn. Mem. Cogn.]* 14:223–239.

Squire, L. R., 1992. Declarative and nondeclarative memory: Multiple brain systems supporting learning and memory. *J. Cognitive Neurosci.* 99:195–231.

Squire, L. R., J. G. Ojemann, F. M. Miezin, S. E. Petersen, T. O. Videen, and M. E. Raichle, 1992. Activation of the hippocampus in normal humans: A functional anatomical study of memory. *Proc. Natl. Acad. Sci. U.S.A.* 89:1837–1841.

Tulving, E., 1985. How many memory systems are there? *Am. Psychol.* 40:385–398.

Tulving, E., and A. J. Flexser, 1992. On the nature of the Tulving-Wiseman function. *Psychol. Rev.* 99:543–546.

Tulving, E., and D. L. Schacter, 1990. Priming and human memory systems. *Science* 247:301–306.

Tulving, E., D. L. Schacter, and H. Stark, 1982. Priming effects in word-fragment completion are independent of recognition memory. *J. Exp. Psychol. [Learn. Mem. Cogn.]* 8:336–342.

Warrington, E. K., and L. Weiskrantz, 1974. The effect of prior learning on subsequent retention in amnesic patients. *Neuropsychologia* 12:419–428.

Weiskrantz, L., 1985. On issues and theories of the human amnesic syndrome. In *Memory Systems of the Human Brain: Animal and Human Cognitive Processes*, N. M. Weinberger, J. L. McGaugh, and G. Lynch, eds. New York: Guilford, pp. 381–415.

Weldon, M. S., and H. L. Roediger, 1987. Altering retrieval demands reverses the picture superiority effect. *Mem. Cognition* 15:269–280.

Witherspoon, D., and M. Moscovitch, 1989. Stochastic independence between two implicit memory tasks. *J. Exp. Psychol. [Learn. Mem. Cogn.]* 15:22–30.

53 Memory, Hippocampus, and Brain Systems

LARRY R. SQUIRE AND BARBARA J. KNOWLTON

ABSTRACT The topic of memory is considered from a combined psychology and neuroscience perspective. Topics discussed include the nature of retrograde amnesia, the link between recall and recognition memory, the relationship between remembering and knowing, episodic and semantic memory, and the contribution of the frontal lobes to memory. In addition, fact-and-event (declarative) memory is contrasted with a collection of nondeclarative memory abilities including priming, skill and habit learning, and the acquisition of category-level knowledge. Finally, the brain systems that underlie different forms of learning and memory are considered.

Multiple forms of memory

A major theme in current studies of both humans and experimental animals is that memory is not a single entity but is composed of separate systems (Weiskrantz, 1990; Squire, 1992; Schacter and Tulving, 1994). The dissociation between declarative (explicit) and nondeclarative (implicit) memory is based on studies of experimental animals as well as amnesic patients and normal subjects showing that fact-and-event memory is distinct from other kinds of memory (skills, habits, and priming). Figure 53.1 illustrates a way of classifying kinds of memory.

Declarative memory refers to memory for facts and events. It is well suited to storing arbitrary associations after a single trial. Nondeclarative memories are generally acquired gradually across multiple trials (There are exceptions such as priming and taste aversion conditioning, which can occur after a single trial). Declarative knowledge is also flexible and can be readily applied to novel contexts. Nondeclarative memory tends to be inflexible, bound to the learning situation, and not readily accessed by response systems that did not participate in the original learning. The most compelling evidence for this property of declarative and nondeclarative memory systems has come from studies of experimental animals (Eichenbaum, Mathews, and Cohen, 1989; Saunders and Weiskrantz, 1989), although there are also some indications that declarative memory in humans is more flexible than nondeclarative memory (Glisky, Schacter, and Tulving, 1986; but see Shimamura and Squire, 1988).

LARRY R. SQUIRE and BARBARA J. KNOWLTON VA Medical Center and Departments of Psychiatry and Neurosciences, University of California, San Diego School of Medicine, San Diego, Calif.

THE DISTINCTION BETWEEN SHORT-TERM AND LONG-TERM MEMORY In human amnesia, short-term (immediate) memory is fully intact (Baddeley and Warrington, 1970; Cave and Squire, 1992b). The distinction between short-term and long-term memory is also present in experimental animals (Kesner and Novak, 1982; Wright et al., 1985; Overman, Ormsby, and Mishkin, 1991; Alvarez-Royo, Zola-Morgan, and Squire, 1992; Alvarez, Zola-Morgan, and Squire, in press; figure 53.2). These findings together demonstrate that the behavioral impairment in experimental animals following damage to the hippocampus and related structures is a memory problem, not an impairment in perception, rule learning, or some other cognitive function. Indeed, all available evidence supports the conclusion that rats, monkeys, and other animals with damage to the hippocampus and related structures provide a good animal model of human amnesia (Zola-Morgan and Squire, 1990; Squire, 1992).

RETROGRADE AMNESIA The brain system that supports declarative memory has only a temporary role in the formation of long-term memory. Retrograde amnesia, the loss of memories that were acquired prior to the onset of amnesia, is usually temporally graded, such that recent memories are lost more easily than remote memories (Ribot, 1881). Retrograde amnesia

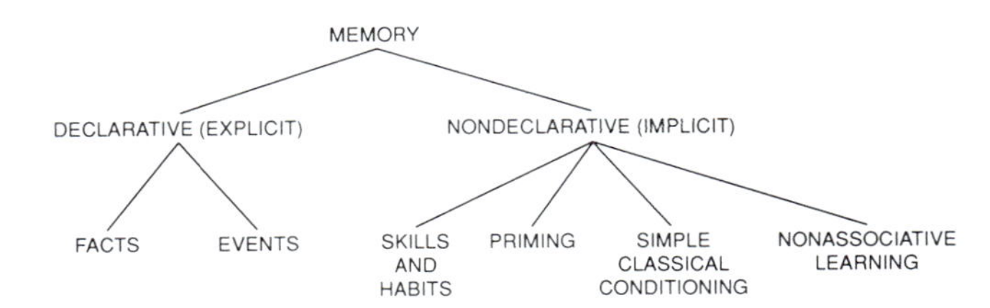

FIGURE 53.1 A taxonomy of long-term memory. (Squire and Zola-Morgan, 1991.)

can sometimes be ungraded and extensive, as in conditions such as encephalitis and head trauma, when damage typically occurs beyond the brain system that supports declarative memory (e.g., Damasio et al., 1985). Nevertheless, in patients with restricted damage within the hippocampal formation, such as patient R. B., retrograde amnesia is brief, perhaps covering a year or two at the most prior to surgery (Zola-Morgan, Squire, and Amaral, 1986). Other patients, who presumably have more extensive damage within the medial temporal lobe, have temporally limited retrograde amnesia that extends back one to two decades (Squire, Haist, and Shimamura, 1989).

Results from experimental animals (Zola-Morgan and Squire, 1990; Kim and Fanselow, 1992; Cho, Beracochea, and Jaffard, 1993) provide evidence for a gradual process of organization and consolidation whereby memory eventually becomes independent of the medial temporal lobe (figure 53.3). The medial temporal lobe is the target of highly processed information originating from a variety of cortical regions, and it returns projections to these same cortical regions. The hippocampal formation may store conjunctions that tie distributed memory storage sites together until more permanent corticocortical connections are formed. Thus, this system may serve to bind together disparate aspects of a memory and distill them into a coherent memory trace that can subsequently be accessed by many routes. Alternatively, current data leave open the possibility that the medial temporal lobe is the exclusive site of long-term memory storage until a cortical representation is fully developed. Computational models of hippocampal-cortical interactions and single-unit studies of the dynamic properties of cortical long-term memory representations will be needed to decide between these alternatives (e.g., see Alvarez and Squire, in press; McClelland, in press).

RECALL AND RECOGNITION MEMORY Amnesic patients perform poorly on conventional memory tasks that assess recall or recognition. Yet, it has also been proposed that improved perceptual fluency (e.g., the phenomenon of priming) might lead to a sense of familiarity and thereby support recognition memory judgments to some extent, independently of declarative

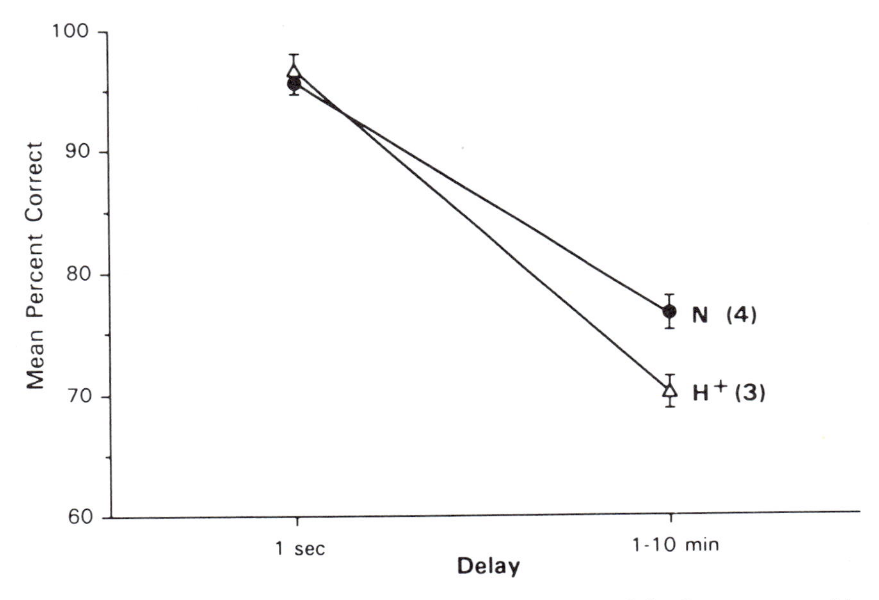

FIGURE 53.2 Percent of correct responses on the delayed nonmatching-to-sample task for 4 normal monkeys (N) and 3 monkeys with damage to the hippocampal formation (H^+). Performance of the 2 groups was identical at the 1-s delay, but differed at longer delays. The delays were presented in a mixed order. (Alvarez, Zola-Morgan, and Squire, in press).

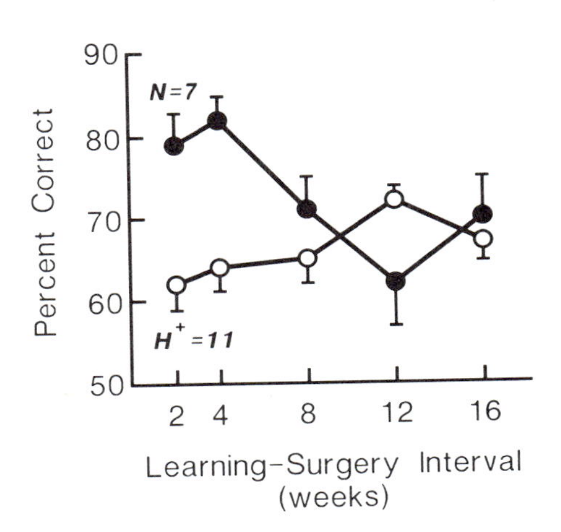

FIGURE 53.3 The effect of hippocampal formation lesions on retrograde amnesia. Monkeys with lesions were impaired in remembering information learned 2 to 4 weeks before surgery, but remembered objects learned long ago as well as normal monkeys. N = 7 normal monkeys; H^{+} = 11 monkeys with damage to the hippocampal formation. Brackets show standard error of the mean (Zola-Morgan and Squire, 1990).

memory (Mandler, 1980; Jacoby and Dallas, 1981). In two studies, amnesic patients performed well on a test of recognition as compared to a test of recall (Hirst et al., 1986; Hirst et al., 1988). However, many of these patients may have had frontal lobe pathology, which impairs the effortful search of memory required for recall (Jetter et al., 1986). In a recent study evaluating recall and recognition performance of amnesic patients over a wide range of retention intervals, recognition was impaired proportionately to recall (Haist, Shimamura, and Squire, 1992). The view that recognition performance derives little, if any, benefit from nondeclarative memory is also supported by findings that amnesic patients can sometimes perform at chance levels on measures of recognition memory at the same time that priming is fully intact (Squire, Shimamura, and Graf, 1985; Cave and Squire, 1992). Although nondeclarative memory may not affect recognition memory judgments in typical recognition tests, it remains possible that item fluency, that is, the process that supports priming, could influence recognition judgments under some circumstances (Johnston, Dark, and Jacoby, 1985; Johnston, Hawley, and Eliott, 1991).

The distinction between remembering and knowing

When an item evokes a conscious recollection including specific information about the item and the learning situation, a subject is said to "remember" (R). When a subject is confident an item is familiar and was seen before, but is unable to remember anything about the item in the learning situation, the subject is said to experience "knowing" (K) (Tulving, 1985). In some respects, the distinction between remembering and knowing is similar to the distinction between declarative and nondeclarative memory, and R and K responses can be dissociated in a number of ways that are reminiscent of that distinction. For example, the frequency of R responses is reduced when items are acquired during divided attention, but K responses are not affected (Gardiner and Parkin, 1990). The divided attention manipulation typically affects declarative memory (e.g., recognition memory) more than it affects nondeclarative memory (e.g., priming).

It is also possible that both remembering and knowing are dependent on the limbic and diencephalic brain structures that support declarative memory, but that remembering depends additionally on other brain systems important for source memory such as the frontal lobes (Schacter, Harbluk, and McLachlan, 1984; Janowsky, Shimamura, and Squire, 1989b). In one study, elderly subjects with age-appropriate memory abilities generated fewer R responses and more K responses than young adults (Parkin and Walter, 1992). The frequency of R responses in the elderly subjects correlated negatively with signs of frontal lobe dysfunction.

In another study, event-related potentials (ERPs) from old items that elicited R responses were similar to ERPs from old items that elicited K responses until 500 ms after each item was presented (Smith, 1993). Yet, items that were endorsed as old (i.e., all the items that received either R or K responses) could be distinguished from items that were endorsed as new beginning about 350 ms after item presentation. Moreover, electrical activity in the hippocampal formation during recognition memory performance appears to be most closely related to task performance during the period 400–500 ms after item presentation (Heit, Smith, and Halgren, 1990; see also Smith, 1993). Smith (1993) suggested that both R and K responses result from a common process of recollection, dependent on declarative memory and the hippocampus and related structures. The distinction between R and K responses arises from a postrecollective process, when subjects attend to the products of their retrieval efforts.

The distinction between episodic and semantic memory

Episodic memory refers to autobiographical memory for events, while semantic memory refers to factual memory (Tulving, 1983). Although both episodic and semantic memory are impaired in amnesia, (Shimamura and Squire, 1987; Gabrieli, Cohen, and Corkin, 1988), the possibility remains that amnesic patients have disproportionately impaired episodic memory. However, it is difficult to compare episodic and semantic memory, because episodic memory is specific to events that cannot be repeated. Accordingly, in amnesic patients, the ability to acquire some semantic memory through repetition will always exceed the ability to acquire episodic memory. Second, depending on how one defines semantic memory, there are domains of semantic memory that are severely affected in amnesia (e.g., the ability to learn new facts), and there are domains of semantic memory that are relatively preserved (e.g., the capacity for the gradual learning of artificial grammars and other abilities; see the section later in this chapter on nondeclarative memory).

In one study, the severely amnesic patient K. C. was able to learn simple sentences despite having virtually no episodic memory (Tulving, Hayman, and MacDonald). This apparent dissociation between episodic and semantic memory may depend on the fact that patient K. C. became amnesic following head trauma, a condition commonly associated with damage to both the frontal lobe and the temporal lobe. Interestingly, a more recent study found that the severely amnesic patient H. M., who had surgical damage to the medial temporal lobe, did not exhibit successful semantic learning, although the testing procedure used for H. M. was similar to the one used for K. C. (Tulving, personal communication, 1993). Thus, frontal lobe damage can impair episodic memory more than semantic memory. Indeed, episodic memory may be similar to (or in some instances identical to) source memory, which has previously been linked to frontal lobe function.

THE FRONTAL LOBES, THE DIENCEPHALON, AND THE MEDIAL TEMPORAL LOBE Patients with lesions involving the frontal lobes have a variety of deficits that affect performance, such as impaired source memory (Schacter, Harbluk, and McLachlan, 1984; Shimamura and Squire, 1987), impaired metamemory, that is, impaired ability to make judgments and predictions about one's own memory ability (Janowsky, Shimamura, and Squire, 1989a), impaired memory for temporal order (Milner, Petrides, and Smith, 1985; Shimamura, Janowsky, and Squire, 1990), and impaired recall abilities (Jetter et al., 1986). Diencephalic amnesic patients with Korsakoff's syndrome typically exhibit frontal lobe damage in addition to medial diencephalic damage. The presentation of amnesia in Korsakoff's syndrome is therefore somewhat different than in amnesia resulting from other etiologies (Janowsky, Shimamura, and Squire, 1989a; Shimamura, Janowsky, and Squire, 1991).

Other than the cognitive deficits attributable to frontal lobe damage, there is striking similarity between diencephalic amnesia and medial temporal lobe amnesia. Both groups have similar forgetting rates within long-term memory (McKee and Squire, 1992) and similar spatial memory abilities (Cave and Squire, 1991). The similarity between diencephalic and medial temporal lobe amnesia presumably reflects the close anatomical connections between the diencephalic midline and the medial temporal lobe, and suggests that these two regions can be considered to belong to a single functional system. The two regions undoubtedly make somewhat different contributions to memory, but from the perspective of behavioral criteria, the similarities are more prominent than the differences.

Nondeclarative memory

PRIMING Priming refers to the increased ability to identify or detect a stimulus as a result of its recent presentation. The first encounter with an item results in a representation of the item, which can then be subsequently accessed more readily than information about stimuli that have not been presented previously. Amnesic patients exhibit intact priming effects (figure 53.4; for a recent review, see Schacter, Chiu, and Ochsner, 1993). It is important to note that intact priming in amnesic patients has been demonstrated for novel materials that have no preexisting representations (see Schacter, Chiu, and Ochsner, 1993; Squire, Knowlton, and Musen, 1993). These results indicate that priming is not derived simply by activating stored memory representations, but rather is based on the sensory-perceptual traces created by stimulus presentation.

Presentation of items can also influence preferences and judgments about the items. For example, both amnesic patients and normal subjects are more likely

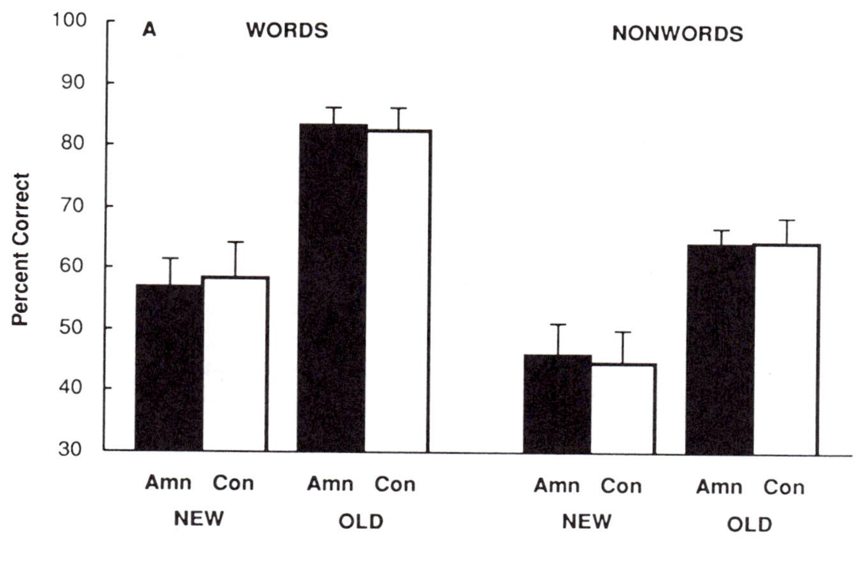

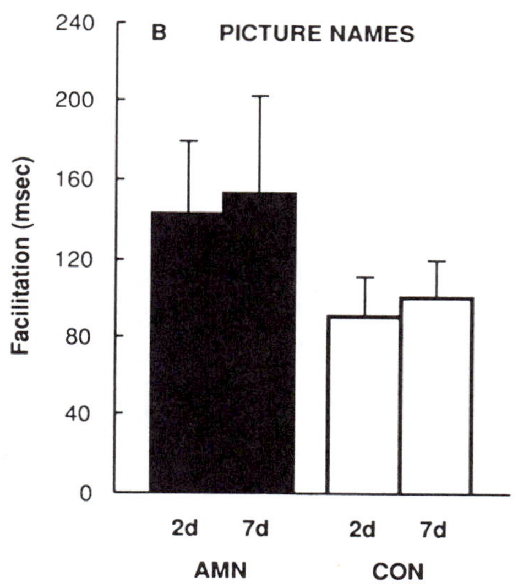

FIGURE 53.4 Intact priming in amnesic patients on two different tests. (A) Percent of words and nonwords correctly identified in a perceptual identification task. Old items had been presented once previously, and the advantage for identifying old items compared to new items indicates priming. (B) Facilitation of picture naming at 2 days (2d) and 7 days (7d) after a single presentation of the pictures. The facilitation score was obtained by subtracting the time required to name 50 old pictures from the time required to name 50 new pictures. Brackets show standard errors of the mean. AMN, amnesic patients; CON, control subjects (Haist et al., 1991; Cave and Squire, 1992a).

to judge a proper name as famous if the name has been presented recently (Neeley and Payne, 1983; Jacoby, Woloshyn, and Kelley, 1989; Squire and McKee, 1992). Normal subjects can suppress this effect in some circumstances because they can draw on declarative memory to recall that the items were just presented (Jacoby et al., 1989). Thus, in one study, only nonfamous names were presented first, then subjects were informed that all the names were nonfamous, then subjects were asked to judge the fame of new famous names together with both old and new nonfamous names. Amnesic patients continued to exhibit a fame judgment bias, but normal subjects did not (Squire and McKee, 1993).

The anatomical locus of priming is probably the neocortex. Studies using positron emission tomography (PET) are consistent with a right posterior neocortical site for word-stem completion priming (Squire et al.,

1992). The finding of a right posterior locus suggests that word-stem completion priming relies importantly on visual, orthographic features of the presented material. Priming across modalities, priming across typefaces, auditory priming, and priming of semantic information presumably depend on other cortical regions.

Although priming can be long-lasting and can result in new representations, priming is nevertheless limited in comparison to declarative memory. While declarative memory is well suited for forming new associations between arbitrary stimuli in a single trial, in nondeclarative memory novel associations are formed more gradually. Thus, implicit learning of novel associations does occur over multiple trials, but one-trial implicit learning of novel associations does not occur readily for either normal subjects or amnesic patients (Musen and Squire, 1993). However, associations that are easily integrated into a single perceptual unit, such as a word and the color in which it is printed, can be learned nondeclaratively in a single trial (Musen and Squire, 1992).

Skills and Habits The learning of skills and habits is largely nondeclarative in some circumstances, as evidenced by the fact that amnesic patients learn at an entirely normal rate (see Squire, Knowlton, and Musen, 1993, for a review). Amnesic patients can learn normally even when the information to be acquired is not exclusively perceptual or motor. In one study, subjects performed a serial reaction-time task in which they responded successively to a sequence of four illuminated spatial locations (Nissen and Bullemer, 1987). The task was to press one of four keys as rapidly as possible as soon as the location above that key was illuminated. Amnesic patients and normal subjects successfully learned a repeating sequence as indicated by decreasing reaction time for key presses as the sequence repeated itself. When the sequence was changed, reaction times increased again. Subjects were able to learn the sequence even when they were judged to have no declarative knowledge of it. Subjects were judged to have no declarative knowledge when they were unable to generate the sequence in subsequent tests, and were unaware that a sequence had been presented (Nissen and Bullemer, 1987; Willingham, Nissen, and Bullemer, 1989).

A recent study has challenged the idea that sequence learning is implicit, on the basis of findings that subjects were able to recognize and reproduce correct sequences when methods were used that were more sensitive than those used previously (Perruchet and Amorim, 1992). It was suggested that subjects do have declarative knowledge of the material, and that a distinction between memory systems is not required by the data. The findings from amnesic patients are particularly useful in this context, because the patients provide a tool for assessing whether declarative memory is only epiphenomenal or whether it is important for task performance. A finding that amnesic patients learn and remember entirely normally provides strong evidence that long-term declarative memory is not needed for performance. It is possible that some declarative knowledge develops during initial learning, and that even in amnesic patients such knowledge is supported by their intact immediate memory capacity, but it is a different matter whether in normal subjects declarative knowledge for what is learned in a task can or does persist within long-term memory once learning is completed. If performance is intact in amnesia at some time after learning, one has grounds for concluding that performance is supported by nondeclarative memory.

Production systems

In some cases, skills can involve more abstract information, and what is learned is neither perceptual nor motor. For example, subjects can gradually learn to control the level of an output variable by manipulating an input variable that relates to the output variable by a simple formula, although they need not acquire much reportable knowledge about the rule (Berry and Broadbent, 1984). Amnesic patients perform as well as normal subjects during the early learning of this task (Squire and Frambach, 1990). With extended training, normal subjects are able to outperform amnesic patients, and normal subjects are also better than amnesic patients at answering questions about task strategy.

Probabilistic Classification Learning In probabilistic category learning, subjects try to predict one of two outcomes based on a set of cues that are probabilistically associated with each outcome. In one such task, a list of one to four symptoms is presented, and

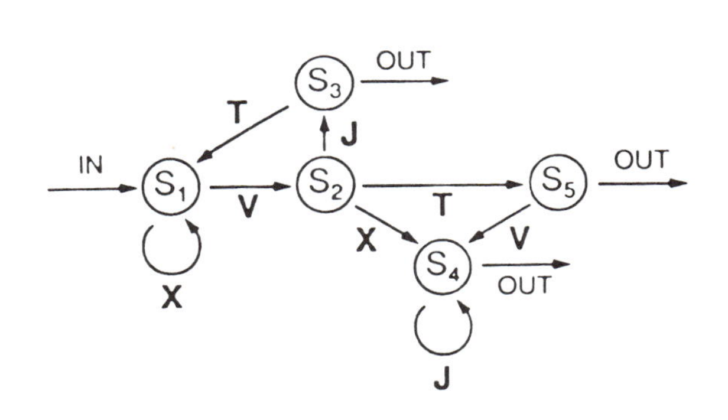

IN
OUT
OUT
OUT
S1
S2
S3
S4
S5
L
L
F
B
Z
F
Z
B

Grammatical	Nongrammatical
XXVT	TVT
XXVXJJ	TXXXVT
VXJJ	VXXXVJ
VTV	VJTVTX

Grammatical	Nongrammatical
BFZBZ	FBZ
LBF	BB
LLBL	ZZB
BZB	LFZBZF

FIGURE 53.5 Two finite-state rule systems used to generate the letter strings of artificial grammars. Examples of grammatical and nongrammatical letter strings are listed below each rule system. (Abrams and Reber, 1989; Knowlton et al., 1992).

each predicts one of two disease outcomes with a particular probability (Gluck and Bower, 1988). This task shares formal aspects with classical conditioning. That is, the separate cues (symptoms) compete for associative strength with the outcome (disease) in much the same way that conditioned stimuli compete for associative strength with the unconditioned stimulus (Gluck and Bower, 1988; Chapman and Robbins, 1990; Shanks, 1991). In three different tasks of probabilistic classification learning, amnesic patients improved their classification performance at the same rate as normal subjects (Knowlton, Squire, and Gluck, submitted). Probabilistic associations may be learned implicitly because information about a single trial is not as useful for performance as information about the probabilistic relationship between cues and outcomes, which necessarily accrues over many trials. In the study, learning occurred at a normal rate for the amnesic patients during approximately the first 50 trials of training. However, with extended training normal subjects surpassed the performance of amnesic patients, presumably because they were able to decipher the task to some extent and to remember some of the relationships explicitly.

ARTIFICIAL GRAMMAR LEARNING In artificial grammar learning, subjects see a series of letter strings generated by a finite-state rule system (figure 53.5). Subjects are told about the underlying rule system only after viewing the letter strings. They are then asked to judge whether new letter strings adhere or do not adhere to the rules. Although normal subjects are not able to report much explicit knowledge about their judgments, they are able to classify new letter strings at a level above chance (see Reber, 1989, for a review). However, it has also been argued that subjects may be using partially valid declarative knowledge of the grammar to make their judgments, and that declarative knowledge about the grammar can be elicited from subjects using sensitive test measures (Perruchet and Pacteau, 1990; Dulany, Carlson, and Dewey, 1984). Recent studies have helped to resolve this debate by showing that amnesic patients are able to make classification judgments as well as normal subjects in an artificial grammar learning task, despite their severe impairment in the ability to recognize the particular letter strings that were used to teach the grammar (Knowlton, Ramus, and Squire, 1992; Knowlton and Squire, 1994; figure 53.6).

PROTOTYPE ABSTRACTION AND CATEGORY LEARNING When subjects see a series of examples belonging to a single category, they can later classify new examples correctly. One possibility is that subjects abstract a

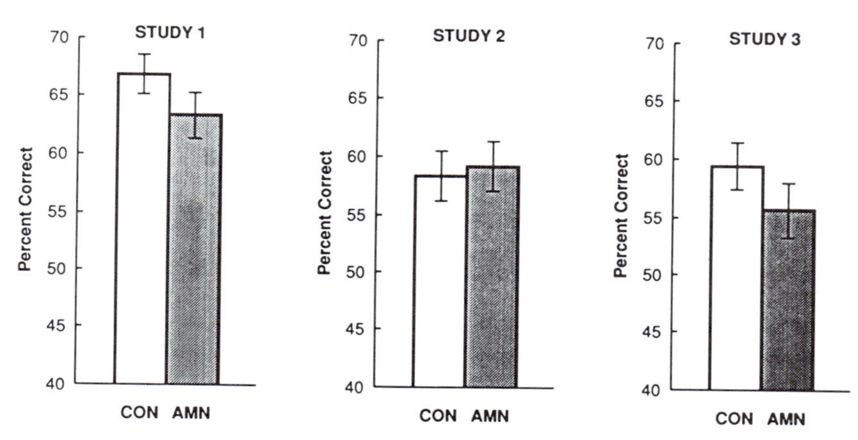

FIGURE 53.6 The results of three separate studies showing normal performance of amnesic patients (AMN) compared to control subjects (CON) on classification tasks based on artificial grammars. Brackets show standard error of the mean. (Knowlton et al., 1992; Knowlton and Squire, 1994).

prototype, or central tendency, from these examples and use the abstracted prototype to classify new items (Posner and Keele, 1968; Rosch, 1973). Alternatively, category judgments may be based on a comparison of test items to examples stored in declarative memory (Medin and Schaffer, 1978; Hintzman, 1986). Studies of amnesic patients should illuminate this issue. Amnesic patients and normal subjects were shown distortions of a prototypic dot pattern during training (Knowlton and Squire, 1993). The two groups performed equivalently on a later classification test, thereby demonstrating that they had abstracted the prototype from the examples. The prototype and low distortions of the prototype were judged to be members of the training category more often than higher distortions (figure 53.7). These results suggest that category-level information is acquired independently of declarative memory for training exemplars. Category-level information might be constructed nondeclaratively (implicitly) either by forming an abstracted prototype or by making comparisons with instances that are stored in implicit memory. In either case, it appears that category-level judgments can be independent of the ability to remember declaratively the particular instances that are encountered during training.

From memory systems to brain systems

Declarative memory is the product of a unique system that is dependent on medial temporal lobe/diencephalic structures, which operate in concert with neocortex (figure 53.8). Studies of nonhuman primates have elucidated the brain structures and connections that support declarative memory (Mishkin, 1982; Squire and Zola-Morgan, 1991). The important structures in the medial temporal lobe are the hippocampus, the entorhinal cortex, the parahippocampal cortex, and the perirhinal cortex. The amygdala is not part of the medial temporal lobe system for declarative memory (Zola-Morgan, Squire, and Amaral, 1989).

Damage to the hippocampal region, caused either by ischemia or radio frequency lesions, resulted in a significant memory impairment. Yet this level of impairment was increased when the area of the damage was systematically enlarged to include, first, the parahippocampal cortex and posterior entorhinal cortex (the H^+ lesion). The impairment associated with an H^+ lesion was increased still further when the H^+ lesion was extended forward to include anterior entorhinal cortex and perirhinal cortex (Zola-Morgan et al., 1993; Zola-Morgan, Squire, and Ramus, in press; figure 53.9).

These findings are consistent with the findings from human amnesia. Patient R. B., who exhibited a significant memory impairment following damage to field CA1 of the hippocampus (Zola-Morgan, Squire, and Amaral, 1986) was not nearly so impaired as patient H. M., who sustained much more extensive medial temporal lobe damage (Scoville and Milner, 1957). Thus, the parahippocampal and perirhinal cortices are not simply conduits for sending information to the hippocampus. Damage to the hippocampal region itself causes a relatively mild level of impairment. The fact that memory impairment increases when the adja-

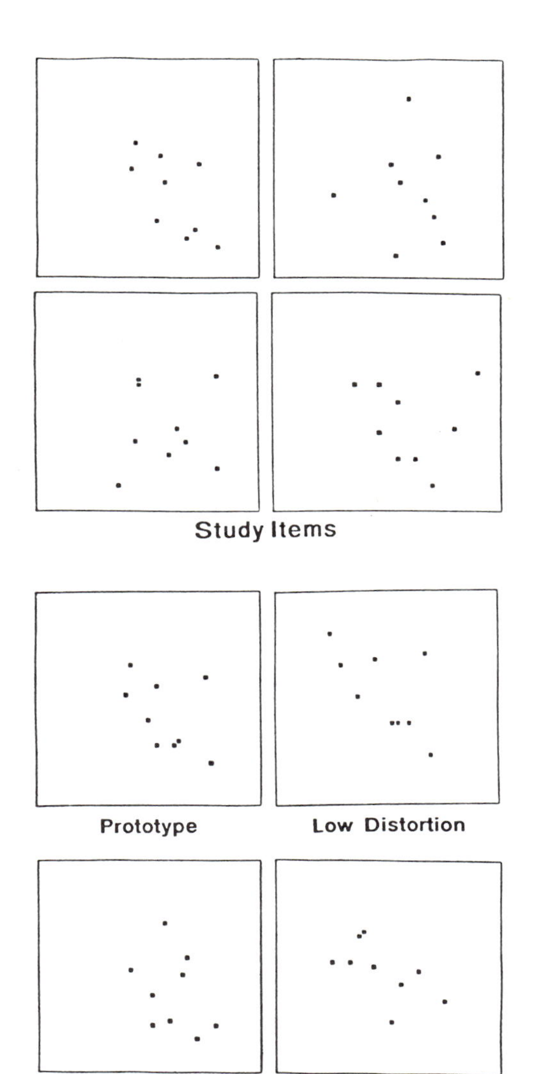

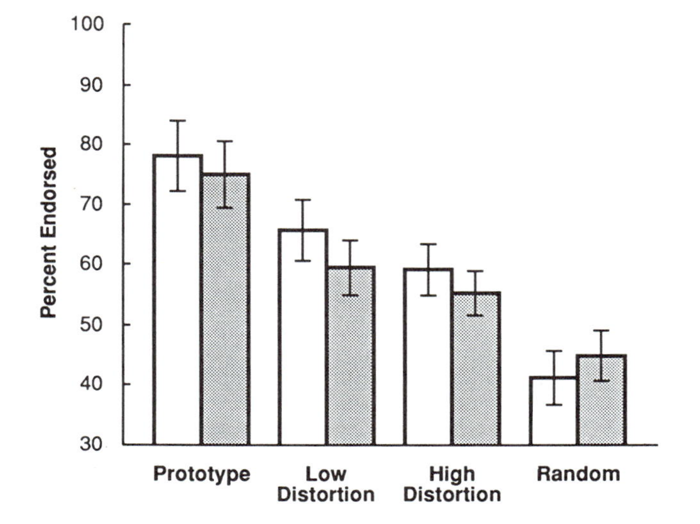

FIGURE 53.7 (Top) Examples of 4 study items and 4 test items used to assess classification learning. The study items are all arithmetic distortions of a prototype dot pattern that subjects do not see. The test items include the prototype pattern, novel distortions of the prototype, and random dot patterns that provide a measure of baseline classification performance. (Bottom) Performance on the dot pattern classification task according to type of test item. Open bars, control subjects; shaded bars, amnesic patients. Brackets show standard error of the mean. (Knowlton and Squire, 1993).

cent cortical regions are damaged indicates that these cortical areas themselves also contribute to memory function.

The information processed by medial temporal lobe structures is also directed to areas in the diencephalon important for declarative memory (Graff-Radford et al., 1990; Zola-Morgan and Squire, 1993). The development of an animal model of alcoholic Korsakoff's syndrome in the rat (Mair et al., 1988) provides a particulary favorable opportunity for investigating the anatomy of diencephalic amnesia.

Anatomical Substrates of Nondeclarative Memory

Brain systems other than the medial temporal lobe and the diencephalic midline are involved in acquiring nondeclarative information. For example, classical conditioning of discrete responses of the skeletal musculature depends on the cerebellum (for a review, see Thompson, 1990), while the conditioning of emotional responses depends on the amygdala (LeDoux, 1987; Davis, 1992). Caudate lesions in rats and monkeys impair the learning of win-stay habits and stimulus response tasks that are insensitive to lesions of the hippocampal formation (Packard, Hirsh, and White, 1989; Wang, Aigner, and Mishkin, 1990).

The neostriatum may also be important for the learning of skills and habits in human subjects. Patients with Huntington's disease were impaired at learning sensorimotor, skill-based tasks (Martone, Butters, and Payne, 1984; Heindel, Butters, and Salmon, 1988; Knopman and Nissen, 1991). Although declarative memory is not normal in these patients, the same patients who performed more poorly than amnesic patients on sensorimotor skill learning tasks performed better than amnesic patients on tests of declarative memory. Patients with Huntington's disease may be impaired on the sensorimotor tasks because they are deficient at forming motor programs. An important

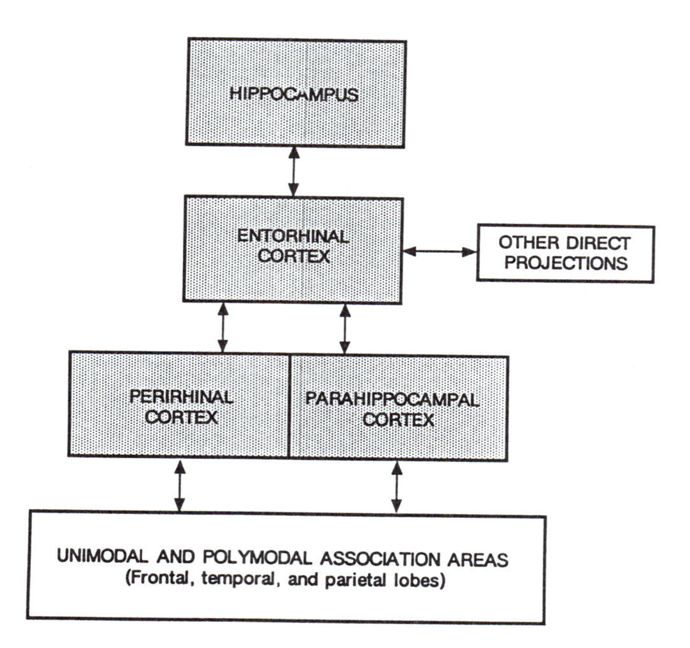

FIGURE 53.8 A schematic view of the structures and connections important for declarative memory. Shaded areas indicate structures within the medial temporal lobe.

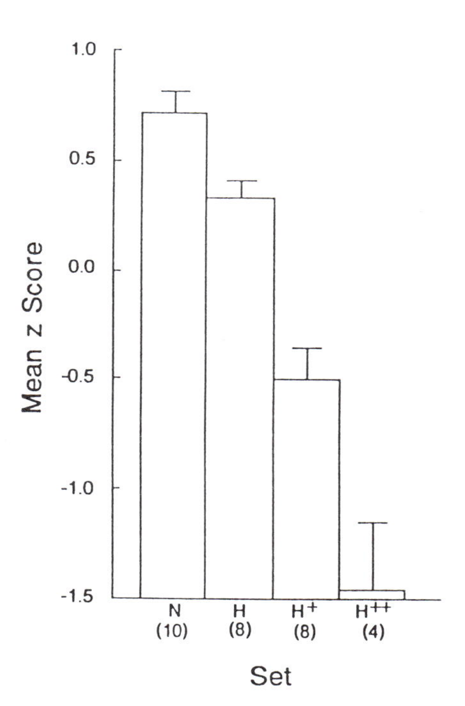

FIGURE 53.9 Mean z scores based on 4 measures of memory for 10 normal monkeys (N), 8 monkeys with damage to the hippocampal region (H), 8 monkeys with damage that also included the adjacent entorhinal and parahippocampal cortices (H^+), and 4 monkeys in which the H^+ lesion was extended forward to include the anterior entorhinal cortex and the perirhinal cortex (H^{++}). As more components of the medial temporal lobe memory system were included in the lesion, the severity of memory impairment increased. Brackets show standard errors of the mean (Zola-Morgan, Squire, and Ramus, in press).

question is whether patients with Huntington's disease would be impaired on the learning of habit-like tasks that do not have a motor component, such as artificial grammar learning or probabilistic classification learning. Alternatively, the neostriatum might not participate in this kind of learning. The processing of exemplars in neocortex might gradually lead to a corti-

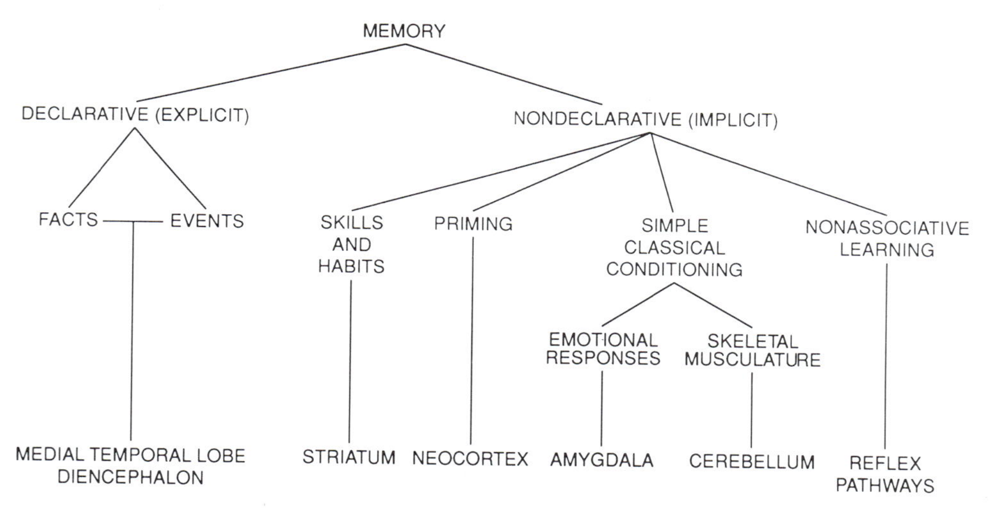

FIGURE 53.10 A taxonomy of long-term memory and associated brain structures.

cal representation of the commonalities between the training items, and the resulting cortical representation could provide a basis for category-level judgments.

It is now possible to link particular brain regions and systems to various kinds of memory (figure 53.10). The next challenge for cognitive neuroscience is to make these links specific. Ultimately, this means identifying where the synaptic changes occur that support different kinds of memory.

REFERENCES

Abrams, M., and A. S. Reber, 1989. Implicit learning in special populations. *J. Psycholinguist. Res.* 17:425–439.

Alvarez, P., and L. R. Squire, in press. Memory consolidation and the medial temporal lobe: A simple network model. *Proc. Natl. Acad. Sci. U.S.A.*

Alvarez, P., S. Zola-Morgan, and L. R. Squire, in press. The animal model of human amnesia: Long-term memory impaired and short-term memory intact. *Proc. Natl. Acad. Sci. U.S.A.*

Alvarez-Royo, P., S. Zola-Morgan, and L. R. Squire, 1992. Impaired long-term memory and spared short-term memory in monkeys with medial temporal lobe lesions: A response to Ringo. *Behav. Brain Res.* 52:1–5.

Baddeley, A. P., and E. K. Warrington, 1970. Amnesia and the distinction between long and short-term memory. *J. Verb. Learn. Verb. Behav.* 9:176–189.

Berry, D., and D. Broadbent, 1984. On the relationship between task performance and associated verbalizable knowledge. *Q. J. Exp. Psychol.* 36A:209–231.

Cave, C. B., and L. R. Squire, 1991. Equivalent impairment of spatial and nonspatial memory following damage to the human hippocampus. *Hippocampus* 1:329–340.

Cave, C., and L. R. Squire, 1992a. Intact and long-lasting repetition priming in amnesia. *J. Exp. Psychol. [Learn. Mem. Cogn.]* 18:509–520.

Cave, C. B., and L. R. Squire, 1992b. Intact verbal and nonverbal short-term memory following damage to the human hippocampus. *Hippocampus* 2:151–164.

Chapman, G. B., and S. J. Robbins, 1990. Cue interaction in human contingency judgment. *Mem. Cognition* 18:537–545.

Cho, Y. H., D. Beracochea, and R. Jaffard, 1993. Extended temporal gradient for the retrograde and anterograde amnesia produced by ibotenate entorhinal cortex lesions in mice. *J. Neurosci.* 13:1759–1766.

Damasio, A. R., P. J. Eslinger, H. Damasio, G. W. Van Hoesen, and S. Cornell, 1985. Multimodal amnesic syndrome following bilateral temporal and basal forebrain damage. *Arch. Neurol.* 42:252–259.

Davis, M., 1992. The role of the amygdala in fear-potentiated startle: Implications for animal models of anxiety. *Trends Pharmacol. Sci.* 13:35–41.

Dulany, D. E., R. A. Carlson, and G. I. Dewey, 1984. A case of syntactical learning and judgment: How conscious and how abstract? *J. Exp. Psychol. [Gen.]* 113:541–555.

Eichenbaum, H., P. Mathews, and N. J. Cohen, 1989. Further studies of hippocampal representation during odor discrimination learning. *Behav. Neurosci.* 103:1207–1216.

Gabrieli, J. D. E., N. J. Cohen, and S. Corkin, 1988. The impaired learning of semantic knowledge following medial temporal-lobe resection. *Brain Cog.* 7:157–177.

Gabrieli, J. D. E., W. Milberg, M. M. Keane, and S. Corkin, 1990. Intact priming of patterns despite impaired memory. *Neuropsychologia* 28:417–427.

Gardiner, J. M., and A. J. Parkin, 1990. Attention and recollective experience in recognition memory. *Mem. Cognition.* 18:579–583.

Glisky, E. L., D. L. Schacter, and E. Tulving, 1986. Learning and retention of computer-related vocabulary in memory-impaired patients: Method of vanishing cues. *J. Clin. Exp. Neuropsychol.* 8:292–312.

Gluck, M. A., and G. H. Bower, 1988. From conditioning to category learning: An adaptive network model. *J. Exp. Psychol. [Gen.]* 117:227–247.

Graff-Radford, N. R., D. Tranel, G. W. Van Hoesen, and J. Brandt, 1990. Diencephalic amnesia. *Brain* 113:1–25.

Haist, F., G. Musen, and L. R. Squire, 1991. Intact priming of words and nonwords in amnesia. *Psychobiology* 19:275–285.

Haist, F., A. P. Shimamura, and L. R. Squire, 1992. On the relationship between recall and recognition memory. *J. Exp. Psychol. [Learn. Mem. Cogn.]* 18:691–702.

Heindel, W. C., N. Butters, and D. P. Salmon, 1988. Impaired learning of a motor skill in patients with Huntington's disease. *Behav. Neurosci.* 102:141–147.

Heit, G., M. E. Smith, and E. Halgren, 1990. Neuronal activity in the human medial temporal lobe during recognition memory. *Brain* 113:1093–1112.

Hintzman, D., 1986. Schema abstraction in a multiple-trace memory model. *Psychol. Rev.* 93:411–428.

Hirst, W., M. K. Johnson, J. J. Kim, E. A. Phelps, G. Risse, and B. T. Volpe, 1986. Recognition and recall in amnesics. *J. Exp. Psychol. [Learn. Mem. Cogn.]* 12:445–451.

Hirst, W., M. K. Johnson, E. A. Phelps, and B. T. Volpe, 1988. More on recognition and recall in amnesics. *J. Exp. Psychol. [Learn. Mem. Cogn.]* 14:758–762.

Jacoby, L. L., 1983. Remembering the data: Analyzing interactive processes in reading. *J. Verbal Learn. Verbal Be.* 22:485–508.

Jacoby, L. L., and M. Dallas, 1981. On the relationship between autobiographical memory and perceptual learning. *J. Exp. Psychol. [Gen.]* 3:306–340.

Jacoby, L. L., C. Kelley, J. Brown, and J. Jasechko, 1989. Becoming famous overnight: Limits on the ability to avoid unconscious influences of the past. *J. Pers. Soc. Psychol.* 56:326–338.

Jacoby, L. L., V. Woloshyn, and C. Kelley, 1989. Becoming famous without being recognized: Unconscious influences of memory produced by dividing attention. *J. Exp. Psychol. [Gen.]* 118:115–125.

Janowsky, J. S., A. P. Shimamura, and L. R. Squire, 1989a. Memory and metamemory: comparisons between patients with frontal lobe lesions and amnesic patients. *Psychobiology*, 17:3–11.

Janowsky, J. S., A. P. Shimamura, and L. R. Squire, 1989b. Source memory impairment in patients with frontal lobe lesions. *Neuropsychologia* 27:1043–1056.

Jetter, W., U. Poser, R. B. Freeman, and J. H. Markowitsch, 1986. A verbal long term memory deficit in frontal lobe damaged patients. *Cortex* 22:229–242.

Johnston, W. A., W. J. Dark, and L. L. Jacoby, 1985. Perceptual fluency and recognition judgments. *J. Exp. Psychol. [Learn. Mem. Cogn.]* 11:3–11.

Johnston, W. A., K. J. Hawley, and M. G. Elliott, 1991. Contribution of perceptual fluency to recognition judgments. *J. Exp. Psychol. [Learn. Mem. Cogn.]* 17:210–223.

Kesner, R. P., and J. M. Novak, 1982. Serial position curve in rats: role of the dorsal hippocampus. *Science* 218:173–175.

Kim, J. J., and M. S. Fanselow, 1992. Modality-specific retrograde amnesia of fear. *Science* 256:675–677.

Knopman, D. S., and M. J. Nissen, 1991. Procedural learning is impaired in Huntington's disease: Evidence from the serial reaction time task. *Neuropsychologia* 29:245–254.

Knowlton, B. J., L. R. Squire, and M. Gluck, Submitted. Probabilistic classification learning in amnesia. *Learning and Memory*.

Knowlton, B. J., S. J. Ramus, and L. R. Squire, 1992. Intact artificial grammar learning in amnesia: Dissociation of classification learning and explicit memory for specific instances. *Psychol. Sci.* 3:172–179.

Knowlton, B. J., and L. R. Squire, 1993. The learning of categories: Parallel brain systems for item memory and category level knowledge. *Science* 262:1747–1749.

Knowlton, B. J., and L. R. Squire, 1994. The information acquired during artificial grammar learning. *J. Exp. Psychol. [Learn. Mem. Cogn.]* 20:79–91.

LeDoux, J. E., 1987. Emotion. In *Handbook of Physiology: The Nervous System*, vol. 5: *Higher Functions of the Nervous System*, J. M. Brookhart and V. B. Mountcastle, eds. Bethesda, Md.: American Physiological Society, pp. 419–460.

Mair, R. G., C. D. Anderson, P. J. Langlais, and W. J. McEntee, 1988. Behavioral impairments, brain lesions and monoaminergic activity in the rat following a bout of thiamine deficiency. *Behav. Brain Res.* 27:223–239.

Mandler, G., 1980. Recognizing: The judgment of previous occurrence. *Psychol. Rev.* 87:252–271.

Martone, M., N. Butters, and P. Payne, 1984. Dissociations between skill learning and verbal recognition in amnesia and dementia. *Arch. Neurol.* 41:965–970.

McClelland, J. L., in press. The organization of memory: A parallel distributed processing perspective. *Rev. Neurol. (Paris)*.

McKee, R. D., and L. R. Squire, 1992. Equivalent forgetting rates in long-term memory for diencephalic and medial temporal lobe amnesia. *J. Neurosci.* 12:3765–3772.

Medin, D. L., and M. M. Schaffer, 1978. Context theory of classification learning. *Psychol. Rev.* 85:207–238.

Milner, B., M. Petrides, and M. L. Smith, 1985. Frontal lobes and the temporal organization of memory. *Hum. Neurobiol.* 4:137–142.

Mishkin, M., 1982. A memory system in the monkey. *Philos. Trans. R. Soc. Lond. [Biol.]* 298:85–92.

Musen, G., and L. R. Squire, 1992. Implicit learning of one-trial color-word "associations" in amnesic patients. *Soc. Neur. Abstr.* 18:386.

Musen, G., and L. R. Squire, 1993. On the implicit learning of novel associations by amnesic patients and normal subjects. *Neuropsychology* 7:119–135.

Neely, J. H., and D. G. Payne, 1983. A direct comparison of recognition failure rates for recallable names in episodic and semantic memory tests. *Mem. Cog.* 11:161–171.

Nissen, M. J., and P. Bullemer, 1987. Attentional requirements of learning: Evidence from performance measures. *Cog. Psychol.* 19:1–32.

Overman, W. H., G. Ormsby, and M. Mishkin, 1991. Picture recognition vs. picture discrimination learning in monkeys with medial temporal removals. *Exp. Brain Res.* 79:18–24.

Packard, M. G., R. Hirsh, and N. M. White, 1989. Differential effects of fornix and caudate nucleus lesions on two radial maze tasks: Evidence for multiple memory systems. *J. Neurosci.* 9:1465–1472.

Parkin, A. J., and B. M. Walter, 1992. Recollective experience, normal aging, and frontal dysfunction. *Psychol. Aging* 7:290–298.

Perruchet, P., and M. Amorim, 1992. Conscious knowledge and changes in performance in sequence learning: Evidence against dissociation. *J. Exp. Psychol. [Learn. Mem. Cogn.]* 18:785–800.

Perruchet, P., and C. Pacteau, 1990. Synthetic grammar learning: Implicit rule abstraction or explicit fragmentary knowledge? *J. Exp. Psychol. [Gen.]* 119:264–275.

Posner, M. I., and S. W. Keele, 1968. On the genesis of abstract ideas. *J. Exp. Psychol.* 77:353–363.

Reber, A. S., 1989. Implicit learning and tacit knowledge. *J. Exp. Psychol. [Gen.]* 118:219–235.

Ribot, T., 1881. *Les maladies de la memoire.* New York: Appleton-Century-Crofts.

Rosch, E. H., 1973. On the internal structure of perceptual and semantic categories. In *Cognitive Development and the Acquisition of Language*, T. E. Moore, ed. New York: Academic Press, pp. 111–144.

Saunders, R. C., and L. Weiskrantz, 1989. The effects of fornix transection and combined fornix transection, mammillary body lesions and hippocampal ablations on object-pair association memory in the rhesus monkey. *Behav. Brain Res.* 35:85–94.

Schacter, D. L., C.-Y. Chiu, and K. N. Ochsner, 1993. Implicit memory: A selective review. *Annu. Rev. Neurosci.* 16:159–182.

Schacter, D. L., J. L. Harbluk, and D. R. McLachlan,

1984. Retrieval without recollection: An experimental analysis of source amnesia. *J. Verb. Learn. Verb. Be.* 23: 593–611.

Schacter, D. L., and E. Tulving, eds., 1994. *Memory Systems 1994.* Cambridge, Mass.: MIT. Press.

Scoville, W. B., and B. Milner, 1957. Loss of Recent memory after bilateral hippocampal lesions. *J. Neurol., Neurosurg. Psychiatry* 20:11–21.

Shanks, D. R., 1991. Categorization by a connectionist network. *J. Exp. Psychol. [Learn. Mem. Cogn.]* 17:433–443.

Shimamura, A. P., J. S. Janowsky, and L. R. Squire, 1990. Memory for the temporal order of events in patients with frontal lobe lesions and amnesic patients. *Neuropsychologia* 28:803–814.

Shimamura, A. P., J. S. Janowsky, and L. R. Squire, 1991. What is the role of frontal lobe damage in memory disorders? In *Frontal Lobe Function and Dysfunction,* H. D. Levin, H. M. Eisenberg, and A. L. Benton, eds. New York: Oxford University Press, pp. 173–195.

Shimamura, A. P., and L. R. Squire, 1987. A neuropsychological study of fact memory and source amnesia. *J. Exp. Psychol. [Learn. Mem. Cogn.]* 13:464–473.

Shimamura, A. P., and L. R. Squire, 1988. Long-term memory in amnesia: Cued recall, recognition memory, and confidence ratings. *J. Exp. Psychol. [Learn. Mem. Cogn.]* 14:763–770.

Smith, M. E., 1993. Neurophysiological manifestations of recollective experience during recognition memory judgments. *J. Cognitive Neurosci.* 5:1–13.

Squire, L. R., 1992. Memory and the hippocampus: A synthesis from findings with rats, monkeys, and humans. *Psychol. Rev.* 99:143–145.

Squire, L. R., 1987. *Memory and Brain.* New York: Oxford University Press.

Squire, L. R., and M. Frambach, 1990. Cognitive skill learning in amnesia. *Psychobiology* 18:109–117.

Squire, L. R., F. Haist, and A. P. Shimamura, 1989. The neurology of memory: Quantitative assessment of retrograde amnesia in two groups of amnesic patients. *J. Neurosci.* 9:828–839.

Squire, L. R., B. Knowlton, and G. Musen, 1993. The structure and organization of memory. *Annu. Rev. Psychol.* 44:453–495.

Squire, L. R., and R. McKee, 1993. Declarative and nondeclarative memory in opposition: When prior events influence amnesic patients more than normal subjects. *Mem. Cognition* 21:424–430.

Squire, L. R., and R. McKee, 1992. The influence of prior events on cognitive judgments in amnesia. *J. Exp. Psychol. [Learn. Mem. Cogn.]* 18:106–115.

Squire, L. R., J. G. Ojemann, F. M. Miezin, S. E. Petersen, T. O. Videen, and M. E. Raichle, 1992. Activation of the hippocampus in normal humans: A functional anatomical study of memory. *Proc. Natl. Acad. Sci. U.S.A.* 89:1837–1841.

Squire, L. R., A. Shimamura, and P. Graf, 1985. Independence of recognition memory and priming effects. A neuropsychological analysis. *J. Exp. Psychol. [Learn. Mem. Cogn.]* 11:37–44.

Squire, L. R., and S. Zola-Morgan, 1991. The medial temporal lobe memory system. *Science* 253:1380–1386.

Thompson, R. F., 1990. Neural mechanisms of classical conditioning in mammals. *Philos. Trans. R. Soc. Lond. [Biol.]* 329:161–170.

Tulving, E., 1983. *Elements of Episodic Memory.* Cambridge: Oxford University Press.

Tulving, E., 1985. How many memory systems are there? *Am. Psychol.* 40:385–398.

Tulving, E., C. A. G. Hayman, and C. A. MacDonald, 1991. Long-lasting perceptual priming and semantic learning in amnesia: A case experiment. *J. Exp. Psychol. [Learn. Mem. Cogn.]* 17:595–617.

Wang, J., T. Aigner, and M. Mishkin, 1990. Effects of neostriatal lesions on visual habit formation in rhesus monkeys. *Soc. Neur. Abstr.* 16:617.

Weiskrantz, L., 1990. Problems of learning and memory: One or multiple memory systems? *Philos. Trans. R. Soc. Lond. [Biol.]* 329:99–108.

Willingham, D. B., M. J. Nissen, and P. Bullemer, 1989. On the development of procedural knowledge. *J. Exp. Psychol. [Learn. Mem. Cogn.]* 15:1047–1060.

Wright, A. A., H. C. Santiago, S. F. Sands, D. F. Kendrick, and R. G. Cook, 1985. Memory processing of serial lists by pigeons, monkeys and people. *Science* 229: 287–289.

Zola-Morgan, S., and L. R. Squire, 1990. The primate hippocampal formation: Evidence for a time-limited role in memory storage. *Science* 250:288–290.

Zola-Morgan, S., and L. R. Squire, 1993. Neuroanatomy of memory. *Annu. Rev. Neurosci.* 16:547–563.

Zola-Morgan, S., L. R. Squire, and D. G. Amaral, 1986. Human amnesia and the medial temporal region: Enduring memory impairment following a bilateral lesion limited to field CAl of the hippocampus. *J. Neurosci.* 6: 2950–2967.

Zola-Morgan, S., L. R. Squire, and D. G. Amaral, 1989. Lesions of the amygdala that spare adjacent cortical regions do not impair memory or excerbate the impairment following lesions of the hippocampal formation. *J. Neurosci.* 9:1922–1936.

Zola-Morgan, S., L. R. Squire, R. P. Clower, and N. L. Rempel, 1993. Damage to the perirhinal cortex exacerbates memory impairment following lesions to the hippocampal formation. *J. Neurosci.* 13:251–265.

Zola-Morgan, S., L. R. Squire, and S. Ramus, in press. Severity of memory impairment in monkeys as a function of locus and extent of damage within the medial temporal lobe memory system. *Hippocampus.*

54 Organization of Memory: Quo Vadis?

ENDEL TULVING

ABSTRACT Research in cognitive psychology and neuropsychology of memory has produced a wealth of data that can be meaningfully ordered with the help of two general classes of concepts—memory processes and memory systems. This chapter proposes a simple model of organization of memory in which cognitive memory systems are related to one another in terms of the principal processes of encoding, storage, and retrieval. The central assumption of this SPI model—serial, parallel, independent—is that the relations among systems are *process specific*: Information is *encoded* into systems *serially*, and encoding in one system is contingent on the successful processing of the information in another system; information is *stored* in different systems in *parallel*; and information from each system can be *retrieved independently* of information in other systems.

Memory is one of Nature's most jealously guarded secrets. At the beginning of the second century of its scientific study, it continues to baffle, frustrate, and mystify those who would explore it. Although we have learned a great deal about memory over the years, it often seems that whenever we discover yet another previously unknown fact about memory, we have succeeded in adding more to what there is to know than to what we do know.

Past research has generated an immense wealth of data, and more is being added every day. In cognitive psychology alone we have a staggering number of findings and facts. Surprisingly perhaps, from the perspectives of some other disciplines, these findings and facts are reliable and robust. The problem of replicability of data hardly ever arises. Over a hundred years of patient and sometimes plodding study we have apparently learned a few useful tricks about how to get Nature to provide reasonably consistent answers to the questions that we are capable of putting to her in the form of experiments. However, our success has been somewhat less remarkable in interpreting and making sense of this abundance of data (cf. Tulving, 1979). There is less agreement among practitioners as to what the findings and facts tell us about the larger picture of memory than about what the findings and facts are that require explanation and understanding. Thus it is that a major challenge facing memory researchers today lies in correcting the imbalance between what we know well and what we know less well, the imbalance between what the facts about memory are and what they mean. How are we to make sense of the data? How are we to order the facts into a more comprehensive totality? What is the story that Nature, through the bits and pieces of empirical facts she has seen fit to share with us, is trying to tell us about the organization of the complex functioning structure that we think of as memory?

This chapter explores these questions and proposes some tentative answers. After a brief historical sketch, it describes a plausible organization of different forms of human memory, or memory systems, and then suggests how the different systems are related to one another in terms of some of the major processes of memory. The major objective of the chapter is to take the first step toward integrating memory processes and memory systems.

A historical sketch

The first era of scientific memory research began with Ebbinghaus (1885) and ended around 1960. This was the long period of "verbal learning." Its emphasis was on experimental design and precise measurement of basic phenomena of learning and forgetting, in normal adults, of serial and paired-associate lists of verbal items. The concept of association, and its single property of "strength," explained most of the known facts to the satisfaction of most practitioners.

ENDEL TULVING Rotman Research Institute of Baycrest Centre, Toronto, Canada

Around 1960 the associative verbal learning framework was largely replaced by the "information processing" paradigm. A wider variety of problems, issues, approaches, methods, and theoretical interpretations were adopted. Paired-associate and serial learning procedures were largely abandoned in favor of free and cued recall, as well as recognition and various kinds of memory judgments—recency, frequency, and the like. Experimental studies of short-term memory led to theoretical distinction between primary (short-term) and secondary (long-term) memory. Units of analysis shifted from lists to single items. Experimenters and theorists began to think of single items as "to-be-remembered events." The analytical distinction between storage and retrieval was translated into experimental paradigms that allowed the separation of the two processes. Influential theoretical concepts such as levels of processing, encoding specificity, and encoding/retrieval interactions emerged during this stage, as did "context" and "context effects." Connections were established between the previously isolated disciplines of cognitive psychology and neuropsychology. The concept of association as the basic theoretical building block was replaced by the concept of multiple processes, among which encoding, storage, and retrieval played a dominant role.

The current era of research, beginning sometime around 1980, can be thought of as cognitive neuroscience of memory. It is characterized by further expansion and liberalization of methods, techniques, and choices of questions and problems. The domain of "memory" has expanded considerably, both horizontally and vertically. The dominant concepts of the era so far have been *priming* and *memory systems*. There has been a steadily growing convergence between cognitive psychology and neuropsychology; interest has deepened in the study of learning and retention in memory-impaired patients; increasingly more attention is being paid to memory across life-span development; theoretically motivated and precisely controlled psychopharmacological studies of memory have appeared on the scene; computer modeling of memory processes has become more and more sophisticated; and the neuroimaging approach to the study of memory is rapidly overcoming its initial difficulties. This is the age of multidisciplinary study of memory.

All three epochs of memory research have contributed to the data and theory. The major theoretical tools that are available today for ordering the data have been developed in more recent times. They can be conveniently divided into two main classes of concepts—*memory processes* and *memory systems* (Tulving, 1991). Memory processes represent a bequest of the information-processing era. Their story has been told in some detail by a number of writers elsewhere (e.g., Crowder, 1976; Eysenck, 1977; Klatzky, 1980; Tulving, 1983). The concept of memory systems is a gift of the cognitive neuroscience of memory (e.g., Shallice, 1979; Warrington, 1979; Tulving, 1983, 1985a; Cohen, 1984; Mishkin, Malamut, and Bachevalier, 1984; Schacter and Moscovitch, 1984; Weinberger, McGaugh, and Lynch, 1985; Weiskrantz, 1987). The time seems ripe now for uniting the two sets of concepts—processes and systems—into a more comprehensive framework. To set the stage for such an understanding, we consider first what we know about different memory systems.

Human memory systems

Our current understanding of the organization of human memory has gradually evolved from various conceptual dichotomies: memory and habit, short-term and long-term memory, episodic and semantic memory, procedural and declarative memory, and the like (e.g., Sherry and Schacter, 1987; Schacter and Tulving, in press). Combining these dichotomies into a more general scheme allows us to identify at least five major categories of human memory, or memory "systems," together with a number of subcategories or subsystems. These systems and subsystems are listed in table 54.1.

The procedural systems are behavioral or cognitive *action* systems, whereas the other four forms in table 54.1 are cognitive *representation* systems. One of the major differences between the two kinds of system lies in the feasibility of characterization of the changes that result from learning or acquisition in a propositional or some other symbolic form: it is possible to do so for the four cognitive systems, but not quite possible for the procedural systems. The operations of procedural memory are expressed in the form of skilled behavioral and cognitive procedures independently of any cognition. Skillful performance of many perceptual-motor and cognitive tasks, such as balancing a stick on one's finger or reading text, are examples of tasks that depend heavily on the procedural memory systems.

The other four systems mediate changes in cognition, or thought. In the course of normal activity of an individual, the computational outputs of the cognitive

TABLE 54.1
Major categories of human learning and memory

System	Other terms	Subsystems	Retrieval
Procedural	Nondeclarative	Motor skills Cognitive skills Simple conditioning Simple associative learning	Implicit
PRS	Priming	Structural description Visual word form Auditory word form	Implicit
Semantic	Generic Factual Knowledge	Spatial Relational	Implicit
Primary	Working Short-term	Visual Auditory	Explicit
Episodic	Personal Autobiographical Event memory		Explicit

memory systems typically guide overt behavior, but such conversion of cognition into behavior is not an obligatory part of memory. Rather it is an optional postretrieval process. The ultimate output of cognitive memory systems is expressed in conscious awareness, which can, but need not, be converted into overt behavior such as verbal expression. Behavioral responses that subjects make in cognitive memory tasks in the laboratory serve merely as reports of cognitive processes.

Perceptual priming—recently reviewed by Schacter, Chiu, and Ochsner (1993) and by Roediger and McDermott (1993), and covered by Schacter's chapter in this volume—is a special form of perceptual learning that is expressed in enhanced identification of objects as structured physical-perceptual entities. A perceptual encounter with an object on one occasion primes or facilitates the perception of the same or a similar object on a subsequent occasion, in the sense that the identification of the object requires less stimulus information or occurs more quickly than it does in the absence of priming. Because perceptual priming represents a rudimentary capability whose biological utility seems to be obvious, it seems reasonable to expect that it is represented across a wide spectrum of species. Nevertheless, priming has not yet been observed or measured in nonhuman animals or in preverbal humans.

Semantic memory makes possible the acquisition and retention of factual information in the broadest sense; the structured representation of this information, semantic knowledge, models the world. Semantic knowledge provides the individual with the necessary material for thought, that is, for cognitive operations on the aspects of the world beyond the reach of immediate perception. The semantic memory systems are not tied either to language or to meaning. The designation *semantic memory* is merely a historical accident, and a better phrase to refer to the same concept might be "general knowledge of the world." It is a conjecture that human semantic memory has evolved from the spatial learning and knowledge of the ancestors of humans.

Primary memory, also referred to as short-term memory or working memory, registers and retains incoming information in a highly accessible form for a short period of time after the input. Primary memory, like other memory systems, is identified through dissociations of its products from those of other systems. It makes possible a lingering impression of the individual's present environment beyond the duration of the physical presence of the stimulus information emanating from the environment. This topic is covered in Baddeley's chapter in the present volume.

Episodic memory enables individuals to remember their personally experienced past, that is, to remember experienced events as embedded in a matrix of other personal happenings in subjective time. It depends on but transcends the range of the capabilities of semantic memory. The most distinctive aspect of episodic memory is the kind of conscious awareness that characterizes recollection of past happenings. This awareness is unique and unmistakably different from the kinds of awareness that accompany perceptual experiences, imagining, dreaming, solving of problems, and retrieval of semantic information. To distinguish the episodic-memory awareness from these other kinds, it has been referred to as autonoetic consciousness (Tulving, 1985b).

Relations among memory systems

Two entries in table 54.1, semantic and episodic memory, are sometimes categorized together as *declarative* (Squire, 1982) or *propositional* (Tulving, 1983) memory, as they share a number of features. Another frequently used distinction is that between implicit and explicit memory (Graf and Schacter, 1985; Schacter, 1987a). These are not memory systems, but forms of *expression* of memory. Implicit memory designates the expression of stored information without awareness of its acquisition coordinates in space and time, that is, expression of what the individual has learned without necessarily remembering how, when, or where the learning occurred. Explicit memory, on the other hand, refers to the expression of what the person is consciously aware of as a personal experience. Retrieval operations in the earlier systems, as shown in table 54.1, can be said to be *implicit*, whereas in the later systems they are *explicit*.

The forms of memory in table 54.1 are listed in order of their assumed emergence, from the earliest to the latest, both with respect to the phylogenetic and ontogenetic development, and with respect to the dependence relations that govern certain aspects of their operations. Thus, procedural forms of learning and memory probably evolved first and develop early in human infants, and episodic memory evolved last and develops later in human children. Learning to make appropriate responses to simple sensory stimuli (procedural memory) has obvious biological utility to organisms at all stages of evolution and development. So does the ability to identify objects in one's environment, and learning to do so quickly and effortlessly (perceptual priming). Forms of learning and memory that evolved later, such as primary or working memory and episodic memory, are not necessary for survival in a relatively simple and stable environment such as that in which the Paleolithic ancestors of humans lived, and in which human infants and amnesic patients live today. Working memory becomes more critical when the demand arises for intra- and interindividual communication—that is, for abstract thought and language. Episodic memory becomes critical as a catalyst for the acquisition of knowledge of the world (semantic memory) through its capability of encoding and storing information about similar events at different times, and for counteracting associative interference (Tulving, 1991, 1993). The ordering of the major memory systems in table 54.1 also reflects the conjectured relations among the systems: Some of the operations of the later ones depend on, and are supported by, the operations of the earlier ones, whereas earlier systems can operate essentially independently of the later ones. We will return to this theme when we discuss the SPI model of organization of memory.

The evidence for the biological and functional separability of the categories and subcategories listed in table 54.1 is still fragmentary, largely indirect, and of variable quality and quantity. Much of the relevant information comes from the study of brain-damaged patients with *selective* memory impairments (Milner, Corkin, and Teuber, 1968; Warrington and Weiskrantz, 1968, 1974). To test the hypothesis that perceptual priming, semantic memory, and episodic memory, like other major categories of memory, represent different neurocognitive systems and subsystems requires the boot-strapping-like delineation and unraveling of the myriad relations among tasks and systems. It represents a formidable scientific challenge. A healthy beginning has been made, however, in the form of experiments conforming to the *task-comparison* methodology (Richardson-Klavehn and Bjork, 1988). Outcomes of different memory tasks that are known or assumed to be differentially weighted by contributions of different systems are systematically compared. Dissociations among these outcomes are regarded as providing support for the hypothesis of separability of systems. Outcomes of tests are said to be dissociated if they differ as a function of an independent variable, or if they differ for different groups of subjects or patients, or for different brain states. Dissociations contrast with *parallel effects*, observations that the manipulation of an independent variable or a treatment produces similar changes in the outcomes of different tasks, or different measures of memory performance.

When a number of different dissociations—yielded by different kinds of subjects, different tasks and situations, and different techniques—are seen as converging on the same classificatory scheme, it becomes reasonable to hypothesize the existence of separate memory systems. Although dissociations of interest are usually observed first at the level of behavior, behavioral data on their own are seldom sufficiently compelling to preclude alternative interpretations of empirical facts. Classification of memory into different systems and subsystems requires a broad-based multidisciplinary approach. Functional analyses must be integrated with relevant neuroanatomical, neurochemical, and neurophysiological methods.

SPI model of organization of memory

The work in the mainstream cognitive psychology of memory during the information-processing era yielded not only a large array of interesting, unexpected, and robust empirical facts about memory. It also produced a remarkable consensus among practitioners regarding the interpretation of phenomena of memory in terms of processes such as encoding, storage, and retrieval, and the interactions among them (Crowder, 1976; Eysenck, 1977; Klatzky, 1980). More recently, systems-based views of memory have been making steady progress. Today our understanding of the relation between working memory and long-term memory (Baddeley, 1986), between episodic and semantic memory (Tulving, 1991, 1993), and between perceptual representation systems (PRS) and episodic memory (Schacter, 1990; Schacter, Chiu, and Ochsner, 1993) is considerably richer and sharper than it was only ten years ago. Given that the usefulness of conceptualizing memory in terms of memory processes and their interactions is almost universally accepted, and that the "structural" approach to memory—through postulation of multiple systems—is gaining ground, the time has now come to tackle the next problem, that of working out the relations between the two major classes of concepts of memory—processes and systems. Is there some systematic way of integrating memory processes and memory systems?

It is possible to propose a simple model, dubbed the SPI model, that may take us a step closer to this objective. SPI stands for *serial, parallel, and independent.* The model's central assumption is that the nature of relations among different cognitive systems is *process specific*: The relations among systems depend on the nature of the processes involved, as follows. (1) Information is *encoded* into systems *serially*, and encoding in one system is contingent on the successful processing of the information in some other system, that is, the output from one system provides the input into another. (2) Information is *stored* in different systems in *parallel*. The information in each system and subsystem, even if it all originates in one and the same act of perception, or "study episode," is different from that in others, its nature being determined by the nature of the original information and the properties of the system. Thus, what appears to be a single act of encoding—a single glance at a visual display, or a single short learning trial—produces multiple mnemonic effects, in different regions of the brain, all existing (i.e., available for potential access) in parallel. Thus, with respect to storage, different systems operate in parallel. (3) Information from each system and subsystem can be *retrieved* without any necessary implications for retrieval of corresponding information in other systems. Thus, with respect to the process of retrieval, different systems are *independent*.

The SPI model holds for *cognitive* memory systems. Its possible extension to noncognitive systems requires further study. The current list of cognitive systems (table 54.1) includes PRS, semantic memory, primary or working memory, and episodic memory. According to the model, when an event such as the presentation of an unfamiliar but meaningful sentence occurs, information about different aspects of the event may be registered in all four systems, or their appropriate subsystems. Information embodying the structural features of constituent stimulus objects (words) is registered in the perceptual representation (word form) systems. In general, the PRS information tells the brain about the kinds of objects that exist in the world. (Priming does not seem to occur for objects that the brain interprets globally but that do not exist in the world (Schacter et al., 1991; Nilsson, Olofsson, and Nyberg, 1992). The products of the processing in PRS can be retrieved, as happens in priming experiments, or they can be forwarded to the semantic systems for more elaborate processing of the relations among the

TABLE 54.2

Process-specific relations among cognitive memory systems

Process	Relation
Encoding	Serial
Storage	Parallel
Retrieval	Independent

words and their meaning. The output of the semantic system tells the brain about the contingencies of the world. This output normally also reaches both the working memory and the episodic memory systems. The former allows further elaboration of the information in terms of various kinds encoding and rehearsal operations; the latter computes the temporal-spatial contextual coordinates of the incoming information in relation to already existing episodic information, or to the self. The SPI scheme provides for the distribution of the information generated by a single event through many regions of the brain, whereby different aspects of the information are coded in their own specific, possibly unique, forms in different regions. We do not yet know whether the information within a given system is tightly localized or more widely distributed.

In summary, the SPI model suggests one possible way of answering some of the questions that can be posed about the relations between and among cognitive memory systems: In what sense are memory systems and subsystems independent? In what sense are they interdependent? How dow they interact? Do they operate in series? Does information enter into, or is it retrieved from, one system through another? Do they operate in parallel? The model proposes that there is no single answer to these questions: *Relations among systems are process specific*. Different systems are dependent on one another in the operations of interpreting, encoding, and initial storing of information. Once encoding has been completed, different kinds of information about the same initial event are held in various systems in parallel, depending upon the nature of the information and the evolved properties of the systems (Tulving, 1984a). Access to different kinds of information about a given event is possible in any system independently of what happens in other systems.

By relating different systems to one another through different processes, the SPI model suggests one way of integrating—bringing into a common reference frame—the concepts of memory processes and memory systems. As an abstract model, however, it specifies neither the nature of particular processes characteristic of different systems nor any neuroanatomical or neurophysiological substrates of the systems. It is compatible with many possible more specific and more concrete cognitive, neuropsychological, and neurocognitive models. For instance, we can replace any one of the abstract systems in the model with a corresponding (known or assumed) neuroanatomical concept—such as the hippocampal structure—without changing the basic logic. The basic assumptions of process-specific relations among systems remain the same: Encoding is serial, storage is parallel, and retrieval can be independent.

The SPI model not only helps us to order known facts, it also makes specific testable predictions. On reflection, the list of such facts and predictions turns out to be reasonably long; we consider a few examples here.

1. The model predicts that double dissociations are possible between tasks whose execution draws heavily on the resources of different systems, but only in post-acquisition situations—that is, in situations involving retrieval of previously stored information. Double dissociations may *not* be possible in comparable situations—same tasks, or using same systems—in which encoding of new information is required. Consider the double dissociation in retrieval of episodic and semantic information. Some brain-damaged patients can retrieve previously acquired semantic-memory information better than they can retrieve previously acquired episodic-memory information (e.g., Cermak and O'Connor, 1983), whereas others show the opposite pattern (e.g., DeRenzi, Liotti, and Nichelli, 1987). Such a double dissociation, however, should not be possible in situations requiring new learning. Indeed, no patients have so far been described who can remember an autobiographical encounter, say, with a celebrity, but who do not know what the celebrity is known for. According to the model such a double dissociation between semantic and episodic memory is not possible, and only single dissociations (impaired episodic memory and preserved semantic memory) can occur.

2. The model allows stochastic independence of retrieval of information representing nominally identical individual items, as revealed by contingency analyses, from PRS and semantic memory (e.g., Tulving, Hayman, and Macdonald, 1991), or PRS and episodic memory (e.g., Tulving, Schacter, and Stark, 1982), while assuming that successful encoding of information into the semantic-memory system is contingent upon successful processing of information *through* the PRS, and encoding of information into the episodic-memory system is contingent upon successful processing of information *through* the semantic system. In the absence of the model, the two sets of findings—serial encoding dependency and stochastic retrieval independence—would appear puzzling.

3. The serial encoding dependency of the SPI model allows acquisition of information by an earlier system (such as semantic) even if later systems (such as episodic) are dysfunctional, but not vice versa. Thus the model accords with experimental and clinical observations that brain-damaged patients suffering from anterograde amnesia may be capable of acquiring new semantic information even when they have no conscious recollection of any learning episodes (e.g., Schacter, Harbluk, and McLachlan, 1984; McAndrews, Glisky, and Schacter, 1987; Shimamura and Squire, 1987; Glisky and Schacter, 1988; for a review see Tulving, Hayman, and Macdonald, 1991). The learning observed in these studies was always slower than that in normal controls, but it did occur. The model's accordance with the data suggesting that semantic memory can be at least partly intact while episodic memory is severely impaired contrasts with some other models that rule out such a possibility. For instance, Squire's (1987, 1992) hippocampally oriented dichotomy between declarative and nondeclarative memory, in which declarative memory is defined in terms of the learner's ability to acquire and store information about general facts and personally experienced events, cannot be readily reconciled with the findings of differential impairment of episodic and semantic memory.

4. The SPI model accords with the many findings of asymmetries in the acquisition and retrieval of information in the brain. For example, it is well known that the limbic areas, including the hippocampal structures, play a crucial role in the acquisition of new information. The evidence comes from many observations that damage to these structures renders such acquisition difficult or impossible. (See Squire and Knowlton's chapter in the present volume.) However, after the initial acquisition, hippocampal patients or animals can frequently retrieve the learned information without difficulty (e.g., Staubli and Lynch, 1987). The famous amnesic patient H. M., whose bilateral medial temporal lobe resection rendered him essentially incapable of learning any new semantic or episodic information, has a largely unimpaired IQ, indicating unhindered access to a great deal of previously acquired information, and he can also produce reports of recollections from his youth (Milner, Corkin, and Teuber, 1968). Similar findings of asymmetry between acquisition and retrieval have been reported from experiments with amnestic drugs, such as benzodiazepines: These drugs impair new learning of semantic and episodic information, but leave retrieval of old learning intact (Lister, 1985; Ghoneim and Mewaldt, 1990).

5. Within the general framework of the SPI model it is possible to assume that the medial temporal lobe and diencephalic structures—damage to which frequently produces global amnesia—are critical for episodic memory insofar as episodic learning depends on the intact semantic systems. It is also possible to assume that other brain regions, including certain prefrontal cortical areas, are involved in the encoding and retrieval of aspects of personal experiences, such as temporal sequencing of separate and otherwise unrelated events (Milner, Petrides, and Smith, 1985; Schacter, 1987b). At Toronto we are currently exploring the hypothesis that the frontal pole regions (Brodmann Area 10) play a special role in episodic memory and autonoetic awareness of the past, and in projection of past experiences into the future in the form of plans and intentions (Ingvar, 1985; Tulving, 1985b; Stuss and Benson, 1986). We think of area 10, in addition to other prefrontal regions, as a promising candidate region in the quest for localization of episodic memory and autonoetic awareness, because, like episodic memory, it has evolved recently, it has appropriate connections to the limbic system, including the amygdala, and its function or functions are largely unknown (Pandya, personal communication, 1993).

Quo vadis?

The current version of the SPI model represents an extension and elaboration of earlier ideas concerning the relations among memory systems, especially the conjecture that, contrary to popular views, episodic memory evolves and develops later than, and in its operations depends upon, semantic memory (Tulving, 1984b, 1985a). It has clear affinities to other theories of organization of memory. Thus, to give just a few of many possible examples, it shares with Weiskrantz (1987) the current listing of five major systems, with Squire (1987) the ideas of earlier (nonhippocampal) and later (hippocampal) systems, with Johnson (1983) the notion of multiple entries of information into different systems, and with Moscovitch (1992) the emphasis on the nature of interrelations among memory processes and components.

With respect to serial processing of information through various systems at encoding, the SPI model is closely related to Lynch and Granger's (1992) "assembly line" model. In the Lynch and Granger model, too, operations occur in a serial fashion, leading, among other things, to "the expectation that late functions can be dissociably removed without affecting early functions, but that damage to early functions will also damage late functions" (1992, 196). Lynch and Granger come to their views on the basis of work on long-term potentiation as a possible storage mechanism of olfactory memory in rats, whereas the (mono) hierarchical features of the SPI model were suggested, among other things, by observed dissociations between forms of awareness associated with retrieval of personal and impersonal information in memory pathology (Tulving, 1985b; Tulving, Hayman, and Macdonald, 1991). Such a convergence of ideas from two rather different starting points may be purely coincidental. It is also possible, however, that the convergence reflects something more than just an accident. It may even point to the future, providing a hint of an answer to the question, Organization of memory: Quo vadis?

Conclusion

The proposal for organization of memory as described is just that, a proposal. The scheme will undoubtedly turn out to be inadequate, deficient, or just plain wrong. It nevertheless provides an explicit starting point for a more systematic pursuit of what is clearly the next problem that needs to be tackled. In science, as in chess, a plan or a theory, even a poor one, is better than no plan or theory at all. The confusion that usually prevails in the absence of a theory is likely to breed only more of the same, whereas an incorrect theory can be always be corrected.

ACKNOWLEDGMENTS The author's research is supported by the Natural Sciences and Engineering Research Council of Canada. The author is grateful to Reza Habib for assistance in the preparation of the chapter.

REFERENCES

BADDELEY, A. D., 1986. *Working Memory*. Oxford: Clarendon.

CERMAK, L. S., and M. O'CONNOR, 1983. The anterograde and retrograde retrieval ability of a patient with amnesia due to encephalitis. *Neuropsychologia*. 21:213–234.

COHEN, N. J., 1984. Preserved learning capacity in amnesia: Evidence for multiple memory systems. In *Neuropsychology of Memory*, L. R. Squire and N. Butters, eds. New York: Guilford, pp. 83–103.

CROWDER, R. G., 1976. *Principles of Learning and Memory*. Hillsdale, N.J.: Erlbaum.

DERENZI, E., M. LIOTTI, and P. NICHELLI, 1987. Semantic amnesia with preservation of autobiographic memory. A case report. *Cortex*. 23:575–597.

EBBINGHAUS, H., 1885. *Über das Gedächtnis*. Leipzig: Duncker and Humblot.

EYSENCK, M. W., 1977. *Human memory: Theory, Research and Individual Differences*. Oxford: Pergamon.

GHONEIM, M. M., and S. P. MEWALDT, 1990. Benzodiazepines and human memory: A review. *Anesthesiology*. 72:926–938.

GLISKY, E. L., and D. L. SCHACTER, 1988. Long-term retention of computer learning by patients with memory disorders. *Neuropsychologia*. 26:173–178.

GRAF, P., and D. L. SCHACTER, 1985. Implicit and explicit memory for new associations in normal and amnesic subjects. *J. Exp. Psychol. [Learn. Mem. Cogn.]* 11:501–518.

HAYMAN, C. A. G., and E. TULVING, 1989a. Contingent dissociation between recognition and fragment completion: The method of triangulation. *J. Exp. Psychol. [Learn. Mem. Cogn.]* 15:228–240.

HAYMAN, C. A. G., and E. TULVING, 1989b. Is priming in fragment completion based on a "traceless" memory system? *J. Exp. Psychol. [Learn. Mem. Cogn.]* 15:941–956.

INGVAR, D., 1985. "Memory of the future": An essay on the temporal organization of conscious awareness. *Hum. Neurobiol.* 4:127–136.

JOHNSON, M. K., 1983. A multiple-entry modular memory system. In *The Psychology of Learning and Motivation: Advances in Research and Theory*, G. H. Bower, ed. New York: Academic Press, pp. 81–123.

KLATZKY, R. L., 1980. *Human Memory: Structures and Processes*. San Francisco: W. H. Freeman.

LISTER, R. G., 1985. The amnesic action of benzodiazepines in man. *Neurosci. Behav. Rev.* 9:87–94.

LYNCH, G., and R. GRANGER, 1992. Variations in synaptic plasticity and types of memory in corticohippocampal networks. *J. Cognitive Neurosci.* 4:189–199.

MCANDREWS, M. P., E. L. GLISKY, and D. L. SCHACTER, 1987. When priming persists: Long-lasting implicit memory for a single episode in amnesic patients. *Neuropsychologia*. 25: 497–506.

MILNER, B., S. CORKIN, and H. L. TEUBER, 1968. Further analysis of the hippocampal amnesic syndrome: 14 year follow-up study of H. M. *Neuropsychologia*. 6:215–234.

MILNER, B., M. PETRIDES, and M. L. SMITH, 1985. Frontal lobes and the temporal organization of memory. *Hum. Neurobiol.* 4:137–142.

MISHKIN, M., B. MALAMUT, and J. BACHEVALIER, 1984. Memories and habits: Two neural systems. In *Neurobiology of Learning and Memory*, G. Lynch, J. L. McGaugh, and N. M. Weinberger, eds. New York: Guilford, pp. 65–88.

MOSCOVITCH, M., 1992. Memory and working-with-memory: A component process model based on modules and central systems. *J. Cognitive Neurosci.* 4:257–267.

NILSSON, L.-G., U. OLOFSSON, and L. NYBERG, 1992. Implicit memory of dynamic information. *Bull. Psychonom Soc.* 30: 265–267.

Richardson-Klavehn, A., and R. A. Bjork, 1988. Measures of memory. *Annu. Rev. Psychol.* 39:475–543.

Roediger, H. L., III, and K. B. McDermott, 1993. Implicit memory in normal human subjects. In *Handbook of Neuropsychology*, vol. 8, F. Boller and J. Grafman, eds. Amsterdam: Elsevier, pp. 63–131.

Schacter, D. L., 1987a. Implicit memory: History and current status. *J. Exp. Psychol. [Learn. Mem. Cogn.]* 13:501–518.

Schacter, D. L., 1987b. Memory, amnesia and frontal lobe dysfunction: A critique and interpretation. *Psychobiology.* 15:21–36.

Schacter, D. L., 1990. Preceptual representation systems and implicit memory: Toward a resolution of the multiple memory systems debate. *Ann. N. Y. Acad. Sci.* 608:543–571.

Schacter, D. L., 1992. Understanding implicit memory: A cognitive neuroscience approach. *Am. Psychol.* 47:559–569.

Schacter, D. L., C. Y. P. Chiu, and K. N. Ochsner, 1993. Implicit memory: A selective review. *Annu. Rev. Neurosci.* 16:159–182.

Schacter, D. L., L. A. Cooper, S. M. Delaney, M. A. Peterson, and M. Tharan, 1991. Implicit memory for possible and impossible objects: constraints on the construction of structural descriptions. *J. Exp. Psychol. [Learn. Mem. Cogn.]* 17:3–19.

Schacter, D. L., J. Harbluk, and D. McLachlan, 1984. Retrieval without recollection: An experimental analysis of source amnesia. *J. Verb. Learn. Verb. Be.* 23:593–611.

Schacter, D. L., and M. Moscovitch, 1984. Infants, amnesics, and dissociable memory systems. In *Infant Memory*, M. Moscovitch, ed. New York: Plenum, pp. 173–216.

Schacter, D. L., and E. Tulving, in press. What are the memory systems of 1994? In *Memory Systems 1994*, D. L. Schacter and E. Tulving, eds. Cambridge, Mass.: MIT Press.

Shallice, T., 1979. Neuropsychological research and the fractionation of memory systems. In *Perspectives in Memory Research*, L.-G. Nilsson, ed. Hillsdale, N.J.: Erlbaum, pp. 257–277.

Sherry, D. F., and D. L. Schacter, 1987. The evolution of multiple memory systems. *Psychol. Rev.* 94:439–454.

Shimumura, A. P., and L. R. Squire, 1987. A neuropsychological study of fact memory and source amnesia. *J. Exp. Psychol. [Learn. Mem. Cogn.]* 13:464–473.

Squire, L. R., 1982. The neuropsychology of human memory. *Annu. Rev. Neurosci.* 5:241–273.

Squire, L. R., 1987. *Memory and Brain.* New York: Oxford University Press.

Squire, L. R., 1992. Declarative and nondeclarative memory: Multiple brain systems supporting learning and memory. *J. Cognitive Neurosci.* 4:232–243.

Staubli, U., and G. Lynch, 1987. Stable hippocampal long-term potentiation elicited by 'theta' pattern stimulation. *Brain Res.* 435:227–234.

Stuss, D. T., and D. F. Benson, 1986. *The Frontal Lobes.* New York: Raven.

Tulving, E., 1979. Memory research: what kind of progress? In *Perspectives on Memory Research: Essays in Honor of Uppsala University's 500th Anniversary*, L.-G. Nilsson, ed. Hillsdale, N.J.: Erlbaum, pp. 19–34.

Tulving, E., 1983. *Elements of Episodic Memory.* Oxford: Clarendon.

Tulving, E., 1984a. Multiple learning and memory systems. In *Psychology in the 1990's*, K. M. J. Lagerspetz and P. Niemi, eds. Amsterdam: North Holland, pp. 163–184.

Tulving, E., 1984b. Relations among components and processes of memory. *Behav. Brain Sci.* 7:257–268.

Tulving, E., 1985a. How many memory systems are there? *Am. Psychol.* 40:385–398.

Tulving, E., 1985b. Memory and consciousness. *Can. Psychol.* 25:1–12.

Tulving, E., 1991. Concepts of human memory. In *Memory: Organization and Locus of Change*, L. Squire, G. Lynch, N. M. Weinberger, and J. L. McGaugh, eds. New York: Oxford University Press, pp. 3–32.

Tulving, E., 1993. What is episodic memory? *Current Directions in Psychological Sciences* 2:67–70.

Tulving, E., C. A. G. Hayman, and C. A. Macdonald, 1991. Long-lasting perceptual priming and semantic learning in amnesia: A case experiment. *J. Exp. Psychol. [Learn. Mem. Cogn.]* 17:595–617.

Tulving, E., and D. L. Schacter, 1990. Priming and human memory systems. *Science.* 247:301–306.

Tulving, E., D. L. Schacter, and H. A. Stark, 1982. Priming effects in word-fragment completion are independent of recognition memory. *J. Exp. Psychol. [Learn. Mem. Cogn.]* 8:352–373.

Warrington, E. K., 1979. Neuropsychological evidence for multiple memory systems. In *Brain and Mind: Ciba Foundation Symposium.* Amsterdam: Excerpta Medica, pp. 153–166.

Warrington, E. K., and L. Weiskrantz, 1968. New method of testing long-term retention with special reference to amnesic patients. *Nature.* 217:972–974.

Warrington, E. K., and L. Weiskrantz, 1974. The effect of prior learning on subsequent retention in amnesic patients. *Neuropsychologia.* 12:419–428.

Weinberger, N. M., J. L. McGaugh, and G. Lynch, 1985. *Memory Systems of the Brain.* New York: Guilford.

Weiskrantz, L., 1987. Neuroanatomy of memory and amnesia: A case for multiple memory systems. *Hum. Neurobiol.* 6:93–105.

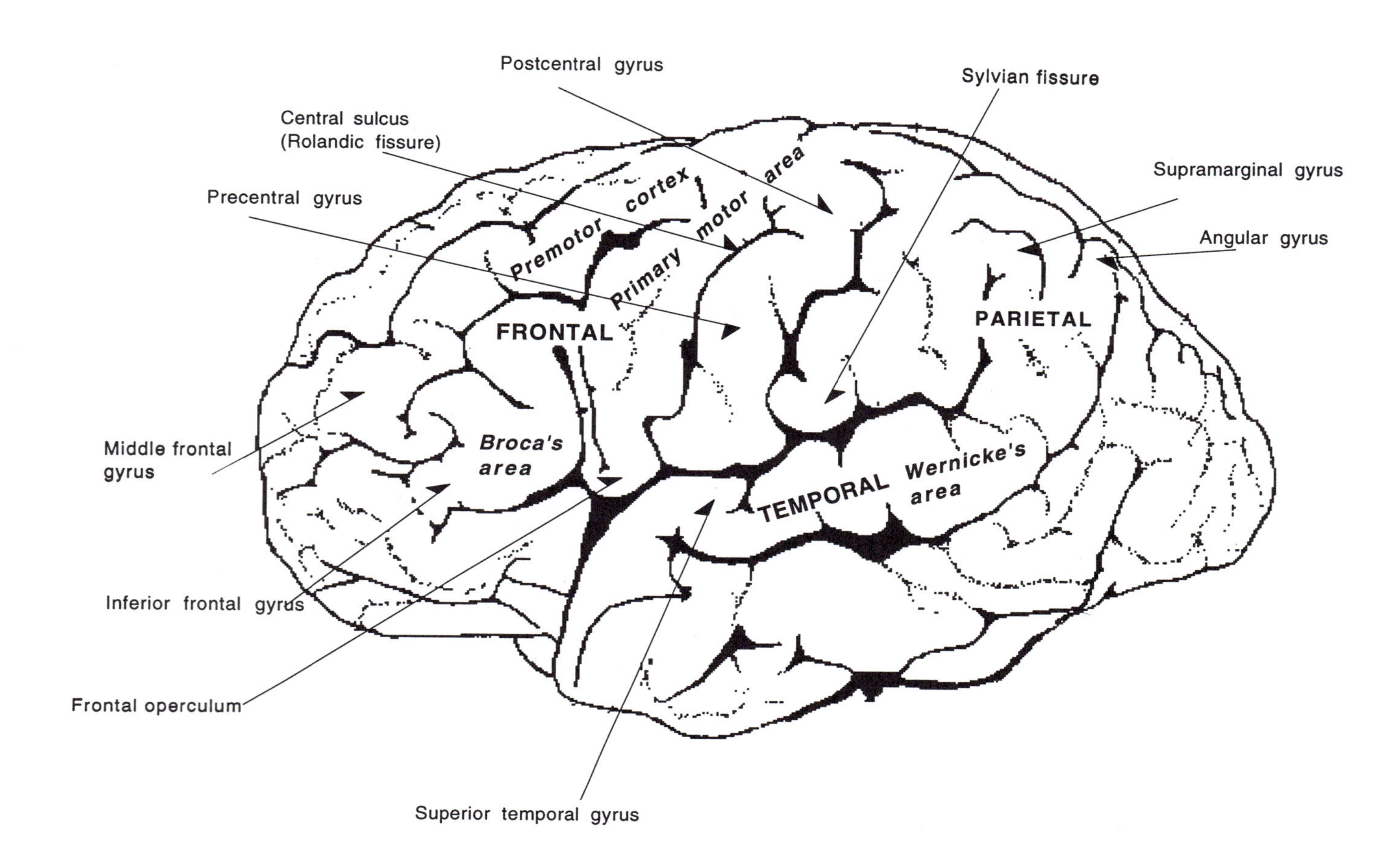
Postcentral gyrus
Sylvian fissure
Central sulcus
(Rolandic fissure)
Supramarginal gyrus
Precentral gyrus
Angular gyrus
Premotor cortex
Primary motor area
FRONTAL
PARIETAL
Middle frontal
gyrus
Broca's
area
TEMPORAL
Wernicke's
area
Inferior frontal gyrus
Frontal operculum
Superior temporal gyrus

VII LANGUAGE

Lateral view of the left cerebral cortex.

Introduction

STEVEN PINKER

LANGUAGE IS A system that allows humans to communicate thoughts by a highly structured stream of sound, or, in signed languages, manual gestures. Research in linguistics and psycholinguistics of the past 35 years has shown that all languages are based on the same complex computational design, which emerges spontaneously in all normal children in all societies (Pinker, 1994). The cognitive neuroscience of language is the study of the neural system that underlies this ability. Thus "language," in this sense of species-wide adaptation for mapping thoughts onto signals, should be distinguished from (a) the thoughts themselves, since thinking surely can go on in the absence of language (e.g., in infants, nonhuman primates, and aphasics); (b) reading and writing, which are recent inventions in human history and must be explicitly taught, with uneven results; (c) the "prescriptive grammar" of schoolmarms and style manuals, which list differences between standard and nonstandard dialects of English, and lay down conventions of written prose.

The design of language is based on two principles. The first is the word, an arbitrary pairing between a sound and a meaning. The word *duck* does not look like a duck, walk like a duck, or quack like a duck, but refers to a duck all the same, because the members of a language community, as children, all memorized the pairing. Young adults are estimated to have at least 60,000 words in their "mental dictionaries." The scientific concept of "word" in the sense of mental dictionary entry is not identical to the common-sense notion of a string of letters separated by blank spaces. On the one hand, idioms like *hit the fan* and *bite the dust* must be memorized entries; on the other, novel forms like *unfaxability* or *out-Nixoned* need not be listed in the mental dictionary but can be created and understood on the fly.

The second principle is grammar. Words can be combined into an bigger words, phrases, and sentences by rules that give a precise meaning to every combination. These rules, the "mental grammar," allow a person to express and understand an infinite number of novel thoughts. Thus we can distinguish between the newsworthy headline *Man Bites Dog* and the everyday *Dog Bites Man* by the positions of the words *man* and *dog* with respect to *bites*. Grammar consists of three components.

Morphology in the system that combines words and bits of words into larger words, as in *fax* + *able* + *ity* or *bite* + *s*. It has two subcomponents: *inflection*, the modification of a word according to its context (as in plural *-s* and past *-ed*), and *derivation*, the creation of new words from old ones (as in suffixes like *-able* and *-ity* and compounds like *toothbrush* and *Nixon-hater*).

Syntax combines words into phrases and sentences. The meaning and well-formedness of sentences do not depend directly on word-by-word ordering but on more complex principles of combination. First, words are interpreted in terms of their grouping into constituents, or "phrase structure"; thus the ordered string of words *On tonight's program, Dr. Ruth will discuss sex with Dick Cavett* has two meanings, depending on whether *Dick Cavett* is grouped with *sex* or with *discuss*. Second, phrases are assigned semantic or "thematic" roles by the verb: in *Man Fears Dog* and *Man Frightens Dog*, the verbs assign opposite roles to the subject and object phrases. Third, there are dependencies, often spanning long distances in a sentence, between a phrase and some other position: In *Which man did the dog bite?* the semantic role played by *Which man* (the bitee) is determined not by its own position but by the empty object position after the word *bite*.

Phonology defines combinations of sounds that conform to a consistent pattern in the language (e.g., the sound of English). It can be divided roughly into *segmental phonology*, the possible combinations of sound units into words (e.g., *blicket* is a possible English word; *ngagak* is not), and *prosody*, the patterns of intonation, stress, and timing that span many sound units. Prosody is used systematically in grammar (e.g., in determining the difference between *blackboard* and *black board*), and also more broadly in communication, as in questions, emphasis, sarcasm, and emotion.

Grammar is merely the code or protocol that defines a consistent signal-to-meaning mapping for some language. Actual production and comprehension in real time require input and output systems: *acoustic phonetics*, the decoding of the sound stream into words during speech perception; *parsing*, the mental grouping of words into phrases and sentences and the interpretation of their meanings during comprehension; *production*, the mental planning of the words and sentences to be uttered; and *articulatory phonetics*, the conversion of words into motor commands during speaking.

The study of the neural basis of language is in flux. Most neurology and psychology textbooks list discrete brain areas with dedicated input-output functions, damage to which causes well-defined syndromes (e.g., Broca's area for speech production, Wernicke's area for speech comprehension). But this century-old model had little empirical support to begin with, and antedated both modern linguistics and modern brain imaging. Because the effects of brain lesions on language are variable from person to person and rarely fit the textbook categories, the model is rejected by all research-oriented neurolinguists.

Aside from a near consensus that language is concentrated in the left perisylvian areas, there is no agreed-upon replacement for the textbook model, and there are no unambiguous anatomical correlates of the linguistic and processing components described above. One possibility is that language is somehow "distributed" all over the left hemisphere. But that conclusion may be premature. Language is probably computed by microcircuity defined at the cellular, subcellular, and synaptic levels, where manipulation and recording in the human brain is currently impossible. Thus there may be discrete language circuits, but current technology would have no way of identifying them, because they might differ in macroscopic size, shape, and location from person to person, have fine-grained, tortuous shapes that never coincide exactly with brain lesions, and compute internal information-processing subroutines rather than overt behavioral tasks like naming or repeating words. As the papers in this section attest, currently there is optimism that advances in static and functional brain imaging, developmental studies, linguistic and psycholinguistic analyses, and computational modeling will soon lead to better understanding of the neural organization of language.

REFERENCES

PINKER, S., 1994. *The Language Instinct.* New York: Morrow.

55 The Cognitive and Neural Bases of Language Acquisition

KARIN STROMSWOLD

ABSTRACT This chapter reviews findings from research on normal language acquisition, learnability theory, developmental and acquired language disorders, and language acquisition after the critical period that indicate that the ability to acquire language is the result of innate brain mechanisms. It is possible that infants' brains are predisposed to perceive categorically stimuli such as phonemes, words, syntactic categories, and phrases, and that this predisposition allows children to acquire language rapidly and with few errors.

Because the ability to learn a language is a uniquely human ability, language acquisition is an important topic in cognitive neuroscience. Perhaps the most fundamental question about language and language acquisition is the extent to which the ability to learn language is innate. Innate abilities often share certain characteristics. If an ability is innate, it usually is present in all normal individuals. Its acquisition tends to be uniform and automatic, with all normal individuals going through the same stages at the same ages, without specific instruction being required. There may be a critical period for successful acquisition. The ability is likely to be functionally and anatomically autonomous or modular with respect to other abilities, and the trait may be heritable.

Although these characteristics are by no means definitive, they can be used to evaluate supposedly innate traits. For example, the ability to walk presumably is innate, and exhibits most of the hallmarks of innate abilities, whereas the ability to knit presumably is not innate and exhibits few of these hallmarks. If children's brains are innately predisposed to learn language, then given adequate exposure to language, all children with normal brains should, without instruction, learn language in a uniform way, just as normal vision develops given adequate exposure to visual stimuli (Hubel and Wiesel, 1970). The ability to acquire language should be functionally and anatomically autonomous or modular from other abilities, and genetic and acquired lesions may specifically impair or spare the ability to learn language. If the ability to learn language is not innate, then instruction may be necessary to learn language, the course of acquisition may vary greatly from person to person (perhaps as a function of the quality of instruction), there should be no critical period for acquisition, and there should be no evidence for the functional or anatomical modularity of language.

KARIN STROMSWOLD Department of Psychology and Center for Cognitive Science, Rutgers University, New Brunswick, N.J.

Language development

LINGUISTICS AND THE UNIVERSAL FEATURES OF LANGUAGE Superficially, learning to talk differs from learning to walk in that children are capable of learning many different languages, but only one basic walk. If human languages are all fundamentally different from one another, it would be impossible for a child to be innately predisposed to be able to learn all human languages. Linguists have discovered that, although some languages seem to be radically different from other languages (e.g., Turkish and English), in essential ways all human languages are remarkably similar to one another (Chomsky, 1981, 1986; Croft, 1990). The principles-and-parameters approach to language argues that all languages share a common universal grammar—a common set of principles—and that any differences that exist among the grammars of languages are attributable to differences in the settings (or values) chosen from a finite and fixed set of parameters, or dimensions, for those principles (Chomsky, 1981, 1986). According to this view, children are born knowing the features that are universal to all languages (universal grammar), and all they must do to learn a particular language is learn the vocabulary and parametric settings of that language.

Uniformity in Language Acquisition The course of language acquisition is remarkably uniform (Brown, 1973). Most children say their first referential words at about 9 to 12 months of age (Morley, 1965), and for the next 6 to 8 months, children continue to acquire single words in a fairly slow fashion until they have acquired approximately 50 words. Most of children's first 50 words are labels for objects such as *cookie*, *mother*, *father*, and *bottle*, with a few action verbs such as *eat*, *come* and *go*, social terms such as *good-bye* and *hello*, and prepositions such as *up* and *down* rounding out the list (Nelson 1973). Once children have acquired 50 words, their vocabularies often increase rapidly in size (e.g., Reznick and Goldfield, 1992), as they acquire between 7 and 9 words a day from ages two to six years (Carey, 1977).

At around 18 to 24 months of age, children begin to combine words and form two-word utterances such as *want cookie*, *play checkers*, and *big drum* (Brown, 1973). During the two-word utterance stage, the vast majority of children's utterances are legitimate portions of sentences in the language they are learning. Thus, in a language such as English that has restricted word order, children will say *want cookie* but not *cookie want* (Brown, 1973), and *he big* but not *big he* (Bloom, 1990). Children gradually begin to use sentences longer than two words, but for several months their speech often lacks phonetically unstressed functional category morphemes such as determiners, auxiliary verbs, and verbal and nominal inflectional endings. Representative utterances during this period include *Sarah want cookie*, *Where Humpty Dumpty go?* and *Adam write pencil*. Because children's speech during this stage is similar to the way adults speak when words are at a premium (e.g., when they send a telegraph), children's early speech is often described as telegraphic (Brown, 1973). Gradually, omissions become rarer until by the time children are between three and four years old, the vast majority of their utterances are completely grammatical (Stromswold, 1990a,b; 1994b).

Most children master the syntax (grammar) of their language in a surprisingly similar manner (see Brown, 1973). For example, Brown (1973) found that all three children he studied acquired 14 grammatical morphemes in essentially the same order. Similarly, all 15 of the children I studied acquired the 20-odd English auxiliary verbs in essentially the same order (Stromswold, 1990a). The order in which these 15 children acquired complex constructions such as questions, negative constructions, passives, datives, exceptional case-marking constructions, embedded sentences, preposition-stranding constructions, causative constructions, small clause constructions, verb-particle constructions, and relative clauses constructions was also extremely regular (Stromswold, 1988, 1989a,b, 1990a,b, 1992, 1994b; Snyder and Stromswold, 1994). Finally, to a remarkable degree both within and across languages, children make certain types of mistakes and not others (see Errors, Instruction, and the Automaticality of Language later in this chapter).

The Acquisition of Syntactic Categories In order to acquire their language, children must not only learn the meanings of words like *cat* and *eat* (for a discussion of word learning see Soja, Carey, and Spelke, 1991), children must also learn that words like *cat* are nouns and words like *eat* are verbs. Learning the categorical membership of words (i.e., which words are nouns, which words are verbs, etc.) is critical because whether a syntactic or morphological rule applies to a particular word depends on the categorical membership of that word and not on the meaning of the word. This can easily be demonstrated with an example. Upon hearing the sentence *Linus cratomizes Lucy*, any speaker of English automatically knows that the nonsense word *cratomize* is a lexical verb. Even without knowing what *cratomize* means, an English-speaker automatically knows, for example, that its progressive form is *cratomizing* and its past tense form is *cratomized*; that *do*-support is required to ask a standard matrix question (e.g., *Did Linus cratomize Lucy?* and not **Cratomized Linus Lucy?*[1] or **Linus cratomized Lucy?*) or to negate an utterance (e.g., *Linus didn't cratomize Lucy*, and not **Linus cratomized not Lucy*, **Linus not cratomized Lucy*, etc.); and that the grammatical subject precedes rather than follows *cratomize* in simple declarative utterances (e.g., *Linus cratomizes Lucy* and not **Cratomizes Linus Lucy*). The fact that English-speakers know the syntactic and morphological behavior of *cratomize* without having the slightest idea what *cratomize* means demonstrates that categorical membership and not meaning determines syntactic and morphological behavior. A central question in the field of language acquisition is how children learn the categorical membership of words. For adults, the answer is simple. Even from the single sentence

Linus cratomizes Lucy, to an adult, *cratomize* is clearly a verb because it appears after the grammatical subject (*Linus*) and before the object (*Lucy*), has the third-person verbal inflection *-s*, and exhibits other verblike properties. The answer is much trickier for children.

How do children learn which words are verbs if they do not know what properties are typical of verbs, and how can they learn the properties of verbs if they do not know which words are verbs? One simple possibility is that every verb in every human language shares some readily accessible property, and children are innately predisposed to look for this property. Unfortunately, there doesn't seem to be any such property. Instead, infants probably must rely on a combination of prosodic cues, semantic cues, and correlational cues to learn which words are nouns and which words are verbs in their language (Pinker, 1987). Infants may use prosodic cues such as changes in fundamental frequency and lengthening to help determine where major phrasal boundaries are, and this plus knowledge of universal properties of phrases (e.g., verbs are contained within verb phrases; sentential phrases contain verbs) could help children learn which words are verbs (Jusczyk et al., 1992). Infants might also set up an enormous correlation matrix in which they record all of the behaviors associated with words, and categories could be the result of children noticing that certain behaviors tend to be correlated. For example, the verb category would result from children noticing that there is a class of words that often end in *-ing*, *-ed*, or *-s*, frequently occur in the middle of sentences, and rarely appear at the beginning of a sentence (see Maratsos and Chalkley, 1981). The problem with a simple, unconstrained correlational account is that there are an infinite number of correlations that children must consider, most of which will never appear in any language (Pinker, 1984, 1987). If infants are born "knowing" that in all languages, objects are expressed by nouns, physical actions are expressed by verbs, and attributes are expressed by adjectives, infants could infer that words referring to physical objects are nouns, words referring to actions are verbs, and words referring to attributes are adjectives. They could learn the properties of nouns and verbs from these semantically prototypical cases, a process often referred to as semantic bootstrapping (see Pinker, 1984, 1987).

The Acquisition of Auxiliary and Lexical Verbs

In a sense, the paradox of syntax acquisition is that unless children basically know what they have to learn before they begin, they cannot succeed in learning the grammar of their language. One might argue that, although it has been demonstrated that children must have innate mechanisms that allow them to learn the categorical membership of words, it is possible that these innate mechanisms are not specifically linguistic. The acquisition of auxiliary verbs and lexical verbs can be used to determine whether children have specifically linguistic innate mechanisms that allow them to acquire language.

The acquisition of English auxiliary and lexical verbs is a particularly good test case because the two types of verbs are so semantically, syntactically, and lexically similar to one another that a learner who has no knowledge about auxiliary and lexical verbs (i.e., a simple correlational learner) is almost certain to confuse the two types of verbs. For example, for many auxiliaries there is a lexical verb counterpart with an extremely similar meaning (e.g., the pairs *can* and *is able to*, *will* and *is going to*, *must* and *have to*). Auxiliary and lexical verbs are syntactically similar in that both types of verbs often take verbal endings, follow subject noun phrases, and lack all of the grammatical properties of nouns, adjectives, and other syntactic categories. Finally, auxiliary and lexical verbs typically are identical forms (e.g., copula and auxiliary forms of *be*, possessive and auxiliary forms of *have*, lexical verb and auxiliary forms of *do*, etc.). The remarkable degree of similarity can be appreciated by comparing pairs of sentences such as *He is sleepy* and *He is sleeping*, *He has cookies* and *He has eaten cookies* and *He does windows* and *He does not do windows*.

The syntactic and morphological behavior of auxiliaries is extremely complex, and there are no obvious nonlinguistic correlates for this behavior to aid in learning (Stromswold, 1990a). Without innate, specifically linguistic mechanisms, it is unclear how children would be able to correctly identify the 100 unique strings of auxiliaries that are acceptable in English from among the over 20! (2.43×10^{18}) unique strings of English auxiliaries (Stromswold, 1990a). Descriptively, the basic restrictions on auxiliaries can be summarized as follows:

AUX →
(Modal) (*have -en*) (progressive *be -ing*) (passive *be -en*)

Any or all of the auxiliaries are optional, but if they are present, they must occur in the above order. In addition, each auxiliary requires that the verb that follows it be of a certain form. Modal auxiliaries (e.g., *can, will,* and *might*) require that the verb that follows be an infinitival form (e.g., *eat*); perfect *have* requires that the verb that follows be a perfect participle (e.g., *eaten*); progressive *be* requires that the verb that follows be a progressive participle (e.g., *eating*); and passive *be* requires that the main verb be a passive participle (e.g., *eaten*). In addition, the first verbal element must be tensed in a matrix clause. Finally, matrix questions and negative statements are formed by inverting or negating the first auxiliary. If no auxiliary is present, *do*-support is required (see Stromswold, 1990a, Stromswold, 1992 for additional restrictions and complications). Lexical and auxiliary verbs pose a serious learnability question (Baker, 1981; Pinker, 1984; Stromswold, 1989a, 1990a, 1992): How can children distinguish between auxiliary and lexical verbs before they learn the behavior of the two types of verbs, and how do children learn the two types of verbs' behaviors before they can distinguish between them?

If children do not distinguish between auxiliary and lexical verbs, they will generalize what they learn about one type of verb to the other type of verb. This will result in rapid learning. It will also lead children to make errors that can only be set right by negative evidence (explicit information that a particular construction is ungrammatical). Unfortunately, parents don't seem to provide negative evidence (Brown and Hanlon, 1970). Thus, if children do not distinguish between auxiliaries and lexical verbs, they are destined to make certain types of inflectional errors (e.g., **I aming go*, **I musts eat*), combination errors involving multiple lexical verbs (e.g., **I hope go Disneyland*), negated lexical verbs (e.g., **I eat not cookies*), lone auxiliaries (e.g., **I must coffee*), and unacceptable combinations of auxiliaries (e.g., **I may should go*). They will also make word order errors, scrambling the order of lexical verbs and auxiliaries (e.g., **I go must*), scrambling the order of auxiliaries (e.g., **He have must gone*), and incorrectly inverting lexical verbs (e.g., *eats he meat?*). If, on the other hand, children have innate predispositions that allow them to distinguish between auxiliary and lexical verbs, they may not make these errors.

In order to test whether English-speaking children distinguish between auxiliary and lexical verbs, I searched the transcripts of 14 children's speech, examining by hand the over 66,000 utterances by these children that contained auxiliaries (Stromswold, 1989a, 1990a). I found that the children acquired the auxiliary system with remarkable speed and accuracy. In fact, I found no clear examples of the types of inflectional errors, combination errors, or word order errors that they would have made had they confused auxiliary and lexical verbs. Thus, children seem to have innate, specifically linguistic mechanisms that allow them to distinguish between auxiliary and lexical verbs.

ERRORS, INSTRUCTION, AND THE AUTOMATICALITY OF LANGUAGE One of the hallmarks of innate abilities is that they can be acquired without explicit instruction. This seems to be true for language. Although parents do correct their children when they make errors that affect the meaning of utterances, parents do not seem to reliably correct children's grammatical errors (Brown and Hanlon, 1970; Marcus, 1993). Furthermore, even when parents *do* try to correct grammatical errors, their efforts are often in vain (McNeill, 1966). Moreover, it is clear that correction is not necessary for lexical and syntactic acquisition, because some children who are unable to speak (and, therefore, cannot be corrected by their parents) have normal receptive language (Stromswold, 1994a). If teaching and correction is necessary for language development, it should not be possible for children to have impaired production and intact comprehension. Recently, I have been studying the language acquisition of a young child who is unable to speak. Despite the fact that he has essentially no expressive language (he can only say a handful of phonemes), his receptive language is completely intact. For example, at age 4 he was able to distinguish between reversible active and passive sentences (correctly distinguishing the meanings conveyed by sentences such as *The dog kicked the cat, The cat kicked the dog, The dog was kicked by the cat,* and *The cat was kicked by the dog*) and to make grammaticality judgments (e.g., correctly recognizing that *What can Cookie Monster eat?* is grammatical, whereas **What Cookie Monster can eat?* is not) (see Stromswold, 1994a).

Children learn language quickly, never making certain types of errors that seem very reasonable to make (e.g., certain types of auxiliary errors). But as Pinker (1989) points out, children are not perfect: They *do* make certain types of errors. They overregularize inflectional endings, saying *eated* for *ate* and *mouses* for *mice* (Pinker, 1989). They make lexical errors, such as sometimes passivizing verbs such as *die* that do not passivize (e.g., *He get died*; from Pinker, 1989). They also make certain types of syntactic errors, such as using *do*-support when it is not required (e.g., *Does it be around it?* and *This doesn't be straight*; Stromswold, 1990a, 1992) and failing to use *do*-support when it is required (e.g., *What she eats?* Stromswold, 1990a, 1990b). What do these errors tell us? First, they confirm that children use language productively and are not merely repeating back what they hear their parents say, because parents do not use these unacceptable forms (Pinker, 1989). These errors may also provide an insight into the peculiarities of languages. For example, children's difficulty with *do*-support suggest that *do*-support is not part of universal grammar, but rather is a peculiar property of English (Stromswold, 1990a,b, 1994b).

Finally, these errors may provide insight into the types of linguistic categories that children are predisposed to acquire. Consider, for example, the finding that children overregularize lexical *be*, *do*, and *have* but they *never* overregularize auxiliary *be*, *do*, and *have* (Stromswold, 1990a, 1994c). The fact that children say sentences like **She beed happy* but not sentences like **She beed smiling* indicates that children not only can distinguish between auxiliary verbs and lexical verbs, as discussed in the previous section, but they treat the two types of verbs differently. What kind of innate learning mechanism could result in children overregularizing lexical verbs but not the homophonous auxiliaries? One possibility is that children have specific, innate learning mechanisms that cause them to treat auxiliaries differently than lexical verbs. Unfortunately, there are problems with this explanation. Athough many languages contain words that are semantically and syntactically similar to English auxiliaries (Steele, 1981), and all languages are capable of making the semantic and syntactic distinctions that in English are made by auxiliaries, some languages either appear not to have auxiliaries (instead making use of inflectional affixes) or not to distinguish between auxiliaries and lexical verbs. Given that not all languages contain auxiliary verbs that can easily be confused with lexical verbs, it seems unlikely that there are innate mechanisms that permit children to distinguish specifically between auxiliary verbs and lexical verbs. In addition, the hypothesis that specific innate mechanisms allow children to distinguish between auxiliaries and lexical verbs has little explanatory power because it explains nothing beyond the phenomena that led us to propose it.

Alternatively, children's ability to distinguish between auxiliary and lexical verbs might reflect a more general ability to distinguish between functional categories (e.g., determiners, auxiliaries, nominal and verbal inflections, pronouns, etc.) and lexical categories (nouns, verbs, adjectives, etc.). Lexical categories are fairly promiscuous: They freely admit new members (e.g., *fax*, *modem*, *email*, etc.) and the grammatical behavior of one member of a lexical category can fairly safely be generalized to another member of the same lexical category. Functional categories are conservative: New members are not welcome and generalizations even within a functional category are very dangerous (see Stromswold, 1990a, 1994c). Innate mechanisms that specifically predispose children to distinguish between lexical and functional categories have a number of advantages over similar mechanisms for auxiliary and lexical verbs. Unlike the distinction between auxiliary and lexical verbs, the distinction between lexical and functional categories is found in all human languages; thus, mechanisms that predispose children to distinguish between lexical categories and functional categories are better candidates a priori for being innate. In addition, research on speech errors (e.g., Garrett, 1976), neologisms (Stromswold, 1994c), parsing (e.g., Morgan and Newport, 1981), linguistic typology (e.g., Croft, 1990), aphasia (e.g., Goodglass, 1976), developmental language disorders (e.g., Guilfoyle, Allen, and Moss, 1991), and event-related potentials (Neville, 1991) all points to the importance of the lexical-functional distinction.

If children have innate mechanisms that predispose them to distinguish between lexical and functional categories, this would help them to distinguish not only between auxiliary and lexical verbs but also between pronouns and nouns, determiners and adjectives, verbal stems and verbal inflections, and other pairs of lexical and functional categories that are hard to distinguish using nonlinguistic knowledge. If children have innate mechanisms that specifically predispose them to distinguish between syntactic categories that allow for free generalization (lexical categories) and those that do not (functional categories), this would explain why children overregularize lexical *be*, *do* and *have* but not auxiliary *be*, *do*, and *have*. It would also help explain why children are able to learn language so rapidly and with so few errors, because such a learning mechanism would permit children to generalize only where it is safe to do so (i.e., within a lexical category).

Computationally, the difference between lexical and functional categories might be expressed as the difference between rule-based and list-based generalizations or, within a connectionist framework, between network architectures that have different degrees and configurations of connectivity (see Stromswold, 1994c).

Language acquisition and brain development

In the section on language development, results of research on normal language acquisition were used to argue that the ability to learn language is the result of innate, language-specific learning mechanisms. This section of the chapter will review neurobiological evidence that supports this theory.

THE DEVELOPMENT OF LANGUAGE REGIONS OF THE BRAIN Contrary to claims by Lenneberg (1967), the language areas of the human brain appear to be anatomically and functionally asymmetrical at or before birth. Anatomically, analyses of fetal brains reveal that the temporal plane is larger in the left hemisphere than in the right hemisphere (Wada, Clarke, and Hamm, 1975). Development of the cortical regions that subserve language in the left hemisphere consistently lags behind the development of the homologous regions in the right hemisphere. The right temporal plane appears during the thirtieth gestational week, whereas the left temporal plane first appears approximately 7–10 days later (Chi, Dooling, and Gilles, 1977). Even in infancy, dendritic development in the region around Broca's area on the left lags behind that found in the homologous region on the right (Scheibel, 1984). Event-related potential (ERP) and dichotic listening experiments suggest that the left hemisphere is differentially sensitive for speech from birth (for a review see Mehler and Christophe, this volume).

Only a few studies have investigated the neural bases of lexical or syntactic abilities in neurologically intact children. For example, Molfese and his colleaguges (Molfese, 1990; Molfese, Morse, and Peters, 1990) taught infants as young as 14 months labels for novel objects and then compared the children's ERPs when the novel objects were paired with correct verbal labels and incorrect verbal labels. A late-occurring response was recorded in the left hemisphere electrode sites when the correct label was given but not when an incorrect label was given. Similarly, an early-occurring response was recorded bilaterally in the frontal electrodes when the correct label was given, but not when an incorrect label was given. In another ERP study, Holcomb, Coffey, and Neville (1992) found no clear evidence prior to age 13 of the normal adult pattern of greater negativity in the left hemisphere than the right hemisphere for semantically plausible sentences (e.g., *We baked cookies in the oven*) and greater negativity in the right hemisphere than the left hemisphere for semantically anomalous sentences (e.g., *Mother wears a ring on her school*). In addition, the negative peak associated with semantical anomalies (the N400) was later and longer in duration for younger subjects than older subjects (Holcomb, Coffey, and Neville, 1992). In summary, although linguistic stimuli evoke similar types of electrical activity in young children's brains as those recorded in adult brains, children's ERPs may not become indistinguishable from adult ERPs until around puberty.

THE MODULARITY OF LANGUAGE ACQUISITION With a few exceptions such as those just mentioned, most of what is known about the relationship between brain development and lexical and syntactic development has come from studying exceptional language acquisition. If, as was argued earlier, language acquisition involves the development of specialized structures and operations that have no counterparts in nonlinguistic domains, then it should be possible for a child to be cognitively intact and linguistically impaired or to be linguistically intact and cognitively impaired. If, on the other hand, language acquisition involves the development of the same general symbolic structures and operations used in other cognitive domains, then dissociation of language and general cognitive development should not be possible. Recent studies suggest that language development is selectively impaired in children with specific language impairment (SLI) and selectively spared in children who suffer from disorders such as Williams syndrome.

Specific language impairment The term "specific language impairment" (SLI) is often used to refer to developmental disorders that are characterized by severe deficits in the production and/or comprehension of language that cannot be explained by hearing loss, mental retardation, motor deficits, neurological or psychiatric disorders, or lack of exposure to language. Because SLI is a diagnosis of exclusion, SLI children are a very heterogeneous group. This heterogeneity can and does affect the outcome of behavioral and neurological studies, with different studies of SLI children frequently reporting different results depending on how SLI subjects were chosen. The exact nature of the deficits underlying SLI remains uncertain; suggestions range from transient, fluctuating hearing loss (Gordon, 1988; but see Bishop and Edmundson, 1986), impairments in auditory sequencing (Efron, 1963), impairments in rapid auditory processing (Tallal and Piercy, 1973), general impairment in representational or symbolic reasoning (Morehead and Ingram, 1973), to specifically linguistic deficits such as the inability to extract linguistic features (Gopnik, 1990a,b). At the neural level, the cause of SLI is also uncertain. Initially, it was theorized that children with SLI had bilateral damage to the perisylvian cortical regions that subserve language in adults (Bishop, 1987). Because SLI is not a fatal disorder and people with SLI have normal life spans, to date only one brain of a possible SLI child has come to autopsy. Postmortem examination of this brain revealed atypical symmetry of the temporal planes and a dysplastic microgyrus on the inferior surface of the left frontal cortex along the inferior surface of the sylvian fissure (Cohen, Campbell, and Yaghmai, 1989), findings similar to those reported in dyslexic brains by Geschwind and Galaburda (1987). Although it is tempting to use the results of this autopsy to argue—as Geschwind and Galaburda (1987) have for dyslexia—that SLI is the result of subtle anomalies in the left perisylvian cortex, the child whose brain was autopsied had a performance IQ of only 74 (and a verbal IQ of 70), and, hence, the anomalies noted on autopsy may have been related to the child's general cognitive impairment rather than to her language impairment.

Computed tomography (CT) and magnetic resonance imaging (MRI) scans of SLI children have failed to reveal the types of gross perisylvian lesions typically found in patients with acquired aphasia (Jernigan et al., 1991; Plante et al., 1991). CT and MRI scans have revealed, however, that the brains of SLI children often do not have the normal pattern of the left temporal plane being larger than the right temporal plane. A number of researchers have also studied the functional characteristics of SLI children's brains. Data from dichotic listening experiments (e.g., Arnold and Schwartz, 1983; Boliek, Bryden, and Obrzut, 1988; Cohen et al., 1991) and ERP experiments (e.g., Dawson et al., 1989) suggest that at least some SLI children have aberrant functional lateralization for language, with language present either bilaterally or predominantly in the right hemisphere. Single photon emission computed tomography (SPECT) studies of normal and language-impaired children have revealed hypoperfusion in the inferior frontal convolution of the left hemisphere (including Broca's area) in two children with isolated expressive language impairment (Denays et al., 1989), hypoperfusion of the left temporoparietal region and the upper and middle regions of the right frontal lobe in 9 of 12 children with expressive and receptive language impairment (Denays et al., 1989), and hypoperfusion in the left temporofrontal region of language-impaired children's brains (Lou, Henriksen, and Bruhn, 1990). In summary, in vivo studies suggest that SLI brains lack the functional and anatomical asymmetries typically associated with language.

Hyperlexia Although mental retardation generally results in depression of language function (Rondal, 1980), researchers have reported that some mentally retarded children have remarkably intact language. This condition, called "hyperlexia," has been reported in some children with hydrocephalus (Swisher and Pinsker, 1971), Turner's syndrome (Yamada and Curtiss, 1981), infantile hypercalcemia or Williams syndrome (Bellugi et al., 1992), and mental retardation of unknown etiology (Yamada, 1990).

Children with Williams syndrome (WS) have particularly extreme dissociation of language and cognitive functions (Bellugi et al., 1992). Hallmarks of WS include microcephaly with a "pixie-like" facial appearance, general mental retardation with IQs typically in the 40s–50s range, delayed onset of expressive language, and "an unusual command of language combined with an unexpectedly polite, open and gentle manner" by early adolescence (Von Armin and Engel, 1964). Although their language is often deviant for their chronological age, Williams syndrome children have larger vocabularies and speak in sentences that are more syntactically and morphologically complex and well formed than do children of equivalent mental ages. In addition, WS children demonstrate good metalinguistic skills such as the ability to recognize an utterance as ungrammatical and to respond in a contextually appropriate manner (Bellugi et al., 1992).

Volumetric analyses of MRI scans indicate that compared to normal brains, cerebral volume and cerebral gray matter of WS brains are significantly reduced in size and the neocerebellar vermal lobules are increased in size, with paleocerebellar vermal regions of low-normal size (Jernigan et al., 1993). Although auditory ERPs for WS adolescents have similar morphology, distribution, sequence, and latency to those found in age-matched controls, WS adolescents display large-amplitude responses even at short interstimulus intervals, suggesting hyperexcitability of auditory mechanisms at the cortical level (Neville, Holcomb, and Mills, 1989). To date no PET or SPECT studies of WS children have been reported. It will be interesting to learn from such studies whether the classically defined language areas in WS brains become hyperperfused in response to linguistic stimuli.

THE GENETIC BASIS OF LANGUAGE If the acquisition of language is the result of specialized structures in the brain that are coded for by information contained in the genetic code (see Pinker and Bloom, 1990), then one might expect to find evidence for the hereditability of language. If, on the other hand, language acquisition is essentially the result of instruction and involves no specifically linguistic structures, there should be no evidence of genetic transmission of language.

Familial aggregation studies Developmental language disorders such as SLI often seem to run in families. Depending on what was counted as evidence of language impairment, studies have revealed that between 30% and 40% of language-impaired children have a family history of language impairment (McCready, 1926; Ingram, 1959; Luchsinger, 1970; Byrne, Willerman, and Ashmore, 1974; Fundudis, Kolvin, and Garside, 1979; Martin, 1981; Robinson, 1987; Tallal, Ross, and Curtiss, 1989a; Tomblin, 1989), whereas only 3% to 14% of non-language-impaired children have a positive family history of language impairment (Fundudis, Kolvin, and Garside, 1979; Bishop and Edmundson, 1986; Neils and Aram, 1986; Tallal, Ross, and Curtiss, 1989a; Tomblin 1989).[2]

Although data on familial aggregation of language disorders suggest that some developmental language disorders have a genetic component, it is possible that children who have language-impaired parents or siblings are more likely to be linguistically impaired themselves because they are exposed to deviant language. If this is the cause of familial aggregation of language impairment, then the most severely linguistically impaired children should come from families with the highest incidence of language impairment. Contrary to this prediction, Byrne, Willerman, and Ashmore (1974) found that children who had profound language impairments were *less* likely to have positive family histories of language impairment than were children who were only moderately language impaired, and Tallal and colleagues (1991) found no differences in the language abilities of children who did and did not have a positive family history of language disorders. Similarly, if familial aggregation is merely the result of a deviant language environment, then within a family with a high incidence of language impairment one would not expect to find any family members with completely normal language. In their study of a single extended family, Gopnik and Crago (1991) found that 13 of the 29 family members had absolutely normal language, despite extensive exposure to deviant language from the 16 family members who were language impaired.

Twin studies The influences of environmental and genetic factors can be teased apart by comparing the concordance rates for language impairment in monozygotic (MZ) and dizygotic (DZ) twins. Because MZ and DZ twins share the same pre- and postnatal environment, if the concordance rate for a particular trait is greater for MZ than DZ twins, this can be taken to reflect the fact that MZ twins share 100% of their genetic material, whereas, on average, DZ twins share only 50% of their genetic material (for a review, see Eldridge, 1983). In a study of articulatory errors in preschool-age children, Locke and Mather (1989) found concordance rates of 82% in MZ twins and 56% in DZ twins. In a study of 32 pairs of single-sex MZ twins and 25 pairs of single sex DZ twins in which at least one member of each set of twins had a spoken language impairment, Lewis and Thompson (1992) found an 86% concordance rate for the MZ twins and a 48% concordance rate for the DZ twins. To date, more twin studies have focused on written language than on spoken language. If reading and spelling disabilities are simply expressions of a more general language impairment (see Geschwind and Galaburda, 1987), then the results of these twin studies of dyslexia can be used to determine whether there is a strong genetic component to developmental language impairments. In a study of reading ability, Bakwin (1973) found an 83% concordance rate for 31 pairs of MZ twins and a 29% concordance rate for 31 pairs of DZ twins. Analyzing a different group of MZ and DZ twins, Stevenson et al. (1987) found evidence of a strong genetic influence on spelling abilities, but only a moderate influence on reading abilities. A multiple regression analysis of the reading abilities of 64 pairs of MZ twins and 55 pairs of DZ twins, in which at least one twin in each pair was dyslexic, suggests that about 30% of dyslexics' reading deficits are the result of heritable factors (DeFries, Fulker, and LaBuda, 1987). In summary, the higher concordance rates for written and spoken language among MZ twins than DZ twins suggests there is a significant genetic component for developmental language disorders.

Modes of transmission If careful pedigree studies of large, multigeneration families reveal that the pattern of family members who are language impaired and non-language-impaired is as would be predicted given a certain type of genetic transmission, this supports the notion that SLI is a genetic disorder. It should be remembered, however, that complex behavioral traits are usually the result of multiple genes interacting in complex ways, with less than 100% penetrance and expressivity of a trait. Thus, the failure to find a simple model of transmission cannot be taken as evidence that a behavioral trait is not genetically transmitted (see Pauls, 1983). Hurst et al. (1990) performed a pedigree analysis on the same family studied by Gopnik (Gopnik, 1990a) and concluded that the mode of transmission in this family was a single autosomal dominant gene with near 100% penetrance. Samples and Lane (1985) performed a similar analysis on a family in which six of six siblings had a severe developmental language disorder and concluded that the mode of transmission in that family was a single autosomal recessive gene. It has been suggested that the sex-ratio data for SLI are most consistent with an autosomal dominant transmission, with greater penetrance through mothers than fathers (see Tallal, Ross, and Curtiss, 1989b). If there are multiple modes of transmission for SLI, as the results of these studies seem to indicate, this suggests that SLI is genetically heterogeneous, just as dyslexia appears to be genetically heterogeneous.

The final—and most definitive—method for determining whether there is a genetic basis for familial language disorders is to determine which gene (or genes) is responsible for the language disorders found in these families. The typical way this is done is to use linkage analysis techniques to compare the genetic material of language-impaired and normal family members and determine how the genetic material of affected family members differs from that of unaffected family members. Linkage analyses performed on 245 members of 19 extended families with a high incidence of dyslexia suggest that dyslexia is a genetically heterogeneous disorder, linked to chromosome 15 in about 30% of the 16 families (Pennington and Smith, 1988). Although there are no such results for SLI, Gopnik (1990a) reports that Pembry is performing linkage analyses in the extended family studied by Gopnik (1990a) and by Hurst et al. (1990).

RECOVERY FROM POSTNATALLY ACQUIRED BRAIN DAMAGE Lesions acquired during infancy typically result in transient, minor linguistic deficits, whereas similar lesions acquired during adulthood typically result in permanent, devastating language impairments (see, for example, Guttmann, 1942; Lenneberg, 1967; and the references cited in this section). The more optimistic prognosis for childhood aphasia may reflect that less neuronal pruning has occurred in young brains (Cowan et al., 1984), and that the creation of new synapses and the reactivation of latent synapses is more likely in younger brains (Huttenlocher, 1979). Language recovery after childhood aphasia typically has been attributed either to recruitment of brain regions that are adjacent to the damaged perisylvian language regions in the left hemisphere or to recruitment of the topographically homologous regions in the undamaged right hemisphere. According to Lenneberg (1967), prior to puberty the right hemisphere can completely take over the language functions of the left hemisphere. The observation that infants and toddlers who undergo complete removal of the left hemisphere acquire or recover near-normal language suggests that the right hemisphere can take over *most* of the language functions of the left hemisphere if the transfer of function happens at an early enough age (Dennis and Kohn, 1975; Dennis and Whitaker, 1976; Dennis, 1980; Rankin, Aram, and Horwitz, 1981; Byrne and Gates, 1987; but see Bishop, 1983, for a critique). Because few studies have examined the linguistic abilities of children who undergo left hemispherectomy during middle childhood, the upper age limit for hemispheric transfer of language is unclear. Right-handed adults who undergo left hemispherectomy typically become globally aphasic with essentially no recovery of language (e.g., Zollinger, 1935; Crockett and Estridge, 1951; Smith, 1966). The observation that a right-handed 10-year-old child (Gardner et al., 1955) and a right-handed 14-year-old child (Hillier, 1954) who underwent left hemispherectomy reportedly suffered from global aphasia with modest recovery of language function suggests that hemispheric transfer of language function is greatly reduced by not completely eliminated by puberty.

Studies that reveal that left-hemisphere lesions are more often associated with (subtle) syntactic deficits than are right-hemisphere lesions (Dennis and Kohn, 1975; Dennis and Whitaker, 1976; Woods and Carey, 1979; Dennis, 1980; Rankin, Aram, and Horwitz, 1981; Kiessling, Denckla, and Carlton, 1983; Aram et al., 1985; Aram, Ekelman and Whitaker, 1986; Byrne and Gates, 1987; Thal et al., 1991) call into question the complete equipotentiality of the right and left hemispheres for language and suggest that regions in the left hemisphere may be uniquely and innately designed to acquire syntax. It should be noted, however, that some studies have not found greater syntactic deficits with left- than right-hemisphere lesions (e.g., Basser, 1962; Feldman et al., 1992; Levy, Amir, and Shalev, 1992). Such studies may have included children whose lesions were smaller (Feldman et al., 1992) or in different locations than those of the children described in studies that have found a hemispheric difference for syntax.

In children who suffer from partial left-hemisphere lesions rather than complete left hemispherectomies, language functions could be assumed by adjacent undamaged tissues within the left hemisphere or by homotopic structures in the intact right hemisphere. Results of Wada tests (in which lateralization of language is determined by testing language function when each hemisphere is temporarily anesthetized) indicate that children with partial left-hemisphere lesions often have language represented bilaterally or in the right hemisphere (Rasmussen and Milner, 1977; Mateer and Dodrill, 1983). However, one ERP study suggests that children with partial left-hemisphere lesions are more likely to have language localized in the left hemisphere than the right hemisphere (Papanicolaou et al., 1990). There are a number of possible reasons for this discrepancy, including differences in the types of linguistic tasks used in the ERP and Wada studies and possible differences in the sizes and sites of left-hemisphere lesions in the children studied. In addition, it is possible that the discrepancy is due to the fact that most of the children in the ERP study acquired their lesions after age four and that it is unclear to what extent any of them ever exhibited any sign of language impairment.

Although there is some disagreement about the details of language recovery after postnatally acquired left-hemisphere lesions, the following generalizations can be made. Behaviorally, the prognosis for recovery of language is generally better for lesions that are acquired at a young age, and syntactic deficits are among the most common persistent deficits. If a lesion is so large that there is little or no undamaged tissue adjacent to the language regions of the left hemisphere, regions of the right hemisphere (presumably homotopic to the left-hemisphere language areas) will be recruited for language. The essentially intact linguistic abilities of children with extensive left-hemisphere lesions are particularly remarkable when contrasted with the markedly impaired linguistic abilities of SLI children who have very minimal evidence of neuropathology on CT or MRI scans. Perhaps the reason for this curious finding is that, although SLI children's brains are not deviant on a macroscopic level, SLI brains may have pervasive, bilateral, microscopic anomalies such that there is no normal tissue that can be recruited for language function. The results of one autopsy support this hypothesis: A boy who suffered a severe cyanotic episode at 10 days of age subsequently suffered from pronounced deficits in language comprehension and expression until his death (from mumps and congenital heart disease) at age 10. Autopsy revealed that the boy had bilateral loss of cortical substance starting at the inferior and posterior margin of the central sulci and extending backward along the course of the insula and sylvian fissures for 8 cm on the left side and 6 cm on the right side (Landau, Goldstein, and Kleffner, 1960). Perhaps the reason this child did not outgrow his language disorder is that, because he had extensive bilateral lesions, there were no appropriate regions that could be recruited for language.

THE CRITICAL PERIOD FOR LANGUAGE ACQUISITION AND NEURONAL MATURATION Although the ability to learn language appears to be innate, exposure to language during childhood is necessary for normal language development, just as the ability to see is innate yet visual stimulation is necessary for normal visual development (Hubel and Wiesel, 1970). The hypothesis that exposure to language must occur by a certain age in order for language to be acquired normally is called the critical period hypothesis. The critical period for language acquisition is generally believed to coincide with the period of great neural plasticity and is often thought to end at or sometime before the onset of puberty (see Lenneberg, 1967).

Wild children Skuse (1984a, 1984b) reviewed nine well-documented cases of children raised in conditions of extreme social and linguistic deprivation for $2\frac{1}{2}$ to 12 years. All of these cases involved grossly impoverished environments, and frequently malnourishment and physical abuse. At the time of discovery, the children ranged in age from $2\frac{1}{2}$ years to $13\frac{1}{2}$ years, had essentially no receptive or expressive language, and were globally retarded in nonlinguistic domains. The six children who eventually acquired normal or near-normal language function were all discovered by age 7 and had no signs of brain damage. Of the three children who remained language impaired, one was discovered at age 5 but had clear evidence of brain damage (Davis, 1940, 1947), and one was discovered at $3\frac{1}{2}$ years of age but had organic abnormalities not attributable to extreme deprivation (Skuse, 1984a). Genie, the third child with persistent linguistic impairments, is remarkable both for having the most prolonged period of deprivation (12 years) and, at almost 14 years of age, for being the oldest when discovered (Curtiss, 1977). Neuropsychological testing suggests that Genie does not have the expected left-hemisphere lateralization for language. Although it is tempting to conclude that Genie's failure to acquire normal language and her anomalous lateralization of language function are both the result of her failure to be exposed to language prior to the onset of puberty, it is possible that cortical anomalies in the left hemisphere are the cause of both her anomolous lateralization and her failure to acquire language (Curtiss, 1977).

Deaf isolates As Curtiss (1977, 1989) points out, it is impossible to be certain that the linguistic impairments observed in children such as Genie are the result of linguistic isolation and not the result of social and physical deprivation and abuse. Curtiss (1989) has described the case of Chelsea, a deaf woman who had essentially no exposure to language until age 32 when she began to receive intensive speech and language therapy. Unlike children such as Genie, Chelsea did not experience any social or physical deprivation. Chelsea's ability to use language (particularly syntax) is at least as impaired as Genie's, an observation consistent with the critical-period hypothesis (Curtiss, 1989). To test whether there is a critical period for first-language acquisition, Newport and her colleagues (Newport, 1990) have studied the signing abilities of deaf people whose first exposure to American Sign Language (ASL) was at birth (native signers), before age 6 (early signers), or after age 12 (late signers). Consistent with the critical-period hypothesis, even after 30 years of using ASL, on tests of morphology and complex syntax, native signers outperform early signers, who in turn outperform late signers (Newport 1990).

Second-language acquisition To test whether there is a critical period for second-language acquisition, Johnson and Newport (1989) studied the English abilities of native speakers of Korean or Chinese who first became immersed in English between the ages of 3 and 39. For subjects who began to learn English before puberty, age of English immersion was highly correlated with proficiency with English syntax and morphology, whereas no significant correlation was found for subjects who began to learn English after puberty (Johnson and Newport, 1989).

Evidence from studies of children such as Genie, deaf isolates, and people who acquire a second language suggests that the ability to acquire language diminishes with age. Consistent with research showing that (essentially) complete language recovery rarely occurs if a left-hemisphere lesion occurs after age 5, and substantial recovery rarely occurs if a lesion is acquired after the onset of puberty, subtle tests of linguistic abilities reveal that native fluency in a language is rarely attained if one's first exposure to that language occurs after early childhood, and that near-native fluency in a language is rarely attained if first exposure occurs after the onset of puberty. This is consistent with Hubel and Wiesel's (1970) finding that normal visual development requires visual stimuli during a critical period of neural development, and suggests that neural fine-tuning is critical to normal language acquisition and that this fine-tuning can only occur with exposure to language during a certain time period.

Summary

Evidence from normal and abnormal language acquisition suggests that innate mechanisms allow children to acquire language. Just as is the case with vision, given adequate, early exposure to language, children's language develops rapidly and with few errors, despite little or no instruction. The brain regions that permit this to occur seem to be functionally and anatomically distinct at birth and may correspond to what linguists refer to as universal grammar. It is possible that the reason that exposure to a particular language must occur during infancy and early childhood in order for that language to be mastered is that the type of neural fine-tuning that is associated with learning the particular parameters of a language must occur while there is a high degree of neural plasticity. The structures and operations that are involved in language seem to be anatomically and functionally modular and apparently do not have nonlinguistic counterparts. One possibility is that children have innate mechanisms that predispose them to perceive categorically such linguistic stimuli as phonemes, words, syntactic categories, and phrases, and that exposure to these types of linguistic stimuli facilitates the neural fine-tuning necessary for normal language acquisition. For example, children might have innate mechanisms that predispose them to assume that certain types of meanings and distinctions are conveyed by morphemes. They also might have innate mechanisms that specifically predispose them to distinguish between syntactic categories that allow for free generalization (lexical categories) and those that do not (functional categories). These innate mechanisms may allow children's brains to solve the otherwise intractable induction problems that permeate language acquisition.

In the future, fine-grained linguistic analyses of the speech of language-impaired children may be used to distinguish between different types of specific language impairment. Linkage studies of SLI may tell us which genes code for the brain structures that are necessary for language acquisition. The exquisite sensitivity of magnetic resonance imaging to white matter–gray matter distinctions means that MRI could be used to look for subtle defects that may be associated with developmental language disorders, such as subtle disorders of neuronal migration, or dysmyelinization (Barkovich and Kjos, 1992; Edelman and Warach, 1993). Furthermore, the correlation between myelinization and development of function (Smith 1981) means that serial MRIs of normal children, SLI children, and Williams syndrome children might shed light on the relationship between brain maturation and normal and abnormal language development. Finally, functional neuroimaging techniques such as ERP, PET, and functional MRI may eventually help answer questions about the neural processes that underlie language and language acquisition in normal children, SLI children, WS children, children with left-hemisphere lesions, and children who are exposed to language after the critical period.

ACKNOWLEDGMENTS Preparation of this chapter was supported by a Merck Foundation Fellowship in the Biology of Developmental Disabilities and also by a faculty fellowship from the Rutgers University Center for Cognitive Science.

NOTES

1. Throughout the paper, ungrammatical sentences are indicated with an asterisk (*).
2. One notable exception to this pattern was a recent study by Whitehurst et al. (1991). Whitehurst et al. may have found no significant difference in the rates of familial language impairment for control children and children with isolated expressive language delay (ELD) because ELD is a distinct entity from language impairments involving comprehension (as the study's authors suggest), or because the transient nature of ELD (Silva, Williams, and McGee, 1987) means that a history of ELD is more likely to go unreported than is a history of comprehension delay.

REFERENCES

Aram, D. M., B. L. Ekelman, D. F. Rose, and H. A. Whitaker, 1985. Verbal and cognitive sequelae of unilateral lesions acquired in early childhood. *J. Clin. Exp. Neuropsychol.* 7:55–78.

Aram, D. M., B. L. Ekelman, and H. A. Whitaker, 1986. Spoken syntax in children with acquired unilateral hemisphere lesions. *Brain Lang.* 27:75–100.

Arnold, G., and S. Schwartz, 1983. Hemispheric lateralization of languge in autistic and aphasic children. *J. Autism Dev. Disord.* 13:129–39.

Baker, C. L., 1981. Learnability and the English auxiliary system. In *The Logical Problem of Language Acquisition*, C. L. Baker and J. J. McCarthy, eds. Cambridge, Mass., MIT Press.

Bakwin, H., 1973. Reading disabilities in twins. *Dev. Med. Child Neurol.* 15:184–187.

Barkovich, A. J., and B. O Kjos, 1992. Grey matter heterotopias: MR characteristics and correlation with developmental and neurological manifestations. *Radiology*, 182: 493–499.

Basser, L. S., 1962. Hemiplegia of early onset and faculty of speech, with special reference to the effects of hemispherectomy. *Brain* 85:427–460.

Bellugi, U., A. Birhle, H. Neville, T. L. Jernigan, S. Doherty, 1992. Language, cognition, and brain organization in a neurodevelopmental disorder. In *Developmental Behavioral Neuroscience*, M. Gunnar and C. Nelson, eds. Hillsdale, N. J.: Erlbaum.

Bishop, D. V. M., 1983. Linguistic impairment after left hemidecortication for infantile hemiplegia: A reappraisal. *Q. J. Exp. Psychol.* 35A:199–207.

Bishop, D. V. M., 1987. The causes of specific developmental language disorder ("developmental dysphasia"). *J. Child Psychol. Psychiatry* 28:1–8.

Bishop, D. V. M., and A. Edmundson, 1986. Is *otitis media* a major cause of specific developmental language disorders? *British J. Disord. Commun.* 21:321–338.

Bloom, P., 1990. Syntactic distinctions in child language. *J. Child Lang.* 17:343–355.

Boliek, C. A., M. P. Bryden, and J. E. Obrzut, 1988. Focused attention and the perception of voicing and place of articulation contrasts with control and learning-disabled children. Paper presented at the 16th Annual Meeting of the International Neuropsychological Society, January 1988.

Brown, R., 1973. *A First Language: The Early Stages*. Cambridge, Mass.: Harvard University Press.

Brown, R., and C. Hanlon, 1970. Derivational complexity and order of acquisition in child speech. In *Cognition and the Development of Language*, J. R. Hayes, ed. New York: Wiley.

Byrne, B. M., L. Willerman, and L. L. Ashmore, 1974. Severe and moderate language impairment: Evidence for distinctive etiologies. *Behav. Genet.* 4:331–345.

Byrne, J. M., and R. D. Gates, 1987. Single-case study of left cerebral hemispherectomy: Development in the first five years of life. *J. Clin. Exp. Neuropsychol.* 9:423–434.

Carey, S., 1977. The child as word learner. *Linguistic Theory and Psychological Reality*, M. Halle, J. Bresnan, and G. A. Miller, eds. Cambridge, Mass.: MIT Press.

Chi, J. G., E. C. Dooling, and F. H. Gilles, 1977. Left-right asymmetries of the temporal speech areas of the human brain. *Arch. Neurol.* 34:346–348.

Chomsky, N., 1981. *Lectures on Government and Binding*. Dordrecht, Holland: Foris.

Chomsky, N., 1986. *Knowledge of Language: Its Nature, Origin and Use*. New York: Praeger.

Cohen, H., C. Gelinas, M. Lassonde, and G. Geoffroy, 1991. Auditory lateralization for speech in language-impaired children. *Brain Lang.* 41:395–401.

Cohen, M., R. Campbell, and F. Yaghmai, 1989. Neuropathological abnormalities in developmental dysphasia. *Ann. Neurol.* 25:567–570.

Cowan, W. M., J. W. Fawcett, D. D. O'Leary, and B. B. Stanfield, 1984. Regressive events in neurogenesis. *Science* 225:1258–1265.

Crockett, H. G., and N. M. Estridge, 1951. Cerebral hemispherectomy. *Bull. Los Angeles Neurol. Soc.* 16:71–87.

Croft, W., 1990. *Typology and Universals*. New York: Cambridge University Press.

Curtiss, S., 1977. *Genie: A Psycholinguistic Study of a Modern Day "Wild Child"*. New York: Academic Press.

Curtiss, S., 1989. The independence and task-specificity of language. In *Interaction in Human Development*, A. Bornstein and J. Bruner, eds. Hillsdale, N.J.: Erlbaum.

Davis, K., 1940. Extreme social isolation of a child. *Am. J. Sociol.* 45:554–565.

Davis, K., 1947. Final note on a case of extreme isolation. *A. J. Sociol.* 52:432–437.

Dawson, G., C. Finley, S. Phillips, and A. Lewy, 1989. A comparison of hemispheric asymmetries in speech-related brain potentials of autistic and dysphasic children. *Brain Lang.* 37:26–41.

DeFries, J. C., D. W. Fulker, and M. C. LaBuda, 1987. Evidence for a genetic aetiology in reading disability of twins. *Nature* 329:537–539.

Denays, R., M. Tondeur, M. Foulon, F. Verstraeten, H. Ham, A. Piepsz, and P. Noel, 1989. Regional brain blood flow in congenital dysphasia studies with technetium-99M HM-PAO SPECT. *J. Nucl. Med.* 30:1825–1829.

Dennis, M., 1980. Capacity and strategy for syntactic comprehension after left or right hemidecortication. *Brain Lang.* 10:287–317.

Dennis, M., and B. Kohn, 1975. Comprehension of syntax in infantile hemiplegics after cerebral hemidecortication: Left hemisphere superiority. *Brain Lang.* 2:475–486.

Dennis, M., and H. A. Whitaker, 1976. Language acquisition following hemi-decortication: Linguistic superiority of the left over the right hemisphere. *Brain Lang.* 3:404–433.

Edelman, R. R., and S. Warach, 1993. Magnetic Resonance Imaging, part 1. *N. England J. Med.* 328:708–716.

Efron, R., 1963. Temporal perception, aphasia, and deja vu. *Brain* 86:403–424.

Eldridge, R., 1983. Twin studies and the etiology of complex neurological disorders. In *Genetic Aspects of Speech and Language Disorders*, C. L. Ludlow and J. A. Cooper, eds. New York: Academic Press.

Feldman, H., A. L. Holland, S. S. Kemp, and J. E. Janosky, 1992. Language development after unilateral brain injury. *Brain Lang.* 42:89–102.

Fundudis, T., I. Kolvin, and G. Garside, 1979. *Speech Retarded and Deaf Children*. London: Academic Press.

Gardner, W. J., L. J. Karnosh, C. C. McClure, and A. K. Gardner, 1955. Residual function following hemispherectomy for tumour and for infantile hemiplegia. *Brain* 78: 487–502.

Garrett, M., 1976. Syntactic processes in sentence production. In *New Approaches to Language Mechanisms*, R. Wales and E. Walker, eds. Amsterdam: North-Holland.

Geschwind, N., and A. Galaburda, 1987. *Cerebral Lateralization: Biological Mechanisms, Associations, and Pathology*. Cambridge, Mass.: MIT Press.

Goodglass, H., 1976. Agrammatism. In *Perspectives in Neurolinguistics and Psycholinguistics*, H. Whitaker and H. Whitaker, eds. New York: Academic Press.

Gopnik, M., 1990a. Feature-blind grammar and dysphasia. *Nature* 344:715.

Gopnik, M., 1990b. Feature blindness: A case study. *Lang. Acquisition* 1:139–164.

Gopnik, M., and M. B. Crago, 1991. Familial aggregation of a developmental language disorder. *Cognition* 39:1–50.

Gordon, A. G., 1988. Some comments on Bishop's annotation 'Developmental dysphasia and otitis media'. *J. Child Psychol. Psychiatry* 29:361–363.

Guilfoyle, E., S. Allen, and S. Moss, 1991. Specific Language Impairment and the maturation of functional categories. Paper presented at the 16th Annual Boston University Conference on Language Development, October 19, 1991.

Guttman, E., 1942. Aphasia in children. *Brain* 65:205–219.

Hillier, W. F., 1954. Total left cerebral hemispherectomy for malignant glioma. *Neurology* 4:718–721.

Holcomb, P. J., S. A. Coffey, and H. J. Neville, 1992. Visual and auditory sentence processing: A developmental analysis using event-related brain-potentials. *Dev. Neuropsychol.* 8:203–241.

Hubel, D., and T. Wiesel, 1970. The period of susceptibility to the physiological effects of unilateral eye closure in kittens. *J. Physiol. (Lond.)* 206:419–436.

Hurst, J. A., M. Baraitser, E. Auger, F. Graham, and S. Norell, 1990. An extended family with a dominantly inherited speech disorder. *Dev. Med. Child Neurol.* 32:347–355.

Huttenlocher, P. R., 1979 Synaptic density in human frontal cortex—Developmental changes and effects of aging. *Brain Res.* 163:195–205.

Ingram, T. T. S., 1959. Specific developmental disorders of speech in childhood. *Brain* 82:450–467.

Jernigan, T. L., U. Bellugi, E. Sowell, S. Doherty, and J. Hesselink, 1993. Cerebral morphological distinctions between Williams and Down Syndromes. *Arch. Neurol.* 50: 186–191.

Jernigan, T. L., J. R. Hesselink, E. Sowell, and P. A. Tallal, 1991. Cerebral structure on Magnetic Resonance Imaging in language-impaired and learning-impaired children. *Arch. Neurol.* 48:539–545.

Johnson, J., and E. Newport, 1989. Critical period effects in second language learning: The influence of maturational state on the acquisition of English as a second language. *Cognitive Psychol.* 21:60–99.

Jusczyk, P. W., K. Hirsch-Pasek, D. Kemler Nelson, and L. J. Kennedy, 1992. Perception of acuoustic correlates of major phrasal units by young infants. *Cognitive Psychol.* 24:252–293.

Kiessling, L. S., M. B. V. Denckla, and M. Carlton, 1983. Evidence for differential hemispheric function in children with hemiplegic cerebral palsy. *Dev. Med. Child Neurol.* 25:727–734.

Landau, W. M., R. Goldstein, and F. R. Kleffner, 1960. Congenital aphasia: A clinicopathological study. *Neurology* 10:915–921.

Lenneberg, E. H., 1967. *Biological Foundations of Language.* New York: Wiley.

Levy, Y., N. Amir, and R. Shalev, 1992. Linguistic development of a child with congenital localised L. H. lesion. *Cognitive Neuropsychol.* 9:1–32.

Lewis, B. A., and L. A. Thompson, 1992. A study of developmental speech and language disorders in twins. *J. Speech Hear. Res.* 35:1086–1094.

Locke, J. L., and P. L. Mather, 1989. Genetic factors in the ontogeny of spoken language: Evidence from monozygotic and dizygotic twins. *J. Child Language*, 16:553–559.

Lou, H. D., L. Henriksen, and P. Bruhn, 1990. Focal cerebral dysfunction in developmental learning disabilities. *Lancet* 335:8–11.

Luchsinger, R., 1970. Inheritance of speech deficits. *Folia Phoniatrica* 22:216–230.

Maratsos, M., and M. Chalkley, 1981. The internal language of childen's syntax: The ontogenesis and representation of syntactic categories. In *Children's Language*, vol. 2, K. Nelson, ed. New York: Gardner.

Marcus, Gary F., 1993. Negative evidence in language acquisition. *Cognition.* 46(1):53–85.

Martin, J. A., 1981. *Voice, Speech and Language in the Child: Development and Disorder.* New York: Springer.

Mateer, C. A., and C. B. Dodrill, 1983. Neuropsychological and linguistic correlates of atypical language lateralization: Evidence from sodium amytal studies. *Hum. Neurobiol.* 2:135–142.

McCready, E. B., 1926. *Am. J. Psychiatry* 6:267.

McNeill, D., 1966. Developmental psycholinguistics. In *The Genesis of Language*, F. Smith and G. Miller, eds. Cambridge, Mass.: MIT Press.

Molfese, D. L., 1990. Auditory evoked responses recorded from 16-month old human infants to words they did and did not know. *Brain Lang.* 36:345–363.

Molfese, D. L., Morse, P. A., and Peters, C. J. (1990). Auditory evoked responses to names for different objects: Cross-modal processing as a basis for infant language acquisition. *Developmental Psychology*, 26(5):780–795.

Morehead, D., and D. Ingram, 1973. The development of base syntax in normal and linguistically deviant children. *J. Speech and Hearing Res.* 16:330–352.

Morgan, J., and E. Newport, 1981. The role of constituent structure in the induction of an artifical language. *J. Verb. Learn. Verb. Be.* 20:67–85.

Morley, M., 1965. *The Development and Disorders of Speech in Children.* Edinburgh: Livingstone.

Neils, J., and D. M. Aram, 1986. Family history of children with developmental language disorders. *Percept. Mot. Skills* 63:655–658.

Nelson, K., 1973. Structure and strategy in learning to talk. *Monog. Soc. Res. Child Dev.* 38.

Neville, H., 1991. Neurobiology of cognitive and language

processing: Effects of early experience. In *Brain Maturation and Cognitive Deveopment*, K. Gibson and A. Petersen, eds. New York: de Gruyter.

Neville, H. J., P. J. Holcomb, and D. M. Mills, 1989. Auditory sensory and language processing in Williams Syndrom: An ERP study. Paper presented at the International Neuropsychological Society, January 1989.

Newport, E., 1990. Maturational constraints on language learning. *Cognitive Sci.* 14:11–28.

Papanicolaou, A. C., A. DiScenna, L. Gillespie, and D. Aram, 1990. Probe-evoked potenital finding following unilateral left-hemisphere lesions in children. *Arch. Neurol.* 47:562–566.

Pauls, D. L., 1983. Genetic analysis of family pedigree data: A review of methodology. In *Genetic Aspects of Speech and Language Disorders*, C. L. Ludlow and J. A. Cooper, eds. New York: Academic Press.

Pennington, B., and S. Smith, 1988. Genetic influences on learning disabilities: An update. *J. Consult. Clin. Psychol.* 56:817–823.

Pinker, S., 1984. *Language Learnability and Language Development.* Cambridge, Mass.: Harvard University Press.

Pinker, S., 1987. The bootstrapping problem in language acquisition. In *Mechanisms of Language Acquisition*, B. MacWhinney, ed. Hillsdale, N.J.: Erlbaum.

Pinker, S., 1989. *Learnability and Cognition: The Acquisition of Argument Structure.* Cambridge, Mass.: MIT Press.

Pinker, S., and P. Bloom, 1990. Natural language and natural selection. *Behav. Brain Sci.* 13:707–784.

Plante, E., L. Swisher, R. Vance, and S. Rapsak, 1991. MRI findings in boys with Specifically Language Impaiment. *Brain Lang.* 41:52–66.

Rankin, J. M., D. M. Aram, and S. J. Horwitz, 1981. Language ability in right and left hemiplegic children. *Brain Lang.* 14:292–306.

Rasmussen, T., and B. Milner, 1977. The role of early left-brain injury in determining lateralization of cerebral speech functions. *Ann. N. Y. Acad. Sci.* 299:335–369.

Reznick, J. S., and B. A. Goldfield, 1992. Rapid change in lexical development in comphrension and production. *Dev. Psychol.* 28:406–413.

Robinson, R. J., 1987. The causes of language disorder: An introduction and overview. In *Proceedings of the First International Symposium on Specific Speech and Language Disorders in Children.* London: AFASIC.

Rondal, J., 1980. Language delay and language difference in moderately and severely retarded children. *Special Education in Canada* 54:27–32.

Samples, J., and V. Lane, 1985. Genetic possibilities in six siblings with specific language learning disorders. *Asha* 27: 27–32.

Scheibel, A. B. (1984). A dendritic correlate of human speech. In N. Geschwind and A. M. Galaburda (Eds.), *Cerebral Dominance: The Biological Foundations* Cambridge, Mass.: Harvard University Press.

Silva, P. A., S. Williams, R. McGee, 1987. Early language delay and later intelligence, reading and behavior problems. *Dev. Med. Child Neurol.* 29:630–640.

Skuse, D. H., 1984a. Extreme deprivation in early childhood I: Diverse outcomes for 3 siblings from an extraordinary family. *J. Child Psychol. Psychiatry* 25:523–541.

Skuse, D. H., 1984b. Extreme deprivation in early childhood II: Theoretical issues and a comparative review. *J. Child Psychology and Psychiatry* 25:543–572.

Smith, A., 1966. Speech and other functions after left dominant hemispherectomy. *Journal of Neurology, Neurosurgery and Psychiatry* 29:467–471.

smith, J. F., 1981. Central nervous system. In *Paediatric Pathology*, C. L. Berry, ed. Berlin: Springer Verlag: 147–148.

Snyder, W., and Stromswold, K. 1994. The structure and acquisition of English dative constructions. Under review.

Soja, N. N., Carey, S., and Spelke, E. S. (1991). Ontological categories guide young children's inductions of word meaning: Object terms and substance terms. *Cognition*, 38:179–211.

Steele, S., 1981. *An Encyclopedia of AUX: A Study in Cross-Linguistic Equivalence.* Cambridge, Mass.: MIT Press.

Stevenson, J., and P. Graham, G. Fredman, and V. McLoughlin, 1987. A twin study of genetic influences on reading and spelling ability and disability. *J. Child Psychol. Psychiatry* 28:229–247.

Stromswold, K., 1988. Linguistic representations of children's *wh*-questions. *Papers and Reports on Child Lang.* 27: 107–114. Stanford: Stanford University.

Stromswold, K., 1989a. How conservative are children? *Papers and Reports on Child Lang.* 28:148–155. Stanford: Stanford University.

Stromswold, K., 1989b. Using naturalistic data: Methodological and theoretical issues (or How to lie with naturalistic data). Paper presented at the 14th Annual Boston University Conference on Langauge Development, October 13–15, 1989.

Stromswold, K., 1990a. Learnability and the acquisition of auxiliaries. Unpublished Ph.D. dissertation. Cambridge, MA: MIT (Available through MIT's Working Papers in Linguistics.)

Stromswold, K., 1990b. The acquisition of language-universal and language-specific aspects of Tense. Paper presented at the 15th Boston University Conference on Language Development, October 19–21, 1990.

Stromswold, K., 1992. Learnability and the acquisition of auxiliary and copula *be*. In *ESCOL* '91. Columbus: Ohio State University.

Stromswold, K., 1994a. Language comprehension without language production: Implications for theories of language acquisition. Paper presented at the 18th Annual Boston University Conference on Language Development. January 1994.

Stromswold, K. 1994b. The nature of children's early grammar: Evidence from inversion errors. Paper presented at the 1994 Linguistic Society of America Conference, January 1994. Boston, Massachusetts.

STROMSWOLD, K., 1994c. Lexical and functional categories in language and language acquisition. Rutgers University Center for Cognitive Science Technical Report.

SWISHER, L. P., and E. J. PINSKER, 1971. The language characteristics of hyperverbal hydrocephalic children. *Dev. Child Neurol.* 13:746–755.

TALLAL, P., and M. PIERCY, 1973. Defects of non-verbal auditory perception in children with developmental dysphasia. *Nature* 241:468–469.

TALLAL, P., R. ROSS, and S. CURTISS, 1989a. Familial aggregation in specific language impairment. *J. Speech Hear. Disord.* 54:167–173.

TALLAL, P., R. ROSS, and S. CURTISS, 1989b. Unexpected sex-ratios in families of language/learning impaired children. *Neuropsychologia* 27:987–998.

TALLAL, P., J. TOWNSEND, S. CURTISS, and B. WULFECK, 1991. Phenotypic profiles of language-impaired children based on genetic/family history. *Brain Lang.* 41:81–95.

THAL, D. J., V. MARCHMAN, J. STILES, D. ARAM, D. TRAUNER, R. NASS, and E. BATES, 1991. Early lexical development in children with focal brain injury. *Brain Lang.* 40:491–527.

TOMBLIN, J. B., 1989. Familial concentrations of developmental language impairment. *J. Speech Hear. Disord.* 54: 287–295.

VON ARMAN, G., and P. ENGEL, 1964. Mental retardation related to hypercalcaemia. *Dev. Med. Child Neurol.* 6:366–377.

WADA, J. A., R. CLARKE, and A. HAMM, 1975. Cerebral hemispheric asymmetry in humans. *Arch. Neurol.* 32:239–246.

WHITEHURST, G. J., D. S. ARNOLD, M. SMITH, J. E. FISCHEL, C. J. LONIGAN, and M. C. VALDEZ-MENCHACHA, 1991. Family history in developmental expressive language delay. *J. Speech Hear. Res.* 34:1150–1157.

WOODS, B. T., and CAREY, S., 1979. Language deficits after apparent clinical recovery from childhood aphasia. *Annals of Neurology*, 6:405–409.

YAMADA, J., 1990. *Laura: A Case for the Modularity of Language.* Cambridge, Mass.: MIT Press.

YAMADA, J., and S. CURTISS, 1981. The relationship between language and cognition in a case of Turner's Syndrome. *UCLA Working Papers in Cognitive Linguistics* 3:93–115.

ZOLLINGER, R., 1935. Removal of left cerebral hemisphere: Report of a case. *Arch. Neurol. Psychiatry* 34:1055–1064.

56 The Cognitive Neuroscience of Syntactic Processing

DAVID CAPLAN

ABSTRACT This chapter reviews the nature of the syntactic structures that serve to relate words in sentences, the processing of those structures, and its neural localization. Syntactic structures are hierarchically organized sets of categories, over which specifically defined relationships serve to determine specific aspects of sentence meaning. Experimental psychological evidence suggests that the recognition and production of these structures is at least partially independent of other linguistic, verbal, and cognitive processes, with respect to both the operations involved and the processing resources they use. Evidence from lesion sites in patients with disorders of syntactic processing, and from functional neuroimaging studies in normal subjects, indicate that syntactic processing occurs in the left perisylvian cortex in right-handed individuals and perhaps is more narrowly localized within a part of Broca's area.

Sentences convey aspects of the structure of events and states in the world. These semantic values are collectively known as the propositional content of a sentence. These values include thematic roles (information about who did what to whom), attribution of modification (which adjectives go with which nouns), the reference of pronouns, and other aspects of meaning. The propositions conveyed by sentences are entered into higher-order structures that constitute the discourse level of linguistic structure (van Dijk and Kintsch, 1983; Grosz, Pollack, and Sidner, 1989). Discourse includes information about the general topic under discussion, the focus of a speaker's attention, the novelty of the information in a given sentence, the temporal order of events, causation, and so on. The relationships between words that are established by the syntactic structure of a sentence allow aspects of propositional meaning to be expressed in different ways as a function of discourse context.

In what follows, I shall briefly describe the nature of syntactic structures, review the evidence for the uniqueness of the cognitive system that processes these representations, and describe the neural systems involved in those processes.

DAVID CAPLAN Neuropsychology Laboratory, Neurology Department, Massachusetts General Hospital, Boston, Mass.

The nature of syntactic structures

Syntactic structures consist of hierarchical groupings of syntactic categories (e.g., noun, verb, preposition, noun phrase, verb phrase, etc.) in which particular categories bear a number of defined relationships to one another. Different aspects of propositional meaning are determined by these different relationships (see figure 56.1). One example of this system—the relationship between syntactic structure, sentential semantic features, and discourse-level semantic features—is illustrated in sentences (1) through (3):

(1) The dog scratched the cat.

(2) The cat_i was scratched [t_i] by the dog.

(3) It was the cat_i that the dog scratched [t_i].

The thematic roles (who does what to whom) in all these sentences are the same: The dog is agent (the element that accomplishes the action), and the cat is theme (the element upon which the action devolves). Sentences (2) and (3) indicate that different syntactic forms, serving different discourse functions such as establishing the topics, foci, and presuppositions of a discourse, can convey the same thematic roles. This expressive power carries with it a need for complexity in syntactic structures. In sentences (2) and (3), Chomsky (1986) postulates a trace of a moved noun phrase in these sentences. This trace, indicated in brackets, is assigned the thematic role that is appropriate for the grammatical role it plays, and it transmits this thematic role to the noun phrase with which it is coindexed.

A certain degree of parsimony is thought to operate across human languages with respect to the nature

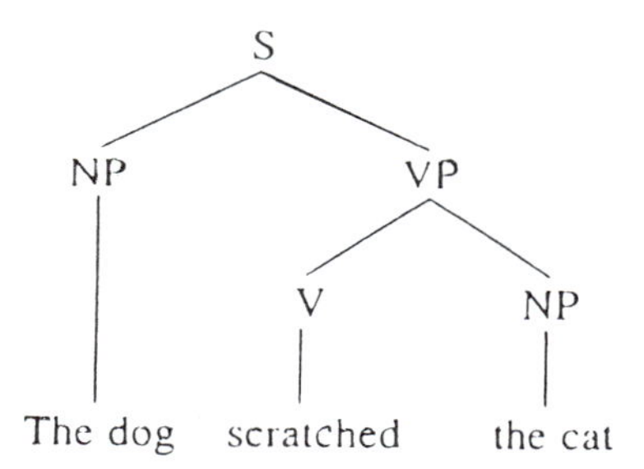

FIGURE 56.1 Syntactic structure of the sentence *The dog scratched the cat*, showing the hierarchical organization of syntactic categories that determines sentence meaning. *The dog* is the subject and agent of *scratched*, and *the cat* is its object and theme.

of syntactic structures. At a sufficiently abstract level, syntactic structures are thought to be identical in all languages (Chomsky has gone so far as to suggest that there is only one human language). In large part because of the universal character of syntactic structures (when characterized at an adequately abstract level of description), Chomsky and others consider that these structures are a cognitive reflection of an innate, specifically human, mental capacity. It would be misleading to represent the above description of syntactic structures as uncontroversial. One group of linguists holds a variety of different positions from Chomsky's regarding the exact nature of these representations (e.g., whether sentences such as (2) and (3) have a trace is debated), but nonetheless maintains that syntactic structures are complex and abstract entities (see Sells, 1985, for discussion). A very different position is taken by researchers who only recognize the existence of very simple syntactic structures, such as the order of syntactic categories, declension, agreement, and other surface-visible features (Bates et al., 1982; Bates and MacWhinney, 1989). In my view, the models developed by this latter group of researchers pose no serious challenge to the models of syntactic structures I have been describing. In the first place, they are simply unable to represent the range of form-meaning pairings found at the level of sentences in natural languages; that is, they fail to constitute an adequate description of language. In addition, they do not deal adequately with a wide variety of facts about language use. Whatever may be the deeper regularities of its structure when examined in detail, the syntax of natural languages is a complex mental object.

Processing syntactic structures

Information-processing models of language can be expressed as flow diagrams (often called functional architectures) that indicate the sequence of operations of the different components that perform a language-related task. One hypothesized processor is devoted to processing syntactic structure. We shall consider the nature of the operations of this hypothesized processor, which bear on the degree to which this processor is specialized (encapsulated and domain specific). We shall begin by presenting the model developed by Frazier (1987a, 1987b, 1989, 1990) as a point of departure.

In Frazier's model, the sentence comprehension process involves a number of independent modules: one that builds phrase structure (the c-structure module), one that assigns co-reference (the binding module), one that assigns thematic roles and predication (the θ-predication module), and one that assigns reference to pronouns and other referential items (the reference module). Evidence for the independence of these components comes from several sources. One source is the existence of preferred interpretations of sentences, as in (4) and (5) below, and garden-path effects in sentences, as in (6) through (8):

(4) The boy hit the girl with a book. (The preferred reading attaches *with a book* to *hit*, not to *the girl*.)

(5) Joyce said that Tom left yesterday. (The preferred reading attaches *yesterday* to *left*, not to *said*.)

(6) Ernie kissed Marcie and her sister laughed. (A temporary misanalysis yields the interpretation that Ernie kissed Marcie and her sister.)

(7) Since Jay always jogs a mile seems like a short distance to him. (A temporary misanalysis yields the interpretation that Jay always jogs a mile.)

(8) The horse raced past the barn fell. (A temporary misanalysis yields the interpretation that the horse raced past the barn.)

Frazier (1987a, 1987b) has argued that these effects reflect basic aspects of how the parser operates. She argued that the parser incorporates each new word in a sentence into a syntactic structure, following principles that minimize the work that it must do. Two of these principles are known as *minimal attachment* and *late closure*. Minimal attachment specifies that the parser does not postulate any potentially unnecessary nodes.

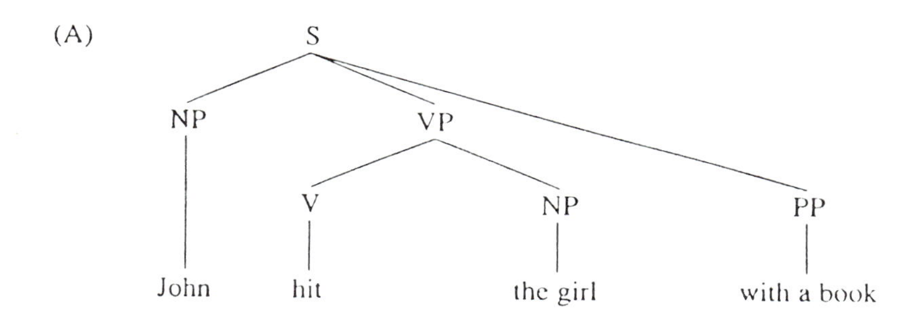

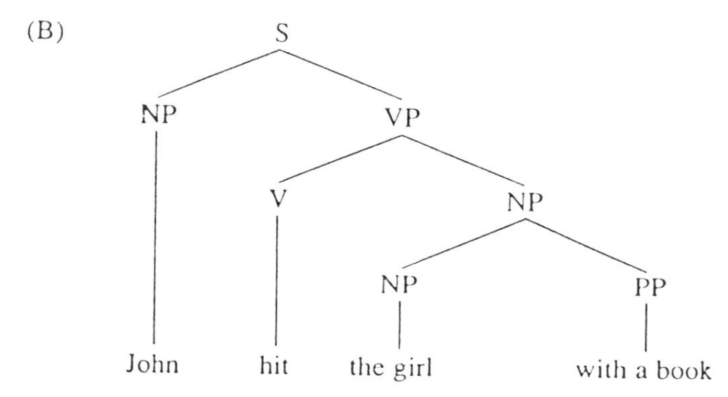

FIGURE 56.2 Preferred and unpreferred syntactic structures of the two meanings of the sentence *The boy hit the girl with a book*. The preferred reading attaches *with a book* to *hit*, and the unpreferred reading attached it to *the girl*. The preferred reading has fewer and less complex nodes that the unpreferred reading.

Late closure specifes that new items are attached to the phrase or clause being processed, if grammatically possible. Minimal attachment is illustrated in figure 56.2, which shows how it leads to the structure (A) for (4), rather than the alternative structure (B).

Models of sentence comprehension differ in important ways with respect to their claims regarding what information is used to determine sentence meaning and how different types of information interact in this process. Models such as Frazier's maintain that the sentence-comprehension process is divided into different types of operations and that each can make use of only certain types of information (Clifton and Ferreira, 1987; Frazier, 1987a, 1987b, 1989; Frazier and Clifton, 1989; Frazier and Rayner, 1982). Models that are slightly more interactive maintain that many different types of information all interact together at the same level of processing to determine the meaning that is assigned to a sentence (Holmes, 1987; but see Ford, Bresnan, and Kaplan, 1982; Mitchell, 1987, 1989; Stowe, 1989; Tanenhaus and Carlson, 1989; Fodor, 1990; Tanenhaus, Garnsey, and Boland, 1990). Highly interactive models take the position that there is no level of representation that is constructed during the sentence-comprehension process that is purely syntactic and that all types of syntactic and semantic information interact to produce the final output of the comprehension process (Bates et al., 1982; Bates, Friederici, and Wulfeck, 1987; Bates and MacWhinney, 1989; Bates, McDonald, and MacWhinney, 1989; MacWhinney, 1989; McClelland, St. John, and Taraban, 1989; McDonald and MacWhinney, 1989). However, the data base upon which most of these interactive theories rests consists largely of results of experiments using paradigms in which problem-solving strategies are very likely to have contributed to the subjects' performances. Thus, these models may be more relevant to problem solving than to sentence comprehension itself (see Gibson, 1992, for additional discussion). Models that recognize the existence of complex syntactic structures and a parser are the only models that deal with many of the psycholinguistic phenomena currently described in the literature, such as the garden-path effects illustrated above.

Some researchers have argued that the effects that we contend are due to the way the parser operates are really due to the role of discourse structure in interpreting sentences. According to these researchers, biases such as those seen in sentence (4) are due to listeners' attempts to achieve "referential success"; that is, to assign all items mentioned in a sentence to specific individuals in a discourse. In sentence (4), for instance, attaching *the book* to *the girl* implies that the context has identified more than one girl and that the sentence is referring to the girl with the book (of the girls mentioned in the context). Since presenting the sentence in isolation does not establish a context with more than one girl, the sentence is preferentially interpreted as attaching *the book* to *hit*. Evidence supporting this analysis would come from demonstrations that context reverses preferences such as those that occur for sentences like (4) when presented in isolation. There is evidence that context changes many of these effects, and does so while sentences are being processed (Altmann and Steedman, 1988). Thus, controversy surrounds the issue of how the syntactic processor works. Though there is good reason to believe that there is a cognitive mechanism that is specialized for the assignment of syntactic structure, how restricted the input to this mechanism is remains unclear.

The neural basis for sentence processing

Correlations between language impairments and lesion sites indicate that the association cortex in the region of the sylvian fissure is responsible for language processing (see Caplan, 1987, for review). This region includes the pars triangularis and opercularis of the third frontal convolution (Broca's area, or Brodmann's areas 45 and 44), the association cortex in the opercular area of the pre- and postcentral gyri, the supramarginal and angular gyri of the parietal lobe (Brodmann areas 39 and 40), the first temporal gyrus from the supramarginal gyrus to a point approximately lateral to Heschl's gyrus (Wernicke's area, or Brodmann areas 41 and 42), and possibly a portion of the adjacent second temporal gyrus. Language processing is restricted to this area in the left hemisphere in up to 98% of right-handed individuals (Milner, 1974) and recruits or is exclusively based in the right hemisphere in up to a third of individuals who are ambidextrous or left-handed (Goodglass and Quadfasel, 1954). The supplementary motor area and several subcortical gray matter structures (the caudate, putamen, parts of the thalamus) may play roles in language processing, but correlational and path analyses of lesion sites in patients with language impairments suggest that most of language processing is cortically based (Metter et al., 1988).

Two general classes of theories have been developed that address the relationship of portions of the perisylvian association cortex to components of the language-processing system, one based on "holist" or distributed views of neural function (Lashley, 1950) and one based on localizationist principles (Geschwind, 1965). The basic tenet of holist or distributed theories of the functional neuroanatomy for language is that language-processing components each involve all or most of the perisylvian association cortex. The existence of specific language-processing deficits following small lesions in this area would appear to make most versions of holist models untenable. Such cases favor localizationist models that maintain that specific types of language-processing operations are carried out in particular parts of this cortical region. Recently, hybrid models have been proposed, according to which specific language-processing operations are carried out in particular locales, which are themselves organized in a distributed fashion (Mesulam, 1990).

Strong evidence supporting the assignment of specific language-processing components to particular neural areas has been difficult to obtain, largely for technical reasons. Series of patients with deficits affecting the same restricted language processor, in whom lesions have been localized on the basis of state-of-the-art neuroimaging, are and will likely remain infrequent because of the difficulties associated with identifying such patients. This is not surprising if we assume that specific psycholinguistic processes are localized in areas of the brain whose blood supply also supports additional regions that carry out other language processes. Electrocortical stimulation and electrocorticography can only be used in subjects with abnormal brains in which reorganization may affect language localization (Ojemann, 1983). Techniques such as ERP recordings, which are proving increasingly sensitive to aspects of syntactic processing in normal subjects (Neville et al., 1991; Osterhout and Holcomb, 1990; see Garrett, this volume, for review), are difficult to use for localization purposes because of the challenges involved in identifying the point sources of scalp-detected electric fields.

The limits of current knowledge regarding the exact localization of specific language processors can be seen in the domain of syntatic processing. Although the discussion of the psychology of sentence processing concentrated on comprehension, we shall review the literature on the neural correlates of impairments of both sentence comprehension and sentence production.

Patients with aphasia have been described who appear to have suffered selective impairments of the ability to construct syntactic structures in sentence production. One group of these patients—so-called Broca's aphasics who have an expressive language disturbance known as agrammatism—tend to simplify syntactic structures and to omit function words (e.g., articles, auxiliary verbs, etc.) and grammatical morphemes (e.g., agreement markers, plural markers, etc.) from their speech. These patients tend to have lesions that include Broca's area. However, the lesions in patients with the larger syndrome of Broca's aphasia often extend well beyond this region (Mohr et al., 1978; Metter et al., 1989). We have studied a series of 20 patients with the agrammatic form of Broca's aphasia (Vanier and Caplan, 1990). Large lesions, affecting the entire territory supplied by the middle cerebral artery, were found in five cases. Middle-sized lesions, resulting from occlusion of several terminal branches of the middle cerebral artery but not the entire territory of the middle cerebral artery, occurred in seven cases. Two of these seven lesions were similar to those described by Mohr and colleagues (1978) in cases of Broca's aphasia and were almost exclusively localized to the frontal and parietal lobes, which receive their blood supply from upper bank of the middle cerebral artery. However, the other five middle-sized lesions affected variable amounts of temporal lobe as well as areas above the sylvian fissure, and one of these cases showed complete sparing of Broca's area. Small lesions, resulting from occlusion of a single terminal branch of the middle cerebral artery, occured in several cortical and subcortical locations. Four small lesions were frontal. One, in a right-hemisphere case (i.e., a crossed dextral aphasic), involved only pars triangularis of the third frontal convolution. One showed the opposite pattern in frontal lobe, involving pars opercularis and the precentral opercular association cortex and sparing pars triangularis. Two small lesions in frontal lobe involved both Broca's area and the precentral opercular cortex. Three small lesions were entirely subcortical, involving the insula, the arcuate fasciculus, and the corona radiata. Finally, one small lesion was located entirely in the inferior parietal and superior temporal region, and spared the frontal lobe completely. Overall, the lesions showed a tendency to involve frontal and insular cortex, but exhibited considerable variability in size and location. It may be that different basic impairments in sentence production all resulted in similar performances on variables measured in the tasks administered in this study. However, on the assumption that agrammatism represents a deficit in a stage in language processing involving the construction of the syntactic form of a sentence (Caplan, 1991), these data are consistent with the view that this stage of sentence planning is accomplished by quite different areas of the association cortex in the perisylvian area in different individuals.

Patients who have difficulty constructing syntactic structures in comprehension tasks and/or in using these structures to determine the propositions expressed in sentences have been described by many researchers (Caramazza and Zurif, 1976; Schwartz, Saffran, and Marin, 1980; Linebarger, Schwartz, and Saffran, 1983; Caplan, Baker, and Dehaut, 1985; Tyler, 1985; Caplan and Futter, 1986; Caplan and Hildebrandt, 1988; Zurif et al., 1993; see Caplan, 1992, for review). Many of these patients are agrammatic Broca's aphasics, a fact that has led several researchers to suggest that Broca's area is responsible for syntactic processing (Mesulam, 1990; Damasio and Damasio, 1992). Grodzinsky (1990) presents a more specific hypothesis regarding the syntactic operations carried out in this brain region. However, as noted above, lesions in patients with Broca's aphasia, and in the subset of Broca's aphasics with expressive agrammatism, often extend well beyond Broca's area (Mohr et al., 1978; Metter et al., 1989) and may even spare this area as judged by CT-scan data (Vanier and Caplan,1990). Tramo, Baynes, and Volpe (1988) documented syntactic comprehension impairments in three patients with Broca's aphasia, of which two had anterior lesions and one had a posterior lesion. In many of the studies cited above, lesion data have been presented in summary form, and it is impossible to know whether the summarized data represent the entirety of the lesion. Some of these patients have also been shown to be able to make judgments regarding the syntactic well-formedness of sentences (Linebarger, Schwartz, and Saffran, 1983; Linebarger, 1990), suggesting that they can construct syntactic structures but not map them them onto sentence meaning (Schwartz et al., 1987). For all these reasons, the hypothesis that Broca's area is the sole area responsible for parsing—or for a particular set of parsing operations—can only be said to receive modest support from the existing data on sentence-comprehension impairments in agrammatic patients.

In addition to these issues about the lesion sites in patients with agrammatism, other data raise questions about the hypothesis that Broca's area is the sole or major site of syntactic processing in sentence comprehension. Many patients with aphasic syndromes other than agrammatic Broca's aphasia also often show impairments of syntactic comprehension (Caplan, Baker, and Dehaut, 1985; Caplan and Hildebrandt, 1988). The locations of lesions in many of these patients are very likely to involve or be restricted to regions other than Broca's area. One study identified the dominant inferior parietal lobe as the location of lesions that produced impairments in sentence comprehension while sparing comprehension of single words (Selnes et al., 1983). However, it is not clear whether the patients in this study had impairments in parsing sentences, or in interpreting them, or both; nor is it clear to what extent disturbances of verbal short-term memory contributed to their sentence-comprehension impairments. Our own work has found no correlation between lesion site within the dominant perisylvian cortex and the severity of a syntactic comprehension deficit (Caplan, Baker, and Dehaut, 1985; Caplan 1987); however, the determination of lesion site was based on a variety of clinical as well as radiological data sources in these studies, and cannot be considered definitive. In general, very few adequate neuroradiological studies of patients who have been shown to have syntactic processing deficits been published.

Overall, though the available evidence suggests that syntactic processing in sentence-comprehension tasks takes place in Broca's area, strong evidence that this is so is still lacking. Whether parsing and sentence interpretation are carried out in different areas of the perisylvian cortex, or whether parsing is carried out in part of the perisylvian cortex and sentence interpretation in another area of the brain, remains unclear.

Virtually all the data relevant to localization of these processes comes from deficit-lesion correlational studies. (Event-related potentials have been recorded in response to processing aspects of sentence meaning (Kutas and Hillyard, 1980) and syntactic form (Neville et al., 1991; Osterhout and Holcomb, 1990), but, as noted above, these studies have not primarily focused on the localization of the neural generators of the ERPs associated with syntactic processing.) Functional neuroimaging has emerged as a potentially powerful new way to provide data regarding the localization of language processors. The basis of the PET approach is to identify local increases in cerebral blood flow (rCBF) that accompany cognitive functions by comparing blood-flow differences in suitably designed cognitive tasks (Peterson et al., 1988; Posner et al., 1988; Peterson et al., 1990). These increases reflect increases in neurophysiological activity associated with performing a cognitive operation. We have recently used this approach to investigate the neural basis of sentence comprehension and, more specifically, the ability to parse syntactic structures in sentence comprehension.

The subjects were six normal male college students between 19 and 28 years old. All were native speakers of English who were strongly right-handed and had no first-degree left-handed relatives. They performed three tasks that required making judgments about whether sentences were acceptable or unacceptable, while PET activity was being measured (table 56.1). In task 1, the subjects were presented sentences with structures known as center-embedded relative clauses. In task 2, they were shown sentences with structures known as right-branching relative clauses. In task 3, subjects were presented acceptable center-embedded and right-branching sentences, and unacceptable sentences containing a pseudo-word that was an orthographically and phonetically possible string in English (e.g., *The video cowned the businessman that bribed the senator*).

Both task 1 and task 2 require a subject to construct the syntactic structure of each sentence and assign its meaning. This process is more difficult in task 1 because of an increase in memory load that is associated with processing center-embedded compared to right-branching relative clause structures. Consistent with this claim, and with other experimental findings, the intrascan mean reaction times were significantly longer in task 1 than in task 2. Subtraction of PET activity associated with task 2 from that associated with task 1 thus provides an indication of the location of the neural system involved in this specific syntactic operation. Task 3 requires subjects to read all the words in a sentence, but not necessarily to assign the sentence's structure or meaning. However, behavioral evidence indicates that subjects did assign syntactic structures in this task: In task 3, subjects took significantly more

Table 56.1
Design of PET syntax experiment

Sentence types in tasks 1–3	
Task 1: Center-embedded sentences	
Acceptable	*The limerick that the boy recited appalled the priest*
Anomalous	*The teenager that the miniskirt wore horrified the mother*
Task 2: Right-branching sentences	
Acceptable	*The biographer omitted the story that insulted the director*
Anomalous	*The woman tipped the hairdresser that pleased the haircut*
Task 3: Sentences with nonwords	
Acceptable	As in tasks 1 and 2
Unacceptable	*The economist predicted the recession that chorried the man*
	The sculpture that the artist exhibited shocked the findle

Language functions assessed in PET subtractions

Comparison	Stimulated state	Baseline state	Cognitive operation
Tasks 1–2	Center-embedded sentences	Right-branching sentences	Syntactic processing
Tasks 1–3	Center-embedded sentences	Sentences with nonwords	Sentential semantic processing
Tasks 2–3	Right-branching sentences	Sentences with nonwords	Sentential semantic processing

time to judge center-embedded sentences than right-branching sentences, indicating that they processed syntactic structures even if the task did not require them to do so. This suggests that the difference between tasks 1 and 2 and task 3 lies in whether subjects interpreted sentences semantically or not, and that subtraction of PET activity associated task 3 from that associated with either task 1 or task 2 may provide an indication of the location of the neural system(s) involved in assigning sentence meaning.

Subtracting PET activity in task 3 from either task 1 or task 2 showed that all left-sided perisylvian language structures increased their blood flow. For the subtraction of task 2 from task 1, a significant increase in rCBF occurred only in pars opercularis of Broca's area. The results are consistent with the view that the process of sentence comprehension involves the left perisylvian association cortex in right-handed individuals. They suggest more precise localization of one aspect of syntactic processing in pars opercularis of Broca's area. This area may be specialized for memory processes associated with such operations.

These results raise a number of neurobiological questions. As noted above, other studies based on deficit-lesion correlations, including our own, have suggested a considerable degree of variability in the localization of syntactic processing within the perisylvian association cortex. The discrepancy between the two sets of results may point to a deeper scientific generalization. The subjects in our PET experiment were carefully selected to represent as likely a population as possible in which to find that an aspect of the sentence comprehension process was localized in a specific brain area. One possibility is that the variability in deficit-lesion correlations found in patients with impairments of (roughly) similar aspects of sentence processing is linked to variability in such factors as the degree of handedness, familial handedness, age, or sex of the patients in the deficit-leison correlational studies. Additional PET studies with different populations may provide data regarding this possibility.

Summary

I have reviewed the nature of syntactic structures, models of their processing, and data regarding their functional neuroanatomy. Syntactic structures are complex mental entities that are similar in all languages and that may be processed by specialized cognitive operators. Data from deficit-lesion correlations in patients with disorders of syntactic processing and from functional neuroimaging in normal subjects indicate that aspects of syntactic processing are very narrowly localized within the brain. The neural network that is involved in syntactic processing appears to have the pars opercularis of the inferior frontal gyrus as its most active cortical region, although it may be that this localization varies somewhat among normal individuals.

ACKNOWLEDGMENTS The work reported here was partially supported by a grant from the National Institute of Deafness and Other Communication Disorders (DC00942).

REFERENCES

ALTMANN, G., and M. J. STEEDMAN, 1988. Interaction with context in human syntactic processing. *Cognition* 30:191–238.

BATES, E., and B. MACWHINNEY, 1989. Functionalism and the competition model. In *A Cross-Linguistic Study of Sentence Processing*, B. MacWhinney and E. Bates, eds. Cambridge: Cambridge University Press, pp. 3–73.

BATES, E., A. FRIEDERICI, and B. WULFECK, 1987. Comprehension in aphasia: A cross-linguistic study. *Brain Lang.* 32:19–67.

BATES, E., J. MCDONALD, and B. MACWHINNEY, 1989. A maximum likelihood procedure for the analysis of group and individual data in aphasia research. Paper presented at The Academy of Aphasia, Nashville, Tenn.

BATES, E., S. MCNEW, B. MACWHINNEY, A. DEVESCOVI, and S. SMITH, 1982. Functional constraints on sentence processing. *Cognition* 11:245–299.

CAPLAN, D., 1987. Discrimination of normal and aphasic subjects on a test of syntactic comprehension. *Neuropsychologia* 25:173–184.

CAPLAN, D., 1991. Agrammatism is a theoretically coherent aphasic category. *Brain Lang.* 40:274–281.

CAPLAN, D., 1992. *Language: Structure, Processing and Disorders.* Cambridge, Mass.: MIT Press, Bradford Books.

CAPLAN, D., and C. FUTTER, 1986. Assignment of thematic roles to nouns in sentence comprehension by an agrammatic patient. *Brain Lang.* 27:117–134.

CAPLAN, D., and N. HILDEBRANDT, 1988. *Disorders of Syntactic Comprehension.* Cambridge, Mass.: MIT Press, Bradford Books.

CAPLAN, D., C. BAKER, and F. DEHAUT, 1985. Syntactic determinants of sentence comprehension in aphasia. *Cognition* 21:117–175.

CARAMAZZA, A., and E. B. ZURIF, 1976. Dissociation of algorithmic and heuristic processes in language comprehension: Evidence from aphasia. *Brain Lang.* 3:572–582.

CHOMSKY, N., 1986. *Knowledge of Language.* New York: Praeger.

CLIFTON, C., and F. FERREIRA, 1987. Modularity in sentence comprehension. In *Modularity in Knowledge Representation and Natural-Language Understanding*, J. L. Garfield, ed. Cambridge, Mass.: MIT Press, pp. 277–290.

DAMASIO, A. R., and H. DAMASIO, 1992. Brain and language. *Sci. Am.* September 1992:89–95.

FODOR, J. A., 1990. *Modularity in sentence processing.* Paper presented at the City University of New York Conference on Sentence Processing.

FORD, M., J. BRESNAN, and R. KAPLAN, 1982. A competence-based theory of syntactic closure. In *The Mental Representation of Grammatical Relations*, J. Bresnan and R. Kaplan, eds. Cambridge, Mass.: MIT Press, pp. 727–796.

FRAZIER, L., 1987a. Sentence processing: A tutorial review. In *Attention and Performance XII: The Psychology of Reading*, M. Coltheart, ed. London: Erlbaum, pp. 559–586.

FRAZIER, L., 1987b. Theories of sentence processing. In *Modularity in Knowledge Representation and Natural-Language Processing*, J. Garfield, ed. Cambridge, Mass.: MIT Press, pp. 291–307.

FRAZIER, L., 1989. Against lexical generation of syntax. In *Lexical Representation and Process*, W. Marslen-Wilson, ed. Cambridge, Mass.: MIT Press, pp. 505–528.

FRAZIER, L., 1990. Exploring the architecture of the language-processing system. In *Cognitive Models of Speech Processing: Psycholinguistic and Computational Perspectives*, G. T. M. Altmann, ed. Cambridge, Mass.: MIT Press Bradford Books, pp. 409–433.

FRAZIER, L., and C. CLIFTON, 1989. Successive cyclicity in the grammar and the parser. *Lang. Cognitive Proc.* 4:93–126.

FRAZIER, L., and K. RAYNER, 1982. Making and correcting errors during sentence comprehension. *Cognitive Psychol.* 14:178–210.

GESCHWIND, N., 1965. Disconnection syndromes in animals and man. *Brain* 88:237–294, 585–644.

GIBSON, E., 1992. On the adequacy of the competition model. *Language* 68:812–830.

GOODGLASS, H., and F. A. QUADFASEL, 1954. Language laterality in left-handed aphasics. *Brain* 77:521–548.

GRODZINSKY, Y., 1990. *Theoretical Perspectives on Language Deficits.* Cambridge, Mass.: MIT Press.

GROSZ, B. J., M. E. POLLACK, and C. L. SIDNER, 1989. Discourse. In *Foundations of Cognitive Science*, M. Posner, ed. Cambridge, Mass.: MIT Press, pp. 437–468.

HOLMES, V. M., 1987. Syntactic parsing: In search of the garden path. In *Attention and Performance XII: The Psychology of Reading*, M. Coltheart, ed. London: Erlbaum, pp. 587–600.

KUTAS, M., and S. A. HILLYARD, 1980. Reading senseless sentences: Brain potentials reflect semantic anomaly. *Science* 207:203–205.

LASHLEY, K. S., 1950. In search of the engram. *Symp. Soc. Exp. Biol.* 4:454–482.

LINEBARGER, M. C., 1990. Neuropsychology of sentence parsing. In *Cognitive Neuropsychology and Neurolinguistics: Advances in Models of Cognitive Function and Impairment*, A. Caramazza, ed. Hillsdale, N.J.: Erlbaum, pp. 55–122.

LINEBARGER, M. C., M. F. SCHWARTZ, and E. M. SAFFRAN, 1983. Sensitivity to grammatical structure in so-called agrammatic aphasics. *Cognition* 13:361–392.

MACWHINNEY, B., 1989. Competition and connectionism. In *A Cross-Linguistic Study of Sentence Processing*, B. MacWhinney and E. Bates, eds. Cambridge: Cambridge University Press, pp. 422–457.

MCCLELLAND, J., and A. KAWAMOTO, 1986. Mechanisms of sentence processing: Assigning role to constituents. In *Parallel Distributed Processing*, J. McClelland and D. Rumelhart eds. Cambridge, Mass.: MIT Press.

MCCLELLAND, J. L., M. ST. JOHN, and R. TARABAN, 1989. Sentence comprehension: A parallel distributed processing approach. *Lang. Cognitive Proc.* 4:287–336.

MCDONALD, J., and B. MACWHINNEY, 1989. Maximum likelihood models for sentence processing. In *A Cross-Linguistic Study of Sentence Processing*, B. MacWhinney and E. Bates, eds. Cambridge: Cambridge University Press, pp. 397–422.

MESULAM, M.-M., 1990. Large-scale neurocognitive networks and distributed processing for attention, language, and memory. *Ann. Neurol.* 28(5):597–613.

METTER, E. J., W. H. RIEGE, W. R. HANSON, C. A. JACKSON, D. KEMPLER, D. VANLANCKER, 1988. Subcortical structures in aphasia: An analysis based on (F-18)-fluorodoxyglucose positron emission tomography, and computed tomography. *Arch. Neurol.* 45:1229–1234.

METTER, E., D. KEMPLER, C. JACKSON, W. HANSON, J. C. MAZZIOTTA, and M. E. PHELPS, 1989. Cerebral Glucose Metabolism in Wernicke's, Broca's, and Conduction Aphasia. *Arch. Neurol.* 46:27–34.

MILNER, B., 1974. Hemispheric Specialization: Its scope and limits. In *The Neurosciences: Third Study Program*, F. O. Schmidt and F. G. Warden, eds. Cambridge, Mass.: MIT Press, pp. 75–89.

MITCHELL, D. C., 1987. Lexical guidance in human parsing: Locus and process characteristics. In *Attention and Performance XII: The Psychology of Reading*, M. Coltheart, ed. London: Erlbaum, pp. 601–618.

MITCHELL, D. C., 1989. Verb-guidance and other lexical effects in parsing. *Lang. Cognitive Proc.* 4:123–154.

MOHR, J. P., M. S. PESSIN, S. FINKELSTEIN, H. FUNKENSTEIN, G. W. DUNCAN, and K. R. DAVIS, 1978. Broca aphasia: Pathologic and clinical. *Neurology* 28:311–324.

NEVILLE, H. J., J. NICOL, A. BARSS, K. FORSTER, and M. GARRETT, 1991. Syntactically based sentence processing classes: Evidence from event-related brain potentials. *J. Cognitive Neurosci.* 3:155–170.

OJEMANN, G., 1983. Brain organization for language from the perspective of electrical stimulation mapping. *Behav. Brain Sci.* 6:189–230.

OSTERHOUT, L., and P. J. HOLCOMB, 1990. Event-related potentials elicited by grammatical anomalies. In *Psychophysiological Brain Research*, C. H. M. Bruina, A. W. K. Gaillard, and A. Kok, eds. Tilburg: Tilburg University Press, pp. 299–302.

PETERSEN, S. E., P. T. FOX, A. Z. SNYDER, and M. E. RAICHLE, 1990. Activation of extrastriate and frontal cortical areas by visual words and word-like stimuli. *Science* 249:1041–1044.

PETERSON, S. E., P. T. FOX, M. I. POSNER, M. MINTUN, and M. E. RAICHLE, 1988. Positron emission tomographic studies of the cortical anatomy of single-word processing. *Nature* 331:585–589.

POSNER, M. I., S. E. PETERSON, P. T. FOX, M. E. RAICHLE, 1988. Localization of cognitive operations in the human brain. *Science* 240:1627–1632.

SCHWARTZ, M. F., M. C. LINEBARGER, E. M. SAFFRAN, and D. S. PATE, 1987. Syntactic transparency and sentence interpretation in aphasia. *Lang. Cognitive Proc.* 2:85–113.

SCHWARTZ, M., E. SAFFRAN, and O. MARIN, 1980. The word order problem in agrammatism I: Comprehension. *Brain Lang.* 10:249–262.

SELLS, P., 1985. *Lectures on Contemporary Syntactic Theories*. Chicago: University of Chicago Press.

SELNES, O. A., D. KNOPMAN, N. NICCUM, A. B. RUBENS, and D. LARSON, 1983. CT scan correlates of auditory comprehension deficits in aphasia: A prospective recovery study. *Neurology* 13:558–566.

STOWE, L. A., 1989. Thematic structures and sentence comprehension. In *Linguistic Structure in Language Processing*, G. Carlson, and M. Tanenhaus, eds. Drodrecht: Kluwer, pp. 319–357.

TANENHAUS, M. K., and G. N. CARLSON, 1989. Lexical structure and language comprehension. In *Lexical Representation and Process*, W. Marslen-Wilson, ed. Cambridge, Mass.: MIT Press, pp. 529–561.

TANENHAUS, M. K., S. M. GARNSEY, and J. BOLAND, 1990. Combinatory lexical information and language comprehension. In *Cognitive Models of Speech Processing: Psycholinguistic and Computational Perspectives*, G. T. M. Altmann, ed. Cambridge, Mass.: MIT Press, Bradford Books, pp. 383–408.

TRAMO, M. J., K. BAYNES, and B. T. VOLPE, 1988. Impaired syntactic comprehension and production in Broca's aphasia: CT lesion localization and recovery patterns. *Neurology* 38:95–98.

TYLER, L., 1985. Real-time comprehension processes in agrammatism: A case study. *Brain Lang.* 26:259–275.

VAN DIJK, T. A., and W. KINTSCH, 1983. *Strategies of Discourse Comprehension*. New York: Academic Press.

VANIER, M., and D. CAPLAN, 1990. CT scan correlates of agrammatism. In *Agrammatic Aphasia*, L. Menn and L. Obler, eds. Amsterdam: Benjamins, pp. 97–114.

ZURIF, E., D. SWINNEY, P. PRATHER, J. SOLOMON, and C. BUSHELL, 1993. An on-line analysis of syntactic processing in Broca's and Wernicke's aphasia. *Brain Lang.* 45:448–464.

57 The Structure of Language Processing: Neuropsychological Evidence

MERRILL GARRETT

ABSTRACT Language use depends on multiple knowledge types; these include systems of specifically linguistic information and systems of nonlinguistic conceptual and perceptual information. Language processing is globally organized into two major performance domains designed for language production and comprehension respectively, within which descriptions of phonological, syntactic, semantic, and discourse structure must be coordinated. The time course and interactions among these structures are of central concern in the development of language-processing theory. Results from recent event-related potential (ERP) studies of sentence processing demonstrate distinct ERP responses to semantic and syntactic processes; multiple, distinct ERP responses to differing types of syntactic and lexical processes; and variation in ERP responses associated with temporal variation of processes in sentence and lexical analysis. These and related findings from brain-imaging studies and from studies of aphasia and other language disturbances support models in which modular processes for the assignment of language structure are filtered in real time by semantic and conceptual constraints.

Analysis of language use must be treated in terms of multiple systems: Production and comprehension processes are distinct but intricately intertwined; both are deployed in time, and both draw on multiple cognitive resources. Caplan's outline of grammar and processor (chapter 56) focuses on syntactic structures and associated parsing algorithms that lie at the heart of any account of language processes. These are, accordingly, principal targets for explanation in any neuropsychological account of language. Rapp and Caramazza (chapter 58) and Blumstein (chapter 59) provide discussions of central aspects of the lexical and sound processing systems that feed the parsing and sentence-construction systems described by Caplan; the three classes of processing activity thus addressed must be integrated in real time to provide the basis for interpretive processing. Discussion of these areas as discrete domains leads naturally enough to the assumption that they are separable processes in the mental activity that underlies language use. The extent to which this is true is an empirical question, at the center of enquiry in behavioral studies of language processing and in the neuroscience of language processes. Though many questions about the detailed architecture of the system are unsettled, the essential structural contrasts of sound, word, sentence, and discourse are not seriously at issue. As we will shortly see, and as the other language reviews in this section attest, cognitive neuroscience–based investigations of language do converge in major respects with such a general decomposition of language activity. The first section of this chapter provides a processing outline of these general architectural features of language systems; the second section treats neuropsychological evidence that bears on relations among lexical, syntactic, and semantic processes.

MERRILL GARRETT Cognitive Science Program and Department of Psychology, University of Arizona, Tucson, Ariz.

Real-time language processing systems

PRODUCTION AND COMPREHENSION MODELS: TWO FACES OF THE LANGUAGE PROCESSOR The primary language performance modes are speaking and listening. Figure 57.1 outlines their major components.

Note the major parallels of organization in the two systems: Both have levels of process that encompass phonology (i.e., segmental sound structure), prosody (i.e., stress and intonational phrasing), lexical structure, syntax, and message-level expressions based on discourse and conceptual structure. Both have two major processing streams—one focused on lexical recovery and the other on building the structures with which

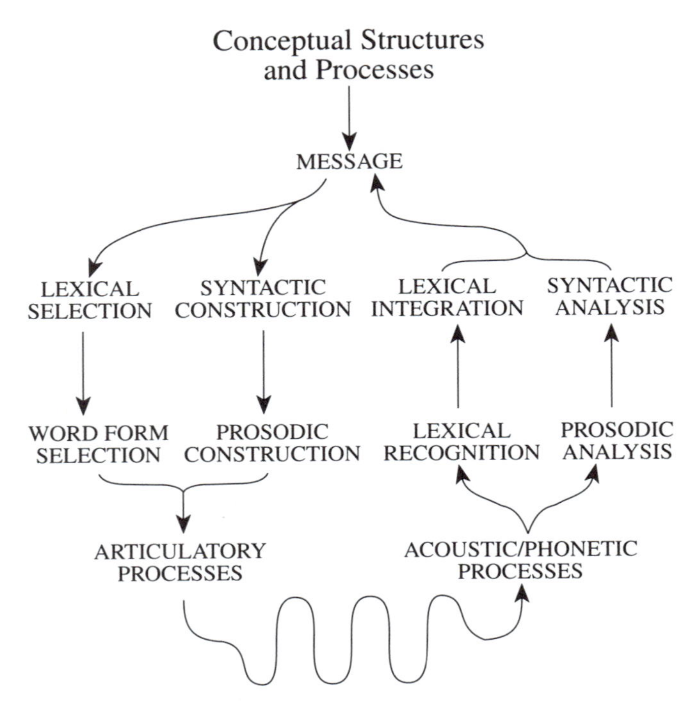

FIGURE 57.1 An outline of information flow for language-production and language-comprehension systems.

lexical elements are integrated. A further general matter is interaction between the two systems. Though not explicitly represented in the diagram, there are clear dependencies between production and comprehension. The comprehension system provides support for production processes: We not only monitor our own speech for reasons of stylistic control, we monitor it for error, and routinely inhibit error outputs during or prior to their overt utterance (Levelt, 1989). A complementary point concerning production-system support for comprehension processes may also hold (see the discussion that follows under "Relations among processing types").

LEXICAL AND STRUCTURAL PROCESSING STREAMS The separation of lexical recovery and phrasal construction processes poses a coordination problem. Accordingly, a significant aspect of language-processing activity revolves around the association of lexical content with structural frameworks of various types. In this connection, note that Caplan's characterization of parsing emphasizes that the association of lexical elements to phrasal environments created by the parser is incremental—that is, lexical input is linked to evolving sentence structure at the earliest feasible moment. Although assimilation of lexical elements to phrasal environments may sometimes be momentarily delayed, most evidence indicates that assignments occur in the real-time frame established by lexical identification of the input. It is important to keep in mind, however, that this still leaves undetermined the nature and time course of interpretive processes that are consequent to phrasal assimilation of a lexical form. In any event, the essential picture is one of incremental parsing decision, with circumscribed occasions for look-ahead or delay when structural ambiguity must be resolved. Similar general observations apply for the production system: Sentence generation is incrementally driven by syntactic information in abstract lexical records ("lemmas"; Kempen and Hoenkamp, 1987).

Figure 57.1 characterizes structural processing as "multilevel." In comprehension, lexical forms are first assimilated to a prosodically motivated phrasal structure. The parsing operations outlined by Caplan in the previous chapter (i.e., assignment of constituent structure, thematic role, predication relations, and coreference) would, in figure 57.1, be represented in the mapping from prosodic structure to messages via intervening representations of syntax and logical form. For production a similar organization holds, with the order of events reversed—syntactic construction precedes phonological interpretation of lemmas and assimilation of the resulting word forms into prosodic structures (see Garrett, in press; Levelt, 1989; and references therein). This description of structural processing incorporates hypotheses about lexical recovery that posit staged retrieval of lexical records for both comprehension and production. For comprehension, staged retrieval holds that the mapping from acoustic-phonetic representations to lexical records is form-driven and not directly dependent on word meaning; some aspects of this issue are discussed later in connection with ambiguity processing. For discussion of corresponding production problems, see Levelt et al., 1991, and Rapp and Caramazza, chapter 58 of this volume.

RELATIONS AMONG PROCESSING TYPES Concurrent processing is assumed for both production and comprehension, that is, phonological, syntactic, and semantic analyses are assumed to be active simultaneously. In general terms, this is taken for granted, but the matter develops force when questions are posed about relations between the levels vis-à-vis particular input elements. Dynamic questions arise as to how closely yoked the processes are across levels, as do questions of how sharply distinguished these broad processing classes are in their real-time exercise. For comprehension, the issue of bottom-up versus top-down effects is a recurrent theme, and the canonical question is how much influence on data-driven processes may derive from higher levels of structure. These issues are complex and empirically vexing, to say the least. See, among many examples, Forster (1979), Marslen-Wilson and Tyler (1980), Fodor (1983), Crain and Steedman (1985), Frazier (1987), McClelland, St. John, and Taraban (1989), and Bock and Kroch (1992), for diverse views. In production, similar issues arise, and questions are posed regarding the role of feedback from lower levels to earlier levels (see Levelt, 1989).

Although there are gray areas, the assumption here is that the overall evidence supports a substantial degree of independence for semantic, syntactic, and phonological domains, and requires a significantly modular language-processing architecture. A central empirical challenge for modular accounts of language comprehension is to maintain plausible, real-time models that are form-driven (i.e., parsing is based only on the serial order of sentence elements classified for their potential syntactic roles), while at the same time providing very rapid and powerful interpretive constraints on sentence processing (e.g., preferred selection of parsing solutions that fit the prevailing discourse context). To insulate basic lexical and syntactic mechanisms from direct influences of semantic interpretation, context effects must be indirectly accounted for by interpretive filtering. Such filtering requires that appropriate interpretive schemas be integrated with each of the main aspects of comprehension structures. The sentence-production system offers one source of such schemas (Forster, 1979): Lexical recognition outputs would be routed to parsing and also to message construction (see figure 57.1) to generate plausible construals of the sentence's lexical content in light of the prevailing discourse context; these filter competing analyses from the comprehension system. By this or other means (see, e.g., Tanenhaus, Carlson, and Trueswell, 1989), top-down constraints on comprehension must integrate with bottom-up constraints by way of filtering that preserves the processing integrity of syntactic and phonological systems.

A variety of detailed considerations of method and theory bear on assessment of the processing problems that have been outlined in this section, most of which we must ignore; the reader is directed to the sources cited for discussion. Our purpose is to relate some recent electrophysiological evidence to the major questions posed in the last two subsections: How do syntactic processes in sentence comprehension relate to their associated semantic interpretation processes, and how are the structures associated with lexical items integrated with sentence and discourse environments?

The neuropsychology of real-time sentence processing

Evidence from imaging studies shows an association of distinct brain regions with the major structural classes of languages (see chapters 56 and 61). The commitment of distinct neural systems for phonological, syntactic, and semantic structures fits the general architecture sketched in the previous section of this chapter. How might the processes associated with those regions interact to support real-time language use? Findings from studies of language disorders and from recent event-related potential (ERP) studies of language bear on this question. Major features of the ERP evidence converge with observations from language pathologies; to provide a background to the ERP discussion, we note some salient observations from the latter area.

Some Neuropsychological Evidence for Relations Between Syntactic and Conceptual Processes

Normal language use integrates language-specific structures with systems of encyclopedic background knowledge, and with the pragmatic knowledge that directs discourse and narrative structure. How tightly tied is this more general knowledge to the function of basic language mechanisms? A variety of formal, experimental, and observational outcomes indicate the possibility of dissociation. For example, reports of hyperlinguistic abilities in cognitively impaired children demonstrate that grave limitations on general intelligence are compatible with the development of language with near-normal range of morphosyntactic structure, linguistic hierarchy and embedding, and movement and ellipsis conventions (Yamada, 1990; Rondal, 1993; see chapter 55). Patients with Alzheimer's disease and certain aphasic patients provide other compelling observations. These patients' comprehension is severely compromised, but they succeed quite well at repetition. And most importantly, they spontaneously correct ungrammatical deviations in number and tense marking (Whitaker, 1975; Martin and Saffran, 1990). In short, absent the direction of significant semantic and discourse competence, complex mechanisms that insure concord for widely separated elements of sentence structure continue to function. The import of this oft-cited performance is highlighted by recent experimental work. That work (e.g., Bock and Eberhard, 1992) explored error-induction procedures for agreement processes in English. Normal speakers occasionally err in structures for which an embedded verb mismatches in number the head noun that governs agreement (e.g., **A man of 320 pounds are too heavy*; The asterisk indicates syntactic deviance). Semantic variables that might have been expected to influence the incidence of such errors were ineffective; only syntactic variation produced error. Thus, although number is a semantic dimension, once its value is set, the mechanism that distributes its marking in sentence form seems not to be directly dependent on nonlinguistic cognitive processes; this is as the observations of Alzheimer's and aphasia patients suggest.

Such findings show that the capacity to construct complex sentence forms in speaking is not to be identified with conceptual competence. And, similarly, the capacity to construct basic sentence representations during listening is dissociable from conceptual processes. Some research on agrammatism in aphasia illustrates this latter point. Clinically, agrammatism is a condition in which verbal output is labored, with sharply reduced incidence of adjectives, verbs, and especially functional vocabulary both bound and free (namely, inflections and closed-class words such as articles, prepositions, conjunctions, quantifiers, complementizers, etc.). Agrammatic speech is sometimes labeled "telegraphic" for this reason. Caramazza and Zurif (1976) advanced the hypothesis that agrammatic speakers might also necessarily be agrammatic comprehenders. This hypothesis was strongly indicated by several reports of agrammatic speakers who failed to understand sentences when interpretation depended on closed-class vocabulary—precisely the vocabulary most strikingly compromised in their speech output. In general, such patients had diminished capacity to deal with syntactically complex sentence structures. Significantly, where syntactic problems arose, these patients were able to rely on intact capacities for semantic inferences based on the lexical content of sentences. So, for example, they could interpret passives like *The flowers were watered by the girl* by plausible construal of the objects and events independent of their syntax, but they failed on passives like *The boy was chased by the girl* where plausible inference is not decisive and syntax is indispensable for accurate response. Distinct interpretive processes were thus postulated: those based on syntactic representations and those with primary reliance on lexical inference. The thesis entertained to account for agrammatic comprehension was that the central syntactic capacity was compromised.

Subsequent investigation showed this latter claim could not be fully sustained, and in the process yielded evidence for a somewhat different picture of the dissociation of parsing capacity and semantic interpretation. Briefly, the evidence showed that some agrammatics could parse sentences they could not comprehend. This emerged when well-formedness judgments were used to test performance (Linebarger et al., 1983). Task performance was not completely normal for all structural types, but accuracy was high for many aspects of structure for which sentence interpretation tests showed poor performance. These results showed that ability to marshal detailed syntactic representations (of the sort required to distinguish well-formed from ill-formed word sequences) should not be identified with the ability to interpret those representations. Moreover, these agrammatic performances demonstrated multiple modes of semantic processing—some dependent on syntactic analysis, and others dependent on the inferential analysis of lexical content. The neurophysiological evidence from ERP studies, to which we now turn, significantly sharpens our understanding of the relations between mechanisms for the analysis of sentence form and mechanisms for sentence interpretation.

Neurophysiological Evidence for Relations Between Syntactic and Semantic Processes In 1980, Kutas and Hilyard correlated a neurophysiological measure of brain activity with an abstract linguistic response and changed the face of neuropsychological enquiry into language processing. The technique involves averaging time segments of electroencephalogram recordings that follow sets of relevantly similar stimuli (e.g., words of a particular semantic or grammatical category); the average waveforms so generated are conventionally referred to as event-related potentials (ERPs). The stable association of such stimuli with the timing of waveforms of particular morphology and distribution (using standard recording sites on the scalp) provides a basis for the characteristic nomenclature for ERP responses (see e.g., Hilyard and Picton, 1987, and chapter 50). We will look at evidence for several distinct ERP responses to aspects of syntactic processing—distinct in two ways: from each other, and from semantic interpretation effects—where the index of semantic interpretation is the well-established N400 response.

Kutas and Hilyard's (1980) work, and subsequently that of other investigators, demonstrated a robust and highly specific ERP response to semantic deviance. Their original report contrasted responses to sentence-final words in frames like *John spread the warm bread with———*." Completions like *socks* were compared with those like *butter*; the former were associated with a substantial negative shift in polarity of the ERP beginning around 300 ms after stimulus onset (visual) and peaking around 400 ms; distribution was posterior, bilateral, and larger over the right hemisphere. Figure 57.2 illustrates such a waveform.

N400 is better described as a plausibility response than a deviance response, for subsequent work showed that this response is graded. Magnitude of N400 responses varies as a function of the rated sensibleness or acceptability of lexical targets in their sentence frames (Kutas and Hilyard, 1984), and seems to reflect the readiness with which lexical items are integrated into an interpretive framework. Semantic deviance is merely the limiting case, not the necessary trigger.

Several extensions of this approach have yielded evidence for diverse lexical-processing effects (see, e.g., Kutas and Van Petten, 1988), and, more recently, for sentence-processing effects. Note that the following report of such studies is qualitative, provides little methodological description, and ignores many details of potential significance. Word-by-word presentations for visual stimuli should be assumed throughout, with some rate variations (but 600 ms/word, with 300 ms exposure and 300 ms blank interval is common). Measurements are reported for exposure of the target words given in italic in the example stimulus sentences; topographical effects are noted only where relevant to the argument.

Work by Garnsey, Tannenhaus, and Chapman (1989) provides a good beginning for discussion. They used N400 effects to investigate the parsing of embedded *wh* questions like (a) and (b) below. Understanding the example in (a) requires knowing that *customers* is object of the verb, as in the declarative (c)—that is, the reader must connect a *wh*-marked element and its interpretive site (sometimes referred to as a gap; marked by _ in the examples).

(a) The businessman knew which customers the secretary *called* _ at home.

(b) The businessman knew which article the secretary *called* _ at home.

(c) The businessman knew that the secretary called the *customers* at home.

(d) The businessman knew that the secretary called the *article* at home.

The researchers presented such sentences visually in sensible versions and in versions with an implausible noun phrase, as in (b) and (d). Results for declaratives showed an N400 for words like *article* in (d), but not for those like *customers* in (c), as the findings of Kutas and Hilyard would lead one to expect. But the same contrast also appeared for the moved noun phrases in (a) and (b) at the gap positions after the verb: An N400 occurred for (b) sentence types but not (a) types. Subjects interpreted the gap following the verb as though the noun were there (for similar claims based on other measures, see chapter 56). These results also indicate that incremental syntactic assignments were made strictly on grounds of the projection of the syntactic information associated with lexical items. Notice that there are semantically acceptable versions of (b) if the material following the gap is varied, for example, *The businessman knew which article the secretary called the library for*. In normal language use, if the parser could suppress semantically implausible bindings and wait for a sensible gap location, the strategy would seldom fail, because most of what we say is semantically well formed. The evidence that there is momentary (usually unconscious) consideration of the inappropriate syntactic analysis fits a filtering model of semantic and syntactic interaction—a point to which we will return.

The experiment by Garnsey, Tannenhaus, and Chapman used semantic interpretation to show when the parser makes certain syntactic assignments. The identification of syntactically specific ERP components is a separate and complementary issue. Kutas and Hilyard did, in fact, address an aspect of this by looking at ERP responses to violations of tense and number agreement in contrast to the N400 effects induced by semantic implausibility. ERPs to agreement violations did not resemble the N400 responses in amplitude or scalp distribution, and the results helped establish the

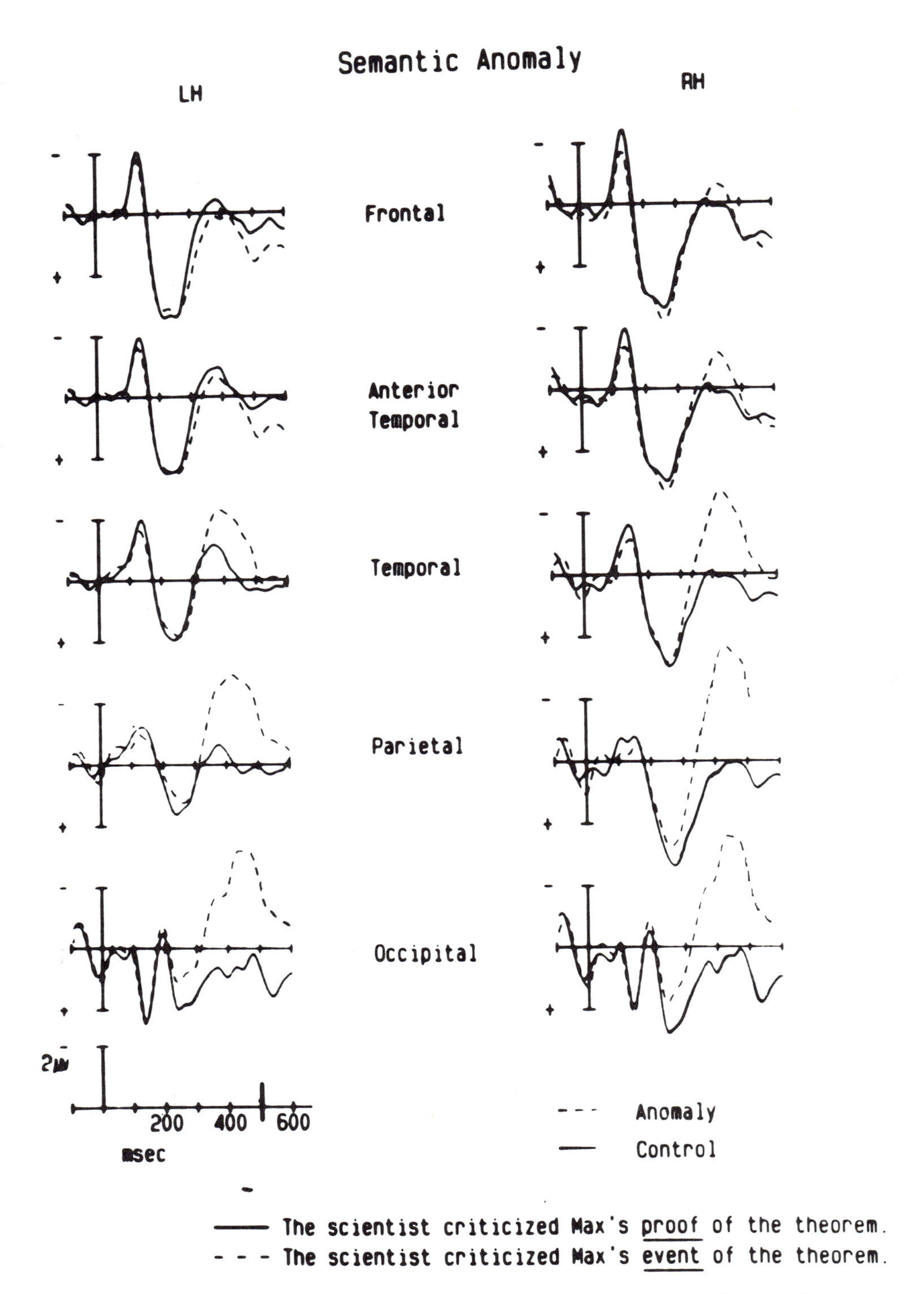

FIGURE 57.2 ERPs at homologous left- and right-hemisphere recording sites for semantic anomalies (from Neville et al., 1991). Responses in dashed lines were elicited by the anomalous words; solid lines are for the corresponding control word in the matched sentences.

N400's specificity to semantic processes (Kutas and Hilyard, 1983). Several recent studies have sought to discover ERP patterns that reflect syntactic integration in the way that the N400 seems to be modulated by interpretive integration. Two components that those studies jointly link to syntactic processing are (1) a sustained and widely distributed positive shift in the 500–900 ms range, hereafter SPS; and (2) a left, anterior, negative shift in the 400–700 ms range, hereafter LAN.

Osterhout and Holcomb (1992) reported an ERP test for aspects of verb subcategorization in which subjects made an acceptability judgment after each visually presented stimulus. Subcategorization constraints express restrictions on the types of structures that can follow particular verbs, as in the contrast between (a) and (b) below:

(a) *The broker hoped to sell the stock.*

(b) **The broker persuaded to sell the stock.*

The experimenters reported a sustained positive shift —which they called the P600 and we are labeling SPS —beginning some 500 ms after the point of the structural violation (the word *to* following *persuade*). See figure 57.3.

Notice that acceptability of the lexical sequence following the two verb types can be reversed by continuing the sentences as in (c) and (d) below——(c) is a reduced version of *The broker who was persuaded to sell the stock . . .*; *hoped* does not passivize:

(c) The broker persuaded to sell the stock was sent to jail.

(d) *The broker hoped to sell the stock was sent to jail.

For these latter versions, Osterhout and Holcomb again observed an SPS at *to* in *persuade* versions like (c), and an additional such response, this time in *hope* versions like (d) at the auxiliary verb *was*. They thus observed complementary SPS effects at successive serial locations. This makes two important points beyond identifying a potential ERP response to syntactic deviation. First, There is a "premature analysis" effect: The first SPS effect in (c) arises from a syntactic option (*persuade* taken as main verb) that is not required by the new sentence environment. But, given incremental processing, it is momentarily embraced—hence, an SPS like the one in (b). Second, an SPS effect is absent for (c) types at the second locus—after *was*, where (d) types did show an SPS. This indicates that the initial inappropriate analysis is rapidly reanalyzed and the correct syntax assigned in (c). An important additional result is that apparent N400 effects occurred at the ends of the deviant sentences. Osterhout and Holcomb interpreted these as indications of delayed semantic integration in the syntactically deviant sentences.

Hagoort, Brown, and Groothusen (1993) report studies with results similar to those of Osterhout and Holcomb, but with additional structural types and significant variation of method. For their tests (done in Dutch), subjects read passively; acceptability judgments were not required. Subcategorization constraints, phrase-structure constraints, and agreement relations were tested. Agreement violations were subject-verb number mismatches; phrasal violations reversed an obligatory adverb-adjective order (translated from the Dutch: **Most of the visitors like the colorful very tulips in Holland*). Subcategorization violations used obligatorily intransitive verbs immediately followed by a noun phrase inflectionally marked as a direct object —for example (translated), **The son of the rich man boasts the car of his father.* ERP results showed an SPS for agreement and phrase-structure violations (see figure 57.4), but not for subcategorization violations. All three violation types produced N400 shifts at sentence-final positions.

A full account of the absent SPS for the subcategorization violations reported by Hagoort, Brown, and Groothusen will require further study. They suggest that the violations were so immediately linked to failure of a semantic prediction from the subcategorization frame that an N400 very likely occurred more or less simultaneously with a positive shift, canceling out evidence for SPS. That suggestion and the claim for SPS as a syntactic reflex are strongly reinforced by a subsequent variation of the same experiment (Hagoort and Brown, in press). The variation substitutes semantically unrelated words for open-class elements in their materials, producing structurally parallel nonsense versions (e.g., adverb-adjective inversions like **The heel tripped over the inhabited rather cat in his pocket*). This experiment displayed the same SPS results as were obtained by Hagoort, Brown, and Groothusen. But, importantly, N400 effects were not observed at ends of sentences; interpretive integration effects were eliminated for nonsense versions. This finding indicates that some aspect of syntactic analysis underlies the SPS pattern.

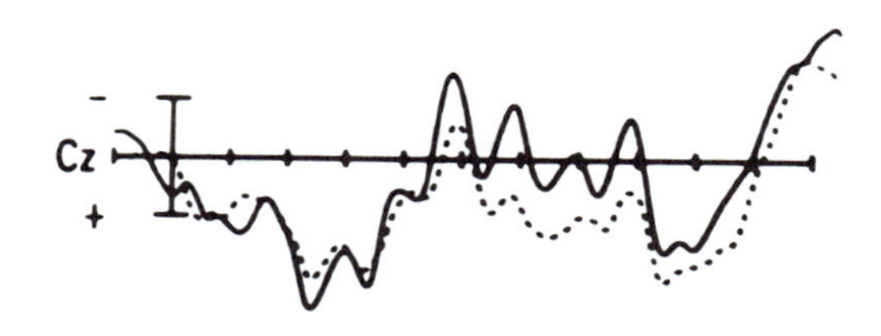

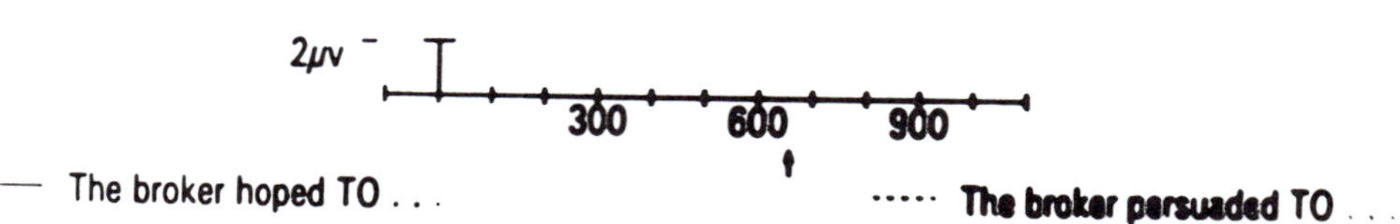

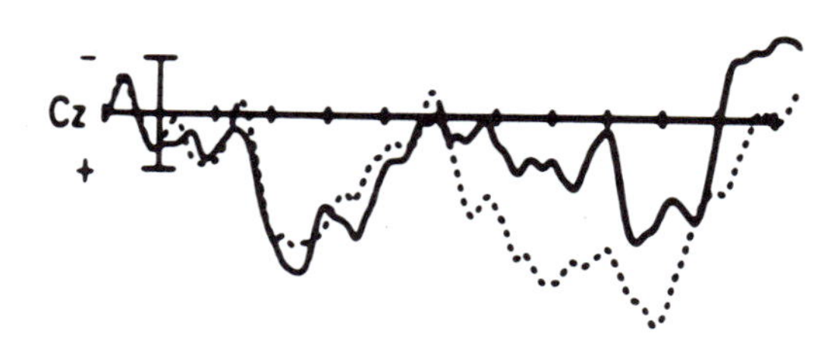

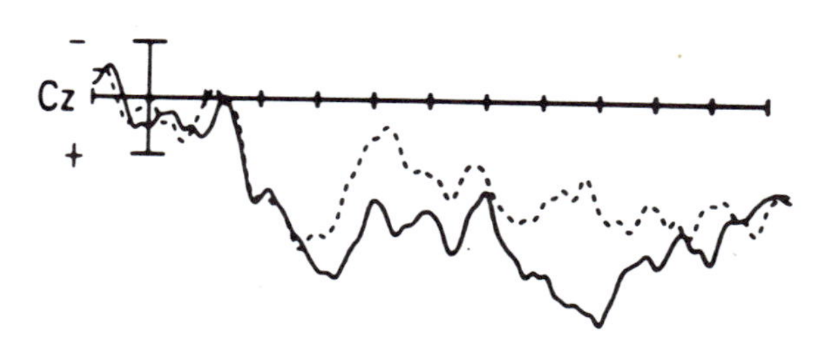

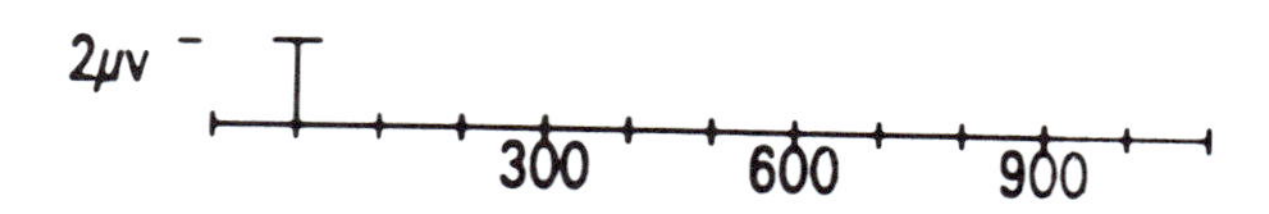

— . . . was sent to JAIL. ····· . . . was sent to JAIL.

FIGURE 57.3 ERPs at midline (from Cz) for the infinitival (top), auxilliary (middle), and sentence-final (bottom) words (from Osterhout and Holcomb, 1992). Responses in dashed lines were elicited by the syntactically deviant words; solid lines are for the corresponding control word in the matched sentences.

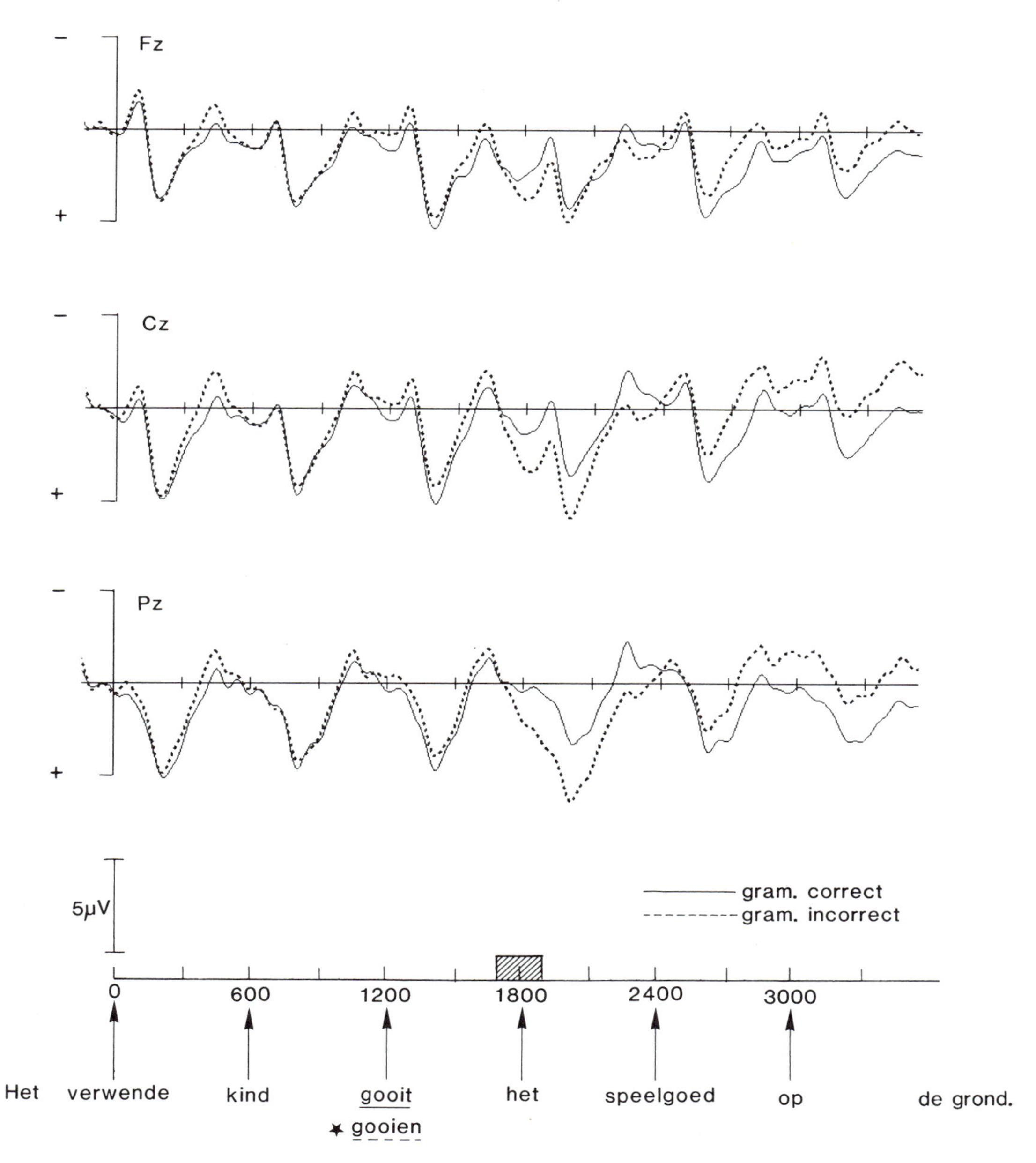

FIGURE 57.4 ERPs at three midline sites for agreement (above) and phrase-structure violations (p. 890) (from Hagoort, Brown, and Groothusen, 1993). Responses in dashed lines were elicited by the syntactically deviant words; solid lines are for the corresponding control word in the matched sentences.

Work by Neville and colleagues (1991) also implicates an SPS effect and introduces evidence for earlier negative shifts, including an LAN effect. The experiment tested phrase-structure constraints, and two classes of movement constraints on noun phrases; semantic violations were also tested in versions matched to the phrase-structure violations; subjects made acceptability judgments for visually presented sentences. Semantic anomalies (see figure 56.2) produced an N400 effect of the amplitude and distribution expected

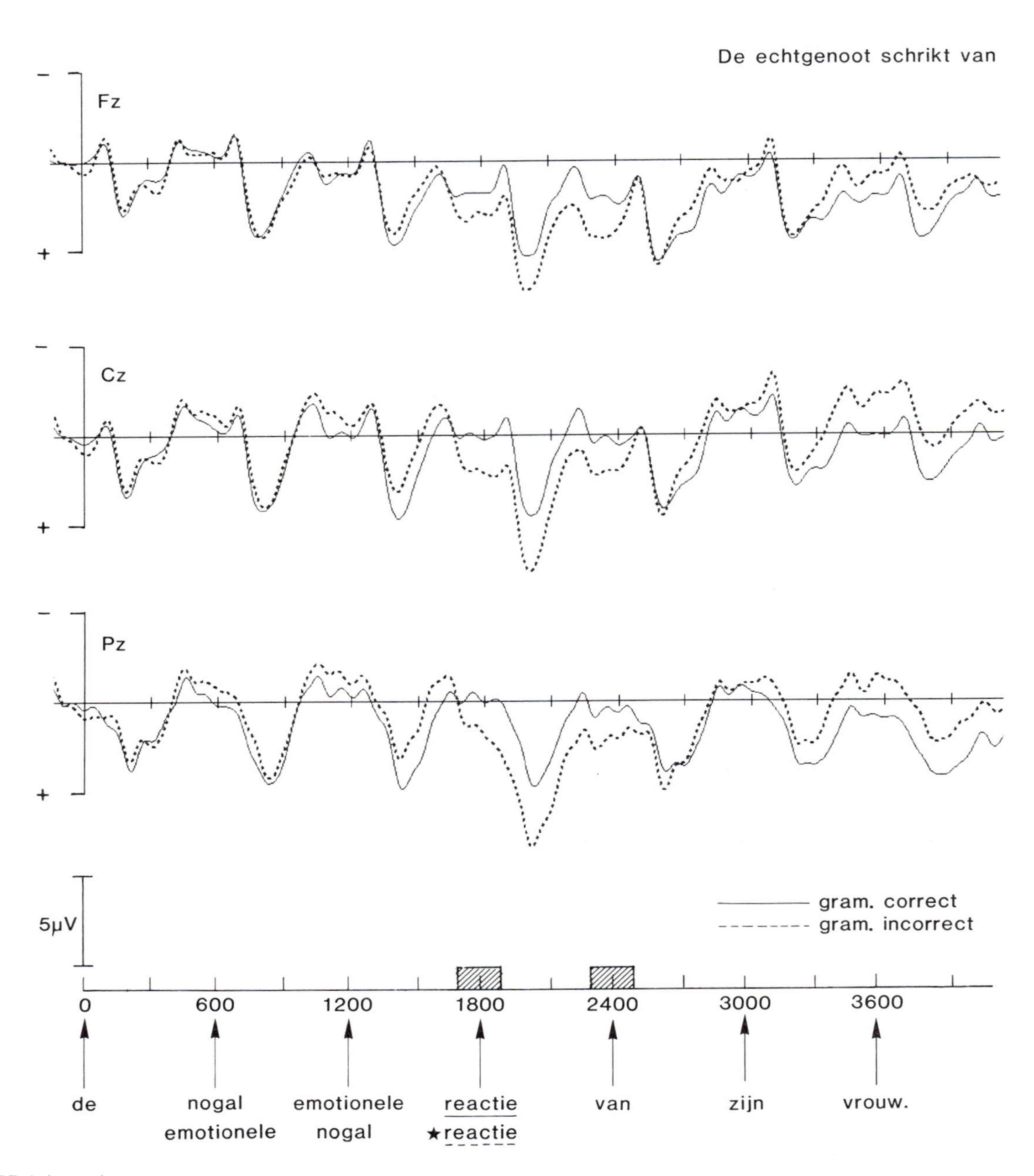

FIGURE 57.4 (cont.)

from many prior experiments. Figures 57.5 and 57.6 show results for two of the deviance types.

The phrase-structure violations (e.g., **The scientist admired Max's of proof the theorem*) produced an early negativity (N125) at anterior sites in the left hemisphere, and a second negative shift, which we are labeling LAN, in the 300–500 ms range (also strongly lateralized to the left hemisphere, maximal in temporal and parietal locations but also manifest at anterior sites); this was followed by a sustained positive shift beginning around 500 ms—similar to the SPS already discussed. Note that the distribution of the LAN is

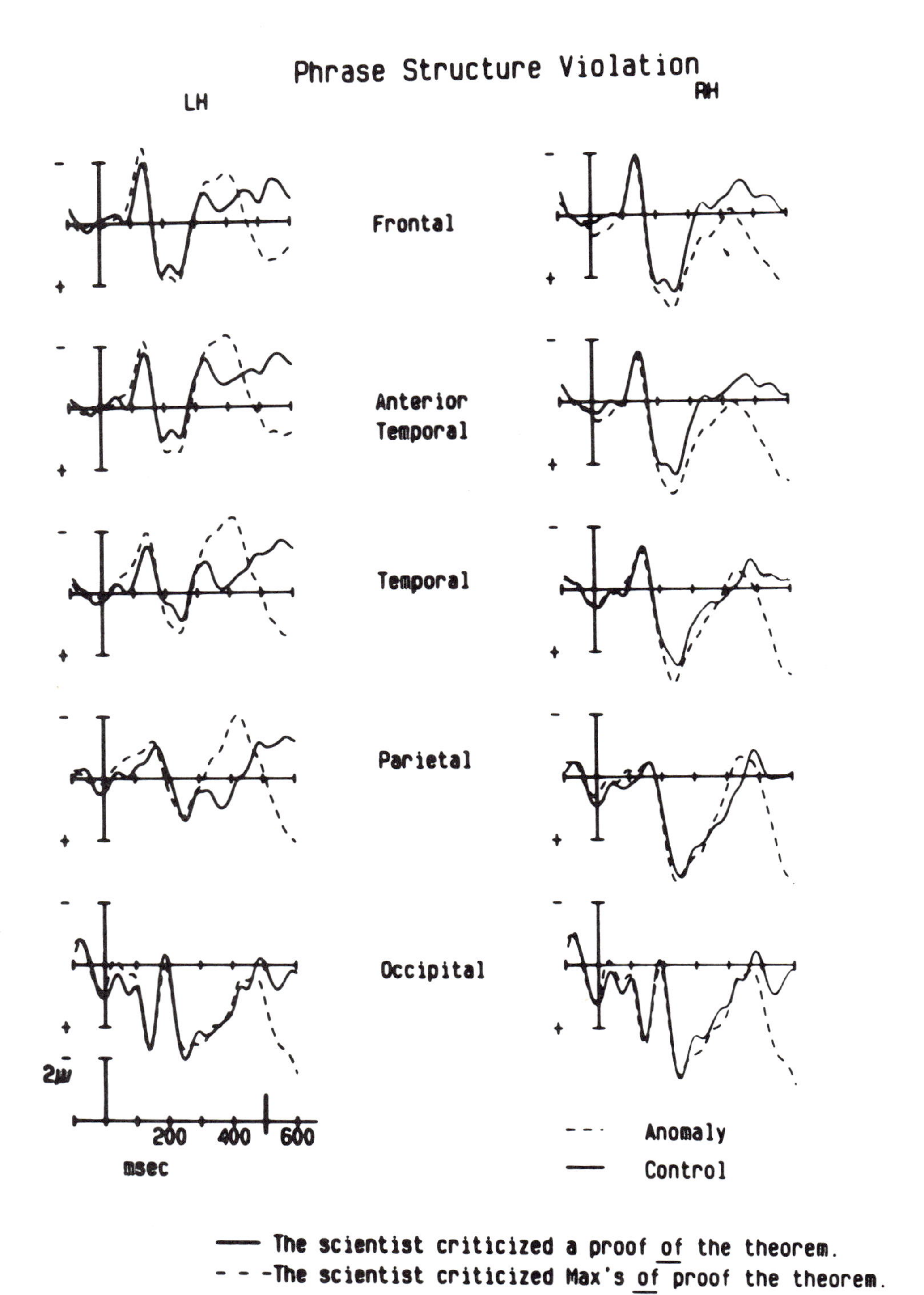

FIGURE 57.5 ERPs at homologous left- and right-hemisphere recording sites for phrase-structure violations (from Neville et al., 1991). Responses in dashed lines were elicited by the sytactically deviant words; solid lines are for the corresponding control word in the matched sentences.

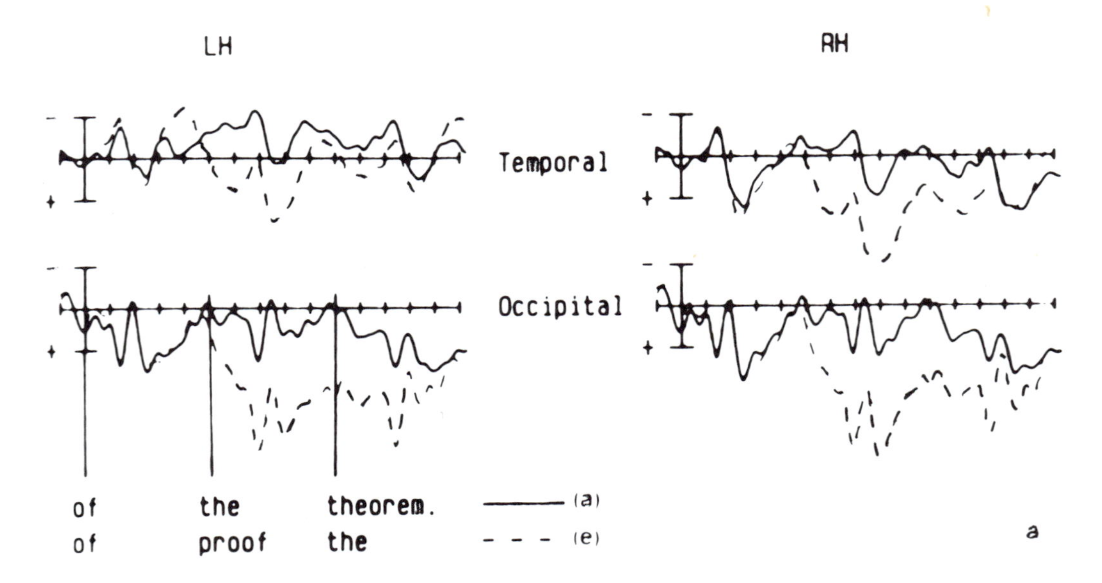

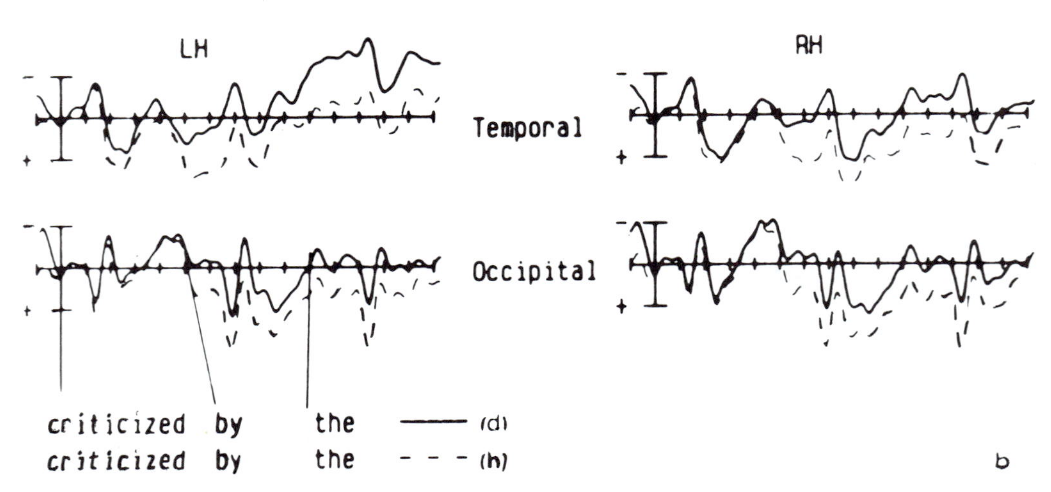

FIGURE 57.6 ERPs at left- and right-hemisphere temporal and occipital recording sites for phrase-structure and subjacency violation comparison words and the two following words (from Neville et al., 1991). Responses in dashed lines were elicited by the deviant words; solid lines are for the corresponding control word in the matched sentences.

complementary to that of the N400: It is strong where N400 is absent, and weak or absent where N400 is strongest.

The movement violations (specificity violations such as **What did the scientist criticize Max's proof of?* and subjacency violations such as **What was a proof of criticized by the scientist?*) differ in their linguistic description (see Neville et al., 1991). The two types of movement violation produced different ERP patterns: The specificity violation produced a negative shift similar to that for phrase-structure violations but without an SPS; the subjacency violation produced an SPS but no negative shift. The characterization of the SPS as a response to syntactic deviance is fit by two of the three conditions,

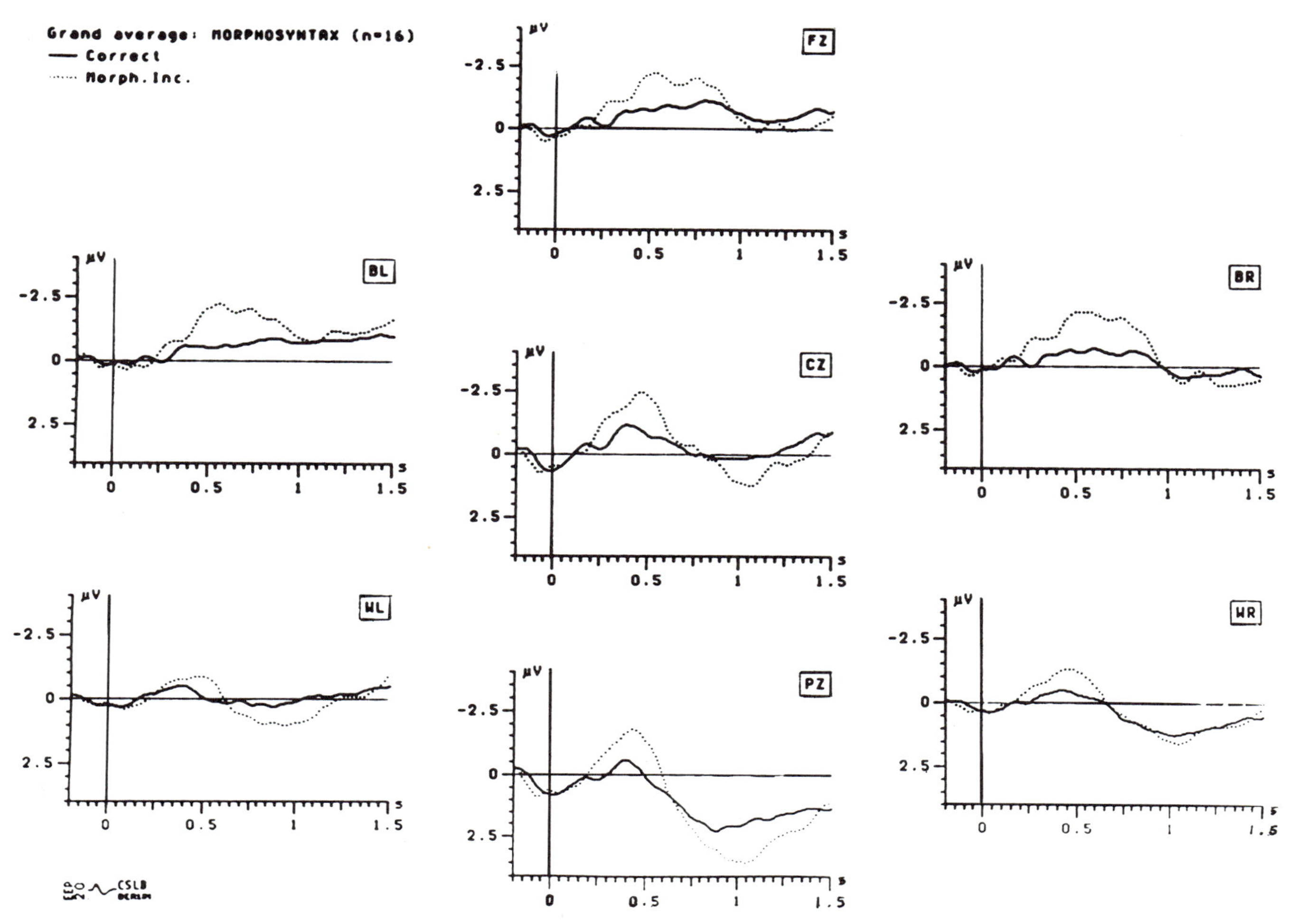

FIGURE 57.7 ERPs at midline and homologous left- and right-hemisphere recording sites for morphosyntactic violations (from Friederici et al., in press). Responses in dashed lines were elicited by the deviant words; solid lines are for the corresponding control word in the matched sentences.

but not by the specificity case. The SPS may be absent in that case for reasons similar to the subcategorization case in Hagoort, Brown, and Groothusen: Specificity violations are easy to understand in spite of the deviance, and an interaction with N400 effects may have occurred because of immediate reinterpretation; also, specificity violations were near ends of sentences and could have interacted with interpretive closure effects.

Two German-language experiments also show LAN effects. Rosler et al. (1993) tested deviant subcategorization structures in a visual sentence-completion paradigm. The test word was a well-formed or an ill-formed continuation of a sentence fragment, for example, *Der Prasident wurde begrusst* ("The President is being greeted") versus **Der Lehrer wurde gefallen* ("The teacher is being fallen"). ERP measures showed a left anterior negativity in the 300–700 ms range that was temporally and topographically somewhat similar to that observed by Neville et al. for phrase-structure and specificity violations. Friederici, Pfeifer, and Hahne (in press) report an auditory test comparing semantic, morphosyntactic, and phrasal violations in a similar paradigm. That test also showed negative shifts at anterior sites for morphological and phrasal violations; see figure 57.7 for one example. Neither study reported an SPS effect; its absence may be linked to the end-of-sentence locus of the violations.

The studies reviewed show syntactically driven ERP responses for several different types of structure. Observations by Kluender and Kutas (1993) reveal a facet of such effects for working memory interactions with sentence processing. Memory overhead is an inescapable consequence of computations that tie noun phrase fillers to gaps. Kluender and Kutas raise the hypothesis that LAN may index processes associated with entering gap fillers in memory and then restoring them to interpretation sites encountered later in the sentence. They provide evidence for this by comparing embedded sentences introduced by a relative pronoun, such as (a) below, with those introduced by complementizers that do not signal an upcoming gap, such as (b).

(a) Have you forgotten who he could coerce ____ into singing songs at . . .

(b) Have you forgotten that he could coerce her into singing songs at . . .

Note that we are dealing here only in contrasts of well-formed structures, so the processes at issue are not dependent on detecting deviant syntax. For the first comparison site (at *he*), the LAN was significantly greater following *wh* pronoun's than following, for example, "that-complementizer" versions (see figure 57.8). An LAN difference was also observed at the second comparison site (*into*), which follows the gap site. This pattern fits the memory hypothesis well.

It is possible that a filler-gap effect might be present in the sustained negativity found by Neville et al. for specificity violations. It is less easy to see how the memory hypothesis applies for their LAN response to phrase-structure violations. The effects observed by Rosler et al. and by Friederici, Pfeifer, and Hahne also do not readily fit a memory account. In brief, LAN effects are likely to have more than one explanation.

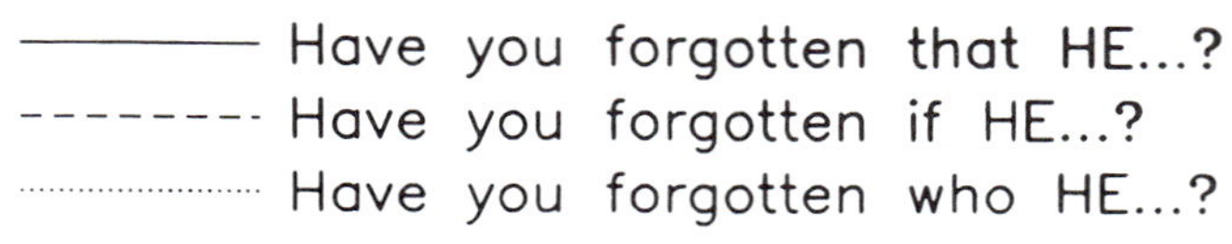

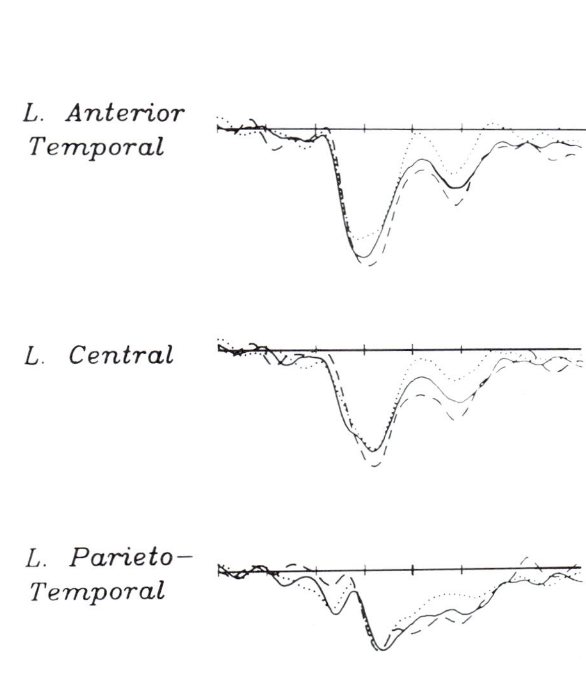

FIGURE 57.8 ERPs at left- and right-hemisphere recording sites for filler-gap words and two complementizer control words (from Kluender and Kutas, 1993). Responses in dotted lines were elicited by the filler-gap words; dashed and solid lines are for the corresponding control words in the matched sentences.

LEXICAL PROCESSES IN SENTENCE CONTEXTS This section moves our focus to semantic and syntactic influences on lexical processes. Lexical ambiguity and grammatical word-class effects are considered.

Lexical ambiguity Recognition processes for lexical ambiguities reflect sentential constraints on lexical access and interpretation. Though there are a number of unsettled issues, it seems clear that analysis of ambiguous words in neutral or weakly constraining contexts is marked by multiple access: Both senses are activated, with priority for the more frequently occurring sense (see, e.g., Simpson and Krueger, 1991). A recent study by Hagoort and Brown (in press) reinforces this and identifies relevant ERP responses for future investigation. When Dutch sentences with an ambiguous word were tested, ERP responses to the ambiguity diverged from those for a control word. The effect was a negative shift with some similarity to LAN effects—though not so sharply focused at anterior left: Effects were most pronounced for anterior left and right, with some posterior left but no significant posterior right effect; left-side responses were more sustained than right. This evidence indicates that encounters with ambiguities in neutral contexts occasion immediate changes in processing profile. What about ambiguities with a prior constraining context? Is the effect modified? Does selection of a contextually preferred sense generally occur without activation of contextually irrelevant senses? Behavioral studies suggest not. Cross-modal priming tasks (see, e.g., Swinney, 1979) have shown effects for both contextually relevant (CR) and contextually irrelevant (CI) senses of ambiguities. In these tasks, subjects listen to sentences containing an ambiguous word while monitoring a visual display for lexical decision targets. Sentence context is manipulated to control ambiguity interpretation, and visual targets semantically related to different senses of an ambiguity are tested for priming from the auditory channel. Several such experiments have shown time-dependent effects: Probes at the ends of ambiguous words show facilitation for both CR and CI senses; probes presented 200–500 ms later show effects only for CR senses (e.g., Seidenberg, et al., 1982). This suggests data-driven access that yields multiple (possibly frequency-ordered) analyses that are rapidly modulated by context to select the appropriate reading.

An ERP version of this kind of test by Van Petten and Kutas (1988) used visual presentation for both the ambiguous sentences and the probe words: Ambiguous words were sentence-final, and were followed by the probe at a 200 or 700 ms interval (onset to onset). A lexical-decision version of the task produced the usual reaction-time profile—both meanings were activated at 200 ms, but only the CR sense was activated at 700 ms. A subsequent ERP version of the experiment compared N400 responses to the probe words (semantic relatedness reduces the magnitude of N400 responses). The 700 ms delay condition showed the selective pattern: Only CR words reduced N400 responses. At the 200 ms delay, an interesting difference between behavioral measures and the ERP index appeared: Both probe types had diminished N400s, but the CR probe diverged from control about 200 ms earlier and to a greater degree than the CI probe (figure 57.9). Van Petten and Kutas examined several explanations for the time course of processing revealed by the ERP measure, and opted for an account that assumes selective access for the CR sense. Activation of the CI sense was attributed to a backward-acting priming process in which the probe for the CI sense interacts with the memory record of the sentence-final ambiguous word, triggering an interpretation that would otherwise not have been activated. This account minimizes the context-free, data-driven aspects of lexical processing and places contextual constraint mechanisms much earlier in the process than most earlier analyses.

The contrast between behavioral and ERP measures raises interesting questions about the generality of backward-acting context mechanisms—questions that we must set aside (see, e.g., Tannenhaus et al., 1988; Van Petten and Kutas, 1991a). But, given Hagoort and Brown's evidence for an immediate reflex of ambiguity, we can ask when prior contextual constraints affect that processing—and, relatedly, how conditions of the ERP test fit the rate of form-driven lexical access processes. In Van Petten and Kutas's short-interval condition, probes occurred 200 ms after ambiguity onset; 250 ms after that, the first ERP evidence of the CR meaning appears, and, a like interval later, that of the CI meaning. Estimates of the rate of ambiguity processing can be fit into this frame. So, for example, ambiguity and control appear to diverge some 200 ms after onset for Hagoort and Brown's ERP study, and a reaction-time study by Zwitserlood (1989) suggests about the same order of effect for auditory processing. Applied to the Van Petten and Kutas experiment, this suggests that processing of an ambiguous word would indeed be well under way or completed by stimulus offset. Intrinsic or contextually derived influences on the process could be established at the point of presen-

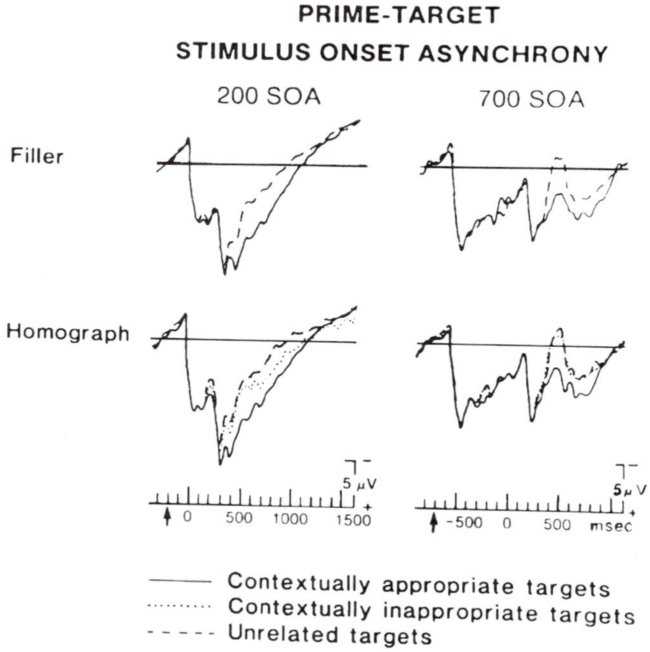

FIGURE 57.9 ERPs at midline for ambiguity probe words at 200 ms and 700 ms test intervals (from Van Petten and Kutas, 1991a). Responses in solid lines were elicited by the contextually appropriate probe words; responses in dotted lines were elicited by the contextually inappropriate probes; dashed lines are for the unrelated control words.

tation of probes and could produce differential effects for CR and CI probes. How does this relate to the effect of data-driven, multiple-access mechanisms? A bit more detail from Zwitserlood's experiment is instructive. She used cross-modal priming (acoustic sentences with visual probes; Dutch language) for words that were not fully ambiguous but were compatible with multiple analysis for the first portion of the acoustic representation of the test word—for example, *Kapitaan* 'captain' and *Kapital* 'capital'—and phonetically identical till the last syllable. Visual probes presented after varying amounts of the acoustic signal representing test words (50–100 ms increments) showed that contextual constraints from the test word's sentence environment were irrelevant to priming at the two earliest testing positions—that is, both meanings facilitated related probes. But, context-specific priming effects did appear before the test word's completion (one test position earlier in constraining sentence contexts than in a context-free control condition). In short, data-driven, multiple-access processes were manifest at early positions, but acoustic and contextual constraint began to effect lexical closure at the ends of words. Such a picture applied to the visual ambiguities suggests that primary access mechanisms may achieve full or partial closure (including some response to context) before significant processing of a following probe word is available. Reopening of lexical processing for the ambiguity could be triggered by presentation of the probe word, and ERP effects would reflect the time course of that process. The timing estimates here are delicate, but future ERP measures should permit resolution of this kind of detailed question about processing at close temporal intervals.

Lexical integration in sentence processing Another significant ERP experiment by Van Petten and Kutas (1991b) bears on incremental sentence processing issues. Subjects read sensible normal prose sentences, syntactically legal but nonsensical sentences (called "syntactic prose" hereafter), and random word strings. ERP measures were averaged for words at matched serial positions across stimuli, enabling the experimenters to look at ERP profiles as words in the sentence or word string were progressively assimilated during reading. They examined several lexical variables; we focus on grammatical category. Grammatical category effects were discussed in terms of a broad classification into open (nouns, verbs, and adjectives) and closed classes (minor grammatical categories).

For open-class vocabulary, N400 responses showed a systematic decline across serial sentence positions for normal prose stimuli; no such decline occurred in for syntactic prose and random word string stimuli. This suggests progressively easier assimilation of lexical content to an interpretation as context builds in sentences. Absence of such an effect in the syntactically organized but semantically incoherent sentences indicates dependence of the effect on interpretation. Closed-class words presented an interesting contrast to open-class effects: Overall, their N400 responses were reduced in amplitude for the normal prose condition compared to the syntactic prose, which was in turn reduced compared to the random word string condition. Open-class words were not similarly reduced in the syntactic prose condition, and Van Petten and Kutas note that this indicates a specific configurational effect in syntactic prose items for processing of closed-class words. Closed-class responses also showed no modification as a function of serial position in sentences, from which Van Petten and Kutas infer that the closed-class constraints were local as well as configural.

Van Petten and Kutas do not endorse a qualitative contrast between open- and closed-class elements to account for these differences. Rather, they associate the performance contrasts in ERP outcomes with correlated variables such as frequency, length, semantic richness, and a variable role in sentence processing. There is good reason to take such a position with respect to N400 indices of closed-class processing, for it is clear that the closed class is not homogeneous, containing as it does a variety of subtypes that relate to language structures in different ways. Put another way, there is more than one open versus closed class distinction. There are (at least) two messages in this observation. The import of one is outlined by Van Petten and Kutas. The other is that since open/closed processing may differ in different parts of the multicomponent language-processing system, observational and experimental results may reflect the sensitivity of our measures to different aspects of the processes. This is not a new problem. To illustrate with one example: Early-stage parsing algorithms may use procedures based on the stress neutralization of closed-class vocabulary in English; this would implicate aspects of syntactic analysis that integrate the closed class under prosodic description. But such a classification would lump together elements with variable relations to the syntax and semantics for levels of sentence structure motivated by configural and logical constraints of interpretation.

Some additional evidence that bears on this question is available from studies by Neville, Mills, and Lawson (1992), who address issues of relations between semantic and syntactic processing in still another way. They investigated the reading of normally hearing and congenitally deaf subjects with ERP measures, and compared the semantic and syntactic processing for reading English text in these two groups by looking at responses to open- and closed-class vocabulary. The latter subjects learned American Sign Language as their first and primary language and acquired their English-language competence "late and imperfectly because of their congenital deafness"—and after, the authors argue, the period in which languages are acquired with native fluency and control. The deaf subjects generally read with significantly less facility than the hearing subjects and displayed significantly less understanding of the syntactic features of English than the hearing group. In their comparison, Neville, Mills, and Lawson found that semantic processes, as indexed by the N400 response to open-class words in semantically implausible materials, were virtually identical for the two populations. But, ERP responses linked to grammatical processes sharply differed: The researchers report an ERP pattern for closed-class words shown only by the hearing subjects. They observed an N280 response, localized to anterior temporal and frontal regions of the left hemisphere, as well as a response like the LAN discussed earlier. They evaluated differences in length and frequency of occurrence for open and closed comparison sets, and concluded that the characteristic responses for the closed class could not plausibly be accounted for in those terms. Such ERP studies that contrast N400 effects with the family of left-dominant negative responses that seem to be linked to syntactic processes may help us to resolve the different roles that closed-class elements play in sentence processes.

Summary: SPS, LAN, and N280

ERP investigations of sentence and lexical processing enrich evaluation of language performance in many detailed ways, most of which have been only partially explored. The initial family of results supports a tightly coupled interplay of form-driven and interpretive processes that have been suggested by a wide range of behavioral tests. The existence of distinguishable ERP reflexes for interpretive and syntactic processes promises a more precise assessment of the core issues of interaction and modular processing, which are the complementary foci of theoretical activity in this domain.

SPS and LAN types of effects appear across a significant range of sentence structures. Both are likely to have multiple causes, and they are subject to interaction with each other, with N400 effects, and with ERP responses to attentional factors related to the timing of parsing and of interpretive processes at different serial locations in sentences. There are inconsistencies without ready accounts for both LAN and SPS effects, namely, their failures to appear in some cases; these might plausibly relate to such interactions. A summary discussion of the sort in this chapter sets aside many differences of detail among experimental data reports that may ultimately prove of importance, and it would be unwise to celebrate the similarities of outcome without recognizing the potential complications presented by unevaluated differences. Nevertheless, the several observations of left anterior focused negative shifts in the 200–700 ms range (including LAN) and those of the more broadly distributed positive shifts in the 500–900 ms range (including SPS) across different task demands with different sentence structures are encouraging. Careful titration of these effects can yield information about the nature and time course of sentence processing that is otherwise hard to come by, particularly as each of these types of effect may be compared with the N400 effects tied to aspects of sentence interpretation.

The still earlier left anterior negative shifts (viz., N125 and N280 effects) are also intriguing, particularly as they may relate to the integration of closed-class vocabulary in parsing activity. The distinction between open- and closed-class vocabulary has a significant record in neuropsychology (in particular, in classical clinical accounts of agrammatism in Broca's aphasia and of paragrammatism in Wernicke's aphasia). These and related observations of normal sentence processing and language development have had a checkered history, and no fully satisfactory account of the special role in language processing of the closed class or of its behavioral reflexes is currently on offer.

In this connection, we should also recall the syntactic prose effects in Hagoort and Brown (in press) and those in Van Petten and Kutas (1991b). Those effects appear in strings for which open-class content is misleading or irrelevant and for which functional vocabulary, bound and free, is the primary vehicle that supports structural integration. Such findings raise the prospect that further careful investigation of this vocabulary contrast, using ERP and related procedures, may well lead to a more satisfactory resolution of a psycholinguistic research problem of enduring interest. In so doing, it is perhaps fitting that the latest in language processing technology, ERP techniques, should be combined with one of hoariest illustrations of the dissociation of syntactic form and semantic content.

ACKNOWLEDGMENTS Preparation of this chapter and partial support for some of the author's work referred to herein was provided by a grant from the McDonnell-Pew Cognitive Neurosciences Program and an NIH grant to the Center for Neurogenics Communications Disorders at University of Arizona.

REFERENCES

BOCK, J. K., and K. EBERHARD, 1992. Meaning, sound and syntax in English number agreement. *Lang. Cognitive Proc.* 8:57–99.

BOCK, J. K., and A. S. KROCH, 1992. The isolability of syntactic processing. In *Linguistic Structure in Language Processing*, G. N. Carlson and M. K. Tanenhaus, eds. Dordrecht: Reidel.

CARAMAZZA, A., and E. ZURIF, 1976. Dissociation of algorithmic and heuristic processes in language comprehension: Evidence from aphasia. *Brain Lang.* 3:572–582.

CRAIN, S., and M. STEEDMAN, 1985. On not being led up the garden path: The use of context by the psychological processor. In *Natural Language Parsing*, D. Dowty, L. Karttunen, and A. Zwicky, eds. Cambridge: Cambridge University Press.

DELL, G., 1986. A spreading activation theory of retrieval in sentence production. *Psychol. Rev.* 93:283–321.

FODOR, J. A., 1983. *The Modularity of Mind.* Cambridge, Mass.: MIT Press.

FORSTER, K. I., 1979. Levels of processing and the structure of the language processor. In *Sentence Processing*, W. Cooper and E. Walker, eds. Englewood, N.J.: Erlbaum.

FRAZIER, L., 1987. Theories of sentence processing. In *Modularity in Knowledge Representation and Natural Language Processing*, J. Garfield, ed. Cambridge, Mass.: MIT Press.

FRIEDERICI, A., E. PFEIFER, and A. HAHNE, (in press) Event related brain potentials during natural speech processing: Effects of semantic, morphological and syntactic violations. *Brain Res. Cogn. Brain Res.*

GARNSEY, S., M. TANNENHAUS, and R. CHAPMAN, 1989. Evoked potentials and the study of sentence comprehension. *J. Psycholinguistic Res.* 18:51–60.

GARRETT, M. (in press) Errors and their relevance for models of language production. In *Handbooks of Linguistics and Communication Science: Linguistic Disorders and Pathologies*, Jurgen Dittman and Gerhard Blanken, eds. New York: De Gruyter.

HAGOORT, P., and C. BROWN, (in press) Brain responses to lexical ambiguity resolution and parsing. *J. Psycholinguistic Res.*

HAGOORT, P., C. BROWN, and J. GROOTHUSEN, 1993. The syntactic positive shift (SPS) as an ERP-measure of syntactic processing. *Lang. Cognitive Proc.*

HILYARD, S., and T. PICTON, 1987. Electrophysiology of cognition. In *Handbook of Physiology.* Section 1, *The Nervous System.* Vol. 5, *Higher Functions of the Brain*, part 2, F. Plum, ed. Bethesda, M.D.: American Physiological Society, pp. 519–584.

KEMPEN, G., and P. HOENKAMP, 1987. Incremental procedural grammar for sentence formulation. *Cognitive Sci.* 11: 201–258.

KLUENDER, R., and M. KUTAS, in press. Bridging the gap: Evidence from ERP's on the processing of unbounded dependencies. *J. Cognitive Neurosc.* 5:196–214.

KUTAS, M., and S. HILYARD, 1980. Reading between the lines: Event related potentials during sentence processing. *Brain Lang.* 11:354–373.

KUTAS, M., and S. HILYARD, 1983. Event related brain potentials to grammatical errors and semantic anomalies. *Mem. Cognition* 11:539–550.

KUTAS, M., and S. HILYARD, 1984. Brain potentials during reading reflect word association and semantic expectation. *Nature* 307:161–163.

KUTAS, M., and C. VAN PETTEN, 1988. Event related brain potential studies of language. In *Advances in Psychophysiology*, vol. 3, P. K. Ackles, J. R. Jennings, and M. G. H. Coles, eds. Greenwich, CT: JAI Press, pp. 139–187.

LEVELT, W. J. M., 1989. *Speaking: From Intention to Articulation.* Cambridge, Mass.: MIT Press.

LEVELT, W., H. SCHRIEFERS, A. MEYERS, T. PECHMAN, D. VORBERG, and J. HAVINGA, 1991. The time course of lexical access in speech production: A study of picture naming. *Psychol. Rev.* 98:122–142.

LINEBARGER, M., M. SCHWARTZ, and E. SAFFRAN, 1983. Sensitivity to grammatical structure in so-called agrammatic aphasics. *Cognition* 13:361–392.

MARSLEN-WILSON, W. D., and L. TYLER, 1980. The temporal structure of spoken language understanding. *Cognition.*

MARTIN, N., and E. SAFFRAN, 1990. Repetition and verbal STM in Transcortical Sensory Aphasia: A case study. *Brain Lang.* 39:254–288.

MCCLELLAND, J., M. ST. JOHN, and R. TARABAN, 1989. Sentence comprehension: A parallel distributed processing approach. *Lang. Cognitive Proc. 4* (special issue on Parsing and Interpretation): 287–336.

NEVILLE, H., D. MILLS, and D. LAWSON, 1992. Fractionating

language: Different neural subsystems with different sensitive periods. *Cerebral Cortex* 2:244–258.

Neville, H., J. Nicol, A. Barss, K. Forster, and M. Garrett, 1991. Syntactically based sentence processing classes: Evidence from event related potentials. *J. Cognitive Neurosci.* 3:151–165.

Osterhout, L., and P. Holcomb, 1992. Event related brain potentials elicited by syntactic anomaly. *J. Mem. Lang.* 31:785–806.

Rondal, J. A., in press. Language systemic specificity in Down syndrome. *Proceedings of the International Symposium on the Specificity of Down Syndrome.*

Rosler, F., A. Friederici, P. Putz, and A. Hahne, 1993. Event related brain potentials while encountering semantic and syntactic constraint constraint violations. *J. Cognitive Neurosci.* 5:345–362.

Seidenberg, M., M. Tannenhaus, J. Leiman, and M. Beinkowski, 1982. Automatic access of the meanings of ambiguous words in context: Some limitations of knowledge-based processing. *Cognitive Psychol.* 14:489–537.

Simpson, G., and M. Krueger, 1991. Selective access of homograph meanings in sentence context. *J. Mem. Lang.* 30:627–643.

Swinney, D., 1979. Lexical access during sentence comprehension: (Re)Consideration of context effects. *J. Verb. Learn. Verb. Be.* 18:645–659.

Tanenhaus, M., C. Burgess, and M. Seidenberg, 1988. Is multiple access an artifact of backward priming? In *Lexical Ambiguity Resolution*, S. Small, G. Cottrell, and M. Tanenhaus, eds. San Mateo, Calif.: Morgan Kaufman.

Tanenhaus, M., G. Carlson, and J. Trueswell, 1989. The role of thematic structures in interpretation and parsing. *Lang. Cognit. Proc.* 4 (special issue on Parsing and Interpretation): 211–334.

Van Petten, C., and M. Kutas, 1988. Ambiguous words in context: An event related potential analysis of the time course of meaning activation. *J. Mem. Lang.* 26:188–208.

Van Petten, C., and M. Kutas, 1991a. Electrophysiological evidence for the flexibility of lexical processing. In *Understanding Word and Sentence*, G. Simpson, ed. Amsterdam: Elsevier, North Holland.

Van Petten, C., and M. Kutas, 1991b. Influences of semantic and syntactic context on open- and closed-class words. *Mem. Cognition* 19:95–112.

Whitaker, H., 1975. A case study of the isolation of the language function. In *Studies in Neurolinguistics*, vol. 2, H. Whitaker and H. A. Whitaker, eds. New York: Academic Press.

Yamada, J. E., 1990. *Laura: A Case for the Modularity of Mind.* Cambridge, Mass.: MIT Press.

Zwitserlood, P., 1989. The locus of the effects of sentential-semantic context in spoken word processing. *Cognition* 32: 25–64.

58 Disorders of Lexical Processing and the Lexicon

BRENDA C. RAPP AND ALFONSO CARAMAZZA

ABSTRACT We review a number of patterns of performance observed in individuals with acquired neurological impairments affecting their abilities to represent or process words. These observations serve as the basis for a detailed proposal regarding the representation of lexical knowledge. In order to account for these various performance patterns we must assume a premorbid lexical system in which knowledge of the meaning, phonology, orthography, and syntax of lexical items consists of internally complex representations that are componentially organized. The distinctions that are revealed by the examination of impaired performance presumably are anatomically/physiologically based and thus serve as a source of information regarding the functional organization of the human brain.

The principal questions with which we will be concerned here pertain to the content and organization of our knowledge of words. Thus, we will be concerned both with trying to discover the specific content of the knowledge that underlies our use of words and with determining how the various aspects of that knowledge are interrelated. We discuss findings that are sufficiently robust and consistent with one another that they must be accounted for by any theory of lexical representation and processing. We will reach the conclusion that the available data motivate a view of lexical representations as internally complex and componentially organized mental objects. We will argue, in fact, that the available data constitute a real challenge for those theories that either deny the existence of lexical and sublexical (e.g., morphemes, syllables) representations or for those that assume that lexical knowledge is organized in an inextricably tangled manner.

BRENDA C. RAPP Department of Cognitive Science, Johns Hopkins University, Baltimore, Md.
ALFONSO CARAMAZZA Department of Psychology, Dartmouth College, Hanover, N.H.

The independent representation of meaning and form

One relatively broad distinction in the representation of lexical knowledge that is revealed by the performance of certain brain-damaged individuals is the distinction between the meaning of a lexical item and its form. Thus, evidence for the preservation of meaning in the presence of the unavailability of form, as well as for the reverse phenomenon, has been well documented.

SPARED KNOWLEDGE OF MEANING As "normal" speakers of the language we have all experienced the feeling of knowing the meaning of a word and yet not knowing its name (see Brown and McNeill, 1966, for discussion of the "tip of the tongue" state). With brain-damaged subjects the classical examples of this phenomenon are the cases of anomia, in which difficulties in retrieving the form corresponding to a concept are apparent in spontaneous conversation as well as in picture naming (see Geschwind, 1967; Goodglass et al., 1976). For example, Kay and Ellis (1987) describe the performance of E. S. T., who suffered left temporal lobe damage as a consequence of a tumor and its excision. E. S. T., when shown a picture of a snowman, was unable to provide the name but responded: "It's cold, it's a ma ... cold ... frozen." Often, limited aspects of the form were available, such as when E. S. T. attempted to name the picture of a stool: /stop/, /stɛp. ... seat, small seat, round seat, sit on the ... sit on the ... /stə/. ...

Interestingly, the difficulty in recovering the phonological form of items can extend to oral reading. In such cases the individual indicates comprehension of the written word by providing adequate definitions and yet is still unable to produce the correct name. Thus, E. S. T., in attempting to read STEAK said "I'm going to eat something ... it's beef ... you can have a /sə/ ... different ... costs more ... a different da di da ... used to know ... in Yankee land ... great big ... it's ridiculous...." The written word SHAMPOO provoked "don't need much, haven't got much on my head ... if I wanted to wash ... /šægən/." Clearly the meanings of the printed words are accessed but their names are not.

The differential availability of meaning and form following brain damage can extend to written forms as well. For example, J. G. who suffered damage to the left temporal-parietal area adjacent to the occipital and temporal horns (Goodman and Caramazza, 1986; see also Beauvois and Derouesne, 1981), when asked to define and spell auditorially presented words, was able to provide excellent definitions, although she was unable to produce the correct written form. Thus, "digit" was defined as "a number" and spelled D-I-D-G-E-T, while a "thief" was identified as a "person who takes things" and spelled T-H-E-F-E." In fact, J. G. was able to provide correct definitions for 98% of the items she was presented, although she was only able to spell 58% of the correctly defined item.[1]

It is important to note that difficulty in producing the name that corresponds to a given meaning would not be particularly surprising if the difficulty were the result of a relatively peripheral deficit, such as one at the level of the articulatory mechanisms for spoken forms or at the level of executing the appropriate motor plans for written forms. If that were the case, one could not speak of a dissociation between knowledge of meaning and knowledge of form, since it is not specifically lexical forms that are affected, but instead processes involved more generally in production (spoken or written). Data regarding such phenomena would not speak to the issue of the representation of *lexical* meanings and forms. Thus, the relevant deficits are those that can be localized to a level of representation or processing that involves word knowledge itself. What is interesting about the cases mentioned above is that although the individuals had no trouble producing spoken or written forms in general, they often did have difficulty in producing the specific lexical forms they were searching for.

Spared Knowledge of Forms Linguists have long argued for the independence of meaning and form at the level of sentential semantics and syntax. Specifically with regard to single-word processing, the apparent selective sparing of the knowledge of lexical forms has been documented in a number of cases of individuals who read or write words that they apparently cannot correctly comprehend. If the words they were able to read or write were words with very predictable letter-sound correspondences (e.g., HOT, MAST) this might not be particularly informative, since good performance with such words could be attributed not to spared lexical knowledge but rather to knowledge of the predictable letter-sound correspondences of the language. This would not be relevant to the issue of lexical knowledge representation because the knowledge of letter-sound regularities is *sub*lexical in that it can be applied equally well to lexical forms (e.g., HAT) or to invented forms (e.g., LAT). Thus the relevant finding is the observation of poor comprehension of correctly read or spelled words whose pronunciations and spellings are not predictable—words such as CAUGHT, DOVE, etc. A number of such cases have been reported both for reading (Schwartz, Saffran, and Marin, 1980; Shallice and Warrington, 1980) and for writing (Hillis and Caramazza, 1991a; Patterson, 1986).

For example, Hillis and Caramazza (1991a) presented the case of J. J. (a patient with damage to the left temporal area and basal ganglia as the result of a cerebral vascular accident, or CVA) who displayed relative preservation of orthographic and phonological forms in the face of impaired comprehension. This was demonstrated when he was asked to define and spell or define and read a number of words. His general ability to define words he could understand was intact ("type" was defined as "tapping a device for numbers and letters"). If one considers his performance only on those words whose spellings or pronunciations are highly irregular, then he was able to read correctly 75% of words he was not able to define correctly. Although his written performance was worse overall, he was nonetheless able to spell correctly 32% of words he was not able to comprehend correctly. Thus, J. J. is an example of these individuals in whom neurological damage has affected the representation of the meanings of words far more than it has affected the knowledge of their written and spoken forms.

The case of J. J. involves the selective preservation of the *production* of spoken and written forms in the absence of comprehension. In addition, there are cases in which words are *recognized* as familiar forms but apparently are not fully understood. Thus, D. R. B. (Franklin, Howard, and Patterson, in press), who suffered from a left middle cerebral artery infarct, was able to discriminate correctly between words of the language and invented wordlike forms whether they were presented visually or auditorially (98% accuracy in both cases). In addition, although he was able to judge whether two written words were synonyms of one another with an accuracy of 95%, for auditorially presented words he was able to judge synonymy with only 61% accuracy.[2] That is, he apparently exhibited difficulty in assigning the correct meaning to words that he was able to recognize as legitimate forms of the language.

In sum, cases in which meaning or form—written or spoken, in comprehension or production—are selectively affected reveal that the names of concepts and their meanings are separable. The findings strongly support the conclusion that the representation of our knowledge of the meanings of words is functionally (and, presumably, anatomically and physiologically) distinct from the representation of their forms.

Interestingly, evidence from PET studies seems to be consistent with this distinction if not with the identification of specific neuroanatomic areas that subserve different components of lexical processing. Thus, Petersen et al., (1989; see also Petersen et al., 1988; Posner et al., 1988) have reported that modality-specific sensory areas are activated by passive word input, whereas anterior inferior frontal cortex is activated by semantic tasks. Although it is not implausible that frontal cortex may play an important role in semantic processing, the evidence from studies of brain-damaged subjects points to a significant role for the temporal and parietal lobes. Thus, for example, patient J. J., who was impaired in lexical-semantic processing, had lesions involving the left temporal lobe and basal ganglia (see Damasio, 1990, for further discussion of this issue).

The relationship between orthographic and phonological forms

In a number of the cases presented above, there are indications that knowledge of the written (orthographic) and spoken (phonological) forms of words may be differentially affected by neurological damage. In this context, Two questions have attracted a great deal of interest: To what extent is our knowledge of the phonological forms of words separable from our knowledge of their orthographic forms? And, more specifically, Can orthographic forms be retrieved without the prior retrieval of phonological forms? These questions arise in response to the claim that orthographic knowledge is fundamentally parasitic upon phonological knowledge. The claim has some empirical motivation (for a review see Van Orden, Pennington, and Stone, 1990) but is fundamentally based on the introspective experience of the primacy of inner speech (in contrast to inner reading and writing) as well as on the fact that, both historically and developmentally, orthography has followed phonology. Regardless of their validity, these observations do not rule out the possibility that the adult knowledge system is organized such that the two information types are fully independent. Thus, we would like to know if one can retrieve the written name that corresponds to a concept without the mediating role of phonology. A schematic depiction of the alternative hypotheses appears in figure 58.1. The evidence from impaired performance that would favor the hypothesis of independence would be the ability of an individual to write the names of words he or she is unable to produce in spoken form.

In fact, striking cases have been described in which written naming abilities are far superior to verbal ones (Ellis, Miller, and Sin, 1983; Lhermitte and Dérouesné, 1974). For example, Lhermitte and Dérouesné (1974) presented two cases in which written picture-naming responses were largely correct in comparison with spoken responses. In these cases spoken errors consisted primarily of well-articulated nonwords, referred to as neologisms. For example, when a subject was asked to repeat and write the French word "couteau," the responses were "mocrida" and C-O-U-T-E-A-U. However, as we argued above, it is crucial that the deficit be localized to the lexical level. In the case of neologisms, more peripheral loci of impairment have not been ruled out, so it can be argued that lexical phonological representations are, in fact, intact and serve as the basis for the retrieval of the correct written forms, but that spoken responses are distorted due to one or more deficits further downstream.

More clearly relevant, therefore, are those cases in which the errors in phonological output are more uncontroversially lexical in origin, such as cases in which semantic errors are made exclusively or primarily in spoken naming (e.g., oral naming of a picture of a brush as "comb") while spelling is intact (at least in terms of lexical access). In such cases it becomes difficult to argue that the semantically related response "comb" serves as the basis for the retrieval of the cor-

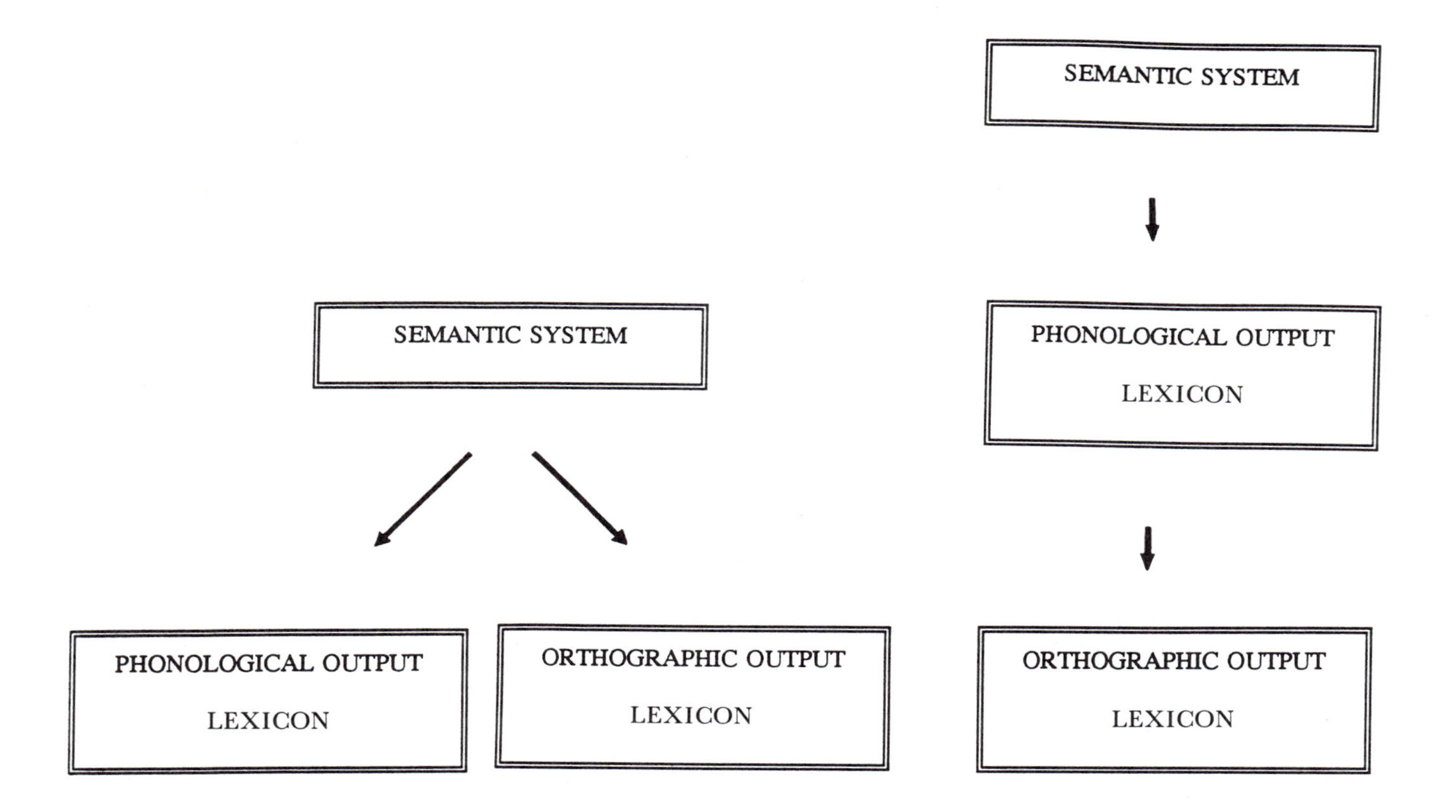

FIGURE 58.1 Alternative hypotheses regarding the relationship between orthographic and phonological knowledge. According to Hypothesis A the two are independent; according to Hypothesis B phonology mediates access to orthographic form.

rect response B-R-U-S-H. Caramazza and Hillis (1990) report the cases of H. W. (with left parietal and occipital damage resulting from various CVAs) and R. G. B. (with left frontoparietal damage as a consequence of an infarct) who made semantic errors exclusively in the spoken modality. These individuals also produced errors in written naming, but these were usually identifiable misspellings of the target, and never semantic errors (e.g., [kangaroo] → K-A-G-O-O or [thumb] → THUB). Importantly, both individuals clearly understood the pictures that they misnamed, as evidenced by the definitions they were able to provide. R. G. B. read RECORDS as "radio" but described the stimulus by saying, "You play 'em on a phonograph ... can also mean notes you take and keep." H. W. produced "bank" for INTEREST and said, "You go to the bank and put it in and you get more money ... not very much now." Given that speech was not generally affected and that semantic processes seemed intact, the impairment that produced the semantic errors in the speech of these individuals was localized to the level of the lexical phonological representations used in output.

The performance of R. G. B. and H. W. is clearly just the sort of pattern that favors the hypothesis that orthographic and phonological forms are independently represented and accessed from lexical meaning representations (figure 58.1). The alternative hypothesis that the retrieval of orthographic forms is mediated by phonological representations founders, because it is difficult to argue that a phonological representation that results in a semantic error in spoken naming can be the basis for the correct or recognizable response in written naming.

If the architecture proposed under hypothesis A (figure 58.1) correctly captures the relationship between semantic, orthographic, and phonological lexical knowledge, then there are additional performance patterns one might expect to observe. For example, one would expect to find individuals who, when asked to provide both spoken and written names on each presentation of a picture, provide different responses in the two modalities. P. W., with anterior parietal and posterior frontal damage from a CVA (Rapp, Benzing, and Caramazza, in preparation), exhibited just this behavior. He not only produced semantic errors in writing and correct responses in spoken naming ([skirt] → S-O-C-K, "skirt") and vice versa ([brush] → B-R-U-S-H, "comb"), but also produced different semantic errors in the two modalities on the very same trial ([knife] → S-P-O-O-N, "fork").

In conclusion, the evidence presented argues strongly for the separate representation of written and spoken lexical forms. Furthermore, the results support the view that phonological and orthographic representations can be independently accessed from meaning representations. This example illustrates how data from impaired performance can be brought to bear on an old and persistent question regarding the representation of lexical knowledge. Once again, the data from PET studies (see, for example, Petersen et al., 1989) are consistent with the basic distinctions drawn here, and the principal areas of activation in processing auditorially and visually presented words involve, respectively, temporoparietal and extrastriate cortex.

The internal organization of meaning and form stores

Up to this point we have motivated three basic distinctions in lexical knowledge types—semantic, orthographic, and phonological—and have presented specific claims regarding the processing relationships among them. This puts us in a position to examine a type of impaired performance that appears to be relevant to the further specification of each of these knowledge types: the category-specific deficits. These are deficits that seem to affect certain categories selectively within one of the lexical knowledge stores, and therefore they are potentially informative regarding more detailed aspects of their organization. For example, there have been reports of category-specific deficits affecting animate but not inanimate concepts, fruits and vegetables but not other foods; nouns but not verbs and vice versa, and so forth. In this section we will examine first those deficits that affect categories of semantic knowledge and then those that affect categories of form knowledge (see also Caramazza et al., in press, for additional commentary).

Category-Specific Deficits of Meaning The modern interest in category-specific deficits dates back at least to the landmark study of Goodglass and colleagues (Goodglass et al., 1976), who examined the production and comprehension abilities of a number of patients for various categories (numbers, letters, colors, objects, and actions) and discovered interesting differences in performance both across patients and across categories. Since then a number of detailed case studies of category-specific deficits have been reported involving selective impairment of categories such as abstract words (Marshall and Newcombe, 1966), concrete words (Warrington, 1975; Warrington and Shallice, 1984), animate objects (Warrington and Shallice, 1984; Warrington and McCarthy, 1987; Hillis and Caramazza, 1991b), fruits and vegetables (Hart, Berndt, and Caramazza, 1985), verbs (Miceli et al., 1984; McCarthy and Warrington, 1985), nouns (Miceli et al., 1984; Zingeser and Berndt, 1988), body parts (Dennis, 1976), geographic names (McKenna and Warrington, 1978), proper names (Semenza and Zettin, 1988, 1989), and colors (Geshwind and Fusillo, 1966). However, the dissociation that has received the most attention (and is perhaps the most widely observed) is that between performance with animate concepts and with inanimate concepts.

Warrington and Shallice (1984) described the performance of two individuals who showed a dramatic difference between their knowledge of inanimate concepts and their impoverished knowledge of animate concepts (table 58.1). Both individuals, who suffered from herpes simplex encephalitis that affected both temporal lobes, were able to provide satisfactory definitions of inanimate concepts such as *tent*, *briefcase*, and *compass*, although they could not properly define animate concepts such as *wasp*, *duck*, and *snail*. Such findings naturally provoke the question, What do such selective patterns of impairment reveal about knowledge

TABLE 58.1

Naming performance of patients reported by Warrington and Shallice

	Animate	Inanimate
J. B. R.	8%	79%
S. B. Y.	0%	52%

J. B. R.	ostrich	"unusual"
	snail	"an insect animal"
	daffodil	"plant"
	tent	"temporary outhouse, living home"
	briefcase	"small case used by students to carry papers"
	compass	"tools for telling direction you are going"
S. B. Y.	wasp	"bird that flies"
	duck	"an animal"
	holly	"what you drink"
	wheelbarrow	"object used by people to take material about"
	submarine	"ship that goes underneath sea"
	umbrella	"object used to protect you from water that comes"

representation and processing? The most straightforward interpretation of the data is that these dissociations reflect quite directly the categories of knowledge represented in the brain, and, therefore, that objects and actions, animate and inanimate categories, and so forth, are, in fact, the basic semantic "kinds" that are neurally implemented. The alternative interpretation is that categories such as these reflect other, more basic properties of brain organization. For example, it has been suggested that the animacy distinction is simply correlated with differences in complexity and familiarity or with differences in the extent to which animate and inanimate concepts are defined by visual-perceptual features. We will examine each of these proposals in turn.

Stewart, Parkin, and Hunkin (1992) as well as Funnel and Sheridan (1992) and more recently Gaffan and Heywood (1993) have proposed that the animacy distinction reflects the fact that animate categories include concepts of greater physical complexity and less familiarity than do the inanimate categories. The proposal, therefore, is that the brain is sensitive to complexity rather than to the animate/inanimate distinction. There are, however, a number of reasons to be confident that observed animacy differences are not necessarily due to these factors. First of all, Sartori, Miozzo, and Job (1993) have documented a case of selective impairment to animate concepts even when items were matched for complexity and familiarity. Secondly, the reverse pattern of impairment, in which performance with animate concepts is superior to that with inanimate concepts, has been documented a number of times (Warrington and McCarthy, 1983, 1987; Hillis and Caramazza, 1991b; Sacchett and Humphreys, 1992). Especially useful for resolving this issue are two cases described by Hillis and Caramazza (1991b) who were studied using the very same stimulus materials and yet revealed complementary patterns of impairment (see figure 58.2)—J. J. exhibited a selective preservation of animal concepts while P. S. exhibited a selective impairment of the animate category. (P. S., as a result of a traumatic brain injury, had sustained extensive damage to the left temporal region and a lesser amount of damage to right temporal and frontal areas; J. J. had damage to the left temporal area and basal ganglia resulting from a CVA.)

It is unlikely that all of the reported dissociations result from sensitivity to complexity and familiarity; therefore, we can assume, at least temporarily, that some of these dissociations actually reflect organizational aspects of the lexical semantic system. The question remains, however, as to how directly one can interpret the dissociations. Warrington and Shallice (1984; see also Farah and McClelland, 1991) have argued that the animate-inanimate distinction reflects the structure of a semantic system that is organized not into categories based on animacy, but rather into modality- or sensory-specific subsystems based on visual, functional, auditory, verbal, and (presumably) olfactory features. Specifically, they have argued that animate concepts rely more heavily on visual features for their definition than do the inanimate concepts, which, in contrast, are defined by their functional properties. According to this view (see figure 58.3), one would expect to observe deficits primarily with items in the animate class if "visual semantics" were affected by neurological damage.

There are, however, a number of reasons to question this interpretation of the observed dissociations. Some of these reasons involve uncertainty regarding the specific content of the proposal: Further clarification is needed regarding what is meant by visual, verbal, and functional semantics. For example, it is not obvious how features as disparate as "king of the jungle" and "lives in Africa" as well as "serves for cutting food" can

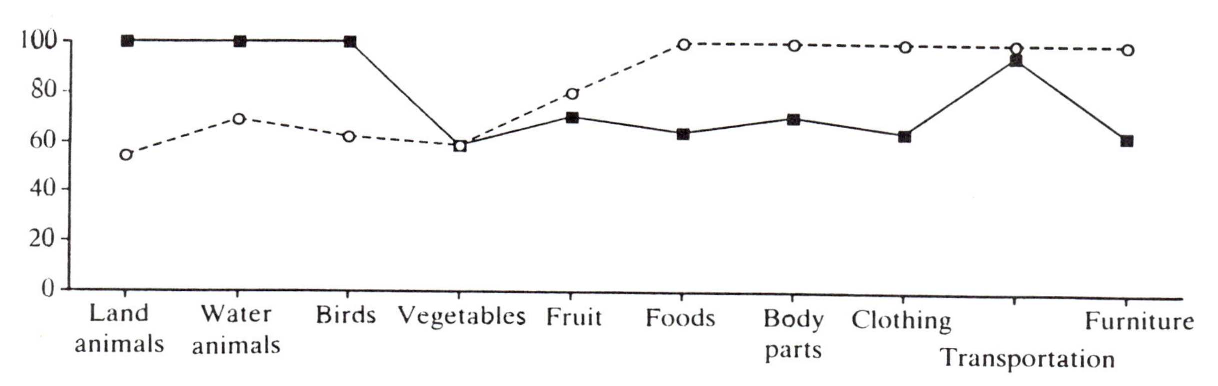

FIGURE 58.2 Category-specifc effects in the oral naming performance of J. J. (solid line) and P. S. (dashed line) (from Hillis and Caramazza, 1991b).

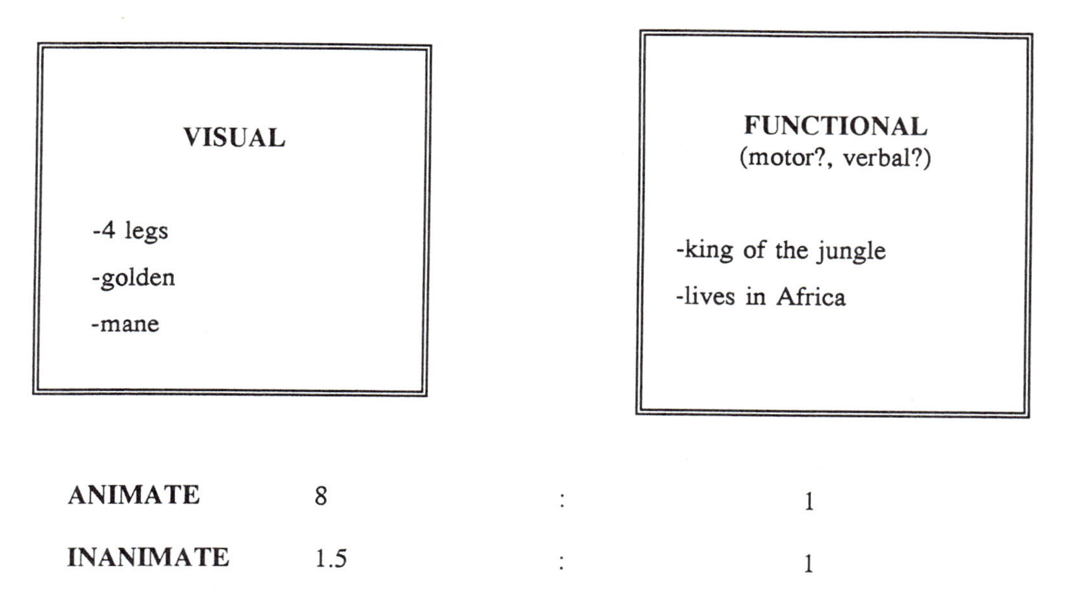

FIGURE 58.3 A schematic depiction of the proposal according to which categorical organization is an emergent property of shared sensory-specific semantic features. Animate and inanimate categories differ in terms of the characteristic ratio of visual to functional features.

be attributed to a system that stores information regarding the functional features of concepts (see Caramazza et al., 1990; Hillis et al., 1990; and Rapp, Hillis, and Caramazza, 1993; but also Shallice, 1993). More immediately problematic, however, is that the proposal does not account for the full range of relevant patterns of impairment that have been reported.

Specifically, it does not seem to follow from the sensory-specific hypothesis that selective islands of preservation or impairment should emerge *within* the animate or inanimate category. In this respect the performance of individuals such as M. D. (Hart, Berndt, and Caramazza, 1985) is problematic. M. D., with damage to the left frontal region and basal ganglia as a result of several CVAs, had particular difficulty with fruits and vegetables, while other food categories, animals, and vehicles were largely unaffected.

Warrington and McCarthy (1987) have suggested that such within-category differences may result from the selective damage to or preservation of specialized channels within the sensory-specific semantic systems. They argue, for example, that fruits and vegetables are distinguished from one another primarily on the basis of their color and, thus, that their selective impairment results from damage to a specialized color channel within the visual semantic store. Despite the surface plausibility of such a proposal, it is inconsistent with a number of facts. First, we are able to identify fruits and vegetables on the basis of shape information alone, indicating that color information cannot be essential to their definition. Furthermore, M. D. had comparable difficulty recognizing fruits and vegetables when they were presented only for tactile exploration, indicating that the impairment was present even when color information was not available. Finally, M. D.'s color-naming skills were intact, indicating no apparent deficit to the representation of color.

Given that these attempts at interpretation of the category-specific semantic deficits have not been entirely successful, we are left with the original question regarding what these striking differences reveal about the representation of lexical semantic knowledge. There are a number of alternatives to be explored. For example, it is possible that the dissociations do not indicate categorical organization of semantic knowledge at all, but rather are a consequence of the fact that the similarity in meaning among different concepts is represented through the actual sharing of featural material. In such a view, fruits and vegetables emerge as a single category because the actual representations of the individual fruits and vegetables are overlapping. In the case of neurological damage, if the subset of semantic features that are involved in the definition of fruits and vegetables happen to be affected, we should expect to find a deficit that apparently selectively affects the fruit-and-vegetable category. It is important to note that according to such a hypothesis there is no reason to assume that semantic features are sensorially organized (Caramazza et al., in press). Thus, shared features of fruits and vegetables may involve information about their relationship to plants or their function as food, for example. The prediction is that items from other categories that rely heavily on such features should also be affected. Unfortunately, this hypothesis has not yet been examined at the level of detail that would be required to evaluate it.

In sum, the category-specific semantic deficits represent intriguing glimpses into the internal organization of the lexical meaning system. Until now, however, they have resisted satisfactory interpretation.

Category-Specific Deficits of Form A number of cases were mentioned above in which specific grammatical categories—nouns, verbs, and function words—appear to be selectively affected. Although most of these cases were not studied so as to determine the level of the functional deficit, there is some evidence that certain of these deficits may reside at the level of the meaning system (see McCarthy and Warrington, 1985), and a strong case can be made for grammatical category impairments that are specific to either the phonological or orthographic lexicons. Deficits of this latter sort are particularly interesting because they are not amenable to explanation under the sensory-specific hypothesis proposed for the category-specific semantic impairments.

Caramazza and Hillis (1991) report the cases of two patients who show selective deficits to the category of verbs in only one modality of output. H. W. produced semantic errors with verbs in oral naming only, and S. J. D. produced semantic errors in naming verbs in the written modality only (figure 58.4). Comparable results were obtained regardless of the modality of the stimulus—in picture-naming tasks as well as in reading and writing to dictation. This pattern of results was also obtained using identical words in noun and verb contexts: There's a *crack* in the mirror./ Don't *crack* the nuts in here. Whereas S. J. D. correctly produced *crack* in the noun context in either written or spoken forms, verbs were produced without semantic errors only in the spoken modality. H. W. showed the opposite modality effect.

A number of characteristics of these results allow us to be confident about the locus of the deficits. The principal one, of course, is the fact that the deficit appears to be specific to one of the form systems, allowing us to rule out a more central or semantic locus of impairment. Furthermore, it is clear that the differences in performance between nouns and verbs cannot be attributed to differences in abstractness, not only because one would expect such differences to manifest themselves regardless of modality of output, but also because abstract nouns and verbs such as *belief*, *mercy*, *achieve*, and *learn* dissociate in the same manner and to the same degree as do the more concrete nouns and verbs. Similarly, it is difficult to argue that the differences result from the extent to which different grammatical categories rely on sensorially defined semantic features, given that there is no reason to think that spoken and written output modalities should be differentially affected. Finally, the use of homonyms makes it impossible to argue that the effects might have been

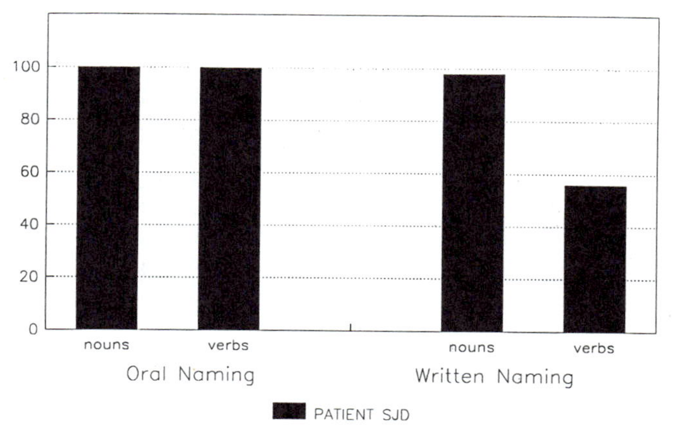

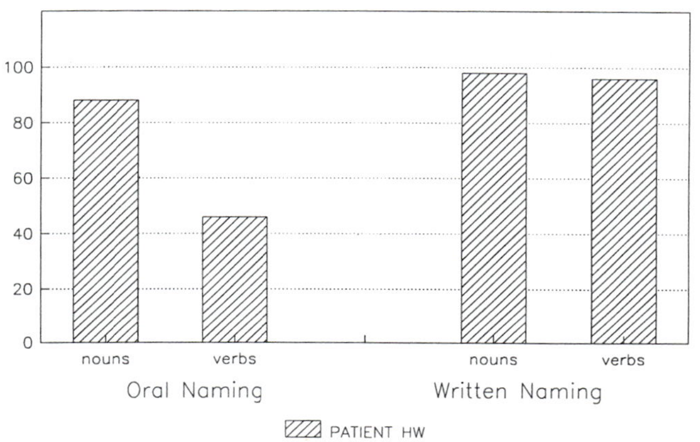

Noun: There's a crack in the mirror.

Verb: Don't crack the nuts in here.

FIGURE 58.4 S. J. D.'s and H. W.'s accuracy in the reading and writing homonyms.

due to differences in the complexity of the stimuli that interact with modality of output.

The hypothesis that the reported grammatical class effects are simply a reflection of the fact that verbs are generally more difficult to produce is ruled out by the reports of patients with the opposite grammatical class effect: poor performance in the oral production of nouns relative to verbs. Of particular interest is the case of E. B. A., who, with left frontal lobe and basal ganglia damage as a result of a CVA, (Hillis and Caramazza, submitted) shows contrasting grammatical class effects in recognition and production of nouns

TABLE 58.2
Accuracy of performance of P. B. S. according to grammatical class and output modality

	Nouns	Function Words
Written picture description	98%	13%
Oral picture description	20%	75%

and verbs: She is impaired in oral production of nouns relative to verbs and in the recognition of written verbs relative to nouns. The double dissociation of grammatical class by modality within a single patient rules out explanations of the grammatical class effect in terms of the relative difficulty of the two classes of words.

A striking example that grammatical category differences are not restricted to nouns and verbs involves the performance of an individual described by Rapp, Benzing, and Caramazza (1992). P. B. S., who had suffered a left middle cerebral artery infarction, showed a double dissociation in his performance with nouns and function words across written and spoken modalities, such that nouns were produced relatively well in writing although function words were not. In the spoken modality, on the other hand, function words were often appropriately produced and nouns were severely affected (table 58.2). Thus, it is not only the case that function words were relatively better preserved than nouns within the spoken modality, but that they were also better preserved than function words in the written modality. Nouns, in the reverse pattern, were almost invariably substituted for by neologisms in spoken output although generally produced correctly in written output.

As in the cases of S. J. D., H. W., and E. B. A., the finding of impaired performance with particular grammatical categories that are specific to one lexical form and not the other allows us to rule out a deficit to a more abstract level of representation that is common to both output modalities. Thus, in the case of P. B. S. we can assume that because function words are appropriately produced in speech they must have been represented appropriately at the sentence-planning level that is common to both written and spoken sentence production. The same can be said for the nouns in writing—the fact that they are produced appropriately in written output indicates that they were specified correctly at a prior level of sentence planning. These cases, therefore, lead to the conclusion that grammatical category is an organizing parameter of the orthographic and phonological form systems. Interestingly, this conclusion is consistent with certain models of sentence production developed on the basis of the speech errors of speakers who have not suffered brain damage (Garrett, 1980, 1982). Differences in localization of processing of noun and verb representations in the brain (although not specific to an output modality) have been suggested (Miceli et al., 1984; Damasio and Tranel, 1993; Caramazza, in press). It would appear that selective difficulties in processing verbs are associated with frontal lobe lesions, whereas selective difficulties in processing nouns are associated with lesions in the temporoparietal area.

In sum, this examination of the category-specific effects of brain damage on lexical processing abilities, although inconclusive regarding the interpretation of the deficits localized to the semantic level of representation, strongly supports the claim that the representation of our knowledge of orthographic and phonological forms is organized such that grammatical categories can be independently affected in one modality and not the other. We will now turn to performance patterns that allow us to examine the internal structure of the lexical knowledge representations themselves.

The structure of phonological and orthographic representations: Morphology

Speakers and writers can be enormously creative in their ability to generate appropriate lexical forms with which they were previously unfamiliar, for example, mutation → mutationless but not unmutation. This ability appears to be a reflection of our knowledge of forms, their meanings, and the possibilities and limitations of their combination. There are a great many theoretical issues related to how best to characterize this knowledge. Among them is the question of whether or not morpheme-sized units of meaning and/or form are actually manipulated either in the use of familiar morphologically complex forms or in the apparently compositional creative process. A number of observations of impaired performance following brain damage are relevant to this issue. These results have been interpreted as indicating that the orthographic and phonological form systems include morpheme-sized units, and thus have been taken as support for a compositional view of form representation (see also Sproat, this volume).

Numerous individuals have been described who make errors in reading and naming tasks in which the relationship between the target word and the error can be considered to be morphological, for example producing "talking" for "talked" (see Coltheart, Patterson, and Marshall, 1980). It has been argued that such errors result from an impairment to morphological processes that are required to correctly assemble morpheme-sized units into words. However, the difficulty in interpreting many of these cases has been that the production of morphological errors is typically accompanied by the erroneous production of semantically and/or visually or phonologically similar words. That is, in addition to errors such as TALKING → "talked," errors involving only a semantic relationship (FLUTE read as "piano") or only a visual/phonological one (SANK → "sand") are typically also present. In such cases it could be argued that the so-called morphological errors are actually semantic or form-related errors, and thus that their presence has no implications for issues regarding the morphological nature of lexical knowledge. That is, it is argued that a semantic deficit that produces responses that are semantically close to the target (FLUTE → "piano") should also result in errors such as WALKING → "walk"; similarly, a deficit at the level of forms that results in the confusion of similar forms (CHAIN → "chair") should also result in CHAINS → "chain." In either case affixed forms could be represented as single units.

In this respect, the case of S. J. D. (Badecker and Caramazza, 1991) is particularly interesting because morphological errors were accompanied only by phonemic errors (e.g., SHRILLY → /šruli/) and, in addition, were produced only in response to affixed stimuli. When reading performance with affixed and unaffixed homophones such as TEAS and TEASE was compared, it was found that morphologically related responses occurred only with the affixed words. The fact that semantic errors were not produced rules out the possibility that the morphological errors were simply a result of semantic proximity; the fact that the morphologically related responses were produced only for the affixed stimuli makes it difficult to argue that morphologically related forms were produced merely as a result of phonemic proximity (presumably, homophonous unaffixed forms such as TEASE and affixed forms such as TEAS have identical phonological neighborhoods). Having established that S. J. D.'s performance appears to be relevant specifically to issues regarding the representation and processing of morphology, it is also important to note that, given S. J. D.'s excellent comprehension of the written stimuli that resulted in morphological errors, the locus of the deficit appears to be at the level of the retrieval of the spoken forms. Particularly informative are the illegal morphological errors that S. J. D. produced; that is, she produced morpheme combinations that do not occur in the language: endangered → endangerous, patronize → patroning. These responses were accompanied by indications that the target word had been correctly comprehended (WRITTEN → "writer, writed, not today, but before"). Data of these sort are certainly difficult to explain without assuming that morpheme-sized units exist at the level of the phonological lexicon and that, in the case of S. J. D., the functional deficit results in the miscombination of these morpheme units. Comparable data and arguments have been presented for morphological composition of orthographic forms (Caramazza, et al., 1985; Badecker, Hillis, and Caramazza, 1990; Badecker, Rapp, and Caramazza, in press).

Concluding remarks

The objective of this chapter has been to illustrate how the performance of brain-damaged individuals can be brought to bear on questions regarding the manner in which knowledge of words is actually represented and organized in the human brain. We have argued that the extent to which observations of impaired performance will be useful for such an endeavor will be determined not only by the coherence and reliability of the data but also by the degree to which data from different brain-damaged individuals and data from other sources converge to produce similar descriptions of lexical knowledge representation and processing. It should be apparent from the cases that have been described that in interpreting the performance of different individuals, speaking different languages and exhibiting different patterns of errors, we have had to make similar assumptions about the structure of the premorbid cognitive system. That is, although no two brain-damaged individuals we have described exhibited precisely the same patterns of impaired performance, in order to account for the systematic aspects of their errors we have come to common conclusions.

The results we have described also have important implications for theories of the functional organization of the human brain. The patterns of performance we have described suggest that distinct neural mechanisms are responsible for the representation and processing of the meanings and the orthographic and phonological forms of words. Furthermore, within such mechanisms further subdivisions must exist for the organization of grammatical categories, and neurally based distinctions must support the internal structure of lexical representations that we have described. Unfortunately, it is the case that stable functional deficits such as those we have described are usually associated with fairly large lesions, which complicates the task of relating specific functional deficits to specific brain mechanisms (Caramazza, in press). Nonetheless, we have attempted to review briefly, whenever possible, studies that have attempted to define the neural mechanisms that are dedicated to the processing of different aspects of lexical knowledge.

In conclusion, the examination of the patterns of ability and inability of brain-damaged individuals to perform different tasks—with different classes of words, involving different modalities of input and output—as well as the examination of the types of errors such individuals produce, has allowed us to develop a fairly detailed proposal regarding the structure of human lexical knowledge representation.[3] The empirical results described here are clearly problematic for those who assume that our lexical knowledge and skills are based on the relationships between *sounds* (or letters) and meanings rather than on the relationships between *lexical forms* (orthographic or phonological) and meanings (Seidenberg and McClelland, 1989). Instead, the picture that emerges is one in which our knowledge of words involves mental entities corresponding to lexical meanings and forms that are internally complex and componentially organized. Proposals such as this one should be useful for the further and more detailed understanding of the functional organization of the human brain and for our ability to make progress in developing increasingly successful therapeutic techniques.

ACKNOWLEDGMENTS We would like to thank our colleagues William Badecker, Argye Hillis, and Gabriele Miceli for discussing their work and sharing their insights with us. We would also like to express our appreciation to Steven Pinker for his comments on an earlier version of this chapter. Finally, we would like to acknowledge the support provided by NIH grants NS22201, DC00366, and DC01423 to Dartmouth College.

NOTES

1. We will use the form "hat" if we are referring to a spoken word; /hæt/ if an actual phonetic transcription is used; H-A-T if we are referring to the spelling of the word; HAT if we are referring to the visual presentation of a word; and [hat] to represent the presentation of a picture.
2. These results represent D. R. B.'s performance with abstract words. Franklin et al. demonstrate a striking dissociation in comprehension in the auditory modality between abstract and concrete words. For our purposes here, however, this further dissociation is not relevant.
3. In this chapter we have focused almost entirely on developing a description of lexical *representations*. Due to space limitations we have not included a discussion of the nature of relevant *processes*.

REFERENCES

Badecker, W., and A. Caramazza, 1991. Morphological composition in the lexical output system. *Cognitive Neuropsych.* 8(5):335–367.

Badecker, W., A. Hillis, and A. Caramazza, 1990. Lexical morphology and its role in the writing process: Evidence from a case of acquired dysgraphia. *Cognition* 34:205–243.

Badecker, W., B. C. Rapp, and A. Caramazza, in press. Lexical morphology and the two orthographic routes. *Cognitive Neuropsych.*

Beauvois, M. F., and J. Dérouesné, 1981. Lexical or orthographic agraphia. *Brain* 104:21–49.

Brown, R., and D. McNeill, 1966. The "tip of the tongue" phenomenon. *J. Verb. Learn. Verb. Be.* 5:325–337.

Caramazza, A. (in press). The representation of lexical knowledge in the brain. In *Decade of the Brain*, vol. 1, R. D. Broadwell, ed. Library of Congress.

Caramazza, A., and A. Hillis, 1990. Where do semantic errors come from? *Cortex* 26:95–122.

Caramazza, A., and A. E. Hillis, 1991. Lexical organization of nouns and verbs in the brain. *Nature* 349:788–790.

Caramazza, A., A. E. Hillis, E. C. Leek, and M. Miozzo, (In press). The organization of lexical knowledge in the brain: Evidence from category-and modality-specific deficits. In *Mapping the Mind: Domain Specificity in Cognition and Culture*. L. Hirschfeld and S. Gelman, eds. Cambridge: Cambridge University Press.

Caramazza, A., A. Hillis, B. C. Rapp, and C. Romani, 1990. The multiple semantics hypothesis: Multiple confusions? *Cognitive Neuropsych.* 7:161–189.

Caramazza, A., G. Miceli, M. C. Silveri, and A. Laudanna, 1985. Reading mechanisms and the organization of the lexicon: Evidence from acquired dyslexia. *Cognitive Neuropsych.* 2:81–114.

Coltheart, M., K. Patterson, and J. Marshall, 1980. *Deep Dyslexia*. London: Routledge and Kegan Paul.

Damasio, A. R., 1990. Category-related recognition defects as a clue to the neural substrates of knowledge. *Trends Neurosci.* 13:95–98.

Damasio, A. R., and D. Tranel, 1993. Verbs and nouns are retrieved from separate neural systems. *Proc. Natl. Acad. Sci. U.S.A.* 90:4957–4960.

Dennis, M., 1976. Dissociated naming and locating of body parts after left anterior temporal lobe resection: An experimental case study. *Brain Lang.* 3:147–163.

Ellis, A. W., D. Miller, and G. Sin, 1983. Wernicke's aphasia and normal language processing: A case study in cognitive neuropsychology. *Cognition* 15:111–114.

Farah, M. J., and J. L. McClelland, 1991. A computational model of semantic memory impairment: Modality-specificity and emergent category-specificity. *J. Exp. Psychol. (Gen.)* 120:339–357.

Franklin, S., D. Howard, and K. Patterson, in press. Abstract word meaning deafness. *Cognitive Neuropsych.*

Funnell, E., and J. Sheridan, 1992. Categories of knowledge? Unfamiliar aspects of living and non-living things. *Cognitive Neuropsych.* 9:135–153.

Gaffan, D., and C. A. Heywood, 1993. A spurious category-specific visual agnosia for living things in normal human and nonhuman primates. *J. Cognitive Neurosci.* 5:118–128.

Garrett, M., 1980. Levels of processing in sentence production. In *Language Production, vol. 1: Speech and Talk*, B. Butterworth, ed. New York: Academic Press.

Garrett, M., 1982. Production of speech: Observations from normal and pathological language use. In *Normality and Pathology in Cognitive Functions*, A. Ellis, ed. London: Academic Press.

Geschwind, N., 1967. The varieties of naming errors. *Cortex* 3:97–112.

Geschwind, N., and M. Fusillo, 1966. Colour naming deficits in association with alexia. *Arch. Neurol.* 15:137–146.

Goodglass, H., E. Kaplan, S. Weintraub, and N. Ackerman, 1976. The "tip of the tongue" phenomenon in aphasia. *Cortex* 12:145–153.

Goodman, R. A., and A. Caramazza, 1986. Aspects of the spelling process: Evidence from a case of acquired dysgraphia. *Lang. Cognitive Proc.* 1(4):263–296.

Hart, J., R. S. Berndt, and A. Caramazza, 1985. Category-specific naming deficit following cerebral infarction. *Nature* 316:439–440.

Hillis, A. E., and A. Caramazza, 1991a. Mechanisms for accessing lexical representations for output: Evidence from a case with category-specific semantic deficit. *Brain Cogn.* 40:106–144.

Hillis, A. E., and A. Caramazza, 1991b. Category specific naming and comprehension impairment: A double dissociation. *Brain* 114:2081–2094.

Hillis, A. E., and A. Caramazza, (submitted). The representation of grammatical knowledge in the brain.

Hillis, A. E., B. C. Rapp, C. Romani, and A. Caramazza, 1990. Selective impairment of semantics in lexical processing. *Cognitive Neuropsych.* 7:191–243.

Kay, J., and A. W. Ellis, 1987. A cognitive neuropsychological case study of anomia: Implications for psychological models of word retrieval. *Brain* 110:613–629.

Lhermitte, F., and J. Derouesné, 1974. Paraphasies et jargonaphasie dans le langage oral avec conservation de langage écrit: Genèse des néologismes. *Rev. Neurol. (Paris)* 130:21–38.

Marshall, J. C., and F. Newcombe, 1966. Syntactic and semantic errors in paralexia. *Neuropsychologia* 4:169–176.

McCarthy, R. A., and E. K. Warrington, 1985. Category specificity in an agrammatic patient: The relative impairment of verbal retrieval and comprehension. *Neuropsychologia* 23:709–723.

McKenna, P., and E. K. Warrington, 1978. Category specific naming preservation: A single case study. *J. Neurol. Neurosurg. Psychiatry* 43:781–788.

Miceli, G., M. C. Silveri, G. Villa, and A. Caramazza, 1984. On the basis for the agrammatic's difficulty in producing main verbs. *Cortex* 20:207–220.

Patterson, K. E., 1986. Lexical but nonsemantic spelling. *Cognitive Neuropsych.* 3:341–367.

Petersen, S. E., P. T. Fox, M. I. Posner, M. Mintum, and M. E. Raichle, 1988. Positron emission tomographic studies of cortical anatomy of single word processing. *Nature* 331:585–589.

Petersen, S. E., P. T. Fox, M. I. Posner, M. Mintum, and M. E. Raichle, 1989. Positron emission tomographic studies of the processing of single words. *J. Cognitive Neurosci.* 1:153–170.

Posner, M. I., S. E. Petersen, P. T. Fox, and M. E. Raichle, 1988. Localization of cognitive operations in the human brain. *Science* 240:1627–1631.

Rapp, B. C., L. Benzing, and A. Caramazza, 1992. The dissociation of grammatical categories in the spoken versus written production of a single patient. Paper presented at the Academy of Aphasia, Toronto, Canada.

Rapp, B., and A. Caramazza (submitted). The dissociation of grammatical categories in the spoken vs. written production of a single patient.

Rapp, B. C., A. E. Hillis, and A. Caramazza, 1993. The role of representations in cognitive theory: More on multiple semantics and the agnosias. *Cognitive Neuropsych.* 10(3): 235–249.

Sacchett, C., and G. W. Humphreys, 1992. Calling a squirrel a squirrel but a canoe a wigwam: Category specific deficits for artefactual objects and body parts. *Cognitive Neuropsych.* 9:73–86.

Sartori, G., M. Miozzo, and R. Job, 1993. Category-specific naming impairment? Yes. *Q. J. Exp. Psychol.* 46A: 489–504.

Schwartz, M. F., E. M. Saffran, and O. S. M. Marin, 1980. Fractionating the reading process in dementia: Evidence for word-specific print-to-sound associations. In *Deep Dyslexia*, M. Coltheart, K. Patterson, and J. C. Marshall, eds. London: Routledge and Kegan Paul.

Seidenberg, M., and J. McClelland, 1989. A distributed, developmental model of visual word recognition and naming. *Psychol. Rev.* 96(4):523–568.

Semenza, C., and M. Zettin, 1988. Generating proper names: A case of selective inability. *Cognitive Neuropsychol.* 5:711–721.

Semenza, C., and M. Zettin, 1989. Evidence from aphasia for the role of proper names as pure referring expressions. *Nature* 342:678–679.

Shallice, T., 1993. Multiple semantics: Whose confusions? *Cognitive Neuropsych.* 10:251–261.

Shallice, T., and E. K. Warrington, 1980. Single and multicomponent central dyslexic syndromes. In *Deep Dyslexia*, M. Coltheart, K. Patterson, and J. Marshall, eds. London: Routledge and Kegan Paul.

Stewart, Parkin, and Hunkin, 1992. Naming impairments following recovery from herpes simplex encephalitis: Category specific? *Q. J. Exp. Psychol. [A]* 44:261–284.

Van Orden, G. C., B. F. Pennington, and G. O. Stone, 1990. Word identification in reading and the promise of subsymbolic psycholinguistics. *Psychol. Rev.* 97:488–522.

Warrington, E. K., 1975. The selective impairment of semantic memory. *Q. J. Exp. Psychol.* 27:635–657.

Warrington, E. K., and R. A. McCarthy, 1983. Category-specific access dysphasia. *Brain* 106:859–878.

Warrington, E. K., and R. A. McCarthy, 1987. Categories of knowledge: Further fractionations and an attempted explanation. *Brain* 110:1273–1296.

Warrington, E. K., and T. Shallice, 1984. Category-specific semantic impairments. *Brain* 107:829–853.

Zingeser, L. B., and R. S. Berndt, 1988. Grammatical class and context effects in a case of pure anomia: Implications for models of language production. *Cognitive Neuropsych.* 5:473–516.

59 The Neurobiology of the Sound Structure of Language

SHEILA E. BLUMSTEIN

ABSTRACT The neurobiology of the sound structure of language is explored by investigating the principles of the sound structure of language, delineating the mechanisms implicated in language use, and specifying the neural instantiation of this system. Converging evidence from clinical investigations of lesion patients and electrophysiological and metabolic studies of both normal and brain-damaged patients is reviewed. Results show that there is a common phonological representation of the sound structure of words that informs both speech production and speech perception, and that this system is broadly distributed in the perisylvian areas of the left hemisphere. Specific neural substrates are identified for the phonetic implementation of speech production. Brain damage in these areas results in an articulatory rather than a linguistic disorder. The evidence challenges the classical view of the organization of the sound structure of language, in which speech production is subserved solely by anterior brain structures and reception is subserved solely by posterior brain structures.

Introduction

Speech is the primary interface with the language system. It serves as the gateway for language production and language reception. And yet, it is clear that speech is not simply a code or cipher in which there is a one-to-one mapping from symbol to representation; rather it has—as does language—its own internal structure, representations, and processes. The nature of the structure is shaped not only by the physiological constraints of the speech apparatus (the vocal tract) in speech production and of the auditory system (the auditory pathway) in speech perception, but also by constraints and principles that are unique to language itself.

It is the goal of this chapter to identify the representations, algorithms, and principles of the sound structure of human language; to delineate the mechanisms implicated in language use, including both speech production and speech reception; and to specify, to the extent possible, the neural instantiation of this system. To that end, converging evidence from a number of fields of research will be used, including clinical/lesion investigations of patients, and electrophysiological and metabolic studies of both normal and brain-damaged patients. To date, most evidence concerning the neurobiology of language derives from studies of patients with lesions. Animal models, unfortunately, provide a limited testing framework to study human speech or language, and the electrophysiological and PET studies with humans focusing particularly on speech are few.

Figure 59.1 shows a working model of the levels of representation and the processes contributing to both speech production and speech perception. A number of observations concerning this model are worth making here. It has been assumed that a single lexicon (words of the language) is shared by the speech production and speech reception mechanisms. That is, words to be produced or perceived ultimately contact a common representation. The nature of that representation is in terms of segments, phonetic features, and rules for their combination that are specific to the sound structure of language. In addition, as the schematic shows, it is assumed that in speech reception, all auditory speech input ultimately accesses the lexicon. Thus, as depicted, there is no separate mechanism for the processing of nonsense syllables independent of the lexicon. A similar assumption is made in speech production. Nonetheless, to effect decoding and encoding of speech requires different mechanisms relating to the auditory system in the former and to the speech apparatus in the latter, and for that reason the interface of these two systems with the lexicon for speech production and speech perception requires a unique set of operations.

SHEILA E. BLUMSTEIN Department of Cognitive and Linguistic Sciences, Brown University, Providence, R.I.

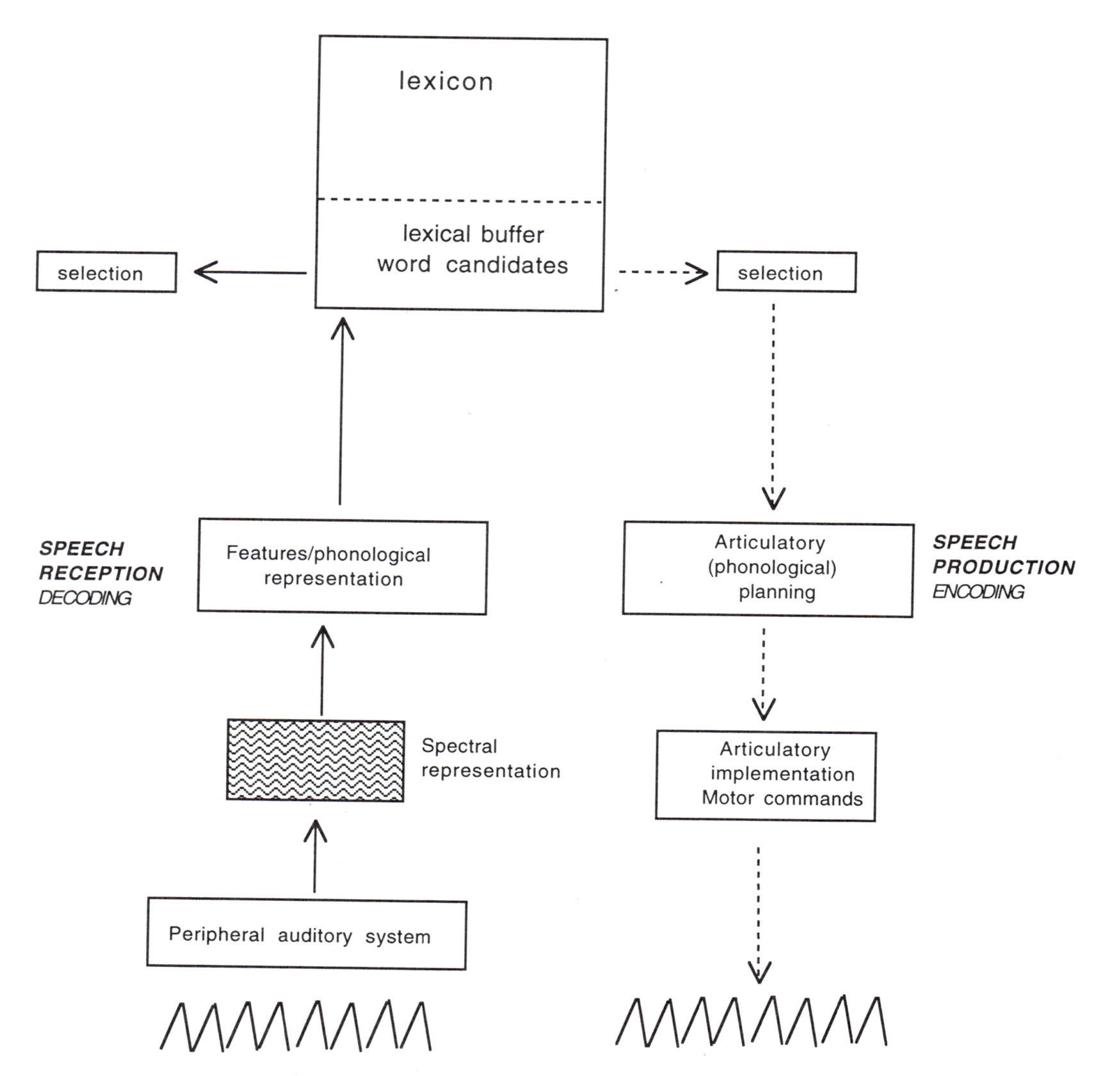

FIGURE 59.1 A model of speech production and speech perception.

CLINICAL AND NEUROLOGICAL BASES OF LANGUAGE DISORDERS IN APHASIA Classical approaches to the clinical and neurological bases of language disorders in adult aphasics have typically characterized the aphasia syndromes in broad anatomical (anterior and posterior) and functional (expressive and receptive) dichotomies (see Geschwind, 1965). The left hemisphere is dominant for most right-handers for language, and the study of the aphasias has focused primarily on patients who have sustained unilateral left-hemisphere brain damage. Anatomically, anterior aphasics have left-hemisphere lesions that are defined as suprasylvian and anterior to the central sulcus, whereas posterior aphasics have lesions defined as subsylvian that are posterior to the central sulcus (see figure 59.2). Functionally, anterior aphasics have greater expressive deficits, whereas posterior aphasics typically have greater receptive deficits. A number of aphasic syndromes are subsumed within each of the anterior and posterior aphasias; each has a distinct language profile and an associated neuroanatomical basis (Geschwind, 1965). The clinical syndromes that characterize these aphasias are based upon a constellation of language abilities and disabilities that are defined in terms of language faculties or functions such as speaking, understanding, reading, and writing (Goodglass and Kaplan, 1972).

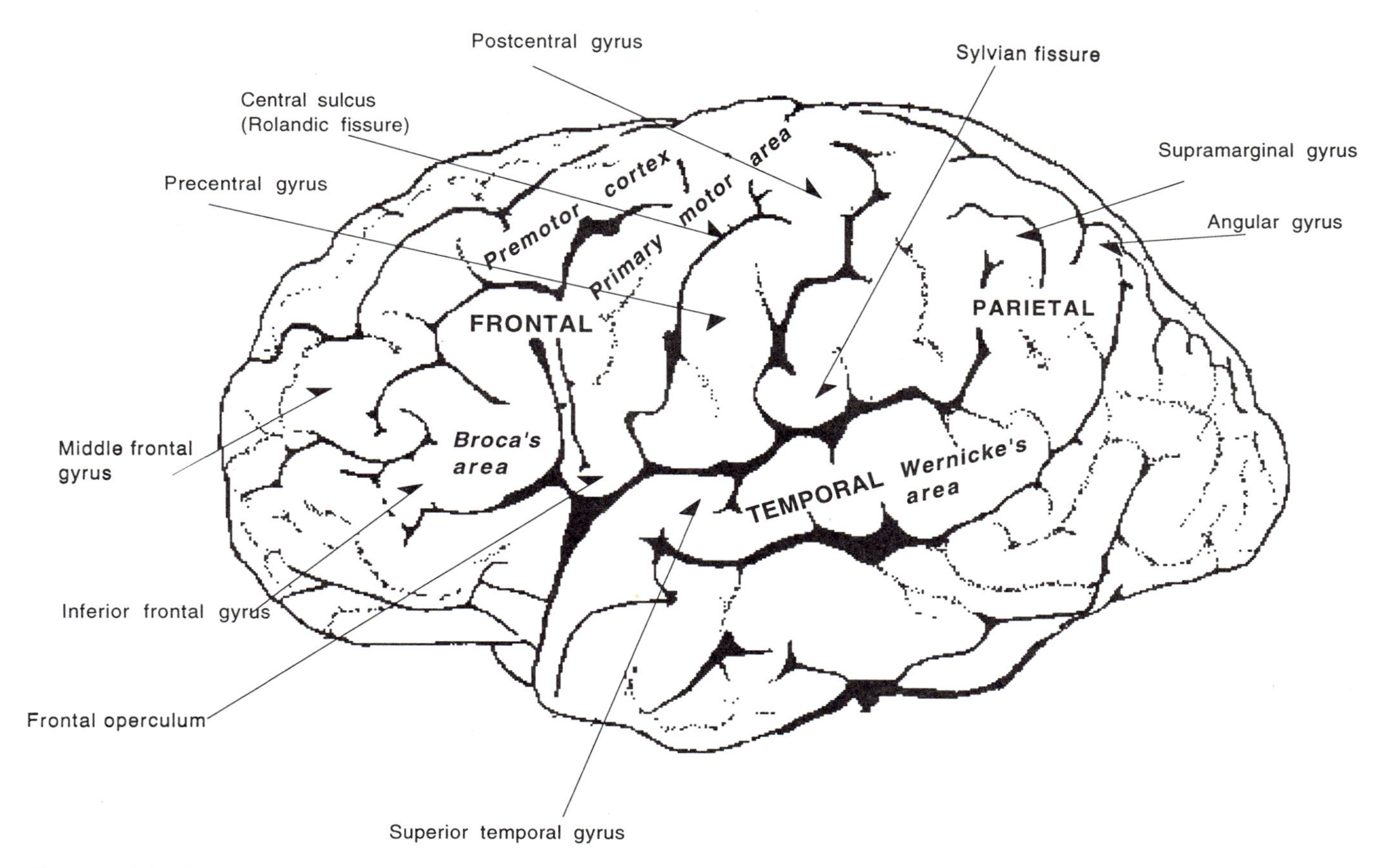

FIGURE 59.2 Lateral view of the left cerebral cortex.

The two aphasia syndromes that best characterize the anterior-posterior and expressive-receptive dichotomy are Broca's aphasia and Wernicke's aphasia. Broca's aphasics show a profound expressive deficit in the face of relatively good auditory language comprehension. Speech output is typically nonfluent in that it is slow, labored, and often dysarthric, and the melody pattern seems flat. Furthermore, speech output is often agrammatic, characterized by the omission of grammatical words such as *the* and *is*, as well as the incorrect usage of inflectional endings. The underlying neuropathology includes the frontal operculum as well as premotor and motor regions posterior and superior to the frontal operculum, extending to the white matter below and including the basal ganglia and insula (Damasio, 1991; see figure 59.2).

Wernicke's aphasics, in contrast, exhibit a dense auditory language comprehension deficit in the context of well-articulated and fluent speech. The speech output of these patients often contains paraphasias—errors in the production of words—including literal paraphasias (sound substitutions, e.g., *boy* → *poy*), verbal paraphasias (word substitutions, e.g. *boy* → *girl*), and, in some cases, neologisms (productions that are possible but nonexistent words in the language, e.g., *toofbay*). In addition, although speech output is fluent and grammatically full, it is often paragrammatic (syntactic phrases are inappropriately juxtaposed), is often empty of semantic content, and is marked with the overuse of high frequency, contentless nouns and verbs such as *thing* and *be*. The neuropathology associated with Wernicke's aphasia involves the posterior region of the left superior temporal gyrus, often extending to the supramarginal and angular gyrus (Damasio, 1991; see figure 59.2).

To a first approximation the anterior-posterior anatomical dichotomy corresponds well with the functional expressive-receptive dichotomy. After all, it is not surprising to find expressive language deficits with damage to the motor areas and receptive language deficits with damage to the auditory association areas. Nevertheless, as we will see, this functional and anatomical dichotomy is not upheld as we consider in detail the neurobiology of the sound structure of language.

Speech production

As figure 59.1 shows, in order to produce a word or group of words, the speaker must select the word candidate(s) from the lexicon; encode the abstract phonological representation of the word in terms of the articulatory parameters required for realizing the phonological properties in the particular context in which they appear (i.e., engage in articulatory planning); and then turn this phonetic string into a set of motor commands or motor programs to the vocal tract. The final speech output must conform to the phonological rules of the language, including the correct production of the sound segments in their phonetic environment; the appropriate stress pattern of the word; and, for groups of words, the appropriate prosodic structure or intonation pattern of the utterance.

Evidence that there are at least two distinct levels of representation, including a phonological representation corresponding to the underlying representation of the lexical item and a phonetic representation corresponding to the articulatory parameters of an utterance, has emerged most clearly in the patterns of speech production deficits in the adult aphasias. In some cases, patients may produce a wrong sound segment, but its phonetic implementation is correct; for example, for *teams* the patient says *keams*. In other cases, patients may produce the correct sound segments but their phonetic implementation is distorted: for *teams* the patient produces an initial /t/ that is overly aspirated. Such a production does not occur in English or in any other language. In order for both types of productions to occur, two different levels of representation—phonological and phonetic—are implicated.

PHONOLOGICAL REPRESENTATIONS: MECHANISMS AND NEUROBIOLOGY To explore the nature of phonological representations, research has focused on the patterns of speech production in aphasic patients. A number of different methodologies have been used to elicit productions from patients, including conversational speech, naming, and repetition. Nonetheless, a consistent finding is that nearly all aphasic patients, regardless of the aphasia syndrome, display impairments implicating a deficit at the phonological level. Critically, the patterns of impairment are similar, suggesting that despite the different underlying neuropathologies, a common mechanism is impaired. The similar patterns of performance are particularly striking given the very different clinical characteristics and neuropathologies of the patients investigated.

The mechanism impaired in the phonological output impairment of all of these patients relates to the selection and/or the organization of the features that make up the candidate lexical entries. In the model for speech production shown in figure 59.1, a word candidate is selected from the lexicon. Production of the word requires that its sound properties (i.e., its segments and features) be specified so that they can be planned for articulation and ultimately translated into neuromuscular commands relating to the speech apparatus. It is generally assumed that the activation of a lexical item excites its phonological representation, defined in terms of nodes or units corresponding to segments and features. A number of models of speech production propose that these word candidates are scanned into a short-term buffer, in order to account for the fact that the ultimate production of a sequence of words or an utterance is influenced not only at the segmental level but also at the prosodic level by the phonological context of neighboring words and ultimately by the syntactic role that the individual lexical item plays in the utterance string (see Levelt, 1989, for discussion). For example, the auxiliary *have* may be reduced in certain syntactic contexts to a /v/ and appended to the preceding word—for example, *I have eaten* may be produced as *I've eaten*.

The selection of sound properties, or phonological planning, can be impaired in a number of ways following brain damage, and such planning impairments may be at the single-word level or may include units larger than individual words. There are four descriptive categories of errors produced by patients, all of which are consistent with the view that they have a selection or phonological planning deficit (Blumstein, 1973). These errors are as follows: phoneme substitution errors in which a phoneme is substituted for another, e.g., *teams* → *keams*; simplification errors in which a phoneme or syllable is deleted, e.g. *green* → *geen*; addition errors in which an extra phoneme or syllable is added to a word, e.g., *see* → *stee*; and environment errors in which the occurrence of a particular phoneme is influenced by the surrounding phonetic context. These errors include metathesis, e.g. *degree* → *gedree*, and progressive and regressive assimilation errors, e.g., *Crete* → *kreke* and *trete*, respectively. The stability of these patterns is evidenced by their occurrence across languages, including French (Bouman and Grunbaum, 1925; Lecours and Lhermitte, 1969), German (Bouman and Grunbaum, 1925; Goldstein, 1948), English (Green, 1969; Blumstein, 1973), Turkish (Peuser and Fittschen, 1977), Russian (Luria, 1966), and Finnish (Niemi, Koivuselka-Sallinen, and Hanninen, 1985).

The patterns of sound substitutions are consistent with the view that the incorrect phonetic features have been selected or activated, but they have been correctly implemented by the articulatory system. Thus, most substitution errors involve a change in the value of a single phonetic feature. For example, the production of *doy* for *toy* reflects a change in the feature of voicing. Relatively few substitution errors produced by aphasic patients involve changes in more than one feature. In addition, the pattern of errors is consistent with the view that the phonetic features are organized in terms of tiers, because sound substitution errors rarely involve more than one tier at a time. Tiers have been defined in phonological theory to reflect the fact that phonetic features correspond to independent articulatory gestures (and consequent acoustic events) such as tongue placement and movement, lip movement, laryngeal activity, and height of the velum. Phoneme substitution errors in aphasia involve almost exclusively feature changes relating to place of articulation (the point in the vocal tract where the constriction takes place; e.g., *teams* → *keams*), voicing (the state of the larynx; e.g. *toy* → *doy*), nasality (the height of the velum; e.g., *nut* → *dut*), and manner of articulation (the nature of the constriction and its release; e.g., *sun* → *tun*)—all changes on a single tier (Blumstein, 1990).

Finally, phonological errors also suggest that the nature of the syllable structure (i.e., the organization of consonants and vowels) of the lexical candidate constrains the type and extent of errors made during the selection process (Blumstein, 1973, 1990). In particular, phoneme substitution errors are more likely to occur for a single consonant than for a part of a consonant cluster; for example, /f/ is more likely to undergo a phoneme substitution error in the word *feet* than in the word *fleet*. Simplification and addition errors are more likely to result in what is considered to be the simplest and thus the canonical syllable structure—consonant-vowel—such that consonants are more likely to be deleted in a word beginning with two consonants (*sky* → *ky*), and consonants are more likely to be added in a word beginning with a vowel (*army* → *jarmy*). And finally, assimilation errors across word boundaries (errors in which one sound segment is influenced by the presence of another occurring either earlier or later in the utterance) preserve the syllable structure relations of the lexical candidates. That is, if the influencing phoneme is at the beginning of the target word, so is the assimilated phoneme: *history books* → *bistory books*. If the influencing phoneme is at the end of the target word, so is the assimilated phoneme: *roast beef* → *roaf beef*. These results show that the syllable structure of a word is represented in the lexicon, and that this information is used in the planning buffer for sentence production. If this were not the case, the syllable constraints shown in the assimilation errors would not occur across word boundaries.

Nevertheless, despite the systematicity and regularity of the phonological errors, the particular occurrence of such an error cannot be predicted. That is, the patient may sometimes make an error on a particular word, and other times produce it correctly. Nor are the errors unidirectional (Blumstein, 1973; Hatfield and Walton, 1975). A voiced-stop consonant may become voiceless, e.g., /d/ → /t/, and a voiceless-stop consonant may become voiced, e.g., /t/ → /d/. Thus, the patterns of errors reflect statistical tendencies. Because patients are able to produce the correct utterances and because their errors are systematic (although not predictable), the error patterns suggest that the underlying phonological representations are intact but that there has been a change in the level of activation of the nodes corresponding to the phonetic representation (i.e., features and syllable structure) or, alternatively, there has been a failure to select the appropriate phonological representation when the lexical candidate was stored in the short-term lexical buffer. Whatever the nature of the underlying deficit, it is clear that virtually all aphasic patients, regardless of lesion site, display phonological output deficits. Importantly, the patterns of impairment among these patients seem to be similar. What does seem to vary by type of aphasia is the severity of the impairment, with Broca's aphasics and anterior aphasics displaying more of such errors than either conduction or Wernicke's aphasics (Blumstein, 1973).

With respect to the neural localization of phonological disorders, there does not seem to be a distinct neural locus. That phonological disorders occur in the context of the aphasias implicates the left perisylvian areas. However, the impairments can occur in temporal, temporoparietal, and anterior areas including Broca's area and the precentral gyrus (see figure 59.2). They almost always involve cortical lesions that extend to subcortical areas, and do not typically occur in the context of subcortical lesions alone, even among the subcortical aphasias (Naeser et al., 1982; Damasio, 1991). Nevertheless, subcortical pathology is typically involved along with the cortical areas typically implicated in the aphasias, particularly if the aphasias are not transient but persist subsequent to the initial insult (usually around 4–6 weeks).

It is worth adding a cautionary note concerning localization of function and, in particular, localization of components of the sound structure of language. First, many neurolinguistic investigations have focused on clinical characteristics of patients as indications of underlying neuropathology rather than on a close mapping of neuroanatomical loci and speech or phonological problems per se. Only in recent years have sophisticated neuroimaging techniques been available for such mapping, and even now these techniques have their limitations. Second, it is difficult to interpret a failure to show a one-to-one mapping of structure to function. It is possible that a failure to show close localization of a so-called function reflects individual differences in the neurological organization of language. Moreover, the processing components that ultimately result in speech output are indeed complex and are inferred through investigation of patients' behavior on different language tasks such as naming and repetition—which in themselves may require different processing components. As a result, it may be difficult to identify a single locus corresponding to phonological output disorders even though there is a hypothetically localized module or component. As John Hughlings Jackson noted, "To locate the damage which destroys speech and to localise speech are two different things" (quoted in Head, 1963, vol. 1, p. 50).

Having said all this, however, the view that speech and language functions are not narrowly localized but rather are more broadly distributed in the perisylvian areas is not new. Proponents of this view include some of the classical aphasiologists including John Hughlings Jackson (see Head, 1963) and A. R. Luria (1966). The fact that phonological disorders are not localizing impairments may not be surprising given that lexical access—in particular, naming—is perhaps the least localizing of language functions. Naming deficits of varying severity emerge in nearly all the aphasias (and also in the dementias). They are also evoked in seizure patients by electrical stimulation of the surface of the cortex throughout the left perisylvian area, extending posteriorly and also anteriorly to the frontal lobe including Broca's area, the inferior frontal lobe immediately adjacent to the face motor cortex, and the left precentral gyrus (excluding the frontal and occipital poles; see figure 59.2) (see also Penfield and Roberts, 1966; Ojemann, 1983).

Phonetic Representations: Mechanisms and Neurobiology As Figure 59.1 shows, subsequent to the selection of a lexical candidate or candidates and the articulatory planning of the utterance, the phonetic string is ultimately converted into a set of motor commands to the articulatory system. A wide range of speech production deficits reflect impairments to the motor commands or motor programs to the vocal tract system. For the purpose of this chapter, we will limit discussion to those phonetic disorders that occur in the context of a language impairment, that is, aphasia.

Whereas the neural basis of phonological representations seems to be diffusely present in the left dominant language hemisphere, the neural basis of phonetic disorders seems to be more localized and to implicate very specific neural structures. A long-observed phenomenon is that anterior aphasics produce phonetic errors. The implied basis for these errors is one of articulatory implementation; that is, the commands to the articulators to encode the word are incorrect, poorly timed, and so forth. A number of studies have explored these phonetic patterns of speech by investigating the acoustic properties or the articulatory parameters underlying the production of particular phonetic dimensions. The dimensions investigated included voicing in stop consonants and fricatives, place of articulation in stop consonants and fricatives, and the nasal and stop manner of articulation. Studies of speech production in anterior patients have shown that these patients have difficulty producing phonetic dimensions that require the timing of two independent articulators. These findings have emerged in the analysis of two phonetic dimensions, voicing and nasality. In the case of the voicing feature, the dimension studied is voice-onset time, that is, the time relationship between the release of a stop consonant and the onset of vocal cord vibration. The production of nasal consonants also requires appropriate timing between two articulators—in this case, the release of the closure in the oral cavity and the opening of the velum.

Results of analyses of the production of the voicing and nasal phonetic dimensions have shown that anterior aphasics evidence significant deficits (Blumstein, et al., 1977; Freeman, Sands, and Harris, 1978; Itoh, Sasanuma, and Ushijima, 1979; Blumstein et al., 1980; Itoh et al., 1980; Itoh et al., 1982; Gandour and Dardarananda, 1984a; Shewan, Leeper, and Booth, 1984). These same patterns emerge across different languages. They occur not only in English and Japanese, for which voice-onset time serves to distinguish two categories of voicing (voiced and voiceless), but also in Thai, for which voice-onset time serves to distinguish three categories of voicing in stop consonants (prevoiced, voiced, and voiceless aspirated).

Nevertheless, these anterior aphasics also show normal patterns of production. The constellation of spared and impaired patterns of articulation suggests that their disorder produces impairments in particular articulatory maneuvers rather than in the articulatory implementation of phonetic features. The evidence comes from studies of voicing in stop consonants. In English, the voicing feature in stop consonants can be cued in several ways. As discussed earlier, voice-onset time provides one measure of voicing for stop consonants occurring in initial position. A second measure is the duration of the vowel preceding a stop consonant. Vowels are short before voiceless stops (e.g., *write*) and long before voiced stops (e.g., *ride*). If patients have a deficit related to the implementation of voicing, then they should display impairments in the production of voice-onset time as well as vowel length preceding voiced and voiceless stop consonants. In contrast, if they have a deficit related to particular articulatory maneuvers, such as the timing of two independent articulators, the production of voice-onset time may be impaired while the production of vowel length may be normal. Results indicate that while these patients show an impairment in the implementation of the voicing phonetic dimension via voice-onset time, they are able to maintain the distinction between voiced and voiceless stops on the basis of the duration of the preceding vowel (Duffy and Gawle, 1984; Baum et al., 1990). Moreover, there is not a systematic relation within the same patient between the ability to realize the voicing dimension by means of voice-onset time and the ability to do so on the basis of vowel duration (Tuller, 1984). Thus, these patients do not have a disorder affecting the articulatory production of the feature voicing, but rather a disorder affecting particular articulatory maneuvers, namely the timing or integration of movements of two independent articulators.

Consistent with this view are the results from acoustic analysis of the production of vowels. Differences among vowel sounds such as /i a u/ are determined acoustically by the frequencies of the first two resonant peaks, called formant frequencies. Analyses of the formant frequencies of spoken vowel utterances show that anterior aphasics, including Broca's aphasics, maintain the formant frequency characteristics of different vowels, despite increased variability in their productions (Ryalls, 1981, 1986, 1987; Kent and Rosenbek, 1983). The production of vowels requires articulatory gestures that are based on the overall shape of the tongue rather than on the coordination of independent articulators.

Although anterior aphasics show a disorder in temporal coordination, their disorder does not reflect a pervasive timing impairment. Fricative durations do not differ significantly from those of normal subjects (Harmes et al., 1984), and the patients maintain the intrinsic duration differences characteristic of fricatives that vary with respect to place of articulation; for example, /s/ and /š/ are longer in duration than /f/ and /θ/ (Baum et al., 1990). Although overall vowel duration is longer for anterior aphasics than for normal subjects (see Ryalls, 1987, for review), these patients do maintain differences in the intrinsic durations of vowels; for example, tense vowels such as /i/ and /e/ are longer than their lax vowel equivalents /I/ and /E/. In addition, Thai-speaking anterior aphasics maintain the contrast between short vowels and long vowels, a contrast that is phonemic in their language (Gandour and Dardarananda, 1984b).

In addition to the impairment in the timing of independent articulators, difficulties for anterior aphasics have also emerged with laryngeal control. The patients have shown impairments in voicing in the production of voiced fricatives (Kent and Rosenbek, 1983; Harmes et al., 1984; Baum et al., 1990), and impairments in voicing that influence the spectral shape associated with place of articulation in stop consonants (Shinn and Blumstein, 1983). Consistent with the findings that anterior aphasics have impairments involving laryngeal control are studies of intonation. Intonation, or the melody of language, is ultimately determined by laryngeal maneuvers. Acoustic analyses of two-word spontaneous speech utterances and reading in Broca's aphasics has shown that although these patients have rudimentary control over some features of prosody in that they maintain a terminal falling fundamental frequency (Cooper et al., 1984), they show a restriction in the fundamental frequency range (Cooper et al., 1984; Ryalls, 1982). Restrictions in fundamental frequency range support the clinical impression that these speakers produce utterances in a monotone or with a flattened intonation.

Kent and Rosenbek (1983) have suggested that the timing problem found in anterior aphasics for individual segments and their underlying features is a manifestation of a broader impairment in the integration of articulatory movements from one phonetic segment to another. The sounds of speech are affected by the phonetic contexts in which they occur. Recent investigations of coarticulation effects in anterior aphasics show that they produce relatively normal anticipatory coarticulation. For example, in producing the syllable /s/, they anticipate the rounded vowel /u/ in the production of the preceding /s/ (Katz, 1988). Nevertheless, they seem to show a delay in the time it takes to produce these effects (Ziegler and von Cramon, 1985, 1986), and they may show some deficiencies in their productions (Tuller and Story, 1986; but see Katz, 1987, for discussion). What these results suggest is that phonological planning is relatively intact and it is the ultimate timing or coordination of the articulatory movements that is impaired. Consistent with this view are results showing that Broca's aphasia patients demonstrate impairments in the complex timing relation between syllables; that is, they do not show the normal decrease in the duration of a root syllable as word length increases (see Baum, 1992), nor do they show a normal ability to increase rate of articulation beyond a certain limit.

Taken together, several conclusions can be reached concerning the nature of the phonetic disorders and their ultimate underlying mechanisms. In particular, the impairment is not a linguistic one in the sense that the patient is unable to implement a particular phonetic feature. Moreover, the patients have not lost the representation for implementation nor the knowledge base for how to implement sounds in context. Rather, particular maneuvers relating to timing of articulators seem to be impaired, ultimately affecting the phonetic realization of some sound segments and some aspects of speech prosody.

Both lesion data from aphasic patients and evoked potential data from normal subjects suggest that there are specific neuroanatomical substrates relating to such phonetic implementation patterns (see figure 59.2). In particular, CT scan correlations with patterns of speech production deficits suggest the involvement of Broca's area (slices B and B/W), and the anterior limb of the internal capsule (including slices B, B/W, and W) (Baum et al., 1990). The lower motor cortex regions for larynx, tongue, and lips (slices W and SM) are also implicated, although less consistently so. The role of the left inferior frontal cortex (including Broca's area) and the precentral gyrus in the production of speech have also been shown in normal subjects (see, e.g., Petersen et al., 1989), and seems to be limited to articulatory maneuvers related to the production of speech sounds (McAdam and Whitaker, 1971, but cf. Grozinger, Kornhuber, and Kriebel, 1977). The supplementary motor area has been identified as well (Petersen et al., 1989). Of particular interest is the fact that the speech mechanism requires the coordination of both sides of the laryngeal/articulatory mechanism, and, in this sense, speech production is bilaterally innervated. Nevertheless, phonetic disorders such as those described in this chapter do not emerge with damage to analogous speech areas in the right hemisphere, suggesting that while both hemispheres may be ultimately involved in the production of speech, the control site for these mechanisms is in the left hemisphere.

While it is not surprising to find that anterior portions of the left hemisphere, particularly those localized in the vicinity of the motor cortex, are implicated in the production of speech, recent results suggest that posterior areas of the brain are also involved. There is no question that phonetic patterns of speech are qualitatively distinct in anterior and posterior aphasics. Posterior aphasics do not display the timing deficits that anterior aphasics manifest in the production of voice-onset time in stop consonants (Blumstein et al., 1980; Hoit-Dalgaard, Murry, and Kopp, 1983; Gandour and Dardarananda, 1984a; Shewan, Leeper, and Booth, 1984; Tuller, 1984) or in the production of nasal consonants (Itoh and Sasanuma, 1983). Nor do they show the impairments in laryngeal control either for the production of voicing or for those articulatory maneuvers requiring the integration of laryngeal movements and movements of the supralaryngeal vocal tract (Shinn and Blumstein, 1983; Baum et al., 1990). Nevertheless, although it is clearly distinct from that of anterior aphasics, posterior patients do display a subtle phonetic impairment. Most typically, they show increased variability in the implementation of a number of phonetic parameters (Ryalls, 1986; Kent and McNeill, 1987), including vowel formant frequencies (Ryalls, 1986) and vowel durations (Tuller, 1984; Ryalls, 1986; Gandour et al., 1992).

These subclinical impairments in speech production found in left-hemisphere posterior aphasics do not emerge in right-hemisphere patients (Mathison, 1992). Thus, the increased variability in posterior aphasics is not due to a so-called brain-damage effect. Rather, these impairments suggest that the speech production system is a network involving both posterior and anterior brain structures. The role of these brain structures in speech production seems to be different, because anterior patients show clear-cut phonetic impairments while posterior patients show subclinical phonetic impairments. Nevertheless, both anterior and posterior structures ultimately contribute to the speech production process.

Electrical stimulation studies have also suggested that the neural basis of speech production involves more than the anterior regions of the left hemisphere. The classic brain stimulation studies of Penfield and Roberts (1966) showed speech interference, described as hesitations and slurring, with stimulation of the motor areas, perisylvian and especially suprasylvian sites, and superior frontal, superior parietal, and superior and middle temporal lobes (see figure 59.2). These findings were replicated by Ojemann (1983). Nevertheless, because it is not clear whether the descriptive category of speech interference defined by Penfield and Roberts and by Ojemann consisted of the same types of phonetic output errors described in the aphasias, it is difficult to determine how these findings map onto those described with brain-damaged patients.

The nature of the posterior mechanism contributing to articulatory implementation is not clear. Two hypotheses may be suggested, but at this point they remain speculative. Posterior fibers project anteriorly to the motor cortex system, and damage to those fibers could affect the speech implementation system. Alternatively, the auditory feedback system normally contributing to the control of the articulatory parameters of speech may be impaired. More research is required to determine the nature of the mechanisms involved, but what is clear is that the traditional dichotomy between production, subserved by anterior brain structures, and perception, subserved solely by posterior structures, is not supported. As we will see, evidence from speech perception provides further support for the view that the language processing system is a complex network with both feedforward and feedback pathways between anterior and posterior portions of the brain.

Speech perception

Because the primary auditory pathway surfaces in Heschl's gyrus within the temporal lobe, it would not be surprising to find that the auditory association areas in the left temporal lobe are actively involved in speech reception. The classical view of the aphasias, in fact, made this claim, and attributed the language comprehension deficit of Wernicke's aphasia to impairments in the "sound images" of words (Geschwind, 1974) or to impairments in "phonemic hearing" (Luria, 1966), or to both. As Luria reasoned in explicating his hypothesis, if patients could not perceive phonological contrasts, then they would be unable to process words appropriately for meaning: *bee* might be misperceived as *pea*, resulting in a severe auditory comprehension disorder.

A review of figure 59.1 also suggests that contact with the lexicon (and, ultimately, with meaning) requires the conversion of the auditory input from the peripheral auditory system to a more abstract feature (phonological) representation. In fact, as illustrated in figure 59.1, the auditory reception of words involves the encoding of the auditory input into a spectral representation based on the extraction of more generalized auditory patterns or properties from the acoustic waveform, the conversion of this spectral representation to a more abstract feature (phonological) representation, and then the selection of a word candidate from a set of potential word candidates sharing phonological properties with the target word.

While speech perception studies with aphasic patients have supported the view that perceptual impairments reflect the misperception of phonetic features (i.e., the more abstract phonological properties of words or word candidates), they do not support the classical hypothesis that speech perception deficits per se underlie the auditory language comprehension impairments of Wernicke's aphasics, nor do they support the proposal that speech perception impairments are restricted to patients with left posterior brain damage, and in particular, with temporal lobe pathology. Let us turn to the evidence.

Phonological Representations in Speech Perception Like production studies with aphasic patients, most studies exploring the role of speech perception deficits in auditory comprehension impairments have focused on the ability of aphasic patients to perceive phonemic or segmental contrasts. Studies on segmental perception have indeed shown that aphasic patients evidence deficits in processing segmental contrasts. These studies have either explored patients' abilities to discriminate pairs of words and nonwords, such as *pear* versus *bear* or *pa* versus *ba*, or they have asked subjects to point to the appropriate word or consonant from an array of phonologically confusable pictures or nonsense syllables. Results show that nearly all aphasic patients show some problems in discriminating phonological contrasts (Blumstein, Baker, and Goodglass, 1977; Jauhiainen and Nuutila, 1977; Miceli et al., 1978, 1980) or in labeling or identifying consonants presented in a consonant-vowel context (Basso, Casati, and Vignolo, 1977; Blumstein et al., 1977; see also Boatman and Gordon, 1993). These problems emerge for the perception of both real words and nonsense syllables. Although there are more errors in the perception of nonsense syllables than real words, the overall patterns of performance are similar and essentially mirror the patterns found in the analysis of phonological errors in speech production. That is, subjects are more likely to make speech perception errors when the test stimuli differ by a single phonetic feature than when they differ by two or more features (Blumstein, Baker, and Goodglass, 1977; Miceli et al., 1978; Baker, Blumstein, and Goodglass, 1981). Among the various types of feature contrasts, the perception of place of articulation is particularly vulnerable (Blumstein, Baker, and Goodglass, 1977; Miceli et al., 1978; Baker, Blumstein, and Goodglass, 1981). Interestingly, similar patterns emerge in normal subjects when they perceive speech under difficult listening conditions (see, e.g., Miller and Nicely, 1955).

Significantly, there does not seem to be a relationship between speech perception abilities and auditory language comprehension. Patients with good auditory comprehension skills have shown impairments in speech processing; conversely, patients with severe auditory language comprehension deficits have shown minimal speech perception deficits (Basso, Casati, and Vignolo, 1977; Blumstein et al., 1977; Jauhiainen and Nuutila, 1977; Miceli et al., 1980; for general discussion see Boller, 1978). These patients have been drawn from a broad range of clinical types and underlying neuropathologies and include Broca's aphasics, Wernicke's aphasics, mixed anterior aphasics, conduction aphasics, and others.

Phonetic Representations in Speech Perception What is not clear from many of the studies exploring the perception of segmental contrasts is whether the failure to perceive such contrasts reflects an impairment in the perception of abstract phonetic features or an impairment relating to the extraction of the acoustic patterns associated with these features. To explore this issue, several studies have investigated the perception of the acoustic parameters associated with phonetic features. To this end, subjects are presented with an acoustic continuum in which the acoustic parameters of the end-point stimuli are appropriate to two different phonetic categories, and the remaining stimuli on the continuum range continuously and parametrically across the values associated with those phonetic segments. When given such stimuli, normal subjects show categorical perception. That is, when asked to label or identify the stimuli, they perceive them as belonging to discrete categories corresponding to the end-point, exemplar stimuli, and they show a sharp change in the identification of the categories usually at a particular stimulus along the continuum; when asked to discriminate the stimuli, they discriminate only those stimuli that they labeled as belonging to two different categories, and fail to discriminate those stimuli that they labeled as belonging to the same phonetic category, even though all of the discrimination pairs vary along the same physical dimension.

The studies exploring categorical perception in aphasia have investigated two phonetic dimensions—voicing (Basso, Casati, and Vignolo, 1977; Blumstein, Baker, and Goodglass, 1977; Gandour and Dardarananda, 1982) and place of articulation in stop consonants (Blumstein et al., 1984). For voicing, the acoustic dimension varied was voice-onset time, and for place of articulation, the dimension varied was the frequency of the formant transitions appropriate for /b d g/, and the presence or absence of a burst preceding the transitions. Results showed that if aphasic patients could perform either of the two tasks (labeling or discrimination), it was the discrimination task. Most importantly, the discrimination functions were generally similar in shape, and the locus of the phonetic boundary was comparable to that in normal subjects, even in those patients who could only discriminate the stimuli.

The facts that no perceptual shifts were obtained for the discrimination and labeling functions for aphasic patients, that the discrimination functions remained stable even in those patients who could not label the stimuli, and that the patients perceived the acoustic dimensions relating to phonetic categories in a fashion similar to normal subjects suggest that aphasic patients do not have a deficit specific to the extraction of the spectral patterns corresponding to the phonetic categories of speech. Rather, their deficit seems to relate to the threshold of activation of the phonetic or phonological representation itself or to its ultimate contact with the lexicon. Consistent with this view is the finding that while patients may show speech perception impairments, their performance is variable; they do not show selective impairments relating to a particular phonetic feature; and the pattern of errors is bidirectional—for example, voiced consonants may be perceived as voiceless, and voiceless consonants may be perceived as voiced. Interestingly, this pattern of results mirrors that found in speech production studies with aphasic patients.

In contrast to the segmental features of speech, the prosodic cues (intonation and stress) are consistently less affected in aphasia. Severely impaired aphasics have been shown to retain some ability to recognize and distinguish the syntactic forms of commands, yes-no questions, and information questions marked only by intonation cues (Green and Boller, 1974), even when they are unable to do so when syntactic forms are marked by lexical and syntactic cues. Nonetheless, as with intonation cues, patients' performance is not completely normal. A number of studies have revealed impairments in the comprehension of lexical and phrasal stress contrasts such as *hótdog* versus *hotdóg*) (Baum, Kelsch-Daniloff, and Daniloff, 1982; Emmorey, 1987), as well as sentential contrasts such as *He fed her dog bíscuits* versus *He fed her dóg biscuits* (Baum et al., 1982). Similar findings emerged for the perception of tone contrasts serving as lexical cues in Thai (Gandour and Daradarananda, 1983) and Chinese (Naeser and Chan, 1980). Importantly, no differences have emerged in any studies between the performance of anterior and posterior aphasics—a finding consistent with the results for the perception of phonemic contrasts.

Speech Perception and Lexical Access While results of speech perception experiments suggest that aphasic patients do not seem to have a selective impairment in processing the segmental structure underlying the auditory properties of speech, several recent studies have suggested that aphasic patients display impairments in the intersection of the sound properties of speech with lexical access. These studies have shown some interesting dichotomies in the performance of Broca's and Wernicke's aphasics that suggest that patients' impairments do not reflect speech perception deficits per se but rather deficits in the interaction of sound structure as it contacts the lexicon (see also Martin et al., 1975; Baker, Blumstein, and Goodglass, 1981).

Milberg, Blumstein, and Dworetzky (1988) explored the extent to which phonological distortions affect semantic facilitation in a lexical decision task. They investigated whether a phonologically distorted prime word such as *gat* or *wat* would affect the amount of semantic facilitation for target words semantically related to the undistorted prime, such as *cat*. Subjects were asked to make a lexical decision on the second item of stimulus pairs in which the first stimulus was semantically related to the second (e.g., *cat–dog*), or alternatively, was systematically changed by one or more features (e.g., *gat–dog*, *wat–dog*). Fluent patients showed priming in all phonologically distorted conditions (e.g., *gat–dog*, *wat–dog*), suggesting a reduced threshold for lexical access. In contrast, anterior aphasics showed priming only in the undistorted, semantically related condition (e.g., *cat–dog*), suggesting an increased threshold for lexical access. These results suggest that impairments in the use of phonological information to access the lexicon can manifest themselves in different ways in aphasic patients in the absence of a deficit in processing the phonological properties of speech themselves.

Similarly, the lexical status of a word affects differentially how aphasic patients perform phonetic categorization. Normal subjects typically show a lexical effect. That is, the locus of a phonetic boundary in an acoustic continuum such as voice onset time changes as a function of the lexical status of the end-point stimuli. When the endpoint /d/ stimulus is a word, such as *dash*, and the endpoint /t/ stimulus is a nonword, such as *tash*, there are more *d* responses along the continuum; in contrast, when the endpoint /t/ stimulus is a word, such as *task*, and the endpoint /d/ stimulus is a nonword, such as *dask*, there are more *t* responses along the continuum (Ganong, 1980). Broca's aphasics show a larger-than-normal lexical effect, placing a greater reliance on the lexical status of the stimulus in making their phonetic decisions than on the perceptual information in the stimulus. In contrast, Wernicke's aphasics do not show a lexical effect, suggesting that lexical information does not influence phonetic categorization, and perhaps that such top-down processing even fails to guide their language performance (Blumstein et al., 1993).

Neurobiology of Speech Perception Overall, the findings from speech perception studies of aphasic patients suggest that the neural basis for speech reception is neurally complex and includes far greater neural involvement than simply the primary auditory areas and auditory association areas in the temporal lobe. While the number of neurophysiological and electrophysiological studies focusing particularly on speech reception are few, their results provide converging evidence consistent with this view. PET studies have shown that the primary auditory cortex is activated in the processing of simple auditory stimuli (Lauter et al., 1985), and that both the primary auditory cortex and the superior temporal gyrus are activated in passive word recognition (Petersen et al., 1988, 1989). These results are not surprising. What is of interest, however, is that other cortical areas seem to be activated as well. For example, Zatorre et al. (1992) showed increased activity in Broca's area near the junction with premotor cortex, as well as in the superior parietal area near the supramarginal gyrus, when subjects were required to make a phonetic decision or phonetic judgment about auditorily presented consonant-vowel-consonant (CVC) stimuli (see figure 59.2). Further, Ojemann (1983) has shown impairments in the ability of patients to identify auditorily presented sound segments embedded in a phonetic context such as /a_ma/ during electrical stimulation to a wide range of cortical areas including the inferior frontal cortex (Ojemann, 1983). The exact role of these anterior areas in speech reception is not clear. However, these areas seem to be a part of a single neural system. Research with monkeys is consistent with this view. Auditory stimulation studies have shown direct ipsilateral projections to prefrontal regions, and ablation studies of these areas have shown diminished auditory discrimination skills (Pandya, Hallett, and Mukherjee, 1969).

The physical properties and complexity of the auditory system also seem to play a role in determining the neural system involved. A number of PET studies have shown that simple auditory stimuli consisting of noise bursts, clicks, or tones produce activation bilaterally in Heschl's gyrus. With the presentation of more complex speech CVC stimuli, both the primary auditory areas and the auditory association areas along the superior temporal gyrus are activated (Zatorre et al., 1992; Petersen et al., 1988, 1989). When subjects are required to focus on the pitch parameters of CVC stimuli rather than on the phonetic identity of the final consonant, greater right-hemisphere activity has been shown. The importance of these results is that the identical auditory stimuli (CVC syllables) produce different neural activity as a function of the task demands on the subject. These results support long-standing findings suggesting that different neural mechanisms are involved in the processing of verbal and nonverbal stimuli (Milner, 1962; Mazziotta et al., 1982), and in particular, in the perception of pitch and phonetic segments (see also Milner, 1962; Wood, Goff, and Day, 1971).

Summary

In this chapter we have attempted to characterize the neurobiology of the sound structure of language. Evidence from both speech production and speech perception suggests that there are various stages of processing leading ultimately to the production or perception of a word or utterance. Nevertheless, there is a common phonological representation of the sound structure of words that informs both speech production and speech perception mechanisms. This representation is in terms of abstract phonetic features relating to segment structure, and also in terms of representations defining syllable structure and prosodic structure. That similar patterns of phonological errors are shown by aphasic patients in both speech production and speech perception regardless of lesion site supports the view that a common phonological representation for words subserves both speech production and speech perception.

Both speech production and speech perception require different neural mechanisms to accomplish their respective goals. Evidence from aphasia supports the distinctions between phonetic and phonological representations in speech production, and between spectral (auditory) and phonological representations in speech perception. Overall, the patterns of errors from aphasic patients, as well as the evidence from brain stimulation studies and PET and evoked potential studies with normal subjects, suggest that the speech production and speech perception systems are distributed networks. There is no evidence supporting selective deficits for particular types of aphasia or for particular lesions sites in either speech production or speech perception; the patterns of errors of aphasic patients suggest inconsistent but rule-governed misproductions or misperceptions; and the neural localization of the processes involved in speech production and speech perception seems to be broadly distributed in the perisylvian area of the left hemisphere, not narrowly localized. The only exception seems to be the evidence of specific neural substrates relating to the phonetic implementation of speech; these areas correspond to the left inferior frontal cortex and the precentral gyrus, particularly for the larynx and the face.

The classical view, in which speech production is subserved by anterior brain structures and reception is subserved by posterior brain structures, is not supported by the data reviewed. The speech production system extends to posterior regions including the temporal and parietal lobes, as evidenced by subclinical phonetic deficits in the speech production of Wernicke's aphasics and by speech output impairments consequent to brain stimulation in posterior cortical areas. The speech perception system extends to anterior regions including the frontal lobe, as evidenced by speech perception impairments in anterior aphasics, speech perception problems consequent to brain stimulation in anterior cortical areas, and increased metabolic activity in anterior portions of the brain when phonetic judgements of auditory stimuli are required. The contributions of posterior areas to speech production and anterior areas to speech perception and the nature of the neural system involved are not fully understood at this time, but provide challenges for future researchers in exploring the neurobiology of the sound structure of language.

ACKNOWLEDGMENTS This research was supported in part by NIH grant DC00314 to Brown University. Many thanks to Steven Pinker for helpful comments on an earlier draft of this chapter.

REFERENCES

Baker, E., S. E. Blumstein, and H. Goodglass, 1981. Interaction between phonological and semantic factors in auditory comprehension. *Neuropsychologia* 19:1–16.

Basso, A., G. Casati, and L. A. Vignolo, 1977. Phonemic identification defects in aphasia. *Cortex* 13:84–95.

Baum, S. R., 1992. The influence of word length on syllable duration in aphasia: Acoustic analyses. *Aphasiology* 6:501–513.

Baum, S. R., S. E. Blumstein, M. A. Naeser, and C. L. Palumbo, 1990. Temporal dimensions of consonant and vowel production: An acoustic and CT scan analysis of aphasic speech. *Brain Lang.* 39:33–56.

Baum, S. R., J. Kelsch-Daniloff, and R. Daniloff, 1982. Sentence comprehension by Broca's aphasics: Effects of some suprasegmental variables. *Brain Lang.* 17:261–271.

Blumstein, S. E., 1973. *A Phonological Investigation of Aphasic Speech*. The Hague: Mouton.

Blumstein, S. E., 1990. Phonological deficits in aphasia: Theoretical perspectives. In *Cognitive Neuropsychology and Neurolinguistics: Advances in Models of Cognitive Function and Impairment*, A. Caramazza, ed. Hillsdale N.J.: Erlbaum.

Blumstein, S. E., E. Baker, and H. Goodglass, 1977. Phonological factors in auditory comprehension in aphasia. *Neuropsychologia* 15:19–30.

Blumstein, S. E., M. Burton, S. Baum, R. Waldstein, and D. Katz, 1993. The role of lexical status on the phonetic categorization of speech in aphasia. *Brain Lang.* in press.

Blumstein, S. E., W. E. Cooper, H. Goodglass, S. Statlender, and J. Gottlieb, 1980. Production deficits in aphasia: A voice-onset time analysis. *Brain Lang.* 9:153–170.

Blumstein, S. E., W. E. Cooper, E. B. Zurif, and A. Caramazza, 1977. The perception and production of voice-onset time in aphasia. *Neuropsychologia* 15:371–383.

Blumstein, S. E., V. C. Tartter, G. Nigro, and S. Statlender, 1984. Acoustic cues for the perception of place of articulation in aphasia. *Brain Lang.* 22:128–149.

Boatman, D., and B. Gordon, 1993. The functional organization of auditory speech processing as revealed by direct cortical electrical interference. manuscript.

Boller, F., 1978. Comprehension disorders in aphasia: A historical overview. *Brain Lang.* 5:149–165.

Bouman, L., and A. Grunbaum, 1925. Experimentell-psychologische Untersuchungen sur Aphasie und Paraphasie. *Z. Gesamte Neurol. Psychiatrie* 96:481–538.

Cooper, W. E., C. Soares, J. Nicol, D. Michelow, and S. Goloskie, 1984. Clausal intonation after unilateral brain damage. *Lang. Speech* 27:17–24.

Damasio, H., 1991. Neuroanatomical correlates of the aphasias. In *Acquired Aphasia*, ed. 2, M. T. Sarno, ed. New York: Academic Press.

Duffy, J., and C. Gawle, 1984. Apraxic speakers' vowel duration in consonant-vowel-consonant syllables. In *Apraxia of Speech*, J. Rosenbek, M. McNeil, and A. Aronson, eds. San Diego, Calif.: College-Hill Press.

Emmorey, K. D., 1987. The neurological substrates for prosodic aspects of speech. *Brain Lang.* 30:305–320.

Freeman, F. J., E. S. Sands, and K. S. Harris, 1978. Temporal coordination of phonation and articulation in a case of verbal apraxia: A voice-onset time analysis. *Brain Lang.* 6:106–111.

Ganong, W. F., 1980. Phonetic categorization in auditory word perception. *J. Exp. Psychol. [Hum. Percept.]* 6:110–125.

Gandour, J., and R. Dardarananda, 1982. Voice onset time in aphasia: Thai, I. Perception. *Brain Lang.* 17:24–33.

Gandour, J., and R. Dardarananda, 1983. Identification of tonal contrasts in Thai aphasic patients. *Brain Lang.* 17:24–33.

Gandour, J., and R. Dardarananda, 1984a. Voice-onset time in aphasia: Thai, II: Production. *Brain Lang.* 18:389–410.

Gandour, J., and R. Dardarananda, 1984b. Prosodic disturbances in aphasia: Vowel length in Thai. *Brain Lang.* 23:177–205.

Gandour, J., S. Ponglorpisit, F. Khunadorn, S. Dechongkit, P. Boongird, and R. Boonklam, 1992. Timing characteristics of speech after brain damage: Vowel length in Thai. *Brain and Lang.* 42:337–345.

Geschwind, N., 1965. Disconnexion syndromes in animals and man. *Brain* 88:237–294, 585–644.

Geschwind, N., 1974. Wernicke's contribution to the study of aphasia. In *Selected Papers on Language and the Brain*, R. S. Cohen and M. W. Wartofsky, eds. Boston Studies in the Philosophy of Science, vol. 16. Dordrecht: Reidel.

Goldstein, K., 1948. *Language and Language Disturbances.* New York: Grune and Stratton.

Goodglass, H., and E. Kaplan, 1972. *The Assessment of Aphasia and Related Disorders.* Philadelphia: Lea and Febiger.

Green, E., 1969. Phonological and grammatical aspects of jargon in an aphasic patient: A case study. *Lang. Speech* 12:103–118.

Green, E., and F. Boller, 1974. Features of auditory comprehension in severely impaired aphasics. *Cortex* 10:133–145.

Grozinger, B., H. H. Kornhuber, and J. Kriebel, 1977. Human cerebral potentials preceding speech production, phonation, and movements of the mouth and tongue, with reference to respiratory and extracerebral potentials. In *Language and Hemispheric Specialization in Man: Cerebral Event-Related Potentials*, J. E. Desmedt, ed. Basel: Karger.

Harmes, S., R. Daniloff, P. Hoffman, J. Lewis, M. Kramer, and R. Absher, 1984. Temporal and articulatory control of fricative articulation by speakers with Broca's aphasia. *J. Phonetics* 12:367–385.

Hatfield, F. M., and K. Walton, 1975. Phonological patterns in a case of aphasia. *Lang. Speech* 18:341–357.

Head, H., 1963. *Aphasia and Kindred Disorders of Speech*, vol. 1. New York: Hafner.

Hoit-Dalgaard, J., T. Murry, and H. Kopp, 1983. Voice onset time production and perception in apraxic patients. *Brain Lang.* 20:329–339.

Itoh, M., and S. Sasanuma, 1983. Velar movements during speech in two Wernicke aphasic patients. *Brain Lang.* 19:283–292.

Itoh, M., S. Sasanuma, H. Hirose, H. Yoshioka, and T. Ushijima, 1980. Abnormal articulatory dynamics in a patient with apraxia of speech. *Brain Lang.* 11:66–75.

Itoh, M., S. Sasanuma, I. Tatsumi, S. Murakami, Y. Fukusako, and T. Suzuki, 1982. Voice onset time characteristics in apraxia of speech. *Brain Lang.* 17:193–210.

Itoh, M., S. Sasanuma, and T. Ushijima, 1979. Velar movements during speech in a patient with apraxia of speech. *Brain Lang.* 7:227–239.

Jauhiainen, T., and A. Nuutila, 1977. Auditory perception of speech and speech sounds in recent and recovered aphasia. *Brain Lang.* 4:572–579.

Katz, W. F., 1988. Anticipatory coarticulatory in aphasia: Acoustic and perceptual data. *Brain Lang.* 35:340–368.

Katz, W., 1987. Anticipatory labial and lingual coarticulation in aphasia. In *Phonetic Approaches in Speech Production in Aphasia and Related Disorders*, J. Ryalls, ed. Boston: College-Hill Press.

Kent, R., and M. McNeill, 1987. Relative timing of sentence repetition in apraxia of speech and conduction aphasia. In *Phonetic Approaches to Speech Production in Aphasia and Related Disorders*, J. Ryalls, ed. Boston: College-Hill Press.

Kent, R., and J. Rosenbek, 1983. Acoustic patterns of apraxia of speech. *J. Speech Hear. Res.* 26:231–248.

Lauter, J., P. Herscovitch, C. Formby, and M. E. Raichle, 1985. Tonotopic organization in human auditory cortex revealed by positron emission tomography. *Hearing Res.* 20:199–205.

Lecours, A. R., and F. Lhermitte, 1969. Phonemic paraphasias: Linguistic structures and tentative hypotheses. *Cortex* 5:193–228.

Levelt, W. J. M., 1989. *Speaking: From Intention to Articulation.* Cambridge, Mass.: MIT Press.

Luria, A. R., 1966. *Higher Cortical Functions in Man.* New York: Basic Books.

Martin, A. D., N. H. Wasserman, L. Gilden, and J. West, 1975. A process model of repetition in aphasia: An investigation of phonological and morphological interactions in aphasic error performance. *Brain Lang.* 2:434–450.

Mathison, H. K., 1992. An acoustic analysis of vowel production in a right hemisphere-damaged population. Unpublished Honors Thesis in Cognitive Science, Brown University.

MAZZIOTTA, J. C., M. E. PHELPS, R. E. CARSON, and D. KUHL, 1982. Tomographic mapping of human cerebral metabolism: Auditory stimulation. *Neurology* 32:921–937.

MCADAM, D. W., and H. A. WHITAKER, 1971. Language production: Electroencephalographic localization in the normal human brain. *Science* 172:499–502.

MICELI, G., C. CALTAGIRONE, G. GAINOTTI, and P. PAYER-RIGO, 1978. Discrimination of voice versus place contrasts in aphasia. *Brain Lang.* 2:434–450.

MICELI, G., G. GAINOTTI, C. CALTAGIRONE, and C. MASULLO, 1980. Some aspects of phonological impairment in aphasia. *Brain Lang.* 11:159–169.

MILBERG, W., S. E. BLUMSTEIN, and B. DWORETZKY, 1988. Phonological processing and lexical access in aphasia. *Brain Lang.* 34:279–293.

MILLER, G. A., and P. E. NICELY, 1955. An analysis of perceptual confusions among some English consonants. *J. Acous. Soc. Am.* 27:338–352.

MILNER, B., 1962. Laterality effects in audition. In *Interhemispheric Relations and Cerebral Dominance*, V. B. Mountcastle, ed. Baltimore: M. D. Johns Hopkins Press.

NAESER, M. A., M. P. ALEXANDER, N. HELM-ESTABROOKS, H. L. LEVINE, S. A. LAUGHLIN, and N. GESCHWIND, 1982. Aphasia with predominantly subcortical lesion sites. *Arch. Neurol.* 39:2–14.

NAESER, M. A., and S. W.-C. CHAN, 1980. Case study of a Chinese aphasic with the Boston Diagnostic Aphasia Examination. *Neuropsychologia* 18:389–410.

NIEMI, J., P. KOIVUSELKA-SALLINEN, and R. HANNINEN, 1985. Phoneme errors in Broca's aphasia: Three Finnish cases. *Brain Lang.* 26:28–48.

OJEMANN, G. A., 1983. Brain organization for language from the perspective of electrical stimulation mapping. *Behav. Brain Sci.* 6:189–230.

PANDYA, D. N., M. HALLETT, and S. K. MUKHERJEE, 1969. Intra- and inter-hemispheric connections of the neocortical auditory system in the rhesus monkey. *Brain Res.* 14: 49–65.

PENFIELD, W., and L. ROBERTS, 1966. *Speech and Brain Mechanisms*. New York: Atheneum.

PETERSEN, S. E., P. T. FOX, M. I. POSNER, M. MINTUN, and M. E. RAICHLE, 1988. Positron emission tomographic studies of the cortical anatomy of single-word processing. *Nature* 331:585–589.

PETERSEN, S. E., P. T. FOX, M. I. POSNER, M. MINTUN, and M. E. RAICHLE, 1989. Positron emission tomographic studies of the processing of single words. *J. Cognitive Neurosci.* 1:153–170.

PEUSER, G., and M. FITTSCHEN, 1977. On the universality of language dissolution: The case of a Turkish aphasic. *Brain Lang.* 4:196–207.

RYALLS, J., 1981. Motor aphasia: Acoustic correlates of phonetic disintegration in vowels. *Neuropsychologia* 20:355–360.

RYALLS, J., 1982. Intonation in Broca's aphasia. *Neuropsychologia* 20:355–360.

RYALLS, J., 1986. An acoustic study of vowel production in aphasia. *Brain Lang.* 29:48–67.

RYALLS, J., 1987. Vowel production in aphasia: Towards an account of the consonant-vowel dissociation. In *Phonetic Approaches to Speech Production in Aphasia and Related Disorders*, J. Ryalls, ed. Boston: College-Hill Press.

SASANUMA, S., I. F. TATSUMI, and H. FUJISAKI, 1976. Discrimination of phonemes and word accent types in Japanese aphasic patients. *International Congress of Logopedics and Phoniatrics* 16:403–408.

SHANKWEILER, D. P., K. S. HARRIS, and M. L. TAYLOR, 1968. Electromyographic study of articulation in aphasia. *Arch. Phys. Med. Rehab.* 49:1–8.

SHEWAN, C. M., H. LEEPER, and J. BOOTH, 1984. An analysis of voice onset time (VOT) in aphasic and normal subjects. In *Apraxia of Speech*, J. Rosenbek, M. McNeill, and A. Aronson, eds. San Diego, Calif.: College-Hill Press.

SHINN, P., and S. E. BLUMSTEIN, 1983. Phonetic disintegration in aphasia: Acoustic analysis of spectral characteristics for place of articulation. *Brain Lang.* 20:90–114.

TULLER, B., 1984. On categorizing aphasic speech errors. *Neuropsychologia* 22:547–557.

TULLER, B., and R. S. STORY, 1986. Co-articulation in aphasic speech. *J. Acoust. Soc. Am.* 80: Suppl. 1, MM17.

WOOD, C., W. R. GOFF, and R. S. DAY, 1971. Auditory evoked potentials during speech perception. *Science* 173: 1248–1251.

ZATORRE, R. J., A. C. EVANS, E. MEYER, and A. GJEDDE, 1992. Lateralization of phonetic and pitch discrimination in speech processing. *Science* 256:846–849.

ZIEGLER, W., and D. VON CRAMON, 1985. Anticipatory coarticulation in a patient with apraxia of speech. *Brain Lang.* 26:117–130.

ZIEGLER, W., and D. VON CRAMON, 1986. Disturbed coarticulation in apraxia of speech: Acoustic evidence. *Brain Lang.* 29:34–47.

60 Computational Interpretations of Neurolinguistic Observations

RICHARD SPROAT

ABSTRACT Neurolinguists and computational linguists share one goal—namely, the understanding of how the human language-processing mechanism works—yet they typically approach that goal in different ways. This chapter examines a selected handful of neurolinguistic results relating to human processing of prosody and morphology, and it suggests for these results interpretations in the context of various recent computational models of language. Two main methodological points are made. First, when one designs a computational model, one must carefully consider the kinds of information that such a model might use to compute its output, and how the modules of the model should interact. Such considerations can prove useful in interpreting certain neurolinguistic data, because in some cases computations that are currently difficult or impossible to do automatically correspond rather nicely to particular deficits evidenced in aphasic patients. Second, if one starts with a neurolinguistic or psycholinguistic model and considers its computational implementation, one may end up with a clearer picture of how the model should work. Occasionally, one may find that the model, despite its psychological appeal, makes no computational sense and is therefore suspect from a computational point of view.

Many neurolinguists and computational linguists share one goal—namely, the understanding of how the human language-processing mechanism works—yet they typically approach that goal in opposite ways. Whereas neurolinguists try to understand normal language function by observing people for whom language function has become somehow abnormal, computational linguists attempt to understand how language works by building models that, ideally, mimic normal human linguistic behavior. Both techniques are crude: In neurolinguistics one is at the mercy of patients' misfortunes and it is virtually impossible to replicate an identical set of conditions in another patient; computational models are usually not good mimics of human performance, and even if they were, it is not always clear what that would mean because the system may be a good mimic only of *what* people do, not of *how* they do it. Nonetheless, it is useful to ask both what neurolinguistics can learn from computational linguistics, and (to a lesser extent) what the latter can learn from the former, and it is these questions that I intend to pursue here.

Specifically, I will discuss computational models of some linguistic phenomena and relate those models to published neurolinguistic or psycholinguistic findings. Space allows for discussion of only a few topics, but I hope that the case studies I have chosen will convince the reader of a couple of points. First, in building a computational model, one is forced to think carefully about the different kinds of information that such a model might use to compute its output, and about how the modules of the model should interact. Such considerations can turn out to be useful in interpreting certain neurolinguistic data: In some cases, computations that are currently difficult or impossible to do automatically correspond rather nicely to particular deficits evidenced in aphasic patients. Second, if one starts with a neurolinguistic or psycholinguistic model and considers its computational implementation, one may end up with a clearer picture of how the model should work: In some cases, one may find that the model, despite its psychological appeal, makes no computational sense and is therefore suspect from a computational point of view.

Computational models of prosody

An examination of text-to-speech will demonstrate some of the practical limitations on building computational models. The problem of text-to-speech can be simply stated (Allen, Hunnicut, and Klatt, 1987; Klatt, 1987): Given a text in some language, convert that text into speech output that sounds as much as possible like a native speaker of that language reading

RICHARD SPROAT Linguistics Research Department, AT&T Bell Laboratories, Murray Hill, N.J.

that text. Crucial components of naturalness in a text-to-speech system are various problems that can be classified under the general rubric of prosody: Appropriate assignment of accents, pitch (F_0) contours, and segmental durations are all important if the resulting speech is to sound natural. In this section I will briefly outline some of the methods for prosody used in the AT&T Bell Laboratories text-to-speech system—which we call TTS (see, e.g., Anderson et al., 1984; Sproat et al., 1992)—and I will discuss some limitations of those methods. It will be shown that these limitations relate in interesting ways to the effects on prosody of various kinds of brain damage. Although I will only discuss the Bell Laboratories system here, this does not affect the point—no text-to-speech system has solved the limitations that I will discuss for TTs.

TONE AND INTONATION Consider first the problem of assigning an intonation contour to a sentence in English. One popular model of English intonation is due to Pierrehumbert 1980, and is the model used in TTS (Anderson et al., 1984). In this model pitch accents are associated to various words in the sentence. In an intonational language like English, we distinguish between stress, which is a property of syllables in a word (in the word *rèlatívity* there are two stressed syllables, *-tiv-* having primary stress and *rel-* having secondary stress), and *accent*, which denotes a salient perturbation in the pitch of the voice that is (generally) associated with the primary stressed syllable of the word. The salient perturbation is one of several varieties of upward or downward movement in pitch, and each of these varieties constitutes an accent type. The simplest pitch accents in English are represented phonologically by an H (High) pitch target (with an upward movement in pitch) and an L (Low) pitch target (a downward movement). More complex pitch accents are also found, for example, the H* + L accent (Pierrehumbert, 1980), whose pitch contour consists of a rise to a high pitch target followed immediately by a fall to a low pitch.

In addition to pitch accents, there are also pitch perturbations that are associated not with words but with particular positions in the prosodic phrase: Boundary tones are associated with phrasal boundaries, and phrase accents with various phrase-internal positions. The different ways in which an English sentence can be rendered intonationally correspond to different combinations of tones in the phonological representation of the sentence. For example, the same sentence can be rendered with a falling intonation—appropriate for a simple declarative—or a rising intonation—appropriate for a simple yes-no question—and these two possibilities correspond to different phonological specifications for the sentence; see figure 60.1. Note that while the accents and tones chosen differ in the two examples, various other linguistic properties remain constant; thus, primary stress is on the second syllable of the word *achar* in both cases.

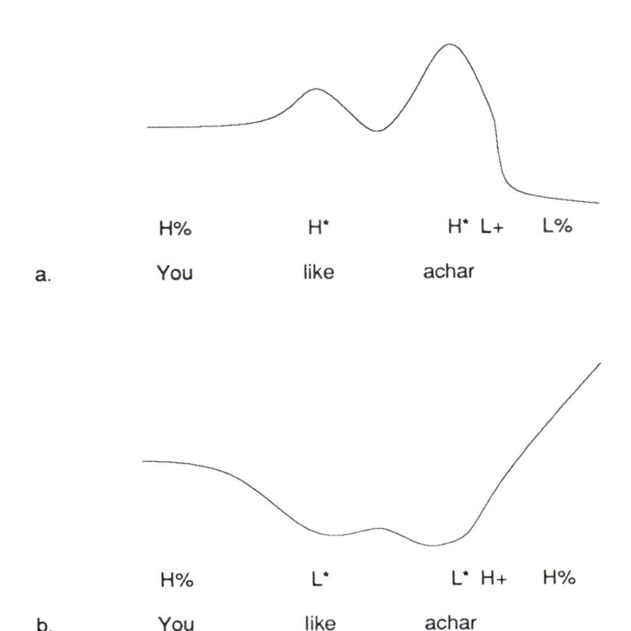

FIGURE 60.1 Tonal specifications for two intonational renditions of the sentence *You like achar.* The first figure represents a falling contour with low final boundary tone (L%), high pitch accents (H*) and a low phrase accent (L+). The second represents a rising contour, with opposite specifications for pitch accents, phrase accent, and final boundary tone.

PROBLEMS FOR TEXT-TO-SPEECH

Accent placement While such a theory of intonation provides a simple model for various attested English intonation patterns, there are numerous problems with its practical application in a text-to-speech system. First, pitch accents must be associated with words, but this raises the problem of how to compute appropriate accenting in the general case. A noun phrase such as *Park Avenue* is often accented on both words, with the more prominent accent on the second word. Yet there are circumstances in which it is appropriate to deaccent the second word:

(1) Drive along MADISON AVENUE, and then turn onto PARK Avenue.

This problem can be solved partly for text-to-speech (Hirschberg, 1993; Sproat, Hirschberg, and Yarowsky, 1992). For example, many function words (e.g., *to*, *the*) rarely receive an accent, and it is possible to achieve good results by never accenting these words. Furthermore, if a word was used in the previous context, subsequent uses of that word will often be deaccented: This simple method will render examples like (1) appropriately. However, consistent with Bolinger's (1972) claim that "accent is predictable if you're a mind-reader", there are still many instances of the creative use of accent placement to convey speakers' intentions that we currently cannot model. As a result, TTS is still far from natural in its performance. Put differently, relatively natural accentuation can currently be achieved only to the extent that accents can be predicted on the basis of purely grammatical or easily computed semanticopragmatic information.

Tonal specification Once one has determined pitch accent placement, there is a further problem of picking appropriate tones to associate to those accents, and also to the phrase accents and boundaries. As we saw in figure 60.1, even with identical accent placement it is possible to have more than one pitch contour for a sentence in English. There are only two basic contours that TTS selects automatically: the simple 'declarative' contour as in figure 60.1a, and simple yes-no question contours as in figure 60.1b. The former is the default contour; the latter is chosen only if it can be determined from text analysis that the sentence is a yes-no question. No attempt is made to assign complex pitch accents automatically (though these can be assigned manually). The use of some complex accents by English-speakers is understood to some extent; see, for example, Ward and Hirschberg (1985). However, we currently have no way to predict their appropriate use on the basis of what we can infer from text.

While the choice of tone associated with accents is a very difficult problem in English, the analogous problem in a tone language like Mandarin is much more straightforward. Unlike in English, tone in Mandarin and other Chinese languages is lexically specified. Each syllable in a Mandarin word has an inherent tone, and one can find sets of words that are distinguished solely on the basis of their lexical tone: Compare *mà* 'scold' with a falling tone (represented phonologically as the sequence HL), *má* 'hemp' (rising tone—LH), and *mǎ* 'horse' (low tone—L). As a result, Mandarin is somewhat easier to synthesize insofar as plausible tonal contours are generally more predictable (simply on the basis of what words are in the sentence) than they are in English.

Phonetic implementation of tone From an abstract representation in terms of a sequence of H and L tones, it is necessary to apply a set of phonetic implementation models in order to arrive at a pitch contour for the sentence. The model used by Anderson, Pierrehumbert, and Liberman (1984) provides a number of parameters that affect the exact outcome of phonetic implementation. Among these parameters are the utterance's pitch range, expressed in terms of the highest and lowest pitches (normally) reachable for that utterance; the prominence of the accents, which determines how high (for an H) or low (for an L) within the pitch range a particular accent will be; and the final lowering factor, which determines, for example, how quickly the final fall of a simple declarative contour will occur. While it is possible to set reasonable values for these parameters automatically, producing contours that often sound plausible, normal human speakers show much finer control over phonetic implementation. For example, particularly high prominences can be used to convey greater-than-normal emphasis on a particular word. Most such subtleties are not automatically derivable from text, and it is thus beyond the capabilities of current TTS to produce anything other than a rather vanilla set of prominences.

DURATION A similar set of problems arises with respect to segmental durations. The duration module of TTS (part of the empirical basis of which is described in van Santen, 1992), provides a set of parametric statistical models that produce reasonable durations for different classes of segments in different contexts. Various factors are known to affect segmental duration. For example, vowels tend to be longer in stressed syllables (the *a* in *BAT* vs. that in *COMbat*), longer still in accented syllables (*large BAT* vs. *BASEBALL bat*), longer in phrase-final position (*I saw the bat* vs. *The bat is over there*), and longer before voiced consonants (*bad* vs. *bat*). Parameters modeling such factors are used to determine durations for segments in TTS. While this approach provides useful models of duration, humans have much more control over segmental duration than we are currently able to impart to TTS. Abnormal duration can be used affectively to convey various kinds of information, including uncertainty or added emphasis. We have no way at present to compute from text the appropriate use of such abnormal duration.

The Effects of Brain Damage on Prosody

English-speaking Broca's aphasics Danly and Shapiro (1982) present a study of the prosodic characteristics of five Broca's aphasic speakers of English. While some aspects of prosody—such as the sentence-final F_0 fall of simple declarative sentences (see figure 60.1a) and F_0 declination (discussed below)—seemed to be preserved for these speakers, others, such as phrase-final segmental lengthening, were impaired.

Consider first segmental lengthening. For the control subjects in Danly and Shapiro's experiment, the mean duration of phrase-final words was 223 ms, as opposed to 169 ms for the same words in phrase-medial position. Surprisingly, the Broca's aphasics consistently showed exactly the opposite effects: In phrase-final position, their average word duration was 277 ms, as compared with 337 ms for the same words in medial position. Danly and Shapiro propose a number of possible causes for this observation, including the hypothesis that the aphasics' syntactic representations are underspecified so as to prevent the triggering of normal phrase-final lengthening. This suggestion is intriguing in light of the text-to-speech model discussed in previous sections. As we noted, it is possible to compute appropriate phrase-final lengthening for text-to-speech, and to that extent phrase-final lengthening can be said to be determined by grammatical factors; what cannot be so easily computed is the affective use of duration, which is not based on grammatical properties of the input. Since Broca's aphasics are known to exhibit various grammatical deficits, it is interesting that they seem to show a deficit of some of the grammatically controlled aspects of phonetic implementation.

Let us turn now to the preservation of the final F_0 fall. In the Pierrehumbert model of intonation (which is not the model assumed by Danly and Shapiro) the final F_0 fall in a declarative sentence is a consequence of a final sequence of H pitch accent, L phrase accent, and L boundary tone; or, in other words, it is a consequence of the choice of tones associated with those accents. As we have seen, TTS automatically assigns this choice of tones in the majority of sentences; but recall that this tonal combination is by no means always appropriate: At best it can be said to be a somewhat unmarked default intonation pattern, one that it is reasonable to have the system use unless there is reason for it to do otherwise. As such, then, the choice of a final F_0 fall is not grammatically determined. Thus, it is perhaps unsurprising that Broca's aphasics are relatively unaffected in their production of this intonation contour: If their deficit affects grammatical processing, then the largely extragrammatical process of choosing an appropriate intonation contour should be unaffected.

Finally, let us consider F_0 declination. F_0 declination denotes the general downtrend that is commonly observed in pitch contours in natural speech. In the Liberman and Pierrehumbert (1984) model (partly incorporated into the intonation model in TTS), one aspect of declination is seen as a consequence of *catathesis*, by which the topline of the pitch range is reset after a sequence of tones of the form H L. Since catathesis depends upon the particular choice of tones, one would expect that declination should be optional; indeed, although declination is commonly observed, it is by no means mandatory. In other words, the occurrence of declination depends in part upon extragrammatical factors and as such is not expected to he affected in Broca's aphasics. There is a complication in that the aphasics' preservation of F_0 declination is only observed for short sentences: For long sentences the effect is lost, and the patients show little or no declination. Danly and Shapiro's closer examination of their data, however, reveals that individual patients actually show a wide range of behaviors; two of the patients show normal or near-normal declination even over long sentences. One patient showed no declination at all, and the two remaining patients showed a small rise in pitch over the span of the sentence. Interestingly, these latter two patients also showed an F_0 rise—not an F_0 declination—in an earlier experiment testing shorter sentences, suggesting that even for shorter sentences the neat picture painted here is not entirely satisfactory for all patients.

But even if the interpretation is not wholly straightforward, I think it has been shown that there is merit in looking at aphasic data in terms of computational models of prosody such as the one used in TTS. Broca's aphasia can be roughly characterized as a disorder of grammar. Consistent with this, Broca's aphasics seem to show prosodic deficits in areas where a text-to-speech system can succeed on the basis of linguistic information derivable from the input text; they seem to show fewer deficits in areas that are arguably extragrammatical and where TTS does not enjoy such success.

Mandarin-speaking Broca's aphasics What would one expect to find if one turned to a tone language like Mandarin? Recall that one aspect of Mandarin TTS is easier than the equivalent aspect for English, namely, the selection of appropriate tonal contours, and that this is because Mandarin tones are determined purely by the grammatical or lexical content of the sentence. Given this, one might expect that Mandarin-speaking Broca's aphasics would show deficits in tonal selection resulting in utterances that have inappropriate tonal contours. This is correct, as was shown by Packard (1986).

Packard's subjects consisted of a group of eight nonfluent Mandarin-speaking aphasics, and an appropriately matched set of normal subjects. Subjects were asked to repeat a series of short words. Their responses were recorded and scored for segmental and tonal errors. One of the recognized symptoms of aphasia is *phonological paraphasia* (see Blumstein, 1973). Phonological paraphasia consists in speech errors in which one or more phonological segments are incorrectly substituted for the correct targets. Thus *cat* might be produced as *gat*. Given that tone is part of the phonological representation of morphemes in Mandarin, it is expected that phonological paraphasias should affect tone just as they affect, say, consonants. As Packard shows, patients' errors in tone production (average 22.4 errors) and errors in consonant production (average 26.9) are roughly on a par, indicating that, in Mandarin, tones are affected just as much as other segments.

Right-hemisphere damage in Taiwanese speakers Ross and colleagues (1992) discuss the effect of right-hemisphere (inferior frontoparietal) damage on the affective use of pitch in a group of speakers of Taiwanese, a Chinese language of the Southern Min group. In Taiwanese, as in Mandarin, the overall shape of the pitch contour is determined in large measure by lexical specification. However, in both languages one can affectively modify various aspects of the prosody, much as in English. Patients and controls were presented with five different affective renditions of a single Taiwanese sentence containing six toned syllables. The renditions were angry, sad, surprised, happy, and neutral. After hearing each rendition, subjects were asked to reproduce what they heard as closely as possible, and these repetitions were recorded. For each of the six syllables for all five renditions, the following measurements were taken: average F_0, initial F_0, slope of F_0, and duration of the syllable (actually, the duration of the F_0 contour associated with the syllable was measured, but this is at least somewhat proportional to the syllable's duration.) Then an "emotional range" was computed for each acoustic measure for each syllable by computing the difference between the highest and the lowest values found among the five renditions of that syllable.

In general, normal controls showed much larger values for the emotional range indices than did the patients. For example, for the fourth syllable of the sentence, which has a high-falling tone, normal subjects' mean emotional range for the slope of F_0 is about 70 semitones per second. The right-hemisphere-damaged patients, on the other hand, showed a mean range of only about 30 semitones per second for that syllable. In other words, in normal subjects some affective conditions occasion a very steep F_0 slope for the fourth syllable of the utterance, whereas other conditions produce a very shallow slope. In contrast, for the patients, the difference between the steepest and the least steep pitch drop was much less severe. The experimenters emphasize that nothing in the patients' tonal production "could be construed as aphasic" (444)—that is, there is nothing grammatically wrong with the patients' tonal competence. Interestingly too, the durational emotional range measure is overwhelmingly the largest for the sentence-final syllable, in patients as well as normals subjects, consistent with the fact that lengthening effects of all kinds are enhanced in the final position. One presumes, therefore, that the patients, like the control subjects, preserve final lengthening. Unlike Broca's aphasics, therefore, these patients have preserved grammatically controlled aspects of prosody; their impairment consists in their inability to affectively vary prosodic factors. In this respect these patients are rather similar to text-to-speech systems, which currently have little ability to compute from text appropriate affective use of prosody.

Computational morphology

The analysis of word structure provides another rich ground for the computational interpretation of psycholinguistic and neurolinguistic analyses. In this section I examine some neurolinguistic and psycholinguistic findings on morphology and propose some computational interpretations.

DO PEOPLE MORPHOLOGICALLY DECOMPOSE WORDS? Computational morphological models assume that part of the analysis of morphologically complex words involves morphological decomposition: In order to analyze a word such as *fishiness*, one often wants to break it into its component morphemes, *fish*, *-y*, and *-ness*. The details of which particular types of morphology are handled vary between different systems, depending partly upon the intended purpose of the program. Some programs may need to analyze the internal structure of compounds like *doghouse*, or deal with derivational morphology such as that found in *fishiness*; others may only handle inflectional morphology, such as noun plurals like *fishes*. But there is generally no disagreement on the necessity of doing some morphological decomposition. I will return momentarily to the practical reasons for that conclusion, but first let us consider the question of what neurolinguists and psycholinguists have to say on the matter.

Models of human lexical access can be distinguished roughly on the basis of whether they assume that people store morphologically complex words as wholes or that we store them in morphologically decomposed form and retrieve them using (some amount of) morphological analysis. Proponents of the former view differ on how they assume the various morphologically nondecomposed words to be stored in the mental lexicon. One of the more oft-cited versions of this approach is work of Lukatela et al. (1980) on Serbo-Croatian noun paradigms. Serbo-Croatian nouns inflect for seven different cases and two numbers. In a lexical-decision task (a task in which the subject is to decide whether a given stimulus is a word), speakers of Serbo-Croatian were presented with nouns in the nominative singular, genitive singular, and instrumental singular forms. According to Lukatela and colleagues (1980), if one takes a simple model of the mental lexicon wherein individual word forms are listed as completely unrelated lexical entries, then one would expect lexical access to be via the (unanalyzed) fully inflected form of the word. Now, Serbo-Croatian noun paradigms are replete with syncretism, whereby several slots in the paradigm may be filled with a single form: The noun *žena* 'woman' appears as *žene* in the genitive singular, nominative plural, and accusative plural. As a consequence the form *žene* is more frequent than the form *žena*—even though the genitive itself is less frequent than the nominative—and thus by the frequency effect (see Bradley, 1978), we would expect faster reaction times for *žene* than for the nominative *žena*. However, the subjects recognized the nominative singular of the word much more rapidly than the other case forms, refuting the simplistic model. But the researchers also cannot accept the view that lexical access for forms like *žena* (nominative) and *žene* (genitive) involves morphological decomposition. Both words are bimorphemic, consisting of a stem (*žen*) and a case affix. If morphological decomposition were necessary to recognize Serbo-Croatian inflected nouns, both nominative and genitive singular feminine nouns ought to require the same amount of processing, and the reaction times should therefore be the same. As an alternative, Lukatela et al. propose the "satellite entries" model, whereby nouns are stored fully formed in clusters, with the nominative in the center of the cluster and the other case forms arranged as satellites around the center of the cluster. Inflected nouns are accessed via the nominative form; if the form being looked up is in the nominative, then there is a direct hit and the lookup succeeds rapidly; if it is another case form, then search continues among the satellites of the cluster, which will in turn mean a longer search. This model, then, apparently predicts the results that the experimenters found.

Yet, there are a couple of problems with the view of Lukatela et al. when it is evaluated from a computational perspective. The first is purely mechanical: For an inflected noun such as *žene* (genitive), how exactly does lookup via the *nominative* form proceed? Evidently *žene* is not the same form as *žena*, so a simple match will not show that the cluster centered at *žena* is the correct cluster to be searching. Fortunately, the two words share the substring *žen* (the stem of the word) and one might therefore imagine that looking up the word *žene* involves comparing its stem with the stem of the nominative entry in the cluster for *žena*. But notice that this requires some amount of morphological analysis—stripping or ignoring the suffix *-e*. In other case forms, even more material must be ignored, as with *dinarima*, dative plural of *dinar* 'money'. So, it is hard to see how the satellite entries model can really function without allowing for *some* morphological decomposition in the lexical access process. The model, while perhaps psychologically appealing, is evidently not well conceived computationally.

More problematic is the assumption that morphologically complex words can generally be expected to be listed in the lexicon. The problem with that view is simply that there are too many words, and it is not practical to assume that one is going to find all of them sitting in the lexicon ready for access as whole words. As I argued in Sproat 1992, this point can be made even in English, which is often claimed to be morphologically impoverished. But the case can be made even more strongly in a heavily inflected language like Serbo-Croatian, where almost any word that is not a function word can appear in many different forms. Recall that Serbo-Croatian nouns are inflected for seven cases and two numbers. Assuming a reasonably sized lexicon of 35,000 nouns, half a million entries would be required to list all of the case and number forms for nouns alone; the situation would be even worse in languages like Turkish or Finnish, whose noun paradigms have even more forms than do those of Serbo-Croatian. Not surprisingly, and especially for highly inflected languages, it has long been recognized in the computational linguistics community that some amount of on-line morphological analysis is necessary if one expects to handle all the words that one will encounter. While it is conceivable that humans are not limited by the same kinds of practical limitations as machines, the onus is on those who would claim that people store all morphologically complex words to propose a viable mechanism for this approach.

Fortunately, I do not think that data reported by Lukatela et al. are at odds with the view that speakers often (if not always) parse morphologically complex words during recognition. To see this, consider that in many languages, including Serbo-Croatian, the nominative singular of a noun is the "citation form": It is the form of the word that one will generally find listed in a printed dictionary. Being the citation form, it is the form that one would expect to see if a noun were presented in isolation, and this is precisely how nouns were presented to subjects in the experiments by Lukatela and colleagues. Since nominative case is expected in this context, one can also assume that it is primed, and that, therefore, subjects would react more quickly to nouns in the nominative merely because nominative case itself is primed by the context. Presumably, if one were to present words in a different context—for example, as the object of a verb—where some other case is appropriate, then one would still expect subjects to react more rapidly to words in the contextually appropriate case—here, something other than the nominative. If this is correct, then nominative case has no special status; it is simply accessed more rapidly in those contexts where nominative is expected. One could then have a model wherein *žena* and *žene* are both morphologically decomposed during lexical access, and still obtain the results observed by Lukatela et al.

FINITE-STATE MORPHOLOGY AND NEUROLINGUISTICS: STORING WORDS IN DECOMPOSED FORM In contrast to the approach of Lukatela et al., other researchers have proposed that at least some words are morphologically decomposed, or at least are stored in morphologically decomposed form. One such recent proposal is advanced by Miceli and Caramazza (1988), who describe the morphological problems of a severely aphasic Italian-speaking patient, F. S., with a large left-hemisphere lesion. Partly as a consequence of his morphological deficit, F. S. performed poorly at word repetition. When presented with a list of inflected verbs and derived nouns and adjectives, F. S. was able to correctly repeat only 33.4% of the inflected words, and the same percentage of the derived words. For the incorrectly repeated inflected words, all of the morphological errors made were errors in inflection. In the case of derived words, most of the errors were errors in inflection, although there were a few errors in derivation as well.

Inflectional errors consisted of substituting a different, though legal, inflectional ending for the target—for example, *legg* + *eva* (read + Imperfect3Sg) 'he was reading' in place of *legg* + *ere* (read + Infinitive) 'to read'; or in other cases substituting impossible inflectional endings—for example, substituting for a target consisting of the the first conjugation infinitive form *aspett* + *are* (wait + Infinitive) 'to wait' a form consisting of the stem plus a third conjugation imperfect suffix **aspett* + *iva* (wait + Imperfect3Sg). Derivational errors consisted in the substitution of derivational endings different from the intended ones, for example, *pitt* + *ore* 'painter' instead of *pitt* + *ura* '(a) painting'.

Miceli and Caramazza draw two conclusions from their results. First, since inflectional morphology is much more prone to error in F. S.'s productions than derivational morphology, they conclude that inflectional and derivational morphology are represented in different components of the morphology, and that the lesion has damaged the inflectional component. This, they suggest, makes their findings consistent with linguistic theories of morphology that advocate a separation of inflectional and derivational morphology (e.g., Anderson, 1982). Second, they argue for a model in which "lexical items are represented in morphologically decomposed form" (p. 55). Their main reason for preferring this to, say, a version of the satellite entries hypothesis is similar to the conclusion arrived at in my earlier discussion of the work of Lukatela et al.—namely, that such a scheme does not obviously allow for novel, morphologically well-formed constructions.

What might it mean computationally to say that words are stored in morphologically decomposed form? Is the lexicon basically a list of entries annotated with their morphological analyses? It may make sense to think of things in another way, whereby the representation that is used to decompose novel word forms is exactly the same representation in which known words may also be stored. As an example, consider the finite-state model of morphology in the KIMMO morphological analyzer (Koskenniemi, 1983) and its derivaties—probably the best-known and most widely used approach to computational morphology at the present time. (Of course, discussion of KIMMO here should not be taken as an endorsement of the psychological reality of this system; see Sproat, 1992.) Monomorphemic entries in the KIMMO system are represented using *letter trees*, or *tries* (Knuth, 1973). Accessing a monomorphemic entry involves starting at the root of the lexicon and traversing the lexicon tree one letter (or phoneme) at a time. If one runs out of input letters

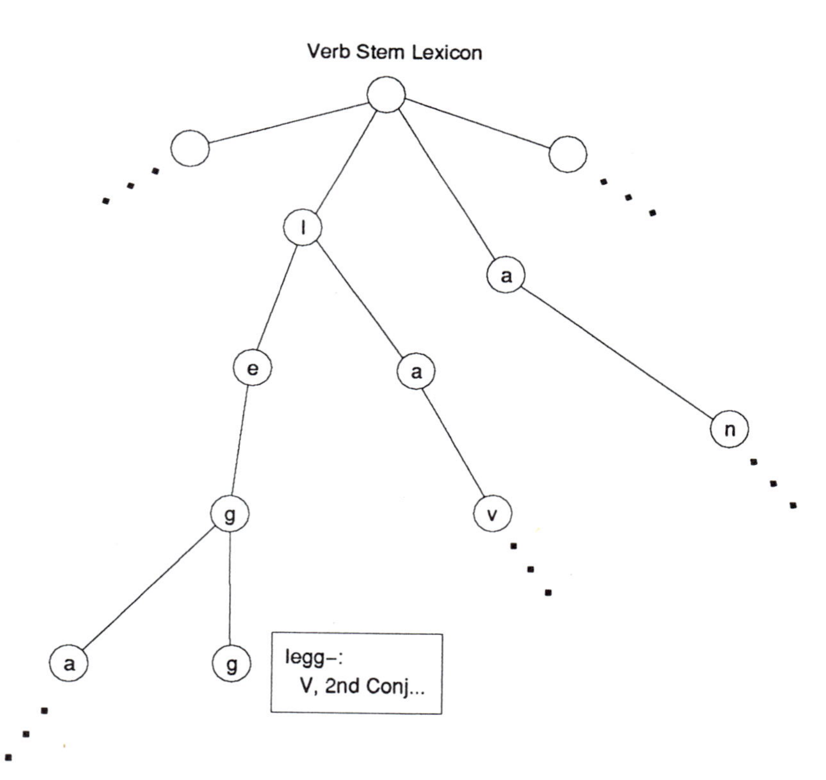

FIGURE 60.2 A partial letter tree for the Italian verb-stem lexicon. *Legg* would be accessed by first matching the *l* of the input to the node labeled *l* in the tree; then input *e* would be matched against node *e*, and so on.

and simultaneously lands on a node that is labeled as a lexical entry, then lexical access is successful; otherwise it fails. The diagram in figure 60.2 shows a letter-tree representation for an Italian verb-stem lexicon, and the steps involved in accessing the stem *legg-* 'read'.

Lexical decomposition within the KIMMO approach is a straightforward extension of this model. Part of the lexical entry for a morpheme is what Koskenniemi terms the continuation pattern, consisting of the names of zero or more other letter-tree lexicons where the lookup can continue. For example, part of the process of looking up the Italian word *legg + eva* '(s)he was reading' involves looking in the verb-stem lexicon and finding the stem *legg-*. In order to proceed from there, one has to see which other (suffix) lexicons may be opened in order to continue the search. *Leggere* is a second-conjugation verb in Italian, so the stem *legg-* would minimally contain the information instructing one to search in the second-conjugation verbal suffix lexicon for continuations of the stem *legg-*. Figure 60.3 shows the lookup of *leggeva* under such a scheme.

The simplistic model that I have just described has at least three properties that are relevant to neuro- and psycholinguistic models of morphology. The first is that, although a fully formed word like *leggeva* is not listed per se, it and all other morphologically well-formed words can be said to be virtually represented in the lexicon (Sproat, 1992, 131). This, I claim, is a reasonable computational interpretation of the notion that the lexicon consists of morphologically decomposed lexical entries. But there is a consequence of this model that may not be so desirable from a psychological point of view, and that is the second property: As it stands, the KIMMO model makes no representational distinction between previously seen morphologically complex words and words that are novel (to the speaker-hearer) though clearly morphologically possible. Within psycholinguistics, it is often assumed that words that have been seen before are represented in the mental lexicon in a way that unseen (though possible) words are not; consider the augmented addressed morphology model (see Caramazza et al., 1988), wherein both the full form of a morphologically complex word (e.g., *walked*) and its constituent parts (*walk*, *-ed*) are active during lexical access, but in the case of normal lexical access of previously seen words, morphological decomposition is generally not performed, and the word is instead accessed via the fully formed entry.

The psychological drawback of finite-state morphological models such as KIMMO could be remedied by assuming that the connections between lexicons have costs associated with them such that the cost is inversely proportional to the frequency of the transition in the data on which the system was trained. Morphologically complex words that have been seen many times would still be represented in morphologically decomposed form, but the cost associated with traversal of the network would be lower than for a fairly or completely novel word. Some sort of trainable weighting scheme is desirable for purely computational reasons anyway. Morphological analyzers such as KIMMO return all legal analyses for a word, but in many applications one does not want all possible analyses, but rather the one that is contextually most likely. Cost schemes, when implemented, have traditionally involved ad hoc hand assignment of weights; for example, the DECOMP morphological analysis module of the the MITalk text-to-speech system (Allen, Hunnicutt, and Klatt, 1987). A more principled, data-driven method is desirable, and recently some promising work along these lines has been reported (Heemskerk, 1993).

The third property of KIMMO-type morphological analysis relates to the main claim of Miceli and Caramazza's (1988) paper, namely that there is a difference between the lexical representation of derivation and inflection. F. S.'s derivational errors were few when compared with his inflectional errors, so whatever is "broken" in F. S.'s morphological performance affects inflectional morphology but has a much less pronounced effect on derivational morphology. The most straightforward explanation is that inflection and derivation are in separate components and that the break has occured within the component associated with inflectional morphology. However, KIMMO-style morphological models—and at least some linguistic models of morphology (see Di Sciullo and Williams, 1987)—make no formal distinction between derivation and inflection. If one wants to represent the morphological decomposition of *pitt + ore* 'painter', one would do it in the KIMMO system in exactly the same way as was outlined for some inflectional cases discussed earlier. But the crux of the matter lies in the conditional in the last sentence: *If* one were to represent the morphological decomposition of *pittore*, then one would do so as described, but would one necessarily want to

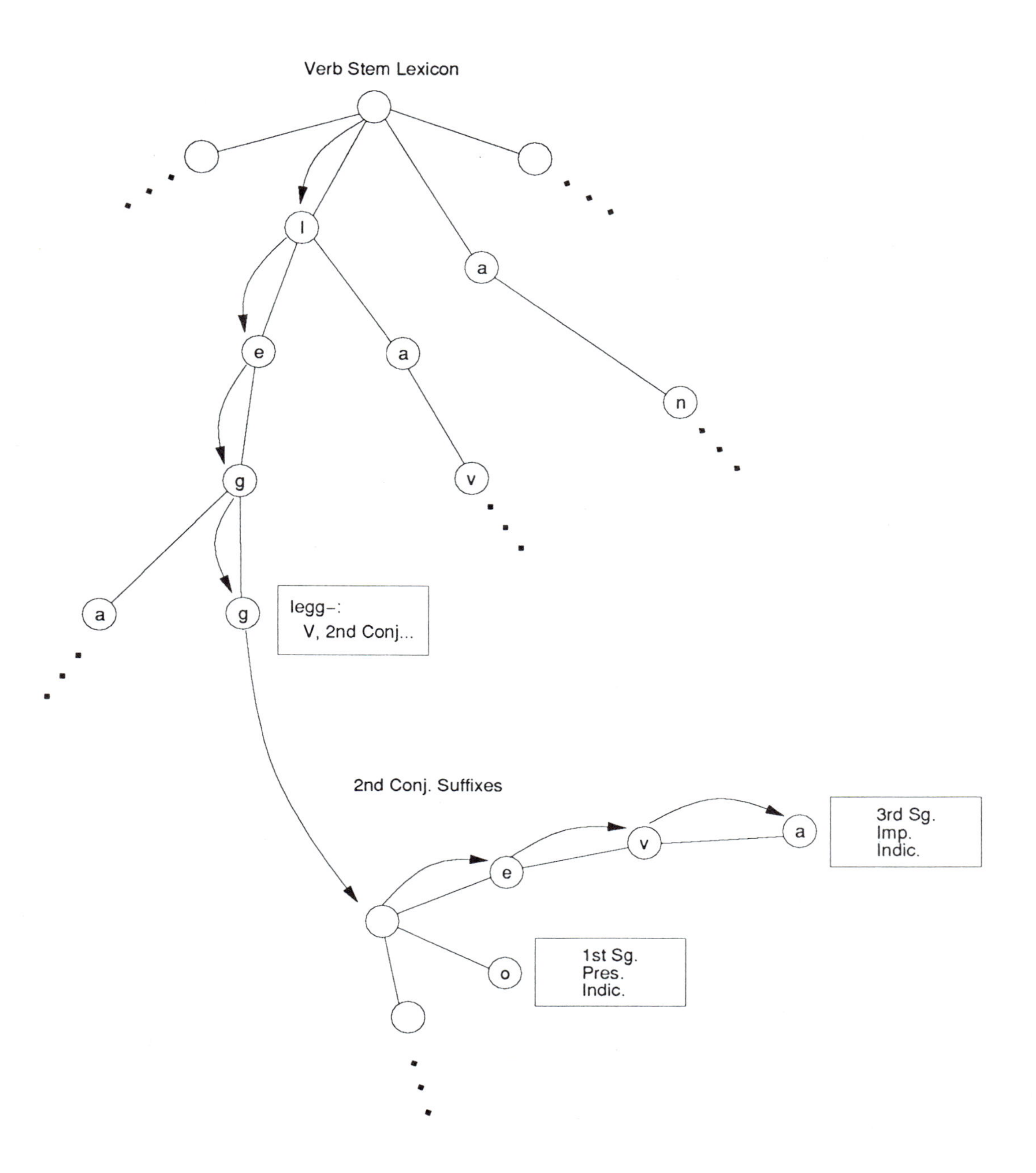

FIGURE 60.3 Lexical access of the Italian inflected verb form *leggeva* '(s)he was reading'.

decompose this word morphologically as part of its normal lexical access? Certainly Miceli and Caramazza are correct in noting that derivational processes must be available to handle novel words. But one characteristic of many (though not all) derivational processes that distinguishes them from many (though not all) inflectional processes is their idiosyncrascy: Simply put, many derived words probably must be listed as wholes in the lexicon for no other reason than that they have acquired excess baggage in the form of idiosyncratic and thus unpredictable interpretations. One English example is *payable*, which usually means "must be paid" rather than "may be paid," the literal meaning. F. S.'s behavior may therefore simply be a consequence of the fact that many derived words really are not actively decomposed, and thus the chance of error of a purely morphological kind is severely reduced. Interestingly, this suggestion makes a different prediction than the one Miceli and Caramazza would appear to be making. Under their proposal, given that derivation and inflection occur in two different boxes of the morphology, then one might expect to find a patient exactly like F. S., except that the break would be in the derivation box. Such a patient's morphological errors would consist mostly of misselection of derivational affixes; there would be very few errors in inflectional morphology. On the alternative view just presented, such a patient is not expected. I am not aware of the existence of precisely that set of symptoms, and thus the view that inflection and derivation are simply two sub-boxes of the morphological component seems unsatisfactory. (One might expect patients with general memory problems, such as Alzheimer's patients, to make many errors in retrieving stored derivational forms, and fewer errors in productive inflected forms. Note, though, that Alzheimer's patients also make errors in irregular inflected forms, such as *caught*. I thank Steve Pinker for this example.)

In any event, I would be surprised if the neurolinguistic distinction between inflection and derivation turned out to be a categorical one: Some derivational affixes—such as *-ness* in English—are much more productive than others—such as *-dom*—and one would expect to find a greater number of errors in the more productive affixes for patients like F. S. A priori classifications, such as inflection versus derivation, should be used with caution.

In this chapter I have examined a few neurolinguistic and psycholinguistic phenomena, and have interpreted them within the context of recent computational work. It should be remembered, of course, that the limitations on computational methods that I have assumed are only *current* limitations: It is entirely possible, for example, that we will eventually be able to build TTS systems that are able to make appropriate affective use of prosody. Be that as it may, it is interesting that properties of current computational systems seem to shed some light on studies of human performance.

ACKNOWLEDGMENTS I wish to thank Steve Pinker, Jerry Packard, and Chilin Shih for useful comments on previous versions of this chapter.

REFERENCES

ALLEN, J., M. S. HUNNICUTT, and D. KLATT, 1987. *From Text to Speech: The MITalk System*. Cambridge: Cambridge University Press.

ANDERSON, M., J. PIERREHUMBERT, and M. LIBERMAN, 1984. Synthesis by rule of English intonation patterns. *Proceedings of the IEEE International Conference on Acoustic, Speech, and Signal Processing*. 2.8.2–2.8.4. New York: IEEE.

ANDERSON, S., 1982. Where's morphology? *Linguistic Inq*. 13: 571–612.

BLUMSTEIN, S., 1973. *A Phonological Investigation of Aphasic Speech*. The Hague: Mouton.

BOLINGER, D., 1972. Accent is predictable (if you're a mind-reader). *Language* 48:633–644.

BRADLEY, D., 1978. Computational distinctions of vocabulary type. Ph.D. diss., Massachusetts Institute of Technology.

CARAMAZZA, A., A. LAUDANNA, and C. ROMANI, 1988. Lexical access and inflectional morphology. *Cognition* 28: 297–332.

DANLY, M., and B. SHAPIRO, 1982. Speech prosody in Broca's aphasia. *Brain Lang*. 16:171–190.

DI SCIULLO, A-M., and E. WILLIAMS, 1987. *On the Definition of Word*. Cambridge, Mass.: MIT Press.

HEEMSKERK, J., 1993. A probabilistic context-free grammar for disambiguation in morphological parsing. *Proceedings of the Sixth European Chapter of the Association for Computational Linguistics*. 183–192.

HIRSCHBERG, J., 1993. Pitch accent in context: Predicting intonational prominence from text. *Artif. Intell*. 63:1–2.

KLATT, D., 1987. Review of text-to-speech conversion for English. *J. Acoust. Soc. Am*. 82:737–793.

KNUTH, D., 1973. *The Art of Computer Programming*, vol. 3. Reading, Mass.: Addison-Wesley.

KOSKENNIEMI, K., 1983. Two-level morphology: A general computational model for word-form recognition and production. Ph.D. diss., University of Helsinki.

LIBERMAN, M., and J. PIERREHUMBERT, 1984. Intonational invariance under changes in pitch range and length. In *Language Sound Structure: Studies in Phonology Presented to Morris Halle*, M. Aronoff and R. Oehrle, eds. Cambridge, Mass.: MIT Press.

LUKATELA, G., B. GLIGORIJEVIĆ, A. KOSTIĆ, and M. TURVEY, 1980. Representation of inflected nouns in the internal lexicon. *Mem. Cognition* 8:415–423.

MICELI, G., and A. CARAMAZZA, 1988. Dissociation of inflectional and derivational morphology. *Brain Lang*. 35:24–65.

PACKARD, J., 1986. Tone production deficits in nonfluent aphasic Chinese speech. *Brain Lang*. 29:212–223.

Pierrehumbert, J., 1980. The Phonology and Phonetics of English Intonation. PhD dissertation, MIT, Cambridge, Mass.

Ross, E., J. Edmondson, G. Seibert, and J.-L. Chan, 1992. Affective exploitation of tone in Taiwanese: An acoustical study of "tone latitude". *J. Phonetics* 20:441–456.

Sproat, R., 1992. *Morphology and Computation.* Cambridge, Mass.: MIT Press.

Sproat, R., J. Hirschberg, and D. Yarowsky, 1992. A corpus-based synthesizer. *Proceedings of the Second International Conference on Speech and Language Processing*, Banff, Alberta: University of Alberta, 563–566.

Van Santen, J., 1992. Contextual effects on vowel duration. *Speech Commun.* 11:513–546.

Ward, G., and J. Hirschberg, 1985. Implicating uncertainty: The pragmatics of fall-rise intonation. *Language* 61:747–776.

61 Maturation and Learning of Language in the First Year of Life

JACQUES MEHLER AND ANNE CHRISTOPHE

ABSTRACT Language, a species-specific aptitude, seems to be acquired by selection from a set of innate dispositions. In the first part of this chapter, we assess the cortical substrate of speech processing. We show that even though, from birth on, speech is preferentially processed in the left hemisphere, the underlying cortical structures continue to specialize over the period of language acquisition. In the remainder of the chapter, we illustrate a number of innate mechanisms that help the baby to learn language. Infants are able to segregate utterances according to their source language; they are able to segment speech streams into basic units that may be different for contrasting languages. Throughout the chapter we focus on the processes that enable the infant to go from universal capacities to language-specific ones. The general hypothesis that acquisition proceeds by selection is supported by data.

What advantages can an infant possibly draw from listening to speech? One could imagine that, at birth, the infant neglects speech, until interest in learning words and the syntactic rules linking words in sentences begins at around 12 months. If this is the case, one might as well ignore what infants do during the first few months of life and start to explain language acquisition at the point when the acquisition of the lexicon begins. Alternatively, one might imagine that the infant, at birth, has to pay attention to incoming speech signals. Language acquisition may begin with the establishment of some important properties of the language, which are reflected in the speech signal. This view presupposes that the infant interacts in a predetermined way with speech stimuli, in much the way other animals interact with signals that are specific to their own species.

JACQUES MEHLER and ANNE CHRISTOPHE Laboratoire de Sciences Cognitives et Psycholinguistique, CNRS-EHESS, Paris, France

We hypothesize that during their first encounters with speech infants determine how many linguistic systems coexist in their surroundings, and moreover that they make the necessary adjustments for the acquisition of the maternal language. Chomsky (1988) has proposed a model of how this happens and addresses the issue of how infants can be set to learn different languages:

> The principles of universal grammar have certain parameters, which can be fixed by experience in one or another way. We may think of the language faculty as a complex and intricate network of some sort associated with a switch box consisting of an array of switches that can be in one of two positions. Unless the switches are set one way or another, the system does not function. When they are set in one of the permissible ways, then the system functions in accordance with its nature, but differently, depending on how the switches are set. The fixed network is the system of principles of universal grammar; the switches are the parameters to be fixed by experience. The data presented to the child learning the language must suffice to set the switches one way or another. When the switches are set, the child has command of a particular language and knows the facts of that language: that a particular expression has a particular meaning and so on.... Language learning, then, is the process of determining the values of the parameters left unspecified by universal grammar, of setting the switches that make the network function.... Beyond that, the language learner must discover the lexical items of the language and their properties.... Language learning is not really something that the child does; it is something that happens to the child placed in an appropriate environment, much as the child's body grows and matures in a predetermined way when provided with appropriate nutrition and environmental stimulation. (Chomsky, 1988, 62–63, 134)

Chomsky and most of his colleagues, even if they believe that the setting of parameters accounts for most aspects of language acquisition, have focused on how it applies to syntax. Our own investigation focuses on how initial adjustments apply to the most external aspect of language, namely, its sound structure. Below, we investigate how infants behave when confronted with speech signals and explore when and how they parse and represent language.

If language is a species-specific attribute, which are the specific neurological structures that insure its emergence in every member of the species? As far back as Gall (1835), neurologists have suspected that most species-specific attributes are the expression of dedicated organs. Although we know this to be the case (see Knudsen and Konishi, 1978; Nottebohm, 1981; Geschwind and Galaburda, 1987, for some abilities) after decades of research, it is still difficult to locate the "language organ" precisely (see Ojemann, 1983).

The search for the neurological structures underlying language is still in progress. Pathologies described by Gopnik and her colleagues (Gopnik, 1990; Gopnik and Crago, 1991), show that certain kinds of developmental dysphasia (difficulties or delays in learning language) run in families with a distribution that is consistent with a single autosomal dominant mutation. Williams syndrome children, described by Bellugi and her colleagues (see, e.g., Bellugi et al., 1992), are children who have some cognitive deficits but relatively spared linguistic abilities. Indeed, they have very poor spatial abilities, and their capacity to do even the simplest arithmetical operations is impaired. In contrast, their syntactic and lexical abilities are exceptionally good. Their behaviors contrast markedly with those of children with Down's syndrome, suggesting that the behavioral differences arise because of genetic differences between these populations. Moreover, it is striking that Curtiss and Yamada (1981) also found children who, despite impressive language abilities, had greatly impaired cognitive abilities. These studies show that language acquisition is marginally related to general intelligence, which would be a necessary condition for learning by inductive inferences and hypothesis testing. Likewise, Klima and Bellugi (1979) and Poizner, Bellugi, and Klima (1987) have shown that deaf children acquire sign language in the manner that hearing children learn to speak. Landau and Gleitman (1985) have shown that blind people will acquire a spoken language despite their major sensory deprivation. Both deaf and blind children can acquire language with astonishing ease. If the surroundings are sufficiently congenial to ease the child's predicament, the road to language can be as swift and smooth as that of any other child. The evidence suggests that language arises because of specific neural structures that are part of the species' endowment. Neuropsychology teaches that language is basically lateralized in the left hemisphere. But do these structures arise from the child's acquisition of language, or are they responsible for the acquisition of language in the first place?

What comes first, language or lateralization?

Lenneberg (1967) argues that at birth the brain is equipotential and that lateralization arises as a by-product of language acquisition. Lenneberg proposes equipotentiality to accommodate the early findings that in younger children, both lesions to the left hemisphere and lesions to the right hemisphere may result in aphasia that is transitory unless the lesions are really very great. When they are not, language tends to be acquired or reacquired a few weeks after it was affected. It is only after puberty that impairments from small focal lesions to the left hemisphere become irreversible, and also that aphasias almost never arise after focal lesions to the right hemisphere. Today, new investigations make it difficult to accept that the infant's brain is nonlateralized at birth. Before presenting infant data it is necessary to certify that speech comprehension is mostly a left-hemisphere aptitude. A new technique, positron emission tomography (PET), makes it possible to study how the cortex processes not only sentences in one's maternal language but also sentences in a language one does not understand.

Most of the classical investigations with PET have focused on cortical activation while listening to a list of words (see Howard et al., 1992) and little, if any, research has focused on the cortical activation that arises while listening to connected speech sounds. The situation has been corrected by Mazoyer et al. (1993), who have used PET combined with sophisticated individual anatomical and brain-imaging techniques to study the activation of cortical areas in normal subjects who listen to connected speech signals. The activation in the left hemisphere increases as the signals get progressively closer to a normal story in the subjects' mother tongue. Color plate 21 reproduces the activation in subjects who were listening to one of five conditions: a story in their native language (French), a list of French words, a story that was syntactically adequate but had no meaning whatsoever (the nouns and verbs of the original story were replaced by randomly chosen words from the same category), a "story" in which the nouns and verbs were nonexistent (but that were phonologically legal in French), and a story in an unknown language, Tamil (recorded by a very proficient French-Tamil bilingual who also recorded the story in French). By and large, this study endorses the view that the left hemisphere is more actively engaged when processing standard connected speech than the right. Of course, our findings suggest that this is so only when one listens to a familiar language. The activation in the left hemisphere observed for French but not for Tamil could reflect the fact that the subjects understand French or the fact that French is their mother tongue. These results address a classical concern, namely, whether left-hemisphere superiority emerges through the learning of language. The brain-imaging results show that adults who listen to speech in a language they do not understand, have a left hemisphere that remains unactivated, as compared to the resting baseline.[1]

Consider what happens in the brains of infants who do not understand either French, Tamil, or, for that matter, any other language. Does this mean that at birth, or shortly thereafter, no functional asymmetry will be observable? Is the brain of the neonate activated in the same way by any speech sound that reaches its ears, and if so, does this activation show a pattern that is characteristic for speech stimuli and only for them? Or could it be that speech sounds, regardless of language, mostly activate the neonate's left hemisphere?

Entus (1977) adapted the dichotic presentation technique (i.e., the simultaneous presentation of two stimuli, one to each ear) and found a right-ear advantage for syllables and a left-ear advantage for musical stimuli in 4-month-old infants. Vargha-Khadem and Corballis (1979) were unable to replicate Entus's finding. Bertoncini et al. (1989) carried out a new investigation with younger infants (4-day-olds) and better stimuli. As can be seen in figure 61.1, Bertoncini et al. found that neonates have a right-ear advantage for syllables and a left-ear advantage for musiclike noises of otherwise similar acoustic characteristics. This result is comparable to that reported by Best (1988). However, Best also reports that 2-month-old infants show a left-ear advantage for musiclike stimuli but no right-ear advantage for syllables. This minor discrepancy is likely to reflect the sensitivity of the methods or the

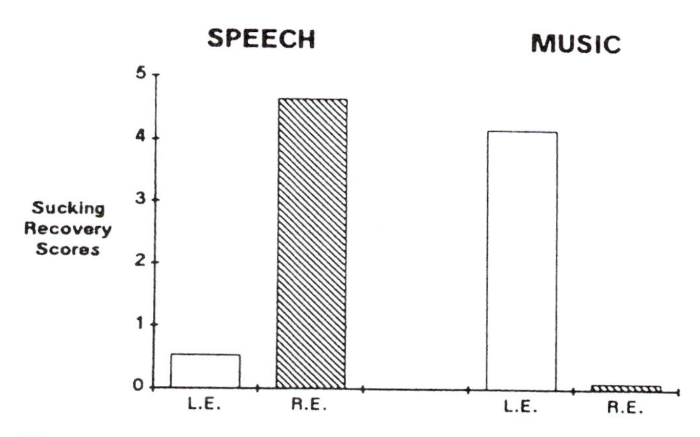

FIGURE 61.1 Sucking recovery scores (average of the first 2 postshift min minus average of the last 2 preshift min) as a function of the ear receiving the change (left ear or right ear) in the speech and music conditions. (From Bertoncini et al., 1989, with permission)

response to the contrasting stimuli used by each group. In any case, both studies agree that very young infants display ear advantages that are asymmetric as a function of stimuli. Today the consensus seems to be that the greater engagement of the left hemisphere with speech originates in our genetic endowment rather than occurring as a side effect of language learning. These findings clarify one of the central issues concerning the biological foundations of language. However, we can go one step further.

We are considering the hypothesis that, at birth, speech stimuli, regardless of the language from which they stem, activate the left hemisphere, whereas later in life, the only speech stimuli that can activate the left hemisphere must either belong to the maternal language or, possibly, to a second language. This raises the possibility that the mother tongue has a special status and that language learning consists not only in the dismissal of acoustic stimuli that are nonlinguistic in nature, but also, possibly, in the dismissal of all speech stimuli that do not belong to one's mother tongue. If so, a related question arises in relation to the representation of speech in bilinguals. This hypothesis, as we shall see later, is compatible with the sparse data now available and seems to mesh well with evidence about the behavioral changes one observes at the end of the first year of life.[2]

In our review of behavioral data about the tuning of infants toward their maternal language phonology, we will examine in turn how infants treat and represent speech at different levels (the utterance, prosodic, prelexical, and segmental levels) and for each level, investigate how infants' performance depends on their maternal language. The evidence we will present suggests that during acquisition the infant not only privileges some properties (possibly setting some phonological parameters) but also learns how to neglect aspects that are not relevant to the language he or she is trying to master.

Discovering the languages of the world

In order to acquire language, a first prerequisite that infants have to meet is that they should be able to sort linguistic stimuli from nonlinguistic ones: They have to attribute meaning and linguistic function to words, not to dog barks or telephone rings. Above, we showed that the infant processes speech-stimuli asymmetrically and in areas that are classically related to speech-processing centers in adults. This suggests that somehow the human brain is capable from birth of sorting stimuli according to whether they are speechlike or not. Moreover, Colombo and Bundy (1983) showed that 4-month-old infants prefer to listen to words than to other sounds. This experiment is suggestive, though not conclusive; indeed, infants' preferences may depend on the choice of the nonspeech stimuli; for example, animal noises may generate more interest than speech. Sturdier evidence has recently been reported by Eimas and Miller (1992): They showed that 4-month-old infants behave as if they experienced duplex perception, a phenomenon that in adults has been taken to mean that speech is processed in specific ways (Liberman, 1982; Whalen and Liberman, 1987). Eimas and Miller suggest that there is a speech mode that is engaged when processing speech stimuli even in 3- to 4-month-old infants.

However, even though there is good evidence that infants can distinguish speech from nonlinguistic stimuli, they still face a momentous problem. We argued in the introduction that the only way infants could possibly acquire language is by making some choices among a set of finite, innately determined possibilities. Indeed, the input with which they are presented is not rich enough to allow them to build the language structure from scratch. But, according to current statistics, most children listen to more than a single language in their immediate surroundings. How can infants successfully converge toward any given language, that is, correctly select the options adequate for this language, if they indiscriminately use sentences from different languages with incompatible option settings? Yet, there seems to be little ground for worry. Indeed, there is strong converging evidence that the child has predispositions that allow it to cope with this problem in a smooth and efficient fashion.

Bahrick and Pickens (1988) have shown that 4-month-olds react to a change in language. When infants were habituated to an English sentence and then switched to a novel English utterance, they showed no recovery of interest, compared to infants that were switched to a (novel) Spanish utterance. Mehler and his colleagues (1988), using many different short utterances instead of only one, found that 4-day-olds tested in a French environment prefer to listen to their maternal language: They suck more when listening to French than when listening to Russian.[3] Moreover, 4-day-olds were able to discriminate the set of Russian utterances from the set of French utterances. Two-month-old American infants also discriminated English from Italian sentences, but in contrast to newborns, they displayed no preference for either English or Italian. Interestingly, the 2-month-old American infants failed to discriminate the Russian and French sentences that French newborns discriminated.

New results show that 4-day-old infants discriminate French from English sentences, corroborating the discrimination part of the previously reported results. However, the preference of newborn infants for their maternal language was not replicated. Mehler et al. (1988) also made the claim that 4-day-olds could not discriminate Italian from English nor English from Italian, two languages unfamiliar to the infants born to French parents. However, our published results show that both experimental groups have a tendency to increase their sucking rate during the postshift phase. The original claim was based upon the fact that neither experimental group exhibited a significant difference to its corresponding control. However, because there is no significant interaction between the groups and their effects, both groups can be considered together. When one assesses jointly the results for both groups, it appears that 4-day-olds discriminate the two unfamiliar languages without any difficulty. In contrast, at two months, American infants do not discriminate two unfamiliar languages (French and Russian), even when one combines the results for both experimental groups. This suggests that while younger infants still work on all the utterances to which they are exposed, the older ones already concentrate on utterances that share a structure corresponding to the maternal language and neglect utterances that do not. If so, one has to conclude that by the age of two months the infant has set the first values to individuate the structure of the maternal language.

What might these first adjustments rely upon? The above studies have shown that the main properties that infants pay attention to are carried by the lower 400 Hz of the spectrum. Indeed, when the utterances are reversed, infants do not respond to the change in language, although they still do when they are exposed to filtered speech that contains only the frequencies below 400 Hz. These observations allow us to conjecture that infants begin by paying attention mainly to the prosodic properties of speech (that is, overall properties of utterances such as intonation and rhythm). This allows them to classify inputs according to whether they are drawn from one or another natural language. Rapidly they extract a representation that captures the prosodic properties that characterize their maternal language. Recently, Moon, Panneton-Cooper, and Fifer (in press) have begun investigating the same question with a new experimental method that also relies on the sucking response, and is explicitly designed to test preference. They observed that infants of Spanish-speaking parents preferred Spanish over English, whereas infants of English-speaking parents showed the reverse pattern of preference. These results suggest that by two days of age infants already have acquired a sensitivity for some regularity of the maternal language.

Lambertz (in preparation) has recently adapted a method originally used to study visual perception (Johnson, Posner, and Rothbart, 1991) to assess 2-month-olds' preference for auditory stimuli. After attracting the infants' gaze to the center of a display by switching on a multicolor moving spiral pattern, Lambertz presents an auditory stimulus through either one of two loudspeakers that are located 30° to the right and to the left of the moving spiral. The experimental measure is the time the infant needs to initiate a visual saccade toward the source of the sound. Lambertz used as auditory stimuli French and English utterances that were less than 3 seconds long and were either intact or low-pass filtered. The results, illustrated in figure 61.2, show that American-born infants start reacting to English utterances significantly faster than they do to French utterances. This result is obtained regardless of whether the utterances are filtered or not. Lambertz's results thus show that 2-month-old infants, like newborns, display a preference for their maternal language when tested under appropriate conditions. Moreover, the fact that the preference holds for filtered speech confirms the fact that infants probably rely on some gross properties of utterances to categorize them (their prosody).

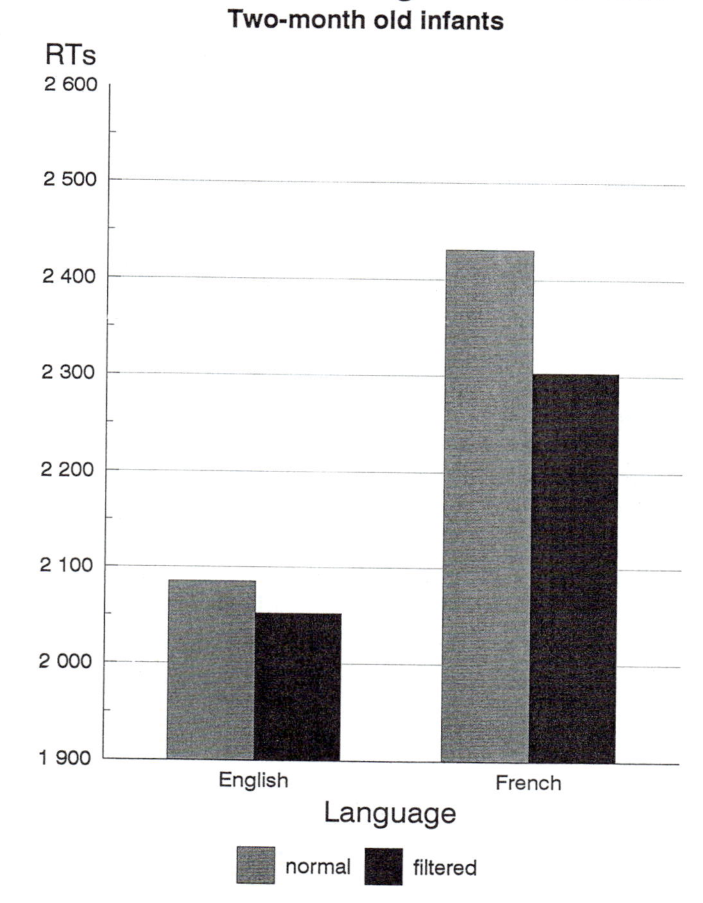

FIGURE 61.2 Mean latencies of initiation of a visual saccade in the direction of the sound for American 2-month-olds listening to French and English sentences that are either filtered or nonfiltered.

The picture that seems to arise from all these studies is that a few days after birth, infants are able to tell apart two different languages, even when neither of them is present in their environment; moreover, they already show a preference for their maternal language. Two months after birth, infants still show a preference for their maternal language (not surprisingly), but they seem to have lost the ability to distinguish two languages that are not familiar to them. These results seem to hold with low-pass-filtered speech stimuli, which indicates that infants rely on prosodic properties of speech to perform this task. This picture, which needs to be made more precise and confirmed, raises a number of captivating questions. For instance, what happens with children raised in bilingual environments? Do they show a preference for the two languages in their environment, or does one of them take precedence over the other? Also, when and how does left-hemisphere lateralization restrict itself to speech stimuli belonging to the maternal language? Last, what happens to deaf children who are not exposed to speech stimuli? Is it the case that signs play a role exactly equivalent to speech, benefiting from a special treatment from birth on, or is it only later, because of the absence of speech, that deaf infants realize that signs are the vehicle of language in their environment? Even though we still have no idea how to answer these questions, our original concern in starting these investigations is satisfactorily addressed: Infants do not get confused when exposed to several languages, because they know how to tell them apart. Even better, they seem already to know, a few days after birth, which language is going to be their maternal language.

Knowing how to distinguish languages is, of course, only the very first step in language acquisition. In the next section, we will turn to the question of how infants segment the continuous speech stream into linguistically relevant units.

How to find the assembling units of language

While speech is a continuous signal, language is discrete; indeed, we compute the meaning of sentences by operating on individual words. There is thus no doubt that upon hearing utterances, speakers of a language have to access individual words. If we now think about the task facing the infant hearing a language, it is clear that acquiring the lexicon is a necessary stage of learning. Indeed, the sounds languages use to refer to things are extremely idiosyncratic. There is nothing *kelef*-ish, *perro*-ish, or *hund*-ish about a dog, no more that there is something very *chien*-ish about it. Moreover, infants should be able to access the same representation regardless of who pronounces a word, the rate with which the word is spoken, whether it is whispered or screamed, said in a happy or sad intonation, and so forth. Mehler, Dupoux, and Segui (1990) have argued, on the basis of the preceding arguments, that infants should have access to a prelexical, normalized representation of speech. In the next section, we will examine what we currently know about such a representation. Moreover, Mehler, Dupoux, and Segui argue that infants should also be able to segment the continuous speech stream into word-sized units, in order to be able to memorize the word forms that will constitute their future lexicon. Two alternatives to this solution have been offered in the literature, both considered implausible by Christophe, Dupoux, and Mehler (1993). The first one states that all words to be learned are first heard in isolation. This implies that speech directed to infants has to be special, since adults rarely utter words in isolation. But then, mothers who—out of belief or distraction—neglected to speak in this special way would be responsible for their children's failure to learn language. We prefer to avoid such an implausible claim and explore alternative solutions. The second alternative (Hayes and Clark, 1970) assumes that infants are endowed with a mechanism that computes transition probabilities between successive prelexical units. In this view, word boundaries appear as troughs in the transitional probability function. However, if such a clustering mechanism might plausibly work on short strings of prelexical units, it is quite implausible for segmenting whole sentences into words, given the size of the lexicon.

Most parties accept that words have to be recovered from the speech stream. The same must be claimed for syntactic constituents. Indeed, infants do at some point have to acquire syntax. Mazuka (1993) has recently made a principled argument to support the view that even syntactic parameters should be set through prosody. She argues that syntactic parameters cannot be set by using the syntactic category of words, because parameters are supposed to be useful precisely to find categories. Instead, syntactic parameters should be set independently of syntactic analysis. Mazuka makes the assumption that prosody may inform the infant not only about where the units are, but also about their dominance relationships. More generally, Mazuka suggests that some syntactic parameters that are relevant for analysis of the fine-grained structure of language might be set by focusing on information available at the utterance level.

Christophe et al. (1994) attempted to assess the hypothesis that infants are able to segment the speech stream into word-sized units. They reasoned that if this hypothesis is correct, then infants should be able to discriminate bisyllabic contexts that contain a word boundary from bisyllabic contexts that do not. Indeed, they showed that four-day-old French infants were able to discriminate consonant-vowel-consonant-vowel strings (CVCVs) spliced out from the middle of a long word (e.g., *mati* in "mathé*mati*cien") from CVCVs spliced out from between two words (e.g., *mati* in "panora*ma ti*betain"). What could have been the basis of discrimination? Christophe and her colleagues measured a significant word-final vowel lengthening, which is not surprising since French is an accent-final language. They also measured a significant word-initial consonant lengthening, which has been observed in many different languages (for English, Umeda, 1976; for Dutch, Quené, 1991; for Czech, Lehiste, 1965; for Estonian, Lehiste, 1966; and for Italian and Swedish, see Vaissière, 1983). The fact that infants are sensitive to word boundaries should be true universally (for all languages of the world), even though different languages may use different boundary cues—indeed, in order to realize which cues signal which boundaries in their particular language, infants should at least perceive all potential cues. Other languages thus have to be investigated before we can conclude that infants are sensitive to word-boundary cues in general. But we already know that infants have the capacity to perceive potential word boundary cues in French.

Jusczyk and his colleagues have conducted a number of experiments directly aimed at evaluating which kind of units infants perceive in the signal. They presented infants with continuous speech samples that had been interrupted either at a boundary between two constituents or within a constituent. They used a modified preference-looking paradigm much like the one developed by Fernald (1985). Studies with 9-month-old, 6-month-old, and even $4\frac{1}{2}$-month-old American infants show that they prefer listening to speech that has been artificially interrupted between two clauses rather than within a clause (see Hirsh-Pasek et al., 1987). Moreover, $4\frac{1}{2}$-month-old American infants who are presented with Polish sentences also show a preference for the speech samples that preserve the clauses (Jusczyk, Kemler Nelson et al., in preparation). The authors suggest that the clause's integrity is signaled by both temporal and frequency cues, and that these cues may well be universal. Indeed, in languages as different as English and Japanese, clause boundaries are marked by a fall in pitch and a lengthening of the last segments of the clause (Fisher and Tokura, in press).

Jusczyk et al. (1992) investigated whether infants were also sensitive to units below the clause. They placed the pause either at the major syntactic boundary, namely between the subject and the verb, or at a minor syntactic boundary, generally between the verb and its complement. They observed that 9-month-olds but not 6-month-olds prefer listening to stories with the pause at the major syntactic break. However, in some conditions this major syntactic break, between subject and verb, does not correspond to a major prosodic break. In particular, when the subject is a pronoun, it tends to cliticize with the following verb (unless it bears emphatic stress); when this happens, subject and verb together form a clitic group, a very cohesive prosodic unit. Recently, Gerken, Jusczyk, and Mandel (1994) carried out an experiment explicitly designed for contrasting syntactic and prosodic units. They used the same stories where either the full noun phrase was repeated for each sentence (e.g., *the caterpillar*), or where it was replaced by a pronoun (*it*). Only in the first case did the major syntactic break between subject and verb correspond to a major prosodic boundary (a phonological phrase boundary). Unsurprisingly, the researchers found that 9-month-old infants were sensitive to prosodic units, but not to syntactic units per se. Thus, 9-month-old infants who heard sentences with lexical noun phrases preferred sentences with a pause inserted between the subject and the verb to the same sentences with the pause inserted between the verb and its complement. However, when the infants were tested with pronoun-subject sentences, they did not show a marked preference for either segmentation.

From this series of studies, one is tempted to conclude that infants perceive sentences as strings of clauses from a very early age. Moreover, they should do so in all languages worldwide, without any need for adaptation or tuning. Only at the age of 9 months is there positive evidence that infants hear phonological phrases. We still do not know at what age infants display sensitivity to clitic groups (formed by the grouping of a content word and its adjacent clitics, such as articles, pronouns, auxiliaries, etc.), but a reasonable bet given the present data would be even later than 9 months. But, as we will see now, there is evidence from other paradigms, using lists of words instead of long samples of continuous speech, that infants are probably sensitive to prosodic units smaller than the phonological phrase before 9 months.

Jusczyk, Friederici, and colleagues (1993) report that 6-month-old American infants show a preference for lists of unfamiliar English words as compared to Norwegian words uttered by a single bilingual speaker. However, these infants show no preference for lists of English as opposed to Dutch words. Moreover, these results hold when the words are low-pass filtered, indicating that the infants relied on the prosodic properties of the words. The authors stress the fact that the prosodic properties of English and Norwegian words are very different, while English and Dutch words are quite similar in their prosody. This study implies that 6-month-old infants already have a notion of what is a legal prosodic unit in their language, even though these prosodic units are only two syllables long.

Jusczyk, Cutler, and Redanz (1993) have raised a similar question in relation to stress. As Cutler and her colleagues have argued, most content words in English start with a strong syllable (Cutler and Carter, 1987), and adult English-speaking listeners make use of this regularity in their processing of continuous speech. Jusczyk, Cutler, and Redanz showed that American 9-month-olds, but not 6-month-olds, show a preference for lists of unfamiliar strong-weak words, as opposed to lists of unfamiliar weak-strong words. Again, the result holds when the words are low-pass filtered. This study may be taken as evidence that at the age of 9 months, infants have an idea of what is the most frequent prosodic structure in their native language. The implication is that before the age of 9 months, infants have some means of segmenting speech into word-sized units, which allows them to make statistics about a sufficient number of such units. Another implication, of course, is that from the age of 9 months infants may take advantage of this regularity in English.

It appears that the age at which sensitivity to smaller-sized prosodic units arises depends on the paradigm chosen. This should not be too surprising. When presented with long samples of continuous speech, infants' attention is likely to be focused on the larger-sized prosodic units (the intonational phrase, the phonological phrase). This would explain why they do not react to the disruption of smaller units such as clitic groups or phonological words. In contrast, presenting infants with lists of words focuses their attention on the smaller units. This may be why the results of the two word-list studies reported suggest sensitivity to smaller prosodic units before the age of 9 months, while the continuous-speech studies did not.

The studies reviewed in this section relate to how the child recovers specific cues to parse sentences, and recovers prosodic, syntactic, and semantic units. However, as we mentioned in the introduction of this section, the infant needs some prelexical, normalized representation of speech to work on. In the next section, we review what we currently know about such representations, with special emphasis on the question of their language specificity.

How is speech represented?

Psychologists and linguists have combined the study of adult competence for language with that of the properties of the human brain that are responsible for its emergence. In other words, they pursue the study of infants' abilities before learning has had the opportunity to leave an imprint, that is the initial state, and also the ability of adults who share a grammatical system, namely, a stable state (see Chomsky, 1965; Mehler and Bever, 1968). One of the major advantages of jointly considering the initial state and the stable state is best illustrated by focusing on how infants discover contrasting language phonologies and on how adults use them. Numerous studies have established that speakers of different languages process speech in ways that are suited to the phonology and prosodic structure of their maternal language. Mehler et al. (1981), Cutler et al. (1983, 1986), and Segui, Dupoux, and Mehler (1990) have shown that native speakers of French and of English rely upon different prelexical units to access the lexicon: French-speakers rely on syllables, whereas English-speakers rely on stressed syllables (see also Cutler and Norris, 1988; Cutler and Butterfield, 1992). More recently, Otake and colleagues (1993) have shown that the *mora* acts as the segmentation unit, or prelexical atom, for speakers of Japanese.[4] The aforementioned work is distinct from more standard conceptions in that it does not regard the phoneme as the necessary and unique device to represent the speech stream, and in its reflection of the notion that languages may use differing prelexical devices to segment and represent the speech stream. Furthermore, these authors argue that natural languages have rhythmic structures that are related to the "timing units": French would be syllable-timed, English stress-timed, and Japanese mora-timed. Do these rhythmical structures play any role during language acquisition?

This is precisely the question that Bertoncini and her colleagues have started to investigate. Bijeljac-Babic, Bertoncini, and Mehler (1993) investigated whether infants rely on a syllable-like notion to organize speech stimuli. Could the infant detect the difference between words with different numbers of syllables? The researchers showed that French 4-day-olds discriminated between bi- and trisyllabic items. Moreover, when the length of the items was equated by means of a speech editor and a special algorithm that leaves the spectral templates unmodified except for duration, the infants were still capable of discriminating the lists. Last but not least, when the number of phonemes changed from four to six, keeping the number of syllables constant, infants did not react to the change in stimulation. These results confirm the hypothesis that the syllable, or a covariant structure, is used to represent speech stimuli. Bertoncini et al. (1988) had already noticed a vowel-centered representation of speech sounds in very young infants. This representation becomes more phoneme-like for the 2-month-olds.

The above results are difficult to assess in the absence of cross-linguistic validation. French, the language of the stimuli used, but also the surrounding language for most of the tested infants, is a syllable-based language (Mehler et al., 1981) but this is not the case for other languages (see Cutler, Norris, and Segui, 1983; Otake et al., 1993). What would infants do if they were tested with Japanese utterances? As mentioned above, Japanese is a mora-based language. Items like *toki* and *tooki* are both bisyllabic, but the first one is bimoraic, while the second one is trimoraic. Bertoncini and her colleagues (in preparation) have explored this issue in detail. In a first experiment it was shown that French 4-day-olds discriminate bi- and trisyllabic items (that were also bi- and trimoraic) spoken by a native Japanese monolingual speaker. Second, the experimenters used lists of stimuli that differed only by the number of moras, but not by the number of syllables (e.g., *toki* and *tooki*). The results suggest that French 4-day-olds are insensitive to this contrast. If this observation is confirmed, one may conclude that the syllable (or something correlated with it) is the universal structure in terms of which speech is represented at birth. Another possible interpretation would be that French newborns are incapable of discriminating morae because they have already realized that French is syllable-based. Given the young age (3 days) of the infants tested, we consider this alternative interpretation implausible. But to reject it definitely, we will have to test Japanese newborns under the same conditions, and show that they behave like French newborns.

In this section we have raised the question of how infants converge toward the rhythmical organization of their maternal language. In one hypothesis, the syllable is the primitive core structure for representing any language. If so, according to the particular language to which one is exposed, one learns to correct this representation and specify whether the syllable is accented or not, whether it has vowel reduction or not, and so forth. In another hypothesis, all of the many organizations displayed in speech signals are computed in parallel by infants. Other components of language determine which organization should be preserved, given the constraints arising from production and lexical considerations.

From prosody to segments

In a series of classic studies, Werker and Tees (1984) showed that younger infants are able to discriminate consonant contrasts that are not valid in their maternal language, even though older infants and adults are not. This decline in the ability to distinguish foreign contrasts starts at 8 months of age. By the time infants are one year old they behave just like adults; that is, they do not react to foreign contrasts. This behavior suggests a critical period and a related neurological rearrangement. Best, McRoberts, and Sithole's (1988) study of click discrimination by English-speaking adults and American-born infants leads to a different interpretation. Zulu clicks are ingressive consonants that are absent from English. Consequently, these speech sounds cannot be assimilated to a contrast used in English. English adults without any experience with Zulu clicks process these as natives do, suggesting that adults do not lose an early ability when they acquire the phonology of their mother tongue. Indeed, Best, McRoberts, and Sithole studied American infants ranging from 6 to 14 months in age with Zulu clicks and with contrasts that do not belong to English but which may be assimilated to phonemes used in that language. Best (1990) replicated Werker's reports of decreasing performance with foreign contrasts that are similar to one in the maternal language, but Best, McRoberts, and Sithole's subjects behaved very differently with Zulu clicks. Zulu clicks are so distinguishable from any native English category that they are not assimilated to categories in English. Werker and Tees's results are then due to assimilation of novel sounds to the categories of one's native language whenever assimilation remains possible.

This new view about the convergence toward the phonology of the maternal language does mean the process is less biological and ultimately unrelated to a biological clock. Whether infants assimilate foreign contrasts to familiar categories or use routines that are compiled for the purpose of production to code the surrounding contrasts, the fact of the matter remains that around 8 to 10 months of age, infants change from a phonetic to a phonemic mode of perceiving consonants (that is, they stop treating speech as consisting of mere sounds, and instead begin to represent only the linguistically relevant contrasts). More recently, Kuhl and colleagues (1992) have shown that 6-month-olds have already acquired some properties that characterize the vowels in their maternal language. From these studies, and from related studies by Werker and Polka (1993), it appears that infants might specify vocalic properties before they specify the properties of the consonants that figure in their native languages.

From the above studies one can draw the obvious conclusion that learning to speak requires the discovery of the segments that are part of the language. But as we saw before, this is far from being all there is to learning one's mother tongue. We saw that the infant has the ability to establish natural boundaries for prosodic structures. One can propose that infants learn the prosodic structure of their language before they can represent the segmental information about the vowels and consonants in that language.

Conclusions

Our investigations have led us to propose a general hypothesis about the relationship between certain areas of the brain and language. We have argued that the child is already functionally lateralized at birth. Later in life, as PET studies show, only utterances that belong to one's own language will increase the activity of the left hemisphere. Hence, one has to conclude that while the brain is initially prepared to be charitable about what will be processed linguistically, it becomes more and more demanding until it ends up accepting as linguistic only those stimuli that are part of the mother tongue (or possibly of another language that has been learned).

In the years to come we need to investigate in greater detail the nature of the units that are used to represent the speech stream in contrasting languages. We have proposed that languages can be classified into families according to whether they rely on the syllable, the foot, the mora, or other units. The data we presented, however, are also compatible with the view that the syllable is the universal atom for representing speech, regardless of language. In learning more about one's language—in particular, its lexicon and the production routines to build sentences—other segments or structures tend to be discovered. Indeed, our preliminary results point in this direction. However, research at this juncture and on this point is so scarce that we can say little before more data become available.

ACKNOWLEDGMENTS The preparation of this chapter was supported by a grant from the Human Frontier Scientific Program, as well as by a grant from the Human Capital Mobility Project. It was also carried out with the help of CNET (Convention 837BD28 00790 9245 LAA/TSS/CMC), and CNRS (ATP "Aspects Cognitifs et Neurobiologiques du Langage"). We would like to thank Luca Bonatti, Teresa Guasti, and Christophe Pallier for their help, comments, and fruitful discussion, as well as Steve Pinker for many thoughtful suggestions on a first version of this chapter.

NOTES

1. Of course, the primary acoustic cortical areas will be bilaterally activated by noise or any other acoustic stimulation.
2. Work by Petitto (1993), and by Poizner, Bellugi, and Klima (1987) suggests that deaf, native speakers of sign language have cortical asymmetries homologous to those of hearing subjects. It seems unlikely to us that signs taken from, say, American Sign Language would spontaneously activate the left hemisphere of newborns. Could it be that in the absence of speech, the left cortical areas devoted to its processing are taken over by other highly structured inputs?
3. The same bilingual speaker pronounced the utterances in both languages, so this result cannot be attributed to potential differences in timbre.
4. A *mora* is a subsyllabic unit that may consist of either a consonant-vowel or vowel syllable, or of the nasal or stop coda of a syllable.

REFERENCES

Bahrick, L. E., and J. N. Pickens, 1988. Classification of bimodal English and Spanish language passages by infants. *Infant Behav. Dev.* 11:277–296.

Bellugi, U., A. Bihrle, H. Neville, T. L. Jernigan, and S. Doherty, 1992. Language cognition and brain organization in a neurodevelopmental disorder. In *Developmental Behavioral Neuroscience*, M. Gunnar and C. Nelson, eds. Hillsdale, N.J.: Erlbaum, pp. 201–232.

Bertoncini, J., C. Floccia, T. Nazzi, K. Miyagishima, and J. Mehler, in preparation. Morae and syllable? Rhythmical basis of speech representation in neonates.

Bertoncini, J., Morais, R. Bijeljac-Babic, P. W. Jusczyk, L. Kennedy, and J. Mehler, 1988. An investigation of young infants' perceptual representations of speech sounds. *J. Exp. Psychol. [Gen.]* 117:21–33.

Bertoncini, J., J. Morais, R. Bijeljac-Babic, S. MacAdams, I. Peretz, and J. Mehler, 1989. Dichotic perception and laterality in neonates. *Brain Cogn.* 37:591–605.

Best, C. T., 1988. The emergence of cerebral asymmetries in early human development: A literature review and a neuroembryological model. In *Brain Lateralization in Children*, D. L. Molfese and S. J. Segalowitz, eds. New York: Guilford.

Best, C. T., 1990. Adult perception of nonnative contrasts differing in assimilation to native phonological categories. *J. Acoust. Soc. Am.* 88:S177.

Best, C. T., G. W. McRoberts, and N. M. Sithole, 1988. Examination of perceptual reorganization for nonnative speech contrasts: Zulu click discrimination by English-speaking adults and infants. *J. Exp. Psychol. [Hum. Percep. Performance.]* 14:345–360.

Bijeljac-Babic, R., J. Bertoncini, and J. Mehler, 1993. How do four-day-old infants categorize multisyllabic utterances? *Dev. Psychol.* 29:711–721.

Chomsky, N., 1965. *Aspects of a Theory of Syntax*. Cambridge, Mass.: MIT Press.

Chomsky, N., 1988. *Language and Problems of Knowledge*. Cambridge, Mass.: MIT Press.

Christophe, A., E. Dupoux, J. Bertoncini, and J. Mehler, 1994. Do infants perceive word boundaries? An empirical study of the bootstrapping of lexical acquisition. *J. Acoust. Soc. Am.* 95:1570–1580.

Christophe, A., E. Dupoux, and J. Mehler, 1993. How do infants extract words from the speech stream? A discussion of the bootstrapping problem for lexical acquisition. In *Proceedings of the 24th Annual Child Language Research Forum*, E. V. Clark, ed. Stanford, CSLI, 209–224.

Colombo, J., and R. S. Bundy, 1983. Infant response to auditory familiarity and novelty. *Infant Behav. Dev.* 6:305–311.

Curtiss, and Yamada, 1981. Selectively intact grammatical development in a retarded child. UCLA Working Papers in Cognitive Linguistics, 3:61–91.

Cutler, A., and S. Butterfield, 1992. Rhythmic cues to speech segmentation: Evidence from juncture misperception. *J. Mem. Lang.* 31:218–236.

Cutler, A., and D. M. Carter, 1987. The predominance of strong initial syllables in the English vocabulary. *Comput. Speech Lang.* 2:133–142.

Cutler, A., and D. Norris, 1988. The role of strong syllables in segmentation for lexical access. *J. Exp. Psychol. [Hum. Percept. Performance.]* 14:113–121.

Cutler, A., J. Mehler, D. Norris, and J. Segui, 1983. A language-specific comprehension strategy. *Nature* 304: 159–160.

Cutler, A., J. Mehler, D. Norris, and J. Segui, 1986. The syllable's differing role in the segmentation of French and English. *J. Mem. Lang.* 25:385–400.

Eimas, P. D., and J. L. Miller, 1992. Organization in the perception of speech by young infants. *Psychol. Sci.* 3:340–345.

Entus, A. K., 1977. Hemispheric asymmetry in processing of dichotically presented speech and nonspeech stimuli by infants. In *Language Development and Neurological Theory*, S. J. Segalowitz and F. A. Gruber, eds. New York: Academic Press, pp. 63–73.

Fernald, A., 1985. Four-month-olds prefer to listen to motherese. *Infant Behav. Dev.* 8:181–195.

Fisher, C., and H. Tokura, in press. Acoustic cues to clause boundaries in speech to infants: Cross-linguistic evidence.

Gall, F. J., 1835. *Works: On the Function of the Brain and Each of its Parts*, vol. 1–6. Boston: March, Capon and Lyon.

Gerken, L., P. W. Jusczyk, and D. Mandel, 1994. When prosody fails to cue syntactic structure: Nine-month-olds' sensitivity to phonological vs syntactic phrases. *Cognition* 51:237–265.

Geschwind, N., and A. Galaburda, 1987. *Cerebral Lateralization.* Cambridge, Mass.: MIT Press.

Gopnik, M., and M. B. Crago, 1991. Familial aggregation of a developmental language disorder. *Cognition* 39:1–50.

Gopnik, M., 1990. Feature-blind grammar and dysphasia. *Nature* 344:615.

Hayes, J. R., and H. H. Clark, 1970. Experiments on the segmentation of an artificial speech analogue. In *Cognition and the development of language*, J. R. Hayes, ed. New York: Wiley.

Hirsh-Pasek, K., D. G. Kemler Nelson, P. W. Jusczyk, K. W. Cassidy, B. Druss, and L. Kennedy, 1987. Clauses are perceptual units for young infants. *Cognition* 26:269–286.

Howard, D., K. Patterson, R. Wise, W. D. Brown, K. Friston, C. Weiller, and R. Frackowiak, 1992. The cortical localization of the lexicons. *Brain* 115:1769–1782.

Johnson, M. H., M. I. Posner, and M. K. Rothbart, 1991. Components of visual orienting in early infancy: Contingency learning, anticipatory looking, and disengaging. *J. Cognitive Neurosci.* 3:335–344.

Jusczyk, P. W., A. Cutler, and N. J. Redanz, 1993. Infants' sensitivity to the predominant stress patterns of English words. *Child Dev.* 64:675–687.

Jusczyk, P. W., A. Friederici, J. Wessels, V. Svenkerud, and A. M. Jusczyk, 1993. Infants' sensitivity to the sound patterns of native language words. *J. Memory Lang.* 32: 402–420.

Jusczyk, P. W., K. Hirsh-Pasek, D. G. Kemler-Nelson, L. J. Kennedy, A. Woodward, and J. Piwoz, 1992. Perception of acoustic correlates of major phrasal units by young infants. *Cognitive Psychol.* 24:252–293.

Jusczyk, P. W., D. G. Kemler Nelson, K. Hirsh-Pasek, and T. Schomberg, in preparation. Perception of acoustic correlates to clausal units in a foreign language by American infants.

Klima, E., and U. Bellugi, 1979. *The Signs of Language.* Cambridge, Mass.: Harvard University Press.

Knudsen, E. I., and M. Konishi, 1978. A neural map of auditory space in the owl. *Science* 200:795–797.

Kuhl, P. K., K. A. Williams, F. Lacerda, K. N. Stevens, and B. Lindblom, 1992. Linguistic experience alters phonetic perception in infants by 6 months of age. *Science* 255:606–608.

Lambertz, in preparation. Language recognition in 2-month-olds.

Landau, B., and L. Gleitman, 1985. *Language and Experience: Evidence from a Blind Child.* Cambridge, Mass.: Harvard University Press.

Lehiste, I., 1965. Juncture. In *Proceedings of the Fifth International Congress of Phonetic Sciences* (Münster 1964). Basel: Karger, 172–200.

Lehiste, I., 1966. *Consonant quantity and phonological units in Estonian* The Hague: Mouton.

Lenneberg, E., 1967. *Biological Foundations of Language.* New York: Wiley.

Liberman, A. M., 1982. On finding that speech is special. *Am. Psychol.* 37:148–167.

Mazoyer, B. M., S. Dehaene, N. Tzourio, V. Frak, N. Murayama, L. Cohen, O. Levrier, G. Salamon, A. Syrota, and J. Mehler, 1993. The cortical representation of Speech. *J. Cognitive Neurosci.* 5:467–479.

Mazuka, R., 1993. How can a grammar parameter be set before the first word? Paper presented at the Signal to Syntax Conference, Brown University, February 1993.

Mehler, J., and T. G. Bever, 1968. The study of competence in cognitive psychology. *Int. J. Psychol.* 3:237–280.

Mehler, J., J. Y. Dommergues, U. Frauenfelder, and J. Segui, 1981. The syllable's role in speech segmentation. *J. Verb. Learn. Verb. Be.* 20:298–305.

Mehler, J., E. Dupoux, and J. Segui, 1990. Constraining models of lexical access: The onset of word recognition. In *Cognitive Models of Speech Processing: Psycholinguistic and Computational Perspectives*, G. Altmann, ed. Cambridge, Mass.: MIT Press, Bradford Books, pp. 236–262.

Mehler, J., P. W. Jusczyk, G. Lambertz, N. Halsted, J. Bertoncini, and C. Amiel-Tison, 1988. A precursor of language acquisition in young infants. *Cognition* 29:143–178.

Moon, C., R. Panneton-Cooper, and W. P. Fifer, 1993. Two-day-olds prefer their native language. *Infant Behav. Dev.* 16:495–500.

Nottebohm, F., 1981. A brain for all seasons: Cyclical anatomical changes in song control nuclei of the canary brain. *Science* 214:1368–1370.

Ojemann, G., 1983. Brain organization for language from the perspective of electrical stimulation mapping. *Behav. Brain Sci.* 6:189–230.

Otake, T., G. Hatano, A. Cutler, and J. Mehler, 1993. Mora or syllable? Speech segmentation in Japanese. *J. Mem. Lang.* 32:258–278.

Petitto, L., 1993. On the ontogenetic requirements for early language acquisition. In *Developmental Neurocognition: Speech and Face Processing in the First Year of Life*, B. de Boysson-Bardies, S. de Schonen, P. Jusczyk, P. McNeilage, and J. Morton, eds. Dordrecht: Kluwer, pp. 365–383.

Poizner, H., U. Bellugi, and E. S. Klima, 1987. *What the Hand Reveals about the Brain.* Cambridge, Mass.: MIT Press.

Quené, H., 1991. Word segmentation in meaningful and nonsense speech. Actes du XIIème Congrès International des Sciences Phonétiques, 19–24 Août 1991, Aix-en-Provence, France.

Segui, J., E. Dupoux, and J. Mehler, 1990. The syllable's role in speech segmentation, phoneme identification, and lexical access. In *Cognitive Models of Speech Processing: Psycholinguistic and Computational Perspectives*, G. Altmann, ed. Cambridge, Mass.: MIT Press, Bradford Books, pp. 263–280.

Umeda, N., 1976. Consonant duration in American English. *J. Acoust. Soc. Am.* 61:3, 846–858.

Vaissière, J., 1983. Language-independant prosodic features. In *Prosody: Models and Measurements*, A. Cutler and D. Ladd, eds. Berlin: Springer-Verlag.

Vargha-Khadem, F., and M. Corballis, 1979. Cerebral asymmetry in infants. *Brain Lang.* 8:1–9.

Werker, J. F., and Polka, 1993. The ontogeny and developmental significance of language-specific phonetic perception. In *Developmental Neurocognition: Speech and Face Processing in the First Year of Life*, B. de Boysson-Bardies, S. de Schonen, P. Jusczyk, P. McNeilage, and J. Morton, eds. Dordrecht: Kluwer, pp. 275–288.

Werker, J. F., and R. C. Tees, 1984. Cross-language speech perception: Evidence for perceptual reorganization during the first year of life. *Infant Behav. Dev.* 7:49–63.

Whalen, D. H., and A. M. Liberman, 1987. Speech perception takes precedence over nonspeech perception. *Science* 237:169–171.

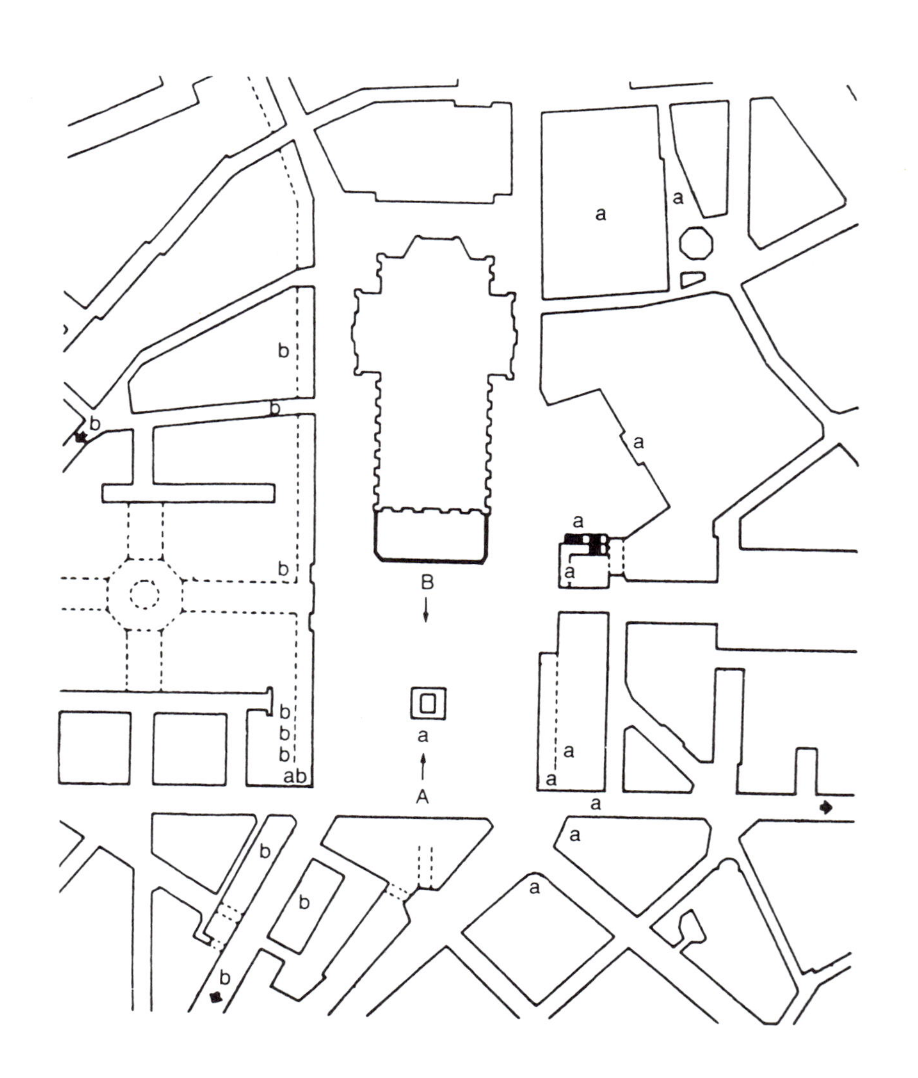
a
a
b
b
b
a
a
b
a
B
b
b
b
a
a
a
ab
A
a
a
b
a
b
b

VIII THOUGHT AND IMAGERY

Map of the Piazza del Duomo in Milan, showing the two positions, A and B, from which patients were asked to imagine viewing the piazza, and the landmarks they recalled from each imagined position, labeled "a" and "b" respectively.

Introduction

STEPHEN M. KOSSLYN

The chapters in this section address some of the most vexing problems in philosophy and psychology, which center on questions about the nature of thinking. The term *thinking* is notoriously vague, but we can all agree that it hinges on two kinds of properties. First, information must be represented internally; and second, that information must be manipulated in order to draw inferences and conclusions. Most of what we mean by "thinking" corresponds to reasoning, to determining what follows from specific circumstances or how to reach a goal given a set of initial conditions. Depending on how information is represented, different types of processing are appropriate to carry out reasoning. Thus, issues about how information is stored in memory become central to our understanding of the nature of thought.

The study of reasoning has had a checkered past. The behaviorists, responding to the inconclusive efforts of the introspectionist psychologists of the day, rejected such questions altogether; they were treated as beyond the purview of science. John Watson claimed that thinking was no more than talking to oneself. But many researchers must have harbored secret doubts about such assertions: We all have had the experience of mulling something over, being puzzled, working through a problem, planning ahead, or just plain daydreaming. Any reasonable person would not deny the existence of such a thing as thought. It's one of the things we do much better than the other animals.

The invention of the digital computer offered a rigorous way to conceive of internal representation and reasoning—namely, as information processing—and cognitive psychology was not long in being born (see Gardner, 1985). Cognitive psychology was a major advance over behaviorism; it brought the mind back to psychology. But cognitive psychology inherited a bias from its behavioral predecessors: It focused almost entirely on behavioral data, such as response times and error rates, to draw inferences about mental events. Moreover, as part of its love affair with the computer metaphor, cognitive psychology likened mental processes to programs in a computer that could be characterized independently of the physical machine on which they ran. This orientation led the field virtually to ignore the brain.

The bias toward focusing on behavioral data and the computer metaphor had some unfortunate drawbacks. For one, it turned out to be extremely difficult to draw inferences about abstract mental activity purely on the basis of behavioral data. Early attempts to characterize the stages of real-time processing, such as that of Sternberg (1969), soon became mired in controversy (e.g., see Townsend, 1974). Much of the field backed away from attempts to characterize the mechanisms underlying cognition, and instead probed the limits and properties of various phenomena (e.g., the circumstances under which priming occurs).

The advent of cognitive neuroscience has once again focused attention on the nature of the real-time processes that underlie mental functioning. It replaced the mind-as-program metaphor with the conception that mental events correspond to brain functions. With the recent availability of sensitive brain-scanning techniques and sophisticated ways of testing brain-damaged patients, we can begin to characterize how the brain gives rise to the mind. The cognitive neuroscience approach is exciting in part because it again leads us to focus on mechanism. In addition, as its name implies, cognitive neuroscience roots cognition in the brain—and in so doing integrates the study of the mind with the rest of the natural sciences.

Imagery was one of the earliest victims of the behaviorist attempts to develop a rigorous and scientific psychology. Hence, it is somewhat ironic, but no accident, that research on mental imagery has been a major focus of cognitive neuroscience studies of mental events. As discussed in the following chapters, imagery turns out to share mechanisms with more basic processes used in like-modality perception, memory, and even motor control. For example, visual mental imagery involves the so-called ventral and dorsal pathways, which are used in perception to encode object properties and spatial properties, respectively. Such high-level mechanisms have been among the first to be studied in some detail. Thus, researchers could use facts about such processing to place constraints on theories of the neural bases of imagery. Chapters by Farah, Kosslyn and Sussman, and Paivio reveal that much has been accomplished in such studies. Farah provides compelling evidence that visual mental imagery does indeed rely on at least some of the same underlying mechanisms as visual perception. Kosslyn and Sussman reinforce this point, and also argue that the motor system is involved in at least some forms of imagery. And Paivio reminds us of the relation between imagery and performance in various memory tasks—empirical and theoretical knowledge that is yet to be incorporated into a complete cognitive neuroscience of imagery and memory. Although the authors have slightly different perspectives, it is clear that there is enormous cohesion in this line of research. Not only are facts about the brain helping to shape theorizing, but they are leading to a convergence of views. This may be more than mere fashion; it is possible that the results of empirical investigations are actually converging—actually coming to characterize aspects of how imagery arises from brain function.

Much of the theorizing in the chapters on imagery is rooted in research with nonhuman primates. As is evident in earlier chapters of this volume, the past decade has seen spectacular progress in understanding the neural bases of vision in monkeys. Because imagery and perception share common mechanisms, researchers have turned to animal models as a source of inspiration for theorizing about the mechanisms underlying imagery. The findings summarized by Graziano and Gross underline the fact that cognitive functions are not unitary and undifferentiated; we should expect that any given cognitive function is accomplished by many networks working in concert. Specifically, we are likely to have a number of different representations of space, each of which would be useful for different types of reasoning.

Smith and Jonides describe another way in which our understanding of cognitive processes is enhanced by drawing on animal models, this time models of working memory (see Goldman-Rakic, 1987). Although there is no guarantee that human and monkey brains operate the same way in any given instance, there is good reason to believe that basic visual, memory, and motor processing is similar. As Smith and Jonides illustrate, research on nonhuman primates can play an invaluable role in inspiring theories of human brain function. Smith and Jonides used positron emission tomography (PET) to study the nature of working memory. Working memory lies at the heart of reasoning processes; it is the mechanism whereby one applies processes to representations of specific circumstances. The limitations of working memory are without question a major determinant of how we reason.

The research and theory on imagery and working memory are important in part because they are building the foundations of theories of reasoning, of thought, per se. Robin and Holyoak illustrate how investigations of thought processes can benefit from the broader framework and perspective formed by more basic research. They develop a novel theory of frontal lobe function, which depends on the idea that specific representations are continually being "bound" together as a prerequisite to reasoning. Before one can reason about the correct course of action, one must understand that a specific object is in a specific relation to time and place. Moreover, one often must understand the relations among features and attributes of objects, things that they have in common and things that are distinct. The process of selecting specific attributes and binding them appropriately presumably depends in part on working memory.

And, finally, the chapter by Johnson-Laird outlines types of representations and processes that are used in the highest forms of reasoning. Johnson-Laird argues that such reasoning relies on the construction and comparison of "mental models," which presumably involve working memory. These models provide a way to determine what is implied by a set of conditions; more difficult problems appear to require that one construct and consider more such models—which presumably would tax working memory, and hence the theory could be tested in part using methods like those of Smith and Jonides. Such models might sometimes involve forming mental images of specific objects or events. Thus, research on mental imagery may come to provide a foundation for some research on reasoning, and studies of the highest form of cognition will indirectly benefit from the research on the most basic perceptual, memory, and motor processes that has advanced research on imagery.

In short, cognitive neuroscience is providing the means for grappling with central problems in the study of the mind. It is clear that such work is just now coming into its own, and the following chapters offer only a taste of things to come.

REFERENCES

GARDNER, H., 1985. *The Mind's New Science: A History of the Cognitive Revolution.* New York: Basic Books.

GOLDMAN-RAKIC, P. S., 1987. Circuitry of primate prefrontal cortex and regulation of behavior by representational knowledge. In F. Plum, vol. ed., and V. B. Mountcastle sec. ed., *Handbook of Physiology.* Sec. 1: *The Nervous System.* Vol. 5: *Higher Functions of the Brain.* Bethesda, Md.: American Physiological Society.

STERNBERG, S., 1969. The discovery of processing stages: Extensions of Donders' method. In *Attention and Performance II*, W. G. Koster, ed. Amsterdam: North-Holland, pp. 276–315.

TOWNSEND, J. T., 1974. Issues and models concerning the processing of a finite number of inputs. In *Human Information Processing: Tutorials in Performance and Cognition*, B. H. Kantrowitz, ed. New York: Wiley.

62 The Neural Bases of Mental Imagery

MARTHA J. FARAH

ABSTRACT What neural events underlie the generation of a visual mental image? This chapter reviews evidence from patients with brain damage and from measurements of regional brain activity in normal subjects. The answer emerging from these studies is that many of the same modality-specific cortical areas used in visual perception are also used in imagery. These areas include spatially mapped regions of occipital cortex. There is also evidence for a distinct imagery mechanism, not used under normal circumstances for perception, which is required for the generation of images from memory. Evidence concerning the localization of this process is not entirely consistent, but shows a trend toward regions of the posterior left hemisphere.

Introduction

The Study of Imagery in Cognitive Psychology The term "mental imagery" has been used in a number of different ways in psychology, from referring to the relatively specific act of forming a "picture in the head" to denoting a more general class of nonverbal thought processes in which spatial representations are actively recalled or manipulated. This chapter is primarily concerned with the first sense of mental imagery, the kind most readers will find themselves experiencing when they attempt to answer the question What color are the stars on the American flag? The distinctive phenomenology of imagery is probabably its most salient characteristic feature as a form of mental representation. Most people report that images seem much like real percepts, although fainter, sketchier, and more effortful to maintain. However, a scientific psychology cannot allow distinctions among different forms of mental representation to rely entirely on phenomenology, and therefore much of the research on imagery in cognitive psychology and neuropsychology has been devoted to characterizing imagery in more objective ways.

To most nonspecialists, it seems unbelievable that the existence of mental images, as a qualitatively different form of mental representation from verbal memory, could be doubted. Yet numerous psychologists did doubt this, in part because of the influence of the computer analogy in psychology and the conviction that cognition is symbol manipulation, and in part based on a preference for the parsimony of having fewer types of mental representations (see, e.g., Pylyshyn, 1973). Early imagery researchers devised many elegant and ingenious experimental paradigms to demonstrate the distinction between imagery and verbal thought, and to characterize imagery in objective information-processing terms (e.g., Paivio, 1971; Shepard, 1978; Kosslyn, 1980). Paivio (this volume) summarizes the research by himself and others on the role of imagery in memory. This research was important not only for our understanding of learning and memory, but also because it demonstrated the separate status of imagery as a distinct "code" or form of representation. Subsequent research by Kosslyn, Shepard, and others focused on the question of *how* imagery differed from verbal thought. This question was the subject of the so-called imagery debate.

The Imagery Debate Starting in the late 1970s, two related issues concerning imagery came into focus. The first was whether mental imagery involved some of the same representations normally used during visual perception, or whether imagery involved only more abstract, postperceptual representations. This issue was discussed most explicitly by Finke (e.g., 1980) and Shepard (e.g., 1978). The second issue was whether mental images had a spatial, or array-like, format, or whether they were propositional in format. Much of Kosslyn's research was aimed at addressing this issue (e.g., Kosslyn, 1980). The two issues are in principle

MARTHA J. FARAH Department of Psychology, University of Pennsylvania, Philadelphia, Pa.

independent, although given the fact that much of visual representation is array-like (e.g., Maunsell and Newsome, 1987), they are closely related.

Although these issues are straightforwardly empirical in nature, they proved difficult to settle using the experimental methods of cognitive psychology. The classic image-scanning experiments of Kosslyn (e.g., Kosslyn, Ball, and Reiser, 1978) will be used to illustrate this point. Kosslyn and associates found that when subjects were instructed to focus their attention on one part of an image and then move it continuously, as quickly as possible, to some other part of the image, the time taken to scan between the two locations was directly proportional to their metric separation, just as if subjects were scanning across a perceived stimulus. This finding follows naturally from the view that images share representations with visual percepts, and that these representations have a spatial format. However, not all psychologists agreed with this interpretation. Various types of alternative explanations were proposed that accounted for the scanning findings, and many others, without hypothesizing shared representations for imagery and perception or a spatial format for imagery. These alternative explanations are discussed in detail by Farah (1988) as motivation for turning to neuropsychological evidence. Two examples will be mentioned briefly here.

Pylyshyn (1981) suggested that subjects in imagery exeriments take their task to be simulating the use of visual-spatial representations, and that with their tacit knowledge of the functioning of their visual system subjects are able to perform this simulation using nonvisual representations. Intons-Peterson (1983) suggested that subjects in imagery experiments may be responding to experimenter expectancies, and she has shown that at least certain aspects of the data in imagery experiments can indeed be shaped by the experimenters' preconceptions. Although these accounts strike many people as less plausible than the hypothesis that images are modality-specific visual representations that are intrinsically spatial in format, it has proven difficult to rule them out. In fact, Anderson (1978) has argued that no behavioral data (i.e., sets of stimulus inputs paired with subjects' reponses to those stimuli and the latencies of the responses) can ever distinguish alternative, nonvisual theories of imagery from the visual-spatial theories. One of the incentives for studying the neuropsychology of mental imagery is that it has the potential to be more decisive on these issues, because it provides more direct evidence on the internal processing stages intervening between stimulus and response in imagery experiments.

The brain bases of mental images: Modality specificity and format

The neuropsychological studies relevant to the two issues of the imagery debate—are images visual, and do they have a spatial format?—include both studies of brain-damaged subjects and psychophysiological measures in normal subjects. These studies will be reviewed very briefly here; a more detailed review of some of this material can be found in Farah, 1988, and Farah, 1989.

STUDIES OF BRAIN-DAMAGED PATIENTS If forming a mental image consists of activating cortical visual representations, then patients with selective impairments of visual perception should manifest corresponding impairments in mental imagery. This is often the case. For example, DeRenzi and Spinnler (1967) investigated various color-related abilities in a large group of unilaterally brain-damaged patients and found an association between impairment on color-vision tasks, such as the Ishihara test of color blindness, and on color-imagery tasks, such as verbally reporting the colors of common objects from memory. Beauvois and Saillant (1985) studied the imagery abilities of a patient with a visual-verbal disconnection syndrome. The patient could perform purely visual color tasks (e.g., matching color samples) and purely verbal color tasks (e.g., answering questions such as "What color is associated with envy?") but could not perform tasks in which a visual representation of color had to be associated with a verbal label (e.g., color naming). On tasks that tested the patient's color imagery ability purely visually, such as selecting the color sample that represented the color of an object depicted in black and white, she did well. However, when the equivalent problems were posed verbally (e.g., "What color is a peach?") she did poorly. In other words, mental images interacted with other visual and verbal task components as if they were visual representations. More recently, De Vreese (1991) reported two cases of color imagery impairment, one of whom had left occipital damage and displayed the same type of visual-verbal disconnection as the patient just described, and the other of whom had bilateral occipital damage and parallel color-perception and color-imagery impairments.

In another early study documenting the relations between imagery and perception, Bisiach and Luzzatti (1978) found that patients with hemispatial neglect for visual stimuli also neglected the contralesional sides of their mental images. Their two right-parietal-damaged patients were asked to imagine a well-known square in Milan, shown in figure 62.1. When they were asked to describe the scene from vantage point A in the figure, they tended to name more landmarks on the east side of the square (marked with lowercase *as* in the figure); that is, they named the landmarks on the right side of the imagined scene. When they were then asked to imagine the square from the opposite vantage point, marked B on the map, they reported many of the landmarks previously omitted (because these were now on the right side of the image) and omitted some of those previously reported.

Levine, Warach, and Farah (1985) studied the roles of the "two cortical visual systems" (Ungerleider and Mishkin, 1982) in mental imagery, with a pair of patients. Case 1 had visual disorientation following bilateral parieto-occipital damage, and case 2 had visual agnosia following bilateral inferior temporal damage. We found that the preserved and impaired aspects of visual imagery paralleled the patients' visual abilities:

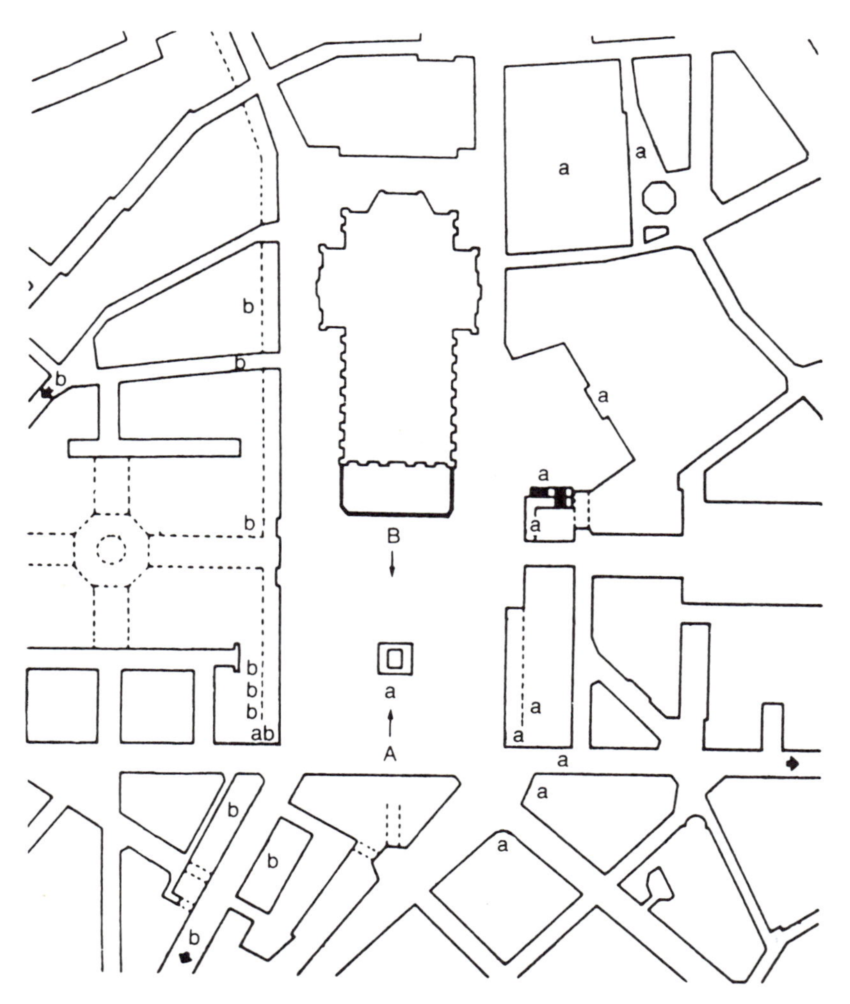

FIGURE 62.1 Map of the Piazza del Duomo in Milan, showing the two positions, A and B, from which patients were asked to imagine viewing the piazza, and the landmarks they recalled from each imagined position, labeled "a" and "b" respectively.

Case 1 could neither localize visual stimuli in space nor accurately describe the locations of familiar objects or landmarks from memory. However, he was good at both perceiving object identity from appearance and describing object appearance from memory. Case 2 was impaired at perceiving object identity from appearance and at describing object appearance from memory, but was good at localizing visual stimuli and at describing their locations from memory.

Farah, Hammond, et al., (1988) carried out more detailed testing on the second patient, L. H. We adapted a large set of experimental paradigms from the cognitive psychology literature that had been used originally to argue for either the visual nature of imagery (i.e., "picture in the head" imagery) or for its more abstract spatial nature. Our contention was that both forms of mental imagery exist, contrary to much of the research in cognitive psychology aimed at deciding which of the two characterizations of imagery is correct. On the basis of the previous study, we conjectured that cognitive psychology's so-called visual imagery tasks would be failed by the patient with the damaged ventral temporo-occipital system, whereas cognitive psychology's so-called spatial imagery tasks would pose no problem for him because of his intact dorsal parieto-occipital system. Figure 62.2 shows the locations of his brain lesions, which spare the dorsal visual system.

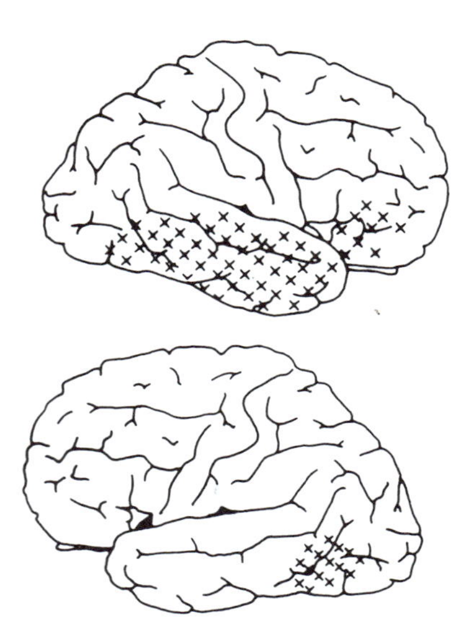

FIGURE 62.2 Diagram showing regions of damage in the brain of case L. H.

The visual imagery tasks included imagining animals and reporting whether they had long or short tails, imagining common objects and reporting their colors, and imagining triads of states within the United States and reporting which two are most similar in outline shape. The spatial imagery tasks included such mental image transformations as mental rotation, scanning, and size scaling, and imagining triads of shapes and reporting which two are closest to one another. As shown in figure 62.3, the patient was impaired relative to control subjects at the visual, pattern, and color imagery tasks, but entirely normal at the spatial imagery tasks.

Although the foregoing studies implicate modality-specific visual representations in imagery, they either are ambiguous as to the level of visual representation involved, or they implicate relatively high-level representation in the temporal and parietal lobes. In a recent study, Farah, Soso, and Dasheiff (1992) examined the role of the occipital lobe in mental imagery. If mental imagery consists of activating relatively early representations in the visual system, at the level of the occipital lobe, then it should be impossible to form images in regions of the visual field that are blind due to occipital lobe destruction. This predicts that patients with homonymous hemianopia should have a smaller maximum image size, or visual angle of the mind's eye. The maximum image size can be estimated using a method developed by Kosslyn (1978), in which subjects imagine walking toward objects of different sizes and report the distance at which the image just fills their mind's eye's visual field and is about to "overflow." The trigonometric relation between the distance, object size, and visual angle can then be used to solve for the visual angle.

We were fortunate to encounter a very high-functioning, educated young woman who could perform the rather demanding task of introspecting on the distance of imagined objects at "overflow." In addition, she could serve as her own control because she was about to undergo unilateral occipital lobe resection for treatment of epilepsy. We found that the size of her biggest possible image was reduced after surgery, as represented in figure 62.4. Furthermore, by measuring maximal image size in the vertical and horizontal dimensions separately, we found that only the horizontal dimension of her imagery field was significantly reduced. These results provide strong evidence for the use of occipital visual representations during imagery.

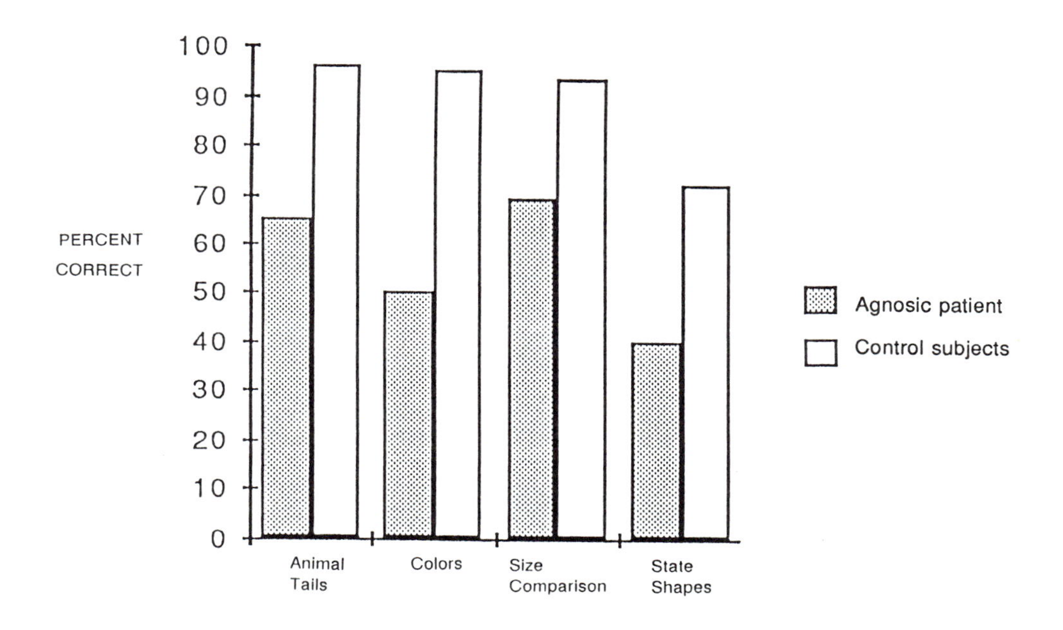

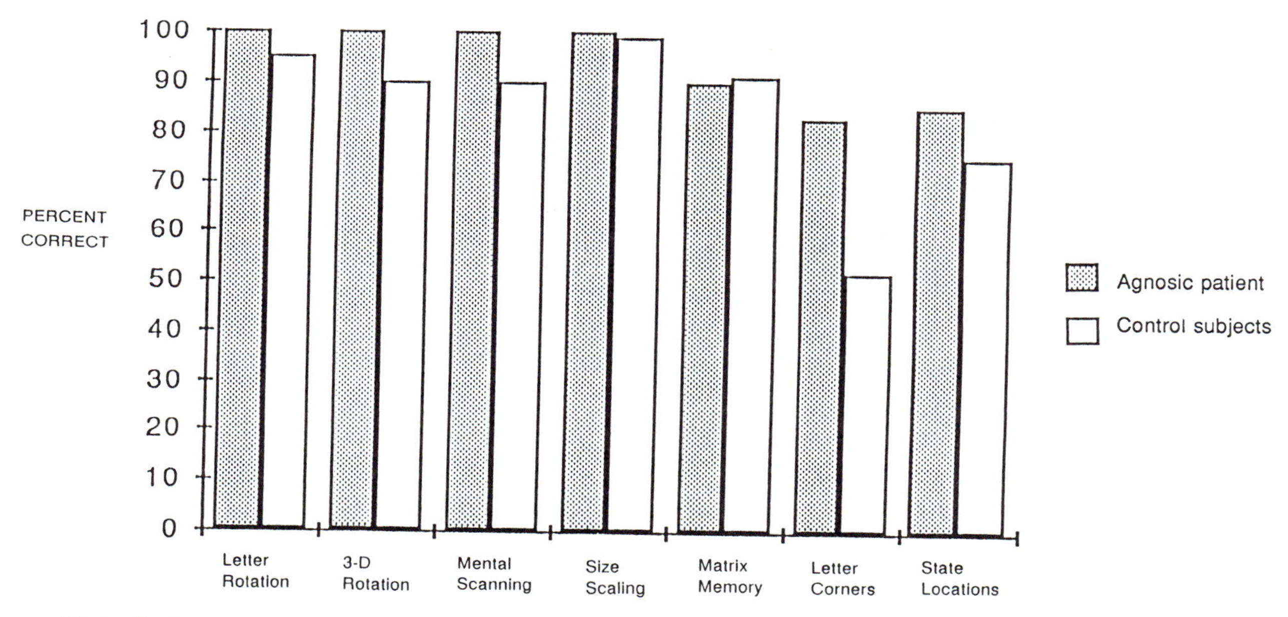

FIGURE 62.3 Performance of case L. H. and control subjects on mental imagery tasks requiring retreival of visual (or "what") and spatial (or "where") information.

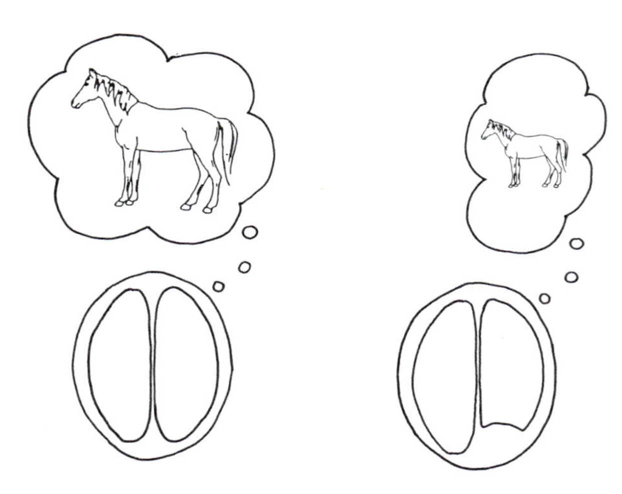

"I can get to within 15 feet of the horse in my imagination before it starts to overflow"

"The horse starts to overflow at an imagined distance of about 35 feet"

FIGURE 62.4 Representation of the effects of occipital lobectomy on the maximal mental image size of case M. G. S.

Although the results from brain-damaged patients are generally consistent with the hypothesis that mental imagery involves representations within the visual system proper—including relatively early representations in the occipital lobe that are known to be spatial in format—there are discrepant findings as well. It has occasionally been noted that an agnosic patient's drawings from memory are satisfactory. Recently, Behrmann, Winocur, and Moscovitch (1992) reported a severely agnosic patient who demonstrated good visual mental imagery abilities. It is difficult to reconcile this observation with the hypothesis that imagery and visual perception share representations, unless one supposes either a very peripheral perceptual abnormality that for some reason interferes more with tests of object recognition than with tests of seemingly lower level visual functions, or an impairment in the unidirectional transformation of relatively early occipital representations into higher-level object representations. In the absence of any independent support for these interpretations, they seem rather unsatisfactory.

To conclude, in most but not all cases of selective visual impairments following damage to the cortical visual system, patients manifest qualitatively similar impairments in mental imagery and perception. This provides some evidence for the hypothesis that imagery and perception share at least some modality-specific cortical representations, and that those representations are specialized for the same kinds of visual or spatial information in both perception and imagery. Let us now turn to a different source of evidence on the relation between imagery and perception.

BRAIN-IMAGING STUDIES IN NORMAL SUBJECTS Starting in the mid-1980s, regional brain activity during mental imagery has been monitored in normal subjects using a variety of techniques, as summarized in table 62.1. An early study relevant to the relation of imagery and perception was reported by Roland and Friberg (1985). They examined patterns of regional blood flow using single photon emisson computed tomography (SPECT) while subjects performed three different cognitive tasks, one of which was to visualize a walk through a familiar neighborhood, making alternate left and right turns. In this task, unlike the other tasks, blood flow indicated activation of the posterior regions of the brain, including visual cortices of the parietal and temporal lobes. Although most of occipital cortex was not monitored in this study, a later positron emisson tomography (PET) study by the same group of researchers also failed to find significant occipital activation during the "mental walk" task (Roland et al., 1987). The most pronounced effects of imagery were seen in posterior parietal cortex, with nonsignificant increases in occipital and temporal areas. These results are therefore consistent with the hypothesis that mental imagery is a function of higher visual cortical areas, but the study failed to implicate early, occipital, areas.

Goldenberg and his colleagues have performed a series of blood-flow studies of mental imagery using SPECT. They inferred which brain areas were activated by mental imagery using very elegant experimental designs in which the imagery task was closely matched with control tasks involving many of the same processing demands, except for the mental imagery per se (e.g., Goldenberg et al., 1987; Goldenberg et al., 1989a; Goldenberg et al., 1989b; Goldenberg et al., 1991; Goldenberg et al., 1992). For example, one imagery task was the memorization of word lists using an imagery mnemonic, and its control task was memorization without imagery (Goldenberg et al., 1987). Another task involved answering questions of equal difficulty, which either required mental imagery (e.g., "What is darker green, grass or a pine tree?) or did not (e.g., "Is the Categorical Imperative an ancient grammatical form?"; Goldenberg et al., 1989a). In all of these studies, visual imagery is found to be associated

Table 62.1

Functional imaging studies of visual mental imagery implicating visual cortical involvement

Authors	Method	Visual Activity	Additional Comments
Roland & Friberg (1985)	SPECT	Parietal, temporal	Most occipital activity not monitored
Roland et al. (1987)	PET	Parietal	
Goldenberg et al. (1987)	SPECT	Occipital, temporal	
Farah, Peronnet, et al. (1988)	ERP	Occipital, temporal	
Farah et al. (1989)	ERP	Occipital, temporal	Time course implicates area 18
Farah & Peronnet (1989)	ERP	Occipital	Temporal not monitored; magnitude predicted by imagery vividness
Goldenberg et al. (1989a)	SPECT	Occipital, temporal	
Goldenberg et al. (1989b)	SPECT	Occipital, temporal	
Uhl et al. (1990)	ERP	Occipital, temporal, parietal	Parietal associated with map images; occipital-temporal with face and color images
Goldenberg et al. (1991)	SPECT	Occipital	
Goldenberg et al. (1992)	SPECT	Occipital, temporal	
Charlot et al. (1992)	SPECT	Occipital, temporal	
Kosslyn et al. (1993)	PET	Occipital, temporal, parietal	Activity in area 17
Le Bihan et al. (1993)	MRI	Occipital	Activity in area 17

with occipital and temporal activation. It is possible that the greater parietal involvement observed by Roland and colleagues (Roland and Friberg, 1985; Roland et al., 1987) is related to the subjects' need to represent spatial aspects of the environment in their mental walk task (cf. the findings of Farah, Hammond, et al., 1988, on dissociable visual and spatial mental imagery).

A recent study by Charlot and colleagues (1992) used SPECT rCBF while subjects generated and scanned images in the classic cognitive psychology image-scanning paradigm developed by Kosslyn, Ball, and Reiser (1978). These authors also found activation of visual association cortex, including occipital cortex.

In general, these findings with SPECT rCBF are consistent with the previous patient work, in that they show visual cortical activity associated with mental imagery. Perhaps of greatest interest, pronounced and consistent activation of occipital cortex was observed in these studies. Thus suggests that imagery involves spatially mapped visual representations.

Our group has used event-related potentials (ERPs) to address the question of whether visual mental imagery has a visual locus in the brain. In one study, Farah, Peronnet, et al. (1988) used ERPs to map out, in space and in time, the interaction between mental imagery and concurrent visual perception. We found that imagery affected the ERP to a visual stimulus early in stimulus processing, within the first 200 ms. This implies that imagery involves visual cortical regions that are normally activated early in visual perception. The visual ERP component that is synchronized with the effect of imagery, the N1, is believed to originate in areas 18 and 19, implying a relatively early extrastriate locus for imagery in the visual system. Interpolated maps of the scalp-recorded ERPs were also consistent with this conclusion.

In a second series of studies, Farah et al. (1989) took a very different approach to localizing imagery in the brain using ERP methods. Rather than observing the interaction between imagery and concurrent perception, we simply asked subjects to generate a mental image from memory, in response to a visually presented word. By subtracting the ERP to the same words when no imagery instructions were given from the ERP when subjects were imaging, we obtained a relatively pure measure of the brain electrical activity that is synchronized with the generation of a mental image. Again, we constructed maps of the scalp distribution of the ERP imagery effect, in order to determine whether the maxima lay over modality-specific visual perceptual areas. Despite the very different ex-

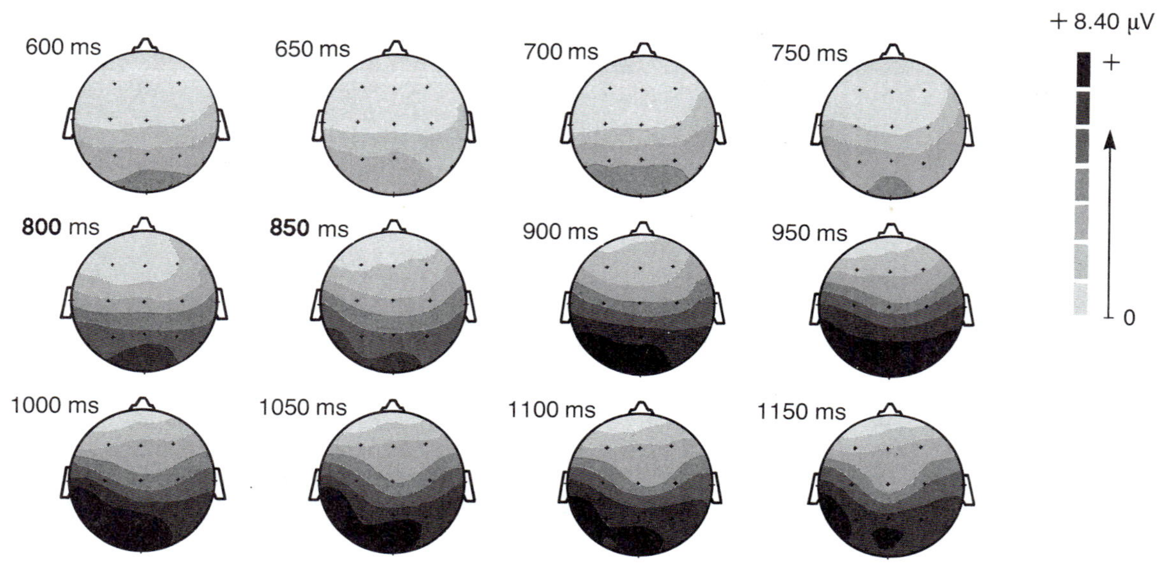

FIGURE 62.5 Scalp distribution of potentials synchronized with the generation of a mental image, in 50 ms time slices.

perimental paradigm, we found a highly similar scalp distribution to that found in the previous experiment, clearly implicating visual areas. Figure 62.5 shows the evolving scalp distribution of subjects' brain electrical activity as they generate images from memory. When the experiment was repeated using auditory word presentation, the same visual scalp topography was obtained. Control experiments showed that the imagery effects in these experiments were not due to the cognitive effort expended by subjects when imaging (as opposed to imagery per se), or to eye movements.

Farah and Peronnet (1989) reported two studies in which subjects who rated their imagery as relatively vivid showed a larger occipital ERP imagery effect when generating images than subjects who claimed to be relatively poor imagers. This result, which we then replicated under slightly different conditions, suggests that some people are more able to efferently activate their visual systems than others, and that such people experience especially vivid imagery.

Uhl and colleagues (1990) used scalp-recorded DC shifts to localize brain activity during imagery for colors, faces, and maps. Following transient positive deflections of the kind observed by Farah et al. (1989), a sustained negative shift was observed over occipital, parietal, and temporal regions of the scalp. Consistent with the different roles of the two cortical visual systems, the effect was maximum over parietal regions during map imagery, and maximum over occipital and temporal regions during face and color imagery.

More recently, Kosslyn et al. (1993) brought the PET method to bear on the localization of imagery. In the first two of their experiments, subjects viewed grids in which block letters were either present or to be imagined, and judged whether an *X* occupying one cell of the grid fell on or off the letter. Comparisons between imagery and relevant baseline conditions showed activation of many brain areas, including occipital visual cortex. Figure 62.6 shows the areas of greatest significant activity when subjects generated images from memory, relative to when they performed an analogous task with perceived stimuli. In a third experiment, subjects generated either large or small images of letters of the alphabet with eyes closed, and the researchers directly compared the two imagery conditions. They found that the large images activated relatively more anterior parts of visual cortex than the small ones, consistent with the known mapping of the visual field onto area 17.

Le Bihan and colleagues (1993) used magnetic resonance imaging (MRI) to investigate the role of visual cortex in mental imagery. Like PET, MRI has excellent spatial resolution; in addition, MRI has excellent temporal resolution. Le Bihan et al. measured regional brain activity as subjects alternately viewed flashing patterns and imagined them. The results shown for one

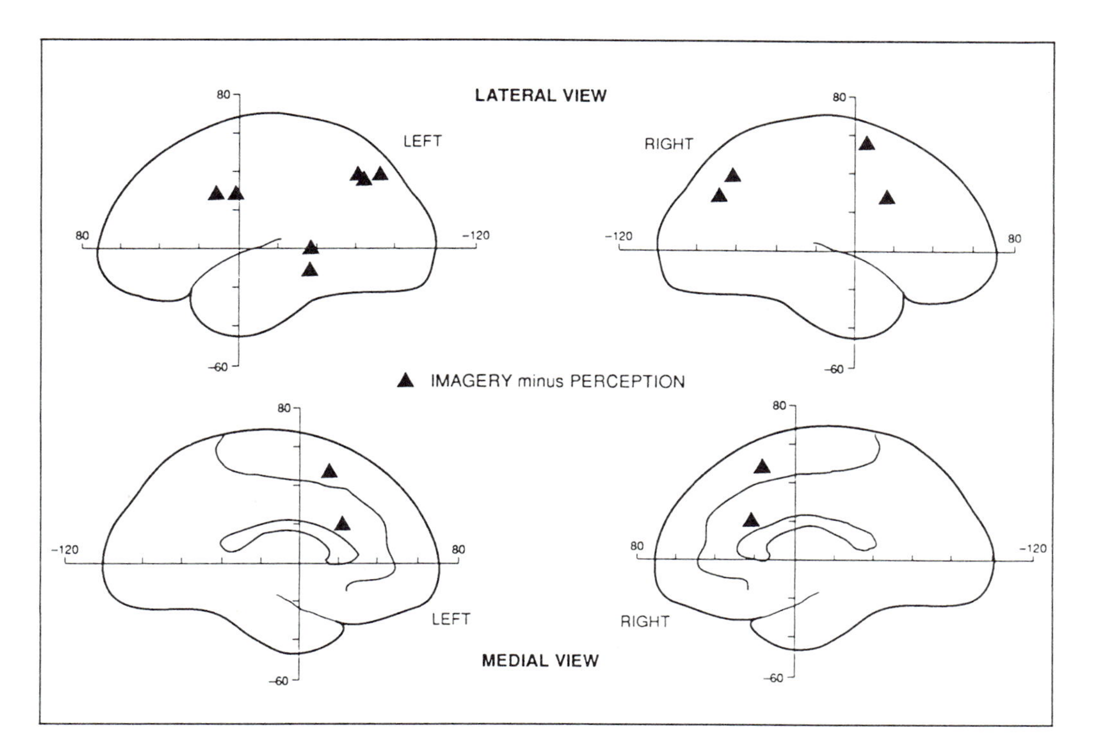

FIGURE 62.6 Areas of significant increases in activity measured by PET during the generation of images from memory, relative to an analogous perceptual task.

subject in figure 62.7 provide a striking demonstration of the involvement of area 17 in mental imagery as well as in perception.

THE IMAGERY DEBATE REVISITED In Anderson's (1978) discussion of the intrinsic ambiguities of behavioral data for resolving the imagery debate, he suggested that physiological data might someday provide a decisive answer. Only fifteen years later, that day appears to have arrived. Although there is much more to find out, and some of our current conclusions will doubtless need to be revised, we can point to a body of converging evidence that supports the modality-specific visual nature of mental images, and suggests that at least some of their neural substrates have a spatial representational format: Damage to visual areas representing such specialized stimulus properties as color, location, and form result in the loss of these properties in mental imagery. Spatially delimited impairments of visual attention and of visual representation are accompanied by corresponding impairments in imagery. Psychophysiological studies using blood-flow imaging methods and ERPs have operationalized imagery in a wide variety of ways, including instructions to take a mental walk, imagery mnemonics, general knowledge questions about the appearances of familiar objects, the effect of imagery on concurrent perception, instructions to image common objects, self-reported individual differences in vividness of imagery, and judgments about imagined letters of the alphabet. Across all of these superficially different tasks, indices of regional brain activity implicate modality-specific visual cortex in mental imagery. Further, the N1 locus of the imagery-perception interaction effect suggests that specifically area 18 is involved, and the PET and MRI results suggest that area 17 may be involved as well.

Image generation

Kosslyn (e.g., 1980) made a distinction between three general components of the imagery system: the long-term visual memories, from which images are generated; the visual buffer, a spatially formatted representational medium that is activated to produce phenomenally experienced images; and an image-generation process, whereby the visual buffer is activated in accordance with memory. The two issues concerning image generation to be discussed here are

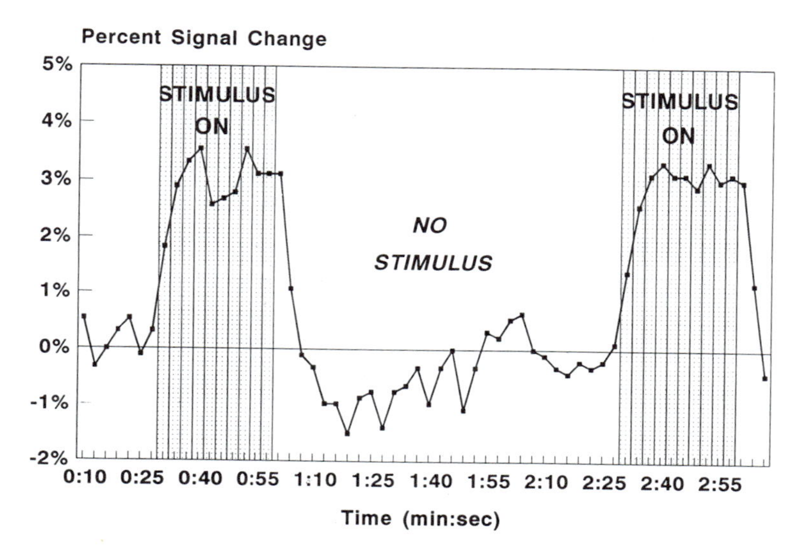

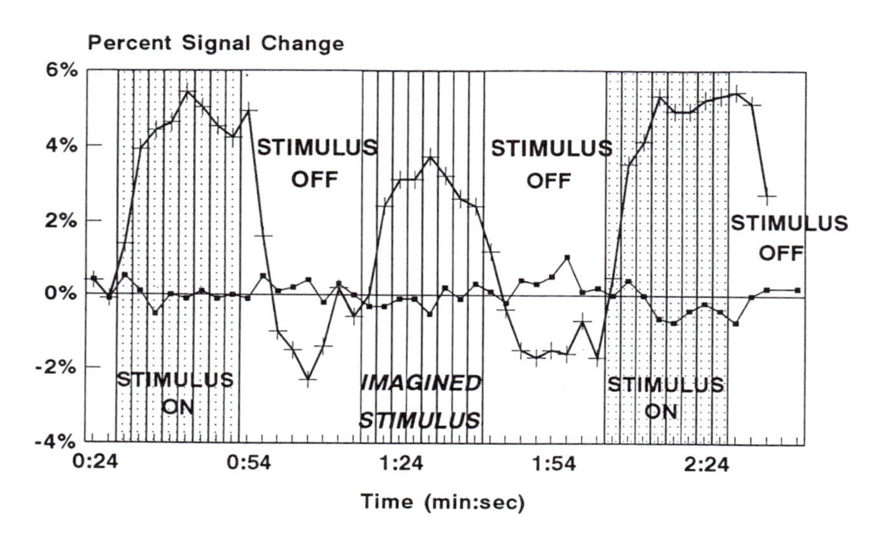

FIGURE 62.7 Plot of activity over time in striate cortex measured by MRI, as a function of the subject's activity. In upper panel, the subject was resting between two real visual stimuli. In lower panel, the subject imagined the stimuli between actual stimulus presentations.

whether image generation is really a distinct and dissociable component from the long- and short-term representational components of imagery, and where mental image generation is localized. Readers are directed to Tippett's (1992) article for an excellent review of the neuropsychology of image generation.

From the perspective of neural network computation, the distinctions proposed by psychologists between representations and processes often appear unnecessary, and one might therefore question the need to hypothesize a separate image generation process. For example, it is sometimes assumed that visual recognition requires a kind of matching process, separate from the visual representations themselves, to compare the representations active in the visual buffer with the corresponding long-term visual memory representations (e.g., Farah, 1984). However, such a process would be superfluous in an associative network in which the representations of visual knowledge at various levels of the visual system, embodied in the connection strengths, would be sufficient to activate the appropriate high-level object representations given an active representation in the visual buffer. Even so, the generation of an image might be expected to require mechanisms beyond those needed for object recognition. It is clear that every time one thinks about an object, one's visual system does not inexorably perform pattern completion and create a mental image. The activation of early, visual buffer representations by active, high-level visual memory representations does not appear to be as automatic as the reverse direction of activation flow. This is consistent with the existence of a component of the cognitive architecture that is needed for image generation but not for visual perception and object recognition.

Farah (1984) reviewed the neurological literature on imagery impairments and identified a set of cases in which perception was grossly intact. In subsequent years, a small number of additional cases of selectively impaired imagery have been reported (e.g., Grossi, Orsini, and Modafferi, 1986; Farah, Levine, and Calvanio, 1988; Riddoch, 1990; Goldenberg, 1992). On the face of things, this seems to imply the existence of a distinct image generation component. However, this conclusion must be viewed as tentative. Many of the cases had subtle visual perceptual impairments, whose role in the imagery impairment cannot be ruled out. Another possibility is a subtle visual memory impairment, as suggested by Goldenberg and Artner (1991). In support of this hypothesis, they showed that a group of subjects selected for left posterior brain damage, who were impaired on image generation tasks, were also impaired at making subtle visual discriminations, for example between a bear with pointy ears and a bear with rounded ears. However, it is possible that when confronted with a pair of such pictures, normal people generate an image of a bear to recall whether bears have pointy or rounded ears.

The localization of mental image generation has been a controversial topic. Although mental imagery was for many years assumed to be a function of the right hemisphere, Ehrlichman and Barrett (1983) pointed out that there was no direct evidence for this assumption. Farah (1984) noted a trend in the cases she reviewed for left posterior damage. Farah, Levine, and Calvanio (1988) suggested that the left temporo-occipital area may be critical. The recent focally damaged cases mentioned above (Grossi, Orsini, and Modafferi; Farah, Hammond, et al., 1988; Riddoch, 1991; Goldenberg, 1992) have supported this suggestion. The rarity of cases of image generation deficit suggests that this function may not be strongly lateralized in most people; however, when impairments are observed after focal unilateral damage, the left or dominant hemisphere is implicated.

Other neuropsychological methods have been brought to bear on the laterality of mental image generation and have revealed the same general trend toward left-hemisphere specialization, although exceptions exist. The ERP experiments on the generation of images described earlier showed greater effects on the left than the right, maximum over the temporo-occipital region. In addition, left temporo-occipital foci of activity have been noted by Goldenberg in several, but not all, of his rCBF studies (Goldenberg et al., 1987, 1989a, 1991, 1992). Charlot et al. (1992) also found left temporo-occipital cortex activated in their rCBF study. Kosslyn and his colleagues (1993) reported two experiments that they classified as requiring subjects to generate images from long-term visual memory. Although one of these experiments was aimed at comparing activation patterns for large and small images, and thus did not include a no-imagery baseline to allow inferences about image generation per se, the other showed greater activity in left temporal and occipital association areas associated with image generation.

Research with split-brain patients has produced variable results, although in all cases the left hemisphere has shown at least an initial or partial superiority for image generation (Farah et al., 1985; Kosslyn et al., 1985; Corballis and Sergent, 1988). Divided-field studies with normal subjects have offered less clear support. For example, Cohen (1975) and Farah (1986) found evidence consistent with a left-hemisphere locus for image generation. In contrast, Sergent (1989) reports finding opposite results in her divided-visual-field experiment. The possibility that there are different types of mental image generation, with differing hemispheric loci, has been proposed by Kosslyn (1988) and deserves further systematic study.

Conclusions

Research on the neural bases of mental imagery is a fairly recent development in neuropsychology, and our conclusions at present can therefore only be tentative. Nevertheless, a reasonably consistent picture seems to be emerging, across a variety of neuropsychological research methods. In this picture, mental imagery is the efferent activation of some subset of the brain's visual areas, subserving the same types of functions (what, where, color, spatial attention, and so on) in imagery and in perception. In addition, evidence from both brain-damaged patients and functional imaging of normal subjects' brains suggests that some of the visual areas shared with imagery include spatially mapped areas of the occipital lobe. This finding is particularly important for resolving the long-standing issue in cognitive psychology of the format of mental images.

The study of image generation has yielded results that are less clear and consistent, concerning both its existence as a distinct component of the cognitive architecture and its neuroanatomic localization. Although a body of evidence favors the existence of an image-generation mechanism localized to the posterior left hemisphere, discrepant data also exist, and further research will be needed to draw any firm conclusions.

ACKNOWLEDGMENTS The writing of this chapter was supported by ONR grant N00014-931-0621, NIMH grant R01 MH48274, and grant 90-36 from the McDonnell-Pew Program in Cognitive Neuroscience.

REFERENCES

Anderson, J. R., 1978. Arguments concerning representation for mental imagery. *Psychol. Rev.* 85:249–277.

Beauvois, M. F., and B. Saillant, 1985. Optic aphasia for colours and colour agnosia: A distinction between visual and visuo-verbal impairments in the processing of colours. *Cognitive Neuropsychol.* 2:1–48.

Behrmann, M., G. Winocur, and M. Moscovitch, 1992. Dissociation between mental imagery and object recognition in a brain-damaged patient. *Nature* 359:636–637.

Bisiach, E., and C. Luzzatti, 1978. Unilateral neglect of representational space. *Cortex*, 14:129–133.

Charlot, V., N. Tzourio, M. Zilbovicius, B. Mazoyer, M. Denis, 1992. Different mental imagery abilities result in different regional cerebral blood flow activation patterns during cognitive tasks. *Neuropsychologia* 30:565–580.

Cohen, G., 1975. Hemispheric differences in the utilization of advance information. In *Attention and Performance* V, P. M. A. Rabbit and S. Dornic, eds. New York: Academic Press.

Corballis, M. C., and J. Sergent, 1988. Imagery in a commissurotomized patient. *Neuropsychologia* 26:13–26.

De Vreese, L. P., 1991. Two systems for colour-naming defects: verbal disconnection vs colour imagery disorder. *Neuropsychologia*, 29:1–18.

DeRenzi, E., and H. Spinnler, 1967. Impaired performance on color tasks in patients with hemispheric lesions. *Cortex* 3:194–217.

Ehrlichman, H., and J. Barrett, 1983. Right hemisphere specialization for mental imagery: A review of the evidence. *Brain Cogn.* 2:39–52.

Farah, M. J., 1984. The neurological basis of mental imagery: A componential analysis. *Cognition* 18:245–272.

Farah, M. J., 1986. The laterality of mental image generation: A test with normal subjects. *Neuropsychologia* 24:541–551.

Farah, M. J., 1988. Is visual imagery really visual? Overlooked evidence from neuropsychology. *Psychol. Rev.* 95: 307–317.

Farah, M. J., 1989. The neural bases of mental imagery. *Trends Neurosci.* 12:395–399.

Farah, M. J., M. S. Gazzaniga, J. D. Holtzman, and S. M. Kosslyn, 1985. A left hemisphere basis for visual mental imagery? *Neuropsychologia* 23:115–118.

Farah, M. J., K. L. Hammond, D. N. Levine, and R. Calvanio, 1988. Visual and spatial mental imagery: Dissociable systems of representation. *Cognitive Psychol.* 20: 439–462.

Farah, M. J., D. N. Levine, and R. Calvanio, 1988. A case study of mental imagery deficit. *Brain Cogn.* 8:147–164.

Farah, M. J., and F. Peronnet, 1989. Event-related potentials in the study of mental imagery. *J. Psychophysiol.* 3:99–109.

Farah, M. J., F. Peronnet, M. A. Gonon, and M. H. Giard, 1988. Electrophysiological evidence for a shared representational medium for visual images and percepts. *J. Exp. Psychol. [Gen.]* 117:248–257.

Farah, M. J., F. Peronnet, L. L. Weisberg, and M. A. Monheit, (1989). Brain activity underlying mental imagery: Event-related potentials during image generation. *J. Cognitive Neurosci.* 1:302–316.

Farah, M. J., M. J. Soso, and R. M. Dasheiff, 1992. The visual angle of the mind's eye before and after unilateral occipital lobectomy. *J. Exp. Psychol. [Hum. Percept.]* 18: 241–246.

Finke, R. A., 1980. Levels of equivalence in imagery and perception. *Psychol. Rev.* 87:113–132.

Goldenberg, G., 1992. Loss of visual imagery and loss of visual knowledge: A case study. *Neuropsychologia* 30:1081–1099.

Goldenberg, G., and C. Artner, 1991. Visual imagery and knowledge about visual appearance in patients with posterior cerebral artery lesions. *Brain Cognition* 15:160–186.

Goldenberg, G., I. Podreka, M. Steiner, and K. Willmes, 1987. Patterns of regional cerebral blood flow related to memorizing of high and low imagery words: An emission computer tomography study. *Neuropsychologia* 25: 473–486.

Goldenberg, G., I. Podreka, M. Steiner, K. Willmes, E. Suess, and L. Deecke, 1989a. Regional cerebral blood flow patterns in visual imagery. *Neuropsychologia* 27:641–664.

Goldenberg, G., I. Podreka, F. Uhl, M. Steiner, K. Willmes, and L. Deecke, 1989b. Cerebral correlates of imagining colours, faces and a map: I. Spect of regional cerebral blood flow. *Neuropsychologia* 27:1315–1328.

Goldenberg, G., I. Podreka, M. Steiner, P. Franzen, and L. Deecke, 1991. Contributions of occipital and temporal brain regions to visual and acoustic imagery: A spect study. *Neuropsychologia* 29:695–702.

Goldenberg, G., M. Steiner, I. Podreka, L. Deecke, 1992. Regional cerebral blood flow patterns related to verification of low- and high-imagery sentences. *Neuropsychologia* 30:581–586.

Grossi, D., A. Orsini, and A. Modafferi, 1986. Visuoimaginal constructional apraxia: On a case of selective deficit of imagery. *Brain Cogn.* 5:255–267.

Intons-Peterson, M. J., 1983. Imagery paradigms: How vulnerable are they to experimenters' expectations? *J. Exp. Psychol. [Hum. Percept.]* 9:394–412.

Kosslyn, S. M., 1978. Measuring the visual angle of the mind's eye. *Cognitive Psychol.* 10:356–389.

Kosslyn, S. M., 1980. *Image and Mind*. Cambridge, Mass.: Harvard University Press.

Kosslyn, S. M., 1988. Aspects of a cognitive neuroscience of mental imagery. *Science* 240:1621–1626.

Kosslyn, S. M., N. M. Alpert, W. L. Thompson, V. Maljkovic, S. Weise, C. F. Chabris, S. E. Hamilton, S. L. Rauch, and F. S. Buonanno, 1993. Visual mental imagery activates topographically organized visual cortex: PET investigations. *J. Cognitive Neurosci.* 5:263–287.

Kosslyn, S. M., T. M. Ball, and B. J. Reiser, 1978. Visual images preserve metric spatial information: Evidence from studies of image scanning. *J. Exp. Psychol. [Hum. Percept.]* 4:47–60.

Kosslyn, S. M., J. D. Holtzman, M. J. Farah, and M. S. Gazzaniga, 1985. A computational analysis of mental image generation: Evidence from functional dissociations in split-brain patients. *J. Exp. Psychol. [Gen.]* 114:311–341.

Le Bihan, D., R. Turner, T. A. Zeffiro, C. A. Cuenod, P. Jezzard, and V. Bonnerot, in press. Activation of human primary visual cortex during visual recall: A magnetic resonance imaging study. *Proc. Nat. Acad. Sci. U.S.A.*

Levine, D. N., J. Warach, and M. J. Farah, 1985. Two visual systems in mental imagery: Dissociation of "What" and "Where" in imagery disorders due to bilateral posterior cerebral lesions. *Neurology* 35:1010–1018.

Maunsell, J. H. R., and W. T. Newsome, 1987. Visual processing in monkey extrastriate cortex. *Annu. Rev. Neurosci.* 10:363–401.

Paivio, A., 1971. *Imagery and Verbal Processes*. New York: Holt, Rinehart, and Winston.

Pylyshyn, Z. W., 1973. What the mind's eye tells the mind's brain: A critique of mental imagery. *Psychol. Bull.* 80:1–24.

Pylyshyn, Z. W., 1981. The imagery debate: Analogue media versus tacit knowledge. *Psychol. Rev.* 88:16–45.

Riddoch, M. J., 1990. Loss of visual imagery: A generation deficit. *Cognitive Neuropsychol.* 7:249–273.

Roland, P. E., L. Eriksson, A. Stone-Elander, and L. Widen, 1987. Does mental activity change the oxidative metabolism of the brain? *J. Neurosci.* 7:2373–2389.

Roland, P. E., and L. Friberg, 1985. Localization of cortical areas activated by thinking. *J. Neurophysiol.* 53:1219–1243.

Sergent, J., 1989. Image generation and processing of generated images in the cerebral hemispheres. *J. Exp. Psychol. [Hum. Percept.]* 15:170–178.

Shepard, R. N., 1978. The mental image. *Am. Psychol.* 33: 125–137.

Tippett, L. J., 1992. The generation of visual images: A review of neuropsychological research and theory. *Psychol. Bull.* 112:415–432.

Uhl, F., G. Goldenberg, W. Lang, G. Lindinger, M. Steiner, and L. Deecke, 1990. Cerebral correlates of imagining colours, faces and a map: II. Negative cortical DC potentials. *Neuropsychologia* 28:81–93.

Ungerleider, L. G., and M. Mishkin, 1982. Two cortical visual systems. In *Analysis of Visual Behavior*, D. J. Ingle, M. A. Goodale, and R. J. W. Mansfield, eds. Cambridge, Mass.: MIT Press.

63 Imagery and Memory

ALLAN PAIVIO

ABSTRACT Thirty years of modern research and theory on imagery and episodic memory are reviewed in two parts. Part 1 covers the first 20 years, during which most of the basic effects of imagery variables were established. Theoretical lines were drawn between such modality-specific approaches as dual-coding theory (DCT) and more abstract (e.g., propositional) views. Part 2 updates the review, focusing on the explanatory adequacy of DCT as compared to approaches that emphasize modality-neutral representations and processes. As in the earlier period, DCT and other modality-specific theories appear to explain the broadest range of memory effects of imagery variables.

Imagery has been controversial throughout much of its psychological history in Western society. Viewed as the essence of memory for centuries following Simonides' invention of the method of loci, it was rejected as an educational memory technique during the Reformation as part of the general iconoclastic movement of the time. Then it fell victim to the behavioristic attack on all mentalistic concepts. Today, this iconoclasm of the mind reappears in at least some computational approaches to cognition. The alternatives to imagery have always been more abstract forms of internal representation, ranging from natural language to formal propositions.

This general dialectic provides the theme for my review of the study of imagery and memory over the last 30 years. Something like the classical view persists among researchers who assume that the underlying representations for imagery are concrete and multimodal, directly reflecting their perceptual origins. Dual-coding theory (DCT; Paivio, 1971, 1986, 1992a) is based on that view. Others have proposed an abstract substrate without rejecting the reality of imagery. Kosslyn (1980), for example, suggested that propositions and computational processes account for the generation and manipulation of images, but the generated images themselves contain "emergent properties" that make them functionally useful. Still others explain imagery effects in terms of amodal structures and processes without relying on computational modeling. Specific alternatives are described in appropriate contexts.

Several important topics are excluded because of space limitations. I have omitted neuropsychological research because the general topic is covered in the chapters by Farah and Kosslyn, and detailed reviews from my own and other theoretical perspectives are available elsewhere (e.g., Paivio and te Linde, 1982; Paivio, 1986; Tippett, 1992; Goldenberg, 1993). I only touch on picture-memory studies, although they buttress theoretical conclusions arising from the more numerous studies of imagery and memory for language (see Paivio, 1986, ch. 8) that are the focus of this chapter. Finally, I do not discuss recent connectionistic approaches to cognition, because they have not yet been applied to the complex memory functions of imagery that have been empirically discovered and explained in noncomputational terms (cf. Paivio, 1992b).

The first part of this chapter reviews the early period of modern research, emphasizing outcomes that have had a lasting influence on subsequent developments. The second part brings the story up to date.

The first 20 years

Major advances in our understanding of imagery and memory occurred over a 20-year period that began in the early 1960s. Imagery was systematically operationalized in terms of measures of the image-evoking capacity of items; imagery instructions and other procedures designed to affect the availability of imagery in memory tasks; and individual differences in imagery abilities. The effects of these variables motivated specific imagery hypotheses as well as dual-coding theory. The following question accompanied these developments: Is imagery *really* the effective empirical variable and the most appropriate explanatory concept, or can the memory effects be better explained in other ways?

ALLAN PAIVIO Department of Psychology, University of Western Ontario, London, Ontario, Canada

Concreteness Effects and the Conceptual Peg Hypothesis The initial studies focused on the effects of noun concreteness in paired-associate learning of adjective-noun and noun-noun pairs (Paivio, 1963, 1965). The pertinent results were that pairs were learned more easily when the nouns were concrete rather than abstract, and that concreteness of the stimulus noun was more helpful than concreteness of the response.

The effects were predicted from the conceptual peg hypothesis, which stated that concrete nouns readily evoke integrated images that provide mediational links to the associated responses. Presentation of a concrete noun as a recall cue reactivates the integrated image, which can be decoded into the appropriate verbal response. The imagery interpretation was supported by rating data that established that concrete words evoke images more easily than abstract words. The stimulus advantage of concrete nouns was shown to be due to their function as retrieval cues during recall rather than to their position in the pair during study (e.g., Yarmey and O'Neill, 1969).

Item recognition and free-recall performances also proved to be superior for concrete nouns compared to abstract nouns (Olver, 1965; Paivio, 1967), although the effects are weaker than in paired-associate recall, presumably because those tasks lack explicit cues for image arousal during recall. The effect in the noncued tasks could be explained by the hypothesis that the probability of verbal-imaginal dual coding is higher for concrete than for abstract words and that the two codes are retrieved independently, with additive effects on memory.

Begg (1972) obtained evidence for both code additivity and imagery integration effects in memory for words and phrases. Concrete materials were free-recalled twice as well as abstract ones, suggesting an additive increment due to imagery. Cuing with one word from a phrase substantially increased recall for the other word, but only for concrete phrases, suggesting that concrete phrases are stored as integrated images and that presentation of one word as a cue redintegrates the entire image, whereas abstract phrases are stored nonintegratively as separate words.

Finally, the concreteness effect is weak or absent in tasks that require memory for the temporal order of items (e.g., Paivio and Csapo, 1969). The DCT explanation is that the imagery system is less effective than the verbal system in sequential processing of discrete items, a controversial view that is justified on balance by the empirical evidence (see Paivio, 1986, 171–175).

Alternative Interpretations of Concreteness Effects Verbal associative meaningfulness (*m*) was the first alternative to imagery as an explanation of concreteness effects. It turned out that controlling *m* did not reduce the positive relation between word imagery and recall, whereas controlling imagery essentially eliminated the effect of *m* (e.g., Paivio, Smythe, and Yuille, 1968). A related explanation based on stimulus-response transition probability (TP) was investigated by Kusyszyn and Paivio (1966) using noun-adjective and adjective-noun pairs in which the nouns were either concrete or abstract and the target responses had either few (high TP) or many (low TP) alternative associates. The relevant outcome was that concreteness and TP had independent and additive effects on recall. We shall see later that TP, now called associative set size, has been used recently to test context-availability explanations of concreteness effects on memory.

Paivio (1965) disconfirmed item distinctiveness as another possible explanation of the concreteness effect in paired-associate learning, leaving image-mediated association as the favored explanation at that time. Subsequent research, however, implicated both distinctiveness and associative processes in the effect (e.g., Anderson et al., 1977; Paivio, 1971, 293–294).

More than 20 potentially confounding attributes of concrete and abstract words were included in a factor analytic study (Paivio, 1968) along with memory scores for the words. The analysis yielded a strong imagery-concreteness factor defined by several rating and imagery–reaction time measures. Memory scores loaded significantly on this factor, most strongly when the words served as stimuli in paired-associate recall. The best single predictor of recall was the rated imagery value of the words.

Coding Instructions The separate and joint effects of imagery and verbal coding instructions generally parallel the effects of word concreteness on memory. Imagery coding of words usually enhances recall relative to control levels in paired-associate learning, presumably because imagery instructions increase the probability of integrative organization and dual coding in the task. Consistent with the organization hypothesis, instructions to construct integrated images of the referents of word pairs produces higher cued recall than instructions to generate separate images for each pair member (Bower, 1970). Moreover, a recall increment from free to cued recall was greater under integrated than separate imagery conditions (Begg, 1973). Code independence was confirmed by additive effects of imaginal and verbal coding in free recall, with imagery contributing more than verbal coding to their combined effect (Paivio, 1975).

Verbal mediational theories could not explain the separate and independent memory contributions of imaginal and verbal coding, and alternative explanations in terms of levels of processing, elaborative encoding, and abstract knowledge structures also proved to be inadequate. For example, Paivio and Lambert (1981) had French-English bilinguals repeat some English words, translate others into French, and image to a third subset. A subsequent incidental memory test showed that recall increased progressively from repeat, to translate, to image conditions. The translation effect suggests that bilingual translation equivalents are stored independently (cf. Paivio and Desrochers, 1980; Paivio, Clark, and Lambert, 1988). The further increase under image coding is problematic for elaboration and levels-of-processing theories because there is no a priori ground for assuming that imaging is any deeper or more elaborate than translating words. Finally, experiments by Keenan and Moore (1979) in response to a critique of "pictorial" images by Neisser and Kerr (1973) reinforced the view that effective memory images are modality specific rather than abstract knowledge structures.

Strategy Reports as Evidence Experimental subjects were often asked about the strategies they used to learn word pairs (see Paivio, 1971). They consistently reported using more imagery than verbal (or other) strategies with concrete words, and the reverse with abstract words. Moreover, frequency of reported imagery correlated substantially with learning scores, especially for concrete words (cf. Richardson, 1978). Consistent with the conceptual peg hypothesis, the correlation between imagery use and paired-associate learning was higher when the scores were computed for items serving as stimuli ($r = .56$) than for the same items when they served as responses ($r = .39$). A causal interpretation of the correlational data was strengthened by the strategy-specific nature of the relations: Subjects were free to report verbal strategies and they often did, but verbal strategies did not correlate well with learning.

Individual Differences Imagery questionnaires and spatial ability tests have long been used to predict memory performance, with mixed results: Sometimes they correlated positively and sometimes not at all with memory performance (e.g., see Paivio, 1971, ch. 14; Ernest, 1977). Thus, individual difference measures diverged from item concreteness, imagery instructions, and strategy reports as defining operations for imagery in memory studies.

Other Modalities of Imagery and Memory Nonvisual perceptual and imagery modalities received some attention in the early research. For example, Paivio, Philipchalk, and Rowe (1975) found that, like pictures, environmental sounds were inferior to words in sequential memory tasks unless the sounds could be easily coded verbally, but sounds were remembered as well as words in item memory tasks. Paivio and Okovita (1971) found that blind people benefited from the auditory but not the visual imagery value of words in paired-associate learning, whereas sighted people benefited relatively more from visual imagery. Similarly, Conlin and Paivio (1975) found that deaf subjects were affected uniquely by the signability of words in paired-associate learning, suggesting that they make use of a gestural code.

Such results motivated an orthogonal model of the relation between verbal-nonverbal symbolic dichotomy and specific sensorimotor modalities (Paivio, 1972): Language and nonverbal objects and events alike can be visual (printed words versus visual objects), auditory (speech versus environmental sounds), or haptic (tactual and motor feedback from the activity of writing versus manipulating objects). Gustatory, olfactory, and affective modalities are entirely nonverbal because we do not construct linguistic symbols from tastes, smells, or emotional experiences. The description refers to sensorimotor modalities but it applies as well to imagery and internalized language. This descriptive scheme has been augmented by assumptions concerning differences in the structural and functional properties of the subsystems (Paivio, 1986).

A Problem for Integrative Imagery Imagery theories could explain most of the pertinent findings that emerged during this period, but one result was problematic for any theory that stressed the unique integrative role of imagery. Marschark and Paivio (1977) found that concrete sentences were generally remembered better than abstract sentences, but, when recall was successful, it was equally integrated or holistic for both types of sentences. This equivalence seemed inconsistent with the DCT hypothesis that imagery enhances integrative memory for concrete materials. The hypothesis would have been upheld had subjects used imagery when they recalled abstract sentences integratively, but subjects reported using imagery only with concrete sentences. Marschark and Paivio concluded that dual coding could account for the concrete-abstract differences in overall recall but that some common mechanism is needed to explain the equivalent integrative memory effects. We shall see later that the problem has since been resolved in a way that is consistent with dual-coding theory.

Progress in the 1980s and beyond

The conceptual gap between abstract representational theories and dual-coding theory was reduced in the 1980s when a number of researchers proposed triple-coding theories that included representations corresponding to the surface features of language and to nonverbal information along with abstract representations (e.g., Potter and Faulconer, 1975; Anderson, 1983; van Dijk and Kintsch, 1983; Snodgrass, 1984). Such theories are more powerful than DCT, but the problem is to specify the conditions under which the different representational levels are used. This has not been fully realized at present, and it suffices to note here that the approaches have generated relevant memory studies in which imagery plays a role (e.g., Perrig and Kintsch, 1985).

Other researchers continue to seek explanations of imagery phenomena in terms of general processes that do not depend on modality-specific representations, while at the same time there is an expanding interest in specific modalities of imagery. Moreover, measures of individual differences continue to be inconsistent as predictors of memory performance. I turn to these developments, beginning with a series of alternatives to imagery theories of memory.

Schematic Knowledge Structures Day and Bellezza (1983) reported results that challenged DCT explanations of the effects of imagery variables on cued recall. Subjects were required to form composite images to word pairs that varied in concreteness and pair relatedness. They rated the vividness of their images and then were unexpectedly given a cued recall test. The critical results were that unrelated concrete pairs were rated lower in imagery but nonetheless were recalled better than the related abstract pairs. Day and Bellezza concluded that the results contradicted DCT and other imagery theories. Their explanation was in terms of "organized generic knowledge structures" (schemata) based on "relations among objects in the physical world rather than their mode of representation in memory" (1983, 256–257).

Paivio, Clark, and Khan (1988) used the same procedures to show that DCT could explain the Day and Bellezza results. Subjects were presented noun pairs that varied in concreteness and relatedness. Different groups rated the pairs on composite imagery or relatedness, and then were unexpectedly given a cued recall test. The crucial result was that superior recall for unrelated concrete pairs occurred under imagery but not relatedness instructions, suggesting that the Day and Bellezza concreteness effect depended on the mode of representation and not simply on generic knowledge about the relations between objects.

Subjects in a second experiment rated imagery value, recalled pairs, and completed a questionnaire that tested for image retrieval. The important result was that composite images were retrieved better given stimuli from unrelated concrete pairs than from related abstract pairs. This was predicted from the conceptual peg hypothesis, which emphasizes the potent effect of stimulus concreteness on image retrieval and recall.

Context Availability The context availability hypothesis emphasizes differential availability of information from prior knowledge rather than imagery to explain concreteness effects (e.g., Kieras, 1978; Schwanenflugel and Shoben, 1983). The relevant studies have generally found that presenting abstract and concrete materials in a supportive context (e.g., in paragraphs or meaningfully related pairs) reduced concrete-abstract differences in recall.

Proponents of the hypothesis have focused on differences in the number of implicit associates that concrete and abstract words can evoke, and contradictory claims are made: Concrete words are said to have a recall advantage either because they have a smaller associative set than abstract words—thereby facilitating access to relevant associates—or because they have larger associative sets, enriching their retrieval context (for a review, see Nelson and Schreiber, 1992).

The debate seems to have been empirically resolved. Nelson and Schreiber (1992) found that ratings of concreteness and associative set size were only slightly correlated, and that the two variables had independent effects in cued recall, so that concrete words were recalled better than abstract words, and words with smaller associative sets better than those with larger sets. Concreteness also facilitated free recall, whereas set size had no effect. Nelson and Schreiber concluded that context availability does not account for concreteness effects, and that the results are consistent instead with several other theories, including DCT.

Recall that the above results were anticipated earlier, in that response set size (transition probability) and noun concreteness had independent effects in cued recall of adjective-noun pairs (Kusyszyn and Paivio, 1966). Moreover, the two variables were factorially independent in a study of noun attributes and recall (Paivio, 1968). Nelson and Schreiber went beyond the early studies, however, in that they used a larger pool of words and additional recall conditions (e.g., intra- and extra-list cues), so that the generality and limiting conditions of the effect of associative set size are now better understood.

Schwanenflugel, heretofore a proponent of the contextual availability hypothesis, has also clarified the issues. Schwanenflugel, Akin, and Luh (1992) equated concrete and abstract words on context availability as measured by ratings. Three free-recall experiments showed that concrete words were recalled better than abstract words in the following conditions: when context availability was controlled, but not when it was confounded with imageability of the words; following ratings of imagery but not ratings of context availability; and by subjects who reported using an imagery strategy, but not by those who did not report imagery. Schwanenflugel, Akin, and Luh concluded that the results supported a strategic imagery view of concreteness effects in free recall, but not context availability or an automatic-imagery interpretation.

Dual-coding theory accommodates and extends the strategic-imagery interpretation. Our earlier research showed that imagery-coding instructions facilitated free recall of concrete words even under incidental learning conditions in which subjects rated ease of image generation to each word and then unexpectedly were asked to recall the words (Paivio, 1975). Thus, it is unlikely that imagery was used as a learning strategy, although it may have been used strategically during retrieval. The DCT position in any case is that recall will be enhanced if subjects happen to remember the images, even if such recollections are not the result of deliberate learning or retrieval strategies.

Sadoski, Goetz, and Fritz (1993) provided evidence on the context availability hypothesis as applied to concrete and abstract sentences and paragraphs that were carefully matched according to several criteria. Other attributes were also measured. Recall was substantially higher for concrete than for abstract materials in both immediate and delayed cued-recall tests. Among the other attributes, familiarity of the sentences' and paragraphs' topics is especially relevant because some versions of the context availability hypothesis (see Schwanflugel, Akin, and Luh, 1992, 97) emphasize availability of prior knowledge to explain the concreteness effect, and familiarity is a measure of such knowledge. Concreteness was far superior to familiarity as a recall predictor. Sadoski, Goetz, and Fritz concluded that their results are best explained by dual-coding theory.

Relational and Distinctiveness Processing Marschark and Hunt (1989) proposed an alternative to DCT and other imagery-based explanations of concreteness effects in paired-associate learning, according to which memory for a response word depends on activation of both relational and distinctive information during pair encoding and subsequent response retrieval. They emphasized that concreteness enhances distinctiveness processing, which will benefit recall provided that relational processing has occurred (p. 711). Thus, the superiority of concrete over abstract pairs should increase under conditions that encourage relational processing. The predicted interactions were generally confirmed: Concreteness effects were greater (a) under cued recall than under free recall, presumably because cuing ensured that encoded relations would be reactivated at retrieval; and (b) when relational encoding was encouraged by use of related pairs and instructions to integrate or associate the pair members during study.

Paivio, Walsh, and Bons (in press) argued that Marschark and Hunt's approach and DCT differ only on one point that has predictive consequences: Whereas Marschark and Hunt asserted that concreteness effects are dependent on relational processing, DCT assumes that concreteness and relational variables are independent. We tested these alternatives by comparing free and cued recall of noun pairs that varied in concreteness and pair relatedness. According to Marschark and Hunt's theory, concreteness effects should be greater for responses from related than unrelated pairs—especially in cued recall, because cuing ensures access to relational information at retrieval (see Marschark and Hunt, 1989, 714). The DCT prediction was that the effects of concreteness and relatedness would be independent and additive.

The results of two experiments generally supported DCT: In experiment 1, concreteness and relatedness were independent and additive in both cued and free recall; in experiment 2, the two variables were independent in cued recall and they interacted reliably in free recall only with subjects and not with items as the random factor. The complete absence of any interaction in cued recall, where it should have emerged most clearly according to Marschark and Hunt, is the strongest support for DCT.

THE IMAGERY INTEGRATION PUZZLE REVISITED The foregoing experiments also addressed the long-standing puzzle concerning imagery and integrative memory (Marschark and Paivio, 1977). The problem was that, although overall recall was better for concrete than for abstract sentences, the proportion of sentences recalled as integrated wholes did not differ for the two types. Paivio, Walsh, and Bons (in press) proposed an explanation derived from the general assumptions of DCT: Verbal associative processes provide an integrative mechanism that is equally available to abstract and concrete materials, whereas imagery provides an additional integrative base that is relatively more available to concrete material.

We tested the hypothesis using the increment from free to cued recall as the principal measure of integrative memory (cf. Begg, 1972). A further measure was the proportion of responses that were recalled together with their stimulus partners in the free-recall condition. Both measures confirmed DCT predictions, most persuasively in that integrative recall scores were high for concrete unrelated pairs as well as for concrete and abstract related pairs, whereas integration was very low for abstract unrelated pairs. In a nutshell, strong verbal associative relations were *necessary* for integrative recall of abstract pairs, whereas high imagery value was *sufficient* for integrative recall of concrete pairs.

OTHER MODALITIES OF IMAGERY Systematic research has begun on the separate and joint contributions to memory of different modalities of imaged or perceptually presented objects. Winnick and Brody (1984) found that words high in both auditory and visual imagery (e.g., *clock*) were recalled no better than words high in only one modality (e.g., *cigar*, *click*), and instructions to image in two modalities was no more effective than instructions to image in only one. Paivio (1989b) obtained similar results along with other findings that prompted two possible explanations.

One possibility is that visual presentation of a word list selectively interfered with recall of words high on visual imagery or both visual and auditory imagery, and of words learned under visual or auditory-visual instructions. A procedure designed to reduce such interference (Paivio, Clark, and Bons, 1993) yielded an additive effect of visual and auditory word imagery, but dual-modality imagery instructions still did not produce better recall than single-modality instructions. A second explanation followed from the observation (Paivio, 1989b) that visual images were often reported by subjects who were instructed to use auditory imagery, whereas auditory images were infrequently reported by subjects instructed to use visual imagery. This asymmetry in cross-modal coding could have obscured any additive effect of audiovisual instructions because single-modality auditory instructions also induced dual coding. The issue remains to be resolved.

Perceptual-memory studies avoid the above problems. Lyman and McDaniel (1990) found that the recall of odors increased significantly when each studied odor was coupled with its name or a source-object picture. Combining all three codes tended to produce a further (but nonsignificant) increase. Thompson and Paivio (in press) studied the joint effects of auditory and visual components of such audiovisual objects as the telephone, by capitalizing on a critical feature of the repetition lag effect in free recall: Successive (zero-lag) repetition of a word or picture results in a less-than-additive recall increment, whereas increasing the number of intervening items between repetitions augments recall to a point that reaches and sometimes exceeds additivity. Earlier DCT research using that procedure demonstrated that picture-name repetitions had additive memory effects even at zero lag (see Paivio, 1975). Thompson and Paivio (1994) accordingly used only zero-lag repetitions of object pictures, related environmental sounds, or picture-sound pairs. The picture-sound condition produced additive recall relative to two presentations of either pictures or sounds, even when the two modalities (e.g., a telephone and its ring) were presented simultaneously.

Other studies have shown that motoric imagery is independent of other modalities. Saltz and Nolan (1981) compared the effects of visual, motoric, and verbal competition on memory for sentences learned using visual imagery, motor enactment, or verbal mnemonic instructions. Recall was disrupted when the modality of the competition task matched the modality of the mnemonic technique but not when the two differed, thereby confirming the mnemonic independence of imaginal and verbal codes and revealing the independence of visual and motor modalities of imagery. A variety of other procedures have supported the distinction between motor and visual imagery in memory (e.g., Hall, 1980; Englekamp, 1988) as well as a more subtle kinematic-motor distinction (Zimmer and Engelkamp, 1985).

Finally, the modality distinctions revealed earlier by the observation that blind people profit from auditory but not visual imagery value of words in associate learning (Paivio and Okovita, 1971) have been challenged by demonstrations of similar memory effects of visual imagery manipulations with blind and sighted subjects (e.g., Zimler and Keenan, 1983). De Beni and Cornoldi (1988; see also Paivio, 1986, 157–158) critically evaluated those studies and then showed that memory performance of blind subjects suffered relative to that of controls when the task required construction of multiple interactive images. Interpretive complexities in this research area remain to be resolved.

Individual Differences Current studies of individual differences in imagery and memory could be guided by general theories (e.g., Kosslyn and Jolicoeur, 1980; Paivio, 1986, ch. 6). From a DCT perspective, for example, consideration could be given to the person's overall ability to use imagery and/or verbal processes during encoding, to organize and transform images (e.g., generating interactive images), to decode images into words referentially during retrieval, and so on. Such component abilities have not been systematically related to memory. Instead, researchers have continued to focus on imagery vividness tests and, occasionally, imagery thinking habits as measured by the Individual Difference Questionnaire (IDQ; Paivio and Harshman, 1983). Spatial ability tests, frequently used in earlier research (see Ernest, 1977), have been infrequent in recent memory studies.

The popular tests continue to be inconsistent as predictors. Thus, scores on Marks's (1973) Vividness of Visual Imagery Questionnaire (VVIQ) correlated positively with memory performance in some studies (e.g., Denis, 1982; Hanggi, 1988) but not in others (e.g., Chara and Hamm, 1989; Cohen and Saslona, 1990). The VVIQ even showed a negative relation between imagery and color memory (e.g., Heuer, Fischman, and Reisberg, 1986), apparently because imagery vividness contributes to false alarms and to response choices more distant from the correct choice than those associated with less vivid images. Relations between imagery scores on the IDQ and memory, which also varied in earlier research (see Paivio, 1971, ch. 14; Ernest, 1977), have recently correlated positively with color recall in a picture-memory test (Cohen and Saslona, 1990) and with short-term memory scores (Tanwar and Malhotra, 1990). The reasons for the variable results remain unclear.

Other individual difference variables have been found to influence the mnemonic effectiveness of imagery. Pressley et al. (1987) related children's learning of concrete sentences under imagery instructions or no-strategy control conditions to measures of their short-term memory and verbal competence. They found that greater short-term memory and verbal competence were associated with better sentence learning in the imagery condition but not in the control condition. Among other things, these findings implicate dual coding in that verbal and imagery skills contributed to concrete sentence learning.

The methodological and conceptual problems in this area have been comprehensively evaluated by Katz (1983). He proposed a model in which past and present environmental influences interact with individual differences in imagery to produce different effects in memory and other cognitive tasks.

Some general conclusions

Our scientific understanding of imagery and memory increased rapidly during the first 20 years of modern research, beginning in the 1960s. Imagery was systematically operationalized, and its effects on episodic memory were distinguished from effects attributable to other variables. The image-evoking value of words, imagery instructions, and imagery strategies were especially predictive of memory for language materials. The justified explanations were that composite images formed during study were redintegrated by high-imagery stimuli during response retrieval in cued-recall tasks; and that images and verbal codes contributed additively to recall in the absence of explicit retrieval cues.

Computational theories assumed alternatively that the internal memory code is amodal and abstract. Unable to accommodate modality-specific aspects of memory, such approaches evolved into theories in which a propositional base was augmented by components corresponding more directly to imagery and language. Other alternatives (e.g., context availability, relational-distinctiveness processing) likewise have not displaced imagery explanations, which are expanding to take account of different sensory modalities of imagery. Yet to be resolved is the continued inconsistency of individual differences in imagery as memory predictors.

ACKNOWLEDGMENTS The author's research and preparation of this chapter were supported by grant OGP0004866 from the Natural Sciences and Engineering Research Council of Canada.

REFERENCES

Anderson, J. R., 1983. *The Architecture of Cognition.* Cambridge, Mass.: Harvard University Press.

Anderson, R. C., E. T. Goetz, J. W. Pichert, and H. M. Halff, 1977. Two faces of the conceptual peg hypothesis. *J. Exp. Psychol. [Hum. Learn. Mem.]* 3:142–149.

Begg, I., 1972. Recall of meaningful phrases. *J. Verb. Learn. Verb. Be.* 11:431–439.

Begg, I., 1973. Imagery and integration in the recall of words. *Can. J. Psychol.* 27:159–167.

Bower, G. H., 1970. Imagery as a relational organizer in associative learning. *J. Verb. Learn. Verb. Be.* 9:529–533.

Chara, P. J., and D. A. Hamm, 1989. An inquiry into the construct validity of the Vividness of Visual Imagery Questionnaire. *Percept. Mot. Skills* 69:127–136.

Cohen, B. H., and M. Saslona, 1990. The advantage of being an habitual visualizer. *J. Mental Imagery* 14:101–112.

Conlin, D., and A. Paivio, 1975. The associative learning of the deaf: The effects of word imagery and signability. *Mem. Cognition* 3:335–340.

Day, J. C., and F. S. Bellezza, 1983. The relation between visual imagery mediators and recall. *Mem. Cognition* 11: 251–257.

De Beni, R., and C. Cornoldi, 1988. Imagery limitations in totally congenitally blind subjects. *J. Exp. Psychol. [Learn. Mem. Cogn.]* 14:650–655.

Denis, M., 1982. On figuration components of mental representations. In *Cognitive Research in Psychology*, F. Klix, J. Hoffmann, and E. van der Meer, eds. Berlin: Verlag der Wissenchaften.

Engelkamp, J., 1988. Images and actions in verbal learning. In *Cognitive and Neuropsychological Approaches to Mental Imagery*, M. Denis, J. Engelkamp, and J. T. E. Richardson, eds. Amsterdam: Martinus Nijhoff.

Ernest, C. H., 1977. Imagery ability and cognition: A critical review. *J. Mental Imagery* 1:181–216.

Goldenberg, G., 1993. The neural basis of mental imagery. *Baillière's Clinical Neurology* 2:265–286.

Hall, C. R., 1980. Imagery for movement. *J. Hum. Movement Studies* 6:252–264.

Hanggi, D., 1988. Differential aspects of visual short- and long-term memory. Third Conference of the European Society for Cognitive Psychology Symposium: Imagery and the processing of visuo-spatial representations (Cambridge, England). *Eur. J. Cognitive Psychol.* 1:285–292.

Heuer, F., D. Fischman, and D. Reisberg, 1986. Why does vivid imagery hurt colour memory? *Can. J. Psychol.* 40: 161–175.

Katz, A., 1983. What does it mean to be a high imager? In *Imagery, Memory, and Cognition: Essays in Honor of Allan Paivio.* J. C. Yuille, ed. Hillsdale, N.J.: Erlbaum.

Keenan, J. M., and R. E. Moore, 1979. Memory for images of concealed objects: A reexamination of Neisser and Kerr. *J. Exp. Psychol. [Hum. Learn. Mem.]* 5:374–385.

Kieras, D., 1978. Beyond pictures and words: Alternative information-processing models for imagery effects in verbal memory. *Psychol. Bull.* 85:532–554.

Kosslyn, S. M., 1980. *Image and Mind.* Cambridge, Mass.: Harvard University Press.

Kosslyn, S. M., and P. Jolicoeur, 1980. A theory-based approach to the study of individual differences in mental imagery. In *Aptitude, Learning and Instruction: Cognitive Processes Analysis of Aptitude*, vol. 1. R. E. Snow, P. A. Federico, and W. E. Montague, eds. Hillsdale, N.J.: Erlbaum.

Kusyszyn, I., and A. Paivio, 1966. Transition probability, word order, and noun abstractness in the learning of adjective-noun paired associates. *J. Exp. Psychol.* 71:800–805.

Lyman, B. J., and M. A. McDaniel, 1990. Memory for odors and odor names: Modalities of elaboration and imagery. *J. Exp. Psychol. [Learn. Mem. Cogn.]* 16:656–664.

Marks, D. F., 1973. Visual imagery differences in the recall of pictures. *B. J. Psychol.* 64:17–24.

Marschark, M., and R. R. Hunt, 1989. A reexamination of the role of imagery in learning and memory. *J. Exp. Psychol. [Learn. Mem. Cogn.]* 15:710–720.

Marschark, M., and A. Paivio, 1977. Integrative processing of concrete and abstract sentences. *J. Verb. Learn. Verb. Be.* 16:217–231.

Neisser, U., and N. Kerr, 1973. Spatial and mnemonic properties of visual images. *Cognitive Psychol.* 5:138–150.

Nelson, D. L., and T. A. Schreiber, 1992. Word concreteness and word structure as independent determinants of recall. *J. Mem. Lang.* 31:237–260.

Olver, M. A., 1965. Abstractness, imagery, and meaningfulness in recognition and free recall. Master's thesis, University of Western Ontario.

Paivio, A., 1963. Learning of adjective-noun paired-associates as a function of adjective-noun word order and noun abstractness. *Can. J. Psychol.* 17:370–379.

Paivio, A., 1965. Abstractness, imagery, and meaningfulness in paired-associate learning. *J. Verb. Learn. Verb. Be.* 4:32–38.

Paivio, A., 1967. Paired-associate learning and free recall of nouns as a function of concreteness, specificity, imagery, and meaningfulness. *Psychol. Rep.* 20:239–245.

Paivio, A., 1968. A factor-analytic study of word attributes and verbal learning. *J. Verb. Learn. Verb. Be.* 7:41–49.

Paivio, A., 1971. *Imagery and Verbal Processes.* New York: Holt, Rinehart, and Winston. (Reprinted 1979, Hillsdale, N.J.: Erlbaum.)

Paivio, A., 1972. Symbolic and sensory modalities of memory. In *Third Western symposium on learning: Cognitive learning.* M. E. Meyer, ed. Bellingham: Western Washington State College.

Paivio, A., 1975. Coding distinctions and repetition effects in memory. In. *The Psychology of Learning and Motivation,* vol. 9, G. H. Bower, ed. New York: Academic Press.

Paivio, A., 1986. *Mental Representations: A Dual-Coding Approach.* New York: Oxford University Press.

Paivio, A., 1989a. A dual coding perspective on imagery and the brain. In *Neuropsychology of Visual Perception,* J. Brown, ed. Hillsdale, N.J.: Erlbaum.

Paivio, A., 1989b. Modalities of imagery and memory. In A. F. Bennett and K. McConkey, eds. *Cognition in Individual and Social Contexts.* Selected/revised papers in vol. 3, 24th International Congress of Psychology. Amsterdam: Elsevier.

Paivio, A., 1992a. Dual coding theory: Retrospect and current status. *Can. J. Psychol.* 45:255–287.

Paivio, A., 1992b. Imagery in the computational age. Symposium presentation, 25th International Congress of Psychology. Abstract in the *Int. J. Psychol.* 3:144.

Paivio, A., J. Clark, and T. Bons, 1993. Visual and auditory imagery effects on recall. Unpublished manuscript.

Paivio, A., J. M. Clark, and M. Khan, 1988. Effects of concreteness and semantic relatedness on composite imagery ratings and cued recall. *Mem. Cognition* 16:422–430.

Paivio, A., J. M. Clark, and W. E. Lambert, 1988. Bilingual dual-coding theory and semantic repetition effects on recall. *J. Exp. Psychol. [Learn. Mem. Cogn.]* 14:163–172.

Paivio, A., and K. Csapo, 1969. Concrete image and verbal memory codes. *J. Exp. Psychol.* 80:279–285.

Paivio, A., and K. Csapo, 1973. Picture superiority in free recall: imagery or dual coding? *Cognitive Psychol.* 5:176–206.

Paivio, A., and A. Desrochers, 1980. A dual-coding approach to bilingual memory. *Can. J. Psychol.* 34:388–399.

Paivio, A., and R. A. Harshman, 1983. Factor analysis of a questionnaire on imagery and verbal habits and skills. *Can. J. Psychol.* 37:461–483.

Paivio, A., and W. Lambert, 1981. Dual coding and bilingual memory. *J. Verb. Learn. Verb. Be.* 20:532–539.

Paivio, A., and H. W. Okovita, 1971. Word imagery modalities and associative learning in blind and sighted subjects. *J. Verb. Learn. Verb. Be.* 10:506–510.

Paivio, A., R. Philipchalk, and E. J. Rowe, 1975. Free and serial recall of pictures, sounds, and words. *Mem. Cognition* 3:586–590.

Paivio, A., P. C. Smythe, and J. C. Yuille, 1968. Imagery versus meaningfulness of nouns in paired-associate learning. *Can. J. Psychol.* 22:427–441.

Paivio, A., and J. te Linde, 1982. Imagery, memory, and the brain. *Can. J. Psychol.* 36:243–272.

Paivio, A., M. Walsh, and T. Bons, in press. Concreteness effects on memory: When and Why? *J. Exp. Psychol. [Learn. Mem. Cogn.]*

Perrig, W., and W. Kintsch, 1985. Propositional and situational representations of text. *J. Mem. Lang.* 24:503–518.

Potter, M. C., and B. A. Faulconer, 1975. Time to understand pictures and words. *Nature* 253:437–438.

Pressley, M., T. Cariglia-Bull, S. Deane, and W. Schneider, 1987. Short-term memory, verbal competence, and age as predictors of imagery instructional effectiveness. *J. Exp. Child Psychol.* 43:194–211.

Richardson, J. T. E., 1978. Reported mediators and individual differences in mental imagery. *Mem. Cognition* 6:376–378.

Sadoski, M., E. T. Goetz, and J. B. Fritz, 1993. Impact of concreteness on comprehensibility, interest, and memory for text: Implications for dual coding theory and text design. *J. Educ. Psychol.* 85:291–304.

Saltz, E., and S. D. Nolan, 1981. Does motoric imagery facilitate memory for sentences? A selective interference test. *J. Verb. Learn. Verb. Be.* 20:322–332.

Schwanenflugel, P. J., C. Akin, and W. Luh, 1992. Context availability and the recall of abstract and concrete words. *Mem. Cognition* 20:96–104.

Schwanenflugel, P. J., and E. J. Shoben, 1983. Differential context effects in the comprehension of abstract and concrete verbal materials. *J. Exp. Psychol. [Learn. Mem. Cogn.]* 9:82–102.

Snodgrass, J. G., 1984. Concepts and their surface representations. *J. Verb. Learn. Verb. Be.* 23:3–22.

Tanwar, U., and D. Malhotra, 1990. Imagery variables in short term memory. *Psychol. Studies* 35:191–196.

Thompson, V. A., and A. Paivio, in press. Memory for pictures and sounds: Independence of auditory and visual codes. *Can. J. Exp. Psychol.*

TIPPETT, L. J., 1992. The generation of visual images: A review of neuropsychological research and theory. *Psychol. Bull.* 112:415–432.

VAN DIJK, T. A., and W. KINTSCH, 1983. *Strategies of Discourse Comprehension.* New York: Academic Press.

WINNICK, W. A., and N. BRODY, 1984. Auditory and visual imagery in free recall. *J. Psychol.* 118:17–29.

YARMEY, A. D., and B. J. O'NEILL, 1969. S-R and R-S paired-associate learning as a function of concreteness, imagery, specificity, and association value. *J. Psychol.* 71:95–109.

ZIMLER, J., and J. M. KEENAN, 1983. Imagery in the congenitally blind: How visual are visual images? *J. Exp. Psychol. [Learn. Mem. Cogn.]* 9:269–282.

ZIMMER, H. D., and J. ENGELKAMP, 1985. An attempt to distinguish between kinematic and motor memory components. *Acta Psychol.* 58:81–106.

64 Relational Complexity and the Functions of Prefrontal Cortex

NINA ROBIN AND KEITH J. HOLYOAK

ABSTRACT We propose a theoretical framework for understanding the core cognitive functions of prefrontal cortex. Our general claim is that the prefrontal cortex is responsible for the creation and maintenance of explicit relational representations that guide thought and action. The framework provides a formal characterization of the types of representations that pose the greatest difficulty for animals with prefrontal damage, an explanation of the close tie between phylogenetic and ontogenetic development of the frontal lobes and cognitive capabilities, and an explanation of the deficits that can result from the use of impoverished representations.

Theoretical perspectives on prefrontal functions

The human prefrontal cortex presents us with a striking paradox. On the one hand it is widely believed to control the highest and most distinctively human forms of thinking; yet at the same time, extensive damage to this area of the brain is apt to yield only modest decrements in traditional measures of intelligence. This paradox offers a clear challenge to cognitive science. An adequate theory of mental representation and processing should be able to shed light on the relationship between high-level thinking and what we call intelligence. Resolution of the paradox may be achieved through the research strategy of cognitive neuroscience, which seeks closer links between models of brain and of cognition. Such links may provide needed constraints on the basic concepts used in theories of high-level cognition.

Despite the fact that patients with prefrontal damage perform in the normal range on standard intelligence tests, they are clearly impaired on a wide variety of cognitive functions, including planning, monitoring and modifying behavior, learning complex tasks, and temporal sequencing. Clinical reports reveal that these frontal patients fail to plan for future events, including social interactions, and have difficulty implementing formerly routine plans such as food shopping. Deficits in formulating strategies and plans have also been revealed in experimental studies of tasks involving hypothesis testing, spatial working memory, construction of categorization schemes, and problem solving. The diversity of deficits after frontal lobe damage has fostered the development of an equally diverse set of theories of frontal lobe functions. Recent theories have emphasized the function of the frontal lobes in executive control, attentional control based on internal representations of context, creation and use of abstract event knowledge, and sequencing of actions. (See Fuster, 1989, for a review.) Each of these perspectives appears to capture important insights into the patterns of deficits that have been observed. It has not been entirely clear, however, how the various approaches relate either to each other or to general theories of cognition. Work on frontal functions highlights a number of basic psychological concepts that have not been clearly defined. For example, what makes a task novel? At one extreme, any two experiences differ at least in terms of their spatiotemporal coordinates, so we could claim that everything is novel. At the other extreme, any two experiences have some similarity, such as "both happened to me" or "both happened in this century," so we could argue that nothing is entirely novel. Similar definitional problems cloud concepts such as complexity and context. How should complexity be measured? How is context different from any other feature of a task situation? And how are these concepts related to one another and to mechanisms for working memory and attentional control?

We believe it is possible to clarify, and perhaps even to answer, some of these questions by introducing a general framework for understanding *relational complexity*. We will argue that the central commonality linking

NINA ROBIN and KEITH J. HOLYOAK Department of Psychology, University of California, Los Angeles, Calif.

the various aspects of the frontal syndrome in humans and other animals involves the explicit representation and processing of relational information. The complexity of relations is closely related both to the abstraction of concepts and to distinct types of similarities between events. The prefrontal cortex appears to be responsible for the creation and maintenance of relational representations that guide thought and action. This basic function depends on a working-memory system that serves to bind elements into relational structures, which can then constrain behavior in a goal-directed manner across time. The framework provides a formal characterization of the types of representations that pose the greatest difficulty for animals with prefrontal damage; an explanation of the close tie between phylogenetic and ontogenetic development of the frontal lobes and cognitive capabilities; and an explanation of the deficits that can result from the use of impoverished representations. We will first introduce the framework, and then illustrate its application in interpreting empirical evidence obtained with humans and other primates.

A framework for defining relational complexity

Our thesis is that the functions of prefrontal cortex are linked to the acquisition and use of explicit relational knowledge in the service of a goal. Explicit knowledge, as we use the term here, is based on slow, effortful, and conscious processing that is directly dependent on working memory, and contrasts with implicit knowledge based on relatively rapid, effortless, and unconscious processing. By explicit *relational* knowledge we mean knowledge that differentiates roles from their fillers, and hence relates the latter to the former. The minimal requirement for inferring a distinction between role and filler is the ability to respond on the basis of a specific dimension of variation. At an earlier level, both phylogenetically and ontologically, an animal will react to individual objects without any explicit dimensional analysis. Work on children's classification reveals that even infants can match one apple to another on the basis of global similarity, but only later do children become capable of matching a red apple to a red block on the basis of their common color and despite differences on other dimensions (see Smith, 1989, for a review). Success at the latter task is indicative of differentiation between a role (e.g., "red thing") and the filler of the role (e.g., a particular red apple), where the filler can be freely varied without altering the child's basic response to the dimension that defines the role. We will refer to the more primitive ability to react to global similarity between situations or objects as *holistic* processing, adopting a term often used in the developmental literature on classification.

Explicit processing of dimensions, in which the frontal lobes appear to play a major role, can involve role-filler relationships at different levels of complexity. Here we will adapt a general taxonomy developed by Halford and his colleagues (Halford and Wilson, 1980; Halford, 1993), which has been related to working-memory requirements of cognitive tasks, cross-species cognitive comparisons, and levels of human cognitive development. Halford (1993) defines relational complexity in terms of the number of independently varying dimensions that must be considered together to generate an appropriate response. Each such dimension can be viewed as a variable argument or "slot" linked to a relational concept (i.e., a predicate, function, or operator); in a specific instantiation of the relational structure, the filler of each such slot is logically bound to it. For example, the structure

red (apple-1)

relates the predicate *red* to a certain red apple, apple-1. Apple-1 serves as the filler of the single argument slot associated with the predicate *red* (equivalent to the role of "red thing"), and thus is bound to the slot in this instantiation of the predicate. More generally, the filler of a slot is a logical constant bound to a variable.

In symbolic knowledge representations of this sort as they are typically used in artificial intelligence, the arguments of a predicate are viewed as a list. It is assumed that the arguments in the list can be processed sequentially, so that complexity would be linear with the number of arguments. However, Halford argues that complexity increases worse than linearly with the number of arguments. In his view, arguments psychologically correspond to dimensions of potential simultaneous variation that jointly determine a response. For example, the one-place predicate *red* picks out a single dimension, corresponding to objects that are red (e.g., apples, fire trucks, sunsets), where the appropriate response to each would be "true"; similarly, the two-place predicate *larger than* ranges over two dimensions corresponding to pairs of objects in which one is larger than the other (e.g., {dog, cat}, {truck, car}); and the three-place predicate *between* ranges over three dimensions corresponding to appropriate triples of elements (e.g., {United States, Canada, Mexico}, as in *The United States is between Canada and Mexico*). The psychological benefit of an n-place predicate (i.e., a predicate with n distinct arguments or roles) is that it codes information about n-way interactions between the dimensions, which can then guide appropriate responses. Each dimension and each interaction between dimensions represents a potential source of variation relevant to determining the appropriate response. Psychologically, higher-level interactions provide a basis for responding to complex combinations of elements by virtue of their roles in the overall relational structure, where the response could not be determined by the individual elements considered in isolation from each other. The psychological cost associated with an n-place predicate is that it requires a representation of n dimensions of variation (i.e., n elements filling roles) that have to be considered simultaneously. Moreover, this representation must represent not only the individual elements, but also the binding of each element into its appropriate role.

To characterize the levels of relational complexity more formally, we will generally speak of functions that map a set of argument slots, each based on a dimension that varies independently of the others, to a response. Thus $f(D_1, D_2, \ldots D_n) \rightarrow R$ signifies that a function f applied to the fillers of n slots based on n dimensions, D, generates a response R, where the arrow represents response generation.[1] A taxonomy of levels of complexity can then be defined as follows:

Level 1. A function of 1 dimension provides an *attribute* mapping.

Level 2. A function of 2 dimensions provides a *relational* mapping.

Level 3. A function of 3 dimensions provides a *system* mapping.

Level 4. A function of 4 dimensions provides a *multiple system* mapping.

This taxonomy generates a hierarchy of abstraction for concepts. Holistic processing, which we assume does not depend on frontal functions, may be considered level 0 in the taxonomy, and does not involve abstraction of specific dimensions or representations of variables. Attribute mappings require abstraction of a single dimension of variation, the minimal requirement for differentiating a role from its filler. A relational mapping depends on the ability to relate two dimensions of variation to each other. A system mapping makes it possible to define a relation between relations (i.e., a higher-order relation). A transitive ordering, for example, requires three elements for which the ordering of two pairs constrains the ordering of the third pair (e.g., *A before B* and *B before C* implies *A before C*, where the implication is based on the relation between two *before* relations).

Because of the cognitive burdens imposed by concepts of high dimensionality, reasoners will often shift representations and strategies to reduce the effective complexity of tasks. Halford (1993) suggests two basic mechanisms for complexity reduction: conceptual chunking and segmentation. Conceptual chunking involves collapsing a multidimensional concept into one based on fewer dimensions. When complex concepts are chunked, immediate access to higher-order interactions between dimensions is lost. A chunk is nonetheless informationally richer than a simple holistic representation, in that a chunk is constructed from a complex relational structure defined over symbols; furthermore, its internal structure is potentially recoverable by "unchunking." We assume that a variety of factors will affect the ease with which chunks can be formed. In particular, forming chunks from individual elements will be difficult when the elements are separated from each other in time or space, or when they do not consistently co-occur (i.e., when the same elements are often rearranged to form different relational combinations, yielding what is sometimes termed a *varied mapping* from inputs to response). As we will see, conditions that impede chunking, and hence increase relational complexity, often appear to increase the dependence of a task on prefrontal cortex.

Segmentation is a processing strategy that reduces task complexity by dividing a task into smaller components that can be processed independently and therefore serially. Often, chunking can be used to facilitate task segmentation. For example, suppose a task requires learning the order of n actions that must be performed sequentially to achieve a goal. If n is large, the complexity would be overwhelming if all the elements were considered in parallel. Suppose, however, that one first learns to perform action A before B, and then chunks this unit of activity, which we can refer to as AB. One can then learn to perform AB before C, chunking this new unit as ABC, and so on until the complete ordering has been mastered. Using such a segmentation strategy (which a benevolent teacher might encourage), the task complexity of learning an ordering (of any number of items) can be reduced to a series of relational mappings.

Judicious use of chunking and segmentation will be essential for performing any task that would otherwise exceed the reasoner's maximum limit on parallel processing of dimensions. Halford argues that in humans this maximum limit increases over the course of cognitive development to a maximum of four. Tasks used to test frontal functions in monkeys can, we believe, be plausibly performed using only relational mappings (coupled with the ability to form chunks and to learn tasks in segments), suggesting that the maximum dimensionality of concepts for nonhuman primates may be two.

Although Halford's taxonomy of relational complexity is purely formal, it suggests important constraints on the cognitive mechanisms required to learn and reason explicitly about relational contingencies. Selective attention is needed in order to focus on specific dimensions of the representation for an object or event, as must be done to move beyond primitive holistic processing. The system requires a working memory capable of processing n dimensions in parallel, where the size of n determines the maximum limit on complexity of individual concepts. This working memory, which can be distinguished from a short-term, declarative store used simply to maintain items over time, is required for relational processing. The working memory must code the bindings of elements into specific roles with respect to relational concepts. To maintain appropriate bindings, some form of attentional control must prevent "cross talk" that could blur the identities and role assignments of the individual elements. The bindings must be dynamic so that intermediate computations can be performed without permanently altering the representations of elements. In addition, task-irrelevant elements of the situation may need to be inhibited in order to prevent interference with the representation of goal-relevant elements. In order to segment a task, it will be necessary to swap dynamically between subsets of goal-relevant elements. Attentional control of the sort associated with executive functions will be required to adjudicate between conflicting responses generated by functions at different complexity levels. That is, a certain element considered in isolation may tend to trigger action A, while that same element in combination with one or more additional elements may trigger the incompatible action B. Often, a function based on an isolated element will trigger a response that is a useful default, but which is contextually inappropriate and hence must be inhibited in favor of a response that is a function of the relations between multiple elements considered together. Attentional control and dynamic binding is thus the key to reducing interference with adaptive problem solving.

Another crucial component of the overall system will be a mechanism for forming conceptual chunks. A conceptual chunk—a unitized representation of a complex relational structure—is closely related to the psychological concept of a schema. Formation of a chunk, or schema, requires a mechanism that can take the transient bindings of elements to roles in working memory and create a structure in long-term memory from which the bindings can be recovered by a mechanism for unchunking.

We propose that the major functions of the prefrontal cortex can be understood as aspects of an overall system for reasoning with and learning about explicit relational concepts. (This is not to deny that this region of the brain performs other functions as well.) Within any mammalian species, we conjecture that tasks requiring the highest complexity level attainable by the species will be maximally dependent on frontal functions. Our framework helps to integrate the various frontal functions that have been the focus of previous theories. Executive control of attention will be required in complex planning and problem solving because these operations intrinsically involve high levels of relational complexity. Contextual elements that need to be considered in combination with other elements will increase relational complexity, and also will require inhibition of responses based on individual elements. Abstract concepts will have relatively high dimensionality, and hence will be more dependent on prefrontal functions. Conditional contingencies based on multiple elements, especially when these are difficult to chunk due to their temporal or spatial separation, will also generate high levels of relational complexity. Once complex relations are chunked and stored in long-term memory, and appropriate task segmentations and specific role bindings have been learned, frontal involvement will be sharply reduced. Because conventional tests of intelligence largely tap previously acquired knowledge rather than reasoning with novel relational concepts, or even reasoning that requires novel bindings of elements to roles in familiar schemas, such tests may often fail to detect the deficits that result from frontal damage.

Evidence of relational processing in prefrontal cortex

In this section we will review evidence that indicates how damage to the prefrontal cortex can impair the ability to create relational and system-level mappings in primates. While our analysis is based on data from both human and nonhuman primates, the application of animal results to humans must be done with several caveats (Fuster, 1989). First, neurological insult seldom respects anatomical boundaries. Tumors and closed head injury rarely produce focal deficits specific to a single functional region. Second, much of the lesion work on animals has involved bilateral ablation, whereas many of the human studies involve unilateral lesions. Thus, deficits may be more severe in animal subjects than in human frontal patients because these patients have the capacity to compensate with their remaining prefrontal cortex. An additional complication hinges on the fact that the prefrontal cortex appears evolutionarily more complex in humans than in nonhuman primates and other mammals. There may thus be important species differences in the complexity of the tasks governed by the prefrontal cortex in humans, as our analysis of relational complexity in fact suggests. Humans have more strategies available to compensate in complex situations.

Sketch of Functional Neuroanatomy Figure 64.1 provides a lateral view of a rhesus monkey brain. The frontal lobes correspond to the region anterior to the central sulcus. The frontal cortex can be divided into the prefrontal, premotor, and the cingulate or limbic cortex. The prefrontal cortex can be circumscribed on the basis of its connectivity to the thalamus: It receives afferent projections from the dorsomedial nucleus of the thalamus (Fuster, 1989). Figure 64.1 shows three major regions of the prefrontal cortex. The *dorsolateral* region, for which the principal sulcus is the most prominent morphological landmark in the rhesus monkey, has many reciprocal connections with the posterior cortex, receiving visual, auditory, and somatosensory information. There are also corticocortical connections with the orbital regions, as well as subcortical connections, particularly to the caudate nucleus. The *periarcuate* region (the arcuate sulcus and its surrounding area) is the caudal-most region of the prefrontal cortex. The *orbitofrontal* region is located ventral to the dorsolateral area, and in the rhesus monkey extends from the ventral lip of the principal sulcus (the only portion visible from the lateral view of figure 64.1) around the orbital convexity to the ventral surface. This area has cortical communication with the lateral areas of the prefrontal cortex as well as the limbic system. The posteromedial areas of the orbitofrontal region include the basal forebrain, which connects to the amygdala and hippocampus.

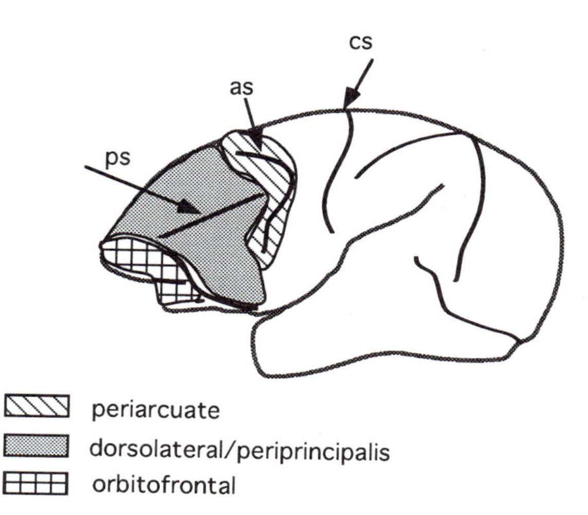

Figure 64.1 Lateral view of the brain of the rhesus monkey: cs, central sulcus; as, arcuate sulcus; ps, principal sulcus.

It should be clear from this description that the prefrontal cortex has reciprocal connections with a wide range of cortical and subcortical structures, including almost all areas of the central nervous system, enabling it to play an integrative role in the control of behavior. It is also clear that the prefrontal cortex is a heterogeneous region, both architecturally and functionally. Damage to different areas of the prefrontal cortex is associated with separable syndromes. In particular, damage to the dorsolateral area, including the principal sulcus and its surrounding area (i.e., the periprincipalis region), is associated with impaired planning, distractibility, and working-memory deficits that interfere with learning contingencies between stimuli and responses separated in time (Fuster, 1989; Goldman-Rakic, 1990; Knight, 1991). The periarcuate region is important for learning conditional contingencies between stimuli and responses, especially in situations that do not involve significant delays (Petrides, 1985; Stamm, 1987). Insult to the orbitofrontal region is associated with loss of inhibitory control over environmentally cued responses, with associated affective disturbances, including personality changes, mood swings, and socially inappropriate behavior resulting from decreased inhibitions (Stuss and Benson, 1984; Fuster, 1989; Knight, 1991).

Phylogenetic and Ontogenetic Trends in Frontal Development Increases in capability to handle relational complexity, both across species and within individuals over the course of development, are correlated with changes in the frontal lobes. This cortical region has been the latest to evolve, and its size and complexity increases more with phylogenetic development than does any other region. Ontogenetically, the prefrontal cortex is one of the last cortical areas to become fully myelinated, a process that may not be complete in humans until the beginning of the second decade; synaptic and neuronal density also increases significantly through childhood (Thatcher, 1991). This late development strongly suggests that the role of the prefrontal cortex in guiding behavior increases throughout development. Moreover, close parallels can be found between patterns of developmental changes in the coordination of electrical activity in the frontal cortex with that in posterior cortical systems, on the one hand, and developmental changes in attentional control and relational processing, on the other (Case, 1992). It is also the case that performance on a variety of tests that reveal frontal deficits in human adults shows pronounced developmental changes over middle to late childhood (Levin et al., 1991).

Although few in number, case studies of the consequences of childhood frontal damage provide additional support for the postulated link between frontal functions and relational learning. These studies reveal a pattern of delayed onset of behavioral deficits, followed by a period of modest progress, which ends with arrested development at adolescence (Ackerly and Benton, 1948, cited in Damasio, 1985; Grattan and Eslinger, 1991). The deficits that arise over the course of childhood include a variety of cognitive deficiencies related to organization and planning, coupled with lack of self-regulation and abnormal social behavior. Damasio (1985, 351) has concluded, "It seems probable that bilateral damage to the frontal lobes in infancy or childhood produces a more devastating effect on personality and cognitive ability than the same amount of damage sustained elsewhere in the brain at any time in the course of development." He particularly notes the striking contrast with cases of early hemispherectomy on either side, which demonstrate near-normal cognitive and social development. If the central function of the frontal lobes is to support learning of complex relational concepts, then it follows that frontal damage will be much more detrimental if it precedes the period when major concepts based on system and multiple-system mappings are normally acquired, which is middle childhood through adolescence. It is also possible that social cognition is especially dependent on the ability to respond to subtle relations between multiple cues, including contextual cues accumulated over time, and hence is particularly vulnerable to early frontal insults.

Working Memory and Integration of Relations across Delays The relational complexity of a task is almost inevitably increased when multiple dimensions of information must be integrated over time. Separation in time will hinder chunking, so the information will need to be processed as distinct but related units, making relational mapping necessary. At the same time, the need to maintain representations of at least one dimension over time until the other relevant information is presented will impose a burden on working memory and attentional control. It is therefore not surprising that one of the most robust findings obtained with monkeys after prefrontal lesions is that they exhibit impaired performance on delay tasks (Jacobson, 1935). Delay tasks all involve responding to a stimulus when there is a delay imposed between the stimulus and the response. The classical delay task involves baiting one of two identical food wells; after a fixed delay, the monkey must choose the baited food well. After a prespecified intertrial interval, the sequence is repeated with a new, randomly chosen food well. Successful performance on this task requires learning a functional rule that might be represented as a function of two variables,

$$\text{same}\ (X, \text{baited}_n) \rightarrow \text{choose}\ X,$$

where X is a food well and baited_n is the location baited on the current trial. For example, if the left food well has been baited, then the animal must choose it. Learning such a rule requires the capability to perform a relational mapping.[2] Performance also depends on the ability to remember the most recently baited location during the delay, and to suppress any interfering memories or response tendencies from prior trials. That is, the critical bindings must be transitory, as the same elements play different roles from trial to trial.

The delayed match-to-sample task is similar to the basic delay task, except the relevant relation is nonspatial. A sample object is first presented, and then, after a delay, two objects are presented, one of which is the same object as the sample. The animal must learn to approach the previously shown object. The rule that must be acquired is,

$$\text{same } (X, \text{sample}_n) \rightarrow \text{choose } X,$$

which has the same logical form as the relational mapping required in the delay task. The primary difference between these two tasks is that in the basic delay task the animal must remember a particular spatial location across a delay, whereas in the delayed match-to-sample task the animal must remember a particular object identity.

Intact monkeys can succeed at both of the above tasks, as well as at more complex variations of each. In a variation of the basic delay task, called a delayed alternation task, the animal must learn to go to the location that was *not* baited on the prior trial. On the first trial, both food wells contain food and the animal is rewarded for approaching either one. On the next trial, the food well that the animal did not choose on the prior trial is baited (without the animal seeing this being done). The rule in a delayed alternation task is roughly,

$$\text{different } (X, \text{baited}_{n-1}) \rightarrow \text{choose } X,$$

Monkeys with lesions to the dorsolateral area, particularly the periprincipalis region, exhibit deficits in all these relational delay tasks, both spatial and nonspatial (Mishkin and Manning, 1978; Stamm, 1987). The magnitude of the deficit is proportional to the length of the delay (Fuster, 1989). Furthermore, lesions of the dorsolateral region do not impair performance on discrimination learning for visual stimuli (Petrides, 1985; Fuster, 1989), spatial stimuli (Passingham, 1985), or stimuli that differ in temporal duration (Goldman-Rakic, 1990).

A study by Rosenkilde, Rosvald, and Mishkin (1981) provides an especially informative contrast. They assessed performance on a temporal discrimination task using monkeys with dorsolateral, orbitofrontal, or periarcuate lesions. The temporal discrimination task involved learning that a 10-second delay signaled that the left food well had been baited, whereas a 30-second delay signaled that the right food well was baited. Monkeys were trained on the task preoperatively, and relearning was assessed approximately two weeks after surgery. Only the orbitofrontal group was impaired on relearning, and these animals exhibited significant perseverative tendencies. The results thus indicated that damage to the principal sulcus does not impair time discrimination per se. In terms of our complexity analysis, the temporal discrimination task used by Rosenkilde, Rosvald, and Mishkin can be represented as two attribute mappings,

$$\text{short (delay interval)} \rightarrow \text{choose left}$$

$$\text{long (delay interval)} \rightarrow \text{choose right,}$$

and hence is representationally simpler than the more complex conditional contingency tasks that require relational mappings. It thus appears that the deficits caused by periprincipalis damage are not tied simply to temporal coding per se; rather, the impairments are specific to tasks that require responses based on representations with at least the complexity of relational mappings that must be maintained in working memory.

Electrophysiological studies complement lesion studies in elucidating the functions of specific cortical regions in delay tasks. Studies based on single-cell recordings indicate that cells in the prefrontal cortex have increased rates of firing to specific task-related features, including the identity of the cue, the response, and the delay (Fuster, 1989, 1993; Goldman-Rakic, 1990). Neurons selectively responsive to stimuli presented in delay tasks have been found throughout the prefrontal cortex, although primarily in the region of the principal sulcus (Yamatani et al., 1990). The wide distribution of these cells may account for the fact that size of lesion is correlated with degree of deficit (Stamm, 1987). Furthermore, in the delayed match-to-sample task, many neurons respond to any object that is the same as the sample just presented, independently of the visual features of the object (Yamatani et al., 1990). Such cells are candidates for the neural basis for the coding of variables and bindings, in that they respond to the current instantiation of a relationship. In general, learning delay tasks based on relations between two dimensions of variation appears to involve changes in neuronal firing patterns in prefrontal cortex, primarily in the dorsolateral but also in the periarcuate and orbitofrontal regions.

Although clear evidence of localization is difficult to obtain with humans, the broad pattern of findings with frontal patients is consistent with involvement of prefrontal cortex in coding contingent relationships between elements separated in time. Patients with frontal lobe damage usually are not significantly more impaired than those with posterior damage on tests of short-term memory for information from visual, auditory, and kinesthetic modalities (Ghent, Mishkin, and Teuber, 1962). These negative findings involve tasks that do not require relational mappings (e.g., recalling the specific orientation of a line after a delay). In contrast, frontal patients are selectively impaired on tasks that involve formulating contingencies based on spatiotemporally distinct events, such as delayed-response and delayed-alternation tasks (Freedman and Oscar-Berman, 1986).

The ability retrospectively or prospectively to sequence actions and events also involves formulating relational and higher-order mappings. When subjects are required to recall the order of recently presented items, patients with frontal lobe damage are impaired in both delayed and immediate recall conditions (Lewinsohn et al., 1972). For example, Lewinsohn and colleagues presented two items sequentially on each trial. Each item had three features that had to be recalled: a picture, a background pattern, and a background color. The background patterns were highly similar across items; in addition, pictures and backgrounds were repeated in different combinations across trials, making chunking difficult. Recall was scored for the specific features of each item as well as for the order of the two items. Frontal patients showed deficits for both immediate and delayed recall.

Because the stimuli were constructed in a manner that would tend to prevent processing the items as chunks, subjects presumably were forced to treat the three features of each item as separate elements. Correct recall of any feature of an item in relation to the corresponding feature in the other item would therefore involve computing at least two relations: the part-whole relation between a feature and the item in which it occurred, and the relation between the order of the two items. For example, suppose I_A and I_B are the two items to be recalled, and f_1 is a feature of I_A. Correct recall of f_1 in its appropriate position would require processing the relations

$$\text{part-of}\ (f_1, I_A)$$

and

$$\text{before}\ (I_A, I_B).$$

Considering these two relations together involves three dimensions of variation (the feature and the two items), and therefore constitutes a system mapping. The level of relational complexity is thus far greater for this ordered recall task than for one that simply requires unordered recall of individual items. Thus for humans, as for monkeys, frontal involvement in temporal and spatial coding is linked to relational complexity.

The ability to sequence events is a crucial component of planning. Shallice (1982) and Owen et al. (1990) found that patients with frontal lobe damage were impaired in planning a sequence of moves that would rearrange an initial pattern of colored beads into a goal state. Efficient performance on this task depends upon the ability to break the task into subgoals and then reach each of the subgoals. The representational complexity of the task depends on the number of subgoals and their relationship to one another. In general, planning a set of actions requires at least relational-level mappings, because the choice of action depends on the relationship between the goal and an available operator.

Learning Conditional Contingencies That Do Not Require Temporal Bridging On the basis of evidence that the deficits resulting from prefrontal lesions are not specific to spatial or visual discriminations, or to delay per se, it has been argued that dorsolateral lesions impair the more general ability to learn relationships between distinct temporal and spatial events (Petrides, 1985, 1987; Fuster, 1989). Even when temporal gaps are not involved, however, it is possible to design tasks that require learning conditional contingencies with the complexity of relational mappings. Petrides (1985) compared monkeys with periarcuate lesions to those with lesions of the principal sulcus region on learning a nontemporal conditional contingency task. On this task, monkeys were rewarded when they opened a lit box in the presence of a toy clown or when they opened an unlit box in the presence of a yellow disc. (Both boxes were always presented.) Monkeys with periarcuate lesions were severely impaired, while monkeys with lesions in the principal sulcus were slower than normal controls but able to learn the task.

However, there were no differences between groups when monkeys were rewarded for opening the lit box and not otherwise.

There is a crucial difference between tasks involving conditional versus nonconditional contingencies in terms of our complexity analysis: The former task depends on forming two relational mappings, whereas the latter depends on only a single attribute mapping (or even on holistic processing). The above conditional task required learning two relational rules,

appear-together (toy clown, lit box) → open lit box,

appear-together (yellow disc, unlit box) → open unlit box.

In contrast, the nonconditional contingency can be learned using a much simpler representation, such as

appear (lit box) → open lit box.

In general, it seems that although deficits in delay tasks are greatest when the periprincipalis region is damaged, deficits in nontemporal conditional contingency learning are greatest after damage to the periarcuate area (Petrides, 1985, 1986; Stamm, 1987; see Petrides, 1987, for a review). It thus seems that both these prefrontal regions play roles in binding elements into relations. The periprincipalis region is especially relevant to the maintenance of relevant dimensions of variation in working memory as relations are dynamically formed, whereas the periarcuate region is central to the formation of relations between elements when temporal gaps are not a major factor.

Similar deficits have been found in human frontal patients when they perform nontemporal tasks that nonetheless involve high relational complexity. For instance, one of the most frequent findings with human patients who have suffered frontal lobe insult is poor ability to categorize stimuli (Milner, 1964; Stuss and Benson, 1984; Owen et al., 1990). Stuss et al. (1983), for example, found that patients with prefrontal leucotomies were impaired on the ability to identify similarities between three of four objects. On each trial, subjects were shown a card with pictures of four objects. Three of the four objects could be classified in two ways on the basis of their values on the dimensions of color, form, size, and orientation. Subjects were first required to identify three similar items, and then to verbally state the reason for the similarity. Next, subjects were required to identify three *other* similar items, and again provide a verbal explanation of the basis for the grouping. The patients were able to point to a correct grouping of the items on the first categorization task, but often gave inconsistent or concrete responses when explaining their decisions. Furthermore, they were markedly impaired in their ability both to create and to explain a second, alternative grouping of the items.

This pattern of performance can be directly related to the complexity of each of the successive tasks. The initial categorization could be done on the basis of holistic processing based on the overall similarities of the items. However, to identify the basis for the similarity requires forming a relational mapping involving a dimension shared by three items but not the fourth. The second categorization task is yet more demanding in that success requires establishing a relational mapping that is different from that used for the first grouping. Representing the difference between two relational mappings has the complexity of a system mapping. Such results support our claim that the frontal lobes play an increasingly large role in cognitive tasks as the relational complexity of the tasks increases, even when temporal bridging is not required.

Inhibition of Interference from Rival Responses

In normal information processing, selection of items or actions is accompanied by inhibition of competing responses. When the contextually appropriate action is based on a high level of relational complexity, it will be necessary to inhibit alternative responses based on simpler representational levels, such as responses to single, isolated elements, as well as any previously established rival responses. Damage to the prefrontal cortex will lead to formation of impoverished representations of relations. This in turn will decrease inhibition of rival responses and hence lead to greater interference.

Such decreased inhibition appears to be selectively associated with damage to the orbitofrontal region. One of the classical paradigms used to measure response interference is the "go, no-go" paradigm, in which success requires learning to make a response to one stimulus but to inhibit that same response to a similar stimulus. Lesions of the orbitofrontal region lead to higher error rates on this task than do lesions of the dorsolateral cortex (Fuster, 1989). Damage to the orbitofrontal region has been shown to lead to increased interference from perseverative responses, an error pattern that has also been observed in delayed match-to-sample tasks with a single pair of stimuli, and in object-alternation tasks (Mishkin and Manning, 1978).

Drewe (1975) found that humans with frontal lobe damage were impaired on learning a go, no-go task. This type of error is also commonly observed in patients' performance on the Wisconsin Card Sorting Test. On this task, the subject must match a set of cards that vary in color, shape, and number to a set of target cards on the basis of experimenter feedback. The task is to match the cards in the deck to the target cards according to one of the three dimensions. After the subject has achieved criterion on one dimension, the experimenter changes the concept, and the subject must learn to categorize along a different dimension based on the feedback. Frontal patients tend to perseverate in their initially learned response pattern after the experimenter has shifted the concept (Milner, 1964; Stuss and Benson, 1984).

Conclusion

Our review of the functions of prefrontal cortex leaves us in broad agreement with the conclusions of Fuster (1989), who argued that converging evidence from lesion studies, single-cell recordings, and electrophysiological measures have shown that the prefrontal cortex subserves three primary functions: maintaining representations of elements in working memory to process cross-temporal relationships; learning conditional contingencies; and providing resistance to interference. Experimental lesion studies with monkeys indicate that these functions have separate anatomical loci: The dorsolateral region acts as a working memory to code relationships that occur across temporal discontinuities; the periarcuate region is central to learning conditional contingencies that do not involve temporal gaps; and the orbitofrontal region is critical to minimizing interference. However, all three loci act together to constrain behavior in a goal-directed manner across time.

Because relational complexity of tasks is maximal when novel combinations of elements must be processed, damage to the prefrontal cortex is especially deleterious when it occurs at a young age, before the major relational schemas required for mature thinking and behavior have been acquired. Once relational schemas have been formed, their subsequent use will be less dependent on prefrontal functions. Thus both humans and other primates are able to perform previously acquired habits, even those that span temporal intervals, despite suffering frontal injury as adults. The relative independence of well-learned responses from frontal control accounts in part for the fact that frontal patients often do not show major impairment on conventional measures of intelligence. In addition, the tester, who guides the patient's attentional focus, may to some degree play the part of the patient's frontal lobes, minimizing interference and distraction (Stuss and Benson, 1984). The testing environment itself can facilitate task segmentation and thus reduce a task's relational complexity. Nonetheless, damage to the prefrontal cortex will continue to produce deficits in tasks involving relations between novel combinations of elements, including tasks in which established schemas must be reinstantiated in new ways.

Our framework for analyzing relational complexity is in general agreement with previous theories of frontal lobe functions, but at the same time may prove useful in developing closer ties between these theories and computational models of thought. The diverse tasks that show impairment after frontal damage—including planning, sequencing of actions, using context to modulate social behavior, learning contingencies between spatiotemporally separate stimuli and responses, and forming flexible categories—all share a common requirement: Independently varying elements must be bound to specific roles with respect to relations. Future work should include task analyses to isolate the representational units required for successful task performance. In addition, theoretical work should be directed at the development of models of relational processing that capture the role of working-memory and attentional constraints.

ACKNOWLEDGMENTS Joaquim Fuster, Jordan Grafman, Graeme Halford, Jim Kroger, and Phil Johnson-Laird provided helpful comments.

NOTES

1. We assume that the generation of the response need not be explicitly represented by the animal; hence, only the arguments that serve as inputs to the function—and not the response produced as its output—are treated as dimensions of variation. This assumption is reflected in our notation: The dimensions are included within parentheses as arguments to the function, and "$\rightarrow R$," which is outside of the parentheses, signifies the mapping from the set of inputs to a response.
2. The required representation may actually be less complex than a full relational mapping in the sense of Halford

(1993), because the two arguments can be assigned separately on the basis of temporal cues. Similar caveats apply to all the two-slot representations suggested for tasks performed by nonhuman primates.

REFERENCES

CASE, R., 1992. The role of the frontal lobes in the regulation of cognitive development. *Brain Cogn.* 20:51–73.

DAMASIO, A. R., 1985. The frontal lobes. In *Clinical Neuropsychology*, ed. 2, K. M. Heilman and E. Valenstein, eds. New York: Oxford University Press, pp. 339–375.

DREWE, E. A., 1975. Go-no go learning after frontal lobe lesions in humans. *Cortex* 11:8–16.

FREEDMAN, M., and M. OSCAR-BERMAN, 1986. Bilateral frontal lobe disease and selective delayed response deficits in humans. *Behav. Neurosci.* 100:337–342.

FUSTER, J. M., 1989. *The Prefrontal Cortex: Anatomy, Physiology, and Neuropsychology of the Frontal Lobe.* New York: Raven.

FUSTER, J. M., 1993. Frontal lobes. *Curr. Opin. Neurobiol.* 3:160–165.

GHENT, L., M. MISHKIN, and H. L. TEUBER, 1962. Short-term memory after frontal lobe lesion in man. *J. Comp. Physiol. Psychol.* 55:705–709.

GOLDMAN-RAKIC, P. S., 1990. Cellular and circuit basis of working memory in prefrontal cortex of nonhuman primates. In *Progress in Brain Research 85: The Prefrontal Cortex: Its Structure, Function and Pathology*, H. B. M. Uylings, C. G. Van Eden, J. P. De Bruin, M. A. Corner, and M. G. Feenstra, eds. New York: Elsevier, pp. 325–336.

GRATTAN, L., and P. ESLINGER, 1991. Frontal lobe damage in children and adults: A comparative review. *Dev. Neuropsychol.* 7:283–326.

HALFORD, G. S., 1993. *Children's Understanding: The Development of Mental Models.* Hillsdale, N.J.: Erlbaum.

HALFORD, G. S., and W. H. WILSON, 1980. A category theory approach to cognitive development. *Cognitive Psychol.* 12:356–411.

JACOBSON, C. F., 1935. Functions of frontal association area in primates. *Arch. Neurol. Psychiatry* 33:558–569.

KNIGHT, R. T., 1991. Evoked potential studies of attention capacity in human frontal lobe lesions. In *Frontal Lobe Function and Dysfunction*, H. S. Levin, H. M. Eisenberg, and A. L. Benton, eds. New York: Oxford University Press, pp. 139–155.

LEVIN, H., K. CULHANE, J. HARTMANN, K. EVANKOVICH, A. MATTSON, H. HARWARD, G. RINGHOLZ, L. EWING-COBBS, and J. FLETCHER, 1991. Developmental changes in performance on tests of purported frontal lobe functioning. *Dev. Neuropsychol.* 7:377–395.

LEWINSOHN, P. M., R. E. ZIELER, J. LIBET, S. EYEBERG, and G. NIELSON, 1972. Short-term memory: A comparison between frontal and nonfrontal right- and left-hemisphere brain-damaged patients. *J. Comp. Physiol. Psychol.* 81:248–255.

MILNER, B., 1964. Some effects of frontal lobectomy in man. In *The Frontal Granular Cortex and Behavior*, W. J. Warren and K. Akert, eds. New York: McGraw-Hill, pp. 313–334.

MISHKIN, M., and F. J. MANNING, 1978. Non-spatial memory after selective prefrontal lesions in monkeys. *Brain Res.* 1443:313–323.

OWEN, A., J. DOWNES, B. SAHAKIAN, C. POLKEY, and T. ROBBINS, 1990. Planning and spatial working memory following frontal lobe lesions in man. *Neuropsychologia* 28: 1021–1034.

PASSINGHAM, R. E., 1985. Prefrontal cortex and the sequencing of movement in monkeys Macaca mulatta. *Neuropsychologia* 23:453–462.

PETRIDES, M., 1985. Deficits in non-spatial conditional associative learning after periarcuate lesions in the monkey. *Behav. Brain Res.* 16:95–101.

PETRIDES, M., 1986. The effects of periarcuate lesions in the monkey on the performance of symmetrically and asymmetrically reinforced visual and auditory go, no-go tasks. *J. Neurosci.* 6:2054–2063.

PETRIDES, M., 1987. Conditional learning and primate cortex. In *The Frontal Lobes Revisited*, E. Perecman, ed. N.J.: Erlbaum, pp. 91–108.

ROSENKILDE, C. E., H. E. ROSVOLD, and M. MISHKIN, 1981. Time discrimination with positional responses after selective prefrontal lesions in monkeys. *Brain Res.* 210:129–144.

SHALLICE, T., 1982. Specific impairments of planning. *Phil. Trans. R. Soc. Lond. [Biol.]* 298:199–209.

SMITH, L. B., 1989. From global similarities to kinds of similarities: The construction of dimensions in development. In *Similarity and Analogical Reasoning*, S. Vosniadou and A. Ortony, eds. Cambridge: Cambridge University Press, pp. 146–178.

STAMM, J. S., 1987. Monkey's delayed response deficit solved. In *The Frontal Lobes Revisited*, E. Perecman, ed. N.J.: Erlbaum, pp. 73–89.

STUSS, D. T., and D. F. BENSON, 1984. Neuropsychological studies of the frontal lobes. *Psychol. Bull.* 95:3–28.

STUSS, D. T., D. F. BENSON, E. F. KAPLAN, W. S. WEIR, M. A. NAESER, I. LIEBERMAN, and D. FERRILL, 1983. The involvement of orbitofrontal cerebrum in cognitive tasks. *Neuropsychologia* 21:235–248.

THATCHER, R. W., 1991. Maturation of the human frontal lobes: Physiological evidence for staging. *Dev. Neuropsychol.* 7:397–419.

YAMATANI, K., O. TAKETOSHI, H. NISHIJO, and A. TAKAKU, 1990. Activity and distribution of learning-related neurons in monkey Macaca fuscata prefrontal cortex. *Behav. Neurosci.* 104:503–531.

65 Mental Models, Deductive Reasoning, and the Brain

PHILIP N. JOHNSON-LAIRD

ABSTRACT This chapter considers the two main approaches to deductive thinking: theories based on formal rules of inference postulate that deduction is a syntactic process akin to a logical proof; the mental model theory postulates that it is a semantic process akin to the search for counterexamples. Experimental evidence bears out the predictions of the model theory: the more models needed for a deduction, the harder it is; erroneous conclusions are consistent with the premises; and general knowledge affects the process of search. Recent neurological evidence bears out, as the model theory predicts, a significant involvement of the right hemisphere in reasoning.

If deduction is a purely verbal process then
it will not be affected by damage to the
right hemisphere.
It *is* affected by such damage.
∴ It is not a purely verbal process.

This argument is an example of a valid deduction: Its conclusion must be true if its premises are true. (They may not be, of course.) Deductive reasoning is under intensive investigation by cognitive scientists, and more is known about it than about any other variety of thinking. The aim of this chapter is to explain its nature and to relate it to the brain. "The cerebral organization of thinking has no history whatsoever," Luria remarked (1973, 323); and Fodor (1983, 119) suggested that nothing can be known about the topic, because thinking does not depend on separate "informationally encapsulated" modules (but cf. Shallice, 1988, 271). Many regions of the brain are likely to underlie it, but as we shall see, a start has been made on the neuropsychology of reasoning.

Many cognitive scientists have argued that deductive reasoning depends on formal rules of inference like those of a logical calculus, and that these unconscious rules are used to derive conclusions from the representations of premises. These "propositional" representations are syntactically structured strings of symbols in a mental language, and the chain of deductive steps is supposedly analogous to a logical proof (see, e.g., the theories of Braine, Reiser, and Rumain, 1984; Osherson, 1974–1976; Rips, 1983). An alternative account postulates a central role for mental models. This account does not reject propositional representations, but it treats them as the input to a process that constructs a mental model corresponding to the situation described by the verbal discourse. The process of deduction—as well as induction and creation (Johnson-Laird, 1993)—is carried out on such models rather than on propositional representations. Models are the natural way in which the human mind constructs reality, conceives alternatives to it, and searches out the consequences of assumptions. They are, as Craik (1943) proposed, the medium of thought. But what *is* a mental model?

The underlying idea is that the understanding of discourse leads to a model of the relevant situation akin to one created by perceiving or imagining events instead of merely being told about them (Johnson-Laird, 1970). Experimental studies have indeed found evidence for both initial propositional representations and mental models (see e.g., Johnson-Laird, 1983; van Dijk and Kintsch, 1983; Garnham, 1987). The same idea has led to the model theory of deductive reasoning. The theory was not cut from whole cloth, but was gradually extended from one domain to another. From a logical standpoint, there are at least four main domains of deduction:

1. Relational inferences based on the logical properties of such relations as *greater than*, *on the right of*, and *after*.
2. Propositional inferences based on negation and on such connectives as *if*, *or*, and *and*.

PHILIP N. JOHNSON-LAIRD Department of Psychology, Princeton University, Princeton, N.J.

3. Syllogisms based on pairs of premises that each contain a single quantifier, such as *all* or *some*.

4. Multiply quantified inferences based on premises containing more than one quantifier, such as *Some pictures by Turner are more valuable than any by any other English painter*.

Logicians have formalized a predicate calculus that covers all four domains and includes the propositional calculus, which deals with inferences based on connectives. The model theory was developed first for relational inferences and syllogisms, and recently for propositional and multiply quantified inferences. In contrast, psychological theories based on formal rules exist for relational and propositional inferences, but not for syllogisms or for multiply quantified inferences.

Theories and evidence have been reviewed in detail elsewhere (Johnson-Laird and Byrne, 1991, 1993; Holyoak and Spellman, 1993). In this chapter, we will stand back from the details and present an integrated account of mental models based on all of this work. We will also bring the story up to date and relate it to the neuropsychology of thinking. The chapter begins with relational inferences and establishes that a model-based system does not require postulates specifying the logical properties of relations. It then shows how models can underlie reasoning with sentential connectives, such as *or*, and quantifiers, such as *all*. Next, it shows how certain sorts of diagrams inspired by the model theory can help reasoners to cope with disjunctions. Finally, it considers the neuropsychological findings, and draws some conclusions about the assumptions underlying mental models.

Relational inferences and emergent logical properties

Consider the following simple inference:

The Turner painting is on the right of the Daumier.
The Corot sketch is on the left of the Daumier.
What follows?

A valid answer is that the Turner painting is on the right of the Corot sketch. Psychological theories based on formal rules of inference (e.g., Hagert, 1984; Ohlsson, 1984) explain the derivation of the answer in terms of a formal proof. It depends on the logical properties of the relations: *on the left of* is the converse of *on the right of*, and both are transitive relations. These properties have to be added to the premises by stating them in so-called meaning postulates, that is, postulates that depend on the meanings of these relations:

For any x, y, if x is on the left of y, then y is on the right of x.
For any x, y, z, if x is on the right of y, and y is on the right of z, then x is on the right of z.

With these postulates the conclusion can be derived:

The Turner painting is on the right of the Corot,

using various rules of inference, including *modus ponens*:

$$\begin{array}{l} \text{if } p \text{ then } q \\ p \\ \therefore q \end{array}$$

where p and q denote any propositions whatsoever. The formal derivation for this simple inference is surprisingly long: It calls for eight steps, but that is the price to be paid for using formal rules.

The theory of mental models takes a different approach. It treats propositional representations as instructions for the construction of models. The meaning of, say, *on the right of* consists in the appropriate increments to the Cartesian coordinates of one object, y, in order to locate another object, x, so that: x is on the right of y. Hence, the propositional representation of the assertion:

The Turner painting is on the right of the Daumier

can be used to construct a spatial model:

$$d \quad t$$

where d denotes the Daumier and t denotes the Turner. The information in the second premise:

The Corot sketch is on the left of the Daumier

can be added to yield:

$$c \quad d \quad t$$

This model supports the conclusion:

The Turner painting is on the right of the Corot sketch.

The conclusion is true in the model, but does it follow validly from the premises? The crucial manipulation to test validity is to search for alternative models of the premises that refute the conclusion. In fact, there are no alternative models of the premises in which the conclusion is false, and so it is valid. The model-based method of reasoning accordingly has no need of meaning postulates or formal rules of inference. The logical properties of a relation, such as its transitivity, are not explicitly represented at all, but emerge from the meaning of the relation when it is put to use in the construction of models. The general procedure of searching for alternative models is used to test validity.

The evidence from three-term series problems, such as the example above, does not suffice to decide between formal rules and mental models. However, studies of two-dimensional spatial reasoning have produced more decisive data (Byrne and Johnson-Laird, 1989). We examined problems of the the following sort:

The cup is on the right of the saucer.
The plate is on the left of the saucer.
The fork is in front of plate.
The spoon is in front of the cup.
What is the relation between the fork and the spoon?

Subjects tend to imagine symmetrical arrangements, and so the description corresponds to a single model:

plate	saucer	cup
fork		spoon

It should be relatively easy to answer that the fork is on the left of the spoon. When the second premise of the problem is changed to

The plate is on the left of the cup

the resulting premises are consistent with at least two distinct models:

plate	saucer	cup		saucer	plate	cup
fork		spoon			fork	spoon

The same relation holds between the fork and the spoon in both models, but the theory predicts that the task should be harder because both models must be constructed in order to test the validity of the answer. The task should be still harder where the correct response can be made only by constructing both models. The description

The cup is on the right of the saucer
The plate is on the left of the cup
The fork is in front of plate
The spoon is in front of saucer

is consistent with two distinct models:

plate	saucer	cup		saucer	plate	cup
fork	spoon			spoon	fork	

that have no relation in common between the fork and the spoon, and so there is no valid answer to the question. Granted that the mind has a limited processing capacity, the model theory predicts the following rank order of increasing difficulty: one-model problems, multiple-model problems with valid answers, and multiple-model problems with no valid answers.

Formal-rule theories need complex meaning postulates to support two-dimensional deductions (Hagert, 1984; Ohlsson, 1984). Whatever rules a theory uses, however, the one-model problem calls for a longer derivation than the multiple-model problem with a valid answer. It is necessary to infer the relation between the plate and the cup for the one-model problem, but there is no need for such a derivation with the multiple-model problem because the relation is directly asserted by the second premise:

The plate is on the left of the cup.

Hence, formal rule theories predict that the one-model problems should be harder than the multiple-model valid problems, which is exactly the opposite prediction to the one made by the model theory.

Our experiments compared the predictions of the two theories (Byrne and Johnson-Laird, 1989). In one experiment, 18 adults carried out four inferences of each of the three sorts, and the percentages of their correct responses were as follows: 70% for the one-model problems, 46% for the multiple-model valid problems, and 8% for the multiple-model problems with no valid conclusion. This robust trend corroborates the model theory but runs counter to the formal-rule theories. The same results have been obtained from analogous problems concerning temporal relations (Schaeken and Johnson-Laird, 1993). Subjects also drew correct conclusions to one-model problems reliably faster than to multiple-model problems.

Models for connectives and quantifiers

What remains to be accounted for are the logical constants—sentential connectives and quantifiers. Some psychological theories postulate formal rules of inference for connectives, but no such theories exist for quantifiers. The model theory, however, proposes an account for both. Connectives call for models of alternative possibilities. A conjunction of the form:

p and q

requires only a single model:

$p \quad q$

where p and q respectively denote the situations described by the two propositions. But an exclusive disjunction such as:

p or else q, but not both

requires two alternative models, which are shown here on separate lines:

$$\begin{array}{ll} p & \\ & q \end{array}$$

A conditional of the form:

If p, then q

calls—at least initially—for one explicit model (of the antecedent and consequent) and one implicit model of an alternative situation:

$$\begin{array}{ll} p & q \\ \multicolumn{2}{c}{\dots} \end{array}$$

The implicit model symbolized by the three dots may subsequently be rendered explicit, but for many inferences the implicit model suffices. We have implemented a computer program (Propsych) that computes the numbers of explicit models required by inferences (see Johnson-Laird, Byrne, and Schaeken, 1992). Consider, for example, the following argument:

Studies have shown that children of people who smoke more than two packs per day have a greater exposure than others to secondhand smoke or a lowered resistance to viral infection. Children exposed to secondhand smoke have an increased risk of lung cancer. Children with lowered resistance to viral infection are harder to treat with chemotherapy. These two factors make for intractable cases of lung cancer. Thus, these children risk contracting untreatable lung cancer.

The first step is to represent the underlying propositional connectives in the premises:

If child of smoker *then* (exposed to smoke *or* lowered resistance).
If exposed to smoke *then* greater risk.
If lowered resistance *then* chemotherapy harder.
If greater risk or chemotherapy harder *then* risk of untreatable cancer.
∴ *If* child of smoker *then* risk of untreatable cancer.

We can then use the Propsych program to work out the total number of explicit models that have to be constructed to carry out the inference. Thus, the first premise calls for two explicit models and one implicit model:

$$\begin{array}{lll} c & & s \\ c & l & \\ \multicolumn{3}{c}{\dots} \end{array}$$

where c denotes a child of a smoker, s denotes exposure to smoke, and l denotes lowered resistance. The second premise calls for the following models:

$$\begin{array}{ll} s & r \\ \multicolumn{2}{c}{\dots} \end{array}$$

where r denotes a greater risk. The principles for combining sets of models are simple: A new model is made, if possible, from each pairwise combination of a model from one set with a model from the other set, according to the principles in Johnson-Laird, Byrne, and Schaeken (1992, 425):

1. If the model in one set is implicit and the model in the other set is implicit, then the result is an implicit model.
2. If the model in one set is implicit but the model in the other set is not, then no new model is formed from them.
3. If the pair of models is inconsistent, that is, one contains the representation of a proposition and the other contains a representation of its negation, then no new model is formed from them.
4. Otherwise the two models are joined together, eliminating any redundancies.

The result of combining the sets of models for the first two premises is, accordingly:

$$\begin{array}{llll} c & & s & r \\ c & l & s & r \\ \multicolumn{4}{c}{\dots} \end{array}$$

Hence, so far, the process of inference has called for the construction of five explicit models. The set of premises as a whole calls for the construction of nine explicit models. Initial models of this sort suffice for all the 61 direct inferences used in a study by Braine, Reiser, and Rumain (1984), and the program was used to count them: They predicted the difficulty of the problems as well as these authors' rule-based theory.

Although many inferences in daily life can be made with such models, sometimes one has to think more carefully and flesh out the models completely. Given the conditional

If there is a triangle then there is a circle,

could there be a circle without a triangle? Presumably so, given one interpretation of the conditional. Could there be a triangle without a circle? Of course not. That would contravene the meaning of the conditional. Hence, as soon as individuals begin to think more closely about the meaning of the conditional, they realize that the explicit model in the following set:

△ ○
...

represents the only possible situation in which a triangle can occur. That is, it must occur with a circle given the truth of the conditional. One way to represent this information is to use a special annotation:

[△] ○
...

where the square brackets indicate that triangles have been exhaustively represented in relation to circles. The procedure for fleshing out models works as follows: When a proposition has been exhaustively represented, its negation is added to any other models; when a proposition has not been exhaustively represented, it and its negation form separate models that replace the implicit model (denoted by three dots). Triangles cannot occur in fleshing out the implicit model above, because they are already exhausted, but their negations can occur with either a circle or its negation. Hence, the result is:

△ ○
¬△ ○
¬△ ¬○

where ¬ is an annotation representing negation. Because there is no longer any implicit model, there is no need for symbols representing exhaustive representations. Exhaustion is thus a device that allows the inferential system to represent certain information implicitly—it can be made explicit, but at the cost of fleshing out the models.

The same principles suffice for the representation of quantifiers. The interpretation of an assertion, such as

All the Frenchmen in the restaurant are gourmets

calls for a model of the following sort, in which each line no longer represents a separate model, but rather a separate individual in one and the same model:

[*f*] *g*
[*f*] *g*
...

where *f* denotes a Frenchman, *g* denotes a gourmet, and the three dots represent implicit individuals. As before, the square brackets indicate an exhaustive representation: The tokens denoting Frenchmen exhaust the set in relation to the set of gourmets. The set of gourmets, however, is not exhaustively represented. Hence, if the implicit individuals are fleshed out explicitly, some of them may be gourmets, but none of them can be Frenchmen unless they are also gourmets.

The information from a second premise, say

Some of the gourmets are wine drinkers

can be added to the model

[*f*] *g* *w*
[*f*] *g*
w
...

This model supports the believable conclusion:

Some of the Frenchmen in the restaurant are wine drinkers

This conclusion is erroneous, though it is drawn by most subjects (see Oakhill, Johnson-Laird, and Garnham, 1989). It is refuted by an alternative model of the premises:

[*f*] *g*
[*f*] *g*
g *w*
w
...

When the second premise is instead:

Some of the gourmets are Italians

the initial model supports the unbelievable conclusion:

Some of the Frenchmen are Italians

and hardly any subjects err now. In other words, reasoners tend to "satisfice" (see Simon, 1959): If they reach a congenial conclusion they tend not to search for alternative models. Satisficing is a frequent cause of everyday disasters, both major and minor. It seems an obvious danger, yet it cannot be predicted by rule theories, which contain no elements corresponding to models of situations.

The model theory generalizes to multiply quantified assertions. For example, the premises:

None of the Avon letters is in the same place as any of the Bury letters
All of the Bury letters are in the same place as all of the Caton letters

yield the valid conclusion:

None of the Avon letters is in the same place as any of the Caton letters.

Granted the following definition of *in the same place as*:

x is in the same place as y = x is in a place that has the same spatial coordinates as those for y

the premises support a model of the state of affairs:

| [a] [a] [a] | [b] [b] [b] [c] [c] [c] |

where the vertical barriers demarcate separate places, and there are arbitrary numbers of individuals of each sort (*a*s denote Avon letters, *b*s denote Bury letters, and *c*s denote Caton letters). This model yields the conclusion:

None of the Avon letters is in the same place as any of the Caton letters.

No alternative model of the premises refutes the conclusion. As the theory predicts, one-model deductions are easier than multiple-model deductions (see Johnson-Laird, Byrne, and Tabossi, 1989).

Diagrams and disjunctions

Formal-rule theories predict that the difficulty of a deduction depends on the length of its derivation; the model theory predicts that it depends on the number of models that have to be constructed. Although some of these predictions run in parallel, there are interesting divergencies between them. According to the model theory, inferences based on exclusive disjunctions (two models) should be easier than inferences based on inclusive disjunctions (three models). Rule theories can accommodate this result by assuming that the rule for exclusive disjunction is easier to use than the rule for inclusive disjunction, but they cannot *predict* the phenomenon. The simplest prediction of the model theory, however, does not require any detailed account of numbers of models: Erroneous conclusions should tend to be consistent with the truth of the premises rather than inconsistent with them, because reasoners will often base their conclusions on only some of the possible models of the premises. Current theories based on formal rules of inference make no predictions about the nature of systematically erroneous conclusions.

Experiments in all the main domains of deduction have corroborated these two predictions of the model theory. Deductions that call for only a single model are reliably easier than those that call for multiple models; and erroneous conclusions tend to be consistent with the premises rather than inconsistent with them. We will illustrate this evidence with some recent studies of so-called double disjunctions (see Johnson-Laird, Byrne, and Schaeken, 1992; Bauer and Johnson-Laird, 1993).

If you wish to experience the phenomenon, then ask yourself what, if anything, follows from these double disjunctive premises:

Raphael is in Tacoma or Jane is in Seattle, or both.
Jane is in Seattle or Paul is in Philadelphia, or both.

Each premise supports three explicit models, and when the information from both premises is combined the result is five distinct models:

[t]	[s]	[p]
[t]	[s]	
	[s]	[p]
	[s]	
[t]		[p]

where *t* denotes Raphael in Tacoma, *s* denotes Jane in Seattle, and *p* denotes Paul in Philadelphia, though the actual models that people construct will probably represent particular individuals in particular cities. The models support the conclusion:

Jane is in Seattle, or Raphael is in Tacoma and Paul is in Philadelphia.

As the model theory predicts, a double disjunction is reliably easier when the disjunctions are exclusive, e.g.:

Raphael is in Tacoma or Jane is in Seattle, but not both.
Jane is in Seattle or Paul is in Philadelphia, but not both.
What follows?

because there are now only two possible models:

[t]		[p]
	[s]	

which support the conclusion:

Jane is in Seattle, or Raphael is in Tacoma and Paul is in Philadelphia.

The problems are difficult, and most errors are consistent with the premises, that is, based on only some of their possible models, and typically on only a single model.

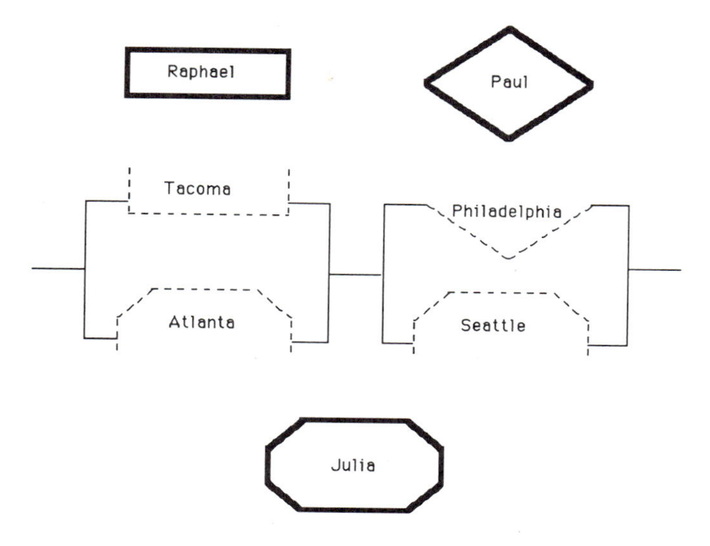

FIGURE 65.1 A diagram presenting a double disjunction problem: An event occurs only if the circuit diagram is completed from left to right by moving the shapes into their congruent positions in the circuit.

The psychological problem of deduction is to keep track of alternative possibilities. One way in which performance can be strikingly improved is to use diagrams rather than verbal premises. Not any sort of diagram will do, however. The evidence suggests that the diagram must use graphical means to make the alternative possibilities more explicit. With diagrams that resemble electrical circuits, such as figure 65.1, subjects drew 30% more valid conclusions than with the equivalent verbal problems (Bauer and Johnson-Laird, 1993).

Reasoning and the brain

The mental model theory is particularly pertinent to the neuropsychology of thinking. Unlike theories that stress that reasoning is a verbal process (Polk and Newell, 1992) or is governed by formal rules (e.g., Rips, 1983), it assumes that a major component of reasoning is nonverbal—that is, the construction of models with a structure corresponding to the structure of situations. Hence, the theory predicts that the right cerebral hemisphere should play a significant part in reasoning (Whitaker et al., 1991).

In general, neuropsychological evidence bears out this prediction. Several studies have shown that damage to the right hemisphere impairs patients' ability to make inferences. Thus, Caramazza et al. (1976) have shown that such patients have problems in deducing the converse of relations. They fail such problems as:

John is taller than Bill.
Who is shorter?

Similarly, Read (1981) found that they are impaired in comparison with normals with such three-terms series problems:

Arthur is taller than Bill.
Bill is taller than Charles.
Who is shortest?

These studies were motivated by the possibility that visual imagery underlies performance (but cf. McDonald and Wales, 1986) and the knowledge that visuospatial thinking appears to depend on the right hemisphere. However, reasoning can also be based on models that have no perceptible correlates. Individuals who are capable reasoners often report that they have not experienced visual imagery, and yet their performance is entirely consistent with the predictions of the model theory: They find multiple-model problems difficult, and their errors are consistent with premises. If the construction of models depends on the right hemisphere, then patients with right-hemisphere damage should find it just as hard to reason about abstract matters as to reason about topics that are easy to visualize. Some neurological studies have examined the ability to make inferences that do not depend on visuospatial thinking. Thus, given the sentences:

Sally approached the movie star with pen and paper in hand.
She was writing an article about famous people's views about nuclear power.

normal individuals are likely to infer that Sally wanted to ask the star about nuclear power. Patients with damage to the right hemisphere, as Brownell et al. (1986) observed, infer that Sally wanted the movie star's autograph. They are misled by the first sentence and cannot make the bridging inference from the second sentence to revise their interpretation. In general, it seems that right-hemisphere damage leads to an inability to get the point of a story, to make implicit inferences establishing coherence, to grasp the force of indirect illocutions such as requests (see e.g., Wapner, Hamby, and Gardner, 1981; Beaman, 1993), although at least one study failed to detect effects of right-hemisphere damage on implicit inferences (Tompkins and Mateer, 1985). What complicates matters is that damage to the right hemisphere can lead to semantic difficulties in the interpretation of words (see, e.g., Joanette and Brownell, 1990), and so in consequence the comprehension of discourse may also be impaired. Conversely, there is also evidence from split-brain patients that either hemisphere is capable of nonverbal reasoning (Zaidel, Zaidel, and Sperry, 1981), though the left hemisphere is superior to the right in problem solving (Gazzaniga, 1992, 103).

The strongest evidence for the model theory's predictions comes from experiments on conditional reasoning carried out by Whitaker and his colleagues. In a study of brain-damaged patients, Whitaker et al. (1991) examined conditional reasoning in two groups. The patients in both groups had undergone a unilateral anterior temporal lobectomy to relieve focal epilepsy, one group to the right hemisphere and the other group to the left hemisphere. Those with right-hemisphere damage were poorer at reasoning from false conditional premises than those with left-hemisphere damage. Thus, given the following conditional:

If it rained the streets will be dry

and the categorical assertion:

It rained,

the right-hemisphere-damaged group had a reliable tendency to conclude:

The streets will be wet.

In other words, these patients were unable to carry through the process of deduction in isolation from their knowledge of reality. In an ingenious study, Savary, Whitaker, and Markovits (1992), have extended this research to normal individuals. They argued that if reasoning depends on a major nonverbal component, then it should interfere more than a verbal memory task with a nonverbal secondary task. The primary task was either reasoning with a conditional problem or memorizing a sentence. While engaged in a primary task, the subjects had to judge whether two shapes were similar (the nonverbal secondary task) or decide whether a visually presented string of letters was a word or not (the verbal secondary task). The experimenters also obtained response times to the two secondary tasks when they were performed alone, and the key comparisons concerned the difference between these control measures and those obtained while the subjects were performing a primary task. The results confirmed the prediction: Reasoning, unlike memorizing a sentence, slowed down the judgments of the similarity of shapes, whereas there was no difference between the two in their effects on lexical decision.

The finding that certain sorts of diagrams can help reasoners (see the previous section) allows investigators to study reasoning without the need for verbal comprehension of premises. A major task for the future is to use brain-scanning techniques (see Kosslyn and Koenig, 1992) to investigate which areas of the brain are active during verbal and diagrammatic reasoning. The model theory predicts that both sorts of reasoning depend on the right hemisphere, and that diagrams should reduce the dependence on the left hemisphere.

Conclusions

The model theory is based on six main assumptions:

1. Entities are represented by tokens in models, their properties by properties of the tokens, and the relations between them by the relations between the tokens.
2. Alternative possibilities are represented by alternative models.
3. Negation is represented by a propositional annotation.
4. Implicit individuals and situations are represented by a propositional annotation that works in concert with an annotation indicating what has been represented exhaustively.
5. To account for counterfactual reasoning or reasoning about what is permissible, the epistemic status of a model can also be represented by a propositional annotation, such as: a model represents a real possibility, a counterfactual state of affairs, a permissible state of affairs, and so on.
6. Reasoning calls for the construction of models of premises, the formulation of conclusions based on them, and a search for alternative models to test validity.

Models based on the first two assumptions represent a class of situations (Barwise, 1993), and models that include propositional annotations can represent a finite set of alternative classes of situations.

The resulting theory makes three principal predictions. First, the greater the number of models called for to make an inference, the harder the task will be. Second, erroneous conclusions will tend to be consistent with the premises rather than inconsistent with them. Third, knowledge can influence the deductive process: Subjects will search more assiduously for alternative models when a putative conclusion is unbelievable than when it is believable. This chapter has illustrated the corroboration of these predictions in experiments from several domains of deduction. The model theory also makes a critical prediction about the role of the cerebral hemispheres in reasoning. As Whitaker et al. (1991) first pointed out, the construction of models is likely to depend on the right hemisphere. Although there is some evidence for this prediction, the crucial experiment has yet to be done. It calls for brain scanning during two sorts of reasoning, verbal and diagrammatic.

ACKNOWLEDGMENTS I am grateful for support from the James S. McDonnell Foundation. I thank Keith Holyoak and Nina Robin for useful criticisms of a draft of this chapter, and Malcolm Bauer and Ruth Byrne for help in carrying out much of the research.

REFERENCES

BARWISE, J., 1993. Everyday reasoning and logical inference. (Commentary on Johnson-Laird and Byrne, 1991) *Behav. Brain Sci.* 16:337–338.

BAUER, M. I., and P. N. JOHNSON-LAIRD, 1993. How diagrams can improve reasoning. *Psychol. Sci.* 4:372–378.

BEAMAN, M., 1993. Semantic processing in the right hemisphere may contribute to drawing inferences from discourse. *Brain Lang.* 44:80–120.

BRAINE, M. D. S., B. J. REISER, and B. RUMAIN, 1984. Some empirical justification for a theory of natural propositional logic. In *The Psychology of Learning and Motivation*, vol. 18. New York: Academic Press.

BROWNELL, H. H., H. H. POTTER, A. M. BIHRLE, and H. GARDNER, 1986. Inference deficits in right brain-damaged patients. *Brain Lang.* 27:310–321.

BYRNE, R. M. J., and P. N. JOHNSON-LAIRD, 1989. Spatial reasoning. *J. Mem. Lang.* 28:564–575.

CARAMAZZA, A., J. GORDON, E. B. ZURIF, and D. DE LUCA, 1976. Right hemispheric damage and verbal problem solving behavior. *Brain Lang.* 3:41–46.

CRAIK, K., 1943. *The Nature of Explanation.* Cambridge: Cambridge University Press.

FODOR, J. A., 1983. *The Modularity of Mind.* Cambridge, Mass.: MIT Press.

GARNHAM, A., 1987. *Mental Models as Representations of Discourse and Text.* Chichester: Ellis Horwood.

GAZZANIGA, M. S., 1992. *Nature's Mind: The Biological Roots of Thinking, Emotions, Sexuality, Language, and Intelligence.* New York: Basic Books.

HAGERT, G., 1984. Modeling mental models: experiments in cognitive modeling of spatial reasoning. In *Advances in Artificial Intelligence*, T. O'Shea, ed. Amsterdam: North-Holland.

HOLYOAK, K. J., and B. A. SPELLMAN, 1993. Thinking. *Ann. Rev. Psychol.* 44:265–315.

JOANETTE, Y., and H. H. BROWNELL, eds., 1990. *Discourse Ability and Brain Damage: Theoretical and Empirical Perspectives.* New York: Springer-Verlag.

JOHNSON-LAIRD, P. N., 1970. The perception and memory of sentences. In *New Horizons in Linguistics*, J. Lyons, ed. Harmondsworth, Middlesex, England: Penguin, pp. 261–270.

JOHNSON-LAIRD, P. N., 1983. *Mental Models: Towards a Cognitive Science of Language, Inference and Consciousness.* Cambridge: Cambridge University Press.

JOHNSON-LAIRD, P. N., 1993. *Human and Machine Thinking.* Hillsdale, N.J.: Erlbaum.

JOHNSON-LAIRD, P. N., and R. M. J. BYRNE, 1991. *Deduction.* Hillsdale, N.J.: Erlbaum.

JOHNSON-LAIRD, P. N., and R. M. J. BYRNE, 1993. Precis of *Deduction*, and authors' response: Mental models or formal rules? *Behav. Brain Sci.* 16:323–333, 368–376.

JOHNSON-LAIRD, P. N., R. M. J. BYRNE, and W. SCHAEKEN, 1992. Propositional reasoning by model. *Psychol. Rev.* 99: 418–439.

JOHNSON-LAIRD, P. N., R. M. J. BYRNE, and P. TABOSSI, 1989. Reasoning by model: The case of multiple quantification. *Psychol. Rev.* 96:658–673.

KOSSLYN, S. M., and O. KOENIG, 1992. *Wet Mind: The New Cognitive Neuroscience.* New York: The Free Press.

LURIA, A. R., 1973. *The Working Brain: An Introduction to Neuropsychology.* Harmondsworth, Middlesex, England: Penguin.

MCDONALD, S., and R. WALES, 1986. An investigation of the ability to process inferences in language following right hemisphere brain damage. *Brain Lang.* 29:68–80.

OAKHILL, J. V., P. N. JOHNSON-LAIRD, and A. GARNHAM, 1989. Believability and syllogistic reasoning. *Cognition* 31: 117–140.

OHLSSON, S., 1984. Induced strategy shifts in spatial reasoning. *Acta Psychol.* 57:46–67.

OSHERSON, D. N., 1974–1976. *Logical Abilities in Children*, vols. 1–4. Hillsdale, N.J.: Erlbaum.

POLK, T. A., and A. NEWELL, 1992. A verbal reasoning theory for categorical syllogisms. Mimeo, Department of Computer Science, Carnegie Mellon University.

READ, D. E., 1981. Solving deductive reasoning problems after unilateral temporal lobotomy. *Brain Lang.* 12:116–127.

RIPS, L. J., 1983. Cognitive processes in propositional reasoning. *Psychol. Rev.* 90:38–71.

SAVARY, F., H. WHITAKER, and H. MARKOVITS, 1992. Counterfactual reasoning deficits in right brain damaged patients. Unpublished paper, Department of Psychology, University of Quebec at Montreal.

Schaeken, W., and P. N. Johnson-Laird, 1993. Mental models and temporal reasoning. Unpublished paper, Department of Psychology, Princeton University.

Shallice, T., 1988. *From Neuropsychology to Mental Structure.* Cambridge: Cambridge University Press.

Simon, H. A., 1959. Theories of decision making in economics and behavioral science. *Am. Econ. Rev.* 49:253–283.

Tompkins, C., and C. A. Mateer, 1985. Right hemisphere appreciation of prosodic and linguistic indications of implicit attitude. *Brain Lang.* 24:185–203.

van Dijk, T. A., and W. Kintsch, 1983. *Strategies of Discourse Comprehension.* New York: Academic Press.

Wapner, W., S. Hamby, and H. Gardner, 1981. The role of the right hemisphere in the apprehension of complex linguistic materials. *Brain Lang.* 14:15–33.

Whitaker, H. A., F. Savary, H. Markovits, C. Grou, and C. M. J. Braun, 1991. Inference deficits after brain damage. Paper presented at the annual INS meeting, San Antonio, Texas.

Zaidel, E., D. W. Zaidel, and R. W. Sperry, 1981. Left and right intelligence: Case studies of Raven's Progressive Matrices following brain bisection and hemi-decortication. *Cortex* 17:167–186.

66 Working Memory in Humans: Neuropsychological Evidence

EDWARD E. SMITH AND JOHN JONIDES

ABSTRACT There is a growing body of neuropsychological evidence supporting the idea of separate working memories for verbal, visual, and spatial information. Much of this research employs the logic of double dissociation and shows, for example, that a lesion in one brain site is associated with an impairment in spatial working memory, whereas a lesion in another site is associated with an impairment in visual working memory. We have extended this double-dissociation logic to PET studies of working memory. PET measurements were taken while subjects were engaged in either a spatial or a visual object task. In the spatial task, three dots were presented briefly, followed by a brief delay, followed by a probe; and subjects had to indicate whether the probe was in the same location as one of the dots. In the object task, two objects were presented briefly, followed by a brief delay, followed by a probe object; and subjects indicated whether the probe object was the same as one of the two preceding objects. The two tasks led to different patterns of activation —the spatial task led to activation in the right-hemisphere occipital, parietal, and prefrontal regions, whereas the object task resulted in activation in left-hemisphere parietal and prefrontal regions. This double dissociation further supports the idea of separate working memories.

Introduction

WORKING MEMORIES The concept of human working memory grows out of two distinct lines of research. One consists of experimental studies of short-term memory, which has been conceptualized as a system that holds limited information for only a brief time until it can be externalized. A prototypical example is retaining a newly learned phone number until it can be dialed. Building on the tradition of such studies, Baddeley and Hitch (1974) were among the first to note that the short-term memory system may be involved in reasoning and comprehension. These researchers demonstrated that requiring people to remember irrelevant material for a brief time while they simultaneously engaged in a reasoning task resulted in a decrement in reasoning. This result led directly to the notion of a *working memory* system involved in reasoning.

A second line of research that also implicates a working memory system involves computational modeling of higher cognitive processes, such as language, reasoning, and problem solving. Here, researchers include in a working-memory system not only information that has to be retrieved imminently, but also the intermediate products of mental computations involved in higher cognitive processes, as well as various representations of the current environment (e.g., Anderson, 1983; Newell, 1990). Empirical support for this computational notion of working memory comes principally from studies showing that variations in the capacity of working memory are correlated with variations in performance on reasoning and language tasks (e.g., Carpenter, Just, and Shell, 1990; Just and Carpenter, 1992).

Whichever tradition of working memory one examines, a basic question one can ask is whether working memory is unitary in character or whether different working memories are employed for different kinds of information. Baddeley (e.g., 1986, 1992) has been a leader in addressing this question, and his research implicates at least two different working memories—one that holds verbal material in a speechlike code, and a second that holds visual or spatial material in some sort of pictorial code. Further research has indicated that there may be yet separate working memories for visual and spatial information (e.g., Farah et al., 1988).[1]

If there are indeed separate buffers for verbal, visual, and spatial information, then one might expect the buffers to be served by different brain structures. This possibility is the major concern of the present chapter, and we note at the outset that it is compatible with two basic facts about the organization of the cerebral cor-

EDWARD E. SMITH and JOHN JONIDES Department of Psychology, University of Michigan, Ann Arbor, Mich.

tex. First, the cortex (and the central nervous system in general) is organized by sensory modality, with different cortical regions serving vision, audition, touch, etc. This fact makes plausible the idea that there are different working memories for information presented visually and for that presented auditorally (see, e.g., Wilson, Skelly, and Goldman-Rakic, 1992, for a similar supposition). Second, the cortical basis of vision seems to comprise two subsystems, a ventral system that mediates perception of object information, and a dorsal system that mediates perception of location (e.g., Ungerleider and Mishkin, 1982). This makes plausible the idea of different working memories for visual information concerned with the identity of objects and for visual information concerned with the spatial location of objects.

This background about working memory leads naturally to the agenda for the remainder of this chapter. In the next section we review some of the relevant evidence for supposing that there are identifiable working memories for visual and spatial information. We emphasize neuropsychological and neurophysiological findings rather than strictly behavioral evidence in this review because evidence that different brain mechanisms are associated with different memory tasks seems to us to be the most telling for the notion of separate buffers. In the third section, we describe studies using positron emission tomography (PET) to provide evidence about the distinctions among working memory systems in normal humans.

Prior Evidence for Multiple Working Memories

Perhaps the most compelling evidence for different working memories relies on the logic of double dissociations: If there is a factor that influences or is associated with performance on task A but not task B, and there is another factor that influences or is associated with performance on task B but not task A, then these two tasks must be mediated by different processing mechanisms. Demonstrating such a pair of factors provides support for the hypothesis of separate systems.

The logic of double dissociation is implemented differently in behavioral, neuropsychological, and neurophysiological studies. In strictly behavioral studies, double dissociations are shown when a behavioral manipulation affects performance on task A but not task B, whereas a second behavioral manipulation affects performance on B but not A. In neuropsychological studies of lesioned animals (most prominently, humans), the critical factors are not behavioral manipulations but rather lesions in two different brain sites. In this case, the experimenter must demonstrate that one lesion affects performance on task A but not B, whereas another lesion affects task B but not A. Obviously, this requires the use of different subjects, each with one type of lesion, rather than allowing the use of a subject as his or her own control as in the strictly behavioral case (and other cases that we shall consider). In typical neurophysiological experiments involving single-cell recordings in nonhumans, the critical factors are neurons in two different brain locations. In this case, one must show that neurons in one location are active during task A but not during B, whereas neurons in a second location show the opposite pattern of activation.

The logic of testing for double dissociations is sufficiently telling that we introduce here its application to research on normal humans using positron emission tomography (PET) and, potentially, other imaging techniques. As in neuropsychological and neurophysiological studies, the relevant factors are again two different brain regions. A double dissociation is revealed if one brain region is active during task A but not B, whereas the other region is active during task B but not A.

Let us now assess the evidence for different working memories by considering double dissociations of the above four types—from behavioral, neuropsychological, neurophysiological, and PET studies. The data from the four lines of research show some convergence.

Behavioral studies There are several behavioral studies that have used double dissociations to establish a distinction between verbal and visuospatial buffers, and between visual and spatial buffers (for reviews, see Baddeley, 1986, 1992). Consider an illustration of this approach from an experiment performed by Logie (1986; see also Baddeley, 1986). In this study, subjects learned a list of digit-word pairs presented auditorally. Half the subjects were trained to learn the list by a visual imagery mnemonic. They first memorized rhymes like "1 is a bun, 2 is a shoe...." Then, when presented a digit-word pair to be memorized (e.g., "2-ashtray"), subjects tried to retrieve the rhyme with the same number ("2 is a shoe"), and generate an image linking the word in the rhyme with the word to be learned (an image linking shoe and ashtray). The remaining subjects were instructed to learn the list by rote rehearsal. Arguably, the imagery mnemonic involves a visual and/or spatial buffer, whereas the rote strategy involves a verbal buffer. While learning the digit-word pairs, distracting stimuli were presented that subjects had to ignore. These distractors could be either pictures or the names of the pictures, with both types of distractors occurring with both methods of learning. The rationale of the experiment (making use of the logic of double dissociations) was that the picture distractors should engage the visual and/or spatial buffer, which would impair imagery-based learning but not rote-based learning; in contrast, the name distractors should engage the verbal buffer, which would impair rote-based but not imagery-based learning. Logie obtained just this double dissociation, thereby supporting the idea of separate buffers for visual and verbal material.

This study illustrates not only the application of the double dissociation logic to behavioral studies of working memory, but also some of the problems that plague these studies. One difficulty is that the main tasks involved—learning by imagery or by rote in this example—are quite complex and may well include unwanted components (e.g., the imagery mnemonic clearly involves some verbal components). This lack of purity can be alleviated by devising simpler and more refined tasks. But other potential problems with this line of research may be inherent in the strictly behavioral methodology. Following an argument articulated by Pylyshyn (1981), subjects who learned by the imagery mnemonic may not have used anything like visual processing, but rather may have relied on nonvisual knowledge *about* visual processes. For example, in the study just described, subjects in the mnemonic condition may not so much have visualized the objects involved, but rather described to themselves the objects' visual appearances. In a related vein, subjects in the imagery-learning condition may merely have been supplying data that conformed to what they took to be the experimenter's expectations (e.g., Intons-Peterson, 1983). Thus, in the study just described, since subjects in the mnemonic condition knew they should be doing something visual, they might have suspected that they were supposed to have trouble when their secondary task was visual, and this suspicion somehow led to the decrement in performance. These kinds of objections arise whenever a researcher proposes that imagery involves perceptual processes, and they are as relevant to studies of working memory as they are to studies of imagery phenomena. Although there are counterarguments to these objections (see Kosslyn, 1981), the case for perceptual processing in the absence of visual stimuli may be made more strongly with neuropsychological data, which are not open to these particular objections (Kosslyn, 1987; Farah, 1988).

Neuropsychological studies Consider evidence for separate visual and spatial buffers that stems from research by Levine, Warach, and Farah (1985). These researchers studied two patients who had different patterns of brain damage. Patient 1 had bilateral temporo-occipital lobe damage, whereas patient 2 had bilateral lesions in the parieto-occipital region. The patients were given two sets of tasks. One set involved "bedside" tests of visual processing and memory that do *not* involve spatial information, and the other set included comparable tests of visual processing and memory that rely heavily on spatial information. Examples of nonspatial, visual tasks include describing an animal solely from imagery, or drawing an animal from memory. An example of a spatial task is the description of some map or route solely from memory. Compared to control subjects, patient 1 (temporo-occipital damage) showed normal to better-than-normal performance on the spatial tasks, but impaired performance on all nonspatial visual tasks; patient 2 (parieto-occipital damage), however, showed the opposite pattern, thus demonstrating a clear double dissociation. Although only two subjects were involved in the actual experiment, Levine, Warach, and Farah (1985) surveyed the relevant literature and found that patients known to have problems in object recognition (presumably a nonspatial visual task) were not particularly likely to have problems in spatial imagery, and patients known to have problems in spatial imagery were unlikely to have problems in object recognition or imagery.

Neurophysiological studies Relevant neurophysiological data come from studies of visual and spatial buffers in monkeys by Goldman-Rakic and her colleagues (e.g., Goldman-Rakic, 1987; Wilson, Scalaidhe, and Goldman-Rakic, 1993). In experiments focusing on spatial memory, monkeys were trained to remember the spatial location of a target stimulus for several seconds before executing a saccade to that location. Activity in some cells in the region of the principal sulcus of dorsolateral prefrontal cortex occurred *only* during a delay period during which the animal was presumably storing the position of the previously presented visual target. In contrast, in experiments in which monkeys were trained to remember a nonspatial visual object, such as a simple pattern or a monkey's face, neural activity during the delay period involved only neurons in another region of prefrontal cortex, inferior to the first region. Thus, activation of neurons in the region of the principal sulcus is associated with storage of spatial information but not nonspatial information, whereas activation of neurons in the inferior convexity region showed the opposite pattern. This line of research reveals evidence of a double dissociation at the level of individual neurons.

PET studies Although neuropsychological and neurophysiological studies converge in that they both show double dissociations between visual and spatial buffers, the two sets of studies implicate different anatomical regions. In the Levine, Warach, and Farah (1985) study, the patient who was impaired only on spatial tasks had a lesion in the parieto-occipital region, suggesting that this region is involved in the functioning of a spatial buffer, whereas the studies of spatial memory in monkeys implicated a region in prefrontal cortex (although Goldman-Rakic and her coworkers argue for parietal involvement in this task as well). This disagreement is more apparent then real, though, once one recognizes a limitation of lesion studies in localizing functions like spatial memory. Specifically, performance on any task, including the spatial memory tasks used by Levine, Warach, and Farah, inevitably involves several processes or functions, and if any one of these is impaired by a lesion, then performance on the entire task will suffer. For example, in the spatial task requiring description of a route from memory, one must first encode the relevant spatial information, then store it, and finally retrieve and utilize it at the time of test. Performance on this task would therefore suffer if there were an impairment just in encoding visual information, and such an impairment would be consistent with a lesion in the parieto-occipital region such as that found in Levine, Warach, and Farah's patient of interest.

To overcome this limitation, one needs data of brain activity that provide a more direct picture of all major anatomical areas that mediate performance in a particular task. Positron emission tomography provides just such data. In doing so, PET studies overcome yet another limitation of studies of human lesions. Findings from studies of lesions are often based on a small number of subjects (e.g., just two patients in the Levine, Warach, and Farah study). The small number is dictated by the availability of appropriate patients or by methodological considerations (e.g., Caramazza, 1986), yet it raises potential problems in generalizing the results of these studies. This limitation is offset in PET studies in which multiple subjects can be tested, as in strictly behavioral experiments. Another difficulty in interpreting lesion data is that recovery of function after brain damage may result in somewhat irregular brain function, again leading to problems of generalization. In PET studies, of course, the brains involved can be normal ones. (See Fox, 1992, for further comparison of PET and lesion data.)

We do not argue that PET studies should replace lesion studies (PET, too, has its own set of limitations), but rather that PET measures can be used as converging evidence to lesion data. In the present context, this means bringing to bear PET measures on the issues of different working-memory buffers. The questions of particular interest are whether PET measures obtained with normal human subjects will yield a double dissociation between spatial and visual working memory tasks, just as neuropsychological and neurophysiological studies have; and whether they will reveal a consistency between the cortical regions activated and those implicated in prior neuropsychological and neurophysiological studies.

Some experimental results

We have performed a series of PET experiments that bear on the issues mentioned above. Since some of the details of these studies appear elsewhere (Jonides et al., 1993; Koeppe et al., 1993), in what follows we focus on a brief presentation of these studies, highlighting those aspects critical to the issue of double dissociations derived from imaging data.

TASKS Two groups of subjects were tested on visual, working-memory tasks while PET measurements were taken; one group was tested on spatial memory and the other on object memory. The spatial-memory task is schematized at the top of figure 66.1. On each trial, subjects began by fixating a cross in the center of a computer screen for 500 ms. The cross was replaced by three randomly arrayed dots on the circumference of an imaginary circle of 14° diameter, centered on the

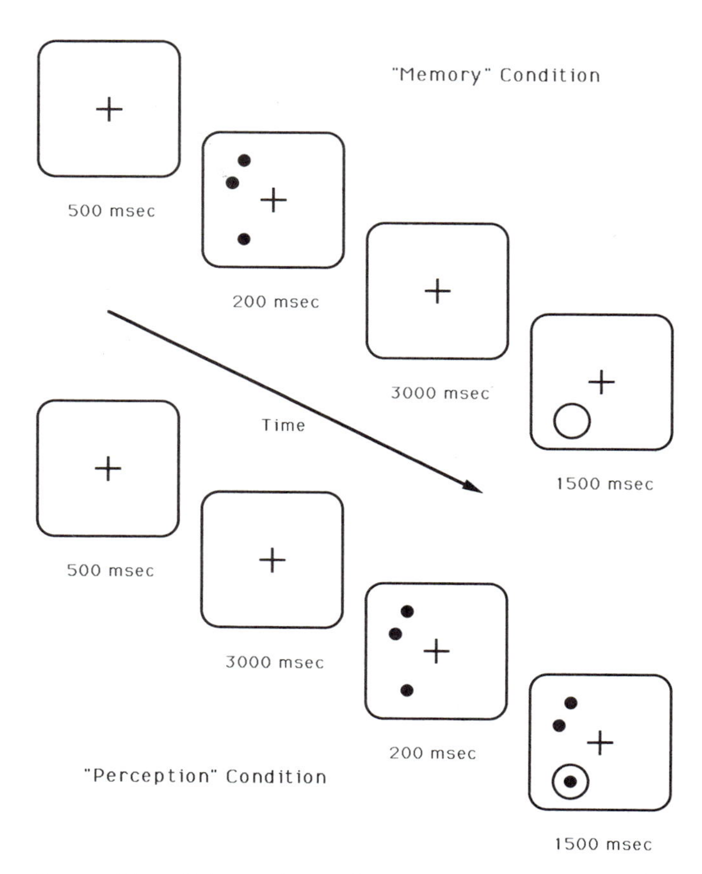

FIGURE 66.1 Schematic drawings of the events on each trial of the spatial memory and spatial perception conditions.

cross. Subjects had to encode the locations of the three dots quickly because the dots remained in view for only 200 ms (too brief a duration for a successful saccade to the dots, on average). Next, the fixation cross reappeared for a retention interval of 3000 ms. The retention interval was followed by a location probe, which consisted of a single outline circle that either encircled the location of one of the previous dots (with probability .50) or not. The subjects' task was to press a response button once or twice to indicate whether or not the probe encircled the location of a target dot; subjects had to make their responses within a 1500 ms interval. The probe circle was either centered directly about the location of a previously presented dot, or it missed the nearest dot location by 15°–40°.

This spatial-memory task requires numerous processes in addition to working memory of spatial information, including maintaining attention, encoding spatial information, and selecting and executing a response. In order to remove the effects of these additional processes from the activation patterns of the PET images, a control condition was needed that included these processes but did not include a memory requirement (following the methodological approach developed by Posner et al., 1988). This spatial-perception control task is schematized at the bottom of figure 66.1. Again, each trial began with a fixation cross, but in this condition the cross remained in view for 3500 ms (the duration of the fixation plus the retention interval in the spatial-memory condition). The three dots were then presented for 200 ms, followed immediately by an interval of 1500 ms during which the three dots and the probe circle were presented simultaneously. As in the memory condition, subjects pressed a response button once or twice to indicate whether or not the probe encircled a dot, in this case based on a display in which the probe and dots were presented for viewing simultaneously so that no memory of dot locations was needed to make a successful response. This spatial-perception control task thus included nearly the same trial events as the spatial-memory task, but with no memory requirement.

The object-memory task and its control condition are schematized in figure 66.2. Consider first the object-memory task illustrated at the top of figure 66.2. It was designed to have a similar structure to the spatial task. Again, a trial began with a fixation cross exposed for 500 ms. The cross was replaced by two objects that were presented on either side of the cross's previous location. These remained in view for 400 ms (This duration was selected on the basis of pilot experimentation that determined the duration needed for accurate encoding of the objects.). The objects were unfamiliar geometric figures, constructed so that each consisted of an outline shape with a second shape embedded within it, with lines connecting the inner and outer shapes. The third trial event consisted of just the fixation cross, for a retention interval of 3000 ms. This was followed by a probe, which consisted of an object located where the fixation cross had been. The probe was either identical to one of the two target objects (with probability .50) or not. The subjects' task was to press a button once or twice to indicate whether or not the probe object was identical to one of the target objects; subjects had to make this response within a 1500 ms interval. As in the spatial-memory task, there was need for a control condition that included all but the memory processes in the main task. This object-perception control task is schematized at the bottom of figure 66.2. It involved approximately the same trial

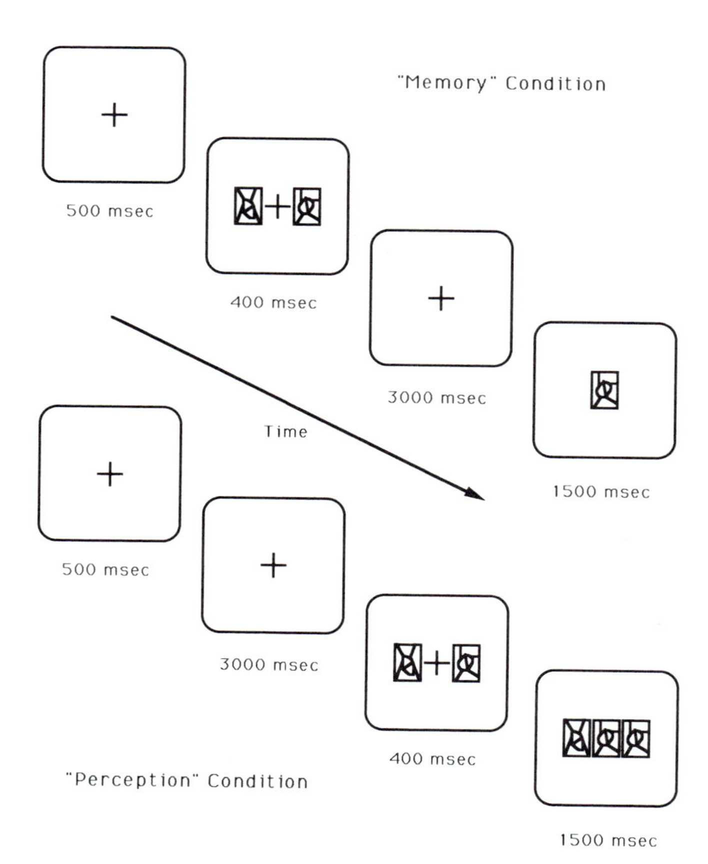

FIGURE 66.2 Schematic drawings of the events on each trial of the object memory and object perception conditions.

events as the memory task, but with the events rearranged so that successful performance did not require any memory for object information. Subjects responded by pressing a key either once or twice if the center object in the last display matched either of the two flanking objects.

One group of 18 subjects was given two blocks of 20 trials of each of the spatial-memory and spatial-perception tasks. (This group was also scanned on two blocks of another condition, not reported here.) The order of administration of these blocks was counterbalanced across subjects. Another group of 12 subjects was tested on the object-memory and object-perception tasks. This group received three blocks of 20 trials for each condition, again counterbalanced for order of conditions across subjects. All subjects were right-handed adults drawn from the University of Michigan community who had volunteered to participate for pay. Subjects were given practice on their respective tasks prior to testing in the experiment proper.

PROTOCOL FOR PET MEASURES Subjects were familiarized with the PET apparatus prior to the experiment proper. Each subject had an intravenous catheter inserted into the left arm to receive the injection of radioactive tracer, and was then positioned in the scanner with tape applied from the head holder to the forehead to constrain head movement. The experimental protocol consisted of six scans. Each scan consisted of 20 trials, with the first 3 trials presented prior to the injection of the radionuclide (the total duration of these 3 trials was approximately 15 s.). Immediately following these trials, a bolus injection of 66 mCi of oxygen 15–labeled water was administered, after which approximately 15 s elapsed before the radionuclide reached the brain. Trials continued to be administered during this interval. Recording of activity began 5 s after the count rate was observed to increase above the background level and continued for 60 s thereafter. Injections for subsequent scans were separated by 14 min intervals, permitting the oxygen 15 to decay to an acceptable background level.

The PET images for each subject were transformed to a stereotaxic coordinate system (Minoshima et al., 1993) and linearly standardized to an atlas brain (Talairach and Tournoux, 1988). After normalizing pixel values for global flow rate differences among scans (Fox et al., 1985), the data were averaged across the subjects in a condition, giving mean and variance values for each condition. The average image for each control condition was subtracted from that of its corresponding memory condition to reveal differences in activation between these conditions. The difference image was analyzed for statistical significance on a pixel-by-pixel basis using t-statistics, followed by a multiple-comparison adjustment based on the Bonferroni method (Worsley et al., 1992). A one-tailed adjusted value of $p < .05$ was used as a criterion for reliability.

RESULTS

Spatial memory Subjects were quite accurate in their behavioral performance on this task, making no errors in the perception condition, and 16% errors in the memory condition. Of these errors, roughly half were false alarms, occurring on those trials in which a probe did not encircle a target location. (The other half, of course, were misses, in which a probe did encircle a target but the subject failed to report this correctly.)

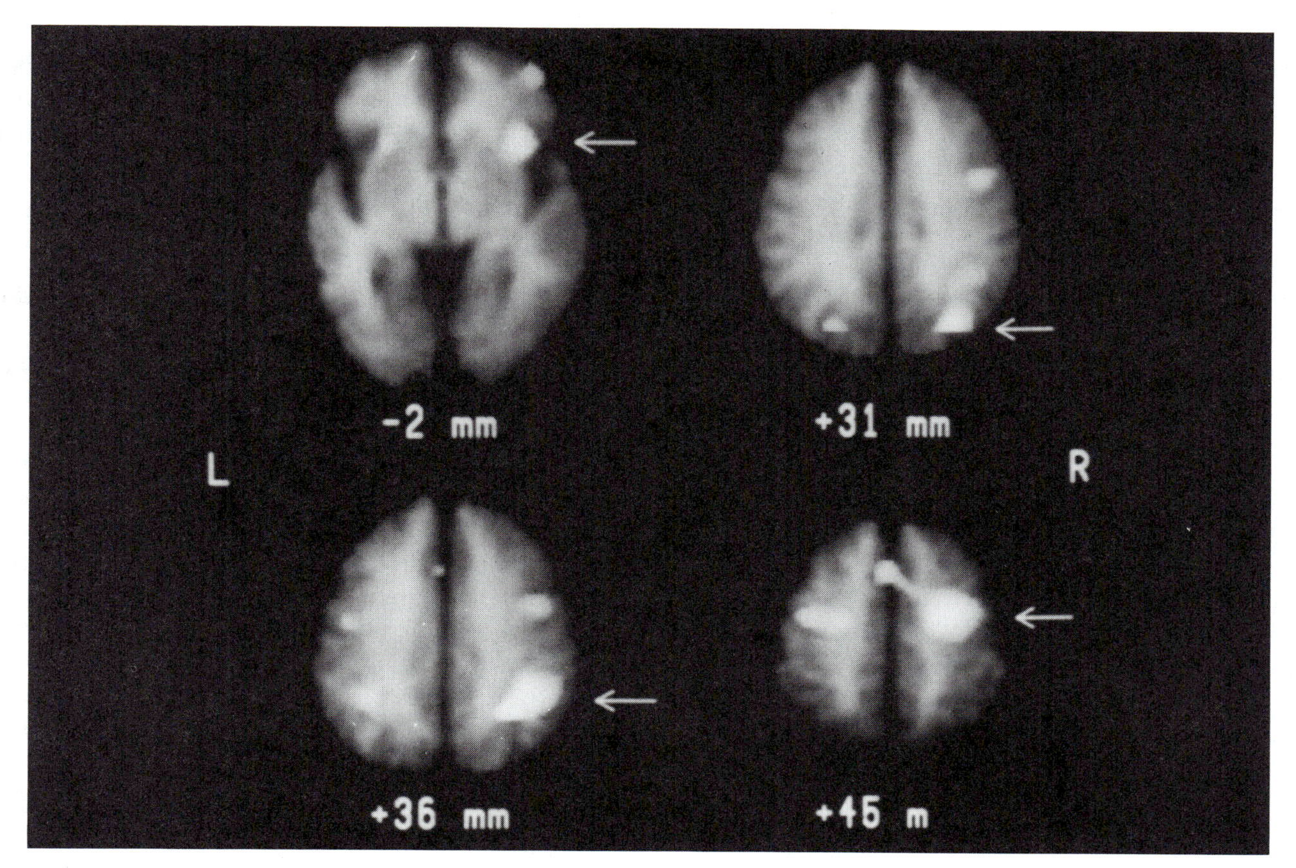

FIGURE 66.3 PET images of the four statistically significant activation sites in the spatial memory task, each superimposed on an MRI image of a composite brain for the purpose of illustrating the anatomical localization of activation foci. The top right and bottom left images include activation foci that are cut off due to the fact that the remaining activation fell outside the field of view of the camera. All four significant foci of activation were in the right hemisphere. Inspecting the images from top to bottom, and then from left to right, the foci of activation were in prefrontal cortex (centered in Brodmann's area 47; stereotaxic coordinates: −35, 19, −2), parietal cortex (centered in Brodmann's area 40; stereotaxic coordinates: −42, −40, 36), occipital cortex (centered in Brodmann's area 19; stereotaxic coordinates: −30, −76, 31), and premotor cortex (centered in Brodmann's area 6; stereotaxic coordinates: −34, −1, 45).

An indication of the fact that subjects used a spatial memory representation to mediate their performance comes from an examination of the false-alarm trials, which can be categorized as ones in which the probe was relatively near a dot location even through it encircled none, or relatively far away. When near, subjects' average error rate was 28.9%, and when far, 7.5%. These values are significantly different ($p < .01$). This is precisely the pattern one would predict if subjects coded the locations spatially: A probe that appeared relatively close to the location of a previous target would be difficult to reject as having encircled that location.

The major difference between the perception and memory conditions was that only the latter required subjects to store location information; hence, differences between these two conditions provide information about the localization of these storage processes. Figure 66.3 presents four brain images showing those brain activation levels resulting from the subtraction of the perception from the memory condition that exceeded a criterion of $p < .05$ before correction for multiple comparisons. The figure includes an arrow beside each image that marks the one site of activation on that image whose significance level exceeded $p < .05$ after correction by the Bonferroni method (Friston et al., 1991). These areas of activation have been superimposed on magnetic resonance images of a composite brain to illustrate the anatomical localization of the foci.

One of the areas of reliable activation is in the prefrontal cortex of the right hemisphere, centered in Brodmann's area 47 approximately, as shown in the upper left image of figure 66.3 (stereotaxic coordinates: −35, 19, −2). This focus is consistent with neurophysiological results described earlier showing single-cell activity in monkeys in a related area, in the vicinity of the principal sulcus (Goldman-Rakic, 1987). The task used in the present research with humans is similar to the one used with monkeys, and so the analogy in locus of brain activity suggests a similarity in the memory processes in these two species.

A second area of activation in our subjects was centered in the parietal cortex, principally in the posterior portion as shown in the lower left image of figure 66.3 (approximately Brodmann's area 40; stereotaxic coordinates: −42, −40, 36). This area has previously been reported to be engaged by visual tasks that require object localization (e.g., Chaffee, Funahashi, and Goldman-Rakic, 1989; Haxby et al., 1991). Because our task requires the localization of target dots in visual space (prior to their storage), it seems reasonable that the posterior parietal area should be active. Note that there is ample evidence of projections from the posterior parietal cortex to the prefrontal area (Goldman-Rakic, 1987), permitting a passage of information about dot location to the area putatively involved in the storage of that information.

A third area of activation, shown in the upper right image of figure 66.3 is in occipital cortex, approximately centered in Brodmann's area 19 (stereotaxic coordinates: −30, −76, 31). This area has been implicated in the creation of images in humans (e.g., Kosslyn et al., 1993), suggesting that image creation is an important process in the present task as well.

The final focus of activation, illustrated in the lower right image of figure 66.3 is in premotor cortex, approximately in the region of Brodmann's area 6 (centered on stereotaxic coordinates −34, −1, 45). Note that activation in this region is probably *not* due to the explicitly motor requirement of the present task (responding with button pushes), because this requirement was identical in the perception and memory conditions and therefore should have been subtracted out of the image in figure 66.3. Research has shown that neurons in the premotor area are involved in the planning of motor responses (Tanji, 1985; Mauritz and Wise, 1986). In our experiment, of course, a manual motor response was required in both perception and memory conditions, so the effect of planning this response should have been removed by subtracting the two conditions. Furthermore, the principal premotor activity revealed in our data is in the right hemisphere, and yet all subjects responded with their right hands. Further investigation of this locus of activation will be required to establish its function.

The statistically significant areas of increased activation shown in figure 66.3 are all located in the right hemisphere. There was increased activation as well in the homologous areas in the left hemisphere (as suggested by an examination of figure 66.3), although activation in none of the areas in the left hemisphere reached statistical significance by our criterion. It may be telling that correlations between accuracy of behavioral performance and right-hemisphere brain activation were consistently higher (averaging $r = .33$) than those between accuracy and left-hemisphere activation (averaging $r = .01$). In fact, the four correlations between accuracy and right-hemisphere activation in the four cortical sites discussed above are all larger than the comparable four correlations between accuracy and left-hemisphere activation. This is an indication that performance in the spatial memory task is mediated more directly by activation in the right than the left hemisphere.

Object memory Accuracy in the object task was quite good, averaging 94% in the object perception condition and 78% in the object memory condition. Note that for both conditions, this is somewhat lower accuracy than in the comparable perception and memory conditions of the spatial task, suggesting that subjects had greater difficulty with the object tasks.

Examination of the activations in the PET images reveals a quite different pattern than that obtained with the spatial task. Figure 66.4 shows subtraction images of the object memory minus the object perception conditions. Once again, the images indicate all areas of activation that exceeded a threshold of $p < .05$; but the area on each image that exceeded this criterion after correction for multiple comparisons is marked with an arrow. There are four statistically significant foci of activation that are indicated: left inferior prefrontal gyrus, left parietal lobule, left inferior temporal gyrus, and anterior cingulate gyrus.

The area of activation in the prefrontal lobe (shown in the upper right image of figure 66.4) was centered in Brodmann's area 6 in the left hemisphere (stereotaxic

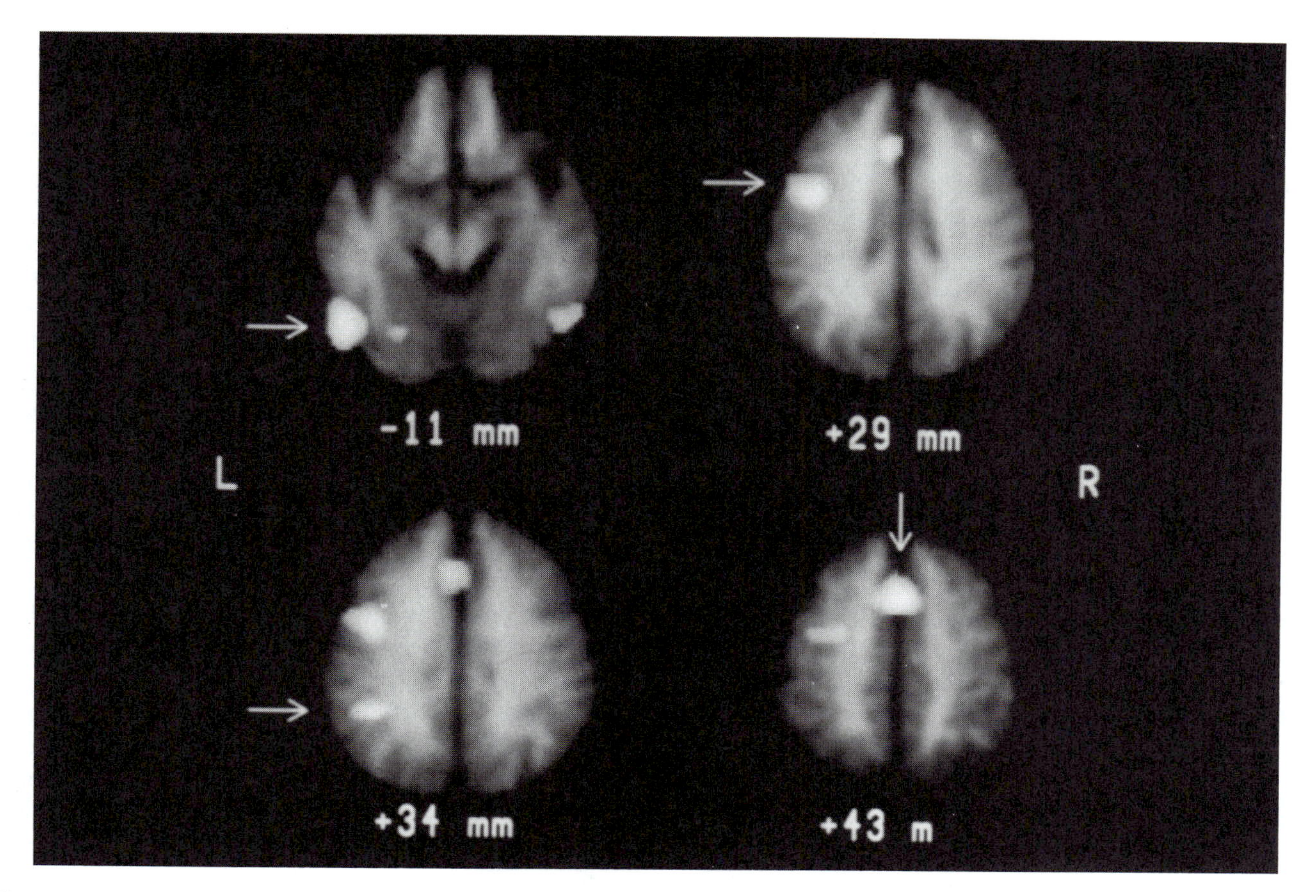

FIGURE 66.4 PET images of the four statistically significant activation sites in the object memory task, each superimposed on an MRI image of a composite brain for the purpose of illustrating the anatomical localization of activation foci. The bottom left image has an activation focus that is cut off due to the fact that the remaining activation fell outside the field of view of the camera. Three of the four significant foci of activation were in the left hemisphere. Inspecting the images from top to bottom, and then from left to right, the foci of activation were in inferior temporal cortex (centered in Brodmann's area 37; stereotaxic coordinates: 48, −58, −11), parietal cortex (centered in Brodmann's area 40; stereotaxic coordinates: 35, −42, 34), prefrontal cortex (centered in Brodmann's area 6; stereotaxic coordinates: 39, 3, 29), and anterior cingulate (centered in Brodmann's area 32; stereotaxic coordinates: −1, 14, 43).

coordinates: 39, 3, 29). Activation of this area has been implicated by others in working memory tasks with verbal materials (Petrides et al., 1993). The activation seen in our data would be consistent with the use of language processes if subjects attempted to name the geometric forms and rehearse these names. Below, we discuss the implication of this site of activation for a model of the circuitry involved in this task.

The site of activation in left posterior parietal cortex shown in the lower left image of figure 66.4 was centered in Brodmann's area 40 (stereotaxic coordinates: 35, −42, 34). Based on studies of brain injuries and their effects on language function, this area of the left hemisphere has been identified as one involved in the development of symbolic representations to stimuli presented to the sensorium. Consequently, activation in this area may be associated with processes that are involved in the creation of symbolic mental representations of the objects in our experiment, representations that are then stored briefly until the probe is presented seconds later.

The activation in left temporal cortex shown in the upper left image of figure 66.4 was located in the vicinity of Brodmann's area 37 (stereotaxic coordinates: 48, −58, −11). By now, it has been well documented that inferior temporal cortex is intimately involved in object recognition (the so-called ventral pathway of the visual system; see, e.g., Ungerleider and Mishkin, 1982; Livingstone and Hubel, 1988; Haxby et al., 1991). Consistent with this view, we interpret the activation in left temporal cortex as an index of processes involved in creating a representation of the target objects, a representation that is then subject to further processing by parietal and frontal areas.

Finally, the anterior cingulate activation shown in the lower right image of figure 66.4 was in the region of Brodmann's area 32 (stereotaxic coordinates: −1, 14, 43). Although there was some indication of anterior cingulate activation in the spatial memory condition as well, it did not approach statistical significance. The anterior cingulate has been implicated in tasks that require selective attention and in tasks that require significant processing capacity (e.g., Pardo et al., 1990). One possible reason for this activation appearing in the object task but not in the spatial task may be that the object task requires more in the way of processing effort than the spatial task. This possibility is suggested by our behavioral data, which showed that accuracy was lower in the object task than the spatial task.

Spatial and object memory compared The patterns of brain activation revealed in this experiment invite several possible models of working-memory processes, all of which are broadly consistent with the modal conceptualization of working memory derived from behavioral experiments (Baddeley, 1992). In detail, moreover, these models begin to suggest circuitry for working memory, as well as differences in this circuitry depending on the sort of representations involved in a task and the processes that operate on the representations. Let us outline some possible models suggested by the PET results in the spatial and object memory tasks and indicate how these models may expand on current conceptualizations of working memory.

One model that would be consistent with the activation data in the spatial memory task posits that internal processes calculate locations of the target dots at the time these targets are presented, using mechanisms in the parietal lobe. This calculation then feeds into the generation of a mental image of the dots in their locations by processes in occipital cortex. An internal image of the dots might correspond to a matrix-like representation that includes all locational information simultaneously, thus permitting emergent properties of the array to present themselves (e.g., the relative locations among dots). Following generation of this image, storage processes of the prefrontal cortex are responsible for maintaining the image in memory during the retention interval. An alternative model is that presentation of the target dots elicits image generation processes directly in the occipital cortex, following which there is a calculation from this image of the dot coordinates by processes in parietal areas. This calculation then results in a memory trace of the target locations that is retained by prefrontal activity. In still another alternative model, image generation and location calculation processes operate in parallel, with either or both contributing to the memory trace that is stored by prefrontal processes.

Data from the object memory task suggest a very different class of models to account for the activation patterns. The following model, for example, is one that would be consistent with these patterns and with previous neuropsychological research. Processes of the inferior temporal lobe begin by creating representations of the two target objects.[2] These representations can then serve as the basis for subjects to create a symbolic code for the objects, using processes in the left parietal lobe. In view of the fact that this area has been shown to be associated with language activity, it seems reasonable to assume that the symbolic code may often be linguistically based. This code is then retained for several seconds using rehearsal processes of the left prefrontal cortex.

Of course, the results of just one experiment do not provide sufficient data to affirm one or another of the illustrative models described above and to rule out alternatives to them that might be equally consistent with the activation patterns obtained. That aside, however, these data clearly show a double dissociation between the object and spatial working-memory tasks involved. At a gross level, one can see in figures 66.3 and 66.4 that the spatial task recruits processes largely in the right hemisphere whereas the object task recruits processes in the left hemisphere. At a finer level of detail, the circuitry suggested for processing in the object task appears to be quite different from that used in the spatial task, hemispheric differences aside. First, the object pathway includes the inferior temporal lobe. There is ample documentation from studies of both monkeys and humans that this site is included in a pathway that is required for object-recognition. In the present experiment, the activation of temporal lobe mechanisms is presumably an indication that subjects engage an object-recognition mechanism as they encounter the target objects in the task. Second, the activation in prefrontal cortex in the object task is in a region superior and posterior to that found for the spatial task. As we suggested above, this may be an indication of the use of rehearsal processes that are tied to a linguistic code for the objects in the task. Whether this is so or not, however, the areas of activation in prefrontal cortex in the two tasks seem not to be homologous. This is consistent with recent evidence found by Wilson, Scalaidhe, and Goldman-Rakic (1993) that spatial and object memory processes in monkeys are mediated by adjacent areas of prefrontal cortex, one area in the vicinity of the principal sulcus and rostral to the arcuate sulcus, and the other in the inferior convexity. These studies, however, did not find evidence of hemispheric differences in the localization of these areas, and this may constitute an important difference in the processing of these tasks by monkeys and by humans.

A final comment is in order concerning the prefrontal cortex and its role in working-memory processes. Both tasks in the present experiment provide data implicating lateral prefrontal cortex in the maintenance of spatial and object information for short periods of time, consistent with recent evidence from studies of monkeys, as discussed above. While it is becoming increasingly clear that prefrontal cortex is involved in working memory, it is not yet clear what its role is. It might be that activity of prefrontal cortex is itself the mechanism of storage for internal codes. It might also be that this activity represents a pointer or index to other circuitry that is responsible for maintaining the actual engram for spatial or object information. Yet another possibility is that prefrontal activity is associated with rehearsal processes that recirculate a representation to maintain its freshness during the retention interval. Depending on whether a representation is linguistic (or otherwise symbolic) or spatial in nature, this rehearsal may take different forms and recruit processes in somewhat different regions of lateral prefrontal cortex. Much remains to be learned about the role that prefrontal cortex plays in a working-memory system.

Whatever the details of models that derive from imaging data, it is clear that a new set of technologies has provided us with a new domain in which to investigate double dissociations in cognitive processing. These data could add importantly to the data from behavioral, neuropsychological, and neurophysiological studies to converge on detailed models of the processes involved in working memory.

ACKNOWLEDGMENTS The research reported here was supported by grants from the Office of Naval Research and the James S. McDonnell Foundation. The work was performed while Smith was supported by the John Simon Guggenheim Foundation and Jonides was supported by a fellowship from the James S. McDonnell Foundation.

NOTES

1. A few words are in order about terminology. Researchers who champion the idea of multiple working memories often have in mind a single system that includes some specialized short-term stores, or *buffers* (e.g., Baddeley, 1986). We hold open the possibility that there may be multiple systems, not just multiple buffers; but since the work that follows does not distinguish between these views we will tend to talk of multiple buffers rather than multiple systems. Another point concerns our use of the terms *verbal*, *visual*, and *spatial*. By a verbal buffer, we mean one that holds verbal material, which would include what Baddeley (e.g., 1986) terms the articulatory loop. By spatial information, we refer to information about the relations among objects in space, regardless of the modality of input of that information; hence, a spatial buffer need not necessarily have visual contents. Finally, often we use the term *visual* to contrast with *spatial*; hence *visual* means visual information that is not about spatial relations.
2. The fact that there is activation in this area after subtraction of the object perception from the object memory condition suggests that subjects in the former condition did not process the figures as objects per se. Rather, they may have simply engaged in some sort of featural comparison of the simultaneously presented target and probe objects.

REFERENCES

ANDERSON, J. R., 1983. *The Architecture of Cognition*. Cambridge, Mass.: Harvard University Press.

BADDELEY, A. D., 1992. Working memory. *Science* 255:566–559.

BADDELEY, A. D., 1986. *Working Memory*. Oxford: Oxford University Press.

BADDELEY, A. D., and G. J. HITCH, 1974. Working memory. In *The Psychology of Learning and Motivation*, vol. 8, G. H. Bower, ed. New York: Academic Press.

CARAMAZZA, A., 1986. On drawing inferences about the structure of normal cognitive systems from the analyses of patterns of impaired performance: The case for single-patient studies. *Brain Cogn.* 5:41–66.

CARPENTER, P. A., M. A. JUST, and P. SHELL, 1990. What one intelligence test measures: A theoretical account of the processing in the Ravens Progressive Matrices Test. *Psychol. Rev.* 97:404–431.

CHAFFEE, M., S. FUNAHASHI, and P. S. GOLDMAN-RAKIC, 1989. Unit activity in the primate posterior parietal cortex during an oculomotor delayed response task. *Soc. Neurosci. Abstr.* 15:786.

Farah, M. J., 1988. Is visual imagery really visual? Overlooked evidence from neuropsychology. *Psychol. Rev.* 95: 307–317.

Farah, M., K. M. Hammond, D. N. Levine, and R. Calvanio, 1988. Visual and spatial mental imagery: Dissociable systems of representation. *Cognitive Psychol.* 20:439–462.

Fox, P. T., 1992. Human brain mapping with positron emission tomography. In *Notes for Neurology of Behavior Course*, M. M. Mesulam, ed. Harvard Medical School.

Fox, P. T., J. M. Fox, M. E. Raichle, and R. M. Burde, 1985. The role of cerebral cortex in the generation of saccadic eye movements: A positron emission tomography study. *J. Neurophysiol.* 54:348–368.

Friston, K. J., C. D. Frith, P. F. Liddle, R. S. J. Frackowiak, 1991. Comparing functional (PET) images: The assessment of significant change. *J. Cereb. Blood Flow Metab.* 11:690–699.

Funahashi, S., C. J. Bruce, and P. S. Goldman-Rakic, 1989. Mnemonic coding of visual space in the monkey's dorsolateral prefrontal cortex. *J. Neurophysiol.* 61:2, 331–349.

Goldman-Rakic, P. S., 1987. In *Handbook of Physiology: The Nervous System*, F. Plum, ed. Bethesda, M.D.: American Physiological Society.

Haxby, J. V., C. L. Grady, B. Horwitz, L. G. Ungerleider, M. Mishkin, R. E. Carson, P. Herscovitch, M. B. Schapiro, and S. I. Rapoport, 1991. Dissociation of object and spatial vision processing pathways in human extrastriate cortex. *Proc. Natl. Acad. Sci. U.S.A.* 88:1621–1625.

Intons-Peterson, M. J., 1983. Imagery paradigms: How vulnerable are they to experimenters' expectations: *J. Exp. Psychol. [Hum. Percept.]* 9:394–412.

Jonides, J., E. E. Smith, R. A. Koeppe, E. Awh, S. Minoshima, and M. A. Mintun, 1993. Spatial working memory in humans as revealed by PET. *Nature* 363:623–625.

Just, M. A., and P. A. Carpenter, 1992. A capacity theory of comprehension: Individual differences in working memory. *Psychol. Rev.* 99:1, 122–149.

Koeppe, R. A., S. Minoshima, J. Jonides, E. E. Smith, E. S. Awh, and M. A. Mintun, 1993. PET studies of working memory in humans. Paper presented at the Society for Nuclear Medicine 40th Annual Meeting, Toronto.

Kosslyn, S. M., 1981. The medium and the message in mental imagery: A theory. *Psychol. Rev.* 88:46–66.

Kosslyn, S. M., 1987. Seeing and imaging in the cerebral hemispheres: A computational approach. *Psychol. Rev.* 94: 148–175.

Kosslyn, S. M., N. M. Alpert, W. L. Thompson, V. Maljkovic, S. B. Weise, C. F. Chabris, S. E. Hamilton, and F. S. Buonanno, 1993. Visual mental imagery activates topographically organized cortex: PET investigations. *J. Cognitive Neurosci.* 5:263–287.

Levine, D. N., J. Warach, and M. J. Farah, 1985. Two visual systems in mental imagery: Dissociation of "what" and "where" in imagery disorders due to bilateral posterior cerebral lesions. *Neurology* 35:1010–1018.

Livingstone, M., and D. Hubel, 1988. Segregation of form, color, movement, and depth: Anatomy, physiology, and perception. *Science* 240:740–750.

Logie, R. H., 1986. Visuo-spatial processing in working memory. *Q. J. Exp. Psychol.* 38:229–247.

Mauritz, K. H., and S. P. Wise, 1986. Premotor cortex of the rhesus monkey: Neuronal activity in anticipation of predictable environmental events. *Exp. Brain Res.* 61:229–244.

Minoshima, S., R. A. Koeppe, M. A. Mintun, K. L. Berger, S. F. Taylor, K. A. Frey, and D. E. Kuhl, 1993. Automated detection of the intercommisural line for stereotactic localization of functional brain images. *J. Nuclear Med.* 34:322–329.

Newell, A., 1990. *Unified Theories of Cognition.* Cambridge, Mass.: Harvard University Press.

Pardo, J. V., P. J. Pardo, K. W. Janer, and M. E. Raichle, 1990. The anterior cingulate cortex mediates processing selection in the Stroop attentional conflict paradigm. *Proc. Natl. Acad. Sci. U.S.A.* 87:256–259.

Petrides, M., B. Alivisatos, A. C. Evans, and E. Meyer, 1993. Dissociation of human mid-dorsolateral from posterior dorsolateral frontal cortex in memory processing. *Proc. Natl. Acad. Sci. U.S.A.* 90:873–877.

Posner, M. I., S. E. Peterson, P. T. Fox, and M. E. Raichle, 1988. Localization of cognitive functions in the human brain. *Science* 240:1627–1631.

Pylyshyn, Z. W., 1981. The imagery debate: Analogue media versus tacit knowledge. *Psychol. Rev.* 88:16–45.

Shallice, T., 1988. *From Neuropsychology to Mental Structure.* Cambridge: Cambridge University Press.

Skelly, J. P., F. A. W. Wilson, and P. S. Goldman-Rakic, 1992. Neurons in the prefrontal cortex of the Macaque selective for faces. *Neurosci. Abstr.* 18:705.

Talairach, J., and P. Tournoux, 1988. *A Co-planar Stereotaxic Atlas of a Human Brain.* Stuttgart: Thieme.

Tanji, J., 1985. Comparison of neuronal activities in monkey supplementary and precentral motor areas. *Behav. Brain Res.* 18:137–142.

Ungerleider, L. G., and M. Mishkin, 1982. Two cortical visual systems. In *Analysis of Visual Behavior*, D. J. Ingle, M. A. Goodale, and R. J. W. Mansfield, eds. Cambridge, Mass.: MIT Press.

Wilson, F. A.W., S. P. O. Scalaidhe, and P. S. Goldman-Rakic, 1993. Dissociation of object and spatial processing domains in primate prefrontal cortex. *Science* 260:1955–1958.

Wilson, F. A. W., J. P. Skelly, and P. S. Goldman-Rakic, 1992. Areal and cellular segregation of spatial and feature processing by prefrontal neurons. *Neurosci. Abstr.* 18:705.

Worsley, K. J., A. C. Evans, S. Marrett, and P. Neelin, 1992. A three-dimensional statistical analysis for CBF activation studies in human brain. *J. Cereb. Blood Flow Metab.* 12:900–918.

67 The Representation of Extrapersonal Space: A Possible Role for Bimodal, Visual-Tactile Neurons

MICHAEL S. A. GRAZIANO AND
CHARLES G. GROSS

ABSTRACT We propose that extrapersonal space is represented in the brain by bimodal, visual-tactile neurons in inferior area 6 in the frontal lobe, area 7b in the parietal lobe, and the putamen. In each of these areas, there are cells that respond to both tactile and visual stimuli. In each area, the tactile receptive fields are arranged to form a somatotopic map. The visual receptive fields are usually adjacent to the tactile ones and extend outward from the skin about 20 cm. Thus each area contains a somatotopically organized map of the visual space that immediately surrounds the body. These three areas are monosynaptically interconnected, and may form a distributed system for representing extrapersonal visual space. For many neurons with tactile receptive fields on the arm or hand, when the arm was moved, the visual receptive field moved with it. Thus, these neurons appear to code the location of visual stimuli in arm-centered coordinates. More generally, we suggest that the bimodal cells represent near extrapersonal space in a body part–centered fashion, rather than in an exclusively head-centered or trunk-centered fashion.

A central issue in cognitive neuroscience is how the brain constructs a stable map of the world. The retinal image of an object moves every time the head or the eyes move, and yet we perceive objects as having stable positions in space. We are able to reach toward objects, saccade to targets, and avoid threatening or looming stimuli. How is the location of these nearby stimuli encoded in the brain?

We propose that in the primate, the visual space near the body—extrapersonal or peripersonal space—is encoded by a system of interconnected brain areas that includes parietal area 7b, the inferior portion of premotor area 6, and the putamen. Neurons in these areas respond to somatosensory stimuli, and have discrete receptive fields that are arranged to form a somatotopic map of the body (Robinson and Burton, 1980a; Hyvarinen, 1981; Crutcher and DeLong, 1984a; Gentilucci et al., 1988). Many of these neurons, particularly in the head and arm portions of the map, are bimodal, responding to visual as well as tactile stimuli (Graziano and Gross, 1992, 1993; Hyvarinen, 1981; Hyvarinen and Poranen, 1974; Leinonen et al., 1979; Leinonen and Nyman, 1979; Rizzolatti et al., 1981b; Robinson and Burton, 1980a, 1980b). For the bimodal neurons, the visual receptive field usually matches the location of the tactile receptive field, and is confined in depth to a region within reach of the animal's arm. Thus these neurons provide a somatotopically organized representation of the visual space near the body.

In this chapter we describe the properties of bimodal, visual-tactile neurons in the putamen, parietal area 7b, and inferior area 6, and suggest how these areas may encode the location of visual stimuli in extrapersonal space.

Neuronal response properties in three bimodal areas

We recorded from single neurons in the putamen, area 7b, and inferior area 6 in macaque monkeys anesthetized with nitrous oxide and immobilized with Pavulon. In each area, we found three types of responsive cells: somatosensory cells, visual cells, and bimodal

MICHAEL S. A. GRAZIANO and CHARLES G. GROSS Department of Psychology, Princeton University, Princeton, N.J.

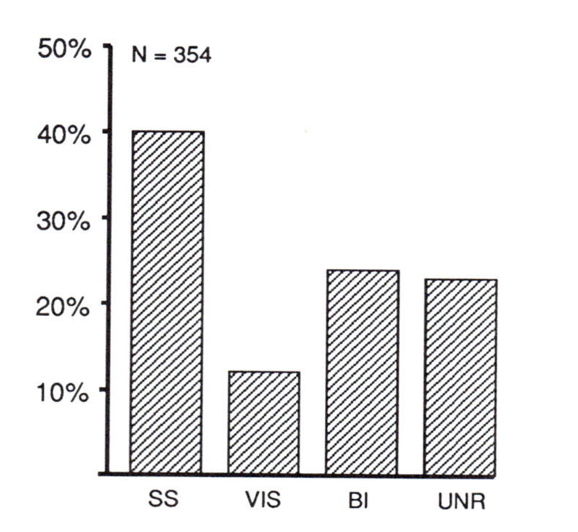

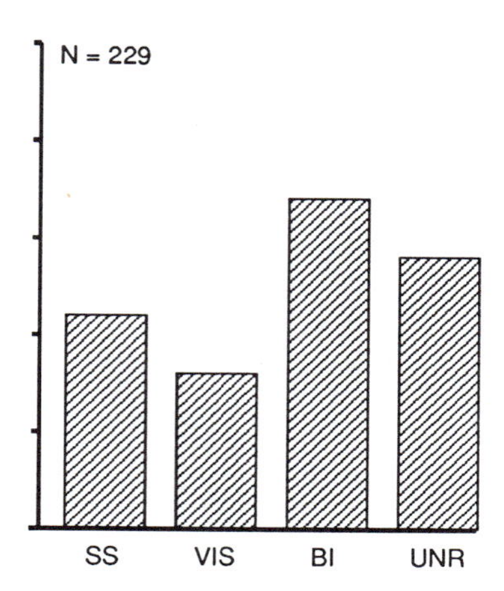

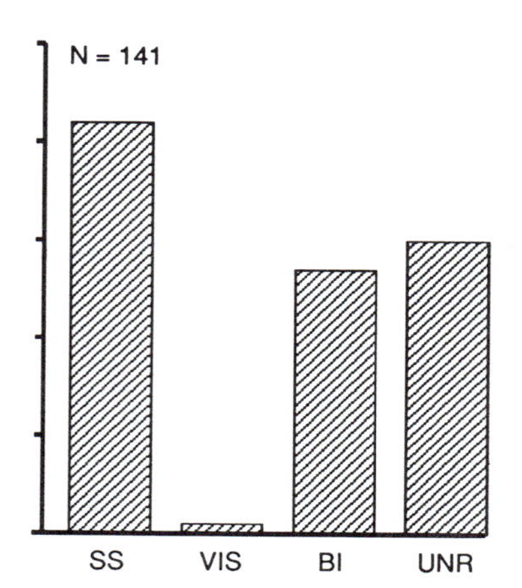

FIGURE 67.1 Proportions of somatosensory cells (SS), visual cells (VIS), bimodal cells (BI), and unresponsive cells (UNR) in the putamen, area 7b, and area 6.

cells. The bimodal cells responded both to visual and to somatosensory stimuli. In this section, we describe the properties of these cells for each of the three brain areas.

PUTAMEN We studied 354 putamen neurons, of which 40% were somatosensory, 12% were visual, 24% were bimodal, and 23% were unresponsive (figure 67.1).

Somtosensory cells Somatosensory responses were studied using manual palpation, manipulation of joints, gentle pressure, and stroking with cotton swabs. Receptive fields were plotted by repeated presentation of the most effective of these stimuli. Neurons were somatotopically organized in a manner similar to that described by Crutcher and DeLong (1984a). On vertical electrode penetrations, the first cells encountered had receptive fields on the tail or the legs. As the electrode moved ventrally, cells had receptive fields on the trunk, then the shoulders and arms, then the face, and finally inside the mouth. Figure 67.2 shows a representative penetration.

Bimodal cells In addition to somatosensory neurons, we found bimodal, visual-somesthetic neurons in the face and arm region of the somatotopic map (eg, figure 67.2, cells 6, 7, and 8). Most bimodal cells (86%) responded to light cutaneous stimulation. Sixty-six percent had somatosensory receptive fields on the face,

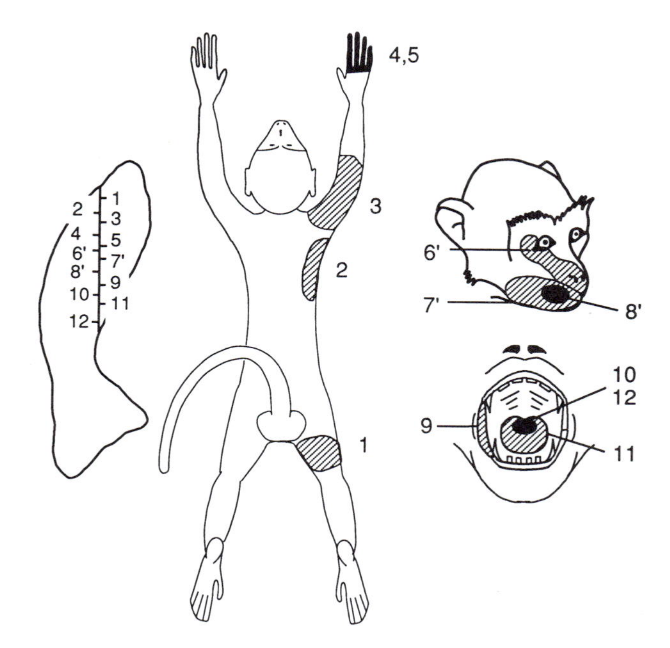

FIGURE 67.2 Somatotopic organization of the putamen. A representative electrode penetration is shown on a coronal section through the putamen, at AP level 14. Receptive field locations for the neurons are shown to the right. Cells indicated with single quotes (eg, 6′) responded to visual as well as tactile stimuli.

29% had receptive fields on the arm, and 5% had receptive fields that encompassed the entire body.

We tested bimodal cells with moving bars of light on a tangent screen. Since cells often appeared to be selective for the depth of the visual stimulus, the screen was placed at various distances from the animal and the lenses were changed to adjust the animal's plane of focus. Many cells did not respond to these projected light stimuli, and only responded to stimuli moving near the animal's face or hands. Cells that preferred small stimuli particularly close to the skin were tested with a cotton swab. The stimulus was moved slowly toward and away from the animal to determine the maximum distance for which a response could be obtained. The dimensions of the responsive region were determined by approaching the animal from various angles.

A typical example of a bimodal cell is shown in figure 67.3. The tactile receptive field was plotted while the animal's eyes were covered. The cell was activated by a light touch to the facial hairs, and the responsive region covered most of the contralateral cheek and the area around the mouth (A and B). However, when the animal's eyes were uncovered, the response began before the stimulus had touched the face. A cotton swab was moved toward the tactile receptive field, and the cell began responding when the stimulus was within about 10 cm of the face (C). We know that this response was not caused by inadvertent tactile stimulation, such as by air movement, because it was eliminated when the eyes were covered (D).

By approaching the tactile receptive field from various angles, we determined the three-dimensional responsive region, which we called the visual receptive field. This responsive region differed from a classical receptive field because it was not only restricted in visual angle, it was also confined in depth. As shown in figure 67.3A, the visual receptive field as thus defined was a solid angle centered at the tactile receptive field and extending out approximately 10 cm. The response was weak and erratic toward the edges of the visual receptive field. The response was better to a stimulus moving toward the face than to a stimulus moving away.

Figure 67.4 shows several more examples of bimodal cells with tactile receptive fields on the face. As in the previous example, these cells responded to touching of the facial hair. They also responded to visual stimuli moving toward the tactile receptive field. For the cells shown in A and B, the visual receptive field extended outward about 10 cm from the tactile receptive field. The cells shown in C, D, and E differed slightly from this basic pattern. The cell in C had a bilateral tactile receptive field, but a contralateral visual receptive

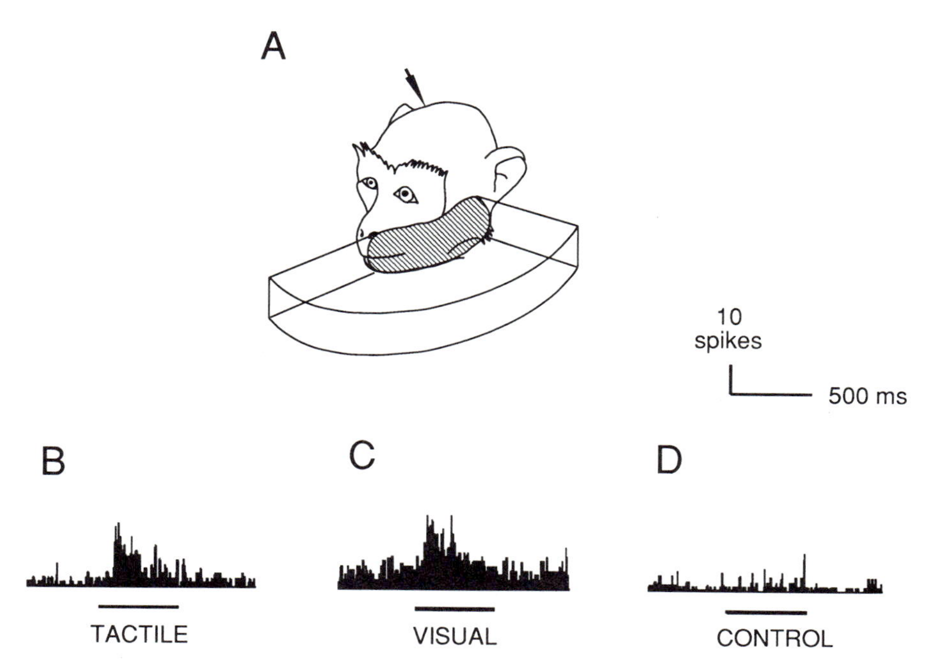

FIGURE 67.3 Poststimulus time histograms, summed over 10 trials, for a typical bimodal putamen cell. (A) The tactile receptive field (stippled) and the visual receptive field (boxed) are in register. The arrow indicates the hemisphere recorded from. (B) Response to a cotton swab touching the face while the eyes are covered. (C) Response to a cotton swab approaching the face within 10 cm while the eyes are open. (D) Same as C, with the eyes covered.

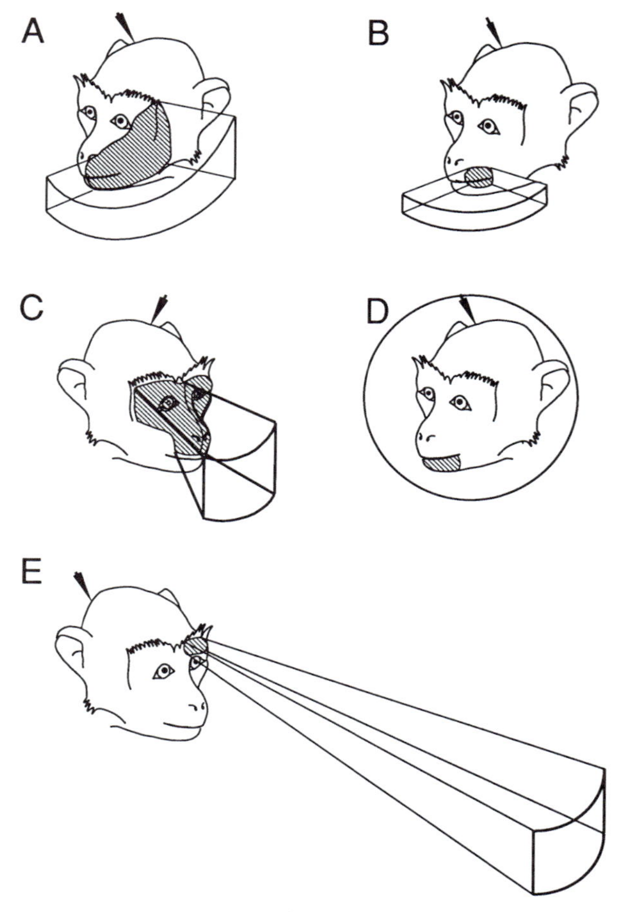

FIGURE 67.4 (A and B) Typical bimodal cells from the putamen, in that the visual and tactile receptive fields correspond and the visual receptive field extends about 10 cm from the face. (C–E) Atypical bimodal cells, because in (C) the tactile receptive field is bilateral and the visual receptive field is contralateral; in (D) the tactile receptive field is confined to the lower jaw but the visual receptive field extends from the face in all directions; and in (E) the visual receptive field extends about 100 cm from the face.

field. Stimuli moving toward the ipsilateral side of the face did not activate the cell, even though touching the ipsilateral side of the face did. The cell in figure 67.4D had a small bilateral tactile receptive field covering the chin, and a visual receptive field covering the entire visual field but extending outward only about 10 cm from the face. Approaching any part of the face, even the upper face, caused a visual response. The cell in E had a tactile receptive field on the contralateral brow, and a visual receptive field that extended out about one meter from the monkey.

Figure 67.5 shows several examples of bimodal cells with tactile receptive fields on the arm. The cells shown in A and B had tactile receptive fields on the contralateral arm and visual receptive fields in the contralateral periphery. Both cells responded to visual stimuli as far away as 1.5 m. The cell shown in figure 67.5C responded to touching of both arms, and the visual receptive field was bilateral. Again, the cell responded to stimuli as far away as 1.5 m.

We were able to characterize the visual receptive field for 48 bimodal cells. Of these, 77% responded best or only to visual stimuli within 20 cm of the skin, while 23% responded to stimuli at greater distances. Cells with tactile receptive fields on the arm generally responded to more distant stimuli than cells with tactile receptive fields on the face.

As illustrated above, for most bimodal cells the location of the tactile receptive field matched the location of the visual receptive field. However, for bimodal cells with tactile receptive fields on the arm, what happens when the arm is moved to a new location? Do the tactile and visual receptive fields become dissociated, or does one receptive field shift in order to remain in register with the other? Figure 67.6 shows the result for two cells. The cell shown in A responded to visual stimuli only when the arm was propped forward into the monkey's field of view. When the arm was tucked back, thus placing the tactile receptive field out of sight, the cell no longer responded to visual stimuli presented anywhere in the visual field. The tactile response, however, was equally good for both arm positions. The cell shown in figure 67.6B had a particularly close match between the tactile and visual receptive fields; the visual receptive field extended 5 cm from the hand. When the arm was moved to different locations within the animal's sight, the visual receptive field also moved to follow the location of the hand. When the hand was placed out of sight, the cell did not respond at all to visual stimuli. Of 25 bimodal cells with tactile receptive fields on the arm, 5 had visual responses that were gated by the position of the arm in this fashion.

Bimodal cells or cells with visual receptive fields near the body have not been reported previously for the putamen. This may be because other groups studying single-unit activity in the putamen used awake animals sitting in chairs (DeLong, 1973; Liles, 1983; Crutcher and DeLong, 1984a, 1984b; Liles, 1985; Liles and Updyke, 1985; Alexander, 1987; Schultz and Romo, 1988). Under these conditions, stimuli moving close to the head or arms would be likely to elicit movements, and any associated neuronal discharges might have

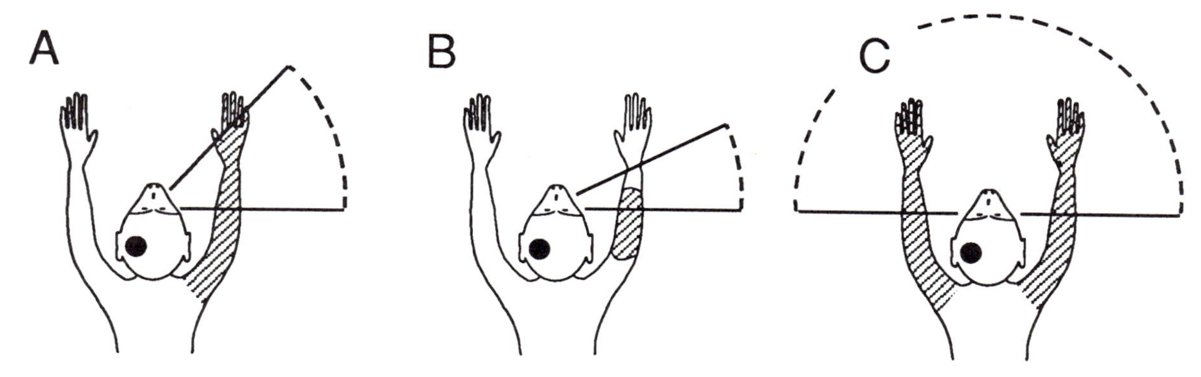

FIGURE 67.5 Bimodal cells from the putamen with tactile receptive fields on the arm. The lines indicate the angles subtended by the visual receptive fields in the horizontal plane. The dashed lines indicate that the receptive fields extend farther than one meter. The stippling shows the tactile receptive fields, and the black circles on the head show the hemisphere recorded from.

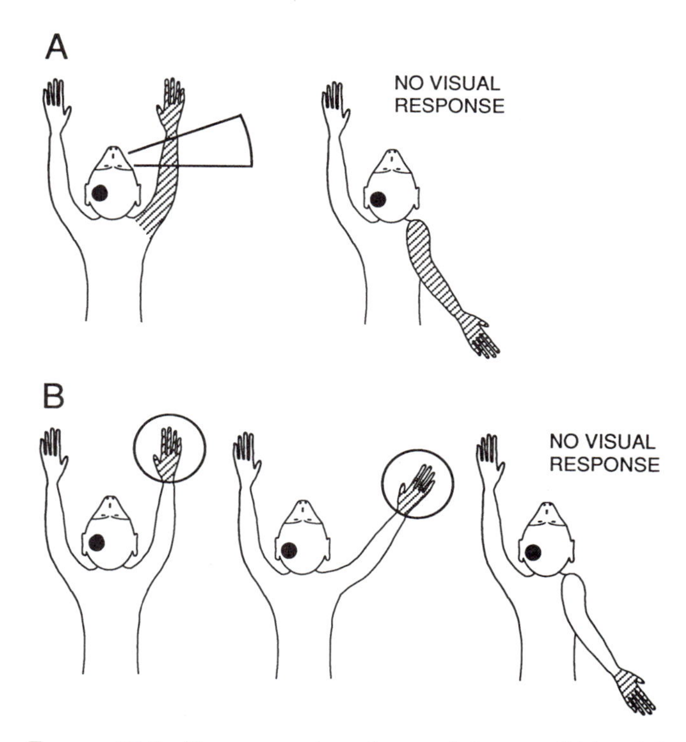

FIGURE 67.6 Two examples of a special type of bimodal arm cell from the putamen. These cells responded visually when the arm was within the monkey's field of view (left), but did not respond when the arm was moved out of view (right). For the cell shown in (B), the visual receptive field moved as the hand moved.

been interpreted as motor or somatosensory responses rather than visual ones.

Visual cells In addition to somatosensory cells and bimodal cells, we also found cells that responded only to visual stimuli. Figure 67.7 shows the response of a visual cell, which was located within the face portion of the somatotopic map. As was the case for many of the bimodal cells, this cell responded to visual stimuli moving toward the face within about 50 cm, but not to stimuli moving away.

PARIETAL AREA 7B We studied 229 neurons in area 7b, of which 22% were somatosensory, 16% were visual, 34% were bimodal, and 28% were unresponsive (see figure 67.1).

Somatosensory cells The somatotopic organization in area 7b is crude, with considerable overlap between the representations of different body parts (Hyvarinen, 1981). However, like Hyvarinen, we found that the representation of the face is generally more anterior than the representation of the arm.

Bimodal cells We found a high proportion of bimodal neurons in area 7b, in agreement with previous reports (Hyvarinen, 1981; Hyvarinen and Poranen, 1974; Leinonen et al., 1979; Leinonen and Nyman, 1979; Robinson and Burton, 1980a, 1980b). As was the case for the putamen, most of these neurons (65%) responded to light cutaneous stimulation. Bimodal cells had somatosensory receptive fields on the face (13%), the arm (48%), both the face and the arm (33%), the chest (2%), and the whole upper body (4%). We obtained visual receptive field plots for 50 bimodal cells. Of these, 42% preferred stimuli out to 20 cm from the animal, 42% preferred stimuli out to one meter, and 16% responded well to stimuli at greater distances.

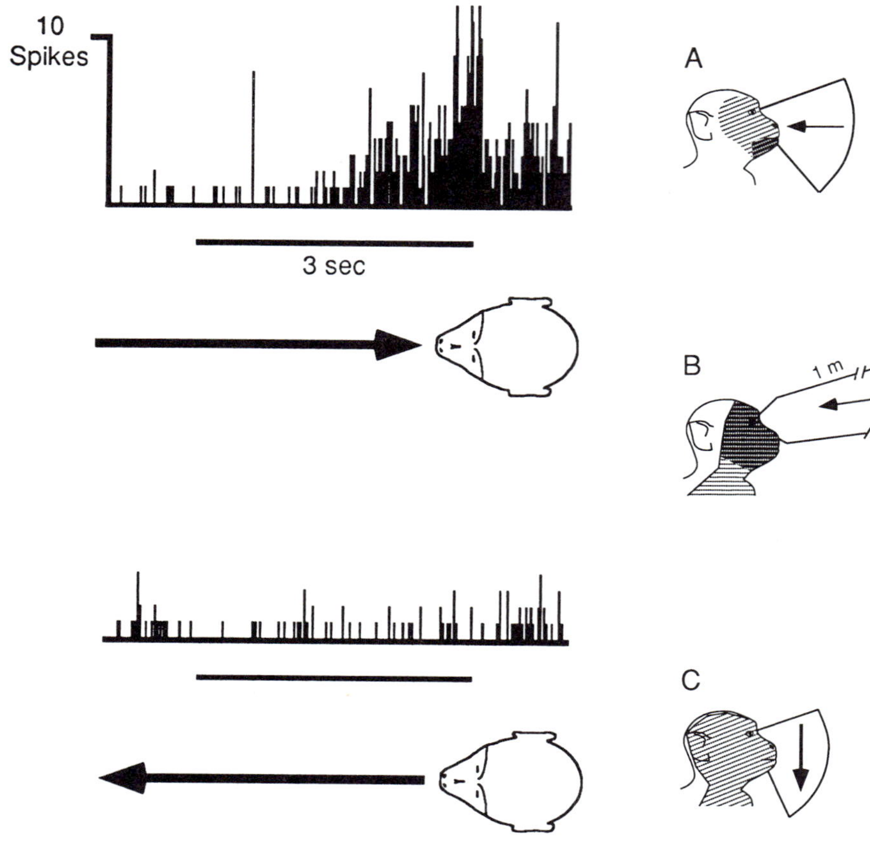

FIGURE 67.7 Response of a visual putamen cell to a sphere 5 cm in diameter moving on a track at 23.3 cm/s toward or away from the face. Far point, 78 cm; near point, 8 cm; duration, 3 s; based on 9 trials with an intertrial interval of 15 s.

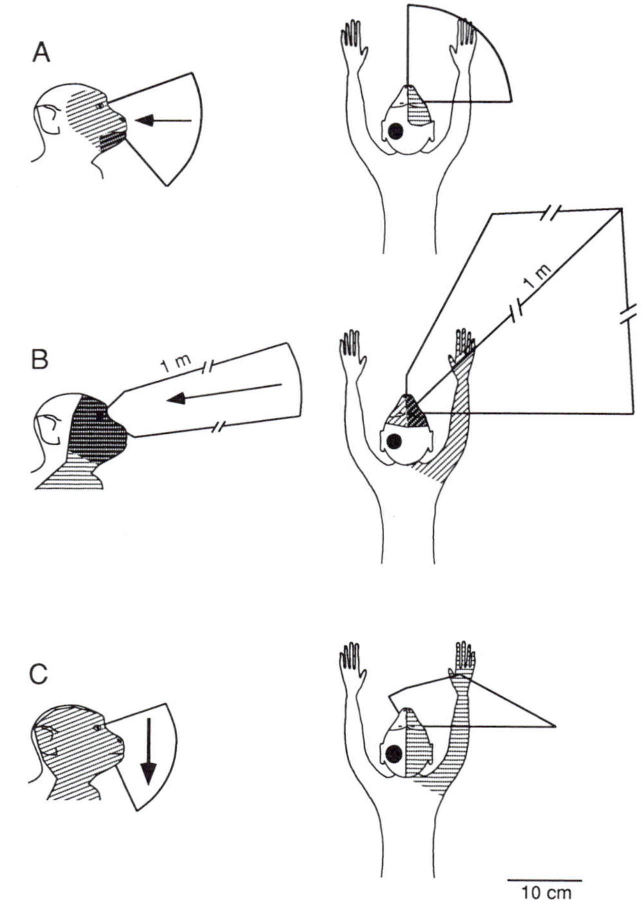

FIGURE 67.8 Three examples of bimodal cells from area 7b. The lines indicate the angles subtended by the visual receptive fields in both the horizontal and vertical planes. Arrows indicate directional selectivity. All lines are drawn to the scale shown at bottom, except for the broken lines in B, which are as labeled. The stippling indicates the tactile receptive fields; the darker stippling indicates regions of strongest response. The black circles on the head indicate the hemisphere recorded from.

Figure 67.8 shows several examples of bimodal responses. The cell shown in A had a tactile receptive field on the contralateral side of the face. It responded best to touching the jaw, and more weakly to touching the cheek or eyebrow. It also responded to visual stimuli within about 15 cm of the face. The visual receptive field was contralateral and mostly in the lower visual field, thus matching the tactile receptive field. The cell only responded to visual stimuli moving inward toward the face, never to stimuli moving outward. The cell shown in B had a large tactile receptive field on the face and the contralateral arm. It also responded to contralateral visual stimuli within about one meter of the animal. Again, the cell responded only to inward motion toward the face, never to outward motion. The cell shown in C also had a contralateral tactile receptive field on the face and the arm. It responded best to visual stimuli in a region within about 20 cm of the animal; however, it also responded weakly to more distant stimuli, as far away as 2 m. This cell responded to visual stimuli moving in any direction within the receptive field, but responded best to downward motion.

Thirty bimodal neurons with tactile receptive fields on the arm were further tested, by moving the arm to different locations. Unlike in the putamen, in all 30 cases the visual response was independent of the position of the arm. Even when the arm was placed entirely out of the animal's view, the visual receptive field remained unchanged. There is a previous report of visual responses in area 7b that change as the arm moves (Leinonen et al., 1979), however no further information, such as receptive field plots, is provided. These results suggest that such cells may be found in area 7b as well as in the putamen, but perhaps in much smaller numbers.

Inferior Premotor Area 6 We recorded from 141 neurons in inferior area 6, of which 42% were somatosensory, 1% were visual, 27% were bimodal, and 30% were unresponsive (see figure 67.1).

Somatosensory cells Neurons in inferior area 6 were somatotopically organized in a manner similar to that described in previous reports (Gentilucci et al., 1988). When electrode penetrations were made in the lateral part of inferior area 6, the tactile receptive fields were located on the face, and when electrode penetrations were made in the medial part, the tactile receptive fields were located on the arm.

Bimodal cells A high proportion of neurons in inferior area 6 were bimodal, in agreement with previous reports (Rizzolatti et al., 1981b). As in the putamen and in area 7b, most of these (79%) responded to light cutaneous stimulation. Bimodal cells had somatosensory receptive fields on the face (24%), the arm (34%), both the face and the arm (29%), the chest (2%), the face and the chest (2%), and the whole upper body (8%). Of cells with sufficiently clearly plotted visual receptive fields, 39% preferred stimuli within 20 cm of the animal, 22% preferred stimuli within 1 m, and 39% responded well to stimuli at greater distances.

Figure 67.9 shows two examples of bimodal neurons with tactile receptive fields on the face. The cell shown in A had a receptive field on both sides of the face and on the ipsilateral shoulder. It responded to visual stimuli within one meter of the animal, and preferred stimuli within 30 cm. The visual receptive field was bilateral and extended farther into the ipsilateral side, thus matching the tactile receptive field. For the cell shown in B, the tactile response was directional. It preferred stimuli that moved across the skin from left to right. The visual response matched the location of the tactile response, and was also directionally selective, from left to right.

Figure 67.10 shows a bimodal cell with a tactile receptive field on the contralateral arm. The visual receptive field was confined to the lower visual field, and the response was strongest on the contralateral side, thus matching the location of the tactile receptive field. However, when the arm was bent back, placing the tactile receptive field out of the animal's field of view, the cell no longer responded to visual stimuli. There are two possible ways that arm position might affect the response of the cell: through proprioceptive feedback or through visual feedback. We tested these alternatives by placing an opaque shield between the face and the arm, thus blocking any visual feedback from the arm. As illustrated in figure 67.10C, when the arm was bent forward, even though it was blocked from view, the cell responded to visual stimuli. When the arm was bent back (figure 67.10D), the visual response disappeared. Therefore, the visual response for this neuron was modulated by proprioceptive feedback about the position of the arm.

Figure 67.11 shows another example of a bimodal neuron with a tactile receptive field on the arm. This neuron responded to visual stimuli at least 2 m from the animal, and preferred movement from left to right.

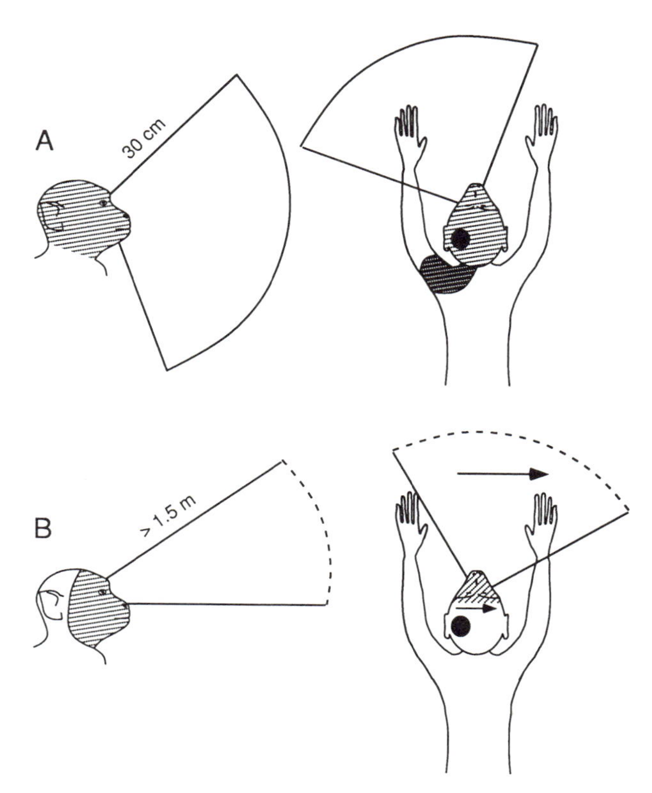

Figure 67.9 Two examples of bimodal cells from inferior area 6. For the cell in (B), both the visual and tactile responses preferred rightward motion.

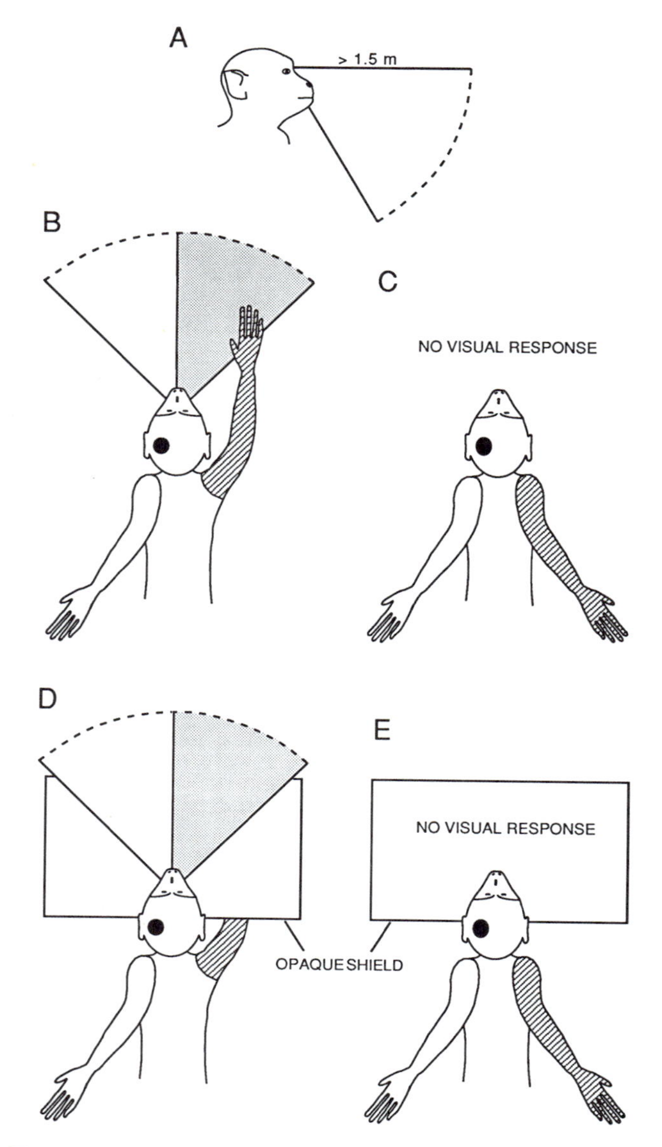

Figure 67.10 An example of a cell from inferior area 6, for which the visual response depended on arm position. (A and B) The visual receptive field was lower field and bilateral, but the response was best contralateral. The tactile receptive field was on the contralateral arm. (C) When the arm was placed out of sight, the visual response disappeared. (D and E) The visual response depended on arm position even when the arm was blocked from view with an opaque shield.

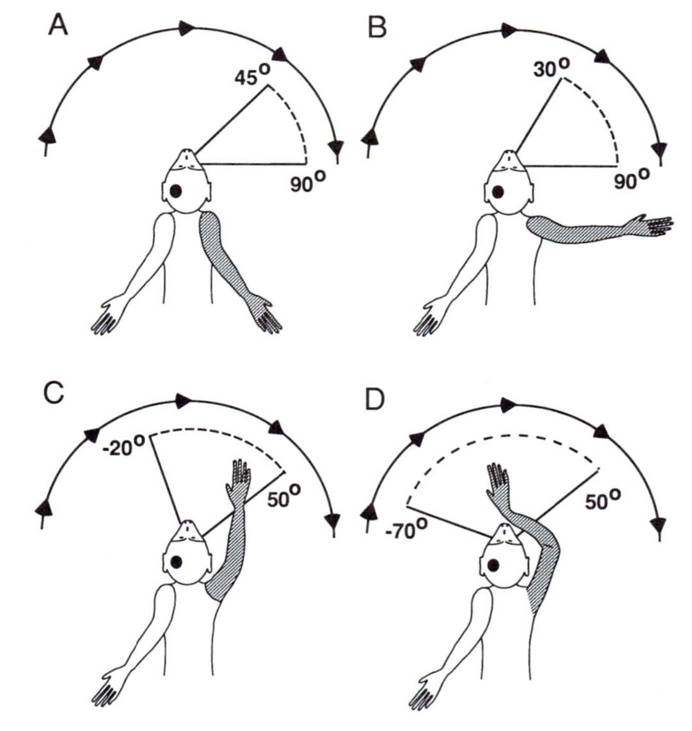

Figure 67.11 An example of a cell from inferior area 6 for which the visual receptive field moved as the arm moved. The stimulus was swept in an arc from left to right, in the cell's preferred direction of motion.

When the arm was bent backward (as in A), the visual response began 45° into the contralateral field and continued to the edge of sight. When the arm was positioned out to the side (as in B), the visual response began closer to the midline, at 30°. When the arm was bent forward (C), the visual response began 20° into the ipsilateral field, and no longer extended to the edge of the contralateral field. Finally, with the hand roughly centered at the nose (D), the visual response began as far as 70° in the ipsilateral field.

In total, 12 bimodal neurons with tactile receptive fields on the arm were tested by placing the arm in different positions, and for 8 of these the visual receptive field moved with the arm. That is, the visual receptive field remained attached to the tactile receptive field. This phenomenon was noted earlier by Rizzolatti and colleagues (personal communication), who studied inferior area 6 in awake monkeys trained to fixate. They also found that for many cells, the visual receptive field remained fixed to the tactile receptive field even when the monkey's eyes moved to a new location (Fogassi et al., 1992; Gentilucci et al., 1983). That is, whether the arm is moved as in our experiment, or the retina is moved as in the experiments by Gentilucci et al. and by Fogassi et al., the visual receptive field adjusts in order to remain attached to the tactile receptive field.

Summary of Bimodal Properties As described above, neurons in the putamen, area 7b, and inferior area 6 have many properties in common. All three areas are somatotopically organized, although the map in area 7b is relatively crude. In addition to the somatosensory neurons, all three areas contain bimodal, visual-tactile neurons. In the putamen, 24% of the cells were bimodal; in area 7b, 34% were bimodal; and 27% of the cells in area 6 were bimodal (see figure 67.1). For these bimodal cells, the visual and tactile receptive fields corresponded, and visual stimuli near the animal drove the cells best. For bimodal cells that had tactile receptive fields on the arm, we tested whether the visual receptive field moved as the arm moved. In the putamen, the visual receptive field was modulated by the position of the arm for 20% of the cells tested. Sixty-seven percent of the cells in inferior area 6 were modulated by arm position. However, in area 7b, none of the 30 cells tested showed any modulation of the visual response by the position of the arm.

There appears to be a fourth brain area with bimodal, visual-tactile responses nearly identical to those in the putamen, area 7b, and area 6. Neurons in the ventral intraparietal area (VIP) respond to tactile stimuli, primarily on the face, and to visual stimuli presented within a few centimeters of the tactile receptive field (Colby and Duhamel, 1991; Colby, Duhamel, and Goldberg, in press; Duhamel, Colby, and Goldberg, 1991). For at least some of these neurons, the visual receptive field appears to be fixed with respect to the face, even when the eyes move to a new location (Colby, Duhamel, and Goldberg, in press). For example, one neuron preferred a stimulus moving toward the chin, but not the forehead; this was so whether the animal's gaze was directed downward or upward. Since VIP has few if any tactile receptive fields on the arm, it may not be possible to test the dependence of the visual receptive field on arm position, as we did for the other three areas.

Could the sensory responses have been motor?

In the awake monkey, cells in the putamen, area 7b, and inferior area 6 respond during voluntary movement (e.g., Hyvarinen, 1981; Crutcher and DeLong, 1984b; Gentilucci et al., 1988; Rizzolatti et al., 1988). Could the responses to visual and tactile stimuli that we observed in these three areas have actually been motor rather than sensory responses, representing the animal's attempt to avoid or to reach for the stimulus? Because the animal was immobilized with Pavulon, such attempts to move could not have been noticed. However, in control tests when the animal was respirated with nitrous oxide and oxygen but not immobilized with Pavulon, there was no obvious motor response to these stimuli. Furthermore, the characteristics of the responses we observed suggest that they are sensory and not motor. As described above, both the tactile and visual responses had discrete receptive fields that varied from one cell to the next. Many cells were directionally selective in the tactile modality, the visual modality, or both. It is difficult to imagine how such stimulus selectivity could have been caused by the animal attempting to move.

Although a motor explanation of the responses we observed is thus inherently implausible, we directly tested the possibility. We recorded in the putamen of an awake monkey whose head was fixed by a head bolt and whose arms were loosely constrained in padded arm rests. Eye position was measured with a scleral search coil, and electrical activity was measured through surface electrodes pasted over various muscles of the upper and lower arm, using electromyography (EMG). First, the animal was trained to fixate an LED during presentation of visual and tactile stimuli. These stimuli included cotton swabs that were brought near the face, shoulders, arms, or hands at various speeds and then touched the skin. After several weeks the animal became so habituated to the situation that it sat quietly and continued to fixate the LED even during presentation of these stimuli.

We then recorded from single neurons in the putamen while simultaneously taking EMG recordings from the arm. As in the anesthetized animals, we found neurons that responded to visual and tactile stimuli, and the location of the visual and tactile receptive fields corresponded. For example, the neuron shown in figure 67.12 had a tactile receptive field on the contralateral arm, and responded to visual stimuli within about 10 cm of the arm. The rasters and histogram in figure 67.12A show the response as the visual stimulus was moved toward the tactile receptive field. An EMG record during one trial is also shown. There is clearly no change in EMG activity during the presentation of the stimulus. By contrast, figure 67.12B shows the EMG activity when the animal was fondling a grape that was

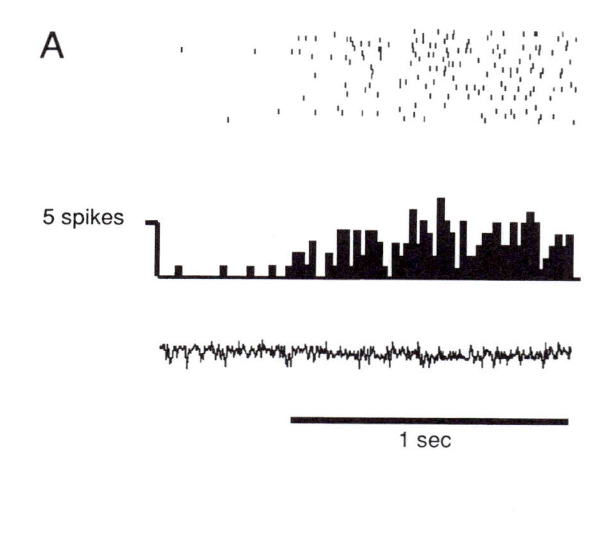

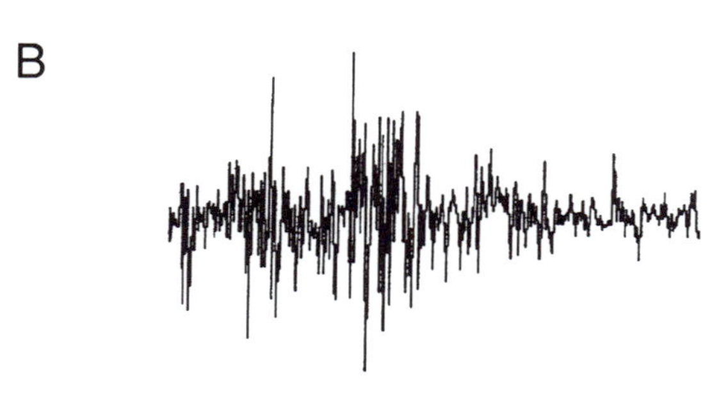

FIGURE 67.12 A bimodal neuron recorded from the putamen of an awake, fixating monkey. (A) The neuronal response, based on 20 trials, as the visual stimulus approached the tactile receptive field on the arm. The EMG trace (palmaris longus muscle), taken from one of the 20 trials, shows that the arm was stationary during stimulus presentation. (B) EMG trace while the animal touched a grape presented near its fingers.

placed near its fingers. The number of cells we have sampled so far is insufficient to assess any possible quantitative differences between bimodal putamen cells in the anesthetized and unanesthetized monkeys. However, these results demonstrate that bimodal responses with corresponding visual and tactile receptive fields occur in awake monkeys, unassociated with arm movements.

An interconnected system of bimodal areas

At the cortical level, the initial convergence of vision and somesthesis appears to occur in the parietal lobe. Somatosensory areas project to the medial bank of the intraparietal sulcus (area MIP) (Jones and Powell, 1970; Vogt and Pandya, 1978), visual areas project to the lateral bank (area LIP) (Selzer and Pandya, 1980; Maunsell and Van Essen, 1983; Ungerleider and Desimone, 1986; Neal, Pearson, and Powell, 1988; Cavada and Goldman-Rakic, 1989a; Boussaoud, Ungerleider, and Desimone, 1990; Baizer, Ungerleider, and Desimone, 1991), and both projections overlap in the fundus (area VIP) (Maunsell and Van Essen, 1983; Ungerleider and Desimone, 1986; Colby and Duhamel, 1991; Duhamel, Colby, and Goldberg, 1991). All three intraparietal areas innervate 7b (Jones and Powell, 1970; Mesulam et al., 1977; Cavada and Goldman-Rakic, 1989a), which also receives other somatosensory input, primarily from the second somatosensory area (SII) (Stanton et al., 1977; Cavada and Goldman-Rakic, 1989a). Inferior area 6 and area 7b are heavily interconnected (Mesulam et al., 1977; Kunzle, 1978; Matelli et al., 1986; Cavada and Goldman-Rakic, 1989b), and both project to the putamen (Kunzle, 1978; Weber and Yin, 1984; Cavada and Goldman-Rakic, 1991; Parthasarathy, Schall, and Graybiel, 1992).

These connections suggest that bimodal responses may be generated in area VIP and area 7b from convergent visual and somesthetic input. Area 7b may then transmit its bimodal properties to inferior area 6 and the putamen, where further processing is done. In the next section we propose that these areas form a system for representing extrapersonal space.

Bimodal cells may code extrapersonal visual space

The putamen, inferior area 6, area 7b, and area VIP form a distributed system of bimodal cells. Each of these areas is somatotopically organized, except perhaps for VIP, where the organization is not known. In each map, neurons in the arm portion have visual receptive fields around the arm, while neurons in the face portion have visual receptive fields around the head. That is, each area contains a somatotopically organized map of the visual space near the animal. We suggest, therefore, that these areas encode near extrapersonal visual space. This view is supported by the results of lesion experiments. Lesions to inferior area 6 impair the ability to localize nearby visual stimuli, but leave intact the ability to localize stimuli that are beyond reaching distance (Rizzolatti, Matelli, and Pavesi, 1983; Rizzolatti and Berti, 1990). Lesions to the parietal lobe cause a whole constellation of spatial deficits, including deficits in processing extrapersonal space. For example, parietal lesions in humans often produce optic ataxia, an inability to judge the locations of stimuli for the purpose of reaching toward them (Newcombe and Ratcliff, 1989).

What spatial coordinate system do these bimodal areas use to encode the location of visual stimuli? Several different visual coordinate systems have been described for other regions of the brain. In the primary visual cortex, visual space is mapped retinocentrically; that is, neurons in V1 have visual receptive fields that are fixed with respect to the retina. In parietal area 7a, the receptive fields are also retinocentric, but the response magnitude is modulated by eye position (Andersen, Essick, and Siegel, 1985). A population of such neurons could encode the location of stimuli with respect to the head, that is, in craniocentric space (Zipser and Andersen, 1988). Finally, visual receptive fields in inferior area 6 remain in the same location even when the eyes move (Fogassi et al., 1992; Gentilucci et al., 1983). That is, they appear to encode space in a way that is explicit at the level of single neurons.

One explanation for these visual receptive fields that remain stationary when the eyes move is that they represent space in craniocentric coordinates (Fogassi et al., 1992). Our own work, however, suggests that the mapping of extrapersonal visual space in area 6 (and in the putamen) is not exclusively craniocentric. In both areas, we found bimodal neurons with visual receptive fields that move with the arm. These cells could encode the location of visual stimuli in arm-centered coordinates. More generally, we suggest that these bimodal areas may encode stimuli in a body part–centered fashion. According to this view, bimodal cells with tactile receptive fields on the face encode the location of stimuli with respect to the head; bimodal cells with tactile receptive fields on the arm encode the location of stimuli with respect to the arm; and bimodal cells with tactile receptive fields on the chest encode the location of stimuli with respect to the trunk. That is, the visual space near the animal is represented as if it were a gelatinous medium surrounding the body that deforms whenever the head rotates or the limbs move. Such a map would give the location of the visual stimulus with respect to the body surface, in somatotopic coordinates.

This hypothesis of body part–centered coordinates for the mapping of near extrapersonal visual space yields the following predictions. If a bimodal cell has a tactile receptive field on the face, then the visual receptive field would move as the head is rotated, but it would not move with the eye or the arm. If a bimodal cell has a tactile receptive field on the chest, then the visual receptive field would move as the trunk is rotated, but not with the head, the eyes, or the arm. We would expect to find neurons with these visual properties in some or all of the bimodal areas described above.

Relationship between sensory and motor properties

Cells in the putamen, area 7b, and inferior area 6 have motor functions as well as sensory functions (e.g., Hyvarinen, 1981; Crutcher and DeLong, 1984b; Gentilucci et al., 1988; Rizzolatti et al., 1988). Indeed, the same neurons often have both sensory and motor activity. These areas are probably best described as sensory-motor interfaces, which help to encode the location of sensory stimuli and to generate the motor responses to those stimuli. Are the sensory and motor responses expressed in a common coordinate system? There is some evidence that this is the case for area 6. Many neurons in inferior area 6 respond when the monkey reaches toward a target (Caminiti, Johnson, and Urbano, 1990). These neurons are broadly tuned to a preferred direction of reach, and this motor field moves as the arm moves, rotating at roughly the same angle that the shoulder has rotated. That is, just as for the visual receptive fields, the motor response fields for arm movements appear to be arm centered. The relevant experiment has not been done for neurons in area 7b or the putamen. However, there is psychophysical evidence from humans that, whatever portion of the brain may control reaching movements, it is done in an arm- or shoulder-centered coordinate system (Soechting and Flanders, 1989a, 1989b).

Another area with both sensory and motor responses in the same neurons is the superior colliculus (for review, see Sparks, 1991; Stein and Meredith, 1993). Neurons in the deep layers of the superior colliculus respond to visual, auditory, and tactile stimuli, and they also respond during saccadic eye movements. Many neurons are multimodal, and in these cases the response fields for the saccadic eye movements and the receptive fields for the different sensory modalities correspond spatially. Exactly how the location of a stimulus is encoded in the colliculus has been an issue of great interest. It appears that saccade targets are encoded by sensory receptive fields that are fixed with respect to the retina. This retinocentric organization was particularly clear in experiments by Jay and Sparks (1987), who recorded from neurons that had auditory receptive fields. When the monkey moved its eyes to different locations, these auditory receptive fields also moved, and thus remained at the same retinal coordinates.

Retinocentric coordinates are particularly appropriate for encoding the location of saccadic targets, because these coordinates give the distance and direction between the fovea and the desired target location. That is, they encode the saccadic "motor error" (Sparks, 1991). Similarly, arm-centered responses are appropriate for reaching toward an object, because they give the motor error between the current arm position and the desired arm position. However, retinocentric and arm-centered coordinate systems are not the only useful choices. When a monkey reaches toward another monkey with its teeth, it must encode the spatial relationship between its mouth and the target. When a soccer player butts a ball with his head, he must encode the changing relationship between the ball and his forehead. When he elbows a fellow player, he must encode the distance and direction between his elbow and the other player's stomach. Indeed, it would be useful to have a visual coordinate frame fixed to every part of the body surface, for the purpose of hitting, grasping, or avoiding visual stimuli in extrapersonal space. We hypothesize that the bimodal portions of the brain provide exactly such a somatotopically organized map of space. The arm and face portion of this map is clearly exaggerated. Indeed, in the putamen, we did not find any visual responses in the leg or trunk portions of the somatotopic map. However, they may exist in much lower proportions. In area 7b and area 6, the leg and trunk representations have not been adequately explored.

Two types of spatial maps have generally been distinguished (e.g., Hein and Jeannerod, 1983; Stiles-Davis, Kritchevsky, and Bellugi, 1988; Paillard, 1991). The first is an egocentric map, in which objects are located with respect to the body—usually with respect to a point in the middle of the forehead. The other is an allocentric map, in which objects are located in a fixed, external frame. The bimodal system proposed in this chapter contains a type of egocentric representation, a body part–centered one rather than a head-centered one.

ACKNOWLEDGMENTS We thank Dr. Hillary Rodman for her help in every stage of these experiments, and Dr. Barry Stein, Dr. Giacomo Rizzolatti, and Dr. Michael Goldberg for their helpful comments. This work was supported by NIMH grant MH-19420.

REFERENCES

Alexander, G. E., 1987. Selective neuronal discharge in monkey putamen reflects intended direction of planned limb movements. *Exp. Brain Res.* 67:623–634.

Andersen, R. A., G. K. Essick, and R. M. Siegel, 1985. Encoding of spatial location by posterior parietal neurons. *Science* 230:456–458.

Baizer, J. S., L. G. Ungerleider, and R. Desimone, 1991. Organization of visual inputs to the inferior temporal and posterior parietal cortex in macaques. *J. Neurosci.* 11:168–190.

Boussaoud, D., L. G. Ungerleider, and R. Desimone, 1990. Pathways for motion analysis: cortical connections of visual areas MST and FST in the macaque. *J. Comp. Neurol.* 296:462–495:

Caminiti, R., P. B. Johnson, and A. Urbano, 1990. Making arm movements within different parts of space: Dynamic aspects in the primate motor cortex. *J. Neurosci.* 10:2039–2058.

Cavada, C., and P. S. Goldman-Rakic, 1989a. Posterior parietal cortex in rhesus monkey: I. Parcellation of areas based on distinctive limbic and sensory corticocortical connections. *J. Comp. Neurol.* 287:393–421.

Cavada, C., and P. S. Goldman-Rakic, 1989b. Posterior parietal cortex in rhesus monkey: II. Evidence for segregated corticocortical networks linking sensory and limbic areas with the frontal lobe. *J. Comp. Neurol.* 287:422–445.

Cavada, C., and P. S. Goldman-Rakic, 1991. Topographic segregation of corticostriatal projections from posterior parietal subdivisions in the macaque monkey. *Neurosci.* 42: 683–696.

Colby, C. L., and J. Duhamel, 1991. Heterogeneity of extrastriate visual areas and multiple parietal areas in the macaque monkey. *Neuropsychologia* 29:517–537.

Colby, C. L., J. Duhamel, and M. E. Goldberg, 1993. The ventral intraparietal area (VIP) of the macaque: Anatomical location and visual response properties. *J. Neurophysiol.* 69:902–914.

Crutcher, M. D., and M. R. DeLong, 1984a. Single cell studies of the primate putamen. I. Functional organization. *Exp. Brain Res.* 53:233–243.

Crutcher, M. D., and M. R. DeLong, 1984b. Single cell studies of the primate putamen. I. Relations to direction of movement and pattern of muscular activity. *Exp. Brain Res.* 53:244–258.

DeLong, M. R., 1973. Putamen: Activity of single units during slow and rapid arm movements. *Science* 179:1240–1242.

Duhamel, J., C. L. Colby, and M. E. Goldberg, 1991.

Congruent representations of visual and somatosensory space in single neurons of monkey ventral intra-parietal cortex (area VIP). In *Brain and Space*, J. Paillard, ed. New York: Oxford University Press, pp. 223–236.

Fogassi, L., V. Gallese, G. di Pellegrino, L. Fadiga, M. Gentilucci, G. Luppino, M. Matelli, A. Pedotti, and G. Rizzolatti, 1992. Space coding by premotor cortex. *Exp. Brain Res.* 89:686–690.

Gentilucci, M., L. Fogassi, G. Luppino, R. Matelli, R. Camarda, and G. Rizzolatti, 1988. Functional organization of inferior area 6 in the macaque monkey: I. Somatotopy and the control of proximal movements. *Exp. Brain Res.* 71:475–490.

Gentilucci, M., C. Scandolara, I. N. Pigarev, and G. Rizzolatti, 1983. Visual responses in the postarcuate cortex (area 6) of the monkey that are independent of eye position. *Exp. Brain Res.* 50:464–468.

Graziano, M. S., and C. G. Gross, 1992. Somatotopically organized maps of near visual space exist. *Behav. Brain Sci.* 15:750.

Graziano, M. S., and C. G. Gross, 1993. A bimodal map of space: Tactile receptive fields in the macaque putamen with correspondiong visual receptive fields. *Exp. Brain Res.* 97:96–109.

Hein, A., and M. Jeannerod, 1983. *Spatially Oriented Behavior.* New York: Springer-Verlag.

Hyvarinen, J., 1981. Regional distribution of functions in parietal association area 7 of the monkey. *Brain Res.* 206: 287–303.

Hyvarinen, J., and A. Poranen, 1974. Function of the parietal associative area 7 as revealed from cellular discharges in alert monkeys. *Brain* 97:673–692.

Jay, M. F., and D. L. Sparks, 1987. Sensorimotor integration in the primate superior colliculus. II. Coordinates of auditory signals. *J. Neurophysiol.* 57:35–55.

Jones, E. G., and T. P. S. Powell, 1970. An anatomical study of converging sensory pathways within the cerebral cortex of the monkey. *Brain* 93:739–820.

Kunzle, H., 1978. An autoradiographic analysis of the efferent connections from premotor and adjacent prefrontal regions (areas 6 and 9) in Macaca fascicularis. *Brain Behav. Evol.* 15:185–234.

Leinonen, L., J. Hyvarinen, G. Nyman, and I. Linnankoski, 1979. I. Functional properties of neurons in the lateral part of associative area 7 in awake monkeys. *Exp. Brain Res.* 34:299–320.

Leinonen, L., and G. Nyman, 1979. II. Functional properties of cells in anterolateral part of area 7 associative face area of awake monkeys. *Exp. Brain Res.* 34:321–333.

Liles, S. L., 1983. Activity of neurons in the putamen associated with wrist movements in the monkey. *Brain Res.* 263:156–161.

Liles, S. L., 1985. Activity of neurons in putamen during active and passive movement of wrist. *J. Neurophysiol.* 53: 217–236.

Liles, S. L., and B. V. Updyke, 1985. Projection of the digit and wrist area of precentral gyrus to the putamen: Relation between topography and physiological properties of neurons in the putamen. *Brain Res.* 339:245–255.

Matelli, M., R. Camarda, M. Glickstein, G. Rizzolatti, 1986. Afferent and efferent projections of the inferior area 6 in the macaque monkey. *J. Comp. Neurol.* 251:281–298.

Maunsell, J. H., and D. C. Van Essen, 1983. The connections of the middle temporal visual area (MT) and their relationship to a cortical hierarchy in the macaque monkey. *J. Neurosci.* 3:2563–2856.

Mesulam, M., G. W. Van Hoesen, D. N. Pandya, and N. Geschwind, 1977. Limbic and sensory connections of the inferior parietal lobule (area PG) in the rhesus monkey: A study with a new method for horseradish peroxidase histochemistry. *Brain Res.* 136:393–414.

Neal, J. W., R. C. A. Pearson, and T. P. S. Powell, 1988. The organization of the cortico-cortical connections between the walls of the lower part of the superior temporal sulcus and the inferior parietal lobule in the monkey. *Brain Res.* 438:351–356.

Newcombe, F., and G. Ratcliff, 1989. Disorders of visuospatial analysis. In *Handbook of Neuropsychology*, vol. 2, F. Boller and J. Grafman, eds. New York: Elsevier, pp. 333–356.

Paillard, J., 1991. *Brain and Space.* New York: Oxford University Press.

Parthasarathy, H. B., J. D. Schall, and A. M. Graybiel, 1992. Distributed but convergent ordering of corticostriatal projections: Analysis of the frontal eye field and the supplementary eye field in the macaque monkey. *J. Neurosci.* 12:4468–4488.

Rizzolatti, G., and A. Berti, 1990. Neglect as a neural representation deficit. *Rev. Neurol. (Paris)* 146:626–634.

Rizzolatti, G., R. Camarda, L. Fogassi, M. Gentilucci, G. Luppino, and M. Matelli, 1988. Functional organization of inferior area 6 in the macaque monkey: II. Area F5 and the control of distal movements. *Exp. Brain Res.* 71: 491–507.

Rizzolatti, G., M. Matelli, G. Pavesi, 1983. Deficits in attention and movement following the removal of postarcuate (area 6) and prearcuate (area 8) cortex in macaque monkey. *Brain* 106:655–673.

Rizzolatti, G., C. Scandolara, M. Matelli, and M. Gentilucci, 1981a. Afferent properties of periarcuate neurons in macaque monkeys: I. Somatosensory resposes. *Behav. Brain Res.* 2:125–146.

Rizzolatti, G., C. Scandolara, M. Matelli, and M. Gentilucci, 1981b. Afferent properties of periarcuate neurons in macaque monkeys: II. Visual resposes. *Behav. Brain Res.* 2:147–163.

Robinson, C. J., and H. Burton, 1980a. Organization of somatosensory receptive fields in cortical areas 7b, retroinsular, postauditory, and granular insula of M. fascicularis. *J. Comp. Neurol.* 192:69–92.

Robinson, C. J., and H. Burton, 1980b. Somatic submodality distribution within the second somatosensory area

(SII), 7b, retroinsular, postauditory, and granular insular cortical areas of M. fascicularis. *J. Comp. Neurol.* 192:93–108.

SCHULTZ, W., and R. ROMO, 1988. Neuronal activity in the monkey striatum during the initiation of movements. *Exp. Brain Res.* 71:431–436.

SELTZER, B., and D. N. PANDYA, 1980. Converging visual and somatic coritcal input to the intraparietal sulcus of the rhesus monkey. *Brain Res.* 192:339–351.

SOECHTING, J. F., and M. FLANDERS, 1989a. Sensorimotor representations for pointing to targets in three-dimensional space. *J. Neurophysiol.* 62:582–594.

SOECHTING, J. F., and M. FLANDERS, 1989b. Errors in pointing are due to approximations in sensorimotor transformations. *J. Neurophysiol.* 62:595–608.

SPARKS, D. L., 1991. The neural encoding of the location of targets for saccadic eye movements. In *Brain and Space*, J. Paillard, ed. New York: Oxford University Press, pp. 3–19.

STANTON, G. B., W. L. R. CRUCE, M. E. GOLDBERG, and D. L. ROBINSON, 1977. Some ipsilateral projections to areas PF and PG of the inferior parietal lobule in monkeys. *Neurosci. Lett.* 6:243–250.

STEIN, B. E., and M. A. MEREDITH, 1993. *The Merging of the Senses*. Cambridge, Mass.: MIT Press.

STILES-DAVIS, J., M. KRITCHEVSKY, and U. BELLUGI, 1988. *Spatial Cognition: Brain Bases and Development*. Hillsdale, N.J.: Erlbaum.

UNGERLEIDER, L. G., and R. DESIMONE, 1986. Cortical connections of visual area MT in the macaque. *J. Comp. Neurol.* 248:190–222.

VOGT, B. A., and D. N. PANDYA, 1978. Cortico-cortical connections of somatic sensory cortex (areas 3, 1 and 2) in the rhesus monkey. *J. Comp. Neurol.* 177:179–192.

WEBER, J. T., and T. C. T. YIN, 1984. Subcortical projections of the inferior parietal cortex (area 7) in the stumptailed monkey. *J. Comp. Neurol.* 224:206–230.

ZIPSER, D., and R. A. ANDERSEN, 1988. A back-propagation programmed network that simulates response properties of a subset of posterior parietal neurons. *Nature* 331: 679–684.

68 Roles of Imagery in Perception: Or, There Is No Such Thing as Immaculate Perception

STEPHEN M. KOSSLYN AND AMY L. SUSSMAN

ABSTRACT This chapter reviews the neural and computational bases of the roles of imagery in perception, and argues that imagery is used to complete fragmented perceptual inputs, to match shape during object recognition, to prime the perceptual system when one expects to see a specific object, and to prime the perceptual system to encode the results of specific movements. According to the present analysis, imagery acts as a bridge not only between perception and memory but also between perception and motor control. The functions of mental imagery in perception and motor control have apparently shaped its properties even when it is used autonomously.

In this chapter we argue that mental imagery plays an integral role in perception, and that it may have evolved primarily for this purpose and later have been recruited into higher cognitive processes. In this sense, imagery is like one's nose: It did not evolve to hold up glasses, but once it was present, it could be used to do so (see Gould and Lewontin, 1979). The idea that higher cognitive processes contribute to perception is not new (e.g., see Neisser, 1967, 1976; Gregory, 1970); however, we have recently learned much about the neuroanatomy and neurophysiology of vision that allows us to develop old ideas in new directions.

We take as our starting point the finding that imagery and perception share common mechanisms (for reviews, see Farah, 1988; Finke, 1989; Kosslyn, 1994). Kosslyn et al. (1993) used positron emission tomography (PET) to provide strong evidence that imagery activates topographically organized parts of visual cortex. For example, in one study they found greater activation in posterior parts of the medial occipital lobe (which represent foveal input during vision) when subjects formed small visual mental images, and greater activation in more anterior parts of the medial occipital lobe (which represent more peripheral input during vision) when subjects formed large visual mental images; these findings were obtained even though the subjects had their eyes closed during the entire task. In this chapter we explore the relationship between patterns of activation in topographically organized cortex that are evoked from memory (mental images) and those that are evoked by on-line input (percepts).

Using imagery in perception

Visual perception must begin with bottom-up processes: We register what we see. However, in many situations, the input activates multiple representations in memory. This may happen when relatively few distinguishing properties are visible, and hence the input places relatively few constraints on what the object must be. For example, if one cannot see the edges that indicate a pen's clip, it will be difficult to use the remaining edges to distinguish between a pen and pencil. This problem is even more difficult when one is trying to identify a specific member of a category, such as one's own pen rather than somebody else's.

To solve such problems, David Lowe implemented a computer vision system that incorporated a second phase of processing. He found that when the stable properties of the input do not strongly implicate a particular object, it is useful to activate a stored "model" of the best-matching shape (Lowe, 1985, 1987a, 1987b). Such models are stored in his system's visual memory (which corresponds to the pattern activation subsystem of Kosslyn et al., 1990). When activated, the

STEPHEN M. KOSSLYN Department of Psychology, Harvard University, Cambridge, Mass.
AMY L. SUSSMAN Department of Psychology, Yale University, New Haven, Conn.

model generates an image of an object in the input array (which corresponds to the visual buffer, in our terms). In Lowe's system, this generated image is then compared to the input image itself; the generated image is adjusted in size and orientation until it makes the best possible match to the input image. If the match is good enough, then the object is recognized. If Lowe is correct, then the ability to generate images is an essential part of our ability to recognize objects.

This sort of image-matching process can also augment the input itself. Recurrent connections allow the states of units farther along in the processing sequence to provide feedback to earlier units. In models of neural networks (e.g., see Rumelhart and McClelland, 1986), this feedback may actually fill in missing elements of the input, a process called vector completion. Imagery may play a similar role: The image provides information that will complete missing parts of the input.

Evidence for Image Matching Cave and Kosslyn (1989) investigated whether an image generated from a stored model is sometimes used to augment the input image during human object recognition. Based on Lowe's ideas, they expected a mental image to be used when the input properties were insufficient for a match to be made easily. Subjects were shown a rectangle superimposed on a diamond; one object was drawn in heavy black lines while the other was drawn in lighter, narrow lines. The subjects' task was to decide whether lines of equal length were used to draw the lighter object (e.g., whether it was a square or nonsquare rectangle), which was partly obscured. On half the trials the lines were of equal length; on half they were not. If imagery would be useful in this situation, Cave and Kosslyn expected the subjects to use an image of the previous target object (it was the most active and hence the most readily available appropriate representation). On 75% of the trials, successive target objects were the same size, but on the remaining 25% of the trials they were of different sizes. If subjects matched an image of the previous stimulus, then on 25% of the trials they would need to adjust the size of the image to fit the input. Previous research had shown that the greater the extent of a size transformation, the more time is required to complete it (e.g., see Sekuler and Nash, 1972; Bundesen and Larsen, 1975, 1981; Larsen and Bundesen, 1978). Finally, on half of each type of trial the target figure was the same object as on the previous trial (e.g., both were diamonds), and on half the two were different. If an image of the previous object was used, then the subjects should be faster when the same object appeared twice in succession than when different objects appeared.

As predicted, when a target object was the same as the previous one, the time to evaluate it increased linearly with the disparity in the sizes of the two objects. The size-scaling function was very similar to that obtained when subjects adjust images, and was much slower than that obtained when subjects adjust the scope of attention (to surround an image in the visual buffer). In contrast, when the second stimulus was a different object, the subjects were slower and the size-scaling function was very similar to that obtained when subjects merely adjusted the scope of attention. These results, then, were just as expected if subjects formed an image of the previous stimulus and used it to recognize successive stimuli; if the image augmented the input, recognition was facilitated, but if it could not supplement the input, the subject had to engage in more complete, slower, bottom-up processing.

Similar results have been obtained when orientation, instead of size, has been varied. Koriat, Norman, and Kimchi (1991) showed subjects letters sequentially at different orientations, and asked them to identify each one by pressing a key. Analogous to Cave and Kosslyn's findings when they examined the effects of size disparity, Koriat, Norman, and Kimchi found that even though the response times increased with greater orientation disparity when the same letter appeared a second time in succession, these response times were faster than the trials in which the letter differed from the preceding one. The representation of the preceding letter may have allowed an image to be generated, which facilitated encoding when that letter appeared again.

This role of imagery in recognition may be counterintuitive because we are not aware of generating images when we look at objects. However, if we usually generate appropriate images, then we should not be aware of them. When the generated image matches the input image, we simply experience seeing the object. This observation suggests a simple mechanism for determining a match: If the generated image strengthens the input to visual memory, then it matched the input image. In this case, the generated image would simply add onto the image from the eyes, producing a greater signal to areas farther downstream. When the generated image does not match the image in the visual buffer, the input to visual memory would become noisier, not stronger.

Neural plausibility

We have suggested that when an input reaching visual memory is not adequate to implicate one particular object during perception, a mental image of the most likely stored representation is projected back into the visual buffer in an attempt to augment the input image. This concept is consistent with the fact that visual areas that send information to higher visual areas generally also receive information from those areas.

Pathways of Image Activation The theory that visual mental images arise via feedback from higher visual centers has been proposed at least since the time of William James (e.g., see James, 1890), and the fact that reciprocal connections are abundant in the visual system lends credence to this idea. However, the idea is not without potential problems. One major difficulty is that the connections that provide feedback (efferent pathways) are not exact reversals of the connections that provide input (afferent pathways; see Rockland and Virga, 1989). The efferent pathways contain more diffuse connections than the afferent pathways. As an example, Zeki and Shipp (1988) point out that area V5 (also known as area MT) receives input from the "thick stripes" of area V2 (a layer of neurons identified through staining; see Livingstone and Hubel, 1988), and sends feedback primarily to these "thick stripes"; however, it also sends feedback to the "thin stripes" and "interstripes" of area V2.

Rockland and Virga (1989) suggest that connections within higher areas might produce such divergence. Indeed, each higher-level neuron might receive input from multiple lower-level neurons, perhaps indirectly via the other neurons within the same area that have lateral connections to one another. Such diffuse efferent connections imply that if information is stored in high-level visual areas, then reconstructing the original image itself will not be trivial.

This problem of diffuse efferent connections may provide a strong hint about the process by which images are created: It may be that the feedback mechanism involves "coarse coding" instead of a point-to-point mapping (e.g., see Hinton, McClelland, and Rumelhart, 1986). A familiar example of coarse coding is the fact that there are only three types of cones in the retina, which are tuned to respond maximally to different wavelengths. Because the cones have relatively broad tuning functions, any given wavelength produces more or less activity in the different cones—and it is this mixture that indicates a specific hue. Analogously, information may be stored according to the relative strengths of input from large regions, and images are reconstructed when the higher-level visual areas send feedback to these overlapping areas in lower-level, topographically mapped regions of cortex.

The concept of coarse coding suggests a possible solution to a related problem: How can a spatial image be created on the basis of nonspatially stored representations? Higher-level visual areas, such as area IT, do not have precise topographic organization (see Felleman and Van Essen, 1991). Therefore, information in higher-level areas cannot be transcribed into images simply by reversing the process by which they were stored. In fact, Douglas and Rockland (1992) have found direct pathways from TE (at the anterior part of the inferior temporal lobe) to V1. Course coding offers a mechanism for taking a nonspatial representation and projecting it back to form a spatial one.

In addition, it is worth noting that imagery may not rely solely on direct corticocortical feedback connections; it may also involve pathways through the thalamus (see Mumford, 1991). Cortical structures send long axons from deep layers to the thalamus. The thalamus, in turn, sends some of this information back to the cortex, ending in layer 4, the cortical layer through which input generally enters. This type of feedback connection differs from the direct corticocortical feedback connections, which generally terminate in superficial layers (1 or 2) or deep layers (5 or 6), rather than in layer 4.

Priming and Imagery The kind of vector completion process we described earlier can occur in advance of seeing an object, which would prime one to encode an expected pattern. Kosslyn, Anderson, and Chiu (in preparation) showed that perception can prime imagery and vice versa. For example, they asked subjects first to examine a list of words and count the number of vowels in each. Subjects then visualized these and other words in lowercase letters, and decided whether the first and last letters were the same height; they performed this task faster for primed words than for novel words. More priming was observed when the subjects visualized the words in the same font in which they had previously seen them. In another experiment subjects were faster to decide whether a picture was a real object or a scrambled-up object if they had formed an image of the object prior to seeing the stimulus than if they formed an image of another object or merely described categories of the proper object.

The kind of mental images that are used in memory retrieval (as when you recall how many windows are in your living room) or in reasoning (as when you think of the best route to the airport from a friend's house) may result from a large amount of priming. Information from higher-level visual areas projected to lower-level ones may not only serve to prime the matching process in visual memory, but it may also cause the threshold of neurons in the visual buffer to lower so much that they begin firing spontaneously. This mechanism may create a pattern of activation in topographically mapped areas, and this pattern is a representation of a mental image.

Image Transformations In Lowe's model, a mental image is adjusted to match the perceptual (input) image. The ability to transform the size, orientation, location, and even the shape of an imaged pattern is an essential aspect not only of using images to augment percepts but also of many other imagery functions. One hint about the processes underlying image transformation was suggested by what may happen when mental rotation ability is lost. Kosslyn, Daly, and Wray (unpublished) tested a brain-damaged patient who had great difficulty in mentally rotating objects; during testing, she continually reached toward the screen and moved her hand as if she were actually turning the stimulus. For at least some types of image transformations, it is possible that the image is transformed via priming from the motor system: One is anticipating what one would see if one performed a specific action (for similar ideas, see, e.g., Weimer, 1977; Finke, 1979, 1989; Annett, 1990).

At least some image transformations, then, may arise from another type of priming: One primes the visual system to encode what one would see after an object was manipulated. This notion allows us to understand a mystery: People require more time to transform imaged objects greater amounts (see Shepard and Cooper, 1982). For example, subjects require more time to imagine rotating an object through more degrees. This is a mystery because there is no constraint in the brain that requires mental images of objects to pass through trajectories. However, limbs must pass through trajectories; thus, if one images what one would see if an object were manipulated, then imaged objects would pass through trajectories.

There is evidence that at least some types of mental transformations involve motor processing. Georgopoulos et al. (1989) found that as a monkey prepared to move its arm in a specific way, the activity of neurons in motor cortex changed in succession along a trajectory, as if anticipating going through the different arm positions. Deutsch et al. (1988) studied brain activation while their human subjects performed a mental rotation task. They found selective activation of frontal and parietal cortex (particularly in the right cerebral hemisphere), regions that are known to be involved in programming and executing actions (see Kosslyn and Koenig, 1992, ch. 7). In addition, Decety, Philippon, and Ingvar (1988) found that the prefrontal cortex, supplementary motor cortex, and parts of the cerebellum were activated both when subjects imagined that they were writing letters and when they actually wrote the letters.

Several studies of the mental rotation of body parts also suggest that motor processing is used in some types of image transformations. Cooper and Shepard (1975) asked subjects to view pairs of drawings of hands and decide whether the hands were the same (e.g., both left hands) or different. The hands were presented at various angles. Response times varied with the positions of the hands, which appeared to reflect in part the physical awkwardness of actually moving one's hands from the given positions. Similarly, Sekiyama (1982, 1983) showed that the subjects were faster when mentally rotating hands in a manageable direction (e.g., rotating the left hand clockwise and rotating the right hand counterclockwise; Parsons, 1987a, b, reports converging results).

Motion-Added versus Motion-Encoded Transformations Consider two possible ways to transform objects in images. First, one could activate stored memories of moving objects. This would be a kind of *motion-encoded* image transformation. For example, if one sees race cars zooming around a track, one can later visualize the race. However, such purely visual motion may often be supplemented by motor processes. We typically move our eyes when examining scenes. This pattern of eye movements may index when a particular encoding was made, and the pattern of eye movements may be stored (see Noton and Stark, 1971). If so, then the eye movements may serve as retrieval cues during imagery, indicating which image segment is the next one in the sequence (see Stark and Ellis, 1981).

The second possible method by which mental images could be transformed involves imposing motion information on an imaged object that was not moving during the initial encoding of the representation. This *motion-added* type of transformation is the kind of imaged motion that is more likely to underlie the kind of processing that Lowe posited. After all, we rarely see an object move in exactly the way that would be necessary to match a previously stored image with perceptual input. When using a mental image to augment input, one may not simply activate a previously stored motion sequence; rather, one may be guided by the shape in the visual buffer. This kind of imaged motion could be produced by incrementally altering the way information in visual memory is mapped to the visual buffer.

Logically, the mapping function from visual memory to the visual buffer could be altered via purely visual processes or via the motor system. We suggested earlier that at least some image transformations are evoked when one anticipates (primes) what one would expect to see if one manipulated an object in a certain way. Processes that program the motor system interact closely with those that subserve visual perception. For example, Prablanc, Pelisson, and Goodale (1986) found that when subjects reach for a target, they are three times more accurate when the target is in view the entire time than when it is removed shortly after the reach is initiated. This demonstrates that visual perception is needed to correct movements on-line, as they are executed. The speed with which on-line movement corrections take place suggests that a rather tight linkage exists between what one expects to see and an error-correction motor output system (for relevant computational models, see Reeke et al., 1989; Mel, 1991; for a review of other such models, see Kosslyn and Koenig, 1992, ch. 7).

Neurons in the posterior parietal lobe play a special role in this process. Andersen (1989) suggests that neurons in the posterior parietal lobe ordinarily have both sensory- and motor-related responses: Neurons that respond when the animal is reaching or planning to reach also have somatosensory inputs, and neurons that respond to smooth pursuit, saccades, or fixations respond to visual stimuli as well (see also, e.g., Taira et al., 1990; Kalaska and Crammond, 1992; Sakata et al., 1992). It is possible that these neurons participate in producing a visual representation that anticipates what one would see after manipulating an object.

Many others have suggested that the anticipated consequences of one's actions have an effect on one's perception. For example, Muensterberg (1914, p. 141) asserted, "We all perceive the world just as far as we are prepared to react to it." It has, in fact, been shown that damage to the posterior parietal lobe can cause human patients (Perenin and Vighetto, 1983) and monkeys (Lamotte and Acuna, 1978) to have difficulty anticipating the consequences of their own reaching behavior, causing them to guide reaching very carefully using visual feedback. Our hypothesis is similar that of Droulez and Berthoz (1990), among many others (e.g., see Neisser, 1976; Hochberg, 1981, 1982; Shepard, 1984; Coren, 1986; Finke, 1989). As predicted by this hypothesis, Droulez and Berthoz observed that the maximum speed with which one can rotate a mental image is similar to the maximum speed at which one can make actual orienting movements. This discovery led them to suggest that mental rotation is computed on the basis of the simulation of an orienting movement. They also suggest that the process that produces these types of simulated movements is the same process that is used in actual motor control.

Imagery and perception revisited

Imagery thus may play an integral role in perception, helping one not only to recognize objects (when they are either static or moving) but also to anticipate the consequences of events. Indeed, Freyd (1987) summarizes a series of findings that suggest that motion is not encoded entirely by bottom-up processes. She and her colleagues have documented a phenomenon they call representational momentum. When one sees a sequence of static images progressing along a trajectory (such as a falling object), one is apt to incorrectly recall the end of the sequence as if the object were farther along in the trajectory than it actually was. This discovery suggests that one mentally extends a known progression, and stores the end result of this anticipation. In fact, Finke and Shyi (1988) found that shifts in memory of this sort are highly correlated with the actual rate at which implied motion is extrapolated along a trajectory. These results support the proposal that images are formed by priming the representation of what one anticipates seeing.

This idea illuminates an interesting result reported by Meador and colleagues (1987). They studied three patients who had right parietal lesions with accompanying unilateral neglect, and found that the patients not only ignored objects to their left in perception, but also neglected objects to the left in their mental images (see also Bisiach and Luzzatti, 1978; Bisiach and Berti, 1988). The experimenters instructed one of the patients to move his head and eyes so that he actually looked into the left half of space when recalling imaged objects. When the subject did this, they found marked improvement in his ability to "see" objects on the left side of his images. Actually moving and anticipating what one would see may have been an especially effective method of forming the images.

The two types of image transformations, motion-added and motion-encoded, may often work together. Perception is not like a camera, or even a movie camera. One's attention wanders, and one often does not study an object carefully. Therefore, one may not have encoded a moving pattern completely, or over a very long period of time; rather, one may have encoded a succession of images, which may or may not be moving (as the philosopher Bergson, 1954, seems to have suggested). In this case, motion-added imagery will supplement motion-encoded imagery.

Conclusions

Many of the properties of imagery appear to mimic those of perception. This is not surprising if imagery arises from the same mechanisms that are used during perception, and in fact plays a critical role in perception itself. If this is true, researchers interested in imagery are fortunate indeed. Perception, with its systematic moorings in observable properties of the world, is much easier to study than a slippery creature like mental imagery.

We have argued further that imagery is a bridge between perception and motor control. If so, then we are in a position to begin to understand central psychological phenomena such as imitation: How is it that observing someone else's behavior allows one to perform that behavior? And why is this performance not perfect? Possible answers to such questions can begin to be formed if one thinks about the role of imagery in programming behavior and about the origins of images in perception (see Kosslyn and Koenig, 1992, chs. 4 and 7).

ACKNOWLEDGMENTS Preparation of this chapter was supported by NSF grant BNS 90 09619, and NINDS grant 2 P01 17778-09. We thank Pierre Jolicoeur for comments on related material, and Gregg DiGirolamo for a careful reading of the manuscript.

REFERENCES

ANDERSEN, R. A., 1989. Visual and eye movement functions of the posterior parietal cortex. *Ann. Rev. Neurosci.* 12:377–403.

ANNETT, J., 1990. Relations between verbal and gestural explanations. In *Cerebral Control of Speech and Limb Movement*, G. E. Hammond, ed. Amsterdam: North-Holland, Elsevier, pp. 327–346.

BERGSON, H., 1954. *Creative Evolution*. London: Macmillan.

BISIACH, E., and A. BERTI, 1988. Hemineglect and mental representation. In *Cognitive and Neuropsychological Approaches to Mental Imagery*, M. Denis, J. Engelkamp, and J. T. E. Richardson, eds. Dordrecht: Martinus Nijhoff.

BISIACH, E., and C. LUZZATTI, 1978. Unilateral neglect of representational space. *Cortex* 14:129–133.

BUNDESEN, C., and A. LARSEN, 1975. Visual transformation of size. *J. Exp. Psychol. [Hum. Percept.]* 1:214–220.

BUNDESEN, C., and A. LARSEN, 1981. Mental transformations of size and orientation. In *Attention and Performance*, vol. 9, A. Baddeley, and J. Long, eds. Hillsdale, N.J.: Erlbaum, pp. 279–294.

CAVE, K. R., and S. M. KOSSLYN, 1989. Varieties of size-specific visual selection. *J. Exp. Psychol. [Gen.]* 118:148–164.

COOPER, L. A., and R. N. SHEPARD, 1975. Mental transformations in the identification of left and right hands. *J. Exp. Psychol. [Hum. Percept.]* 1:48–56.

COREN, S., 1986. An efferent component in the visual perception of direction and extent. *Psychol. Rev.* 93:391–410.

DECETY, J., B. PHILIPPON, and D. H. INGVAR, 1988. rCBF landscapes during motor performance and motor ideation of a graphic gesture. *Eur. Arch. Psychiatry Neurol. Sci.* 238:33–38.

DEUTSCH, G., W. T. BOURBON, A. C. PAPANICOLAOU, and H. M. EISENBERG, 1988. Visuospatial experiments compared via activation of regional cerebral blood flow. *Neuropsychologia* 26:445–452.

DOUGLAS, K. L., and K. S. ROCKLAND, 1992. Extensive visual feedback connections from ventral inferotemporal cortex. *Soc. Neurosci. Abs.* 18:390.

DROULEZ, J., and A. BERTHOZ, 1990. The concept of dynamic memory in sensorimotor control. In *Freedom to Move: Dissolving Boundaries in Motor Control*, D. R. Humphrey and H. J. Freund, eds. Chichester, England: Wiley.

FARAH, M. J., 1988. Is visual imagery really visual? Overlooked evidence from neuropsychology. *Psychol. Rev.* 95:307–317.

Felleman, D. J., and D. C. Van Essen, 1991. Distributed hierarchical processing in the primate cerebral cortex. *Cerebral Cortex* 1:1–47.

Finke, R. A., 1979. The functional equivalence of mental images and errors of movement. *Cognitive Psychol.* 11:235–264.

Finke, R. A., 1989. *Principles of Mental Imagery*. Cambridge, Mass.: MIT Press.

Finke, R. A., and G. C.-W. Shyi, 1988. Mental extrapolation and representational momentum for complex implied motions. *J. Exp. Psychol. [Learn. Mem. Cogn.]* 14:112–120.

Freyd, J. J., 1987. Dynamic mental representations. *Psychol. Rev.* 94:427–438.

Georgopoulos, A. P., J. T. Lurito, M. Petrides, A. B. Schwartz, and J. T. Massey, 1989. Mental rotation of the neuronal population vector. *Science* 24:234–236.

Gould, S. J., and R. C. Lewontin, 1979. The spandrels of San Marco and the Panglossian paradigm: A critique of the adaptationist programme. *Proc. R. Soc. Lond. [Biol.]* 205:581–598.

Gregory, R. L., 1970. *The Intelligent Eye*. London: Weidenfeld and Nicholson.

Hinton, G. E., J. L. McClelland, and D. E. Rumelhart, 1986. Distributed processing. In *Parallel Distributed Processing: Explorations in the Microstructure of Cognition*. Vol. 1, *Foundations*, D. E. Rumelhart, and J. L. McClelland, eds. Cambridge, Mass.: MIT Press, pp. 77–109.

Hochberg, J., 1981. On cognition in perception: Perceptual coupling and unconscious inference. *Cognition* 10:127–134.

Hochberg, J., 1982. How big is a stimulus? In *Organization and Representation in Perception*, J. Beck, ed. Hillsdale, N.J.: Erlbaum, pp. 191–217.

James, W., 1890. *Principles of Psychology*. New York: Holt.

Kalaska, J. F., and D. J. Crammond, 1992. Cerebral cortical mechanisms of reaching movements. *Science* 255:1517–1522.

Koriat, A., J. Norman, and R. Kimchi, 1991. Recognition of rotated letters: Extracting invariance across successive and simultaneous stimuli. *J. Exp. Psychol. [Hum. Percept.]* 17:444–457.

Kosslyn, S. M., 1994. *Image and Brain: The Resolution of the Imagery Debate*. Cambridge, Mass.: MIT Press.

Kosslyn, S. M., N. M. Alpert, W. L. Thompson, V. Maljkovic, S. B. Weiss, C. F. Chabris, S. E. Hamilton, S. L. Rauch, and F. S. Buonanno, 1993. Visual mental imagery activates topographically organized visual cortex: PET investigations. *J. Cognitive Neurosci.* 5:263–287.

Kosslyn, S. M., R. A. Flynn, J. B. Amsterdam, and G. Wang, 1990. Components of high-level vision: A cognitive neuroscience analysis and accounts of neurological syndromes. *Cognition* 34:203–277.

Kosslyn, S. M., and O. Koenig, 1992. *Wet Mind: The New Cognitive Neuroscience*. New York: Free Press.

Lamotte, R. H., and C. Acuna, 1978. Defects in accuracy of reaching after removal of posterior parietal cortex in monkeys. *Brain Res.* 13:309–326.

Larsen, A., and C. Bundesen, 1978. Size scaling in visual pattern recognition. *J. Exp. Psychol. [Hum. Percept.]* 4:1–20.

Livingstone, M., and D. Hubel, 1988. Segregation of form, color, movement, and depth: Anatomy, physiology, and perception. *Science* 240:740–749.

Lowe, D. G., 1985. *Perceptual Organization and Visual Recognition*. Boston: Kluwer.

Lowe, D. G., 1987a. Three-dimensional object recognition from single two-dimensional images. *Artif. Intel.* 31:355–395.

Lowe, D. G., 1987b. The viewpoint consistency constraint. *Int. J. Comput. Vis.* 1:57–72.

Meador, K. J., D. W. Loring, D. Bowers, and K. M. Heilman, 1987. Remote memory and neglect syndrome. *Neurology* 37:522–526.

Mel, B. W., 1991. A connectionist model may shed light on neural mechanisms for visually guided reaching. *J. Cognitive Neurosci.* 3:273–292.

Muensterberg, H., 1914. *Psychology, General and Applied*. New York: Appleton.

Mumford, D., 1991. On the computational architecture of the neocortex: I. The role of the thalamo-cortical loop. *Biol. Cybern.* 65:135–145.

Neisser, U., 1967. *Cognitive Psychology*. New York: Appleton.

Neisser, U., 1976. *Cognition and Reality*. San Francisco: W. H. Freeman.

Noton, D., and L. Stark, 1971. Scanpaths in saccadic eye movements while viewing and recognizing patterns. *Vision Res.* 11:929–942.

Parsons, L. M., 1987a. Imagined spatial transformations of one's body. *J. Exp. Psychol. [Gen.]* 116:172–191.

Parsons, L. M., 1987b. Imagined spatial transformations of one's hands and feet. *Cognitive Psychol.* 19:178–241.

Perenin, M. T., and A. Vighetto, 1983. Optic ataxia: A specific disorder in visuomotor coordination. In *Spatially Oriented Behavior*, A. Hein, and M. Jeannerod, eds. New York: Springer-Verlag, pp. 305–326.

Prablanc, C., D. Pelisson, and M. A. Goodale, 1986. Visual control of reaching movements without vision of the limb. *Exp. Brain Res.* 62:293–302.

Reeke, G. N., Jr., L. H. Finkel, O. Sporns, and G. M. Edelman, 1989. Synthetic neural modeling: A multilevel approach to the analysis of brain complexity. In *Signal and Sense: Local and Global Order in Perceptual Maps*, G. M. Edelman, W. E. Gall, and W. M. Cowan, eds. New York: Wiley.

Rockland, K. S., and A. Virga, 1989. Terminal arbors of individual "feedback" axons projecting from area V2 to V1 in the macaque monkey: A study using immunohistochemistry of anterogradely transported *phaseolus vulgaris*-leucoagglutinin. *J. Comp. Neurol.* 285:54–72.

Rumelhart, D. E., and J. L. McClelland, eds., 1986. *Parallel Distributed Processing: Explorations in the Microstructure of Cognition*. Vol. 1, *Foundations*. Cambridge, Mass. MIT Press.

Sakata, H., M. Taira, S. Mine, and A. Murata, 1992. The hand-movement-related neurons of the posterior parietal cortex of the monkey: Their role in the visual guidance of hand movement. In *Control of Arm Movement in Space: Neurophysiological and Computational Approaches*, R. Caminiti, P. Johnson, and Y. Burnod, eds. New York: Springer-Verlag.

SEKIYAMA, K., 1982. Kinesthetic aspects of mental representations in the identification of left and right hands. *Percept. Psychophys*. 32:89–95.

SEKIYAMA, K., 1983. Mental and physical movements of hands: Kinesthetic information preserved in representational systems. *Jpn. Psychol. Res*. 25:95–102.

SEKULER, R., and D. NASH, 1972. Speed of size scaling in human vision. *Psychon. Sci*. 27:93–94.

SHEPARD, R. N., 1984. Ecological constraints on internal representation: Resonant kinematics of perceiving, imagining, thinking, and dreaming. *Psychol. Rev*. 91:417–447.

SHEPARD, R. N., and L. A. COOPER, 1982. *Mental Images and Their Transformations*. Cambridge, Mass.: MIT Press.

STARK, L., and S. ELLIS, 1981. Scanpaths revisited: Cognitive models direct active looking. In *Cognition and Visual Perception*, D. Fisher, R. Monty, and J. Senders, eds. Hillsdale, N.J.: Erlbaum, pp. 193–226.

TAIRA, M., S. MINE, A. P. GEORGOPOULOS, A. MURATA, and H. SAKATA, 1990. Parietal cortex neurons of the monkey related to the visual guidance of hand movement. *Exp. Brain Res*. 83:29–36.

WEIMER, W. B., 1977. A conceptual framework for cognitive psychology: Motor theories of the mind. In *Perceiving, Acting and Knowing: Towards Ecological Psychology*, R. Shaw and J. Bransford, eds. Hillsdale, N.J.: Erlbaum, pp. 267–311.

ZEKI, S., and S. SHIPP, 1988. The functional logic of cortical connections. *Nature* 335:311–317.

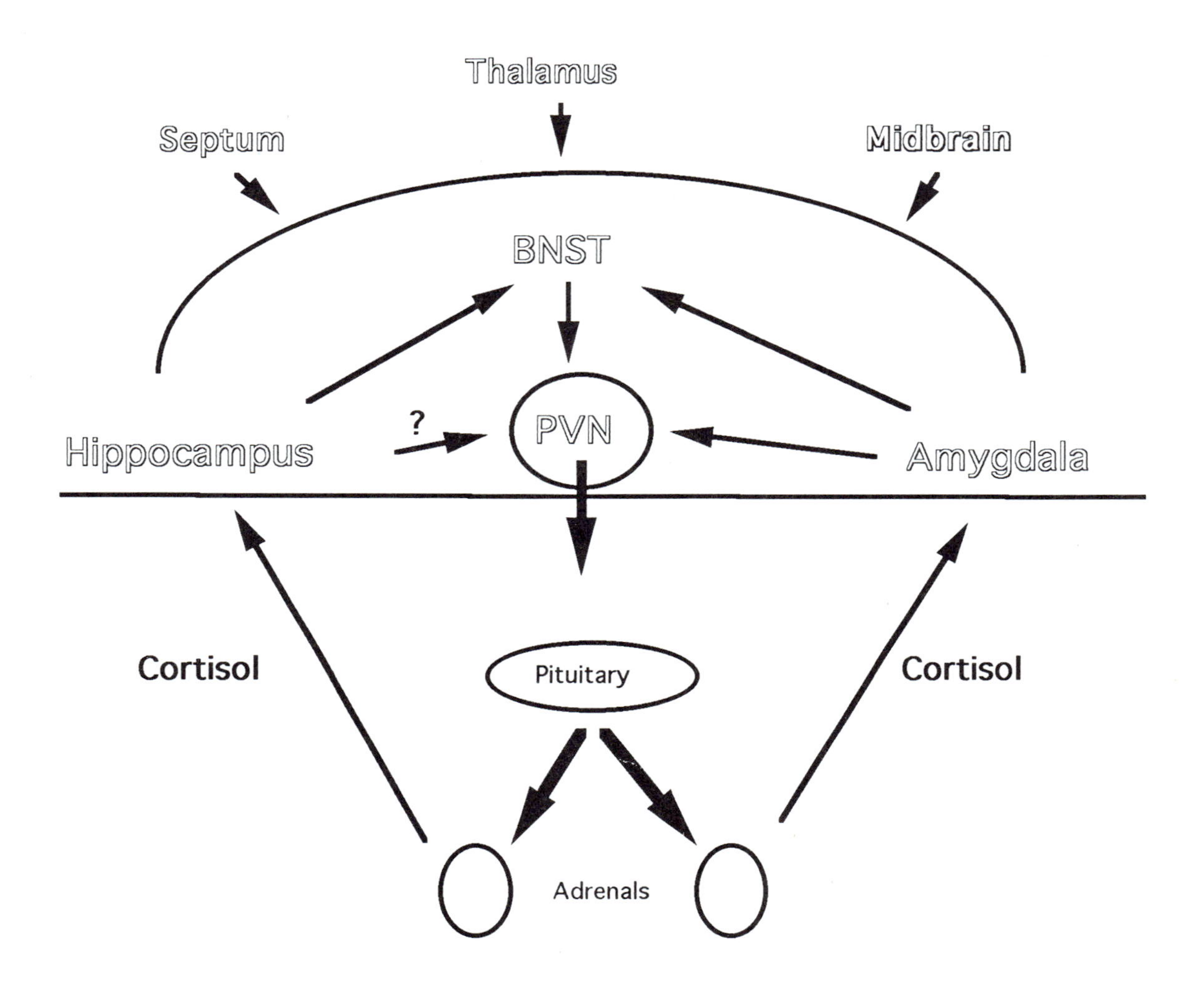

Thalamus
Septum
Midbrain
BNST
?
PVN
Hippocampus
Amygdala
Cortisol
Pituitary
Cortisol
Adrenals

IX EMOTION

Role of limbic brain structures in the regulation of hypothalamo-pituitary-adrenal activity. Arc at top indicates that exact inputs of thalamus, septum, and midbrain are not specified. Below the arc, the input of the hippocampus and amygdala are believed to involve the bed nucleus of the stria terminalis as a way station; but there are also direct inputs to the paraventricular nucleus (PVN), which is the site of production of the releasing factor CRH and of vasopressin (AVP).

Introduction

JOSEPH E. LEDOUX

EMOTION WAS A major concern of early neuroscientists working around the turn of the century, and continued to flourish as a research area through midcentury. However, research on the relation between brain and mind has, in recent years, been more concerned with cognition than with emotion. There is, in fact, no cohesive group of neuroscientists who explicitly identify themselves as primarily researching brain mechanisms of emotion. This is in part related to trends and fashions in science and the resulting funding priorities—cognitive research is more popular and more heavily funded than emotion research. It is probably also related in part to the perceived difficulty in studying a process so complex and so subjective as emotion. In fact, however, emotion is inherently no more complex or subjective than other processes that have been extensively studied, such as memory and perception.

In spite of the relative neglect of emotion by neuroscientists in the recent past, there is a growing interest in the problem of emotion and its neural organization. This is related to the long-standing and widening recognition that mind and cognition are not the same (e.g., Neisser, 1967; Hilgard, 1980), as well as the renewed sense that emotion is itself an interesting brain function (and a much less crowded research area). It is also related to the fact that discoveries about the brain mechanisms of emotion have implications for understanding and treating the various emotional disorders that affect people. The chapters in this section illustrate the breadth of the research and theorizing that is being done today. With the cognitive revolution and cognitive neuroscience now well in place, emotional neuroscience may be the next frontier of research on the relation between mind and brain.

The first chapter, by LeDoux, outlines what is known about the neural systems mediating emotion. LeDoux argues that we are not ready for global theories of emotion and its relation to brain mechanisms and instead need to work on specific, experimentally tractable aspects of the problem. He uses learned fear as a model system for exploring brain mechanisms of emotion and discusses the implications for understanding how emotional behavior is controlled and how subjective emotional states of consciousness arise. The centerpiece of this emotion system is the amygdala.

Bloom takes on the problem of identifying some of the cellular mechanisms that might underlie emotional processes. He surveys some general principles of cellular functioning, particularly mechanisms of interneuronal signaling and transduction, and then discusses how these mechanisms might be operative in specific circuits (especially within the amygdala) that have been implicated in emotional processes through behavioral studies. Implications for emotion and health are also considered.

Rolls presents a theory of emotion based on his research on the neurophysiology of emotional processes in primates. The theory, in brief, is that emotions are states produced by reinforcing stimuli. Unit recording studies aimed at identifying sites in the brain involved in processing reinforcing stimuli are described. The research points to the amygdala and the orbitofrontal cortex as key areas in the brain's system for computing the reinforcing value of stimuli.

Weinberger also utilizes single-unit recording procedures to study emotional processes. His work is focused on determining how emotional events can alter the processing of sensory stimuli. Specifically, he shows that processing of sensory signals is altered when those signals occur in temporal association with aversive events. The retuning is specific to the signal and appears to be related to interactions between thalamocortical and cholinergic pathways in the cortex. The amygdala is postulated to play a key role as the source of activation of the cholinergic inputs to the cortex.

Brothers takes the neurophysiology of emotion to the more complex level of social interactions. She argues that social responses involve brain areas that have links with effector systems, on the one hand, and access to highly processed multisensory activity, on the other. She describes several brain regions that meet these criteria, including the amygdala, orbital cortex, and anterior cingulate cortex. She then suggests how these brain systems might mediate the perception of the intentions of others and considers the implications of her ideas for understanding autism.

The topic of stress is the focus of McEwen's chapter. He examines individual differences in the physiological response to psychological stress, and the neural systems involved in processing stress signals and in responding to the physiological response to stress. The adverse effects of stress on brain and body are discussed.

Halgren and Marinkovic's research on depth recordings from epileptic patients is the basis for their chapter on human emotional neurophysiology. They postulate the existence of an emotional system consisting of an orienting complex, event integration, and response selection. A cortico-limbic-brainstem network is described as the neural instantiation of these processes.

Drevets and Raichle present their findings on the use of PET imaging to study emotion-related processes in the brains of humans. The chapter focuses on their work on familial pure depressive disease. Comparisons are made between depressed and normative states, depressed and remitted states, and untreated depressed versus treated states. Additionally, comparisons are made between depressed subjects and normal subjects in whom sadness is experimentally induced. Depression and anxiety are also compared.

The neuropsychology of anxiety and schizophrenia is the topic of Gray's chapter. He describes his limbic system–basal ganglia theory of these processes. These systems are said to be involved in the attainment of goals. The limbic system (hippocampus and amygdala) is particularly involved in the recognition of goals and the evaluation of outcomes of action, while the basal ganglia are concerned with the execution of motor programs. The model is applied to anxiety and schizophrenia.

REFERENCES

HILGARD, E. R., 1980. The trilogy of mind: Cognition, affection, and conaton. *J. History Behav. Sci.* 16:107–117.

NEISSER, U., 1967. *Cognitive Psychology*. New York: Appleton-Century-Crofts.

69 In Search of an Emotional System in the Brain: Leaping from Fear to Emotion and Consciousness

JOSEPH E. LEDOUX

ABSTRACT Efforts to understand the psychology and biology of emotion have resulted in only limited success. It is proposed that emotion be investigated using a model systems approach, whereby specific, experimentally tractable aspects of emotion are studied. This approach is outlined in detail for the emotion fear, the psychology and biology of which is understood better than that of any other emotion. The implications of research on fear for our understanding of the more general concept of emotion and for our understanding of the relationship between emotion and consciousness are considered.

The first step in a neurobiological approach to any mental or behavioral function is to identify the neural system that mediates the function in question. For many years, it seemed that this problem was solved for emotion by the existence of the limbic system theory of emotion (MacLean, 1952). However, in recent years, the limbic system theory of emotion has been called into question (e.g., LeDoux, 1991), as has the anatomical concept of a limbic system (Brodal, 1982; Kotter and Meyer, 1992). The neural system of emotion needs to be reexamined.

Unfortunately, when we attempt to relate emotion to brain we quickly run into a problem. There is no universally agreed upon definition or theory of emotion that delimits emotional phenomena in a way that is useful for relating those phenomena to neural systems. If we cannot define emotion, how can we hope to identify the emotion system or systems in the brain?

JOSEPH LEDOUX Center for Neural Science, New York University, New York, N.Y.

Emotion researchers might profit from a lesson learned by investigators attempting to understand the neural basis of memory. For many years, memory was conceived of as a unitary phenomenon mediated by a single neural system. This led to the search for *the* engram (Lashley, 1950), a kind of mnemonic holy grail. Behavioral procedures used to examine memory were applied indiscriminately, for they were thought to all provide equal access to the "memory" system. This is now known to be completely wrong. There are multiple kinds of memory, each mediated by different neural systems, and behavioral tests of memory must be tailored to the specific kind of memory under investigation. Today, research on the neural basis of memory is centered around specific model systems of memory, each attempting to elucidate how a particular kind of memory function is organized in the brain. In retrospect, it is clear why there were so many contradictory findings about what memory is and how it is organized in the brain, for diverse phenomena were being treated as if they were related, but they were not.

In research and theorizing on emotion, global questions about emotion are often the main concern. Findings on fear are integrated with findings on love, aggression, and empathy in efforts to explain emotion. This approach has led to some progress but has on the whole been disappointing in explaining what emotion is and how it is represented in the brain. Following the example from memory research, it might be useful to focus on specific, well-defined, and experimentally addressable aspects of emotion, through the use of model systems of emotion, rather than on the global concept of emotion. This might allow progress on important issues that otherwise might be left unaddressed, but, in the end, it may also help to illuminate the more global problem.

In the following, my aim is to describe research on the neural basis of fear, and especially the neural basis of conditioned or learned fear. I first defend fear as a model emotion, and describe and rationalize fear conditioning as a model systems approach to the study of fear. I then discuss the current understanding of the neural system of fear conditioning, and compare fear conditioning and its neural circuitry to other neurobehavioral approaches to studying fear. Finally, I consider the implications of research on fear conditioning for understanding the more general concept of emotion and the relationship between emotion and consciousness.

Fear as a model emotion

Fear is an emotion that is particularly important to human and animal existence. It is a normal reaction to threatening events and is a common part of life. A number of psychiatric conditions that afflict humans represent disorders of fear regulation, including anxiety, phobia, panic, and posttraumatic stress disorders. Although there is considerable disagreement among emotion taxonomists as to which emotions make up a core set of so-called basic emotions, most if not all emotion lists include fear. Fear is also particularly amenable to laboratory experimentation (it can readily be elicited, and a number of quantitative measures of fear can be made). Fear is probably the emotion that has been studied the most, both by psychologists and by neuroscientists. If through studies of fear we were able to improve our understanding of fear alone, rather than of emotion in general, the more modest goal would itself be an important achievement.

Fear conditioning as an experimental approach to fear

A commonly used technique for studying fear involves the procedures of aversive classical conditioning, otherwise known as fear conditioning. For example, if a tone is presented to a rat several times in association with an aversive event, such as a mild electric shock to the feet, the rat will, upon exposure to the tone in the absence of the shock, exhibit characteristic signs of fear: The rat will freeze and its blood pressure and heart rate will be elevated in very stereotypical ways. The tone is referred to a conditioned stimulus (CS), the shock as an unconditioned stimulus (US), and the behavioral and autonomic expressions as conditioned responses (CRs).

It is important to emphasize that the CRs are not themselves learned. The CRs elicited by CSs are often hard-wired (genetically specified) responses that occur whenever the animal is threatened. For example, laboratory-bred rats who have never seen a cat will freeze if they encounter a cat (Blanchard and Blanchard, 1972), just as they freeze when they hear a tone that has been paired with shock (Bouton and Bolles, 1980; LeDoux, 1990). Fear conditioning is stimulus learning, not response learning. That is, through fear conditioning, novel environmental events can gain access to neural circuits that control hard-wired responses. The hard-wired responses are naturally activated by stimuli that have acquired meaning through evolutionary learning by the species rather than through individual learning. Fear conditioning opens up this channel of evolutionarily perfected responsivity to new environmental events.

Conditioned fear learning occurs quickly (in as little as a single trial) and is very difficult to extinguish. Moreover, since the stimuli used (brief tones or flashing lights) are not likely to have significance to the subject outside of the experimental setting, the subject's prior history with the stimulus is not a factor that greatly affects the reaction that is expressed as a result of conditioning. These qualities (rapid learning, enduring effects, and emotional neutrality of the conditioned stimuli) make fear conditioning an especially powerful technique for studying the neural basis of emotion and emotional learning and memory in animals (e.g., Kapp, Pascoe, and Bixler, 1984; Gentile et al., 1986; Davis, Hitchcock, and Rosen, 1987; LeDoux, 1987, 1990, 1992a; Kapp et al., 1990; Davis, 1992; Kapp et al., 1992; Kimand Fanselow, 1992) as well as in humans (e.g., Öhman, Erixon, and Lofberg, 1975).

Neural systems and cellular mechanisms of fear conditioning

Studies conducted over the past 15 or so years have made significant progress in identifying the neural circuits and some of the cellular mechanisms involved in fear conditioning in mammals. The work has involved several different experimental preparations and conditioned response measures: heart rate responses in restrained rabbits, blood pressure and freezing responses in freely behaving rats, and potentiation of startle responses by conditioned fear stimuli in rats (for reviews, see Kapp et al., 1990; LeDoux, 1990, 1992a; Davis, 1992a). In spite of this diversity, one fact has emerged as unequivocal: The amygdala is an essential part of the fear-conditioning circuitry. Below, the neural circuitry of fear conditioning is described, and then some mechanisms that might contribute to fear learning through this circuitry are discussed.

Neural Circuitry of Fear Conditioning The contribution of the amygdala to fear conditioning became apparent in the late 1970s when Kapp and colleagues demonstrated that lesions of the central nucleus of the amygdala (ACe) interfered with the conditioning of fear reactions, as measured by heart-rate CRs (see Kapp et al., 1984). Subsequent work in each of the other model systems described above implicated ACe in fear conditioning (see Kapp et al., 1990; LeDoux, 1990; Davis, 1992a). Further, projections from ACe to different target regions in the brain stem were shown to be involved in the expression of conditioned responses through different response modalities: Projections to the dorsal motor nucleus of the vagus are involved in conditioned bradycardia (Kapp et al., 1990); projections to lateral hypothalamus, which projects to the tonic vasomotor center in the rostral ventral medulla, in conditioned blood pressure increases (LeDoux, Farb, and Ruggiero, 1990); projections to the central gray in conditioned freezing (LeDoux, Farb, and Ruggiero, 1990); projections to the reticulopontis caudalis in potentiation of startle (Davis, 1992); projections to the bed nucleus of the stria terminalis, which projects to the endocrine control regions of the paraventricular and supraoptic nuclei of the hypothalamus, in conditioned release of "stress hormones" from the pituitary gland (Gray et al., in press). The central nucleus is thus a key interface with motor systems involved in the expression of conditioned fear reactions through various response modalities, some of which are depticted in figure 69.1.

An important question raised by these observations is, How does CS information reach ACe and thereby activate these varied response systems? The amygdala receives inputs from cortical association areas associated with the major sensory modalities (reviewed by Amaral et al., 1992; Turner, Mishkin, and Knapp, 1980). While removal of the auditory cortex does not interfere with the conditioning of fear reactions to auditory stimuli, damage to the medial geniculate body or inferior colliculus completely abolishes this capacity (LeDoux, Sakaguchi, and Reis, 1984). The medial geniculate body also has direct projections to the amygdala (LeDoux, Ruggiero, and Reis, 1985), and interruption of these connections interferes with conditioning (summarized in LeDoux et al., 1990). The lateral nucleus of the amygdala (AL) is the main recipient of acoustic inputs from the auditory thalamus, and selective destruction of this region interferes with conditioning (see LeDoux, Farb, and Ruggiero, 1990; LeDoux, 1992a). Although AL has only sparse if any direct connections with ACe, it projects to the basolateral and basomedial nuclei of the amygdala (Stefanacci et al., 1992), both of which project to ACe (Amaral et al., 1992).

Together, the various findings described thus far provide a description of neural pathways that are capable of bringing auditory CS inputs from the auditory thalamus to AL, from AL to ACe by way of intra-amygdala connections, and from ACe to various brainstem areas involved in the control of specific CRs (figure 69.2). This is thus a complete circuit description, from sensory to motor neurons, of a fear-conditioning circuit involved when an auditory stimulus is paired with foot shock. Conditioning with visual stimuli involves a similar, though less clearly defined, set of connections (LeDoux, 1990; Davis, 1992b).

Several additional points should be considered. First, thalamic projections to the amygdala are sufficient to mediate simple fear conditioning (one tone paired with a shock), but so are cortical projections (Romanski and LeDoux, 1992). Second, if the conditioning paradigm involves discrimination training, where one CS is paired with shock and a different one is not, cortical lesions interfere with conditioning (Jarrell et al., 1987). However, the lesions do not eliminate the presence of a response to the CS but instead result in responses to both the CS+ and CS− (this suggests that thalamic projections to amygdala, the main pathways left when the cortex is damaged, are unable to distinguish the two stimuli, and thus both stimuli elicit CRs). AL receives inputs from cortical sensory association regions as well as from sensory thalamus, and thus is likely to be the amygdala interface with both thalamic and cortical sensory systems (see LeDoux, 1992a). Third, when rats are conditioned

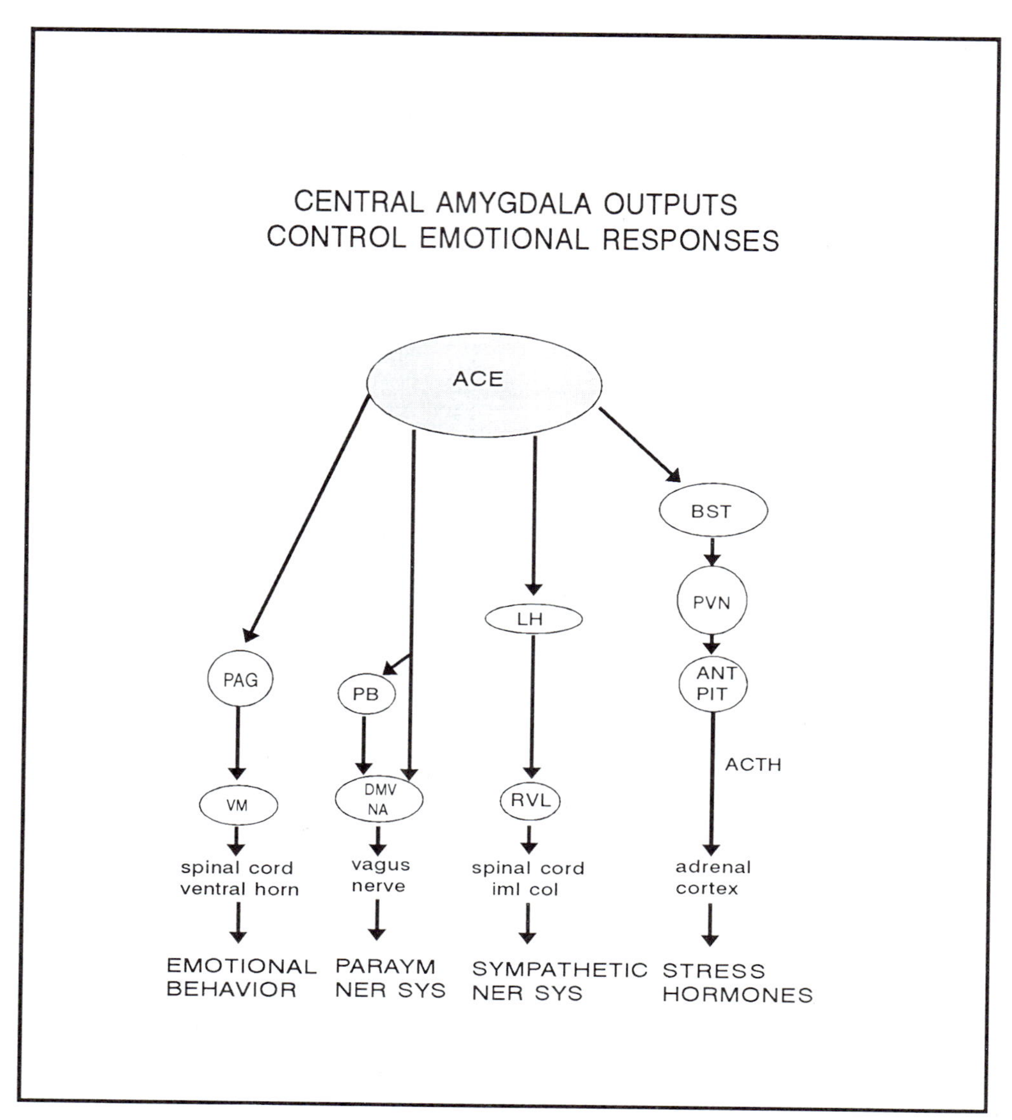

FIGURE 69.1 Some of the efferent projections of the central amygdaloid nucleus that have been implicated in the control of emotional responses elicited by conditioned fear stimuli are illustrated. ACE, central nucleus of the amygdala; ANT PIT, anterior lobe of the pituitary gland; BST, bed nucleus of the stria terminalis; DMV, dorsal motor nucleus of the vagus; iml col, intermediolateral column of the spinal cord; LH, lateral nucleus of the hypothalamus; NA, nucleus ambiguous; PAG, periaquiductal gray matter; PB, parabrachial nucleus; RVL, rostral ventral lateral medulla; VM, ventral medulla.

they develop fear reactions to the chamber in which the shocks occur, as well as the to CS itself. Lesions of the hippocampus (and amygdala) interfere with the conditioning of fear to the chamber (contextual conditioning), but lesions of the hippocampus have no effect on conditioning to the CS (Kim and Fanselow, 1992; Phillips and LeDoux, 1992a). The hippocampal formation, by way of the CA1 and subiculum areas, also projects to AL (Phillips and LeDoux, 1992b). These connections may mediate contextual conditioning, allowing complex stimulus representations to gain access to fear responses controlled through the amygdala. Fourth, extinction of learned fear appears to be mediated by cortical projections to the amgydala (LeDoux,

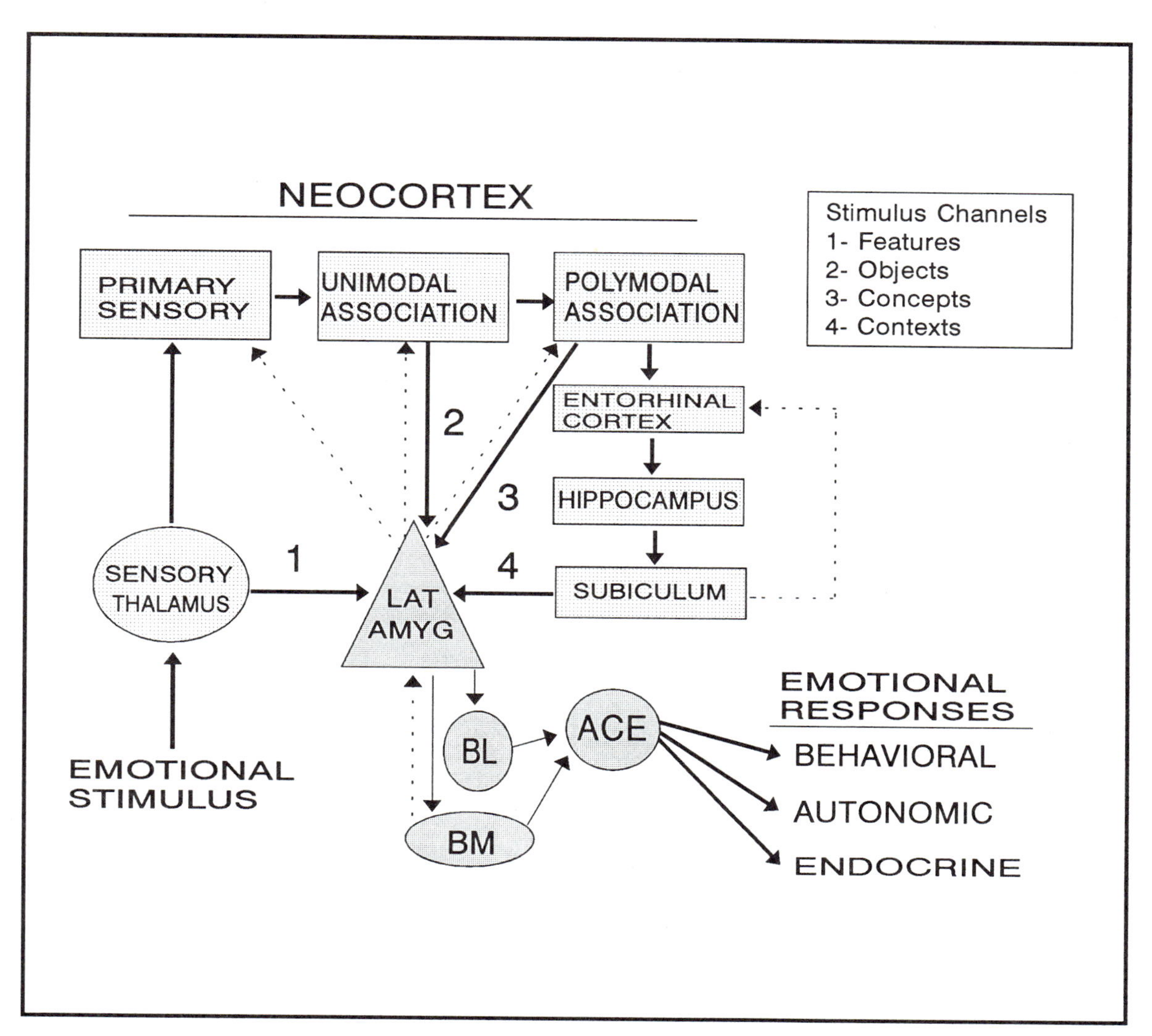

FIGURE 69.2 Amygdala pathways and fear conditioning. The lateral nucleus of the amygdala (LAT AMYG) receives inputs from sensory processing areas in the thalamus (1) and neocortex (2), as well as inputs from higher-order association regions in the neocortex (3) and hippocampal formation (4). During fear conditioning, the amygdala may process information in parallel from these various channels. Fear conditioning can be mediated by pathway 1 or 2 when a simple, undiscriminated cue serves as the CS, whereas pathway 2 is required when closely related stimuli must be discriminated. Pathway 4 is involved in the conditioning of fear responses to contextual stimuli. Pathway 3, from the medial prefrontal cortex to the amygdala, appears to play a role in extinction. Within the amgydala, information is relayed from the lateral nucleus to the central nucleus (ACE) by way of intermediate connections in the basolateral (BL) and basomedial (BM) nuclei. Outputs of ACE are involved in the control of specific emotional responses.

Romanski, and Xagoraris, 1989)—specifically, projections from the medial prefrontal area (Morgan, Romanski, and LeDoux, 1993).

Thus, the amygdala is a key structure in fear conditioning, regardless of the nature of the information that enters into the association with the US. AL, in particular, appears to be especially important as the sensory interface of the amygdala, receiving crude stimulus information from the thalamus, perceptual information from the cortex, and higher-order information from the hippocampal formation (figure 69.2). Through connections of AL to ACe, each of these forms of stimulus information can gain access to emotional response mechanisms.

Cellular Mechanisms of Fear Conditioning The neural modifications underlying fear conditioning are likely to occur in synaptic connections of the CS processing pathway. The US somehow changes the strength of connections in the pathway, allowing the CS to have an amplified effect. If this view of learning is correct, elucidation of how the brain processes auditory CS information before conditioning should provide a strong foundation upon which to assess changes in CS processing induced by the US. In recent studies, we have therefore examined the responses of neurons in AL, the sensory interface of the amygdala, to auditory stimulation (Bordi and LeDoux, 1992, 1993; Romanski et al., 1993). AL cells have short latency responses to auditory stimuli. The latencies can be explained only in terms of direct transmission from the thalamus, since the earliest latencies in AL are about the same as the earliest latencies in auditory cortex. Second, some cells habituate quickly to repetition of a stimulus, but respond afresh if the stimulus is changed. The habituation can last for minutes, suggesting that the cells retain (remember) some specific information about the stimulus. Third, other cells respond to repeated stimulation, and some of these are fairly narrowly tuned to specific frequencies of tones. Fourth, when tuned, AL cells almost exclusively prefer high-frequency stimulation (16–30 kHz tones). Fifth, AL cells have relatively high thresholds (about 30–50 dB above the primary auditory system). Finally, almost every cell that responded to auditory stimuli also responded to somatosensory stimuli. Some of these auditory response properties are illustrated in figure 69.3.

These findings suggest insights into the neural organization of fear and fear learning. First, habituating cells in AL are novelty detectors. Novel stimuli that have no consequence are not fertile emotional stimuli, and AL learns to ignore them. Fear conditioning might alter this tendency to habituate, allowing the cell to continue to respond to significant information that is repeated. Second, some AL cells are hard-wired to respond to loud stimuli. Loudness is a clue to distance, and things that are close are more likely to be dangerous than things that are far away. This is a useful, albeit primitive, computation for a fear system to perform. Fear conditioning may change the threshold for these cells, allowing less intense but meaningful stimuli to activate them. Third, the preference for high-frequency stimuli is interesting in light of recent studies showing that when rats are threatened they tend to emit warning calls in the same frequency range (Blanchard et al., 1991). Other species may have similar stimulus-specific responses. For example, the sound of a baby crying is believed to be an innate releaser in humans, and it is conceivable that spectral components embedded in baby cries might have a particularly strong representation in the human AL. Several studies conducted in humans have shown that so-called prepared stimuli produce particularly strong conditioned responses (Öhman, Erixon, and Lofberg, 1975). Fear conditioning might therefore modify the ease with which prepared stimuli gain access to response systems through the amygdala. Finally, the fact that almost all of the AL cells that respond to sounds also respond to somatosensory stimuli suggests that almost all of the cells in this region are potentially modified by the pairing of stimuli from these two modalities in conditioning studies.

The actual mechanisms through which fear conditioning or any other form of learning takes place have not been clearly elucidated in vertebrate species. However, the most popular ideas about how learning might take place involve a phenomenon known as long-term potentiation (LTP). LTP is induced by giving tetanizing high-frequency electrical stimulation to a pathway and recording a long-lasting increase in synaptic efficacy as a result of the stimulation. Synaptic efficacy is measured by testing the magnitude of the postsynaptic response to some stimulus before and after giving the tetanizing stimulation. LTP has been most extensively studied in the hippocampal formation, but occurs in a variety of other areas of the brain as well. LTP is attractive as a model of learning because it is an experience-dependent form of synaptic plasticity, it involves cooperative activity of afferents, and it is specific to the synapses activated during the tetanization (e.g., Brown et al., 1988).

The cellular basis of LTP in the hippocampus has been worked out in exquisite detail (see Brown et al., 1988; Madison, Malenka, and Nicoll, 1991). Although there are several forms of LTP in hippocampus, the one that has attracted the most interest as a model of associative learning is referred to as NMDA-dependent LTP. During tetanization, presynaptically released glutamate (Glu) binds to NMDA and non-NMDA excitatory amino acid receptors, but is prevented from opening the NMDA channel by a magnesium block at resting membrane potentials. The binding to non-NMDA receptors depolarizes the membrane, removes

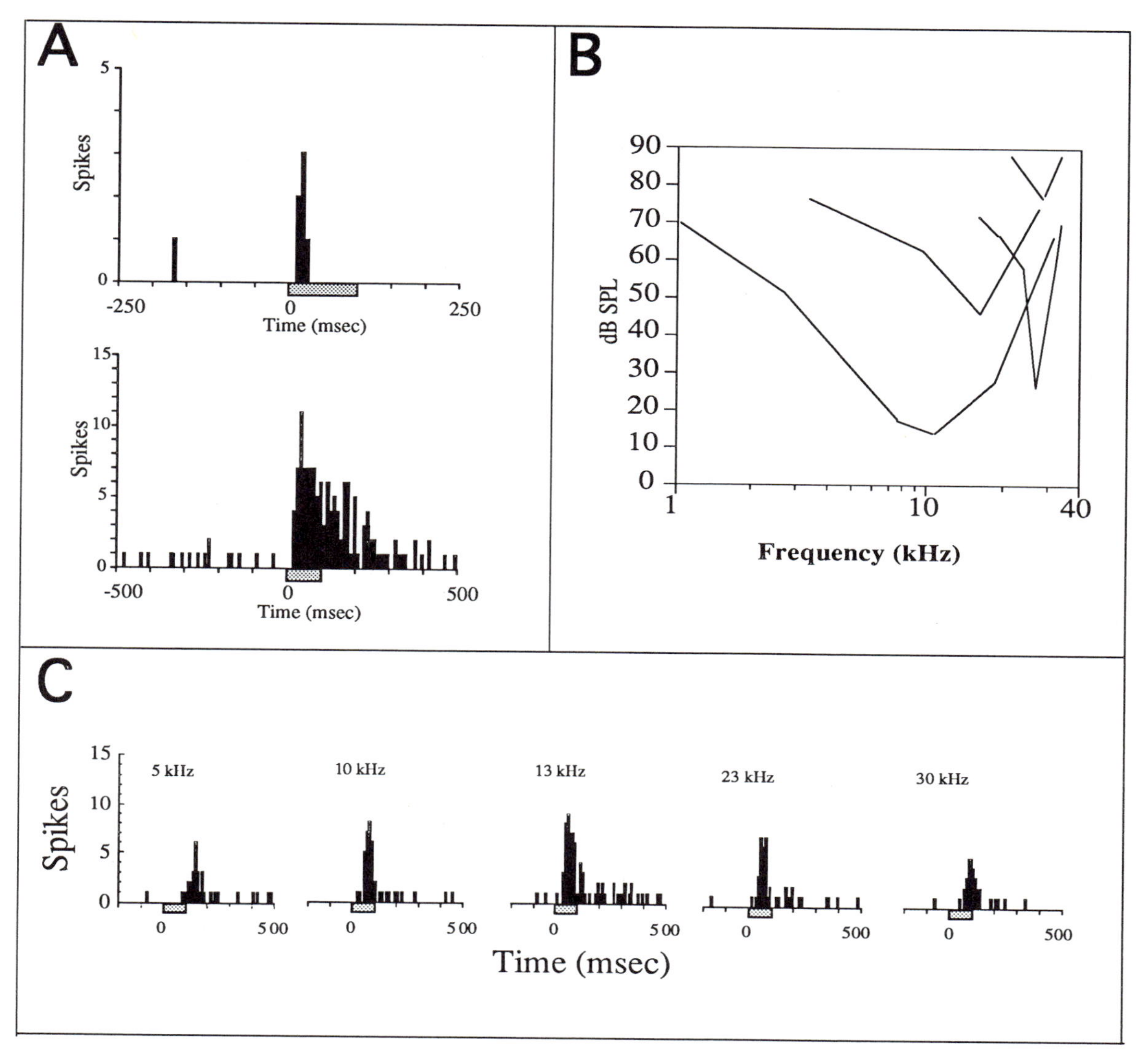

FIGURE 69.3 Auditory processing in lateral amygdala. (A) Typical, short-latency responses of lateral amygdala cells to auditory stimuli (bursts of white noise). (B) Frequency-threshold (tuning) curves of lateral amygdala cells. Note the relatively high intensity thresholds of the cells, the preference for higher frequencies, and the fairly broad bandwidth of responsivity, especially at higher intensities. (C) Histograms of responses summed over 10 sweeps of pure tones of different frequencies at about 30 dB above threshold, illustrating the preference of this cell for frequencies around 13 kHz. (Based on findings reported in Bordi and LeDoux, 1992, and Bordi et al., 1993).

the block, and allows Glu bound to NMDA receptors to open the NMDA channel. As a result, calcium enters the postsynaptic cell; this is believed to be an important step in triggering LTP.

LTP has been induced in the amygdala via tetanizing stimulations to the thalamo-amygdala projection (Clugnet and LeDoux, 1990; Rogan and LeDoux, 1993) (figure 69.4). The contribution of transmitter receptors has not been thoroughly analyzed yet. However, the cells of origin and preterminal axons of the pathway contain Glu (LeDoux, Farb, and Milner, 1991; Farb et al., 1992), and simultaneous blockade of NMDA and non-NMDA receptors interferes with synaptic responses elicited by thalamic stimulation (figure 69.5). While one study suggests that NMDA receptors are not involved in the LTP induced in amygdala by stimulation of the external capsule (Chapman and Bellavance, 1992), this conclusion is

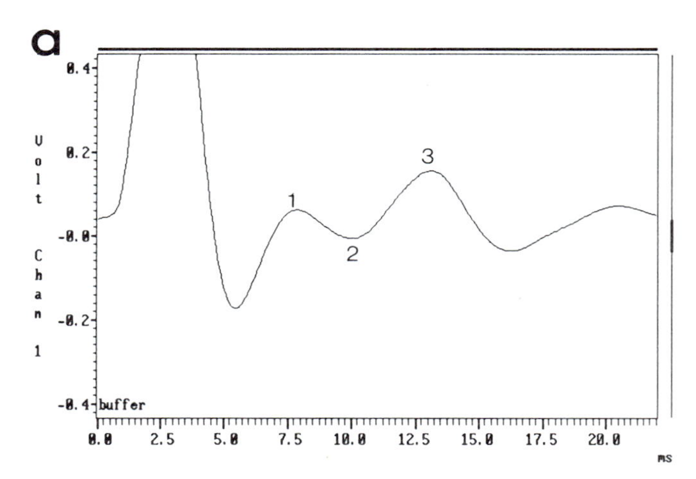

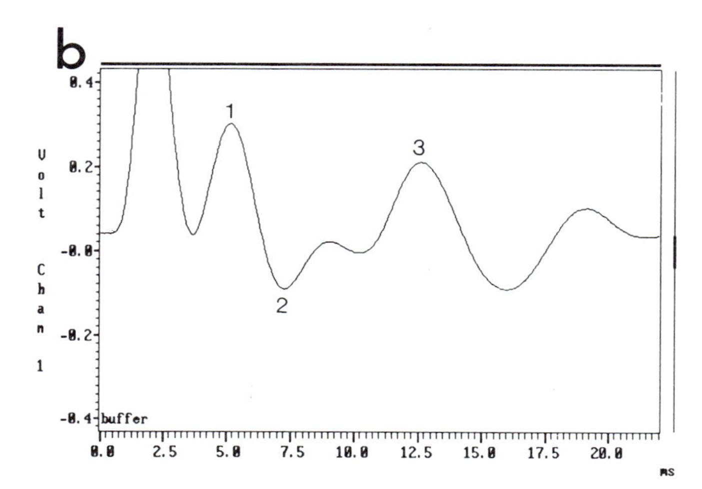

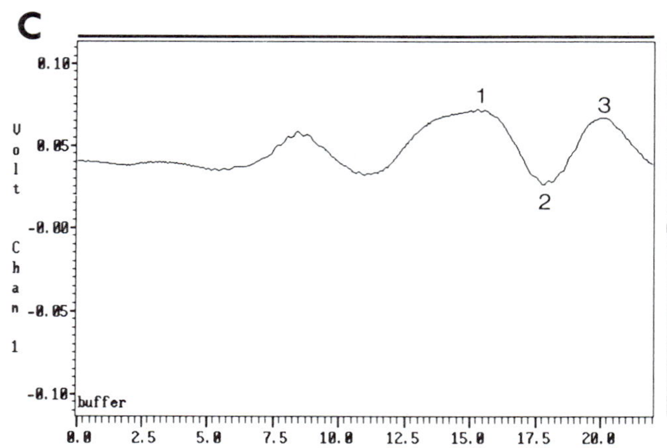

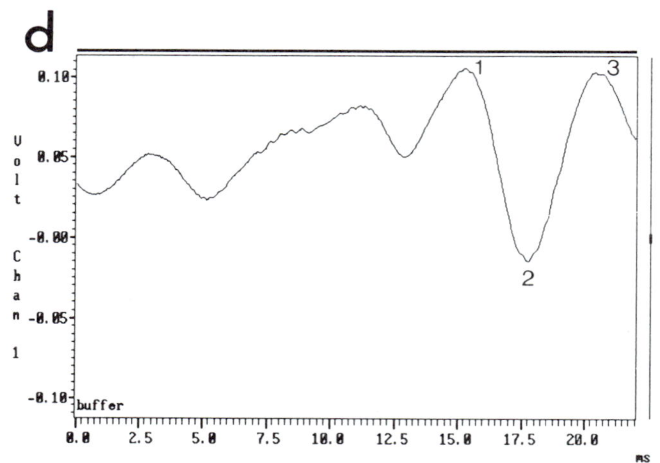

FIGURE 69.4 LTP in lateral amygdala. (a) Field potential elicited in lateral amgydala by stimulation of the medial geniculate body. (b) Field potential elicited by the same stimulus as in (a) but about one hour after delivery of tetanizing electrical stimuli to the pathway. (c) Field potential elicited in lateral amygdala by peripheral auditory stimulation. (d) Field potential elicited by the same stimulus as in (c) but about one hour after delivery of tetanizing electrical stimuli to the pathway. The increase in the amplitude of the field potential at measurement point 2, relative to point 1, in panel (b) compared to (a) illustrates thalamo-amgydala LTP. The increase at point 2, relative to point 1, in panel (d) compared to panel (c) shows that natural (auditory) stimuli can act on tetanized synapses. (Based on Rogan and LeDoux, 1993)

premature; even if it is later determined to be warranted for external capsule LTP, it does not directly apply to thalamo-amygdala LTP (see LeDoux, 1992a).

An important step toward establishing a relationship between amygdala plasticity and memory has been taken by studies showing that blockade of NMDA receptors in the lateral and basolateral amygdala interferes with fear conditioning (Miserendino et al., 1990). However, even if NMDA receptors turn out to be involved in amygdala LTP and in fear conditioning, it would remain possible that these receptors are used quite differently, and independently, during natural learning and during LTP induction.

Other approaches to fear

A number of other behavioral tasks have been used to study fear and related aversive reactions (see Gray, 1987; Handley, 1991; Everitt and Robbins, 1992;

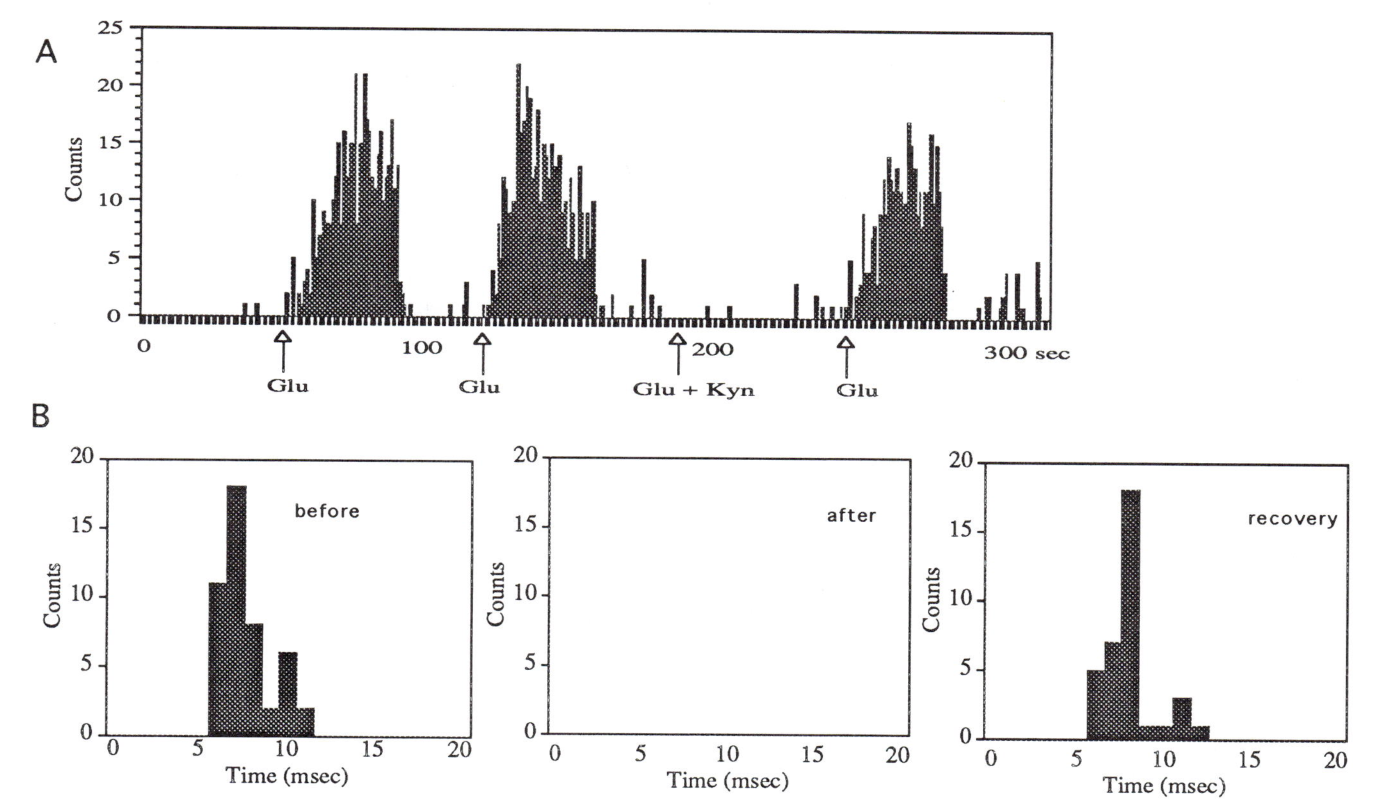

FIGURE 69.5 Excitatory amino acid transmission in thalamo-amgydala pathways. (A) Rate histogram showing excitation of a single cell in lateral amygdala to iontophoresis of glutamate (Glu) and the blockade of this evoked excitation by application of kynurenic acid (Kyn), a broad-spectrum antagonist of excitatory amino acid receptors. (B) Poststimulus histograms showing responses elicited in lateral amygdala by electrical stimulation of the medial geniculate body before and after iontophoresis of Kyn and after recovery. (Based on unpublised findings obtained by M. C. Clugnet, R. Phillips, and J. E. LeDoux, 1991)

Panksepp et al., 1992). Some of these involve new learning (such as passive and active avoidance tasks), whereas the contribution of new learning to others is minimal (open field test, response to novel situations, elevated maze performance, social interaction test). Relative to conditioned fear, each of these tasks represents a more complex methodological approach to the identification of neural systems. In fear conditioning, a discrete eliciting stimulus is used, and a stereotyped fear reaction is directly elicited by the CS. With the other fear tests, the eliciting stimulus is usually unclear, and the assessment of fear usually occurs indirectly, as it depends on the measurement of an instrumental behavioral response (which may consist of the withholding of a response, in passive-avoidance conditioning or performing a response in active avoidance).

The amygdala has been implicated in most tests of fear. However, not all variations of every task have proven to be sensitive to amygdala damage. There are many ways to structure any of these tasks, and it is important to ask, for every task and every variation of the task, whether it is truly a measure of fear. Further, for every brain lesion that interferes with performance on a potential measure of fear, it is important to distinguish the effects of the lesion on the fear part of the task, as opposed to the more general cognitive, behavioral, or motivational aspects of the task.

Some measures of fear have implicated the hippocampus and related structures such as the septum. Evidence of this type has been used to build a septohippocampal theory of fear and anxiety (Gray, 1987). This well-developed theory accounts for much of the data on the effects of septohippocampal damage and on the effects of antianxiety drugs. However, it is problematic in several respects (reviewed in LeDoux, 1987, 1992b). One problem involves its failure to account for the role of the amygdala in fear and anxiety, as revealed by studies of fear conditioning. Another problem has to do with the difficulty in distinguishing the contribution of the hippocampus to the fear part of tasks as opposed to the stimulus representation part of the task. Extrapolating from studies of fear conditioning described above, we might conclude that the hippocampus is mainly involved in fear when the task used to assess fear requires the processing of complex stimulus representations. Considerably more work is needed to resolve differences between the amygdala and septo-hippocampal theories of fear and anxiety.

From animal fear to human emotion

By focusing on conditioned fear responses in animals as a model system of emotion, we run the risk that our results will be limited in several ways. Three of these will be described.

One limitation, that our results might be applicable only to fear and not to other emotions, is not very troubling. If we could truly nail down neural mechanisms of an emotion as important as fear, that itself would be an important achievement. In any event, it seems that at least some of the findings will apply to emotions other than fear. For example, we have identified the lateral nucleus of the amygdala as the amygdala interface with the sensory world in fear conditioning. Since this structure is the primary location through which sensory information reaches the amygdala, it is likely to serve as a general sensory interface for any emotional function involving the amygdala. And since many of the emotional processes that have been studied involve the amygdala to some degree (see Aggleton and Mishkin, 1986; Everitt and Robbins, 1992; LeDoux, 1992b; Ono, 1992; Rolls, 1992), the identification of the amygdala's sensory interace for fear is an important observation with considerable generality.

A second limitation, that our results might apply only to animals, is unlikely to be a problem. There is good evidence for the conservation of fear (and some other emotions) across mammalian species (and possibly across some nonmammals as well). There are common modes of fear expression in animals and humans. The universality of the so-called fight-or-flight response and its physiological manifestations is an example. Another example is the fact that "freezing" is a commonly expressed response to threats in most if not all mammals (including humans). In addition, the amygdala is a critical part of the circuitry involved in fear in rats, as well as in nonhuman primates (e.g., Aggleton and Mishkin, 1986; Ono and Nishijo, 1992). Although it has been difficult to study the contribution of the amygdala to emotion in humans, due to ethical considerations and due to the lack of a sufficient population of patients with brain damage restricted to the amygdala, there is some evidence for the existence of similar mechanisms in humans (Halgren, 1992). Further, the fact that fear responses in animals can be used very effectively as a basis for developing drugs that relieve fear and anxiety in humans further supports the conservation of fear mechanisms across species. Conservation across species is less clear for other emotions, but this may be related to the fact that less work has been done with other emotions than with fear.

The third possible limitation, that our results will shed light on emotional responses but not on the true emotion problem—subjective emotional experience—requires some discussion. Subjective emotional experience, though important, is but one aspect of emotion that needs to be understood, and it is not necessarily the aspect that must be (or can be) understood first.

Consider the following proposal about the subjective state we call fear. Fear is a state of consciousness that can (but does not necessarily) result when the defense system of the brain (the system that detects threats and organizes responses to threats) is activated. The extent to which any given animal experiences fear when the neural system that processes threats is activated is directly related to the extent to which that animal's species possesses a well-developed capacity for conscious awareness. For species in which consciousness is robust, such as humans, subjective fear states can (but do not necessarily) occur when this system is activated. For species in which consciousness is less robust or nonexistent, subjective fear is less robust or nonexistent. The function of the defense system is not to produce fear but to cope efficiently with threats. This proposal makes no claims about how extensive consciousness is in specific nonhuman animals, but simply suggests that the extent to which subjective fear occurs is dependent upon the extent to which consciousness is present. A similar conclusion could be drawn for other instances of the various subjective states we call emotions.

If we want to understand how the subjective state, fear, comes about, we have to examine how information about the state of the defense system comes to be represented as conscious content. This means that we need to know as much as possible about how the defense system processes threats and, on the basis of such processing, produces protective responses—for these central and peripheral processes are the raw materials out of which fear is constructed in consciousness. To the extent that different subjective emotions result from activity in different systems that may process information and produce responses in different ways, it may be necessary to examine each emotion and its underlying system separately. This task may not be as onerous as it seems, as there are likely to be groups or families of emotions (Ekman, 1992) that share similar features, functional rules, and neural mechanisms.

Emotion and consciousness

Finally, we come to the question of how emotional information comes to be represented in consciousness. This is the question that has most preoccupied emotion scientists, who have typically treated emotions are special kinds of conscious experiences. I suggest that there is nothing special about subjective emotional states as states of consciousness—that humans have one mechanism of consciousness, the contents of which can be occupied by mundane events or by highly charged emotions. Because emotions are particularly intense and long-lasting states of consciousness (due to their association with central nervous system arousal and peripheral autonomic and endocrine activation), they feel different. But these differences are more likely due to the kind of information that is fed into consciousness during emotional states than to differences in the nature of the conscious mechanism itself in emotional and nonemotional states. If this is true, it is understandable why the field of emotion has had so much trouble in solving the problem of emotion—it has set as its goal the task of understanding consciousness.

The ideas outlined above suggest that there is much to be done in addition to solving the problem of consciousness, which is a general problem for the field of psychology and not a burden that rests entirely on the shoulders of emotion scientists and theorists. The study of visual perception progressed in leaps and bounds once researchers realized that they could study how the brain processes color, form, motion, and other aspects of visual perception without having to first solve the problem of how the brain experiences sensory qualia. The same is true of the field of emotion, which should be particularly concerned with understanding how various specialized brain systems process information and produce the responses that are characteristic of various classes of emotions before attempting to explain how emotions result from these processes. Since emotions are determined by the specific activity of the specialized systems, studying the specialized systems may be the only way to understand the subjective correlates that result.

This is not meant to diminish the importance of subjective emotional states. Once these occur, the mind is changed in significant ways. They dominate consciousness until they dissipate. They are powerful motivators of instrumental actions that aim to terminate or prolong the state. When we are unable effectively to cope with subjective emotional states, psychopathology can result. Paradoxically, we may need to step back from the emotions themselves in order to understand them better. That is, until we understand the underlying systems that generate specific subjective emotional states, we will not get very far in understanding how emotions arise.

Conclusion

Emotion has proven to be a slippery concept for both psychologists and neuroscientists. After much research and debate, there is still no consensus as to exactly what emotion is. Nevertheless, considerable progress has been made in understanding specific, clear instances of emotion, such as fear. In general, we may be better off pursuing these experimentally tractable aspects of emotion using model emotion systems than attempting to capture the holy grail of "emotion". The model systems approach sidesteps the important question of what emotion is, but in so doing allows us to make progress in understanding some fundamental aspects of emotion, and may someday also help us understand the essence of emotion as well.

ACKNOWLEDGMENTS This chapter was written while the author was supported by USPHS grants R37MH38774, R01MH46516, and K02MH00956, and NSF grant IBN9209646.

REFERENCES

AGGLETON, J. P., and M. MISHKIN, 1986. The amygdala: Sensory gateway to the emotions. In *Emotion: Theory, Research and Experience*, vol. 3, R. Plutchik and H. Kellerman, eds. Orlando, Fla.: Academic Press, pp. 281–299.

AMARAL, D. G., J. L. PRICE, A. PITKÄNEN, and S. T.

Carmichael, 1992. Anatomical organization of the primate amygdaloid complex. In *The Amygdala: Neurobiological Aspects of Emotion, Memory, and Mental Dysfunction*, J. P. Aggleton, ed. New York: Wiley, pp. 1–66.

Blanchard, D. C., and R. J. Blanchard, 1972. Innate and conditioned reactions to threat in rats with amygdaloid lesions. *J. Comp. Physiol. Psychol.* 81(2):281–290.

Blanchard, R. J., S. Weiss, R. Agullana, T. Flores, and D. C. Blanchard, 1991. Antipredator ultrasounds: Sex differences and drug effects. *Neurosci. Abstr.* 17.

Bordi, F., and J. LeDoux, 1992. Sensory tuning beyond the sensory system: An initial analysis of auditory properties of neurons in the lateral amygdaloid nucleus and overlying areas of the striatum. *J. Neurosci.* 12(7):2493–2503.

Bordi, F., J. E. LeDoux, M. C. Clugnet, and C. Pavlides, 1993. Single unit activity in the lateral nucleus of the amygdala and overlying areas of the striatum in freely-behaving rats: Rates, discharge patterns, and responses to acoustic stimuli. *Behav. Neurosci.* 107:757–769.

Bouton, M. E., and R. C. Bolles, 1980. Conditioned fear assessed by freezing and by the suppression of three different baselines. *Anim. Learn. Behav.* 8:429–434.

Brodal, A., 1982. *Neurological Anatomy*. New York: Oxford University Press.

Brown, T. H., P. F. Chapman, E. W. Kairiss, and C. L. Keenan, 1988. Long-term synaptic potentiation. *Science* 242:724–728.

Chapman, P. F., and L. L. Bellavance, 1992. NMDA receptor-independent LTP in the amygdala. *Synapse* 11: 310–318.

Clugnet, M. C., and J. E. LeDoux, 1990. Synaptic plasticity in fear conditioning circuits: Induction of LTP in the lateral nucleus of the amygdala by stimulation of the medial geniculate body. *J. Neurosci.* 10:2818–2824.

Davis, M., 1992a. The role of the amygdala in fear-potentiated startle: Implications for animal models of anxiety. *Trends Pharmacol. Sci.* 13:35–41.

Davis, M., 1992b. The role of the amygdala in conditioned fear. In *The Amygdala: Neurobiological Aspects of Emotion, Memory, and Mental Dysfunction*, J. P. Aggleton, ed. New York: Wiley-Liss, pp. 255–306.

Davis, M., J. M. Hitchcock, and J. B. Rosen, 1987. Anxiety and the amygdala: Pharmacological and anatomical analysis of the fear-potentiated startle paradigm. In *The Psychology of Learning and Motivation*, vol. 21, G. H. Bower, ed. San Diego: Academic Press, pp. 263–305.

Eichenbaum, H., 1992. The hippocampal system and declarative memory in animals. *J. Cog. Neurosci.* 4(3):217–231.

Ekman, P., 1992. An overview of emotion. *Paper presented at a conference on affective neuroscience, Nov. 8–10, 1992, Washington, D.C., sponsored by the McArthur Foundation and the McDonnell Foundation.*

Everitt, B. J., and T. W. Robbins, 1992. Amygdala-ventral striatal interactions and reward-related processes. In *The Amygdala: Neurobiological Aspects of Emotion, Memory, and Mental Dysfunction*, J. P. Aggleton, ed. New York: Wiley, pp. 401–429.

Farb, C., C. Aoki, T. Milner, T. Kaneko, and J. LeDoux, 1992. Glutamate immunoreactive terminals in the lateral amygdaloid nucleus: A possible substrate for emotional memory. *Brain Res.* 593:145–158.

Gentile, C. G., T. W. Jarrell, A. Teich, P. M. McCabe, and N. Schneiderman, 1986. The role of amygdaloid central nucleus in the retention of differential Pavlovian conditioning of bradycardia in rabbits. *Behav. Brain Res.* 20: 263–273.

Gray, J. A., 1987. *The Psychology of Fear and Stress*. New York: Oxford University Press.

Gray, T. S., R. A. Piechowski, J. M. Yracheta, P. A. Rittenhouse, C. L. Betha, and L. D. van der Kar, in press. Ibotenic acid lesions in the bed nucleus of the stria terminalis attenuate conditioned stress induced increases in prolactin, ACTH, and corticosterone.

Halgren, E., 1992. Emotional neurophysiology of the amygdala within the context of human cognition. In *The Amygdala: Neurobiological Aspects of Emotion, Memory, and Mental Dysfunction*, J. Aggleton, ed. New York: Wiley, pp. 191–228.

Handley, S. L., 1991. Serotonin in animal models of anxiety: The importance of stimulus and response. In *Serotonin, Sleep, and Mental Disorder*, C. Idzikowski and A. Cowan, eds. Oxford: Wrightson/Blackwell, pp. 89–115.

Jarrell, T. W., C. G. Gentile, L. M. Romanski, P. M. McCabe, and N. Schneiderman, 1987. Involvement of cortical and thalamic auditory regions in retention of differential bradycardia conditioning to acoustic conditioned stimulii in rabbits. *Brain Res.* 412:285–294.

Kapp, B. S., J. P. Pascoe, and M. A. Bixler, 1984. The amygdala: A neuroanatomical systems approach to its contributions to aversive conditioning. In *Neuropsychology of Memory*, N. Butters and L. R. Squire, eds. New York: Guilford, pp. 473–488.

Kapp, B. S., P. J. Whalen, W. F. Supple, and J. P. Pascoe, 1992. Amygdaloid contributions to conditioned arousal and sensory information processing. In *The Amygdala: Neurobiological Aspects of Emotion, Memory, and Mental Dysfunction*, J. P. Aggleton, ed. New York: Wiley, pp. 229–254.

Kapp, B. S., A. Wilson, J. Pascoe, W. Supple, and P. J. Whalen, 1990. A neuroanatomical systems analysis of conditioned bradycardia in the rabbbit. In *Learning and Computational Neuroscience: Foundations of Adaptive Networks*, M. Gabriel and J. Moore, eds. Cambridge, Mass.: MIT Press, pp. 53–90.

Kim, J. J., and M. S. Fanselow, 1992. Modality-specific retrograde amnesia of fear. *Science* 256:675–677.

Kotter, R., and N. Meyer, 1992. The limbic system: A review of its empirical foundation. *Behav. Brain Res.* 52: 105–127.

Lashley, K. S., 1950. In search of the engram. *Symp. Soc. Exp. Biol.* 4:454–482.

LeDoux, J. E., 1987. Emotion. In *Handbook of Physiology*. Sec. 1, *The Nervous System*. Vol. 5, *Higher Functions of the Brain*, part 2, V. B. Mountcastle, ed. Bethesda, Md.: American Physiological Society, pp. 419–460.

LeDoux, J. E., 1990. Information flow from sensation to emotion: Plasticity in the neural computation of stimulus value. In *Learning and Computational Neuroscience: Foundations of Adaptive Networks*, M. Gabriel and J. Moore, eds. Cambridge, Mass.: MIT Press, pp. 3–52.

LeDoux, J. E., 1991. Emotion and the Limbic System Concept. *Concepts Neurosci.* 2:169–199.

LeDoux, J. E., 1992a. Brain mechanisms of emotion and emotional learning. *Curr. Opin. Neurobiol.* 2:191–198.

LeDoux, J. E., 1992b. Emotion and the amygdala. In *The Amygdala: Neurobiological Aspects of Emotion, Memory, and Mental Dysfunction*, J. Aggleton, ed. New York: Wiley, pp. 339–352.

LeDoux, J. E., P. Cicchetti, A. Xagoraris, and L. M. Romanski, 1990. The lateral amygdaloid nucleus: Sensory interface of the amygdala in fear conditioning. *J. Neurosci.* 10:1062–1069.

LeDoux, J. E., and C. R. Farb, 1991. Neurons of the acoustic thalamus that project to the amygdala contain glutamate. *Neurosci. Lett.* 134:145–149.

LeDoux, J. E., C. R. Farb, and T. A. Milner, 1991. Ultrastructure and synaptic associations of auditory thalamo-amygdala projections in the rat. *Exp. Brain Res.* 85:577–586.

LeDoux, J. E., C. F. Farb, and D. A. Ruggiero, 1990. Topographic organization of neurons in the acoustic thalamus that project to the amygdala. *J. Neurosci.* 10:1043–1054.

LeDoux, J. E., L. M. Romanski, and A. E. Xagoraris, 1989. Indelibility of subcortical emotional memories. *J. Cog. Neurosci.* 1:238–243.

LeDoux, J. E., D. A. Ruggiero, and D. J. Reis, 1985. Projections to the subcortical forebrain from anatomically defined regions of the medial geniculate body in the rat. *J. Comp. Neurol.* 242:182–213.

LeDoux, J. E., A. Sakaguchi, and D. J. Reis, 1984. Subcortical efferent projections of the medial geniculate nucleus mediate emotional responses conditioned by acoustic stimuli. *J. Neurosci.* 4(3):683–698.

MacLean, P. D., 1952. Some psychiatric implications of physiological studies on frontotemporal portion of limbic system (visceral brain). *Electroencephalogr. Clin. Neurophysiol.* 4:407–418.

Madison, D. V., R. C. Malenka, and R. A. Nicoll, 1991. Mechanisms underlying long-term potentiation of synaptic transmission. *Annu. Rev. Neurosci.* 14:379–397.

Miserendino, M. J. D., C. B. Sananes, K. R. Melia, and M. Davis, 1990. Blocking of acquisition but not expression of conditioned fear-potentiated startle by NMDA antagonists in the amygdala. *Nature* 345:716–718.

Morgan, M. A., L. M. Romanski, and J. E. LeDoux, 1993. Extinction of emotional learning: Contribution of medial prefrontal cortex. *Neurosci. Lett.* 163:109–113.

O'Keefe, J., and L. Nadel, 1978. *The Hippocampus as a Cognitive Map*, Oxford: Clarendon Press.

Öhman, A., G. Erixon, and I. Lofberg, 1975. Phobias and preparedness: Phobic versus neural pictures as conditioned stimuli for human autonomic responses. *J. Abnorm. Psychol.* 84:41–45.

Ono, T., and H. Nishijo, 1992. Neurophysiological basis of the Klüver-Bucy syndrome: Responses of monkey amygdaloid neurons to biologically significant objects. In *The Amygdala: Neurobiological Aspects of Emotion, Memory, and Mental Dysfunction*, J. P. Aggleton, ed. New York: Wiley, pp. 167–190.

Panksepp, J., D. S. Sacks, L. J. Crepau, and B. B. Abbot, 1991. The psycho- and neurobiology of fear systems in the brain. In *Fear, Avoidance, and Phobias*, M. R. Denny, ed. Hillsdale, N.J.: Erlbaum, pp. 7–59.

Phillips, R. G., and J. E. LeDoux, 1992a. Differential contribution of amygdala and hippocampus to cued and contextual fear conditioning. *Behav. Neurosci.* 106:274–285.

Phillips, R. G., and J. E. LeDoux, 1992b. Overlapping and divergent projections of CA1 and the ventral subiculum to the amygdala. *Soc. Neurosci. Abstr.* 18:518.

Rogan, M., and J. E. LeDoux, 1993. Long-term increases in auditory-evoked responses accompany tetanically-induced LTP in the thalamo-amgydala pathway. *Soc. Neurosci. Abstr.* 19:1227.

Rolls, E. T., 1992. Neurophysiology and functions of the primate amygdala. In *The Amygdala: Neurobiological Aspects of Emotion, Memory, and Mental Dysfunction*, J. P. Aggleton, ed. New York: Wiley, pp. 143–165.

Romanski, L. M., M. C. Clugnet, F. Bordi, and J. E. LeDoux, 1993. Somatosensory and auditory convergence in the lateral nucleus of the amygdala. *Behav. Neurosci.* 107:444–450.

Romanski, L. M., and J. E. LeDoux, 1992. Equipotentiality of thalamo-amygdala and thalamo-cortico-amygdala projections as auditory conditioned sitmulus pathways. *J. Neurosci.* 12:4501–4509.

Stefanacci, L., C. R. Farb, A. Pitkanen, G. Go, J. E. LeDoux, and D. G. Amaral, 1992. Projections from the lateral nucleus to the basal nucleus of the amygdala: a light and electron microscopic PHA-L study in the rat. *J. Comp. Neurol.* 323:2–17.

Turner, B. H., M. Mishkin, and M. Knapp, 1980. Organization of the amygdalopetal projections from modality-specific cortical association areas in the monkey. *J. Comp. Neurol.* 191:515–543.

70 Cellular Mechanisms Active in Emotion

FLOYD E. BLOOM

ABSTRACT This chapter examines the mechanisms of intercellular communication pertinent to the neuroscience of emotion. Recent clinical and preclinical studies have advanced the characterization of interneuronal signals and their mechanisms of transduction relative to the neural circuitry implicated in the expression and regulation of emotion. Within the context of the neuronal circuitry demonstrated by behavioral strategies to be critical for emotion, we consider the neurotransmitters employed by these circuits, and the interactive transductive mechanisms, established for certain combinations of neurotransmitters at other locations, that may also be operating at these emotion-regulating locations. The behavioral studies have emphasized the pivotal role of the amygdaloid complex, and in particular its central nucleus, in at least the experimental analysis of the emotion of fear. The shared mechanisms of transduction used in these regulatory systems may also permit emotional events to influence or be influenced by activity in the immune or endocrine systems.

Hierarchical levels of research in neurosciences

The strategies used to analyze the neuroscientific substrates of emotion are molecular, cellular, and behavioral. These three terms constitute the minimal dissection of a complex hierarchical ensemble that we have previously described to epitomize the principal methods of neuroscience research (see Bloom, 1988). The intensively exploited molecular level has been the traditional focus for characterizing the action of drugs that alter behavior. Molecular discoveries provide biochemical probes for identifying the appropriate neuronal sites, their mediative mechanisms, and the pharmacological tools to verify the working hypotheses of other strategies.

Research at the cellular level determines which specific neurons and which of their most proximate synaptic connections may mediate a behavior or the effects of a given drug. The cellular level of research into the basis of emotion exploits both molecular and behavioral leads to determine the most likely brain sites at which behavioral changes pertinent to emotion can be analyzed (see Bloom, 1988; Koob and Bloom, 1988, for references). Inferences as to the locus of the cells or cell systems central to experimental analysis of emotion have also been drawn from studies involving lesions or stimulations of specific brain sites (see LeDoux, this volume).

Research on emotion at the behavioral level (see Gray, 1982, and this volume; LeDoux, 1987, and this volume; also see Davis, 1989; Davis 1992a; Everitt and Robbins, 1992) centers on the integrative phenomena that link populations of neurons into extended specialized circuits, ensembles, or more pervasively distributed systems that integrate the physiological expression of emotions. Our inferences from such physiological changes (heart rate, respiration, locomotion, etc.) that the states experienced by animals are equivalent to the emotional states that are experienced by humans and accompanied by the same sorts of physiological changes are continuously tested by clinical observations on the neurochemical and neuropharmacological correlates of emotional diseases—in particular, major depression (Siever et al., 1986; Coccaro et al., 1989; Gorman et al., 1989; Kupfer, 1989) and the anxiety disorders (Gorman et al., 1989; Grillon et al., 1991)—as well as by the emotional consequences of the self-administration of stimulant or depressant drugs (see Koob and Bloom, 1988; Koob, Wall, and Bloom, 1989).

Neurotransmitters: General considerations

As originally conceived, central neurotransmitters operated uniformly, exciting or inhibiting their postsynaptic targets via receptors that activated passive ionic conductances. As research has progressed, the

FLOYD E. BLOOM Department of Neuropharmacology, The Scripps Research Institute, La Jolla, Calif.

list of transmitter substances has expanded from simple amino acids and amines to include a host of neuronally derived peptides. Moreover, the range of actions attributable to neurotransmitters has also expanded. Instead of being limited to receptors that open ionic channels to permit the passive flow of ions into or out of neurons, current concepts of transmitter actions include a variety of intramembranous and intracytoplasmic second messengers that can regulate both active and passive ionic conductances (Siggins and Gruol, 1986; Bloom, 1988, 1990a; Shepherd, 1990).

Neurotransmitter Organization There are three major chemical classes of neurotransmitters: the *amino acid transmitters*, of which glutamate and aspartate are recognized as the major excitatory transmitting signals, and gamma-aminobutyric acid (GABA) and glycine as the major inhibitory transmitters; the *aminergic transmitters* (acetylcholine, epinephrine, norepinephrine, dopamine, serotonin, and histamine), and the literally dozens of *peptides* (see Bloom, 1988, 1990b; Cooper, Bloom, and Roth, 1991). A revolutionary finding has emerged here in concepts of brain system interactions (see Hökfelt, Fuxe, and Pernow, 1987): It now seems that neuropeptides are almost certainly never the sole signal secreted by a central neuron that contains such a ligand, but rather represent one of two or more potentially secreted signals, the other of which may be an amino acid or an amine (at intrasynaptic terminal concentrations a thousand times higher) or a second or third peptide. It is also likely, but not yet definitively established, that purines, lipids, and steroids made within neurons may play important auxiliary roles in intercellular transmission in the nervous system.

Synaptic Interactions Classically, the electrophysiological events underlying synaptic transmission have been divided into two categories, *excitation* and *inhibition* (see Bloom, 1990a; Cooper, Bloom, and Roth, 1991). In both cases, the classical examples of transmission at the neuromuscular junction and at autonomic ganglia characterized the process of transmission as though it were synonymous with enhancement of ionic conductance, leading to the end event of excitation (by depolarizing ion flow by cations into the cell) or inhibition (by cationic, largely K, efflux, or anionic chloride influx). However, in the 1970s, both vertebrate and invertebrate nervous systems offered many examples of nonclassical effects of transmitters, in which transmitter actions, and the effects of the pathways that contained these transmitters, were associated with diminished ion flow. Often termed "modulatory," these effects have also been held to represent the major means by which transmitter actions of pathways converging on common targets can interact to provide for signal integration (see Bloom, 1990; Cooper, 1991, for additional references).

Implications for mechanisms of transmission underlying emotion

Traditionally, behaviorally oriented neuroscience has stretched across a broad chasm between short-term events of synaptic transmission and longer-term events in behavioral contexts, with often vaguely defined changes in the short-term events suggested as being responsible for consequent behavioral plasticity. At present, a widely accepted concept, that of activity-dependent modification of both structure and function from the synaptic to the behavioral levels is serving to make these biobehavioral connections more amenable to constrained experimental hypothesis generation and testing. The classical experiments of Hubel and Wiesel on the developmental irregularities of binocularly sensitive neurons within the visual cortex following unilateral visual deprivation established that normal neuronal activity was essential for the formation of appropriate neuronal connections even in the absence of direct experimental damage to the developing brain. Subsequent studies have revised those initial concepts of activity-dependent neuronal plasticity to include the modification of synaptic innervation in the adult animal. Under certain experimental conditions, such plasticity can be described as "massive reorganization" even in the brains of adult primates (Garraghty, Pons, and Kass, 1990; Pons et al., 1991; Wiesel, Anderson, and Katz, 1991). Given that adult neurons and their circuits may be capable of exuberant degrees of structural synaptic modification, under the control of locally derived "growth factors" and their receptors, and driven by synaptic activity, it is only a modest conceptual extension to relate this body of data to another extremely active element of modern cellular and molecular neuroscience, namely the elucidation of molecular synaptic mechanisms for enduring activity-dependent modifications.

At present, simple forms of behavior modification such as habituation and sensitization have been dissected into their molecular mechanisms through studies carried out in marine mollusks, particularly the *Aplysia* (Kandel, 1989; Siegelbaum and Kandel, 1991; Bailey and Kandel, 1993; Hawkins, Kandel, and Siegelbaum, 1993). These studies suggest that a specific form of activity-dependent interneuronal plasticity, namely long-term potentiation, can serve as a model both for classical memory and learning and for the more elusive phenomena by which emotional contexts may be linked to specific environmentally triggered responses.

Long-term potentiation as a model for expressing conditioned emotion

In the mammalian brain, the most widely pursued model of synaptic plasticity, long-term potentiation (LTP), has concentrated on the long-lasting (ten minutes to days, depending on the conditions) enhancement in synaptic transmission, measured as increased amplitudes of excitatory postsynaptic potentials or the currents generated by these potentials in specific circuits after high-frequency, high-intensity activation of other discrete paths, mainly within the hippocampal formation. This model system illustrates one means by which separate neurocircuits mediated by different transmitter systems converging on common target neurons can have their effects integrated in an enduring manner. In this way, the effects of a neutral conditional stimulus can be more tightly linked to an unconditional expression system such as the physiological signs of pain- or fear-induced responses. LeDoux (1992) has suggested that LTP-like events between neocortex and amygdala could underlie certain forms of emotional learning and other conditional associations mediated by the amygdala.

Originally, LTP was described with macroelectrode recordings in the hippocampus of intact animals as a means to link learning and cellular modifications (see Bliss, 1973; Brown, 1988) for intervals of days to weeks. However, most recent work has concentrated on slice preparations of the hippocampus or neocortex (see Nicoll, Kauer, and Malenka, 1988; Kauer et al., 1990; Zalutsky and Nicoll, 1990). Long-term potentiation has also been observed within the hippocampal formation at the synapses between the dentate granule cells and the targets of their mossy fiber synapses, at the CA3 pyramidal neurons, and also between the Schaffer collaterals of CA3 pyramidal neurons, as well as in associational fibers from the contralateral hippocampus to CA1 pyramidal neurons (Nicoll, Kauer, and Malenka, 1988; Nicoll, Malenka, and Kauer, 1989; Madison, Malenka, and Nicoll, 1991). Although neuropeptides are unquestionably also present in these circuits, their role in the modulatory process remains unclear. LTP has also been demonstrated within the lateral nucleus of the amygdala from stimulation of the medial geniculate, a critical relay site in the pathways responsible for acoustically conditioned fear (see LeDoux, 1987; Clugnet and LeDoux, 1990; Clugnet, LeDoux, and Morrison, 1990; LeDoux, 1992).

Antagonists of the NMDA subtype of Glu receptor will prevent LTP induction without interfering with already-potentiated transmission or with normal low-frequency transmission in the pathway under study. According to current interpretations (see Madison, 1991), the NMDA receptor couples the depolarization with increased Ca^{++} entry either to depolarize small dendritic domains that are the site of convergent afferents in the studied pathway or to enhance transmitter release presynaptically. In either case—and perhaps in both cases—the overall effect is to increase subsequent transmission in the tetanized pathway. Once potentiation is established, its maintenance appears to depend on postsynaptic protein phosphorylations through protein kinase C (Linden and Aryeh, 1989).

It is unclear at this time whether the molecular events underlying LTP are presynaptic, postsynaptic, or both (see Madison, 1991, 1993), or whether other neurotransmitter systems operating in these fields of the hippocampal formation can also play a role in the modulation of these phenomena (see Dutar and Nicoll, 1988; Kauer, Malenka, and Nicoll, 1988; Dutar and Nicoll, 1989; Kauer et al., 1990). An interesting facet of this body of research has been the implication that a short-lived gaseous mediator, nitric oxide (NO), may be released from the postsynaptic elements undergoing Glu-mediated LTP, through a Ca-dependent activation of nitric oxide synthase, and passively released from the postsynaptic membrane to regulate the presynaptic release of neurotransmitter (see Gally et al., 1990; Madison, Malenka, and Nicoll, 1991; O'Dell et al., 1991; Madison, 1993). NO, like adrenal and gonadal steroids (McEwen, 1991, and this volume) may be viewed as additional chemical agents for regulation of neuronal activity, and may offer a ground on which "volume transmission" events (See Agnati, Bjelke, and Fuxe, 1992) may become physiologically significant.

Résumé of circuits and transmitters implicated in emotion

Recent reviews (Davis, 1992a, 1992b; LeDoux, 1992, and this volume) strongly support the working conclusion that the amygdaloid complex is a major brain site at which fear-inducing conditional associations can be acquired and can elicit a physiological expression of fear. The classic literature reviewed by these authors suggested that lesions of the amygdala, and especially the lateral and central nuclei, can inhibit both unconditioned fear reactions and fear elicited by conditional associations, both in terms of locomotor attempts to flee, and in terms of the physiological and endocrine sequelae. Direct microinjections into the amygdala of drugs whose effects are mediated through specific neurotransmitter systems has implicated opioid peptides (see Kapp et al., 1990) neuropeptide Y, or NPY (Heilig et al., 1993), and GABA (Hodges, Green, and Glenn, 1987) in the suppression of conditioned aversive or anxiety effects; these effects of NPY were evoked selectively from the central nucleus. In contrast, microinjections of antagonists for the 5-HT_3 receptor subtype will elicit direct anxiolytic effects and block some of the aversive physiological effects of withdrawal from drugs on which the animals had developed dependency (Costall et al., 1990); although other sites, especially the nucleus accumbens, seem to be the sites at which the aversive effects of precipitated withdrawal arise (Koob, Wall, and Bloom, 1989).

Davis's 1992 analysis of the transmitter systems emphasizes the role of the central monoamine systems mainly in the expression of the events downstream of the amygdala (e.g., see Davis, 1992a, fig. 2). However, other evidence, mainly from clinical examinations of the indirect manifestations of major depression and anxiety, suggests a role for both noradrenergic and serotonergic deficiency as a basis for the mood changes, in that drugs that deplete or block their function can lead to depressed moods as well as to decreased levels of arousal (see Siever et al., 1986; Yudofsky, 1992). Ascending influence of pontine monoaminergic systems to undefined forebrain targets has also been implicated in the sensitization of patients with panic disorder to their attacks: Drugs (yohimbine or clonidine) or natural factors (the peptide corticotropin releasing factor, or CRF; Valentino and Foote, 1987, 1988) that enhance locus coeruleus activity lead to signs of anxiety, whereas drugs that reduce locus coeruleus activity have the reverse effect (see Gorman et al., 1989, for discussion). Examination of the immunocytochemical maps within the regions of the amygdala that are critical for the acquisition and display of conditioned emotional responses (see Price, Russchen, and Amaral, 1987) reveals that dopaminergic and noradrenergic afferents are prominent in the central nucleus, and especially its medial component, while 5-HT afferents are more ubiquitous throughout the amygdaloid complex, and cholinergic afferents are excluded from the central nucleus and prominent only in the posterior portion of the basal nucleus. Within the regions of the central nucleus receiving the amines, there is a reasonable overlap: Convergent afferents contain the neuropeptides neurotensin, somatostatin, vasoactive intestinal polypeptide, and the proenkephalin derived opioid peptides. Somatostatin, neurotensin, and CRF, as well as substance P, are also found within neurons of the nucleus, and could play a role in the convergent afferent information as well as in the efferent steps.

EXAMPLES OF CONVERGENT HETEROSYNAPTIC INTERACTIONS UNDERLYING EMOTION Given that the precise circuits and transmitters responsible for the enduring associations responsible for auditory or visual conditioned fear or startle responses remain to be defined, the application of the preceding principles of signal transduction to the behavioral expression of emotion obviously cannot be satisfactory. The sorts of pertinent integrative synaptic mechanisms that may mediate emotional events can be visualized in two recent examples from our collaborators.

Vasoactive intestinal polypeptide (VIP) and norepinephrine (NE) in rat cerebral cortex VIP and norepinephrine are both regarded as likely neurotransmitters in cerebral cortex, where they subserve a mutual coregulation of intracellular cAMP levels at the level of their identified convergent cellular target, the neocortical pyramidal neurons. VIP neurons are intrinsic, bipolar, radially oriented, intracortical neurons whose radial structure spans the entire thickness of the cortex with approximately 1 VIP neuron for every 30-μ-wide cylinder of cortical thickness; the NA innervation arises only from locus coeruleus and innervates a broad expanse of cortex in the horizontal plane. Identified cortical pyramidal neurons are depressed in spontaneous firing by iontophoresis of either NA or cyclic AMP. Magistretti and colleagues (see Magistretti, 1990, for references) demonstrated that VIP and NA can act synergistically to increase cyclic AMP in cerebral cortex. Physiological tests of VIP and NA interactions on rat cortical neurons are significant: Iontophoretic application of VIP during subthreshold NA administration causes pronounced inhibitions of cellular discharge regardless of the effect of VIP prior to NA.

The biochemical and physiological synergism of VIP and NA for cAMP synthesis was blocked by phentolamine, an alpha adrenergic receptor antagonist, and mimicked by phenylephrine, an α receptor agonist (Ferron, Siggins, and Bloom, 1985; Magistretti, 1990). Subsequent studies by Magistretti and collaborators (see Magistretti, 1990) revealed that the α_1 augmentation involves the formation of an unknown arachidonic acid metabolite as deduced by a dose-dependent blockade of this augmentation by cyclo-oxygenase inhibition, and its restoration by submicromolar concentrations of prostaglandins $F2_a$ and $E2_a$. These data provide insight into the phenomenon of conditional responsiveness: the effectiveness of VIP as a regulator of the properties of cortical pyramidal neurons depends on the context or conditions provided by concurrent or preexisting responses to NA. Conversely, the concurrent or preexisting actions of NA in this spatial response domain can "condition" the pyramidal neurons to express a more robust response to VIP, and thus to enhance signal-to-noise levels as observed in earlier descriptions of intracortical NA actions. If we apply this concept of conditional transmitter operations to the present context of transmitter system interactions engaged in emotion, then it would be predicted that the effectiveness within cortex of the VIP-local effects would be vastly different if they were initiated in the time over which neocortical noradrenergic arousal was in effect.

Interactions between somatostatin and acetylcholine: An alternative interactive mechanistic sequence Similar interactive data have emerged from the interactions of the peptide somatostatin (SS14) with the more classical transmitter acetylcholine within the hippocampal formation. Our studies (Mancillas, Siggins, and Bloom, 1986) confirmed that SS14, when applied alone, directly depresses spontaneous discharge rate, and its effects have slow onset and offset. However, when tested concurrently with either brief pulses or small amounts of ACh continuously leaked from an iontophoretic pipette, SS14 caused a dose-dependent enhancement of facilitations induced by ACh but not of those produced by glutamate. The molecular mechanisms responsible for the observed interactions between SS14 and ACh have recently been significantly extended through the work of Siggins and colleagues (Moore et al., 1988; Schweitzer, Madamba, and Siggins, 1990; Schweitzer et al., 1993). In current-clamp recordings in hippocampus and nucleus tractus solitarius, SS14 hyperpolarized the neurons and reduced their spontaneous activity. Under voltage-clamp conditions, a notable action ascribed to ACh at muscarinic receptors had been to block the "slow, inward, relaxation," or M-current, a slowly inactivating outward K^+ current (Siggins, 1986; Schweitzer et al., 1993). In contrast to the ionic effects attributed to ACh on this M-current, SS14 *increased* the amplitude of the M-current, in some cases more than fourfold. Subsequently, Siggins and colleagues have demonstrated that opioid peptides can produce a similar effect to SS14 through opiate receptors and not through somatostatin receptors. Most recently (Schweitzer et al., 1993) Siggins and colleagues have provided evidence that the SS14-induced electrophysiological effects are mediated by arachidonic acid, in a manner similar to the effects of the α-adrenergic potentiation of VIP noted above: Effects in modifying the M-current appear to be due to an as-yet-unspecified metabolite of arachidonic acid, while a direct hyperpolarizing effect may be directly caused by arachidonic acid. As might be predicted from the juxtaposition of these two series of interactive events, others have described an arachidonic acid–mediated enhancement by SS14 of the effects of α-adrenergic responses, although on cultured striatal astrocytes rather than hippocampal neurons (Marin et al., 1991).

This dynamic, opposing regulatory interaction of SS14 and ACh is of considerable interest in characterizing the sorts of effects that peptide-signaling molecules may elicit from neurons, either as transmitters released from convergent pathways, or as a means to enhance the effectiveness of a single pathway under conditions of differing afferent activity, as for example those events taking place within the amygdala or in which amygdala output is relayed cholinergically to somatostatin-enriched layers of neocortex. Furthermore, SS14 coexists in synaptic terminals with gamma-aminobutyric acid, and the GABA-mediated postsynaptic responses may be key to the effectiveness of anxiety-reducing treatments.

Some concluding ideas: Emotion and health

I propose that these examples of conditional signaling mechanisms may well set the scales on which other spatial and temporal signals between neurons are ultimately measured as purposeful. In the present contexts, such interactions might represent means by which arousal, attention, and fear can be functionally associated both with environmental signals and with the locomotor or other effector emotional responses elicited. It is becoming increasingly evident that the vocabulary of neuronal communication is much larger and more complex than previously conceived. The results reviewed here suggest that deciphering the grammar of neuronal communication that underlies the acquisition and expression of emotion will require careful evaluation of the interactions of neurotransmitters. Suffice it to suggest that the engagement of the CRF systems arising within the amygdaloid complex, and perhaps also the activation of the hypothalamic CRF system to influence the pontine visceral targets of these neurons, could in addition alter the set points of the hypothalamic-pituitary-adrenal axis as well. If this occurs opportunities emerge for the wholesale realignment of integrative adaptive systems outside of the brain, including the endocrine and immune systems, whose dysfunctions are frequently ascribed to emotional antecedents. Since the healthy brain is in general self-regulating and highly homeostatic, a failure to achieve the necessary homeostatic feedback could set the stage for prolonged hyper- or hyporesponsiveness to emotion-eliciting stimuli or emotion-sensitive responses. As the technological advances of functional imaging as well as cellular-level neuropharmacology continue, direct experimental evaluation of such ideas will soon be at hand, changing what is currently near-justifiable speculation into testable hypothesis.

REFERENCES

Agnati, L. F., B. Bjelke, and K. Fuxe, 1992. Volume transmission in the brain. *Am. Sci.* 80:362–373.

Bailey, C. H., and E. R. Kandel, 1993. Structural changes accompanying memory storage. *Annu. Rev. Physiol.* 55: 397–426.

Bliss, T. V. P., 1973. Long-lasting potentiation of synaptic transmission in the dentate area of the anesthetized rabbit following stimulation of the perforant path. *J. Physiol.* 232:331–356.

Bloom, F. E., 1988. Neurotransmitters: Past, present, and future directions. *Faseb. J.* 2:32–41.

Bloom, F. E., 1990a. Neurohumoral transmission and the central nervous system. In *The Pharmacological Basis of Therapeutics*, ed. 8, A. G. Gilman, T. W. Rall, A. S. Nies, and P. Tayler, eds. New York: Pergamon, pp. 244–268.

Bloom, F. E., 1990b. Neuropeptides in the nineties. In *Neuropeptides*, T. W. Schwartz, L. M. Hilsted, and J. F. Rehfeld, eds. Copenhagen: Benzon Foundation, pp. 1–24.

Brown, T. H., P. F. Chapman, E. W. Kairiss, and C. L. Keenan, 1988. Long term synaptic potentiation. *Science* 242:724–728.

Clugnet, M. C., and J. E. LeDoux, 1990. Synaptic plasticity in fear conditioning circuits: Induction of LTP in the lateral nucleus of the amygdala by stimulation of the medial geniculate body. *J. Neurosci.* 10:2818–2814.

Clugnet, M. C., J. E. LeDoux, and S. F. Morrison, 1990. Unit responses evoked in the amygdala and striatum by electrical stimulation of the medial geniculate body. *J. Neurosci.* 10:1055–1061.

Coccaro, E. F., L. J. Siever, H. M. Klar, G. Maurer, and K. Cochrane, 1989. Serotonergic studies in patients with affective and personality disorders. *Arch. Gen. Psychiatry* 46:587–599.

Cooper, J. R., F. E. Bloom, and R. H. Roth, 1991. *Biochemical Basis of Neuropharmacology*. New York: Oxford University Press.

Costall, B., B. J. Jones, M. E. Kelley, R. J. Naylor, E. S. Onaivi, and M. B. Tyers, 1990. Sites of action of ondansetron to inhibit withdrawal from drugs of abuse. *Pharmacol. Biochem. Behav.* 36:97–104.

Davis, M., 1989. Neural systems involved in fear-potentiated startle. *Ann. N. Y. Acad. Sci.* 563:165–183.

Davis, M., 1992a. The role of the amygdala in fear and anxiety. *Annu. Rev. Neurosci.* 15:353–375.

Davis, M., 1992b. The role of the amygdala in fear-potentiated startle: Implications for animal models of anxiety. *Trends Pharmacol. Sci.* 13:35–41.

Dutar, P., and R. A. Nicoll, 1988. Stimulation of phosphatidylinositol (PI) turnover may mediate the muscarinic suppression of the M-current in hippocampal pyramidal cells. *Neurosci. Lett.* 85:89–94.

Dutar, P., and R. A. Nicoll, 1989. Pharmacological characterization of muscarinic responses in rat hippocampal pyramidal cells. *Experientia Suppl.* 57:68–76.

Everitt, B. J., and T. W. Robbins, 1992. Neurochemically defined arousal systems. In *The Amygdala: Neurobiological Aspects of Emotion, Memory, and Mental Dysfunction*, J. P. Aggleton, ed. New York: Wiley, pp. 138–149.

Ferron, A., G. R. Siggins, and F. E. Bloom, 1985. Vasoactive intestinal polypeptide acts synergistically with norepinephrine to depress spontaneous discharge rate in cerebral cortical neurons. *Proc. Natl. Acad. Sci. U.S.A.* 82: 8810–8812.

Gally, J. A., A. R. Montague, J. Reeker, G. N., and G. M. Edelman, 1990. The NO hypothesis: Possible effects of a short lived, rapidly diffusible signal in the development and function of the nervous system. *Proc. Natl. Acad. Sci. U.S.A.* 87:3547–3551.

Garraghty, P. E., T. P. Pons, and J. H. Kaas, 1990. Ablations of areas 3a and 3b of monkey somatosensory cortex abolish cutaneous responsivity in area 1. *Brain Res.* 528: 165–169.

Gorman, J. M., M. R. Liebowitz, A. J. Fyer, and J. Stein, 1989. A neuroanatomical hypothesis for panic disorder. [See comments] *Am. J. Psychiatry* 146:148–161.

Gray, J., 1982. The neuropsychology of anxiety: An enquiry into the functions of the septo-hippocampal system. *Behav. Brain Sci.* 5:469–532.

Grillon, C., R. Ameli, S. W. Woods, K. Merikangas, and M. Davis, 1991. Fear-potentiated startle in humans: Effects of anticipatory anxiety on the acoustic blink reflex. *Psychophysiology* 28:588–595.

Hawkins, R. D., E. R. Kandel, and S. A. Siegelbaum, 1993. Learning to modulate transmitter release: Themes and variations in synaptic plasticity. *Annu. Rev. Neurosci.* 16:625–665.

Heilig, M., S. McLeod, M. Brot, S. C. Heinrichs, F. Menzaghi, G. F. Koob, and K. T. Britton, 1993. Anxiolytic-like action of neuropeptide Y: Mediation by Y1 receptors in amygdala, and dissociation from food intake effects. *Neuropsychopharmacology* 8:357–363.

Hodges, H., S. Green, and B. Glenn, 1987. Evidence that the amygdala is involved in benzodiazepine and serotonergic effects on punished responding but not on discrimination. *Psychopharmacology* 92:491–504.

Hökfelt, T., K. Fuxe, and B. Pernow, 1987. *Coexistence of Neuronal Messengers: New Principle in Chemical Transmission.* Amsterdam: Elsevier.

Kandel, E. R., 1989. Genes, nerve cells, and the remembrance of things past. *J. Neuropsychiatry Clin. Neurosci.* 1: 103–125.

Kapp, B. S., A. Wilson, J. P. Pascoe, W. F. Supple, and P. J. Whalen, 1990. A neuroanatomical systems analysis of conditioned bradycardia in the rabbit. In *Neurcomputation and Learning: Foundations of Adaptive Networks,* M. Gabriel and J. Moore, eds. New York: Bradford Books, pp. 135–158.

Kauer, J. A., R. C. Malenka, and R. A. Nicoll, 1988. A persistent postsynaptic modification mediates long-term potentiation in the hippocampus. *Neuron* 1:911–917.

Kauer, J. A., R. C. Malenka, D. J. Perkel, and R. A. Nicoll, 1990. Postsynaptic mechanisms involved in long-term potentiation. *Adv. Exp. Med. Biol.* 268:291–299.

Koob, G. F., and F. E. Bloom, 1988. Cellular and molecular mechanisms of drug dependence. *Science* 242:715–723.

Koob, G. F., T. L. Wall, and F. E. Bloom, 1989. Nucleus accumbens as a substrate for the aversive stimulus effects of opiate withdrawal. *Psychopharmacology [Berlin]* 98:530–534.

Kupfer, D. J., 1989. Neurophysiological factors in depression: New perspectives. *Eur. Arch. Psychiatry Clin. Neurosci.* 238:251–258.

LeDoux, J. E., 1987. Emotion. In *Handbook of Neurophysiology.* Sec. 1, *The Nervous System.* Vol. 5, *Higher Functions of the Brain,* part 2. V. B. Mountcastle, ed. Bethesda, Md.: American Physiological Society, pp. 419–459.

LeDoux, J. E., 1992. Brain mechanisms of emotion and emotional learning. *Curr. Opin. Neurobiol.* 2:191–197.

Linden, D. J., and R. Aryeh, 1989. The role of protein kinase C in long term potentiation: A testable model. *Brain Res. Brain Res. Rev.* 14:179–296.

Madison, D. V., 1993. Pass the nitric oxide. *Proc. Natl. Acad. Sci. U.S.A.* 90:4329–4331.

Madison, D. V., R. C. Malenka, and R. A. Nicoll, 1991. Mechanisms underlying long-term potentiation of synaptic transmission. *Annu. Rev. Neurosci.* 14:379–397.

Madison, D. V., 1991. LTP, post or pre? A look at the evidence for the locus of long-term potentiation. *The New Biol.* 3:549–557.

Magistretti, P. J., 1990. VIP neurons in the cerebral cortex. *Trends Pharmacol. Sci.* 11:250–254.

Magistretti, P. J., and J. H. Morrison, 1988. Noradrenaline- and vasoactive intestinal peptide–containing neuronal systems in neocortex: Functional convergence with contrasting morphology. *Neurosci.* 24:367–378.

Mancillas, J. R., G. R. Siggins, and F. E. Bloom, 1986. Somatostatin selectively enhances acetylcholine-induced excitations. *Proc. Natl. Acad. Sci. U.S.A.* 83:7518–7521.

Marin, P., J. C. DeLumeau, M. Tence, J. Cordier, J. Glowinski, and J. Premont, 1991. Somatostatin potentiates the a1-adrenergic activation of phospholipase C in striatal astrocytes through a mechanism involving arachidonic acid and glutamate. *Proc. Natl. Acad. Sci. U.S.A.* 88:9016–9020.

McEwen, B. S., 1991. Non-genomic and genomic effects of steroids on neural activity. *Trends Pharmacol. Sci.* 12:141–147.

Moore, S. D., S. G. Madamba, M. Joels, and G. R. Siggins, 1988. Somatostatin augments the M-current in hippocampal neurons. *Science* 239:278–280.

Nicoll, R. A., J. A. Kauer, and R. C. Malenka, 1988. The current excitement in long-term potentiation. *Neuron* 1: 97–103.

Nicoll, R. A., R. C. Malenka, and J. A. Kauer, 1989. The role of calcium in long-term potentiation. *Ann. N. Y. Acad. Sci.* 568:166–170.

O'Dell, T. J., R. D. Hawkins, E. R. Kandel, and O. Arancio, 1991. Tests of the role of two diffusible substances in long-term potentiation: Evidence for nitric oxide as a possible early retrograde messenger. *Proc. Natl. Acad. Sci. U.S.A.* 88:11285–11289.

Paul, S. M., and R. H. Purdy, 1992. Neuroactive steroids. *Faseb. J.* 6:2311–2322.

Pons, T. P., P. E. Garraghty, A. K. Ommaya, J. H. Kaas, E. Taub, and M. Mishkin, 1991. Massive cortical reorganization after sensory deafferentation in adult macaques. *Science* 252:1857–1860.

Price, J. L., F. T. Russchen, and D. G. Amaral, 1987. The limbic region. II: The amygdaloid complex. In *Handbook of Chemical Neuroanatomy*, vol. 5, A. Björklund, T. Hökfelt, and L. W. Swanson, eds. Amsterdam: Elsevier, pp. 279–388.

Purdy, R. H., P. H. Moore, Jr., A. L. Morrow, and S. M. Paul, 1992. Neurosteroids and $GABA_A$ receptor function. *Adv. Biochem. Psychopharmacol.* 47:87–92.

Schweitzer, P., S. Madamba, J. Champagnat, and G. R. Siggins, 1993. Somatostatin inhibition of hippocampal CA1 pyramidal neurons by arachidonic acid and its metabolites. *J. Neurosci.* 13:2033–2049.

Schweitzer, P., S. Madamba, and G. R. Siggins, 1990 Arachidonic acid metabolites as mediators of somatostatin-induced increase of neuronal M-current. *Nature* 346:464–467.

Shepherd, G. M., 1990. *The Synaptic Organization of the Brain.* New York: Oxford University Press.

Siegelbaum, S. A., and E. R. Kandel, 1991. Learning-related synaptic plasticity: LTP and LTD. *Curr. Opin. Neurobiol.* 1:113–20.

Siever, L. J., T. W. Uhde, D. C. Jimerson, C. R. Lake, I. J. Kopin, and D. L. Murphy, 1986. Indices of noradrenergic output in depression. *Psychiatry Res.* 19:59–73.

Siggins, G. R., and D. L. Gruol, 1986. In *Mechanisms of Transmitter Action in the Vertebrate Nervous System: Integrative Systems of the Reticular Core*, F. E. Bloom, ed. Bethesda, Md.: American Physiological Society, pp. 1–114.

Valentino, R. J., and S. L. Foote, 1987. Corticotropin-releasing factor disrupts sensory responses of brain noradrenergic neurons. *Neuroendocrinology* 45:28–36.

Valentino, R. J., and S. L. Foote, 1988. Corticotropin-releasing hormone increases tonic but not sensory-evoked activity of noradrenergic locus coeruleus neurons in unanesthetized rats. *J. Neurosci.* 8:1016–1025.

Wiesel, T. N., D. Anderson, and L. Katz, eds., 1991. *Neural Development: Discussions in Neuroscience.* Amsterdam: Elsevier.

Yudofsky, S. C., 1992. Beta-blockers and depression. The clinician's dilemma (editorial). *JAMA* 267:1826–1827.

Zalutsky, R. A., and R. A. Nicoll, 1990. Comparison of two forms of long-term potentiation in single hippocampal neurons. *Science* 248:1619–1624.

71 Retuning the Brain by Fear Conditioning

NORMAN M. WEINBERGER

ABSTRACT Fear conditioning produces highly specific retuning of receptive fields in the auditory cortex; responses to the conditioned stimulus frequency are increased whereas responses to other frequencies are decreased, shifting tuning toward or even to the conditioned stimulus frequency. This receptive field plasticity is associative and highly specific, is established rapidly, lasts indefinitely, and can be expressed under general anesthesia. A model of associative retuning in the auditory cortex, involving the convergence of lemniscal and nonlemniscal thalamic auditory inputs plus muscarinic mechanisms, has found recent support. We conclude that emotional learning involves an enduring change in sensory cortical tuning.

Emotion and its neural substrates are essential to integrated, adaptive behavior. Fear, an important, usually acquired, emotional state or response to a potentially damaging or life-threatening event, has been studied extensively with the use of Pavlovian classical defensive conditioning paradigms (hereafter called, simply, conditioning). Fear of a specific stimulus is elicited by and associated with the conditioned stimulus (CS) during the interval between its onset and the onset of an aversive unconditioned stimulus (US) (e.g., Weinberger, 1965). Behaviorally, conditioned fear to a stimulus (e.g., tone) generally is evident as rapidly acquired conditioned responses (CRs) such as interruption of ongoing behavior ("freezing") and the elicitation of a variety of conditioned autonomic responses (Lennartz and Weinberger, 1992a).

Neurobiological studies have identified the amygdaloid complex as critical for fear conditioning, particularly as the site of integration leading to the behavioral expression of fear (LeDoux, 1990). Our complementary line of inquiry seeks an understanding of the acquisition and representation of sensory stimuli as they acquire emotional valence, rather than the circuitry that produces behavioral conditioned responses. Fundamental to this approach is the concept that acquired *information* can be employed to subserve cognitive processes and adaptive behavior for an indefinite future.

NORMAN M. WEINBERGER Department of Psychobiology, Center for the Neurobiology of Learning and Memory, University of California, Irvine, Calif.

Approach

To understand the neural bases of learning and memory for fear and other states and responses, it is necessary to understand the principles and mechanisms by which information is acquired and stored in the brain. An extensive literature has established that conditioning involves the plasticity of neuronal responses to the CS in many brain systems. In general, conditioning causes associatively dependent increased neuronal responses to the CS, including the sensory neocortex of its modality (Weinberger and Diamond, 1987).

However, these findings do not directly address the issue of *information processing*, but rather document changes in excitability as indexed by responsivity, which could be either specific to the CS or general to other stimuli. This distinction is of critical importance. If learning specifically alters the processing of information about the CS, then potentiation would be restricted mainly to the frequency of the CS. On the other hand, if learning generally increases responsivity to acoustic stimuli, then responses to both the CS and non-CS frequencies would be enhanced. This issue cannot be resolved in standard conditioning experiments because both specific informational and general excitability mechanisms would produce response facilitation to the CS on training trials.

It is important to understand that an increase in general excitability due to association is different from an increase in general excitability caused by nonassociative processes such as sensitization. This distinction has been overlooked. But associative learning could facilitate the processing of a class of stimuli, such as acoustic frequencies, rather than a specific member of the class, specifically the tonal frequency that is the CS. Although such general learning may be adaptive, it does not involve a modification of the way in which specific information about the CS itself is represented and stored. In short, to establish that learning specifically alters the processing of information about the CS, it is necessary *but not sufficient* to demonstrate that the plasticity is due to association.

A sensory system perspective, together with a learning perspective, provides a way to distinguish between specific informational and general excitability processes. The sensory perspective highlights a fundamental construct of sensory physiology—the *receptive field.* The receptive field of a cell is delineated by the set of sensory stimuli that affect the activity of that cell. For example, there are receptive fields for the orientation of lines in the visual cortex, for specific sites of stimulation of the skin in the somatosensory system, and for sound frequencies in the auditory system.

Learning experiences that produce increased responses to a tonal conditioned stimulus could change receptive fields in either of two ways: a general increase in response across the receptive field, or a highly specific change, perhaps centered on the CS. The former would reflect an associative, that is, learned, but *general* increase in neuronal excitability. In contrast, the latter would indicate a change in the *processing of information* about the acquired behavioral meaning of the CS, for example, a *specific* modification of its coding and representation (figure 71.1).

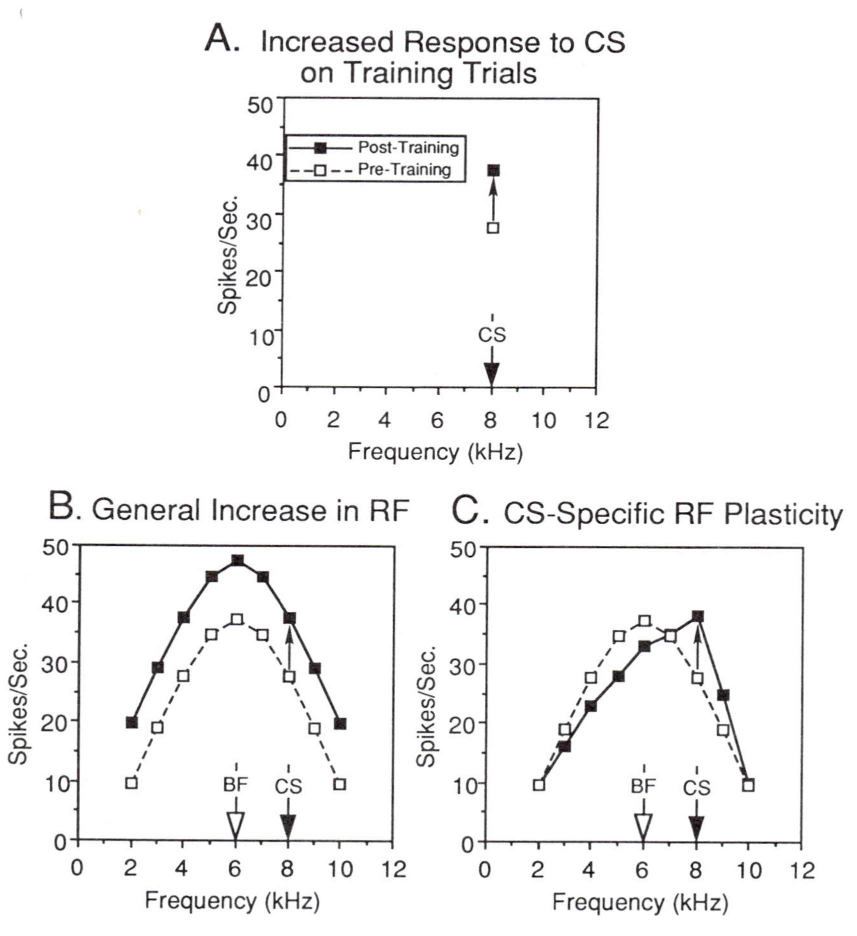

FIGURE 71.1 The application of receptive field (RF) analysis to learning. (A) Increased response to an acoustic conditioned stimulus (CS) due to training (e.g., classical conditioning). The increased response could be due either to a general increase in response across the receptive field (B) or to a CS-specific change in which response to the CS is increased while responses to other frequencies are increased less, unchanged, or decreased (C).

Receptive field plasticity in fear conditioning

Studies of receptive field plasticity in behavioral learning were first reported by Diamond and Weinberger (1989; 1986), for the secondary auditory cortex of the cat. Subsequent experiments have concerned the tonotopic auditory cortex and the medial geniculate complex of the guinea pig. To date, classical conditioning, sensitization, and habituation have been studied. The findings demonstrate that conditioning specifically alters the processing of information about the conditioned stimulus. In contrast, sensitization training increases excitability in a non-frequency-specific manner. Habituation alters information processing specifically about the repeated stimulus, producing a frequency-specific decrease in response. Brief summaries are provided in the following sections.

RECEPTIVE FIELD PLASTICITY IN GUINEA PIG AUDITORY CORTEX Classical conditioning produces receptive field plasticity in the primary auditory cortex of waking, adult guinea pigs (Bakin and Weinberger, 1990). Recordings from layers V and VI revealed that conditioning induces specific receptive field plasticity with *facilitation at the frequency of the CS and depression of responses to other frequencies, including that of the pretraining best frequency*. Moreover, frequency tuning could be *shifted* to the CS frequency (figure 71.2A). This receptive field plasticity was highly specific (figure 71.2B1). Additionally, receptive field plasticity specific to the CS was maintained at 24 hours posttraining and was often more pronounced at this time than immediately after training. Thus, receptive fields in primary auditory cortex are not fixed in the adult animal, but can be specifically modified rapidly by associative processes. The overall effect is to potentiate responses to the frequency of the conditioned stimulus and depress responses to the pretraining best frequency and other frequencies.

Subjects trained in a sensitization paradigm never developed CS-specific receptive field plasticity, thus demonstrating the associative basis of the retuning, but rather developed a general increase in response across the receptive field (figure 71.2B2) (Bakin and Weinberger, 1990). Sensitization training with a visual stimulus (flashing light) also produces a general increase in response across the frequency domain (figure 71.2B2; Bakin, Lepan, and Weinberger, 1992), showing that sensitization effects are not based on auditory experience. (Hereafter, the phrase "receptive field plasticity" refers strictly to CS-specific modifications of receptive fields, not to the general, transmodal effects of sensitization training.)

HABITUATION PRODUCES FREQUENCY-SPECIFIC DECREMENTS IN RECEPTIVE FIELDS Habituation completes the triumvirate of basic learning, with conditioning and sensitization. We studied habituation by presenting tones of a given frequency repeatedly and noting the decrement in neuronal discharges in the auditory cortex of the guinea pig. Comparison of receptive fields obtained before and after this repetition-induced decrement revealed a specific effect. Responses to the repeated frequency were greatly decreased (an average of 74% of control). In contrast, responses to other frequencies were unchanged. This receptive field plasticity was highly specific, with frequencies as close as 0.125 octaves from the repeated frequency showing no significant decrement (Condon and Weinberger, 1991; figure 71.2B3).

RECEPTIVE FIELD PLASTICITY IN DISCRIMINATION LEARNING Receptive field plasticity developed in two-tone discrimination learning: Responses to the CS+ were increased, responses to the CS− and other non-CS+ stimuli were depressed and tuning shifted toward or to the CS+ frequency. Moreover, receptive field plasticity developed across tone intensity (figure 71.3; Edeline and Weinberger, 1993). The degree of frequency specificity of discrimination was very high; responses to frequencies as close as 0.10 octaves from the frequency of the CS+ (smallest distance used) were either not facilitated or reduced.

RAPID DEVELOPMENT OF FREQUENCY-SPECIFIC RECEPTIVE FIELD PLASTICITY Fear conditioning is known to be established very quickly, in as few as 5–10 trials (for review, see Lennartz and Weinberger, 1992a). Our previous studies in the guinea pig used 30 CS-US pairing trials and receptive field plasticity was evident immediately thereafter. To determine whether this is the lower limit for the rate of development of receptive field plasticity, we obtained receptive fields after 5, 15, and 30 trials and after a 1 hour retention period, using our standard cardiac conditioning paradigm. Receptive field plasticity developed very rapidly, in most cases after only 5 trials. Of particular interest, increased response to the conditioned stimulus frequency appears to develop simultaneously, rather than succes-

A.

1. Pre-Conditioning RF

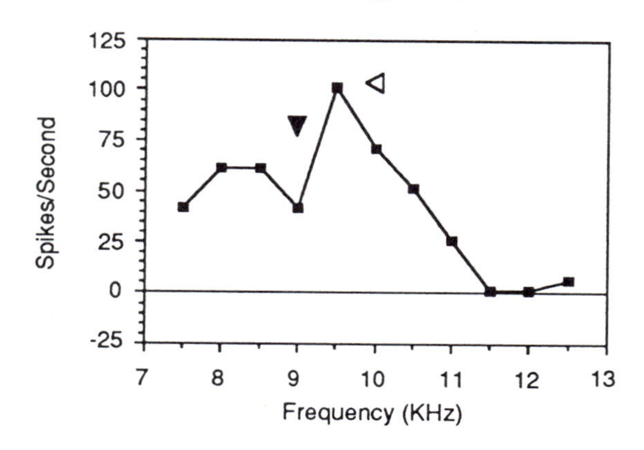

2. One Hour Post-Condit. RF

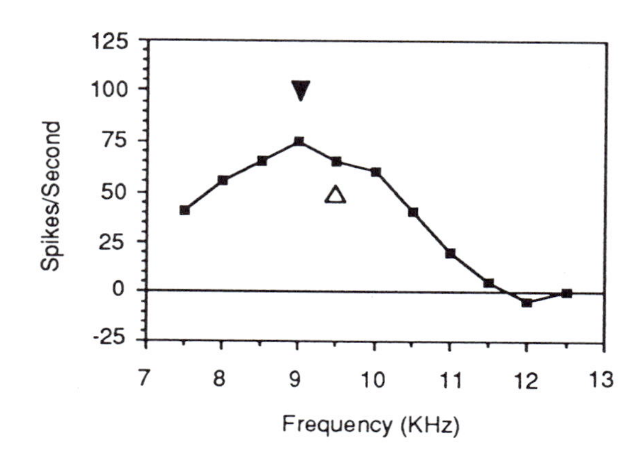

3. One Hour Post minus Pre-RF

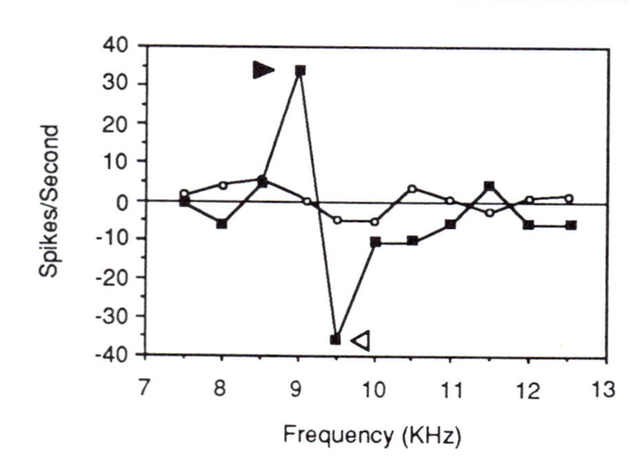

B.

1. Conditioning Modification of RF

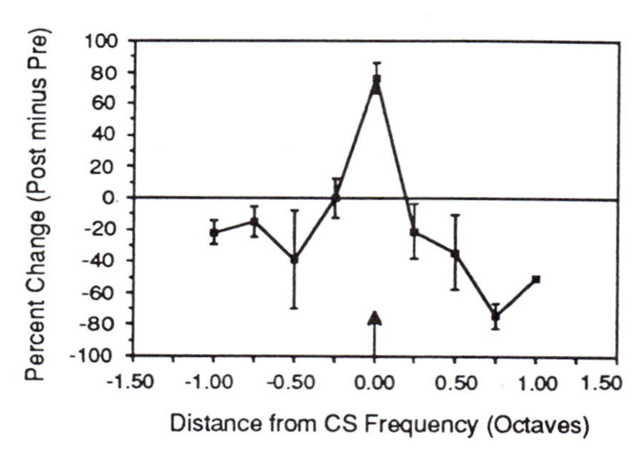

2. Auditory vs. Visual Sensitization

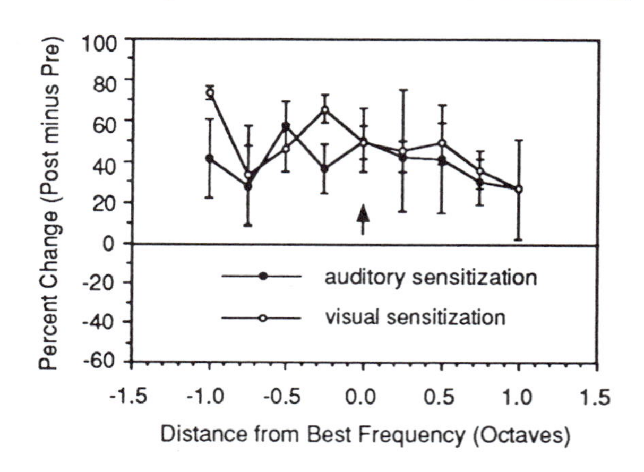

3. Habituation Modification of RF

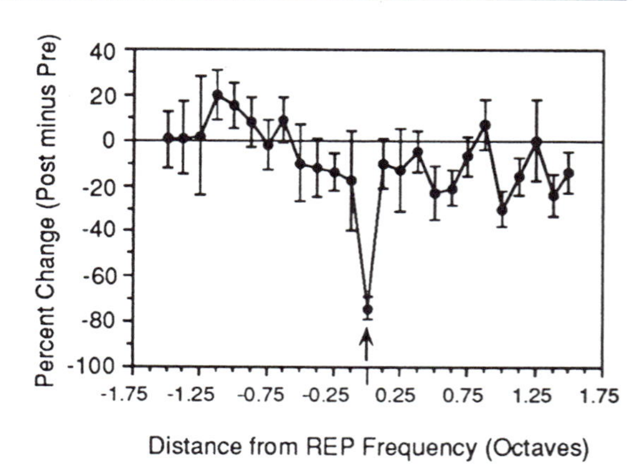

sively, with decreased response to the pretraining best frequency (figure 71.4) (Edeline, Pham, and Weinberger, 1993). Thus, receptive field plasticity develops very rapidly, as does fear conditioning.

FIGURE 71.2 Modification of frequency receptive fields (RFs) in the primary auditory cortex of the guinea pig. (A and B1) Classical conditioning specifically modifies frequency receptive fields. Illustrated is a case in which the conditoned stimulus (CS) frequency became the best frequency. (A1) Preconditioning, the quantified receptive field (tone-evoked response minus pretone background activity as a function of frequency) exhibited a best frequency (BF) of 9.5 kHz (open arrowhead). The animal then underwent a single session of 30 trials of classical conditioning (tone followed by shock) in which the frequency of the conditioned stimulus was 9.0 kHz (closed arrowhead); the subject developed behavioral fear conditioning within 10 trials (not shown). (A2) One hour postconditioning,, the response to the BF (and many other frequencies) decreased, whereas the response to the CS frequency increased, such that it became the new BF; i.e., frequency tuning shifted to the CS frequency. (A3) The RF difference function reveals a maximal increase at the CS frequency and maximal decrease at the pretraining best frequency; open circles show no systematic effect on spontaneous activity. (B1) Group average RF difference functions for classical conditioning ($n = 7$) reveals a highly specific increase in the receptive field such that response is increased only to the frequency of the CS, whereas the responses to almost all other frequencies are decreased ("side-band suppression"). Here and B2 and B3, the abscissa is in octaves from a reference frequency; in this case of conditioning, the reference frequency is the CS frequency. (B2) Average RF difference functions for auditory sensitization training (tone and shock unpaired, $n = 7$) reveal that only pairing produces CS-specific RF plasticity; sensitization training produces only a general increase in response across the receptive field (closed circles). Moreover, this sensitization effect is not even modality specific; when sensitization training is performed with light and shock unpaired ($n = 4$), the RF difference functions for acoustic frequency also develop a general increase in response (open circles). Data have been normalized for response magnitude and frequency response range. The x axis is centered around the best frequency. (B3) Acoustic habituation (repeated presentation of one frequency, 1/s for 15–30 min) produces a frequency-specific decrease in receptive fields; shown is the group average RF difference function (posthabituation minus prehabituation, $n = 26$). Changes in receptive fields are expressed as octave distance from the repeated frequency (REP). Note the decrease of 74% at the REP frequency with much smaller decreases at adjacent frequencies, showing a high degree of frequency specificity. The bandwidth of the REP decrease is very narrow, about 0.125 octaves at 50% of the decrement (37%).

LONG-TERM RETENTION OF RECEPTIVE FIELD PLASTICITY Previous findings showed that receptive field plasticity in the ACx is maintained at least 24 hours, the longest retention interval tested (Bakin and Weinberger, 1990). To determine the duration of retention, guinea pigs underwent a single session of conditioning (30 trials), and receptive fields were determined on the preceding day and at various retention intervals from 1 hour up to 8 weeks posttraining (Weinberger, Javid, and Lepan, 1993). Naturally, subjects were awake during training. However, were determined while they were under general anesthesia (sodium pentobarbital or ketamine). Frequency-specific receptive field plasticity was present 1 hour after training and for periods of up to 8 weeks, the longest retention period tested (figure 71.5).

Thus, receptive field plasticity is retained indefinitely. Moreover, it is sufficiently robust to be expressed under deep general anesthesia (both pentobarbital and ketamine). This finding also rules out the possibility that receptive field plasticity is due to selective arousal or attention to the CS frequency during receptive field determination (and presumably selective "de-arousal" to the pretraining best frequency and other frequencies that show decreased responses; see also Lennartz and Weinberger, 1992b). These results have implications for sensory physiology experiments, which generally use anesthetized animals. Our findings suggest that the learning experiences of an animal, although not explicitly acknowledged or assessed, may be reflected in recordings obtained under anesthesia.

SUMMARY In summary, receptive field analysis reveals that *conditioning has specific effects on information processing, whereas sensitization generally increases excitability.* The effects may be conveniently summarized as analagous to changing stations on a radio (retuning) versus increasing the volume (gain). Importantly, receptive field plasticity in conditioning is associative, is highly frequency specific and discriminative, develops rapidly, and appears to last indefinitely. These findings may help explain the results of studies that have found conditioning induced increased metabolic activity for the representation of the CS frequency band in the ACx (Gonzalez-Lima, 1992). Overall, the receptive field and metabolic findings are mutually supportive, strengthening the conclusion that fear conditioning involves a *specific modification of the processing, representation,*

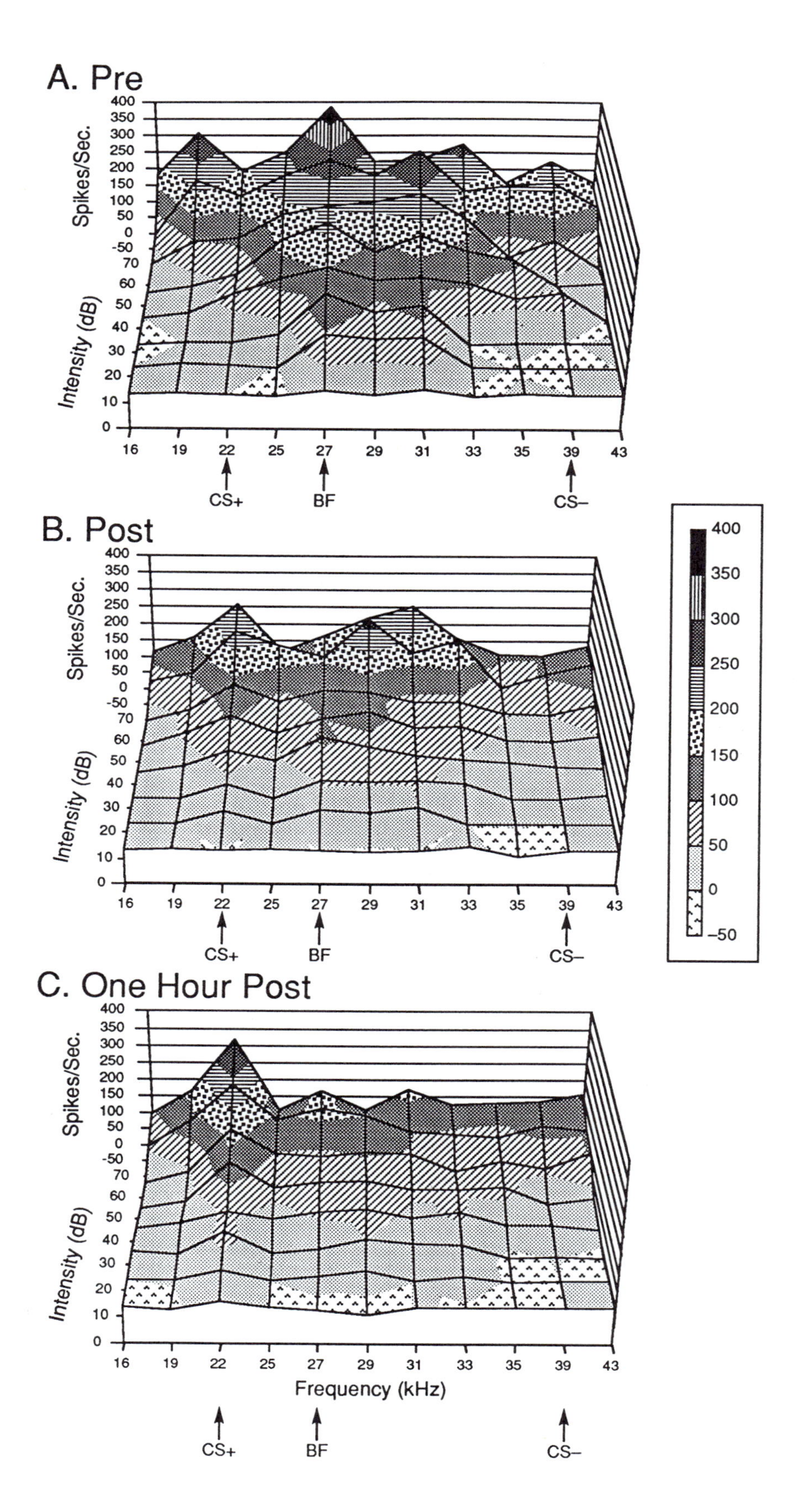
A. Pre
B. Post
C. One Hour Post
Spikes/Sec.
Intensity (dB)
Frequency (kHz)
CS+
BF
CS–
400
350
300
250
200
150
100
50
0
–50

and possibly even the storage of information concerning acoustic conditioned stimuli.

Toward mechanisms of receptive field plasticity in fear conditioning

We briefly review here evidence that various components of the thalamic auditory system play different roles in ACx receptive field plasticity and in the behavioral expression of conditioned fear. We then review findings that implicate the basal forebrain cholinergic system in receptive field plasticity. We conclude with a résumé of a model that integrates the various findings and has received empirical support.

MEDIAL GENICULATE NUCLEUS The medial geniculate nucleus provides direct auditory input to the primary auditory cortex from two nuclei that are morphologically and physiologically distinct and are differentially plastic during learning. The ventral medial geniculate (MGv), which is the "classical" precisely tuned lemniscal nucleus, *develops no change* in responses to an acoustic CS during fear conditioning. In contrast, the magnocellular medial geniculate (MGm), is the broadly tuned "lemniscal-adjunct" input and *rapidly develops discharge plasticity*, as observed in cluster recordings in the rabbit, cat, and rat (reviewed in Weinberger and Diamond, 1987).

Anatomical and physiological findings support the convergence of acoustic (i.e., CS) and spinothalamic (i.e., US) input in the MGm (Jones, 1985). Thus, it could be the initial site of plasticity in acoustic fear conditioning and could play a major role in the development of receptive field plasticity in the ACx (see the next section, "An integrated model"). Therefore, we investigated whether synaptic plasticity can be rapidly induced in the MGm, and whether microstimulation in this area can substitute for a standard peripheral shock US to support behavioral conditioning.

FIGURE 71.3 Retuning to the CS+ frequency, across intensity in two-tone discrimination training. The data are shown on a 3-D graph where the x axis is frequency, the y axis is response magnitude and the z axis is the intensity (in 10-dB steps). Pretraining (A), the best frequency (BF) was 27 kHz across intensities, the CS+ was in a "valley" at 22 kHz, and the CS− frequency was a secondary peak at 39 kHz. Immediately posttraining (B), there were pronounced decreased responses at the pretraining BF across intensity, while increased excitatory responses were observed at the CS+ frequency (22 kHz) and also at 31 kHz and 33 kHz. There were also decreased responses at the CS−. One hr posttraining (C), the only peak of strong excitation was at the CS+ frequency and this was present across intensities. Note that the selective increase at the CS+ frequency can be observed across intensity both posttraining and 1 hr posttraining.

Long-term potentiation in the magnocellular medial geniculate nucleus (MGm) Brief high-frequency stimulation of the brachium of the inferior colliculus produces long-term potentiation seen in field potentials in the MGm (Gerren and Weinberger, 1983) and single neurons (Weinberger, 1982).

Stimulation of the medial geniculate as an unconditioned stimulus The MGm projects to the amygdala as well as to auditory cortex. Lesions near or in the MGm prevent fear conditioning to sounds (LeDoux, 1990). This could be caused either by disruption of auditory input to the amygdala, which might be the main site of CS-US convergence, or by disruption of projected plasticity to the amygdala from the MGm, which also could be a site of plasticity. To clarify this issue, we substituted electrical microstimulation of the medial geniculate complex (including the posterior intralaminar nucleus, PIN) for the US in a cardiac conditioning experiment. Cardiac conditioning was established with stimulation of the PIN, but stimulation of other divisions of the medial geniculate, including the MGm, was ineffective, as was unpaired (sensitization) training (Cruikshank, Edeline, and Weinberger, 1992; see figure 71.6). These results are consistent with the view that the PIN is involved with the behavioral expression of learned fear, perhaps as a site of CS-US convergence. They also require distinguishing between two medial geniculate components; the PIN, which is linked to the behavioral expression of learning via the amygdala, and the MGm proper, which remains the best candidate for a critical role in the production of receptive field plasticity in the auditory cortex.

Receptive field plasticity in the medial geniculate nucleus Despite these findings, one could argue that ACx receptive field plasticity is "projected" directly from the medial geniculate because receptive fields had not been obtained from this nucleus. To resolve this issue, we determined receptive fields in the MGv and MGm before and after cardiac conditioning in the guinea pig.

Within the MGv, some weak and transient receptive field plasticity developed (figure 71.7A; Edeline and

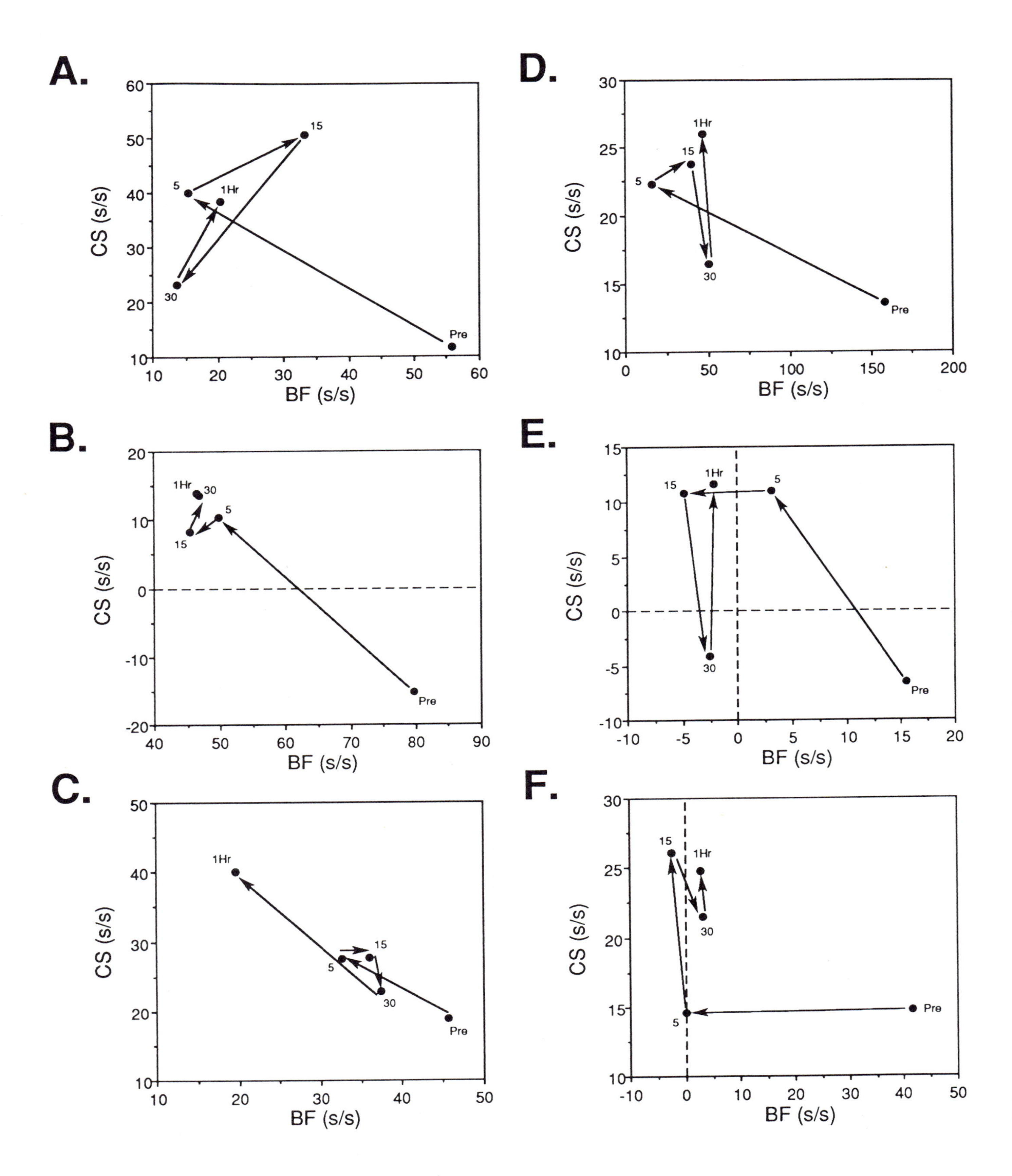
A.
CS (s/s)
BF (s/s)
Pre
5
15
30
1Hr
B.
CS (s/s)
BF (s/s)
Pre
5
15
30
1Hr
C.
CS (s/s)
BF (s/s)
Pre
5
15
30
1Hr
D.
CS (s/s)
BF (s/s)
Pre
5
15
30
1Hr
E.
CS (s/s)
BF (s/s)
Pre
5
15
30
1Hr
F.
CS (s/s)
BF (s/s)
Pre
5
15
30
1Hr

Weinberger, 1991). Robust receptive field plasticity developed in the MGm, which contains two types of tuning: narrowly tuned cells, as is the case for all MGv neurons, and very broadly tuned cells. Receptive field plasticity of narrowly tuned cells was restricted and transient, like that found in the MGv. The receptive fields of most MGm cells are exceptionally broad and multipeaked both before and after training. These broadly tuned cells in the MGm developed stronger receptive field plasticity (Edeline and Weinberger, 1992; figure 71.7B).

Is receptive field plasticity in the auditory cortex generated within the cortex itself to any extent, or is it due to the "projection" of receptive field plasticity from the medial geniculate? The weak and transient nature of receptive field plasticity in the MGv rules out this nucleus as the only source of cortical plasticity. The MGm also is unlikely to be the only source of cortical receptive field plasticity because its narrowly tuned cells have the same characteristics as MGv cells, and the receptive fields of its broadly tuned cells are much broader and more complex than those found in the cortex. Therefore, receptive field plasticity in the auditory cortex cannot easily be explained as having developed exclusively in the medial geniculate nucleus and then been "projected" to the auditory cortex.

FIGURE 71.4 Vector diagrams illustrating the development of changes of response to the pretraining best frequency (BF) and to the conditioned stimulus (CS) frequency. The rate of discharge (spikes/s) for these two frequencies is shown for each of the receptive fields: pretraining, 5, 15, and 30 trials, and 1 hr posttraining (arrows connect successive time periods). Ninety percent of recordings meeting the frequency-specific criterion were characterized by a simultaneous decreased response to the BF and increased response to the CS; vectors for this relationship move from the lower right to the upper left parts of the diagrams. Various cases are illustrated in A through E. (A) Opposite changes for the BF and CS after 5 trials, which varied somewhat thereafter but were maintained at 15 and 30 trials and at the 1 hr time point. (B) Opposite changes for the BF and CS after 5 trials with minimal variation at 15 and 30 trials and no change from 30 trials to 1 hr. Note that the pretraining response to the CS frequency was suppressive but that this became excitatory after only 5 trials. (C) Opposite changes at the BF and CS at 5 trials, with small variations at 15 and 30 trials, followed by a very large change at 1 hr ("incubation"). (D) Rapid opposite changes at the BF and CS after 5 trials with some "regression" of these effects at 15 and particularly at 30 trials but nonetheless a maintained effect 1 hr later. (E) Example of opposite changes to the CS frequency and BF after 5 trials, with loss of response to the CS at 30 trials but recovery of the increased response to the CS at 1 hr posttraining. Note that pretraining, the response to the BF was excitatory and the response to the CS frequency was inhibitory; nonetheless, 1 hr posttraining, the signs had reversed: Responses to the CS became excitatory and responses to the BF became inhibitory. (F) An unusual case of sequential rather than simultaneous changes. The response to the BF decreased after 5 trials without an increase to the CS frequency, and then the response to the CS frequency increased after 15 trials, and both changes were thereafter maintained.

Cholinergic Mechanisms in Frequency-Specific Receptive Field Plasticity

Effects of iontophoresis of agents on receptive fields Cholinergic agonists (acetylcholine, ACh, or the specific muscarinic agonist acetyl-beta-methacholine (MCh)) applied to neurons in the primary auditory cortex of the waking cat produced *frequency-specific changes* in receptive fields, including tuning shifts with decreased response to the best frequency and increased responses to an adjacent frequency. These effects could be blocked by atropine (McKenna, Ashe, and Weinberger, 1989). Endogenous ACh also produces tuning changes in the ACx; iontophoretic or micropressure application of anticholinesterases produced tuning shifts that were similar to those obtained with muscarinic agonists (Ashe, McKenna, and Weinberger, 1989). The observations indicate that endogenous ACh, acting at muscarinic receptors, can produce organized modification of receptive fields.

Frequency-specific modification of receptive fields by tone-ACh pairing If ACh released in ACx during training trials participates in the induction of CS specific receptive field plasticity, then pairing a tone with direct iontophoretic application of ACh in place of a US might produce such plasticity. A test of this hypothesis was positive: After such pairing, the major change in tuning was at the frequency of the paired tone, with adjacent tones showing opposite effects (figure 71.8). The pairing effects could be blocked with atropine (Metherate and Weinberger, 1990). Thus, the presence of increased levels of ACh at or near a cortical neuron during a period of excitation with an acoustic stimulus results in plasticity that can be specific to the paired tone. This is consistent with the possibility that cortical ACh is involved in the receptive field plasticity that is induced by conditioning.

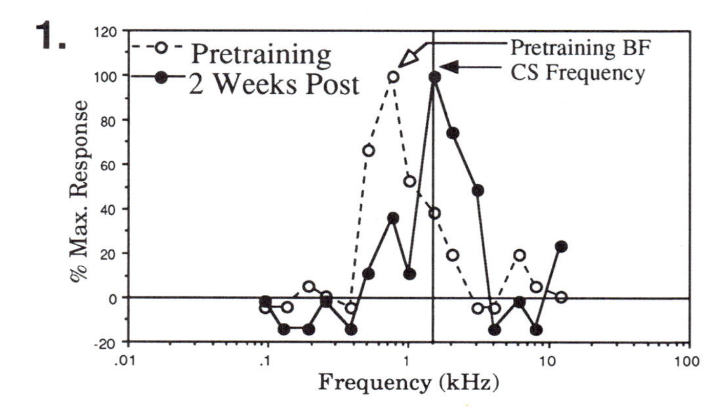

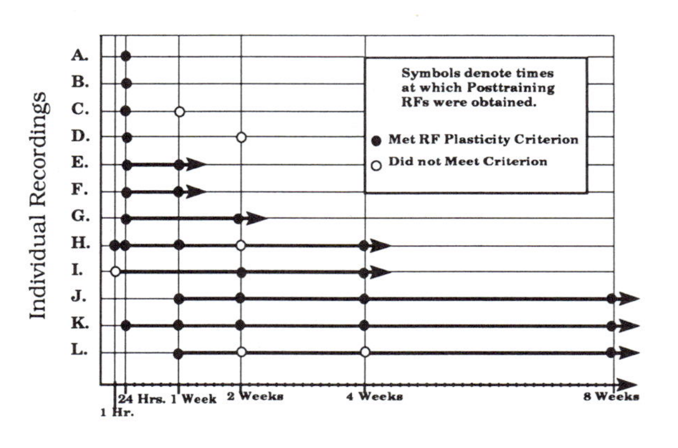

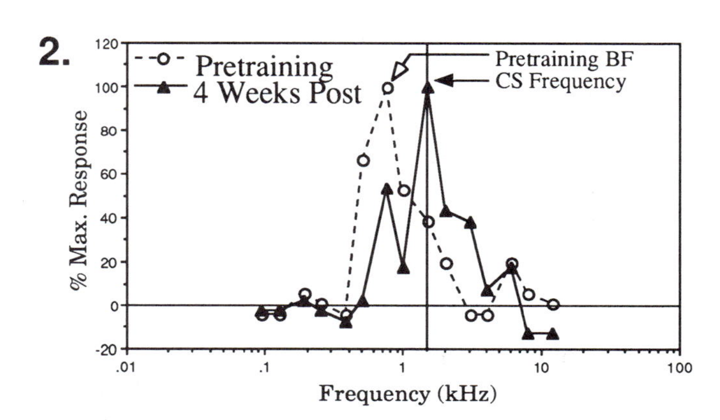

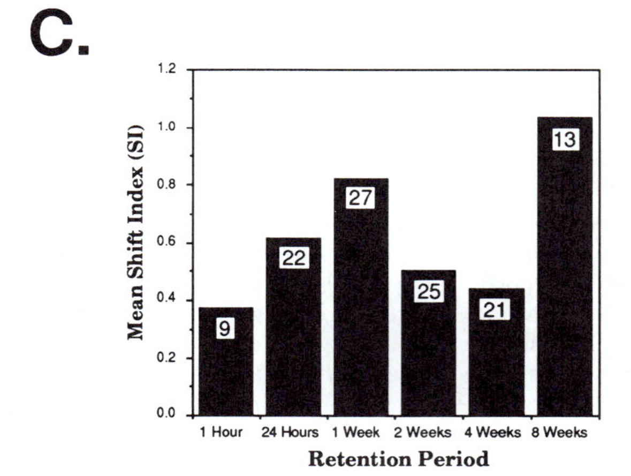

FIGURE 71.5 Long-term retention of conditioned stimulus–specific specific receptive field (RF) plasticity. (A) Frequency receptive fields (60 dB) illustrating long-term retention for four weeks and complete shift of tuning. Pretraining best frequency (BF) = 0.75 kHz, conditioned stimulus (CS) = 1.5 kHz. Note the shift of tuning so that the CS frequency became the new BF, shown here at two (A1) and four (A2) weeks posttraining, the last recording available for this subject. Subject was trained while awake, but RFs were obtained while the subject was under ketamine anesthesia. (B) Group summary for long-term retention of RF plasticity under anesthesia for all subjects. Recordings A and B did not yield data after the 24 hr retention period; recordings C and D did not show retention after 24 hrs. All other recordings exhibited retention of frequency-specific plasticity at the last retention interval for which acceptable recordings could be obtained. Recordings H and L failed to meet a statistical criterion during one or more intermediate retention sessions, and recording I was not tested at 24 hrs or 1 week, but showed RF plasticity at 2 and 4 weeks. (C) Group data showing average amount of shift of the BF toward the CS frequency at retention intervals of 1 hour to 8 weeks. A shift index (SI) of 1.0 indicates a shift from the BF to the CS; a shift of 0.5 denotes a shift 50% of the frequency distance to the CS frequency. Note that tuning shifts are generally 50% or more and are close to 100% at 8 weeks. The numbers denote the number of tuning curves for each average SI.

An integrative model of learning-induced receptive field plasticity in ACx

A preliminary model that incorporates the major features of receptive field plasticity in the ACx and its possible substrates has helped to organize the findings and has received experimental support (Weinberger et al., 1990a; Weinberger, Ashe, and Edeline, in press). Briefly, we hypothesize that three systems that project to the ACx act synergistically to shift tuning to or toward the CS, by strengthening cortical synapses for this frequency and by weakening synapses for other

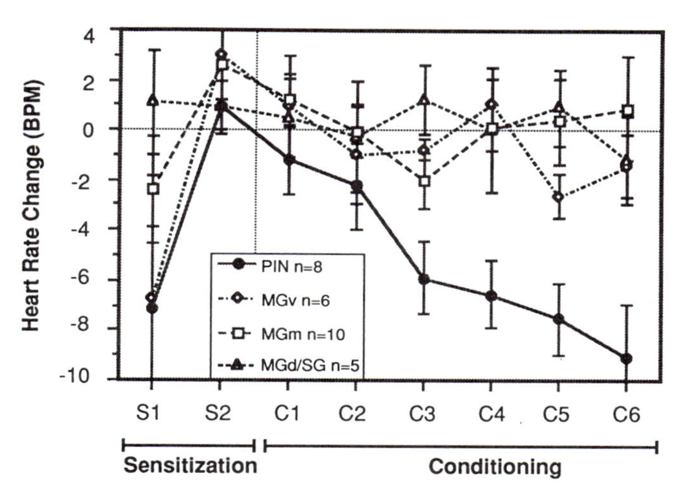

FIGURE 71.6 The effects of stimulation within the medial geniculate nucleus used as the unconditioned stimulus to establish cardiac conditioning. Shown are subgroup learning curves according to the MG subdivision in which the stimulating electrode was located. PIN, posterior intralaminar nucleus; MGm, medial division of the MG; MGv, ventral division of the MG; MGd/SG, dorsal division of the MG and suprageniculate. MGd and SG are combined into one group. Ordinate represents the magnitude of the heart rate response during the tone conditioned stimulus (CS), measured as change from baseline in beats per minute (BPM). Abscissa represents trial blocks across the training session. The sensitization phase (S1 and S2) consists of explicitly unpaired presentations of the conditioned and unconditioned stimuli. C1 through C6 compose the classical conditioning phase. For the conditioning phase, each point represents the mean conditioned response (CR) for the group indicated, averaged over ten trials. In the sensitization phase the trials were divided into 2 blocks of five in order to illustrate the initial orienting response to the tone (see S1) and its subsequent decrement (S2). Error bars indicate standard error. Note the rapid and robust development of cardiac responses during the conditioning phase for the PIN group, while the other 3 anatomical groups have near-zero conditioned responses.

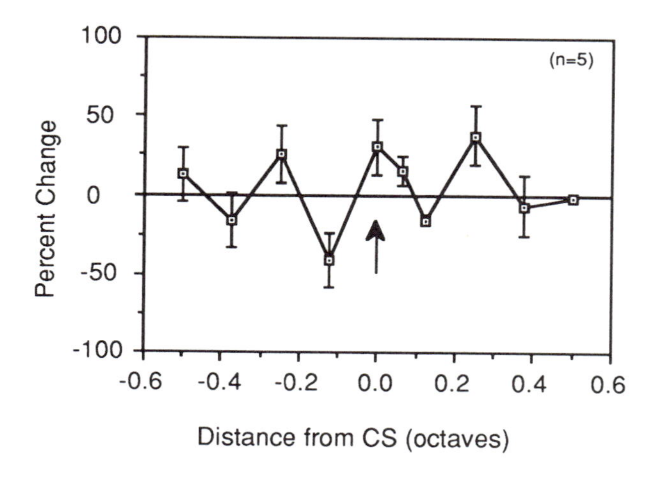

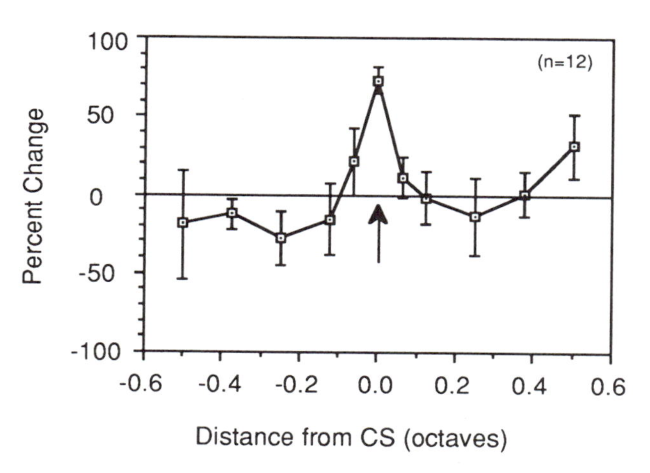

FIGURE 71.7 The effects of classical conditioning on receptive fields in the medial geniculate nucleus, the thalamic source of input to the auditory cortex. (A) Group receptive field (RF) difference functions for the ventral (lemniscal) medial geniculate nucleus, 1 hr postconditioning. There was no systematic change in receptive fields. (B) Group data for the RF plasticity in the magnocellular (nonlemniscal) medial geniculate nucleus, 1 hr postconditioning. Note the CS-specific changes in the receptive field, with a 73% increase at the CS frequency; the bandwidth of this increase at 50% of maximum was less than ±0.1 octaves, and side-band suppression at more distant frequencies is evident.

frequencies. These systems are: (1) the auditory lemniscal non-plastic line from the MGv; (2) the auditory nonlemniscal plastic path from the MGm; and (3) neuromodulatory, providing nonauditory, cholinergic influences across the auditory cortex from the nucleus basalis (NBM) (figure 71.9).

During CS-US pairing trials, the MGv provides unaltered, detailed frequency input to layer IV of the auditory cortex. In contrast, the MGm, which receives input from both the CS and US, is thought to be the first site of associative plasticity. However, its broad tuning and complex response properties provide little if any detailed frequency information to the auditory cortex. Rather, it projects an increased response to the

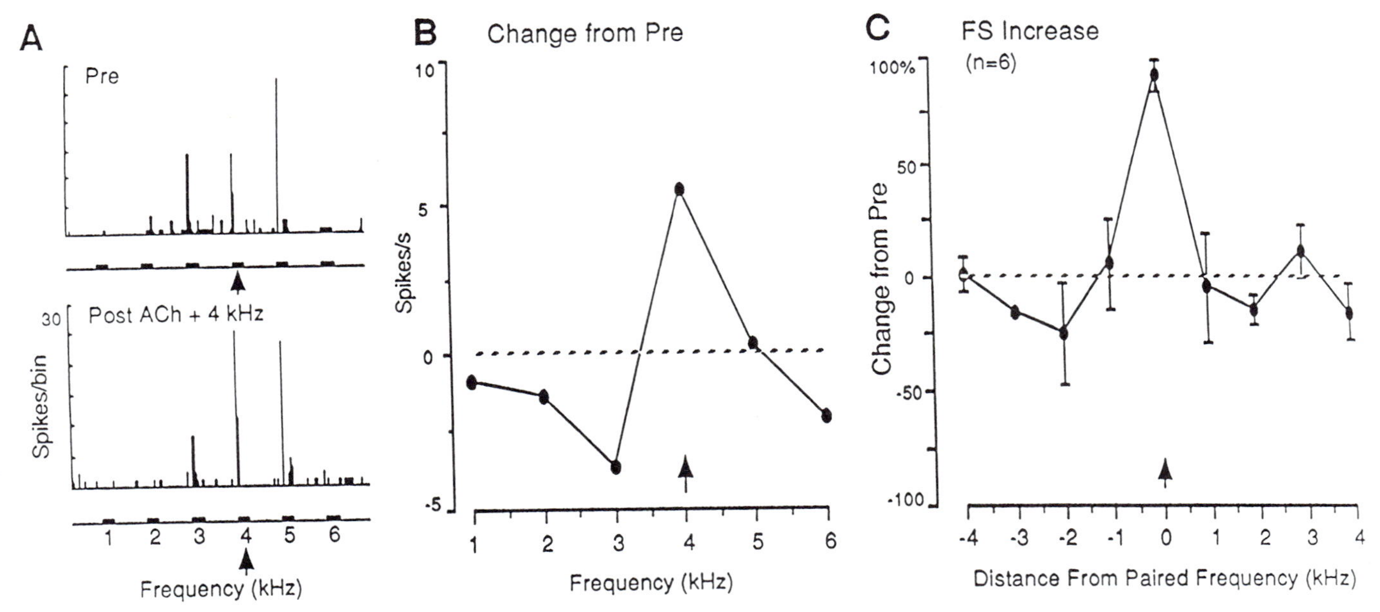

FIGURE 71.8 Cholinergic "substitution" for an unconditioned stimulus. CS-specific RF plasticity due to pairing a tone with the administration of ACh (no behavioral training). (A) Response to tones (80 dB) prior to pairing of ACh (30 nA) with 4 kHz is shown in histogram on top. Bottom histogram depicts response after the ACh-tone pairing. Histogram bins 10 ms. (B) Quantification of the receptive field change shows that the ACh treatment facilitated the response to the frequency of the paired tone (arrow), whereas responses to adjacent frequencies were unchanged or reduced. (C) Mean (±SE) receptive field changes of six cells that displayed CS-specific RF plasticity. Frequency of paired tone is at 0 (arrow), and altered responses to adjacent tones are plotted on either side in 1 kHz steps. Note the high degree of specificity of the pairing effect. FS, frequency specific.

CS to the apical dendrites in layers I and II (perhaps also to basal dendrites) of pyramidal cells in lower layers. The MGm/PIN also provides information on the CS-US association to the amygdala, which initiates autonomic and certain somatic conditioned responses and also initiates an increased release of acetylcholine (ACh) via its projections to the nucleus basalis. The release of ACh amplifies the input from the MGm on the apical dendrites of pyramidal cells, by increasing membrane resistance without altering membrane potential. This is thought to produce a widespread enhancement of postsynaptic activation during training trials. Via extended Hebbian rules (further explicated later), MGv effects on cortical pyramidal cells are strengthened for the CS frequency, because it is the only frequency active during CS presentation, so that CS input axons are active at the time that postsynaptic cells are excited by the MGm and nucleus basalis inputs. This sequence of hypothesized increased strength of CS synapses in the MGm and in the auditory cortex is summarized in figure 71.10.

Synapses for non-CS frequencies in the auditory cortex are thought to be weakened, as follows. During training trials, non-CS inputs to the auditory cortex (and also to the MGm) are inactive because only the CS frequency is presented during paired conditioning trials in which the conditoned stimulus is followed by the unconditioned stimulus. Nonetheless, postsynaptic pyramidal cells in the auditory cortex are excited by the MGm and nucleus basalis influences, as explained above. Via modified Hebbian rules, this "mismatch" results in a decrease in synaptic strength for non-CS frequency synapses. The overall result is seen in post-training cortical receptive fields as increased responses to the CS frequency and decreased responses to many other frequencies (figure 71.11A).

In summary, the hypothesis is that synapses are first changed in the MGm and next changed in the synapses between layer IV cells (which receive frequency-specific input from the MGv) and layer V and VI cortical pyramidal cells. The cholinergic input is thought to modulate the amount of synaptic change

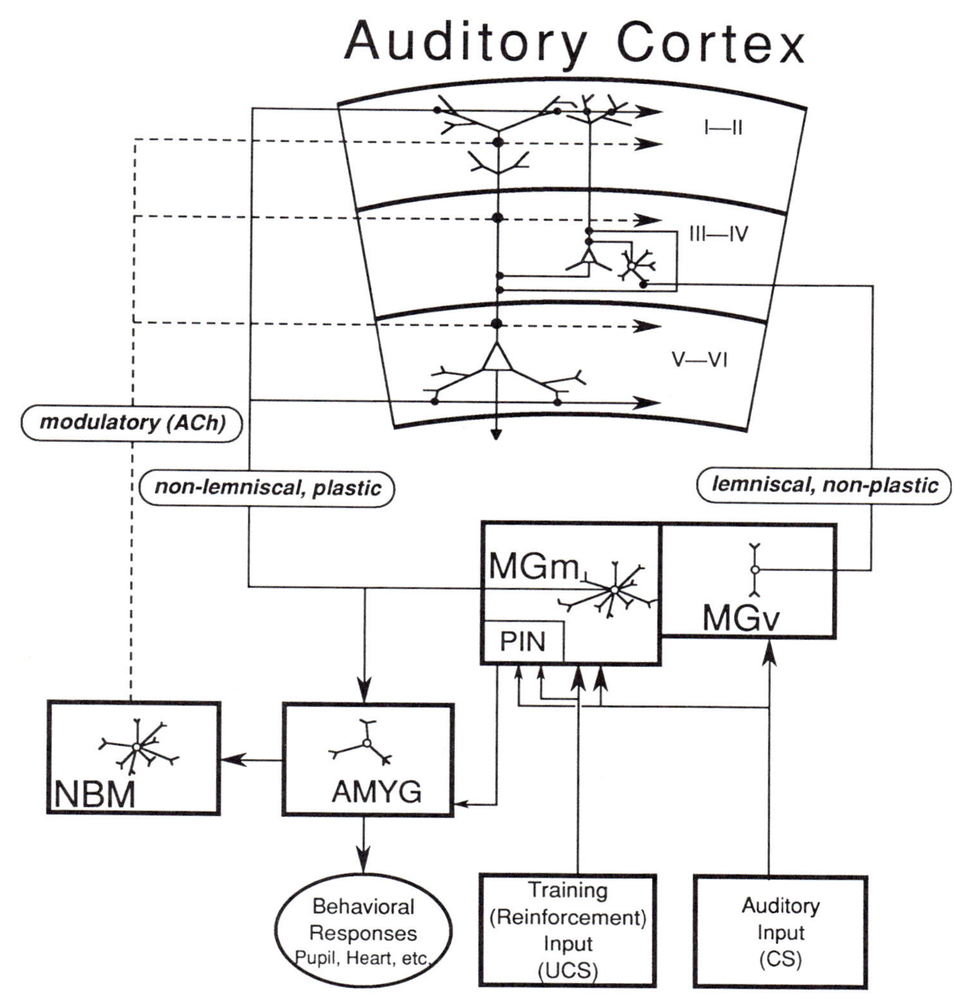

FIGURE 71.9 Diagram showing the major components of the model of conditioning-induced, rapidly developing auditory plasticity and behavioral conditioned responses. The mechanisms for associative CS-specific receptive field plasticity in the auditory cortex are based on the convergence of three subcortical systems at the auditory cortex, which (1) provide detailed frequency information ("lemniscal, non-plastic"), (2) indicate the behavioral significance of a current acoustic stimulus ("non-lemniscal, plastic"), and (3) produce neuromodulation based on the importance of the current auditory stimulus ("modulatory (ACh)"). See text and figure 71.10 for detailed explanation. AMYG, amygdala; MGm, magnocellular medial geniculate nucleus; MGv, ventral medial geniculate nucleus; NBM, nucleus basalis of Meynert; PIN, posterior intralaminar nucleus. Roman numerals refer to cortical laminar zones.

rather than to be essential for such changes. Thus, the MGm input itself may produce plasticity, but the strength and the duration of the plasticity may depend upon the muscarinic effect of the nucleus basalis on the cortex. Also, the cortical domain of ACh is very extensive compared to that of the MGm, which is confined to auditory cortical fields. Thus, the nucleus basalis effect may extend across the cortical mantle, serving to "bind" together the plasticities of various cortical fields, given that learning is likely to transcend the sensory modality of the conditioned stimulus, including information such as where and when conditioning occurred.

The operation of simple extended Hebbian rules is summarized in figure 71.11B. There are four possible combinations of active or inactive states of pre- or postsynaptic elements. Note that two of these combinations could in principle account for the strengthening of CS synapses (labeled 1 in the figure) and the weakening

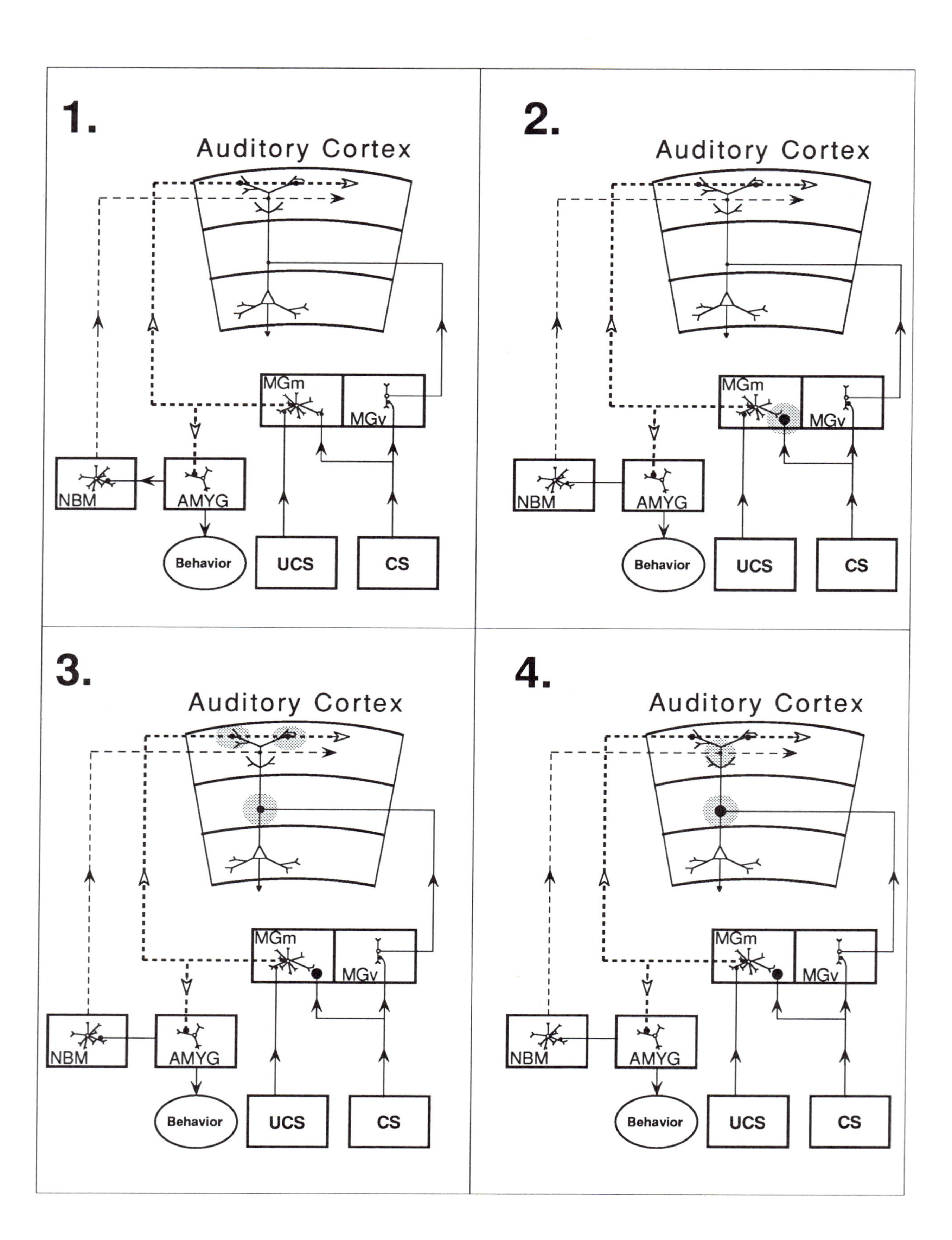

1.
Auditory Cortex
MGm
MGv
NBM
AMYG
Behavior
UCS
CS
2.
Auditory Cortex
MGm
MGv
NBM
AMYG
Behavior
UCS
CS
3.
Auditory Cortex
MGm
MGv
NBM
AMYG
Behavior
UCS
CS
4.
Auditory Cortex
MGm
MGv
NBM
AMYG
Behavior
UCS
CS

of non-CS synapses (2) during classical conditioning. Moreover, a third combination, specifically, presynaptic active input with a nonactive postsynaptic element (3), could account for a frequency-specific decrement in habituation to a repeated tone, as summarized previously (Condon and Weinberger, 1991). Although this is likely to be an oversimplified account of specific receptive field plasticity in conditioning and habituation, the extended Hebbian rules have unexpectedly great explanatory power.

Experimental Support The model explicitly predicts that receptive field plasticity should develop across the frequency representation such that the CS frequency should have an expanded representation while other frequencies should have a reduced representation in the tonotopic frequency map of the cortex (Weinberger et al., 1990a). Direct support for this prediction has recently been reported by the laboratory of M. Merzenich, in which owl monkeys were trained in a frequency discrimination task. Extensive mapping of the frequency representation revealed that, compared to control conditions, the area of representation of the frequency of the CS+ tone was significantly increased (Recanzone, Schreiner, and Merzenich, 1993).

Also, the model posits that ACh acting at muscarinic synapses should potentiate the response of cortical neurons to ascending auditory input. This effect has been found in two types of studies. Stimulation of the nucleus basalis does facilitate responses in the auditory cortex that are elicited by electrical stimulation of the medial geniculate nucleus. Moreover, this facilitation is blocked by atropine (Metherate and Ashe, 1991). Stimulation of the nucleus basalis produces long-duration facilitation of neuronal responses to tones in auditory cortex of the waking rat. The facilitation was found in the ipsilateral but not the contralateral auditory cortex and was blocked by atropine (Hennevin et al., 1992).

A third type of support has been summarized above. Pairing a tone with the application of ACh to the ACx can produce frequency-specific receptive field plasticity that is similar to that induced by conditioning and is blocked by atropine (Metherate and Weinberger, 1990; see also Lin and Phillis, 1991).

FIGURE 71.10 Sequence of hypothesized increases in CS synaptic strength during a conditioned stimulus (CS)–unconditioned stimulus (UCS) trial, in the magnocellular medial geniculate nucleus (MGm) and primary auditory cortex (ACx). Synaptic strengthening is denoted by increased size of "synaptic dots" and location/time of action is denoted by a gray circle. (1) Pretrial status. (2) Initial change in the MGm due to CS-UCS convergence. (3) Immediate subsequent increased strength of the input from the ventral medial geniculate nucleus to pyramidal cells due to convergence of unchanged CS frequency input from the MGv with increased output from the MGm to apical dendrites. The MGm output also activates the amygdala, initiating behavioral conditioned responses and also activation of the nucleus basalis. (4) Additional strengthening of the CS synapses from the MGv to pyramidal cells due to the release of acetylcholine from the nucleus basalis, acting at muscarinic synapses, on apical dendrites of pyramidal cells.

Problems and future directions

Learning-induced receptive field plasticity is important in that it distinguishes between excitability and information in the neurophysiology of conditioning; provides findings that fulfill neuronal criteria for behaviorally demonstrated modification of stimulus representation; and establishes that both of these fundamental processes involve sensory neocortex. These findings may provide a neurobiological basis for the important finding from purely behavioral studies that the representation of a conditioned stimulus is changed by learning (e.g., Holland, 1990). The discovery and characterization of specific, associative receptive field plasticity in the auditory cortex leads in two directions: reductionistic, to the search for its mechanisms; and synthetic, to the determination of its role in the storage of information and the later use of that information in cognitive processes and behavior.

Mechanisms Regarding mechanisms, an immediate problem is that the architecture of the neocortex is considerably more complex than that of, for example, the hippocampus. The remarkable accomplishments in understanding synaptic plasticity in the hippocampus depended upon knowledge of the involved circuitries; physiological characterization of its components; and analysis of the processing "flow" among its components. The same stages of inquiry are required for the neocortex. The extensive laminar organization of the neocortex has long been recognized as its distinctive feature, and one that must be fundamental to its processing, and quite likely to information storage and cognition. In view of the period of about two decades

A. Possible Changes in Synaptic Strength Underlying Receptive Field Plasticity

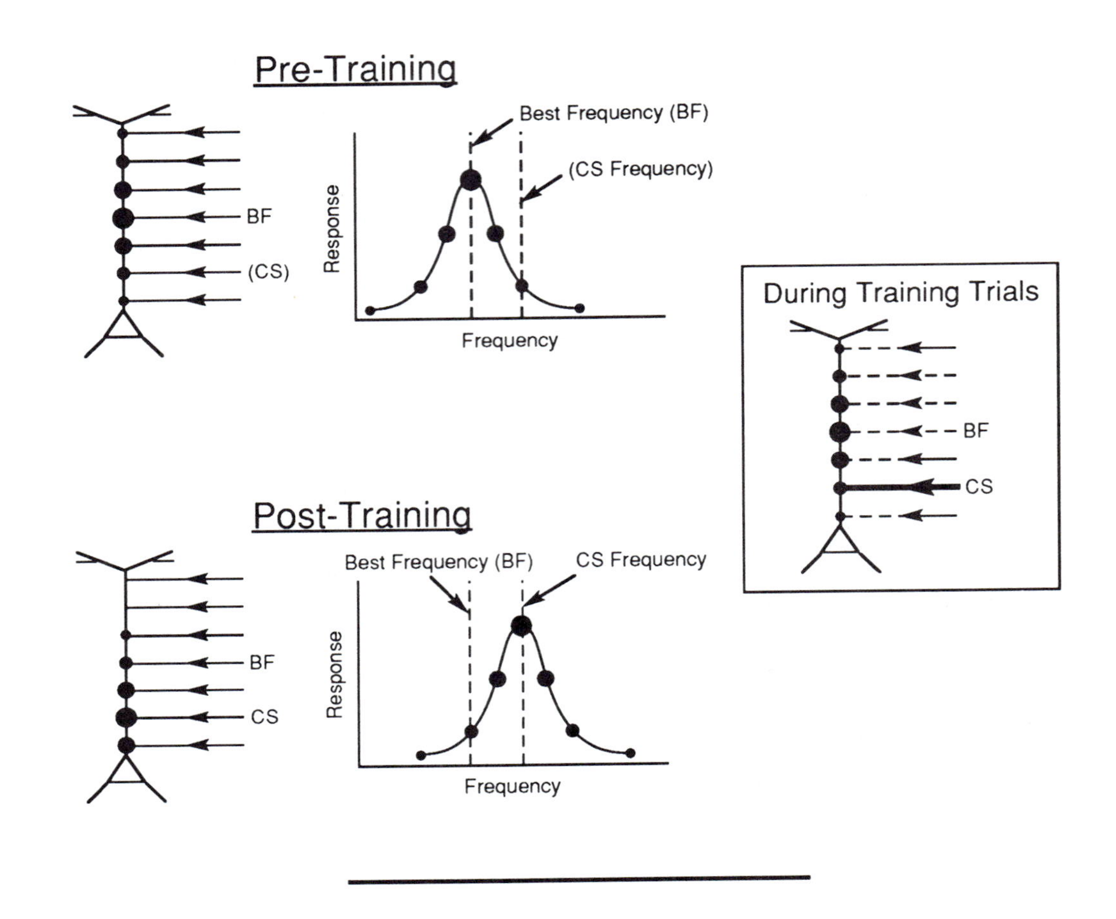

B. Hebbian Rules for Changing Synaptic Strength

		Pre-Synaptic Input Active?	
		Yes	No
Post-Synaptic Cell Depolarized?	Yes	1 *strengthen* (CS, conditioning)	2 *weaken* (non-CS, conditioning))
	No	3 *weaken* (CS, habituation)	4 *no change* (non-CS, Habituation)

of intense work on hippocampal physiology and plasticity, which is not yet complete, it appears that we may not be able to expect equivalent success in the neocortex in a shorter period. A fundamental difference between the hippocampus and the neocortex is that the former appears to be necessary for long-term memory for certain types of learning, but is not the site of long-term storage. In contrast, it seems very likely that the neocortex is an important site for storage of such memories. There is reason to believe that the storage of information may occur in cortex where it is processed (Squire, 1987). Thus, in addition to understanding cortical function in emotional and other types of learning, it will be necessary to comprehend the relationships between the neocortex and other structures, including the hippocampus.

There are also formidable technical problems. Chief among these is the continuing dearth of routine ways to determine the functional organization of ensembles of neurons. We must bear in mind that all techniques both provide insights into brain function and also limit views.

On the positive side, within the past few years, there has been a pronounced increase of research on plasticity in the cerebral cortex. Moreover, many sensory neurophysiology laboratories are well equipped to study receptive fields. The addition of learning treatments to sensory experiments is not an insurmountable problem. Thus, a greater focus on receptive field plasticity and learning, both emotional and otherwise, may well be forthcoming. Additionally, the approach of determining the effects of behavioral experiences on receptive fields need not be limited either to elementary sensory dimensions or to classical conditioning. Rather, receptive field analysis could be applied to any behavioral treatment or controlled cerebral process and to complex sensory events, provided that a systematic relationship can be shown between a controlled input dimension and some property of cellular response.

FIGURE 71.11 Application of extended Hebbian rules to receptive field (RF) plasticity for classical conditioning and habituation. (A) Possible changes in synaptic strength underlying receptive field plasticity during classical conditioning. Diagrams show inputs to a pyramidal cell of varying synaptic strength (denoted by size of "synaptic dots") from various frequencies that produce the excitatory receptive field of the target cell, and the resultant RFs based on these synaptic strengths for pretraining and posttraining. The inset shows that during training trials, only the conditioned stimulus (CS) frequency is present (heavy arrow). The effect of classical conditioning is hypothesized to increase synaptic strength to the CS frequency and to decrease synaptic strength to most other frequencies. (B) Hebbian rules for changing synaptic strength. Conditioning: During each trial, the correlation of active states of the CS input and the postsynaptic cell would strengthen the CS synapses (1) while the discorrelation caused by absence of activity in non-CS inputs would weaken synapses for non-CS frequencies (2). Habituation: During repeated presentation of a CS alone, the activation of the CS input would be discorrelated with the state of the postsynaptic cell because of the absence of unconditioned stimulus effects, which are necessary to activate this cell (3), i.e. either via the projection from the MGm or from the nucleus basalis, or both. There should be no change in the strength of non-CS synapses because of the absence of both acoustic input and activation (4).

FUNCTIONS OF RECEPTIVE FIELD PLASTICITY With reference to the roles of learning-induced receptive field plasticity, it would be premature to conclude that there is yet an acceptable understanding of its functions, for several reasons.

First, the findings are relatively recent, and, although they appear to be yielding a reasonably coherent picture, they are incomplete. It will be necessary to more fully determine the behavioral circumstances under which receptive field plasticity develops. Such studies would involve both more complex aspects of classical conditioning and also other behavioral situations.

Second, although receptive field plasticity satisfies several criteria for the storage of information (rapid development, specificity, and indefinite retention), there is not yet a generally accepted, incisive set of criteria for the neurophysiological or other functional identification of a memory.

Third, neurobiological approaches to brain function remain unable to "read out content," being limited largely to statements regarding the form, distribution, amount, and change in neural excitability or responsivity.

Fourth, the Zeitgeist in much behavioral neurophysiology continues to remain in the long shadow of Descartes, that is, in the search for stimulus-response circuitry. While this is of importance, most learning produces the enduring storage of information that can be used not only for immediate conditioned responses but also in unanticipated ways in the future. Contemporary understanding of classical conditioning shows that it includes cognitive functions. However, these cognitive functions continue to be ignored in neurobiology, due in part to the apparent simplicity of the procedures in classical conditioning. These cognitive aspects of conditioning make it more difficult to predict a future behavioral outcome from a present change in receptive fields than to predict an immediate, specific, motor conditioned reflex from a change in a low-order synaptic chain.

Overall, it is important to acknowledge the relatively primitive state of current research with respect to both conceptual and technical problems. The acquisition, storage, and ultimate use of information for the production of behavior constitutes a central problem of neuroscience, if not *the* central problem. It involves all levels of organization of brain structure and function, from the molecule through cognition and goal-directed behavior. The novel finding that emotional learning, specifically fear conditioning, involves highly specific modifications of receptive fields for auditory frequencies may have its greatest significance by redirecting attention, thought, and research in neuroscience.

ACKNOWLEDGMENTS This research was supported by the following grants: MH 11250, 51342, 11095, 22712, 05440, 14599, 05424, 05440; BNS 76-81924; NS 16108; ONR N00014-84-K-0391, N00014-87-K-0043, N00014-91-J-1193; DARPA N00014-89-J-3178 to the author; and an unrestricted grant from the Monsanto Company. I wish to give special acknowledgment to John Ashe, Jon Bakin, Scott Cruikshank, Jean-Marc Edeline, Roxanna Javid, Robert Lennartz, Tom McKenna, Raju Metherate, and Dave South, and to gratefully acknowledge the technical support of Ann Markham, Manprit Dhillon, Duc Pham, Branko Lepan, Phuc Pham, Thu Huynh. I thank Gabriel Hui for computer programming and support, and Jacquie Weinberger. I also wish to give special thanks to Jeff Winer for his advice on the morphological organization of the auditory system, and to the late Howard Schneiderman.

REFERENCES

Ashe, J. H., T. M. McKenna, and N. M. Weinberger, 1989. Cholinergic modulation of frequency receptive fields in auditory cortex: II. Frequency-specific effects of anticholinesterases provide evidence for a modulatory action of endogenous ACh. *Synapse* 4:44–54.

Bakin, J. S., and N. M. Weinberger, 1990. Classical conditioning induces CS-specific receptive field plasticity in the auditory cortex of the guinea pig. *Brain Res.* 536:271–286.

Bakin, J. S., B. Lepan, and N. M. Weinberger, 1992. Sensitization induced receptive field plasticity in the auditory cortex is independent of CS-modality. *Brain Res.* 577:226–235.

Condon, C. D., and N. M. Weinberger, 1991. Habituation produces frequency-specific plasticity of receptive fields in the auditory cortex. *Behav. Neurosci.* 105:416–430.

Cruikshank, S. J., J-M Edeline, and N. M. Weinberger, 1992. Stimulation at a site of auditory-somatosensory convergence in the medial geniculate nucleus is an effective unconditioned stimulus for fear conditioning. *Behav. Neurosci.* 106:471–483.

Diamond, D. M., and N. M. Weinberger, 1986. Classical conditioning rapidly induces specific changes in frequency receptive fields of single neurons in secondary and ventral ectosylvian auditory cortical fields. *Brain Res.* 372:357–360.

Diamond, D. M., and N. M. Weinberger, 1989. Role of context in the expression of learning-induced plasticity of single neurons in auditory cortex. *Behav. Neurosci.* 103: 471–494.

Edeline, J.-M., and N. M. Weinberger, 1991. Thalamic short term plasticity in the auditory system: Associative retuning of receptive fields in the ventral medial geniculate body. *Behav. Neurosci.* 105:618–639.

Edeline, J.-M., and N. M. Weinberger, 1992. Associative retuning in the thalamic source of input to the amygdala and auditory cortex: Receptive field plasticity in the medial division of the medial geniculate body. *Behav. Neurosci.* 106:81–105.

Edeline, J.-M., and N. M. Weinberger, 1993. Receptive field plasticity in the auditory cortex during frequency discrimination training: Selective retuning independent of task difficulty. *Behav. Neurosci.* 107:82–103.

Edeline, J.-M., P. Pham, and N. M. Weinberger, 1993. Rapid development of learning-induced receptive field plasticity in the auditory cortex. *Behav. Neurosci.* 107:539–551.

Gerren, R., and N. M. Weinberger, 1983. Long term potentiation in the magnocellular medial geniculate nucleus of the anesthetized cat. *Brain Res.* 265:138–142.

Gonzalez-Lima, F., 1992. Brain imaging of auditory learning functions in rats: Studies with fluorodeoxyglucose auto radiography and cytochrome oxidase histochemistry. In *Advances in Metabolic Mapping Techniques for Brain Imaging of Behavioral and Learning Functions*, F. Gonzalez-Lima, T. Finkenstadt, and H. Scheich, eds. Boston: Dordrecht, pp. 39–109.

Hennevin, E., J.-M. Edeline, B. Hars, and C. Maho, 1992. Basal forebrain (BF) stimulation enhances tone-evoked responses in the auditory cortex of awake rats. *Fifth Conference on the Neurobiology of Learning and Memory* 5:40.

Holland, P. C., 1990. Forms of memory in Pavlovian conditioning. In *Brain Organization and Memory: Cells, Systems, and Circuits*, J. L. McGaugh, N. M. Weinberger, and G. Lynch, eds. New York: Oxford University Press, pp. 78–105.

Jones, E. G., 1985. *The Thalamus*. New York: Plenum Press.

LeDoux, J. E., 1990. Information flow from sensation to emotion: Plasticity in the neural computation of stimulus value. In *Learning and Computational Neuroscience: Foundation of adaptive networks*, M. Gabriel and J. Moore, eds. Cambridge, Mass.: MIT Press, Bradford Books, pp. 3–51.

Lennartz, R. C., and N. M. Weinberger, 1992a. Analysis of response systems in Pavlovian conditioning reveals rapidly vs slowly acquired conditioned responses: Support for two factors, implications for neurobiology. *Psychobiology* 20:93–119.

Lennartz, R. C., and N. M. Weinberger, 1992b. Frequency-specific receptive field plasticity in the medial geniculate body induced by Pavlovian fear conditioning is expressed in the anesthetized brain. *Behav. Neurosci.* 106: 484–497.

Lin, Y., and J. W. Phillis, 1991. Muscarinic agonist-mediated induction of long-term potentiation in rat cerebral cortex. *Brain Res.* 551:342–345.

McKenna, T. M., J. H. Ashe, and N. M. Weinberger, 1989. Cholinergic modulation of frequency receptive fields in auditory cortex: I. Frequency-specific effects of muscarinic agonists. *Synapse* 4:30–43.

Metherate, R., and J. H. Ashe, 1991. Basal forebrain stimulation modifies auditory cortex responsiveness by an action at muscarinic receptors. *Brain Res.* 559:163–167.

Metherate, R., and N. M. Weinberger, 1990. Cholinergic modulation of responses to single tones produces tone-specific receptive field alterations in cat auditory cortex. *Synapse* 6:133–145.

Recanzone, G. H., C. E. Schreiner, and M. M. Merzenich, 1993. Plasticity in the frequency representation of primary auditory cortex following discrimination training in adult owl monkeys. *J. Neurosci.* 13:87–103.

Scheich, H., and C. Simonis, 1991. Conditioning changes frequency representation in gerbil auditory cortex. *Soc. Neurosci. Abstr.* 17:450.

Squire, L., 1987. *Memory and Brain.* New York: Oxford University Press.

Weinberger, N. M., 1965. Effect of detainment on extinction of avoidance responses. *J. Comp. Physiol. Psychol.* 60: 135–138.

Weinberger, N. M., 1982. Sensory plasticity and learning: The magnocellular medial geniculate nucleus of the auditory system. In *Conditioning: Representation of Involved Neural Function*, C. D. Woody, ed. New York: Plenum Press, pp. 697–710.

Weinberger, N. M., and D. M. Diamond, 1987. Physiological plasticity of single neurons in auditory cortex: Rapid induction by learning. *Prog. Neurobiol.* 29:1–55.

Weinberger, N. M., J. H. Ashe, D. M. Diamond, R. Metherate, T. M. McKenna, and J. S. Bakin, 1990. Retuning auditory cortex by learning: a preliminary model of receptive field plasticity. *Concepts Neurosci.* 1:91–132.

Weinberger, N. M., J. H. Ashe, and J.-M. Edeline, 1994. Learning-induced receptive field plasticity in the auditory cortex: Specificity of information storage. In *The Memory System of the Brain*, J. Delacour ed. Singapore: World Scientific Publishing.

Weinberger, N. M., R. Javid, and B. Lepan, 1993. Long term retention of learning-induced receptive field plasticity in auditory cortex. *Proc. Natl. Acad. Sci. U.S.A.* 90: 2394–2398.

72 A Theory of Emotion and Consciousness, and Its Application to Understanding the Neural Basis of Emotion

EDMUND T. ROLLS

ABSTRACT It is shown that emotions can usefully be considered as states produced by reinforcing stimuli. The ways in which a wide variety of emotions can be produced, and the functions of emotion, are considered. There is evidence from single-neuron recording and the effects of lesions that the amygdala is involved in the formation of stimulus-reinforcement associations, and that the orbitofrontal cortex is involved with correcting behavioral responses when these are no longer appropriate because of changes in previous reinforcement contingencies. Both structures receive information about primary (unlearned) reinforcers such as taste, smell, and touch, and also about potential learned reinforcers such as visual and auditory stimuli. It is suggested that these brain structures are important in emotion because they are involved in processing primary reinforcers, and particularly because they are involved in stimulus-reinforcement association learning. The information that reaches the amygdala for these functions includes information about faces, which activates a population of neurons in the amygdala and parts of the temporal lobe visual cortex specialized to respond to faces. These regions of visual cortex may be involved in social and emotional responses to faces.

In order to understand the neural basis of emotions, it is helpful to consider what emotions are, and their functions. First, therefore, a brief foundation is provided. A more complete account can be found elsewhere (Rolls, 1990, 1992b). Then the neural bases of emotion are considered. Particular attention is paid to research with nonhuman primates, partly because the developments in primates of the structure and connections of neural systems involved in emotion such as the amygdala and orbitofrontal cortex make studies in primates particularly important for understanding emotion in humans.

EDMUND T. ROLLS Department of Experimental Psychology, University of Oxford, Oxford, England

A theory of emotion

Emotions can usefully be defined as states produced by instrumental reinforcing stimuli (see Rolls, 1990, and earlier work by Millenson, 1967; Weiskrantz, 1968; Gray, 1975, 1987). (Instrumental reinforcers are stimuli that, if their occurrence, termination, or omission is made contingent upon the making of a response, alter the probability of the future emission of that response.) Some stimuli are unlearned reinforcers (e.g., the taste of food if the animal is hungry, or pain), while others may become reinforcing by learning because of their association with such primary reinforcers, thereby becoming secondary reinforcers. This type of learning may thus be called stimulus-reinforcement association, and probably occurs via the process of classical conditioning. If a reinforcer increases the probability of the emission of a response on which it is contingent, it is said to be a positive reinforcer, or reward; if it decreases the probability of such a response it is a negative reinforcer, or punishment. For example, fear is an emotional state that might be produced by a sound (the conditioned stimulus) that has previously been associated with a shock (the primary reinforcer).

The converse reinforcement contingencies produce the opposite effects on behavior. The omission or termination of a positive reinforcer (extinction and time

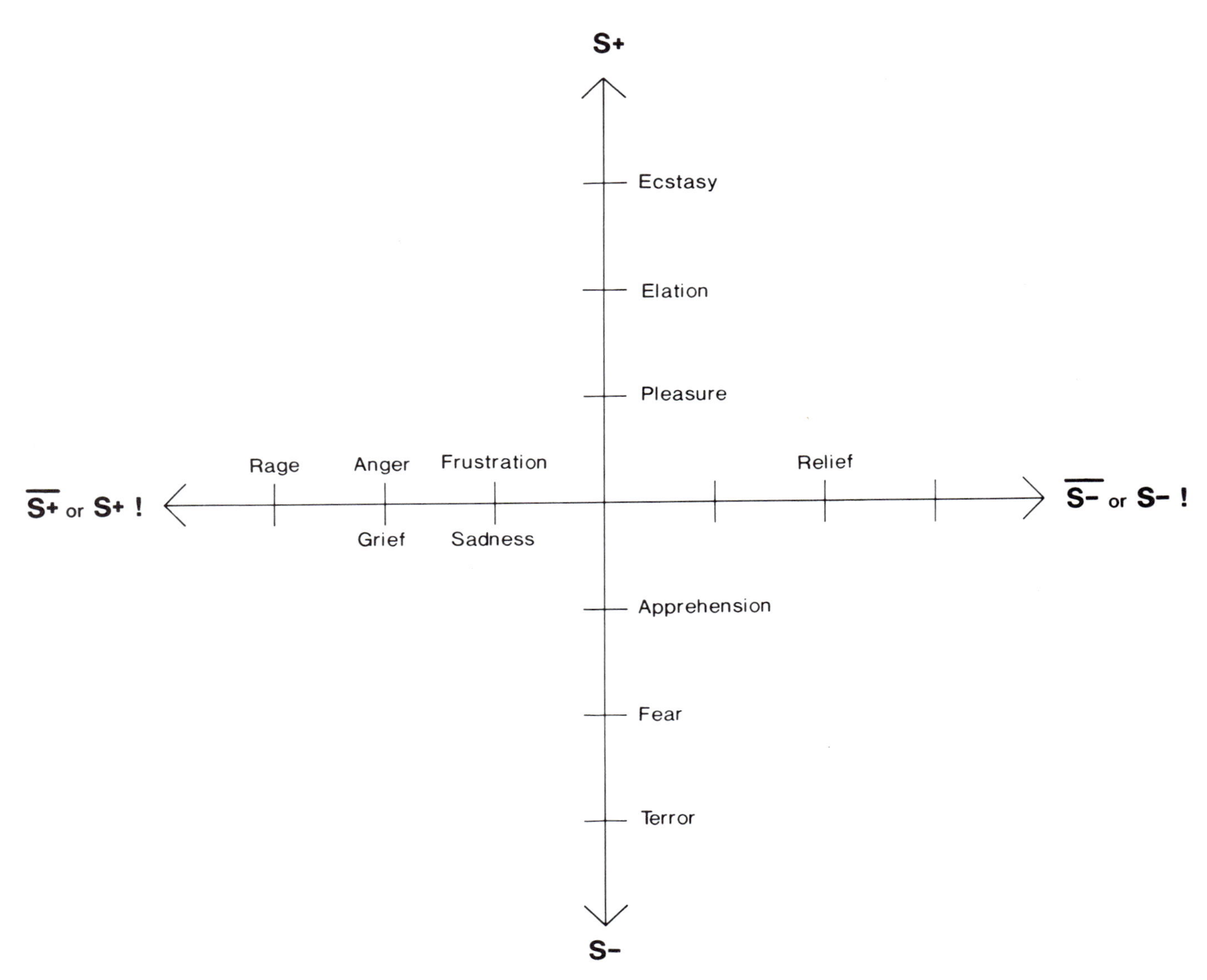

FIGURE 72.1 Some of the emotions associated with different reinforcement contingencies are indicated. Intensity increases away from the center of the diagram, on a continuous scale. The classification scheme created by the different reinforcement contingencies consists of (1) the presentation of a positive reinforcer (S+), (2) the presentation of a negative reinforcer (S−), (3) the omission of a positive reinforcer ($\overline{S+}$) or the termination of a positive reinforcer (S+!), and (4) the omission of a negative reinforcer ($\overline{S-}$) or the termination of a negative reinforcer (S−!).

out, respectively; sometimes described as punishment), decrease the probability of responses. Responses followed by the omission or termination of a negative reinforcer increase in probability, this pair of negative reinforcement operations being termed, respectively, active avoidance and escape (see Gray, 1975, and Mackintosh, 1983). The different emotions can be described and classified according to whether the reinforcer is positive or negative, and by the reinforcement contingency. An outline of such a classification scheme, elaborated more precisely by Rolls (1990), is shown in figure 72.1.

The mechanisms described here are not limited in terms of the range of emotions for which they can account. Some of the factors that enable a very wide range of human emotions to be analyzed with this foundation are elaborated elsewhere (Rolls, 1990), and include the following:

1. The reinforcement contingency (see figure 72.1).

2. The intensity of the reinforcer (see figure 72.1).

3. Any environmental stimulus might have a number of different reinforcement associations. For example, a stimulus might be associated with both the presentation of a reward and the presentation of a punishment, allowing states such as conflict and guilt to arise.

4. Emotions elicited by stimuli associated with different primary reinforcers will be different.

5. Emotions elicited by different secondary reinforcing stimuli will be different from each other (even if the primary reinforcer is similar).

6. The emotion elicited can depend on whether an active or passive behavioral response is possible. For example, if an active behavioral response can occur to the omission of a positive reinforcer, then anger might be produced, but if only passive behavior is possible, then sadness, depression, or grief might occur.

By combining these six factors, it is possible to account for a very wide range of emotions (see Rolls, 1990). Several other points are worth noting: (1) Emotions can be produced by the recall of reinforcing events as well as by external reinforcing stimuli; (2) cognitive processing (whether conscious or not) is important in many emotions, because very complex cognitive processing may be required to determine whether environmental events are reinforcing or not; (3) emotions normally consist of the cognitive processing that determines the reinforcing valence of the stimulus, and the elicited mood change if the valence is positive or negative; and (4) stability of mood implies that absolute levels of reinforcement must be represented over moderately long time spans by the firing of mood-related neurons—a difficult operation that may contribute to seemingly spontaneous mood swings, depression that occurs without a clear external cause, and the multiplicity of hormonal and transmitter systems that seem to be involved in the control of mood (see also Rolls, 1990).

The key point of this introduction is that in order to understand the neural bases of emotion, we must consider brain mechanisms that are involved in reward and punishment, and in learning in which environmental stimuli are associated—or are no longer associated—with rewards and punishments. Before considering this, it will be useful to summarize the functions of emotions, because these functions are important for understanding the output systems with which brain mechanisms involved in emotion must interface. The functions, described more fully elsewhere (Rolls, 1990), can be summarized as follows:

1. The elicitation of autonomic responses (e.g., change in heart rate) and endocrine responses (e.g., the release of adrenalin). These prepare the body for action. There are output pathways from the amygdala and orbitofrontal cortex, directly and via the hypothalamus, to the brainstem autonomic nuclei.

2. Flexibility of behavioral responses to reinforcing stimuli. Emotional (and motivational) states allow a simple interface between sensory inputs and motor outputs, because only the valence of the stimulus to which attention is being paid need be passed to the motor system, rather than a full representation of the sensory world. In addition, when a stimulus in the environment elicits an emotional state (because it is a primary reinforcer or because of classical conditioning), we can flexibly choose any appropriate instrumental response to obtain the reward or avoid the punishment. This is more flexible than simply learning a fixed behavioral response to a stimulus (see Gray, 1975; Rolls, 1990). Pathways from the amygdala and orbitofrontal cortex to the striatum are implicated in these functions.

3. Emotion is motivating. For example, fear learned by stimulus-reinforcement association formation provides the motivation for actions performed to avoid noxious stimuli.

4. Communication. For example, a monkey may communicate its emotional state to others, by making an open-mouth threat to indicate the extent to which it is willing to compete for resources, and this may influence the behavior of other animals. There are neural systems in the amygdala and overlying temporal cortical visual areas that are specialized for the face-related aspects of this processing.

5. Social bonding. Examples of this are the emotions associated with the attachment of the parents to their young, and the attachment of the young to their parents (see Dawkins, 1989).

6. The current mood state can affect the cognitive evaluation of events or memories (see Blaney, 1986), and this may have the function of facilitating continuity in the interpretation of the reinforcing value of events in the environment. A hypothesis on the neural

pathways that implement this function is presented later, in the section "Effects of emotions on cognitive processing."

7. Emotion may facilitate the storage of memories. One way in which this occurs is that episodic memory (i.e., one's memory of particular episodes) is facilitated by emotional states. This may be advantageous in that storing many details of the prevailing situation when a strong reinforcer is delivered may be useful in generating appropriate behavior in similar situations in the future. This function may be implemented by the relatively nonspecific projecting systems to the cerebral cortex and hippocampus, including the cholinergic pathways in the basal forebrain and medial septum, and the ascending noradrenergic pathways (Wilson and Rolls, 1990a, 1990b; Treves and Rolls, 1994). A second way in which emotion may affect the storage of memories is that the current emotional state may be stored with episodic memories, providing a mechanism for the current emotional state to influence which memories are recalled. A third way in which emotion may affect the storage of memories is by guiding the cerebral cortex in the representations of the world that it sets up. For example, in the visual system, it may be useful to build perceptual representations or analyzers that are different from each other if they are associated with different reinforcers, and to be less likely to build them if they have no association with reinforcement. Ways in which backprojections from parts of the brain important in emotion (such as the amygdala) to parts of the cerebral cortex could perform this function are discussed by Rolls (1989a, 1990, 1992b).

8. By enduring for minutes or longer after a reinforcing stimulus has occurred, emotion may help to produce persistent motivation and direction of behavior.

9. Emotion may trigger the recall of memories stored in neocortical representations. Amygdala backprojections to the cortex could perform this function for emotion in a way analogous to that in which the hippocampus could implement the retrieval in the neocortex of recent (episodic) memories (Treves and Rolls, 1994).

The neural bases of emotion

Some of the principal brain regions implicated in emotion will now be considered, including the amygdala, orbitofrontal cortex, and basal forebrain. Some of these brain regions are indicated in figures 72.2 and 72.3. Particular attention is paid to the functions of these regions in primates, for in primates the neocortex undergoes great development and provides major inputs to these regions, in some cases to parts of these structures thought not to be present in nonprimates. An example of this is the projection from the primate neocortex in the anterior part of the temporal lobe to the basal accessory nucleus of the amygdala (see the following section).

The Amygdala

Connections of the amygdala The amygdala in the primate receives massive projections from the overlying temporal lobe cortex (see Van Hoesen, 1981; Amaral et al., 1992). In the monkey these come to overlapping but partly separate regions of the lateral and basal amygdala from the inferior temporal visual cortex, the superior temporal auditory cortex, the cortex of the temporal pole, and the cortex in the superior temporal sulcus. These inputs thus come from the higher stages of sensory processing in the visual and auditory modalities, and not from early cortical processing areas. The amygdala also receives from the part of the insula that receives projections from the somatosensory cortical areas (Mesulam and Mufson, 1982). It also receives projections from the posterior orbitofrontal cortex (see figure 72.2, areas 12 and 13). Subcortical inputs to the amygdala include projections from the midline thalamic nuclei, from the subiculum and CA1 parts of the hippocampal formation, from the hypothalamus and substantia innominata, from the nucleus of the solitary tract (which receives gustatory and visceral inputs), and from olfactory structures (Amaral et al., 1992). Although there are some inputs from early on in some sensory pathways—for example, auditory inputs from the medial geniculate nucleus (LeDoux, 1987, 1992)—this route is unlikely to be involved in most emotions, for which cortical analysis of the stimulus is likely to be required. Emotions are usually elicited by environmental stimuli (including other organisms) analyzed to the object level and not to retinal arrays of spots or pure tones. Consistent with this view that neural systems involved in emotion in primates generally receive from sensory systems where analysis targets the identity of the stimulus rather than its emotional significance or hedonic value, neurons that have responses related to the association with reinforcement of visual stimuli are

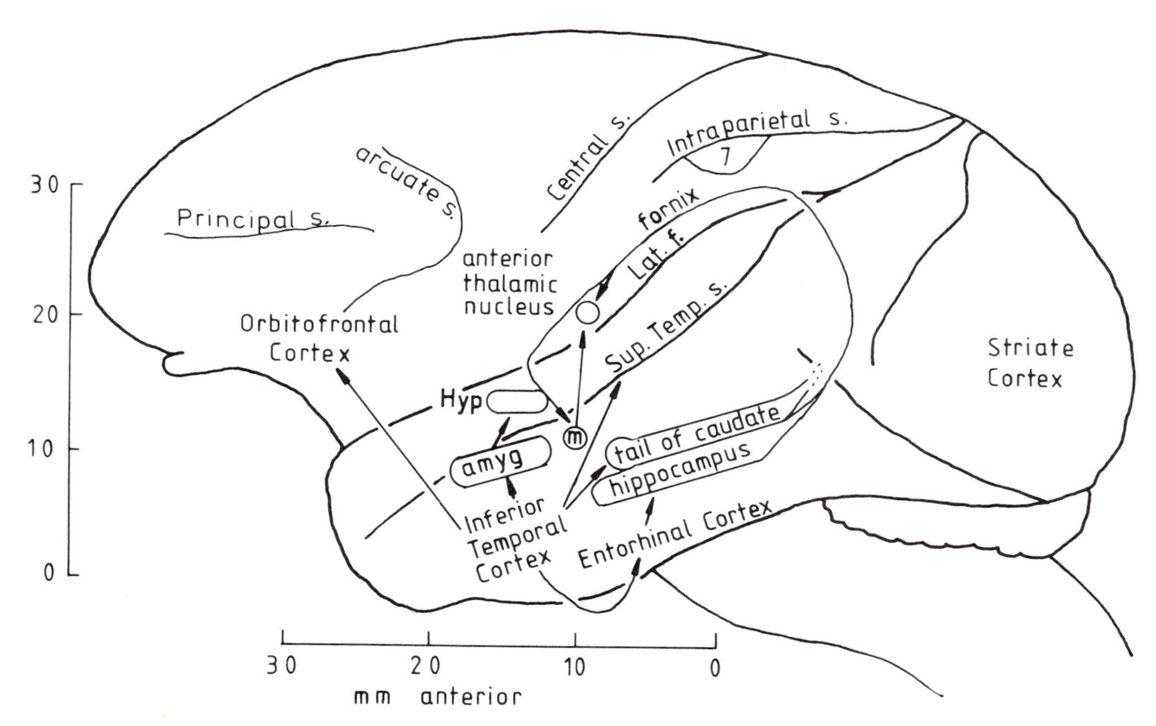

FIGURE 72.2 Some of the pathways described in the text are shown on this lateral view of the rhesus monkey brain. amyg, amygdala; central s, central sulcus; Hyp, hypothalamus/substantia innominata/basal forebrain; Lat f, lateral (or sylvian) fissure; m, mammillary body; Sup Temp s, superior temporal sulcus; 7, posterior parietal cortex (area 7).

not found in the inferior temporal visual cortex (Rolls, Judge, and Sanghera, 1977); whereas such neurons are found in the amygdala and orbitofrontal cortex (see below; cf. figures 72.2 and 72.3). Similarly, processing in the taste system of primates up to and including the primary taste cortex reflects the identity of the stimulus, whereas its hedonic value as influenced by hunger is reflected in the responses of neurons in the secondary taste cortex (Rolls, 1989b, 1994b; see figure 72.3). The outputs of the amygdala (Amaral et al., 1992) include the well-known projections to the hypothalamus, from the lateral amygdala via the ventral amygdalofugal pathway to the lateral hypothalamus; and from the medial amygdala, which is relatively small in the primate, via the stria terminalis to the medial hypothalamus. The ventral amygdalofugal pathway includes some long descending fibers that project to the autonomic centers in the medulla oblongata, and provide a route for cortically processed signals to reach the brain stem. Another interesting output of the amygdala is to the ventral striatum (Heimer, Switzer, and Van Hoesen, 1982) including the nucleus accumbens, for via this route information processed in the amygdala could gain access to the basal ganglia and thus influence motor output. (The output of the amygdala also reaches more dorsal parts of the striatum.) The amygdala also projects to the medial part of the mediodorsal nucleus of the thalamus, which projects to the orbitofrontal cortex and provides the amygdala with another output. In addition, the amygdala has direct projections back to many areas of the temporal, orbitofrontal, and insular cortices from which it receives inputs (Amaral et al., 1992). It is suggested elsewhere (Rolls, 1989a,c) that the functions of these backprojections include the guidance of information representation and storage in the neocortex, as well as recall (when this is related to reinforcing stimuli). Another interesting set of output pathways of the amygdala projects to the entorhinal cortex, which provides the major input to the hippocampus and dentate gyrus, and to the ventral subiculum (Amaral et al., 1992).

These anatomical connections of the amygdala indicate that it is placed to receive highly processed information from the cortex and to influence motor systems, autonomic systems, some of the cortical areas from which it receives inputs, and other limbic areas. The

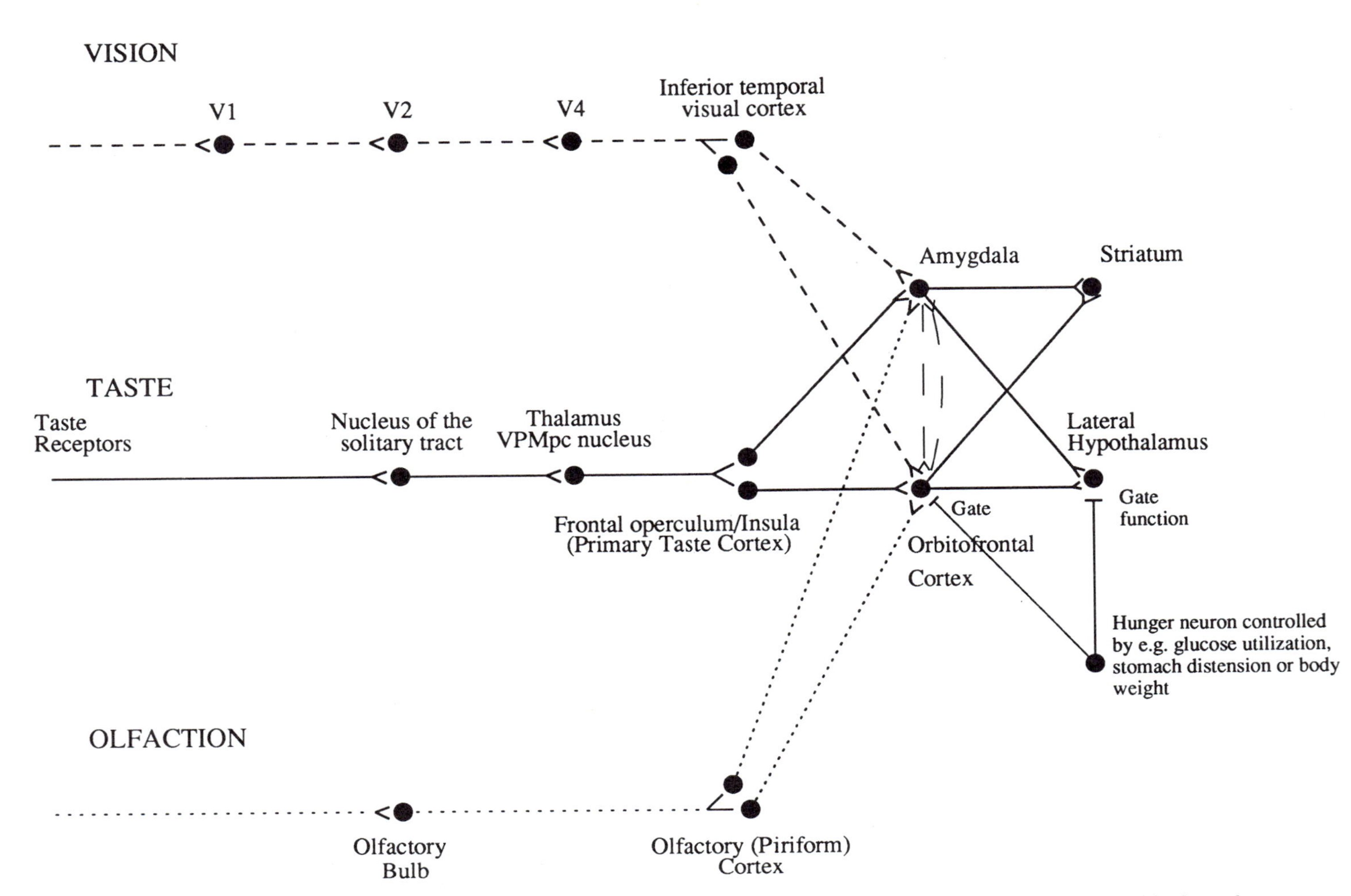

FIGURE 72.3 Diagrammatic representation of some of the connections described in the text. V1, striate visual cortex. V2 and V4, cortical visual areas. In primates, sensory analysis proceeds in the visual system as far as the inferior temporal cortex and the primary gustatory cortex; beyond these areas, in, for example, the amygdala and orbitofrontal cortex, the hedonic value of the stimuli is represented, as well as whether they are reinforcing or are associated with reinforcement (see text).

functions mediated through these connections will now be considered, using information available from the effects of damage to the amygdala and from the activity of neurons in the amygdala.

Effects of amygdala lesions Bilateral removal of the amygdala in monkeys produces tameness, a lack of emotional responsiveness, excessive examination of objects, often with the mouth, and eating of previously rejected items such as meat (Weiskrantz, 1956). These behavioral changes compose much of the Klüver-Bucy syndrome, which is produced in monkeys by bilateral anterior temporal lobectomy (Klüver and Bucy, 1939). In analyses of the bases of these behavioral changes, it has been observed that there are deficits in learning to associate stimuli with primary reinforcement, including both punishments (Weiskrantz, 1956) and rewards (Jones and Mishkin, 1972; Gaffan, 1992; Aggleton, 1993). The association-learning deficit is present when the associations must be learned from a previously neutral stimulus (e.g., the sight of an object) to a primary reinforcing stimulus (such as the taste of food). The impairment is not found when the association learning is between a visual stimulus and an auditory stimulus that is already a secondary reinforcer (because of prior pairing with food). This demonstrates that the amygdala is involved in the learning of associations between neutral stimuli and primary (but not secondary) reinforcers (see Gaffan and Harrison, 1987; Gaffan, Gaffan, and Harrison, 1988, 1989; Baylis and Gaffan, 1991; Gaffan, 1992). Further evidence linking the amygdala to reinforcement mechanisms is that monkeys will work in order to obtain electrical stimulation of the amygdala, and that single neurons in the amygdala are activated by brain-stimulation reward produced at a number of different brain sites (Rolls, 1975; Rolls, Burton, and Mora, 1980). The symptoms of the Kluver-Bucy syndrome, including the emotional changes, could be a result of this type of deficit in learning stimulus-reinforcement associations (Jones and Mishkin, 1972; Aggleton and Passingham, 1981; Mishkin and Aggleton, 1981; Rolls, 1986a,b, 1990, 1992b). For example, the tameness, the hypoemotionality, the increased orality, and the altered responses to food might arise because of damage to the normal mechanism by which stimuli become associated with reward or punishment.

The amygdala is well placed anatomically for learning associations between objects and primary reinforcers, for it receives inputs from the higher parts of the visual system and from systems that process primary reinforcers such as taste, smell, and touch (see figure 72.3). The association learning in the amygdala may be implemented by Hebb-modifiable synapses from visual and auditory neurons onto neurons receiving inputs from taste, olfactory, or somatosensory primary reinforcers. Consistent with this, Davis (1992) has found that at least one type of associative learning in the amygdala can be blocked by local application to the amygdala of an NMDA receptor blocker. Once the association has been learned, the outputs from the amygdala could be driven by the conditioned as well as the unconditioned stimuli. In line with this, LeDoux et al., (1988) were able to show that lesions of the lateral hypothalamus (which receives from the central nucleus of the amygdala) blocked conditioned heart-rate (autonomic) responses. Lesions of the central gray of the midbrain (which also receives from the central nucleus of the amygdala) blocked the conditioned freezing but not the conditioned autonomic response to the aversive conditioned stimulus. Further, Cador, Robbins, and Everitt (1989) obtained evidence consistent with the hypothesis that the learned incentive (conditioned reinforcing) effects of previously neutral stimuli paired with rewards are mediated by the amygdala acting through the ventral striatum: Amphetamine injections into the ventral striatum enhanced the effects of a conditioned reinforcing stimulus only if the amygdala was intact (see also Everitt and Robbins, 1992).

Thus there is much evidence that the amygdala is involved in responses made to stimuli associated with primary reinforcement. There is evidence that it may also be involved in determining whether novel stimuli are approached, for monkeys with amygdala lesions place novel foods and nonfood objects in their mouths, and rats with amygdala lesions have decreased neophobia, more quickly accepting new foods (Rolls and Rolls, 1973; see also Dunn and Everitt, 1988; Rolls, 1992b; Wilson and Rolls, 1993, and submitted).

Neuronal activity in the primate amygdala to reinforcing stimuli Recordings from single neurons in the amygdala of the monkey have shown that some neurons do respond to visual stimuli, consistent with the inputs from the temporal lobe visual cortex (Sanghera, Rolls, and Roper-Hall, 1979). Other neurons responded to auditory, gustatory, olfactory, or somatosensory stimuli, or in relation to movements. In tests of whether the neurons responded on the basis of the association of stimuli with reinforcement, it was found that approximately 20% of the neurons with visual responses had responses that occurred primarily to stimuli associated with reinforcement, for example to food and to a range of stimuli the monkey had learned signified food in a visual discrimination task (Sanghera, Rolls, and Ropper-Hall, 1979; Rolls, 1981b; Wilson and Rolls, 1993a,b). However, none of these neurons (in contrast to some neurons in the hypothalamus and orbitofrontal cortex described later) responded exclusively to rewarded stimuli, in that all responded at least partly to one or more neutral, novel, or aversive stimuli. Neurons with responses that are probably similar to these have also been described by Ono et al. (1980), and by Nishijo, Ono, and Nishino (1988; see Ono and Nishijo, 1992).

The degree to which the responses of these amygdala neurons are associated with reinforcement has been assessed in learning tasks. When the association between a visual stimulus and reinforcement was altered by reversal (so that the visual stimulus formerly associated with juice reward became associated with aversive saline and vice versa), it was found that 10 of 11 neurons did not reverse their responses, and for the other neuron the evidence was not clear (Sanghera, Rolls, and Roper-Hall, 1979; Wilson and Rolls, 1993, submitted; see also Rolls, 1992b). On the other hand, in a simpler relearning situation in which salt was added to a piece of food such as watermelon, the responses of 4 amygdala neurons to the sight of the watermelon diminished (Nishijo, Ono, and Nishino (1988). More investigations are needed to show the extent to which amygdala neurons do alter their activity flexibly and rapidly in relearning tests such as these (see Rolls, 1992b). What has been found is that neurons in the orbitofrontal cortex show very rapid reversal of their responses in visual discrimination reversal, and it therefore seems likely that the orbitofrontal cortex is especially involved when repeated relearning and reassessment of stimulus-reinforcement associations is required, as described later, rather than initial learning, in which the amygdala may be involved.

Responses of amygdala neurons to novel stimuli that are reinforcing As described above, some of the amygdala neurons that responded to rewarding visual stimuli also responded to some other stimuli that were not associated with reward. Wilson and Rolls (Rolls, 1992b) discovered a possible reason for this. They showed that the neurons with reward-related responses also responded to relatively novel visual stimuli, for example, in visual recognition memory tasks. When monkeys are given such relatively novel stimuli outside the task, they will reach out for and explore the objects, and in this respect the novel stimuli are reinforcing. Repeated presentation of the stimuli results in habituation of the neuronal response and of behavioral approach, if the stimuli are not associated with primary reinforcement. It is thus suggested that the amygdala neurons described operate as filters, providing an output if a stimulus is associated with a positive reinforcer or is positively reinforcing because of its relative novelty. The functions of this output may be to influence the interest shown in a stimulus, whether it is approached or avoided, whether an affective response to it occurs, and whether a representation of the stimulus is made or maintained via an action mediated through either the basal forebrain nucleus of Meynert or the backprojections to the cerebral cortex (Rolls, 1987, 1989a, 1990, 1992b).

It is an important adaptation to the environment to explore relatively novel objects or situations, for in this way advantage due to gene inheritance can become expressed and selected for. This function appears to be implemented in the amygdala in this way. Lesions of the amygdala impair the operation of this mechanism, in that objects are approached and explored indiscriminately, relatively independently of whether they are associated with positive or negative reinforcement, or of whether they are novel or familiar.

Neuronal responses to faces in the primate amygdala Another interesting group of neurons in the amygdala responds primarily to faces (Rolls, 1981b; Leonard et al., 1985). Each of these neurons responds to some but not all of a set of faces, and thus an ensemble of these neurons could convey information about the identity of the faces. These neurons are found especially in the basal accessory nucleus of the amygdala (Leonard et al., 1985), a part of the amygdala that develops markedly in primates (Amaral et al., 1992). This is probably part of a system that has evolved for the rapid and reliable identification of individuals from their faces, and of facial expressions, because of the importance of such identification in primate social behavior (Rolls, 1981b, 1984, 1990, 1992a,b,c; Perrett and Rolls, 1982; Leonard et al., 1985). The amygdala may allow neurons that reflect the social and emotional significance of faces to be formed by associations between face representations received from the temporal cortical areas and information about primary reinforcers received from, for example, the somatosensory system (via the insula; Mesulam and Mufson, 1982), and the gustatory system (via, for example, the orbitofrontal cortex; see figure 72.3) (Rolls, 1984, 1987, 1990, 1992a,b,c). Indeed, it is suggested that the tameness of the Klüver-Bucy syndrome, and the inability of amygdalectomized monkeys to interact normally in a social group (Kling and Steklis, 1976), arise because of damage to this system that is specialized for processing faces.

Cortical cells found in certain of the temporal lobe regions (e.g., TEa, TEm and TPO; Baylis, Rolls, and Leonard, 1987) that project into the amygdala have properties that would enable them to provide useful inputs to such an associative mechanism in the amygdala (see Rolls, 1987, 1989a, 1990, 1992a,b,c). These cortical neurons in many cases use ensemble encoding with sparse distributed representations (Baylis, Rolls, and Leonard, 1985; Rolls and Treves, 1990). This type of tuning is appropriate for input to an associative memory such as that believed to be implemented in the amygdala; it allows many memories to be stored, and allows generalization and graceful degradation (Rolls, 1987, 1989a; Rolls and Treves, 1990). There is evidence that the responses of some of these neurons are altered by experience so that new stimuli become incorporated into the network (Rolls et al., 1989). Competition may play a role in the self-organization of such networks (Rolls, 1989a,c,d). The representation that is built in temporal cortical areas shows considerable size, contrast, spatial frequency, and translation invariance (Rolls and Baylis, 1986; Rolls, 1992a,c). This is useful for an associative mechanism that must associate individuals (independently of the exact retinal position, size, etc.) with reinforcement. A further advantage of this function is that for some of the cortical neurons the representation that is built is object-based rather than viewer-centered; that is, the firing is relatively independent of viewing angle (Hasselmo et al., 1989).

In addition to the population of neurons that codes for face identity, which tend to have object-based representations and are in areas TEa and TEm on the ventral bank of the superior temporal sulcus, there is a separate population in the cortex in the superior temporal sulcus (e.g., area TPO) which conveys information about facial expression (Hasselmo, Rolls, and Baylis, 1989). Some of the neurons in this region tend to have view-based representations (so that information is conveyed, for example, about whether the face is looking at one or is looking away), and may respond to moving faces and to facial gesture (Hasselmo et al., 1989; Rolls, 1992a,c). Information about both face identity and facial expression is important in social and emotional responses to other primates, which must be based on who the individual is as well as on the facial expression or gesture being made. One output from the amygdala for this information is probably via the ventral striatum, for a small population of neurons has been found in the ventral striatum with responses selective for faces (Rolls and Williams, 1987; Williams et al., 1993).

The evidence described above implicates the amygdala in the learning of associations between stimuli and reinforcement. This means that it must be important in learned emotional responses, and at least part of the importance of the amygdala in emotion appears to be that it is involved in this type of emotional learning.

THE ORBITOFRONTAL CORTEX The orbitofrontal cortex (see figures 72.2 and 72.3) receives inputs via the mediodorsal nucleus of the thalamus, pars magnocellularis, which itself receives afferents from temporal lobe structures such as the prepiriform (olfactory) cortex, the amygdala, and the inferior temporal cortex (Krettek and Price, 1974, 1977). Another set of inputs reaches the orbitofrontal cortex directly from the inferior temporal cortex, the cortex in the superior temporal sulcus, the temporal pole, and the amygdala (Jones and Powell, 1970; Barbas, 1988; Seltzer and Pandya, 1989); a third set reaches it from the primary taste cortex in the insula and frontal operculum (Rolls, 1989b, 1994) and the primary olfactory (piriform) cortex (Price et al., 1991; see figure 72.3); and a fourth set makes up the at least partly dopaminergic projection from the ventral tegmental area. The orbitofrontal cortex projects back to temporal lobe areas such as the inferior temporal cortex, and, in addition, to the entorhinal cortex and the cingulate cortex (Van Hoesen, Pandya, and Butters, 1975). The orbitofrontal cortex also projects to the preoptic region and lateral hypothalamus, to the ventral tegmental area (Nauta, 1964), and to the head of the caudate nucleus (Kemp and Powell, 1970).

Damage to the caudal orbitofrontal cortex in the monkey produces emotional changes. These include decreased aggression to humans and to stimuli such as a snake and a doll, and a reduced tendency to reject foods such as meat (Butter, McDonald and Snyder, 1969; Butter, Snyder, and McDonald, 1970; Butter and Snyder, 1972). In the human, euphoria, irresponsibility, and lack of affect can follow frontal lobe damage (see Kolb and Whishaw, 1990). These changes following frontal lobe damage may be related to a failure to react normally to and learn from nonreward in a number of different situations. This failure is evident as a tendency to respond when responses are inappropriate, that is, no longer rewarded. For example, monkeys with orbitofrontal damage are impaired on go/no-go task performance, in that they go on the no-go trials (Iversen and Mishkin, 1970); in an object-reversal task, in that they respond to the object that was formerly rewarded with food; and in extinction, in that they continue to respond to an object that is no longer rewarded (Butter, 1969; Jones and Mishkin, 1972; see Rolls, 1990). The visual discrimination learning deficit shown by monkeys with orbitofrontal cortex damage (Jones and Mishkin, 1972; Baylis and Gaffan, 1991) may be due to the tendency of these monkeys not to withold responses to nonrewarded stimuli (Jones and Mishkin, 1972).

The hypothesis that the orbitofrontal cortex is involved in correcting responses made to stimuli previously associated with reinforcement has been investigated via recordings from single neurons in the orbitofrontal cortex made while monkeys performed tasks known to be impaired by damage to the orbitofrontal cortex (Thorpe, Rolls, and Maddison, 1983). It has been found that one class of neurons in the orbitofrontal cortex of the monkey responds in certain nonreward situations (Thorpe, Rolls, and Maddison, 1983). For example, some neurons responded in extinction, immediately after a lick had been made to a visual stimulus that had previously been associated with fruit juice reward, and other neurons responded in a reversal task, immediately after the monkey had responded to the previously rewarded visual stimulus but had obtained punishment rather than reward. Another class of orbitofrontal neurons responded to particular visual stimuli only if they were associated with reward, and these neurons showed one-trial stimulus-reinforcement association reversal. Another class of neurons conveyed information about whether a reward had been given, responding, for example, to the taste of sucrose, or, for other neurons, to the taste of saline (Thorpe, Rolls, and Maddison, 1983). These orbitofrontal neurons with gustatory responses have now been analyzed further (Rolls, 1989b, 1994), and can be tuned quite finely to gustatory stimuli such as a sweet taste (Rolls, 1989b; Rolls, Yaxley, and Sienkiewicz, 1990). Moreover, their activity is related to reward, in that those that respond to the taste of food do so only if the monkey is hungry (Rolls, Sienkiewicz, and Yaxley, 1989). These orbitofrontal neurons receive their input from the primary gustatory cortex, in the frontal operculum (Scott et al., 1986; Rolls, 1989b). Moreover, in part of this orbitofrontal region, some neurons are bimodal, combining taste and olfactory inputs (Rolls, 1989b, 1994b).

It is suggested that these types of information are represented in the responses of orbitofrontal neurons because they are part of a mechanism that evaluates whether a reward is expected and generates a mismatch (evident as a firing of the nonreward neurons) if reward is not obtained when it is expected (Thorpe, Rolls, and Maddison, 1983; Rolls, 1990). These neuronal responses provide further evidence that the orbitofrontal cortex is involved in emotional responses—particularly when these involve correcting previously learned reinforcement contingencies—in situations that include those usually described as involving frustration.

A number of the symptoms of frontal lobe damage in humans appear to be related to this function of altering behavior when particular responses become inappropriate. Thus, humans with frontal lobe damage can show impairments in a number of tasks in which an alteration of behavioral strategy is required in response to a change in environmental reinforcement contingencies (see Goodglass and Kaplan, 1979; Jouandet and Gazzaniga, 1979; Kolb and Whishaw, 1990). For example, Milner (1963) showed that on the Wisconsin Card Sorting Task (in which cards are to be sorted according to the color, shape, or number of items on each card depending on whether the examiner says "right" or "wrong" to each placement), frontal patients had difficulty either in determining the first sorting principle or in shifting to a second principle when required to. Also, in stylus mazes, frontal patients have difficulty in changing direction when a sound indicates that they have left the correct path (see Milner, 1982). In both types of test, frontal patients may be able to verbalize the correct rules and yet be unable to correct their behavioral sets or strategies appropriately. Some of the personality changes that can result from frontal lobe damage may be related to a similar type of dysfunction. For example, the euphoria, irresponsibility, lack of affect, and lack of concern for the present or future that sometimes follow frontal lobe damage (see Hecaen and Albert, 1978) may also be related to a dysfunction in altering behavior appropriately in response to a change in reinforcement contingencies (Rolls, 1990).

Basal Forebrain and Hypothalamus There are neurons in the lateral hypothalamus and basal forebrain of monkeys that respond to visual stimuli associated with rewards such as food (Rolls, 1975, 1981a, 1982, 1986a,b,c, 1990, 1993; Burton, Rolls, and Mora, 1976; Mora, Rolls, and Burton, 1976; Rolls, Burton, and Mora, 1976; Wilson and Rolls, 1990a), or with punishment (Rolls, Sanghera, and Roper-Hall, 1979). These neurons with reinforcement-related activity may receive from the amygdala and orbitofrontal cortex, and produce autonomic responses to emotional stimuli, via pathways that descend from the hypothalamus toward the brainstem autonomic motor nuclei (see Rolls, 1990, 1992b). It is also possible that these outputs could influence emotional behavior, through, for example, the connections from the hypothalamus to the amygdala (Aggleton, Burton, and Passingham, 1980), to the substantia nigra (Nauta and Domesick, 1978), or even to the neocortex (Divac,

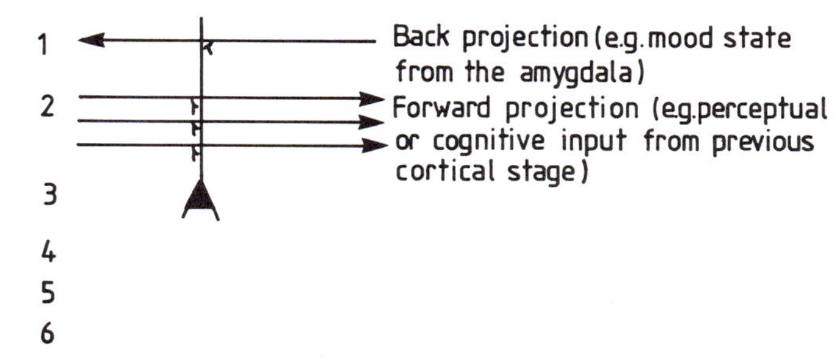

FIGURE 72.4 Pyramidal cells in, for example, layers 2 and 3 of the temporal lobe association cortex receive forward inputs from preceding cortical stages of processing, and also backprojections from the amygdala. It is suggested that the backprojections from the amygdala make modifiable synapses on the apical dendrites of cortical pyramidal cells during learning when amygdala neurons are active in relation to a mood state; and that the backprojections from the amygdala via these modified synapses allow mood state to influence later cognitive processing, for example by facilitating some perceptual representations.

1975; Kievit and Kuypers, 1975). Indeed, it is suggested that the latter projection, by releasing acetylcholine in the cerebral cortex when emotional stimuli (including reinforcing and novel stimuli) are seen, provides one way in which emotion can influence the storage of memories in the cerebral cortex (Rolls, 1987; Wilson and Rolls, 1990a,b).

EFFECTS OF EMOTION ON COGNITIVE PROCESSING Emotional states may affect whether or how strongly memories are stored using the basal forebrain memory strobe; may be stored as part of many memories, using, for example, amygdala-entorhinal inputs to the hippocampus; and may influence both the recall of such memories and the operation of cognitive processing, by amygdalocortical backprojections (figure 72.4; see Rolls, 1990, 1992b; Treves and Rolls, 1994).

Consciousness and emotion

It might be possible to build a computer that would perform the functions of emotions that have been discussed, and yet we might not want to ascribe to it emotional feelings. This point raises the general issue of consciousness and its functions. Because the topic of emotional feelings is of considerable interest, one view on consciousness, influenced by contemporary cognitive neuroscience, is outlined next. However, this view is only preliminary, and theories of consciousness are likely to undergo considerable development.

A starting point is that many actions can be performed relatively automatically, without apparent conscious intervention. An example sometimes given is driving a car. Such actions could involve control of behavior by brain systems that are old in evolutionary terms, such as the basal ganglia. It is of interest that the basal ganglia (and cerebellum) do not have backprojection systems to most of the parts of the cerebral cortex from which they receive inputs (see, e.g., Rolls and Johnstone, 1992; Rolls, 1994a). In contrast, parts of the brain such as the hippocampus and the amygdala, which are involved in functions such as episodic memory and emotion, respectively, about which we can make (verbal) declarations, do have major backprojection systems to the high parts of the cerebral cortex from which they receive forward projections (Rolls, 1992b; Treves and Rolls, 1994). It may be that evolutionarily newer parts of the brain, such as the language areas and parts of the prefrontal cortex, are involved in an alternative type of control of behavior, in which actions can be planned with the use of a (language) system that allows relatively arbitrary (syntactic) manipulation of semantic entities.

The general view that there are many routes to behavioral output is supported by the evidence that there are many input systems to the basal ganglia (from almost all areas of the cerebral cortex), and that neuronal activity in each part of the striatum reflects the activity in the overlying cortical area (Rolls and Johnstone, 1992; Rolls, 1994a). The evidence is consistent with the possibility that different cortical areas, each specialized for a different type of computation, have their outputs directed to the basal ganglia, which then select the strongest input and map this into action (via outputs directed, for example, to the premotor cortex) (Rolls and Johnstone, 1992). Within this scheme, the language areas would offer one of many routes to action—a route particularly suited to planning actions, because of the syntactic manipulation of semantic entities that may make long-term planning possible. Consistent with the hypothesis of multiple routes to action, only some of which utilize language, is the evidence that split-brain patients may not be aware of actions being performed by the nondominant hemisphere (Gazzaniga and LeDoux, 1978). Also consistent with multiple routes to action, including nonverbal ones, patients with focal brain damage, for example to the prefrontal cortex, may emit actions, yet comment verbally that they should not be performing those actions.

In both split-brain patients and those with focal brain damage, confabulation may occur, in that a patient's verbal account of why an action was performed may not be related at all to the environmental event that actually triggered the action (Gazzaniga and LeDoux, 1978). It is possible that sometimes in normal humans when actions are initiated as a result of processing in a specialized brain region, the language system subsequently elaborates a coherent account of why that action was performed. This would be consistent with a general view of brain evolution that holds that as areas of the cortex evolve they are laid on top of existing circuitry connecting inputs to outputs, and that each level in this hierarchy of input-output pathways may control behavior according to the specialized function it can perform. (It is of interest that mathematicians may get a hunch that something is correct, yet not be able to verbalize why. They may then resort to formal, more serial and language-like theorems to prove the case, and these seem to require more conscious processing. This is a further indication of a close association between linguistic processing and consciousness. The linguistic processing need not, as in reading, involve an inner articulatory loop.)

We may next examine some of the advantages and behavioral functions that language, present as the most recently added layer to the above system, would confer. One major advantage would be the ability to plan actions through many potential stages and to evaluate the consequences of those actions without having to perform the actions. For this, the ability to form propositional statements, and to perform syntactic operations on the semantic representations of states in the world, would be important. Also important in this system would be the ability to have second-order thoughts (e.g., I think that he thinks that ...), as this would allow much better modeling and prediction of others' behavior, and therefore of planning, particularly planning when it involves others. This capability for higher-order thoughts would enable reflection on past events, which would also be useful in planning. In contrast, nonlinguistic behavior would be driven by learned reinforcement associations, learned rules, and so forth, but not by flexible planning for many steps ahead involving a model of the world including others' behavior. (For an earlier view that is close to this part of the argument, see Humphrey, 1980.) (The examples of behavior from nonhumans that may reflect planning may reflect much more limited and inflexible planning. For example, the dance of the honeybee to signal to other bees the location of food may be said to reflect planning, but the symbol manipulation is not arbitrary. There are likely to be interesting examples of nonhuman primate behavior, perhaps in the great apes, that reflect the evolution of an arbitrary symbol manipulation system that could be useful for flexible planning (cf. Cheney and Seyfarth, 1990.)

It is next suggested that this arbitrary symbol manipulation—using important aspects of language processing, and used for planning, but not used in initiating all types of behavior—is what consciousness is about. Indeed, consciousness may *be* the state when this type of processing is being performed. This is consistent with the points made above that the brain systems that are required for consciousness and language are either very similar or the same. According to this explanation, the feeling of anything is the state that is present when linguistic processing, involving thoughts of the second or higher order, is being performed (Rosenthal, 1990, 1993; cf. Dennett, 1991).

This analysis does not yet give an account for qualia, for example why the color red feels red, or why a rewarding touch produces an emotional feeling of pleasure. The view I suggest on qualia is as follows. Information processing in and from our sensory systems (e.g., the sight of the color red) may be relevant to planning actions using language and to the conscious processing thereby implied. Given that these inputs must be represented in the system that plans, we may ask whether it is more likely that we would be conscious of them or that we would not. I suggest that it would be a very specialized system that would allow such sensory and emotional inputs to be part of (verbal) planning, and yet remain unconscious. It seems to be much more parsimonious to hold that we would be conscious of those inputs because they would be being used in one type of processing just described, and this is what I propose.

The explanation of emotional feelings or qualia suggested by this discussion is that they should be felt as conscious because they enter into a specialized linguistic symbol manipulation system that has evolved for flexible planning of actions. It would require a very special machine to enable this processing—which is conscious by its nature—to occur without the sensory and emotional inputs becoming felt qualia. The qualia are thus accounted for by the evolution of the linguistic system for flexible planning. This account implies that it may be especially animals with a higher-order belief and thought system and with linguistic symbol manipulation that have qualia. It may be that much nonhuman animal behavior, provided that it does not require flexible linguistic planning, takes place according to reinforcement-guided behavior and rule following. Such behaviors might appear very similar to human behavior performed in similar circumstances, but would not imply qualia. It would be primarily by virtue of flexible, linguistic, planning behavior that humans would be different from other animals, and would have elaborated qualia.

It is of interest to comment on how the evolution of a system for flexible planning might affect emotions. Consider grief, which may occur when a reward is terminated and no immediate action is possible. It may be adaptive by leading to a cessation of the formerly rewarded behavior, thus facilitating the possible identification of other positive reinforcers in the environment. In humans, grief may be particularly potent because it becomes represented in a system that can plan ahead and understand the enduring implications of the loss.

This account of consciousness also leads to a suggestion about the processing that underlies the feeling of free will. Free will would in this scheme involve the use of language to check many moves ahead on a number of possible series of actions and their outcomes, and with this information to make a choice from the likely outcomes. (If, in contrast, choices were made only on the basis of the reinforcement value of immediately available stimuli, without the arbitrary syntactic symbol manipulation made possible by language, then the choice strategy would be much more limited, and we might not want to use the term *free will*, as all the consequences of the possible actions would not have been computed.) In the operation of such a free will system, the uncertainties introduced by the limited information available about the likely outcomes of series of actions, and the inability to use optimal algorithms when combining conditional probabilities, would be much more important factors than whether or not the brain operates deterministically. (The operation of brain machinery must be relatively deterministic, for it has evolved to provide reliable outputs for given inputs.)

This chapter has presented some ideas on why we have consciousness and are conscious of emotional feelings. It is likely that theories of consciousness will continue to undergo rapid development, and current theories should not be taken to have practical implications.

ACKNOWLEDGMENTS The author has worked on some of the experiments described here with G. C. Baylis, M. J. Burton, M. E. Hasselmo, C. M. Leonard, F. Mora, D. I. Perrett, M. K. Sanghera, T. R. Scott, S. J. Thorpe, and F. A. W. Wilson. Their collaboration, and helpful discussions with or communications from M. Davies and C. C. W. Taylor (Corpus Christi College, Oxford), and M. S. Dawkins, are gratefully acknowledged. Some of the research described was supported by the Medical Research Council.

REFERENCES

AGGLETON, J. P., 1993. The contribution of the amygdala to normal and abnormal emotional states. *Trends Neurosci.* 16:328–333.

AGGLETON, J. P., M. J. BURTON, and R. E. PASSINGHAM, 1980. Cortical and subcortical afferents to the amygdala in the rhesus monkey (Macaca mulatta). *Brain Res.* 190: 347–368.

AGGLETON, J. P., and R. E. PASSINGHAM, 1981. Syndrome produced by lesions of the amygdala in monkeys (Macaca mulatta). *J. Comp. Physiol.* 95:961–977.

AMARAL, D. G., J. L. PRICE, A. PITKANEN, and S. T. CARMICHAEL, 1992. Anatomical organization of the primate amygdaloid complex. In *The Amygdala*, J. P. Aggleton, ed. New York: Wiley, pp. 1–66.

BARBAS, H., 1988. Anatomic organization of basoventral and mediodorsal visual recipient prefrontal regions in the rhesus monkey. *J. Comp. Neurol.* 276:313–42.

BAYLIS, L. L., and D. GAFFAN, 1991. Amygdalectomy and ventromedial prefrontal ablation produce similar deficits in food choice and in simple object discrimination learning for an unseen reward. *Exp. Brain Res.* 86:617–622.

BAYLIS, G. C., E. T. ROLLS, and C. M. LEONARD, 1985. Selectivity between faces in the responses of a population of neurons in the cortex in the superior temporal sulcus of the monkey. *Brain Res.* 342:91–102.

BAYLIS, G. C., E. T. ROLLS, and C. M. LEONARD, 1987. Functional subdivisions of temporal lobe neocortex. *J. Neurosci.* 7:330–342.

BLANEY, P. H., 1986. Affect and memory: A review. *Psych. Bull.* 99:229–246.

BURTON, M. J., E. T. ROLLS, and F. MORA, 1976. Visual responses of hypothalamic neurones. *Brain Res.* 107:215–216.

BUTTER, C. M., 1969. Perseveration in extinction and in discrimination reversal tasks following selective prefrontal ablations in Macaca mulatta. *Physiol. Behav.* 4:163–171.

BUTTER, C. M., J. A. McDONALD, and D. R. SNYDER, 1969. Orality, preference behavior, and reinforcement value of non-food objects in monkeys with orbital frontal lesions. *Science* 164:1306–1307.

Butter, C. M., D. R. Snyder, and J. A. McDonald, 1970. Effects of orbitofrontal lesions on aversive and aggressive behaviors in rhesus monkeys. *J. Comp. Physiol. Psychol.* 72: 132–144.

Butter, C. M., and D. R. Snyder, 1972. Alterations in aversive and aggressive behaviors following orbitofrontal lesions in rhesus monkeys. *Acta Neurobiol. Exp.* 32:525–565.

Cador, M., T. W. Robbins, and B. J. Everitt, 1989. Involvement of the amygdala in stimulus-reward associations: interaction with the ventral striatum. *Neurosci.* 30: 77–86.

Cheney, D. L., and R. M. Seyfarth, 1990. *How Monkeys See the World.* Chicago: University of Chicago Press.

Davis, M., 1992. The role of the amygdala in conditioned fear. In *The Amygdala*, J. P. Aggleton, ed. New York: Wiley, pp. 255–305.

Dawkins, R., 1989. *The Selfish Gene*, ed. 2. Oxford: Oxford University Press.

Dennett, D. C., 1991. *Consciousness Explained.* London: Penguin.

Divac, I., 1975. Magnocellular nuclei of the basal forebrain project to neocortex, brain stem, and olfactory bulb: Review of some functional correlates. *Brain Res.* 93:385–398.

Dunn, L. T., and B. J. Everitt, 1988. Double dissociations of the effects of amygdala and insular cortex lesions on conditioned taste aversion, passive avoidance, and neophobia in the rat using the excitotoxin ibotenic acid. *Behav. Neurosci.* 102:3–23.

Everitt, B. J., and T. W. Robbins, 1992. Amygdala-ventral striatal interactions and reward-related processes. In *The Amygdala*, J. P. Aggleton, ed. New York: Wiley, pp. 401–430.

Gaffan, D., 1992. Amygdala and the memory of reward. In *The Amygdala*, J. P. Aggleton, ed. New York: Wiley, pp. 471–483.

Gaffan, E. A., D. Gaffan, and S. Harrison, 1988. Disconnection of the amygdala from visual association cortex impairs visual reward-association learning in monkeys. *J. Neurosci.* 8:3144–3150.

Gaffan, D., E. A. Gaffan, and S. Harrison, 1989. Visual-visual associative learning and reward-associative learning in monkeys: The role of the amygdala. *J. Neurosci.* 9:558–564.

Gaffan, D., and S. Harrison, 1987. Amygdalectomy and disconnection in visual learning for auditory secondary reinforcement by monkeys. *J. Neurosci.* 7:2285–2292.

Gazzaniga, M. S., and J. LeDoux, 1978. *The Integrated Mind.* New York: Plenum.

Goodglass, H., and E. Kaplan, 1979. Assessment of cognitive deficit in brain-injured patient. In *Handbook of Behavioral Neurobiology*. Vol. 2, *Neuropsychology*. M. S. Gazzaniga, ed. New York: Plenum, pp. 3–22.

Gray, J. A., 1975. *Elements of a Two-Process Theory of Learning.* London: Academic Press.

Gray, J. A., 1987. *The Psychology of Fear and Stress*, ed. 2. Cambridge: Cambridge University Press.

Hasselmo, M. E., E. T. Rolls, and G. C. Baylis, 1989. The role of expression and identity in the face-selective responses of neurons in the temporal visual cortex of the monkey. *Behav. Brain Res.* 32:203–218.

Hasselmo, M. E., E. T. Rolls, G. C. Baylis, and V. Nalwa, 1989. Object-centered encoding by face-selective neurons in the cortex in the superior temporal sulcus of the monkey. *Exp. Brain Res.* 75:417–429.

Hecaen, H., and M. L. Albert, 1978. *Human Neuropsychology*. New York: Wiley.

Heimer, L., R. D. Switzer, and G. W. Van Hoesen, 1982. Ventral striatum and ventral pallidum: Components of the motor system? *Trends Neurosci.* 5:83–87.

Humphrey, N. K., 1980. Nature's psychologists. In *Consciousness and the Physical World*, B. D. Josephson and V. S. Ramachandran, eds. Oxford: Pergamon, pp. 57–80.

Iversen, S. D., and M. Mishkin, 1970. Perseverative interference in monkey following selective lesions of the inferior prefrontal convexity. *Exp. Brain Res.* 11:376–386.

Jones, B., and M. Mishkin, 1972. Limbic lesions and the problem of stimulus-reinforcement associations. *Exp. Neurol.* 36:362–377.

Jones, E. G., and T. P. S. Powell, 1970. An anatomical study of converging sensory pathways within the cerebral cortex of the monkey. *Brain* 93:793–820.

Jouandet, M., and M. S. Gazzaniga, 1979. The frontal lobes. In *Handbook of Behavioral Neurobiology*. Vol. 2, *Neuropsychology* M. S. Gazzaniga, ed. New York: Plenum, pp. 25–59.

Kemp, J. M., and T. P. S. Powell, 1970. The cortico-striate projections in the monkey. *Brain* 93:525–546.

Kievit, J., and H. G. J. M. Kuypers, 1975. Subcortical afferents to the frontal lobe in the rhesus monkey studied by means of retrograde horseradish peroxidase transport. *Brain Res.* 85:261–266.

Kling, A., and H. D. Steklis, 1976. A neural substrate for affiliative behavior in nonhuman primates. *Brain Behav. Evol.* 13:216–238.

Klüver, H., and P. C. Bucy, 1939. Preliminary analysis of functions of the temporal lobes in monkeys. *Arch. Neurol. Psychiatry* 42:979–1000.

Kolb, B., and I. Q. Whishaw, 1990. *Fundamentals of Human Neuropsychology*, ed. 2. New York: Freeman.

Krettek, J. E., and J. L. Price, 1974. A direct input from the amygdala to the thalamus and the cerebral cortex. *Brain Res.* 67:169–174.

Krettek, J. E., and J. L. Price, 1977. The cortical projections of the mediodorsal nucleus and adjacent thalamic nuclei in the rat. *J. Comp. Neurol.* 171:157–192.

LeDoux, J. E., 1987. Emotion. In *Handbook of Physiology*. Sec. 1, *The Nervous System*. Vol. 5, *Higher Functions of the Brain*, part 2, V. B. Mouncastle, ed. Bethesda, Md.: American Physiological Society, pp. 419–459.

LeDoux, J. E., 1992. Emotion and the amygdala. In *The Amygdala*, J. P. Aggleton, ed. New York: Wiley, pp. 339–351.

LeDoux, J. E., J. Iwata, P. Cicchetti, and D. J. Reis, 1988. Different projections of the central amygdaloid nucleus mediate autonomic and behavioral correlates of conditioned fear. *J. Neurosci.* 8:2517–2529.

Leonard, C. M., E. T. Rolls, F. A. W. Wilson, and G. C. Baylis, 1985. Neurons in the amygdala of the monkey with responses selective for faces. *Behav. Brain Res.* 15:159–176.

Mackintosh, N. J., 1983. *Conditioning and Associative Learning.* Oxford: Oxford University Press.

Mesulam, M.-M., and E. J. Mufson, 1982. Insula of the old world monkey: III. Efferent cortical output and comments on function. *J. Comp. Neurol.* 212:38–52.

Millenson, J. R., 1967. *Principles of Behavioral Analysis.* New York: Macmillan.

Milner, B., 1963. Effects of different brain lesions on card sorting. *Arch. Neurol.* 9:90–100.

Milner, B., 1982. Some cognitive effects of frontal-lobe lesions in man. *Philo. Trans. R. S. Lond. [Biol.]* 298:211–226.

Mishkin, M., and J. Aggleton, 1981. Multiple functional contributions of the amygdala in the monkey. In *The Amygdaloid Complex*, Y. Ben-Ari, ed. Amsterdam: Elsevier, pp. 409–420.

Mora, F., E. T. Rolls, and M. J. Burton, 1976. Modulation during learning of the responses of neurones in the lateral hypothalamus to the sight of food. *Exp. Neurol.* 53:508–519.

Nauta, W. J. H., 1964. Some efferent connections of the prefrontal cortex in the monkey. In *The Frontal Granular Cortex and Behavior*, J. M. Warren and K. Akert, eds. New York: McGraw Hill, pp. 397–407.

Nauta, W. J. H., and V. B. Domesick, 1978. Crossroads of limbic and striatal circuitry: Hypothalamo-nigral connections. In *Limbic Mechanisms*, K. E. Livingston and O. Hornykiewicz, eds. New York: Plenum, pp. 75–93.

Nishijo, H., T. Ono, and H. Nishino, 1988. Single neuron responses in amygdala of alert monkey during complex sensory stimulation with affective significance. *J. Neurosci.* 8:3570–3583.

Ono, T., H. Nishino, K. Sasaki, M. Fukuda, and K. Muramoto, 1980. Role of the lateral hypothalamus and amygdala in feeding behavior. *Brain Res. Bull. Suppl.* 4, 5:143–149.

Ono, T., and H. Nishijo, 1992. In *The Amygdala*, J. P. Aggleton, ed. Chichester: Wiley, pp. 167–190.

Perrett, D. I., and E. T. Rolls, 1982. Neural mechanisms underlying the visual analysis of faces. In *Advances in Vertebrate Neuroethology*, J.-P. Ewert, R. R. Capranica and D. J. Ingle, eds. New York: Plenum.

Perrett, D. I., E. T. Rolls, and W. Caan, 1982. Visual neurons responsive to faces in the monkey temporal cortex. *Exp. Brain Res.* 47:329–342.

Price, J. L., S. T. Carmichael, K. M. Carnes, M.-C. Clugnet, and M. Kuroda, 1991. Olfactory input to the prefrontal cortex. In *Olfaction: A Model System for Computational Neuroscience*, J. L. Davis and H. Eichenbaum, eds. Cambridge, Mass.: MIT Press, pp. 101–120.

Rolls, E. T., 1975. *The Brain and Reward.* Oxford: Pergamon.

Rolls, E. T., 1981a. Processing beyond the inferior temporal visual cortex related to feeding, memory, and striatal function. In *Brain Mechanisms of Sensation*, Y. Katsuki, R. Norgren and M. Sato, eds. New York: Wiley, pp. 241–269.

Rolls, E. T., 1981b. Responses of amygdaloid neurons in the primate. In *The Amygdaloid Complex*, Y. Ben-Ari, ed. Amsterdam: Elsevier, pp. 383–393.

Rolls, E. T., 1982. Feeding and reward. In *The Neural Basis of Feeding and Reward*, D. Novin and B. G. Hoebel, eds. Brunswick, Maine: Haer Institute for Electrophysiological Research.

Rolls, E. T., 1984. Neurons in the cortex of the temporal lobe and in the amygdala of the monkey with responses selective for faces. *Hum. Neurobiol.* 3:209–222.

Rolls, E. T., 1986a. A theory of emotion, and its application to understanding the neural basis of emotion. In *Emotions: Neural and Chemical Control*, Y. Oomura, ed. Tokyo: Japan Scientific Societies Press; Basel: Karger, pp. 325–344.

Rolls, E. T., 1986b. Neural systems involved in emotion in primates. In *Emotion: Theory, Research, and Experience.* Vol. 3, *Biological Foundations of Emotion*, R. Plutchik and H. Kellerman, eds. New York: Academic Press, pp. 125–143.

Rolls, E. T., 1986c. Neuronal activity related to the control of feeding. In *Feeding Behavior: Neural and Humoral Controls*, R. C. Ritter, S. Ritter, and C. D. Barnes, eds. New York: Academic Press, pp. 163–190.

Rolls, E. T., 1987. Information representation, processing and storage in the brain: Analysis at the single neuron level. In *The Neural and Molecular Bases of Learning*, J.-P. Changeux and M. Konishi, eds. Chichester, England: Wiley, pp. 503–540.

Rolls, E. T., 1989a. Functions of neuronal networks in the hippocampus and neocortex in memory. In *Neural Models of Plasticity: Experimental and Theoretical Approaches*, J. H. Byrne and W. O. Berry, eds. San Diego: Academic Press, pp. 240–265.

Rolls, E. T., 1989b. Information processing in the taste system of primates. *J. Exp. Biol.* 146:141–164.

Rolls, E. T., 1989c. The representation and storage of information in neuronal networks in the primate cerebral cortex and hippocampus. In *The Computing Neuron*, R. Durbin, C. Miall, and G. Mitchison, eds. Wokingham, England: Addison-Wesley, pp. 125–159.

Rolls, E. T., 1989d. Functions of neuronal networks in the hippocampus and cerebral cortex in memory. In *Models of Brain Function*, R. M. J. Cotterill, ed. Cambridge: Cambridge University Press, pp. 15–33.

Rolls, E. T., 1990. A theory of emotion, and its application to understanding the neural basis of emotion. *Cognition Emotion* 4:161–190.

Rolls, E. T., 1992a. Neurophysiological mechanisms underlying face processing within and beyond the temporal cortical visual areas. *Philo. Trans. R. Soc. Lond. [Biol.]* 335:11–21.

Rolls, E. T., 1992b. Neurophysiology and functions of the primate amygdala. In *The Amygdala*, J. P. Aggleton, ed. New York: Wiley, pp. 143–165.

Rolls, E. T., 1992c. The processing of face information in the primate temporal lobe. In *Processing Images of Faces*, V.

Bruce and M. Burton, eds. Norwood, N.J.: Ablex, pp. 41–68.

Rolls, E. T., 1993. Neurophysiology of feeding in primates. In *Neurophysiology of Ingestion*, D. A. Booth, ed. Oxford: Pergamon Press, pp. 137–169.

Rolls, E. T., 1994a. Neurophysiology and cognitive functions of the striatum. *Rev. Neurol.*, in press.

Rolls, E. T., 1994b. Central taste anatomy and physiology. In *Handbook of Clinical Olfaction and Gustation*, R. L. Doty, ed. Dekker: New York.

Rolls, E. T., and G. C. Baylis, 1986. Size and contrast have only small effects on the responses to faces of neurons in the cortex of the superior temporal sulcus of the monkey. *Exp. Brain Res.* 65:38–48.

Rolls, E. T., G. C. Baylis, M. E. Hasselmo, and V. Nalwa, 1989. The effect of learning on the face-selective responses of neurons in the cortex in the superior temporal sulcus of the monkey. *Exp. Brain Res.* 76:153–164.

Rolls, E. T., M. J. Burton, and F. Mora, 1976. Hypothalamic neuronal responses associated with the sight of food. *Brain Res.* 111:53–66.

Rolls, E. T., M. J. Burton, and F. Mora, 1980. Neurophysiological analysis of brain-stimulation reward in the monkey. *Brain Res.* 194:339–357.

Rolls, E. T., and S. Johnstone, 1992. Neurophysiological analysis of striatal function. In *Neuropsychological Disorders Associated with Subcortical Lesions*, G. Vallar, S. F. Cappa, and C. W. Wallesch, eds. Oxford: Oxford University Press, pp. 61–97.

Rolls, E. T., S. J. Judge, and M. Sanghera, 1977. Activity of neurones in the inferotemporal cortex of the alert monkey. *Brain Res.* 130:229–238.

Rolls, E. T., and B. J. Rolls, 1973. Altered food preferences after lesions in the basolateral region of the amygdala in the rat. *J. Comp. Physiol. Psychol.* 83:248–259.

Rolls, E. T., M. K. Sanghera, and A. Roper-Hall, 1979. The latency of activation of neurons in the lateral hypothalamus and substantia innominata during feeding in the monkey. *Brain Res.* 194:121–135.

Rolls, E. T., Z. J. Sienkiewicz, and S. Yaxley, 1989. Hunger modulates the responses to gustatory stimuli of single neurons in the caudolateral orbitofrontal cortex of the macaque monkey. *Eur. J. Neurosci.* 1:53–60.

Rolls, E. T., and A. Treves, 1990. The relative advantages of sparse versus distributed encoding for associative neuronal networks in the brain. *Network* 1:407–421.

Rolls, E. T., and G. V. Williams, 1987. Sensory and movement-related neuronal activity in different regions of the primate striatum. In *Basal Ganglia and Behavior: Sensory Aspects and Motor Functioning*, J. S. Schneider and T. I. Lidsky., eds. Bern: Hans Huber, pp. 37–59.

Rolls, E. T., S. Yaxley, and Z. J. Sienkiewicz, 1990. Gustatory responses of single neurons in the orbitofrontal cortex of the macaque monkey. *J. Neurophysiol.* 64:1055–1066.

Rosenthal, D., 1990. A theory of consciousness. ZIF Report No. 40, Zentrum fur Interdisziplinaire Forschung, Bielefeld, Germany.

Rosenthal, D. M., 1993. Thinking that one thinks. In *Consciousness*, M. Davies and G. W. Humphreys, eds. Oxford: Blackwell, pp. 197–223.

Sanghera, M. K., E. T. Rolls, and A. Roper-Hall, 1979. Visual responses of neurons in the dorsolateral amygdala of the alert monkey. *Exp. Neurol.* 63:610–626.

Scott, T. R., S. Yaxley, Z. J. Sienkiewicz, and E. T. Rolls, 1986. Gustatory responses in the frontal opercular cortex of the alert cynomolgus monkey. *J. Neurophysiol.* 56:876–890.

Schwaber, J. S., B. S. Kapp, G. A. Higgins, and P. R. Rapp, 1982. Amygdaloid and basal forebrain direct connections with the nucleus of the solitary tract and the dorsal motor nucleus. *J. Neurosci.* 2:1424–1438.

Seltzer, B., and D. N. Pandya, 1989. Frontal lobe connections of the superior temporal sulcus in the rhesus monkey. *J. Comp. Neurol.* 281:97–113.

Thorpe, S. J., E. T. Rolls, and S. Maddison, 1983. Neuronal activity in the orbitofrontal cortex of the behaving monkey. *Exp. Brain Res.* 49:93–115.

Treves, A., and E. T. Rolls, 1994. A computational analysis of the role of the hippocampus in memory. *Hippocampus*, in press.

Van Hoesen, G. W., 1981. The differential distribution, diversity and sprouting of cortical projections to the amygdala in the rhesus monkey. In *The Amygdaloid Complex*, Y. Ben-Ari, ed. Amsterdam: Elsevier, pp. 77–90.

Van Hoesen, G. W., D. N. Pandya, and N. Butters, 1975. Some connections of the entorhinal (area 28) and perirhinal (area 35) cortices in the monkey: II. Frontal lobe afferents. *Brain Res.* 95:25–38.

Weiskrantz, L., 1956. Behavioral changes associated with ablation of the amygdaloid complex in monkeys. *J. Comp. Physiol. Psychol.* 49:381–391.

Weiskrantz, L., 1968. Emotion. In *Analysis of Behavioral Change*, L. Weiskrantz, ed. New York and London: Harper and Row, pp. 50–90.

Wilson, F. A. W., and E. T. Rolls, 1990a. Neuronal responses related to reinforcement in the primate basal forebrain. *Brain Res.* 502:213–231.

Wilson, F. A. W., and E. T. Rolls, 1990b. Neuronal responses related to the novelty and familiarity of visual stimuli in the substantia innominata, diagonal band of Broca and periventricular region of the primate. *Exp. Brain Res.* 80:104–120.

Wilson, F. A. W., and E. T. Rolls, 1993. The effects of stimulus novelty and familiarity on neuronal activity in the amygdala of monkeys performing recognition memory tasks. *Exp. Brain Res.* 93:367–382.

Wilson, F. A. W., and E. T. Rolls, submitted. The primate amygdala and reinforcement: A dissociation between rule-based and associatively-mediated memory revealed in amygdala neuronal activity.

Williams, G. V., E. T. Rolls, C. M. Leonard, and C. Stern, 1993. Neuronal responses in the ventral striatum of the behaving monkey. *Behav. Brain Res.* 55:243–252.

73 Neurophysiology of the Perception of Intentions by Primates

LESLIE BROTHERS

ABSTRACT This chapter presents an emerging view of the brain basis of perception of others' intentions. The central concept of links between sensory representation of social events and somatic response is introduced, together with empirical evidence for the role of particular neural structures in mediating social perception. The role of somatosensory signals in assigning motivational significance to current social experience is considered. The final section addresses the absence of "theory of mind" in autism.

General considerations for an anatomy of social response

Social responses result from activation of distributed neural subsystems consisting of effectors (motor, autonomic, and endocrine) linked with representations of external events involving conspecifics. While component features of the social environment—such as the flash of canine teeth in a threat display or the color of an estrous swelling—may be simple, social *events* derive their specific meanings from complex combinations of features. A nearby animal's direction and speed of movement, its facial expression, its vocalization, contextual facts such as preceding events, the roles and identities of other individuals in the vicinity, all combine to form unique events with unique implications for the observer. Therefore, neural subsystems that process such events so as to produce appropriate responses should have the following connectivities: intimate links with effector structures on the one hand, and access to highly processed, multimodal sensory activity on the other. A number of brain regions meet these criteria to greater or lesser degrees. Examples include the amygdala, posterior orbitofrontal cortex, and anterior cingulate cortex. In this section, each region is defined and its inputs from sensory cortices and outputs to effector centers are reviewed. Although interconnections among these structures, and outputs to cortical regions are undoubtedly important, they are omitted. A caveat: While the amygdala, orbitofrontal cortex, and anterior cingulate cortex are treated here as single entities, they are in fact composed of multiple subdivisions with different connections, many of which are incompletely understood.

The primate amygdala consists of the lateral, basal, basal accessory, anterior cortical, medial, posterior central, and intercalated nuclei; the nucleus of the lateral olfactory tract, periamygdaloid cortex, anterior amygdaloid area, and the amygdalohippocampal area. The connections of the primate amygdala have recently been reviewed in detail by Amaral et al. (1992), from whom the following selective summary is taken. The nuclei of the amygdala receive unimodal sensory input from higher-level visual and auditory areas, from the olfactory bulb and olfactory (piriform) cortex, and from somatosensory areas via the caudal insula. Polymodal sensory afferents to the amygdala arise from perirhinal, parahippocampal, and temporal polar cortices, and from the dorsal bank of the superior temporal sulcus (STS). The amygdala projects directly to the striatum, the hypothalamus, and brainstem centers. Thus, it is linked directly with motor, endocrine, and autonomic effector systems on the output side, and with unimodal and polymodal sensory regions on the input side.

Although the orbitofrontal cortex is a heterogeneous structure whose subdivisions have different connections, all receive inputs from the amygdala and from the mediodorsal nucleus, pars magnocellularis, of the thalamus (Fuster, 1989). The posterior orbitofrontal

LESLIE BROTHERS Office of Research, Sepulveda Veterans Administration Medical Center, Sepulveda, Calif.

cortex is here taken to include areas 12, 13, and 14 of Walker (1940). Like the amygdala, posterior orbitofrontal cortex receives high-level sensory input and has direct connections to effector structures. It receives visual input from the rostral inferior temporal cortex (Whitlock and Nauta, 1956; Barbas, 1988), auditory input from the rostral superior temporal cortex (Barbas, 1988), somatosensory input from the operculum and insula (Barbas, 1988), somatosensory and visual projections from posterior parietal and opercular cortex (Barbas, 1988; Cavada and Goldman-Rakic, 1989; Preuss and Goldman-Rakic, 1991), olfactory input from olfactory cortex (Powell, Cowan, and Raisman, 1965), and polymodal input from the STS (Barbas, 1988). As reviewed by Fuster (1989, 21), it sends projections directly to the hypothalamus as well as to the mesencephalon and pons. In addition, it projects to the basal ganglia (Selemon and Goldman-Rakic, 1985). The posterior orbitofrontal cortex, then, receives high-level unimodal and polymodal sensory input and projects directly to endocrine, autonomic, and motor areas.

The anterior cingulate includes area 24 and prelimbic area 32. Subdivisions of the anterior cingulate receive inputs from rostral auditory association cortex (Vogt and Pandya, 1987) and from polymodal areas including TG, TPO, TF, TH, and insula (Vogt and Pandya, 1987; Seltzer and Pandya, 1989). No direct outputs to the hypothalamus or brainstem autonomic centers have been described; various divisions of the anterior cingulate, however, send efferents to motor areas at a number of levels, including the primary and supplementary motor cortex (Van Hoesen, Benjamin, and Afifi, 1981; Morecraft and Van Hoesen, 1992), basal ganglia (Selemon and Goldman-Rakic, 1985), spinal cord (Dum and Strick, 1991), and pontine gray matter (Vilensky and Van Hoesen, 1981). Thus, while the particulars differ, the amygdala, posterior orbitofrontal cortex, and anterior cingulate cortex all mediate directly between higher-order sensory regions on the one hand and effector regions on the other. As we shall see below, each is implicated in social responsiveness.

Social animals need to assign distinct meanings to social events that may be rather similar in their sensory aspects. Neural systems mediating social response must produce more than general response tendencies to a few simple classes of external events (such as, e.g., startle reaction to loud noises, or withdrawal to suddenly looming objects). The putative distributed neural subsystems must combine discriminable, significant social situations encoded in sensory cortices with patterns of effector activity that may yield gross responses such as rage or flight, or with subtler and possibly elaborate patterns that produce changes in respiration, bloodflow distribution, or gonadotropin release. Links between complex sensory events and specific responses are not restricted to the domain of social events, but they are particularly useful for generating socially intelligent responses. Leaving aside other purposes that might be served by regions with such connectivity, we will now proceed to consider ensembles that simultaneously encode the sensory aspects of discrete social situations and set into motion the relevant responses. We can call such a distributed ensemble a *social situation representation/response*, henceforth, SSR/R.

Human SSR/Rs are universal

The complexities of social life call for differentiated and varied responses in ever-changing environments consisting of mates, offspring, allies, rivals, leaders, ingroups, bullies, and friends. Judging by the present complex social lives of apes and ourselves, such environments must have existed before the common ancestor of apes and humans evolved, and thus have been in place long enough to have shaped the cognition of hominids. The cognition in question takes the form of a sort of dictionary of innate social "feelings" (which can be read as "dispositions to act"; Brothers, 1990). While systems for conceptualizing feelings are not culturally universal (Wierzbicka, 1992), there are social situations that are universally articulable. Thus, when informants do not have a word for a feeling they wish to describe—or when they do not share a common language with an interlocutor, as happens when anthropologists travel among other cultures—they have recourse to describing the social situation in which the feeling occurs, counting on the other person's empathy to create the feeling as he imagines the recounted situation. The success of this strategy is one line of evidence for the universality of social situations and the feelings to which they give rise. We shall see instances of this tactic later in the chapter.

A second line of evidence for universality comes from the diversity of sources that report on identical situations. Let us now look at some examples of these. A Canadian patient (upon stimulation of the anterior temporal lobe) told his doctor he had a feeling, which he could not name, "like you are demanding ... as if I were guilty of some form of tardiness" (Gloor, 1986, 165). Ifaluk Pacific islanders have a word for this feeling, *metagu*. Anthropologists, who have devoted considerable attention to the concept, gloss it as an anxious or fearful experience engendered in a person who commits a social transgression and is thereby the object of moral anger on the part of someone of higher social status (Lutz, 1988). An example of another SSR/R that appears in many cultures is the negative sense of being socially shunned or excluded (Gruter and Masters, 1986). Also familiar to a Western reader are descriptions of situations causing *hakaika* in Marquesas islanders: For example, a bridegroom was *hakaika* at the thought of accepting an offer of wedding photographs, believing he might be seen as showing off. His discomfort was explained to the anthropologist by another Marquesan as follows: "It's *hakaika* because there are many people. People are *hakaika* at public Scripture recitals and drink two whiskeys to get up the energy to recite. They aren't skilled at that activity, so they are *hakaika*" (Kirkpatrick, 1985, 81).

"Self-conscious" might be the closest English translation for *hakaika*; in any event, as Kirkpatrick indicates, the term "points to a highly generalized sensitivity to others' opinions." A final example is *awumbuk*, "which does not have an encompassing gloss in English" (Fajans, 1985, 380). Compared to the preceding examples, *awumbuk* is a more time-extensive reaction to a social situation. According to Fajans, it is a lassitude that people feel after the departure of visitors, friends, or relatives who have resided with them. The symptoms of *awumbuk* are tiredness (sleeping late in the morning, and an inability to get started), lack of success in activities (failure to find game or to get the garden weeded), and some degree of boredom. The description is strikingly similar to those given in a very different context, Western psychoanalysis, by some patients following ruptures of interpersonal ties. A patient's reactions to episodes of depression in his wife, or to physical separations from her, are described as follows: "In these reactive states he would feel 'defeated,' find it difficult to get out of bed, and experience a 'black cloud' descending over him, along with a nearly complete loss of motivation" (Stolorow, Brandchaft, and Atwood, 1987, 96). The similarity of this description to that of *awumbuk* suggests that the presumed pathology resides in the intensity, but not the content, of the man's reaction.

The brain's potentiality for setting up SSR/R ensembles must be innate, but the content of realized SSR/Rs cannot be entirely genetically specified. Instead, as is the case for language and other learned behavior, the capacity is experience-expectant, and therefore subject to cultural and familial influence. What is entirely culturally specified, and not at all biological, is the assignment of particular linguistic labels (e.g., pride, tenderness, resentment) to feelings stimulated by social situations.

Evidence from the brain

We now turn to brain regions possessing the connectivity essential to SSR/R ensembles. This section surveys lesion and stimulation studies in the primate amygdala, orbitofrontal cortex and anterior cingulate with respect to social behavior. (For a review of single unit data, see Brothers, 1992). It concludes with some comments on the neural basis of failures in social judgment by patients with orbitofrontal lesions.

Lesions The Klüver-Bucy syndrome (Klüver and Bucy, 1939), which consists of tameness, "psychic blindness," and changes in ingestive and sexual behavior, is produced in monkeys by temporal lobe lesions, specifically, lesions of the amygdala and medial temporal polar cortex (Weiskrantz, 1956). The social effects of amygdala lesions in nonhuman primates have been reviewed by Kling and Brothers (1992). In adults, such lesions result in decreased social interaction and loss of rank that are permanent, in contrast to temporary but recoverable changes seen in lesioned infants. Deterioration in maternal behavior is produced by amygdala lesions in some species. Most species display changes in sexual behavior, of which hypersexuality is the commonest. Aggleton (1992), citing Jacobson (1986) and Tranel and Hyman (1990), notes that amygdala lesions in humans are associated with difficulties in face recognition, but states that whether these are due to a more general deficit in visuospatial memory remains an open question. Of interest is the behavior of a woman with congenital bilateral calcification of the amygdala, who is unusually forward with experimenters

(Nahm, personal communication) and has a "tendency to be somewhat coquettish and disinhibited ... and often makes mildly inappropriate sexual remarks" (Tranel and Hyman, 1990).

Lesions of orbitofrontal cortex in rhesus monkeys (*Macaca mulatta*) produce aversion to social contact and a decrease in aggression (Butter, Snyder, and MacDonald, 1970; Butter and Snyder, 1972). Diminished social contact in orbitofrontal operates, measured as decreased frequency and duration of grooming bouts and decreased time spent in spatial proximity to others, is also seen in lesioned vervet monkeys (*Cercopithecus aethiops*). Alterations in previous social roles are observed in these subjects: For example, female operates rebuff juveniles (Raleigh, 1976; Raleigh and Steklis, 1981). In humans, orbitofrontal lesions result in facetiousness without pleasure, social disinhibition, and "acquired sociopathy" (Damasio and Van Hoesen, 1983; Eslinger and Damasio 1985; Damasio, Tranel, and Damasio, 1990). The syndrome characterized by Damasio and his colleagues as "acquired sociopathy" is exemplified by patient E. V. R. Prior to surgery for an orbitofrontal meningioma at age 35, the patient was an exemplary father, businessman, and member of his community. After the surgery, the patient retained a high level of intelligence, as demonstrated on numerous diagnostic tests, but became completely unable to make appropriate decisions regarding work, relationships, and daily living. He had a striking inability to assess the trustworthiness of new associates, and entered into partnerships with persons of doubtful character, which resulted in bankruptcy. Further study revealed that E. V. R. was able to solve difficult social problems, including sophisticated ethical problems, when presented with these verbally (Saver and Damasio, 1991). Findings regarding deficient autonomic activity in this patient, and Damasio's interpretation of these findings, are discussed at the end of this section.

Lesions of the anterior cingulate in humans are generally the result of vascular accidents, tumors, or psychosurgery. The results of psychosurgery will not be reviewed here, because their interpretation is clouded by the fact that the patients have preexisting psychiatric difficulties. Akinetic mutism, that is, lack of voluntary speech and movement in the presence of apparent alertness, has been associated with lesions of the anterior cingulate region (Barris and Schuman, 1953). Jürgens and von Cramon (1982) reported on a case of akinetic mutism with damage to anterior cingulate and neighboring regions. Although the patient recovered his capacity for spontaneous speech, emotional intonation remained permanently defective; together with the results of monkey studies, this led the authors to conclude that "the anterior cingulate cortex is involved in the volitional control of emotional vocal utterances." In a review, Damasio and Van Hoesen wrote, "It is unquestionable that bilateral damage to the cingulate causes a state of disturbed affect, in which expression and experience of emotion are precluded, and any attempt to communicate is curtailed" (1983, 95). They reported on the subjective experience of a woman patient who had mutism as a result of cingulate damage: Upon recovery she said that during her mute phase "nothing mattered" and therefore she had had "nothing to say." Using neuropsychological tests of attention in a patient before and after cingulotomy, Janer and Pardo (1991) concluded that destruction of the anterior cingulate impairs selective attention. The contrast between this view of anterior cingulate function and the preceding one shows that approaches to brain function that are framed in terms of nonsocial cognition elicit different results than those framed in terms of social interaction.

STIMULATION Stimulation of the amygdala in animals produces behavior that appears fearful, such as piloerection and attempts to escape. Rage and attack behaviors, components of copulation sequences, and vocalizations are also produced by amygdala stimulation (see Kling and Brothers, 1992, for review). Stimulation of the amygdala in human patients has been reviewed by Halgren (1992) and Gloor (1992). In contrast to the case of nonhuman primates, sexual automatisms are almost never the result of amygdala stimulation, although psychosexual experiences may occur in women. Elementary experiences of fear are common, as are autonomic effects and sensations referable to the viscera. The most striking effects of amygdala stimulation are mnemonic episodes of an autobiographical nature. Of all brain structures that have been subjected to systematic stimulation, including temporal isocortex and hippocampus, the amygdala produces mnemonic episodes most reliably, in the absence of afterdischarge spread to neighboring regions (Gloor, 1992).

Gloor noted that these mnemonic episodes are "most often set in what might be called a 'social context' in the broadest sense of the term, or better perhaps in an 'ethological context'.... They frequently bear some relationship with familiar situations or situations with an affective meaning. They frequently touch on some aspect of [the patient's] relationship with other people, either known specifically to him or not" (1986, 165).

Here are two examples of such episodes. (The second, which involves hippocampal stimulation, is included because the hippocampus and the amygdala are intimately interconnected and, as the example shows, may both participate in generating responses to social situations.)

> Upon stimulating his left amygdala at 1mA, he had a feeling "as if I were not belonging here," which he likened to being at a party and not being welcome.

> Right hippocampal stimulation at 3 mA induced anxiety and guilt, "... like you are demanding to hand in a report that was due 2 weeks ago ... as if I were guilty of some form of tardiness." (Gloor, 1986, 164)

Note that the latter experience has been referred to above as equivalent to *metagu* in Pacific islander society. As noted in Brothers (1990) there are two interesting aspects to these reports. One is that the reported experiences involved actions, attitudes, or intentions attributed to others, perceived by the subjects to be directed at themselves; and they were, simultaneously, feelings (dispositions) on the part of the subjects. This illustrates the dual aspect of the SSR/R, and is evidence for the role of structures such as the amygdala in SSR/Rs. In real-life experience, as opposed to the stimulation setting, the high-level sensory representations of the others would be specific and spelled out, because of activation of widespread areas of sensory cortex. A second interesting aspect is that when the subjects found that they could not name the feelings, they had recourse to describing a social situation, leaving it to the examiner to derive the feeling from the situation, as in the example of the Marquesan and the anthropologist cited earlier.

Stimulation of orbitofrontal cortex in monkeys results in respiratory inhibition and changes in blood pressure (Kaada, 1951). Data on stimulation in this region in humans are not available. Stimulation of the anterior cingulate produces vocalizations in monkeys (Kaada, 1951; Robinson, 1967; Jürgens and Ploog, 1970) and natural-appearing movements that are probably due to the intimate tie between the cingulate and motor regions (Luppino et al., 1991; Van Hoesen, Benjamin, and Afifi, 1981). Similarly, in humans, stimulation produces natural-appearing movements of the hands and mouth, such as rubbing, scratching, or squeezing movements of the hands, puckering of the lips, and movements of the tongue (Talairach et al., 1973). In this study, a tendency to euphoria was noted in response to stimulation, as was an instance of playful behavior directed toward the experimenter. Sexual automatisms may result from seizure activity in the anterior cingulate (Delgado-Escueta et al., 1987).

The Role of the Soma: A debate Before concluding this section on correlates of social function in brain activity, it is appropriate to analyze the findings and theories of Damasio, Tranel, and Damasio (1990, 1991) regarding the social deficits of patient E. V. R., described above in the section on orbitofrontal lesions. Damasio found that while patients such as E. V. R. were able to produce appropriate responses to verbally presented social situations, they were unable to make correct choices in real-life situations. Damasio and his colleagues tested E. V. R. and other patients with similar pathology for their ability to produce autonomic responses to visually presented social stimuli. While elementary stimuli such as loud noises produced normal autonomic responses, target pictures such as scenes of social disaster, mutilation, and nudity did not. The researchers concluded that the deficit in social response in these patients arose from their inability to link current perceptions with somatic states linked to punishment and reward that were previously mobilized by similar situations. Such states could take the form of visceral and skeletal muscular activity, which would normally be signaled to somatosensory and higher-order cortices, thence to be experienced, via limbic mediation, as positive or negative feelings. More usually in adults, however, prior experience would have established central representations of somatic states in somatosensory cortices. Since these cortices possess reciprocal links with structures such as orbitofrontal cortex and amygdala, reencountering a situation that had already been associated with a somatic state would evoke the relevant somatic representation centrally (Tranel and Damasio, 1993; Damasio, personal communication). The destruction of the orbitofrontal link, with the consequent failure to regenerate either actual somatic states or their central representations, deprives the patient of somatic markers to guide behavior in the current situation.

The primary difference between this account and an SSR/R account of E. V. R.'s difficulties is that the somatic marker account is dichotomous, whereas the SSR/R account is pluralistic. That is, Damasio's accounts considered the somatic portion of a social experience simply to signal reward or punishment, while SRR/R holds that (numerous, discriminable) social situations are linked with (numerous, discriminable) patterns of effector activity in autonomic, endocrine, and motor structures of the brain. The significant complexity of efferent activity patterns in the SSR/R model is captured by Gloor's imagery regarding the amygdala, namely, that it "plays upon the autonomic-neuroendocrine keyboard" (1992, 511). While there is at present no empirical evidence that unequivocally supports either the dichotomous or the "keyboard" account, the evidence from amygdala stimulation, at least, is in favor of specific and variegated motivational states (e.g., guilt, feeling ostracized) rather than simply reward or punishment.

The target scenes presented to the orbitofrontal patients were complex, requiring a high level of sensory processing, in contrast to the loud noise that provoked autonomic activity. According to SSR/R, structures such as amygdala, orbitofrontal cortex, and to a lesser extent, the anterior cingulate, specialize in linking highly processed sensory material to effector structures; indeed, an appropriate response to very complex stimuli might require the orchestrated activity of all these regions. Neural activity provoked by simpler stimuli such as loud noises might be relayed from the thalamus (LeDoux, 1992), bypassing higher-order cortices, and pass in parallel through several linking areas to arrive at effector structures. Such activity would be robust to damage in one or several linking areas. In the case of E. V. R., higher-order sensory activity evoked by the target pictures in the experiment, or by real-life social situations, is not linked to effectors such as autonomic activity or other complex somatic dispositions to act.

The SSR/R interpretation is indirectly supported by a more recent study (Tranel and Damasio, 1993). In a seeming paradox, patient Boswell, who has sustained damage not only to his posterior orbitofrontal cortices bilaterally, but to many other structures besides, has been able to form what appears to be a conditioned association, accompanied by autonomic changes, to target pictures of faces. The discrepancy between the tasks presented to E. V. R. and to Boswell clarifies the issue. Whereas the orbitofrontal-lesioned patients were asked to view scenes whose significance was mediated via social understanding (e.g., scenes of disasters, mutilations, and nudity), which therefore require considerable processing, Boswell was asked to view stimuli that had simply been linked with positive or negative experiences for him. The former, more demanding task would critically require the integration of widespread cortical activity; it is also sensitive to the intactness of the orbitofrontal cortices. The latter task requires no such interpretive judgment, and the autonomic response appears despite widespread brain damage.

Theory of mind

"Theory of mind" is a phrase coined by Premack (Premack and Woodruff, 1978; Premack, 1988) and used both in the study of primates "mentalizing" their fellows, and to describe the deficit presumed essential to the pathology of autism (Baron-Cohen, 1990; Frith, Morton, and Leslie, 1991). The capacity to represent others' intentions does more, evolutionarily speaking, than simply permit its possessor to engage in the social gambits of Machiavellian primate societies: It facilitates social learning and thus the spread of technology. This is because a technical innovation is likely to be understood only if the *intention* of the inventor is grasped by the observer of the new method. Accordingly, Tomasello, Kruger, and Ratner (1993) explain the greater cognitive abilities of captive over wild chimpanzees as being due to enhancement of their latent capacity for taking on the perspective of intentional agents. The enhancement arises from frequent experiences of joint attention, with their human caregivers, to objects and activities.

The ontogeny of a theory of mind in normal children is by now well described (Astington, Harris, and Olson, 1988). The child's construction of other minds, in the standard account, is concerned mainly with others' attention, beliefs, and false beliefs. In contrast to this "cold" theory of mind is an SSR/R-based account called the "hot" theory of mind (Brothers and Ring, 1992). In this account, the perception of an attitude that has significance for the perceiving subject (such as, a belittling attitude on the part of another that produces feelings of shame in the subject) brings together the complex sensory representation and the effector response as described in the first two sections of this chapter.

A "hot" theory of mind constructs others' meaningful intentions and their evaluative attitudes. Evaluative attitudes (such as surprise, pleasure, or fear) can be read from the face in expressive movements occurring primarily in the regions around the eyes and mouth (Fridlund, 1991). Evaluative signals, in combination with direction of attention as indicated, for example, by gaze, are used by the observer to create descriptions of intention and disposition. Normal children use "social referencing" in ambiguous situations to discern the evaluative attitudes of caregivers toward novel stimuli (Campos and Sternberg, 1981); autistic children do not (Sigman et al., 1992). Furthermore, autistic people have difficulty analyzing affective information in other settings (Hobson, 1986; Hobson, Ouston, and Lee, 1988). For this reason, difficulty in "primary affective relatedness" is held by Hobson to be the basic defect of autism (Hobson, 1993). Such a defect could result from failure to attend to or process the eye region of the face for affective signals, or to translate such signals, via SSR/R ensembles, into subjective feelings. Indeed, autistic children have difficulty understanding the feelings generated by social situations (Yirmiya et al., 1992). The hypothesis that autistic deficits arise from a developmental failure to attend to and process the eye region of faces is currently under investigation (Baron-Cohen and Cross, 1992; Baron-Cohen et al., in press).

ACKNOWLEDGMENTS The author thanks F. Wilson and J. L. Locke for helpful discussions, and T. Preuss, S. Baron-Cohen and J. LeDoux for valuable comments on the manuscript. The author's research is supported by the Department of Veterans Affairs.

REFERENCES

Aggleton, J. P., 1992. The functional effects of amygdala lesions in humans: A comparison with findings from monkeys. In *The Amygdala: Neurobiological Aspects of Emotion, Memory, and Mental Dysfunction*, J. P. Aggleton, ed. New York: Wiley, pp. 485–503.

Amaral, D., J. L. Price, A. Pitkanen, and S. T. Carmichael, 1992. Anatomical organization of the primate amygdaloid complex. In *The Amygdala: Neurobiological Aspects of Emotion, Memory, and Mental Dysfunction*, J. P. Aggleton, ed. New York: Wiley, pp. 1–66.

Astington, J., P. Harris, and D. Olson, eds., 1988. *Developing Theories of Mind.* Cambridge: Cambridge University Press.

Barbas, H., 1988. Anatomic organization of basoventral and mediodorsal visual recipient prefrontal regions in the rhesus monkey. *J. Comp. Neurol.* 276:313–342.

Baron-Cohen, S., 1990. Autism: A specific cognitive disorder of "mind-blindness". *Int. Rev. Psychiatry.* 2:81–90.

Baron-Cohen, S., R. Campbell, A. Karmiloff-Smith, P. Grant, and J. Walker, 1993. Are children with autism blind to the significance of the eyes? Unpublished manuscript, Institute of Psychiatry, London.

Baron-Cohen, S., and P. Cross, 1992. Reading the eyes: Evidence for the role of perception in the development of a theory of mind. *Mind Lang.* 7:172–186.

Barris, R., and H. Schuman, 1953. Bilateral anterior cingulate gyrus lesions: Syndrome of the anterior cingulate gyri. *Neurology* 3:44–52.

Brothers, L., 1990. The social brain: A project for integrating primate behavior and neurophysiology in a new domain. *Concepts Neurosci.* 1(1):27–51.

Brothers, L., 1992. Perception of social acts in primates: Cognition and neurobiology, *Semin. Neurosci.* 4:409–414.

Brothers, L., and B. Ring, 1992. A neuroethological framework for the representation of minds. *J. Cognitive Neurosci.* 4(2):107–118.

Butter, C., and D. Snyder, 1972. Alterations in aversive and aggressive behaviors following orbital frontal lesions in rhesus monkeys. *Acta Neurobiol. Exp.* 32:525–565.

Butter, C., D. Snyder, and J. MacDonald, 1970. Effects of orbital frontal lesions on aversive and aggressive behaviors in rhesus monkeys. *J. Comp. Physiol. Psychol.* 72:132–144.

Campos, J., and C. Sternberg, 1981. Perception, appraisal and emotion: The onset of social referencing. In *Infant Social Cognition: Empirical and Theoretical Considerations*, M. Lamb and L. Sherrod, eds. Hillsdale, N. J.: Erlbaum, pp. 273–314.

Cavada, C., and P. S. Goldman-Rakic, 1989. Posterior parietal cortex in rhesus monkey: I. Parcellation of areas based on distinctive limbic and sensory corticocortical connections. *J. Comp. Neurol.* 287:393–421.

Damasio, A., D. Tranel, and H. Damasio, 1990. Individuals with sociopathic behavior caused by frontal damage fail to respond autonomically to social stimuli. *Behav. Brain Res.* 41:81–94.

Damasio, A., D. Tranel, and H. Damasio, 1991. Somatic markers and the guidance of behavior: Theory and preliminary testing. In *Frontal Lobe Function and Dysfunction*, H. Levin, H. Eisenberg, and A. Benton, eds. New York: Oxford University Press, pp. 217–229.

Damasio, A., and G. Van Hoesen, 1983. Emotional disturbances associated with focal lesions of the limbic frontal lobe. In *Neuropsychology of Human Emotion*, K. Heilman and P. Satz, eds. New York: Guilford, pp. 85–110.

Delgado-Escueta, A., B. Swartz, H. Maldonado, G. Walsh, R. Rand, and E. Halgren, 1987. Complex partial seizures of frontal lobe origin. In *Presurgical Evaluation of Epileptics*, H. Wieser and C. Elger, eds. Berlin: Springer, pp. 268–299.

Dum, R., and P. Strick, 1991. The origin of corticospinal projections from the premotor areas in the frontal lobe. *J. Neurosci.* 11(3):667–689.

Eslinger, P. J., and A. R. Damasio, 1985. Severe disturbance of higher cognition after bilateral frontal lobe ablations: Patient EVR. *Neurology* 35:1731–1741.

Fajans, J., 1985. The person in social context: The social character of Baining "Psychology." In *Person, Self, and*

Experience: Exploring Pacific Ethnopsychologies, G. M. White and J. Kirkpatrick, eds. Berkeley: University of California Press, pp. 367–397.

Fridlund, A., 1991. Evolution and facial action in reflex, social motive, and paralanguage. *Biol. Psychol.* 32:3–100.

Frith, U., J. Morton, and A. M. Leslie, 1991. The cognitive basis of a biological disorder: Autism. *Trends Neurosci.* 14(10):433–438.

Fuster, J. M., 1989. *The Prefrontal Cortex: Anatomy, Physiology, and Neuropsychology of the Frontal Lobe.* New York: Raven.

Gloor, P., 1986. The role of the human limbic system in perception, memory and affect: Lessons from temporal lobe epilepsy. In *The Limbic System: Functional Organization and Clinical Disorders*, B. K. Doane and K. E. Livingston, eds. New York: Raven, pp. 159–169.

Gloor, P., 1992. Role of the amygdala in temporal lobe epilepsy. In *The Amygdala: Neurobiological Aspects of Emotion, Memory, and Mental Dysfunction*, J. P. Aggleton, ed. New York: Wiley, pp. 505–538.

Gruter, M., and R. Masters, eds., 1986. *Ostracism: A Social and Biological Phenomenon.* New York: Elsevier.

Halgren, E., 1992. Emotional neurophysiology of the amygdala within the context of human cognition. In *The Amygdala: Neurobiological Aspects of Emotion, Memory, and Mental Dysfunction*, J. P. Aggleton, ed. New York: Wiley, pp. 191–228.

Hobson, R., 1986. The autistic child's appraisal of expressions of emotion, *J. Child Psychol. Psychiatry* 27:321–342.

Hobson, R., 1993. Understanding persons: The role of affect. In *Understanding Other Minds: Perspectives from Autism*, S. Baron-Cohen, H. Tager-Flusberg, and D. J. Cohen, eds. Oxford: Oxford University Press, pp. 204–227.

Hobson, R., J. Ouston, and A. Lee, 1988. What's in a face? The case of autism. *Br. J. Psychol.* 79:441–453.

Jacobson, R., 1986. Disorders of facial recognition, social behavior and affect after combined bilateral amygdalotomy and subcaudate tractotomy: A clinical and experimental study. *Psychol. Med.* 16:439–450.

Janer, K., and J. Pardo, 1991. Deficits in selective attention following bilateral anterior cingulotomy. *J. Cognitive Neurosci.* 3:231–241.

Jürgens, U., and D. Ploog, 1970. Cerebral representation of vocalization in the squirrel monkey. *Exp. Brain Res.* 10:532–554.

Jürgens, U., and D. von Cramon, 1982. On the role of the anterior cingulate cortex in phonation: A case report. *Brain Lang.* 15:234–248.

Kaada, B., 1951. Somato-motor, autonomic and electrocorticographic responses to electrical stimulation of "rhinencephalic" and other structures in primates, cat and dog. *Acta Physiol. Scand.* 24:1–285.

Kirkpatrick, J., 1985. Some Marquesan understandings of action and identity. In *Person, Self, and Experience: Exploring Pacific Ethnopsychologies*, G. M. White and J. Kirkpatrick, eds. Berkeley: University of California Press, pp. 80–120.

Kling, A., and L. Brothers, 1992. The amygdala and social behavior. In *The Amygdala: Neurobiological Aspects of Emotion, Memory, and Mental Dysfunction*, J. P. Aggleton, ed. New York: Wiley, pp. 353–377.

Klüver, H., and P. C. Bucy, 1939. Preliminary analysis of functions of the temporal lobes in monkeys. *Arch. Neurol. Psychiatry.* 42:797–1000.

LeDoux, J., 1992. Emotion and the amygdala. In *The Amygdala: Neurobiological Aspects of Emotion, Memory, and Mental Dysfunction*, J. P. Aggleton, ed. New York: Wiley, pp. 339–351.

Luppino, G., M. Matelli, R. Camarda, V. Gallese, and G. Rizzolatti, 1991. Multiple representations of body movements in mesial area 6 and the adjacent cingulate cortex: An intracortical microstimulation study in the macaque monkey, *J. Comp. Neurol.* 311:463–482.

Lutz, C., 1988. *Unnatural Emotions: Everyday Sentiments on a Micronesian Atoll and Their Challenge to Western Theory.* Chicago: University of Chicago Press.

Morecraft, R. J., and G. Van Hoesen, 1992. Cingulate input to the primary and supplementary motor cortices in the rhesus monkey: Evidence for somatotopy in areas 24c and 23c. *J. Comp. Neurol.* 322:471–489.

Powell, T. P. S., W. M. Cowan, and G. Raisman, 1965. The olfactory connections. *J. Anat.* 99:791–813.

Premack, D., 1988. "Does the chimpanzee have a theory of mind?" revisited. In *Machiavellian Intelligence: Social Expertise and the Evolution of Intellect in Monkeys, Apes and Humans*, R. Byrne and A. Whiten, eds. Oxford: Clarendon, pp. 160–179.

Premack, D., and G. Woodruff, 1978. Does the chimpanzee have a theory of mind? *Behav. Brain Sci.* 1:515–526.

Preuss, T., and P. S. Goldman-Rakic, 1991. Ipsilateral cortical connections of granular frontal cortex in the strepsirhine primate *Galago*, with comparative comments on anthropoid primates. *J. Comp. Neurol.* 310:507–549.

Raleigh, M. J., 1976. Unpublished doctoral dissertation, University of California, Berkeley.

Raleigh, M. J., and H. D. Steklis, 1981. Effects of orbital frontal and temporal neocortical lesions on affiliative behavior of vervet monkeys. *Exp. Neurol.* 73:378–389.

Robinson, B., 1967. Vocalization evoked from forebrain in *Macaca mulatta*. *Physiol. Behav.* 2:345–354.

Saver, J. L., and A. R. Damasio, 1991. Preserved access and processing of social knowledge in a patient with acquired sociopathy due to ventromedial frontal damage. *Neuropsychologia* 29:1241–1249.

Selemon, L. D., and P. Goldman-Rakic, 1985. Longitudinal topography and interdigitation of corticostriatal projections in the rhesus monkey. *J. Neurosci.* 5(3):776–794.

Seltzer, B., and D. Pandya, 1989. Frontal lobe connections of the superior temporal sulcus in the rhesus monkey. *J. Comp. Neurol.* 281:97–113.

Sigman, M. D., C. Kasari, J.-H. Kwon, and N. Yirmiya, 1992. Responses to the negative emotions of others by autistic, mentally retarded, and normal children. *Child Dev.* 63:796–807.

Stolorow, R., B. Brandchaft, and G. Atwood, 1987. *Psychoanalytic Treatment: An Intersubjective Approach.* Hillsdale, N. J.: The Analytic Press.

Talairach, J., J. Bancaud, S. Geier, M. Bordas-Ferrer, A. Bonis, G. Szikla, and M. Rusu, 1973. The cingulate gyrus and human behavior. *Electroencephalogr. Clin. Neurophysiol.* 34:45–52.

Tomasello, M., A. C. Kruger, and H. H. Ratner, 1993. Cultural learning. *Behav. Brain Sci.* 16:495–552.

Tranel, D., and A. Damasio, 1993. The covert learning of affective valence does not require structures in hippocampal system or amygdala. *J. Cognitive Neurosci.* 5:79–88.

Tranel, D., and B. Hyman, 1990. Neuropsychological correlates of bilateral amygdala damage. *Arch. Neurol.* 47: 349–355.

Van Hoesen, G., D. Benjamin, and A. Afifi, 1981. Limbic cortical input to area 6 in the monkey. *Anat. Rec.* 199:262–263.

Vilensky, J., and G. Van Hoesen, 1981. Corticopontine projections from the cingulate cortex in the rhesus monkey. *Brain Res.* 205:391–395.

Vogt, B., and D. Pandya, 1987. Cingulate cortex of the Rhesus monkey: II. Cortical afferents. *J. Comp. Neurol.* 262:271–289.

Walker, E., 1940. A cytoarchitectural study of the prefrontal area of the macaque monkey. *J. Comp. Neurol.* 98:59–86.

Weiskrantz, L., 1956. Behavioral changes associated with ablation of the amygdaloid complex in monkeys. *J. Comp. Physiol. Psychol.* 4:381–391.

Whitlock, D. G., and W. J. H. Nauta, 1956. Subcortical projections from the temporal neocortex in the *Macaca mulatta*. *J. Comp. Neurol.* 106:183–212.

Wierzbicka, A., 1992. Talking about emotions: Semantics, culture, and cognition. *Cognition Emotion* 6:285–319.

Yirmiya, N., M. Sigman, C. Kasari, and P. Mundy, 1992. Empathy and cognition in high functioning children with autism. *Child Dev.* 663:150–160.

74 Stressful Experience, Brain, and Emotions: Developmental, Genetic, and Hormonal Influences

BRUCE S. MCEWEN

ABSTRACT Genetic constitution and experience interact to determine brain function, just as they influence other organs and systems of the body. Emotions are the product of an individual's own processing of occurrences on the basis of his or her own prior history and biology, and an emotional response activates neural and neuroendocrine effector systems and leads to a variety of short and long-term consequences that may or may not result in disease. The adaptive response of the organism has been described in terms of *homeostasis*, and by a newer term, *allostasis*, meaning "stability through anticipatory change"; the long-term consequences of continued demand on the physiologic response are referred to as *allostatic load.*

Hormones of the thyroid, gonads, and adrenals have important effects on the developing and adult brain. They play an important role in developmental events that program the response of the adult brain, and they also mediate structural and neurochemical changes that modify how the adult brain responds to ongoing experience. The hippocampus is an important brain structure in cognition, and it also plays a role in anxiety; the hippocampal formation is sensitive to circulating adrenal steroids, which produce both short-term and long-term effects. Adrenal steroids also counterregulate various neurotransmitter systems, including the very important CRH system of the brain, which is involved as a mediator of arousing and anxiogenic effects of stress. The amygdala also plays an important role in governing adrenal steroid secretion during stress as well as in mediating behaviors related to fear and anxiety; however, the role of the amygdala as a target of both CRH and adrenal steroid actions is less well studied. The participation of neural and neuroendocrine mechanisms in pathophysiological responses to "stressors" is discussed in terms of the model of allostatic load. The importance of cognitive processing of information about a potential stressor is emphasized as a factor that determines whether or not the response of the individual is "stressful." However, the brain mechanisms underlying that processing are influenced by their intrinsic connectivity and neurochemistry, which are the product of genetic predisposition, developmental influences, and prior experiences and neuroendocrine influences.

BRUCE S. MCEWEN Laboratory of Neuroendocrinology, Rockefeller University, New York, N.Y.

Stressful experiences occur frequently, and yet stress is a very difficult concept to define and quantify psychologically and physiologically. This is because stressors are heterogenous, and also because the individual responses to a potentially stressful situation are highly variable. In fact, there are major differences in how individuals perceive and respond to challenges, as well as differences in the physiological capacities and limitations of individuals to respond to potentially stressful situations. Natural disasters, physical trauma, human-caused disasters such as accidents and rape, and life events such as divorce, job loss, and death of a loved one are well recognized as producing stress in virtually all people, but even these stressors do not have uniform effects in terms of affecting health, for example (Osterweis, Solomon, and Green, 1984; Weiner, 1992).

Individual differences in responding to challenge are a product of genetics, developmental influences, and experience. This chapter discusses environmental factors that help to bring out individual differences in the neural as well as behavioral and physiological responses to a stressful experience. We shall first discuss why there are individual differences in response to stress, noting a well-recognized behavioral model that emphasizes cognitive processing of and response to potentially stressful experiences. Then we shall describe the biological mediators of the response to stressors, and finally, we shall discuss possible developmental determinants of the behavioral and neural mechanisms underlying the stress response.

We shall next turn to the brain and the endocrine system. After noting the involvement of two limbic brain regions, the hippocampus and the amygdala, in the stress response, we shall discuss the neuroendocrine response to stress and the role of adrenal steroid hormones in the adaptive as well as maladaptive aspects of the response to stress. After summarizing the important, stabilizing effects of adrenal steroids on behavioral consequences of psychological stress, we shall switch to a discussion of the links between stress and disease, in which both acute and chronic stress play a role, and to which wear and tear produced by the body's own adaptive mechanisms to prolonged stress appears to be an important contributor.

Individual differences in response to psychological stress

Stressful experiences activate autonomic and neuroendocrine responses, which originally evolved to enable vertebrates to adapt and survive while either fleeing from predators or fighting members of their own species in competition for mates, food, or territory. The problem is that, for humans, most stressful experiences do not involve overt fight-or-flight responses; rather, they are usually manifestations of individual reactions to interpersonal relations and life events. In other words, humans are particularly vulnerable to suffer from psychological stressors. Thus the endocrine and neural activity resulting from the stress response leads to consequences in humans other than those they were intended to meet in animals that are actively fighting or fleeing. They have the potential to cause harm, including gastrointestinal disturbances and cardiovascular hyperactivity as well as anxiety and depressive illness (Chrousos and Gold, 1992; Weiner, 1992).

However, there are enormous individual differences in the response to potentially stressful situations. Genetic predisposition is an important factor (figure 74.1), and yet genetics alone does not explain individual differences, because there are only 40% to 60% concordances among identical twins for diseases such as Alzheimer's disease (Davies, 1986), atherosclerosis (Berg, 1983), schizophrenia (Plomin, 1990) and type I diabetes (Barnett, Leslie, and Pyke, 1981); and only 19% concordances for asthma (Mrazek and Klinnert, 1991). Thus, environmental factors in the broadest sense, meaning experiences from conception onward, play an important role in determining the expression of disease-related genetic traits. To help understand why some individuals are more susceptible to stressful events, we must consider some of the multiple factors operating under conditions of stress.

Behavioral Aspect On the behavioral side, the most important factor is how the individual interprets and reacts to a challenge (Lazarus, 1966). (See figure 74.1.) The reaction to potentially stressful stimuli is determined, in part, by the social context in which they occur and the social status of the individual (e.g., whether he or she is dominant or subordinate). Moreover, the effect of the potential stress-evoking situation on the nervous system is determined in part by the genetic predisposition, stage of biological development, and sex of the individual, and also by past learned experiences. In light of these factors, the potentially stressful stimulus is perceived as a threat or not perceived as a threat. If it is a threat, the source of the threat is either known or not known. If the source is not known, the individual becomes highly vigilant and remains physiologically aroused until the decision can be made that the situation is, or is not, threatening. If the source is known, then the question is whether or not there is a coping response available. If no response is available, then the response may be one of helplessness or hopelessness, with altered physiological responses.

If a response is available, it may be a low-cost response and therefore would not be classified as stressful. Or it may be a high-cost response such as aggression; or a response involving thrill-seeking or risk-taking behavior, like smoking or drinking or driving recklessly, that is itself dangerous to health. For example, alcohol is both used and abused as a means of coping with stress (Sher, 1987). However, the short-term reduction of stress has its own psychological and biological costs, namely, that increasing alcohol consumption can result in increased stress due to negative social consequences of drinking, as well as feelings of worry and guilt, and adverse physical consequences—which themselves may serve as triggers for anxiety reactions (Kushner, Sher, and Beitman, 1990). In turn, these stress-inducing consequences of repeated alcohol consumption may lead to increased drinking, which will inevitably entail greater overall stress and deterioration of health.

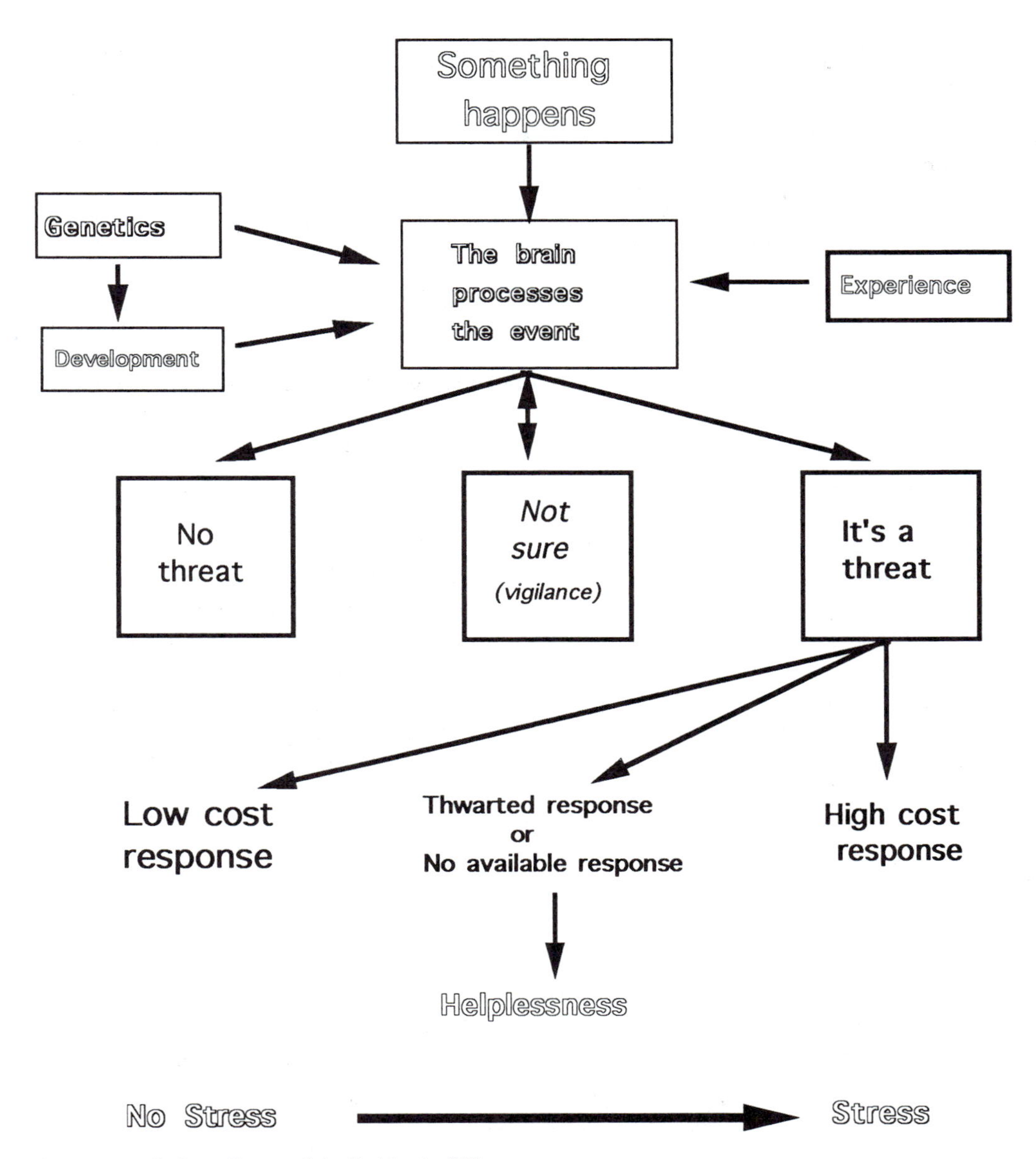

FIGURE 74.1 Diagram of the effects of individual differences in processing of information about a potential stressor on the behavioral response.

Alternatively, the available response can be blocked or thwarted, and this may lead to helplessness, frustration, displaced aggression, and various high-cost responses. Perceptions of threat of response choices are also central aspects of the assessment of risk and of decisions pertaining to health-damaging and health-promoting behaviors, including compliance or noncompliance with medical treatments (Redelmaier, Rozin, and Kahneman, 1993; Horwitz and Horwitz, 1993).

BIOLOGICAL COMPONENT OF RESPONSE TO STRESSFUL EXPERIENCES The biological side of the stress response is also very important, and the qualitative and quantitative nature of the biological stress response is dependent on the outcome of the behavioral assessment of the risk or threat represented by a particular situation. The biological side of the stress response can be divided into three components: mediators, effectors, and disease outcomes (figure 74.2). The mediators of the response to stressful stimuli are the brain, the autonomic

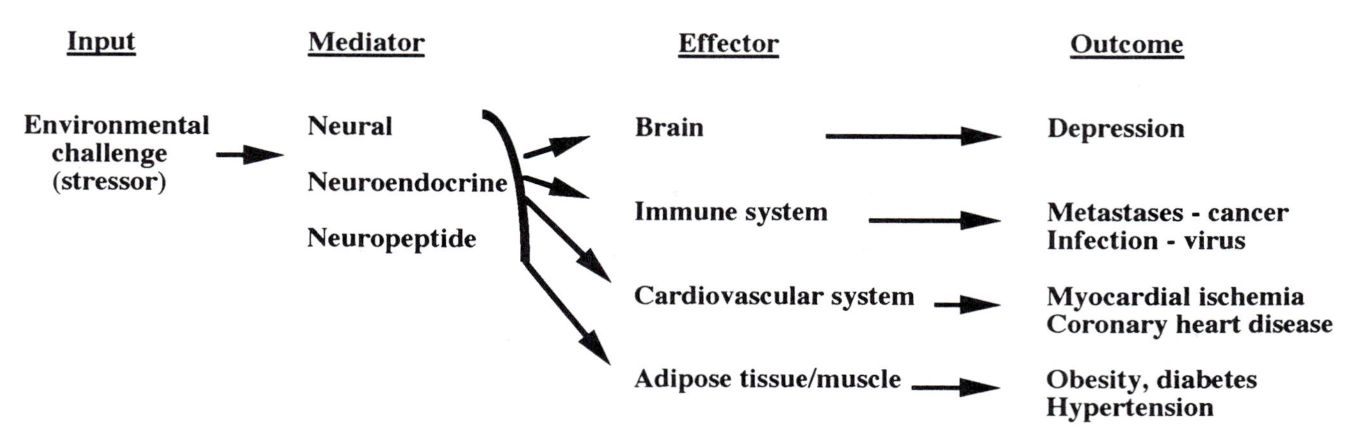

FIGURE 74.2 Diagram of the relationships between mediators, effectors, and disease end points.

nervous system, and the neuroendocrine system. These are affected by the persistence of stress and by the triggering effects of acute stress. The effectors are the brain, the immune system, the cardiovascular systems, and the adipose tissue and muscle. Compromising the immune system can exacerbate the metastatic spread of cancer and of viral infections (Ben-Eliyahu et al., 1991; Cohen, Tyrrell, and Smith, 1991; Kiecolt-Glaser and Glaser, 1991); whereas, in the cardiovascular system, plaque formation leads to atherosclerosis, and plaque rupture and platelet aggregation result in myocardial infarction and often in sudden death (Muller and Tofler, 1990). Adipose tissue and muscle system changes, especially fat deposition, are associated with

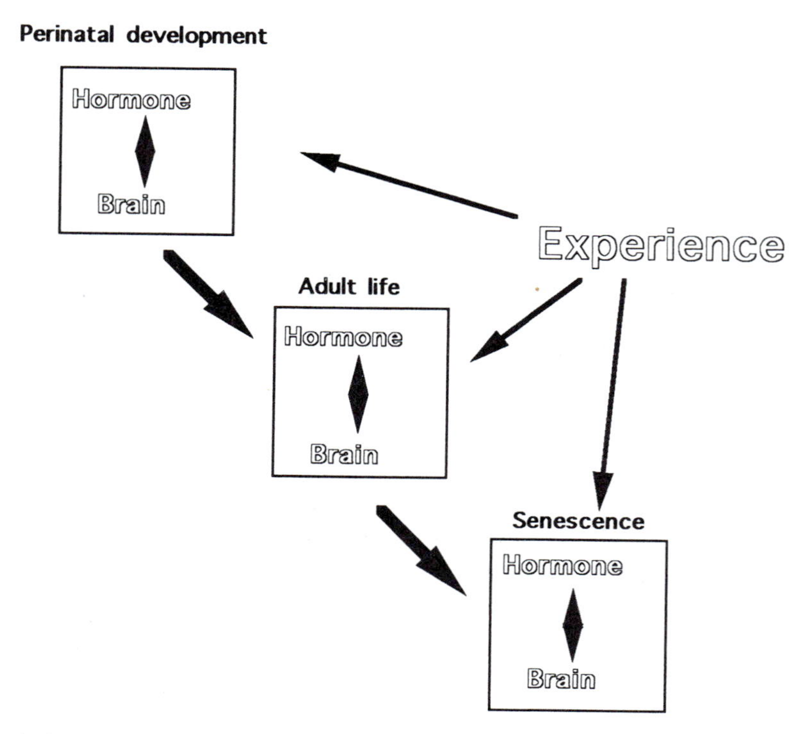

FIGURE 74.3 Diagram of the cascading relationship between hormone-brain interactions and experience at various stages of development.

diabetes and can lead to coronary heart disease. In fact, persistently elevated cortisol secretion promotes elevated insulin secretion, and the combined elevation of these two hormones may accelerate atherosclerosis (Brindley and Rolland, 1989). These are examples of the hidden price of the body's response to stressful challenges, to which we shall return later in our discussion of stress and disease.

Developmental Influences We have seen that the environment, in the broadest sense, plays an important role in determining the expression of genetically determined predisposition to diseases. This means that events from conception onward may have some influence in the expression or nonexpression of phenotypic traits of disease or normal function (figure 74.3). What are some of the developmental influences that may affect emotionality and contribute either to resiliency or to risk of stress-related pathology?

Studies of rats subjected to random prenatal sound stress have indicated that emotionality in adult life is increased as a result of such exposure (Fride and Weinstock, 1984; Fride et al., 1986). Handling of prenatally stressed animals as neonates wipes out the prenatal stress effect and results in normal or subnormal levels of emotionality (Wakschlak and Weinstock, 1990). Postnatal handling, which involves separating the pups from the mother for 10 minutes per day on postnatal days 1 through 14, results in lower emotionality and greater exporation of novel situations. Hypothalamo-pituitary-adrenal reactivity is reduced by neonatal handling; and with this reduction comes a reduction in the rate of aging, as manifested by age-related loss of neurons in the hippocampus and age-related deficits in spatial learning (Meaney et al., 1991).

These two examples out of a vast literature on developmental influences of stress highlight two extremes of outcomes, namely, increased and decreased levels of emotionality that are long-lasting in adult life. The implications of the slower aging of the hippocampus in the less emotional animals are that developmental events help to determine the later life course. However, the example is incomplete: It is not yet known whether rats that have been prenatally stressed and are more emotional actually undergo accelerated aging on the parameters indicated above. Future work must address this important question. The consequences of stress- and age-related decline in hippocampal function will be discussed below.

Brain structures and processes implicated in response to psychological stressors

Before further discussing the stress response, it is important to identify and describe some of the brain regions that play a major role in processing and responding to psychological stressors. The hippocampus and amygdala are two regions of the limbic system of the brain that are intimately involved in the generation of anxiety and learning associated with fear, and they are also important control centers for autonomic and neuroendocrine responses to stressful experiences. One of the features of anxiety and depression is the elevation of adrenal steroid secretion; the neural control of this secretion and the effects of this secretion on the brain and body are the two focal points of this section. The hippocampus and amygdala contain receptors for adrenal steroid hormones and are thus important targets for hormonal feedback. This feedback regulates the neural circuitry through which aversive stimuli are processed and lead to learned behaviors.

Fear and Anxiety The central nucleus of the amygdala is an important neural control center for fear and conditioned fear (Aggleton, 1992). It has projections to autonomic, respiratory, cardiovascular, facial, and neuroendocrine control centers in the brain that affect facial expressions, influence respiratory, cardiovascular, gastrointestinal, and adrenocortical activity, alter social interactions, and cause behavioral arousal and vigilance.

Lesions of the central nucleus of the amygdala block fear conditioning but not the display of previously conditioned aversive responses (Aggleton, 1992; see also Davis, 1992; Phillips and LeDoux, 1992). Whereas lesions of the amygdala interfere with fear conditioning in general, lesions of the hippocampal formation interfere with fear conditioning in a context of multiple sensory cues but not when a single cue is paired with a foot shock (Phillips and LeDoux, 1992). In other words, the role of the hippocampus is defined more by its involvement in processing the spatial and temporal aspects of a changing environment (Eichenbaum and Otto, 1992) than by a direct involvement in processing fear-producing stimuli. However, both the amygdala and the hippocampus play an important role in controlling the secretion of adrenal steroids and responding to these hormones (McEwen, 1977; see figure 74.4).

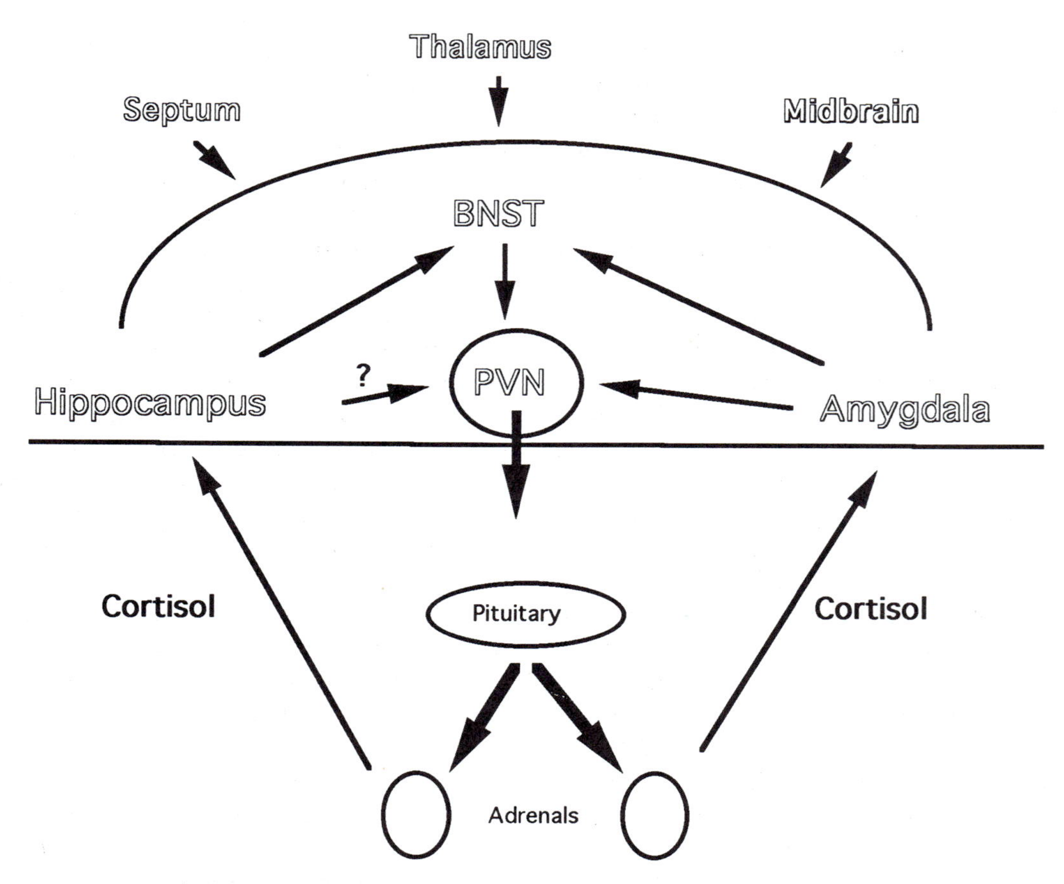

FIGURE 74.4 Role of limbic brain structures in the regulation of hypothalamo-pituitary-adrenal activity. Arc at top indicates that exact inputs of thalamus, septum, and midbrain are not specified. Below the arc, the input of the hippocampus and amygdala are believed to involve the bed nucleus of the stria terminalis as a way station; but there are also direct inputs to the paraventricular nucleus (PVN), which is the site of production of the releasing factor CRH and of vasopressin (AVP).

ADRENAL STEROID STRESS RESPONSE Whereas the hippocampus has a generally inhibitory role in the control of hypothalamo-pituitary-adrenal (HPA) activity (Jacobson and Sapolsky, 1991), the central nucleus of the amygdala is implicated in the stimulation of HPA activity by stressors such as restraint as well as by conditioned stressors (Van de Kar et al., 1991; Roozendaal, Koolhass, and Bohus, 1992). Hyperactivity of the amygdala may play a role in the frequent elevation of cortisol levels seen in endogenous depressive illness (Drevets and Raichle, 1992). The central nucleus of the amygdala has a direct projection to the paraventricular nucleus (PVN), in which neuron cell bodies containing the corticotrophin-releasing hormone (CRF) and vasopressin and oxytocin are found; these substances trigger release of pituitary ACTH (Gray, Charney, and Magnusson, 1989). The central amygdaloid nucleus also has a projection to the bed nucleus of the stria terminalis (Sun, Roberts, and Cassell, 1991). The hippocampus via the corticohypothalamic tract from the ventral subiculum has both a direct projection to the paraventricular nucleus (Silverman, Hoffman, and Zimmerman, 1981; Herman et al., 1989) and a projection to the bed nucleus of the stria terminalis, which sends projections to the PVN that influence HPA activity (Cullinan, Herman, and Watson, 1991). In this way, the hippocampus and amygdala have independent effects on the PVN as well as convergent and conflicting influences that may well be sorted out at the level of the bed nucleus (see figure 74.4).

ADRENAL STEROID FEEDBACK ON THE BRAIN IN CONJUNCTION WITH STRESS One reason for believing that adrenal steroid feedback on the brain has an important influence on how the body and brain handle stressful experiences is that excess or insufficiency of adrenal steroids leads to emotional instability and cognitive deficits (von Zerssen, 1976; Murphy, 1991). However, the actions of adrenal steroids are biphasic, in that low to moderate levels protect the brain from anxiety. In animal experiments, adrenal steroids are anxiolytic when given to adrenalectomized animals (Weiss et al., 1970; File, Vellucci, and Wendlandt, 1979), and they protect rats from developing learned helplessness when subjected to unavoidable foot shock (Edwards et al., 1990).

The role of adrenal steroid feedback in the aftermath of stress is described by the term *counterregulation*, which refers to the negative regulation of various neurochemical systems that are positively regulated, or turned on, by stressors (figure 74.5). Counterregulation has the net effect of keeping at least some of the individual neurochemical signals from going out of control as a result of repeated stress. For example, noradrenergic activity increases during stress, and repeated stress elevates the biosynthetic capacity to make noradrenaline (Nissenbaum et al., 1991) as well as inducing tyrosine hydroxylase and its mRNA (Mussachio et al., 1969; Richard et al., 1988; Angulo et al., 1991). Adrenal steroids counterregulate and reduce the response to noradrenaline, reducing noradrenaline-stimulated cAMP formation and calcium-calmodulin adenylate cyclase activity in cerebral cortex (Mobley, Manier, and Sulser, 1983; Stone et al., 1987; Gannon and McEwen, 1990), as well as noradrenaline-stimulated phosphoinositol hydrolysis in hippocampus (Kolasa, Song, and Jope, 1992). Other examples of this type shown in figure 74.5 include reductions of CRF biosynthesis and release in the paraventricular nuclei of hypothalamus. As shown in figure 74.5, the serotonergic system presents a more complex picture, in that 5-HT1A receptors are down-regulated by elevated glucocorticoids in the hippocampus, but 5-HT2 receptors are up-regulated in cerebral cortex by elevated glucocorticoids (Mendelson and McEwen, 1990; Kuroda et al., 1992). The balance between 5-HT2 and S-HT1A receptors is considered important in maintaining resilience in affective state, and a relative increase in 5-HT2 versus 5-HT1A sensitivity may tip the balance toward depression (Deakin and Graeff, 1991).

Adrenal Steroid Effects on the Body and Brain During the Diurnal Cycle There is a delicate balance in which the level and duration of adrenal steroid secretion may be critical to whether the brain is protected or destabilized as far as its emotional reactivity. The secretion of adrenal steroids is under complex control of a number of neural systems, and it is necessary to understand something about these control

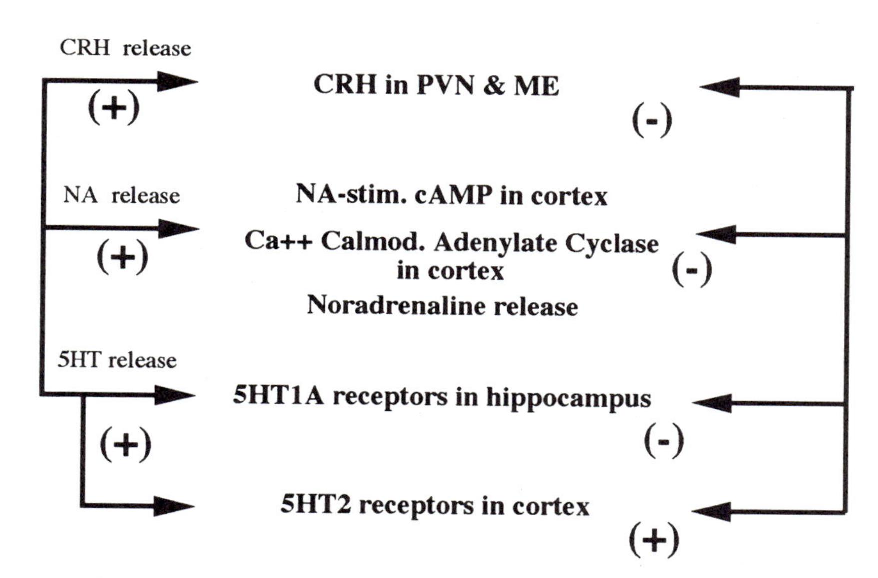

Figure 74.5 Diagram of interactions between stress and the feedback actions of adrenal steroids that mediate or, in some cases, counterregulate the neural response to stressful experiences.

mechanisms. The HPA axis is governed by two independent biological clocks, the circadian clock and a food-entrainable oscillator, and by stressful stimuli (McEwen et al., 1992). Adrenal steroids have important actions on the nervous system during the diurnal and food-entrainable rhythms and in the aftermath of stress (McEwen et al., 1992). The role of adrenal steroid feedback in the two types of rhythms is to coordinate neural and metabolic responses with the cyclic environment to which the animal is entrained, for example, the day-night cycle or the availability of food at specific times.

Adrenal Steroid Effects in Relation to Cognitive Function One effect of adrenal steroid feedback may be to alter the response characteristics of the hippocampus in terms of its susceptibility to the process of long-term potentiation, believed to be a cellular component of a memory mechanism. The peak of sensitivity of the hippocampus to LTP is during the waking period in both rats and squirrel monkeys; in both species the waking period follows the daily rise in adrenal glucocorticoid secretion (Barnes, McNaughton, and Goddard, 1977). Adrenalectomy of rats abolishes the diurnal elevation of LTP sensitivity during the waking period, suggesting that adrenal secretions play a coordinating role (Dana and Martinez, 1984). New data indicate that two types of adrenal steroid receptors in hippocampus play an important role in mediating biphasic effects of glucocorticoids on LTP: High-affinity type I receptors respond to rising basal levels of glucocorticoids during the sleeping period to facilitate the LTP response, whereas lower-affinity type II receptors respond to stress levels of glucocorticoids and inhibit the LTP response (Joels and DeKloet, 1991; Diamond et al., 1992; Pavlides et al., 1993).

Elevated levels of glucocorticoids have long been associated with depression and impaired cognitive function (Starkman and Schteingart, 1981; Wolkowitz et al., 1990; Martignoni et al., 1992). A recent study suggests that there may be some reduction in hippocampal volume in some hypercortisolemic Cushing's patients that is associated with impaired cognitive function (Starkman et al., 1992). Experiments on rats, summarized below, indicate that high levels of glucocorticoids and repeated stress causes atrophy of hippocampal neurons, as well as death of hippocampal neurons if the stress is prolonged or severe. Thus there are at least three mechanisms by which adrenal steroids may adversely affect cognitive performance: through their absence, as in the LTP discussion above; through acute elevation into the stress range, with an inhibition of LTP; and through prolonged elevation leading to neuronal atrophy or outright neuronal loss.

What are the adverse consequences of stress on the brain and body?

Allostasis and Allostatic Load Having spent time on some important issues surrounding the concept of stress, including brain regions, neuroendocrine processes, and behavioral mechanisms, it is time to return to a discussion of stress as it has evolved historically. Stress is often referred to as a threat to homeostasis, either real or implied; this concept has evolved through the work and writings of various investigators beginning with Hans Selye (Weiner, 1992). Homeostasis is a useful concept (Bernard, 1878; Cannon, 1929), but it is not without its limitations. Walter Cannon is credited not only for his input on homeostasis but also for coining the term *wisdom of the body* in referring to the generally beneficial effects of maintaining homeostasis. But homeostasis, in the sense of constancy, does not adequately describe normal physiology, in which blood pressure, heart rate, endocrine output, and neural activity are continually changing—from sleeping to waking, in response to external factors, and in anticipation of coming events. At each time of the daily cycle, these parameters are maintained within an operating range in response to environmental challenges. The operating range, and the ability of the body to increase or decrease vital functions to a new level within that range upon challenge, particularly in anticipation of a challenge, has been defined as *allostasis*, or "stability through change" by Sterling and Eyer (1988). The operating range for most physiological systems is larger in health than in disease, and it is larger in younger compared to older individuals (Lipsitz and Goldberger, 1992; Nabiloff et al., 1991). Exceeding this range can lead to disaster, as is the case when exertion leads to a myocardial infarction (Muller and Toffler, 1990).

When the body is placed under increased demand, the increased activity of neural and hormonal systems places a load on organs and tissues that is often hidden: Think of two equally heavy weights on a seesaw, compared to two much ligher weights on the same seesaw (figure 74.6). Although in both situations the system is in balance, the seesaw itself experiences greater strain

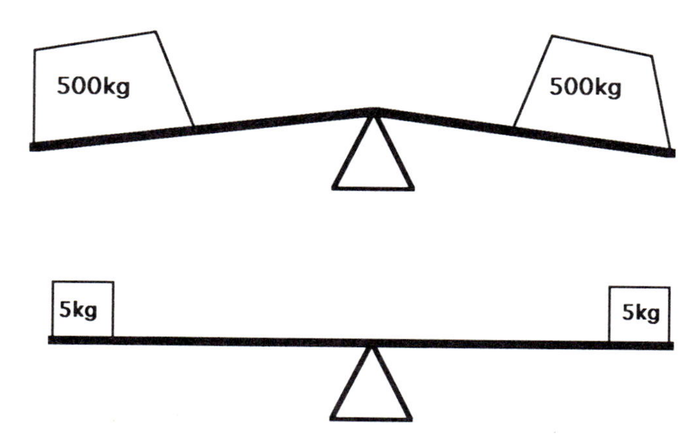

FIGURE 74.6 Representation of the concept of allostatic load, as the hidden price of the response to stress, in terms of small or large weights balancing a seesaw.

with the two heavy weights. The strain on the body produced by the elevated activity of physiological systems under challenge, as well as by repeated fluctuations in physiological response, places a strain on these systems (Sterling and Eyer, 1981). Extended over long periods, this strain can lead to wear and tear on a number of organs and tissues, and can predispose the organism to disease. We have termed this condition *allostatic load* (McEwen and Stellar, 1993). For example, elevated glucocorticoids alter the activity of a number of systems in the body, for example, increasing insulin secretion in response to elevated blood glucose levels that result from elevated glucocorticoid secretion. The combination of elevated glucocorticoids and insulin predisposes the body to deposit fat and atherosclerotic plaques (Brindley and Rolland, 1989). This is an excellent example of the failure of the "wisdom of the body," and it is also an example of allostatic load. Of course, genetic factors, such as the susceptibility to diabetes and obesity, as well as diet in this particular example, are very important in determining the degree to which this load leads to pathology. Moreover, psychological stress, as in an unstable social situation, is compounded by the diet consumed: A diet rich in fats exacerbates the pathological results in terms of atherosclerotic plaques (Manuck et al., 1988).

ALLOSTATIC LOAD, STRESS, DEPRESSION, AND CORONARY HEART DISEASE An example of allostatic load is the increased level of stress associated with job insecurity and work load. It was shown a number of years ago that cyclic occupational stress, as exemplified by the seasonal work load of tax accountants, exacerbates risk factors for coronary heart disease, for example, increasing serum cholesterol and decreasing blood clotting time (Friedman, Rosenman, and Carroll, 1958). A more recent study found positive correlations between perceived job instability and work load and LDL/HDL ratios, consistent with a role of stress in atherogenesis (Siegrist et al., 1988).

A similar type of chronic stress, associated with an allostatic load, may be occurring in endogenous depressive illness, in which cortisol levels are elevated and there is evidence for disregulation and hyperactivity of the central noradrenergic system (Gold, Goodwin, and Chrousos, 1988a, 1988b). Depression is associated with an increased risk of mortality in a nursing home population (Rovner et al., 1991). Specific disease susceptibilities include insulin resistance (Winokur et al., 1988) and an increased incidence of ischemic heart disease (Anda et al., 1993). Both of these are likely consequences of elevated glucocorticoid secretion in depression, according to the allostatic load model as it applies to stress effects on atherosclerosis, obesity, and diabetes.

ALLOSTATIC LOAD AND AUTONOMIC HYPERREACTIVITY Another example of chronic stress in a human population was described in studies of the residents of Three Mile Island, which revealed elevated urinary catecholamines and impairment of concentration and also disturbances of sleep, although there was no data concerning pathology resulting from the increased stress (Davidson, Fleming, and Baum, 1987). Yet another important example of chronic internal stress stemming from previous experience is posttraumatic stress disorder, involving hypervigilance and often anger and hostility (McFall et al., 1990) as well as elevated catecholamines (Mason et al., 1988). As in the Three Mile Island study, the long-term health consequences of posttraumatic stress disorder have yet to be investigated.

What is the connection between stress and disease?

QUALIFICATIONS One of the most difficult questions to answer is whether there is any causal link between stressful experiences and the occurrence of disease. Evidence for effects of stress on health involve a retrospective assessment of stressful life experiences, and their correlation with the occurrence of autonomic, cardiovascular, gastrointestinal, and immune system pathol-

Table 74.1
Possible links between stress and disease

Disease	Reference
Asthma	Mrazek and Klinnert, 1991
Diabetes	Surwit, Ross, and Feingloss, 1991 Cox and Gonder-Fredrick, 1991 Hagglof et al., 1991
Ulceration of GI tract	Weiner, 1992
Myocardial infarction	Muller and Tofler, 1990; McEwen and Stellar, 1993
Immune related:	
Grave's disease	Winsa et al., 1991
Type I diabetes	Hagglof et al., 1991
Tumor metastasis	Spiegel et al., 1989; Ben-Eliyahu et al., 1991
Viral infections	Cohen, Tyrrell, and Smith, 1991
Rheumatoid arthritis	Weiner, 1991

ogy, as well as mental disorders (see McEwen and Stellar, 1993). For a number of diseases, there is evidence that acute or chronic stress contributes significantly as a risk factor to the expression of diseases that are already present (table 74.1). The contributions of stress may be small but they are nevertheless important. For example, for various forms of inflammatory bowel disease, major life stress events were the most significant correlates of disease activity, even though only 7% of the variance was uniquely attributable to stress (Duffy et al., 1991). Another important determiner of the efficacy of stress on a disease process is the existence of a prior vulnerability. For example, the effects of stress on cardiac pathology depend on organ vulnerability (Tapp and Natelson, 1988); that is, the ability of stress to promote acute heart failure is dependent on cardiac pathology, which may in turn be influenced by the consequences of allostatic load.

It should be emphasized that not all environmental input is necessarily damaging; rather, as indicated by the model in figure 74.1, the interpretation of the event by the individual is crucial to the biological response. For example, the immune system is responsive to beneficial behavioral manipulations as well as to suppressive effects of stress. On the one hand, supportive group therapy was reported to double the survival time for metastatic breast cancer patients, and this survival outlasted the end of intervention (Spiegel et al., 1989). On the other hand, psychological stress has been reported to increase susceptibility to the common cold, elevating infection rates from 74% to 90% and clinical colds from 27% to 47% (Cohen, Tyrrell, and Smith, 1991). This is consistent with earlier studies that had shown increased incidence of mononucleosis with examination-related stress in medical students (Kiecolt-Glaser and Glaser, 1991).

In contrast to acute stress effects, longer-term effects on diseases related to the immune system have not been as easy to document. One exception is a recent study of type I diabetes in children, which found that stressful life events stemming from actual or threatened losses within the family and occurring at age 5 to 9 years significantly increased the relative risk; this was so even after the investigators had standardized for possible confounding factors such as age, sex, and family socioeconomic status (Hagglof et al., 1991). Another interesting example is that increased frequency of negative life events have been associated with newly diagnosed Grave's disease in adults, indicating a possible interaction between hereditary factors and stress (Winsa et al., 1991). Furthermore, psychosocial influences on another autoimmune disease, rheumatoid arthritis, are strongly suggestive but are confounded—as in the case of asthma (Mrazek and Klinnert, 1991)—by the heterogeneity of the disease (Weiner, 1991). Personality features, such as the ability to express anger and irritation, and stressful life events were implicated as risk factors in women suffering from rheumatoid arthritis in whom there was not a family history of the disease (Weiner, 1991).

It is widely accepted that stressful experiences exacerbate existing depressive illness (Anisman and Zacharko, 1982). According to the behavioral model presented above, predisposition for depression may be one of the biasing factors in the perception that a particular event leads to a high-cost behavioral or physiological response. On the other hand, it is much more controversial whether stressful life events precipitate depression, although, if so, there are other predisposing factors leading to the disorder as well (Anisman and Zacharko, 1982). Included among the predisposing factors for depression are death of a parent before the age of 17 (see Anisman and Zacharko, 1982). Thus the story for depression resembles that for the other diseases described above, in which no clear or overwhelming role can be assigned to stressful experiences and yet very few people would deny that life stress plays a significant role along with other factors.

Compounding the problem of establishing a link between stressful experience and disease are sex differences in disease onset. Major psychiatric illnesses show a varied pattern of sex differences, with men showing more antisocial behavior and substance abuse, whereas women are more susceptible to anxiety disorders and depression (Regier et al., 1988). Mechanisms of pain sensitivity also differ between the sexes, judging from recent studies on mice (Mogil et al., 1993). In female mice, estradiol induced a state of insensitivity to blocking actions of an NMDA antagonist, dizocilpine, on stress-induced analgesia, whereas male rats were not responsive in this way to estradiol.

Mechanisms Involved in Determining Pathophysiological Response As noted earlier, individual personality traits are very important features of increased or decreased vulnerability to stressful experience, because such traits act as a filter for external events and determine to what degree an experience will act as a stressor (figure 74.1; Lazarus, 1966; Matthews, 1988). Moreover, behavior resulting from processing of a potentially stressful experience may itself contribute to a more or less stressful outcome and to increased or decreased risk of disease or damage through further health-damaging or health-promoting behaviors.

One factor behind individual behavioral responses to internal or external events is the individual's neurochemical phenotype, which can bias the way the nervous system interprets and responds to challenge. For example, low serotonin levels have been linked to hostility, increased alcohol intake, and violent behavior including suicide (Ballenger et al., 1979; Brown et al., 1982; Mann et al., 1989). Animal studies show that brain serotonin levels are low in aggressive strains and high in more docile and domesticated strains (Popova et al., 1990). In addition, low serotonin levels disinhibit sympathetic activity and reduce parasympathetic activity, and lead to increased blood pressure surges that accompany angry and hostile responses (Saxena and Villalon, 1990). Such an excess of sympathetic reactivity increases vulnerability to adverse consequences such as myocardial infarctions and asthmatic attacks.

In terms of physiological response to challenge, we have seen that the body is not always wise in its homeostatic or allostatic control mechanisms and that compensatory changes that are designed to correct a temporary load on the body can, if prolonged, lead the body down a pathway to disease. We have defined this state as allostatic load. Chronic stress can have long-term effects on susceptiblity to diseases that are the result of sustained endocrine imbalance. During sustained elevations of cortisol, caused by high-fat diet or stress, insulin secretion increases to counteract insulin insensitivity produced by the elevated cortisol levels (Brindley and Rolland, 1989). Thus, even while the seesaw, representing allostatic load, is in balance, and the body is apparently coping in the short term with the stressor that has led to the elevated cortisol, the long-term elevations of both cortisol and insulin favor hyperlipidemia and accelerate atherogenesis (Brindley and Rolland, 1989). For example, the stress of an unstable social environment increases atherosclerosis in dominant monkeys fed a moderately atherogenic diet (Kaplan et al., 1991). These are also the conditions that can precipitate or exacerbate diabetes (Cox and Gonder-Frederick, 1991; Surwit, Ross, and Feingloss, 1991).

Allostatic Load and Wear and Tear on the Hippocampus The brain has a very limited capacity to repair itself after damage, and, as it declines in its function after being compromised, the endocrine system and the rest of the body are affected. As depicted in figure 74.5, adrenal steroid secretion is involved in adaptation to stress and in counterregulation, in many ways protecting the brain from its own primary neurochemical responses to environmental challenge (Chrousos and Gold, 1992; McEwen et al., 1992). Yet, we have also noted that adrenal steroids participate—paradoxically—in damaging effects of long-term environmental challenge on the brain, especially the hippocampal formation. This has been revealed by studies in rats and monkeys of the effects of aging and chronic social stress (Sapolsky, Krey, and McEwen, 1986; Uno et al., 1989, 1991; Kerr et al., 1991), and it is another example of allostatic load. Thus, evidence is mounting that the nervous system is subject to wear and tear as a result of stressful experiences. Damage or destruction of neurons can impair brain function and compromise physiological control mechanisms. Glucocorticoid secretion together with neural activity is believed to be responsible for this wear and tear on neuronal structure; but the cellular mechanism is not known, although it may be similar to that which is involved in ischemic damage following stroke (Sapolsky, Krey, and McEwen, 1986; Uno et al., 1989, 1991; Kerr et al., 1991). The functional consequences of such damage include disregulation of the hypothalamo-pituitary-adrenal axis and cognitive impairment resulting from damage to hippocampus (Jacobson and Sapolsky, 1991). The hippocampus and amygdala contain intracellular receptors for adrenal steroids that mediate effects on gene expression and lead to long-term changes in neural structure and neurochemistry (McEwen et al., 1992). Type I and type II adrenal steroid receptors coexist in many hippocampal neurons, and undoubtedly also in neurons of the amygdala; they constitute a two-level recognition system for circulating glucocorticoids, which handles both the cyclic variation and the stress-induced elevation. Type I and type II receptors have different functions. In the hippocampus, for example, type I receptors mediate adrenal steroid effects to stabilize the dentate gyrus neuronal population, suppressing both neuronal death and neurogenesis in this population of neurons, both of which increase after adrenalectomy. On the other hand, type II receptors mediate destructive effects of glucocorticoids on pyramidal neurons (McEwen et al., 1992). On a different time scale, as we have seen, type I and type II receptors mediate biphasic effects of adrenal steroids on long-term potentiation.

Repeated stress also has morphological consquences in the brain. In the hippocampus, repeated restraint stress of rats leads over a number of weeks to atrophy of dendrites of the CA3 pyramidal neurons of Ammons horn (Watanabe, Gould, Cameron, et al., 1992; Watanabe, Gould, Daniels, et al., 1992; Watanabe, Gould, and McEwen, 1992). More severe stress in rats, such as cold swim, and social stress in tree shrews and vervet monkeys leads to neuronal death in the CA3 region (Uno et al., 1989, 1991; Mizoguchi et al., 1992). Treatment with corticosterone mimics the effect of restraint stress on dendritic atrophy; yet, we have found that the effect is prevented by phenytoin, an inhibitor of excitatory amino acid release and action, and by tianeptine, an atypical tricyclic antidepressant that facilitates reuptake of serotonin (Watanabe, Gould, Cameron, et al., 1992; Watanabe, Gould, Daniels, et al., 1992). Prevention of the glucocorticoid stress response using the adrenal steroid synthesis inhibitor cyanoketone also prevents this atrophy of CA3 pyramidal neurons (Magarinos and McEwen, 1993). It remains to be seen what is the causal link between the atrophy of dendrites and the death of pyramidal neurons in hippocampus. On the one hand, dendritic atrophy may be an initial stage in the degenerative process; on the other hand, it may be a protective mechanism intended to reduce the cytotoxic actions of severe stress (Magarinos and McEwen, 1993).

Amygdala Much less is known so far about the amygdala. In the central nucleus of the amygdala, type II receptors are found in neurons that contain CRF, met-enkephalin, neurotensin, and somatostatin (Honkaniemi et al., 1992), although it is not yet clear if glucocorticoids regulate expression of any of these peptides in these neurons. There have been few other studies to date of the actions of adrenal steroids on neurons of the amygdala, and much is yet to be learned about the effects of adrenal steroids on their neurochemistry as well as their morphology, because it is conceivable that effects occur in amygdala that are similar to the stress effects that have been found in the hippocampus.

Immune System Another important area of interaction between behavioral, neural, and endocrine factors is the immune system (figure 74.7). Current research is emphasizing the important regulatory interactions between the immune, nervous, and endocrine systems. Acute and chronic modulation of immune system function is mediated by the autonomic nervous system, by the hypothalamo-pituitary-adrenocortical axis, and by a variety of regulatory peptides and pituitary hormones; and the immune system releases chemical messengers that affect the nervous system (Dunn, 1988). The movement (known as trafficking) of immune cells to or away from sites of infection is an important aspect of this regulation (Cupps and Fauci, 1982), along with modulation of cellular and humoral immunity. Adrenal steroids play an important role in trafficking of immune cells (Cupps and Fauci, 1982), and they increase humoral immunity while reducing the cellular immune response (Mason, 1991). Sympathetic nervous system innervation is also an important factor in modulating the immune system, and neuroactive peptides such as substance P and vasoactive intestinal peptide have immunoregulatory effects (Dunn, 1988; Livnat et al., 1985; Rabin, Cunnick, and Lysle, 1990). Natural killer cells and cytotoxic T cells play a pivotal role in defending against viral infections, and stressors have been shown to influence the time course of a viral infection as well as both the cellular and humoral responses to the virus (Sheridan et al., 1991). Both adrenal steroids and neural input are believed to

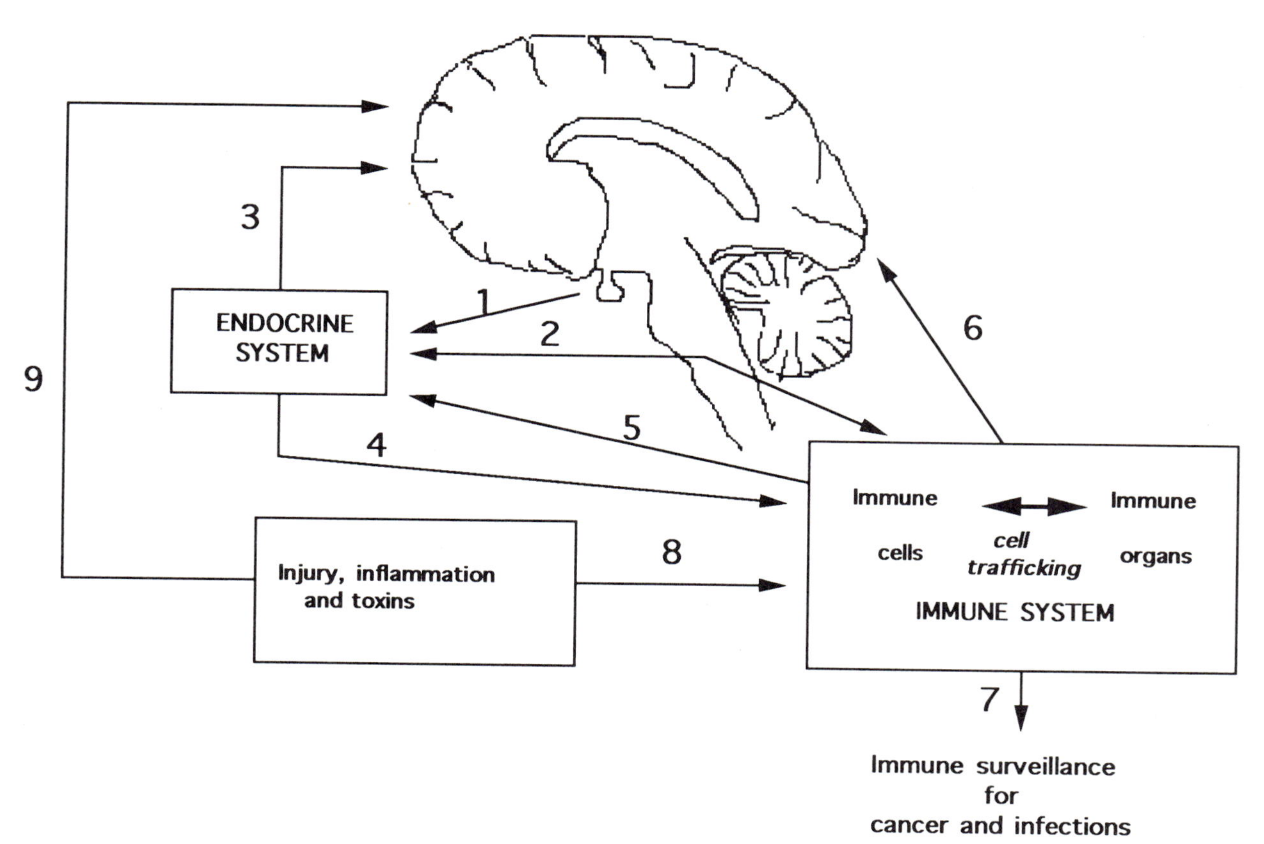

FIGURE 74.7 The brain and the endocrine and immune systems have complex interactions in response to pathogens and tumor cells. (1) Neuroendocrine products control endocrine function. (2) Neural activity also regulates endocrine function, as well as affecting the immune system. (3) Hormones affect the brain and pituitary. (4) Hormones also affect immune cells and organs. (5) Immune-system messengers (e.g., cytokines, thymic hormones) affect endocrine function. (6) Immune-system messengers have direct and indirect effects on pituitary and brain. (7) Immune function, which includes the movement or "trafficking" of immune cells to various tissues and organs through the circulation and lymphatic system, performs surveillance against tumor cells and pathogens as well as foreign cells and substances. (8) Injury, inflammation, and toxins from infection stimulate immunity as well as cytokine productions. (9) Injury, inflammation, and toxins also signal the brain and pituitary (Dunn, 1998; Plata-Salaman, 1992). Reprinted from McEwen and Stellar, 1993, by permission.

be involved in such stress effects on immunity (Rabin, Cunnick, and Lysle, 1990). Likewise, in autoimmune disorders, the adrenal steroids play an important suppressive role, whereas sympathetic neural activity has a complex modulatory role through a variety of adrenergic receptor types (Rabin, Cunnick, and Lysle, 1990; Leonard et al., 1991).

Natural killer cells play a key role in defending against tumor cells (Robertson and Ritz, 1990). An unexpected feature of acute stress is its efficacy in enhancing tumor metastasis in rats. A recent animal study (Ben-Eliyahu et al., 1991) is especially important because it represents a first step in establishing that the magnitude of reductions in natural killer cell activity produced by stress can be sufficient to allow the metastasis. Very few other studies have attempted to show whether a stress-induced change in relevant immune parameters is of sufficient magnitude to lead to a particular health outcome.

DEPRESSIVE ILLNESS With respect to depressive illness, the anxiolytic and counterregulatory effects of adrenal steroids summarized in figure 74.5 appear to stabilize the nervous system and protect it from falling prey to depressive illness. The main evidence for this statement comes from animal experimentation, in which learned helplessness is elicited more frequently in adrenalectomized rats, whereas adrenal steroid replacement therapy reduces the incidence of learned helplessness (Edwards et al., 1990). Adrenal steroids also have anxiolytic effects when given to adrenalectomized rats (Weiss et al., 1970; File, Vellucci, and Wendlandt, 1979); and yet, adrenal steroid excess is also associated with dysphoria and symptoms of depression (von Zerssen, 1976; Murphy, 1991). One way in which this may occur, noted above and in figure 74.5, involves the dual effects of glucocorticoids to increase 5-HT2 receptors and decrease 5-HT1A receptors (Deakins and Graeff, 1991).

When depressive illness occurs, it can be surmised that the normal balance of neurochemical systems has broken down, and that the counterregulatory effects of adrenal steroids are no longer able to keep these systems operating within their normal limits (see Drevets and Raichle, this volume). At such a time, antidepressant treatments are called for, and these may take the form of electroshock therapy or use of a variety of tricyclic and nontricyclic drugs with diverse neuropharmacological profiles, such as fluoxetine or imipramine, which inhibit serotonin uptake (Langer et al., 1980; Richelson and Pfenning, 1984), and tianeptine, which facilitates serotonin uptake (Menini, Mocaer, and Garattini, 1987; Fattacini et al., 1990). Where do such antidepressant treatments act? Because their mechanisms of action are so diverse, it is unlikely that they act in any single neural site or by any one mechanism. In fact, the multiplicity of treatments and the variety of counterbalanced neurochemical systems (see figure 74.5) would argue for multiple causes of depressive illness. Nevertheless, some of the most compelling evidence has implicated the amygdala as a primary site of disturbances in familial unipolar depressive illness (Drevets et al., 1992) as well as a likely site of action of some antidepressant drugs in rats (Duncan et al., 1986; Ordway et al., 1991; Drevets and Raichle, 1992).

The role of the hippocampus as a site of action of antidepressant treatments is less clear. This brain region does not appear to be a primary site of processing of fear and anxiety, but rather a site of integration and initial processing of learning. Phillips and LeDoux (1992) have noted that the hippocampus is implicated in the learning that is linked to aversive events, when the cues associated with them are in a context along with other cues and some discrimination is required along with temporal and spatial processing. Thus, the conditions in which fear-provoking connections are learned, under the biasing influence of depressive illness or severe anxiety, are most likely to be complex enough to involve the hippocampal formation. It remains to be seen whether the ameliorative effects of antidepressant treatments may also involve the hippocampus as well as brains areas like the cerebral cortex, where other types of cognitive processing and information storage also take place.

Conclusions

Stressful experiences are very much an individual matter, involving the interpretation of events and the choosing of responses, as well as the inherent capacities of an individual organism to respond to challenge. Limbic brain structures such as the amygdala and the hippocampus appear to play an important role in the processing of experiences, from both an emotional and a cognitive standpoint. These brain structures also help to determine the individual response, both in terms of overt behavior and in terms of autonomic and neuroendocrine reactions. Furthermore, the magnitude and duration of the neural and hormonal responses to experiences that are stressful have consequences for the functioning of organs and tissues.

According to the concept of allostasis, the arousal in anticipation of or in response to challenges to the organism results, at least temporarily, in a degree of stability in coping with the stressor; but, according to the concept of allostatic load, there is a price—often hidden—to the prolongation of the arousal and maintenance of neural and hormonal activity, in that pathological changes in organs may be accelerated. The resulting pathology limits the organs' ability to handle acute stressful challenges. For example, in the case of the cardiovascular system, acute stress may result in a myocardial infarction, made more likely, in part, by atherosclerotic plaque formation having been exacerbated by a combination of stressful experiences and diet.

Thus the link between stressful experiences, emotions, and disease involves a lifelong process of interaction between genetic constitution, experiences, and learning, in which neuroendocrine and neural mechanisms play a critical role in laying the groundwork for susceptibility or resilience to challenge. One of the surprising findings is the degree to which socioeconomic status is a major factor in the nature of these associations (Adler et al., 1993), and understanding the mechanisms underlying the linkage between stress and socioeconomic status is one of the major challenges that lies ahead.

ACKNOWLEDGMENTS The author acknowledges with deep appreciation the many contributions of fellow members of the Health and Behavior Network of the MacArthur Foundation: Judith Rodin, Nancy Adler, Grace Castellazzo, Ralph Horwitz, Solomon Katz, Robert Lawrence, Karen Matthews, Paul Rosin, Eliot Stellar, and Terence Wilson. Related research in the author's laboratory is supported by NIH grant MH41256.

REFERENCES

ADLER, N., W. T. BOYCE, M. A. CHESNEY, S. FOLKMAN, and S. L. SYME, 1993. Socioeconomic inequalities in health: No easy solution. *JAMA* 269:3140–3145.

AGGLETON, J. P., 1992. *The Amygdala*. New York: Wiley.

ANDA, R., D. WILLIAMSON, D. JONES, C. MACERA, E. EAKER, A. GLASSMAN, and J. MARKS, 1993. Depressed affect, hopelessness and the risk of ischemic heart disease in a cohort of U. S. adult.

ANGULO, J., D. PRINTZ, M. LEDOUX, and B. S. MCEWEN, 1991. Isolation stress increases tyrosine hydroxylase mRNA in the locus coeruleus and midbrain and decreases proenkephalin mRNA in the striatum and nucleus accumbens. *Mol. Brain Res.* 11:301–308.

ANISMAN, H., and R. M. ZACHARKO, 1982. Depression: The predisposing influence of stress. *Behav. Brain Sci.* 5:89–137.

BALLENGER, J., F. K. GOODWIN, L. F. MAJOR, and G. L. BROWN, 1979. Alcohol and central serotonin metabolism in man. *Arch. Gen. Psychiat.* 36:224–227.

BARNES, C., B. MCNAUGHTON, and G. GODDARD, 1977. Circadian rhythms of synaptic excitability in rat and monkey central nervous system. *Science* 197:91–92.

BARNETT, A. H., R. LESLIE, and D. PYKE, 1981. Diabetes in identical twins: A study of 200 pairs. *Diabetologia* 20:87–93.

BEN-ELIYAHU, S., R. YIRMIYA, J. LIEBESKIND, A. N. TAYLOR, and R. P. GALE, 1991. Stress increases metastatic spread of a mammary tumor in rats: Evidence for mediation by the immune system. *Brain Behav. Immun.* 5:193–205.

BERG, K., 1983. Genetics of coronary heart disease. *Prog. Med. Genet.* 5:35–90.

BERNARD, C., 1878. *Les phenomenes de la vie*, vol. 1. Paris: Librarie J-B Bailliere et Fils.

BRINDLEY, D., and Y. ROLLAND, 1989. Possible connections between stress, diabetes, obesity, hypertension and altered lipoprotein metabolism that may result in atherosclerosis. *Clin. Sci.* 77:453–461.

BROWN, G. L., M. H. EBERT, D. C. GOYER, D. C. JIMERSON, W. J. KLEIN, W. E. BUNNEY, and F. K. GOODWIN, 1982. Aggression, suicide and serotonin: Relationships to CSF amine metabolites. *Am. J. Psychiatry* 139:741–746.

CANNON, W., 1929. The wisdom of the body. *Physiol. Rev.* 9:399–431.

CHROUSOS, G., and P. GOLD, 1992. The concepts of stress and stress system disorders: Overview of physical and behavioral homeostasis. *JAMA* 267:1244–1252.

COHEN, S., D. TYRRELL, and A. SMITH, 1991. Psychological stress and susceptibility to the common cold. *N. England J. Med.* 325:606–612.

COX, D. J., L. A. GONDER-FREDERICK, 1991. The role of stress in diabetes mellitus. In *Stress, Coping and Disease*, P. McCabe, N. Schneidermann, T. M. Field, and J. S. Skyler, eds. Hillsdale, N.J.: Erlbaum, pp. 118–134.

CULLINAN, W., J. HERMAN, and S. WATSON, 1991. Morphological evidence for hippocampal interaction with the hypothalamic paraventricular nucleus. *Soc. Neurosci. Abstr.* 17:395.2.

CUPPS, T., and A. FAUCI, 1982. Corticosteroid-mediated immunoregulation in man. *Immunol. Rev.* 65:134–140.

DANA, R., and J. MARTINEZ, 1984. Effect of adrenalectomy on the circadian rhythm of LTP. *Brain Res.* 308:392–395.

DAVIDSON, L. M., R. FLEMING, and A. BAUM, 1987. Chronic stress, catecholamines, and sleep disturbances at Three Mile Island. *J. Hum. Stress* 13:75–83.

DAVIES, P., 1986. The genetics of Alzheimer's disease: A review and discussion of the implications. *Neurobiol. Aging* 7:459–466.

DAVIS, M., 1992. The role of the amygdala in fear and anxiety. *Ann. Rev. Neurosci.* 15:353–375.

DEAKIN, J. F. W., and F. G. GRAEFF, 1991. Critique: 5-HT and mechanisms of defence. *J. Psychopharm.* 5:305–313.

DIAMOND, D. M., M. C. BENNET, M. FLESHNER, and G. M. ROSE, 1992. Inverted-U relationship between the level of peripheral corticosterone and the magnitude of hippocampal primed burst potentiation. *Hippocampus* 2:421–430.

DREVETS, W. C., and M. E. RAICHLE, 1992. Neuroanatomical circuits in depression: Implications for treatment mechanisms. *Psychopharm. Bull.* 28:261–274.

DREVETS, W. C., T. O. VIDEEN, J. L. PRICE, S. H. PRESKORN, S. T. CARMICHAEL, and M. E. RAICHLE, 1992. A functional anatomical study of unipolar depression. *J. Neurosci.* 12:3628–3641.

DUFFY, L. C., M. A. ZIELEZNY, J. R. MARSHALL, J. E. BYERS, M. M. WEISER, J. F. PHILLIPS, B. M. CALKINS, P. L. OGRA, and S. GRAHAM, 1991. Relevance of major stress events as an indicator of disease activity prevalance in inflammatory bowel disease. *Behav. Med.* 17:101–110.

DUNCAN, G. E., G. R. BREESE, H. CRISWELL, W. E. STUMPF, R. A. MUELLER, and J. B. COVEY, 1986. Effects of antidepressant drugs injected into the amygdala on behavioral responses of rats in the forced swim test. *J. Pharmacol. Exp. Ther.* 238:758–762.

DUNN, A., 1988. Nervous system–immune system interactions: An overview. *J. Recept. Res.* 8:589–607.

EDWARDS, E., K. HARKINS, G. WRIGHT, and F. HENN, 1990. Effects of bilateral adrenalectomy on the induction of learned helplessness behavior. *Neuropsychopharmacology* 3:109–114.

EICHENBAUM, H., and T. OTTO, 1992. The hippocampus: What does it do? *Behav. Neural Biol.* 57:2–36.

FATTACINI, C. M., F. BOLANOS-JIMINEZ, H. GOZLAN, and

M. HAMON, 1990. Tianeptine stimulates uptake of 5-hydroxytryptamine in vivo in the rat brain. *Neuropharmacology* 90:1–8.

FILE, S., A. VELLUCCI, and S. WENDLANDT, 1979. Corticosterone: An anxiogenic or anxiolytic agent? *J. Pharm. Pharmacol.* 31:300–305.

FRIDE, E., Y. DAN, J. FELDON, G. HALEVY, and M. WEINSTOCK, 1986. Effects of prenatal stress on vulnerability to stress in prepubertal and adult rats. *Physiol. Behav.* 37:681–687.

FRIDE, E., and M. WEINSTOCK, 1984. The effects of prenatal exposure to predictable or unpredictable stress on early development in the rat. *Dev. Psychobiol.* 17:651–660.

FRIEDMAN, M., R. H. ROSENMAN, and V. CARROLL, 1958. Changes in the serum cholesterol and blood clotting time in men subjected to cyclic variations of occupational stress. *Circulation.* 17:852–861.

GANNON, M., and B. S. MCEWEN, 1990. Calmodulin involvement in stress- and corticosterone-induced down-regulation of cyclic AMP generating systems in rat brain. *J. Neurochem.* 55:276–284.

GOLD, P., F. GOODWIN, and G. CHROUSOS, 1988a. Clinical and biochemical manifestations of depression (part 1). *N. England J. Med.* 319:348–353.

GOLD, P., F. GOODWIN, and G. CHROUSOS, 1988b. Clinical and biochemical manifestations of depression (part 2). *N. England J. Med.* 319:413–420.

GRAY, T. S., M. E. CARNEY, and D. J. MAGNUSSON, 1989. Direct projections from the central amygdaloid nucleus to the hypothalamic paraventricular nucleus: possible role in stress-induced adrenocorticotrophin release. *Neuroendocrinology* 50:433–446.

HAGGLOF, B., L. BLOM, G. DAHLQUIST, G. LONNBERG, and B. SAHLIN, 1991. The Swedish childhood diabetes study: Indications of severe psychological stress as a risk factor for Type I (insulin-dependent) diabetes mellitus in childhood. *Diabetologia* 34:579–583.

HERMAN, J., M. SCHAFER, E. YOUNG, and S. WATSON, 1989. Evidence for hippocampal regulation of neuroendocrine neurons of the hypothalamo-pituitary-adrenocortical axis. *J. Neurosci.* 9:3072–3082.

HONKANIEMI, J., M. PELTO-HUIKKO, L. RECHART, J. ISOLA, A. LAMMI, K. FUXE, J.-A. GUSTAFSSON, C. WIKSTROM, and T. HOKFELT, 1992. Colocalization of peptide and glucocorticoid receptor immunoreactivities in rat central amygdaloid nucleus. *Neuroendocrinology* 55:451–459.

JACOBSON, L., and R. SAPOLSKY, 1991. The role of the hippocampus in feedback regulation of the hypothalamic-pituitary-adrenocortical axis. *Endocr. Rev.* 12:118–134.

JOELS, M., and E. R. DEKLOET, 1991. Control of neuronal excitablity by corticosteroid hormones. *Trends Neurosci.* 15: 25–30.

HORWITZ, R., and K. HORWITZ, 1993. Adherence to treatment and health outcomes. *Arch. Intern. Med.* 153:1863–1868.

KAPLAN, J. R., M. R. ADAMS, T. B. CLARKSON, S. B. MANUCK, and C. A. SHIVERY, 1991. Social behavior and gender in biomedical investigations using monkeys: studies in atherogenesis. *Lab Anim. Sci.* 41:1–9.

KERR, S., L. CAMPBELL, M. APPLEGATE, A. BRODISH, and P. LANDFIELD, 1991. Chronic stress-induced acceleration of electrophysiologic and morphometric biomarkers of hippocampal aging. *J. Neurosci.* 11:1316–1324.

KIECOLT-GLASER, J., and R. GLASER, 1991. Stress and immune function in humans. In *Psychoneuroimmunology*, R. Ader, O. L. Felten, and N. Cohen, eds. New York: Academic Press, pp. 849–867.

KOLASA, K., L. SONG, and R. S. JOPE, 1992. Adrenalectomy increases phosphoinositide hydrolysis induced by norepinephrine or excitatory amino acids in rat hippocampal slices. *Brain Res.* 579:128–134.

KURODA, Y., M. MIKUNI, T. OGAWA, and K. TAKAHASHI, 1992. Effect of ACTH, adrenalectomy and the combination treatment on the density of 5-HT2 receptor binding sites in neocortex of rat forebrain and 5-HT2 receptor-mediated wet dog shake behaviors. *Psychopharmacology* 108: 27–32.

KUSHNER, M. G., J. SHER, and B. D. BEITMAN, 1990. The relation between alcohol problems and the anxiety disorders. *Am. J. Psychiatry* 147:685–695.

LANGER, S. Z., C. MORET, R. RAISMAN, M. L. DUBUCOVICH, and M. BRILEY, 1980. High affinity [3H]-imipramine binding in rat hypothalamus: association with uptake of serotonin but not of norepinephrine. *Science* 210:1133–1135.

LAZARUS, R. S., 1966. *Psychological Stress and the Coping Process.* New York: McGraw-Hill.

LEONARD, J. P., F. J. MACKENZIE, H. A. PATEL, and M. L. CUZNER, 1991. Hypothalamic noradrenergic pathways exert an influence on neuroendocrine and clinical status in experimental autoimmune encephalomyelitis. *Brain Behav. Immun.* 5:328–338.

LIPSITZ, L. A., and A. L. GOLDBERGER, 1992. Loss of "complexity" and aging. *JAMA* 13:1806–1809.

LIVNAT, S., S. FELTEN, S. CARLSON, D. BELLINGER, and D. FELTEN, 1985. Involvement of peripheral and central catecholamine systems in neural-immune interactions. *J. Neuroimmunology* 10:5–30.

MAGARINOS, A. M., and B. S. MCEWEN, 1993. Blockade of glucocorticoid synthesis by cyanoketone prevents chronic stress-induced dendritic atrophy of hippocampal neurons in the rat. *Abstr. Soc. Neurosci.* 19:71.2, P168.

MANN, J., P. M. MARZUK, V. ARANGO, P. A. MCBRIDE, A. C. LEON, and H. TIERNEY, 1989. Neurochemical studies of violent and nonviolent suicide. *Psychopharmacol. Bull.* 25:407–413.

MANUCK, S. B., J. R. KAPLAN, M. R. ADAMS, and T. B. CLARKSON, 1988. Studies of psychosocial influences on coronary artery atherosclerosis in cynomolgus monkeys. *Health Psychol.* 7:113–124.

MARTIGNONI, E., A. COSTA, E. SINFORIANI, A. LIUZZI, P. CHIODINI, M. MAURI, G. BONO, and G. NAPPI, 1992. The brain as a target for adrenocortical steroids: Cognitive implications. *Psychoneuroendocrinology* 17:343–354.

Mason, D., 1991. Genetic variation in the stress response: Susceptibility to experimental allergic encephalomyelitis and implications for human inflammatory disease. *Immunol. Today* 12:57–60.

Mason, J., E. Giller, T. Kosten, and L. Harkness, 1988. Elevation of urinary norepinephrine/cortisol ratio in post-traumatic stress disorder. *J. Nerv. Ment. Dis.* 176:498–502.

Matthews, K. A., 1988. Coronary heart disease and type A behaviors: Update on and alternative to the Booth-Kewley and Friedman (1987) quantitative review. *Psychol. Bull.* 104:373–380.

McEwen, B. S., 1977. Adrenal steroid feedback on neuroendocrine tissues. *Ann. N.Y. Acad. Sci.* 297:568–579.

McEwen, B. S., J. Angulo, H. Cameron, H. Chao, D. Daniels, M. Gannon, E. Gould, S. Mendelson, R. Sakai, R. Spencer, and C. Woolley, 1992. Paradoxical effects of adrenal steroids on the brain: Protection versus degeneration. *Biol. Psychiatry* 31:177–199.

McDaniel, J. S., and C. Nemeroff, 1993. Psychological distress and depression in the cancer patient: Diagnostic, biological and treatment considerations. In *Clinical Oncology*, M. D. Abeloff, J. O. Armitage, A. S. Lighter, and J. E. Niederhuber, eds.

McEwen, B. S., and E. Stellar, 1993. Stress and the individual: Mechanisms leading to disease. *Arch. Intern. Med.* 153:2093–2101.

McFall, M., M. Murburg, G. Ko, and R. Veith, 1990. Autonomic responses to stress in Vietnam combat veterans with posttraumatic stress disorder. *Biol. Psychiatry* 27: 1165–1175.

Meaney, M., J. Mitchell, D. Aitken, S. Bhat Agar, S. Bodnoff, L. Ivy, and A. Sarriev, 1991. The effects of neonatal handling on the development of the adrenocortical response to stress: Implications for neuropathology and cognitive deficits later in life. *Psychoneuroendocrinology* 16: 85–103.

Mendelson, S. D., and B. S. McEwen, 1990. Adrenalectomy increases the density of 5-HT1A receptors in rat hippocampus. *Neuroendocrinology Lett.* 12:353.

Menini, T., E. Mocaer, and S. Garattini, 1987. Tianeptine, a selective enhancer of serotonin uptake in rat brain. *Naunyn Schmiederbergs Arch. Pharmacol.* 336:478–482.

Mizoguchi, K., T. Kunishita, D.-H. Chui, and T. Tabira, 1992. Stress induces neuronal death in the hippocampus of castrated rats. *Neurosci. Lett.* 138:157–160.

Mobley, P., D. Manier, and F. Sulser, 1983. Norepinephrine-sensitive 8adenylate cyclase system in rat brain: role of adrenal corticosteroids. *J. Pharmacol. Exp. Ther.* 226:71–77.

Mogil, J., W. Sternberg, B. Kest, P. Marek, and J. Liebeskind, 1993. Sex differences in the antagonism of swim stress-induced analgesia: Effects of gonadectomy and estrogen replacement. *Pain* 53:17–25.

Mrazek, D., and M. Klinnert, 1991. Asthma: Psychoneuroimmunologic considerations. In *Psychoneuroimmunology*, R. Ader, D. I. Felten, and N. Cohen, eds. New York: Academic Press, pp. 1013–1033.

Muller, J., and G. Tofler, eds., 1990. A symposium: Triggering and circadian variation of onset of acute cardiovascular disease. *Am. J. Cardiol.* 66:1G–70G.

Murphy, B. E. P., 1991. Steroids and depression. *J. Steroid Biochem. Mol. Biol.* 38:537–559.

Mussachio, J., L. Julou, S. Kety, and J. Glowinski, 1969. Increase in rat brain tyrosine hydroxylase produced by electroconvulsive shock. *Proc. Natl. Acad. Sci. U.S.A.* 63: 1117–1119.

Nabiloff, B. D., D. Benton, and G. F. Solomon, J. E. Morley, J. L. Fahey, E. T. Bloom, T. Makinodau, and S. L. Gilmore, 1991. Immunological changes in young and old adults during brief laboratory stress. *Psychosom. Med.* 53:121–132.

Nissenbaum, L. K., M. J. Zigmond, A. F. Sved, and E. D. Abercrombie, 1991. Prior exposure to chronic stress results in enhanced synthesis and release of hippocampal norepinephrine in reponse to a novel stressor. *J. Neurosci.* 11:1478–1484.

Ordway, G. A., C. Gambarana, S. M. Tejani-Butt, P. Areso, M. Hauptmann, and A. Frazer, 1991. Preferential reduction of binding of 125I-iodopindolol to beta-1 adrenoceptors in the amygdala of rat after antidepressant treatments. *J. Pharm. Exp. Ther.* 257:681–690.

Osterweiss, M., F. Solomon, and M. Green, eds., 1984. *Bereavement: Reactions, Consequences and Care.* Washington: National Academy Press, pp. 312.

Pavlides, C., Y. Watanabe, A. M. Magarinos, and B. S. McEwen, in preparation. Opposing role of Type I and Type II adrenal steroid receptors in hippocampal long-term potentiation.

Phillips, R. G., and J. E. LeDoux, 1992. Differential contribution of amygdala and hippocampus to cued and contextual fear conditioning. *Behav. Neurosci.* 2:274–285.

Plata-Salaman, C. R., 1992. Immunoregulators in the nervous system. *Neurosci. Biobehav. Rev.* 15:185–215.

Plomin, R., 1990. The role of inheritance in behavior. *Science* 48:183–188.

Popova, N., A. Kulikov, E. Nikulina, E. Kozlachkova, and G. Maslova, 1990. Serotonin metabolism and serotonergic receptors in Norway rats selected for low aggressiveness to man. *Aggressive Behav.* [Russia] 17:207–213.

Rabin, B., J. Cunnick, and D. Lysle, 1990. Stress-induced alteration of immune function. *Prog. Neuroendoimmunol.* 3: 116–124.

Redelmaier, D., P. Rozin, and D. Kahneman, 1993. Understanding patients' decisions: Cognitive and emotional perspectives. *JAMA* 270:72–76.

Regier, D. A., J. H. Boyd, J. D. Burke, D. S. Rae, J. K. Myers, M. Kramer, L. N. Robbins, L. K. George, M. Karno, and B. Z. Locke, 1988. One-month prevalence of mental disorders in the United States. *Arch. Gen. Psychiatry* 45:977–986.

Richard, F., N. Faucon-Biguet, R. Labatut, D. Rollet, J. Mallet, and M. Buda, 1988. Modulation of tyrosine hydroxylase gene expression in rat brain and adrenals by exposure to cold. *J. Neurosci. Res.* 20:32–37.

Richelson, E., and M. Pfenning, 1984. Blockade by antidepressants and related compounds of biogenic amine uptake into rat brain synaptosomes: Most antidepressants selectively block norepinephrine uptake. *Eur. J. Pharmacol.* 104:277–286.

Robertson, M. J., and J. Ritz, 1990. Biology and clinical relevance of human natural killer cells. *Blood* 76:2421–2438.

Roozendaal, B., J. M. Koolhaas, and B. Bohus, 1992. Central amygdaloid involvement in neuroendocrine correlates of conditioned stress responses. *J. Neuroendocrinology* 4:483–489.

Rovner, B., P. German, L. Brant, R. Clark, L. Burton, and M. Folstein, 1991. Depression and mortality in nursing homes. *JAMA* 27:993–996.

Sapolsky, R., L. Krey, and B. S. McEwen, 1986. Neuroendocrinology of stress and aging: The glucocorticoid cascade hypothesis. *Endocr. Rev.* 7:284–301.

Saxena, P. R., and C. M. Villalon, 1990. Cardiovascular effects of serotonin agonists and antagonists. *J. Cardiovasc. Pharmacol.* 7:S17–S34.

Sher, K. J., 1987. Stress response dampening. In *Psychological Theories of Drinking and Alcoholism*, H. T. Bland and K. E. Leonard, eds. New York: Guilford, pp. 227–271.

Sheridan, J., N. Feng, R. Bonneau, C. Allen, B. Huneycutt, and R. Glaser, 1991. Restraint stress differentially affects anti-viral cellular and humoral immune responses in mice. *J. Neuroimmunol.* 31:245–255.

Siegrist, J., H. Matschinger, P. Cremer, and D. Seidel, 1988. Atherogenic risk in men suffering from occupational stress. *Atherosclerosis* 69:211–218.

Silverman, A. J., D. K. Hoffman, and E. A. Zimmerman, 1981. The descending afferent connections of the paraventricular nucleus of the hypothalamus (PVN) *Brain Res. Bull.* 6:47–61.

Spiegel, D., H. Kraemer, J. Bloom, and E. Gottheil, 1989. Effect of psychosocial treatment on survival of patients with metastatic breast cancer. *The Lancet* 2:888–891.

Starkman, M., S. Gebarski, S. Berent, and D. Schteingart, 1992. Hippocampal formation volume, memory dysfunction and cortisol levels in patients with Cushing's syndrome. *Biol. Psychiatry* 32:756–765.

Starkman, M., and D. Schteingart, 1981. Neuropsychiatric manifestations of patients with Cushing's syndrome. *Arch. Intern. Med.* 141:215–219.

Sterling, P., and J. Eyer, 1981. Biological basis of stress-related mortality. *Soc. Sci. Med. [E]* 15:13–42.

Sterling, P., and J. Eyer, 1988. Allostasis: A new paradigm to explain arousal pathology. In *Handbook of Life Stress, Cognition and Health*, J. Fisher and J. Reason, eds. London: Wiley, pp. 629–649.

Stone, E., B. S. McEwen, A. S. Herrera, and K. Carr, 1987. Regulation of alpha and beta components of noradrenergic cyclic AMP responses in cortical slices. *Eur. J. Pharmacol.* 141:347–356.

Sun, N., L. Roberts, and M. D. Cassell, 1991. Rat central amygdaloid nucleus projections to the bed nucleus of the stria terminalis. *Brain Res. Bull.* 27:651–662.

Surwit, R. S., S. L. Ross, and M. N. Feingloss, 1991. Stress, behavior, and glucose control in diabetes mellitus. In *Stress, Coping and Disease*, P. McCabe, N. Schneidermann, T. M. Field, and J. S. Skyler, eds. Hillsdale, N.J.: Erlbaum, pp. 97–117.

Tapp, W. N., and B. H. Natelson, 1988. Consequences of stress: A muliplicative function of health status. *Faseb J.* 2:2268–2271.

Uno, H., G. Flugge, C. Thieme, O. Johren, and E. Fuchs, 1991. Degeneration of the hippocampal pyramidal neurons in the socially stressed tree shrew. *Abstr. Soc. Neurosci.* 17:52.20, p. 128.

Uno, H., T. Ross, J. Else, M. Suleman, and R. Sapolsky, 1989. Hippocampal damage associated with prolonged and fatal stress in primates. *J. Neurosci.* 9:1705–1711.

Van de Kar, L. D., R. A. Piechowski, P. A. Rittenhouse, and T. S. Gray, 1991. Amygdaloid lesions: Differential effect on conditioned stress and immobilization-induced increases in corticosterone and renin secretion. *Neuroendocrinology* 54:89–95.

Von Zerssen, D., 1976. Mood and behavioral change under corticosteroid therapy. In *Psychotropic Actions of Hormones*, T. Util, G. Landahn, and J. Herrmann, eds. New York: Spectrum, pp. 195–222.

Wakshlak, A., and M. Weinstock, 1990. Neonatal handling reverses behavioral abnormalities induced in rats by prenatal stress. *Physiol. Behav.* 48:289–292.

Watanabe, Y., E. Gould, H. Cameron, D. Daniels, and B. S. McEwen, 1992. Phenytoin prevents stress- and corticosterone-induced atrophy of CA3 pyramidal neurons. *Hippocampus* 2:431–436.

Watanabe, Y., E. Gould, D. Daniels, H. Cameron, and B. S. McEwen, 1992. Tianeptine attenuates stress-induced morphological changes in the hippocampus. *Eur. J. Pharm.* 222:157–162.

Watanabe, Y., E. Gould, and B. S. McEwen, 1992. Stress induces atrophy of apical dendrites of hippocampal CA3 pyramidal neurons. *Brain Res.* 588:341–345.

Weiner, H., 1991. Social and psychobiological factors in autoimmune disease. In *Psychoneuroimmunology*, R. Ader, O. L. Felten, and N. Cohen, eds. New York: Academic Press, pp. 995–1011.

Weiner, H., 1992. *Perturbing the Organism: The Biology of Stressful Experience*. Chicago: University of Chicago Press.

Weiss, J. M., B. S. McEwen, M. Silva, and M. Kalkut, 1970. Pituitary-adrenal alterations and fear responding. *Am. J. Physiol.* 218:864–868.

Winokur, A., G. Maislin, J. Phillips, and J. Amsterdam, 1988. Insulin resistance after oral glucose tolerance testing in patients with major depression. *Am. J. Psychiatry* 145: 325–330.

Winsa, B., H. O. Adami, and R. Bergstrom, A. Gamstedt, P. A. Dahlberg, V. Adamson, R. Jansson, and A. Karlsson, 1991. Stressful life events and Graves' disease. *Lancet* 338:1475–1479.

Wolkowitz, O. M., V. I. Reus, H. Weingartner, K. Thompson, A Breier, A. Doran, D. Rubinow, and D. Pickar, 1990. Cognitive effects of corticosteroids. *Am. J. Psychiatry* 147:1297–1303.

Wolkowitz, O. M., H. Weingartner, K. Thompson, D. Pickar, S. M. Paul, and D. W. Hommer, 1987. Diazepam-induced amnesia: A neuropharmacological model of an organic amnestic syndrome. *Am. J. Psychiatry* 144:25–29.

75 Neurophysiological Networks Integrating Human Emotions

ERIC HALGREN AND KSENIJA MARINKOVIC

ABSTRACT Cortico-limbic-brainstem networks integrate visceral, cognitive, and contextual information in humans to produce emotional feelings and behaviors. The *orienting complex* prepares the organism to process (via arousal and directed attention) and react to (via generalized viscero- and skeletomotor activation) biologically prepotent stimuli. Such stimuli evoke a series of field potentials (N2/P3a/SW) from a distributed frontoparietal circuit subserving attention. Electrical stimulation of the limbic parts of this circuit (rectal and cingulate gyri) evoke visceromotor responses. Emotional input to cognitive *event integration* occurs during the field potentials N4/P3b, generated in high-level sensory and multimodal association cortices and, predominantly, in the hippocampal and amygdala formations and lateral orbital cortex. Medial temporal units respond to both visceral and abstract cognitive information, and medial temporal stimulation evokes viscerosensations, emotionally eloquent hallucinations, and emotional feelings. Emotional influences on voluntary *response selection* may occur during the *readiness potential*, generated in the premotor and central cingulate cortices. Stimulation of the central cingulate cortex evokes partially organized movements. Sustained contextual information, reflected in the frontal contingent negative variation (CNV), provides a long-lasting influence unifying the phasic responses over time. Thus, limbic and perilimbic cortices contribute emotional guidance during successive neurophysiological stages underlying orienting, cognitive integration, voluntary action, and sustained context.

The neurophysiological mechanisms of human emotion encompass a vast range of phenomena, from fixed action patterns triggered by biologically imperative stimuli to deeply considered voluntary acts fulfilling life goals. Such processes that integrate and guide the totality of behavior are necessarily complex. One source of this complexity is the fact that different levels of the human nervous system are organized according to different principles, have access to different sources of information, and control different aspects of behavior. Thus, the motivations that serve biological ends may arise in a system that has direct access to visceral state but no direct means to influence volition and awareness. Conversely, motivations that arise in intellectual projections of possible future scenarios must find means to influence visceral output as well as general cortical tonus. Perhaps most crucially, emotion must include the mechanisms that resolve conflicts between these levels of motivation, such as those that arise, for example, during a religious fast.

In short, while emotion always seems to function to provide a directed integration of the entire organism, both the goals and manifestations of this integration may be extremely diverse. This diversity is reflected in the methods, goals, and terminology used to investigate human emotion. One research approach has concentrated on the psychophysiological manifestations of emotion, especially during visceral conditioning and the orienting response. Another focuses on the cognitive context of emotion, measuring motivation, volition, and interpretation. A third, speaking more directly to clinical concerns, focuses on the personality. In order to relate the discoveries of these distinct research approaches to the neural basis of human emotion, we begin this review by offering a simple classification of emotional processes, according to the time scale at which they operate. We distinguish four successive but overlapping stages in the emotional reaction to a stimulus: (1) the orienting complex, (2) emotional event integration, (3) response selection, and (4) sustained emotional context.

The orienting complex (corresponding to the orienting response in its broadest sense) consists of the orientation of attention toward a potentially important event and the mobilization of resources to cope with it. At its most extreme this reaction is exemplified by the

ERIC HALGREN and KSENIJA MARINKOVIC Institut National de la Santé et de la Recherche Médicale, Rennes, France; Department of Psychiatry and Brain Research Institute, University of California at Los Angeles; (E. H.) Regional Epilepsy Center, Wadsworth Veterans Medical Center, Los Angeles, Calif.

startle response. While the orienting complex is automatic and thus preconscious in itself, it facilitates encoding of the eliciting event for awareness in emotional event integration.

During emotional event integration, the neurally encoded stimulus is integrated with semantic associates and other information from long-term semantic memory, relevant events from declarative memory, and the current cognitive and emotional context. This integration is necessary and sufficient for awareness. It results in an affective coloring and cognitive interpretation of even the most mundane events. The event, as encoded during event integration, is itself integrated with the context for action, resulting in voluntary (conscious) acts. This phase—response selection—also integrates a multitude of influences and thus will reflect a continuum of motivations from the routine to the passionate.

Both orienting complexes and event integrations are heavily influenced by the internal context, that is, the sustained neurophysiological background to phasic events—the subject's mood. Conversely, this context only becomes manifest in directly observable behavior through its effects on orienting complexes and event integrations. More permanent influences (corresponding to the temperament or personality) probably reflect hormonal or structural influences that are beyond the scope of this chapter.

Each stage—orienting complex, event integration, response selection, and internal context—can be associated with a distinct neural substrate. For each type of emotional process, an underlying neural network is suggested and supporting evidence is presented from recording, lesion, and stimulation studies in humans.

Orienting complex

STARTLE: GENETICALLY PROGRAMMED AND BRAINSTEM ORGANIZED The orienting complex is a mobilization of cerebral and somatic resources in order to effectively cope with a biologically important event. Two aspects of this reaction can be distinguished, which mobilize systems for, respectively, perceiving and comprehending the event, and taking physical action. The essential nature of this reaction can be seen vividly in its exaggerated form, the startle response, an involuntary, genetically programmed response to sudden sensory events that signal a possible impending biological emergency. For example, a loud abrupt sound will elicit arousal if the organism is somnolent or interruption of current cognitive processing if alert; orientation of attention and of the sense organs toward the apparent source of the sound; and distribution of sensory information regarding the stimulus to widespread cerebral areas for processing. In addition, the sound will evoke a multifaceted somatic response that prepares the body for action or for withstanding injury, including generalized sympathetic arousal, increased skeletal muscular tonus, and hormonal changes (Ohman, 1987). Animal studies have shown that the startle as well as other forms of the orienting complex are neuronally integrated in the midbrain and perhaps the hypothalamus (Jordan, 1990; Klemm and Vertes, 1990). Apparently the same organization obtains in humans, given that the orienting complex is preserved in anencephalic infants (Tuber et al., 1980), and that stimulation of the human central gray and lateral hypothalamus results in sympathetic arousal (Iacono, Blaine, and Nashold, 1982).

FOREBRAIN IMPLEMENTATION AND CONTROL Brainstem output during the orienting complex is apparent in autonomic activity (especially electrodermal; Ohman, 1987), whereas the forebrain's engagement in orienting can be measured as event-related potentials (ERPs). Startling stimuli evoke a large N1c component, and changes in auditory stimulus trains automatically evoke an MMN (Naatanen and Gaillard, 1983). The most characteristic response to orienting stimuli is the N2/P3a/SW complex, recorded over the frontocentral scalp with peaks at about 200, 280, and 350 ms after stimulus onset (figure 75.1; Halgren, 1990b; Marinkovic, in preparation). Across trials, the N2/P3a size is correlated with electrodermal responses (Lyytinen and Naatanen, 1987), and conforms well to habituation-dishabituation orienting complex parameters (Rockstroh and Elbert, 1990). However, the autonomic orienting complex and the N2/P3a can be dissociated on the basis of details of their reaction to repetition or to changes in the task relevance of the stimuli (Naatanen and Gaillard, 1983; Rosler, Hasselmann, and Sojka, 1987). These differences suggest that the N2/P3a/SW and the classical orienting complex may reflect the afferent (preparation-to-process) and efferent (preparation-to-respond) functions of the orienting complex, respectively.

Slightly before the scalp-recorded N2/P3a/SW, peaks with similar waveforms and task correlates are generated in widespread cortical areas, especially in

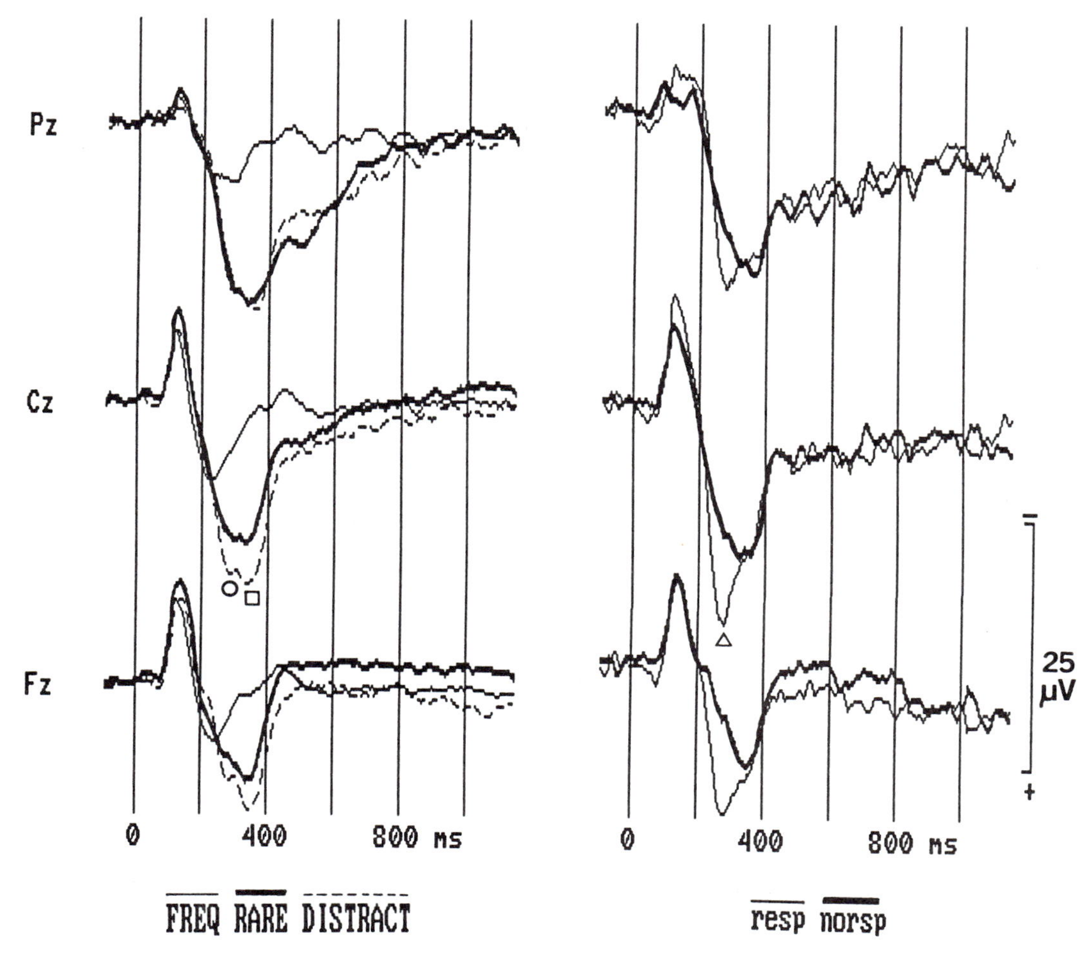

FIGURE 75.1 (Left) Brain potentials recorded from the human scalp during an auditory discrimination task. Interspersed among frequent tones (thin lines), are rare target tones to which the subject pressed a key (thick lines), and occasional unique, strange, distracting sounds to which the subject made no response (dashed lines). Both targets and distractors evoked a large P3 component, with peaks at 285 ms (circle; termed P3a) and 330 ms (square; termed P3b). Depth recordings identify distinct generating systems for the P3a (including cingulate and rectal g.) and the P3b (including hippocampal formation and lateral orbital cortex). (Right) Correlation of the P3a with the autonomic orienting complex. The brain potentials evoked by rare distractor sounds were divided into two groups, according to whether the sounds did (thin line) or did not (thick line) also evoke an electrodermal response. Trials with autonomic responses also evoke a large P3a, especially at frontal (Fz) and central (Cz) electrode sites (triangle). This is consistent with other studies identifying the P3a as the manifestation of the orienting complex in the forebrain. In contrast, the P3b seems to envelope the closure of the intentional event-encoding process. (From Marinkovic, in preparation)

the anterior cingulate gyrus (plus its posterior and inferior extensions, posterior cingulate gyrus and gyrus rectus; see figure 75.2), the inferior parietal lobule (in particular the supramarginal gyrus), and the region of area 46 (in the inferior frontal gyrus, pars triangularis) (Halgren et al., submitted; Wood and McCarthy, 1985; Alain et al., 1989; Smith et al., 1990). This neural circuit also includes a brainstem modulator or trigger, inasmuch as unilateral unusual stimuli still evoke part of the ipsilateral brain P3 after complete section of the forebrain commissures (Kutas et al., 1990).

The same parietal-cingulate-dorsolateral prefrontal circuit that generates the N2/P3a/SW has also been identified by metabolic and lesion studies in humans as a functional network for directed attention (see table 75.1 and chapter 39). Furthermore, electrical stimulation of the limbic parts of this circuit (i.e., the posterior orbital cortex; Livingston et al., 1948, and the anterior

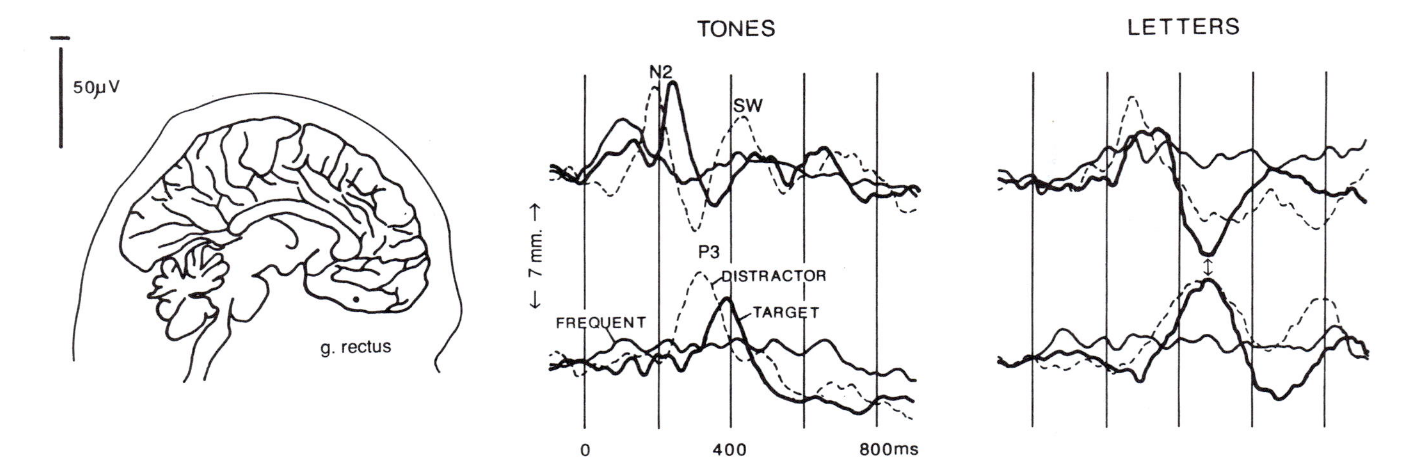

FIGURE 75.2 Generation of the P3a in the human rectal g. (medial orbitofrontal cortex) during the orienting complex. Averaged field potentials from two sites are presented. The location of the most medial site (bottom traces) is indicated at right on a tracing of the sagittal MRI. The upper traces were recorded simultaneously from a site 7 mm directly lateral. In the task at left, 50 ms TONES were presented every 1600 ms. Most of the tones were low-pitched (FREQUENT, thin lines). Interspersed randomly on 11% of the trials was a rising tone indicating that the subject should press a key (TARGET, thick lines). On another 11% of the trials were nontarget (DISTRACTOR) sounds, each unique (dashed lines). In the task at the right (LETTERS), visual symbols were presented for 200 ms every 1600 ms, consisting of rare targets (*) and distractors (letters) interspersed randomly with frequents (x). In the upper traces, a negative (upward)–positive–negative sequence corresponding to the N2/P3a/SW is evoked by rare target and distractor auditory and visual stimuli. In the lower traces, the P3a can be seen to invert to negative polarity, suggesting a local generator. The first vertical line indicates stimulus onset, with successive lines every 200 ms.

cingulate gyrus apparently adjacent to the corpus callosum, area 33; Pool and Ransohoff, 1949) can evoke sympathetic arousal, the efferent limb of the orienting complex.

Thus, during orienting the cortical and limbic areas critical for directed attention are activated more or less synchronously in a series of neurophysiological stages. The task correlates and timing of these stages suggest that they provide an envelope for the integrative construction of an initial rapid neural representation of the imperative stimulus. The prominent involvement of limbic sites in this network suggests a mechanism whereby autonomic aspects of the orienting complex can be integrated with attentional aspects. Furthermore, via this network the orienting complex can be evoked or modulated by cortically recognized events, and not simply by genetically programmed stimuli. Indeed, in humans the learned orienting complex seems to occur only when the subject becomes conscious of the contingency between the conditioned and unconditioned stimuli (Ohman, 1987). The prefrontal P3a has a shorter latency than the posterior parietal (Halgren et al., submitted), and prefrontal lesions eliminate the P3a over both anterior and posterior cortex (Knight, 1984), suggesting that the prefrontal cortex plays a leading role in organizing the cortical orienting complex in humans.

LIMITS OF THE ORIENTING COMPLEX Since the orienting complex functions to reorient behavior and cognition toward significant stimuli, it embodies the essence of emotion, as defined by those theories that consider emotion to be a generalized arousal that disrupts ongoing behavior (e.g., Plutchik, 1980). However, for other theorists (Panksepp, 1982), the orienting complex is too brief and nonspecific to be considered as more than a larval emotion, or preemotional response, constituting the primary emotional appraisal that may then lead to secondary appraisal and true emotion (Ohman, 1987). That is, the orienting complex is not able to organize a high-level, complex response, nor can a high-level, fine analysis of the event occur before the orienting complex is evoked. It is preconscious, in the temporal as well as the hierarchical sense, often occuring precisely in situations that do not allow a fine analysis before some response is demanded. In humans, the function of this level seems to be in facilitating the later occurrence of a more cognitive evaluation and volitional reaction, both by mobilizing resources (attentional, visceral, and somatomotor) and by imposing upon the organism an imperative to attend and respond.

Emotional event integration

Cognition and Emotion In everyday life, emotions are experienced much more frequently as integral parts of complex events—involving the meaning of such stimuli as words, faces, or scenes—than as automatic responses to loud noises. Similarly, emotion is more likely to become manifest as a complex, intentional, goal-directed sequence of movements—such as deciding to undergo elective surgery, or driving to a lover's house—than as global sympathetic arousal.

In contrast to the orienting complex, in which there is little or no distinction between the generalized responses evoked by different stimuli, at the level of event integration the different emotions have distinct subjective qualities and objective consequences. The distinctiveness of different emotions cannot be due to their autonomic concomitants, because these concomitants are very similar for different emotions and only occur *after* the feeling has been defined (Ohman, 1987). Consequently, cognition must play a crucial role in assigning emotional specificity, which reflects mainly the psychosocial context in which the autonomic arousal occurs (Mandler, 1975). Yet emotions have a long evolutionary history and therefore must have a genetic basis (Panksepp, 1982). The direction of experience and behavior toward the biological goals of survival and reproduction is clearly too important to be left entirely to either the cognitive (aware) or the visceral (reflexive) brain systems: The crucial essence of emotion is how the contributions of both are effectively integrated.

Stimulation-Evoked Emotions, Viscerosensations, and Emotionally Symbolic Memories In humans, accurate emotional judgments require the contributions of several brain regions, including the frontal, temporal, and parietal lobes, bilaterally (Kolb and Taylor, 1990; Heilman and Bowers, 1992). The regions where electrical stimulation may give rise to subjective emotional feelings tend to be more localized, but nonetheless include all of the limbic system. Among the variety of emotions evoked by human amygdala and hippocampal stimulation, fear is by far the most common (Halgren et al., 1978; Gloor, 1991; Halgren, 1991a). In contrast, stimulation of the anterior cingulate gyrus has been reported to give rise to about equal numbers of positive and negative emotions (Meyer et al., 1973; Bancaud et al., 1976; Laitinen, 1979). Similarly, stimulation of callosal fibers believed to interconnect the posterior orbital and anterior cingulate (area 32) cortices (Barbas and Pandya, 1984) results in mainly pleasant emotions (Laitinen, 1979). Stimulation in the septal area may also result in pleasurable feelings (Heath, John, and Fontana, 1968; Obrador, Delgado, and Martin-Rodriguez, 1973). Finally, uncontrolled laughing may occur with epileptic seizures arising in the anterior hypothalamic-preoptic area (Gaggero et al., 1991). Despite reports of uncontrolled rage after electrical stimulation of the human amygdala, this response is extremely rare, and appears usually or always to reflect a defensive reaction in a highly confused state rather than a well-directed or intentional attack. Conversely, hostile behavior may be decreased by amygdala lesions, but clear effects are observed only in retarded subjects (for review see Halgren, 1991a).

As might be expected given the strong link between subjective feelings and visceral sensations, stimulation of essentially the same anatomical areas gives rise to both types of mental phenomena (Halgren and Chauvel, 1993). The most common visceral sensation evoked from the medial temporal lobe is the epigastric sensation, which typically rises from the stomach up the chest to the throat and head (Halgren et al., 1978). Intragastric recordings during epigastric sensations indicate that they are true hallucinations, rather than the indirect results of evoked gastric movements (Van Buren, 1963). Conversely, bilateral removal of the medial temporal lobe in one case eliminated the subjective sense of hunger or thirst (Hebben et al., 1985).

Stimulation of the hippocampus and amygdala can also evoke respiratory and cardiovascular phenomena (Halgren et al., 1978). Evidence that these phenomena may be viscerosensory is found in the fact that a large proportion of medial temporal lobe neurons are highly sensitive to respiratory and cardiac cycles, as well as to blood gas levels (Halgren, Babb, and Crandall, 1977; Frysinger and Harper, 1989). Although paraesthesias of the external genitalia are evoked in the basoposterior paracentral lobule (i.e., the medial parietal lobe), sexual feelings per se appear to involve the insula and amygdala (Stoffels et al., 1980). Electrical stimulation

Table 75.1

Neural systems for integrating emotion into orienting, encoding, acting, and maintaining

	Orienting Complex	Event Encoding
Function	Respond to biologically imperative stimuli Prepare to process Alerting Direct attention Distribute information Prepare to react: Autonomic arousal	Encode event for cognition Cognitive Semantic associations Individual identificiation Emotional Emotional associations Psychosocial context
Structures	Brain stem Alerting Autonomic activation Attentional circuit Neocortical (parietofrontal) Orient attention Distribute information Limbic (cingulate and rectal gyri) Integrate hypothalamic and cortical components	Neocortical Fusiform gyrus: Object-encoding Superior temporal sulcus: Semantic association Limbic Amygdala: Emotional association Hippocampus: Emotional memories Lateral orbitofrontal: Emotional context
Lesions	Brain stem: Death Limbic: Dyscontrol of attention/emotion Neocortical: Neglect	Neocortical: Aphasias, agnosias Limbic Amygdala: Less emotionality? Hippocampus: Amnesia Lateral orbitofrontal: Poor judgment
Stimulation	Brain stem: Strong universal autonomic Limbic: Contextual autonomic	Limbic Emotions, viscerosensations Emotionally eloquent memories
Evoked potentials	Components N2/P3a/SW From 130 to 400 ms post–stimulus onset Evoked by unusual stimuli regardless of attention or meaning	Components N4/P3b From 260 to 800 ms post–stimulus onset Evoked by words, faces, etc. if attended and processed for meaning Modulated by ease of cognitive integration with information-specific unit activity

of the amygdala, hippocampal formation, or superior temporal gyrus may also evoke intense memory- or dreamlike hallucinations (Penfield and Perot, 1963; Halgren et al., 1978; Gloor, 1991; Bancaud et al., 1994). Although they are vivid and detailed, the subject knows that these hallucinations are not current reality. In a minority of patients, the particular experience selected appears to symbolize current psychodynamic concerns (Halgren and Chauvel, 1993).

All three phenomena evoked by limbic stimulation (viscerosensory hallucinations, emotional feelings, and emotionally symbolic hallucinated images) can provide a signal to awareness as to the biological significance of an event (Halgren and Chauvel, 1993). The subjective experience of emotion may be considered to be composed of such communications from the unconscious to the conscious brain. Anatomically, this would correspond to the fact that the tertiary sensory and supramodal association cortices receive input not only from sensory cortices but also from the amygdala and hippocampal formations (Amaral and Insausti, 1990). The implication, that experience is constructed by association cortex from limbic as well as sensory input, is supported by recordings in the human brain during event encoding.

Neurophysiological Stages in Event Encoding

Anatomical and physiological studies in monkeys have defined the successive stages of object processing in the visual system (see chapter 24). In humans also, evoked-

Response Choice	Sustained Context
Integrate encoded event with response templates Select and organize movement	Maintain information in active memory Contextual Stimulus
Neocortical Premotor: Movement sequencing Precentral: Movement command Limbic Middle cingulate gyrus: Integrate emotion with movement Supplementary motor cortex: Movement initiation	Brain stem: Set processing mode Neocortical Dorsolateral prefrontal: Cognitive context Ventrolateral prefrontal: Socioemotional context Premotor: Response set Posterior: Stimulus characteristics and identity
Limbic Middle cingulate gyrus: Akinetic mutism Supplementary motor cortex (global) Neocortical Premotor: Uncoordination Precentral: Paralysis (focal)	Neocortical Ventrolateral prefrontal: Contextually inappropriate behavior Premotor: Impulsive behavior
Middle cingulate gyrus: Complex adaptive movement sequences Supplementary motor cortex: Postural movements	Prefrontal: Confusion, forced thoughts
Component readiness potential Begins from 3000 to 300 ms premovement onset	Component contingent negative variation Between related stimuli

potential components reveal successive stages of face and word encoding that begin in the primary sensory cortex, pass through levels of association cortex, and rapidly arrive at multimodal and limbic structures in all lobes (table 75.1). A face-specific potential at 180 ms appears to be generated initially in the basal temporo-occipital cortex (areas 19/37, perhaps corresponding to basal V4), and then transmitted widely to superior temporal sulcal, parietotemporal, and dorsolateral prefrontal cortices (figure 75.3; Halgren et al., 1994a). Words evoke potentials at about 220 ms in the general region of the angular gyrus and Broca's area. Following these specific processing potentials are more widespread negativities that culminate in the well-studied N4 (or N400), and P3b (or P300, or late positive component).

Apparently locally generated N4/P3b's have been recorded in several structures, including the superior temporal sulcus, posterior parietal cortex, supramarginal and cingulate gyri, and area 46 in the dorsolateral prefrontal cortex (Halgren et al., 1994a). Large N4s are also generated in fusiform gyrus, where they may be specific to words or faces. It is striking, however, that the largest N4/P3b generators are in limbic structures (figure 75.4): the medial temporal lobe, where the N4 appears to be generated in the hippocampus, parahippocampal gyrus, and periamygdaloid region and the P3b is probably generated in the hippocampus proper (Stapleton and Halgren, 1987; McCarthy et al., 1989); and the lateral orbital cortex (Halgren et al., 1994b).

The cognitive task correlates, timing, and anatomi-

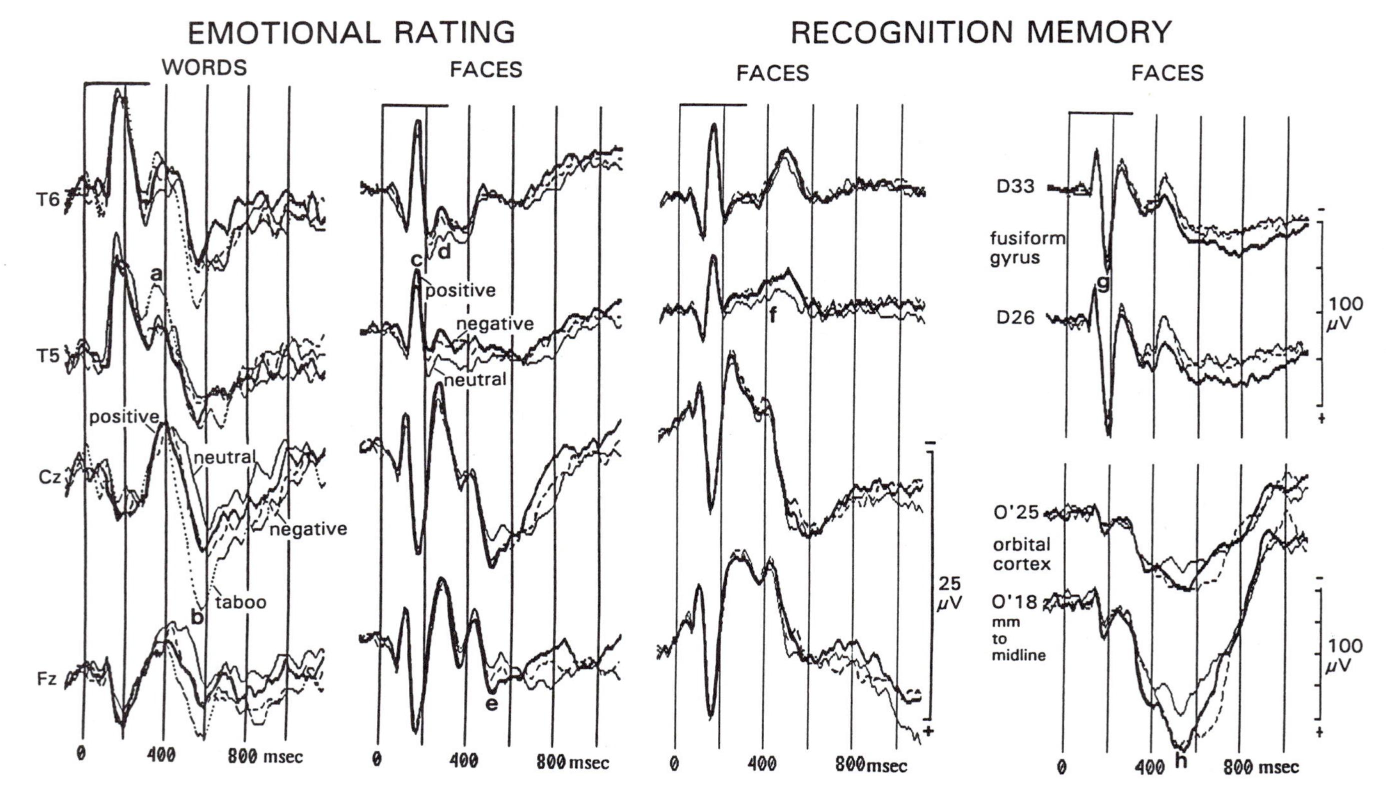

FIGURE 75.3 Potentials recorded from the human scalp (left 3 columns), and depth (fourth column), in response to faces and words expressing different emotions. In the emotional rating task, single words, or faces of unfamiliar people, were presented, and the subject moved a joystick to the left or right (indicating "very negative" to "very positive"). Preliminary data obtained in the emotional word rating task show a large difference in waveforms evoked by emotionally neutral versus taboo words, with negative or positive valence words evoking waveforms midway between neutral and taboo words at frontocentral sites. The taboo words evoked an early negativity over the left posterior temporal scalp (T5), beginning at 280 ms (a). The difference between emotionally charged and neutral words is maximal over the central midline scalp (Cz), where it appears as an increased P3b beginning about 400 ms after stimulus onset (b). The earliest significant waveform differences with respect to emotional expressions of faces appear at temporo-occipital sites and persist there for several hundred ms. The N175 component is significantly smaller in response to both emotional expressions when compared to neutral expression (c). This difference persists as a negativity between 200 and 400 ms to emotional expressions at temporo-occipital sites (d). The only significant differentiation among waveforms evoked by all three emotions appears frontocentrally in the 400–600ms latency range (e). In the recognition memory task, the subject pressed a key when a face that had previously been presented reappeared. Significant effects of emotional expression in this task were limited to the temporo-occipital sites. These effects were observed most clearly in area measures between 200 and 600 ms after stimulus onset (f). No difference between potentials evoked by positive and negative expressions was observed in this task. Thus, distinct brain potentials to different facial expressions were evoked only in the task where emotional judgments were explicitly required. Direct intracerebral recordings during the recognition task reveal a large focal potential to faces in the fusiform g. peaking at about 180 ms after stimulus onset (g). Like the component of similar latency recorded on the scalp over temporo-occipital cortex, the amplitude of this component changed significantly between different emotions. Emotionally expressive faces also evoke a significantly larger late positivity in the left orbitofrontal cortex (h), with a peak latency (520 ms) similar to that of the scalp P3b (e). (From Marinkovic et al., in preparation)

cal extent of the N4 suggest that it embodies the global integration of the stimulus with the current cognitive context, in order to neurally encode the event. This hypothesis is supported by the following experimental findings:

1. The N4 is evoked only by stimuli (in any modality) that are potentially meaningful within a broad semantic system, such as words or faces—that is, precisely those stimuli that must be integrated with the context in order for their meaning to be determined

(Halgren, 1990a). (The N2 plays an analogous role for simple sensory stimuli.)

2. A large number of conditions modulate N4 amplitude to a given stimulus (e.g., sentence context, previous presentation of a semantic associate, truth of the completed event, lexical frequency, presence of the stimulus in primary or recent memory), and the size of the N4 is decreased when this information facilitates integration of the N4 within the cognitive context (Kutas and Van Petten, 1988).

3. The modulating stimuli may be in a different modality (e.g., auditory versus visual), or in a different knowledge domain (e.g., words versus faces), suggesting that the modulated network has access to all of these types of information and/or encodes information at a level deeper than modality or knowledge domain (Domalski, Smith, and Halgren, 1991).

4. Hippocampal formation and amygdala unit-firing during the N4 shows specificity both for the stimulus (i.e., a particular word or face) and for the context in which it is presented (Heit, Smith, and Halgren, 1988).

5. The N4 is the first evoked-potential component to show clearly these sensitivities to meaningfulness, specificity, and context, and is the last to occur before the response must be specified (Halgren, 1990a).

6. The global integration of an event is considered to be the essence of controlled or conscious processing, and the N2/N4/P3b has been associated with such processes via self-report and latency (see Halgren, 1993; Halgren and Chauvel, 1993, for review).

7. The extensive anatomical distribution of putative N4 generators in most or all supramodal association cortex areas provides a sufficiently broad neural substrate to encode all of the knowledge domains that contribute to an encoded event. Again, the triggering of the N4 is routed through the brain stem, inasmuch as the N4 is still evoked bilaterally after presentation of words to only the left hemisphere in patients with com-

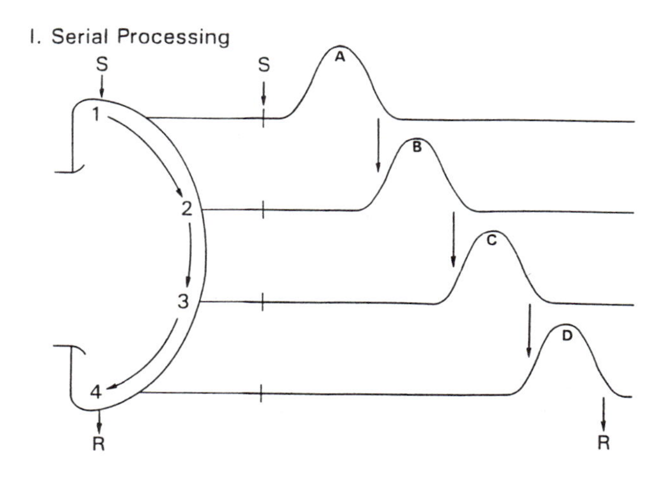

FIGURE 75.4 Cognitive evoked potential (EP) components as integrative stages of information processing in humans. A priori, one could imagine that information processing between stimulus (S) and response (R) is serial (top), with successive EP components (A, B, C, D, E) passing through successive anatomical structures (1, 2, 3, 4). Alternatively (middle), after the activation arrives in each successive structure, it could could persist indefinitely, but independently of other activated structures. What is observed is closer to the third possibility (bottom): Although the earlier stages of processing may be somewhat localized to sensory or sensory association cortex (A, B, C), the later stages of contextual integration (D, E) engage, more or less synchronously, multiple sensory association, motor association, multimodal association, prefrontal, and limbic structures (bottom: Parallel interdependent processing).

plete section of the forebrain commissures (Kutas, Hillyard, and Gazzaniga, 1988).

The P3b follows the N4 in these paradigms, and tends to be modulated by the same conditions that modulate the N4, but in the opposite direction (Halgren, 1990a). Indirect evidence suggests that the P3b represents the second phase of the same cognitive contextual integration process as the N4. The first (N4) phase would provide relative excitation to facilitate the spread of information from its sensory and memory entry points into the cognitive network, and the second (P3b) phase would provide both immediate recurrent inhibition to prevent this spread from recruiting spurious elements, and delayed recurrent inhibition to disrupt recurrent excitatory loops and thus permit the evolution of new networks (Read, Nenov, and Halgren, 1994). Consistent with this model, behavioral responses indicating successful event encoding are issued at the latency of the N4-to-P3b transition (McCarthy and Donchin, 1981).

Integration of Emotional Information In summary, the N4/P3b is generated in multiple association cortex and limbic areas, and has cognitive correlates and other characteristics that suggest it embodies contextual integration. The specific contribution of each area during the N4/P3b can be inferred from the effects of stimulating or lesioning it, or from the behavioral correlates of its unit activity (figure 75.4). In particular, the data reviewed above and in chapters 69, 72, and 73 would suggest that the amygdala contributes an emotional evaluation of the presented stimulus, and the lateral orbitofrontal cortex provides a psychosocial context (Halgren and Chauvel, 1993).

Thus, the fact that the limbic system is heavily activated beginning with the early stages of event encoding implies that emotional information is integrated in very early stages of event encoding. The prominent limbic activity beginning 120 ms after stimulus onset permits limbic input to shape the content of the encoded experience rather than simply to react to its content. Thus, the emotional evaluation of events occurs synchronously with their cognitive evaluation, prior to any conclusion having been obtained from that processing (see chapter 69; Plutchik, 1980; Panksepp, 1982; Ohman, 1987). Conceivably, this would allow the limbic system to contribute to the myriad psychological defense mechanisms (repression, denial, undoing, projection, displacement, etc.) that may distort or eliminate the conscious experience of an emotionally significant event (Brenner, 1974). When sensory input is absent (e.g., during dreams), event integration would be dominated by limbic input, resulting in emotionally eloquent hallucinations.

These inferences are further supported by the potentials evoked by faces showing differing emotional expression (figure 75.3; Marinkovic et al., submitted). Facial expressions provide a rapid, emotion-specific, and largely innate system for the reliable interpersonal communication of emotional state (Izard and Buechler, 1979; Ekman, 1982). Thus, beginning with the initial face-specific component peaking at 180 ms after the face is presented, emotionally expressive faces evoke significantly different scalp potentials in comparison to unexpressive faces. Anatomically, such differences are seen as early in the visual system as the fusiform gyrus, but are much larger in later processing stages, especially in the lateral orbitofrontal cortex during the N4/P3b. Similar differences in scalp potentials are observed to emotionally significant words. These data are consistent with other studies that have generally found that, of stimuli that are perceived but otherwise equated, the scalp P3b is larger to stimuli with greater affective valence (Johnston, Burleson, and Miller, 1987; but see Vanderploeg, Brown, and Marsh, 1987).

Emotion in volition

By definition, an emotion motivates, and usually what it motivates is an action. When the action is volitional, its selection and production must integrate the conclusion of the event-encoding process with the context for action. Key structures in the volitional origins of action are the supplementary motor cortex and the underlying central cingulate gyrus, where lesions can abolish voluntary movement (see Vogt, Finch, and Olson, 1992, for review) and which are activated metabolically during the planning of voluntary movements (Roland, 1985). Electrical stimulation of the supplementary motor cortex can evoke either movement arrest or coordinated, stereotypical "postural" turning movements (Penfield and Welch, 1951; Talairach and Bancaud, 1966). Low-level electrical stimulation of the middle part of area 24 in the cingulate gyrus evokes "highly integrated types of motor behavior which are sometimes very well adapted to the situation, including sucking, nibbling, licking and tactile exploration of the body and surrounding space" (Talairach et al., 1973).

The involvement of these structures in the formulation of voluntary movements can be monitored with the readiness potential (also known as Bereitschaftspotential), a broad negativity beginning 300 to 3000 ms before voluntary movements. Field potentials and/or unit activity correlated with the readiness potential in humans suggests generators in the precentral, premotor, supplementary motor, and central cingulate cortices (Halgren et al., 1994b; Heit, Smith, and Halgren, 1990; Groll-Knapp et al., 1980) and possibly other subcortical areas (Halgren, 1991b), with thalamic modulatory influences (Halgren, 1990b; Raeva, 1986). In simple tasks, the readiness potential appears to have more restricted generators (Neshige, Luders, and Shibasaki, 1988).

Following the same logic as for the cognitive evoked-potential components discussed previously, it may be proposed that the emotional coloring of actions is contributed during the readiness potential to the progressively defined movement (table 75.1). It is possible that the middle cingulate gyrus plays a crucial role in contributing this information, receiving input from anterior cingulate, medial temporal, orbitofrontal, and dorsolateral frontal areas, and projecting to the supplementary motor cortex (Van Hoesen, Morecraft, and Vogt, 1993). Lesions that affect these areas result in a reduction in the emotional coloring of facial expressions and speech (Kolb and Taylor, 1990; Heilman and Bowers, 1992).

Sustained emotional context

Orientation, event encoding, and response choice all occur against the background of sustained neural activity embodying the current schema of the world and of possible actions to be taken within it. Sustained specific firing has been noted in the primate frontal lobe, where different aspects of the current cognitive context appear to be held in different regions (for example, sensory and spatial information in area 46), response mappings appear to be held in premotor cortices (Goldman-Rakic, 1987; Fuster, 1989), and lesion evidence suggests that the socioemotional context may be maintained in orbitofrontal neuronal firing (see chapter 73). Although such sustained responses have not been reported, phasic responses by orbitofrontal neurons to events that fail to match contextually established expectancies are described in chapter 72. Deficient socioemotional context would explain how orbitofrontal lesions produce impulsivity, irresponsibility, and social inappropriateness (Stuss, Gow, and Hetherington, 1992): In the absence of the socioemotional input to the N2/P3a/SW network, attention might be oriented toward irrelevant or inappropriate items; without the socioemotional input to the encoding of events during the N4/P3b, their emotional dimension may be unappreciated; and without a socioemotional input during the readiness potential, response choices could be impulsive and unrelated to the social situation.

This hypothesis can be tested because the sustained firing of frontal neurons induces a current that ultimately causes the CNV (contingent negative variation). Intracranial recordings have found evidence for CNV generation in multiple regions, especially the prefrontal cortex (Papakostopoulos, Cooper, and Crow, 1976; Groll-Knapp et al., 1980). The CNV is strongly modulated by a cholinergic pathway ascending from the brain stem (see Halgren, 1990b, for review). As predicted, the level of the CNV has been found to have a strong effect on the P3 (Rockstroh et al., 1992). This level, in turn, is strongly affected by the level of anxiety or arousal (Tecce and Cole, 1976; Proulx and Picton, 1984), as well as the presence of psychopathology, or even of normal personality variations (Rockstroh et al., 1982).

From physiology to personality

At the psychological level, sustained firing may correspond to a temporary mood, goal, or conviction. Neurobiological influences of longer duration such as hormonal levels, excitability changes, or structural modifications, may underlie such psychological constructs as constitution or temperament. These influences provide a coherence to behavior over time, becoming manifest in the emotions of the moment through their influences on the neurophysiological processes embodying orienting, event integration, volition, and context. As a whole, such influences may be considered to constitute the personality.

There are complex but nonetheless convincing differences in the orienting complex correlated with different personality types (Eysenck, 1981; Cahill and Polich, 1992). Furthermore, lesions of the limbic areas where the N2/P3a/SW is generated may ameliorate the excessive fixation of attention by emotion in obsessional disorders and chronic pain (Teuber, Corkin, and Twitchell, 1977; Laitinen, 1979). Similarly, the P3b is abnormal in schizophrenia (Faux et al., 1987; Pritchard, 1986; Mirsky and Duncan, 1986), and such patients may have abnormalities in the medial temporal limbic structures active during event encoding (see chapters 69, 74, 76, and 77; Halgren and Chauvel, 1993).

Conclusion

In mammals, basic emotional patterns are encoded in the brain stem, but are elaborated and controlled by various limbic structures (see chapter 69). Modern data from humans support this general view, but put greater emphasis on the integration of the neocortex in emotion via the limbic structures' important, direct, and continuous role in cognition (figure 75.5). These forebrain contributions can be differentiated according to their temporal stages and core neural substrates: orienting in the cingulate gyrus and associated prefrontal and posterior parietal cortices; event encoding in the hippocampus and amygdala and associated inferotemporal and lateral-orbital cortices; response choice in the central cingulate gyrus and associated supplementary and premotor cortices; sustained context predominantly in prefrontal areas. In addition to its core substrate, each stage is triggered and modulated by brainstem circuits, and secondarily activates much of limbic and association cortex.

Thus, emotion as a directed integration of the entire organism toward a biological goal is implemented in each of these stages via the simultaneous interaction of neural networks spanning brainstem, limbic, and neocortical structures. Yet, mechanisms exist whereby different levels of the neuraxis can respond independently to biological imperatives: the brain stem by triggering fixed action patterns in response to genetically defined releasing stimuli; the forebrain by implementing different action schemata with respect to a complex world model. In both unconscious reflexes and cold calculations, emotion is absent. Rather, emotional feelings are the subjective aspect of a nervous system in which these disparate neural levels are functionally linked. Thus, emotion can be thought of as a state of communication between brain stem and forebrain in which their distinct resources are mobilized in response to an event recognized as important by either.

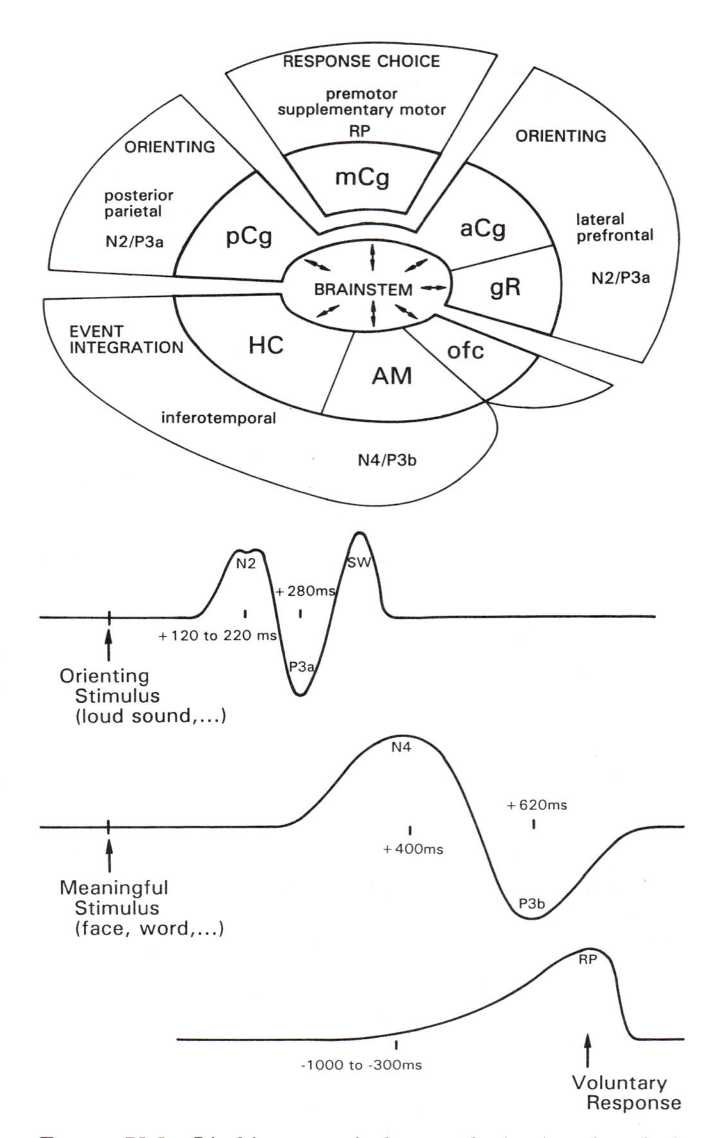

FIGURE 75.5 Limbic-neocortical networks (top) and evoked potential components (bottom) embodying different functional systems that integrate cognitive with visceral aspects of emotion in humans. Biologically imperative stimuli evoke the orienting complex and the N2/P3a evoked potential components, generated in limbic (gR, gyrus rectus; aCg, anterior cingulate gyrus; pCg, posterior cingulate gyrus) and neocortical sites (including sMg, supramarginal gyrus, and a46, area 46). During the N4/P3b, event integration takes place, integrating attended semantic stimuli with the cognitive and emotional context in limbic (AmF, amygdaloid formation; HCF, hippocampal formation; lOFc, lateral orbitofrontal cortex) and neocortical sites (including Fg, fusiform gyrus, and sTs, superior temporal sulcus). This may be followed by neurophysiological activity underlying response choice, the readiness potential (RP), generated in limbic (mCg, middle cingulate gyrus) and neocortical sites (preC, precentral cortex; preM, premotor cortex; and sMc, supplementary motor cortex).

ACKNOWLEDGMENTS We thank Patrick Chauvel, Michael Smith, Gary Heit, June Stapleton, Jeffrey Clarke, Patrick Baudena, Milan Marinkovic, and Tracy Gigg for contributing to the research results reported here. The authors' research is supported by the USPHS (NS18741), the U.S. Department of Veterans Affairs, the French MRT and INSERM, and NATO.

REFERENCES

ALAIN, C., F. RICHER, A. ACHIM, and J. M. SAINTHILAIRE, 1989. Human intracerebral potentials associated with target, novel and omitted auditory stimuli. *Brain. Topogr.* 1:237–245.

AMARAL, D. G., and R. INSAUSTI, 1990. The hippocampal formation. In *The Human Nervous System*, G. Paxinos, ed. New York: Academic Press, pp. 711–755.

BANCAUD, J., F. BRUNET-BOURGIN, P. CHAUVEL, and E. HALGREN, 1994. Anatomical origin of déjà vu and vivid 'memories' in human temporal lobe epilepsy. *Brain* 117:71–90.

BANCAUD, J., J. TALAIRACH, S. GEIER, A. BONIS, S. TROTTIER, and M. MANRIQUE, 1976. Manifestations comportementales induites par la stimulation électrique du gyrus cingulaire antérieur chez l'homme. *Rev. Neurol. [Paris]* 132:705.

BARBAS, H., and D. N. PANDYA, 1984. Topography of commissural fibers of the prefrontal cortex in the rhesus monkey. *Exp. Brain. Res.* 55:187–191.

BRENNER, C., 1974. On the nature and development of affects: A unified theory. *Psychoanal. Q.* 43:532–556.

CAHILL, J. M., and J. POLICH, 1992. P300, probability, and introverted/extroverted personality types. *Biol. Psychol.* 33: 23–35.

DOMALSKI, P., M. E. SMITH, and E. HALGREN, 1991. Cross-modal repetition effects on the N4. *Psychol. Sci.* 2:173–178.

EKMAN, P., 1982. *Emotion in the Human Face.* Cambridge: Cambridge University.

EYSENCK, H. J., ed., 1981. *A Model for Personality.* Berlin: Springer.

FAUX, S. F., M. W. TORELLO, R. W. MCCARLEY, M. E. SHENTON, and F. H. DUFFY, 1987. P300 topographic alterations in schizophrenia: A replication study. *Current Trends in Event Related Potential Research* (*EEG* Suppl. 40), R. Johnson, J. W. Rohrbaugh, and R. Parasuraman, eds. Amsterdam: Elsevier, pp. 688–694.

FRYSINGER, R. C., and R. M. HARPER, 1989. Cardiac and respiratory correlations with unit discharge in human amygdala and hippocampus. *Electroencephalogr. Clin. Neurophysiol.* 72:463–470.

FUSTER, J. M., 1989. *The Prefrontal Cortex: Anatomy, Physiology and Neuropsychology of the Frontal Lobe.* New York: Raven.

GAGGERO, R., M. P. BAGLIETTO, F. BORAGNO, M. ODDONE, P. TOMA, and M. DE NEGRI, 1991. Epilepsy with laughing seizures, hypothalamic hamartoma and precocious puberty: Contributions of MRI, computed EEG topography (CET) and ambulatory EEG (A-EEG). *Minerva Pediatrica* 43:801–810.

GLOOR, P., 1991. Neurobiological substrates of ictal behavioral changes. In *Neurobehavioral Problems in Epilepsy.* Vol. 55, *Advances in Neurology*, D. B. Smith, D. M. Treiman, and M. R. Trimble, eds. New York: Raven, pp. 1–34.

GOLDMAN-RAKIC, P. S., 1987. Circuitry of the prefrontal cortex: Short-term memory and the regulation of behavior by representational knowledge. In *Handbook of Physiology.* Sec. 1, *The Nervous System.* Vol. 5, *Higher Functions of the Brain*, part 1, V. B. Mountcastle, ed. Bethesda, Md.: American Physiological Society, pp. 373–417.

GROLL-KNAPP, E., J. GANGLBERGER, M. HAIDER, and H. SCHMID, 1980. Stereoelectroencephalographic studies on event-related slow potentials in the human brain. In *Electroencephalography and Clinical Neurophysiology*, H. Lechner and A. Aranibar, eds. Amsterdam: Excepta Medica, pp. 746–760.

HALGREN, E., 1990a. Insights from evoked potentials into the neuropsychological mechanisms of reading. In *Neurobiology of Cognition*, A. Scheibel and A. Weschsler, eds. New York: Guilford, pp. 103–150.

HALGREN, E., 1990b. Evoked potentials. In *Neuromethods.* Vol. 15, *Neurophysiological Techniques: Applications to Neural Systems*, A. A. Boulton, G. Baker, and C. Vanderwolf, eds. Clifton, N. J.: Humana, pp. 147–275.

HALGREN, E., 1991a. Emotional neurophysiology of the amygdala within the context of human cognition. In *The Amygdala: Neurobiological Aspects of Emotion, Memory, and Mental Dysfunction*, J. P. Aggleton, ed. New York: Wiley, pp. 191–228.

HALGREN, E., 1991b. Firing of human hippocampal units in relation to voluntary movements. *Hippocampus* 1:153–161.

HALGREN, E., 1993. Physiological integration of the declarative memory system. In *The Memory System of the Brain*, J. Delacour, ed. New York: World Scientific, pp. 69–155.

HALGREN, E., P. BAUDENA, G. HEIT, J. CLARKE, and K. MARINKOVIC, 1994a. Spatio-temporal stages in face and word processing. 1. Depth-recorded potentials in the human occipital, temporal and parietal lobes. *J. Physiol. (Paris)* 88:1–50.

HALGREN, E., P. BAUDENA, G. HEIT, J. CLARKE, K. MARINKOVIC, and P. CHAUVEL, 1994b. Spatio-temporal stages in face and word processing. 2. Depth-recorded potentials in the human frontal and Rolandic cortices. *J. Physiol. (Paris)* 88:51–80.

HALGREN, E., T. L. BABB, and P. H. CRANDALL, 1977. Responses of human limbic neurons to induced changes in blood gases. *Brain. Res.* 132:43–63.

HALGREN, E., and P. CHAUVEL, 1993. Experiential phenomena evoked by human brain electrical stimulation. *Adv. Neurol.* 65:87–104.

HALGREN, E., R. D. WALTER, D. G. CHERLOW, and P. H. CRANDALL, 1978. Mental phenomena evoked by electrical stimulation of the human hippocampal formation and amygdala. *Brain* 101:83–117.

HEATH, R. G., S. B. JOHN, and C. J. FONTANA, 1968. The pleasure response: Studies by stereotaxic techniques in patients. In *Computers and Electronic Devices in Psychiatry*, N. Kline and E. Laska, eds. New York: Grune and Stratton, pp. 178–179.

Hebben, N., S. Corkin, H. Eichenbaum, and K. Shedlack, 1985. Diminished ability to interpret and report internal states after bilateral medial temporal resection: Case H. M. *Behav. Neurosci.* 99:1031–1039.

Heilman, K. M., and D. Bowers, 1992. Neuropsychological studies of emotional changes induced by right and left hemispheric lesions. In *Psychological and Biological Approaches to Emotion*, N. L. Stein, B. Leventhal, and T. Trabasso, eds. Hillsdale,N.J.: Erlbaum, pp. 97–113.

Heit, G., M. E. Smith, and E. Halgren, 1988. Neural encoding of individual words and faces by the human hippocampus and amygdala. *Nature* 333:773–775.

Heit, G., M. E. Smith, and E. Halgren, 1990. Neuronal activity in the human medial temporal lobe during recognition memory. *Brain* 113:1093–1112.

Iacono, R. P., S. Blaine, and S. Nashold, Jr., 1982. Mental and behavioral effects of brain stem and hypothalamic stimulation in man. *Hum. Neurobiol.* 1:273–279.

Izard, C. E., and S. Buechler, 1979. Aspects of consciousness and personality in terms of differential emotions theory. In *Emotion: Theory, Research and Experience*, R. Plutchik and H. Kellerman, eds. New York: Academic Press, pp. 349–366.

Johnston, V. S., M. H. Burleson, and D. R. Miller, 1987. Emotional value and late positive components of ERPs. In *Current Trends in Event-Related Potential Research* (*EEG* suppl. 40), R. Johneson, J. W. Rohrbaugh, and R. Parasuraman, eds. Amsterdam: Elsevier, pp. 198–203.

Jordan, D., 1990. Autonomic changes in affective behavior. In *Central Regulation of Autonomic Functions*, A. D. Loewy and K. M. Spyer, eds. New York: Oxford University Press.

Klemm, W. R., and R. P. Vertes, eds., 1990. *Brainstem Mechanisms of Behavior*. New York: Wiley.

Knight, R. T., 1984. Decreased response to novel stimuli after prefrontal lesions in man. *Electroencephalogr. Clin. Neurophysiol.* 59:9–20.

Kolb, B., and L. Taylor, 1990. Neocortical substrates of emotional behavior. In *Psychological and Biological Approaches to Emotion*, N. L. Stein, B. Leventhal, and T. Trabasso, eds. Hillsdale, N.J.: Erlbaum, pp. 115–144.

Kutas, M., S. A. Hillyard, and M. S. Gazzaniga, 1988. Processing of semantic anomaly by right and left hemispheres of commissurotomy patients: Evidence from event-related brain potentials. *Brain* 111:553–576.

Kutas, M., S. A. Hillyard, B. T. Volpe, and M. S. Gazzaniga, 1990. Late positive event-related potentials after commissural section in humans. *J. Cognitive Neurosci.* 2:258–271.

Kutas, M., and C. Van Petten, 1988. Event-related brain potential studies of language. In *Advances in Psychophysiology*, P. K. Ackles, J. R. Jennings, and M. G. H. Coles, eds. Greenwich, Conn.: JAI Press, pp. 139–187.

Laitinen, L. V., 1979. Emotional responses to subcortical electrical stimulation in psychiatric patients. *Clin. Neurol. Neurosurg.* 81:148–157.

Livingston, R. B., W. P. Chapman, K. E. Livingston, and L. Kraintz, 1948. Stimulation of orbital surface of man prior to frontal lobotomy. *Res. Publ. Assoc. Res. Nerv. Ment. Dis.* 27:421–432.

Lyytinen, H., and R. Naatanen, 1987. Autonomic and ERP responses to deviant stimuli: Analysis of covariation. In *Current Trends in Event-Related Potential Research* (*EEG* suppl. 40), R. Johnson, J. W. Rohrbaugh, and R. Parasuraman, eds. Amsterdam: Elsevier, pp. 108–117.

Mandler, G., 1975. *Mind and Emotion*. New York: Wiley.

McCarthy, G., and E. Donchin, 1981. A metric for thought: A comparison of P300 latency and reaction time. *Science* 211:77–80.

McCarthy, G., C. C. Wood, P. D. Williamson, and D. D. Spencer, 1989. Task-dependent field potentials in human hippocampal formation. *J. Neurosci.* 9:4253–4268.

Meyer, G., M. McElhaney, W. Martin, and C. P. McGraw, 1973. Stereotactic cingulotomy with results of acute stimulation and serial psychological testing. In *Surgical Approaches in Psychiatry*, L. V. Laitinen and K. E. Livingston, eds. Lancaster: Medical and Technical Publishing, pp. 39.

Mirsky, A. F., and C. C. Duncan, 1986. Etiology and expression of schizophrenia: Neurobiological and psychosocial factors. *Annu. Rev. Psychol.* 37:291–319.

Naatanen, R., and A. W. K. Gaillard, 1983. The orienting reflex and the N2 deflection of the ERP. In *Tutorials in Event Related Potential Research: Endogenous Components*, A. W. K. Gaillard and W. Ritter, eds. Amsterdam: Elsevier, pp. 119–141.

Neshige, R., H. Luders, and H. Shibasaki, 1988. Recording of movement-related potentials from scalp and cortex in man. *Brain* 111:719–736.

Obrador, S., J. M. R. Delgado, and J. G. Martin-Rodriguez, 1973. Emotional areas of the human brain and their therapeutical stimulation. In *Cerebral Localization*, K. J. Zuelch, ed. Berlin: Springer, pp. 171–183.

Ohman, A., 1987. The psychophysiology of emotion: An evolutionary-cognitive perspective. In *Advances in Psychophysiology: A Research Annual*, vol. 2, P. K. Ackles, J. R. Jennings, and M. G. H. Coles, eds. Greenwich, Conn.: JAI, pp. 79–127.

Panksepp, J., 1982. Toward a general psychobiological theory of emotion. *Behav. Brain. Sci.* 5:407–422.

Papakostopoulos, D., R. Cooper, and H. J. Crow, 1976. Electrocorticographic studies of the contingent negative variation and "P300" in man. In *The Responsive Brain*, W. C. McCallum and J. R. Knot, eds. Bristol: Wright, pp. 201–204.

Penfield, W. P., and P. Perot, 1963. The brain's record of auditory and visual experience: A final summary and discussion. *Brain* 86:595–696.

Penfield, W., and K. Welch, 1951. Supplementary motor area of the cerebral cortex. *Arch. Neurol. Psychiatry* 66:289–317.

Plutchik, R., 1980. *Emotion: A Psychoevolutionary Synthesis*. New York: Harper and Row.

Pool, J. L., and J. Ransohoff, 1949. Autonomic effects of stimulating the rostral portion of cingulate gyri in man. *J. Neurophysiol.* 12:385–392.

Pritchard, W. S., 1986. Cognitive event-related potential correlates of schizophrenia. *Psychol. Bull.* 100:43–66.

Proulx, G. B., and T. W. Picton, 1984. The effects of anxiety and expectancy on the CNV. In *Brain and Information: Event-Related Potentials* (*Annals of the N.Y. Academy of Sciences*, vol. 425), R. Karrer, J. Cohen, and P. Tueting, eds. New York: New York Academy of Sciences, pp. 617–628.

Raeva, S., 1986. Localization in human thalamus of units triggered during "verbal commands," voluntary movements and tremor. *Electroencephalogr. Clin. Neurophysiol.* 63: 160–173.

Read, W., V. I. Nenov, and E. Halgren, 1994. The role of inhibition in memory-retrieval by hippocampal area CA3. *Neurosci. Biobehav. Rev.* 18:55–68.

Rockstroh, B., and T. Elbert, 1990. On the relations between event-related potentials and autonomic responses: Conceptualization within a feedback loop framework. In *Event-Related Brain Potentials: Issues and Applications*, J. W. Rohrbaugh, R. Parasuraman, and R. Johnson, Jr., eds. New York: Oxford University Press, pp. 90–108.

Rockstroh, B., T. Elbert, N. Birbaumer, and W. Lutzenberger, 1982. *Slow Brain Potentials and Behavior*. Baltimore, M.D.: Urban and Schwartzenberg.

Rockstroh, B., M. Müller, M. Wagner, R. Cohen, and T. Elbert, 1992. Probing the functional brain state during P300 and CNV. *EPIC* [Hungary] 134.

Roland, P. E., 1985. Cortical organization of voluntary behavior in man. *Hum. Neurobiol.* 4:155–167.

Rosler, F., D. Hasselmann, and B. Sojka, 1987. Central and peripheral correlates of orienting and habituation. In *Current Trends in Event-Related Potential Research* (*EEG* suppl. 40), R. Jr. Johnson, J. W. Rohrbaugh, and R. Parasuraman, eds. Amsterdam: Elsevier, pp. 366–372.

Smith, M. E., E. Halgren, M. Sokolik, P. Baudena, A. Mussolino, C. Liegeois-Chauvel, and P. Chauvel, 1990. The intracranial topography of the P3 event-related potential elicited during auditory oddball. *Electroencephalogr. Clin. Neurophysiol.* 76:235–248.

Stapleton, J. M., and E. Halgren, 1987. Endogenous potentials evoked in simple cognitive tasks: Depth components and task correlates. *Electroencephalogr. Clin. Neurophysiol.* 67:44–52.

Stoffels, C., C. Munari, A. Bonis, J. Bancaud, and J. Talairach, 1980. Manifestations genitales et "sexuelles" lors des crises epileptiques partielles chez l'homme. *Rev. Electroencephalogr. Neurophysiol. Clin.* 10:386–392.

Stuss, D. T., C. A. Gow, and C. R. Hetherington, 1992. "No longer Gage": Frontal lobe dysfunction and emotional changes. *J. Consult. Clin. Psychol.* 60:349–359.

Talairach, J., and J. Bancaud, 1966. The supplementary motor area in man. *Int. J. Neurosci.* 5:330–347.

Talairach, J., J. Bancaud, S. Geier, M. Bordas-Ferrer, A. Borris, G. Szikla, and M. Buser, 1973. The cingulate gyrus and human behavior. *Electroencephalogr. Clin. Neurophysiol.* 34:45–52.

Tecce, J. J., and J. O. Cole, 1976. The distraction-arousal hypothesis, CNV, and schizophrenia. In *Behavior Control and Modification of Physiological Activity*, D. I. Mostofsky, ed. Englewood Cliffs, N.J.: Prentice-Hall, pp. 162–219.

Teuber, H.-L., S. H. Corkin, and T. L. Twitchell, 1977. Study of cingulotomy in man: A summary. In *Neurosurgical Treatment in Psychiatry, Pain and Epilepsy*, W. H. Sweet, S. Obrador, and J. G. Martin-Rodriguez, eds. Baltimore, Md.: University Park Press, pp. 355–362.

Tuber, D. S., G. G. Berntson, D. S. Bachman, and J. N. Allen, 1980. Associative learning in premature hydranencephalic and normal twins. *Science* 210:1035–1037.

Van Buren, J. M., 1963. The abdominal aura: A study of abdominal sensation occurring in epilepsy and produced by depth stimulation. *Electroencephalogr. Clin. Neurophysiol.* 15:1–19.

Vanderploeg, R. D., W. S. Brown, and F. T. Marsh, 1987. Judgements of emotion in words and face: ERP correlates. *Int. J. Psychophysiol.* 5:193–205.

Van Hoesen, G. W., R. J. Morecraft, and B. A. Vogt, 1993. Connections of the monkey cingulate cortex. In *Neurobiology of Cingulate Cortex and Limbic Thalamus*, B. A. Vogt and M. Gabriel, eds. Boston: Birkhäuser.

Vogt, B. A., D. M. Finch, and C. R. Olson, 1992. Functional heterogeneity in cingulate cortex: The anterior executive and posterior evaluative regions. *Cerebral Cortex* 2:435–443.

Wood, C. C., and G. McCarthy, 1985. A possible frontal lobe contribution to scalp P300. *Soc. Neurosci. Abstr.* 11: 879–870.

76 Positron Emission Tomographic Imaging Studies of Human Emotional Disorders

WAYNE C. DREVETS AND MARCUS E. RAICHLE

ABSTRACT Positron emission tomography (PET) imaging produces three-dimensional pictures of brain function in terms of regional blood flow and metabolism. Since PET provides this information noninvasively, it is being used in humans to delineate the functional anatomy of various mental processes, including those related to emotion. One approach to identifying the anatomical correlates of emotion has been to compare PET scans of subjects with emotional disorders with scans of control subjects.

To illustrate the types of experimental strategies that are useful in PET investigations of emotional disorders, we highlight our studies of familial pure depressive disease (FPDD). This endeavor has included comparisons between the depressed and the normative states, the depressed and the remitted states, and the untreated-depressed and the treated states. In addition, we compare the results in FPDD with image data from other subtypes of major depression and from a state of induced sadness in normal subjects. Finally, we contrast the functional anatomical correlates of major depression with those of anxiety disorders. The findings in these pathologic emotional states converge with data obtained using other techniques to implicate neural circuits involving limbic and limbic-related structures in emotion.

While animal research has provided important insights into the neural substrates of emotional behavior and autonomic expression, relatively little is known about the anatomical correlates of emotional experience. This is partly because inquests into emotional experience are dependent upon human research, since only humans directly communicate the subjective experience of emotion. The recent development of functional brain imaging technologies such as positron emission tomography (PET) has vastly improved our capabilities for noninvasively studying the distributed neural networks that support the mental operations related to emotion. Compared to techniques involving lesion effects, electroencephalography, and neuropsychological test performance, PET affords superior spatial resolution (≈ 5 mm) and the ability to image multiple structures functioning simultaneously (with temporal resolution of 40 s).

WAYNE C. DREVETS Department of Psychiatry, Washington University School of Medicine, St. Louis, Mo.
MARCUS E. RAICHLE Department of Neurology and Neurological Surgery (Neurology), Division of Radiation Sciences, Mallinckrodt Institute of Radiology, and the McDonnell Center for the Studies of Higher Brain Function, Washington University School of Medicine, St. Louis, Mo.

The relationship between brain function and regional blood flow and metabolism

A tight coupling exists between blood flow, metabolism, and functional activity within the brain (Raichle, 1987). During physiologic activation alterations in neuronal activity are associated with rises in local blood flow and metabolism that follow electrocortical activity by ≈ 0.5 s. By acquiring measurements of regional blood flow or metabolism, PET produces detailed pictures of local neuronal work. The ability to noninvasively image regional blood flow and metabolism provides unprecedented opportunities to map the functional anatomy underlying a variety of brain activities. Functional imaging techniques have been used to identify the anatomical correlates of motor, somatosensory, visual, auditory, linguistic, and purely cognitive tasks using strategies that compare dynamic images obtained during an experimental task with images acquired during a control task (Raichle, 1987, 1990). Such paradigms should eventually prove fruitful in elucidating the functional anatomy of the experience, evaluation, and expression of human emotion.

However, imaging paradigms that employ emotional induction techniques to study emotion in normal subjects suffer from several limitations. The stimulus variables associated with emotional induction techniques are difficult to control, and the magnitude of the emotional response is both variable across subjects and difficult to assess objectively. The stability of the response also varies, as induced emotions fluctuate during the measurement period or may occur too transiently to result in measurable blood-flow changes. Finally, the quality of emotional responses vary, as emotional induction techniques typically produce clusters of emotions rather than isolated emotional states, and such clusters differ between subjects.

One approach for investigating the correlates of emotion is to image subjects suffering from mood or anxiety disorders. In these conditions the emotional state is relatively stable over the duration of the study and is unambiguous in its quality and magnitude. In some cases, well-characterized experimental techniques exist for precipitating or exacerbating these pathologic emotional states (e.g., chemical challenges that induce panic attacks in subjects with panic disorder). Nevertheless, interpreting the results of such comparisons is complicated, since blood-flow or metabolic differences between patients and controls may describe either the pathophysiologic abnormalities associated with the genetic or environmental predisposition to a disorder, or the state-dependent changes that reflect the cognitive, emotional, or behavioral manifestations of the illness. One means of distinguishing the physiologic correlates of symptoms from the pathophysiologic changes that underlie the tendency to develop the emotional disorder involves the differentiation of state-related differences (present only when the subject is symptomatic) from trait differences (present whether the subject is symptomatic or asymptomatic). Since most mood and anxiety disorders follow an episodic course, characterized by periods of illness followed by return to the premorbid baseline, this can be accomplished by scanning subjects in both symptomatic and asymptomatic phases. Complementary information can be obtained by imaging patients before and after treatment, or by imaging normal subjects during experimentally induced emotional states.

PET imaging in pathological emotional states: Major depression

The studies of major depression performed in our PET laboratory illustrate the application of these experimental strategies in investigating emotional disorders. Major depression is characterized by recurrent episodes of the major depressive syndrome that are separated by periods of normal mood. The major depressive syndrome consists of a persistent negative emotional state accompanied by changes in energy, sleep, appetite, psychomotor activity, and thought content. Rather than simply consisting of sadness, the mood change may be described as an emotional "pain" or an inability to experience pleasure (anhedonia). In contrast to a normal reactive low mood, major depression does not simply arise in response to medical or psychosocial stressors, and remits with antidepressant pharmacotherapy (Goodwin and Jamison, 1990).

A variety of genetic, neurochemical, neuroendocrine, neuropsychological, and other biological data suggest but do not define a neurobiologic basis for major depression (Rush et al., 1991). This evidence also indicates that major depression probably encompasses a heterogeneous group of disorders associated with multiple pathophysiologic states (Winokur, 1982; Rice et al., 1987; Schatzberg et al., 1989; Goodwin and Jamison, 1990; Moldin, Reich, and Rice, 1991). Major depression may thus be associated with an assortment of different PET images as different subtypes are considered. For example, two major categories of primary major depression are recognized ("primary" designates that the depressive syndrome antedated other medical or psychiatric illnesses that might be associated with it; Winokur, 1986): bipolar, in which both manic and depressive episodes occur, and unipolar, in which only depressive episodes occur. Unipolar and bipolar depressives appear neurochemically, pharmacologically, and genetically distinct (Rice et al., 1987; Schatzberg et al., 1989; Goodwin and Jamison, 1990; Moldin, Reich, and Rice, 1991). Consistent with these differences, unipolar and bipolar depressives differ in their respective circulatory and metabolic correlates, as described in the following section.

Within primary unipolar depression additional heterogeneity exists, though controversy remains as to subdivisions therein. We employ a classification for unipolar depression that capitalizes on the heritability of affective disorders, and identifies subtypes based upon family history (Winokur, 1982). Of these subtypes, familial pure depressive disease (FPDD, defined as primary unipolar depression in an individual who has a parent, sibling, or offspring with primary unipolar depression but no family history of alcoholism, antisocial personality, or mania) has been most consistently associated with biologic markers (Winokur, 1982).

THE FUNCTIONAL ANATOMY OF FAMILIAL PURE DEPRESSIVE DISEASE Using PET measurements of regional blood flow we distinguished state and trait abnormalities in FPDD by imaging subjects in the depressed and the remitted (asymptomatic) phases. We obtained evidence that flow is increased in the left prefrontal cortex and the left amygdala in FPDD (Drevets, Videen, et al., 1992; see color plates 22 and 23). In the left prefrontal cortex, blood flow was elevated in the depressed but not the remitted phase (see color plate 24). In contrast, left amygdala flow appeared increased in both phases. These data suggest that FPDD is associated with an abnormality in the left prefrontal cortex related to the state of being depressed, and possibly with an abnormality in the amygdala related to the trait of being susceptible to depression, though further assessment is needed to confirm the latter hypothesis.

The affected area in the left prefrontal cortex included the ventrolateral prefrontal cortex, the frontal polar cortex, the medial orbital cortex, and the pregenual portion of the anterior cingulate gyrus (see color plate 22 and figures 76.1). The left ventrolateral prefrontal cortex included the caudal part of Brodmann's area 11 (comparable to Walker's area 12), and areas 45 and 47. The frontal polar region involved Brodmann's area 10, and the cingulate region appeared to involve the pregenual part of area 32. A fourth area of increased activity in the prefrontal cortex was identified in the right frontal polar cortex (see color plate 22). In a subsequent comparison the depressed group was also found to have increased blood flow in the left medial thalamus (see color plate 25) and in the vicinity of the right hippocampal formation or posterior thalamus. In addition, blood flow was decreased in the depressives relative to controls in the medial caudate bilaterally (see color plate 26), the right midtemporal cortex, and the parieto-occipital cortex bilaterally (see color plate 27).

Although other studies of depression have not been restricted to FPDD, some similarities exist between our study and other studies of primary unipolar depressives. Other functional imaging studies of unipolar depression also identified increased metabolism in the ventral prefrontal cortex (Baxter et al., 1985; Uytdenhoef et al., 1983; Buchsbaum et al., 1986) and decreased metabolism in the caudate (Baxter et al., 1985). In contrast to the findings in the ventral prefrontal cortex, decreased flow and metabolism have been reported in the dorsolateral prefrontal cortex by Baxter et al. (1989) and Bench et al. (1992), although the latter study was confounded by medication effects. However, recent reports of decreased frontal lobe size in postmortem and volumetric MRI studies of major depression render these findings of decreased frontal flow and metabolism uninterpretable (Bowen et al., 1989; Coffey et al., 1993). This is because PET data are affected by the inclusion of metabolically inactive cerebrospinal fluid spaces resulting from atrophy (known as the "partial volume effect") such that reduced frontal size could appear as decreased frontal blood flow or metabolism (Mazziotta et al., 1981). Since decreased frontal volume could have reduced the magnitude of our findings in the prefrontal cortex, we are repeating our studies of major depression using methods that correct PET measurements for volumetric differences (Videen et al., 1988).

The areas where we found decreased activity, in the right medial temporal gyrus and the parieto-occipital cortex, were similar to areas where decreased blood flow and metabolism were previously reported in depression (Uytdenhoef et al., 1983; Post et al., 1987). These regions are generally associated with sensory processing or attention (Posner and Petersen, 1990), and the significance of decreased blood flow in such areas is unclear. Dysfunction in these areas during depression had previously been hypothesized based upon neuropsychological test performance abnormalities. For example, right temporal lobe dysfunction was suggested as an explanation for the consistent finding of reduced left-ear auditory sensitivity in depressed subjects (e.g., Yozawitz et al., 1979). In addition, dysfunction in the temporal or occipital neocortex was hypothesized as a cause for sensory memory deficits involving modality-specific failures of preattentive information registration in depression (Sackeim and Steif, 1988). Perhaps in major depression, the neural systems involved in processing exteroceptive information and maintaining alertness to the external environment are inhibited in favor of systems involved in processing emotion or negative thoughts (color plate 27).

Comparison of regional blood flow changes in depression, induced sadness, and language generation

To contrast the functional anatomy of major depression with that of normal sadness, we compared the location of regional abnormalities in the depressed phase of FPDD with blood-flow changes in psychiatrically well subjects as they contemplated sad thoughts or memories to self-induce a sad mood (Drevets, Spitznagel, et al., 1992; Pardo, Pardo, and Raichle, 1993). During induced sadness, blood flow increased in part of the left ventrolateral prefrontal cortex where flow had been elevated in FPDD (figure 76.1).

However, increased flow was also observed in overlapping portions of the left ventrolateral prefrontal cortex in a language generation experiment, implying that activation of the left ventrolateral prefrontal cortex is not specific to the emotional state of the subjects in major depression or induced sadness (figure 76.1). In this language study, subjects were presented nouns (for 150 ms) and asked to generate related verbs (Petersen et al., 1989). Although PET cannot determine whether the same neuronal populations within the ventrolateral prefrontal cortex are involved in each of the three conditions (major depression, induced sadness, and verb generation), it would appear that rather than serving as an emotion-specific area, the left ventrolateral prefrontal cortex may subserve a more general function in cognitive processing. The specificity for each condition may instead be imparted by the distributed networks involving multiple, simultaneously functioning neural structures that are distinctive for each state.

For example, within the prefrontal cortex, areas outside the ventrolateral prefrontal cortex were activated in each of the three conditions. In the language task, blood flow increased the left dorsolateral prefrontal cortex (Petersen et al., 1989). Flow was unchanged in this area in induced sadness and major depression. In contrast, blood flow was increased in the medial orbital

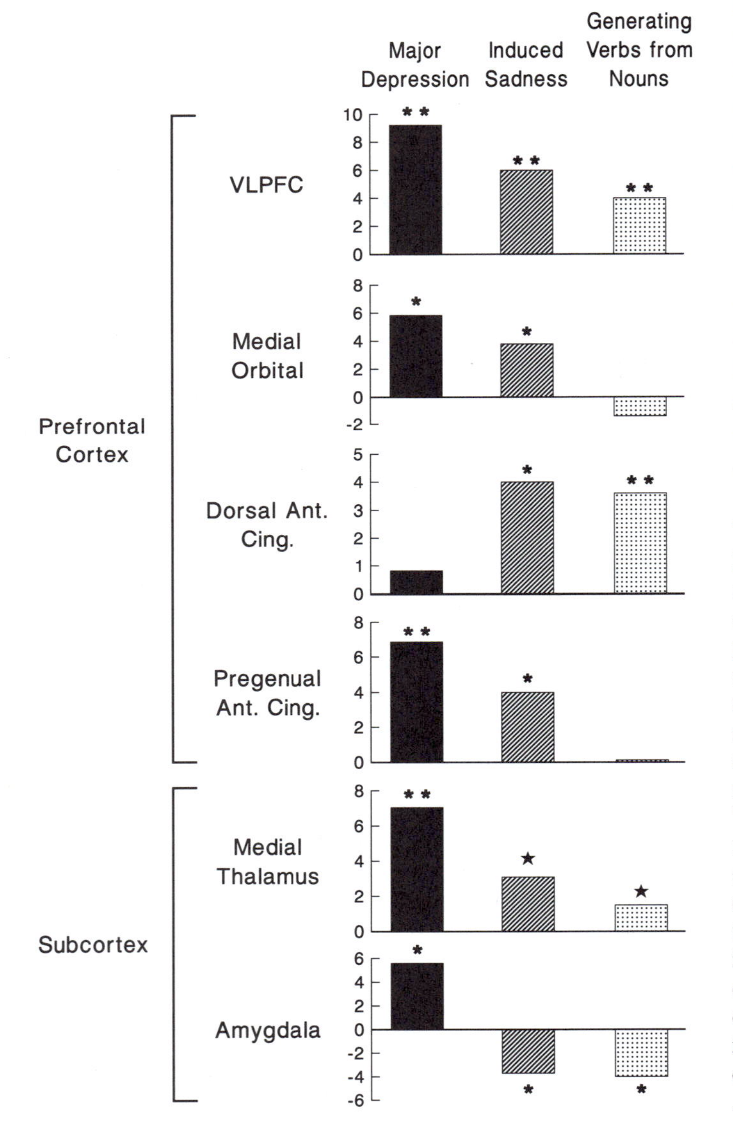

Figure 76.1 Regional blood-flow changes in limbic-thalamo-cortical structures during major depression, induced sadness, and verb generation. Image data are expressed as percent difference in normalized blood flow in the depressed subjects relative to normal subjects (solid bars), in normal subjects during experimentally induced sadness relative to themselves in the resting condition (hatched bars), and in normal subjects generating verbs from related nouns relative to themselves while reading nouns (stippled bars). While the regional data are not directly comparable across the three data sets, they illustrate the general regions that are mutually or uniquely activated (or in the case of the amygdala, deactivated) in these diverse conditions. Such information may provide preliminary insights into the distributed functional neural networks that impart the specificity of each condition. VLPFC, ventrolateral prefrontal cortex; ant. cing., anterior cingulate cortex. The significance of the differences between each experimental group and its respective control group is indicated as follows: ★ = $p < .10$, * = $p < .05$, ** = $p < .01$.

cortex in the depressed phase of FPDD and in the sadness induction task, but not in the verb generation task (figure 76.1). Flow changes in the anterior cingulate cortex also differed between the emotional and verb generation conditions. During verb generation and sadness induction, flow increased in the portion of the anterior cingulate located dorsocaudal to the genu of the corpus callosum (Petersen et al., 1989). Flow was unchanged in this portion of the anterior cingulate in major depression. In contrast, blood flow was increased in the pregenual portion of the anterior cingulate gyrus in both emotional states, but not in the verb generation task (Drevets, Videen, et al., 1992; Drevets et al., in preparation). In the subcortex, all three conditions resulted in increased blood flow in the left medial thalamus. In the amygdala, however, blood flow was increased only in the depressed subjects with FPDD. In contrast, blood flow significantly decreased in the amygdala during both the verb generation and sadness induction tasks (figure 76.1; Drevets, Spitznagel, et al., 1992).

Functional Correlations Increased blood flow in the amygdala and the prefrontal cortex in the depressed phase of FPDD may represent either physiologic concomitants of their roles in mediating the cognitive, emotional, or behavioral symptoms of depression, or disrupted physiology associated with the pathophysiologic defect causing such symptoms. Although it cannot yet be decided which interpretation is correct, determining the state versus trait nature of these abnormalities begins to address this question. Since the abnormality in the left prefrontal cortex appeared to be a state marker in FPDD, it presumably reflected the physiology underlying the expression of at least part of the depressive syndrome. In contrast, if the abnormality in the amygdala is a trait marker, it may represent a more fundamental aspect of the disease process and the vulnerability to major depression.

The Amygdala The amygdala appears to play a major role in assigning affective significance to experiential stimuli by forming associations between stimuli and reinforcement (see chapters 69 and 72). It is conceivable that dysfunction of the amygdala's role in evaluating affective significance could yield a state where negative affective labels are inappropriately assigned to all stimuli, resulting in depressed mood, or where positive affective labels are not assigned to any stimuli, resulting in anhedonia. Our data suggested that increased blood flow in the amygdala reflects a trait abnormality in FPDD, and that the magnitude of this abnormality correlates with the severity of the depressive symptoms (Drevets, Videen, et al., 1992). Conversely, the decrement in amygdala blood flow during antidepressant drug treatment correlated with improvement in depressive symptoms (Drevets et al., 1993).

The Prefrontal Cortex The left prefrontal cortical abnormality included lateral, anterior, and medial components (color plate 22). The lateral component contained much of the left ventrolateral prefrontal cortex, and the medial component involved the medial orbital cortex and the pregenual portion of the anterior cingulate gyrus. All of these regions are heavily interconnected with the ipsilateral amygdala (Amaral and Price, 1984).

These areas of the prefrontal cortex have been postulated to participate in the association of concept with emotion and action (Collins, 1988). In the orbital cortex nearly one-half of all cells demonstrate altered firing rates during the delay period between stimulus and response, and posttrial activity is related to the presence or absence of reward (Rosenkilde, Baver, and Ruster, 1981). Damasio, Tranel, and Damasio, (1990) have reported that humans with destruction of the ventral prefrontal cortex lose the experience and expression of emotion related to concepts that would ordinarily evoke emotion. The prefrontal cortex in general appears to be concerned with the use of short-term, representational memory, such that behavior is guided by representations of a stimulus instead of the stimulus itself (Goldman-Rakic, 1987). In the ventral prefrontal cortex this representational memory may specifically pertain to emotion.

In contrast to the ventral prefrontal (orbital) region, the ventrolateral prefrontal cortex may be more generally involved in representational memory and less specifically involved in emotion. As described above, blood flow increased in the left ventrolateral prefrontal cortex during verb generation as well as during the contemplation of sad thoughts and the depressed phase of FPDD (figure 76.1). A common feature of the verb generation and sadness induction tasks was the process of associating thoughts or emotions with a previous thought. Depressed subjects may also be engaged in a similar associative operation, as they experience incessant negative thoughts concerning personal failure, death, or interpersonal loss. All three conditions (verb generation, contemplation of sad thoughts, and depressive ruminations) involve making associations while information is held in representational memory: semantic associations in the case of the language task, and emotional associations in the case of the negative thoughts in major depression and the sad thoughts task. Increased flow in the left ventrolateral prefrontal cortex in depression may thus reflect a physiologic correlate of depressive ruminations. Consistent with this hypothesis, left frontal EEG activity in depressives correlates with ruminative ideation, but not with depressed mood (Perris and Monakhov, 1979).

Dysfunction of the left ventrolateral prefrontal cortex may also be associated with a perseverative state involving melancholic thoughts or emotions. Monkeys with surgical lesions of this region demonstrate perseverative interference, characterized by difficulty in learning to withhold responses to nonrewarding stimuli, presumably because of interference from a strong, competing response tendency (Iversen and Mishkin, 1970). Likewise humans with lesions of the prefrontal cortex exhibit difficulty shifting intellectual strategies in response to changing demands (i.e., they perseverate in strategies that become inappropriate; Goldman-Rakic, 1987). Depressed patients with FPDD describe their ruminations as intrusive and exceedingly difficult to discontinue. However, the ruminations cease with antidepressant pharmacotherapy (blood flow in the left ventrolateral prefrontal cortex concomitantly decreases during effective antidepressant drug treatment; Drevets et al., 1993). It might be hypothesized that impaired modulation of the ventrolateral prefrontal cortex in patients with FPDD interferes with their ability to shift emotional or cognitive sets, such that they maintain negative thought patterns or moods.

In the medial prefrontal cortex, the anterior cingulate cortex has also been implicated in emotion (LeDoux, 1987). The anterior cingulate appears to participate in the motivational aspects of stimulus coding, as suggested by reinforcement associated with appetitive rewards, affective responses to noxious stimuli, and significance coding (Vogt, 1993). Bilateral cingulate ablation has been associated with akinetic mutism, characterized by the absence of motivation to behave or speak (Damasio, 1983). In contrast, more limited surgical lesions within the anterior cingulate ameliorate severe depressive and anxiety syndromes (Balantine et al., 1987).

The pregenual portion of the anterior cingulate cortex may be most specifically involved in emotion. Functional heterogeneity within the cingulate was originally suggested by the results of electrical stimulation, which elicited arousal and heightened attention, simple motor movements, or emotional changes such as fear, anguish, sadness, or euphoria depending upon which portion of the cingulate was stimulated (Talairach et al., 1973; Laitinen, 1979; Damasio and Van Hoesen, 1983). In PET studies of language and attention, blood flow increases have been identified in the portion of the anterior cingulate caudal and dorsal to the genu of the corpus callosum (Petersen et al., 1989; Pardo et al., 1990; Posner and Petersen, 1990; Corbetta et al., 1991). The role of this area of the cingulate was thought be related to attention and selection for action (Posner and Petersen, 1990; Corbetta et al., 1991), which is consistent with the observations of increased blood flow in this area in the verb generation and sadness induction tasks, as both tasks involved attentive selection of specific cognitive operations (figure 76.1). In contrast, in the pregenual portion of the anterior cingulate cortex, blood flow increased in major depression and induced sadness but not in the language or attentional tasks (figure 76.1).

The anatomical circuitry of FPDD

Local changes in blood flow and metabolism generally reflect the energy utilization associated with synaptic activity, rather than the electrophysiologic activity of cell bodies (Raichle, 1987). Since synaptic activity may originate from cell bodies distantly located from observed changes as well as from local intercellular communication, consideration of the neural circuitry involving the site of a blood-flow difference becomes imperative in interpreting PET image data. In our study of FPDD, the findings of increased blood flow in the left prefrontal cortex, the left amygdala, and the left medial thalamus, and decreased blood flow in the caudate, appear to corroborate a variety of indirect evidence indicating that circuits involving the prefrontal cortex and related parts of the striatum, pallidum, and thalamus are involved in the pathophysiology of depression (Swerdlow and Koob, 1987; McHugh, 1989).

Our findings suggest specifically that two interconnected circuits are involved in the pathophysiology of FPDD: a limbic-thalamo-cortical circuit involving the amygdala, the mediodorsal nucleus of the thalamus (in the medial thalamus), and the ventrolateral and medial prefrontal cortex; and a limbic-striatal-pallidal-thalamic (LSPT) circuit involving related parts of the striatum and the ventral pallidum as well as the components of the other circuit (figure 76.2). The first of these circuits can be conceptualized as an excitatory triangular circuit in which the amygdala and the prefrontal cortical regions are interconnected by excitatory projections with each other and with the mediodorsal nucleus (Amaral and Price, 1984; Kuroda and Price, 1991). Through these connections the amygdala is in a position both to directly activate the prefrontal cortex and to modulate the reciprocal interaction between the prefrontal cortex and the mediodorsal nucleus.

The LSPT circuit constitutes a disinhibitory side-loop between the amygdala or prefrontal cortex and the mediodorsal nuclues. The amygdala and the prefrontal cortex send excitatory projections to overlapping parts of the ventromedial caudate and nucleus accumbens (Fuller, Russchen, and Price, 1987). This part of the striatum sends an inhibitory projection to the ventral pallidum (Graybiel, 1990) which in turn sends GABA-ergic, inhibitory fibers to the mediodorsal nucleus (Kuroda and Price, 1991). Because the pallidal neurons have relatively high spontaneous firing rates (DeLong, 1972), activity in the prefrontal cortex or amygdala that activates the striatum and in turn inhibits the ventral pallidum would release the mediodorsal nucleus from an inhibitory pallidal influence. Thus, if the amygdala is abnormally active in depression, it can potentially produce an episode of abnormal activity in the prefrontal cortex and mediodorsal nucleus both directly and through the striatum and pallidum.

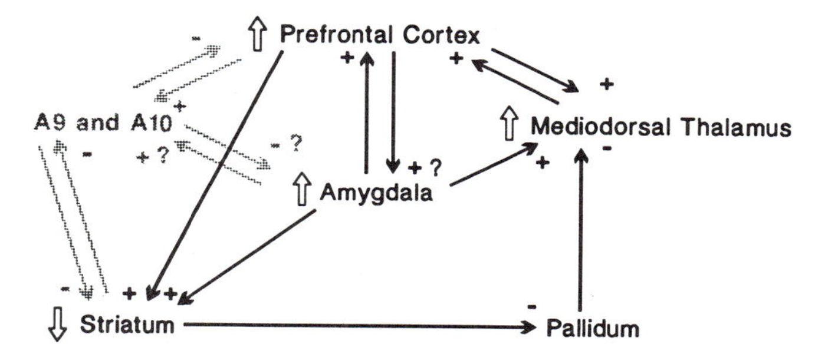

FIGURE 76.2 Neuroanatomical circuits hypothesized to participate in the functional anatomy of FPDD. Regions containing blood-flow differences have adjacent open arrows that indicate the direction of change in flow in the depressives relative to the controls. The regions' monosynaptic connections with each other are illustrated (closed arrows), with (+) indicating excitatory and (−) indicating inhibitory projections, and with (?) indicating where experimental evidence is limited. The portions of the prefrontal cortex referred to involve primarily the ventrolateral and medial prefrontal cortex. The parts of the striatum under consideration are the ventral medial caudate and nucleus accumbens, which particularly project to the ventral pallidum (Nauta and Domesick, 1984). The major dopaminergic projections from the substantia nigra (A9) and the ventral tegmental area (A10) to these structures are illustrated with the shaded tone. (Reprinted from Drevets, Videen, et al., 1992, by permission.)

One theory regarding the pathophysiology of major depression holds that stress-induced long-term potentiative phenomena related to kindling or behavioral sensitization occur in the amygdala and lead to recurrent spontaneous episodes of depression (Post, 1992). Since these phenomena represent permanent changes in neuronal sensitivity, this hypothesis appears compatible with our observation that activity in the amygdala is abnormally increased in remitted subjects with FPDD who are no longer taking antidepressant drugs (Drevets, Videen, et al., 1992). We more recently reported that antidepressant drug treatment profoundly decreases blood flow in the amygdala (Drevets et al., 1993). Our data are thus intriguing in light of Post's hypothesis (1992) that long-term prophylactic treatment may be indicated to prevent abnormal amygdala activity from reemerging and inducing recurrences of major depression.

In a potentially complementary hypothesis regarding the pathophysiology of major depression, Swerdlow and Koob (1987) proposed a neural model of depression involving the LSPT circuit that is also compatible with our data, and may additionally explain the observation of decreased blood flow in the caudate. They hypothesized that decreased dopaminergic transmission into the ventromedial striatum (Nauta and Domesick, 1984) enhances reverberatory activity between the prefrontal cortex, amygdala, and mediodorsal nucleus, which results in the perseveration of a fixed pattern of cortical activity manifested by the emotional, cognitive, and motor processes of depression. Dopaminergic projections from the substantia nigra and ventral tegmental area provide an important inhibitory or modulatory input into the striatum, amygdala, and prefrontal cortex (figure 76.2; Graybiel, 1990; Rowlands and Roberts, 1980; Thierry et al., 1988). Thus, the effect of mesostriatal dopamine (DA) deficiency would be to increase striatal output, thereby inhibiting the pallidum and disinhibiting the mediodorsal nucleus (Wooten and Collins, 1981). This, together with the decrease in direct DA effects in the amygdala and prefrontal cortex, would increase activity in the limbic-thalamo-cortical circuit, potentially yielding the blood-flow increases we found (figure 76.2). Moreover, if mesostriatal dopaminergic transmission is reduced, decreased synaptic activity at striatal DA receptors could appear as decreased blood flow in the caudate (e.g., Brown and Wolfson, 1978; see color plate 26).

The limbic-striatal-pallidal-thalamic circuit and other depressive subtypes

Lesions involving the parts of the prefrontal cortex that participate in these circuits (i.e., tumors or infarctions) and diseases of the basal ganglia (e.g., Parkinson's or Huntington's disease) are associated with elevated rates of major depression (Mayeux, 1983; Jeste, Lohr, and Goodwin, 1988; Starkstein and Robinson, 1989). These associations suggest that dysfunction at multiple points within the LSPT circuit may give rise to the major depressive syndrome. Moreover, because these conditions affect this system in different ways, it may also be suggested that imbalances within these circuits, rather than overall increased or decreased synaptic activity within a particular structure, may produce the depressive syndrome.

For example, in Parkinson's disease increased striatal-pallidal transmission results from the loss of the inhibitory effects of nigrostriatal dopaminergic neurons, whereas in Huntington's disease, decreased striatal-pallidal transmission results from the degeneration of striatal neurons (Rougemont et al., 1984; Young et al., 1986). Both diseases are associated with a four- to tenfold increase in the risk for developing major depression relative to other similarly disabling illnesses. The fact that imbalances arising at multiple points within these circuits may produce mood disturbances resembles the case for movement disorders, which result from disrupted modulation at various points in the motor circuit (which involves the primary and supplementary motor cortices, the dorsolateral striatum, the dorsal pallidum, and the motor thalamus) (DeLong 1972).

PET Findings in Bipolar Depression If diverse types of modulatory dysfunction within the same circuit can result in mood disturbances, then different functional anatomical correlates reflected by distinct PET images could be associated with the major depressive syndrome. For example, in contrast to the PET findings in primary unipolar depressives, bipolar depressives generally display decreased blood flow and metabolism in the prefrontal cortex and normal caudate metabolism (Baxter et al., 1985; Buchsbaum et al., 1986; Cohen et al., 1989; Delvenne et al., 1990). In our own investigations of bipolar depressives, blood flow was decreased in the prelimbic portion of the anterior cingulate gyrus in the ventromedial prefrontal cortex, and the rostroventral thalamus, areas where flow had been abnormally increased in the unipolar depressives with FPDD (Drevets and Raichle, in preparation). Moreover, discriminant analysis of the covariance of blood flow in the left prefrontal cortex, amygdala, caudate, and medial thalamus sensitively distinguished FPDD subjects from both control and bipolar subjects (Drevets, Spitznagel, et al., 1992). The observation that in the ventral prefrontal cortex both increased (in unipolar depression) and decreased (in bipolar depression) blood flow or metabolism are associated with major depression is consistent with the hypothesis that abnormal modulation within the LSPT circuit, rather than increased or decreased activity in any single structure, may produce the major depressive syndrome.

Distinct pathophysiologies within the LSPT circuit could also account for the clinical differences between unipolar and bipolar depression. This would again reflect the case in the motor circuit, where opposite effects upon modulation at a given limb of the circuit yield distinct movement disorders. For example, in Parkinson's disease striatal-pallidal transmission is increased; the resulting abnormal movements consist of tremor, bradykinesia, and cogwheel-rigidity; and the course of the induced depression is always unipolar. In contrast, in Huntington's disease striatal-pallidal transmission is decreased; the movements are choreiform or choreoathetoid; and the course of the secondary mood disorder is often bipolar (Mayeux, 1983).

PET Findings in Other Depressive Subtypes Imaging studies performed in other major depressive subtypes have also emphasized metabolic abnormalities in limbic-cortical regions. In depressives with seasonal affective disorder, Cohen et al. (1992) reported increased medial orbital metabolism and greater metabolism in the left compared with the right midprefrontal cortex, similar to our findings in FPDD (Drevets, Videen, et al., 1992). In contrast, subjects with induced depressive syndromes associated with partial-complex seizure disorder or basal ganglia disorders have demonstrated decreased blood flow in the ventral prefrontal cortex compared with nondepressed controls with the same disorders, similar to the findings in bipolar depression (Mayberg et al., 1990, 1992; Bromfield et al., 1992). Mayberg et al. (1990) also reported that caudate metabolism was decreased in depressed relative to nondepressed patients with Parkinson's disease, and that metabolism in the thalamus was decreased in depressed relative to nondepressed patients with Huntington's disease (Mayberg et al., 1992).

PET imaging in anxiety disorders

Anxiety states may also involve the limbic prefrontal cortex and the amygdala. Reivich, Gur, and Alavi (1983) reported a positive correlation between anxiety in normal subjects and metabolism in the posterior orbital and middle frontal cortices. In subjects with panic disorder imaged between panic attacks (when anticipatory anxiety predominates), blood flow and metabolism were elevated in the right ventral prefrontal cortex (Drevets et al., submitted; Nordahl et al., 1989). Right prefrontal activity decreased in these subjects during antianxiety pharmacotherapy (Drevets et al., submitted; Nordahl et al., 1991).

In the right amygdala, flow increased during anxiety induced in normals by the threat of a painful electric stimulus (Raichle, 1990). In contrast, blood-flow changes in the amygdala have not been observed in panic disorder (Reiman et al., 1986, 1989; Nordahl et al., 1990; Benkelfat et al., 1991).

Limbic-Striatal-Pallidal-Thalamic Circuit Involvement in Obsessive-Compulsive Disorder Obsessive-compulsive disorder (OCD) is an anxiety disorder characterized by persistent intrusive thoughts (obsessions) and behaviors (compulsions) that are perceived as illogical but are maintained because attempts to resist them result in overwhelming anxiety. Similarities between OCD and major depression exist, as the negative ruminations of depression possess an intrusive, obsessive nature, and the pharmacologic and neurosurgical interventions that ameliorate OCD also effectively treat major depression (Rassmussen and Eisen, 1990).

Similar to PET findings in primary unipolar depression, PET studies of OCD have generally found elevated metabolism in the orbital and medial frontal cortices (for review see Insel, 1992). As in the case of depression, activity in this region decreases during antidepressant drug treatment. In contrast to PET studies of unipolar depression, Baxter et al. (1987) found that caudate metabolism is increased in OCD. Baxter et al. (1992) also demonstrated that caudate metabolism decreases during either pharmacologic or behavioral treatment. In a study of secondary OCD, La Plane and colleagues (1989) imaged subjects with necrosis of the globus pallidus associated with induced obsessions. Metabolism was decreased in the prefrontal cortex in these patients. This resembles the aforementioned literature in depression, where primary unipolar depression was associated with increased orbital blood flow or metabolism, but secondary depression induced by basal ganglia disorders (involving the striatum) was associated with decreased orbital metabolism.

The phenomenologic, clinical, and physiologic similarities between OCD and unipolar depression, which are otherwise distinct in their course, prognosis, genetics, and neurochemical concomitants, suggest that these disorders may be associated with different pathophysiologies that overlap in their involvement of the LSPT circuit (Insel et al., 1992). The LSPT circuit in general has been implicated in the organization of internally guided behavior toward a reward, switching of response strategies, and habit formation (e.g., Stern and Passingham, 1990). The orbital cortex in particular appears to participate in correcting behavioral responses that become inappropriate as reinforcement contingencies change (Rolls, this volume). Both unipolar depression and OCD are characterized by nonrewarding cognitive and behavioral sets that are maintained because the ability to switch sets is impaired. OCD is characterized by persistent, intrusive, nonrewarding thoughts and behaviors with inability to switch to goal-oriented sets; and depression involves persistent painful thoughts and inability to convert them to positive or rewarding thoughts. Perhaps dysfunction within the LSPT circuit, or more specifically in the orbital cortex, accounts for the inability to inhibit unrewarded thoughts and behaviors in these disorders.

ACKNOWLEDGMENTS The authors' research is supported by NIH grants HL 13851 and NS 06833, NIMH Scientist Development Award for Clinicians MH 00928-02 (Drevets), the MacArthur Foundation, and the Charles A. Dana Foundation.

REFERENCES

AMARAL, D. G., and J. L. PRICE, 1984. Amygdalocortical projections in the monkey. *J. Comp. Neurol.* 230:465–496.

BALLANTINE, JR., H. T., A. J. BOUCKOMS, E. K. THOMAS, and I. E. GIRIUNAS, 1987. Treatment of psychiatric illness by stereotactic cingulotomy. *Biol. Psychiatry* 22:807–819.

BAXTER, L. R., M. E. PHELPS, J. C. MAZZIOTTA, J. M. SCHWARTZ, R. H. GERNER, C. E. SELIN, and R. M. SUMIDA, 1985. Cerebral metabolic rates for glucose in mood disorders. *Arch. Gen. Psychiatry* 42:441.

BAXTER, L. R., M. E. PHELPS, J. C. MAZZIOTTA, B. H. GUZE, J. M. SCHWARTZ, and C. E. SELIN, 1987. Local cerebral glucose metabolic rates in obsessive-compulsive disorder: A comparison with rates in unipolar depression and in normal controls. *Arch. Gen. Psychiatry* 44:211–218.

BAXTER, L. R., J. M. SCHWARTZ, M. E. PHELPS, J. C. MAZZIOTTA, B. H. GUZE, C. E. SELIN, R. H. GERNER, and R. M. SUMIDA, 1989. Reduction of prefrontal cortex glucose metabolism common to three types of depression. *Arch. Gen. Psychiatry* 46:243.

BAXTER, L. R., J. M. SCHWARTZ, K. S. BERGMAN, M. P. SZUBA, B. H. GUZE, J. C. MAZZIOTTA, A. ALAZRAKI, C. E. SELIN, H. K. FERNG, P. MUNFORD, and M. E. PHELPS, 1992 . Caudate glucose metabolic rate changes with both drug and behavior therapy for obsessive-compulsive disorder. *Arch. Gen. Psychiatry* 49:681–689.

BENCH, C. J., K. J. FRISTON, R. G. BROWN, L. C. SCOTT, R. S. J. FRACKOWIAK, and R. J. DOLAN, 1992. The anatomy of melancholia: Focal abnormalities of cerebral blood flow in major depression. *Psychol. Med.* 22:607–615.

BENKELFAT, C., J. BRADWEJN, E. MEYER, M. ELLENBOGEN, S. MILOT, A. GJEDDE, and A. EVANS, 1991. Neuroanatomical correlates of CCK_4-induced panic attacks in normal volunteers. *Neuropsychopharmacology Abstr.* 30:151.

BOWEN, D. M., A. NAJLERAHIM, A. W. PROCTER, P. T. FRANCIS, and E. MURPHY, 1989. Circumscribed changes of the cerebral cortex in neuropsychiatric disorders of later life. *Proc. Natl. Acad. Sci. U.S.A.* 86:9504j–9508.

BROMFIELD, E. B., L. ALTSHULER, D. B. LEIDERMAN, M. BALISH, T. A. KETTER, O. DEVINSKY, R. M. POST, and W. H. THEODORE, 1992. Cerebral metabolism and depression in patients with complex partial seizures. *Arch. Neurol.* 49:617–623.

BROWN, L. L., and L. I. WOLFSON, 1978. Apomorphine increases glucose utilization in the substantia nigra, subthalamic and corpus striatum of rat. *Brain Res.* 140:188–193.

BUCHSBAUM, M. S., J. WU, L. E. DELISI, H. HOLCOMB, R. KESSLER, J. JOHNSON, A. C. KING, E. HAZLETT, K. LANGSTON, and R. M. POST, 1986. Frontal cortex and basal ganglia metabolic rates assessed by positron emission tomography with [^{18}F]2-deoxyglucose in affective illness. *J. Affective Disord.* 10:137–152.

COFFEY, C. E., W. E. WILKINSON, R. D. WEINER, I. A. PARASHOS, W. T. DJANG, M. C. WEBB, G. S. FIGIEL, and C. E. SPRITZER, 1993. Quantitative cerebral anatomy in depression—a controlled magnetic resonance imaging study. *Arch. Gen. Psychiatry* 50:7–16

COHEN, R. M., M. GROSS, T. E. NORDAHL, W. E. SEMPLE, N. ROSENTHAL, 1992. Preliminary data on the metabolic brain pattern of patients with winter seasonal affective disorder. *Arch. Gen. Psychiatry* 49:545–552.

COHEN, R. M., W. E. SEMPLE, M. D. GROSS, T. E. NORDAHL, A. C. KING, D. PICKAR, and R. M. POST, 1989. Evidence for common alterations in cerebral glucose metabolism in major affective disorders and schizophrenia. *Neuropsychopharmacology* 2:241–254.

COLLINS, R. C., 1988. Prefrontal-limbic systems: Evolving clinical concepts. In *Advances in Contemporary Neurology*, F. Plum, ed. Philadelphia: Davis, pp. 185–204.

CORBETTA, M., F. M. MIEZIN, S. DOBMEYER, G. L. SHULMAN, and S. E. PETERSON, 1991. Selective and divided attention during visual discriminations of shape, color, and speed: Functional anatomy by positron emission tomography. *J. Neurosci.* 11:2383–2402.

DAMASIO, A. R., D. TRANEL, H. DAMASIO, 1990. Individuals with sociopathic behavior caused by frontal damage fail to respond autonomically to social stimuli. *Behav. Brain Res.* 41:81–94.

DAMASIO, A. R., G. W. VAN HOESEN, 1983. Emotional disturbance associated with focal lesions of the limbic frontal lobe. In *Neuropyschology of Human Emotion*, K. Heilman and P. Satz, eds. New York: Guilford, pp. 85–110.

DELONG, M. R., 1972. Activity of basal ganglia neurons during movement. *Brain Res.* 40:127–135.

DELVENNE, V., F. DELECLUSE, P. P. HUBAIN, A. SCHOUTENS, V. DEMAERTELAER, and J. MENDLEWICZ, 1990. Regional cerebral blood flow in patients with affective disorders. *Br. J. Psychiatry* 157:359–365.

DREVETS, W. C., and M. E. RAICHLE, 1992. Neuroanatomical circuits in depression: Implications for treatment mechanisms. *Psychopharmacol. Bull.* 28:261–274.

DREVETS, W. C., E. M. REIMAN, A. M. MACLEOD, E. ROBINS, and M. E. RAICHLE, submitted. Cerebral blood flow changes in panic disorder during alprazolam treatment

DREVETS, W. C., E. L. SPITZNAGEL, A. K. MACLEOD, and M. E. RAICHLE, 1992. Discriminatory capability of PET measurements of regional blood flow in familial pure depressive disease. Abstract presented at Society of Neuroscience Annual Meeting, Anaheim, Calif.

DREVETS, W. C., T. O. VIDEEN, J. L. PRICE, S. H. PRESKORN, S. T. CARMICHAEL, and M. E. RAICHLE, 1992. A functional anatomical study of unipolar depression. *J. Neurosci.* 12:3628–3641.

Drevets, W., T. Videen, A. MacLeod, and M. Raichle, 1993. Regional blood flow changes during antidepressant treatment. *Neurosci. Abstr.* 19:7.

Fuller, T. A., F. T. Russchen, and J. L. Price, 1987. Sources of presumptive glutaminergic/aspartergic afferents to the rat ventral striatopallidal region. *J. Comp. Neurol.* 258:317.

Goldman-Rakic, P. S., 1987. Circuitry of primate prefrontal cortex and regulation of behavior by representational memory. In *Handbook of Physiology*. Sec. 1, *The Nervous System*. Vol. 5, *Higher Functions of the Brain*, part 1, V. B. Mountcastle, ed. Bethesda, Md.: American Physiological Society, pp. 373–417.

Goodwin, F. K., and K. R. Jamison, eds., 1990. *Manic-Depressive Illness*. New York: Oxford University Press.

Graybiel, A. M., 1990. Neurotransmitters and neuromodulators in the basal ganglia. *Trends Neurosci.* 13:244–254.

Insel, T. R., 1992. Toward a neuroanatomy of obsessive-compulsive disorder. *Arch. Gen. Psychiatry* 49:739–744.

Iverson, S. D., and M. Mishkin, 1970. Perserverative interference in monkeys following selective lesions of the inferior prefrontal convexity. *Exp. Brain Res.* 11:376–386.

Jeste, D. V., J. B. Lohr, and F. K. Goodwin, 1988. Neuroanatomical studies of major affective disorders: A review and suggestions for further research. *Br. J. Psychiatry* 153: 444–459.

Kuroda, M., and J. L. Price, 1991. Synaptic organization of projections from basal forebrain structures to the mediodorsal nucleus of the rat. *J. Comp. Neurol.* 303:513.

Laitinen, L. V., 1979. Emotional responses to subcortical electrical stimulation in psychiatric patients. *Clin. Neurol. Neurosurg.* 81–3:148–157.

LaPlane, D., M. Levasseur, B. Pillon, B. Dubois, M. Baulac, B. Mazoyer, S. Tran Dinh, G. Sette, F. Danze, and J. C. Baron, 1989. Obsessive-compulsive and other behavioural changes with bilateral basal ganglia lesions. *Brain* 112:699–725.

LeDoux, J. E., 1987. Emotion. In *Handbook of Physiology*. Sec. 1, *The Nervous System*. Vol. 5, *Higher Functions of the Brain*, part 1, V. B. Mountcastle, ed. Bethesda, Md.: American Physiological Society, pp. 419–459.

Mayberg, H. S., S. E. Starkstein, B. Sadzot, T. Preziosi, P. L. Andrezejewski, R. F. Dannals, H. N. Wagner, and R. G. Robinson, 1990. Selective hypometabolism in the inferior frontal lobe in depressed patients with Parkinson's disease. *Ann. Neurol.* 28:57–64.

Mayberg, H. S., S. E. Starkstein, C. E. Peyser, J. Brandt, R. F. Dannals, and S. E. Folstein, 1992. Paralimbic frontal lobe hypometabolism in depression associated with Huntington's disease. *Neurology* 42:1791–1797.

Mayeux, R., 1983. Emotional changes associated with basal ganglia disorders. In *Neuropsychology of Human Emotion*, K. M. Heilman and P. Satz, eds. New York: Guilford, pp. 141–164.

Mazziotta, J. C., M. E. Phelps, D. Plummer, and D. E. Kuhl, 1981. Quantitation in positron emission computed tomography: 5. Physical-Anatomical Effects. *J. Comput. Assist. Tomogr.* 5:734–743.

McHugh, P. R., 1989. The neuropsychiatry of basal ganglia disorders. *Neuropsychiatry Neuropsychol. Behav. Neurol.* 2: 239–247.

Moldin, S. O., T. Reich, and J. P. Rice, 1991. Current perspectives on the genetics of unipolar depression. *Behav. Genet.* 21:211–242.

Nauta, W. J. H., and V. Domesick, 1984. Afferent and efferent relationships of the basal ganglia. In *Function of the Basal Ganglia*. CIBA Foundation Symposium 107. London, Pitman, pp. 3–29.

Nordahl, T. E., W. E. Semple, M. Gross, T. A. Mellman, M. B. Stein, P. Goyer, A. C. King, T. W. Uhde, and R. M. Cohen, 1990. Cerebral glucose metabolic differences in patients with panic disorder. *Neuropsychopharmacology* 3: 261–272.

Pardo, J. V., P. J. Pardo, K. W. Janer, and M. E. Raichle, 1990. The anterior cingulate cortex mediates processing selection in the Stroop attentional conflict paradigm. *Proc. Natl. Acad. Sci. U.S.A.* 87:256–259.

Pardo, J. V., P. J. Pardo, and M. E. Raichle, 1993. Neural correlates of self-induced dysphoria. *Am. J. Psychiatry* 150:713–719.

Perris, C., and K. Monakhov, 1979. Depressive symptomatology and systemic structural analysis of the EEG. In *Hemisphere Asymmetries of Function in Psychopathology*, J. Gruzelier and P. Flor-Henry, eds. Amsterdam: Elsevier, pp. 223–236.

Petersen, S. E., P. T. Fox, M. I. Posner, M. A. Mintun, and M. E. Raichle, 1989. Positron emission tomographic studies of the processing of single words. *J. Cognitive Neurosci.* 1:153–170.

Posner, M. I., and S. E. Petersen, 1990. The attention system of the human brain. *Annu. Rev. Neurosci.* 13:25–42.

Post, R. M., 1992. Transduction of psychosocial stress into the neurobiology of recurrent affective disorder. *Am. J. Psychiatry* 149:999–1010.

Post, R. M., L. E. DeLisi, H. H. Holcomb, T. W. Uhde, R. Cohen, and M. S. Buchsbaum, 1987. Glucose utilization in the temporal cortex of affectively ill patients: positron emission tomography. *Biol. Pychiatry* 22:545–553.

Post, R. M., S. R. B. Weiss, T. Ketter, P. J. Pazzaglia, and K. Denicoff, 1991. Impact of affective illness on gene expression: rationale for long-term prophylaxis. *Eur. J. Neuropsychopharmacology* 1:214–216.

Price, J. L., F. T. Russchen, and D. G. Amaral, 1987. The limbic region: II. The amygdaloid complex. In *Handbook of Chemical Neuroanatomy*, vol. 5, A. Bjorklund, T. Hokflet, and L. W. Swanson, eds. New York: Elsevier, pp. 279–388.

Raichle, M. E., 1987. Circulatory and metabolic correlates of brain function in normal humans. In *Handbook of Physiology*. Sec. 1, *The Nervous System*. Vol. 5, *Higher Functions of the Brain*, part 2, V. B. Mountcastle, ed. Bethesda, Md.: American Physiological Society, pp. 643–674.

Raichle, M. E., 1990. Exploring the mind with dynamic imaging. *Semin. Neurosci.* 2:307–315.

Rassmussen, S. A., and J. L. Eisen, 1990. Epidemiology and clinical features of obsessive-compulsive disorders. In *Obsessive-Compulsive Disorders: Theory and Management*, M. A. Jenike, L. Baer, and W. E. Minichiello, eds. Chicago: Year Book Medical Publishers, pp. 10–27.

Reiman, E. M., M. E. Raichle, E. Robins, F. K. Butler, P. Herscovitch, P. Fox, and J. Perlmutter, 1986. The application of positron emission tomography to the study of panic disorder. *Am. J. Psychiatry* 143:469–477.

Reiman, E. M., M. E. Raichle, E. Robins, M. A. Mintun, M. J. Fusselman, P. T. Fox, J. L. Price, and K. Hackman, 1989. Neuroanatomical correlates of a lactate-induced panic attack. *Arch. Gen. Psychiatry* 46:493–500.

Reivich, E. M., R. Gur, and A. Alavi, 1983. Positron emission tomographic studies of sensory stimuli, cognitive processes and anxiety. *Hum. Neurobiol.* 2:25–33.

Rice, J., T. Reich, N. Andreasen, J. Endicott, M. Van Eerdewegh, R. Fishman, R. Herschfeld, and G. Klearman, 1987. The familial transmission of bipolar illness. *Arch. Gen. Psychiatry* 44:441–447.

Rosenkilde, C. E., R. H. Bauer, and J. M. Fuster, 1981. Single cell activity in ventral prefrontal cortex of behaving monkeys. *Brain Res.* 209:375–394.

Rougemont, D., J. C. Baron, P. Collard, P. Bustany, D. Comar, and Y. Agid, 1984. Local cerebral glucose utilisation in treated and untreated patients with Parkinson's disease. *J. Neurol. Neurosurg. Psychiatry* 47:824–830.

Rowlands, G. J., and P. J. Roberts, 1980. Specific calcium-dependent release of endogenous glutamate from rat striatum is reduced by destruction of the cortico-striatal tract. *Exp. Brain Res.* 39:239–240.

Rush, A. J., J. W. Cain, J. Raese, R. S. Stewart, D. A. Waller, and J. R. Debus, 1991. Neurobiological bases for psychiatric disorders. In *Comprehensive Neurology*, R. N. Rosenberg, ed. New York: Raven, pp. 555–603.

Sackheim, H. A., and B. L. Steif, 1988. Neurophysiology of depression and mania. In *Depression and Mania*, A. Georgotas and R. Cancro, eds. New York: Elsevier, pp. 265–289.

Schatzberg, A. F., J. A. Samson, K. L. Bloomingdale, P. J. Orsulak, B. Gerson, P. P. Kizuka, J. O. Cole, and J. J. Schildkraut, 1989. Toward a biochemical classification of depressive disorders. *Arch. Gen. Psychiatry* 46: 260–268.

Selemon, L. D., and P. S. Goldman-Rakic, 1985. Longitudinal topography and interdigitation of corticostriatal projection in the rhesus monkey. *J. Neurosci.* 5:776–794.

Starkstein, S. E., and R. G. Robinson, 1989. Affective disorders and cerebral vascular disease. *Br. J. Psychiatry* 154: 170–182.

Stern, C. F., and R. E. Passingham, 1990. The nucleus accumbens and the organization of behavioral sequences in monkeys (*macaca fascicularis*). *Soc. Neurosci. Abst.* 15: 1244.

Swerdlow, N. R., and G. F. Koob, 1987. Dopamine, schizophrenia, mania and depression: Toward a unified hypothesis of cortico-striato-pallido-thalamic function. *Behav. Brain Sci.* 10:197–245.

Talairach, J., J. Bancaud, S. Geier, M. Bordas-Ferrer, A. Bonis, G. Szikla, and M. Rusu, 1973. The cingulate gyrus and human behaviour. *Electroencephalogr. Clin. Neurophysiol.* 34:45–52.

Thierry, A. M., J. Mantz, C. Milla, and J. Glowinski, 1988. Influence of the mesocortical/prefrontal dopamine neurons on their target cells. In *The Mesocorticolimbic Dopamine System*, P. W. Kalivas and C. B. Nemeroff, eds. New York: New York Academy of Sciences, pp. 101–111.

Uytdenhoef, P., P. Portelange, J. Jacquy, G. Charles, P. Linkowski, and J. Mendlewicz, 1983. Regional cerebral flow and lateralized hemispheric dysfunction in depression. *Br. J. Psychiatry* 143:128–132.

Videen, T. O., J. S. Perlmutter, M. A. Mintun, and M. E. Raichle, 1988. Regional correction of positron emission tomography data for the effects of cerebral atrophy. *J. Cereb. Blood Flow Metab.* 8:662–670.

Vogt, B. A., 1993. The structural organization of cingulate cortex: Area, neurons and somatodendritic receptors. In *The Neurobiology of Cingulate Cortex and Limbic Thalamus*, B. A. Vogt and M. Gabriel, eds. Boston: Birkhauser.

Vogt, B. A., and M. Gabriel, 1993. *The Neurobiology of Cingulate Cortex and Limbic Thalamus*. Boston: Birkhauser.

Winokur, G., 1982. The development and validity of familial subtypes in primary unipolar depression. *Pharmacopsychiatry* 15:142–146.

Winokur, G., 1986. Unipolar depression. In *The Medical Basis of Psychiatry*, G. Winokur and P. Clayton, eds. Philadelphia: Saunders, pp. 60–79.

Wooten, G. F., and R. C. Collins, 1981. Metabolic effects of unilateral lesion of the substantia nigra. *J. Neurosci.* 1:285–291.

Young, A. B., J. B. Penney, S. Starosta-Rubinstein, D. S. Markel, S. Berent, B. Giordani, R. Ehrenkaufer, D. Jewett, and R. Hichwa, 1986. PET scan investigations of Huntington's disease: Cerebral metabolic correlates of neurological features and functional decline. *Ann. Neurol.* 20:296–303.

Yozawitz, A., G. Bruder, S. Sutton, L. Sharpe, B. Gurland, J. Fleiss, and L. Costa, 1979. Dichotic perception: Evidence for right hemisphere dysfunction in affective psychosis. *Br. J. Psychiatry* 135:224–237.

77 A Model of the Limbic System and Basal Ganglia: Applications to Anxiety and Schizophrenia

JEFFREY A. GRAY

ABSTRACT The chapter presents a model of the behavioral functions and information processing discharged jointly by the limbic system (especially the hippocampal formation and amygdala) and the basal ganglia (both the dorsal and ventral striatal systems). In general terms, the limbic system plus basal ganglia act as a mechanism for the attainment of goals. The sensory aspects of this overall goal-direction function (recognition of goals and evaluation of the outcomes of action) are dealt with in the limbic system; the motor aspects (establishment and execution of motor programs) in the basal ganglia. The model is applied to an understanding of the neuropsychology of anxiety and schizophrenia, especially the positive symptoms of the latter.

The model of the limbic system plus basal ganglia (LSBG) presented here occupies three closely interrelated levels: behavioral, neural, and cognitive. Previous descriptions of portions of the model (e.g., Gray, 1982a, 1982b, in press; Gray, Feldon, et al., 1991; Gray, Hemsley, et al., 1991; Gray and Rawlins, 1986) have preserved the separation between these three levels of analysis; here, however, less attention is paid to these distinctions. The data on which the model is based are drawn from a wide variety of empirical results and have been summarized and reviewed elsewhere (Gray, 1977, 1982a, 1987; Gray and McNaughton, 1983; Gray, Feldon, et al., 1991). The anatomical regions and interconnections to which the model refers are set out schematically in figures 77.1 and 77.2. These diagrams include the structures familiarly clustered together under the terms *limbic system* and *basal ganglia*. The anatomical justification for these groupings, and for the separation of the limbic system and basal ganglia from other subsystems of the brain, is controversial. Resolution of this controversy, in my view, will require a clear understanding of the specific functions served by these systems (cf. the visual system, defined as it is at least as much by its function in the processing of visual information as by its anatomical interconnectivity). This chapter represents an attempt to contribute to such a functional understanding of the LSBG.

JEFFREY A. GRAY Department of Psychology, Institute of Psychiatry, London, England

The function of the limbic system and basal ganglia: A hypothesis

Let us start with a question of great generality: What function does the limbic system–basal ganglia (LSBG) serve? The answer proposed is that the LSBG is a mechanism for the attainment of goals. The sensory aspects of this overall function of goal direction (recognition of goals and evaluation of the outcomes of action) are dealt with in the limbic system; the motor aspects (establishment and execution of goal-directed motor programs) in the basal ganglia. To carry out the goal-direction function, a number of subsidiary functions must be executed and coordinated. A likely list of such subfunctions, and of the major regions of the LSBG most concerned with them, is as follows.

GOAL SETTING First, goals have to be recognized as goals. The final biological goals of action (positive reinforcers, or rewards) are, of course, innately determined (food, water, etc). An animal cannot, however, wait until one of these materializes and provides innately recognizable sensory stimulation. It must get to the place (in space and time) where such a goal is to be found; and, to do that, it needs to establish a series of linked subgoals that will permit it to achieve this approach behavior. Setting up such a series of linked

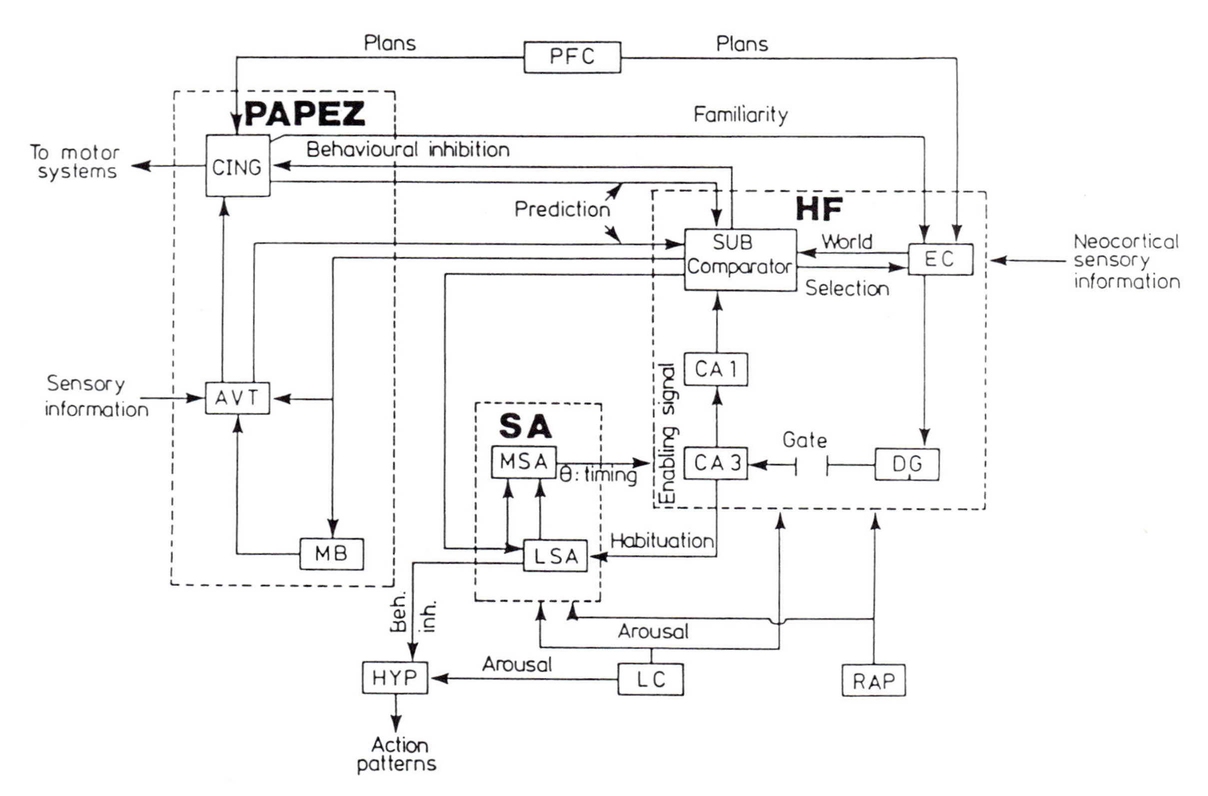

FIGURE 77.1 The septohippocampal system: The three major building blocks are shown in heavy print: HF, the hippocampal formation, made up of the entorhinal cortex, EC, the dentate gyrus, DG, CA3, CA1, and the subicular area, SUB; SA, the septal area, containing the medial and lateral septal areas, MSA and LSA; and the Papez circuit, which receives projections from and returns them to the subicular area via the mammillary bodies, MB, anteroventral thalamus, AVT, and cingulate cortex, CING. Other structures shown are the hypothalamus, HYP, the locus coeruleus, LC, the raphe nuclei, RAP, and the prefrontal cortex, PFC. Arrows show direction of projection; the projection from SUB to MSA lacks anatomical confirmation. Words in lower case show postulated functions; beh. inhib., behavioral inhibition. (From Gray, 1982b.)

subgoals depends upon the process described in animal learning theory as establishing a goal gradient. This process consists of the formation of Pavlovian associations between initially neutral stimuli, or cues, and innate positive reinforcers—the cues now becoming secondary positive reinforcers—followed by the formation of further associations between other cues and those already established as secondary reinforcers (Deutsch, 1964; Gray, 1975). In addition to learning about the spatiotemporal location of desired goals in this way, an animal must also learn about undesirable outcomes (negative reinforcers, or punishments), such as pain or proximity to a predator. This is achieved by a similar process of repeated primary and secondary Pavlovian conditioning, leading to the formation of linked series of secondary negative reinforcers. There is much evidence (LeDoux, 1987; Rolls, 1990) that a key role is played in this process of cue-reinforcer learning, for both positive and negative reinforcement, by neurons in the amygdala.

GOAL ATTAINMENT Once a cue-reinforcer association has been formed (and this can happen very quickly, often in only a single trial), the animal is in a position to do something about the cue: approach it (where the term *approach* includes any behavior that increases proximity in space and time to its occurrence) if it is a secondary positive reinforcer; or avoid it (performing any behavior that decreases proximity in space and time to its occurrence) if it is a secondary negative reinforcer. However, the complexities of the natural environment are such that, normally, a whole chain of linked secondary reinforcers will be required for effective action. The information concerning this chain,

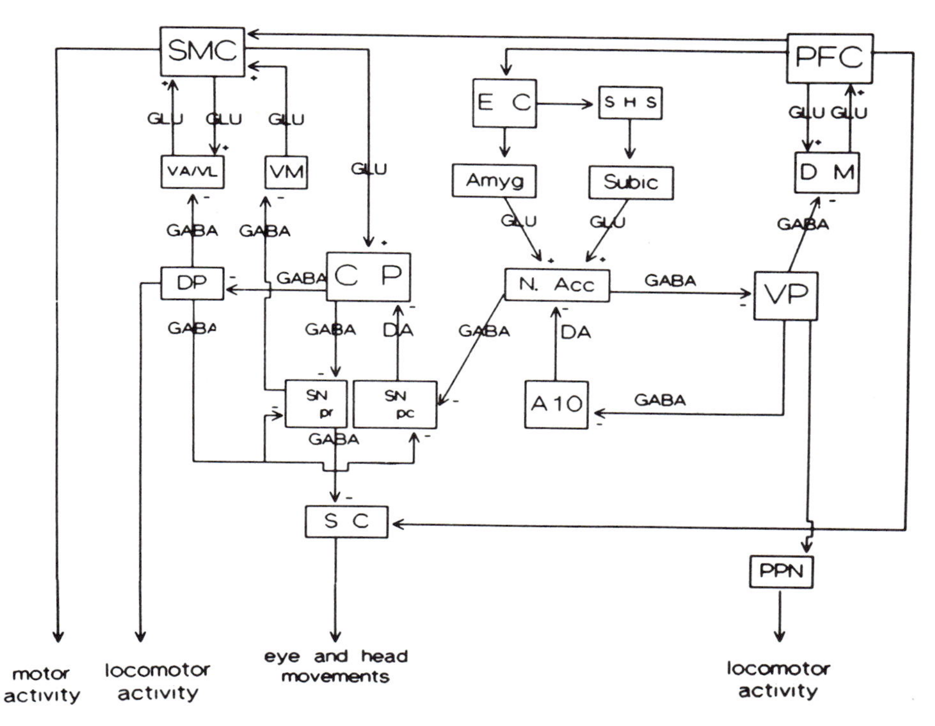

FIGURE 77.2 The basal ganglia and their connections with the limbic system. Structures: SMC, sensorimotor cortex; PFC, prefrontal cortex; EC, entorhinal cortex; SHS, septo-hippocampal system; Subic, subicular area; Amyg, amygdala; VA/VL, nucleus (n.) ventralis anterior and ventralis lateralis thalami; VM, n. ventralis medialis thalami; DM, dorsalis medialis thalami; DP, dorsal pallidum; VP, ventral pallidum; CP, caudate-putamen; N. Acc, n. accumbens; SNpr, substantia nigra, pars reticulata; SNpc, substantia nigra, pars compacta; A 10, n. A 10 in ventral tegmental area; SC, superior colliculus; PPN, penduculopontine nucleus. Transmitters: GLU, glutamate; DA, dopamine; GABA, gamma-aminobutyric acid. (From Gray, Feldon, et al., 1991.)

therefore, must be transmitted from the amygdala, where it is initially established, to motor systems in the basal ganglia. This step appears to be accomplished by the projection from the amygdala to the ventral striatum, or nucleus (n.) accumbens (Rolls and Williams, 1987; Gray, Feldon, et al., 1991). The latter structure has been recognized for some time as a key node in the interface between the limbic system and the basal ganglia (e.g., Mogenson and Nielsen, 1984). There is evidence from single-unit recording studies that accumbal neurons do indeed receive information about associations between cues and positive reinforcers (Rolls and Williams, 1987), as well as evidence for accumbal release of dopamine in close association with rewarded behavior (Fibiger and Phillips, 1988; Hernandez and Hoebel, 1988; Young, Joseph, and Gray, 1992). We have in addition recently demonstrated that cues associated with foot shock elicit conditioned dopamine release in n. accumbens (Young, Joseph, and Gray, 1993); thus n. accumbens receives information about secondary negative as well as secondary positive reinforcement.

That neurons receive a certain class of information does not indicate what they do with it. We have proposed that n. accumbens uses information about cue-reinforcer associations to establish and run the sequences of motor steps that are required to reach specific goals; but that the detailed sensorimotor content of each step is contained in the dorsal striatal system, which links the caudate putamen to sensory and motor cortices, to nuclei ventralis anterior and ventralis lateralis of the thalamus, and to the dorsal pallidum (Gray, Feldon, et al., 1991). To use a computer analogy, n. accumbens holds a list of steps making up a given motor program and is able to switch through the list in an appropriate order; but, in order to retrieve the specific content of each step, it must call up the appropriate subroutine by way of its connections to the dorsal striatal system. Drawing upon previous suggestions (Oades, 1985; Swerdlow and Koob, 1987), Gray,

Feldon, and colleagues (1991) further proposed that switching from one step to the next in a motor program is achieved by the intra-accumbal release of dopamine at terminals projecting from n. A10 in the ventral tegmental area. Swerdlow and Koob (1987) have presented a detailed analysis of the way in which the circuitry linking n. accumbens to the limbic cortex (prefrontal and cingulate areas), to the dorsomedial nucleus of the thalamus, and to the ventral pallidum, would allow activation of the A10 dopaminergic fibers by outputs from n. accumbens itself to achieve this effect (see figure 77.3).

What about the detailed sensorimotor content of the motor steps, as executed by the dorsal striatal system? Rolls and Williams (1987) have used the anatomical organization of this system, together with a general theory of random associative networks, to outline a mechanism by which assemblies of cells with the appropriate connections to motor outputs could be selected. In brief, these authors consider sets of Spiny I striatal cells (the major, GABAergic, output from the caudate putamen), which, because of the particular pattern of connections that they possess, would receive inputs from both (1) neurons that respond to environmental cues associated with positive reinforcers, and (2) other neurons that fire when the animal makes a movement that happens to affect the occurrence of this reinforcer. They show how such cells might initially respond only to the conjunction of cue and movement, but could come eventually to be activated by the cue alone, and so to participate in the production of the appropriate movement, given the cue. If we assume that neurons in set (1) receive information from n. accumbens (indirectly, e.g., by way of the dorsomedial nucleus of the thalamus and the prefrontal cortex), Rolls and Williams's proposal provides a mechanism by which the list of motor steps held in n. accumbens can be translated into a sequence of detailed sensorimotor steps in the caudate putamen and its associated thalamic, pallidal, and cortical connections.

Goal Monitoring In these ways, then, motor programs directed toward goals (primary and secondary positive reinforcers) can be established and executed. The next requirement is that the outcomes of each program be monitored, in order to ensure that the intended goals are indeed achieved. The model supposes that this monitoring or *comparator* process is mediated by the septohippocampal system and the associated Papez circuit, that is, the loop from the subiculum (the major output station for the septohippocampal system) via the mammillary bodies, anteroventral thalamus,

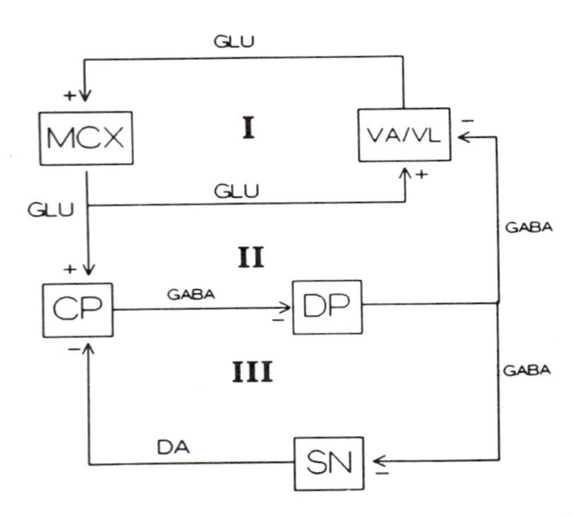

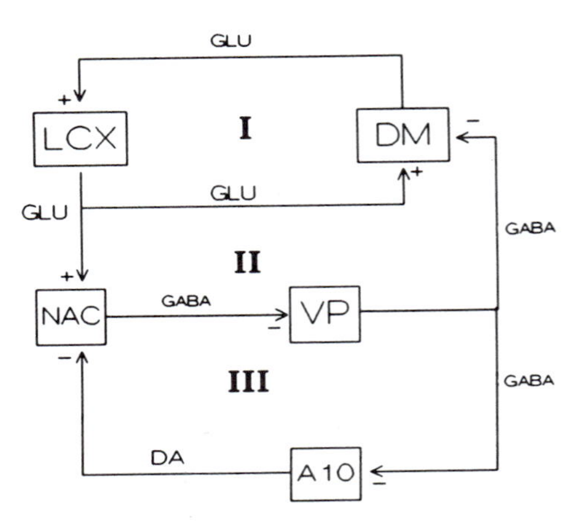

Figure 77.3 *Left* The caudate motor system: nonlimbic cortico-striato-pallido-thalamic-midbrain circuitry. MCX, motor and sensorimotor cortex; VA/VL, ventral anterior and ventrolateral thalamic nuclei; CP, caudate putamen (dorsal striatum); DP, dorsal pallidum; SN, substantia nigra.

Right The accumbens motor system: limbic cortico-striato-pallido-thalamic-midbrain circuitry. LCX, limbic cortex, including prefrontal and cingulate areas; DM, dorsomedial thalamic nucleus; NAC, nucleus accumbens (ventral striatum); VP, ventral pallidum; A 10, dopaminergic nucleus A 10 in the ventral tegmental area.

GLU, GABA and DA, the neurotransmitters glutamate, gamma-aminobutyric acid, and dopamine. +−, excitation and inhibition; I, II, III, feedback loops, the first two positive, the third negative. (Based on Swerdlow and Koob, 1987.)

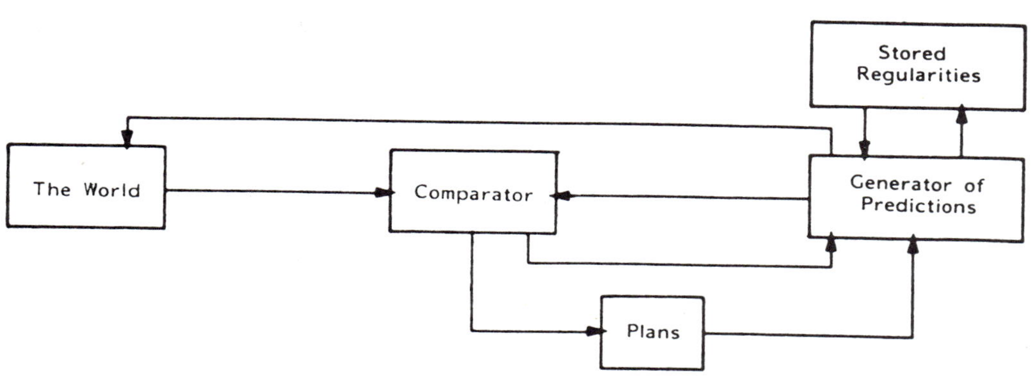

FIGURE 77.4 Information processing required for the comparator function of the septohippocampal system

and cingulate cortex back to the subiculum. The information-processing functions required for such a comparator to work are illustrated in figure 77.4.

Information about the current state of the animal's world is first analyzed in the sensory systems of the neocortex and then fed via the temporal lobe (more specifically, the entorhinal cortex) to the hippocampal formation. The information received by the hippocampal formation in this way has already been heavily processed, being certainly multimodal and probably highly abstract. O'Keefe and Nadel's (1978) influential hypothesis that this information consists of descriptions of spatial locations within a mapping system is supported by much evidence; but it is clear that the hippocampal formation also handles other, nonspatial kinds of information (Gray, 1982a; Rawlins, Lyford, and Seferiades, 1991). Whatever their exact form, these descriptions of the current state of the world must be compared to predicted states of the world. As shown in figure 77.4, the making of such predictions requires the following kinds of data: (1) the last verified current state of the world; (2) the next step in the current motor program; (3) access to stored regularities describing associations between states of the world resembling the last current one and other succeeding states of the world (i.e., stimulus-stimulus associations of the kind set up by Pavlovian conditioning); and (4) access to stored regularities describing associations between the current step in the motor program and succeeding states of the world (i.e., response-stimulus associations of the kind set up by instrumental conditioning).

The model supposes that the circuit responsible for the making of predictions is the Papez loop (subicular area–mammillary bodies–anteroventral thalamus–cingulate cortex–subicular area), and that the actual comparison process is accomplished by subicular neurons. Thus, the last verified current state of the world is coded by subicular neurons at the outset of a cycle around the Papez circuit; a description of the next step in the current motor program is supplied by way of the projection from frontal to cingulate cortex, the frontal cortex itself receiving information about the list of steps in a motor program from n. accumbens via the dorsomedial nucleus of the thalamus; stimulus-stimulus and response-stimulus regularities are stored in the temporal lobe and accessed by way of the projections from the subicular area and the frontal cortex, respectively, to the entorhinal cortex. Finally, timing of cycles around this circuitry is accomplished by the hippocampal theta rhythm (approximately 6–12 Hz, resulting in a quantized timing unit of about 0.1 s).

At the end of each such predictive cycle, the subicular neurons responsible for the comparison process make a match-mismatch decision with regard to (1) the input representing the current state of the world derived from neocortical sensory analysis, and (2) the input representing the predicted state of the world derived from the Papez predictive circuit. A match decision is followed by initiation of the next predictive cycle coupled with the next analysis of the current state of the world. This analysis is biased toward features that will enter the next prediction, a biasing achieved by feedback from the subicular area to the entorhinal cortex. If, however, there is a mismatch decision, or if the predicted state of the world includes negative reinforcers, the current motor program is interrupted.

Match decisions must be communicated to the motor programming system in order to confirm that the last intended step in the current program has been successfully completed. This is accomplished by way of the projection from the subiculum to n. accumbens. This projection terminates upon accumbal GABAergic Spiny I output neurons, which also receive dopaminergic inputs from A10 (Totterdell and Smith, 1989), in the same general caudomedial region where fibers from the amygdala reach n. accumbens (Phillipson and Griffith, 1985). The model therefore supposes that a match output from the subiculum terminates the current step in the motor program, permitting the amygdaloid projection, in conjunction with dopaminergic afferents, to switch in the accumbal output neurons corresponding to the next step. Weiner (1991) has proposed a detailed mechanism, utilizing the accumbens projection to the substantia nigra and the circuitry illustrated in figure 77.3 (taken from Swerdlow and Koob, 1987), by which such switching between steps within the accumbens can be transmitted to the caudate system coding for the detailed sensorimotor content of each step.

In this way, then, the model attempts to give a general account of how the LSBG is able to establish, run, and monitor goal-directed motor programs, although there are a number of features of the model not touched upon here (see Gray, 1982a, 1982b; Gray and Rawlins, 1986; Gray, Feldon, et al., 1991; Gray, Hemsley, et al., 1991).

Applications of the model

The model has developed gradually in a series of attempts to understand the brain functions that underlie particular psychological phenomena. In the second half of this chapter I shall briefly review its application to two of these phenomena.

THE NEUROPSYCHOLOGY OF ANXIETY The hypothesis that the septohippocampal system and its associated Papez circuit discharge a general comparator function, comparing actual and predicted states of the world, was first developed in the context of an account of anxiety. This account was based in the first instance on an analysis of the behavioral effects in experimental animals of antianxiety, or anxiolytic, drugs, including principally benzodiazepines, barbiturates, and ethanol. A review of some 400 studies of this kind in species ranging from goldfish to chimpanzees led to the conclusions summarized in figure 77.5 (Gray, 1977). This figure proposes that anxiety reflects activity in a behavioral inhibition system (BIS) that responds to the threat of punishment, the omission of anticipated reward ("frustrative nonreward"; Amsel, 1992), or extreme novelty by the inhibition of ongoing behavior, increased readiness for vigorous action, and increased attention to environmental cues; and that anxiolytic drugs exert their effects by reducing activity in the BIS. This proposal—or other similar formulations—has been widely accepted. Much more controversial has been the answer to the question, What brain system(s) mediate anxiety?

To a large extent the answer given to this question depends upon the starting point chosen for an experimental analysis of aversively motivated behavior in animals. If one starts from the anxiolytic drugs, one is led to the septohippocampal system by the considerable degree of similarity that exists between, on the one hand, the profile of behavioral change produced

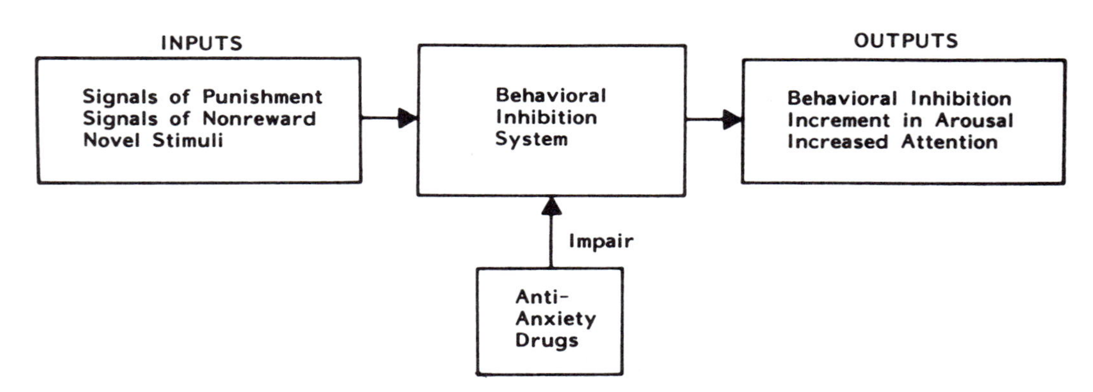

FIGURE 77.5 The behavioral inhibition system (BIS) as defined by its inputs and outputs.

by these agents and, on the other, the profile seen after lesions to either the septal area or the hippocampal formation (Gray, 1982a, 1982b; Gray and McNaughton, 1983). Alternative starting points lead to different destinations. A favorite has been the fight-or-flight behavior elicited by unconditioned punishment and frustrative nonreward; since this type of behavior can also be elicited by electrical stimulation of a range of sites in the amygdala, medial hypothalamus, and central gray of the midbrain, it is naturally a system comprising these regions that is seen as the neural substrate of anxiety (Panksepp, 1982). A second alternative lies in the formation of Pavlovian conditioned fear responses; LeDoux (1987) has marshaled a considerable body of evidence, much of it from his own laboratory, implicating the amygdala in conditioning of this kind.

There are strong reasons why these different flowers cannot be allowed to bloom together—at least, not all under the same name, whether it be *anxiety* or something else. Chief among these reasons is that there is a double dissociation between responses to unconditioned punishment and nonreward, on the one hand, and responses to conditioned stimuli signaling these events, on the other. This dissociation is observed behaviorally (unconditioned stimuli typically provoke vigorous locomotor activity, whereas conditioned stimuli typically provoke behavioral inhibition or "freezing"), pharmacologically (anxiolytics do not generally reduce responses to painful stimuli, except at high doses, and analgesics do not generally reduce responses to conditioned stimuli associated with pain), and neurally (e.g., fight-or-flight behavior is reduced by amygdaloid lesions, and responses to conditioned frustrative stimuli are reduced by septal and hippocampal lesions); for review, see Gray (1982a). A second, though less firmly established, relevant distinction is between fear conditioning (that is, the establishment of a Pavlovian conditioned response using a painful unconditioned stimulus) and the responses elicited by a conditioned fear stimulus after conditioning. Anxiolytic drugs appear to have little if any effect upon the process of formation of the conditioned response, but they weaken the responses elicited by the conditioned stimulus (Gray, 1977, 1982a). Thus, a choice has to be made: One can identify anxiety with the processes antagonized by the anxiolytic drugs, in which case it is the septohippocampal system that is nodal; one can identify anxiety with the process of fear conditioning, in which case it is the amygdala; or one can identify it with fight-or-flight behavior, in which case it is a system involving the amygdala, medial hypothalamus, and central gray—Graeff's (1987) "behavioral aversive system," or Gray's (1987) "fight/flight system" (FFS).

This, of course, is a choice of names only (Gray, 1991). Substantively, there is no reason why all these systems should not coexist, each performing its own function, as supposed, for example, by Graeff (1987) and Gray (1987). In the long run, relatively neutral names, or indeed abbreviations such as BIS or FFS, are to be preferred to terms like *anxiety*, with their burden of surplus meaning. However, if a choice has to be made, there is good reason to equate the term *anxiety* to *activity in the BIS*. Anxiolytics, such as diazepam or chlordiazepoxide, when administered to relevant patient groups, reliably elicit lower ratings and self-ratings of "anxiety" (Rickels, 1978); thus, in the construction of an animal model of anxiety, these compounds benefit from greater face validity than other possible starting points. On this assumption, therefore, the neural substrate of anxiety becomes the one that mediates activity in the BIS. (A complementary suggestion is that activity in the FFS underlies the human emotions of panic and/or rage. This proposal is consistent with evidence that anxiety and panic are susceptible to different drug treatments; Klein, 1981; Graeff, 1987; Gray, 1987.)

As already indicated, lesion evidence implicates the septohippocampal system as a nodal point in the substrate of the behavioral inhibition system. Given the behavioral model of the input-output relations that define the BIS (figure 77.5), the neural systems that instantiate it must be capable of detecting threat (i.e., stimuli associated with punishment or with nonreward) or novelty (i.e., the occurrence of an unexpected stimulus or the nonoccurrence of one that is expected). This role is consistent with the predictive and comparator functions allocated to the septohippocampal system and its associated Papez circuit, and indeed motivated this allocation in the first place (Gray, 1982a, b). Information about threat can be seen as reaching the septohippocampal system in two mutually compatible ways. First, in agreement with LeDoux's (1987) analysis, the formation of the cue-reinforcement associations that underlie fear conditioning takes place in the amygdala; this information can be relayed to the hippocampal formation via the entorhinal cortex (Braak and Braak, 1992). Second, information about predicted outcomes of current motor programs, including the possibility that these will include aversive events, can be relayed to the hippocampal formation via the prefrontal-entorhinal connection (the prefrontal cortex itself receiving information about motor programs via the dorsomedial thalamus). (Important roles are also played by the ascending noradrenergic and serotonergic systems, but there is no space to consider them here; see Gray, 1982a, b.)

Once threat or novelty is detected, the outputs of the BIS (figure 77.5) must be operated. The behavioral inhibition output proper requires inhibition of any ongoing motor programs. The evidence from lesion studies that the hippocampal formation plays a major role in such inhibition of motor programs is very strong (Gray and McNaughton, 1983; Jarrard et al., 1986), although the neural substrate of the "freezing" posture that often accompanies such behavioral inhibition appears to be principally mediated by lower structures in the central gray of the midbrain (Fanselow, 1991), perhaps under amygdaloid control (LeDoux, 1987). It is still unclear, however, which pathway mediates this hippocampal control over motor programs. The projection from the subicular area to n. accumbens appears not to be involved, since section of this pathway (Rawlins et al., 1989) fails to reproduce the increase in resistance to extinction of rewarded running that is characteristic of large hippocampal lesions (Gray and McNaughton, 1983; Jarrard et al., 1986). The most likely alternative is the projection from the subicular region to the cingulate cortex (Gray, 1982a; Gabriel, Sparenborg, and Stolar, 1987). If this hypothesis is correct, the hippocampal formation would be able to facilitate the continuation of motor programs by way of its output to n. accumbens or, alternatively, to interrupt them by way of its output to the cingulate cortex.

The attentional output of the BIS requires increased attention to environmental cues, and especially those that are novel or associated with the threat that has brought the ongoing motor program to a halt. This can be achieved via two complementary routes. First, neocortical sensory analysis can be influenced by way of the subicular projection to the entorhinal cortex and from there to the sensory cortices. Second, the output to n. accumbens can be utilized to influence accumbal projections to two regions controlling exploratory behavior: the pedunculopontine nucleus, part of the mesencephalic motor region concerned with exploratory locomotion (Yang and Mogenson, 1987); and the superior colliculus (Williams and Faull, 1988), a structure known to be of importance in the control of head and eye movements and visual attention (Dean and Redgrave, 1984; Wurtz and Albano, 1980).

Finally, the arousal output of the BIS provides a general facilitation of motor behavior, so that whatever motor program is engaged following the interruption (including the original, interrupted one) occurs with greater-than-normal vigor. This output appears to be discharged by the ascending noradrenergic and serotonergic pathways. In general, activation of these pathways appears to increase the capacity of the organs they innervate to process other neural messages arriving at the same time via more specific, point-to-point afferents (e.g., Segal, 1977). Thus, the monoaminergic inputs to the hippocampal formation increase threat-related cognitive processing while simultaneously increasing the readiness for prompt and vigorous motor behavior by way of parallel projections to, for example, n. accumbens and the caudate-putamen. In this way, these pathways act as a general alarm system, as proposed by Redmond (1979) for the noradrenergic pathway originating in the locus coeruleus.

This model of the neuropsychology of anxiety was based entirely on experiments with animals. How does it fare as an account of the symptoms observed in the major anxiety disorders, such as agoraphobia, social phobia, and anxiety state (Gray, 1982a)? Such symptoms fall into three classes: autonomic (e.g., respiratory and cardiac changes, sweating), which will not be considered here (see Redmond, 1979; Graeff, 1987); behavioral (chiefly, phobic avoidance); and cognitive (e.g., worry, obsessional rumination). The behavioral symptoms of anxiety may be analyzed simply as reflecting the behavioral inhibition output of the BIS, giving rise to phobic avoidance in response to threat-related stimuli. The cognitive symptoms may be analyzed as a search for such stimuli (experienced as worry or, in the extreme, obsessional rumination). This would be mediated by way of the hippocampal outputs (via the subicular area) to the entorhinal cortex, and so to neocortical sensory systems; and to n. accumbens, and so to exploratory systems (mesencephalic locomotor region, superior colliculus), as outlined above. In addition, however, it is necessary to postulate further mechanisms to permit, at the human level, the control of such search processes by symbolic threats formulated linguistically (e.g., the threat that one may fail an examination). This mechanism can plausibly (Gray, 1982a) be found in the pathways linking cortical language areas to the prefrontal cortex and thence both to motor programming circuits in the basal ganglia and to the hippocampal formation via the entorhinal cortex.

THE NEUROPSYCHOLOGY OF SCHIZOPHRENIA The analysis of pathological anxiety in the preceding section treats this condition as overactivity in a normally organized brain. In contrast, it is now generally agreed that schizophrenia reflects a structural *dis*organization in the brain, though one of as-yet-uncertain etiology. The neuropathology of the schizophrenic brain has been widely described; it involves both loss and abnormalities of packaging of neurons, especially in the temporal and frontal neocortex and in the hippocampal formation and amygdala (for references, see Gray, Feldon, et al., 1991). This neuropathology is difficult to reconcile with the leading neurochemical hypothesis of schizophrenia, namely, that it reflects overactivity in one or another (probably the mesolimbic) ascending dopaminergic pathway. This hypothesis is based upon the drugs (indirect dopamine agonists) that give rise to or exacerbate psychotic symptoms, and those (dopamine receptor blockers) that are able to reduce these symptoms; but schizophrenic neuropathology does not usually extend to either the cell-body region (A10, in the ventral tegmental area) or the principal terminal region (n. accumbens) of the mesolimbic dopaminergic pathway. Some pathology is found in terminal regions (in parts of the frontal and temporal neocortex) of the mesocortical dopaminergic pathway, which also originates in A10—but not as a prominent feature of the schizophrenic brain.

In an effort to combine these neuropathological and neuropharmacological data, we have proposed (Gray, Feldon, et al., 1991) that the structural basis of schizophrenia is to be found in an abnormality in the connections between the limbic forebrain (especially the hippocampal formation, via the subicular area) and the basal ganglia (especially n. accumbens). Neurochemically, it is suggested that this structural abnormality is equivalent to an increase in dopaminergic transmission in n. accumbens (as we have seen, the subiculo-accumbens projection terminates in the same region, and probably on the same neurons, as the projection from n. A10; Totterdell and Smith, 1989). Psychologically, given the general model of limbic–basal ganglia interactions outlined above, an interruption in the subiculo-accumbens projection would have the effect that steps in a motor program should fail to receive the confirmatory messages signaling the occurrence of the expected outcome of each step. By the same token, outcomes that would have been expected to occur, given normal information processing, should appear to be novel and so provoke continuing exploratory behavior. Hemsley (1987) has shown how such a failure to utilize past regularities of experience in the control of current behavior and perception could account for the bizarre positive symptoms (Crow, 1980) of schizophrenia, such as delusional beliefs or the capturing of attention by apparently trivial stimuli. Experimental investigations of this account, using tests of selective attention based upon animal learning theory—latent inhibition (Lubow, 1989) and Kamin's (1968) blocking effect—with rats, normal human subjects, and acute and chronic schizophrenics, have provided generally supportive results (for references, see Gray, Feldon, et al., 1991; and, more recently, Gray, Hemsley, and Gray, 1992; Gray et al., 1992; Jones, Gray, and Hemsley, 1992; Young, Joseph, and Gray, 1993). (It may be possible to develop a similar, and complementary, account based upon disruption in the projection to n. accumbens from the amygdala, but this remains to be done; Gray, Feldon, et al., 1991.)

An alternative hypothesis (Frith, 1987) to account for the symptoms of schizophrenia calls upon the same general machinery but proposes that the abnormal connection lies between the prefrontal cortex and the septohippocampal comparator system, that is, in the prefrontal connections to the cingulate and/or entorhinal cortices (figure 77.1). According to the general model developed above, these projections provide information about intended motor programs to the comparator system. Thus an interruption in their functioning would have the consequence that outcomes of such programs would appear to be unexpected and unintended. Frith (1987, 1992) shows in detail how such an abnormality in information processing could give rise to such positive psychotic symptoms as hallucinations, thought insertion, and delusions of alien control. This hypothesis is consistent with the proposal that a disruption in the subiculo-accumbens projection forms the neural substrate of schizophrenia. The pathology observed in the schizophrenic hippocampal formation and parahippocampal gyrus may indicate impaired output from prefrontal to entorhinal cortex (as may the pathology in the frontal cortex itself), impaired output from subiculum to n. accumbens, or both. Another relevant model is that of Weinberger (1987). In line with evidence for lowered functional activity, detected in neuroimaging studies, in the schizophrenic prefrontal cortex, Weinberger sees the primary neural basis of schizophrenia as lying in an *underactive* mesocortical dopaminergic innervation of the dorsolateral prefrontal cortex, thought to lead (as shown in experiments in the rat; Pycock, Kerwin, and Carter, 1980) to increased mesolimbic dopaminergic activity. The pathways responsible for such compensatory changes in dopamine release in different terminal regions of the mesocortical and mesolimbic projections are at present unknown. One possibility is that they include frontotemporal connections, followed by the same subiculo-accumbens projection emphasized in the model of Gray, Feldon, et al. (1991); in that case, Weinberger's (1987) hypothesis is consistent with both this model and the circuitry proposed by Frith (1987).

Conclusion

The model of the functions of the limbic system and basal ganglia presented above is one of wide generality that is undoubtedly capable of application in contexts other than those of anxiety and schizophrenia (see, e.g., Gray, 1993b; Gray and Rawlins, 1986). One particularly interesting possibility is that the model may throw light on the nature of the contents of consciousness, although profound theoretical and possibly philosophical problems have first to be resolved (Gray, 1993a, submitted); both anxiety and acute schizophrenic disturbance, of course, have profound effects upon conscious experience. At present, the model is hampered by being couched in verbal terms, which rob it of precision and leave open the possibility that it contains internal contradictions. However, we have recently made a start upon the construction of a neural-network computer simulation of part of the model (Schmajuk, Lam, and Gray, in preparation). This simulation appears to be able to account for many of the detailed features of a behavioral phenomenon—latent inhibition—that is critical for the application of the model to acute schizophrenia (Gray, Feldon, et al., 1991). Thus, if it achieves nothing else, the model illustrates the possibility in principle of constructing an integrated theory of brain function that straddles structural, physiological, computational, behavioral, and experiential levels.

REFERENCES

AMSEL, A., 1992. *Frustration Theory*. Cambridge: Cambridge University Press.

BRAAK, H., and E. BRAAK, 1992. The human entorhinal cortex: normal morphology and lamina-specific pathology in various diseases. *Neurosci. Res.* 15:6–31.

CROW, T. J., 1980. Positive and negative schizophrenic symptoms and the role of dopamine. *Br. J. Psychiatry* 137: 383–386.

DEAN, P., and P. REDGRAVE, 1984. Superior colliculus and visual neglect in rat and hamster: III. Functional implications. *Brain Res. Rev.* 8:155–163.

DEUTSCH, J. A., 1964. *The Structural Basis of Behaviour*. Cambridge: Cambridge University Press.

FANSELOW, M. S., 1991. The midbrain periaqueductal gray as a coordinator of action in response to fear and anxiety. In *The Midbrain Periaqueductal Grey Matter: Functional, Anatomical and Immunohistochemical Organization*, A. Depaulis and R. Bandler, eds. New York: Plenum.

FIBIGER, H. C., and A. G. PHILLIPS, 1988. Mesocorticolimbic dopamine systems and reward. In *The Mesolimbic Dopamine System*, P. W. Kalivas and C. B. Nemeroff, eds. Bethesda, Md.: New York Academy of Sciences, pp. 206–215.

FRITH, C. D., 1987. The positive and negative symptoms of schizophrenia reflect impairments in the perception and initiation of action. *Psychol. Med.* 17:631–648.

FRITH, C. D., 1992. *The Cognitive Neuropsychology of Schizophrenia*. Hove: Erlbaum.

GABRIEL, M., S. P. SPARENBORG, and N. STOLAR, 1987. Hippocampal control of cingulate cortical and anterior thalamic information processing during learning in rabbits. *Exp. Brain Res.* 67:131–152.

GRAEFF, F. G., 1987. The anti-aversive action of drugs. In *Advances in Behavioral Pharmacology*, vol. 6, T. Thompson, P. B. Dews, and J. Barrett, eds. Hillsdale, N.J.: Erlbaum.

GRAY, J. A., 1975. *Elements of a Two-Process Theory of Learning*. London: Academic Press.

GRAY, J. A., 1977. Drug effects on fear and frustration: possible limbic site of action of minor tranquillizers. In *Handbook of Psychopharmacology*, vol. 8, L. L. Iversen, S. D. Iversen, and S. H. Snyder, eds. New York: Plenum, pp. 433–529.

GRAY, J. A., 1982a. *The Neuropsychology of Anxiety: An Enquiry into the Functions of the Septo-Hippocampal System*. Oxford: Oxford University Press.

GRAY, J. A., 1982b. Précis of "The Neuropsychology of Anxiety: An enquiry into the functions of the septo-hippocampal system." *Behav. Brain Sci.* 5:469–484.

GRAY, J. A., 1987. *The Psychology of Fear and Stress*, ed. 2. Cambridge: Cambridge University Press.

GRAY, J. A., 1991. Fear, panic, and anxiety: What's in a name? *Psychol. Inquiry* 2:77–78.

Gray, J. A., 1993a. Consciousness, schizophrenia and scientific theory. In *Experimental and Theoretical Studies of Consciousness*, J. Marsh, ed. Ciba Foundation Symposium 174. Chichester, England: Wiley, pp. 263–281.

Gray, J. A., 1993b. The neuropsychology of the emotions: Framework for a taxonomy of psychiatric disorders. In *Emotions: Essays on Emotion Theory*, S. van Goozen, ed. Hillsdale, N.J.: Erlbaum, pp. 29–59.

Gray, J. A., in press. The contents of consciousness: A neuropsychological conjecture. *Behav. Brain Sci.*

Gray, J. A., J. Feldon, J. N. P. Rawlins, D. R. Hemsley, and A. D. Smith, 1991. The neuropsychology of schizophrenia. *Behav. Brain Sci.* 14:1–20.

Gray, J. A., D. R. Hemsley, J. Feldon, N. G. Gray, and J. N. P. Rawlins, 1991. Schiz bits: Misses, mysteries and hits. *Behav. Brain Sci.* 14:56–84.

Gray, J. A., and N. McNaughton, 1983. Comparison between the behavioural effects of septal and hippocampal lesions: A review. *Neurosci. Biobehav. Rev.* 5:109–132.

Gray, J. A., and J. N. P. Rawlins, 1986. Comparator and buffer memory: An attempt to integrate two models of hippocampal function. In *The Hippocampus*, vol. 4, R. L. Isaacson and K. H. Pribram, eds. New York: Plenum, pp. 159–201.

Gray, N. S., D. R. Hemsley, and J. A. Gray, 1992. Abolition of latent inhibition in acute, but not chronic, schizophrenics. *Neurol. Psychiatry Brain Res.* 1:83–89.

Gray, N. S., A. D. Pickering, D. R. Hemsley, S. Dawling, and J. A. Gray, 1992. Abolition of latent inhibition by a single 5 mg dose of *d*-amphetamine in man. *Psychopharmacology* 107:425–430.

Hemsley, D. R., 1987. An experimental psychological model for schizophrenia. In *Search for the Causes of Schizophrenia*, H. Hafner, W. F. Gattaz, and W. Janzavik, eds. Stuttgart: Springer, pp. 179–188.

Hernandez, L., and B. G. Hoebel, 1988. Food reward and cocaine increase extracellular dopamine in the nucleus accumbens as measured by microdialysis. *Life Sci.* 42: 1705–1712.

Jarrard, L. E., J. Feldon, J. N. P. Rawlins, J. D. Sinden, and J. A. Gray, 1986. The effects of intrahippocampal ibotenate on resistance to extinction after continuous or partial reinforcement. *Exp. Brain Res.* 61:519–530.

Jones, S. H., J. A. Gray, and D. R. Hemsley, 1992. Loss of the Kamin blocking effect in acute but not chronic schizophrenics. *Biol. Psychiatry* 32:739–755.

Kamin, L. J., 1968. "Attention-like" processes in classical conditioning. In *Miami Symposium on the Prediction of Behavior*, M. R. Jones, ed. Miami, Fla.: University of Miami Press, pp. 9–31.

Klein, D. F., 1981. Anxiety re-conceptualized. In *Anxiety: New Research and Changing Concepts*, D. F. Klein and J. Rabkin, eds. New York: Raven, pp. 235–263.

LeDoux, J. E., 1987. Emotion. In *Handbook of Physiology*. Sec. 1, *The Nervous System*. Vol. 5, *Higher Functions of the Brain*, V. Mountcastle, ed. Bethesda, Md.: American Physiological Society, pp. 419–459.

Lubow, R. E., 1989. *Latent Inhibition and Conditioned Attention Theory*. Cambridge: Cambridge University Press.

Mogenson, G. J., and M. Nielsen, 1984. A study of the contribution of hippocampal-accumbens-subpallidal projections to locomotor activity. *Behav. Neural Biol.* 42: 52–60.

Oades, R. D., 1985. The role of NA in tuning and DA in switching between signals in the CNS. *Neurosci. Biobehav. Rev.* 9:261–282.

O'Keefe, J., and L. Nadel, 1978. *The Hippocampus as a Cognitive Map*. Oxford: Clarendon.

Panksepp, J., 1982. Towards a general psychobiological theory of emotions. *Behav. Brain Sci.* 5:407–422.

Phillipson, O. T., and A. C. Griffiths, 1985. The topographical order of inputs to nucleus accumbens in the rat. *Neurosci.* 16:275–296.

Pycock, C. J., R. W. Kerwin, and C. J. Carter, 1980. Effect of lesion of cortical dopamine terminals on subcortical dopamine in rats. *Nature* 286:74–77.

Rawlins, J. N. P., J. Feldon, J. Tonkiss, and P. J. Coffey, 1989. The role of subicular output in the development of the partial reinforcement extinction effect. *Exp. Brain Res.* 77:153–160.

Rawlins, J. N. P., G. Lyford, and A. Seferiades, 1991. Does it still make sense to develop nonspatial theories of hippocampal function? *Hippocampus* 1:283–286.

Redmond, Jr., D. E., 1979. New and old evidence for the involvement of a brain norepinephrine system in anxiety. In *Phenomenology and Treatment of Anxiety*, W. G. Fann, I. Karacan, A. D. Pokorny, and R. L. Williams, eds. New York: Spectrum, pp. 153–203.

Rickels, K., 1978. Use of anti-anxiety agents in anxious outpatients. *Psychopharmacology* 58:1–17.

Rolls, E. T., 1990. A theory of emotion, and its application to understanding the neural basis of emotion. In *Psychobiological Aspects of Relationships between Emotion and Cognition* (special issue of *Cognition and Emotion*), J. A. Gray, ed. Hillsdale, N.J.: Erlbaum, pp. 161–190.

Rolls, E. T., and G. V. Williams, 1987. Sensory and movement related neuronal activity in different regions of the primate striatum. In *Basal Ganglia and Behavior: Sensory Aspects and Motor Functioning*, J. S. Schneider and T. I. Kidsky, eds. Berne: Hans Huber, pp. 37–59.

Schmajuk, N. A., Y.-W. Lam, and J. A. Gray, in preparation. Latent inhibition: A neural network approach.

Segal, M., 1977. The effects of brainstem priming stimulation on interhemispheric hippocampal responses in the awake rat. *Exp. Brain Res.* 28:529–541.

Swerdlow, N. R., and G. F. Koob, 1987. Dopamine, schizophrenia, mania and depression: toward a unified hypothesis of cortico-striato-pallidothalamic function. *Behav. Brain Sci.* 10:197–245.

Totterdell, S., and A. D. Smith, 1989. Convergence of hippocampal and dopaminergic input onto identified neurons in the nucleus accumbens of the rat. *J. Chem. Neuroanat.* 2:285–298.

Weinberger, D. R., 1987. Implications of normal brain

development for the pathogenesis of schizophrenia. *Arch. Gen. Psychiatry* 44:660–670.

Weiner, I., 1991. The accumbens-substantia nigra pathway, mismatch and amphetamine. *Behav. Brain Sci.* 14:54–55.

Williams, M. N., and R. L. M. Faull, 1988. The nicrotectal projection and tectospinal neurons in the rat: A light and electron microscopic study demonstrating a monosynaptic nigral input to identified tectospinal neurons. *Neurosci.* 25:533–562.

Wurtz, R. H., and J. E. Albano, 1980. Visual-motor function of the primate superior colliculus. *Ann. Rev. Neurosci.* 3:189–226.

Yang, C. R., and G. J. Mogenson, 1987. Hippocampal signal transmission to the pedunculopontine nucleus and its regulation by dopamine D2 receptors in the nucleus accumbens: An electrophysiological and behavioural study. *Neurosci.* 23:1041–1055.

Young, A. M. J., M. H. Joseph, and J. A. Gray, 1992. Increased dopamine release *in vivo* in nucleus accumbens and caudate nucleus of the rat during drinking: A microdialysis study. *Neurosci.* 48:871–876.

Young, A. M. J., M. H. Joseph, and J. A. Gray, 1993. Latent inhibition of conditioned dopamine release in rat nucleus accumbens. *Neurosci.* 54:5–9.

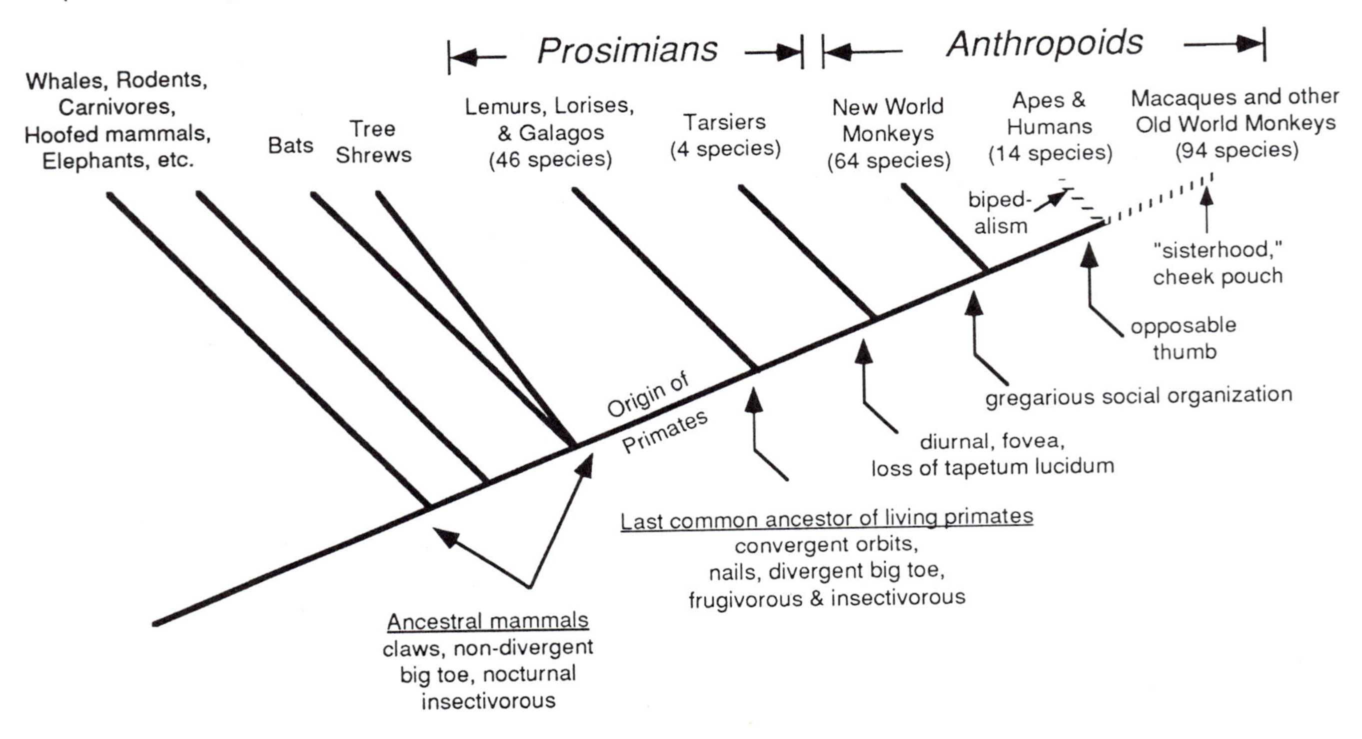
Other Mammals
(3500+ living species)
Primates
(222 living species)
Prosimians
Anthropoids
Whales, Rodents, Carnivores, Hoofed mammals, Elephants, etc.
Bats
Tree Shrews
Lemurs, Lorises, & Galagos (46 species)
Tarsiers (4 species)
New World Monkeys (64 species)
Apes & Humans (14 species)
Macaques and other Old World Monkeys (94 species)
biped-alism
"sisterhood," cheek pouch
opposable thumb
gregarious social organization
diurnal, fovea, loss of tapetum lucidum
Origin of Primates
Last common ancestor of living primates
convergent orbits, nails, divergent big toe, frugivorous & insectivorous
Ancestral mammals
claws, non-divergent big toe, nocturnal insectivorous

X EVOLUTIONARY PERSPECTIVES

The location of the ventral premotor area (PMV) in nonhuman primates compared to that of Brodmann's area 44 (the posterior part of Broca's area) in humans. PMV has been identified in prosimians, New World monkeys, and Old World monkeys, as a discrete region that represents forelimb and orofacial movements and is located immediately anterior to the inferior part of primary motor cortex (M1) (Preuss, 1993). Brodmann's area 44 occupies the same location with respect to M1 in humans. Recent metabolic and stimulation studies in humans suggest that Broca's area is involved in the control of forelimb and orofacial movements.

Introduction

JOHN TOOBY AND LEDA COSMIDES

EVOLUTIONARY biology has a great deal to offer cognitive neuroscience. Because human and nonhuman brains are evolved systems, they are organized according to an underlying evolutionary logic. By knowing what adaptive problems a species faced during its evolutionary history, researchers can gain insight into the functional circuitry of its neural architecture (see the chapters in this section by Daly and Wilson, Gallistel, and Gaulin). Furthermore, many neuroscience methods involve taking observations about nonhuman brains and using them to make inferences about the human brain. But the validity of these inferences rests on facts and principles derived from evolutionary biology; unless these principles are carefully applied, such inferences go awry (see Preuss's chapter, this section). A familiarity with the basics of modern evolutionary biology is, therefore, an important working tool for cognitive neuroscientists. Yet many of those basic elements—such as the difference between phylogenetic and adaptationist approaches—are relatively unknown in the cognitive and neural sciences.

Among brain scientists, there are fundamental differences in what an evolutionary perspective is thought to entail. Most neuroscientists take an evolutionary perspective to mean a *phylogenetic* perspective: the investigation of continuities from species to species implied by the inheritance of homologous features from common ancestors. Others take an "evolutionary perspective" to mean an *adaptationist* perspective: the investigation of evolved functional design. This often entails the functional analysis of niche-differentiated cognitive and neural machinery that is unique to the species being investigated.

One can see the interplay of these sentiments in the controversy over ape language and its implications for primate neuroanatomy. Half the scientific audience cheers for the apes, expecting them to duplicate human linguistic feats; the other half is confident that the apes' linguistic abilities will prove very limited. Scientists with a phylogenetic perspective form the apes' cheering section: They reason that if a human can learn a language, then our nearest relatives should be able to do so as well; after all, common descent ensures that their neuroanatomy will be similar to our own. Scientists with an adapationist perspective are skeptics in the ape language controversy. They (correctly, in our view) see the acquisition of language as a species-specific computational ability, requiring highly complex and specialized cognitive and neural mechanisms. Other primates, who were not selected to participate in a linguistic form of communication, are unlikely to share these brain mechanisms.

Within neuroscience, the adaptationist approach has largely been ignored because many neuroscientists have implicitly adopted a biologically obsolete linear model of phylogenetic continuity. According to this view, the human brain is qualitatively similar to the brains of other primates and, for that matter, of other mammals; though differing in scale, they share a single basic plan. But if brain architectures are fundamentally the same between species, then how could they contain functional specializations for solving adaptive problems that are specific to each species?

The ethical barrier to using invasive techniques to investigate humans has inclined many researchers toward the linear phylogenetic view, by making it convenient to assume and to believe that there are few qualitative differences between the human brain and the "primate brain" or even the "mammalian brain." If this assumption of phylogenetic continuity were true, then the careful study of a few species such as rats and rhesus monkeys would be sufficient to deduce the important features of human neural organization. Modern adaptationists, in contrast, recognize that the design of each species is unique in many ways, and that assumptions based on direct leaps from one species to another will often be wrong. Natural selection tends to specialize and improve the design of circuits that solve important adaptive problems, and each species has faced a somewhat different array of adaptive problems over the course of its evolutionary history. Neuroscientists who take an adaptationist perspective have found strong evidence of functional circuitry that differs from species to species (for examples, see the chapters in this section by Daly and Wilson, Gallistel, Gaulin, and Preuss).

A modern evolutionary framework resolves the apparent tension between phylogenetic and adaptationist approaches. Animals from different species are similar to each other in neural architecture because of inheritance from common ancestors, the same selection pressures operating on different species, or both. Animals from different species differ in neural architecture because of independent descent, the operation of different selection pressures on different species, or random divergence. Adaptationist and phylogenetic analyses are complementary components of an integrated evolutionary approach, and both have value. But their relative validity depends on which brain mechanism is being investigated, and at what level of resolution (for extended discussion, see Tooby and Cosmides, 1989).

At the grossest level of brain anatomy, continuity will seem largely vindicated. More differences will appear, however, as more species are examined and as brain structures are more finely mapped. Although these differences will become more apparent the finer the scale of investigation, there is no reason to think of them as *functionally* "fine" or minor. Hands, anteater claws, seal flippers, bat wings, horse hooves, tiger paws, and elephant legs are all homologous structures sharing a certain basic architecture across mammals. But at a finer level of description they are all very different, and these differences are often related to functional problems that are specific to each group. Neural differences between mammals do not seem so dramatic until one considers that behavioral differences are the most logical assay of differences in brain function. Behavioral differences between bats, chimpanzees, mole rats, meerkats, humans, whales, leopards, macaques, and tamarins are not mild variations of a single model; they reflect highly diversified computational organizations.

The anatomical differences that create diversified cognitive mechanisms are more likely to be found in the details of circuit architecture than in gross anatomy. This is due to the fact that the output of complex, dynamical systems (such as brains) can be dramatically changed by only minor changes in internal structure. Because natural selection shapes brain structure based on its effects on behavior, the behavioral output of the neural architecture can readily be shaped by adaptive demands over evolutionary time, even though the modification of the neural substrate necessary to create adaptively major changes may be anatomically minor. Thus, the nature of complex neural design makes the search for continuities among animal species (including humans) helpful and illuminating at a gross neuroanatomical level, where structural homologies are easily recognizable and readily traceable through related species. Equally, cellular-level processes seem to be conserved across species, making cross-species inferences highly reliable for this class of features. Between these two extreme scales, however, continuity breaks down. The computational properties of neural circuits depends critically upon the exact arrangement of microlevel elements. At this level, an adaptationist perspective—more specifically, the careful analysis of specific adaptive information-processing problems—will be necessary to understand neural structure in functional terms and at its most detailed level. Even more obviously, phylogenetic continuity is powerless to explain or to illuminate the zoologically unique features of the cognitive or neural architecture of any species, and humans exhibit many zoologically unique properties. To understand many aspects of human brain function (such as language, tool use, imitation, and coalition formation), it will be necessary to consider the niche-differentiating, species-specific selection pressures hominids encountered during human evolutionary history (e.g., Tooby and DeVore, 1987). To elucidate the human cognitive architecture and how it is physically implemented, adaptationist analyses are therefore essential.

The news that human brains cannot be thought of as "scaled-up" versions of macaque brains has been unwelcome and resisted by many neuroscientists. But it shouldn't be. Not only can more sophisticated versions of phylogenetic inference be practiced (see Preuss's chapter), but by trading in an obsolete model of linear continuity, neuroscientists will get an expanded set of inferential tools in exchange. Indeed, one of the most powerful tools a cognitive neuroscientist can have is a sophisticated appreciation for adaptation-driven phylogenetic diversity. Gaulin's work on spatial cognition describes how sibling species that have been subjected to different selection pressures can be used to test hypotheses about the relationship between neural structures and cognitive functions. Moreover, functional analyses of adaptive problems can be used to discover neurocognitive systems that were previously unknown (see, e.g., the chapters in this section by Cosmides and Tooby, Daly and Wilson, Gallistel, and Gaulin). Evolutionary biology replaces a linear scale of nature with a rich matrix of relationships between species, selection pressures, behaviors, and neural structures. This matrix permits a far greater variety of methods for testing hypotheses, based upon the genuine diversity of structure-function relationships. As the contributions to this section should make clear, cognitive neuroscientists can find a wealth of new approaches to their research problems by exploring the numerous connections between their field and evolutionary biology.

REFERENCES

Tooby, J., and L. Cosmides, 1989. Adaptation versus phylogeny: The role of animal psychology in the study of human behavior. *Int. J. Comp. Psychol.* 2:175–188.

Tooby, J., and I. DeVore, 1987. The reconstruction of hominid behavioral evolution through strategic modeling. In *Primate Models of Hominid Behavior*, W. Kinzey, ed. New York: SUNY Press, pp. 183–237.

78 Mapping the Evolved Functional Organization of Mind and Brain

JOHN TOOBY AND LEDA COSMIDES

ABSTRACT The human brain is a biological system produced by the evolutionary process, and so cognitive neuroscience is itself a branch of modern evolutionary biology. Accordingly, cognitive neuroscientists can benefit by acquiring a professional knowledge of the recent technical advances made in evolutionary biology and by applying them to their research. Useful tools include the biologically rigorous concept of function that is appropriate to neural and cognitive systems; a growing list of the specialized functions the human brain evolved to perform; and criteria for distinguishing the narrowly functional aspects of the neural and cognitive architecture that are responsible for the brain's organization from the much larger set of properties that are by-products or noise. With such tools, researchers can construct biologically meaningful experimental tasks and stimuli. These are more likely to activate the large array of functionally dedicated mechanisms that constitute the core of human brain function, but which are at present largely unstudied.

Nothing in biology makes sense except in the light of evolution.
—*T. Dobzhansky*

It is the theory which decides what we can observe.
—*Einstein*

Seeing with new eyes: Toward an evolutionarily informed cognitive neuroscience

The task of cognitive neuroscience is to map the information-processing structure of the human mind, and to discover how this computational organization is implemented in the physical organization of the brain. The central impediment to progress is obvious: The human brain is, by many orders of magnitude, the most complex system humans have yet investigated. Purely as a physical system, the vast intricacy of chemical and electrical interactions among roughly one hundred billion neurons defeats any straightforward attempt to build a comprehensive model, as one might attempt to do with particle collisions, geological processes, protein folding, or host-parasite interactions. At present, the underlying logic of the system seems lost among the torrent of observations that have been accumulated to date, and obscured by the inherent complexity of the system.

Historically, however, well-established theories from one discipline have functioned as organs of perception for others. They allow new relationships to be observed and make visible elegant systems of organization that had previously eluded detection. It seems worth exploring whether the same could be true for the brain sciences.

In fact, the brain is more than a physical system: It is both a computational system and an evolved biological system. Although cognitive neuroscience began with the recognition that studying the brain as a computational system would offer important new insights, the field has so far failed to take equal advantage of the fact that the brain is an evolved system as well. Indeed, the brain is a computational system that was organized and specifically designed to solve a narrowly identifiable set of biological information-processing problems. For this reason, evolutionary biology can supply a key missing element in the cognitive neuroscience research program: a list of the native information-processing functions that the human brain was built to execute. Our computational architecture evolved its distinctive sets of structured information-processing relationships as devices or modules to perform this particular targeted set of adaptive functions. In turn, our neural architecture evolved its distinctive physical configu-

JOHN TOOBY Department of Anthropology, and LEDA COSMIDES Department of Psychology, Center for Evolutionary Psychology, University of California, Santa Barbara, Calif.

ration because it brought these targeted sets of functional information-processing relationships into existence. By providing the functional engineering specifications to which human brains were built to conform, evolutionary biology can help researchers to isolate, identify, activate, and map the important functional aspects of the cognitive architecture, aspects that would otherwise be lost among the myriad irrelevant phenomena in which they are embedded. The resulting maps of the computational structure of each device will then allow researchers to isolate, identify and map the functional aspects of the neural architecture. The biologically implausible view that the brain is a general-purpose information-processing system provides little guidance for research in cognitive neuroscience. In contrast, an evolutionary approach allows cognitive neuroscientists to apply a sophisticated body of new knowledge to their problems. In short, because theories and principled systems of knowledge can function as organs of perception, the incorporation of a modern evolutionary framework into cognitive neuroscience may allow the community to detect ordered relationships in phenomena that otherwise seem too complex to be understood.

Over the last 30 years, evolutionary biology has made a number of important advances that have not yet diffused into allied disciplines such as the cognitive and neural sciences. These advances constitute a potent set of new principles relevant to dissecting and understanding the phenomena studied by cognitive neuroscientists (Tooby and Cosmides, 1992). Central to these advances is the modern technical theory of evolution. This consists of the logically derivable set of causal principles that necessarily govern the dynamics of reproducing systems. These principles account for the properties that reproducing systems cumulatively acquire over successive generations. The explicit identification of this core logic has allowed the biological community to develop an increasingly comprehensive set of principles about what kinds of features can and do become incorporated into the designs of reproducing systems down their chains of descent, and what kinds of features do not (Hamilton, 1964, 1972; Maynard Smith, 1964, 1982; Williams, 1966; Dawkins, 1976, 1982, 1986; Cosmides and Tooby 1981; Tooby, 1982). This set of principles has been tested, validated, and enriched through its integration with functional and comparative anatomy, biogeography, genetics, immunology, embryology, behavioral ecology, and a number of other disciplines.

Just as the fields of electrical and mechanical engineering summarize our knowledge of principles that govern the design of human-built machines, the field of evolutionary biology summarizes our knowledge of the engineering principles that govern the design of organisms, which can be thought of as machines built by the evolutionary process (for overviews, see Dawkins, 1976, 1982, 1986; Daly and Wilson, 1984; Krebs and Davies, 1987). Modern evolutionary biology constitutes, in effect, a foundational organism design theory, whose principles can be used to fit together research findings into coherent models of specific cognitive and neural mechanisms.

First principles: Reproduction, feedback, and the antientropic construction of organic design

Within an evolutionary framework, an organism is describable as a self-reproducing machine, and the defining property of life is the presence in a system of devices or organization that cause the system to construct new and similarly reproducing systems. From this defining property of self-reproduction, the entire deductive structure of modern Darwinism logically follows (Dawkins, 1976; Williams, 1985; Tooby and Cosmides, 1990b). Because the replication of the design of the parental machine is not always error-free, randomly modified designs (i.e., mutants) are introduced into populations of reproducers. Because such machines are highly organized so that they cause the otherwise improbable outcome of constructing offspring machines, the great majority of random modifications will interfere with the complex sequence of actions necessary for self-reproduction. Consequently, such modified designs will tend to remove themselves from the population—a case of negative feedback.

However, a small residual subset of design modifications will, by chance, happen to constitute improvements in the design's machinery for causing its own reproduction. Such improved designs (by definition) cause their own frequency to increase in the population—a case of positive feedback. This increase continues until (usually) such modified designs outreproduce and thereby replace all alternative designs in the population, leading to a new species-standard design. After such an event, the population of reproducing machines is different from the ancestral population: The population- or species-standard design has taken a step "uphill" toward a greater degree of functional organization for reproduction. This spontaneous feedback process—natural selection—is the only known process by which functional organization emerges naturally in the world, without intelligent design and intervention. Hence, all naturally occurring functional organization in organisms must be ascribed to its operation and must be consistent with its principles.

Over the long run, down chains of descent, this feedback cycle pushes the design of a species stepwise "uphill" toward arrangements of elements that are increasingly improbably well organized to cause their own reproduction in the environment the species evolved in. Because the reproductive fates of the inherited traits that coexist in the same organism are linked together, traits will be selected to enhance each other's functionality (but see Cosmides and Tooby, 1981; Tooby and Cosmides, 1990b, for the relevant genetic analysis and qualifications). Consequently, accumulating design features will tend to fit themselves together sequentially into increasingly functionally elaborated machines for reproduction, composed of constituent mechanisms—called adaptations—that solve problems whose solutions either are necessary for reproduction or increase its likelihood (Darwin, 1859; Williams, 1966, 1985; Dawkins, 1986; Thornhill, 1990; Tooby and Cosmides, 1990b). Significantly, in species like humans, genetic processes insure that complex adaptations are virtually always species-typical (unlike nonfunctional aspects of the system)—so the functional aspects of the architecture will tend to be genetically universal (Tooby and Cosmides, 1990b).[1] This means that any complex device that cognitive neuroscientists find should be universal, at least at the genetic level.

Because design features are embodied in organisms, they can, generally speaking, propagate themselves in only two ways: by solving problems that will increase the probability that the organism they are situated in will produce offspring, or by solving problems that will increase the probability that the organism's kin will produce offspring (Williams and Williams, 1957; Hamilton, 1964; however, see Cosmides and Tooby, 1981, for intragenomic methods). An individual's relatives, by virtue of having descended from a common ancestor, have an increased likelihood of having the same design feature as compared to other conspecifics, so their increased reproduction will tend to increase the frequency of the design feature. Accordingly, design features that promote both direct reproduction and kin reproduction, and that make efficient trade-offs between the two, will replace those that do not. To put this in standard biological terminology, design features are selected for to the extent that they promote their inclusive fitness (Hamilton, 1964).

Although largely unknown to cognitive neuroscientists, the promotion of inclusive fitness is the ultimate functional product of all evolved cognitive devices. That is, design changes were incorporated into the neural architecture of a species to the extent that they stably promoted inclusive fitness in the past, and were discarded to the extent that they did not. The human brain, to the extent that it is organized to do anything functional at all, will be organized to construct the information, make the decisions, and generate the behavior that would have tended to promote inclusive fitness in the ancestral environments and behavioral contexts of Pleistocene hunter-gatherers. (The preagricultural world of hunter-gatherers is the appropriate ancestral context, because natural selection operates far too slowly to have built complex information-processing adaptations to the post-hunter-gatherer world of the last few thousand years.) Although there are an infinite number of other standards of functionality one could sensibly have for various purposes (e.g., getting an individual to read English or to avoid shouting inappropriately), this biological standard is the only standard of functionality that is relevant to analyzing why the human brain architecture is organized in one fashion rather than another.

There is, however, a second family of evolutionary processes, in addition to selection, by which mutations can become incorporated into species-typical designs: chance processes. For example, the sheer impact of many random accidents may cumulatively propel a useless mutation upward in frequency until it crowds out all alternative design features from the population. Clearly, the presence of such a trait in the architecture is not explained by the (nonexistent) functional consequences it had over many generations on the design's reproduction; as a result, chance-injected traits will not tend to be coordinated with the rest of the organism's architecture in a functional way.

Although such chance events play a restricted role in evolution and explain the existence and distribution of many simple and trivial properties, organisms are not primarily chance agglomerations of stray properties. In the first place, reproduction is a highly improbable outcome in the absence of functional machinery designed to bring it about, and only designs that retain all of the necessary functional organization avoid being selected out. Secondly, to the extent that a mutation has a significant systematic impact on the functional organization leading to reproduction, selection will act on it. For this reason, the significant and consequential aspects of organismic architectures are organized primarily by natural selection. Reciprocally, those modifications that are so minor that their consequences are negligible on reproduction are invisible to selection and are therefore not organized by it. Thus, chance properties do indeed drift through the standard designs of species in a random way, but they are unable to account for the complex, organized design in organisms and are, correspondingly, usually peripheralized into those aspects of the system that do not make a significant impact on its functional operation (Tooby and Cosmides, 1990a,b, 1992). Random walks do not systematically build intricate and improbably functional arrangements such as the visual system, the language faculty, face-recognition competences, emotion-recognition modules, food-aversion circuits, cheater-detection devices, or motor control.

Brains are composed primarily of adaptive problem-solving devices

The foregoing leads to the most important point for cognitive neuroscientists to abstract from modern evolutionary biology: Although not everything in the design of organisms is the product of selection, all complex functional organization is. This is because the only known cause of and explanation for complex functional design in organic systems is natural selection: It is the single natural hill-climbing process that pushes designs through state space toward increasingly well organized—and otherwise improbable—functional arrangements (Williams, 1966, 1985; Dawkins, 1986; Tooby and Cosmides, 1990a,b, 1992; Pinker and Bloom, 1992). Specifically, this means that all of the functional organization present in the human brain was built by natural selection during our evolutionary history. Indeed, selection can account only for functionality of a very narrow kind: approximately, those design features organized to promote the reproduction of an individual and his or her relatives (Williams, 1966; Dawkins, 1986). Fortunately for the modern theory of evolution, the only naturally occurring, complex functionality that has ever been documented in undomesticated plants, animals, or other organisms is functionality of just this kind, along with its derivatives and by-products. This has several important implications for cognitive neuroscientists.

First, whenever one finds functional organization built into our cognitive or neural architecture, one is looking at adaptations—devices that acquired their distinctive organization from natural selection acting on our hunter-gatherer or more distant primate ancestors. Reciprocally, when one is searching for intelligible functional organization underlying a set of cognitive or neural phenomena, one is far more likely to discover it by using an adaptationist framework for organizing observations, because adaptive organization is the only kind of *functional* organization that is there to be found.

Second, because the reliably developing mechanisms (i.e., modules, circuits, functionally isolable units, mental organs, or computational devices) that cognitive neuroscientists study are evolved adaptations, all of the biological principles that apply to adaptations apply to cognitive devices. Thus cognitive neuroscience and evolutionary biology are connected in the most direct possible way. This conclusion should be a welcome one, because it is the logical doorway through which a very extensive body of new expertise and principles can be applied to cognitive neuroscience, stringently constraining the range of valid hypotheses about the functions and structures of cognitive mechanisms. Because cognitive neuroscientists are usually studying adaptations and their effects, they can supplement their present research methods with carefully derived adaptationist analytic tools (e.g., Shepard, 1981, 1984, 1987a, 1987b; Marr, 1982; Freyd, 1987; Leslie, 1987, 1988; Sherry and Schacter, 1987; Cosmides, 1989; Gallistel, 1990; Ramachadran, 1990; Pinker, 1991; Cosmides and Tooby, 1992; Jackendoff, 1992; Pinker and Bloom, 1992; Baron-Cohen, 1994).

Third, our cognitive architectures are designed to incorporate only the precise, narrow, and strange kinds of functional organization that natural selection spontaneously builds, rather than any other kind of functional organization. What this means is that the problems our cognitive devices are designed to solve do not reflect the problems our modern life experiences lead us to see as normal, such as reading, driving cars, working for large organizations, reading insurance forms, learning the oboe, or playing Go. Instead, they are the ancient and seemingly esoteric problems that our hunter-gatherer ancestors encountered generation after generation over hominid evolution. These include such problems as foraging, kin recognition, "mind reading" (i.e., inferring the motives, intentions, and knowledge of others based on their situation, history, and behavior), engaging in social exchange, avoiding incest, choosing mates, interpreting threats, recognizing emotions, caring for children, regulating immune function, and so on, as well as the already well-known problems involved in perception, language acquisition, and motor control. For biological reasons discussed elsewhere, such devices should be far more numerous and far more content-specialized than is usually appreciated even by cognitive scientists sympathetic to modular approaches (for a review of the issues, see Cosmides and Tooby, 1987, 1994, this volume; Tooby and Cosmides, 1992). From an evolutionary perspective, the human cognitive architecture is far more likely to resemble a confederation of hundreds or thousands of functionally dedicated computers, designed to solve problems endemic to the Pleistocene, than it is to resemble a single general-purpose computer equipped with a small number of general-purpose procedures such as association formation, categorization, or production-rule formation (Cosmides and Tooby, 1987; Tooby and Cosmides, 1992; see also Gallistel, 1990, this volume). Although our architectures may be capable of other kinds of functionality or activities (e.g., weaving, playing pianos), these are incidental byproducts of selection for our Pleistocene competences—just as a machine built to be a hair dryer can, incidentally, dehydrate fruit or electrocute. But it will be difficult to make sense of our cognitive mechanisms if one attempts to interpret them as devices designed to perform functions that were not selectively important for our hunter-gatherer ancestors, or if one fails to consider the functions they were designed to accomplish (for the importance of functional analysis, see chapter 79).

Fourth, evolutionary biology gives the concept of function a specific and rigorous content that is otherwise lacking, and imposes strict rules on its use. Although many cognitive scientists are unaware of it, every time the function of a computational device is discussed, this automatically invokes a biological concept with a very specific and narrow technical meaning (Williams, 1966; Dawkins, 1986; Tooby and Cosmides, 1990a, 1992). As discussed, it is only the narrow biological meaning that is relevant in explaining why a system is structured as it is—that is, what specific consequences of a design feature caused it to be propagated and made species-standard within ancestrally structured environments. So, not only are the problems that our devices were designed to solve esoteric to modern sensibilities, but the standards that define what counts as functional solutions to these problems are evolutionary standards, and hence odd and nonintuitive as well. Cognitive neuroscientists need to recognize that in explaining or exploring the reliably developing organization of a cognitive device, the "function" of a design refers solely to how it systematically caused its own propagation in ancestral environments. It does not validly refer to any of the various intuitive or folk definitions of function such as "contributing to the attainment of the individual's goals," "contributing to one's well-being," "contributing to society," or even "making a valid inference." These other kinds of usefulness may or may not exist as side effects of a given evolved design, but they can play no role in explaining how such designs came into existence or why they have the organization that they do. The only kind of functional organization that has been built into our cognitive architectures—and hence that researchers should spend their time looking for—is the kind that matches this peculiar biological standard of functionality: enhancing propagation in ancestral environments. The fact that sexual jealousy, for example, has no truth-value, and may not contribute to any individual's well-being or to any positive social good, is irrelevant to why the cognitive mechanisms that reliably produce it under certain limited conditions became part of our species-typical computational architecture (Daly, Wilson, and Weghorst, 1982). In short, the technical criteria that define what solutions our cognitive devices are designed to produce—that is, what counts as functional design and successful processing—are evolutionary in nature, and usually cannot be supplied by simply consulting common sense.

Fifth, the standard of parsimony imported from physics, from traditional philosophy of science, or from habits of economical programming is inappropriate and misleading in biology and, hence, in neuroscience and cognitive science, which study biological systems. The evolutionary process never starts with a clean workboard, has no foresight, and incorporates new features solely on the basis of whether they lead to systematically enhanced propagation. Enhanced functionality in a complexly structured series of environments is the only criterion for organizing designs. Indeed, when one examines the brain, one sees an amazingly heterogeneous physical structure. A correct theory of evolved cognitive functions should be no less complex and heterogeneous than the evolved physical structure itself, and should map onto the heterogeneous set of recurring adaptive tasks faced by hominid foragers over evolutionary time. Indeed, analyses of the adaptive problems that humans and other animals must regularly have solved over evolutionary time to remain in the world suggests that the mind contains a far greater number of functional specializations than has traditionally been supposed (for discussion, see Cosmides and Tooby, 1987, 1993; Symons, 1987; Tooby and Cosmides, 1992).

Sixth, understanding the neural organization of the brain depends on understanding the functional organization of its cognitive devices. The brain originally came into existence, and accumulated its particular set of design features, only because these features functionally contributed to the organism's propagation. This contribution—that is, the evolutionary function of the brain—is obviously the adaptive regulation of behavior and physiology on the basis of information derived from the body and from the environment. The brain performs no significant mechanical, metabolic, or chemical service for the organism; its function is purely informational, computational, and regulatory in nature. Because the function of the brain is informational in nature, its precise functional organization can be described accurately only in a language that is capable of expressing its informational functions—that is, in cognitive terms, rather than in cellular, anatomical, or chemical terms. Cognitive investigations are not some soft, optional activity that goes on only until the "real" neural analysis can be performed. Instead, the mapping of the computational adaptations of the brain is an unavoidable and indispensable step in the neuroscience research enterprise; it must proceed in tandem with neural investigations, and indeed will provide one of the primary frameworks necessary for organizing the body of neuroscience results.

The reasons why are straightforward. Natural selection operating on hominids in complexly structured ancestral environments posed adaptive information-processing problems, such as effective foraging, object recognition, motivational allocation, contagion avoidance, and so on. These recurrent problems selected for specialized devices that could solve these information-processing problems—that is, cognitive or computational devices. That in turn selected for those precise physical arrangements of cells (and modifications of the internal organization of cells) that could embody the particular computational relationships that reliably solved those adaptive problems. Natural selection retained neural structures on the basis of their ability to create adaptively organized relationships between information and behavior (e.g., the sight of a predator activates inference procedures that cause the organism to hide or flee) or between information and physiology (e.g., the sight of a predator increases the organism's heart rate in preparation for flight). Thus, it is the information-processing structure of the human psychological architecture that has been functionally organized by natural selection, and the neural structures and processes have been organized insofar as they physically realize this cognitive organization. Brains exist and have the structure that they do because of the computational requirements imposed by selection on our ancestors. The adaptive structure of our computational devices provides a skeleton around which a modern understanding of our neural architecture will be constructed.

That is why cognitive neuroscience is pivotal to the progress of the brain sciences. There are an astronomical number of physical interactions and relationships in the brain, and blind empiricism rapidly drowns itself among the deluge of manic and enigmatic measurements. The same is true at the cognitive level: The blind empiricist will drown in the sea of irrelevant phenomena that our computational devices can generate—everything from writing theology or dancing the limbo to calling for the restoration of the Plantagenets to the throne of England. Fortunately, however, evolutionary biology, behavioral ecology, and hunter-gatherer studies can be used to identify and supply descriptions of the recurrent adaptive problems humans faced during their evolution. Supplemented with this knowledge, cognitive research techniques can abstract out of the welter of human cognitive performance a series of maps of the functional information-processing relationships that constitute our computational devices and that evolved to solve this particular set of problems: our cognitive architecture. With these computational maps in hand, we can navigate the ocean of physical relationships in the brain, abstracting out that exact and minute subset that implement those information-processing relationships. It is only those relationships that explain the existence and functional organization of the system. The immense number of other physical relationships in the brain are incidental by-products of those narrow aspects that implement the functional computational architecture. Consequently, an adaptationist inventory and functional mapping of our cognitive devices can provide the essential theoretical guidance for neuroscientists, allowing them to home in on these narrow but meaningful aspects of neural organization, and to distinguish them from the sea of irrelevant neural phenomena.

Brain architectures consist of adaptations, by-products, and random effects

To understand the human computational or neural architecture (or that of any living species) is a problem in reverse engineering: We have working exemplars of the design in front of us, but we must organize our observations of these exemplars into a systematic functional and causal description of the design. One can describe and decompose brains into properties according to any of an infinite set of alternative systems, and hence there are an indefinitely large number of cognitive and neural phenomena that could potentially be defined and measured. However, describing and investigating the architecture in terms of its adaptations is a useful place to begin, because (1) the adaptations are the cause of the system's organization (the reason for the system's existence); (2) organisms, properly described, consist largely of collections of adaptations (evolved problem-solvers); (3) all of the complex, functionally organized subsystems in the architecture are adaptations; (4) an adaptationist frame of reference allows cognitive neuroscientists to apply to their research problems the formidable array of knowledge that evolutionary biologists have accumulated about adaptations; and (5) such a frame of reference permits the construction of economical and principled models of the important features of the system, models in which the wealth of varied phenomena fall into intelligible, functional, and predictable patterns. As Ernst Mayr put it, summarizing the historical record, "the adaptationist question, 'What is the function of a given structure or organ?' has been for centuries the basis for every advance in physiology" (1983, 32). It should prove no less productive for cognitive neuroscientists.

Indeed, all of the inherited design features of organisms can be partitioned into three adaptively defined categories: adaptations (often, though not always, complex), the by-products or concomitants of adaptations, and random effects. Chance and selection, the two components of the evolutionary process, explain different types of design properties in organisms, and all aspects of design must be attributed to one of these two forces. The conspicuously distinctive cumulative impacts of chance and selection allow the development of rigorous standards of evidence for recognizing and establishing the existence of adaptations and distinguishing them from the nonadaptive aspects of organisms caused by the nonselectionist mechanisms of evolutionary change (Williams, 1966, 1985; Tooby and Cosmides, 1990b; Thornhill, 1991; Pinker and Bloom, 1992; Symons, 1992).

Adaptations are systems of properties (called mechanisms) crafted by natural selection to solve the specific problems posed by the regularities of the physical, chemical, developmental, ecological, demographic, social, and informational environments encountered by ancestral populations during the course of a species' or population's evolution (figure 78.1; for other discussions of adaptation, see Williams, 1966, 1985; Dawkins, 1986; Symons, 1989, 1992; Thornhill, 1990; Tooby and Cosmides, 1990a, 1992; Pinker and Bloom, 1992). Adaptations are recognizable by "evidence of special design" (Williams, 1966); that is, certain features of the evolved species-typical design of an organism are recognized "as components of some special problem-solving machinery" (Williams, 1985, 1). Moreover, they are so well organized and represent such good engineering solutions to adaptive problems that a chance coordination between problem and solution is effectively ruled out as a counterhypothesis. Standards for recognizing special design include whether the problem solved by the structure is an evolutionarily long-standing adaptive problem, and such factors as economy, efficiency, complexity, precision, specialization, and reliability, which, like a key fitting a lock, render the design too good a solution to a defined adaptive problem to be coincidence (Williams, 1966). Like most other methods of empirical hypothesis-

The formal properties of an adaptation

An adaptation is:
(1) A cross-generationally recurring set of characteristics of the phenotype,
(2) that is reliably manufactured over the developmental life history of the organism
(3) according to instructions contained in its genetic specification,
(4) in interaction with stable and recurring features of the environment (i.e., it reliably develops normally when exposed to normal ontogenetic environments),
(5) whose genetic basis became established and organized in the species (or population) over evolutionary time, because
(6) the set of characteristics systematically interacted with stable and recurring features of the ancestral environment (the "adaptive problem")
(7) in a way that systematically promoted the propagation of the genetic basis of the set of characteristics better than the alternative designs existing in the population during the period of selection. This promotion virtually always takes place through enhancing the reproduction of the individual bearing the set of characteristics, or the reproduction of the relatives of that individual.

FIGURE 78.1 The most fundamental analytic tool for organizing observations about a species' functional architecture is the definition of an adaptation. In order to function, adaptations evolve such that their causal properties rely on and exploit these stable and enduring statistical and structural regularities in the world, and in other parts of the organism. Things worth noticing include the fact that an adaptation (such as teeth or breasts) can develop at any time during the life cycle, and need not be present at birth; that an adaptation can express itself differently in different environments (e.g., speaking English, speaking Tagalog); that an adaptation is not just any individually beneficial trait, but one built over evolutionary time and expressed in many individuals; that it may not be producing functional outcomes now (e.g., agoraphobia), but was needed to function well in ancestral environments; and finally, that an adaptation is the product of gene-environment interaction, like every other aspect of the phenotype; however, unlike many other phenotypic properties, it is the result of the interaction of the species-standard set of genes with those aspects of the environment that were present and relevant during the evolution of the species. (For a more extensive definition of the concept of adaptation, see Tooby and Cosmides, 1990a, 1992.)

testing, the demonstration that something is an adaptation is always, at core, a probability assessment concerning how likely a set of events is to have arisen by chance. Such assessments are made by investigating whether there is a highly nonrandom coordination between the recurring properties of the phenotype and the structured properties of the adaptive problem, in a way that meshed to promote fitness (genetic propagation) in ancestral environments (Tooby and Cosmides, 1990a, 1992). For example, the lens, pupil, iris, retina, visual cortex, and so on are too well coordinated both with each other and with features of the world, such as the properties of light, optics, geometry, and the reflectant properties of surfaces, to have co-occurred by chance. In short, like the functional aspects of any other engineered system (e.g., the electron gun in a television), they are recognizable by their organized and functional relationships to the rest of the design and to the structure of the world.

In contrast, concomitants or by-products of adaptations are those properties of the phenotype that do not contribute to functional design per se, but that happen to be coupled to properties that are. They were, consequently, dragged along into the species-typical architecture because of selection for the functional design features to which they are linked. For example, bones are adaptations but the fact that they are white is an incidental by-product. Bones were selected to include calcium because it conferred hardness and rigidity to the structure (and was dietarily available), and it simply happens that alkaline earth metals appear white in many compounds, including the insoluble calcium salts that are a constituent of bone. From the viewpoint of functional design, by-products are the result of chance, in the sense that the process that led to their incorporation into the design was blind to their consequences (assuming the consequences were not negative). They are distinguishable from adaptations by the absence of complexly arranged functional consequences (e.g., the whiteness of bone does nothing for the vertebrate). In general, by-products will be far less informative as a focus of study than adaptations, because they are consequences and not causes of the organization of the system. (Hence they are functionally arbitrary and unregulated, and may, for example, vary capriciously between individuals.) Unfortunately, unless researchers actively seek to study organisms in terms of their adaptations, they will usually end up measuring and investigating arbitrary and random admixtures of functional and functionless aspects of organisms—and this hampers the discovery of the underlying organization of a biological system. We do not yet know, for example, which exact aspects of the neuron are relevant to its function and which are by-products, thus many computational neuroscientists are stuck using a model of the neuron that is inaccurate and outdated by decades.

Finally, of course, entropic effects of many types are always acting to introduce functional disorder into the design of organisms. Traits introduced by accident or by evolutionary random walks are recognizable by the lack of coordination they produce within the architecture, or between the architecture and the environment, as well as by the fact that they frequently cause uncalibrated variation between individuals. Examples of such entropic processes include genetic mutation, recent change in ancestrally stable environmental features, and developmentally anomalous circumstances.

How well engineered are adaptations?

The design of our cognitive and neural mechanisms should reflect the structure of the adaptive problems our ancestors faced only to the extent that natural selection is an effective process. Is it one? How well or poorly engineered are adaptations? Some researchers have argued that evolution produces mostly inept designs, because selection does not produce perfect optimality (Gould and Lewontin, 1979). In fact, evolutionary biologists since Darwin have been well aware that selection does not produce perfect designs (Darwin, 1859; Williams, 1966; Dawkins, 1976, 1982, 1986; for a recent convert from the position that organisms are optimally designed to the more traditional adaptationist position, see Lewontin, 1967, 1978; see Dawkins, 1982, for an extensive discussion of the many processes that prevent selection from reaching perfect optimality). Still, because natural selection is a hill-climbing process that tends to choose the best of the variant designs that actually appear, and because of the immense numbers of alternatives that appear over the vast expanse of evolutionary time, natural selection tends to cause the accumulation of superlatively well engineered functional designs.

Empirical confirmation can be gained by comparing how well evolved devices and human-engineered devices perform on evolutionarily recurrent adaptive problems (as opposed to arbitrary, artificial modern tasks, such as chess). For example, the claim that language competence is a simple and poorly engineered adaptation cannot be taken seriously, given the total amount of time, engineering, and genius that has gone into the still unsuccessful effort to produce artificial systems that can remotely approach—let alone equal—human speech perception, comprehension, acquisition, and production (Pinker and Bloom, 1992).

Even more strikingly, the visual system is composed of collections of cognitive adaptations that are well-engineered products of the evolutionary process, and while they may not be "perfect" or "optimal"—however these somewhat vague concepts may be interpreted—they are far better at vision than any human-engineered system yet developed. Wherever the standard of biological functionality can be clearly defined—from semantic induction to capturing solar energy, to object recognition, to color constancy, to echolocation, to relevant problem-solving generalization, to chemical recognition (olfaction), to mimicry, to scene analysis, to chemical synthesis—evolved adaptations are at least as good as, and usually strikingly better than, human-engineered systems, in those rare situations when humans can build systems that can accomplish these tasks at all. It seems reasonable to insist that before a system is criticized as being poorly designed, the critic ought to be able to construct a better alternative—a requirement, it need hardly be pointed out, that has never been met by anyone who has argued that adaptations are poorly designed. Thus, while adaptations are certainly suboptimal in some ultimate sense, it is an empirically demonstrable fact that the short-run constraints on selective optimization do not prevent the emergence of superlatively organized computational adaptations in brains. Indeed, aside from the exotic nature of the problems the brain is designed to solve, it is exactly this sheer functional intricacy that makes our architecture so difficult to reverse-engineer and to understand.

Cognitive adaptations reflect the structure of the adaptive problem and the ancestral world

Looking at known adaptations not only can tell us about the overall engineering quality of evolved computational devices, but also can inform us about the general character of cognitive adaptations and adaptive problems. For example, hundreds of vision researchers, working over decades, have been mapping the exquisitely structured set of information-processing adaptations involved in vision. As Marr (1982) put it, the evolutionary function of vision is scene analysis—the reconstruction of models of real-world conditions from a two-dimensional visual array. As more and more functional subcomponents are explored, and as artificial intelligence researchers try to duplicate vision in computational systems attached to electronic cameras, four things have become clear (Marr, 1982; Poggio, Torre, and Koch, 1985). The first is that the magnitude of the computational problem posed by scene analysis is immensely greater than anyone had suspected before trying to duplicate it. Even something so seemingly simple as perceiving the same object as having the same color at different times of day turns out to require intensely specialized and complex computational machinery, because the spectral distribution of light reflected by the object changes widely with changes in natural illumination (see, e.g., Shepard, 1992). The second conclusion is that, as discussed, our visual system is a very well engineered set of cognitive adaptations, capable of recovering far more sophisticated information from two-dimensional light arrays than the best of the artificially engineered systems developed so far. The third is that successful vision requires specialized neural circuits or computational machinery designed particularly for solving the adaptive problem of scene analysis (Marr, 1982). And the fourth is that scene analysis is an unsolvable computational problem unless the design features of this specialized machinery "assume" that objects and events in the world manifest many specific regularities—that is, unless the cognitive procedures embody a complementary structure that reflects the problem-relevant parts of the world (Shepard, 1981, 1984, 1987a; Marr, 1982, Poggio, Torre, and Koch, 1985).

These four lessons—complexity of the adaptive information-processing problem, well-engineered problem-solving machinery as the evolved solution, specialization of the problem-solving machinery to fit the particular nature of the problem, and the requirement that the machinery embody substantial and contentful innate knowledge about the adaptive problem—recur throughout the study of the evolved computational subcomponents of our cognitive architecture (Cosmides and Tooby, 1987, 1992; Tooby and Cosmides, 1990a, 1990b; on language, see Chomsky, 1975, and Pinker, 1989, 1991; on vision, see Marr, 1982, and Poggio, Torre, and Koch, 1985). Well-studied adaptations overwhelmingly achieve their functional outcomes because they display an intricately engineered coordination between their specialized design features and the detailed structure of the task and the task environment. Like a code that has been torn in two and given to separate couriers, the two halves (the structure of the mechanism and the structure of the task) must be put together to be understood. In order to function, adaptations evolve such that their causal properties rely on and exploit these stable and enduring statistical and structural regularities in the world. Thus, to map the structures of our cognitive devices, we need to understand the structures of the problems they solve, and the problem-relevant parts of the hunter-gatherer world. If studying face recognition mechanisms, one must study the recurrent structure of faces. If studying social cognition, one must study the recurrent structure of hunter-gatherer social life. For vision the problems are not so very different for a modern scientist and a Pleistocene hunter-gatherer, so the folk notions of function that perception researchers use are not a serious problem. But the more one strays away from low-level perception, the more one needs to know about human behavioral ecology and the structure of the ancestral world.

Experimenting with ancestrally valid tasks and stimuli

Although bringing cognitive neuroscience current with modern evolutionary biology offers many new research tools (see, e.g., Preuss, this volume), we have out of necessity limited discussion to only one: an evolutionary functionalist research strategy (see chapter 79 and Tooby and Cosmides, 1992, for a description; for examples, see Barkow, Cosmides, and Tooby, 1992, and chapters 80, 83, and 84). The adoption of such an approach will modify research practice in many ways. Perhaps most significantly, researchers will no longer have to operate purely by intuition or guesswork in deciding what kinds of tasks and stimuli to expose subjects to. Using knowledge from evolutionary biology, behavioral ecology, animal behavior, and hunter-gatherer studies, they can construct ancestrally or adaptively valid stimuli and tasks. These are stimuli that would have had adaptive significance in ancestral environments, and tasks that resemble (at least in some ways) the adaptive problems our ancestors would have been selected for the ability to solve.

The currently widespread practice of using only arbitrary stimuli of no adaptive significance (e.g., lists of random words, colored geometric shapes), or abstract experimental tasks of unknown relevance to Pleistocene life has sharply limited what researchers have and can observe about our evolved computational devices. This is because the adaptive specializations that (arguably) constitute the majority of our neural architecture were designed to remain dormant until triggered by cues of the adaptively significant situations they are designed to handle. The Wundtian and British empiricist methodological assumption that complex stimuli, behaviors, representations, and competences are compounded out of simple ones has now been empirically falsified in scores of cases (see, e.g., Gallistel, 1990), and so restricting experimentation to such stimuli and tasks simply limits what researchers can find to an impoverished and unrepresentative set of phenomena. In contrast, experimenters who use more biologically meaningful stimuli have had far better luck, as the collapse of behaviorism and its replacement by modern behavioral ecology have shown in the study of animal behavior. To take our own research as only one example of its applicability to humans, effective mechanisms for Bayesian inference—undetected by 20 years of previous research using "modern" tasks and data formats—were activated by exposing subjects to information formatted in a way in which hunter-gatherers would have encountered it (Cosmides and Tooby, in press). Equally, when subjects were given ancestrally valid social inference tasks (cheater detection, threat interpretation), previously unobserved adaptive reasoning specializations were activated, guiding subjects to act according to evolutionarily predicted but otherwise unexpected patterns (Cosmides, 1989; Cosmides and Tooby, 1992).

Everyone accepts that one cannot study human language specializations by exposing subjects to meaningless sounds: The acoustic stimuli must contain the subtle, precise, high-level relationships that make sound language. Similarly, in order to move on to the study of other complex cognitive devices, researchers should expose their subjects to stimuli that contain the subtle, ancestrally valid relationships relevant to the diverse functions of those devices. In such an expanded research program, experimental stimuli and tasks would involve constituents like faces, smiles, expressions of disgust, foods, the depiction of socially significant situations, sexual attractiveness, habitat quality cues, animals, navigational problems, cues of kinship, rage displays, cues of contagion, motivational cues, distressed children, species-typical body language, rigid object mechanics, plants, predators, and other functional elements that would have been part of ancestral hunter-gatherer life. Investigations would look for functional subsystems that not only deal with such low-level and broadly functional competences as perception, attention, memory, and motor control, but also address higher-level ancestrally valid competences as well—mechanisms such as eye-direction detectors (Baron-Cohen, 1994), face recognizers (e.g., Johnson and Morton, 1991), food-memory subsystems (e.g., Hart, Berndt, and Caramazza, 1985), person-specific memory, child-care motivators, sexual jealousy modules, and so on.

Although these proposals to look for scores of content-sensitive circuits and domain-specific specializations will strike many as bizarre and even preposterous, they are well grounded in modern biology, and we think that in a decade or so they will look tame. If cognitive neuroscience is anything like investigations in modularist cognitive psychology and in modern animal behavior, researchers will be rewarded with the materialization of a rich array of functionally patterned phenomena: phenomena that have not been observed so far because the relevant mechanisms have not been activated in the laboratory by exposure to ecologically appropriate stimuli. Although the functions of most brain structures are still largely unknown, pursuing such research directions may begin to populate the empty regions of our maps of the brain with circuit diagrams of discrete, functionally intelligible computational devices.

NOTE

1. The genes underlying complex adaptations cannot vary substantially between individuals, because if they did, then the obligatory genetic shuffling that takes place during sexual reproduction would, in the offspring generation, break apart the complex adaptations that had existed in the parents. All of the genetic subcomponents necessary to build the complex adaptation would rarely reappear together in the same individual if they were not being reliably supplied by both parents in all matings (for a discussion of the genetics of sexual recombination, species-typical adaptive design, and individual differences, see Tooby, 1982; Tooby and Cosmides, 1990b).

REFERENCES

Barkow, J., L. Cosmides, and J. Tooby, eds., 1992. *The Adapted Mind: Evolutionary Psychology and the Generation of Culture.* New York: Oxford University Press.

BARON-COHEN, S., 1994. The eye-direction detector: A case for evolutionary psychology. In *Joint-Attention: Its Origins and Role in Development*, C. Moore and P. Dunham, eds. New Jersey: Erlbaum.

CHOMSKY, N., 1975. *Reflections on Language*. New York: Random House.

COSMIDES, L., 1989. The logic of social exchange: Has natural selection shaped how humans reason? Studies with the Wason selection task. *Cognition* 31:187–276.

COSMIDES, L., and J. TOOBY, 1981. Cytoplasmic inheritance and intragenomic conflict. *J. Theor. Biol.* 89:83–129.

COSMIDES, L., and J. TOOBY, 1987. From evolution to behavior: Evolutionary psychology as the missing link. In *The Latest on the Best: Essays on Evolution and Optimality*, J. Dupre, ed. Cambridge, Mass.: The MIT Press, pp. 277–306.

COSMIDES, L., and J. TOOBY, 1992. Cognitive adaptations for social exchange. In *The Adapted Mind: Evolutionary Psychology and the Generation of Culture*, J. Barkow, L. Cosmides, and J. Tooby, eds. New York: Oxford University Press.

COSMIDES, L., and J. TOOBY, 1994. From evolution to adaptations to behavior: Toward an integrated evolutionary psychology. In *Biological Perspectives on Motivated and Cognitive Activities*, Roderick Wong, ed. Norwood, N. J.: Ablex.

COSMIDES, L., and J. TOOBY, in press. Are humans good intuitive statisticians after all? Rethinking some conclusions of the literature on judgment under uncertainty. *Cognition*.

COSMIDES, L., J. TOOBY, and J. BARKOW, 1992. Evolutionary psychology and conceptual integration. In *The Adapted Mind: Evolutionary Psychology and the Generation of Culture*, J. Barkow, L. Cosmides, and J. Tooby, eds. New York: Oxford University Press.

DALY, M., and M. WILSON, 1984. *Sex, Evolution and Behavior*. 2 ed. Boston: Willard Grant.

DALY, M., M. WILSON, and S. J. WEGHORST, 1982. Male sexual jealousy. *Ethol. Sociobiology* 3:11–27.

DARWIN, C., 1859. *On the Origin of Species*. London: Murray; reprinted, Cambridge, Mass.: Harvard University Press.

DAWKINS, R., 1976. *The Selfish Gene*. New York: Oxford University Press.

DAWKINS, R., 1982. *The Extended Phenotype*. San Francisco: Freeman.

DAWKINS, R., 1986. *The Blind Watchmaker*. New York: Norton.

FREYD, J. J., 1987. Dynamic mental representations. *Psychol. Rev.* 94:427–438.

GALLISTEL, C. R., 1990. *The Organization of Learning*. Cambridge, Mass.: MIT Press.

GOULD, S. J., and R. C. LEWONTIN, 1979. The spandrels of San Marco and the Panglossian program: A critique of the adaptationist programme. *Proc. R. Soc. Lond.* 205:281–288.

HAMILTON, W. D., 1964. The genetical evolution of social behavior. *J. Theor. Biol.* 7:1–52.

HAMILTON, W. D., 1972. Altruism and related phenomena, mainly in social insects. *Annu. Rev. Ecol. Syst.* 3:193–232.

HART, JR., J., R. S. BERNDT, and A. CARAMAZZA, 1985. Category-specific naming deficit following cerebral infarction. *Nature* 316:439–440.

JACKENDOFF, R., 1992. *Languages of the Mind*. Cambridge, Mass.: MIT Press, Bradford Books.

JOHNSON, M. H., and J. MORTON, 1991. *Biology and Cognitive Development: The Case of Face Recognition*. Cambridge, Mass.: Blackwell.

KREBS, J. R., and N. B. DAVIES, 1987. *An Introduction to Behavioural Ecology*. Oxford: Blackwell.

LESLIE, A. M., 1987. Pretense and representation: The origins of "theory of mind". *Psychol. Rev.* 94:412–426.

LESLIE, A. M., 1988. The necessity of illusion: Perception and thought in infancy. In *Thought Without Language*, L. Weiskrantz, ed. Oxford: Clarendon, pp. 185–210.

LEWONTIN, R. C., 1967. Spoken remark in *Mathematical Challenges to the Neo-Darwinian Interpretation of Evolution*. (P. S. Moorhead and M. Kaplan, eds). *Wistar Institute Symposium Monograph* 5:79.

LEWONTIN, R. C., 1978. Adaptation. *Sci. Am.* 239:157–169.

MARR, D., 1982. *Vision: A Computational Investigation into the Human Representation and Processing of Visual Information*. San Francisco: Freeman.

MAYNARD SMITH, J., 1964. Group selection and kin selection. *Nature* 20:1145–1147.

MAYNARD SMITH, J., 1982. *Evolution and the Theory of Games*. Cambridge: Cambridge University Press.

MAYR, E., 1983. How to carry out the adaptationist program. *The Am. Nat.* 121:324–334.

PINKER, S., 1989. *Learnability and Cognition: The Acquisition of Argument Structure*. Cambridge, Mass.: MIT Press.

PINKER, S., 1991. Rules of language. *Science* 253:530–535.

PINKER, S., and P. BLOOM, 1992. Natural language and natural selection. Reprinted in *The Adapted Mind: Evolutionary Psychology and the Generation of Culture*, J. Barkow, L. Cosmides, and J. Tooby, eds. New York: Oxford University Press, pp. 451–493.

POGGIO, T., V. TORRE, and C. KOCH, 1985. Computational vision and regularization theory. *Nature* 317:314–319.

RAMACHADRAN, V. S., 1990. Visual perception in people and machines. In *AI and the Eye*, A. Blake and T. Troscianko, eds. New York: Wiley.

SHEPARD, R. N., 1981. Psychophysical complementarity. In *Perceptual Organization*, M. Kubovy and J. R. Pomerantz, eds. Hillsdale, N.J.: Erlbaum, pp. 279–341.

SHEPARD, R. N., 1984. Ecological constraints on internal representation: Resonant kinematics of perceiving, imagining, thinking, and dreaming. *Psychol. Rev.* 91:417–447.

SHEPARD, R. N., 1987a. Evolution of a mesh between principles of the mind and regularities of the world. In *The Latest on the Best: Essays on Evolution and Optimality*, J. Dupre, ed. Cambridge, Mass.: The MIT Press.

SHEPARD, R. N., 1987b. Towards a universal law of generalization for psychological science. *Science* 237:1317–1323.

SHEPARD, R. N., 1992. The perceptual organization of colors: An adaptation to regularities of the terrestrial world? In *The Adapted Mind: Evolutionary Psychology and the Genera-*

tion of Culture, J. Barkow, L. Cosmides, and J. Tooby, eds. New York: Oxford University Press, pp. 495–532.

SHERRY, D. F., and D. L. SCHACTER, 1987. The evolution of multiple memory systems. *Psychol. Rev.* 94:439–454.

SYMONS, D., 1987. If we're all Darwinians, what's the fuss about? In *Sociobiology and Psychology*, C. B. Crawford, M. F. Smith, and D. L. Krebs, eds. Hillsdale, N.J.: Erlbaum, pp. 121–146.

SYMONS, D., 1989. A critique of Darwinian anthropology. *Ethol. Sociobiol.* 10:131–144.

SYMONS, D., 1992. On the use and misuse of Darwinism in the study of human behavior. In *The Adapted Mind: Evolutionary Psychology and the Generation of Culture*, J. Barkow, L. Cosmides, and J. Tooby, eds. New York: Oxford University Press.

THORNHILL, R., 1990. The study of adaptation. In *Interpretation and Explanation in the Study of Animal Behavior*, Vol. 2, M. Bekoff and D. Jamieson, eds. Boulder, Colo.: Westview, pp. 31–62.

TOOBY, J., 1982. Pathogens, polymorphism and the evolution of sex. *J. Theor. Biol.* 97:557–576.

TOOBY, J., and L. COSMIDES, 1990a. On the universality of human nature and the uniqueness of the individual: The role of genetics and adaptation. *J. Pers.* 58:17–67.

TOOBY, J., and L. COSMIDES, 1990b. The past explains the present: Emotional adaptations and the structure of ancestral environments. *Ethol. Sociobiol.* 11:375–424.

TOOBY, J., and L. COSMIDES, 1992. The psychological foundations of culture. In *The Adapted Mind: Evolutionary Psychology and the Generation of Culture*, J. Barkow, L. Cosmides, and J. Tooby, eds. New York: Oxford University Press, pp. 19–136.

WILLIAMS, G. C., 1966. *Adaptation and Natural Selection: A Critique of Some Current Evolutionary Thought*. Princeton: Princeton University Press.

WILLIAMS, G. C., 1957. A defense of reductionism in evolutionary biology. *Oxf. Surv. Evol. Biol.* 2:1–27.

WILLIAMS, G. C., and D. C. WILLIAMS, 1957. Natural selection of individually harmful social adaptations among sibs with special reference to social insects. *Evolution* 17:249–253.

79 From Function to Structure: The Role of Evolutionary Biology and Computational Theories in Cognitive Neuroscience

LEDA COSMIDES AND JOHN TOOBY

ABSTRACT The cognitive neuroscience of central processes is currently a mystery. The brain is a vast and complex collection of functionally integrated circuits. Recognizing that natural selection engineers a fit between structure and function is the key to isolating these circuits. Neural circuits were designed to solve adaptive problems. If one can define an adaptive problem closely enough, one can see which circuits have a structural design that is capable of solving that problem. Evolutionary biologists have developed a series of sophisticated models of adaptive problems. Some of these models analyze constraints on the evolution of the cognitive processes that govern social behavior: cooperation, threat, courtship, kin-directed assistance, and so on. These forms of social behavior are generated by complex computational machinery. To discover the functional architecture of this machinery, cognitive neuroscientists will need the powerful inferential tools that evolutionary biology provides, including its well-defined theories of adaptive function.

The cognitive sciences have been conducted as if Darwin never lived. Their goal is to isolate functionally integrated subunits of the brain and determine how they work. Yet most cognitive scientists pursue that goal without any clear notion of what "function" means in biology. When a neural circuit is discovered, very few researchers ask what its adaptive function is. Even fewer use theories of adaptive function as tools for discovering heretofore unknown neural systems. Indeed, many people in our field think that theories of adaptive function are an explanatory luxury—fanciful, unfalsifiable speculations that one indulges in at the end of a project, after the hard work of figuring out the structure of a circuit has been done.

LEDA COSMIDES Department of Psychology, and JOHN TOOBY Department of Anthropology, Center for Evolutionary Psychology, University of California, Santa Barbara, Calif.

In this chapter, we will argue that theories of adaptive function are not a luxury. They are a necessity, crucial to the future development of cognitive neuroscience. Without them, cognitive neuroscientists will not know what to look for and will not know how to interpret their results. As a result, they will be unable to isolate functionally integrated subunits of the brain.

Explanation and discovery in cognitive neuroscience

> [t]rying to understand perception by studying only neurons is like trying to understand bird flight by studying only feathers: it just cannot be done. In order to understand bird flight, we have to understand aerodynamics; only then do the structure of feathers and the different shapes of birds' wings make sense. (Marr, 1982, 27)

David Marr developed a general explanatory system for cognitive science that is much cited but rarely applied. His three-level system applies to any device that processes information—a calculator, a cash register, a television, a computer, a brain. It is based on the following observations:

1. Information-processing devices are designed to solve problems.
2. They solve problems by virtue of their structure.

3. Hence, to explain the structure of a device, one needs to know

a. *what* problem it was designed to solve, and

b. *why* it was designed to solve that problem and not some other one.

In other words, one must develop a task analysis of the problem, or what Marr called a *computational theory*. Knowing the physical structure of a cognitive device and the information-processing program realized by that structure is not enough. For human-made artifacts and biological systems, form follows function. The physical structure is there because it embodies a set of programs; the programs are there because they solve a particular problem. A computational theory specifies what that problem is and why there is a device to solve it. It specifies the *function* of an information-processing device. Marr felt that the computational theory was the most important and the most neglected level of explanation in the cognitive sciences.

This functional level of explanation has not been neglected in the biological sciences, however, because it is essential for understanding how natural selection designs organisms (for background, see chapter 78). An organism's phenotypic structure can be thought of as a collection of design features—of machines, such as the eye or liver. A design feature can cause its own spread by solving adaptive problems—problems, such as detecting predators or detoxifying poisons, that recur over many generations and whose solution tends to promote reproduction. Natural selection is a feedback process that "chooses" among alternative designs on the basis of how well they function. By selecting designs on the basis of how well they solve adaptive problems, this process engineers a tight fit between the function of a device and its structure. To understand this causal relationship, biologists had to develop a theoretical vocabulary that distinguishes between structure and function. Marr's computational theory is a functional level of explanation that corresponds roughly to what biologists refer to as the "ultimate" or "functional" explanation of a phenotypic structure.

Even though there is a close causal relationship between the function of an information-processing device and its structure, a computational theory of a device does not uniquely specify its structure. This is because there are many ways to skin a cat. More precisely:

4. Many different information-processing programs can solve the same problem. These programs may differ in how they represent information, in the processes whereby they transform input into output, or both. So knowing the goal of a computation does not uniquely determine the design of the program that realizes that goal in the device under consideration.

5. Many different physical systems—from neurons in a brain to silicon chips in a computer—can implement the same program.[1] So knowing the structure of a program does not uniquely determine the properties of the physical system that implements it. Moreover, the same physical system can implement many programs, so knowing the physical properties of a system cannot tell one which programs it implements (table 79.1)

A computational theory defines what problem the device solves and why it solves it, but it does not specify *how* this is accomplished; theories about programs and their physical substrate specify how the device solves the problem. Each explanatory level addresses a different question. To understand an information-processing device completely, Marr argued, one needs explanations on all three levels: computational theory, programming, and hardware (see table 79.1).

A computational theory of function is more than an explanatory luxury, however. It is an essential tool for discovery in the cognitive and neural sciences. Whether a mechanism was designed by natural selection or

TABLE 79.1

Three levels at which any machine carrying out an information-processing task must be understood

1. *Computational theory:*
What is the goal of the computation, why is it appropriate, and what is the logic of the strategy by which it can be carried out?

2. *Representation and algorithm:*
How can this computational theory be implemented? In particular, what is the representation for the input and output, and what is the algorithm for the transformation?

3. *Hardware implementation:*
How can the representation and algorithm be realized physically?

In evolutionary biology:
Explanations at the level of the computational theory are called *ultimate*-level explanations.
Explanations at the level of representation and algorithm, or at the level of hardware implementation, are called *proximate*-level explanations.

From Marr, 1982, 25.

by the intentional actions of a human engineer, one can count on there being a close causal relationship between its structure and its function. A theory of function may not determine a program's structure uniquely, but it reduces the number of possibilities to an empirically manageable number. Task demands radically constrain the range of possible solutions; consequently, very few cognitive programs are capable of solving any given adaptive problem. By developing a careful task analysis of an information-processing problem, one can vastly simplify the empirical search for the cognitive program that solves it. And once that program has been identified, it is easy to develop clinical tests that will target its neural basis.

It is currently fashionable to think that the findings of neuroscience will eventually place strong constraints on theory formation at the cognitive level. In this view, once we know enough about the properties of neurons, neurotransmitters, and cellular development, figuring out what cognitive programs the human mind contains will become a trivial task. This cannot be true. There are millions of animal species on earth, each with a different set of cognitive programs. *The same basic neural tissue embodies all of these programs.* Facts about the properties of neurons, neurotransmitters, and cellular development cannot tell one which of these millions of programs the human mind contains.

The cognitive structure of an information-processing device "depends more upon the computational problems that have to be solved than upon the particular hardware in which their solutions are implemented" (Marr, 1982, 27). In other words, knowing *what* and *why* allows one to generate focused hypotheses about *how*. To figure out how the mind works, cognitive neuroscientists will need to know what problems our cognitive and neural mechanisms were designed to solve.

Beyond intuition: How to build a computational theory

To illustrate the notion of a computational theory, Marr asks us to consider the *what* and *why* of a cash register at a checkout counter in a grocery store. We know the *what* of a cash register: It adds numbers. Addition is an operation that maps pairs of numbers onto single numbers, and it has certain abstract properties, such as commutativity and associativity (table 79.2). How the addition is accomplished is quite irrelevant: Any set of representations and algorithms that satisfies these abstract constraints will do. The input to the cash register is prices, which are represented by numbers. To compute a final bill, the cash register adds these numbers together. That's the *what.*

But *why* was the cash register designed to add the

TABLE 79.2
Why cash registers add

Rules defining addition	Rules governing social exchange in a supermarket
There is a unique element, "zero"; Adding zero has no effect: $2 + 0 = 2$	If you buy nothing, it should cost you nothing; and buying nothing and something should cost the same as buying just the something. (The rules of zero)
Commutativity: $(2 + 3) = (3 + 2) = 5$	The order in which goods are presented to the cashier should not affect the total. (Commutativity)
Associativity: $(2 + 3) + 4 = 2 + (3 + 4)$	Arranging the goods into two piles and paying for each pile separately should not affected the total amount you pay. (Associativity; the basic operation for combining prices)
Each number has a unique inverse that when added to the number gives zero: $2 + (-2) = 0$	If you buy an item and then return it for a refund, your total expenditure should be zero. (Inverses)

Adapted from Marr, 1982, 22–23.

prices of each item? Why not multiply them together, or subtract the price of each item from 100? According to Marr, "the reason is that the rules we *intuitively feel to be appropriate* for combining the individual prices in fact define the mathematical operation of addition" (p. 22, emphasis added). He formulates these intuitive rules as a series of constraints on how prices should be combined when people exchange money for goods, then shows that these constraints map directly onto those that define addition (see table 79.2). On this view, cash registers were designed to add because addition is the mathematical operation that realizes the constraints on buying and selling that our intuitions deem appropriate. Other mathematical operations are inappropriate because they violate these intuitions; for example, if the cash register substracted each price from 100, the more goods you chose the less you would pay—and if you chose enough goods, the store would pay *you*.

In this particular example, the buck stopped at intuition. But it shouldn't. Our intuitions are produced by the human brain, an information-processing device that was designed by the evolutionary process. To discover the structure of the brain, one needs to know *what* problems it was designed to solve and *why* it was designed to solve those problems rather than some others. In other words, one needs to ask the same questions of the brain as one would of the cash register. Cognitive science is the study of the design of minds, regardless of their origin. Cognitive neuroscience is the study of the design of minds that were produced by the evolutionary process. Evolution produced the what, and evolutionary biology is the study of why. Most cognitive neuroscientists know this. What they don't yet know is that understanding the evolutionary process can bring the architecture of the mind into sharper relief. For biological systems, the nature of the designer carries implications for the nature of the design.

The brain can process information because it contains complex neural circuits that are functionally organized. The only component of the evolutionary process that can build complex structures that are functionally organized is natural selection. And the only kind of problems that natural selection can build complexly organized structures for solving are adaptive problems (Williams, 1966; Dawkins, 1986; Tooby and Cosmides, 1990, 1992, this volume). Bearing this in mind, let us consider the source of Marr's intuitions about the cash register. Buying food at a grocery store is a form of social exchange—cooperation between two or more individuals for mutual benefit. The adaptive problems that arise when individuals engage in this form of cooperation have constituted a long-enduring selection pressure on the hominid line. Paleoanthropological evidence indicates that social exchange extends back at least two million years in the human line, and the fact that social exchange exists in some of our primate cousins suggests that it may be even more ancient than that. It is exactly the kind of problem that selection can build cognitive mechanisms for solving.

Social exchange is not a recent cultural invention, like writing, yam cultivation, or computer programming; if it were, one would expect to find evidence of its having one or several points of origin, of its having spread by contact, and of its being extremely elaborated in some cultures and absent in others. But its distribution does not fit this pattern. Social exchange is both universal and highly elaborated across human cultures, presenting itself in many forms: reciprocal gift-giving, food sharing, market pricing, and so on (Cosmides and Tooby, 1992; Fiske, 1992). It is an ancient, pervasive, and central part of human social life.

The computational mechanisms that give rise to social exchange behavior in a species must satisfy certain *evolvability constraints*. Selection cannot construct mechanisms in any species—including humans—that systematically violate these constraints. In evolutionary biology, researchers such as Robert Trivers, W. D. Hamilton, and Robert Axelrod have explored constraints on the evolution of social exchange using game theory, modeling it as a repeated Prisoner's Dilemma. These analyses have turned up a number of important features of this adaptive problem, a crucial one being that social exchange cannot evolve in a species unless individuals have some means of detecting individuals who cheat and excluding them from future interactions (e.g., Williams & Williams, 1957; Trivers, 1971; Axelrod and Hamilton, 1981; Axelrod, 1984; Boyd, 1988).

Behavioral ecologists have used these constraints on the evolution of social exchange to build computational theories of this adaptive problem—theories of what and why. These theories have provided a principled basis for generating hypotheses about the phenotypic design of mechanisms that generate social exchange in a variety of species. They spotlight design features that any cognitive program capable of solving this adaptive problem must have. By cataloging these design features, animal behavior researchers were able to look for—and discover—previously unknown aspects of the psychology of social exchange in species from chimpanzees, baboons and vervets to vampire bats and hermaphroditic coral-reef fish (e.g., Smuts, 1986; de Waal and Luttrell, 1988; Fischer, 1988; Wilkinson, 1988, 1990). This research strategy has been successful for a very simple reason: Very few cognitive programs satisfy the evolvability constraints for social exchange. If a species engages in this behavior (and not all do), then its cognitive architecture must contain one of these programs.

In our own species, social exchange is a universal, species-typical trait with a long evolutionary history. We have strong and cross-culturally reliable intuitions about how this form of cooperation should be conducted, which arise in the absence of any explicit instruction (Cosmides and Tooby, 1992; Fiske, 1992). In developing his computational theory of the cash register—a tool used in social exchange—David Marr was consulting these deep human intuitions.[2]

From these facts, we can deduce that the human cognitive architecture contains programs that satisfy the evolvability constraints for social exchange. As cognitive scientists, we should be able to specify what rules govern human behavior in this domain, and why we humans reliably develop circuits that embody these rules rather than others. In other words, we should be able to develop a computational theory of the organic information-processing device that governs social exchange in humans.

The empirical advantages of using evolutionary biology to develop computational theories of adaptive problems had already been amply demonstrated in the study of animal minds (e.g., Gould, 1982; Krebs and Davies, 1987; Gallistel, 1990; Real, 1991). We wanted to test its utility for studying the human mind. A powerful way of doing this would be to use an evolutionarily derived computational theory to discover cognitive mechanisms whose existence no one had previously suspected. By using evolvability constraints, we developed a computational theory of social exchange (Cosmides, 1985; Cosmides and Tooby, 1989). It suggested that the cognitive processes that govern human reasoning might have a number of design features specialized for reasoning about social exchange—what Gallistel (this volume; also Rozin, 1976) calls *adaptive specializations*.

The goal of our research is to recover, out of carefully designed experimental studies, high-resolution "maps" of the intricate mechanisms that collectively constitute the human mind. Our evolutionarily derived computational theory of social exchange has been allowing us to do that. It led us to predict a large number of design features in advance—features that no one was looking for and that most of our colleagues thought were outlandish. Experimental tests have confirmed the presence of all the design features that have been tested for so far. Those design features that have been tested and confirmed are listed in table 79.3, along with the alternative by-product hypotheses that we and our colleagues have eliminated. So far, no known theory invoking general-purpose cognitive processes has been able to explain the very precise and unique pattern of data that tests like these have generated. The data are best explained by the hypothesis that humans reliably develop circuits that are complexly specialized for reasoning about reciprocal social interactions. Parallel lines of investigation indicate that humans have also evolved additional, differently structured circuits that are specialized for reasoning about aggressive threats and protection from hazards (e.g., Manktelow and Over, 1990; Tooby and Cosmides, 1989). We are now planning clinical tests to find the neural basis for these mechanisms. By studying patient populations with autism and other neurological impairments of social cognition, we should be able to see whether dissociations occur along the fracture lines suggested by our various computational theories. (For a description of the relevant social exchange experiments, see Cosmides, 1985, 1989; Cosmides and Tooby, 1992; Gigerenzer and Hug, 1992.)

Since Marr, cognitive scientists have become familiar with the notion of developing computational theories to study perception and language, but the notion that one can develop computational theories to study the information-processing devices that give rise to social behavior is still quite alien. Yet some of the most important adaptive problems our ancestors had to solve involved negotiating the social world, and some of the best work in evolutionary biology is devoted to analyzing constraints on the evolution of mechanisms that solve these problems. There are many reasons for the neglect of these topics in the study of humans, but the primary one is that cognitive scientists have been relying on their intuitions for hypotheses rather than asking themselves what kind of problems the mind was designed to solve. Evolutionary biology addresses that question. Consequently, evolutionary

Table 79.3

*Reasoning about social exchange: Evidence of special design**

The following design features were predicted and found:	A number of by-product hypotheses were empirically eliminated. It was shown that:
The algorithms governing reasoning about social contracts operate even in unfamiliar situations.	Familiarity cannot explain the social contract effect.
The definition of cheating that these algorithms embody depends on one's perspective.	It is not the case that social contract content merely facilitates the application of the rules of inference of the propositional calculus.
They are just as good at computing the cost-benefit representation of a social contract from the perspective of one party as from the perspective of another.	Social contract content does not merely "afford" clear thinking.
They embody implicational procedures specified by the computational theory.	Permission schema theory cannot explain the social contract effect; in other words, application of a generalized deontic logic cannot explain the results.
They include inference procedures specialized for cheater detection.	It is not the case that any problem involving payoffs will elicit the detection of violations.
Their cheater-detection procedures cannot detect violations of social contracts that do not correspond to cheating.	
They do not include altruist detection procedures.	
They cannot operate so as to detect cheaters unless the rule has been assigned the cost-benefit representation of a social contract.	

* To show that an aspect of the phenotype is an adaptation to perform a particular function, one must show that it is particularly well designed for performing that function, and that it cannot be better explained as a by-product of some other adaptation or physical law.

biology places important constraints on theory formation in cognitive neuroscience, constraints from which one can build computational theories of adaptive information-processing problems.

Organism design theory

Knowing that the circuitry of the human mind was designed by the evolutionary process tells us something centrally illuminating: that, aside from those properties acquired by chance or imposed by engineering constraints, the mind consists of a set of information-processing circuits that were designed by natural selection to solve those adaptive problems that our hunter-gatherer ancestors faced generation after generation (see chapter 78). The better we understand the evolutionary process, adaptive problems, and ancestral life, the more intelligently we can explore and map the intricacies of the human mind.

Figuring out the structure of an organism is an exercise in reverse engineering; the field of evolutionary biology summarizes our knowledge of the engineering principles that govern the design of organisms. Taken together, these principles constitute an *organism design theory*. A major activity of evolutionary biologists is the exploration and definition of adaptive problems. By combining results derived from mathematical modeling, comparative studies, behavioral ecology, paleoanthropology, and other fields, evolutionary biologists try to identify what problems the mind was designed to solve and why it was designed to solve those problems rather than some other ones. In other words, they explore exactly those questions that Marr argued were essential for developing computational theories of adaptive information-processing problems.

Computational theories address what and why, but because there are multiple ways of achieving any solution, they are not sufficient to specify how. But the more closely one can define what and why—the more one can constrain what would count as a solution—the more clearly we can see which hypotheses about mechanisms are viable and which are not. The more constraints one can discover, the more the field of possible solutions is narrowed, and the more one can concentrate empirical efforts on discriminating between viable hypotheses.

Natural selection is capable of producing only certain kinds of designs: designs that have promoted their own reproduction in past environments. It constrains what counts as an adaptive problem, and therefore narrows the field of possible solutions. In evolutionary analyses, cognitive scientists will discover a rich and surprisingly powerful source of constraints from which precise computational theories can be built. Indeed, these analyses provide the only source of constraints for the cognitive processes that govern human social behavior. Table 79.4 lists families of constraints that cognitive scientists could be using, but are not.

We would like to illustrate this point with an extended example involving social behavior. Consider Hamilton's rule, which describes the selection pressures operating on mechanisms that generate behaviors that have a reproductive impact on an organism and its kin (Hamilton, 1964). The rule defines (in part) what counts as biologically successful outcomes in these kinds of situations. These outcomes often cannot be reached unless specific information is obtained and processed by the organism.

In the simplest case of two individuals, a mechanism that produces acts of assistance has an evolutionary advantage over alternative mechanisms if it reliably causes individual i to help relative j whenenever $C_i < r_{i,j}B_j$. In this equation, C_i is the cost to i of rendering an act of assistance to j, measured in terms of foregone reproduction; B_j is the benefit to j of receiving that act of assistance, measured in terms of enhanced reproduction; and $r_{i,j}$ is the probability that a randomly sampled gene will be present at the same locus in the relative due to joint inheritance from a common ancestor.

Other things being equal, the more closely the behaviors produced by cognitive mechanisms conform to Hamilton's rule, the more strongly those mechanisms will be selected for. A design feature that systematically caused an individual to help more than this—or less than this—would be selected against.

This means that the cognitive programs of an organism that confers benefits on kin cannot violate Hamilton's rule. Cognitive programs that systematically violate this constraint cannot be selected for. Cognitive programs that satisfy this constraint can be selected for. A species may lack the ability to confer benefits on kin, but if it has such an ability, then it has it by virtue of cognitive programs that produce behavior that respects this constraint. Hamilton's rule is completely general: It is inherent in the dynamics of natural selection, true of any species on any planet at any time. One can call theoretical constraints of this kind *evolvability constraints*; they specify the class of mechanisms that can, in principle, evolve (Tooby and Cosmides, 1992; Cosmides and Tooby, 1994). The evolvability con-

TABLE 79.4

Evolutionary biology provides constraints from which computational theories of adaptive information-processing problems can be built

To build a computational theory, one must answer two questions:

1. What is the adaptive problem?

2. What information would have been available in ancestral environments for solving it?

Some sources of constraints

1. More precise definition of Marr's "goal" of processing that is appropriate to evolved (as opposed to artificial) information-processing systems

2. Game-theoretic models of the dynamics of natural selection (e.g., kin selection, Prisoner's Dilemma, and cooperation—particularly useful for analysis of cognitive mechanisms responsible for social behavior)

3. Evolvability constraints: Can a design with properties X, Y, and Z evolve, or would it have been selected out by alternative designs with different properties? (i.e., does the design represent an evolutionarily stable strategy?—related to point 2)

4. Hunter-gatherer studies and paleoanthropology—source of information about the environmental background against which our cognitive architecture evolved (Information that is present now may not have been present then, and vice versa.)

5. Studies of the algorithms and representations whereby other animals solve the same adaptive problem (These will sometimes be the same, sometimes different.)

straints for one adaptive problem usually differ from those for another.

Under many ecological conditions, this selection pressure defines an information-processing problem for whose solution organisms will be selected to evolve mechanisms. Hamilton's rule answers the three questions that Marr said a computational theory of an information-processing problem should answer: It identifies the goal of a computation, why it is relevant, and the logic of the strategy by which it can be carried out (Marr, 1982, 25; see table 79.1).

Using this description of an adaptive problem as a starting point, one can immediately begin to define the cognitive subtasks that would have to be addressed by any set of mechanisms capable of producing behavior that conforms to this rule. What information-processing mechanisms evolved to reliably identify relatives, for example? What criteria and procedures to do they embody? That is, do these mechanisms define an individual as a sibling if that individual (a) was nursed by the same female who nursed you, (b) resided in close contact with you during your first three years of life, or (c) looks or smells similar to your mother, within a certain error tolerance? What kind of information is processed to estimate $r_{i,j}$, the degree of relatedness? Under ancestral conditions, did siblings and cousins coreside, such that one might expect the evolution of mechanisms that discriminate between the two? After all, $r_{i,\text{full sib}} = 4r_{i,\text{first cousin}}$. What kind of mechanisms would have been capable of estimating the magnitudes of the consequences of specific actions on one's own and on others' reproduction? (For example, the estimation procedures of vampire bats could be tied directly to volume of regurgitated blood fed to a relative, as this is the only form of help they give.) What kinds of decision rules combine these various pieces of information to produce behavior that conforms to Hamilton's rule? And so on.

This example highlights several points about the connection between evolutionary biology and the cognitive sciences:

1. *Knowledge drawn from evolutionary biology can be used to discover previously unknown functional organization in our cognitive architecture.* Hamilton's rule is not intuitively obvious; researchers would not look for cognitive mechanisms that are well designed for producing behavior that conforms to this rule unless they had already heard of it. After Hamilton's rule had been formulated, behavioral ecologists began to discover psychological mechanisms that embodied it in many nonhuman animals (Krebs and Davies, 1987). Unguided empiricism is unlikely to uncover a mechanism that is well designed to solve a problem of this kind.

2. *By using the definition of an adaptive problem, one can easily generate hypotheses about the design features of information-processing mechanisms, even when these mechanisms are designed to produce social behavior.* Knowing the definition of the problem allows one to break it down into cognitive subtasks, such as kin recognition, kin categorization, and cost-benefit estimation, in the same way that knowing that the adaptive function of the visual system is scene analysis allows one to identify subtasks such as depth perception and color constancy.

3. *Knowing the ancestral conditions under which a species evolved can suggest fruitful hypotheses about design features of the cognitive adaptations that solve the problem.* For example, the key task in developing a computational theory of kin identification is identifying cues that would have been reliably correlated with kinship in ancestral environments without also being correlated with lack of kinship. If there are no such cues, then a kin identification mechanism cannot be selected for. If there are several possible cues, then empirical tests are the only way to determine which one(s) the system uses. Even so, considering what kind of information was available simplifies the task immensely: Coresidence is a reliable cue of sibhood in some species, but other cues would have to be picked up and processed in a species in which siblings and cousins coreside.

4. *Knowing about ancestral conditions can help one avoid conceptual wrong turns in the interpretation of data.* The cue "looks like me" is not a good candidate cue for kin identification, because our hunter-gatherer ancestors did not have mirrors. It therefore would have been difficult to form an accurate template of one's own face for comparison. If one were to find data suggesting that this cue is used, one should consider conducting tests to see whether this is an incidental correlation caused by the use of a more likely cue, such as "looks like my mother."

5. *A computational theory built from evolutionary constraints can provide a standard of good design.* A design for solving this adaptive problem can be evaluated by determining how closely it produces behavior that tracks Hamilton's rule. Standards of good design are an essential tool for cognitive scientists because they allow one to determine whether a hypothesized mechanism is capa-

ble of solving the adaptive problem in question and to decide whether that mechanism would have done a better job under ancestral conditions than alternative designs.

Some programs are not capable of solving a particular problem. Hypotheses that propose such programs should be eliminated from consideration. Cognitive scientists have developed powerful methods for determining whether a program is capable of solving a problem, but these methods can be used only if one has a detailed computational theory defining what the problem is. Two particularly powerful methods are as follows:

a. *Computational modeling.* One can implement the program on a computer, run the program, and see what happens.

b. *Solvability analysis.* Theoretical analyses can sometimes reveal that a proposed program is incapable of solving a problem. These analyses can be formal or informal. The learnability analyses used in developmental psycholinguistics are of both varieties (Pinker, 1979, 1984; Wexler and Culicover, 1980). The problem in question is how a child learns the grammar of his or her native language, given the information present in the child's environment. Mathematical or logical theorems can sometimes be used to prove that programs with certain formal properties are incapable of solving this problem. Informally, a grammar-learning program that works only if the child gets negative feedback about grammatical errors can be eliminated from consideration if one can show that the necessary feedback information is absent from the child's environment.

The use of these powerful methods has been largely restricted to the study of vision and language, where cognitive scientists have developed computational theories. But these methods can be applied to many other adaptive problems—including ones involving social behavior—if evolutionary analyses are used to develop comptuational theories of them. For example, because Hamilton's rule provides a standard of good design, it can be used to evaluate the popular assumption that "central" processes in humans are general purpose and content-free (e.g., Fodor, 1983).

Content-free systems are limited to knowing what can be validly derived by general processes from perceptual information. Imagine, then, a content-free architecture situated in an ancestral hunter-gatherer. When the individual with this architecture sees a relative, there is nothing in the stimulus array that tells her how much she should help that relative. And there is no consequence that she can observe that tells her whether, from a fitness point of view, she helped too much, not enough, or just the right amount, where "the right amount" is defined by $C_{ego} < r_{ego,j}B_j$. Ancestral environments lack the information necessary for inducing this rule ontogenetically (as do modern ones, for that matter). Even worse, the correct rule cannot be learned from others: An implication of Hamilton's rule is that selection will design circuits that motivate kin to socialize a child into behaving in ways that are contrary to the very rule that the child must induce (Trivers, 1974).

By developing a computational theory based on Hamilton's rule, one can easily see that a content-free architecture fails even an informal solvability test for this adaptive problem. And, because Hamilton's rule defines a particularly strong selection pressure, the content-free architecture also fails an evolvability test (Tooby and Cosmides, 1992; Cosmides and Tooby, 1994).

6. *Insights from evolutionary biology can bring functional organization into clear focus at the cognitive level, but not at the neurobiological level.* Hamilton's rule immediately suggests hypotheses about the functional organization of mechanisms described in information-processing terms, but it tells one very little about the neurobiology that implements these mechanisms—it cannot be straightforwardly related to hypotheses about brain chemistry or neuroanatomy. However, once one knows the properties of the cognitive mechanisms that solve this adaptive problem, it should be far easier to discover the structure of the neural mechanisms that implement them (see Tooby and Cosmides, 1992, and chapter 78). The key to finding functional organization at the neural level is finding functional organization at the cognitive level.

Hamilton's rule is a rich source of constraints from which to build computational theories of the adaptive problems associated with kin-directed social behavior. But it is not unique in this regard. When mathematical game theory was incorporated into evolutionary analyses, it became clear that natural selection constrains which kinds of circuits can evolve. For many domains of human activity, evolutionary biology can be used to determine what kind of circuits would have been quickly selected out, and what kind were likely to have become universal and species-typical. For this reason, knowledge of natural selection and of the ancestral environments in which it operated can be used to create computational theories of adaptive information-processing problems. Evolutionary biology provides a principled way of deciding what domains are likely to have associated modules[3] or mental organs—it allows one to pinpoint adaptive problems that the human mind must be able to solve with special efficiency, and it suggests design features that any mechanism capable of solving these problems must have. Of equal importance, evolutionary biology provides the definition of successful processing that is most relevant to the study of *biological* information processing systems: It gives technical content to the concept of function, telling the psychologist what adaptive goals our cognitive mechanisms must be able to accomplish.

The approach employed by Marr and others—developing computational theories of a problem defined in functional terms—has been very successful, especially in the field of perception, where the function or goal of successful processing is intuitively obvious. But for most kinds of adaptive problems (and, therefore, for most of our cognitive mechanisms), function is far from obvious, and intuition uninformed by modern biology is unreliable or misleading. In social cognition, for example, what constitutes adaptive or functional reasoning is a sophisticated biological problem in itself, and is not susceptible to impressionistic, ad hoc theorizing. There exists no domain-general standard for adaptation or successful processing, therefore functionality must be assessed through reference to evolutionary biology, adaptive problem by adaptive problem.

Fortunately, over the last 30 years, there have been rapid advances in the technical theory of adaptation. There are now a series of sophisticated models of what constitutes adaptive behavior in different domains of human life, especially those that involve social behavior. It is therefore possible to develop, out of particular areas of evolutionary biology, computational theories of the specialized cognitive abilities that were necessary for adaptive conduct in humans.

Conclusion

Textbooks in psychology are organized according to a folk-psychological categorization of mechanisms: attention, memory, reasoning, learning. In contrast, textbooks in evolutionary biology and behavioral ecology are organized according to adaptive problems: foraging (hunting and gathering), predator avoidance, resource competition, fighting, coalitional aggression, dominance and status, inbreeding avoidance, sexual attraction, courtship, pair-bond formation, trade-offs between mating effort and parenting effort, mating system, sexual conflict, paternity uncertainty and sexual jealousy, parental investment, discriminative parental care, reciprocal altruism, kin altruism, cooperative hunting, signaling and communication, navigation, habitat selection. Behavioral ecologists and evolutionary biologists have created a library of sophisticated models of the selection pressures, strategies, and trade-offs that characterize these adaptive problems.

Which model is applicable for a given species depends on certain key life-history parameters. Findings from paleoanthropology, hunter-gatherer archeology, and studies of the ways of life of modern hunter-gatherer populations locate humans in this theoretical landscape by filling in the critical parameter values. Ancestral hominids were savannah-living primates; omnivores, exposed to a wide variety of plant toxins and having a sexual division of labor between hunting and gathering; mammals with altricial young, long periods of biparental investment in offspring, pair-bonds, and an extended period of physiologically obligatory female investment in pregnancy and lactation. They were a long-lived, low-fecundity species in which variance in male reproductive success was higher than variance in female reproductive success. They lived in small, nomadic, kin-based bands of perhaps 50 to 100; they would rarely have seen more than 1000 people at one time; they had little opportunity to store provisions for the future; they engaged in cooperative hunting, defense, and aggressive coalitions; they made tools and engaged in extensive amounts of cooperative reciprocation; they were vulnerable to a large variety of parasites and pathogens. When these parameters are combined with formal models from evolutionary biology and behavioral ecology, a reasonably consistent picture of ancestral life begins to appear (e.g., Tooby and DeVore, 1987). In this picture, the adaptive problems posed by social life loom large. Most of these are characterized by strict evolvability constraints, which could only be satisfied by cognitive programs that are specialized for reasoning about the social world. This suggests that our evolved mental architecture contains a large and intricate "faculty" of social cognition (Brothers, 1990; Cosmides and Tooby, 1992; Fiske, 1992; Jackendoff, 1992). Yet virtually no work in cognitive neuroscience is devoted to looking for dissociations between different forms of social reasoning, or between social reasoning and other cognitive functions. The work on autism as a neurological impairment of a "theory of mind" module is a notable and very successful exception (e.g., Baron-Cohen, Leslie, and Frith, 1985; Frith, 1989; Leslie, 1987.)

Textbooks in evolutionary biology are organized according to adaptive problems because these are the only problems that selection can build mechanisms for solving. Textbooks in behavioral ecology are organized according to adaptive problems because circuits that are functionally specialized for solving these problems have been found in species after species. No less should be true of humans. To find such circuits, however, cognitive neuroscientists will need the powerful inferential tools that evolutionary biology provides.

Through the computational theory, evolutionary biology allows the matching of algorithm to adaptive problem: Evolutionary biology defines information-processing problems that the mind must be able to solve, and the task of cognitive neuroscience is to uncover the nature of the algorithms that solve them. The brain's microcircuitry was designed to implement these algorithms, so a map of their cognitive structure can be used to bring order out of chaos at the neural level.

Atheoretical approaches will not suffice—a random stroll through hypothesis space will not allow one to distinguish figure from ground in a complex system. To isolate a functionally integrated mechanism within a complex system, one needs a theory of what function that mechanism was designed to perform. Sophisticated theories of adaptive function are therefore essential if cognitive neuroscience is to flourish.

ACKNOWLEDGMENTS We thank Steve Pinker and Mike Gazzaniga for valuable discussions of the issues discussed in this chapter. We are grateful to the McDonnell Foundation for financial support, and also for NSF grant BNS9157-449 to John Tooby.

NOTES

1. For example, consider the fact that certain text editing programs, such as WordStar, have been implemented on machines with different hardware architectures. The program is the same, in the sense that functional relationships among representations are preserved. The same inputs produce the same outputs: ^G always erases a letter, ^KV always moves a block, and so on.
2. Had Marr known about the importance of cheating in evolutionary analyses of social exchange, he might have been able to understand other features of the cash register as well. Most cash registers have anticheating devices: cash drawers that lock until a new set of prices is punched in, two rolls of tape that keep track of transactions (one is for the customer; the other rolls into an inaccessible place in the cash register, preventing the clerk from altering the totals to match the amount of cash in the drawer). In a way akin to the evolutionary process, as more sophisticated technologies become available and cheap, one might expect the anticheating design features of cash registers to become more sophisticated as well.
3. We do not mean "modules" in Fodor's sense; his criteria do not lay appropriate emphasis on functional organization for solving adaptive problems.

REFERENCES

Axelrod, R., 1984. *The Evolution of Cooperation.* New York: Basic Books.

Axelrod, R., and W. D. Hamilton, 1981. The evolution of cooperation. *Science* 211:1390–1396.

Baron-Cohen, S., A. Leslie, and U. Frith, 1985. Does the autistic child have a "theory of mind"? *Cognition* 21:37–46.

Boyd, R., 1988. Is the repeated prisoner's dilemma a good model of reciprocal altruism? *Ethol. Sociobiol.* 9:211–222.

Brothers, L., 1990. The social brain: A project for integrating primate behavior and neurophysiology in a new domain. *Concepts Neurosci.* 1:27–51.

Cheney, D. L., and R. Seyfarth, 1990. *How Monkeys See the World.* Chicago: University of Chicago Press.

Cosmides, L., 1985. Deduction or Darwinian algorithms? An explanation of the "elusive" content effect on the Wason selection task. Ph.D. diss. Harvard University. University Microfilms #86-02206.

Cosmides, L., 1989. The logic of social exchange: Has natural selection shaped how humans reason? Studies with the Wason selection task. *Cognition* 31:187–276.

Cosmides, L., and J. Tooby, 1989. Evolutionary psychology and the generation of culture, Part II. Case study: A computational theory of social exchange. *Ethol. Sociobiol.* 10: 51–97.

Cosmides, L., and J. Tooby, 1992. Cognitive adaptations for social exchange. In *The Adapted Mind: Evolutionary Psychology and the Generation of Culture*, J. Barkow, L. Cosmides, and J. Tooby, eds. New York: Oxford University Press.

Cosmides, L., and J. Tooby, 1994. Origins of domain specificity: The evolution of functional organization. In *Mapping the Mind: Domain Specificity in Cognition and Culture*, L. Hirschfeld and S. Gelman, eds. New York: Cambridge University Press.

Dawkins, R., 1986. *The Blind Watchmaker.* New York: Norton.

De Waal, F. B. M., and L. M. Luttrell, 1988. Mechanisms of social reciprocity in three primate species: Symmetrical relationship characteristics or cognition? *Ethol. Sociobiol.* 9:101–118.

Fischer, E. A., 1988. Simultaneous hermaphroditism, tit-for-tat, and the evolutionary stability of social systems. *Ethol. Sociobiol.* 9:119–136.

Fiske, A. P., 1992. *Structures of Social Life: The Four Elementary Forms of Human Relations*. New York: Free Press.

Fodor, J. A., 1983. *The Modularity of Mind*. Cambridge, Mass.: MIT Press.

Frith, U., 1989. *Autism: Explaining the Enigma*. Oxford: Blackwell.

Gallistel, C. R., 1990. *The Organization of Learning*. Cambridge, Mass.: MIT Press.

Gigerenzer, G., and K. Hug, 1992. Domain-specific reasoning: Social contracts, cheating and perspective change. *Cognition* 43:127–171.

Gould, J. L., 1982. *Ethology: The Mechanisms and Evolution of Behavior*. New York: Norton.

Hamilton, W. D., 1964. The genetical evolution of social behaviour. I, II. *J. Theor. Biol.* 7:1–52.

Jackendoff, R., 1992. *Languages of the Mind*. Cambridge, Mass.: MIT Press.

Krebs, J. R., and N. B. Davies, 1987. *An Introduction to Behavioural Ecology*. Oxford: Blackwell.

Leslie, A. M., 1987. Pretense and representation: The origins of "theory of mind". *Psychol. Rev.* 94:412–426.

Manktelow, K. I., and D. E. Over, 1990. Deontic thought and the selection task. In *Lines of Thinking*, vol. 1, K. J. Gilhooly, M. T. G. Keane, R. H. Logie, and G. Erdos, eds. New York: Wiley.

Marr, D., 1982. *Vision: A Computational Investigation into the Human Representation and Processing of Visual Information*. San Francisco: Freeman.

Pinker, S., 1979. Formal models of language learning. *Cognition* 7:217–283.

Pinker, S., 1984. *Language Learnability and Language Development*. Cambridge, Mass.: Harvard University Press.

Pinker, S., 1994. *The Language Instinct*. New York: Morrow.

Real, L. A., 1991. Animal choice behavior and the evolution of cognitive architecture. *Science* 253:980–986.

Rozin, P., 1976. The evolution of intelligence and access to the cognitive unconscious. In *Progress in Psychobiology and Physiological Psychology*, J. M. Sprague and A. N. Epstein, eds. New York: Academic Press.

Smuts, B., 1986. *Sex and Friendship in Baboons*. Hawthorne, N.Y.: Aldine.

Tooby, J., and L. Cosmides, 1990. The past explains the present: Emotional adaptations and the structure of ancestral environments. *Ethol. Sociobiol.* 11:375–424.

Tooby, J., and L. Cosmides, 1992. The psychological foundations of culture. In *The Adapted Mind: Evolutionary Psychology and the Generation of Culture*, J. Barkow, L. Cosmides, and J. Tooby, eds. New York: Oxford University Press.

Tooby, J., and L. Cosmides, 1989. The logic of threat. Human Behavior and Evolution Society, Evanston, IL.

Tooby, J., and I. DeVore, 1987. The reconstruction of hominid behavioral evolution through strategic modeling. In *Primate Models of Hominid Behavior*, W. Kinzey, ed. New York: SUNY Press.

Trivers, R. L., 1971. The evolution of reciprocal altruism. *Q. Rev. Biol.* 46:35–57.

Trivers, R. L., 1974. Parent-offspring conflict. *Am. Zool.* 14:249–264.

Wexler, K., and P. Culicover, 1980. *Formal Principles of Language Acquisition*. Cambridge, Mass.: MIT Press.

Wilkinson, G. S., 1988. Reciprocal altruism in bats and other mammals. *Ethol. Sociobiol.* 9:85–100.

Wilkinson, G. S., 1990. Food sharing in vampire bats. *Sci. Am.* February: 76–82.

Williams, G. C., 1966. *Adaptation and Natural Selection: A Critique of Some Current Evolutionary Thought*. Princeton: Princeton University Press.

Williams, G. C., and D. C. Williams, 1957. Natural selection of individually harmful social adaptations among sibs with special reference to social insects. *Evolution* 11:32–39.

80 Does Evolutionary Theory Predict Sex Differences in the Brain?

STEVEN J. C. GAULIN

ABSTRACT A general model is outlined that explains why, and thus predicts when, evolution will produce different phenotypes in the two sexes. In this process—termed sexual selection—the key independent variable is reproductive rate; for the sex with the higher reproductive rate mating opportunities will often limit reproduction, whereas this will seldom be the case for the more slowly reproducing sex. As a consequence, the faster reproducing sex will evolve a wide array of traits that function to increase mating opportunities. Such evolved sex differences will typically be under the ontogenetic control of sex hormones. Cognitive traits and their neuroanatomical bases are likely targets of sexual selection. To exemplify how sexual selection might operate on a cognitive trait, a testable hypothesis is developed to explain the evolution of sex differences in spatial ability. Comparative data from mammals and birds are reviewed in evaluating the hypothesis.

Why are there any sex differences?

As it is now understood, Darwin's (1859) theory of evolution by natural selection explains how the continual sorting of random genetic variation results in the assembly of designs for reproduction (Fisher, 1958; Williams, 1966; Dawkins, 1976, 1982). Those genetic variants that effect a statistical advantage in the rate at which they are transmitted to the next generation—for example, by increasing the reproductive output of the bodies in which they reside—invariably become more common. Each species tends to evolve a somewhat different design for reproduction because its particular ecological niche presents a unique set of design problems and constraints. In other words, the matter of which genetic variants are statistically advantaged is context dependent.

STEVEN J. C. GAULIN Department of Anthropology, University of Pittsburgh, Pittsburgh, Pa.

Thus, Darwin's theory explains why different species come to possess different adaptations. But what the original theory of natural selection does not fully explain is why there would be multiple designs for reproduction within any species. Clearly we could expect some transient variability: New variation is always arising and selection will need time to sort it. But why would a single species contain a bimodal distribution of variation with two distinct and relatively stable types, such as females and males? Within any species, the sexes typically exploit the same environments, exploit the same foods, and are exploited by the same predators and parasites. Beyond the genitalia, why should males and females be built differently by evolution?

Darwin correctly saw that sex differences are not wholly arbitrary. One sex, often but not always males, tends to possess anatomical or behavioral features that appear useless or harmful in avoiding the ecological threats of starvation, hungry predators, and the like. He pointed to traits like antlers, crests, flamboyant coloration, elaborate movements and vocalizations. Why should "ecologically useless" traits like these ever evolve, and if they do, why should their evolution be restricted to just one sex? In general, these ecologically useless traits are functional in a narrow context: They are used to compete for and court members of the opposite sex. Darwin's (1871) theory of sexual selection argued that traits conferring an advantage in this mating context would spread just as certainly as traits that provided an advantage in food finding, digestive efficiency, or predator avoidance. While some modern evolutionists may debate whether sexual and natural selection are different processes, the distinction persists because it is analytically useful.

Selection is the generic term that is taken to summarize the notion that a trait will spread if it is transmitted to subsequent generations at a higher rate than its alternatives. Darwin used the term *natural selection* to denote the spread of traits conferring a wide range of ecological advantages, and reserved *sexual selection* to emphasize that a trait can spread even if its benefits are narrowly restricted to the mating context. Traits that enhance same-sex competitiveness and traits that enhance cross-sex attractiveness are typical results of sexual selection. The distinction between natural and sexual selection makes it apparent how "ecologically useless" traits can evolve, and it similarly suggests why they might be restricted to a single sex: The ecological problems that confront females and males are largely similar whereas the mating problems they face can be quite different.[1]

In order to develop the theory of sexual selection, let us temporarily restrict the discussion to mammals. In this group of animals neither sex can reproduce alone, so males and females might seem to be equally dependent on access to members of the opposite sex. To see that this is not necessarily so, consider the following comparison between a female and a male deer. If the female has just successfully mated and conceived her maximum litter, then subsequent copulations will add nothing to her reproductive output until she has weaned her just-conceived litter. In contrast, a male deer who has just mated could increase his total output of offspring by additional copulations, provided they are with previously unfertilized females. In species like deer, males can benefit from extensive mating in a way that females cannot. From a male point of view, copulatory opportunities limit reproductive output, whereas additional matings have little or no impact on female reproductive output.

While female and male deer would be constrained by similar ecological problems generally, they would be subject to different selection pressures in the mating context. In cases like this, any trait that permits a male to secure a disproportionate share of the available matings would be positively selected. A trait with similar effects on females would yield no reproductive advantage. This kind of imbalance is what drives sexual selection. It leads to the evolutionary spread of traits useful in competition for mates, but only in the sex with the higher potential reproductive rate (Clutton-Brock and Vincent, 1991). Differences in reproductive rate often arise because one sex makes a greater physiological commitment to rearing the offspring than does the other (Williams, 1966; Trivers, 1972), and this is clearly the case in deer and many other mammals. In any particular species, the facts of reproduction determine the relative reproductive rates of the two sexes.

Males produce small sex cells and females large ones. This much is definitional, and it may produce a fundamental bias in the reproductive rates of males and females, since small gametes are presumably cheaper and might therefore be produced more rapidly. But this bias is not without exception. The theory of sexual selection does not predict the evolution of flamboyant, sexually competitive traits in males. It predicts the evolution of these traits in the "fast" sex, that is, the sex with the potentially higher reproductive rate. Thus, important test cases for the theory would include those species where females have higher reproductive rates than males, species where males are the "slow" sex.

Such cases have long been known in various groups of animals including birds, amphibians, fish, and insects (Williams, 1966; Trivers, 1972). To take one example, among at least two species of phalarope (tundra-breeding shorebirds) males incubate the eggs and tend the young alone. Soon after copulation the female lays several eggs in the male's nest and then abandons him to seek additional partners. Females can lay several clutches of eggs per season, but males can rear only one. Thus, in these species female reproductive output depends more on the availability of mates than does male reproductive output. Females are the fast sex: The facts of reproduction allow successful females to produce more offspring than do successful males. As predicted, females are larger and more brightly colored than males; they are more aggressive among themselves than are males, and they take the active role in courtship. Clutton-Brock and Vincent (1991) have recently shown that phalaropes are not unique. In general, males seem to have been the principal targets of sexual selection in that they alone have evolved sexually competitive traits. But where females have higher reproductive rates than males, then females, not males, bear these stigmata of sexual selection.

So far we have discussed the two ends of a continuum, a continuum measuring the relative reproductive rates of females and males. At one end (represented by deer and many other mammals), males can reproduce more rapidly; at the other (represented by phalaropes), females have higher reproductive rates. Another logical possibility is that the sexes could have equal reproductive rates. This is generally the case in *monog-*

amous species, where a single male and female form a more or less exclusive mating bond and opportunities for either to reproduce outside the pair are few. Monogamous mating habits might evolve in particular species because they confront special circumstances. For example, the offspring may survive poorly without biparental care, making the desertion tactic, used by male deer and female phalaropes, unsuccessful.

If all reproduction were within monogamous pairs, neither sex could accelerate its reproductive rate without also accelerating the reproductive rate of the other by the same amount. In such cases the consequent identity in male and female reproductive rates would eliminate sexual selection, and sex differences would therefore be minimal. Of course monogamy is often imperfect, and, to the extent that one sex can elevate its reproductive rate independently of the other, sexual selection will begin to operate and sex differences will begin to evolve. We use the term *polygyny* to refer to those cases where, by taking additional mates before their current partners have finished reproducing, males can attain a higher reproductive rate. Deer are thus polygynous. Similarly, the term *polyandry* is used to denote those cases where the taking of multiple partners elevates female reproductive rate. Phalaropes are polyandrous. If a species deviates from perfect monogamy, it deviates either toward polygyny or toward polyandry. Under polygyny sexual selection tends to produce flamboyant males, under polyandry, flamboyant females.

Do sex differences have a genetic basis?

The modern, rigorous way to state Darwin's theory requires a focus on the causes of variation. Chance errors in gene copying produce a steady supply of new alleles. Over the course of development, these alleles may influence the outward traits of the individuals who carry them. The traits that are sorted by selection, both natural and sexual, are such allelic effects. An allele will spread if its effects cause it to be copied at a higher rate than any of its allelic alternatives. Prospectively we can say that those genes that most favorably affect their own copying rates will become common. Retrospectively we can say that the traits of existing organisms are simply the allelic effects that led to maximal rates of gene copying in the past. We have seen that allelic effects that increase copulatory rates would be beneficial in the fast sex but would confer little if any advantage in the slow sex. Does this mean that certain alleles will spread in one sex and different alleles will spread in the other? A comparison between the evolution of species differences and the evolution of sex differences will help us answer this question. First let us consider species differences.

A particular allele might have positive effects in one environment and negative effects in another. If these two environments are the niches of two different species, then the populations are free to follow their own separate evolutionary paths. Thus, the allele in question is expected to increase in frequency in the former case and decrease in the latter; species become different as a consequence of such genetic differentiation. The point is that the outcome of selection in one species is independent of the outcome of selection in the other. Now consider sex differences. Suppose the two environments mentioned above are sexual environments—for example, the fast sex environment, where access to additional mates augments gene copying rates, and the slow sex environment, where it does not. Again imagine a particular allele with positive effects in one environment and negative effects in the other. An example might be an allele that produces brighter coloration and hence makes the individual more conspicuous. Being more conspicuous probably attracts more potential mates and more potential predators, and the net balance of these two results could vary with sex (Zahavi, 1975; Moller, 1989). For members of the fast sex, additional matings could more than compensate for a higher risk of predation, but since members of the slow sex gain little from additional matings, the net result of brighter coloration might be negative for them.

In such cases one might conclude that the bright-coloration allele will spread in the fast sex and be eliminated in the slow sex. Such an outcome would be rather unusual because, within any species, males and females intimately share virtually all of their genes. Mothers pass their genes to both sons and daughters; sons inherit theirs from both sexes of parent. If bright coloration confers an advantage among members of the fast sex, then the bright-coloration allele will begin to spread, and it will ordinarily be spread to both sexes of offspring. In any single generation, the brightest members of the fast sex will have elevated reproductive success. In contrast, the brightest members of the slow sex will have depressed reproductive success, but this negative selection will not eliminate the bright-coloration allele because, immediately, the brightest members of the fast sex replace it by producing sons and daughters in roughly equal numbers. We seem to have a paradox. How can sex differences evolve if, every generation, both sexes of offspring inherit the alleles that benefit only one sex of parent? Will sexual selection produce a stalemate, with the traits of males and females each compromised for the benefit of the other? This idea may be partially correct, but sex differences are too widespread and conspicuous for this to be the whole answer.

Because of their linkage through reproduction, the two sexes of a single species are not free to differentiate genetically in the way two species can.[2] Sex differences could, however, evolve if alleles had sex-specific effects (Lande, 1980). For example, we suggested that bright coloration might have costs (increased attention from predators) and benefits (increased attention from the opposite sex), and that the fast sex was differentially able to exploit the benefits. Suppose the allele(s) responsible for bright coloration were expressed primarily in members of the fast sex and repressed in the slow sex. In this case the allele would be free to spread because it would not be inflicting costs on members of the slow sex. Moreover, sex differences would appear because of the differential expression in males and females of alleles that are carried by both sexes.

This perspective on the evolution of sex differences can be expressed another way. Over the eons, the typical gene finds itself in approximately equal numbers of male and female bodies. Sometimes, this is not a problem. But when one sex has the potential to reproduce more rapidly than the other, the very traits that most benefit one sex may be harmful to the other. In such cases, the genes that are most likely to be retained by selection are the genes that are capable of constructing good solutions to two different problems: the problem of making a fit female and the problem of making a fit male. The only way a gene could accomplish both tasks is by adjusting its phenotypic effects in response to the sex of its current body.

There might seem to be an obvious alternative way for sex differences to evolve. Not every piece of the genome is shared between the sexes; for example, in mammals, only males possess Y chromosomes. This allows for the possibility of genetic differentiation between the sexes, equivalent to genetic differentiation between species. Genes on the Y could spread without any automatic consequences for gene frequency in females. The reason that this route to sex differences seems relatively unimportant is that the Y appears to be largely inert (Cavalli-Sforza and Bodmer, 1971); few regions on it are transcribed to messenger RNA, and thus the Y has minimal scope to shape the phenotype. Hamilton (1967) and Tooby (1982) have given arguments to explain the evolution of Y-chromosome inertness, but that is another matter. Although the general inertness of the Y prevents it from being the direct cause of most sex differences, it does contain an important piece of the evolutionary puzzle. If sex differences are the result of differential expression of autosomal genes (genes not on the sex chromosomes), any such differential expression requires a control mechanism, a modulator, because autosomal genes must be instructed about the sex of their current body. The Y-resident gene called testes determining factor or TDF (Page, 1986; Vergnaud et al., 1986; Grumbach and Conte, 1992) seems to provide this instruction. TDF activates autosomal genes that in turn cause the fetal gonads to differentiate into testes. The testes produce high levels of androgens that influence the expression of many other autosomal genes. Of course TDF is absent in normal females; thus the fetal gonads become ovaries and a different hormonal regime carries a different message to the autosomes. In this way, genes that only one sex carries (genes on the Y) produce a signal of sexual identity for the rest of the genotype. Against this reliable background signal, selection has apparently constructed a wide array of modulatory mechanisms that restrict the expression of sexually competitive traits to the sex where they will produce a net benefit.

Do sex differences have a genetic basis? Yes, but one that is rather more complicated than the genetic basis of species differences. All of this suggests that sex differences probably evolve much more slowly, or, more precisely, are more resistant to selection, than species differences (Lande, 1980; Rogers and Mukherjee, 1992). For this reason, the presence of adaptive sex differences suggests a history of especially strong selection and thus offers relatively powerful evidence on the adaptive function of the traits in question.

Sexual selection for spatial ability

Just like any other organ or anatomical structure, the brain is a potential target of sexually differentiated selection pressures. One should expect to find adaptive sex differences in the brain whenever the two sexes differ in terms of the adaptive problems they have faced over the evolutionary history of the species; circuits governing aggression, mate choice, and parental care are likely candidates. This section will illustrate the evolutionary approach to sex differences by focusing on a domain of possible interest to cognitive neuroscientists. I have selected the set of cognitive skills conventionally referred to as spatial ability and will explore whether sex differences in spatial learning might be a consequence of sexual selection. Sex differences in this cognitive domain have been relatively well studied for both humans (Harris, 1978; McGee, 1982; Rosenthal and Rubin, 1982; Linn and Petersen, 1985; Halpern, 1986) and laboratory rodents (Barrett and Ray, 1970; Joseph, 1979; Beatty 1984; Mishima et al., 1986). A sexual-selection analysis of this sex difference will require a working hypothesis about how spatial ability could affect copulatory success.

The first assumption I will suggest is that the set of processing functions commonly called spatial ability is an evolved navigational adaptation. Animals that are better able to acquire and manipulate accurate data on the distribution of rewards and risks in their environments will have outreproduced their less able fellows and therefore will have transmitted the genes underlying these abilities to their descendants. Typically the females and males of a single species exploit the environment in the similar ways, confronting essentially similar distributions of reward and risk. Thus the

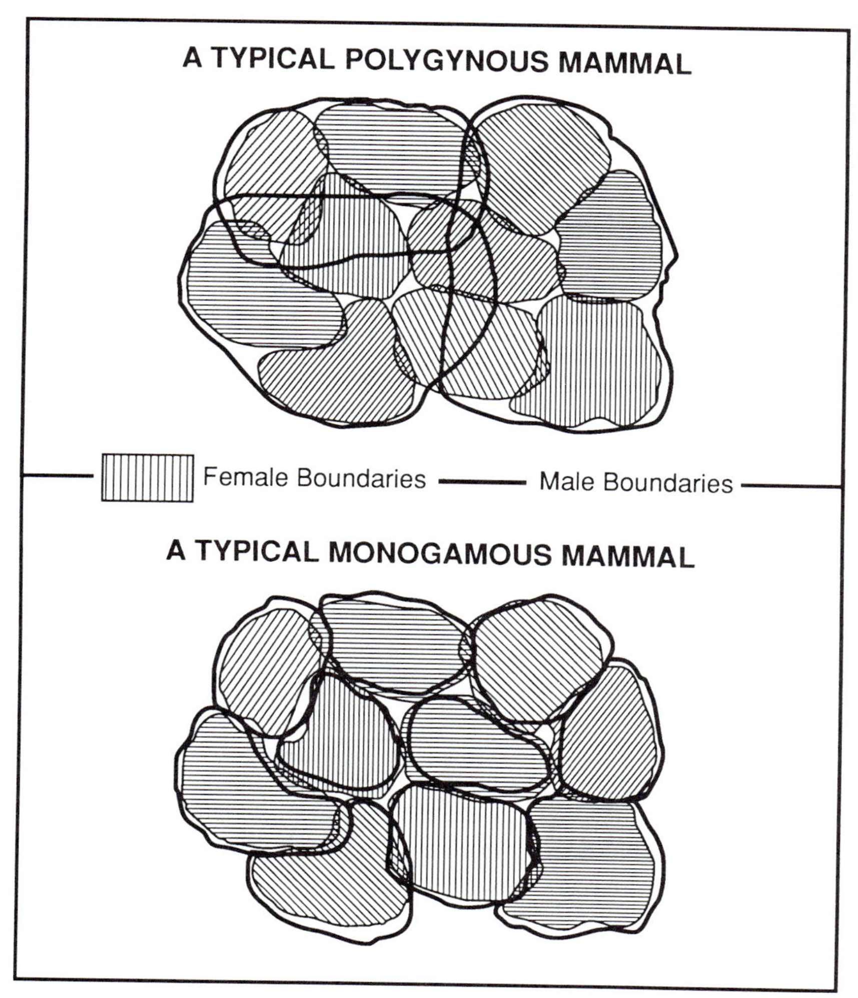

FIGURE 80.1 In many polygynous species male home ranges greatly exceed female home ranges, whereas in monogamous species male and female partners tend to have isomorphic ranges.

two sexes will be subject to similar natural selection for spatial ability. But how will sexual selection affect the evolution of spatial skills?

As we have seen, this depends on the reproductive rates of males and females, which in turn depend on the circumstances of reproduction. In monogamous species, male and female reproductive rates are tightly linked, neither member of a pair being able to accelerate its reproduction beyond that of its partner. In contrast, males of polygynous species have the potential to reproduce more rapidly than females and they therefore experience a scarcity of females that is not reciprocal. Under these circumstances, information on the location and reproductive status of females is probably more useful to competing males than is parallel data on males to competing females. In such cases, the ability to travel widely without disorientation would have a greater impact on male than female fitness.

Field data support this theoretical argument. The way that animals use space depends, of course, on the distribution of available foods, potential predators, and the like, but it also depends on their mating system (Brown, 1966). In almost all monogamous species of birds and mammals, the male and female members of a mated pair maintain close spatial associations and move together through a shared home range (e.g., Chivers, 1974; Kleiman, 1977). Many polygynous species present a sharp contrast to this pattern. Here, the distribution of females may not differ from the monogamous pattern, where each female exploits a relatively exclusive range; however, male home ranges are often much larger than female home ranges. Data from several polygynous species suggest that males who contact more females tend to leave more offspring than do more sedentary males (Rodman, 1984; Kawata, 1988). Figure 80.1 illustrates the relationship between mating system and sex-specific ranging patterns.

Of course, increases in range size and the requisite underlying spatial abilities are probably not free. Thus the second assumption is that there will be energetic costs associated with patrolling a larger range and greater exposure to predators as well. Increases in spatial ability probably entail both neural and metabolic costs. But such costs do not preclude a sexual-selection analysis; they are an integral part of it. Sex differences typically evolve because sexually competitive traits are costly, and provide outweighing benefits only in the faster sex. In this context our theory suggests that the genes fostering range expansion and the supporting spatial skills would probably have persisted only if they produced sex-limited effects.

These ideas can be expressed in terms of our modulatory model of sexually selected traits. In polygynous species, the beneficial effects of range expansion are restricted to males; hence any allele that fosters range expansion will be subject to positive selection if its effects are limited to males. Such sex-dependent modulation is not predicted in perfectly monogamous species because there neither sex can benefit differentially from increased exposure to prospective mates. Thus polygyny is expected to foster the evolution of sexually dimorphic ranging patterns and spatial skills, whereas monogamy will foster isomorphic male and female ranges and spatial abilities.

Testing the model

For the evolutionist, the simplest approach to testing this model is to evaluate how accurately it predicts the cross-species distribution of relevant sex differences. The central prediction is that sex differences in spatial ability and in ranging should covary. Any such covariation will constitute stronger evidence if it derives from closely related species. Distantly related species usually possess many different adaptations, so any difference in the particular traits of interest will be difficult to attribute to a single cause. In contrast, closely related species conserve many of the same adaptations from their recent common ancestor. Thus, if we predict that, for example, mating systems shape ranging patterns, it will be most informative if we examine the ranging patterns of closely related species, species that are otherwise very similar but that nevertheless differ in their mating systems. We have found this sort of useful comparison among wild rodents of the genus *Microtus*, commonly called voles or, in North America, meadow mice.

Rodents were desirable subjects for this research because a wide range of spatial tasks has been developed for laboratory rats and mice. Of course, if one is interested in evolutionary effects it is better to avoid products of artificial selection (animals long subject to controlled breeding programs), because in such cases there is no good way to estimate how much natural and sexual selection has been undone by human intervention. Fortunately, the movement patterns of wild voles can be studied in their natural habitats, they adapt fairly well to spatial testing in the laboratory and, most importantly, there is considerable variation in their naturally occurring mating systems (Wolff, 1985).

When viewed in terms of the preexisting psychological literature, the sexual-selection model of sex differences makes some surprising and some unsurprising predictions. All of the mammalian species previously studied—rats, mice, and people—show reliable sex differences on spatial tasks. The model predicts such differences for polygynous species, and all three—rats, mice, and people—can at least be said to have a recent evolutionary history of polygyny (Dewsbury, 1981; Murdock, 1986). Indeed, any random collection of mammalian species would be predicted to show the same pattern because some 95% of mammalian species are polygynous (Alcock, 1993). A demonstration that sex differences in spatial ability occur in other polygynous mammals would of course be useful, but the more novel predictions of the model concern monogamous species. These species are predicted to lack sex differences in spatial ability.

Our strategy was to contrast monogamous and polygynous species of voles. Meadow voles (*Microtus pennsylvanicus*) are polygynous and, as such, have served as a sort of control species. Unless meadow voles showed sex differences, no pattern of results for monogamous species could be interpreted as strong support for the model. We chose pine voles (*M. pinetorum*) and prairie voles (*M. ochrogaster*), two monogamous species in the same genus, as providing the most appropriate contrasts. Recent advances in radiotelemetry allowed assessment of their ranging patterns in the wild, and standard tests were used to measure their spatial performance in the laboratory.

In the wild, meadow voles showed strong sex differences in ranging, whereas neither pine nor prairie voles, the two monogamous species, ever showed statistically significant sex differences in range size (Gaulin and FitzGerald, 1986, 1988, 1989). These conclusions hold regardless of the time scale (day, months) over which range sizes are computed (table 80.1). The ranging behavior of meadow voles is complicated by developmental and seasonal variation but fits expectations from sexual selection (Gaulin and FitzGerald, 1988, 1989). This polygynous species shows no sex differences in ranging among sexually immature individuals or during the winter months when no breeding occurs. However, during the breeding season, adult male ranges expand dramatically compared to those of adult females (see figure 80.2). It seems likely that range expansion is costly and should therefore be avoided in the absence of compensatory benefits. There is presumably no compensatory benefit to either sex in monogamous species. In polygynous species males could benefit from range expansion, but only upon the attainment of sexual maturity and only if local females are in a reproductive phase. Overall,

TABLE 80.1

Range sizes of breeding voles, by sex and species

	N	30-day range ($m^2 \pm$ SE)	48-hour range ($m^2 \pm$ SE)	24-hour range ($m^2 \pm$ SE)
Pine voles				
Males	10		37.6 ± 6.2	
Females	9		40.8 ± 4.5	
Prairie voles				
Males	21	277.6 ± 63.7		77.0 ± 14.4
Females	26	183.2 ± 24.0		56.4 ± 7.6
Meadow voles				
Males	9		623.3 ± 163.7	
Females	12		209.4 ± 48.5	
Meadow voles				
Males	21	694.6 ± 211.5		198.6 ± 41.9
Females	22	157.4 ± 26.1		64.0 ± 10.4

Data from Gaulin and FitzGerald (1986, 1989).

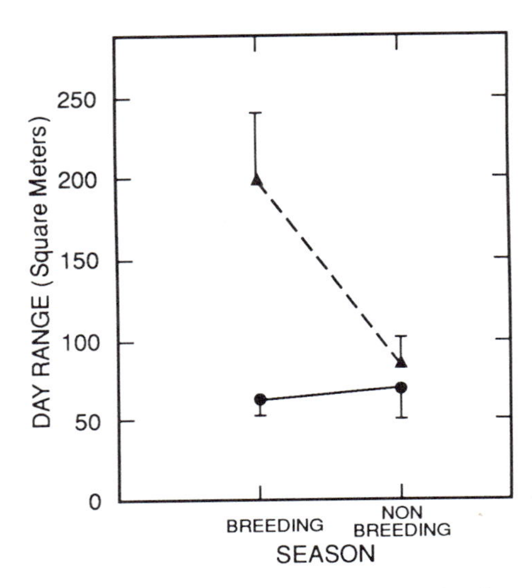

FIGURE 80.2 Day ranges for meadow voles by sex and season. Male means are indicated by triangles, female means by circles (data from Gaulin and FitzGerald, 1989).

these field data suggest that ranging patterns are strongly shaped by sexual selection. The next question is, do spatial abilities covary with ranging behavior?

Over a number of years we have used eight mazes of two different types to test the spatial ability of meadow, pine, and prairie voles under laboratory conditions. Many of the subjects tested in the mazes were individuals whose movements had been previously telemetered in the field; the remainder derived from the same local populations. Our first study (Gaulin and FitzGerald, 1986) involved a comparison of pine and meadow voles' performance on a place-learning task (i.e., a sunburst maze). Later studies (Gaulin and FitzGerald, 1989; Gaulin, FitzGerald, and Wartell, 1990) have focused on comparisons of prairie and meadow voles on a series of seven route-learning tasks (i.e., symmetrical mazes). Throughout this research the stable finding has been that the monogamous species—prairie and pine voles—systematically lack sex differences in maze performance, regardless of maze type, testing conditions, or season (table 80.2, figure 80.3). The polygynous species, meadow voles, were tested under the same conditions. In this species males make significantly fewer errors than females as long as the subjects have been trapped during the breeding season. Even in this polygynous species, nonbreeding animals show minimal sex differences in both range size (see above) and maze performance (figure 80.4).

TABLE 80.2

Median ranks of voles in study 1 on the place-learning task, by sex, species

	N	Meian Rank	Mann-Whitney *U*-Test
Pine voles			
Males	8	10.5	n.s.
Females	13	11.0	
Meadow voles			
Males	9	7.0	$p < .025$
Females	11	15.0	

For each species, the *U*-test evaluates the significance of the sex difference in rank. Data from figure 3 of Gaulin and FitzGerald (1986).

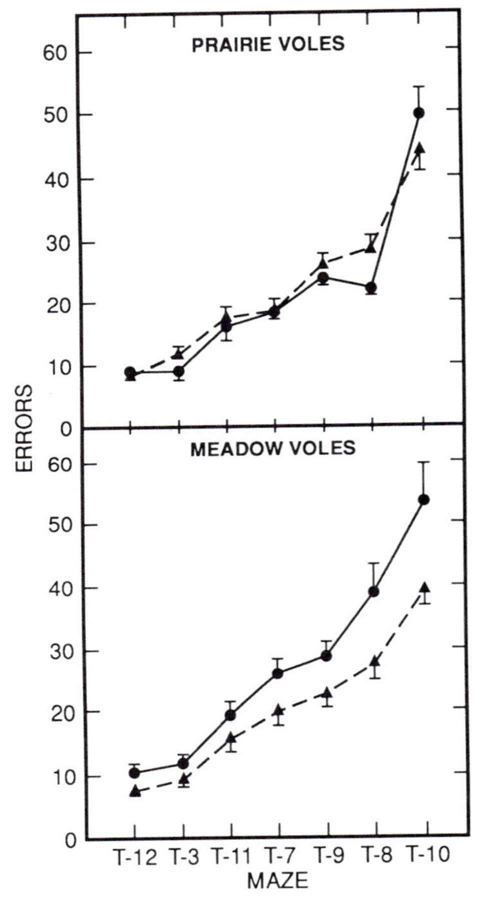

FIGURE 80.3 Symmetrical maze performance by sex and species. Male means are indicated by triangles, female means by circles (data from Gaulin and FitzGerald 1989).

The seasonal dependence of the sex difference in maze performance has recently been replicated for the montane vole (*M. montanus*), another polygynous species (Jacobs, Gaulin, and Rowsemitt, in preparation).

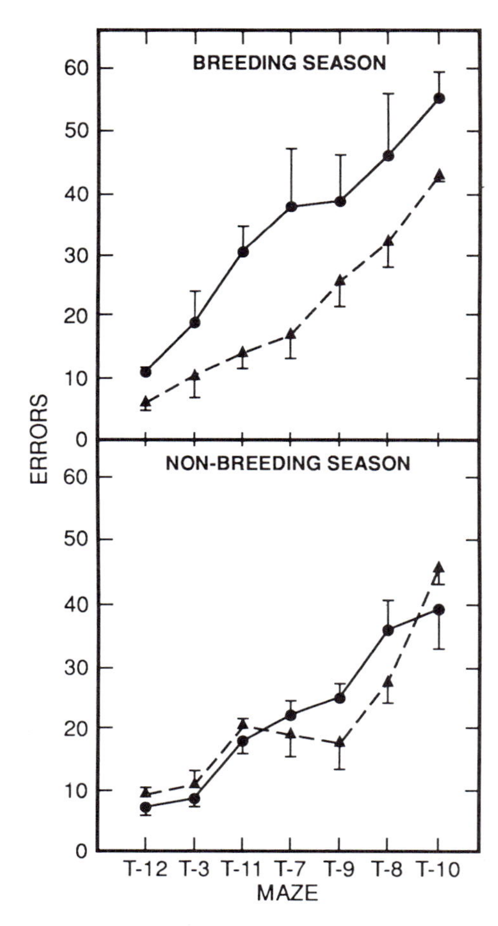

FIGURE 80.4 Spatial performance of meadow voles by sex and season. Male means are indicated by triangles, female means by circles (data from Jacobs and Gaulin, submitted).

It has sometimes been suggested that sex differences in spatial ability are due to sex differences in spatial experience. The credibility of this idea is seriously undermined by the fact that laboratory rodents reliably show sex differences in maze performance despite rearing conditions that are highly controlled and not differentiated by sex. Nevertheless, we felt some obligation to explore this possibility in voles because the predictions of the experience model closely match those of the sexual selection model. The experience hypothesis would predict no sex difference in maze performance for pine and prairie voles because, in each of these species, males and females would have had similar spatial experience in the wild. Thus when tested in our laboratory, they should have shown no sex difference in maze performance. However, in the polygynous species (meadow voles), adult breeding males and females would have had very different amounts of spatial experience in the wild due to their dramatically different range sizes. Meadow voles would then have been expected to show marked sex differences in maze performance. Of course this is precisely what we found.

The experience hypothesis predicts an ontogenetic effect of range size and the sexual selection hypothesis predicts an evolutionary effect of range size. The only way to discriminate between these two ideas is to manipulate experience experimentally. In essence, the experience hypothesis posits that so-called sex differences are really not what they appear, but they are instead just experience differences. Thus, differences in spatial experience should produce differences in maze performance independent of sex. We selected voles trapped in the wild to represent the spatially experienced condition and their first-generation, laboratory-reared offspring, confined to small mouse cages from birth, to represent the spatially deprived condition. These groups differed in "range size" (parental home range versus area of the rearing cages) by a factor of 4000, certainly much greater than any naturally occurring sex difference in home range size. Nevertheless, this extreme manipulation of experience had no effect on maze performance; the spatially deprived group performed just as well as did their wild-caught parents (Gaulin and Wartell, 1990). Ontogenetic effects seem to be minimal over a very wide range of spatial experience.

A neuroanatomical basis for sex differences in spatial ability

A variety of kinds of evidence, deriving from studies of birds and mammals (O'Keefe et al., 1975; Morris et al., 1982; O'Keefe and Speakman, 1987; Sherry and Vaccarino, 1989; Krebs et al., 1989; Sherry et al., 1989; Wehner, Sleight, and Upchurch, 1990) suggests that the hippocampus is critically involved in spatial processing (Sherry, Jacobs, and Gaulin, 1992). If this is so, selection for improved spatial ability should have produced adaptive changes in the hippocampus. The evidence suggests that it has. For example, three groups of Holarctic birds—the tits, nuthatches, and jays—include species that cache food items in scattered locations for subsequent use. Reliance on this spatially demanding foraging tactic has apparently led to evolutionary changes in the hippocampus. In both North America and Europe, birds that store food have significantly larger hippocampi, measured as a proportion of telencephalon, than do similarly sized birds that do not store food (Krebs et al., 1989; Sherry et al., 1989).

Parallel data exist for rodents. Many species of kangaroo rats (*Dipodomys*) store seeds, their principal food, but storage patterns differ by species. Merriam's kangaroo rat (*D. merriami*) uses a scatter-hoarding technique, with many separate caches, whereas the bannertail (*D. spectabilis*) stores all its seed in its home burrow. Thus the bannertail faces a considerably simpler task in locating its seeds and, as expected, the hippocampus is significantly smaller, relative to total brain size, in this species (Jacobs and Spencer, in press).

On the basis of the above comparisons, we might suggest that spatial ability and hippocampal size represent the behavioral and neuroanatomical components of a navigational adaptation. Given that natural selection produces species differences much more readily than sexual selection produces sex differences, a useful test of the hypothesized functional link between navigational demands, spatial ability, and hippocampal size can be devised with voles. Again we expect covariation, and again the data bear out this prediction. In meadow voles, adult, breeding males travel more widely than females and perform better on laboratory measures of spatial ability, whereas in pine voles, neither sex difference is observed. The allocation of brain tissue to hippocampus echoes these patterns (table 80.3). In meadow voles males have significantly larger hippocampi, relative to total brain size, than do females, whereas in pine voles there is no statistically significant sex difference in the size of this brain structure (Jacobs et al., 1990).

The comparative work on birds and other rodents is at least indirectly supportive of the idea that both natural and sexual selection shape hippocampal anatomy. All of the food-caching birds are monogamous. In these species, there are no conspicuous sex differences in space use and ranging, and both sexes employ the food-caching strategy. For none of these food-caching birds are there reports of sex differences in hippocampal size (Krebs et al., 1989; Sherry et al., 1989). In contrast, both species of kangaroo rats are polygynous and, during the breeding season, males range widely in search of more sedentary females. As expected, in both species of kangaroo rats, males have larger hippocampi than do females (Jacobs and Spencer, in press).

Table 80.3
Comparisons of hippocampal (HP) size, by sex, species

		HP Volume (mm^3)	HP Volume × 100 / Brain Volume
Pine Voles			
Males	10	22.89 ± 0.95	4.39 ± 0.15
Females	10	22.42 ± 0.74	4.57 ± 0.23
Meadow Voles			
Males	10	31.42 ± 0.69	4.93 ± 0.07
Females	10	28.22 ± 0.89	4.50 ± 0.12

Data from figure 1 of Jacobs et al. (1990); in that original presentation, the two right-hand graphs are mislabeled on the y-axis: "x 100" should be deleted.

The model outlined in this chapter emphasizes that sex differences are not, fundamentally, a consequence of maleness and femaleness. Rather, they evolve in response to the particular reproductive problems and conditions that have faced each sex over the species' evolutionary history. In principle, traits that we humans think of as masculine might well characterize females in species subject to a different sexual selection regime. In this regard, an especially powerful test of the sexual selection model would focus on species where the circumstances of reproduction pose greater navigational problems for females than for males. In these species, females should perform better on laboratory measures of spatial ability and should have evolved larger hippocampi. Such a species has been identified (Sherry et al., 1993).

The brown-headed cowbird (*Molothrus ater*) is a brood parasite. This means that cowbirds do not rear their own young. Instead, female cowbirds lay their eggs in the nests of host species who unwittingly rear the cowbirds' offspring along with their own. Of course, natural selection has favored discriminatory abilities among the hosts. Some of these abilities depend on timing, such that if the cowbird lays her egg too early or too late in the host's laying cycle, the host will be able to recognize it as foreign and remove it. Thus female cowbirds must search widely for host nests, and must be able to relocate these nests periodically to assess whether they are at an appropriate stage to be parasitized. Male cowbirds neither locate nor monitor host nests. There are no other described sex differences in feeding behavior or ecology in this species. Thus brown-headed cowbirds represent a case where the navigational demands on females seem to be greater than those on males.

All described cowbird species seem to be brood parasites, their closest nonparasitic relatives being other icterine blackbirds, such as the red-winged blackbird (*Agelaius phoeniceus*) and the common grackle (*Quiscalus quiscula*). Neither the redwing nor the grackle is characterized by sex differences in feeding behavior or ecology. Both species breed in relatively dense colonies; males and females do much of their foraging outside the breeding area in mixed-sex flocks. Thus, in these two nonparasitic blackbirds, neither sex should experience a special demand for navigational skills, as females do in cowbirds. Data on hippocampal size in these three species conforms to our expectations. Controlling for the size of the telencephalon, neither redwings nor grackles show a sex difference in hippocampal size; among cowbirds, however, females have significantly larger hippocampi (Sherry et al., 1993). It would now be very desirable to have laboratory measures of spatial performance for males and females of these three species.

Hormonal influences on spatial ability

Here I want to make a general point by emphasizing that the ontogenetic components of our sexual selection model seem to fit the available data. Sex differences are not ordinarily the consequence of genetic differences between females and males. Instead they arise from the differential expression of relatively similar genotypes in male and female bodies. The most successful genes are the ones that can build a fit female and a fit male, and the most reliable signal these genes have as to which they are currently building is the local hormone regime. Thus it is not surprising that gonadal hormones orchestrate the ontogeny of sex differences in spatial ability in all the species for which we have data (Gaulin and Hoffman, 1988). Unfortunately, no direct hormonal data are yet available for the voles, kangaroo rats, and birds that have informed our evolutionary analyses. Systematic studies of hormonal influences on spatial ability have been restricted to laboratory rodents and humans. Most of the documented effects are developmental (i.e., organizational), but there are some indications of activational effects as well.

Clearly, an evolutionary view of sex differences does not require that they be rigidly present from birth. Instead, for reasons of efficiency we should expect to find evolved mechanisms that orchestrate the expression of these sex differences at appropriate times, typically in conjunction with the attainment of sexual maturity or the onset of seasonal breeding. In laboratory rodents, sex-typical patterns of maze performance can be completely reversed by early administration of appropriate hormones (e.g., Dawson, Cheung, and Lau, 1973, 1975; Stewart, Skvarenina, and Pottier, 1975; Joseph, Hess, and Birecree, 1978; Williams, Barnett, and Meck, 1990). Males normally perform better on spatial tasks; thus many workers have assumed, and demonstrated, that testosterone enhances performance. The situation may be more complicated developmentally. The local aromatization of particular androgens to estradiol seems to govern sexual differentiation of the rat CNS (e.g., McEwen et al., 1977). Thus, Williams has argued that spatial ability could also be elevated by early CNS exposure to androgen-derived estradiol (Williams, Barnett, and Meck, 1990) and that the hippocampus is responsive to these estrogenic influences (Williams and Meck, 1991). Only indirect evidence is available for wild rodents. In voles, significant variation in both range size and spatial performance is at least temporally associated with hormonally mediated (Seabloom, 1985) changes in breeding status.

In humans, men are normally exposed to higher levels of androgens during development than are women but these two data points are not very informative by themselves; patients who manifest unusual hormonal regimes have provided the most valuable sorts of evidence on the relationship between gonadal hormones and spatial ability. In Turner's syndrome (XO) the gonads remain undifferentiated; the result is a phenotypic female who was both androgen- and estrogen-deprived during fetal development. These women show a special deficit in spatial processing; their spatial ability is significantly below their verbal ability and significantly below the spatial ability of normal control females (Money, 1963; Alexander, Walker, and Money, 1964; Garron, 1977; Buchsbaum and Henkin, 1980; Gordon and Galatzer, 1980; Rovet and Netley, 1982). Congenital adrenal hyperplasia (CAH) provides an interesting contrast to Turner's syndrome. In CAH females, endocrine malfunction leads to elevated androgen levels during fetal development; on spatial tests, CAH women score significantly above the mean of matched normal females (Perlman, 1973; Resnick et al., 1986).

Males also show relevant variation; males experiencing low androgen levels in utero (e.g., those with Klinefelter's syndrome and idiopathic hypogonadotropic hypogonadism) can be compared with normal males. The former typically have significantly depressed spatial ability (Bobrow, Money, and Lewis, 1971; Buchsbaum and Henkin, 1980; Nyborg and Nielsen, 1981; Hier and Crowley, 1982). The results for both men and women agree with the experimental data from laboratory rodents: Early exposure to androgen (in some form) elevates subsequent spatial ability.

Summary and conclusions

The evolutionary perspective explains why there are sex differences by explaining both their adaptive function and their ontogenetic basis. Moreover, the evolutionary perspective points to appropriate outcrops for studying sex differences because, by identifying their adaptive function, it permits *a priori* predictions about their cross-species distribution. This is a valuable research-design tool, whether the questions are endocrinological, neurobiological, or psychological, because it identifies natural experimental and control groups.

Viewed as components of evolved male and female reproductive strategies, some kinds of sex differences are expected to be functionally linked. Certain sets of traits work well for the fast sex, others work well for the slow sex; some arbitrary mix from the two lists would almost certainly lead to lower reproductive success in either sex. When we are studying the ontogenetic basis of sex differences, we are studying evolved developmental programs. In terms of both efficiency and error reduction, functionally linked traits should also be developmentally linked. This suggests that once we have unlocked one aspect of sexual differentiation we may have a powerful insight into many others as well.

Comparison is one of the evolutionist's favorite tools for studying adaptation. When such comparisons are made across distantly related taxa, the effects of divergent evolutionary histories and independently evolved adaptations cloud the analysis. Because evolution is largely constrained to build males and females out of the same genes, sex differences provide the ultimate controlled comparisons; when we compare the males and females of a single species the confounding effects of evolutionary divergence are minimal. Similarly, because of the reproductive linkage between male and female genotypes, sex differences are more resistant to selection than are, for example, species differences. For this reason, the existence of marked sex differences implies a history of quite potent selection, and thus constitutes especially relevant data in support of an adaptive argument.

In this chapter, sex differences in spatial ability have been used to exemplify the evolutionary approach. Current sexual selection theory has been outlined and used to construct an explanatory model of this sex difference. The model suggests how spatial ability might function in various reproductive contexts. The predictions of this model (table 80.4) fit well with available natural historical data on navigational challenges

TABLE 80.4
Summary of predictions of tests

Predictions (numbered) Test (lettered)	Supported?
1. Sex differences in reproductive opportunities drive sex differences in ranging patterns.	
a. Range size is unrelated to sex in monogamous species.	yes
b. Range size is dependent on both sex and breeding status in polygynous species.	yes
2. Sex differences in spatial ability are an evolved adaptation to reproductively motivated sex differences in ranging patterns.	
a. Spatial ability is unrelated to sex in monogamous species.	yes
b. Spatial ability is dependent on both sex and breeding status in polygynous species.	yes
3. Sex differences in spatial ability are ontogenetically shaped by reliable predictors of future reproductive opportunities.	
a. Sex-specific patterns of spatial ability are relatively unresponsive to spatial experience.	yes
b. Sex differences in spatial ability are ontogenetically dependent on early hormone regimes.	yes
4. Sex differences in the brain evolve in response to sex- and species-specific cognitive demands.	
a. No sex differences are found in the hippocampus of monogamous species.	yes
b. Males have larger hippocampi in polygynous species.	yes
c. Females have larger hippocampi in a brood-parasitic species.	yes

faced by females and males in the wild, with laboratory data on sex differences in spatial performance, and with neuroanatomical data on sexual dimorphism in a brain structure—the hippocampus—thought to be critically involved in the processing of spatial information. What is known about steroid influences on the development of this sex difference also matches expectations from the evolutionary model.

NOTES

1. In fact, sex differences are expected to evolve whenever the selection pressures on males and females differ. While many such cases will involve adaptations for mating, they are not logically restricted to this context. Profet's (1992) discussion of the teratogenic risks to fetuses provides an example of a selection pressure that is restricted to females but has little to do with obtaining mates or excluding sexual competitors.
2. Tooby and Cosmides (1990) have shown that this argument is in fact more general: In species where genetic recombination occurs, polygenetically regulated complex adaptations require a high level of genetic uniformity in the population as a whole.

REFERENCES

Alcock, J., 1993. *Animal Behavior: An Evolutionary Approach*, ed. 5. Sunderland, Mass.: Sinauer.

Alexander, D., H. T. Walker, and J. Money, 1964. Studies in directional sense. *Arch. Gen. Psychiatry* 10:337–339.

Barrett, R. J., and O. S. Ray, 1970. Behavior in the open field, Lashley III maze, shuttle box, and Sidman avoidance as a function of strain, sex, and age. *Dev. Psychol.* 3:73–77.

Beatty, W. W., 1984. Hormonal organization of sex differences in play fighting and spatial behavior. In *Sex Differences in the Brain*, G. J. De Vries, J. P. C. De Bruin, H. B. M. Uylings, and M. A. Corner, eds. Amsterdam: Elsevier, pp. 315–330.

Bobrow, A. A., J. Money, and V. G. Lewis, 1971. Delayed puberty, eroticism and sense of smell: A psychological study of hypogonadotropism, osmatic and anosmatic (Krallmann's syndrome). *Arch. Sex. Behav.* 1:329–344.

Brown, L. E., 1966. Home range and movement in small mammals. *Symp. Zool. Soc. Lond.* 18:111–142.

Buchsbaum, M. S., and R. I. Henkin, 1980. Perceptual abnormalities in patients with chromatin negative gonadal dysgenesis and hypogonadotropic hypogonadism. *Int. J. Neurosci.* 11:201–209.

Cavalli-Sforza, L. L., and W. F. Bodmer, 1971. *The Genetics of Human Populations.* San Francisco: Freeman.

Chivers, D. J., 1974. *The Siamang in Malaya.* Basel: Karger.

Clutton-Brock, T. H., and A. J. C. Vincent, 1991. Sexual selection and the potential reproductive rates of males and females. *Nature* 351:58–60.

Darwin, C., 1859. *On the Origin of Species.* London: Murray.

Darwin, C., 1871. *The Descent of Man, and Selection in Relation to Sex.* London: Murray.

Dawkins, R., 1976. *The Selfish Gene.* Oxford: Oxford University Press.

Dawkins, R., 1982. *The Extended Phenotype.* Oxford: Oxford University Press.

Dawson, J. L. M., Y. M. Cheung, and R. T. S. Lau, 1973. Effects of neonatal sex hormones on sex-based cognitive abilities in the white rat. *Psychologia* 16:17–24.

Dawson, J. L. M., Y. M. Cheung, and R. T. S. Lau, 1975. Developmental effects of neonatal sex hormones on spatial and activity skills n the white rat. *Biol. Psychol.* 3:213–229.

Dewsbury, D. A., 1981. An exercise in the prediction of monogamy in the field from laboratory data on 42 species of muroid rodents. *Biologist* 63:138–162.

Fisher, R. A., 1958. *The Genetical Theory of Natural Selection.* New York: Dover.

Garron, D., 1977. Intelligence among persons with Turner's syndrome. *Behav. Genet.* 7:105–127.

Gaulin, S. J. C., and R. W. FitzGerald, 1986. Sex differences in spatial ability: An evolutionary hypothesis and test. *Am. Nat.* 127:74–88.

Gaulin, S. J. C., and R. W. FitzGerald, 1988. Home-range size as a predictor of mating systems in *Microtus. J. Mammal.* 69:311–319.

Gaulin, S. J. C., and R. W. FitzGerald, 1989. Sexual selection for spatial-learning ability. *Anim. Behav.* 37:322–331.

Gaulin, S. J. C., R. W. FitzGerald, and M. S. Wartell, 1990. Sex differences in spatial ability and activity in two vole species (*Microtus ochrogaster* and *M. pennsylvanicus*). *J. Comp. Psychol.* 104:88–93.

Gaulin, S. J. C., and H. A. Hoffman, 1988. Evolution and development of sex differences in spatial ability. In *Human Reproductive Behaviour: A Darwinian Perspective*, L. L. Betzig, M. Borgerhoff Mulder, and P. W. Turke, eds. Cambridge: Cambridge University Press, pp. 129–152.

Gaulin, S. J. C., and M. S. Wartell, 1990. Effects of experience and motivation on symmetrical-maze performance in the prairie vole (*Microtus ochrogaster*). *J. Comp. Psychol.* 104:183–189.

Gordon, H. W., and A. Galatzer, 1980. Cerebral organization in patients with gonadal dysgenesis. *Psychoneuroendocrinology* 5:235–244.

Grumbach, M. M., and F. A. Conte, 1992. Disorders of sexual differentiation. In *Williams Textbook of Endocrinology*, J. Wilson and D. Foster, eds. Philadelphia: Saunders, pp. 853–952.

Halpern, D. F., 1986. *Sex Differences in Cognitive Abilities.* Hillsdale, N.J.: Erlbaum.

Hamilton, W. D., 1967. Extraordinary sex ratios. *Science* 156:477–488.

Harris, L. J., 1978. Sex differences in spatial ability: Possible environmental, genetic, and neurological factors. In

Asymmetrical Function of the Brain, M. Kinsbourne, ed. New York: Cambridge University Press, pp. 405–522.

HIER, D. B., and W. F. CROWLEY, 1982. Spatial ability in androgen-deficient men. *N. England J. Med.* 306:1202–1205.

JACOBS, L. F., S. J. C. GAULIN, and C. N. ROWSEMITT, submitted. Seasonal modulation of sex differences in spatial learning ability in two species of voles (*Microtus pennsylvanicus* and *M. montanus*). *J. Comp. Psychol.*

JACOBS, L. F., S. J. C. GAULIN, D. F. SHERRY, and G. E. HOFFMAN, 1990. Evolution of spatial cognition: Sex-specific patterns of spatial behavior predict hippocampal size. *Proc. Natl. Acad. Sci. U.S.A.* 87:6349–6352.

JACOBS, L. F., and W. D. SPENCER, in press. Natural space-use patterns and hippocampal size in kangaroo rats. *Brain Behav. Evol.*

JOSEPH, R., 1979. Effects of rearing and sex on maze learning and competitive exploration in rats. *J. Psychol.* 101:37–43.

JOSEPH, R., S. HESS, and E. BIRECREE, 1978. Effects of hormone manipulation and exploration on sex differences in maze learning. *Behav. Biol.* 24:264–277.

KAWATA, M., 1988. Mating success, spatial organization, and male characteristics in experimental field populations of the red-backed vole *Clethrionomys rufocanus bedfordiae*. *J. Anim. Ecol.* 57:217–235.

KLEIMAN, D. G., 1977. Monogamy in mammals. *Q. Rev. Biol.* 52:39–69.

KREBS, J., D. F. SHERRY, S. D. HEALY, V. H. PERRY, and A. L. VACCARINO, 1989. Hippocampal specialization of food-storing birds. *Proc. Natl. Acad. Sci. U.S.A.* 86:1388–1392.

LANDE, R., 1980. Sexual dimorphism, sexual selection, and adaptation in polygenic characters. *Evolution* 34:292–305.

LINN, M. C., and A. C. PETERSEN, 1985. Emergence and characterization of sex differences in spatial ability: A meta-analysis. *Child Dev.* 56:1479–1498.

MCEWEN, B. S., I. LIEBERBURG, C. CHAPTAL, and L. C. KREY, 1977. Aromatization: Important for sexual differentiation of the rat brain. *Horm. Behav.* 9:249–263.

MCGEE, M. G., 1982. Spatial abilities: The influence of genetic factors. In *Spatial Abilities: Developmental and Physiological Foundations*, M. Potegal, ed. New York: Academic Press, pp. 199–222.

MISHIMA, N., F. HIGASHITANI, K. TERAOKA, and R. YOSHIOKA, 1986. Sex differences in appetitive learning of mice. *Physiol. Behav.* 37:263–268.

MOLLER, A. P., 1989. Viability costs of tail ornaments in a swallow. *Nature* 332:640–642.

MONEY, J., 1963. Cytogenic and psychosexual incongruities with a note on space-form blindness. *J. Psychiatry* 119: 820–827.

MORRIS, R. G. M., P. GARRUD, J. N. P. RAWLINS, and J. O'KEEFE, 1982. Place navigation impaired in rats with hippocampal lesions. *Nature* 297:681–683.

MURDOCK, G. P., 1986. Ethnographic atlas. *World Cultures* 2(4).

NYBORG, H., and J. NIELSEN, 1981. Spatial ability of men with karyotype 47,XXY, 47,XYY, or normal controls. In *Human Behavior and Genetics*, W. Schmid, and J. Nielsen, eds. Amsterdam: Elsevier pp. 85–106.

O'KEEFE, J., L. NADEL, S. KEIGHTLY, and D. KILL, 1975. Fornix lesions selectively abolish place learning in the rat. *Exp. Neurol.* 48:152–166.

O'KEEFE, J., and A. SPEAKMAN, 1987. Single unit activity in the rat hippocampus during a spatial memory task. *Exp. Brain Res.* 68:1–27.

PAGE, D. C., 1986. Sex reversal: Deletion mapping of male-determining function of the human Y chromosome. *Cold Spring Harb. Symp. Quant. Biol.* 51:229–235.

PERLMAN, S. M., 1973. Cognitive abilities of children with hormone abnormalities: Screening by psycho-educational tests. *J. Learn. Disabil.* 6:21–29.

PROFET, M., 1992. Pregnancy sickness as adaptation: A deterrent to the ingestion of teratogens. In *The Adapted Mind*, J. H. Barkow, L. Cosmides, and J. Tooby, eds. New York: Oxford University Press, pp. 327–365.

RESNICK, S. M., S. A. BERENBAUM, I. I. GOTTESMAN, and T. J. BOUCHARD, 1986. Early hormonal influences on cognitive functioning in congenital adrenal hyperplasia. *Dev. Psychol.* 22:191–198.

RODMAN, P. S., 1984. Foraging and social systems of orangutans and chimpanzees. In *Adaptations for Foraging in Nonhuman Primates*. P. S. Rodman and J. G. H. Cant, eds. New York: Columbia University Press, pp. 134–160.

ROGERS, A. R., and MUKHERJEE, A., 1992. Quantitative genetics of sexual dimorphism in human body size. *Evolution* 46:226–234.

ROSENTHAL, R., and D. B. RUBIN, 1982. Further meta-analytic procedures for assessing cognitive gender differences. *J. Educ. Psychol.* 74:708–712.

ROVET, J., and C. NETLEY, 1982. Processing deficits in Turner's syndrome. *Dev. Psychol.* 18:77–94.

SEABLOOM, R., 1985. Endocrinology. In *Biology of New World Microtus*, R. H. Tamarin, ed. Stillwater, Okla.: American Society of Mammalogists Special Publications, pp. 685–724.

SHERRY, D. F., M. R. L. FORBES, M. KHURGEL, and G. O. IVY, 1993. Greater hippocampal size in females of the brood parasitic brown-headed cowbird. *Proc. Natl. Acad. Sci. U.S.A.* 90:7839–7843.

SHERRY, D. F., L. F. JACOBS, and S. J. C. GAULIN, 1992. Spatial memory and adaptive specialization of the hippocampus. *Trend. Neurosci.* 15:298–303.

SHERRY, D. F., and A. L. VACCARINO, 1989. Hippocampus and memory for food caches in black-capped chickadees. *Behav. Neurosci.* 103:308–318.

SHERRY, D. F., A. L. VACCARINO, K. BUCKENHAM, and R. HERZ, 1989. The hippocampal complex of food-storing birds. *Brain Behav. Evol.* 34:308–317.

STEWART, J., A. SKVARENINA, and J. POTTIER, 1975. Effects of neonatal androgens on open-field behavior and maze learning in the prepubescent and adult rat. *Physiol. Behav.* 14:291–295.

TOOBY, J., 1982. Pathogens, polymorphism and the evolution of sex. *J. Theor. Biol.* 97:557–576.

Tooby, J., and L. Cosmides, 1990. On the universality of human nature and the uniqueness of the individual: The role of genetics and adaptation. *J. Pers.* 58:17–67.

Trivers, R. L., 1972. Parental investment and sexual selection. In *Sexual Selection and the Descent of Man: 1871–1971*, B. G. Campbell, ed. Chicago: Aldine, pp. 136–179.

Vergnaud, G., D. C. Page, M.-C. Simmler, L. Brown, F. Rouyer, B. Noel, D. Botstein, A. de la Chapelle, and J. Weissenbach, 1986. A deletion map of the human Y chromosome based on DNA hybridization. *Am. J. Hum. Genet.* 38:109–124.

Wehner, J. M., S. Sleight, and M. Upchurch, 1990. Hippocampal protein kinase C activity is reduced in poor spatial learners. *Brain Res.* 523:181–187.

Williams, C. L., A. L. Barnett, and W. H. Meck, 1990. Organizational effects of early gonadal secretions on sexual differentiation in spatial memory. *Behav. Neurosci.* 104: 84–97.

Williams, C. L., and W. H. Meck, 1991. The organizational effects of gonadal steroids on sexually dimorphic spatial ability. *Psychoneuroendocrinology* 16:155–176.

Williams, G. C., 1966. *Adaptation and Natural Selection.* Princeton: Princeton University Press.

Wolff, J. O., 1985. *Behavioral Biology of New World Microtus*, R. H. Tamarin, ed. Stillwater, Okla.: American Society of Mammalogists Special Publications, pp. 340–372.

Zahavi, A., 1975. Mate selection:—A selection for a handicap. *J. Theor. Biol.* 53:205–214.

81 The Argument from Animals to Humans in Cognitive Neuroscience

TODD M. PREUSS

ABSTRACT Neuroscientists make inferences about the human brain by studying nonhuman species, an enterprise that depends on assumptions about the nature of evolution. Traditionally, many neuroscientists have supposed that all mammals possess variants of the same brain which differ only in size and degree of elaboration. Under this model, the brains of nonhuman species can be treated as simplified versions or models of the human brain. However, there is evidence that mammalian cerebral organization is much more variable than is commonly acknowledged. The diversity of mammalian brain organization implies that neuroscientists can make better inferences about human brain organization by comparing multiple species chosen based on their evolutionary relationships to humans, than by studying individual "model" or "representative" species. The existence of neural diversity also suggests that nonhuman species have evolved cognitive specializations that are absent in humans.

During the past two decades, neuroscientists have developed a host of new techniques for studying the organization of the nervous system at a level of detail and precision scarcely imaginable by earlier generations of researchers. However, many of the most useful and informative techniques, including those used to trace neural connections, require invasive or terminal procedures. For this reason, they are used to study nonhuman species almost exclusively, and provide no direct information about the human nervous system. As a result, much of what neuroscientists believe to be true about the human brain, particularly about its connectional and areal organization, is based on inference or extrapolation from studies of other species.

From an evolutionary standpoint, it is reasonable to expect that we can learn much about humans by studying other species, especially closely related species. After all, humans share a long history of common ancestry with other primates and with mammals generally. Evolution necessarily entails change, however, and change poses a problem for those who would extrapolate findings from one species to another. Neuroscientists rarely confront this challenge directly. It is entirely commonplace to read reports of studies of a single primate species, usually the rhesus macaque, that purport to be studies of "the monkey" or "the primate." This practice discourages critical evaluation of evolutionary differences. Furthermore, since humans are primates, it tempts one to conclude that what is true of "the primate" will also be true of humans. The problem is that, strictly speaking, there is no such thing as "the primate" or "the monkey." Rather, there are approximately 200 living species of primates (Fleagle, 1988; Nowak, 1991), representing several distinct phyletic groups: prosimians, New World monkeys, Old World monkeys, and hominoids (apes and humans). Unless evolution has produced very few changes in neural organization, we have no prior grounds to conclude that what is true of any single primate species—rhesus macaque, squirrel monkey, owl monkey, or whatever—is true of primates generally, or of humans in particular.

Furthermore, if it is the case that evolution has produced so few differences between the brains of humans and other primates that we can comfortably ignore them, why should we suppose there is anything unusual about primates? Perhaps the features of cerebral organization found in humans evolved early in mammalian history. If so, the choice of species to be studied becomes largely a matter of cost and convenience, and it would be difficult to defend the study of anything other than rats. Indeed, Kolb and colleagues have vig-

TODD M. PREUSS Department of Psychology, Vanderbilt University, Nashville, Tenn.

orously defended the status of rats as "representative" mammals, and regard rat cortex as a good model system for understanding the structure and function of the human cortex, even the higher-order association regions (Kolb and Tees, 1990; Kolb and Whishaw, 1990). People who study primates may scoff at this, but if one accepts that evolution produced important differences between rodents and primates, one must acknowledge that it could have produced important differences between rhesus macaques and humans as well.

The purpose of this chapter is to consider how ideas about evolution affect the practice of neuroscience. First, I will contrast the modern conception of evolution, which embraces both similarities and differences, with the traditional view adopted by many neuroscientists, which emphasizes similarity and continuity. Next, I will review some of the evidence for phyletic differences in mammalian cerebral organization. I will then propose procedures for making inferences about human neural organization from the study of animals that are grounded in modern evolutionary principles, and thus do not rely on the traditional (and unwarranted) assumption that all primates or all mammals possess, in a fundamental sense, the same brain. Finally, I will consider how the evolutionary model advocated here can provide new perspectives on human cognition and its relationship to other varieties of animal cognition.

The evolutionary tradition in neuroscience

To illustrate the potential pitfalls involved in making inferences about the human brain from studies of macaques or other putative model species, imagine trying to make inferences about the *non*neural characteristics of humans by studying macaques (Fleagle, 1988; Richard, 1985). To be sure, we would get some things right. Humans, like macaques, have eyes set together in the front of the face rather than on the sides of the head, have dexterous hands with opposable thumbs and digits tipped with nails instead of claws, and live in complex societies. Unfortunately, we would also make many mistakes. For example, we would conclude that humans walk on all fours, have a tail, and possess a thick coat of fur. We would infer that humans possess a pouch of tissue in the cheek, where food can be hidden from higher-ranking individuals (figure 81.1). Furthermore, we would conclude that human social groups are centered on a stable core of closely related adult females, with males leaving their natal groups when they reach puberty. Besides leading us to ascribe erroneous characteristics to humans, the study of macaques would provide us no information about the unique evolutionary specializations of humans: bipedalism, functional hairlessness, unique hand muscles, a new layer of fatty tissue, and language, among others (Aiello and Dean, 1990; Napier, 1993).

FIGURE 81.1 A rhesus monkey (*Macaca mulatta*), marked for identification as part of a field study, showing cheek pouches distended with food. Cheek pouches are evolutionary specializations of the Old World monkey subfamily Cercopithecinae, to which macaques belong. (Courtesy of A. Richard)

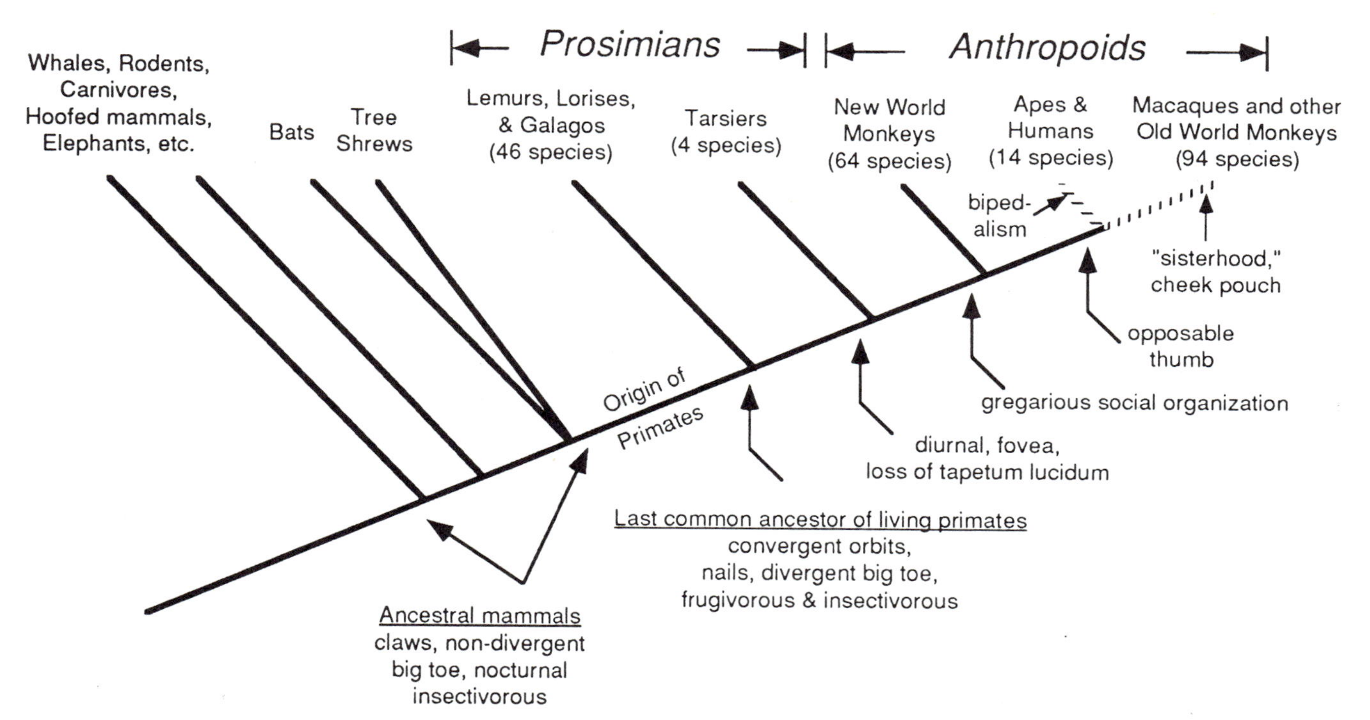

FIGURE 81.2 An evolutionary tree, showing the relationship of primates to other mammals, and the relationships among primates. The origins of evolutionary specializations distinctive of different groups, inferred from comparative and paleontological studies, are mapped onto the tree diagram. So, for example, the origin of primates was marked by the evolution of convergent orbits and the modification of claws into nails, among other changes. The features are shared by later primates. Humans and macaques differ due to evolutionary events that occurred subsequent to the divergence of their respective lineages (marked with horizontal and vertical bars, respectively). The number of species of primates and other mammals is derived from Nowak (1991).

In extrapolating from macaques to humans, why would we get some features right and others wrong? The answer lies in the geometry of evolution. Evolution is like a branching tree (figure 81.2). Every species is a composite of characteristics that evolved at different points in its ancestry; closely related species share more features, distant relatives fewer features. Macaques and humans share a long history of common ancestry. This is reflected in the characteristics they have in common, which are the characteristics we correctly attribute to humans based on the study of macaques. However, the macaque and human lineages separated about 25 million years ago, and since that time each has evolved its own unique features: the cheek pouch and "sisterhood" social organization in the case of macaques, bipedalism and language in the case of humans. It is features such as these, the evolutionary specializations of particular taxonomic groups, that confound attempts to make extrapolations from one group to another. When we consider a particular feature of macaque neural organization, how are we to know whether this is a feature present in humans rather than some unique product of macaque evolution? How can we tell whether we are studying the neural analogues of the opposable thumb and frontated orbits, rather than something akin to a cheek pouch? And how can we ever hope to understand what is distinctive about the human brain by studying macaques?

Given the diversifying nature of phyletic change, it is difficult to imagine a modern evolutionary biologist endorsing the use of macaques (much less rats or cats) as model humans. Why are neuroscientists not troubled by this procedure? The answer seems to be that neuroscientists, for the most part, do not view brain evolution as treelike. There is a tradition in neuroscience and psychology that holds that the important events in brain evolution involved mainly progressive increases in the size and "differentiation" of the brain and its components, without fundamental changes in basic structures and functions (in mammals, at least). That is to say, the pattern of brain evolution has been likened to a unitary scale or ladder rather than to a branching tree.

The view that evolution is like a scale is certainly not unique to neuroscientists: this was the predominant view of evolution among biologists from Darwin's era until well into this century, and it remains the predominant view in our culture at large. Nevertheless, modern evolutionary biologists are inclined to view the phylogenetic scale as a relic of the nineteenth century, with its emphasis on the march of Progress and the inevitable ascent of Man (Richards, 1987, 1992). Faced with the fact that lineages tend to evolve distinctive specializations, and unable to document a unitary, progressive trend in the history of life, evolutionary biologists consider the branching tree a more suitable metaphor for evolution than the scale (Gould, 1989; Williams, 1966, 1992).

In continuing to embrace the phylogenetic scale, therefore, the neurobehavioral sciences are out of step with modern evolutionary biology (Hodos and Campbell, 1969; Kaas, 1987; Povinelli, 1993; Preuss, 1993). How are we to understand this adherence to an otherwise discredited idea? Without question, the scale holds particular appeal for students of the brain and cognition. The scale ranks animals from lower to higher, and few of us would question the placement of humans at the top, at least with regard to cognition. Yet even as it affirms the special status of humans, the scale metaphor suggests a unity among animal brains, and so provides a rationale for extrapolating from animals to humans. For the scale permits change only within very narrow limits: Along the scale, change is strictly cumulative, so that brains may become improved, added to, and enlarged, but they do not *diverge*. Thus, lower forms can serve as simplified models of higher forms. The appeal of the phylogenetic scale may also reflect the fact that neurobiology has deep roots in a particular anatomical doctrine closely linked to the phylogenetic scale. According to this doctrine, known as typological or idealistic morphology, the member species of each major taxonomic group are regarded as manifestations or elaborations of an ideal, inherent form or archetype (Richards, 1992). Under this theory, the goal of comparative anatomy was to glimpse the common archetype through the haze of variations, the different expressions of the archetype presented by actual organisms.

Typological morphology was the dominant anatomical doctrine of the eighteenth and nineteenth centuries, predating the theory of evolution. The idea that differences between related species are not fundamental—that variations represent different expressions of a shared type—was held by many early evolutionists, including Darwin. This is evident in Darwin's insistence that humans possess no structures that are not also present in other animals, and by his emphasis on continuity between living forms in both physical and mental characteristics (Darwin, 1871, chap. 1). For example, in the *Descent of Man*, Darwin asserted that "the difference in mind between man and the higher animals, great as it is, is certainly one of degree and not of kind" (Darwin, 1871, 105). The only uniquely human characteristic he acknowledged (grudgingly, it seems) is language. For Darwin, human evolution was largely a process of improving upon faculties present in other animals, rather than adding new ones (Povinelli, 1993; Preuss, 1993; Richards, 1987).

Evolutionary biologists today take a different view of the evolutionary process: The changes organisms undergo in evolution represent departures or deviations from an ancestral condition, rather than different expressions of a common type (e.g., Williams, 1992). Yet, neurobiologists embraced Darwin's narrow doctrine of evolution, along with its typological foundations, early in the development of their discipline; and the influence of typological morphology continues to this day. For example, Cajal (1904) maintained that structures corresponding to the higher-order cortical areas of humans are present in other mammals, although smaller than in humans. Later, Elliot Smith (1924) and his student Le Gros Clark (1959), both central figures in the development of primate neuroanatomy, acknowledged that the number of cortical areas increased in evolution. However, they rationalized these changes as improvements or refinements of common brain structures rather than as the evolution of new structures. Ebbesson's (1980) idea that brain evolution is largely a matter of "parcellation"—the progressive segregation of elements originally mixed in one structure into multiple daughter structures—also belongs within this tradition: The brain is enlarged, and the parts rearranged, but nothing new appears.

Recently, Kolb and his colleagues have developed a framework for thinking about brain evolution and its relevance to animal models of human neuropsychology (Kolb and Whishaw, 1983, 1990; Kolb and Tees, 1990). Their approach is very much in the traditional vein, in that they acknowledge that mammalian species differ in brain size and in the number of cortical areas while denying that such differences are fundamental or qualitative. They specifically deny that evolution has produced new connectional or functional systems: the major features of cortical organization are present throughout the class Mammalia and are therefore said to be "class common." They conclude, "There is no strong evidence for unique brain-behavior relations in any species within the class Mammalia, including *Homo sapiens*" (Kolb and Whishaw, 1990, 110). From this perspective, it follows that the rat can serve as a "representative mammal" for the purposes of generalizing from animals to humans, even regarding such functionally higher-order regions as prefrontal and parietal cortex (Kolb and Tees, 1990).

The diversity of cortical structure

I now turn to the evidence for variation in neural organization among mammals, focusing on the cortex and related structures. The fact of variation is not at issue: No one suggests that all mammalian brains are exactly the same. What is at issue is the nature of variation. Is it the case, as many neuroscientists have supposed, that brains vary only by degrees from a common mammalian type, such that the study of a few species would lead to an accurate picture of all other species, including humans? Or is there evidence of more substantial differences? A review of the literature suggests that mammalian brains display a number of remarkable variations, at several levels of organization. (For further discussion of neural diversity, see Kaas, 1987; Kaas and Preuss, 1993; and Preuss, 1993.)

HISTOLOGY Ever since techniques were first developed to stain nerve cell bodies, biologists have been struck by the distinctive, laminated appearance of neocortex. Figure 81.3A presents a Nissl-stained section through the primary visual area (V1, area 17) of an owl monkey. The laminated appearance of the cortex reflects in part the segregation of cell types into different strata, for example, the concentration of small, granular cells in layer IV and of larger, pyramidal cells in layers III and V. (Six main layers are usually distinguished in neocortex.) Lamination is developed to an extreme degree in the visual cortex of anthropoid primates, but similar (if less vivid) patterns can be discerned in cortical areas in almost all mammals that have been examined (Brodmann, 1909). But there is at least one outstanding departure from this pattern among mammals—the cortex of cetaceans (whales and dolphins). Figure 81.3B shows a Nissl-stained section through the visual cortex of a bottlenose dolphin. Clearly, this is cortex with a difference. Layer I is enormously thick in these animals, compared to the other layers. No granular layer IV can be distinguished in adult dolphins, although granule cells are present in other cortical layers. Moreover, Golgi studies indicate that the pyramidal cells of dolphin visual cortex include a variety of unusual morphologies (Garey, Winkelmann, and Brauer, 1985; Glezer, Jacobs, and Morgane, 1988). These characteristics are found throughout the cortical mantle of cetaceans that have been studied to date. Some workers believe that the distinctive characteristics of cetaceans represent the retention of a primitive stage in the evolution of mammalian cortex (Glezer, Jacobs, and Morgane, 1988), while others believe they constitute evolutionary specializations (Johnson, 1988; Preuss, 1993). In either case, the unusual character of cetacean cortex is very interesting, in view of the reputedly high intelligence of these animals: If cetaceans are indeed intelligent in a manner recognizable to humans, their intelligence rests on a very different neural foundation than does ours.

Cetaceans pose a challenge for the view that there is a "basic uniformity in structure of the neocortex" across mammals (Rockel, Hiorns, and Powell, 1980). Although cetaceans are probably an extreme case, there is evidence of more subtle variations in the laminar and cellular organization of cortex between mammalian species and between cortical areas in the same species (Beaulieu and Collonier, 1989; Beaulieu, 1993).

CONNECTIVITY Although it is now widely accepted that the number of cortical areas differs among mammalian groups, it is not as commonly recognized that the connectional and functional relationships between areas also vary. Even the relationship between primary

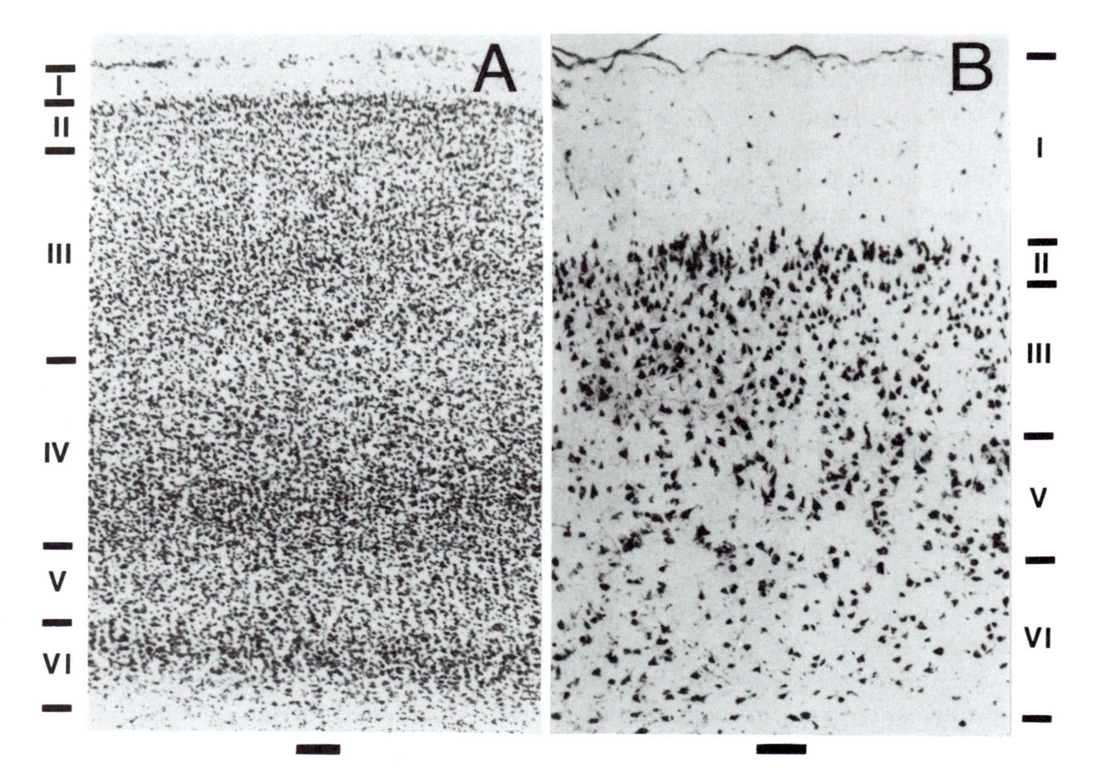

FIGURE 81.3 The primary visual cortex (V1) of an owl monkey (A) and a bottlenose dolphin (B), stained with cresyl violet. Owl monkey V1 is highly stratified: Note for instance the concentration of small (granular) cells in layer IV and the concentration of large, pyramidal cells in the deepest part of layer III. Dolphin visual cortex displays a very different pattern of lamination: Layer I is extremely thick, layer IV is absent, and stratification is generally less marked. The apparent difference in overall cell density between these two sections may be due in part to differences in histological processing. (A) A 40 μm thick frozen section. (B) A 20 μm thick section from a paraffin-embedded brain. The larger average cell size in (B) presumably reflects the fact that dolphins are much larger-bodied animals than owl monkeys. Scale bars represent 100 μm. (B modified from Garey, Winkelmann, and Brauer, 1985 and reproduced courtesy of Lawrence Garey)

sensory areas and secondary areas may change. For example, inactivation of V1 by cooling or lesion in anthropoid primates produces functional deactivation of the immediately adjoining visual area, V2, and other extrastriate visual areas (Rocha-Miranda et al., 1975; Schiller and Malpeli, 1977; Kaas and Krubitzer, 1992). Inactivation of V1 in cats has a less profound effect on the visually driven activity of neurons in extrastriate visual areas (e.g., Sherk, 1978; Guido, Tong, and Spear, 1991), presumably because in cats (unlike in primates), there are significant projections from the lateral geniculate nucleus to visual areas beyond V1 (Rosenquist, 1985). A similar evolutionary "rewiring" of thalamocortical and corticocortical connections in the somatosensory system of primates has also been documented (Garraghty et al., 1991).

The cortical connections of rats and primates appear to differ in several respects. For example, the primary motor area (M1) of rats is connected with orbital cortex (Reep, Goodwin, and Corwin, 1990; Paperna and Malach, 1991); despite intensive study of M1 in nonhuman primates, no orbital connections have been reported (e.g., Stepniewska, Preuss, and Kaas, 1993). The rat primary visual area (V1) also differs from that of primates and carnivores, having direct projections to medial limbic and possibly frontal cortex (Vogt and Miller, 1983; Reep, Goodwin, and Corwin, 1990; Paperna and Malach, 1991); visual-limbic connections have also been described in tree shrews (Sesma, Casagrande, and Kaas, 1984).

The carnivore literature provides two interesting examples of species differences. In rats and macaques, it is well established that the amygdala sends strong projections to the thalamic mediodorsal nucleus (MD), specifically to its medial part (Krettek and Price, 1977; Russchen, Amaral, and Price, 1987), which projects in turn to medial frontal and orbitofrontal cortex. However, studies with comparable techniques indicate that the amygdala does not project significantly to MD in cats (Krettek and Price, 1977; Velayos and Reinoso-Suárez, 1985). In cats, as in primates, there is a region of cortex anterior to the somatic motor region from which eye movements can be elicited by electrical stimulation; this is usually called the frontal eye field, although this region is actually composed of multiple areas in both primates and cats (Schlag and Schlag-Rey, 1987; Stepniewska, Preuss, and Kaas, 1993). While it is tempting to view this as a case of class-common organization, the connectional evidence suggests otherwise. In nonhuman primates, the frontal oculomotor areas receive their main cortical inputs from parietal and temporal areas located at the fringe of the extrastriate visual region (Huerta, Krubitzer, and Kaas, 1987). In cats, by contrast, the main input to frontal oculomotor cortex arises from cortex near the insula, at a distance from the main extrastriate zone, a region that includes the anterior ectosylvian visual area (EVA) (Nakai, Tamai, and Miyashita, 1987; Olson and Graybiel, 1987). EVA is unlike any visual area known in primates, being separated from the rest of extrastriate visual cortex by intervening territories of auditory and somatosensory cortex, and having unusual receptive field properties (Olson and Graybiel, 1987; Mucke et al., 1982). It seems likely that EVA is a cat area that has no homologue in primates. So, although both primates and carnivores possess frontal oculomotor cortex, they differ with respect to the major cortical connections of the region. This suggests that one or more of the frontal oculomotor fields of primates and carnivores are the products of convergent evolution and are therefore not homologous. There is evidence of other differences in frontal lobe organization between primates and nonprimates (Preuss and Goldman-Rakic, 1991a, 1991b; Preuss, 1993).

Modular and Laminar Organization In cortical areas, the terminal fields of input fibers and the cells of origin of outputs are typically segregated in different laminae, and are sometimes also segregated into repeating "columns" or "modules" oriented orthogonally to the laminae. These patterns of segregation vary across mammalian groups. The best-known examples of modular segregation are the ocular dominance columns of primary visual cortex. In most mammals, dense projections from the lateral geniculate nucleus of the thalamus terminate in layers III and IV of cortical area V1; these projections relay visual information from the left and right eyes to V1 on each side of the brain. In some primate species, inputs from the left and right eye are segregated into alternating bands within layers III and IV, known as ocular dominance columns. The degree of segregation varies considerably among primates, however, and some New World monkeys exhibit virtually no segregation at all (reviewed by Florence, Conley, and Casagrande, 1986). Ocular dominance columns are absent in other mammals that have been examined, with the exception of carnivores (reviewed by Casagrande and Kaas, in press). Ocular dominance columns probably evolved independently (convergently) in primates and carnivores. One reason for concluding this is that carnivores, which possess columns, are distantly related to primates, whereas tree shrews, which are thought to be closely related to primates (Novacek, 1992), lack columns (Hubel, 1975). However, tree shrews offer an interesting twist to the ocular dominance story. Although they lack ocular dominance columns, inputs from each eye are differentially distributed within layers III and IV of the primary visual area. Specifically, inputs from both eyes converge in the upper and lower tiers of layer IV, but inputs from the contralateral eye have additional terminations in the middle tier of layer IV and in layer III (Hubel, 1975). This laminar arrangement of ocular inputs is unique among mammals studied to date.

Area V1 of primates exhibits another form of modular organization. As shown in figure 81.4, sections cut parallel to the cortical surface and stained for the metabolic enzyme cytochrome oxidase (CO), exhibit a regular pattern of darkly stained spots, termed "puffs" or "blobs." CO blobs have been described in at least seven genera of anthropoid primates (including *Homo*) as well as in a number of prosimians, but have not been clearly demonstrated in any nonprimate mammal (Horton, 1984; Kaas and Preuss, 1993; Preuss, Beck, and Kaas, 1993; Casagrande and Kaas, in press). These results are consistent with the suggestion of Horton (1984) that CO blobs are an evolutionary specialization of primate V1. CO blobs are thought to

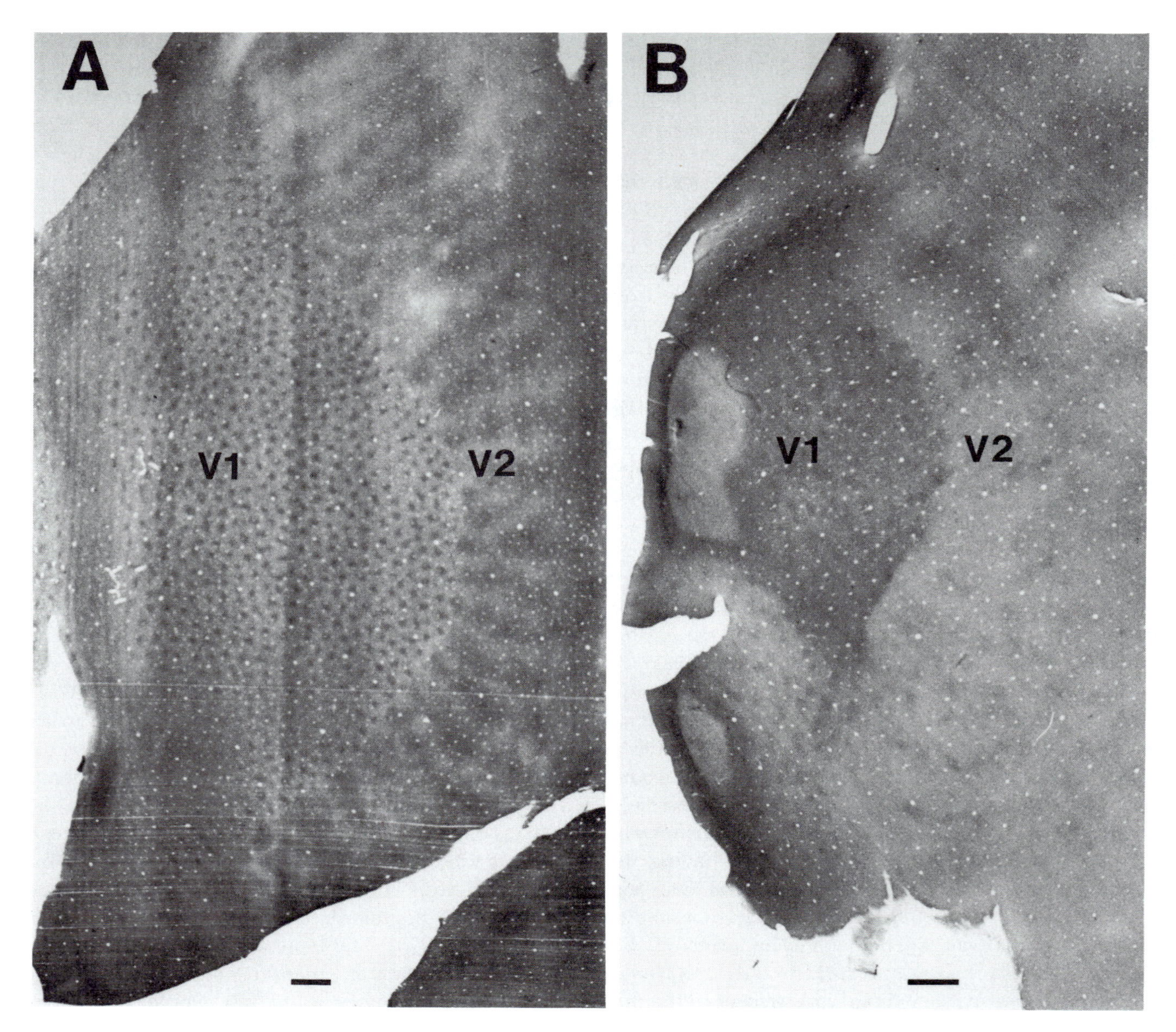

FIGURE 81.4 Modular organization of areas V1 and V2 in (A) a squirrel monkey (*Saimiri sciureus*), a diurnal, anthropoid primate, and (B) a galago (*Galago garnetti*), a nocturnal, prosimian primate. In both cases, the cortex was separated from the underlying white matter, flattened, and sectioned parallel to the cortical surface. Sections were then stained for cytochrome oxidase, a metabolic enzyme. Many small, dark, cytochrome oxidase–rich "blobs" are visible in area V1 of both animals. Squirrel monkeys and other anthropoid primates also possess well-marked cytochrome oxidase–dense stripes in area V2, whereas stripes are very poorly developed in V2 of galagos and other prosimian primates. Scales bars: 1 mm.

constitute a specialized processing channel within the visual cortex, a point I will consider in a later section.

NEUROCHEMICAL DISTRIBUTION It is axiomatic among cell biologists that macromolecular pathways and systems have been highly conserved during the evolution of eukaryotic organisms. This conservatism is reflected in the presence of a common set of neurotransmitter, neuromodulator, and receptor molecules across a spectrum of invertebrate and vertebrate groups. Nevertheless, there are phyletic differences even at this level of organization, particularly with regard to the biochemical phenotypes of specific classes of neurons and the distribution of specific transmitter- and modulator-containing axons within the cortex.

One of the best-studied examples is the differential distribution of dopamine (DA)-containing axons in the frontal cortex of rats and macaques (Berger, Gaspar, and Verney, 1991; Berger, 1992). Dopaminergic neurons, with cell bodies located in the substantia nigra and ventral tegmental area of the midbrain, send strong projections to the frontal cortex, where they appear to modulate the responses of target neurons to other afferents. In rats, DA-containing fibers are concentrated in the medial frontal and orbitofrontal regions, with relatively weak projections to the primary motor cortex (M1) and supplementary motor area. Macaques and humans also have DA projections to medial and orbital cortex, but in addition exhibit very dense DA innervation of M1 and the supplementary motor area, with fibers distributed across all layers of cortex (see also Williams and Goldman-Rakic, 1993). In rat motor cortex, by contrast, DA-containing fibers are essentially restricted to layers V and VI. In rats, furthermore, the DA-containing fibers of frontal cortex also contain neurotensin, whereas in macaques, rats, DA and neurotensin are localized in different fibers, which project to different laminae (Berger, Gaspar, and Verney, 1991).

To take another example, there is now good evidence that rats possess a population of intrinsic neurons in cerebral cortex that contain acetylcholine (ACh); these have been identified using several different antibodies raised against choline acetyltransferase, the synthetic enzyme that is a definitive marker for ACh (Houser et al., 1985). Studies employing similar techniques provide no evidence of ACh-containing cortical neurons in macaques (Lewis, 1991), humans (Mesulam and Geula, 1991), or cats (Kimura et al., 1981).

From nonhumans to humans

The diversity of mammalian neurological organization cannot be denied or dismissed as trivial. We need to consider how the fact of diversity affects the practice of neuroscience. If we cannot simply extrapolate results from nonhumans to humans, how are we to advance our understanding of the human brain?

We can, of course, carry out more intensive studies of human beings, as suggested by Crick and Jones (1993). The problem, as Crick and Jones recognize, is that the techniques currently available for studying the connectional organization of the human brain are markedly inferior to those available for studying nonhumans. Certainly, innovations that improve our ability to study the human brain directly are only to be welcomed. For the immediate future, however, our methods for studying animals will remain superior to our methods for studying people. A second approach, to be advocated here, seeks to improve the inferences we make about humans from animal studies. One major goal of this approach is to determine those features of nonhuman neurological organization that are likely to be present in humans, as determined through a program of comparative research, guided by our understanding of evolutionary relationships. This approach promises fruitful interactions with more direct studies of human organization. Another goal of this approach is to understand in what respects the brains of species differ from each other, and how and why those differences evolved. Of particular interest and importance are those characteristics that distinguish humans from other animals.

The manner in which comparative studies can yield inferences about humans, and the potential for interaction between human and animal studies, are illustrated by recent investigations of extrastriate visual cortex (see also Kaas, 1993). This region has been extensively studied in Old World monkeys, mainly macaques, which have been taken as models for human organization. Crick and Jones (1993) have criticized this approach, arguing that macaque studies can provide only a working hypothesis about human organization, about which we have little direct knowledge. Humans and macaques presumably differ in ways we cannot determine at present. This is a reasonable expectation on evolutionary grounds, but it must to be recognized that our knowledge of visual organization in nonhuman primates is not restricted to Old World monkeys: New World monkeys and prosimians have been studied in detail as well. One can identify a number of visual areas, based on similarities in architectonics, location, connections, and physiological properties, that are present in species belonging to each of these major primate groups (reviewed by Kaas, 1993; Krubitzer and Kaas, 1993; Preuss, Beck, and Kaas, 1993). As shown in figure 81.5A, the shared areas include the primary visual area (V1), second visual area (V2), dorsolateral area (DL; also known as V4), middle temporal area (MT; also known as V5), and the dorsome-

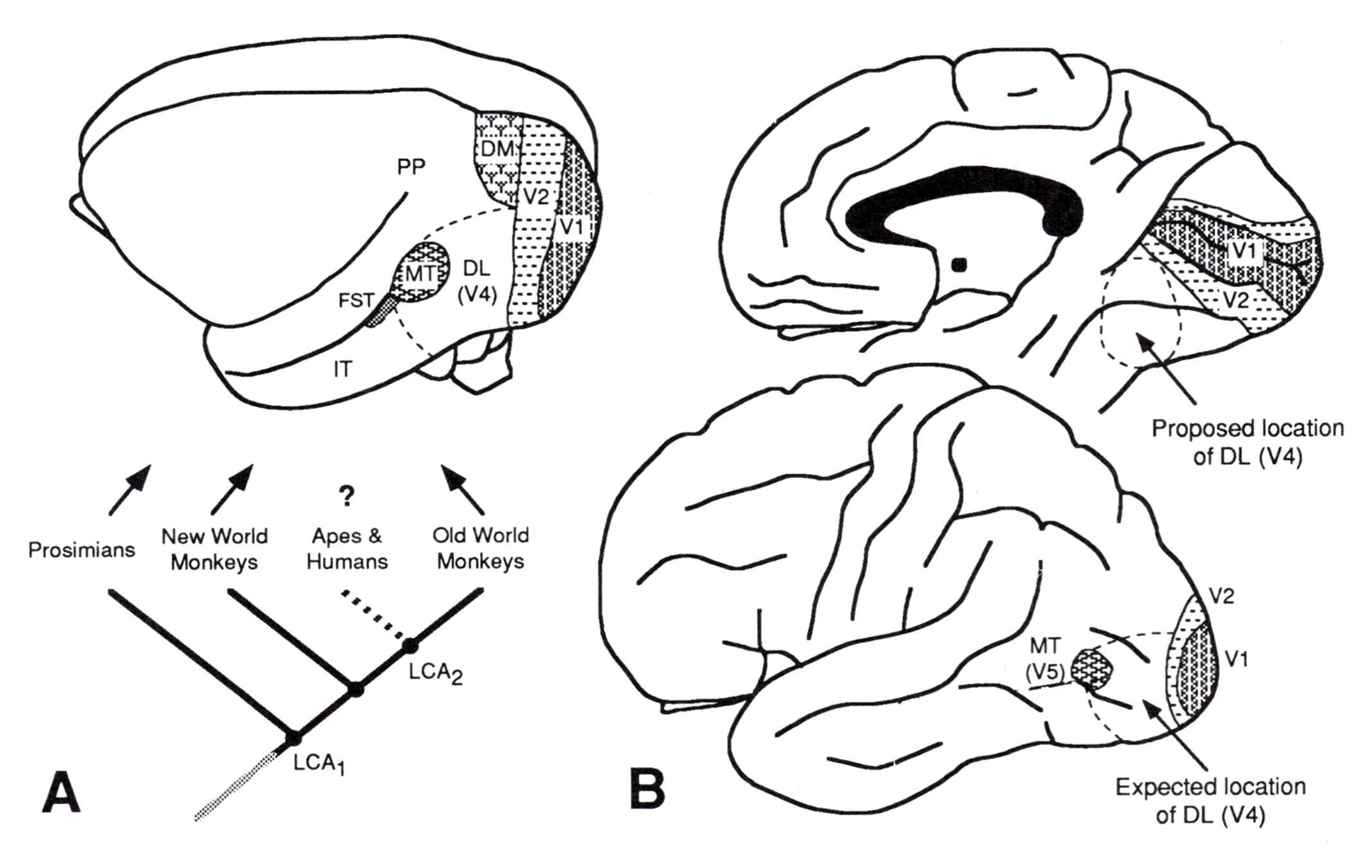

FIGURE 81.5 The organization of extrastriate visual cortex in humans and nonhuman primates. (A) Visual areas shared by nonhuman primates. Studies of prosimians, New World monkeys, and Old World monkeys have made it possible to identify a set of visual areas common to these groups, depicted in the figurine at the top. In addition to the primary visual cortex (V1; striate cortex), the common areas include the second visual area (V2), dorsomedial area (DM), dorsolateral area (DL; and known as V4), middle temporal area (MT; also known as V5), and the area of the fundus of the superior temporal sulcus (FST). There is also evidence for multiple divisions of posterior parietal (PP) and inferotemporal (IT) cortex in these groups. Given the similarities among prosimians, New World monkeys, and Old World monkeys, it is likely that these visual areas were present in the last common ancestor of the living primates (LCA_1), as indicated in the tree diagram at the bottom. The phylogenetic distribution of visual areas also implies that these areas were present in the last common ancestor of apes, humans, and Old World monkeys (LCA_2). (B) The organization of visual cortex in humans, based on recent architectonic and in vivo imaging studies (modified from Kaas, 1992). The area denoted as MT in humans is very densely myelinated and is reponsive to moving stimuli, as is MT in other primates. The location of human DL (V4) is controversial. Zeki et al. (1991) have identified a region of cortex on the ventromedial surface of the occipital lobe that is metabolically active while subjects view colored stimuli; they regard this as the homologue of DL (V4) of nonhuman primates, an area they consider to be the color-processing area. By contrast, Kaas (1992) suggests that in humans DL (V4) should be located the lateral surface of the occipital lobe, between V2 and MT, as it is in all other primates examined.

dial area (DM). There is also evidence that all major primate groups possess posterior parietal and inferotemporal cortex, although much remains to be learned about the organization of these regions, especially in New World monkeys and prosimians. While there are doubtless additional visual areas, these are the areas which, based on current evidence, appear to be shared by the major primate groups and thus are likely to have been present in ancestral primates.

The fact that we can identify a common pattern of extrastriate organization, shared across a wide variety of nonhuman primates, suggests that human extrastriate cortex is organized in similar fashion. This organization may have been modified during human evolution, but the common pattern of primate organization—inferred from studies using multiple techniques and diverse species—provides a solid foundation from which to pursue human investigations. Efforts are currently being made to identify homologues of extrastriate areas in humans, using evidence from architectonics, location, and function (Kaas, 1992). For example, there is good evidence that humans possess a homologue of MT, located anteriorly to V1 and V2 in the lateral occipitotemporal region (figure 81.5B), as one would expect based on comparative studies (Clarke and Miklossy, 1990; Zeki et al., 1991). By contrast, the location of area V4 is controversial. Zeki and colleagues (1991) argue that this area is located in the inferior and medial part of the occipital lobe, because they regard V4 as the cortical color center, and the inferomedial region is activated metabolically when subjects view colored stimuli. In response, Kaas (1992) has noted that if Zeki and colleagues are correct, human V4 differs in its location from that of all other primates studied. In nonhuman primates, V4 or DL is located on the lateral surface, between area MT and the foveal representation of V2. While it is possible that evolution has "relocated" V4 in humans, the spatial arrangement of anatomical structures tends to be highly conserved in evolution (Darwin, 1859), a principle illustrated nicely by the stability of extrastriate organization across nonhuman primates. The comparative evidence thus suggests that the cortical region activated by colored stimuli in the study by Zeki and colleagues (1991) corresponds to some visual area other than V4, or perhaps to a limited portion of V4.

It is common to hear neuroscientists assert that they are not really interested in comparative issues; what they care about is *function*. Yet a comparative perspective can provide unique insights on function. Consider the cytochrome oxidase–rich "blobs" in the primary visual area of primates (figure 81.4). Physiological investigations in Old World and New World monkeys have shown that blobs contain a higher proportion of color-opponent cells than surrounding cortex (Livingstone and Hubel, 1984). Blobs have thus come to be regarded as components of a specialized color-processing "channel" within the visual system—as indeed they may be, in diurnal primates such as macaques and squirrel monkeys (figure 81.4A). However, the comparative evidence puts the relationship between blobs and color vision in a different light (as it were). In addition to diurnal primates, which have well-developed color vision, CO blobs are present in such nocturnal primates as galagos (figure 81.4B), lorises, and owl monkeys (Horton, 1984; Livingstone and Hubel, 1984; Preuss, Beck, and Kaas, 1993), in which color vision is not well developed. What is more, because nocturnality is probably the ancestral condition for primates (Fleagle, 1988), it is likely that blobs originally evolved in animals with poor color vision. For this reason, researchers have considered other possible functions of blobs. Along with color-opponent cells, blobs contain broadband, brightness-selective cells (Livingstone and Hubel, 1984), prompting the suggestion that in primates generally, blobs constitute part of a perceptual brightness constancy system (Allman and Zucker, 1990). Furthermore, Allman and Zucker propose that when diurnal primates evolved, the brightness constancy system was modified to accommodate color-specific brightness differences. That is, they propose that evolution constructed primate color vision using structures that originally subserved other aspects of vision.

The foregoing example illustrates an important evolutionary principle. Evolution is a tinkerer, building new structures and systems by modifying existing structures, rather than by designing them from scratch (Simpson, 1967; Jacob, 1982). If we ask, Why is this system constructed in this way? we must consider not only what the system does but also where it came from. And that is a question about evolution, to be addressed through comparative studies.

It is because evolution adapts old structures to new ends that we may find studies of nonhuman species helpful in understanding the structural basis of even uniquely human functions. Consider the case of language. It has long been argued that the ventral premotor area (PMV) of nonhuman primates resembles Broca's area in cytoarchitecture and location (e.g., (Bonin, 1944; figure 81.6). Recent studies provide additional evidence that Broca's area and PMV are homologous. Connectional and microstimulation studies in nonhuman primates indicate that PMV represents forelimb and orofacial movements (Preuss, 1993; Stepniewska, Preuss, and Kaas, 1993). Metabolic and surface stimulation studies in humans suggest that Broca's area also represents nonlinguistic forelimb and orofacial movements (Fox et al., 1988; see also Roland et al., 1980; Uematsu et al., 1992). Results such as these have

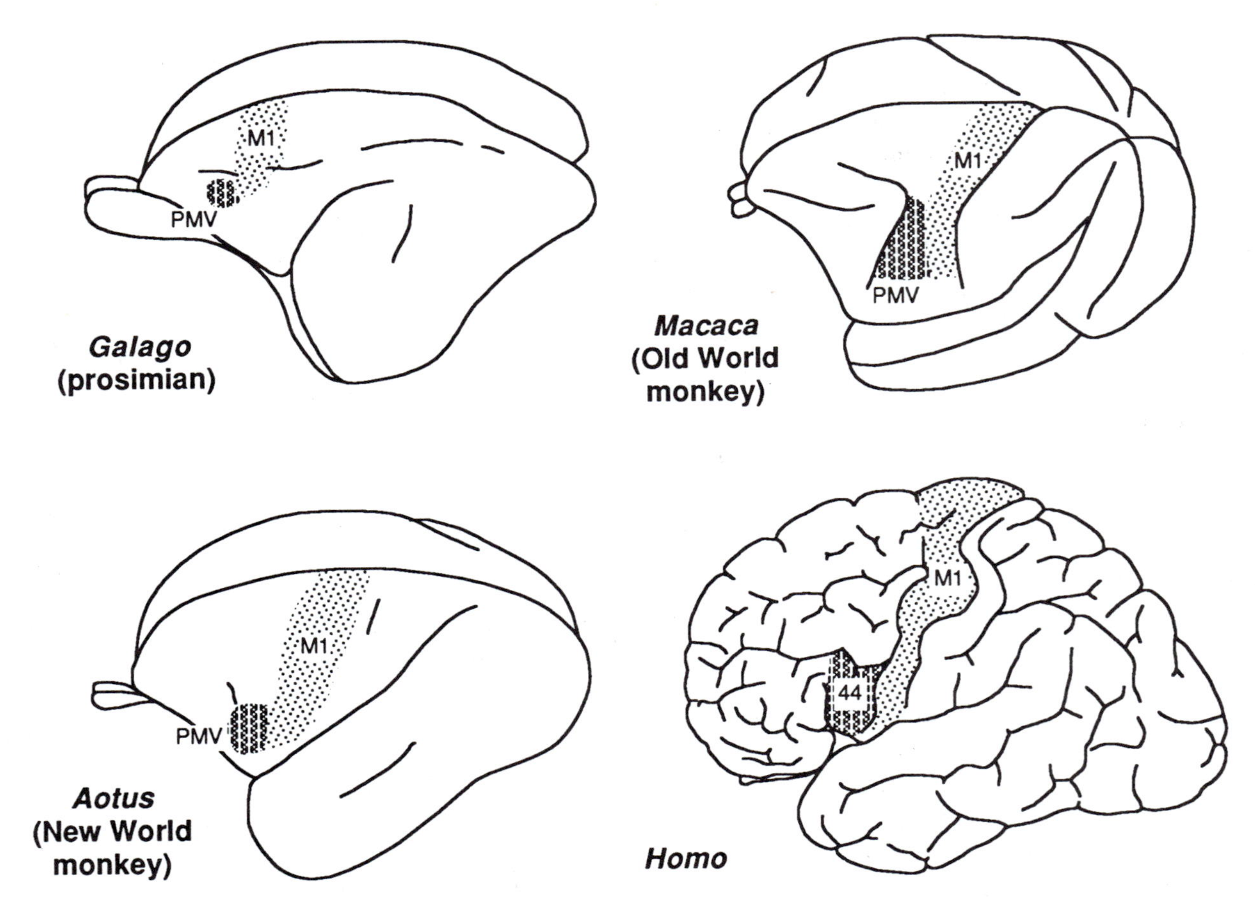

FIGURE 81.6 The location of the ventral premotor area (PMV) in nonhuman primates compared to that of Brodmann's area 44 (the posterior part of Broca's area) in humans. PMV has been identified in prosimians, New World monkeys, and Old World monkeys, as a discrete region that represents forelimb and orofacial movements and is located immediately anterior to the inferior part of primary motor cortex (M1) (Preuss, 1993). Brodmann's area 44 occupies the same location with respect to M1 in humans. Recent metabolic and stimulation studies in humans suggest that Broca's area is involved in the control of forelimb and orofacial movements.

led Fox and colleagues (1988) to suggest that Broca's area is a general premotor area without specific linguistic functions. Alternatively, it might be the case that Broca's area has both linguistic and somatic motor functions, and that evolution constructed the neural systems of language in part by "recruiting" existing motor areas, as Bonin (1944) suggested. The latter hypothesis suggests why language can be conveyed as naturally and fully with manual signs as with speech (Klima and Bellugi, 1979).

The foregoing examples demonstrate how we can make reasonable inferences about the structure of the human brain by comparing the neural organization of animals closely related to humans. This procedure does not require that the animals we compare be alike in *all* features of organization, merely that they possess *some* features in common by virtue of shared ancestry. Studies of nonhuman primates are of particular relevance for identifying likely features of human organization, for these are the animals with which humans share the longest period of common descent.

The matter of inferring human brain structure represents only one aspect of comparative neuroscience, a specific aspect that is best addressed by studying animals closely related to humans. There are other neurobiological issues of a more general nature, though still pertinent to understanding humans, for which different comparative strategies are appropriate. For example, if we want to know why information processing in the human visual system is compartmentalized into blobs and columns, it is useful to consider the range of circumstances under which compartments develop and the structural and functional consequences of compartmentalized processing. Such an inquiry properly extends to whatever species and systems exhibit compartmentalization, from the barrel fields of rat somatosensory cortex (Kaas, Merzenich, and Killackey, 1983) to the optic tectum of three-eyed frogs, with its artificially induced visual columns (Constantine-Paton, 1982).

A diversity of minds

Providing the basis for making inferences about humans from the study of other species is just one way that evolutionary ideas can inform cognitive neuroscience. Indeed, if we accept the modern metaphor of evolution as a branching tree, and if we accept that neural organization varies among mammals, we are led to a new and challenging conception of the relationship between human minds and animal minds, one that is fundamentally different from that suggested by the scale metaphor. Under the older view of life represented by the phylogenetic scale, humans stand at the zenith of a continuum of mental development, and other beings are ranked according to a human standard. The modern conception of evolution as a branching tree removes this standard, so that nonhuman species are seen not as steps on the ladder to humanity but as alternative outcomes of the evolutionary process. Under this view, the human mind represents not the highest expression of a common animal mind, but rather one mind among many.

What are the consequences of acknowledging a diversity of minds? For one thing, it suggests new questions about the human brain and cognition. If the human mind is one evolutionary outcome among many, we must ask: Why this outcome and not some other? What specific cognitive capacities were selected for in human evolution? How were the components of neural and cognitive systems present in our primate ancestors modified to produce new systems in humans? Questions like these make little sense under the phylogenetic scale, in which the emergence of the human mind is the almost inevitable outcome of a general process of improvement. And these are fundamental questions about human nature rarely asked by neuroscientists.

To view evolution as treelike is to raise new questions about other animals as well. We should expect that nonhuman species are in some respects truly different from humans, with neural systems and functional capacities that humans lack. Understanding what other animals are like, when they are not like humans, is perhaps the most profound and intriguing challenge faced by cognitive neuroscientists.

ACKNOWLEDGMENTS The author would like to thank Michael Gazzaniga, Leda Cosmides, John Tooby, Jon Kaas, Mary Anne Case, and Sherre Florence for their support and assistance in preparing this chapter. Conversations with Patricia Gaspar, Jon Kaas, and Frederick Szalay were instrumental in developing the ideas presented here. The author's research is supported in part by the McDonnell-Pew Program in Cognitive Neuroscience.

REFERENCES

AIELLO, L., and C. DEAN, 1990. *Human Evolutionary Anatomy.* London: Academic Press.

ALLMAN, J., and S. ZUCKER, 1990. Cytochrome oxidase and functional coding in primate striate cortex: A hypothesis. *Cold Spring Harb. Symp. Quant. Biol.* 55:979–982.

BEAULIEU, C., 1993. Numerical data on neocortical neurons in adult rat, with special reference to the GABA population. *Brain Res.* 609:284–292.

BEAULIEU, C., and M. COLLONIER, 1989. Number of neurons in individual laminae of areas 3B, 4γ, and 6aα of the cat cerebral cortex: A comparison with major visual areas. *J. Comp. Neurol.* 279:228–234.

BERGER, B., 1992. Dopaminergic innervation of the frontal cerebral cortex: Evolutionary trends and functional implications. In *Frontal Lobe Seizures and Epilepsies*, P. Chauvel, A. V. Delgado-Escuerta, E. Halgren, and J. Bancaud, eds. New York: Raven, pp. 525–544.

BERGER, B., P. GASPAR, and C. VERNEY, 1991. Dopaminergic innervation of the cerebral cortex: Unexpected differences between rodent and primate. *Trends. Neurosci.* 14: 21–27.

BONIN, G. VON, 1944. The architecture. In *The Precentral Motor Cortex.* P. C. Bucy, ed. Urbana: University of Illinois Press, pp. 7–82.

BRODMANN, K., 1909. *Vergleichende Lokalisationslehre der Grosshirnrhinde.* Leipzig: Barth.

CAJAL, S. RAMON Y, 1904. *Textura del Sistema Nervioso del Hombre y de los Vertebrados*, vol. 2. Madrid: N. Moya. English translation in *Cajal on the Cerebral Cortex*, J. DeFelipe and E. G. Jones, eds., 1988. New York: Oxford University Press, pp. 465–490.

CASAGRANDE, V. A., and J. H. KAAS, in press. The afferent, intrinsic, and efferent connections of primary visual cortex in primates. In *Cerebral Cortex*. Vol. 10, *Primary Visual Cortex in Primates*, A. Peters and K. Rockland, eds. New York: Plenum.

CLARKE, S., and J. MIKLOSSY, 1990. Occipital cortex in man: Organization of callosal connections, related myelo- and cytoarchitecture, and putative boundaries of functional visual areas. *J. Comp. Neurol.* 298:188–214.

CONSTANTINE-PATON, M., 1982. The retinotectal hookup: The process of neural mapping. In *Developmental Order:*

Its Origin and Regulation, S. Subtelny, ed. New York: Liss, pp. 317–349.

Crick, F., and E. G. Jones, 1993. Backwardness of human neuroanatomy. *Nature* 361:109–110.

Darwin, C., 1859. *On the Origin of Species*. London: Murray; reprinted, 1984, Cambridge, Mass.: Harvard University Press.

Darwin, C., 1871. *The Descent of Man, and Selection in Relation to Sex*. London: Murray; reprinted, 1981, Princeton: Princeton University Press.

Ebbesson, S. O. E., 1980. The parcellation theory and its relation to interspecific variability in brain organization, evolutionary and ontogenetic development, and neuronal plasticity. *Cell. Tissue Res.* 213:179–212.

Elliot Smith, G., 1924. *The Evolution of Man: Essays*. London: Oxford University Press.

Fleagle, J. G., 1988. *Primate Adaptation and Evolution*. San Diego: Academic Press.

Florence, S. L., M. Conley, and V. A. Casagrande, 1986. Ocular dominance columns and retinal projections in New World spider monkeys (*Ateles ater*). *J. Comp. Neurol.* 243:234–248.

Fox, P., S. Petersen, M. Posner, and M. Raichle, 1988. Is Broca's area language specific? *Neurology* 38 (suppl. 1): 172.

Garey, L. J., E. Winkelmann, and K. Brauer, 1985. Golgi and Nissl studies of the visual cortex of the bottlenose dolphin. *J. Comp. Neurol.* 240:305–321.

Garraghty, P. E., S. L. Florence, W. N. Tenhula, and J. H. Kaas, 1991. Parallel thalamic activation of the first and second somatosensory areas in prosimian primates and tree shrews. *J. Comp. Neurol.* 311:289–299.

Glezer, I. I., M. S. Jacobs, and P. J. Morgane, 1988. The "initial brain" concept and its implications for brain evolution in Cetacea. *Behav. Brain Sci.* 11:75–116.

Gould, S. J., 1989. *Wonderful Life*. New York: Norton.

Guido, W., L. Tong, and P. D. Spear, 1991. Afferent bases of spatial- and temporal-frequency processing by neurons in the cat's posteromedial lateral suprasylvian cortex: Effects of removing areas 17, 18, 19. *J. Neurophysiol.* 64: 1636–1651.

Hodos, W., and C. B. G. Campbell, 1969. *Scala naturae*: Why there is no theory in comparative psychology. *Psychol. Rev.* 76:337–350.

Horton, J. C., 1984. Cytochrome oxidase patches: A new cytoarchitectonic feature of monkey visual cortex. *Philos. Trans. R. Soc. Lond. [Biol.]* 304:199–253.

Houser, C. R., G. D. Crawford, P. M. Salvaterra, and J. E. Vaughn, 1985. Immunocytochemical localization of choline acetyltransferase in rat cerebral cortex: A study of cholinergic neurons and synapses. *J. Comp. Neurol.* 234: 17–34.

Hubel, D. H., 1975. An autoradiographic study of the retino-cortical projections in the tree shrew (*Tupaia glis*). *Brain Res.* 96:41–50.

Huerta, M. F., L. A. Krubitzer, and J. H. Kaas, 1987. Frontal eye field as defined by intracortical microstimulation in squirrel monkeys, owl monkeys, and macaque monkeys. II. Cortical connections. *J. Comp. Neurol.* 265:332–361.

Jacob, F., 1982. *The Possible and the Actual*. New York: Pantheon Books.

Johnson, J. I., 1988. Whose brain is initial-like? *Behav. Brain Sci.* 11:96.

Kaas, J. H., 1987. The organization and evolution of neocortex. In *Higher Brain Function: Recent Explorations of the Brain's Emergent Properties*, S. P. Wise, ed. New York: Wiley, pp. 347–378.

Kaas, J. H., 1992. Do humans see what monkeys see? *Trends. Neurosci.* 15:1–3.

Kaas, J. H., 1993. The organization of visual cortex in primates: Problems, conclusions, and the use of comparative studies in understanding the human brain. In *Functional Organization of the Human Visual System*, B. Gulyas, D. Ottoson, and P. E. Roland, eds. Oxford: Pergamon, pp. 1–11.

Kaas, J. H., and L. A. Krubitzer, 1992. Area 17 lesions deactivate area MT in owl monkeys. *Vis. Neurosci.* 9:399–407.

Kaas, J. H., M. M. Merzenich, and H. P. Killackey, 1983. The reorganization of somatosensory cortex following peripheral nerve damage in adult and developing mammals. *Annu Rev. Neurosci.* 6:325–356.

Kaas, J. H., and T. M. Preuss, 1993. Archontan affinities as reflected in the visual system. In *Mammal Phylogeny: Placentals*, F. S. Szalay, M. J. Novacek, and M. C. McKenna, eds. New York: Springer, pp. 115–128.

Kimura, H., P. L. McGeer, J. H. Peng, and E. G. McGeer, 1981. The central cholinergic system studied by choline acetyltransferase immunohistochemistry. *J. Comp. Neurol.* 200:151–201.

Klima, E. S., and U. Bellugi, 1979. *Signs of Language*. Cambridge, Mass.: Harvard University Press.

Kolb, B., and R. C. Tees, 1990. The rat as a model of cortical function. In *The Cerebral Cortex of the Rat*, B. Kolb and R. C. Tees, eds. Cambridge, Mass.: MIT Press, pp. 3–17.

Kolb, B., and I. Q. Whishaw, 1983. Problems and principles underlying interspecies comparisons. In *Behavioral Approaches to Brain Research*, T. E. Robinson, ed. New York: Oxford University Press, pp. 237–263.

Kolb, B., and I. Q. Whishaw, 1990. *Fundamentals of Human Neuropsychology*, ed. 3. New York: Freeman.

Krettek, J. E., and J. L. Price, 1977. Projections from the amygdaloid complex to the cerebral cortex and thalamus in the rat and cat. *J. Comp. Neurol.* 172:687–722.

Krubitzer, L. A., and J. H. Kaas, 1993. The dorsomedial visual area (DM) of owl monkeys: Connections, myeloarchitecture, and homologies in other primates. *J. Comp. Neurol.* 334:497–528.

Le Gros Clark, W. E., 1959. *The Antecedents of Man*. Edinburgh: Edinburgh University Press.

Lewis, D. A., 1991. Distribution of choline acetyltransferase-immunoreactive axons in monkey frontal cortex. *Neurosci.* 40:363–374.

Livingstone, M. S., and D. H. Hubel, 1984. Anatomy and

physiology of a color system in the primate visual cortex. *J. Neurosci.* 4:309–356.

Mesulam, M.-M., and C. Geula, 1991. Acetylcholinesterase-rich neurons of the human cerebral cortex: Cytoarchitectonic and ontogenetic patterns of distribution. *J. Comp. Neurol.* 306:193–220.

Mucke, L., M. Norita, G. Benedek, and O. Creutzfeldt, 1982. Physiologic and anatomic investigation of a visual cortical area situated in the ventral bank of the anterior ectosylvian sulcus of the cat. *Exp. Brain Res.* 46:1–11.

Nakai, M., Y. Tamai, and E. Miyashita, 1987. Corticocortical connections of frontal oculomotor areas in the cat. *Brain Res.* 414:91–98.

Napier, J., 1993. *Hands*, R. H. Tuttle ed. Princeton: Princeton University Press.

Novacek, M. J., 1992. Mammalian phylogeny: Shaking the tree. *Nature* 356:121–125.

Nowak, R. M., 1991. *Walker's Mammals of the World*, vol. 1, ed. 5. Baltimore, Md.: Johns Hopkins University.

Olson, C. R., and A. M. Graybiel, 1987. Ectosylvian visual area of the cat: Location, retinotopic organization, and connections. *J. Comp. Neurol.* 261:277–294.

Paperna, T., and R. Malach, 1991. Patterns of sensory intermodality relationships in the cerebral cortex of the rat. *J. Comp. Neurol.* 308:432–456.

Povinelli, D., 1993. Reconstructing the evolution of mind. *Am. Psychol.* 48:493–509.

Preuss, T. M., 1993. The role of the neurosciences in primate evolutionary biology: Historical commentary and prospectus. In *Primates and their Relatives in Phylogenetic Perspective*, R. D. E. MacPhee, ed. New York: Plenum Press, pp. 333–362.

Preuss, T. M., P. D. Beck, and J. H. Kaas, 1993. Areal, modular, and connectional organization of visual cortex in a prosimian primate, the slow loris (*Nycticebus coucang*). *Brain Behav. Evol.* 42:321–335.

Preuss, T. M., and P. S. Goldman-Rakic, 1991a. Myelo- and cytoarchitecture of the granular frontal cortex and surrounding regions in the strepsirhine primate *Galago* and the anthropoid primate *Macaca*. *J. Comp. Neurol.* 310:429–474.

Preuss, T. M., and P. S. Goldman-Rakic, 1991b. Ipsilateral cortical connections of granular frontal cortex in the strepsirhine primate *Galago*, with comparative comments on anthropoid primates. *J. Comp. Neurol.* 310:507–549.

Reep, R. L., G. S. Goodwin, and J. V. Corwin, 1990. Topographic organization in the corticocortical connections of medial agranular cortex in rats. *J. Comp. Neurol.* 294:262–280.

Richard, A. F., 1985. *Primates in Nature*. New York: Freeman.

Richards, R. J., 1987. *Darwin and the Emergence of Evolutionary Theories of Mind and Behavior*. Chicago: University of Chicago Press.

Richards, R. J., 1992. *The Meaning of Evolution: The Morphological Construction and Ideological Reconstruction of Darwin's Theory*. Chicago: University of Chicago Press.

Rocha-Miranda, C., D. Bender, C. G. Gross, and M. Mishkin, 1975. Visual activation of neurons in inferotemporal cortex depends on striate cortex and the forebrain commissure. *J. Neurophysiol.* 38:475–491.

Rockel, A. J., R. W. Hiorns, and T. P. S. Powell, 1980. The basic uniformity of structure of the neocortex. *Brain* 103:221–224.

Roland, P., E. Skinhøj, N. A. Lassen, and B. Larsen, 1980. Different cortical areas in man in organization of voluntary movements in extrapersonal space. *J. Neurophysiol.* 43:137–150.

Rosenquist, A. C., 1985. Connections of visual cortical areas in the cat. In *Cerebral Cortex*. Vol. 3, *Visual Cortex*, A. Peters and E. G. Jones, eds. New York: Plenum, pp. 81–117.

Russchen, F. T., D. G. Amaral, and J. L. Price, 1987. The afferent input to the magnocellular division of the mediodorsal thalamic nucleus in the monkey, *Macaca fascicularis*. *J. Comp. Neurol.* 256:175–210.

Schiller, P. H., and J. G. Malpeli, 1977. The effect of striate cortex cooling on area 18 in the monkey. *Brain Res.* 126:366–369.

Schlag, J., and M. Schlag-Rey, 1987. Evidence for a supplementary eye field. *J. Neurophysiol.* 57:179–200.

Sesma, M. A., V. A. Casagrande, and J. H. Kaas, 1984. Cortical connections of area 17 in tree shrews. *J. Comp. Neurol.* 230:337–351.

Sherk, H., 1978. Area 18 responses in cat during reversible inactivation of area 17. *J. Neurophysiol.* 41:204–215.

Simpson, G. G., 1967. *The Meaning of Evolution*, revised ed. New Haven: Yale University Press.

Stepniewska, I., T. M. Preuss, and J. H. Kaas, 1993. Architectonics, somatotopic organization, and ipsilateral cortical connections of the primary motor area (M1) of owl monkeys. *J. Comp. Neurol.* 330:238–271.

Uematsu, S., R. Lesser, B. Gordon, K. Hara, G. L. Krauss, E. I. Vining, and R. W. Webber, 1992. Motor and sensory cortex in humans: Topography studied with chronic subdural stimulation. *Neurosurgery* 31:59–72.

Velayos, J. L., and F. Reinoso-Suárez, 1985. Prosencephalic afferents to the mediodorsal thalamic nucleus. *J. Comp. Neurol.* 242:161–181.

Vogt, B. A., and M. W. Miller, 1983. Cortical connections between rat cingulate cortex and visual, motor, and postsubicular cortices. *J. Comp. Neurol.* 216:192–210.

Williams, G. C., 1966. *Adaptation and Natural Selection: A Critique of Some Current Evolutionary Thought*. Princeton: Princeton University Press.

Williams, G. C., 1992. *Natural Selection: Domains, Levels, and Challenges*. New York: Oxford University Press.

Williams, S. M., and P. S. Goldman-Rakic, 1993. Characteristics of the dopaminergic innervation of the primate frontal cortex using a dopamine-specific antibody. *Cerebral Cortex* 3:199–222.

Zeki, S., J. D. G. Watson, C. J. Lueck, K. J. Friston, C. Kennard, and R. Frackowiak, 1991. A direct demonstration of functional specialization in human visual cortex. *J. Neurosci.* 11:641–649.

82 Evolution of the Human Brain: A Neuroanatomical Perspective

HERBERT P. KILLACKEY

ABSTRACT This chapter focuses on several aspects of the evolution of the human brain; namely, its large size, the multiplicity of neocortical areas, and the possibility that some of these areas may be unique to the human brain. The biological context within which the evolution of the human brain occurred is outlined. The general organization of mammalian neocortex is discussed, and it is proposed that two developmental mechanisms (afferent specification and cortical exuberance) play a role in subdividing the neocortex of all mammals into anatomically and functionally discrete areas.

The enlarged neocortex is the most conspicuous feature of the human cerebral hemispheres. The neocortex is the part of the brain that is most closely associated with human cognitive function. Thus, it is the neocortex that will be the focus of this enquiry. My point of view is that of a neuroanatomist whose research interest is the organization and development of the mammalian neocortex with an emphasis on the somatosensory cortex of the rat. While this may seem a peculiar background from which to approach questions concerning the evolutionary processes that have shaped the organization of the human brain, I would submit that it may be an asset. First, until quite recently most modern techniques for study of the structure and function of the nervous system have been invasive and for obvious reasons have not been applied to the human species. Consequently, we know less about the organization of the human neocortex than we do about neocortical organization in common laboratory species. Further, it is information from these as well as a number of other species that must be examined from a comparative perspective if we are to understand the selective pressures and mechanisms that formed the human brain and how brain structures evolved. Second, with a few notable exceptions, much of the literature on the evolution of the human brain ignores all but the grossest structure of the brain. This is understandable given that the fossil record is mute with regard to most aspects of brain morphology other than size. Nevertheless, there is much from modern developmental and comparative neurobiology that can be incorporated into our attempts to understand the origins of the human brain but is often omitted save at the most superficial level.

The human brain, like any other bit of biological tissue, is the product of evolutionary processes acting over very long periods of time. While the human brain is both relatively large and exceedingly complex, it is similar in its organization to those of all other mammalian species, and it plays a similar role in the adaptation of the human species to its specific environmental niche. However, the evolution of the brain did not proceed by the simple addition of parts to the brain of a preexisting species. The human brain is not that of an ancestral ape with the addition of a bit of frontal lobe and some language areas. While this is easy to state, it is much more difficult to articulate the subtle organizational changes that must have taken place during the evolution of the human brain. It now seems fairly clear that neither new neuronal types nor new neuronal transmitters are to be found in the human brain. What appears to be of particular significance is the presence of neocortical areas not found in other species. The emergence of any new cortical area implies the formation of a complex set of afferent and efferent connections with a number of other cortical areas as well as with a number of subcortical regions.

It should also be clearly stated at the outset that any attempt to understand the evolution of the human brain at this point in time will have limited success. First, our understanding of the organization and function of the neocortex in any given mammalian species is at a very early stage. Second, although the broad outlines of the phyletic history of our species have been

HERBERT P. KILLACKEY Department of Psychobiology, University of California, Irvine, Calif.

determined, many significant questions remain unanswered. Thus, any attempt to address the evolution of the human brain must be more conjectural then authoritative. Consequently, what follows is offered with a caveat: We are members of a species that possesses not only an extreme curiosity about all aspects of the world and our place in it, but also the ability to invent plausible scenarios to satisfy that curiosity in the absence of the necessary empirical data.

A brief outline of the human pedigree

Life has existed on this planet for over 3 billion years, and animals have existed for over 600 million years of that time. Of the vertebrates, the reptiles appeared on earth approximately 350 million years ago and the first mammals are only half as old. The placental mammals are a still more recent innovation and as a group began their rise to prominence approximately 65 million years ago, after the extinction of the large reptiles. Among the major biological innovations possessed by the mammals were homeothermy, a differentiated set of teeth that allowed a wider variety of food resources to be exploited, and an enlarged brain, which resulted in increased behavioral flexibility. The record of the primate lineage also extends back to the initial radiation of placental mammals. The initial primates predated on insects in an arboreal environment that probably required a number of both sensory and locomotor adaptive specializations (Cartmill, 1974). Later primate adaptations include those associated with a diurnal and terrestrial environment. The primate fossil record contains a number of gaps, and tracing the descent of living primates except in the most general terms is very difficult (Martin, 1993).

The broad outline of the evolution of our species is covered in detail in several recent texts (Campbell, 1985; Klein, 1989). I would stress the provisional and idiosyncratic nature of my interpretation of this outline and the fact that many points are still openly debated by authorities in the field (Wood, 1992). Between 10 and 5 million years ago, as the environment in eastern Africa became nonwooded and open, one of the late Miocene terrestrial apes began to alter its quadrupedal gait toward a bipedal gait. This had several possible advantages: energy efficiency, freeing the forelimbs for other functions, and raising the eyes for a better view over savanna grasses. This was the beginning of the hominid family. The oldest known hominid is *Australopithecus afarensis*, which dates to approximately 4 million years ago. This species, which possessed a smaller braincase than later hominids, was clearly a bipedal animal, although the skeletal remains suggest that it still possessed considerable tree-climbing ability. Tool use by *A. afarensis* has not been directly documented. It is presently thought that this species gave rise to two lineages. One led to several other species of australopithecines and eventually became extinct; the second one led to the genus *Homo* in ways that are not yet entirely clear. The first species of this genus, *Homo habilis*, inhabited east Africa from 2.5 million to 1.5 million years ago. This species clearly possessed a larger braincase than the australopithecines and has left copious evidence of stone tool–making ability. It in turn was succeeded by *Homo erectus*, which is believed to be the only hominid species that existed from 1.5 million to half a million years ago. The braincase of *Homo erectus* was still larger than that of its predecessors, and it produced more sophisticated tools than did earlier hominids. *Homo erectus* was probably the first hominid to control fire and was definitely the first to expand beyond Africa into Europe and Asia. *Homo erectus* was in turn succeeded by *Homo sapiens*, in a process that was underway approximately 300,000 years ago. This later event is currently the subject of intense study. The major question concerns whether the emergence of *Homo sapiens* was a unique speciation event that occurred in Africa or an event that occurred over a broader geographic front in a more gradual fashion. Finally, anatomically modern humans appear in the fossil record from approximately 30,000 years ago, and in Europe replaced an earlier *Homo* population, the Neanderthals.

The brain from the outside

Measurement and analysis of brain size provides one important approach to the study of the evolution of the brain. Obviously, brain size can be measured directly from the brains of recent species. It can also be measured relatively accurately from a mold or cast of the inner skull made by filling the cranial cavity with a substance such as latex. In many species, particularly mammals, the brain is tightly packed within the skull and the shape of the brain's external surface is impressed onto the inner cranial surface. Such an endocast can also result from the filling of a skull in the wild by sand or debris that later fossilizes. In any case, endocasts provide a means to measure the size of the brains of many extant and extinct species. This approach has been used by Jerison (1973, 1991) to document the overall size of the brain and its expansion from the time of the appearance of reptiles in the fossil record until the present. Brain size by itself is a complicated statistic that is related to body size among other things. Jerison, in his analysis, attempts to control for body size by comparing data from a given species with an empirically determined average mammalian value. This allows him to compare the size of the brain of any given species with that of any other regardless of their respective size. When species that are similar in body size differ in brain size, this difference in brain size is attributed to the degree of encephalization in the two species. Encephalization is defined as the fraction of gross brain size that represents neuronal processing capacity that is not related to body size. Total neuronal capacity can be thought of as consisting of two parts. One is related to general bodily functions and the sensory and motor requirements of a particular species and thus is relatively closely tied to body size. The second, the encephalization component, is related to aspects of neuronal control that are not directly related to body size—memory, for example. From this perspective, the human brain is more encephalized than the brains of all other primates, which in turn are more encephalized than the brains of most other mammals, which in turn are more encephalized than those of reptiles. This is the sense in which one can say that the human brain is over three times larger than would be expected in a primate of our body size (Passingham, 1982).

The enlargement of the human brain cannot be viewed in isolation from the species' mammalian heritage. The degree of encephalization in the earliest mammalian brain available from the fossil record (approximately 175 million years old) is about four times that of a reptile of the same body size. Jerison (1973) has suggested that this increase in encephalization was a result of a shift from the daylight vision–based niche of many reptiles, which involved midbrain neuronal processing, to the nocturnal niche of the early mammals, which relied more on the auditory system and enlisted forebrain neuronal processing systems as well as the brain stem and midbrain. In the reptilian visual system a considerable degree of processing is carried out peripherally at the retina itself. This is not true of the auditory system, which is characterized by a low number of peripheral processing units, and hence by a greater degree of central processing that is carried out at both the midbrain and forebrain level. There was very little change in the degree of mammalian encephalization over the next 100 million years and until the major mammalian radiation after the age of the dinosaurs. However, it should be noted that the degree of encephalization in mammals of that time is roughly the same as that found in several living mammalian species, for example, the Virginia opossum and the hedgehog.

The initial increased encephalization in primates over basal mammalian levels is most likely related to both visual and locomotor adaptations associated with the arboreal niche. The frontally placed eyes and the related development of stereoscopic vision, which confer a great advantage for the arboreal life-style, would have made increased demands on central neuronal processing capabilities. In primates, as in mammals in general, these demands were met by the further expansion of the forebrain rather than through an elaboration of the reptilian midbrain visual centers. Similarly, a higher degree of precision and rapid control of musculature is necessary to successfully locomote in an arboreal environment. In terms of food sources, many later primate species focused on plant foods such as fruit. Such food sources are unevenly distributed throughout the environment and available only during limited times. This might have been another source of selective pressure for brain enlargement, due to the memory required to forage successfully in such an environment. This factor might also have served as an impetus for the redevelopment of color vision in primates, a feature found in reptiles but believed to have been lost in early nocturnal mammals. At least one cortical area in the rhesus monkey appears to be devoted largely to the processing of color information (Zeki, 1993).

The foregoing has placed an emphasis on food sources as a central feature of an ecological niche and one that may have influenced neocortical expansion. There do indeed appear to be dietary correlates of encephalization. For example, Harvey and Krebs (1990) report that leaf-eating species of primates have smaller brains relative to body size than do closely related species that do not feed on leaves. An associated question concerns the increased energy demands a larger brain places on an organism. It has been reported that the metabolic demands of the central nervous systems of both the rhesus monkey and humans are considerably higher than those of most vertebrate nervous systems (Mink, Blumenschine, and Adams, 1981). These authors report that whereas between 2% and 8% of the basal metabolism of most vertebrate species is attributable to the central nervous system, the comparable figures for the rhesus monkey and humans are 13% and 20%, respectively.

The degree of encephalization among the earliest hominid, *A. afarensis*, is no greater than would be expected in an ape of similar size (an inference based on comparison with modern apes). Given that *A. afarensis* was fully bipedal, the adoption of this unique style of locomotion is not correlated with any detectable increase in brain size. As the body size of later hominid species are all roughly the same, we can equate differences in brain size directly with encephalization. Two and a half million years ago, *H. habilis* possessed a brain that was roughly a third larger than that of *A. afarensis* and half the size of that of anatomically modern humans. The brain of *H. erectus* was further enlarged and intermediate in size between *H. habilis* and *H. sapiens*. This increase in brain size has been attributed by different authors to a wide range of factors, most of which must be regarded as somewhat speculative. As an example, I will detail a scenario recently put forth by Jerison (1991). He places emphasis on a social primate adapted to a savannalike environment shifting from the mainly vegative diet typical of most primates to a more carnivorous diet. By analogy to social carnivores, Jerison speculates that it would be necessary for such primates to develop a way to both mark and map their extensive territory. Jerison suggests that hominids accomplished the marking and knowing of a territory with their neocortex, utilizing an auditory-vocal system. In this scenario, a cognitive system for knowing a territory could also be exploited for communicative purposes. This in turn suggests that our use of language as a communication system differs from vocal communication systems in other primates, which are usually restricted to vocal releasers of an instinctive behavior. Language may be regarded as an extension of our sensory systems, the majority of which are arranged in the neocortex as maps of some aspect of the outside world. Perhaps language is an alternative form of brain mapping that applies to time as well as space and can be directly shared with other members of our species.

Another aspect of the externally visible brain is the complex arrangement of cortical gyri and sulci that are characteristic of many larger brains. These are usually regarded as a mechanical means for increasing the surface area of the neocortex within the confines of a bony, rigid skull. In humans, approximately two-thirds of the cortical surface area is buried within the sulci, whereas in apes the comparable figure is about a quarter. The developmental forces that sculpt sulci and gyri are poorly understood, and the relation of these structures to the underlying functional and morphological areas of the neocortex is not at all straightforward. Indeed, there can be considerable variation from specimen to specimen of the same species (Welker, 1990). Thus, sulcal patterns provide little if any information about the underlying anatomical and functional organization of the neocortex. It should also be noted that sulci in particular are not entities but spaces defined by surrounding tissue. As negatives, they have no function, and, strictly speaking, they should not be compared within a species or homologized across species. This should be kept in mind when evaluating studies that purport to provide evidence of the existence of a given cortical area or that draw functional conclusions on the basis of sulcal patterns in extinct species, particularly given that sulcal patterns in endocasts are often little more than very faint impressions. In this context, a quote from Tobias (1987) is particularly apt. He states, "The recognition of specific cerebral gyri and sulci from their impressions on an endocast is a taxing, often subjective and even invidious undertaking which arouses much argumentation." This caveat also applies to studies that attempt to draw conclusions about the size of functional areas in one hemisphere versus the other in the human brain on the basis of interhemispheric differences in sulcal patterns. For example, the planum temporale—a gross anatomical descriptor for a region on the surface of the temporal lobe near the end of the sylvian fissure—may reliably vary in size between the two hemispheres in the human (Geshwind and Levitsky, 1968). However, it is not clear whether this variance is in any way related to differences in language function between the two hemispheres or is a consequence of the differential expression of mechanical factors that contribute to patterns of sulci and gyri.

Inside the brain

Traditionally, the neocortex is regarded as a six-layered structure, although this is a convenient fiction and a greater or lesser number of layers are found in some areas. The neocortex is a mammalian innovation thought to have derived from the bilaminate general dorsal cortex of the reptile forebrain. Marin-Padilla (1988) has suggested that the remnant of this reptilian structure is to be found in the embryonic primordial plexiform layer of the mammalian brain. This is a transient structure that is thought to function as a scaffolding for the cortical plate. The cortical plate later develops into the six-layered neocortex. The "uniqueness" of the mammalian neocortex has been questioned by Karten (1991; see also Karten and Shimizu, 1989). The major difference in telencephalic organization between birds and mammals, according to Karten, is that in birds, telencephalic circuits are organized in a nucleate fashion rather than a laminar one. A major implication of this hypothesis is that the unique aspect of the mammalian telencephalon is not neocortex per se but its laminated organization. This raises the question of the functional significance of lamination, which is found in the brains of all vertebrates. Two prominent examples are the cerebellum and the tectum. My hypothesis is that lamination, particularly in the case of the neocortex, is a straightforward way of bringing multiple sets of inputs into contact with multiple sets of outputs. Within the neocortex, a single input has access to multiple output channels. For example, a thalamic input to the middle cortical layers of a primary sensory cortex might result in direct activation of neurons that project to other cortical areas, the brain stem, or more caudal brain regions, and back to the thalamus. Conversely, and as a result of the dendritic organization of cortical pyramidal neurons, a single output can be contacted by multiple inputs. An added feature of neocortex is that the remaining dimension is largely organized in a topographic fashion, providing a ready referent to the external world. These features must have greatly enhanced the integrative capabilities of the telencephalon. Further, given that the neocortex is composed of repeated processing modules, it is a structure that can be readily molded into processing systems (cortical areas) that are adapted to specific niches. However, it remains a puzzle why lamination did not also occur in the telencephalon of birds, which have elegantly laminated visual midbrains. Perhaps other factors such as forward eye placement, coupled with aerodynamic considerations, placed constraints on the bony skull that favored a nucleated rather than a laminated telencephalon.

In the horizontal dimension, the cytoarchitectonic heterogeneity of the neocortex was noted by early neuroanatomists, and indeed this heterogeneity provided the first basis for the regional subdivision of the neocortex. The basic premise behind the subdividing of neocortex is the delineation of the relationship between structure and function that is a common focus of all neuroanatomical investigations. Later functional studies provided support for the subdivision of the neocortex into anatomically distinct areas along its horizontal dimension. Thus, a cortical area can be defined as an anatomically distinct region of the neocortex that carries out a specific function. In general, it has been easiest to define cortical areas that are most closely associated with the initial stages of sensory processing, both because these areas seem to have more distinct anatomical properties and because their functional properties are more amenable to experimental analysis.

The second major dimension of the neocortex is the vertical one. While laminar differences in cell type and density were noted by early investigators, the full significance of laminar organization as a reflection of the segregation of both afferent inputs and efferent outputs was not fully realized until the introduction of modern neuroanatomical techniques for tracing neural pathways. Generally speaking, the neocortex can be subdivided in the vertical dimension into three broad domains. The fourth cortical layer, or the granular layer, is the chief receptive layer, and as such is the major target of thalamic inputs and, in many cases, inputs from other areas of the cortex. The supragranular layers (layers III and II) are the chief source of efferents to other cortical areas, and the infragranular layers (layers V and VI) are the chief source of efferent output from the neocortex to other portions of the brain.

It is often suggested that one way the neocortex has expanded is by the addition of new functional areas. The cytoarchitectonic map of the human neocortex most commonly referred to is that of Brodmann (1909), and he subdivided the human neocortex into 52 different areas. How does this compare to the number of cortical areas in other mammalian species? Zilles, in a recent atlas, subdivides the rat neocortex into approximately 20 different areas; and a smaller but closely related species, the mouse, has approximately the same number of cortical areas. However, it should be kept in mind that subdividing the neocortex simply on the basis of cytoarchitectonic differences that are detectable in Nissl-stained material may not provide an accurate assessment of the number of functionally distinct cortical areas in a given species. For example, Zilles defines several different parts (face, forepaw, and hindpaw representations) of the rat's primary somatosensory cortex as cortical areas, and consequently probably overestimates the number of distinct functional areas in the rat brain. On the other hand, the macaque monkey neocortex at latest count contains 25 cortical areas solely devoted to the processing of visual information and another 7 areas in which visual information is also processed along with other information (Van Essen, Anderson, and Felleman, 1992).

Van Essen, Anderson, and Felleman (1992) incidentally illustrate in their first figure a number of other cortical areas in the neocortex of the macaque in addition to the visual areas. A simple tally of all the cortical areas illustrated in this figure leads to the conclusion that the neocortex of the macaque contains more than 52, the number of cortical areas identified by Brodmann in the human neocortex. Does the macaque neocortex contain more cortical areas than the human neocortex? Probably not, but it should be noted that a cortical area in the sense employed by these authors is defined on a combination of anatomical and physiological criteria that are considerably more stringent than the criteria utilized to subdivide human neocortex to date. This would suggest that many functional cortical areas remain to be uncovered in the human brain. This is almost a certainty, as human neocortex has not been explored with many of the techniques currently available to the neurobiologist.

The language areas of the human brain are most likely examples of newly evolved areas. However, it should be kept in mind that neither of the classical language areas, Broca's area and Wernicke's area, are cortical areas in the strict sense in which the term *area* is used by an neuroanatomist. For example, they are not defined according to the same strict and multiple criteria that are employed in defining primary visual cortex (area 17), and each includes more than one cytoarchitectonically distinct area. At best, the terms *Broca's area* and *Wernicke's area* define relatively broad neocortical regions that, when damaged, result in somewhat variable disorders in the realm of language. Modern cognitive neuropsychological research is both redefining the nature of the deficit that is thought to result from damage to these classical areas and broadening the definition of the cortical areas involved in language processing (Caramazza, 1988). For example, Caramazza appears to regard language processing as being distributed over much larger regions of the brain than the classical language areas. The fact that he uses the term *perisylvian region* as an anatomical descriptor rather than referring to poorly defined areas exemplifies this approach.

Deacon (1990) has injected a note of caution into the discussion of increase in the number of neocortical areas. He has pointed out that most discussions of this topic are confounded by the fact that small animal brains in general contain few neocortical areas while larger animal brains contain more areas. Further, primates in general are relatively large animals compared to the baseline mammals with which we tend to compare them. Thus, an increase in the number of neocortical areas may be an allometric correlate of a larger body size rather than a correlate of encephalization. A small mammal with an increased number of neocortical areas relative to other small mammals of roughly the same size would provide evidence that this is not necessarily the case. At least superficially, the insectivorous echolocating bat seems to have more areas devoted to auditory processing than other small mammals of roughly the same size (Suga, 1990). While this can be interpreted as positive evidence on the point, inferences should be made with caution, as the auditory system of most other small mammals have not been examined with the same degree of precision as the bat's auditory system. Nevertheless, auditory cortex in the insectivorous echolocating bat is also an excellent example of how the neocortical areas that evolve in a given species are "unique" in the sense that they are a response to that species' specific ecological adaptations.

One additional aspect of the vertical organization of the neocortex should be noted. A number of investigators have suggested that a fundamental feature of the neocortex is its "columnar" organization (see Mountcastle, 1978). A "column" has been regarded as both the elemental functional unit and building block of the neocortex. From this perspective, a neocortical area may be regarded as a collection of individual cortical columns that share a common set of afferent and efferent connections. Rockel, Hiorns, and Powell (1980) document a marked uniformity in the basic vertical organization of the neocortex across a diverse group of mammalian species. These authors report that the absolute number of neurons in a small volume of neocortex extending from the pial surface to the underlying white matter, in rough terms a neocortical column, is relatively invariant both across cortical areas and species. This invariance in neuronal number is present despite both the cytoarchitectonic and functional differences that characterize different cortical areas and the approximately threefold difference in cortical thickness across species. With the exception of primary visual cortex in primates, all of the cortical areas in all species contain approximately 110 neurons. The authors interpret their results as evidence that it is the areal or horizontal dimension of neocortex that undergoes a major increase while the number of neurons within its depths at a given point remains relatively constant. This study suggests that with the exception of primate visual cortex, roughly the same number of neurons compose the fundamental cortical processing unit in all species and cortical areas. Thus, the basic cortical processing unit is quite conservative and has remained relatively unchanged in the course of mammalian evolution. It also suggests that cortical expansion has occurred largely by the addition of cortical processing units. Thus, a larger brain contains many more cortical processing units than a smaller brain, but the basic unit is the same in both cases. (The vertical dimension of neocortex also exhibits some change. The two- or threefold increase in cortical thickness seen in primates at least partially reflects increases in both dendritic length and spines and the richness of the axonal strata that contact cortical neurons, but this cannot account for the major expansion of the neocortex. However, it certainly would result in a significant increase in processing capabilities within a cortical column.)

We must next ask how the brain of an evolving species adds cortical processing units and how these units are conscripted into the cortical areas that are part of that species' adaptation to its niche. This is done within the framework of the following assumptions. One, cortical areas that are closely tied to peripheral sensory and motor structures are the first cortical areas to form and seem to be common to all mammals. Two, all cortical areas are composed of the same basic unit. Three, the basic unit of cortical organization is a vertically organized small group of cells that form a processing unit or column. Four, the column has a finite processing capability, which, when exceeded, leads to evolutionary pressures for the addition of processing units. Five, new processing units can either be incorporated into existing cortical areas as necessary to adapt to changes in peripheral surfaces, or, more importantly, they can be formed into new cortical areas when new processing needs emerge. Six, new cortical areas are composed of new processing units that are added in series with the preexisting units at the border of a preexisting cortical area. They are connected with those of the preexisting cortical area by intrahemispheric connections that are by necessity topographically organized. These assumptions envision a rather straightforward and flexible mechanism for both the formation of cortical areas that may be unique to a given species and the addition of new cortical areas in an expanding neocortex.

Developmental mechanisms of neocortical expansion and specification

In a previous publication (Killackey, 1990), I suggested that knowledge of the development of the nervous system could contribute to our understanding of how the brain evolved, for it is during development that the processes that mold the mature brain operate. The first order of business in producing an expanded neocortex would simply be to produce more neurons. The neuronal proliferative events in neocortex are believed to be similar in all mammalian species (Rakic, 1988). Germinal cell precursors of neocortical neurons proliferate along the ventricular walls of the telencephalon and then migrate to their final position in the neocortex. They do this in an orderly fashion such that neocortical neurons in the deeper cortical layers take up their final position before neurons that are generated later take up their position. Thus, there is an overall inside-out gradient in the formation of the neocortex. This process continues in a wavelike fashion across the telencephalon, and neocortical columns are added in a fashion that is roughly analogous to the growth rings of a tree (Smart, 1983). Prolonging the period of neurogenesis would allow more waves of neurons to be added, and a larger neocortex would result. Neurons that compose the mouse neocortex are generated over a 5-day period, while those of the human neocortex are generated over an 80-day period (Rakic, 1988). If the proliferative cycle time is roughly the same in the two species, the added cycles could easily produce the extra number of neurons in the human brain. Given that the estimated number of neurons in a mouse brain is 10^7 (Schuz and Palm, 1989) and that the usual estimate for the number of neurons in the human neocortex is 10^{11}, it would take roughly 13 additional proliferative cycles to produce the human neocortex.

It appears that although some defining features of the neocortex are determined during the proliferative period, others are not. Recently, Walsh and Cepko (1992) have reported that clonally related neurons can be dispersed over a wide area of neocortex and not restricted to a single cytoarchitectonic area. The significance of this finding is that it suggests that the specification of neocortical areas occurs after neurogenesis within the forming neocortex and not at the ventricular zone. As a first approximation, I would suggest that factors intrinsic to the neocortex are sufficient to produce a six-layered neocortex. However, such a default cortex is relatively unspecified, and it is factors extrinsic to the neocortex that bring about further differentiation. I would now like to briefly turn to the processes that may play a role in specification of cortical areas.

Two mechanisms that may play a role in subdividing the neocortex into the morphological and functional areas characteristic of the adult are afferent specification and cortical exuberance. Afferent specification assumes that the major afferent input to a neural structure is one major factor that plays a role in the ontogenetic determination of a neural target's structural organization. In terms of neocortex, the dorsal thalamus provides the major afferent input, and hence would be expected to play the major extrinsic role in the guidance of its organization. However, it should be kept in mind that the dorsal thalamus is only the penultimate link in a sequence of neural connections which begins with receptor surfaces at the periphery. Thus, for sensory systems it is ultimately the peripheral receptor surfaces acting through the immediate agency of the dorsal thalamus which play a role in neocortical organization including the formation of cytoarchitectonic borders. Thalamic input reaches the developing neocortex relatively early during the period when neurons are migrating from the ventricular surface to the neocortex and well before there is any sign of areal differentiation within the neocortex.. The process of afferent specification probably occurs during the initial invasion of neocortex by thalamic afferents and is the net result of interactions between thalamic fibers with each other as well as with their target neocortical tissue.

Afferent specification also interacts with the second process, cortical exuberance to sculpt distributions of supragranular and infragranular projection neurons. Cortical exuberance, a term introduced by Innocenti, Fiore, and Caminiti (1977), refers to the fact that neocortical projection neurons initially send processes to multiple target areas and later eliminate processes, rather than eliminate whole cells as in the case of neuronal cell death. In the case of supragranular cortical projection neurons processes are extended to the area of potential targets within the ipsilateral hemisphere, as well as across the corpus callosum to the contralateral hemisphere. Similarly, neurons of infragranular layer V initially send processes to multiple targets such as the superior colliculus and the spinal cord. In both cases, processes to one target are later lost seemingly on the basis of information provided by afferent input to the locus of the projection neurons cell body. Evidence for cortical exuberance has been found in a wide range of mammalian species including the macaque monkey. It has not been reported to occur in nonmammalian species and for the most part it appears to be restricted to neocortex. This would not be unexpected in light of presumed selection pressures to increase neocortical size. Cortical exuberance can be viewed as an adaptive modification of the more general phenomena, developmental neuronal death, common in the central nervous system of all vertebrate species.

Human neocortical expansion

It is my opinion that the developmental events discussed in the last section are a general feature of mammalian brain development and that they provide a reasonable biological framework for understanding the evolution and expansion of the human brain. I would emphasize once again that the evolution of the human brain should not be viewed in isolation. Expansion of the neocortex is not a unique feature of the human brain but a continuation of processes which are detectable in the first mammals, a prominent feature of the primate brain in general and is a characteristic of all species of the genus *Homo*. Exactly what feature or combination of features of the adaptive niche occupied by hominids drove the more recent expansion remains speculative.

A key to where human neocortex has expanded is the suggestion by Deacon (1990) that the primary sensory and motor areas of the human brain are considerably smaller than would be predicted for a primate brain of its size. The inference to be made from this is that it is the cortical areas beyond the primary areas that have expanded the most in the human brain. Indeed, these areas clearly compose the greatest part of the human neocortex. While these areas have traditionally been referred to as "association areas," recent research such as that carried out in the rhesus monkey visual system demonstrates that this is a gross oversimplification. The name *association area* is more the reflection of a theoretical bias and a paucity of experimental evidence than an accurate reflection of the functional role of the cortical areas so named. For example, as mentioned above, the monkey cortex contains a large number of areas devoted to the processing of different and specific aspects of vision, and many of these areas would once have been referred to as association areas. These visual areas form a distributed hierarchical network of cortical areas that are interconnected and contain several separate processing streams (Van Essen, Anderson, and Felleman, 1992). These separate processing streams have been demonstrated to extend into the frontal lobes where they each target distinct areas of prefrontal cortex (Wilson, Scalaidhe, and Goldman-Rakic, 1993), suggesting that similar organizational principles apply to neocortical areas in the rhesus monkey frontal lobes. Recent experiments with noninvasive imaging techniques are beginning to sketch the broad outlines of a similar cortical organization in human neocortex (Kosslyn et al., 1993; Jonides et al., 1993).

I would speculate that afferent specification and cortical exuberance have also played a major role in the evolution of the human neocortex. In general, thalamocortical projections are densest in primary cortical areas, and I would assume that thalamocortical afferents play their major role in these areas. In these areas, thalamocortical afferents also assert an influence on exuberant cortical projection neurons. One form of this interaction is the determination of whether a given cortical projection neuron projects ipsilaterally or contralaterally. A dense thalamic input to an area seems to be correlated with a preponderance of ipsilateral cortical projection neurons. These ipsilateral cortical projection neurons may in turn function as a dominant input that colonizes and specifies cortical areas surrounding primary sensory cortices that are not as densely invaded by a thalamic input. A set of such projection neurons originating from a common primary cortical area would possess an ordered spatial topographic map, which it would in turn impose on its cortical target tissue. Further, if such a projection were formed on a near neighbor–to–near neighbor basis, the mirror-image relationships that normally exist between adjoining and related cortical areas would result. In this simplistic view, the process of specification may be somewhat hierarchical.

The specification of neocortex begins with thalamic input specifying a primary area. In cascading fashion, ipsilateral corticocortical projections originating in the primary cortical area specify the next cortical area, which in turn specifies the next area, and so on. Perhaps in this fashion cortical areas that are furthest removed from the primary areas are specified by multiple inputs. It is also likely that thalamic nuclei such as the pulvinar and the dorsomedial nucleus, which are particularly well developed in humans and are related to neocortex beyond the primary areas, may play some role in the specification process. Although thalamic input from sensory systems have been the focus of this discussion, this is chiefly because it is these systems that lend themselves most readily to experimentation, and hence have been most closely examined. There is no reason that other well-developed thalamic inputs could not play a role in the process.

Concluding ruminations

In the foregoing, I have focused on several aspects of the evolution of the human brain, namely, its large size, the multiplicity of neocortical areas, and the possibility that some of these areas may be unique to the human brain. I have in the fashion of a naive reductionist attempted to explain these in terms of our current biological knowledge, and the attempt, while somewhat speculative, is a reasonable start. However, there is one other aspect of the human brain that should be mentioned, which, although well documented, is poorly understood. This is the functional asymmetry of the human neocortex in the cognitive sphere. This seems to me to be a distinctly human trait. While a number of investigators have attempted to link both anatomical and functional asymmetries that occur in other species with those seen in humans, my opinion is that most of these attempts seem to be forced and rather weak (for recent reviews of this literature, see Bradshaw, 1988). In my opinion, the gap between the functional asymmetries in human neocortex and the symmetries in other species is of roughly the same size as the gap between human language and the communication skills of other species. At one level, the human cortical asymmetry demonstrates how the same neuronal machinery can be applied to solving problems in very different domains; yet the adaptive significance of this trait is unclear. While one could imagine it to be a means of increasing the processing capacity of the neocortex without further increases in its size, this does not appear to be the case. Toth (1985) finds evidence in the tool making of *Homo habilis* that the same proportion of this species was right-handed as is true for modern humans. If one assumes the same high correlation between handedness and functional cortical asymmetry that is found in modern humans, this implies that the neocortex of *Homo habilis* was functionally asymmetric. Given that the size of the brain in this hominid species was only half that found in modern humans, it is unlikely that functional asymmetry is a response to size limitations. Perhaps the functional asymmetry was a necessary part of the cognitive system that Jerison (1991) speculates evolved for marking and mapping a territory. Although this is sheer speculation, the presence of functional asymmetry in an ancestral species once again reinforces the notion that the human brain can only be understood within a larger biological context.

REFERENCES

Bradshaw, J. L., 1988. The evolution of human lateral asymmetries: New evidence and second thoughts. *J. Hum. Evol.* 17:615–637.

Brodmann, K., 1909. *Vergleichende Lokalisationslehre der Grosshirnrinde.* Leipzig: Barth.

Campbell, B. C., 1985. *Human Evolution.* ed. 3 New York: Aldine.

Caramazza, A., 1988. Some aspects of language processing revealed through the analysis of acquired aphasia: The lexical system. *Ann. Rev. Neurosci.* 11:395–421.

Cartmill, M., 1974. Rethinking primate origins. *Science* 184: 436–443.

Deacon, T. W., 1990. Rethinking mammalian brain evolution. *Am. Zool.* 30:629–705.

Geshwind, N., and W. Levitsky, 1968. Human brain: Left-right asymmetries in temporal speech region. *Science* 161: 186–187.

Harvey, P. H., and J. R. Krebs, 1990. Comparing brains. *Science* 249:140–149.

Innocenti, G. M., L. Fiore, and R. Caminiti, 1977. Exuberant projection into the corpus callosum from the visual cortex of newborn cats. *Neurosci. Lett.* 4:237–242.

Jerison, H. J., 1973 *The Evolution of the Brain and Intelligence.* New York: Academic Press.

Jerison, H. J., 1991. *Brain Size and the Evolution of Mind.* New York: American Museum of Natural History.

Jonides, J., E. E. Smith, R. A. Koeppe, E. Awh, S. Minoshima, and M. A. Mintum, 1993. Spatial working memory in humans as revealed by PET. *Nature* 363:623–626.

Karten, H. J., 1991. Homology and evolutionary origins of the "neocortex." *Brain Behav. Evol.* 38:264–272.

Karten, H. J., and T. Shimizu, 1989. The origins of neocortex: Connections and lamination as distinct events in evolution. *J. Cognitive Neurosci.* 1:291–301.

Klein, R. G., 1989. *The Human Career.* Chicago: University of Chicago Press.

Killackey, H. P., 1990. Neocortical expansion: An attempt towards relating phylogeny and ontogeny. *J. Cognitive Neurosci.* 2:1–17.

Kosslyn, S. M., N. M. Alpert, W. L. Thompson, V. Maljkovic, S. B. Weise, C. F. Chabris, S. E. Hamilton, S. L. Rauch, and F. S. Buonanno, 1993. Visual mental imagery activates topographically organized visual cortex: PET investigations. *J. Cognitive Neurosci.* 5:263–287.

Marin-Padilla, M., 1988. Early ontogenesis of the human cerebral cortex. In *Cerebral Cortex*, vol. 7, A. Peters and E. G. Jones, eds. New York: Plenum, pp. 1–34.

Martin, R. D., 1993. Primate origins: Plugging the gaps. *Nature* 363:223–234.

Mink, J. W., R. J. Blumenschine, and D. B. Adams, 1981. Ratio of central nervous system to body metabolism in vertebrates: its constancy and functional basis. *Am. J. Physiol.* 241:R203–R212.

Mountcastle, V. B., 1978. An organizing principle for cerebral function: The unit module and the distributed system. In *The Mindful Brain*, G. M. Edelman, ed. Cambridge Mass.: MIT Press, pp. 7–50.

Passingham, R. E., 1982. *The Human Primate.* San Francisco: Freeman.

Rakic, P., 1988. Specification of cerebral cortical areas. *Science* 241:170–176.

Rockel, A. J., R. W. Hiorns, and T. P. S. Powell, 1980. The basic uniformity in structure of the neocortex. *Brain* 103:221–244.

Smart, I. H. M., 1983. Three dimensional growth of the mouse isocortex. *J. Anat.* 137:683–694.

Schuz, A., and G. Palm, 1989. Density of neurons and synapses in the cerebral cortex of the mouse. *J. Comp. Neurol.* 286:442–455.

Suga, N., 1990. Biosonar and neural computation in bats. *Sci. Am.* 262:60–68.

Tobias, T. B., 1987. The brain of *Homo habilis*: A new level of organization in cerebral evolution. *J. Hum. Evol.* 16: 741–761.

Toth, N., 1985. Archaeological evidence for preferential right handedness in the lower and middle Pleistocene and its possible implications. *J. Hum Evol.* 14:607–614.

Van Essen, D. C., C. H. Anderson, and D. J. Felleman, 1992. Information processing in the primate visual system: An integrated systems perspective. *Science* 255:419–423.

Walsh, C., and C. L. Cepko, 1992. Widespread dispersion

of neuronal clones across functional regions of the cerebral cortex. *Science* 255:434–440.

Welker, W., 1990. Why does cerebral cortex fissure and fold? In *Cerebral Cortex*, vol. 8, E.G. Jones and A. Peters, eds. New York: Plenum, pp. 3–136.

Wilson, F. A. W., S. P. O. Scalaidhe, and P. S. Goldman-Rakic, 1993. Dissociation of object and spatial processing domains in primate prefrontal cortex. *Science* 260:1955–1958.

Wood, B., 1992. Origin and evolution of the genus *Homo*. *Nature* 355:783–790.

Zeki, S., 1993. *A Vision of the Brain*. Oxford: Blackwell.

Zilles, K., 1985. *The Cortex of the Rat*. Berlin: Springer.

83 The Replacement of General-Purpose Theories with Adaptive Specializations

C. R. GALLISTEL

ABSTRACT Associative theories of learning assume a general-purpose learning process whose structure does not reflect the demands of a particular learning problem. By contrast, implicit in the studies of learning conducted by zoologists and, recently, by some experimental psychologists, is the assumption that learning mechanisms are computationally specialized for solving particular kinds of problems. Models of the latter kind have begun to be applied even to the results from experiments on classical and instrumental conditioning. These models imply that to understand learning neurobiologically, we must discover the cellular mechanisms by which the nervous system stores and retrieves the values of variables and carries out the elementary computational operations (the elementary operations of arithmetic and logic), as well as the circuits that implement special-purpose computations using these universal elements of computation.

Theories of learning are and always have been predominantly associative theories. However, in the study of animal learning, where these theories have historically been most dominant, a different conception is gaining ground. Whereas associative theories have their historical roots in the empiricist philosophy of mind, the alternative conception has its roots in evolutionary biology, more particularly in zoology, that is, in the study of the natural history of animal behavior and of the mechanisms that enable animals to cope with the challenges posed by their habits of life.

Associative theories of learning assume a basic learning mechanism, or, in any event, a modest number of learning mechanisms. These mechanisms are distinguished by their properties—for example, whether or not they depend on temporal pairing—not by the particular kind of problem their special structure enables them to solve. Indeed, people doing neural net modeling, which is currently the most widespread form of associative theorizing, are often at pains to point out that the network has solved a problem in the absence of an initial structure tailored to the solution of that problem (e.g., Becker and Hinton, 1992). The alternative conceptualization, by contrast, takes for granted that biological mechanisms are hierarchically nested adaptive specializations, each mechanism constituting a particular solution to a particular problem. The foliated structure of the lung reflects its role as the organ of gas exchange, and so does the specialized structure of the tissue that lines it. The structure of the hemoglobin molecule reflects its function as an oxygen carrier. The structure of the rhodopsin molecule reflects its function as a photon-activated enzyme. One cannot use a hemoglobin molecule as the first molecular stage in light transduction and one cannot use a rhodopsin molecule as an oxygen carrier, any more than one can see with an ear or hear with an eye. Adaptive specialization of mechanism is so ubiquitous and so obvious in biology, at every level of analysis, and for every kind of function, that no one thinks it necessary to call attention to it as a general principle about biological mechanisms.

In this light, it is odd but true that it is radically at variance with most past and contemporary theorizing about learning to assume that learning mechanisms are adaptively specialized for the solution of particular kinds of problems. In this chapter, I first describe the essential features of associationist theories, then contrast them with the essential features of the models of learning that take adaptive specialization for granted. I also point out the implications of the adaptive specialization view for research on the neural mechanisms of learning and memory.

C. R. GALLISTEL Department of Psychology, University of California, Los Angeles

Associations

Associative theories of learning take the mechanism that alters the strengths of associations as the basic learning mechanism and the associations themselves as the elements of memory. An association is a connection between two units of mental or neural activity (two ideas, two neurons, two nodes in a neural net, etc.). The associative connection arises either because the two units have often been active at nearly the same time (the temporal pairing of activation) or through the repeated operation of a feedback mechanism that is activated by errors in the output and adjusts associative strengths to reduce the error. In traditional animal learning theory, the process that forms associations by the first mechanism (temporal pairing of activation) is called the classical or Pavlovian conditioning process, while the process that forms associations through the agency of error-correcting feedback is called the instrumental or operant conditioning process. In neural net modeling, the first process is called an unsupervised learning mechanism (e.g., Becker and Hinton, 1992) in contrast to the more common second mechanism, which requires a supervisor or teacher that knows the correct output. The correct output is required for the error-actuated backpropagation algorithm, which plays the role of the reinforcement process, selectively strengthening those connections that lead to correct outputs. In theories of instrumental learning (Hull, 1943), the error-correcting feedback comes from the rewards generated by effective acts.

The associative link serves as the pathway by which the activation of one unit leads to the activation (excitation) or inactivation (inhibition) of another unit. It is a conducting link in the circuitry by which inputs generate outputs. An associative bond is not a symbol; its strength does not specify the value of a quantifiable fact about the world. From the strength of an association, one cannot in general infer, for example, the distance between two points, or the angle formed by three points, or the position of a point in a constellation of points, or the duration of a temporal interval, or the number of items in an array, or the number of times one stimulus has been paired with another.

In neural net modeling, the pattern of strengths across a large number of associative connections is sometimes taken to specify, or at least correspond to, the value of a real-world variable. But even when patterns of synaptic strengths are imagined to encode the values of external variables, it is rarely proposed that the network performs elementary computational operations (addition, multiplication, ordination) with the values thus specified. Neural net modeling, like most previous associationist theorizing, is by and large antirepresentational; it denies that mental functioning is mediated by symbols and processes that operate with symbols. It argues for a subsymbolic theory of information processing in the brain (Smolensky, 1986).

An implicit assumption in the associative conceptual framework is that, from the standpoint of the basic or elementary learning mechanism, learning problems have the same structure. This assumption is manifest in many ways, for example, in the assumption that the same neural net can solve quite unrelated problems (at least when supplied with certain special-purpose input-conditioning mechanisms). "We've got the solution, now tell us the problem," is a harsh but not entirely inaccurate characterization of the neural net approach to modeling the mechanisms that mediate perception and learning.

In neurobiology, the general-purpose nature of the associative process is manifest in the assumption that the temporal-pairing version of associative theory captures the essence of the learning mechanism. This assumption was reflected in a comment made in the course of a recent seminar, where we were discussing an experiment that employed the Morris water maze (Morris, 1981) to test the hypothesis that long-term potentiation is important in memory formation. One participant objected that the water maze was a poor choice of task because it is difficult to say what the conditioned stimulus (CS) is. This objection takes it for granted that in every learning situation there really is a CS, a stimulus that is motivationally neutral and that is temporally paired with a motivationally significant stimulus (called the unconditioned stimulus, or US) to produce the strengthening of an association.

In the context of contemporary experimental work on the cellular-level mechanism of learning, it is entirely reasonable to object to the use of spatial tasks on the grounds that it is difficult to identify the conditioned stimulus. Most such work takes the process of association formation in a classical conditioning task as the paradigm within which to investigate the basic cellular mechanisms of learning. Moreover, it is often assumed that features of the behavioral task known to be crucial for the appearance of a conditioned response, such as the forward temporal pairing of the CS and

US, are reflected in the properties of the cellular-level mechanism (e.g., Hawkins and Kandel, 1984; Gluck and Thompson, 1987). That is, it is assumed that a change in associative strength is a change in synaptic connectivity, and that the molecular cascade that mediates this change is triggered by the forward temporal pairing of the CS and US signals at the site of the neuronal change. If this ubiquitous assumption is correct, then CS-US pairing is, indeed, the essential driving factor in learning, and we should use tasks where we can identify and control this crucial variable.

From a zoological perspective, however, the assumption implicit in the above objection is bizarre, because it directly contradicts the assumption of adaptive specialization. Zoologists have studied the role that learning plays in the solution to many different problems confronted by widely differing species of animals. In most of these cases, it is impossible to identify a CS and a US whose temporal pairing produces the learning. The conceptual framework that seems obvious and natural in the analysis of classical conditioning seems inapplicable to these other instances of learning. Moreover, a conceptual framework that seems appropriate for analyzing one of these zoological learning phenomena usually seems inappropriate when applied to another. The learning studied by zoologists has a particularity and peculiarity that discourages any thought of applying to one problem an analysis appropriate to another.

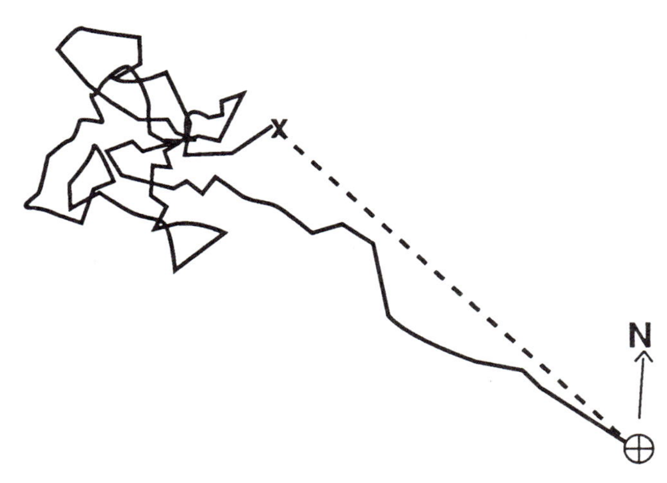

FIGURE 83.1 Track of a foraging ant. The outward (searching) journey is the solid line. The ant found food at X. Its homeward run is the broken line. (Redrawn from Harkness and Maroudas, 1985; used by permission)

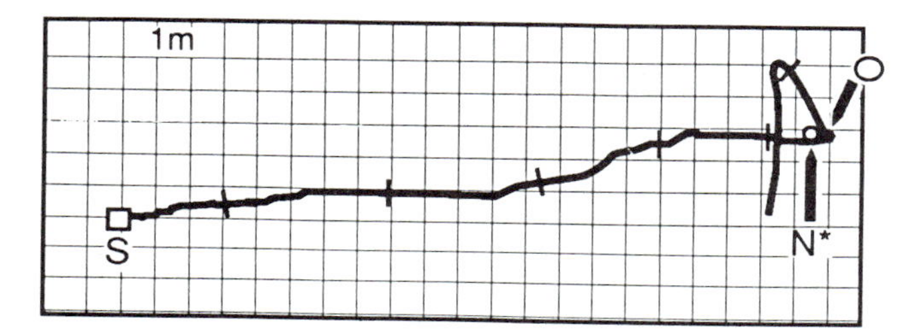

FIGURE 83.2 The homeward run of an ant displaced into unfamiliar territory on which the experimenters had previously marked off a 1 m gridwork. S, the point of release; N*, the location of the fictive nest (where the ant's nest would have been had the ant not been displaced); O, the point at which the ant terminated its nestward run and entered a search pattern. (Redrawn from Wehner and Srinivasan, 1981; used by permission)

The natural history of learning

LEARNING BY PATH INTEGRATION Figure 83.1 shows the track of a long-legged, fast-moving desert ant, *Cataglyphis bicolor*, foraging for and finding a morsel of food on the hot plain of the Tunisian desert. On the outward leg of its journey (solid tracing), it twists and turns this way and that searching for the carcass of an insect that has succumbed to the heat. When it finally finds one, it bites off a chunk, turns, and runs more or less straight toward its nest, a hole 1 mm in diameter, which may be as much as 50 m away. Its ability to orient homeward demonstrably depends on information it acquires during the outward journey. If the ant is deprived of this information by being picked up as it emerges from the nest and transported to an arbitrary point in the vicinity of its nest, it wanders in circles and makes it back to the nest only after a long time, if at all (Wehner and Flatt, 1972). A simple experiment (Wehner and Srinivasan, 1981) reveals the nature of the learning mechanism that acquires the requisite positional information. When the ant turns toward its nest, it is picked up and transported more than half a kilometer across the desert, where it is released to run across a large grid the experimenter has marked in advance on the desert floor. The grid on the desert floor enables the experimenter to trace the ant's course on a graph-paper grid at a reduction of 100 : 1 (figure 83.2). Although the ant is now in territory it has never seen and never traversed, it runs in a direction that lies within a degree or two of the compass direction of its nest from the site where it was picked up (the direction it would have run had it not been displaced). It runs in a straight line for a distance slightly longer than the distance of its nest from the point where it was picked up, then abruptly stops its straight run and begins a systematic search for the nonexistent nest.

The ant's ability to run over unfamiliar terrain a course whose direction and distance equal the direction and distance of its nest from the point of its capture implies that its navigation is based on some form of path integration or dead reckoning. Path integration is the integration of the velocity vector with respect to time to obtain the position vector, or some discrete equivalent of this computation. The discrete equivalent in traditional marine navigation is to record the direction and speed of travel (the velocity) at intervals, multiply each recorded velocity by the interval since the previous recording to get interval-by-interval displacements (e.g., making 5 knots on a northeast course for half an hour puts the ship 2.5 nautical miles northeast of where it was), and sum the successive displacements (changes in position) to get the net change in position. These running sums of the longitudinal and latitudinal displacements are the deduced reckoning of the ship's position.

For this computation to be possible, the ant's nervous system must have elements capable of preserving the value of a variable over time. The essence of the ongoing summation that underlies path integration is the adding of values that specify the most recent displacement to the value that specifies the cumulative prior displacement. To do this one must somehow be able to hold in memory the value of the sum, and must be able to add to that value. Path integration is a process that computes and stores values that specify a quantifiable objective fact about the world—the ant's direction and distance from its nest. The use made of these values, if any, is determined by the decision process controlling behavior at the moment. The integration of velocity to obtain position goes on throughout the ant's journey, but only when the decision to turn for home is made do the positional values stored in the dead-reckoning integrator lead to the well-oriented straight runs for a predetermined distance shown in figures 83.1 and 83.2. Thus, the position-specifying values are not conducting links in the circuitry that links inputs to outputs. Rather, they specify for various decision processes the acquired information the decision process needs to specify the behavioral output.

The values accumulated in the integrator at important points in the forager's journey may also be stored for later use for purposes other than returning to the nest. For example, the ant stores the positional coordinates of the location at which it finds the carcass, so that it can return directly there on its next foraging trip. We infer this because if we set up a food source in a fixed location, the ants soon come directly to it from their nest (Wehner and Srinivasan, 1981). These fast-moving desert ants do not lay an odor trail while returning to the nest, so they can return directly to a food source only if they record its position.

The foraging bee gives us even more direct evidence that it has stored the positional coordinates of a food source for later use. When it returns to its hive, it performs a waggle dance. It runs in a figure-eight path, waggling when it runs the central segment where the two loops join. Other foragers follow behind the dancing bee to learn the location of the food. The direction of this waggling run relative to the gravitational vertical specifies the direction of the food source relative to the sun, while the number of waggles specifies the approximate distance of the source from the hive (Frisch, 1967). In this way, the bee communicates by means of its behavior the position-specifying values computed and saved by its path-integrating mechanism.

If the adaptive specialization of the computations that mediate information acquisition is taken for granted, then the only plausible elementary mechanisms that might be common to learning mechanisms in general are the mechanisms of storage and the mechanisms that carry out the primitive elements of all computations—the operations of arithmetic and logic. Path integration is a simple, primitive, and ubiquitous learning mechanism. Path integration is a notably simple and widespread neural computation that makes elementary use of this reduced instruction set, that is, of the primitive computational operations of multiplying, adding, dividing, subtracting, storing, and retrieving. It ought therefore to be of central interest to the large community of scientists interested in discovering the cellular-level mechanisms that make learning and higher cognitive function possible.

The idea that path integration is a paradigmatic learning process strikes many people as perverse because they feel that in some sense path integration isn't learning at all. Many will agree, at least for the sake of argument, that learning is the process or processes by which we acquire knowledge. When shown figures 83.1 and 83.2, they will also agree that the foraging ant knows the direction and distance of its nest from wherever it happens to be. Its knowledge of the direction and distance back to its nest is certainly not innate. When, however, it is explained that the ant acquires this knowledge by path integration, they begin casting about for a new definition of learning, one that will exclude path integration. The reasons they offer for wanting to find a definition of learning that excludes path integration are revelatory. Often among the first reasons offered is that path integration depends on an innate special-purpose mechanism, dedicated to doing that and only that. In short, it depends on the operation of an adaptive specialization. This is a cogent argument only if we believe that there is a general-purpose learning mechanism, a mechanism that is not specialized to solve one particular information-acquisition problem.

Another reason offered for excluding path integration is that the ant does not have to experience its outward journey repeatedly before it knows the way home. A variant of this argument is that the ant does not get better at running home with repeated practice. This objection is a reification of the associative assumption that learning involves the gradual strengthening of something. The conditioned response in classical conditioning gets stronger as the number of selective pairings of the conditioned and unconditioned stimuli increases, but the only apparent justification for making strengthening through repetition part of the definition of learning is the conviction that there is a general-purpose learning mechanism and that the classical conditioning paradigm captures its essence. If we think that path integration captures the only essence of learning that is there to be captured, then we are not going to make strengthening by repetition part of the definition of learning.

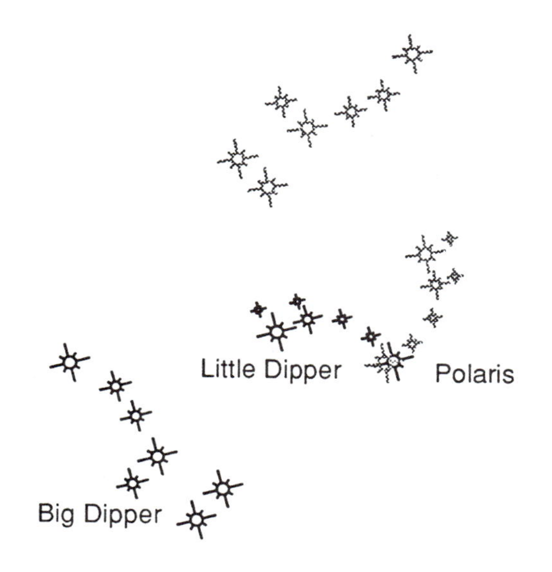

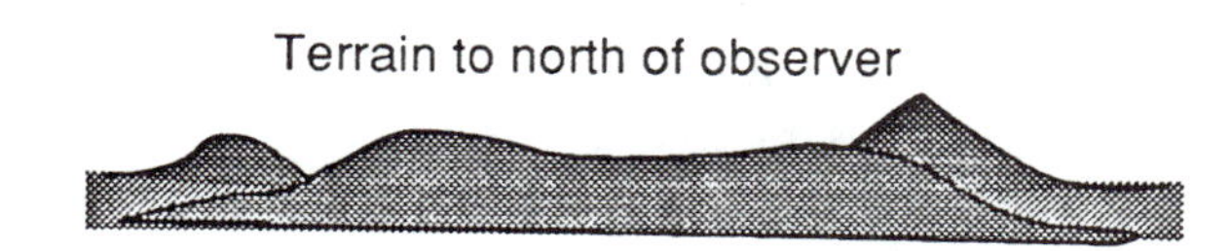

FIGURE 83.3 The Big and Little Dippers (two of the most prominent circumpolar constellations) as seen at two different times (faint, about 9:00 P.M.; dark, about 4:00 A.M.) on a spring night from temperate latitudes in the Northern Hemisphere. By observing the slow wheeling of the star patterns, the nestling migratory bird learns the location of the celestial pole (presently very near Polaris).

LEARNING BY OBSERVING THE NIGHT SKY ROTATE Many migratory birds fly at night. Those that do maintain their compass orientation by reference to the circumpolar constellations, whenever some or all of these are visible. The circumpolar constellations are the star patterns—for example, the Big and Little Dippers; see figure 83.3—that define the celestial pole, the point in the sky toward which the earth's axis of rotation points. The earth's axis of rotation wobbles markedly, in a 27,000-year cycle called the precession of the equinoxes, so the location of the celestial pole within the pattern of circumpolar stars varies greatly on a time scale that is short from an evolutionary perspective. Migratory birds are not born knowing the location of the celestial pole in the night sky; they learn it while they are nestlings and incapable of flight. (Emlen, 1969a; Able and Bingham, 1987). They learn it by observing the rotation of the night sky. From a prolonged and steady gaze at the slowly rotating sky or from separate, widely spaced observations of the changing positions of the salient star clusters in the sky, the nestling bird extracts and stores the location of the center of rotation. Months later, by observing any of several different circumpolar star clusters, the migrating bird can adjust its orientation so as to keep its back to, for example, the north celestial pole. In the spring, a change in the bird's hormonal condition alters a parameter in the decision process that uses the stored information about the location of the celestial pole to keep its migratory flight oriented. Now, it orients toward rather than away from the celestial pole (Emlen, 1969b; Martin and Meier, 1973).

Although we do not know the appropriate computational description of the process by which the nestling bird extracts the center of rotation of the night sky, few would be tempted to try to explain this learning by path integration. And, treating this learning in terms of the temporal pairing of CSs and USs appears just as unpromising. As with other forms of spatial learning, it is difficult to say what the CS is. It is equally difficult to identify the other crucial components in the Pavlovian analysis of learning, the unconditioned stimulus, the unconditioned response, and the conditioned response. Orienting to the south is not an unconditioned response to Polaris. We have to do here with another adaptive specialization. Like the path-integration mechanism, this mechanism is designed to compute and store a particular set of quantitative facts about the world—the values of variables that specify the position of the celestial pole in the pattern of dots defined by the circumpolar stars. The path-integration problem and the pole-localization problem differ with regard to what is specified by the values to be computed and the nature of the data from which they must computed. These differences appear to demand corresponding differences in the computational structure of the requisite learning mechanisms, just as the differences between sound and light demand both differences in the requisite sensory organs and differences in the neural mechanisms that compute similar quantitative facts about the world (for example, the angular position of objects) from these dissimilar input modalities (Mogdnas and Knudsen, 1992).

Learning by Computing Relative Reward Abundance As already noted, the waggle dance of the returned forager bee directly indicates the direction and distance values computed by its path-integration mechanism. It is relatively rare to find a behavioral output that reflects so directly the values computed by a learning mechanism, but there are other instances of this, one of which is matching behavior. Matching behavior is seen in foraging animals that are simultaneously exploiting more than one food patch. They apportion their time among the patches in proportion to the relative abundance of food (Smith and Dawkins, 1971). If, for example, two experimenters stand 30

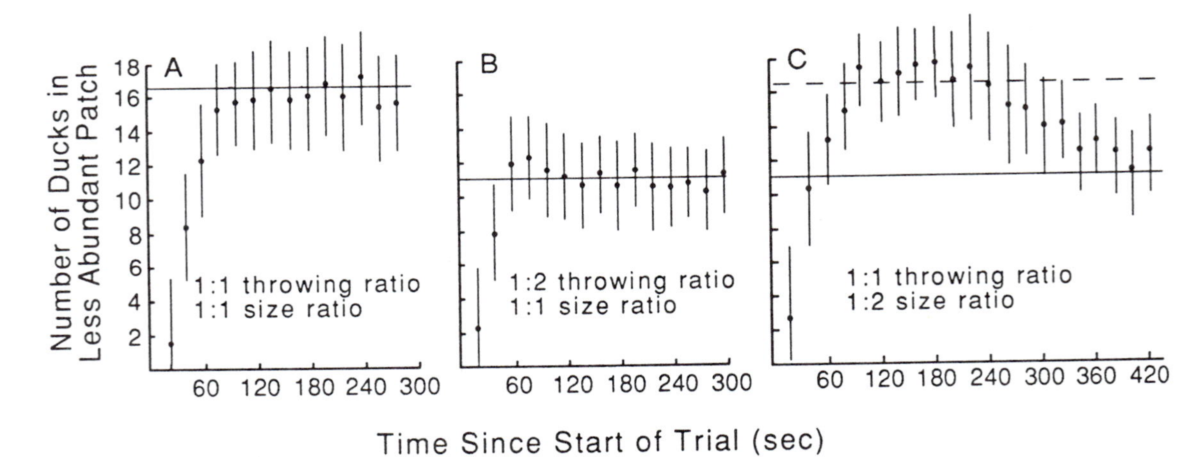

Figure 83.4 Matching in a flock of 33 ducks. The number of ducks in front of the less abundant provider as a function of time since the onset of throwing bread chunks. Points indicate means; vertical bars indicate standard deviations. (A) Both slower and faster providers throwing chunks of equal size (2 g) at same rate (12/min). Solid horizontal line indicates number expected if ducks divide between providers in proportion to the relative abundance of bread provided. Data from 29 trials. (B) Slower provider throws at half the rate of faster provider (6/min vs. 12/min); equal size chunks. Data from 24 trials. (C) Both providers throwing at same rate but more abundant provider throwing chunks twice as big. Horizontal dashed line indicates matching based on rate alone; horizontal solid line indicates matching based on relative abundance $[(M_1/I_1)/(M_2/I_2)]$, where M is reward magnitude (chunk size) and I is the interreward interval. Data from 14 trials. (Adapted from Harper, 1982, 576, 577, 582, by permission)

yards apart on the bank of a pond and throw chunks of bread onto the water, the ducks in an overwintering flock divide themselves 50–50 between the experimenters when they throw equal-sized chunks equally often (figure 83.4A). But when one experimenter throws twice as often as the other, they divide 2 : 1 in favor of the "food patch" that is twice as rich (figure 83.4B). And when the experimenters throw equally often but one throws chunks twice as big, the ducks initially divide themselves in accord simply with the more readily perceptible relative rates of throwing. However, they soon correct for the more difficult to detect difference in chunk size. Within 10 minutes, they are divided between the experimenters in accord with the relative abundance of bread (figure 83.4C).

The abundance of bread in a patch is the magnitude of a chunk divided by the interval between chunks, that is, the ratio M/I, where M specifies the size of a chunk and I specifies the interval between successive throws. Thus, the relative abundance is the ratio of two ratios. The matching law says that the ratio of the times an animal allots to two patches matches the ratio of the food abundances. Thus, the observable behavior output ratio directly reflects a ratio that the brain has computed from its experience of the magnitudes of the pieces of bread and the intervals between throws.

But does an account of matching behavior require that we assume the computation by the brain of the relative abundance ratio? The most common models of the process that generates matching behavior assume that this process does not involve the computation of this ratio—indeed, that it does not even involve the formation within the nervous system of quantities (strengths of associative bonds) that are proportionate to the abundances themselves. Most accounts of matching behavior are similar to Herrnstein's melioration model, which can be understood in terms of the relative strengths of the associations between the patches and food (Herrnstein and Vaughan, 1980; Vaughan, 1985). These associative strengths are not readily thought of as specifying experienced patch profitability, because they are nonlinearly related to patch profitability.

The melioration model assumes that the relative amount of time allocated to a patch is determined by the relative strength of its association with food. The strength of this association is determined by the patch's experienced profitability, which is the amount of food the animal has obtained per unit of its behavioral investment in the patch (for example, per unit of time it has spent in the patch or per response it has made to a key). Note that we have now switched to the instrumental form of associative learning theory; associations are strengthened by reinforcement, not by temporal pairing. Vaughan (1985), however, assumes that this associative process is mathematically analogous to the process described by Rescorla and Wagner (1972) for classical conditioning. It has an upper asymptote—a maximum possible strength of association between a patch and food—and the strength of the association between a patch and food is a monotonically increasing but decelerating function of the experienced profitability of a patch. In other words, the strength of the association is not proportionate to the experienced profitability. The strength of the association between the patch and food gives only an ordinal representation of experienced profitability.

The melioration model assumes that when an animal experiences a difference in the profitability of two patches, it forms a stronger association between the more profitable patch and food. The inequality of associative strengths causes it to increase its behavioral investment in the more profitable patch (spend more time there). Because the amount of food obtainable from a patch is limited (by, for example, the experimenter's throwing rate), the increase in the number of rewards the animal obtains from a patch is not proportionate to the increase in its investment. The experienced profitability of a patch decreases as the animal increases its investment. Conversely, the more animals neglect the patch where food comes along less often, the more likely it is that an already thrown piece of bread will be there for the taking; hence the more profitable a visit to that patch becomes. In short, under the conditions where matching behavior is observed, there is a negative feedback relation between investment in a patch and its experienced profitability. It can be shown that the equilibrium state for this negative feedback system occurs when the investment ratio matches the relative rates at which food becomes available (Vaughan, 1985). At equilibrium, the experienced profitabilities are equal, hence the strengths of the associations between the patches and food are equal, hence there is no tendency for the animal to change its relative investment (time allocation).

The importance of this associative model of matching behavior in the present discussion is twofold. First, the model contains no structure specific to this problem. It simply applies common assumptions about the association-forming process to the analysis of choice behavior under the conditions that yield matching behavior. Second, it is not in a serious sense a computational model because there are no arithmetic operations with quantities that specify anything more than ordinal properties of the experienced world; there is no representation of magnitude or interreward intervals, no computation of a quantity that specifies food abundance (morsel magnitude divided by the interreward interval), nor any computation of relative food abundance (the ratio of food abundances).

There is, however, a serious problem with this model, which is evident from inspection of figure 83.4. The model assumes that matching behavior is driven by the return an animal experiences on its behavioral investment. This means that matching cannot become established before the animal has experienced a return from both patches, that is, until it has spent time in both patches and obtained food from both patches. Moreover, the model requires that the animal experience the effect of varying its investment on the return that it gets. It is the variation in return as a function of variation in investment that eventually brings the animal's investment ratio into equilibrium with the food abundance ratio. The problem is that matching behavior appears much more rapidly than these assumptions would predict.

In figure 83.4B, the flock of ducks distributed itself in accord with the throwing ratio within 1 minute after the onset of throwing. There were 33 ducks in the flock. During the one minute it took for matching to become established, almost half the ducks did not get a single chunk from either patch, and few if any ducks got a chunk from both patches. Matching was already established before any duck had acquired experience of the profitability of both patches. The kinetics of the matching seen in figure 83.4—the extremely rapid establishment of a matching distribution following the onset of differential provisioning—implies that what the ducks do is based on the relative abundances of the food they observe in the two patches and not on the returns they experience from their behavioral investments in the patches.

It might be objected that the group situation is too complicated a situation in which to estimate the kinetics of matching behavior in individual animals. However, two recent experiments have demonstrated equally rapid kinetics in individual animals showing matching behavior in an operant task with concurrent variable-interval schedules (Dreyfus, 1991; Mark and Gallistel, 1994). In these experiments, a single animal (a pigeon in the Dreyfus experiment, a rat in the Mark and Gallistel experiment) works in an experimental chamber with two manipulanda (keys or levers). When a key or lever is depressed, it delivers reward (food in the Dreyfus experiment, brain stimulation reward in the Mark and Gallistel experiment) on a variable-interval schedule. A variable-interval (VI) schedule is a Poisson or random-rate process; it schedules another reward (arms the key or lever) at a variable interval after the previous reward is collected. Only the expectation (average value) of this interval is fixed. Any one actual interval is determined by a process with some constant probability of arming the manipulandum in any one second. The lower this arming probability, the greater the average interval between rewards (the VI interval). With a VI 64 s schedule, there is on average a 64 s interval between the collection of the last reward and the availability of the next reward on that lever or key.

By setting up different VI schedules for the two keys or levers, the experimenter controls the relative rates at which the animal may obtain rewards from the two manipulanda. Thus, this operant task is analogous to the more natural situation of the duck experiment in figure 83.4 in that reward is delivered at two different rates in two different locations. The principal differences are that the animal subjects are alone, not in groups, and that the time at which the next reward will come along can only be predicted on average; it is impossible to anticipate the actual moment when a lever will be armed.

In the Mark and Gallistel (1994) experiment, there were two levers, each in its own alcove. Halfway through each experimental session, there was a signaled reversal in the relative rates of reward, from 4 : 1 in favor of one lever to 4 : 1 in favor of the other. Mark and Gallistel also varied the overall reward density (the number of rewards per minute received from the two levers combined) by a factor of 16—from 1 to 16 rewards per minute. They plotted the rat's time-allocation ratio (the relative amount of time allocated to the two levers) in successive narrow windows, and they also plotted the relative rates of reward experienced during these same time windows, that is, the number of rewards received on one lever divided by the number received on the other. They used windows equal to twice the expected interreward interval on the leaner schedule. Because relative time allocation and relative rate of reward were cumulated in such narrow windows, the numbers of rewards per window were small. Most often only one or two rewards had been received on the leaner schedule. Not uncommonly, none had been received, in which case the relative rate of reward was undefined for that window.

When Mark and Gallistel plotted the rat's window-by-window relative time allocation and its window-by-window relative rate of experienced reward on the same graphs, they were startled to see that the relative time allocation was closely tracking the large random variations in the relative rate of reward (figure 83.5). Cross-correlation and regression analyses confirmed that the rat's time-allocation ratio in a given window

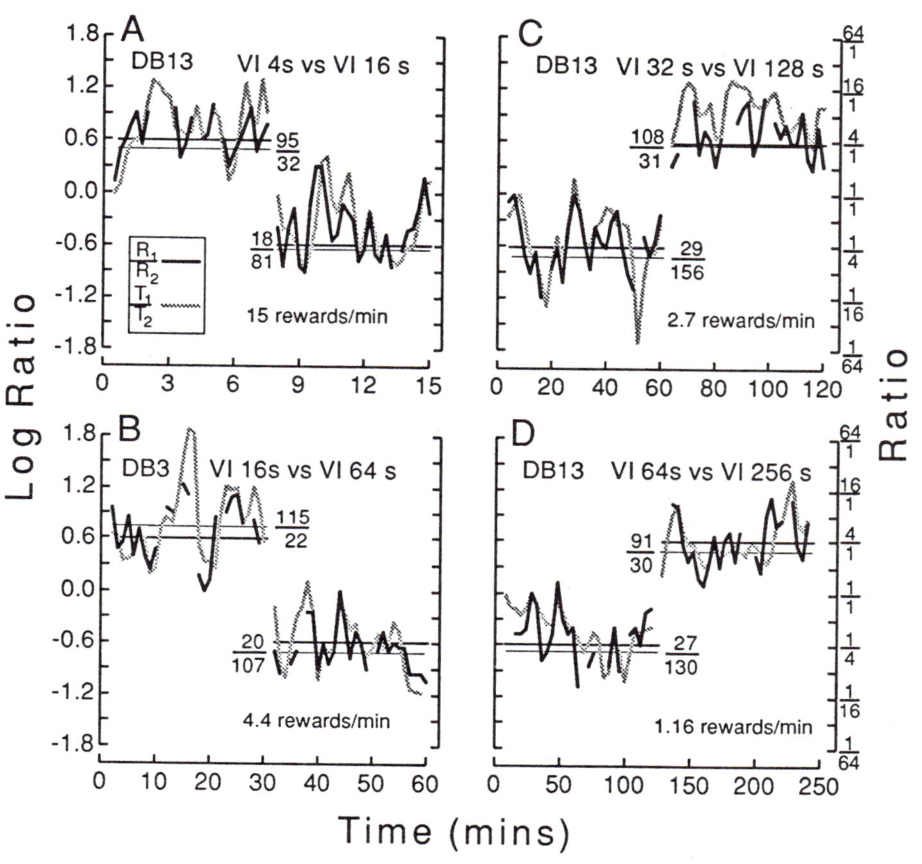

FIGURE 83.5 Representative plots of log reward ratios and log time-allocation ratios in windows equal to twice the expected interreward interval on the leaner schedule. (Windows for successive points overlap by half a window.) The relative (programmed) rates of reward reversed halfway through each session, from 4 : 1 in favor of one lever to 4 : 1 in favor of the other. Note that the time-allocation ratios reversed within the first window after this reversal in the relative rates of reward. Note also that the time-allocation ratio tracks the swings in the reward ratio. The gaps in the solid line are at windows in which no reward was received from the leaner schedule (hence the reward ratio was undefined). The heavy horizontal lines indicate the matching ratio for the programmed rates of reward; the ligher horizontal lines indicate the matching ratio for the ratio actually experienced (as determined from the total rewards received on each side during each half session, which are given beside these matching lines). (A) The highest-density condition. On the richer schedule, the programmed interreward interval was 4 s, while on the leaner schedule it was 16 s. The experienced reward density from the two schedules combined is given in the lower right corner of the panel. (B) Fourfold lower overall reward density. Note 4-fold increase in duration of session. (C) An 8-fold lowering of reward density. (D) A 16-fold lower reward density. In this condition, the average interreward interval on the richer schedule was slightly longer than 1 min; on the leaner schedule, it was slightly longer than 4 min. Note that both the response to the midsession reversal and the tracking of the random swings in the reward ratio are independent of overall reward density. (From Mark and Gallistel, 1994, by permission)

was well predicted by the experienced reward ratio in that same window and that the relative rates of reward in earlier windows had negligible impact on the time-allocation ratio in the most recent window.

The extent to which the rat's time-allocation ratio tracked the reward ratio was independent of the overall reward density. The rat did this as accurately when rewards were coming at the rate of about 16 per minute on the two schedules combined as when they were coming at only about 1 per minute on the two schedules combined. Given that the rats were adjusting so rapidly to the wide, unsignaled, and unpredictable swings in the relative rate of reward, it is no surprise that when Mark and Gallistel looked at the adjustment to the programmed reversal in the relative rates of reward, they found that the reversal was nearly complete within the span of one expected interreward interval on the leaner schedule.

The relative rate of reward is the ratio of two rate estimates. The smallest amount of time in which it is possible in principle to estimate the rate of reward provided by a Poisson process is one interreward interval. If an animal tracked varying rates of payoff as closely as possible, the best it could do would be to use as its momentary estimate of the relative rate of reward the ratio of the most recent interreward intervals on each side. There are wide random variations in the durations of the individual interreward intervals in a Poisson process. Hence, the ratio of two such intervals will itself show wide swings; estimates of the ratio of the rates of payoff based on a sample of only one interreward interval from each process often vary from the ratio of the expected values (the ratio of the averages) by an order of magnitude. These wide variations are completely unpredictable. The only way that the rat's time-allocation ratio can track these wide, unpredictable swings in the relative rate of reward is if the process that determines how the animal allocates its time only looks at the most recent one or two interreward intervals on each side. Thus, we conclude that the animal's choice behavior is determined by the ratio of the most recent interreward intervals on each side.

The assumption that the animal's behavior is driven by the (internally computed) ratio of the most recent interreward interval on each side also explains why its ability to track apparent fluctuations in the relative rate of return is independent of the overall reward density, that is, of the average time it must wait before either schedule generates a reward. The animal is estimating rates by measuring the time intervals between events. This mechanism, unlike the associative averaging mechanism assumed by the melioration model, works regardless of the average interval between rewards.

Thus, experiments that determine the kinetics of matching behavior with individual animals confirm what one sees in figure 83.4: The adjustment to varying rates of reward occurs much too rapidly for this adjustment to be driven by the kind of averaging mechanism envisaged by the melioration model. Like other associative models of matching behavior, this model assumes that the strengths of the associations between each lever (or patch) and food are exponentially relaxing running averages over many previous rewards (Killeen, 1981; Lea and Dow, 1984). These models predict a sluggish adjustment to changes in the relative rates of reward, whereas observed behavior, under both group conditions and individual conditions, demonstrates adjustments whose rapidity is at or close to the theoretical maximum. The only models that can predict adjustments this rapid are models in which a special-purpose learning mechanism computes the ratio of two interevent intervals.

Classical conditioning as an adaptive specialization

I have given a greatly curtailed catalog of special-purpose learning mechanisms. The catalog could be enlarged indefinitely. I have not, for example, described the mechanism by which animals learn the ephemeris function, the position of the sun above the horizon (the solar azimuth) as a function of the time of day, so that they can use the sun to determine their heading (Gould, 1984). Nor have I described the mechanism by which the immature barn owl adjusts its auditory map of the angular positions of sound sources to align it with its visual map (Knudsen and Knudsen, 1990; Mogdnas and Knudsen, 1992). Nor have I described the complex mechanisms by which some nestling and fledgling songbirds learn features of the song they will sing as adults (Marler, 1991b). But the general point is clear; all of these instances of learning are mediated by processes whose structure adapts them to solve one particular kind of problem. They are not general-purpose learning mechanisms; they are adaptive specializations. They are what Marler (1991a) has called instincts to learn. The basic nature of their specialization is computational; they are designed to compute and store certain kinds of information from certain kinds of data. This information is then used by various decision processes to control behavior. The decision processes themselves often perform further computations on the previously learned values in the course of determining the behavioral output.

But what about the learning that occurs during classical (Pavlovian) and instrumental (operant) conditioning experiments? Can that, too, be seen as a product of adaptively specialized learning mechanisms? If so, what is the domain or class of computational problem to which these mechanisms are adapted? Recently, some of the most successful and mathematically simple models of the conditioning mechanism have been based on the assumption that conditioning is mediated by mechanisms that record the durations of temporal intervals (Gibbon et al., 1988; Gibbon and Balsam, 1981; Gallistel, 1990, 1992). These models also assume that the decision processes that translate the remembered values of temporal intervals into conditioned responding perform elementary arithmetic operations with the values of these stored intervals.

Timing models of conditioning implicitly or explicitly assume that the conditioning process is specialized for the solution of problems in multivariate, nonstationary time series analysis (Gallistel, 1992). In particular, they specify computational mechanisms by which the brain can determine which conditioned stimuli predict what interevent intervals in situations like blocking (Kamin, 1969), background conditioning (Rescorla, 1968), and pseudodiscrimination (Wagner et al., 1968). In these experimental paradigms, the animal sorts out which of several CSs actually affects the rate of US occurrence (the expected intervals between occurrences of a motivationally significant US, such as food or foot shock). That is, the animal solves the problem at the heart of multivariate statistical analysis, the problem of figuring out what predicts what. These timing models of conditioning also specify computational mechanisms by which the animal may recognize changes in the US-US interval predicted by a given CS (nonstationarity), and computational mechanisms by which the animal assesses the uncertainties inherent in estimating population parameters (such as the expected US-US interval) from small samples. In short, like computational models of dead reckoning, these models describe the computational processes that adapt the conditioning mechanism to its particular domain.

The discussion of the mechanisms underlying matching behavior, which has been intensively studied by both zoologists interested in foraging behavior and operant psychologists interested in choice processes, is only one indication of the power of the timing models when applied to data from classical and instrumental conditioning experiments. For example, these models give a more straightforward explanation of the effects of background conditioning (Rescorla, 1968), which are crucial in modern analyses of conditioning. When USs are given in the presence of the background alone (that is presented to the animal while it is in the apparatus but when there is no explicit CS), the animal associates these USs with the background, and this association may block the learning that would otherwise occur when the US is paired with an explicit CS (i.e., a transient tone). The associative analysis of this important phenomenon requires an internal trial clock (Rescorla and Wagner, 1972), for which there is no evidence. The interval-timing analysis (Gibbon and Balsam, 1981; Gallistel, 1990) simply assumes that the animal learns the inter US interval to be expected in the apparatus itself and does not exhibit a conditioned response to another CS if the inter US interval in the presence of that CS (and, of course, the background) is no different from the inter US interval predicted by the apparatus (background).

In associative theories of classical conditioning, it is assumed that the temporal interval between CS onset and US onset determines the rate of conditioning, but it has been shown that there is no effect of the CS-US interval on the rate of conditioning if the US-US interval is adjusted in proportion to the change in CS-US interval (Gibbon et al., 1977). Put another way, the rate of conditioning is invariant under scalar transformation of the temporal intervals in the training protocol. Explaining this scalar invariance is a major challenge to associative models, but scalar invariance in the rate of conditioning is a straightforward prediction of the timing models (Gibbon and Balsam, 1981; Gallistel, 1990, 1992).

Making quantitative predictions about the effects of partial reinforcement on the rate of conditioning is another major challenge to associative models. Timing models, on the other hand, predict that partial reinforcement will have a scalar effect on the rate of conditioning. A scalar effect of partial reinforcement on the rate of conditioning is what is observed, at least in autoshaping paradigms (Gibbon et al., 1980). That is, when the US occurs during only one in every 10 occurrences of the CS, it takes 10 times as many CS occurrences to produce the same strength of conditioned responding.

Partial reinforcement makes conditioning slower, but it also makes extinction slower. Explaining the retarding effect of partial reinforcement on the rate of extinction is a problem that has long bedeviled associative models. Why should the history of conditioning determine how many nonreinforcements it takes to weaken the association a given amount? On the other hand, timing models predict that the partial reinforcement will increase the trials necessary to produce extinction by a factor proportionate to the thinning of the reinforcement schedule (Gallistel, 1992), which is borne out by the empirical results, at least in autoshaping experiments (Gibbon et al., 1980). Finally, timing models predict that increasing the intertrial interval, which greatly increases the rate of initial conditioning, will have no effect on the rate of extinction. Again, this is the empirical result (Gibbon et al., 1980).

In short, models that assume adaptively specialized rather than general-purpose learning mechanisms are now being applied to the prediction and explanation of well-established findings in the classical and instrumental conditioning literature. This suggests that learning mechanisms, like other biological mechanisms, will invariably exhibit adaptive specialization. We should no more expect to find a general-purpose learning mechanism than we should expect to find a general-purpose sensory organ.

An increasing number of cognitive competencies are now known to be of a special-purpose nature. Moreover, learning phenomena commonly interpreted as supporting a general-purpose associative analysis of learning—classical and instrumental conditioning—are now better explained as mediated by adaptively specialized learning mechanisms. The mechanism that mediates classical and instrumental conditioning is specialized for multivariate, nonstationary, small-sample time series analysis. This specialization makes conditioning much more "cognitive" than we thought. As Rescorla (1988) has pointed out, modern experiments have shown that conditioning is "not what you think it is." But the specialization that makes the conditioning mechanism so much more intelligent than we thought it was also makes it hopelessly ill suited to other learning tasks, such as path integration or learning the center of rotation of the night sky.

Despite long-standing and deeply entrenched views to the contrary, the brain can no longer be viewed as an amorphous plastic tissue that acquires its distinctive competencies from the environment acting on general-purpose cellular-level learning mechanisms. Cognitive neuroscientists, as they trace out the functional circuitry of the brain, should be prepared to identify adaptive specializations as the most likely functional units they will find. At the circuit level, special-purpose circuitry is to be expected everywhere in the brain, just as it is now routinely expected in the analysis of sensory and motor function. At the cellular level, the only processes likely to be universal are the elementary computational processes for manipulating neural signals in accord with the laws of arithmetic and logic and for storing and retrieving the values of variables.

REFERENCES

Able, K. P., and V. P. Bingham, 1987. The development of orientation and navigation behavior in birds. *Q. Rev. Biol.* 62:1–29.

Becker, S., and G. E. Hinton, 1992. Self-organizing neural network that discovers surfaces in random-dot stereograms. *Nature* 355:161–163.

Dreyfus, L. R., 1991. Local shifts in relative reinforcement rate and time allocation on concurrent schedules. *J. Exp. Psychol. [Anim. Behav. Proc.]* 17:486–502.

Emlen, S. T., 1969a. The development of migratory orientation in young Indigo Buntings. *Living Bird* 8:113–126.

Emlen, S. T., 1969b. Bird migration: Influence of physiological state upon celestial orientation. *Science* 165:716–718.

Frisch, K. V., 1967. *The Dance-Language and Orientation of Bees*. Cambridge, Mass.: Harvard University Press.

Gallistel, C. R., 1990. *The Organization of Learning*. Cambridge, Mass.: MIT Press.

Gallistel, C. R., 1992. Classical conditioning as an adaptive specialization: A computational model. In *The Psychology of Learning and Motivation: Advances in Research and Theory*, D. L. Medin, ed. New York: Academic Press, pp. 35–67.

Gibbon, J., M. D. Baldock, C. M. Locurto, L. Gold, and H. S. Terrace, 1977. Trial and intertrial durations in autoshaping. *J. Exp. Psychol. [Anim. Behav. Proc.]* 3:264–284.

Gibbon, J., and P. Balsam, 1981. Spreading associations in time. In *Autoshaping and Conditioning Theory*, C. M. Locurto, H. S. Terrace, and J. Gibbon, eds. New York: Academic Press, pp. 219–253.

Gibbon, J., R. M. Church, S. Fairhurst, and A. Kacelnik, 1988. Scalar expectancy theory and choice between delayed rewards. *Psychol. Rev.* 95:102–114.

Gibbon, J., L. Farrell, C. M. Locurto, H. J. Duncan, and H. S. Terrace, 1980. Partial reinforcement in autoshaping with pigeons. *Anim. Learn. Behav.* 8:45–59.

GLUCK, M. A., and R. F. THOMPSON, 1987. Modeling the neural substrates of associative learning and memory: A computational approach. *Psychol. Rev.* 94(2):176–191.

GOULD, J. L., 1984. Processing of sun-azimuth information by bees. *Anim. Behav.* 32:149–152.

HARKNESS, R. D., and N. G. MAROUDAS, 1985. Central place foraging by an ant (Cataglyphis bicolor Fab.): A model of searching. *Anim. Behav.* 33:916–928.

HARPER, D. G. C., 1982. Competitive foraging in mallards: Ideal free ducks. *Anim. Behav.* 30:575–584.

HAWKINS, R. D., and E. R. KANDEL, 1984. Is there a cell-biological alphabet for simple forms of learning? *Psychol. Rev.* 91:375–391.

HERRNSTEIN, R. J., and W. J. VAUGHAN, 1980. Melioration and behavioral allocation. In *Limits to Action: The Allocation of Individual Behavior*, J. E. R. Staddon, ed. New York: Academic Press, pp. 143–176.

HULL, C. L., 1943. *Principles of Behavior*. New York: Appleton-Century-Crofts.

KAMIN, L. J., 1969. Predictability, surprise, attention, and conditioning. In *Punishment and Aversive Behavior*, B. A. Campbell and R. M. Church, eds. New York: Appleton-Century-Crofts, pp. 276–296.

KILLEEN, P. R., 1981. Averaging theory. In *Quantification of Steady-State Operant Behavior*, C. M. Bradshaw, E. Szabadi, and C. F. Lowe, eds. Amsterdam: Elsevier, pp. 231–259.

KNUDSEN, E. I., and P. F. KNUDSEN, 1990. Sensitive and critical periods for visual calibration of sound localization by barn owls. *J. Neurosci.* 10(1):222–232.

LEA, S. E. G., and S. M. DOW, 1984. The integration of reinforcements over time. In *Timing and Time Perception*, J. Gibbon and L. Allan, eds. New York: Annals of the New York Academy of Sciences, pp. 269–277.

MARK, T. A., and C. R. GALLISTEL, 1994. The kinetics of matching. *J. Exp. Psychol. [Anim. Behav. Proc.]* 20:1–17.

MARLER, P., 1991a. The instinct to learn. In *The Epigenesis of Mind*, S. Carey and R. Gelman, eds. Hillsdale, N.J.: Erlbaum, pp. 37–66.

MARLER, P., 1991b. Song-learning behavior: The interface with neuroethology. *Trend. Neurosci.* 14(5):199–205.

MARTIN, D. D., and A. H. MEIER, 1973. Temporal synergism of corticosterone and prolactin in regulating orientation in the migratory White-throated Sparrow (*Zonotrichia albicollis*). *Condor* 75:369–374.

MOGDNAS, J., and E. I. KNUDSEN, 1992. Adaptive adjustment of unit tuning to sound localization cues in response to monaural occlusion in developing owl optic tectum. *J. Neurosci.* 12(9):3473–3484.

MORRIS, R. G. M., 1981. Spatial localization does not require the presence of local cues. *Learn. Motiv.* 12:239–260.

RESCORLA, R. A., 1968. Probability of shock in the presence and absence of CS in fear conditioning. *J. Comp. Physiol. Psychol.* 66(1):1–5.

RESCORLA, R. A., 1988. Pavlovian conditioning: It's not what you think it is. *Am. Psychol.* 43:151–160.

RESCORLA, R. A., and A. R. WAGNER, 1972. A theory of Pavlovian conditioning: Variations in the effectiveness of reinforcement and nonreinforcement. In *Classical Conditioning II*, A. H. Black and W. F. Prokasy, eds. New York: Appleton-Century-Crofts, pp. 64–99.

SMITH, J. N. M., and R. DAWKINS, 1971. The hunting behavior of individual Great Tits in relation to spatial variations in their food density. *Anim. Behav.* 19:695–706.

SMOLENSKY, P., 1986. Information processing in dynamical systems: foundations of harmony theory. In *Parallel Distributed Processing: Foundations*, D. E. Rumelhart and J. L. McClelland, eds. Cambridge, Mass.: MIT Press, pp. 194–281.

VAUGHAN, W. J., 1985. Choice: A local analysis. *J. Exp. Anal. Behav.* 43:383–405.

WAGNER, A. R., F. A. LOGAN, K. HABERLANDT, and T. PRICE, 1968. Stimulus selection in animal discrimination learning. *J. Exp. Psychol.* 76(2):171–180.

WEHNER, R., and M. V. SRINIVASAN, 1981. Searching behavior of desert ants, genus *Cataglyphis* (*Formicidae*, Hymenoptera). *J. Comp. Physiol.* 142:315–338.

WEHNER, R., and I. FLATT, 1972. The visual orientation of desert ants, *Cataglyphis bicolor*, by means of territorial cues. In *Information Processing in the Visual System of Arthropods*, R. Wehner, ed. New York: Springer, pp. 295–302.

84 Discriminative Parental Solicitude and the Relevance of Evolutionary Models to the Analysis of Motivational Systems

MARTIN DALY AND MARGO WILSON

ABSTRACT Behavioral biologists customarily distinguish between proximate causal analyses of the mechanisms underlying action, and ultimate (selectionist) analyses, which invoke the adaptive functions (fitness-promoting consequences) for which behavior has evolved. Cognitive neuroscientists, operating in the proximate mode, have paid little attention to selectionist theories. However, proximate causal research is inevitably guided in part by implicit assumptions about adaptive function; and selectionist theories, which make such assumptions explicit and develop their implications, can help generate novel, testable proximate causal hypotheses.

This thesis is illustrated with theory and research on parenting. Selectionist models suggest certain variables (certainty of parenthood, offspring quality, opportunity costs) that evolved parental motivational systems may be expected to track, sometimes providing considerable detail about the form of expected functional relationships. Recent studies demonstrate the utility of these models in the search for cognitive and neural mechanisms.

Why do female rodents become aggressive when lactating? A psychophysiological approach to this question might entail exploring the roles of particular hormonal regimens (e.g., Mayer, Monroy, and Rosenblatt, 1990), sensory inputs (e.g., Stern and Kolunie, 1991), and brain structures (e.g., Hansen et al., 1991). A rather different approach entails asking such questions as whether maternal aggression is specifically directed against genuine threats to the pups, and whether it is effective in protecting them (e.g., Elwood, Nesbitt, and Kennedy, 1990).

Evolutionists commonly refer to psychophysiology's explorations as *proximate causal* analyses, while the second approach, characteristic of behavioral ecology, concerns *adaptive function* or *ultimate causation*. At least since Tinbergen (1963a), animal behaviorists have painstakingly distinguished these two modes of explanation, and rightly so, since many fruitless controversies have been fueled by incomprehension of the distinction. It does not follow, however, that proximate and ultimate analyses can or should be pursued in isolation from one another. The adaptationist theoretical approach of behavioral ecology has much to offer researchers engaged in proximate causal analysis of behavioral control mechanisms. That, in a nutshell, is the thesis of this chapter.

Some writers dismiss adaptationist theorizing as unscientific speculation, but although it is true that ultimate causation hypotheses do not generally submit to experimental testing as straightforwardly as proximate causal hypotheses, this dismissive stance is counterproductive. The most resolutely mechanistic physiologist relies on complex assumptions about adaptive function, so proscription on adaptationist theorizing amounts to insistence that these assumptions remain inexplicit and unexamined. As Mayr (1983) has stressed, every important discovery in physiology and other proximate causal fields of biology has been predicated upon the researchers' interpretations of the func-

MARTIN DALY and MARGO WILSON Department of Psychology, McMaster University, Hamilton, Ontario, Canada tional significance and adaptive design features of the systems under study. Contemporary gut research, for example, is predicated on the understanding that extracting nutrients from ingesta is what the gut is for. This seems obvious and not at all speculative, but the gut's function was not always obvious, and serious investigation of gut physiology was scarcely possible until it *became* obvious. Uncontroversial orienting assumptions about adaptive function are founded in hard-won knowledge; only after basic adaptive functions are correctly apprehended can the research enterprise bloom. There could be no neuroscience, for example, until the relatively recent discovery that information processing is what nerve tissue is for.

Unfortunately, the assumptions about adaptive function that guide research programs in cognitive neuroscience and related disciplines are not always sound. Neither are they always explicit, and this hinders critical scrutiny of them. Making notions of adaptive function explicit and exploring their implications is precisely what the evolutionary models of behavioral ecology are meant to do, and their implications about potentially fruitful directions for proximate causal research can be both straightforward and novel. We shall illustrate this thesis with examples from the domain of parental motivation and behavior.

Evolution by selection and the proximate-ultimate distinction

The adaptive complexity of living things was once the most compelling reason to believe that supernatural powers intervene in our world. Darwin and Wallace (1858) destroyed this theological "argument from design" by discovering a natural process that produces adaptive complexity without intelligence or intention: the continual generation of heritable variation in the characteristics of individuals, followed by a nonrandom differential survival and reproductive success of the variants. Darwin called the latter differentials "natural selection."

Darwin's theory implied that the adaptive function of all traits is ultimately reproductive. Traits proliferate because they contribute to the relative reproductive success of their bearers compared to other members of the same populations. "Survival value" is a popular way of referring to adaptive significance, but it is misleading because personal survival is not the criterion of adaptation. Selection favors whatever traits enhance the proportional representation of their carriers' alleles in future gene pools. It follows that creatures have evolved to enter willingly into life-threatening contests for mating opportunities, to deplete bodily reserves to nourish dependent offspring, to allocate benefits discriminatively with respect to closeness of relationship, and in general to expend their very lives in the pursuit of genetic posterity (fitness).

Darwin clearly understood that natural selection is primarily a process of competition within species. Nevertheless, for a century after him, most behavioral biologists routinely misunderstood this point, blithely imagining that selection equips conspecific animals with the shared purpose of perpetuating their species. This "greater-goodism" (Cronin, 1992) cannot easily be reconciled with the orthodox neo-Darwinian conception of adaptive function as effective contribution to the competitive ascendancy of one's genotypic elements over their alleles. But for decades no one seemed to notice. Greater-goodism dominated discussions of the evolution of social phenomena until demolished by Hamilton (1964) and Williams (1966), and it continues to sow confusion in fields untouched by Darwinism.

Greater-goodism illustrates how unexamined false assumptions can impede research. Konrad Lorenz, for example, so thoroughly and uncritically accepted it as to assert that the "aim of aggression" is never lethal (1966, 38), a claim he insisted he had derived from "objective observation of animals." Although Lorenz was familiar with a literature containing many field observations of fatal fights, he dismissed all such reports as instances of pathology or "mishap." Lethality was expurgated from Lorenz's analysis of aggression and became invisible to his readers. Indeed, he has been widely cited as having documented the sublethal restraint of animal aggression in nature and the unique murderousness of humankind, notions wildly at odds with actual field observations of animal conflict.

Niko Tinbergen, who shared the 1973 Nobel prize with Lorenz and Karl von Frisch for their roles in establishing ethology as a science, paid more careful attention to the problem of transforming ideas about adaptive significance into explicit, testable hypotheses. He championed the view that explanations in terms of proximate cause and adaptive function are equally valid, distinct, and complementary (Tinbergen, 1963a), and in so doing he helped to found the approach of modern behavioral ecology. Tinbergen (1963b) asked, for example, why nesting birds carry the eggshells away after their young hatch, his premise being that selection would have eliminated such behavior unless it had fitness benefits sufficient to offset the costs in time, energy, and temporary absence from the vulnerable hatchlings. So he devised experiments to test whether eggshell removal served a sanitary function, disposed of dangerous sharp objects, or made nests less conspicuous to predators; only the latter function was supported.

Extensive formulation and testing of explicit models of adaptive function proliferated after Tinbergen's early efforts, and in the initial flowering of this approach, its practitioners sometimes declared their autonomy from the enterprise of characterizing proximate causal mechanisms. According to the introduction to an excellent sociobiology textbook, for example, "data from studies of proximate causation usually have only limited value for understanding ultimate causation, and vice versa" (Wittenberger, 1981, 4). Efforts to divorce ultimate and proximate analyses are futile, however, as may be illustrated by consideration of a classic problem in optimal foraging theory.

How should a forager exploit resources distributed in depletable patches? An elegant ultimate causation theory maintains that the forager should leave partially depleted patches to seek fresh ones when the instantaneous rate of food-getting from the present patch equals the highest gross rate of return that can be attained over the total time that is spent both foraging within patches and traveling between them or searching for them (Charnov, 1976). *How* the animal should assess a patch's instantaneous rate of yield, induce the habitat-specific mean interpatch travel time, et cetera, are proximate details that are outside the theory's purview. But trouble begins when we try to decide what would constitute a potentially falsifying test of the theory. Foragers do not always perform perfectly, so how is one to decide whether errors reflect a mere shortfall of information to guide behavioral decisions or a more basic flaw in the optimality analysis? This question inevitably led to hypotheses about information processing and decision rules (e.g., Green, 1984). What information is available to the animal, and how should it be processed to estimate relevant parameters of the situation and make adaptive behavioral decisions? How would the answer change if increased memory loads and computational demands were treated as costs? Thus, optimal foraging theorists have come to address issues like the optimal investment of time or effort in information gathering (Stephens and Krebs, 1986), and the form of optimal forgetting functions for obsolescing information (Healy, 1992). As these issues have been raised and addressed, the general question has been subtly transformed into one that is neither simply adaptationist nor simply proximate, but an amalgam of both: What would an optimal cognitive program for solving this patch-foraging problem look like, and how do actual cognitive programs compare with that theoretical ideal?

Charnov's (1976) patch-foraging theory never was a theory of how animals will behave. It was a task analysis: a theoretical characterization of the essential features of an adaptive problem confronting animals that forage for patchily distributed foods. Natural selection may be expected to have equipped animals with solutions to problems like this, and these evolved solutions are often most usefully described at a cognitive level, as algorithms for information processing and behavioral decision making (Cosmides and Tooby, 1987). Optimal foraging is not unusual in this regard; cognitive formulations are increasingly prominent in other subfields of behavioral ecology and sociobiology, for similar reasons (e.g., Hepper, 1991; Davies, 1992). What, after all, is it that selectionist models and "ultimate explanations" purport to predict and explain? Nothing more nor less than the organization of evolved proximate causal structures.

In attempting to free physiology from vitalism, Claude Bernard (1865) maintained that although "the nature of our mind leads us to seek the essence or the *why* of things ... experience soon teaches us that we cannot get beyond the *how*, i.e., beyond the immediate cause or the necessary conditions of phenomena" (p. 80). Bernard remained oblivious to Darwinism until his death in 1878 (Olmsted, 1938), and so have many of his intellectual descendants for another century. To this day, there are physiological psychologists who cite Bernard's dictum with approval, and to whom it would be anathema to suggest, for example, that predation risk reduction is in any sense the reason why a gull removes eggshells from the vicinity of its nest.

Sober (1983) has proposed that psychology's antipathy to explanations in terms of adaptive function derives from a message inferred from the victory of Newton's blind physics over Aristotle's teleology, namely that "a science progresses by replacing teleological concepts with ones that are untainted by ideas of goals, plans and purposes" (p. 115). This stance, Sober continues, "received further impetus from the Darwinian revolution in biology," because Darwin replaced a purposeful creator with a purposeless mechanism. But if the Darwinian revolution truly contributed to psychology's naive emulation of physics, then that influence entailed a great irony. By providing a thoroughly materialistic explanation for the previously incomprehensible fact that living things have "goals, plans and purposes" instantiated in their structures, Darwin's discovery actually rendered obsolete the sort of doctrinaire antagonism to purposelike concepts exemplified by Bernard's dictum. The creative feedback process of selection justifies invoking the consequences of biological phenomena as part of their explanation: What they achieve is, in a very real sense, why they exist.

Sober is certainly correct in claiming that psychology has been lukewarm about concepts that smack of teleology, but the Darwinian revolution has had little overt relevance to the debate. Those, like Tolman (1932), who rebelled against the doctrinaire exclusion of purposive concepts, were inspired more by the manifest goal-directedness of their subject animals than by an appreciation of the efficacy of natural selection. And when "goals and plans" resurfaced during the cognitive revolution against behaviorism (Miller, Galanter, and Pribram, 1960), they were inspired more by the growing sophistication of cybernetic devices than by an understanding of evolutionary adaptation. The result is that many cognitive scientists have continued to operate with only a superficial understanding of what the psyche is organized to achieve (Barkow, Cosmides, and Tooby, 1992).

Symons (1987) has argued that psychology's failure to exploit evolutionary thinking has impeded progress only in certain limited domains. Researchers in areas like sensation, perception, memory, and motor control rely on complex assumptions about the purposes of the mechanisms they study; indeed, the very delineation and naming of a mechanism for study typically entails parsing the psyche into low-level tasks such as the maintenance of perceptual constancies. These scientists make progress when their functional parsings of the psyche are sound (Tooby and Cosmides, 1992), but according to Symons (1987), sound functional parsings are often attainable without recourse to the modern gene-selectionist theory of evolution:

> Selectional thinking sheds little light on perceptual-constancy mechanisms because an ideal design for such a mechanism probably would be the same whether the mechanism's ultimate goal was to promote the survival of genes, individual human bodies, or *Homo sapiens*; for precisely the same reason, selectional thinking sheds little light on organismic goals as vague as *not being hungry* or *not being frightened*. It is only when *it really matters* that the brain/mind was designed to promote the survival of genes ... that psychology is likely to benefit significantly from Darwin's view of life. (Symons, 1987, 130)

Symons then argues that the domain within which "it really matters" that psychological mechanisms have evolved to promote genetic posterity is that of the "mechanisms of feeling," especially social and sexual feeling.

Although it may indeed be the case that explicit evolutionizing would contribute little to analysis of a perceptual constancy mechanism, Symons's argument greatly understates the breadth of psychological domains within which "selectional thinking" is relevant. His own choice of examples—"organismic goals as vague as *not being hungry* or *not being frightened*"—may be used to illustrate the point.

Feeding research is founded in a sound conception of immediate adaptive function: The controls of feeding behavior are organized to extract energy and nutrients from foodstuffs in accord with organismic needs. But if researchers imagine that mere energy balance or survival is the sole criterion of functionality, and fail to recognize that these are subsidiary goals in a more complex motivational structure that functions to promote gene replication, then their analyses of how organisms go about "not being hungry" will suffer. Consider, for example, the fact that broody hens, who are cryptic while incubating their eggs in nests on the ground, experience a programmed decline of 18% in target body weight during incubation (figure 84.1). This radical modulation of the state of "not being hungry" functions to minimize exposure of the nest to predation when the hen gets up to feed, and it does so at considerable cost to her own bodily condition and survival prospects. This is one of many examples of adaptive anorexias (Mrosovsky and Sherry, 1980), in which the mechanisms determining an animal's inclination to feed are sensitive not only to internal energy reserves but also to cues of the likely fitness costs of taking time out from other adaptive activities.

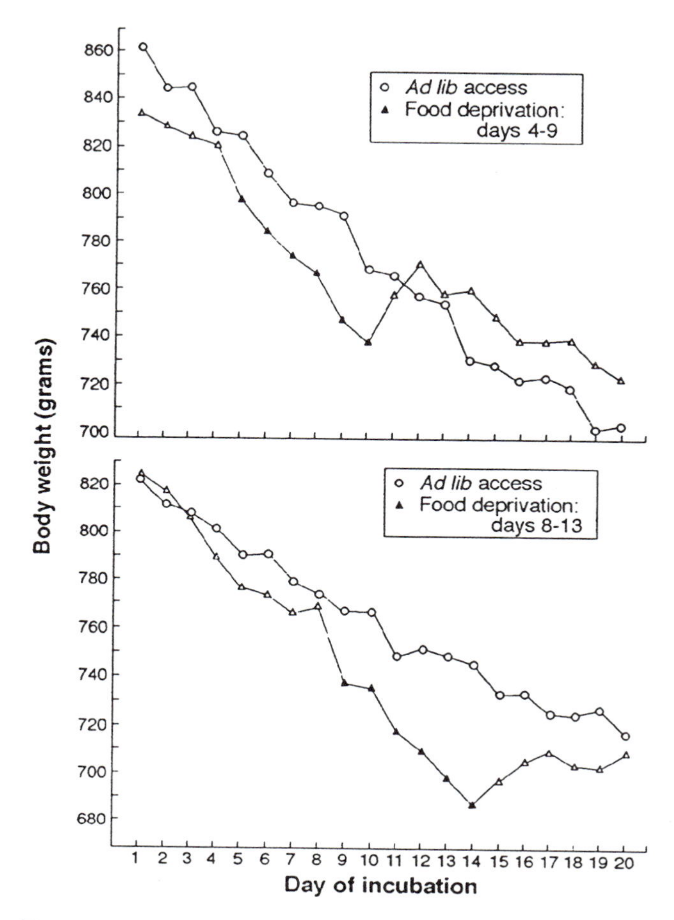

FIGURE 84.1 Adaptive anorexia. Broody jungle fowl hens with ad lib food access lose weight steadily over the 20-day incubation period. If totally deprived of food for 6 days beginning on day 4 (top panel) or day 8 (bottom panel), the birds lose weight more rapidly, then regain weight for 2 or 3 days when food is reinstated, in effect returning to their "programmed" trajectory of weight loss. (From Sherry, Mrosovsky, and Hogan, 1980, figures 6 and 7.)

Similarly, the "vague" organismic goal of "not being frightened" has complex causality that is unlikely to be elucidated without recourse to selectional thinking. A male stickleback fish guarding his nest full of eggs, for example, will stand his ground against an approaching predator longer, and dart at the predator more bravely, the more eggs he has in the nest (Pressley, 1981). In effect, the greater fitness value of a larger brood elevates the statistical probability of death that the little fish is prepared to accept. One correlate of brood size, which might be the cue modulating fear versus bravery in this case, is carbon dioxide production by the eggs, and if so, then it is likely that this cue will prove to mitigate fearfulness only in egg-guarding males. One could never understand (and would be unlikely to discover) such contextual variation in the controls of fearfulness without the basic Darwinian insight that even personal survival is a subordinate objective to that of genetic posterity.

Thus, although the functional design of mechanisms subserving ends like perceptual constancy may indeed be studied and explained without recourse to Darwinism, as Symons (1987) maintains, it seems that almost any "motivational" mechanism will be misunderstood until it is scrutinized in the light of selection theory. This proposition applies to hunger, to fear, and to most of the subject matter of cognitive neuroscience and psychology, including even much of perception. Organisms have not been designed by selection merely to maintain energy balance, repair their tissues, and outlive their fellows. They have been designed by selection to replicate their genes, largely though not exclusively by the debilitating and risky enterprises of sexual reproduction and parental investment.

Parental motivation and discriminative solicitude

Parenting is a prominent component of the behavioral repertoires of many animals, including people. Throughout human prehistory, most women (and perhaps even men) devoted the majority of their waking hours to foraging for, educating, guarding, and otherwise nurturing their young. Yet psychologists have had remarkably little to say about the sources of variation in parental efforts and inclinations. When we prepared a paper on the topic for the 1987 Nebraska Symposium on Motivation (Daly and Wilson, 1988c), for example, we noted that the 34 previous volumes in the series contained not a single paragraph specifically concerned with parental motivation. It seems that psychologists have shied away from this important domain of motivation for want of a theoretical framework from which to approach it. The requisite framework is of course the Darwinian view of behavioral control systems as having been organized by a history of selection to promote fitness.

Parental behavior has obvious, direct links to fitness: Offspring are the vehicles of parental fitness. However, not all offspring are equally capable of translating parental nurture into increments in the long-term survival of parental genetic materials. It follows that selection favors discriminative mechanisms of parental psychology: mechanisms that allocate "parental investment" (Trivers, 1972) with sensitivity to available cues of the statistically expected consequences for parental fitness. For example, selection will favor preferential investment in one's own young as opposed to the young of others, in viable young as opposed to lost causes, and in needy young as opposed to those for whom the investments would be superfluous (Wilson and Daly, 1993).

Parental investments take various forms including direct transfers of heat and nutrients, foraging and food delivery, and protection (Clutton-Brock, 1991). Different forms of parental investment may be temporally disjunct and hence causally distinct, as when lactation succeeds gestation, but the functional commonality among diverse investments provides a rationale for expecting that there will often be some commonality of causation as well. Any offspring whose characteristics make it a good bet to yield fitness for one sort of parental investment will usually be a good bet for other sorts, too. Doubt that a particular youngster is indeed the parent's own, for example, reduces the expected parental fitness payoffs of both feeding the youngster and defending it against predators. Divestment from lost causes should similarly apply in parallel to all manner of parental investments. Thus we may expect parental motivational systems to contain processes and structures that function as if mediated by a unitary parameter of offspring-specific parental love or solicitude, which is influenced by a variety of parental, offspring, and situational cues of fitness value (i.e., of the offspring-specific expected contribution to parental fitness), and which influences in its turn a variety of parental activities.

Even offspring-specific parental investments cannot be adaptively dispensed solely on the basis of cues of offspring fitness value, however. The smaller or younger of two siblings might profit more from a food delivery, for example, and hence be the preferred recipient for such a parental investment, even though the larger or older sibling has the higher fitness value and would be the preferred recipient of parental defense in an antipredator context. Parental solicitude may be expected to be complexly contingent upon variable attributes of the parent, the young, and the situation, because both expected fitness and the expected impact of a given parental investment on expected fitness are contingent upon these variable attributes. In particular, parental solicitude can be predicted to vary adaptively in relation to (1) phenotypic and situational cues affording information about the certainty of parenthood (whether the offspring is indeed the parent's own); (2) phenotypic and situational cues affording information about the offspring's reproductive value (expected future fitness); and (3) the fitness value of the available alternatives to present parental investment.

Certainty of Parenthood An offspring's expected contribution to parental fitness is the product of its reproductive value and its relatedness (r) to the putative parent. In the case of outbred sexually produced offspring, $r = 0.5$, but this value is in effect probabilistically reduced in the case of uncertain parentage.

Indiscriminate allocation of parental benefits without regard to cues of actual parentage would be an evolutionary anomaly. Consider a famous allegation of just such indiscriminacy. Mexican free-tailed bats roost in dark caves in aggregations that can number in the millions. Within hours of giving birth to her single pup, the female leaves it hanging in a crèche while she goes foraging. The crèche may contain several thousand infants per square meter, and they sometimes crawl a meter or more between nursing bouts. Noting these facts, and having demonstrated both that pups would attach to any female held near them and that the females would not then remove the pups, Davis, Herreid, and Short (1962) concluded that female Mexican free-tailed bats act as an anonymous "dairy herd."

This conclusion demands our skepticism, because it is not plausible that lactation could be evolutionarily stable (Maynard Smith, 1976) in such a case. The nursing bat incurs both energetic depletion and predation risk in order to deliver 16% of her body weight in milk each day. If milk were truly a communal resource, selection would surely favor the female who deposits her pup in the care of the dairy herd, dries up, and opts out. Doubting the dairy herd theory for this reason, McCracken (1984) genotyped mothers and infants in the field, and found that while some mismatches indeed occurred, 83% of mothers were actually feeding their own pups. A 17% incidence of nursing nonrelatives represents a substantial failing of discriminative parental investment, but not such an egregious failing as to select against lactation (Beecher, 1991).

The adaptationist expectation of offspring-specific parental solicitude stands in opposition to a prevalent conception of mammalian maternal motivation.

Rosenblatt (1990), Pryce (1992), and many others treat maternalness as a state that is nonspecific with respect to its object. This conception is certainly not a generally applicable one, and its popularity appears to be attributable to happenstance: Maternal solicitude is indeed remarkably indiscriminate in the laboratory rat, and this species has dominated research for no other reason than its convenience. However, the rat's relative imperviousness to the individuality of young turns out to be a peculiarity of a minority of mammals, with a particular ecology.

Mexican free-tailed bats search out, recognize, and selectively nurture their pups in the free-for-all of the crèche (McCracken and Gustin, 1992). Seals that deliver and nurse their pups in close proximity attack unrelated pups who try to suckle, even as they nurse their own (e.g., Trillmich, 1981). Hoofed mammals who raise precocious young in mobile herds do likewise (e.g., Poindron and Le Neindre, 1980). Rats are different: They seem oblivious to the own-versus-alien distinction, and blithely give suck to whatever pups they find in their nests, including even those of other species. Why? Rat pups, unlike bats, seals, and goats, are immobile and sequestered in defended burrows, with the result that mixing of young does not occur in the absence of experimental intervention. The rat mother in nature dispenses nurture selectively to her own young, using her nest site as the cue by which she recognizes them. Moreover, when the growing pups of burrow-dwelling rodents become mobile so that mixing of youngsters is an imminent possibility, mothers then come to recognize their pups as individuals and will no longer accept fosterings (e.g., Holmes, 1984).

The risk that unrelated young will elicit misdirected parental investment varies even among closely related species. Bank and cliff swallows, for example, nest colonially, whereas the closely related rough-winged and barn swallows nest more dispersedly. Newly flying young sometimes return to the wrong nest in the two colonial species, but seldom or never in the dispersed species; moreover, fledged young who are still being fed by their parents aggregate in crèches in the colonial species but not in the dispersed. Thus, the demand for parental discrimination of own versus alien is clearly greater in the colonial species, and Beecher (1990) reports that they indeed recognize their own offspring by voice whereas the dispersed species do not.

Beecher and colleagues have furthermore predicted and demonstrated that selection for offspring recognition in colonial swallows has affected the attributes of both the chicks and their parents. The calls of colonial species chicks are intrinsically more discriminable than those of dispersed species chicks, as shown both by informational analyses of the physical properties of the calls (Beecher, 1988) and by superior discrimination of the calls of colonial species chicks by adults of either nesting type, as well as by other animal species (Loesche et al., 1991; Loesche, Beecher, and Stoddard, 1992). Adaptation on the parental side is indicated by the fact that adult cliff swallows (colonial) outperform adult barn swallows (dispersed) on these tasks, even when the calls to be distinguished are those of barn swallows (Loesche et al., 1991). Selection will not always favor such complementarity of parental and offspring adaptations, however, and Beecher (1988, 1991) has further discussed the circumstances, such as orphaning, in which there may be simultaneous selection in favor of parental discrimination capability but against the evolution of distinctive "signatures" in the young.

As noted above, the physiology of mammalian mothering has been studied primarily in rats and other burrow-dwelling rodents, who, like the dispersed-nesting swallows, have not experienced a history of selection for rapid discriminative attachment to their own young. There is one intensively studied mammalian species, however, that has experienced precisely such selection: the sheep. An individualized bond between ewe and lamb is typically established within three hours of parturition (Poindron and Le Neindre, 1980). Many elements of the complex neuroanatomy and chemistry of maternal motivation are similar in sheep and rats, and perhaps even across the class Mammalia (e.g., Kendrick et al., 1992). Other things vary among species, such as the specific central effects of oxytocin (Lévy et al., 1992), which plays a taxonomically broad role in parturition but has more species-specific effects on sexual receptivity (hardly surprising when one considers that rats, for example, become sexually receptive after giving birth whereas sheep do not).

Most interesting in the present context are the ewe's adaptations to the specific problem of individualized maternal responsiveness. Kendrick, Lévy, and Keverne (1992) found that single neurons in the mitral cell layer of the olfactory bulb never responded preferentially to lamb odors in late pregnancy, but that more than half had switched to doing so soon after birth,

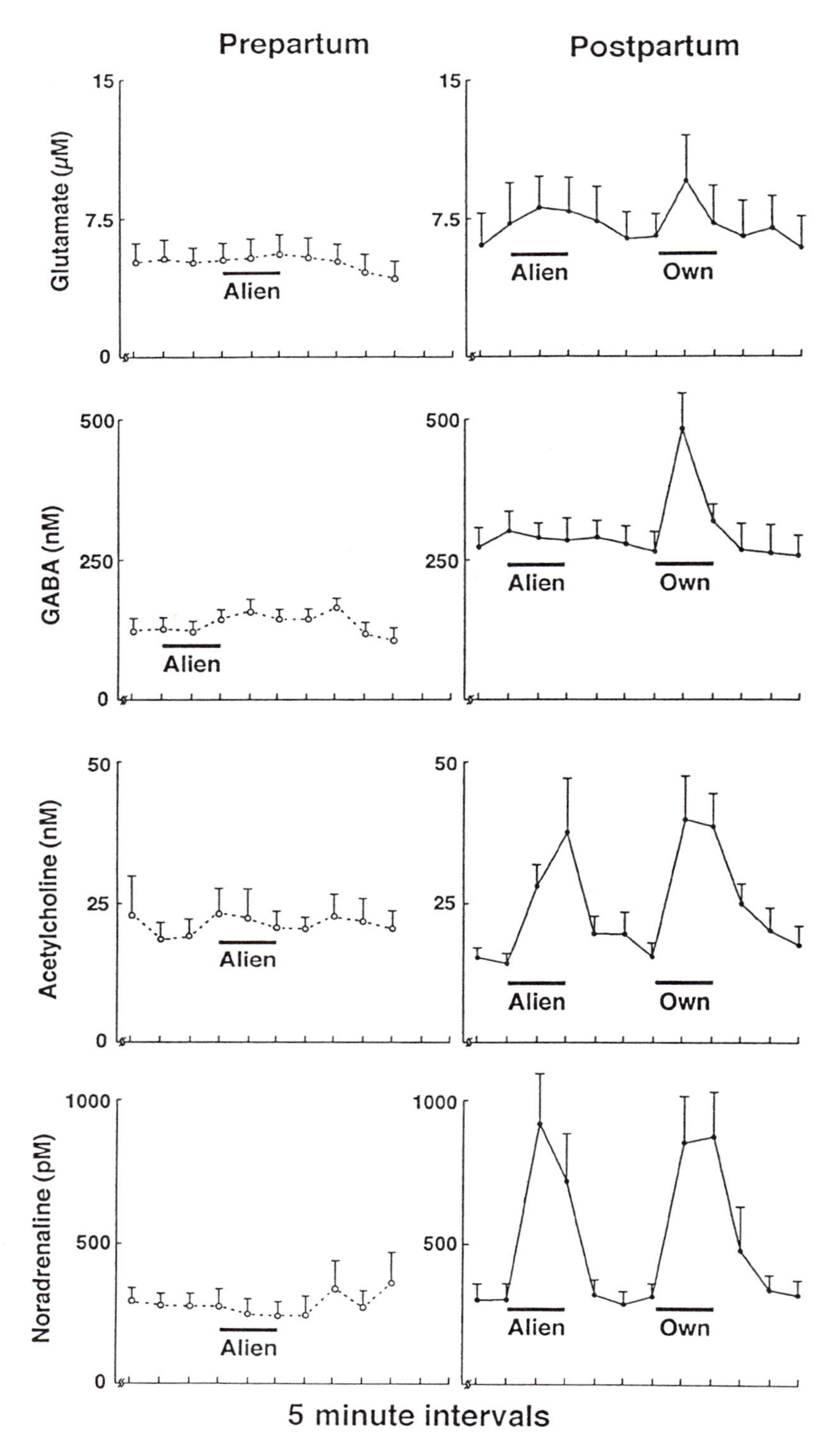

FIGURE 84.2 Discriminative response to own offspring on the day of birth. Concentrations of glutamate, GABA, acetylcholine, and norepinephrine (mean ± SEM) in microdialysis samples taken at 5 min intervals from the olfactory bulbs of nine sheep during 10 min exposures to lamb odors (black bars), within 24 hrs prepartum and postpartum. Although postpartum data are portrayed in the order alien lamb, own lamb, order of presentation was in fact randomized. (From Kendrick, Lévy, and Keverne, 1992, figure 2.)

mainly at the expense of responsiveness to food; 70% of these cells responded indiscriminately to the odors of any lamb, but the remaining 30% responded selectively to the particular lamb with which the ewe had formed a selective bond at birth. Other experiments measuring concentrations of neurotransmitters in specific cells of the olfactory bulbs revealed substantial changes in the release of olfactory bulb neurotransmitters immediately after birth (figure 84.2). Within hours of birth, lamb odors elicit major acetylcholine (ACh) and norepinephrine responses (figure 84.2C and 84.2D) in the centrifugal projections relevant to the storage of olfactory information, where the same odors had elicited no response 24 hours earlier. Moreover, although these ACh and norepinephrine responses are indiscriminate with respect to lamb identity, only the ewe's own lamb elicits release of the excitatory amino acid glutamate from mitral cells (figure 84.2A) and the release of the inhibitory amino acid gamma-aminobutyric acid (GABA; figure 84.2B) from olfactory bulb interneurons that affect the activity of mitral cells. Kendrick, Lévy, and Keverne (1992) conclude that these electrophysiological and neurochemical changes in the ewe's olfactory bulbs after birth represent an adaptive specialization for the task of individualized bonding. Presumably no such discriminative responses occur in the rat brain, and no one would have thought to look for them in sheep, either, without an adaptationist appreciation of the demand for rapid selective maternal bonding in precocious, gregarious animals.

Paternal solicitude, where it exists, is another matter. In animals with internal fertilization, males typically incur greater uncertainty about parenthood than females, and the attendant risk of misdirected paternal care is presumably at least part of the reason why parental care is female dominated where fertilization is internal, in contrast to externally fertilizing fish and amphibians (Ridley, 1978). When females mate with two or more males in a single fertile period, the ensuing "sperm competition" (Parker, 1970) is a potent selective force, affecting attributes of both parties (Smith, 1984; Birkhead and Møller, 1992; Baker and Bellis, 1993). One consequence is the evolution of "mate guarding" by males (Birkhead and Møller, 1992; Wilson and Daly, 1992). In species in which males make postzygotic investments in their putative offspring, the stakes rise: Losing a fertilization to a rival entails continuing penalties if the cuckolded male persists in playing father, so the selective premium on mate guarding rises.

If mate-guarding fails, a male can contain the damage by adjusting his paternal efforts in relation to probabilistic cues of paternity. At least two classes of cues are available: those indicative of his share of mating access to the female when she conceived, and similarities or differences in the attributes of offspring and their possible sires. In some bird species with biparental care, a significant amount of the variance among males in their rates of feeding hatchlings is attributable to variations either in the actual rate of extrapair copulation by their mates or in the time that females spent outside the males' surveillance when fertile weeks earlier (figure 84.3). This is not always the case, however, and Whittingham, Taylor, and Robertson (1992) and Westneat and Sherman (1993) consider possible reasons for species differences, such as when seasonally constrained breeding makes feeding his nestlings a

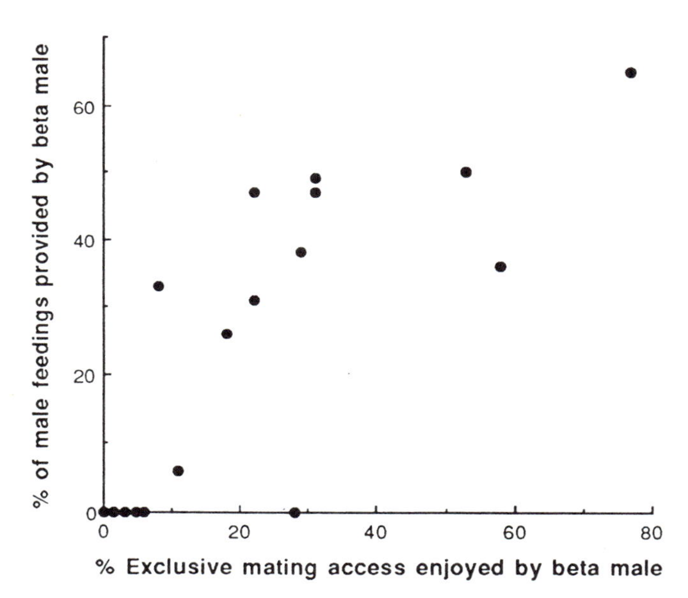

FIGURE 84.3 Modulated feeding effort in response to paternity probability cues. In polyandrous trios of dunnocks (*Prunella modularis*), the female mates both with the alpha (dominant) male and with the beta (subordinate) male when she can escape the alpha's attention. Weeks later, when eggs hatch and nestlings are being fed, the proportion of paternal feedings provided by the beta male closely tracks his proportionate share of mating access. Each point represents one of 17 such trios, observed during both the mating and provisioning phases of the nesting cycle. (From Davies et al., 1992, figure 1c.)

male's best or only option across a wide range of values of paternity probability.

Surprisingly, there is no evidence that nonhuman males ever use offspring phenotypes as paternity cues. Human males certainly do, however (Daly and Wilson, 1988c). One corollary is intense interest in the resemblances of newborn babies to putative paternal kin, and relative noninterest in their resemblances to maternal kin (Daly and Wilson, 1982; Regalski and Gaulin, 1993). A possible adaptation in human mothers is strong motivation to discover paternal features in their babies (Daly and Wilson, 1982; Robson and Kumar, 1980); pregnant women's fantasies suggest that this motive begins to be felt before birth (Leifer, 1977). What has yet to be investigated is to what degree, if at all, human paternal affection and investment are affected by resemblance. There is much anecdotal evidence of complete divestment in response to phenotypic evidence of nonpaternity (Daly and Wilson, 1988c), but no systematic study has been made of the phenomenon. Of more general interest, perhaps, is the likelihood that even in the absence of conscious paternity doubt, the psychological mechanisms affecting parental affection may be sexually differentiated, with resemblance to self relevant to fathers but not mothers (Daly and Wilson, 1981).

A rather obvious implication of the notion that selection favors parental discriminativeness is that parental inclinations may fail when adults find themselves in loco parentis to unrelated young (Rohwer, 1986). Remarkably, this possibility seems not to have occurred to social and medical scientists seeking the sources of variable child abuse risk, until Wilson, Daly, and Weghorst (1980) showed that stepchildren are vastly overrepresented as victims. Differential risk is especially large when the criterion of abuse is unequivocal and extreme, that is, in cases of child homicide (figure 84.4). Excess risk to stepchildren cannot be attributed to poverty, coresidency from birth, maternal age, brood size, incidental traits of persons who remarry, or any other suggested confounding (Daly and Wilson, 1985; Wilson and Daly, 1987; Flinn, 1988). Much converging evidence indicates that violence against stepchildren is simply one extreme reflection of a large difference in the (undoubtedly overlapping) distributions of genetic parental and stepparental affection (Wilson and Daly, 1987; Daly and Wilson, 1991, 1993). Although the evidence is scanty, adoptive parenthood appears to be much less problematic than stepparenthood. One possible interpretation is that the contemporary Western practice of adoption by nonrelatives is an evolutionary novelty against which the evolved parental psyche has

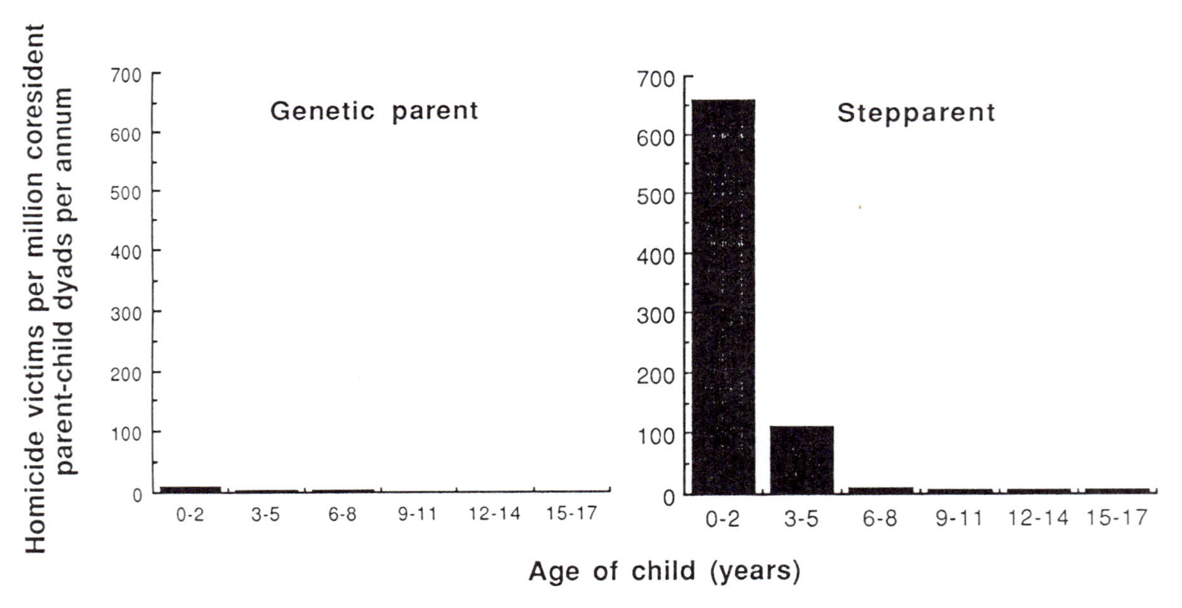

Figure 84.4 Rates of filicide, according to age of victim, by genetic parents ($N = 341$ victims) versus stepparents ($N = 67$ victims) in Canada. Data represent all filicide cases known to Canadian police forces between 1974 and 1983. (From Daly and Wilson, 1988a, figure 1.)

no specific defenses, so that the psychology of genetic parenthood is more readily engaged than in the case of stepparenthood, which has been a recurring adaptive problem because of parental mortality and/or marital dissolution for as long as hominids have formed biparental unions.

In some animals (though clearly not in people), males routinely dispose of their predecessors' young when they can gain earlier use of their new mates' maternal investments for the benefit of their own young by so doing (Hausfater and Hrdy, 1984). This antithesis of parental solicitude must also be adaptively discriminative with respect to probability of paternity, and a diversity of mechanisms apparently exists. As with modulations of positive investments, there is no evidence that males of any nonhuman species use offspring phenotype as a paternity cue in infanticidal decisions. In some species, such as gerbils, males selectively spare young associating with or carrying odor cues of females with whom the male has mated (Elwood and Kennedy, 1993). Others, such as mice (vom Saal, 1985) rely less on the individual identities of past mates than on a remarkable timing heuristic. Intravaginal ejaculation is the necessary and sufficient condition for male mice tested with stimulus pups to switch from infanticidal to parental responses. However, the switch does not typically occur for many days after mating, until shortly before the male's own pups would be born (vom Saal, 1985). In effect, the male clocks the female's pregnancy, without cues from her and by a distinct neural timing mechanism; the duration of pregnancy remains constant in real time when day length is artificially modified, but the male's "pregnancy clock" counts light-dark cycles instead (Perrigo, Bryant, and vom Saal, 1990). As in the case of the specificity of sheep mitral cell responses, this sophisticated physiological mechanism was discovered as a direct result of adaptationist theorizing.

Reproductive Value of the Young An act of parental investment like food delivery cannot be considered a straightforward index of discriminative parental solicitude. The offspring receiving more parental investment in a delimited observation period is not necessarily the offspring that the parent values more. An older offspring, for example, may be better able to feed itself and hence less in need of parental feeding than a younger sibling, though the older has higher reproductive value (age-, sex- and phenotype-specific expected future fitness), and hence higher fitness value to the parent. Where we might expect the parent to reveal its greater valuation of the older offspring is when confronted with the choice of saving only one. This choice is hard to operationalize in nature, but a related paradigm has become popular for measuring variations in parental valuation of offspring. The method is that described earlier with reference to the male stickleback's greater valuation of a larger brood: How much risk to self is the parent prepared to accept to defend its helpless young from a predator dangerous to both?

An adaptive parental psyche may be expected to tolerate more risk to self in defense of young of greater reproductive value. One determinant of reproductive value is brood size. Another is offspring age, since the reproductive value of immature animals increases over time at least until maturity, by simple virtue of surviving successive periods of potential prereproductive mortality. Studies of nestbound young who remain helpless unless defended are especially germane here, since they avoid confoundings of offspring reproductive value with changes in self-defensive or escape capabilities; many such studies have been conducted with birds and fishes, and the general result, albeit with many complications, is that parental defense indeed increases as offspring reproductive value increases (Montgomerie and Weatherhead, 1988; Redondo, 1989).

Maternal aggression in laboratory rodents is presumably an analogous manifestation of parental readiness to defend the young, and Maestripieri and Alleva (1991) have shown that it varies with brood size as predicted from reproductive value theory. Oddly, however, maternal aggression does not appear to increase as helpless young age, and the adaptive significance of the time course of maternal aggression in these species remains obscure. One possibility is that the changes are adapted to changing threats from infanticidal conspecifics of both sexes rather than to predators (Daly, 1990).

In the human case, parentally perpetrated infanticide can be treated as a reverse assay of parental solicitude for which it has some of the same advantages as parental defense. Any factor that may be expected to influence parental investment allocations should also be relevant to the likelihood of lethal divestment, regardless of whether infanticide is a fitness-promoting adaptation or an incidental and maladaptive epiphenomenon of parental unconcern with the offspring's welfare. According to a systematic review of rationales for infanticide in nonstate societies where it is not criminalized, human infanticide is primarily a response to cues of low infant reproductive value, namely bad circumstances such as famine at the time of the birth or defects in the child itself (Daly and Wilson, 1984); the principal rationale for infanticide outside this category is dubious or inappropriate paternity.

Human infanticide also provides an instance of the hypothetical test of parental solicitude suggested above: Which do you save when one must be sacrificed? The answer is that mothers confronted with this dilemma save the older child, whose reproductive value is usually greater. A common rationale for infanticide is maternal overburdening when the birth interval is too short; nowhere do people solve this problem by disposing of the toddler (Daly and Wilson, 1984).

In the absence of mishap, a child's reproductive value increases steadily from birth until at least puberty. With modern medicine, the early increase is muted by declines in infant and juvenile mortality, but where mortality and fertility are closer to the levels that must have prevailed for most of human history, the prepubertal increase in reproductive value is not trivial. We would thus expect parental feelings to have evolved such that parents will seem to value offspring increasingly with age, and we might therefore expect to see an age-related decrease in the likelihood of lapses of parental solicitude. Increased parental solicitude with offspring age may be difficult to detect because the offspring's dependence is waning at the same time, but parental valuation of the young can again be assayed by the parent's declining willingness to tolerate or expose the young to lethal risk. One apparent manifestation of such an age-related change is a monotonic decrease in the risk of filicide (Daly and Wilson, 1988a, 1988b), which continues to near zero as the offspring approaches maximal reproductive value in young adulthood. Not merely the direction of change, but the specific shape of the age-related filicide curve tracks ancestral reproductive value schedules remarkably well (Daly and Wilson, 1988b). It is especially striking that children become increasingly immune from parental lethal action as they mature, since this maturation entails increasing competitiveness in their interactions with nonrelatives, and an increasing overall risk of becoming involved in lethal interpersonal conflict, both as killer and as victim (Wilson and Daly, 1985; Daly and Wilson, 1990).

Alternatives to Present Parental Investment

The final class of determinants of variable parental response to be considered is alternative uses of parental resources. Polygynous red-winged blackbirds that have sired young in different nests commonly help provision only one brood, at least for a period of days, and they correctly prefer the brood where their efforts best promote their fitness (Yasukawa, Leanza, and King, 1993). Broods that were neglected when there was a more profitable option become effective elicitors of paternal investment when that more profitable option disappears.

Less obvious than this sort of parental allocation problem is the trade-off between parental investment and other fitness-promoting activities, such as the pursuit of additional matings, or the use of available physiological resources for growth and tissue repair with expected fitness benefits in the future. As parents age and senesce, their own residual reproductive value declines, and future alternatives may deserve less weight in present parental decision making. In the parental defense paradigm discussed above, there is some evidence that parents accept greater risk as their own reproductive value declines (Montgomerie and Weatherhead, 1988; Thornhill, 1989), although effects are less clear than those of offspring reproductive value. Efforts to show effects of parental reproductive value on parental investment decisions have been somewhat bedeviled by the fact that aging animals change in other possibly relevant ways as their reproductive value declines. Greater parental effort with age could be confused, for example, with effects of experience that make the parent more effective without really incurring greater risk to self or otherwise investing more; however, parental experience effects do not seem to explain away increases in parental effort with age in jungle fowl (Thornhill, 1989) or California gulls (Pugesek, 1983, 1987).

Human females have an unusually discrete end to the potentially reproductive life span, and so we might expect a woman to exhibit a decrease, with age, in any tendency to devalue a present offspring in terms of its compromising effects on her future. It follows that the risk of maternally perpetrated infanticide might decline as a function of maternal age, and so it does

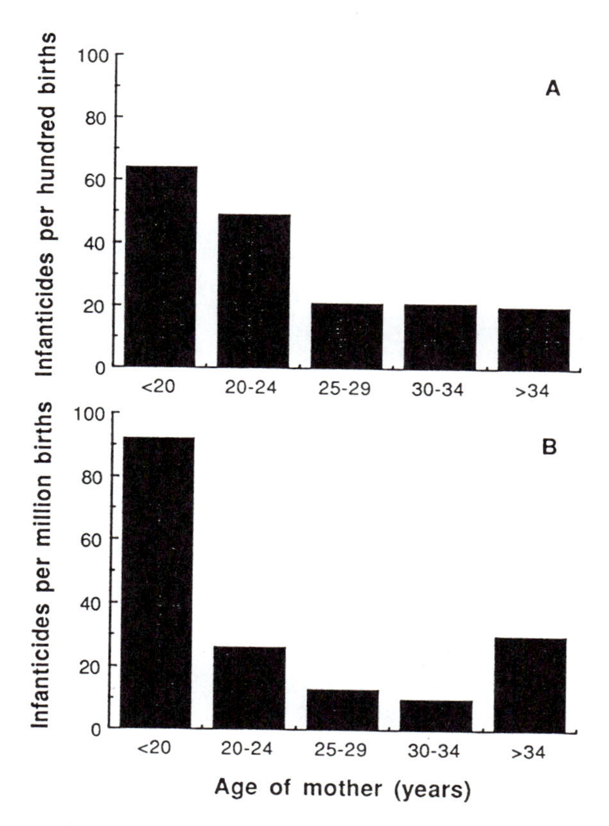

FIGURE 84.5 Rates of infanticide by mothers as a function of maternal age, among (A) Ayoreo Indians of South America ($N = 54$) and (B) Canadians ($N = 87$). (From Daly and Wilson, 1988a, figure 3, with data in (A) from Bugos and McCarthy, 1984.)

(figure 84.5). We would expect that many of the variables that are relevant to changes in maternal solicitude (or lapses therein, as in the case of maternal filicides) should be similarly relevant for fathers. However, women's reproductive life spans end before those of men, so the utility of alternative reproductive efforts declines more steeply as a function of age for women than for men. Moreover, dependent children impose different opportunity costs on mothers and on fathers: A nursing infant constrains mother's immediate alternative reproductive prospects much more than father's, and the magnitude of this differential impact upon mother versus father declines with time since birth. These considerations suggest that a mother's valuation of a child relative to her valuation of herself is likely to rise more steeply with time since the child's birth than is the corresponding quantity for the father. If filicides constitute a sort of reverse assay of parental solicitude, it follows that filicide rate should decline more steeply for mothers than for fathers, and it does (Daly and Wilson, 1988b).

A three-stage theory of maternal bonding

The idea that variations in parental solicitude are the outputs of motivational systems designed by selection to modulate parental efforts has implications for the controversial topic of maternal bonding. The theoretical considerations reviewed above suggest that the development of child-specific parental love is likely to involve at least three separable processes with different time courses: (1) an initial assessment of the newborn's fitness prospects on the basis of both its quality and the situation; (2) discriminative attachment to the baby as an individual; and (3) a much more prolonged and gradual deepening of individualized love.

INITIAL CUES OF NEWBORN'S FITNESS PROSPECTS The first process to be expected is an assessment, immediately after birth, of the child and of how its qualities and present circumstances combine to predict its prospects. Conspicuous signs of low viability increase maternal inclination to divest; in the modern West, births of children with major defects commonly evoke immediate shocked rejection by the parents (e.g., Drotar et al., 1975), a rejection that would have led to quick abandonment in ancestral settings (Dickeman, 1975). Where genetic counseling and termination of pregnancy are available, parents commonly want to abort seriously defective fetuses (e.g., Leschot, Verjaal, and Treffers, 1985). Beliefs that deformed infants are ghosts or demons (or the progeny thereof) are found sporadically throughout the world and are invoked to justify infanticide. Analogous superstitious allegations about well-formed, healthy babies are essentially nonexistent, implying that so-called superstition cannot be dismissed as ignorant foolishness, but functions instead as an ideological buttress of difficult but functional choices (Daly and Wilson, 1988a).

Conspicuously defective newborns are rare, of course, but maternal responsiveness in the immediate postpartum period also varies with subtle cues of the infant's quality and health (e.g., Mann, 1992). Small, premature babies incur increased risk of abandonment or abuse (e.g., Hunter et al., 1978) and when such babies are likely to die, parents may distance themselves emotionally and fail to participate in the infant's hospital care (Newman, 1980). Emotional distancing has also been described among impoverished Brazilian mothers of weak, sickly infants who are expected to die (Scheper-Hughes, 1985). It should be noted that low birth weight in North America is associated with low socioeconomic status, maternal youth, large family size, and close birth spacing (e.g., Zuckerman et al., 1984), but size and vigor of a newborn may be salient cues modulating the development of parental solicitude. In an observational study of low-birth-weight twins, the healthier twin was more effective in eliciting maternal responsiveness at 8 months of age, and factors such as duration of postpartum separation or the incidence of positive interactions with the mother did not account for the mother's differential treatment of the twins (Mann, 1992).

Within the first few hours after birth, healthy human infants exhibit a precocious social responsiveness—eye contact and selective attention to maternal speech—which may be an adaptation for advertising quality and eliciting maternal commitment during this assessment phase. If circumstances are dicey and the mother is in any way ambivalent, poor responsiveness might tip the scales toward disinclination to raise the child.

There has been considerable controversy about the effects of mother-infant contact in the immediate postpartum period on the quality of the mother-child bond (e.g., Klaus and Kennell, 1976; Herbert, Sluckin, and Sluckin, 1982). Effects of circumstantial variables other than the contact itself have been taken as evidence against a specialized bonding adaptation, but from a selectionist perspective, it would not be surprising to discover that extra postpartum contact has little ameliorative effect on mothers when circumstances such as poverty, lack of paternal support, and other indices of maternal overburdening cue poor fitness prospects. More work is needed to assess whether situational and other variables (such as the mother's residual reproductive value as measured by her age) interact with and modify the effects of particular postpartum experiences. Moreover, modern medical techniques of fetal assessment can provide mothers with information relevant to the hypothesized assessment phase of bonding even before birth, with effects on the time course of "maternal-fetal bonding" as assessed by questionnaire (e.g., Caccia et al., 1991); it is an open question whether prenatally and postnatally received information on offspring quality have essentially similar impact or are instead processed differently as a result of hormonal or other events surrounding parturition.

Many new mothers experience a brief period of the "blues" within the first few postpartum days (e.g., Cutrona, 1982; Hopkins, Marcus, and Campbell, 1984). A lesser but considerable number experience a more debilitating postpartum depression. Such depression is apparently especially likely when the mother is young, single, at odds with the father, or otherwise lacking social support (e.g., Cox, Connor, and Kendell, 1982; Cutrona, 1982; Hopkins et al., 1984), and when the infant is suffering from poor health (e.g., Blumberg, 1980; Grossman, Eichler, and Winickoff, 1980). These circumstances are very similar to the infanticide circumstances described in the ethnographic literature (Daly and Wilson, 1984). Women suffering from extreme postpartum depression are sometimes characterized by clinicians as delusional, but the typical content of the "delusions" seems not at all fantastic: concern about inability to care for the baby, fear of not having enough love for the baby, and guilt aroused by infanticidal thoughts (e.g., Herzog and Detre, 1976).

Discriminative Bonding to Own Offspring Parents are highly sensitive to their babies' distinctive features, recognizing them by voice (e.g., Formby, 1967) and by smell (Porter, 1991) with only minimal exposure. Some have implied that these abilities represent psychological adaptations for discriminative bonding, but, of course, people are very good at recognizing individual faces generally (see *J. Cognitive Neurosci.* 3[1], 1991), and persons other than mothers may be just as good at discriminating babies by smell (Porter et al., 1986). Whether there is a specific heightened postpartum infant recognition ability is still to be determined.

Rather than having merely to recognize her own baby, the "task" confronting the new mother is to develop an individualized commitment to it, such that she is emotionally prepared to invest heavily in its welfare without being at the same time vulnerable to parasitism by children generally. Many new mothers report an initial feeling of indifference to their babies (perhaps reflecting the initial assessment phase as well as the lack of individuation), but very few feel the same way by one week postpartum (e.g., Robson and Kumar, 1980). After having had close contact with their infants over the first few days, mothers commonly report developing a feeling that their baby is uniquely wonderful (e.g., Klaus and Kennell, 1976).

A Gradual Deepening of Parental Love The third predictable process of parental attachment is a much more gradual one: The strength of parental love may be expected to grow with the child's increasing reproductive value, especially over the first few years when there is the steepest increase in that value. Fleming et al. (1990) analyzed the content of women's utterances at intervals over 16 postpartum months, and found that the mothers talked more and more positively about their infants over time; the effect was not merely due to changes in maternal condition or situation, as the same measures with respect to self and to husband did not exhibit similar trends. Postpartum growth in the salience and importance of the infant was also reflected by "an increasingly large proportion of women reporting such things as feelings of closeness to their infants, being pleased with their infants' development, or enjoying child-care activities" (p. 141).

The information that parents garner from their continued monitoring of offspring quality should affect the depth and time course of their love and commitment, especially while infant mortality risk remains high. Since parental effort is a resource to be invested, not squandered, chronic changes in the infants' responsiveness and robustness, consequent upon the effects of malnutrition, dehydration, and pathogens, can be expected to dampen parental love, in spite of the infant's greater need. In many societies, newborn babies are not immediately named or officially acknowledged by the community, a practice more or less explicitly linked to their uncertain future. Naming bestows personhood and facilitates the individuation of affection. (Indirect evidence for this claim can be found in observations that naming children after relatives is effective in inspiring namesake investment and inheritance; Smith, 1977; Furstenberg and Talvitie, 1980.) The postpartum delay in recognizing infants' personhood corresponds to a period of high mortality risk, perhaps with the effect of facilitating difficult decisions of divestment and lessening the emotional pain should the infant die (Mull and Mull, 1988; Scheper-Hughes, 1985). It is something of a cliché to claim that the valuation of children is a recent Western cultural invention, introducing tales of child brutalization and parental indifference in history and in other societies as support; those making this argument fail to appreciate that seeming callousness is an understandable response to circumstances that make children poor prospects for survival and reproduction, and that the same mothers who seem indifferent to the plight of one child in one context can be profoundly nurturant to others born in more auspicious circumstances (e.g., Bugos and McCarthy, 1984).

Concluding remarks

Evolutionary theory is not a substitute for proximate causal analysis, but a valuable aid thereto: Understanding how selection operates and what behavioral control mechanisms have been designed to achieve affords innumerable hints to their probable organization. Improved maternal efficacy over the life span, to take one example, has routinely been assumed to reflect the acquisition of skills and/or knowledge, and the "immaturity" of young mothers. However, selectional thinking suggests that such changes over the reproductive life span may often reflect adaptive changes in maternal inclinations as maternal reproductive value and opportunities change. This alternative view suggests many possible lines of research about the contingent modulation of parental solicitude.

Selectional thinking provides reason to believe that the individualistic focus of cognitive neuroscientists will illuminate family relations better than more sociological approaches. The popular focus on families as "systems" and their members as components thereof cannot be correct, insofar as it elevates the so-called system's objectives above those of its actors and ignores the fact that family members are agents with only partially congruent interests. A quarter century of criticism of greater-goodism in biology has clarified why individual organisms are the appropriate level in the hierarchy of life at which to impute integrated agendas, and why the analogizing of larger groups to self-interested individuals typically fails.

REFERENCES

Baker, R. R., and M. A. Bellis, 1993. Human sperm competition: Ejaculate manipulation by females and a function for the female orgasm. *Anim. Behav.* 46:887–909.

Barkow, J. H., L. Cosmides, and J. Tooby, eds., 1992. *The Adapted mind: Evolutionary Psychology and the Generation of Culture.* New York: Oxford University Press.

Beecher, M. D., 1988. Kin recognition in birds. *Behav. Genet.* 18:465–482.

Beecher, M. D., 1990. The evolution of parent-offspring recognition in swallows. In *Contemporary Issues in Compara-*

tive Psychology, D. A. Dewsbury, ed. Sunderland, Mass.: Sinauer, pp. 360–380.

Beecher, M. D., 1991. Successes and failures of parent-offspring recognition in animals. In *Kin Recognition*, P. G. Hepper, ed. Cambridge: Cambridge University Press, pp. 94–124.

Bernard, C., 1865. *Introduction à l'étude de la médecine expérimentale*. Paris: Baillière. (Quoted from the translation by H. C. Greene. New York: Dover, 1957.)

Birkhead, T. R., and A. P. Møller, 1992. *Sperm Competition in Birds*. London: Academic Press.

Blumberg, N. J., 1980. Effects of neonatal risk, maternal attitude, and cognitive style on early postpartum adjustment. *J. Abnorm. Psychol.* 89:139–150.

Bugos, P. E., and L. M. McCarthy, 1984. Ayoreo infanticide: A case study. In *Infanticide: Comparative and Evolutionary Perspectives*, G. Hausfater and S. B. Hrdy, eds. New York: Aldine, pp. 503–520.

Caccia, N., J. M. Johnson, G. E. Robinson, and T. Barna, 1991. Impact of prenatal testing on maternal-fetal bonding: Chorionic villus sampling versus amniocentesis. *Am. J. Obstet. Gynecol.* 165:1122–1125.

Charnov, E. L., 1976. Optimal foraging: The marginal value theorem. *Theor. Popul. Biol.* 9:129–136.

Clutton-Brock, T. H., 1991. *The Evolution of Parental Care*. Princeton: Princeton University Press.

Cosmides, L., and J. Tooby, 1987. From evolution to behavior: Evolutionary psychology as the missing link. In *The Latest on the Best: Essays on Evolution and Optimality*, J. Dupré, ed. Cambridge, Mass.: MIT Press, pp. 277–306.

Cox, J. L., Y. Connor, and R. E. Kendell, 1982. Prospective study of the psychiatric disorders of childbirth. *Br. J. Psychiatry* 140:111–117.

Cronin, H., 1992. *The Ant and the Peacock*, Cambridge: Cambridge University Press.

Cutrona, C. E., 1982. Nonpsychotic postpartum depression: A review of recent research. *Clin. Psychol. Rev.* 2:487–503.

Daly, M., 1990. Evolutionary theory and parental motives. In *Mammalian Parenting Biochemical, Neurobiological and Behavioral Determinants*, N. A. Krasnegor and R. Bridges, eds. New York: Oxford University Press, pp. 25–39.

Daly, M., and M. I. Wilson, 1981. Child maltreatment from a sociobiological perspective. *New Dir. Child Dev.* 11: 93–112.

Daly, M., and M. I. Wilson, 1982. Whom are newborn babies said to resemble? *Ethol. Sociobiol.* 3:69–78.

Daly, M., and M. Wilson, 1984. A sociobiological analysis of human infanticide. In *Infanticide: Comparative and Evolutionary Perspectives*, G. Hausfater and S. B. Hrdy, eds. New York: Aldine, pp. 487–502.

Daly, M., and M. Wilson, 1985. Child abuse and other risks of not living with both parents. *Ethol. Sociobiol.* 6:197–210.

Daly, M., and M. Wilson, 1988a. Evolutionary social psychology and family homicide. *Science* 242:519–524.

Daly, M., and M. Wilson, 1988b. *Homicide*. New York: Aldine.

Daly, M., and M. Wilson, 1988c. The Darwinian psychology of discriminative parental solicitude. *Nebr. Symp. Motiv.* 35:91–144.

Daly, M., and M. Wilson, 1990. Killing the competition. *Hum. Nature* 1:83–109.

Daly, M., and M. Wilson, 1991. A reply to Gelles: Stepchildren are disproportionately abused, and diverse forms of violence can share causal factors. *Hum. Nature* 2:419–426.

Daly, M., and M. Wilson, 1993. Stepparenthood and the evolved psychology of discriminative parental solicitude. In *Infanticide and Parental Care*, S. Parmigiami and F. vom Saal, eds. London: Harwood, pp. 121–134.

Darwin, C., and A. R. Wallace, 1858. *Evolution by Natural Selection*. Reprinted, 1958, London: Cambridge University Press.

Davies, N. B., 1992. *Dunnock Behaviour and Social Evolution*. Oxford University Press.

Davies, N. B., B. J. Hatchwell, T. Robson, and T. Burke, 1992. Paternity and parental effort in dunnocks *Prunella modularis*: How good are male chick-feeding rules? *Anim. Behav.* 43:729–745.

Davis, R. B., C. F. Herreid, and H. L. Short, 1962. Mexican free-tailed bats in Texas. *Ecol. Monogr.* 32:311–346.

Dickeman, M., 1975. Demographic consequences of infanticide in man. *Annu. Rev. Ecol. Syst.* 6:107–137.

Drotar, D., A. Baskiewicz, N. Irvin, J. Kennell, and M. Klaus, 1975. The adaptation of parents to the birth of an infant with a congenital malformation: A hypothetical model. *Pediatrics* 56:710–717.

Elwood, R. W., and H. F. Kennedy, 1993. Selective allocation of parental and infanticidal responses in rodents: a review of mechanisms. In *Infanticide and Parental Care*, S. Parmigiami and F. vom Saal, eds. London: Harwood Press.

Elwood, R. W., A. A. Nesbitt, and H. F. Kennedy, 1990. Maternal aggression in response to the risk of infanticide by male mice, *Mus domesticus*. *Anim. Behav.* 40:1080–1086.

Fleming, A. S., D. N. Ruble, G. L. Flett, and V. Van Wagner, 1990. Adjustment in first-time mothers: Changes in mood and mood content during the early postpartum months. *Dev. Psychol.* 26:137–143.

Flinn, M. V., 1988. Step- and genetic parent/offspring relationships in a Caribbean village. *Ethol. Sociobiol.* 9:335–369.

Formby, D., 1967. Maternal recognition of infant's cry. *Dev. Med. Child Neurol.* 9:293–298.

Furstenberg, F. F., and K. G. Talvitie, 1980. Children's names and paternal claims. *J. Fam. Issues* 1:31–57.

Green, R. F., 1984. Stopping rules for optimal foragers. *Am. Nat.* 123:30–40.

Grossman, F. K., L. S. Eichler, and S. A. Winickoff, 1980. *Pregnancy, Birth, and Parenthood*. San Francisco: Jossey-Bass.

Hamilton, W. D., 1964. The genetical evolution of social behaviour: I and II. *J. Theor. Biol.* 7:1–52.

Hansen, S., C. Harthorn, E. Waslin, L. Lofberg, and K. Svensson, 1991. Mesotelencephalic dopamine system and reproductive behavior in the female rat: Effects of ventral tegmental 6-hydroxydopamine lesions on maternal and sexual responsiveness. *Behav. Neurosci.* 105:588–598.

HAUSFATER, G., and S. B. HRDY, eds., 1984. *Infanticide: Comparative and Evolutionary Perspectives.* New York: Aldine.

HEALY, S., 1992. Optimal memory: Toward an evolutionary ecology of animal cognition. *Trend. Ecol. Evol.* 7:399–400.

HEPPER, P. G., ed., 1991. *Kin Recognition.* Cambridge: Cambridge University Press.

HERBERT, M., W. SLUCKIN, and A. SLUCKIN, 1982. Mother-to-infant "bonding"? *J. Child Psychol. Psychiat.* 23:205–221.

HERZOG, A., and T. DETRE, 1976. Psychotic reactions associated with childbirth. *Dis. Nerv. Syst.* 37:229–235.

HOLMES, W. G., 1984. Ontogeny of dam-young recognition in captive Belding's ground squirrels. *J. Comp. Psychol.* 98:246–256.

HOPKINS, J., M. MARCUS, and S. B. CAMPBELL, 1984. Postpartum depression: A critical review. *Psyhol. Bull.* 95:498–515.

HUNTER, R. S., N. KILSTROM, E. N. LODA, and F. KRAYBILL, 1978. Antecedents of child abuse and neglect in premature infants: A prospective study in a newborn intensive care unit. *Pediatrics* 61:629–635.

KENDRICK, K. M., E. B. KEVERNE, M. R. HINTON, and J. R. GOODY, 1992. Oxytocin, amino acide and monoamine release in the region of the medial preoptic area and bed nucleus of the stria terminalis of the sheep during parturition and suckling. *Brain Res.* 569:199–209.

KENDRICK, K. M., F. LÉVY, and E. B. KEVERNE, 1992. Changes in the sensory processing of olfactory signals induced by birth in sheep. *Science* 256:833–836.

KLAUS, M. H., and J. H. KENNELL, 1976. *Maternal-Infant Bonding.* St. Louis: Mosby.

LEIFER, M., 1977. Psychological changes accompanying pregnancy and motherhood. *Genet. Psychol. Monogr.* 95:55–96.

LESCHOT, N. J., M. VERJAAL, and P. E. TREFFERS, 1985. A critical analysis of 75 therapeutic abortions. *Early Hum. Dev.* 10:287–293.

LÉVY, F., K. M. KENDRICK, E. B. KEVERNE, V. PIKETTY, and P. POINDRON, 1992. Intracerebral oxytocin is important for the onset of maternal behavior in inexperienced ewes delivered under peridural anesthesia. *Behav. Neurosci.* 106: 427–432.

LOESCHE, P., M. D. BEECHER, and P. K. STODDARD, 1992. Perception of cliff swallow calls by birds (*Hirundo pyrrhonota* and *Sturnus vulgaris*) and humans (*Homo sapiens*). *J. Comp. Psychol.* 106:239–247.

LOESCHE, P., P. K. STODDARD, B. J. HIGGINS, and M. D. BEECHER, 1991. Signature versus perceptual adaptations for individual vocal recognition in swallows. *Behaviour* 118: 15–25.

LORENZ, K. Z., 1966. *On Aggression.* New York: Harcourt, Brace, and World.

MAESTRIPIERI, D., and E. ALLEVA, 1991. Litter defense and parental investment allocation in house mice. *Behav. Process* 23:223–230.

MANN, J., 1992. Nurturance or negligence: Maternal psychology and behavioral preference among preterm twins. In *The Adapted Mind,* J. Barkow, L. Cosmides, and J. Tooby, eds. New York: Oxford University Press, pp. 367–390.

MAYER, A. D., M. A. MONROY, and J. S. ROSENBLATT, 1990. Prolonged estrogen-progesterone treatment of nonpregnant ovariectomized rats: Factors stimulating home-cage and maternal aggression and short-latency maternal behavior. *Horm. Behav.* 24:342–364.

MAYNARD SMITH, J., 1976. Evolution and the theory of games. *Am. Sci.* 64:41–45.

MAYR, E., 1983. How to carry out the adaptationist program? *Am. Nat.* 121:324–334.

MCCRACKEN, G. F., 1984. Communal nursing in Mexican free-tailed bat maternity colonies. *Science* 223:1090–1091.

MCCRACKEN, G. F., and M. K. GUSTIN, 1992. Nursing behavior in Mexican free-tailed bat maternity colonies. *Ethology* 89:305–321.

MILLER, G. A., E. GALANTER, and K. PRIBRAM, 1960. *Plans and the Structure of Behavior.* New York: Holt, Rinehart, and Winston.

MONTGOMERIE, R. D., and P. J. WEATHERHEAD, 1988. Risks and rewards of nest defence by parent birds. *Q. Rev. Biol.* 63:167–187.

MROSOVSKY, N., and D. F. SHERRY, 1980. Animal anorexias. *Science* 207:837–842.

MULL, D. S., and J. D. MULL, 1988. Infanticide among the Tarahumara of the Mexican Sierra Madre. In *Child Survival: Anthropological Perspectives on the Treatment and Maltreatment of Children,* N. Scheper-Hughes, ed. Boston: Reidel.

NEWMAN, L. F., 1980. Parents' perceptions of their low birth weight infants. *Pediatrician* 9:182–190.

OLMSTED, J. M. D., 1938. *Claude Bernard, Physiologist.* London: Harper.

PARKER, G. A., 1970. Sperm competition and its evolutionary consequences in the insects. *Biol. Rev.* 45:525–567.

PERRIGO, G., W. C. BRYANT, and F. S. VOM SAAL, 1990. A unique neural timing system prevents male mice from harming their own offspring. *Anim. Behav.* 39:535–539.

POINDRON, P., and P. LE NEINDRE, 1980. Endocrine and sensory regulation of maternal behavior in the ewe. *Adv. Stud. Behav.* 11:75–119.

PORTER, R. H., 1991. Mutual mother-infant recognition in humans. In *Kin Recognition,* P. G. Hepper, ed. Cambridge: Cambridge University Press, pp. 413–432.

PORTER, R. H., R. D. BALOGH, J. M. CERNOCH, and C. FRANCHI, 1986. Recognition of kin through characteristic body odors. *Chem. Senses* 11:389–395.

PRESSLEY, P. H., 1981. Parental effort and the evolution of nest-guarding tactics in the threespine stickleback, *Gasterosteus aculeatus* L. *Evolution* 35:282–295.

PRYCE, C. R., 1992. A comparative systems model of the regulation of maternal motivation in mammals. *Anim. Behav.* 43:417–441.

PUGESEK, B. H., 1983. The relationship between parental age and reproductive effort in the California gull (*Larus californicus*). *Behav. Ecol. Sociobiol.* 13:161–171.

Pugesek, B. H., 1987. Age-specific survivorship in relation to clutch size and fledging success in California gulls. *Behav. Ecol. Sociobiol.* 21:217–221.

Redondo, T., 1989. Avian nest defence: Theoretical models and evidence. *Behaviour* 111:161–195.

Regalski, J. M., and S. J. C. Gaulin, 1993. Whom are Mexican infants said to resemble? Monitoring and fostering paternal confidence in the Yucatan. *Ethol. Sociobiol.* 14:97–113.

Ridley, M., 1978. Paternal care. *Anim. Behav.* 26:904–932.

Robson, K. M., and R. Kumar, 1980. Delayed onset of maternal affection after childbirth. *Br. J. Psychiatry* 136: 347–353.

Rohwer, S., 1986. Selection for adoption versus infanticide by replacement "mates" in birds. *Curr. Ornithol.* 3:353–395.

Rosenblatt, J. S., 1990. Landmarks in the physiological study of maternal behavior with special reference to the rat. In *Mammalian Parenting: Biochemical, Neurobiological and Behavioral Determinants*, N. A. Krasnegor and R. Bridges, eds. New York: Oxford University Press, pp. 40–60.

Scheper-Hughes, N., 1985. Culture, scarcity, and maternal thinking: Maternal detachment and infant survival in a Brazilian shantytown. *Ethos* 13:291–317.

Sherry, D. F., N. Mrosovsky, and J. A. Hogan, 1980. Weight loss and anorexia during incubation in birds. *J. Comp. Physiol. Psychol.* 94:89–98.

Smith, D. S., 1977. Child-naming patterns and family structure change: Hingham, Massachusetts 1640–1880. *Newberry Papers in Family and Community History*, paper 76–75.

Smith, R. L., ed., 1984. *Sperm Competition and the Evolution of Animal Mating Systems*. Orlando, Fla.: Academic Press.

Sober, E., 1983. Mentalism and behaviorism in comparative psychology. In *Comparing Behavior*, D. W. Rajecki, ed. Hillsdale, N.J.: Erlbaum, pp. 113–142.

Stephens, D. W., and J. R. Krebs, 1986. *Foraging Theory*. Princeton: Princeton University Press.

Stern, J. M., and J. M. Kolunie, 1991. Trigeminal lesions and maternal behavior in Norway rats: I. Effects of cutaneous rostral snout denervation on maintenance of nurturance and maternal aggression. *Behav. Neurosci.* 105:984–997.

Symons, D., 1987. If we're all Darwinians, what's the fuss about? In *Sociobiology and Psychology*, C. Crawford, M. Smith, and D. Krebs, eds. Hillsdale, N.J.: Erlbaum, pp. 121–146.

Thornhill, R., 1989. Nest defense by red jungle fowl (*Gallus gallus spadiceus*) hens: The roles of renesting potential, parental experience and brood reproductive value. *Ethology* 83:31–42.

Tinbergen, N., 1963a. On aims and methods of ethology. *Z. Tierpsychol.* 20:410–433.

Tinbergen, N., 1963b. The shell menace. *Nat. Hist.* 72:28–35.

Tolman, E. C., 1932. *Purposive Behavior in Animals and Men*. New York: Century.

Tooby, J., and L. Cosmides, 1992. The psychological foundations of culture. In *The Adapted Mind: Evolutionary Psychology and the Generation of Culture*, J. H. Barkow, L. Cosmides and J. Tooby, eds. New York: Oxford University Press, pp. 19–136.

Trillmich, F., 1981. Mutual mother-pup recognition in Galapagos fur seals and sea lions: Cues used and functional significance. *Behaviour* 78:21–42.

Trivers, R. L., 1972. Parental investment and sexual selection. In *Sexual Selection and the Descent of Man 1871–1971*, B. Campbell, ed. Chicago: Aldine.

vom Saal, F. S., 1985. Time-contingent change in infanticide and parental behavior induced by ejaculation in male mice. *Physiol. Behav.* 34:7–15.

Westneat, D. F., and P. W. Sherman, 1993. Parentage and the evolution of parental behavior. *Behav. Ecol.* 4:66–77.

Whittingham, L. A., P. D. Taylor, and R. J. Robertson, 1992. Confidence of paternity and male parental care. *Am. Nat.* 139:1115–1125.

Williams, G. C., 1966. *Adaptation and Natural Selection*. Princeton: Princeton University Press.

Wilson, M., M. Daly, 1985. Competitiveness, risk-taking and violence: The young male syndrome. *Ethol. Sociobiol.* 6:59–73.

Wilson, M., and M. Daly, 1987. Risk of maltreatment of children living with stepparents. In *Child Abuse and Neglect: Biosocial Dimensions*, R. J. Gelles and J. B. Lancaster, eds. Hawthorne, N.Y.: Aldine, pp. 215–232.

Wilson, M., and M. Daly, 1992. The man who mistook his wife for a chattel. In *The Adapted Mind: Evolutionary Psychology and the Generation of Culture*, J. H. Barkow, L. Cosmides, and J. Tooby, eds. New York: Oxford University Press, pp. 289–322.

Wilson, M., and M. Daly, 1993. The psychology of parenting in evolutionary perspective and the case of human filicide. In *Infanticide and Parental Care*, S. Parmigiami and F. vom Saal, eds. London: Harwood Press, pp. 73–104.

Wilson, M. I., M. Daly, and S. J. Weghorst, 1980. Household composition and the risk of child abuse and neglect. *J. Biosoc. Sci.* 12:333–340.

Wittenberger, J. F., 1981. *Animal Social Behavior*. Boston: Duxbury.

Yasukawa, K., F. Leanza, and C. D. King, 1993. An observational and brood-exchange study of paternal provisioning in the red-winged blackbird, *Agelaius phoeniceus*. *Behav. Ecol.* 4:78–82.

Zuckerman, B. S., D. K. Walker, D. A. Frank, C. Chase, and B. Hamburg, 1984. Adolescent pregnancy: Biobehavioral determinants of outcome. *Pediatrics* 105:857–863.

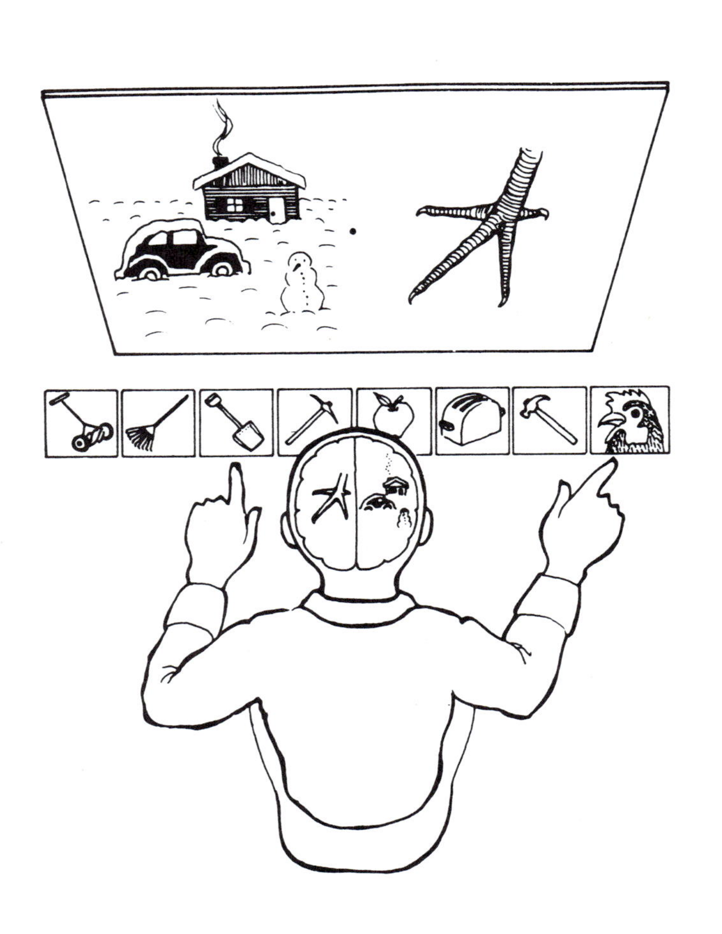

XI
CONSCIOUSNESS

The illustration depicts how each hemisphere can be examined and studied separately following brain bisection.

Introduction

DANIEL L. SCHACTER

HUMAN consciousness is just about the last surviving mystery.
—Daniel C. Dennett, 1991

Every period in the relatively brief history of neuroscience has considered that it had a special, sometimes even ultimate, insight into the neural mechanisms of conscious awareness.
—Lawrence Weiskrantz, 1991

Mention the term *consciousness* to a cognitive neuroscientist, and you will probably elicit one of two very different reactions. Some will likely shrug or groan, mumbling uncomfortably that consciousness is a construct that we do not yet know how to approach sensibly or to investigate productively. Dennett (1991, 21) acknowledged the vexing "mystery" of consciousness: "A mystery is a phenomenon that people don't know how to think about—yet." Others, however, will probably respond with enthusiasm, even excitement, citing the latest experimental dissociations between conscious and unconscious processes in neuropsychology or electrophysiology as evidence that cognitive neuroscience is on the verge of finally cracking the riddle of consciousness. Weiskrantz (1991) has noted that this enthusiasm is not without historical precedent.

Consciousness is a particularly challenging topic for cognitive neuroscience. The challenge begins with the fact that most investigators have a difficult time agreeing on an adequate definition of the term (e.g., Wilkes, 1988). Contemporary treatments of consciousness often begin with an acknowledgment that although we all have a subjective sense of what we mean when we use the expression, satisfactory formal definitions are difficult to come by. Indeed, a number of writers have argued that the term *consciousness* is simply too coarse to be useful theoretically. Tulving (1985), for example, distinguished among three types of consciousness: anoetic (nonknowing), which entails simple awareness of external stimuli; noetic (knowing), which involves awareness of symbolic representations of the world; and autonoetic (self-knowing) consciousness, which involves awareness of self and personal experience extended in time. Farthing (1992) has offered a distinction between primary consciousness—simple perceptual awareness of extemal and internal stimuli—and reflective consciousness—"thoughts about one's own conscious experiences per se" (1992, 13). Natsoulas (1978) has distinguished seven different ways in which the term *consciousness* has been used, and numerous other distinctions among forms or types of consciousness could be cited (cf. Marcel and Bisiach, 1988; Milner and Rugg, 1991).

Definitional problems notwithstanding, cognitive neuroscientists have approached phenomena of consciousness from a variety of perspectives. During the 1950s and 1960s, the discovery of rapid eye movement sleep (Aserinsky and Kleitman, 1953) led to excitement about possibilities for understanding the neurophysiological basis of "states" of consciousness. A new and startling perspective on consciousness was provided by the observations of commissurotomy or split-brain patients that began to appear in the 1960s and 1970s. Led by such investigators as Sperry (1966), Bogen (1969), and Gazzaniga (1970), studies of split-brain patients produced striking observations that suggested the possible existence of independent systems of consciousness in each hemisphere. These observations provided fertile ground for theorizing about consciousness—much of it rather speculative—in neuroscience, psychology, and philosophy (cf. Popper and Eccles, 1977; Puccetti, 1981; Springer and Deutsch, 1985).

While much of the initial excitement surrounding split-brain studies developed during the 1960s and early 1970s, a different sort of phenomenon began to capture the attention of cognitive neuroscientists during the late 1970s and 1980s: demonstrations that various kinds of brain-damaged patients exhibit preserved access to nonconscious or implicit knowledge despite a profound impairment of conscious or explicit knowledge. Perhaps the best known and most arresting example is that of *blindsight*, where patients with lesions to striate cortex who deny conscious perception of visual stimuli nonetheless can "guess" their location and other attributes (e.g., Weiskrantz, 1986). Similarly, amnesic patients who lack explicit or conscious memory for their recent experiences can exhibit nonconscious or implicit memory for aspects of those experiences, as exemplified by such phenomena as priming and skill learning (Schacter, 1987). Similar kinds of dissociations have been observed in patients with aphasia, alexia, and unilateral neglect, among others, and have led to the discovery of analogous phenomena in normal subjects (for review, see Schacter, 1992). These dissociations have led to a variety of proposals concerning the nature, function, and neural basis of consciousness (see Milner and Rugg, 1991).

The 1980s also witnessed renewed attention to another neuropsychological phenomenon with important implications for thinking about consciousness: unawareness of deficit, or anosognosia. The observation that some brain-damaged patients claim to be entirely unaware of the existence of deficits that are all too obvious to others was first reported in the late nineteenth and early twentieth centuries. Unfortunately, implications of the phenomenon for understanding the nature of conscious experience were not pursued systematically, perhaps because of the prevalence of psychodynamic approaches to the issue (for historical review, see McGlynn and Schacter, 1989). However, stimulated largely by the pioneering research of Bisiach and his colleagues concerning anosognosia in neglect patients (Bisiach and Geminiani, 1991), the central importance of anosognosia and related phenomena for theories of consciousness has come to be more widely appreciated (see Prigatano and Schacter, 1991, for an overview of contemporary approaches).

The chapters in this section provide a broad overview of approaches to consciousness in contemporary cognitive neuroscience that summarize in some depth the foregoing areas of investigation, and point the way to future research. The first three chapters focus on general conceptual and theoretical issues. Farber and Churchland provide a philosophical treatment of various constructs that are central to discussions of consciousness, and develop conceptual perspectives that should prove quite stimulating for cognitive neuroscientists. Hirst summarizes empirical and theoretical approaches to consciousness that have emerged from cognitive science and delineates their significance for cognitive neuroscience. Kinsbourne outlines a theoretical model of consciousness in which he argues for the viability of a model in which multiple representations are active in parallel, and rejects serial models in which a sequence of information-processing stages culminate in a privileged consciousness center.

The subsequent chapters also address general conceptual issues in the context of more focused discussions of specific phenomena that are critical to the analysis of consciousness. Bisiach and Berti discuss unilateral neglect and related impairments of spatial representation with respect to their implications for understanding consciousness. Moscovitch considers the relation between memory and consciousness, analyzing dissociations between implicit and explicit forms of memory within the context of a model that draws a sharp distinction between modules and central systems. Knight and Grabowecky focus on the role of frontal regions in conscious experience, drawing on both neuropsychological and electrophysiological observations to formulate a general perspective on the issue.

Hobson and Stickgold consider problems of consciousness with respect to sleep and alterations in arousal. They approach issues of sleep and consciousness from the perspective of contemporary cognitive neuroscience, suggesting explicit linkages between underlying neurophysiology and conscious experience during sleep. Gazzaniga summarizes and integrates numerous observations concerning consciousness and the cerebral hernispheres, and considers phenomena observed in split-brain patients with respect to his notion of a left-hemisphere "interpreter" that plays a key role in generating aspects of conscious experience.

Are scientists of the 1990s any closer to understanding consciousness than those of preceding decades? Only time will tell, but the ferment of activity surrounding the topic can at least assure us that the age-old enigma of consciousness will continue to occupy cognitive neuroscientists well into the twenty-first century.

REFERENCES

Aserinsky, E., and N. Kleitman, 1953. Regularly occurring periods of ocular motility and concomitant phenomena during sleep. *Science* 118:361–375.

Bisiach, E., and G. Geminiani, 1991. Anosognosia related to hemiplegia and hemianopia. In *Awareness of Deficit after Brain Injury*, G. P. Prigatano and D. L. Schacter, eds. New York: Oxford University Press, pp. 17–39.

Bogen, J. E., 1969. The other side of the brain: An appositional mind. *Bull. Los Angeles Neurolo. Soc.* 34:135–162.

Dennett, D. C., 1991. *Consciousness Explained.* Boston: Little, Brown.

Farthing, G. W., 1992. *The Psychology of Consciousness.* Englewood Cliffs, N.J: Prentice Hall.

Gazzaniga, M. S., 1970. *The Bisected Brain.* New York: Appleton-Century-Crofts.

Marcel, A. J., and E. Bisiach, eds., 1988. *Consciousness in Contemporary Science.* New York: Oxford University Press.

McGlynn, S. M., and D. L. Schacter, 1989. Unawareness of deficits in neuropsychological syndromes. *J. Clin. Exp. Neuropsychol.* 11:143–205.

Milner, A. D., and M. D. Rugg, eds. 1991. *The Neuropsychology of Consciousness.* San Diego: Academic Press.

Natsoulas, T., 1978. Consciousness. *Am. Psychol.* 33:906–914.

Popper, K. R., and J. C. Eccles, 1977. *The Self and Its Brain: An Argument for Interactionism.* Berlin: Springer.

Prigatano, G. P., and D. L. Schacter, eds., 1991. *Awareness of Deficit after Brain Injury.* New York: Oxford University Press.

Puccetti, R., 1981. The case for mental duality: Evidence from split brain data and other considerations. *Behav. Brain Sci.* 4:93–123.

Schacter, D. L., 1987. Memory, amnesia, and frontal lobe dysfunction. *Psychobiology* 15:21–36.

Schacter, D. L., 1992. Understanding implicit memory: A cognitive neuroscience approach. *Am. Psychol.* 47:559–569.

Sperry, R. W., 1966. Brain bisection and mechanisms of consciousness. In *Brain and Conscious Experience*, J. C. Eccles, ed. New York: Springer, pp. 298–313.

Springer, S. P., and G. Deutsch, 1985. *Left Brain, Right Brain.* New York: Freeman.

Tulving, E., 1985. How many memory systems are there? *Am. Psychol.* 40:385–398.

Weiskrantz, L., 1986. *Blindsight: A Case Study and Implications.* Oxford: Clarendon.

Weiskrantz, L., 1991. Introduction: Dissociated issues. In *The Neuropsychology of Consciousness*, A. D. Milner and M. D. Rugg, eds. San Diego: Academic Press, pp. 1–10.

Wilkes, K. V., 1988. —— yìshì, duh, um, and consciousness. In *Consciousness in Contemporary Science*, A. J. Maucel and E. Bisiach eds. New York: Oxford University Press.

85 Consciousness and the Neurosciences: Philosophical and Theoretical Issues

ILYA B. FARBER AND PATRICIA S. CHURCHLAND

ABSTRACT In this chapter we suggest that *consciousness*, while admittedly a fragmented and vague concept, should not presently be subjected to more precise definition. Rather, following the traditional path of conceptual development in science, the concept should be allowed to coevolve with the growing body of relevant empirical data and techniques. In this spirit, we present and discuss examples drawn from several different (and possibly incompatible) conceptions of consciousness, classifying them into three general aspects: conscious awareness, higher faculties, and conscious states. This particular division is largely arbitrary, and is not intended as a prediction or recommendation about the structure of future theories of consciousness; its purpose is simply to highlight some of the specific philosophical and scientific challenges that are raised by the various conceptions of consciousness, and to provide some conceptual resources for forestalling the confusion that can result when one word is used to refer to so many different ideas. In closing, we present some of the methods for approaching consciousness that have recently been developed by neuroscientists, and discuss the general prospects for neuroscientific explanation of conscious phenomena.

What are we trying to explain?

There is a powerful temptation to deal with the vagueness and obscurity of "consciousness" by proposing that we set aside empirical inquiry and theory building until we have determined quite precisely the proper definition of the term and the scope of the concept. After all, it is clear that, even just within science, many people mean different things when they talk about consciousness, and using the same word for so many different things (some of which will be discussed in detail below) is bound to produce a certain amount of confusion and pointless argumentation. The natural suggestion, then, is that we should stop and make sure we've got our terms straight before we go on building and testing the theories that use them.

This idea, while profoundly seductive, is both misguided and potentially hazardous. The first part of this chapter will be devoted to showing why this is so, and to motivating an alternative approach that is more sensitive to the ways in which scientific categories traditionally develop. Subsequent sections will discuss various aspects of what we traditionally call consciousness, and the final section will address the prospects for a neuroscientific understanding of these phenomena.

Prior to the emergence of a reasonably well-developed theory, the relevant phenomena do not classify themselves together for the scientists' convenience. The world does not come to us prepackaged; there may be some structure in it, and from our perspective further structure is imposed by the constitution of our sensory and cognitive systems; but even within these constraints, there are still many different ways in which we could group the things around us.

In the early days of our acquaintance with a given range of phenomena, we have little to rely on except for gross physical and behavioral similarity. Especially favored are properties relevant to our interactions with the things involved: plants that are edible or poisonous, and animals that are docile or dangerous, are more likely to be grouped with others that share those properties. Highly unusual or distinctive features —flight, the production of heat or light—also exert a strong influence on pre- or protoscientific schemes of categorization.

Developed sciences tend not to use such criteria. As understanding improves, the classifications get re-

ILYA B. FARBER and PATRICIA S. CHURCHLAND Department of Philosophy, University of California at San Diego, La Jolla, Calif.

drawn according to different sorts of principles. In developed sciences, properties such as *edible* or *predatory* are not necessarily left out, but they no longer serve as the bases for our taxonomies. In the modern zoological taxonomy, such physically and behaviorally disparate animals as the rock hyrax and the elephant are grouped together by virtue of genetic and evolutionary criteria; likewise, we place in different categories genetically disparate animals that, due to convergent evolution, may have very similar forms and behaviors. The history of the term *fire* provides another striking example of category shift: The term was once used to classify not only burning carbon-stuffs, but also activity on the sun and various stars (actually nuclear fusion), lightning (actually electrically induced incandescence), the northern lights (actually spectral emission), and the flash of fireflies (actually phosphorescence). In our modern conceptual scheme, since none of these things involves oxidation, none belongs to the same class as wood fires. Moreover, some processes that turned out to belong to the oxidation class—rusting, tarnishing, and metabolism—were not originally considered to share anything with burning, since felt heat was taken to be an essential feature of the class.

So why is it that science tends to reject simple, obviously useful criteria in favor of others that are arcane and of little apparent relevance to daily life? One possible answer is that the resultant categories more accurately reflect the structure of reality itself—that they map more closely onto "natural kinds," or, as it is sometimes said, carve nature at her joints. In less metaphysically loaded terms, however, it can simply be said that the resultant categories ultimately turn out to be more useful in the production of sophisticated and powerful scientific theories, for a number of reasons:

1. Increases in insight and control tend to cut across the old categories. For example, the knowledge gained from research into the chemical processes involved in fire was (ultimately) relevant to scientists' understanding of rusting and metabolism, but of little help to investigators of fireflies or the Northern Lights.

2. The old categories do not have as much deep structure, and thus place limits on useful definitional precision. Terms like *predator* and *weed* will always admit of gray areas, and while sharp dividing lines can be constructed, they are bound to be arbitrary and not very useful.

3. Since they have their bases in a common set of basic sciences, modern categories enable scientists to connect and unify their understanding in a way that primitive categories do not. The contrast here is sharpest with consciousness itself—the grand challenge in philosophy of mind has long been simply to find a *place* for it in our scientific framework, a way to relate it to anything else that we study. The dearth of satisfying solutions has led many (including John Searle, whose argument will be discussed later) to conclude that consciousness is somehow irredeemably special and different; but a more optimistic inference is that the concept is too closely tied to an archaic worldview, and just needs some reworking before it is ready for the scientific "prime time."[1]

A natural consequence of this criterial shift is that *theories* about something, and *definitions* as to what in the world counts as that thing, evolve together, hand in hand. It is only with the deeper insights provided by theoretical advances that the most useful groupings begin to be discernible, and with each regrouping of the phenomena new theoretical advances become possible. In fact, firm, explicit definitions become available only fairly late in the game; they may be useful for teaching the mature and well-established parts of a discipline to students, but play little role in directing the course of science itself.[2] Thus the danger of the define-first-and-hypothesize-later approach suggested above: During the tentative, pioneering period in which we first attempt to come to theoretical grips with a phenomenon, such a process encourages us to stipulate rather than to examine, to settle prematurely on criteria that are familiar and readily available rather than to strive for the definition that will most benefit scientific progress and unification. Rather than an exercise in wordplay, what is really needed is the familiar scientific bootstrap operation in which genuine theory and increasing definitional precision coevolve (P. S. Churchland, 1983). The words will come to have a more precise meaning as they are more deeply embedded within the framework of empirical theory; beyond this, any additional enforced precision would only be arbitrary and spurious.[3]

While *consciousness* may at present lack definitional precision, it is rich in examples, and some light may be shed on the concept itself by focusing on these phenomena and trying to discern what, if anything, they have in common. The hope here is not that some single property will be found that unifies everything we currently want to call *consciousness*; as happened with *fire*, our current intuitive classification will be redrawn as a more scientifically embedded theory emerges, and new theories may pull together what now seem to be diverse phenomena by virtue of some shared explanation that is presently unavailable to us. Rather, it will be enough to discern a few *subcategories* or *aspects* of consciousness, grouping together phenomena that have similar characteristics and that present similar problems for the theorist.

This sort of subdivision can serve at least two purposes. First, it makes it easier for theorists to avoid at least some of the aforementioned "confusion and pointless argumentation" by specifying which *aspects* of consciousness a given theory is meant to account for. And second, it brings into sharper relief the particular issues and problems that must be faced by anyone who seeks to explain a given class of conscious phenomena. The next three sections of this chapter are organized with these functions in mind: Each is devoted to a particular category of conscious phenomena, starting with an examination of some of the major examples and ending with a discussion of the theoretical challenges that those examples pose.[4]

Conscious awareness

In this sense, consciousness is a property of the relation between an observer and a phenomenon; some*one* is consciously aware of some*thing*. The major subcategories here are variations on just what that thing is.

Sensory awareness. This would include awareness of most of the sorts of things we call stimuli. We tend to speak more or less interchangeably of being aware of states of the world and being aware of the sensations that inform us of those states; but we can also include hallucinations and modality-specific imagery, which carry no direct information about the outside world.

Generalized awareness. One can be aware of inner states with no clear link to any modality: fatigue, dizziness, anxiousness, comfort, hunger, and so on. From the perspective of science, we may understand these to be derived in whole or in part from more specific states of the body (muscle tension, movement of fluid in the ears), but we don't *experience* them as such. This category could also include awareness of temporal durations (how long ago one turned the oven on); of spatial layout, both immediate and extended; and perhaps even of self, as a thing that changes somewhat but endures through time.

Metacognitive awareness. There are all sorts of things that one can be aware of in the realm of one's own cognition. The most common examples have to do with what one does and does not know: "I am aware that I know the maiden name of my paternal grandmother, and that I don't know how many first cousins I have." It's also possible to become conscious of what one is now thinking,[5] or even of the train of thoughts (now past) that led up to the present one.

Conscious recall. We can become consciously aware of events that occurred earlier. Sometimes this involves mental imagery of some kind, though it need not; and the recall may or may not be prompted or assisted by current sensory experience.

Awareness has proven to be the aspect of consciousness that lends itself most readily to scientific study, for a number of reasons. First, it allows of a fairly satisfying operational definition: Barring deception, neurological dysfunction, or other such interferences, we can say that a human subject counts as being consciously aware of something if and only if he or she can report it. This makes testing much easier, and allows the scientist to bypass the problem of subjectivity (discussed later, in the section on states of consciousness). On the down side, it also makes it possible indefinitely to postpone the question of nonhuman consciousness, since the operational definition has no obvious extension to animals that cannot talk.

Second, awareness has a good contrast class. Implicit in the very term *conscious perception* is the idea that we can and do have unconscious perception. One of cognitive science's central insights, present in early psychophysics but not fully appreciated until this century, is that our thoughts, decisions, and behavior are often influenced by things of which we are not aware. Stimuli that someone doesn't consciously notice—tachistoscopically presented words, the color of a room, the dilation of another person's pupils—can nonetheless have profound effects on one's mood and behavior. The same can be said for generalized awareness (which explains the adage that one should never go grocery shopping while hungry), though the distinction between *unconsciously perceiving* such a state and merely *being in it* is not so clear.

For metacognition and memory, the procedural/declarative distinction (Squire and Cohen, 1984) cuts along similar lines; we can demonstrate knowledge and cognitive skills (or lack of such) without knowing that we possess them. For example, some densely amnesic patients are able briskly and correctly to solve a puzzle on which they have long practiced, even though they cannot recall ever having seen it before; normal subjects can be influenced by priming effects of words they do not recall seeing; and it is not uncommon for patients with brain damage to be unaware of even the most striking of deficits, such as word-salad aphasia, cortical blindness, or hemiparesis.

And third, experimenters have a fairly good repertoire of tools for directing and controlling awareness. At minimum, one can simply tell a subject to pay attention to something, and/or assign a task that requires the subject consciously to process the desired phenomena. One can exploit known predispositions, such as the tendency to notice flashy, changing stimuli over dull, constant ones; and by overloading the subject's attentive resources ("count your heartbeats, press the left button if you see a spelling error in a green or blue word, and press the right button whenever you hear the middle-pitch note . . . ") one can drastically reduce the chances that a task-irrelevant stimulus will be noticed.

Together, these factors make it possible to mount relatively controlled and precise experiments that target awareness. A review of the cognitive neuroscience literature testifies to this fact; awareness is by far the most studied aspect of consciousness, so much so that the two terms often come to seem synonymous. And here we see one of the dangers of the coevolutionary approach to theory making: If one of the phenomena in our early, intuitive classification of some domain turns out to be a more tractable and rewarding subject of scientific study, it will tend to be elevated in theoretical importance, and its less tractable brethren will tend to be marginalized. Inevitably, the choice of research projects in science will be to a large degree technique-driven. This certainly leads to an efficient allocation of effort and resources, but we must also remember that one of the purposes of scientific theorizing is to illuminate the objects of everyday experience—and awareness does not even begin to exhaust the catalog of experiences in which consciousness plays a part.

Higher faculties

In its standard sense, the term *consciousness* carries an implication of agency and control, which unconscious but animate entities—insects, zombies, industrial robots, pick your favorite example—lack. The significance of this sort of consciousness is recognized in many disciplines. In the philosophical and political literature, agency is often seen as equivalent to, or at least necessary for, free will and moral rights and responsibilities; while in robotics and artifical intelligence *control* denotes certain functions that are vital for the coordination and efficient functioning of a complex information-processing system. If consciousness is to fulfill either of these roles, it cannot be simply an aware but passive reception of information; it must involve a repertoire of *faculties*, of things that a creature can do, or at least cause to be done, by virtue of that creature's being conscious.

The list of such faculties[6] could be quite long, and would depend at least in part on one's choice of current theories of mental function; but there are a few standard examples that will serve to illustrate the general boundaries of the class.

Attention. This is the way consciousness influences what it will and will not become aware of. Given the more or less serial and remarkably slow character of conscious information processing, it is essential that only a few things be impinging on consciousness at any given time. As discussed above, certain types of stimuli are automatically granted high priority; but we can also *decide* to focus our attention on some particular sensation, modality, or external object, or to withdraw from reality—usually by rolling our eyes up and to the side—and reserve our awareness for internal phenomena.

Reasoning. Reasoning is the classic example of a higher function. Skirting the historical baggage about pure thought and eternal ideas, we can simply say that reasoning is a type of information processing that operates at a high level of abstraction. Once one knows algebra (a perennial favorite example), one can apply it to all sorts of different objects, or even just go through the motions *in crania* without applying them to any real objects at all. This is the essence of *formal* operation; as long as one keeps one's symbols straight and applies the rules properly (that is, as long as one gets the *syntax* right) it doesn't really matter what, if anything, the symbols refer to (what the *semantics* are). In contrast, a conditioned aversion is perhaps the ultimate nonrational mental process: It is modality specific, functions only in the presence of a particular stimulus, and always follows the same sequence (e.g., if you hear the bell, you cringe).

Self-Control This is consciousness acting as the arbiter for internal disputes. Most traditional examples involve the imposition of reason and moral beliefs on base, physical impulses, but this faculty can be seen in a much more general and pragmatic way. It is readily apparent that much decision making is done unconsciously and in parallel, with many different factors contributing to the final outcomes (more on this below); decisions can be, and usually are, made while one's conscious attention is somewhere else entirely. But in a parallel decision-making system of this sort, there needs to be some way of resolving cases where different factors strongly suggest conflicting paths of action. Things might sort themselves out eventually, but standing around indecisively is evolutionarily costly, and one conscious decision can resolve things immediately and perhaps lay down a habit of future unconscious action.

The scientific status of these faculties, and of the general notion of consciousness-as-agency, is much more problematic than that of awareness. To some extent, they are holdovers from older models of mind that presumed that most mental "work" was done within consciousness.[7] There is a growing consensus across the cognitive sciences that our conscious experience of decision making and control only poorly reflects the way things actually get done; the actual mechanisms seem to be more parallel, and to act in a manner that is more or less inaccessible to consciousness (Gazzaniga, 1985; Minsky, 1986; Lycan, 1987; Dennett, 1991). The theories currently on offer vary widely in their accounts of how these mechanisms are constituted, how "smart" the elements of the parallel system can each be, and how their activities are orchestrated to produce speedy, decisive behavior; but the theories are united in denying that consciousness is as unitary, irreducible, and aloof as discussions about faculties or free will typically assume.

What, then, can be said about the persistent intuition that we really are, so to speak, the captains of our own ships? One possibility is that it is just an illusion. According to Gazzaniga, and to some extent Dennett, when we experience ourselves as having made a conscious decision, what we are in fact getting is a "story" reconstructed from the (external and internal) evidence. Implausible as this may seem at first blush, it does have considerable empirical support. Dennett (1991) discusses some puzzling phenomena that seem to support such a conclusion, and Nisbett and Wilson (1977) have performed a number of striking experiments that suggest that "introspectively" produced stories about the reasons behind one's own choices are heavily influenced by external evidence, and closely resemble those that an outside observer would construct.[8] But what could possibly be the use of such a "reconstructive consciousness," or of the illusion of unity it creates; why should the brain bother to create it at all? If this is all there is to it, why be conscious?

One answer might be that consciousness is just an easy way to enhance the coordination of diverse mental/neural processes. Access to a common autobiographical story and a "sense of self" might, for example, make it easier for subsystems to cooperate and to reach compatible decisions. But it is also intriguing that many of the apparent characteristics of conscious mental processes are radically unlike those of unconscious processes, or for that matter of neural processes in general: Conscious thought is comparatively slow, serial, and abstract; it deals with only a few objects at a time; its contents are readily translated into a communicable form (i.e., language);[9] and its storage and processing limits can be overcome by the use of external objects such as books, calculators, maps, and word-processing programs.

To those familiar with the connectionist/symbolic rift in the field of artifical intelligence (AI), these properties are quite suggestive. They begin to make it look as though conscious thought is the eminently parallel brain's attempt to *act like* a traditional, symbolic system.[10] This would explain why conscious thought is so slow and circumscribed; neural nets simply aren't very good at that sort of thing, and it takes them a lot of work to do what a digital computer does quite easily and naturally (Baars 1988). At the same time, it makes sense of our lack of conscious access to so much of our own mental functioning. Most of the work, including the "implementation" of conscious functions, will still be carried out in the usual neural fashion, via highly distributed networks performing all sorts of recurrent nonlinear operations on huge activation vectors, and this sort of activity is notoriously hard to describe in a neat symbolic way (as can be attested to by any neural network theorist who has had the experience of staring at the weight and activation matrices of a net and trying to figure out what's going on).

The most recent version of this sort of theory is that propounded by Dennett in his book *Consciousness Ex-*

plained (1991). He suggests that each brain is running a socially conditioned "virtual machine," whose operations and contents constitute a person's conscious actions and awareness. The introspective experience of symbolic, linguistic mentality can thus be likened to the user illusion presented by a home computer that is acting like a word processor, calculator, piano, fighter plane, or whatever.[11]

If one assumes that all this has some basis in the evolutionary history and structure of the brain, there is potential for a scientific investigation into the difference between operations inside and outside of the virtual machine (Farber, 1991). Dennett, however, contends that the installation of the machinery of consciousness is strictly cultural—following Dawkins (1989), he makes use of the notion of self-perpetuating units of cultural transmission, called *memes*—and that we would therefore be mistaken to look for a neural-structural basis for consciousness. This claim puts his otherwise interesting theory on shaky footing with the cognitive sciences. First, it is unclear what in the brain of a preconscious child would enable these memes to take root and so drastically restructure the child's cognitive mechanisms; it seems the child must somehow unconsciously comprehend and implement instructions contained in overheard adult dialogue and narrations from storybooks, or implicit in adult behavior, and Dennett gives us no indication as to how this would work.

But more importantly, Dennett is quite up-front in claiming that the learned rather than genetic character of the virtual machine makes it unlikely that there are useful neural correlates to be found.[12] Though the functions are all ultimately produced by the brain, they may be produced in different ways by different brains, and may be implemented quite diffusely across the cortex. Thus, despite his frequent use of examples from cognitive neuroscience, the theory essentially denies neuroscientists a place in the study of consciousness; as discussed in his chapter on the "heterophenomenological method," his position is that the most effective way to study consciousness will be simply to talk to subjects and perform various psychophysical experiments on them. Needless to say, the present authors find such a conclusion unduly pessimistic and exclusionary; and in the section on prospects for a neuroscience of cognition, we will present a few examples of the sorts of research that do indeed promise to shed light on the neural basis of consciousness.

States of consciousness

This final category covers the aspects of consciousness that are both the most commonsensical and, in many ways, the most problematic. The word *consciousness* is often used to refer to a *state*, which can be present to varying degrees, and which in some sense embodies what is going on in a person's mind. Consciousness is treated both as a single state that can be present to varying degrees, and as a collection of states that come and go separately, as follows.

Conscious versus unconscious. When people have just been hit on the head and are now lying immobile, it is usually safe to infer that they are *un*conscious—that they have no awareness, are not engaging in reasoning or self-control, do not feel any emotions, and in short are enjoying no more of a mental life than do plants and insects. But at the same time, these observational criteria are defeasible; we are willing to admit that someone can be completely immobilized and still be perfectly conscious. Interestingly, both scientists and lay people alike feel more confident in making such judgments if they have reliable information about the subject's brain state, even if only in the form of activity on an EEG.

General modulations. We can also detect other, less drastic states, such as wakefulness, surprise, anger, and the like. Unlike unconsciousness, these states (supposedly) have internal feels, which allow us to identify them in ourselves without examining our own behavior.[13] Much philosophical hay has been made of the point that we only experience these feels in our own case, and hence seem only marginally justified in assuming that others feel the same feels that we do. As we learn more about the way the brain produces these feels, this sort of worry is bound to become less plausible; but in the meantime, it (along with its cousin, in the next paragraph) funds a considerable amount of opposition to physical, scientific accounts of the mind.

Qualia or "Raw Feels." Quale is a philosopher's term of art for the raw *feel* of an experience, as divorced from any behavioral or cognitive effects the experience might have. So, the story goes, if one sees something red, one may acquire a new belief ("There's red here") and perhaps alter one's behavior in some way, but one also experiences a unique feel that is similar to that of seeing reddish-purple or reddish-orange and unlike that of sticking a hand in ice water. This leads naturally to an even uglier version of the above problem, since these qualia don't have even the erratic behavioral manifestations of, say, emotional states. It seems that there is no way to check on anyone else's qualia, so it is always, in principle, possible that they are radically different from our own, or even missing entirely.

The problem that has been mounting throughout these examples has two classical formulations—the Other Minds Problem and the Subjectivity Problem. The Other Minds Problem is this: How do we know that other people have conscious minds (as opposed to being zombies or robots), and that those minds are anything like our own? Perhaps the most common response to this problem, and one that is not entirely without merit, is a hearty "Who cares!" of the same sort that's offered in response to solipsism ("How do I know you aren't all just figments of my imagination?") and Hume's problem ("how do I know that all the regularities I've observed won't fail tomorrow?"). The very fact that we get along just fine without answers to these crushing questions can be taken as a hint that the questions themselves are ill posed. A more up-front defense can be constructed, however, by pointing out that we assign mental states to others as part of a general (if only implicit) theory of their functioning, and that the burden of proof for mental states should thus be no stronger than that applied to unobservable theoretical entities in other sciences. We treat such postulated entities as "real," not because we've had some sort of direct personal contact with them, but because they form the basis of theories that allow us successfully to explain and predict the relevant phenomena (in this case, the phenomena of human behavior).

For the specific case of qualia, a few philosophers have suggested ways to meet the problem head-on, either by denying the very existence of qualia as conventionally construed (Dennett, 1988), or by showing in a general way how they might be identified with patterns of neuronal activity (P. M. Churchland, 1986). The latter strategy is reductionist, in the sense that it opts for the possibility that feels can be given an explanation in neurobiological terms. It also points the way to a general means of resolving the "seen from the inside/described from the outside" problem: If one can discover a sufficiently rich physical description, one whose internal relations and dynamics closely mirror those found in the mental or qualitative realm, *that's enough* by the standards of physical science—as seen in past reductions, such as that of temperature to mean molecular kinetic energy—and no further proof of the "absolute necessity" of the correlation need be given.

The second version of the problem—the Subjectivity Problem—is more abstract and subtle, but perhaps of more direct interest to the cognitive sciences. The above states were described as having internal, subjective feels. How, it is asked, can we possibly describe these feels within science? If they are strictly subjective, only available to one person, not measurable by any instruments, how can we incorporate them into our objective science?

In its most extreme form, this problem has been taken as demonstrating that a strictly objective science of the mind is impossible. In his recent book *The Rediscovery of the Mind*, John Searle (1992) devotes a chapter to his claim that subjectivity is beyond the descriptive resources of objective science as we now conceive it. It follows from this that we can never have an adequate theory of consciousness unless we allow irreducibly subjective elements ("feels" of some sort or other) as basic objects in our science, right alongside the atoms, force fields, and so forth. His argument goes like this:

1. Something cannot be part of an objective science unless it is in principle observable.
2. Our notion of observation presupposes the distinction between the observation—the creation of a subjective representation—and the thing observed.
3. We can never make this distinction for consciousness, since its workings are accessible only to itself.
4. Therefore, we can never construct a purely objective account of consciousness.

And the moral, as Searle puts it, is that we should "stop [trying to 'picture subjectivity as part of our world view'] altogether at this point and just acknowledge the facts. The facts are that biological processes produce conscious mental phenomena, and these are irreducibly subjective" (pp. 97–99).

This is an interesting argument, to which no one-paragraph summary will do justice. Fortunately, it doesn't take a detailed treatment to show that Searle is making a mistake in telling neuroscientists that an objective science of the mind is a mare's nest. We can begin by noting that this argument bears a striking similarity to a host of apparent paradoxes that have been advanced as "proof" that there are metaphysical limits on what can be discovered or accounted for by science. Kant, for example, thought it was a priori certain (provable by reason alone) that space could not be discovered to be non-Euclidean, and that it made no sense for the universe to have a beginning or for space to be bounded. All of these claims, springing from the seemingly unassailable source of pure reason, turned out to be false. Likewise, before Gödel, the possibility of a dissociation between theorem truth and theorem provability seemed ruled out as paradoxical —as utterly illogical. Until Einstein, the reasonableness of the idea of absolute simultaneity, and the nonsensicalness of relative simultaneity, seemed as obvious as anything could be.

"Obvious" is the problem here. No argument created by a human about real-world phenomena is ever truly a priori; of necessity, any such argument will utilize terms and concepts that have empirical origins and that reflect the arguer's conceptual framework. Thus, what seems at one stage in our knowledge to be an a priori limitation on what can be discovered often turns out to be a limitation based on ignorance.[14] There was much about universes, space, and time that had yet to be discovered when Kant thought about them—and the Newtonian theory he was working with was highly sophisticated by comparison to our current understanding of consciousness.

One can already discern the beginnings of possible fractures in Searle's a priori truths: For example, does consciousness really have to be an undifferentiated blob, such that "any introspection I have of my own conscious state is itself that conscious state," (p. 97) or could a sophisticated theory of consciousness make subtler distinctions than that? Searle's frequent dependence on "obvious facts" to establish his ideas about consciousness, here and throughout the book,[15] is misleading. It's obvious that there's *something* like consciousness going on, but the same cannot be said for the particular *properties* of consciousness as we now describe it, and we should certainly be wary of any argument that relies on assumptions about those selfsame properties in telling us what we can and cannot ever find out about them.

Prospects for a neuroscience of consciousness

As we have seen, some philosophers are pessimistic about the ability of neuroscience to explain conscious phenomena. Even though the specific arguments for such positions may be unconvincing, it is nevertheless reasonable to concede the abstract point that we do not know how far science can take us. For all we know (in our current state of profound ignorance), it might be that problems now unseen will ultimately frustrate neurobiological approaches to consciousness. This concession, however, is of the vacuous "we-might-all-be-dead-tomorrow" variety, and as such it lacks a single identifiable methodological consequence. It does not, for example, mean that we should give up neuroscientific exploration of the various conscious phenomena, or that projects studying the neurobiology of visual awareness or working memory or attention are a waste of time. It does not entail that any specific ongoing projects should be abandoned, or that any particular phenomenon should be given up on as permanently inscrutable. It should be evident, even to the die-hard pessimist, that much can still be learned about the mind via neuroscience. Unless and until science has hit the wall, so to speak, it makes sense to keep trying to make progress.

Moreover, it is useful to remember that scientific discoveries typically provoke a recharacterization of the problems and a recharting of the logical terrain, along with a host of new experimental questions. In consequence, what seems muddy and intractable at an early stage may well become much clearer and more approachable in the context of an advanced, experimentally grounded understanding. In addition, the advent of new techniques, such as functional magnetic resonance imaging (MRI) and magnetic electroencephalography (MEG), can open up whole new domains of study, expanding the list of phenomena that can be considered experimentally approachable.

This doesn't mean that the difficulties should be denied or downplayed; neuroscience *is* hard (Crick, 1994; Churchland and Sejnowski, 1992), and the brain might be more complicated than it is smart (Mark M. Churchland, personal communication). The pragmatic response to the pessimist is simply that progress is clearly still possible, and progress changes the look of the problem. Because science regularly and *unpredictably* overturns deep-seated, heartfelt, and plain-as-plants verities about what we shall never, never understand, the most practical plan is to avoid prohibitive speculation on the ultimate limits of our understanding and to work instead on specific problems, seeing how much progress we in fact make and in what directions our discoveries take us.

There are a number of such promising entry points

into the various issues concerning consciousness, from which specific neurobiological and psychological questions can be asked. For example, as discussed earlier in the section on conscious awareness, one basic expectation is that there are discoverable differences in the brain between the following two conditions: (1) A stimulus is presented and the subject is aware of it; and (2) a stimulus is presented and the subject is not aware of it.

The working hypothesis is, quite simply, that it should be possible to discover what those differences are. The next step will be to figure out how such data fit into what else is known about neuronal function, connectivity patterns, the pharmacological milieu, the functional role of particular brain regions, behavioral data on perception, learning, memory, feelings, attention, dreaming, and so forth. Coevolution of understanding of what is going on at many different levels of organization, from molecules to neurons to networks to behavior, is the benchmark of modern neuroscience.

This sort of multilevel approach is already being applied to mental functions that bear on consciousness. Working memory, for example, has been the focus of intense research by Patricia Goldman-Rakic (1987) and her colleagues. Studying macaques, they found that lesions to the dorsolateral frontal region of cortex specifically impair working memory for spatial location. In the task, the monkey sees a light come on at a specific location, and has to remember its location long enough to be able to report it via eye movements once the end of the delay period (varying from one second to tens of seconds) is signaled.

At the single-cell level, Goldman-Rakic and colleagues find a topographic organization of cells responsive to perception of the light in particular locations. The breakthrough relevant to working memory is that they also find a population of spatially mapped cells active only during the delay period—that is, during the period when the light is no longer visible and the monkey has to keep the light location in memory while waiting for a cue to produce behavior. When "delay-over" is cued, activity in these "delay" cells falls off, while activity in a distinct, third population of cells begins. Despite the brevity of this summary, which omits the anatomical and physiological data, the basic point emerges quite clearly: The Goldman-Rakic laboratory has discovered what appears to be a neuronal correlate of working memory for particular visual-spatial locations.

Holding something in mind while waiting for the right moment to act involves aspects of awareness, attention, and conscious control, and the Goldman-Rakic work begins to let us see what sort of brain mechanisms might subserve these functions. Obviously, Goldman-Rakic has not yet explained how working memory works; we still have a long way to go on that subject. But all the customary cautions notwithstanding, the fact is that this research constitutes a major experimental and conceptual advance, and one directly relevant to phenomena involving consciousness.

Binocular rivalry turns out to be another rewarding entry route into the domain of consciousness. Here too, productive research has the virtue of interlocking data from both behavioral and cellular levels of organization (Logothetis and Schall, 1989). The standard setup involves constructing opposing visual stimuli, such as vertical and horizontal bars (or upgoing and downgoing gratings), and presenting stimuli so that each eye is continuously receiving a stimulus that rivals the image presented to the other eye.

In such "rivalrous" conditions, human subjects report that they see first one (e.g., horizontal), then the other (e.g., vertical), then the first, and so on in an alternating fashion, with a frequency of about 1 Hz. In other words, though the rivalrous stimuli are continuously present, the perception fluctuates regularly. Logothetis and his colleagues use the rivalry paradigm to ask the following question: Given that there are tuned cells in visual cortical area MT that continuously respond to the continuous stimuli (e.g., upgoing gratings), is there also a population of tuned cells whose activity fluctuates concordantly (at about 1 Hz) with the subject's fluctuating perception? In other words, are there cells tuned to upgoing movement whose activity is confined to those periods when the subject *consciously perceives* such movement?

The answer is yes, where a behavioral response is used as an index of what the animal sees. That is, as expected, a subpopulation of "upgoing" cells is continuously active during the presentation of upgoing stimuli to one eye and downgoing stimuli to the other eye. The breakthrough, however, consists in the discovery that another population of "upgoing" cells is active only when the subject *perceives* upward motion.

Observing the customary cautions and sidelining many important details, we can sum up the crucial finding: Logothetis and colleagues find some cells whose activity profile parallels the subject's *visual ex-*

perience, as opposed to the mere presence of the visual stimulus. Obviously, this research alone does not provide a neurobiological explanation of visual awareness. It is, however, a substantial foot in a most intriguing door. Among other things, it provokes a range of new questions bearing on visual awareness, and it suggests that exploring the relations between working memory and visual perception may be rewarding. The results also confer on this research a kind of bellwether status, inasmuch as they represent the sort of direct correlation between identifiable features of brain activity and conscious experience that antireductionist theories such as Searle's and Dennett's would suggest we are unlikely to find.

These are merely two examples; interesting work is being done on many other aspects of consciousness, including, of course, attention (Desimone and Ungerleider, 1989; Maunsell et al., 1991).[16] Here too, questions are framed and data are interpreted against a background of psychological experiments on attentional phenomena, as well as lesion studies on humans and other animals. It is not yet understood how attention connects to working memory and visual awareness, nor how these interact with more general conscious states; but there is no reason to assume that these problems are any less tractable than the ones on which these researchers are already making such impressive progress.

Concluding remarks

If there is a unifying purpose to the foregoing, it is that of clearing ground for the development and acceptance of a neuroscientific account of the various phenomena we collect under the term *consciousness*. Considering the surprising resistance that neuroscientific approaches to consciousness encounter in the scientific, philosophical, and lay communities, one must devote a certain amount of time to proving that such an account is desirable and possible. Far more interesting, however, are the questions neuroscientists must pose for themselves: What exactly is to be explained? How does one apply the scientific method to something so nebulous and ill defined? What counts as a successful explanation? And what is it about consciousness that makes even the most tentative of advances so exciting?

As seen in the previous section, methodologies are already developing for the study of certain aspects of consciousness. Each of the cases described involved searching for neural correlates of particular phenomena that are normally described in terms of conscious experience. Rather than addressing the whole grand, multifarious concept all at once, scientists such as Goldman-Rakic and Logothetis are finding tractable, well-defined subquestions, which increase the likelihood of their achieving results that are unambiguous and easily integrated with the rest of science. In addition, the specificity of their claims prevents the talking at cross purposes that can result from overly general use of the word *consciousness*. Ultimately, this approach also squares with the moral about conceptual evolution that was developed in the first section of the chapter: Since the findings can be described without essential reference to consciousness, they do not commit us to any particular conception of it, and hence can be used as data in the gradual process of developing a conception that is both scientifically grounded and philosophically satisfying.

NOTES

1. None of this implies that the old categories need be discarded entirely; they may still be quite useful, especially where the more sophisticated forms of description are not needed or are not available. Ultimately, one would hope that the scientific categories would come to be reflected in daily use (which would make tough "everyday" questions susceptible to scientific inquiry), but even in the most established sciences this is happening slowly at best.
2. Though popularized in modern times by T. S. Kuhn, this point was recognized at least as early as 1830; in their *Treatise on Mechanics*, Henry Kater and Dionysius Lardner follow their discussion of Newton's laws and definitions with this disclaimer: "We have noticed these formularies more from a respect for the authorities by which they have been adopted, than from any persuasion of their utility. Their full import cannot be comprehended until nearly the whole of elementary mechanics has been acquired, and then all such summaries become useless."
3. Recognizing that significant definitional precision *accompanies* rather than *precedes* empirical discovery, Dennett (in conversation) mocks the define-the-words-first strategy as "the heartbreak of premature definition."
4. The authors claim no special status for the divisions offered here; there are doubtless other, equally valid ways of carving up the domain. In particular, since these categories were chosen to reflect the structure of our *current* (effectively prescientific) ways of talking about consciousness, they should not be taken as predictions about, or suggestions for, the shape that more sophisticated theories might take.

5. It should be noted that, at least with our current vocabulary for describing these things, this seems to lead to problems about whether the "self" that is being consciously aware is really the same as the "self" that is doing the thinking; William James puzzled over this at length, and there is a good discussion of it in Lyons (1986).
6. Our choice of terms here is deliberately archaic; using *faculties* instead of, say, *abilities* or *mechanisms* highlights the fact that we are dealing with a set of intuitions that are for the most part very old, poorly understood, and probably in for a good deal of revision.
7. It is interesting that this supposition, so universal and innocuous just a short time ago, is almost inconceivable to modern theorists in artificial intelligence or neuroscience.
8. For example, in one experiment subjects were asked to rate the emotional tone of a selection from a novel; some groups read versions from which certain passages had been deleted. They were then asked to describe how the passages affected their decisions, or, for the deletion groups, to *predict* what effect they would have had if present. All the groups answered similarly, and in a way that assigned considerable importance to the passages, despite the fact that none of the passages exhibited any significant effect on actual responses. While not conclusive, the data strongly suggest that the subjects who did read the passage were in some sense "predicting" too, rather than consulting a privileged introspective history.
9. Some claim that this is because the basic structure of thought is linguistic (as in Fodor's "language of thought" hypothesis). The neuroscientific data have not been kind to this theory, but there is no reason to doubt that language and consciousness exerted considerable influence on each other's evolution, and continue to interact within development.
10. This description is admittedly a bit backward, since the symbolic systems themselves—AI programs, computer architectures, logics—were modeled after the perceived symbolic character of thought. Dennett (1991, 215) has a good discussion of this point with respect to the von Neumann computer architecture.
11. The idea here is that, at the level of the actual operations of the computer's central processing unit, there are no pieces of paper, hammers and strings, or ailerons; but for the user's benefit, the computer *acts as if* there were. This illusion is most complete when the computer is emulating something else with very similar input and output capabilities (e.g., another computer), so that there is no easy way to tell whether one is dealing with an emulation or the real thing.
12. Dennett 1991, p. 219: "[S]uccessful installation [of human consciousness] is determined by myriad microsettings in the brain, which means that its functionally important features are very likely to be invisible to neuroanatomical scrutiny in spite of the extreme salience of the effects."
13. There are, however, good reasons for being skeptical about this, which will be discussed later.
14. Searle seems to acknowledge something like this point in a critique of Thomas Nagel: "[Consciousness] seems mysterious because we do not know how the system of neurophysiology/consciousness works, and an adequate knowledge of how it works would remove the mystery." (Searle 1992, 102)
15. See especially pp. 3, 30, 48, 51, 52, 60, and 247.
16. As in the above cases, the research strategy for investigating the neurobiology of attention involves exploring neuronal properties in the brain of an awake monkey trained to make a behavioral response indexed to a mental state, in this case a shift in attention.

REFERENCES

Baars, B., 1988. *A Cognitive Theory of Consciousness.* Cambridge: Cambridge University Press.

Churchland, P. M., 1986. Some reductive strategies in cognitive neurobiology. *Mind* 379:95. Reprinted in *A Neurocomputational Perspective.* Cambridge, Mass.: MIT Press.

Churchland, P. S., 1983. Consciousness: The transmutation of a concept. *Pacific Philos. Q.* 64:80–95.

Churchland, P. S., and T. Sejnowski, 1992. *Computational Neuroscience.* Cambridge, Mass.: MIT Press.

Crick, F., 1994. *The Astonishing Hypothesis: The Scientific Search for the Soul.* New York: Macmillan.

Dawkins, R., 1989. *The Selfish Gene,* New Edition. Oxford: Oxford University Press.

Dennett, D., 1988. Quining qualia. In *Consciousness in Contemporary Science,* A. Marcel and E. Bisiach, eds. New York: Oxford University Press, pp. 42–77.

Dennett, D., 1991. *Consciousness Explained.* Boston, Mass.: Little, Brown and Company.

Desimone, R., and L. Ungerleider, 1989. Neural mechanisms of visual processing in monkeys. In *Handbook of Neuropsychology,* vol. 2, F. Boller and J. Grafman, eds. Amsterdam: Elsevier, pp. 267–299.

Farber, I., 1991. Symbolic consciousness, connectionist brain: Architectural clues from introspection and cognitive science. Unpublished manuscript.

Fodor, J., 1975. *The Language of Thought.* New York: Crowell.

Gazzaniga, M., 1985. *The Social Brain: Discovering the Networks of the Mind.* New York: Basic Books.

Goldman-Rakic, P., 1987. Circuitry of the pre-frontal cortex and the regulation of behavior by representational memory. In *Higher Cortical Function: Handbook of Physiology,* F. Blum and V. Mountcastle, eds. Washington, D.C.: American Physiological Society, pp. 373–417.

Logothetis, N., and J. Schall, 1989. Neuronal correlates of subjective visual perception. *Science* 245:761–763.

Lycan, W., 1987. *Consciousness.* Cambridge, Mass.: MIT Press.

Lyons, W., 1986. *The Disappearance of Introspection.* Cambridge, Mass.: MIT Press.

Maunsell, J., G. Sclar, T. Nealey, and D. DePriest, 1991. Extraretinal representations in area V4 in the macaque monkey. *Vis. Neurosci.* 7:561–573.

Minsky, M., 1986. *The Society of Mind.* New York: Simon and Schuster.

Nisbett, R., and T. Wilson, 1977. Telling more than we can know: Verbal reports on mental processes. *Psychol. Rev.* 84(3).

Searle, J., 1992. *The Rediscovery of the Mind.* Cambridge, Mass.: MIT Press.

Squire, L. R., and N. Cohen, 1984. Human memory and amnesia. In *Neurobiology of Learning and Memory*, G. Lynch, J. L. McGaugh, and N. M. Weinberger, eds. New York: Guilford Press, pp. 3–64.

86 Cognitive Aspects of Consciousness

WILLIAM HIRST

ABSTRACT Cognitive scientists have addressed several issues concerning the nature of consciousness that can be addressed empirically. This paper reviews several of them: (1) Can people be aware of processes as well as products of processes? (2) What are the limits on conscious experience? (3) Do people go beyond the information given in the stimulus array? (4) What and how much can be processed unconsciously? (5) What guides unconscious processing? (6) What are the effects of the unconscious on conscious behavior? Throughout this review, I argue that it is better to view attention as a collection of processes rather than as a distinct faculty or mechanism.

In this chapter I examine six questions about consciousness for which cognitive psychologists have supplied at least partial answers, answers built around issues of processing, representation, and task demands. Each section of the chapter focuses on one of these questions. I should make clear from the start what I mean when I use the word *consciousness*. I am referring to the awareness people have of mental objects, be they percepts, images, or feelings. Although some might contend that people can be aware of something before reflecting on it—that is, they can be aware of something without realizing it—as I am using the term, to be *aware* entails being able to report this awareness. People are not merely aware of a percept, image, or feeling; they also are aware that they are experiencing it. They know that *they* are seeing, imagining, and feeling. Any act of consciousness involves self-awareness as well as awareness of the world, mental images, or feelings. How a person reports this awareness can vary, but in each case, in order to say that a person is conscious of a mountain, a crashing wave, or desire, the person must acknowledge that he or she sees a mountain, imagines a crashing wave, or feels desire. The six questions addressed herein concern this sense of consciousness. (Although I use *attending* and *being conscious* or *being aware* interchangeably in this chapter, I recognize that these terms, while closely related, do not capture the same phenomena. To attend to something is to bring it into the focus of conscious awareness, so the term implies that one is conscious or aware. On the other hand, one can be conscious of things that one is not necessarily attending to, such as objects on the periphery of one's consciousness.)

WILLIAM HIRST Department of Psychology, New School for Social Research, New York, N.Y.

Can people be aware of processes as well as products of processes?

According to Nisbett and Wilson (1977), people are aware of content, not process. For Nisbett and Wilson, the products of perceptual processing, mnemonic retrieving, and imaging—the produced percepts, memory beliefs, and images—form the landscape of conscious experience. Of course, people can report on processing, but as Nisbett and Wilson cautioned, these introspective reports on internal processing are often merely constructions after the fact—theories about their processing—rather than on-line conscious awareness of processing.

For instance, Nisbett and his colleagues asked subjects to memorize word pairs that might prime responses in a subsequent word-association task. The experimenters asked subjects to study *ocean–moon*, anticipating that the pair might bias them to respond "Tide" on a free association task when given *detergent*. The experimenters were effective in creating the expected bias; subjects were twice as likely to respond with the desired response when exposed to the relevant word pair than when not exposed. However, subjects almost never explained their behavior by referring to the memorization task. They opined that "Tide is the best-known detergent" or "My mother uses Tide." Their mistaken reports simply did not reflect the underlying processing producing this behavior.

Sometimes, however, introspective reports may capture what is going on internally. For certain tasks, cognitive psychologists treat subjects' introspections about internal processing as primary data. In their study of human problem solving, Newell and Simon (1972) employed protocol analysis as their basic investigative tool, for instance, asking subjects to speak aloud their thoughts while solving the puzzle the Tower of Hanoi. Subjects' articulation of their line of reasoning proved to be incredibly perspicacious, providing enough insight in the underlying processing to allow Newell and Simon to construct a working computer program simulating the subjects' behavior.

Why are the subjects seemingly aware of internal processing in the Newell and Simon study, but not in the studies cited by Nisbett and Wilson? According to Ericsson and Simon (1984), the verbal reports of cognitive processes in a protocol are based on a subset of the information in short-term memory and long-term memory, the kind of information Nisbett and Wilson claimed people have access to. Nisbett and Wilson maintained that people have access to their focus of attention, their current sensations, their emotions and evaluations, and their plans. Subjects in Newell and Simon's study reported on these features of their consciousness and the maneuvers they undertook to change these features. Their subjects did not, however, report on the processes that underlie their attention, sensations, emotions, evaluations, and plans. To the extent, then, that people are accurately aware of processing, these processes are probably transformations of information already in consciousness. As far as I can tell, no one has claimed that people can be aware of the processes by which information first springs into consciousness.

Are there limits on conscious experience?

Conscious awareness of the external world does not mirror the world. People can fail to be conscious of much that impinges on the sensorium. Some limits arise for what might be called trivial reasons, so-called data limitations (Norman and Bobrow, 1975). People cannot see light with wavelengths above a certain frequency or hear sounds below a certain decibel level. These limitations usually reflect restrictions imposed by our sensory organs. At least from the perspective of a cognitive psychologist, the more interesting restrictions on conscious awareness go beyond data limitations. Some scholars have claimed that people are conscious of only one thing at a time—or, as William James (1890) put it, "not easily more than one, unless the processes are very habitual; but then two, or even three" (p. 409). Whatever consciousness is, it cannot hold on to many things at a time.

One could conceive of this limit as an inherent characteristic of consciousness. From this perspective, human consciousness cannot function like a split screen, with, say, two simultaneous messages appearing with equal clarity and detail. People must trade off attention to one message for attention to another (Norman and Bobrow, 1975). On the other hand, one could trace the limit to the processing underlying conscious perception rather than to an inherent characteristic of consciousness itself. That is, limits may exist on the means by which products are constructed rather than on the size of the stage on which these products are displayed. From this perspective, how much or what people can be simultaneously conscious of would depend on the processing demands and the repertoire of mental skills.

The current evidence supports a processing explanation. For instance, the ability to attend to two messages simultaneously depends not just on how much one must keep in mind at one time, but also on how hard it is to keep the two messages internally segregated (Hirst and Kalmar, 1987). The human mind may be a parallel processor par excellence as long as concurrent streams of processing remain separate. When the output at any stage in one stream of processing serves as the input at some stage in the other stream of processing, cross talk or a breakdown in one or both streams of processing could result. Consequently, how much one can simultaneously attend to should depend on the modality of the stimulus and the mode of response, among other factors. Indeed, it is easier to hold in consciousness one auditory and one visual message than either two auditory messages or two visual messages (Wickens, 1980, 1984).

Multiple-resource models of attention provide an alternative to this segregation model. According to Wickens (1980, 1984), a small number of distinct processing resources may exist. Resources can be thought of as reservoirs of fuel that drive processing, or as collections of processing mechanisms. One resource "pool" might be limited to processing visual material and another to processing auditory messages. The multiple-resource explanation, however, cannot elegantly account for why it is easier to attend to one string of animal terms and one string of vegetable terms than either to two strings of animal terms or to two strings of vegetable terms (Hirst and Kalmar, 1987). From a multiple-resource perspective, this latter finding suggests that separate resources exist for every semantic distinction, clearly an untenable notion. The segregation approach offers a more parsimonious explanation, with semantically distinct concurrent messages providing a basis for segregation.

Limits on conscious experience cannot be accounted for solely by appeals to segregation. Limits will also vary as people organize incoming information. For instance, in parallel search tasks, in which subjects scan a long list of letter strings for target letters, one might attribute the observed "pop-out" effect to subjects' ability to structure target letters into larger units—for instance, to organize into a larger unit the targets *B* and *P* because they possess curvilinear shapes (Neisser, Novick, and Lazar, 1963; Neisser, 1964; Gibson and Yonas, 1966; see also Schneider and Shiffrin, 1977; Shiffrin and Schneider, 1977).

People also can overcome limitations by automatizing what initially required attention (Schneider and Shiffrin, 1977; Shiffrin and Schneider, 1977; see also Cheng, 1985). Definitions of automaticity vary (see Kahneman and Treisman, 1984, for a review). Automatic processing usually does not interfere with concurrent processing and does not require effort or intentionality (Shiffrin and Schneider, 1977; Hasher and Zacks, 1979; but see Hanson and Hirst, 1989). For our purposes here, automaticity transforms processes that once occurred consciously and effortfully into ones that occur effortlessly and without awareness or consciousness. Target detection in parallel search tasks may at first involve conscious search, but with sufficient practice, the search becomes automatic, outside conscious experience (Schneider and Shiffrin, 1977). A beginning pianist may initially consciously read a note and then press the correct key, but the expert smoothly and without any conscious awareness simply plays what is written on the page (Sudnow, 1978). Education plays a part in transforming conscious, effortful mental activity to automatic activity.

A full account of automaticity should bear on issues about the nature of consciousness. In the current debate of automaticity, some scholars argue that automatic and effortful processing differ qualitatively (Shiffrin and Schneider, 1977); that the two differ only quantitatively, that is, that automatic processing is effortful processing "sped up" (Posner and Snyder, 1975); that automatic processing reflects changes in the organizational structure of the processing (Cheng, 1985); or that automaticity arises from multiple representations of instances of past experience (Logan, 1988). These differing positions have their counterparts in discussions of consciousness. Thus, a quantitative difference in automatic and effortful processing might suggest that consciousness may differ only quantitatively from unconsciousness. However, it makes little sense to pursue the implications of a quantitative difference in conscious and unconscious processing if the automaticity issue comes down solidly for a qualitative difference; and, to some extent, psychologists have resisted studying the connection between automaticity and consciousness until they better understand the phenomenon of automaticity. Whatever the exact relation between automaticity and consciousness, the importance of automaticity to an understanding of consciousness is clear: How much a person can be conscious of at one time can change as processes become automatic.

Finally, people can expand their capacity either by increasing their effort (Kahneman, 1973) or by improving their perceptual and attentional skills (Hirst, 1986). One could trace some benefit of improved skills to increases in automaticity or organization, but sometimes the improvement cannot be straightforwardly traced to either of these factors. Studies on simultaneous reading and writing suggest that practice may actually increase mental capacity, just as exercise increases muscle strength (Hirst et al., 1980).

In each of these cases, how much is held in consciousness depends on how information is processed. People can see, hear, and sense more or less of the world depending on how they organize it; their repertoire of automatic skills; the modality, semantic characteristics, and mode of response; the amount of effort they are exerting; and their level of perceptual and attentional skills. As people's organizational and perceptual skills change, as effortful mental acts become automatic, as people put more effort into a task, limits on conscious experience can change. What an individual could previously discern only in a sequence of conscious images becomes available in a single image. When learning to identify a painting by Klee, I may initially see first the distinctive coloring and then, separately, the primitive line drawing. Yet with practice, I begin to see the painting as a unit, lines and colors combined to give the canvas its Klee quality. Consciousness is limited by flexible constraints on processing rather than by an inherent, fixed limit on conscious experience.

Does consciousness go beyond the information given?

Conscious awareness of the external world may fail to mirror the world not only because of limits on how much can be processed at the same time, but also because it imports the characteristics of mental processing to the conscious experience. Again, consider perception, although a similar argument could be made for memory and imagination. People do not just see what is out there in the world. They go beyond the information given; they see, hear, touch, and generally are conscious of nonexistent stimuli. This phenomenon is well documented. In studies of the phoneme restoration effect, an experimenter replaced a phoneme in an auditory message with white noise (Warren, 1970). Subjects reported hearing the missing phoneme, with the white noise floating above the message. They were conscious of something not there. Similarly, people do not see the blind spot in their vision (Hochberg, 1964). They fill the gap and have no sense that they are blind to a small spot near the center of their visual field.

If people fill gaps as they perceive, then consciousness reflects this construction, not what is in the world. This point may seem obvious, but it cannot be overemphasized. Just as perceptual processing is not passive, so consciousness is not a passive reflection of the world. People may build the content of their consciousness out of the elements of the world, but they clearly go beyond this information. Although people sometimes actually hallucinate, mostly the world constrains the perceptual content of consciousness.

In treating consciousness as the product of mental processing, I am, of course, finessing a central problem. Most models of cognition involve a chain of discrete unconscious processes, each with its own input and output. Only the final result becomes conscious. For each process in this chain, there is presumably a product, or output, that serves as the input to the next process in the chain. Why does the final output serve as the content of consciousness and not the intervening outputs? How can one determine which output is the conscious one?

I am not sure how to answer these questions. The simplest answer, and perhaps the correct one, is that the chain of processing is built that way. Although infants come into the world with many innate abilities, they also must acquire many perceptual skills (Gibson, 1969). For each new perceptual skill, they must learn, among myriad things, to produce a conscious experience consistent with what impinges on the sensorium. People are said to have learned to discriminate wines when, after considerable hours of pleasant practice, they taste the difference between a St. Emillion and a Pauliac. They previously did not have the conscious experience of the wines differing in taste, and they would claim to have learned the appropriate enological skills only when their conscious experiences differ as the wines differ. In their struggle to acquire the appropriate underlying processing, people seek to change their conscious experience. The skill can be said to be learned only when newly acquired processing produces the desired conscious experience. Consequently, the final output of a chain of processing serves as the content of consciousness in part because the processing was acquired to produce this output.

What or how much can be processed unconsciously?

Almost any discussion of consciousness inevitably turns to the unconscious. The distinction between consciousness and the unconscious usually is associated with Freud, but can be traced back to Leibnitz, or perhaps farther (Ellenberger, 1970). There is little doubt that unconscious processing exists. The processes transforming stimulation on the retina into the percept of a glass remain beyond introspection, yet most cognitive psychologists would claim that they exist (Kihlstrom, 1987). Determining the character of this cognitive unconscious represents a major endeavor for cognitive science. Disagreements exist over the level of analysis of incoming information possible without consciousness. A low, or shallow, level of analysis would involve the processing of physical features of the stimulus, for example, the shape of the letters in the word *bottle*. That is, a person may be able to process the physical, even the orthographic, features of the word, but fail to identify the word as having the meaning "bottle." A higher, or deeper, level of analysis might entail the processing of the meaning of the word.

Or consider face perception. A person may be able to recognize a picture as depicting the face of an elderly woman with blue eyes, blond hair, and pug nose, yet she may fail to identify the picture as one of her mother. She accomplishes the low level of analysis of detecting the features, but not the deeper level of analysis of identifying the person depicted in the picture.

Is the unconscious capable of deep analysis (Dixon, 1981; Loftus and Klinger, 1992)? The Freudian unconscious is "hot and wet," (Kihlstom, 1987) with dynamic conflicts between deeply meaningful forces creating at times a bubbling cauldron of psychic pressure. This unconscious clearly is capable of deep analysis. Some cognitive psychologists strip this characterization of much of its affective valence, yet still assign to the unconscious the ability to perform of a deep level of analysis (Marcel, 1983a, b; Greenwald, 1992). Other cognitive psychologists attribute only shallow processing to the unconscious (e.g., Broadbent, 1958). For them, information impinging on the sensorium can be processed at a meaningful level only if it emerges into consciousness. Psychologists have explored the depth of unconscious processing using three experimental paradigms: subliminal perception, dichotic listening, and neuropsychological investigation.

Starting with Poetzl (1917) and achieving full steam with the work of Klein (1954) and his associates, researchers have examined whether subliminally presented stimuli affect subsequent judgments or behaviors. If the subthreshold stimuli affect subsequent judgments, then subjects must have unconsciously processed the material. For instance, Eagle (1959), working in the tradition of George Klein, flashed subliminally a picture depicting either a boy angrily throwing a birthday cake or the same boy pleasantly presenting the cake, and then followed it with a superthreshold presentation of the boy adopting a neutral posture. Subjects judged the personality of the boy in the superthreshold presentation. Although subjects claimed not to see the first picture, its content biased their judgment of second picture.

The Eagle study, as well as related efforts (see Ericksen, 1956, for a review), suggest that people can meaningfully and affectively process a subliminally presented picture. Yet replication is difficult (Ericksen, 1956). Critics wondered whether subjects were indeed unaware of the subliminal material. For instance, subjects may not see the picture in its entirety, but may see fragments of the picture that themselves could bias the subsequent judgment. They may see angular line segments in the picture of the angry boy and curved lines in the picture of the pleasant boy without seeing the picture as a whole or processing its meaning. The perception of these fragments may not be sufficient to exceed subjects' criteria to say they saw the picture, but may still be sufficient to influence the interpretation of the superthreshold picture. In addition, subjects may even see the subliminally presented picture for a few milliseconds, but their memory for this brief sighting may quickly fade. Consequently, they may fail to report seeing the pictures, but the picture may nevertheless have left enough of a trace to bias the interpretation of superthreshold picture. Whatever the possible confounds, the subliminal perception effect could be traced to some level of awareness. This awareness may be enough to bias their judgments about the superthreshold presentation, but not enough to support recognition of the subliminally presented material.

Recent work has tried to avoid such problems, with only limited success. For instance, Marcel (1983a, b) flashed at subthreshold speeds either a word or a blank field, followed by a mask. The mask was a cross patch of letters designed to prevent the perseverance of the word. One of two tasks followed the mask. In the detection task, subjects indicated whether a blank field or a word had been flashed. In the lexical decision task, subjects saw a string of letters and specified whether the string formed a word. Response times were recorded. The experiment was designed so that in some trials, the tachistoscopically presented words were semantically related to the subsequent letter strings. If the tachistoscopically presented word was processed to at least a lexical level, then its presence should facilitate lexical decision. On the other hand, if the tachistoscopically presented word was not processed deeply, then its presence should have no effect on lexical decision.

The results supported the claim for deep unconscious processing. Subjects performed at chance levels in the detection task, indicating that they did not see the tachistoscopically presented word. They did, however, demonstrate a clear priming effect: Lexical decision was significantly faster when the lexical string was semantically related to the tachistoscopically presented word than when it was unrelated.

Marcel's results depend on the criteria he used to define the perceptual threshold. In their replication of Marcel, Chessman, and Merikle (1986) distinguished between subjective and objective criteria. Subjective thresholds reflect subjects' reports as to whether they saw something on the screen. Objective thresholds rely on subjects' forced-choice responses. Subjective thresholds are usually higher than objective thresholds. In Chessman and Merikle (1984), subjects correctly detected 66% of the time stimuli that they claimed not to see (chance would have predicted 25%). Subjects may fail to "see" the stimuli—to be consciously aware of it—yet they may respond on a forced-choice examination better than one would predict from chance.

Many researchers have found priming effects similar to those reported by Marcel using subjective thresholds (Fowler et al., 1981; Chessman and Merikle, 1984, 1986; Merikle, 1992). On the other hand, reliable subliminal priming effects remain elusive using an objective threshold (Chessman and Merikle, 1984; Merikle, 1992). The subliminal priming effect may not be limited to single words; subliminal activation has been reports for sentences such as *Mommy and I are one*. This weak but reproducible sentence effect must be approached cautiously, however, because it may reduce to a single-word effect. Sentences other than *Mommy and I are one*, but including the word *Mommy* or *one* can yield effects similar to, albeit weaker than, the original sentence (Silverman and Weinberger, 1985).

A full interpretation of this pattern of results depends critically on how one interprets the two measures of threshold. In both threshold measures, subjects must decide whether they saw something, yet the criteria for making a positive response probably differs. Conscious awareness is not all-or-none, but a matter of degree. One need only think about the objects in the periphery of vision to see how difficult it is to determine whether one is conscious of something. Anyone who has experienced fast tachistoscopic presentation knows how "iffy" the conscious awareness of stimuli can be.

In order to decide whether they actually "saw" something in such "iffy" situations—that is, to decide whether they had the phenomenological feel of seeing—people may establish a stricter criteria than if they merely must choose between "present" and "absent." If the signal is too weak, they may report that they do not see the object, yet still be reluctant to say that nothing was there in a forced-choice examination (Holender, 1986).

There may be, then, a degree of conscious awareness bounded by objective and subjective thresholds. In this realm of conscious experience, subjects may be aware of something—a brief sighting that leaves an impoverished memory trace, fragments of stimuli, or only a weak signal. This limited awareness may be enough to affect subsequent judgments about superthreshold material. For instance, a brief but forgotten sighting may activate the lexical node associated with the subthreshold stimulus. This activation may in turn be strong enough to prime the superthreshold lexical string.

Clearly, one cannot ask, as I did at the beginning of this section, what or how much can be processes unconsciously. This question treats the conscious and unconscious dichotomy as sharp and definite. Rather, if unconsciousness grades into consciousness, a more appropriate question would probe the levels of analysis at various degrees of consciousness. When the most conservative threshold of conscious experience is used—an objective threshold—evidence for meaningful unconscious processing remains elusive. Only if one probes people in the twilight of consciousness between objective and subjective thresholds does evidence of meaningful processing appear. I would maintain that the level of analysis that occurs unconsciously is shallow. When the unconscious is defined using objective criteria, then the level of analysis is limited to the activation of physical features (Chessman and Merikle, 1984, 1986). Only as people begin to be consciously aware—as a subjective threshold permits—does meaningful processing begin to be possible (Marcel, 1983a,b).

The results from the dichotic listening literature tell a similar story. In these experiments, subjects hear two simultaneously presented messages and shadow one of the two messages. Although some dichotic listening experiments support the claim that people can meaningfully process unattended material (Lewis, 1970; Corteen and Dunn, 1972; Corteen and Wood, 1972; Lackner and Garrett, 1973; MacKay, 1973), attempts to replicate these experiments have routinely failed (Treisman, Nauires, and Green, 1974; Newstead and Dennis, 1979; Wardlaw and Kroll, 1976). In other words, the dichotic listening experiments do not offer support for claims of deep unconscious processing.

Studies of the performance of brain-damaged patients do not remarkably change the unfolding story (see Schacter, 1992, for a more complete review). In many instances, subjects seem to be unaware of salient aspects of their environment, but show in some indirect way that they have processed the material. A careful examination of this so-called unconscious processing reveals that analysis rarely goes beyond the physical level (see Bisiach and Berti, this volume). With prosopagnosia, patients demonstrate a galvanic skin response to pictures of familiar faces, but cannot identify the person (see Young and De Haan, 1992, for a review). Patients may be reacting to the physical features of the stimulus. There is no evidence that they unconsciously identify the face in the picture.

Similarly, blindsighted patients report not seeing objects in front them. They can nevertheless point to them (Weiskrantz, 1986). Again, blindsight appears to involve the unconscious processing of physical features of the stimuli—such as the spatial location of the object—but not deeper processing, such as that necessary for discrimination and identification.

Finally, there is the phenomenon of extinction (Volpe, LeDoux, and Gazzaniga, 1979). In a test of extinction, examiners expose patients to stimuli in both visual hemifields either simultaneously or consecutively. With consecutive presentation, patients report seeing both stimuli. However, when the stimuli are flashed simultaneously, the patients can "see" only one of the stimuli, yet they can report correctly whether the two stimuli are the same with an accuracy above that expected from chance alone. They do not need to process the stimuli beyond its physical features to make this judgment.

All three lines of evidence, then, tell a strikingly consistent tale: Unconscious processing unquestionably occurs, yet it analyzes incoming information at a shallow level unless the product of the process emerges into at least the periphery of consciousness.

What guides unconscious processing?

Human information processing could be governed by the goals people adopt. The goals people seek to accomplish—the tasks at hand—could direct what conscious and unconscious processes are undertaken. They could halt a particular line of processing as soon as it is no longer pertinent to the task (Norman, 1968), processing information with a low pertinence shallowly and reserving deep processing for information with the highest pertinence. Information that emerged into consciousness, then, would be both deeply processed and highly relevant to the goals of the organism.

On the other hand, people may process everything that impinges upon the sensorium to a deep level—to what Marcel (1983b) called "the highest and most abstract level" (p. 244)—and then select from this processed information what they want to place into consciousness. From this perspective, what people process unconsciously is not identical with or, in some cases, even closely related to what they are conscious of. The goals people have do not guide unconscious processing, only which product of unconscious processing is placed into consciousness.

As my review to this point indicates, the evidence supports a goal-directed view of unconscious processing. Not everything that impinges on the sensorium is fully processed. Moreover, work on subliminal perception and dichotic listening and neuropsychological investigation suggest that meaningful processing emerges only as consciousness emerges. There is no compelling evidence of meaningful unconscious processing when the strictest criteria for unconsciousness are used.

People do not process everything to the "highest and most abstract level" before they select what will emerge into consciousness. Humans are goal-directed creatures, and what is processed unconsciously will depend on how pertinent it is to accomplishing a goal. Most lines of processing may be safely rejected based on physical features alone. People will whittle down parallel streams of processing based on myriad features, but cannot process something at the "highest and most abstract level" unless it is going to emerge into at least the periphery of consciousness.

Does the unconscious affect conscious behavior?

Freud argued strenuously that unconscious dynamics shape conscious thought and behavior, and that the content of the unconscious has long-term effects on behavior. Again, one need not accept the complex interplay between the unconscious and consciousness envisioned by Freud to recognize that the content of the unconscious may have a long-term yet unverbalizable influence on conscious behavior.

At least two factors appear to govern the impact of the unconscious on conscious behavior: task demands, and the representation people form of the incoming material. In order to understand the importance of these two factors, consider the distinction between *mem-*

ory images and *memory beliefs* (Hirst, 1990; also Johnson and Hirst, 1991; Johnson, Hashtroudi, and Linsday, 1993). When remembering a previously studied word —GARAGE, for instance—an image of the word may come to mind. This *memory image* may be vivid and detailed, or faint and fragmented. It may be elaborated, with information about the spatiotemporal context in which the word occurred, the orthography of the word, the color of the background, and so on, or it may be impoverished, with the only available information being the word itself. People can construct from this memory image a *memory belief* that the image refers to an past event or learned fact. This belief is not an inherent quality of memory, but an adscititious feature. It is constructed, for instance, when the image contains spatial-temporal information, is vivid and rapidly retrieved, and follows from other recollections (Kelley and Lindsay, 1992; Manier and Hirst, in preparation; see also Whittlesea, Jacoby, and Girard, 1990).

Consequently, when a person does not form a rich representation of a past event or the task does not demand that a rich representation be retrieved, it is possible to form a memory image without necessarily forming a memory belief. Past experience, then, could influence present behavior without people being aware of the source of the influence. My contention is that the effect of the unconscious on conscious behavior can be understood in these terms. I will examine several ways the unconscious can have an impact on conscious behavior.

First, can subliminally presented material have a long-term influence on behavior (an effect I will call *subliminal learning*)? Subliminally presented material clearly has an immediate effect on behavior. In the Marcel studies, for example, the influence of the subliminal prime on lexical decision lasted at least a few seconds. The *mere exposure effect* offers an example of long-term subliminal learning: A brief exposure to an item will increase a subject's preference for that item, not just for a few seconds but for minutes and more (Kunst-Wilson and Zajonc, 1980). This exposure need not be conscious (Kunst-Wilson and Zajonc, 1980; Mandler, 1985; Mandler, Nakamura, and Van Zandt, 1987; Bornstein, 1993). Researchers several times flashed visual figures, such as an irregular octagons, fast enough so that subjects could not subsequently recognize the figures. When given a preference test after the material was presented, subjects choose the subliminally presented figure more often than one might expect from chance. Subjects in this experiment formed a memory that could alter preference judgment but could not support recognition.

I want to make two observations about the mere exposure effect. First, once again, an effect is found in the twilight of conscious awareness. Unlike in the subliminal priming work, subjects in this paradigm are not even held to the standard of a subjective threshold. Rather their awareness of the item must be only faint enough to ensure that they cannot later recognize the item. This criterion is presumably at or slightly above the subjective threshold. Under presentation conditions meeting this new criterion, people appear able to form memories.

Second, the mere exposure effect can be understood by assuming that if people are not aware of an item, they will not encode its spatiotemporal context or details about its physical features, relate it to other material, or more generally elaborate upon it; rather the representation will be decontextualized and unelaborated. It is amazing enough that subjects encode item information about material they do not see. More elaborate encoding would require people to go beyond item encoding to relate the item to a spatiotemporal context in which they are not even aware the item is placed. This elaborate processing would require the kind of deep analysis that I argue simply does not occur unconsciously. The automatic encoding of spatial-temporal information observed by Hasher and Zacks (1979) and Manier and Hirst (in preparation) involved superthreshold presentation; it does not speak to encoding under subliminal presentation conditions.

Subjects in a mere exposure experiment, then, are probably forming decontextualized, unelaborated representations. One may not need a vivid, contextualized, and elaborated image to bias a preference judgment—even a faint signal may be enough to alter preferences. On the other hand, it is unlikely that one would construct a memory belief from a decontextualized, unelaborated representation. The subliminal presentation allows one to produce a mnemonic representation rich enough to alter preferences, but so impoverished as to deny recognition.

Sometimes, people may clearly see stimuli, but fail to notice features or relations between or within the stimuli. Subliminal perception is not involved. People claim not to notice that which, if it were pointed out to them, they could unquestionably see. Do subjects in such circumstances form memories of these unnoticed relations, through what I will call *implicit learning*? Are their subsequent judgments or behaviors affected by an implicit knowledge of these features or relations?

Several lines of experimentation have putatively documented implicit learning. In studies of artificial grammars, subjects appear to learn grammatical rules generating structured strings of letters, even though they cannot articulate these rules (Reber, 1967, 1989; Reber and Allen, 1978). In studies of problem solving, subjects appear to learn underlying principles governing a transportation system without being to articulate these principles (Broadbent, Fitzgerald, and Broadbent, 1986; Berry and Broadbent, 1988).

Lewicki, Hill, and Bizet (1980) and Lewicki, Hill, and Czyewska (1992; see also Bargh, 1993) demonstrated the implicit learning of contingencies in social judgment tasks. They presented subjects with pictures of women with short or long hair. Subjects heard short stories about each woman, indicating that they were either kind or capable. Personality traits and hair length were contingent; for example, long-haired women were always kind. Following this presentation, subjects judged the traits of unfamiliar women in a new set of pictures. Although subjects did not notice the relation between hair length and traits, they nevertheless used this association in making subsequent judgments.

And Manier and Hirst (in prep.) showed that subjects can utilize information about an unnoticed contingency between the spatial location of a word on a CRT screen and its membership in one of four semantic categories to guide recognition. Subjects were not consciously using the contingency when making a positive recognition judgment, inasmuch as they were not even aware that the contingency existed. Nevertheless, its unconscious influence led to a conscious sense of recognition, even for semantically related distractors.

Again, I have two points to make about this line of research. First, many researchers conclude from this research that people can implicitly learn complex rules (Reber, 1989), yet in order to make successful judgments in these studies, subjects may need to learn only simple contingencies. In the Lewicki studies, there were one-to-one correspondences between hair length and personality; in the Broadbent work, subjects may merely have learned simple up-down relations between the four variables. And in the Manier study, subjects' judgment relied on simple associations between spatial locations and the targets. Even the artificial grammar studies may reduce to the learning of simple associations. In some instances, subjects must learn only bigrams from the strings to distinguish accurately between grammatical and nongrammatical strings (Dulany, Carlson, and Dewey, 1984; Perruchet, Gallego, and Savy, 1990; Perruchet and Pacteau, 1990). Subjects in the artificial grammar studies may not be able to verbalize the grammatical rules, then, because they have not learned them, implicitly *or* explicitly.

Second, a careful examination of the representations formed in these experiments could account for the pattern of results. In the Lewicki and Broadbent experiments, subjects recognized the material that they were shown, but could articulate any of the contingencies. Their recognition of the woman, her hair length, the values for the variables, and the letter strings is not surprising. After all, they clearly noticed this material in the study phases, and consequently one might expect a contextualized and elaborated representation of this material, a representation that could support recognition.

Their inability to verbalize the contingencies also is not surprising. The representation of the unnoticed contingency in the Lewicki and Broadbent studies probably remains decontextualized and unelaborated. This impoverished representation may not support recognition, let alone recall, yet it might be embellished enough to support a shift in social perception or a change in guessing a value. The case of implicit learning, then, is similar to that of subliminal learning. Subjects' performance on the direct and indirect memory tests probably reflects their inability to build a contextualized and elaborated representation.

Of course, subjects may both be aware of the stimulus material at presentation and remember it later on a memory test. In this instance, I will claim that the material is *explicitly learned. Implicit memory* refers to those instances in which the explicitly learned material affects subsequent behavior without the subjects being aware that they are remembering the material. Moscovitch (in chapter 89) surveys the literature in detail. Suffice it to say that many researchers maintain that at least two different kinds of memories exist: implicit memory, in which perceptual representations are stored and retrieval occurs without awareness, and explicit memory, in which more elaborated representations are stored and retrieval occurs with awareness (see Tulving, 1983; Schacter, 1987, 1989; Tulving and Schacter, 1990; Squire, 1992). Interest here focuses not on the claim that two separable memory systems exist, but on the degree of awareness involved in these two kinds of memory.

What are subjects aware of in direct memory tests that they are unaware of in indirect memory tests? The distinction between memory image and memory belief will help answer this question. Consider an experiment in which subjects study the word *garage* and then are asked either to recall the word that began with *gar* (cued recall) or to say the first word that comes to mind (stem completion). In both tasks, subjects make the same response—they bring to mind the same memory image, *garage*. In the cued-recall task, however, subjects are also aware of the source of their response—that the word *garage* occurred in the studied list. They possess what I call a memory belief. As we have already seen, memory beliefs need not accompany the memory image produced in a stem completion task. Subjects often report in stem-completion tasks that they do not know why the word *garage* sprung to mind.

A difference in task demands, not representation, could account for the phenomenological difference in these two experimental conditions (Roediger, 1990). Both tasks demand that subjects retrieve a memory image. In the cued-recall condition, subjects also must retrieve information about the source of the memory image—that it occurred in the list. In the stem-completion task, they must retrieve only the memory image. It is not that the retrieval in the stem-completion task is unconscious, as Greenwald (1992) has recently written, or that the memory itself is unconscious. Rather, subjects may be unaware of the source of their memory image and consequently fail to construct a memory belief.

Conclusion

Are there empirically addressable yet meaningful questions about consciousness? The answer should be clear at this point. I have raised six questions:

1. Can people be aware of processes as well as the products of these processes? The answer depends on how we define *process*. In most cases, the transformations of the content of short-term and long-term memory are consciously available and can be elicited through protocols, as Newell and Simon (1972) have amply demonstrated. The processes producing this content are rarely, if ever, accessible to conscious awareness.

2. What are the limits on conscious experience? Consciousness itself may not be limited. The amount of information held in consciousness at any moment will vary as a function of the nature of the material, the degree to which people can segregate competing messages, their perceptual and attentional skills, the effort they put into the task at hand, and the automatic nature of the underlying processes. Consequently, limits on conscious experience reflect underlying processing rather the amount of information the stage lights of consciousness can illuminate at the same time.

3. Is conscious awareness a direct reflection of the world? A clear *no*. Not only do limits on consciousness force people to select what they will attend to, but the constructive nature of mental processing allows people to go beyond the information given and be aware of things that are not in the stimulus. How much of consciousness reflects this construction—and hence is open to social influence—and how much depends on encapsulated, modular processing (Fodor, 1983) is not clear.

4 and 5. What and how much is processed unconsciously? The answer seems to be that it depends on how we define *unconscious*. If strict criteria are adopted, then unconscious processes seem limited to a low level of analysis. However, if one extends the discussion to the twilight of consciousness—that "iffy" arena between absolutely not seeing something and definitely seeing it, for example—then meaningful processing becomes a possibility. Meaningful processing seems possible only if one has some degree of conscious experience, vague as the experience might be.

6. What is the effect of unconscious material on conscious behavior? Since Freud's introduction of the concept of repression, the effects of the unconscious on the conscious have proved enduringly fascinating. We do not need as complex a notion as repression to account for the evidence (Holmes, 1974). The answer to the question breaks into two components: representation and task demands. People encode events to varying levels of depth. In many tasks involving subliminal or implicit learning, limits on the depth at which people can unconsciously process incoming information ensure that the information is represented shallowly. This shallow representation may be so impoverished that it cannot support recognition. It may, however, affect a variety of other tasks, including social judgments and preference. Other tasks involve explicit learning, with

subjects often forming contextualized and elaborated representations of the material. Here the representation can support recognition and the construction of memory beliefs. Furthermore, many tasks, such as priming, do not require subjects to retrieve the information necessary for the construction of a memory belief. The effect here on behavior would be outside subjects' awareness—not because of an impoverished representation but because the task does not demand that they retrieve the kind of information needed to construct a memory belief.

Psychologists can begin to draw from these sets of findings at least a sketch of the dynamics of the unconscious and conscious. The sketch focuses on processing, representation, and task demands. Consciousness is limited because of the processing producing the content of consciousness, not by an inherent characteristic of consciousness itself. The shallow processing of the unconscious limits the effect of unconscious processing on conscious behavior. Explicitly learned material affects behavior outside the awareness of the individual when the task demands do not require people to retrieve the kind of information necessary for the construction of memory beliefs.

This emphasis on representation, processing, and task demands permits the psychologist to address many issues of consciousness, such as the ones raised in this chapter. It does not, however, address many mysteries. Why must consciousness be limited? Why cannot meaning be assigned to material unconsciously? Why be conscious at all? And finally, how does consciousness get its feel, its quality of being conscious? These are intriguing questions—and I suspect that the present approach cannot hope to address them.

ACKNOWLEDGMENTS This research was supported by grants from the National Institutes of Health RO1-MH42064 and the McDonnell-Pew Program in Cognitive Neuroscience.

REFERENCES

Bargh, J. A., 1993. Being unaware of the stimulus versus unaware of how it is interpreted: Why subliminality *per se* does not matter to social psychology. In *Perception without Awareness*, R. F. Bornstein and T. S. Pittman, eds. New York: Guilford Press.

Berry, D. C., and D. E. Broadbent, 1988. Interactive tasks and the inplicit-explicit distinction. *Br J. Exp. Psychol.* 79: 251–272.

Bornstein, R. F., 1993. Subliminal mere exposure effects. In *Perception without Awareness*, R. F. Bornstein and T. S. Pittman, eds. New York: Guilford Press.

Broadbent, D. E., 1958. *Perception and Communication.* London: Pergamon.

Broadbent, D. E., P. Fitzgerald, and M. H. P. Broadbent, 1986. Implicit and explicit knowledge in the control of complex systems. *J. Exp. Psychol. Monogr.* 80(2, pt. 2): 1–17.

Cheesman, J., and P. M. Merikle, 1984. Priming with and without awareness. *Percept. Psychophys.* 36:387–395.

Cheesman, J., and P. M. Merikle, 1986. Distinguishing conscious from unconscious perceptual processes. *Can. J. Psychol.* 40:343–367.

Cheng, P. W., 1985. Restructuring versus automaticity: Alternative accounts of skill acquisition. *Psychol. Review.* 92:414–423.

Churchland, P. S., 1986. *Neurophilosophy.* Cambridge, Mass.: MIT Press.

Corteen, R. S., and D. Dunn, 1972. Shock-associated words in a nonattended message: A test for momentary awareness. *J. Exp. Psychol.* 94:308–314.

Corteen, R. S., and B. Wood, 1972. Autonomic responses to shock-associated words in an unattended channel. *J. Exp. Psychol.* 94:308.

Dixon, N. F., 1981. *Preconscious Processing.* Chichester, England: Wiley.

Dulany, D. E., A. Carlson, and G. I. Dewey, 1984. A case of syntactical learning and judgment: How conscious and how abstract? *J. Exp. Psychol. [Gen.]* 113:541–555.

Eagle, M., 1959. The effects of subliminal stimuli of aggressive content upon conscious cognition. *J. Pers.* 27:578–600.

Ellenberger, H., 1970. *The Discovery of the Unconscious: The History and Evolution of Dynamic Psychology.* New York: Basic Books.

Ericksen, C. W., 1956. Subception: Fact or artifact. *Psychol. Rev.* 63:74–80.

Ericsson, K. A., and H. A. Simon, 1984. *Protocol analysis.* Cambridge, Mass.: MIT Press.

Fodor, J. A., 1983. *The Modularity of Mind: An Essay in Faculty Psychology.* Cambridge, Mass.: MIT Press.

Fowler, C. A., G. Wolford, R. Slade, and L. Tassinary, 1981. Lexical access with and without awareness. *J. Exp. Psychol. [Gen.]* 110:341–362.

Gibson, E. J., 1969. *Principles of Perceptual Learning and Development.* New York: Appleton-Century-Crofts.

Gibson, E. J., and A. A. Yonas, 1966. A developmental study of the effects of visual search behavior. *Percept. Psychophys.* 1:169–171.

Greenwald, A. G., 1992. New Look 3: Unconscious cognition reclaimed. *Am. Psychol.* 47:766–779.

Hanson, C., and W. Hirst, 1989. Representation of events: A study of orientation, recognition and recall. *J. Exp. Psychol. [Gen.]* 118:136–147.

Hasher, L., and R. T. Zacks, 1979. Automatic and effortful processing in memory. *J. Exp. Psychol. [Gen.]* 108:365–388.

Hirst, W., 1986. The psychology of attention. In *Mind and brain: Dialogues in Cognitive Neuroscience*, J. E. LeDoux and W. Hirst, eds. New York: Cambridge University Press.

Hirst, W., 1990. On consciousness, recall, recognition, and the architecture of memory. In *Implicit Memory*, K. Kirsner, S. Lewandowsky, and J. C. Dunn, eds. Hillsdale, N.J.: Erlbaum.

Hirst, W., and D. Kalmar, 1987. Characterizing attentional resources. *J. Exp. Psychol. [Gen.]* 116:68–81.

Hirst, W., E. S. Spelke, C. C. Reaves, G. Caharack, and U. Neisser, 1980. Dividing attention without alternation or automaticity. *J. Exp. Psychol. [Gen.]* 109:98–117.

Hochberg, J. E., 1964. *Perception.* Englewood Cliffs, N.J.: Prentice-Hall.

Holender, D., 1986. Semantic activation without conscious identification in dichotic listening, parafoveal vision, and visual masking: A survey and appraisal. *Behav. Brain Sci.* 9:1–23.

Holmes, D., 1974. Investigations of repression: Differential recall of material experimentally or naturally associated with ego threat. *Psychol. Bull.* 81:632–653.

James, W., 1890. *The Principles of Psychology*. New York: Dover, 1950.

Johnson, M. K., S. Hashtroudi, and S. Lindsay, 1993. Source monitoring. *Psychol. Bull.* 11:1–26.

Johnson, M. K., and W. Hirst, 1991. Processing subsystems of memory. In *Cognitive Neuroscience*, H. J. Weingartner and R. Lister, ed. New York: Oxford University Press.

Johnston, W. A., and V. J. Darks, 1986. Selective attention. *Ann. Rev. Psych.* 37:43–75.

Kahneman, D., 1973. *Attention and Effort*. Englewood Cliff, N.J.: Prentice-Hall.

Kahneman, D., and A. M. Treisman, 1984. Changing views of attention and automaticity. In *Varieties of Attention*, R. Parasuraman and R. Davies, eds. New York: Academic Press.

Kelley, C. M., and D. S. Lindsay, 1992. Remembering mistaken for knowing: Ease of retrieval as a basis for confidence in answers to general knowledge questions. Manuscript, University of Western Ontario.

Kihlstrom, J. F., 1987. The cognitive unconscious. *Science* 237:1445–1452.

Klein, G. S., 1954. Need and regulation. In *Nebraska Symposium on Motivation*, M. R. Jones, ed. Lincoln: University of Nebraska Press, pp. 224–277.

Kunst-Wilson, W. R., and R. B. Zajonc, 1980. Affective discrimination of stimuli that cannot be recognized. *Science* 207:557–558.

Lackner, J., and M. Garrett, 1973. Resolving ambiguity: Effects of biasing context in the unattended ear. *Cognition* 1:359–372.

Lewicki, P., T. Hill, and E. Bizet, 1988. Acquisition of procedural knowledge about a pattern of stimuli that cannot be articulated. *Cognitive Psychol.* 20:24–37.

Lewicki, P., T. Hill, and M. Czyewska, 1992. Nonconscious acquisition of information. *Am. Psychol.* 47:792–801.

Lewis, J. L., 1970. Semantic processing of unattended messages using dichotic listening. *J. Exp. Psychol.* 85:225–228.

Loftus, E. F., and M. R. Klinger, 1992. Is the unconscious smart or dumb? *Am. Psychol.* 47:761–765.

Logan, G. D., 1988. Towards an instance theory of automatization. *Psychol. Rev.* 95:492–527.

Mack, A., B. Tang, R. Tuma, and S. Kahn, 1992. Perceptual organization and attention. *Cognitive Psychol.* 24:475–501.

MacKay, D. G., 1973. Aspects of a theory of comprehension, memory, and attention. *Q. J. Exp. Psychol.* 25:22–40.

Mandler, G., 1985. *Cognitive psychology: An Essay in Cognitive Science.* Hillsdale, N.J.: Erlbaum.

Mandler, G., Y. Nakamura, and B. J. S. Van Zandt, 1987. Nonspecific effects of exposure on stimuli that cannot be recognized. *J. Exp. Psychol. [Learn. Mem. Cogn.]* 13:646–648.

Manier, D., and W. Hirst, in preparation. On the origins of memory.

Marcel, A., 1983a. Conscious and unconscious perception: Experiments on visual masking and word recognition. *Cognitive Psychol.* 15:197–237.

Marcel, A., 1983b. Conscious and unconscious perception: An approach to the relations betwen phenomenal experience and perceptual processes. *Cognitive Psychol.* 15:238–300.

Merikle, P. M., 1992. Perception without awareness: Critical issues. *Am. Psychol.* 47:792–795.

Neisser, U., 1964. Visual search. *Sci. Am.* 210:94–101.

Neisser, U., R. Novick, and R. Lazar, 1963. Searching for ten targets simultaneously. *Percept. Mot. Skills.* 17:955–961.

Newell, A., and H. A. Simon, 1972. *Human Problem Solving*. Englewood Cliffs, N.J.: Prentice-Hall.

Newstead, S. E., and I. Dennis, 1979. Lexical and grammatical processing of unshadowed messages: A reexamination of the MacKay effect. *Q. J. Exp. Psychol.* 31:447–488.

Nisbett, R. E., and T. D. Wilson, 1977. Telling more than we can know: Verbal reports on mental processes. *Psychol. Rev.* 84:231–259.

Norman, D. A., 1968. Toward a theory of memory and attention. *Psychol. Rev.* 75:522–536.

Norman, D. A., and D. G. Bobrow, 1975. On data-limited and resouce-limited processes. *Cognitive Psychol.* 7:44–64.

Perruchet, P., J. Gallego, and I. Savy, 1990. A critical reappraisal of the evidence for uncosncious abstaction of deterministic rules in complex experimental situations. *Cognitive Psychol.* 22:493–516.

Perruchet, P., and C. Pacteau, 1990. Synthetic grammar learning: Implicit rule abstraction or explicit fragmentary knowledge. *J. Exp. Psychol. [Gen.]* 119:264–275.

Poetzl, O., 1917. The relationship between experimentally induced dream images and indirect vision. In *Psychological Issues*, J. Wolff, D. Rapaport, and S. H. Annin, eds. and trans. 2 (3, Monog. 7): 41–120.

Posner, M. I., and C. R. R. Synder, 1975. Attention and cognitive control. In *Information Processing and Cognition: The Loyola Symposium*, R. L. Solso, ed. Hillsdale, N.J.: Erlbaum.

Reber, A. S., 1967. Implicit learning of artificial grammars. *J. Verb. Learn. Verb. Be.* 6:855–863.

Reber, A. S., 1989. Implicit learning and tacit knowledge. *J. Exp. Psychol. [Gen.]* 118:219–235.

Reber, A. S., and R. Allen, 1978. Analogy and abstraction strategies in synthetic grammar learning: A functional interpretation. *Cognition* 6:189–221.

Roediger, H. L., III, 1990. Implicit memory: Retention without remembering. *Am. Psychol.* 45:1043–1056.

Schacter, D. L., 1987. Implicit memory: History and current status. *J. Exp. Psychol. [Learn. Mem. Cogn.]* 13:501–518.

Schacter, D. L., 1989. On the relation between memory and consciousness: Dissociable interactions and conscious experience. In *Varieties of Memory and Consciousness: Essays in Honor of Endel Tulving*, H. L. Roediger, III and F. I. M. Craik, eds. Hillsdale, N.J.: Erlbaum.

Schacter, D. L., 1992. Implicit knowledge: New perspectives on unconscious processes. *Proc. Natl. Acad. Sci. U.S.A.* 89:11113–11117.

Schneider, W. E., and R. M. Shiffrin, 1977. Controlled and automatic human information processing: I. Detection, search, and attention. *Psychol. Rev.* 84:1–66.

Shiffrin, R. M., and W. E. Schneider, 1977. Controlled and automatic human information processing: II. Perceptual learning, automatic attending, and a general theory. *Psychol. Rev.* 84:128–190.

Silverman, L. H., and J. Weinberger, 1985. Mommy and I are one: Implications for psychotherapy. *Am. Psychol.* 40:1296–1308.

Squire, L. R., 1992. Memory and the hippocampus: A synthesis from findings with rats, monkeys, and humans. *Psychol. Rev.* 99:195–231.

Sudnow, D., 1978. *Ways of the Hand*. New York: Bantam.

Treisman, A. M., R. Nauires, and J. Green, 1974. Semantic processing in dichotic listening: A replication. *Mem. Cognition* 2:641–646.

Tulving, E., 1983. *Elements of Episodic Memory*. New York: Oxford University Press.

Tulving, E., and D. L. Schacter, 1990. Priming and human memory systems. *Science* 247:301–306.

Volpe, B. T., J. E. LeDoux, and M. S. Gazzaniga, 1979. Information processing in an "extinguished" visual field. *Nature* 282:722–724.

Wardlaw, K. A., and N. E. Kroll, 1976. Autonomic responses to shock-associated words in a non-attended message: A failure to replicate. *J. Exp. Psychol. [Hum. Percept.]* 2:357–360.

Warren, R. M., 1970. Perceptual restoration of missing speech sounds. *Science* 167:392–393.

Watson, J. B., 1924. *Behaviorism*. New York: Norton.

Weiskrantz, L., 1986. *Blindsight: A Case Study and Implications*. Oxford, England: Clarendon.

Whittlesea, B. W. A., L. L. Jacoby, and K. A. Girard, 1990. Illusions of immediate memory: Evidence of an attributional basis for feelings of familiarity and perceptual quality. *J. Mem. and Lang.* 29:716–732.

Wickens, C., 1980. The structure of attentional processes. In *Attention and Performance*, vol. 8, R. S. Nickerson, ed. Hillsdale, N.J.: Erlbaum.

Wickens, C., 1984. Processing resources in attention. In *Varieties of Attention*, R. Parasuraman and D. R. Davies, eds. Orlando, Fla.: Academic Press.

Young, A. W., and E. H. F. De Haan, 1992. In *The Neuropsychology of Consciousness*, A. D. Milner and M. D. Rugg, eds. San Diego: Academic Press.

87 Models of Consciousness: Serial or Parallel in the Brain?

MARCEL KINSBOURNE

ABSTRACT The traditional, serial hierarchical centered-brain model conflicts with findings in brain structure, physiology, and focal lesion effects on behavior. An alternative model is proposed, in which multiple representations are active in parallel. Some are bound by bidirectional corticocortical interaction into a *dominant focus*, the moment-by-moment composition of which determines the contents of consciousness. Phenomena of unilateral neglect of space and of person, as well as of blindsight, are shown to be compatible with the model.

Two problems about consciousness

Brain activity giving rise to conscious experience raises two separable challenges: One is to characterize the neural activity that underlies consciousness in terms of the dynamic properties and topographic distribution of this neural activity. The other is to explain how this neural activity engenders phenomenal experience, relative to which there are two competing positions.

IS NEURAL ACTIVITY TRANSDUCED INTO CONSCIOUSNESS? Some think that awareness is generated by, but different in nature from, neural activity (a distinct "information bearing medium" Mangan, 1993). This putative transformation of neural activity into awareness must involve formidable complexities and even a mathematics or physics that is currently (Penrose, 1989; Searle, 1992) if not permanently (McGinn, 1991) beyond our grasp. This last-frontier mystique of consciousness shows how greatly these theorists value their subjectivity, but it lacks objective support. What form might such support take?

If consciousness is a separate product, it should be separable from the neural activity of the brain. Consciousness should be demonstrable in isolation, and brain activity and information processing should proceed unaltered in the absence of consciousness. But neither phenomenon has been demonstrated. This encourages the more parsimonious view that consciousness is emergent from the neural activity without intervening transduction: It is the subjective aspect of the underlying neuronal activity; subjectivity refers to what it is like for the brain to be in relevant functional states. (Many would find this counterintuitive, but intuitions are data to be explained, not conclusions to be respected. There is no more justification for using intuition to judge theory in behavioral science than in physics; see Dennett, 1991). This directs discussion to the functional states in question. What is the neuropsychology of consciousness?

ARE WE CONSCIOUS OF PROCESSING OR OF REPRESENTATIONS? It is customary to refer to processing that generates representations of information and of intentions in the brain. Is consciousness the actual functioning of our brains, or only the results of this functioning? Lashley (1958) pointed out that "no activity of mind is ever conscious—experience clearly gives no clue as to the means by which it is organized" (p. 4). Thus conscious contents are the products of preconscious processing. At any time they embody the state changes that have occurred. The processing is opaque to awareness (Prinz, 1992). What, then, are the necessary and sufficient conditions for any item of information that is represented in terms of neuronal activity to contribute to the construction of the phenomenal present?

IS THERE SOMETHING SPECIAL ABOUT CONSCIOUS REPRESENTATIONS? In an information processing sequence, each output serves as input to the next processor. Not all these representations contribute to consciousness. To do so, do the representations have to be of a special kind, enjoy special relationships, or be located in a special place in the brain?

MARCEL KINSBOURNE Center for Cognitive Studies, Tufts University, Medford, Mass.

Type of representation An obvious approach to characterizing representations that contribute to consciousness is top-down: Hypothesize the adaptive function of consciousness, and then arrive at the set of representations that would suffice to accomplish that purpose. This tactic usually targets effortful attentional processes, and ends up with far too restricted a set of represented experiences, such as verbalizable features at the focus of attention (e.g., Johnson-Laird, 1983). It ignores less articulated global feelings such as pain or panic, and perception in a more dispersed or global mode. An example of the latter is panoramic vision, in which there is immediate (i.e., parallel) acquisition of general textural information, and the observer has a sense of clarity and richness of content, which can only be verbalized after a succession of fixations that enable serial focal analysis. A listing of what *can* enter consciousness might suggest that almost anything that can be represented can be experienced. Conversely, content of which one is usually aware can also control behavior outside awareness (i.e., subliminally).

Because the contents of consciousness are so diverse, they may not all be adapted to the same specific purpose. They may instead characterize a particular stage in the information-processing sequence, or they may share some adaptively important contextual attribute (or both).

Stage of representation Jackendoff (1987) observed that neither the most elementary representations (Marr's primal sketch, phonological distinctive features) nor the most abstract (3-D representation, meaning, syntax) contribute their contents to awareness. When viewing a scene, we are not aware of the as-yet-unorganized elements (features) that the brain is extracting from visual input. Nor is speech initially a potpourri of as-yet-unorganized auditory distinctive features. Conversely, when speaking, we do not first represent in awareness the semantic "formal elements" or the syntactic organization of what we intend to say. So, according to Jackendoff, consciousness is embodied in "intermediary representations," not in representations either at the highest abstract level or at the low level of sensory primitives.

Are lower-order representations really precluded from awareness? In the unfolding of a percept, it is unknown, and probably unknowable, whether early representations were really preconscious, or conscious but already forgotten at the time of report. This problem is pervasive when two or more sensory events occur in very rapid succession (Dennett and Kinsbourne, 1992). As the neural activity "settles," its previous instantiations prove evanescent.

Early representations may not be preserved in awareness simply because they are too short-lived, regardless of their stage in processing per se. But if a processing sequence is interrupted by a focal brain lesion, a representation that normally does not access awareness may do so, generating appearances that normal people never experience. For instance, patients recovering from cortical blindness have unusual experiences that could be thus explained (Poppelreuter, 1923). Patients with phantom limbs that they include in intended actions experience the represented intentions as movements in the absent limb, because they are not supplanted by the realized movement.

Are higher-order representations really excluded from awareness? The alleged end points of input processing (e.g., 3-D representation) may in fact be fully represented, but implicitly as knowledge about the objects and relationships in the represented scene. Unlike the scene itself, knowledge about it becomes explicit only when called for. But the scene itself is organized by prior knowledge about its constituents, as is the decoding of a verbal utterance (Liberman and Studdert-Kennedy, 1978). This is an interactive (heterarchical) rather than a strictly serial process. The knowledge about the input is not represented separately in awareness, but is inherent in the relationships that form between the input's representation and other representations that coexist (see integrated field model, below).

Abstract representations thought to precede speech subdivide into semantic and syntactic categories. The semantics of an utterance may be represented in awareness in the pervasive though unanalyzed feeling that one knows what one is about to say. Again, that ill-differentiated awareness is supplanted by the explicit content of the utterance that follows. If the syntactical elements into which utterances can be analyzed have psychological reality, then again they are supplanted by the elaborated utterance that immediately follows, and we cannot know whether they were transiently in consciousness. To be a viable candidate for consciousness, a representation has to persist for a minimum time. Some persisting representations become mutually entrained to give rise to the experience of objects and scenes. Representations are not entrained if they lack intensity (are subthreshold), are incomplete, are too short-lived, or even because they make incompatible demands on output mechanisms.

The field of conscious awareness imposes no restrictions on the types of representations that contribute to it. Probably all categories of representation are sometimes in consciousness and sometimes excluded from it. According to the integrated cortical field model, representations do not change character to become conscious, nor are their contents shuttled to some privileged place. Whether they are conscious depends on their interrelationships.

Integrated field model of awareness

What does distinguish information that is in awareness from the same information when it is not? The integrated field model (Kinsbourne, 1988) supposes that this attribute is contextual, consisting of a relationship to other content that is simultaneously represented in awareness. Conversely, when sensory input is processed outside awareness, it may modify (i.e., prime) response predispositions in interesting ways, but that processing is unrelated to anything else that is currently happening to the individual. It is not incorporated into the phenomenal here-and-now, and therefore cannot be recovered by event memory. Nor can it serve as a cue to what else was happening at the time.

Any experience is inevitably in a context; detailed and rich, or minimal and confined to the fact that it is happening "to me, here and now." It is part of an event, a personal experience, potentially retrieved by event (episodic) memory. This experiential binding gives information acquired in awareness the additional potential to be related to whatever else was happening at the time to or around the observer. The adaptive implications of rare combinations of contingencies can therefore be explored without multitrial learning. Having its contents contribute to awareness does not preclude a representation from modifying response predispositions. But it also enables the individual to evaluate a signal in context, in case its adaptive implication is context dependent.

If awareness stands out not by what it is but by the company it keeps, then we should, in conceptualizing its brain basis, look for corresponding relational design characteristics in the cerebral neuronal architecture that accomodates the neural representations that underlie awareness.

The model of the centered brain

Ever since its mid-nineteenth-century inception, the dominant model in neuropsychology has been the centered brain. Extrapolating the communication channels that converge from the body periphery on the brain, this model postulates continuing convergence within the brain substance, in service of multimodal integration into objects (each uniquely coded by a "grandmother cell"). This hierarchically serial organization culminates in some ill-specified place, the consciousness module, where the by now elaborately preprocessed information purportedly enters consciousness. Suitably informed, the consciousness module arrives at some appropriate decision, which is implemented by the reverse sequence of messages, becoming ever more fragmentary and dispersed as they recede from the apex of the hierarchy.

As little as the centered brain model has to commend it in any subarea of neuropsychology (Kinsbourne, 1982), it is particularly inadequate as a basis for consciousness. Dennett and Kinsbourne (1992) epitomized the centered brain vis-à-vis consciousness by the metaphorical "Cartesian Theater" along the lines of Descartes's concept of the role of the pineal gland. At some privileged locus, the brain provides a display for the viewing pleasure of a homunculus, who is both experiencer and agent. In that article we examined one consequence—that events should necessarily be experienced in the temporal sequence in which they cross a neural "finish line" for theatric screening—and showed that this conflicts with well-established time-related phenomena in perception.

Evidence Against the Centered Brain A centered cortical neuroanatomy, with converging unidirectional channels linking node with node at successively higher levels of integration is not to be found. Instead, there is a diffuse neural network, the neuropil, that is continuous in the cortical gray matter. Many neurons do not fire in the sense of generating action potentials, but instead propagate slow-wave graded potentials (Bullock, 1981). Patches of neuropil are interconnected by corticosubcortical loops or by U-shaped corticocortical connections. No progression in level or hierarchy is indicated by the anatomy, and U-shaped projections are fully reciprocal, with as extensive and specific an innervation in one direction as in the other. There is no point where it all comes together, strategically located at the apex of a hierarchy.

Corresponding to this massively parallel (and heterarchical) anatomy is a parallel neuropsychology. No focal cortical lesion ablates awareness and intention, leaving nonconscious processing intact. Focal cortical lesions impair specific abilities. They do not render a person unconscious. Correspondingly, when either hemisphere is briefly inactivated by intracarotid amobarbital, the other suffices to enable the patient to respond to simple verbal commands. Neurophysiology has resoundingly confirmed the long-standing localizationist viewpoint that the cerebral neural network is differentiated in its various parts so as to implement different mental operations. But it has not found that information represented at two (or more) points is integrated at yet another point. Metabolic studies carried out during mental activity reveal a distribution of activation that encompasses the areas that correspond to the ingredients of the task, but no separate area of confluence of these ingredients (Lassen and Roland, 1983). Separate dorsal and ventral "streams" of visual information processing (Ungerleider and Mishkin, 1982) nowhere merge. No centered locus of activation characterizes all tasks that are consciously performed or indicates that consciousness has been entered.

The parallel brain

In contrast to the exclusively connected centered brain, the brain conceptualized as processing in parallel offers scope for the lateral interaction of unrelated processors. In normative studies, activating one processor exerts systematic effects (priming or interfering) on other, functionally unrelated systems. These effects are predictable, based on the functional cerebral distance principle (e.g., Kinsbourne, 1970, 1972, 1973, 1975; Kinsbourne and Hicks, 1978; Kinsbourne and Byrd, 1985). This "penetrability" of modules to each others' influences is consistent with a parallel neuronal architecture. According to the centered brain model, different, separately analyzed aspects of a stimulus are bound by conjoint representation at a node downstream in the brain. Abandoning this model implies binding at a distance, probably by repetitive interaction between complementary representations. How this might work is only beginning to be elucidated. Nonetheless, the parallel model, though as yet incompletely specified, does probe brain organization from a more realistic perspective.

How, then, can we arrive at a plausible neural basis for consciousness within the framework of a massively parallel brain, or at least, cerebral cortex?

Dominant Focus In order to contribute to awareness, representations must be adequately activated. What happens when they are? I have suggested (Kinsbourne, 1988) that at any time in the awake individual's brain there is a dominant focus of patterned neural activity that underlies the phenomenal experience of that moment, the momentarily dominant "draft" of the "multiple drafts" that in rapid succession constitute the ostensibly continuous "stream" of ever-changing consciousness (Dennett and Kinsbourne, 1992). Whatever subset of currently active cell assemblies participates in this dominant focus determines what content is represented in awareness and is related to the self (and later available to recollection). From moment to moment the composition of the dominant focus changes, as representations become bound to it while others break away. Subsidiary clusters of entrained representations may coexist, candidates for future inclusion in the dominant focus. Thus subjectivity depends on patterned neural activity that is widely and variably dispersed over the cerebral cortex.

Mechanisms of Binding into the Dominant Focus One way in which representations might bind is if the respective cell assemblies fire synchronously, or perhaps enter into joint oscillation of firing (von der Malsburg, 1977; Gray and Singer, 1989). This entrainment is perhaps mediated by plastic synapses such as those carrying NMDA receptors (Flohr, 1991). Certain high-frequency rhythms recordable on the EEG may reflect ongoing binding (Loring and Sheer, 1984). This account of binding certainly raises major issues, such as how an attribute can bind simultaneously with other attributes of the same object, with other objects into coordinated displacement, with other coexisting objects into a scene, and with the state (physical, emotional, expectancy) of the observer. Does volume oscillation at multiple simultaneous frequencies implement this nested complex of relationships?

Another issue is the representation of temporal order of succession. If simultaneity is applied to the process of binding, the brain cannot at the same time code temporal order in terms of relative time of arrival of stimuli (at least in the microtime of fractions of a second). Dennett and Kinsbourne (1992) suggest that the brain organizes the onrush of inputs by "microtakings," which may violate the strict physical succession of stimulation in favor of a best-fit organization that takes into account content and experience. Obviously, most of this remains to be worked out. Here I restrict myself to observing that synchronous firing and co-oscillation are consistent with the bidirectionality of corticocortical connections. Any successful account of consciousness must explain how cell assemblies widely scattered in the cerebral cortex can contribute to consciousness.

Evidence from syndromes of selective unawareness

There is no focal lesion that generally strips the subjective aspect from cognitive processing. But there are partial syndromes of unawareness in which abilities are preserved that are normally associated with consciousness (Schacter, 1992), such as blindsight (Weiskrantz et al., 1974).

Blindsight Poppel, Held, and Frost (1973) reported four patients who could, by pointing, localize stimuli that they could not see. An intensive research effort has dramatically increased the inventory of what blindsight patients can accomplish in forced-choice situations, that is, when they are told to respond as best they can although they cannot see what they are responding to (Weiskrantz, 1986). Pollatsek, Rayner, and Collins (1984) reported analogous observations in normal people. During a saccade to a peripheral target that is not recognizable until it is fixated, the target is changed. On account of saccadic suppression, the subject does not observe the change as it occurs. Subsequently, the subject is unaware that the target has changed. Nonetheless, in a forced-choice paradigm subjects are significantly successful in judging that a change has occurred. So, at a level of detail sufficient to make the conscious discrimination of a change, the change information is processed without benefit of awareness. But it is what the blindsight patients (and normals under saccadic suppression) do *not* do that indicates what subjects gains when the contents of their representations are conscious. Blindsight patients do not relate the information in the blind field to the summed experience of the moment. Their responding without awareness is a fragmentary activity that they would not have initiated spontaneously.

If the blindsight patient is unaware of what she can do, the neglect patient is unaware of what she cannot do.

Unilateral Neglect of Space

Neglect as directionality-disinhibited orienting The unilateral neglect syndrome is a selective disorder of awareness that gives rise to strikingly disordered behavior. Superficially characterized, the patient has lost awareness of the left side of space and/or his body (though individual patients differ considerably in the relative involvement of different modalities of input and in motor involvement). He neither experiences input from the left nor orients or acts toward that side. Equally important, the patient is unaware that he is not using left-sided information and therefore does not attempt to compensate. Using the centered brain as a model, one would postulate that modules for attention to left-sided information have been inactivated by the (right parietal) injury and remark upon, but not explain, the anosognosia for that loss. Screens in the Cartesian Theater that should depict information originating from the left are blank. But an analysis of neglect phenomenology undermines any such assumption.

Rather than a focal loss from the cognitive repertoire, what is observed is the overcontrol of behavior by the residual intact brain—specifically by an intact but disinhibited facility that implements orientation to the right (Kinsbourne, 1970, 1987), without opposition from the lesioned opponent facility that previously swung orientation to the left. The obverse of the failure to attend to the left is an excess of attending to the right extreme of a display ("a heightened compulsive response to ipsilateral stimuli," in Sprague's 1966 cat model). The patient's gaze swings toward the rightmost visible feature, regardless of instruction or the requirements of the task. Contrary to the Western left-to-right reading habit, the patient scans a letter sequence right to left. In speed of reaction to the rightmost of a series of targets, the neglect patient even outdoes the normal control (reviewed by Kinsbourne, 1993). Once focusing on a right-sided stimulus, the patient finds it hard to disengage toward the left (Posner et al., 1984).

Not only does the neglect patient orient to the right end of things, but he relies excessively upon right-located features. Shown a series of letters that could have served as the end of several words, he completes the word backward to its beginning as though a whole word had been presented, for example ging to singing (Kinsbourne and Warrington, 1962, and recent studies of "neglect dyslexia," e.g., Riddoch, 1990). Similarly he overinterprets contours that could be the right sides of forms, reporting that he saw the complete form (Warrington, 1962). When sketching from memory, he regards the sketch as complete although he has not sketched in its left side. The brain does not register the fact that a part is missing, because there are no analyzers that signal "nothing there" or "something missing." It generates a complete percept by virtue of the "imaginative completion of incomplete figures" (Poppelreuter, 1917). As normal people experience shapes and contours as continuous across the blind spot, many hemianopics systematically "complete" figures that overlap their blind visual field. That is, they report having seen the complete shape, even thought they could not have seen the part that was presented to the blind field. Indeed, such a patient will regularly perceive as complete a figure that is objectively incomplete, so long as the missing part impinged on defective visual field. The missing part is "assumed" to be present, and it is reported as being seen (Kinsbourne and Warrington, 1962; Warrington, 1962). Is this filling in beyond the information given accomplished by activating the same visual cortical neurons that would have been active had the inferred contours actually been presented? That is hardly possible, when the corresponding area of cortex is inactivated or disconnected by disease. Therefore the completion type of filling in must represent decision based on expectancy. The expectancy in turn is based on what else is happening at the same time, and on previous experience of similar situations. This bias in evaluation indicates a higher-order defect than would be expected from a mere failure to obtain complete information, and it parallels striking aspects of the patient's attitude toward his body and the ambient space.

Awareness of deficit and the representation of knowledge in neglect The patient with neglect shows no sign that she is aware of the degree to which the information available to her consciousness is restricted and incomplete. In contrast to the patient with a sensory deficit, who is fully aware of having lost sensory input and attempts compensatory strategies, she behaves as if her experience were unchanged from the premorbid state, and explicitly denies the deficit. It follows that she must be accomplishing satisfactory matches between input and memory representations, just as before the brain disease began. Given that what the patient perceives is restricted to the right extreme of the form, for such a fragment to achieve a complete match the patient's memory representation must be equally fragmented. The representational bias in neglect involves remembered information. The patient therefore draws conclusions (unconsciously) that are actually not justified based on the limited amount of information that is available. But the resulting experience, veridical or not, seems normal to her. In view of this, the patient cannot agree with the clinician's bleak appraisal of her perceptual capabilities.

That memory representations can be subject to neglect was elegantly illustrated by Bisiach and Luzzatti (1978). They asked two patients to describe a scene well known to them first from one imagined vantage point, then from the opposite view. In each case the patient described what he would have seen to his right, had he been there in person. The landmarks omitted from one perspective were described from the other, and vice versa. It follows that the complete scene was potentially available, but only the imaged right side could be brought into awareness. From the evidence that follows, we conclude that the left part of the representation was insufficiently activated to become integrated into the conscious field. The lowered level of activation could suffice, however, to influence semantic memory, as indicated by priming effects that have been reported following left-sided input of which the patient remained unaware (Kinsbourne and Warrington, 1962; Volpe, LeDoux, and Gazzaniga, 1979; Marshall and Halligan, 1988; McGlinchy-Berroth et al., 1993).

The effect on neglect of nonspecific lateral ascending activation An early empirical finding by Silberpfennig (1949) reveals that underactivation is the basis of the representational impairment in neglect. He found that caloric stimulation of the inner ear (cold contralateral or warm ipsilateral to the lesion) could temporarily correct the neglect and restore normal perception. This finding has recently been confirmed for unilateral neglect both of space (Rubens, 1985) and of person (Vallar et al., 1990). Evidently the vestibular stimulation activated the damaged hemisphere and restored its ability to act as an effective opponent processor in equilibrating representations across the lateral plane. The representations themselves must have been intact, that is, available, if not accessible. For complete access into awareness (i.e., binding to the other activated representations that subserved the awareness of the moment) they required the added induced hemisphere activation. Soon after the stimulation is over, the patient reverts to the previous state of neglect, unaware that only moments earlier there had been available to him a more complete worldview. I suggest that this is because he is again unable mentally to represent the spatial domain to which he had just had access. One cannot experience the lack of information in a domain that one cannot represent.

Implications for the location of conscious representations It is clear that unilateral neglect is not the disconnection of input from a consciousness center further downstream. It is hard to conceive of projections dedicated to transmitting information about the right end of things or the left end of things. Instead, it is activation that maintains representations in awareness. Whether a representation contributes to the neglecter's awareness does not depend on the representation itself, but on whether other representations are formed of input to its right. In neglect, a structure that assists in maintaining an even distribution of activation across the representation is disabled. Recognition depends on a match between input and memory representation, and a laterally depleted representation can still execute a match based on right-located features. Information to the left is not observed, and, if absent, neither is its absence noted. Unattended parts of objects and scenes are "filled in."

Unilateral Neglect of Person and Autotopagnosia A long-standing (centered brain) view of the body image—the collective awareness of the body parts—is that it is based on a neural representation somewhere in the cerebral cortex, distinct from the somatosensory representation of the same parts. An extensive lesion might eliminate the body image completely without causing sensory loss. This would cause complete unawareness of the body as a spatially articulated structure (autotopagnosia). More restricted lesions would inactivate portions of the body image representation, causing unawareness of the corresponding body parts (local autotopagnosia). The total set of local autotopagnosias should, like jigsaw puzzle pieces, superimpose on a general autotopagnosia. But the clinical literature has uncovered no such set of deficits. Insofar as general autotopagnosia exists, it is a category-specific conceptual disorder (Semenza and Goodglass, 1985) rather than a body-image disorder in the sense of impaired awareness. As for partial autotopagnosias, left neglect of the body is the only face-valid candidate. However, unawareness of left body parts, notably the left hand, is better understood as an attentional bias away from the left extreme of a lateral somatosensory coordinate (Kinsbourne, in press). It resolves, like left neglect of space, during unilateral vestibular stimulation by caloric irrigation (Bisiach, Rusconi, and Vallar, 1991).

Contrary to the classical notion, one cannot attend at one and the same time to the whole body (any more than to the whole visual field). When one invests somatosensory attention in a body part, the rest of the body is experienced as an undifferentiated background, rather like the visual periphery is experienced during focal visual attention. One can no more suffer from a local unawareness of a body part than from an unawareness of a patch in the visual field. Again, the findings favor a laterally interacting set of representations rather than a hierarchy of representation. Somatosensory attention, like visual attention, is controlled by laterally interacting opponent processors.

Conclusions

The neuropsychological evidence is consistent with a model of consciousness as arising in active representations that are dispersed, at times widely, across the cerebral cortex. The representations contribute to consciousness by providing context for each other (an integrated field). The neuropsychological foundation of consciousness will not be found in a dedicated brain module, but rather in the interaction, in a manner to be clarified, of representations whose defining characteristics are beginning to be understood.

REFERENCES

Bisiach, E., and C. Luzzatti, 1978. Unilateral neglect of representational space. *Cortex* 14:129–133.

Bisiach, E., M. L. Rusconi, and G. Vallar, 1991. Remission of somalophrenic delusions through vestibular stimulation. *Neuropsychologia* 29:1029–1032.

Bullock, T. H., 1981. Spikeless neurons: Where do we go from here? In *Neurons without Impulses*, A. Roberts, and B. M. H. Bush, eds. Cambridge: Cambridge University Press, pp. 269–284.

Dennett, D. C., 1991. *Consciousness Explained.* Boston: Little, Brown.

Dennett, D. C., and M. Kinsbourne, 1992. Time and the observer: The where and when of consciousness in the brain. *Behav. Brain Sci.* 15:183–247.

Flohr, H., 1991. Brain processes and phenomenal consciousness. *Theory Psychol.* 1:245–262.

Gray, C. M., and W. Singer, 1989. Stimulus-specific neuronal oscillations in orientation columns of cat visual cortex. *Proc. Natl. Acad. Sci. U.S.A.* 86:1689–1702.

Jackendoff, R., 1987. *Consciousness and the Computational Mind.* Cambridge, Mass.: MIT Press.

Johnson-Laird, P., 1983. *Mental Models: Towards a Cognitive Science of Language, Inference, and Consciousness.* Cambridge: Cambridge University Press.

Kinsbourne, M., 1970. The cerebral basis of lateral asymmetries in attention. *Acta Psychologica* 33:193–201.

Kinsbourne, M., 1972. Eye and head turning indicate cerebral lateralization. *Science* 176:539–541.

Kinsbourne, M., 1973. The control of attention by interaction between the cerebral hemispheres. In *Attention and Performance IV*, S. Kornblum, ed. New York: Academic Press, pp. 239–256.

Kinsbourne, M., 1975. The mechanism of hemispheric control of the lateral gradient of attention. In *Attention and Performance V*, P. M. A. Rabbitt and S. Dornic, eds. London: Academic Press, pp. 81–97.

Kinsbourne, M., 1982. Hemispheric specialization and the growth of human understanding. *Am. Psychol.* 37:411–420.

Kinsbourne, M., 1987. Mechanisms of unilateral neglect. In *Neurophysiological and Neuropsychological Aspects of Spatial Neglect*, M. Jeannerod, ed. Amsterdam: Elsevier, pp. 69–86.

Kinsbourne, M., 1988. Integrated field theory of consciousness. In *Consciousness in Contemporary Science*, A. J. Marcel and E. Bisiach, eds. New York: Oxford University Press, pp. 239–256.

Kinsbourne, M., 1993. Orientational bias model of unilateral neglect: Evidence from attentional gradients within hemispace. In *Unilateral Neglect: Clinical and Experimental Studies*, I. H. Robertson and J. C. Marshall, eds. New York: Erlbaum.

Kinsbourne, M., in press. Awareness of one's own body: An attentional theory. In *The Body and the Self*, J. Bermudez and A. J. Marcel, eds. Cambridge, Mass.: MIT Press.

Kinsbourne, M., and M. Byrd, 1985. Word load and visual hemifield shape recognition: Priming and interference effects. In *Mechanisms of Attention: Attention and Performance XI*, M. I. Posner and O. S. M. Marin, eds. Hillsdale, N.J.: Erlbaum, pp. 529–543.

Kinsbourne, M., and R. E. Hicks, 1978. Functional cerebral space: A model for overflow, transfer and interference effects in human performance: A tutorial review. In *Attention and Performance XI: Mechanisms of Attention*, J. Requin, ed. Hillsdale, N.J.: Erlbaum, pp. 345–362.

Kinsbourne, M., and E. K. Warrington, 1962. A variety of reading disability associated with right hemisphere lesions. *J. Neurol. Neurosurg. Psychiatry* 25:339–344.

Lashley, K., 1958. Cerebral organization and behavior. In *The Brain and Human Behavior*, Research Publications, Association for Research on Nervous and Mental Disorders, No. 36, H. C. Solomon, S. Cobb, and W. Penfield, eds. Baltimore, Md.: Williams and Wilkins, pp. 1–18.

Lassen, N. A., and P. E. Roland, 1983. Localization of cognitive function with cerebral blood flow. In *Localization in Neuropsychology*, A. Kertesz, ed. New York: Academic Press, pp. 141–152.

Liberman, A., and M. Studdert-Kennedy, 1978. Phonetic perception. In *Handbook of Sensory Physiology*, vol. 8, *Perception*, R. Held, H. Leibowitz, and H.-L. Teuber, eds. Heidelberg: Springer.

Loring, D. W., and D. E. Sheer, 1984. Laterality of 40 Hz EEG and EMG during cognitive performance. *Psychophysiology* 21:34–38.

Malsburg, C., von der, 1977. Self-organization of orientation-sensitive cells in striate cortex. *Kybernetik* 14:85–199.

Mangan, B., 1993. Dennett, consciousness and the sorrows of functionalism. *Consciousness Cognition* 2:1–17.

Marshall, J. C., and P. W. Halligan, 1988. Blindsight and insight in visuospatial neglect. *Nature* 336:766–767.

McGinn, C., 1991. *The Problem of Consciousness.* Oxford: Blackwell.

McGlinchy-Berroth, R., W. P. Milberg, M. Verfaillie, M. Alexander, and P. Kilduff, 1993. Semantic processing in the neglected visual field: Evidence from a lexical decision task. *Cognitive Neuropsychol.* 10:79–108.

Penrose, R., 1989. *The Emperor's New Mind: Concerning Computers, Minds and the Laws of Physics.* Oxford: Oxford University Press.

Pollatsek, A., K. Rayner, and W. E. Collins, 1984. Integrating pictorial information across eye movements. *J. Exp. Psychol. [Gen.]* 113:426–442.

Poppel, E., R. Held, and D. Frost, 1973. Residual function after brain wounds involving the central visual pathways in man. *Nature* 243:295–296.

Poppelreuter, W., 1917. *Disturbances of Lower and Higher Visual Capacities Caused by Occipital Damage.* Leipzig: Voss. Translation by J. Zihl. Oxford: Clarendon Press, 1990.

Poppelreuter, W., 1923. Zur Psychologie und Pathologie der optischen Wahrnehmung. *Z. Gesamte Neurol. Psychiatrie* 83:26–152.

Posner, M. I., J. A. Walker, F. J. Friedrich, and R. D. Rafal, 1984. Effects of parietal injury on covert orienting of visual attention. *J. Neurosci.* 7:1863–1974.

Prinz, W., 1992. Why don't we perceive our brain states? *Eur. J. Cognitive Psychol.* 4:1–20.

Riddoch, M. J., ed., 1990. Neglect and the peripheral dyslexias. *Cognitive Neuropsychol.* 7:369–554.

Rubens, A. B., 1985. Caloric stimulation and unilateral visual neglect. *Neurology* 35:1019–1024.

Schacter, D. L., 1992. Implicit knowledge: New perspectives on unconscious processes. *Proc. Natl. Acad. Sci. U.S.A.* 89:11113–11117.

Searle, J. R., 1992. *The Rediscovery of the Mind*. Cambridge, Mass.: MIT Press.

Semenza, C., and H. Goodglass, 1985. Localization of body parts in brain injured subjects. *Neuropsychologia* 23:161–175.

Silberpfennig, J., 1949. Contributions to the problem of eye movements: III. Disturbance of ocular movements with pseudo hemianopsia in frontal tumors. *Confinia Neurologica* 4:1–13.

Sprague, J. M., 1966. Interactions of cortex and superior colliculus in mediation of visually guided behavior in the cat. *Science* 153:1544–1547.

Ungerleider, L. G., and M. Mishkin, 1982. Two cortical visual systems. In *Analysis of Visual Behavior*, D. J. Ingle, M. A. Goodale, and R. J. W. Mansfield, eds. Cambridge, Mass.: MIT Press, pp. 549–586.

Vallar, G., R. Sterzi, G. Bottini, S. Cappa, and M. L. Rusconi, 1990. Temporary remission of left hemianesthesia after vestibular stimulation. A sensory neglect phenomenon. *Cortex* 26:123–131.

Volpe, B. T., J. E. LeDoux, and M. S. Gazzaniga, 1979. Information processing of visual stimuli in an extinguished field. *Nature* 282:722–724.

Warrington, E. K., 1962. The completion of visual forms across hemianopic visual field defects. *J. Neurol. Neurosurg. Psychiatry* 25:208–217.

Weiskrantz, L., 1986. *Blindsight: A Case Study and Implications*. Oxford: Oxford University Press.

Weiskrantz, L., E. K. Warrington, M. D. Sanders, and J. Marshall, 1974. Visual capacity in the hemianopic field following a restricted occipital ablation. *Brain* 97:709–728.

88 Consciousness in Dyschiria

EDOARDO BISIACH AND ANNA BERTI

ABSTRACT This chapter reviews evidence for nonconscious mental representation in dyschiria (unilateral neglect and related disorders). Instances of dissociation of consciousness relative to the representation of one's own body and environment and of dissociation between awareness of the *what* and *where* of that representation are reported and discussed. A model is outlined that can collect the reviewed symptoms into a unitary syndrome of representational impairment confined to the side of egocentric space opposite to the side of a brain lesion. The chapter ends with some remarks about the current status of concepts such as *representation* and *consciousness*.

Dyschiria (from the Greek *chéir* 'hand') is a representational disorder confined to the side of space opposite to the side of a brain lesion involving specific neural circuits (see Bisiach and Berti, 1987, for review). It is generally believed to be more frequent and severe after lesions of the right hemisphere. Its symptoms may be defective and, less frequently, productive. Defective symptoms are designated by the collective term *unilateral neglect*. In most severe instances, patients behave as if the contralesional side of egocentric space were neither perceived nor conceived. By *egocentric space* we broadly refer to a polar coordinate system the origin of which is the subject him- or herself; further analysis of this concept may be found elsewhere (Bisiach, 1993). Productive phenomena are delusions confined to the side of egocentric space contralateral to the lesion. These delusions are known under the collective term *somatoparaphrenia* (Gerstmann, 1942), because they are almost always related to one side of the body, or part of it, and range from simple disavowal of ownership of body parts to bizarre ascriptions and metamorphic perception by which a patient may, for example, claim that a scaffolding has been substituted for the left half of his body, as in Ehrenwald's case (quoted by Hécaen et al., 1954). The spatial frame of dyschiria is manifold, since it comprises retinotopic, somatotopic, and object-centered components; within these frames, symptoms show a left-right gradient of severity that, as initially claimed by Kinsbourne (1970), is much more likely to be monotonic than stepwise (see Bisiach, 1993, for further discussion of the spatial features of the syndrome).

Since Dunn's pioneering observation (1845), accumulating neuropsychological experience in different areas has shown that mental contents of which patients are apparently unaware may be inferred from indications ranging from elementary physiological reactions, such as galvanic skin reaction, to intentional behavior, such as responses to questionnaires (see Milner and Rugg, 1991, for review). The study of dyschiria has since the beginning contributed in a substantial way to the discovery of nonconscious mental representation, although this has only recently come to be fully appreciated. Even more crucial has been its contribution in demonstrating the dissociability of consciousness and suggesting the diachronic structure of conscious representation—that is, the timing of recursive processes that are likely to underlie phenomenal experiences corresponding, for example, to the paradoxical behavior shown by our patients E. B. and F. S. (discussed in the next two sections, respectively). Furthermore, the study of dyschiria plays a unique role in elucidating the spatial constraints to which conscious representation is submitted.

We will first review evidence concerning nonconscious representation in dyschiria. By *representation* we only refer to sensorially driven or internally generated mental episodes, not to the long-term (quiescent) memories underlying them. Therefore, we do not refer to, for example, inactive knowledge that can or cannot access consciousness depending on the spatial location it must occupy when summoned into a representational event (Bisiach and Luzzatti, 1978).

Then, we will show how consciousness relative to a certain content may dissociate depending on response

---

EDOARDO BISIACH Dipartimento di Psicologia Generale, Università di Padova, Padova, Italy, and
ANNA BERTI Istituto di Fisiologia Umana, Università di Parma, Parma, Italy

characteristics and make some remarks about the dissociation between representation and representational attitudes such as beliefs and desires. We will also deal with the dissociation between consciousness of somatic or extrasomatic states of affairs and consciousness of their spatial location.

Finally, we will comment on empirical data and draw some conclusions about the distributed character of consciousness and the articulation inherent in the phenomenal self.

Nonconscious representation in dyschiria

Somatic Domain Nonconscious veridical information about the patient's contralesional body parts was first noticed by Anton (1899), who called it *dunkle Kenntnis* (dim knowledge). Left hemiplegic patients may pay no attention to the left side of their body, deny its inability to move, and assert that nothing is the matter with their physical state in general; nevertheless, they do not remonstrate to being bedridden and show total lack of intentionality with regard to the affected limbs: They do not engage in bimanual activities, nor express desires the satisfaction of which would require use of those limbs. (Other patients may show the converse behavior: They acknowledge hemiplegia but are unappreciative of its consequences.) The adequacy of nonconscious representation of the actual conditions of plegic limbs is evidenced by recent observations made by Marcel and Tegnér (1993). These authors asked hemiplegic patients to rate on a 10-point scale their own presumed ability to carry out a number of activities involving the use of limbs of both sides, such as tying knots, clapping hands, climbing a ladder, etc., and, a few moments later, to rate what they thought would have been the examiner's ability with respect to those same activities had he been affected by exactly the same impairments as theirs. Anosognosic patients who rated their own ability as totally unimpaired were often found to judge that the experimenter's ability would be reduced or even totally abolished.

Extrasomatic Domain Evidence for the possibility of nonconscious perception of the contralesional side of visual stimuli can be traced back to a study in which Kinsbourne and Warrington (1962) found that in neglect dyslexia the length of miscompleted words uttered by patients tended to increase with the length of word stimuli.

Volpe, LeDoux, and Gazzaniga (1979) asked four right-parietal lesion patients with no hemianopia but showing left extinction when two stimuli were simultaneously presented, one in their left and one in their right visual hemifield, to report whether two words or line-drawn pictures exposed for 150 ms, one on each side of the fixation point, were the same or different. Two patients claimed that the task was "silly," since there were no stimuli in the left visual field with which to compare the stimuli they could perceive in their right visual field; nonetheless, forced same/different guesses were perfectly (patient 1), or nearly perfectly accurate (patient 2).

Some years later, Bisiach, Meregalli, and Berti (1990) described the case of a patient, E. B., who neglected or miscompleted the left side of 10-letter words and pronounceable nonwords she had to read, but was able correctly to report all letters from items in which the left side of words had been substituted with a nonpronounceable string of letters. E. B. thus showed what could be called late selection for *in*attention. A similar dissociation was found by Hillis and Caramazza (1990) in their neglect patient R. B, who showed momentary awareness of graphemes on the left side of words he would then neglect or miscomplete: On being presented with the word *village*, for example, he would first spell it correctly and then misread it as *mileage*.

Further suggestion of nonconscious perception of the neglected side of visual stimuli came from Marshall and Halligan's observation (1988) of a right-brain-damaged patient who, though resolutely denying any differences between the line-drawn picture of a house and the same picture with added red flames on the left side, consistently chose the former in a series of trials in which she had to point to the house she would prefer to live in. The results of a similar investigation by Bisiach and Rusconi (1990), however, showed that behavior such as that observed in Marshall and Halligan's experiment may not necessarily imply nonconscious processing up to the level of meaning extraction; two patients of theirs, indeed, consistently preferred the burning house. Nonconscious perception up to the level of meaning extraction was instead convincingly demonstrated by Berti and Rizzolatti (1992).

With the aim of replicating and extending the results obtained by Volpe, LeDoux, and Gazzaniga (1979), Berti and Rizzolatti devised a priming experiment. In two conditions, respectively "highly congruent" and "congruent," a 200 ms presentation of the picture of an animal or a fruit in the left (contralesional) visual field of right-brain-damaged patients was followed 200 ms later by right-visual-field presentation of another picture that matched the former, either being identical or belonging to the same category (animal or fruit). In the third, "noncongruent," condition the two pictures belonged to two different categories. Patients had to press one or the other of two keys according to whether the picture that appeared to the right of fixation was of an animal or a fruit. Five patients who fully denied the occurrence of stimuli in the left visual field had significantly shorter reaction times to the right-field stimuli in the highly congruent and congruent conditions (743 and 750 ms, respectively) than in the noncongruent condition (855 ms).

Vallar, Rusconi, and Bisiach (1994) asked five left-neglect patients to give same/different responses to pairs of line-drawn pictures of animals. The two pictures were shown one above the other; different pairs were composed by a drawing of an animal and a drawing in which the left half of that animal had been substituted by the left half of a different animal. All patients misjudged different pairs as "same." However, when asked which of the two drawings more properly corresponded to the name of the animal depicted in the nonchimeric picture, three patients made correct choices and noticed the noncongruous half of one of the drawings in different pairs, but one patient, G. S., made correct choices without noticing the presence of the chimeric pictures. The remaining patient, R. G., gave paradoxical responses: In most trials with different pairs (16 of 20; $p < .05$, binomial test) he chose the chimeric animal and justified his choice by mentioning (nonexisting) differences between the two *right* halves of the drawings in the misjudged pairs. The lack of any spontaneous comment by G. S. on her choices makes it very unlikely that they were based on overt recognition of chimeric drawings. Conscious identification, however, can be confidently ruled out in patient R. G., whose paradoxical responses were based on the very details he did not overtly identify. The nature of these responses was tentatively interpreted by the experimenters in terms of "avoidance" (Denny-Brown and Chambers, 1958).

Làdavas, Paladini, and Cubelli (in preparation) demonstrated, in a patient suffering from left neglect, nonconscious semantic priming during a lexical decision task. The patient had to press a bar as soon as a word was projected in his right (ipsilesional) visual field, while no response was required for nonwords. Right visual field stimuli were preceded by short exposure in the left visual field of words that, in *go* trials, were either semantically related or unrelated to the target word. Responses to words preceded by semantically related primes were significantly faster than responses to words preceded by unrelated primes. In a forced-choice detection task with short exposure of single words in the left visual field and yes-or-no responses, the patient's performance did not exceed chance level.

Summing up, the results of the experiments by Volpe, LeDoux, and Gazzaniga (1979), Marshall and Halligan (1988), and Bisiach and Rusconi (1990) might have revealed only nonconscious processing of visual information below the level of meaning extraction. Neglect and miscompletion of words and pronounceable nonwords contrasting with normal report of nonpronounceable nonwords by patient E. B. (Bisiach, Meregalli, and Berti, 1990) showed only that the patient had implicitly recognized the lawful intercorrelation among unreported graphemes. The findings by Berti and Rizzolatti (1992), Vallar, Rusconi, and Bisiach (1994), and Làdavas, Paladini, and Cubelli (in preparation), by contrast, strongly suggest that processing of neglected visual stimuli can be carried out until the level of meaning extraction and nonetheless fail to reach the threshold of consciousness. Converging evidence is provided by what Làdavas, Bosinelli, and Bisiach (1993) have found with the Rorschach test in a left hemineglect patient (discussed in the section on dissociation between *what* and *where*).

Dissociation of consciousness depending on response characteristics

SOMATIC DOMAIN The term *response* is used here in a very broad sense, including spontaneous activity of any kind prompted by, or conditional to, conscious or nonconscious knowledge of the current state of contralesional body parts. The phenomena referred to in the previous section by the term *dim knowledge* of hemiplegia, therefore, are themselves the result of a dissociation between verbal denial of paralysis and its implicit behavioral acknowledgment. The converse dissociation may be observed in patients who verbally admit being paralyzed—either spontaneously or because they have been told they are—and nonetheless try to rise from bed and walk, or are ready to engage in bimanual activities; such as knitting, that are obviously precluded to them. A dissociation of consciousness is also evident in the above-mentioned observations made by Marcel and Tegnér (1993). These authors have found that patients who firmly deny hemiplegia may give positive answers to questions such as whether their contralesional limbs are "lazy" or "naughty." Inconsistent conscious self-evaluation may also be revealed by the sensitivity of denial of illness in general, and in particular of hemiplegia, to interpersonal factors. Jaffe and Slote (1958), in fact, showed that two biased questionnaires, one maximizing and the other minimizing the implications of the illness, elicited opposite patterns of reaction from their patients, in compliance with the examiner's prearranged attitude.

Extrasomatic Domain Dissociations of awareness depending on response factors are even more clear-cut in neglect of outer space. Detection of contralesional visual stimuli has been found to be more impaired in right-brain-damaged patients—both in conditions of single and of double simultaneous stimulation—for manual than for verbal responses (Bisiach, Vallar, and Geminiani, 1989). In an experiment on line bisection in which left neglect patients had to mark the line midpoint by displacing a pointer mounted on a pulley device, neglect was found to be more pronounced when the pointer moved in the same direction as the hand movement than when it moved in the opposite direction (Bisiach et al., 1990).

Even more impressive is the reversal of perceptual awareness demonstrated by Tegnér and Levander (1991) by means of a modified version of Albert's (1973) line cancellation task. After having been given the canonical form of the test, in which they had to cross 40 short lines drawn at random on a sheet of paper, neglect patients were tested again in a condition in which the sheet of paper and the performing hand were concealed by a bench, and cancellation was guided by a left-right inverted image of the visual array, provided by two mirrors joined to form a 90° vertical dihedron. In this condition, some patients still crossed out lines they saw on the right side—although in order to do that their hand had to cross the body's sagittal midplane and work on the left side of the sheet of paper—and neglected lines reflected by the left side of the mirror device. These patients, therefore, showed the same kind of perceptual unawareness in both conditions: The left side was visually ignored. By contrast, other patients crossed out, in the mirror-reversed condition, only lines lying on the right side of the sheet of paper, although those lines appeared on the side of the visual array—the left side—they had neglected in the canonical condition of the task. Some of them spontaneously claimed to see no lines on the right!

In the first type of neglect thus individuated by Tegnér and Levander, therefore, lack of consciousness of what lies on one side of the perceptual array seems to depend on contralesional deficit of visual exploration, whereas in the second type it is likely to be subordinate to a disorder of manual exploration of the contralesional side of egocentric space, that is, to what Heilman and colleagues (1985) called directional hypokinesia. What is relevant here is the apparent dramatic dissociation found in the second type of neglect between consciousness and nonconsciousness of visual information on one side of space by contrasting congruent and noncongruent visuomanual exploration.

Further experimental manipulation has provided additional insight into the dynamics underlying this dissociation. It has in fact been shown that the two types of neglect individuated by means of the mirror-reversed line cancellation task can be further differentiated according to whether they are rigidly determined or reversible into one another by forcing patients to direct attention onto the neglected side (Bisiach, 1994). After having been tested on the mirror-reversed version of the task, patients were tested again in a constrained condition in which they had to start from the side they had only a moment before neglected in the unconstrained mirror-reversed condition. Some of them immediately relapsed into the type of behavior they had shown in the prior condition. Much more interesting is the fact that some patients who had complied with the instructions and crossed out lines on the previously neglected side, instead of extending their search to what had been the favored side in the unconstrained condition, stopped near the midline as if totally unaware of stimuli of which they had been perfectly conscious a few instants earlier.

A further example of response-dependent dissociation of consciousness is the following (Bisiach, Berti, and Vallar, 1985). Patients suffering from left visuospatial neglect were placed in front of a panel with two rows of light-emitting diodes, one green and one red.

They were instructed to respond to the flashing of any diode by naming its color (first experiment) or by pressing a response key of the same color that simultaneously lighted on one side of the panel while the key of the wrong color lighted on the other side (second experiment). The stimulus, therefore, could appear on the left or the right side with respect to the visual fixation point in the middle of the panel, while the appropriate response key (in the second experiment) lighted either on the same side as the stimulus or on the opposite side. In the first experiment, patients were always aware of right-side stimuli. In the second experiment, patients were aware of right-side stimuli on virtually all trials in which the appropriate response key was on the right side. But on trials in which pressing of the left-side key was required, they often showed incorrect reactions—pressing either the key on the right side, which was unlit but of the correct color, or the lit key of the wrong color—or no reaction at all, sometimes spontaneously denying (patient F. S.) that any stimulus had occurred in the ipsilesional (unaffected) visual field. Peculiar to this experiment were the facts that the crucial variable—the side of the manual response—was not known to the subject before the occurrence of the stimulus, and that, on trials in which the subject denied the presence of the stimulus, gating of "normal" sensory information out of consciousness originated *after* the stages of color identification and initiation of the search for the corresponding response key. This may be taken as evidence for diachronic mechanisms underlying a certain kind of dissociation of consciousness, namely, the dissociation between consciousness of a stimulus when its sensory processing is part of a full stimulus-response cycle and lack of consciousness of an identical stimulus when the cycle is interrupted (i.e., when access to response is witheld).

Dissociation between the what *and* where *of conscious representation*

Somatic Domain It has long been known that patients with unilateral brain lesions may report elementary contralesional stimulations as felt on the opposite (ipsilesional) side of somatic or extrasomatic space, usually on a mirror-symmetrical point. The phenomenon was first described by Obersteiner (1882), who termed it *allochiria.* Since then, it has been frequently reported and found to occur in different sensory modalities: somatosensory, visual, auditory, and even olfactory (see Bisiach and Vallar, 1988, for a concise review). Transposition from the contra- to the ipsilesional side of the body of more complex percepts was observed by Brain (1941) in three patients. "In such patients," he wrote, "not only may a stimulus applied to the affected half of the body be felt on the normal side, but *voluntary* movement *spontaneously* initiated may feel as if carried out on the normal side" (p. 265; italics added). This is probably the phenomenon to which Jones had made allusion under the term *motorische Allochirie* (1910, 146). In some instances "the whole affected half of the body may feel transposed" (Brain, 1941, 265). Further examples of massive allochiria in the somatic domain are the transposition of pain from the contralesional to the ipsilesional side of the body—for example, in the case of lower limb phlebitis—and the complaint of motor impairment of the *unaffected* limbs in association with denial of ownership of contralesional limbs observed in patient L. A.-O. (described in Bisiach and Geminiani, 1990).

Transposition of conscious mental contents from the affected to the unaffected side of the body has been interpreted by Brain (1941) in terms of disordered body schema: "It appears that severe damage to the scheme for one half of the body causes events occurring in that half, if perceived at all, to be related in consciousness to the surviving scheme representing the normal half" (p. 265). Although this interpretation is likely to be to a large extent true, no matter how problematical the concept of body schema might be, it must be further worked out to account for instances in which sensorially driven or internally generated representation is transposed from the unaffected to the affected limbs. Critchley (1953, 91), for example, noted that "patients may feel stereognosic phantom sensations in the affected hand when an object is held in the normal hand." He also cited the phenomenon observed by Allen (1928) in one of his patients: While looking for a match after having put a cigarette between his lips he suddenly felt the phantom sensation of a box of matches in his affected hand.

It must be noted that transposition of conscious somatosensory events from one side of the body to the opposite cannot be explained on the basis of hardwired mechanisms of sensory transduction. This kind of explanation is overruled not only by exceptional instances such as that described by Allen, but also by the fact that the occurrence of allochiria in the affected limbs increases according to a proximal-to-distal gradient, as is the case with phantom-limb phenomena in amputees and (most likely) with anosognosic and somatoparaphrenic phenomena, whereas the gradient of bilateral projection of somatosensory pathways to the cortex is exactly the opposite.

Extrasomatic Domain Although outside the somatic domain clear instances of allochiria have until recently been found to occur almost exclusively with respect to elementary stimuli, there is fresh evidence of side-to-side transposition of complex visual percepts in patients suffering from left hemineglect. Halligan, Marshall, and Wade (1992), for example, described a patient who accurately copied the picture of a butterfly with a missing left wing; however, in copying the complete picture of a similar butterfly, she omitted the left wing but transposed some of its details onto the right one.

In investigating the performance of neglect patients on the Rorschach test for different purposes, Làdavas, Bosinelli, and Bisiach (1993) came across a similar finding. One of their patients had first been asked to interpret the 10 inkblot figures of the test in the canonical condition, then to interpret the same figures after removal of the left half of each of them and, finally to interpret five chimeric figures obtained by juxtaposing the left half of figures III, IX, VIII, II, and X to the right half of figures V, I, VI, IV, and VII, respectively. The patient had been asked, as usual, to point with her finger to the part of the figure of which she was giving an interpretation. After that, she had to trace the whole contour of each figure. Two aspects of her behavior are relevant here. First, The patient gave global responses (i.e., responses based on the whole configuration) to 3 of the 10 original figures (figures IV, V, and VI), but no global responses to halved or chimeric figures. She always pointed only to the entire right half of the figure when she gave global responses, or to right-side details when responses were not global; furthermore, she always accurately traced the right outline of original, halved, and chimeric configurations and the left (straight vertical) outline of halved configurations, while drawing a vertical line at the middle of complete and chimeric configurations. It can therefore be inferred that global responses were based on nonconscious perception of the left half of original figures and not on a phenomenon of perceptual completion—which, if present, should also have occurred with halved and chimeric figures.

Second, The patient gave allochiric-reduplicative responses to 5 of the 10 original figures (figures I, II, III, VII, and VIII) and to one of the chimeric figures (figure II left–IV right). For example, she said "two monkey" while pointing at, and tracing, only the right-side black silhouette in figure III. Likewise, she interpreted the chimeric configuration obtained by juxtaposition of the left and right halves of figures II and IV, respectively, as "Two bears that have fought; there is blood and this is their hair," but only pointed at, and traced the contour of, the right half of the figure. (In conformity with current policy relative to the reproduction of the Rorschach test, no figure is given here.) No reduplicative responses were given to halved figures. On Tegnér and Levander's (1991) mirror-reversed cancellation task, the patient was found to be affected by left visuo-exploratory neglect, which militates against an interpretation of her behavior in terms of manual directional hypokinesia.

Further evidence of left-to-right transposition, and reduplication of content on the right side of visuospatial patterns, is the observation made by Vallar, Rusconi, and Bisiach (1994) in patient R. G. during the experiments mentioned earlier, in the section on nonconscious representation. While tracing the head of a kangaroo on a deer-left/kangaroo-right chimera, R. G. said: "Here one could draw a deer." Likewise, while tracing the head of a swan on a shark-left/swan-right chimera, he said: "Here one could draw a shark." In both cases the patient had not recognized the chimeric nature of the drawing.

Implications and conclusions

Some years ago, we outlined a model aimed at demonstrating that closely related dysfunctions of a single neural mechanism may lead to either defective or productive phenomena of misrepresentation on the contralesional side of egocentric space (Bisiach, Meregalli, and Berti, 1990). The model was held to support our claim that these phenomena should be considered components of a more general representational syndrome that we called, after Zingerle (1913), dyschiria. Our claim was based empirically on factors such as the frequent association of defective and productive phenomena of unilateral misrepresentation and the fact that both kinds of phenomena may undergo temporary

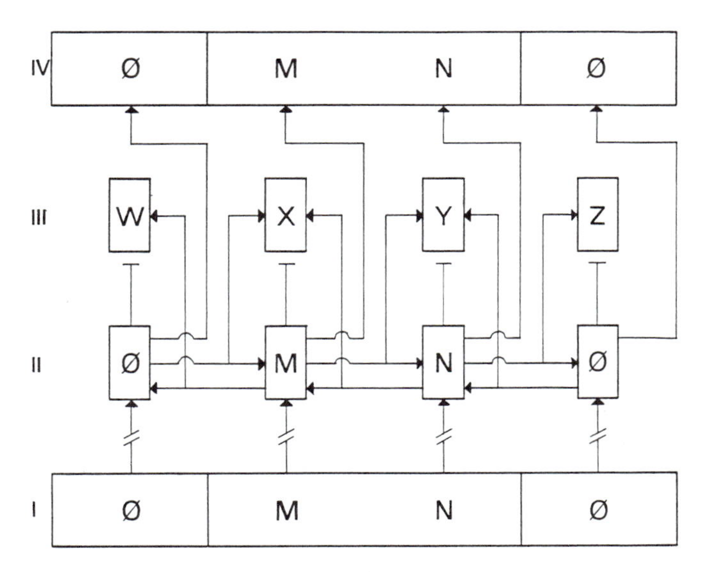

FIGURE 88.1 Model showing the emergence of a visuospatial percept (Reproduced by permission from Bisiach, Meregalli, and Berti, 1990)

remission during vestibular stimulation (see Bisiach, 1993, for review). Although originally unconcerned with the phenomena surveyed in this chapter, our model can easily integrate them into the pathophysiology of dyschiria.

Figure 88.1 outlines the model in its intact structure; diagrams showing its dysfunctions may be found elsewhere (Bisiach, Meregalli, and Berti, 1990). Layer I is the terminal of a sensory transducer carrying information about a left-to-right elongated object MN and its flanks 00. It projects to layer II, a sensorially driven topological representational network in which the content of layer I is analyzed and synthesized. Layer II is composed of a horizontal array of interconnected sets of cell assemblies firing in response to the message carried by the sensory transducer. The output of these cell assemblies, layer IV, is a veridical, topological, conscious representation of sensory input. Layer III is constituted by an array of internally driven, topologically organized cell assemblies the firing of which corresponds to autochthonous (i.e., not sensorially driven) representation. Layer III cell assemblies accept excitatory input (arrows) from heterotopic, but inhibitory input (T ends) from homotopic layer II cell assemblies; the latter insures that in normal waking conditions the product of autochthonous representation does not reach layer IV, or is tagged as imaginary. Horizontal connections (between adjoining cell assemblies in layer II, and between heterotopic cell assemblies of layers II and III) inform the receiver about the current status of the sender. Except for abnormal conditions, the information they carry influences the firing of target cell assemblies without shaping them in the form of a displaced duplicate of the transmitting cell assembly. Their function is the mutual integration of spatially contiguous contents and the implementation of a *local* system of cross-expectations, such as those giving rise to different kinds of perceptual and conceptual completion.

More or less complete inactivation of one side of layers II and III gives rise to unilateral lack of representation at the level of layer IV (unilateral neglect). Reduction or abolition of activity confined to one side of layer II frees from inhibition homotopic layer III cell assemblies, thus releasing phenomena of unilateral, nonveridical, autochthonous representation that may show through layer IV in the form of delusional contents (such as somatoparaphrenia).

Phenomena of contralesional nonconscious sensorially driven representation such as those reviewed earlier in the chapter may be held to depend on hypoactivation of layer II cell assemblies on the affected side, such that they are still sufficient to inhibit homotopic layer III cell assemblies and to give rise to contentful cognitive activity. This activity, though not reaching the level of processing required for its becoming conscious (layer IV), may nonetheless affect behavior, for example, by causing deportments such as those ascribed by Anton to "dim knowledge" of illness. Pathological conditions apt to interrupt the normal (homotopic) flow of information processing on one side may favor its shunt to the opposite side through horizontal connections in layers II and III and give rise to phenomena of allochiria and reduplication.

The model would obviously need a great deal of refinement in order to get closer to the actual machinery whose dysfunction results in the various phenomena of dyschiria. There are aspects it does not address, such as the extinction of contralateral stimuli in conditions of double simultaneous stimulation, the manifold spatial frames revealed by recent investigation of neglect patients, the issue of the left-right gradient of neglect, and so forth. Furthermore, it refers to a single sensory analyzer and therefore disregards the problem of supramodal determinants of dyschiria (if any). Finally, it totally ignores the issue of (in a very broad sense) premotor factors, such as those suggested by the dissociations of consciousness depending on response modality. These dissociations demonstrate that layer IV cannot be meant as a flat interface between input and output processes, as a literal reading of the model might suggest (see Bisiach, 1994, for further discussion of this point). In other words, these dissociations show that consciousness is not *diachronically* monolithic, in the sense that, as suggested by our patient F. S.'s denial of visual stimuli to which he was not able to give manual responses, it may be subject to reentrant influences (Edelman, 1989) from output-related events on earlier, input-related, processing stages. The basic implication of our model, on the other hand, is that consciousness is also not *synchronically* monolithic, in the sense that, at any given time, disorders of conscious attributes of mental representation may be spatially circumscribed: A patient may in fact be aware of one side of an object or of the whole environment, while the representation of the opposite side—though sometimes available to high-level cognitive processing—remains nonconscious.

There are two main points that remain deeply unclear. One concerns the status of mental contents that, although (apparently) nonconscious, can be shown to affect behavior, for example, in the clinical and experimental situations reviewed in the section on nonconscious representation. To what extent can we talk of nonconscious *representation* in those instances? The impossibility of a clear-cut answer to this question is inherited from the insufficient elucidation of the concept of representation itself. On the one hand, one would hesitate to claim that eye-blinking conditioned to a tone is due to a *representation* of the forthcoming unconditioned stimulus; on the other hand, mental contents such as those giving rise to semantic priming (Berti and Rizzolatti, 1992), or the nonconscious knowledge of illness that shows, in anosognosic patients, through the questionnaire devised by Marcel and Tegnér (1993), seem to be very close to what we would unhesitatingly call representation. In which respect—apart from not being conscious—do these mental contents differ from the cognitive representation?

The other point is even more problematical. Sensory information of which patients seem to be unaware may be integrated in intentional actions such as the consistent choice of one of two allegedly identical stimuli in Marshall and Halligan's and similar experiments. As aptly pointed out by Marcel (1988, 146–147), it is true that patients do not, as a rule, initiate any action spontaneously intended toward objects of which they only have this kind of nonconscious information. There are, however, exceptions: Our patient L. A.-O., on being asked to pick up objects aligned from left to right in front of her, started collecting them from the rightmost one, but once arrived about the midline of the row pushed the remainder leftward while saying: "There aren't any more" (Bisiach and Geminiani, 1991). Would we confidently assert that an action of that kind was executed with no consciousness whatsoever of what was explicitly denied? This and similar paradoxes arise from the fact that we often treat other people's consciousness as a hypothetical construct. First-person phenomenal experience, however, is *not* a construct (see Bisiach, 1988, 1991). Therefore, whatever might be the extent to which in normal conditions we can legitimately accept another person's attestations about his or her phenomenal experience, we must be extremely cautious in dealing with anomalous cases.

There may well be instances in which phenomenal experience of denied sensory information is indistinguishable from lack of any experience, for example, in experiments such as those done by Volpe, LeDoux, and Gazzaniga (1979) and by Berti and Rizzolatti (1992), but perhaps this is not the most significant point for the understanding of consciousness. Much more significant are probably the cases in which consciousness appears to be dissociated. These cases suggest that the crucial contraposition is not between consciousness and nonconsciousness, but between *co-consciousness* of a person's mental states and events and its breakdown (by which different mental states or events, though being individually endowed with phenomenal quality, are kept separate from one another within the stream of consciousness). This leads to what for many might be the counterintuitive conclusion that consciousness is not the exclusive prerogative of an indivisible central executive component of the mind and that the self is in fact a "virtual captain of the crew" (Dennett, 1991, 228), whose composite structure is dramatically evident in the many bizarre phenomena of disordered representation and delusional belief relative to the side of egocentric space opposite to the side of a brain lesion.

ACKNOWLEDGMENTS Preparation of this chapter has been supported by MURST and CNR grants to the first author. We are indebted to the philosopher Susan Hurley for the term *co-conscious* and its implications with respect to the phenomena we have reviewed and discussed. We are also grateful to Joe Bogen and David Galin for their helpful comments on our first draft.

REFERENCES

Albert, M. L., 1973. A simple test of visual neglect. *Neurology* 23:658–664.

Allen, I. M., 1928. Unusual sensory phenomena following removal of a tumour of the sensory cortex. *J. Neurol. Psychopathol.* 4:133–145.

Anton, G., 1899. Ueber die Selbstwahrnehmung der Herderkrankungen des Gehirns durch den Kranken bei Rindenblindheit und Rindentaubheit. *Arch. Psychiatr. Nervenkrank.* 32:86–127.

Berti, A., and G. Rizzolatti, 1992. Visual processing without awareness: Evidence from unilateral neglect. *J. Cognitive Neurosci.* 4:345–351.

Bisiach, E., 1988. The (haunted) brain and consciousness. In *Consciousness in Contemporary Science*, A. J. Marcel and E. Bisiach, eds. Oxford: Oxford University Press, pp. 101–120.

Bisiach, E., 1991. Understanding consciousness: Clues from unilateral neglect and related disorders. In *The Neuropsychology of Consciousness*, A. D. Milner and M. D. Rugg eds. London: Academic Press, pp. 113–137.

Bisiach, E., 1993. Mental representation in unilateral neglect and related disorders: The Twentieth Sir Frederic Bartlett Lecture. *Q. J. Exp. Psychol.* 46A:435–461.

Bisiach, E., 1994. Perception and action in space representation: Evidence from unilateral neglect. In *International Perspectives on Psychological Science. Vol. 2: The State of the Art*, P. Bertelson, P. Eelen, and G. d'Ydewalle, eds. Hillsdale, N.J.: Erlbaum, pp. 51–66.

Bisiach, E., and A. Berti, 1987. Dyschiria: An attempt at its systemic explanation. In *Neuropysiological and Neuropsychological Aspects of Spatial Neglect*, M. Jeannerod, ed. Amsterdam: North-Holland, pp. 183–201.

Bisiach, E., A. Berti, and G. Vallar, 1985. Analogical and logical disorders underlying unilateral neglect of space. In *Attention and Performance*, vol. 9, M. I. Posner and O. S. M. Marin, eds. Hillsdale, N.J.: Erlbaum, pp. 239–246.

Bisiach, E., and G. Geminiani, 1991. Anosognosia related to hemiplegia and hemianopia. In *Awareness of Deficit after Brain Injury*, G. P. Prigatano and D. L. Schacter, eds. New York: Oxford University Press, pp. 17–39.

Bisiach, E., G. Geminiani, A. Berti, and M. L. Rusconi, 1990. Perceptual and premotor factors of unilateral neglect. *Neurology* 40:1278–1281.

Bisiach, E., and C. Luzzatti, 1978. Unilateral neglect of representational space. *Cortex* 14:129–133.

Bisiach, E., S. Meregalli, and A. Berti, 1990. Mechanisms of production control and belief fixation in human visuospatial processing: Clinical evidence from unilateral neglect and misrepresentation. In *Quantitative Analyses of Behavior*. Vol. 9, *Computational and Clinical Approaches to Pattern Recognition and Concept Formation*, M. L. Commons, R. G. Herrnstein, S. M. Kosslyn, and D. B. Mumford, eds. Hillsdale, N.J.: Erlbaum, pp. 3–21.

Bisiach, E., and M. L. Rusconi, 1990. Break-down of perceptual awareness in unilateral neglect. *Cortex* 26:643–649.

Bisiach, E., and G. Vallar, 1988. Hemineglect in humans. In *Handbook of Neuropsychology*, vol. 1, F. Boller and J. Grafman, eds. Amsterdam: Elsevier, pp. 195–222.

Bisiach, E., G. Vallar, and G. Geminiani, 1989. Influence of response modality on perceptual awareness of contralesional visual stimuli. *Brain* 112:1627–1636.

Brain, W. R., 1941. Visual disorientation with special reference to lesions of the right cerebral hemisphere. *Brain* 64: 244–272.

Critchley, M., 1953. *The Parietal Lobes.* London: Hafner.

Dennett, D. C., 1991. *Consciousness Explained.* Boston: Little, Brown.

Denny-Brown, D., and R. A. Chambers, 1958. The parietal lobe and behavior. *Res. Publ. Assoc. Res. Nerv. Ment. Dis.* 36:35–117.

Dunn, R., 1845. Case of suspension of the mental faculties, of the powers of speech, and special senses. *Lancet* 2:536–538, 588–590.

Edelman, G. M., 1989. *The Remembered Present.* New York: Basic Books.

Gerstmann, J., 1942. Problem of imperception of disease and of impaired body territories with organic lesions. *Arch. Neurol. Psychiatry* 48:890–913.

Halligan, P. W., J. C. Marshall, and D. T. Wade, 1992. Left on the right: Allochiria in a case of left visuo-spatial neglect. *J. Neurol. Neurosurg. Psychiatry* 55:717–719.

Hécaen, H., J. de Ajuriaguerra, L. Le Guillant, and R. Angelergues, 1954. Delire centré sur un membre-phantome chez une hémiplégique gauche par lésion vasculaire avec anosognosie. *Evolution Psychiatrique* 19:273–279.

Heilman, K. M., D. Bowers, H. B. Coslett, H. Whelan, and R. T. Watson, 1985. Directional hypokinesia. *Neurology* 35:855–859.

Hillis, A., and A. Caramazza, 1990. The effects of attentional deficits on reading and spelling. In *Cognitive Neuropsychology and Neurolinguistics: Advances in Models of Cognitive Function and Impairment*, A. Caramazza, ed. Hillsdale, N.J.: Erlbaum, pp. 211–275.

Jaffe, J., and W. H. Slote, 1958. Interpersonal factors in denial of illness. *Arch. Neurol. Psychiatry* 80:653–656.

Jones, E., 1910. Die Pathologie der Dyschirie. *J. Psychol. Neurol.* 15:145–183.

Kinsbourne, M., 1970. A model for the mechanism of unilateral neglect of space. *T. Am. Neurol. Assoc.* 95:143–146.

Kinsbourne, M., and E. K. Warrington, 1962. A variety of reading disability associated with right hemisphere lesions. *J. Neurol. Neurosurg. Psychiatry* 25:339–344.

Làdavas, E., M. Bosinelli, and E. Bisiach, 1993. Non-conscious perception of neglected stimuli: Evidence from Rorschach's test. Poster presented at the 11th European Workshop on Cognitive Neuropsychology, Bressanone, 24–26 January.

Làdavas, F., R. Paladini, and R. Cubelli, 1993. Implicit associative priming in a patient with left visual neglect. *Neuropsychologia* 32:1307–1320.

Marcel, A. J., 1988. Phenomenal experience and functionalism. In *Consciousness in Contemporary Science*, A. J. Marcel and E. Bisiach, eds. Oxford: Oxford University Press, pp. 121–158.

Marcel, A. J., and R. Tegnér, 1993. Knowing one's plegia. Paper read at the 11th European Workshop on Cognitive Neuropsychology, Bressanone, 24–26 January.

Marshall, J. C., and P. W. Halligan, 1988. Blindsight and insight in visuo-spatial neglect. *Nature* 336:766–767.

Milner, D. A., and M. D. Rugg, eds. 1991. *The Neuropsychology of Consciousness*. London: Academic Press.

Obersteiner, H., 1882. On allochiria: A peculiar sensory disorder. *Brain* 4:153–163.

Tegnér, R., and M. Levander, 1991. Through a looking glass: A new technique to demonstrate directional hypokinesia in unilateral neglect. *Brain* 114:1943–1951.

Vallar, G., M. L. Rusconi, and E. Bisiach, 1994. Awareness of contralesional information in unilateral neglect: Effects of verbal cueing, tracing and vestibular stimulation. In *Attention and Performance XV*, M. Moscovitch and C. Umiltà, eds. Cambridge, Mass.: MIT Press.

Volpe, B. T., J. E. LeDoux, and M. S. Gazzaniga, 1979. Information processing in an "extinguished" visual field. *Nature* 282:722–724.

Zingerle, H., 1913. Ueber Stoerungen der Wahrnehmung des eigenen Koerpers bei organischen Gehirnerkrankungen. *Monatsschr. Psychiatr. Neurol.* 34:13–36.

89 Models of Consciousness and Memory

MORRIS MOSCOVITCH

ABSTRACT Why are only some memories accompanied by a conscious awareness that what has been recovered has been experienced before? Following a brief historical overview and a critique of the current literature, an answer to the question is proposed. The explanation is derived from a neuropsychological model based on modules and central systems. According to the model, the hippocampus and related memory structures bind or integrate into a memory trace the neural elements that mediate a conscious experience. In this way, consciousness is bound to other (information-containing) elements of the memory trace and becomes an intrinsic property of it. At retrieval, consciousness is recovered along with the other elements. This *recovered consciousness* is the signal or marker that confers on conscious memory the felt experience of recollection. Memory mediated by nonhippocampal structures, such as those in the posterior neocortex, cannot have consciousness bound in the trace by virtue of the organization of the mediating structures. Those structures can be modified by experience and thereby influence behavior but without giving rise to conscious recollection. Neuropsychological evidence is provided to support these ideas.

This chapter addresses a fundamental question about memory: Why are only some forms of memory accompanied by conscious awareness? Specifically, why is conscious recollection a necessary component of traditional tests of memory, like recall and recognition, but not of other tests, like direct or repetition priming in perceptual identification or completion? I begin with a brief historical survey, and then examine current models and theories of memory to see how the relation of consciousness to memory is treated by them. Of those models in which consciousness plays a significant role, I distinguish between two types: *content-based models* in which consciousness is associated, or even equated, with a particular content of memory, such as its temporal-contextual aspect; and *attribute-based models* that treat consciousness as a separate attribute or property of memory that is not linked necessarily to a particular kind of content. A solution to the puzzle posed at the beginning of the chapter is then derived from a neuropsychological model of memory based on modules and central systems that has elements in common with both types of models.

MORRIS MOSCOVITCH Psychology Department, University of Toronto, Mississauga, Ontario, and Rotman Research Institute of Baycrest Centre, North York, Ontario

Definitions and assumptions

Consciousness has a variety of meanings. What I mean by consciousness in this chapter is phenomenological awareness. In other words, I use *consciousness* in the ordinary language sense of being aware *of* something. One is *conscious of* or aware of something when a verbal or nonverbal description can be provided of the object of that awareness or a voluntary response can be made that comments on it. With regard to memory, to be conscious implies being aware of mental experience as a memory rather than as a percept, thought, or image without the sense that any of them had occurred before. Also, I wish to focus on those instances in which awareness of memory has an immediate, noninferential quality in much the same way that we have an immediate awareness of a perceived object. I wish to exclude from this discussion those cases where awareness of memory is inferred—"I must have experienced this before, otherwise how could I account for my behavior, feelings, knowledge?" To be sure, sometimes conscious recollection is based on inference and sometimes it may be false. But not all, or even most, memories are of this sort. If they were, we would all behave like some patients with organic amnesia who lack "raw" memories of recently experienced events but whose ability to make memory inferences or attributions on the basis of current knowledge is intact.

Finally, I wish to assert that there is a component to this awareness that goes beyond the ability to respond or react to stimuli differentially on the basis of informational content. For this last point, there is some empirical evidence from patients with cortical blindness, agnosia, dyslexia, and neglect. There are now numerous reports that patients can respond differentially to familiar and unfamiliar faces, can pick up information about stimuli in the neglected field or blind field, and can glean lexical information about words without being conscious of the very information to which they have responded (see Schacter, McAndrews, and Moscovitch, 1988). I do not wish to imply by this that there is no relationship between consciousness and content of the experienced event, but rather that conscious awareness of an event is an aspect of that experience that may be separable from content. Indeed, I am willing to go so far as to say that unless conscious awareness accompanies the experienced event, one has not truly experienced the event in question. Thus, patients with blindsight may respond correctly to different types of visual input in the blind field, but on questioning they deny having had a visual experience, or at least one that was related in a meaningful way to their performance (Weiskrantz, 1988). The same can be said of the behavior of the other types of patients whose various disorders have been mentioned. This idea that consciousness is a component or aspect of experienced events that is distinct from the informational content of the event, though it may be intimately related to it, is critical to the argument I develop about the relation between consciousness and memory.

Although I have not provided a precise definition of consciousness, I think the sense I have conveyed of my use of the term, and the empirical justification for it, provides an appropriate starting point for discussion and research. As Weiskrantz (1988) reminded us, "Definitions and precise theoretical constructs are the final product, not the starting point of inquiry" (p. 183).

Tests of memory can be classified as conscious or unconscious. Because the terms *conscious* and *unconscious* carry heavy theoretical implications, particularly with regard to the role such terms play in psychodynamic theories, psychologists have adopted the more neutral, operational terms *explicit* and *implicit*, or direct or indirect, to refer to the different types of tests. (Graf and Schacter, 1985; Richardson-Klavehn and Bjork, 1989). Explicit or direct tests of memory depend on the subject's ability to reflect consciously on the past in order to respond correctly and to recognize the recovered information as a memory. In contrast, on implicit tests, examiners assess memory indirectly by observing whether an encounter with a target alters the subject's performance without asking the subject to recollect the target consciously. For example, rather than asking the subject to recall or recognize a particular word, either of which is an explicit test, the experimenter simply asks the subject to read or perceive a word. If the speed of reading or accuracy of perceptual identification is different for previously studied than for nonstudied words, then the experimenter can infer that memory for the word was retained. The benefit accrued to repeated over new material is known as direct priming, or repetition priming. If designed properly, the experiment can be conducted so that subjects not only are unaware that they have studied some of the words but also are unaware that memory is being tested.

Historical overview

The distinction between conscious and unconscious processes is not a new one in psychology. By the end of the nineteenth century, it figured prominently in the ideas of Sigmund Freud and William James, and even entered into the writings of Hermann Ebbinghaus, three major theorists who were otherwise quite different from each other in their concerns, methodology, and theories. And in all cases, it was with respect to memory that the distinction between conscious and unconscious processes was most prominent. This should not surprise us because memory, more than any other psychological process, invites speculation about the role of conscious and unconscious processes. Except for memories that occupy our immediate awareness, all other ones lie dormant, to be revived only as the occasion demands or to influence our behavior without our awareness. What is surprising is that mainstream experimental psychology, but not psychodynamic theory, virtually banned consciousness from its domain for about 60 years of this century, and, in the process, replaced the study of memory with that of learning.

It is most interesting to note that most of the tests used during this period would now be classified as explicit. Few implicit tests of memory were administered. Thus, by their choice of tests, investigators inadvertently conceded that consciousness plays a role in memory even while publicly denying that consciousness was a useful construct, or even that it existed. Only when the concept of consciousness is admitted into psychology can unconscious tests of memory be discovered or be appreciated for what they are.

As information-processing, cognitive, and physiological theories began replacing learning theory in the 1950s and 1960s, discussions of the role of consciousness in behavior slowly crept into the literature. A number of discoveries called attention to the role of consciousness in memory (see Mandler, 1989). In studying memory in normal people, several investigators noted that a single encounter with a stimulus was sufficient to alter its identifiability on subsequent tests of perception, reading, lexical decision, or completion. (for historical review, see Schacter, 1987; for review of current literature, see Roediger and McDermott, 1993). More importantly, this improved performance was independent of the subject's ability to recollect having encountered the stimulus previously. These findings indicated that information about a stimulus could be registered, retained, and recovered implicitly and that conscious recollection is a feature only of explicit tests of memory.

The idea that memory is divisible into fundamentally different forms received added impetus from investigators working on the neural basis of memory in mammals. With their focus on the hippocampus, these investigators noted that only some types of memory were affected by hippocampal lesions, leading them to propose that fundamentally different forms of memory were mediated by hippocampal and nonhippocampal structures (see reviews in O'Keefe and Nadel, 1978; Nadel, 1992). Although few of these investigators were concerned with consciousness (O'Keefe, 1985, is an exception), their studies prepared the field for a profound reinterpretation of organic amnesia in humans from a syndrome that was considered to involve almost all memory functions to one that is now believed to affect only some.

The greatest impact on attitudes regarding the importance of consciousness to memory came from studies of patients with severe anterograde amnesia. It was discovered that even profoundly amnesic people could acquire sensorimotor skills (Milner, 1966; Corkin, 1968) and even recover information about specific items, such as words or pictures, if appropriate cues were provided (Warrington and Weiskrantz, 1973). The theoretical significance of these spared memory functions in amnesic patients was not fully appreciated until the late 1970s and early 1980s.

Rozin and Diamond (Rozin, 1976a; Diamond and Rozin, 1984) were the first to suggest that, unlike normal people, amnesic patients did not respond to the cue by consciously recollecting the target; instead, they suggested that amnesics provided the first response that the cue brought to mind. As a result, the amnesics did not experience a sense of familiarity, a phenomenological awareness, that they had encountered the item in the past. Indeed, on explicit tests of recognition, amnesic patients could not choose as targets the very items (e.g., words) that these cues elicited as responses (Warrington and Weiskrantz, 1973; Graf and Schacter, 1985). Pioneering studies by Cohen and Squire (1980) and Graf, Mandler, and Haden (1982) were instrumental in calling attention to the link between preserved memory in amnesia and the performance of normal people on implicit tests of memory.

On the basis of these studies, and the many others that they inspired, amnesia is now viewed as a selective rather than global memory impairment. How best to characterize the impairment or define the domain that is affected is currently being debated (see Schacter and Tulving, 1992). I favor the view that amnesia is a disorder that affects only the conscious recollection of postmorbid events (Moscovitch, 1982; Moscovitch, Winocur, and McLachlan, 1986).

The status of consciousness in models of memory

Although it is now commonplace to distinguish between explicit and implicit tests on the basis of the demands they make on conscious awareness of the target as a memory, it is surprising that only a handful of investigators have assigned a primary role to consciousness in their theories and models.

PROCESSING THEORIES Proponents of processing theories argue that the dissociations observed between performance on explicit and implicit tests arise because the two involve different processes. The transfer-appropriate processing theory (see Roediger, 1990) states that memory performance depends on the overlap in processing operations performed at encoding and at retrieval. Typically, explicit tests are conceptually driven and demand semantic processing whereas implicit tests are data-driven tests that emphasize perceptual processes. Conceptually driven implicit tests have also been devised, and performance on them is dissociable from performance on explicit tests both in normal people and in amnesics (see Moscovitch et al., 1993). Consciousness plays no role in this theory. The failure to consider consciousness may yet prove to be a critical drawback. If performance on explicit tests of memory is conceptually driven, why should dissociations be found between them and conceptually driven implicit tests unless the explicit tests recruit additional (consciousness-related?) processes that are not used on conceptual implicit tests?

The relation between consciousness and memory is a central concern, however, for other processing theorists, like Jacoby (1991). In recent years, Jacoby developed techniques that could be used to determine the relative contributions of conscious and nonconscious processes to performance on memory tests. It is telling, however, that the contrastive terms *controlled* versus *automatic* are coming to replace "memory with and without awareness" in his papers. This shift in terminology is indicative of a change in attitude that is reflected also in Jacoby's theoretical work. There is some discussion of the conditions that promote conscious and nonconscious use of memory, but little space is devoted to considering why consciousness accompanies some types of memory but not others. Control may be a diagnostic marker of memory with awareness, but it is not to be equated with it.

Systems Theories In contrast to processing-based theories of memory, multiple systems theories posit that performance on different types of memory tests is mediated by different, distinct memory systems that are distinguished from one another on the basis of structure and function. Systems theorists, however, often are not much better than their processing counterparts in assigning a significant role for consciousness in their work, especially if they work primarily, or exclusively, with nonhuman species (for notable exceptions see Rozin, 1976a, 1976b; O'Keefe, 1985; Weiskrantz, 1988). Although many acknowledge that conscious recollection is associated with one system but not the other, few make any further attempt to develop a theory of memory based on that distinction. Instead, they focus on content, on the type of information each system is believed to process. Researchers working primarily with humans contrast context-dependent versus context-free memory (Winocur and Kinsbourne, 1978; Mayes, 1992); declarative versus procedural or nondeclarative (Cohen and Squire, 1980; Squire, 1992); episodic versus semantic (Cermak, 1984).

Those working on animal models prefer other dichotomies: spatial (locale) versus nonspatial (taxon) (O'Keefe and Nadel, 1978; Nadel, 1992); configural versus simple associations (Rudy and Sutherland, 1992); relational versus nonrelational associations (Eichenbaum, 1992); or dichotomies that are a combination of content and process such as working versus reference memory (Olton, Becker, and Handelmann, 1979) or memories versus habits (Mishkin and Appenzeler, 1987) (for critique, see Moscovitch, in press). Then there are researchers who simply avoid these dichotomous distinctions and adopt a far more pragmatic, functional neuroanatomic approach (Butters, Heindel, and Salmon, 1990).

The Toronto Group Among the few people who have made consciousness an integral part of their theory of human memory are those who belong to what I shall call the Toronto group. Among its members I include Tulving, Schacter, Graf, and myself as well as Gardiner (Gardiner and Java, 1994) who frequently visits Toronto. Also, many of the opponents of the particular approach associated with this group, such as Craik, Jacoby, Kolers, Mandler, and Roediger were either based at the University of Toronto or had strong affiliations with it.[1] Because I am most familiar with the work of the Toronto group, and because I think the relationship between memory and consciousness is spelled out most clearly in their writings, I will deal primarily with their work. First, it is necessary to provide some background information about different types of explicit and implicit tests of memory.

Types of memory tests

Perceptual Item-Specific Tests On the basis of studies conducted in the last decade, it is generally conceded that there are a variety of different types of implicit tests, each making different cognitive demands, each being mediated by different neural substrates, and, perhaps, each involving conscious and nonconscious processes to different degrees (for a taxonomy of different implicit tests, see Moscovitch, 1992; Squire, 1992). Because to deal adequately with all of them in this chapter is impossible, I focus only on *perceptual, item-specific* tests of memory because probably more is known about this type than any other and because they are the most obvious implicit counterpart to explicit tests. An item-specific test is considered to be perceptual if the study material is reinstated in whole or in part at test and identification of the target or some aspect of it is required.

Associative and Strategic Explicit Tests of Episodic Memory There are also a variety of different types of explicit tests, but for the purposes of this chapter I wish to distinguish between two of them: associative/cue dependent (or ecphoric) and strategic. Associative episodic memory tests are those in which the cue is sufficient to elicit a memory that the subject consciously apprehends as one. When given the cue "Have you visited Paris?" or "Have you met Donald Hebb?" or "Have you had dinner yet?" the answer is elicited effortlessly. It is the episodic equivalent of responding with "Guildenstern" to the cue "Rosencrantz" on a semantic memory task. This associative episodic memory process, I will argue, is at the core of conscious recollection and is mediated by the hippocampus and related structures. On the other hand, on *strategic* tests, the cue does not elicit the target memory automatically but provides the starting point of a memory search that has elements in common with problem solving. Such strategic processes are often initiated by questions that require the reinstatement of a particular spatial and temporal context such as "Where were you in mid-August two summers ago?" or "Can you recall the first list you learned today?" Arriving at a proper solution to these memory problems involves, among other things, providing cues to the component mediating associative memory, evaluating the response that is elicited, comparing it with other knowledge, using that information to construct or select another cue for delivery to the associative component, and repeating the process until a satisfactory outcome is obtained. In contrast to the associative process, the strategic one is effortful and under voluntary control. The prefrontal cortex is among the critical structures involved.

Properties of Perceptual Repetition Priming Effects There is general consensus that performance on perceptual item-specific tests is dissociable from performance on explicit tests for the very same items in both normal and amnesic patients. Perceptual repetition-priming effects have been reported for items that were already familiar to the individual such as words, faces, and pictures, as well as for novel, sometimes meaningless material, such as two-dimensional (e.g., Musen and Squire, 1992) and three-dimensional drawings of possible objects (Schacter, Cooper, and Delaney, 1990), for nonwords (Musen and Squire, 1991), and for unfamiliar faces (Ellis et al., 1987).

Perceptual repetition-priming effects are now believed to have the following properties, many of which distinguish them from variables that affect performance on explicit tests (for review and references, see Moscovitch, Vriezen, and Goshen-Gottstein, 1993; Roediger and McDermott, 1993). They are *modality* and *format* specific, the rule of thumb being that repetition-priming effects can tolerate changes in surface features so long as the structurally invariant properties of the stimulus are similar at study and at test. Thus, priming effects are affected little by changes in size or reflection (or typeface, for words) but are diminished noticeably by a change from pictorial to written format. Level-of-processing manipulations also have much less influence on perceptual repetition priming than on explicit tests of memory. Repetition-priming effects, some lasting for months, can be obtained after a single, brief exposure to the stimulus and, in many cases, the effect is not increased by multiple presentations (for recent evidence and references, see Bentin and Moscovitch, 1988; Sloman et al., 1988; Challis and Sidhu, 1993).

A theory of perceptual repetition priming based on perceptual input modules or representation systems

The evidence from psychological studies of normal and amnesic people led some of us in the Toronto group along with collaborators Lynn Cooper and Carlo Umiltà (but see also Jackson and Morton, 1984; Kirsner and Dunn, 1985) to propose that performance on perceptual implicit tests is mediated by the very structures involved in picking up stimulus information and transforming it into structural, presemantic representations (Moscovitch, 1989; 1992; Schacter, 1989, 1992; Tulving and Schacter, 1990; Moscovitch and Umiltà, 1991). These networks of structures are termed *perceptual representation systems* by Schacter and Tulving (figure 89.1), and perceptual input modules by Moscovitch and Umiltà (figure 89.2). Each subsystem or module is domain specific in that it can process only one type of information. The visual word-form module or system processes written words, the face-recognition module processes faces, and so on. Being presemantic, the modules or systems represent information at a structural or perceptual level, not a conceptual one.

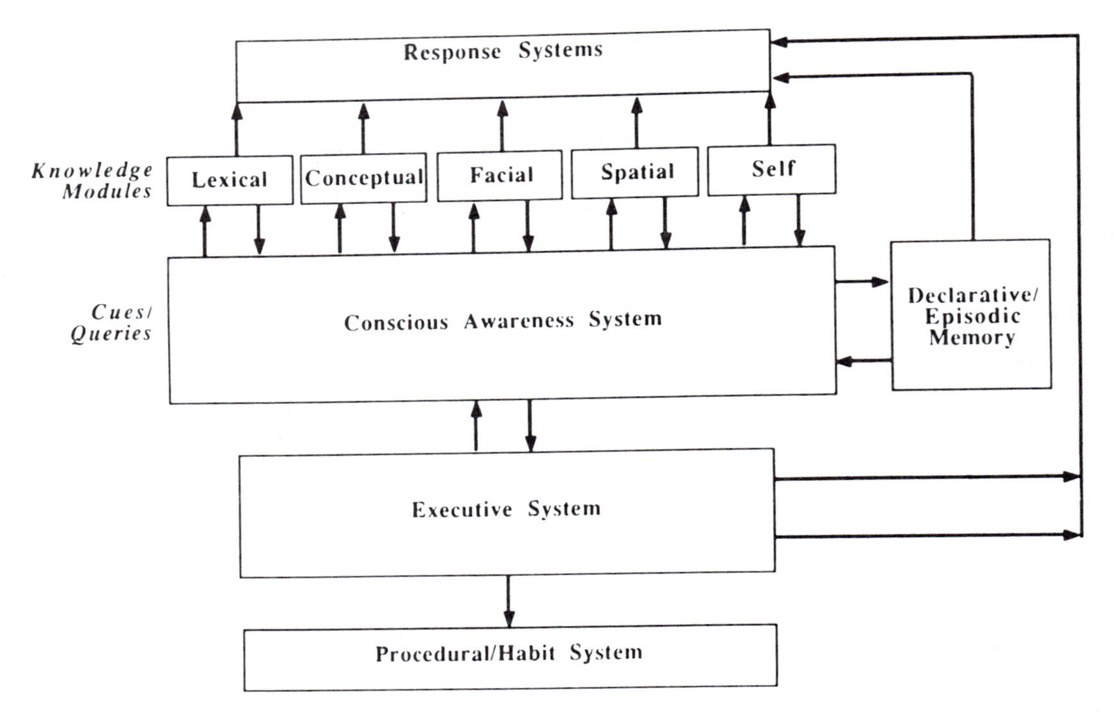

FIGURE 89.1 A schematic depiction of DICE: dissociable interactions and conscious experience. (From Schacter, 1989)

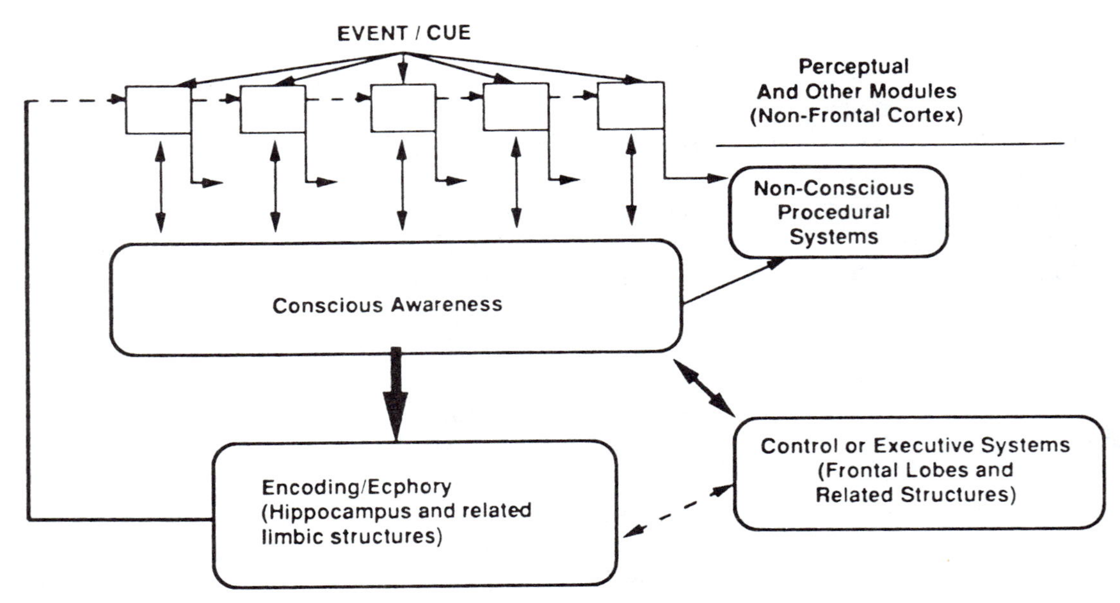

FIGURE 89.2 A neuropsychological model of memory (from Moscovitch, 1989). The dashed lines indicate that the interaction is optional. The cortical modules that interact with the hippocampal system will vary depending on the information that is available to consciousness when the event is initially experienced and when it is being remembered.

Evidence from agnosia, dyslexia, dementia, and amnesia, as well as from PET studies in normal people, indicates that modules and perceptual representation systems are not just hypothetical constructs but are localizable to structures in the posterior neocortex, in what Luria (1966) called secondary-zone structures (for review see Schacter, 1992; Moscovitch, Vriezen, and Goshen-Gottstein, 1993).

In the course of processing information, the systems or modules are modified, leaving behind a perceptual record (Kirsner and Dunn, 1985) or engram of the stimulating event. The altered neuronal circuitry that underlies the record preserves information about the stimulating event and enables subsequent, related events to be processed and identified more quickly and accurately. This reactivation of perceptual records is the basis for repetition-priming effects.

As a result, repetition-priming effects are sensitive to manipulations of perceptual features, especially those related to invariant structural properties of stimuli. Semantic variables have little influence on repetition priming. Because perceptual modules and perceptual representation systems are necessary for the structural identification of objects, sounds, words, and so forth, they must of necessity be modifiable by experience to represent the myriad of items we encounter in our lives. This quality is manifested in the rapid registration of novel information that accounts for evidence of repetition-priming effects for unfamiliar items.

The term *registration* refers to the neocortical process of forming lasting perceptual records. I have proposed this term to distinguish this process from other ones involved in forming long-term memory traces that typically mediate performance on explicit tests and for which the generic term *consolidation* has traditionally been used (Moscovitch, 1992).

As figures 89.1 and 89.2 indicate, perceptual input modules or perceptual representation systems operate outside of conscious awareness. Their output can be delivered to semantic systems for interpretation, and to consciousness (or a conscious awareness system), resulting in conscious apprehension of the perceptual event. The output can also be delivered to procedural or action systems, where they may influence behavior without concomitant subjective awareness of the precipitating stimulus event.

Modules and Central Systems Because perceptual input modules are a critical part of my model (figure 89.2), understanding what is meant by modules is necessary for a proper appreciation of the model and what it entails. Umiltà and I (Moscovitch and Umiltà, 1990, 1991) have a restrictive definition of modules that is based on a modified version of Fodor's (1983) conception of modularity. Modules are computational devices that have propositional content and that satisfy all of the following three criteria: domain specificity; informational encapsulation or cognitive impenetrability; and shallow output. *Domain specificity* entails that the input each module can accept for processing is restricted or circumscribed to a particular type. For example, face-recognition modules process information only about faces but not about other visual stimuli. *Informational encapsulation* or *cognitive impenetrability* implies that modules are largely resistant to top-down cognitive influences and are cognitively impenetrable to probes of their content or operation. In other words, higher-order knowledge, motivations, and expectations have little effect on how modules operate. Also, there is no conscious access to the internal workings or contents of modules—only to their output, which is shallow. An output is *shallow* if it is not semantically interpreted but is confined to domain-specific features. The output cannot convey any information either about how the output was derived or about the relation of that output to general knowledge about the world. In short, the output is both presemantic and ahistorical in that interlevel representations that led to the shallow output are not available for conscious inspection. Thus, for example, the shallow output of an object-recognition module (or structural representation system) conveys information about the object's structural features but nothing about its membership in particular semantic categories or about the processing history that led to that output. Evidence reviewed elsewhere (Moscovitch and Umiltà, 1990, 1991) from studies of normal and brain-damaged people strongly suggests that perceptual representation systems or perceptual input modules satisfy the criteria of modularity.

In contrast to modules, *central systems* are "slow, deep, global rather than local, largely under voluntary (or, as one says, "executive") control.... Above all, they are paradigmatically unencapsulated; the higher the cognitive process, the more it turns on the integration of information across superfically dissimilar domains" (Fodor, 1985, 4). In short, they satisfy none of the criteria of modularity, at least at a global level of analysis (for further discussion and provisos, see Moscovitch and Umiltà, 1990, and references cited therein).

Consciousness and perceptual repetition-priming effects

Armed with these data and theories about perceptual, item-specific implicit tests, we can return to the discussion of consciousness. The question can now be rephrased, Why should the memory that is registered by perceptual representation systems not be made conscious? Why can't perceptual records preserve their characteristics, yet still give rise to memory with conscious awareness when they are reactivated? There seems nothing contradictory about this, yet it does not occur.

Process-, Content-, and Attribute-Based Explanations of Consciousness and Memory According to Schacter (1989), explicit remembering involves gaining access not only to consciousness but to the context in which the target item was embedded. For any mental event to become conscious it must first gain access to the conscious awareness system (CAS) (figure 89.1). When a record in the perceptual representation system is activated, it gives rise to the conscious experience of a percept, not of a memory. Why? Because "access of an activated representation to CAS does not provide any contextual information about the occurrence of a recent event, and therefore does not provide a basis for explicit remembering" (p. 367). Yet Schacter's own work on source amnesia or forgetting contradicts this statement: Normal people and amnesic patients can have explicit memory for the target without memory for the context (Schacter, Harbluk, and McLachlan, 1984). Moreover, performance even on implicit tests is influenced by context, as shown in recent studies for newly formed associations between two words (Graf and Schacter, 1985; Moscovitch, Winocur, and McLachlan, 1986) or between a rich environmental context and the words studied in it (Graf, 1993) (see Goshen-Gottstein and Moscovitch, 1992; Graf, 1993). Consciousness, not context, is the critical element of explicit remembering.

A plausible, process-based explanation would be that conscious recollection is typically conceptually driven whereas repetition priming is not; once semantics is recruited, consciousness is conferred on the semantically encoded memories. This explanation fails because division along the conceptual-perceptual dimension does not always distinguish explicit from implicit tests. Some explicit tests are perceptually driven, just as some implicit tests are conceptually driven (Blaxton, 1989). Moreover, performance on conceptual implicit tests is dissociable from performance on explicit tests in normal and amnesic people (Tulving and Schacter 1990; Moscovitch, Vriezen, and Goshen-Gottstein, 1993; Roediger and McDermott, 1993). What distinguishes performance on the two types of tests is that only in the explicit case is the information that is recovered accompanied by conscious awareness of a memory.

In contrast to these process- and content-based explanations, Tulving (1985) holds that different kinds of consciousness are inherent properties or attributes of specific memory systems. *Autonoetic* (knowing with the self in it) consciousness is "correlated with episodic memory. It is necessary for the remembering of personally experienced events. . . . It is autonoetic consciousness that confers the phenomenal flavour to the remembering of past events, the flavour that distinguishes remembering from other kinds of awareness such as those characterizing perceiving, thinking, imagining or dreaming" (p. 3). According to Tulving, the system mediating performance on perceptual and conceptual item-specific implicit tests is characterized by anoetic and noetic consciousness, respectively. Tulving's proposal is not consistent with the idea that all systems or modules feed into a common consciousness system. Because Schacter believes they do, the memories must be distinguished on the basis of content—that is, of some properties other than consciousness—but we have seen that this is not always the case in principle or in fact. Tulving's proposal, though appealing on some grounds, lacks any principled rationale for assigning different types of consciousness to different systems. Tulving (1985, p. 5) answers the question with which this section began by definition. "Autonoetic consciousness is a *necessary correlate* of episodic memory . . . there is no such thing as 'remembering without awareness.' Organisms can behave and learn without (autonoetic) awareness, but they cannot *remember* without awareness." But why? What is it about these systems that makes them that way?

Why Repetition Priming Is Not Accompanied by Conscious Awareness of the Target as a Memory

The answer, I believe, lies in combining aspects of Tulving's and Schacter's explanations. Rather than being dichotomous, their explanations can become complementary. As Schacter asserts, processes that are confined to perceptual modules do not give rise to conscious experience (see also Moscovitch, 1989). Some interaction with other functional systems is necessary for phenomenal awareness to occur. Yet whether that interaction gives rise to a sense of familiarity, to a recollection of the past, or to a thought or percept without that sense of pastness (or feeling of familiarity) is not simply a function of contacting or failing to contact the conscious awareness system. Instead, I argue, it is also a function of the properties of the perceptual modules themselves. By virtue of their being modular, perceptual representation systems cannot, in principle, give rise to the feeling of familiarity.

If it is indeed the case that repetition-priming effects, both at encoding and at reactivation, are mediated by perceptual input modules as my version of the Toronto group's theory asserts, then it follows directly from the modules' adherence to modularity criteria that reactivation of perceptual records cannot lead to conscious remembering or even to an output that carries with it a sense of familiarity. If the output of perceptual modules is truly shallow, that is, presemantic and ahistorical, all that can be delivered to consciousness or the conscious awareness system is information at the level at which the module processes it and commensurate with the representations it forms. Typically, it is information about the structural, perceptual features of the processed stimulus. What is delivered to consciousness, and to semantic systems, is a percept stripped of meaning and history, though it may be contextually bound at the perceptual level. With repetition, the percept is delivered more accurately, efficiently, and quickly. But what is delivered is not a memory, not even an impoverished one without context. Even supposing that modules retain, but do not deliver, some historical information about the record's antecedents, we cannot penetrate the module to gain conscious access to that information. Consequently, on repetition priming we are aware of having a perceptual experience not a mnemonic one.

Jacoby (1983) has suggested that the experience of perceptual fluency associated with the more rapid and accurate identification of repeated, as opposed to new, stimuli can give rise to a feeling that the stimulus is familiar. As Jacoby has been careful to indicate, this sense of familiarity is an attribution, an inference, based on perceptual judgments. It should not be confused with the idea that a conscious sense of familiarity or pastness is associated inextricably with the information that the shallow output of the module conveys.

Evidence That Conscious Awareness of Target Information Is Not a Necessary Condition for Acquisition and Recovery of That Information on Perceptual Implicit Tests According to our models (see figures 89.1 and 89.2) and to the argument I have developed, the locus of perceptual repetition-priming effects is at a level *before* the information gains access to consciousness. Although under normal circumstances neurologically intact people consciously apprehend the target information at study and at test, much of the important business for repetition priming occurs before that apprehension takes place. The prediction, therefore, is that perceptual repetition-priming effects should be observed in patients whose input modules are sufficiently intact to pick up the relevant information—whether or not the patient is consciously aware of that information. Conversely, repetition-priming effects should be absent in patients whose modules are so damaged that the information cannot be picked up even without awareness.

There is good evidence that normal perceptual repetition-priming effects can be obtained in patients with a variety of disorders that spare the sensory or parasensory areas of posterior neocortex where perceptual input modules are believed to reside. Thus, amnesic patients, patients with Alzheimer's disease, and patients with Huntington's or Parkinson's disease, all display normal perceptual repetition-priming effects although their performance on explicit or other implicit tests of memory is often compromised (see Butters, Heindel, and Salmon, 1990; Keane et al., 1991; Moscovitch, Vriezen, and Goshen-Gottstein, 1993). Functional neuroimaging studies of repetition priming in normal people have also yielded encouraging results indicating that repetition priming for words implicates a region in the extrastriate cortex that corresponds to the word-form system (Squire et al., 1992). Because the region was on the right rather than the left hemisphere, these results should be treated with caution until further work confirms them.

More important to our case is evidence from patients with agnosia and dyslexia, and from normal people, showing that perception without awareness is sufficient to support memory without awareness. There are now a number of reports of agnosic patients who show domain-specific, perceptual impairments on explicit tests, that is, on tests involving conscious awareness, but whose performance on corresponding implicit tests of perception is relatively spared (see Schacter, McAndrews, and Moscovitch, 1988). Thus, prosopagnosic patients fail to identify faces on explicit tests but not on implicit ones; letter-by-letter readers cannot read words explicitly, yet they indicate on implicit tests that they can extract lexical, and some form of semantic, information from the very same words. These behaviors suggest that their modules should be sufficiently intact to support priming, a prediction confirmed by the observations of spared repetition-priming effects in these same patients. Significantly, their conscious memory for the very same items is severely reduced. This point will become critical in our discussion of consciousness and explicit tests of memory. In prosopagnosic patients whose module is so damaged that even implicit knowledge of faces is absent, repetition-priming effects are also not observed, as our model predicts (for review, see Moscovitch, Vriezen, and Goshen-Gottstein, 1993).

Also consistent with our model and argument is evidence of substantial and long-lasting perceptual repetition-priming effects in studies of normal people in which the stimulus is so degraded that the subject often is not aware of it and explicit recognition approaches chance (see Moscovitch and Bentin, 1993, and references cited therein). Reducing awareness of the target by diverting attention away from it with a secondary, concurrent task also has relatively little effect on performance on implicit, in comparison to explicit, tests of memory for the target (Eich, 1984; Jacoby, Woloshyn, and Kelley, 1989; Parkin, Reid, and Russo, 1990). Repetition-priming effects have even been obtained for material presented while subjects were anesthetized, when consciousness presumably is reduced to a minimum (Kihlstrom and Couture, 1992).

Consciousness and explicit tests of memory

We can now turn to the related question, Why should some material that is retained be accompanied when it is recovered by an immediate, conscious sense that it has been experienced before? In other words, what accounts for the fundamental distinction between memory with and without awareness?

Tulving's position that conscious awareness is a property of the system that supports remembering is appealing, but he provides no explanation of how consciousness comes to be a property of that system. By referring to my and Umiltà's (Moscovitch and Umiltà, 1990, 1991) framework of how modules and central systems interact in memory (Moscovitch, 1992), I will try to sketch a theory of how consciousness comes to be a property of remembering.

Recovered consciousness: A theory of conscious recollection and hippocampal function

The central idea is that the hippocampal complex acts as an associative, episodic memory module that mandatorily picks up information that is consciously apprehended, and only that information. I will develop this idea to argue that as a result consciousness becomes a primitive attribute or inherent property of memories that are recollected. Because I am more comfortable using neuropsychological terminology, I will refer to this module as the hippocampal component and hope that the reader will overlook any category errors that my use of this terminology entails. Those errors should disappear if the term *associative memory module* is substituted for *hippocampal component*.

The hippocampal component consists of a variety of structures in the medial temporal lobes and diencephalon that form a circuit. In addition to the hippocampus, the structures involved are the parahippocampal gyrus, the subiculum, the entorhinal and perirhinal cortices, the mammilary bodies and dorsomedial nuclei of the thalamus, the cingulate cortex, the fornix, and fiber tracts connecting the structures, and, perhaps, other structures such as the septum. Each of these structures makes a special, perhaps unique, contribution to the operation of the entire complex. Because we know very little about that, I refer to the entire complex by the name of its most conspicuous structure, the hippocampus.

Perceptual input modules and interpretative central systems deliver their output to consciousness or CAS (see Moscovitch and Umiltà, 1990, 1991, for a discussion on the interrelationship of working memory and consciousness). Once in consciousness, the information is mandatorily picked up by the hippocampal component in much the way that perceptual modules obligatorily pick up input in their domain (figure 89.2).

Let us assume that there are neural correlates of being consciously aware of something. That is, when we are aware of something, the pattern of neural firing (and maybe the type and number of neurons involved) is different from that which occurs when we are not aware. The hippocampal component, I propose, is a module whose specific domain is information that is consciously apprehended. Conscious apprehension is not all-or-none. To the extent that an event is not apprehended, it is not picked up by the hippocampal component. The evidence that unattended or poorly attended information is poorly recollected on explicit tests of memory, which presumably involve the hippocampus, is overwhelming and need not be reviewed here (see Moscovitch, Vriezen, and Goshen-Gottstein, 1993; Roediger and McDermott, 1993). What is equally critical for the model I have proposed, is that performance on implicit tests is far less affected by attentional manipulations (see above).

Using reciprocal pathways that connect parts of the hippocampal complex to the cortex, the hippocampal component binds or integrates into a *memory trace* the neural elements that mediate the conscious experience. Those include the records of the modules and central systems whose output contributed to the conscious experience as well as whatever elements made the experience conscious. Simultaneously, a file entry or index is created within the hippocampal complex via which the trace can be reactivated (Teyler and DiScenna, 1986). Alternatively, the trace itself may be accessed directly with the hippocampus playing the necessary but temporary role of keeping elements of the trace bound together. Because the events encoded by the hippocampal component were consciously experienced, the information that made the experience conscious is also part of the memory trace. In this way, consciousness is bound to other aspects of the event and becomes an intrinsic property of the memory trace.

COHESION AND CONSOLIDATION I refer to the hippocampal binding of elements into a memory trace as *cohesion*. Once bound, a slower process ensues that makes the memory trace permanent. This is *consolidation*. When consolidation is complete, the hippocampus is no longer necessary for recovery of memories. (see Moscovitch, 1994, for further discussion).

RECOLLECTION In order for an event to be recollected consciously, a memory trace must be reactivated. This occurs when an external or internally generated cue interacts with the memory trace, a process called *ecphory* by Semon (1921, cited in Tulving, 1983).

Previously (Moscovitch, 1989, 1992; Moscovitch and Umiltà, 1990, 1991), I wrote that the product of that ecphoric process is delivered to consciousness (Moscovitch, 1989, 1992; Moscovitch and Umilta, 1990). If consciousness is an intrinsic property of the memory trace, there is also a sense in which the product of the ecphoric process becomes conscious, as if ecphory enabled that which was dormant to become active. Consciousness is recovered along with the other elements of the memory trace.

It is this recovery of a trace imbued with consciousness that accounts for the felt experience of remembering. Put another way, it is the aura of consciousness surrounding a recovered trace that makes it feel familiar and immediately recognizable as something that has previously been experienced; in short, that makes it a memory.[2]

Recovered consciousness is the signal or marker that distinguishes memories from thoughts and percepts involving *on-line consciousness*. On-line conscious awareness may accompany the acquisition of information by perceptual input modules or perceptual representation systems as their shallow output becomes accessible. The information represented in the module itself remains preconscious (see figures 89.1 and 89.2). Consciousness, therefore, is not part of the perceptual record whose reactivation leads to repetition-priming effects.

Peter Milner (1989) suggested something similar when he proposed that an event that activates the hippocampus acquires a familiarity attribute or feature that is incorporated into the cell assembly for that event. When that cell assembly is reactivated, the familiarity feature becomes part of our experience of the recovered event in much the same way that we experience color or motion as qualifying features of an event if their neural substrates were active as the event was perceived.

The view that I have advanced resembles Tulving's (1985) in that consciousness is held to be an inherent property of remembering. It differs from his in two ways. One is that it provides an explanation of how consciousness came to be part of remembering, whereas Tulving assumed it was part of remembering as a first principle. Second, it does not hold that the consciousness associated with remembering is fundamentally different in kind from the consciousness associated with perception or thought. Rather, on-line consciousness associated with perception and thought (experience) is incorporated into a memory trace and becomes recovered consciousness at retrieval.

Benefits and Costs of Modular Memory Components Elsewhere I have argued and presented evidence in support of the idea that the hippocampus is a memory module that satisfies the criteria of modularity (Moscovitch and Umilta, 1991; Moscovitch, 1992). Here I wish only to pursue some of the implications that this idea has for the organization of memory and consciousness.

Like other modular processes, ecphoric (automatic retrieval) processes (Tulving, 1983), once initiated, are rapid, obligatory, informationally encapsulated, and cognitively impenetrable. The same is true of the initial formation and encoding of memory traces. We are aware only of the input to the hippocampal component and the memory trace that shallow output from it makes available. Thus, we remember countless daily events without intending to remember them: Memories may "pop" into mind much as preattentive perceptual stimuli "pop out" of their background.

This analogy of memory with perception is appropriate insofar as aspects of both are modular. Just as it would be maladaptive to have a perceptual system that is too much under our control and subject to our motivations and expectancies, so it would not be useful to have a memory system that relies on our intentions to remember. Because most often we do not know in advance what will be worth committing to memory, it is important to have a system that is capable of encoding and storing information automatically, as a natural consequence of apprehending the material consciously. Moreover, since events unfold at their own pace, most of them would not be encoded by the time we could determine that they were worth remembering.

The central idea of the levels-of-processing framework (Craik and Lockhart, 1972), that remembering is a natural by-product of cognition, follows directly from this view that the hippocampal component is modular. What determines what is remembered is not the intention to remember as such, but the extent to which events are attended and information from them is processed to a deep level and properly organized. Paying close attention to the target and encoding it semantically makes it distinctive and its memory traces more easily retrievable (Moscovitch and Craik, 1976). Once events are fully apprehended in consciousness, they are obligatorily picked up and encoded by the hippocampus. Conversely, without a hippocampal component, no lasting memory traces can be formed and consciously recollected no matter how deeply information is processed (Cermak and Reale, 1978).

An additional benefit of an automatic hippocampal component is that it does not draw cognitive resources away from other activities. If committing something to memory always required additional effort beyond that involved in apprehending the relevant information, it would lead to a peculiar trade-off: resources allocated to attention and comprehension would be unavailable for memory, and vice versa. We would remember well only those items that were processed poorly, an unacceptable and counterfactual condition.

At retrieval, conscious recollection is constantly occurring. It provides us with a sense of familiarity and continuity in daily life that we take for granted. Amnesic patients comment that they always feel as if they have just awakened in a strange place or from a dream. Conscious recollection becomes effortful and demands our attention when we wish to reflect on the memories that have emerged or when we wish to initiate strategies to recover memories (see below).

The cost of being modular is that the hippocampal component lacks the "intelligence" for self-organization, strategic intervention, and monitoring. Events are encoded only by simple contiguity and by associations that memory traces form with each other and with cues. Hippocampally activated memories lack temporal organization, or what I have termed historical context (Moscovitch, 1994). The hippocampal component responds reflexively to cues; if cues are initially ineffective, it cannot conduct a memory search or monitor the ecphoric output to determine whether the recovered memories are veridical or even plausible. The job of organizing the input, devising retrieval strategies, verifying the output, and placing it in proper historical context (temporal order), is left to pre- and postecphoric extrahippocampal processes, probably involving the frontal lobes. It is because its organization and retrieval method is associative, and cue dependent, that we refer to the hippocampal component as an associative episodic memory component, terms that also describe the explicit memory tests that involve it.

Memories of which we are not conscious are memories revealed only in action and elicited only by external stimulation. Consciousness may be necessary to allow us to have memories without acting on them and to retrieve them without the need of an external stimulus cue. In this way, we can think about and manipulate memories until we are ready to act. Just as perceptual modules function "to represent the external world and make it accessible to thought" (Fodor, 1983, 40), so the associative memory module is necessary to represent the internally experienced world and make it available to thought. Consciousness serves as the vehicle for that process. We now turn to the role of extrahippocampal, strategic processes in remembering.

Working with memory: The role of the frontal lobes in strategic explicit tests of memory

The prefrontal cortex is a large, heterogeneous structure consisting of a number of distinct areas, each with its own projections to and from other brain regions and each having presumably different functions (Pandya and Barnes, 1987). Despite the evidence for localization of function among regions of prefrontal cortex, Umiltà and I (Moscovitch and Umiltà, 1990, 1991) argued that they are central system structures that contribute to performance on strategic, explicit tests of memory.

The frontal lobes are prototypical, organizational structures that are critical for selecting and implementing encoding strategies that organize the input to the hippocampal component and the output from it; for evaluating that shallow output and determining its correct temporal sequence and spatial context with respect to other events; and for using the resulting information either to guide further mnemonic searches, to direct thought, or to plan future action. In short, the frontal lobes are necessary for converting remembering from a stupid, reflexive act triggered by a cue to an intelligent, reflective, goal-directed activity that is under conscious, voluntary control. When one is trying to place a person who looks familiar or to determine one's whereabouts last July, the appropriate memory does not emerge automatically but must be ferreted out, often laboriously, by retrieval strategies of which we are conscious and that probably rely on the frontal lobes for their execution.

Memory disorders following frontal lesions are not related to deficits in storage and retention, which are hippocampal functions. Instead, they are associated with impaired organizational and strategic processes that, at the extreme, result in confabulations in which accurately remembered elements of one event are combined with those of another without regard to their internal consistency or plausibility (Moscovitch, 1989). As befits a central system structure for which the criterion of domain specificity does not apply, the frontal lobe's function with respect to memory is similar to its function in other domains. The frontal lobes organize the raw material that is made available by other modules and central systems. The frontal lobes' representations are available to conscious inspection, and the output is deep. If the hippocampal complex can be considered to consist of "raw memory" structures, then the frontal lobes are "working-with-memory" structures that operate on the input to the hippocampal component and the output from it (Moscovitch and Winocur, 1992). It is here that our knowledge, expectations, fears, and desires influence what is remembered.

Consciousness and memory: content, process, or intrinsic property (attribute)?

I have proposed a theoretical account of performance on implicit and explicit tests of memory. By referring to a theory of memory based on modules and central systems, I have tried to indicate how memory with and without conscious awareness emerges from the basic operations of modules and central systems. Consciousness is an inherent property of the memory trace, being bound to it along with other aspects of the experienced event by the hippocampus and structures related to it. It is these structures that support performance on associative, explicit episodic memory tests. At retrieval, what is recovered is a consciousness-imbued memory trace that accounts for our conscious recollection. "Consciousness" is recovered along with other aspects of the experienced event. This follows from the encoding specificity principle (Tulving, 1983): Put bluntly, *consciousness in, consciousness out*. The frontal lobes act as working-with-memory structures that organize the input to the hippocampal component and the output from it. Performance on perceptual implicit tests, on the other hand, is mediated by perceptual input modules whose input is domain-specific stimulus information and whose output is a structural representation of the stimulus. On such tests, the subject experiences the target as a percept, at best a fluent percept, but not as a remembered stimulus.

In the course of this chapter I have favored the view that consciousness as an intrinsic property or attribute of memory, and not informational content, is the distinguishing feature of explicit remembering. The content view states that it is the rich context (Schacter, 1989; Mayes, 1992) or the configural (Sutherland and Rudy, 1989) or relational (Eichenbaum, 1992) properties of the memory trace that allows for the conscious experience of recollection. In the one case, consciousness is itself a property, in the other, it is a product or outcome of other factors.

Having stated that I favor the "attribute" position, I now want to indicate clearly that this position does not preclude the possibility that the content of conscious memories is different from that of nonconscious ones. Indeed, it may be the case that our brains have evolved in such a way that consciousness is associated with collections of complex, cross-modal representations (Baars, 1988). The point I wish to make is that issues regarding consciousness cannot be reduced to ones concerned only with the content of the conscious experience.

Other investigators substitute process-laden words such as *controlled* versus *automatic* to distinguish between conscious and nonconscious memory. This idea has been very useful when implemented as an experimental procedure (see especially Jacoby, 1991). Our performance on memory tests is determined by the contribution of conscious and nonconscious memory components. Since no memory test is process pure, it is important to establish techniques to distinguish between conscious and nonconscious memories. It is also important, however, not to confuse consciousness itself with processes associated with consciousness, or with those that consciousness makes possible. Consciousness is the phenomenal awareness that accompanies some mental activities. With respect to remembering, and perhaps with respect to no other function, consciousness is also an inherent property of the very object of our apprehension.

ACKNOWLEDGMENTS Work on this chapter was supported by the Natural Sciences and Engineering Research Council of Canada, grant A8347 and an Ontario Mental Health Foundation Research Associateship. I thank Marlene Behrmann, Peter Milner, Dan Schacter, and Tim Shallice for their insightful comments.

NOTES

1. Like good rock bands, it is hard to keep the members of psychology departments together. Mandler is at University of California, San Diego. Schacter is now at Harvard, by way of Arizona, and Graf is at the University of British Columbia. Kolers is deceased.
2. To reiterate, the consciousness to which I refer is not self-consciousness but rather a consciousness or awareness that is quite primitive and continuously present and is an integral aspect of our experiences. It is the type of consciousness that distinguishes the visual experience of someone with intact vision from that of someone with "blindsight," who does not experience sight but who can respond, without awareness, to visual stimuli. It is this "consciousness" that is bound in the memory trace and is recovered on remembering.

REFERENCES

Baars, B. J., 1988. *A Cognitive Theory of Consciousness*. New York: Cambridge University Press.

Bentin, S., and M. Moscovitch, 1988. The time course of repetition effects for words and unfamiliar faces. *J. Exp. Psychol. Gen.* 117:148–160.

Blaxton, T. A., 1989. Investigating dissociations among memory measures: Support for a transfer appropriate processing framework. *J. Exp. Psychol. [Learn. Mem. Cogn.]* 15:657–668.

Butters, N., W. C. Heindel, and D. P. Salmon, 1990. Dissociation of implicit memory in dementia: Neurological impliciations. *Bull. Psychonom. Soc.* 28:359–366.

Cermak, L. S., 1984. The episodic-semantic distinction in amnesia. In *The Neuropsychology of Memory*. L. R. Squire and N. Butters, eds. New York: Guilford.

Cermak, L. S., and L. Reale, 1978. Depth of processing and retention of words by alcoholic Korsakoff patients. *J. Exp. Psychol. [Hum. Learn. Mem.]* 4:165–174.

Challis, B. H., and R. Sidhu, 1993. Dissociative effect of massed repetition on implict and explicit measures of memory. *J. Exp. Psychol. [Learn. Mem. Cogn.]* 19:115–127.

Cohen, N., and L. R. Squire, 1980. Preserved learning and retention of pattern analyzing skill in amnesia: Dissociation of knowing how and knowing that. *Science* 210:207–210.

Corkin, S., 1968. Acquisition of motor skill after bilateral medial temporal-lobe excision. *Neuropsychologia* 6:255–265.

Craik, F. I. M., and R. S. Lockhart, 1972. Levels of processing: A framework for memory research. *J. Verb. Learn. Verb. Beh.* 11:671–684.

Diamond, R., and P. Rozin, 1984. Activation of existing memories in anterograde amnesia. *J. Abnorm. Psychol.* 93: 98–105.

Eich, E., 1984. Memory for unattended events: Remembering with and without awareness. *Mem. Cognition.* 12: 105–111.

Eichenbaum, H., 1992. The hippocampal system and declarative memory in animals. *J. Cognitive Neurosci.* 4:217–231.

Ellis, A. W., A. W. Young, B. M. Flude, and D. C. Hay, 1987. Repetition priming of face recognition. *Q. J. Exp. Psychol. [A]* 39:193–210.

Fodor, J., 1983. *The Modularity of Mind.* Cambridge, Mass.: MIT Press, Bradford Books.

Fodor, J., 1985. Multiple book review of *The modularity of mind. Behav. Brain Sci.* 8:1–42.

Gardiner, J. M., and R. I. Java, 1994. Recognizing and remembering. In *Theories of Memory*, A. Collins, M. A. Conway, S. E. Gathercole, and P. E. Morris, eds. Hillsdale, N.J.: Erlbaum.

Goshen-Gottstein, Y., and M. Moscovitch, 1992. Repetition priming effects for pre-existing and novel associations between words. Paper presented at the Psychonomic Society, St. Louis, Mo.

Graf, P., 1993. Ten years of research on implicit memory. In *Attention and Performance XV: Conscious and Nonconscious Information Processing*, C. Umiltà and M. Moscovitch, eds. Cambridge, Mass.: MIT Press.

Graf, P., G. Mandler, and P. E. Haden, 1982. Simulating amnesic symptoms in normal subjects. *Science* 218:1243–1244.

Graf, P., and D. L. Schacter, 1985. Implicit and explicit memory for new associations in normal and amnesic subjects. *J. Exp. Psychol. [Learn. Mem. Cogn.]* 11:501–518.

Jackson, A., and J. Morton, 1984. Facilitation of auditory word recognition. *Mem. Cogn.* 12:568–574.

Jacoby, L. L., 1983. Perceptual enhancement: Persistent effects of an experience. *J. Exp. Psychol. [Learn. Mem. Cogn.]* 9:21–38.

Jacoby, L. L., 1991. A process dissociation framework: Separating automatic from intentional uses of memory. *J. Mem. Lang.* 30:513–541.

Jacoby, L. L., V. Woloshyn, and C. M. Kelley, 1989. Becoming famous without being recognized: Unconscious influences of memory produced by dividing attention. *J. Exp. Psychol. [Gen.]* 118:115–125.

Keane, M. M., J. D. E. Gabrieli, A. C. Fennema, J. H. Growdon, and S. Corkin, 1991. Evidence for a dissociation between perceptual and conceptual priming in Alzheimer's disease. *Behav. Neurosci.* 105:326–342.

Kihlstrom, J. F., and L. J. Couture, 1992. Awareness and information processing in general anaesthesia. *J. Psychopharm.* 6:410–417.

Kirsner, K., and D. Dunn, 1985. The perceptual record: A common factor in repetition priming and attribute retention. In *Attention and Performance XI*, M. I. Posner and O. S. M. Marin, eds. Hillsdale, N.J.: Erlbaum.

Luria, A. R., 1966. *Higher Cortical Functions in Man.* New York: Basic Books.

Mandler, G., 1989. Memory: Conscious and unconscious. In *Memory: Interdisciplinary Approaches.* P. R. Solomon, G. R. Goethals, C. M. Kelley, and B. R. Stephens, eds. New York: Springer.

Mayes, A. R., 1992. Automatic memory processes in amnesia: how are they mediated? In *The Neuropsychology of Consciousness*, A. D. Milner and M. D. Rugg, eds. New York: Academic Press.

Milner, B., 1966. Amnesia following operation on the temporal lobes. In *Amnesia*, C. W. M. Whitty and O. L. Zangwill, eds. London: Butterworth, pp. 109–133.

Milner, P. M., 1989. A cell assembly theory of hippocampal amnesia. *Neuropsychologia* 27:31–40.

Mishkin, M., and T. Appenzeller, 1987. Anatomy of memory. *Sci. Am.* 256:80–89.

Moscovitch, M., 1982. Multiple dissociations of function in amnesia. In *Human Memory and Amnesia*, L. S. Cermak, ed., Hillsdale, N.J.: Erlbaum, pp. 337–370.

Moscovitch, M., 1989. Confabulation and the frontal system: Strategic vs. associative retrieval in neuropsychological theories of memory. In *Varieties of Memory and Consciousness: Essays in Honor of Endel Tulving*, H. L. Roediger, III and F. I. M. Craik, eds. Hillsdale, N.J.: Erlbaum.

Moscovitch, M., 1992. Memory and working-with-memory: A component process model based on modules and central systems. *J. Cognitive Neurosci.* 4:257–267.

Moscovitch, M., 1994. Memory and working-with-memory: Evaluation of a component process model and comparison with other models. In *Memory Systems*, 1994 D. L. Schacter and E. Tulving, eds. Cambridge, Mass.: MIT Press.

Moscovitch, M., and S. Bentin, 1993. The fate of repetition effects when recognition approaches chance. *J. Exp. Psychol. [Learn. Mem. Cogn.]* 19:148–158.

Moscovitch, M., and F. I. M. Craik, 1976. Depth of processing, retrieval cues and uniqueness of encoding as factors in recall. *J. Verb. Learn. Verb. Behav.* 15:447–458.

Moscovitch, M., and C. Umiltà, 1990. Modularity and neuropsychology: Implications for the organization of attention and memory in normal and brain-damaged people. In *Modular Processes in Dementia*, M. E. Schwartz, ed. Cambridge, Mass.: MIT Press.

Moscovitch, M., and C. Umiltà, 1991. Conscious and nonconscious aspects of memory: A neuropsychological framework of modules and central systems. In *Perspectives on Cognitive Neuroscience*, R. G. Lister and H. J. Weingartner, eds. Oxford: Oxford University Press.

Moscovitch, M., E. Vriezen, and Y. Goshen-Gottstein, 1993. Implicit tests of memory in patients with focal lesions and degenerative brain disorders. In *Handbook of Neuropsychology*, H. Spinnler and F. Boller, eds. Amsterdam: Elsevier.

Moscovitch, M., and G. Winocur, 1992. The neuropsychology of memory and aging. In *The Handbook of Aging and Cognition*, T. A. Salthouse and F. I. M. Craik, eds. Hillsdale, N.J.: Erlbaum.

Moscovitch, M., G. Winocur, and D. McLachlan, 1986. Memory as assessed by recognition and reading time in normal and memory impaired people with Alzheimer's disease and other neurological disorders. *J. Exp. Psychol. [Gen.]* 115:331–347.

Musen, G., and L. R. Squire, 1991. Normal acquisition

of novel information in amnesia. *J. Exp. Psychol. [Learn. Mem. Cogn.]* 17:1095–1104.

MUSEN, G., and L. R. SQUIRE, 1992. Nonverbal priming in amnesia. *Mem. Cogn.* 20:441–448.

MUSEN, G., and L. R. SQUIRE, 1993. On the implicit learning of novel associations by amnesic patients and normal subjects. *Neuropsychology*. 7:119–135.

NADEL, L., 1992. Multiple memory systems: What and why. *J. Cognitive Neurosci.* 4:179–188.

O'KEEFE, J., 1985. Is consciousness the gateway to the hippocampal cognitive map: A speculative essay on the neural basis of mind. In *Brain and Mind*, D. Oakley, ed. London: Methuen.

O'KEEFE, J., and L. NADEL, 1978. *The Hippocampus as a Cognitive Map*. Oxford: Clarendon.

OLTON, D. S., J. T. BECKER, and G. E. HANDELMANN, 1979. Hippocampus, space and memory. *Behav. Brain Sci.* 2:313–365.

PANDYA, D., and C. L. BARNES, 1987. Architecture and connections of the frontal lobe. In *The Frontal Lobes Revisited*, E. Perecman, ed. New York: IRBN.

PARKIN, A. J., T. REID, and R. RUSSO, 1990. On the differential nature of implicit and explicit memory. *Mem. Cognition* 18:307–314.

RICHARDSON-KLAVEHN, A., and R. A. BJORK, 1988. Measures of memory. *Annu. Rev. Psychol.* 39:475–543.

ROEDIGER, H. L., 1990. Implicit memory: Retention without remembering. *Am. Psychol.* 45:1043–1056.

ROEDIGER, H. L., III, and K. B. MCDERMOTT, 1993. Implicit memory in normal human subjects. In *Handbook of Neuropsychology*, F. Boller and J. Grafman, eds. Amsterdam: Elsevier, pp. 63–131.

ROZIN, P., 1976a. The psychobiological approach to human memory. In *Neural Mechanisms of Learning and Memory*, M. R. Rosenzweig, and E. L. Bennett, eds. Cambridge, Mass.: MIT Press, pp. 3–46.

ROZIN, P., 1976b. The evaluation of intelligence and access to the cognitive unconscious. In *Progress in Psychobiology and Physiological Psychology*, J. Sprague and A. Epstein, eds. New York: Academic Press.

RUDY, J. W., and R. J. SUTHERLAND, 1992. Configural and elemental associations and the memory coherence problem. *J. Cognitive Neurosci.* 4:208–216.

SCHACTER, D. L., 1987. Implicit memory: History and current status. *J. Exp. Psychol. [Learn. Mem. Cogn.]* 13:501–518.

SCHACTER, D. L., 1989. On the relation between memory and consciousness: Dissociable interactions and conscious experience. In *Varieties of Memory and Consciousness: Essays in Honour of Endel Tulving*, H. L. Roediger, III and F. I. M. Craik, eds. Hillsdale, N.J.: Erlbaum.

SCHACTER, D. L., 1992. Priming and multiple memory systems: Perceptual mechanisms of implicit memory. *J. Cogn. Neurosci.* 4:244–256.

SCHACTER, D. L., L. A. COOPER, and S. M. DCLANEY, 1990. Implicit memory for unfamiliar objects depends on access to structural descriptions. *Journal of Experimental Psychology: General* 119:5–24.

SCHACTER, D. L., J. L. HARBLUK, and D. R. MCLACHLAN, 1984. Retrieval without recollection: An experimental analysis of source amnesia. *J. Verb. Learn. Verb. Behav.* 23:593–611.

SCHACTER, D. L., M. P. MCANDREWS, and M. MOSCOVITCH, 1988. Access to consciousness: Dissociations between implicit and explicit knowledge in neuropsychological syndromes. In *Thought without Language*, L. Weiskrantz, ed. Oxford: Oxford University Press, pp. 242–278.

SCHACTER, D. L., and E. TULVING, eds., 1992. Special issue on memory systems. *J. Cognitive Neurosci.* 4(3).

SLOMAN, S. A., C. A. G. HAYMAN, N. OHTA, J. LAW, and E. TULVING, 1988. Forgetting in primed fragment completion. *J. Exp. Psychol. [Learn. Mem. Cogn.]* 14:223–239.

SQUIRE, L. R., 1992. Declarative and nondeclarative memory: Multiple brain systems supporting learning and memory. *J. Cognitive Neurosci.* 4:232–243.

SQUIRE, L. R., J. OJEMANN, F. MIEZIN, S. PETERSEN, T. VIDEEN, and M. RAICHLE, 1992. Activation of the hippocampus in normal humans: A functional anatomical study of memory. *Proc. Natl. Acad. Sci. U.S.A.* 89:1837–1841.

SUTHERLAND, R. J., and J. W. RUDY, 1989. Configural association theory: The role of the hippocampal formation in learning, memory, and amnesia. *Psychobiology* 17:129–144.

TEYLER, T. J., and P. DISCENNA, 1986. The hippocampal memory indexing theory. *Behav. Neurosci.* 100:147–154.

TULVING, E., 1983. *Elements of Episodic Memory*. Oxford: Clarendon.

TULVING, E., 1985. Memory and consciousness. *Can. Psychol.* 25:1–12.

TULVING, E., and D. L. SCHACTER, 1990. Priming and human memory systems. *Science* 247:301–306.

WARRINGTON, E. K., and L. WEISKRANTZ, 1973. An analysis of short-term and long-term memory deficits in man. In *The Physiological Basis of Memory*, J. A. Deutsch, ed. New York: Academic Press.

WEISKRANTZ, L., 1988. Some contributions of neuropsychology of vision and memory to the problems of consciousness. In *Consciousness in Contemporary Science*, A. J. Marcel and E. Bisiach, eds. Oxford: Clarendon.

WINOCUR, G., and M. KINSBOURNE, 1978. Contextual cuing as an aid to Korsakoff amnesics. *Neuropsychologia* 16:671–682.

90 Escape from Linear Time: Prefrontal Cortex and Conscious Experience

ROBERT T. KNIGHT AND MARCIA GRABOWECKY

ABSTRACT Patients with prefrontal damage suffer from deficits in inhibitory control and novelty detection, which lead to distractibility, a noisy internal milieu, impairments in attention, and poor temporal coding. These deficits cumulate to produce further problems, including stimulus-bound behavior, reduced decision confidence, perseveration, poor planning and memory organization, difficulty in generating novel ideas, and deficits in generating and evaluating counterfactual scenarios. Modulation of internal and external events makes it possible to remove oneself from the present and construct alternative interpretations of past, present, and future events. In neurologically intact human subjects this "off-line" ability to consciously evaluate and adjust behavior is proposed to rely on prefrontal cortex.

Insights into the brain mechanisms of conscious experience are provided by the study of neurological patients. Patients with prefrontal damage exhibit deficits in behaviors that are crucial for normal conscious experience. Problems with inhibitory control of external sensory inputs and internal cognitive processes, coupled with abnormalities in the detection of novel events, lead to a cascade of behavioral deficits. This chapter reviews neuropsychological and neurophysiological evidence linking prefrontal cortex to these deficits and to conscious experience.

The prefrontal cortex can be divided into three major regions. Damage to each region produces a different set of behavioral manifestations. This discussion of the prefrontal cortex and consciousness will be limited to the role of dorsolateral prefrontal cortex and will not address the contributions of orbitofrontal or mesial regions. Damage to dorsolateral prefrontal cortex (Brodmann's areas 6, 8, 9, 10, 44, 45, and 46) results in a range of cognitive disturbances. In early disease from tumors or degenerative disorders that spare language cortex, subtle deficits in creativity and mental flexibility can be observed. When unilateral disease progresses or becomes bilateral, pronounced behavioral problems emerge. Deficits in planning, temporal coding, metamemory, judgment, and insight predominate. Attention is invariably impaired in advanced disease. Acute infarcts involving the precentral branch of the middle cerebral artery are associated with a transient syndrome of global confusion and attention abnormalities. Right-sided infarcts involving rostral area 46 and the frontal eye field can result in contralateral hemispatial neglect. In cases of advanced bilateral dorsolateral prefrontal damage, perseveration and frontal release signs such as snout, grasp, and palmomental reflexes are usually present.

An important consequence of prefrontal damage is a decreased confidence in many aspects of behavior. Prefrontal patients are uncertain about their ability to correctly perform a variety of behaviors despite objective evidence of correct performance. The prefrontal patient's inability to selectively inhibit unwanted input results in a noisy internal milieu, which, coupled with an inability to detect deviance, hampers the ability to code the beginning and ending of discrete events and tag them with appropriate spatiotemporal information. Lacking the ability to maintain a temporal stream and reach decisions with confidence, the prefrontal patient is left without a coherent past or future and is locked into an uncertain present. In advanced prefrontal disease this deficit is clinically observable as perseveration and stimulus-bound behavior.

The capacity to monitor and adjust one's behavior confidently and be aware of one's consciousness is a key

ROBERT T. KNIGHT and MARCIA GRABOWECKY Department of Neurology, Center for Neuroscience, University of California, Davis, Calif.

facet of human cognition. Evidence from neurological patients suggests that this capacity is dependent on the dorsolateral prefrontal cortex. Cognitive neuropsychological and neurophysiological findings that support this proposed role of dorsolateral prefrontal cortex in consciousness will be reviewed in turn.

Cognitive neuropsychological studies

The behavioral changes that arise from damage to the prefrontal cortex are notoriously difficult to capture with standardized neuropsychological tests. Patients with large prefrontal lesions can perform within the normal range on tests of memory, intelligence, and other cognitive functions. Even the Wisconsin Card Sorting Test (WCST) and other tests that are believed to be especially sensitive to frontal damage sometimes fail to discriminate patients with frontal lesions from normals or those with lesions in other regions (Eslinger and Damasio, 1985; Grafman, Jonas, and Salazar, 1990). At the same time, patients with prefrontal cortex damage may be quite impaired in their daily lives. How can this paradox of good performance on standardized tests be reconciled with impaired functioning in daily life?

Theoretical Framework It is proposed that two classes of cognitive operations, which will be referred to as *simulation* and *reality checking* are selectively impaired after dorsolateral prefrontal damage. Simulation refers to the process of generating internal models of external reality. These models may represent an accurate past or an alternative past, present, or future, and include models of the environment, of other people, and of the self. Reality checking refers to processes that monitor information sources to represent accurately their spatiotemporal context. These monitoring processes are critical for discriminating between simulations of alternate possibilities and veridical models of the world. Simulation and reality checking can be considered supervisory (Shallice, 1988), or executive functions (Milner and Petrides, 1984; Stuss and Benson, 1986; Baddeley and Wilson, 1988). These functions are essential for behavior to be integrated, coherent, and contextually appropriate, features that may be necessary for conscious experience (Allport, 1988). Simulation and reality checking are prerequisites for permitting actions to be dissociated from current environmental constraints while maintaining awareness of those constraints. A patient who cannot simulate alternatives to a situation becomes "stimulus bound" (Luria, 1966) and is incapable of responding flexibly. Without reality checking, a patient cannot discriminate between internally generated possibilities and the model of the external world as it currently exists. Simulation and reality checking work together, allowing humans to simulate manipulations of the external environment, evaluate the consequences of those manipulations, and act on the results of the simulations with a richer knowledge of potential outcomes.

The frontal lobes have been proposed to be involved in decision making, goal-directed behavior, planning, and monitoring behavior (see Stuss, Eskes, and Foster, 1994, for review). Stuss and colleagues suggest that behavior is controlled by a hierarchical process with three levels of activity. The first includes sensory-perceptual mechanisms that are located in posterior regions of the brain and tend to be modality specific. The second level, executive functions, receives input from the sensory-perceptual level and is responsible for the direction of lower-level subsystems toward achievement of a goal. Control by the executive level is effortful in novel situations, but given sufficient practice executive functions may become automatized. The highest level in this framework is that of consciousness and self-awareness (Stuss, 1991). Inputs to the level of self-awareness may include representations of the options generated by simulation processes operating at the executive level.

Simulation and Counterfactuals Simulation processes have been studied extensively in normal populations (Tversky and Kahneman, 1973; Kahneman and Tversky, 1982; Kahneman and Miller, 1986; Kahneman and Varey, 1990). Judgments and decisions in any situation are arrived at as a consequence of the evaluation of a set of internally generated alternatives. These are called counterfactual scenarios because they are an alternative reality to the one experienced. Counterfactual alternatives are omnipresent in normal human cognition, and form a large and important part of our conscious awareness. The generation of counterfactuals occurs in both emotional and more cognitive behavior. Counterfactual expressions occur often in everyday life, especially in situations involving regret or grief. For example, a distraught parent may say, "If only I had not given my son the keys to the car, the accident would not have occurred." Such expressions reflect internally generated alternatives that may assume a high degree of realism to the person generating them. Of course, counterfactual alternatives occur in less emotionally weighted situations as well. When one thinks, "If I had ordered the pasta with white sauce instead of marinara, this stain would be less obvious," a counterfactual scenario has been generated.

According to Kahneman and Miller (1986), all events are compared to counterfactual alternatives. Events that recruit few alternatives, perhaps because they match the norm of similar events, are judged to have been predictable and unavoidable. Thus, even unanticipated events are seen as normal if they do not cause the generation of many counterfactual alternatives. It is proposed that patients with prefrontal damage are impaired in their ability to generate and evaluate counterfactuals, and that this problem may stem from a dysfunction in the automatic detection of novelty. Because prefrontal patients do not respond appropriately to novel events they generate fewer counterfactual alternative realities. Counterfactuals are constructed to compare what happened with what could have happened. Without such simulations it is difficult to avoid making the same mistakes over and over again.

Impairments in simulation processes Stimulus-bound behavior is often noted in patients with dorsolateral prefrontal lesions and may be the result of a deficit in simulation (Luria, 1966; Lhermitte, 1986). For example, objects placed in front of a prefrontal patient are picked up and used (utilization behavior) without the patient being asked to do so. In addition, behavior of the experimenter may be imitated, even when this behavior is bizarre and socially inappropriate (Lhermitte, Pillon, and Serdaru, 1986). In short, patients with frontal lesions appear excessively bound by environmental cues. Patients with prefrontal cortex lesions have also been described as lacking insight and foresight; as incapable of planning either for the near or distant future; and as deficient in creativity (Hebb and Penfield, 1940; Ackerly and Benton, 1947; Damasio, 1985). This set of abnormal behaviors may be a consequence of a deficit in the ability to simulate alternatives.

Deficits in estimation and frequency judgments Patients with frontal lesions are impaired at estimation problems. Smith and Milner (1984) showed that patients with focal frontal lesions performed poorly at a task involving estimating the price of common objects. Shallice and Evans (1978) had frontal patients make estimates of the size of well-known objects for which no ready size information was available, such as the length of the average man's spine. Frontal patients were seriously impaired at this task, at times giving wildly inaccurate and even impossible answers. To perform estimation tasks, subjects must recruit appropriate exemplars, generate category norms, and then select an appropriate size or price judgment.

Temporal coding abnormalities Patients with frontal lobe lesions are impaired in tasks involving temporal ordering, such as the sequencing of recent or remote events (Milner, Petrides, and Smith, 1985; Moscovich, 1989; Shimamura, Janowsky, and Squire, 1990; McAndrews and Milner, 1991). In addition, these patients are impaired in making recency judgments (Milner, 1971; Milner, Petrides, and Smith, 1985), a process that may rely on the correct temporal coding of events. Self-ordered pointing, a task in which the patient must remember the order in which objects have been indicated, is also impaired in patients with frontal lesions (Petrides and Milner, 1982). Patients with extensive frontal lesions also have little concern for either the past or the future (Goldstein, 1944; Ackerly and Benton, 1947).

Reality Checking The term *reality checking* refers to those aspects of monitoring the external world that have been called reality testing when they concern the present, and reality monitoring when they concern the past. Reality checking includes both an awareness of the difference between an internally generated alternate reality and a current reality, and the maintenance of a true past in the presence of counterfactual alternatives. Reality checking is essential for carrying out simulation processes without compromising the ability to respond to the objective environment. Simulation processes generate an alternate reality that must be evaluated in relation to its divergence from the current reality. This is necessary for generation of a plan of action to bring the current reality and the desired reality into accord. Without the ability to maintain an up-to-date model of external reality, this process will fail. If the internal model of an alternate reality is confused with the model of the true external reality, the process will also fail.

Reality monitoring Memories are created both for events experienced in the world and for simulated events. These two sources of memories must be treated differently to be used effectively. Given that both internal and external events create memory representations, what cues differentiate our internal models of reality from our internal simulations of reality?

Johnson and Raye (1981) studied normal subjects' ability to discriminate between memories of external events and of internally generated events. Memories of external events tend to be more detailed and have more spatial and temporal contextual information. Internally generated memories tend to be abstract and schematic, lacking in detail. These two memory representations form overlapping populations, and similar internal and external events may become confused. Reality checking involves a continual assessment of the relationship between behavior and the environment. The detection of novelty is essential to this process so that changes in the environment can be incorporated into a representation of the external world. As an individual acts on the environment, the consequences of the action must be incorporated into existing plans. If the environment deviates from expectations, plans must be reassessed.

Source memory and confabulation Patients with damage to prefrontal cortex show a disproportionate impairment in memory for the source of information (Schacter, Harbluk, and McLachlan, 1984; Shimamura and Squire, 1987; Janowsky, Shimamura, and Squire, 1989). Factual information is correctly recalled, but the spatiotemporal context in which the information was acquired is forgotten.

An interesting subgroup of amnesics are those who confabulate. Not all amnesics confabulate, but those who do are likely to have damage to the prefrontal cortex (Mayes, 1988; Moscovitch, 1989). For patients who confabulate, plausible memories are not discriminated from implausible ones. This may reflect a reality monitoring deficit. In combination with a deficit in the temporal coding of events, these impaired processes lead to the selection of one alternative from among many others retrieved in a memory search, but the selection is made with little confidence, and many errors occur.

Cognitive neurophysiological studies

Prefrontal damage in humans results in abnormalities in attention and early memory mechanisms manifesting in the initial 20–500 ms after sensory stimulation (Knight, 1994). Disinhibition of neuronal activity in primary sensory cortex is observable 20 ms after sensory stimulation as an increase in amplitude of primary sensory cortical evoked responses. This chronic leakage of irrelevant sensory inputs may contribute to the distractibility observed in prefrontal patients. Sustained attention is also impaired after prefrontal damage, leading to reduced discrimination abilities and distractibility when temporal discontinuities require bridging. Sensory detection requires both phasic attention and comparison of the stimulus to an internal memory template. P300 amplitudes predict long-term memory for the stimulus and may measure engagement of neocortical working memory processes preceding long-term encoding. P3b may match external reality to an internal model whereas P3a registers deviations from this model. P3a responses are reduced by prefrontal damage.

Inhibitory Control Prefrontal cortex has a net inhibitory output to subcortical (Edinger, Siegel, and Troiano, 1975) and cortical regions (Alexander, Newman, and Symmes, 1976; Skinner and Yingling, 1977). A prefrontal-thalamic gating mechanism has been reported that provides for modality-specific suppression of sensory input to primary cortical regions. Blockade of this prefrontal-thalamic mechanism results in increased amplitudes of primary sensory cortex evoked responses (Skinner and Yingling, 1977; Yingling and Skinner, 1977). This prefrontal-thalamic system provides a mechanism for inhibitory control of early sensory inputs, which is important for intermodality suppression of irrelevant inputs.

Prominent features of prefrontal damage, including altered attention ability and perseveration, may be linked to problems in inhibitory control (Lhermitte, 1986; Lhermitte, Pillon, and Serdaru, 1986). For example, inability to suppress previous incorrect responses may underlie the poor performance of prefrontal subjects on the Wisconsin Card Sorting Task and on Stroop phenomena (Shimamura et al., 1992).

Experiments were conducted to assess whether deficits in early sensory inhibitory control occur after human prefrontal damage. Irrelevant auditory and somatosensory stimuli were delivered to patients with

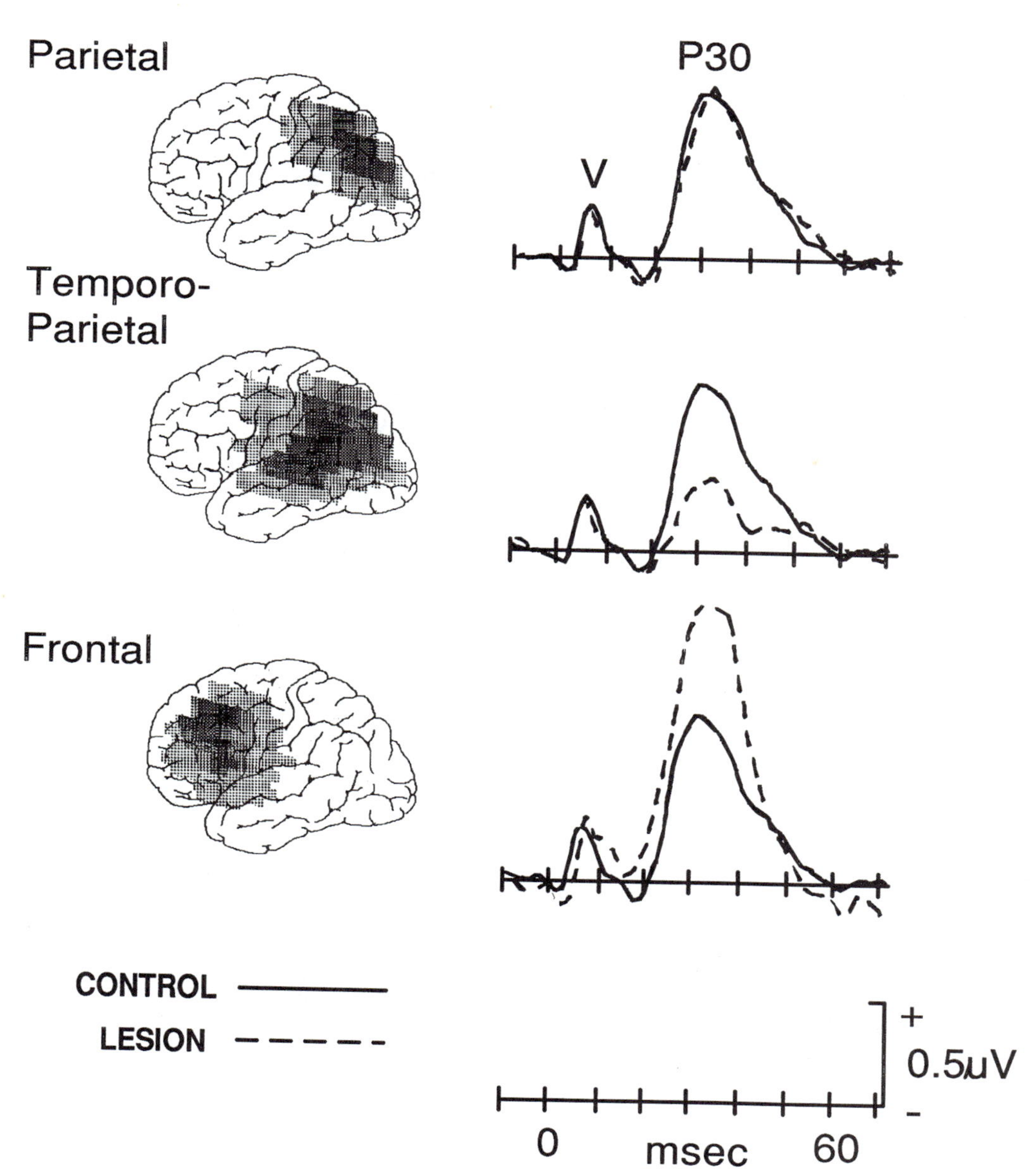

FIGURE 90.1 Auditory evoked responses generated in the inferior colliculus (wave V) and the primary auditory cortex (P30) are shown for controls (solid line) and patients (dashed line) with focal damage in the temporal-parietal junction (top, $N = 13$), lateral parietal cortex (middle, $N = 8$), or dorsolateral prefrontal cortex (bottom, $N = 13$). Reconstructions of the extent of damage in each patient group are shown on the left. Stimuli were clicks delivered at a rate of 13/s at intensity levels of 50 dB HL. Unilateral damage in the temporal-parietal junction extending into primary auditory cortex reduces P30 responses. Lateral parietal damage sparing primary auditory cortex has no effect on P30 responses. Dorsolateral prefrontal damage results in normal collicular potentials but an enhanced P30 primary cortical response.

discrete damage to dorsolateral prefrontal cortex and to patients with comparably sized lesions in the temporal-parietal junction or the lateral parietal cortex. Evoked responses generated in primary auditory and somatosensory cortices were recorded in these patients and in age-matched controls (see figures 90.1 and 90.2).

Posterior lesions invading primary cortical regions reduced evoked responses. Lesions in posterior association cortex sparing primary sensory regions had no

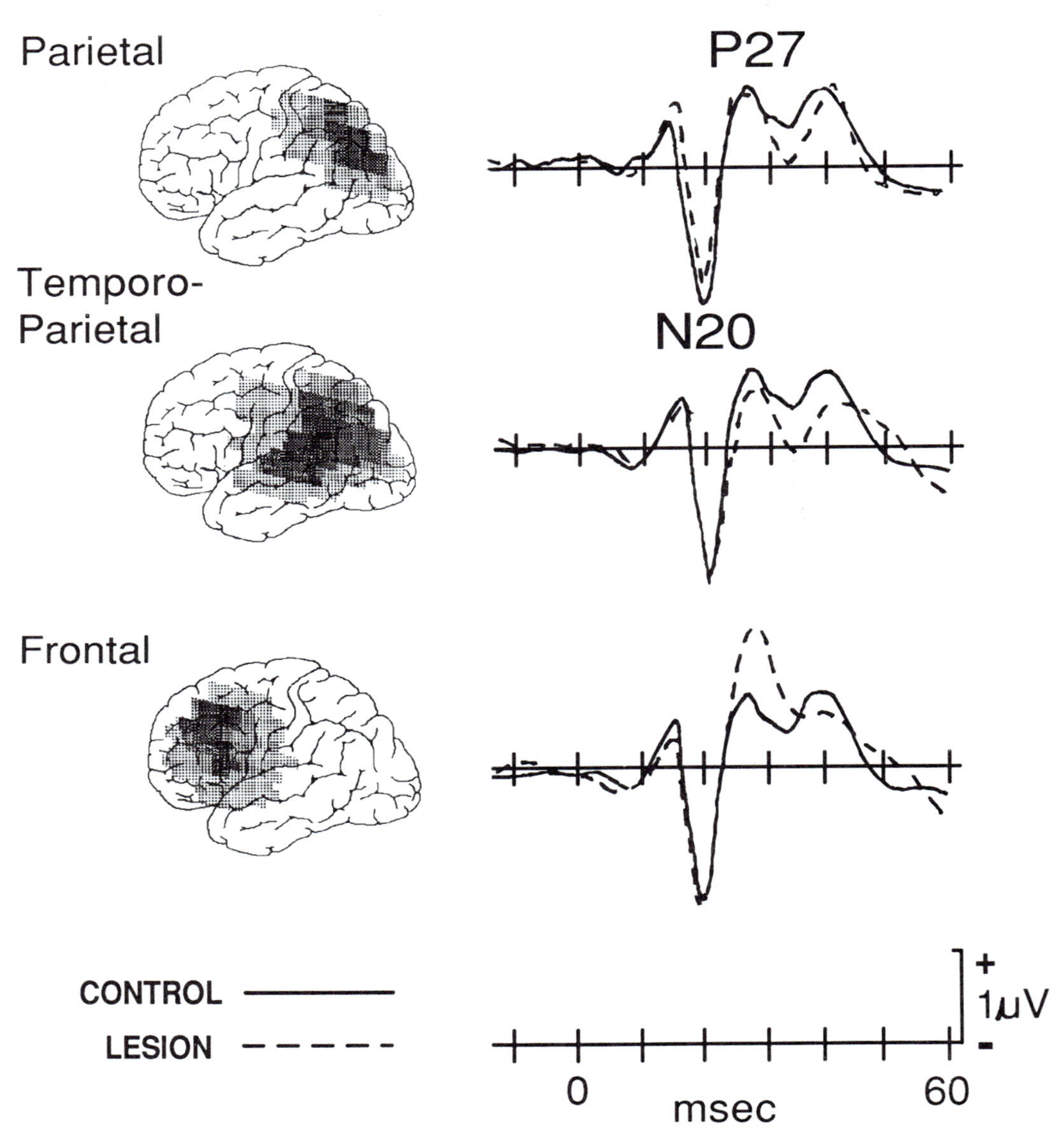

FIGURE 90.2 Somatosensory evoked responses were recorded from area 3b (N20) and areas 1 and 2 (P26). Stimuli were square-wave pulses of 0.15 ms duration delivered to the median nerve at the wrist. Stimulus intensity was set at 10% above opponens twitch threshold and stimuli were delivered at a rate of 3/s. Responses are shown from controls and the same patient groups represented in figure 90.1. Damage in cortical regions sparing primary somatosensory cortex (temporal-parietal $N = 12$, parietal $N = 8$) had no effect on the N20 or earlier spinal cord potentials. Prefrontal damage ($N = 11$), resulted in a selective increase in the amplitude of the P26 response.

effects on the primary cortical evoked responses. Conversely, prefrontal damage resulted in a disinhibition of both the primary auditory and somatosensory evoked responses generated from 20–40 ms poststimulation. This effect was measured as an amplitude enhancement of primary cortical responses (Knight, Scabini, and Woods, 1989; Yamaguchi and Knight, 1990). Spinal cord and brainstem potentials were unaffected, indicating that the loss of sensory control was due to abnormalities in either a prefrontal-thalamic or a direct prefrontal-sensory cortex mechanism. Chronic disinhibition of sensory inputs can have severe behavioral sequelae. For instance, inability to suppress irrelevant inputs has been shown to decrease attention capacity and habituate the orienting response (Sokolov, 1963; Knight, 1984).

Attention Abnormalities Focused attention to tones in one ear results in an enhancement of evoked potentials to all stimuli in that ear (Hillyard et al., 1973). This effect onsets by 25 ms after stimulation, indicating that humans are able to exert attention effects on inputs to primary cortical regions (McCallum et al., 1983; Woldorff and Hillyard, 1991). This physiological evidence supports an early sensory filtering mechanism (Broadbent, 1958; Treisman, 1960).

Auditory selective attention is impaired in patients with unilateral damage in left or right dorsolateral pre-

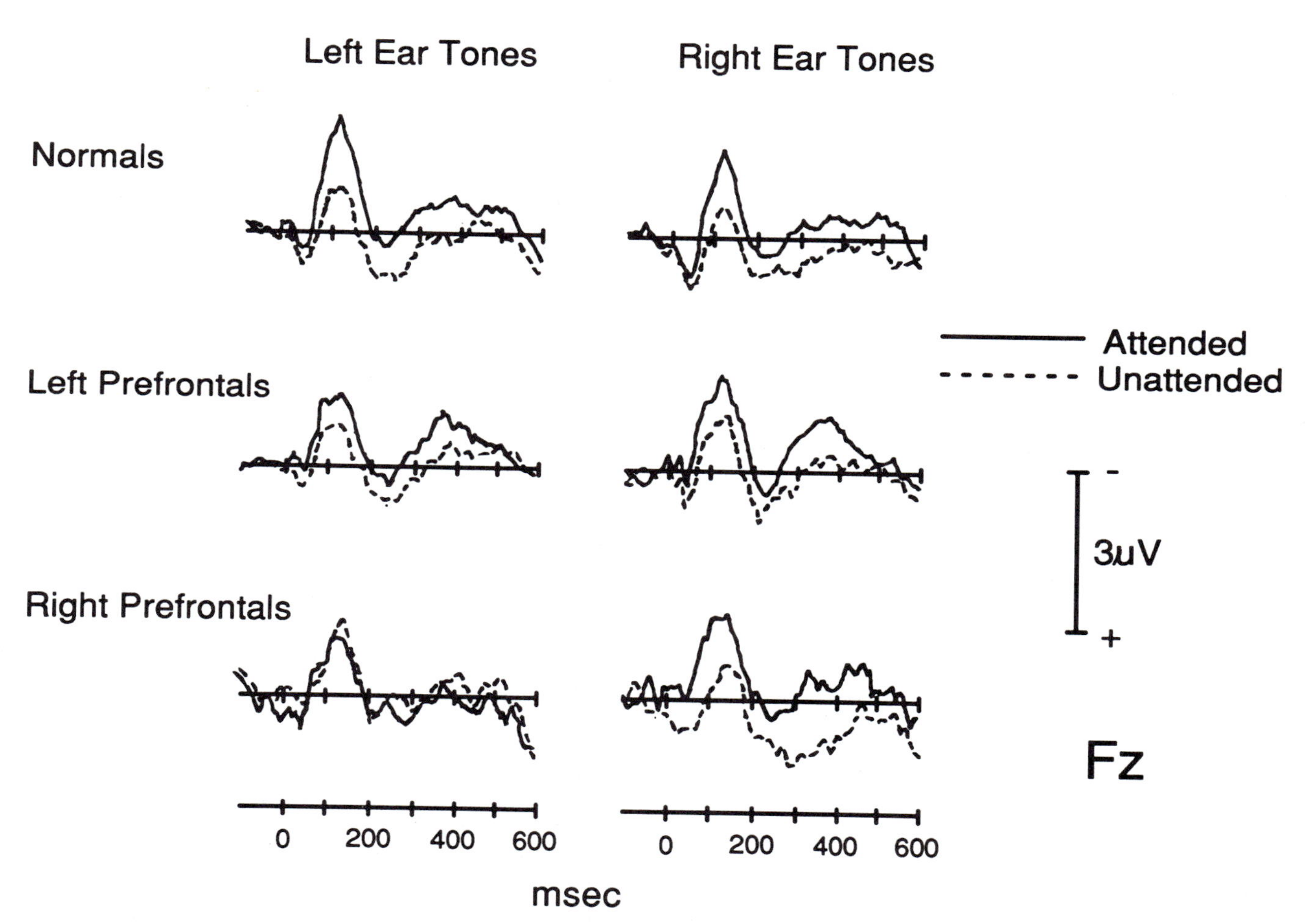

Figure 90.3 Attention effects from an auditory dichotic experiment are shown for age-matched control subjects (top), patients with unilateral left prefrontal damage (middle), and patients with right prefrontal damage (bottom). Controls generate enhanced amplitude of all attended stimuli (Nd, solid line). Patients with left prefrontal damage show diminished attention effects in both the left and right ear. Right prefrontal damage results in an inability to generate attention effects in the left ear contralateral to damage.

frontal cortex. Damage in these subjects is due to infarction of the precentral branch of the middle cerebral artery resulting in damage centered in areas 9 and 46, which are the human analogues of the sulcus principalis region in monkeys. In dichotic listening tasks, normal subjects generate an enhanced negativity to all stimuli in an attended channel. This negativity onsets from 25 to 50 ms after delivery of an attended stimulus and has been referred to as the processing negativity or Nd wave in the ERP literature. Control subjects generated comparable selective attention effects (Nd) for left- and right-ear stimulation. Left prefrontal patients generated reduced attention effects (Nd potential) in both ears. Right prefrontal patients have absent Nd potentials for left-ear stimuli (see figure 90.3; Knight et al., 1981). Patients with posterior association cortex lesions involving temporal-parietal areas were studied using this same paradigm. Comparable attention deficits were observed for left and right posterior lesions, suggesting that posterior cortical areas may not be asymmetrically organized for auditory selective attention (Woods, Knight, and Scabini, 1993).

Electrophysiological results in prefrontal patients parallel clinical observations in the neglect syndrome, supporting a right frontal dominance in attention allocation (Mesulam, 1981). One hypothesis is that the left frontal lobe is capable of allocating attention only to the contralateral right hemispace whereas the right frontal lobe can allocate attention to both the contralateral and ipsilateral hemispace. Enlarged right prefrontal cortex may provide the underlying anatomical substrate for this attention asymmetry (Wada, Clarke, and Hamm, 1975).

Whereas normal subjects generate attention effects at interstimulus intervals varying from 200–400 ms, attention abnormalities were maximal in the prefrontal patients at interstimulus intervals longer than 300 milliseconds (Woods and Knight, 1986). Fuster has proposed that the prefrontal cortex is crucial for bridging temporal discontinuities in the environment (Fuster, 1980). This "synthetic temporal function" could be more critical for long than for short interstimulus intervals. Another possibility is that at longer interstimulus intervals prefrontal subjects are more likely to be distracted by irrelevant stimuli. Distractibility is reported to be a prominent behavioral feature of prefrontal lesioned animals and humans (Bartus and Levere, 1977; Milner, 1982). For instance, delayed-response deficits in monkeys with prefrontal lesions (Jacobsen, 1935) are influenced by distractibility in the delay interval (Brutkowski, 1965). Electrophysiological attention effects in humans are also influenced by distractibility. The presentation of an irrelevant stimulus leads to a decrement in the response to a subsequent stimulus in prefrontal patients, but not in normal subjects. This effect is particularly pronounced in the ear contralateral to a prefrontal lesion at long interstimulus intervals. Attention performance is improved in prefrontal patients if no irrelevant stimuli are present, and sensory gating of nonattended inputs is not required.

Orientation and Memory The P300 component, first reported in 1965 (Desmedt, Debecker, and Manil, 1965; Sutton et al., 1965), is generated in all sensory modalities after detection of a potentially significant environmental event. P300 potentials peak in amplitude between 300 and 600 ms after sensory stimulation. The earlier phase of the P300 (P3a) is maximal over prefrontal regions and peaks in amplitude about 50 ms prior to parietal P300 activity (P3b). The P300 is not a unitary brain electrical response. Scalp recordings (Courchesne, Hillyard, and Galambos, 1975; Squires, Squires, and Hillyard, 1975; Yamaguchi and Knight, 1991a), intracranial recordings from depth electrodes (McCarthy et al., 1989; Smith et al., 1990), and the effects of focal brain lesions (Knight et al., 1989; Yamaguchi and Knight, 1991b) have provided evidence that field potentials generated by neural activity in prefrontal cortex, the temporal-parietal junction, and the hippocampal formation sum to produce the scalp-recorded P300.

The P300 phenomenon has received considerable attention since it has been linked to both orientation and memory mechanisms (Karis, Fabiani, and Donchin, 1984; Fabiani, Karis, and Donchin, 1986; Paller, Kutas, and Mayes, 1987). P300-like potentials have been reported in a variety of mammalian species including rats, cats, and monkeys (see Paller, 1994, for a review). Its ubiquitous occurrence across species indicates that the P300 phenomenon may represent activity of a basic neural system involved in the early detection and encoding of sensory stimuli. Subcomponents of the P300 may index engagement of working memory mechanisms in humans. For instance, prefrontal-hippocampal-dependent early latency P300 activity is linked to detection of perturbations from an ongoing memory template (Knight 1991a, 1991b). Ruchkin and colleagues have provided evidence that longer-latency P300 responses index activity in phonological and visuospatial systems of working memory (Ruchkin et al., 1990; Ruchkin et al., 1992).

Novelty has prominent effects on both the latency and the scalp distribution of P300 responses. Predictable and task-relevant stimuli generate small prefrontal P3a's and large parietal P3b's. Presentation of unexpected and novel stimuli results in a differential increase in prefrontal P3a amplitude. Prefrontal P300 responses (P3a) are generated about 50 ms prior to the parietal P300 (P3b). Damage to various brain regions

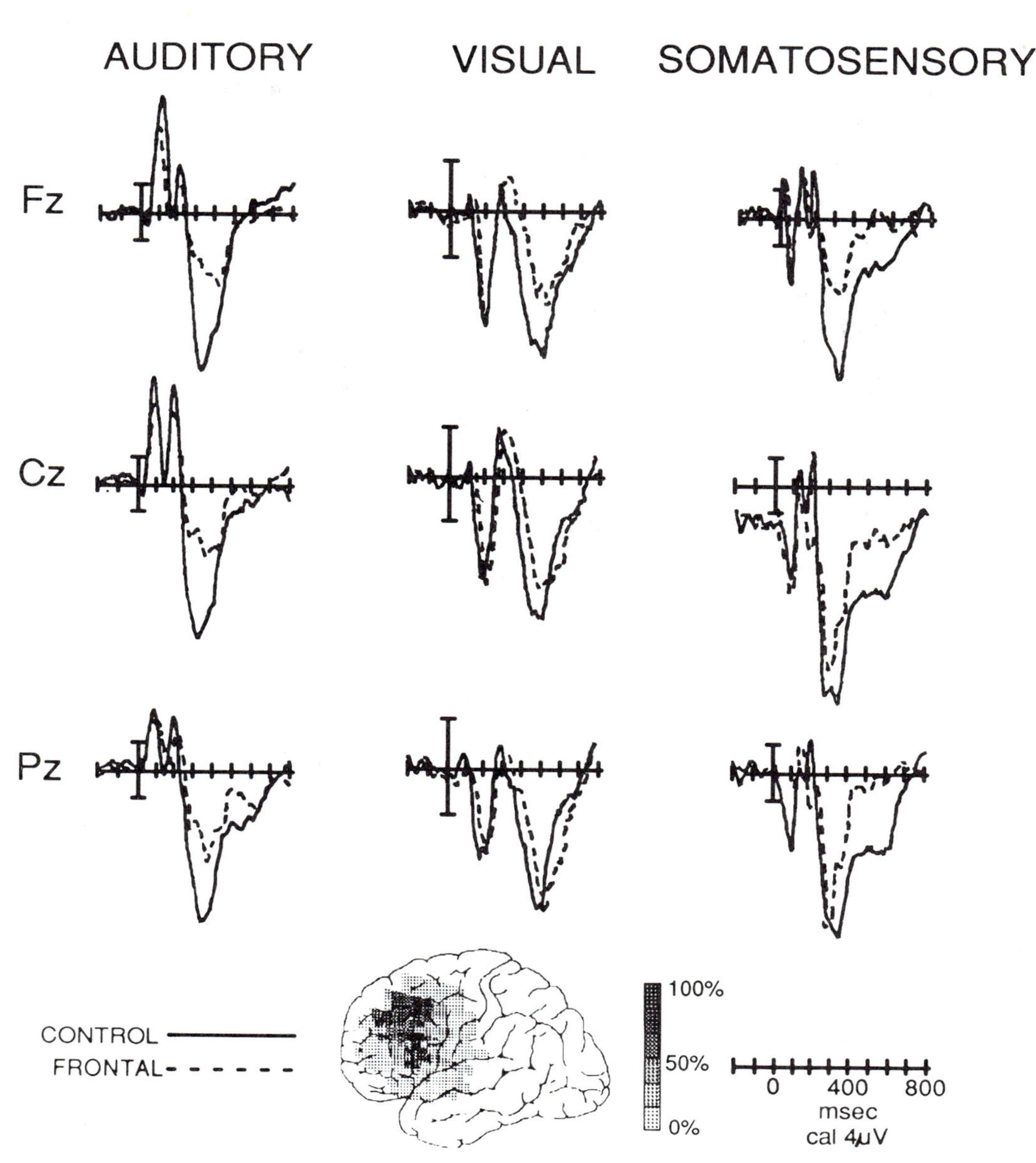

FIGURE 90.4 P3a responses recorded from controls (solid lines) and patients with unilateral damage in the dorsolateral prefrontal cortex (dashed lines, $N = 13$). P3a potentials were recorded to unexpected and novel auditory, somatosensory, and visual stimuli. Damage in the patients was due to infarction of the precentral branch of the middle cerebral artery. Both left and right prefrontal lesions are reflected onto the left side of the averaged lesion shown in the figure (from Knight, 1991a).

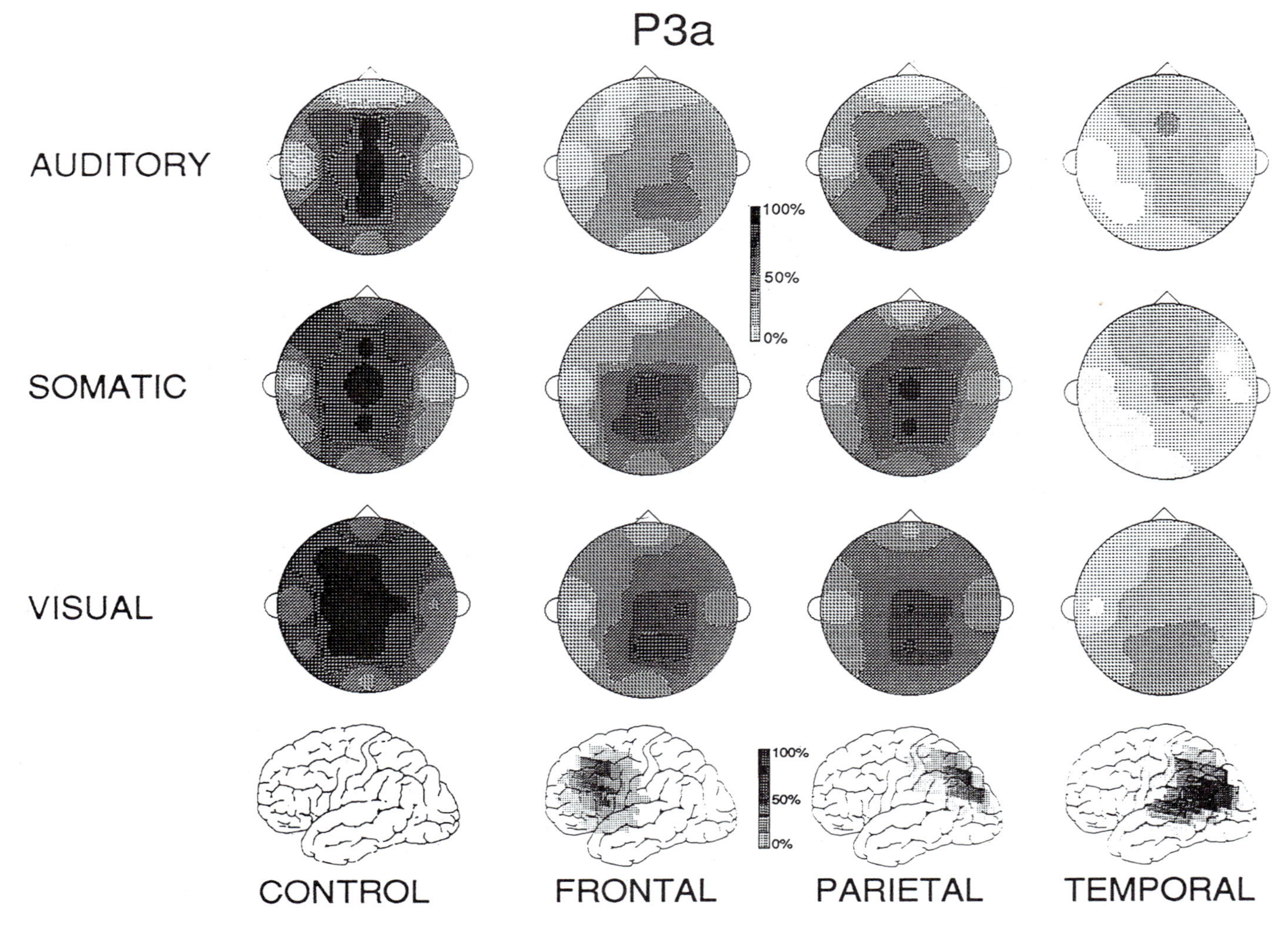

FIGURE 90.5 Scalp voltage distributions for P3a responses to novel auditory, somatosensory, and visual stimuli are shown for controls and patients with damage in the dorsolateral prefrontal cortex ($N = 13$), the temporal-parietal junction ($N = 13$), and the lateral parietal lobe ($N = 8$). The averaged lesions for the groups are represented on the bottom of the figure. Both P3a and P3b responses are reduced in patients with temporal-parietal junction damage. P3a responses are more reduced than P3b responses in the prefrontal patients. Reductions are recorded over the entire lesioned hemisphere in the prefrontal-damaged patients (left side of voltage maps).

has been shown to produce different patterns of P3a and P3b abnormalities. For example, unilateral damage in the temporal-parietal junction results in multimodal P300 reductions throughout the 300–600 ms poststimulation interval (Knight et al., 1989; Knight, 1990; Yamaguchi and Knight, 1991b, 1992). In contrast, prefrontal damage results in reduced P3a responses to novel stimuli with reductions observed throughout the lesioned hemisphere. Comparable P3a decrements have been observed in the auditory (Knight, 1984), visual (Knight, 1990), and somatosensory modalities in prefrontal-damaged humans (Yamaguchi and Knight, 1991b; see figures 90.4 and 90.5). Novelty P3a decreases were more severe after right prefrontal damage (Scabini, 1992). These findings converge with clinical observations and animal experimentation supporting a critical role of prefrontal structures in the detection of novel stimuli (Kimble, Bagshaw, and Pribram, 1965; Luria and Homskaya, 1970).

Consciousness and prefrontal cortex

Prefrontal-damaged patients lack inhibitory control over internal processes, are stimulus bound, and have little confidence in their ability to interact with the environment. Although memory functions and intelligence appear to be intact, these individuals function poorly in their daily lives. The case of W. R. illustrates this paradox.

CASE REPORT: PATIENT W. R. W. R., a 31-year-old lawyer, presented to the neurology clinic with family concern over his lack of interest in important life events. When questioned by the neurologist (R.T.K.) as to why he was in the clinic, the patient stated that he had "lost his ego." His difficulties began four years previously when he had a tonic-clonic seizure after staying up all night and drinking large amounts of coffee while studying for midterm exams in his final year of law school. An extensive neurological evaluation conducted at the National Institutes of Health, including EEG, CT scan, and PET scan, was all unremarkable. The diagnosis of generalized seizure disorder exacerbated by sleep deprivation was made, and the patient was placed on an anticonvulsant. W. R. graduated from law school but did not enter a practice because he was unable to decide where to take the bar exam. Over the next year he worked as a tennis instructor in Florida. He then broke off a two-year relationship and moved to California to live near his brother. His brother reported that W. R. was indecisive, procrastinated in carrying out planned activities, and was becoming progressively isolated from family and friends. The family attributed these problems to a "mid-life crisis." Four months prior to neurological consultation, W. R.'s mother died. At the funeral and during the time around it the family noted that he expressed no grief regarding his mother's death. The family decided to have the patient reevaluated. General neurological exam was unremarkable except for a mild snout reflex. W. R. was pleasant but somewhat indifferent to the situation. His speech was mildly tangential. On questioning about his mother's death, W. R. confirmed that he did not feel any strong emotions, either about his mother's death or about his current problem. The patient's brother mentioned that W. R. "had never lost it" emotionally during the previous week, at which point W. R. immediately interjected, "and I'm not trying not to lose it." Regarding his mother's death, he stated "I don't feel grief; I don't know if that's bad or good." These statements were emphatic, but expressed in a somewhat jocular fashion (witzelsucht). W. R. was asked about changes in his personality. He struggled for some minutes to describe changes he had noticed, but did not manage to identify any. He stated, "Being inside, I can't see it as clear." He was distractible and perseverative, frequently reverting to a prior discussion of tennis, and repeating phrases such as "yellow comes to mind" in response to queries of his memory. When asked about either the past or the future, his responses were schematic and stereotyped. He lacked any plans for the future, initiated no future-oriented actions, and stated, "It didn't matter that much, it never bothered me" that he never began to practice law. A CT scan revealed a left dorsolateral prefrontal astrocytoma that had infiltrated through the corpus callosum into the right frontal lobe. After discussion of the serious nature of the diagnosis, W. R. remained indifferent. The family members were distressed by the gravity of the situation and showed appropriate anxiety and sadness. Interestingly, they noted that their sadness was alleviated when in the presence of W. R.

W. R. remained an intelligent and articulate individual in spite of his frontal deficits, yet he was unable to carry on the activities to make him a fully functioning member of society. His behavior was completely constrained by his current circumstances. His jocularity was a genuine reaction to the social situation of the moment, and was not influenced by the larger context of his recent diagnosis. He appeared to have difficulty with source monitoring, and had little confidence in his answers to memory queries, complicated by frequent intrusions from internal events. Most marked in his behavior was a complete absence of counterfactual expressions. In particular, W. R. expressed no counterfactual emotions. He seemed unable to feel grief or regret, nor was he bothered by their absence even though he was aware of his brother's concern over his lack of emotion. These observations are not unique to W. R. and suggest that damage in dorsolateral prefrontal cortex leads to deficits in both simulation and reality monitoring, processes that are essential for the normal planning and decision-making functions necessary for consciousness.

Conclusions

These proposals are similar to those of Nauta (1971). Based on connectivity of the prefrontal cortex, Nauta suggested that this region was ideally situated to generate and evaluate internal models of action. It is proposed that in addition to this generation function the

Table 90.1

A cascade of prefrontal deficits

Primary	Secondary	Tertiary
Deficits in inhibitory control	Distractibility	Stimulus-bound behavior
Deficits in detecting novelty	Noisy internal milieu	Reduced decision confidence
	Impaired sustained and phasic attention	Perseveration
	Poor temporal coding of events	Poor planning and memory organization
		Trouble generating novel ideas
		Trouble ordering past, present, and future
		Impaired reality checking
		Deficits in generating and evaluating counterfactural scenarios

prefrontal cortex is crucial for detecting changes in the external environment and for discriminating internally and externally derived models of the world. These functions occupy a major portion of conscious awareness, including rumination on the past, speculation about the future, and daydreams about a different present.

In a similar vein, Oatley (1988) suggested that consciousness has four subcomponents. Two involve functions akin to reality checking. These include consciousness of sensation and the conscious monitoring of behavior. Another subcomponent of consciousness involves the internal simulation of future plans, where the consequences of future actions can be anticipated and evaluated. This form of consciousness allows the cognitive system to communicate with itself (cf. Baars, 1983). A final mode of consciousness allows the continuous monitoring of behavior against a model of the self, to insure integrity of behavior over time.

This chapter has described a cascade of deficits that result from damage to dorsolateral prefrontal cortex. Primary deficits in inhibitory control and in the detection of novelty result in a series of deficits in cognitive and metacognitive behaviors. These deficits are summarized in table 90.1.

Many aspects of conscious awareness are altered in patients with prefrontal damage. Awareness of the sensory world, and of the apparent stream of internal and external events, is impaired by deficits in novelty detection. Changes in the world, internal or external, may not be noticed in a noisy internal milieu. These deficits contribute to impaired reality monitoring and to a subsequent lack of confidence in behavior. An inability to bridge temporal gaps and to temporally sequence internal events, together with deficits in inhibitory control systems, leads to an impairment in the ability to generate coherent representations of alternate or counterfactual realities. While damage to the prefrontal cortex does not eliminate the experience of consciousness, it ensnares it in a noisy and temporally constrained state, locking the patient into the immediate space and time with little ability to escape.

ACKNOWLEDGMENTS Special thanks to Clay C. Clayworth for technical assistance in all phases of the work, and thanks to Ken Paller for helpful comments and discussion. Supported by NINDS Javits Award NS21135 to Knight, the Veterans Administration Medical Research Service, and the McDonnell-Pew Charitable Trust.

REFERENCES

Ackerly, S. S., and A. L. Benton, 1947. Report of a case of bilateral frontal lobe defect. *Res. Publ. Assoc. Res. Nerv. Ment. Dis.* 27:479–504.

Alexander, G. E., J. D. Newman, and D. Symmes, 1976. Convergence of prefrontal and acoustic inputs upon neurons in the superior temporal gyrus of the awake squirrel monkey. *Brain Res.* 116:334–338.

Allport, A., 1988. What concept of consciousness? In *Consciousness in Contemporary Science*, A. J. Marcel and E. Bisiach, eds. Oxford: Clarendon, 159–182.

Baars, B. J., 1983. Conscious contents provide the nervous system with coherent global information. In *Consciousness and Self-Regulation*, vol. 3, R. J. Davidson, G. E. Schwartz, and D. Shapiro, eds. New York: Plenum.

BADDELEY, A., and B. WILSON, 1988. Frontal amnesia and the dysexecutive syndrome. *Brain Cogn.* 7:212–230.

BARTUS, R. T., and T. E. LEVERE, 1977. Frontal decortication in Rhesus monkeys. A test of the interference hypothesis. *Brain Res.* 119:233–248.

BROADBENT, D. E., 1958. *Perception and Communication*. London: Pergamon.

BRUTKOWSKI, S., 1965. Functions of prefrontal cortex in animals. *Physiol. Rev.* 45:721–746.

COURCHESNE, E., S. A. HILLYARD, and R. GALAMBOS, 1975. Stimulus novelty, task relevance, and the visual evoked potential in man. *Electroencephalogr. Clin. Neurophysiol.* 39: 131–143.

DAMASIO, A. R., 1985. The frontal lobes. In *Clinical Neuropsychology*, K. M. Heilman and E. Valenstein, eds. New York: Oxford University Press, pp. 339–375.

DESMEDT, J. E., J. DEBECKER, and J. MANIL, 1965. Mise en evidence d'un signe electrique cerebral associe a la detection par le sujet d'un stimulus sensoriel tactile. *Bull. Acad. R. Med. Belgique* 5:887–936.

EDINGER, H. M., A. SIEGEL, and R. TROIANO, 1975. Effect of stimulation of prefrontal cortex and amygdala on diencephalic neurons. *Brain Res.* 97:17–31.

ESLINGER, P. J., and A. R. DAMASIO, 1985. Severe disturbance of higher cognition after bilateral frontal lobe ablation: Patient EVR. *Neurology* 35:1731–1741.

FABIANI, M., D. KARIS, and E. DONCHIN, 1986. P300 and recall in an incidental learning paradigm. *Psychophysiology* 23:298–300.

FUSTER, J. M., 1980. *The Prefrontal Cortex*. New York: Raven.

GOLDSTEIN, K., 1944. Mental changes due to frontal lobe damage. *J. Psychol.* 17:187–208.

GRAFMAN, J., B. JONAS, and A. SALAZAR, 1990. Wisconsin Card Sorting Test performance based on location and size of neuroanatomical lesion in Vietnam veterans with penetrating head injury. *Percep. Mot. Skills* 71:1120–1122.

HEBB, D. O., and W. PENFIELD, 1940. Human behavior after extensive bilateral removals from the frontal lobes. *Arch. Neurol. Psychiatry* 4:421–438.

HILLYARD, S. A., R. F. HINK, U. L. SCHWENT, and T. W. PICTON, 1973. Electrical signs of selective attention in the human brain. *Science* 182:177–180.

JACOBSEN, C. F., 1935. Functions of frontal association areas in primates. *Arch. Neurol. Psychiatry* 33:58–569.

JANOWSKY, J. S., A. P. SHIMAMURA, and L. R. SQUIRE, 1989. Source memory impairment in patients with frontal lobe lesions. *Neuropsychologia* 27:1043–1056.

JOHNSON, M. K., and C. L. RAYE, 1981. Reality Monitoring. *Psychol. Rev.* 88(1):67–85.

KAHNEMAN, D., and D. T. MILLER, 1986. Norm theory: Comparing reality to its alternatives. *Psychol. Rev.* 93(2): 136–153.

KAHNEMAN, D., and A. TVERSKY, 1982. The simulation heuristic. In *Judgment under uncertainty: Heuristics and Biases*, D. Kahneman, P. Slovic, and A. Tversky, eds. New York: Cambridge University Press, pp. 201–208.

KAHNEMAN, D., and C. A. VAREY, 1990. Propensities and counterfactuals: The loser that almost won. *J. Pers. Soc. Psychol.* 59(6):1101–1110.

KARIS, D., M. FABIANI, and E. DONCHIN, 1984. "P300" and memory: Individual differences in the Von Restorff effect. *Cognitive Psychol.* 16:177–216.

KIMBLE, D. P., M. H. BAGSHAW, and K. H. PRIBRAM, 1965. The GSR of monkeys during orienting and habituation after selective partial ablations of the cingulate and frontal cortex. *Neuropsychol* 3:121–128.

KNIGHT, R. T., 1984. Decreased response to novel stimuli after prefrontal lesions in man. *Electroencephalogr. Clin. Neurophysiol.* 59:9–20.

KNIGHT, R. T., 1990. ERPS in patients with focal brain lesions. *Electroencephalogr. Clin. Neurophysiol. Abstr.* 75:72.

KNIGHT, R. T., 1991a. Evoked potential studies of attention capacity in human frontal lobe lesions. In *Frontal Lobe Function and Dysfunction*, H. Levin, H. Eisenberg, and F. Benton, eds. London: Oxford University Press, pp. 139–153.

KNIGHT, R. T., 1991b. Effects of hippocampal lesions on the human P300. *Soc. Neurosci. Abstr.* 17:657.

KNIGHT, R. T., 1994. Attention regulation and human prefrontal cortex. In *Motor and Cognitive Functions of the Prefrontal Cortex: Research and Perspectives in Neurosciences*, A. M. Thierry, J. Glowinski, P. Goldman-Rakic, and Y. Christen, eds. Berlin: Springer-Verlag, pp. 160–173.

KNIGHT, R. T., S. A. HILLYARD, D. L. WOODS, and H. J. NEVILLE, 1981. The effects of frontal cortex lesions on event-related potentials during auditory selective attention. *Electroencephalogr. Clin. Neurophysiol.* 52:571–582.

KNIGHT, R. T., D. SCABINI, and D. L. WOODS, 1989. Prefrontal cortex gating of auditory transmission in humans. *Brain Res.* 504:338–342.

KNIGHT, R. T., D. SCABINI, D. L. WOODS, and C. C. CLAYWORTH, 1989. Contribution of the temporal-parietal junction to the auditory P3. *Brain Res.* 502:109–116.

LHERMITTE, F., 1986. Human autonomy and the frontal lobes. Part II: Patient behavior in complex and social situations: The "environmental dependency syndrome". *Ann. Neurol.* 19:335–343.

LHERMITTE, F., B. PILLON, and M. SERDARU, 1986. Human anatomy and the frontal lobes. Part I: Imitation and utilization behavior: A neuropsychological study of 75 patients. *Ann. Neurol.* 19:326–334.

LURIA, A. R., 1966. *Higher Cortical Functions in Man*. New York: Basic Books.

LURIA, A. R., and E. D. HOMSKAYA, 1970. Frontal lobes and the regulation of arousal process. In *Attention: Contemporary Theory and Analysis*, D. I. Mostofsky, ed. New York: Appleton-Century-Crofts, pp. 303–330.

MAYES, A. R., 1988. *Human Organic Memory Disorders*. New York: Cambridge University Press.

MCANDREWS, M. P., and B. MILNER, 1991. The frontal cortex and memory for temporal order. *Neuropsychologia* 29(9):849–859.

MCCALLUM, W. C., S. H. CURRY, R. COOPER, P. V. POCOCK, and D. PAPAKOSTOPOULOS, 1983. Brain event-related potentials as indicators of early selective processes in auditory target localization. *Psychophysiology* 20:1–17.

McCarthy, G., C. C. Woods, P. D. Williamson, and D. D. Spencer, 1989. Task-dependent field potentials in human hippocampal formation. *J. Neurosci.* 9:4253–4268.

Mesulam, M. M., 1981. A cortical network for directed attention and unilateral neglect. *Ann. Neurol.* 10:309–325.

Milner, B., 1971. Interhemispheric differences in the localization of psychological processes in man. *Br. Med. Bull.* 27:272–277.

Milner, B., 1982. Some cognitive effects of frontal lesions in man. *Philo. Trans. R. Soc. Lond.* 298:211–226.

Milner, B., and M. Petrides, 1984. Behavioural effects of frontal-lobe lesions in man. *Trends Neurosci.* 7:403–407.

Milner, B., M. Petrides, and M. L. Smith, 1985. Frontal lobes and the temporal organization of memory. *Hum. Neurobiol.* 4:137–142.

Moscovitch, M., 1989. Confabulation and the frontal systems: Strategic versus associative retrieval in neuropsychological theories of memory. In *Varieties of Memory and Consciousness: Essays in Honour of Endel Tulving*, H. L. Roedigger, III, and F. I. M. Craik, eds. Hillsdale, N.J.: Erlbaum, pp. 133–160.

Nauta, W. J. H., 1971. The problem of the frontal lobe: A reinterpretation. *J. Psychiat. Res.* 8:167–187.

Oatley, K., 1988. On changing one's mind: A possible function of consciousness. In *Consciousness in Contemporary Science*, A. J. Marcel and E. Bisiach, eds. Oxford: Clarendon, pp. 369–389.

Paller, K. A., 1994. The neural substrates of cognitive event-related potentials: A review of animal models of P3. In *Cognitive Electrophysiology: ERPs in Basic and Clinical Research*, H. J. Heinze, T. F. Munte, and G. R. Mangun, eds. Boston: Birkhauser, pp. 300–333.

Paller, K. A., M. Kutas, and A. R. Mayes, 1987. Neural correlates of encoding in an incidental learning paradigm. *Electroencephalog. Clin. Neurophysiol.* 55:417–426.

Petrides, M., and B. Milner, 1982. Deficits on subject-ordered tasks after frontal- and temporal-lobe lesions in man. *Neuropsychologia* 20:249–262.

Ruchkin, D. S., R. Johnson, Jr., H. Canoune, and W. Ritter, 1990. Short-term memory storage and retention: An event-related brain potential study. *Electroencephalog. Clin. Neurophysiol.* 76:419–439.

Ruchkin, D. S., R. Johnson, Jr., J. Grafman, H. Canoune, and W. Ritter, 1992. Distinctions and similarities among working memory processes: An event-related potential study. *Cognitive Brain Res.* 1:53–66.

Scabini, D., 1992. Contribution of anterior and posterior association cortices to the human P300 cognitive event related potential. Ph. D. diss., University of California, Davis.

Schacter, D. L., 1992. Implicit knowledge: New perspectives on unconscious processes. *Proc. Natl. Acad. Sci. U.S.A.* 89:11113–11117.

Schacter, D. L., J. L. Harbluk, and D. R. McLachlan, 1984. Retrieval without recollection: An experimental analysis of source amnesia. *J. Verb. Learn. Verb. Be.* 23: 593–611.

Shallice, T., 1988. *From Neuropsychology to Mental Structure.* Cambridge: Cambridge University Press.

Shallice, T., and M. E. Evans, 1978. The involvement of the frontal lobes in cognitive estimation. *Cortex* 14:294–303.

Shimamura, A. P., F. B. Gershberg, P. J. Jurica, J. A. Mangels, and R. T. Knight, 1992. Intact implicit memory in patients with focal frontal lobe lesions. *Neuropsychology* 30:931–937.

Shimamura, A. P., J. S. Janowsky, and L. R. Squire, 1990. Memory for the temporal order of events in patients with frontal lobe lesions and amnesic patients. *Neuropsychologia* 28:803–813.

Shimamura, A. P., and L. R. Squire, 1987. A neuropsychological study of fact memory and source amnesia. *J. Exp. Psychol. [Learn. Mem. Cogn.]* 13:464–473.

Skinner, J. E., and C. D. Yingling, 1977. Central gating mechanisms that regulate event-related potentials and behavior. In *Progress Clinical Neurophysiology*, vol. 1, J. E. Desmedt, ed. Basel: Karger, pp. 30–69.

Smith, M. E., E. Halgren, M. E Sokolik, P. Baudena, C. Liegeois-Chauvel, A. Musolino, and P. Chauvel, 1990. The intra-cranial topography of the P3 event-related potential elicited during auditory oddball. *Electroencephalogr. Clin. Neurophysiol.* 76:235–248.

Smith, M. L., and B. Milner, 1984. Differential effects of frontal lobe lesions on cognitive estimation and spatial memory. *Neuropsychologia* 22:697–705.

Sokolov, E. N., 1963. Higher nervous functions: The orienting reflex. *Ann. Rev. Physiol.* 25:545–580.

Squires, N., K. Squires, and S. A. Hillyard, 1975. Two varieties of long-latency positive waves evoked by unpredictable auditory stimuli in man. *Electroencephalog. Clin. Neurophysiol.* 38:387–401.

Stuss, D. T., 1991. Self, awareness, and the frontal lobes: A neuropsychological perspective. In *The Self: Interdisciplinary Approaches*, J. Strauss and G. R. Goethals, eds. New York: Springer, pp. 255–278.

Stuss, D. T., and D. F. Benson, 1986. *The Frontal Lobes.* New York: Raven.

Stuss, D. T., G. A. Eskes, and J. K. Foster, 1994. Experimental neuropsychological studies of frontal lobe functions. In *Handbook of Neuropsychology*, vol. 9, F. Bollen and J. Grafman, eds. Amsterdam: Elsevier.

Sutton, S., M. Baren, J. Zubin, and E. R. John, 1965. Evoked potentials correlates of stimulus uncertainty. *Science* 150:1187–1188.

Treisman, A. M., 1960. Contextual cues in selective listening. *Q. J. Exp. Psychol.* 12:242–248.

Tversky, A., and D. Kahneman, 1973. Availability: A heuristic for judging frequency and probability. *Cognitive Psychol.* 5:207–232.

Wada, J. A., R. Clarke, and A. Hamm, 1975. Cerebral hemispheric asymmetry in humans. *Arch. Neurol.* 32:239–246.

WOLDORFF, M. G., and S. A. HILLYARD, 1991. Modulation of early auditory processing during selective listening to rapidly presented tones. *Electroencephalogr. Clin. Neurophysiol.* 79:170–191.

WOODS, D. L., and R. T. KNIGHT, 1986. Electrophysiological evidence of increased distractibility after dorsolateral prefrontal lesions. *Neurol.* 36:212–216.

WOODS, D. L., R. T. KNIGHT, and D. SCABINI, 1993. Anatomical substrates of auditory selective attention: Behavioral and electrophysiological effects of temporal and parietal lesions. *Cognitive Brain Res.* 1:227–240.

YAMAGUCHI, S., and R. T. KNIGHT, 1990. Gating of somatosensory inputs by human prefrontal cortex. *Brain Res.* 521: 281–288.

YAMAGUCHI, S., and R. T. KNIGHT, 1991a. P300 generation by novel somatosensory stimuli. *Electroencephalogr. Clin. Neurophysiol.* 78:50–55.

YAMAGUCHI, S., and R. T. KNIGHT, 1991b. Anterior and posterior association cortex contributions to the somatosensory P300. *J. Neurosci.* 11 (7):2039–2054.

YAMAGUCHI, S., and R. T. KNIGHT, 1992. Effects of temporal-parietal lesions on the somatosensory P3 to lower limb stimulation. *Electroencephalogr. Clin. Neurophysiol.* 84:139–148.

YINGLING, C. D., and J. E. SKINNER, 1977. Gating of thalamic input to cerebral cortex by nucleus reticularis thalami. In *Progress Clinical Neurophysiology*, vol. 1, J. E. Desmedt, ed. Basel: Karger, pp. 70–96.

91 The Conscious State Paradigm: A Neurocognitive Approach to Waking, Sleeping, and Dreaming

J. ALLAN HOBSON AND ROBERT STICKGOLD

ABSTRACT One approach to the neurocognitive study of consciousness is to track the changes in conscious states that normally accompany the human wake, sleep, and dream cycle. When distinctive changes in conscious states are identified and documented, their brain basis can be sought at the cellular and molecular level in an animal model. For example, internally generated visual perception is intensified during dreaming. Simultaneously, cognition becomes bizarre and memory is markedly impaired. We suggest that the bizarre cognition and amnesia of REM sleep result from the withdrawal of aminergic modulation of the forebrain. Similarly, the heightened, often bizarre imagery of REM sleep results from the activation of brainstem cholinergic systems, which send chaotic stimuli to the forebrain. In this chapter we show how studies of animal physiology, human cognitive performance and dream reports can be combined to synthesize a neurocognitive model of brain/mind states.

Philosophical and experimental psychological approaches have failed in their efforts to clearly define, let alone to explain, human consciousness. As this volume attests, there has recently been a renewed interest in this ancient and obdurate problem, inspired by the recent rapid growth in the power of the cognitive and brain sciences. Two recurrent strategies characterize an impressive number of these new studies. At one extreme is a global strategy that regards consciousness as an entity localizable to some particular brain region. At the opposite extreme is an analytic strategy that examines operationally defined aspects of conscious experience and then attempts to explain them at the cellular or molecular level. While the first of these strategies is plagued by vague and imprecise conclusions, the latter approach often isolates pieces of the process that are not easily reassembled to give a picture of the whole. Yet these strategies remain popular, in large part because there is as yet no clear method of mapping directly from cellular and molecular processes all the way to the phenomenon of consciousness.

Is there a middle way between the Scylla of superficial generality and the Charybdis of irrelevant detail? This chapter explores the possibility that global aspects of consciousness can be meaningfully correlated with precisely specified microscopic neurobiological mechanisms by mapping *global changes* in conscious states onto *global changes* in brain states. In the transition from waking to dreaming, so dramatic and so synchronous are the alterations in mind and brain states as to argue persuasively that specific, identified brain processes can simultaneously affect the entire constellation of psychological functions that together constitute consciousness.

Prominent among the psychological functions that shift dramatically in the transitions from waking first to sleeping and then to dreaming are perception, memory, orientation, attention, and emotion. It is becoming increasingly clear that a series of events in the brain stem initiates changes in the global physiology and chemistry of the limbic system and cortex, which in turn produce profound alterations in these components of our conscious experience. By studying the physiology and phenomenology of waking, sleeping, and dreaming, and by elucidating the mechanisms that

J. ALLAN HOBSON and ROBERT STICKGOLD Laboratory of Neurophysiology, Department of Psychiatry, Harvard Medical School, Boston, Mass.

shift the brain and mind from one of these states to another, we can learn much about the control and alteration of consciousness. And by providing techniques for analyzing and altering conscious states, such studies further provide us with powerful experimental tools with which we can learn more about consciousness itself. We call this set of facts, assumptions, and strategies the *conscious state paradigm*.

Toward a neurocognitive definition of consciousness

If we are to study the waking, sleeping, and dreaming states from the perspective of consciousness, we need a working definition of this phenomenon. Since our approach to sleep and dreams is a decidedly neurocognitive one, we have developed a neurocognitive definition of consciousness. From this perspective, consciousness is our integrated awareness of the external world, of our bodies, and of our mental processes, and it is dependent on an integrated set of higher-order representations of both internal and external perceptions. It depends not only upon primary representations of the world, as in simple visual perception, but on representations of these representations, such as our identification of a visually perceived object or our awareness of its beauty.

From a neurobiological perspective, this definition of consciousness implies that the nature and richness of an individual's conscious experience depends in large part upon the size and organization of higher-order neuronal networks. A complex, hierarchically organized brain is clearly required to produce the extensive set of perceptual and higher-order representations of our multifaceted experiences. And these representations must be integrated if a unified consciousness is to appear.

From a phylogenetic and ontogenetic point of view, this definition admits of a gradualism whereby an individual or a species could experience a more or less complex and a more or less abstract kind of consciousness depending upon the number and size of higher-order neuronal networks that could be allocated to the functions of consciousness. The greater the size and number of these networks, the greater the integrative challenge in order for experiential unity of consciousness to be achieved, but the greater the reward as well.

The conscious states paradigm, which holds that an individual's experience of consciousness varies over the course of the day as the functional organization of the brain's higher-order networks changes, is a potentially powerful tool for the investigation of the organization of these networks. We begin to develop this strategy by simply tracking how the components or modules of conscious experience change when we first fall asleep; then focus more sharply on our experience of consciousness when we later enter REM sleep and begin to dream. Our basic assumption is that the phenomenological changes that differentiate, say, dreaming from waking, can be understood at the level of the brain. By detailing the specific physiological and biochemical changes that underlie this stereotyped shift in conscious state, we hope to deduce some of the organizational rules controlling the large and populous neuronal networks that together generate full and unified conscious experience. And by observing how consciousness breaks down with these natural changes in brain state, we may also be able to deduce some of the causes of the diminished conscious unity experienced in mental illness.

Sleep and consciousness

The normal, spontaneous, and radical shifts in cognitive processing that occur over the wake-sleep cycle provide a unique opportunity to study some of the basic rules organizing cognition. This opportunity is enhanced by the availability of animal models of the neurobiological substrate of these cognitive states (Dement, 1958; Jouvet and Michel, 1959; Jouvet, 1962). Over the last two decades, our laboratory has also been elucidating the physiology of waking and of the REM and Non-REM sleep states (Hobson and Steriade, 1986; Hobson et al., 1987; Hobson, 1988, 1989) (figure 91.1). Our general strategy is to examine brain and cognitive functions at a formal and global level. For example, in relation to perception, we have explored the hypothesis that ponto-geniculo-occipital (PGO) waves, which are absent or powerfully suppressed in waking, constitute an internal stimulus source for the visual system in REM (rapid eye movement) sleep, thus explaining the vivid visual imagery of our dreams (Mamelak and Hobson, 1989a; Kahn and Hobson, 1991a, 1991b). We further hypothesize that the memory loss in REM sleep may be a consequence of aminergic demodulation of the brain, thus explaining dream amnesia. We have recently begun a detailed

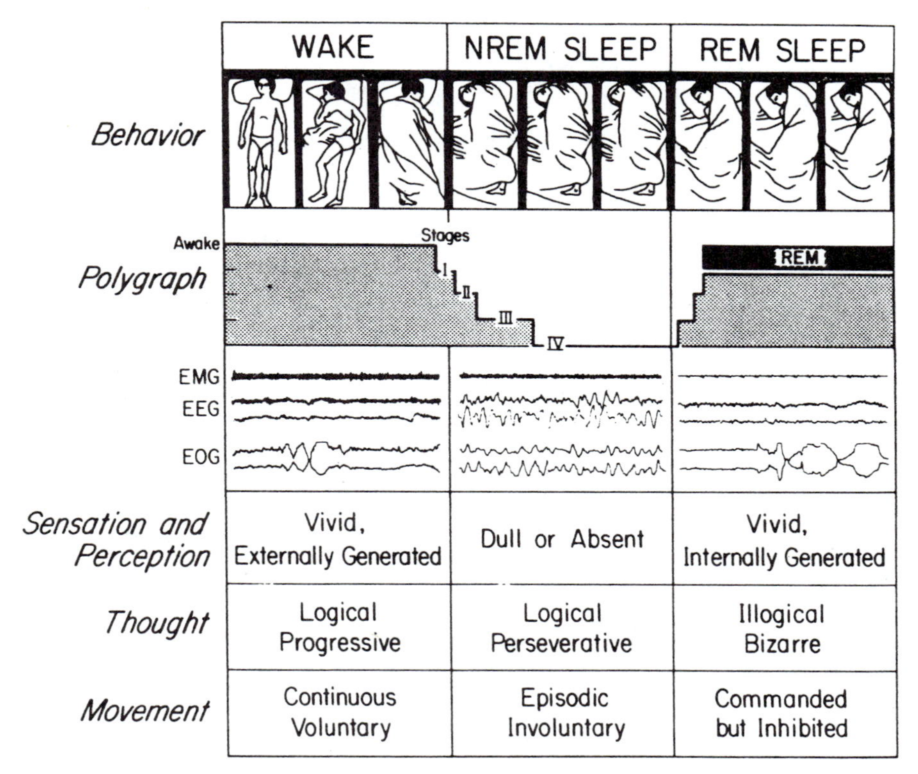

FIGURE 91.1 Behavioral states in humans. States of waking, non-REM (NREM) sleep, and REM sleep have behavioral, polygraphic, and psychological manifestations. In the behavior channel, posture shifts (detectable by time-lapse photography or video) can occur during waking and in concert with phase changes of sleep cycle. Two different mechanisms account for sleep immobility: disfacilitation (during stages I–IV of NREM sleep) and inhibition (during REM sleep). In dreams we imagine that we move but we do not. Sequence of these stages are represented in the polygraph channel. Sample tracings of three variables used to distinguish state are also shown: electromyogram (EMG), which is highest in waking, intermediate in NREM sleep, and lowest in REM sleep; and electroencephalogram (EEG) and electro-oculogram (EOG), which are both activated in waking and REM sleep and inactivated in NREM sleep. Each sample record is 20 s. The three lower channels describe other subjective and objective state variables. (From J. A. Hobson and M. Steriade, 1986)

and quantitative characterization of cognitive functions that change in parallel with the succession of waking and sleep stages. Our overall strategy is to look at low-level cognitive functions using specific behavioral tests (such as perceptual cuing and semantic priming) and high-level cognitive functions (by analyzing reports of sleep mentation).

It is important to recognize that we use the phrase *state-dependent* differently from scientists interested in state-dependent learning, which focuses on the fact that recall of previously learned material depends on the states in which the learning and testing occur. In our usage specific *aspects* of cognitive processes are seen as state-dependent. For example, memory appears to be impaired in REM sleep so that it is difficult to recall mental content that arises in that state; we call this memory failure a state-dependent amnesia.

ORIENTATION AND THE ORIENTING RESPONSE Two of the specific cognitive functions that we would like to investigate are orientation (in the mental status sense of knowing time, place, and person) and orienting (in the attentional sense of selecting the most germane channels or sectors of input to process perceptually). In addition to the phenomenologically strong interaction of these two processes in creating both the anchor and the compass of cognition, we have three empirical reasons to believe that these two processes are tightly linked to one another in a highly state-dependent manner. The first is that they both change in REM sleep such that in our dreams we constantly reorient in both the mental status sense (characters and settings change) and the perceptual sense (in a kaleidoscopically shifting sequence of scenes we look around and search in vain for a stable set of referents); the second is that these two functions change together in organic delirium and Korsakoff's psychosis; the third is that the onset of delirium in alcohol withdrawal is associated with a sudden and marked increase in REM sleep. For us, then, dreaming is a state-dependent delirium.

In developing our conceptual framework and research strategies, we follow Farah (1989) in drawing a distinction between spatial localization of a stimulus (which we call orienting) and the subsequent recognition and evaluation of that stimulus (which is dependent on subjective *orientation*). Since orienting is, by definition, the redirection of attention, we have begun to use the perceptual cuing task to obtain objective evidence of state-dependent changes in orienting. Similarly, since orientation defines the subject's perceptual reference frame, we look for state-dependent changes in performance on the semantic priming test as evidence of changes in the content of and ease of access to associational knowledge networks. In addition, we analyze discontinuities in objects, characters, and locations found in sleep and waking mentation reports to measure orientational stability as it is experienced subjectively.

Methodological Issues The study of sleep mentation suffers from several major limitations (Arkin, Antrobus, and Ellman, 1978) which must be overcome or minimized. First, no behavioral tests of cognition are possible during sleep, except insofar as the behavior involves automatic functions, such as heart rate, PGO activity (McCarley, Winkleman, and Duffy, 1983), or rapid eye movements (Roffwarg et al., 1962). Second, since sleep mentation is not usually under conscious control, it is impossible to manipulate the content of the cognition, eliminating many types of exploration (but see LaBerge, 1985). Third, reports of sleep cognition are suspect, because they are subjective and because recall of sleep cognition is so short-lived and uncertain (Badia, 1991). Fourth, while the sleep laboratory setting allows accurate assessment of brain states, reports gathered in this setting are more constricted than those that can be gathered in more familiar home settings. Finally, the high cost and discomfort of the sleep lab preclude long-term studies that could establish a large data base from individual subjects in longitudinal designs, studies that could be sensitive to the normal vicissitudes of the life cycle and to experimental interventions.

Before presenting the new methods that we have designed to overcome these problems, we review and critique what we call the cross-sectional paradigm of the early psychophysiological era of sleep research (1953–1975). We then discuss the development of the basic neurobiology of sleep (1959 to the present) and explain why we see this rich and solid data as a base for our more global brain-mind mapping effort. Our attempt to develop a scientific approach to the phenomenology of conscious states by quantifying formal cognitive aspects of subjective reports is illustrated in a series of new studies on dream bizarreness. Finally, we present our rationale for a neurocognitive program that is designed to provide a bridge between the basic neurobiology and the phenomenology of wake-sleep states.

Psychophysiology of Waking, Sleeping and Dreaming The discovery of REM sleep and its correlation with dreaming (Aserinsky and Kleitman, 1953) opened a new era of research in the relation of brain to mind. In the early days of the human sleep-dream laboratory, much attention was paid to the specificity or lack thereof of the REM-dream correlation using the newly available sleep laboratory paradigm (Arkin, Antrobus, and Ellman, 1978). Normal subjects, usually students, were awakened from either the non-REM or REM phase of sleep and asked to report their recollection of any mental experience preceding the awakening. The main conclusions of this cross-sectional normative paradigm were that reports from REM sleep awakenings were typically longer, more perceptually vivid, more motorically animated, and more emotionally charged than the non-REM reports; and that non-REM reports tended to be more thoughtlike and contained more representations of current concerns than did REM sleep reports. It seemed most likely that the activation level of the brain was the determinant of these observed differences. This hypothesis was supported by the observation that all measures of dream intensity peaked during the eye movement clusters within REM periods. Attempts to further detail the psychophysiology of dreaming (Roffwarg et al., 1962) and to understand the functional benefits of REM sleep for cognition (Dement, 1960) were impeded by replication failures (Moscowitz and Berger, 1969; Kales et al., 1970). It is our view that the ensuing controversies (Herman, 1989; Antrobus, 1991) regarding these two fundamental issues will be resolved only by the introduction of new concepts and methods (Rechtschaffen et al., 1989), including those of the cognitive sciences.

Numerous studies have investigated cognitive functioning in the postawakening (hypnopompic) period. The general phenomenon of impaired orientation and processing of inputs immediately following awakening has been called sleep inertia (Lubin et al., 1976). Decrements in memory (Stones, 1977) and cognitive performance (Lavie and Sutter, 1975) have been reported with differential effects following awakenings from REM and non-REM sleep. Other studies (for review see Dinges, 1990) point out that both the time of night and the duration of the prior period of wakefulness strongly affect such hypnopompic cognitive performance. All told, these studies point to the existence of sleep inertia and state carryover, and the importance of controlling for time of night and requiring that subjects have normal sleep on the night prior to testing.

Neurobiology of Waking, Sleeping, and Dreaming

The discovery of the ubiquity of REM sleep in mammals (Dement, 1958; Jouvet and Michel, 1959; Jouvet, 1962) gave the brain side of the brain-mind state question an animal model. While animal studies showed that potent and widespread activation of the brain did occur in REM sleep, it soon became clear that Moruzzi and Magoun's concept of a brainstem reticular activating system (Moruzzi and Magoun, 1949) required extension and modification to account for the differences between the EEG activation of waking and that of REM sleep. This was provided by the discovery of the chemically specific neuromodulatory subsystems of the brain stem and of their differential activity in waking (noradrenergic and serotonergic systems on, cholinergic system damped) and REM sleep (noradrenergic and serotonergic systems off, cholinergic system undamped). The resulting model of reciprocal interaction (Hobson, McCarley, and Wysinki, 1975; McCarley and Hobson, 1975) provided a theoretical framework for experimental interventions at the cellular and molecular level that have vindicated the notion that waking and dreaming are at opposite ends of an aminergic-cholinergic neuromodulatory continuum on which non-REM sleep holds an intermediate position.

The articulation of the reciprocal interaction model and the emergence of a wealth of detail regarding the functional reorganization of the visual system in sleep suggested a new conceptual approach to the science of brain-mind state determination (McCarley and Hobson, 1977). First expressed as the activation-synthesis hypothesis of dreaming (Hobson and McCarley, 1977), this new global brain-state to mind-state mapping effort has dealt with the contributions to mental state differentiation of the changes in brain neuromodulation (from aminergic to cholinergic) and of changes in stimulus origin (from external to internal). More specifically, we proposed that the state-dependent changes in thinking and memory were a function of changes in ratios of aminergic to cholinergic neurotransmitters (Flicker, McCarley, and Hobson, 1981).

To illustrate our way of thinking in more detail, let us consider the mechanism and functional consequences of the shift in visual system input source from the retina (in waking) to the brain stem (in REM sleep) (figure 91.2). This shift in input source is associated with the aminergic demodulation and disinhibition of the visual cortex, the lateral geniculate body, and the brainstem oculomotor networks, which occur because noradrenergic and serotonergic neurons stop firing. As a result of disinhibition, the peribrachial cholinergic neurons become hyperexcitable and fire in clusters, initiating the phasic activation of the geniculate body and visual cortex, which are recordable in REM sleep as PGO waves and which correlate with rapid eye movements. In both cats and humans, this cholinergically mediated stimulation conveys to the visual system information about the direction of the eye movements, which in REM sleep become completely uncoupled from external stimulus control (see Callaway et al., 1987; Hobson, 1990, for details and references). The net result of this shift is an electrically activated visual system that is both aminergically demodulated and cholinergically self-stimulated-by signals arising in the brain stem that convey information about eye-movement direction. These changes could determine such cognitive features of dreaming as the formed visual imagery; the frequent reorientation, especially that seen in complete scene shifts; and the loss of volition and of the voluntary control of internal attention that robs dreaming of self-reflective awareness, of deliberate thought, and of guided action.

When the brain changes state, as from REM sleep to waking, it is typical to observe that the polygraphic

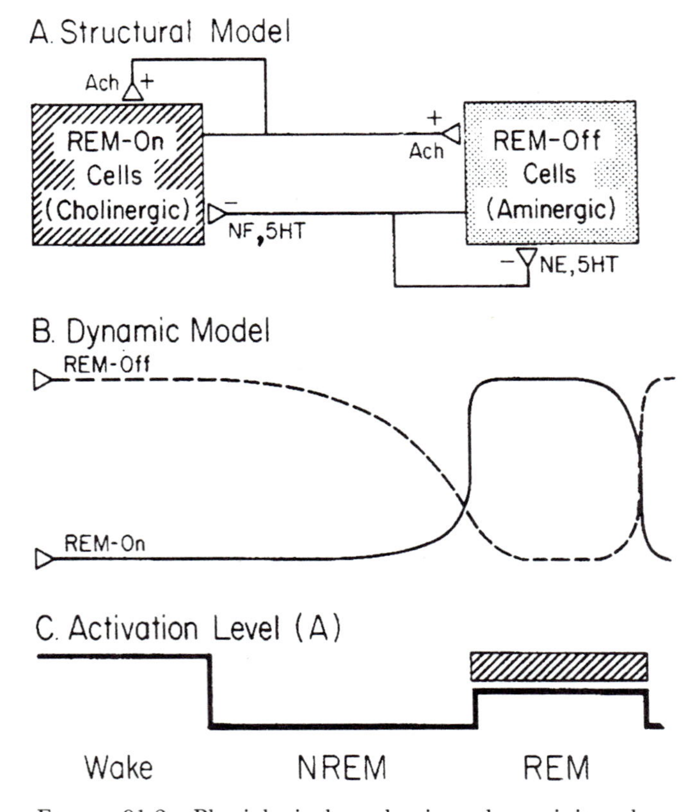

FIGURE 91.2 Physiological mechanisms determining alterations in activation level. (A) Structural model of reciprocal interaction. REM-on cells of the pontine reticular formation are cholinoceptively excited and/or cholinergically excitatory (Ach +) at their synaptic endings. Pontine REM-off cells are noradrenergically (NE) or serotonergically (5HT) inhibitory (−) at their synapses. (B) Dynamic model. During waking the pontine aminergic system is tonically activated and inhibits the pontine cholinergic system. During NREM sleep, aminergic inhibition gradually wanes and cholinergic excitation reciprocally waxes. At REM sleep onset, aminergic inhibition is shut off and cholinergic excitation reaches its high point. (C) Activation level. As a consequence of the interplay of the neuronal systems shown in A and B, the net activation level of the brain is at equally high levels in waking and in REM sleep and at about half this peak level in NREM sleep. (From Hobson, 1990, by permission.)

state parameters do not change simultaneously or instantaneously. This gradualism is relevant to the state carryover phenomenon that can be objectified by neurocognitive testing. At the cellular and molecular levels there is further evidence that the immediate post-REM waking period is a hybrid state with important REM residua due to a lag in the rise in the brain's concentration of norepinephrine and serotonin after activation of the cell bodies in the brain stem (Hobson and Steriade, 1986). While these slowly conducting, widely branching neuronal systems are well designed to perform the neurohormonal task of stabilizing state parameters, they are not characterized by abrupt and synchronous steplike changes in output. Subjectively, we notice this as the difficulty with orientation and attention on arousal from sleep. Depending on the time of night and the stage from which we are aroused, these carryover phenomena may last several minutes. Evidence of more specific state carryover from REM to waking is experienced in sleep paralysis, persistent anxiety, and the hypnopompic hallucinations that we sometimes experience upon awakening from intense dreams. Thus, the switch from sleep to waking may be better characterized as a gradual transition than an instantaneous event.

The phenomenology of sleep and dreaming

It is an indication of the confusion surrounding the definition of consciousness that many researchers still disagree on whether people are conscious or unconscious while sleeping. From our perspective, the presence in sleep of thoughts, images, and emotions is sufficient to characterize this sleep as altered consciousness rather than unconsciousness. What then are the formal differences between waking and sleeping consciousness? Behaviorally, there is an almost complete cessation of responses to external stimuli, and little evidence that subjects are normally aware of such stimuli. There are also qualitative changes in the nature of mentation, and it is these changes on which we have focused our attention. Normally, such studies are done in the sleep laboratory, where polysomnography allows the precise division of the sleep period into stages and substages (see figure 91.1 and our earlier discussion of sleep laboratory studies).

Our research has been aimed at rigorously and quantitatively measuring attributes of sleep mentation within the context of our neurophysiological model. For our purposes, the sleep lab is most valuable for distinguishing REM from non-REM sleep (classic dream imagery is consistently found to be more robust in REM sleep than in non-REM sleep), and sleep from wake. But these distinctions can now be made easily in the home, obviating the expense, discomfort, and inconvenience of the sleep laboratory. Furthermore, it is easier to collect large numbers of reports from subjects in the home, and the familiar surroundings result in reports that are less constrained and more naturalistic. Through the development of a new technology for monitoring sleep stages in the home and of new methodologies for analyzing narrative reports of sleep mentation, we have extended our understanding of sleep mentation and discovered possible links to the underlying neurophysiology of sleep.

Home-Based Sleep Studies Using the Nightcap Data collection in a natural home setting has been achieved using our portable Nightcap recording system (Mamelak and Hobson, 1989b). It combines the advantages of home-based, or free-ranging, subjective sleep-state monitoring with objective verification of brain state. By monitoring postural shifts (which demarcate human sleep at transitions between non-REM, REM, and wake states) and eye movements, this simple, portable sleep-state detector can score wake, REM, and non-REM sleep with high reliability (Mamelak and Hobson, 1989b; Ajilore et al., 1993). Eye movements are monitored by an adhesive-backed piezoelectric film that is attached to the upper eyelid.

The Nightcap contains the sensor transducers, a microprocessor chip, and a 32 Kbyte memory chip mounted in a small (12 cm × 7 cm × 2 cm) case that can be placed on the bed next to the subject or under the pillow, and can record up to 30 nights of data. This approach has already demonstrated sensitivity to subjective estimates of goodness of sleep (Hobson, Spagna, and Malenka, 1978; Ajilore et al., 1993), which might in turn be correlated with measurable aspects of cognitive capability. The Nightcap can also be used in conjunction with a Macintosh computer, which can then identify sleep states on a real-time basis and perform awakenings on a predesigned schedule. The same computer can then be programmed to test for state-dependent changes in cognition (see our later discussion of semantic priming and perceptual cuing).

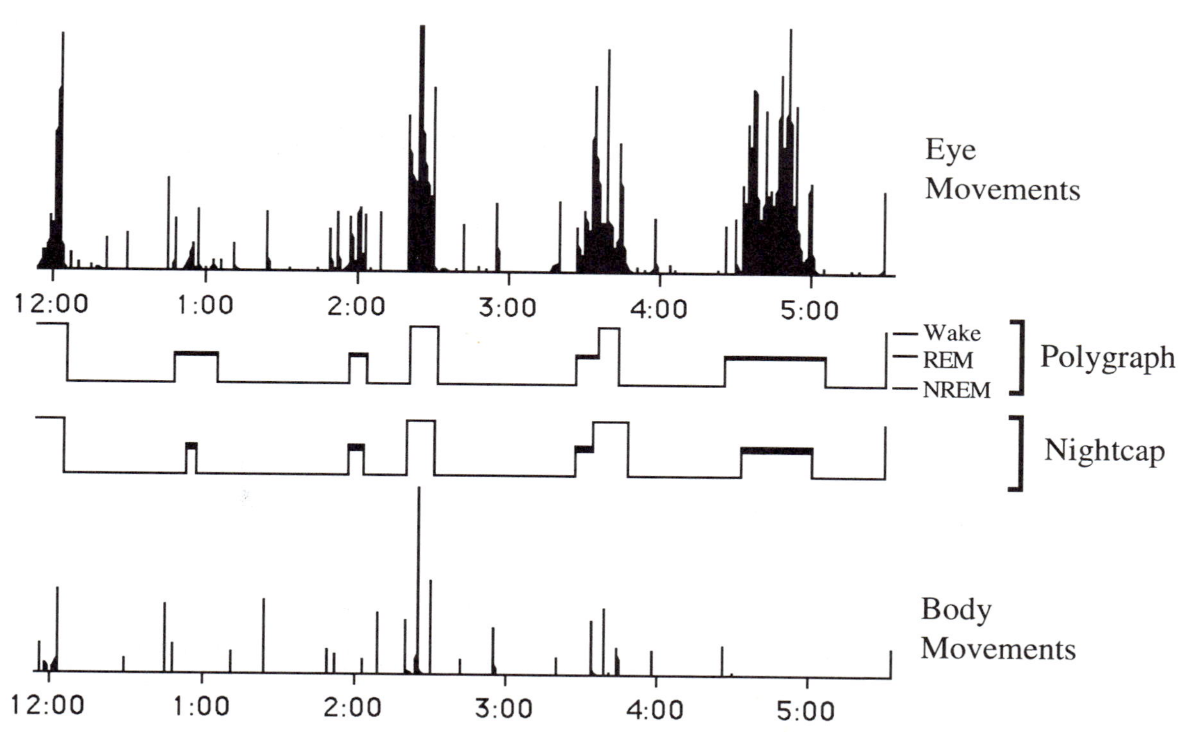

Figure 91.3 Sample Nightcap output and analysis. Top trace, histogram plot of Nightcap-detected eye movements; second trace, hypnogram representing the manually scored polysomnograph record; third trace, hypnogram of computer-scored Nightcap data; fourth trace, histogram plotting Nightcap-detected body movements. On all hypnograms, the top level represents wake, the second level represents REM, and the third level is non-REM. Periodic body movements are not scored. The lower axis indicates the time of night.

The example shown in figure 91.3 shows data from a subject who spent a night in the sleep laboratory wearing both the Nightcap and the standard array of electrodes for polysomnography. The hand-scoring of the polygraph record is displayed along with the Nightcap data. The Nightcap determination of sleep onset was within a minute of that made by the polysomnograph. All REM and non-REM periods, as well as two nocturnal awakenings, were accurately identified by automated computer analysis of the Nightcap data. During 30 nights of simultaneous Nightcap and polygraphic recording, 87% of all minutes were scored the same by the two methods. Agreement on individual nights varied from 76% to 97%.

Dreams Reported after Spontaneous Awakenings

An example of the use of the Nightcap is a study of spontaneously recalled dreams. In this study, 11 subjects each wore the Nightcap at home for 10 consecutive nights, dictating dream reports (or noting the absence of dreaming) after each awakening. During the 110 nights of recording, 239 reports were obtained, 194 of which could be unambiguously classified as to the sleep stage preceding the awakening.

Our first question was whether or not home-based mentation reports would show the same increases in duration and intensity during REM that have been reported in the laboratory. In fact, we found that report length differences were even greater in the home-based sample. The median length for REM reports was 148 words compared to only 21 words for non-REM reports, a ratio of 7 to 1. Similarly, the mean lengths were 317 words and 65 words, a ratio of 5 to 1 ($t = 4.98$, $df = 148$, $p < .001$). Almost half of the REM reports (44%) were over 200 words long, while only 10% of the non-REM reports exceeded that length ($\chi^2 = 20.3$, $df = 1$, $p < .001$). In addition, 24% of the REM reports were longer than 500 words, but only one of the non-REM reports ($\chi^2 = 16.0$, $df = 1$, $p < .001$) was that long.

Because we allowed the subjects to sleep and wake ad libitum, the 194 reports were naturally distributed

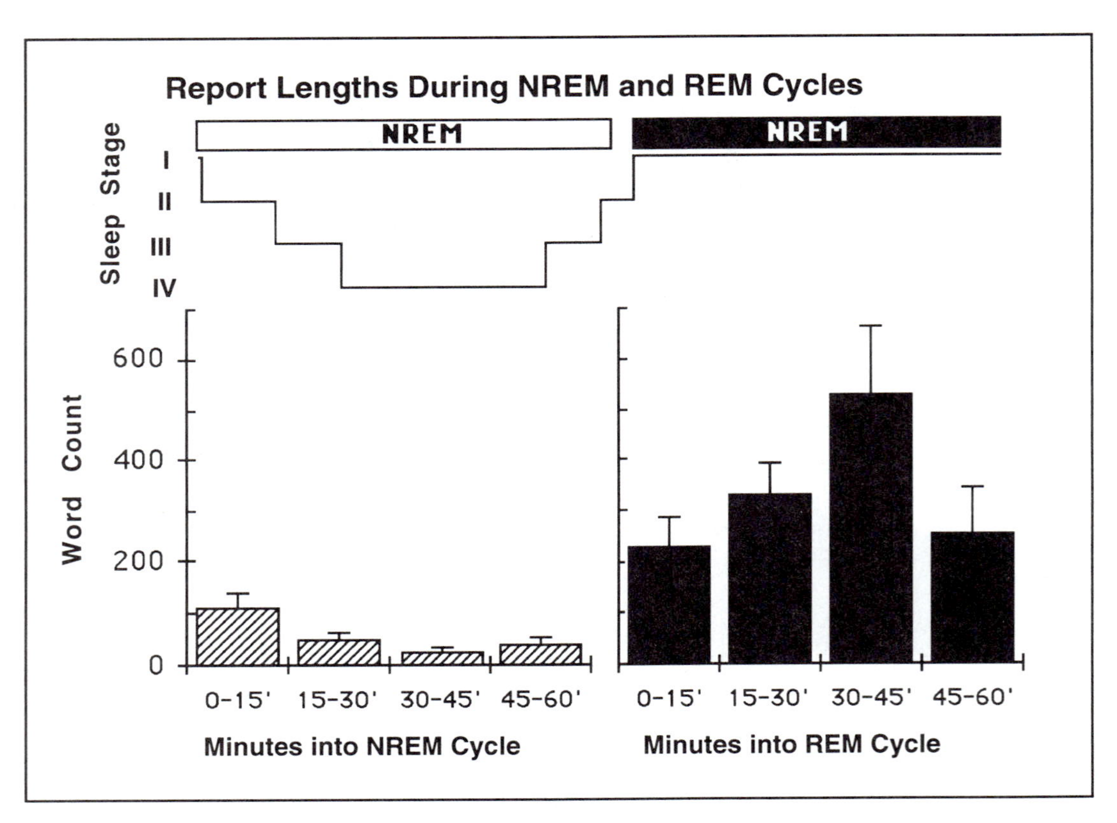

Figure 91.4 Top, schematic representation of normal sleep stage cycle; bottom, temporal distribution of report lengths in non-REM and REM cycles. Times indicate minutes after the start of a non-REM (left) or REM (right) period. Error bars indicate SEM for each sample.

over a wide range of elapsed times in REM and non-REM sleep. This allowed us to correlate report properites with phases of the sleep cycle. When the average length of dream reports was plotted as a function of time into a REM or non-REM cycle at which awakening occurred, a clear cycle was observed (figure 91.4). Non-REM reports were longest during the first 15 minutes after the end of a REM cycle, whereas REM reports were longest 30–45 minutes into a REM cycle. Thus it is clear that sleep mentation tracks the cortical activation level quite well in non-REM sleep, but that within REM, mentation has a more complex relationship to brain physiology, in which increasing REM density and greater REM duration are possible correlates of the increases in word count.

This study is a good example of the power of the Nightcap in studies of sleep mentation. In addition to being collected more rapidly and more easily, these reports are also significantly longer than those collected in the laboratory setting. When John Antrobus collected REM reports from 72 subjects in the sleep laboratory, the longest of the 72 REM reports was 254 words (Antrobus, personal communication). In contrast, 33 of our REM reports (38%) were over 250 words in length. It is our belief that the methodology that we have employed involving home monitoring provides a more accurate sample of normal sleep mentation.

Dream Bizarreness In addition to improving the methodology for sleep mentation sampling, we have also developed new analytic techniques aimed at providing more reliable, objective measures of dream content. Our recent work has focused primarily on the characterization of dream bizarreness. The foundation for this work was laid in the activation-synthesis model of dreaming (Hobson and McCarley, 1977), which proposed that dream bizarreness was a product of REM state physiology. Subsequently, we have worked to quantify dream bizarreness (McCarley and Hoffman, 1981; Hobson et al., 1987; Mamelak and Hobson, 1989a; Williams et al., 1992).

We have outlined three qualitative characteristics of dream bizarreness: incongruity (mismatching features of characters, objects, actions, or settings), discontinuity (sudden changes in these features, resulting in interruptions of orientational stability), and cognitive uncertainty (explicit vagueness) (Hobson et al., 1987). Williams and colleagues (1992) further demonstrated that discontinuity chiefly affects three loci of dreams—character, object, and plot—and comparison of home dream reports with waking fantasies showed that bizarreness is significantly more common in dream reports than in reports of fantasies. Discontinuity was the most state-specific class of bizarreness, being six times more frequent in dreams than in fantasies.

Dream bizarreness has been described and measured by others using various descriptive scales (e.g., Snyder, 1970; Giora, 1981; Hunt, 1989; Antrobus, 1991; Bonato et al., 1991). These scales generally examine improbable and impossible events, transformations, and scene shifts. While some have focused on scene shifts (e.g., Mamelak and Hobson, 1989a, 1989b; Antrobus, 1991), none have examined discontinuity in detail and few have divided it into specific classes (e.g., McCarley and Hoffman, 1981; Williams et al., 1992).

We have defined cognitive discontinuity as "a rapid transition from one thought, action, image, or setting to a completely unrelated one" (Mamelak and Hobson, 1989a). To explain the dramatic discontinuities seen in dream scene shifts we have proposed that chaotically generated internal signals, generated by the activated brain in REM sleep, result in state-related bifurcations in neuronal firing patterns. Attempts by higher cortical systems to integrate the signals into the ongoing dream plot result in bizarre discontinuities. We have postulated that the underlying changes in brain physiology are twofold. First, phasic bursts of activity in the form of ponto-geniculo-occiptal waves provide the chaotic internal signals (Calloway et al., 1987; Kahn and Hobson, 1991a). Second, the chronic absence of aminergic modulation of cortical systems decreases the stability of neural networks and increases the probability of bifurcations in firing patterns (Mamelak and Hobson, 1989a). In a recent theoretical paper, the dynamic interaction between chaos and self-organization has been applied to the activation-synthesis model of the dreaming brain (Kahn and Hobson, 1991a, 1991b).

It is the powerful link between the basic neurophysiology of the brain and the formal characteristics of mental experience that has encouraged us to study dream bizarreness as an embodiment of state-dependent aspects of consciousness and cognition. Specifically, we wish to ask how the top-down influences of higher cortical association nets interact with chaotic, bottom-up brainstem signals in producing the imagery found in dreams. As a first step, we have developed analytic techniques that allow us to objectively characterize dream discontinuities. Our results suggest that character and object transformations follow specific rules, but that these rules do not apply to coarser-grained scene and plot transformations. These bizarreness transformations appear to reflect orientational instabilities whose varying degrees of intensity can be directly and accurately assessed using a graph theory technique (as discussed in a subsequent section of this chapter).

Object and Character Transformations If bizarre discontinuities result from attempts by higher cortical centers to incorporate chaotic brainstem signals into ongoing dreams, how successful are these higher centers at creating cognitive order out of neuronal chaos? Analyzing reports of transformations in dreams can provide insights into the range of semantic networks activated during REM sleep. It was our initial belief that the integration was largely unsuccessful, with the transformed object being largely unrelated to the pretransformation object. But two studies have convinced us that there is a sort of transformative grammar that constrains the types of transformations that can occur. In both studies, transformations of inanimate objects and characters were studied. Consider the following examples.

Table 91.1 shows the combined results of these studies. In a total of 453 home dream reports collected from 45 subjects, 44 characters and inanimate objects were found to undergo transformations (Rittenhouse, Stickgold, and Hobson, 1991, 1994). Of these transformations, 80% were characterized as "intra-class transformations," consisting of one inanimate object transforming into another or one character into another. The remaining 20% involved transformation of inanimate objects into animate objects (e.g., a rope into a snake) and characters transforming into animate objects (e.g., a man into a shark). No transformations were observed where the object transformed was an animate object. More importantly, no transformations of inanimate objects into characters—or vice versa—were observed. These findings suggest that while objects and characters can go through physically impossible transformations, some abstract categorical features of the original items are usually maintained across the transformation.

Assessing Associative Relationships between Paired Cognates We have recently begun an attempt to determine whether object transformations follow even stricter rules (Rittenhouse et al., 1994). Examples such as a rope turning into a snake suggest that there is a strong thematic continuity from the initial object to the final object. In the past, no objective manner of studying this question has been available. While one could ask judges to rate the "coherence" of the initial object–final object pair, such judgments are subjective post hoc analyses, unavoidably colored by one's knowledge that the transformation did occur and one's own beliefs about such coherence. This problem of subjective text analysis has plagued all attempts at objective dream interpretation, and is a topic of active concern in areas of philosophy, linguistics, and literary criticism as well (see Eco, 1992).

In one recent study, we have eliminated this subjective, post hoc paradigm for text analysis, and have developed a pair-matching protocol. Eleven object transformations were identified in a set of dream reports. In each case, an initial object (labeled A1 through A11 for the 11 transformations) was described as suddenly changing into another object (B1 through B11). We then put A1 through A11 in one column on a score sheet, and B1 through B11 (in a randomized order) in a second column. If there is no identifiable connection between the initial and final objects of a

Table 91.1
Object and character transformations in dreams

		Transformed form		
		Inanimate object	Animate object	Character
Transformed into	Inanimate	21	—	—
	Animate object	2	—	7
	Character	—	—	14

Values indicate number of instances identified for each category of transformation.

transformation, then judges should be unable to pair items in column A with their matches in column B. Random guessing (the null hypothesis) would yield one correct match out of 11 (9%). On the other hand, if there are identifiable relationships between the initial and final objects, then judges could conceivably match as many as all of the 11 pairs (100%).

When 7 judges attempted to match the 11 pairs, 78% of the attempted matchings were correct, with 6 of the 11 pairs correctly identified by 6 of the 7 judges ($p < .0001$ for each pair), 4 of the 11 pairs correctly identified by 5 of the 7 judges ($p < .002$ for each pair), and 1 pair correctly identified by 4 of the 7 judges ($p < .02$). Thus, each of the 11 pairs was scored correctly at a greater than random rate. In other words, transformed objects bear an easily identifiable relationship to the object from which they were transformed.

Dream Plot Transformations A more dramatic and radical type of bizarre discontinuity occurs in dreams when there is a scene and plot shift. Scene and plot discontinuities are characterized by, respectively, an abrupt change in location and an abrupt change in thoughts, emotions, and actions, not obviously linked to the previous portion of the report. A subset of these reports can be identified that have "complete discontinuities," defined by the following criteria: the location of the events changes; no character (except the dreamer) is reported at both locations; no object is reported at both locations; and no reference to one location, its characters, or its objects is made at the other location, In a recent study (Stickgold, Pace-Schott, and Hobson, 1994a), a total of 25 (18%) of the 139 reports collected from 16 subjects, had scene or plot discontinuities. In 14 of these reports, from 7 students, the discontinuities were considered complete.

As in the case of the character and object discontinuities, we want to determine whether the pretransformation scene and plot constrain the transformed scene and plot. Put another way, does the overall dream retain a thematic coherence despite the discontinuity of scene and plot? Again, we sought to determine objectively whether there was any identifiable continuity between the two portions of the dream (Stickgold, Rittenhouse, and Hobson, 1991, 1994b). To answer this question, 14 dream reports with scene and plot discontinuities were collected. Seven of these reports were cut in half at the discontinuities, and spliced together with segments from other reports. The other 7 were left intact. The 14 spliced and intact dreams were placed in a random order, and judges were asked to determine which reports were spliced and which were intact.

To our surprise, the 13 judges performed only slightly better than chance, scoring reports correctly on only 59% of attempts (108 out of 182 correct; $p < .01$, binomial test, single-tailed). Individual judges showed accuracies of 43% to 79%, and none of the judges scored significantly better than chance ($p > .2$, binomial test, single-tailed, for each judge). While several reports showed higher than chance scores, it is clear that at least two-thirds to three-quarters of the reports lacked sufficient continuity to be reliably identified as intact or spliced.

This study and the pair-matching study described above, represent a new and powerful means of analyzing texts for coherence, and for the first time make the objective study of this important feature of dream reports possible. By comparing these features in the wake and sleep states, we expect to add important information on the state-dependent aspects of consciousness and cognition.

Orientation Mapping Not all shifts in orientation seen in dream reports are bizarre. Indeed, most are normal and appropriate to the narrative evolution of the dream plot. Because dreaming is a state of consciousness that proceeds independently of sensory input and with altered neuromodulation of the brain, it is of great interest to examine the nature of orientational and attentional shifts in dream reports. For example, do dream reports show more frequent shifts in orientation than reports of waking experience, or are the shifts more dramatic? To answer these questions, we have begun to develop a method of mapping orientation shifts, using a mathematical technique known as graph theory (Sutton, Beis, and Trainor, 1988; Sutton, Mamelak, and Hobson, 1991; Sutton et al., 1994a,b). In this approach, we create graphical representations of plot and scene developments within narrative reports (Sutton et al., 1994a,b). Spatial features and interrelationships between characters, objects, and locations are formally mapped from the narrative description onto a treelike diagram, termed a hierarchy graph. Figure 91.5 shows the graphical representation of a portion of the following dream report:

Last night I was back in the *house* where I grew up, but the house was very different in the decoration and size. My

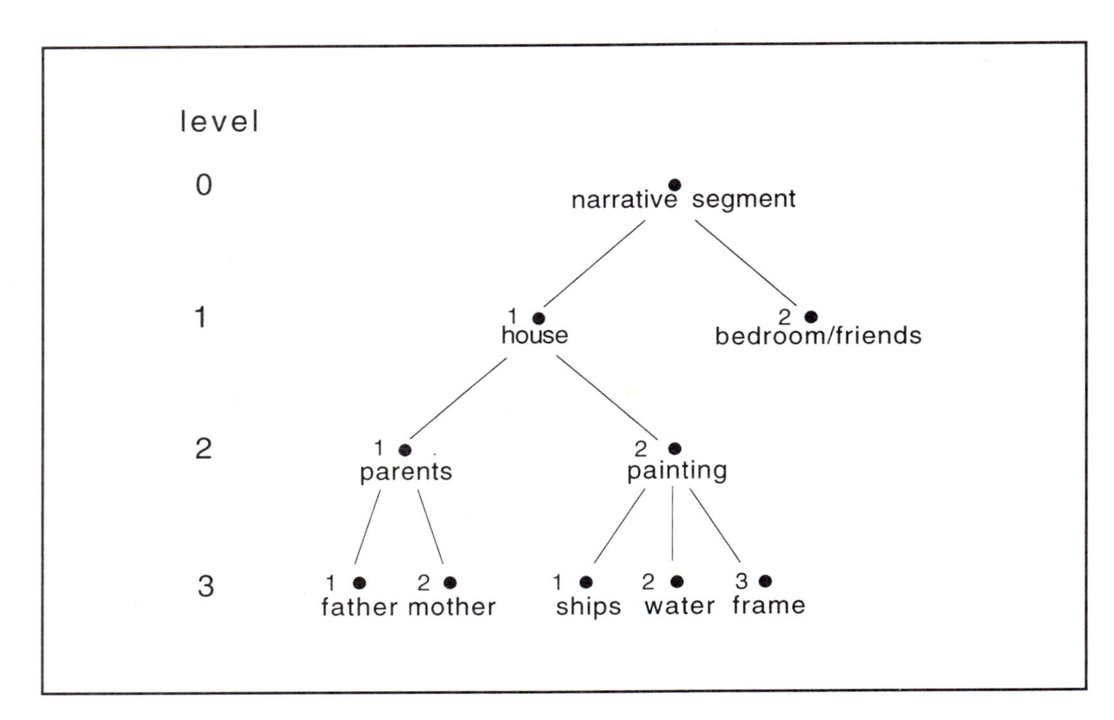

FIGURE 91.5 Hierarchy graph of example narrative. Each level of the graph is marked, and within each level the nodes are labeled numerically from left to right. A unique address label is assigned to each node. The label is determined by the pathway from the root node to the node in question. For example, the node representing *frame* has the address label (1, 2, 3).

parents were fighting about a *painting* which my *mother* said my *father* ruined when he added two *ships* to the painting. The painting was very vivid in my memory when I awoke. The ships were black and red among the deep blue *water* surrounded by a gold painted *frame*. In the same dream I was in my *bedroom*, which was not my bedroom at home, when my *friends* from work came into my room. We all sat under the canopy on my bed. . . .

Features are assigned nodal positions within the graph and are also assigned numerical values based solely on their position within the graph structure. The narrative is then transformed into a sequence of nodal transitions and each transition is assigned a length, d, equal to the weighted sum of the values of the intervening nodes. From the neurocognitive viewpoint, a significant aspect of this approach is that it captures and measures shifts in the dreamer's visual attention as dream consciousness unfolds, without reference to external stimuli.

Using this methodology, a sample of dreams has been mapped onto hierarchy graphs and each dream reduced to a sequence of transitions, each with its own length, d. When occurrences of subject-reported emotion were correlated with these transitions, 88% (35 out of 40) of transitions with $d \geq 1$ had associated emotion, whereas only 25% (26 out of 105) of transitions with $d < 1$ were associated with emotion. These preliminary results suggest that plot transitions associated with high d-values are correlated with emotional experiences in dreams (Sutton et al., 1994b). It is our hope that such mathematical reductions of state-dependent cognitive structures can be of more general use in quantifying cognitive processes.

A neurocognitive approach

Many of the sleep and dream scientists who have continued their research despite the decline of the original sleep and dream lab paradigm now share the goal of establishing a cognitive neuroscience of brain-mind states. Thus, for example, John Antrobus (1991) is modeling mentation across states, using a nonquantitative parallel processing approach that is informed by neurobiology as well as by the data of the sleep lab. Antrobus has also introduced important new controls of the wake state conditions he compares with sleep (see Reinsel, Antrobus, and Wollman, 1991).

Inspired by the tonic-phasic model of Molinari and Foulkes (1969) as well as by the activation-synthesis model, Martin Seligman and Amy Yellen (1987) have added consideration of emotional evaluation to the concepts of primary visual activation and secondary cognitive elaboration. The result is a more comprehensive and more descriptive approach to some of the very same dream features that interested Freud but which obviate the ad hoc psychology of psychoanalytic dream theory. Even David Foulkes, while eschewing any reference to neuroscience, emphasizes the importance of the cognitive approach (Foulkes, 1985).

Semantic Priming and Perceptual Cuing A major aim of our current work is to develop a cognitive methodology that will prove sensitive to changes in conscious state and their functional significance. Recognizing the many limitations of the self-report method, we hope to enrich and solidify our assessment of state-dependent aspects of cognition by using behavioral techniques. Because the waking and dreaming states differ attentionally as well as orientationally, we have begun to test aspects of this cognitive function using semantic priming (Meyer, Schvaneveldt, and Ruddy, 1974; Neely, 1977; Farah, 1989) and perceptual cuing (Posner, 1980) tasks. In our preliminary studies, surprisingly strong carryover effects have been demonstrated with both tests, indicating that the underlying brain state from before the awakening influences subsequent performance. Reaction times are generally longer for all experimental conditions following arousal from sleep. However, subjects show enhanced accuracy in distinguishing words and nonwords when performing the priming task following arousal from REM (Spitzer et al., 1991). Similarly, REM sleep arousals are followed by a trend toward reduced costs in performance on the cuing test (Doricchi et al., 1991a,b).

The theoretical reasons for beginning our behavioral studies with tests of semantic priming and perceptual cuing are based on the supposition that both are sensitive to changes in state-dependent aspects of orienting and orientation. The perceptual cuing task tests the responsiveness of the posterior orienting system, while the semantic priming task tests those more frontal networks (Posner et al., 1988; Posner and Peterson, 1990) that are involved in orientation in the mental status sense. Because these two tests have specifically localized cortical correlates in PET studies (Posner et al., 1988), the chances of developing a hypothesis-driven approach to state-dependent aspects of cognition, using PET and related imagery techniques, are augmented.

Mediation of Cognitive Test Differences We believe that both the posterior orienting and frontal orientation systems encompass the same networks that we have shown to be affected by both aminergic demodulation and cholinergic autostimulation (by eye movement direction signals) in REM sleep. This leads to the hypothesis that the state-dependent changes in orientation that we have observed may be related to specifiable changes in the brain's orienting control system. Insofar as semantic priming and perceptual cuing share common attentional mechanisms, we might expect to see similar state-dependent effects for the two tests. To the degree to which they also activate different cognitive modules (e.g., location vs. identification), as suggested by Farah (1989), we might expect differential effects from state changes. We schematize the interrelationships among these approaches in figure 91.6.

The semantic priming test has an additional advantage: It has both an automatic aspect (by which a related prime results in a faster response to the target word) and an attentional aspect (by which an unrelated word prime results in a slower response than a nonword prime). As a result, state-dependent differences would be expected if REM sleep increases the activation of associative networks or decreases the effort required to shift attention. Thus, this test might both measure and dissociate the effects of state upon memory and attentional mechanisms.

Schvaneveldt and McDonald (1981) have suggested that lexical decisions involve two stages, an early semantic-encoding stage and a late word-comparison stage. Several workers (Samuel, 1981; Schvaneveldt and McDonald, 1981; Johnston and Hale, 1984) have argued that priming of the first of these stages is accomplished by an increase in bias (β in signal-processing terminology), while priming of the later stage involves an increase in sensitivity (the d' parameter of signal processing). Using variations of the semantic-priming paradigm presented below, state-dependent aspects of the early and late stages could be isolated and identified. These differences are relevant to studies of sleep mentation, since preliminary analyses of bizarre uncertainties (Hobson et al., 1987) indicate severe abnormalities in object and character identification suggestive of dramatic shifts in bias.

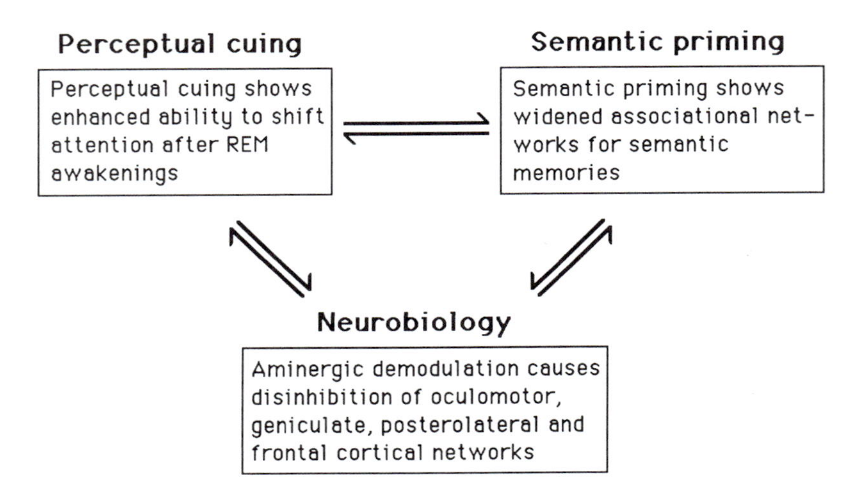

FIGURE 91.6 Paradigm for awakening studies of semantic priming and perceptual cuing.

We attach special importance to our preliminary observations with these two cognitive tests for several reasons. One is that state-of-the-art models for the brain basis of attention (Posner and Peterson, 1990) posit a noradrenergic modulatory influence upon the anterior (frontal) and posterior (parieto-occipital) cortical subsystems of attention. Because our model specifies marked state-dependent differentiation in noradrenergic input to the cortex, it may become possible to demonstrate common elements in the two models. We note that a leading hypothesis of how norepinephrine may modulate the cortex, so as to facilitate attention, holds that it increases the signal-to-noise ratio (Foote, Bloom, and Aston-Jones, 1983). Because we have already begun to explore the effects of changes in noradrenergic modulation upon the behavior of neural nets and have demonstrated decreases in signal-to-noise ratios in simulated REM sleep (Mamelak and Hobson, 1989a), we can offer realistic suggestions about how the state-dependent changes in cognition may be mediated at the neuronal level. More important, we believe our approach may ultimately lead us to hypothesis-testing experiments on the effects of various sleep states on attentional function (and vice versa).

Summary and conclusions

Because we can record brain activity at the level of cells and molecules (in animals) and measure concomitant cognitive functions in humans, it is both feasible and desirable to study the neurocognitive differences between waking, sleeping, and dreaming. By focusing upon the precise ways in which these states change, it is possible to develop a preliminary mapping between microscopic and ubiquitous neurobiological events and consciousness. We call our strategies and findings the conscious state paradigm.

To obtain more naturalistic phenomenological data on the nature of conscious experience we have developed the Nightcap home-based sleep recording system, which reliably detects wake, non-REM, and REM sleep using eye and head movement parameters. The Nightcap can now be interfaced with a Macintosh computer, which can analyze the sleep data, perform awakenings, and even run cognitive tasks following programmed arousals. When subjects awaken spontaneously at home, they are often able to give more detailed descriptions of their antecedent conscious experiences then when awakened in the sleep lab. Using data obtained in this way we have begun an intensive effort to identify and quantify such distinctive aspects of dream cognition as discontinuity and incongruity of persons, objects, settings, and actions. A novel method of narrative graphing has been developed to quantify orientational instability, and should prove valuable to the study of other conscious states.

Preliminary results suggest that performance on both the semantic priming and perceptual cuing tests are dramatically state dependent. On the priming task, subjects made fewer errors on arousal from REM then during either presleep, wake, or non-REM trials. And despite the fact that reaction times are generally longer for perceptual cuing following arousal from all stages of sleep, the post-REM trials show a trend toward reduced costs. The main obstacle still to be overcome in these studies is sleep inertia. (the difficulty that subjects experience in giving reports or performing tests on arousal from sleep).

Taken together, our new results suggest that compared to waking, the REM-sleep or dreaming brain is disinhibited, thus enhancing associative memory and attentional shifts. The molecular basis of the cognitive changes may be cholinergic activation and aminergic demodulation. The behavioral studies are complemented by quantitative analysis of report content from subjects who awaken spontaneously from sleep at home. The data indicate that while subjects frequently and rapidly reorient their visual attention while dreaming, they simultaneously experience a relaxation of the rules of associative memory. We conclude that consciousness depends upon the chemical microclimate of neuronal networks throughout the forebrain. Particularly critical is the balance between aminergic and cholinergic neuromodulators secreted in the forebrain by neurons whose cell bodies are in the pontine brain stem.

ACKNOWLEDGMENTS Supported by grants from the National Institutes of Health (MH 13-923 and MH 48-832) and the Mind-Body Network of the MacArthur Foundation.

REFERENCES

AJILORE, O., R. STICKGOLD, C. RITTENHOUSE, and J. A. HOBSON, 1993. Assessment of good and poor sleepers using the Nightcap home-based sleep monitor. *Sleep Res.* 22.

ANTROBUS, J., 1991. The neurocognition of sleep mentation: Rapid eye movements, visual imagery, and dreaming. In *Sleep and Cognition*, R. Bootzin, J. Kihlstrom, and D. Schacter, eds. Washington D.C.: American Psychological Association, pp. 3–24.

ARKIN, A., J. ANTROBUS, and S. ELLMAN, eds. 1978. *The Mind in Sleep*. Hillsdale, N.J.: Erlbaum.

ASERINSKY, E., and N. KLEITMAN, 1953. Regularly occurring periods of ocular motility and concomitant phenomena during sleep. *Science* 118:361–375.

BADIA, P., 1991. Memories in sleep: Old and new. In *Sleep and Cognition*, R. Bootzin, J. Kihlstrom, D. Schacter, eds. Washington D.C.: American Psychological Association, pp. 67–76.

BONATO, R. A., A. R. MOFFITT, R. F. HOFFMANN, M. A. CUDDY, and F. L. WIMMER, 1991. Bizarreness in dreams and nightmares. *Dreaming* 1:1, 63–61.

CALLAWAY, C. W., R. LYDIC, H. A. BAGHDOYAN, and J. A. HOBSON, 1987. Ponto-geniculo-occipital waves: Spontaneous visual system activation occurring in REM sleep. *Cell. Mol. Neurobiol.* 7:105–149.

DEMENT, W., 1958. The occurrence of low voltage, fast, electroencephalogram patterns during behavioral sleep in the cat. *EEG Clin. Neurophysiol.* 10:291–296.

DEMENT, W., 1960. Dream deprivation. *Science* 131:1705.

DINGES, D. F., 1990. Are you awake? Cognitive performance and reverie during the hypnopompic state. In *Sleep and Cognition*, R. Bootzin, J. Kihlstrom, and D. Schacter, eds. Washington D.C.: American Psychological Association, pp. 159–178.

DORICCHI, F., C. IPPOLITI, P. BRAIBANTI, C. VIOLANI, J. A. HOBSON, and M. BERTINI, 1991. Behavioral states and selective visual attention: preliminary evidence for a deficit in maintainance of attentional focus upon REM awakenings. *Sleep Res.* 20:140.

ECO, U., 1992. Overinterpreting Texts. In *Interpretation and Overinterpretation*, S. Collini, ed. Cambridge: Cambridge University Press.

FARAH, M. J., 1989. Semantic and perceptual priming: how similar are the underlying mechanisms? *J. Exp. Psychol.* 15:188–194.

FLICKER, C., R. W. MCCARLEY, and J. A. HOBSON, 1981. Aminergic neurons: State control and plasticity in three model systems. *Cell. Mol. Neurobiol.* 1:123–166.

FOOTE, S. L., F. E. BLOOM, and G. ASTON-JONES, 1983. Nucleus locus ceruleus: New evidence of anatomical and physiological specificity. *Physiol. Rev.* 63:844–914.

FOULKES, D., 1985. *Dreaming: A Cognitive-Psychological Analysis*. Hillsdale, N.J.: Erlbaum.

GIORA, Z., 1981. Dream styles and the psychology of dreaming. *Psychoanal. Contemporary Thought*, 4:291–381.

HERMAN, J., 1989. Transmutative and reproductive properties of dreams: Evidence for cortical modulation of brain stem generators. In *The Neuropsychology of Dreaming Sleep*, J. Antrobus, M. Bertini, eds. Hillsdale, N.J.: Erlbaum.

HOBSON, J. A., 1988. *The Dreaming Brain*. New York: Basic Books.

HOBSON, J. A., 1989. *Sleep*. New York: Scientific American Library.

HOBSON, J. A., 1990. Activation, input source, and modulation: A neurocognitive model of the state of the brain-mind. In *Sleep and Cognition*, R. Bootzin, J. Kihlstrom, and D. Schacter, eds. Washington, D.C.: American Psychological Association, pp. 25–40.

HOBSON, J. A., S. A. HOFFMAN, R. HELFAND, and D. KOSTNER, 1987. Dream bizarreness and the activation-synthesis hypothesis. *Hum. Neurobiol* 6:157–164.

HOBSON, J. A., and R. W. MCCARLEY, 1977. The brain as a dream state generator: An activation-synthesis hypothesis of the dream process. *Am. J. Psychiatry* 134:1335–1348.

HOBSON, J. A., R. W. MCCARLEY, and P. W. WYZINKI, 1975. Sleep cycle oscillation: reciprocal discharge by two brainstem neuronal groups. *Science* 189:55–58.

HOBSON, J. A., T. SPAGNA, and R. MALENKA, 1978. Ethology of sleep studied with time-lapse photography: Postural immobility and sleep-cycle phase in humans. *Science* 201: 1251–1253.

Hobson, J. A., and M. Steriade, 1986. The neuronal basis of behavioral state control. In *Handbook of Physiology*. Sec. 1, *The Nervous System*. Vol. 4, *Intrinsic Regulatory Systems of the Brain*, V. B. Mountcastle, ed., Md.: American Physiological Society, pp. 701–823.

Hunt, H., 1989. *The Multiplicity of Dreams*. New Haven: Yale University Press.

Johnston, J. E., and B. L. Hale, 1984. The influence of prior context on word identification: Bias and sensitivity effects. In *Attention and Performance* X, H. Bouma, and D. Bouwhis, eds. Hillsdale, N.J.: Erlbaum, pp. 243–255.

Jouvet, M., 1962. Recherche sur les structures nerveuses et les mechanismes responsables des differentes phases du sommeil physiologique. *Arch. Ital. Biol.* 100:125–206.

Jouvet, M., and F. Michel, 1959. Correlation electromyographiques du sommeil chez le chat decortique et mesencephalique chronique. *C. R. Soc. Biol.* 153:422–425.

Kahn, D., and J. A. Hobson, 1991a. Pontogeniculooccipital (PGO) waves produce noise-induced transitions in REM sleep. *Sleep Res.* 20:142.

Kahn, D., and J. A. Hobson, 1991b. Self-organization theory of dreaming. *Sleep Res.* 20:141.

Kales, A., T. Tan, E. Kollar, P. Naitoh, T. A. Preston, and E. J. Malmstrom, 1970. Sleep patterns following 205 hours of sleep deprivation. *Psychosom. Med.* 32(2):189–200.

LaBerge, S., 1985. *Lucid Dreaming*. Los Angeles: Tarcher.

Lavie, P., and D. Sutter, 1975. Differential responding to the beta movement after waking from REM and non-REM sleep. *Am. J. Psychol.* 88:595–603.

Lubin, A., D. Hord, M. L. Tracy, and L. C. Johnson, 1976. Effects of exercise, bedrest and napping on performance decrement during 40 hours. *Psychophysiology* 13: 334–339.

Mamelak, A. N., and J. A. Hobson, 1989a. Dream bizarreness as the cognitive correlate of altered neuronal behavior in REM sleep. *J. Cognitive Neurosci.* 1:201–22.

Mamelak, A. N., and J. A. Hobson, 1989b. Nightcap: A home-based sleep monitoring system. *Sleep* 12:157–166.

McCarley, R. W., and J. A. Hobson, 1975. Neuronal excitability modulation over the sleep cycle: A structural and mathematical model. *Science* 189:58–60.

McCarley, R. W., and J. A. Hobson, 1977. The neurobiological origins of psychoanalytic theory. *Am. J. Psychiatry* 134:1211–1221.

McCarley, R. W., and E. Hoffman, 1981. REM sleep dreams and the activation-synthesis hypothesis. *Am. J. Psychiatry* 138:7.

McCarley, R. W., J. W. Winkelman, and F. H. Duffy, 1983. Human cerebral potentials associated with REM sleep rapid eye movements: Links to PGO waves and waking potentials. *Brain Res.* 274:359–364.

Meyer, D. E., R. W. Schvaneveldt, and M. G. Ruddy, 1974. Functions of graphemic and phonemic codes in visual word-recognition. *Mem. Cognition* 2:309–321.

Molinari, S., and D. Foulkes, 1969. Tonic and phasic events during sleep: Psychological correlates and implications. *Percept. Mot. Skills* 29:343–368.

Moruzzi, G., and H. W. Magoun, 1949. Brainstem reticular formation and activation of the EEG. *Electroencephalogr. Clin. Neurophysiol.* 1:455–473.

Moskowitz, E., and R. J. Berger, 1969. Rapid eye movements and dream imagery: Are they related? *Nature* 224(5219):613–614.

Neely, J. H., 1977. Semantic priming and retrieval from lexical memory: Roles of inhibitionless spreading activation and limited-capacity attention. *J. Exp. Psychol. [Gen.]* 106:226–254.

Posner, M. I., 1980. Orienting of attention. *Q. J. Exp. Psychol.* 32:3–25.

Posner, M. I., and S. E. Petersen, 1990. The attention system of the human brain. *Annu. Rev. Neurosci.* 13:25–42.

Posner, M. I., S. E. Petersen, P. T. Fox, and M. E. Raichle, 1988. Localization of cognitive operations in the human brain. *Science* 240:1627–1631.

Rechtschaffen, A., B. M. Bergmann, C. A. Everson, C. A. Kushida, and M. A. Gilliland, 1989. Sleep deprivation in the rat: X. Integration and discussion of the findings. *Sleep* 12:68–87.

Reinsel, R., J. Antrobus, and M. Wollman, 1991. Bizarreness in waking and sleep mentation: Waking, REM-non-REM, and phasic-tonic differences in bizarreness. In *The Neuropsychology of Dreaming*, J. Antrobus and M. Bertini, eds. Hillsdale, N.J.: Erlbaum.

Rittenhouse, C., O. Ajilore, R. Stickgold, and J. A. Hobson, 1993. State dependent aspects of attention: Differences in reaction time on an exogenous cuing task. *Sleep Res. Abstr.* 22:96.

Rittenhouse, C., D. Broadley, R. Stickgold, and J. A. Hobson, 1993. Increased semantic priming upon awakenings from REM sleep. *Sleep Res. Abstr.* 22:97.

Rittenhouse, C., R. Stickgold, and J. A. Hobson, 1991. Dream discontinuity: Insertion, transformation, and removal. *Sleep Res.* 20:149.

Rittenhouse, C., R. Stickgold, and J. A. Hobson, 1994. Constraint on the transformation of characters, objects, and settings in dream reports. *Consciousness Cognition* 3:100.

Roffwarg, H. P., W. Dement, J. Muzio, and C. Fisher, 1962. Dream imagery: Relationship to rapid eye movements of sleep. *Arch. Gen. Psychiatry* 7:235–258.

Samuel, A. G., 1981. Phonemic restoration: Insights from a new methodology. *J. Exp. Psychol. [Gen.]* 110:474–494.

Schvaneveldt, R. W., and J. E. McDonald, 1981. Semantic context and the encoding of words: Evidence for two modes of stimulus analysis. *J. Exp. Psychol. [Hum. Percept.]* 7:673–687.

Seligman, M. E. P., and A. Yellen, 1987. What is a dream? *Behav. Res. Ther.* 25:1–24.

Snyder, F., 1970. The phenomenology of dreaming. In *The Psychodynamic Implications of the Physiological Studies on Dreams*, L. Madow and L. Snow, eds. Springfield, Ill.: Thomas, pp. 124–151.

Spitzer, M., A. Mamelak, R. Stickgold, J. Williams, W. Koutstall, C. Rittenhouse, B. A. Maher, and J. A. Hobson, 1991. Semantic priming in a lexical decision task on awakenings from REM-sleep: Evidence for a disinhibited semantic network. *Sleep Res.* 20:131.

Stickgold, R., E. Pace-Schott, and J. A. Hobson, 1994a. A New paradigm for dream research: Mentation reports following spontaneous arousal from REM and NREM sleep recorded in a home setting. *Consciousness Cognition* 3:16.

Stickgold, R., C. Rittenhouse, and J. A. Hobson, 1991. Dream coherence: Is beauty in the eye of the beholder? *Sleep Res.* 20:150.

Stickgold, R. A., C. D. Rittenhouse, and J. A. Hobson, 1994b. Dream splicing: A new technique for assessing thematic coherence in subjective reports of mental activity. *Consciousness Cognition* 3:114.

Stones, M. J., 1977. Memory performance after arousal from different sleep stages. *Br. J. Psychol.* 68:177–181.

Sutton, J. P., J. S. Beis, and L. E. H. Trainor, 1988. Hierarchical model of memory and memory loss. *J. Physiol. [Math. Gen.]* 21:4443–4454.

Sutton, J. P., A. N. Mamelak, and J. A. Hobson, 1991. Neural network model of state-dependent sequencing. *Sleep Res.* 20:151.

Sutton, J. P., C. D. Rittenhouse, E. Pace-Schott, R. Stickgold, and J. A. Hobson, 1994a. A new approach to dream bizarreness: Graphing continuity and discontinuity of visual attention in narrative reports. *Consciousness Cognition* 3:61

Sutton, J. P., C. D. Rittenhouse, E. Pace-Schott, J. Merritt, R. Stickgold, and J. A. Hobson, 1994b. Emotion and visual imagery in dream reports: A narrative graphing approach. *Consciousness Cognition* 3:89.

Williams, J., J. Merritt, C. Rittenhouse, and J. A. Hobson, 1992. Bizarreness in dreams and fantasies: Implications for the activation-synthesis hypothesis. *Consciousness Cognition* 1:172–185.

92 Consciousness and the Cerebral Hemispheres

MICHAEL S. GAZZANIGA

ABSTRACT Split-brain subjects have been studied extensively over the past 35 years. Results from these studies have illuminated the nature of human consciousness, with each decade of research revealing new insights. In the following chapter, this work is reviewed and considered in the modern light of cognitive neuroscience.

Thirty-two years ago, I was examining a war veteran who was about to undergo cerebral commissurotomy to control his capricious epilepsy. He was a robust and charming man and one who instilled respect, especially in a young, green graduate student like myself. W. J. was shortly to be operated on and have his corpus callosum and anterior commissure sectioned by Dr. P. J. Vogel. Joseph Bogen, a resident at the time, was the driving force behind the medical idea. Bogen had critically reviewed an earlier literature and had determined that the surgery might well have beneficial effects for the control of W. J.'s epilepsy. My chore was to determine if his preoperative brain and his postoperative brain—which had been divided in two—behaved any differently. Could it be that a disconnected right hemisphere was as conscious as a disconnected left hemisphere? Could it be that a state of co-consciousness could be produced in a human being? What would it mean if any of these proved to be so?

The original studies on humans were carried out in a special context. The immediately preceding years had found that a similar surgery rendered cats, monkeys, and chimpanzees special creatures. Study after study in these surgically altered animals showed them to be unique. Information trained to one side of the brain remained isolated to that half brain. It was as if severing the great cerebral commissure that tied the left and right cortices together produced an animal with two minds. Mind left didn't know the information from mind right and vice versa. In a long series of studies by Myers and Sperry (see Sperry, 1961), it had been established that the two cerebral hemispheres worked quite independently following commissurotomy, and the researchers described such animals as having split brains. The idea that this might be the case for humans seemed bizarre.

MICHAEL S. GAZZANIGA Center for Neuroscience, University of California, Davis, Calif.

Preoperatively, W. J. could name stimuli presented to either visual field or placed in either hand (Gazzaniga, Bogen, and Sperry, 1961). With his eyes closed he could understand any command and carry it out with either hand. In short, he revealed himself to be entirely normal. The stage was set for understanding what might happen to him following the disconnection of his two cerebral hemispheres. There were a number of claims that nothing would happen. Several papers by A. J. Akelaitis (1941), who had studied other patients with similar surgeries, suggested that callosum-sectioned humans behaved no differently than normals. This finding was always a puzzle and one that Karl Lashley seized on to push his idea of mass action and equipotentiality of the cerebral cortex. Discrete circuits of the brain were not important, in this view—only cortical mass.

When W. J. returned for testing, I experienced one of those unforgettable moments in life. First, and to no one's surprise, information presented to his left hemisphere was normally named and described. Then came the critical test. What would happen when information was flashed to his nondominant and isolated right hemisphere? The earlier Akelaitis work on humans predicted that nothing would happen since the corpus callosum played no essential role in the interhemispheric integration of cerebral information. On the other hand, the earlier animal work predicted that some difficulties would emerge. I first flashed a simple light. W. J. said he didn't see anything, even though his left hand responded to the stimulus by pushing a reaction-time Morse code key. This was the beginning of the idea that splitting the human brain produced two separate conscious systems. It was a revolutionary idea, and over 30 years later, it is one that still needs study and clarification.

It is curious that the problem of consciousness is studied even though there is no general agreement about what is meant by the term. If you asked 20 students of the problem to finish the sentence, "Consciousness is ...", 20 different definitions would result. Yet, most of us know what is really meant by the term. It refers to that subjective state we all possess when awake. In short, it refers to our feelings about our mental capacities and functions.

When the first split-brain patients were studied, we avoided confronting the essential definitional question and instead chose to measure the separate capacities of each half brain. Following my initial encounter with W. J., I have spent the last 30 years trying to characterize the nature of conscious mechanisms in these patients. It is not an easy chore. In what follows I attempt to track the major findings in split-brain research that relate to the problem of consciousness. What becomes apparent is that our understanding of the problem of consciousness is continually evolving, and with each new decade, new dimensions to the problem are presented.

The first decade: Basic principles

W. J., after his split-brain surgery and like all subsequent right-handed cases of relevance, normally named and described information presented to his left speaking hemisphere. What was surprising was his seeming lack of response to stimuli presented to his surgically isolated right hemisphere. It was as if he were blind to visual stimuli presented to the left of fixation. Yet, through a series of tests we devised (Gazzaniga et al., 1962; Gazzaniga et al., 1965), it became obvious that while the left talking hemisphere could not report on these stimuli, the right hemisphere, with its ability to control the manual responses of the left hand, could easily react to a simple visual stimulus.

An early conclusion about these phenomena suggested that dividing the hemispheres in order to control intractable epilepsy left each half brain behaving independently of the other. Information experienced by one side seemed unavailable to the other. Moreover, each half brain seemed specialized for particular kinds of mental activity. The left was superior for language while the right seemed more able to carry out visual-spatial tasks. We had separated structures with specific and complex functions.

The capacities that were demonstrated from the left hemisphere were no surprise. However, when the first patients were able to read from the right hemisphere and were able to take that information and choose between alternatives, the case for a double conscious system seemed strong indeed. We could even provoke the right hemisphere in the sphere of emotions. We carried out dozens upon dozens of studies over five years on these patients, and all pointed to this dramatic state of affairs (Sperry et al., 1969; Gazzaniga, 1970). After separating the human cerebral hemispheres, each half brain seemed to work and function outside the conscious realm of the other. Each could learn, remember, emote, and carry out planned activities.

Although the observations were dramatic, we were still not close to capturing the essential nature of these patients' dual consciousness. We had not pushed the upper limits of the mental capacities of the right hemisphere. We also had not seen many patients and had not had the opportunity to see the rich variation in right-hemisphere capacity that has since been discovered through the study of more split-brain patients. But most importantly, we really had not addressed the question of what is consciousness. After all, demonstrating that a seemingly bilateral symmetrical brain can be divided, thereby producing two conscious systems, merely leaves us with two systems we do not understand instead of just one! It does not advance our attack on what is meant by the experience of being conscious.

The second decade: Origins of modular concepts and the interpreter

With time, more patients and more studies, the original picture came to be modified in the early seventies. Instead of attacking the question of the nature of consciousness, the field drifted for awhile into thinking about different kinds of consciousness. The notion that mind left dealt with the world differently than mind right seemed to be the major conclusion of studies during this era. While interesting, this characterization of how each hemisphere processed information again begged the question of what consciousness actually was and how the brain enabled it to be experienced.

In many ways the work in the early part of the decade proved misleading. After the interesting report on how chimeric stimuli found split-brain patients preferring the right hemisphere for gestalt stimuli and the left hemisphere for analytic tasks, the field briefly took on a new color (Levy, Trevarthen, and Sperry, 1972). It wasn't so much that there were separate conscious systems following commissure section, it was that each hemisphere possessed a different cognitive style. This characterization was short-lived in the scientific community but long-lived in the public mind. The original paper that triggered this distinction of left-brain versus right-brain thinking used stimuli that were already well known to elicit preferred-hemisphere functioning in a chimeric format. Demonstration through another stimulus preparation medium that the left hemisphere preferred language-based stimuli and the right preferred faces was a result that was to be predicted from earlier neurologic work.

The mid-seventies found a number of reports being published that emphasized an additional feature of right-hemisphere specialization. Milner and Taylor reported a superior performance in the right hemisphere on nonverbal tactile stimuli (Milner and Taylor, 1972). LeDoux, Wilson and Gazzaniga found the manipulations of a stimulus to be critical in bringing out right-hemisphere superiorities (LeDoux, Wilson, and Gazzaniga, 1976). In the block design test of the WAIS, for example, right-hemisphere superiority was revealed only when subjects manipulated the blocks. If the same task was a match-to-sample task, the superiority effects disappeared.

While on the one hand the new observations were challenging to the simple view of hemispheric functioning and ideas about dual consciousness, the new conceptual framework was even more challenging to existing concepts about the unity of conscious experience. In brief, the new view suggested the brain was organized in a modular fashion with multiple subsystems active at all levels of the nervous system and each subsystem processing data outside the realm of conscious awareness (Gazzaniga, 1976; Gazzaniga and LeDoux, 1976). These modular systems were fully capable of producing behaviors, mood changes, and cognitive activity. These activities were in turn monitored and synthesized by a special system in the left hemisphere that I call the interpreter.

We first revealed the interpreter using a simultaneous concept test. The patient is shown two pictures, one exclusively to the left hemisphere and one exclusively to the right, and is asked to choose from an array of pictures placed in full view in front of him the ones associated with the pictures lateralized to the left and right brain. In one example of this kind of test, a picture of a chicken claw was flashed to the left hemisphere and a picture of a snow scene to the right hemisphere. Of the array of pictures placed in front of the subject, the obviously correct association is a chicken for the chicken claw and a shovel for the snow scene. Split-brain subject P. S., responded by choosing the shovel with the left hand and the chicken with the right. When asked why he chose these items, his left hemisphere replied, "Oh, that's simple. The chicken claw goes with the chicken, and you need a shovel to clean out the chicken shed." Here, the left brain, observing the left hand's response, interprets that response according to a context consistent with its sphere of knowledge—one that does not include information about the left hemifield snow scene.

Another example of this phenomenon of the left brain interpreting actions produced by the disconnected right brain involves lateralizing a written command, such as "laugh," to the right hemisphere by tachistoscopically presenting it to the left visual field. After the stimulus is presented, the patient laughs, and when asked why, says, "You guys come up and test us every month. What a way to make a living!" In still another example, if the command "walk" is flashed to the right hemisphere, the patient will typically stand up from the chair and begin to take leave from the testing van. When asked where he or she is going, the subject's left brain says, "I'm going into the house to get a Coke." However this type of test is manipulated, it always yields the same kind of result (Gazzaniga, 1983).

There are many ways to influence the left-brain interpreter. As already mentioned, we wanted to know whether or not the emotional response to stimuli presented to one half brain would have an effect on the affective tone of the other half brain. In this particular study, we showed under lateralized stimulus presentation procedures a series of film vignettes that included either violent or calm sequences. We showed that the emotional valence of the stimulus crossed over from the right to the left hemisphere. The left hemisphere remains unaware of the content that produced the emotional change, but it experiences and must deal with the emotion and give it an interpretation.

It is interesting to note that while the patients possess at least some understanding of their surgery, they never say things like, "Well, I chose this because I have a split brain and the information went to the right, nonverbal hemisphere." Even patients who have higher IQs than P. S., based on IQ testing, view their responses as behaviors emanating from their own volitional selves, and as a result they incorporate these behaviors into a theory to explain why they behave as they do. Certainly, one can imagine that at some future point a patient might be studied who chooses not to interpret such behaviors because of an overlying psychological structure that prevents the response. Or, one can imagine a patient learning by rote, as it were, what a "split brain" is all about and why, therefore, a certain behavior most likely occurred. With that set the patient might not offer such an explanation.

There are occasions where a patient, having trouble controlling his left arm due to a transient state of dyspraxia, will tend to write off anything it does under the direction of the right brain, thereby making the foregoing test inappropriate for demonstrating the interpretation phenomenon. In such situations, a single set of pictures is presented and only one hand is allowed to make the response. Thus, in this test the word *pink* is flashed to the right hemisphere and the word *bottle* is flashed to the left. Placed in front of the patient are pictures of at least 10 bottles of different colors and shapes. When this test was run on split-brain patient J. W., on a particular day when he kept saying that his left hand did what it wanted to do, he immediately pointed to the pink bottle with the right hand. When asked why he had done this, J. W. said, "Pink is a nice color."

The modular organization of the human brain has now been well established. The functioning modules do have some kind of physical instantiation, but the brain sciences are not yet able to specify the nature of the actual neural networks involved. What is clear is that they operate largely outside the realm of awareness and announce their computational products to various executive systems that result in behavior or cognitive states. Catching up with all this parallel and constant activity seems to be a function of the left hemisphere's interpreter module. The interpreter is the system of primary importance to the human brain. It is what allows for the formation of beliefs, which in turn are mental constructs that free us from simply responding to stimulus-response aspects of everyday life. In many ways it is the system that provides the story line or narrative of our lives. Yet, it would not appear to be the system that provides the heat, the stuff, the feelings about our thoughts. I will have much more to say about this property of the left hemisphere later in the chapter.

The third decade: Variations in patterns

Starting in the eighties, major advances were made in understanding the split-brain syndrome and how it could vary from one individual to another. It became apparent with the reporting of more cases that there was considerable variation in the pattern of hemisphere specialization (Gazzaniga, 1988). What was so striking in these new studies was that the personal states of mind felt by the patients were similiar and seemed unchanged from their preoperative feelings of mind. In other words, even though the distribution of specialized functions varied, the felt state of mind was no different.

New questions were also being asked about the actual nature of lateral specialization. While there was some new evidence that lateral specialization phenomena seen in humans were also seen in nonhuman primates (Hamilton and Vermiere, 1988), other studies began to examine whether the so-called specializations were more apparent than real (Gazzaniga, 1987; Gazzaniga and Smylie, 1984; Hellige, 1993). There was also work supporting certain specializations; new studies demonstrated hemisphere specialization for the control of facial expressions (Gazzaniga and Smylie, 1990) and visual imagery (Kosslyn et al., 1986). As this continuing line of research unfolds, it becomes clear that determining the actual possible uniqueness of right-hemisphere states of mind will be a difficult task.

Finally, there was increasing interest in unconscious processes and the role subcortical processes may play for such activities (Volpe, LeDoux, and Gazzaniga, 1981; Weiskrantz, 1990). Some studies tried to assert that the split brain in humans was not so split and that in fact there were common subcortical integrating mechanisms for high-level information (Sergent, 1990).

Sorting out all of these issues affects how the split-brain human illuminates the problem of consciousness per se. It is important to determine whether or not there are actual structural differences in how each brain processes information. Equally important is how the variations seen in cortical organization affect the felt state of personal consciousness. And finally, it is important to determine if one must appeal to subcortical systems when examining the mechanisms by which the brain enables unconscious processing.

Pattern Variations in Cortical Specialization Over the past 30 years of split-brain research it has been difficult to isolate what costs to cognition might be incurred by having the human cerebrum divided in two. Many earlier studies have shown there is no change in response reaction time for simple discriminations (Gazzaniga and Sperry, 1966), in the capacity to form hypotheses (LeDoux et al., 1977), and in verbal IQ (Campbell, Bogen, and Smith, 1981). There have been some reports that negative effects can be registered on memory function (Zaidel and Sperry, 1974) while others have not confirmed this finding (LeDoux et al., 1977). There are data suggesting that hemispheric disconnection actually allows each half brain to function without perceptual interference from the other, thereby conferring on the whole brain a supernormal capacity to apprehend perceptual information (Holtzman and Gazzaniga, 1985). Although most prior studies have been carried out in the context that each half brain is a functioning, independent system that operates no differently when separated than when connected, new studies are beginning to challenge this original view. The old assumption was based on the kind of behavioral profile seen in the split-brain cases who possess language in each hemisphere. In this small and highly select group of subjects, each hemisphere seems capable of responding in its own way to a wide variety of stimuli.

But then there are the other cases in whom right-hemisphere performance after surgery was poor to nonexistent. The question becomes whether such patients possess right-hemisphere skills at all, or are merely "locked in" after disconnection from the dominant left half brain. The idea is that the perceptual capacity and engrams are there, but there is no capacity to operate on the perceptual capacity. Consider subject E. B.

Prior to split-brain surgery, E. B. was tested on a number of tests including the nonsense wire figure test of Milner and Taylor (1972). It is believed that this task taps into right-hemisphere specialized systems, and case E. B. was able to perform the task with either hand when the objects were presented out of view. Her intact callosum, it would appear, assisted in distributing the information arriving in her left brain from the right hand over to the specialized system in the right hemisphere. Or that, at least, is how we have come to think about such processes.

After the posterior half of the callosum was cut, E. B. was unable to name objects placed in the left hand, in typical split-brain fashion. The fibers crucial for the interhemispheric transfer of tactile information had been severed, and as a result, what the right hand knew the left did not. E. B. also proved to be a patient without right-hemisphere language. While she was able to find points of stimulation on the left hand by touching them with the left thumb, thereby demonstrating good right-hemisphere cortical somatosensory function as already described, she was unable to retrieve an object named by the examiner with the left hand. This task is easily managed by patients with right-hemisphere language. Most importantly, however, E. B. could no longer perform the wire figure task with either hand.

Since E. B. could perform the task preoperatively, it seems clear that the right hemisphere had the capacity to contribute to solving this kind of task when it was connected to the left. Disconnected from the left, it appeared unable to function. Performance on this task with both hands was greatly impaired. As already mentioned, this kind of finding suggests that the left may normally contribute certain executive functions to specialized systems in the right brain. What was thought to be one integrated system—that is, the capacity to carry out these nonverbal tasks—is actually the product of the interaction of at least two systems, each of which can be located in a different brain area.

The evidence to date suggests that dissociable factors are active in what appear to be unified mental activities. In short, one can begin to envision that there are something like executive controllers active in manipulating the data of specialized processing systems. These controllers normally tend to be lateralized in the left brain, and when the right brain becomes isolated from their influence, the specific functions of the right brain become hard to detect when the right hemisphere is tested alone.

And yet the impact of all the newly discovered variation in left-hemisphere organization is virtually nil on the patient's own sense of consciousness. If consciousness reflects the felt state about specialized capacities, a neural system like the left hemisphere can be aware of only those specialized properties in its possession. If a feature of cognitive life is not housed in its networks, it cannot sense its absence. It is data like these that suggest to me that consciousness is best thought of as an instinct. It is not learned, not dependent on neural mass. It is a property of a neural network.

Discovering Unconscious Processes It was also during the eighties that the phenomenon of blindsight took center stage. Ever since the original report of blindsight, philosophers, psychologists, and neuroscientists have been fascinated with the phenomenon (see Weiskrantz, 1990). The original reports were intriguing and sensible. Patients suffering lesions of visual cortex were nonetheless able to carry out various visually triggered tasks. More interesting, these activities went on outside their realm of consciousness. Patients seemed to deny that they were able to do a task, and were reported to say that if they were at least aware of responding, they were responding only at chance. In short, it looked as though the great silent dimension of cognitive life had been tapped. The unconscious was now explorable in scientific terms. In addition to the excitement the work provoked within the visual sciences, it looked as though subcortical and parallel pathways and centers could now be examined in the human brain. It was also the case that a large nonhuman primate literature had developed on the subject. Monkeys with occipital lesions were reported to be able not only to localize objects in space but also to carry out pattern discriminations.

In the context of split-brain research, we were excited by these reports. The idea of the interpreter also implies that there are vast systems working outside the realm of conscious awareness. The functions of specific modules need not be monitored at the level of consciousness, only their products or outputs. Blindsight phenomena offered a window into these unconscious processes, and we began our own studies in this context (Holtzman, 1984).

As the first reports came out, my own lab began to examine related issues. We first wondered whether or not there might be interhemispheric interactions in patients with their cerebrums divided. Patients with the corpus callosum sectioned provided a rich opportunity for examining how subcortical systems that had been posited as possible brain centers managing blindsight phenomena might interact. Indeed, our first studies on interhemispheric interactions in the domain of spatial attention clearly showed that subcortical mechanisms were active in exchanging information (Holtzman et al., 1981). In the early eighties, we also reported on some preliminary studies suggesting that we could obtain interhemispheric semantic priming (Sidtis and Gazzaniga, 1983). Finally, in examining patients with neglect, we were able to show that information presented in the neglected half field could be used in a same-different discrimination even though the subject denied its presentation (Volpe et al., 1979). Finally, we had observed a curious finding in one of our split-brain subjects, which we thought about dubbing "blindspeech." Case J. W. was able to name (speak) about one of two stimuli projected to the right hemisphere from his disconnected left hemisphere. Curiously, the left hemisphere, even though it was producing a correct spoken response, could not use the information about which it was speaking in a simple match-to-sample task. All of this made sense, especially in the light of blindsight phenomena. In short, and in retrospect, it hardly seemed surprising to discover that information not accessible to conscious processes could nonetheless influence behavior and cognition. And, given the many subcortical networks and their interhemispheric connections, the behavioral results could rest on a plausible anatomy.

As our studies progressed, however, it became apparent that our original suggestion that higher-order information might interact following split-brain surgery was in error. Not only were we unable to reproduce the original findings, but we have been unable to see any sort of interhemispheric interactions, even using stabilized images (Seymour et al., 1993). Additionally, beginning in the early 1980s and continuing until the present, our laboratory has examined patients with occipital lesions (Holtzman, 1984; Fendrich et al., 1992). Our overall findings suggest that when residual visual function is seen, it is due to spared visual cortex, not secondary visual pathways and subcortical processing centers. These findings have us reassessing the significance of so-called blindsight phenomena for understanding mechanisms of consciousness. This is not to say, of course, that there are not vast numbers of mental computations going on outside of conscious awareness. It is probably the case that most of our mental life requires unconscious processes. There is, however, no reason to think those processes are not cortical as well. We need not appeal to the subcortex for the total management of such activities.

The fourth decade: Establishing the evolutionary context

It was not until late in the third decade that I began to think about the problem of consciousness from an evolutionary perspective. My evolving view on the matter has left me convinced that a key to our understanding of consciousness is to place the phenomenon in an evolutionary perspective. As I do that, certain truths emerge for me that argue that, at its core, human consciousness is a *feeling* about specialized capacities.

Through all these decades of split-brain research we have been faced with one unalterable fact. Disconnecting the two cerebral hemispheres, which leaves one half of the cortex no longer interacting in a direct way the other half of the cortex, does not typically disrupt the cognitive or verbal intelligence of these patients. The left dominant hemisphere remains the major force in our cognitive lives, and that force is sustained, it would seem, not by the whole cortex but by specialized circuits within the left hemisphere. In short, the inordinately large human brain does not take on its unique capacities simply by virtue of its size.

With the realization that it is the accumulation of specialized circuits that account for the unique human experience, I have come to look upon the problem of consciousness from a particular perspective. This perspective comes from both the realization that the human brain is a constellation of specialized circuits and the fact that throughout normal aging, our sense of being conscious never changes (Gazzaniga, 1992). When these views are taken together, it becomes evident that what we mean when we talk about consciousness is the feelings we have about our specialized capacities. We have feelings about objects we see, hear, and feel. We have feelings about our capacity to think, to use language, to apprehend faces. In other words, consciousness is not another system. It reflects the affective component of specialized systems that have evolved to enable human cognitive processes. With the human inferential system in place, which seems to be limited to the left hemisphere (Gazzaniga and Smylie, 1984), we have a system that empowers all sorts of mental activity. And again, our consciousness about those mental activities is related to our capacity to assign feelings to those activities. That is what distinguishes us from the electronic artifacts that surround modern humans.

Left and right hemisphere consciousness

Of course, viewing consciousness as a feeling about specialized abilities would lead to the prediction that the quality of consciousness emanating from each hemisphere should differ radically. While left-hemisphere consciousness would reflect what we mean by normal conscious experience, right-hemisphere consciousness would vary as a function of the kinds of specialized circuits that a given half brain may possess. Mind left, with its complex cognitive machinery, can distinguish between the states of, say, sorrow and pity, and appreciates the feelings associated with each state. The right hemisphere does not have the cognitive apparatus for such distinctions, and as a consequence has a reduced state of awareness. Consider the following examples of reduced capacity in right hemispheres and the implications such reduced capacity would have for states of consciousness.

Split-brain patients without right-hemisphere language have a limited capacity for responsiveness to patterned stimuli, which ranges from none to the capacity to make simple matching judgments above chance. In the split-brain patients with the capacity to make perceptual judgments not involving language, there was no ability to make a simple same/difference judgment within the right brain when both the sample and match were lateralized simultaneously. In other words, when a judgment of sameness was required for two simultaneously presented figures, the right hemisphere failed. This profile is commonly seen in patients of all kinds, including patients of similar and sometimes greater overall intelligence than those who possess some right-hemisphere language.

This minimal profile of capacity stands in marked contrast to the patients with right-hemisphere language. The right brain of these patients is responsive, and their overall capacity to respond to both language and nonlanguage stimuli has been well cataloged and reported. In the East Coast series of patients we studied, such cases included J. W., who understood language and who had a rich right-brain lexicon as assessed by the Peabody Picture Vocabulary Test as well as other special tests. At the same time, J. W., until recently, could not access speech from the right hemisphere. Studies on V. P. and P. S. revealed that these patients were able to both understand language and speak from each half brain. Would this extra skill lend a greater capacity to right brain's ability to think, that is to say, to interpret the events of the world?

It turns out that the right hemispheres of both patient groups are poor at making simple inferences. For example, when shown two pictures, one after the other, such as a picture of a match and a picture of a woodpile, the right hemisphere cannot combine the two elements into a causal relation and choose the proper result, that is, a picture of a burning woodpile as opposed to a picture of a woodpile and a set of matches. In other testing, simple words were serially presented to the right brain. The task was to infer the causal relation that obtains between the two lexical elements and pick the answer from a list of six possible answers printed and in full view of the subject. A typical trial would consist of words like *pin* and *finger* being flashed to the right brain with the correct answer being *bleed.* While the right hemisphere could always find a close lexical associate of each of the words used when tested separately, it could not make the inference that *pin* and *finger* should result in the answer *bleed.*

In this light, it is hard to imagine that the left and right hemispheres have similar conscious experiences. The right cannot make inferences, and as a consequence is extremely limited in what it can have feelings about. It seems to deal mainly with raw experience in an unembellished way. The left hemisphere, on the other hand, is constantly, almost reflexively, labeling experiences, making inferences as to cause, and carrying out a host of other cognitive activities. In recent studies it has been observed that the left brain carries out visual search tasks in a "smart" way whereas the right hemisphere performs in a poor way (Kingstone, Enns, and Gazzaniga, 1993). Everywhere we turn, we see these clues. The left hemisphere is busy differentiating the world whereas the right is simply monitoring the world (Mangun et al., 1993).

Summary and conclusions

The past 30 years have seen a marked change in how the state of consciousness is viewed, as the result of considering patients with split brains. After marveling at the dramatic breakdown in the transfer of information between the two hemispheres, researchers attempted to characterize the cognitive styles of the two hemispheres. The overpopularized versions of mind left versus mind right soon gave way to yet another realization: The left brain possessed the major cognitive apparatus driving normal cognition. These findings support the view that our complex mental capacities are the products of discrete, specialized circuits in the brain, not the products of a general computational capacity supported by the human brain's entire cortical mantle. In this light, it is argued that understanding what consciousness *is* means understanding that the phenomenon of consciousness consists of feelings we have about these specialized capacities. It is not something separate from these capacities.

In this light we can see that consciousness can be viewed as an instinct. It is certainly not learned. It is the efficient way the brain puts heat to our cold specialized capacities. It is nothing more and nothing less. It is what makes it all worthwhile.

ACKNOWLEDGMENTS Research for this chapter was supported by NIH grants NINDS 5 R01, NS22626-09, NINDS 5 P01, and NS1778-012, and the James S. McDonnell Foundation.

REFERENCES

AKELAITIS, A. J., 1941. Studies on the corpus callosum: Higher visual functions in each homonymous visual field following complete section of corpus callosum. *Arch. Neurol. Psychiatry* 45:788.

BOGEN, J. E., E. D. FISHER, and P. J. VOGEL, 1965. Cerebral commissurotomy: A second case report. *JAMA* 194:1328–1329.

CAMPBELL, A. L., J. E. BOGEN, and A. SMITH, 1981. Disorganization and reorganization of cognitive and sensorimotor functions in cerebral commissurotomy: Compensatory roles of the forebrain commissures and cerebral hemispheres in man. *Brain* 104:493–511.

FENDRICH, R., C. M. WESSINGER, and M. S. GAZZANIGA, 1992. Residual vision in a scotoma: Implications for blindsight. *Science* 258:1489–1491.

GAZZANIGA, M. S., 1970. *The Bisected Brain.* New York: Appleton-Century-Crofts.

GAZZANIGA, M. S., 1978. On dividing the self: Speculations from brain research. In *Neurology*, W. A. de Hartog Jager et al., eds. Amsterdam: Excerpta Medica, pp. 233–244.

GAZZANIGA, M. S., 1983. Right hemisphere language: A twenty year perspective. *Am. Psychol.* 38:525–537.

GAZZANIGA, M. S., 1985. *The Social Brain.* New York: Basic Books.

GAZZANIGA, M. S., 1987. Cognitive and neurological aspects of hemispheric disconnection in the human brain. *Monograph of the FESN Geneva.* IV(4).

Gazzaniga, M. S., 1992. *Nature's Mind.* New York: Basic Books.

Gazzaniga, M. S., J. E. Bogen, and R. W. Sperry, 1962. Some functional effects of sectioning the cerebral commissures in man. *Proc. Natl. Acad. Sci. U.S.A.* 48:1765–1769.

Gazzaniga, M. S., J. E. Bogen, and R. W. Sperry, 1963. Laterality effects in somesthesis following cerebral commissurotomy in man. *Neuropsychologia* 1:209–215.

Gazzaniga, M. S., J. E. Bogen, and R. W. Sperry, 1965. Observations on visual perception after disconnection of the cerebral hemispheres in man. *Brain* 88:221–236.

Gazzaniga, M. S., and S. A. Hillyard, 1971. Language and speech capacity of the right hemisphere. *Neuropsychologia* 9:273–280.

Gazzaniga, M. S., J. D. Holtzman, and C. S. Smylie, 1987. Speech without conscious awareness. *Neurology* 35: 682–685.

Gazzaniga, M. S., and J. E. LeDoux, 1978. *The Integrated Mind.* New York: Plenum Press.

Gazzaniga, M. S., and C. S. Smylie, 1984. Dissociation of language and cognition: A psychological profile of two disconnected right hemispheres. *Brain* 107:145–153.

Gazzaniga, M. S., C. S. Smylie, K. Baynes, W. Hirst, and C. McCleary, 1984. Profiles of right hemisphere language and speech following brain bisection. *Brain Language* 22:206–220.

Gazzaniga, M. S., and R. W. Sperry, 1966. Simultaneous double discrimination response following brain bisection. *Psychon. Sci.* 4:261–262.

Gazzaniga, M. S., and R. W. Sperry, 1967. Language after section of the cerebral commissures. *Brain* 90:131–148.

Hamilton, C. R., and B. A. Vermeire, 1988. Complementary lateralization in monkeys. *Science* 242:1691–1694.

Hellige, J. B., 1993. *Hemispheric Asymmetry: What's Right and What's Left.* Cambridge, Mass.: Harvard University Press.

Holtzman, J. D., 1983. Blindsight following corpus commissurotomy and occipital lobe lesion. Paper presented at the *International Neuropsychological Symposium*, Rethymnon, Crete.

Holtzman, J. D., 1984. Interactions between cortical and subcortical visual areas: Evidence from human commissurotomy patients. *Vis. Res.* 24:801–813.

Holtzman, J. D., and M. S. Gazzaniga, 1982. Dual task interactions due exclusively to limits in processing resources. *Science* 218:1325–1327.

Holtzman, J. D., and M. S. Gazzaniga, 1985. Enhanced dual task performance following callosal commissurotomy in humans. *Neuropsychologia* 23:315–321.

Holtzman, J. D., J. J. Sidtis, B. T. Volpe, D. H. Wilson, and M. S. Gazzaniga, 1981. Dissociation of spatial information for stimulus localization and the control of attention. *Brain* 104:861–872.

Kingstone, A., J. Enns, and M. S. Gazzaniga, 1993. Abstract, Society for Neuroscience.

Kosslyn, S. M., J. D. Holtzman, M. J. Farah, and M. S. Gazzaniga, 1985. A computational analysis of mental image generation: Evidence from functional dissociations in split-brain patients. *J. Exp. Psychol. [Gen.]* 114:311–341.

LeDoux, J. E., G. Risse, S. Springer, D. H. Wilson, and M. S. Gazzaniga, 1977. Cognition and commissurotomy. *Brain* 100:87–104.

LeDoux, J. E., D. H. Wilson, and M. S. Gazzaniga, 1977. Manipulo-spatial aspects of cerebral lateralization: Clues to the origin of lateralization. *Neuropsychologia* 15:743–750.

Levy, J., C. B. Trevarthen, and R. W. Sperry, 1972. Perception of bilateral chimeric figures following hemispheric deconnection. *Brain* 95:61–78.

Mangun, G. R., R. Plager, W. Loftus, S. A. Hillyard, S. J. Luck, T. Handy, V. Clark, and M. S. Gazzaniga, 1994. Monitoring the visual world: Hemispheric asymmetries and subcortical processes in attention. *J. Cogn. Neurosci.*, in press.

Marcel, A. J., 1983. Conscious and unconscious perception: An approach to the relations between phenomenal experience and perceptual processes. *Cogn. Psychol.* 15: 238–300.

Milner, B., and L. Taylor, 1972. Right hemisphere superiority in tactile pattern-recognition after cerebral commissurotomy. Evidence for nonverbal memory. *Neuropsychologia* 10:1–15.

Nass, R. D., and M. S. Gazzaniga, 1987. Lateralization and specialization in human central nervous system. In *Handbook of Physiology* Sec. 1., *The Nervous System.* Vol. 5, parts 1 and 2, *Higher Functions of the Brain*, F. Plum, ed. Bethesda, Md.: The American Physiological Society, pp. 701–761.

Poppel, E., R. Held, and D. Frost, 1973. Residual visual capacities in a case of cortical blindness. *Cortex* 10:605–612.

Sergent, J., 1990. Furtive incursions into bicameral minds. *Brain* 113:537–568.

Seymour, S. E., P. A. Reuter-Lorenz, and M. S. Gazzaniga, 1994. The disconnection syndrome: Basic findings reaffirmed. *Brain*, in press.

Sidtis, J. J., and M. S. Gazzaniga, 1983. Competence versus performance after callosal section: Looks can be deceiving. In *Cerebral Hemisphere Asymmetry: Method, Theory, and Application*, J. B. Hellige, ed. New York: Praeger Scientific Press.

Sidtis, J. J., B. T. Volpe, J. D. Holtzman, D. H. Wilson, and M. S. Gazzaniga, 1981. Cognitive interaction after staged callosal section: Evidence for a transfer of semantic activation. *Science* 212:344–346.

Sperry, R. W., 1961. Cerebral organization and behavior. *Science* 133:1749–1757.

Sperry, R. W., 1968. Mental unity following surgical disconnection of the cerebral hemispheres. *The Harvey Lecture Series 62.* New York: Academic Press, pp. 293–323.

Sperry, R. W., M. S. Gazzaniga, and J. E. Bogen, 1969. Interhemispheric relationships: The neocortical commissures; syndromes of hemisphere disconnection. In *Handbook of Clinical Neurology*, vol. 4, P. J. Vinken and G. W. Bruyn, eds. Amsterdam: North-Holland and New York: Wiley, pp. 273–290.

Volpe, B. T., J. E. LeDoux, and M. S. Gazzaniga, 1979. Information processing of visual stimuli in an extinguished field. *Nature* 282:344–6.

Weiskrantz, L., 1990. The Ferrier Lecture, 1989. Outlooks for blindsight: Explicit methodologies for implicit processes. *Proc. R. Soc. Lond. [Biol.]* 239:247–278.

Weiskrantz, L., E. K. Warrington, M. D. Sanders, and J. Marshall, 1974. Visual capacity in the hemianopic field following a restricted occipital ablation. *Brain* 97:709–728.

Zaidel, D., and R. W. Sperry, 1974. Memory impairment after commissurotomy in man. *Brain* 97:263–272.

Zaidel, E., 1990. Language functions in the two hemispheres following complete cerebral commissurotomy and hemispherectomy. In *Handbook of Neuropsychology*, vol. 4, F. Boller and J. Grafman, eds. Amsterdam: Elsevier.

CONTRIBUTORS

ANDERSEN, RICHARD A. Division of Biology, California Institute of Technology, Pasadena, California

BADDELEY, ALAN Medical Research Council Applied Psychology Unit, Cambridge, England

BAILEY, CRAIG H. Center for Neurobiology and Behavior, College of Physicians and Surgeons, Columbia University, New York, New York

BAILLARGEON, RENÉE Department of Psychology, University of Illinois at Urbana-Champaign, Champaign, Illinois

BARLOW, HORACE Department of Physiology, University of Cambridge, Cambridge, England

BERTI, ANNA Institute of Human Physiology, University of Parma, Parma, Italy

BISIACH, EDOARDO Department of General Psychology, University of Padua, Padua, Italy

BIZZI, EMILIO Department of Brain and Cognitive Sciences, Massachusetts Institute of Technology, Cambridge, Massachusetts

BLACK, IRA B. Department of Neurosciences and Cell Biology, UMDNJ-Robert Wood Johnson Medical School, Piscataway, New Jersey

BLAKEMORE, COLIN University Laboratory of Physiology, University of Oxford, Oxford, England

BLOOM, FLOYD E. Department of Neuropharmacology, The Scripps Research Institute, La Jolla, California

BLUMSTEIN, SHEILA E. Department of Cognitive and Linguistic Sciences, Brown University, Providence, Rhode Island

BRITTEN, KENNETH H. Department of Neurobiology, Stanford University School of Medicine, Stanford, California

BROTHERS, LESLIE Center for Neuroscience, University of California, Davis, Davis, California

CAPLAN, DAVID Neuropsychology Laboratory, Massachusetts General Hospital, Boston, Massachusetts

CARAMAZZA, ALFONSO Department of Psychology, Dartmouth College, Hanover, New Hampshire

CARLILE, SIMON Department of Physiology, University of Sydney, Sydney, New South Wales, Australia

CHALUPA, LEO M. Department of Psychology, University of California, Davis, Davis, California

CHELAZZI, LEONARDO Laboratory of Neuropsychology, National Institute of Mental Health, Bethesda, Maryland

CHRISTOPHE, ANNE Laboratory of Cognitive Sciences and Psycholinguistics, CNRS-EHESS, Paris, France

CHURCHLAND, PATRICIA S. Department of Philosophy, University of California, San Diego, La Jolla, California

COSMIDES, LEDA Department of Psychology, Center for Evolutionary Psychology, University of California, Santa Barbara, California

CUMMING, BRUCE G. University Laboratory of Physiology, University of Oxford, Oxford, England

DALY, MARTIN Department of Psychology, McMaster University, Hamilton, Ontario, Canada

DESIMONE, ROBERT Laboratory of Neuropsychology, National Institute of Mental Health, Bethesda, Maryland

DEYOE, EDGAR A. Department of Cell Biology, Medical College of Wisconsin, Milwaukee, Wisconsin

DREVETS, WAYNE C. Department of Psychiatry, Washington University School of Medicine, St. Louis, Missouri

DUNCAN, JOHN MRC Applied Psychology Unit, Cambridge, England

EVERITT, BARRY J. Department of Anatomy, University of Cambridge, Cambridge, England

FARAH, MARTHA J. Department of Psychology, University of Pennsylvania, Philadelphia, Pennsylvania

FARBER, ILYA B. Department of Philosophy, University of California, San Diego, La Jolla, California

FERRERA, VINCENT P. Department of Physiology, University of California, San Francisco, San Francisco, California

GALLISTEL, C. R. Department of Psychology, University of California, Los Angeles, Los Angeles, California

GARRETT, MERRILL Cognitive Science Program, Department of Psychology, University of Arizona, Tucson, Arizona

GAULIN, STEVEN J. C. Department of Anthropology, University of Pittsburgh, Pittsburgh, Pennsylvania

GAZZANIGA, MICHAEL S. Center for Neuroscience, University of California, Davis, Davis, California

GEORGOPOULOS, APOSTOLOS P. Brain Sciences Center, VA Medical Center, Departments of Physiology and Neurology, University of Minnesota Medical School, Minneapolis, Minnesota

GHEZ, CLAUDE Center for Neurobiology and Behavior, New York State Psychiatric Institute, College of Physicians and Surgeons, Columbia University, New York, New York

GHILARDI, MARIA FELICE Center for Neurobiology and Behavior, New York State Psychiatric Institute, College of Physicians and Surgeons, Columbia University, New York, New York

GILBERT, CHARLES D. The Rockefeller University, New York, New York

GORDON, JAMES Department of Psychology, Hunter College, New York, New York

GRABOWECKY, MARCIA Center for Neuroscience, University of California, Davis, Davis, California

GRAY, JEFFREY A. Department of Psychology, Institute of Psychiatry, London, England

GRAZIANO, MICHAEL S. A. Department of Psychology, Princeton University, Princeton, New Jersey

GROH, JENNIFER M. Department of Psychology, University of Pennsylvania, Philadelphia, Pennsylvania

GROSS, CHARLES G. Department of Psychology, Princeton University, Princeton, New Jersey

GULYÁS, BALÁZS Laboratory of Brain Research and Positron Emission Tomography, Nobel Institute for Neurophysiology, Karolinska Institute, Stockholm, Sweden

HALGREN, ERIC Neurological Clinic, Institut National de la Santé et de la Recherche Médicale, Rennes, France

HILLYARD, STEVEN A. Department of Neurosciences, University of California, San Diego, La Jolla, California

HIRST, WILLIAM Department of Psychology, New School for Social Research, New York, New York

HOBSON, J. ALLAN Department of Psychiatry, Harvard Medical School, Boston, Massachusetts

HOLYOAK, KEITH J. Department of Psychology, University of California, Los Angeles, Los Angeles, California

HSIAO, STEVEN S. Philip Bard Laboratories of Neurophysiology, Department of Neuroscience, Johns Hopkins University School of Medicine, Baltimore, Maryland

HURLBERT, ANYA C. Department of Physiology, University of Newcastle-on-Tyne Medical School, Newcastle-on-Tyne, England

JOHNSON, KENNETH O. Philip Bard Laboratories of Neurophysiology, Department of Neuroscience, Johns Hopkins University School of Medicine, Baltimore, Maryland

JOHNSON, MARK H. Department of Psychology, Carnegie Mellon University, Pittsburgh, Pennsylvania

JOHNSON-LAIRD, PHILIP N. Department of Psychology, Princeton University, Princeton, New Jersey

JOHNSTON, ELIZABETH B. Sarah Lawrence College, Bronxville, New York

JONIDES, JOHN Department of Psychology, University of Michigan, Ann Arbor, Michigan

JORDAN, MICHAEL I. Department of Brain and Cognitive Sciences, Massachusetts Institute of Technology, Cambridge, Massachusetts

KAAS, JON H. Department of Psychology, Vanderbilt University, Nashville, Tennessee

KANDEL, ERIC R. Center for Neurobiology and Behavior, Howard Hughes Medical Center, Columbia University, New York, New York

KAWASHIMA, RYUTA Division of Human Brain Research, Karolinska Institute, Stockholm, Sweden

KILLACKEY, HERBERT P. Department of Psychobiology, University of California, Irvine, California

KING, ANDREW J. University Laboratory of Physiology, University of Oxford, Oxford, England

KINSBOURNE, MARCEL Center for Cognitive Studies, Tufts University, Medford, Massachusetts

KNIERIM, JAMES J. Arizona Research Laboratories, Division of Neural Systems, Memory and Aging, University of Arizona, Tucson, Arizona

KNIGHT, ROBERT T. Center for Neuroscience, University of California, Davis, Davis, California

KNOWLTON, BARBARA J. Department of Neuroscience, VA Medical Center, San Diego, California

KONISHI, MASAKAZU Department of Biology, California Institute of Technology, Pasadena, California

KOSSLYN, STEPHEN M. Department of Psychology, Harvard University, Cambridge, Massachusetts

LABERGE, DAVID Department of Cognitive Science, University of California, Irvine, Irvine, California

LEDOUX, JOSEPH E. Center for Neural Science, New York University, New York, New York

LEVITT, PAT Department of Neuroscience and Cell Biology, UMDNJ-Robert Wood Johnson Medical School, Piscataway, New Jersey

LUCK, STEVEN J. Department of Neurosciences, University of California, San Diego, La Jolla, California

LUESCHOW, ANDREAS Laboratory of Neuropsychology, National Institute of Mental Health, Bethesda, Maryland

MANGUN, GEORGE R. Center for Neuroscience, University of California, Davis, Davis, California

MARINKOVIC, KSENIJA Neurological Clinic, Institut National de la Santé et de la Recherche Médicale, Rennes, France

MARKOWITSCH, HANS J. Physiological Psychology, University of Bielefeld, Bielefeld, Germany

MAUNSELL, JOHN H. R. Division of Neuroscience, Baylor College of Medicine, Houston, Texas

MCEWEN, BRUCE S. Laboratory of Endocrinology, Rockefeller University, New York, New York

MCNAUGHTON, BRUCE Arizona Reseach Labs Division of Neural Systems, University of Arizona, Tucson, Arizona

MEHLER, JACQUES Laboratory of Cognitive Science and Psycholinguistics, CNRS-EHESS, Paris, France

MEREDITH, M. ALEX Department of Anatomy, Medical College of Virginia, Richmond, Virginia

MERZENICH, MICHAEL Department of Otolaryngology and Physiology, University of California, San Francisco, School of Medicine, San Francisco, California

MILLER, EARL K. Laboratory of Neuropsychology, National Institute of Mental Health, Bethesda, Maryland

MISHKIN, MORTIMER Laboratory of Neuropsychology, National Institute of Mental Health, Bethesda, Maryland

MOSCOVITCH, MORRIS Rotman Research Institute of Baycrest Centre, Department of Psychology, University of Toronto, Toronto, Ontario, Canada

MOVSHON, J. ANTHONY Center for Neural Science, New York University, New York, New York

MUNOZ, DOUGLAS P. Department of Physiology, Queens University, Kingston, Ontario, Canada

MUSSA-IVALDI, FERDINANDO A. Department of Brain and Cognitive Sciences, Massachusetts Institute of Technology, Cambridge, Massachusetts

NEVILLE, HELEN J. Neuropsychology Laboratory, The Salk Institute, San Diego, California

NEWSOME, WILLIAM T. Department of Neurobiology, Stanford University School of Medicine, Stanford, California

NOWLAN, STEVEN J. Computational Neuroscience Laboratory, The Salk Institute, San Diego, California

O'Sullivan, Brendan Division of Human Brain Research, Karolinska Institute, Stockholm, Sweden

Paivio, Allan Department of Psychology, University of Western Ontario, London, Ontario, Canada

Parker, Andrew J. University Laboratory of Physiology, University of Oxford, Oxford, England

Pinker, Steven Center for Cognitive Science, Massachusetts Institute of Technology, Cambridge, Massachusetts

Posner, Michael I. Department of Psychology, University of Oregon, Eugene, Oregon

Premack, Ann James Laboratory of Developmental Psychobiology, National Center for Scientific Research, Paris, France

Premack, David Laboratory of Developmental Psychobiology, National Center for Scientific Research, Paris, France

Preuss, Todd M. Department of Psychology, Vanderbilt University, Nashville, Tennessee

Rafal, Robert Department of Neurology, VA Medical Center, Martinez, California

Raichle, Marcus E. Department of Neurology and Radiology, Washington University School of Medicine, St. Louis, Missouri

Rakic, Pasko Section of Neuroanatomy, Yale University School of Medicine, New Haven, Connecticut

Rapp, Brenda C. Department of Cognitive Science, Johns Hopkins University, Baltimore, Maryland

Robbins, Trevor W. Department of Experimental Psychology, University of Cambridge, Cambridge, England

Robertson, Lynn Department of Neurology, UCD School of Medicine, VA Medical Center, Martinez, California

Robin, Nina Department of Psychology, University of California, Los Angeles, Los Angeles, California

Roe, Anna W. Division of Neuroscience, Baylor College of Medicine, Houston, Texas

Roland, Per E. Division of Human Brain Research, Department of Neuroscience, Karolinska Institute, Stockholm, Sweden

Rolls, Edmund T. Department of Experimental Psychology, University of Oxford, Oxford, England

Rugg, Michael D. Wellcome Brain Research Group, School of Psychology, University of Saint Andrews, Saint Andrews, Scotland

Sainburg, Robert Center for Neurobiology and Behavior, Columbia University, New York, New York

Schacter, Daniel L. Department of Psychology, Harvard University, Cambridge, Massachusetts

Sejnowski, Terrence J. Computational Neurobiology Laboratory, The Salk Institute, San Diego, California

Shadlen, Michael N. Department of Neurobiology, Stanford University Medical Center, Stanford, California

Shapley, Robert Center for Neural Science, New York University, New York, New York

Shepherd, Gordon M. Section of Neurobiology, Yale University School of Medicine, New Haven, Connecticut

Shimamura, Arthur P. Department of Psychology, University of California, Berkeley, Berkeley, California

Singer, Wolf Department of Neurophysiology, Max Planck Institute, Frankfurt, Germany

Smith, Edward E. Department of Psychology, University of Michigan, Ann Arbor, Michigan

Sparks, David L. Department of Psychology, University of Pennsylvania, Philadelphia, Pennsylvania

Spelke, Elizabeth S. Department of Psychology, Cornell University, Ithaca, New York

Sproat, Richard Linguistics Research Department, AT&T Bell Laboratories, Murray Hill, New Jersey

Squire, Larry R. VA Medical Center, Department of Psychiatry, University of California, San Diego, School of Medicine, La Jolla, California

Stein, Barry E. Department of Neurobiology and Anatomy, Bowman Gray School of Medicine, Wake Forest University, Winston-Salem, North Carolina

Stickgold, Robert Laboratory of Neurophysiology, Department of Psychiatry, Harvard Medical School, Boston, Massachusetts

Stromswold, Karin Department of Psychology, Center for Cognitive Science, Rutgers University, New Brunswick, New Jersey

Stryker, Michael Department of Physiology, University of California, San Francisco, San Francisco, California

Suga, Nobuo Department of Biology, Washington University, St. Louis, Missouri

Sussman, Amy L. Department of Psychology, Yale University, New Haven, Connecticut

Tooby, John Department of Anthropology, Center for Evolutionary Psychology, University of California, Santa Barbara, Santa Barbara, California

Ts'o, Daniel Y. Division of Neuroscience, Baylor College of Medicine, Houston, Texas

Tulving, Endel Rotman Research Institute of Baycrest Centre, University of Toronto, Toronto, Canada

Twombly, I. Alexander Philip Bard Laboratories of Neurophysiology, Department of Neuroscience, Johns Hopkins University School of Medicine, Baltimore, Maryland

Ullman, Shimon Department of Applied Mathematics and Computer Science, Weizmann Institute of Science, Rehovot, Israel, and Department of Brain and Cognitive Sciences, Massachusetts Institute of Technology, Cambridge, Massachusetts

Vallbo, Åke B. Department of Physiology, University of Göteborg, Göteborg, Sweden

Van Essen, David C. Department of Anatomy and Neurobiology, Washington University School of Medicine, St. Louis, Missouri

Vishton, Peter Department of Psychology, Cornell University, Ithaca, New York

von der Heydt, Rüdiger Philip Bard Laboratories of Neurophysiology, Department of Neuroscience, Johns Hopkins University School of Medicine, Baltimore, Maryland von Hofsten, Claes Department of Psychology, Umeå University, Umeå, Sweden

Wallace, Mark T. Department of Neurobiology and Anatomy, Bowman Gray School of Medicine, Wake Forest University, Winston-Salem, North Carolina

Weinberger, Norman M. Department of Psychobiology, Center for the Neurobiology of Learning and Memory, University of California, Irvine, Irvine, California

Wilson, Margo Department of Psychology, McMaster University, Hamilton, Ontario, Canada

Wilson, Matthew A. Division of Neural Systems, University of Arizona, Tucson, Arizona

Woldorff, Marty G. Research Imaging Center, University of Texas Health Sciences Center, San Antonio, Texas

Wurtz, Robert H. Laboratory of Sensorimotor Research, National Eye Institute, Bethesda, Maryland

Young, Malcolm P. University Laboratory of Physiology, University of Oxford, Oxford, England

Zohary, Ehud Department of Neurobiology, Stanford University School of Medicine, Stanford, California

INDEX

Note: Page numbers in *italics* refer to figures, those followed by "t" refer to tables, and those followed by "n" refer to notes.

B

E

F

G

H

I

M

N

S

T

Z